# BMW
# Z3

## Service Manual

**1.9, 2.3, 2.5i, 2.8, 3.0i, 3.2**

**Z3 Roadster, Z3 Coupe, M Roadster, M Coupe**

**1996, 1997, 1998, 1999, 2000, 2001, 2002**

[B] BentleyPublishers®
.com

**BENTLEY PUBLISHERS®** | Automotive Reference™

Bentley Publishers, a division of Robert Bentley, Inc.
1734 Massachusetts Avenue
Cambridge, MA 02138 USA
800-423-4595 / 617-547-4170

Information that makes the difference®

BentleyPublishers®.com

**Technical contact information**

We welcome your feedback. Please submit corrections and additions to our BMW technical discussion forum at:

http://www.BentleyPublishers.com

**Updates and corrections**

We will evaluate submissions and post appropriate editorial changes online as updates or tech discussion. Appropriate updates and corrections will be added to the book in future printings. Check for updates and corrections for this book before beginning work on your vehicle. See the following web address for additional information:

http://www.BentleyPublishers.com/updates/

**WARNING—Important Safety Notice**

Do not use this manual for repairs unless you are familiar with automotive repair procedures and safe workshop practices. This manual illustrates the workshop procedures for some maintenance and service work. It is not a substitute for full and up-to-date information from the vehicle manufacturer or for proper training as an automotive technician. Note that it is not possible for us to anticipate all of the ways or conditions under which vehicles may be serviced or to provide cautions as to all of the possible hazards that may result.

We have endeavored to ensure the accuracy of the information in this manual. Please note, however, that considering the vast quantity and the complexity of the service information involved, we cannot warrant the accuracy or completeness of the information contained in this manual.

**FOR THESE REASONS, NEITHER THE PUBLISHER NOR THE AUTHOR MAKES ANY WARRANTIES, EXPRESS OR IMPLIED, THAT THE INFORMATION IN THIS MANUAL IS FREE OF ERRORS OR OMISSIONS, AND WE EXPRESSLY DISCLAIM THE IMPLIED WARRANTIES OF MERCHANTABILITY AND OF FITNESS FOR A PARTICULAR PURPOSE, EVEN IF THE PUBLISHER OR AUTHOR HAVE BEEN ADVISED OF A PARTICULAR PURPOSE, AND EVEN IF A PARTICULAR PURPOSE IS INDICATED IN THE MANUAL. THE PUBLISHER AND AUTHOR ALSO DISCLAIM ALL LIABILITY FOR DIRECT, INDIRECT, INCIDENTAL OR CONSEQUENTIAL DAMAGES THAT RESULT FROM ANY USE OF THE EXAMPLES, INSTRUCTIONS OR OTHER INFORMATION IN THIS MANUAL. IN NO EVENT SHALL OUR LIABILITY WHETHER IN TORT, CONTRACT OR OTHERWISE EXCEED THE COST OF THIS MANUAL.**

Before attempting any work on your BMW, read **001 GENERAL WARNINGS and CAUTIONS** and any **WARNING** or **CAUTION** that accompanies a procedure in the service manual. Review the **WARNINGS** and **CAUTIONS** each time you prepare to work on your BMW.

Your common sense and good judgment are crucial to safe and successful service work. Read procedures through before starting them. Think about whether the condition of your car, your level of mechanical skill, or your level of reading comprehension might result in or contribute in some way to an occurrence which might cause you injury, damage your car, or result in an unsafe repair. If you have doubts for these or other reasons about your ability to perform safe repair work on your car, have the work done at an authorized BMW dealer or other qualified shop.

Part numbers listed in this manual are for identification purposes only, not for ordering. Always check with your authorized BMW dealer to verify part numbers and availability before beginning service work that may require new parts.

Special tools required to perform certain service operations are identified in the manual and are recommended for use. Use of tools other than those recommended in this manual may be detrimental to the car's safe operation as well as the safety of the person servicing the car.

The vehicle manufacturer will continue to issue service information updates and parts retrofits after the editorial closing of this manual. Some of these updates and retrofits will apply to procedures and specifications in this manual. We regret that we cannot supply updates to purchasers of this manual.

This manual is prepared, published and distributed by Bentley Publishers, 1734 Massachusetts Avenue, Cambridge, Massachusetts 02138 USA. All information contained in this manual is based on the information available to the publisher at the time of editorial closing. BMW has not reviewed and does not vouch for the accuracy or completeness of the technical specifications and work procedures described and given in this manual.



**ISBN 978-0-8376-1617-9** *Editorial closing 02 / 2005 Job code: BZ02-08

The Library of Congress has cataloged the paperback version as follows:

BMW Z3 : service manual : 1.9, 2.3, 2.5i, 2.8, 3.0i, 3.2 Z3 roadster, Z3 coupe, M roadster, M coupe, 1996, 1997, 1998, 1999, 2000, 2001, 2002.
p. cm.
ISBN 978-0-8376-1250-8 (alk. paper)
1. BMW Z3 automobile--Maintenance and repair--Handbooks, manuals, etc. I. Robert Bentley, Inc.

TL215.B2528B59 2005
629.28'722--dc22

2004029283



Manufactured in the United States of America

202607R001

# Foreword

For the BMW owner with automotive mechanical skills, and for independent service professionals, this manual includes maintenance and repair procedures the BMW Z3 is likely to need during its service life, including many procedures and specifications that were available in an authorized BMW dealer service department at the time this manual was prepared. The BMW owner with no intention of working on his or her car will find that owning and referring to this manual will make it possible to be better informed and to more knowledgeably discuss repairs with a professional automotive technician.

For those intending to do maintenance and repair on their BMW, it is essential that safety equipment be used and safety precautions observed when working on your vehicle. A minimum safety equipment list includes hand protection, eye protection, and a fire extinguisher. A selection of good quality hand tools is also needed. This includes a torque wrench to ensure that fasteners are tightened in accordance with specifications. In some cases, the text refers to special tools that are recommended or required to accomplish adjustments or repairs. These tools are often identified by their BMW special tool number and illustrated.

### Disclaimer

We have endeavored to ensure the accuracy of the information in this manual. When the vast array of data presented in the manual is taken into account, however, no claim to infallibility can be made. We therefore cannot be responsible for the result of any errors that may have crept into the text. Please also read **WARNING–Important Safety Notice** on the copyright page.

Prior to starting a repair procedure, read the procedure, **001 Warnings and Cautions** and the warnings and cautions that accompany the procedure. Reading a procedure before beginning work helps you determine in advance the need for specific skills, identify hazards, prepare for appropriate capture and handling of hazardous materials, and the need for particular tools and replacement parts such as gaskets.

Bentley Publishers encourages comments from the readers of this manual with regard to errors, and/or suggestions for improvement of our product. These communications have been and will be carefully considered in the preparation of this and other manuals. If you identify inconsistencies in the manual, you may have found an error. Please contact the publisher and we will endeavor to post applicable corrections on our website. Posted corrections (errata) should be reviewed before beginning work. Please see the following web address:

**http://www.BentleyPublishers.com/updates/**

BMW continues to issue service information and parts retrofits after the editorial closing of this manual. Some of this updated information may apply to procedures and specifications in this manual. For the latest information, please see the following web address:

**http://www.bmwtechinfo.com**

BMW offers extensive warranties, especially on components of the fuel delivery and emission control systems. Therefore, before deciding to repair a BMW that may be covered wholly or in part by any warranties issued by BMW of North America, LLC, consult your authorized BMW dealer. You may find that the dealer can make the repair either free or at minimum cost. Regardless of its age, or whether it is under warranty, your BMW is both an easy car to service and an easy car to get serviced. So if at any time a repair is needed that you feel is too difficult to do yourself, a trained BMW technician is ready to do the job for you.

**Bentley Publishers**

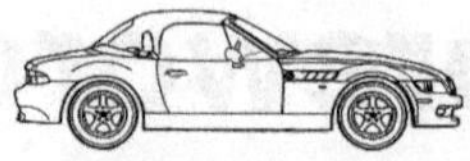

E36/7

# Vehicle Identification and VIN Decoder

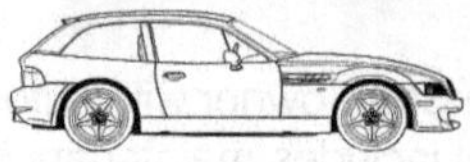

E36/8

This service manual covers 1996 through 2002 BMW Z3 models (Z3 Roadster, Z3 Coupe, M Roadster and M Coupe), and is specifically designed to cover only those models built for sale in the United States.

BMW, and most accessories and parts sellers specializing in BMWs, refer to the Z3 family of cars covered by this manual as the E36/7 (Roadster) and the E36/8 (Coupe) platform. E36 is BMW's internal code for the basic platform shared by all of the cars covered by this manual. The 7 and 8 are a further classification within the E36 family. Except for the engine and a few other components, the models covered by this manual share many of same components such as suspension parts, body panels and the interior pieces. When ordering parts, especially body parts, knowing the platform code may be helpful. Engine codes for Z3 models are listed in **100 Engine–General**.

Some of the information in this manual applies only to cars of a particular model year or range of years. For example, "1999 m.y." refers to the 1999 model year. The model year does not necessarily match the calendar year in which the car was manufactured or sold. To be sure of the model year of a particular car, check the vehicle identification number (VIN) on the car.

The VIN is a unique sequence of 17 characters assigned by BMW to identify each individual car. 4USCN33492LK51496 is an example. Each of the 17 letters and numbers indicates certain facts about the car and its manufacture.

Your BMW's VIN can be found on the top of the dashboard, on the driver's side where the number can be seen through the windshield. The 10th character is the model year code. The letters "I", "O", "Q" and "U" are not used for model year designation, for example, "T" for 1996 m.y., "V" for 1997 m.y., etc. This manual covers BMW Z3 models for the model years 1996 through 2002. The table below explains some of the various codes in the VIN for models covered by this manual.

| VIN Position | Description | Decoding Information | |
|---|---|---|---|
| 1-3 | Manufacturer | WBA | BMW, AG. Munich, Germany |
| | | 4US | BMW, AG. Munich, Germany |
| | | WBS | BMW Motorsport, Gmbh. Munich, Germany |
| | | 5UM | BMW Motorsport, Gmbh. Munich, Germany |
| 4-7 | Vehicle description (body and engine) | CH73 | Z3 roadster, 1.9l (M44) |
| | | CH33 | Z3 roadster/coupe, 2.8l (M52 TU) |
| | | CH93 | Z3 roadster, 2.3l (M52 TU) |
| | | CJ33 | Z3 roadster, 2.8l (M52) |
| | | CK53 | Z3 roadster/coupe, 2.8l (M52 TU) |
| | | CK73 | Z3 coupe 3.0l (M54) |
| | | CK93 | M roadster, 3.2l (S52) |
| | | CL93 | M roadster, 3.2l (S54) |
| | | CM93 | M coupe, 3.2l (S52) |
| | | CN33 | Z3 roadster, 2.5l (M54) |
| | | CN53 | Z3 roadster, 3.0l (M54) |
| | | CN93 | M coupe, 3.2l (S54) |
| 8 | Restraint system | 2 | Manual seat belts with dual SRS airbags |
| | | 3 | Manual seat belts with dual SRS airbags |
| | | 4 | Manual belts with driver and passenger Front and side airbags |
| 9 | Check Digit | | 0-9 or X, calculated by NHTSA |
| 10 | Model Year | T | 1996 |
| | | V | 1997 |
| | | W | 1998 |
| | | X | 1999 |
| | | Y | 2000 |
| | | 1 | 2001 |
| | | 2 | 2002 |
| 11 | Assembly Plant | L | Spartanburg, South Carolina |
| 12-17 | Serial Number | | Sequential production number for specific vehicle |

# 001 General Warnings and Cautions



## PLEASE READ THESE WARNINGS AND CAUTIONS BEFORE PROCEEDING WITH MAINTENANCE AND REPAIR WORK.

### WARNINGS—
### *See also* CAUTIONS

- Some repairs may be beyond your capability. If you lack the skills, tools and equipment, or a suitable workplace for any procedure described in this manual, we suggest you leave such repairs to an authorized BMW dealer service department or other qualified shop.
- Do not reuse any fasteners that are worn or deformed. Many fasteners are designed to be used only once and become unreliable and may fail when used a second time. This includes, but is not limited to, nuts, bolts, washers, self-locking nuts or bolts, circlips and cotter pins. Always replace these fasteners with new parts.
- Never work under a lifted car unless it is solidly supported on stands designed for the purpose. Do not support a car on cinder blocks, hollow tiles or other props that may crumble under continuous load. Never work under a car that is supported solely by a jack. Never work under the car while the engine is running.
- If you are going to work under a car on the ground, make sure that the ground is level. Block the wheels to keep the car from rolling. Disconnect the battery negative (–) terminal to prevent others from starting the car while you are under it.
- Never run the engine in the work area unless it is well-ventilated. Exhaust should be vented to the outside. Carbon monoxide (CO) in the exhaust kills.
- Remove rings, bracelets and other jewelry so that they cannot cause electrical shorts, get caught in running machinery, or be crushed by heavy parts.
- Tie long hair behind your head. Do not wear a necktie, a scarf, loose clothing, or a necklace when you work near machine tools or running engines. If your hair, clothing, or jewelry were to get caught in the machinery, severe injury could result.
- Do not attempt to work on your car if you do not feel well. You increase the danger of injury to yourself and others if you are tired, upset or have taken medication or any other substance that may keep you from being fully alert.
- Illuminate your work area adequately but safely. Use a portable safety light for working inside or under the car. Make sure the bulb is enclosed by a wire cage. The hot filament of an accidentally broken bulb can ignite spilled fuel, vapors or oil.
- Catch draining fuel, oil, or brake fluid in suitable containers. Do not use food or beverage containers that might mislead someone into consuming the contents. Store flammable fluids away from fire hazards. Wipe up spills at once, but do not store the oily rags, which can ignite and burn spontaneously.
- Always observe good workshop practices. Wear goggles when you operate machine tools or work with battery acid. Wear gloves or other protective clothing whenever the job requires working with harmful substances.
- Greases, lubricants and other automotive chemicals contain toxic substances, many of which are absorbed directly through the skin. Read the manufacturer's instructions and warnings carefully. Use hand and eye protection. Avoid direct skin contact.
- Disconnect the battery negative (–) terminal whenever you work on the fuel system or the electrical system. Do not smoke or work near heaters or other fire hazards. Keep an approved fire extinguisher handy.
- Friction materials (such as brake pads or shoes or clutch discs) contain asbestos fibers or other friction materials. Do not create dust by grinding, sanding, or by cleaning with compressed air. Avoid breathing dust. Breathing any friction material dust can lead to serious diseases and may result in death.
- Batteries give off explosive hydrogen gas during charging. Keep sparks, lighted matches and open flame away from the top of the battery. If hydrogen gas escaping from the cap vents is ignited, it will ignite gas trapped in the cells and cause the battery to explode.
- Connect and disconnect battery cables, jumper cables or a battery charger only with the ignition switched off. Do not disconnect the battery while the engine is running.
- Do not quick-charge the battery (for boost starting) for longer than one minute. Wait at least one minute before boosting the battery a second time.
- Do not allow battery charging voltage to exceed 16.5 volts. If the battery begins producing gas or boiling violently, reduce the charging rate. Boosting a sulfated battery at a high charging rate can cause an explosion.
- The air-conditioning system is filled with chemical refrigerant, which is hazardous. The A/C system should be serviced only by trained technicians using approved refrigerant recovery / recycling equipment, trained in related safety precautions, and familiar with regulations governing the discharging and disposal of automotive chemical refrigerants.
- Do not expose any part of the A/C system to high temperatures such as open flame. Heat increases system pressure and excessive heat may cause the system to burst.

***Continued on next page***

## WARNINGS (continued)

- Some aerosol tire inflators are highly flammable. Be extremely cautious when repairing a tire that may have been inflated using an aerosol tire inflator. Keep sparks, open flame or other sources of ignition away from the tire repair area. Inflate and deflate the tire at least four times before breaking the bead from the rim. Completely remove the tire from the rim before attempting any repair.
- Cars covered by this manual are equipped with a supplemental restraint system (SRS), that automatically deploys airbags and pyrotechnic seat belt tensioners in case of a frontal or side impact. These are explosive devices. Handled improperly or without adequate safeguards, they can be accidently activated and cause serious injury.
- The ignition system produces high voltages that can be fatal. Avoid contact with exposed terminals and use extreme care when working on a car with the engine running or the ignition switched on.
- Place jack stands only at locations specified by the manufacturer. The jack supplied with the vehicle is intended for tire changes only. Use a heavy duty floor jack to lift vehicle before installing jack stands. See **010 General**.
- Battery acid (electrolyte) can cause severe burns. Flush contact area with water and seek medical attention.
- Aerosol cleaners and solvents may contain hazardous or deadly vapors and are highly flammable. Use only in a well ventilated area. Do not use on hot surfaces (engines, brakes, etc.).
- Do not remove coolant reservoir or radiator cap with the engine hot. Burns or engine damage may result.
- Fuel is highly flammable. When working around fuel, do not smoke or work near heaters or other fire hazards. Keep an approved fire extinguisher handy.
- The fuel system is designed to retain pressure even when the ignition is off. When working with the fuel system, loosen the fuel lines slowly to allow the residual pressure to dissipate gradually. Take precautions to avoid spraying fuel.
- Never lay tools or parts in the engine compartment or on top of the battery. They may fall into confined spaces and be difficult to retrieve, become caught in belts or other rotating parts when the engine is started, or cause electrical shorts and damage to the electrical system.

## CAUTIONS—
## *See also* WARNINGS

- If you lack the skills, tools and equipment, or a suitable workshop for any procedure described in this manual, leave such repairs to an authorized BMW dealer or other qualified shop.
- BMW is constantly improving its cars and sometimes these changes, both in parts and specifications, are made applicable to earlier models. Therefore, part numbers listed in this manual are for reference only. Always check with your authorized BMW dealer parts department for the latest information.
- Before starting a job, make certain that you have all the necessary tools and parts on hand. Read all the instructions thoroughly, and do not attempt shortcuts. Use tools appropriate to the work and use only replacement parts meeting BMW specifications.
- Use pneumatic and electric tools only to loosen threaded parts and fasteners. Never use these tools to tighten fasteners, especially on light alloy parts. Always use a torque wrench to tighten fasteners to the tightening torque specification listed.
- Be mindful of the environment and ecology. Before you drain the crankcase, find out the proper way to dispose of the oil. Do not pour oil onto the ground, down a drain, or into a stream, pond or lake. Dispose of waste in accordance with federal, state and local laws.
- The control module for the anti-lock brake system (ABS) cannot withstand temperatures from a paint-drying booth or a heat lamp in excess of 203°F (95°C). Do not subject to temperatures in excess of 185°F (85°C) for more than two hours.
- Before doing any electrical welding on cars equipped with ABS, disconnect the battery negative (–) terminal (ground strap) and the ABS control module connector.
- Always make sure ignition is OFF before disconnecting battery.
- Label battery cables before disconnecting. On some models, battery cables are not color coded.
- Disconnecting the battery may erase fault code(s) stored in control module memory. Using special BMW diagnostic equipment, check for fault codes prior to disconnecting the battery cables. If the malfunction indicator light (MIL) is illuminated, see **OBD On-Board Diagnostics** at the back of this manual. This light is identified as the Check Engine or Service Engine Soon light. If any other system faults have been detected (indicated by an illuminated warning light), see an authorized BMW dealer.
- If a normal or rapid charger is used to charge battery, disconnect and remove the battery from the vehicle in order to avoid damaging the vehicle.
- Do not quick-charge the battery (for boost starting) for longer than one minute. Wait at least one minute before boosting the battery a second time.
- Connect and disconnect a battery charger only with the battery charger switched off.
- Sealed or "maintenance free" batteries should be slow-charged only, at an amperage rate that is approximately 10% of the battery's ampere-hour (Ah) rating.
- Do not allow battery charging voltage to exceed 16.5 volts. If the battery begins producing gas or boiling violently, reduce the charging rate. Boosting a sulfated battery at a high charging rate can cause an explosion.

# 010 General





**TABLE**

## GENERAL

This section is intended to help the do-it-yourselfer get started. Tips on workshop practices, basic tools, and a quick reference guide to emergencies can be found here.

> ***WARNING —***
> *Do not use this manual unless you are familiar with basic automotive repair procedures and safe workshop practices. This manual illustrates the workshop procedures required for most service work; it is not a substitute for full and up-to-date information from the vehicle manufacturer or for proper training as an automotive technician. Note that it is not possible for us to anticipate all of the ways or conditions under which vehicles may be serviced or to provide cautions as to all of the possible hazards that may result.*

## HOW TO USE THIS MANUAL

> ***WARNING —***
> *Your common sense and good judgment are crucial to safe and successful service work. Read procedures through before starting them. Think about whether the condition of your car, your level of mechanical skill, or your level of reading comprehension might result in or contribute in some way to an occurrence that might cause you injury, damage your car, or result in an unsafe repair. If you have doubts for these or other reasons about your ability to perform safe repair work on your car, have the work done at an authorized BMW dealer or other qualified shop.*

The manual is divided into ten sections:

- **0 General Data and Maintenance**
- **1 Engine**
- **2 Transmission**
- **3 Suspension, Steering and Brakes**
- **4 Body**
- **5 Body Equipment**
- **6 Electrical System**
- **7 Equipment and Accessories**
- **ELE Electrical Wiring Diagrams**
- **OBD On-Board Diagnostics**

**0 General Data and Maintenance** covers warnings and cautions (**Repair Group 001**), general vehicle information (**Repair Group 010**) as well as the recommended maintenance schedules and service procedures to perform BMW scheduled maintenance work (**Repair Group 020**).

The next seven sections (**Repair Group 1** through **Repair Group 7**) are repair based and organized by three digit repair groups. Most major sections begin with a General repair group, e.g. **100 Engine–General.** These "00" (double zero) groups are mostly descriptive in nature, covering topics such as theory of operation and troubleshooting. The remainder of the repair groups contain the service and repair information.

The last two major sections, **ELE** and **OBD**, contain detailed electrical wiring schematics and scan tool codes.

A master listing of the individual repair groups can be found on the inside front cover. A comprehensive index can be found at the back of the manual.

### Warnings, cautions and notes

Throughout this manual are many passages with the headings **WARNING**, **CAUTION**, or **NOTE**. These very important headings have different meanings.

> ***WARNING—***
> *The text under this heading warns of unsafe practices that are very likely to cause injury, either by direct threat to the person(s) performing the work or by increased risk of accident or mechanical failure while driving.*

> ***CAUTION—***
> *A caution calls attention to important precautions to be observed during the repair work that will help prevent accidentally damaging the car or its parts.*

***NOTE—***
*A note contains helpful information, tips that will help in doing a better job and completing it more easily.*

Please read every **WARNING**, **CAUTION**, and **NOTE** at the front of the manual and as they appear in repair procedures. They are very important. Read them before you begin any maintenance or repair job.

## GETTING STARTED

Most of the necessary maintenance and minor repair that an automobile will need can be done with ordinary tools. Below is some important information on how to work safely, a discussion of what tools will be needed and how to use them.

### Safety

Although an automobile presents many hazards, common sense and good equipment can help ensure safety. Many accidents happen because of carelessness. Pay attention and stick to safety rules as explained in this manual.

## Lifting the car

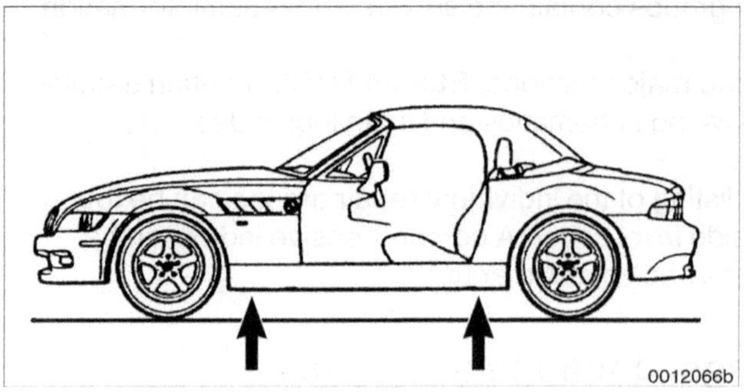


The proper jacking points should be used to raise the car safely and avoid damage. The jack supplied with the car can only be used at the four side points (**arrows**)—just behind the front wheels or just in front of the rear wheels.

> ***WARNING —***
> - *Never work under a lifted car unless it is solidly supported on jack stands that are intended for that purpose.*
> - *When raising the car using a floor jack or a hydraulic lift, carefully position the jack pad to prevent damaging the car body. Plastic pads are provided for this purpose by the manufacturer at the jacking points.*
> - *Watch the jack closely. Make sure it stays stable and does not shift or tilt. As the car is raised, it may roll slightly and the jack may shift.*

## Raising car safely

- Park car on flat, level surface.
- If changing a tire, loosen lug bolts before rasing car. See **Changing a tire** in this repair group.

- Place jack into position. Make sure jack is resting on flat, solid ground. Use a board or other support to provide a firm surface for the jack, if necessary.
- Raise car slowly while constantly checking position of jack and car.
- Once car is raised, block wheel that is opposite and farthest from jack to prevent car from unexpectedly rolling.

> ***WARNING —***
> - *Do not rely on the transmission or the emergency brake to keep the car from rolling. They are not a substitute for positively blocking the opposite wheel.*
> - *Never work under a car that is supported only by a jack. Use jack stands that are designed to support the car. See* **Tools** *in this repair group.*

## Working under car safely

- Disconnect negative (–) cable from battery so that no one can start car. Let others know what you will be doing.

> ***CAUTION—***
> *Prior to disconnecting the battery, read the battery disconnection cautions in* **001 General Warnings and Cautions**.

- Raise car slowly as described above.
- Use at least two jack stands to support car. Use jack stands designed for the purpose of supporting a car. For more information on jack stands, see **Tools** in this repair group.

> ***WARNING —***
> - *A jack is a temporary lifting device and should not be used alone to support the car while you are under it.*
> - *Do not use wood, concrete blocks, or bricks to support a car. Wood may split. Blocks or bricks, while strong, are not designed for that kind of load, and may break or collapse.*

- Place jack stands on firm, solid surface. If necessary, use a flat board or similar solid object to provide a firm footing.
- Lower car slowly until its weight is fully supported by jack stands. Watch to make sure that the jack stands do not tip or lean as the car settles on them.
- Observe all jacking precautions again when raising car to remove jack stands.

## ADVICE FOR THE BEGINNER

The tips in the paragraphs that follow are general advice to help any do-it-yourself BMW owner perform repairs and maintenance tasks more easily and more professionally.

### Planning ahead

To prevent getting in too deep, know what the whole job requires before starting. Read the procedure thoroughly, from beginning to end, in order to know just what to expect and what parts will have to be replaced.

### Cleanliness

Keeping things organized, neat, and clean is essential to doing a good job. When working under the hood, fender covers will protect the finish from scratches and other damage. Make sure the car finish is clean so that dirt under the cover does not scratch the finish.

Any repair job will be less troublesome if the parts are clean. For cleaning old parts, there are many solvents and parts cleaners commercially available.

For cleaning parts prior to assembly, commercially available aerosol cans of parts cleaner or brake cleaner are handy to use, and the cleaner will evaporate completely.

> ***WARNING —***
> *Most solvents used for cleaning parts are highly flammable as well as toxic, especially in aerosol form. Use with extreme care. Do not smoke. Do not use these products indoors or near any source of heat, spark or flame.*

## Non-reusable fasteners

Many fasteners used on the cars covered by this manual must be replaced with new ones once they are removed. These include but are not limited to: bolts, nuts (self-locking, nylock, etc.), cotter pins, studs, brake fittings, roll pins, clips and washers. Genuine BMW parts should be the only replacement parts used for this purpose.

Some bolts are designed to stretch during assembly and are permanently altered rendering them unreliable once removed. These are known as torque-to-yield fasteners. Always replace fasteners where instructed to do so. Failure to replace these fasteners could cause vehicle damage and personal injury. See an authorized BMW dealer for applications and ordering information.

## Tightening fasteners

When tightening the bolts or nuts that attach a component, it is always good practice to tighten the bolts gradually and evenly to avoid misalignment or over stressing any one portion of the component. For components sealed with gaskets, this method helps to ensure that the gasket will seal properly.

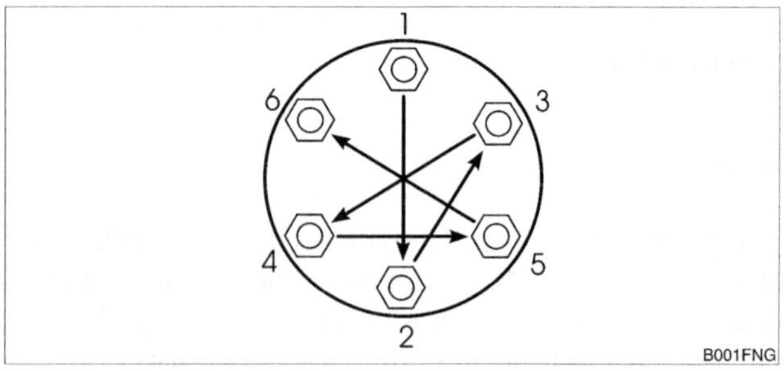


Where there are several fasteners, tighten them in a sequence alternating between opposite sides of the component. Repeat the sequence until all the bolts are evenly tightened to the proper specification.

For some repairs a specific tightening sequence is necessary, or a particular order of assembly is required. Such special conditions are noted in the text, and the necessary sequence is described or illustrated. Where no specific torque is listed, **Table a** can be used as a general guide for tightening fasteners.

> ***WARNING —***
> ***Table a** is a general reference only. The values listed in the table are not intended to be used as a substitute for torques specifically called out in the text.*

***NOTE—***
- *Metric bolt classes or grades are marked on the bolt head.*
- *Do not confuse wrench size with bolt diameter. For a listing of the common wrenches used on various bolt diameters, see* **Basic tool requirements** *in this repair group.*

**Table a. General bolt tightening torques in Nm (max. permissible)**

| Bolt diameter | Bolt Class (according to DIN 267) | | | | | |
|---|---|---|---|---|---|---|
| | **5.6** | **5.8** | **6.8** | **8.8** | **10.9** | **12.9** |
| **M5** | 2.5 | 3.5 | 4.5 | 6 | 8 | 10 |
| **M6** | 4.5 | 6 | 7.5 | 10 | 14 | 17 |
| **M8** | 11 | 15 | 18 | 24 | 34 | 40 |
| **M10** | 23 | 30 | 36 | 47 | 66 | 79 |
| **M12** | 39 | 52 | 62 | 82 | 115 | 140 |
| **M14** | 62 | 82 | 98 | 130 | 180 | 220 |
| **M16** | 94 | 126 | 150 | 200 | 280 | 340 |
| **M18** | 130 | 174 | 210 | 280 | 390 | 470 |

## Gaskets and seals

Gaskets are designed to crush and become thinner as the mating parts are bolted together. Once a gasket has been used, it is no longer capable of making as good a seal as when new, and is much more likely to leak. For this reason, gaskets should not be reused.

Some gaskets—such as head gaskets—are directional. Make sure that these are installed correctly. This same logic applies to any part used for sealing, including rubber O-rings and copper sealing washers.

In places where a shaft must pass through a housing, flexible lip seals are used to keep the lubricating oil or grease from leaking out past the rotating shaft. Seals should never be re-used once they have been removed. When removing a seal, be careful not to damage the metal surfaces.

The key to seal installation is to get the seal in straight without damaging it. Use a seal driver that is the same diameter as the seal housing to gently and evenly install into place. If a proper size seal driver is not available, a socket of the right size will do.

Some seals are directional and special installation instructions apply. Make sure a seal is installed with the lip facing the correct way. Note the installation direction of the old seal before removing it.

## Electrical testing

Many electrical problems can be understood and solved with only a little fundamental knowledge of how electrical circuits function.

Electric current only flows in a complete circuit. To operate, every electrical device in the car requires a complete circuit including a voltage source and a path to ground. The positive (+) side of the battery is the original voltage source, and ground is any return path to the negative (–) side of the battery, whether through the wiring harness or the car body. Except for portions of the charging system, all electrical current in the car is direct current (DC) and flows from positive (+) to negative (–).

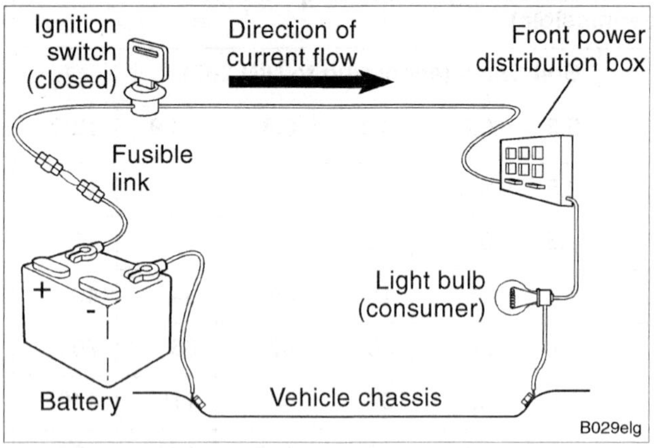


Switches are used to turn components on or off by completing or interrupting the circuit. A switch is "open" when the circuit is interrupted, and "closed" when the circuit is completed. See **600 Electrical System–General** for electrical troubleshooting.

## Wire repairs

Repairs to a wiring harness require special care to make the repair permanent. The wire ends must be clean. If frayed or otherwise damaged, cut off the end. If the wire is too short, splice in a new piece of wire of the same size and make two connections.

Use connectors that are designed for the purpose. Crimped-on or soldered-on connectors are best. Crimp connectors and special crimping pliers are widely available. If soldering, use needlenose pliers to hold the wire near the solder joint and create a "heat dam". This keeps the heat and the solder from traveling up the wire. Always use a solder made specifically for electrical work (rosin core).

***NOTE—***
*Twisting wires together to make a repair is not recommended. Corrosion and vibration will eventually spoil the connection and may lead to irreparable damage to sensitive electronic components.*

Insulate the finished connection. Electronics stores can supply heat-shrinkable insulating tubing that can be placed onto the wire before connecting, slid over the finished joint, and shrunk to a tight fit with a heat gun or hair dryer. The next best alternative is electrical tape. Make sure the wire is clean and free of solder flux or other contamination. Wrap the joint tightly to seal out moisture. See **600 Electrical System–General** for more information.

## BUYING PARTS

Many of the maintenance and repair tasks in this manual call for the installation of new parts, or the use of new gaskets and other materials when reinstalling parts. Most often, the parts that will be needed should be on hand before beginning the job. Read the introductory text and the complete procedure to determine which parts will be needed.

For some bigger jobs, partial disassembly and inspection are required to determine a complete parts list. Read the procedure carefully and, if necessary, make other arrangements to get the necessary parts while your car is disassembled.

### Genuine BMW parts

Genuine BMW replacement parts from an authorized BMW dealer are designed and manufactured to the same high standards as the original parts. They will be the correct material, manufactured to the same specifications, and guaranteed to fit and work as intended by the engineers who designed the car. Some genuine BMW parts have a limited warranty.

Many independent repair shops make a point of using genuine BMW parts, even though they may at times be more expensive. They know the value of doing the job right with the right parts. Parts from other sources can be as good, particularly if manufactured by one of BMWs original equipment suppliers, but it is often difficult to know.

BMW is constantly updating and improving their cars, often making improvements during a given model year. BMW may recommend a newer, improved part as a replacement, and your authorized dealer's parts department will know about it and provide it. The BMW parts organization is best equipped to deal with any BMW parts needs.

### Non-returnable parts

Some parts cannot be returned, even for credit. The best example is electrical parts, which are almost universally considered non-returnable. Buy electrical parts carefully, and be as sure as possible that a replacement is needed, especially for expensive parts such as electronic control units. It may be wise to let an authorized BMW dealer or other qualified shop confirm your diagnosis before replacing an expensive non-returnable part.

### Information you need to know

**Model**. When ordering parts it is important that you know the correct model designation for your car. Models covered in this manual are the Z3/M Roadster and Z3/M Coupe from 1996 to 2002 in both 4-cylinder and 6-cylinder configurations.

**Model year** is not necessarily the same as date of manufacture or date of sale. A 1999 model may have been manufactured in late 1998, and perhaps not sold until early 2000. It is still a 1999 model. Model years covered by this manual are 1996 to 2002.

**Date of manufacture** information is necessary when ordering replacement parts or determining if any of the warranty recalls are applicable to your car. The label on the driver's door below the door latch will specify the month and year that the car was built.

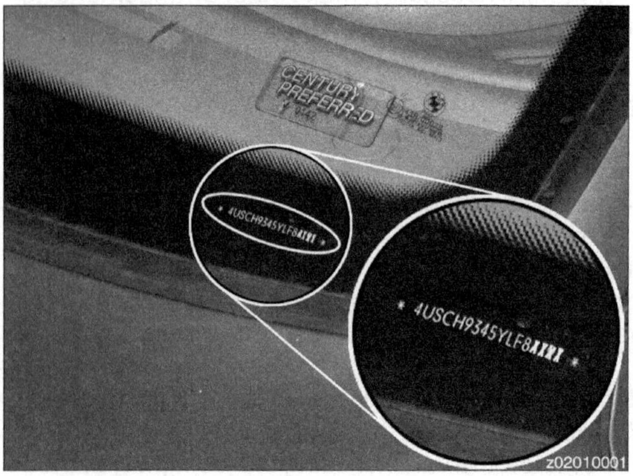


Vehicle Identification Number (VIN)** is a combination of letters and numbers that identify the particular car. The VIN appears on the state registration document, and on the car itself. One location is in the lower left corner of the windshield (**inset**).

**Engine code**. Cars covered in this manual are powered by 4-cylinder (1996 only) and several 6-cylinder engines. For information on engine codes and engine applications, see **100 Engine–General**.

**Transmission code**. The transmission type with its identifying code may be important when buying clutch parts, seals, gaskets, and other transmission-related parts. For information on transmission codes and applications, see **200 Transmission–General**.

## Service

BMW dealers are uniquely qualified to provide service for BMW cars. Their authorized relationship with the large BMW service organization means that they are constantly receiving special tools and equipment, together with the latest and most accurate repair information.

The BMW dealer's service technicians are highly trained and very capable. Authorized BMW dealers are committed to supporting the BMW product. On the other hand, there are many independent shops that specialize in BMW service and are capable of doing high quality repair work. Checking with other BMW owners for recommendations on service facilities is a good way to learn of reputable BMW shops in your area.

## Tools

Most maintenance can be accomplished with a small selection of the right tools. Tools range in quality from inexpensive junk, which may break at first use, to very expensive and well-made tools for the professional. The best tools for most do-it-yourself BMW owners lie somewhere in between.

Many reputable tool manufacturers offer good quality, moderately priced tools with a lifetime guarantee. These are your best buy. They cost a little more, but they are good quality tools that will do what is expected of them. Sears' Craftsman® line is one such source of good quality tools.

Some of the repairs covered in this manual require the use of special tools, such as a custom puller or specialized electrical test equipment. These special tools are called out in the text and can be purchased through an authorized BMW dealer. As an alternative, some special tools mentioned may be purchased from the following tool manufacturers and/or distributors:

- Assenmacher Specialty Tools
  6440 Odell Place, Boulder, CO 80301
  303-530-2424
  http://www.asttool.com
- Baum Tools Unlimited, Inc.
  P.O. Box 5867, Sarasota, FL 34277-5867
  800-848-6657
  http://www.baumtools.com
- Diagnos.co.uk Ltd
  Cuxham Road, Watlington, Oxon, OX49 5NB, UK
  +44 (0)1491 614300
  http://www.autologic-diagnos.com
- Schley Products, Inc.
  5350 E. Hunter Ave., Anaheim Hills, CA 92807
  714-693-7666
  http://www.sptool.com
- Zelenda Machine and Tool Corp.
  65-60 Austin Street, Forest Hills, NY 11374-4695
  888-892-8348, 718-896-2288
  http://www.zelenda.com

## Basic tool requirements

The basic hand tools described below can be used to accomplish most of the simple maintenance and repair tasks.

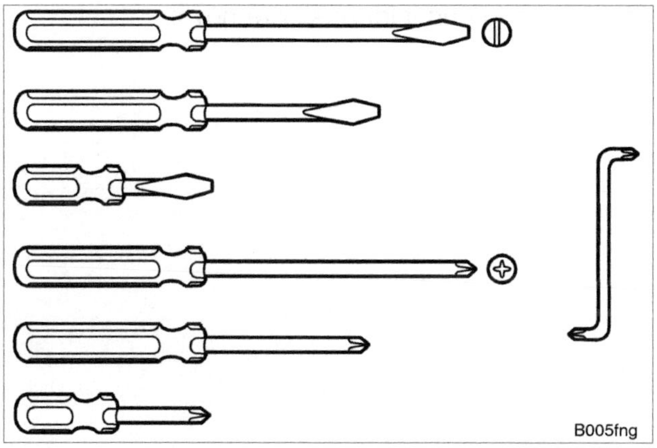
B005fng

**Screwdrivers**. The common flat-blade type and the Phillips type will handle many of the screws used on BMWs. Two or three different sizes of each type will be required, since a screwdriver of the wrong size will damage the screw head.

A complete set of screwdrivers should also include Torx® type screwdrivers.

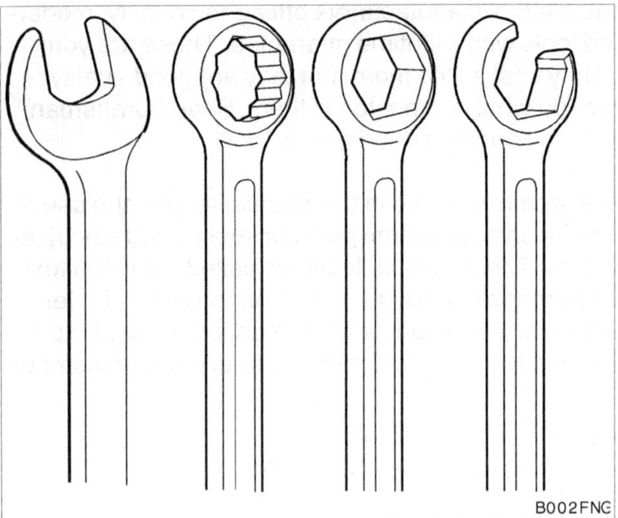
B002FNG

**Wrenches** come in different styles, including open-end, 12-point box-end, 6-point box-end and flare nut. The basic open-end wrench is the most widely used, but grips on only two sides. The box-end wrench has better grip on all six sides of a nut or bolt.

A 12-point box-end can loosen a nut or bolt where there is less room for movement, while a 6-point box-end provides better grip. For hex fasteners on fluid lines, like brake lines and fuel lines, a flare-nut wrench offers the advantages of a box-end wrench with a slot that allows it to fit over the line.

The combination wrench is the most universal. It has one open-end and one box-end. 10 mm and 13 mm wrenches are the most common sizes needed. A more complete set of wrenches would include 6 mm through 19 mm sizes.

**Sockets** are used with a ratchet handle for speed and convenience and can be combined with extensions and universal joints (swivels) to reach fasteners more easily. The most common drive sizes are ¼ in., ⅜ in., and ½ in.

Sockets come in 6-point and 12-point styles. The 6-point offers a better grip on tight nuts and bolts. 6 mm to 19 mm sockets are the most needed sizes. Below is a list of typical bolt diameters and the corresponding wrench sizes.

| Common bolt diameters and wrench sizes | |
|---|---|
| M5 | 8 mm |
| M6 | 10 mm |
| M8 | 12 mm or 13 mm |
| M10 | 17 mm |
| M12 | 19 mm |
| M14 | 22 mm |

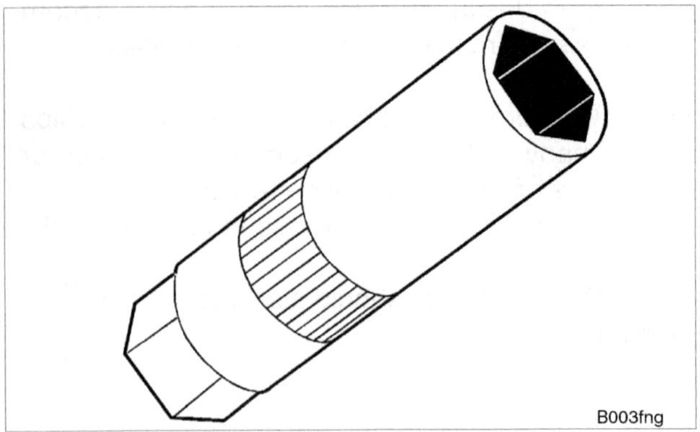
B003fng

**Spark plug socket.** A special socket for spark plugs is the correct size, is deep enough to accommodate a spark plug's length, and includes a rubber insert to both protect the spark plug from damage and grip it for easier removal.

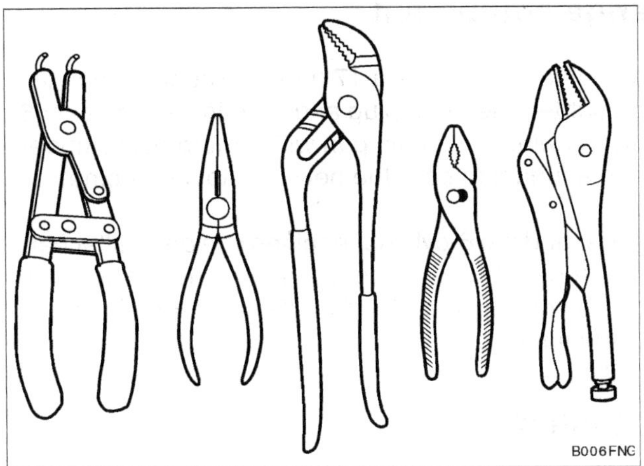
B006FNC

**Pliers**. There are many types of pliers including snap-ring, needlenose, adjustable-joint, slip-joint (Channellock®), and locking (Vise-Grip®) most of which are used for holding irregular objects, bending, or crimping. Some have special applications.

Needlenose pliers are used for gripping small and poorly accessible objects, and are useful for wiring and other electrical work. Locking pliers such as the Vise-Grip® are useful because of their tight grip.

Snap-ring and circlip pliers with special tipped jaws are used to remove and install snap-rings or circlips. Channellock® or water pump pliers have adjustable jaws that can be quickly changed to match the size of the object being held to give greater leverage.

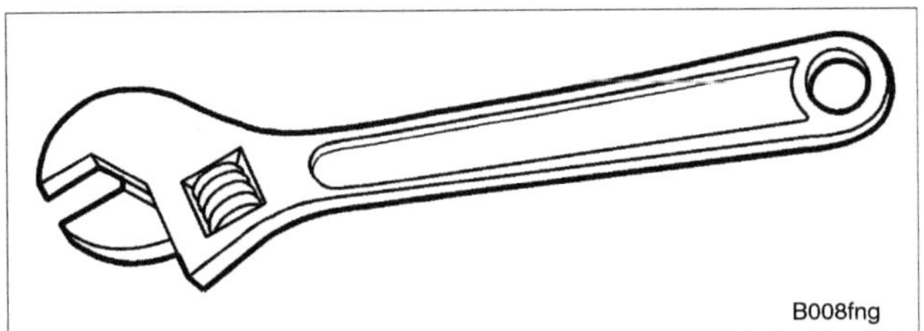
B008fng

An adjustable wrench can be a useful addition to a small tool kit. It can substitute in a pinch if two wrenches of the same size are needed to remove a nut and bolt. Use extra care with adjustable wrenches, as they tend to loosen, slip, and damage fasteners.

Compared to a wrench of the correct size, an adjustable wrench is always second best. They should only be used when the correct size wrench is not available. Choose one of average size range, about 6 to 8 inches in length.

## Jack stands

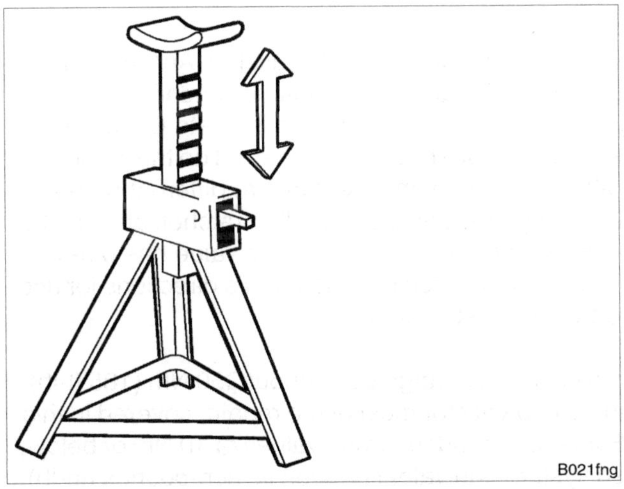
B021fng

Strong jack stands are extremely important for any work that is done under the car. Use only jack stands that are designed for the purpose. Blocks of wood, concrete, bricks, etc. are not safe or suitable substitutes.

> ***WARNING—***
> *A jack should be used only to raise the vehicle and should not be used to support the car for a long period. Always use jack stands to support a raised vehicle.*

Jack stands are available in several styles. The best ones are made of heavy material for strength, have a wide base for stability, and are equipped to positively lock in their raised positions. Get the best ones available.

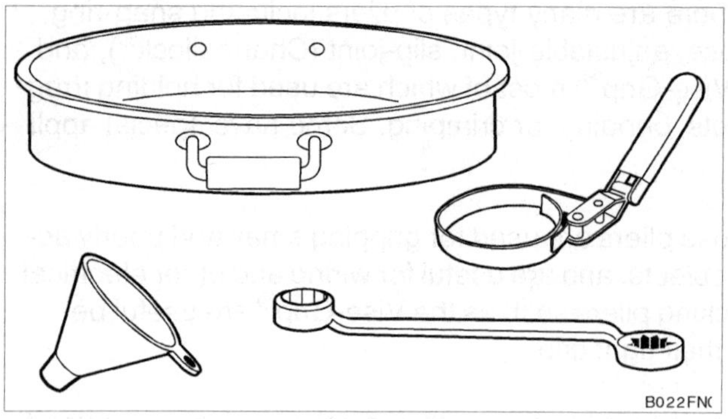


## Oil change equipment

Changing engine oil requires a 17 mm socket or wrench to loosen and tighten the drain plug and a drain pan (at least 8 qt. capacity). A wide, low drain pan will fit more easily under the car. Use a funnel to pour the new oil into the engine.

The 4-cylinder and the 6-cylinder engines use an oil filter canister with a disposable filter cartridge insert. Therefore, an oil filter wrench is not required. See **020 Maintenance** for oil change instructions.

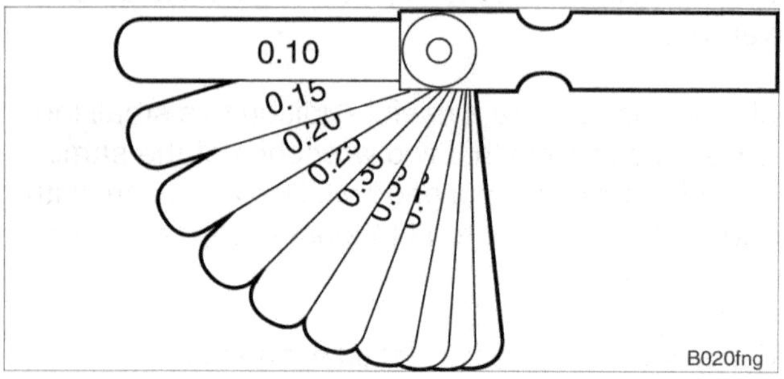


## Feeler gauges

Feeler gauges are thin metal strips of precise thickness, used to measure small clearances. They are normally available as a set, covering a range of sizes.

## Torque wrench

A torque wrench is used to precisely tighten threaded fasteners to a predetermined value. Many of the repair procedures in this manual include BMW-specified torque values in Newton-meters (Nm) and the equivalent values in foot-pounds (ft-lb).

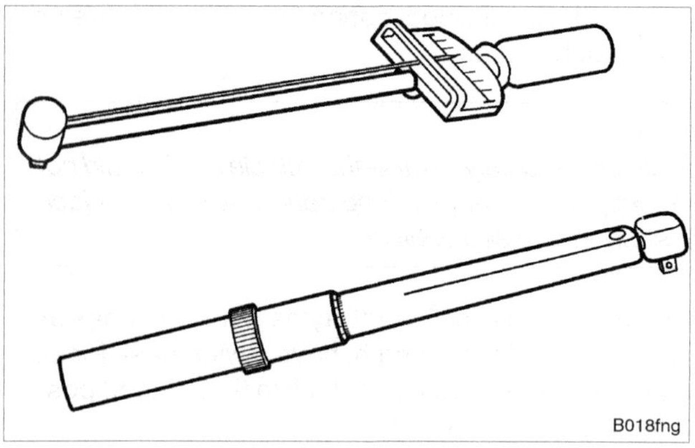


Several types of torque wrench are available. An inexpensive beam-type (top) is adequate but must be read visually. A ratchet-type (bottom) can be preset to indicate (click) when the torque value has been reached. They all do the same job, but offer different convenience features at different prices. The most convenient ones have a built-in ratchet, and can be preset to indicate when a specific torque value has been reached. Follow the wrench manufacturer's directions for use to achieve the greatest accuracy.

A torque wrench with a range up to about 250 Nm (185 ft-lb) has adequate capacity for most of the repairs covered in this manual. For recommended torque values of 10 Nm or below, the English system equivalent is given in inch-pounds (in-lb). These small values may be most easily reached using a torque wrench calibrated in inch-pounds. To convert inch-pounds to foot-pounds, divide by 12.

0012223

## Digital multimeter

< Many of the electrical tests in this manual call for the measurement of resistance (ohms) or voltage values. For safe and accurate tests of sensitive electronic components and systems, a multimeter or digital volt/ohmmeter (DVOM) with high input impedance (at least 10,000 Ω) should be used. Some meters have automotive functions such as dwell and pulse width that are useful for troubleshooting ignition and fuel injection problems.

> ***CAUTION—***
> *Vehicle electronic systems may be damaged by the high current draw of a test light with a normal incandescent bulb. As a general rule, use a high impedance digital multimeter or an LED test light for all electrical testing.*

## BMW special tools

Many repairs covered in this manual call for the use of BMW special tools. This, however, does not automatically mean that the job is too complicated or out of reach of the do-it-yourselfer.

Many of the BMW special tools mentioned in this manual are simply the best thing to use to do the job correctly. In these cases, the tool is identified with a BMW part number. See your authorized BMW dealer parts department for information on how to order special tools.

There are some jobs for which expensive special tools are essential, and not a cost-effective purchase for one-time repair by the do-it-yourself owner. This manual includes such repairs for the benefit of those with the necessary experience and access to tools. For the do-it-yourselfer, the need for special tools is noted in the text, and whether or not BMW dealer service is recommended.

# EMERGENCIES

## Changing a tire

– Stop car on as flat a surface as possible, in a place where you can be easily seen by other drivers. Avoid stopping just over the crest of a hill.

- Turn on emergency flashers and set out flares or emergency markers well behind car. Block wheel diagonally opposite to the one being changed. Passengers should get out of car and stand well away from road.

> ***WARNING —***
> *If a tire goes flat while driving, pull well off the road. Changing a tire on a busy street or highway is very dangerous. If necessary, drive a short distance on the flat tire to get to a safe place. It is much better to ruin a tire or rim than to risk being hit.*

- Take jack and tools from tool area beneath trunk mat. Remove spare tire from tire storage tray.

- Loosen wheel bolts while car is on ground, but leave them a little snug.

0011038

- ◄ Place jack in lifting point nearest wheel being changed. Use a board to provide a firm footing for jack if ground is soft. Raise car only far enough so that wheel is fully off ground and then remove wheel nuts and wheel.

- Install spare wheel. Install wheel nuts and tighten them hand tight using lug nut wrench.

- Lower car. With all wheels on ground, fully tighten nuts in a crisscross pattern. Torque wheel nuts when installing wheel. Check inflation pressure of spare tire.

| Tightening torque | |
|---|---|
| Road wheel to hub | 120 ± 10 Nm (90 ± 7 ft-lb) |

## Spare tire, removing

- Make sure car is parked on flat, level surface away from traffic and other hazards.

- Apply parking brake. Place selector lever in PARK on automatic transmission cars or engage first gear on manual transmission cars.

- Open trunk and remove mat from trunk of floor.

- Remove car jack from under trunk mat.

- Remove first aid kit from console tray in right side (as viewed from rear) of trunk. Then unscrew threaded fastener at bottom of tray and remove tray completely.

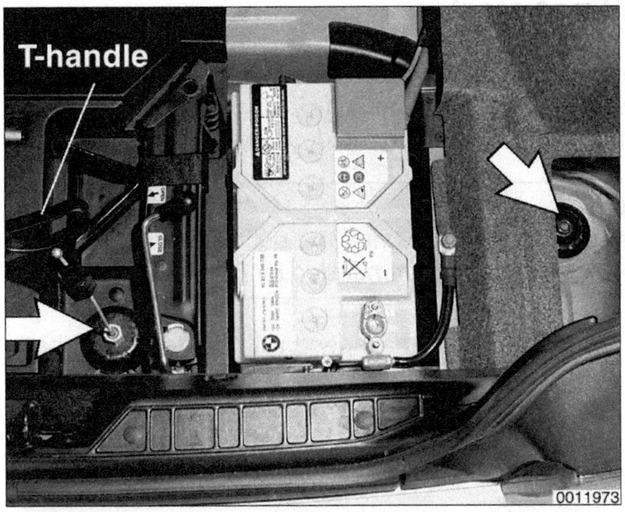


– Remove spark plug wrench from tool tray. Place wrench on nut located below console tray.

***NOTE—***
*If necessary, use the wheel centering pin from the tool tray and insert it through the end of wrench to help loosen nut.*

< Remove T-handle from tool tray. Slide T-handle down cable and onto nut at base of cable.

- Fully loosen and remove spare tire storage tray mounting nuts (**arrows**).

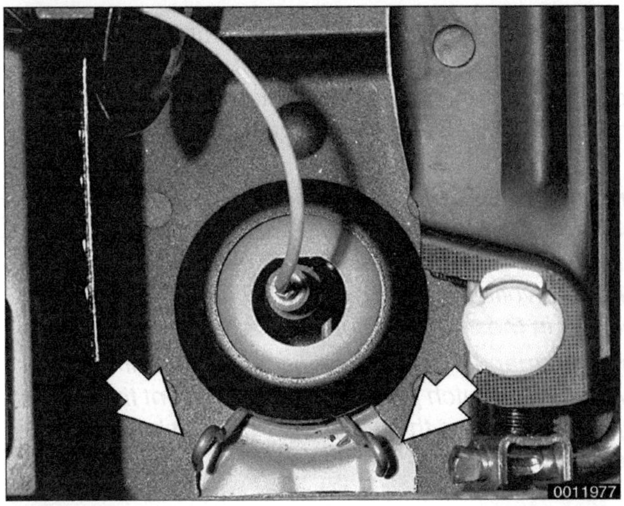


< While pulling T-handle upward, squeeze retaining clips (**arrows**) and lower spare tire to ground.

***CAUTION—***
*Do not place feet under rear of vehicle as the spare tire may drop down while being lowered.*

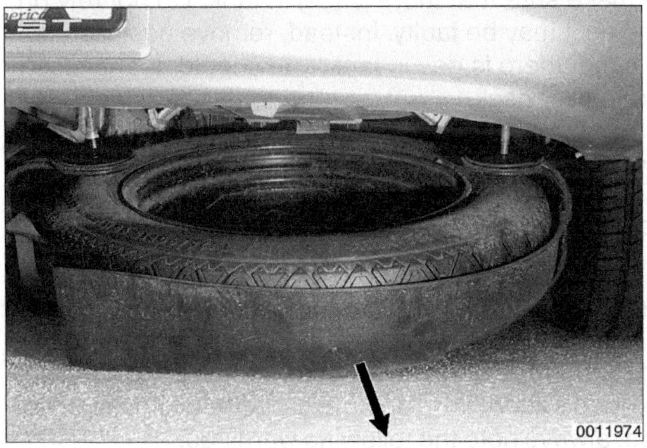


< Pull storage tray with spare tire to rear of car.

– Lift spare tire out of storage tray and unscrew valve extension. Place valve cap on spare tire.

– When remounting spare tire storage tray, pull up on T-handle until retaining clip catches. Then tighten mounting nuts.

***WARNING—***
*Do not place the defective wheel in the spare tire storage tray.*

## Car will not start

- If engine turns over slowly or not at all, especially on cold mornings, battery may not be sufficiently charged. Jump-start from another car battery as described below.

> ***CAUTION—***
> *On cars with manual transmission, push starting (or tow starting) a car is not recommended by BMW.*

- If engine turns over at normal speed with starter motor, battery and starter are fine.

- If starter will not operate see **121 Battery, Starter, Alternator** for starter troubleshooting information.

***NOTE—***
- *A factory-installed electronic immobilization system, referred to as EWS, is used in all Z3 cars. This system prevents operation of the starter if an incorrectly coded ignition key is used. See* **515 Central Locking and Anti-theft**.
- *On cars with automatic transmission, the transmission gear position switch signals EWS to prevent the engine from starting in gear positions other than PARK or NEUTRAL. If voltage is not present at starter terminal* ***50*** *with the key in the start position, check these components.*
- *On cars with manual transmissions, a starter immobilization switch at the clutch pedal is used to prevent the starter from operating unless the clutch pedal is pushed fully to the floor.*

- Check to make sure that there is fuel in tank. Do not rely on fuel gauge as it may be faulty. Instead, remove gas filler cap and rock car. If there is gas in tank, you should hear a sloshing sound at filler neck.

***NOTE—***
*See* **OBD On-Board Diagnostics** *for information on diagnostic trouble codes (DTCs).*

## Jump starting

Cars with discharged or dead batteries can be jump started using the good battery from another car. When jump starting the engine, always note the following warnings.

- Place cars close together, but do not allow them to touch each other.
  - Turn off engine of car with good battery.
  - Turn off ignition switch of car with discharged battery.

***NOTE—***
- *The positive battery post is marked with a plus (+) sign.*

- *In 1996 models the battery is located in the right rear of the engine compartment. In 1997 and later models, the battery is in the trunk.*

> ***WARNING —***
> - *Battery acid (electrolyte) can cause severe burns, and will damage the car and clothing. If electrolyte is spilled, wash the surface with large quantities of water. If it gets into eyes, flush them with water for several minutes and call a doctor.*
> - *Batteries produce explosive and noxious gasses. Keep sparks and flames away. Do not smoke near batteries.*
> - *Do not jump-start the engine if you suspect that the battery is frozen. Trapped gas may explode. Allow the battery to thaw first.*
> - *Do not quick-charge the battery (for boost starting) for longer than one minute, and do not exceed 16.5 volts at the battery with the boosting cables attached. Wait at least one minute before boosting the battery a second time.*

◄ Connect one end of positive (+) cable to positive (+) post of good battery. Remove cover from positive (+) junction post (**A**) in engine compartment and connect other end of the positive (+) cable to junction post.

– Connect one end of negative (–) cable to negative (–) battery post of good battery. Connect opposite end of negative cable (–) to ground lug (–) in engine compartment (**B**).

– Start car with good battery and run engine at about 2,000 rpm, then start car with dead battery.

– With engine at idle, switch on blower motor and rear window defogger to avoid a damaging voltage surge from alternator.

– Carefully disconnect jumper cables, starting with negative cable on ground lug. Turn all electrical consumers off.

***NOTE —***
*Run engine for at least an hour to recharge the battery.*

## Indicator and warning lights

Many of the vehicle systems are self-monitored both statically and dynamically while driving. Generally, a red warning lamp that comes on during driving should be considered serious. If you cannot immediately determine the seriousness of the warning light, stop the vehicle in a safe place and turn the engine off as soon as possible. Consult the owner's manual in the glove compartment for additional information on the warning lamp and the recommended action.

If the malfunction indicator lamp (MIL) comes on or flashes, it indicates that an emissions-related fault has occurred.

***NOTE—***
*On model years 1996 through 2000 cars, the MIL is labeled Check Engine. On model year 2001 and later cars, the MIL is labeled Service Engine Soon.*

Problems with an oxygen sensor, a fuel injector or the fuel tank evaporative control system can cause vehicle emissions to exceed specified limits. This causes the MIL to be turned on. The vehicle can be safely driven with the light on, although the emission systems should be checked as soon as possible. See **OBD On-Board Diagnostics** for more information on the MIL and the on-board diagnostic system.

## Towing

The cars covered by this manual should be towed with a tow truck using wheel lift or flat bed equipment. Do not tow the car on all four wheels except for very short distances to move it to a safe place.

***CAUTION—***
*Do not tow with sling-type equipment. The front spoilers and bumper covers may sustain damage.*

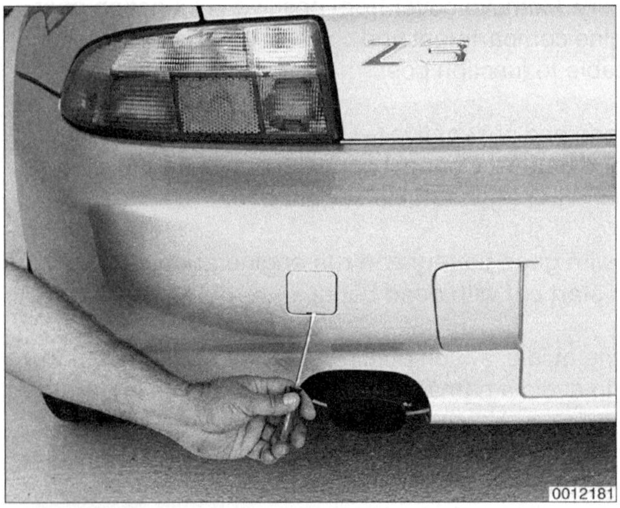

0012181

To access threaded towing eye socket, pry open trim on front or rear bumper.

***NOTE—***
*A towing eye is provided in the luggage compartment tool kit. The towing eye can be screwed into the front or rear bumper.*

0012182

Install towing eye into threaded hole.

– Standard transmission: A maximum towing distance of 20 miles is acceptable with the rear wheels on the ground and the transmission out of gear. If the car needs to be towed further, have the rear wheels placed on dollies.

– Automatic transmission: If absolutely necessary, car can be towed with rear wheels on ground, but the tow should not exceed 25 miles (40 km), at speeds at or below 30 mph (48 km/h).

- Be sure transmission fluid has been topped off before starting tow.
- Always tow car with transmission lever in N (neutral). If tow must exceed 25 miles (40 km), add one quart of ATF to transmission, or better yet, remove driveshaft.
- Be sure to drain or pump out added fluid once tow has been completed.

***NOTE—***
- *ATF draining and filling is covered in* **240 Automatic Transmission**. *Be sure to read the filling and draining procedure, as special service equipment is required to check the fluid level.*
- *Driveshaft removal is covered in* **260 Driveshaft**.

## Spare parts kit

Carrying a basic set of spare parts can prevent a minor breakdown from turning into a major annoyance. Many of the following items won't allow you to do major repair work on the car, but they will help in the event of the failure of something that can disable the car or compromise its safety.

**Spare parts kit – basic**
- Poly ribbed drive belt(s)
- Engine oil (one or two quarts)
- Engine coolant (1 gallon of premixed 50/50 BMW anti-freeze and water)
- Fuse assortment (7.5A, 10A, 15A, 20A, 30A, 50A)
- Radiator hoses (upper and lower)

**Spare parts kit – additional contents**
- Exterior lighting bulbs (headlight, brake light, turn signal, and taillight)
- Wiper blades
- Brake fluid (new unopened bottle, DOT 4 specification)
- Main relay for DME system

# 020 Maintenance





**TABLES**

## GENERAL

The information given in this repair group includes the routine checks and maintenance steps that are both required by BMW under the terms of the vehicle warranty protection and recommended by BMW to ensure long and reliable vehicle operation.

***NOTE—***
*Aside from keeping your car in the best possible condition, proper maintenance plays a role in maintaining full protection under BMW's new car warranty coverage. If in doubt about the terms and conditions of your car's warranty, an authorized BMW dealer should be able to explain them.*

### Special tools

◄ Service indicator resetting tool (Peake Research)

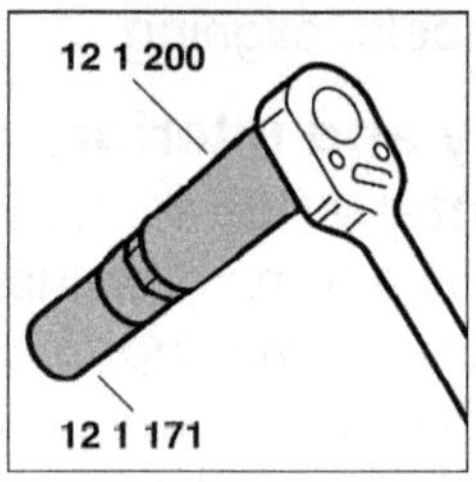


◄ Spark plug socket set (BMW special tools 12 1 171 / 12 1 200)

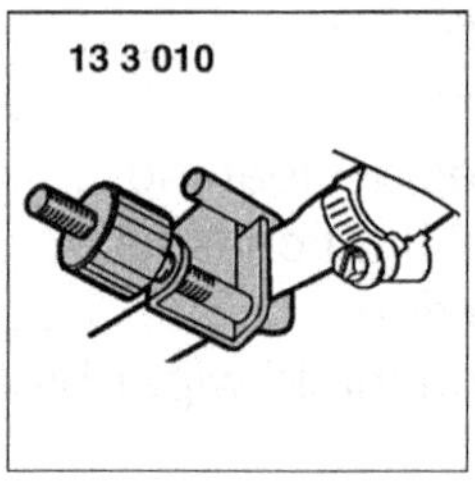


◄ Pinch clamp (BMW special tool 13 3 010)

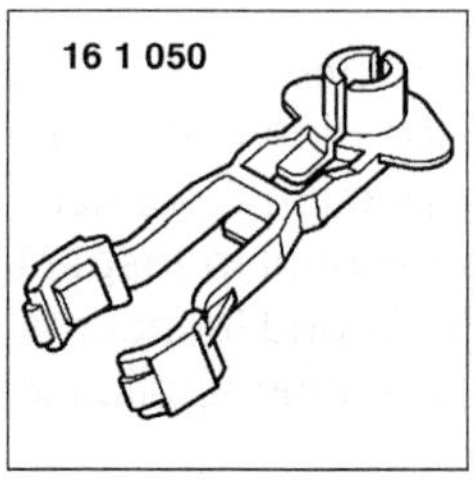


◄ Fuel line fitting tool (BMW special tool 16 1 050)

## Service Interval Indicator (SII)

***NOTE—***
*For 1999 and later Z3 cars, but not M version cars, BMW suggests an extended oil change interval of approximately 15,000 miles, depending on engine operating conditions. The introduction of BMW High Performance Synthetic Oil has made this longer interval possible.*

**1996 - 6 / 2000**. BMW's Service Interval Indicator (SII) calculates maintenance intervals based on elapsed mileage, engine speed, engine temperature, number of starts, length of trips and the amount of fuel used.

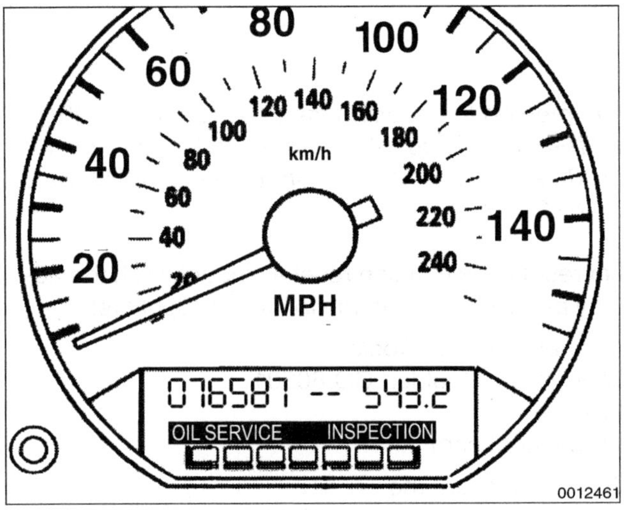


The SII consists of multiple light emitting diodes (LEDs)—typically five green, one yellow, and one red—as well as OIL SERVICE and INSPECTION indicators.

When the ignition is turned on, the green LEDs illuminate. They go out when the engine is started. After the SII is reset, all green LEDs will be illuminated with the key on. As the car is driven in normal use, fewer and fewer green LEDs will be illuminated before start-up, indicating that the next maintenance interval is approaching.

When the car has accumulated sufficient use to require the next maintenance, a yellow LED will illuminate along with either the OIL SERVICE indicator or the INSPECTION indicator. These LEDs will stay on after the engine is started. If maintenance service is delayed, the red LED will also illuminate as a reminder that maintenance service is overdue.

The SII signals the need for routine maintenance:

- Engine oil and oil filter change, after the engine has been warmed up.
- BMW-recommended additional maintenance as listed in **Table a** through **Table c**.

***NOTE—***
- *BMW oil service requirements are approximately equivalent to the maintenance that other European manufacturers specify at intervals with a maximum of every 15,000 miles or 12 months.*
- *An OIL SERVICE interval is always followed by an INSPECTION interval, which is then followed by an OIL SERVICE interval, and so on.*

After the service is performed the SII can be reset to zero.

The SII displays the mileage remaining before service is due. The type of service (OIL SERVICE or INSPECTION) is also illuminated in the display.

0011910

## Service Interval Indicator (SII), resetting

When the specified maintenance has been carried out, reset SII memory.

The SII may be reset using the BMW service and scan tool, or a specialty reset tool from an aftermarket supplier. Plug the tool (**arrow**) into the 20-pin DLC (rear engine compartment).

> ***CAUTION—***
> *Follow the manufacturer's directions when resetting the SII. If the reset procedures are done incorrectly, the reset tool or the electronic Service Interval Indicator may be damaged.*

***NOTE—***
*An alternate method for resetting the SII is to use the OBD II connector (16-pin DLC) adapter harness (BMW part no. 90 88 6 614 300). Also needed is the SII reset tool (BMW part no. 88 88 6 621 110).*

Aftermarket reset tools that can reset all models with the 20-pin DLC may be available from the following suppliers:

- Assenmacher Specialty Tools
  6440 Odell Place, Boulder, CO 80301
  303-530-2424
  http://www.asttool.com
- Baum Tools Unlimited, Inc.
  P.O. Box 5867, Sarasota, FL 34277-5867
  800-848-6657
  http://www.baumtools.com
- Peake Research, Automotive Products Division
  P.O. Box 28776, San Jose, CA 95159
  408-369-0406
  http://www.peakeresearch.com

## Inspection I and Inspection II

If the Service Interval Indicator (SII) signals the need for more comprehensive maintenance and inspection, there are two sets of inspection requirements. These inspections alternate throughout a car's maintenance history. If the last inspection interval was Inspection I, the next inspection interval (following an oil service) will be Inspection II, the next after that will be Inspection I, and so on.

Inspection I tasks are listed in **Table b**. Inspection II includes all of the tasks from Inspection I with additional Inspection II tasks. A listing of Inspection II tasks additional to Inspection I is in **Table c**.

***NOTE—***
*BMW Inspection I and Inspection II requirements are approximately equivalent to the maintenance requirements that other European manufacturers specify. Inspection I is normally due at intervals with a maximum of 30,000 miles or 24 months. Inspection II is normally due at intervals with a maximum of 60,000 miles or 48 months.*

## MAINTENANCE TABLES

The intervals for most tasks listed in the maintenance table below are determined by the SII.

Except where noted, the maintenance items listed apply to all models and model years covered by this manual. The text in the "additional repair information" column refers to repair groups in this manual.

***NOTE—***
*The Inspection II table is supplemental to the Inspection I table.*

**Table a. Oil service**

| Maintenance item | Tools required | New parts required | Warm engine required | Dealer service recommended | Additional repair information |
|---|---|---|---|---|---|
| **Engine compartment maintenance** | | | | | |
| Change oil and oil filter. | * | * | * | | 020 |
| Reset Service Interval Indicator (SII). | * | | | | 020 |
| **Under car maintenance** | | | | | |
| Check steering rack and linkages for leaks and excessive play. | | | | | 320 |
| Check overall thickness of front and rear brake pads. If replacement is necessary: Examine brake disc surfaces. Clean brake pad contact points in calipers. Grease wheel centering hubs (alloy wheels only). Check thickness of parking brake linings only when replacing rear brake pads. Check operation of parking brake and adjust as necessary. | * | | | | 340 |
| Check and adjust tire pressures, including spare. | * | | | | 020 |
| **Body and interior maintenance** | | | | | |
| Check operation of exterior and interior lights, including glove compartment light, engine compartment light, trunk light, instrument panel lights, horns, headlight flasher and dimmer switch. Check headlight and driving light aiming. | | | | | 630 |
| Check function/condition of seat belts. | | | | | 720 |
| Check windshield washer fluid level and concentration, add as necessary. Check operation of washer system. Check condition of wiper blades. | | * | | | 611 |
| **Road test** | | | | | |
| Check braking performance, steering, heating and ventilation, manual or automatic transmission, and mirrors. | | | | | |

| Table b. Inspection I service | | | | | |
|---|---|---|---|---|---|
| **Maintenance item** | **Tools required** | **New parts required** | **Warm engine required** | **Dealer service recommended** | **Additional repair information** |
| **Engine compartment maintenance** | | | | | |
| Read out diagnostic trouble codes (DTCs). | * | | | | **OBD** |
| Check and adjust valve clearance (S54 engine only). | * | * | * | * | **116** |
| Check engine coolant system/heater hose connections for leaks. Check coolant level and antifreeze protection level. Add coolant as necessary. | * | | | | **170** |
| Lubricate accelerator linkage. | | | | | **020** |
| Inspect battery, test for efficiency (1996 models only; for 1997 and later models, battery is in trunk). | * | | | | **121** |
| Check brake/clutch fluid level. | | | | | **340** |
| Check windshield washer fluid level and antifreeze protection. Add washer fluid as necessary. | * | | | | **611** |
| Reset Service Interval Indicator (SII). | * | | | | **020** |
| **Under car maintenance** | | | | | |
| Change engine oil and filter. | * | * | * | | **020** |
| Check transmission and differential for external leaks. | | | | | **230**<br>**240** |
| Check CV joint boots for damage or leaks. | | | | | **331** |
| Visually check fuel tank, fuel lines and connections for leaks. | | | | | **160** |
| Check condition, position, and mounting of exhaust system. Inspect for leaks. | | | | | **180** |
| Check power steering system for leaks. Check power steering fluid level and adjust if necessary. | | | | | **320** |
| Check steering rack and tie rods for tightness. Check condition of ball joints, steering linkage and steering shaft joints. | | | | | **320** |
| Check overall thickness of front and rear brake pads using special BMW tool. If replacement is necessary: Examine brake disc surfaces. Clean brake pad contact points in calipers. Grease wheel centering hubs (alloy wheels only). Check thickness of parking brake linings only when replacing rear brake pads. Check operation of parking brake and adjust as necessary. | * | | | | **340** |
| Check front control arm bushings for damage or wear. | | | | | **310** |
| Check brake system connections and lines for leaks, damage and incorrect positioning. | | | | | **340** |
| Check parking brake actuator. Adjust if necessary. | * | | | | **340** |

| Table b. Inspection I service | | | | | |
|---|---|---|---|---|---|
| **Maintenance item** | **Tools required** | **New parts required** | **Warm engine required** | **Dealer service recommended** | **Additional repair information** |
| Check all tire pressures (including spare) and correct if necessary. Check condition of tires (outer tread surfaces), tread wear pattern. In case of uneven tread wear, perform a wheel alignment. | * | | | | 020 |
| Inspect entire body according to terms of rust perforation limited warranty. (Must be performed at least every two years.) | | | | | |
| **Body/electrical** | | | | | |
| Check A/C operation. | | | | | 640 |
| Check operation of headlights, parking lights, back-up lights, license plate lights, interior lights, glove compartment light, engine compartment light, trunk light, turn signals, emergency flashers, stop lights, horns, headlight flasher and dimmer switch. Adjust headlight aim as necessary. | | | | | 630 |
| Check instrument panel and dashboard lights. | | | | | 620 |
| Check wipers and windshield washer system. Check aim of washer jets and adjust if necessary. | | | | | 611 |
| Check condition and function of seat belts. | | | | | 720 |
| Visually examine all airbag units for torn covers, obvious damage or attachment of decals, decorations or accessories. | | | | | 721 |
| Check central locking system and double lock. Tighten door and door lock striker fasteners. | | | | | 515 |
| Replace batteries for alarm remote controls in all master keys (if applicable). | * | * | | | 515 |
| Check operation and condition of all door, hood and trunk latches. Lubricate hinges and latches. | | | | | 410<br>515 |
| Check A/C blower and rear window defroster operation. | | | | | 640 |
| Activate automatic roll-over protection system via diagnostic link. Note: first remove hardtop or lower convertible top. | * | | | * | 541 |
| Check all warning/indicator lights, Check Control. | | | | | 620 |
| Check operation of rear view mirrors. | | | | | |
| **Road test** | | | | | |
| Check braking performance, steering, heating, A/C operation. Check manual transmission, clutch operation or automatic transmission operation. | | | | | |

**Table c. Inspection II service** (includes all inspection items listed in **Inspection I**)

| **Maintenance item** | **Tools required** | **New parts required** | **Warm engine required** | **Dealer service recommended** | **Additional repair information** |
|---|---|---|---|---|---|
| Inspect drive belt condition and tension. | * | * | | | **020** |
| Inspect body for anticorrosion warranty (every year). | * | | | | |
| Replace brake fluid every 2 years (time interval begins from vehicle production date). | * | * | | * | **340** |
| Z3 models: Replace coolant every 4 years (time interval begins from vehicle production date).<br>M version: Replace coolant every 3 years (time interval begins from vehicle production date). | * | * | | | **170** |
| Replace oxygen sensor<br>1997 - 1998 every 100,000 miles<br>1999-2002 every 120,000 miles | * | * | | | **180** |
| Replace spark plugs<br>1997- 1998, all M models - every 60,000 miles<br>1999-2002 - every 100,000 miles | * | * | | | **020** |
| Change transmission fluid. | * | * | * | * | **230**<br>**240** |
| Check clutch plate for wear. | | | | | **210** |
| Replace fuel filter. | * | * | | | **020** |
| Replace air filter element. | | * | | | **020** |
| Inspect front and rear suspension, including front and rear wheel bearings. | | | | | **310**<br>**330** |
| Tighten A/C compressor mounting bolts. | * | | | | **640** |

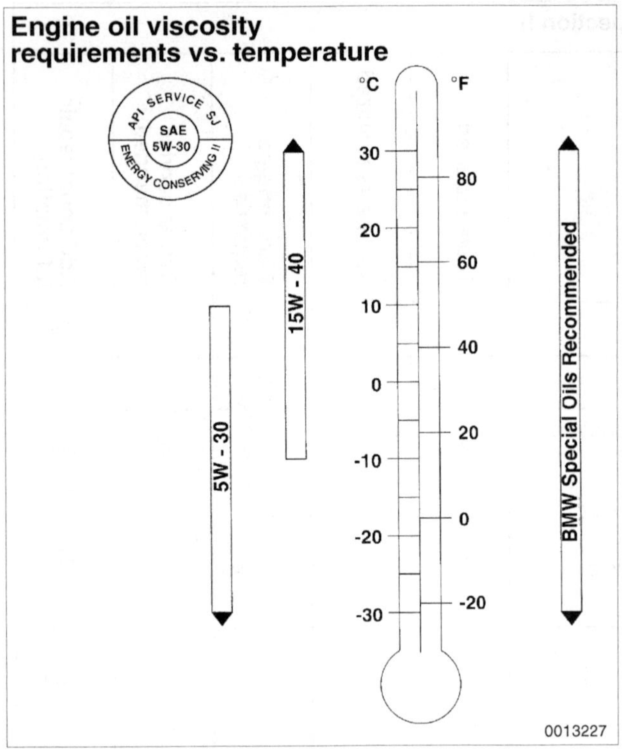


## FLUIDS AND LUBRICANTS

The illustration shows engine oil viscosity (SAE grade) vs. operating temperature range for the BMW engines covered in this manual.

Fluid and lubricant capacities and specifications for Z3 cars are listed in **Table d**.

> ***WARNING —***
> *The use of fluids that do not meet BMW specifications may impair performance and reliability, and may void warranty coverage.*

**Table d. Fluids and lubricants**

| **Fluid** | **Approximate capacity** | **Specification** |
|---|---|---|
| **Engine oil with filter change** | | |
| Z3 Coupe / Roadster<br>• 4-cylinder engine<br>• 6-cylinder engine: M52, M52 TU or M54 engine | <br>5.0 liters (5.30 US qt)<br>6.5 liters (6.9 US qt) | <br>Synthetic oil with<br>API rating SH or higher |
| M Coupe / Roadster<br>• 1998 - 2000: S52 engine<br>• 2001 - 2002: S54 engine | <br>6.0 liters (6.3 US qt)<br>5.5 liters (5.8 US qt) | <br>API SH/CD<br>API SH/CE |
| **Manual transmission oil** | | |
| Getrag S5D 250G<br>ZF S5D 320Z | 1.1 liters (1.15 US qt)<br>1.3 liters (1.37 US qt) | Automatic transmission fluid (ATF). See **230 Manual Transmission.**<br>BMW 83 22 9 408 942 (MTF-LT-1) |
| **Automatic transmission fluid (ATF)** (drain and fill) (additional fluid required when installing a dry torque converter) | | |
| GM A4S 270R / 310R<br>GM A5S 360R<br>GM A5S 390R | 3.3 liters (3.5 US qt)<br>4.0 liters (4.2 US qt)<br>4.0 liters (4.2 US qt) | Dexron III® ATF<br>BMW 83 22 9 407 807<br>(Exxon LT 71141) BMW 83 22 0 024 359<br>(Texaco ETL 8072B) |
| **Differential** | | |
| Z3 Coupe / Roadster<br>• 4-cylinder engine<br>• 6-cylinder engine | <br>1.1 liters (1.2 US qt)<br>1.7 liters (1.8 US qt) | BMW SAF-XO<br>(w/o limited slip)<br>BMW SAF-XJ (with limited slip) |
| M Coupe / Roadster | 1.4 liter (1.5 US qt.) | Hypoid oil (check with BMW dealer) |
| **Power steering fluid** | | |
| All models | Sealed (no drain plug) | Dexron III® ATF |
| **Brake fluid** | | |
| All models | - | SAE DOT4 low viscosity |
| **Engine coolant** | | |
| 4-cylinder engine<br>6-cylinder engine | 6.6 liters (6.9 US qt)<br>10.5 liters (11.0 US qt) | 50% BMW antifreeze/ 50% distilled water |

## Brake fluid

Brake fluid absorbs moisture easily, and moisture in the fluid affects brake performance and reliability. This is why brake fluid should be flushed from the system every two years. When replacing or adding brake fluid, use only new fluid from previously unopened container. Do not use brake fluid that has been bled from the system, even if it is brand new. Use only DOT 4 brake fluid.

***NOTE—***
*See* **340 Brakes** *for more brake fluid information.*

## Cooling system

For engine coolant, BMW recommends a 50/50 mixture of distilled water and phosphate/nitrate free antifreeze containing ethylene glycol. Antifreeze raises the boiling point and lowers the freezing point of the coolant. It also contains additives that help prevent cooling system corrosion.

Model with translucent expansion tank:

- Always check coolant level with engine cold.
- Coolant level should be at maximum mark on expansion tank.

Model with built in coolant level sensor:

- Remove radiator cap with engine cold.
- When upper mark on float is level with top of filler neck, coolant is at minimum allowable level.
- When lower mark on float is level with top of filler neck, coolant is at maximum level.

Routine cooling system maintenance consists of maintaining the coolant level and inspecting hoses. Because the coolant's anticorrosion and antifreeze additives gradually lose their effectiveness, replacement of the coolant every four years is recommended.

Make sure hose connections are tight and dry. Coolant seepage indicates either that the hose clamp is loose, that the hose is damaged, or that the connection is dirty or corroded. Dried coolant has a chalky appearance. Hoses should be firm and springy. Replace any hose that is cracked, that has become soft and limp, or has been contaminated by oil.

> ***CAUTION—***
> *Use only BMW approved phosphate-free antifreeze when filling the cooling system. Use of antifreeze containing phosphates is considered to be harmful to the cooling system.*

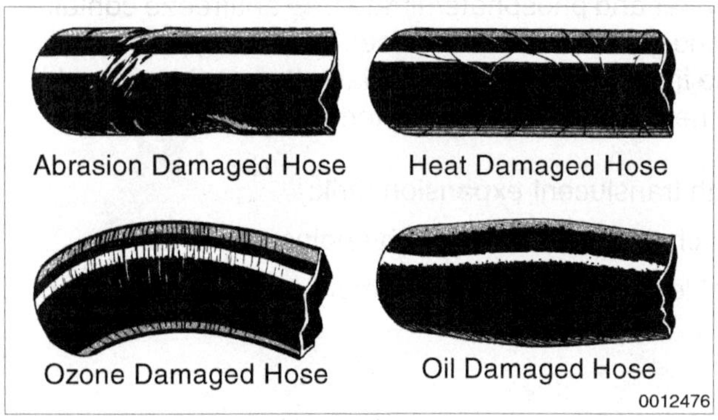


As a preventive measure, replace cooling system hoses every four years. The illustration shows examples of damage to coolant hoses. Any of conditions shown is cause for replacement. *(Courtesy of Gates Rubber Company, Inc.)*

> ***WARNING —***
> *Always allow the engine to cool down before attempting inspection or repairs of the cooling system. Hot coolant scalds.*

## Differential gear oil

Z3 differentials for 1996 - 1997 and M version differentials for 1998 - 2002 are filled or topped up with Hypoid oil.

For Z3 differentials from 1998 on, BMW recommends using only specially formulated synthetic "lifetime" gear oils (SAF-XO), available through an authorized BMW dealer parts department.

For additional information on lubricants that may be compatible, contact an authorized BMW dealer service department.

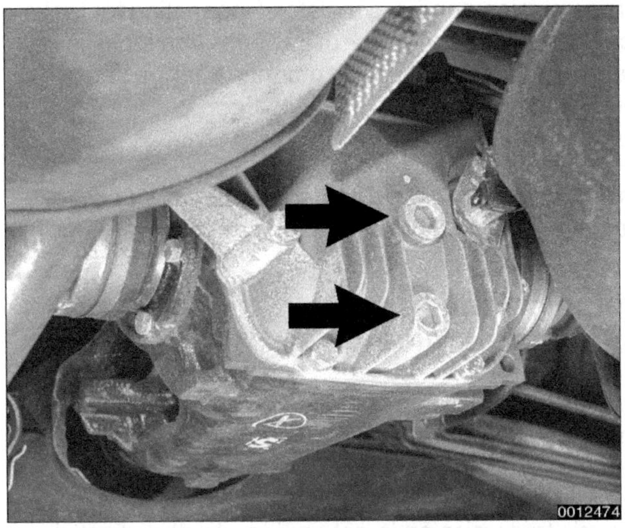


◄ Check differential oil level at filler plug (**upper arrow**).

– Check lubricant level with car on a level surface:
  - Remove oil filler plug.
  - Level is correct when fluid just reaches edge of filler hole.
  - If necessary, top up fluid.
  - Install and tighten oil filler plug when oil level is correct.

*NOTE—*
- *Use a 14 mm or 17 mm Allen socket to remove the fill or drain plug.*
- *Make sure the car is level when checking fluid.*

| Tightening torque | |
|---|---|
| Rear differential filler plug to housing | 70 Nm (52 ft-lb) |

## Power steering fluid

The power steering fluid is Dexron III® ATF, or equivalent. The system is permanently filled and does not have a drain. Routinely adding ATF is not required unless the system is leaking.

◄ To check power steering fluid level in fluid reservoir:
- Park car on level ground with engine off.
- Level is correct if it is between **MIN** and **MAX** marks on dipstick.
- If level is below **MIN** mark, start engine and add fluid to reservoir to bring level up.
- Stop engine and recheck level.
- Hand-tighten reservoir cap.

| Power steering fluid | |
|---|---|
| Recommended fluid | Dexron III® ATF |

## Transmission fluid (manual transmission)

The manual transmission in Z3 models is filled with lifetime oil that ordinarily does not need to be changed.

> *CAUTION—*
> *Do not use multiviscosity engine oils in the manual transmission. Use of such an oil could shorten the service life of the transmission.*

◄ Check manual transmission oil level at transmission filler plug (**arrow**). Make sure car is on level surface.

Manual transmission service consists of inspecting for leaks and checking the fluid. Evidence of transmission leaks is likely to be seen around the driveshaft mounting flange and at the bottom of the bellhousing.

***NOTE—***
*Transmission fluid level checking and replacement procedures are covered in* **230 Manual Transmission.**

### Transmission fluid (automatic transmission)

The automatic transmissions installed in the Z3 models are filled with special fluids, depending on transmission type and model year.

The automatic transmission is not equipped with a dipstick. Checking the ATF level is a complicated procedure which includes measuring and maintaining a specified ATF temperature during the checking procedure.

***NOTE—***
- *The transmission lubricant type can be found on the "typeplate" on the side of the transmission. See* **200 Transmission–General** *and* **240 Automatic Transmission**.
- *Consult an authorized BMW dealer for alternate fluid use and the most-up-to-date information regarding transmission operating fluids.*
- *For more complete ATF service information, including checking ATF level and ATF filter replacement procedures, see* **240 Automatic Transmission**.

## ENGINE OIL SERVICE

BMW recommends the use of BMW High Performance Synthetic Oil in most Roadster and Coupe applications.

A special oil is needed for 2001 - 2002 M Roadster / Coupe models with the S54 engine. For those models BMW recommends Castrol RS Synthetic Oil (also called Castrol TWS Motorsport Synthetic Oil).

| **Oil recommendation** | |
|---|---|
| 2001 - 2002 M Roadster / Coupe (S54 engine) | 10W-60 |
| All other models | 5W-30 |

***NOTE—***
- *Do not use engine oil additives when using BMW High Performance Synthetic engine oil.*
- *The information contained here is as accurate as possible at the time of publication. Consult an authorized BMW dealer for the latest information on factory-recommended maintenance.*

## Engine oil, checking level

◄ Engine oil level is checked with a dipstick (**arrow**) in engine block.

- Check oil level with car on a level surface, after engine has been stopped for at least a few minutes.
- Check level by pulling out dipstick and wiping it clean. Re-insert it all the way and withdraw it again.
- Oil level is correct if it is between two marks near end of stick.

◄ Add oil through filler cap (**arrow**) on top of cylinder head. Add only amount needed to bring oil level to **MAX** mark on dipstick, using an oil of correct viscosity and grade. Too much oil can be just as harmful as too little.

## Engine oil and filter, changing

A complete oil change requires new oil, a new oil filter insert kit, and a new drain plug sealing washer. The tools needed, a 17 mm drain plug socket or box wrench and a drain pan (8 - 10 US qt. capacity), are described in **010 General**.

> ***CAUTION—***
> *If using a "fast-lube" service facility for oil changes, make sure the technician hand-starts and torques the engine oil drain plug using hand-tools. Power tools can strip the threads of the plug and/or the oil pan.*

– Run engine for a few minutes to warm engine oil. Shut engine off.

– With car on level ground, place drain pan under oil drain plug.

◄ Remove filter housing cover:

- Loosen oil filter housing cover (**arrow**) and allow oil to drain back into oil pan.
- Remove cover and filter cartridge and discard any O-rings.

***NOTE—***
*Observe location of sealing O-rings.*

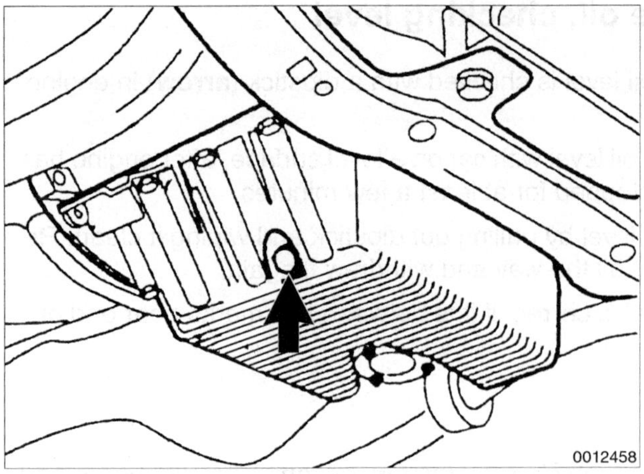


Using a socket or box wrench, loosen drain plug at oil drain pan. Remove plug by hand and let oil drain into pan.

> ***CAUTION—***
> *Pull the loose plug away from the hole quickly to avoid being scalded by hot oil. It will run out quickly when the plug is removed. If possible, use gloves to protect your hands.*

– When oil flow has diminished to an occasional drip, reinstall drain plug with a new metal sealing washer and torque plug.

| Tightening torque | |
|---|---|
| Engine oil drain plug (M12 bolt) | 25 Nm (18 ft-lb) |

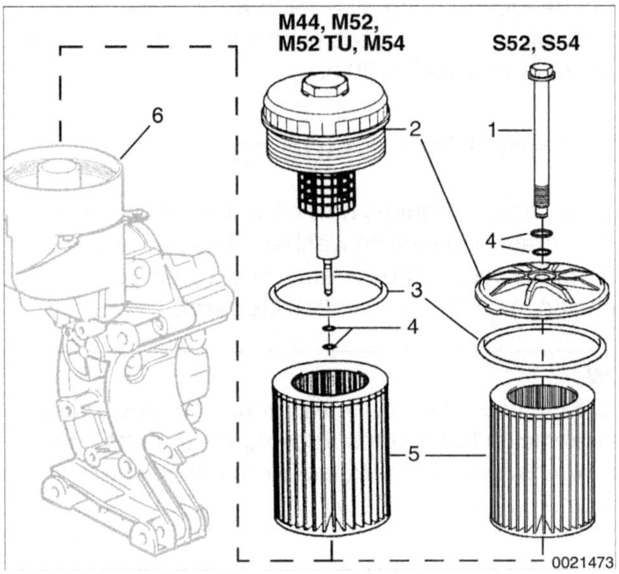


Oil filter components:

1. Filter cover bolt (M10) (tighten to 33 Nm (24 ft-lb))
2. Filter housing cover (tighten to 25 Nm (18 ft-lb))
3. O-ring (always replace)
4. O-rings (always replace)
5. Filter element
6. Filter housing

– Working at oil filter housing:
  - Lubricate new sealing O-rings and mount on cover.
  - Install new filter element.
  - Reinstall housing cover and tighten.

| Tightening torques | |
|---|---|
| Cover bolt to filter housing | 33 Nm (24 ft-lb) |
| Cover to filter housing | 25 Nm (18 ft-lb) |

– Refill crankcase with oil. Approximate oil capacity is listed in **Table d**. Use dipstick to check correct oil level.

– Start engine and check that oil pressure warning light immediately goes out.

– Allow engine to run for a few minutes to circulate new oil, then check for leaks at drain plug and oil filter. Stop engine and recheck oil level.

## ENGINE COMPARTMENT MAINTENANCE

The information under this heading describes routine maintenance, other than oil change, done in the engine compartment. Information on oil change is given earlier in this group.

To perform these operations, it is not necessary for the car to be raised and supported off the ground.

## Accelerator linkage

The accelerator and throttle linkage should be lubricated periodically. Use a general purpose oil on the joints and bearings of the linkage. Use a multipurpose grease on the bearing points of the throttle plate.

## Air filter, replacing

The specified replacement intervals for the air filter are based on normal use. If the car is operated primarily in dusty conditions, the air filter should be serviced more frequently.

< 4-cylinder engine:

- Separate upper air filter housing section from lower section by unfastening spring clips.
- Remove air filter element from housing.
- Wipe inside of air filter housing clean and install new element.
- Reinstall upper air filter housing, making sure that two halves of filter housing are mated correctly. Refasten spring-clips.

< 6-cylinder engine:

- Working at air intake housing, release clips on air filter sliding tray.
- Pull out tray and remove filter element.
- Replace filter element and slide tray back into place.

## Drive belt service

Drive belts and pulleys transfer power from the engine crankshaft to various accessories. Depending on the model and year, Z3 cars use a variety of drive belt configurations:

- The front belt only turns the A/C compressor.
- The rear, longer belt activates the alternator, coolant pump and power steering.

Belt tensioners are used to keep the belts tensioned properly. Unless a tensioner mechanism malfunctions, the ribbed belts do not require tension adjustment. On the other hand, V-belts, used in earlier models, require more frequent attention and, if necessary, tensioning or replacement.

Inspect drive belts with the engine off. If a belt shows signs of wear, cracking, glazing, or missing sections, it should be replaced immediately. To reduce the chance of belt failure while driving, replacement of the belts every four years is recommended.

***NOTE—***
- *Engine compartment layout varies. It may be necessary to remove the air filter housing, radiator shroud or engine cooling fan to gain access to the drive belts.*
- *Always mark a belt with direction of rotation, so that it can be reinstalled in its original orientation.*
- *When belts are replaced with new ones, store the old set in the luggage compartment for emergency use.*

## Drive belts, removing and installing (4-cylinder engine)

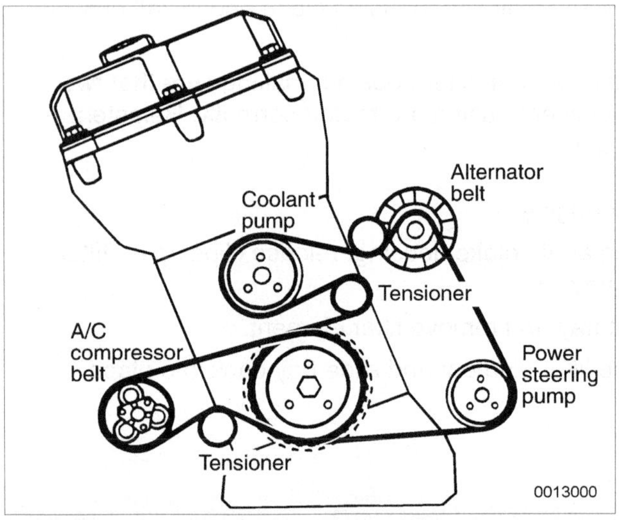


M44 engine (4-cylinder):
- A/C compressor belt is V-belt style.
- Alternator belt is ribbed style.

– Remove splash shield from under engine.

Use screwdriver to pry off A/C belt tensioner trim cap.

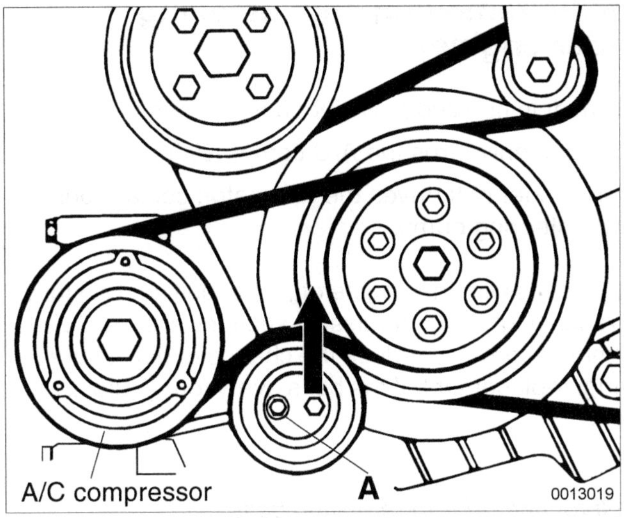


Loosen A/C belt tensioner bolt (**A**) under trim cap, then remove belt from A/C compressor.

- When reinstalling, use long-handled wrench to rotate tensioner counterclockwise (direction of arrow) until proper V-belt tension is attained.

Release alternator drive belt:

- Use thin screwdriver to pry off tensioner cover.
- Use socket wrench and handle to lever spring-loaded tensioner counterclockwise.
- Slip belt off pulleys.

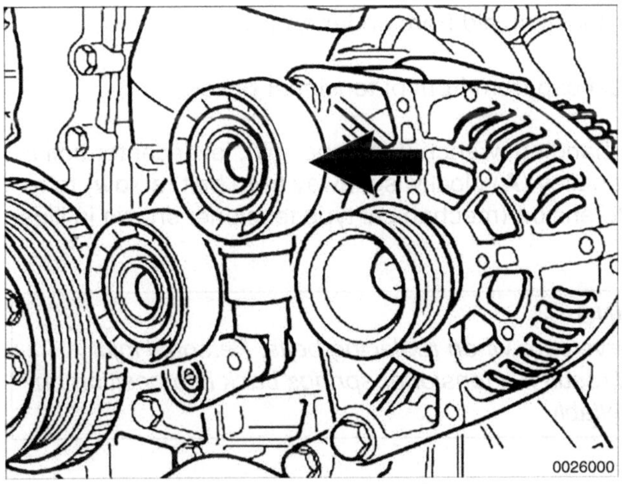


If there is an additional idler pulley (**arrow**) that the alternator belt passes over, be sure to note the routing of the belt.

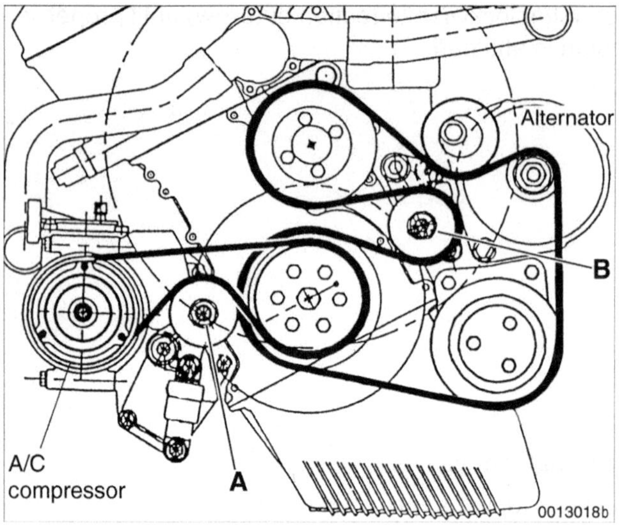


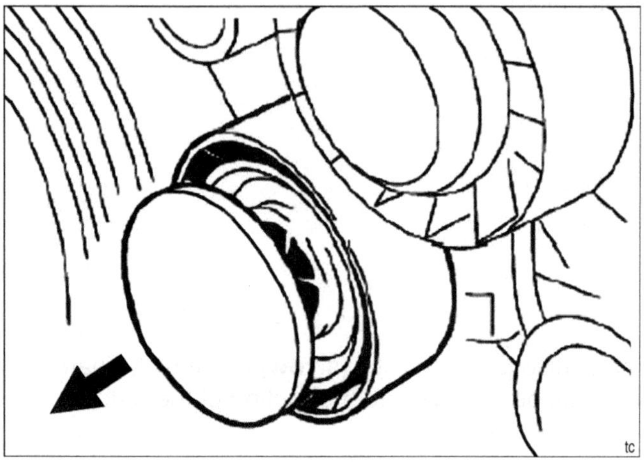


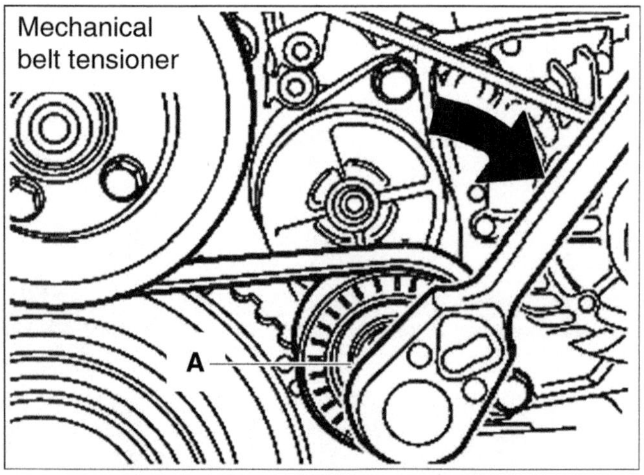


## Drive belts, removing and installing (6-cylinder engine)

6-cylinder models use two poly-ribbed (serpentine) belts:

- The outer belt only turns the A/C compressor.
- The inner, longer belt drives the alternator, coolant pump and power steering pump.

Inspect drive belts with the engine off. If the belt shows signs of wear, cracking, glazing, or missing sections, it should be replaced immediately. To reduce the chance of belt failure while driving, replacement of the belts every four years is recommended.

***NOTE—***
- *Note belt routing prior to removing belts.*
- *When belts are replaced with new ones, keep the old set for emergency use.*

**To remove drive belts**:

To remove the inner (alternator) belt, the outer (A/C compressor) belt must first be removed.

Two types of automatic belt tensioners are used to keep the belts tensioned properly; a hydraulic version and a mechanical version.The procedure for releasing the tension on the belt is similar for both types.

***CAUTION—***
- *Mark drive belt rotation direction if removing and reusing an old belt.*

– Remove splash shield from under engine.

Pry off dust cap from front of tensioner pulley.

Use long-handled wrench (8 mm hex bit), lever center bolt in tensioner pulley (**A**) clockwise, away from belt (**arrow**) to release belt tension (mechanical type tensioner shown in illustration).

***WARNING—***
- *Observe care when replacing belts. Personal injury could result if a tensioner springs back into position uncontrollably.*

***NOTE—***
- *Mechanical belt tensioners have a cast hex boss on the body of the tensioner. Alternatively, use a 17 mm wrench on the hex boss to release belt tension.*

– With tension released, slide belt off pulleys.

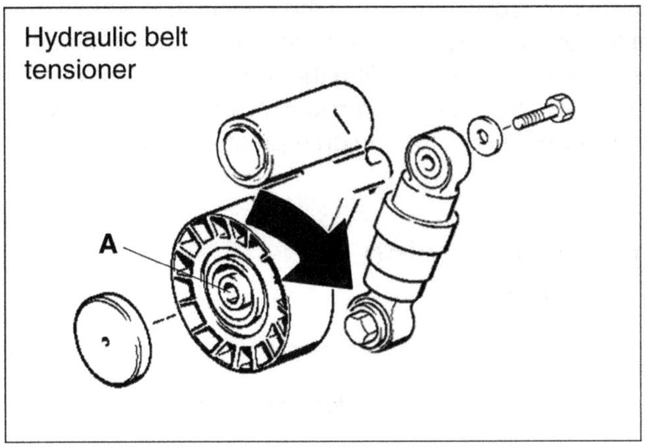


## Idle speed

Engine idle speed can change due to a number of factors, including normal wear. The idle speed in Z3 models is electronically adaptive and non-adjustable.

The S54 engine in 2001-2002 M Coupe / Roadster models uses six individual throttle housings connected via a link rod. Individual idle stop screws on each housing are factory sealed to prevent tampering. Idle speed adjustment is rarely necessary.

No repairs can be carried out to the throttle intake manifold/throttle assembly unless a dedicated BMW scan tool or a compatible aftermarket tool is used to access fault codes. For additional information on engine management and idle problems, see **130 Fuel Injection**. Also see **On-Board Diagnostics** at rear of manual for details of electronic fault finding.

## Oxygen sensors

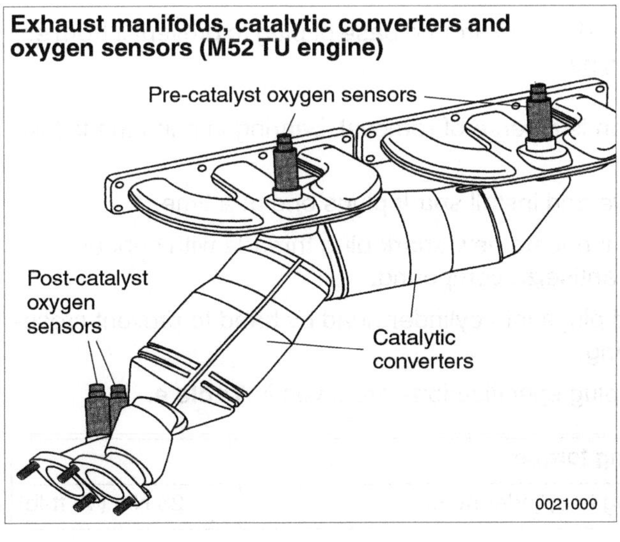


The engine management system in Z3 models is equipped with multiple oxygen sensors. A regulating sensor is mounted before each catalytic converter and a monitoring sensor downstream of each converter. The regulating sensor monitors engine combustion efficiency and helps to control the fuel injection system and exhaust emissions. The monitoring sensor is used by the on-board diagnostic (OBD II) system to monitor the function of the catalytic converter.

OBD II enhanced emission standards require the engine control module (ECM) to monitor the oxygen content in the exhaust both before and after the catalytic converter. This allows for tighter control of tailpipe emissions and also allows the ECM to diagnose converter problems. If the ECM detects that catalytic converter or oxygen sensor efficiency has degraded past a certain pre-programmed limit, it will turn on the malfunction indicator light (MIL), and store a diagnostic trouble code (DTC) in the ECM. See **OBD On-Board Diagnostics**.

Replacement of oxygen sensors at the specified intervals ensures that the engine and emission control system continue to operate as designed. Extending the replacement interval may void the emission control warranty coverage. See **180 Exhaust System** for information on replacing the oxygen sensors.

## Spark plugs, replacing (4-cylinder engine)

> ***WARNING—***
> *To avoid personal injury, be sure the engine is cold before starting the procedure.*

◀ Turn plastic retainers on spark plug cover at top of engine 90°. Remove cover.

◀ Use special spark plug wire removal tool (**arrow**) to gently pull spark plug wire from spark plug.

***NOTE—***
*Removal tool is attached to rear of cylinder head beneath engine cover.*

– Blow away dirt from around base of plug to prevent it from entering engine.

– Installation is reverse of removal, bearing in mind the following:
  - Remove and install spark plugs one at a time.
  - Lightly lubricate new spark plug threads with copper-based antiseize compound.
  - Thread plugs into cylinder head by hand to prevent cross-threading.
  - Spark plug specifications are given in **Table e**.

| **Tightening torque** | |
|---|---|
| Spark plug to cylinder head | 25 Nm (18 ft-lb) |

## Spark plugs, replacing (M52, S52, M52 TU, M54 engine)

6-cylinder engines use a "coil-over" configuration, with one ignition coil above each spark plug.

> ***WARNING —***
> *To avoid personal injury, be sure the engine is cold before starting the procedure.*

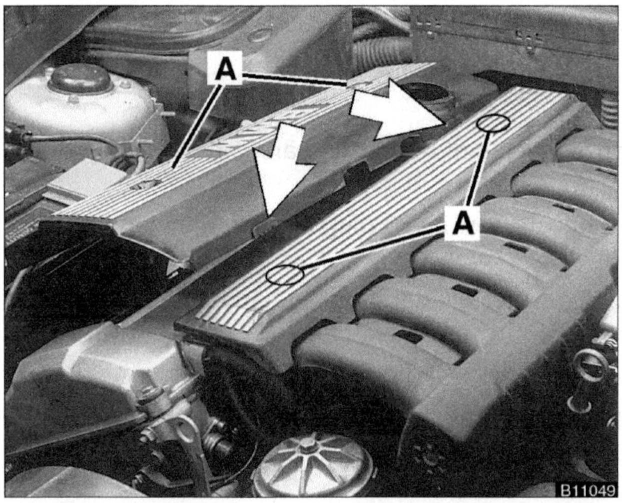


1997 - 1998 Z3, 1998 - 2000 M version:

- Remove top engine covers by prying out small plastic trim covers (**A**) and removing cover retaining nuts.
- Make sure rubber insulators (**arrows**) on either side of large cover do not fall off during cover removal.

1999 - 2002 Z3:

- Remove oil filler cap.
- Remove plastic trim covers (**arrows**).
- Remove cover hold down bolts.
- Remove ignition coil cover.

Remove ignition coil grounding harnesses (**arrows**).

Pull up on spring clips to disconnect ignition coil harness connectors.

- Remove coil mounting bolts. Remove coils.

– Remove spark plugs.

– Installation is reverse of removal, bearing in mind the following:

- Lightly lubricate new spark plug threads with copper-based antiseize compound.
- Thread plugs into cylinder head by hand to prevent cross-threading.
- Be sure to reinstall coil grounding harnesses.
- Spark plug specifications are given in **Table e**.

| **Tightening torque** | |
|---|---|
| Spark plug to cylinder head | 25 Nm (18 ft-lb) |

***NOTE—***
*The spark plug gap in the recommended spark plugs is not adjustable.*

## Spark plugs, replacing (S54 engine)

The S54 engine in 2001 - 2002 M version cars uses a "coil-over" configuration, with one ignition coil above each spark plug.

***WARNING—***
*To avoid personal injury, be sure the engine is cold before starting the procedure.*

– Make sure ignition is OFF. Remove ignition key.

– Working at rear of engine compartment, remove cable duct from bulkhead.

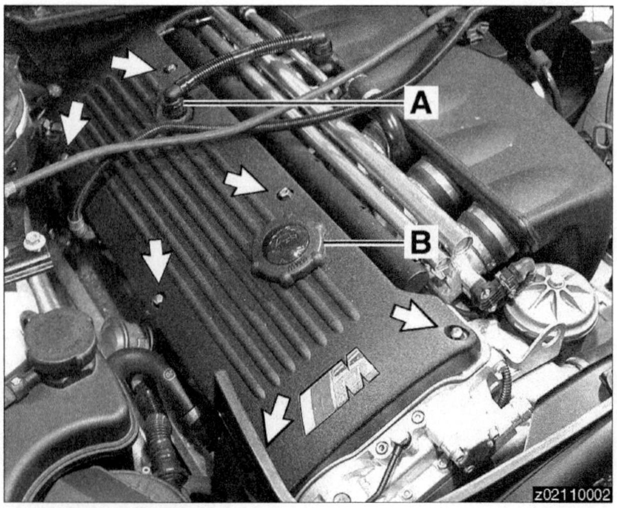


Remove cylinder head top cover:

- Disconnect vent hose (**A**).
- Remove oil filler cover (**B**).
- Remove cover hold-down bolts (**arrows**) and lift off cover.

***NOTE—***
*Remove oil filler cap to allow removal of plastic cover. Be sure to replace it on the oil filler immediately to prevent debris from entering the engine.*

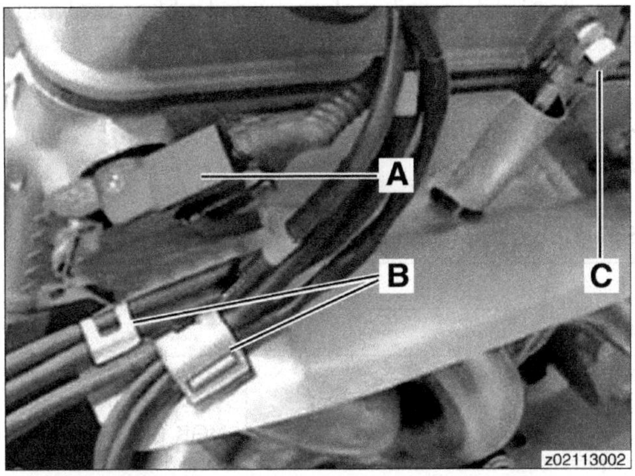


◀ Working at right rear of cylinder head:

- Detach exhaust camshaft position sensor electrical harness connector (**A**).
- Release oxygen sensor wiring harnesses from exhaust heat shield brackets (**B**).
- Remove oil return line banjo bolt (**C**).

> ***CAUTION—***
> *Make sure that sealing ring between banjo bolt and cylinder head does not drop down side of engine.*

– Detach ignition coils grounding strap at right side of cylinder head.

◀ Working above cylinder head cover:

- Carefully unclip knock sensor harness loom at brackets (**arrows**).
- Swing loom off studs and lay aside.

◀ Pull up on ignition coil harness lock to disengage wiring loom from coil.

– Remove ignition coils from spark plugs.

– Remove spark plugs.

– Installation is reverse of removal, bearing in mind the following:

- Lightly lubricate new spark plug threads with copper-based antiseize compound.
- Thread plugs into cylinder head by hand to prevent cross-threading.
- Spark plug specifications are given in **Table e**.

| **Tightening torque** | |
|---|---|
| Spark plug to cylinder head | 25 Nm (18 ft-lb) |

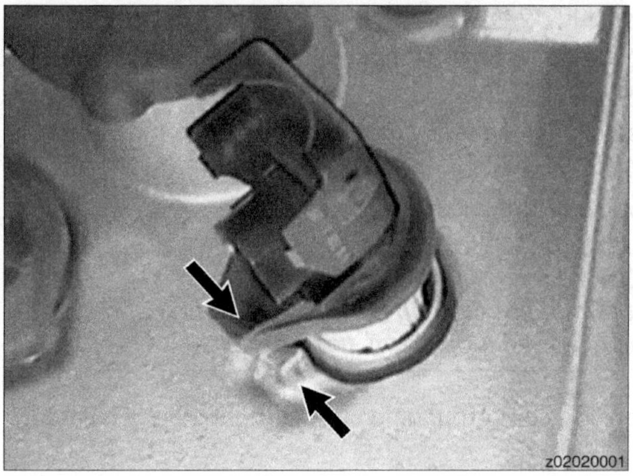
z02020001

When reinstalling ignition coil, make sure coil rubber seal and cylinder head cover lug (**arrows**) fit snugly together.

***NOTE —***
*The spark plug gap in the recommended spark plugs is not adjustable.*

| Table e. Spark plug application | |
|---|---|
| **Engine** | **Spark plug** |
| **4-cylinder** | |
| Z3 Roadster 1.9 (M44 engine) | Bosch F7 LDCR<br>NGK BKR 6 EK<br>NGK BKR 6 EQUP |
| **6-cylinder** | |
| 1997<br>Z3 Roadster 2.8 (M52) | Bosch F8 LDCR<br>NGK BKR 6 EQUP |
| 1998 - 2002<br>Z3 Coupe / Roadster 2.5, 2.5i, 2.8, 3.0i<br>(M52, M52 TU, M54 engine) | Bosch F7 LDCR<br>NGK BKR6 EQUP |
| 1998 - 2000<br>M Coupe / Roadster 3.2 (S52 engine) | Bosch FGR 8KQC |
| 2001 - 2002<br>M Coupe / Roadster 3.2 (S54 engine) | NGK DCPR 8EKP |

## Other Mechanical Maintenance

### Battery, checking and cleaning

***NOTE—***
*In 1996 cars the battery is located in the right rear of the engine compartment. In 1997 and later cars the battery is located in the trunk, under the floor mat.*

Simple maintenance of the battery and its terminal connections ensures maximum starting performance, especially in winter when colder temperatures reduce battery power.

– Make sure battery cables are tight and terminals, cable clamps and battery case are free of the white deposits that indicate corrosion and acid salts. Even a thin layer of dust containing conductive acid salts can cause battery discharge.

– To remove battery corrosion:

- Disconnect battery cables. Disconnect negative (–) cable first.

***CAUTION—***
*Prior to disconnecting the battery, read the battery disconnection cautions given in* **001 General Warnings and Cautions**.

- Remove battery from trunk or engine compartment.
- Clean terminal posts and cable clamps with a wire brush.
- Clean main chassis ground terminal next to battery. Corrosion can be washed away with a baking soda and water solution that neutralizes acid. Apply solution carefully, though, since it also neutralizes acid inside battery.
- Reconnect cable clamps, positive (+) cable first.
- Lightly coat outside of terminals, hold-down screws, and clamps with petroleum jelly, grease, or a commercial battery terminal corrosion inhibitor.

***WARNING—***
*Battery acid is extremely dangerous. Take care to keep it from contacting eyes, skin, or clothing. Wear eye protection. Extinguish all smoking materials and do not work near any open flames.*

– In batteries that have removable access plugs to electrolyte chambers, maintain battery electrolyte at the correct level, approximately 5 mm (¼ in) above top of battery plates or to top of the indicator marks (if applicable).

- Battery plates and the indicator marks can be seen once filler caps are removed.
- If electrolyte level is low, replenish it by adding distilled water only.

***NOTE —***
*The original electrolyte in a maintenance-free battery will normally last the entire service life of the battery under moderate climate conditions.*

## Battery, replacing

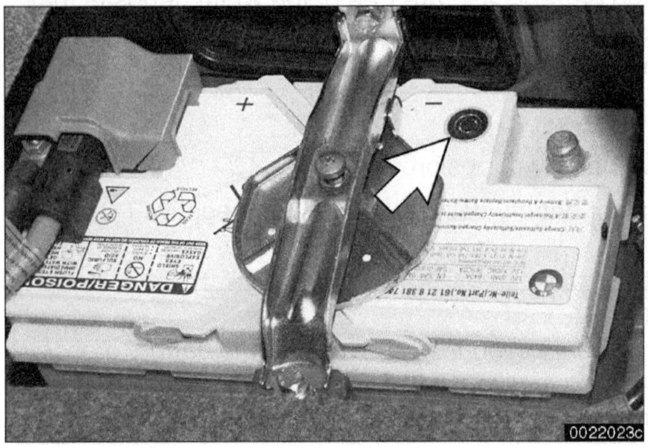

◄ The modern BMW battery is equipped with a built-in hydrometer "magic eye" (**arrow**). Battery condition is determined by the color of the eye:

- Green: Adequate charge
- Black: Inadequate charge; recharge
- Yellow: Defective battery; replace

Batteries are rated by ampere hours (Ah), the number of hours a specific current drain can be sustained before complete discharge, or by cold cranking amperes (CCA), the number of amps available to crank the engine in cold weather conditions. In general, replacement batteries should always be rated equal or higher than the original battery.

***CAUTION—***
*Prior to disconnecting the battery, read the battery disconnection cautions given in* **001 General Warnings and Cautions**.

The battery is held in place by a single hand screw and plate. A secure battery hold-down is important in order to prevent vibrations and road shock from damaging the battery.

***NOTE —***
- *Always disconnect the negative (–) cable first, and connect it last. While changing the battery, clean away any corrosion in or around the battery tray.*
- *More battery and charging system information is in* **121 Battery, Starter, Alternator**.

## Brake pad/rotor wear, checking

Z3 cars are fitted with disc brakes at all four wheels. Although the brakes are equipped with a brake pad warning system, the system only monitors one wheel per axle. Check pad thickness whenever the wheels are off or brake work is being done.

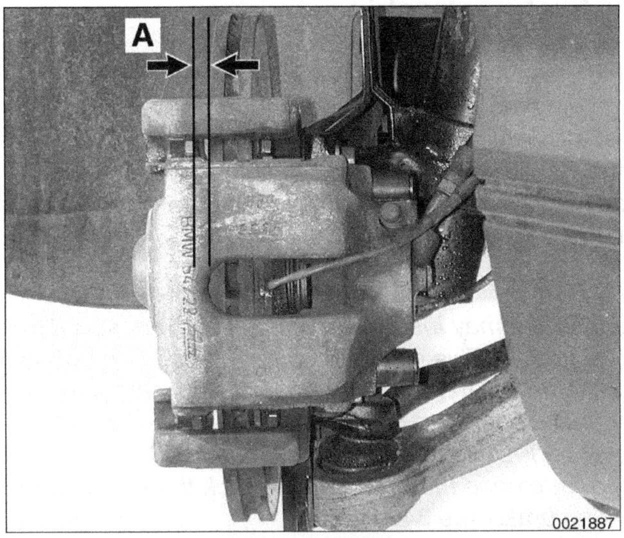


Disc brake pad wear can be checked through opening in caliper:

- Measure distance (**A**) of brake pad "ear" to brake rotor. Compare to specification below.

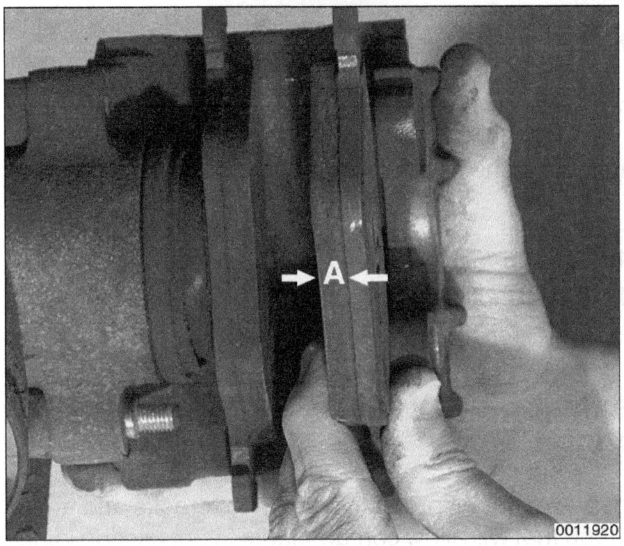


Unbolt caliper from steering arm to properly inspect:

- Brake pad thickness (**A**)
- Brake rotors
- Condition of caliper seal
- Condition of caliper slider bolts

***NOTE—***
*Brake caliper removal and installation procedures are given in* **340 Brakes**.

| **Brake pad lining minimum thickness (A)** | |
|---|---|
| Front or rear pad | 3.0 mm (0.12 in) |

## Brake system, inspecting

Routine maintenance of the brake system includes maintaining the brake fluid in the reservoir, checking brake pads for wear, checking parking brake function, and inspecting the system for fluid leaks or other damage:

- Check that brake hoses are correctly routed to avoid chafing or kinking.
- Inspect unions and brake calipers for signs of fluid leaks.
- Inspect rigid lines for corrosion, dents, or other damage.
- Inspect flexible hoses for cracking.
- Replace faulty hoses or lines.

## Parking brake, checking

The parking brake system is independent of the main braking system and may require periodic adjustment depending on use. Adjust the parking brake if the brake lever can be pulled up more than 8 clicks. Check that the cable moves freely. A description of the parking brake and parking brake adjustment can be found in **340 Brakes**.

***NOTE—***
*The parking brake may lose some of its effectiveness if it is not used frequently. This is due to corrosion build-up on the parking brake drum. To remove corrosion, apply the parking brake just until it begins to grip, then pull the lever up one more stop (click). Drive the car approximately 400 meters (1,300 ft.) and release the brake. To recheck the adjustment of the parking brake see* **340 Brakes**.

## Clutch fluid, checking

The hydraulic clutch and the brake system share the same reservoir and the same brake fluid. Check clutch fluid and brake fluid level at the same time.

***NOTE—***
- *See* **340 Brakes** *for clutch fluid information.*
- *See* **210 Clutch** *for information on the clutch and the hydraulic clutch operating system.*

## Drive axle joint (CV joint) boots, inspecting

Closely inspect CV joint protective boots for cracks and any other damage that will allow contaminants to get into the joint. If a rubber boots fails, the water and dirt that enter the joint will quickly damage it.

***NOTE—***
*Replacement of CV joint boots and inspection of the joints are described in* **330 Rear Suspension**.

## Exhaust system, inspecting

Exhaust system life varies widely according to driving habits and environmental conditions. If short-distance driving predominates, the moisture and condensation in the system will not fully dry out. This will lead to early corrosion damage and more frequent replacement.

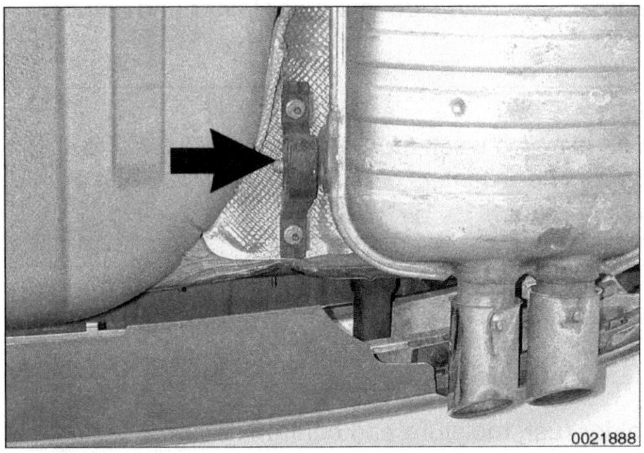


Scheduled maintenance of the exhaust system is limited to inspection:

- Check to see that all the hangers (**arrow**) are in place and properly supporting the system and that the system does not strike the body.
- Check for restrictions due to dents or kinks.
- Check for weakness or perforation due to rust.

***NOTE —***
*Alignment of the system and the location of the hangers are described in* **180 Exhaust System**.

## Fuel filter, replacing

The fuel filter (**arrow**) is located beneath the center of the car, approximately under the driver's seat. Remove protective cover to access filter.

– 2001 - 2002 Z3 Coupe / Roadster cars with M54 engine (DME MS 43.0) are equipped with a fuel filter that has a built in fuel pressure regulator.

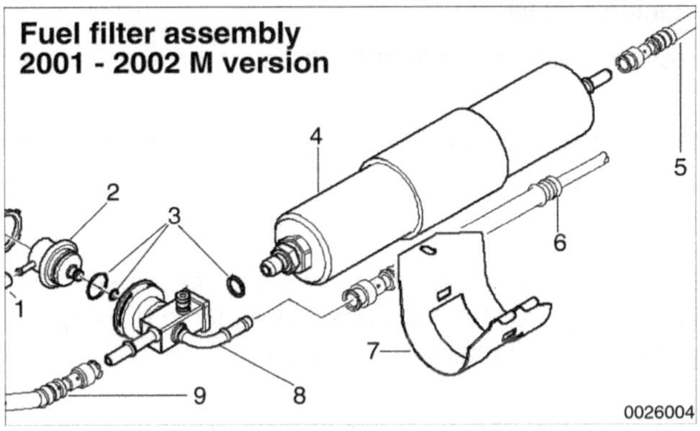


2001 - 2002 M Coupe / Roadster fuel filter components:

1. Vacuum line
2. Fuel pressure regulator
3. Sealing O-rings
4. Fuel filter
5. Fuel line from tank
6. Fuel return line
7. Support bracket
8. Pressure regulator mounting
9. Fuel line to engine

– 2001 - 2002 M Coupe / Roadster cars with S54 B32 engine are equipped with a fuel filter that is attached at the front to the fuel pressure regulator mounting.

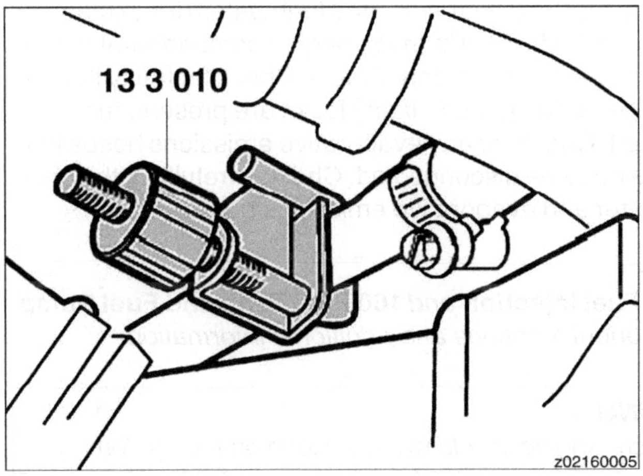


Use BMW special tool 13 3 010 or equivalent to clamp filter supply and return hoses to lessen fuel spillage.

– Drain fuel filter from inlet side into a container and inspect drained fuel.

> ***WARNING —***
> *Fuel will be expelled when fuel line is disconnected. Do not smoke or work near heaters or other fire hazards. Keep a fire extinguisher handy.*

> ***CAUTION—***
> *Clean thoroughly around the filter connections before removing them.*

– Check fuel for rust, moisture and contamination.

– Detach vacuum line from filter or pressure regulator, if applicable.

– Loosen filter support bracket and hose clamps on either end of filter.

***NOTE —***
*Use BMW special tool 161 0 50, or equivalent to separate fuel lines.*

– 2001 - 2002 M version: Remove fuel filter/fuel pressure regulator assembly, then separate filter from regulator on bench.

– When replacing fuel filter:
  - Note arrow or markings indicating direction of flow on new filter.
  - Install new filter using new O-rings (where applicable) and new hose clamps.

***NOTE —***
*When installing the fuel filter cover, take care to reinstall the foam rubber seal in front of the cover correctly to prevent flooding of the filter with rain splash water.*

## Fuel tank and fuel lines, inspecting

Inspect the fuel tank, fuel lines, and fuel system for damage or leaks. Check for fuel leaks in the engine compartment or fuel odors in the passenger compartment. Check for faulty flexible fuel lines by bending them. If any leaks are present, fuel will be expelled. Check for any evaporative emissions hoses that may have become disconnected. Check carefully at the charcoal canister and evaporative emissions purge system.

***NOTE —***
*See* **130 Fuel Injection** *and* **160 Fuel Tank and Fuel Pump** *for component locations and additional information.*

> ***WARNING —***
> *Make sure engine is cold when working on fuel system.*

## Suspension, front, inspecting

Inspection of the front suspension and steering includes a check of all moving parts for wear and excessive play. Inspect ball joint and tie-rod rubber seals and boots for cracks or tears that allow the entry of dirt, water, and other contaminants. See **310 Front Suspension**.

## Suspension, rear, inspecting

Inspect rear suspension bushings for wear. Rear shock absorbers should be checked for fluid leaks. Inspect the rear springs for cracks or road damage.

Differential and drive axle service consists of checking and changing the gear oil, inspecting for leaks, and checking the drive axle rubber boots (CV joint boots) for damage.

The areas where leaks are most likely to occur are around the drive shaft and drive axle mounting flanges. See **330 Rear Suspension** and **331 Differential**.

## Tires, checking inflation pressure

Correct tire pressures are important to handling and stability, fuel economy, and tire wear. Tire pressures change with temperature. Check tire pressures often during seasonal temperature changes. Correct inflation pressures can be found on the driver's door pillar and in the owner's manual. Note that tire pressures should be higher when the car is more heavily loaded.

> ***WARNING—***
> *Do not inflate any tire to a pressure higher than the tire's maximum inflation pressure listed on the sidewall. Use care when adding air to warm tires. Warm tire pressures can increase as much as 4 psi (0.3 bar) over their cold pressures.*

## Tires, rotating

BMW does not recommend tire rotation. Due to the car's suspension design, the front tires begin to wear first at the outer shoulder and the rear tires begin to wear first at the middle of the tread or inner shoulder. Rotating the tires may adversely affect road handling and tire grip.

## Wheels, aligning

BMW recommends checking the front and rear alignment once a year and whenever new tires are installed.

***NOTE—***
*See* **320 Steering and Wheel Alignment** *for alignment specifications.*

# Body and Interior Maintenance

## Body and hinges, lubricating

- Lubricate door locks and lock cylinders with an oil that contains graphite.
- Lubricate body and door hinges, hood latch, and door check rods with SAE 30 or SAE 40 engine oil.
- Lubricate seat runners with multipurpose grease.
- Lubricate hood release cables.

Do not apply oil to rubber parts. If door weather-strips are sticking, lubricate them with silicone spray or talcum powder.

Use winter lock de-icer sprays sparingly. Alcohol in de-icer will wash grease out of lock assemblies and cause locks to corrode internally or become difficult to operate.

## Exterior washing

The longer dirt is left on the paint, the greater the risk of damaging the glossy finish, either by scratching or by the chemical effect dirt particles may have on the painted surface.

Do not wash the car in direct sunlight. If the engine hood is warm, allow it to cool. Beads of water not only leave spots when dried rapidly by the sun or heat from the engine, but also can act as small magnifying glasses and burn spots into the finish. Wash the car with a mixture of lukewarm water and a car wash product. Rinse using plenty of clear water. Wipe the body dry with a soft cloth towel or chamois to prevent water-spotting.

## Interior care

Dirt spots can usually be removed with lukewarm soapy water or a dry foam cleaner. Use spot remover for grease and oil spots. Do not pour the liquid directly on the carpet or fabric, but dampen a clean cloth and rub carefully, starting at the edge of the spot and working inward. Do not use gasoline, naptha, or other flammable substances.

## Leather upholstery and trim

Clean leather upholstery and trim periodically using a slightly damp cotton or wool cloth. The idea is to get rid of the dirt in the creases and pores that can cause brittleness and premature aging. On heavily soiled areas, use a mild detergent (such as Woolite®) or other specially formulated leather cleaners. Use two tablespoons to one quart of cold water. Dry the trim and upholstery completely using a soft cloth. Regular use of a good quality leather conditioner will reduce drying and cracking of the leather.

## Polishing

Use paint polish only if the finish assumes a dull look after long service. Polish can be used to remove tar spots and tarnish, but afterwards apply a coat of wax to protect the clean finish. Do not use abrasive polish or cleaners on aluminum trim or accessories.

## Seat belts

Dirt and other abrasive particles will damage seat belt webbing. If it is necessary to clean seat belts, use a mild soap solution. Avoid bleach and other strong cleaning agents which may weaken the belt webbing.

***WARNING —***
*Do not clean the seat belt webbing using dry cleaning or other chemicals. Allow wet belts to dry before allowing them to retract.*

Inspect condition of the belt webbing and function of retractor mechanisms.

***NOTE —***
*See* **720 Seat Belts** *for seat belt inspection information.*

## Special cleaning

Tar spots can be removed with a bug and tar remover. Never use gasoline, kerosene, nail polish remover, or other unsuitable solvents. Insect spots also respond to tar remover. A bit of baking soda dissolved in the wash water will facilitate their removal. This method can also be used to remove spotting from tree sap.

## Washing chassis

Periodically wash underside of car, especially in winter, to prevent accumulation of road salt and rust. The best time to wash the underside is just after the car has been driven in wet conditions. Spray the chassis with a powerful jet of water. Commercial or self-service car washes may not be best for this, as they may recycle the salt-contaminated water.

## Waxing

For a long-lasting, protective and glossy finish, apply a hard wax after the car has been washed and dried. Use carnauba or synthetic based products. Waxing is not needed after every washing. You can tell when waxing is required by looking at the finish when it is wet. If the water coats the paint in smooth sheets instead of forming beads that roll off, a new coat of wax is needed. Do not apply wax to black trim pieces, rubber or other plastic parts.

## Windshield wiper blade maintenance

Common problems with the windshield wipers include streaking or sheeting, water drops after wiping and blade chatter. Streaking is usually caused when wiper blades are coated with road film or car wash wax. Clean the blades using soapy water. If cleaning the blades does not cure the problem, replace them. BMW recommends replacing the wiper blades twice a year, before and after the cold season.

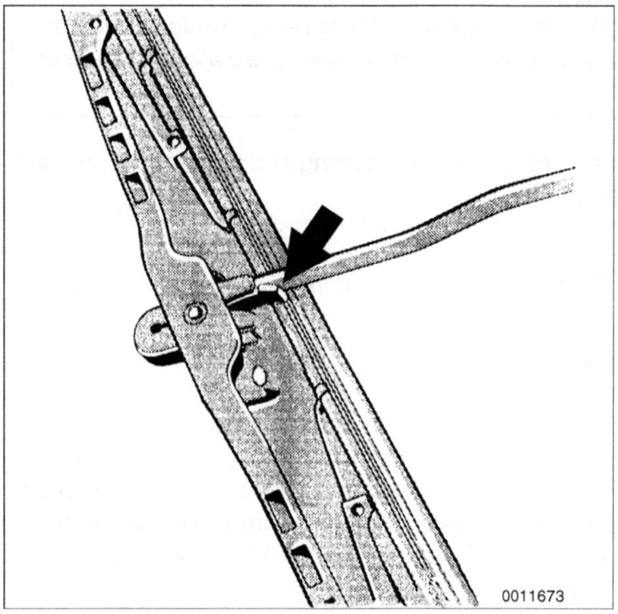


To replace wiper blade, depress retaining tab (**arrow**) and slide blade out of arm.

On older cars, check the wiper arm tension spring that holds the wiper to the glass. Replace the wiper arm if the springs are weak.

Drops that remain behind on the windshield after wiping are caused by oil, road film, or diesel exhaust coating the windshield. Use an alcohol or ammonia solution or a non-abrasive cleanser to clean the windshield.

Wiper blade chatter may be caused by dirty or worn blades, by a dirty windshield, or by bent or twisted wiper arms. Clean the blades and windshield as described above. Adjust the wiper arm so that there is even pressure along the blade, and so that the blade is perpendicular to the windshield at rest. Lubricate the wiper linkage with a light oil. The linkage is located under the hood on the driver's side. If the problem persists, replace the blades.

***NOTE —***
*See* **611 Wipers and Washers** *for more information.*

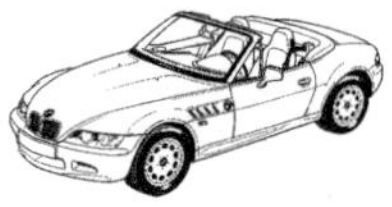

# 100 Engine–General



**TABLES**

## General

This section covers system descriptions and general information for engines and engine management systems in Z3 Roadster, Z3 Coupe, M Roadster and M Coupe models. Also covered is basic engine troubleshooting.

For specific repair procedures, refer to the appropriate repair group:

- **110 Engine Removal and Installation**
- **113 Cylinder Head**
- **116 Cylinder Head and Valvetrain**
- **117 Camshaft Timing Chains**
- **119 Lubrication System**
- **120 Ignition System**
- **130 Fuel Injection**
- **170 Radiator and Cooling System**

***NOTE—***
*M Coupe and M Roadster models for model years 2001 and 2002 are equipped with a completely redesigned inline 6-cylinder engine, the S54 B32. Engine mechanical and engine management topics for this new engine are presented in a separate heading in this repair group.*

## Engine mechanical

### 4-cylinder engine

**M44 engine**

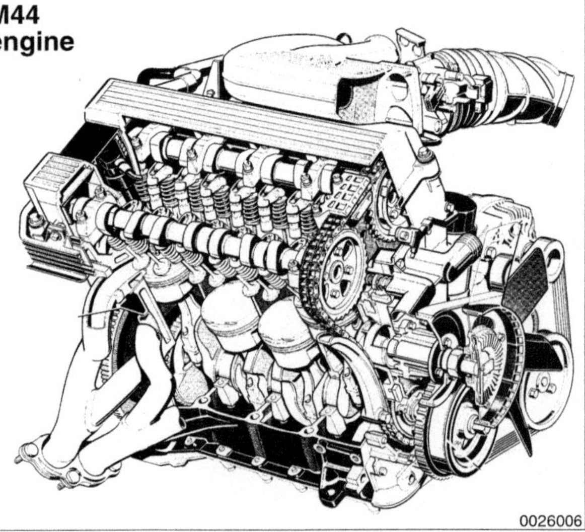

0026006

The 4-cylinder M44 engine cylinder block is cast iron with integral cylinders. The cylinders are exposed on all sides to circulating coolant.

This engine was offered in Z3 cars from the introduction year, 1996, through 1998.

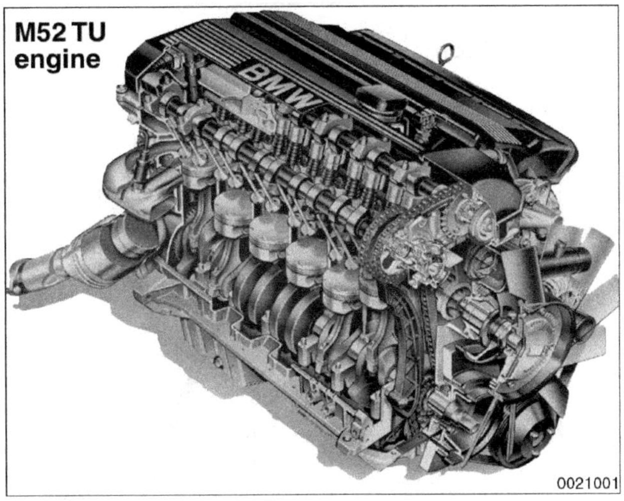


## 6-cylinder engines

Starting with model year 1997, Z3 and M version cars were also equipped with a variety of 6-cylinder engines. See **Table a**.

The cylinder block in all Z3 6-cylinder models is cast aluminum alloy ($Al_9Cu_3$) with cast iron cylinder liners. The cylinders are exposed on all sides to circulating coolant.

The 3.2 liter engine in M versions has a cast iron block for reasons of compactness and strength.

- Steel cylinder liners take up extra space. To keep the overall length of the 6-cylinder engine as small as possible, an all-cast iron block was deemed desirable.
- The 3.2 liter engines develop as much as 100 hp / liter. The S54 B32 engine (2001 - 2002 models) runs at a compression ratio of 11.5 : 1. Cast iron can best withstand the stresses of high cylinder pressures and high piston speed at maximum rpm.

**Table a. Engine applications**

| Model, year | Engine code | No. of cylinders | Displacement cc (cu. in.) | Bore / stroke | Compression ratio | Torque lb-ft/rpm | Horsepower hp / rpm | Engine management |
|---|---|---|---|---|---|---|---|---|
| **Z3 Roadster 1.9** 1996 - 1998 | M44 B19 | 4 | 1894 (115.6) | | 10.0 : 1 | 133 @ 4,300 | 138 @ 6,000 | Bosch DME M5.2 |
| **Z3 Roadster 2.3** 1999 - 2000 | M52 TU B25 | 6 | 2494 (152.2) | 84 mm (3.307 in)/ 75 mm (2.953 in) | 10.5 : 1 | 181 @ 3,500 | 170 @ 5,500 | Siemens DME MS 42.1 |
| **Z3 Roadster 2.5i** 2001 - 2002 | M54 B25 | 6 | 2494 (152.2) | 84 mm (3.307 in)/ 75 mm (2.953 in) | 10.5 : 1 | 175 @ 3,500 | 184 @ 6,000 | Siemens DME MS 43.0 |
| **Z3 Roadster / Coupe 2.8** | | | | | | | | |
| 1997 - 1998 | M52 B28 | 6 | 2793 (170.4) | 84 mm (3.307 in)/ 84 mm (3.307 in) | 10.2 : 1 | 203 @ 3950 | 189 @ 5,300 | Siemens DME MS 41.1 |
| 1999 - 2000 | M52 TU B28 | 6 | 2793 (170.4) | 84 mm (3.307 in)/ 84 mm (3.307 in) | 10.2 : 1 | 206 @ 3,500 | 193 @ 5,500 | Siemens DME MS 42.1 |
| **Z3 Roadster / Coupe 3.0i** | | | | | | | | |
| 2001 - 2002 | M54 B30 | 6 | 2979 (181.8) | 84 mm (3.307 in)/ 89.6 mm (3.528 in) | 10.2 : 1 | 214 @ 3,500 | 225 @ 5,900 | Siemens DME MS 43.0 |
| **M Roadster / Coupe** | | | | | | | | |
| 1998 - 2000 | S52 B32 | 6 | 3152 (192) | 86.4 mm (3.402 in)/ 89.6 mm (3.528 in) | 10.5 : 1 | 236 @ 3,800 | 240 @ 6,000 | Siemens DME MS 41.1 |
| 2001 - 2002 | S54 B32 | 6 | 3246 (198) | 87.0 mm (3.425 in)/ 91.0 mm (3.583 in) | 11.5 : 1 | 251 @ 4,900 | 315 @ 7,400 | Siemens DME MS S54 |

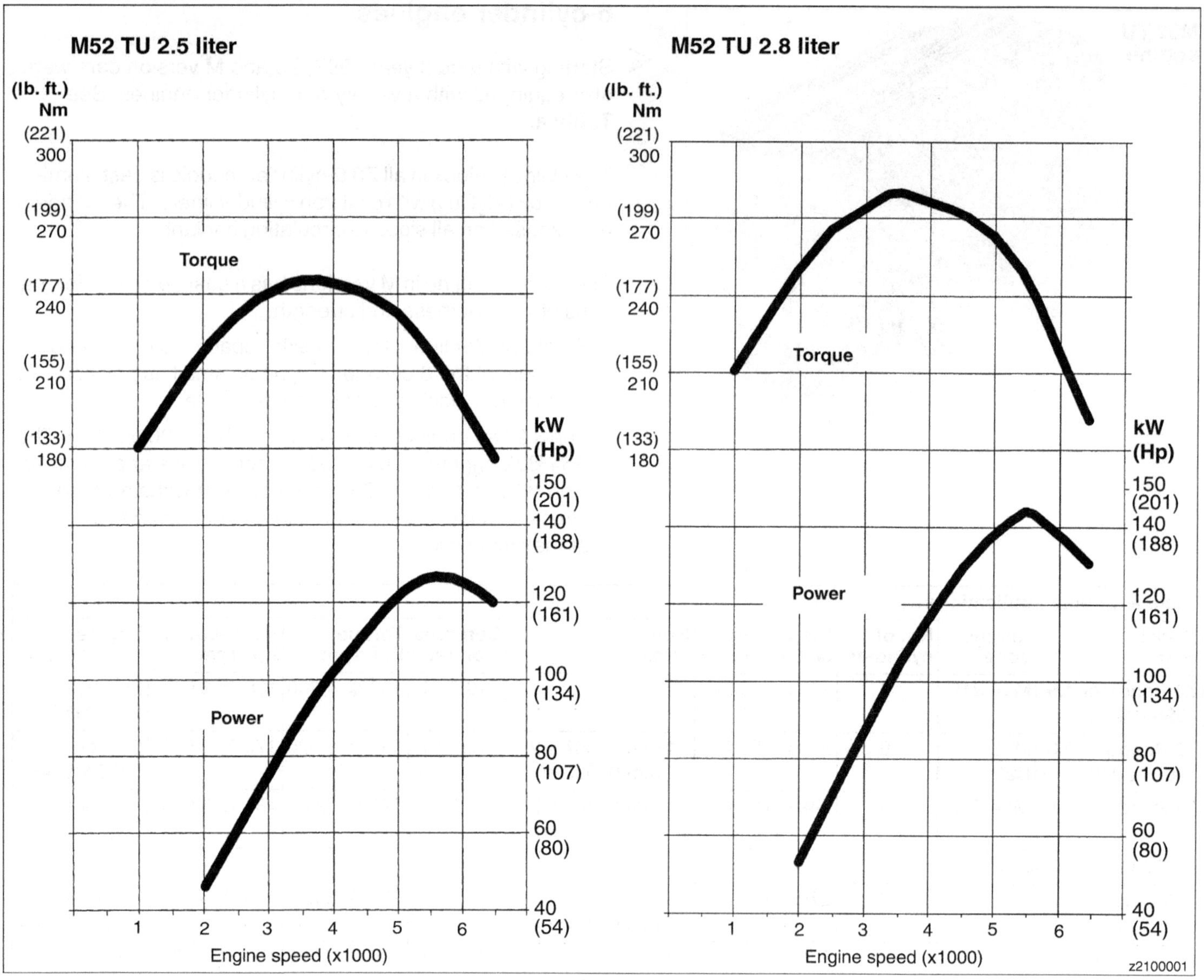


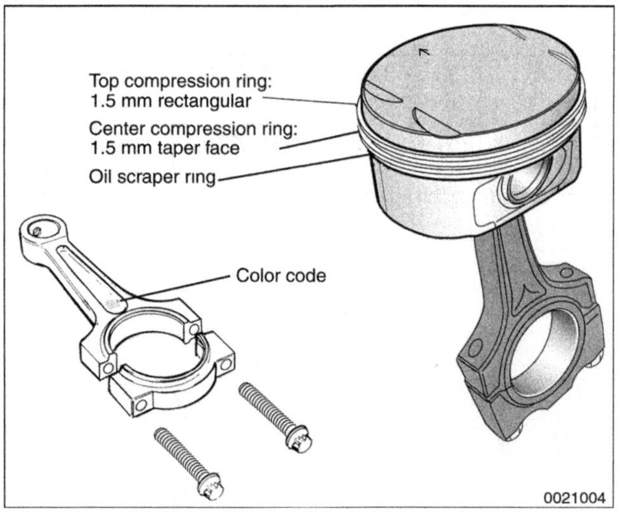


## Crankshaft

The counterweighted crankshaft rotates in replaceable split-shell main bearings. Oilways drilled into the crankshaft provide bearing lubrication. Oil seals pressed into alloy seal housings are installed at both ends of the crankshaft.

The crankshaft in the 2.5 liter engine is cast iron. The 2.8 and 3.0 liter engines use a forged steel crankshaft to accommodate the higher torque.

## Connecting rods and pistons

The forged steel connecting rods use replaceable split-shell bearings at the crankshaft end and solid bushings at the piston pin end.

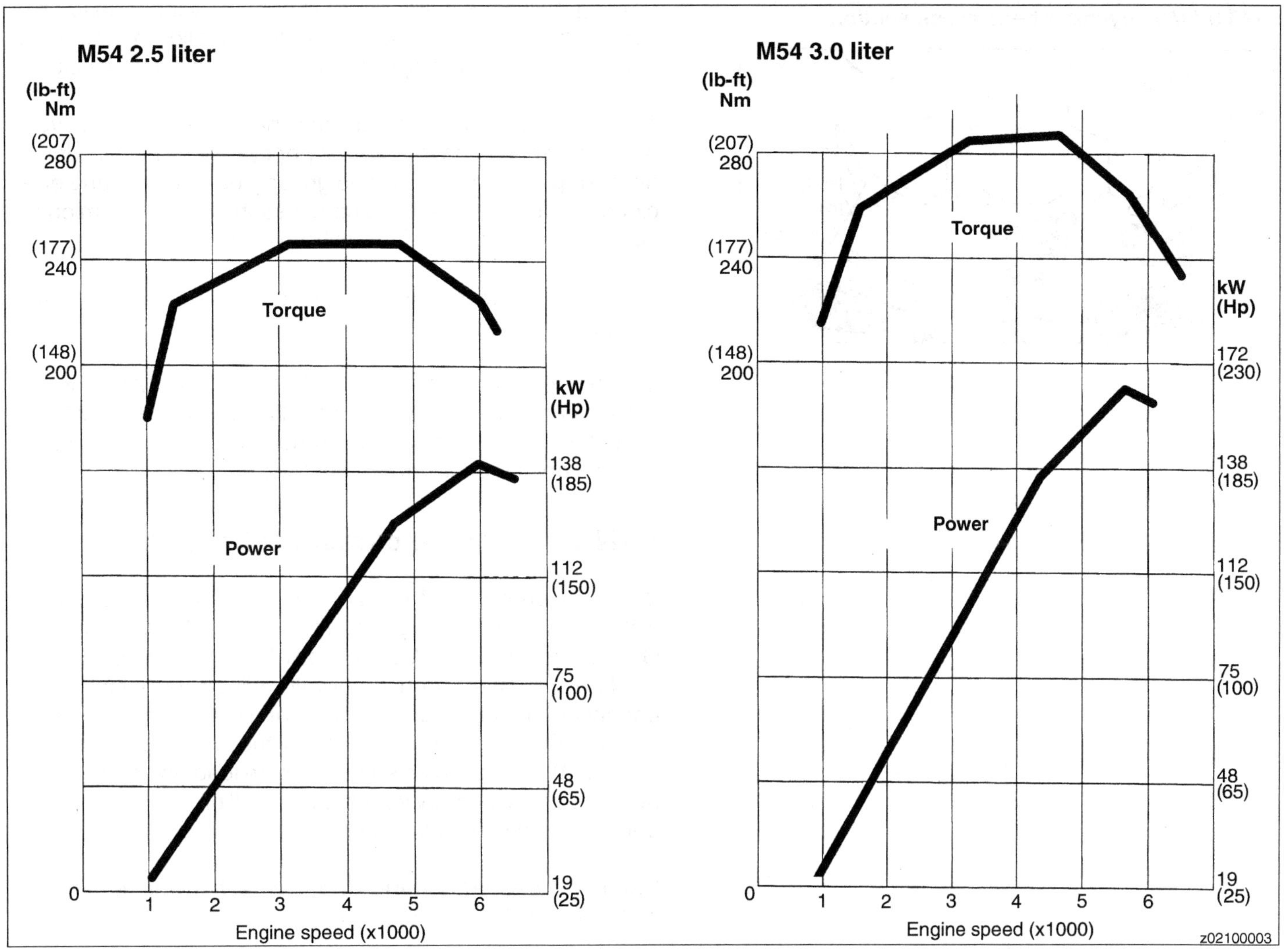


The pistons are of the three-ring type with two upper compression rings and a lower one-piece oil scraper ring. Full-floating piston pins are retained with circlips.

The 2.8 and 3.0 liter engines use a graphite coating on the piston skirts to reduce friction and noise.

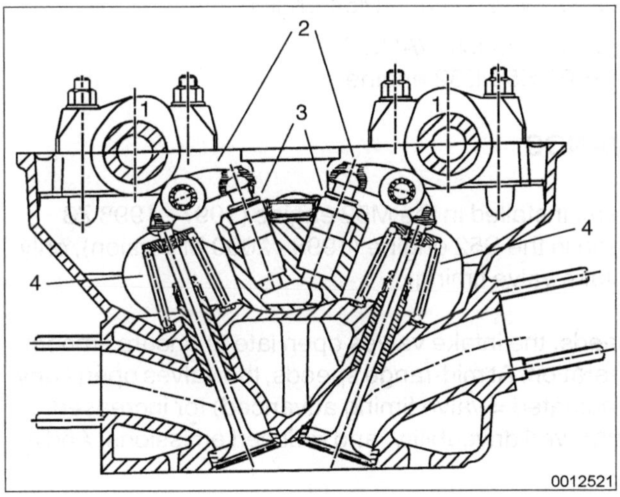


## Cylinder head and valvetrain

In the M44 engine, hydraulic valve adjusters (HVAs) are used in combination with rocker arms to actuate the valves. HVAs have the same function as hydraulic lifters.

1. Camshafts
2. Rocker arms
3. Hydraulic valve adjusters (HVA)
4. Valve and conical valve spring

M52 TU / M54 cylinder head cross section

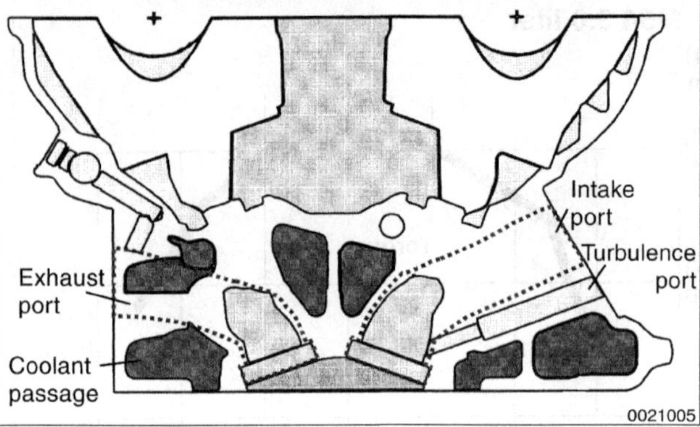


Cooling passages in the M52 TU or M54 cylinder head are designed for optimum coolant circulation, allowing the head to operate at lower temperatures than the cylinder block.

The aluminum cylinder head uses chain-driven double overhead camshafts and four valves per cylinder. The cylinder head employs a crossflow design for greater power and efficiency. Intake air enters the combustion chamber from one side while exhaust gasses exit from the other.

Oilways in the head provide lubrication for the camshafts and valvetrain.

In all models except the 2001 - 2002 M Coupe / M Roadster, valve clearance is set by self-adjusting (zero-lash) hydraulic lifters for zero valve clearance, reduced valve noise and the elimination of routine valve adjustment.

## VANOS (variable camshaft timing)

6-cylinder engines in Z3 and M version cars are equipped with variable valve timing system known as VANOS (from the German words *VAriable NOckenwellen Steuerung*). The VANOS system electrohydraulically adjusts valve timing for enhanced mid-range performance and improved emissions. The VANOS system is controlled by the engine control module (ECM), using engine speed, engine load and engine temperature as the primary inputs. Engine oil pressure is used for powering camshaft adjustment.

See **117 Camshaft Timing Chain** for testing and repair information on the VANOS system.

Three different systems for dynamically adjusting camshaft timing are used in Z3 and M version cars:

- Single VANOS:
  -1997 - 1998 M52 engine
  -1998 - 2000 S52 engine
- Double VANOS:
  -1999 - 2002 M52 TU and M54 engine
- High pressure double VANOS:
  -2001 - 2002 S54 B32 engine

### Single VANOS

This system, installed in the M52 engine (1997 - 1998 Z3 models) and in the S52 engine (1998 - 2000 M version), only adjusts intake valve timing.

At low speeds, the intake valves open late to ensure smooth engine operation. At mid-range speeds, the valves open early (VANOS actuated = valve timing advanced) for increased torque, improved driveability, and reduced emissions. And at high speeds, the valves again open late for optimum power and performance.

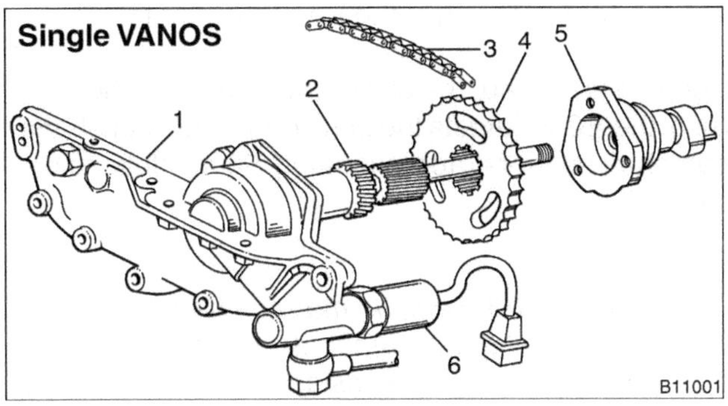

**Single VANOS**

The main components of the single VANOS system are the piston housing with integral valve and solenoid, and the modified intake camshaft and sprocket assembly.

1. VANOS unit with gear cup
2. Helical gear
3. Secondary timing chain
4. Intake camshaft sprocket
5. Intake camshaft
6. VANOS solenoid

When the VANOS solenoid is actuated, oil pressure is directed to the front side of the gear cup piston. This forces the helical gear into the camshaft to advance intake valve timing.

When the engine is running, the piston housing is supplied with pressurized engine oil via the solenoid-actuated spool valve. Depending on the position of the spool valve, oil is directed to either the front or back side of the gear cup piston.

When the solenoid is in the OFF position, engine oil is directed to the back side of the piston. This holds the gear cup forward and valve timing is maintained at the normal (late) position. When the solenoid is energized, oil pressure is directed to the front side of the piston. This in turn moves the gear cup further into the camshaft secondary drive, causing the camshaft to advance 12.5°. The helical gears are cut so that forward motion of the gear cup is translated into rotational motion of the camshaft.

The single VANOS system varies intake camshaft timing in an "all on or all off" manner.

## Double VANOS

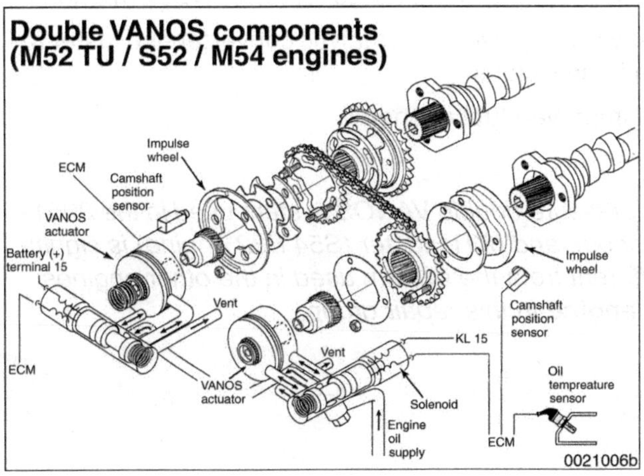

**Double VANOS components (M52 TU / S52 / M54 engines)**

Double VANOS, used in Z3 Roadster and Coupe engines for model years 1999 - 2002, operates independently on both intake and exhaust sides. In this system, the camshafts are infinitely adjustable within the mechanical travel limits of the camshaft drive gears. The engine control module (ECM) controls double VANOS operation.

Engine oil pressure is used to position the VANOS actuators. Oil pressure is fed from the engine oil pump up to the VANOS solenoids; oil drains back to the sump as the camshafts are adjusted during engine operation.

With the engine off, the base setting of the camshafts is as follows:

- Intake camshaft: Retarded
- Exhaust camshaft: Advanced

This is also the fail safe position in the event of an electronic control failure. Both camshafts are held in these positions by oil pressure from the engine oil pump. The exhaust camshaft is held additionally by a spring in the VANOS actuator.

When the engine is started, the camshafts remain in these positions until the ECM detects the positions of the camshafts from the camshaft sensors (approximately 50 revolutions or 2 - 5 seconds).

***NOTE—***
*Vehicles with DME MS 43.0 engine management are equipped with static Hall-effect camshaft sensors so that camshaft positions are recognized as soon as the ignition is turned on, before the engine is started.*

Once the camshaft positions are recognized, the ECM makes an initial camshaft timing adjustment based on oil temperature, oil pressure and engine RPM. Following this initial setting, engine RPM, throttle position signal, intake air temperature and engine coolant temperature are used to adjust camshaft timing.

***NOTE—***
- *VANOS is deactivated, resulting in engine power loss, if the oil is extremely hot.*
- *In case of high oil viscosity, a VANOS fault may be set in the ECM.*

When the ECM detects that the camshafts are in the desired position, the solenoids are modulated at 100 - 220 Hz, maintaining oil pressure on both sides of the actuators to maintain camshaft timing.

In addition to increasing engine power output, double VANOS offers the following advantages:

- Increased torque at lower and medium RPM ranges
- More efficient combustion and improved idle quality
- Internal EGR in part-load range for lower $NO_x$ emissions
- Quicker warm-up cycle for catalytic converter and faster reduction in emissions
- Overall improved fuel economy

***NOTE—***
*The high pressure double VANOS system used in the 2001 - 2002 M Coupe and M Roadster (S54 B32 engine) is significantly different from the system used in the other engines. See* **S54 engine** *in this repair group.*

## Lubrication system

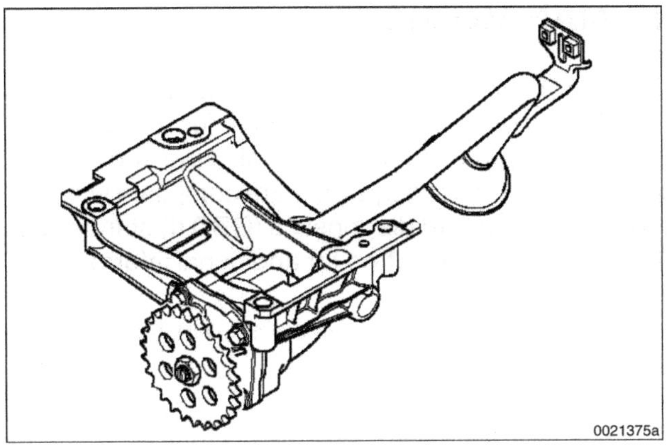


The lubrication system is pressurized whenever the engine is running. The oil pump draws oil through a pickup in the bottom of the oil pan, then forces it through a replaceable oil filter and into the engine oil passages.

The chain-driven oil pump is bolted to the bottom of the cylinder block inside the oil pan. A pressure relief valve limits the maximum system pressure. A bypass valve prevents the oil filter from bursting and insures engine lubrication should the filter become plugged. See **119 Lubrication System** for additional information.

## Cooling system

Whenever the engine is running, a coolant pump circulates coolant through the engine and, if the heater control valve is open, through the heater core in the passenger compartment. The coolant absorbs excess heat and carries it to the radiator where it is transferred into the passing airstream. A thermostat controls the flow of coolant through the radiator based on engine temperature.

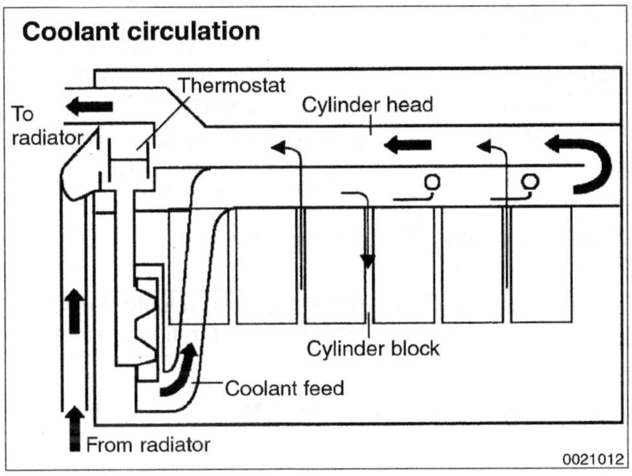


The M52 TU and M54 engine cooling system is designed so that coolant flows directly from the coolant pump to the cylinder head. The coolant is fed from the coolant pump through a cast coolant feed passage to the rear of the cylinder head. From there it flows forward to the thermostat housing, radiator and heater valve. The coolant passages in the cylinder block are only connected to the coolant supply through the holes in the head gasket. A reduced volume of coolant flows through the cylinder block.

starting with 1999 models, Z3 Roadster and Coupe models are equipped with an ECM-controlled electric cooling fan. The traditional BMW belt-driven viscous clutch fan is used for supplemental cooling. See **170 Radiator and Cooling System** for specific configuration and application information.

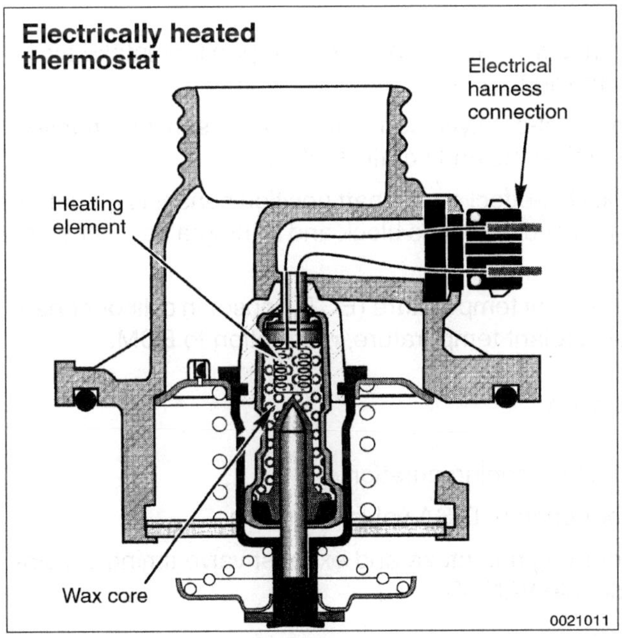


In cars with the M52 TU or M54 engine, an electrically heated thermostat controls the flow of coolant through the radiator based on coolant and outside temperature, engine load and driving speed, rather than just coolant temperature.

The ECM-controlled heated thermostat allows the engine to be operated at higher controlled temperatures during low and part throttle. This optimizes operating temperatures in both cylinder head and block, reduces friction and thereby fuel consumption.

The cooling system is designed to reduce cylinder head operating temperature, resulting in increased torque due to improved volumetric efficiency, and to raise cylinder block operating temperature.

## Engine management

The digital motor electronics (DME) engine management systems used in Z3 and M version cars combine fuel injection, ignition and other functions under the control of the engine control module (ECM). These systems are compliant with second generation on-board diagnostics (OBD II) standards. See **OBD On-Board Diagnostics** for additional information.

Engine driveability troubleshooting is covered in **130 Fuel Injection**.

**Table b. Engine management system applications**

| Year: engine | System | Identifying features |
|---|---|---|
| **Z3** | | |
| 1996 - 1998: M44 | Bosch DME M5.2 | DISA<br>Auxiliary throttle housing (ADS) |
| 1997 - 1998: M52 | Siemens DME MS 41.1 | Single VANOS<br>Auxiliary throttle housing (ADS) |
| 1999 - 2000: M52 TU B25<br>M52 TU B28 | Siemens DME MS 42.1 | Double VANOS<br>Resonance/turbulence intake<br>Motor driven throttle (MDK) |
| 2001: M54 B25<br>M54 B30 | Siemens DME MS 43.0 | Double VANOS<br>Resonance/turbulence intake<br>Electronic throttle (EDK) |
| **M version** | | |
| 1998 - 2000: S52 | Siemens DME MS 41.1 | Single VANOS |
| 2001 - 2002 S54 B32 | Siemens DME MS S54 | High pressure double VANOS<br>6 individual throttle assemblies |

### Engine control module (ECM)

#### ECM inputs

- Hot film mass air flow (MAF) sensor monitors intake air with no moving parts.
- Pre- and post-catalytic converter oxygen sensors monitor engine emissions and catalyst efficiency.
- Dynamic Hall-effect crankshaft position sensor is mounted at flywheel end of engine block and is integral to misfire detection.
- Engine coolant temperature (ECT) sensor in cylinder head supplies coolant temperature information to ECM.

#### ECM outputs

- Fuel injector opening duration
- 4-cylinder engine: DISA valve operation
- 6-cylinder engine: Intake and exhaust valve timing via single or double VANOS

## Siemens DME MS 42.0

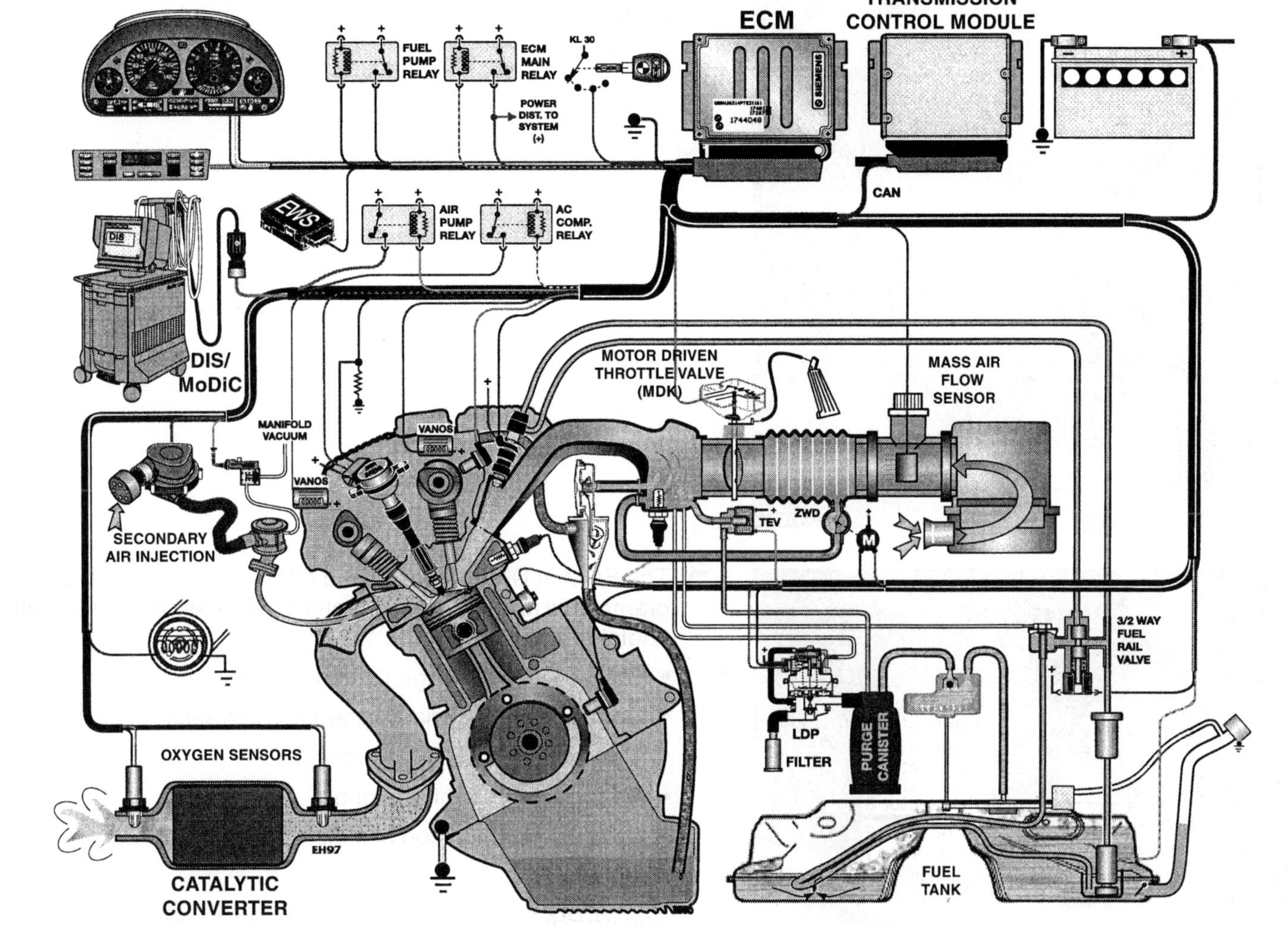


0021025

- Idle speed control valve
- Fuel pressure via 3/2 way valve monitors running losses (evaporative losses), as applicable
- Secondary air injection to reduce HC and CO, as applicable
- Maximum vehicle speed and cruise control, as applicable
- Radiator cooling fan and heated coolant thermostat (DME MS 42.1 / MS 43.0)
- Motor-driven throttle valve (MDK) (DME MS 42.1)
- Electronic throttle valve (EDK) (DME MS 43.0)
- Intake manifold resonance valve (DME MS 42.1 / MS 43.0)
- Multiple spark ignition system (DME MS 42.1 / MS 43.0)
- Fuel purge valve and fuel tank leakage diagnosis pump (LDP or DMTL) (DME MS 42.1 / MS 43.0 / MS S54)

The engine control module (ECM) (**arrow**) is mounted in the electronics box (E-box) at the right rear of the engine compartment.

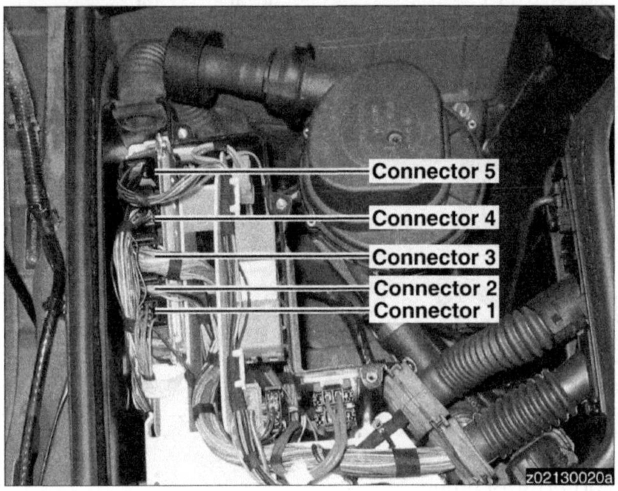


The ECM in DME MS 42.1 and MS 43.0 cars is flash-programmable and features 5 electrical harness connectors with a total of 134 pins.

- **Connector 1:** Voltages and grounds
- **Connector 2:** Ancillary signals (oxygen sensors, etc.)
- **Connector 3:** Engine signals
- **Connector 4:** Vehicle signals
- **Connector 5:** Ignition signals

***NOTE—***
*The EEPROM (chip) in the DME MS 42.1 and MS 43.0 ECM is coded to the vehicle. It can be flashed (reprogrammed) up to 13 times.*

## Fuel metering

The ECM meters pressurized fuel by changing the opening time (pulse width) of the fuel injectors. The exact amount of fuel injected is determined by the amount of time the injectors are open. To ensure that injector pulse width is the only factor that determines fuel metering, fuel pump pressure is maintained by a pressure regulator. The injectors are mounted to a common fuel rail.

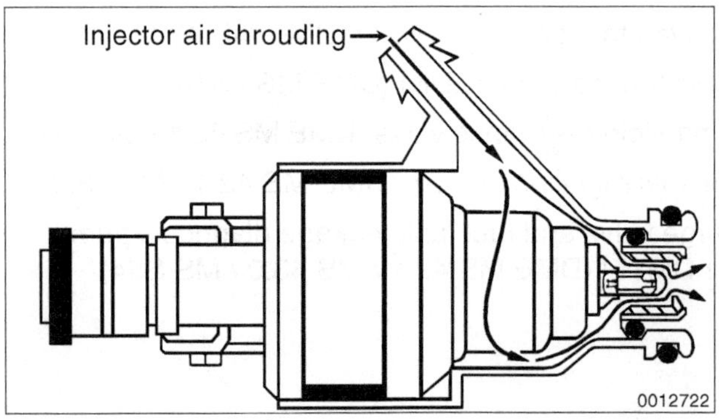


Fuel injectors in the 4-cylinder engine (M44) are air-cooled in order to keep the injector tip from clogging. Each injector is connected to a common air supply.

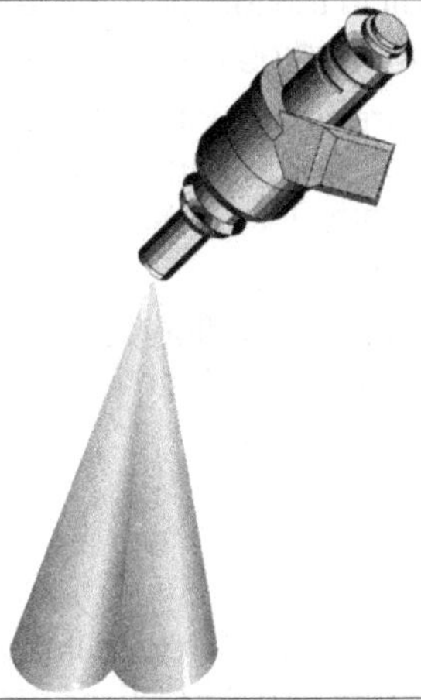
0021010

In cars with DME MS 42.1 and MS 43.0, the fuel-injectors inject fuel at an angle in a dual cone spray pattern.

The ECM monitors engine operating conditions to determine injector opening duration. Each injector can be individually controlled for cylinder selective fuel trim.

## Air intake

Air entering the engine passes through a pleated paper air filter in the air cleaner. Intake air mass is then measured by a mass air flow sensor. A reference current is used to heat a thin film in the sensor when the engine is running. The current needed to hold the temperature of the film constant is the basis of the electronically converted voltage measurement corresponding to the mass of the intake air.

## Idle speed control

Idle speed is electronically controlled via the idle speed control valve by bypassing varying amounts of air around the closed throttle valve. Idle speed is not adjustable. The ECM determines idle speed by controlling a dual-winding rotary idle control valve. The basic functions and parameters of idle speed control are as follows:

- Control of cold air intake volume. For example, at air temperatures below 0°C (32°F), electronic throttle control (MDK, EDK) is also opened during idle (DME MS 42.1, MS 43.0).
- Smooth idle speeds regardless of load and inputs
- Smooth transition from acceleration to deceleration

In cars with DME MS 42.1 and MS 43.0, idle speed stabilization is active during the following conditions:

- Engine warm up
- Heating or A/C activation
- Drive gear selected (automatic transmission)
- Varying cooling fan speeds

If the ECM detects a fault in the idle control valve, it increases or decreases air flow, depending on the nature of the fault:

- If the fault causes decreased air flow (idle control valve closed), the electronic throttle control (MDK or EDK) compensates to maintain idle. The EML lamp illuminates to inform the driver of a fault.
- If the fault causes increased air flow (actuator failed open), VANOS and knock control are deactivated. This reduces engine performance noticeably.

## Throttle control

In order to integrate the driver's wish with the requirements of the traction control systems (ABS / ASC or ABS / DSC) and cruise control, DME systems feature electronic throttle control.

### Auxiliary throttle control system (ADS)

All 4-cylinder models and 6-cylinder models with DME MS 41.1 engine management utilize a conventional throttle assembly with an accelerator cable. In front of the cable controlled throttle in the incoming airstream is a second throttle plate controlled by the ADS actuator. This motor is in turn controlled by the traction control (ASC / DSC) module.

The ADS throttle control assembly can close down the throttle, thus reducing engine torque and assisting in reduced traction situations, but it cannot open the throttle, as that is controlled by the throttle plate attached to the accelerator ca-

**Auxiliary throttle control system (ADS)**

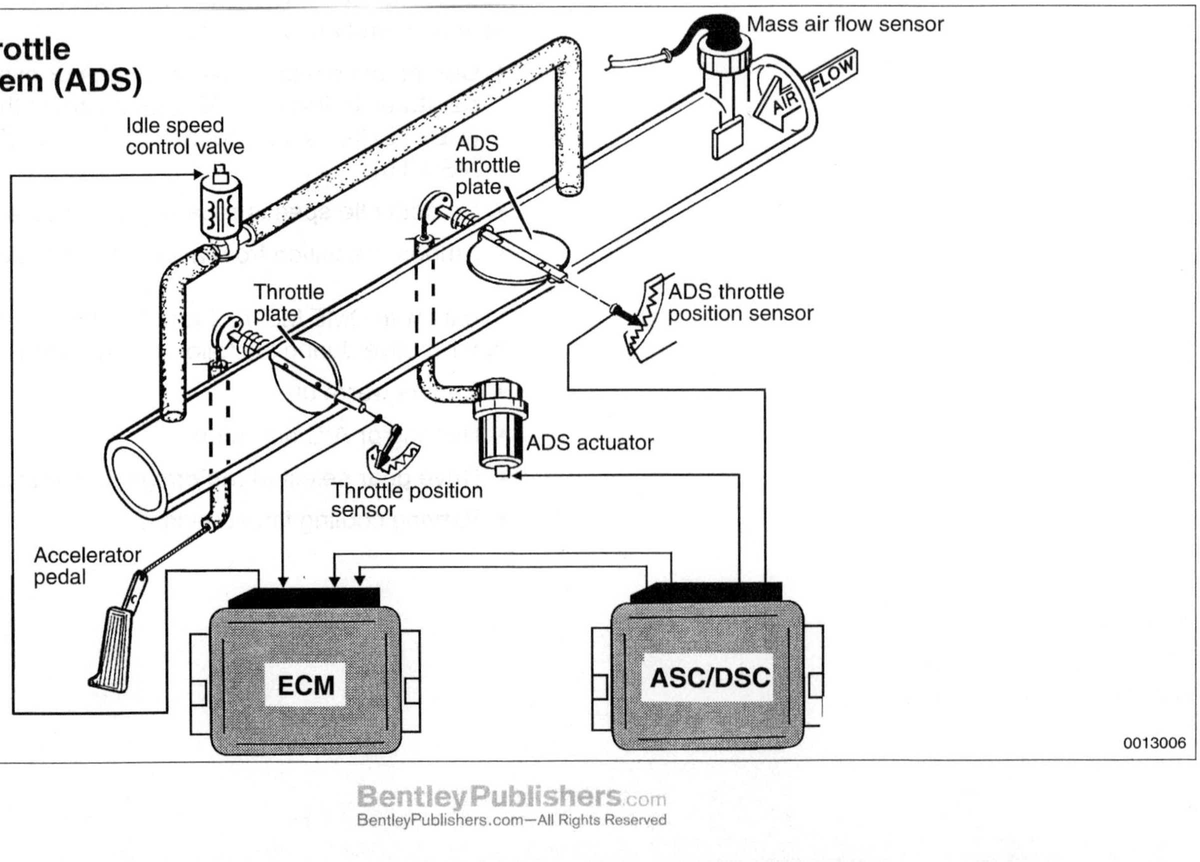

ble. The fail-safe position for the ADS throttle plate is wide open.

Further information about the traction control system can be found in **300 Suspension, Steering and Brakes–General.**

### Motor driven throttle control (MDK)

**MDK throttle assembly**

< DME MS 42.1 engine management incorporates a throttle valve actuator (MDK) at the throttle assembly. Identify MDK as follows:

- Throttle cable is used to actuate accelerator pedal position potentiometers.
- Throttle cable also serves as backup to open throttle plate (full control) when MDK system is in fail-safe mode.

### Electronic throttle control (EDK)

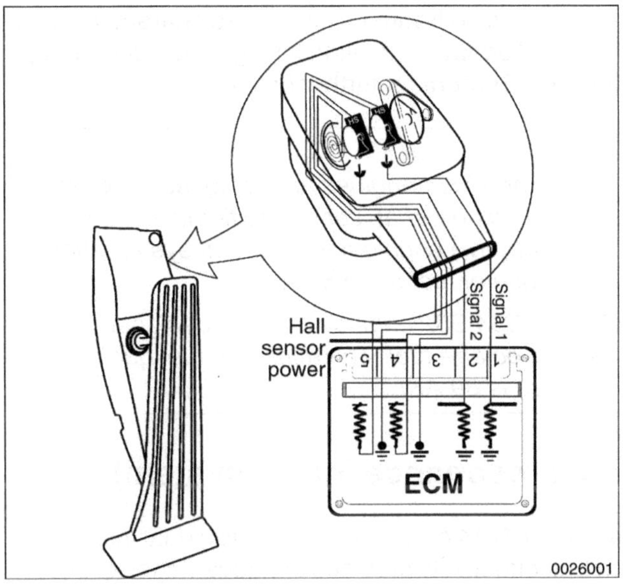


< DME MS 43.0 uses a drive-by wire throttle system referred to as EDK, identified as follows:

- Accelerator position potentiometers (PWG) are located in accelerator pedal assembly.
- EDK is fully electronic. Accelerator pedal is not mechanically linked to EDK throttle housing.

In both DME MS 42.1 and MS 43.0 systems the throttle valve is electronically operated for precise throttle operation, OBD II compliance for fault monitoring, and ASC / DSC and cruise control. This integrated electronic throttle reduces additional control modules, wiring, and sensors. The electronic throttle housing is not adjustable.

***NOTE—***
*The electronic throttle system used in 2001 - 2002 M version (S54 B32 engine) is known as EDR. See* **S54 engine management** *in this repair group.*

## Ignition function

In all models a distributorless ignition system with individual ignition coils for each cylinder is employed.

M44 (4-cylinder) engine: Ignition coil pack is mounted to the right front strut tower in the engine compartment. The coil pack integrates four individual coils into a single unit. The individual coils are not replaceable separately.

6-cylinder engines: Six individual ignition coils are located directly above the six spark plugs. The coils are mounted over the spark plugs and connect to the plugs via a short lead.

The DME engine management systems in 6-cylinder cars use multiple spark ignition to reduce emissions and extend spark plug life.

## Knock control

Knock (detonation) sensors monitor and control potentially damaging ignition knock through the ECM. The knock sensors function like microphones and are able to convert mechanical vibration (knock) into electrical signals. The ECM is programmed to react to frequencies that are characteristic of engine knock and adapt the ignition timing point accordingly. See **120 Ignition System** for further details.

***CAUTION—***
- *The M52 TU and M54 engines are designed to operate best with fuel octane of at least 91 anti-knock index (AKI). The adaptive engine management system, however, allows use of fuel rated 87 AKI.*
- *The S52 and S54 B32 engines are also designed to operate with 91 octane fuel. The minimum fuel quality for these engines is 89 AKI.*

## DISA (dual resonance intake system)

DISA, used on the M44 4-cylinder engine, is a dual intake runner system that effectively provides the advantages of both short and long intake runners within the same engine. For best performance, long intake runners are most beneficial at low- and mid-engine speeds (below 4,200 rpm), and short intake runners enhance torque at high engine speeds (above 4,200 rpm).

***NOTE—***
*The term DISA comes from the German words* DIfferenzierte SAuganlage, *and can roughly be translated as a 'differing intake manifold configuration'.*

The DISA solenoid valve is controlled by the Bosch M5.2 engine control module (ECM), using engine speed as the primary input. The main components of the system are the modified twin-section intake manifold with change-over valve, the twinbarrel throttle body, and the electrical/pneumatic actuating components.

The change-over valve is held in the normally open position by a spring. This allows for an open valve in the event of system failure. During low and mid-range engine speeds, the ECM supplies power to the solenoid valve, which in turn allows vacuum to be applied to the vacuum diaphragm. This causes the valve to close (long-pipe configuration). When engine speed reaches approximately 4,800 rpm, the ECM electrically signals the solenoid valve and the valve opens, creating the short pipe configuration.

**Manifold construction.** The intake manifold is a two-piece metal construction, with a pair of runners in the top section and four runners in the lower section. A changeover valve (butterfly valve) is installed in the lower section, enabling the DISA solenoid to isolate one pair of runners from the other pair.

**DISA components**

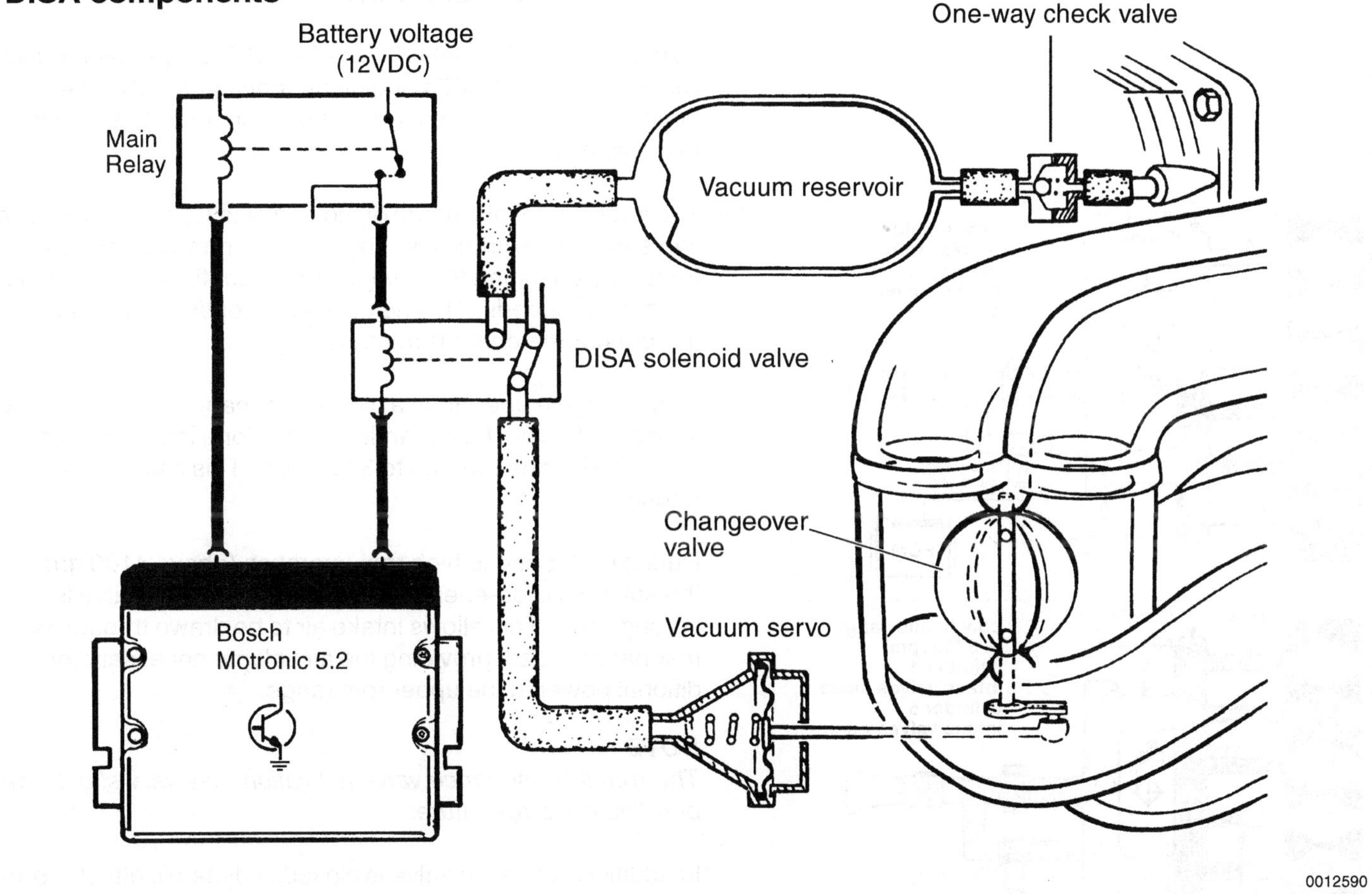

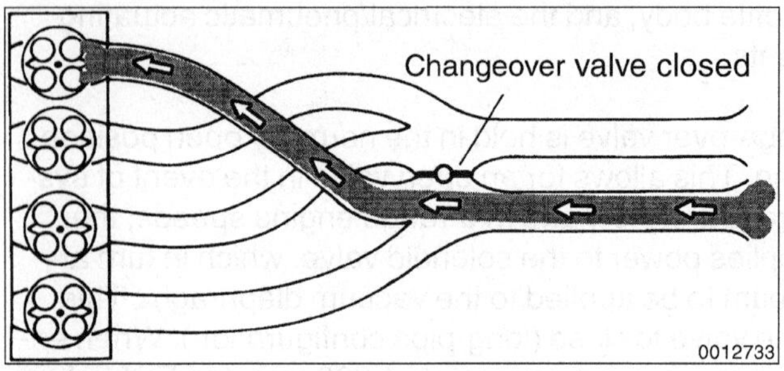


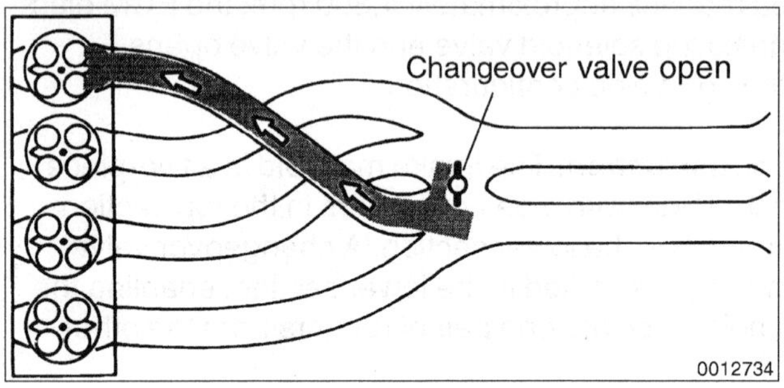


**Operation.** With the DISA changeover valve closed, the pipes in the top half of the manifold act together with the ram air pipes in the lower half to produce a single, long air intake pipe for each cylinder. The column of air oscillating in this combined pipe significantly increases engine torque in the medium rpm range.

Above approximately 4,800 rpm, the changeover valve between the intake air pipes for the two cylinder groups is opened. The shorter pipes in the lower manifold section now become the main suppliers of ram air to the cylinders, yielding greater power at the upper end of the engine rpm range.

**Control components.** The DISA changeover valve is actuated electro-pneumatically via the engine control module (ECM). The valve begins to open as engine speed rises above 4,840 rpm and closes below 4,760 rpm. The action of the valve is deliberately delayed to prevent it from opening and closing repeatedly within a short time.

## Resonance / turbulence intake system

Beginning with the 1999 Z3 Roadster 2.3, 6-cylinder models with the M52 TU or M54 engine are equipped with a variable intake runner system known as the resonance / turbulence intake system.

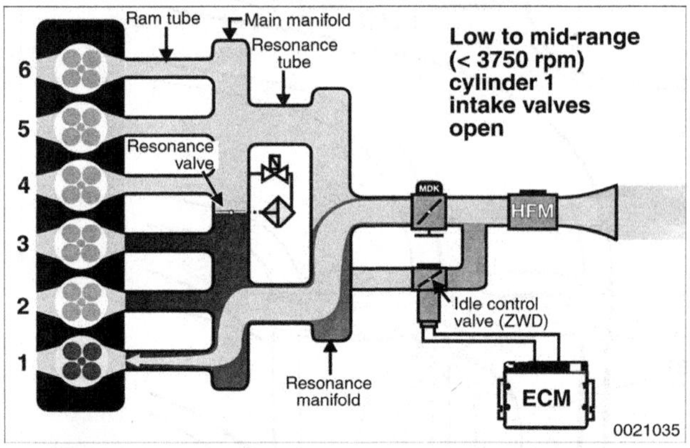


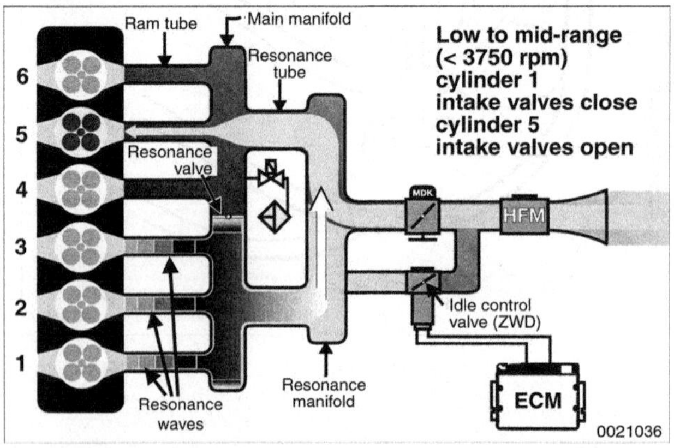


The intake manifold, made of molded plastic, is configured as two sets of three runners. This design enhances low end torque by changing the intake air flow configuration for varying engine speeds. This helps achieve optimum torque throughout the entire rpm range.

During engine operation, a closed resonance valve gives the intake air charge the dynamic effect of long intake runners at low to mid-range rpm (up to 3750 rpm). This helps increase torque.

During mid-range to high rpm operation (above 4100 rpm), the solenoid is de-energized and the resonance valve is sprung open. This allows intake air to be drawn through both resonance tubes, providing the air volume necessary for additional power at the upper rpm range.

***NOTE—***
*The rpm for resonance valve activation may vary slightly depending on temperature.*

In addition, when the valve is closed, a dynamic effect is produced. For example, as intake air is flowing into cylinder 1, the intake valves close. This blocks the onrushing air. The cylinder 1 air flow rebounds backward (resonance back pulse) to fill cylinder 5. The resonance wave along with the intake velocity enhances cylinder filling.

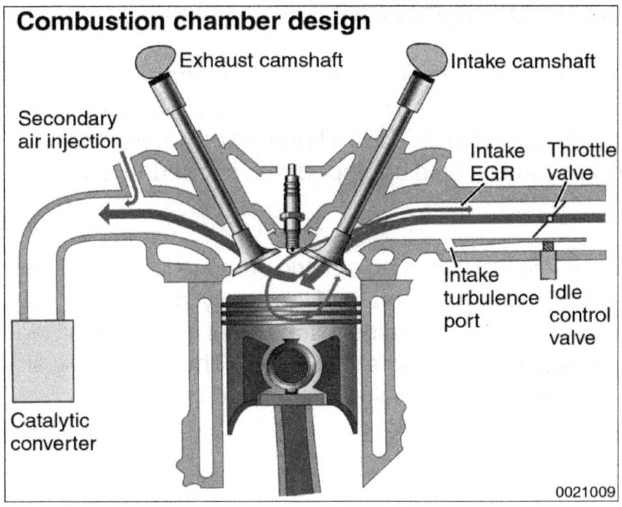


The intake manifold includes intake turbulence ports. The 5.5 mm (0.217 in) turbulence ports channel idle and low speed air directly from the idle speed control valve to one intake valve of each cylinder.

Routing intake air to one intake valve per cylinder causes the air charge to swirl in the cylinder. Together with the high flow rate of intake air across the small (5.5 mm) port, intake fluctuations are reduced for more stable combustion.

## Exhaust system and catalytic converters

OBD II enhanced emission standards require the engine control module (ECM) to monitor the oxygen content in the exhaust both before and after the catalytic converter. This allows for tighter control of tail pipe emissions and also allows the ECM to monitor catalytic converter problems and report catalyst-damaging faults. If the ECM detects that catalytic converter or oxygen sensor efficiency has degraded past a certain preprogrammed limit, it turns on the malfunction indicator light (MIL), and stores a diagnostic trouble code (DTC) in the ECM.

4-cylinder Z3 cars are built with a conventional exhaust manifold. The two oxygen sensors are installed ahead and behind the catalytic converter, which is underneath the car in the middle exhaust pipe.

1997 - 1998 6-cylinder (M52) engines are equipped with two intake manifolds leading to a dual middle exhaust pipe. The pipe connects to the catalytic converter at a Y-junction. The two oxygen sensors are ahead of and behind the catalyst.

In 1999 - 2000 M version (S52 engine), a pair of precatalyst oxygen sensors are installed in the exhaust manifolds. A pair of post-catalyst oxygen sensors are in the pipes leading out of the dual catalytic converters.

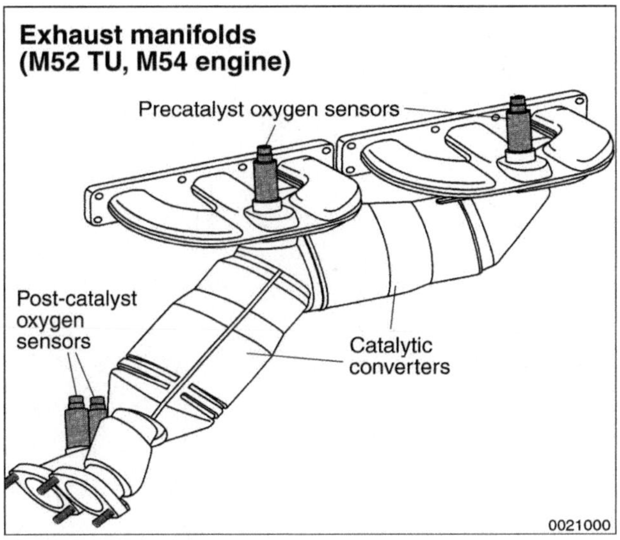


In 1998 - 2002 Z3 models, (M52 TU, M54 engine), each exhaust manifold incorporates a catalytic converter. The converters are mounted close to the engine for faster heat up. This allows the engine management system to switch to closed loop operation faster. There is an oxygen sensor before each catalyst and one after each catalyst.

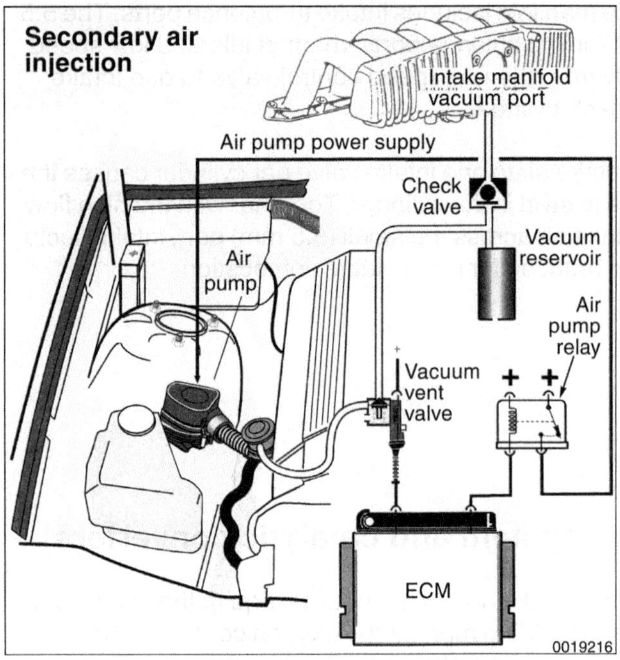


## Secondary air injection

6-cylinder engines are equipped with a secondary air injection system. In addition, beginning in January 1997, the M44 engine was classified as a TLEV (Transitional Low Emission Vehicle) engine and is also equipped with secondary air injection.

Secondary air injection uses an electric pump to force fresh air into the exhaust system upstream of the catalytic converter during engine warm-up. By providing extra oxygen to the unburned fuel in the exhaust, emissions are reduced. The secondary air injection pump stops within a specified timed interval.

## S54 ENGINE

The S54 B32 engine in the M Coupe / M Roadster models shares the basic architecture of other BMW inline 6-cylinder engines.

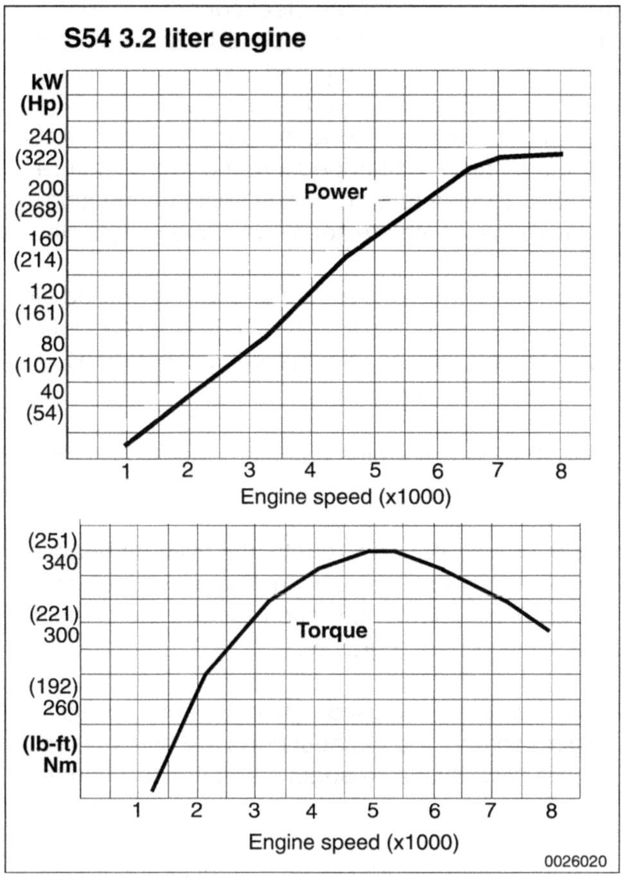


Virtually all of its engineering details, however, are oriented to producing high power and torque output.

***NOTE —***
*Application information and specifications for Z3 and M version engines are in* **Table a**.

## S54 engine mechanical

**Engine block**. The S54 engine block, like the S52 in M version cars from previous years, is made of cast iron for the same reasons of compactness and structural strength.

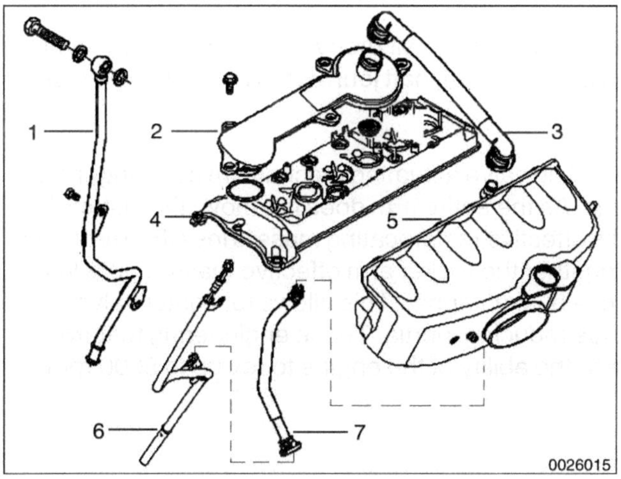


**Crankcase ventilation**. The S54 features a non-pressurized sealed crankcase ventilation system for blow-by vapors. The crankcase blow-by vapors are purged by intake manifold vacuum. The vapors are drawn from the cylinder head cover (**4**) through the inlet of an oil separator (**2**). A labyrinth within separates the oil from the vapors, and the condensed oil returns to the oil pan through return lines (**1** and **7**). The vapors exit the oil separator through an outlet hose (**3**) to the intake manifold (**5**) and are inducted into the combustion chambers.

When the engine is running, intake manifold vacuum closes a check valve in the left side return line (**7**). When the engine is not running, the check valve opens. This allows any condensates that have collected in the intake manifold to drain back to the oil pan through the dipstick tube (**6**).

**Crankshaft**. To achieve the high revving capabilities of the S54 engine (8000 rpm limit), a forged nitro-carbonized steel crankshaft provides strength and rigidity.

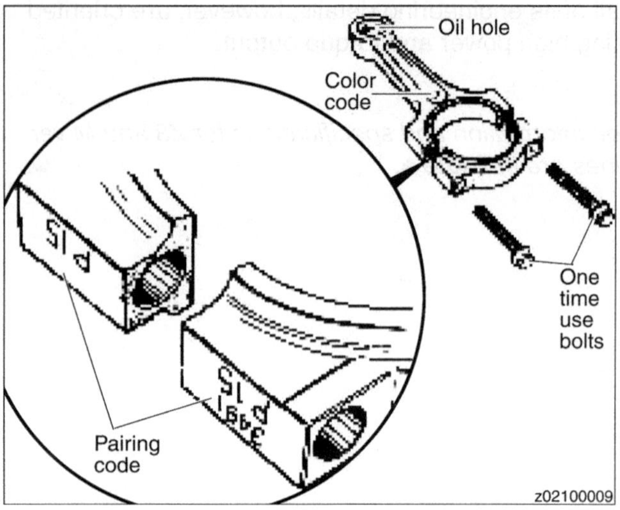


**Connecting rods**. Cracked forged steel connecting rods eliminate the need for bolt sleeves, thus reducing reciprocating weight. The connecting rod big end is fractured and split off, leaving rough surfaces on both the cap and the rod. Centering of the cap on the rod is done through the structure of the split. Pairing codes are stamped into the rod to ensure proper assembly. To maintain proper balance, connecting rods are replaced in a complete set of the same weight class.

**Pistons**. Piston skirts are constructed of graphite coated aluminum. Pistons are cooled by oil spray nozzles that are bolted into the cylinder block. The nozzles deliver a constant oil spray to the underside of the pistons.

**Timing chain**. The double timing chain is hydraulically tensioned and needs no periodic adjustment or replacement.

**Cylinder head**. For increased strength, the cylinder head is manufactured as a single aluminum casting which includes the camshaft carriers. This construction saves 29 lb., helping to lower the vehicle center of gravity.

For smooth airflow, the combustion chambers and intake ports are fully machined; the exhaust ports are partially machined. For durability, the valve seats are of especially hard steel.

A 3-layer stainless steel cylinder head gasket ensures effective sealing of the cylinder head to the engine block.

**Camshafts**. The hollow cast iron overhead camshafts are strengthened by heat-treating the journals and camshaft lobes. The duration and lift (12 mm) of the lobes are the same on both camshafts. The camshafts are not interchangeable.

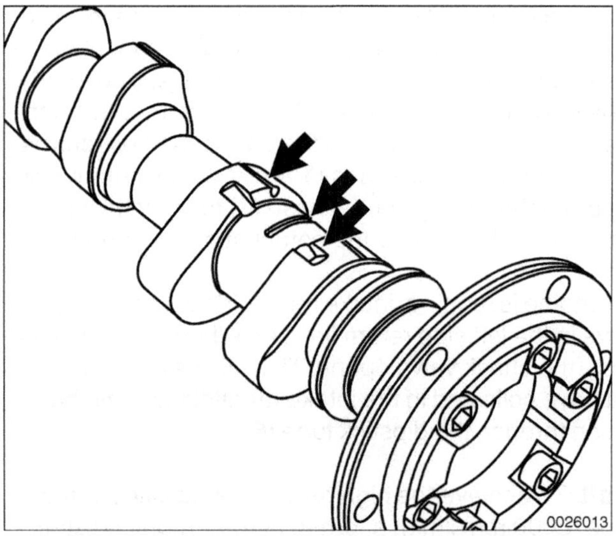


The camshaft lobes have oil grooves (**arrows**) that provide lubrication from the camshaft journals to the lobes and rocker arms.

**Valve train**. The valve actuating mechanism uses finger-type rocker arms. As the entire arm does not move the distance of valve lift, its effective reciprocating mass is less than its actual mass. Altogether the rocker arm effective mass is 30% less than that of a bucket tappet. This allows for lighter valve springs, thus reducing inertia. These engineering features contribute to the ability of the engine to rev up to 8000 rpm.

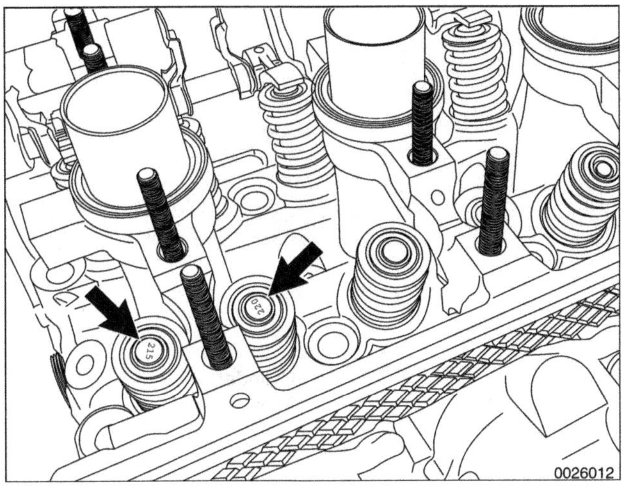

0026012

The valve actuation system requires periodic inspection and, if necessary, adjustment of valve lash using different thicknesses of shims (**arrows**). Camshaft removal is not required for valve lash adjustment. See **116 Cylinder Head and Valvetrain.**

The intake and exhaust valves are lightweight in design to reduce reciprocating mass. The exhaust valve stems are sodium-filled to enhance cooling.

Two progressively wound valve springs are used at each valve.

**High pressure double VANOS.** The S54 engine has stepless double VANOS, which varies valve timing on both the intake and exhaust camshafts. This system varies valve timing over a wide range and contributes to the engine's high power output.

| **High pressure VANOS range** | |
|---|---|
| Total camshaft adjustment range<br>• Intake<br>• Exhaust | <br>60°<br>46° |
| Default camshaft position without VANOS<br>• Intake<br>• Exhaust | <br>130° spread angle<br>83° spread angle |

VANOS is hydraulically actuated via electronic controls. The VANOS system has its own radial-piston oil pump. This pump is part of the exhaust camshaft's VANOS mechanism, and produces up to 1740 psi (approximately 100 bar) of oil pressure. The high pressure enables the system to vary valve timing quickly at very high rpm.

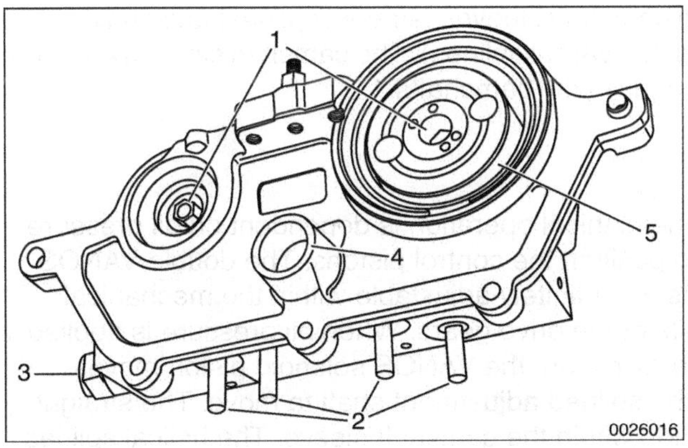

0026016

The VANOS unit is mounted to the front of the cylinder head and contains:

1. Hydraulically actuated mechanical sprocket drives
2. Electronically controlled oil pressure regulating solenoids
3. Pressure regulating valve
4. Inlet oil supply pressure reducing valve
5. Radial piston high pressure output pump driven by exhaust camshaft

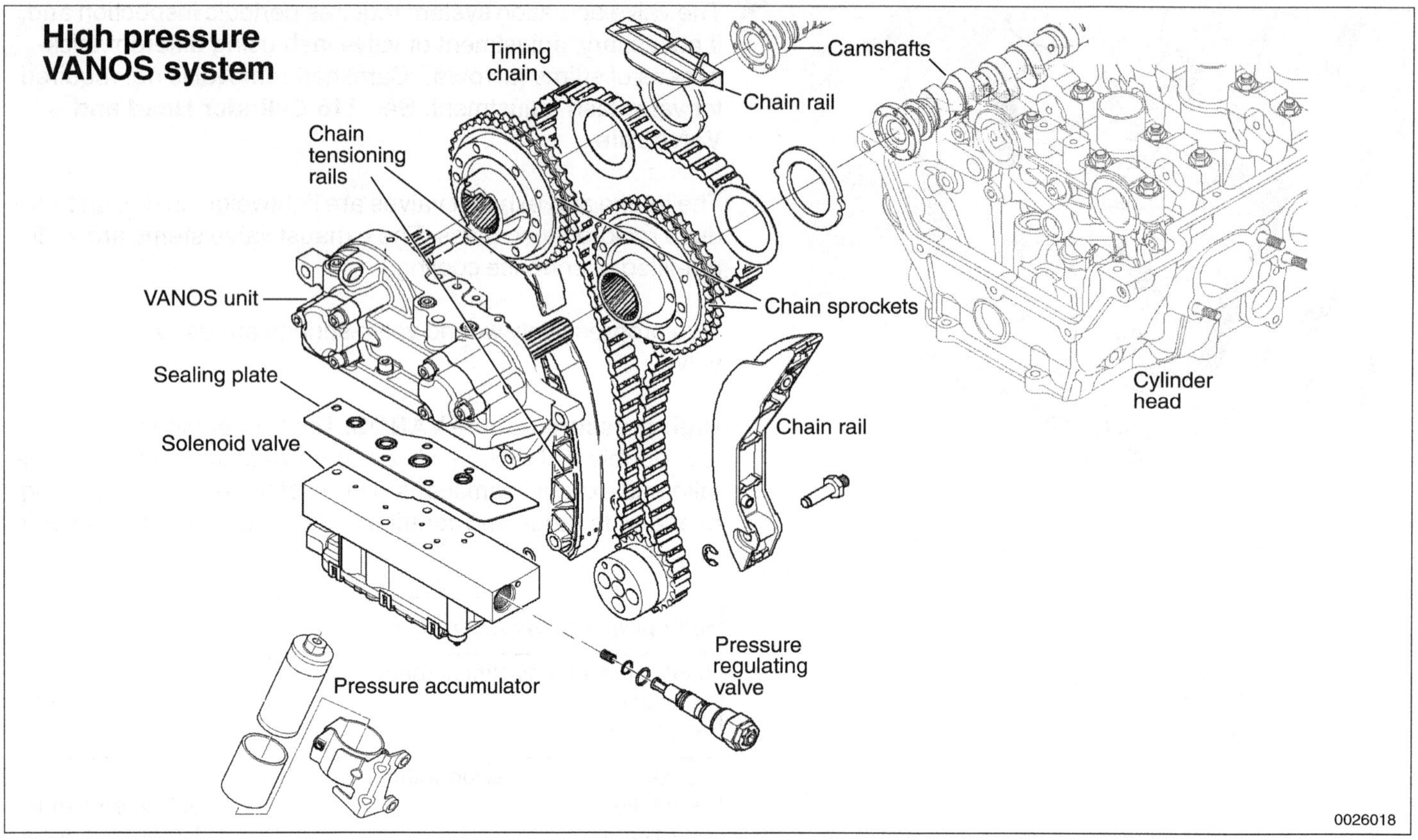


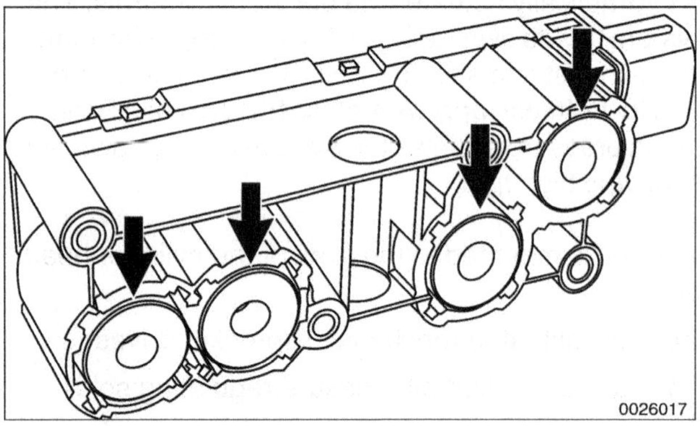


The VANOS solenoid electrical assembly (shown removed from the VANOS unit) contains four solenoids (**arrows**). Two ECM-controlled solenoids are required for each camshaft, one for advancing and one for retarding the camshaft timing.

VANOS adjustment shafts contain two sets of splines that engage with camshaft sleeves (straight splines) and chain sprockets (helical cut splines). The camshaft sleeves are bolted to the ends of the camshafts.

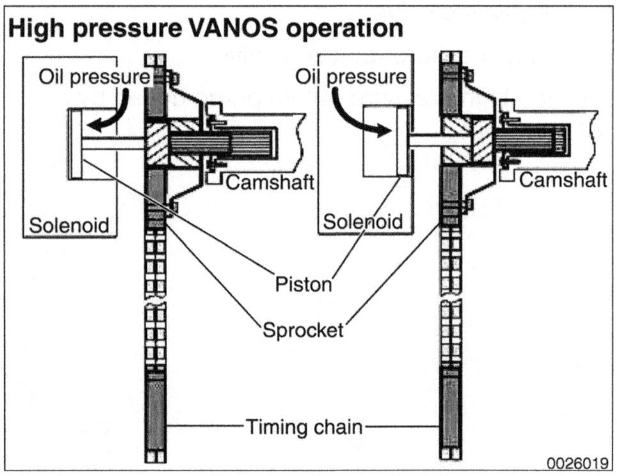


VANOS mechanical operation is dependent on oil pressure applied to position the control pistons. The double VANOS camshafts are infinitely adjustable within the mechanical travel limits of the drive gears. When oil pressure is applied to the control piston, the VANOS solenoid piston moves, causing the splined adjustment shaft to move. The straight splines slide within the camshaft sleeve. The helical splines rotate the camshaft drive sprocket, changing sprocket position in relation to camshaft position. This advances or retards camshaft timing.

Oil is supplied from the main oil galley through the front of the cylinder head to the inlet pressure reducing valve. The pressure reducing valve supplies oil to the radial piston high pressure pump located between the cylinder head and the

VANOS unit. The valve ensures the oil pressure supply to the VANOS pump is 0.5 bar regardless of the varying pressure from the main oil pressure galley. The pressure reducing valve is pressed into the VANOS unit and secured by an O-ring. The pressure regulating valve is mounted in the VANOS unit and regulates the pressure produced by the radial piston high pressure pump.

The pressure accumulator ensures that there is a sufficient volume of oil under pressure to adjust the camshafts under all engine operating conditions. The accumulator is nitrogen charged and is located on the exhaust side of the engine behind the A/C compressor. It is connected to the VANOS unit by a high pressure line.

**Lubrication system**. A semi-dry-sump system helps ensure adequate lubrication under high cornering, acceleration and braking loads. Particularly in hard cornering to the left, it is critical to achieve effective return of oil to the sump.

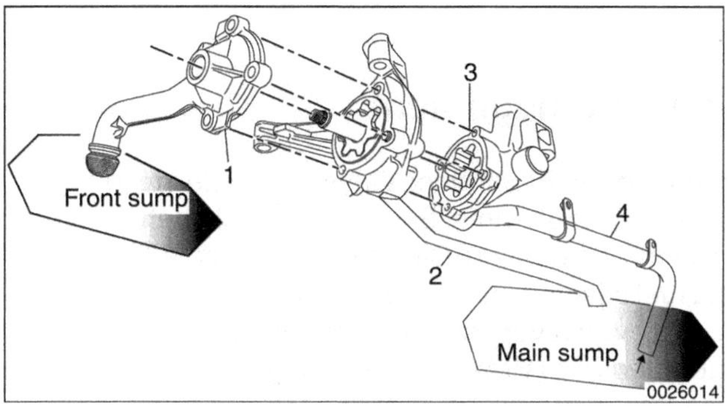


Integrated into the gear-type pressure pump is a scavenging pump (**1**) that collects oil from the right side of the small forward oil sump and pumps it back into the main, larger rear sump through a small pipe (**2**). The main oil pump (**3**) draws oil from the main sump through a large pipe (**4**) and supplies the main oil galley in the engine block.

The rear sump is almost completely closed off from the rest of the system, and thus able to hold the oil necessary for the required lubrication throughout the engine. Specific return passages are also incorporated into the intake (left) side of the engine to ensure ideal oil flow under all conditions.

The graphite-coated aluminum pistons are cooled by oil spray. Each valvetrain rocker arm is sprayed with oil just as it is about to be loaded by the camshaft lobe.

A thermal sender monitors both oil level and temperature and an oil temperature gauge is included in the instrument console.

**S54 B32 engine coolant pump/ thermostat assembly**

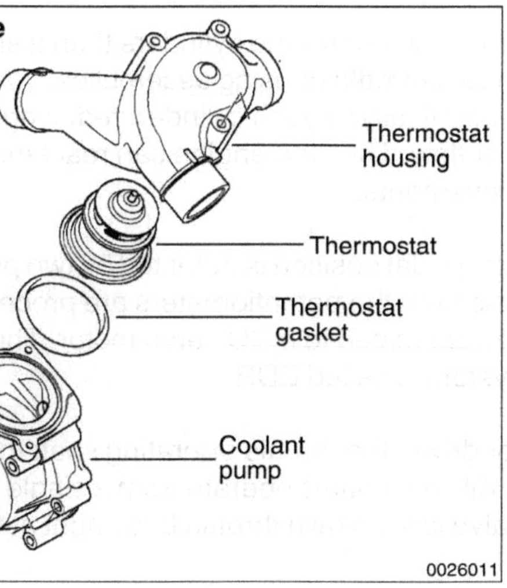


**Cooling system**. The cylinder head incorporates crossflow cooling, promoting consistent temperatures from the front to the rear of the head and helping to minimize distortion and wear under the extreme heat such a high performance engine develops when its full power is being exploited.

Crossflow cooling requires high coolant flow. To this end a coolant pump with contoured vanes is used. To control production costs, each coolant pump vane is produced as a small plastic casting, pressed into a plastic rotor and then welded into place. This impeller boosts coolant pump efficiency significantly.

Also boosting efficiency is a ring-type thermostat, bolted directly into the coolant pump housing, which imposes less resistance to coolant flow than conventional plate thermostats. The 80°C (176°F) thermostat is not electrically heated.

2001 - 2002 M version exhaust system

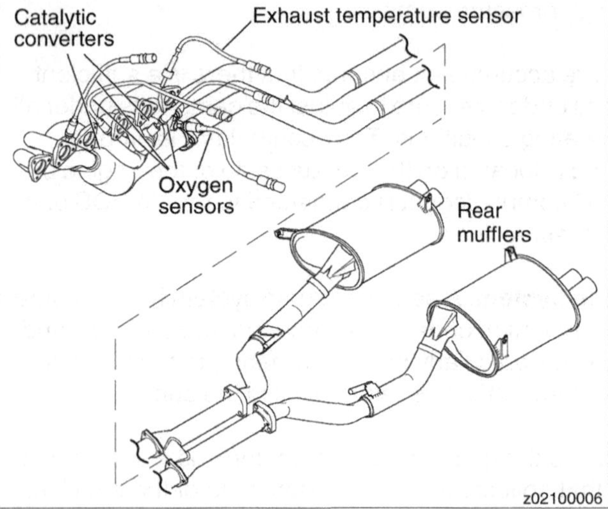


**Exhaust system**. For the 2001 - 2002 M models a new, relatively free flowing exhaust system has been developed.

Two elaborate, snaking stainless headers, each serving three cylinders, incorporate the catalytic converters and include threaded bosses for the 4 oxygen sensors and the exhaust temperature sensor. The headers are formed under high pressure with water inside them. This ensures even distribution of the forming pressure and makes it possible to manufacture the headers with a consistent wall thickness of 1 mm (about 0.04 in). This not only helps to save weight, but also speeds up engine and exhaust system warm-up. Each header is a single piece instead of several welded pieces.

## S54 engine management

**Engine electronics**. Electronics plays a key role in the ability of the S54 to reach high rpms. Every 6° of crankshaft rotation, the engine control module (ECM) calculates and adjusts the ignition and fuel injection at each cylinder individually. Ignition takes place through a very small-diameter pencil coil at each cylinder, adapted from BMW's V-12 racing engine.

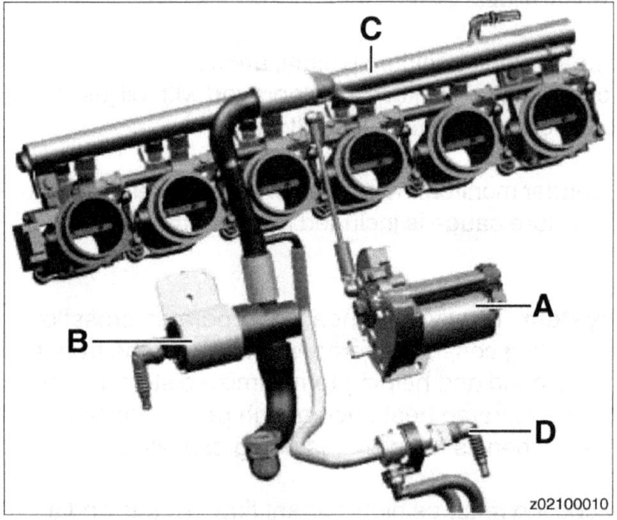


**Intake air system**. The S54 engine uses six individual throttle housings operated by the following components:

- Electronic throttle control actuator (**A**)
- Idle air actuator (**B**)
- External air distribution pipe (**C**) to the individual throttle housings
- Evaporative emissions (fuel tank ventilation) valve (**D**)

Positioned much nearer the cylinders than a single throttle can be, the six throttle housing assemblies bring atmospheric pressure practically to each cylinder, reducing the lagtime inherent in airflow. Thus the engine can react more quickly to throttle movements.

Accelerator pedal position is detected by two potentiometers. The signals from the potentiometers are processed by the ECM and transmitted to a DC servo motor. This electronic throttle system is called EDR.

The motor drives the throttle operating shaft through a tiny gearbox. All six throttles operate from a single shaft. Each throttle valve is in its own throttle body right at the intake port.

Upstream of the throttle bodies are the six intake trumpets, made of weight efficient fiberglass thermoplastic; their diameter-to-length ratio was calibrated with a computer program of BMW's Formula 1 Motorsport department. The trumpets are laser-welded into the induction plenum of the same composite material to form a single assembly.

Together with the stepless VANOS, this induction system contributes to the engine's breathing and fuel / air processing capabilities.

**Knock sensors**. Three knock sensor units are installed on the intake side of the cylinder head.

**Catalytic converters and oxygen sensors.** Each exhaust header includes a catalytic converter. This puts the catalysts closer to the engine, improving emission control when the engine is started from cold and meeting stringent U.S. regulations in this regard. Four oxygen sensors are employed, two pre- and two post-catalyst. The engine meets U.S. LEV (low emissions vehicle) standards.

# 110 Engine Removal and Installation

## GENERAL

Engine removal and installation procedures are covered in this repair group.

Remove the intake manifold and transmission prior to engine assembly removal. Procedures for removing the intake manifold and throttle assembly are given in **113 Cylinder Head Removal and Installation**. Transmission removal and installation procedures are covered in **230 Manual Transmission** or **240 Automatic Transmission**.

For additional procedures required during engine removal, refer to the following repair groups:

- **020 Maintenance**
- **121 Battery, Alternator, Starter**
- **170 Radiator and Cooling System**
- **180 Exhaust System**
- **410 Fenders, Engine Hood**

### Special tools

Special tools are required for engine removal and installation. Be sure to have the necessary equipment on hand before starting the job.

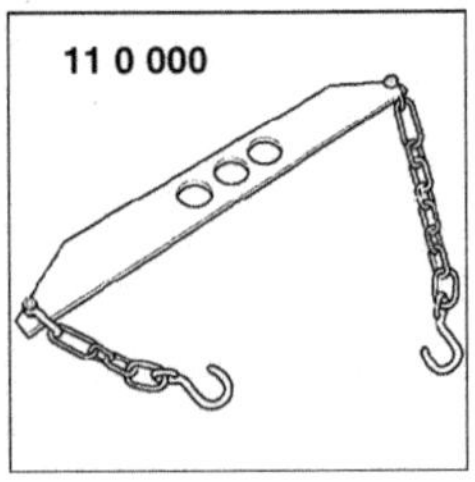


Engine lifting bracket
(BMW special tool 11 0 000)

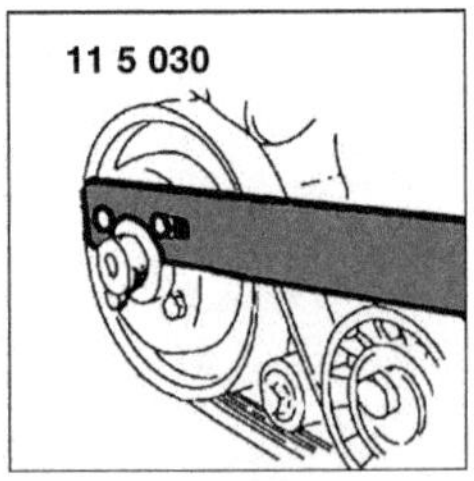


Cooling fan counterhold tool
(BMW special tool 11 5 030)

Cooling fan wrench
(BMW special tool 11 5 040)

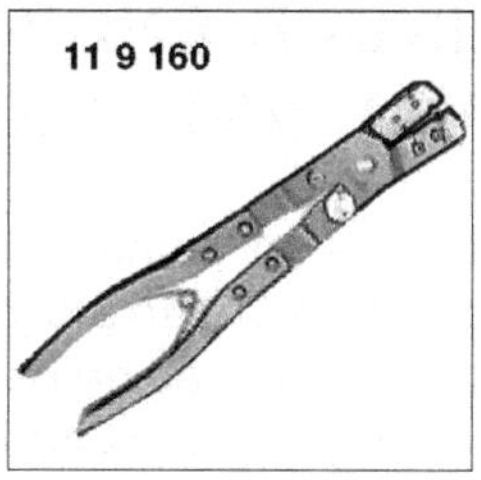


Hose clamp pliers
(BMW special tool 11 9 160)

Pry tool for injector rail
(BMW special tool 12 1 120)

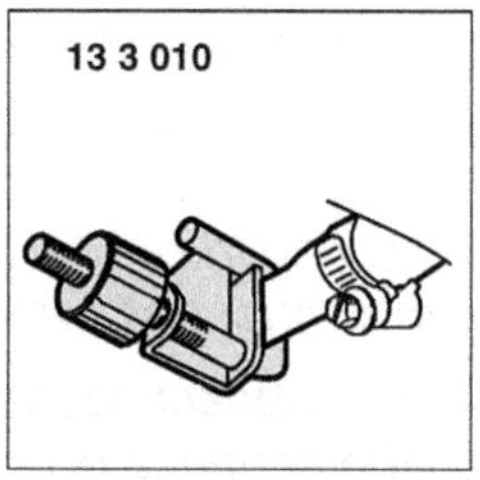


Fuel line clamp
(BMW special tool 13 3 010)

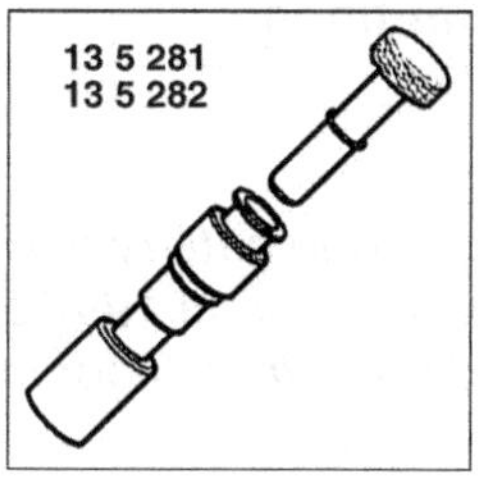


Fuel line plugs
(BMW special tool 13 5 281 / 13 5 282)

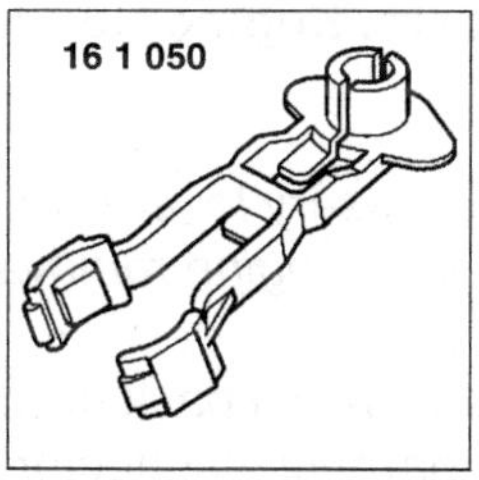


Fuel line removal tool
(BMW special tool 16 1 050)

***CAUTION—***

- *Disconnecting the battery may erase fault code(s) stored in memory. Check for fault codes prior to disconnecting the battery cables.*
- *If the MIL (malfunction indicator light) is illuminated, see* **OBD On-Board Diagnostics** *for DME fault code information.*
- *If any other system faults have been detected (indicated by an illuminated warning light), see the appropriate repair group in this manual or an authorized BMW dealer. Additional systems with self-diagnostic capabilities include:*
  *-ABS (antilock brakes)*
  *-automatic transmission*
  *-SRS (airbags)*
  *-ASC / DSC (traction and stability control)*
  *-EWS (electronic immobilizer)*

## 4-CYLINDER ENGINE REMOVAL AND INSTALLATION

Be sure to cover all painted surfaces before beginning the removal procedure. As an aid to installation, label all components, wires and hoses before removing them. Do not reuse gaskets, O-rings or seals during reassembly.

> ***WARNING —***
> *To avoid personal injury, be sure the engine is cold before beginning the engine removal procedure.*

### Engine, removing and installing (4-cylinder engine)

– Disconnect negative (–) battery cable.

> ***CAUTION—***
> - *Disconnecting the battery cables may erase fault codes stored in control unit memory, as well as radio presets.*
> - *Prior to disconnecting the battery, read the battery disconnection cautions given in* **001 General Warnings and Cautions**.

***NOTE —***
*In 1996 cars, the battery is located in the right rear of the engine compartment. In 1997 and later cars, the battery is located under the trunk floor mat.*

– Remove engine hood or place hood in service position. See **410 Fenders, Engine Hood**.

***NOTE —***
*It is not necessary to remove the engine hood, but it is helpful and will make engine removal and installation easier.*

– Unbolt ignition coil pack from right strut tower. Disconnect electrical harness connector from coil pack and place coil pack on engine.

– Raise vehicle and support safely.

> ***WARNING —***
> *Make sure the car is stable and well supported at all times. Use a professional automotive lift or jack stands designed for the purpose. A floor jack is not adequate support.*

– Remove splash guard under engine.

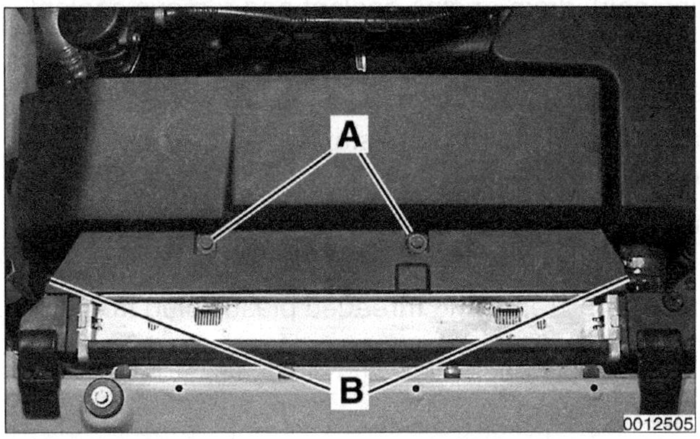


◀ Remove engine intake air duct mounting screws (**A**) from top of radiator.

- Detach plastic rivets (**B**) at sides of radiator.
- Disconnect air duct and remove from engine compartment.

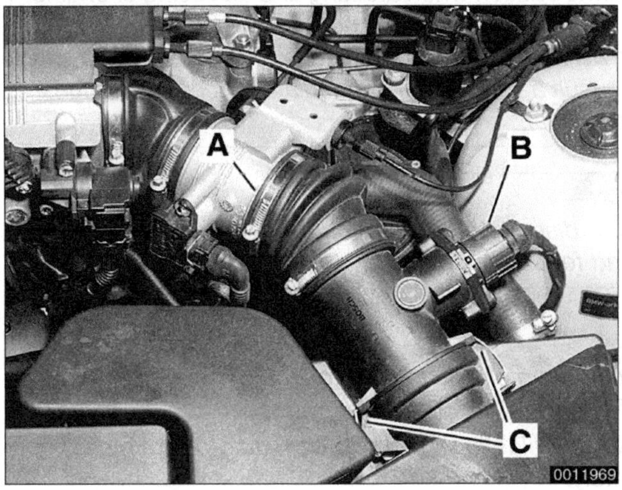


◀ Unbolt and remove complete air cleaner housing with mass air flow sensor.

- Remove intake air duct hose clamp (**A**).
- Detach mass air flow sensor electrical harness connector (**B**).
- Unclip mass air flow sensor retaining clips (**C**).

– Detach harness connectors and throttle cable(s) from throttle housing.

– Label and disconnect fuel lines. See **160 Fuel Tank and Fuel Pump**.

> ***WARNING—***
> - *Fuel may be expelled under pressure. Do not smoke or work near heaters or other fire hazards. Keep a fire extinguisher handy. Before disconnecting fuel hoses, wrap a cloth around fuel hoses to absorb any leaking fuel. Plug all open fuel lines.*
> - *Always unscrew the fuel tank cap to release pressure in the tank before working on the tank or lines.*

– Remove upper and lower intake manifolds. See **113 Cylinder Head Removal and Installation**.

> ***CAUTION—***
> *Stuff clean rags into the open intake ports to prevent any parts from falling into the engine intake.*

– Detach engine wiring harness connectors at following components:

- Starter motor
- Alternator
- Ignition coils
- Fuel injectors
- Temperature sensors
- Crankshaft and camshaft position sensors

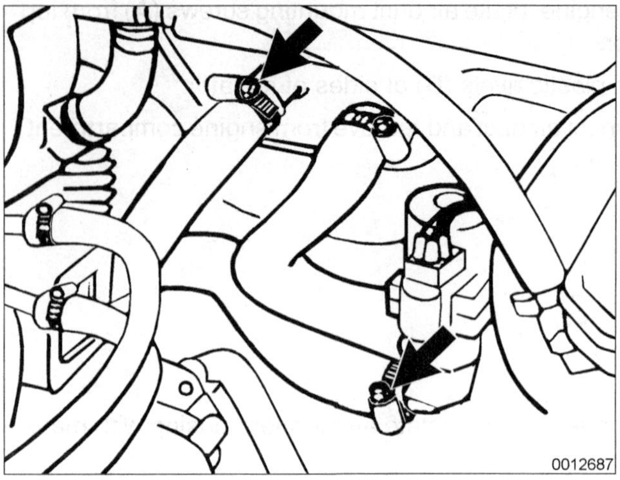


With engine cold, drain engine coolant and remove coolant hoses:

- Drain engine block by removing block drain.
- Disconnect hoses from thermostat housing in front of cylinder head.
- Disconnect hose clamps (**arrows**) at rear of engine and at heater valve. Pull off hoses.
- Drain radiator by removing threaded plastic plug from lower left corner.

> ***WARNING —***
> *Use extreme caution when draining and disposing of engine coolant. Coolant is poisonous and lethal to humans and pets. Pets are attracted to coolant because of its sweet smell and taste. Consult a veterinarian immediately if coolant is ingested by an animal.*

> ***CAUTION—***
> *Catch and dispose of drained coolant according to local, state, and federal laws.*

***NOTE —***
- *The block drain plug is located on the exhaust side toward rear of engine.*
- *Remove tiny plastic lock clip to pull radiator drain plug out completely.*

– Remove radiator cooling fan and radiator. See **170 Radiator and Cooling System**.

– Remove drive belts. Mark belt direction of rotation if belts are to be reused. See **020 Maintenance**.

– Unbolt power steering fluid reservoir from its mounting bracket. Use stiff wire to hang reservoir to one side. Do not disconnect fluid lines.

– Remove power steering pump from its mounting bracket. Use stiff wire to hang pump from body. Do not disconnect fluid lines.

Remove A/C compressor mounting bolts (**arrows**).

- Do not disconnect any refrigerant lines.
- Move A/C compressor out of the way without distorting or damaging refrigerant lines. Support compressor by hanging it from chassis using stiff wire.

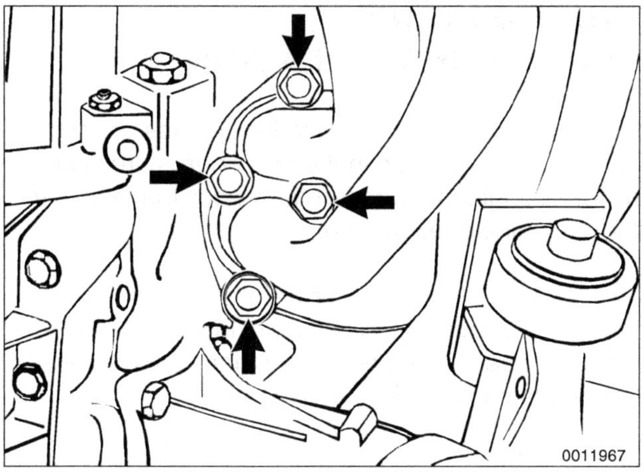


◀ Remove front exhaust pipe mounting nuts (**arrows**) from exhaust manifold. Remove complete exhaust system. See **180 Exhaust System**.

– Remove transmission from car. See **230 Manual Transmission** or **240 Automatic Transmission**.

***NOTE —***
*On vehicle with automatic transmission: Remove front and rear brackets holding transmission cooler lines to engine carrier and oil pan.*

– Install engine lifting device and raise engine slightly.

– Detach right side engine ground strap at engine mount.

– Remove nuts from left and right engine mounts.

– Carefully raise engine out of car, checking for any wiring, fuel lines or mechanical parts that might become snagged as engine is removed.

– Installation is reverse of removal.
  - Be sure drive belts engage pulley grooves correctly.
  - Change engine oil and filter and check all other fluid levels. See **020 Maintenance**.
  - Refill and bleed cooling system. See **170 Radiator and Cooling System**.
  - Install front exhaust pipe using new gaskets. Use copper antiseize paste on threads. See **180 Exhaust System**.

| **Tightening torques** | |
|---|---|
| Coolant drain plug to cylinder block (M14 x 1.5) | 25 Nm (18 ft-lb) |
| Exhaust manifold to cylinder head (replace nuts, use copper antiseize paste))<br>• M6 nut<br>• M7 nut | <br><br>10 Nm (7 ft-lb)<br>20 Nm (15 ft-lb) |
| Exhaust pipe to exhaust manifold (M10) | 45 Nm (33 ft-lb) |
| Intake manifold to cylinder head<br>• M6 nut<br>• M7 nut | <br>10 Nm (7 ft-lb)<br>15 Nm (11 ft-lb) |
| Radiator cooling fan to coolant pump | 40 Nm (30 ft-lb) |
| Radiator drain screw to radiator | 2.5 Nm (22 in-lb) |
| Wiring to alternator<br>• Terminal D+ (M6 nut)<br>• Terminal B+ (M8 nut) | <br>7 Nm (5 ft-lb)<br>13 Nm (10 ft-lb) |
| Wiring to starter<br>• M6 nut<br>• M8 nut | <br>5 Nm (44 in-lb)<br>11 Nm (8 ft-lb) |

## 6-CYLINDER ENGINE REMOVAL AND INSTALLATION

Engine removal procedures for 6-cylinder engines are covered in this section. Most steps in the procedures are similar or the same for all engines. Specific differences that apply are noted at the beginning of each step.

Be sure to cover all painted surfaces before beginning the removal procedure. As an aid to installation, label all components, wires and hoses before removing them. Do not reuse gaskets, O-rings or seals during reassembly.

> ***WARNING —***
> *Due to risk of personal injury, be sure the engine is cold before beginning the engine removal procedure.*

### Engine, removing and installing (6-cylinder engine)

– Disconnect negative (–) battery cable.

> ***CAUTION—***
> - *Disconnecting the battery cables may erase fault codes stored in control unit memory, as well as radio presets.*
> - *Prior to disconnecting the battery, read the battery disconnection cautions given in* **001 General Warnings and Cautions**.

– Remove engine hood or place hood in service position. See **410 Fenders, Engine Hood**.

***NOTE —***
*It is not necessary to remove the engine hood, but it is helpful and will make engine removal and installation easier.*

– Raise vehicle and support safely.

> ***WARNING —***
> *Make sure the car is stable and well supported at all times. Use a professional automotive lift or jack stands designed for the purpose. A floor jack is not adequate support.*

– Remove splash guard under engine.

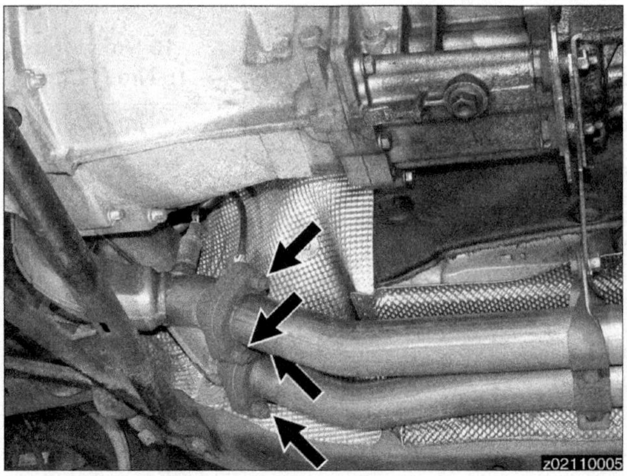


< Remove front exhaust pipe mounting nuts (**arrows**). Remove complete exhaust system. See **180 Exhaust System**.

– Remove transmission from car. See **230 Manual Transmission** or **240 Automatic Transmission**.

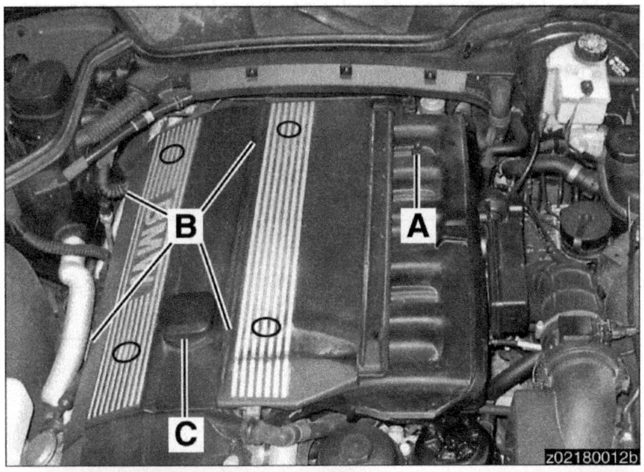


◀ M52, S52, M52 TU or M54 engine: Working at top of engine:

- If applicable: Loosen positive battery cable connection (**A**) and push connector down under intake manifold.
- Remove trim covers (**black ovals**) from engine plastic covers. Remove cover hold down fasteners and lift off covers.
- Make sure rubber insulators (**B**) on either side of large cover do not fall off during cover removal.
- Remove oil filler cap (**C**) to allow removal of large plastic cover. Be sure to replace it on the oil filler immediately to prevent debris from entering the engine.

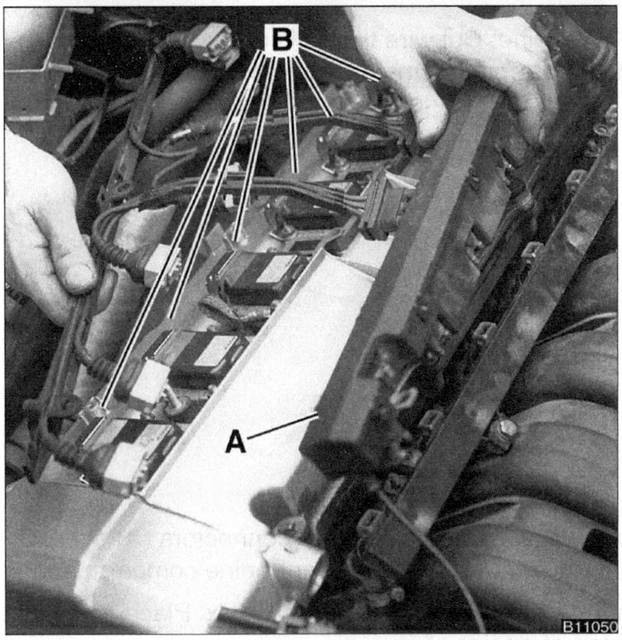


◀ M52 or S52 engine: Working at top of engine:

- Remove hold-down fasteners from fuel injector harness (**A**).
- Disconnect ignition coil harness connectors (**B**) from coils.
- Lift off both harnesses and place at base of windshield.

– Label and disconnect fuel lines. See **160 Fuel Tank and Fuel Pump**.

> ***WARNING—***
> - *Fuel may be expelled under pressure. Do not smoke or work near heaters or other fire hazards. Keep a fire extinguisher handy. Before disconnecting fuel hoses, wrap a cloth around fuel hoses to absorb any leaking fuel. Plug all open fuel lines.*
> - *Always unscrew the fuel tank cap to release pressure in the tank before working on the tank or lines.*

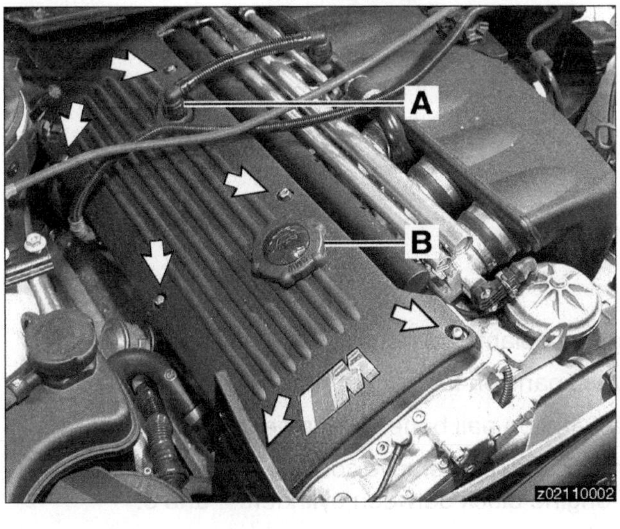


◀ S54 engine: remove cylinder head top cover:

- Disconnect vent hose (**A**).
- Remove oil filler cover (**B**).
- Remove cover hold-down bolts (**arrows**) and lift off cover.

S54 engine: Working at rear of cylinder head cover, separate oxygen sensor connectors (**arrows**). Thread harnesses off top of cylinder head and lay aside.

– Remove intake manifold and throttle housing. See **113 Cylinder Head Removal and Installation**.

> ***CAUTION—***
> *Stuff clean rags into the open intake ports to prevent any parts from falling into the engine intake.*

M52 or S52 engine: Cut wire ties holding engine harness to bulkhead and disconnect main engine harness connectors (**arrows**) at left rear of engine compartment.

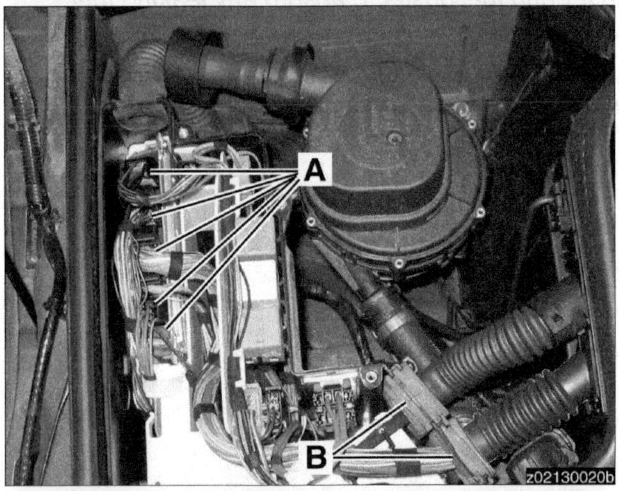


M52 TU, M54 or S54 engine:

- Disconnect engine wiring harness connectors (**A**) at DME control module at E-box, right rear engine compartment.
- Slide harness boots (**B**) off edge of E-box. Place connectors on top of engine.

***NOTE—***
*Remove engine electrical harness with engine.*

Drain engine coolant and remove coolant hoses.

- Remove expansion tank cap on radiator.
- Place a 3-gallon pail beneath engine to capture coolant.
- M52 or S52 engine: Remove coolant drain plug on exhaust side of engine block between cylinders 4 and 5.
- M52 TU, M54 or S54 engine: Remove coolant drain plug on exhaust side of engine block at cylinder 2 (**arrow**).

◄ Drain radiator into a 3-gallon pail by removing plastic drain plug (**arrow**) at base of radiator completely.

> ***WARNING—***
> *Use extreme caution when draining and disposing of engine coolant. Coolant is poisonous and lethal to humans and pets. Pets are attracted to coolant because of its sweet smell and taste. Consult a veterinarian immediately if coolant is ingested by an animal.*

> ***CAUTION—***
> *Catch and dispose of drained coolant according to local, state, and federal laws.*

***NOTE—***
*The radiator drain plug shape and design varies. However, it is always found on the left (driver) side of the radiator.*

– Remove radiator cooling fan and radiator. See **170 Radiator and Cooling System**.

> ***CAUTION—***
> *32 mm radiator fan mounting nut has left hand threads.*

– S54 engine: Remove engine oil cooler.

– Automatic transmission model: Detach transmission cooler lines from radiator, remove brackets holding lines to side of engine, and store lines in a clean environment.

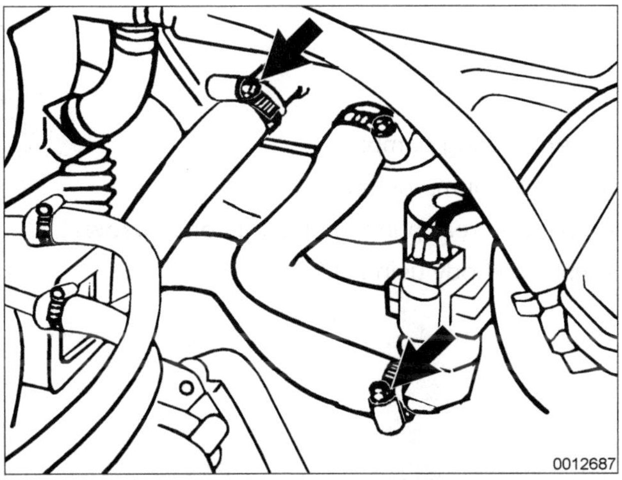


◄ M52 or S52 engine: Disconnect hose clamps (**arrows**) at rear of engine and at heater valve. Pull off hoses.

◄ M52 TU or M54 engine: Pull out quick-release circlip (**arrow**) to detach coolant hose at left rear of engine above starter.

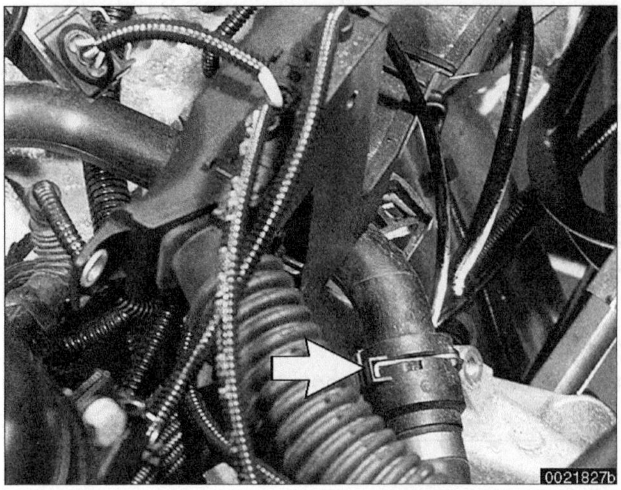


M52 TU or M54 engine: Pull out quick-release circlip (**arrow**) to detach coolant hose at left front of engine above engine mount.

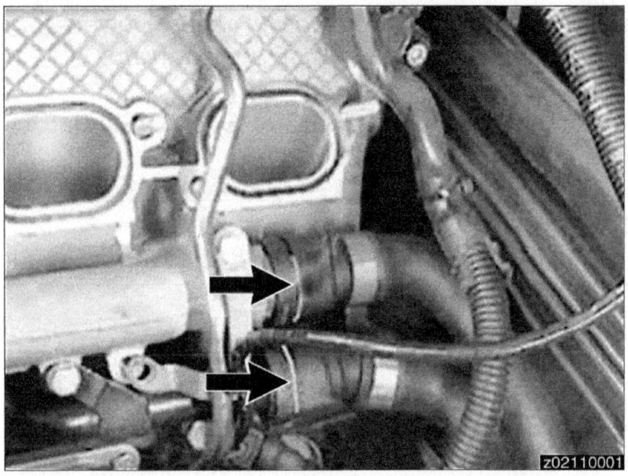


S54 engine: Pull out quick-release circlips (**arrows**) to detach coolant hoses at left rear of engine below cylinder head intake ports.

– Remove drive belts. Mark belt direction of rotation if belts are to be reused. See **020 Maintenance**.

– Unbolt power steering fluid reservoir from its mounting bracket. Use stiff wire to hang reservoir to one side. Do not disconnect fluid lines.

Remove power steering pump:

- Remove steering pump pulley.
- Remove front and rear pump mounting bolts (**arrows**).
- Hang pump from body using stiff wire.

Remove A/C compressor mounting bolts (**arrows**).

- Do not disconnect any refrigerant lines.
- Move A/C compressor out of the way without distorting or damaging refrigerant lines. Support compressor by hanging it from chassis using stiff wire.

– Remove windshield and headlight washer reservoir:

- Remove bolt on top of reservoir.
- Lift tank and disconnect electrical connections to pump and to washer fluid level sensor.
- Disconnect hoses to windshield washer and headlight washer (if applicable). Tilt reservoir to prevent fluid from leaking out.

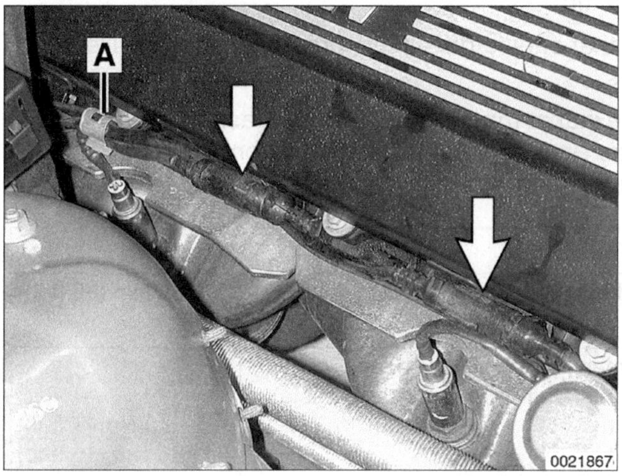


M52 TU or M54 engine: Separate oxygen sensor electrical harness connectors (**arrows**) and mounting clip (**A**) on right side of engine.

– Detach exhaust manifolds and catalytic converters (if applicable) from cylinder head and set aside.

> ***CAUTION—***
> *M52 TU, M54 or S54 engine: Store exhaust manifolds in a safe place to avoid damage to oxygen sensors.*

Remove chassis ground strap (**arrow**) at right engine mounting pedestal.

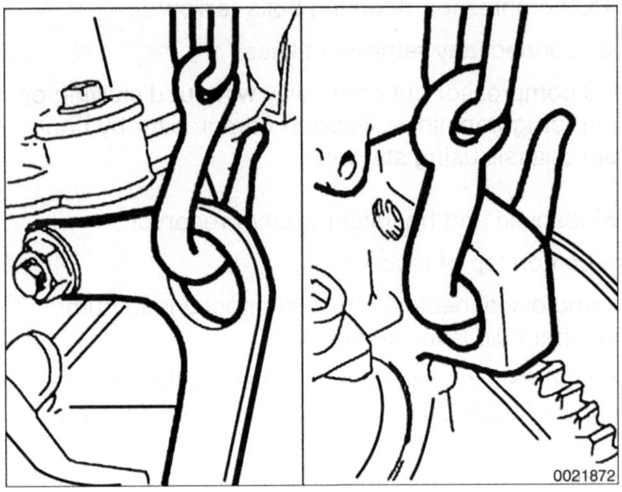


- Install engine lifting device (BMW 11 0 000 or equivalent) to front and rear engine supports and raise engine until its weight is supported. Detach left and right engine mounts.

– Carefully raise engine out of car, checking for any wiring, fuel lines or mechanical parts that might become snagged as engine is removed.

– Installation is reverse of removal, noting the following:
  - Replace all gaskets, O-rings and seals.
  - Change engine oil and filter and check all other fluid levels. See **020 Maintenance**.
  - Refill and bleed cooling system. See **170 Radiator and Cooling System**.
  - To facilitate reassembly, coat mass air flow sensor O-ring seal with acid-free grease.
  - When reattaching throttle housing harness connector, be sure to tighten connector until arrows on connector and plug line up.
  - Check that engine drive belts properly engage pulley grooves.
  - Install exhaust manifolds using new gaskets and self-locking nuts. Use copper antiseize paste on threads. See **180 Exhaust System**.

| **Tightening torques** | |
|---|---|
| Coolant drain plug to cylinder block (M14 x 1.5) | 25 Nm (18 ft-lb) |
| Exhaust manifold to cylinder head (replace nuts, use copper antiseize paste)<br>• M6 nut<br>• M7 nut | <br><br>10 Nm (7 ft-lb)<br>20 Nm (15 ft-lb) |
| Exhaust pipe to exhaust manifold (M10) | 45 Nm (33 ft-lb) |
| Intake manifold to cylinder head<br>• M6 nut<br>• M7 nut | <br>10 Nm (7 ft-lb)<br>15 Nm (11 ft-lb) |
| Radiator cooling fan to coolant pump | 40 Nm (30 ft-lb) |
| Radiator drain screw to radiator | 2.5 Nm (22 in-lb) |
| Wiring to alternator<br>• Terminal D+ (M6 nut)<br>• Terminal B+ (M8 nut) | <br>7 Nm (5 ft-lb)<br>13 Nm (10 ft-lb) |
| Wiring to starter<br>• M6 nut<br>• M8 nut | <br>5 Nm (44 in-lb)<br>11 Nm (8 ft-lb) |

# 113 Cylinder Head Removal and Installation

## GENERAL

This group covers cylinder head removal and installation as well as cylinder head and valve diagnosis procedures. For the procedures given in this repair group, it is assumed that the engine is installed in the engine bay.

Engines are referred to by engine code. If necessary, see **100 Engine–General** for engine code and application information.

In order to remove the cylinder head from the 6-cylinder engine block, remove the VANOS control unit from the cylinder head first. In addition, when working on the M52TU or M54 engine, remove the camshafts. Refer to the following groups:

- **116 Cylinder Head and Valvetrain** for cylinder head and valvetrain disassembly and reconditioning information
- **117 Camshaft Timing Chains** for VANOS repair information

***NOTE—***
- *Cylinder head removal for the S54 engine used in 2001 - 2002 M version is not included in this repair group. S54 cylinder head removal and installation is a highly complex job and is outside the scope of this manual.*
- *If a cylinder head gasket problem is suspected, a compression test or leak-down test usually detects the fault. See* **Diagnostic testing** *in this repair group.*

### Special tools

Special BMW service tools are required to properly remove and install the cylinder head on engines covered by this manual. Special tools are used to time the valvetrain to the crankshaft and to remove the VANOS control unit, camshafts and Torx (E12) head bolts.

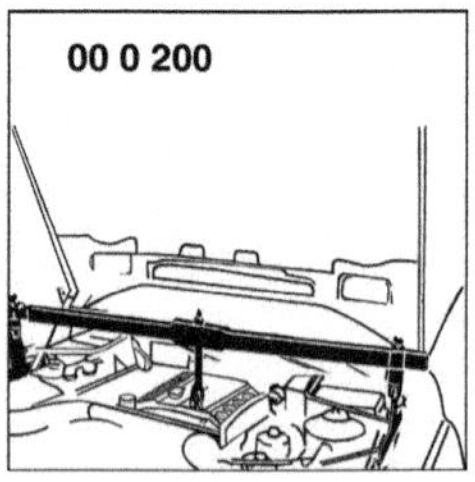


Engine lifting tools
(BMW special tool 00 0 200)

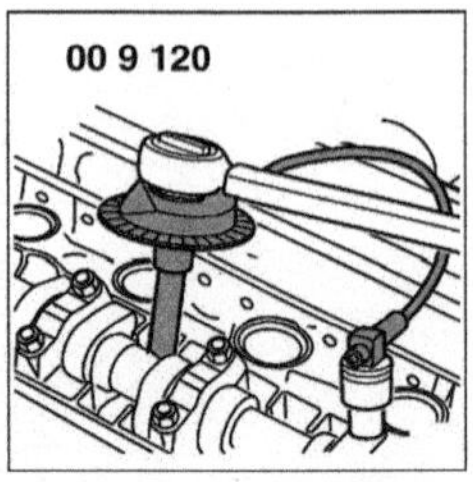


Rotary angle dial gauge
(BMW special tool 00 9 120)

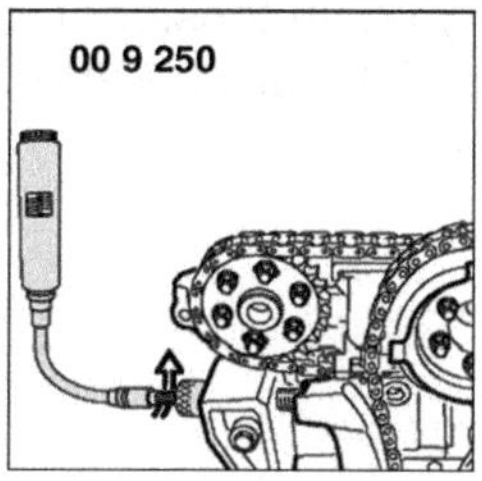


Torque wrench with flex extension
(BMW special tool 00 9 250)

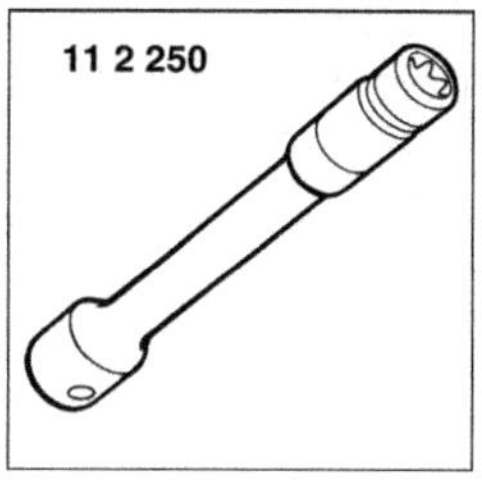


E12 Torx socket
(BMW special tool 11 2 250)

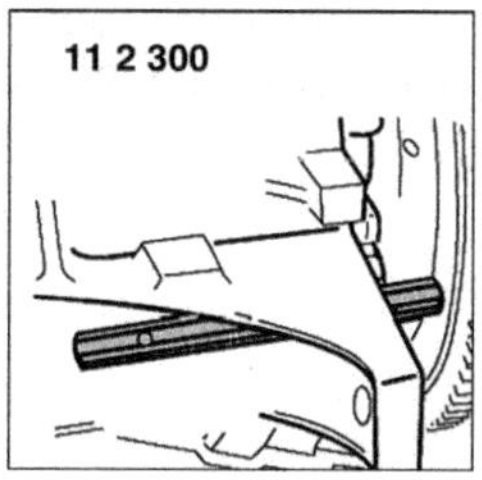


Crankshaft locking tool
(BMW special tool 11 2 300)

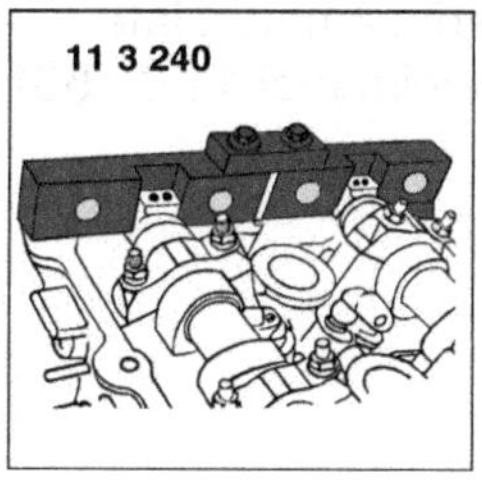


Camshaft locking tool
(BMW special tool 11 3 240)

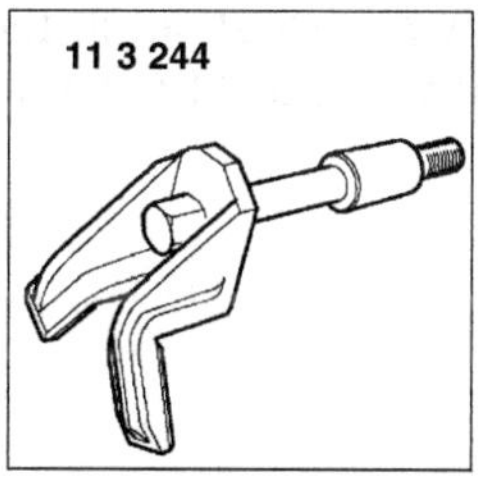


Camshaft locking tool bracket
(BMW special tool 11 3 244)

Hydraulic lifter retaining tool
(BMW special tool 11 3 250)

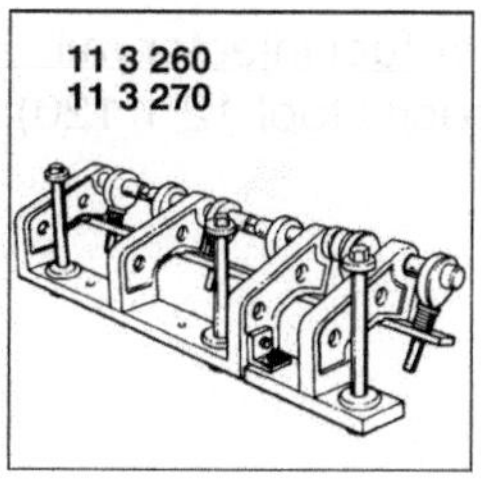


Camshaft bearing cap removal tool
(BMW special tools 11 3 260 / 11 3 270)

◀ Secondary chain tensioner locking pin
(BMW special tool 11 3 290)

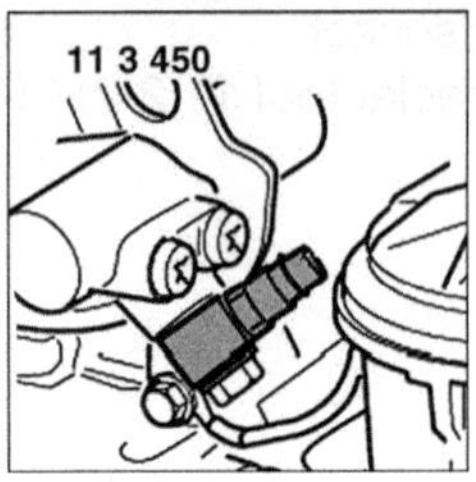


◀ Air line adapter
(BMW special tool 11 3 450)

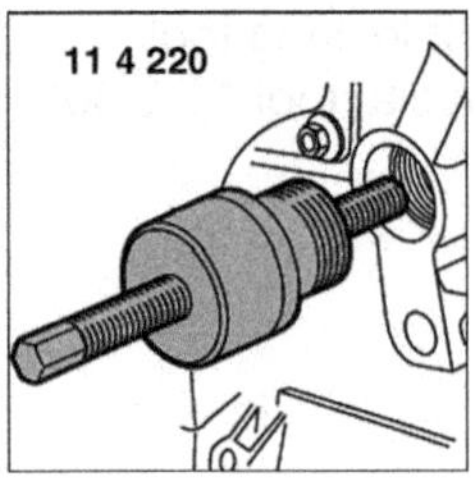


◀ Timing chain tensioner tool
(BMW special tool 11 4 220)

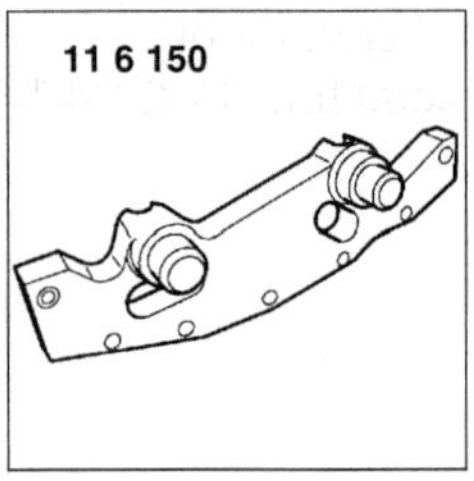


◀ VANOS adjustment plate
(BMW special tool 11 6 150)

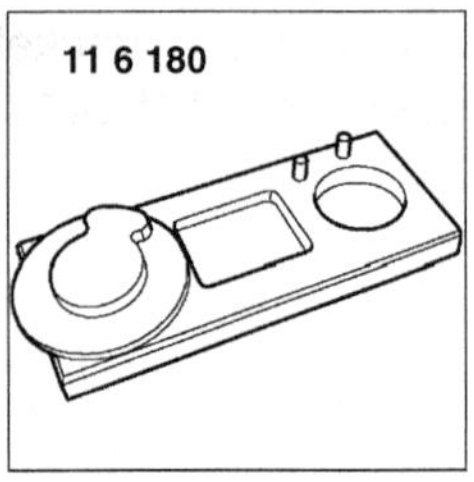


◀ Secondary sprocket setup tool
(BMW special tool 11 6 180)

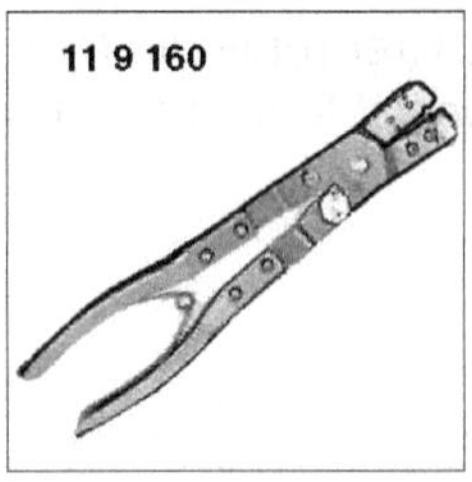


◀ Hose clamp pliers
(BMW special tool 11 9 160)

◀ Pry tool for fuel injector rail
(BMW special tool 12 1 120)

# Diagnostic Testing

## Cylinder compression, checking

A compression gauge is needed to make a compression test. For accurate testing, be sure that the battery and starter are capable of cranking the engine at least 300 rpm and that the engine is at normal operating temperature.

***NOTE—***
*Performing a compression test may cause a fault to set in the engine control module (ECM) and may illuminate the malfunction indicator light (MIL). The light can only be turned out using either BMW special service scan tools or an equivalent aftermarket scan tool. Disconnecting the battery does not erase the fault memory nor turn out the light. See* **OBD On-Board Diagnostics**.

– Disable ignition system by removing DME main relay. See **610 Component Locations** for relay location.

***WARNING—***
- *The ignition system produces high voltages that can be fatal. Avoid contact with exposed terminals and use extreme caution when working on a car with the ignition switched on or the engine running.*
- *Do not touch or disconnect ignition components while the engine is running or being cranked by the starter.*

***CAUTION—***
*Failure to remove the DME main relay or attempting to disable the fuel and ignition systems by other methods may result in damage to the engine control module (ECM).*

– Remove cylinder head top plastic cover.

< 4-cylinder model:
- Slide special tool (**arrow**) over spark plug wire. Use screwdriver for added leverage to pull straight up.
- Repeat for all cylinders.

***NOTE—***
*Special tool for spark plug connector removal is stored underneath cylinder head top plastic cover.*

6-cylinder model:

- Pull up on spring clip to disconnect ignition coil harness connector at each cylinder.
- Remove coil grounding straps.
- Remove coils.

> ***CAUTION—***
> *Note location of all grounds. Failure to reinstall grounds can result in permanent damage to engine control module or ignition system components.*

– Remove spark plugs from all cylinders.

***NOTE—***
*Check the spark plugs for oil deposits that may indicate poor cylinder sealing, then set them aside in order. Reinstall used spark plugs in the same cylinders from which they were removed.*

Install compression gauge in first cylinder spark plug hole, tight enough to form a good seal.

– With parking brake set, transmission in PARK or NEUTRAL, and accelerator pedal pressed to floor, crank engine with starter. Record highest value indicated by gauge.

***NOTE—***
- *The compression gauge reading should increase with each compression stroke and reach near its maximum reading in 4 - 6 strokes.*
- *All cylinders should reach maximum compression in the same number of strokes. If a cylinder needs significantly more strokes to reach maximum compression, there is a problem.*

– Release pressure at compression gauge valve, then remove gauge from spark plug hole. Repeat test for each cylinder and compare results with values given in **Compression specifications**.

| **Table a. Compression specifications** | |
|---|---|
| Minimum compression<br>• 4-cyl., M52, S52, M52 TU, M54<br>• S54 | <br>10 - 11 bar (145 - 160 psi)<br>11 - 12 bar (160 - 174 psi) |
| Maximum difference between cylinders (all engines) | 0.5 bar (7 psi) |

- Interpret compression readings as follows:
  - Low compression: Poorly sealed combustion chamber
  - Relatively even pressures that are below specifications: Worn piston rings and/or cylinder walls
  - Erratic values: Valve leakage
  - Dramatic differences between cylinders: Failed head gasket, burned valve, or broken piston ring

- If readings are within specifications, reinstall spark plugs and ignition coils.

***NOTE—***
*Reinstall used spark plugs in the same cylinders from which they were removed.*

| **Tightening torque** | |
|---|---|
| Spark plug to cylinder head | 25 Nm (18 ft-lb) |
| Ignition coil to cylinder head | 10 Nm (7 ft-lb) |

- Remainder of installation is reverse of removal. Be sure to reinstall all wires disconnected during test, especially ground wires at coils and cylinder head cover (where applicable).

## Wet compression test

To further analyze the source of poor compression, a wet compression test is the next step.

- Repeat compression test, but this time squirt a teaspoon of oil into each cylinder. The oil temporarily helps the seal between piston rings and cylinder wall, practically eliminating leakage past rings for a short time.

- If this test yields a higher reading than “dry” compression test, there is probably leakage between piston rings and cylinder walls, due either to wear or to broken piston rings.

- Little or no change in compression reading indicates other leakage, probably from valves.

## Cylinder leak-down test

The most conclusive diagnosis of low compression symptoms requires a cylinder leak-down test. Using a special tester and compressed air, each cylinder, in turn, is pressurized. The rate at which the air leaks out of the cylinder, as well as where the air leaks out, can accurately pinpoint the magnitude and location of the leakage.

Before attempting any repair that requires major engine disassembly, use a leak-down test to confirm low compression.

## CYLINDER HEAD COVER REMOVAL AND INSTALLATION

You must remove the cylinder head cover to remove the cylinder head.

### Cylinder head cover, removing and installing (4-cylinder engine)

> ***WARNING —***
> *To avoid personal injury, be sure the engine is cold before starting the procedure.*

– Make sure ignition is OFF. Remove ignition key.

◄ Disconnect engine breather hose (**A**) from left side of cylinder head cover.

– Detach wiring harness duct (**B**) from brackets at rear bulkhead and pull complete duct forward and up to allow access to rear of cylinder head cover. Remove duct brackets from bulkhead.

– Detach ignition coil pack (**C**) from right strut tower.

– Remove spark plugs and spark plug wire loom.
  - Remove plastic cover from top of cylinder head.
  - Disconnect spark plug wires from spark plugs and remove spark plugs.
  - Unbolt spark plug cable harness and heat shield from right side of cylinder head cover and set aside.

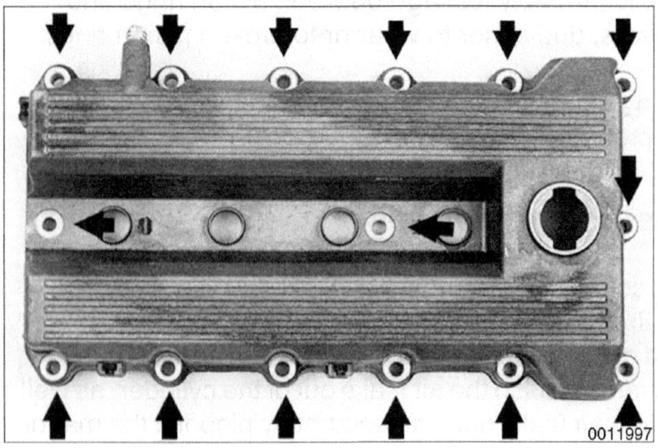


◄ Unscrew and remove cylinder head cover cap screws (**arrows**) and remove cover.

> ***CAUTION—***
> *Two of the cover hold-down screws are under the spark plug cable duct.*

***NOTE —***
*Make note of the arrangement of cap screws, washers and rubber grommets holding the cylinder head cover in place.*

Inspect camshafts oil supply tube seal (**arrow**) at front left of cylinder head. Replace if rubber is cracked or brittle.

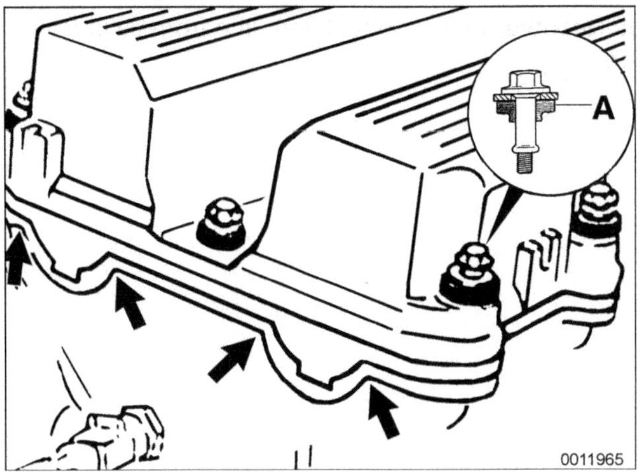


Installation is reverse of removal. Keep in mind the following:

- Make sure sealing surfaces are clean and free of old gasket material.
- Replace cylinder head cover gasket, as well as rubber grommets and seals as necessary.
- Use a small amount of 3-Bond®1209 adhesive or equivalent at corners (**arrows**) of half-moon cut-outs at back of cylinder head.
- Inspect cylinder head cover mounting screw assembly rubber seals (**A**). Replace if rubber is cracked or brittle.

| **Tightening torque** | |
|---|---|
| Cylinder head cover to cylinder head (M6) | 10 Nm (7 ft-lb) |

## Cylinder head cover, removing and installing (M52, S52, M52 TU, or M54 engine)

> ***WARNING—***
> *To avoid personal injury, be sure the engine is cold before starting the procedure.*

***NOTE—***
*The illustrations in this procedure depict an M52 TU (6-cylinder) engine installed in an E46 (3-series) chassis. Z3 models with M52, S52, M52 TU or M54 engines are similar.*

– Make sure ignition is OFF. Remove ignition key.

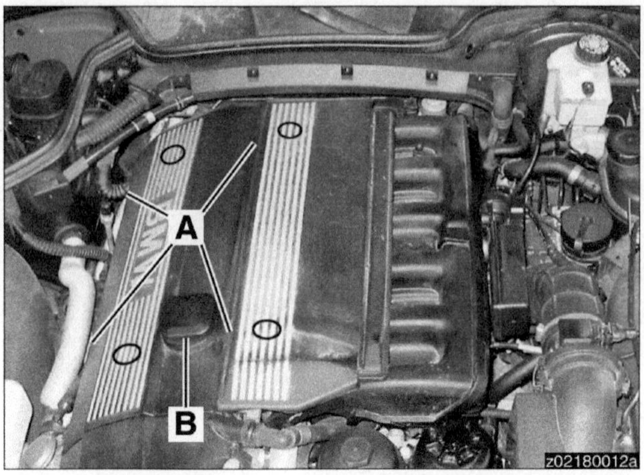


< Working at top of engine:

- Remove trim covers (**black circles**) from engine plastic covers. Remove cover hold down fasteners and lift off covers.
- Make sure rubber insulators (**A**) on either side of large cover do not fall off during cover removal.
- Remove oil filler cap (**B**) to allow removal of large plastic cover. Be sure to replace it on the oil filler immediately to prevent debris from entering the engine.

< Pinch clips (**arrows**) to disconnect vent line from cylinder head cover.

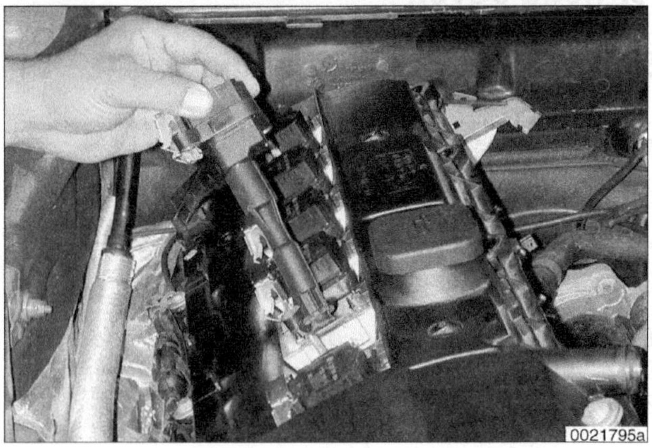


< Remove ignition coils:

- Disconnect ignition coil harness connectors.
- Remove coil mounting fasteners.
- Remove coils.
- Remove ground straps.
- Set coil harness to side of engine compartment.

***CAUTION—***
*Note location of all ground wires. Failure to reinstall grounds can result in permanent damage to engine control module or ignition system components.*

– Remove cylinder head cover mounting fasteners and remove cylinder head cover.

***NOTE—***
*In order to reinstall cylinder head cover mounting bolt insulators and gaskets in their original locations, make note of their arrangement during removal.*

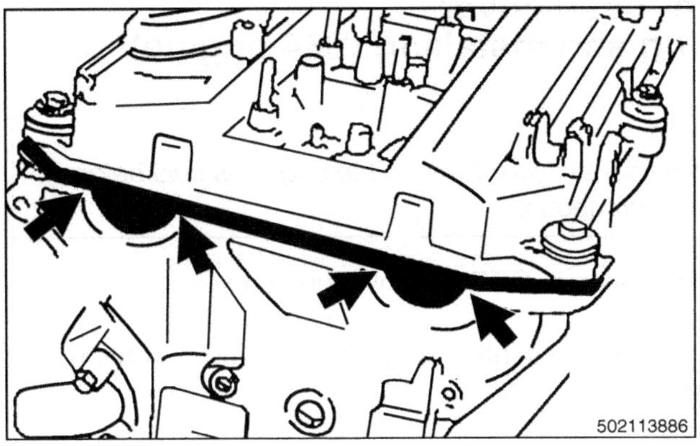


Installation is reverse of removal. Keep in mind the following:

- Make sure sealing surfaces are clean and free of old gasket material.
- Replace cylinder head cover gasket, as well as rubber insulators and seals as necessary.
- Use a small amount of 3-Bond®1209 adhesive or equivalent at corners (**arrows**) of half-moon cut-outs at back of cylinder head.
- Seat gasket and seal corners in front of cylinder head at VANOS unit.
- Inspect cylinder head cover mounting screw assembly rubber seals. Replace if rubber is cracked or brittle.

| **Tightening torque** | |
|---|---|
| Cylinder head cover to cylinder head (M6) | 10 Nm (7 ft-lb) |

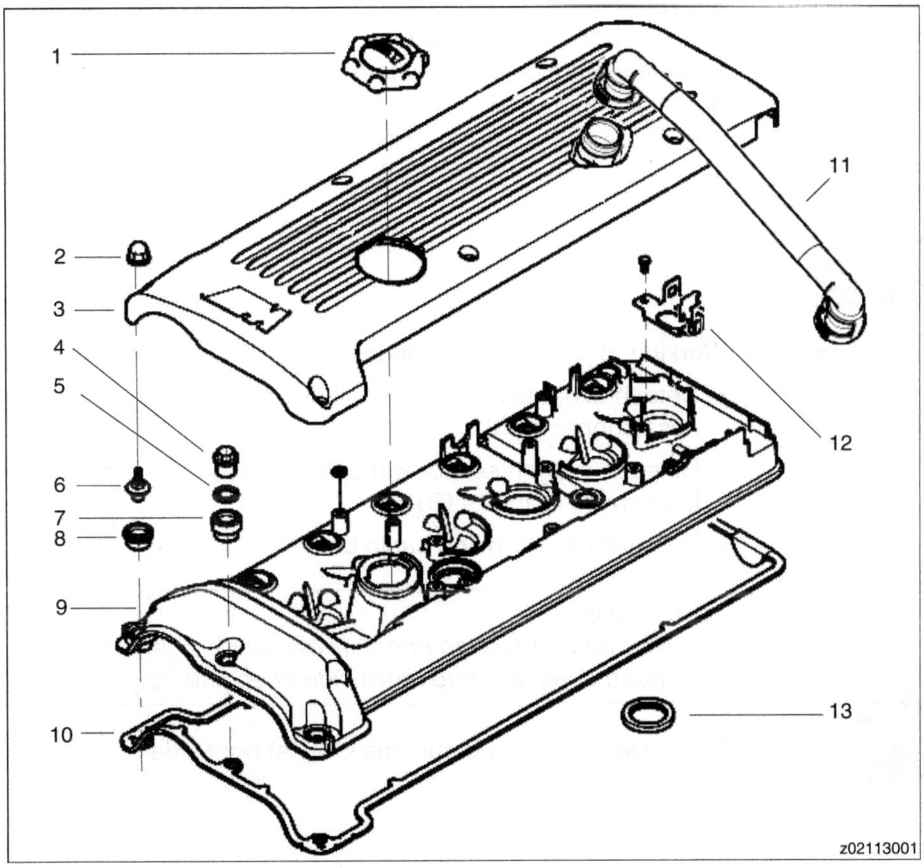


## Cylinder head cover components (S54 engine)

1. **Oil filler cap**
2. **Cap nut (M6)**
3. **Top engine plastic cover**
4. **Cap nut (M6)**
5. **Washer**
6. **Shoulder stud**
7. **Rubber insulator**
8. **Rubber insulator**
9. **Cylinder head cover**
10. **Cylinder head cover gasket**
11. **Engine vent hose**
12. **Oxygen sensor connector bracket**
13. **Spark plug well seal**

## Cylinder head cover, removing and installing (S54 engine)

> ***WARNING —***
> *To avoid personal injury, be sure the engine is cold before starting the procedure.*

– Make sure ignition is OFF. Remove ignition key.

– Working at rear of engine compartment, remove cable duct from bulkhead.

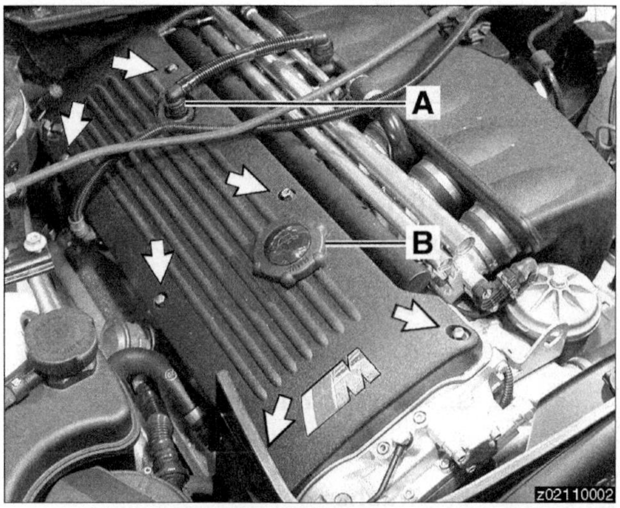


◄ Remove cylinder head top cover:

- Disconnect vent hose (**A**).
- Remove oil filler cover (**B**).
- Remove cover hold-down bolts (**arrows**) and lift off cover.

***NOTE —***
*Remove oil filler cap to allow removal of plastic cover. Be sure to replace it on the oil filler immediately to prevent debris from entering the engine.*

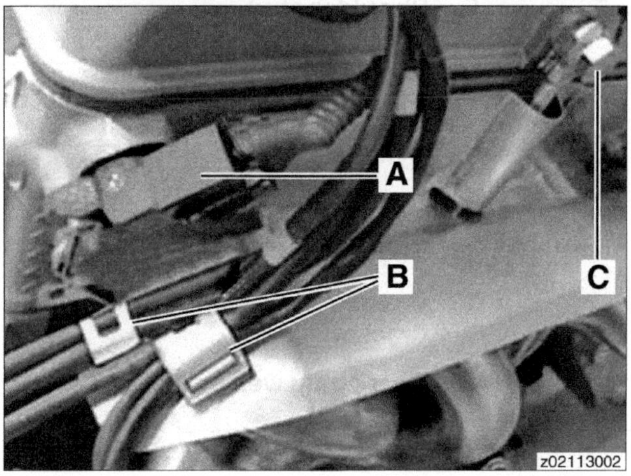


◄ Working at right rear of cylinder head:

- Detach exhaust camshaft position sensor electrical harness connector (**A**).
- Release oxygen sensor wiring harnesses from exhaust heat shield brackets (**B**).
- Remove oil return line banjo bolt (**C**).

> ***CAUTION—***
> *Make sure that sealing ring between banjo bolt and cylinder head does not drop down side of engine.*

– Detach ignition coils grounding strap at right side of cylinder head.

Working at rear of cylinder head cover:

- Match mark oxygen sensor connectors (**arrows**) for correct reassembly.
- Separate connectors.
- Thread harnesses off top of cylinder head and lay aside.

Working above cylinder head cover:

- Carefully unclip knock sensor harness loom at brackets (**arrows**).
- Swing loom off studs and lay aside.

Disconnect and remove ignition coils. See **120 Ignition System**.

- Remove ignition coil harness loom fasteners (**arrows**).
- Lay ignition coil harnesses to one side.

- Remove cylinder head cover mounting fasteners (**arrows**) and remove cylinder head cover.

***NOTE—***
*In order to reinstall cylinder head cover mounting bolt insulators and gaskets in their original locations, make note of their arrangement during removal.*

- Remove old gasket material from sealing faces.
- Use a small amount of 3-Bond®1209 adhesive or equivalent on left and right sides at transition between cylinder head and VANOS adjustment unit.
- Replace spark plug well sealing O-rings.
- Lay new cylinder head cover gasket on cylinder head. Note that gasket is secured with molded rubber guide pins on cylinder head and on VANOS adjustment unit.

***CAUTION—***
- *The rubber guide pins may buckle very slightly during installation. Make sure that guide pins mate exactly with the bore holes.*
- *Make sure gasket is seated correctly on front and rear of cylinder head.*

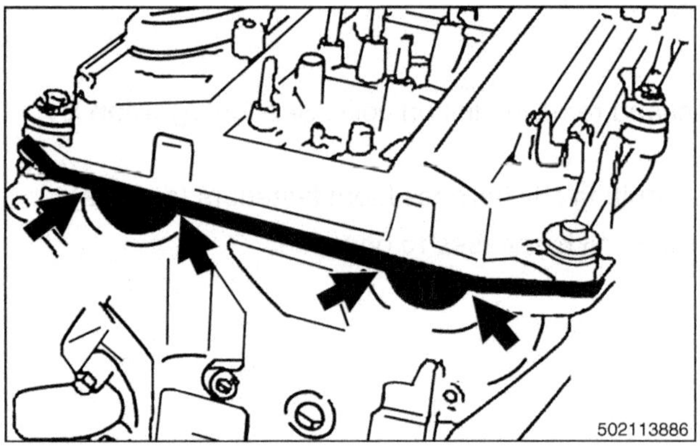


- Remainder of installation is reverse of removal. Keep in mind the following:
  - Inspect cylinder head cover mounting screw assembly rubber insulators and seals. Replace if rubber is cracked or brittle.
  - Use a small amount of 3-Bond®1209 adhesive or equivalent at corners (**arrows**) of half-moon cut-outs at back of cylinder head.
  - Fit cylinder head cover mounting cap nuts finger tight. Then tighten in diagonal sequence from inside to outside.
  - Replace oil return line banjo bolt sealing rings.

| **Tightening torque** | |
|---|---|
| Cylinder head cover to cylinder head (M6) | 10 Nm (7 ft-lb) |

***NOTE—***
- *The oxygen sensor connectors for cylinders 1 -3 and 4 - 6 are different. Use match marks made previously to assemble.*
- *Ensure oxygen sensor, ignition coil and knock sensor electrical harnesses are routed correctly.*

# INTAKE MANIFOLD REMOVAL AND INSTALLATION

You must remove the intake manifold to remove the engine assembly or the cylinder head.

## Intake manifold, removing and installing (4-cylinder engine)

> ***WARNING—***
> *To avoid personal injury, be sure the engine is cold before starting the procedure.*

– Disconnect negative (–) battery cable.

> ***CAUTION—***
> *Prior to disconnecting the battery, read the battery disconnection cautions given in* **001 General Warnings and Cautions**.

***NOTE—***
*On 1996 cars, the battery is located in the right rear of the engine compartment. On 1997 and later cars, the battery is located under the trunk floor mat.*

– Disconnect injector air shrouding hose from fitting at center of upper intake manifold.

– Remove splash guard under engine.

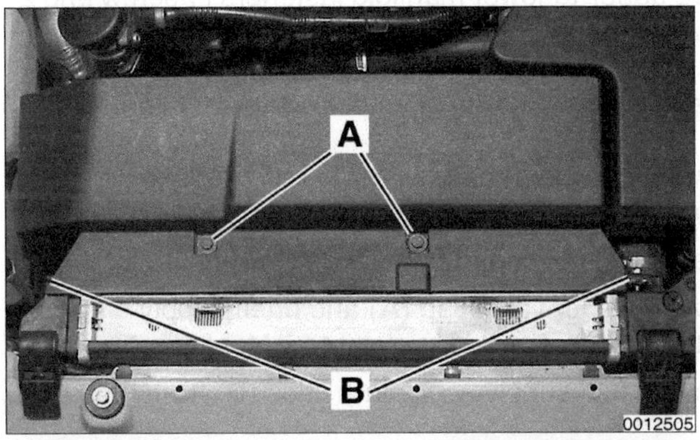


◄ Remove engine intake air duct mounting screws (**A**) from top of radiator.

- Detach plastic rivets (**B**) at sides of radiator.
- Disconnect air duct and remove from engine compartment.

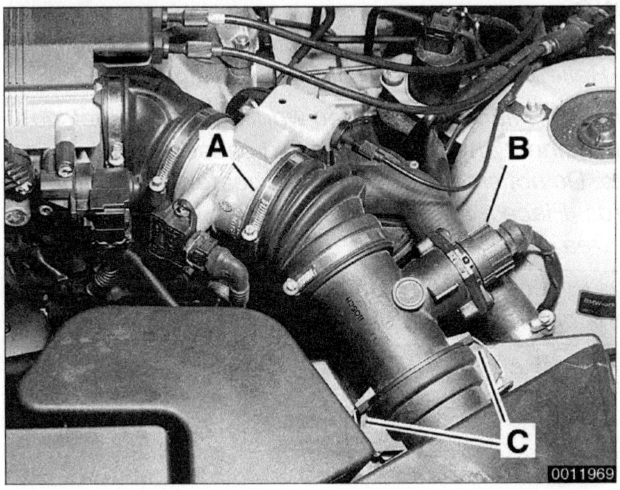


◄ Unbolt and remove complete air cleaner housing with mass air flow sensor.

- Remove intake air duct hose clamp (**A**).
- Detach mass air flow sensor electrical connector (**B**).
- Unclip mass air flow sensor retaining clips (**C**).

– Disconnect intake duct from throttle housing.

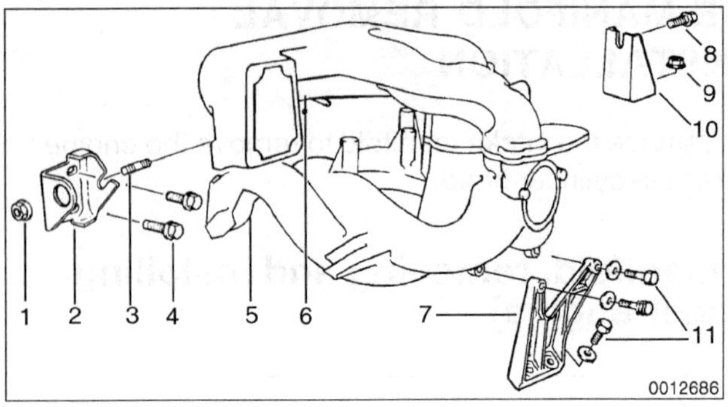


Intake manifold components (4-cylinder engine):

1. Nut (M7)
2. Support shackle
3. Stud (M7)
4. Bolt (M8)
5. Lower intake manifold
6. Upper intake manifold
7. Y-support
8. Bolt (M8)
9. Nut (M7)
10. Support bracket
11. Bolt (M8)

– Remove upper intake manifold:
  - Disconnect throttle cable from throttle housing.
  - Remove manifold support brackets at upper manifold.
  - Disconnect all harness connectors from throttle housing and intake manifold.
  - Disconnect vacuum hose fitting from brake booster on rear bulkhead. Plug opening in booster.
  - Disconnect fuel tank vent hose from vent valve on lower side of manifold.
  - Remove upper-to-lower manifold fasteners (1 screw and 2 nuts) and separate upper manifold from lower manifold.

– Disconnect oil dipstick tube bracket from lower manifold.

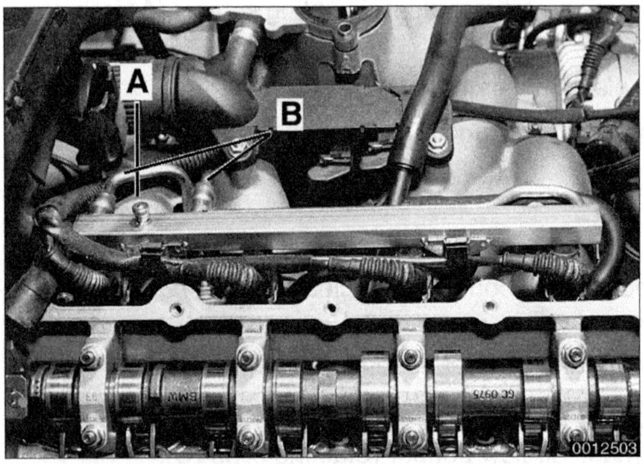


Relieve fuel pressure in fuel rail using compressed air (maximum 3 bar pressure) at air line fitting on fuel rail.

- Unscrew fuel tank cap to release pressure in tank.
- Remove shræder valve cap (**A**) and briefly apply air pressure at valve to force fuel back into fuel tank.
- Disconnect fuel supply and return lines (**B**) from fuel rail.

> ***WARNING—***
> - *The fuel system retains fuel pressure when the engine is off. Relieve line pressure before opening fuel lines.*
> - *Fuel is highly flammable. When working around fuel, do not disconnect any wires that could cause electrical sparks. Do not smoke or work near heaters or other fire hazards. Place a fire extinguisher in the vicinity of the work area.*
> - *Always unscrew the fuel tank cap to release pressure in the tank before working on the tank or lines.*

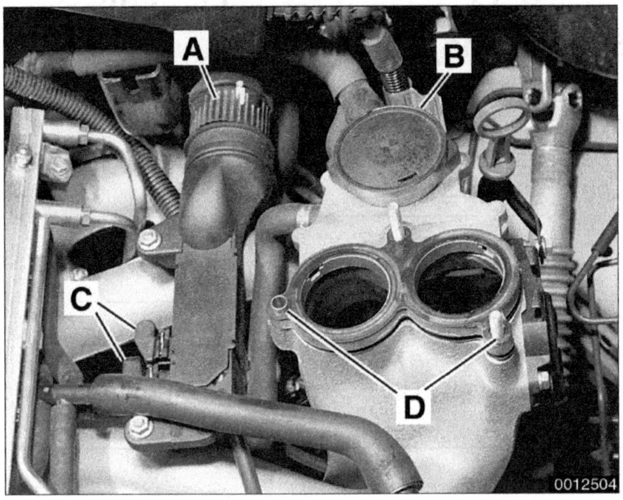


Prepare lower intake manifold for removal:

- Disconnect main engine wiring harness connector (**A**).
- Disconnect coolant hoses from crankcase vent valve (**B**). Remove vent valve from manifold.
- Label and disconnect harness connectors for crankshaft and camshaft sensors (**C**).
- Note locating dowels (**D**) on lower manifold mounting studs. Remove and set aside dowels before removing lower manifold.

– Remove lower intake manifold:

- Remove Y-support from underside of manifold.
- Unbolt alternator and starter support brackets.
- Remove five M7 nuts holding manifold to cylinder head and pull manifold back until harness connectors for oil pressure switch and coolant temperature sensor are accessible. Label and disconnect connectors, then remove manifold and set aside.

***CAUTION—***
*Stuff clean rags into the open intake ports to prevent any parts from falling into the engine intake.*

– Installation is reverse of removal. Take note of the following:

- Make sure sealing surfaces are clean and free of old gasket material.
- Install new intake manifold gasket.
- Reinstall harness connectors for oil pressure switch and coolant temperature sensor before placing manifold flush against cylinder head.
- Tighten manifold mounting nuts in stages.
- When installing lower intake manifold, make sure hollow locating bushings are on two lower studs.
- If necessary, adjust accelerator cable.

| **Tightening torques** | |
|---|---|
| Lower manifold to cylinder head | 15 Nm (11 ft-lb) |
| Lower manifold to upper manifold | 15 Nm (11 ft-lb) |
| Manifold support to manifold | 15 Nm (11 ft-lb) |

## Intake manifold, removing and installing (M52, S52 engine)

> ***WARNING —***
> *To avoid personal injury, be sure the engine is cold before starting the procedure.*

- Disconnect negative (–) battery cable in trunk.

> ***CAUTION—***
> *Prior to disconnecting the battery, read the battery disconnection cautions given in* **001 General Warnings and Cautions**.

***NOTE —***
*On 1996 cars, the battery is located in the right rear of the engine compartment. On 1997 and later cars, the battery is located under the trunk floor mat.*

- Remove fresh air plenum from rear of engine compartment.
- Disconnect vacuum hose from brake system booster. Plug openings in hose and booster.

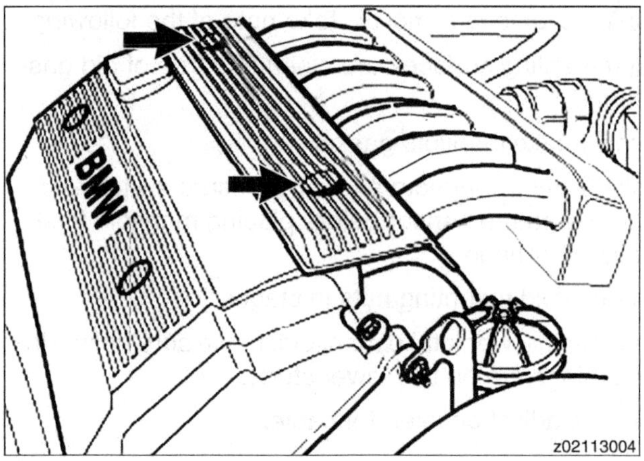


◄ Remove fuel injector rail cover hold down fasteners under small plastic trim covers (**arrows**). Remove cover.

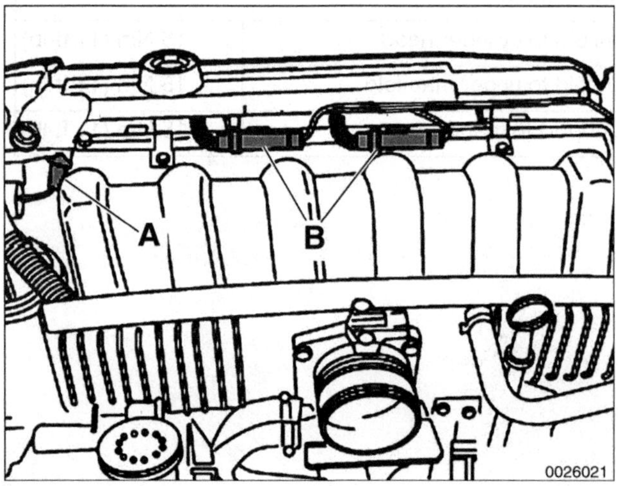


◄ Detach electrical harness connectors next to cylinder head:

- VANOS solenoid connector (**A**)
- Oxygen sensor connectors (**B**)

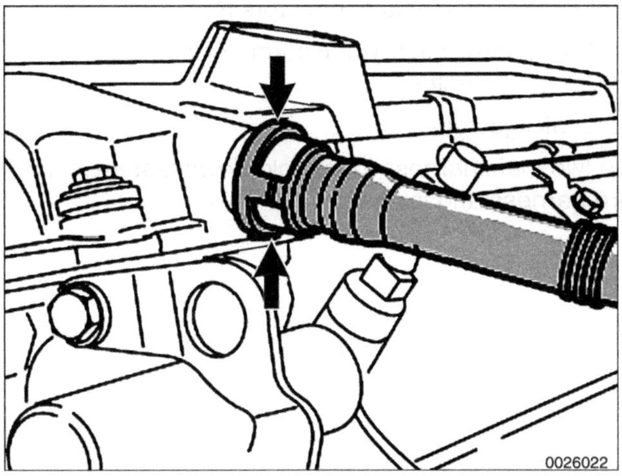


Detach engine vent hose from cylinder head cover by pinching spring clip (**arrows**).

If equipped: Remove schræder valve cap (**arrow**) from air connection on fuel rail. Blow fuel back through supply line using a brief burst of compressed air (maximum of 3 bar or 44 psi).

> ***WARNING—***
> - *The fuel system retains fuel pressure when the engine is off. Relieve line pressure before opening fuel lines.*
> - *Fuel is highly flammable. When working around fuel, do not disconnect any wires that could cause electrical sparks. Do not smoke or work near heaters or other fire hazards. Place a fire extinguisher in the vicinity of the work area.*
> - *Always unscrew the fuel tank cap to release pressure in the tank before working on the tank or lines.*

Working at rear of intake manifold, label and then remove fuel supply and fuel return lines. Disconnect fuel lines from support bracket on manifold.

***NOTE—***
*A special fitting is used to retain the fuel lines to the fuel rail. Use BMW special tool 16 1 050 or equivalent to compress the locking clip inside the end of the fuel line fittings to release the line.*

◄ Detach fuel injector electrical connectors from injectors:

- Use small screwdriver to pry one corner of wire lock clip on fuel injector 1 connector.
- Repeat for all injectors.
- Remove two main harness loom hold-down fasteners. Lift off entire harness and lay aside.

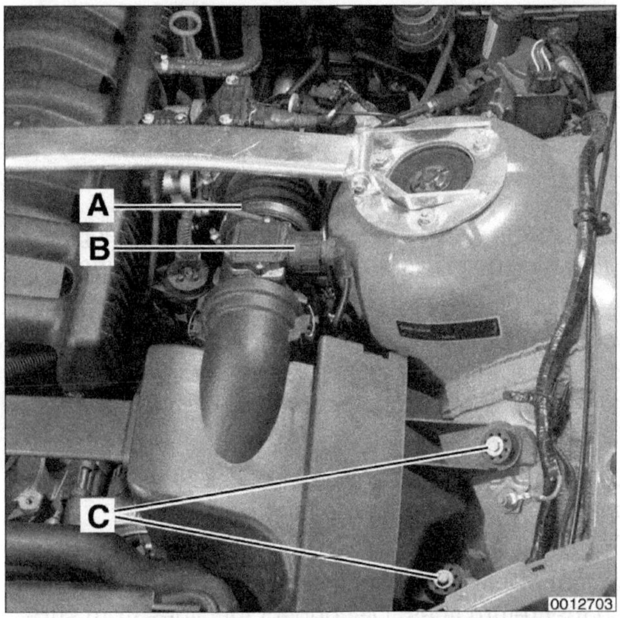


◄ Unbolt and remove complete air cleaner housing, disconnecting and labeling connectors, ducting and hoses as necessary:

- Air duct hose clamp (**A**)
- Mass air flow sensor connector (**B**)
- Air cleaner housing mounting bolts (**C**)

***NOTE—***
*If equipped: Remove strut tower brace for better access to top of engine.*

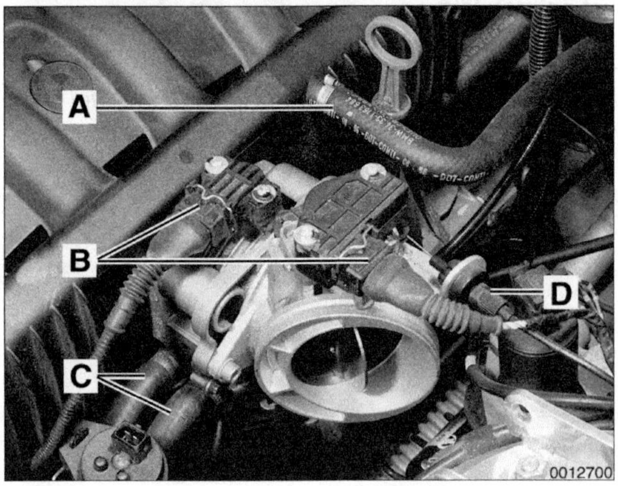


◄ If removing throttle housing, label and disconnect hoses, cables, and connectors from throttle housing and intake manifold:

- Brake booster hose (**A**)
- Electrical harness connectors (**B**)
- Coolant hoses (**C**)
- Throttle cable (**D**)

***CAUTION—***
*When disconnecting coolant hoses, be prepared for dripping coolant. Immediately plug hoses and throttle assembly coolant fittings.*

***NOTE—***
*If you are only removing the intake manifold or cylinder head, leave coolant hoses and throttle cable attached to the throttle assembly.*

– Unbolt throttle housing from intake manifold and set aside.

– Unbolt dipstick guide tube from intake manifold. Detach oil return hose at base of dipstick guide tube.

– Unbolt coolant hose retaining bracket underneath manifold.

– Detach idle speed control valve and intake temperature sensor electrical harness sensors.

– Disconnect connector at fuel tank breather valve near throttle assembly. Pull valve up to remove from bracket. Press down bracket to remove breather hose.

∢ Remove intake manifold from engine:

- Remove seven mounting nuts from above and two mounting bracket bolts from below (**arrows**).
- Check for wiring harnesses, fuel lines or mechanical components that might become snagged as manifold is removed.

> ***CAUTION—***
> *Stuff clean rags into the open intake ports to prevent any parts from falling into the engine intake.*

– Installation is reverse of removal.

- Use new fuel injector seals, hose clamps and wire ties.
- Carefully check intake manifold gaskets and replace if necessary.

| **Tightening torque** | |
|---|---|
| Intake manifold to cylinder head (M7 nut) | 15 ± 2 Nm (11 ± 1 ft-lb) |

## Intake manifold, removing and installing (M52 TU, M54 engine)

> ***NOTE—***
> *The illustrations in this procedure depict an M52 TU (6-cylinder) engine installed in an E46 (3-series) chassis. Z3 models with M52 TU or M54 engine are similar.*

> ***WARNING—***
> *To avoid personal injury, be sure the engine is cold before starting the procedure.*

– Disconnect negative (–) battery cable in trunk.

> ***CAUTION—***
> *Read the battery disconnection cautions given in* **001 General Warnings and Cautions**.

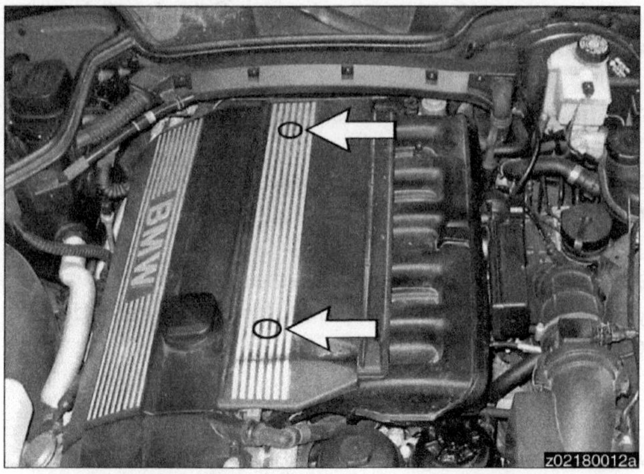


Remove intake manifold cover:

- Remove plastic trim covers (**arrows**).
- Remove cover hold down fasteners and lift off cover.

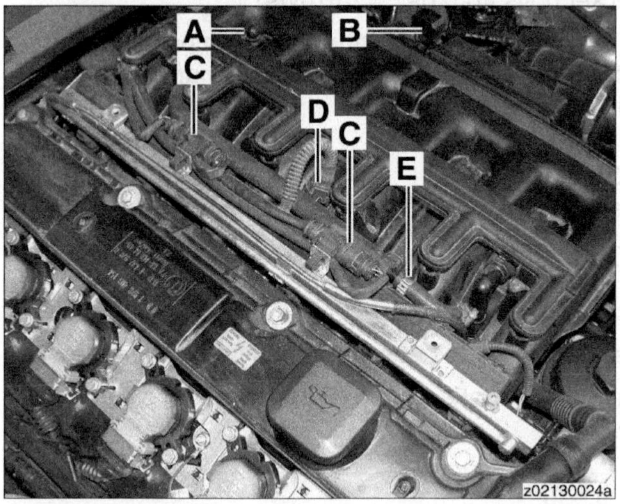


Working above engine, detach the following:

- **A** Positive engine lead at B+ terminal
- **B** Resonance valve electrical connector
- **C** Oxygen sensor connectors
- **D** Electrical connector for intake air temperature sensor
- **E** Manifold vacuum line

> ***CAUTION—***
> *Be sure to mark oxygen sensor connectors so that they can be reassembled as before.*

Working at left side of cylinder head:

- Detach engine vent hose from cylinder head cover by pinching spring clips (**arrows**).
- Detach VANOS solenoid electrical connector (**A**).

Detach fuel injector electrical connectors from injectors:

- Use small screwdriver to pry one corner of wire lock clip on fuel injector 1 connector.
- Repeat for all injectors.
- Remove two main harness hold-down fasteners. Lift off entire harness and lay aside.

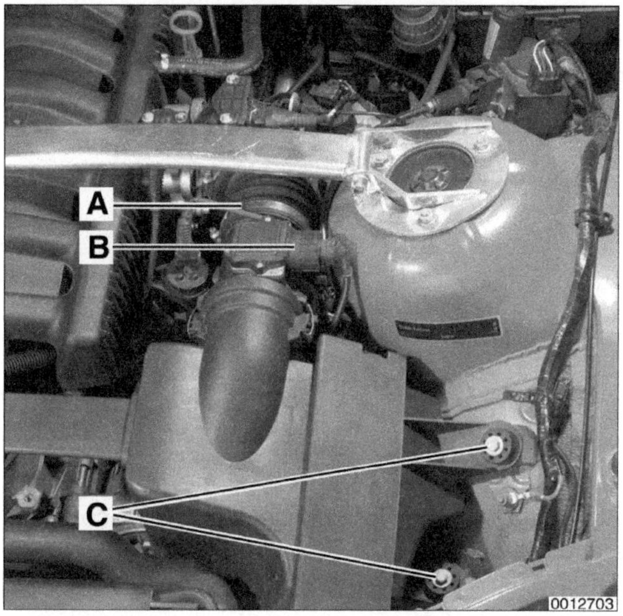


Unbolt and remove complete air cleaner housing, disconnecting and labeling connectors, ducting and hoses as necessary:

- Air duct hose clamp (**A**)
- Mass air flow sensor connector (**B**)
- Air cleaner housing mounting bolts (**C**)

***NOTE—***
*If equipped: Remove strut tower brace for better access to top of engine.*

**Mass air flow sensor ducts (M52 TU, M54 engine)**

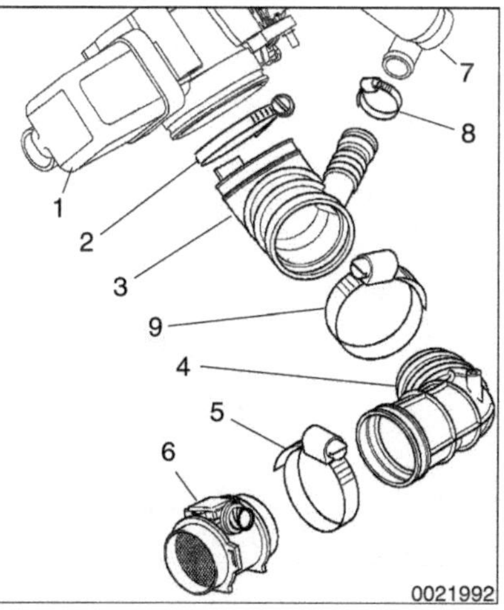


Loosen clamps **2** and **8** and remove mass air flow sensor and air ducts.

1. Throttle assembly
2. Hose clamp 77 - 84 mm
3. Y-duct
4. Air duct
5. Hose clamp 83 - 90 mm
6. Mass air flow sensor
7. Idle control valve
8. Hose clamp 28 - 33 mm
9. Hose clamp 77 - 84 mm

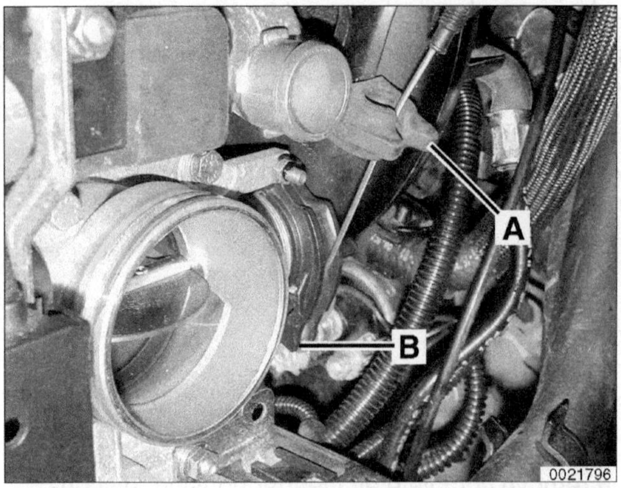


M52 TU engine: Working under intake manifold, pull throttle cable upwards out of rubber retainer (**A**) and unhook ball end of cable (**B**) from throttle actuator.

***NOTE—***
*M54 engine does not utilize throttle cable.*

Remove fasteners (**arrows**) retaining wiring conduit to throttle housing.

Working at throttle housing: Turn connector (**arrow**) counterclockwise and remove.

◀ Disconnect electrical connector (**arrow**) at idle speed control valve.

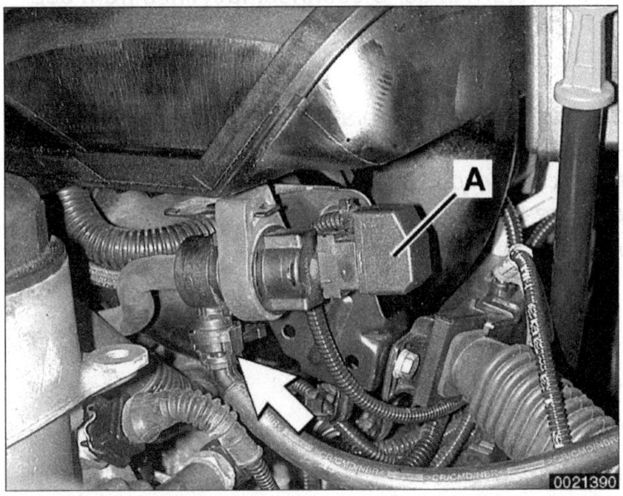


◀ Detach electrical connector (**A**) at fuel tank venting valve. Disconnect hose at quick disconnect fitting (**arrow**).

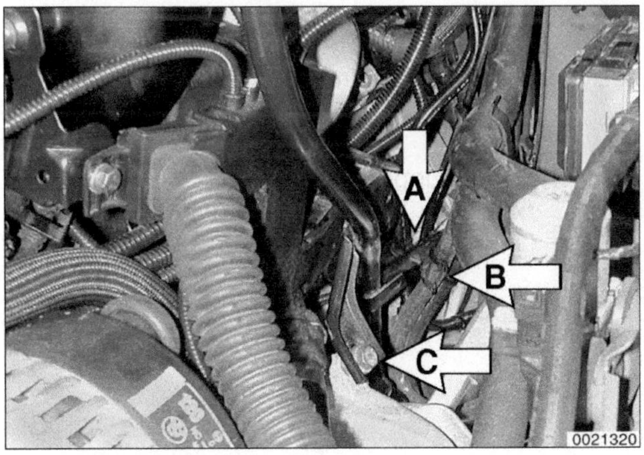


◀ Remove dipstick guide tube:

- Detach harness brackets from tube (**A**).
- Detach fuel lines from tube (**B**).
- Remove lower guide tube mounting bolt (**C**). Pull out dip stick guide tube.

< Remove schræder valve cap (**arrow**) from air connection on fuel rail. Blow fuel back through supply line using a brief burst of compressed air (maximum of 3 bar or 44 psi).

***WARNING—***
- *Fuel in fuel line is under pressure (approx. 3 - 5 bar or 45 - 75 psi) and may be expelled under pressure. Do not smoke or work near heaters or other fire hazards. Keep a fire extinguisher handy. Before disconnecting fuel hoses, wrap a cloth around fuel hoses to absorb any leaking fuel. Catch and dispose of escaped fuel. Plug all open fuel lines.*
- *Always unscrew the fuel tank cap to release pressure in the tank before working on the tank or lines.*

< Working at rear of intake manifold, label and then remove fuel supply and fuel return lines. Disconnect fuel lines from support bracket on manifold.

***NOTE—***
- *A special fitting is used to retain the fuel lines to the fuel rail. Use BMW special tool 16 1 050 or equivalent to compress the locking clip inside the end of the fuel line fittings to release the line.*
- *M54 engine: There is only one fuel line.*

< Working underneath car, remove lower intake manifold support mounting bolt (**arrow**), located adjacent to left engine mount.

– Detach knock sensor connector from bracket at base of manifold.

– Remove manifold lower support nuts.

◄ Remove intake manifold mounting nuts (**arrows**).

– Remove intake manifold from cylinder head while carefully checking for electrical connections or hoses that might become snagged.

> ***CAUTION—***
> *Stuff clean rags into open intake ports to prevent any parts from falling into the engine intake.*

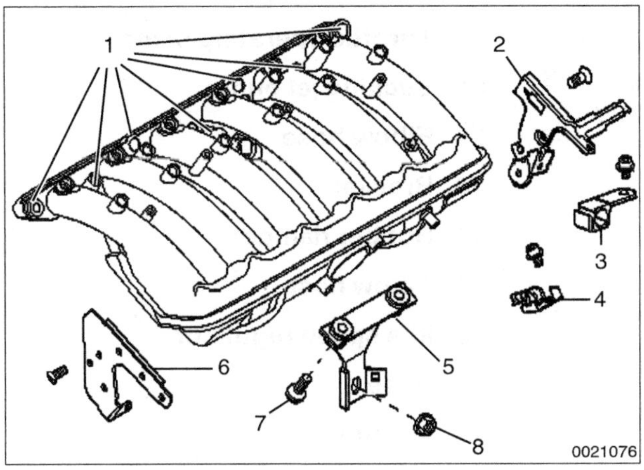


◄ Note intake manifold brackets and mounts:

1. Manifold mounting nuts
   -tighten to 15 Nm (11 ft-lb)
2. Fuel pipe bracket
3. Fuel pipe bracket
4. Vacuum pump bracket
5. Manifold mounting bracket
6. Tank venting valve bracket
7. Mounting bracket to manifold bolt (M6)
   -tighten to 10 Nm (7 ft-lb)
8. Mounting bracket to engine block nut (M10)
   -tighten to 47 Nm (33 ft-lb)

– Intake manifold installation is reverse of removal.

- Use new fuel injector seals.
- Carefully check intake manifold gaskets and replace if necessary.
- Check condition of rubber dampers at throttle housing and idle speed control valve. Replace as necessary.
- Inspect O-ring seal between mass air flow sensor and air filter housing. To facilitate reassembly, coat seal with acid-free grease.

> ***CAUTION—***
> *When reattaching throttle assembly harness connector, make sure arrows on connector and plug line up when fully tightened.*

| **Tightening torques** | |
|---|---|
| Intake manifold to cylinder head (M7 nut) | 15 Nm (11 ft-lb) |
| Mounting bracket to engine block (M10) | 47 Nm (33 ft-lb) |
| Mounting bracket to intake manifold (M6) | 10 Nm (7 ft-lb) |

S54 intake manifold and throttle assemblies

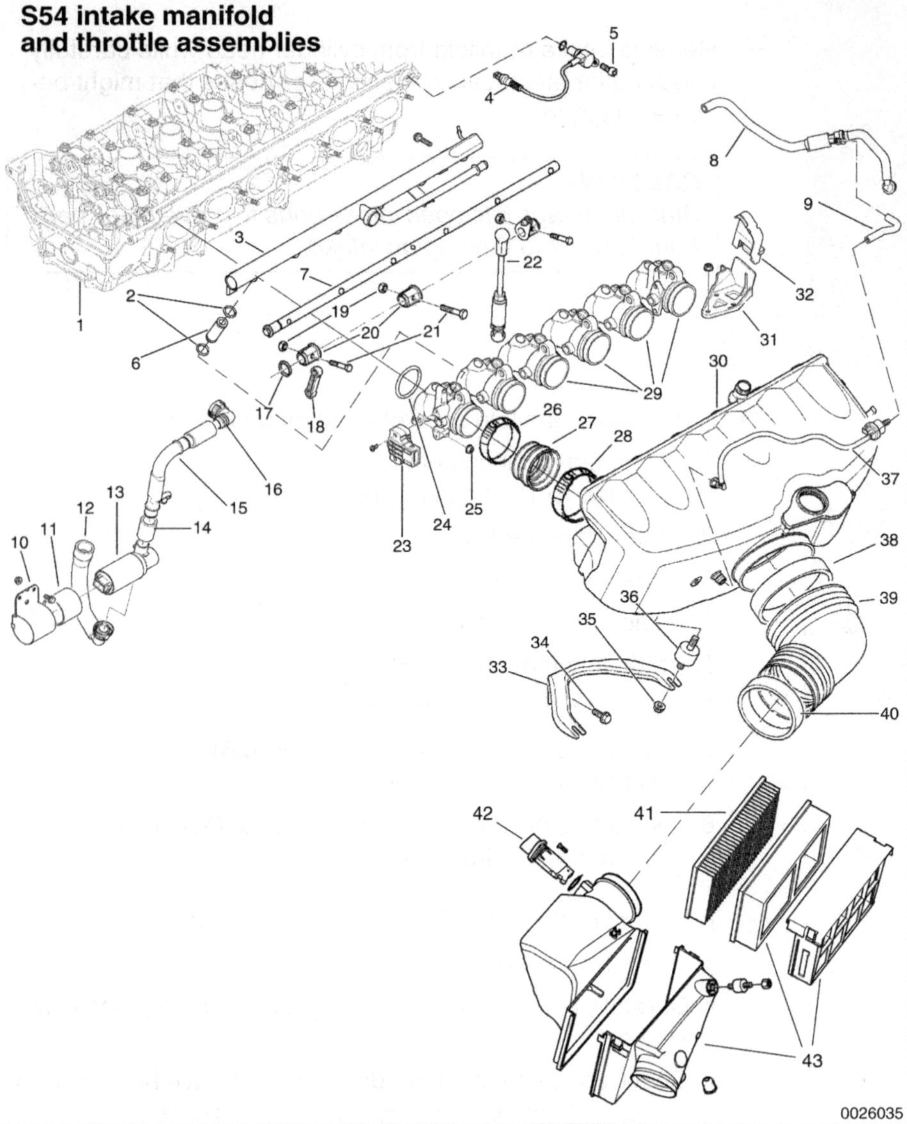


## Air filter, intake manifold and throttle assemblies (S54 engine)

1. Cylinder head
2. Sealing O-ring
3. Injector air distribution pipe
4. Intake camshaft position sensor
5. Bolt (M6)
   - tighten to 10 Nm (7 ft-lb)
6. Connection pipe
7. Throttle operating shaft
8. Sucking jet pump
9. Elbow hose
10. Bracket
11. Rubber damper
12. Elbow hose
13. Idle speed regulator
14. Hose
15. Air tube
16. Quick release coupling
17. Circlip
18. Pull rod
19. Nut (M5)
20. Operating lever
21. Bolt (M5)
22. Main throttle pull rod
23. Throttle position sensor
24. Intake manifold sealing O-ring (gasket)
25. Nut (M7)
   - tighten to 15 Nm (11 ft-lb)
26. Clamp 63 mm
27. Coupling hose
28. Clamp
29. Throttle housings
30. Intake manifold
31. Lower cover
32. Upper cover

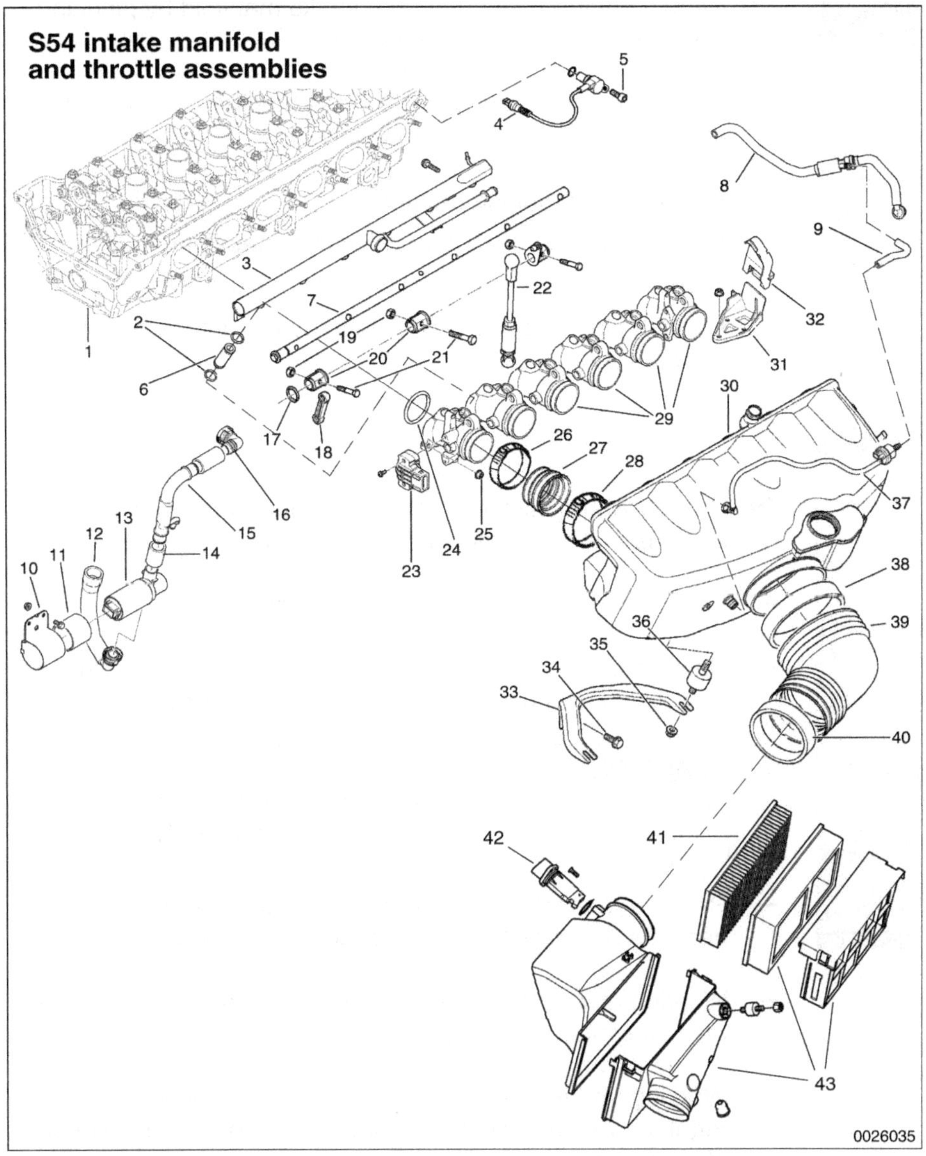


## Air filter, intake manifold and throttle assemblies (S54 engine) (continued)

**33. Intake manifold support bracket**

**34. Bolt (M8)**

**35. Nut (M6)**

**36. Rubber mount**

**37. Duct hose with shut off valve**

**38. Gasket**

**39. Coupling hose**

**40. Gasket**

**41. Air filter element**

**42. Mass air flow sensor**

**43. Air filter housing components**

## Intake manifold and throttle housings, removing and installing (S54 engine)

> ***WARNING—***
> *To avoid personal injury, be sure the engine is cold before starting the procedure.*

– Disconnect negative (–) battery cable in trunk.

> ***CAUTION—***
> *read the battery disconnection cautions given in* **001 General Warnings and Cautions**.

– Remove air filter housing. See **Air filter, intake manifold and throttle assemblies (S54 engine)** in this repair group.

– Remove fresh air plenum from rear of engine compartment.

- Detach engine vent hose from intake manifold by pinching sides of plastic lock clip (**arrows**).
- Detach positive battery cable, hose brackets, vacuum lines, and oil dipstick tube from manifold.
- Working under intake manifold:
  - Detach fuel tank venting valve from manifold.
  - Loosen mounting bracket nuts at front and rear of manifold.
  - Pinch plastic lock clip to detach hose from bottom of manifold.

- Use BMW special tool 11 9 160 or equivalent hose clamp pliers to release clamps at throttle housings.
- Detach crankcase breather return hose.
- Raise intake air manifold slightly and detach from throttle assemblies. Detach vacuum line and wiring harness.

> ***CAUTION—***
> *Do not pull off intake manifold without detaching vacuum line and wiring harness on the reverse side of manifold.*

- Remove manifold while carefully checking for electrical connections or hoses that might become snagged.
- Working at injector air distribution pipe, pinch lock clip to detach idle speed regulator hose.
- Working underneath throttle housings, remove idle speed regulator mounting nuts. Remove regulator and bracket and place aside.
- Detach vacuum lines and electrical harness connectors from throttle housings and fuel injectors.

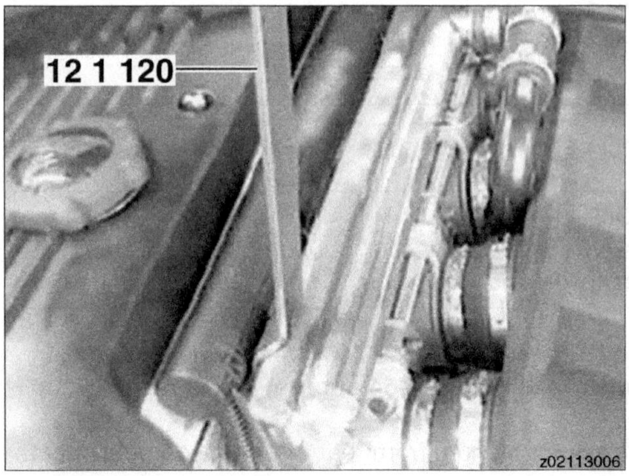


- Working from front to rear of engine (injector 1 - 6), use BMW special tool 12 1 120 or equivalent to pry off fuel injector electrical connector strip.

> ***CAUTION—***
> *Connector strip may break if it is removed without special tool 12 1 120.*

z02113007

◄ Pry off fuel line retainer (**arrow**) at fuel rail.

> ***WARNING —***
> - *Fuel in fuel line is under pressure (approx. 3 - 5 bar or 45 - 75 psi) and may be expelled under pressure. Do not smoke or work near heaters or other fire hazards. Keep a fire extinguisher handy. Before disconnecting fuel hoses, wrap a cloth around fuel hoses to absorb any leaking fuel. Catch and dispose of escaped fuel. Plug all open fuel lines.*
> - *Always unscrew the fuel tank cap to release pressure in the tank before working on the tank or lines.*

z02113008

◄ Remove retaining bolts (**arrow**) at front and rear of fuel rail.

- Detach fuel rail from fuel injectors, air pipe and throttle housings. Remove fuel rail from engine compartment and store in safe place.
- Remove intake camshaft position sensor from left rear of cylinder head.

z02113009

◄ Detach main throttle pull rod from center throttle operating lever (**arrow**).

- Unbolt and remove throttle housings while carefully checking for electrical connections or hoses that might become snagged.
- Clean sealing surfaces. Note correct position of adapter sleeves.
- Detach fuel rail from fuel injectors airpipe and throttle housings. Remove fuel rail from engine compartment and store in safe place.

– On installation:
- Clean all sealing surfaces.
- Replace throttle housing sealing O-rings.
- Be sure throttle housing dowel sleeves fit into cylinder head correctly.
- Check fuel injector O-ring seals. Replace as necessary. Lubricate with SAE 90 oil before installing.

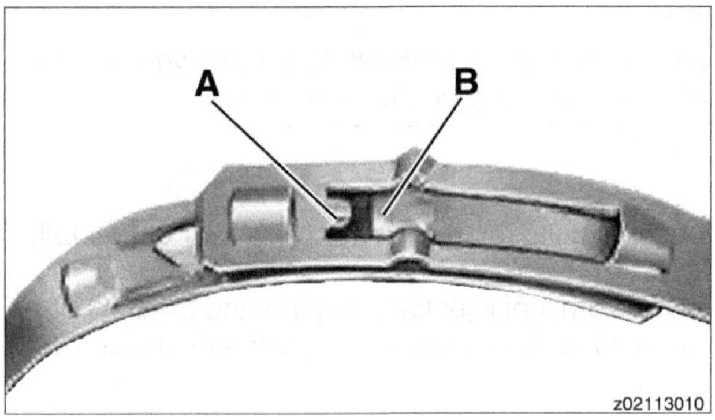


Replace intake manifold to throttle housing coupling clamps.
- Use BMW special tool 11 9 160 to install clamps. Press hose clamp together until lock (**A**) engages catch (**B**).
- Slide preassembled clamps onto ducts prior to final installation.

> ***CAUTION—***
> *Use only original clamps. Use of non-original hose clamps may result in interference with the throttle linkage and cause the throttle to jam or stick in the open position.*

– After assembly:
- Check throttle position settings. See **130 Fuel Injection**.
- Test throttle system with DISplus, GT1, MoDiC or equivalent scan tool.

| **Tightening torques** | |
|---|---|
| Intake camshaft position sensor to cylinder head (M6) | 10 Nm (7 ft-lb) |
| Throttle housing to cylinder head (M7) | 15 ± 2 Nm (11 ± 1 ft-lb) |

## CYLINDER HEAD REMOVAL AND INSTALLATION (4-CYLINDER ENGINE)

Cylinder head removal and installation is a complicated repair. Timing chain removal and installation require special tools. Read the entire procedure before beginning repairs.

### Cylinder head, removing (4-cylinder engine)

> ***WARNING—***
> *Due to risk of personal injury, be sure the engine is cold before beginning the removal procedure.*

> ***CAUTION—***
> *Cover all painted surfaces before beginning the removal procedure. As an aid to installation, label all components, wires, and hoses before removing them. Do not reuse gaskets, O-rings or seals during reassembly.*

– Disconnect negative (–) battery cable.

> ***CAUTION—***
> *Prior to disconnecting the battery, read the battery disconnection cautions given in* **001 General Warnings and Cautions***.*

***NOTE—***
*On 1996 cars, the battery is mounted in the rear of the engine compartment. On 1997 and later cars, the battery is mounted below the trunk floor covering.*

– Raise vehicle and support safely.

> ***WARNING—***
> *Make sure the car is stable and well supported at all times. Use a professional automotive lift or jack stands designed for the purpose. A floor jack is not adequate support.*

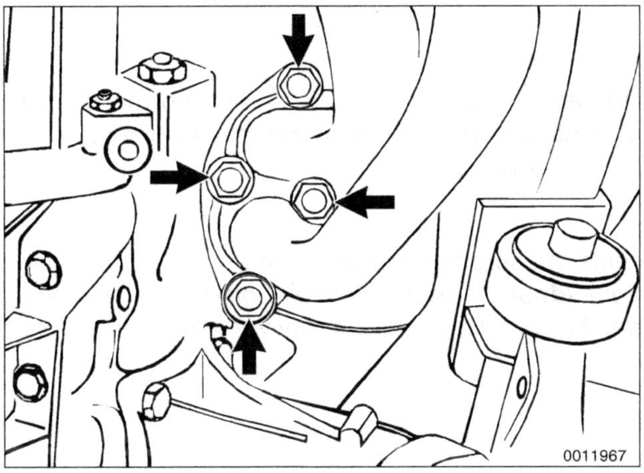


◄ Working underneath vehicle, remove front exhaust pipe mounting nuts (**arrows**) and detach pipe from exhaust manifold.

– Disconnect fuel lines and remove air filter housing and intake manifold. See **Intake manifold, removing and installing (4-cylinder engine)** in this repair group.

> ***WARNING—***
> *Fuel is highly flammable. When working around fuel, do not disconnect any wires that could cause electrical sparks. Do not smoke or work near heaters or other fire hazards. Place a fire extinguisher in the vicinity of the work area.*

> ***CAUTION—***
> *Stuff clean rags into the open intake ports to prevent any parts from falling into the engine.*

– With engine cold, drain coolant from engine block and radiator. See **170 Radiator and Cooling System**.

***NOTE—***
- *On M44 engine, the block drain is on the exhaust side of the engine, near cylinder 4.*
- *The radiator drain plug is on the bottom of the radiator on the driver's side. Remove small plastic lock clip and pull radiator drain plug out completely.*

– Disconnect coolant hoses from thermostat housing in front of cylinder head. Unbolt thermostat housing from cylinder head.

– Disconnect coolant hoses from flange on rear of cylinder head.

– Remove cylinder head cover. See **Cylinder head cover, removing and installing (4-cylinder engine)** in this repair group.

– Disconnect main engine wiring connector. Label and disconnect connectors for crankshaft and camshaft sensors.

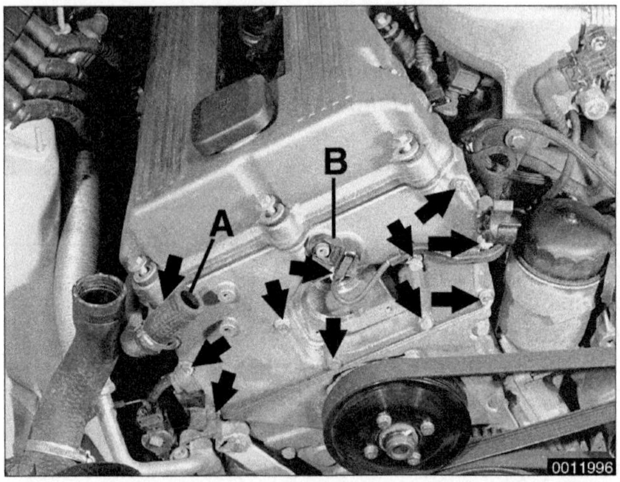


Working at front of cylinder head:

- Remove secondary air injection check-valve from duct (**A**) and lay aside.
- Remove camshaft sensor (**B**) from upper timing chain cover.
- Remove upper timing chain cover mounting bolts (**arrows**). Remove cover.

Set engine to approximate TDC by rotating in normal operating direction until camshaft lobes at cylinder 1 point up and face each other (**arrows**).

***NOTE—***
*At TDC, camshaft sprocket marks are lined up with vertical axis of engine (**dashed lines**). As the engine is installed at a slight angle, the sprocket marks are not vertical relative to the ground.*

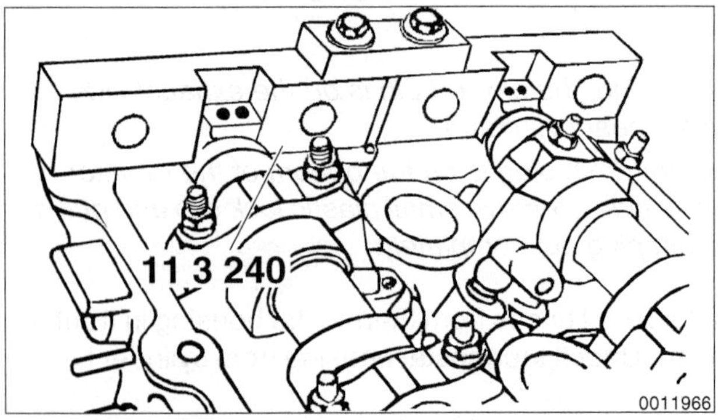


Lock camshafts at TDC by mounting BMW special tool 11 3 240 over square ends of camshafts.

> ***CAUTION—***
> *Lock camshafts in TDC position using the special service tool before removing the timing chain. The tool holds the camshafts to prevent valve contact with pistons.*

***NOTE—***
*Check that tool is squarely seated on cylinder head gasket surface. If necessary, rotate camshafts slightly using an open-end wrench at hex on camshafts.*

∢ Remove hydraulic chain tensioner from right side of cylinder head.

– Remove mounting bolts from left and right camshaft sprockets.

∢ Working at timing chain:

- Remove locking bolt (5 mm Allen wrench) from left chain guide adjuster (**arrow**). Loosen slotted adjusting sleeve approximately two turns.
- Remove camshaft sprockets from camshafts.

***NOTE—***
- *Note orientation of sensor wheel.*
- *It is not necessary to remove the cam sprockets from the chain unless these parts are being replaced. Wire tie each sprocket to the chain to keep the sprockets timed to each other. Keeping the sprockets timed simplifies installation.*

– Hold chain under tension. Using crankshaft vibration damper bolt, turn engine opposite its normal direction of rotation (counterclockwise as viewed from the front) approximately 45°. This prevents accidental contact between valves and pistons.

∢ Using Torx E12 socket, remove cylinder head bolts in sequence indicated. Remove cylinder head and head gasket. Discard head bolts.

## Cylinder head, installing (4-cylinder engine)

– Clean mating surfaces of cylinder head, engine block and timing chain cover. Remove foreign matter and liquid from bolt holes.

> ***CAUTION—***
> *Do not use a metal scraper or wire brush to clean the aluminum cylinder head or pistons. If necessary, use a hard wooden or plastic scraper. Also available are abrasive discs to be used in conjunction with an electric drill. Be sure to use the correct disc for the type of metal being cleaned.*

– Check gasket surface of cylinder head and engine block for warpage using a straight edge.

– Place new cylinder head gasket on surface of engine block. Check that two locating dowels are correctly positioned in block and are not damaged.

***NOTE—***
*Make sure the word OBEN, printed on the gasket, is facing up.*

– Place new rubber profile gasket in groove of front timing cover.

– With camshafts locked in TDC position with BMW special tool 11 3 240, set cylinder head in position.

> ***CAUTION—***
> - *Lock camshafts in TDC position at the rear of the cylinder head with BMW special tool 11 3 240 before installing the cylinder head.*
> - *Make sure that the crankshaft, which had been rotated approximately 45° opposite the direction of engine rotation, is still in that position.*

– Lightly lubricate new cylinder head bolts. Loosely install bolts and their washers, then thread them in until they are finger tight.

> ***CAUTION—***
> *Do not reuse Torx-type cylinder head bolts. They are stretch-type bolts. Always replace after loosening.*

***NOTE—***
*Check that all washers for the cylinder head bolts are in place before installing the bolts.*

◄ Tighten cylinder head bolts in correct sequence, as shown.

◄ Tighten cylinder head bolts in three stages as listed below. For stage 2 and stage 3, use BMW special tool 11 2 110 or suitable protractor to tighten bolts to specified torque angle.

| Tightening torque | |
|---|---|
| Cylinder head to engine block (Torx E12 M10 bolts):<br>• Stage 1<br>• Stage 2<br>• Stage 3 | <br><br>30 Nm (22 ft-lb)<br>+90°<br>+90° |

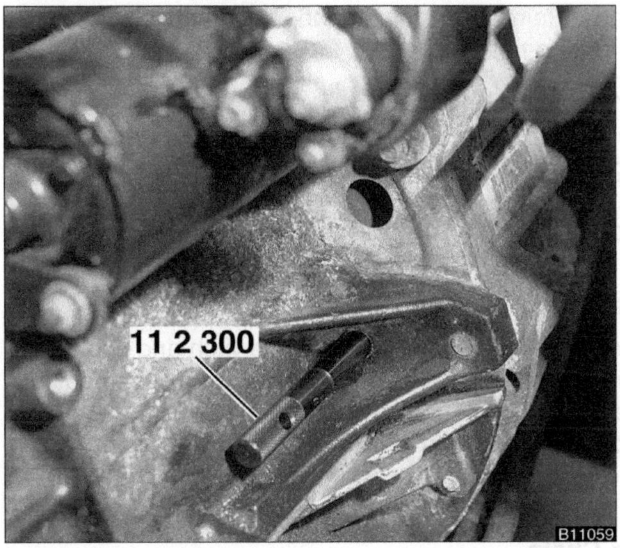


◄ Carefully rotate crankshaft in direction of engine rotation and lock it at TDC: Insert BMW special tool 11 2 300 through transmission bellhousing and into flywheel or drive plate bore.

– Turn left chain guide adjusting sleeve in until it just contacts cylinder head. Insert chain guide retaining screw and tighten.

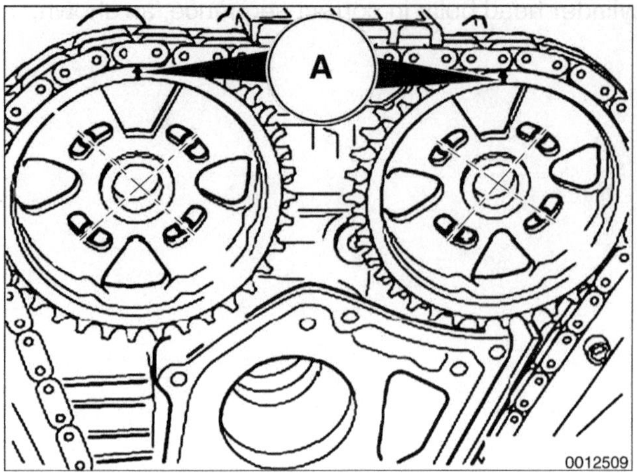


Place chain sprockets with chains on camshafts:

- Marks (**A**) on sprockets point up.
- Tapped camshaft bores are centered to slots in sprockets (**dashed lines**).

– Place sensor wheel on intake camshaft sprocket so that arrow on sensor wheel points up.

– Install and hand-tighten sprocket mounting bolts.

– Install hydraulic chain tensioner to cylinder head. Use a new sealing washer on tensioner when installing.

| Tightening torque | |
|---|---|
| Chain tensioner to cylinder head | 40 Nm (30 ft-lb) |

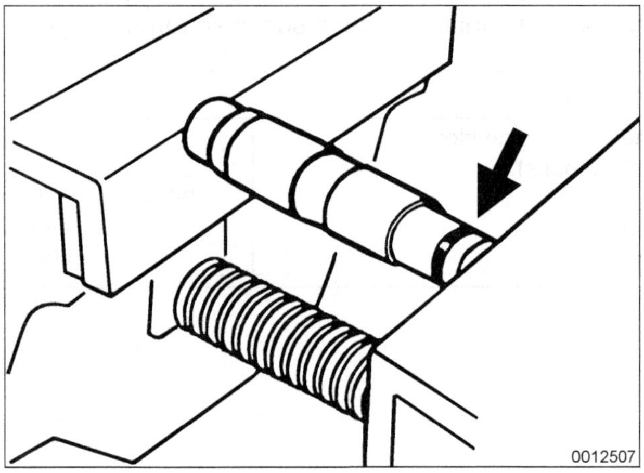


If reusing old chain tensioner:

- First clamp tensioner piston in a vice (soft jaws) and squeeze oil from it.
- Slowly compress piston in vise up to end circlip (**arrow**).
- Repeat this procedure twice to ensure all oil is expelled.

– Check that chain is free of slack. Tighten down sprocket mounting bolts.

| Tightening torque | |
|---|---|
| Chain sprocket to camshaft flange (M7) | 15 Nm (11 ft-lb) |

– Remove crankshaft locking tool from transmission bellhousing and camshaft locking tool from rear of cylinder head.

– Turn engine over by hand in direction of rotation at least two full rotations. Reinstall camshaft and crankshaft locking tools to check engine timing.

- Remove locking tools if no faults are found.
- If necessary, reset camshaft timing. See **117 Camshaft Timing Chains**.

– Installation of remaining parts is reverse of removal. Note the following:

- Make sure sealing surfaces are clean.
- Reinstall connectors for oil pressure switch and coolant temperature sensor before placing manifold flush against cylinder head. Tighten manifold nuts in stages. See **Intake manifold, removing and installing (4-cylinder engine)** in this repair group.
- Install upper timing chain cover. See **117 Camshaft Timing Chain**.
- Reinstall cylinder head cover and ignition coils. See **Cylinder head cover, removing and installing (4-cylinder engine)** in this repair group.

- Refill cooling system. See **170 Radiator and Cooling System**.
- Change engine oil and filter. See **020 Maintenance**.
- If necessary, adjust accelerator cable.

| Tightening torques | |
|---|---|
| Coolant drain plug to engine block | 40 Nm (30 ft-lb) |
| Lower intake manifold to cylinder head | 15 Nm (11 ft-lb) |
| Lower intake manifold to upper intake manifold | 15 Nm (11 ft-lb) |
| Manifold support to manifold | 15 Nm (11 ft-lb) |
| Radiator cooling fan to coolant pump | 40 Nm (30 ft-lb) |
| Radiator drain screw to radiator | 2.5 Nm (22 in-lb) |

## CYLINDER HEAD REMOVAL AND INSTALLATION (M52, S52 ENGINE)

Cylinder head removal and installation is a complicated repair procedure. VANOS and timing chain removal and installation require special tools. Read the entire procedure before beginning the repair.

### Cylinder head, removing (M52, S52 engine)

> ***WARNING —***
> *Due to risk of personal injury, be sure the engine is cold before beginning the removal procedure.*

> ***CAUTION—***
> *Cover all painted surfaces before beginning the removal procedure. As an aid to installation, label all components, wires, and hoses before removing them. Do not reuse gaskets, O-rings or seals during reassembly.*

– Disconnect negative (–) battery cable.

> ***CAUTION—***
> *Prior to disconnecting the battery, read the battery disconnection cautions given in* **001 General Warnings and Cautions**.

– Raise vehicle and support safely.

> ***WARNING —***
> *Make sure the car is stable and well supported at all times. Use a professional automotive lift or jack stands designed for the purpose. A floor jack is not adequate support.*

– Remove front exhaust pipes from manifolds. Discard nuts.

– Disconnect fuel lines and remove air filter housing and intake manifold. See **Intake manifold, removing and installing (M52, S52 engine)** in this repair group.

> ***WARNING —***
> *Fuel is highly flammable. When working around fuel, do not disconnect any wires that could cause electrical sparks. Do not smoke or work near heaters or other fire hazards. Place a fire extinguisher in the vicinity of the work area.*

> ***CAUTION—***
> *Stuff clean rags into the open intake ports to prevent any parts from falling into the engine.*

– Remove secondary air injection check-valve from front of engine and lay aside.

– With engine cold, drain engine coolant and remove coolant hoses from thermostat housing. Drain engine block by removing block drain. See **170 Radiator and Cooling System**.

***NOTE —***
*The block drain is on the exhaust side of the engine, near cylinder 5.*

– Remove radiator cooling fan and fan shroud. See **170 Radiator and Cooling System**.

***NOTE —***
*The radiator cooling fan nut (32 mm wrench) has left-hand threads.*

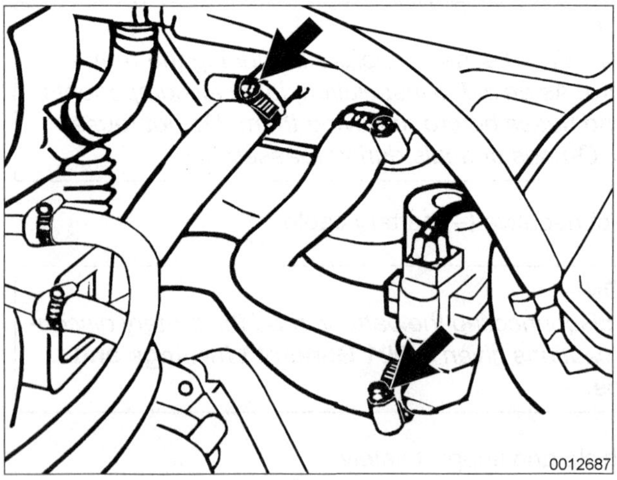


◄ Loosen heater hose clamps (**arrows**) at rear of cylinder head. Detach hoses.

– Remove alternator cooling duct from rear of alternator.

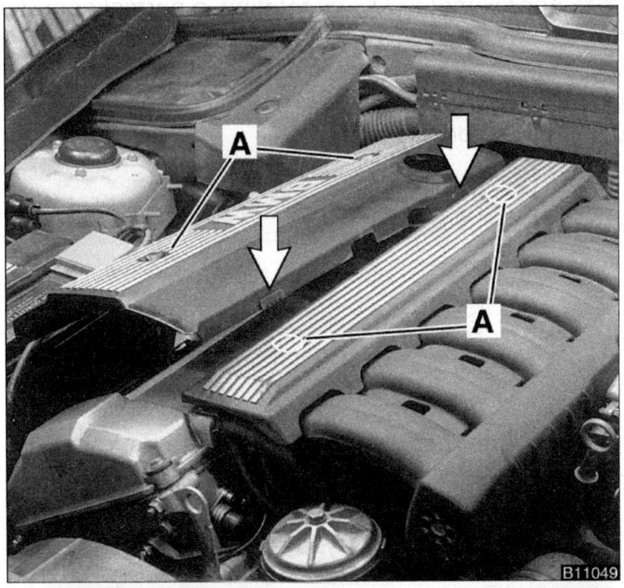


Remove plastic cylinder head covers from above fuel injectors and ignition coils.

- Pry out trim plugs (**A**) and remove retaining fasteners.
- Pull larger cover forward when removing.
- Make sure rubber insulators (**arrows**) on either side of large cover do not fall off during cover removal.

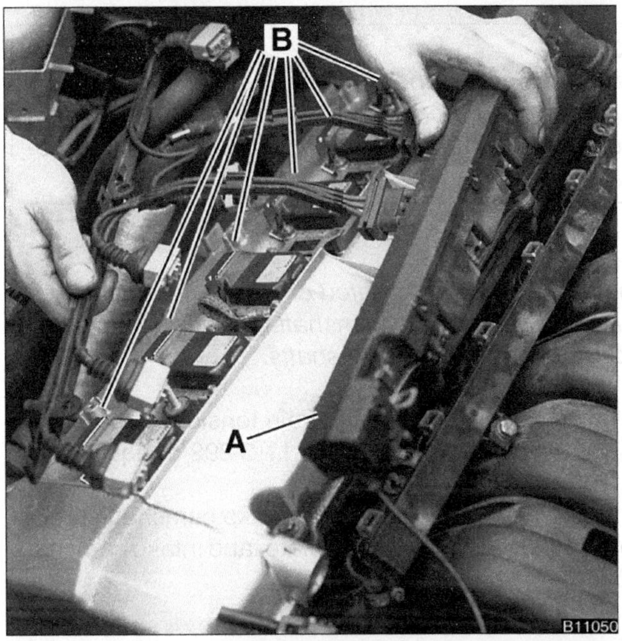


Working above cylinder head cover:

- Remove mounting fasteners from fuel injection harness (**A**). Pry back injector connector retaining clips and lift entire harness off engine.
- Detach ignition coil electrical connectors (**B**).
- Lift both harnesses off engine and lay aside.

– Remove cylinder head cover mounting bolts and remove cylinder head cover. See **Cylinder head cover, removing and installing (M52, S52, M52 TU, or M54 engine)** in this repair group.

***NOTE—***
*Make note of the arrangement of cylinder head cover mounting bolt insulators and gaskets during removal. Reinstall in the same order.*

Remove oil baffle cover from above intake camshaft.

– Disconnect large hose from brake system vacuum booster. Plug opening in vacuum booster.

– Remove VANOS unit from front of cylinder head. See **117 Camshaft Timing Chains**.

◄ Set engine to approximate TDC by rotating in normal operating direction until **0/T** mark (0° TDC) on front lower timing chain cover is aligned with vibration damper. In this position, camshaft lobes at cylinder 1 point up and face each other. Arrows on camshaft sprockets point up.

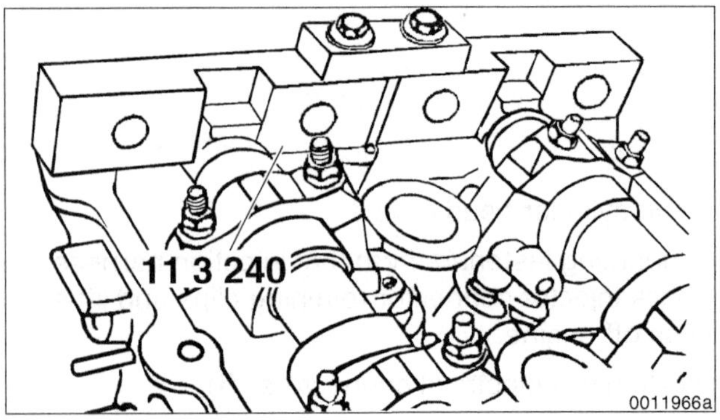


– Remove three cylinder head cover mounting studs from across rear of head.

◄ Lock camshafts at TDC by mounting BMW special tool 11 3 240 over square ends of camshafts.

> ***CAUTION—***
> *Lock camshafts in TDC position using the special service tool, or equivalent, before removing the timing chain. The tool holds the camshafts to prevent valve contact with pistons.*

***NOTE—***
*Check that tool is squarely seated on cylinder head gasket surface. If necessary, rotate camshafts slightly using an open-end wrench at hex on camshafts.*

◄ Press down on secondary timing chain tensioner and lock it down by inserting BMW special tool 11 3 292 or equivalent.

– Remove mounting nuts from front of intake camshaft sprocket. Remove thrust washers, spring plate, and intake camshaft sprocket.

***NOTE—***
*Make note of the thrust washer positions. The thicker (0.4 mm or 0.016 in) washer is mounted outermost.*

– Remove mounting nuts from front of exhaust camshaft sprocket. Remove secondary sprockets together with secondary timing chain from front of camshafts.

> ***CAUTION—***
> *Do not allow crankshaft to rotate once the camshaft timing chain is removed. The pistons can contact the valves.*

***NOTE—***
*It is not necessary to remove the cam sprockets from the chain unless these parts are being replaced. Wire tie each sprocket to the chain to keep the sprockets timed to each other. Keeping the sprockets timed simplifies installation.*

– Remove secondary timing chain tensioner from top of cylinder head.

◄ Remove primary timing chain tensioner (**arrow**) from side of timing chain cover.

> ***CAUTION—***
> *Remove tensioner slowly. The tensioner is under spring pressure.*

◄ Use Torx E8 socket to remove timing chain guide mounting bolts (**arrows**). Remove guide.

– Lift primary chain sprocket off exhaust camshaft. Remove chain from sprocket.

– Hold chain under tension. Using crankshaft vibration damper bolt, turn engine opposite its normal direction of rotation (counterclockwise as viewed from the front) approx. 30° before TDC. This prevents accidental contact between valves and pistons.

– Tie chain up using stiff wire.

◄ Use Torx E8 socket to remove cylinder head to lower timing chain cover mounting bolts (**arrows**).

< Using BMW special tool 11 2 250, loosen cylinder head bolts in several stages in sequence illustrated.

***NOTE—***
*BMW special tool 11 2 250 is a ½-inch drive thin-walled Torx E12 socket and extension. The cylinder head bolts are recessed into the head below the camshaft towers with little working space.*

– Remove cylinder head and discard head bolts.

## Cylinder head, installing (M52, S52 engine)

***CAUTION—***
*If the camshafts are removed and reinstalled while the cylinder head is off the engine block, a minimum waiting time, as listed be low, is required before reinstalling the cylinder head. When the camshaft is removed, the hydraulic lifters can expand. This expansion causes increased valve lift, resulting in open valves and possible damage from contact with piston tops.*

| **Cylinder head installation waiting times** | |
|---|---|
| At or above 68°F (20°C) | 10 minutes |
| 50° - 68°F (10° -20°C) | 30 minutes |
| 32° - 50°F (0° - 10°C) | 75 minutes |

– Clean mating surfaces of cylinder head, engine block and timing chain cover. Remove foreign matter and liquid from bolt holes.

***CAUTION—***
*Do not use a metal scraper or wire brush to clean the aluminum cylinder head or pistons. If necessary, use a hard wooden or plastic scraper. Also available are abrasive discs to be used in conjunction with an electric drill. Be sure to use the correct disc for the type of metal being cleaned.*

– Check gasket surface of cylinder head and engine block for warpage using a straight edge.

- M52: If cylinder head is warped, machine 0.3 mm (0.012 in) off milled surface.
- S52: Do not machine cylinder head. Replace warped head.

– Place new cylinder head gasket on engine block. Check that two locating dowels are correctly positioned in block and are not damaged.

***NOTE—***
- *If the cylinder head is machined, use a special 0.3 mm (0.012 in) thicker gasket, available from an authorized BMW dealer.*
- *Make sure the word "OBEN", printed on the gasket, is facing up. The cylinder head gasket fits correctly in only one orientation.*

– With camshafts locked in TDC position with BMW special tool 11 3 240, set cylinder head in position.

***CAUTION—***
- *Lock camshafts in TDC position at the rear of the cylinder head with BMW special tool 11 3 240 before installing the cylinder head.*
- *Make sure that the crankshaft, which had been rotated approximately 30° opposite the direction of engine rotation, is still in that position.*
- *Do not reuse Torx-type cylinder head bolts. They are stretch-type bolts. Always replace after loosening.*

– Lightly lubricate new cylinder head bolts. Loosely install bolts and their washers, then thread them in until they are finger tight.

***NOTE—***
- *Check that all washers for the cylinder head bolts are in place before installing the bolts. Some of the washers may be staked to the cylinder head.*
- *Different length cylinder head bolts are used dependent on type of engine block:*
  *-M52 engine (aluminum block): M10 x 110 mm bolt*
  *-S52 engine (cast iron block): M10 x 95 mm bolt*

◄ Tighten cylinder head bolts in correct sequence, as illustrated.

Tighten cylinder head bolts in three stages as listed below. For stage 2 and stage 3, use BMW special tool 11 2 110 or suitable protractor to tighten bolts to specified torque angle.

| Tightening torque | |
|---|---|
| Cylinder head to engine block (Torx E12 M10 bolts):<br>• Stage 1<br>• Stage 2<br>• Stage 3 | <br><br>30 Nm (22 ft-lb)<br>+90°<br>+90° |

– Install and tighten cylinder head to lower timing chain cover bolts.

| Tightening torque | |
|---|---|
| Cylinder head to lower timing chain cover | 10 Nm (7 ft-lb) |

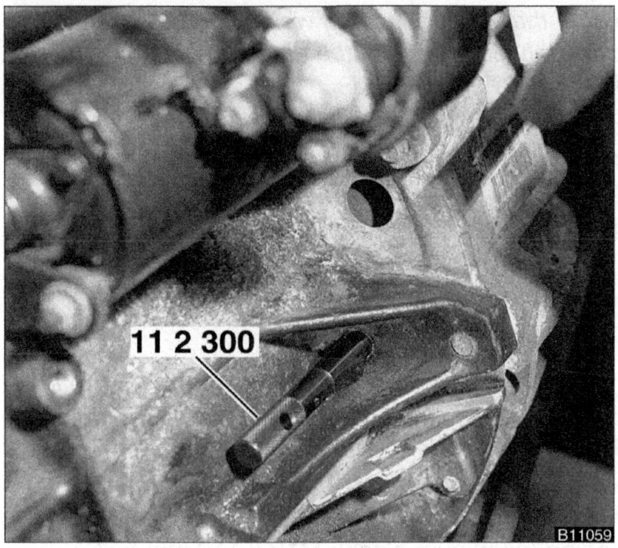


Carefully rotate crankshaft in direction of engine rotation and lock it at TDC: Insert BMW special tool 11 2 300 through transmission bellhousing and into flywheel or drive plate bore.

Mount primary chain sprocket to primary chain so that arrow on sprocket faces up. Mount sprocket to exhaust camshaft so camshaft tapped bores (**arrows**) are on left side of sprocket slotted holes.

– Install timing chain guide and secondary chain tensioner. Tighten mounting bolts.

Install secondary chain sprockets and chain so that arrows on sprockets are pointing up. Center camshaft threaded bores to sprocket slotted holes.

> ***CAUTION—***
> *Mount the intake camshaft sprocket so that collar side of sprocket faces camshaft.*

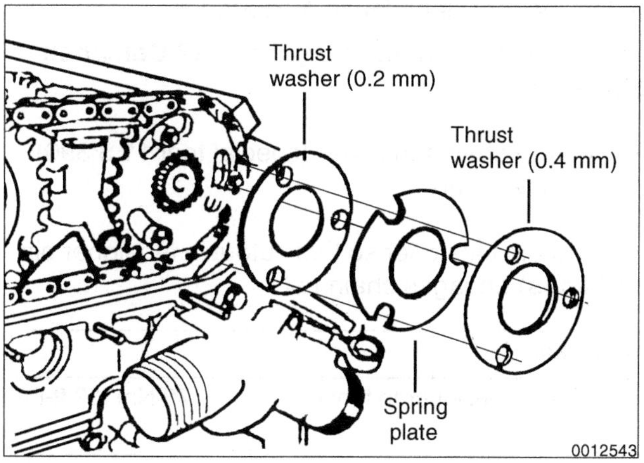


Install thrust washers and spring plate to intake camshaft sprocket.

– Install and tighten exhaust camshaft sprocket mounting bolts finger tight only. Tighten intake camshaft sprocket mounting nuts to specified torque.

| **Tightening torque** | |
|---|---|
| Intake camshaft sprocket to intake camshaft (M6) | 10 Nm (7 ft-lb) |

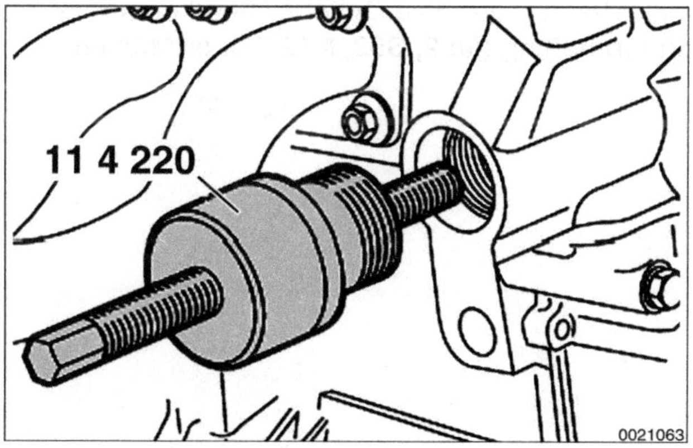


Insert BMW special tool 11 3 390 into primary timing chain tensioner sleeve. Thread tool in finger tight until slack is removed from primary chain. Check that chain is free of slack.

***NOTE—***
*BMW special tool 11 3 390 is a temporary primary chain tensioner and simulates the function of the tensioner.*

– Install VANOS control unit. See **117 Camshaft Timing Chains**.

> ***CAUTION—***
> *Special installation instructions apply when mounting the VANOS control unit to the cylinder head. Be sure to follow the procedure found in* **117 Camshaft Timing Chains**.

– Release secondary chain tensioner slowly.

Preload primary chain tensioning rail:

- Tighten adjusting screw on BMW special tool 11 4 220.

| Tightening torque | |
|---|---|
| Primary chain tensioner preload | 1.3 Nm (11 in-lb) |

– Tighten exhaust camshaft sprocket bolts in two stages.

| Tightening torque | |
|---|---|
| Exhaust camshaft sprocket bolts (M7)<br>• Stage 1<br>• Stage 2 | <br>5 Nm (44 in-lb)<br>22 Nm (17 ft-lb) |

– Remove crankshaft and camshaft locking tools.

– Turn engine over by hand in direction of rotation at least two full rotations. Reinstall camshaft and crankshaft locking tools to check engine timing.

- Remove locking tools if no faults are found.
- If necessary, reset camshaft timing. See **117 Camshaft Timing Chains**.

– Remove temporary chain tensioner (special tool 11 3 390) from side of cylinder head.

Install primary chain tensioner so that cutout in tensioner piston (**dashed lines**) engages chain rail.

| Tightening torque | |
|---|---|
| Primary timing chain tensioner to tensioner sleeve | 70 Nm (52 ft-lb) |

– Install cylinder head cover. See **Cylinder head cover, removing and installing (M52, S52, M52 TU, or M54 engine)**.

– Install front exhaust pipes to manifolds using new nuts. Coat manifold studs with copper paste prior to installing nuts.

| Tightening torques | |
|---|---|
| Exhaust pipe to exhaust manifold (replace nuts)<br>• Stage 1<br>• Stage 2 | <br><br>35 Nm (26 ft-lb)<br>55 Nm (41 ft-lb) |

– Installation of remaining parts is reverse of removal.

- Install and tighten all grounds.
- Refill and bleed cooling system. See **170 Radiator and Cooling System**.
- Change engine oil and filter. See **020 Maintenance**.
- Check adjustment of accelerator cable.

| Tightening torques | |
|---|---|
| Coolant drain plug to engine block | 40 Nm (30 ft-lb) |
| Intake manifold to cylinder head | 15 Nm (11 ft-lb) |
| Radiator cooling fan to coolant pump | 40 Nm (30 ft-lb) |
| Radiator drain screw to radiator | 2.5 Nm (22 in-lb) |

## CYLINDER HEAD REMOVAL AND INSTALLATION (M52 TU, M54 ENGINE)

Cylinder head removal and installation is a complicated repair procedure. VANOS, timing chains and camshafts must be removed first. This requires special tools. Read the entire procedure before beginning the repair.

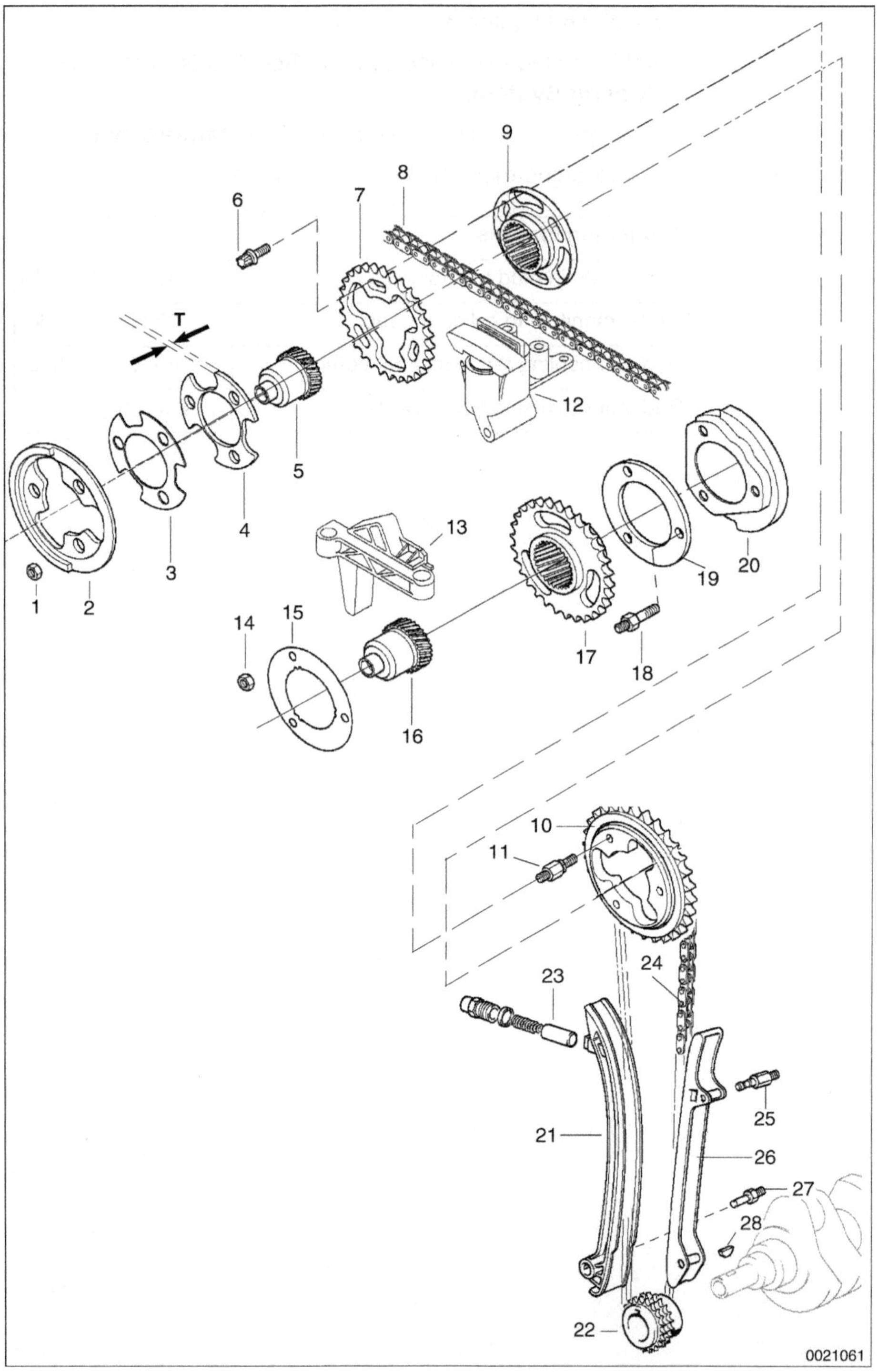


### Timing chain components (M52 TU, M54 engine)

1. **Impulse wheel mounting nut**
2. **Impulse wheel (exhaust)**
3. **Spring plate**
4. **Thrust spacer (T = 3.5 mm)**
5. **Splined shaft**
6. **Torx screw**
7. **Exhaust secondary sprocket**
8. **Secondary timing chain**
9. **Splined sleeve**
10. **Primary sprocket**
11. **Threaded locating stud**
12. **Secondary chain tensioner**
13. **Secondary chain lower guide**
14. **Sprocket mounting nut**
15. **Spring plate**
16. **Splined shaft**
17. **Intake camshaft sprocket**
18. **Locating stud**
19. **Thrust spacer**
20. **Impulse wheel (intake)**
21. **Chain tensioner rail**
22. **Crankshaft sprocket**
23. **Primary chain tensioner**
24. **Primary chain**
25. **Locating stud**
26. **Guide rail**
27. **Locating stud**
28. **Woodruff key**

## Cylinder head, removing (M52 TU, M54 engine)

> ***WARNING—***
> *Due to risk of personal injury, be sure the engine is cold before beginning the removal procedure.*

> ***CAUTION—***
> *Cover all painted surfaces before beginning the removal procedure. As an aid to installation, label all components, wires, and hoses before removing them. Do not reuse gaskets, O-rings or seals during reassembly.*

***NOTE—***
*The illustrations used in this procedure depict an M52 TU engine installed in an E46 (3-series) chassis. Z3 models with M52 TU or M54 engine are similar.*

– Disconnect negative (–) battery cable.

> ***CAUTION—***
> *Prior to disconnecting the battery, read the battery disconnection cautions given in* **001 General Warnings and Cautions**.

– Raise vehicle and support safely.

> ***WARNING—***
> *Make sure the car is stable and well supported at all times. Use a professional automotive lift or jack stands designed for the purpose. A floor jack is not adequate support.*

– Remove splash shield from under engine.

– Remove cylinder head cover. See **Cylinder head cover, removing and installing (M52, S52, M52 TU, or M54 engine)** in this repair group.

– Remove spark plugs.

– Disconnect fuel lines and remove air filter housing and intake manifold. See **Intake manifold, removing and installing (M52 TU, M54 engine)** in this repair group.

> ***WARNING—***
> *Fuel is highly flammable. Place a fire extinguisher in the vicinity of the work area.*

> ***CAUTION—***
> - *Stuff clean rags into the open intake ports to prevent any parts from falling into the engine.*
> - *Mark oxygen sensor connectors before disassembly.*

– Remove exhaust manifolds:
  - Remove windshield washer fluid reservoir.
  - Detach exhaust pipes from manifolds.
  - Using BMW special tool 00 0 200 or equivalent engine lifting harness, raise engine approx. 5 mm (¼ in), taking care to not damage lines, cables or hoses.
  - Remove right engine mount and bracket.
  - Remove oxygen sensors from exhaust manifolds.
  - Remove front exhaust manifold with catalytic converter; then remove rear manifold.
  - See **180 Exhaust System** for additional details.

– Reinstall engine mount and bracket. Remove special tool 00 0 200.

◀ Drain engine coolant from engine block:
  - Remove expansion tank cap on radiator.
  - Place 3-gallon pail underneath engine to capture coolant.
  - Remove coolant drain plug located on exhaust side of cylinder 2 on engine block (**arrow**).

◀ Drain radiator into a 3-gallon pail by removing plastic drain plug (**arrow**) at base of radiator completely.

> ***WARNING —***
> *Use extreme caution when draining and disposing of engine coolant. Coolant is poisonous and lethal to humans and pets. Pets are attracted to coolant because of its sweet smell and taste. Consult a veterinarian immediately if coolant is ingested by an animal.*

> ***CAUTION—***
> *Catch and dispose of drained coolant according to local, state, and federal laws.*

***NOTE—***
*The radiator drain plug shape and design varies. However, it is always found on the left (driver) side of the radiator.*

◄ Remove hoses from thermostat housing by releasing locking clips (**arrows**).

– Disconnect electrical harness at thermostat housing.

– Remove radiator cooling fan and shroud. See **170 Radiator and Cooling System**.

> ***CAUTION—***
> *On cars with viscous clutch cooling fan, radiator fan mounting nut (32 mm) has left-hand threads.*

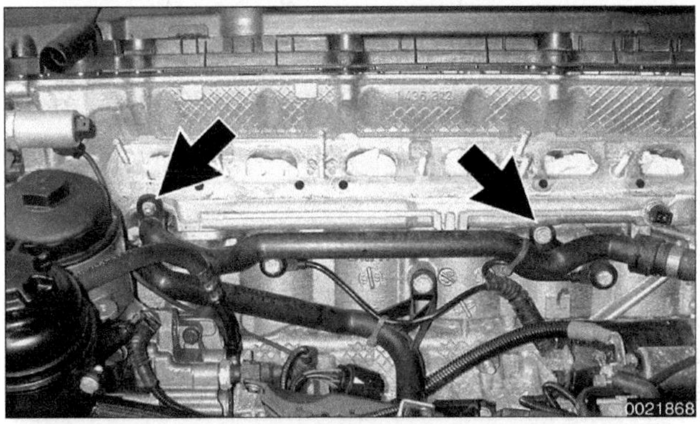


◄ Remove fasteners (**arrows**) from heater bypass tube and set tube aside, leaving heater hose connected.

– Remove secondary air pump:
  - Disconnect hose at one-way valve.
  - Remove bolts at air pump support bracket.
  - Lift up pump and disconnect electrical harness from bottom of pump. Remove pump.

◄ Remove oil baffle cover from above intake camshaft.

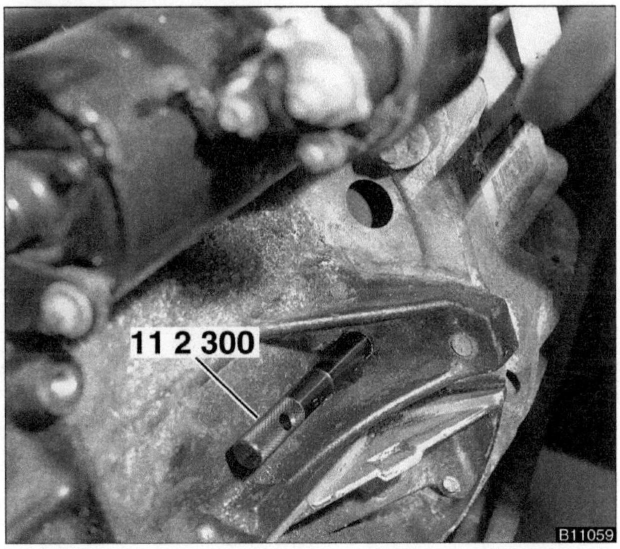


◄ Remove sealing plug from special tool bore on lower left side of engine block near flywheel. Secure crankshaft in TDC position with BMW special tool 11 2 300.

< Unscrew and remove two cylinder head cover studs (**arrows**) at rear of cylinder head.

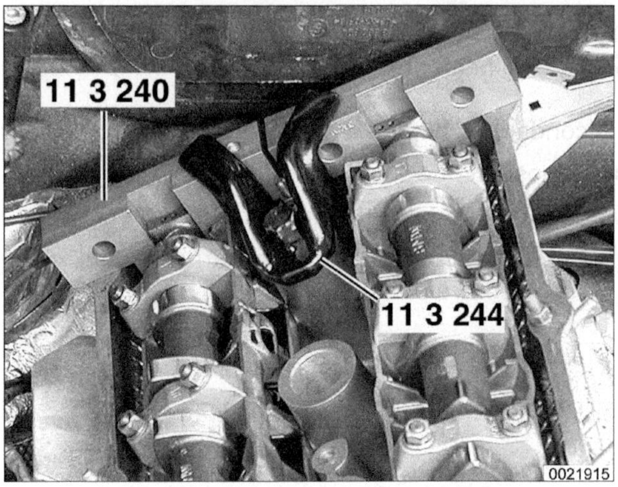


< Secure camshafts in TDC position using BMW special tools 11 3 240 and 11 3 244.

– Remove VANOS unit from front of cylinder head. See **117 Camshaft Timing Chains**.

> ***CAUTION—***
> *Do not rotate the engine with VANOS unit removed. The splined shaft on the intake camshaft might slip out of the VANOS spline teeth, causing the intake cam to no longer be positively connected to the timing chain, allowing for possible piston/valve interference.*

***NOTE—***
*The VANOS unit contains a small quantity of oil. Place shop towels to catch oil as unit is removed or tilted.*

< Remove primary camshaft chain tensioner (**arrow**).

> ***CAUTION—***
> *Primary camshaft chain tensioning piston is under spring pressure.*

Press down on secondary chain tensioner and lock into place using BMW special tool 11 3 292 or a similar size pin.

Remove exhaust camshaft impulse wheel mounting nuts (**arrows**). Remove impulse wheel (**A**).

Remove spring plate (**A**).

Remove intake camshaft sprocket mounting nuts (**arrows**) and remove spring plate (labelled **FRONT**).

Remove torx screws from exhaust camshaft sprocket (**arrows**).

Lift off exhaust and intake sprockets together with secondary chain, thrust spacer (**A**) from exhaust camshaft, and splined shaft (**B**) from intake camshaft.

> ***CAUTION—***
> *Splined shafts share the same part number for both intake and exhaust camshafts. Remove and mark used splined shafts in order and reinstall in original locations.*

Remove exhaust camshaft splined sleeve (**A**) and shaft (**B**).

Remove secondary chain tensioner mounting bolts (**arrows**). Remove tensioner while keeping retaining pin in place.

Remove primary chain sprocket mounting studs (**arrows**) on exhaust camshaft.

Lift primary chain sprocket off exhaust camshaft. Remove sprocket from chain.

< Hang timing chain on end of exhaust camshaft.

< Remove locating studs (**arrows**) from intake camshaft. Lift off intake camshaft thrust spacer (**A**) and impulse wheel (**B**).

< Do not remove end spline retaining screws from camshafts (**arrows**).

< Remove cylinder head cover mounting studs (**arrows**) from center of cylinder head.

– Remove flywheel locking tool from transmission bellhousing so that crankshaft is no longer secured.

< Lift primary chain and hold under tension, then rotate engine against direction of rotation (counterclockwise) approximately 30°. This prevents accidental contact between valves and pistons.

- Remove BMW special tools 11 3 240 and 11 3 244 from rear of cylinder head.

< Remove intake camshaft retaining nuts (**arrows**) at bearing cap 1.

> ***CAUTION—***
> *Intake camshaft bearing cap 1 is centered with adapter sleeves. To avoid camshaft damage, remove before any other caps.*

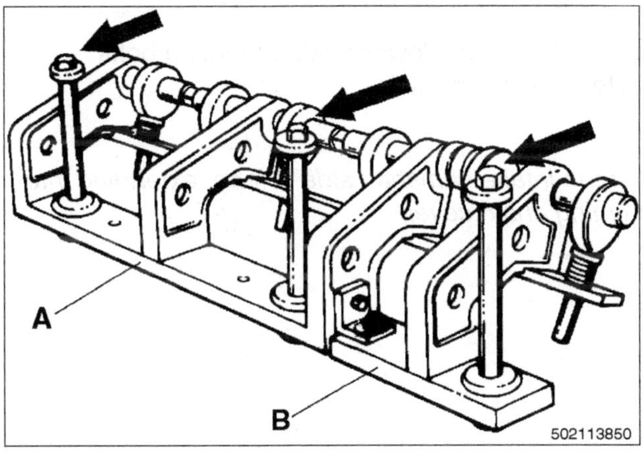


< Fit BMW special tools 11 3 260 (**A**) and 11 3 270 (**B**) to cylinder head and screw long bolts (**arrows**) into spark plug threads. Tensioners aligns with the bearing caps on intake camshaft.

> ***CAUTION—***
> *Do not over-torque bolts into spark plug holes.*

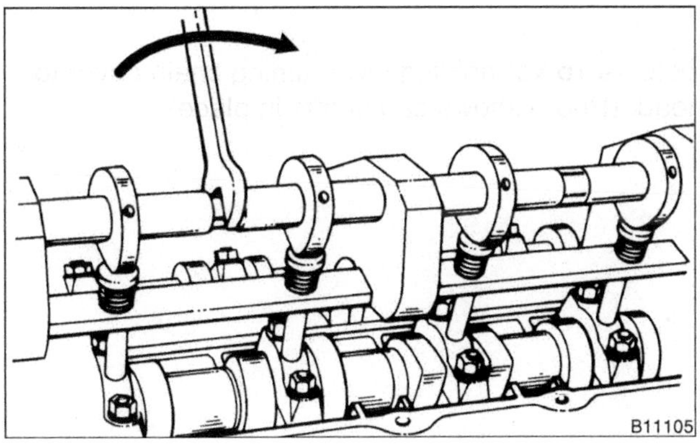


< Turn eccentric shaft of special tool to pretension intake camshaft bearing caps. Remove nuts on remaining bearing caps.

- Release tension on eccentric shaft and remove BMW special tools 11 3 260 and 11 3 270. Remove bearing caps and set aside in order. Remove intake camshaft and store safely.

- Repeat procedure for exhaust camshaft.

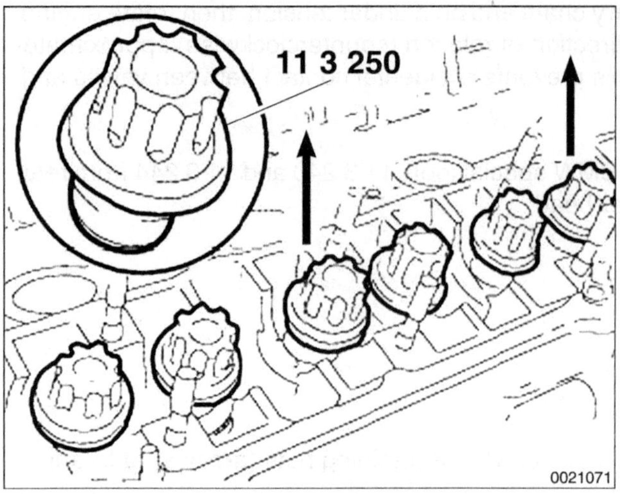


Secure hydraulic lifters in lifter bores using BMW special tool 11 3 250, or remove lifters using a magnetic pick-up tool. With lifters secure or removed, lift out camshaft bearing carriers from cylinder head.

> ***CAUTION—***
> - *Do not let the hydraulic lifters fall out as the camshaft carrier is removed.*
> - *Store hydraulic lifters upright.*
> - *Reinstall used hydraulic lifters into original lifter bores.*

Remove intake camshaft position sensor (**arrow**) from side of cylinder head.

Remove secondary chain lower guide mounting bolts (**arrows**). Remove chain guide.

***NOTE—***
*The bolt on the intake camshaft side is long, extending into the lower timing chain cover.*

Remove bolts (**arrows**) holding lower timing chain cover to cylinder head. (Photo shows camshafts in place).

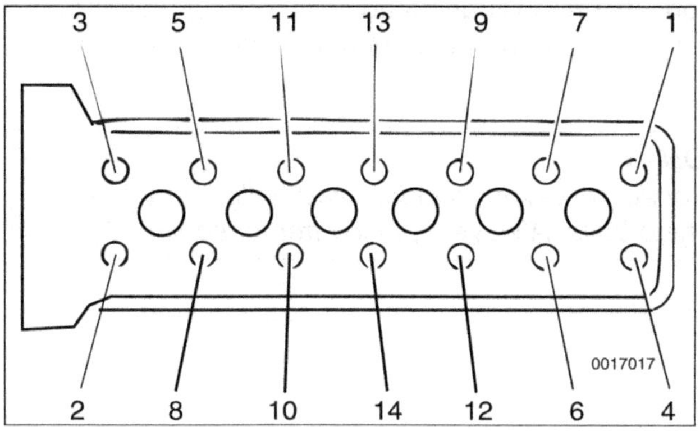


◀ Using BMW special tool 11 2 250 or equivalent, loosen cylinder head bolts in several stages in sequence shown. Discard head bolts.

***NOTE—***
*BMW special tool 11 2 250 is a thin-walled Torx E12 socket with an extended reach. The cylinder head bolts are recessed into the head below the camshaft towers with little working space.*

– Lift off cylinder head. Refer to **116 Cylinder Head and Valvetrain** to evaluate cylinder head.

## Cylinder head, installing (M52 TU, M54 engine)

– Clean mating surfaces of cylinder head, engine block and timing chain cover. Remove foreign matter and liquid from bolt holes.

***CAUTION—***
*Do not use a metal scraper or wire brush to clean the aluminum cylinder head or pistons. If necessary, use a hard wooden or plastic scraper. Also available are abrasive discs to be used in conjunction with an electric drill. Be sure to use the correct disc for the type of metal being cleaned.*

– Check gasket surface of cylinder head and engine block for warpage using a straight edge. If cylinder head is warped, machine 0.3 mm (0.012 in) off milled surface.

***NOTE—***
*If the cylinder head is machined, use a special 0.3 mm (0.012 in) thicker cylinder head gasket, available from an authorized BMW dealer.*

– Place new cylinder head gasket on engine block, guiding primary chain through cylinder head opening. Check that two locating dowels are correctly positioned in block and are not damaged.

***NOTE—***
*Make sure the word "OBEN", printed on the gasket, is facing up. The cylinder head gasket fits correctly in only one orientation.*

– Lightly lubricate new cylinder head bolts. Loosely install bolts and their washers, then thread them in until they are finger tight.

> ***CAUTION—***
> *Do not reuse Torx-type cylinder head bolts. They are stretch-type bolts. Always replace after loosening.*

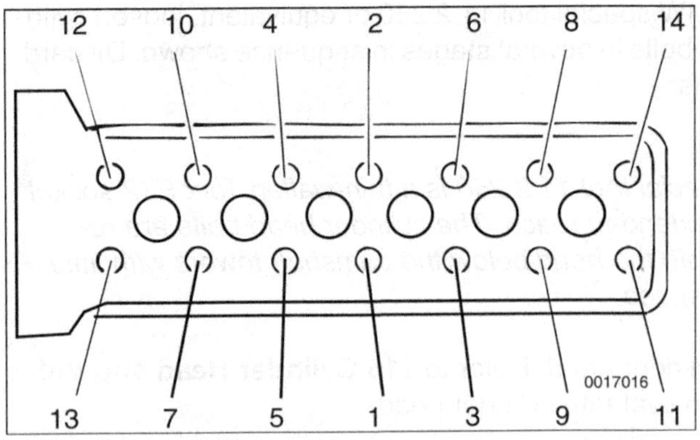


◄ Tighten cylinder head bolts in correct sequence, as illustrated.

◄ Tighten cylinder head bolts in three stages as listed below. For stage 2 and stage 3, use BMW special tool 11 2 110 or suitable protractor to tighten bolts to specified torque angle.

| Tightening torque | |
|---|---|
| Cylinder head to engine block (Torx E12 M10 bolts):<br>• Stage 1<br>• Stage 2<br>• Stage 3 | <br><br>40 Nm (30 ft-lb)<br>+90°<br>+90° |

◄ Install lower secondary chain guide. Tighten Torx bolts (**arrows**) to specifications.

***NOTE—***
*The bolt on the intake camshaft side is long, extending into the lower timing chain cover.*

| Tightening torque | |
|---|---|
| Secondary chain guide to cylinder head | 10 Nm (7 ft-lb) |

Install intake camshaft position sensor (**arrow**).

Install and tighten cylinder head-to-lower timing chain cover bolts (**arrows**). (Photo shows camshafts in place.)

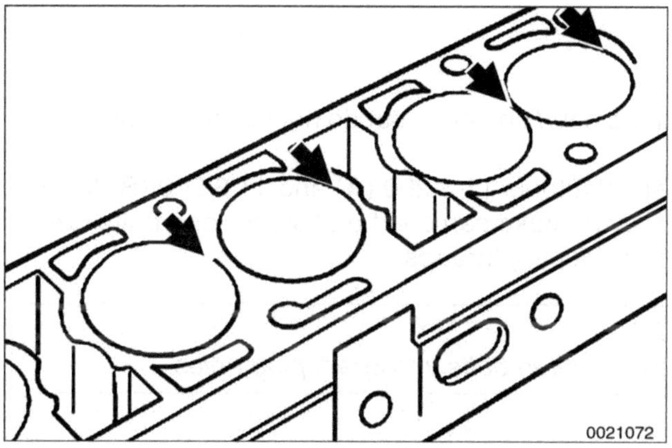


Before installing camshaft carriers, examine bearing points (**arrows**) on hydraulic lifter bores for signs of wear.

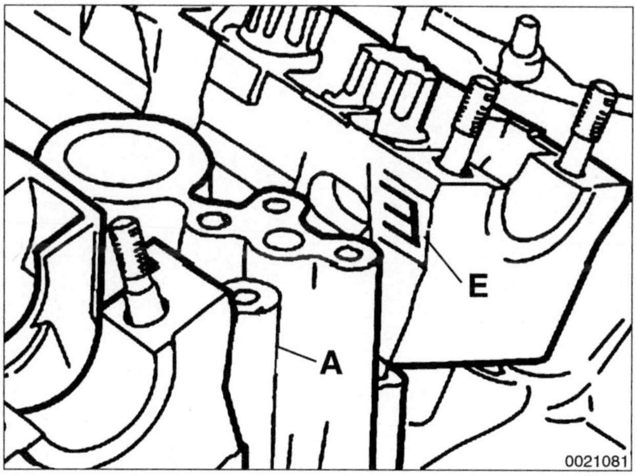


Install camshaft carriers with hydraulic lifters into cylinder head. Note marks on carriers: (**E)** for intake side and (**A)** for exhaust side.

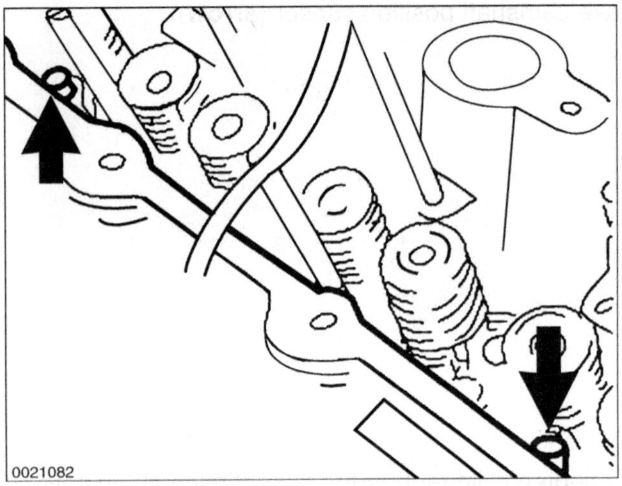


◀ Center camshaft carrier on pins (**arrows**) at bearing positions 2 and 7.

– Lift timing chain and place exhaust camshaft on exhaust camshaft carrier. Place intake camshaft on camshaft carrier.

> ***CAUTION—***
> *Be sure that crankshaft is still positioned at least 30° back from TDC.*

◀ Rotate camshafts so that intake and exhaust lobes for cylinder 1 face each other (**arrows**).

– Place bearing caps on cams, but do not install retaining nuts.

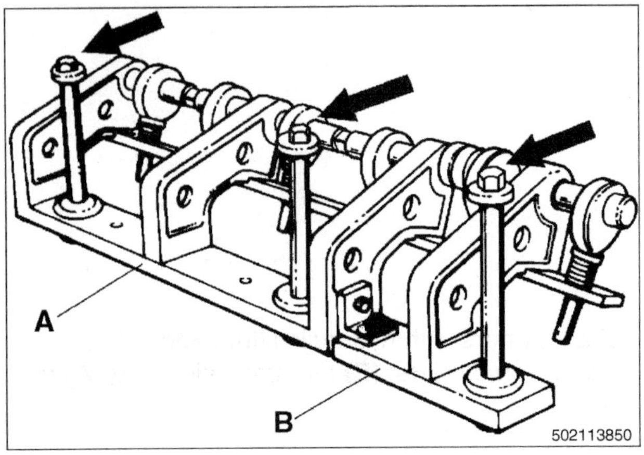


◀ Fit BMW special tools 11 3 260 (**A**) and 11 3 270 (**B**) to cylinder head over intake camshaft and screw long bolts (**arrows**) into spark plug threads.

> ***CAUTION—***
> *Do not over-tighten bolts into spark plug holes.*

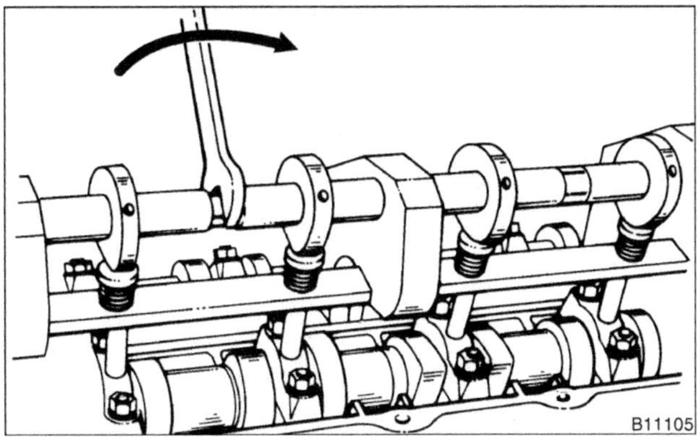


Turn eccentric shaft of special tool to pretension bearing caps. Install and torque nuts on intake camshaft bearing caps.

| Tightening torque | |
|---|---|
| Camshaft bearing cap to cylinder head (M7) | 14 Nm (10 ft-lb) |

– Release tension on eccentric shaft and remove BMW special tool 11 3 260 and 11 3 270.

– Repeat procedure for exhaust camshaft.

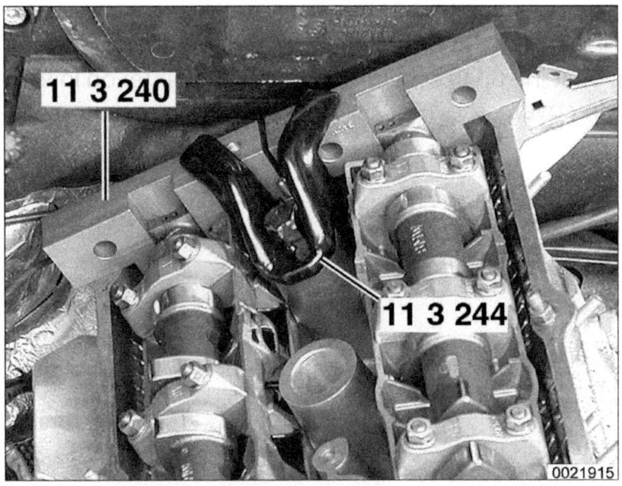


Secure camshafts in TDC position using BMW special tools 11 3 240 and 11 3 244.

If necessary, turn camshaft so that special tools are squarely seated on cylinder head.

> ***CAUTION—***
> *When the camshafts are removed and reinstalled, a minimum waiting time, as listed be low, is required before returning the crankshaft to TDC. When the camshaft is removed, the hydraulic lifters can expand. This expansion causes increased valve lift, resulting in open valves and possible damage from contact with piston tops.*

– Before the next step, observe waiting time for lifter bleed down.

| Camshaft installation waiting times | |
|---|---|
| At or above 68°F (20°C) | 4 minutes |
| 50° - 68°F (10° -20°C) | 11 minutes |
| 32° - 50°F (0° - 10°C) | 30 minutes |

***NOTE—***
*When reinstalling camshaft timing chains, refer to* **Timing chain components (M52 TU, M54 engine)** *in this repair group.*

Lift primary timing chain and hold under tension.

While maintaining tension on timing chain, rotate crankshaft from 30° before TDC in direction of rotation up to cylinder 1 TDC position. Line up **OIT** on front pulley with pointer on lower timing chain cover.

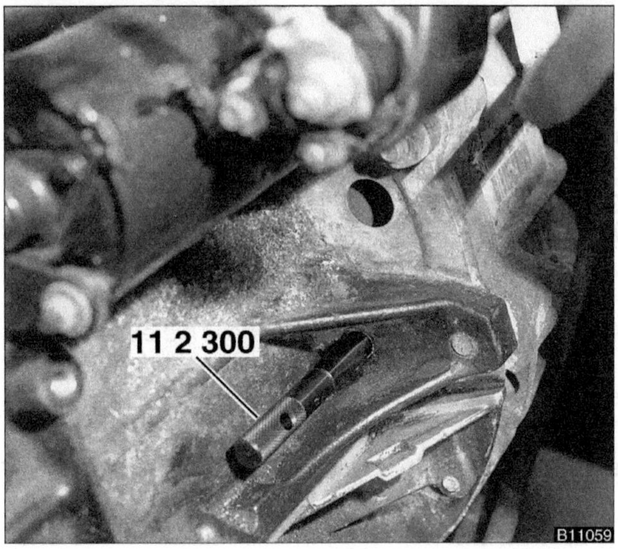


Secure crankshaft in TDC position with BMW special tool 11 2 300.

Slide impulse wheel on intake camshaft, aligning boss with raised portion on camshaft (**arrow**).

Fit thrust spacer to intake camshaft and tighten down with threaded locating studs (place longer threaded portion of studs into camshaft).

| **Tightening torque** | |
|---|---|
| Impulse wheel studs to intake camshaft (M7) | 20 Nm (15 ft-lb) |

Fit top sprocket to primary timing chain and install on exhaust camshaft so that pointer on sprocket lines up with cylinder head sealing surface (**dashed line**).

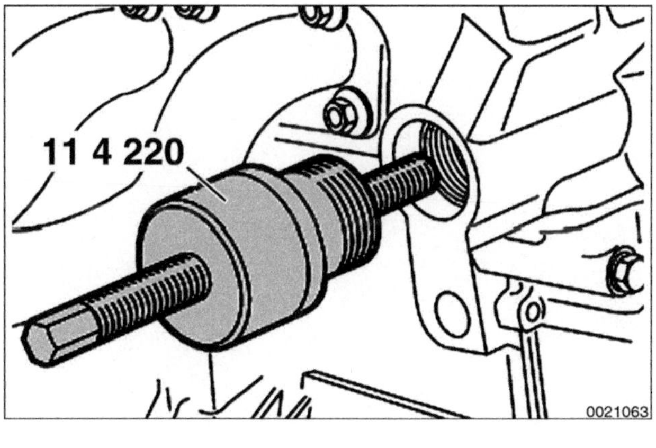


Insert BMW special tool 11 4 220 into primary timing chain tensioning piston bore and bring adjustment screw into contact with tensioning rail, but do not pretension timing chain.

***NOTE—***
*BMW special tool 11 4 220 is a temporary primary chain tensioner and simulates the function of the tensioner.*

– Recheck that arrow on top primary sprocket is aligned with upper edge of cylinder head. Reposition sprocket if necessary.

< Insert and tighten down threaded locating studs (**arrows**) in end of exhaust camshaft.

| Tightening torque | |
|---|---|
| Exhaust camshaft locating stud | 20 Nm (15 ft-lb) |

< Install secondary chain tensioner on cylinder head (**arrows**). Keep tensioner compressed using BMW special tool 11 3 292 or suitable pin.

< Fit exhaust camshaft splined sleeve. Confirm that gap in sleeve splines aligns with corresponding gap in camshaft splines (**arrows**).

Slide splined shaft on exhaust camshaft. Be sure that locating tooth of shaft (**arrow**) fits into spline gaps of camshaft and splined sleeve.

- Slide splined shaft in further until three small slots on splined sleeve are centered on three threaded holes in primary chain sprocket.

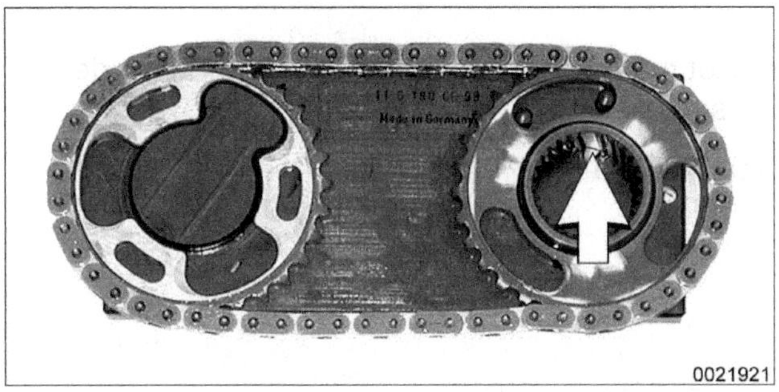


Place intake and exhaust sprockets in BMW special tool 11 6 180. Position spline gap on intake sprocket (**arrow**) as shown and place secondary chain on sprockets.

Remove chain and sprockets from tool and slide sprockets on camshafts. Confirm that gap in intake sprocket splines lines up with corresponding gap in camshaft splines (**arrow**).

> ***CAUTION—***
> *Do not alter position of sprockets with respect to chain when removing from special tool 11 6 180.*

Slide splined shaft on intake camshaft until approx. 1 mm (0.04 in.) of splines (**arrows**) are visible. Confirm that locating tooth of shaft fits into spline gaps on camshaft and sprocket.

⮜ Install intake camshaft spring plate so that **FRONT** mark is visible. Install mounting nuts (**arrows**) finger tight.

⮜ Insert sprocket mounting bolts (**arrows**) on exhaust side camshaft assembly.

- Initially tighten to approx. 5 Nm (44 in-lb) and then back off by half a turn.

⮜ Fit thrust spacer (**A**) on exhaust camshaft.

⮜ Install spring plate (**A**) to exhaust camshaft. Make sure that **F** mark is visible.

***NOTE—***
*If **F** mark is no longer visible, install spring plate so that convex side points forward (toward front of car).*

⮜ Install exhaust camshaft impulse wheel, aligning pointer with top edge of cylinder head (**dashed line**). Install mounting nuts (**arrows**) finger tight.

⮜ Pull out exhaust camshaft splined shaft to stop.

– Press down on secondary chain tensioner and remove tensioner lock-down tool.

Preload primary chain:

- Tighten adjusting screw on BMW special tool 11 4 220 to specified torque.
- Check that chain is free of slack.

| Tightening torque | |
|---|---|
| Primary chain tensioner preload | 0.7 Nm (6 in-lb) |

***NOTE—***
*BMW special tool 11 3 390 is a temporary primary chain tensioner and simulates the function of the tensioner.*

Preload exhaust camshaft spring plate by pressing on impulse wheel while tightening mounting nuts (**arrows**) finger tight.

Install BMW special tool 11 6 150 (VANOS setup bracket) to front of cylinder head timing case. Install nuts (**arrows**) finger tight, then tighten down uniformly until special tool is in full contact with cylinder head.

***CAUTION—***
*Make sure all gasket material is removed from face of cylinder head. Clean sealing face and keep free of oil. If any foreign material is present on the sealing surface, the camshaft timing is incorrect.*

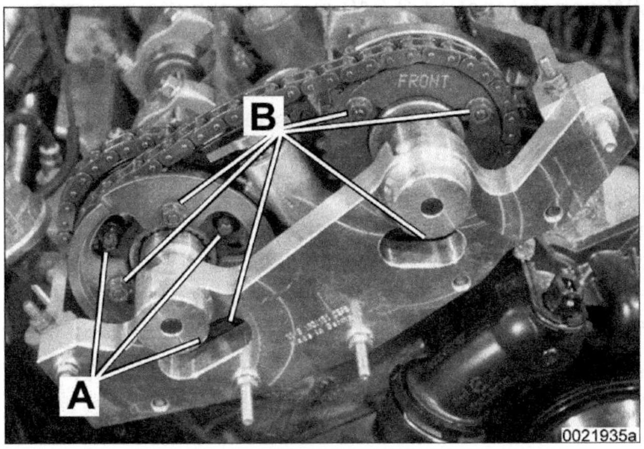


◄ Secure camshaft sprockets and impulse wheels:

- Tighten mounting screws (**A**) on exhaust camshaft impulse wheel to approx. 5 Nm (44 in-lb).
- Tighten mounting nuts (**B**) on exhaust and intake sprocket assemblies to approx. 5 Nm (44 in-lb).
- Torque down mounting screws (**A**) and nuts (**B**) to final specifications.

| **Tightening torques** | |
|---|---|
| Sprocket assembly to camshaft (initial torque) | 5 Nm (44 in-lb) |
| Sprocket assembly wheel to camshaft (final torque)<br>• M7 Torx screws (**A**)<br>• M6 mounting nut (**B**) | <br>20 Nm (15 ft-lb)<br>10 Nm (7 ft-lb) |

– Remove flywheel locking tool from transmission bellhousing so that crankshaft is no longer secured.

– Remove camshaft locking tools from cylinder head.

◄ Turn engine over twice in normal direction of rotation so that cylinder 1 intake and exhaust lobes (**arrows**) face each other.

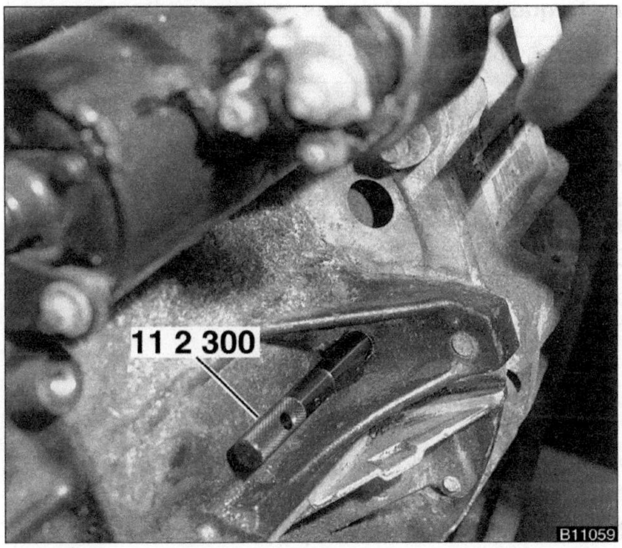


◄ Secure crankshaft in TDC position with BMW special tool 11 2 300.

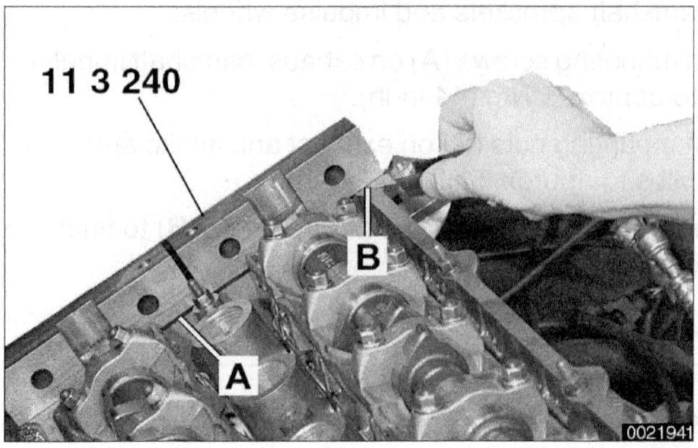


Place BMW special tool 11 3 240 over camshafts and measure clearances to cylinder head top surface.

- If exhaust side of tool is not flush with head (gap **A**), camshaft timing is incorrect. Reset camshaft timing as described in **117 Camshaft Timing Chains**.
- Normally, due to flexible sprocket design, VANOS unit tolerances and play in VANOS splines, when camshaft timing is set correctly, intake side of special tool 11 3 240 may be up to 1 mm (0.04 in) above cylinder head top surface (gap **B**). If so, reassemble engine.
- Otherwise, reset camshaft timing as described in **117 Camshaft Timing Chains.**

– Remove BMW special tool 11 6 150 (VANOS setup bracket) from front of cylinder head.

– Install VANOS control unit. See **117 Camshaft Timing Chains**.

***CAUTION—***
*Special installation instructions apply when mounting the VANOS control unit to the cylinder head. Be sure to follow the procedure found in* **117 Camshaft Timing Chains**.

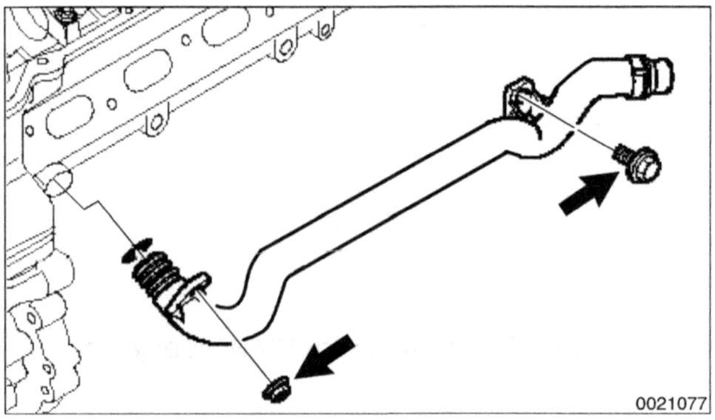


Install coolant pipe at base of cylinder head and tighten fasteners (**arrows**).

***NOTE—***
*Use new sealing O-ring on coolant pipe.*

– Install VANOS oil line banjo bolt with new seals. Attach oil line to VANOS unit.

| Tightening torque | |
|---|---|
| Oil line to VANOS unit (banjo bolt) | 32 Nm (24 ft-lb) |

– Remove BMW special tool 11 4 220 and reinstall chain tensioner cylinder.

| Tightening torque | |
|---|---|
| Primary chain tensioner cylinder to cylinder head | 70 Nm (52 ft-lb) |

– Remove flywheel locking tool from transmission bellhousing. Replace special tool bore sealing plug.

– Remove camshaft locking tools from cylinder head.

– Install intake camshaft cover and cylinder head cover. See **Cylinder head cover, removing and installing (M52, S52, M52 TU, or M54 engine)** in this repair group. Use new gaskets, rubber insulators and seals as necessary.

| Tightening torque | |
|---|---|
| Cylinder head cover to cylinder head (M6) | 10 Nm (7 ft-lb) |

– Install exhaust manifolds using new gaskets and nuts. Coat manifold studs with copper paste prior to installing nuts. Reinstall oxygen sensors.

| Tightening torques | |
|---|---|
| Exhaust manifold to cylinder head (M7) | 20 Nm (15 ft-lb) |
| Oxygen sensor to exhaust manifold | 50 Nm (37 ft-lb) |

– Install electrical connectors for oil pressure sender and coolant temperature sensor before installing intake manifold.

– Installation of intake manifold, throttle housing and air filter assembly is reverse of removal. See **Intake manifold, removing and installing (M52 TU, M54 engine)** in this repair group.
  - Use new fuel injector seals.
  - Carefully check intake manifold gasket and replace if necessary.
  - Inspect O-ring seal between mass air flow sensor and air filter housing. To facilitate reassembly, coat seal with acid-free grease.

***CAUTION—***
*When reattaching throttle assembly harness connector, make sure arrows on connector and plug line up when fully tightened.*

– Installation of remaining parts is reverse of removal. Note the following:

- Reinstall engine cooling fan. Reinstall and adjust engine drive belt(s). See **020 Maintenance**.
- Refill and bleed cooling system. See **170 Radiator and Cooling System**.
- Change engine oil and filter. see **020 Maintenance**.
- If necessary, adjust accelerator cable.
- Reconnect battery.

> ***CAUTION—***
> *To prevent damage to engine electronic systems, be sure to install all ground wires previously removed, including the ground wires for the ignition coils.*

| **Tightening torques** | |
|---|---|
| Coolant drain plug to engine block | 25 Nm (18 ft-lb) |
| Radiator cooling fan to coolant pump | 40 Nm (30 ft-lb) |
| Radiator drain screw to radiator | 2.5 Nm (22 in-lb) |
| Spark plug to cylinder head | 25 Nm (18 ft-lb) |

# 116 Cylinder Head and Valvetrain





**TABLES**

## General

This repair group covers cylinder head and valvetrain service and repair. Most of the repairs described here require that the cylinder head first be removed. See **113 Cylinder Head Removal and Installation**.

***NOTE—***
- *Camshaft removal for the S54 engine used in 2001 - 2002 M version is not included in the repair group. S54 camshaft removal and installation is a highly complex job and is outside the scope of this manual.*
- *If the cylinder head requires significant reconditioning work, consider purchasing a remanufactured cylinder head from an authorized BMW dealer.*
- *For timing chain and VANOS repair information, see* **117 Camshaft Timing Chain**.

### Special tools

Special tools are required for most cylinder head service, disassembly and reconditioning described in this repair group. Be sure to read each procedure thoroughly before starting a job to determine which tools and equipment are necessary.

Micrometer

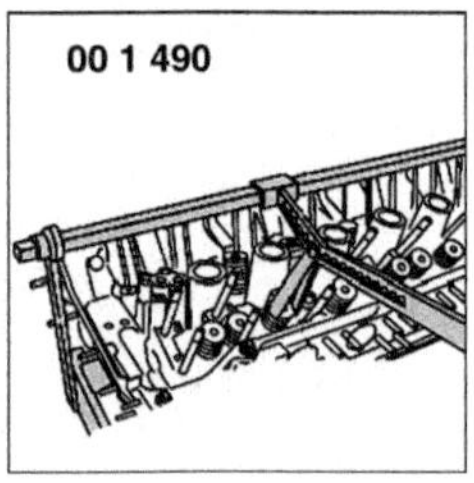


Cylinder head jig
(BMW special tool 00 1 490)

Valve seat grinder
(BMW special tool 00 3 520)

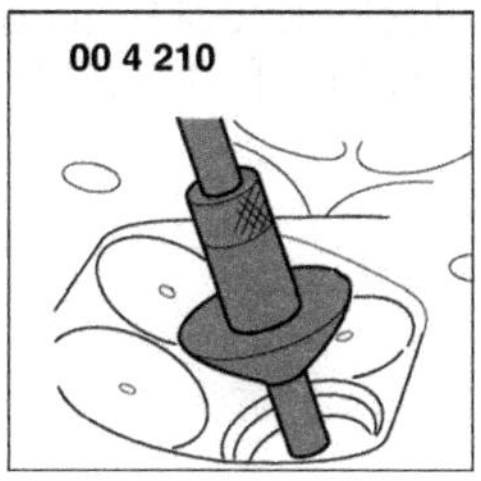


◀ Valve guide reamer (BMW special tool 00 4 210)

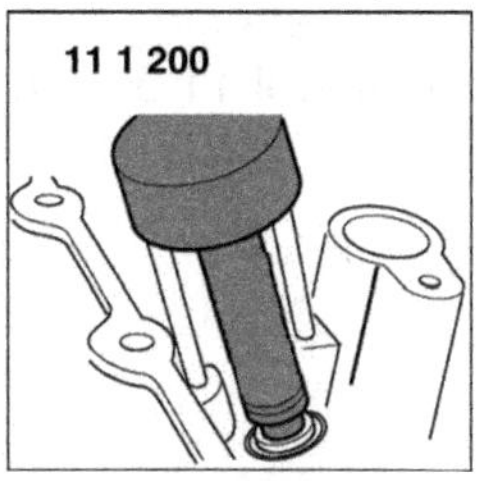


◀ Valve stem seal driver (BMW special tool 11 1 200)

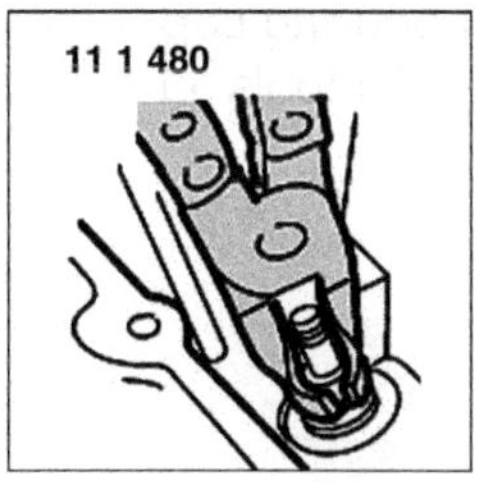


◀ Valve stem seal puller (BMW special tool 11 1 480)

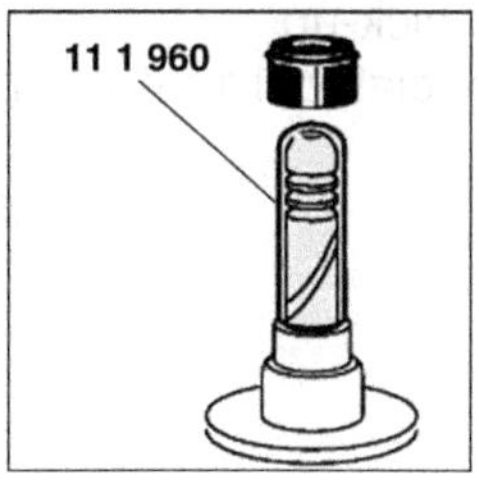


◀ Valve stem seal installation guide (BMW special tool 11 1 960)

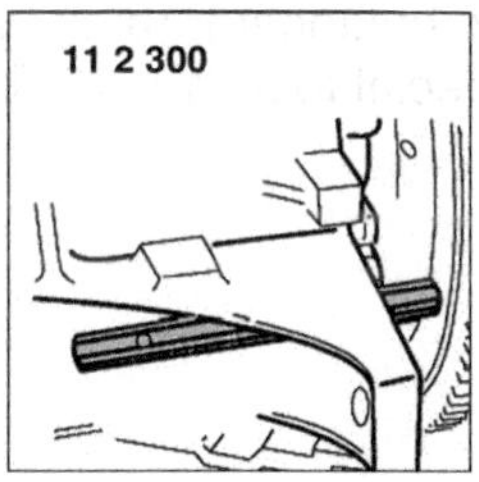


◀ Crankshaft locking tool (BMW special tool 11 2 300)

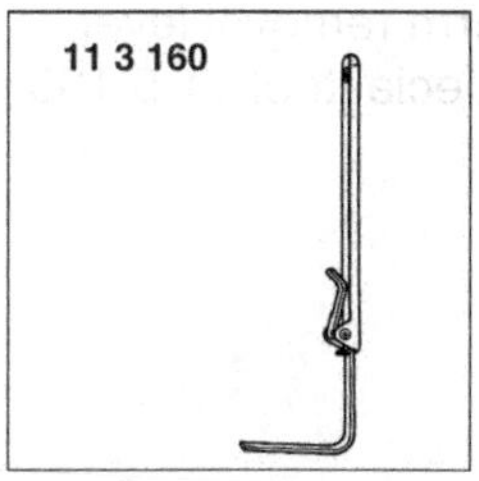


◀ Feeler gauge holder (BMW special tool 113 160)

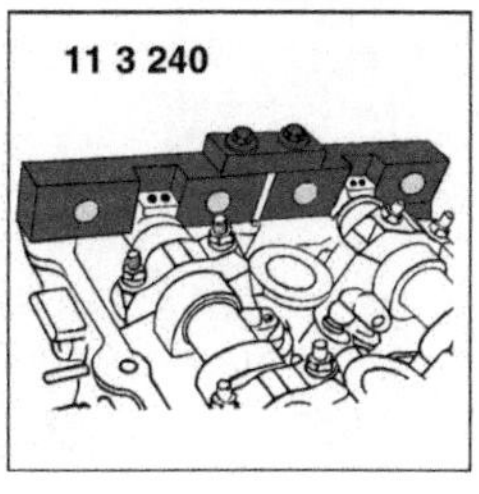


◀ Camshaft locking tool (BMW special tool 11 3 240)

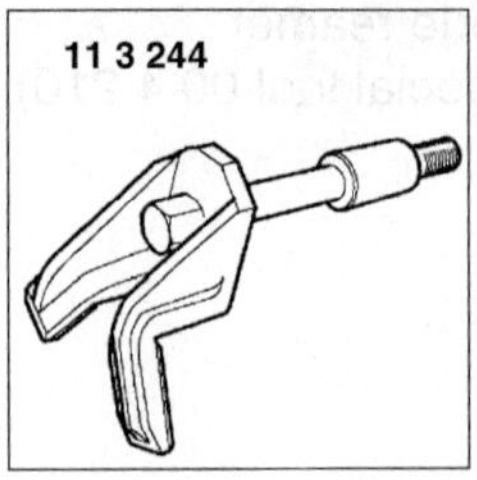


Camshaft locking tool bracket
(BMW special tool 11 3 244)

Hydraulic lifter retaining tool
(BMW special tool 11 3 250)

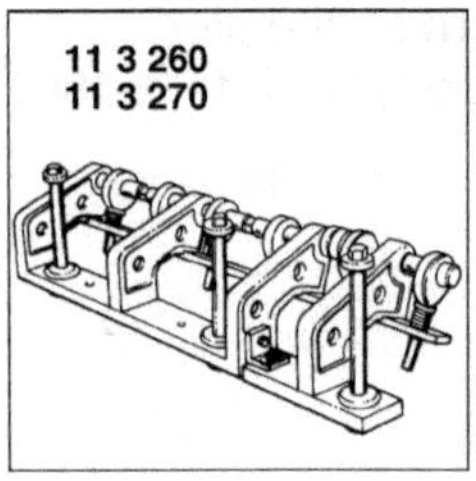


Camshaft bearing cap removal tool
(BMW special tools 11 3 260 / 11 3 270)

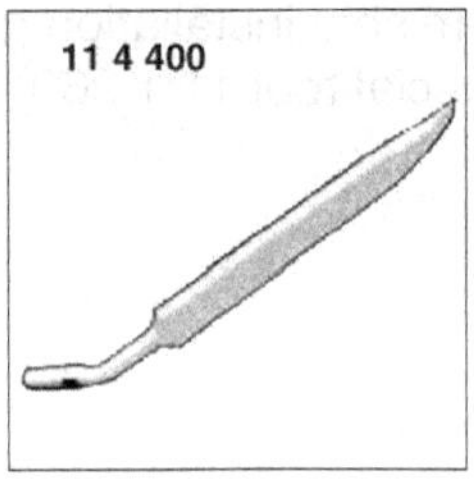


Magnetic pick-up
(BMW special tool 11 4 400)

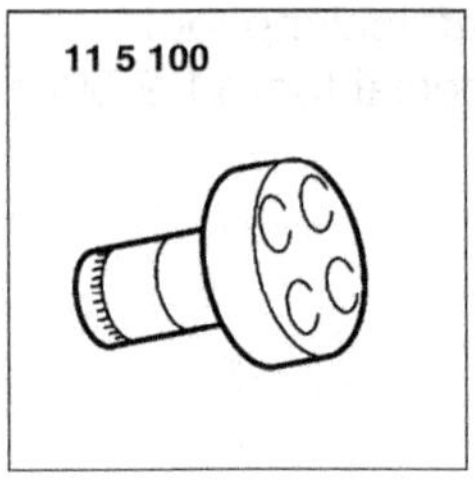


Harmonic balancer turning tool
(BMW special tool 11 5 100)

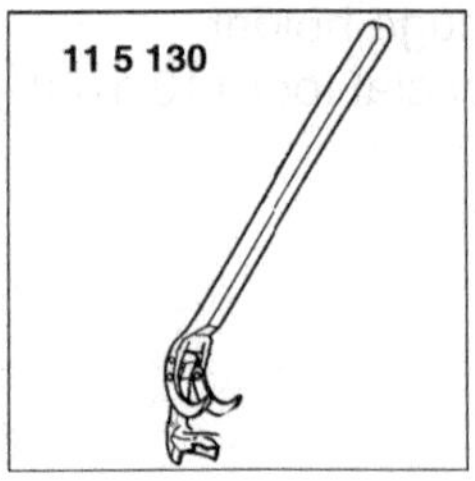


Rocker arm removal lever
(BMW special tool 11 5 130)

## Engine applications

The information given in this repair group is organized according to engine code. **Table a** lists the engines used in the 1996 - 2002 Z3 and M Coupe / Roadster cars. For additional engine application information, see **100 Engine—General**.

**Table a. Engine application codes**

| Engine code | Model (model year) |
|---|---|
| 4-cylinder | |
| M44 B19 | Z3 Roadster 1.9 (1996 - 1998) |
| 6-cylinder | |
| M52 B28 | Z3 Coupe / Roadster 2.8 (1997 - 1998) |
| M52 TU B25 | Z3 Coupe / Roadster 2.3 (1999 - 2000) |
| M52 TU B28 | Z3 Coupe / Roadster 2.8 (1999 - 2000) |
| M54 B25 | Z3 Coupe / Roadster 2.5i (2001 - 2002) |
| M54 B30 | Z3 Coupe / Roadster 3.0i (2001 - 2002) |
| S52 B32 | M Coupe / Roadster (1998 - 2000) |
| S54 B32 | M Coupe / Roadster (2001 - 2002)) |

## Hydraulic valve adjusters (HVA) (4-cylinder engine)

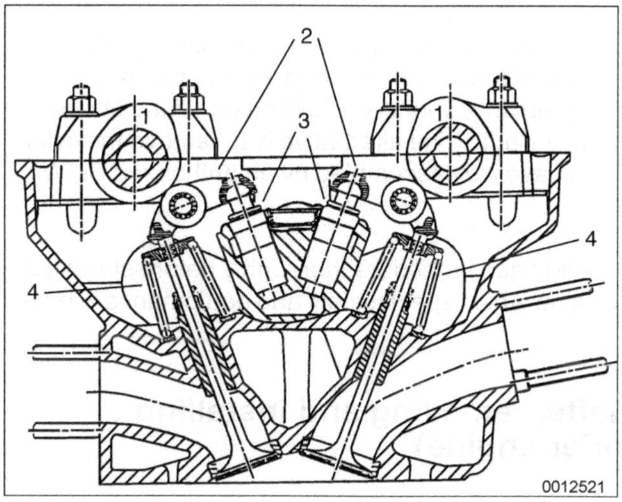


The 4-cylinder engine uses hydraulic valve adjusters (HVA) to keep valve clearances within a limited working range. HVAs are sealed units and require no maintenance.

1. Camshafts
2. Rocker arms
3. Hydraulic valve adjusters (HVA)
4. Valve and conical valve spring

Under some circumstances, such as a cold start, the hydraulic units may become noisy. Hydraulic valve adjuster noise is usually a high-pitched tapping or chattering noise. In most instances, this is considered normal as long as the noise goes away in a few minutes (maximum 20 minutes). If the noise does not go away, either the hydraulic unit is faulty or the oil pressure to the lifter(s) is low.

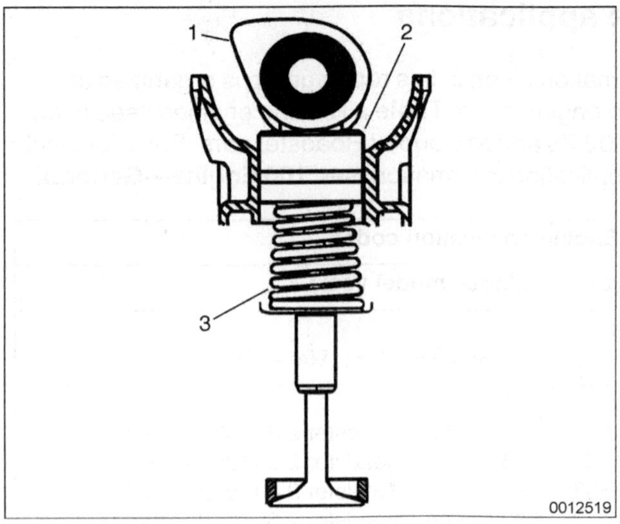


### Hydraulic cam followers (M52, S52, M52 TU or M54 engine)

6-cylinder engines except the S54 engine in 2001 - 2002 M Coupe / Roadster use self-adjusting hydraulic cam followers to keep valve clearances within a limited working range. Cam followers are sealed units and require no maintenance.

1. Section view of camshaft
2. Hydraulic cam follower
3. Valve with conical valve spring

Under some circumstances, such as a cold start, the cam followers may become noisy. Hydraulic cam follower noise is usually a high-pitched tapping or chattering noise. In most instances, this is considered normal as long as the noise goes away in a few minutes (maximum 20 minutes). If the noise does not go away, either the follower is faulty or the oil pressure to the follower is low. Hydraulic cam follower replacement requires camshaft removal.

## CYLINDER HEAD DISASSEMBLY AND ASSEMBLY

Some of these procedures recommend the use of BMW special tools. Many of these tools are expensive and are only available through an authorized BMW dealer. If the special tools are not available, one alternative is to remove the cylinder head and have it disassembled by an authorized BMW dealer.

Be sure to read each procedure thoroughly before starting a job to determine which special tools and equipment are necessary.

### Camshafts, removing and installing (4-cylinder engine)

Camshafts on the 4-cylinder (M44) engine can be removed with the cylinder head installed on the engine. The first step is to remove the cylinder head cover.

> ***CAUTION—***
> *Be sure to use BMW special tool 11 5 130 to remove camshaft rocker arms prior to camshaft removal.*

– Remove cylinder head cover. See **113 Cylinder Head Removal and Installation.**

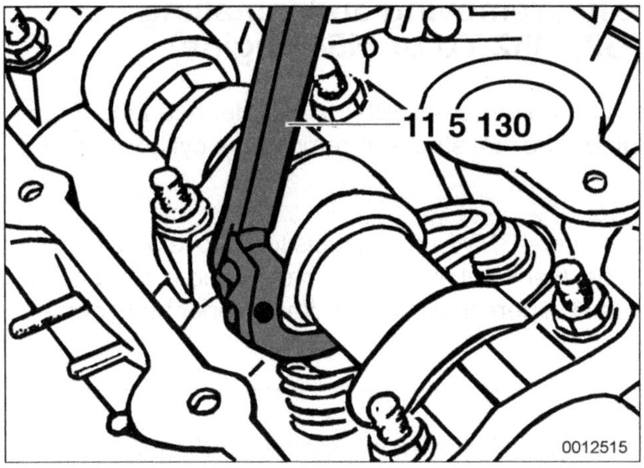


< Using BMW special tool 11 5 130, depress each valve and valve spring enough to remove its rocker arm.

- For each cylinder, be sure camshaft lobes are pointing up before using tool.
- Rotate engine to bring next camshaft lobe pair to vertical position, then remove corresponding rocker arms.
- Place removed rocker arms in order so that they can be reinstalled in the same locations.

– Remove hydraulic chain tensioner from side of cylinder head. Remove timing chain sprockets and timing chains from camshafts. Remove upper timing chain guide at front of cylinder head. See **117 Camshaft Timing Chain**.

– Remove camshaft bearing caps and remove camshaft(s).

– Check all components for wear. See **Cylinder head and valvetrain reconditioning** in this repair group.

– Camshaft installation is reverse of removal.

- Lubricate camshaft(s) prior to installation.
- Assemble timing chain sprockets and tensioner and set valve timing correctly before attempting to reinstall rocker arms. See **117 Camshaft Timing Chain**.
- Make sure all parts are reinstalled in their original positions.

***NOTE—***
- *The camshaft bearing caps are coded with a letter and a cylinder number. For example, "A1" is the exhaust camshaft bearing cap for cylinder 1 and "E1" is the intake camshaft bearing cap for cylinder 1. Similarly, the camshafts are stamped with a code letter: "E" for intake and "A" for exhaust.*
- *"E" is for Einlaß, (German for intake) and "A" is for Auspuff, (German for exhaust).*
- *Cylinder 1 is at the front of the engine.*

| Tightening torque | |
|---|---|
| Camshaft bearing cap to cylinder head (M7) | 15 Nm (11 ft-lb) |

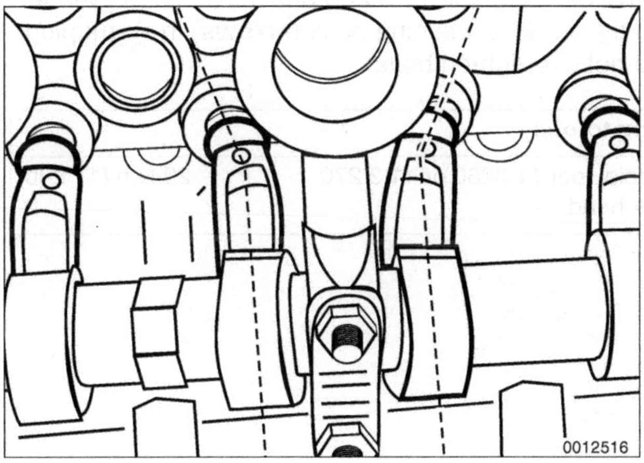


< When installing rocker arms, note that each rocker arm angles away from corresponding spark plug hole.

- Make sure camshaft lobe is pointed up when installing rocker arm.

– Reinstall cylinder head cover.

| Tightening torque | |
|---|---|
| Cylinder head cover to cylinder head (M6) | 10 Nm (7 ft-lb) |

## Camshafts, removing and installing (M52, S52, M52 TU or M54 engine)

Camshafts can be removed with the cylinder head mounted on the engine. The first step is to remove the cylinder head cover.

Always use special BMW service tools to remove and install the camshafts. Removal of the camshafts without the special tools can result in personal injury and/or cylinder head damage.

> ***CAUTION—***
> *The common practice of evenly loosening the camshaft bearing cap nuts is NOT an acceptable method of removing camshafts. Be sure to read the procedure through before starting the job.*

– Remove cylinder head cover. See **113 Cylinder Head Removal and Installation**.

– Remove spark plugs.

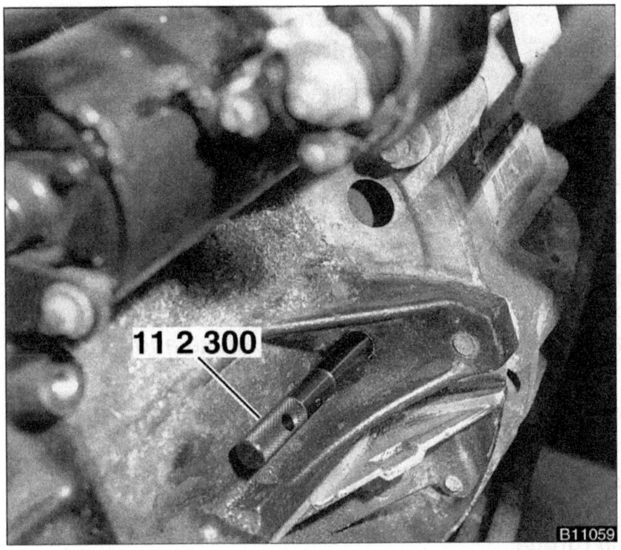


◄ Set engine to TDC and lock into position by installing BMW special tool 11 2 300 through bellhousing and into bore in flywheel or drive plate.

– Remove VANOS control unit from cylinder head. Remove timing chain sprockets from camshafts. See **117 Camshaft Timing Chains**.

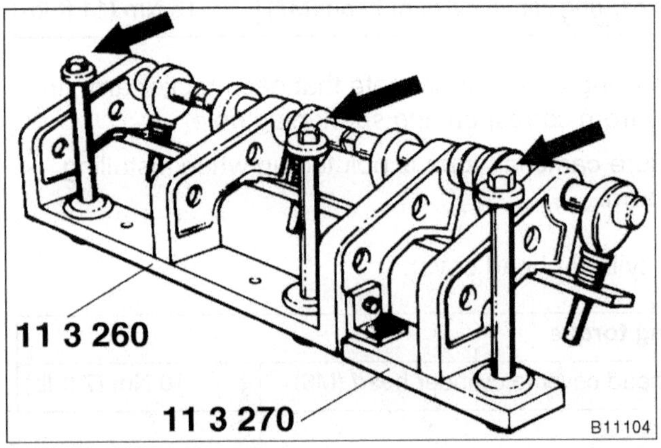


◄ Install BMW special tools 11 3 260 and 11 3 270 to cylinder head and tighten tool retaining bolts (**arrows**) in spark plug holes. Do not overtighten bolts.

| Tightening torque | |
|---|---|
| BMW special tool 11 3260 or 11 3 270 to cylinder head | 23 Nm (17 ft-lb) |

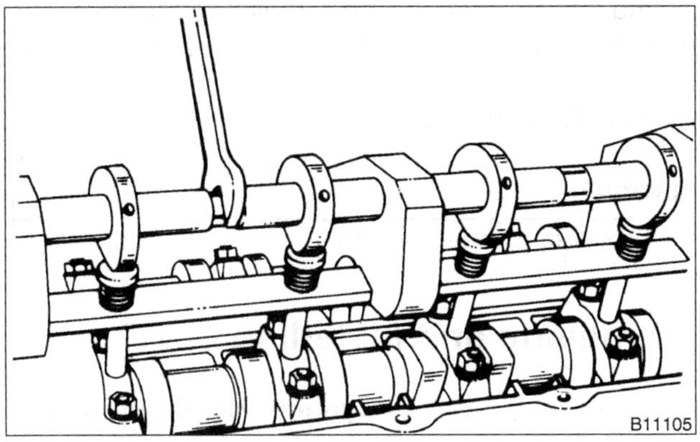


◄ Turn eccentric shaft on special tool until pressure is applied to camshaft bearing caps.

– Remove bearing cap mounting nuts.

– Slowly release pressure on camshaft bearing caps by turning eccentric shaft. Remove bearing caps and camshaft. Repeat procedure on remaining camshaft.

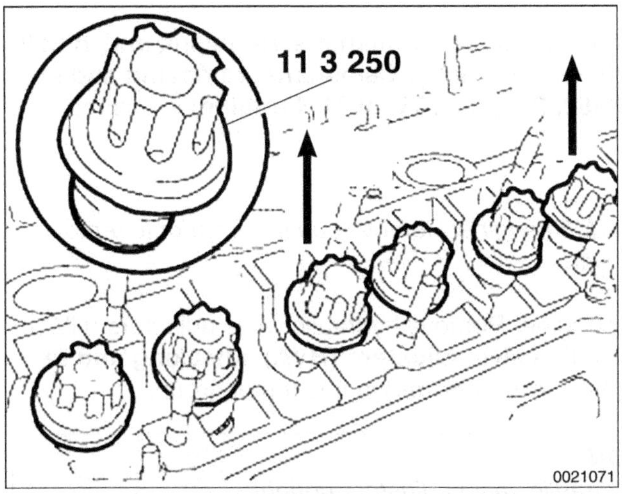


◄ Secure hydraulic lifters in lifter bores using BMW special tool 11 3 250, or remove lifters using a magnetic pick-up tool. With lifters secure or removed, lift out camshaft bearing carriers from cylinder head.

> ***CAUTION—***
> - *Do not let the hydraulic lifters fall out as the camshaft carrier is removed.*
> - *Store hydraulic lifters upright.*
> - *Reinstall used hydraulic lifters into original lifter bores.*

– Check all components for wear. See **Cylinder head and valvetrain reconditioning** in this repair group.

◄ Camshaft installation is reverse of removal.

- Lubricate camshafts prior to installation.
- Position camshafts so that cylinder 1 lobes (**arrows**) point up and toward each other.
- Install camshafts and camshaft bearing caps. Start camshaft bearing nuts by hand, then use special tools 11 3 260 and 11 3 270 to compress valve springs. Torque bearing nuts.
- Make sure all parts are reinstalled in their original positions.

***NOTE—***
- *The camshaft bearing caps are coded with a letter and a cylinder number. For example, "A1" is the exhaust camshaft bearing cap for cylinder 1 and "E1" is the intake camshaft bearing cap for cylinder 1. Similarly, the camshafts are stamped with a code letter: "E" for intake and "A" for exhaust.*
- *"E" is for Einlaß, (German for intake) and "A" is for Auspuff, (German for exhaust).*
- *Cylinder 1 is at front of engine.*

| **Tightening torque** | |
|---|---|
| Camshaft bearing cap to cylinder head (M7) | 15 Nm (11 ft-lb) |

– Reinstall timing chain, sprockets and VANOS control unit. Reset camshaft timing. See **117 Camshaft Timing Chains**.

– Reinstall cylinder head cover.

| Tightening torque | |
|---|---|
| Cylinder head cover to cylinder head (M6) | 10 Nm (7 ft-lb) |

## Valves, removing and installing

– Remove cylinder head. See **113 Cylinder Head Removal and Installation**.

– Remove camshafts. See **Camshafts, removing and installing (4-cylinder engine)** or **Camshafts, removing and installing (M52, S52, M52 TU or M54 engine)** in this repair group.

***NOTE—***
*M52 TU or M54 engine: Remove camshafts prior to removing cylinder head.*

– 4-cylinder engine: Remove hydraulic valve adjusters (HVA). See **Hydraulic valve adjusters (HVA), checking and replacing (4-cylinder engine)** in this repair group.

***CAUTION—***
*Store hydraulic valve adjusters upright.*

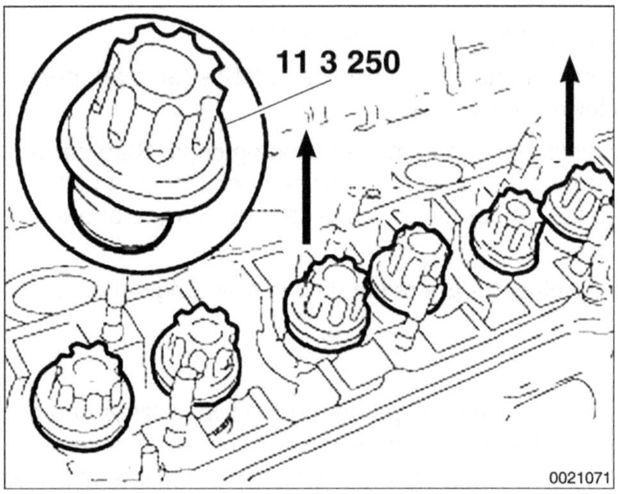


6-cylinder engine: Secure hydraulic lifters in lifter bores using BMW special tool 11 3 250, or remove lifters using a magnetic pick-up tool. With lifters secure or removed, lift out camshaft bearing carriers from cylinder head.

***CAUTION—***
- *Do not let the hydraulic lifters fall out as the camshaft carrier is removed.*
- *Store hydraulic lifters upright.*
- *Reinstall used hydraulic lifters into original lifter bores.*

– Remove valves using a valve spring compressor.

***NOTE—***
- *4-cylinder engine has recessed valve stems. This requires the use of a special valve spring compressor tool.*
- *Label each valve assembly as it is removed so it can be installed in its original position.*

– Remove and discard valve stem oil seals from valve guides.

– See **Cylinder head and valvetrain reconditioning** in this repair group for additional repair information.

– Installation is reverse of removal. Remember to:
  - Install new valve stem oil seals. See **Valve stem oil seals** in this repair group.
  - Reinstall exhaust manifold(s) with new gaskets. Use Molycote® HSC or equivalent anti-seize thread compound on fasteners.

| Tightening torque | |
|---|---|
| Exhaust manifold to cylinder head | 20 Nm (15 ft-lb) |

## Hydraulic valve adjusters (HVA), checking and replacing (4-cylinder engine)

– Run engine until it reaches normal operating temperature.

– Turn engine off and remove top engine cover, spark plugs, and cylinder head cover. See **113 Cylinder Head Removal and Installation**.

– Use a plastic or wooden stick to press down on top of rocker arm end (side away from camshaft). If any play can be detected, replace faulty HVA.

***NOTE —***
*When checking the HVA, makes sure the camshaft lobe is facing up so that there is no valve spring pressure on the adjuster.*

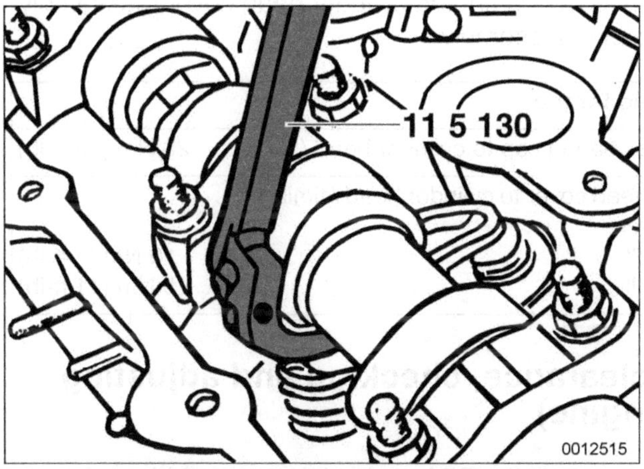


◄ To remove an HVA, turn engine over until camshaft lobe points straight up. Using BMW special tool 11 5 130, depress each valve and valve spring enough to remove its rocker arm.

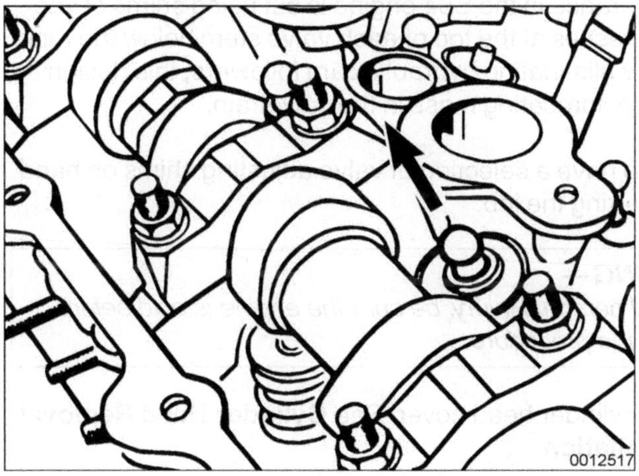


◄ Pull hydraulic valve adjuster straight out of cylinder head (**arrow**).

– Installation is reverse of removal. Be sure to inspect HVA bore in cylinder head for wear or scoring.

| Tightening torques | |
|---|---|
| Cylinder head cover to cylinder head / timing chain cover:<br>• M6 bolt<br>• M7 bolt | <br><br>10 Nm (7 ft-lb)<br>15 Nm (11 ft-lb) |

## Hydraulic cam followers, checking and replacing (M52, S52, M52 TU or M54 engine)

– Run engine until it reaches normal operating temperature.

– Turn engine off and remove top engine covers, ignition coils, and cylinder head covers. See **113 Cylinder Head Removal and Installation**.

< Use a plastic or wooden stick to press down on top of cam follower. If there is any noticeable clearance, replace faulty follower.

***NOTE—***
*When checking a cam follower, make sure the corresponding camshaft lobe is facing up so that there is no valve spring pressure on the follower.*

– To replace a cam follower, remove appropriate camshaft. See **Camshafts, removing and installing (M52, S52, M52 TU or M54 engine)** in this repair group.

– Once camshaft is out of the way, withdraw faulty cam follower and replace with new one. Be sure to inspect cam follower bores for wear and scoring.

– Camshaft, timing chain and cylinder head cover reassembly is reverse of disassembly.

| **Tightening torque** | |
|---|---|
| Camshaft bearing cap to cylinder head (M7) | 20 Nm (15 ft-lb) |
| Cylinder head cover to cylinder head / timing chain cover:<br>• M6 bolt<br>• M7 bolt | <br><br>10 Nm (7 ft-lb)<br>15 Nm (11 ft-lb) |

## Valve clearance, checking and adjusting (S54 engine)

Valve clearance in the S54 engine is set using shims of various thicknesses at the top of each valve stem below the rocker arm. By eliminating hydraulic cam followers, this design reduces reciprocating mass in the valvetrain.

Be sure to have a selection of valve adjusting shims on hand before starting the job.

> ***WARNING—***
> *To avoid personal injury, be sure the engine is cold before starting the procedure.*

– Remove cylinder head cover. See **Cylinder Head Removal and Installation**.

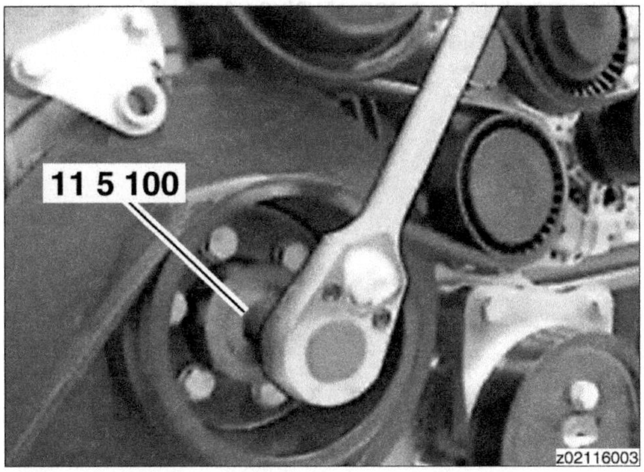


– Remove fan clutch with fan impeller and fan cowl. See **170 Radiator and Cooling System**.

– Remove spark plugs. See **120 Ignition System**.

< Fit BMW special tool 11 5 100 on 4 vibration damper bolts and turn engine over (in normal direction of rotation) until camshaft lobes at cylinder 1 are pointing up.

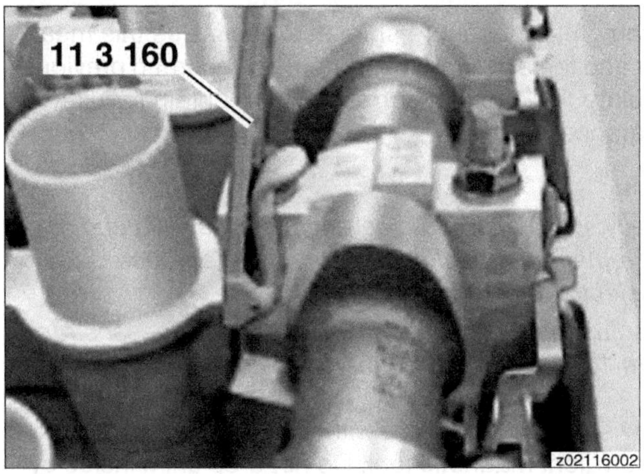


< With camshaft lobes pointing up, measure valve clearance with feeler gauge (BMW special tool 11 3 160 or equivalent).

– Record measured values. Valve clearance specifications are listed in **Table b**. If clearance is not within specified range, adjust as described below.

| Table b. S54 valve clearance (measure at max. temp. of 95°F) | |
|---|---|
| **Valve** | **Clearance** |
| Intake | 0.18 - 0.23 mm (0.0070 - 0.0090 in) |
| Exhaust | 0.28 - 0.33 mm (0.0110 - 0.0130 in) |

– Place clean cloth around valve to be adjusted. Use lint free rags or tape to seal off cylinder head openings including oil return and vent holes, opening to timing case cover at cylinder 1 and spark plug holes.

< Remove retaining clip from rocker arm shaft by pulling straight up (**arrow**).

> ***CAUTION—***
> *Avoid marring or damaging cylinder head gasket surface.*

– Carefully slide rocker arm to one side.

> ***CAUTION—***
> *Valve adjusting shim can easily fall off valve when sliding rocker arm.*

< Remove adjusting shim from top of valve using BMW special tool 11 4 400 or equivalent magnetic tool.

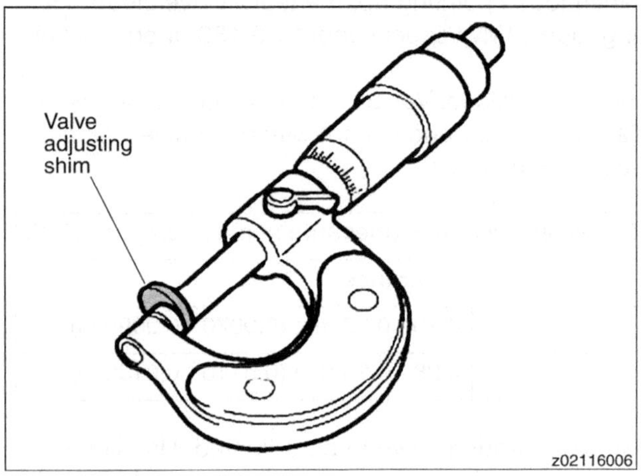


< Use micrometer to measure shim thickness. Use this measurement and feeler gauge measurement recorded previously to calculate thickness of correct shim to obtain required valve clearance.

***NOTE—***
- *Generally, the measured clearance is excessive and a thicker replacement shim is required. If clearance is too tight, select a thinner shim.*
- *Valve adjustment plates are available in thicknesses ranging from 1.72 mm to 2.60 mm, in 0.04 mm increments.*

– Install new shim to valve stem with BMW special tool 11 4 400 or equivalent magnetic tool. Do not drop shim inside cylinder head.

– Carefully push rocker arm back in place. Check for correct shim position. If necessary, use mirror to check correct positioning of components.

***CAUTION—***
*Valve adjusting shim can easily fall off valve when sliding rocker arm.*

– Reinstall rocker arm retaining clip. Retaining clip must snap into place on rocker arm shaft.

***CAUTION—***
*Avoid marring or damaging cylinder head gasket surface.*

– Recheck valve clearance using feeler gauge.

– Repeat checking and adjustment sequence for remainder of valves.

– Engine reassembly is reverse of disassembly.

## CYLINDER HEAD AND VALVETRAIN RECONDITIONING

This section provides the specifications and special reconditioning information necessary to repair the cylinder heads covered by this manual.

If machine shop services are not readily available, one alternative is to install a remanufactured cylinder head from an authorized BMW dealer.

### Cylinder head

– Clean disassembled head before inspecting.

– Inspect cylinder head for cracks. If a cracked cylinder head is suspected and no cracks are detected through visual inspection, have head further tested for cracks by an authorized BMW dealer. Replace cracked cylinder head.

– Check valve guides and valve seats for wear before machining warped head.

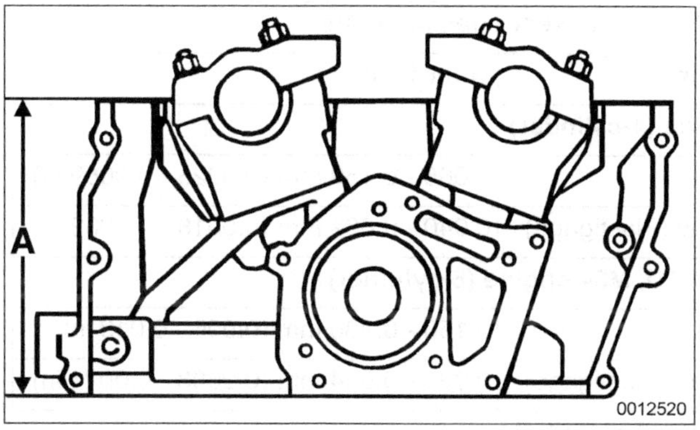


Measure total cylinder head height (**A**) and compare to minimum resurfacing height specifications (**Table c**). If unit does not meet minimum height dimension after machining, replace cylinder head.

| Table c. Cylinder head height | | |
|---|---|---|
| **Engine** | **New** | **Minimum permissible height (dimension A)** |
| M44 | 140.0 mm (5.512 in) | 139.55 mm (5.494 in) |
| M52, M52 TU, M54 | 140.0 mm (5.512 in) | 139.7 mm (5.500 in) |
| S52 | 140.0 mm (5.512 in) | Do not machine. Replace warped cylinder head. |
| S54 | 146.0 ± 0.05 mm (5.748 ± 0.002 in) | |

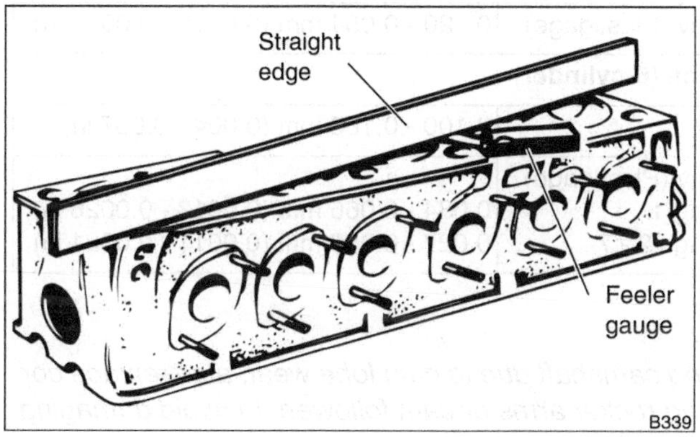


Use high quality straight edge and feeler gauge to check warpage:

- Cylinder head gasket surface
- Cylinder block gasket surface
- Camshaft carrier bearing surface

– Machine or replace components which exceed warpage limits given in **Table d**.

| Table d. Cylinder head warpage limits | |
|---|---|
| **Component** | **Limit** |
| Cylinder head | 0.05 mm (0.002 in) |
| Camshaft carrier bearing surface | 0.05 mm (0.002 in) |

– M52, M52 TU or M54 engine:
  - If cylinder head is warped, machine 0.3 mm (0.012 in) off milled surface.
  - If further machining is required, replace cylinder head. Removing more than this amount reduces size of combustion chamber and may adversely affect engine performance.
  - A 0.3 mm thicker gasket is available from an authorized BMW parts department.

– S52 or S54 engine: Do not machine. Replace warped cylinder head.

## Camshafts

Camshaft wear is usually caused by insufficient lubrication.

– Inspect removed camshaft lobes and journals for wear. Camshaft wear specifications are given in **Table e**.

< M44 engine: check oil sprayer tube and seal (**arrow**). A clogged sprayer tube or a defective oil seal in cylinder head reduces oil flow to camshaft.

**Table e. Camshaft wear specifications**

| Parameter | Wear limits |
|---|---|
| **M44 engine (4-cylinder)** | |
| Axial play | 0.065 - 0.150 mm (0.0025 - 0.0059 in) |
| Radial play (Plastigage) | 0.040 - 0.082 mm (0.0016 - 0.0032 in) |
| **M52, M52 TU, M54 engine (6-cylinder)** | |
| Axial play | 0.150 - 0.330 mm (0.006 - 0.013 in) |
| Radial play (Plastigage) | 0.020 - 0.054 mm (0.0008 - 0.0021 in) |
| **S52 engine (6-cylinder)** | |
| Axial play<br>• Intake camshaft<br>• Exhaust camshaft | <br>0.080 - 0.148 mm (0.003 - 0.006 in)<br>0.080 - 0.167 mm (0.003 - 0.007 in) |
| Radial play (Plastigage) | 0.020 - 0.054 mm (0.0008 - 0.0021 in) |
| **S54 engine (6-cylinder)** | |
| Axial play | 0.100 - 0.183 mm (0.004 - 0.007 in) |
| Radial play (Plastigage)<br>• Bearing 1<br>• Bearings 2 - 7 | <br>0.034 - 0.066 mm (0.0013 - 0.0026 in)<br>0.027 - 0.053 mm (0.0011 - 0.0021 in) |

***NOTE—***
*If replacing camshaft due to cam lobe wear, also replace corresponding rocker arms or cam followers to avoid damaging the new camshaft.*

## Valve guides

◄ Check valve guide wear using a new valve.

- Insert new valve until stem end is flush with end of guide (**white arrow**).
- Rock valve from side to side (**black arrows**). Note dial indicator readings.
- Compare to specification in **Table f**.

If valve guide is excessively worn, ream it to accept oversized valve stems.

**Table f. Valve guide specifications**

| Parameter | Specification |
|---|---|
| Valve guide wear, maximum (valve tilt measured with new valve) | 0.5 mm (0.020 in) |
| Valve guide inside diameter, installed (tolerance per ISO allowance H7)<br>• Standard<br>• Oversize 1<br>• Oversize 2 | <br><br>6.0 mm (0.236 in)<br>6.1 mm (0.240 in)<br>6.2 mm (0.244 in) |

***NOTE—***
- *International Organization for Standardization (ISO) tolerance classes are listed in* **Table f**. *ISO allowances are based on nominal sizes and are used to determine proper fit. Most machine shops have this information readily available.*
- *BMW supplies oversize valves. See* **Valves** *for additional information.*
- *BMW does not supply valve guides as replacement parts.*
- *Replacement valve guides may be available through aftermarket suppliers. Valve guide replacement requires special tools and a press. Heat the cylinder head and chill valve guides when replacing guides.*

## Valves

Valve specifications are listed in **Table g**. Remove carbon deposits from valves using a wire brush or wire wheel.

| Table g. Valve specification | |
|---|---|
| **Parameter** | **Specifications** |
| **Valve head diameter** | |
| M44, M52, M52 TU or M54 engine:<br>• Intake<br>• Exhaust | <br><br>33.0 mm (1.299 in)<br>30.5 mm (1.201 in) |
| S52 engine:<br>• Intake<br>• Exhaust | <br>30.0 mm (1.181 in)<br>27.0 mm (1.063 in) |
| S54 engine:<br>• Intake<br>• Exhaust | <br>35.0 mm (1.378 in)<br>30.5 mm (1.201 in) |
| **Valve stem diameter** | |
| Standard:<br>• Intake<br>• Exhaust | <br>5.975 - 0.015 mm (0.235 - 0.0006 in)<br>5.960 - 0.015 mm (0.2346 - 0.0006 in) |
| Oversize 1:<br>• Intake<br>• Exhaust | <br>6.075 - 0.015 mm (0.239 - 0.0006 in)<br>6.060 - 0.015 mm (0.239 - 0.0006 in) |
| Oversize 2:<br>• Intake<br>• Exhaust | <br>6.175 - 0.015 mm (0.243 - 0.0006 in)<br>6.160 - 0.015 mm (0.243 - 0.0006 in) |

## Valves, leak test

– Remove cylinder head and place on workbench. See **113 Cylinder Head Removal and Installation**.

– Remove camshafts and rocker arm assemblies. See **Camshafts, removing and installing (4-cylinder engine)** or **Camshafts, removing and installing (M52, S52, M52 TU or M54 engine)** in this repair group.

– Install valve assemblies and spark plugs in each cylinder.

– With combustion chamber facing upward, fill each chamber with a thin non-flammable liquid, such as parts cleaning fluid.

– After fifteen minutes, check fluid level. If fluid level in any chamber drops, valves for that cylinder are not sealing properly.

## Valve stem oil seals

Valve stem oil seals prevent excess oil from entering combustion chambers. Excessive oil consumption and smoke from exhaust immediately after starting and during deceleration indicate faulty valve stem oil seals.

***NOTE—***
*Do not reuse valve stem oil seals. If the valves are removed, install new valve stem oil seals.*

– Disassemble cylinder head fully. See **Camshafts, removing and installing** and **Valves, removing and installing** in this repair group.

Use BMW special tool 11 1 480 or equivalent to remove recessed valve stem oil seals.

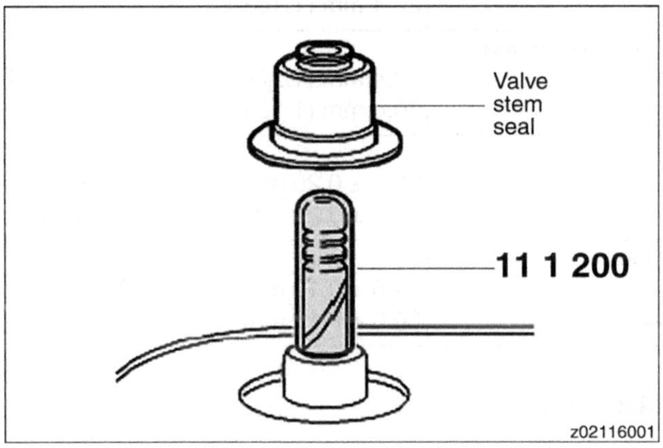


Install valve stem oil seals using BMW special tool 11 1 200 or equivalent.

- Lubricate new seals and install them using hand pressure only.
- Be sure to install valve spring seat before installing seal.

## Valve seats

Resurface valve seats whenever new valves or valve guides are installed. Cutters are required to resurface the seats. Always check valves for leaks after reconditioning a valve seat. See **Valves, leak test** in this repair group.

See **Table h** for valve seat dimensions.

***NOTE—***
*Standard size replacement valve seats are not available from BMW. Replacement valve seats are only available from BMW in 0.4 mm oversize (oversized in both height and diameter). The manufacturer does not provide specifications for valve seat replacement for the engines covered by this manual.*

**Table h. Valve seat dimensions**

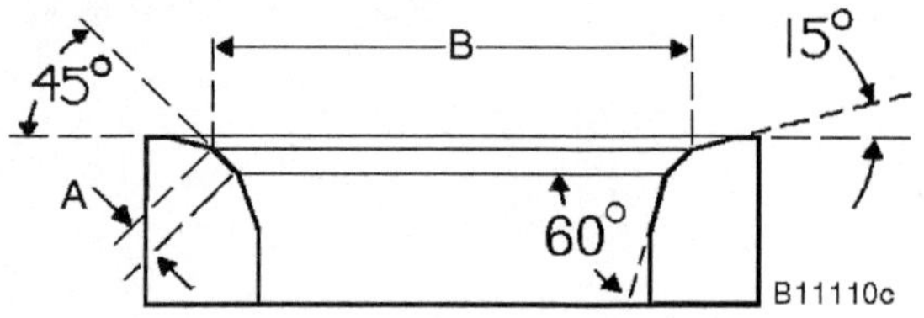


| Parameter | Specification |
|---|---|
| Valve seat angle | 45° |
| Correction angle, outside | 15° |
| Correction angle, inside | 60° |
| **Valve seat width (A)** | |
| M44, M52, S52, M52 TU or M54 engine | 1.65 ± 0.25 mm (0.065 ± 0.014 in) |
| S54 engine<br>• Intake<br>• Exhaust | <br>1.0 + 0.1 mm (0.040 + 0.004 in)<br>1.2 + 0.1 mm (0.047 + 0.004 in) |
| **Valve seat outside dia. (B)** | |
| M44 engine<br>• Intake<br>• Exhaust | <br>32.4 mm (1.276 in)<br>30.1 mm (1.185 in) |
| M52 engine<br>• Intake<br>• Exhaust | <br>32.4 mm (1.276 in)<br>29.4 mm (1.157 in) |
| M52 TU or M54 engine<br>• Intake<br>• Exhaust | <br>32.4 mm (1.276 in)<br>30.0 mm (1.181 in) |
| S52 engine<br>• Intake<br>• Exhaust | <br>29.4 ± 0.20 mm (1.157 ± 0.008 in)<br>26.4 ± 0.20 mm (1.039 ± 0.008 in) |
| S54 engine<br>• Intake<br>• Exhaust | <br>34.5 + 0.10 mm (1.358 + 0.004 in)<br>30.0 + 0.10 mm (1.181 + 0.004 in) |

## Valve springs

While cylinder head is disassembled, check conical valve springs for fatigue.

– Line up springs in a row.

– Place straight edge across top of springs.

– Replace any spring that is significantly shorter than others.

***NOTE —***
*Valve spring specifications and wear limits are not available from BMW.*

# 117 Camshaft Timing Chains

## GENERAL

This repair group covers timing chain repair information as well as camshaft timing and VANOS removal and installation procedures.

The information given in this repair group assumes that the engine is installed in the engine bay and that the cylinder head cover is off the engine. See **113 Cylinder Head Removal and Installation** for cylinder head cover removal procedures.

In the procedures given in this repair group, engines are referred to by engine code. If necessary, see **100 Engine–General** for engine code and application information.

Timing chains are lubricated by engine oil and do not require maintenance. A worn timing chain and sprockets can lead to noisy operation and erratic valve timing. A faulty tensioner can also cause timing chain noises.

***NOTE—***
*Timing chain and VANOS service for the S54 engine used in 2001 - 2002 M version cars is a highly complex job and is outside the scope of this manual.*

### Special tools

Special BMW service tools are required to time the valvetrain to the crankshaft and to remove the VANOS control unit. Be sure to read the procedures through before starting work.

00 9 250

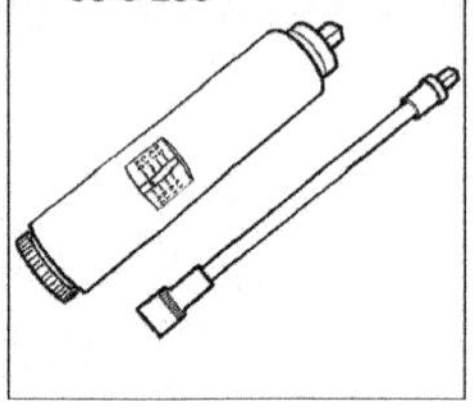

◀ Torque wrench with flex extension (BMW special tool 00 9 250)

11 0 280

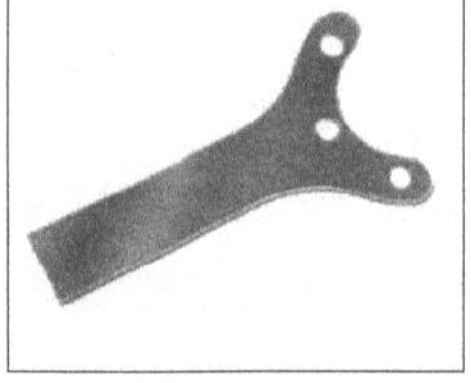

◀ S54 engine vibration damper counterhold tool (BMW special tool 11 0 280)

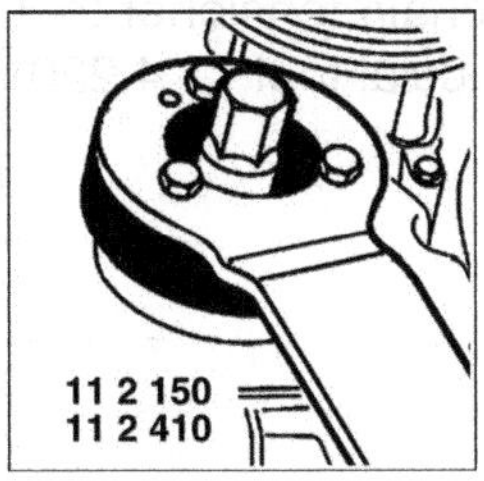


Crankshaft hub locking tools
(BMW special tool 11 2 150 /11 2 410)

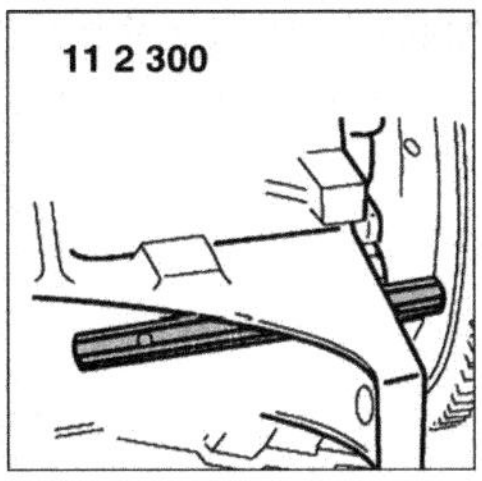


Crankshaft locking tool
(BMW special tool 11 2 300)

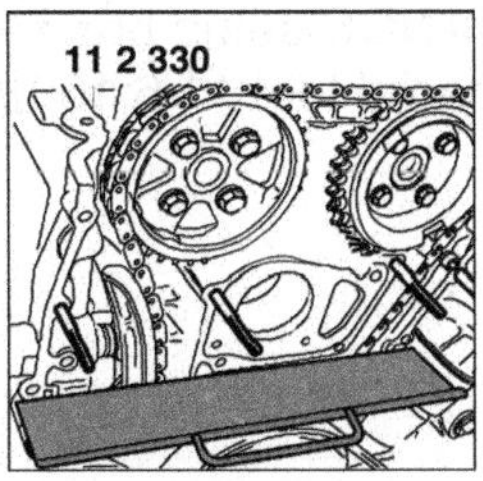


Profile gasket seating tool (4-cylinder engine)
(BMW special tool 11 2 330)

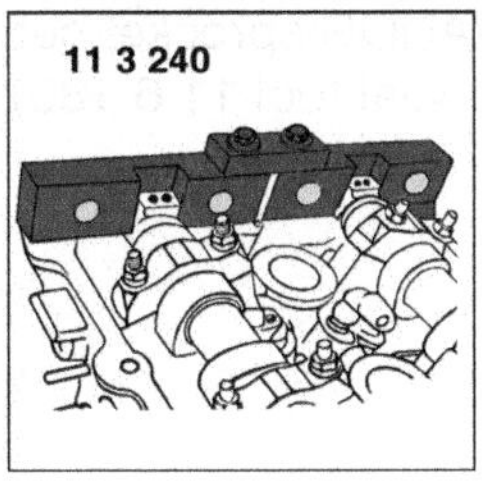


Camshaft locking tool
(BMW special tool 11 3 240)

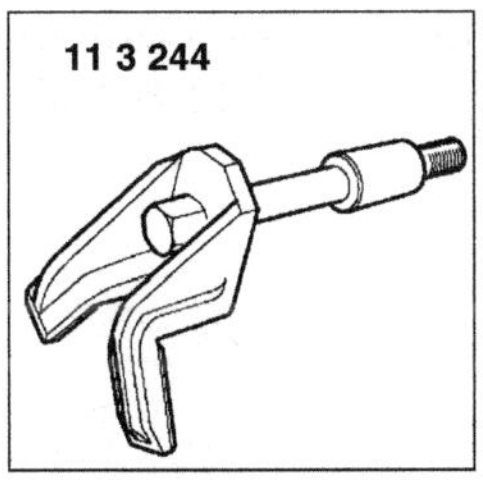


Camshaft locking tool bracket
(BMW special tool 11 3 244)

Secondary chain tensioner locking pin
(BMW special tool 11 3 290)

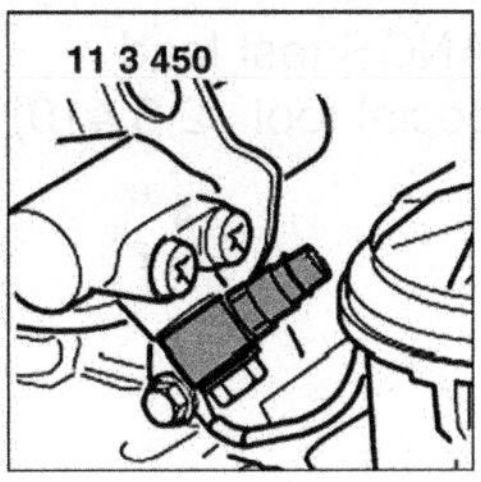


Air line adapter
(BMW special tool 11 3 450)

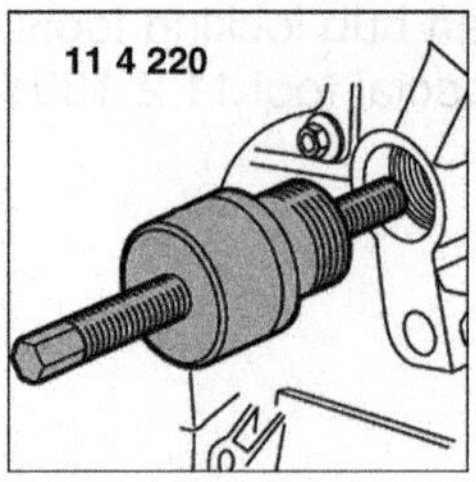


Primary chain tensioner tool (BMW special tool 11 4 220)

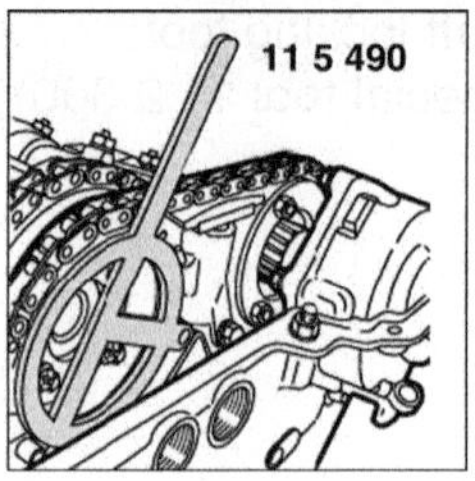


Single VANOS sprocket spanner (BMW special tool 11 5 490)

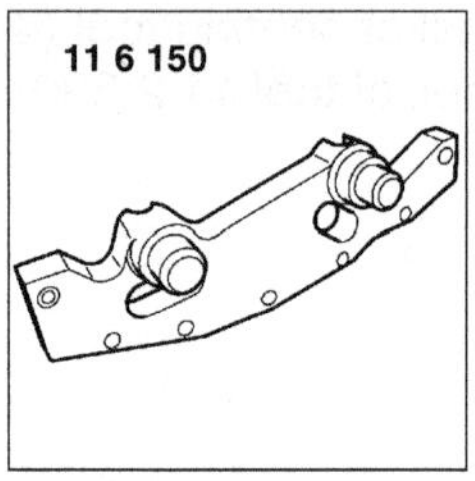


Double VANOS setup bracket (BMW special tool 11 6 150)

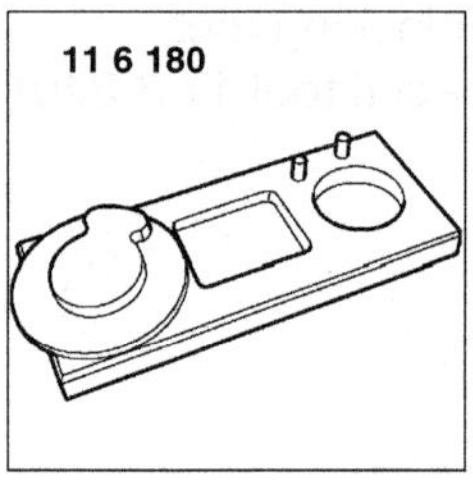


Double VANOS sprocket setup tool (BMW special tool 11 6 180)

Crankshaft vibration damper hub locking tool (M54 engine) (BMW special tool 11 8 190 /11 8 200)

Crankshaft vibration damper hub puller (BMW special tool 11 8 210)

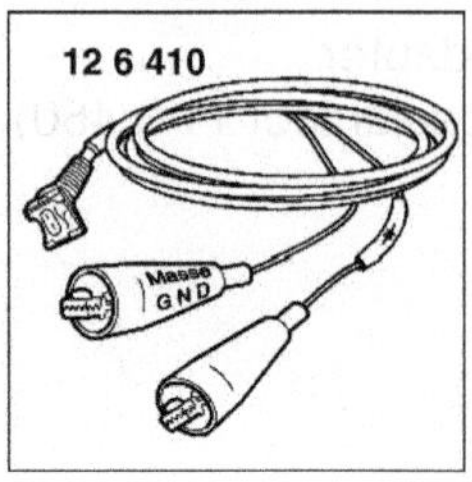


Single VANOS test lead (BMW special tool 12 6 410)

## VIBRATION DAMPER REMOVAL AND INSTALLATION

The vibration damper and front pulley are bolted to the front of the crankshaft. Always replace the vibration damper mounting bolt after removal. Use special tools to counterhold crankshaft hub or pulley while loosening the mounting bolt.

> ***CAUTION—***
> *Do not use flywheel locking tool to hold crankshaft stationary to loosen or tighten vibration damper mounting bolt. Use only the recommended special tools, or equivalent tools for counterholding the hub or pulley.*

### Vibration damper, removing and installing (4-cylinder engine)

– Raise front of car and support in a safe manner.

> ***CAUTION—***
> *Make sure the car is stable and well supported at all times. Use a professional automotive lift or jack stands designed for the purpose. A floor jack is not adequate support.*

– Remove splash guard under engine.

– Remove engine drive belts. See **020 Maintenance**.

***NOTE—***
*Mark direction of belt rotation if reusing belts.*

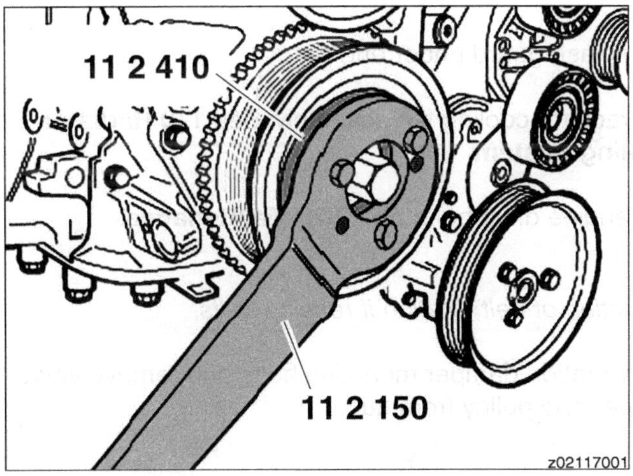


◄ Use BMW special tools 11 2 150 and 11 2 410 (or equivalent) to counterhold vibration damper while loosening mounting bolt.

> ***CAUTION—***
> - *Make sure an assistant holds special tool so it does not slip.*
> - *If an assistant is not available to grip the tool, remove engine components which are in the way. Rest handle of tool on front subframe. Protect subframe from damage with a piece of wood.*
> - *Do not use flywheel locking tool to hold crankshaft stationary to loosen or tighten vibration damper mounting bolt. Use only the recommended special tools, or equivalent tools for counterholding the pulley.*

– Remove vibration damper mounting bolt and discard.

– Slide vibration damper off crankshaft.

– Inspect front crankshaft seal for leaks. Seal replacement is covered in **119 Lubrication System**.

- Installation is reverse of removal:
  - Align vibration damper groove to crankshaft woodruff key.
  - Replace mounting bolt.

> ***CAUTION—***
> *Inspect woodruff key for damage before installing damper.*

- Use BMW special tools 11 2 150 and 11 2 410 to counterhold vibration damper while tightening mounting bolt.

> ***CAUTION—***
> *Do not use flywheel locking tool to hold crankshaft stationary to loosen or tighten vibration damper mounting bolt.*

| **Tightening torque** | |
|---|---|
| Vibration damper to crankshaft (M16 x 1.5) (replace bolt) | 330 Nm (244 ft-lb) |

## Vibration damper, removing and installing (M52, S52 engine)

M52 and S52 engines use a 2-piece crankshaft vibration damper and hub assembly.

- Raise front of car and support in a safe manner.

> ***CAUTION—***
> *Make sure the car is stable and well supported at all times. Use a professional automotive lift or jack stands designed for the purpose. A floor jack is not adequate support.*

- Remove splash guard under engine.
- Remove radiator cooling fan assembly. See **170 Radiator and Cooling System**.
- Remove engine drive belts. See **020 Maintenance**.

  ***NOTE—***
  *Mark direction of belt rotation if reusing belts.*

- Remove vibration damper mounting bolts and remove vibration damper and pulley from hub.

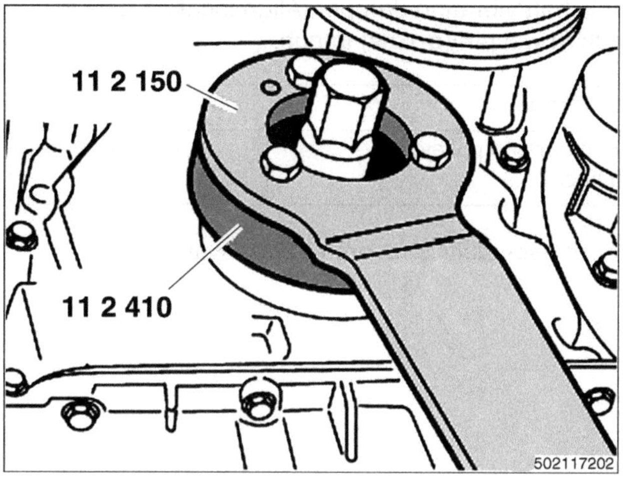


Use BMW special tools 11 2 150 and 11 2 410 (or equivalent) to counterhold vibration damper hub while loosening hub mounting bolt.

> ***CAUTION—***
> - *Make sure an assistant holds special tool so it does not slip.*
> - *If an assistant is not available to grip the tool, remove engine components which are in the way. Rest handle of tool on front subframe. Protect subframe from damage with a piece of wood.*
> - *Do not use flywheel locking tool to hold crankshaft stationary to loosen or tighten crankshaft hub mounting bolt. Use only the recommended special tools, or equivalent hub holding tool.*

– Remove vibration damper hub mounting bolt and discard.

– Slide hub off crankshaft.

– Inspect front crankshaft seal for leaks. Seal replacement is covered in **119 Lubrication System**.

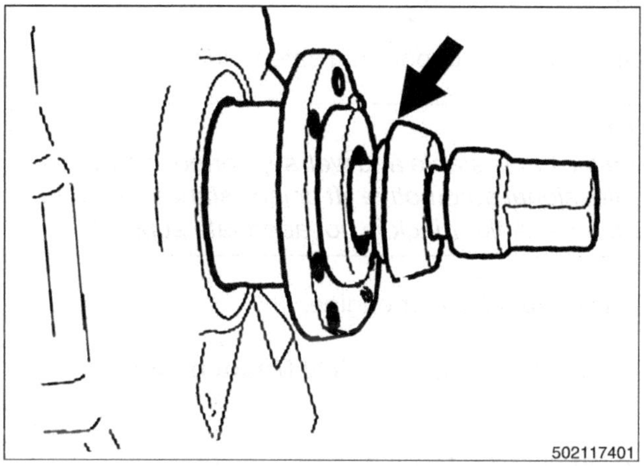


When reinstalling:

- Align vibration damper hub groove to crankshaft woodruff key.
- Insert new mounting bolt with washer shoulder (**arrow**) facing hub.

> ***CAUTION—***
> *Inspect woodruff key for damage before installing hub.*

– Use BMW special tools 11 2 150 and 11 2 410 to counterhold hub while tightening mounting bolt.

> ***CAUTION—***
> *Do not use flywheel locking tool to hold crankshaft stationary to loosen or tighten vibration damper mounting bolt. Use only the recommended special tools, or equivalent tools for counterholding the hub or pulley.*

| **Tightening torque** | |
|---|---|
| Vibration damper hub to crankshaft (use new bolt) | 410 Nm (300 ft-lb) |

◀ Install crankshaft vibration damper. Align locating hole in damper with dowel pin (**arrow**) in hub.

| Tightening torque | |
|---|---|
| Vibration damper and pulley to crankshaft hub | 22 Nm (17 ft-lb) |

– Remainder of assembly is reverse of removal.

## Vibration damper, removing and installing (M52 TU, M54 engine)

M52 TU and M54 engines use a 1-piece crankshaft vibration damper, integrated with the hub.

– Raise front of car and support in a safe manner.

> ***CAUTION—***
> *Make sure the car is stable and well supported at all times. Use a professional automotive lift or jack stands designed for the purpose. A floor jack is not adequate support.*

– Remove splash guard under engine.

– Remove engine drive belts. See **020 Maintenance**.

***NOTE—***
*Mark direction of belt rotation if reusing belts.*

– Remove radiator cooling fan assembly. See **170 Radiator and Cooling System**.

◀ Bolt BMW special tools 11 8 190 and 11 8 200 together. Insert fingers of special tool 11 8 200 in vibration damper openings. Use this tool setup to counterhold vibration damper hub while loosening vibration damper mounting bolt (**arrow**).

> ***CAUTION—***
> - *Make sure an assistant holds special tool.*
> - *If assistant is not available to grip the tool, rest handle of tool on front subframe. Protect subframe from damage with a piece of wood.*
> - *Do not use flywheel locking tool to hold crankshaft stationary to loosen or tighten vibration damper mounting bolt.*

– Remove vibration damper mounting bolt and discard.

– Bolt BMW special tools 11 8 210 and 11 8 200 together. Insert fingers of special tool 11 8 200 in vibration damper openings. Use this tool set up to pull vibration damper off crankshaft.

> ***CAUTION—***
> - *Avoid damaging radiator.*
> - *If using a different puller than the BMW special tools, be careful to avoid damaging crankshaft threads.*

– Inspect front crankshaft seal for leaks. Seal replacement is covered in **119 Lubrication System**.

– When reinstalling:
  - Align vibration damper groove to crankshaft woodruff key.
  - Insert new vibration damper mounting bolt and washer.

> ***CAUTION—***
> *Inspect woodruff key for damage before installing vibration damper.*

– Use BMW special tools 111 8 190 and 11 8 200 to counterhold vibration damper while tightening mounting bolt.

> ***CAUTION—***
> *Do not use flywheel locking tool to hold crankshaft stationary to loosen or tighten vibration damper mounting bolt. Use only the recommended special tools, or equivalent tools for counterholding the pulley.*

| Tightening torque | |
|---|---|
| Vibration damper to crankshaft (use new bolt) | 410 Nm (300 ft-lb) |

## Vibration damper, removing and installing (S54 engine)

- Raise front of car and support in a safe manner.

> ***CAUTION—***
> *Make sure the car is stable and well supported at all times. Use a professional automotive lift or jack stands designed for the purpose. A floor jack is not adequate support.*

- Remove splash guard under engine.
- Remove engine drive belts. See **020 Maintenance**.

***NOTE—***
*Mark direction of belt rotation if reusing belts.*

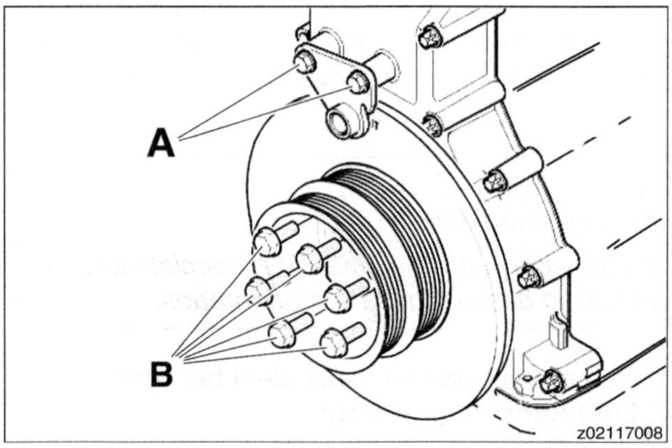


Working at front of engine:

- Remove TDC bracket mounting bolts (**A**) and bracket.
- Remove pulley mounting bolts (**B**) and remove pulley.

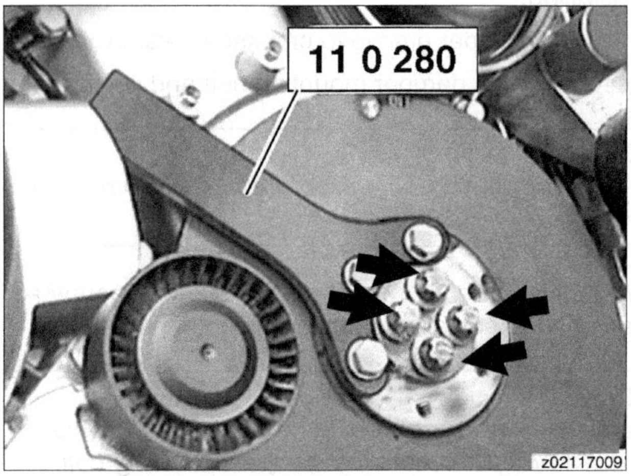


Secure crankshaft hub using BMW special tool 11 0 280. Loosen vibration damper mounting bolts (**arrows**). Remove vibration damper by lifting out from below first (to clear coolant pump pulley).

***NOTE—***
*Special tool 11 0 280 rests on A/C compressor bracket when loosening center bolts.*

> ***CAUTION—***
> *Use only the recommended special tools, or equivalent hub holding tool to loosen bolts.*

- Inspect front crankshaft seal for leaks. Seal replacement is covered in **119 Lubrication System**.
- When reinstalling vibration damper to crankshaft, make sure locating hole in damper is aligned to dowel in crankshaft hub. Install new mounting bolts finger tight.

> ***CAUTION—***
> *Always use new stretch bolts when installing vibration damper.*

– Torque vibration damper mounting bolts while counterholding crankshaft with BMW special tool 11 0 280.

| Tightening torques | |
|---|---|
| Vibration damper hub to crankshaft (M12) (use new bolts):<br>• Stage 1<br>• Stage 2 (angle torque)<br>• Stage 3 (angle torque) | <br><br>60 Nm (44 ft-lb)<br>50°<br>50° |

– Install pulley and TDC bracket. Make sure dowels for TDC bracket are correctly located in timing chain cover.

| Tightening torques | |
|---|---|
| Pulley to vibration damper:<br>• M8 10.9<br>• M8 8.8 | <br>34 Nm (25 ft-lb)<br>22 Nm (16 ft-lb) |

## VANOS (VARIABLE CAMSHAFT TIMING)

Z3 and M cars with 6-cylinder engine are equipped with VANOS (variable camshaft timing). Two versions of VANOS are used in cars covered by this manual:

- Single VANOS: 1997 - 1998 M52 engine, 1998 - 2000 S52 engine
- Double VANOS: 1999 - 2002 M52 TU or M54 engine

***NOTE—***
*2001 - 2002 M version cars with S54 engine are equipped with a high pressure VANOS system. Servicing this system is beyond the scope of this manual.*

### Single VANOS (M52, S52 engine)

Single VANOS in M52 or S52 engine is actuated by engine oil pressure and controlled by the Siemens MS 41.x engine management system. Single VANOS dynamically adjusts intake camshaft timing based on engine load, engine speed and engine temperature.

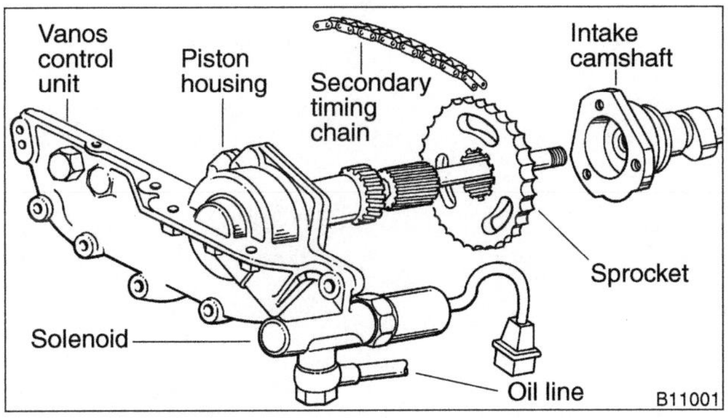


◄ The main components of the single VANOS system are the control unit with piston housing and integral solenoid, and the modified intake camshaft.

When the engine is running, the piston housing is supplied with pressurized engine oil. At idle, the solenoid is in the OFF position (de-energized) and valve timing is maintained in the normal position. When the solenoid is energized, the gear cup piston moves forward to advance the camshaft by 12.5°.

## Single VANOS, testing

> ***CAUTION—***
> *Special BMW service tools are required to check and repair the VANOS system. Read the procedure through before beginning the job.*

There are 3 special tools required to check single VANOS operation:

**Electrical test lead** (BMW special tool 12 6 410). Use to power VANOS solenoid, simulating ground signal from DME control module.

**Air line adapter** (BMW special tool 11 3 450). Substitute for oil supply pipe fitting to simulate oil pressure.

**Crankshaft TDC locking tool** (BMW special tool 11 2 300). Position and lock crankshaft at TDC for cylinder 1.

– Remove cylinder head cover. See **113 Cylinder Head Removal and Installation**.

– Unclip and remove oil baffle cover above intake camshaft.

Set engine to approximate TDC by rotating in normal operating direction until camshaft lobes at cylinder 1 (**arrows**) point up and face each other.

Set engine to TDC by aligning **0IT** mark (0°TDC) on crankshaft vibration damper with cast boss on lower timing chain cover.

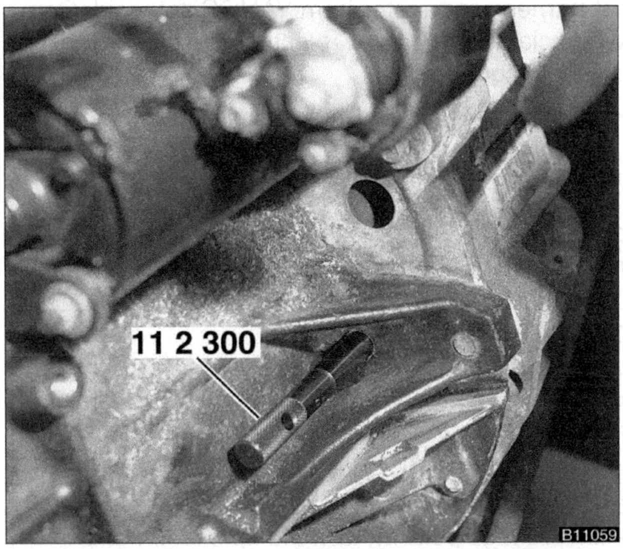


< Lock crankshaft in position by inserting BMW special tool 11 2 300 through transmission bellhousing and into hole in flywheel or drive plate.

***NOTE—***
*Confirm that the locking tool has been correctly installed by trying to rotate the crankshaft.*

– Remove oil line fitting from VANOS control unit.

***NOTE—***
*Wrap the VANOS oil line fitting with a shop rag to absorb leaking oil.*

< Using hollow bolt and seals from oil supply line, install BMW special tool 11 3 450 (air line adapter) and connect a supply of compressed air (30 - 115 psi) to fitting.

< Measure and record dimension **A**, distance between trigger plate edge and side of secondary timing chain tensioner. This is VANOS reference measurement.

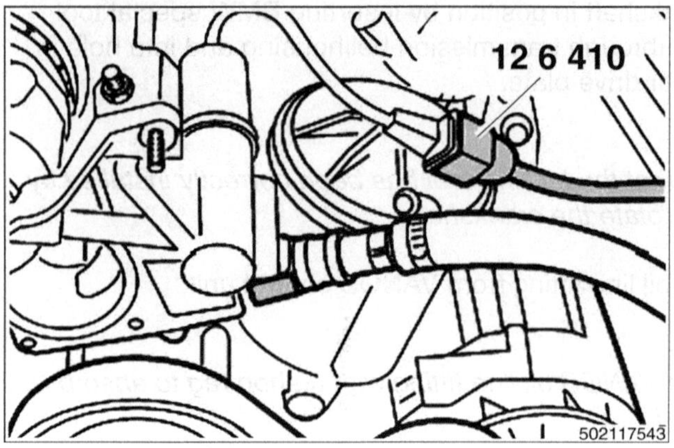


Disconnect harness connector from VANOS solenoid. Connect BMW special tool 12 6 410 (electrical test lead) to VANOS solenoid connector.

- Connect positive (+) test lead to positive (+) terminal of battery.
- Connect negative (–) test lead to chassis ground.
- With compressed air still attached, solenoid should audibly click and intake camshaft should advance.

> **_CAUTION—_**
> - *Be sure to connect the test harness polarity correctly. If the polarity is reversed, the internal diode in the VANOS solenoid will be destroyed. Although the solenoid will still function with a faulty diode, a fault code may be set in the ECM memory.*
> - *To confirm voltage polarity, turn the ignition on and check for positive (+) battery voltage at the terminal corresponding to the red/white wire in the main harness connector. Check that the positive terminal (+) in the main connector corresponds to the same terminal in the solenoid connector that is being connected to the battery.*

With solenoid actuated, measure and record dimension **B** between trigger plate edge and side of secondary timing chain tensioner.

| **VANOS control travel** | |
|---|---|
| Dimension **B** - dimension **A** | min 8.5 mm (0.3346 in) |

– If any faults are found, check VANOS solenoid by removing it from control unit.

- Check that solenoid and control unit plunger move freely. If plunger is sticking, replace complete control unit.
- Solenoid is available as a replacement part. Replace if necessary.

– If VANOS system does not advance correctly and no other faults can be found, VANOS control unit may have been installed incorrectly after other repairs. Remove and reinstall control unit. See **Single VANOS control unit, removing (M52, S52 engine)** and **Single VANOS control unit, installing (M52, S52 engine)** in this repair group.

– Reassemble engine:
  - Remove crankshaft locking tool.
  - Remove test harness and reconnect solenoid electrical harness connector.
  - Remove air line adapter and install oil supply line using new seals.
  - Reinstall cylinder head cover.

| Tightening torques | |
|---|---|
| Cylinder head cover to cylinder head (M6) | 10 Nm (7 ft-lb) |
| Oil supply line to VANOS control unit | 32 Nm (24 ft-lb) |
| Solenoid to VANOS control unit | 30 Nm (22 ft-lb) |

## Double VANOS (M52 TU, M54 engine)

The M52 TU and M54 6-cylinder engines are equipped with double VANOS. This is a variable camshaft timing system for both the intake and exhaust camshafts.

**Double VANOS components**

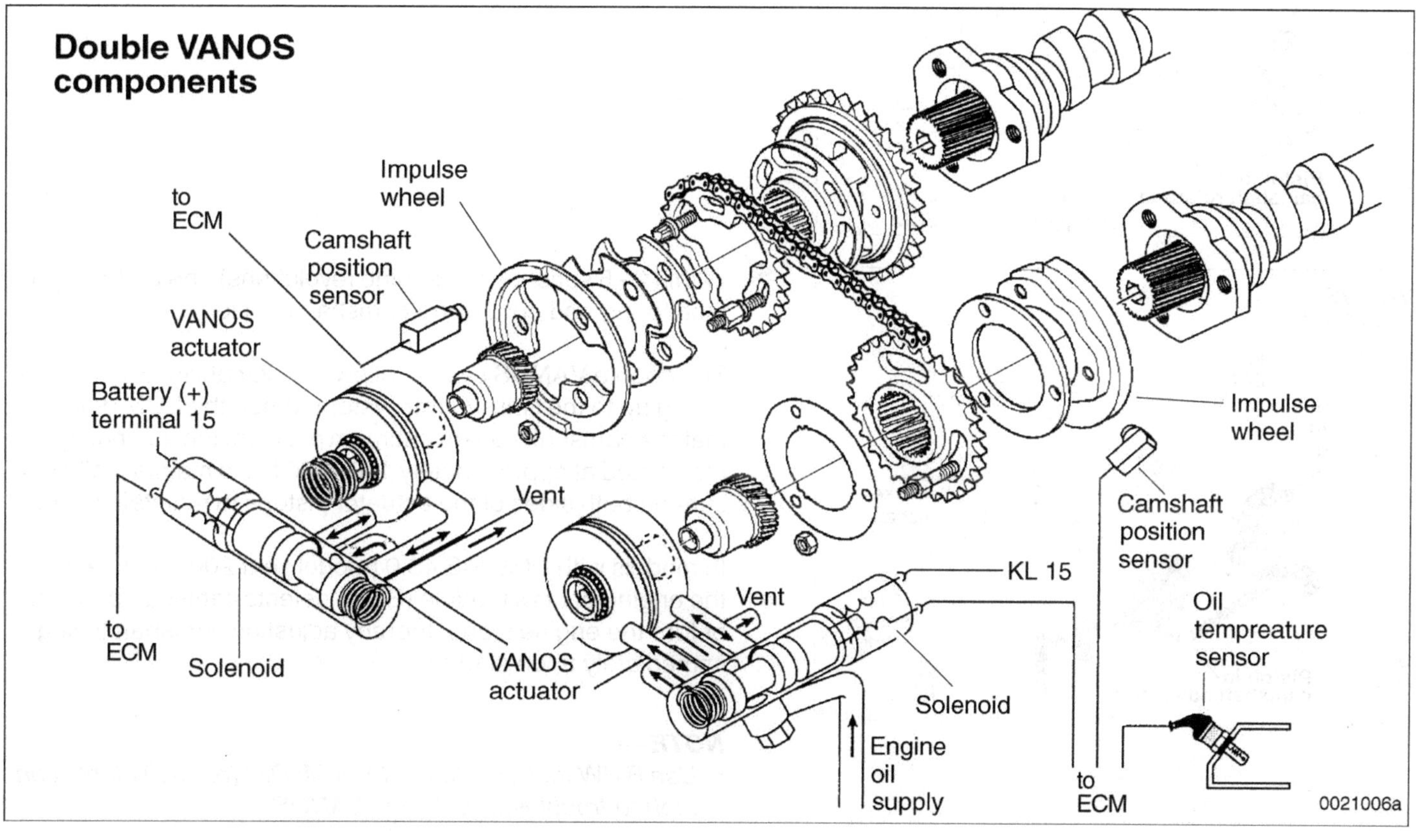

The main components of double VANOS are:

- Intake and exhaust camshafts with helical spline inserts.
- Camshaft sprockets with adjustable gears.
- VANOS actuators (two position piston housing with internal/external helical spline cup).
- Three-way solenoid switching valves.
- Camshaft position impulse wheels.
- Hall effect camshaft position sensors.

Camshaft position is varied by the engine control module (ECM) based on engine rpm, throttle position signal, intake air temperature and coolant temperature.

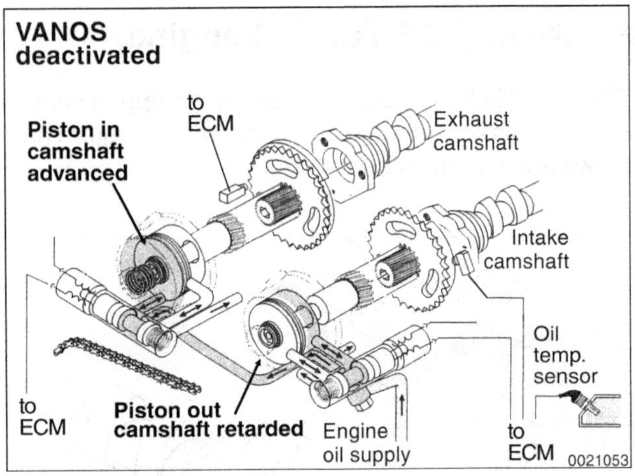


When engine is started, camshafts are in deactivated position:

- Intake camshaft is held in RETARDED position by oil pressure.
- Exhaust camshaft is held in ADVANCED position by preload spring and oil pressure.

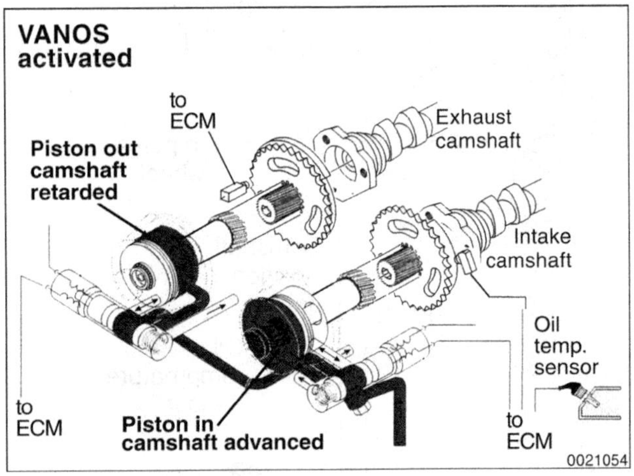


Within 2 - 5 seconds (50 engine revolutions), the ECM begins monitoring and controlling camshaft positions.

The double VANOS system allows full variability of camshaft timing up to the limits of the system. When the ECM detects that the camshafts are in optimum position, the solenoids are modulated at approximately 100 - 220 Hz to maintain oil pressure on both sides of the actuator pistons to maintain timing.

In models with DME MS 43.0 (model year 2001 and newer), the engine control module (ECM) detects camshaft position before the engine starts, thereby adjusting camshaft timing immediately upon start-up.

***NOTE—***
- *Use BMW tool DISplus, GT1 or MoDiC (or equivalent scan tool) to troubleshoot double VANOS.*
- *Diagnostic trouble codes (DTCs) pertaining to VANOS are listed in the table* **Double VANOS fault codes**. *See also* **OBD On-Board Diagnostics**.
- *Elevated oil temperatures can cause VANOS to deactivate. Oil that is too thick (high viscosity) may cause a DTC to be set in the ECM. If VANOS is deactivated (limp-home mode), there will be a noticeable loss of power.*

| Double VANOS fault codes | | |
|---|---|---|
| BMW code | P-code | Fault description |
| 19 | P1529 | VANOS solenoid valve activation, exhaust |
| 21 | P1525 | VANOS solenoid valve activation, intake |
| 103 | P1519 | VANOS faulty reference value, intake |
| 104 | P1520 | VANOS faulty reference value, exhaust |
| 105 | P1522 | VANOS stuck (bank 1), intake |
| 106 | P1523 | VANOS stuck (bank 2), exhaust |

## CAMSHAFT TIMING CHAIN REMOVAL AND INSTALLATION (4-CYLINDER ENGINE)

0011961

◄ The M44 engine uses a single, double-row chain to drive the dual overhead camshafts.

Camshaft timing chain service and repairs requires the removal of front crankshaft vibration damper and timing chain housing covers. Be sure to read the procedures through before starting work.

> ***WARNING—***
> *To avoid personal injury, be sure the engine is cold before starting any of these procedure.*

### Timing chain housing cover, upper, removing and installing (4-cylinder engine)

– Remove cylinder head cover. See **113 Cylinder Head Removal and Installation**.

– With engine cold, drain coolant from engine block and radiator. See **170 Radiator and Cooling System**.

***NOTE—***
- *On M44 engine, the block drain is on the exhaust side of the engine, near cylinder 4.*
- *The radiator drain plug is on the bottom of the radiator on the driver's side. Remove small plastic lock clip and pull radiator drain plug out completely.*

– Disconnect hoses from coolant thermostat housing in front of cylinder head. Unbolt thermostat housing from cylinder head.

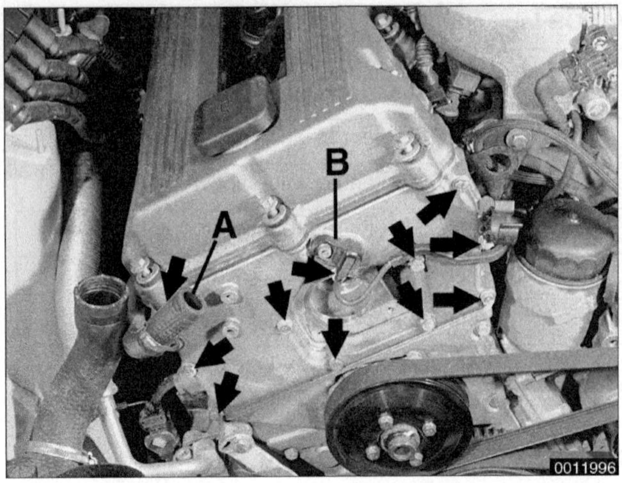


Working at top front of cylinder head:

- Detach secondary air valve duct (**A**).
- Remove camshaft position sensor (**B**).
- Remove upper timing chain cover mounting bolts (**arrows**).
- Remove cover.

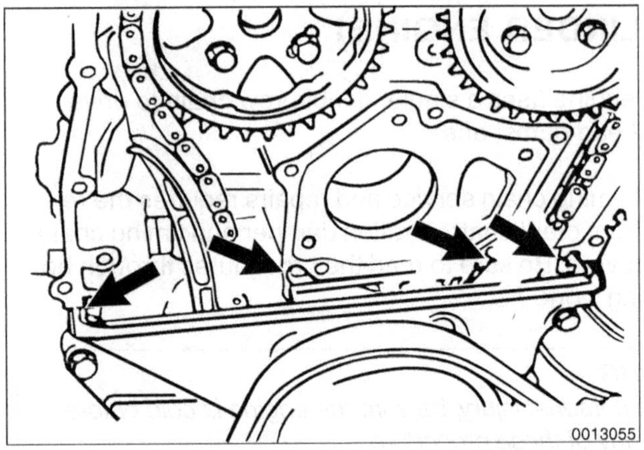


Cut off and remove profile seal (rubber sealing strip between lower and upper timing covers) at corners (**arrows**).

***NOTE—***
*Separate seals for the top timing housing cover are available as spare parts.*

– Prior to reinstallation:

- Clean profile seal groove and sealing surfaces.
- Seat new rubber-profile seal in groove.

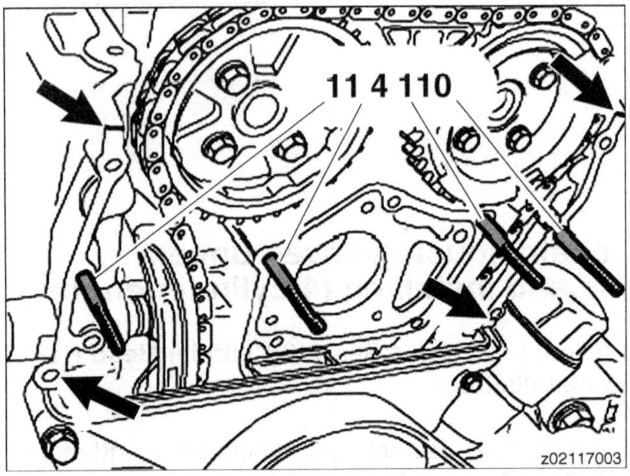


Install BMW special tool 11 4 110 (M6 stud set).

- Apply 3-Bond®1209 or equivalent sealer thinly and evenly to joint corners where upper chain cover meets cylinder head and cylinder block (**arrows**).

***NOTE—***
*BMW special tool 11 4 110 is a set of studs. When installed in upper timing cover mounting bores, they allow cover to be installed loosely so that cylinder head cover can be installed over it. Torquing the cylinder head cover compresses the profile gasket and provides a proper seal between the upper and lower timing chain covers.*

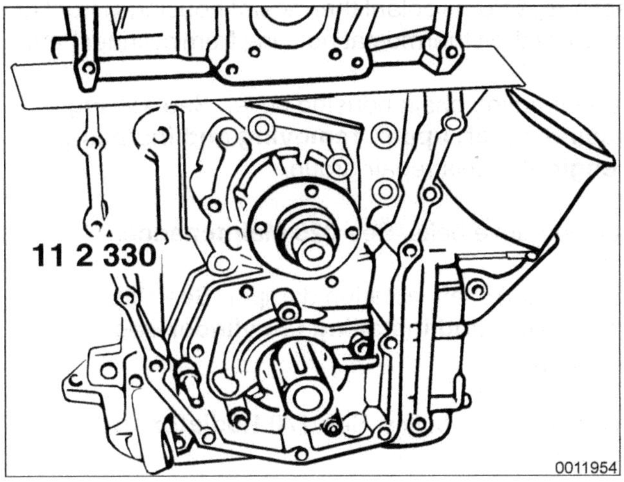


◄ Install upper timing chain cover:

- Apply a thin coating of grease to top of rubber profile gasket and both sides of BMW special tool 11 2 330 or equivalent.
- Attach new gaskets to timing cover with light dabs of gasket sealant. Hold gaskets in place using cover mounting bolts.
- Place cover into position by sliding over special tool 11 2 330 and onto studs (special tool 11 4 110).

***NOTE—***
*A piece of thin sheet metal may be substituted for BMW special tool 11 2 330.*

– Snug down cover mounting bolts finger tight. Pull out special tool 11 2 330.

– Remove camshaft oil supply tube seal at right end of cylinder head

– Temporarily install and torque down cylinder head cover without gasket to fully seat upper timing chain cover. Use large washers under cylinder head cover mounting bolts.

- Fully tighten front timing chain cover mounting bolts.
- Remove special tool 11 4 110 studs. Install and torque remaining front cover mounting bolts.
- Remove cylinder head cover. Reinstall oil supply tube seal. Reinstall cylinder head cover with gasket.
- Recheck torque on timing chain cover bolts.

| **Tightening torques** | |
|---|---|
| Cylinder head cover to cylinder head or upper timing chain cover (M6) | 10 Nm (7 ft-lb) |
| Upper timing chain cover to cylinder head (M6) | 10 Nm (7 ft-lb) |

## Timing chain housing cover, lower, removing and installing (4-cylinder engine)

– Remove cylinder head cover. See **113 Cylinder Head Removal and Installation.**

– With engine cold, drain coolant from engine block and radiator. See **170 Radiator and Cooling System**.

***NOTE—***
- *On M44 engine, the block drain is on the exhaust side of the engine, near cylinder 4.*
- *The radiator drain plug is on the bottom of the radiator on the driver's side. Remove small plastic lock clip and pull radiator drain plug out completely.*

– Disconnect hoses from coolant thermostat housing in front of cylinder head. Unbolt thermostat housing from cylinder head.

– Remove upper timing chain housing cover. See **Timing chain housing cover, upper, removing and installing (4-cylinder engine)** in this repair group.

– Remove engine drive belts. See **020 Maintenance**.

– Remove front crankshaft vibration damper. See **Vibration damper, removing and installing (4-cylinder engine)** in this repair group.

– Remove coolant pump pulley. See **170 Radiator and Cooling System**.

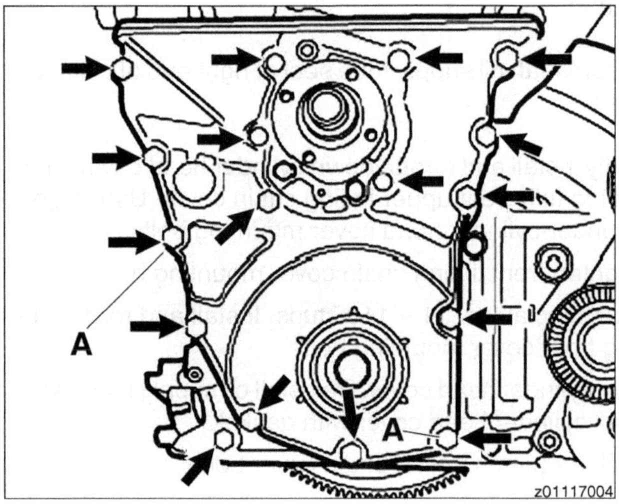


◄ Remove lower cover mounting bolts (**arrows**) and carefully lift cover off.

***NOTE—***
*Use drift to loosen cover at hollow locating dowel sleeves (**A**).*

– Clean sealing surfaces and gasket grooves.

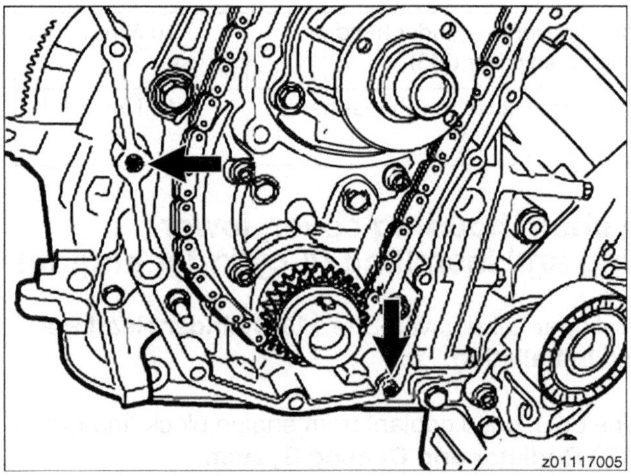


◄ Make sure dowel sleeves (**arrows**) are straight and undamaged.

– If necessary, replace crankshaft front seal. See **119 Lubrication System**.

– Reinstall cover with new gasket.

– Install mounting bolts and initially tighten to approx. 5 Nm (44 in-lb).

– Fully tighten all bolts in alternate sequence.

| **Tightening torques** | |
|---|---|
| Lower timing chain cover to engine block<br>• M6 bolt<br>• M8 bolt | <br>10 Nm (7 ft-lb)<br>22 Nm (16 ft-lb) |

– Recheck torque on mounting bolts.

– Reassemble engine.

## Timing chain, removing (4-cylinder engine)

Special BMW service tools are needed for this procedure. The special tools assure proper timing of the valvetrain, as there are no precise marks to set the timing on the camshafts. Read the procedure through before beginning the job.

– Disconnect negative (–) battery cable.

> ***CAUTION—***
> *Disconnecting the battery cables may erase fault codes stored in control module memory.*

– With engine cold, drain coolant from engine block and radiator. See **170 Radiator and Cooling System**.

***NOTE—***
- *On M44 engine, the block drain is on the exhaust side of the engine, near cylinder 4.*
- *The radiator drain plug is on the bottom of the radiator on the driver's side. Remove small plastic lock clip and pull radiator drain plug out completely.*

– Remove engine drive belts. See **020 Maintenance**.

– Remove radiator cooling fan assembly. See **170 Radiator and Cooling System**.

– Remove cylinder head cover. See **113 Cylinder Head Removal and Installation**.

– Remove spark plugs.

– Remove upper timing chain housing cover. See **Timing chain housing cover, upper, removing and installing (4-cylinder engine)** in this repair group.

< Set engine to approximate TDC by rotating in normal operating direction until camshaft lobes at cylinder 1 point up and face each other.

***NOTE—***
*At TDC, camshaft sprocket marks are lined up with vertical axis of engine (**dotted lines**). As the engine is installed at a slight angle, the sprocket marks are not vertical relative to the ground.*

– Remove front crankshaft vibration damper. See **Vibration damper, removing and installing (4-cylinder engine)** in this repair group.

– Remove coolant pump pulley. See **170 Radiator and Cooling System**.

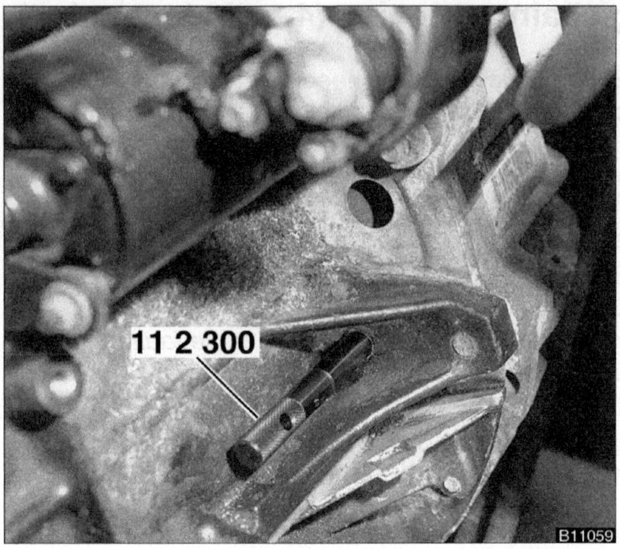


– Remove lower timing chain housing cover. See **Timing chain housing cover, lower, removing and installing (4-cylinder engine)** in this repair group.

< Lock crankshaft in TDC position by inserting BMW special tool no. 11 2 300 through bore and into hole in flywheel or drive plate.

***NOTE—***
- *If necessary, remove plug from locating bore in lower edge of bell housing.*
- *Confirm that the locking tool has been correctly installed by trying to rotate the crankshaft.*

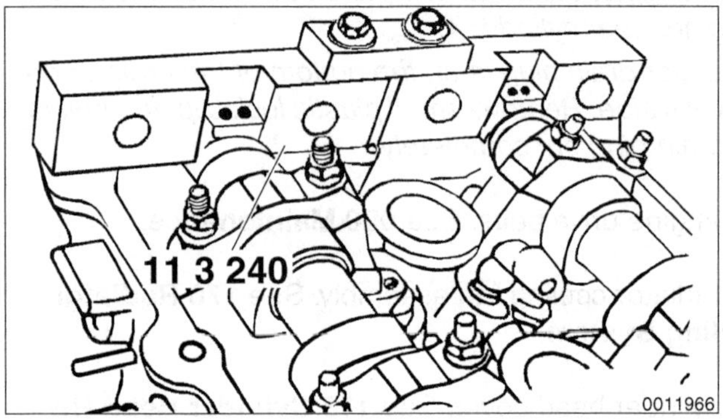


< Lock camshafts at TDC by mounting BMW special tool 11 3 240 over square ends of camshafts.

> ***CAUTION—***
> *Lock camshafts in TDC position using the special service tool, or equivalent, before removing the timing chain. The tool holds the camshafts to prevent valve contact with pistons.*

***NOTE—***
*Check that tool is squarely seated on cylinder head gasket surface. If necessary, rotate camshafts slightly using an open-end wrench at hex on camshafts.*

< Remove hydraulic chain tensioner from right side of cylinder head.

– Remove mounting bolts from intake and exhaust camshaft sprockets. Then remove camshaft sprockets and sensor wheels from camshafts together with chain and crankshaft sprocket.

> ***CAUTION—***
> - *Do not allow crankshaft woodruff key to drop out.*
> - *Do not allow the crankshaft to rotate once the timing chain is removed. The pistons can contact the valves.*

***NOTE—***
*Do not remove cam sprockets from the chain unless these parts are being replaced. Wire tie each sprocket to the chain to keep the sprockets timed to each other. Keeping the sprockets timed will simplify installation.*

## Timing chain, installing (4-cylinder engine)

– Prior to reinstallation:
  - Inspect sprockets for wear or damage.
  - Inspect chain guide and tensioner rails for grooves caused by chain contact.
  - Replace worn parts.
  - If any of the sprockets are being replaced due to wear, replace chain as well.

– Make sure camshafts and crankshaft are locked in TDC position.

– Install timing chain to crankshaft sprocket, then slide sprocket on crankshaft while aligning woodruff key.

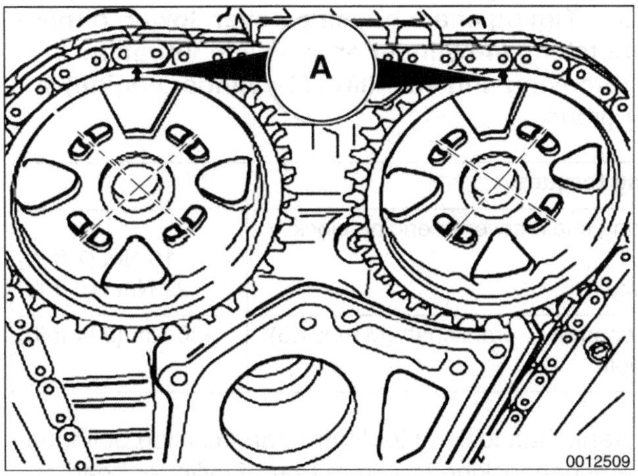


◀ Place chain sprockets with chains on camshafts:
  - Marks (**A**) on sprockets point up.
  - Tapped camshaft bores are centered to slots in sprockets (**dotted lines**).

– Place sensor wheel on intake camshaft sprocket so that arrow on sensor wheel points up.

– Install and hand-tighten sprocket mounting bolts.

◀ Remove locking bolt (5 mm Allen wrench) from left chain guide adjuster (**arrow**). Loosen slotted adjusting sleeve approximately two turns.

– Install hydraulic chain tensioner to lower timing cover. Use a new sealing washer on tensioner when installing.

| Tightening torque | |
|---|---|
| Chain tensioner to cylinder head | 40 Nm (30 ft-lb) |

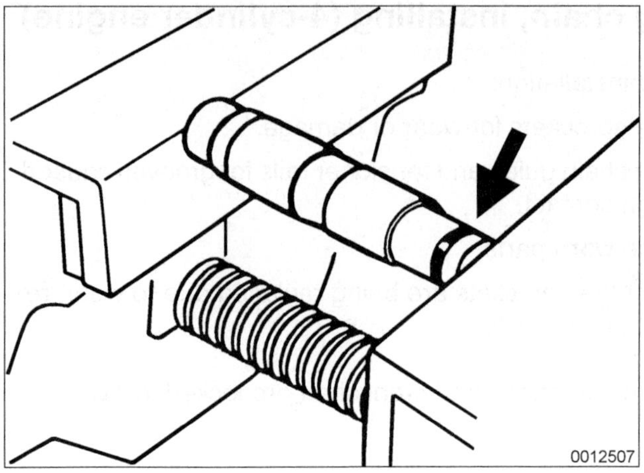
0012507

If reusing old chain tensioner:

- First clamp tensioner piston in a vice (soft jaws) and squeeze oil from it.
- Slowly compress piston in vise up to end circlip (**arrow**).
- Repeat this procedure twice to ensure all oil is expelled.

– Check that chain is free of slack. Tighten down sprocket mounting bolts.

| Tightening torque | |
|---|---|
| Chain sprocket to camshaft flange (M7) | 15 Nm (11 ft-lb) |

– Turn left chain guide adjusting sleeve in until it just contacts cylinder head. Insert chain guide retaining screw and tighten.

– Reinstall lower timing chain housing cover and vibration damper. See **Timing chain housing cover, lower, removing and installing (4-cylinder engine)** and **Vibration damper, removing and installing (4-cylinder engine)** in this repair group.

| Tightening torques | |
|---|---|
| Lower timing chain cover to engine block<br>• M6 bolt<br>• M8 bolt | <br>10 Nm (7 ft-lb)<br>22 Nm (16 ft-lb) |
| Vibration damper to crankshaft (M16 x 1.5) (replace bolt) | 330 Nm (244 ft-lb) |

– Remove crankshaft locking tool from transmission bellhousing and camshaft locking tool from rear of cylinder head.

– Turn engine over by hand in direction of rotation at least two full rotations. Reinstall camshaft and crankshaft locking tools to check engine timing. Remove locking tools if no faults are found.

– Reinstall upper timing chain housing cover. See **Timing chain housing cover, upper, removing and installing (4-cylinder engine)** in this repair group.

| Tightening torque | |
|---|---|
| Upper timing chain cover to cylinder head (M6) | 10 Nm (7 ft-lb) |

– Reinstall cylinder head cover. Check condition of camshaft oil supply tube seal and replace if necessary.

| Tightening torque | |
|---|---|
| Cylinder head cover to cylinder head or upper timing chain cover (M6) | 10 Nm (7 ft-lb) |

– Installation of the remaining parts is reverse of removal.
  - When installing thermostat, make sure arrow or vent hole faces up.
  - Refill and bleed cooling system. See **170 Radiator and Cooling System**.

| Tightening torque | |
|---|---|
| Coolant drain plug to cylinder block | 40 Nm (30 ft-lb) |
| Radiator cooling fan to coolant pump | 40 Nm (30 ft-lb) |
| Radiator drain screw to radiator | 2.5 Nm (22 in-lb) |

## Camshaft timing, checking and adjusting (4-cylinder engine)

– Remove radiator cooling fan assembly. See **170 Radiator and Cooling System**.

– Remove cylinder head cover. See **113 Cylinder Head Removal and Installation**.

– Remove spark plugs.

◄ Set engine to approximate TDC by rotating in normal operating direction until camshaft lobes at cylinder 1 point up and face each other.

***NOTE—***
*At TDC, camshaft sprocket marks are lined up with vertical axis of engine (**dotted lines**). As the engine is installed at a slight angle, the sprocket marks are not vertical relative to the ground.*

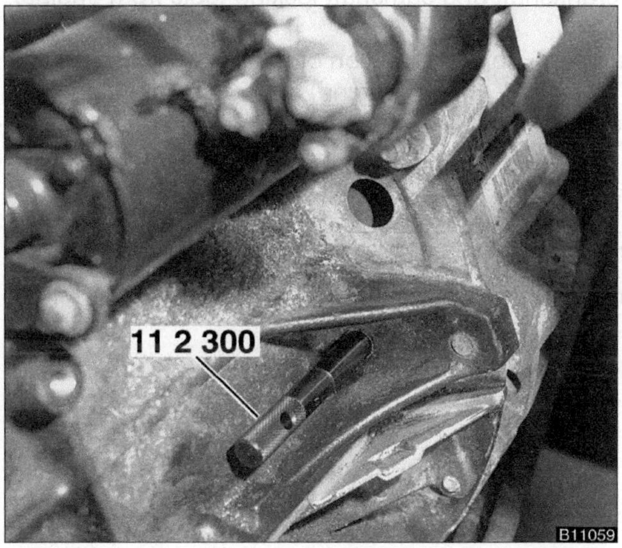


◄ Lock crankshaft in TDC position by inserting BMW special tool 11 2 300 through bore and into hole in flywheel or drive plate.

***NOTE—***
- *If necessary, remove plug from locating bore in lower edge of bell housing.*
- *Confirm that the locking tool has been correctly installed by trying to rotate the crankshaft.*

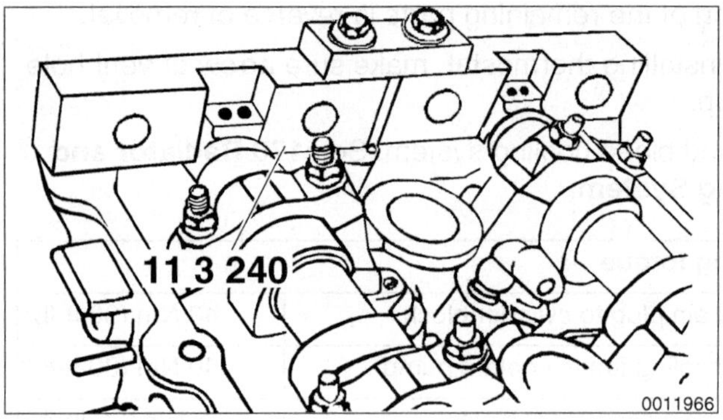


Lock camshafts at TDC by mounting BMW special tool 11 3 240 over square ends of camshafts. Check that tool is squarely seated on cylinder head gasket surface. If necessary, rotate camshafts slightly using an open-end wrench at hex on camshafts.

- If tools fits, camshaft timing is correct.
- If camshaft tool does not fit over camshafts with crankshaft at TDC, follow procedure below to adjust camshaft timing.
- Remove crankshaft and camshaft locking tools from engine.
- Rotate engine in normal direction of rotation until camshaft lobes on cylinder 4 (rear of engine) are up and pointing at each other.
- Loosen two bolts on each camshaft sprocket. Do not remove bolts.
- Rotate engine and lock crankshaft in TDC position by inserting BMW special tool 11 2 300 through bore and into hole in flywheel or drive plate.

***NOTE—***
*Confirm that the locking tool has been correctly installed by trying to rotate the crankshaft.*

- Loosen remaining two bolts on each camshaft sprocket.
- Place BMW special tool 11 3 240 over camshaft ends and make sure it seats flush against cylinder head top surface.
- Tighten two bolts on each camshafts sprocket.

| **Tightening torque** | |
|---|---|
| Chain sprocket to camshaft flange (M7) | 15 Nm (11 ft-lb) |

- Remove crankshaft and camshaft locking tools from engine.
- Rotate engine in normal direction of rotation until camshaft lobes on cylinder 4 (rear of engine) are up and pointing at each other.
- Torque remaining camshaft sprocket bolts.

| **Tightening torque** | |
|---|---|
| Chain sprocket to camshaft flange (M7) | 15 Nm (11 ft-lb) |

- Recheck camshaft timing. If correct, reassemble engine.

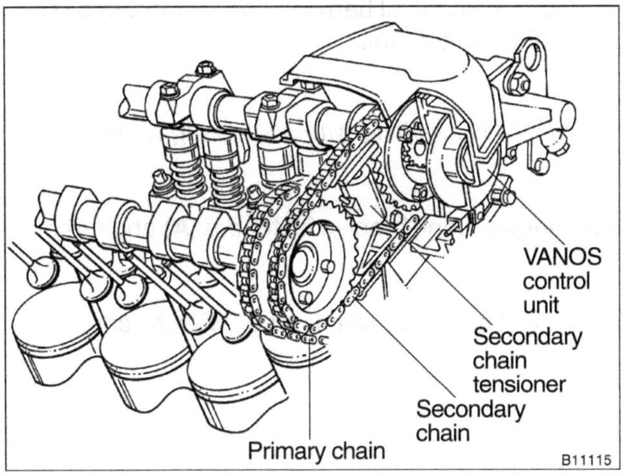


# CAMSHAFT TIMING CHAINS REMOVAL AND INSTALLATION (M52, S52 ENGINE)

M52 or S52 engine uses a pair of chains to drive the camshafts. A primary chain is driven off the crankshaft and drives the primary sprocket on the exhaust camshaft. The secondary chain drives the intake camshaft. Each chain is tensioned by a hydraulic tensioner.

Camshaft timing chain service and repairs requires the removal of front crankshaft vibration damper, VANOS control unit, engine oil pan and timing chain housing cover. Be sure to read the procedures through before starting work.

> ***WARNING —***
> *To avoid personal injury, be sure the engine is cold before starting any of these procedure.*

## Single VANOS control unit, removing (M52, S52 engine)

> ***CAUTION—***
> - *Special BMW service tools are required to remove and install the VANOS control unit. Read the procedure through before starting the job.*
> - *If the camshafts are not properly timed, the pistons can contact the valves.*

– Disconnect negative (-) cable from battery.

> ***CAUTION—***
> - *Disconnecting the battery may erase fault code(s) stored in control module memory. Prior to disconnecting the battery, check for fault codes using special BMW diagnostic equipment.*
> - *Prior to disconnecting the battery, read the battery disconnection cautions given in* **001 General Warnings and Cautions**.

– Remove viscous clutch cooling fan. See **170 Radiator and Cooling System**.

> ***CAUTION—***
> *The radiator cooling fan nut (32 mm wrench) has left- hand threads.*

– Remove engine covers and cylinder head cover. See **113 Cylinder Head Removal and Installation.**

– Unclip and remove oil baffle cover above intake camshaft.

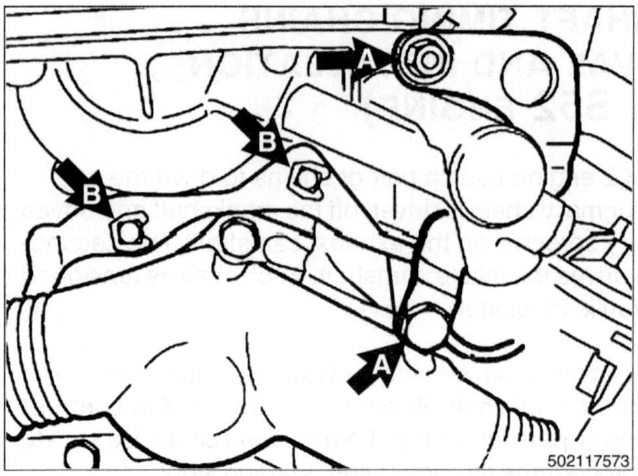


- Disconnect VANOS electrical harness connector and oil line fitting from VANOS control unit.

***NOTE—***
*Wrap the VANOS oil line fitting with a shop rag to absorb leaking oil.*

- Remove front engine lifting bracket fasteners (**A**) and remove bracket.

- Remove cable duct clips (**B**) from front of VANOS control unit.

- Set engine to approximate TDC by rotating in normal operating direction until camshaft lobes at cylinder 1 (**arrows**) point up and face each other.

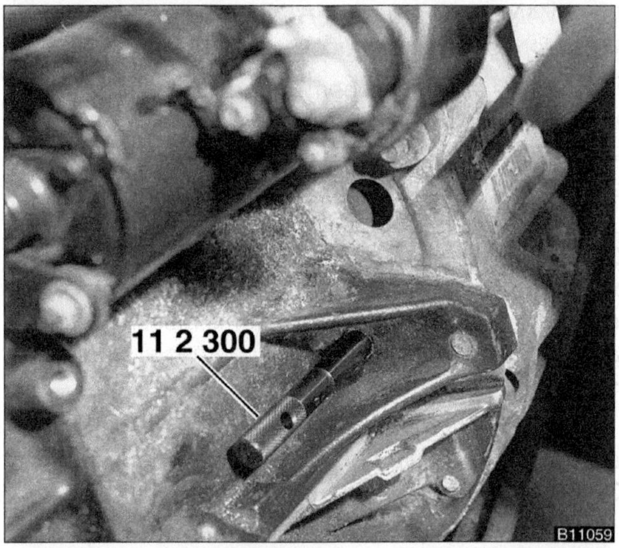


- Lock crankshaft in position by inserting BMW special tool 11 2 300 through transmission bellhousing and into hole in flywheel or drive plate.

***NOTE—***
*Confirm that the locking tool has been correctly installed by trying to rotate the crankshaft.*

- Unscrew and remove cylinder head cover studs (**arrows**) at rear of cylinder head.

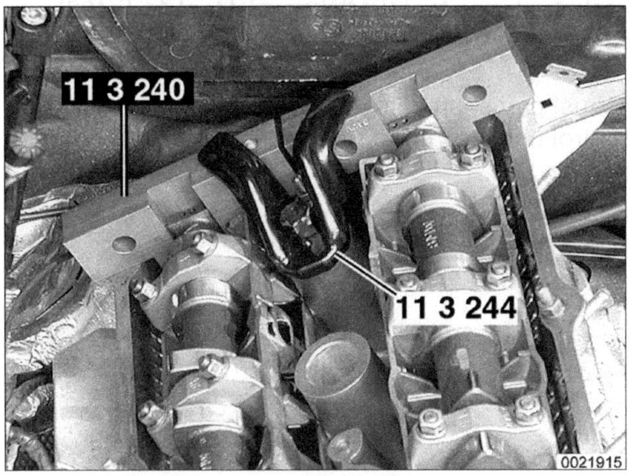


Secure camshafts in TDC position using BMW special tool set 11 3 240.

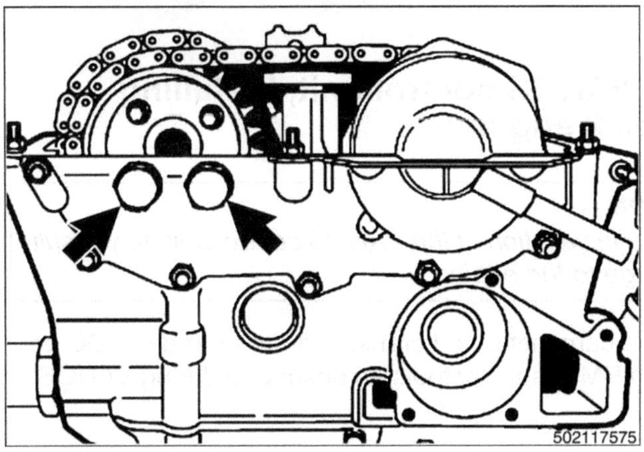


Remove access plugs (**arrows**) from VANOS control unit.

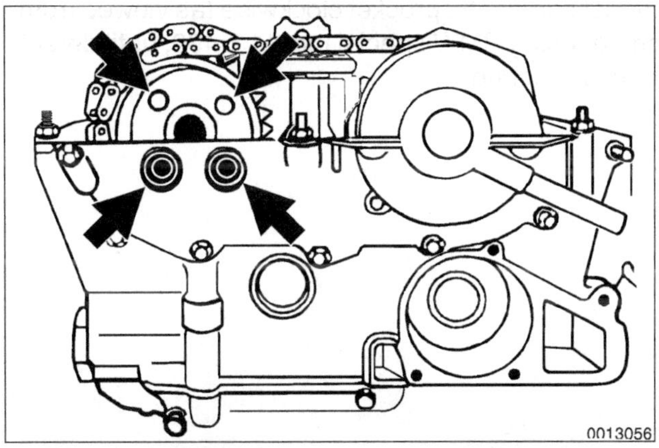


Loosen exhaust camshaft sprocket mounting bolts (**arrows**) (Torx E10).

Press down on secondary chain tensioner and lock into place using BMW special tool 11 3 292 or a similar size pin.

– Remove VANOS control unit mounting fasteners at front of cylinder head.

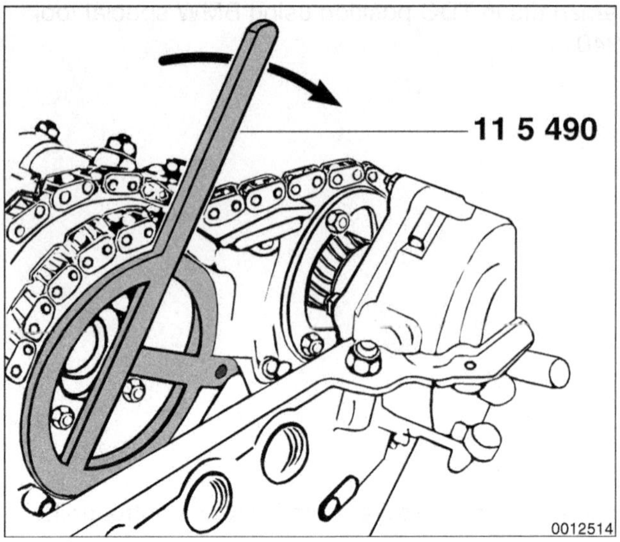


Install sprocket rotating tool (BMW special tool 11 5 490) on exhaust camshaft.

- Rotate exhaust camshaft sprocket clockwise while sliding VANOS control unit forward to remove from cylinder head.

## Single VANOS control unit, installing (M52, S52 engine)

> ***CAUTION—***
> *Incorrect installation of the VANOS control unit may result in damage to the engine.*

- Make sure camshafts and crankshaft are locked in TDC position with BMW special tools as described during VANOS removal.

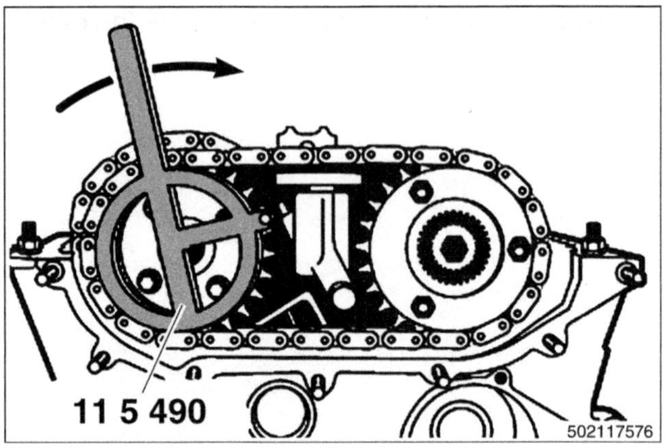


Rotate exhaust camshaft sprocket clockwise (as viewed from front of engine) using BMW special tool 11 5 490, until detent position stop is reached.

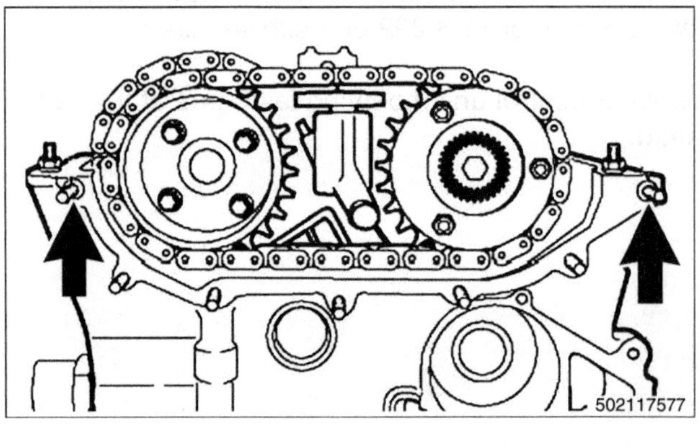


Apply a small amount of liquid gasket sealer (3-Bond 1209® or equivalent) to VANOS housing alignment dowels on front of cylinder head. Install a new gasket.

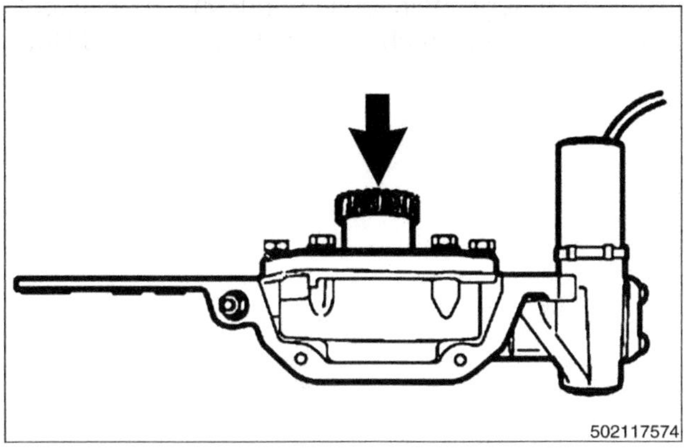


◄ Push VANOS piston gear cup (**arrow**) into VANOS housing until it bottoms out. Position VANOS control unit in front of camshafts, but do not engage splines on gear cup.

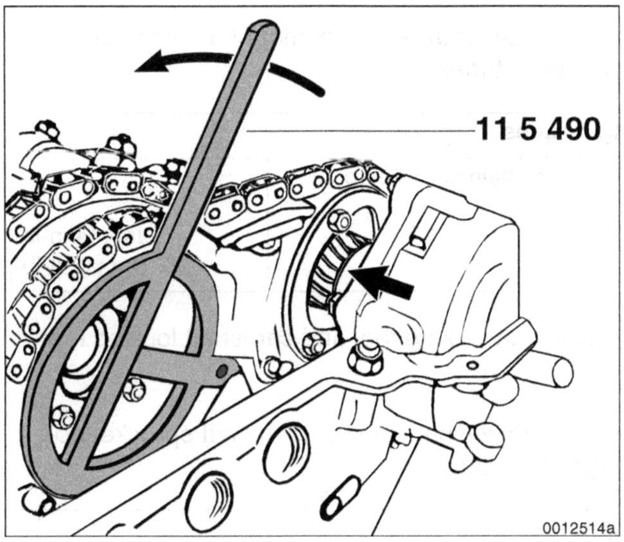


◄ Install BMW special tool 11 5 490 on exhaust camshaft.

– Insert the helical-toothed splined shaft (**arrow**) into the helical-toothed sprocket while rotating gears and chain counterclockwise with special tool.

> ***CAUTION—***
> *Ensure that the first available tooth engages. It is important that the exhaust camshaft sprocket be turned only the minimum amount necessary to engage the gear cup. If this procedure is done incorrectly, the VANOS system may not fully function.*

– Install and tighten down VANOS control unit fasteners.

| **Tightening torques** | |
|---|---|
| VANOS control unit to cylinder head<br>• M6 nuts<br>• M8 bolts | <br>10 Nm (7 ft-lb)<br>22 Nm (17 ft-lb) |

– Press down on secondary timing chain tensioner and remove lock tool.

◄ Remove primary timing chain tensioner (**arrow**) at right side of timing chain cover.

> ***CAUTION—***
> *Timing chain tensioner is under spring tension.*

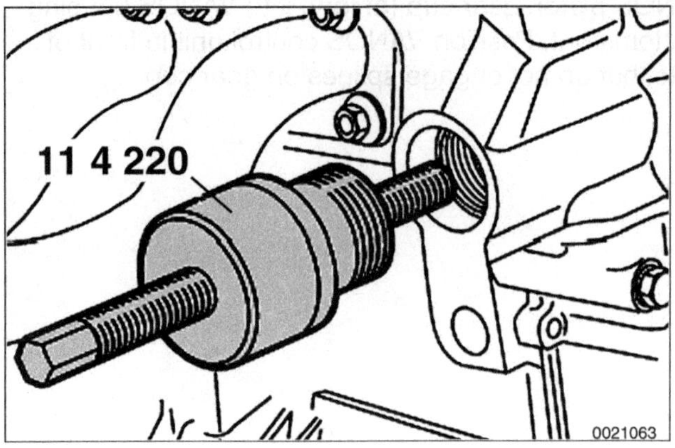


Install adjustable primary chain tensioner (BMW special tool 11 4 220) and preload timing chain by tightening to 1.3 Nm (11 in-lb).

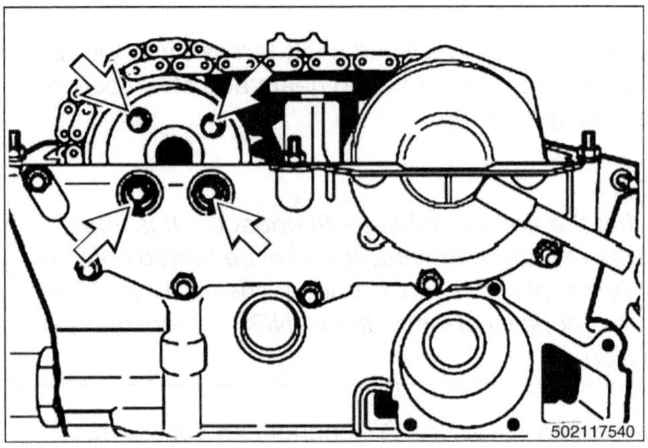


Tighten exhaust camshaft sprocket mounting bolts (**arrows**) alternately in two stages.

| **Tightening torques** | |
|---|---|
| Secondary timing chain sprocket to exhaust camshaft<br>• Stage 1<br>• Stage 2 | <br><br>5 Nm (44 in-lb)<br>22 Nm (17 ft-lb) |

- Remove crankshaft locking tool and camshaft locking tool set.

- Using new sealing washers, install camshaft sprocket bolt access plugs and tighten.

| **Tightening torque** | |
|---|---|
| Access plugs to VANOS control unit | 50 Nm (37 ft-lb) |

- Remove timing chain tensioning tool from side of cylinder head. Install primary chain tensioner with new sealing washer and tighten.

| **Tightening torque** | |
|---|---|
| Primary timing chain tensioner plug to tensioner sleeve | 70 Nm (52 ft-lb) |

- Check VANOS adjustment. See **Single VANOS, testing** in this repair group.

## Timing chains, removing (M52, S52 engine)

Camshaft timing chain removal requires engine oil pan removal. See **119 Lubrication System**.

Special BMW service tools are needed for this procedure. These tools assure proper timing of the valvetrain, as there are no precise timing marks to set the timing on the camshafts. Read the procedure through before beginning the job.

> ***CAUTION—***
> *If the camshafts are not properly timed, the pistons can contact the valves.*

– Disconnect negative (–) battery cable.

> ***CAUTION—***
> *Disconnecting the battery cables may erase fault codes stored in control unit memory.*

– Raise front of car and support in a safe manner.

> ***CAUTION—***
> *Make sure the car is stable and well supported at all times. Use a professional automotive lift or jack stands designed for the purpose. A floor jack is not adequate support.*

– With engine cold, drain engine coolant. Drain engine block by removing block drain. See **170 Radiator and Cooling System**.

– Remove engine drive belts. See **020 Maintenance**.

– Remove radiator cooling fan, fan shroud, and radiator. See **170 Radiator and Cooling System**.

***NOTE—***
*The radiator cooling fan nut (32 mm wrench) has left-hand threads.*

– Drain engine oil and remove oil pan. See **119 Lubrication System**.

– Remove alternator cooling air duct.

– Remove two large coolant hoses from thermostat housing. Remove engine lifting bracket and thermostat housing from front of cylinder head.

– Remove cylinder head cover. See **113 Cylinder Head Removal and Installation**.

– Unclip and remove baffle cover above intake camshaft.

– Remove coolant pump pulley. See **170 Radiator and Cooling System**.

– With crankshaft at TDC position, remove crankshaft vibration damper. See **Vibration damper, removing and installing (M52, S52 engine)**.

< Remove primary timing chain tensioner (**arrow**) at right side of cylinder head.

> ***CAUTION—***
> *Timing chain tensioner is under spring tension.*

< Press down on secondary chain tensioner and lock into place using BMW special tool 11 3 292 or a similar size pin.

– Remove VANOS control unit. See **Single VANOS control unit, removing (M52, S52 engine)** in this repair group.

– Remove intake and exhaust camshaft sprocket mounting bolts. Remove sprockets together with secondary chain.

< Remove secondary timing chain tensioner mounting bolts (**arrows**) and remove tensioner.

– Remove primary timing chain sprocket from exhaust camshaft.

– Remove primary sprocket from chain. Hang chain from end of exhaust camshaft.

Remove secondary chain guide bolts (**arrows**) and remove guide.

Remove lower timing chain cover to cylinder head bolts (**arrows**).

– Remove engine oil pan. See **119 Lubrication System**.

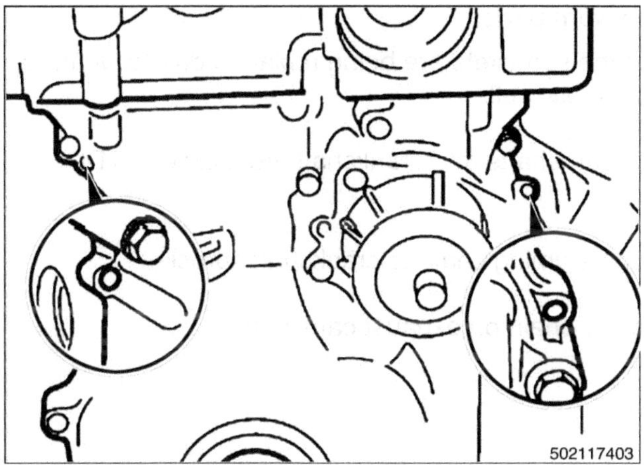


Using a 5 mm or smaller drift, drive two locating dowels (**insets**) in lower timing cover toward rear of car.

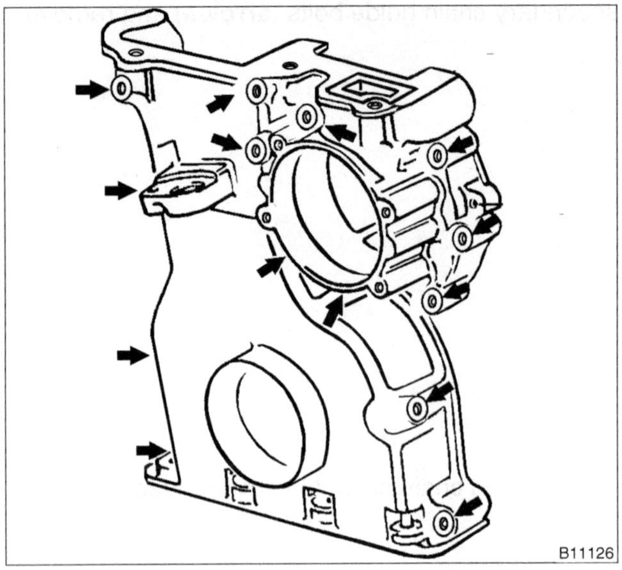


< Remove timing cover mounting bolts (**arrows**).

- Carefully slide cover with coolant pump forward to remove.

> ***CAUTION—***
> - *Use care when removing the cover from the cylinder head gasket. If the cover is stuck, use a sharp knife to separate it from the head gasket.*
> - *If cylinder head gasket is damaged, remove cylinder head and replace head gasket. See* **113 Cylinder Head Removal and Installation**.

***NOTE—***
*For clarity, front timing cover is illustrated with coolant pump removed.*

– Lift primary chain off crankshaft sprocket.

## Timing chains, installing (M52, S52 engine)

– Prior to reinstallation:

- Inspect sprockets for wear or damage.
- Inspect chain guide and tensioner rails for grooves caused by chain contact.
- Replace worn parts.
- If any of the sprockets are being replaced due to wear, replace chain as well.

– Make sure camshafts and crankshaft are locked in TDC position.

– Install primary timing chain to crankshaft sprocket.

< Hang primary chain on exhaust camshaft.

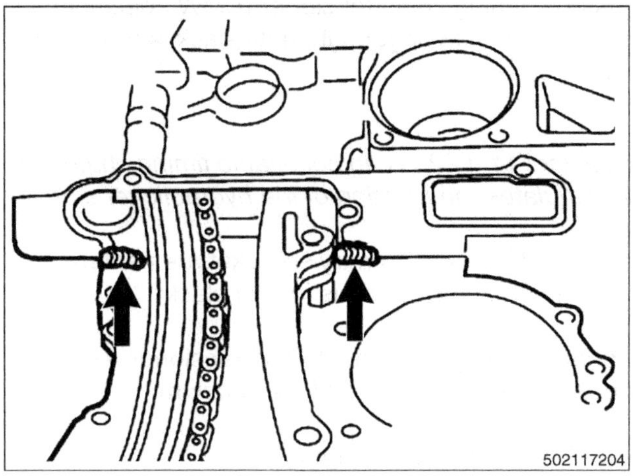


Apply a small bead of silicon sealer (3-Bond 1209® or equivalent) to corners (**arrows**) where timing cover meets cylinder head.

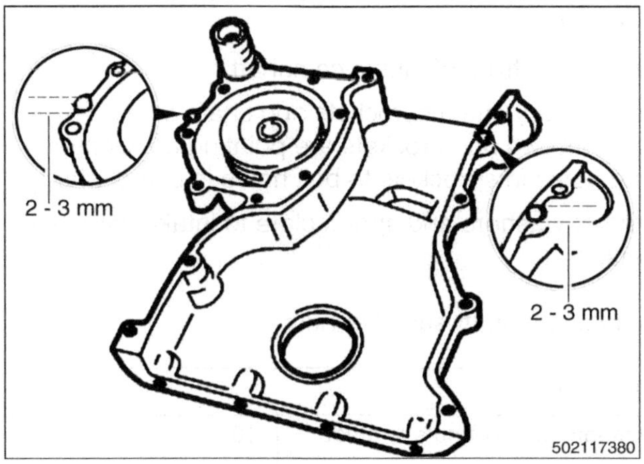


Remove timing cover locating dowels from front of engine block. Tap dowels into cover until they protrude 2 to 3 mm from sealing surface.

– Using new gaskets, install lower timing chain cover:
  - Tap cover into position to engage dowels.
  - Install timing chain cover fasteners and initially tighten to 5 Nm (44 in-lb).
  - Drive dowels in flush to front of cover.
  - Tighten cover mounting bolts alternately and in stages.
  - Install cylinder head to timing chain cover fasteners and tighten.

| **Tightening torques** | |
|---|---|
| Lower timing chain cover to engine block<br>• M7<br>• M8 | <br>15 Nm (11 ft-lb)<br>20 Nm (15 ft-lb) |

– Replace front crankshaft radial seal. See **119 Lubrication System**.

– Reinstall engine oil pan using a new gasket.

Mount exhaust camshaft primary timing chain sprocket to primary chain so that arrow on sprocket faces up. Mount sprocket to camshaft so that top camshaft tapped bores (**arrows**) are on left side of sprocket slotted holes.

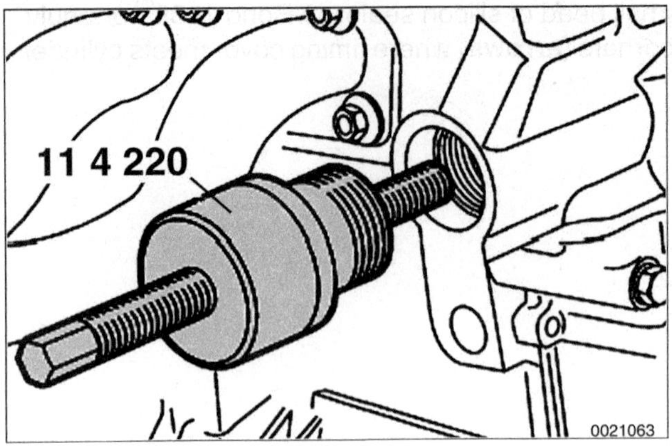


◀ Install adjustable timing chain tensioner (BMW special tool 11 4 220). Thread tool in finger tight until slack is removed from chain.

***NOTE—***
*BMW special tool 11 4 220 is an adjustable timing chain tensioner that simulates the function of the hydraulic tensioner.*

– Check that slotted holes in primary sprocket are centered to bolt holes in camshaft and chain is free of slack.

– Install secondary timing chain guide and tensioner. Tighten mounting bolts (tensioner remains locked down).

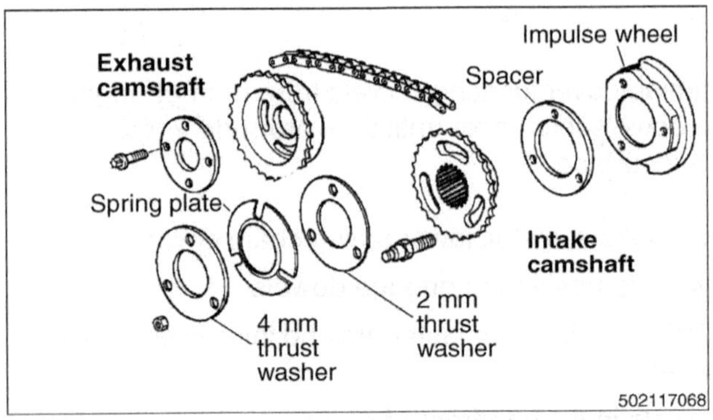


◀ Install VANOS spacers and washers:

- Install cylinder identification impulse wheel to front of intake camshaft.
- Install spacer to front of intake camshaft.
- Install secondary chain sprockets and chain as an assembly so that arrows on sprockets are pointing up. Center mounting holes in sprockets to bolt holes in camshafts.
- Install thrust washers and spring plate to intake camshaft sprockets.

– Tighten intake camshaft sprocket nuts.

| **Tightening torque** | |
|---|---|
| Intake camshaft sprocket nuts to studs (M6) | 10 ± 1 Nm (7 ± 1 ft-lb) |

– Install VANOS control unit, see **Single VANOS control unit, installing (M52, S52 engine)**.

| **Tightening torques** | |
|---|---|
| VANOS control unit to cylinder head<br>• M6 nuts<br>• M8 bolts | <br>10 Nm (7 ft-lb)<br>22 Nm (17 ft-lb) |

– Unlock and release secondary chain tensioner.

◀ Preload primary chain tensioning rail:

- Tighten adjusting screw on BMW special tool 11 4 220.

| **Tightening torque** | |
|---|---|
| Primary chain tensioner preload | 1.3 Nm (11 in-lb) |

– Tighten exhaust camshaft sprocket bolts in two stages.

| **Tightening torque** | |
|---|---|
| Exhaust camshaft sprocket bolts (M7)<br>• Stage 1<br>• Stage 2 | <br>5 Nm (44 in-lb)<br>22 Nm (17 ft-lb) |

– Remove crankshaft and camshaft locking tools.

– Reinstall crankshaft vibration damper and hub.

| Tightening torques | |
|---|---|
| Vibration damper and pulley to crankshaft hub | 22 Nm (17 ft-lb) |
| Vibration damper hub to crankshaft (use new bolt) | 410 Nm (300 ft-lb) |

– Turn engine over by hand in direction of rotation at least two full rotations. Reinstall camshaft and crankshaft locking tools to recheck engine timing. Remove locking tools if no faults are found.

– Remove BMW special tool 11 3 390 (dummy tensioner) from side of cylinder head.

◄ Install primary chain tensioner so that cutout in tensioner piston (**dashed lines**) engages chain rail.

| Tightening torque | |
|---|---|
| Primary camshaft timing chain tensioner to tensioner sleeve | 70 Nm (52 ft-lb) |

– Using new sealing washers, install camshaft sprocket bolt access plugs and tighten.

| Tightening torque | |
|---|---|
| Access plugs to VANOS control unit | 50 Nm (37 ft-lb) |

– Install cylinder head cover mounting studs, camshaft cover, and cylinder head cover.

| Tightening torque | |
|---|---|
| Cylinder head cover to cylinder head (M6) | 10 Nm (7 ft-lb) |

– Installation of remaining parts is reverse of removal. Note the following:

- When installing thermostat, make sure arrow or vent hole faces up.
- Reattach and torque ground wires at cylinder head cover mounting studs and at front of cylinder head and thermostat housing, where applicable.
- Refill and bleed cooling system. See **170 Radiator and Cooling System**.
- Fill engine with oil and install new oil filter. See **020 Maintenance**.

| **Tightening torques** | |
|---|---|
| Coolant drain plug to cylinder block | 40 Nm (30 ft-lb) |
| Radiator cooling fan to coolant pump | 40 Nm (30 ft-lb) |
| Radiator drain screw to radiator | 2.5 Nm (22 in-lb) |
| Thermostat housing to cylinder head | 10 Nm (7 ft-lb) |
| VANOS oil supply pipe to VANOS control unit (banjo bolt) | 32 Nm (24 ft-lb) |

## CAMSHAFT TIMING CHAINS REMOVAL AND INSTALLATION (M52 TU, M54 ENGINE)

z02117007

M52 TU and M54 engines use a pair of chains to drive the camshafts. A primary chain is driven off the crankshaft and drives the primary sprocket on the exhaust camshaft. The secondary chain drives the intake camshaft. Each chain is tensioned by a hydraulic tensioner.

Intake and exhaust camshaft timing is variable and is controlled by the engine control module (ECM). This system is known as double VANOS.

Camshaft timing chain service and repairs requires the removal of front crankshaft vibration damper, double VANOS control unit, engine oil pan and timing chain housing cover. Be sure to read the procedures through before starting work.

> ***WARNING —***
> *To avoid personal injury, be sure the engine is cold before starting any of these procedure.*

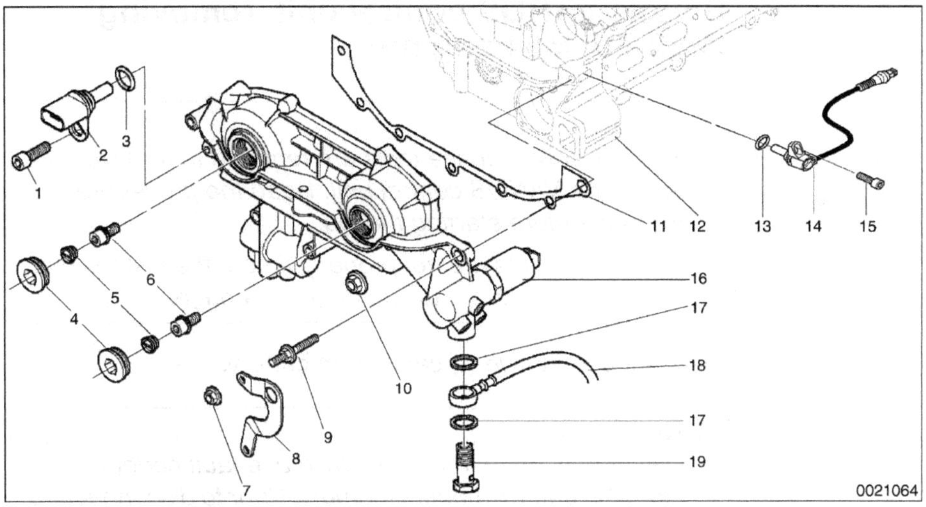


## Double VANOS control unit components

1. **Bolt (M6)**
2. **Exhaust camshaft sensor**
3. **Sealing ring**
4. **Camshaft end sealing plug**
   - tighten to 50 Nm (37 ft-lb)
5. **Camshaft seal cap**
6. **VANOS hydraulic piston set screw**
   - Left hand thread
     tighten to 10 Nm (7 ft-lb)
7. **Nut (M7)**
   - tighten to 14 Nm (10 ft-lb)
8. **Engine lifting hook**
9. **Stud (M7)**
10. **Nut (M6)**
    - tighten to 10 Nm (7 ft-lb)
11. **Gasket**
12. **Cylinder head**
13. **Sealing ring**
14. **Intake camshaft sensor**
15. **Bolt (M6)**
16. **VANOS control unit**
17. **Copper sealing ring**
18. **VANOS oil feed line**
19. **Banjo bolt**
    - tighten to 32 Nm (24 ft-lb)

## Double VANOS control unit, removing (M52 TU, M54 engine)

> ***CAUTION—***
> - *Special BMW service tools are required to remove and install the VANOS control unit. Read the procedure through before starting the job.*
> - *If the camshafts are not properly timed, the pistons can contact the valves causing engine damage.*

– Disconnect negative (-) cable from battery.

> ***CAUTION—***
> - *Disconnecting the battery may erase fault code(s) stored in control module memory. Prior to disconnecting the battery, check for fault codes using special BMW diagnostic equipment.*
> - *Prior to disconnecting the battery, read the battery disconnection cautions given in* **001 General Warnings and Cautions**.

– Remove air filter housing. See **130 Fuel Injection**.

– Remove engine drive belt. See **020 Maintenance**.

– Remove viscous clutch cooling fan. See **170 Radiator and Cooling System**.

> ***CAUTION—***
> *The radiator cooling fan nut (32 mm wrench) has left- hand threads.*

– Remove engine covers and cylinder head cover. See **113 Cylinder Head Removal and Installation**.

– Remove spark plugs.

– Unclip and remove oil baffle cover above intake camshaft.

– Disconnect oil line fitting from VANOS control unit.

***NOTE—***
*Wrap the oil line fitting with a shop rag to absorb leaking oil.*

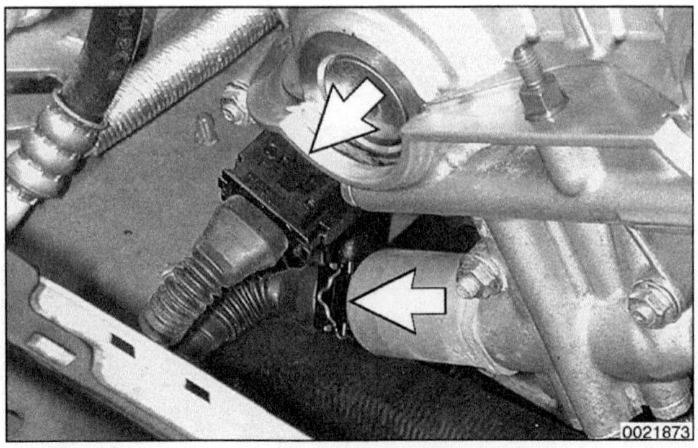


◄ Disconnect electrical connections from exhaust camshaft sensor and exhaust camshaft VANOS solenoid (**arrows**).

◀ Using banjo bolt from VANOS control unit oil line, install test fitting (BMW special tool 11 3 450) to VANOS control unit.

◀ Cover oil hole in VANOS control unit (**arrow**) with shop towel to capture oil which will spray when compressed air line is connected.

– Connect compressed air line to air fitting. Apply air pressure set to 2 - 8 bar (30 - 110 psi).

◀ With compressed air line connected, turn engine at least twice in direction of rotation until cylinder 1 intake and exhaust camshaft lobes (**arrows**) point up and face each other.

> ***CAUTION—***
> *Do not rotate engine counterclockwise to reach the top dead center position. Instead, complete another two complete rotations.*

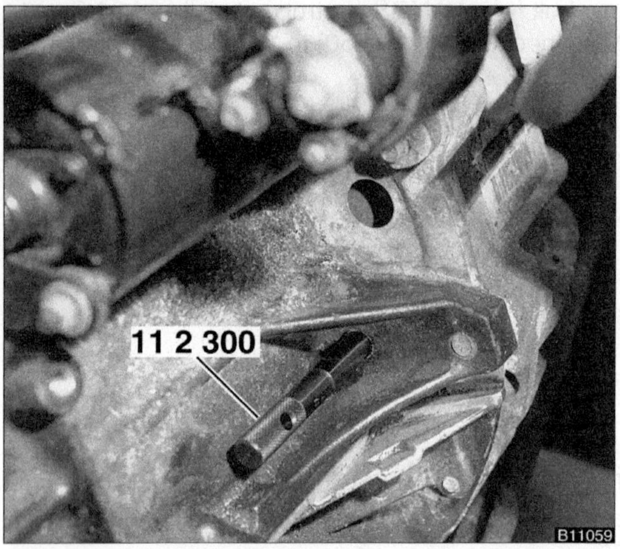


Remove sealing plug from bore under starter. Lock crankshaft in TDC position with BMW special tool 11 2 300 (**arrow**).

Unscrew and remove threaded studs (**arrows**) at rear of cylinder head.

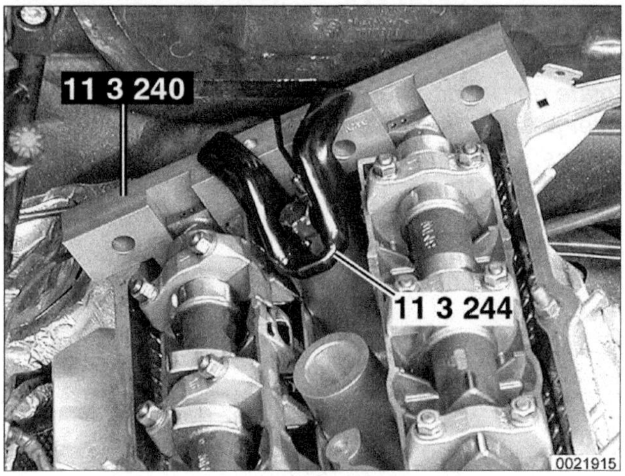


Secure camshafts in TDC position using BMW special tool set 11 3 240.

- Disconnect compressed air line, leaving compressed air fitting attached to VANOS control unit.

***NOTE—***
*Oil drains from pressure line. Have a container and rags ready.*

Unscrew sealing plugs (**arrows**) from VANOS control unit.

***NOTE—***
*Oil drains from sealing plug bores. Have a container and rags ready.*

Pull sealing caps straight out of VANOS control unit with BMW special tool 11 6 170, or with short needle-nosed pliers.

***NOTE—***
*Additional oil may drain from VANOS control unit.*

Remove set screws on ends of intake and exhaust camshafts.

***NOTE—***
*Set screws have left hand threads. Unscrew clockwise.*

Remove VANOS unit:

- Remove fasteners from engine support eye.
- Remove VANOS mounting nuts (**arrows**) from cylinder head. Slide VANOS control unit and metal gasket off.

*NOTE—*
- *The VANOS control unit contains residual oil. Place shop towels underneath when removing.*
- *If the double VANOS control unit is replaced, or if repair operations are completed that may change camshaft timing, recheck and adjust camshaft timing. See* **Camshaft timing, checking and adjusting (M52 TU, M54 engine)**.

*CAUTION—*
*Do not crank the engine with VANOS unit removed. Damage to engine may result from piston / valve interference.*

## Double VANOS control unit, installing (M52 TU, M54 engine)

Clean contact edges of cylinder head face and double VANOS control unit and apply a thin coat of sealing compound (3-Bond 1209® or equivalent) to surfaces.

*CAUTION—*
- *Make sure all gasket material is removed from face of cylinder head. Clean sealing face and keep free of oil.*
- *Check locating dowel (**A**) and dowel sleeve (**B**) at top of cylinder head for damage or incorrect installation.*

*NOTE—*
- *If the double VANOS control unit is being replaced, be sure to check and adjust camshaft timing. See* **Camshaft timing, checking and adjusting (M52 TU, M54 engine)**.
- *If work being carried out has no effect on camshaft timing, it is not necessary to recheck the timing. However, check VANOS operation using a compatible scan tool once the engine is running.*

– Using new gasket, install VANOS control unit to cylinder head.

- Reinstall engine support eye.

| Tightening torques | |
|---|---|
| VANOS control unit to cylinder head<br>• M6 nut<br>• M7 nut | <br>10 Nm (7 ft-lb)<br>14 Nm (10 ft-lb) |

Insert and tighten VANOS hydraulic piston set screws on intake and exhaust camshafts.

*NOTE—*
*Set screws have left hand thread. Tighten counterclockwise.*

| Tightening torque | |
|---|---|
| Hydraulic piston to splined shaft (set screw) | 10 Nm (7 ft-lb) |

Replace sealing caps inside VANOS control unit with BMW special tool 11 6 170, or short needle-nosed pliers.

Insert and secure VANOS sealing plugs (**arrows**), using new sealing O-rings.

| Tightening torque | |
|---|---|
| Sealing plug to VANOS control unit | 50 Nm (37 ft-lb) |

- Remove compressed air fitting from VANOS control unit.
- Fit VANOS oil line banjo bolt with new seals. Attach oil line to VANOS control unit.

| Tightening torque | |
|---|---|
| Oil line to VANOS control unit (banjo bolt) | 32 Nm (24 ft-lb) |

- Connect electrical harness connectors to camshaft position sensors and VANOS solenoid valves.
- Remove BMW special locking tools from rear of cylinder head.
- Remove crankshaft locking tool and replace sealing plug.

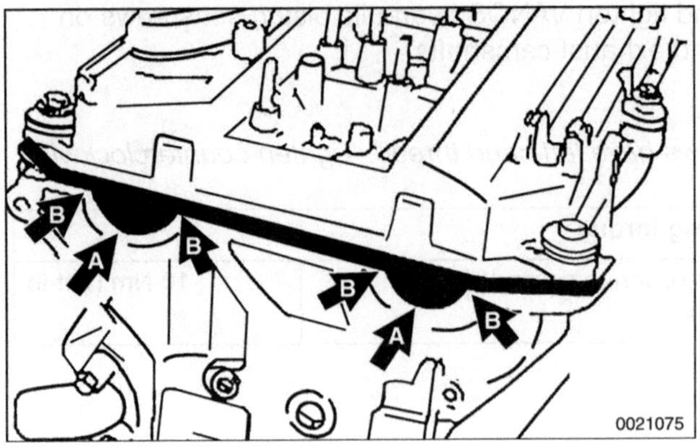


Install intake camshaft baffle and cylinder head cover.

- Check for correct seating of half-moon seals (**A**) in back of cylinder head cover.
- Use a small amount of 3-Bond 1209® or equivalent sealant at corners (**B**) of half-moon cutouts.
- Similarly, seat gasket and seal corners in front of cylinder head, at VANOS control unit.

| **Tightening torque** | |
|---|---|
| Cylinder head cover to cylinder head (M6) | 10 Nm (7 ft-lb) |

– Remainder of installation is reverse of removal.

| **Tightening torques** | |
|---|---|
| Radiator fan (viscous clutch) to coolant pump | 40 Nm (30 ft-lb) |
| Spark plug to cylinder head | 25 Nm (18 ft-lb) |

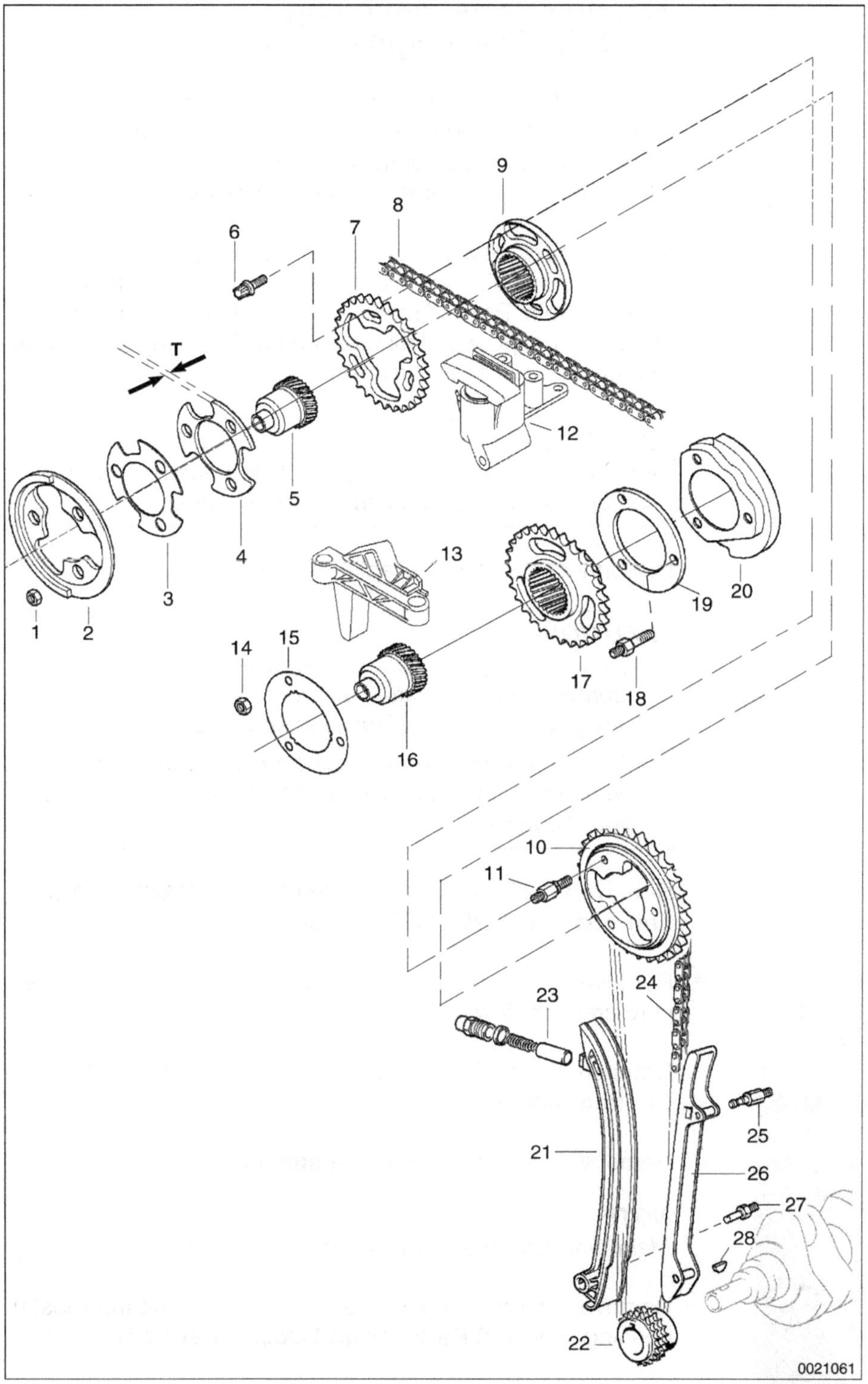


## Timing chain components (M52 TU, M54 engine)

1. **Impulse wheel mounting nut**
2. **Exhaust camshaft impulse wheel**
3. **Spring plate**
4. **Thrust spacer**
   - thickness (**T**) = 3.5 mm
5. **Splined shaft**
6. **Torx screw**
7. **Exhaust secondary sprocket**
8. **Secondary timing chain**
9. **Splined sleeve**
10. **Primary sprocket**
11. **Threaded locating stud**
12. **Secondary chain tensioner**
13. **Secondary chain lower guide**
14. **Sprocket mounting nut**
15. **Spring plate**
16. **Splined shaft**
17. **Intake camshaft sprocket**
18. **Locating stud**
19. **Thrust spacer**
20. **Intake camshaft impulse wheel**
21. **Chain tensioner rail**
22. **Crankshaft sprocket**
23. **Primary chain tensioner**
24. **Primary chain**
25. **Locating stud**
26. **Guide rail**
27. **Locating stud**
28. **Woodruff key**

## Timing chains, removing (M52 TU, M54 engine)

Camshaft timing chain removal requires that the oil pan be removed. See **119 Lubrication System**. Also required is removal of the VANOS control unit and disassembly of the intake and exhaust camshaft sprocket assemblies.

Special BMW service tools are needed for timing chain removal and installation. These tools assure proper timing of the valvetrain. Precise marks to set the timing on the camshafts are not provided for reassembly. Read the procedures through before beginning the job.

> ***CAUTION—***
> *If the camshafts are not properly timed, the pistons can contact the valves, causing engine damage.*

– Disconnect negative (-) cable from battery.

> ***CAUTION—***
> - *Disconnecting the battery may erase fault code(s) stored in control module memory. Check for fault codes using special BMW diagnostic equipment.*
> - *Prior to disconnecting the battery, read the battery disconnection cautions given in* **001 General Warnings and Cautions**.

– Remove VANOS control unit. See **Double VANOS control unit, removing (M52 TU, M54 engine)**.

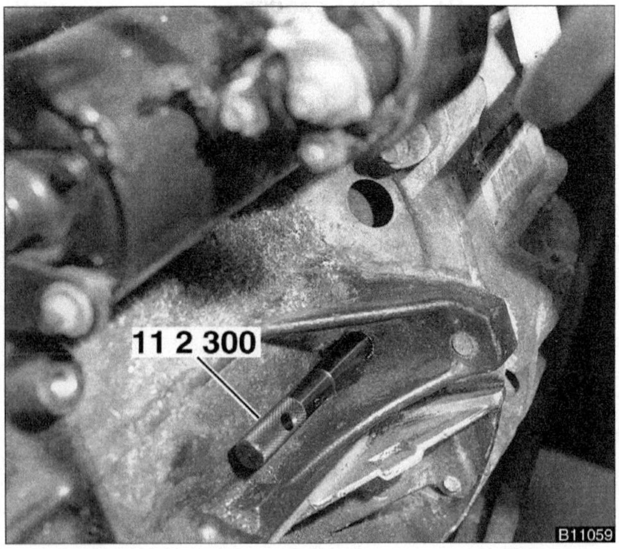


◄ Remove sealing plug from crankcase below starter and secure crankshaft with BMW special tool 11 2 300.

– Drain engine cooling system. See **170 Radiator and Cooling System**.

– Remove drive belts. See **020 Maintenance**.

***NOTE—***
*Mark direction of belt rotation if reusing belt.*

– Remove belt driven cooling fan, fan shroud, and thermostat housing. See **170 Radiator and Cooling System**.

– Remove coolant pump pulley.

– Drain engine oil and remove oil pan as described in **119 Lubrication System**.

◀ Remove primary camshaft chain tensioner (**arrow**) from timing chain cover.

> ***CAUTION—***
> *Primary camshaft chain tensioning piston is under spring pressure.*

◀ Press down on secondary chain tensioner and lock into place using BMW special tool 11 3 292 or a similar size wire.

◀ Remove exhaust camshaft impulse wheel mounting nuts (**arrows**). Remove impulse wheel (**A**).

◀ Remove spring plate (**A**).

Remove intake camshaft sprocket mounting nuts (**arrows**) and remove spring plate (labelled **FRONT**).

Remove torx screws from exhaust camshaft sprocket (**arrows**).

Lift off exhaust and intake sprockets with secondary chain, thrust spacer (**A**) and splined shaft (**B**).

***CAUTION—***
*Splined shafts for both intake and exhaust camshafts share the same part number. Remove and mark used splined shafts in order to reinstall in original locations.*

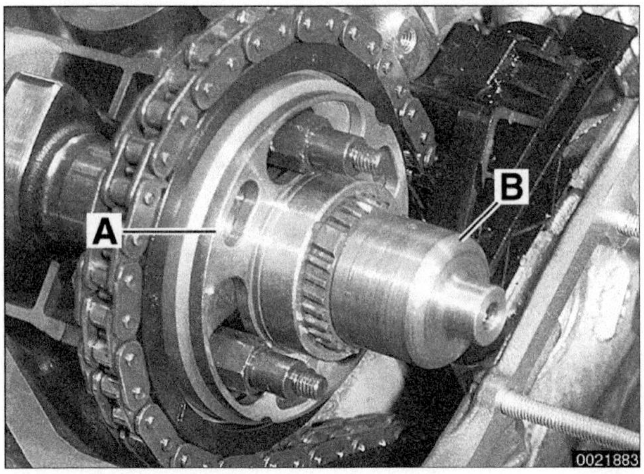


Remove exhaust camshaft splined sleeve (**A**) and shaft (**B**).

◄ Remove secondary chain tensioner mounting bolts (**arrows**). Remove tensioner while keeping locking pin in place.

◄ Remove primary chain sprocket mounting studs (**arrows**) on exhaust camshaft.

◄ Lift primary chain sprocket off exhaust camshaft. Remove sprocket from chain.

◀ Hang timing chain on exhaust camshaft end.

◀ Working in cylinder head cavity, remove timing chain cover bolts (**arrows**).

◀ Remove secondary chain lower guide bolts (**arrows**). Remove chain guide.

***NOTE—***
*Bolt on intake camshaft side is long and extends into timing chain cover.*

– Use BMW special tools 11 8 190 and 11 8 200 to counterhold crankshaft vibration damper while removing vibration damper mounting bolt. See **Vibration damper, removing and installing (M52 TU, M54 engine)** in this repair group.

***CAUTION—***
- *Do not use flywheel locking tool to hold crankshaft stationary to loosen or tighten vibration damper mounting bolt. Use only the special tools specified, or equivalent hub holding tool.*
- *Do not allow crankshaft to rotate while timing chains are loosened or removed. The pistons can contact the valves, causing engine damage.*

***NOTE—***
*The crankshaft vibration damper mounting bolt is tightened to a torque of 410 Nm (300 ft-lb).*

– Use BMW special tools 11 8 210 and 11 8 200 to pull vibration damper off crankshaft.

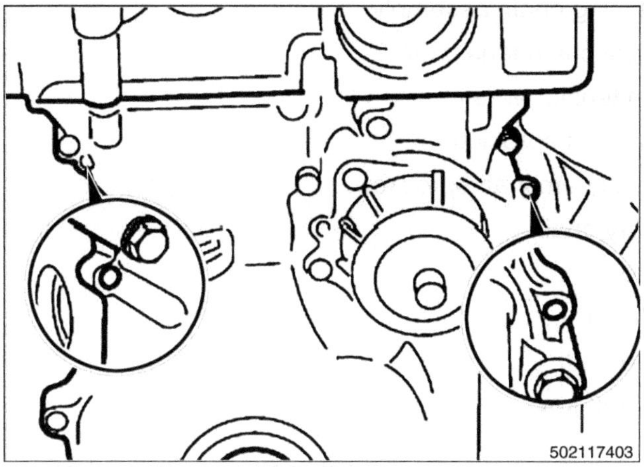


Using a 5 mm or smaller drift, drive two locating dowels (**insets**) in lower timing cover toward rear of car.

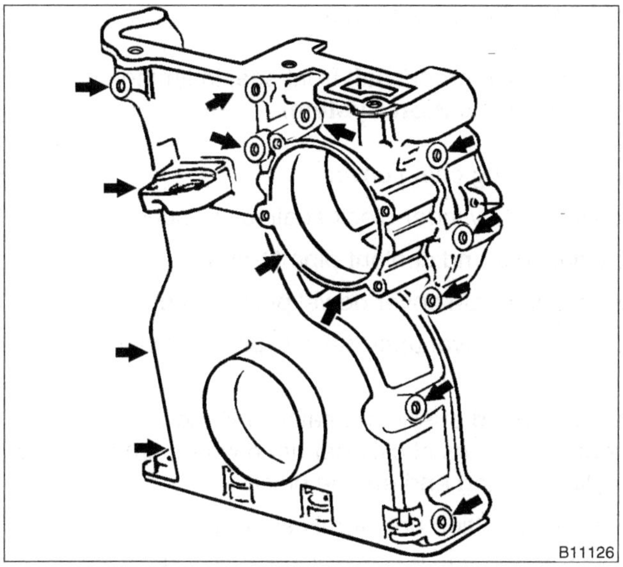


Remove timing cover mounting bolts (**arrows**).

- Carefully slide cover with coolant pump forward to remove.

> ***CAUTION—***
> - *Use care when removing the cover from the cylinder head gasket. If the cover is stuck, use a sharp knife to separate it from the head gasket.*
> - *If cylinder head gasket is damaged, remove cylinder head and replace head gasket. See* **113 Cylinder Head Removal and Installation**.
> - *Use care when detaching front timing cover from coolant pipe at rear of cover.*

***NOTE—***
*For clarity, front timing cover is illustrated with coolant pump removed.*

– Push primary chain tensioner guide rail aside and remove chain.

## Timing chains, installing (M52 TU and M54 engines)

> ***CAUTION—***
> *The procedure outlined below assumes that the camshafts are locked in the TDC installation position with special locking tools as shown earlier.*

– Inspect timing chain sprockets. Inspect guide rail and tensioner rail for damage. Replace any part that is worn or damaged.

***NOTE—***
*If any sprockets are being replaced due to wear, also replace chain.*

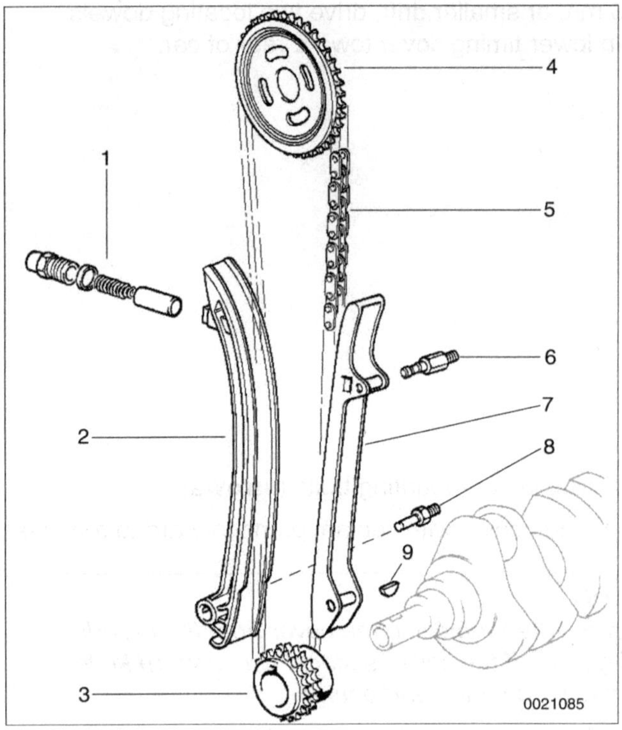


Primary timing chain assembly
1. Primary chain tensioner
2. Chain tensioner rail
3. Crankshaft sprocket
4. Exhaust camshaft sprocket
5. Primary chain
6. Anchor bolt
7. Guide rail
8. Anchor bolt
9. Woodruff key

– Install primary timing chain to crankshaft sprocket and hang upper end from exhaust camshaft.

– If necessary, replace crankshaft front seal in timing chain cover. See **119 Lubrication System**.

– To install lower timing chain cover:
  - Clean cover and cylinder block sealing surfaces.
  - Use new gaskets and coolant pipe O-ring.
  - Remove timing chain cover dowel pins from engine block.
  - Drive dowels into cover until they just protrude slightly from sealing surface.
  - Apply a small bead of silicon sealer (3-Bond 1209® or equivalent) to corners of cylinder head where timing cover meets cylinder head and engine block.
  - Tap cover into position until firmly seated.
  - Install all bolts hand tight, including two Torx bolts from above.
  - Install secondary chain lower guide.
  - Tighten cover mounting bolts alternately and in stages.

| Tightening torque | |
|---|---|
| Lower timing cover to cylinder block (M6) | 10 Nm (7 ft-lb) |

– Install oil pan with new gasket. See **119 Lubrication System**.

| Tightening torques | |
|---|---|
| Oil pan to engine block (M6)<br>• 8.8 grade<br>• 10.9 grade | <br>10 Nm (7 ft-lb)<br>12 Nm (9 ft-lb) |

– Reinstall crankshaft vibration damper. See **Vibration damper, removing and installing (M52 TU, M54 engine)** in this repair group.

> ***CAUTION—***
> - *Do not use BMW special tool 11 2 300 to hold crankshaft stationary to tighten vibration damper mounting bolt. Use only the special tools specified, or equivalent hub holding tool.*
> - *Have a second person hold special tool 11 8 190 while tightening vibration damper mounting bolt.*

| **Tightening torque** | |
|---|---|
| Vibration damper hub to crankshaft (use new bolt) | 410 Nm (300 ft-lb) |

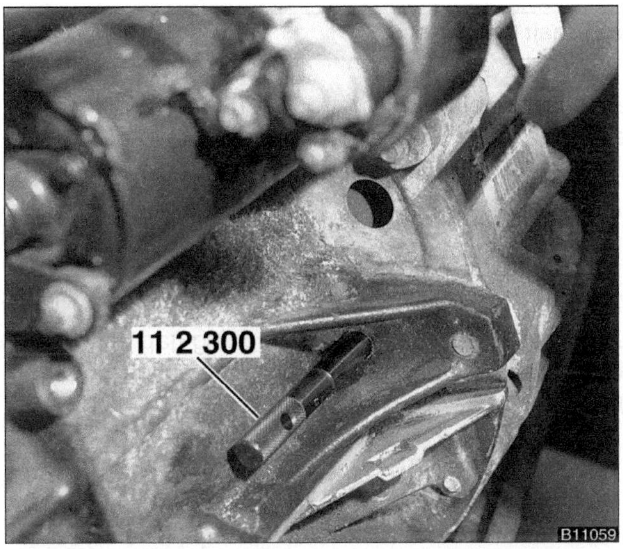


◄ Remove crankshaft locking tool (BMW special tool 11 2 300).

◄ Fit top sprocket to primary timing chain and install on exhaust camshaft so that pointer on sprocket lines up with cylinder head sealing surface (**dashed line**).

– Position sprocket so that chain slack is on tensioner rail side.

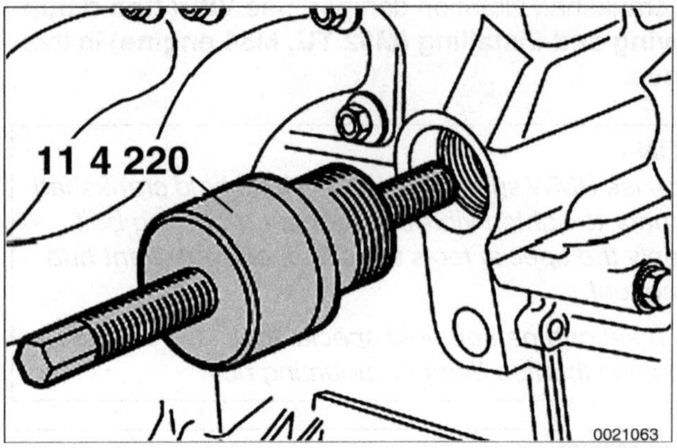


Insert BMW special tool 11 4 220 in cylinder head and screw in adjustment screw by hand just until it contacts tensioning rail and chain slack is removed.

*NOTE—*
*BMW special tool 11 4 220 is an adjustable primary chain tensioner that simulates the function of the tensioner.*

– Make sure arrow on top primary sprocket is still lined up with upper edge of cylinder head. Reposition sprocket if necessary.

Install and tighten down threaded locating studs (**arrows**) in end of exhaust camshaft.

| **Tightening torque** | |
|---|---|
| Exhaust camshaft locating stud | 20 Nm (15 ft-lb) |

– Install secondary chain tensioner on cylinder head. Make sure tensioner is locked in compressed position.

Fit exhaust camshaft splined sleeve, aligning gap in sleeve splines with corresponding gap on camshaft splines (**arrows**).

Slide splined shaft onto exhaust camshaft, aligning larger tooth with corresponding gap of splined sleeve (**arrows**).

- Slide splined shaft in further until 3 small or elongated slots in splined sleeve are centered to 3 threaded holes in sprocket.

Place intake and exhaust sprockets in BMW special tool 11 6 180. Position spline gap on intake sprocket (**arrow**) as shown and place secondary chain on sprockets.

Carefully remove chain and sprockets from tool and slide onto camshafts. Align gap in intake sprocket splines with corresponding gap in camshaft splines (**arrow**).

> ***CAUTION—***
> *Do not alter position of sprockets with respect to chain when removing from special tool 11 6 180.*

Slide splined shaft onto intake camshaft until approx. 1 mm (0.04 in.) of splines can still be seen.

Install intake camshaft spring plate so that FRONT mark is visible. Install mounting nuts (**arrows**) finger tight.

Working at exhaust side, insert sprocket mounting Torx screws into threaded holes (**arrows**).

- Initially tighten screws to approx. 5 Nm (44 in-lb), then back off one-half turn.

Fit thrust spacer (**A**) on exhaust camshaft.

Install spring plate (**A**) to exhaust camshaft. Make sure that **F** mark is visible.

***NOTE—***
*If* **F** *mark is no longer visible, install spring plate as shown in* **inset**.

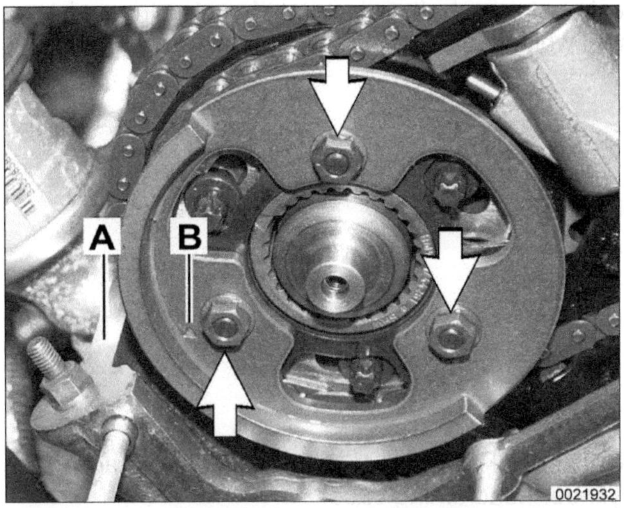


Install exhaust camshaft impulse wheel, aligning mark (**B**) with top edge of cylinder head (**A**). Install mounting nuts (**arrows**) finger tight.

Pull out exhaust splined shaft until it stops.

– Press down on secondary chain tensioner and remove tensioner lock-down tool.

Preload primary chain:

- Tighten adjusting screw on BMW special tool 11 4 220 to specified torque.

| Tightening torque | |
|---|---|
| Primary chain tensioner preload | 0.7 Nm (6 in-lb) |

Preload exhaust camshaft spring plate by pressing on impulse wheel while tightening mounting nuts (**arrows**) finger tight.

Install BMW special tool 11 6 150 (VANOS setup bracket) to front of cylinder head timing case.

- Tighten tool mounting nuts (**arrows**) by hand, then tighten down uniformly until special tool contact cylinder head firmly.

> ***CAUTION—***
> *Make sure all gasket material is removed from face of cylinder head. Clean sealing face and keep free of oil.*

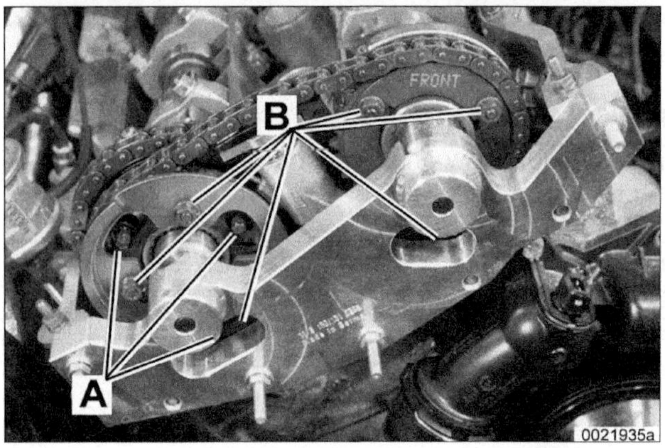


Secure camshaft sprockets and impulse wheels:

- Tighten mounting screws (**A**) on exhaust camshaft impulse wheel to approx. 5 Nm (44 in-lb).
- Tighten mounting nuts (**B**) on exhaust and intake sprocket assemblies to approx. 5 Nm (44 in-lb).
- Torque down mounting screws (**A**) and nuts (**B**) to final specifications.

| **Tightening torques** | |
|---|---|
| Sprocket assembly to camshaft initial torque | 5 Nm (44 in-lb) |
| Sprocket assembly to camshaft<br>• M7 Torx screw (**A**)<br>• M6 mounting nut (**B**) | <br>20 Nm (15 ft-lb)<br>10 Nm (7 ft-lb) |

– Remove flywheel locking tool from transmission bellhousing so that crankshaft is no longer secured.

– Remove camshaft locking tools from rear of camshafts.

Crank engine over by hand twice in direction of rotation until cylinder 1 intake and exhaust camshaft lobes (**arrows**) face each other.

– Secure crankshaft in TDC position with BMW special tool 11 2 300.

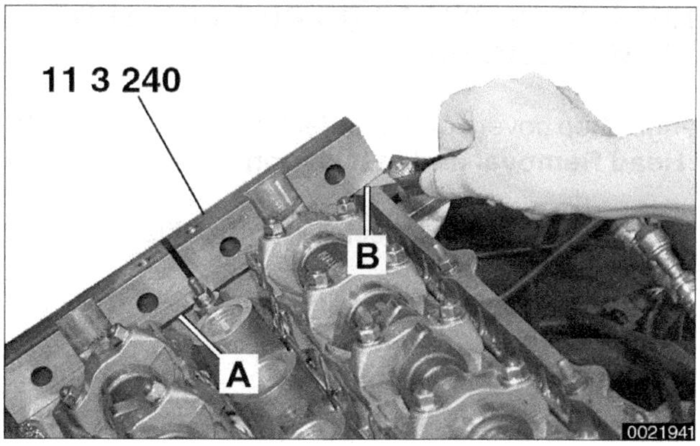


Place BMW special tool 11 3 240 over camshaft ends and measure clearance of tool to cylinder head surface.

***NOTE—***

- *If the exhaust side of the tool is not flush with the head (gap* **A***), retime the camshafts as described later in this repair group.*
- *Due to flexible sprocket design and VANOS control unit tolerances and play, the intake side of special tool 11 3 240 may be up to 1 mm (0.04 in) above the surface of the cylinder head (gap* **B***). This is normal. Otherwise, retime the camshafts as described in* **Camshaft timing, checking and adjusting (M52 TU, M54 engine)**.

– Remove VANOS setup bracket from front of cylinder head.

– Install VANOS control unit. See **Double VANOS control unit, installing (M52 TU, M54 engine)**.

– Install intake camshaft oil baffle.

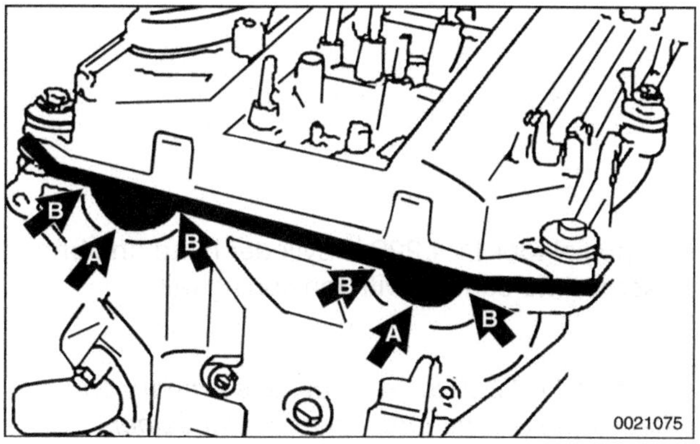


Install cylinder head cover:

- Check for correct seating of half-moon seals (**A**) in back of cylinder head cover.
- Use a small amount of 3-Bond 1209® or equivalent sealant at corners (**B**) of half-moon cutouts. Seal corners in front of cylinder head at VANOS control unit.

| Tightening torque | |
|---|---|
| Cylinder head cover to cylinder head | 10 Nm (7 ft-lb) |

– Remainder of reassembly is reverse of removal, noting the following:

- Secure all coolant hoses, thermostat housing, engine coolant drains.
- Install new engine oil filter and oil. See **119 Lubrication System**.
- Refill and bleed cooling system. See **170 Radiator and Cooling system**.
- Use scan tool to check VANOS operation with engine idling.

| Tightening torques | |
|---|---|
| Coolant drain plug to cylinder block | 25 Nm (18 ft-lb) |
| Radiator cooling fan (viscous clutch) to coolant pump | 40 Nm (30 ft-lb) |
| Radiator drain screw to radiator | 2.5 Nm (22 in-lb) |
| Vibration damper hub to crankshaft (use new bolt) | 410 Nm (300 ft-lb) |

## Camshaft timing, checking and adjusting (M52 TU, M54 engine)

– Remove engine top covers and cylinder head cover. See **113 Cylinder Head Removal and Installation**.

– Remove intake camshaft oil deflector plastic baffle.

– Remove engine drive belts and cooling fan. See **170 Radiator and Cooling System**.

– Remove spark plugs.

◀ Remove primary camshaft chain tensioner cylinder (**arrow**).

> ***CAUTION—***
> *Primary camshaft chain tensioning piston is under spring pressure.*

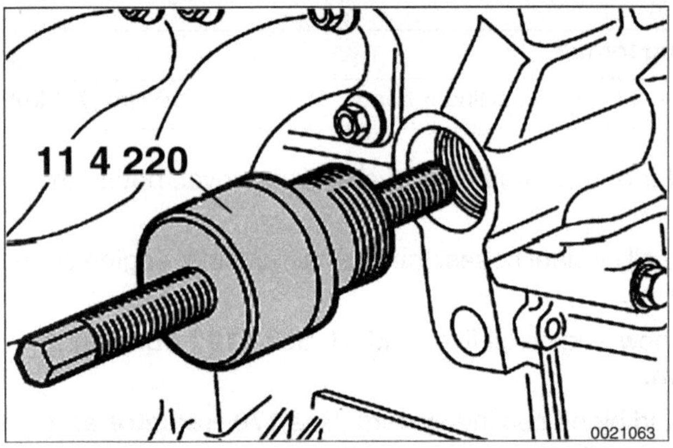


◀ Insert BMW special tool 11 4 220 in cylinder head and bring adjustment screw into contact with tensioning rail.

***NOTE—***
*BMW special tool 11 4 220 is an adjustable primary chain tensioner that simulates the function of the tensioner.*

◀ Preload primary chain tensioner rail:

- Tighten adjusting screw on BMW special tool 11 4 220.

| Tightening torque | |
|---|---|
| Primary chain tensioner preload | 0.7 Nm (6 in-lb) |

Unscrew oil pressure pipe from VANOS control unit. Remove banjo bolt from oil pressure line and install BMW special tool 11 3 450.

> ***CAUTION—***
> *Cover top of VANOS control unit with lint-free shop cloth. Attachment of compressed air line will force oil to spray out of bore.*

– Connect compressed air line to air fitting. Apply air pressure set to 2 - 8 bar (30 - 110 psi).

With compressed air line connected, turn engine at least twice in direction of rotation until cylinder 1 intake and exhaust camshaft lobes (**arrows**) face each other.

> ***CAUTION—***
> *Do not rotate engine counterclockwise to reach the top dead center position. Instead, complete another two complete rotations.*

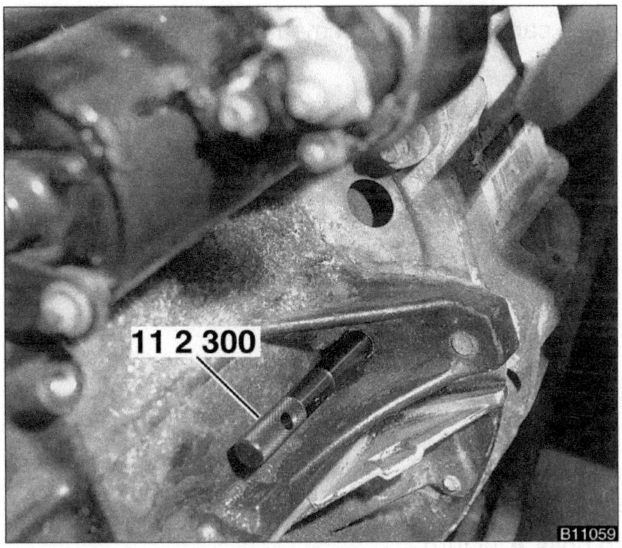


Remove sealing plug from bore at bellhousing. Secure crankshaft in TDC position with BMW special tool 11 2 300 (**arrow**).

– Remove cylinder head cover studs at rear of cylinder head.

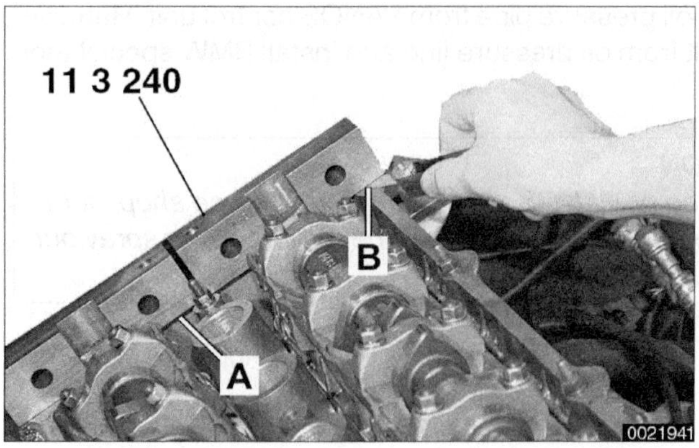


Place BMW special tool 11 3 240 over camshafts ends and measure clearance of tool to cylinder head surface.

***NOTE—***
- *If the exhaust side of the tool (gap* **A***) is not flush with the head, adjust camshaft timing as described below.*
- *Due to flexible sprocket design and VANOS control unit tolerances and play, the intake side of special tool 11 3 240 (gap* **B***) may be up to 1 mm (0.04 in) above the surface of the cylinder head. This is normal. Reassemble engine. Otherwise, adjust camshaft timing as described below.*

– Remove double VANOS control unit. See **Double VANOS control unit, removing (M52 TU, M54 engine)** in this repair group.

Press down on secondary chain tensioner and lock into place using BMW special tool 11 3 292 or equivalent pin or drift.

Loosen intake camshaft sprocket mounting nuts (**arrows**) by 1 turn.

Loosen exhaust camshaft sprocket mounting fasteners:
- Release bolts **A** ½ turn.
- Release nuts **B** 2 turns.

◀ Pull out intake camshaft splined shaft out so that approx. 1 mm (0.04 in) of splines (**arrows**) can be seen.

◀ Pull out exhaust camshaft splined shaft to stop.

– Make sure camshafts are secured in TDC position using BMW special tools 11 3 240 and 11 3 244.

◀ Make sure chain tensioning rail is preloaded.

| Tightening torque | |
|---|---|
| Primary chain tensioner preload | 0.7 Nm (6 in-lb) |

◀ Preload exhaust camshaft spring plate by pressing on impulse wheel. Tighten mounting nuts (**arrows**) by hand. Do not tighten fully.

Install BMW special tool 11 6 150 (VANOS setup bracket) to front of cylinder head timing case. Tighten nuts (**arrows**) by hand, and then tighten down uniformly until special tool contacts cylinder head firmly.

> ***CAUTION—***
> *Make sure all gasket material is removed from face of cylinder head. Clean sealing face and keep free of oil.*

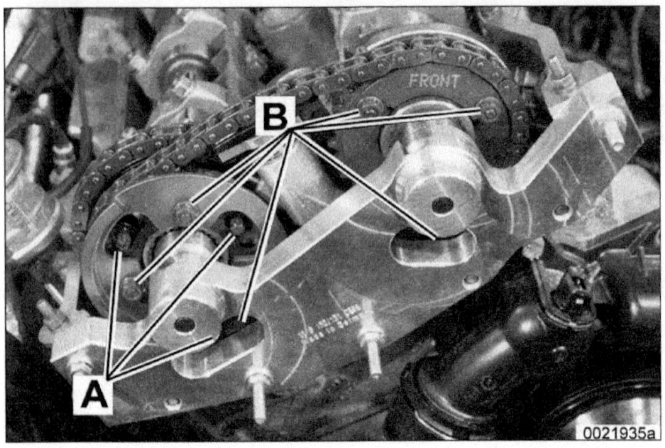


Secure camshaft impulse sprockets and wheels:

- Pretighten Torx screws (**A**) on exhaust camshaft impulse wheel to approx. 5 Nm (44 in-lb).
- Pretighten mounting nuts (**B**) on exhaust and intake sprocket assemblies to approx. 5 Nm (44 in-lb).
- Torque down Torx screws (**A**) and nuts (**B**) to final specifications.

| **Tightening torques** | |
|---|---|
| Sprocket assembly to camshaft initial torque | 5 Nm (44 in-lb) |
| Sprocket assembly to camshaft<br>• M7 Torx screw (**A**)<br>• M6 mounting nut (**B**) | <br>20 Nm (15 ft-lb)<br>10 Nm (7 ft-lb) |

– Remove flywheel locking tool from transmission bellhousing so that crankshaft is no longer locked.

– Remove camshaft locking tools from rear of cylinder head.

– Crank engine over twice by hand in direction of rotation until cylinder 1 intake and exhaust camshaft lobes face each other again.

– Secure crankshaft with BMW special tool 11 2 300.

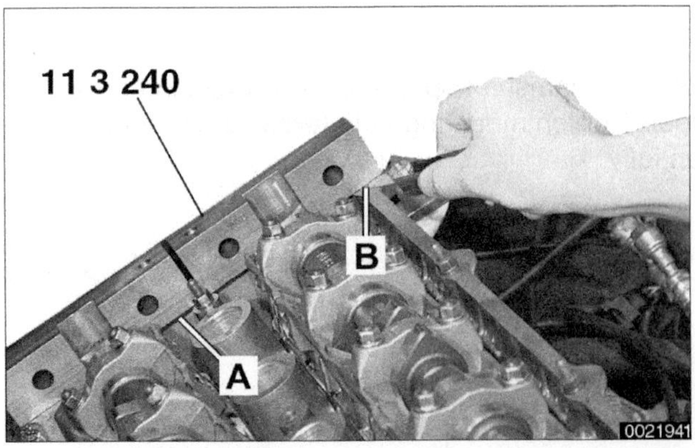


Place BMW special tool 11 3 240 over camshaft ends and measure clearance of tool to cylinder head surface.

***NOTE—***
- *If the exhaust side of the tool (gap **A**) is not flush with the head, camshaft timing is incorrect. Repeat camshaft timing procedure.*
- *Due to flexible sprocket design and VANOS control unit tolerances and play, the intake side of special tool 11 3 240 (gap **B**) may be up to 1 mm (0.04 in) above the surface of the cylinder head. This is normal. Otherwise, repeat camshaft timing procedure.*

– Remove camshaft locking tools from rear of camshafts.

– Remove BMW special tool 11 4 220 (adjustable primary chain tensioner). Reinstall primary chain tensioner.

| Tightening torque | |
|---|---|
| Primary chain tensioner to timing chain cover (M26 x 1.5) | 70 Nm (52 ft-lb) |

– Remove crankshaft locking tool from transmission bellhousing. Reinstall sealing plug.

– Remove VANOS setup bracket from front of cylinder head.

– Install VANOS control unit as described earlier in this group.

– Remove compressed air line adapter (BMW special tool 11 3 450) from VANOS control unit.

– Fit VANOS oil line banjo bolt with new seals. Attach oil line to VANOS control unit.

| Tightening torque | |
|---|---|
| Oil line to VANOS control unit (banjo bolt) | 32 Nm (24 ft-lb) |

– Remainder of installation is reverse of removal. Note the following:

- Install spark plugs.
- Install intake camshaft plastic baffle prior to installing cylinder head cover. Install ignition coil packs and connect electrical harness connectors.
- Install radiator cooling fan.
- Install top engine covers.

| Tightening torques | |
|---|---|
| Radiator cooling fan (viscous clutch) to coolant pump | 40 Nm (30 ft-lb) |
| Cylinder head cover to cylinder head | 10 Nm (7 ft-lb) |
| Spark plug to cylinder head | 25 Nm (18 ft-lb) |

# 119 Lubrication System



**TABLES**

## General

This repair group covers crankshaft seal replacement, oil pan removal and oil pump replacement. Some lubrication system troubleshooting information is provided as well.

In the procedures given in this repair group, engines are referred to by engine code. If necessary, see **100 Engine–General** for engine code and application information.

### Special tools

To access the oil pan, the engine needs to be properly supported while the front suspension subframe is lowered.

A standard oil pressure gauge may be used for measuring oil pressure.

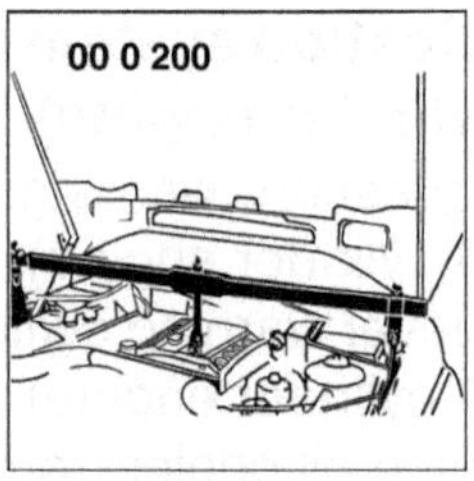


Engine support bracket
(BMW special tool set 00 0 200)

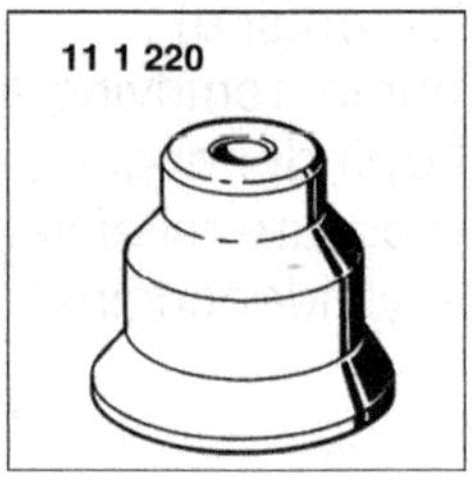


Crankshaft front seal guide, S54 engine
(BMW special tool 11 1 220)

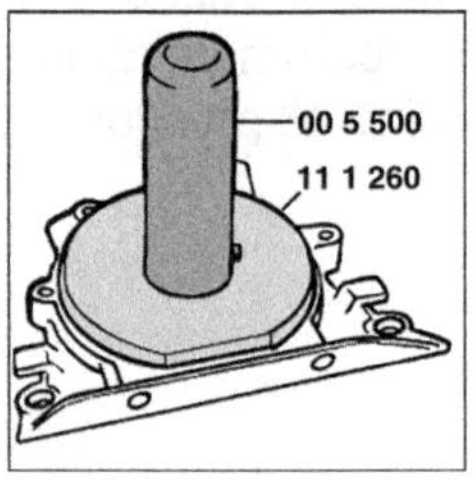


Crankshaft rear main seal driver, 6-cylinder engine
(BMW special tools 00 5 500, 11 1 260)

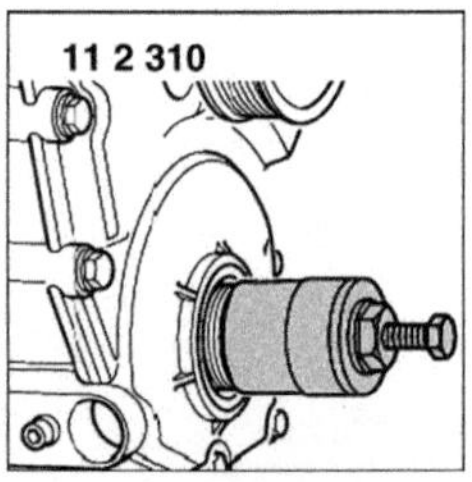


Crankshaft front seal extractor, 4-cylinder engine
(BMW special tool 11 2 310)

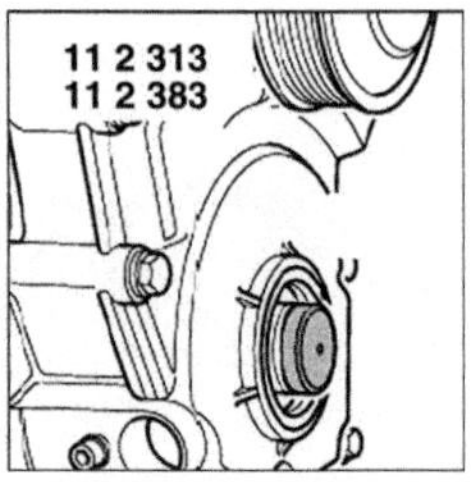


Crankshaft end protector
(BMW special tool 11 2 313 or 11 2 383)

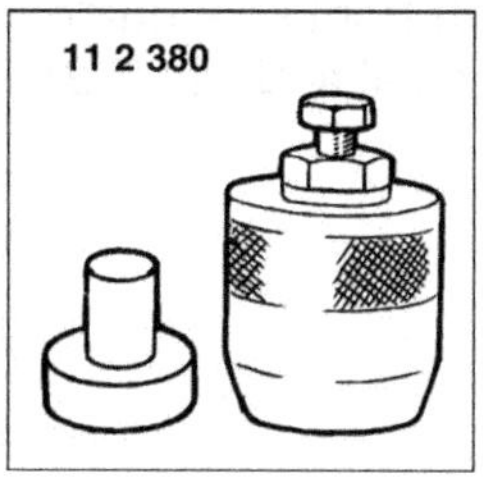


◀ Crankshaft front seal extractor, 6-cylinder engine (BMW special tool 11 2 380)

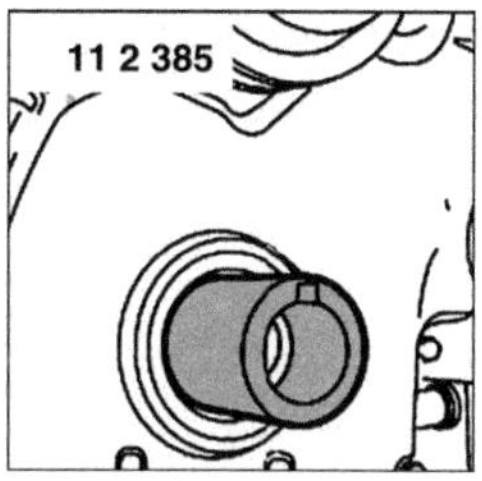


◀ Crankshaft front seal guide, 6-cylinder engine (BMW special tool 11 2 385)

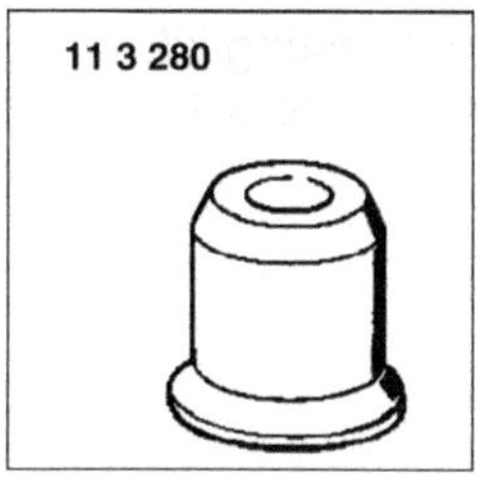


◀ Crankshaft front seal installer, 6-cylinder engine (BMW special tool 11 3 280)

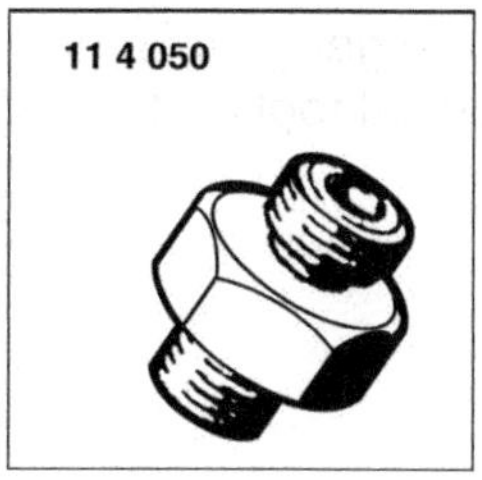


◀ Adapter for BMW oil pressure gauge (BMW special tool 11 4 050)

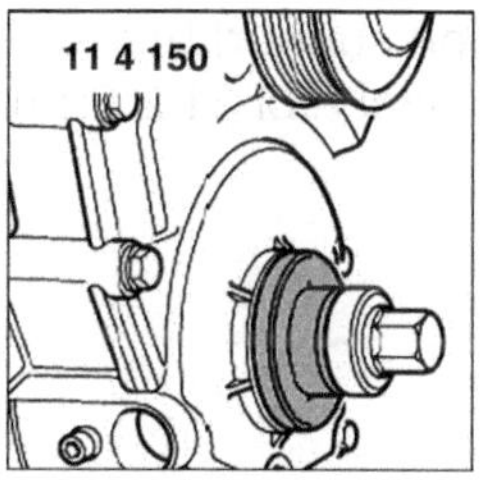


◀ Crankshaft front seal installer, 4-cylinder engine (BMW special tool 11 4 150)

◀ Adapter for BMW oil pressure gauge (BMW special tool 11 4 390)

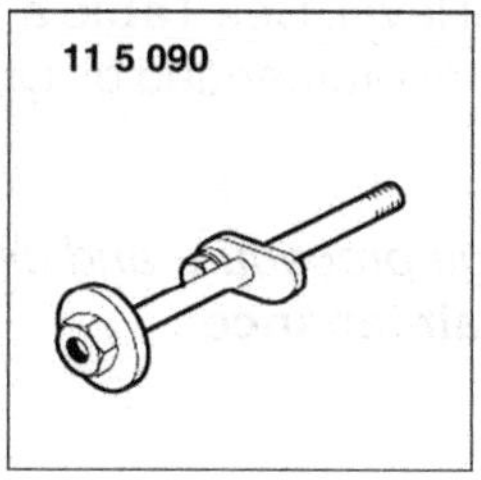


◀ Crankshaft front seal installation fixture, S54 engine- (BMW special tool 11 5 090)

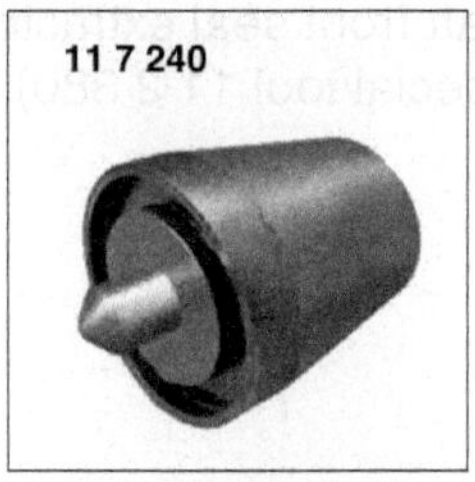


Crankshaft front seal guide, S54 engine (BMW special tools 11 7 240)

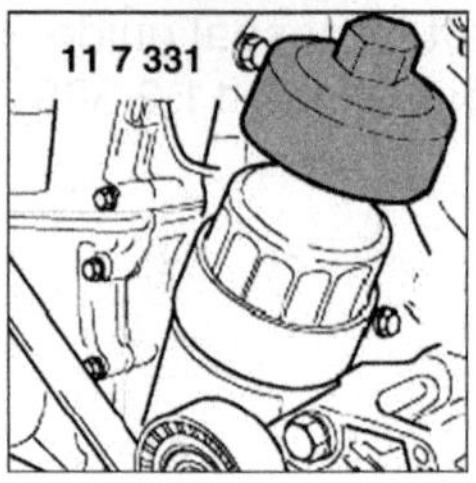


Oil filter cover removal tool (BMW special tool 11 7 331)

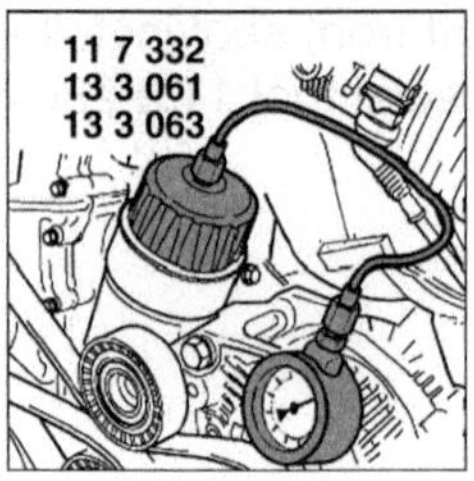


Oil pressure testing kit (BMW special tools 11 7 332, 13 3 061, 13 3 063)

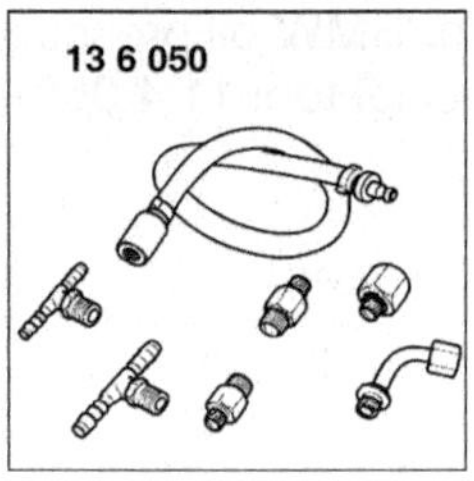


Oil pressure gauge fittings (BMW special tool kit 13 6 050)

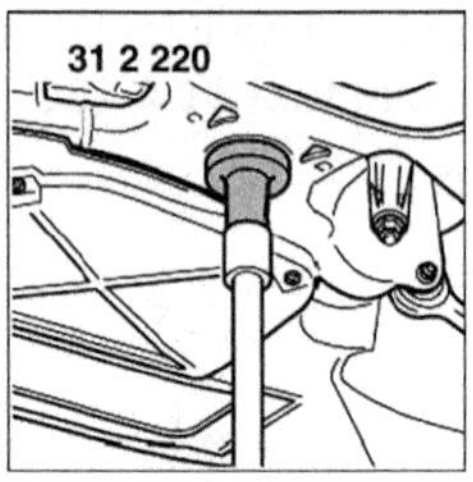


Suspension support bracket (BMW special tool 31 2 220)

## Engine lubrication

Engine oil pressure is generated by a pump which picks up oil from the oil sump. Oil is then pumped through a pressure relief valve and through oil galleys to the oil filter and engine components which need lubrication.

For engine longevity, use the correct type of oil and maintain correct oil level. See **Table a** for engine oil capacities and **Table b** for recommended oil types.

***NOTE—***
*Oil change procedure and oil filter replacement are covered in* **020 Maintenance**.

| Table a. Engine oil capacity (includes oil filter replacement) | |
|---|---|
| 4-cylinder: M44 engine | 5.0 liters (5.3 US qt) |
| 6-cylinder:<br>• M52, M52 TU or M54 engine<br>• S52 engine<br>• S54 engine | <br>6.5 liters (6.9 US qt)<br>6.0 liters (6.3 US qt)<br>5.5 liters (5.8 US qt) |

| Table b. Oil recommendation | |
|---|---|
| 2001 - 2002 M Roadster / Coupe (S54 engine) | Castrol RS 10W-60 (Castrol TWS Motorsport 10W-60) |
| All other models | BMW high performance synthetic 5W-30 |

Two different oil pump systems are used in Z3 and M version cars:

- **4-cylinder engine:** Oil pump is integral with the lower timing chain housing. The oil pump internal gears are driven directly by the crankshaft.
- **6-cylinder engine:** Oil pump is bolted to the bottom of the engine, inside the oil pan. It is chain driven off the front of the crankshaft.

## Engine lubrication warning system

The oil pressure warning system consists of an oil pressure switch mounted in the oil circuit and an instrument panel warning light. Other lubrication system safety features include:

- Dynamic oil level warning
- Engine oil temperature monitoring (S54 engine)
- Oil filter bypass to provide lubrication should the oil filter become clogged
- Oil pump pressure relief valve to prevent excessive system pressure

***NOTE—***
*The engine management system in 1999 - 2000 Z3 cars (DME MS 42.1) and 2001 - 2002 M version cars (DME MS S54) monitors engine oil temperature and signals the instrument cluster to illuminate warning LEDs in case of lubrication system or cooling system malfunction.*

## TROUBLESHOOTING

When the ignition is turned on, the oil pressure warning light comes on. When the engine is started and oil pressure rises slightly, the oil pressure switch opens and the warning light goes out.

> ***CAUTION—***
> *If the red oil pressure warning comes on or flashes on while driving, always assume that the oil pressure is low.*

Before starting to troubleshoot, make sure of the following:

- Fresh or clean oil in engine
- Correct oil level

### Oil pressure, checking

To perform an oil pressure test, BMW specifies special tools that attach to the top of the oil filter housing. See **Special tools** in this repair group.

It is possible to use standard automotive oil pressure testing equipment attached to the oil pressure switch port located at the base of the oil filter housing. Access to this port, however, is extremely restricted, specially in the 4-cylinder and the S54 engine.

If testing shows low oil pressure, one or more of the following conditions may be indicated:

- Worn or faulty oil pump or faulty pump pressure relief valve
- Worn or damaged engine bearings
- Severe engine wear

Any of these conditions indicates the need for major repairs.

#### Oil pressure test, BMW method

– M44, M52, M52 TU or M54 engine: Use BMW special tool 11 7 331 to loosen oil filter cover. Allow engine oil to drain back down into oil pan, then remove oil filter cover.

***NOTE—***
- *Thoroughly clean around oil filter cover before loosening.*
- *Be prepared to catch leaking oil with a shop towel.*

◄ S52 or S54 engine: Loosen center bolt (**arrow**) from oil filter cover. Allow engine oil to drain back down into oil pan, then remove long bolt from cover. Remove cover.

***NOTE—***
- *Thoroughly clean around oil filter cover bolt before loosening.*
- *Be prepared to catch leaking oil with a shop towel.*

– M44, M52, M52 TU or M54 engine:
  - Install BMW special tool 11 7 332 (oil filter cover adapter).
  - Connect BMW special tools 13 3 063 (connector hose) and 13 3 061 (pressure gauge) to oil filter cover adapter.

| Tightening torque | |
|---|---|
| BMW special tool 11 7 332 to oil filter housing | 25 Nm (18 ft-lb) |

– S52 or S54 engine:
  - Install adapter bolt, BMW special tool 11 4 390, in oil filter cover. Install cover and tighten adapter.
  - Connect BMW special tools 13 3 063 (connector hose) and 13 3 061 (pressure gauge) to adapter bolt.

– Start engine and allow to reach operating temperature. Check oil pressure both cold and hot. See **Table c**.

– When finished, remove pressure gauge and adapters.

– M44, M52, M52 TU or M4 engine: Reinstall oil filter cover. Make sure oil filter O-ring seals are seated properly.

| Tightening torque | |
|---|---|
| Oil filter cover to housing | 25 Nm (18 ft-lb) |

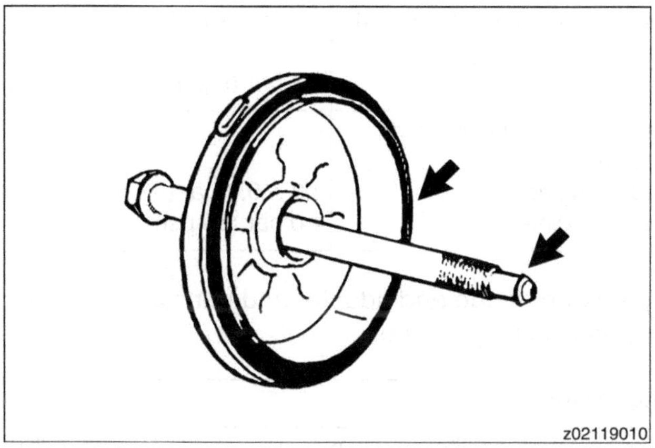
z02119010

◄ S52, S54 engine:
  - Check oil filter cover sealing O-rings (**arrows**). Replace if necessary.
  - Torque down filter cover.

| Tightening torque | |
|---|---|
| Oil filter cover bolt to housing | 33 Nm (24 ft-lb) |

– Check and top off engine oil as necessary.

### Oil pressure test, conventional method

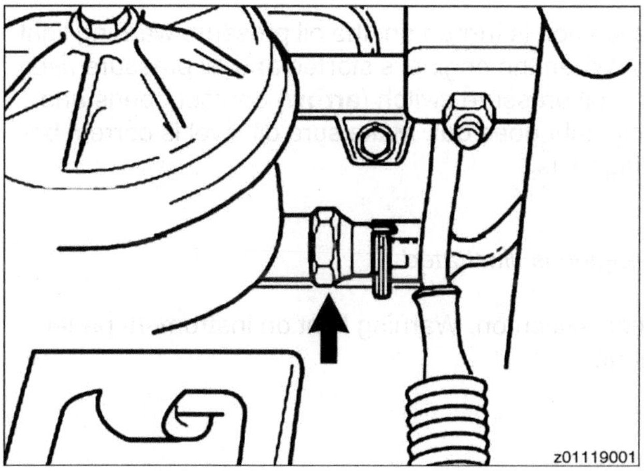
z01119001

◄ 4-cylinder engine: Oil pressure switch (**arrow**) is located at base of oil filter housing, underneath intake manifold.

– Remove engine air intake duct for access.

6-cylinder engine: Oil pressure switch (**arrow**) is located underneath intake manifold. (Intake manifold removed for purpose of illustration.)

- Loosen oil filter cover in order to allow engine oil to drain back down into oil pan.
- Disconnect electrical connector from oil pressure switch and remove switch.

***NOTE*—**
- *Thoroughly clean around the oil pressure switch before removing it.*
- *Be prepared to catch leaking oil with a shop towel.*

- Using adapter (BMW special tool 11 4 050) attach pressure gauge in place of oil pressure switch.
- Start engine and allow to reach operating temperature. Check oil pressure both cold and hot. See **Table c**.

| **Table c. Engine oil pressure** | |
|---|---|
| **Condition** | **Pressure** |
| Idle<br>• M44, M52, S52, M52 TU or M54 engine<br>• S54 engine | <br>0.5 bar (7 psi)<br><br>0.7 bar (9 psi) |
| Regulated oil pressure (elevated engine speed)<br>• 4-cylinder engine<br>• 6-cylinder engine | <br><br>4.3 ± 0.2 bar (63 ± 3 psi)<br>4.0 bar (59 psi) |

- Remove pressure gauge and adapter and reinstall pressure switch with new sealing washer.

| **Tightening torque** | |
|---|---|
| Oil pressure switch to oil filter housing | 27 Nm (20 ft-lb) |

## Oil pressure warning system, testing

When the ignition is turned on, the oil pressure warning light comes on. When the engine is started and oil pressure rises slightly, the oil pressure switch (**arrow**) contact opens and the warning light goes out. Make sure oil level is correct before starting tests.

***NOTE*—**
*M52 TU engine is illustrated.*

- Turn ignition switch on. Warning light on instrument panel must light up.

– Remove connector from oil pressure switch. Warning light on instrument panel must go out.

> ***CAUTION—***
> *Running the engine with the oil pressure switch disconnected may set a fault code (DTC).*

– If light does not go out, wiring to switch is most likely grounded somewhere between switch terminal and warning light. See **ELE Electrical Wiring Diagrams**.

**Oil pressure switch circuit**

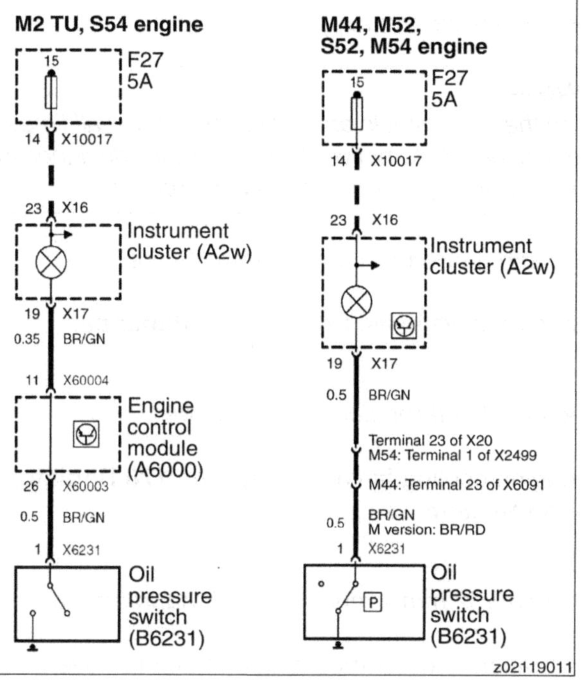


◀ The oil pressure warning switch circuit path depends on engine application.

– If warning light does not illuminate when ignition is on, remove connector from oil pressure switch and use a jumper wire to ground connector terminal to a clean metal surface. Oil pressure warning light should come on.

***NOTE—***
*If there are two wires to the oil pressure switch, ground the brown / green wire that carries the signal to the warning light.*

– If warning light comes on, check switch as described below. If warning light does not come on, wiring to instrument cluster or to light itself is faulty.

– To test switch, connect an ohmmeter between terminal in switch body and ground.

- With engine off, there should be continuity.
- With engine running, oil pressure should open switch and there should be no continuity.
- Replace a faulty switch.

> ***CAUTION—***
> *Keep in mind that low oil pressure may prevent the switch from turning the light out. If the light remains on while the engine is running, check oil pressure. See* **Oil pressure, checking** *in this repair group. Do not drive the car until the problem is corrected. The engine may be severely damaged.*

## CRANKSHAFT SEALS

Special BMW service tools are recommended to extract and to press in front or rear crankshaft seals. See **Special tools** in this repair group. Some commonly available seal removal and installation tools may be substituted. Read the procedures through before starting the work.

Crankshaft front seal replacement requires the removal of the engine drive belts and the vibration damper / crankshaft pulley assembly. For access reasons and to protect the radiator from damage, remove radiator before starting to work on the front seal. **See 170 Radiator and Cooling System.**

Crankshaft rear main seal replacement requires removal of the transmission and flywheel. See **230 Manual Transmission** or **240 Automatic Transmission.**

## Crankshaft front seal, replacing (4-cylinder engine)

The crankshaft front seal is located in the lower timing cover, behind the vibration damper.

– Raise front of vehicle and support safely.

> ***WARNING —***
> *Make sure the car is stable and well supported at all times. Use a professional automotive lift or jack stands designed for the purpose. A floor jack is not adequate support.*

– Remove splash guard from underneath engine.

– Remove engine drive belts. See **020 Maintenance.**

***NOTE —***
*Mark direction of belt rotation if reusing belts.*

– Remove engine cooling fan assembly. See **170 Radiator and Cooling System.**

***NOTE —***
*Cooling fan nut (32 mm wrench) has left-hand threads.*

– Remove vibration damper mounting bolt and remove vibration damper and pulley from crankshaft. See **117 Camshaft Timing Chains.**

> ***CAUTION—***
> *Do not use flywheel locking tool to hold crankshaft stationary to loosen or tighten vibration damper mounting bolt. Use the recommended special tools specified in* **117 Camshaft Timing Chains**, *or equivalent tools for counterholding the pulley.*

– Place protective cap (BMW special tool 11 2 313) on end of crankshaft.

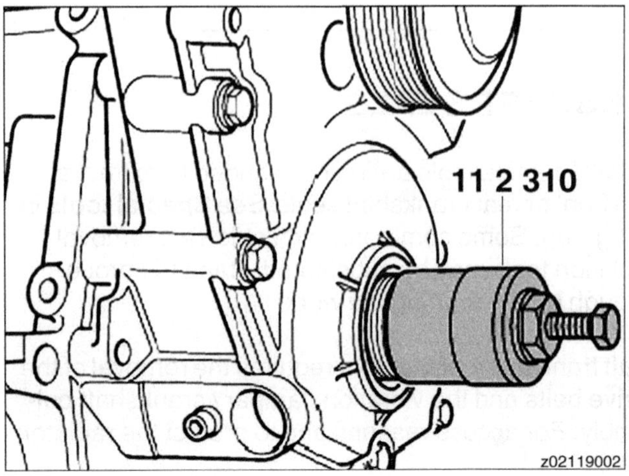


◄ Place oil seal puller (BMW special tool 11 2 310) over end of crankshaft and screw in body of puller until it contacts seal. Tighten center bolt to draw seal out of cover.

– Coat new seal with oil and position in timing chain cover. Use BMW special tool 11 4 150 and vibration damper mounting bolt to draw seal in flush with timing case cover.

– Remount vibration damper using new bolt. Counterhold damper with appropriate tools while tightening mounting bolt. See **113 Camshaft Timing Chains**.

> ***CAUTION—***
> - *Inspect crankshaft woodruff key for damage before installing vibration damper.*
> - *Do not use flywheel locking tool to hold crankshaft stationary to tighten vibration damper mounting bolt.*

| Tightening torque | |
|---|---|
| Vibration damper to crankshaft (M16 x 1.5) (replace bolt) | 330 Nm (244 ft-lb) |

– Installation is reverse of removal, noting the following:
- Install drive belts. See **020 Maintenance**.
- Install cooling fan. See **170 Radiator and Cooling System**.

## Crankshaft front seal, replacing (M52, S52, M52 TU or M54 engine)

The crankshaft front seal is located in the lower timing cover, behind the vibration damper.

– Raise front of vehicle and support safely.

> ***WARNING—***
> *Make sure the car is stable and well supported at all times. Use a professional automotive lift or jack stands designed for the purpose. A floor jack is not adequate support.*

– Remove splash guard from underneath engine.

– Remove engine drive belts. See **020 Maintenance**.

***NOTE—***
*Mark direction of belt rotation if reusing belts.*

– Remove engine cooling fan assembly. See **170 Radiator and Cooling System**.

***NOTE—***
*Cooling fan nut (32 mm wrench) has left-hand threads.*

– M52 or S52 engine (2-piece vibration damper assembly): Remove vibration damper mounting bolts and separate damper from crankshaft hub.

– Using special tools to counterhold crankshaft hub (M52, S52 engine) or vibration damper (M52 TU, M54 engine), remove mounting bolt from crankshaft. Remove hub or damper from crankshaft. See **117 Camshaft Timing Chains**.

> ***CAUTION—***
> *Do not use flywheel locking tool to hold crankshaft stationary to loosen or tighten vibration damper mounting bolt. Use the recommended special tools specified in* **117 Camshaft Timing Chains**, *or equivalent tools for counterholding the pulley.*

– Place protective cap (BMW special tool 11 2 383) on end of crankshaft. Mount seal guide (BMW special tool 11 2 385) over end of crankshaft. Be sure to align groove with keyway on crankshaft.

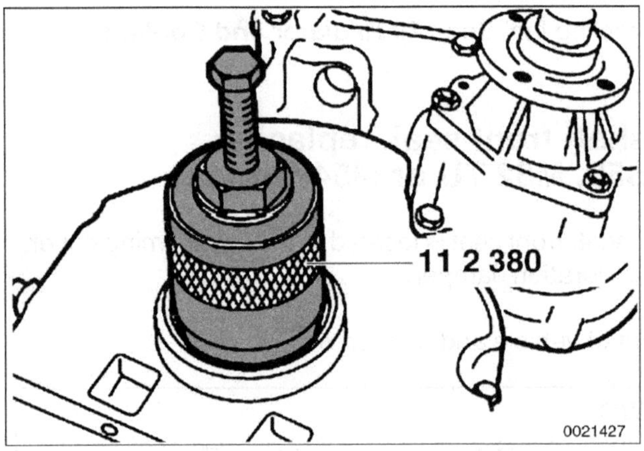


◄ Place oil seal puller (BMW special tool 11 2 380) and screw in body of puller until it contacts seal. Tighten screw in center of tool to draw seal out of cover.

– Coat new seal with oil and position in timing chain cover. Use BMW special tool 11 3 280 and crankshaft center bolt to draw seal in flush with timing case cover.

– Reinstall crankshaft hub or vibration damper using new mounting bolt. Counterhold with appropriate tools while tightening mounting bolt. See **113 Camshaft Timing Chains**.

> ***CAUTION—***
> - *Inspect crankshaft woodruff key for damage before installing hub or vibration damper.*
> - *Do not use flywheel locking tool to hold crankshaft stationary to tighten mounting bolt.*

| **Tightening torque** | |
|---|---|
| Vibration damper or hub to crankshaft (replace bolt) | 410 Nm (302 ft-lb) |

– Installation is reverse of removal, noting the following:

- Install drive belts. See **020 Maintenance**.
- Install cooling fan. See **170 Radiator and Cooling System**.

## Crankshaft front seal, replacing (S54 engine)

The crankshaft front seal is located in the lower timing cover, behind the vibration damper.

– Raise front of vehicle and support safely.

> ***WARNING —***
> *Make sure the car is stable and well supported at all times. Use a professional automotive lift or jack stands designed for the purpose. A floor jack is not adequate support.*

– Remove splash guard from underneath engine.

– Remove engine drive belts. See **020 Maintenance**.

***NOTE —***
*Mark direction of belt rotation if reusing belts.*

– Remove engine cooling fan assembly. See **170 Radiator and Cooling System**.

***NOTE —***
*Cooling fan nut (32 mm wrench) has left-hand threads.*

– Remove TDC bracket. Separate crankshaft pulley from vibration damper. Using special tools to counterhold crankshaft hub, remove vibration damper mounting bolts from crankshaft. Remove damper from crankshaft hub. See **117 Camshaft Timing Chains**.

> ***CAUTION—***
> *Do not use flywheel locking tool to hold crankshaft stationary to loosen or tighten vibration damper mounting bolt. Use the recommended special tools specified in* **117 Camshaft Timing Chains**, *or equivalent tools for counterholding the pulley.*

◄ Place BMW special tool 11 2 384 into crankshaft center.

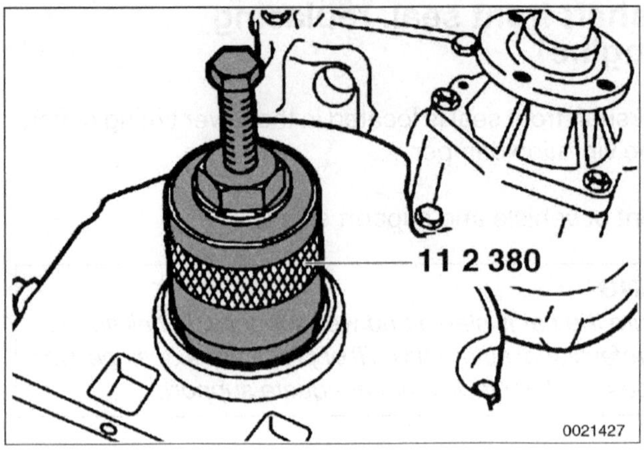


Install BMW special tool 11 2 380 over crankshaft, and screw in until it makes firm contact with seal. Tighten screw in end of
11 2 380 to draw out seal.

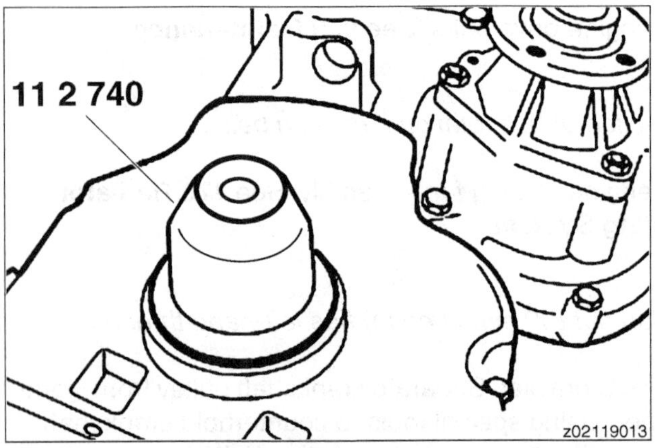


Coat new seal with oil. Place BMW special tool 11 2 740 on crankshaft to protect seal during installation. Slide on new seal until it is seated on timing chain cover. Remove special tool 11 2 740.

> ***CAUTION—***
> *Do not touch sealing lips on new seal with your fingers.*

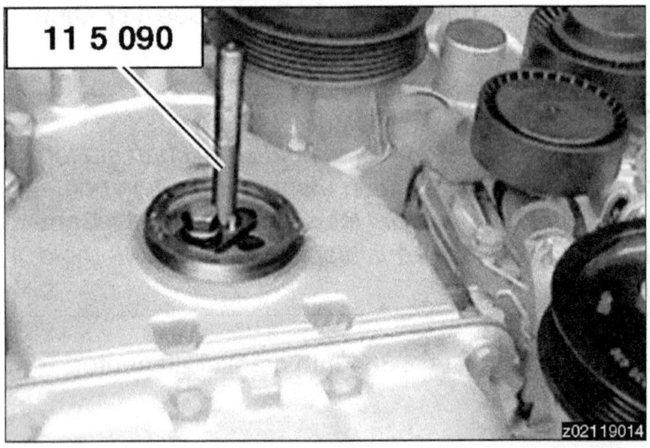


Fasten special tool 11 5 090 to crankshaft.

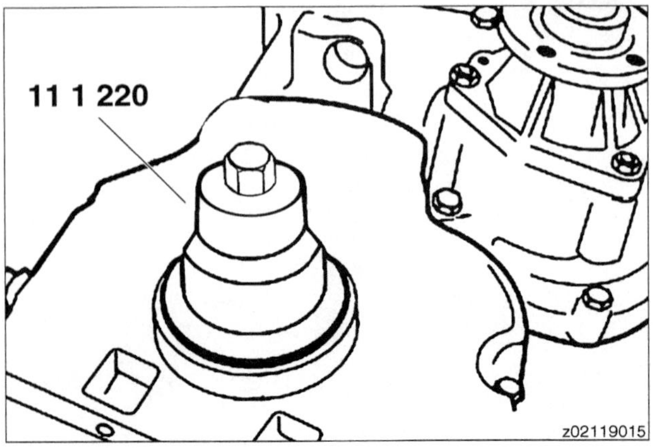


Install special tool 11 1 220 and draw seal in flush with timing case cover.

– Install vibration damper to crankshaft, making sure to align locating hole in damper to dowel in crankshaft hub. Install new bolts finger tight.

> ***CAUTION—***
> *Use new stretch bolts when installing vibration damper.*

– Torque vibration damper mounting bolts while counterholding crankshaft with special tool. See **117 Camshaft Timing Chains**.

| Tightening torques | |
|---|---|
| Vibration damper to crankshaft (use new bolts):<br>• Stage 1<br>• Stage 2 (angle torque)<br>• Stage 3 (angle torque) | <br><br>60 Nm (44 ft-lb)<br>50°<br>50° |

– Install pulley and TDC bracket. Make sure dowels for TDC bracket are correctly located in timing chain cover.

| Tightening torques | |
|---|---|
| Pulley to vibration damper (M8)<br>• 10.9 grade<br>• 8.8 grade | <br>34 Nm (25 ft-lb)<br>22 Nm (16 ft-lb) |

## Crankshaft rear main seal, replacing

The crankshaft rear main seal is located behind the flywheel at the rear end of the engine.

– Remove transmission. See **230 Manual Transmission** or **240 Automatic Transmission**.

– Manual transmission vehicle: Remove clutch pressure plate and disc. See **210 Clutch**.

– Remove flywheel or torque plate.

– Drain engine oil.

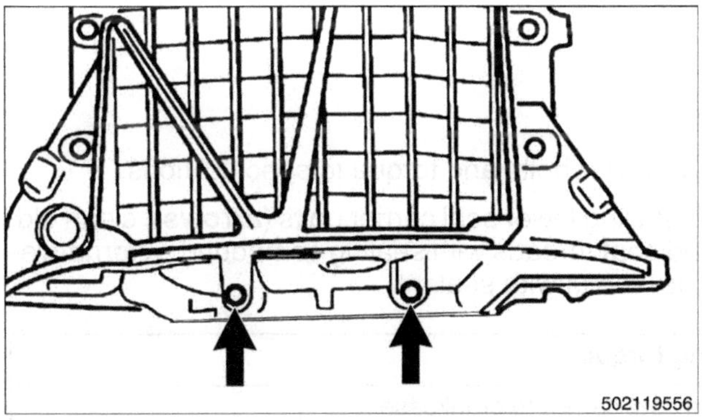

502119556

◄ Remove rear main seal carrier bolts (**arrows**) from oil pan. Loosen all oil pan bolts.

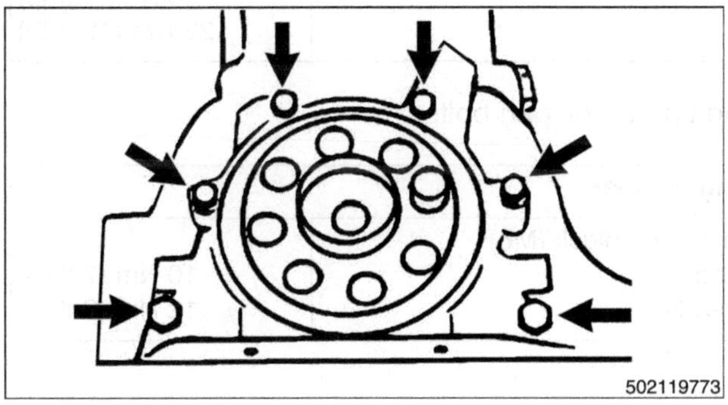

502119773

◄ Remove remaining seal carrier bolts (**arrows**) and carefully remove seal carrier without damaging oil pan gasket.

***CAUTION—***
*After removing seal carrier, check oil pan gasket for damage. If gasket is damaged during removal, remove oil pan and replace gasket. See* **Engine oil pan removal and installation** *in this repair group.*

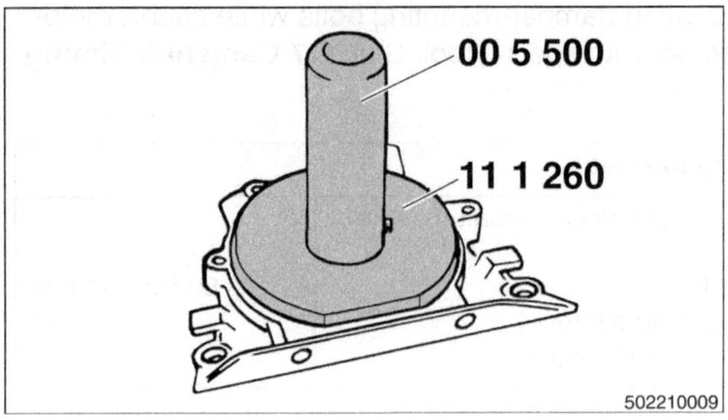

502210009

Remove old seal and install new seal into seal carrier using BMW special tools 11 1 260 and 00 5 500.

*NOTE—*
- *A redesigned rear main seal is used starting with 4 / 1998 production, and is the only version available from BMW as a replacement. The new seal no longer uses a spring to support the sealing lip. Do not kink or damage the sealing lip. Do not touch with fingers.*
- *BMW offers the rear main seal carrier and seal in a kit. A plastic support bushing, required for sliding the seal over the crankshaft, is included with the kit.*

***CAUTION—***
*Do not use BMW special tool 11 2 213 (installation guide) with the new style seal. Use only the special tool supplied with the seal kit.*

– Before reinstalling seal carrier to crankcase:
- Check locating dowels for damage.
- Replace gasket between rear main seal carrier and crankcase.
- Apply thin coat of 3-Bond® 1209 or equivalent sealant to oil pan gasket sealing surfaces.
- Lubricate crankshaft where it contacts rear main seal.

Z02119003

Using plastic support bushing (**arrow**) that comes with seal kit as an installation guide, fit seal over end of crankshaft. Use care to push seal carrier and support bushing on straight, without tilting sideways. Remove support bushing.

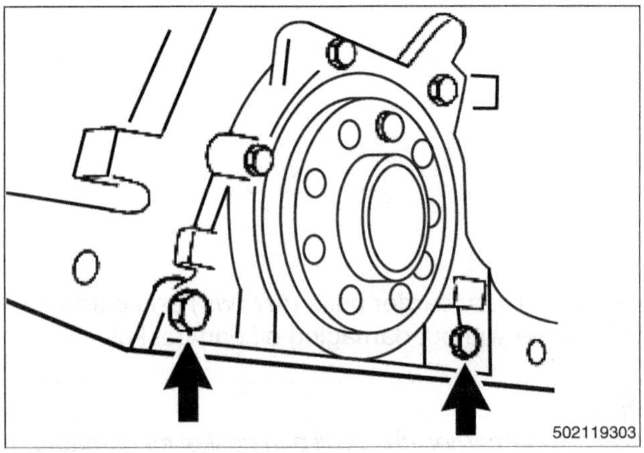
502119303

Install seal carrier bolts and torque to specifications.
- When installing lower seal carrier bolts (**arrows**), clean old bolts and coat threads with sealing compound. Alternatively, use new bolts with sealant.

| **Tightening torques** | |
|---|---|
| Rear main seal carrier to crankcase<br>• M6<br>• M8 | <br>10 Nm (7 ft-lb)<br>22 Nm (16 ft-lb) |

– Install and tighten oil pan bolts.

| **Tightening torques** | |
|---|---|
| Oil pan to engine block (M6)<br>• 8.8 grade<br>• 10.9 grade | <br>10 Nm (7 ft-lb)<br>12 Nm (9 ft-lb) |

– Remainder of assembly is reverse of disassembly.
  - Install flywheel, clutch or torque converter and transmission.
  - Fill engine with oil.
  - Run engine and check for leaks.

| Tightening torques | |
|---|---|
| Flywheel or torque plate to crankshaft (replace bolts) | 120 Nm (90 ft-lb) |
| Oil drain plug to oil pan (M12) | 25 Nm (18 ft-lb) |

## ENGINE OIL PAN REMOVAL AND INSTALLATION

In all models, engine oil pan removal involves suspending the engine from above and lowering the front subframe. Be sure to read the procedures through before starting work.

### Oil pan, removing and installing (4-cylinder engine)

– Place steering wheel in straight ahead position. Turn off ignition and remove ignition key.

– Raise engine hood and secure in service position. See **410 Fenders, Engine Hood**.

– Raise front of vehicle and support safely.

> ***WARNING—***
> *Make sure the car is stable and well supported at all times. Use a professional automotive lift or jack stands designed for the purpose. A floor jack is not adequate support.*

– Remove splash guard from underneath engine.

– Drain engine oil. See **020 Maintenance**.

– Detach vacuum hose adapter from vacuum brake booster at rear of engine compartment.

◄ Remove oil dipstick guide tube mounting nut (**arrow**) and pull guide tube from oil pan.

***NOTE—***
*The guide tube is sealed in the pan with an O-ring. Check that the O-ring comes out with the tube. Use a new O-ring when installing the tube.*

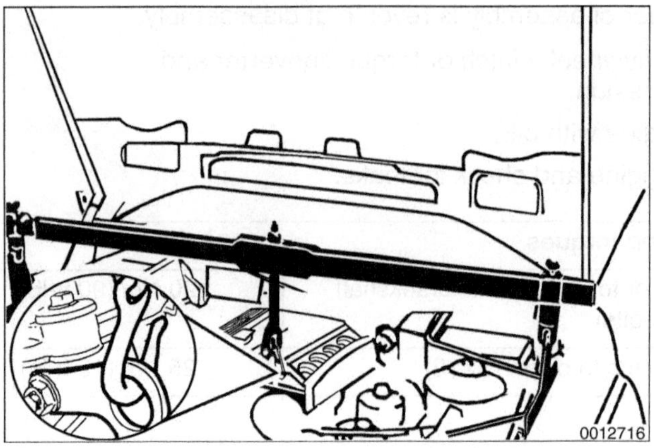
0012716

Install engine support equipment (BMW special tool set 00 0 200 or equivalent) at front engine lifting point and lift approximately 5 mm (¼ inch) until engine weight is supported.

0022004

Working underneath car, remove pinch bolt (**arrow**) and separate steering column shaft from steering rack at universal joint. Matchmark steering column shaft joint to steering rack spindle. See **320 Steering and Wheel Alignment**.

> ***CAUTION—***
> *In order to avoid the need for front-end realignment, do not unbolt power steering rack from front subframe.*

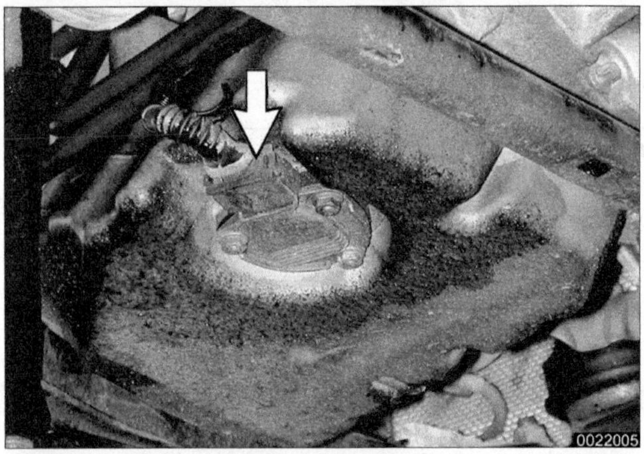
0022005

Working underneath engine oil pan, remove oil level warning switch (**arrow**) located at bottom of oil pan.

***NOTE—***
*Be prepared to catch dripping oil.*

0011960

Working underneath front of car, at left and right sides:

- Remove control arm bushing carrier bolts (**A**) from body.
- If applicable: Remove stabilizer bar link fastener (**B**) and detach link from control arm.
- Remove front crossbrace mounting bolts (**C**). Place brace aside.

– Support front subframe from below using BMW special tool 31 2 220 or equivalent on appropriate jacking equipment.

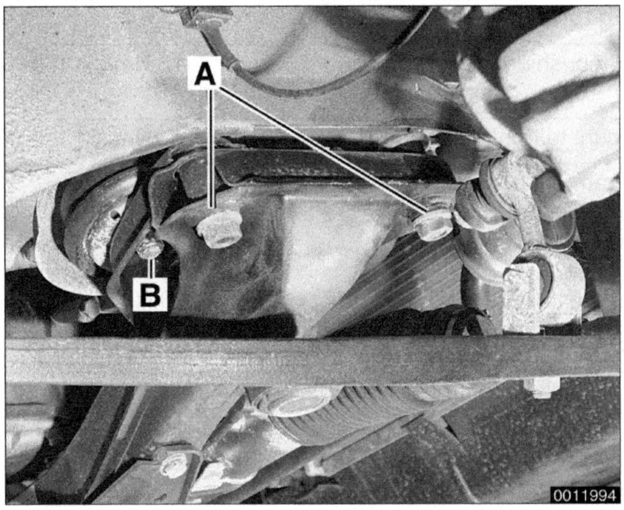


◀ Working at front suspension subframe:

- Remove subframe mounting bolts (**A**) from left and right sides.
- Remove lower engine mount nuts (**B**).
- Detach left and right stabilizer bar anchors from frame rails.
- Unbolt and detach control arms from front subframe. Push control arms to side.
- Lower subframe as far as possible.

– Remove clamping brackets holding fuel lines to oil pan.

– Automatic transmission model: Detach ATF cooler line brackets from oil pan.

– Remove oil pan screws. Lower and remove oil pan toward rear.

> ***CAUTION—***
> *Never pry the oil pan loose.*

***NOTE—***
*If the oil pan does not separate easily from the engine cylinder block, make sure all mounting bolts have been removed. If necessary, a few taps with a rubber mallet should break it free.*

– Thoroughly clean all old gasket material from mating surfaces. Install a new oil pan gasket.

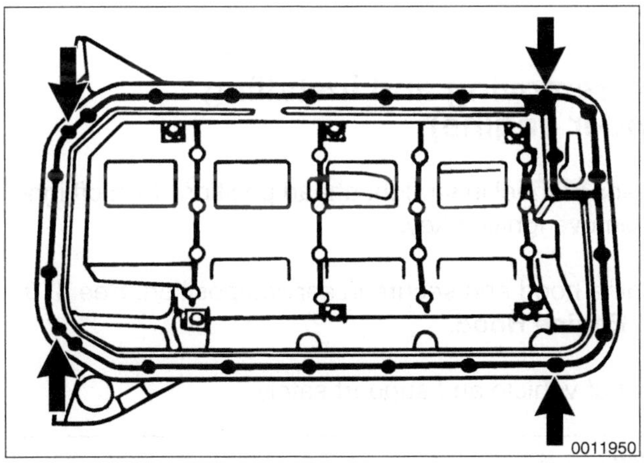


◀ Apply a bead of non-hardening sealer (3-Bond® 1209 or equivalent) to oil pan gasket directly below joints for end cover and front timing case cover (**arrows**). Make sure bead of sealant is 3 mm wide by 2 mm high.

– When installing oil pan to engine, tighten forward screws first, then tighten rear.

| Tightening torques | |
|---|---|
| Oil pan to engine block (M6)<br>• 8.8 grade<br>• 10.9 grade | <br>10 Nm (7 ft-lb)<br>12 Nm (9 ft-lb) |

– Remainder of installation is reverse of removal.

> ***WARNING—***
> *Use new bolts when mounting the subframe to the body.*

***NOTE—***
*The oil pickup tube is attached to the oil pan using self-tapping screws. For this reason no threading for the pickup is present on the new oil pan. Replace the pickup O-ring seal and use old screws to reattach the pickup to the pan.*

| Tightening torques | |
|---|---|
| Control arm bushing carrier to body (M10 bolt) | 42 Nm (31 ft-lb) |
| Engine mount to front subframe (M10 nut) | 42 Nm (31 ft-lb) |
| Front subframe to body rail (M12, replace bolts)<br>• 8.8 grade<br>• 10.9 grade<br>• 12.9 grade | <br><br>77 Nm (57 ft-lb)<br>105 Nm (77 ft-lb)<br>110 Nm (81 ft-lb) |
| Oil drain plug to oil pan (M12) | 25 Nm (18 ft-lb) |
| Stabilizer bar link to control arm (M10) | 42 Nm (31 lb-ft) |

– Center steering spindle to steering rack before installing steering column shaft. See **320 Steering and Wheel Alignment** for specific installation markings and procedures.

| Tightening torques | |
|---|---|
| Steering column universal joint to steering rack spindle (M8, replace pinch bolt) | 19 Nm (14 ft-lb) |

– Fill engine with oil as described in **020 Maintenance**. After adding engine oil, start and run engine. Raise engine speed to 2,500 rpm until oil pressure warning light goes out (about 5 seconds).

***NOTE—***
- *See* **Table a** *and* **Table b** *in this repair group for engine oil capacities and specifications.*
- *BMW does not specify a front end alignment following this procedure.*

## Oil pan, removing and installing (6-cylinder engine)

– Place steering wheel in straight ahead position. Turn off ignition and remove ignition key.

– Raise engine hood and secure in service position. See **410 Fenders, Engine Hood**.

– Raise front of vehicle and support safely.

> ***WARNING—***
> *Make sure the car is stable and well supported at all times. Use a professional automotive lift or jack stands designed for the purpose. A floor jack is not adequate support.*

– Remove splash guard from underneath engine.

– Drain engine oil. See **020 Maintenance**.

– Remove air filter housing and ducts. See **130 Fuel Injection**.

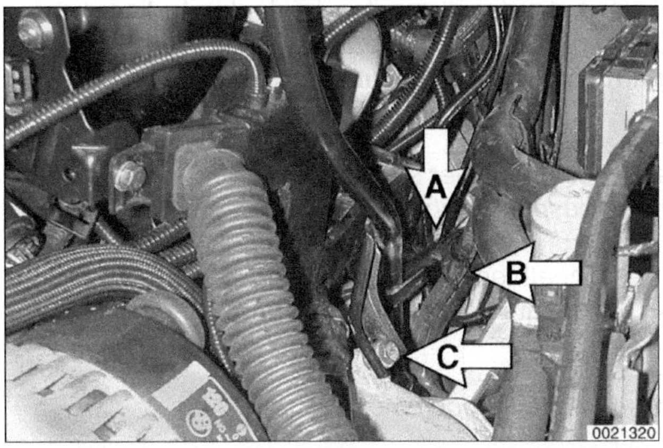


M52, S52, M2 TU or M54 engine: Remove oil dipstick guide tube:

- Disconnect fuel lines and wiring harness from bracket (**A**).
- Disconnect oil separator hose (**B**) from base of guide tube.
- Remove mounting bolt (**C**) from left engine mounting bracket and remove tube from oil pan.

***NOTE—***
*The dipstick guide tube is sealed in the block using an O-ring. Check that the O-ring comes out with the tube.*

S54 engine: Working at left side of engine:

- Remove dipstick guide tube mounting nut (**arrow**) at side of intake manifold.
- Remove dipstick guide tube bracket mounting bolt at top of oil pan. Separate guide tube from oil pan.
- Disconnect oil separator hose at top of oil pan (quick-disconnect fitting).
- Unclip wiring harness secured to top of oil pan.

***NOTE—***
*The dipstick guide tube is sealed in the block using an O-ring. Check that the O-ring comes out with the tube.*

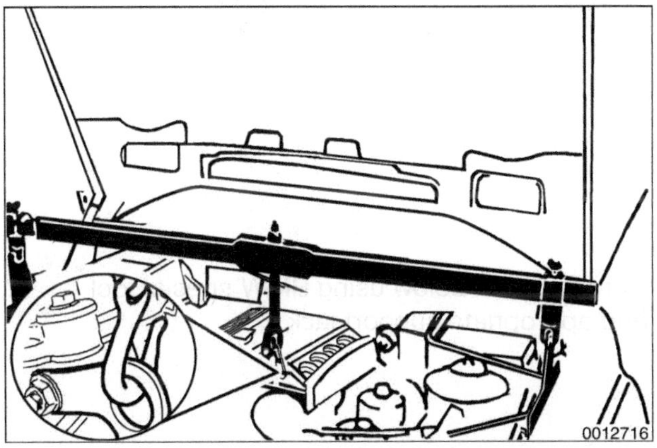


Install engine support equipment (BMW special tool set 00 0 200 or equivalent) at front engine lifting point and lift approximately 5 mm (¼ inch) until engine weight is supported.

Working underneath car, remove pinch bolt (**arrow**) and separate steering column shaft from steering rack at universal joint. Matchmark steering column shaft joint to steering rack spindle. See **320 Steering and Wheel Alignment**.

***CAUTION—***
*In order to avoid the need for front-end realignment, do not unbolt power steering rack from front subframe.*

– Remove power steering pump, leaving hydraulic lines connected to pump. Suspend pump using strong wire. See **320 Steering and Wheel Alignment**.

***NOTE—***
*For better access to power steering assembly, unbolt fluid reservoir and suspend with wire.*

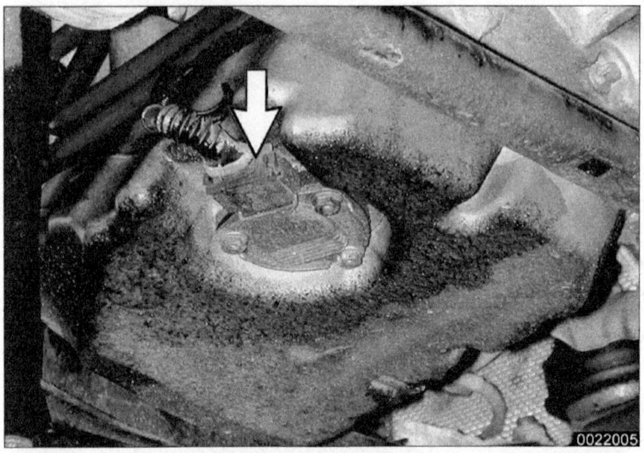


< Working underneath engine oil pan, remove oil level warning switch (**arrow**) located at bottom of oil pan.

***NOTE—***
*Be prepared to catch dripping oil.*

< Working underneath front of car, at left and right sides:

- Remove control arm bushing carrier bolts (**A**) from body.
- If applicable: Remove stabilizer bar link fastener (**B**) and detach link from control arm.
- Remove front crossbrace mounting bolts (**C**). Place brace aside.

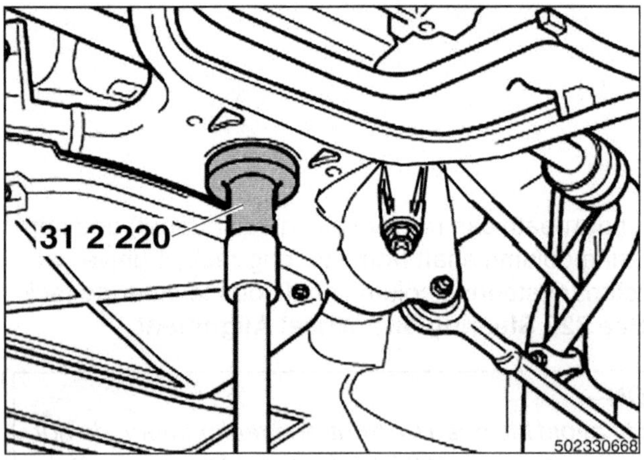


< Support subframe from below using BMW special tool 31 2 220 and appropriate support jack.

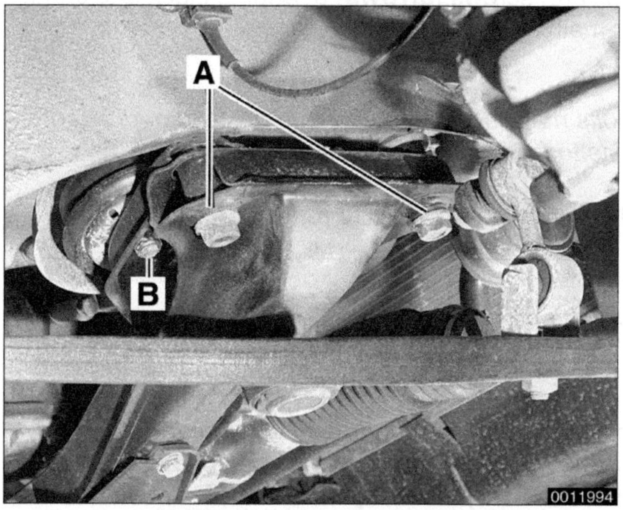


◄ Working at front suspension subframe:

- Remove subframe mounting bolts (**A**) from left and right sides.
- Remove lower engine mount nuts (**B**).
- Detach left and right stabilizer bar anchors from frame rails.
- Unbolt and detach control arms from front subframe. Push control arms to side.
- Lower subframe as far as possible.

– Remove clamping brackets holding fuel lines to oil pan.

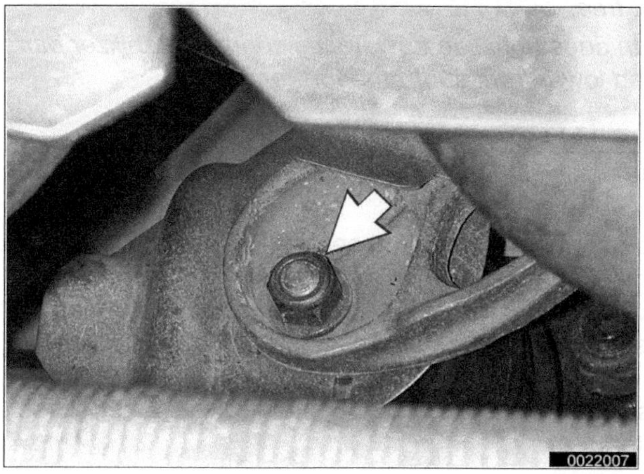


◄ Loosen upper engine mount nuts (**arrow**) on right and left.

– Automatic transmission model: Detach ATF cooler line brackets from engine oil pan.

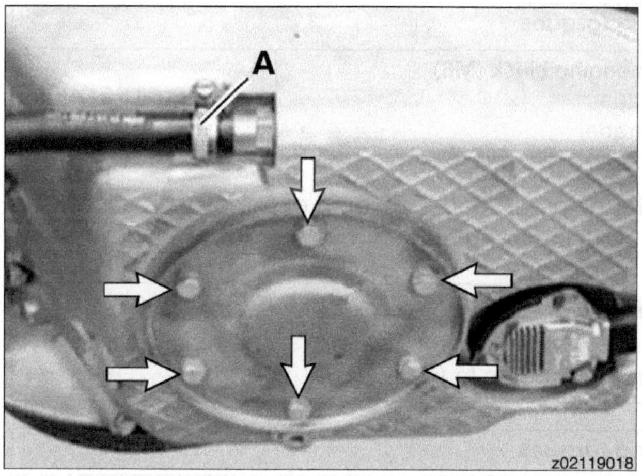


◄ S54 engine: working underneath oil pan:

- Detach return oil line (**A**) from oil pan.
- Remove sump cover fasteners (**arrows**). Remove cover.

***NOTE—***
*Be prepared to catch dripping oil.*

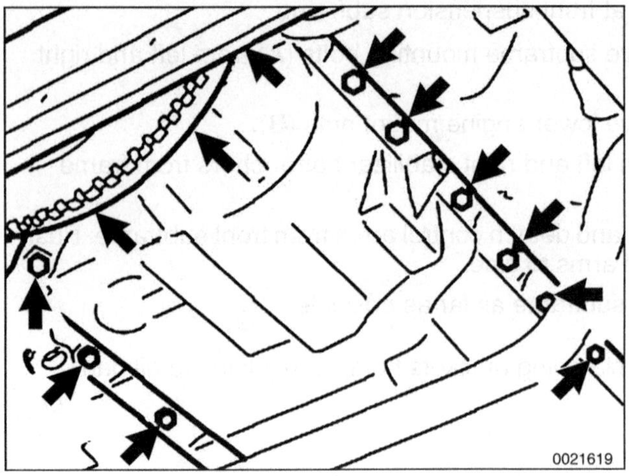


- Remove oil pan screws (**arrows**) at engine block and crankshaft rear main seal carrier.

*NOTE—*
*Oil pan bolts may vary in length. Note position of different length bolts to facilitate reinstallation.*

- Lower oil pan and pull forward to remove.

> ***CAUTION—***
> *Never pry the oil pan loose.*

*NOTE—*
- *If the oil pan does not separate easily from the engine cylinder block, make sure all mounting bolts have been removed. If necessary, a few taps with a rubber mallet should break it free.*
- *If oil pan does not slide forward, disconnect stabilizer bar links and lower front subframe further.*

- Thoroughly clean all old gasket material from mating surfaces. Install a new oil pan gasket.

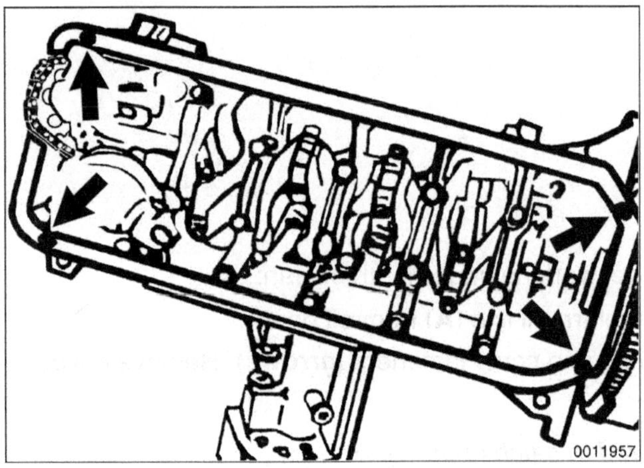


- Apply a bead of non-hardening sealer (3-Bond® 1209 or equivalent) to oil pan gasket directly below joints for end cover and front timing case cover (**arrows**). Make sure bead of sealant is 3 mm wide by 2 mm high.

- When installing oil pan to engine, tighten forward screws first, then tighten rear.

| **Tightening torques** | |
|---|---|
| Oil pan to engine block (M6)<br>• 8.8 grade<br>• 10.9 grade | <br>10 Nm (7 ft-lb)<br>12 Nm (9 ft-lb) |

- Remainder of installation is reverse of removal. Be sure to use new sealing O-ring at engine dipstick tube.

> ***WARNING—***
> *Always use new self-locking fasteners when mounting suspension components to the body.*

| **Tightening torques** | |
|---|---|
| Control arm bushing carrier to body (M10 bolt) | 42 Nm (31 ft-lb) |
| Engine mount to front subframe (M10 nut) | 42 Nm (31 ft-lb) |
| Front subframe to body rail (M12, replace bolts)<br>• 8.8 grade<br>• 10.9 grade<br>• 12.9 grade | <br><br>77 Nm (57 ft-lb)<br>105 Nm (77 ft-lb)<br>110 Nm (81 ft-lb) |

| Tightening torques | |
|---|---|
| Oil drain plug to oil pan (M12) | 25 Nm (18 ft-lb) |
| Stabilizer bar link to control arm (M10) | 42 Nm (31 ft-lb) |

– Center steering spindle to steering rack before installing steering column shaft. See **320 Steering and Wheel Alignment** for specific installation markings and procedures.

| Tightening torques | |
|---|---|
| Steering column universal joint to steering rack spindle (M8, replace pinch bolt) | 19 Nm (14 ft-lb) |

– Fill engine with oil as described in **020 Maintenance**. After adding engine oil, start and run engine. Raise engine speed to 2,500 rpm until oil pressure warning light goes out (about 5 seconds).

***NOTE—***
- *See* **Table a** *and* **Table b** *in this repair group for engine oil capacities and specifications.*
- *S54 engine: A small quantity of oil, which otherwise would stay inside the oil pan during regular oil and filter service, is drained when the pan is removed. Compensate for this with an additional 1.0 liter / 1.1 US qt of oil (for a total of 6.5 liters / 6.9 US qt) when refilling.*
- *BMW does not specify a front end alignment following this procedure.*

## LUBRICATION SYSTEM COMPONENT REPLACEMENT (4-CYLINDER ENGINE)

### Oil pressure warning switch, replacing (4-cylinder engine)

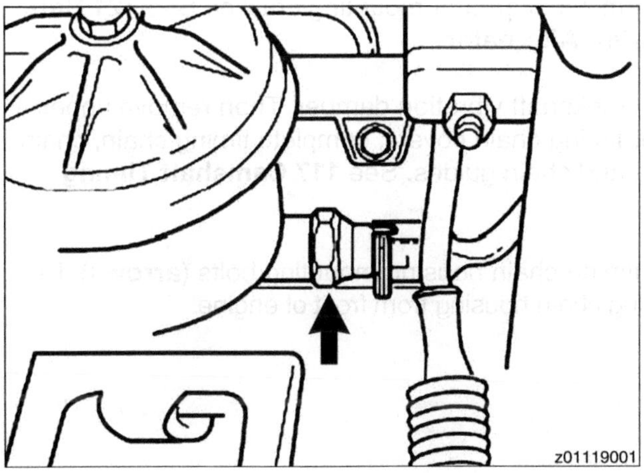
z01119001

◄ Oil pressure switch (**arrow**) is located at base of oil filter housing, underneath intake manifold.

– Remove engine air intake duct for access.

– Unscrew oil filter cover in order to allow engine oil to drain back down into oil pan.

– Disconnect electrical connector from oil pressure switch and remove switch.

***NOTE—***
- *Thoroughly clean around the oil pressure switch before removing it.*
- *Be prepared to catch leaking oil with a shop towel.*

– Install new pressure switch with new sealing washer.

| Tightening torque | |
|---|---|
| Oil pressure switch to oil filter housing | 27 Nm (20 ft-lb) |

## Oil pump, removing and installing (4-cylinder engine)

Oil pressure in the M44 engine is generated by a pump mounted to the rear of the timing chain housing. The pump is gear-driven off the front of the crankshaft.

Oil pump removal requires lowering of the front suspension and removal of the oil pan, as well as removal of the timing chain covers, complete timing chain assembly and timing chain housing. Special service tools are needed for these procedures. Read the procedures through before starting work.

– Disconnect negative (–) battery cable from battery.

> ***CAUTION—***
> *Prior to disconnecting the battery, read the battery disconnection cautions given in* **001 General Warnings and Cautions**.

– Loosen oil filter cover to allow engine oil to drain into oil pan. Reinstall oil filter cover.

– Drain engine oil. See **020 Maintenance**.

– With engine cold, drain coolant. See **170 Radiator and Cooling System**.

– Remove oil pan. See **Oil pan, removing and installing (4-cylinder engine)** in this repair group.

– Remove alternator. Unbolt power steering pump from bracket, then remove alternator mounting bracket. See **121 Battery, Starter, Alternator**.

– Remove crankshaft vibration damper. Then remove upper and lower timing chain covers, complete timing chain, chain sprockets and chain guides. See **117 Camshaft Timing Chain.**

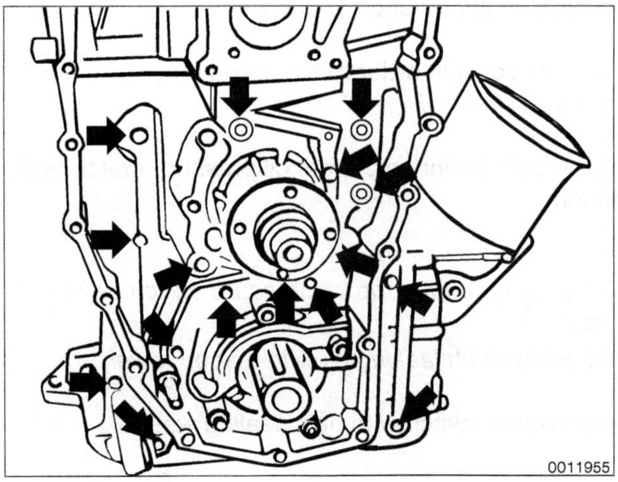


◄ Remove timing chain housing mounting bolts (**arrows**). Remove timing chain housing from front of engine.

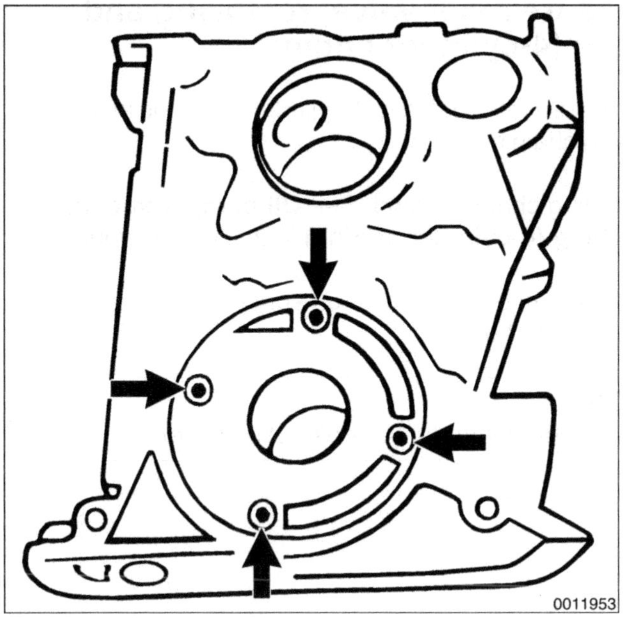


◄ Remove oil pump cover mounting screws (**arrows**) on rear of timing chain housing.

- Remove oil pump gears.
- Remove oil pressure relief valve from top of housing. See **Oil pressure relief valve, removing and installing (4-cylinder engine)**.
- Inspect engine oil pump gears and oil pump housing walls for wear or scoring. Replace components as necessary.

– Installation is reverse of removal.

- Replace all gaskets and O-rings.
- Install a new rubber seal (profile gasket) between top of timing chain housing and cylinder head.
- Be sure to thoroughly clean sealing surfaces.
- Use 3-Bond 1209® or equivalent sealant on both sides of seal before installing.

– Reinstall timing chain. See **117 Camshaft Timing Chains**.

– Reinstall timing housing covers.

> ***CAUTION—***
> *Be sure to protect the profile gasket by using BMW special tool 11 2 330 or equivalent. The full procedure for upper and lower timing chain housing cover installation is given in* **117 Camshaft Timing Chains**.

| **Tightening torques** | |
|---|---|
| Alternator to alternator bracket | 43 Nm (32 ft-lb) |
| Oil filter cover to housing | 25 Nm (18 ft-lb) |
| Oil pan to engine block (M6 bolt)<br>• 8.8 grade<br>• 10.9 grade | <br>10 Nm (7 ft-lb)<br>12 Nm (9 ft-lb-lb) |
| Oil pump cover to timing chain housing (M6) | 10 Nm (7 ft-lb) |
| Thermostat housing to upper timing chain cover (M6) | 10 Nm (7 ft-lb) |
| Timing chain housing to engine block or cylinder head<br>• M6<br>• M8 | <br><br>10 Nm (7 ft-lb)<br>22 Nm (16 ft-lb) |

– Fill engine with oil.

– Fill and bleed cooling system. See **170 Radiator and Cooling System**.

## Oil pressure relief valve, removing and installing (4-cylinder engine)

The oil pressure relief valve in 4-cylinder engine is at the top of the timing chain housing.

– Remove timing chain housing. See **Oil pump, removing and installing (4-cylinder engine)** in this repair group.

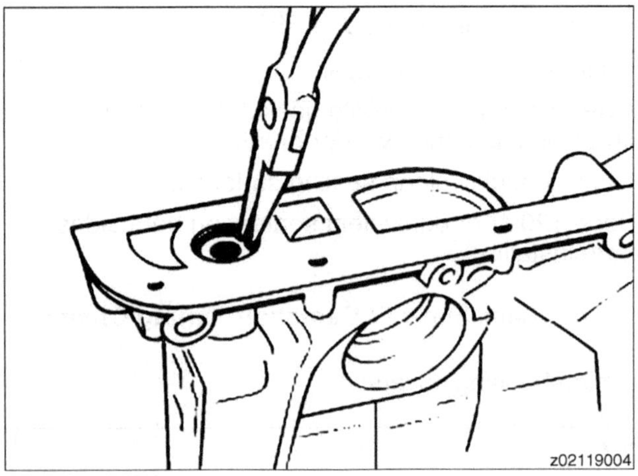


< Use suitable drift to press down on spacer. Remove circlip with needle-nosed pliers.

> ***CAUTION—***
> *There is strong spring force behind the spacer.*

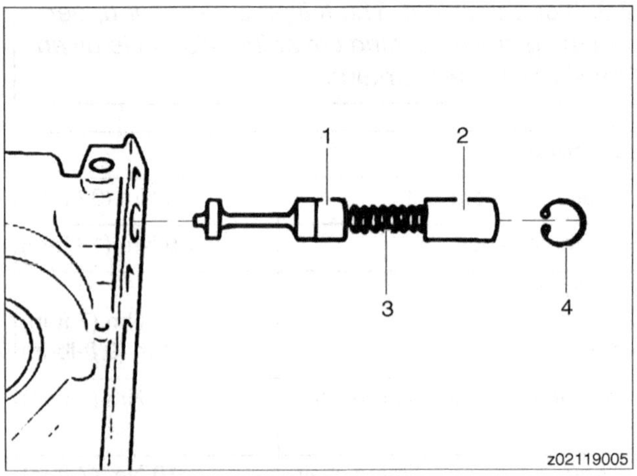


< Remove pressure relief valve and inspect for wear.

1. Control plunger
2. Spacer
3. Spring
4. Circlip

– Assembly is reverse of disassembly.

# LUBRICATION SYSTEM COMPONENT REPLACEMENT (6-CYLINDER ENGINE)

## Oil pressure warning switch, replacing (6-cylinder engine)

< Oil pressure switch (**arrow**) is located underneath intake manifold. (Intake manifold removed for purpose of illustration.)

– Unscrew oil filter cover in order to allow engine oil to drain back down into oil pan.

– Disconnect electrical connector from oil pressure switch and remove switch.

***NOTE—***
- *Thoroughly clean around the oil pressure switch before removing it.*
- *Be prepared to catch leaking oil with a shop towel.*

– Install new pressure switch with new sealing washer.

| Tightening torque | |
|---|---|
| Oil pressure switch to oil filter housing | 27 Nm (20 ft-lb) |

## Oil level warning switch, replacing (6-cylinder engine)

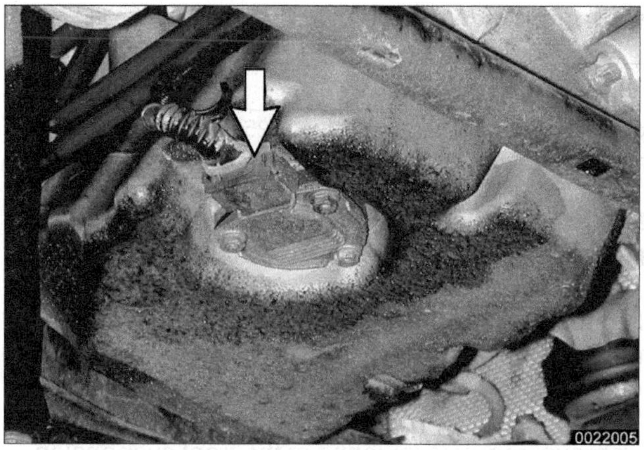
0022005

< Working underneath engine oil pan, drain oil, then remove oil level warning switch (**arrow**) located at bottom of oil pan.

– Reinstall switch using new sealing O-ring.

– Be sure to refill engine and check for leaks after work is completed.

## Engine oil cooler, removing and installing (S54 engine)

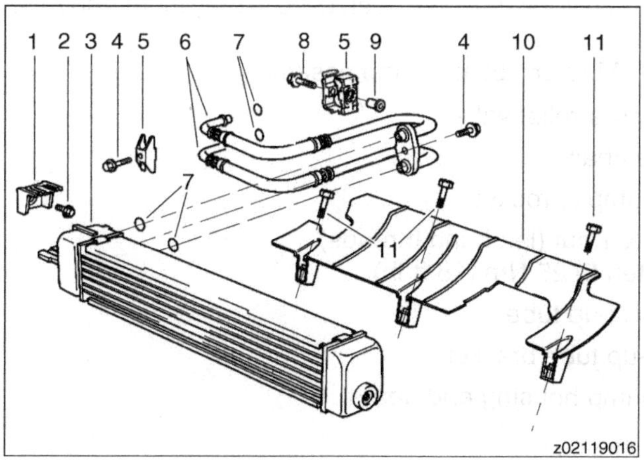

z02119016

< The engine oil cooler is mounted directly below the radiator.

1. Oil cooler mounting bracket
2. M6 bolt
3. Oil cooler
4. M8 bolt
5. Oil pipe mounting bracket
6. Oil pipes
7. O-rings
8. M6 bolt
9. Sleeve
10. Air duct
11. Sheet metal screw

- Raise front of vehicle and support safely.

> ***WARNING —***
> *Make sure the car is stable and well supported at all times. Use a professional automotive lift or jack stands designed for the purpose. A floor jack is not adequate support.*

- Remove splash guard from underneath engine.
- Drain engine oil. See **020 Maintenance**.
- Detach oil cooler lines from oil cooler.
- Remove engine oil cooler carrier mounting screws. Pull oil cooler with carrier straight off.
- Installation is reverse of removal. Replace O-rings at oil line fitting. Install and torque drain plug. Fill engine with oil.

| Tightening torque | |
|---|---|
| Oil drain plug to oil pan (M12) | 25 Nm (18 ft-lb) |

## Oil pump, removing and installing (6-cylinder engine)

The 6-cylinder engine oil pump is located on the underside of the engine block, inside the oil pan. It is driven by the crankshaft via chain and sprocket assembly.

Oil pump removal requires lowering of the front suspension and removal of the oil pan. Read the procedure through before starting work.

Four oil pump versions have been used in 6-cylinder engines:

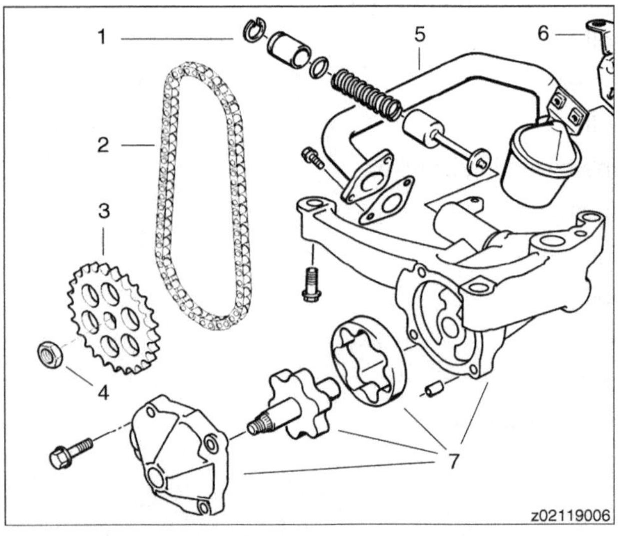


Version 1, M52 or S52 oil pump assembly:

1. Pressure relief valve
2. Drive chain
3. Oil pump sprocket
4. M10 x 1 nut (left hand threads) -tighten to 25 Nm (18 ft-lb)
5. Oil pick-up tube
6. Pick-up tube bracket
7. Oil pump housing and gears

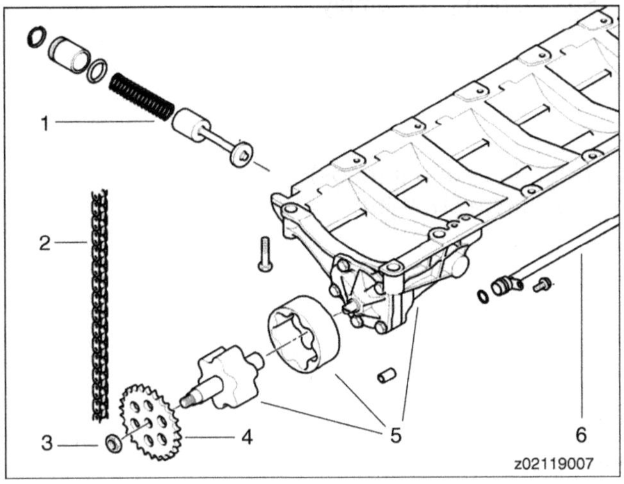


Version 2, M52 TU oil pump assembly:

1. Pressure relief valve
2. Drive chain
3. M10 x 1 nut (left hand threads)
   -tighten to 25 Nm (18 ft-lb)
4. Oil pump sprocket
5. Oil pump housing and gears
6. Oil pick-up tube

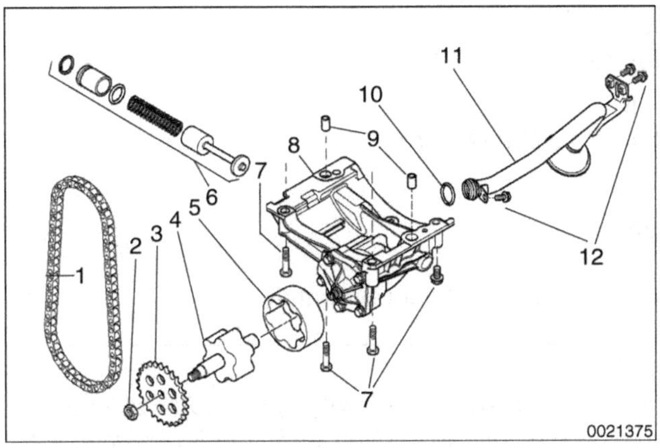


Version 3, M54 oil pump assembly:

1. Drive chain
2. M10 x 1 nut (left hand threads)
   -tighten to 25 Nm (18 ft-lb)
3. Oil pump sprocket
4. Inner pump rotor
5. Outer pump rotor
6. Pressure relief valve
7. M8 bolt
   -tighten to 23 Nm (17 ft-lb)
8. Oil pump housing
9. Locating dowels
10. Pick-up tube sealing O-ring
11. Pick-up tube
12. Pickup tube mounting bolts (M6)
    -tighten to 10 Nm (7 ft-lb)

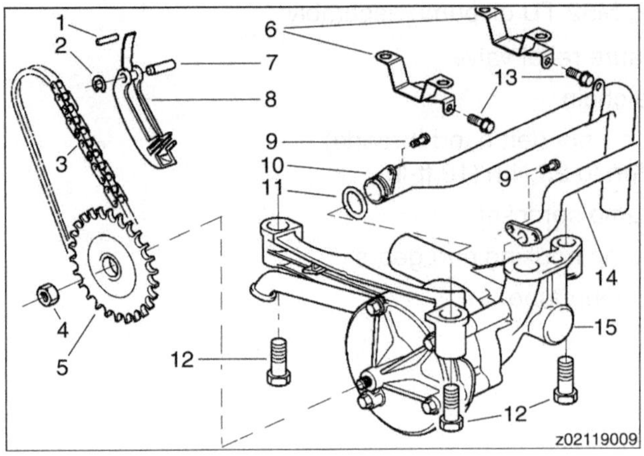


◄ Version 4, S54 oil pump assembly:

1. Dowel
2. Circlip
3. Drive chain
4. M10 x 1 nut (left hand threads) -tighten to 25 Nm (18 ft-lb)
5. Drive sprocket
6. Oil pipe bracket
7. Bearing bolt
8. Chain tensioner
9. M5 bolt
10. Pick-up tube
11. O-ring
12. M8 bolt -tighten to 23 Nm (17 ft-lb)
13. M6 bolt -tighten to 10 Nm (7 ft-lb)
14. Return pipe
15. Oil pump

– Raise front of vehicle and support safely.

> ***WARNING —***
> *Make sure the car is stable and well supported at all times. Use a professional automotive lift or jack stands designed for the purpose. A floor jack is not adequate support.*

– Loosen oil filter cover to allow engine oil to drain into oil pan. Reinstall oil filter cover.

– Drain engine oil. See **020 Maintenance**.

– Lower front suspension and remove oil pan. See **Oil pan, removing and installing (6-cylinder engine)**.

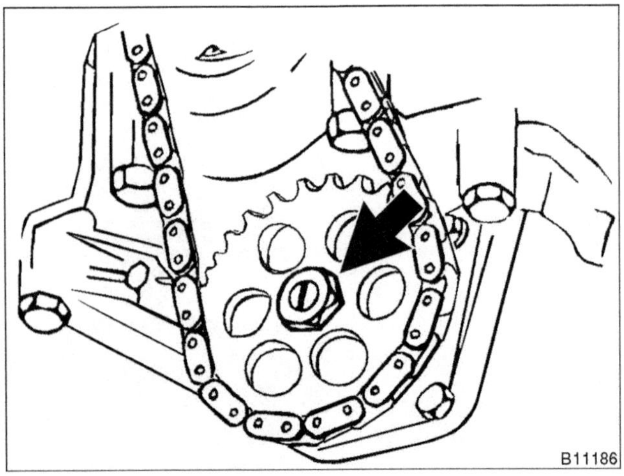


◄ Remove oil pump sprocket mounting nut (**arrow**).

> ***CAUTION—***
> *Oil pump sprocket nut has left hand threads. Turn clockwise to loosen.*

S54 engine: When loosening sprocket nut, protect plastic chain tensioner from damage by removing chain slack: Wrap approx. 7 inches of soft wire (**arrow**) around chain to keep it away from tensioner.

- Lift sprocket off pump together with drive chain.

- S54 engine: Push back on chain tensioner to allow sprocket to disengage from pump.

- Unbolt oil pump from engine block.
  - M52 or S52 engine: Remove oil pump with pick-up tube. Pick-up tube is bolted to side of oil pump.
  - M52 TU engine: Remove pick-up tube, scraper tray and oil pump as a unit.
  - M54 engine: Remove oil pump pick-up tube bolts and remove pick-up tube before removing pump.
  - S54 engine: Remove oil pickup and return tubes before removing pump.
  - Remove mounting bolts from oil pump. Withdraw pump.

***NOTE—***
*Record location of spacers between pump and engine block.*

- Remove cover from oil pump and check for wear or scoring. Spin oil pump shaft and check that gears turn smoothly. Replace pump if gears spin with difficulty or any wear is present.

- Inspect oil pressure relief valve. See **Oil pressure relief valve (6-cylinder engine)** in this repair group.

- Installation is reverse of removal, noting the following:
  - Inspect pump locating dowels for damage and correct positioning.
  - Replace oil pick-up and return tube sealing O-rings (as applicable).
  - Align sprocket splines to oil pump shaft splines before tightening sprocket nut.

***CAUTION—***
*Oil pump sprocket nut has left hand threads. Turn counterclockwise to tighten.*

| Tightening torques | |
|---|---|
| Oil drain plug to oil pan | 25 Nm (18 ft-lb) |
| Oil pan to engine block (M6)<br>• 8.8 grade<br>• 10.9 grade | <br>10 Nm (7 ft-lb)<br>12 Nm (9 ft-lb) |
| Oil pump sprocket to oil pump shaft (M10 x 1) (left hand thread) | 25 Nm (18 ft-lb) |
| Oil pump to crankcase (M8) | 23 Nm (17 ft-lb) |
| Oil pump pickup tube to crankcase (M6) | 10 Nm (7 ft-lb) |

- Fill engine with oil as described in **020 Maintenance**. After adding engine oil, start and run engine. Raise engine speed to 2,500 rpm until oil pressure warning light goes out (about 5 seconds).

***NOTE—***
- *See* **Table a** *and* **Table b** *in this repair group for engine oil capacities and specifications.*
- *S54 engine: A small quantity of oil, which otherwise would stay inside the oil pan during regular oil and filter service, is drained when the pan is removed. Compensate for this with an additional 1.0 liter / 1.1 US qt of oil (for a total of 6.5 liters / 6.9 US qt) when refilling.*

## Oil pressure relief valve (6-cylinder engine)

Remove oil pump. See **Oil pump, removing and installing (6-cylinder engine)**. The oil pump pressure relief valve is held in the side of the oil pump with a circlip.

**Oil pressure relief valve assembly (6-cylinder engine)**

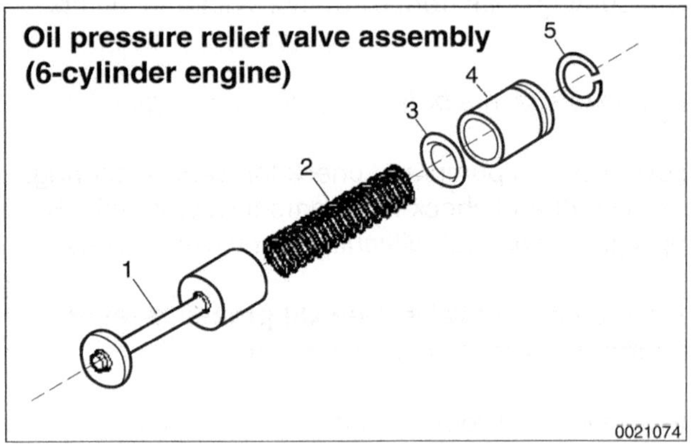


Inspect oil pump relief valve components and bore for scoring or other damage.

1. Control plunger
2. Spring
3. Sealing O-ring
4. Sleeve
5. Circlip

- Install using a new circlip.

# 120 Ignition System





**TABLES**

## General

This repair group covers component troubleshooting and replacement information for the ignition system.

Engines are referred to by engine code. See **Table a** for engine code and application data. See **100 Engine–General** for more information.

When diagnosing engine management problems, including on-board diagnostics (OBD II) fault code analysis, also refer to these repair groups:

- **130 Fuel Injection**
- **600 Electrical System–General**
- **ELE Electrical Wiring Diagrams**
- **OBD On-Board Diagnostics**

***NOTE—***
*Spark plug replacement is covered in* **020 Maintenance**.

### Special tools

Special test equipment is recommended for system diagnosis and testing of the ignition system.

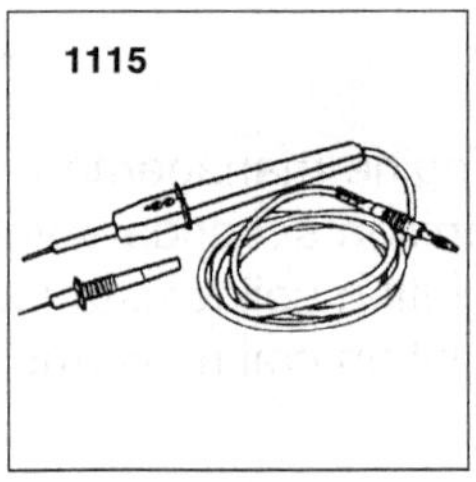


◄ LED test light
(Baum tool 1115)

◄ Automotive digital multimeter

> ***CAUTION—***
> *Only use a digital multimeter when testing automotive electrical components.*

◄ Pry tool for fuel injector rail
(BMW special tool 12 1 120)

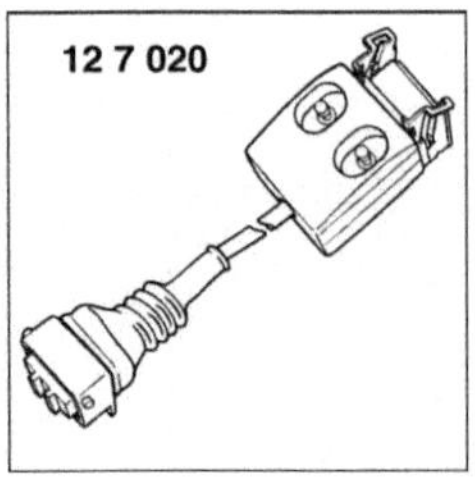


◄ Primary voltage test harness
(BMW special tool 12 7 020)

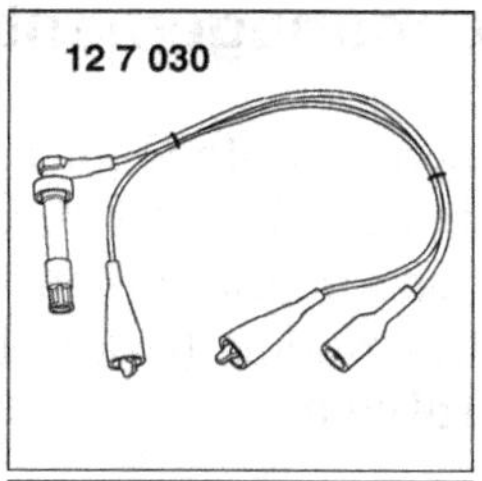


Secondary voltage test harness (BMW special tool 12 7 030)

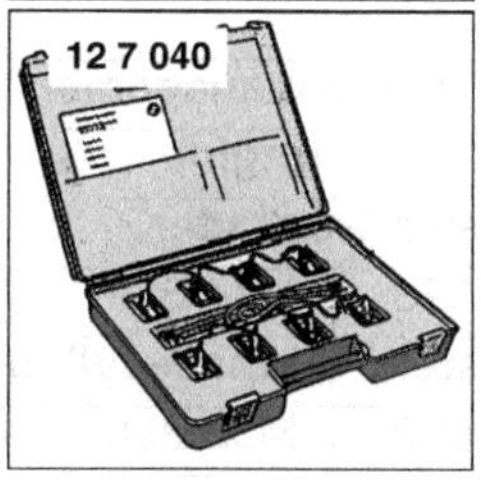


Ignition coil test adapters (BMW special tool 12 7 040)

## Warnings and cautions

The ignition system produces dangerous high voltage. In addition, sensitive electronic components can be damaged if proper precautions are not adhered to.

***WARNING—***
- *Do not touch or disconnect any cables from the coils while the engine is running or being cranked by the starter.*
- *The ignition system produces high voltages that can be fatal. Avoid contact with exposed terminals. Use extreme caution when working on a car with the ignition switched on or the engine running.*
- *Connect and disconnect the DME system wiring and test equipment leads only when the ignition is OFF.*
- *Before operating the starter without starting the engine (for example when making a compression test) always disable the ignition. See* **Disabling ignition system** *in this repair group.*

***CAUTION—***
- *Do not attempt to disable the ignition by disconnecting the coils from the spark plugs.*
- *Connect or disconnect ignition system wires, multiple wire connectors and ignition test equipment leads only while the ignition is OFF. Switch multimeter functions or measurement ranges only with the test probes disconnected.*
- *Do not disconnect the battery while the engine is running.*
- *Use a high impedance digital multimeter for all voltage and resistance tests.*
- *Use an LED test light in place of an incandescent-type test lamp.*
- *In general, make test connections only as specified by BMW, as described in this manual, or as described by the instrument manufacturer.*

## Engine management

The digital motor electronics (DME) engine management systems used in Z3 and M version cars combine fuel injection, ignition and other functions under the control of the engine control module (ECM). DME systems applications are given in **Table a**.

| Table a. Engine management system applications | | |
|---|---|---|
| **Year** | **Engine** | **DME system** |
| **Z3 Roadster 4-cylinder** | | |
| 1996 - 1998 | M44 B19 | Bosch DME M5.2 |
| **Z3 Roadster / Coupe 6-cylinder** | | |
| 1997 - 1998 | M52 B28 | Siemens DME MS 41.1 |
| 1999 - 2000 | M52 TU B25<br>M52 TU B28 | Siemens DME MS 42.1 |
| 2001 - 2002 | M54 B25<br>M54 B30 | Siemens DME MS 43.0 |
| **M Roadster / Coupe 6-cylinder** | | |
| 1998 - 2000 | S52 B32 | Siemens DME MS 41.1 |
| 2001 - 2002 | S54 B32 | Siemens DME MS S54 |

Second generation on-board diagnostics (OBD II) is incorporated into the engine management systems used on the cars covered by this manual. Use a BMW-specific electronic scan tool, such as DISplus, GT1 or MoDiC, or a generic OBD II scan tool to access diagnostic trouble codes (DTCs). DTCs can help pinpoint ignition and other engine management problems.

Additional information about DTCs and engine management system electronic system diagnosis is provided in **OBD On-Board Diagnostics.**

## Ignition system

All engines use a distributorless ignition system with individual ignition coils for each cylinder. There is no distributor cap or ignition rotor.

## Ignition coils

4-cylinder engine: Ignition coils are located in external housing to right of engine, connected to spark plugs via conventional spark plug wires. Cylinders are numbered from front to rear.

M52, S52, M52 TU or M54 engine (6-cylinder): Ignition coils are arranged directly above spark plugs.

- Coils are under engine top plastic cover.
- Coils are mounted to cylinder using M6 bolts.
- Each coil is connected to corresponding spark plug via short lead.
- Coils are connected to engine control module (ECM) via large electrical harness.
- Cylinders are numbered from front to rear.

S54 engine (6-cylinder): Ignition coils are arranged directly above spark plugs, similar to other 6-cylinder engines.

- Coils are pressed into cylinder head.

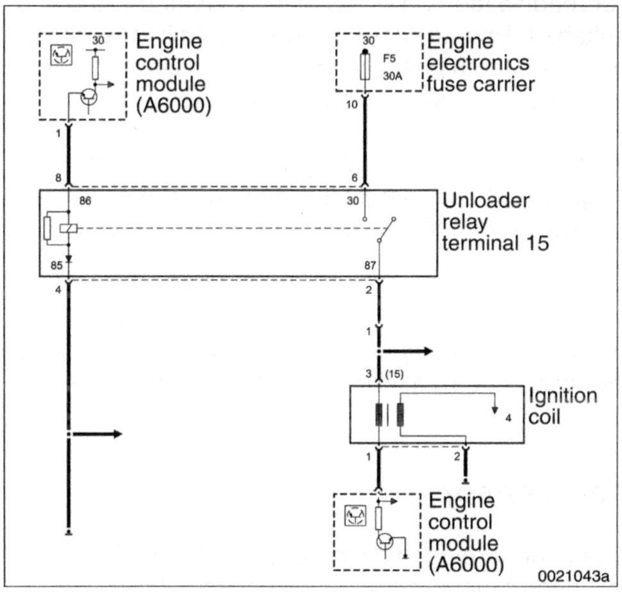


Each coil is controlled and monitored by the ECM on a cylinder-by-cylinder basis.

***NOTE—***
*Schematic illustrates ignition coil wiring for M52 TU engine with DME MS 43.0.*

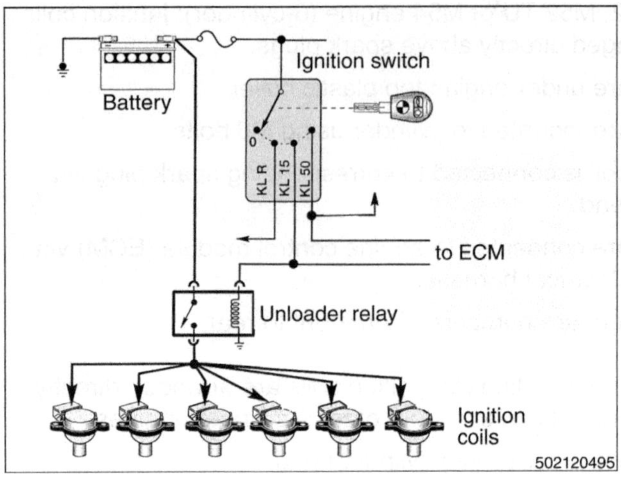


Power (+) from ignition switch is supplied to ignition coils via an unloader relay. Coils are actuated on negative side by ECM.

*NOTE—*
*Diagram is for illustrative purposes only. See* **ELE Electrical Wiring Diagrams** *for wiring information for a specific engine management system.*

## Crankshaft sensor

If the DME control module (ECM) does not receive a crankshaft position signal during cranking the engine does not start.

On both 4- and 6-cylinder engines, the crankshaft sensor is mounted in the left rear side of the engine block.

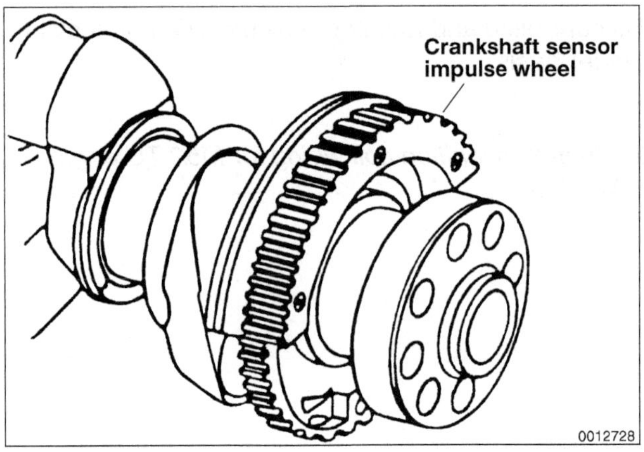


The sensor reads a toothed wheel that is mounted to the end of the crankshaft. Engine speed is determined by the rate at which the wheel's teeth pass the sensor. Crankshaft position is determined by the missing-teeth gap on the wheel.

If the OBD II system misfire detection protocol detects a catalyst damaging fault due to a malfunction in crankshaft sensor components, the malfunction indicator light (MIL) is illuminated.

*NOTE—*
*If the crankshaft sensor impulse wheel is damaged, disassemble the engine down to the crankshaft to replace the wheel.*

## Camshaft sensors

Camshaft sensors are used by the DME for VANOS control, sequential firing of fuel injectors and knock control.

The camshaft sensor reads an impulse wheel mounted to the camshaft. There is an intake camshaft sensor on the M44 (4-cylinder), M52 or S52 (6-cylinder) engine. M52 TU, M54 or S54 engine is equipped with 2 camshaft sensors, one each for the intake and the exhaust camshafts.

## Knock sensors

0022502c

Knock sensors (**arrows**) monitor the combustion chamber for engine-damaging knock. All 4-cylinder models and six-cylinder models with M52, S52, M52 TU or M54 engine use two sensors bolted to the left side of the engine block, underneath the intake manifold. Each sensor monitors multiple cylinders.

6-cylinder models with S54 engine use 3 knock sensors bolted to the left side of the engine block, underneath the intake manifold.

## Ignition timing

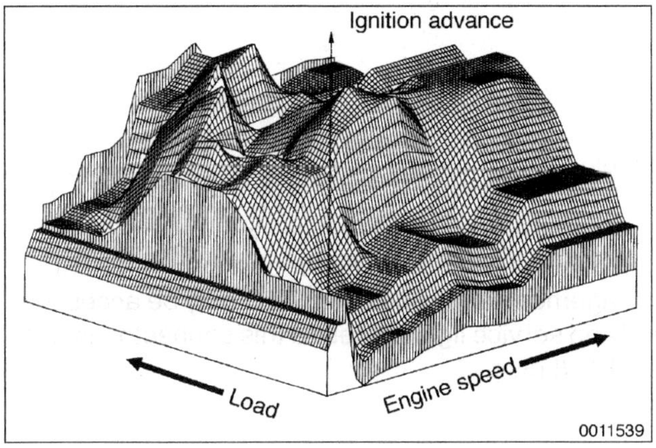

0011539

Ignition timing is electronically mapped and not adjustable. The ECM uses engine load, engine speed, coolant temperature, knock detection, and intake air temperature as the basic inputs for timing control. A three dimensional ignition timing map, similar to the one shown, is digitally stored in the ECM.

The initial ignition point is determined by the camshaft and crankshaft sensors during cranking. Once the engine is running, timing is continually adjusted based on operating conditions.

## IGNITION SYSTEM TROUBLESHOOTING

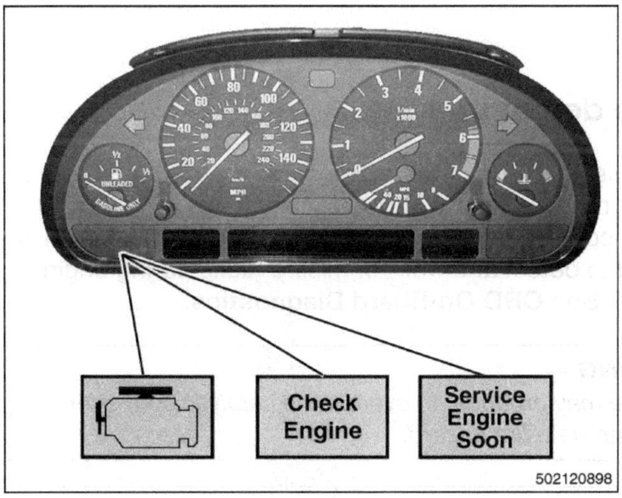

502120898

The malfunction indicator light (MIL) illuminates if an emissions related fault is detected.

***NOTE—***
*BMW models covered in this manual are equipped with one of three versions of MIL (as shown).*

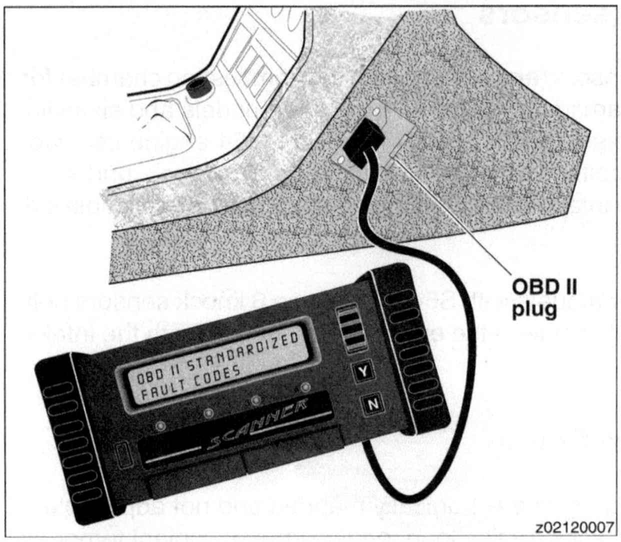


To begin ignition system troubleshooting, attach a diagnostic scan tool (DISplus, GT1, MoDiC or other BMW-compatible tool) to OBD II plug on right side of center console.

Interrogate engine control module (ECM) for stored faults. The OBD II system detects ignition system misfire and other related faults and stores a diagnostic trouble code (DTC) in the ECM along with other pertinent fault information.

Additional OBD II information, including a DTC lookup table, can be found in **OBD On-Board Diagnostics**.

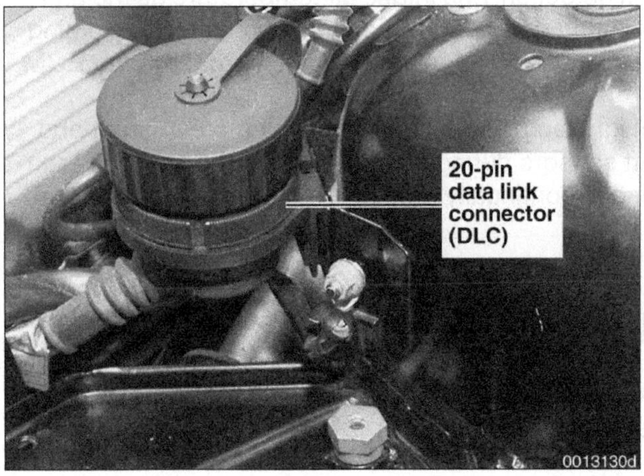


The OBD II plug is mandated by federal law from 1996 on and is found in all Z3 and M version cars. However, BMW cars produced through 6 / 2000 are also equipped with a 20 pin data link connector (DLC) in the engine compartment. Engine management faults and other DTC may be accessed and cleared and service lights reset via this connector as well as via the OBD II plug.

## Misfire detection

Engine misfire in one or more cylinders, may be caused by a variety of malfunctions in various subsystems. The OBD II system incorporated into the engine management system is designed to detect and warn of misfire faults during engine operation. See **OBD On-Board Diagnostics**.

***WARNING —***
*A misfire may produce an overheated catalytic converter, which can be a fire hazard.*

***CAUTION—***
*Ignition misfires can cause high hydrocarbon exhaust emissions and catalytic converter damage. For this reason, if a severe misfire is detected, the fuel injector for the affected cylinder is turned OFF by the ECM and the MIL is illuminated.*

## Oscilloscope diagnostic diagrams

Engine misfire may be caused by a faulty coil. One way to diagnose a faulty coil is to use an oscilloscope to analyze spark quality with the engine running.

Use the oscilloscope diagrams and refer to **Table b** for help in diagnosing some common ignition system problems.

***NOTE—***
*Note the length of sparking period and the up-down voltage attenuations. A short sparking period and/or an irregular/low voltage attenuation usually indicates a faulty coil.*

**Table b. Ignition secondary voltage diagnostics**

| Parameter | Secondary voltage low | Secondary voltage high |
|---|---|---|
| Spark plug electrode gap | Too small | Too big |
| Spark plug electrode condition | | Worn/burnt |
| Spark plug electrode temperature | Too high | Too low |
| Engine compression | Too low | Too high |
| Spark plug wires | | Faulty |
| Fuel air mixture | | Too lean |

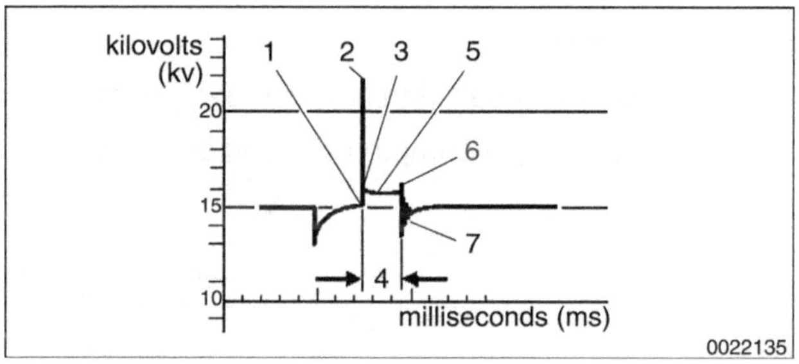


Normal scope trace of spark at idle

1. Start of ignition voltage spike
2. Ignition voltage spike
3. Combustion voltage
4. Period of combustion
5. Combustion curve characteristics
6. Start of voltage decay
7. Termination oscillations

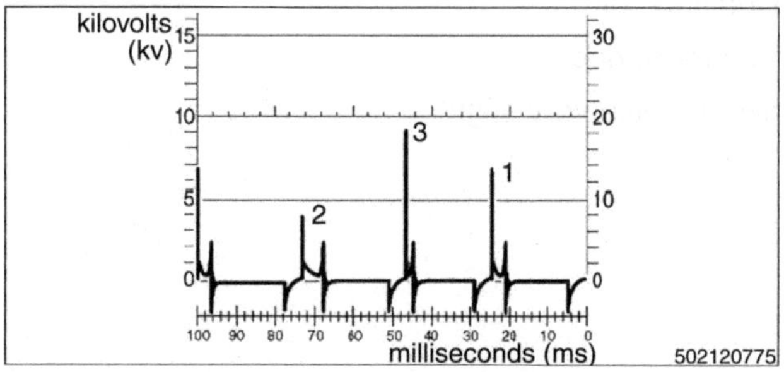


Ignition spark at idle

1. Normal ignition voltage peak (good spark plug)
2. Low voltage peak = closed plug gap
3. High voltage peak = open plug gap

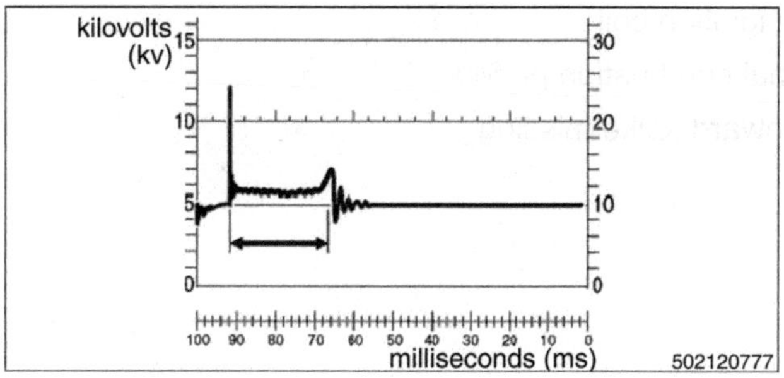


Long combustion period = small spark plug gap

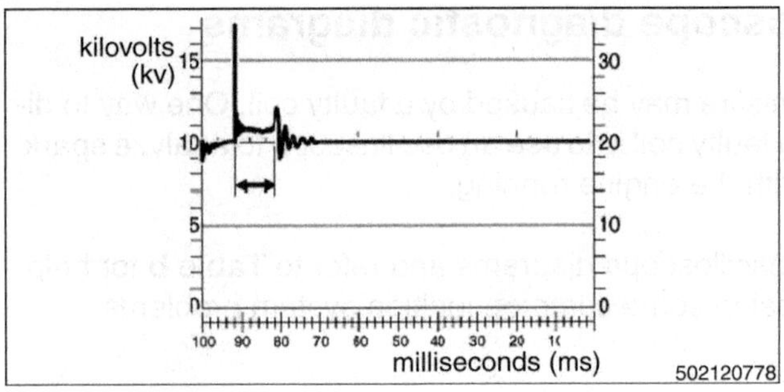


Short combustion period = large spark plug gap

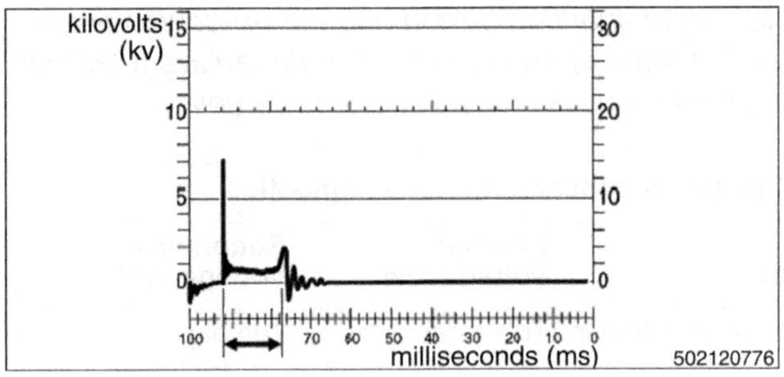


Normal combustion period at idle (single cylinder)

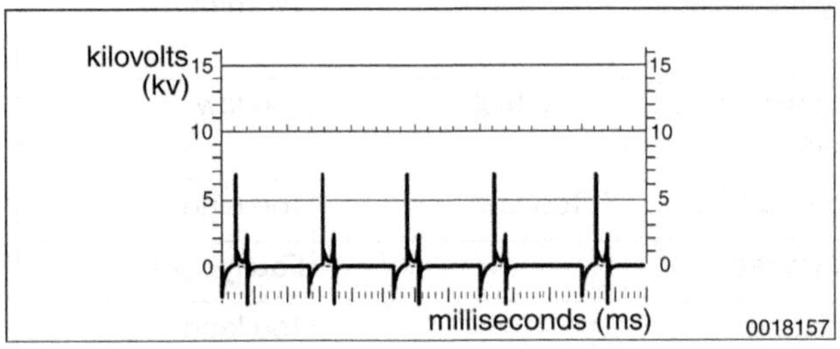


Normal oscilloscope pattern at idle (multiple cylinders)

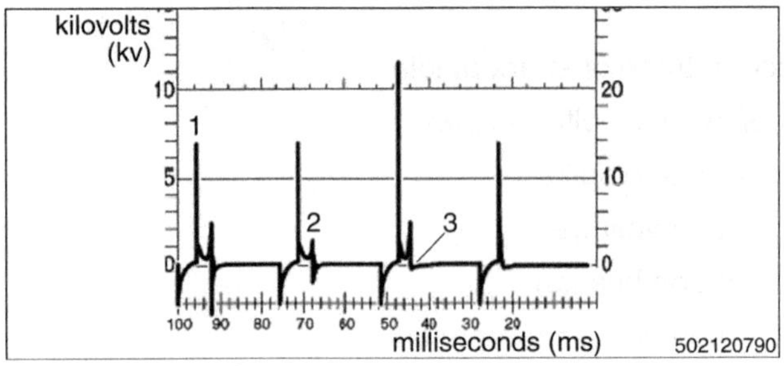


Ignition voltage peaks at idle

1. Normal ignition peaks
2. Downward spike shortened (ignition coil defective)
3. Downward spike missing completely (ignition coil defective)

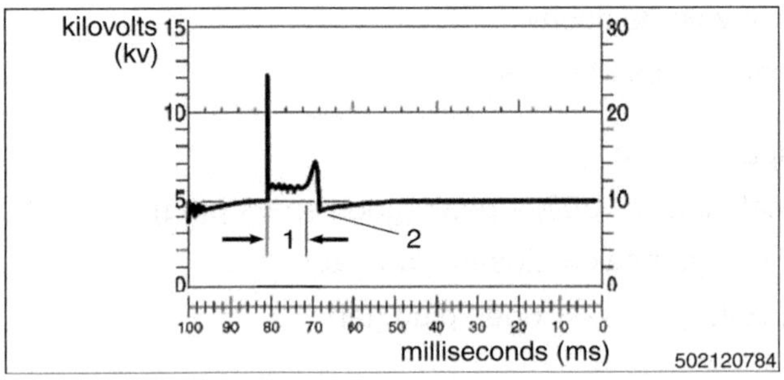


Defective ignition coil

1. Short spark period
2. Downward spike very slight

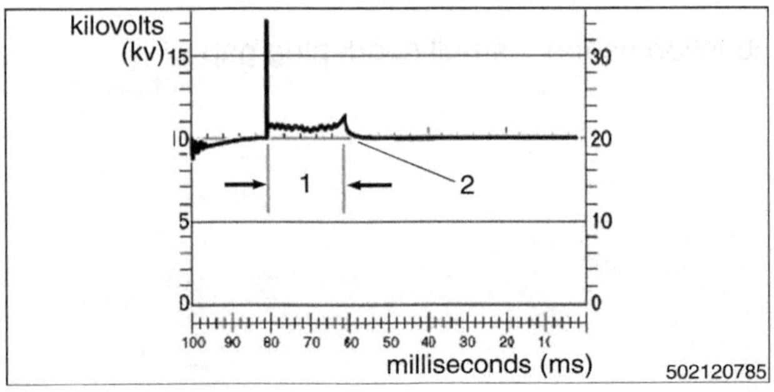


Defective ignition coil

1. Normal combustion period
2. Downward spike missing

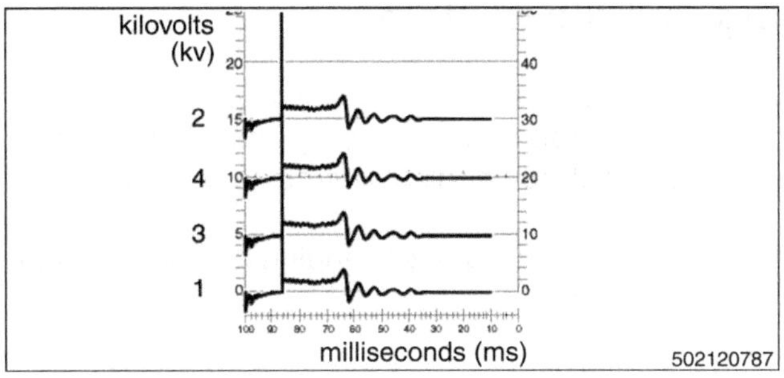


Normal secondary voltage patterns (4–cylinder pattern)

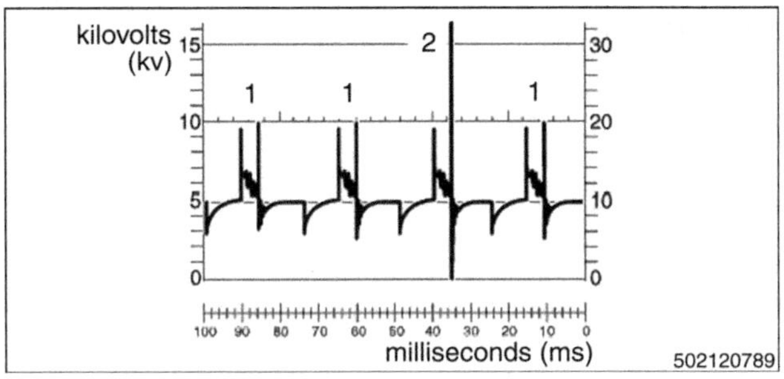


Ignition voltage peaks in response to sudden acceleration load:

1. Normal ignition pattern. Beginning of dying out pattern not much higher than ignition voltage peak
2. Beginning of dying out pattern considerably higher than ignition voltage peak. Fault in injection system:
   -Lean fuel mixture
   -Defective fuel injector
   -Low compression in cylinder

## IGNITION SYSTEM SERVICE

### Ignition firing order

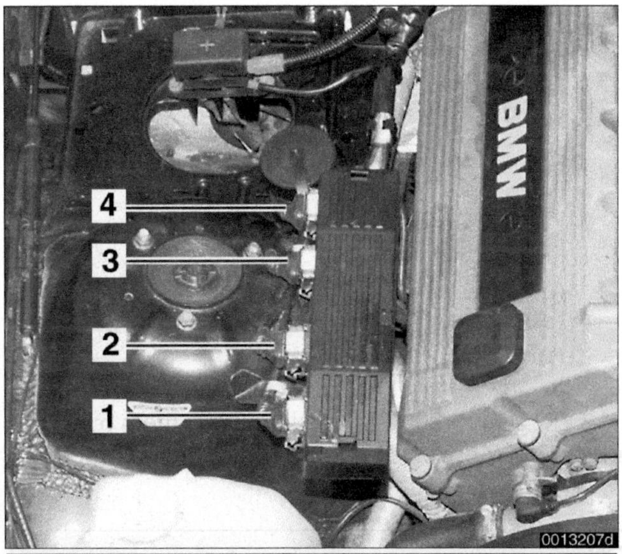


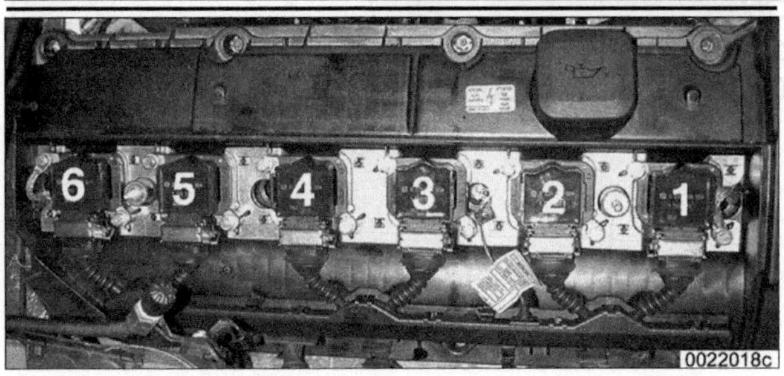


Cylinder 1 is located at the front on all engines.

| Ignition firing order | |
|---|---|
| 4-cylinder engine (M44) | 1 - 3 - 4 - 2 |
| 6-cylinder engine (M52, S52, M52 TU, M54, S54) | 1 - 5 - 3 - 6 - 2 - 4 |

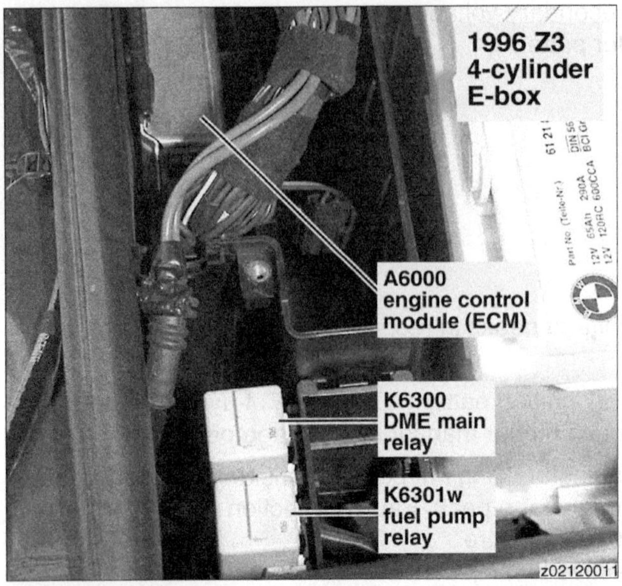


## Disabling ignition system

The ignition system operates in a lethal voltage range. Any time engine service or repair work is being done that requires the ignition to be switched on, disable ignition system.

– Remove DME main relay in right rear engine compartment to disable ignition system.

◄ 1996 4-cylinder model: DME main relay location at right rear of engine compartment.

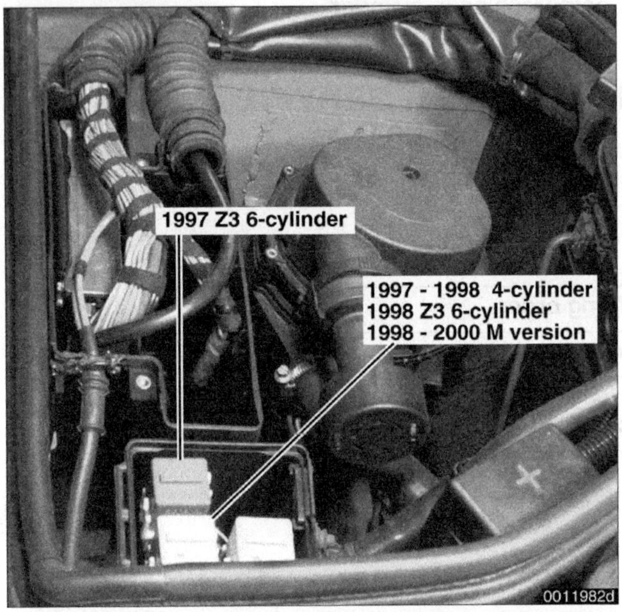


◄ 1997 - 1998 Z3, 1998 - 2000 M version: DME main relay location at right rear of engine compartment.

◄ 1999 - 2002 Z3, 2001 - 2002 M version: DME main relay is located under E-box cover, at right rear of engine compartment.

> ***CAUTION—***
> *Relay locations vary. Use care when identifying relays. See* **610 Electrical Component Locations**.

## Checking for spark

> ***CAUTION—***
> *If a spark test is done incorrectly, damage to the engine control module (ECM) or the ignition coils may result.*

***NOTE—***
*Spark plug replacement is covered in* **020 Maintenance**.

Checking for spark is difficult on engines with distributorless ignition systems.

Try removing the spark plugs and inspecting for differences among them. A poor-firing plug may be wet with fuel and/or black and sooty, but not always. If a coil is not operating, the engine management system disables the fuel injector to that cylinder. The key is to look for differences between cylinders.

## Ignition coil, testing and replacing (4-cylinder engine)

> ***CAUTION—***
> *Use a digital multimeter for the following tests.*

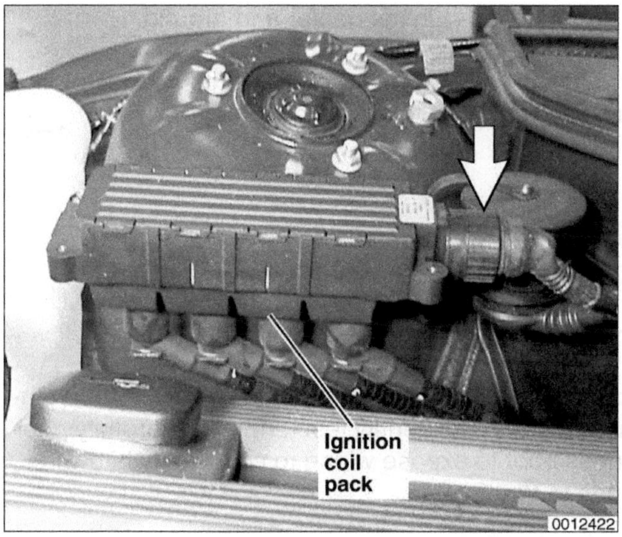


◄ Working at right side of engine, disconnect main harness connector (**arrow**) from coil pack.

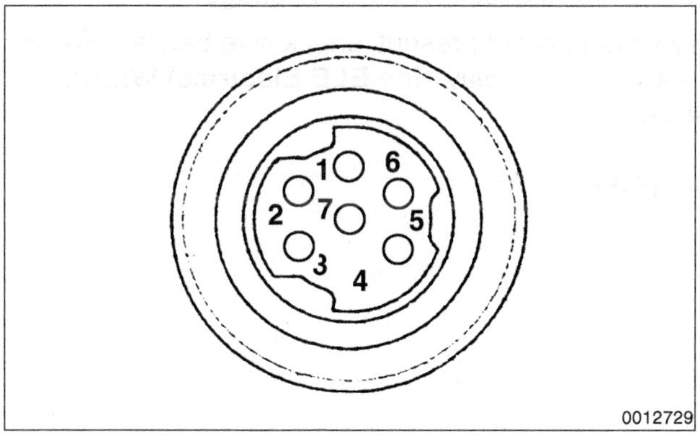


◄ Connect voltmeter between terminal **2** (15) in connector and ground. Turn ignition on and check for battery voltage.

- If battery voltage is not present, check wire between terminal 15 and ignition switch. See **ELE Electrical Wiring Diagrams**.

– Turn ignition off.

– Use an ohmmeter to test coil primary resistance. Compare to values in **Table c**.

| Table c. Ignition coil resistance (M44 engine) | | |
|---|---|---|
| **Circuit** | **Terminals** | **Resistance** |
| Coil #1 primary | **1** (–) and **2**(+) | 0.4 - 0.8 Ω |
| Coil #2 primary | **6** (–) and **2**(+) | 0.4 - 0.8 Ω |
| Coil #3 primary | **7** (–) and **2**(+) | 0.4 - 0.8 Ω |
| Coil #4 primary | **5** (–) and **2**(+) | 0.4 - 0.8 Ω |
| Coil secondary | N.A. | N.A. |

***NOTE—***
*If a single coil is faulty, replace the complete coil pack. Coils are not available individually.*

## Ignition coil, testing and replacing (6-cylinder engine)

***CAUTION—***
*Use a digital multimeter for the following tests.*

– Remove plastic engine cover from top of cylinder head cover.

– Disconnect harness connector from coil. Connect multimeter between terminal 15 (+) in connector and ground.

502120021c

◄ Terminal identification is typically molded into connector. In photo, terminals from left to right are: 1, 4a, 15.

z02113012a

◄ S54 engine: Coil connector and coil differ in design from other 6-cylinder models. To identify test terminals, carefully peel back harness boot to expose wiring to coil. Red-white wire corresponds to terminal 15.

– Turn ignition ON and check for battery voltage.
  - If battery voltage is not present, check wire between terminal 15 and unloader relay. See **ELE Electrical Wiring Diagrams**.

– Turn ignition OFF.

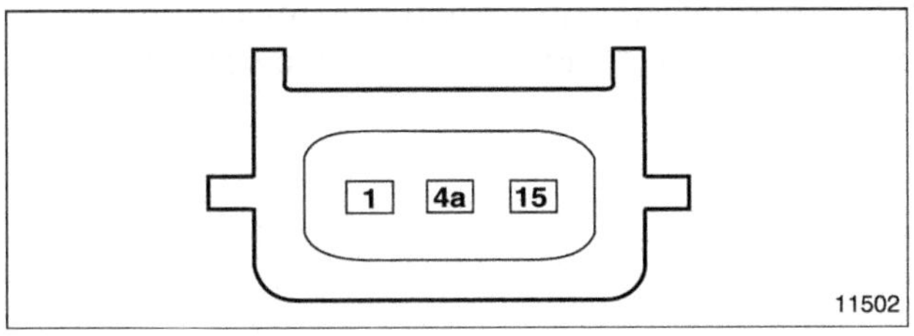


Use a multimeter at coil connectors and pins to test:

- Coil primary resistance
- Spark plug connector resistance
- Compare to values in **Table d**.

***NOTE—***
*The information in the table applies to M52 TU engine. Coil specifications for other engines is not available from BMW. Use this information as a guide when testing coils on engines other than the M52 TU.*

| Table d. Ignition coil resistance (M52 TU engine) | | |
|---|---|---|
| **Component** | **Terminals** | **Resistance** |
| Coil primary resistance | **1** (-) and **15** (+) | 0.8 Ω |
| Coil secondary resistance | | N/A |
| Spark plug connector:<br>• Bosch<br>• Beru | | <br>1.0 kΩ ± 20%<br>1.8 kΩ ± 20% |

– Replace ignition coil or spark plug connector if resistance readings do not meet specifications.

Remove coil and inspect coil housing for hairline cracks or leaking fluid. Replace cracked or leaking coil.

***NOTE—***
*A leaky coil may indicate a faulty engine control module (ECM). Check ECM before installing a new coil.*

> ***CAUTION—***
> *Note location of coil ground straps before coil removal; re-install in the same locations.*

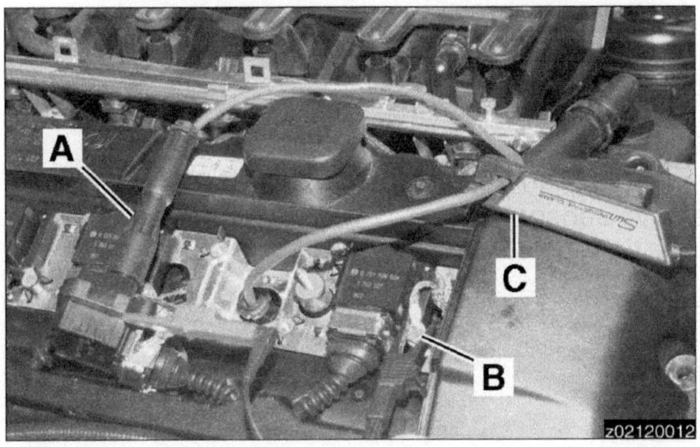


Oscilloscope coil test:

- Remove coil from cylinder head.
- Attach BMW special tool 12 7 030 to coil (**A**).
- Connect ground jumper (**B**) between coil mounting point and engine ground point.
- Clip oscilloscope pick-up lead (**C**) around ignition lead.
- Start engine. Compare 'scope pattern with diagrams in **Oscilloscope diagnostic diagrams** in this repair group.

***NOTE—***
*When replacing ignition coils, ensure that the replacement parts are from the same manufacturer with the same part/code numbers, or replace all coils together.*

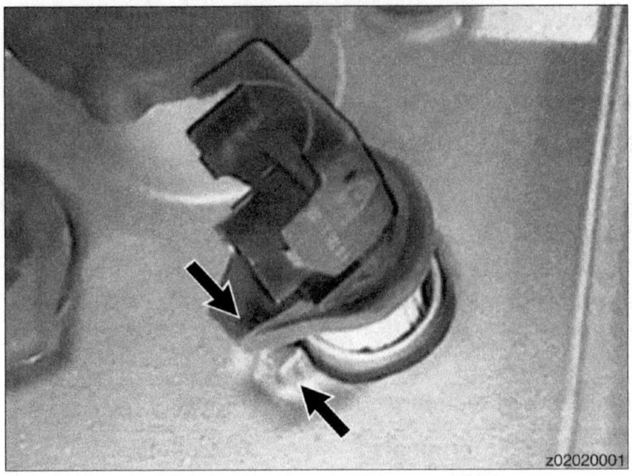

S54 engine: When reinstalling ignition coil, make sure coil rubber seal and cylinder head cover lug (**arrows**) fit snugly together.

## CRANKSHAFT SENSOR REMOVAL AND INSTALLATION

### Crankshaft sensor, replacing (4-cylinder engine)

Crankshaft sensor in 4-cylinder (M44) engine is located on left rear of engine block, underneath intake manifold.

- Make sure ignition is OFF.
- Remove upper section of intake manifold. See **113 Cylinder Head**.

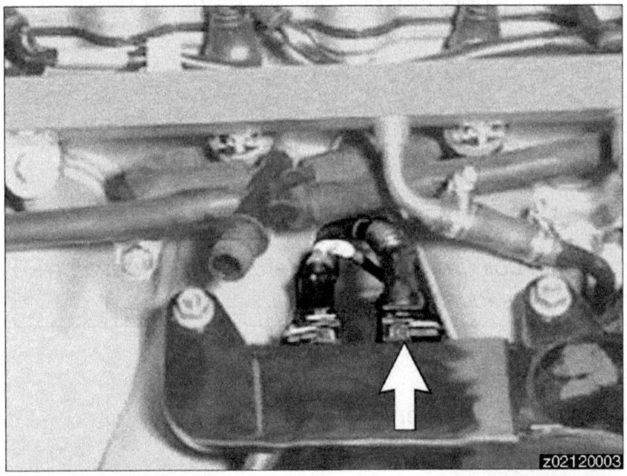

Working at left side of engine, detach crankshaft sensor electrical connector (**arrow**).

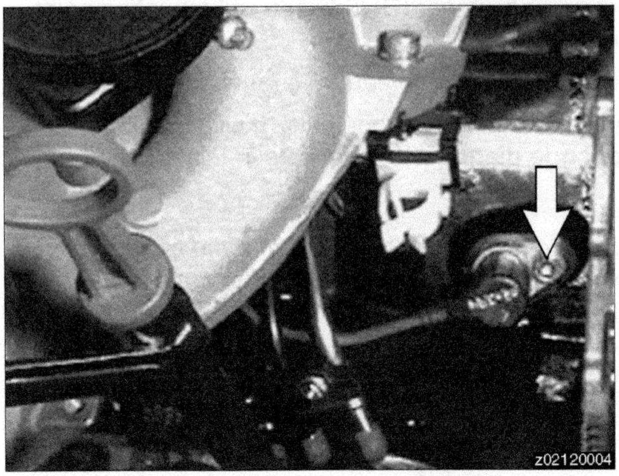

Remove sensor mounting bolt (**arrow**) at left rear of engine block. Remove sensor.

- Installation is reverse of removal.
  - Use new sealing O-ring when installing sensor.
  - Replace self-locking mounting bolt.
  - Be sure wiring is rerouted in same orientation.

| Tightening torque | |
|---|---|
| Crankshaft sensor to engine block (M6, replace bolt) | 10 Nm (7 ft-lb) |

- Check for fault codes and clear and reset DME control module memory.

## Crankshaft sensor, replacing (6-cylinder engine)

Crankshaft sensor in 6-cylinder (M52, S52, M52 TU, M54 or S54) engine is located on left rear of engine block, underneath intake manifold and starter motor.

- Make sure ignition is OFF.
- Raise car and support in a safe manner.

> ***WARNING —***
> *Make sure the car is stable and well supported at all times. Use a professional automotive lift or jack stands designed for the purpose. A floor jack is not adequate support.*

- Remove splash guard under engine.

◀ Working underneath starter:
  - Cut off wire tie securing crankshaft sensor harness connector (**arrow**) to sensor.
  - Disconnect harness connector.

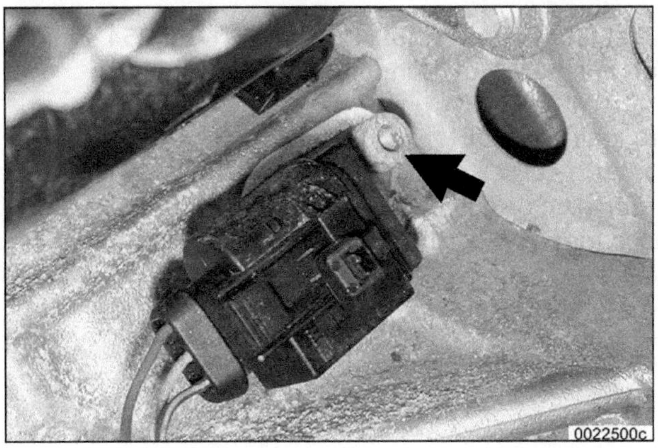


◀ Remove sensor mounting fastener (**arrow**) and remove sensor from cylinder block.

- Installation is reverse of removal.
  - Use new sealing O-ring when installing sensor.
  - Replace self-locking mounting bolt.
  - Be sure wiring is rerouted in same orientation.

| Tightening torque | |
|---|---|
| Crankshaft sensor to engine block (M6, replace bolt) | 10 Nm (7 ft-lb) |

- Check for fault codes and clear and reset DME control module memory.

# Camshaft Sensor Removal and Installation

## Intake camshaft sensor, replacing (4-cylinder engine)

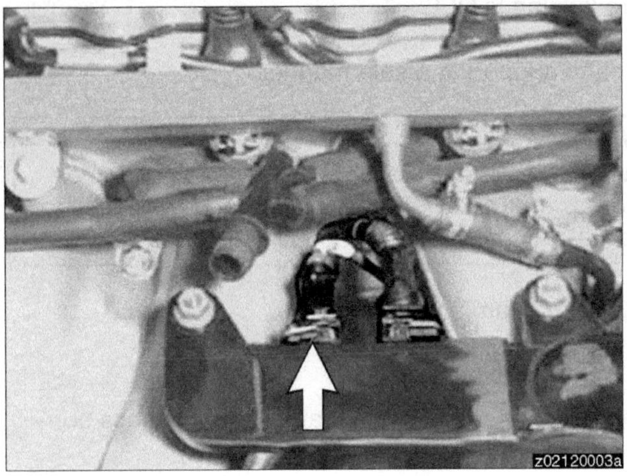


Intake camshaft sensor on 4-cylinder (M44) engine is mounted on upper camshaft timing chain cover.

- Make sure ignition is OFF.
- Remove upper section of intake manifold. See **113 Cylinder Head**.
- Remove camshaft sensor mounting bolt (**arrow**). Remove sensor from timing chain cover.
- Detach camshaft sensor harness from left front of cylinder head.
- Working at left side of engine, detach camshaft sensor electrical connector (**arrow**).
- Installation is reverse of removal. Remember to:
  - Use new sealing O-ring when installing sensor.
  - Be sure harness is rerouted in same orientation as before.

| Tightening torque | |
|---|---|
| Camshaft sensor to upper timing chain cover (M6) | 10 Nm (7 ft-lb) |

- Check for fault codes and clear and reset DME control module memory.

## Intake camshaft sensor, replacing (M52, S52, M52 TU or M54 engine)

Intake camshaft sensor on M52, S52, M52 TU or M54 engine is mounted at left side of cylinder head near front.

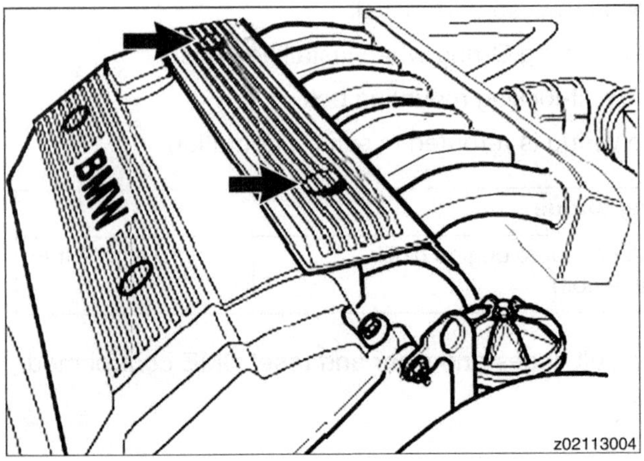


- Make sure ignition is OFF.
- Remove fuel injector rail cover hold down fasteners under small plastic trim covers (**arrows**). Remove cover.
- Remove air filter housing and ducts.

Working at left side of cylinder head:

- Detach engine vent hose from cylinder head cover by pinching spring clips (**arrows**).
- Detach VANOS solenoid electrical connector (**A**).
- Unscrew VANOS solenoid.
- If necessary, unfasten oil supply line on VANOS adjustment unit.

***NOTE—***
*Use shop rags to soak up leaking oil.*

Remove camshaft sensor mounting bolt. Remove sensor (**arrow**) from cylinder head.

– Reach under intake manifold to disconnect camshaft sensor harness.

***NOTE—***
*Prior to removal, attach a stiff piece of wire to connector to aid in proper routing of harness during reinstallation.*

– Installation is reverse of removal. Remember to:

- Replace camshaft sensor mounting bolt.
- Use new sealing O-rings when installing sensor, VANOS solenoid and VANOS oil supply line.
- Be sure sensor harness is rerouted as before.

| **Tightening torques** | |
|---|---|
| Intake camshaft sensor to cylinder head (M6, replace bolt) | 10 Nm (7 ft-lb) |
| Oil supply line to VANOS unit (banjo bolt) | 32 Nm (24 ft-lb) |
| VANOS solenoid to VANOS unit | 30 Nm (22 ft-lb) |

– Check for fault codes and clear and reset DME control module memory.

## Exhaust camshaft sensor, replacing (M52 TU, M54 engine)

Exhaust camshaft sensor on M52 TU or M54 engine is mounted at right side of cylinder head near front.

– Make sure ignition is OFF.

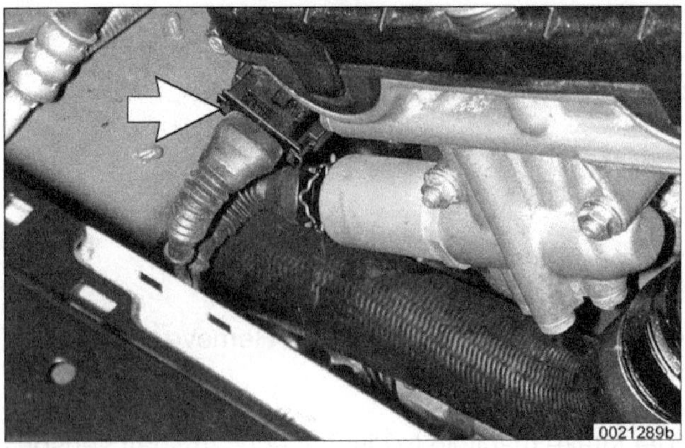


< Working at front of engine on exhaust (right) side, detach exhaust camshaft sensor electrical harness connector (**arrow**).

– Remove sensor mounting bolt and remove sensor.

– Installation is reverse of removal.
- Replace camshaft sensor mounting bolt.
- Use new sealing O-ring when installing sensor.

| **Tightening torque** | |
|---|---|
| Exhaust camshaft sensor to cylinder head (M6, replace bolt) | 10 Nm (7 ft-lb) |

– Check for fault codes and clear and reset DME control module memory.

## Intake camshaft sensor, replacing (S54 engine)

Intake camshaft sensor on S54 engine is mounted at left side of cylinder head near rear. Replacement requires intake manifold removal.

– Make sure ignition is OFF.

– Remove intake manifold. See **113 Cylinder Head Removal and Installation**.

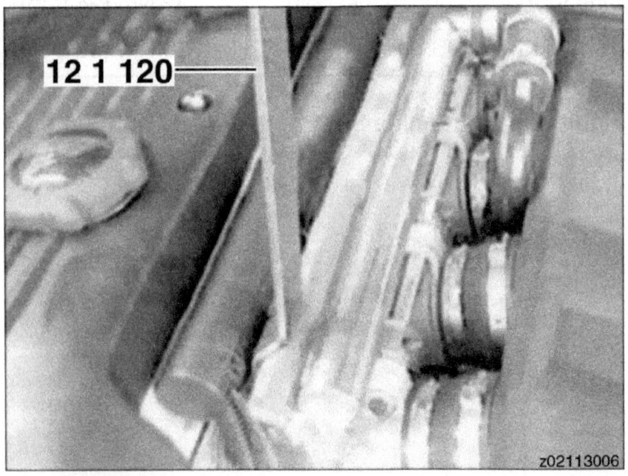


< Working from outer ends and moving towards center, carefully pry off fuel injector electrical connector strip. Use BMW special tool 12 1 120 or equivalent.

> ***CAUTION—***
> *Connector strip may break if it is removed without special tool 12 1 120.*

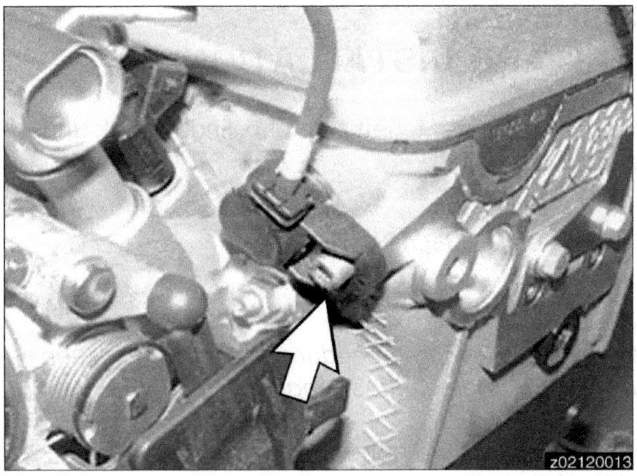


◄ Remove camshaft sensor mounting bolt (**arrow**) from left rear of cylinder head.

- Disconnect camshaft sensor harness.
- Remove sensor from cylinder head.

– Installation is reverse of removal. Remember to:

- Use new self-locking bolt and new O-ring when installing sensor.
- Be sure wiring is rerouted in same orientation.

> ***CAUTION—***
> *When reinstalling intake manifold, use only original clamps. Use of non-original hose clamps may result in interference with the throttle linkage and cause the throttle to jam or stick in the open position.*

| Tightening torque | |
|---|---|
| Intake camshaft sensor to cylinder head | 10 Nm (7 ft-lb) |

– Check for fault codes and clear and reset DME control module memory.

## Exhaust camshaft sensor, replacing (S54 engine)

Exhaust camshaft sensor on S54 engine is mounted at right side of cylinder head near rear.

– Make sure ignition is OFF.

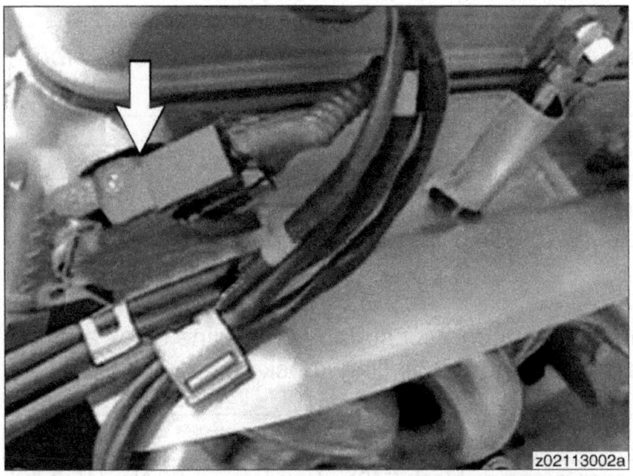


◄ Working at rear of engine on exhaust (right) side, detach exhaust camshaft position sensor electrical harness connector (**arrow**).

– Remove camshaft sensor mounting bolt and remove sensor from cylinder head.

– Installation is reverse of removal. Use new self-locking bolt and sealing O-ring.

| Tightening torque | |
|---|---|
| Exhaust camshaft sensor to cylinder head | 10 Nm (7 ft-lb) |

– Check for fault codes and clear and reset DME control module memory.

# KNOCK SENSOR REMOVAL AND INSTALLATION

Knock sensors are bolted to left side of cylinder block under intake manifold.

## Knock sensors, replacing

***NOTE—***
*The photos in the following procedure illustrate knock sensors in an M52 TU engine. Other engines are similar.*

– 4-cylinder engine: Remove upper intake manifold. See **113 Cylinder Head Removal and Installation**.

– 6-cylinder engine: Remove intake manifold. See **113 Cylinder Head Removal and Installation**.

◄ S54 engine: Remove idle speed control valve mounting nuts (**arrows**) and set valve aside.

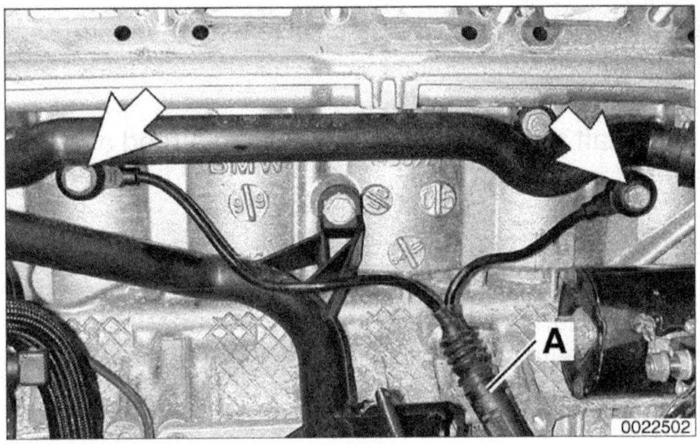


◄ Disconnect knock sensor electrical harness connector (**A**) on left side of engine cylinder block.

- Remove knock sensor mounting bolts (**arrows**) on side of cylinder block.
- Remove sensors.

> ***CAUTION—***
> *Note the installed angle of the knock sensor on the block before removing it. Reinstall the sensor in the same position. Be sure to use a torque wrench when tightening the sensor mounting bolt.*

***NOTE—***
*M52 TU engine with two knock sensors illustrated. Other engines are similar. S54 engine is equipped with 3 knock sensors.*

◄ Clean knock sensor contact surface on engine block and sensor (**arrows**) before installing knock sensor.

| Tightening torque | |
|---|---|
| Knock sensor to engine block (M8) | 20 Nm (15 ft-lb) |

– Check for fault codes and clear and reset DME control module memory.

# 121 Battery, Starter, Alternator





**TABLES**

## GENERAL

This repair group covers battery, starter, alternator and associated components of the electrical system.

Procedures in this section are grouped by engine code. See **100 Engine–General** for application information.

Charging system and starting system troubleshooting is in **Table a**. For additional information, see the following repair groups:

- **020 Maintenance** for drive belt service
- **600 Electrical System–General**
- **ELE Electrical Wiring Diagrams**

### Special tools

◀ Automotive digital multimeter

> ***CAUTION—***
> *Only use a digital multimeter when testing automotive electrical components.*

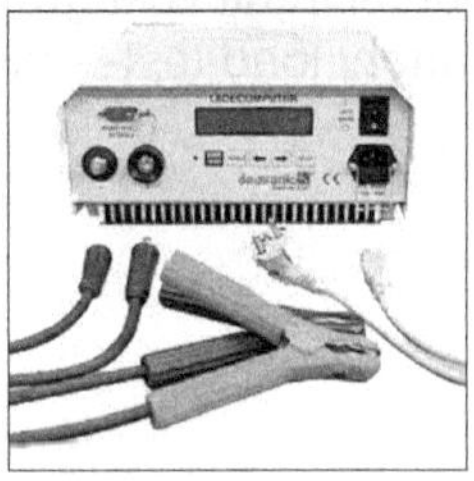

◀ Electronic battery charger (Deutronic DBL 1000-14)

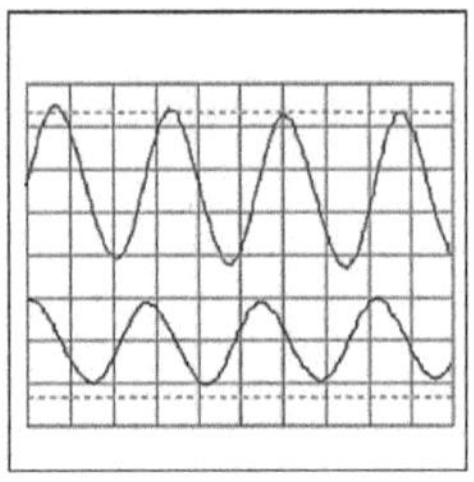

◀ Oscilloscope

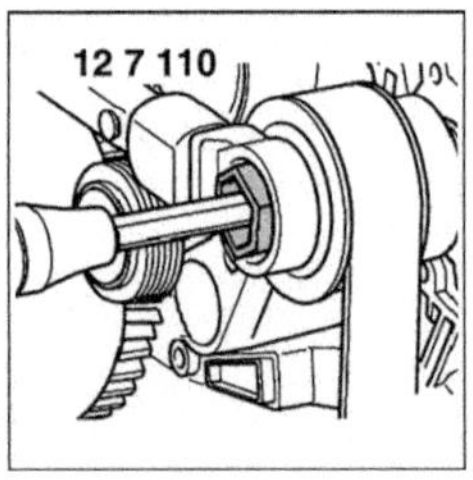


◀ Alternator pulley nut removal tool (BMW special too 12 7 110)

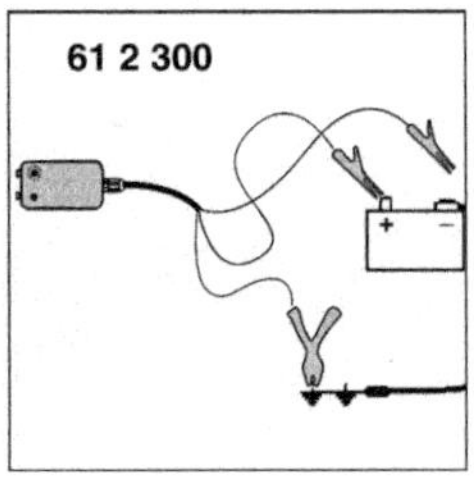


◀ Closed-circuit current measurement adapter (BMW special tool 61 2 300)

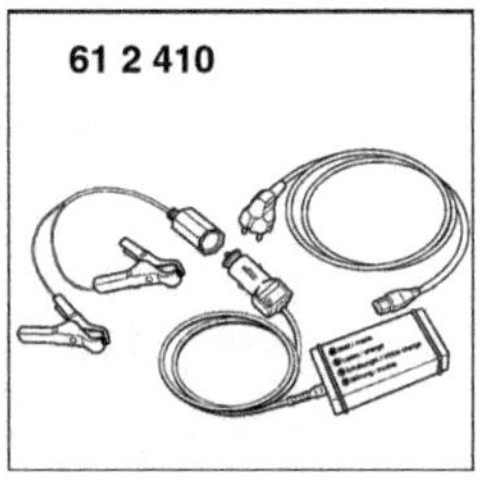


Cigarette lighter trickle charger (BMW special tool 61 2 410)

## Engine electrical system

The alternator and starter are wired directly to the battery. To prevent accidental shorts that might blow a fuse or damage wires and electrical components, always disconnect the negative (-) battery lead before working on the electrical system.

Several versions of alternators, voltage regulators, starters and batteries have been used in BMW models covered by this manual. Replace components according to the original equipment specification. When in doubt, consult an authorized BMW parts department.

## Battery safety terminal (SBK)

Z3 and M version models from 4 / 1999 are equipped with a battery safety terminal (SBK), installed at the battery positive pole. This feature and a comprehensive electrical fusing system aim to minimize the danger of short circuits in case of a severe accident.

The vehicle electrical system is divided into the starter-alternator circuit and the vehicle system supply circuit. The starter and the alternator supply circuit, since it must carry very high current when the starter is engaged and is permanently subjected to voltage, is protected by SBK.

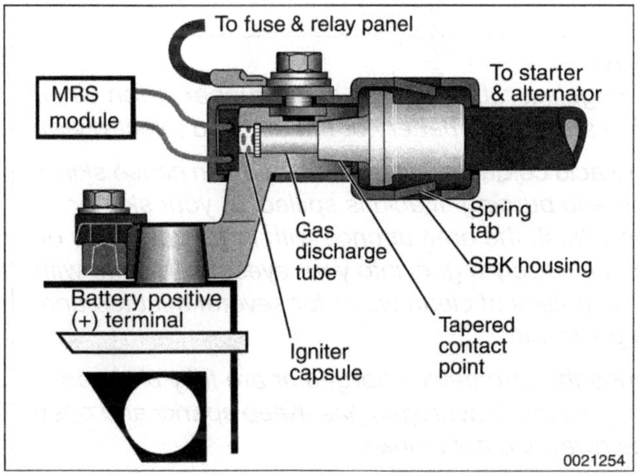


The multiple restraint system (MRS) module controls the SBK. In case of an accident, the MRS module fires the encapsulated pyrotechnic device in the SBK, disconnecting power to the starter and alternator, but maintaining power to the exterior lights and interior of the vehicle.

For additional MRS information, see **721 Airbag System (SRS)**.

## Troubleshooting

Tests for individual electrical system components are described under different components in this repair group. **Table a** gives some general troubleshooting ideas.

**Table a. Battery, starter and alternator troubleshooting**

| Symptom | Probable cause | Corrective action |
|---|---|---|
| Engine does not crank. | Fault in immobilizer system (EWS). | Try another ignition key. If problem persists, contact your authorized dealer. See **515 Central Locking and Anti-theft**. |
| Engine cranks slowly or not at all, solenoid clicks when starter is operated. | Battery cables loose, dirty or corroded. | Clean or replace cables. See **020 Maintenance**. |
| | Battery discharged. | Charge battery and test. Replace if necessary. See **020 Maintenance**. |
| | Battery to body ground cable loose, dirty or corroded. | Inspect ground cable. Clean, tighten or replace if necessary. |
| | Inadequate voltage at starter motor terminal 30. | Check connections, test for voltage at starter. Test for voltage at automatic transmission range switch or starter immobilization switch at clutch pedal. See **250 Gearshift Linkage** or **612 Switches**. |
| | Starter motor or solenoid faulty. | Test starter. See **Starter troubleshooting** in this repair group. |
| Battery does not stay charged more than a few days. | Short circuit draining battery. | Test for excessive current drain with everything electrical off. See **Closed-circuit current measurement** in this repair group. |
| | Short driving trips and high electrical drain on charging system. | Evaluate driving style. Where possible, reduce electrical consumption when making short trips. |
| | Engine drive belt loose, worn, damaged. | Inspect or replace belt. See **020 Maintenance**. |
| | Battery faulty. | Test battery and replace if necessary. See **Battery testing** in this repair group. |
| | Battery cables loose, dirty or corroded. | Clean or replace cables. See **020 Maintenance**. |
| | Alternator or voltage regulator faulty. | Test alternator and voltage regulator. See **Charging system troubleshooting** in this repair group. |

## Warnings and cautions

***WARNING—***

- *Wear goggles, rubber gloves and a rubber apron when working around the battery or battery acid (electrolyte).*
- *Battery acid contains sulfuric acid and can cause skin irritation and burning. If acid is spilled on your skin or clothing, flush the area at once with large quantities of water. If electrolyte gets into your eyes, flush them with large quantities of clean water for several minutes and see a physician.*
- *Batteries that are being charged or are fully charged give off explosive hydrogen gas. Keep sparks and open flames away. Do not smoke.*

> ***CAUTION—***
> - *Only use a digital multimeter when testing automotive electrical components.*
> - *Prior to disconnecting the battery, read the battery disconnection cautions in* **001 General Warnings and Cautions**.
> - *Before disconnecting the battery, switch the ignition OFF. Otherwise, diagnostic troubles codes (DTCs) may be set in some electronic control modules.*
> - *Disconnecting the battery cables may erase DTCs stored in ECM memory.*
> - *Disconnecting the battery erases the radio presets. Therefore, note stored stations and restore them after connecting the battery.*
> - *Stored settings of the on-board computer and clock are also lost when the battery is disconnected.*
> - *Always disconnect the negative (–) battery cable first and reconnect it last. Cover the battery post with an insulating material whenever the cable is removed.*
> - *Do not disconnect battery, alternator or starter wires while the engine is running.*
> - *Never reverse the battery cables. Even a momentary wrong connection can damage the alternator and other electrical components.*
> - *Do not depend on the color of insulation to tell battery positive and negative cables apart. Label cables before removing.*

## BATTERY

***NOTE—***
*Battery replacement and basic battery service is covered in* **020 Maintenance**.

In 1996 Z3 roadster models, the battery is located in the right rear of the engine compartment.

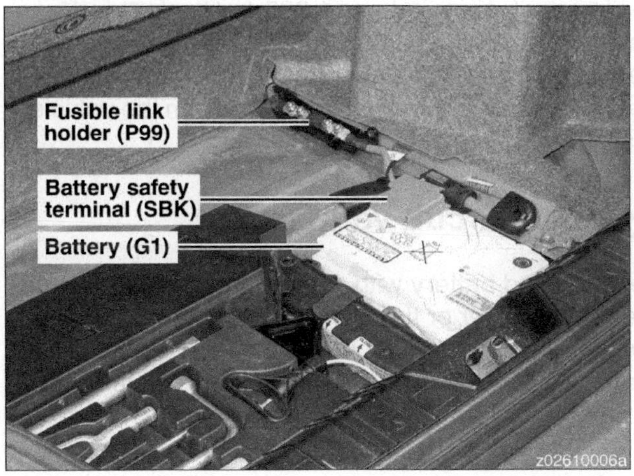


In later models the battery is mounted in the floor of the trunk, concealed under a trim panel.

***NOTE—***
*2000 Z3 Roadster trunk is illustrated. Other models are similar.*

Battery capacity is determined by the amount of current needed to start the vehicle and by the amount of current consumed by the electrical system. This varies from model to model.

BMW batteries are rated by ampere / hours (Ah) and cold cranking amps (CCA) rating, listed on the battery. The Ah rating is determined by the average amount of current the battery can deliver over time without dropping below a specified voltage. The CCA is determined by the battery's ability to deliver starting current at 0°F (-18°C) without dropping below a specified voltage. Always replace the battery with one having the same or higher ratings.

| Recommended battery rating | |
|---|---|
| Z3 / M Coupe or Roadster | 12 volt / 70 Ah |

## Battery testing

Battery testing determines the state of battery charge. For conventional or low-maintenance batteries, the most common method of testing the battery is that of checking the specific gravity of the electrolyte using a hydrometer. Before testing the battery, check that the cables are tight and free of corrosion.

## Hydrometer testing

Before hydrometer testing, load the battery with 15 amperes for one minute. If the battery is installed in the vehicle, this can be done by turning on the headlights without the engine running. The state of battery charge based on electrolyte specific gravity values are given in **Table b**.

**Table b. Electrolyte specific gravity *vs.* battery charge**

| Specific gravity at 80°F (27°C) | State of charge |
|---|---|
| 1.265 | Fully charged |
| 1.225 | 75% charged |
| 1.190 | 50% charged |
| 1.155 | 25% charged |
| 1.120 | Fully discharged |

The hydrometer indicates the specific gravity (or density) of the electrolyte. The denser the concentration of sulfuric acid in the electrolyte, the higher the state of charge.

If the specific gravity is at or above 1.225, but the battery lacks power for starting, determine the battery's service condition with a load voltage test. If the average specific gravity of the six cells is below 1.225, remove the battery from the vehicle and recharge.

## Battery open-circuit voltage test

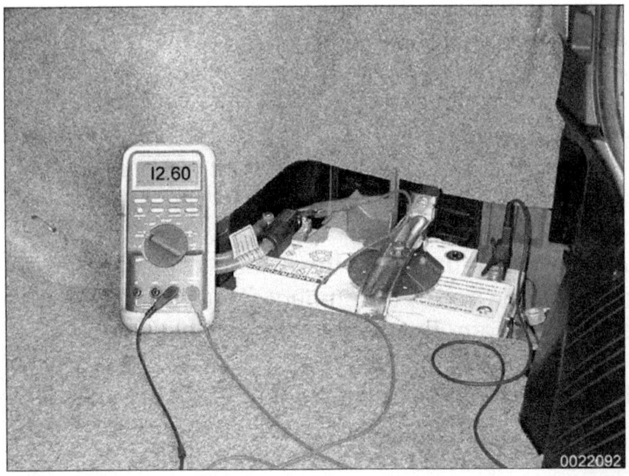


Before testing, load battery with 15 amperes for one minute with battery load-tester or turn on headlights for about one minute without engine running. Connect digital voltmeter across battery terminals. Open-circuit voltage levels are given in **Table c**.

– If open-circuit voltage is OK but battery still lacks power for starting, perform a load voltage test.

– If open-circuit voltage is below 12.4 vdc, recharge battery and retest.

| Table c. Open-circuit voltage *vs.* battery charge | |
|---|---|
| **Open-circuit voltage** | **State of charge** |
| 12.6 vdc or more | Fully charged |
| 12.4 vdc | 75% charged |
| 12.2 vdc | 50% charged |
| 12.0 vdc | 25% charged |
| 11.7 vdc or less | Fully discharged |

## Battery load voltage test

A battery load tester is required for a load voltage test. The test is made by applying a high resistive load to the battery terminals and then measuring battery voltage. The battery should be fully charged for the most accurate results. The battery cables must be disconnected before making the test. Replace the battery if the voltage is below that listed in **Table d**.

> ***WARNING —***
> *Always wear protective goggles and clothing when performing a load test.*

| Table d. Battery load test—minimum voltage | |
|---|---|
| **Ambient temperature** | **Voltage*** |
| 27°C (80°F) | 9.6 vdc |
| 16°C (60°F) | 9.5 vdc |
| 4°C (40°F) | 9.3 vdc |
| -7°C (20°F) | 8.9 vdc |
| -18° C (0° F) | 8.5 vdc |
| * Measure after applying a 200 amp load for 15 seconds. | |

## Closed-circuit current measurement

If the vehicle battery is discharged for an unknown reason, perform a closed-circuit current measurement as the first test. For diagnostic purposes, it is important to not disconnect the battery. Disconnecting the battery may reset a faulty control unit, preventing proper diagnosis.

The following test is generally completed overnight using a multimeter with a recording function and a closed circuit current measurement adapter (BMW special tool 61 2 300 or equivalent).

– Check that battery voltage is 12 vdc or higher. If lower, recharge battery.

– Switch off electrical consumers (interior lights, telephone, aftermarket equipment, etc.).

– Open trunk and remove battery trim panel. Use screwdriver to lock latch on trunk lid or hatch, simulating closed trunk. Close other doors.

***NOTE—***
*1996 model: Battery is in engine compartment. Open engine hood and use screwdriver to lock latch, simulating closed hood.*

– Open driver's door, then close it, simulating driver entering vehicle.

– Switch ignition to RUN position for at least five seconds, then turn ignition to OFF position. This simulates driving vehicle.

– Open and close driver's door, simulating driver leaving vehicle.

– Lock car and arm alarm.

– 1997 and later models: Wait at least 16 minutes for electrical consumer sleep mode to set in.

***NOTE—***
- *Starting with 1997 models, the consumer sleep mode function interrupts battery voltage to vehicle circuits, preventing inadvertent battery drain if one of the consumers remains activated. See* **600 Electrical System–General** *for more details.*
- *Automatic transmission cars: amber shifter LED goes out when consumer sleep mode sets in.*

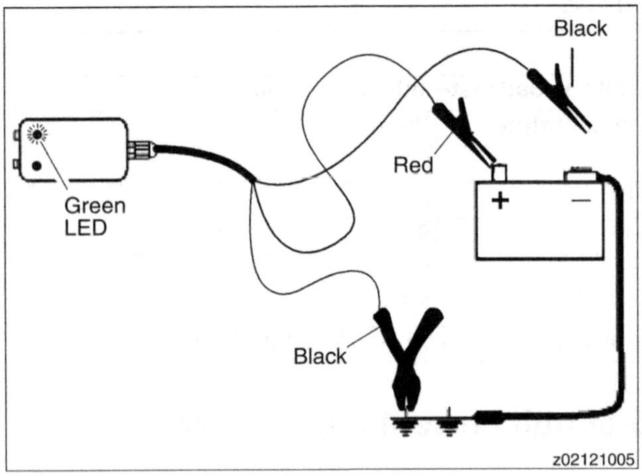


◄ Attach closed-circuit current measurement adapter (BMW special tool 61 2 300):
- Connect red clamp to positive battery terminal.
- Connect large black clamp to chassis ground point.

***NOTE—***
*Green LED lights up to indicate correct adapter installation.*

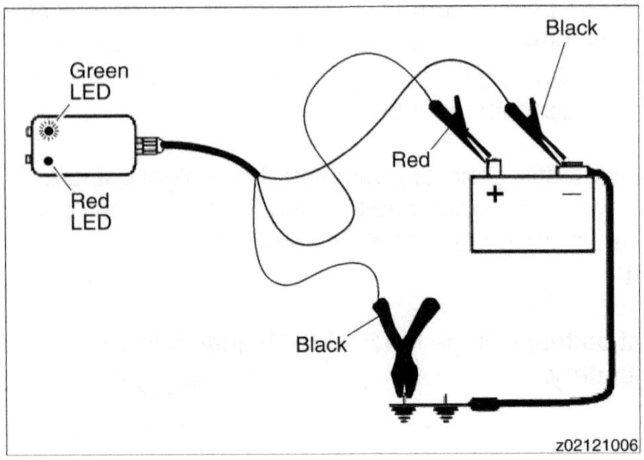


◄ Connect small black clamp to negative battery terminal. Green LED light goes out.

***NOTE—***
*Any electrical overloading is indicated at this point by lighting the red LED. If the red LED lights up, recheck the adapter connections.*

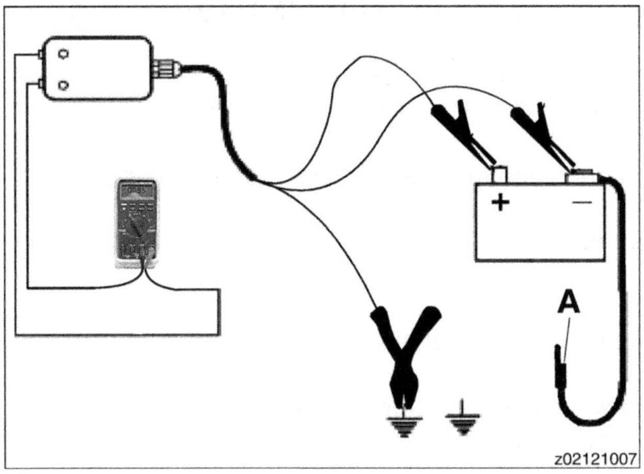


Set multimeter to 2 - 5 A measuring range and connect to adapter.

***NOTE—***
- *Set multimeter to record average readings.*
- *If possible, turn off powersaving features of multimeter which prevent a long term (overnight) test by automatically powering down meter.*

– Disconnect battery ground strap (**A**) at chassis mounting point.

– Record closed-circuit current using multimeter.

– After measuring current, reconnect battery ground before removing large tester clamp. Switch ignition ON for at least 5 seconds and then OFF again to prevent setting faults in control modules.

– Investigate readings over 30 mA.

***NOTE—***
*Momentary spikes of closed-circuit current may be recorded intermittently due to use of remote-control keys in other vehicles nearby or other radio transmitters. Such spikes can be ignored as a cause of battery discharge.*

## Battery, charging

Discharged batteries can be recharged using a battery charger. Remove the battery from the vehicle during charging.

Prolonged charging causes electrolyte evaporation to a level that can damage the battery. It is best to use a low-current charger (6 amperes or less) to prevent battery damage caused by overheating.

***WARNING—***
*Hydrogen gas given off by the battery during charging is explosive. Do not smoke. Keep open flames away from the top of the battery, and prevent electrical sparks by turning the battery charger OFF before connecting or disconnecting it.*

***CAUTION—***
- *Battery electrolyte (sulfuric acid) can damage the car. If electrolyte is spilled, clean the area with a solution of baking soda and water.*
- *Always allow a frozen battery to thaw before attempting to recharge it.*
- *Always disconnect both battery cables and remove battery from vehicle during battery charging. Do not exceed 16.5 charging voltage at the battery.*

## Battery safety terminal (SBK), replacing

If the battery safety terminal (SBK) has been triggered, investigate and correct the cause prior to replacement.

- Working in right side of trunk, remove battery trim panel.
- Disconnect negative (–) battery terminal and cover with insulator.
- Disconnect positive (+) cable from battery.
- Release positive (+) cable from cable brackets on trunk side wall. Lay new repair cable parallel to positive cable.
- Mark off length of new cable on old cable. Allow 10 cm (2.5 in) of slack.
- Saw through old cable.

> ***CAUTION—***
> *Do not use bolt cutters or similar tools to cut through cable. A cable end that has been squashed flat may not fit into the clamping sleeve of the new cable.*

- Strip approximately 15 mm (0.6 in) of insulation from cable end.

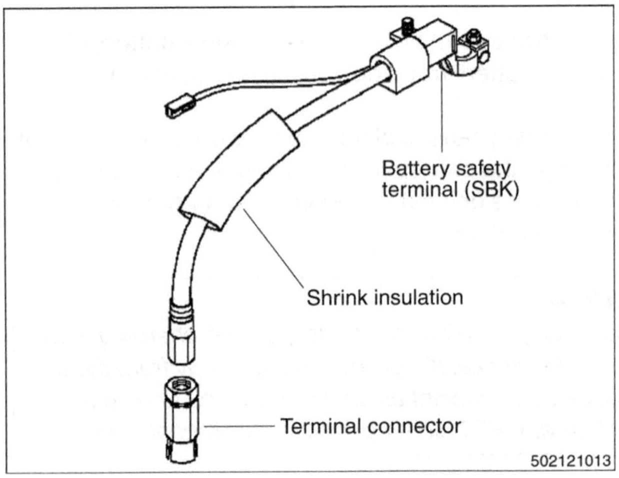


- Push terminal connector of new cable over stripped end of old cable. Tighten clamp.
- Slip shrink-fit insulation over positive cable. Attach new safety harness to end of positive cable. Shrink insulation with hot air blower.
- Reinstall cable in cable bracket and attach to battery.
- Reattach negative battery cable. Reinstall trunk trim.
- Use BMW scan tool DISplus, GT1 or MoDiC (or equivalent) to read out airbag control module (SRS) fault memory. Correct faults, then clear fault memory.

502121025

## ALTERNATOR (GENERATOR)

Several versions of alternators have been used in Z3 and M version cars. Representative wiring schematics on this page depict two of the major types of charging system circuitry used. Refer to **ELE Electrical Wiring Diagrams** for further details.

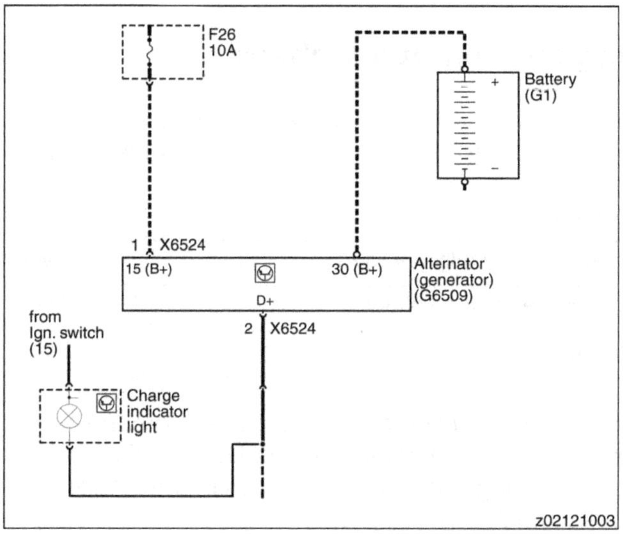


z02121003

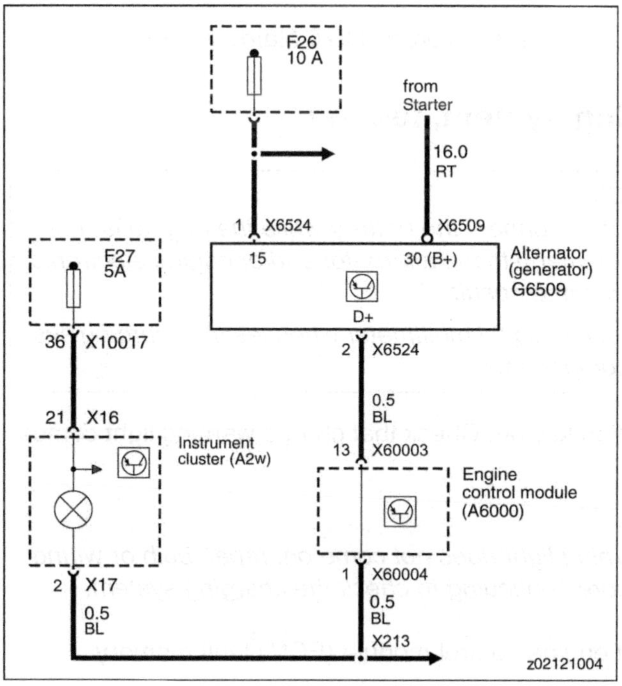


z02121004

## Charging system troubleshooting

Charging system diagnostics requires special test equipment. If the test equipment is not available, charging system fault diagnosis can be performed by an authorized BMW dealer or other qualified repair shop. See **Table a** for general electrical component troubleshooting.

Before checking the alternator, make sure the battery is fully charged and capable of holding a charge. Check that the battery terminals are clean and tight and the alternator drive belt is properly tensioned and not severely worn.

***NOTE—***
*The alternator requires up to 15 second to reach full output.*

## Charging system quick-check

Use a digital multimeter to measure voltage across the battery terminals with key off and then again with engine running.

| Battery voltage | |
|---|---|
| Engine at rest, battery fully charged | 12.6 vdc |
| Engine running, regulated voltage | 13.2 - 14.5 vdc |
| Faulty regulator | >14.8 vdc |

Check for clean and tight battery cables. Check ground cable running from negative (–) battery terminal to chassis and ground cable running from engine to chassis. Check alternator drive belt condition and tension.

***NOTE—***
*Drive belt service is covered in* **020 Maintenance**.

## Charging system, testing

***CAUTION—***
- *Do not disconnect the battery while the engine is running. Damage to the alternator and/or engine electronic systems may result.*
- *Only use a digital multimeter when testing charging system components.*

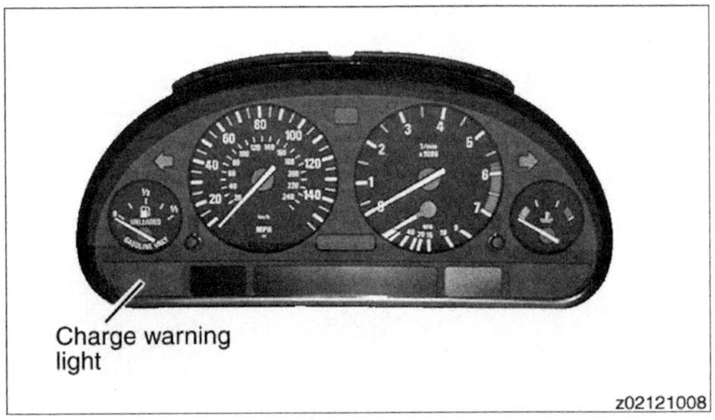


Turn ignition key on. Check that charge warning light comes on.

***NOTE—***
*If the warning light does not come on, repair bulb or wiring faults before continuing to check the charging system.*

– Read out engine control module (ECM) fault memory.

- Connect BMW diagnostic scan tool or oscilloscope to check alternator function.

***NOTE—***
*An alternator output pattern other than "normal" indicates that the alternator needs rebuilding or replacement.*

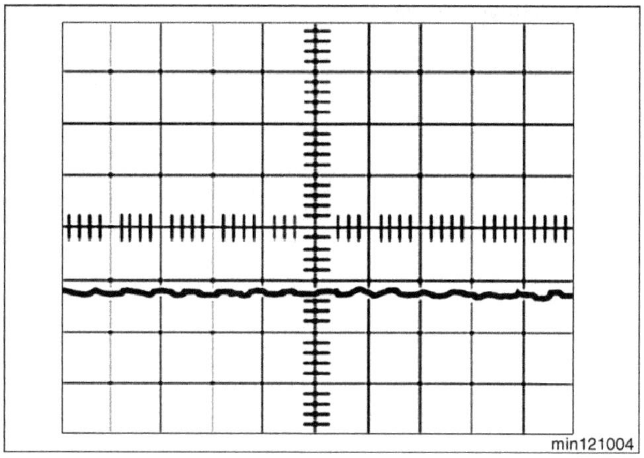
min121004

◄ Normal alternator

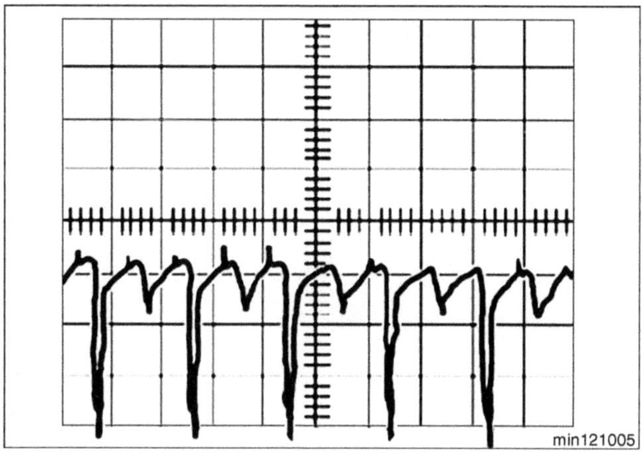
min121005

◄ One phase interrupted

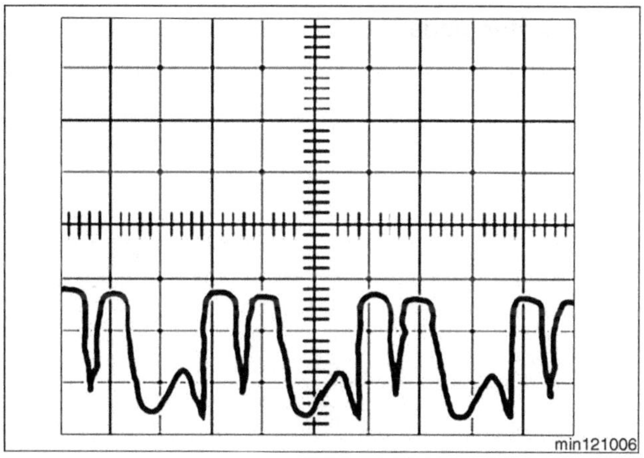
min121006

◄ Interturn (short circuit in rotor windings) fault

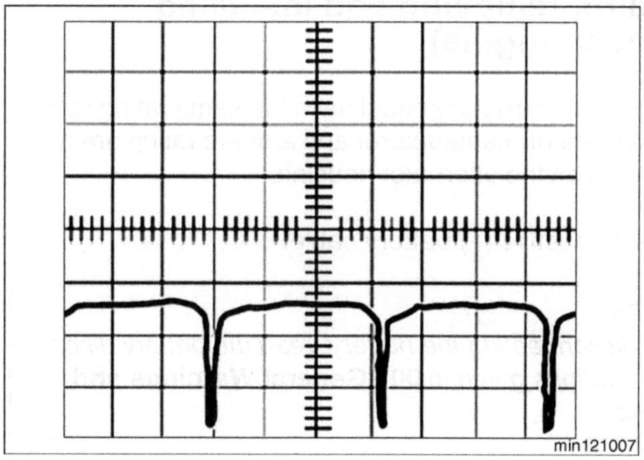
min121007

◄ Open circuit in negative diode

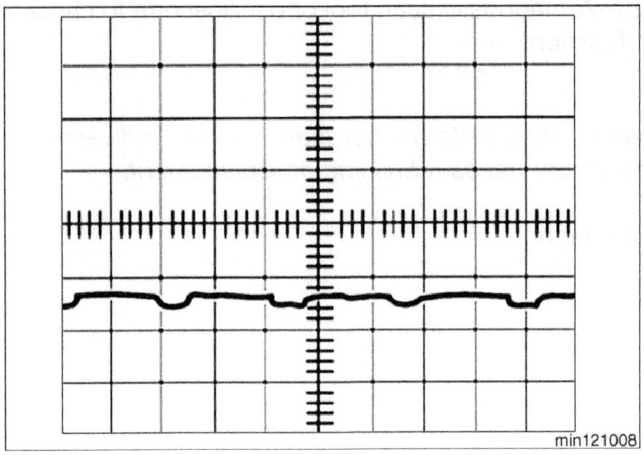


Short circuit in positive diode

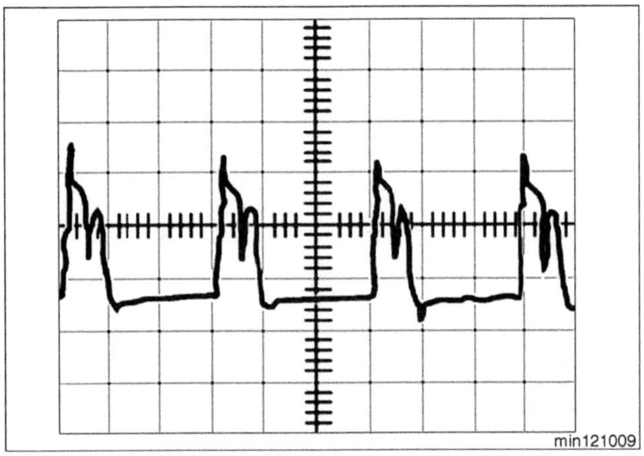


Open circuit in positive diode

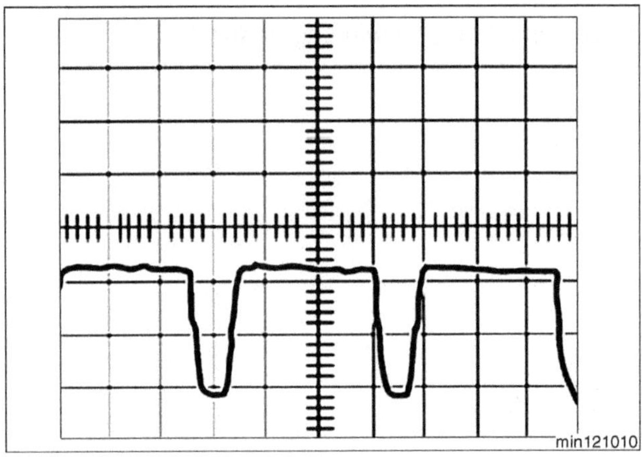


Open circuit in exciter diode

## Alternator, removing and installing (4-cylinder engine)

A replacement alternator should have the same rating as the original. Alternator manufacturer and ampere rating are normally marked on the alternator housing.

– Disconnect negative (–) battery cable.

> ***CAUTION—***
> *Prior to disconnecting the battery, read the battery disconnection cautions given in* **001 General Warnings and Cautions**.

– Remove air cleaner upper section with air flow sensor.

– Remove terminal cover from rear of alternator. Disconnect wiring.

– Remove drive belt from alternator pulley. See **020 Maintenance**.

***NOTE—***
*If reusing drive belt, mark direction of rotation on belt before removing.*

– Engine with hydraulic belt tensioner: Remove tensioner idler pulley (upper roller) from alternator bracket.

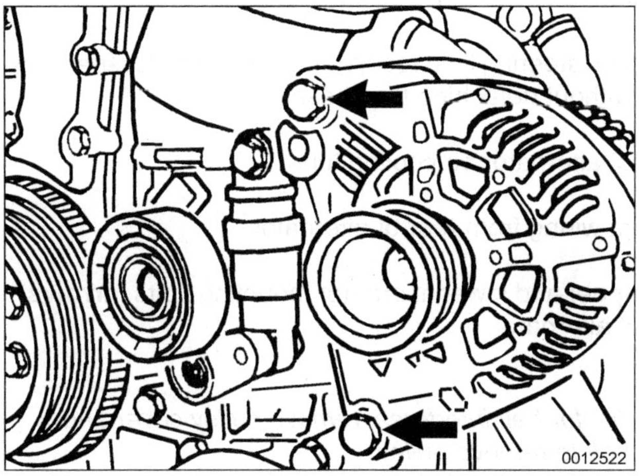


◄ Remove upper and lower alternator mounting bolts (**arrows**) and lift out alternator.

– Installation is reverse of removal. Install drive belt as described in **020 Maintenance**.

| **Tightening torques** | |
|---|---|
| Alternator to mounting bracket:<br>• M8 bolt<br>• M10 bolt | <br>21 Nm (16 ft-lb)<br>43 Nm (32 ft-lb) |
| Pulley to alternator (M16 nut) | 70 Nm (52 ft-lb) |
| Wiring terminals to alternator<br>• B+ wire to alternator (M8 nut)<br>• D+ wire to alternator (M6 nut) | <br>14 Nm (10 ft-lb)<br>6 Nm (53 in-lb) |

## Alternator, removing and installing (6-cylinder engine)

A replacement alternator should have the same rating as the original. Alternator manufacturer and ampere rating are normally marked on the alternator housing.

– Disconnect negative (–) battery cable.

***CAUTION—***
*Prior to disconnecting the battery, read the battery disconnection cautions given in* **001 General Warnings and Cautions**.

– Remove complete air filter housing and ducts. See **130 Fuel Injection**.

– Remove alternator drive belt. See belt removal procedures in **020 Maintenance**.

– Remove radiator cooling fan and fan shroud. See **170 Radiator and Cooling System**.

*NOTE—*
*The radiator cooling fan nut (32 mm wrench) has left-hand threads.*

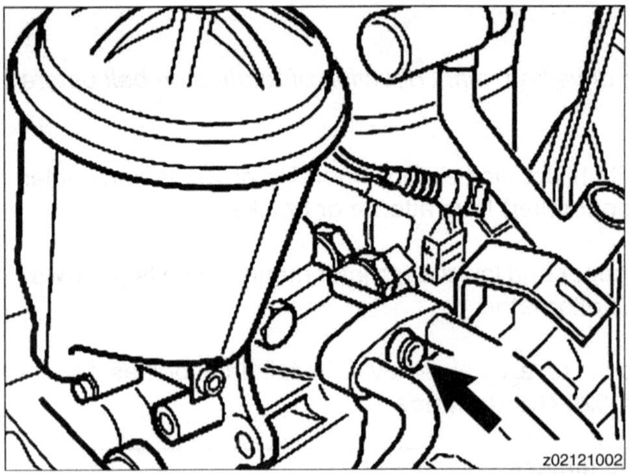


S54 engine:

- Loosen oil filter cover center bolt to allow engine oil to drain back down into oil pan.
- Remove oil cooler lines mounting bolt (**arrow**) from oil filter housing. Tie oil lines to side.

*NOTE—*
*Be prepared to catch leaking oil with a shop towel.*

– Unbolt power steering fluid reservoir and bracket. Suspend to one side without detaching fluid hoses.

– Disconnect alternator air cooling tube.

– Disconnect wiring from rear of alternator.

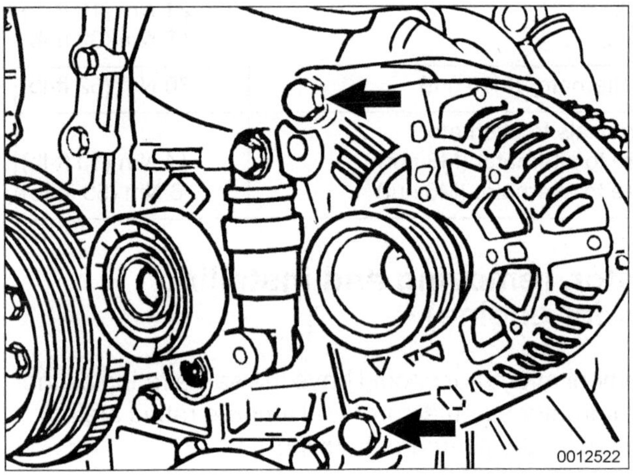


Remove upper and lower alternator mounting bolts (**arrows**) and lift out alternator.

*NOTE—*
*If idler pulley for belt interferes, remove pulley cover and retaining bolt to remove pulley.*

– Installation is reverse of removal.

| Tightening torques | |
|---|---|
| Alternator to mounting bracket (M10 bolt) | 43 Nm (32 ft-lb) |
| Belt tensioner to engine block<br>M8 (8.8 grade) | 22 Nm (16 ft-lb) |
| Fan clutch to coolant pump<br>• with BMW special tool 11 5 040<br>• without BMW special tool 11 5 040 | <br>30 Nm (22 ft-lb)<br>40 Nm (29 ft-lb) |
| Pulley to alternator (M16 nut) | 70 Nm (52 ft-lb) |
| Wiring terminals to alternator<br>• B+ wire to alternator (M8 nut)<br>• D+ wire to alternator (M6 nut) | <br>14 Nm (10 ft-lb)<br>6 Nm (53 in-lb) |

## Voltage regulator, removing and installing

– Disconnect negative (–) battery cable.

> ***CAUTION—***
> *Prior to disconnecting the battery, read the battery disconnection cautions given in* **001 General Warnings and Cautions**.

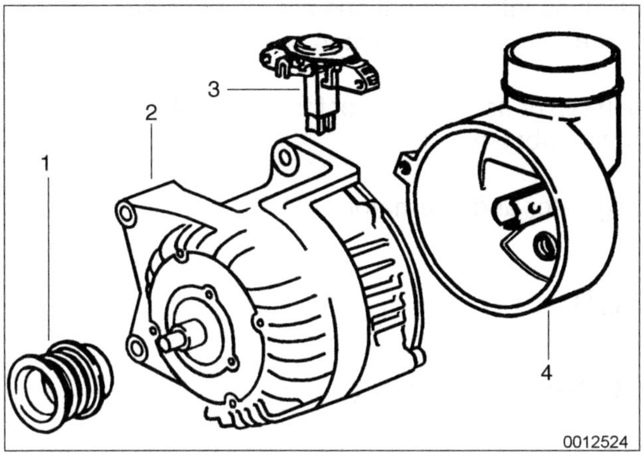


– Remove alternator. See **Alternator, removing and installing** in this repair group.

◀ Remove cooling duct cover from rear of alternator.

1. Alternator drive pulley
2. Alternator
3. Voltage regulator and brush holder
4. Cooling duct cover

– Remove voltage regulator mounting screws and remove regulator from alternator.

– Installation is reverse of removal. Clean brush contact surfaces in alternator and check brush length. See **Alternator brushes, inspecting and replacing** in this repair group.

***NOTE—***
*Bosch alternator is illustrated. Valeo alternator is similar.*

| Tightening torques | |
|---|---|
| Regulator to alternator<br>• M4<br>• M5 | <br>2 Nm (17 in-lb)<br>4 Nm (35 in-lb) |

## Alternator brushes, inspecting and replacing

Regulator brushes are not available as replacement parts from BMW. Replacement brushes may be available from aftermarket sources.

– Remove voltage regulator. See **Voltage regulator, removing and installing** in this repair group.

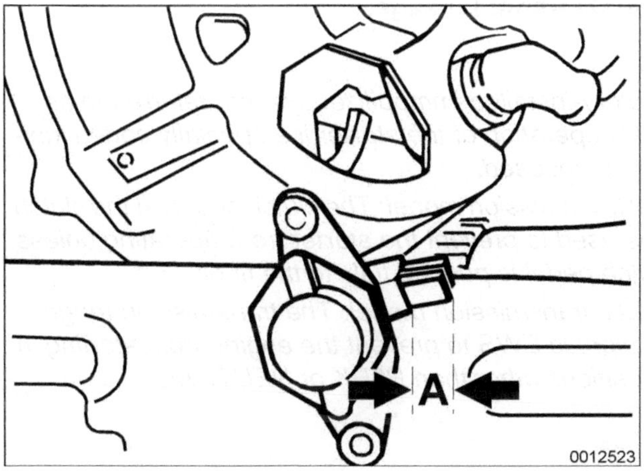


◀ Clean brush contact surfaces and measure brush protrusion.

| Voltage regulator brush protrusion | |
|---|---|
| Dimension **A** (minimum) | 5 mm (¼ in) |

– If brushes are worn or damaged, carefully and as quickly as possible unsolder brush lead from brush holder terminal, withdrawing brush from holder at same time.

***CAUTION—***
*Overheating of the solder joints may cause damage to the voltage regulator electronic components.*

– Remove traces of solder from brush holder terminal.

– Fit spring into brush holder and insert new brush.

– Guide brush lead into terminal and solder into place. Check for free movement of brushes when solder cools.

– Check brush slip rings in alternator for wear. Lightly clean slip rings using fine abrasive cloth.

– Reinstall regulator and alternator.

## STARTER

### Starter troubleshooting

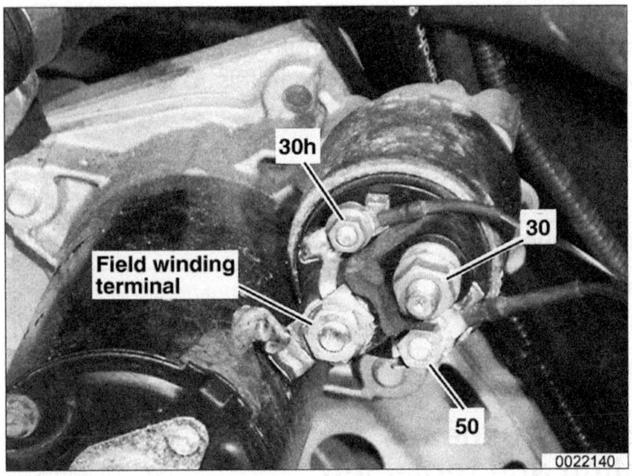


< Typical starter wiring terminal identification is shown. Large wire at terminal 30 is direct battery voltage. Smaller wire at terminal 50 operates starter solenoid via ignition switch.

– If starter turns engine slowly when ignition is in start position:
  - Check battery state of charge.
  - Inspect starter wires, terminals and ground connections for good contact. In particular, make sure ground connections between battery, body and engine are completely clean and tight.
  - If no faults are found, starter may be faulty and should be replaced.

– If starter fails to operate:
  - Check EWS (electronic immobilizer system). Try another ignition key. If no faults can be found, have the EWS system checked using BMW scan tool equipment. See **515 Central Locking and Anti-theft**.
  - Manual transmission: Check clutch switch. See **612 Switches**.
  - Automatic transmission: Check transmission range switch. See **250 Gearshift Linkage**.

***NOTE —***
- *The factory-installed immobilizer system, referred to as EWS, Rs operation of the starter if a specially coded ignition key is not used.*
- *Manual transmission model: The clutch switch at the clutch pedal is used to prevent the starter from operating unless the clutch pedal is pushed fully to the floor.*
- *Automatic transmission model: The transmission range switch signals EWS to prevent the engine from starting in gear positions other than PARK or NEUTRAL.*

– With key in START position, check for battery voltage at terminal 50 of starter motor.
  - If voltage is not present, check wiring between ignition switch and starter terminal.
  - Check EWS system and other sources of potential interruption to solenoid power, such as an aftermarket anti-theft system.
  - If voltage is present and no other visible wiring faults can be found, problem is most likely in starter motor.

## Starter, removing and installing (4-cylinder engine)

– Disconnect negative (–) battery cable.

> ***CAUTION—***
> *Prior to disconnecting the battery, read the battery disconnection cautions given in* **001 General Warnings and Cautions**.

– Raise car and support in a safe manner.

> ***WARNING —***
> *Make sure the car is stable and well supported at all times. Use a professional automotive lift or jack stands designed for the purpose. A floor jack is not adequate support.*

– Remove guide tube for oil dipstick.

– Disconnect wiring from starter.

– Remove top and bottom starter bolts. Remove top bolt from above and bottom bolt from below.

– Pull starter downward and rotate until solenoid is at top, then remove from car.

– Check starter pinion gear and flywheel teeth for damage.

– Installation is reverse of removal.

| **Tightening torques** | |
|---|---|
| Starter to engine (M10) | 47 Nm (35 ft-lb) |
| Wire terminals to starter<br>• M5 nut<br>• Terminal 50 (M6 nut)<br>• Terminal 30 (M8 nut) | <br>5 Nm (44 in-lb)<br>7 Nm (62 in-lb)<br>13 Nm (10 ft-lb) |

## Starter, removing and installing (M52 or S52 engine with manual transmission)

The starter on M52 or S52 engine with manual transmission is removed from below. Access to starter bolts is difficult. Be sure to have the proper tools on hand before proceeding.

***NOTE —***
*Access to the starter wiring is very limited. For better access, remove the intake manifold.*

– Disconnect negative (–) battery cable.

> ***CAUTION—***
> *Prior to disconnecting the battery, read the battery disconnection cautions given in* **001 General Warnings and Cautions**.

– Raise car and support in a safe manner.

> ***WARNING —***
> *Make sure the car is stable and well supported at all times. Use a professional automotive lift or jack stands designed for the purpose. A floor jack is not adequate support.*

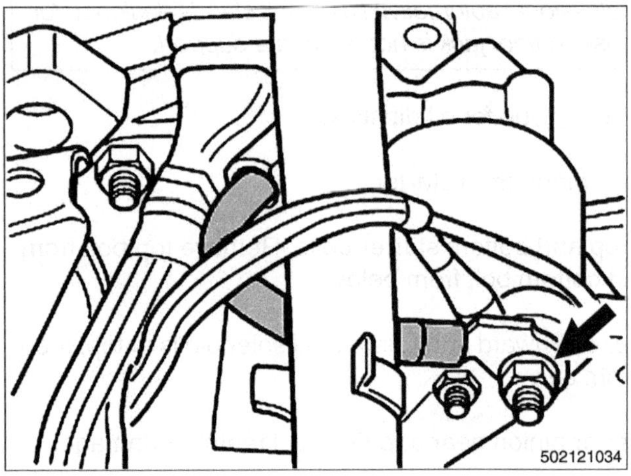


– Working underneath car, remove body crossbrace under rear of engine.

– Detach fuel lines and harness connectors from retaining brackets, as necessary.

◄ Remove fasteners (**arrow**) and disconnect wiring from starter.

- Note routing for cables. Routing varies depending on model and transmission.

– Unbolt starter support bracket from starter and engine block.

– Using a long extension and universal joint from output end of transmission, remove bolts fastening starter to transmission bell housing and engine block.

– Remove starter from below.

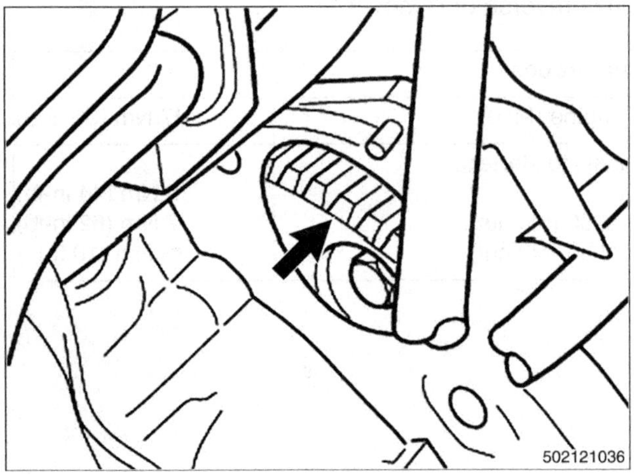


◄ Check starter pinion gear and flywheel teeth (**arrow**) for damage.

– Installation is reverse of removal.
  - Make sure starter electrical wires do not chafe.

| Tightening torques | |
|---|---|
| Starter to engine (M10) | 47 Nm (35 ft-lb) |
| Starter support bracket to starter | 5 Nm (44 in-lb) |
| Starter support bracket to engine | 47 Nm (35 ft-lb) |
| Wire terminals to starter<br>• M5 nut<br>• Terminal 50 (M6 nut)<br>• Terminal 30 (M8 nut) | <br>5 Nm (44 in-lb)<br>7 Nm (62 in-lb)<br>13 Nm (10 ft-lb) |

## Starter, removing and installing (M52 engine with automatic transmission)

The starter on M52 engine with automatic transmission is removed from above. It is necessary to remove the intake manifold and disconnect a number of electrical harness connectors.

– Disconnect negative (–) battery cable.

> ***CAUTION—***
> *Prior to disconnecting the battery, read the battery disconnection cautions given in* **001 General Warnings and Cautions**.

– Remove engine plastic covers and air filter housing. Disconnect fuel lines and remove intake manifold. See **113 Cylinder Head Removal and Installation**.

> ***WARNING —***
> *When disconnecting fuel lines, fuel may be expelled under pressure. Do not smoke or work near heaters or other fire hazards. Keep a fire extinguisher handy. Before disconnecting fuel hoses, wrap a cloth around fuel hoses to absorb any leaking fuel. Plug all open fuel lines.*

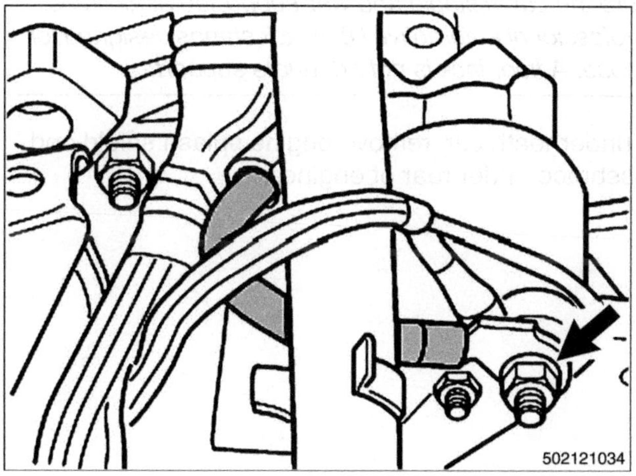


◀ Remove fasteners (**arrow**) and disconnect wiring from starter.
  - Note routing for cables. Routing varies depending on model and transmission.

– Raise car and support in a safe manner.

> ***WARNING —***
> *Make sure the car is stable and well supported at all times. Use a professional automotive lift or jack stands designed for the purpose. A floor jack is not adequate support.*

– Unbolt starter support bracket from starter and engine block.

– Using a long extension and universal joint from output end of transmission, remove bolts fastening starter to transmission bell housing and engine block.

– Remove starter from above.

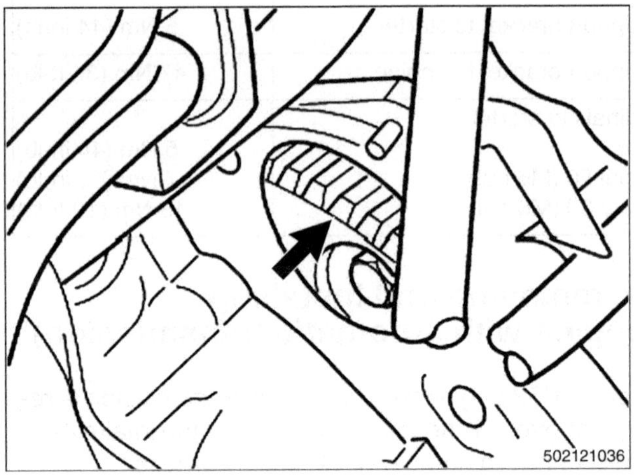


◄ Check starter pinion gear and flywheel teeth (**arrow**) for damage.

– Installation is reverse of removal.

- Make sure starter electrical wires do not chafe.

| **Tightening torques** | |
|---|---|
| Starter to engine (M10) | 47 Nm (35 ft-lb) |
| Starter support bracket to starter | 5 Nm (44 in-lb) |
| Starter support bracket to engine | 47 Nm (35 ft-lb) |
| Wire terminals to starter<br>• M5 nut<br>• Terminal 50 (M6 nut)<br>• Terminal 30 (M8 nut) | <br>5 Nm (44 in-lb)<br>7 Nm (62 in-lb)<br>13 Nm (10 ft-lb) |

## Starter, removing and installing (M52 TU, M54 or S54 engine)

The starter on M52 TU, M54 or S54 engine is removed from below. Access to starter bolts is difficult. Be sure to have the proper tools on hand before proceeding.

– Disconnect negative (–) battery cable.

> ***CAUTION—***
> *Prior to disconnecting the battery, read the battery disconnection cautions given in* **001 General Warnings and Cautions**.

– Raise car and support in a safe manner.

> ***WARNING —***
> *Make sure the car is stable and well supported at all times. Use a professional automotive lift or jack stands designed for the purpose. A floor jack is not adequate support.*

– Working underneath car, remove engine splash shield and body crossbrace under rear of engine.

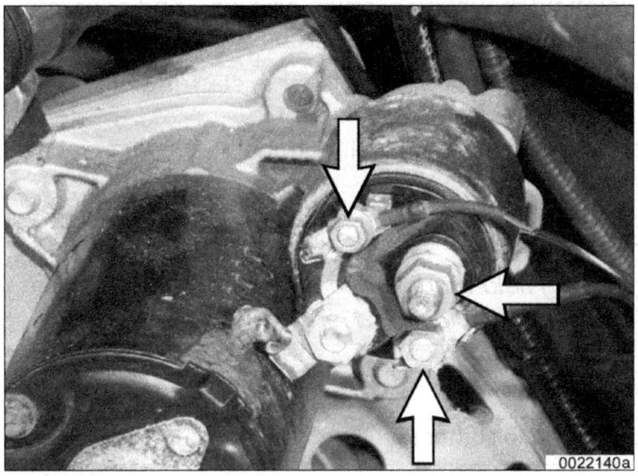

0022140a

◄ Working under intake manifold, remove starter wiring fasteners (**arrows**) and detach harness.

– Starter upper bolt: Reach down from above to loosen and remove. If necessary, detach wiring loom at upper engine compartment bulkhead, then loosen and push aside heater valve for better access to starter bolt.

– Starter lower bolt: Working underneath car, use long extension and universal joint from output end of transmission to loosen and remove.

– Remove starter from below.

> ***CAUTION—***
> *Make sure fuel lines and connectors are not damaged as starter is removed.*

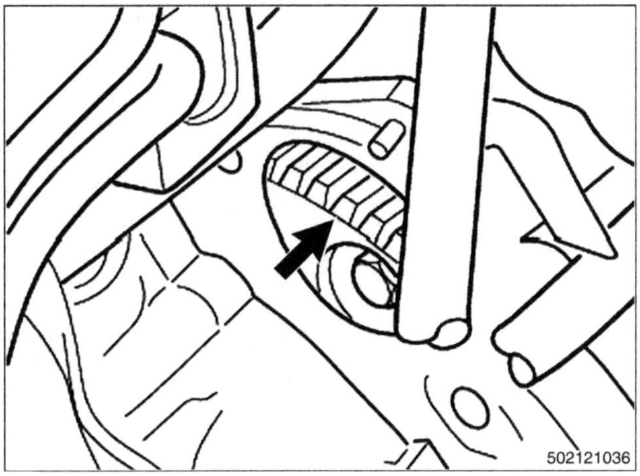

502121036

◄ Check starter pinion gear and flywheel teeth (**arrow**) for damage.

– Installation is reverse of removal.

- Make sure starter electrical wires do not chafe.

| Tightening torques | |
|---|---|
| Starter to engine (M10) | 47 Nm (35 ft-lb) |
| Wire terminals to starter<br>• M5<br>• Terminal 50 (M6)<br>• Terminal 30 (M8) | <br>5 Nm (44 in-lb)<br>7 Nm (62 in-lb)<br>13 Nm (10 ft-lb) |

## Starter solenoid, removing and installing

– Remove starter. See **Starter, removing and installing** in this repair group.

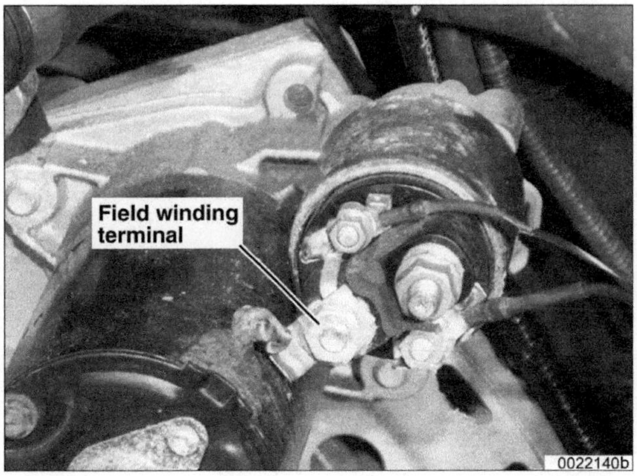


0022140b

◄ Disconnect field winding strap between starter motor and solenoid switch.

***NOTE—***
*If the field winding strap is damaged or burned, a new or rebuilt starter motor is needed.*

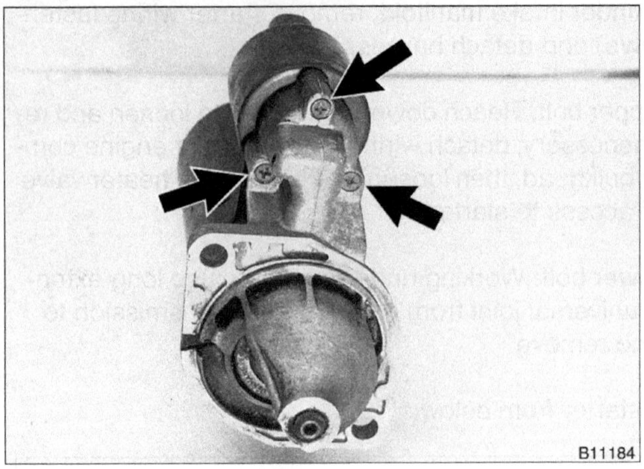


Remove solenoid mounting screws (**arrows**), and separate solenoid from starter.

- Unhook solenoid plunger from lever in starter drive.

– Installation is reverse of removal. Lubricate solenoid piston with light grease.

> ***CAUTION—***
> *When installing field winding strap to starter, position it so that it does not contact the starter body.*

| **Tightening torque** | |
|---|---|
| Field winding strap to starter (M8) | 13 Nm (10 ft-lb) |

# 130 Fuel Injection



**TABLES**

## General

This repair group covers service and repair of the Bosch and Siemens engine management systems used in Z3 and M Coupe and Roadster models.

Procedures in this section are grouped by engine and engine management (DME) code:

- 4-cylinder engine (DME M5.2): Start on page 130-9
- M52, S52 engine (DME MS 41.1): Start on page 130-23
- M52 TU, M54 engine (DME MS 42.1, MS 43.0): Start on page 130-39
- S54 engine (DME MS S54): start on page 130-59

See **Table a** for engine code and DME application data. See **100 Engine–General** for more information.

See the following repair groups for additional engine management service and repair information:

- **020 Maintenance** for oil change, air and fuel filter replacement, spark plug replacement
- **113 Cylinder Head** for intake manifold removal and installation
- **120 Ignition System** for crankshaft sensor, camshaft sensors, knock sensors replacement
- **160 Fuel Tank and Fuel Pump** for fuel system service and troubleshooting, fuel pressure measurement, evaporative system description
- **180 Exhaust System** for oxygen sensor replacement
- **610 Electrical Component Locations** for fuses, relays, control modules
- **ELE Electrical Wiring Diagrams** for wiring and component identification
- **OBD On-Board Diagnostics** for diagnostic trouble codes (DTCs), readiness codes, emissions controls description

### Tools and test equipment

If the equipment required to do the job is unavailable, leave these repairs to an authorized BMW dealer or other qualified BMW repair facility.

Automotive digital multimeter

> ***CAUTION—***
> *Only use a digital multimeter when testing automotive electrical components.*

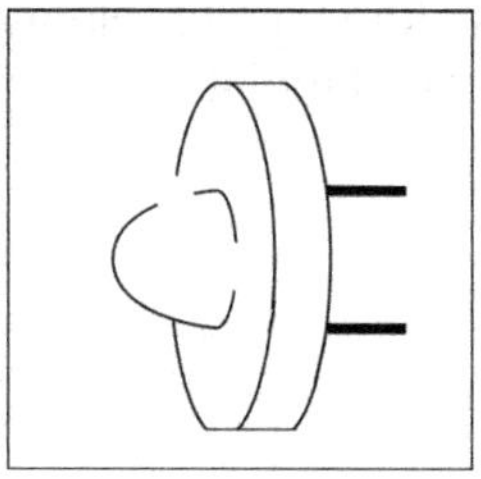

Low current test light ("noid")
(Baum tool 1115a)

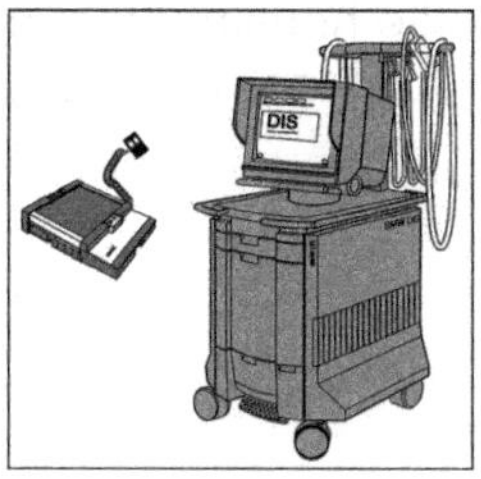

Factory DTC scan tool
(BMW DISplus / GT1 / MoDiC III)

Oxygen sensor tester
(BMW special tool 11 7 450)

Pry tool for fuel injector rail (S54 engine)
(BMW special tool 12 1 120)

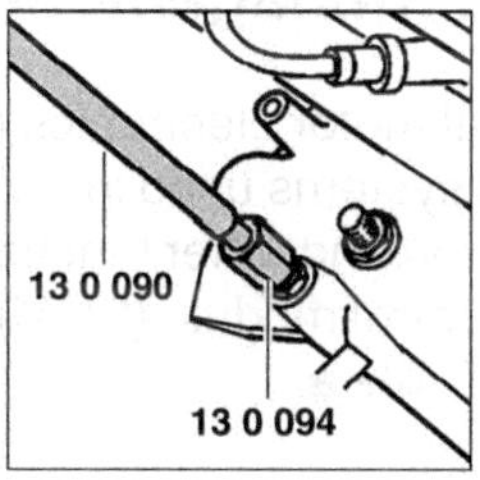


Exhaust gas probe
(BMW special tool kit 13 0 090)

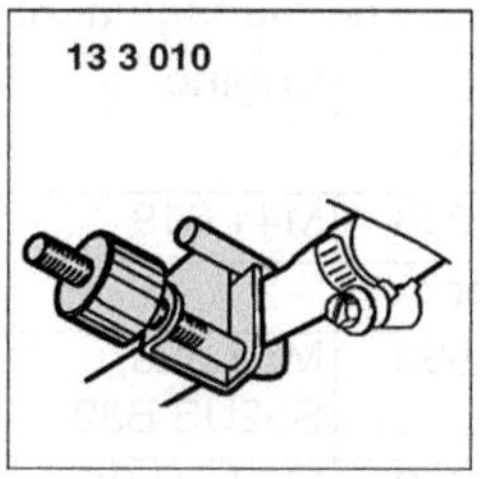


Pinch clamp
(BMW special tool 13 3 010)

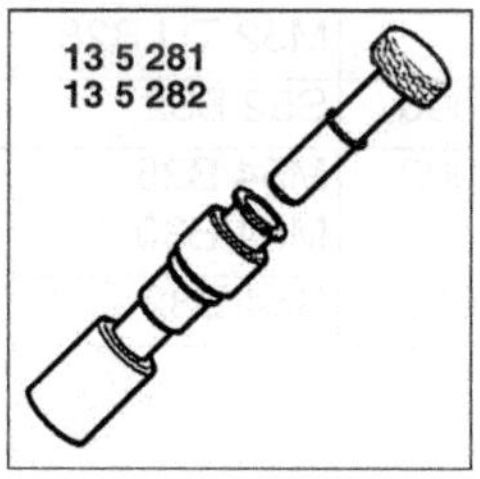


Fuel line plugs
(BMW special tools 13 5 281, 13 5 282)

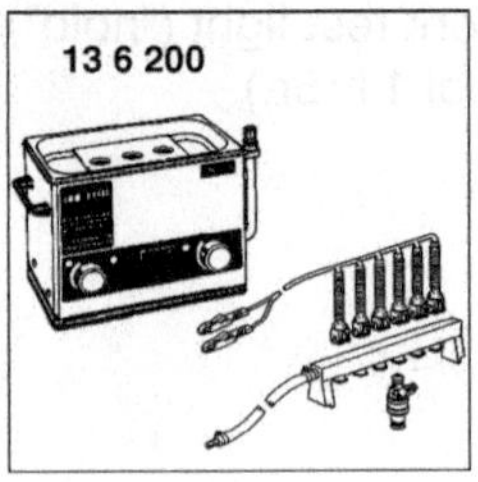


Ultrasonic fuel injector cleaner (BMW special tool kit 13 6 200)

EDK (throttle housing) test adapter (BMW special tool 13 8 030)

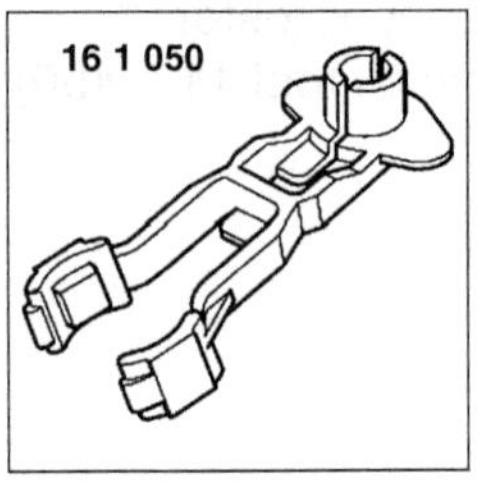


Fuel line fitting tool (BMW special tool 16 1 050)

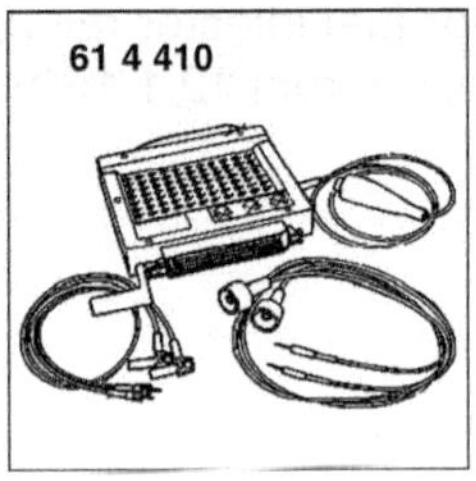


Control module break-out adapter (88 pin) (BMW special tool kit 61 4 410)

## Engine management

The digital motor electronics (DME or Motronic) engine management systems used in Z3 and M cars combine fuel injection, ignition and other functions under the control of the engine control module (ECM). DME system applications are given in **Table a.**

| **Table a. Engine management system applications** | | |
|---|---|---|
| **Year** | **Engine** | **DME system** |
| **4-cylinder** | | |
| 1996 - 1998 | M44 B19 | Bosch DME M5.2 |
| **6-cylinder** | | |
| 1997 - 1998 | M52 B28, S52US B32 | Siemens DME MS 41.1 |
| 1999 - 2000 | M52 TU B25, M52 TU B28 | Siemens DME MS 42.1 |
| 1999 - 2000 | S52 B32 | Siemens DME MS 41.1 |
| 2001 - 2002 | M54 B25, M54 B30 | Siemens DME MS 43.0 |
| 2001 - 2002 | S54 B32 | Siemens DME S54 |

Second generation on-board diagnostics (OBD II) is incorporated into the engine management systems used on the cars covered by this manual. Use a BMW-specific electronic scan tool, such as DISplus, GT1 or MoDiC, or a generic OBD II scan tool to access diagnostic trouble codes (DTCs). DTCs can help pinpoint ignition and other engine management problems.

## Engine management system troubleshooting

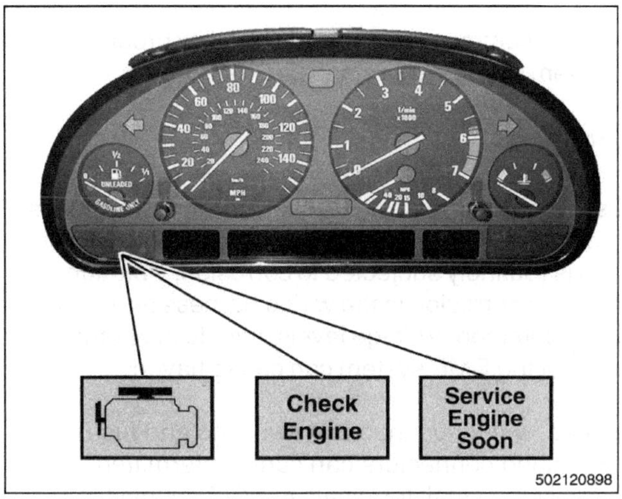

502120898

◄ If vehicle emission levels exceed 1.5 times Federally mandated criteria, the OBD II system illuminates a malfunction indicator light (MIL) in the instrument cluster.

***NOTE—***
*BMW models covered in this manual are equipped with one of three versions of MIL (as shown).*

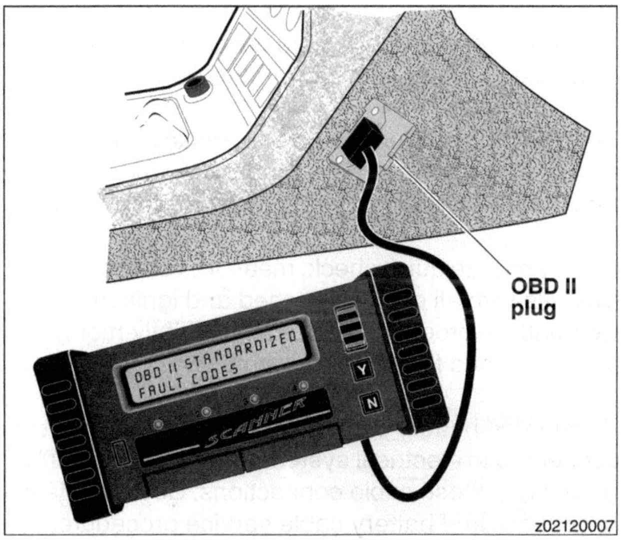

z02120007

◄ To begin engine management system troubleshooting, connect a diagnostic scan tool (DISplus, GT1, MoDiC or other BMW-compatible tool) to OBD II plug on right side of center console.

Interrogate engine control module (ECM) for stored faults. The OBD II system detects ignition and fuel injection faults and stores a diagnostic trouble code (DTC) in the ECM along with other pertinent fault information.

Additional OBD II information, including a DTC lookup table, can be found in **OBD On-Board Diagnostics**.

***NOTE—***
- *Professional diagnostic scan tools available at the time of this printing include the BMW factory tools DISplus, GT1, MoDiC and a small number of aftermarket BMW-specific tools. See* **Tools** *in* **010 General**.
- *In addition to the BMW-specific professional line of scan tools, there are inexpensive generic OBD II scan tool software programs and handheld units available. Although these tools have limited capabilities as compared to the dedicated tools, they are powerful diagnostic tools.*
- *For the DIY owner, a simple aftermarket DTC reader is available through http://www.peakeresearch.com. This tool is capable of checking for DTCs as well as turning off the illuminated MIL and resetting the service indicator lights.*

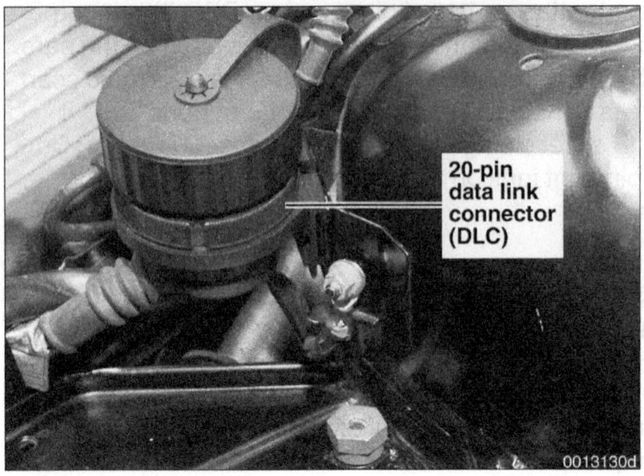


In addition to the OBD II plug, BMW cars produced through 6 / 2000 are also equipped with a 20-pin data link connector (DLC) in the engine compartment. Engine management faults and other DTC may be accessed and cleared and service lights reset via this connector as well as via the OBD II plug.

Additional OBD II information is in **OBD On-Board Diagnostics**.

## Driveability troubleshooting

Two of the most common causes of driveability problems are incorrect system voltage levels and poor grounds.

### System voltage

The DME system operates at low voltage and current levels, making it sensitive to small increases in resistance. The electrical system is routinely subjected to corrosion, vibration and wear, so faults or corrosion in the wiring harness and connectors are not uncommon. Voltage levels outside the normal operating range of the DME system can cause havoc.

– Visually inspect wiring, connectors, switches and fuses. Loose or damaged connectors can cause intermittent problems, especially the small terminals in ECM connectors.

– Disconnect wiring harness connectors to check for corrosion, and use electrical cleaning spray to remove contaminants.

– When troubleshooting an illuminated MIL, make sure battery is fully charged and capable of delivering all its power to electrical system. If battery is old (4 years or more), replace it.

– To make battery charge quick check, measure voltage across battery terminals with all cables attached and ignition off. A fully charged battery produces 12.6 vdc or slightly more, compared to 12.15 vdc for a battery with a 25% charge.

– Even a fully charged battery cannot deliver power unless it is properly connected to electrical system. Check battery terminals for corrosion or loose cable connections. See **020 Maintenance** for battery and battery cable service procedures.

– If battery does not maintain proper voltage, charging system may be at fault. See **121 Battery, Starter, Alternator**.

– If battery cable connection has no visible faults but is still suspect, measure voltage drop across connection.

- Large drop indicates excessive resistance, indicating that connection is corroded, dirty, or damaged.
- Clean or repair connection and retest.

### Engine management system grounds

Good grounds are critical to proper DME operation.

– If a ground connection has no visible faults but is still suspect, measure voltage drop across connection.
  - A large voltage drop indicates high resistance, indicating that connection is corroded, dirty or damaged.
  - Clean or repair connection and retest.

*NOTE—*
- *For ground positions, see* **610 Electrical Component Locations**.
- *For instructions on conducting a voltage drop test and other general electrical troubleshooting information, see* **600 Electrical System–General**.

## Exhaust emissions, testing

– Test exhaust emissions at the tailpipe(s) using an exhaust gas analyzer. Ensure engine is at normal operating temperature.

| Engine normal operating temperature | |
|---|---|
| Oil temperature | >60°C (140°F) |

– Check that engine is mechanically functioning correctly and there are no faults with ignition system.

– Attach DISplus, GT1, MoDiC or equivalent scan tool to OBD II port or DLC. See **Engine management system troubleshooting** in this repair group.

– Check for correct idle speed.

– Check ECM fault memory.

– Insert test probe in tailpipe and test emissions.

◄ If engine is equipped with precatalyst exhaust ports (**arrows**) in exhaust manifold(s), remove plugs and attach test probes (BMW special tool kit 13 0 090 or equivalent).

*NOTE—*
*Switch off shop exhaust ventilation system during tests.*

– If oxygen sensor fault is indicated by scan tool, use oxygen sensor test equipment (BMW special tool kit 11 7 450 or equivalent) to continue testing.

– Problems encountered during emissions test and possible solutions are shown in **Table b**.

| Table b. Engine emissions troubleshooting | |
|---|---|
| **Problem** | **Possible solutions** |
| Emissions at idle too high | Test and clean fuel injectors nozzles. Use BMW special tool 13 6 200 to clean nozzles. |
| | Test fuel pressure. See **160 Fuel Tank and Fuel Pump**. |
| | Coolant temperature too low. Check coolant temperature. |
| | Check engine coolant temperature (ECT) sensor output. |
| | Check for intake air leaks. |
| | Check idle speed control valve. |
| Emissions at idle too low | Check idle speed control valve hoses and connections. |
| | Check for intake air leaks. |

## Warnings and cautions

For personal safety, as well as the protection of sensitive electronic components, adhere to these warnings and cautions.

***WARNING —***
- *Gasoline is highly flammable and its vapors are explosive. Do not smoke or work on a car near heaters or other fire hazards when diagnosing and repairing fuel system problems. Have a fire extinguisher available in case of an emergency.*
- *Renew fuel system hoses, clamps and sealing O-rings any time they are removed.*
- *When working on an open fuel system, wear suitable hand protection, as prolonged contact with fuel can cause illnesses and skin disorders.*
- *The ignition system produces high voltages that can be fatal. Avoid contact with exposed terminals. Use extreme caution when working on a car with the ignition switched on or the engine running.*
- *Disable the ignition if operating the starter without starting the engine. See* **120 Ignition System**.

> ***CAUTION—***
> - *Disconnecting the battery may erase fault code(s) stored in memory. Check for fault codes prior to disconnecting the battery cables. Read the battery disconnection cautions in* **001 General Warnings and Cautions**.
> - *Connect and disconnect the DME system wiring and test equipment leads only when the ignition is switched off.*
> - *Wait at least 1 minute after turning off the ignition before removing the engine control module (ECM) connector. If the connector is removed before this time, residual power in the system relay may damage the control module.*
> - *Fuel system cleaners and other chemical additives other than those specifically recommended by BMW may damage catalytic converters, oxygen sensors or other fuel supply components.*
> - *Do not connect any test equipment that delivers a 12-volt power supply to terminal 15 (+) of the ignition coil. The current flow may damage the ECM. In general, connect test equipment only as specified by BMW or the equipment maker.*
> - *Only use a digital multimeter for electrical tests. Only use an LED test light for quick tests.*

## FUEL SUPPLY

Fuel supply to the injection system is provided by a high pressure electric pump mounted inside the fuel tank. The pump is powered via the fuel pump relay. If the relay is faulty, the fuel pump will not run and the engine will not start.

***NOTE—***
- *Fuel delivery volume and fuel pressure are critical to proper engine operation. Fuel pressure directly influences fuel delivery.*
- *Fuel pump relay, fuel pressure and fuel volume tests are given in* **160 Fuel Tank and Fuel Pump**.

### Relieving system fuel pressure

The fuel system retains fuel pressure in the system when the engine is off. To prevent fuel from spraying on a hot engine, relieve system fuel pressure before disconnecting fuel lines:

- Release fuel tank cap.
- Wrap a shop towel around a fuel line fitting and loosen or disconnect the fitting.

Cleanliness is essential when working with fuel components. Thoroughly clean the unions before disconnecting fuel lines.

## Secondary Air Injection

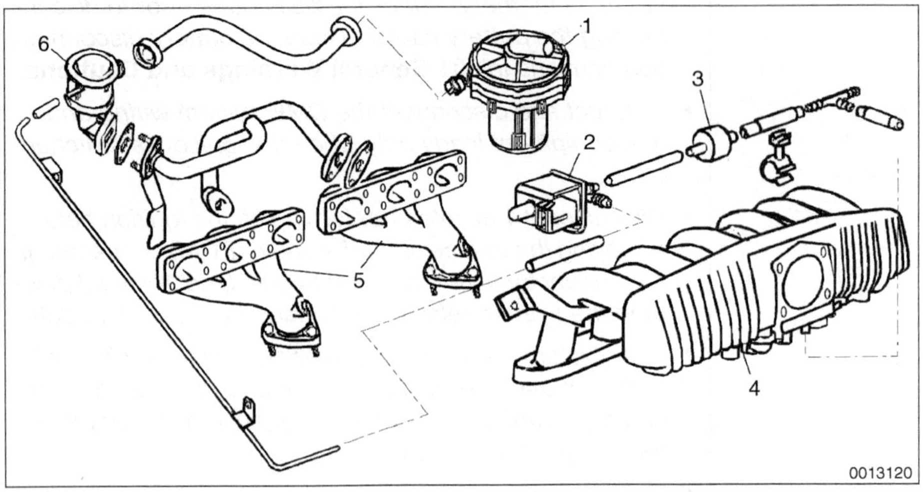


### Secondary air injection components

1. **Secondary air pump in right rear engine compartment**
2. **Secondary air vacuum solenoid (on intake manifold)**
3. **Secondary air check valve**
4. **Intake manifold**
5. **Exhaust manifolds**
6. **Diverter valve (at exhaust manifold)**

***NOTE—***
*Illustration is applicable to 6-cylinder models. Secondary air injection components in 4-cylinder models are similar.*

### Secondary air injection, principles of operation

The secondary air system pumps ambient air into the exhaust stream after a cold engine start to reduce the warm-up time of the catalytic converters and to reduce HC and CO emissions. The ECM controls and monitors the secondary air injection system. The secondary air pump operates at a start temperature of between 10°C (50°F) and 40°C (104°F) and continues to operate for a maximum of 2 minutes at idle speed.

The electric air pump draws in ambient air and supplies it to the diverter valve. The vacuum vent valve (solenoid) switches vacuum to open the diverter valve when activated by the ECM. Once opened, the diverter valve allows air to be pumped into the exhaust stream.

### Secondary air injection applications

Z3 and M version cars with 6-cylinder engine are equipped with secondary air injection.

In addition, beginning with models produced from 1 / 1997, the M44 (4-cylinder) engine was upgraded to transitional low emissions vehicle (TLEV) status for purposes of Federal emission control compliance. This required installation of secondary air injection components in 4-cylinder models.

## Bosch DME M5.2 (4-cylinder engine)

When engine management system faults are detected, a diagnostic trouble code (DTC) is stored in the engine control module (ECM). The MIL also illuminates if an emissions-related fault is detected.

When faults arise, or if the malfunction indicator lamp (MIL) is illuminated, begin troubleshooting by connecting BMW service tester DISplus, MoDiC, GT1 or equivalent scan tool. The capabilities of OBD II software has the potential to save hours of diagnostic time and to help avoid incorrect component replacement and possible damage to system components.

***CAUTION—***
- *The tests in this section may set fault codes (DTCs) in the ECM and illuminate the MIL. After all testing is completed, access and clear DTC fault memory using an OBD II or BMW compatible scan tool. See* **OBD On-Board Diagnostics**.
- *Only use a digital multimeter for electrical tests.*
- *Relay positions can vary. Be sure to confirm relay location and function by identifying the wiring in the socket using the wiring diagrams found in* **ELE Electrical Wiring Diagrams**.

***NOTE—***
- *OBD II fault memory (including an illuminated MIL) can only be reset using the special scan tool. Removing the connector from the ECM or disconnecting the battery will not erase the fault memory.*
- *The BMW-dedicated 20-pin diagnostic link connector (DLC) is in the right rear of the engine compartment. The 20-pin DLC is discontinued on models after 6 / 2000. The 16-pin OBD II diagnostic connector is located inside the car on the right side of the center console. See* **Engine management system troubleshooting** *in this repair group.*

### Air filter housing and ducts, removing and installing (4-cylinder engine)

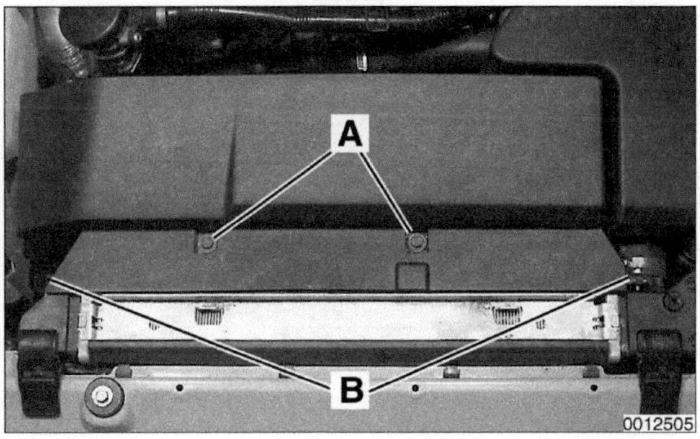


Remove engine intake air duct mounting screws (**A**) from top of radiator.

- Detach plastic rivets (**B**) at sides of radiator.
- Disconnect radiator air shroud and remove from engine compartment.

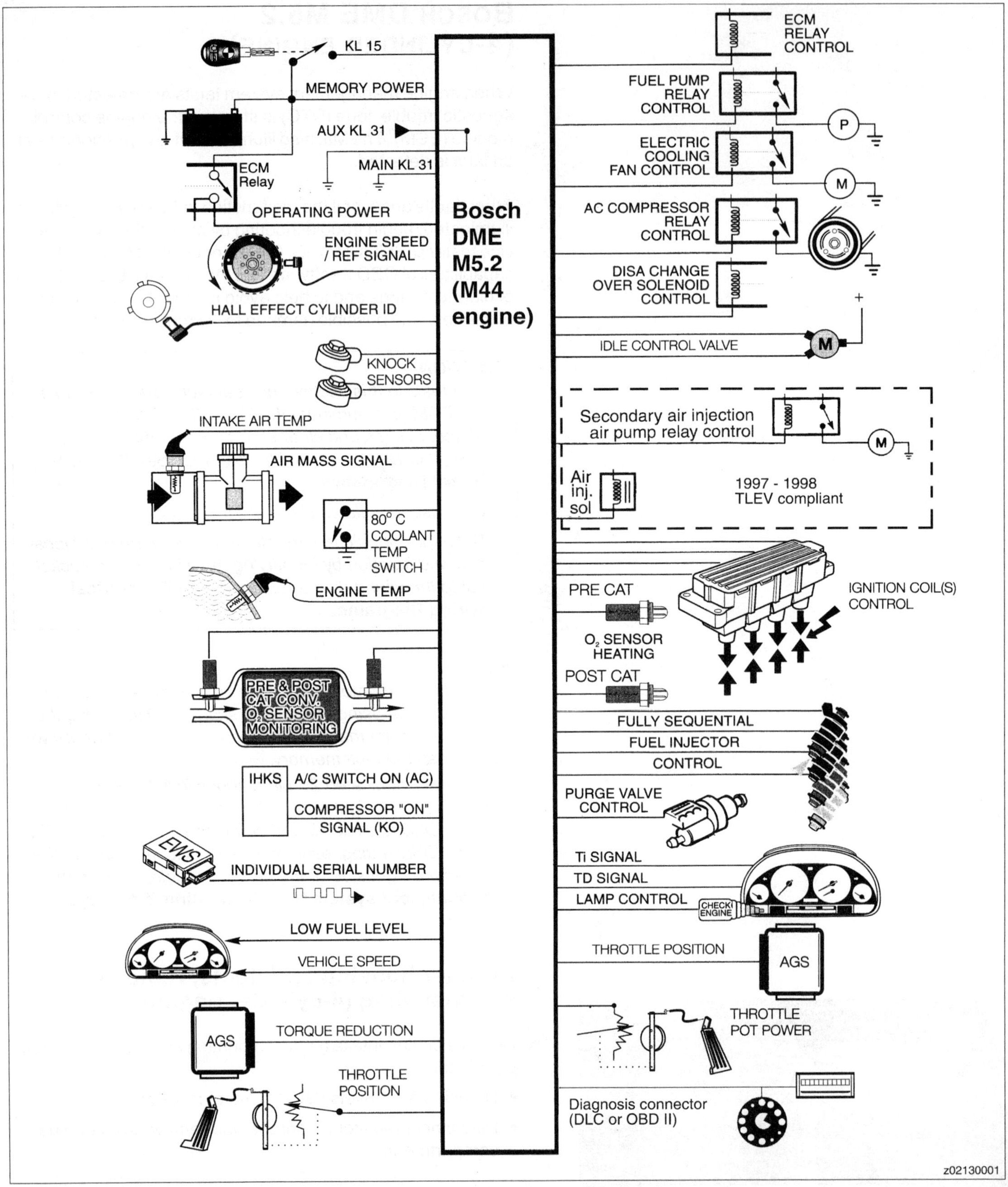
KL 15
MEMORY POWER
AUX KL 31
MAIN KL 31
ECM Relay
OPERATING POWER
ENGINE SPEED / REF SIGNAL
HALL EFFECT CYLINDER ID
KNOCK SENSORS
INTAKE AIR TEMP
AIR MASS SIGNAL
80° C COOLANT TEMP SWITCH
ENGINE TEMP
PRE & POST CAT CONV. O2 SENSOR MONITORING
IHKS
A/C SWITCH ON (AC)
COMPRESSOR "ON" SIGNAL (KO)
EWS
INDIVIDUAL SERIAL NUMBER
LOW FUEL LEVEL
VEHICLE SPEED
AGS
TORQUE REDUCTION
THROTTLE POSITION
Bosch DME M5.2 (M44 engine)
ECM RELAY CONTROL
FUEL PUMP RELAY CONTROL
P
ELECTRIC COOLING FAN CONTROL
M
AC COMPRESSOR RELAY CONTROL
DISA CHANGE OVER SOLENOID CONTROL
+
IDLE CONTROL VALVE
M
Secondary air injection air pump relay control
M
Air inj. sol
1997 - 1998 TLEV compliant
PRE CAT
O2 SENSOR HEATING
POST CAT
IGNITION COIL(S) CONTROL
FULLY SEQUENTIAL FUEL INJECTOR CONTROL
PURGE VALVE CONTROL
Ti SIGNAL
TD SIGNAL
LAMP CONTROL
CHECK ENGINE
THROTTLE POSITION
AGS
THROTTLE POT POWER
Diagnosis connector (DLC or OBD II)
z02130001

Unbolt and remove complete air cleaner housing with mass air flow sensor.

- Detach intake air temperature sensor connector at top air filter housing cover.
- Loosen intake air duct hose clamp (**A**) at auxiliary throttle housing.
- Detach mass air flow (MAF) sensor electrical connector (**B**).

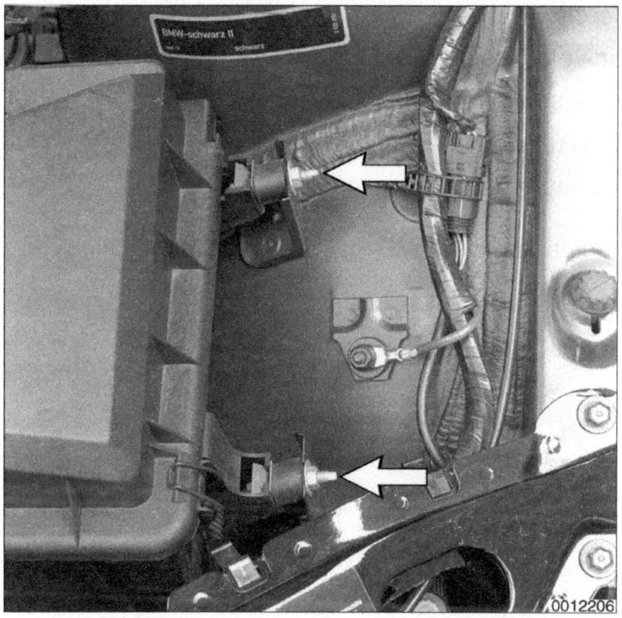


Loosen air filter housing mounting nuts (**arrows**) and lift out housing complete with MAF sensor and ducts.

– After installing, check for fault codes and clear and reset ECM memory.

## Auxiliary throttle housing, removing (4-cylinder engine)

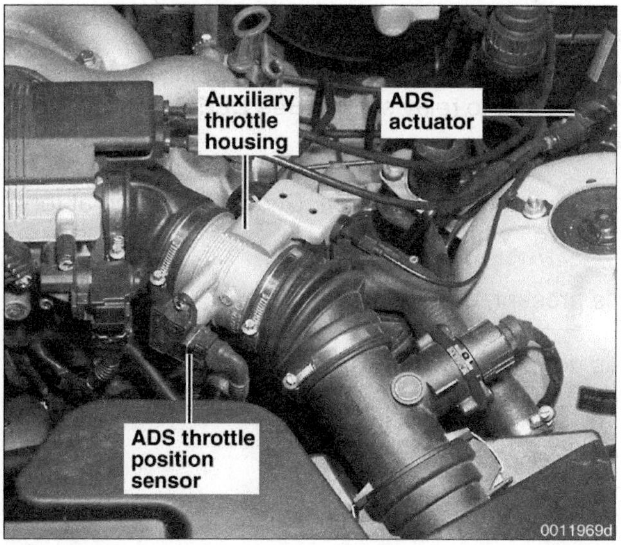


The auxiliary throttle housing is controlled by the ADS actuator. This is part of the traction control system (ABS / ASC or ABS / DSC).

– To remove auxiliary throttle housing:
- Detach ADS actuator Bowden cable.
- Disconnect ADS throttle position sensor electrical connector.
- Loosen air duct clamps and lift out throttle housing.

– Check for fault codes and clear and reset ECM memory.

## DME main relay, testing (4-cylinder engine)

The engine control module (ECM) is energized via the DME main relay and supplies battery positive (B+) power to many of the engine management components and subsystems. If this relay is faulty, the engine will not start.

> ***CAUTION—***
> *Relay positions can vary. Be sure to confirm relay position by identifying the wiring in the socket using the wiring diagrams in* **ELE Electrical Wiring Diagrams**.

The DME main relay powers:

- ECM
- Ignition coils
- Fuel injectors
- Idle air control valve
- Evaporative emissions valve
- Hot film mass air flow (MAF) sensor
- Fuel pump relay
- Oxygen sensor heaters

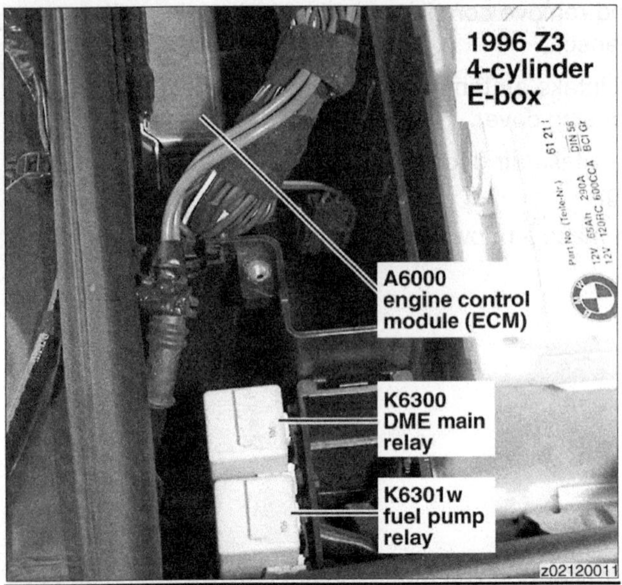

With ignition off, remove DME main relay (K6300) in electronics box (E-box) at right rear of engine compartment.

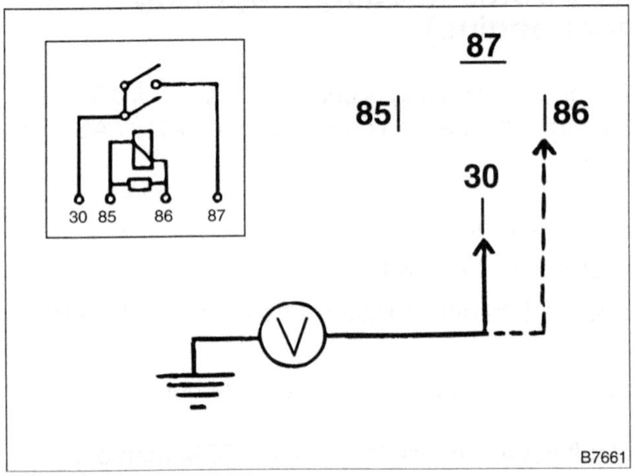


Check for voltage at terminals **30** and **86** of relay socket.

- If battery voltage is present continue testing.
- If battery voltage is not present at either point, check large red wires leading to relay socket. See **ELE Electrical Wiring Diagrams**.

– Reinstall relay and turn ignition on. Check for ground at terminal **85 (**brown / white wire).

– If ground is present, continue testing.

– If ground is not present, signal from ECM is missing. Check wire between ECM and relay.

– If no faults are found:
  - Check ECM ground.
  - If no ground fault is found, ECM may be defective.

– With ignition on and relay installed, check for battery voltage at terminal **87**.

– If battery voltage is present, relay has energized and is functioning correctly.

– If battery voltage is not present and all earlier tests are OK, relay is faulty. Replace.

– When finished testing, check for fault codes and clear and reset ECM memory.

## Engine coolant temperature (ECT) sensor, replacing (4-cylinder engine)

0011984

◄ The ECT sensor is located on the left front of the engine above the oil filter.

The ECT sensor is a dual temperature sensor. One circuit provides coolant temperature information to the ECM while the other circuit provides coolant temperature information to the instrument cluster.

The ECM determines the correct ignition timing and air / fuel mixture required for the engine by monitoring a reference voltage (5 vdc) to the sensor. The voltage drop across the sensor varies as the coolant temperature (sensor resistance) changes.

***NOTE —***
*The ECT sensor is an NTC (negative temperature coefficient) type sensor. In other words, as the temperature rises, resistance through the sensor decreases.*

If the ETC sensor input is faulty or not plausible, the MIL is illuminated when OBD II fault criteria are exceeded. Under these circumstances, the ECM assumes a substitute temperature value (80°C / 176°F) to maintain engine operation. The ignition timing is set to a safely conservative basic setting.

– Disconnect ECT sensor harness.

***WARNING —***
*Due to risk of personal injury, be sure the engine is cold before working on ECT sensor.*

– Unscrew ECT sensor from cylinder head and install new sensor. Be prepared to catch dripping coolant.
  - Use new copper sealing washer when installing sensor.
  - Replace any lost coolant.

| Tightening torque | |
|---|---|
| ECT sensor to cylinder head | 13 Nm (10 ft-lb) |

– Check for fault codes and clear and reset ECM memory.

## Fuel rail and injectors, removing (4-cylinder engine)

> ***WARNING—***
> - *When opening fuel lines, be prepared for fuel to be expelled under pressure (approx. 3 - 5 bar or 45 - 75 psi).*
> - *Do not smoke or work near heaters or other fire hazards. Keep a fire extinguisher handy.*
> - *Before disconnecting fuel hoses, wrap a cloth around fuel hoses to absorb any leaking fuel.*
> - *Catch and dispose of escaped fuel.*
> - *Plug all open fuel lines.*
> - *Unscrew the fuel tank cap to release pressure in the tank before working on the tank or lines.*

> ***CAUTION—***
> *Cleanliness is essential when working with fuel circuit components. Thoroughly clean the unions before disconnecting fuel lines.*

– Disconnect negative (–) battery cable.

> ***CAUTION—***
> *Prior to disconnecting the battery, read the battery disconnection cautions given in* **001 General Warnings and Cautions**.

– Loosen fuel tank cap to release pressure in tank.

– Remove upper section of intake manifold. See **113 Cylinder Head**.

– Remove air shrouding hose for fuel injectors.

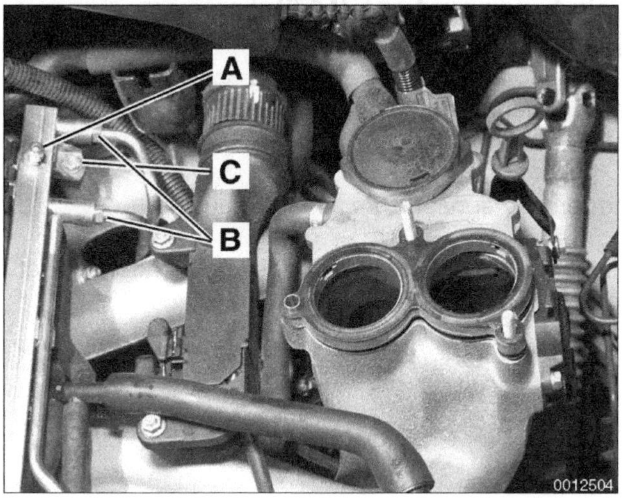


Working at fuel rail:

- Unscrew Schræder valve cap (**A**). Use brief burst of compressed air (max. 44 psi or bar) at valve to blow fuel back into fuel tank.
- Fuel line connections (**B**) are high-pressure threaded fittings. Disconnect fuel supply and return line from fuel rail.
- Remove fuel rail mounting bolts (**C**). (Only one of two mounting bolts shown in photograph.)

> ***WARNING —***
> *The fuel system is designed to retain pressure even when the ignition is off. When working with the fuel system, loosen the fuel lines slowly to allow residual fuel pressure to dissipate gradually. Avoid spraying fuel.*

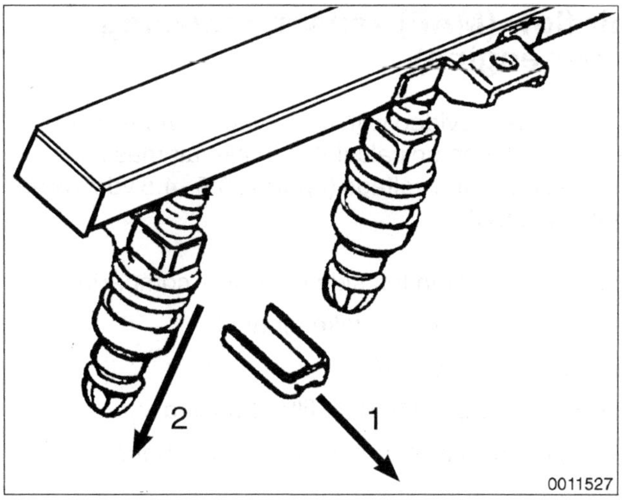


Pull up on fuel rail and remove rail with injectors.

- Pry retaining clip from injector (**1**).
- Pull injector from rail (**2**).

– Installation is reverse of removal. Remember to:

- Fit new sealing O-rings when installing injectors. For ease of installation, lightly lubricate O-rings with assembly lubricant.
- Check that injector electrical connections are correctly fitted and that injectors are fully seated prior to installing fuel rail mounting bolts.
- Replace wire ties.

| Tightening torque | |
|---|---|
| Fuel rail to cylinder head | 10 Nm (7 ft-lb) |

– Check for fault codes and clear and reset ECM memory.

## Idle speed control valve, replacing (4-cylinder engine)

The idle speed valve on the M44 engine is located on the left side of the throttle housing.

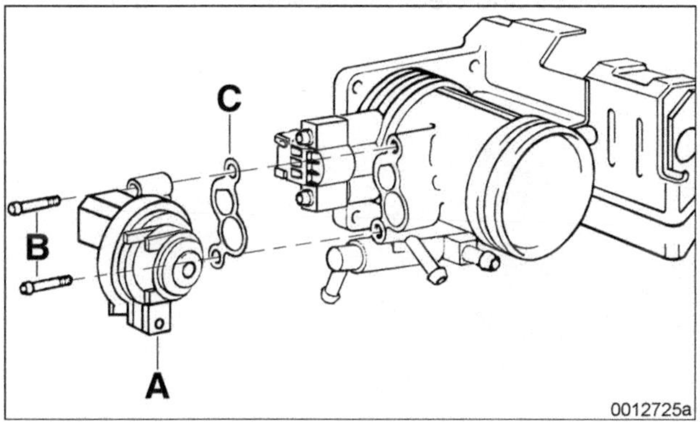


Working at throttle housing:

- Disconnect connector (**A**) from idle valve.
- Remove mounting screws (**B**).
- Pull idle speed control valve from its fittings.

– Installation is reverse of removal. Always replace gasket (**C**).

***NOTE —***
*For about 10 minutes you may notice poor driveability after installing a replacement idle valve.*

– Check for fault codes and clear and reset ECM memory.

## Intake air temperature (IAT) sensor, replacing (4-cylinder engine)

- Disconnect connector from intake air temperature sensor (**arrow**) at air filter housing.
- Unclip and remove sensor from air filter housing.
- After reinstallation, check for fault codes and clear and reset ECM memory.

## Mass air flow (MAF) sensor, replacing (4-cylinder engine)

The MAF sensor is provided with operating power from the DME main relay. Based on calculated intake air mass, the MAF generates a varying voltage (approx. 0.5 - 4.5 vdc) as an input signal to the ECM.

- Working at MAF sensor in left side of engine compartment:
  - Loosen hose clamp (**A**) at intake air duct.
  - Detach sensor connector (**B**).
  - Detach two large clips (**C**) at air filter housing
  - Work MAF sensor out of rubber seal in filter housing.
- If MAF sensor is to be reused, visually inspect sensor for damaged, missing or blocked screen. (Screen affects air flow calibration.)
- Installation is reverse of removal.
  - For ease of installation, lubricate large rubber seal in air filter housing with silicone lubricant or equivalent.
  - Carefully check that all hoses and seals are seated properly.
- Check for fault codes and clear and reset ECM memory.

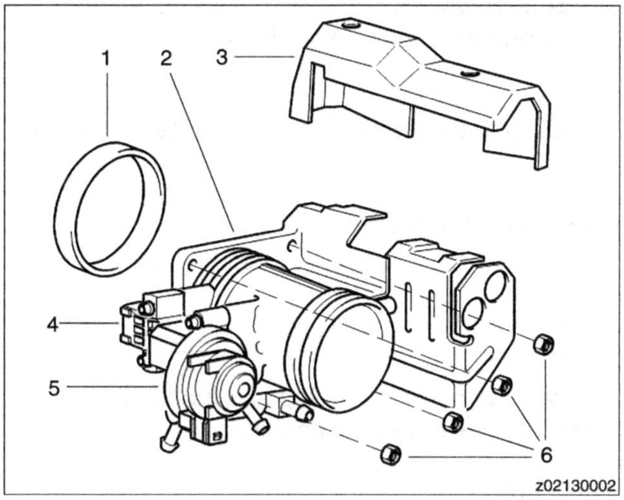


## Throttle housing, replacing (4-cylinder engine)

M44 throttle housing components:

1. Sealing O-ring (profile gasket)
2. Throttle housing assembly
3. Bowden cable cover
4. Throttle position sensor (TPS)
5. Idler speed control valve
6. M6 nut

– Remove air filter housing and ducts. See **Air filter housing and ducts, removing and installing (4-cylinder engine)** in this repair group.

– Remove Bowden cable cover.

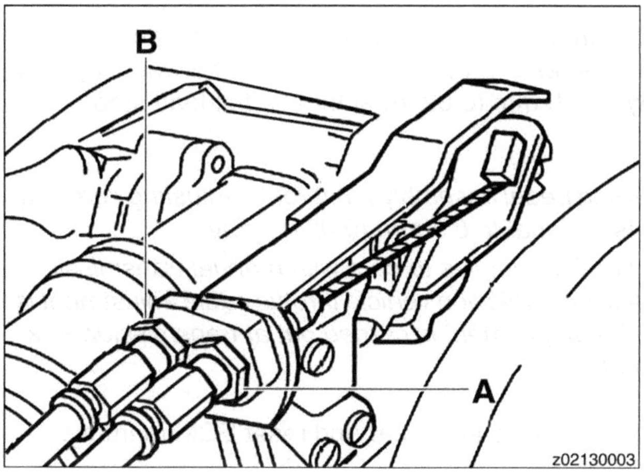


Unscrew and disconnect throttle cable(s):

- **A** to accelerator pedal
- **B** to cruise control (if applicable)

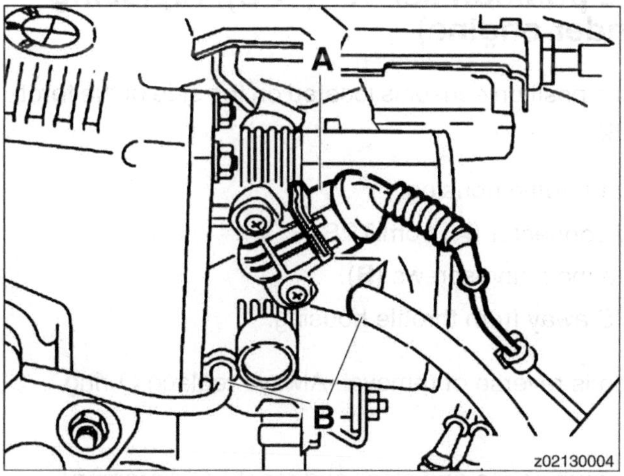


Working at throttle housing:

- Detach throttle position sensor (TPS) connector (**A**).
- Detach vacuum hoses (**B**).

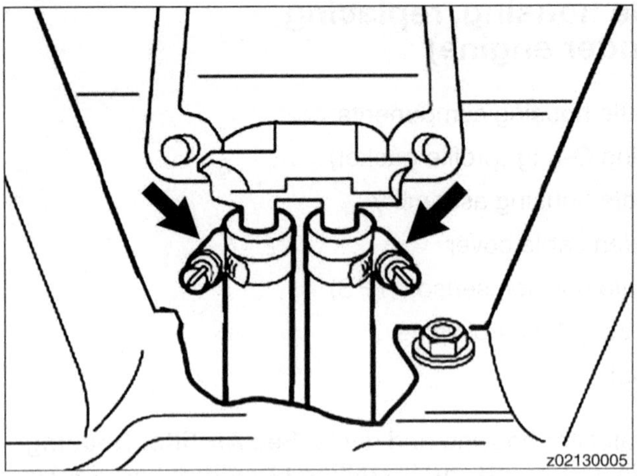
z02130005

Loosen clamps (**arrows**) and detach coolant hoses from throttle housing preheater base. Plug hoses.

***NOTE —***
*Be prepared to catch the small amount of coolant which escapes.*

– Remove M6 mounting nuts and lift off throttle housing.

| ***CAUTION—*** |
|---|
| *Handle throttle housing carefully.* |

– Installation is reverse of removal. Replace rubber O-ring (profile gasket) at base of throttle housing.

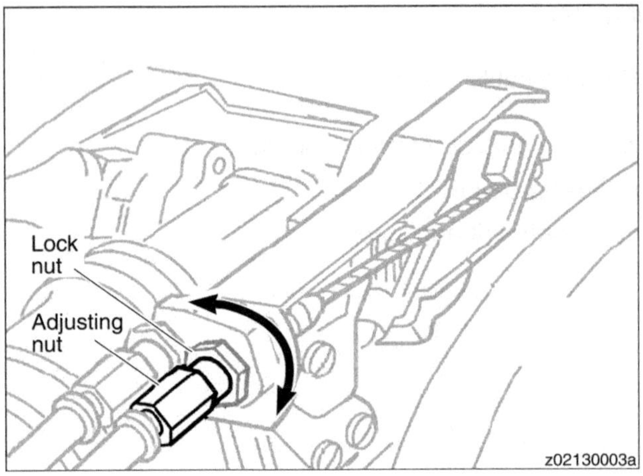

z02130003a

Adjust throttle cable:

- With accelerator pedal unpressed and throttle plate fully closed, turn Bowden cable adjusting nut so that there is no tension on cable. Tighten lock nut.
- Turn adjustment screw on full-throttle stop of pedal assembly so that it rests against accelerator pedal when there is still play of 0.5 mm (0.02 in) at throttle full-load stop.

***NOTE —***
- *Unscrewing pedal assembly full-throttle adjusting screw by 1.5 turns is equal to 0.5 mm (0.02 in) play.*
- *This procedure applies to cars with manual transmission. Automatic transmission vehicle throttle cable adjustment is similar, but adjustment is carried out at transmission kick down switch.*

– Check for fault codes and clear and reset ECM memory.

## Throttle position sensor (TPS), replacing (4-cylinder engine)

The throttle position sensor is located on the side of the throttle housing.

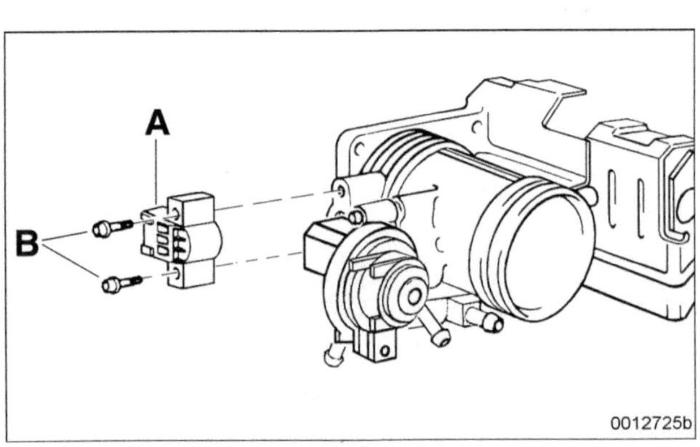

0012725b

Working at throttle housing:

- Unplug connector (**A**) from TPS.
- Remove mounting screws (**B**).
- Pull TPS away from throttle housing.

– Installation is reverse of removal. Always replace O-ring seals.

– Check for fault codes and clear and reset ECM memory.

***NOTE —***
*On cars with traction control, do not confuse the throttle position sensor on the main throttle body with the throttle position switch on the ADS (auxiliary) throttle housing.*

## ECM pin assignments (Bosch DME M5.2)

The engine control module (ECM) is located in the right rear of the engine compartment in the E-box.

ECM pin assignments for Bosch DME M5.2 are given in **Table c**. This information can be helpful when diagnosing faults to or from the ECM.

Generally, absence of voltage or continuity means there is a wiring or connector problem. Test results with incorrect values do not necessarily mean that a component is faulty. Check for loose, broken or corroded connections and wiring before replacing components. For engine management system electrical schematics, see **ELE Electrical Wiring Diagrams**.

***CAUTION—***
- *Always wait at least one minute after turning off the ignition before removing the connector from the engine control module (ECM). If the connector is removed before this time, residual power in the system relay may damage the ECM.*
- *Always connect or disconnect the control module connector and meter probes with the ignition off.*

When making checks at the ECM, use a break-out box to allow tests to be made with the connector attached to the ECM. This prevents damage to the small terminals in the connector. As an alternative, the connector housing can be separated so that electrical checks can be made from the back of the connector. ECM pin numbering, as seen from back of ECM connector, is illustrated.

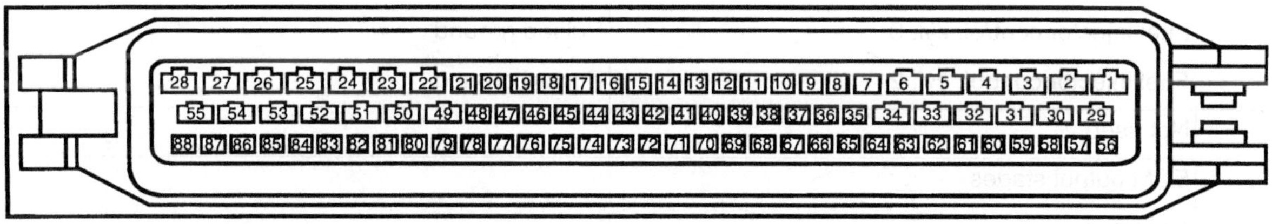

0013211

| Table c. Bosch DME M5.2 ECM pin assignments | | | |
|---|---|---|---|
| **Pin** | **Type** | **Component or function** | **Note** |
| 1 | ground | Oxygen sensor (post-catalyst) heater | Switched ground |
| 2 | output | Idle speed control valve | Pulsed ground, close signal (see also pin 29) |
| 3 | output | Fuel injector control, cyl. 1 | Pulsed ground |
| 4 | output | Fuel injector control, cyl. 4 | Pulsed ground |
| 5 | - | Not used | |
| 6 | ground | Fuel injector output stage | |
| 7 | - | Not used | |

| Table c. Bosch DME M5.2 ECM pin assignments (continued) | | | |
|---|---|---|---|
| **Pin** | **Type** | **Component or function** | **Note** |
| 8 | output | MIL | Instrument cluster |
| 9 | - | Not used | |
| 10 | input | Electronic immobilizer module (EWS) | Activation |
| 11 | output | Automatic climate control (IHKA) | Evaporator control |
| 12 | - | Not used | |
| 13 | - | Not used | |
| 14 | - | Not used | |
| 15 | - | Not used | |
| 16 | input | Intake air temperature (IAT) sensor | Signal |
| 17 | input | Mass air flow (MAF) sensor | Signal |
| 18 | - | Not used | |
| 19 | input | Oxygen sensor (precatalyst) | Signal |
| 20 | output | Crankshaft sensor | Control |
| 21 | input | Camshaft sensor | Signal |
| 22 | output | Ignition coil control, cyl. 3 | Primary signal |
| 23 | output | Ignition coil control, cyl. 4 | Primary signal |
| 24 | - | Not used | |
| 25 | - | Not used | |
| 26 | input | Power supply, terminal 30 | Battery voltage (B+) at all times |
| 27 | input | DME main relay terminal 85 | Activation |
| 28 | ground | Ground | ECM and sensor shielding |
| 29 | output | Idle speed control valve | Pulsed ground open signal (see also pin 2) |
| 30 | ground | Oxygen sensor (precatalyst) heater | Switched ground |
| 31 | output | Fuel injector control, cyl. 3 | Pulsed ground |
| 32 | output | Fuel injector control, cyl. 2 | Pulsed ground |
| 33 | - | Not used | |
| 34 | ground | ECM output stages | |
| 35 | - | Not used | |
| 36 | input | A/C compressor relay | Activation |
| 37 | - | Not used | |
| 38 | - | Not used | |
| 39 | - | Not used | |
| 40 | input | Knock sensor 2 (cyl. 3, 4) | Signal |
| 41 | - | Not used | |
| 42 | input | Vehicle speed | From instrument cluster |
| 43 | - | Not used | |
| 44 | output | Throttle position sensor (TPS) | Reference signal |

| Table c. Bosch DME M5.2 ECM pin assignments (continued) | | | |
|---|---|---|---|
| **Pin** | **Type** | **Component or function** | **Note** |
| 45 | output | Mass air flow (MAF) sensor | Signal |
| 46 | ground | Oxygen sensors | Reference ground |
| 47 | - | Not used | |
| 48 | - | Not used | |
| 49 | output | Ignition coil control, cyl. 1 | Primary signal |
| 50 | output | Ignition coil control, cyl. 2 | Primary signal |
| 51 | - | Not used | |
| 52 | - | Not used | |
| 53 | input | Throttle position sensor | Signal |
| 54 | input | Terminal 87, DME main relay | Power |
| 55 | ground | ECM | |
| 56 | input | Terminal 15 | Switched power |
| 57 | output | Fan relay, normal speed | Engine cooling fan activation, manual transmission |
| 58 | - | Not used | |
| 59 | - | Not used | |
| 60 | input | Data link connector (DLC), pin 18 | Programming voltage |
| 61 | output | Evaporative emission valve | Activation |
| 62 | | Not used | |
| 63 | output | Fuel pump relay | Activation |
| 64 | - | Not used | |
| 65 | - | Not used | |
| 66 | - | Not used | |
| 67 | - | Not used | |
| 68 | input | Double temperature cooling fan switch (manual transmission) | Signal above 80°C |
| 69 | input | Automatic climate control (IHKA) | |
| 70 | input | Knock sensor 1 (cyl. 1, 2) | Signal |
| 71 | ground | Analog signals and knock sensors | |
| 72 | - | Not used | |
| 73 | - | Not used | |
| 74 | input | Engine coolant temperature (ECT) sensor | Signal |
| 75 | - | Not used | |
| 76 | - | Not used | |
| 77 | input | Oxygen sensor (post-catalyst) | Signal |
| 78 | input | Crankshaft sensor | Signal |
| 79 | input | ABS / ASC or ABS / DSC control module | |
| 80 | input | Engine speed sensor | Signal |

| Table c. Bosch DME M5.2 ECM pin assignments (continued) | | | |
|---|---|---|---|
| Pin | Type | Component or function | Note |
| 81 | - | Not used | |
| 82 | - | Not used | |
| 83 | input | On-board computer | Terminal 4 |
| 84 | - | Not used | |
| 85 | - | Not used | |
| 86 | - | Not used | |
| 87 | input | Diagnostic connector (RxD) | Receive signal to pin 15, data link connector (DLC) |
| 88 | output | Diagnostic connector (TxD) | Transmit signal to pin 17, data link connector (DLC) |

## SIEMENS DME MS 41.1 (M52, S52 ENGINE)

When engine management system faults are detected, a diagnostic trouble code (DTC) is stored in the engine control module (ECM). The MIL also illuminates if an emissions-related fault is detected.

When faults arise, or if the malfunction indicator lamp (MIL) is illuminated, begin troubleshooting by connecting BMW service tester DISplus, MoDiC, GT1 or equivalent scan tool. The capabilities of OBD II software has the potential to save hours of diagnostic time and to help avoid incorrect component replacement and possible damage to system components.

***CAUTION—***
- *The tests in this section may set fault codes (DTCs) in the ECM and illuminate the MIL. After all testing is completed, access and clear DTC fault memory using an OBD II or BMW compatible scan tool. See* **OBD On-Board Diagnostics**.
- *Only use a digital multimeter for electrical tests.*
- *Relay positions can vary. Be sure to confirm relay location and function by identifying the wiring in the socket using the wiring diagrams found in* **ELE Electrical Wiring Diagrams**.

***NOTE—***
- *OBD II fault memory (including an illuminated MIL) can only be reset using the scan tool. Removing the connector from the ECM or disconnecting the battery will not erase the fault memory.*
- *For location of the 16-pin OBD II connector or the 20-pin diagnostic link connector (DLC), see* **Engine management system troubleshooting** *in this repair group.*

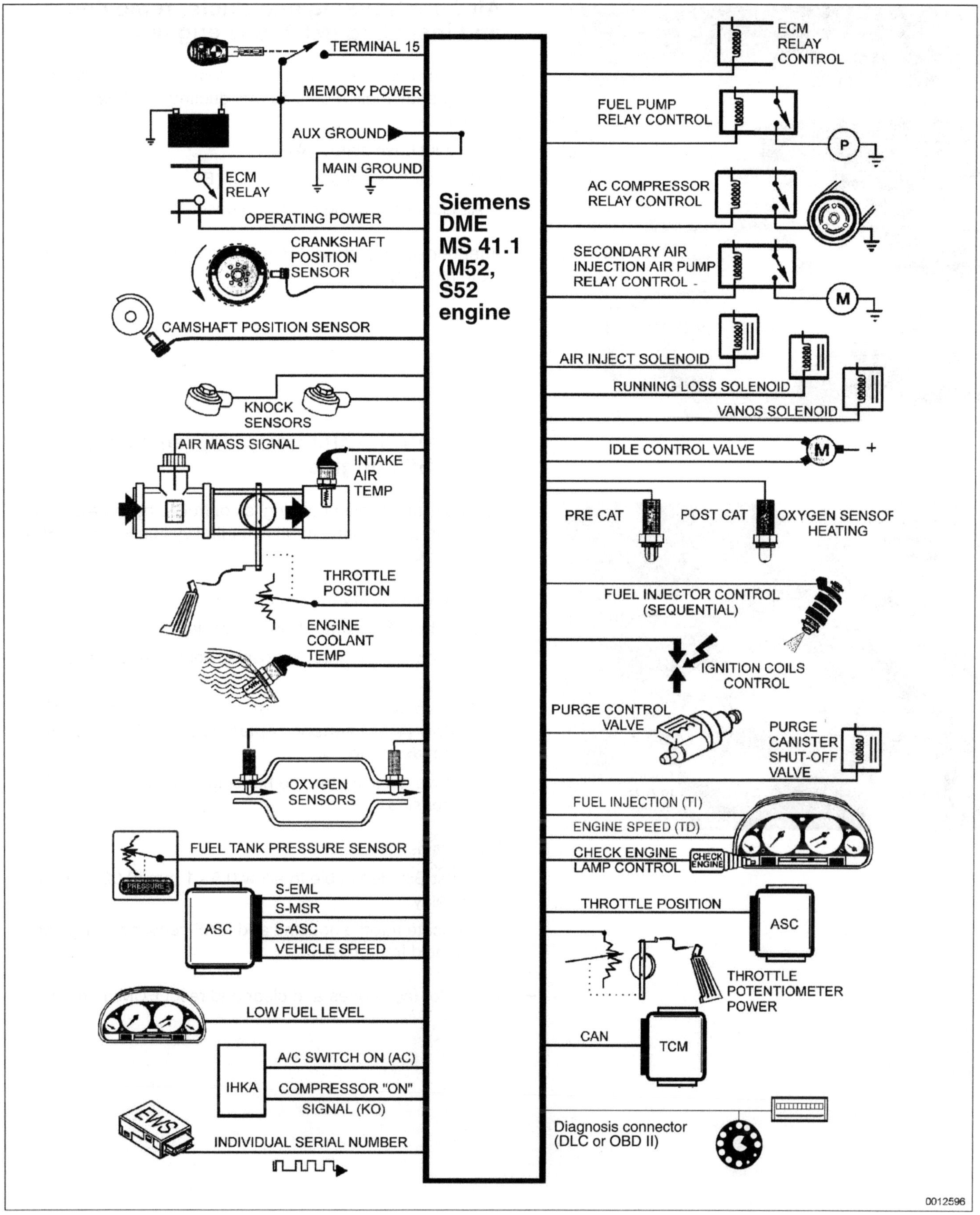

TERMINAL 15
MEMORY POWER
AUX GROUND
MAIN GROUND
ECM RELAY
OPERATING POWER
CRANKSHAFT POSITION SENSOR
CAMSHAFT POSITION SENSOR
KNOCK SENSORS
AIR MASS SIGNAL
INTAKE AIR TEMP
THROTTLE POSITION
ENGINE COOLANT TEMP
OXYGEN SENSORS
FUEL TANK PRESSURE SENSOR
PRESSURE
ASC
S-EML
S-MSR
S-ASC
VEHICLE SPEED
LOW FUEL LEVEL
IHKA
A/C SWITCH ON (AC)
COMPRESSOR "ON" SIGNAL (KO)
EWS
INDIVIDUAL SERIAL NUMBER
Siemens DME MS 41.1 (M52, S52 engine
ECM RELAY CONTROL
FUEL PUMP RELAY CONTROL
P
AC COMPRESSOR RELAY CONTROL
SECONDARY AIR INJECTION AIR PUMP RELAY CONTROL
M
AIR INJECT SOLENOID
RUNNING LOSS SOLENOID
VANOS SOLENOID
IDLE CONTROL VALVE
M
+
PRE CAT
POST CAT
OXYGEN SENSOR HEATING
FUEL INJECTOR CONTROL (SEQUENTIAL)
IGNITION COILS CONTROL
PURGE CONTROL VALVE
PURGE CANISTER SHUT-OFF VALVE
FUEL INJECTION (TI)
ENGINE SPEED (TD)
CHECK ENGINE LAMP CONTROL
CHECK ENGINE
THROTTLE POSITION
ASC
THROTTLE POTENTIOMETER POWER
CAN
TCM
Diagnosis connector (DLC or OBD II)
0012596

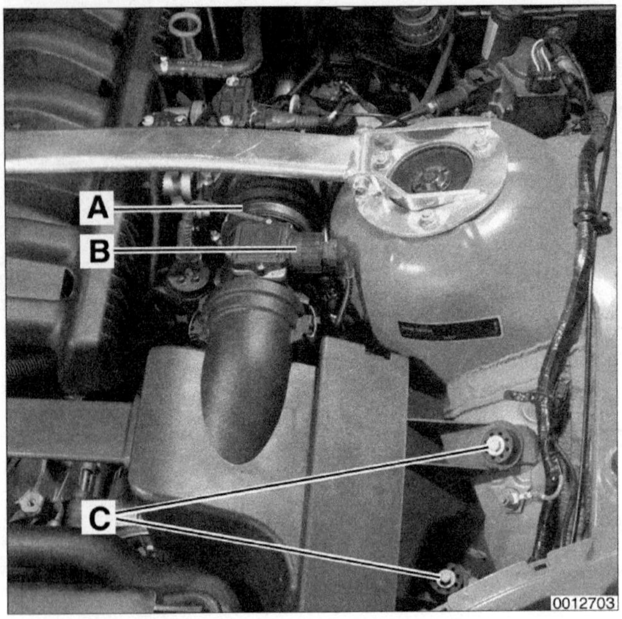


## Air filter housing and ducts, removing and installing (M52, S52 engine)

< Unbolt and remove complete air cleaner housing, disconnecting and labeling connectors, ducting and hoses as necessary:

- Air duct hose clamp (**A**)
- Mass air flow sensor connector (**B**)
- Air cleaner housing mounting bolts (**C**)

– After reinstalling, check for fault codes and clear and reset ECM memory.

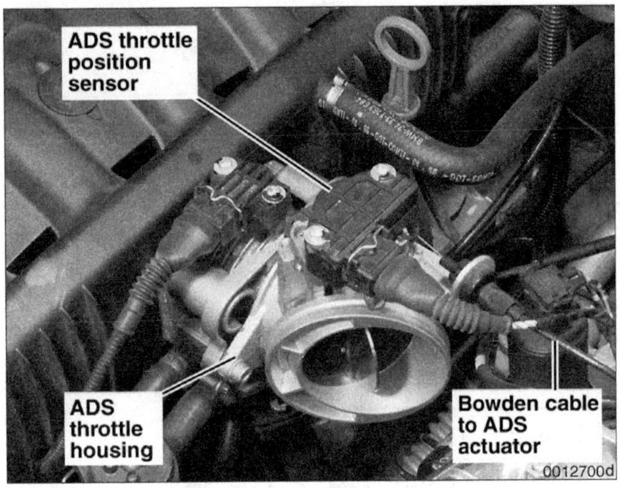


## Auxiliary throttle housing, removing (M52, S52 engine)

< The auxiliary throttle housing is controlled by the ADS actuator. This is part of the traction control system (ABS / ASC or ABS / DSC).

– To remove auxiliary throttle housing:

- Detach ADS actuator Bowden cable.
- Disconnect ADS throttle position sensor electrical connector.
- Remove mass air flow sensor and intake duct. See **Mass air flow (MAF) sensor, replacing (M52, S52 engine)** in this repair group.
- Remove auxiliary throttle housing mounting bolts and lift out housing.

– When reinstalling:

- Adjust Bowden cable to allow 0.5 - 1 mm (0.02 - 0.04 in) free play.
- Lubricate throttle housing and MAF sensor sealing rings with acid free grease.

– Check for fault codes and clear and reset ECM memory.

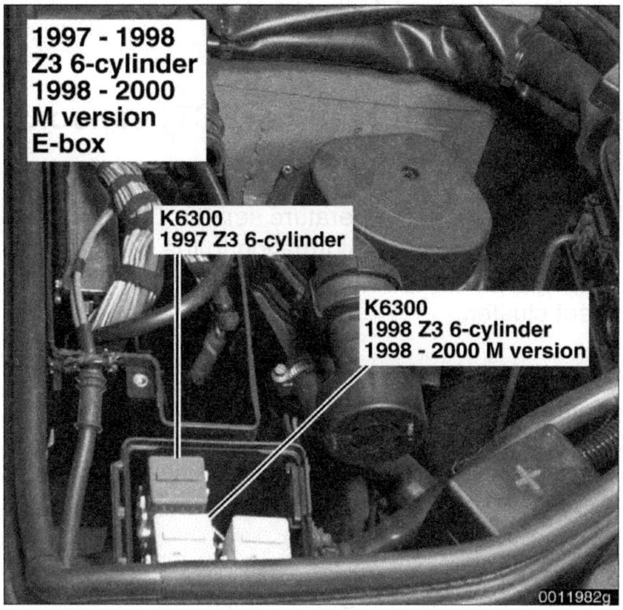


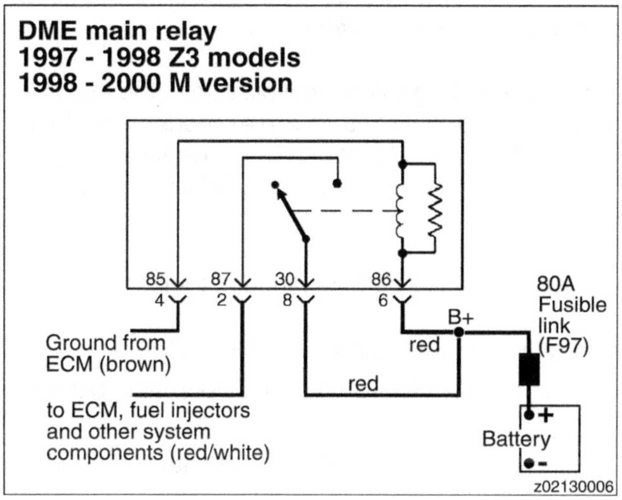


## DME main relay, testing (M52, S52 engine)

The engine control module (ECM) is energized via the DME main relay and supplies battery positive (B+) power to many of the engine management components and subsystems. If this relay is faulty, the engine will not start.

With ignition off, remove DME main relay (K6300) in electronics box (E-box) at right rear of engine compartment.

Check for voltage at terminals **6** and **8** of relay socket (correspond to terminals **30** and **86** on relay).

- If battery voltage is present continue testing.
- If battery voltage is not present at either point, check the following:
  - Large red wires in relay socket
  - B+ (battery positive junction) in right rear of engine compartment
  - 80 amp fusible link (F97) in trunk near battery
  - See **ELE Electrical Wiring Diagrams** for more details
- Check for ground at relay socket **4** (corresponds to terminal **85** on relay).
- If ground is present, continue testing.
- If ground is not present, signal from ECM is missing. Check wire between ECM and relay.
- If no faults are found:
  - Check ECM grounding.
  - ECM may be defective.
- With ignition on and relay installed, check for battery voltage at relay socket **2** (corresponds to terminal **87** on relay).
- If battery voltage is present, relay has energized and is functioning correctly.
- If battery voltage is not present and all earlier tests are OK, relay is faulty. Replace.

– When finished testing, check for fault codes and clear and reset ECM memory.

## Engine coolant temperature (ECT) sensor, replacing (M52, S52 engine)

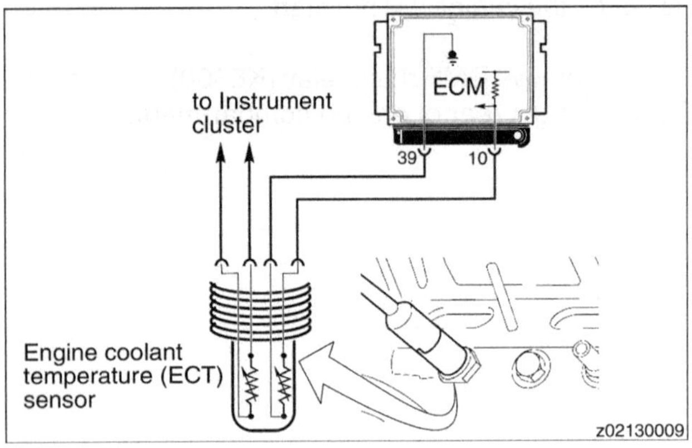


The ECT sensor is a dual temperature sensor. One circuit provides coolant temperature information to the ECM while the other circuit provides coolant temperature information to the instrument cluster.

The ECM determines the correct ignition timing and air / fuel mixture required for the engine by monitoring an reference voltage (5 vdc) to the sensor. Voltage drop across the sensor varies as the coolant temperature (sensor resistance) changes.

***NOTE—***
*The ECT sensor is an NTC (negative temperature coefficient) type sensor. In other words, as the temperature rises, resistance through the sensor decreases.*

If the ECT sensor input is faulty or not plausible, the MIL is illuminated when OBD II fault criteria are exceeded. The ECM assumes a substitute value (80°C / 176°F) to maintain engine operation. The ignition timing is set to a safely conservative basic setting.

The ECT sensor is located below cylinder 1 intake port.

To access engine coolant temperature sensor:

- Remove plastic cover (**A**) from top of intake manifold runners.
- Unclip crankcase vent hose (**B**) from cylinder head cover.

***WARNING—***
*Due to risk of personal injury, be sure the engine is cold before working on ECT sensor removal.*

– Working under front of intake manifold, disconnect ECT harness.

***CAUTION—***
*Be sure ignition is OFF before disconnecting or reconnecting DME system components.*

– Unscrew ECT sensor from cylinder head. Be prepared to catch dripping coolant.

- Installation is reverse of removal:
    - Use new copper sealing washer when installing sensor.
    - Replace any engine coolant lost.

| Tightening torque | |
|---|---|
| Temperature sensor to cylinder head | 13 Nm (10 ft-lb) |

- Check for fault codes and clear and reset ECM memory.

## Fuel rail and injectors, removing (M52, S52 engine)

Fuel injectors are electrically controlled solenoid valves that provide precisely metered and atomized fuel into the engine intake ports.

Injectors receive voltage from the DME main relay. The engine control module (ECM) controls injector opening by activating the ground circuit for the injector valve solenoids. The ECM varies the duration (in milliseconds) of injector opening to regulate air / fuel ratio.

> ***WARNING —***
> - *When opening fuel lines, be prepared for fuel to be expelled under pressure (approx. 3 - 5 bar or 45 - 75 psi).*
> - *Do not smoke or work near heaters or other fire hazards. Keep a fire extinguisher handy.*
> - *Before disconnecting fuel hoses, wrap a cloth around fuel hoses to absorb any leaking fuel.*
> - *Catch and dispose of escaped fuel.*
> - *Plug all open fuel lines.*
> - *Unscrew the fuel tank cap to release pressure in the tank before working on the tank or lines.*

> ***CAUTION—***
> *Cleanliness is essential when working with fuel circuit components. Thoroughly clean the unions before disconnecting fuel lines.*

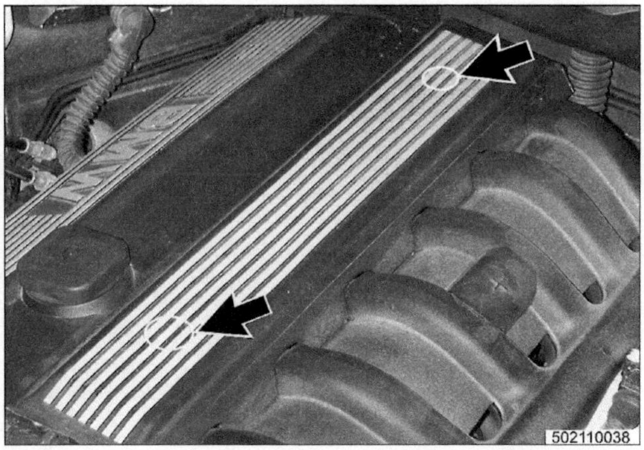
502110038

◄ To access injectors, first remove injector cover by removing plastic trim plugs (**arrows**) and removing fasteners under them. Lift off cover.

- For a quick check of the injectors, run the engine and touch each injector with a screwdriver or stethoscope. You should feel a vibration or hear a buzzing. Turn off engine.

- Loosen fuel tank cap to release pressure in tank.

- Disconnect negative (-) battery cable.

> ***CAUTION—***
> *Prior to disconnecting the battery, read the battery disconnection cautions given in* **001 General Warnings and Cautions**.

◄ Disconnect oxygen sensor connectors (**arrows**) along left side of cylinder head cover.

***NOTE—***
*Mark connectors for later reassembly.*

◄ Working at left side of cylinder head:

- Detach engine vent hose from cylinder head cover by pinching spring clips (**arrows**).
- Detach VANOS solenoid electrical connector (**A**).

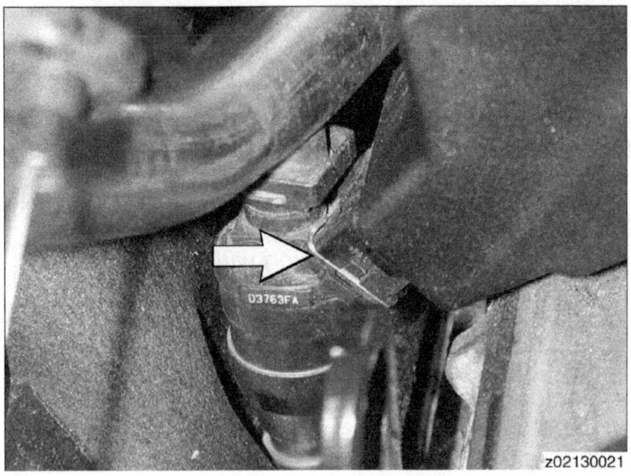


◄ Detach electrical harness from injectors by prying off locking clip (**arrow**) at each injector.

- Remove harness duct mounting fasteners.
- Lift harness off injectors and lay aside.

> ***CAUTION—***
> *Be sure ignition is OFF before disconnecting or reconnecting DME system components.*

Working at rear of manifold, label and disconnect fuel supply and fuel return lines from fuel rail.

> ***WARNING —***
> - *Clamp off the fuel lines and then wrap a clean shop towel around the lines before removing them. Residual fuel pressure is present in the lines.*
> - *Fuel is discharged. Do not disconnect any wires that could cause electrical sparks. Do not smoke or work near heaters or other fire hazards. Keep an approved fire extinguisher handy.*
> - *Loosen fuel tank cap to release pressure.*

***NOTE —***
*A special fitting is used to retain the fuel lines to the fuel rail. Use BMW special tool 16 1 050 or equivalent to compress the locking clip inside the end of the fuel line fittings to release the line.*

– Remove fuel rail mounting fasteners. Pull up on fuel rail to remove fuel injectors from intake manifold.

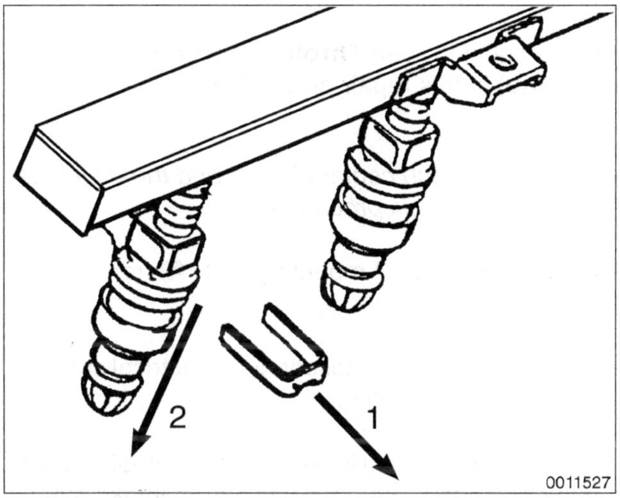


Remove individual injectors:

- Pry retaining clip from injector (**1**).
- Pull injector from rail (**2**).

– Installation is reverse of removal. Remember to:

- Fit new sealing O-rings when installing injectors. For ease of installation, lightly lubricate O-rings with assembly lubricant.
- Check that injector electrical connections are correctly fitted and that injectors are fully seated prior to installing fuel rail mounting bolts.
- Check fuel line rubber O-rings and replace as necessary. For ease of installation, lightly lubricate O-rings with assembly lubricant.
- Replace wire ties.

| **Tightening torque** | |
|---|---|
| Fuel rail to cylinder head | 10 Nm (7 ft-lb) |

– Check for fault codes and clear and reset ECM memory.

## Idle speed control valve, replacing (M52, S52 engine)

***NOTE—***
*If experiencing problems with vehicle idle, be sure to check throttle position sensor.*

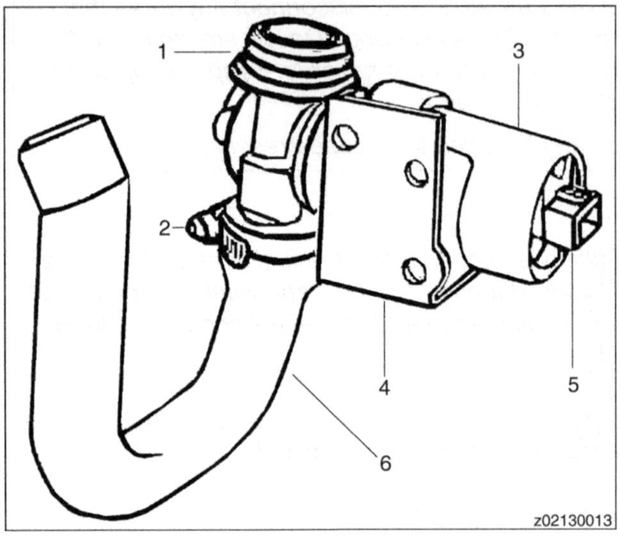


◀ The idle speed control valve regulates idle speed by redirecting air around the throttle valve. It is mounted on the underside of the intake manifold adjacent to the dipstick tube bracket.

1. Vacuum seal to intake manifold
2. Hose clamp
3. Idle speed control valve
4. Mounting bracket
5. Electrical connector
6. Vacuum hose to intake duct

– Remove complete air filter housing and ducts. See **Air filter housing and ducts, removing and installing (M52, S52 engine)** in this repair group.

– Remove throttle housing. See **Throttle housing, replacing (M52, S52 engine)** in this repair group. Set housing aside.

***NOTE—***
*There is no need to detach coolant hoses and throttle cable from throttle housing for this procedure.*

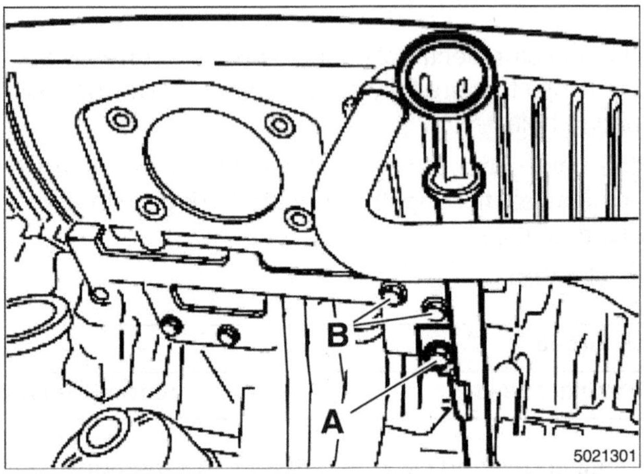


◀ Remove dipstick tube bracket mounting bolt (**A**) from intake manifold.

– Working under intake manifold, disconnect electrical connector from idle speed control valve.

– Remove two bolts (**B**) retaining idle speed control valve to underside of intake manifold. Remove valve from intake manifold. Loosen hose clamp and disconnect hose.

– Installation is reverse of removal. Use new intake manifold vacuum seal when installing. Coat seal with anti-friction sealant.

– Check engine idle speed.

– Check for fault codes and clear and reset ECM memory.

## Intake air temperature (IAT) sensor, replacing (M52, S52 engine)

The intake air temperature (IAT) sensor signal is used by the ECM to monitor intake air temperature for fuel and ignition adaptation. The sensor is located in the intake manifold behind the throttle housing.

The ECM monitors a reference voltage (5 vdc) to the sensor. The voltage drop across the sensor varies as the intake air temperature (sensor resistance) changes.

***NOTE—***
*The IAT sensor is an NTC (negative temperature coefficient) type sensor. In other words, as the temperature rises, resistance through the sensor decreases.*

If IAT sensor input is not plausible, a fault code is set. The MIL is illuminated when OBD II fault criteria are exceeded. The ECM then operates the engine using only the engine coolant temperature (ECT) sensor input.

- Remove complete air filter housing and ducts. See **Air filter housing and ducts, removing and installing (M52, S52 engine)** in this repair group.
- Remove throttle housing from manifold. See **Throttle housing, replacing (M52, S52 engine)**. Set housing aside.

***NOTE—***
*There is no need to detach coolant hoses and throttle cable from throttle housing for this procedure.*

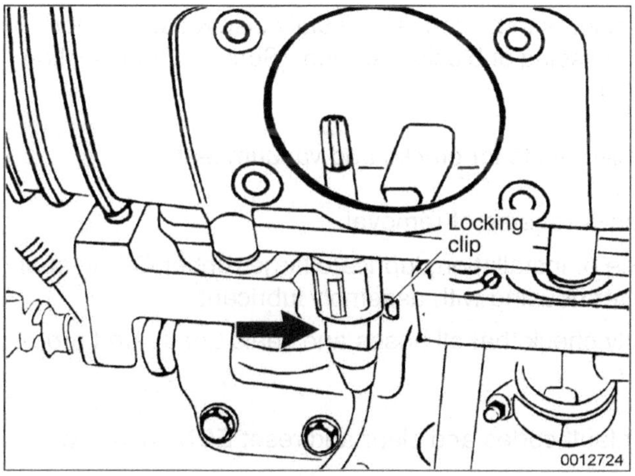


< Remove electrical connector (**arrow**) from IAT sensor.

- Press locking clip to detach temperature sensor and remove from intake manifold.
- Installation is reverse of removal. Use a new sealing ring at intake manifold.
- Check for fault codes and clear and reset ECM memory.

## Mass air flow (MAF) sensor, replacing (M52, S52 engine)

The MAF sensor is provided with electrical power from the DME main relay. Based on calculated intake air mass, the MAF modulates a varying voltage (approx. 0.5 - 4.5 vdc) as an input signal to the ECM.

The DME main relay provides voltage to the MAF sensor to heat a thin metal film to a predetermined temperature. As air flows through the MAF sensor, the heated film is cooled. To maintain the film at a constant temperature, additional heating current is necessary. It is this additional current that is the basis for the input signal.

If there is no output signal from the MAF sensor, the ECM operates the engine using throttle position and engine rpm inputs. A faulty MAF sensor illuminates the MIL.

A faulty MAF sensor can produce the following problems:

- Difficult to restart when engine is hot.
- Engine starts then stalls.
- Engine starts and runs only with accelerator pedal depressed.

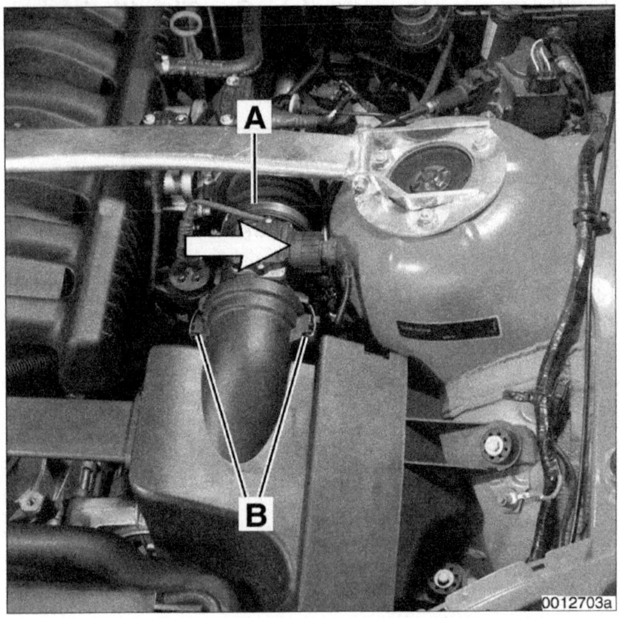


◄ Working at MAF sensor in left side of engine compartment, detach sensor connector (**arrow**).

- Loosen intake boot clamp (**A**).
- Release clips from air filter housing (**B**).
- Remove MAF sensor.

– If MAF sensor is to be reused, visually inspect sensor for damaged, missing or blocked screen. (Screen affects air flow calibration.)

– Check intake ducts for cracks and vacuum leaks.

– Installation is reverse of removal.

- For ease of installation, lubricate large rubber O-ring seal in air filter housing with assembly lubricant.
- Carefully check that all hoses and seals are seated properly.

– Check for fault codes and clear and reset ECM memory.

## Throttle housing, replacing (M52, S52 engine)

– Remove complete air filter housing and ducts. See **Air filter housing and ducts, removing and installing (M52, S52 engine)** in this repair group.

– If necessary, unbolt and remove auxiliary throttle housing. See **Auxiliary throttle housing, removing (M52, S52 engine)** in this repair group.

– Working at throttle assembly, disconnect throttle cable and electrical connector(s).

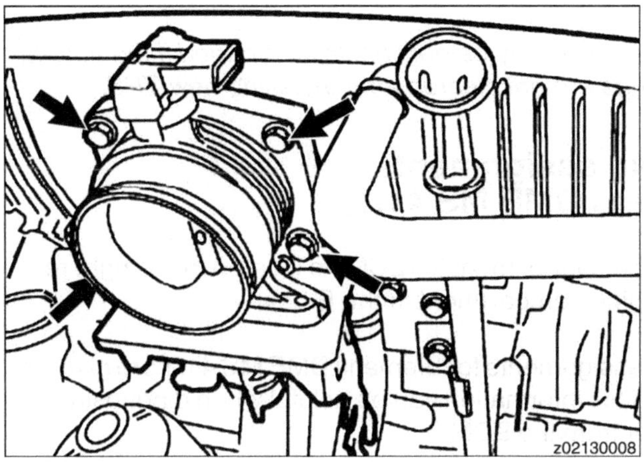


– Loosen clamps and detach coolant hoses from throttle housing preheater base. Plug hoses.

*NOTE—*
*Be prepared to catch the small amount of coolant which escapes.*

< Remove throttle housing mounting bolts (**arrows**). Remove housing and gasket from intake manifold.

– Installation is reverse of removal. Be sure to replace gasket.

– Check for fault codes and clear and reset ECM memory.

## Throttle position sensor (TPS), replacing (M52, S52 engine)

The throttle position sensor (TPS) is located on top of the throttle valve housing and provides the ECM with a voltage value that represents throttle position (angle) and rate of movement.

The TPS is non-adjustable. If it is replaced, the adapted fuel trim settings must be cleared from the ECM using a BMW-compatible scan tool.

If the TPS signal is faulty or not plausible, a fault code is stored and the MIL is illuminated when OBD II fault criteria are exceeded. The ECM still operates the engine using the mass air flow (MAF) sensor and engine rpm inputs.

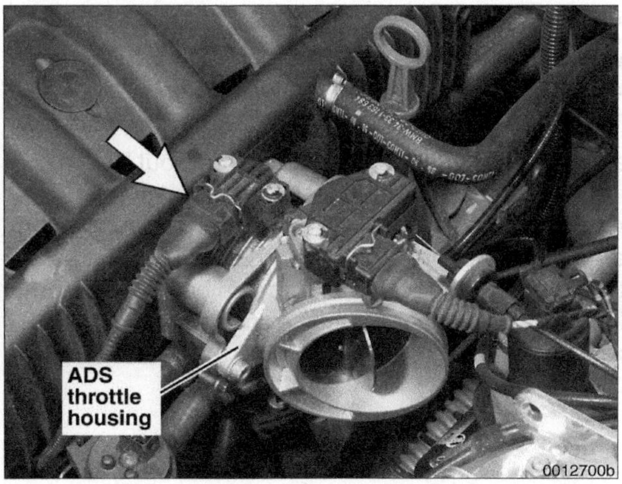


< Working at top of throttle housing, unplug connector **(arrow)** from sensor.

***CAUTION—***
*Be sure ignition is OFF before disconnecting or reconnecting DME system components.*

*NOTE—*
*On cars with traction control, do not confuse the throttle position sensor on the main throttle body with the throttle position sensor on the auxiliary throttle housing.*

– Remove two mounting screws holding sensor to throttle housing.

– Installation is reverse of removal.

– Check for fault codes and clear adaptation settings from ECM using BMW-compatible scan tool.

*NOTE—*
*Poor idle quality may be noticeable for a period if the adaptation settings are not cleared after installing new TPS.*

## ECM pin assignments (Siemens DME MS 41.1)

The engine control module (ECM) is located in the right rear of the engine compartment in the E-box.

ECM pin assignments for Siemens DME MS 41.1 are given in **Table d**. This information can be helpful when diagnosing faults to or from the ECM.

Generally, absence of voltage or continuity means there is a wiring or connector problem. Test results with incorrect values do not necessarily mean that a component is faulty. Check for loose, broken or corroded connections and wiring before replacing components. For engine management system electrical schematics, see **ELE Electrical Wiring Diagrams**.

*CAUTION—*
- *Always wait at least 40 seconds after turning off the ignition before removing the connector from the engine control module (ECM). If the connector is removed before this time, residual power in the system relay may damage the ECM.*
- *Always connect or disconnect the control module connector and meter probes with the ignition off.*

When making checks at the ECM, use a break-out box to allow tests to be made with the connector attached to the ECM. This prevents damage to the small terminals in the connector. As an alternative, the connector housing can be separated so that electrical checks can be made from the back of the connector. ECM pin numbering, as seen from back of ECM connector, is illustrated.

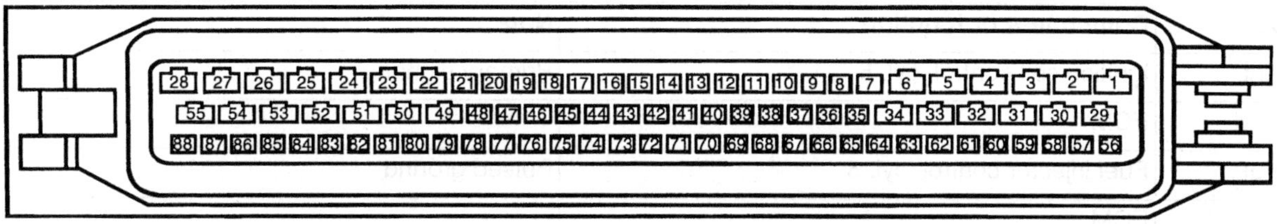


0013211

**Table d. Siemens DME MS 41.1 ECM pin assignments**

| Pin | Type | Component or function | Note |
|---|---|---|---|
| 1 | output | Ignition coil control, cyl. 2 | Primary signal |
| 2 | output | Ignition coil control, cyl. 4 | Primary signal |
| 3 | output | Ignition coil control, cyl. 6 | Primary signal |
| 4 | ground | ECM | |
| 5 | output | Fuel injector control, cyl. 2 | Pulsed ground |
| 6 | output | Fuel injector control, cyl. 1 | Pulsed ground |
| 7 | ground | Mass air flow (MAF) sensor | Signal |
| 8 | input | Mass air flow (MAF) sensor | Signal |
| 9 | output | Instrument cluster | Fuel consumption signal |
| 10 | input | Engine coolant temperature (ECT) sensor | Signal |
| 11 | - | Not used | |
| 12 | input | Throttle position sensor (TPS) | Signal |
| 13 | - | Not used | |
| 14 | input | Intake air temperature sensor (IAT) | Signal |
| 15 | input | ABS / ASC module | Wheel speed signal, right rear |
| 16 | input | A/C refrigerant pressure switch | |
| 17 | - | Not used | |
| 18 | input | Electronic immobilizer (EWS) | Activation |
| 19 | input | Automatic climate control (IHKA) | A/C ON signal |
| 20 | output | Instrument cluster | MIL |
| 21 | output | Camshaft actuator control (VANOS) | Switched ground |
| 22 | output | Fuel injector control, cyl. 3 | Pulsed ground |
| 23 | output | Fuel injector control, cyl. 6 | Pulsed ground |
| 24 | output | Fuel injector control, cyl. 4 | Pulsed ground |
| 25 | ground | Oxygen sensor heater (precatalyst, bank 1) | |
| 26 | input | Power supply (terminal 30) | Battery voltage (B+) at all times |
| 27 | output | Idle speed control valve | Pulsed ground-open signal (see pin 53) |
| 28 | ground | ECM | |
| 29 | output | Ignition coil control, cyl. 1 | Primary signal |
| 30 | output | Ignition coil control, cyl. 3 | Primary signal |

| Table d. Siemens DME MS 41.1 ECM pin assignments (continued) | | | |
|---|---|---|---|
| **Pin** | **Type** | **Component or function** | **Note** |
| 31 | output | Ignition coil control, cyl. 5 | Primary signal |
| 32 | ground | ECM | |
| 33 | output | Fuel injector control, cyl. 5 | Pulsed ground |
| 34 | ground | ECM | |
| 35 | output | Secondary air injection | Activation |
| 36 | output | Instrument cluster | Engine speed signal |
| 37 | - | Not used | |
| 38 | ground | Knock sensor | Shielding |
| 39 | ground | Intake air temp (IAT) sensor and engine coolant temp (ECT) sensor | |
| 40 | ground | Crankshaft sensor, fuel tank pressure sensor | |
| 41 | ground | Camshaft sensor | Shielding |
| 42 | ground | Throttle position sensor (TPS) | |
| 43 | ground | Camshaft sensor | Activation |
| 44 | output | Throttle position sensor (TPS), fuel tank pressure sensor | Reference voltage (5 vdc) |
| 45 | output | ABS / ASC module | Throttle position signal |
| 46 | output | Instrument cluster | Fuel reserve signal |
| 47 | - | Not used | |
| 48 | output | Crankshaft sensor (Hall effect) | Control |
| 49 | input | Terminal 15 | Switched power |
| 50 | output | Running losses solenoid | |
| 51 | output | Fuel tank venting valve | |
| 52 | - | Not used | |
| 53 | output | Idle speed control valve | Pulsed ground signal (see pin 26) |
| 54 | input | DME system power supply | Battery voltage from DME main relay (terminal 87) |
| 55 | ground | Oxygen sensor heater (precatalyst, bank 2) | M Roadster |
| 56 | ground | Ignition circuit | Monitoring signal |
| 57 | output | Knock sensor (cyl. 1 - 3) | |
| 58 | input | Knock sensor (cyl. 1 - 3) | |
| 59 | output | Knock sensor (cyl 4 - 6) | |
| 60 | input/output | Data link connector (DLC) | Diagnostic signal to pin 18 |
| 61 | ground | Oxygen sensor heater (post-catalyst, bank 2) | M Roadster |
| 62 | output | Secondary air injection | |
| 63 | input | Knock sensor (cyl. 4 - 6) | |
| 64 | input | Camshaft sensor | |
| 65 | input | Camshaft sensor | |
| 66 | - | Not used | |

| Table d. Siemens DME MS 41.1 ECM pin assignments (continued) | | | |
|---|---|---|---|
| **Pin** | **Type** | **Component or function** | **Note** |
| 67 | ground | Oxygen sensors (precatalyst) | Signal ground |
| 68 | output | Evaporative purge valve control | Pulsed ground with engine at normal temperature with varying engine load |
| 69 | output | Fuel pump relay control | Energized with engine running |
| 70 | ground | Oxygen sensor (precatalyst, bank 2) | Signal (M Roadster) |
| 71 | ground | Oxygen sensors (post-catalyst) | Signal |
| 72 | ground | Oxygen sensor (post-catalyst, bank 2) | Signal (M Roadster) |
| 73 | input | DME main relay, terminal 85 | Control |
| 74 | output | A/C compressor relay | Activation |
| 75 | input | Oxygen sensor (precatalyst) | Signal |
| 76 | input | Oxygen sensor (precatalyst, bank 2) | Signal (M Roadster) |
| 77 | input | Oxygen sensor (post-catalyst) | Signal |
| 78 | input | Oxygen sensor signal (post-catalyst, bank 2) | Signal (M Roadster) |
| 79 | ground | Oxygen sensor heater (post-catalyst) | |
| 80 | input | ABS / ASC module | Ignition delay signal |
| 81 | input | ABS / ASC module | Ignition timing signal |
| 82 | input | ABS / ASC module | Increase idle speed signal |
| 83 | input | Crankshaft sensor (Hall effect) | Signal |
| 84 | - | Not used | |
| 85 | input/output | CAN-bus low | Automatic transmission control module (TCM) |
| 86 | input/output | CAN-bus high | Automatic transmission control module (TCM) |
| 87 | input | Power supply | Battery voltage from DME main relay, terminal 87 |
| 88 | input output | Diagnostic connector (TxD) | Signal to pin 17, data link connector (DLC) |

## SIEMENS DME MS 42.1 AND MS 43.0 (M52 TU, M54 ENGINE)

When engine management system faults are detected, a diagnostic trouble code (DTC) is stored in the engine control module (ECM). The MIL also illuminates if an emissions-related fault is detected.

When faults arise, or if the malfunction indicator lamp (MIL) is illuminated, begin troubleshooting by connecting BMW service tester DISplus, MoDiC, GT1 or equivalent scan tool. The capabilities of OBD II software has the potential to save hours of diagnostic time and to help avoid incorrect component replacement and possible damage to system components.

***CAUTION—***
- *The tests in this section may set fault codes (DTCs) in the ECM and illuminate the MIL. After all testing is completed, access and clear DTC fault memory using an OBD II or BMW compatible scan tool. See* **OBD On-Board Diagnostics**.
- *Only use a digital multimeter for electrical tests.*
- *Relay positions can vary. Be sure to confirm relay location and function by identifying the wiring in the socket using the wiring diagrams found in* **ELE Electrical Wiring Diagrams**.

***NOTE—***
- *OBD II fault memory (including an illuminated MIL) can only be reset using the scan tool. Removing the connector from the ECM or disconnecting the battery will not erase the fault memory.*
- *The BMW-dedicated 20-pin diagnostic link connector (DLC) is in the right rear of the engine compartment. The 20-pin DLC is discontinued on models after 6 / 2000. The 16-pin OBD II diagnostic connector is located inside the car on the right side of the center console. See* **Engine management system troubleshooting** *in this repair group.*

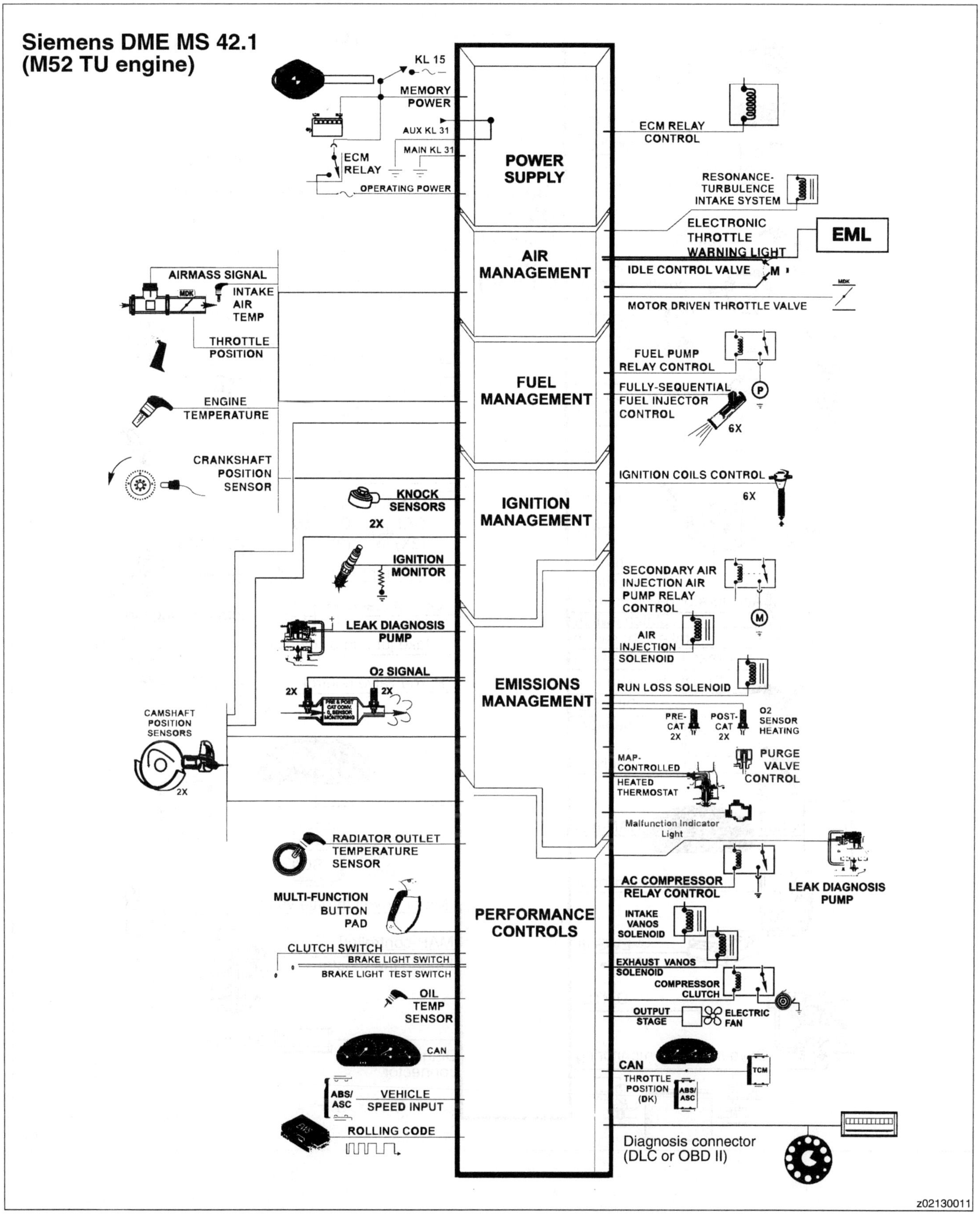
Siemens DME MS 42.1
(M52 TU engine)
KL 15
MEMORY POWER
AUX KL 31
MAIN KL 31
ECM RELAY
OPERATING POWER
POWER SUPPLY
AIR MANAGEMENT
FUEL MANAGEMENT
IGNITION MANAGEMENT
EMISSIONS MANAGEMENT
PERFORMANCE CONTROLS
ECM RELAY CONTROL
RESONANCE-TURBULENCE INTAKE SYSTEM
ELECTRONIC THROTTLE WARNING LIGHT
EML
IDLE CONTROL VALVE
MOTOR DRIVEN THROTTLE VALVE
MDK
AIRMASS SIGNAL
INTAKE AIR TEMP
THROTTLE POSITION
ENGINE TEMPERATURE
CRANKSHAFT POSITION SENSOR
KNOCK SENSORS
2X
FUEL PUMP RELAY CONTROL
FULLY-SEQUENTIAL FUEL INJECTOR CONTROL
6X
IGNITION COILS CONTROL
6X
IGNITION MONITOR
LEAK DIAGNOSIS PUMP
O2 SIGNAL
2X
2X
CAMSHAFT POSITION SENSORS
2X
SECONDARY AIR INJECTION AIR PUMP RELAY CONTROL
AIR INJECTION SOLENOID
RUN LOSS SOLENOID
PRE-CAT 2X
POST-CAT 2X
O2 SENSOR HEATING
MAP-CONTROLLED HEATED THERMOSTAT
PURGE VALVE CONTROL
Malfunction Indicator Light
LEAK DIAGNOSIS PUMP
RADIATOR OUTLET TEMPERATURE SENSOR
MULTI-FUNCTION BUTTON PAD
CLUTCH SWITCH
BRAKE LIGHT SWITCH
BRAKE LIGHT TEST SWITCH
OIL TEMP SENSOR
CAN
ABS/ASC
VEHICLE SPEED INPUT
ROLLING CODE
AC COMPRESSOR RELAY CONTROL
INTAKE VANOS SOLENOID
EXHAUST VANOS SOLENOID
COMPRESSOR CLUTCH
OUTPUT STAGE
ELECTRIC FAN
CAN
TCM
THROTTLE POSITION (DK)
ABS/ASC
Diagnosis connector (DLC or OBD II)
z02130011

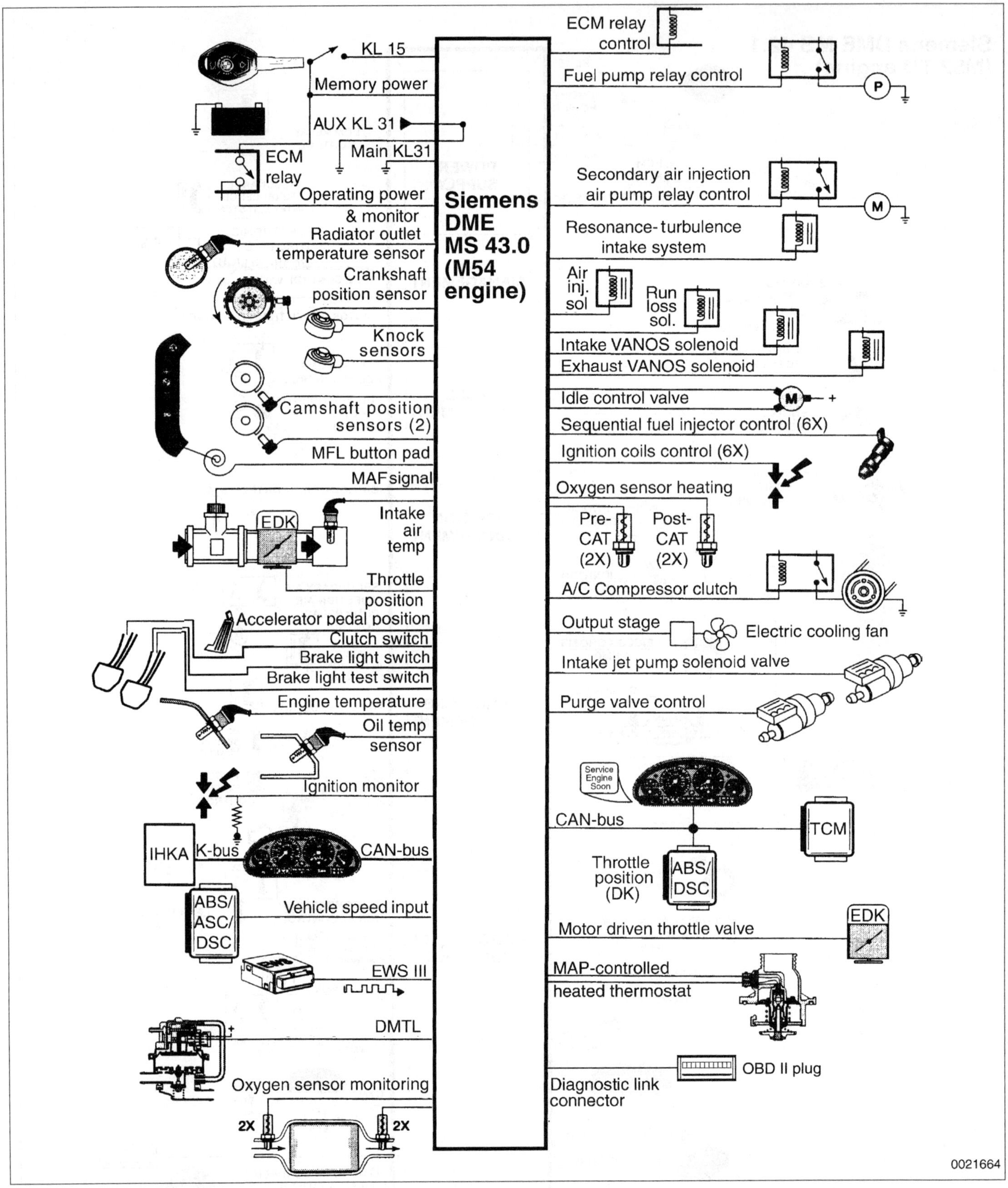
Siemens DME MS 43.0 (M54 engine)
KL 15
Memory power
AUX KL 31
Main KL31
ECM relay
Operating power & monitor
Radiator outlet temperature sensor
Crankshaft position sensor
Knock sensors
Camshaft position sensors (2)
MFL button pad
MAF signal
EDK
Intake air temp
Throttle position
Accelerator pedal position
Clutch switch
Brake light switch
Brake light test switch
Engine temperature
Oil temp sensor
Ignition monitor
IHKA
K-bus
CAN-bus
ABS/ ASC/ DSC
Vehicle speed input
EWS III
DMTL
Oxygen sensor monitoring
2X
2X
ECM relay control
Fuel pump relay control
P
Secondary air injection air pump relay control
M
Resonance-turbulence intake system
Air inj. sol
Run loss sol.
Intake VANOS solenoid
Exhaust VANOS solenoid
Idle control valve
M
Sequential fuel injector control (6X)
Ignition coils control (6X)
Oxygen sensor heating
Pre-CAT (2X)
Post-CAT (2X)
A/C Compressor clutch
Output stage
Electric cooling fan
Intake jet pump solenoid valve
Purge valve control
Service Engine Soon
CAN-bus
TCM
Throttle position (DK)
ABS/ DSC
EDK
Motor driven throttle valve
MAP-controlled heated thermostat
OBD II plug
Diagnostic link connector
0021664

## Air filter housing and ducts, removing and installing (M52 TU, M54 engine)

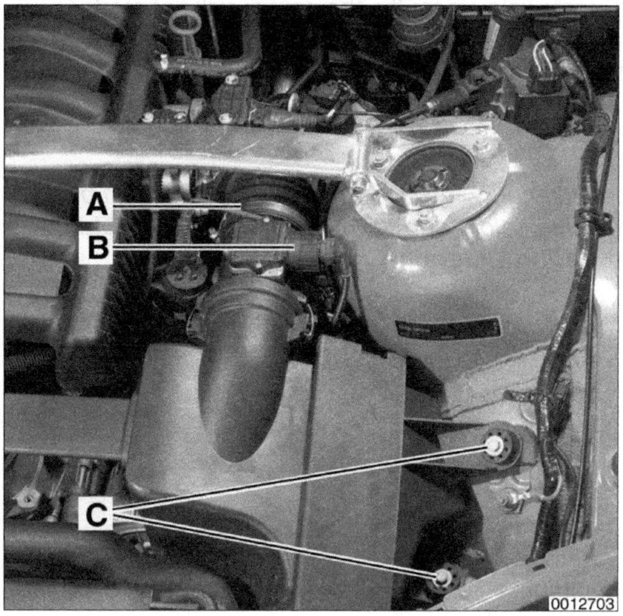


Unbolt and remove complete air cleaner housing, disconnecting and labeling connectors, ducting and hoses as necessary:

- Air duct hose clamp (**A**)
- Mass air flow sensor connector (**B**)
- Air cleaner housing mounting bolts (**C**)

**Mass air flow sensor ducts (M52 TU, M54 engine)**

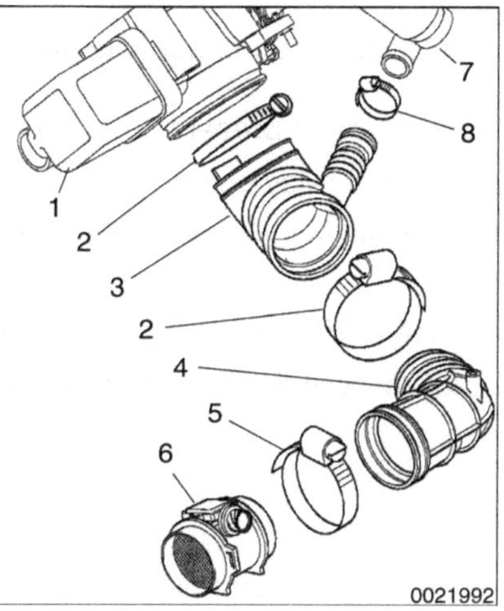


Loosen clamps **2** and **8** and remove mass air flow sensor and air ducts.

1. Throttle assembly
2. Hose clamp 77 - 84 mm
3. Y-duct
4. Air duct
5. Hose clamp 83 - 90 mm
6. Mass air flow sensor
7. Idle control valve
8. Hose clamp 28 - 33 mm

– After reinstalling, check for fault codes and clear and reset ECM memory.

## DME main relay, testing (M52 TU, M54 engine)

The DME main relay is energized via the ECM and supplies battery positive (B+) power to the following:

- ECM
- Ignition coils
- Fuel injectors
- Fuel tank leak detection pump (LDP or DM-TL)
- Idle air control valve
- Camshaft sensors
- Evaporative emissions valve
- Hot film mass air flow (MAF) sensor
- Fuel pump relay
- Oxygen sensor heaters

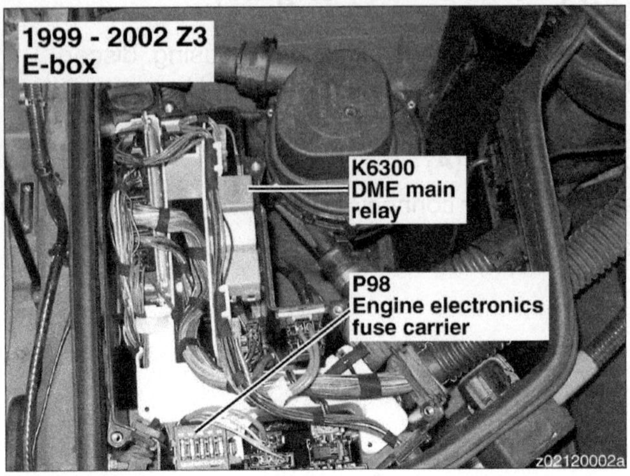


If the DME main relay is faulty, the engine will not start.

With ignition off, remove DME main relay (K6300) in electronics box (E-box) at right rear of engine compartment.

*NOTE—*
*Relay locations can vary. Confirm relay identification by matching wiring colors and terminal numbers. See* **ELE Electrical Wiring Diagrams**.

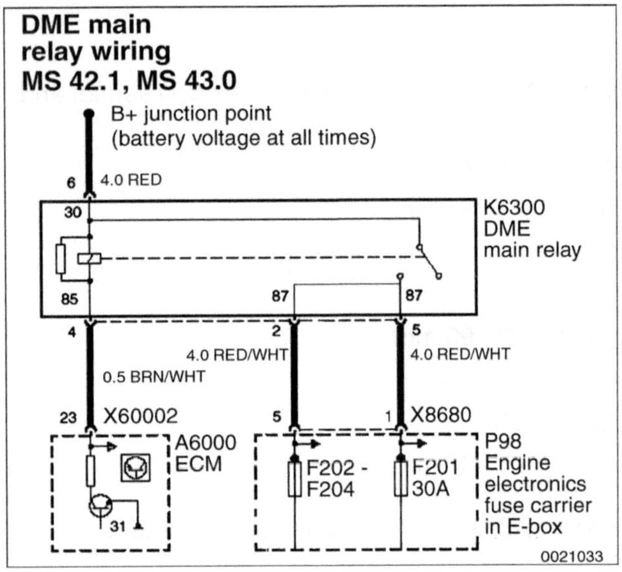


Check for voltage at terminal **6** of relay socket (corresponds to terminal 30 on relay.

- If battery voltage is present, continue testing.
- If battery voltage is not present, check the following:
  - Large red wire in relay socket
  - B+ (battery positive junction) in right rear of engine compartment
  - 200 amp fuse (F100) in trunk near battery
  - See **ELE Electrical Wiring Diagrams** for more details
- Check for ground at relay socket **4** (corresponds to terminal **85** on relay).
- If ground is present, continue testing.
- If ground is not present, signal from ECM is missing. Check wire between ECM and relay.
- If no faults are found:
  - Check ECM grounding.
  - ECM may be defective.
- With ignition on and relay installed, check for battery voltage at relay sockets **2** and **5** (correspond to terminals **87** on relay).

*NOTE—*
*In 1999 and 2000 models, there is only one terminal 87.*

- If battery voltage is present, relay has energized and is functioning correctly.
- If battery voltage is not present and all earlier tests are OK, relay is faulty. Replace.

– If no faults are found during relay testing but power is not reaching ECM or other components, check fuses in the engine electronics fuse carrier P98 in E-box.

– When finished testing, check for fault codes and clear and reset ECM memory.

## Engine coolant temperature (ECT) sensor, replacing (M52 TU, M54 engine)

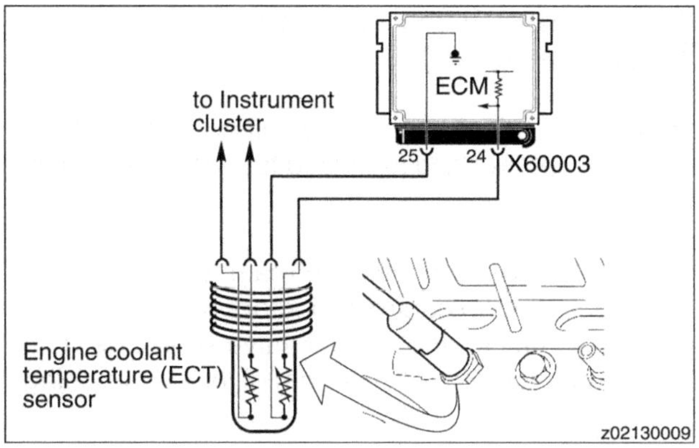


The ECT sensor is a dual temperature sensor. One circuit provides coolant temperature information to the ECM while the other circuit provides coolant temperature information to the instrument cluster.

The ECM determines the correct ignition timing and air/fuel mixture required for the engine temperature by monitoring a reference voltage (5 vdc) to the sensor. The voltage drop across the sensor varies as the coolant temperature (sensor resistance) changes.

***NOTE—***
*The ECT sensor is an NTC (negative temperature coefficient) type sensor. In other words, as the temperature rises, resistance through the sensor decreases.*

If the ECT sensor input is faulty or not plausible, the MIL is illuminated when OBD II fault criteria are exceeded. The ECM assumes a substitute value (80°C / 176°F) to maintain engine operation. The ignition timing is set to a safely conservative basic setting.

The ECT sensor is located below cylinder 6 intake port at the rear of the engine.

– Remove intake manifold. See **113 Cylinder Head**.

***WARNING—***
- *To remove the intake manifold, disconnect fuel lines from fuel rail.*
- *Do not smoke or work near heaters or other fire hazards. Keep a fire extinguisher handy. Before disconnecting fuel hoses, wrap a cloth around fuel hoses to absorb any leaking fuel. Catch and dispose of escaped fuel. Plug all open fuel lines.*

Disconnect ECT sensor connector (**arrow**).

– Unscrew ECT sensor from cylinder head. Be prepared to catch small amount of coolant.

– Installation is reverse of removal:
  - Use new copper sealing washer when installing sensor.
  - Replace any lost coolant.

| Tightening torque | |
|---|---|
| Temperature sensor to cylinder head | 13 Nm (10 ft-lb) |

– Check for fault codes and clear and reset ECM memory.

## Fuel rail and injectors, removing (M52 TU, S54 engine)

Fuel injectors are electrically controlled solenoid valves that provide precisely metered and atomized fuel into the engine intake ports.

Injectors receive voltage from the DME main relay. The engine control module (ECM) controls injector opening by activating the ground circuit for the injector valve solenoids. The ECM varies the duration (in milliseconds) of injector opening to regulate air / fuel ratio.

> ***WARNING—***
> - *When opening fuel lines, be prepared for fuel to be expelled under pressure (approx. 3 - 5 bar or 45 - 75 psi).*
> - *Do not smoke or work near heaters or other fire hazards. Keep a fire extinguisher handy.*
> - *Before disconnecting fuel hoses, wrap a cloth around fuel hoses to absorb any leaking fuel.*
> - *Catch and dispose of escaped fuel.*
> - *Plug all open fuel lines.*
> - *Unscrew the fuel tank cap to release pressure in the tank before working on the tank or lines.*

> ***CAUTION—***
> *Cleanliness is essential when working with fuel circuit components. Thoroughly clean the unions before disconnecting fuel lines.*

◄ To access injectors, first remove injector cover by removing plastic trim plugs (**arrows**) and removing fasteners under them. Lift off cover.

– For a quick check of the injectors, run the engine and touch each injector with a screwdriver or stethoscope. You should feel a vibration or hear a buzzing. Turn off engine.

– Disconnect negative (–) battery cable.

> ***CAUTION—***
> *Prior to disconnecting the battery, read the battery disconnection cautions given in* **001 General Warnings and Cautions**.

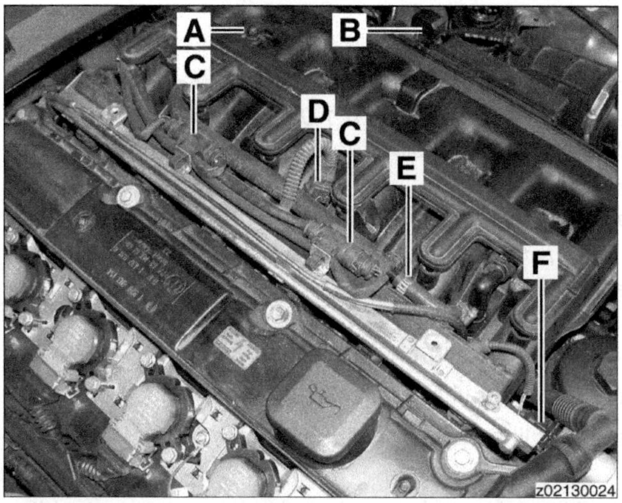


◄ Working above engine, detach the following:

- **A** Positive engine lead at B+ terminal
- **B** Resonance valve electrical connector
- **C** Oxygen sensor connectors
- **D** Electrical connector for intake air temperature sensor
- **E** Manifold vacuum line
- **F** VANOS solenoid electrical connector

> ***CAUTION—***
> *Be sure to mark oxygen sensor connectors so that they can be reassembled as before.*

◄ Disconnect fuel injector electrical connectors from injectors:

- Use small screwdriver to pry one corner of wire lock clip on fuel injector 1 connector.
- Repeat for all injectors.
- Lift off connector loom and set aside.

◄ Unscrew schræder valve cap (**arrow**) from fuel rail.

- Unscrew fuel tank cap to release pressure.
- Using a brief burst of compressed air (maximum of 3 bar or 43.5 psi) blow fuel back into fuel tank.

Working at rear of manifold, label and disconnect fuel supply and fuel return lines from fuel rail.

> ***WARNING —***
> *Clamp off the fuel lines and then wrap a clean shop towel around the lines before removing them. Residual fuel pressure is present in the lines.*

***NOTE —***
- *A special fitting is used to retain the fuel line(s) to the fuel rail. Use BMW special tool 16 1 050 or equivalent to compress the locking clip inside the end of the fuel line fitting to release the line.*
- *If necessary, access fuel line(s) from underneath vehicle.*
- *There is no return fuel line in models with DME MS43.0 (M54 engine).*

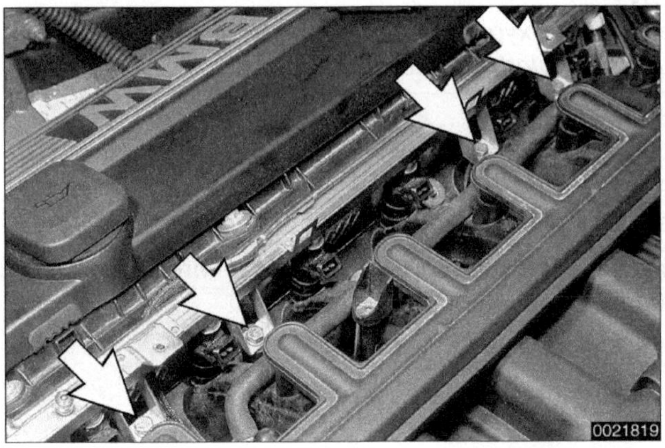


Remove fuel rail mounting bolts (**arrows**).
- Carefully pry fuel rail and injectors off manifold.
- Separate fuel line support bracket at rear of intake manifold.
- Guide fuel line(s) out of rear of engine compartment while lifting fuel rail out.

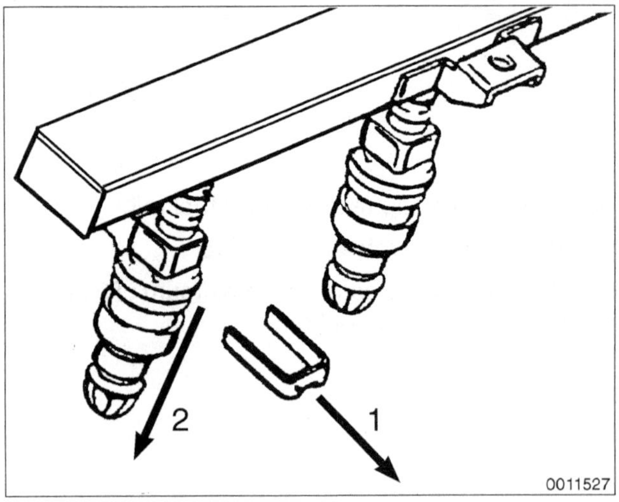


Remove individual injectors:
- Pry retaining clip from injector (**1**).
- Pull injector from rail (**2**).

– Installation is reverse of removal. Remember to:
- Fit new sealing O-rings when installing injectors. For ease of installation, lightly lubricate O-rings with assembly lubricant.
- Check that injector electrical connections are correctly fitted and that injectors are fully seated prior to installing fuel rail mounting bolts.
- Check fuel line rubber O-rings and replace as necessary. For ease of installation, lightly lubricate O-rings with assembly lubricant.
- Replace wire ties.

| **Tightening torque** | |
|---|---|
| Fuel rail to cylinder head | 10 Nm (7 ft-lb) |

– Check for fault codes and clear and reset ECM memory.

## Idle speed control valve, replacing (M52 TU, M54 engine)

The idle speed control valve, mounted directly above the throttle housing, regulates idle speed by redirecting air around the throttle valve. The valve is supplied with battery voltage from the DME main relay. The valve incorporates a two-coil rotary actuator. The ground sides of each coil are pulsed simultaneously by the ECM. The duty cycle of each circuit is varied to achieve the required idle speed.

A faulty idle speed control valve sets a fault code and the MIL is illuminated when OBD II fault criteria are exceeded.

– Remove air filter housing and ducts. See **Air filter housing and ducts, removing and installing (M52 TU, M54 engine)** in this repair group.

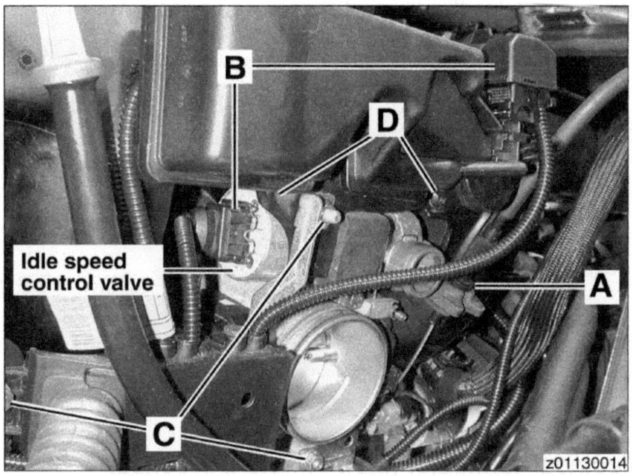


To remove idle speed control valve:

- M52 TU engine: Detach throttle cable (**A**).
- Detach electrical connectors (**B**) at idle speed control valve and intake manifold resonance valve.
- Remove wiring harness housing mounting fasteners (**C**). Pull housing away from throttle valve.
- Remove idle speed control valve bracket mounting bolts (**D**). Remove idle speed control valve with attached bracket.

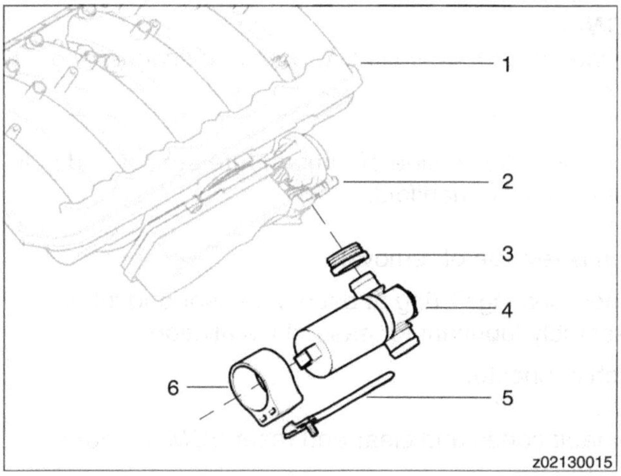


When reinstalling, replace idle valve vacuum seal. Coat seal on inside with assembly lubricant, then install seal in manifold. Press valve into place.

1. Intake manifold
2. Intake manifold resonance valve
3. Vacuum seal
4. Idle control valve
5. Mounting bracket
6. Rubber buffer

– Reattach wiring harnesses and throttle cable (if applicable). Reinstall air filter assembly and ducts.

– Check for fault codes and clear and reset ECM memory.

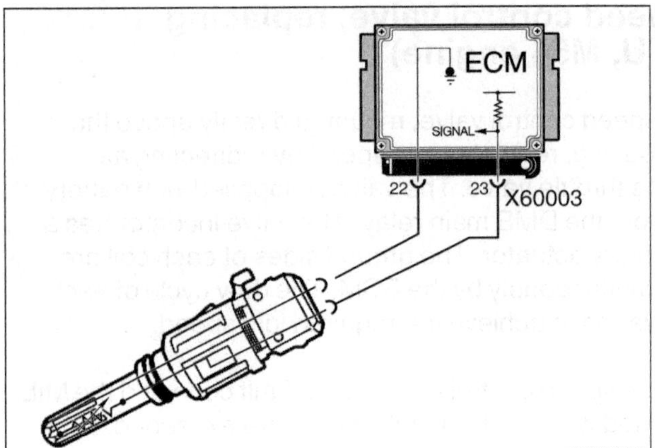

z02130016

## Intake air temperature (IAT) sensor, replacing (M52 TU, M54 engine)

The intake air temperature (IAT) sensor, located at the top of the intake manifold, adapts (fine tunes) the fuel mixture and engine timing based on varying intake air temperatures.

If the intake air temperature signal is not plausible, a fault code is set and the MIL is illuminated when OBD II fault criteria are exceeded. The ECM then operates the engine using the ECT sensor signal input as a back up.

***NOTE—***
*The IAT sensor is an NTC (negative temperature coefficient) type sensor. In other words, as the temperature rises, resistance through the sensor decreases.*

z02180012

To access IAT sensor, remove plastic trim plugs (**arrows**), remove fasteners under them, and remove cover from above fuel injectors.

z02130024b

Detach electrical connector (**arrow**) from IAT sensor.

***CAUTION—***
*Be sure the ignition switch is OFF before disconnecting the IAT sensor.*

- Depress locking clip on side of temperature sensor and pull straight out of intake manifold.
- Installation is reverse of removal.
  - Use a new sealing O-ring at base of sensor and lubricate with assembly lubricant for ease of installation.
  - Reattach connector.
- Check for fault codes and clear and reset ECM memory.

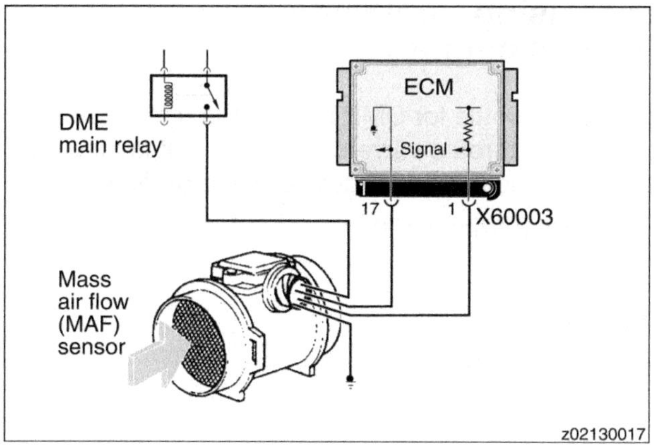


## Mass air flow (MAF) sensor, replacing (M52 TU, M54 engine)

The hot film mass air flow (MAF) sensor sends a varying voltage (approx. 0.5 - 4.5 vdc) to the ECM representing the measured intake air. The air mass input signal is produced electronically using an electrically heated metal film in the air flow stream.

The DME main relay provides the operating voltage to the MAF sensor. As air flows through the sensor, the film is cooled. To maintain the film at a constant temperature, additional current is necessary. It is this additional current that is the basis for the input signal.

If there is no output signal from the MAF sensor, the ECM operates the engine using throttle position and engine rpm inputs. A faulty MAF sensor illuminates the MIL.

A faulty MAF sensor can produce the following problems:

- Difficult to restart when engine is hot
- Engine starts then stalls
- Engine starts and runs only with accelerator pedal depressed

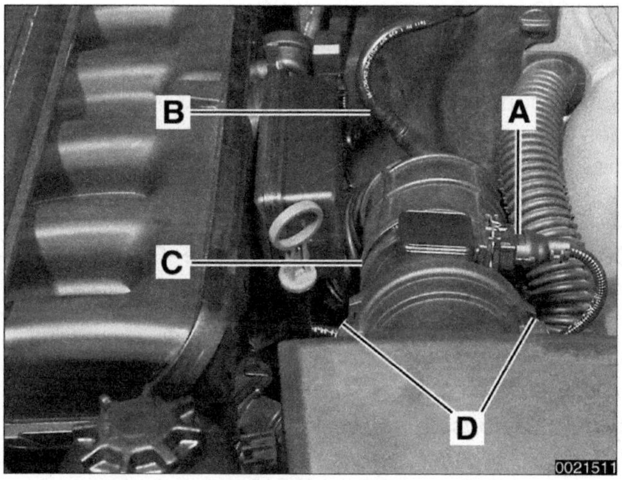


Disconnect electrical connector (**A**) from MAF sensor.

> ***CAUTION—***
> *Be sure ignition is OFF before disconnecting or reconnecting DME system components.*

- Detach brake booster vacuum line (**B**).
- Loosen air duct clamp (**C**).
- Release clips (**D**) from air filter housing.
- Remove MAF sensor.

– If MAF sensor is to be reused, visually inspect sensor for damaged, missing or blocked screen. (Screen affects air flow calibration.)

– Check intake ducts for cracks and vacuum leaks.

– Installation is reverse of removal.

- For ease of installation, lubricate large rubber O-ring seal in air filter housing with assembly lubricant.
- Carefully check that all hoses and seals are seated properly.

– Check for fault codes and clear and reset ECM memory.

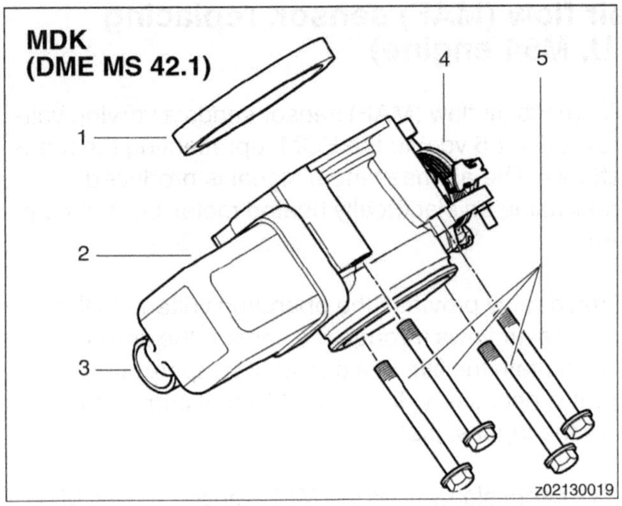


## Throttle housing, replacing (M52 TU, M54 engine)

◀ The motor driven throttle for DME MS 42.1 is referred to as MDK (motor driven throttle valve). The MDK unit is electronically operated by the ECM to regulate intake air flow.

1. Profile gasket (O-ring seal)
2. MDK unit
3. Electrical connector
4. Throttle cable bellcrank
5. M6 bolts

DME MS 42.1 uses a throttle cable from the accelerator pedal to the MDK unit. The throttle cable is connected to the pedal position sensor (PWG) in the MDK unit. This sensor communicates pedal position to the ECM, which in turn electronically regulates throttle plate position. The MDK throttle cable bellcrank is linked to the throttle shaft by springs to allow intervention by the ABS / ASC or ABS / DSC control module (via the ECM) to override the driver's accelerator pedal position.

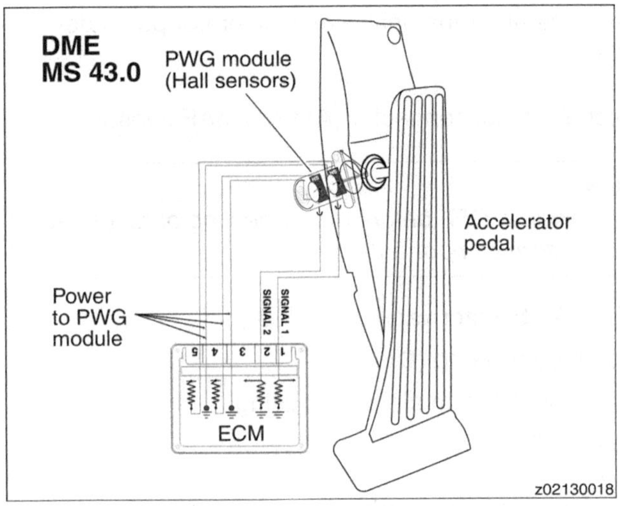


◀ The electronic throttle valve assembly for the MS43.0 system is referred to as EDK (electronic throttle valve). There is no mechanical (cable) link between the accelerator pedal and the throttle plate.

The accelerator pedal module (PWG) at the pedal assembly communicates pedal position directly to the ECM. The module provides two variable voltage signals (via two Hall sensors) to the ECM for pedal position and rate of movement.

***NOTE—***

- *See* **300 Suspension, Steering and Brake–General** *and* **340 Brakes** *for further details about ABS system intervention.*
- *The throttle housing unit is nonadjustable. If found to be faulty, replace as a complete unit.*
- *A faulty throttle housing unit resets the adaptation values stored in the ECM. After replacing the throttle housing, use BMW scan tool DISplus, GT1, MoDiC or equivalent to reset adaptation values.*

> ***CAUTION—***
> *If the adaptation process is not completed correctly, the engine will not start.*

– Disconnect battery negative (-) cable in luggage compartment.

> ***CAUTION—***
> *Prior to disconnecting the battery, read the battery disconnection cautions in* **001 General Warnings and Cautions**.

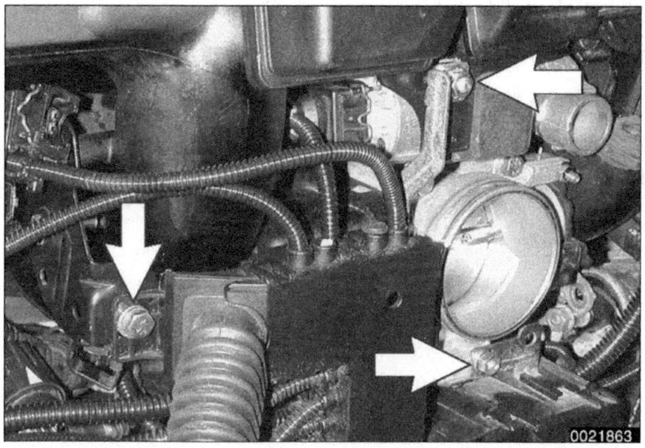


– Remove complete air filter housing. See **Air filter housing and ducts, removing and installing (M52 TU, M54 engine)** in this repair group.

– DME MS42.1: Pull throttle cable upwards out of rubber retainer and unhook ball end of cable from MDK throttle cable bellcrank.

Remove wiring harness housing mounting fasteners (**arrows**) and pull housing away from throttle assembly.

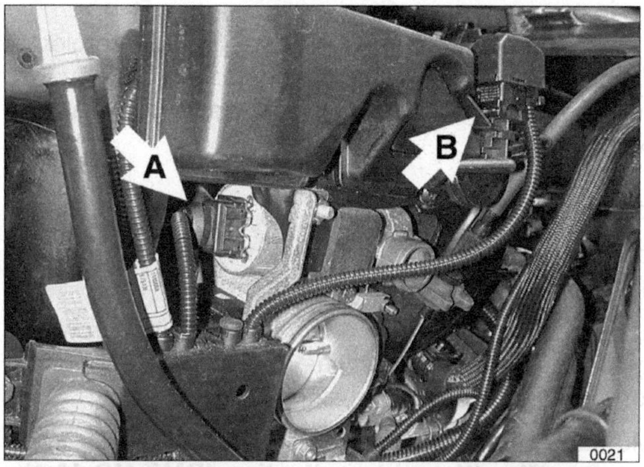


Working at side of intake manifold, disconnect electrical connectors:

- Idle control valve (**A**)
- Intake manifold resonance valve (**B**)

– Disconnect electrical connectors at oil pressure sender and oil temperature sender at base of oil filter housing.

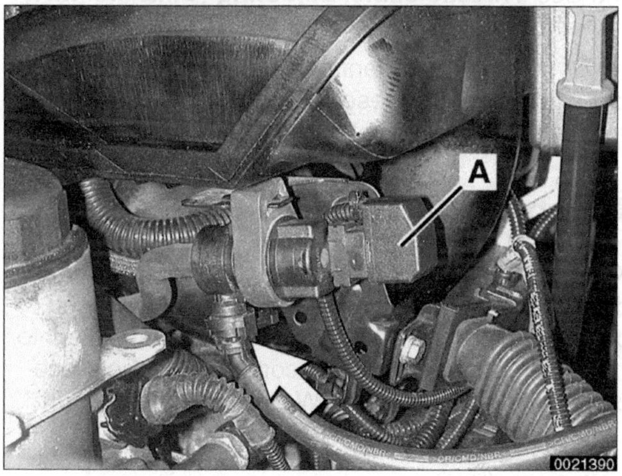


If necessary, disconnect electrical connector at fuel tank venting valve (**A**). Disconnect hose at quick disconnect fitting (**arrow**).

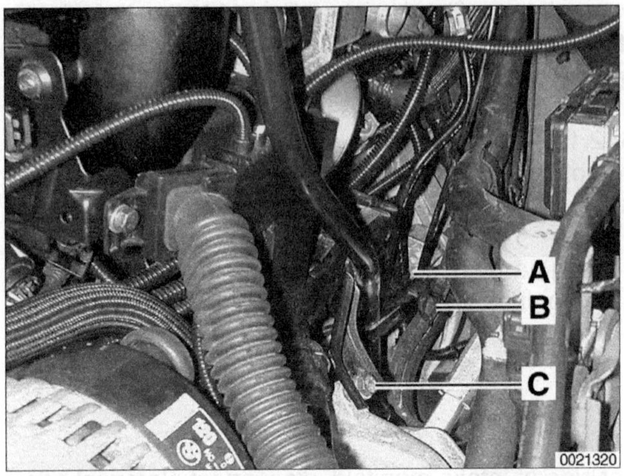


Remove dipstick guide tube:

- Disconnect wiring harness brackets from tube (**A**).
- Unclip fuel lines from dipstick tube (**B**).
- Remove lower guide tube mounting bolt (**C**). Pull out dipstick guide tube.

Working at throttle housing, turn harness plug counterclockwise and disconnect from throttle connector (**A**). Remove throttle housing mounting fasteners (**arrows**) and pull assembly off manifold.

– Installation is reverse of removal.

- Replace profile gasket (sealing O-ring) between throttle assembly and intake manifold.
- Reattach throttle assembly connector with care. Arrow on fully tightened connector must line up with corresponding arrow on throttle assembly housing.
- DME MS 42.1: Reattach throttle cable and check cable adjustment. See **Throttle cable, adjusting (DME MS 42.1)** in this repair group.
- Connect BMW test equipment or equivalent. Check for fault codes and clear and reset ECM memory. Reset throttle plate adaptation values following on-screen directions.

## Throttle cable, adjusting (DME MS 42.1)

The MDK throttle assembly on MS 42.1 cars uses a throttle cable. Adjust cable any time it is disconnected or replaced. This procedure requires the factory scan tool or equivalent to read and adjust throttle angle. It is recommended the procedure be done by an authorized BMW dealer service department.

Basic adjustment:

- Make sure accelerator pedal and accelerator cable bellcrank at MDK unit are in idle position.
- Adjust cable by turning knurled knob on end of throttle cable sheath at MDK unit. Set cable play (**A**).

| **Throttle cable adjustment** | |
|---|---|
| Max. cable play (**A**) | 3 mm (0.12 in) |

– Full throttle adjustment (manual transmission):
  - With ignition key ON, press pedal fully to stop. Check throttle angle using scan tool. Adjust pedal stop underneath accelerator pedal, if necessary.

| Throttle angle specification (manual transmission) | |
|---|---|
| Full throttle | 80° - 84° |

– Full throttle adjustment (automatic transmission):
  - With ignition key on, press accelerator until pedal touches kickdown detent. Check throttle angle. Press accelerator pedal down beyond kickdown detent. Check throttle angle. Adjust kickdown detent if necessary and recheck.
  - Press accelerator down fully against stop. Check throttle angle.
  - Use scan tool diagnostic program or road test vehicle to confirm transmission downshifts when kickdown is actuated.

| Throttle angle specification (automatic transmission) | |
|---|---|
| Kickdown detent contacted | 76° - 80° |
| Kickdown range | 84° - 88° |
| Full-load downshift | 100% |

## Crankcase breather valve, removing and installing (M54 engine)

Engine breather hoses become brittle over time. Replace all hoses along with crankcase breather valve when removing.

– Disconnect battery negative (-) cable in cargo compartment.

> ***CAUTION—***
> - *Prior to disconnecting the battery, read the battery disconnection cautions in* **001 Warnings and Cautions**.

– Remove throttle housing. See **Throttle housing, replacing (6-cylinder engine)** in this repair group.

◄ Working at left side of cylinder head, detach engine vent hose from cylinder head cover by pinching spring clips (**arrows**).

Working between cylinder 1 and cylinder 2 intake manifold runners, detach engine vent hose from crankcase breather hose by pinching spring clips (**arrows**).

Working between cylinder 1 and cylinder 2 intake manifold runners, detach engine vent hose from intake manifold by pinching spring clips (**arrows**).

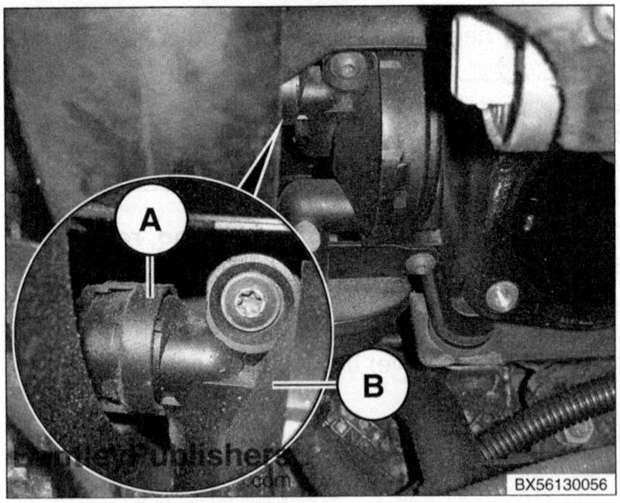


Working under center of intake manifold, use a small pick or flat blade screwdriver to detach breather hose (**A**) from crankcase breather valve (**B**).

< Working under center of intake manifold:

- Detach engine vent hose (**A**) from crankcase breather by pinching spring clips (**arrows**).
- Detach crankcase breather drain hose (**B**) from crankcase breather by pinching spring clips (**arrows**).

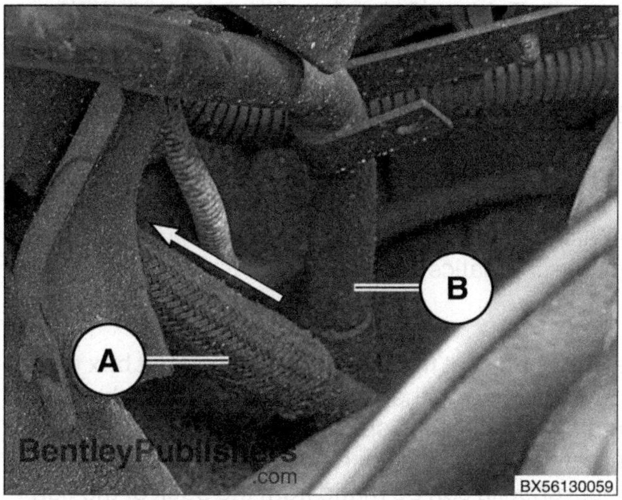


< Remove crankcase breather drain hose (**A**) from dipstick tube (**B**).

– Check that dipstick tube drain for crankcase breather valve is clear of obstructions.

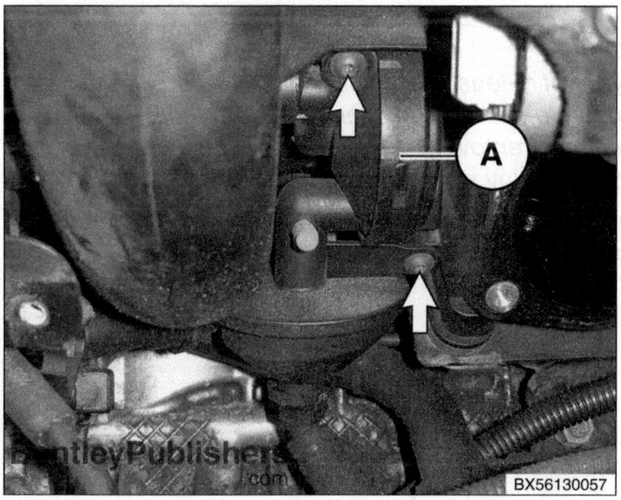


< Remove crankcase breather fasteners (**arrows**) and remove crankcase breather valve (**A**) from intake manifold.

– Installation is reverse of removal. Remember to:

- Check with parts supplier for updated or revised part numbers.
- Replace engine vent hoses with new.
- Listen for an audible click when installing engine vent hoses.

## ECM pin assignments (Siemens DME MS 42.1, MS 43.0)

The engine control module (ECM) is located in the right rear of the engine compartment in the E-box.

The ECM has 5 electrical connectors with a total of 134 pins:

- Connector 1 (X60001): 9 pins, power and grounds
- Connector 2 (X60002): 24 pins, ancillary signals (oxygen sensors, CAN-bus, etc.)
- Connector 3 (X60003): 52 pins, engine signals
- Connector 4 (X60004): 40 pins, vehicle signals
- Connector 5 (X60005): 9 pins, ignition signals

ECM pin assignments for Siemens DME MS 42.1 and MS 43.0 are given in **Table e**. This information can be helpful when diagnosing faults to or from the ECM.

Generally, absence of voltage or continuity means there is a wiring or connector problem. Test results with incorrect values do not necessarily mean that a component is faulty. Check for loose, broken or corroded connections and wiring before replacing components. For engine management system electrical schematics, see **ELE Electrical Wiring Diagrams**.

When making checks at the ECM, use a break-out box to allow tests to be made with the connector attached to the ECM. This prevents damage to the small terminals in the connector. As an alternative, the connector housing can be separated so that electrical checks can be made from the back of the connector.

***CAUTION—***
- *Always wait at least one minute after turning off the ignition before removing the connector from the ECM. If the connector is removed before this time, residual power in the system relay may damage the ECM.*
- *Always connect or disconnect the control module connector and meter probes with the ignition off.*

| Table e. Siemens DME MS 42.1 and MS 43.0 ECM pin assignments | | | |
|---|---|---|---|
| **Connector X60001 9-pin (power, grounds)** | | | |
| **Pin** | **Type** | **Component or function** | **Note** |
| 1 | output | Unloader relay terminal 15 | |
| 2 | - | Not used | |
| 3 | - | Not used | |
| 4 | ground | ECM | |
| 5 | ground | ECM | |
| 6 | ground | ECM | |
| 7 | input | Power supply (terminal 30) | Battery voltage (B+) at all times |
| 8 | input | Fuse carrier, engine electronics | Voltage supply |
| 9 | input | Fuse carrier, engine electronics | Voltage supply |

| Connector X60002 24-pin (ancillary signals) | | | |
|---|---|---|---|
| **Pin** | **Type** | **Component or function** | **Note** |
| 1 | ground | Oxygen sensor heater (precatalyst, bank 1) | |
| 2 | - | Not used | |
| 3 | input/output | CAN-bus low | Automatic transmission control module (TCM) |
| 4 | input/output | CAN-bus high | Automatic transmission control module (TCM) |
| 5 | - | Not used | |
| 6 | input/output | Data link signal (TxD) | Automatic transmission control module (TCM) |
| 7 | ground | Oxygen sensor heater (post-catalyst, bank 1) | |
| 8 | - | Not used | |
| 9 | - | Not used | |
| 10 | - | Not used | |
| 11 | - | Not used | |
| 12 | input | 3/2-way valve, running losses (DME MS 42.1) | Signal |
| 13 | input | Oxygen sensor heater (precatalyst, bank 2) | |
| 14 | input | Oxygen sensor (precatalyst, bank 1) | Signal |
| 15 | input | Oxygen sensor (precatalyst, bank 2) | Signal |
| 16 | input | Oxygen sensor (post-catalyst, bank 1) | Signal |
| 17 | - | Not used | |
| 18 | input | Oxygen sensor (post-catalyst, bank 2) | Signal |
| 19 | ground | Oxygen sensor heater (post-catalyst, bank 2) | |
| 20 | ground | Oxygen sensor heater (precatalyst, bank 1) | |
| 21 | ground | Oxygen sensor heater (precatalyst, bank 2) | |
| 22 | ground | Oxygen sensor heater (post-catalyst, bank 1) | |
| 23 | input | DME system power supply | Battery voltage from DME main relay (terminal 4) |
| 24 | ground | Oxygen sensor heater (post-catalyst, bank 2) | |

| Connector X60003 52-pin (engine signals) | | | |
|---|---|---|---|
| **Pin** | **Type** | **Component or function** | **Note** |
| 1 | input | Mass air flow (MAF) sensor | Signal |
| 2 | input | Exhaust camshaft sensor | Signal |
| 3 | - | Not used | |
| 4 | output | Throttle valve (MDK)(DME MS 42.1) | Throttle position signal 1 |
| 5 | input | Intake camshaft sensor | Signal |
| 6 | - | Not used | |
| 7 | output | Throttle valve (MDK)(DME MS 42.1) | Throttle position signal 2 |
| 8 | input | Crankshaft sensor | Signal |
| 9 | input | Throttle valve (MDK)(DME MS 42.1) | Throttle position signal 2 |
| 10 | input | Throttle valve (MDK)(DME MS 42.1) | Throttle position signal 2 |
| 11 | - | Not used | |
| 12 | input | Starter | Signal |
| 13 | input | Alternator, battery charge indicator light | Signal |
| 14 | ground | Throttle valve (MDK)(DME MS 42.1) | |
| 15 | ground | Exhaust camshaft sensor | |
| 16 | input | Throttle valve (MDK)(DME MS 42.1) | Throttle position signal 1 |
| 17 | ground | Mass air flow (MAF) sensor | |
| 18 | ground | Intake camshaft sensor | |
| 19 | input | Throttle valve (MDK)(DME MS 42.1) | Throttle position signal 1 |
| 20 | ground | Throttle valve (MDK)(DME MS 42.1) | |
| 21 | ground | Crankshaft position sensor | |
| 22 | input | Intake air temperature (IAT) sensor | Signal |
| 23 | ground | Intake air temperature (IAT) sensor | |
| 24 | input | Engine coolant temperature (ECT) sensor | Signal |
| 25 | ground | Engine coolant temperature (ECT) sensor | |
| 26 | input | Engine oil pressure sensor | Signal |
| 27 | input | Engine oil temperature sensor | Signal |
| 28 | ground | Engine oil pressure and temperature sensor | |
| 29 | input | Knock sensor, cyl. 1 - 3 | |
| 30 | input | Knock sensor, cyl. 1 - 3 | |
| 31 | input | Knock sensor, cyl. 4 - 6 | |
| 32 | input | Knock sensor, cyl. 4 - 6 | |
| 33 | output | Fuel injector control, cyl. 1 | Pulsed ground |
| 34 | output | Fuel injector control, cyl. 2 | Pulsed ground |
| 35 | output | Fuel injector control, cyl. 3 | Pulsed ground |
| 36 | output | Fuel injector control, cyl. 4 | Pulsed ground |
| 37 | output | Fuel injector control, cyl. 5 | Pulsed ground |

| Connector X60003 52-pin (engine signals) (continued) | | | |
|---|---|---|---|
| **Pin** | **Type** | **Component or function** | **Note** |
| 38 | output | Fuel injector control, cyl. 6 | Pulsed ground |
| 39 | input | Oil level sensor | Signal |
| 40 | output | VANOS solenoid, intake camshaft | |
| 41 | output | VANOS solenoid, exhaust camshaft | |
| 42 | output | Fuel tank venting valve | |
| 43 | output | Cruise control | |
| 44 | output | Cruise control | |
| 45 | output | Engine coolant thermostat, characteristic map | Activation |
| 46 | output | Idle speed control valve | Close signal |
| 47 | output | Idle speed control valve | Open signal |
| 48 | ground | Knock sensor shield | |
| 49 | output | Resonance valve intake system | Activation |
| 50 | - | Not used | |
| 51 | - | Not used | |
| 52 | output | Secondary air injection pump valve | Activation |

| Connector X60004 40-pin (vehicle signals) | | | |
|---|---|---|---|
| **Pin** | **Type** | **Component or function** | **Note** |
| 1 | output | Instrument cluster | Charge warning light |
| 2 | output | Feedback signal, engine start | General module (ZKE IV) |
| 3 | output | Secondary air injection pump relay | Activation |
| 4 | - | Not used | |
| 5 | - | Not used | |
| 6 | - | Not used | |
| 7 | ground | Pedal position sensor (DME MS 43.0) | |
| 8 | input | Pedal position sensor 1 (DME MS 43.0) | Signal |
| 9 | output | Pedal position sensor (DME MS 43.0) | Voltage supply |
| 10 | output | Fuel pump relay | Activation |
| 11 | output | Instrument cluster (DME MS 42.1) | Oil pressure signal |
| 12 | ground | Pedal position sensor (DME MS 43.0) | |
| 13 | input | Pedal position sensor 2 (DME MS 43.0) | Signal |
| 14 | output | Pedal position sensor (DME MS 43.0) | Voltage supply |
| 15 | - | Not used | |
| 16 | - | Not used | |
| 17 | output | Crankshaft speed signal | OBD II connector or 20-pin DLC |
| 18 | - | Not used | |
| 19 | - | Not used | |

| Connector X60004 40-pin (vehicle signals) (continued) | | | |
|---|---|---|---|
| **Pin** | **Type** | **Component or function** | **Note** |
| 20 | output | Diagnostic module tank leak (DM-TL) detection (DME MS 43.0) | Activation |
| 21 | output | Oil level sensor (DME MS 42.1) | Signal |
| 22 | input | Processed wheel speed, right rear | Signal from ABS / ASC or ABS / DSC module |
| 23 | input | Clutch pedal position switch | Signal |
| 24 | input | Brake light switch | Signal |
| 25 | - | Not used | |
| 26 | input | Terminal 15, fuse F26 | Switched power |
| 27 | input | Cruise control | Signal |
| 28 | input | Brake light switch | Signal |
| 29 | output | A/C compressor relay | Activation |
| 30 | output | Fuel tank leakage diagnosis pump (LDP or DM-TL) | Activation |
| 31 | - | Not used | |
| 32 | input/output | Diagnosis link (TxD) | Signal to OBD II plug or 20-pin DLC |
| 33 | input | Electronic immobilization (EWS) | Code signal |
| 34 | input | Leakage diagnosis pump (LDP) (DME MS 42.1) | Signal |
| 35 | ground | Cruise control | |
| 36 | input/output | CAN-bus high | Signal |
| 37 | input/output | CAN-bus low | Signal |
| 38 | - | Not used | |
| 39 | - | Not used | |
| 40 | - | Not used | |

| Connector X60001 9-pin (ignition signals) | | | |
|---|---|---|---|
| **Pin** | **Type** | **Component or function** | **Note** |
| 1 | output | Ignition coil control, cyl. 3 | Primary signal |
| 2 | output | Ignition coil control, cyl. 2 | Primary signal |
| 3 | output | Ignition coil control, cyl. 1 | Primary signal |
| 4 | - | Not used | |
| 5 | ground | ECM ground | |
| 6 | ground | ECM ground | ground connector |
| 7 | output | Ignition coil control, cyl. 6 | Primary signal |
| 8 | output | Ignition coil control, cyl. 5 | Primary signal |
| 9 | output | Ignition coil control, cyl. 4 | Primary signal |

## Siemens DME MS S54 (S54 engine)

When engine management system faults are detected, a diagnostic trouble code (DTC) is stored in the engine control module (ECM). The MIL also illuminates if an emissions-related fault is detected.

When faults arise, or if the malfunction indicator lamp (MIL) is illuminated, begin troubleshooting by connecting BMW service tester DISplus, MoDiC, GT1 or equivalent scan tool. The capabilities of OBD II software has the potential to save hours of diagnostic time and to help avoid incorrect component replacement and possible damage to system components.

***CAUTION—***
- *After all repairs are completed, access and clear DTC fault memory using an OBD II or BMW compatible scan tool. See* **OBD On-Board Diagnostics**.
- *Only use a digital multimeter for electrical tests.*
- *Relay positions can vary. Be sure to confirm relay location and function by identifying the wiring in the socket using the wiring diagrams found in* **ELE Electrical Wiring Diagrams**.

***NOTE—***
- *OBD II fault memory (including an illuminated MIL) can only be reset using the scan tool. Removing the connector from the ECM or disconnecting the battery will not erase the fault memory.*
- *The 16-pin OBD II diagnostic connector is located inside the car on the right side of the center console. See* **Engine management system troubleshooting** *in this repair group.*

### DME MS S54 component replacement

– Prior to disconnecting fuel injection components, make sure ignition is OFF.

– When finished with repairs, check for fault codes and clear and reset ECM memory.

#### Air filter housing and ducts

– Disconnect mass air flow sensor connector, loosen or cut hose clamps, loosen air filter housing mounting fasteners and lift out housing. See **113 Cylinder Head Removal and Installation**.

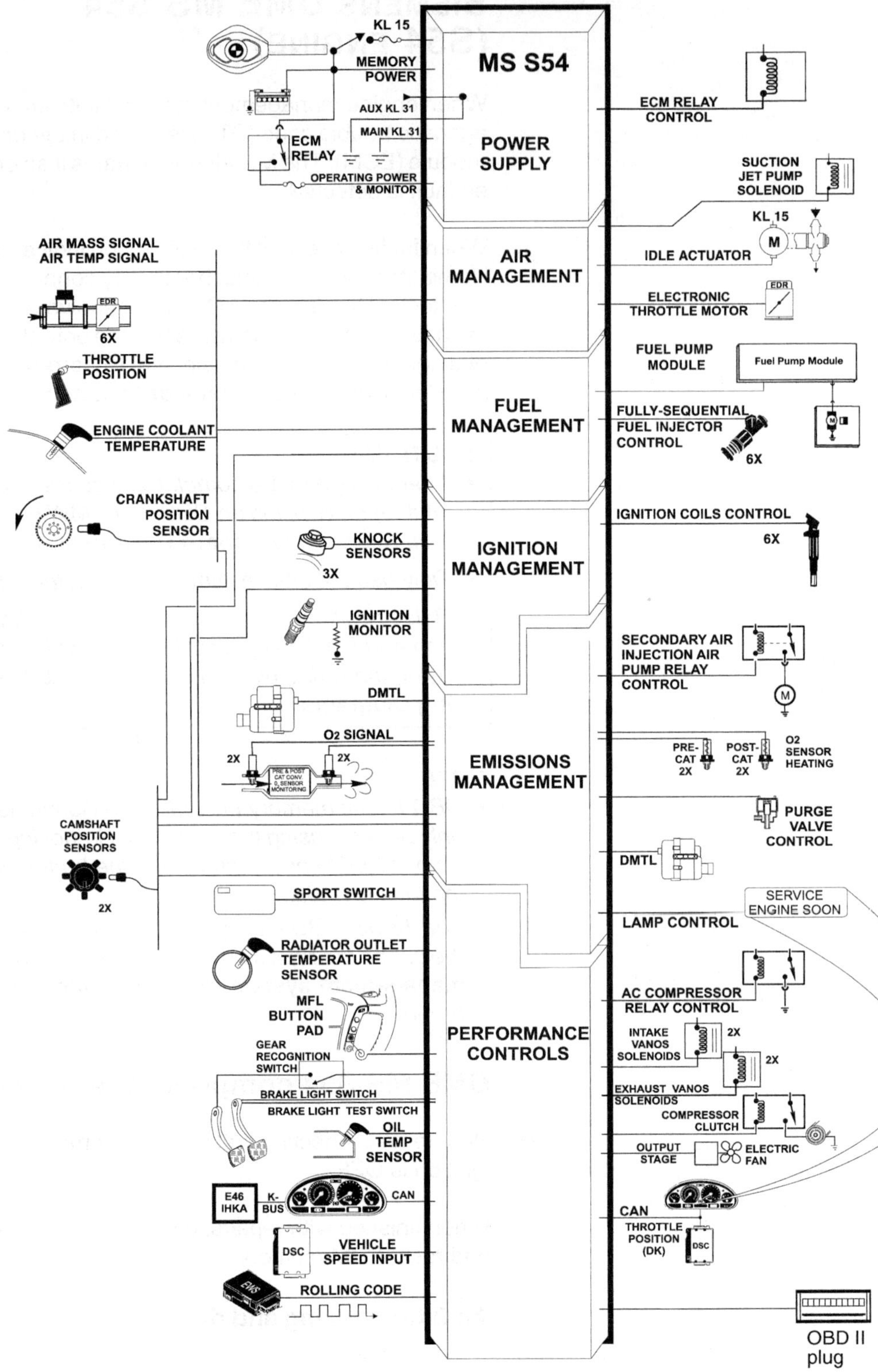

MS S54
KL 15
MEMORY POWER
AUX KL 31
MAIN KL 31
ECM RELAY
OPERATING POWER & MONITOR
POWER SUPPLY
ECM RELAY CONTROL
SUCTION JET PUMP SOLENOID
AIR MASS SIGNAL AIR TEMP SIGNAL
EDR
6X
THROTTLE POSITION
AIR MANAGEMENT
IDLE ACTUATOR
ELECTRONIC THROTTLE MOTOR
FUEL PUMP MODULE
Fuel Pump Module
ENGINE COOLANT TEMPERATURE
FUEL MANAGEMENT
FULLY-SEQUENTIAL FUEL INJECTOR CONTROL
CRANKSHAFT POSITION SENSOR
KNOCK SENSORS
3X
IGNITION MANAGEMENT
IGNITION COILS CONTROL
IGNITION MONITOR
SECONDARY AIR INJECTION AIR PUMP RELAY CONTROL
DMTL
O2 SIGNAL
2X
EMISSIONS MANAGEMENT
PRE-CAT 2X
POST-CAT 2X
O2 SENSOR HEATING
PURGE VALVE CONTROL
CAMSHAFT POSITION SENSORS
SPORT SWITCH
SERVICE ENGINE SOON
LAMP CONTROL
RADIATOR OUTLET TEMPERATURE SENSOR
MFL BUTTON PAD
GEAR RECOGNITION SWITCH
PERFORMANCE CONTROLS
AC COMPRESSOR RELAY CONTROL
INTAKE VANOS SOLENOIDS
EXHAUST VANOS SOLENOIDS
COMPRESSOR CLUTCH
BRAKE LIGHT SWITCH
BRAKE LIGHT TEST SWITCH
OIL TEMP SENSOR
OUTPUT STAGE
ELECTRIC FAN
E46 IHKA
K-BUS
CAN
THROTTLE POSITION (DK)
DSC
VEHICLE SPEED INPUT
EWS
ROLLING CODE
OBD II plug


z02130023

### DME main relay

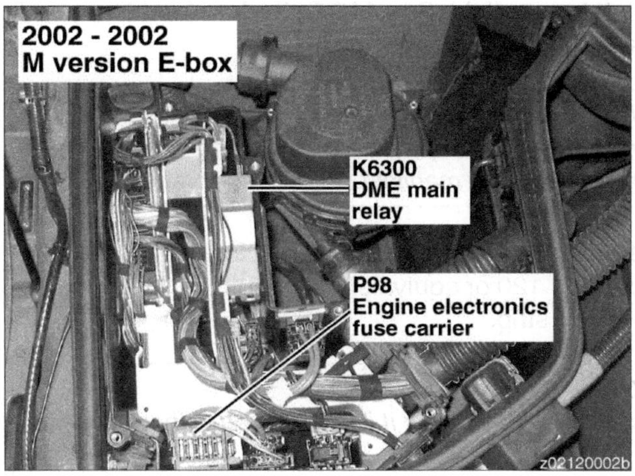


DME main relay (K6300) is located in E-box at right rear of engine compartment.

### Engine coolant temperature (ECT) sensor

Engine coolant temperature (ECT) sensor is located on left side of cylinder head, below cylinder 1 intake runner.

– Remove intake manifold. See **113 Cylinder Head Removal and Installation**.

> ***WARNING*** —
> *Due to risk of personal injury, be sure the engine is cold before working on intake manifold.*

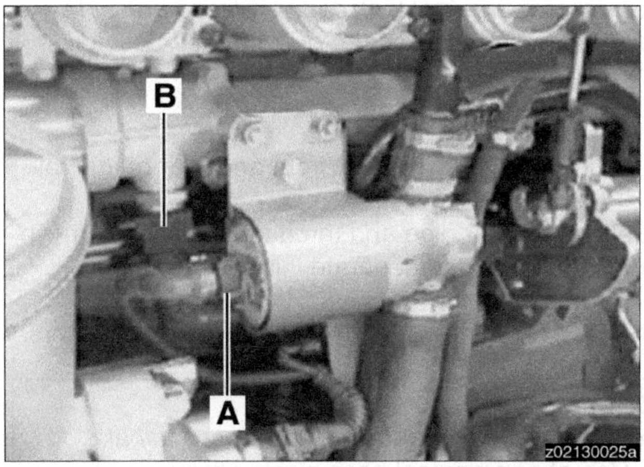


Detach idle speed control valve electrical connector (**A**).

– Detach ECT sensor electrical connector (**B**) and remove sensor. Be prepared to catch small amount of coolant.

– Installation is reverse of removal:
- Use new copper sealing washer when installing sensor.
- Replace any lost coolant.

| **Tightening torque** | |
|---|---|
| Temperature sensor to cylinder head | 13 Nm (10 ft-lb) |

## Fuel rail and injectors

– Remove intake manifold. See **113 Cylinder Head Removal and Installation**.

> ***WARNING —***
> *Due to risk of personal injury, be sure the engine is cold before working on intake manifold.*

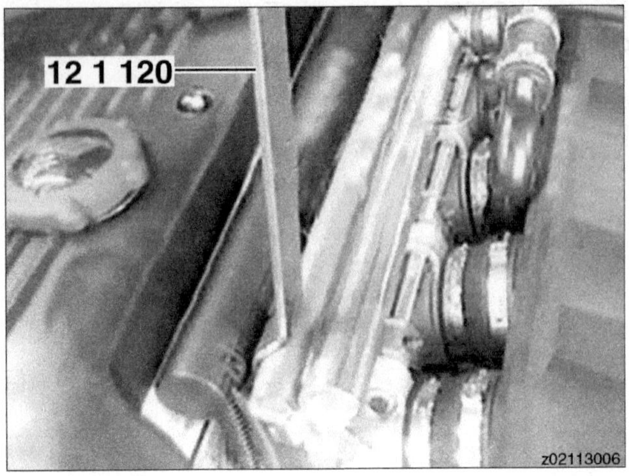

z02113006

< Working from front to rear of engine (injector 1 - 6), use BMW special tool 12 1 120 or equivalent to pry off fuel injector electrical connector strip.

> ***CAUTION—***
> *Connector strip may break if it is removed without special tool 12 1 120*

z02113007

< Pry off fuel line retainer (**arrow**) at fuel rail. Separate fuel line from rail.

> ***WARNING —***
> - *Fuel in fuel line is under pressure (approx. 3 - 5 bar or 45 - 75 psi) and may be expelled under pressure. Do not smoke or work near heaters or other fire hazards. Keep a fire extinguisher handy. Before disconnecting fuel hoses, wrap a cloth around fuel hoses to absorb any leaking fuel. Catch and dispose of escaped fuel. Plug all open fuel lines.*
> - *Always unscrew the fuel tank cap to release pressure in the tank before working on the tank or lines.*

z02113008

< Remove retaining bolts (**arrow**) at front and rear of fuel rail.

– Detach fuel rail from air pipe and throttle housings. Remove fuel rail and injectors, then separate injectors from rail.

– Be sure to use new O-ring seals when reinstalling injectors.

> ***CAUTION—***
> *When reinstalling intake manifold, use only original clamps. Use of non-original hose clamps may result in interference with the throttle linkage and cause the throttle to jam or stick in the open position.*

### Idle speed control valve

– Remove intake manifold. See **113 Cylinder Head Removal and Installation**.

> ***WARNING —***
> *Due to risk of personal injury, be sure the engine is cold before working on intake manifold.*

Working at idle speed control valve:

- Loosen or cut idle speed control valve air duct clamps (**A**) and detach ducts.
- Detach electrical connector (**B**).
- Remove idle speed control valve bracket fasteners (**arrows**) and remove valve.

– When reinstalling:

- Replace rubber damper on idle speed valve, if necessary.
- Use new hose clamps.

### Intake air temperature (IAT) sensor

Intake air temperature (IAT) sensor is integral with mass air flow sensor.

### Mass air flow (MAF) sensor

Working at mass air flow (MAF) sensor, located at air filter intake duct:

- Detach electrical connector (**A**).
- Remove sensor mounting screws (**arrows**). Slide sensor out of duct.

***NOTE —***
*E46 M3 components illustrated. M version Z3 components are similar.*

– When reinstalling, be sure to replace sealing O-ring.

### Throttle housings

6 throttle housings are used, connected by a link rod. See **113 Cylinder Head Removal and Installation** for exploded view of air intake ducts, intake manifold and throttle housings.

## ECM pin assignments (Siemens DME MS S54)

The engine control module (ECM) is located in the right rear of the engine compartment in the E-box.

The ECM has 5 electrical connectors with a total of 134 pins:

- Connector 1 (X60001): 9 pins, power and grounds
- Connector 2 (X60002): 24 pins, ancillary signals (oxygen sensors, CAN-bus, etc.)
- Connector 3 (X60003): 52 pins, engine signals
- Connector 4 (X60004): 40 pins, vehicle signals
- Connector 5 (X60005): 9 pins, ignition signals

ECM pin assignments for Siemens DME MS S54 are given in **Table f**. This information can be helpful when diagnosing faults to or from the ECM.

Generally, absence of voltage or continuity means there is a wiring or connector problem. Test results with incorrect values do not necessarily mean that a component is faulty. Check for loose, broken or corroded connections and wiring before replacing components. For engine management system electrical schematics, see **ELE Electrical Wiring Diagrams**.

When making checks at the ECM, use a break-out box to allow tests to be made with the connector attached to the ECM. This prevents damage to the small terminals in the connector. As an alternative, the connector housing can be separated so that electrical checks can be made from the back of the connector.

***CAUTION—***
- *Always wait at least one minute after turning off the ignition before removing the connector from the ECM. If the connector is removed before this time, residual power in the system relay may damage the ECM.*
- *Always connect or disconnect the control module connector and meter probes with the ignition off.*

| Table f. Siemens DME MS S54 ECM pin assignments | | | |
|---|---|---|---|
| **Connector X60001 9-pin (power, grounds)** | | | |
| **Pin** | **Type** | **Component or function** | **Note** |
| 1 | output | Unloader relay terminal 15 | |
| 2 | output | Throttle actuator (EDR) | Power supply |
| 3 | - | Not used | |
| 4 | ground | ECM | |
| 5 | ground | ECM | |
| 6 | ground | ECM | |
| 7 | input | Power supply, B+ at E-box | Battery voltage at all times |
| 8 | input | Fuse carrier, engine electronics fuse F202 | Voltage supply |
| 9 | ground | Throttle actuator (EDR) | |

| Connector X60002 24-pin (ancillary signals) | | | |
|---|---|---|---|
| **Pin** | **Type** | **Component or function** | **Note** |
| 1 | ground | Oxygen sensor heater (precatalyst, bank 1) | |
| 2 | - | Not used | |
| 3 | - | Not used | |
| 4 | - | Not used | |
| 5 | ground | Oxygen sensors, exhaust temperature sensor | |
| 6 | - | Not used | |
| 7 | ground | Oxygen sensor heater (post-catalyst, bank 1) | |
| 8 | - | Not used | |
| 9 | - | Not used | |
| 10 | input | Exhaust temperature sensor | |
| 11 | output | Fuel pump relay | Signal |
| 12 | output | E-box fan | Signal |
| 13 | ground | Oxygen sensor heater (precatalyst, bank 2) | |
| 14 | input | Oxygen sensor (precatalyst, bank 1) | Signal |
| 15 | input | Oxygen sensor (precatalyst, bank 2) | Signal |
| 16 | input | Oxygen sensor (post-catalyst, bank 1) | Signal |
| 17 | - | Not used | |
| 18 | input | Oxygen sensor (post-catalyst, bank 2) | Signal |
| 19 | ground | Oxygen sensor heater (post-catalyst, bank 2) | |
| 20 | input | Clutch switch | |
| 21 | - | Not used | |
| 22 | - | Not used | |
| 23 | output | DME main relay | Signal |
| 24 | - | Not used | |

| Connector X60003 52-pin (engine signals) | | | |
|---|---|---|---|
| **Pin** | **Type** | **Component or function** | **Note** |
| 1 | input | Mass air flow (MAF) sensor | Signal |
| 2 | input | Exhaust camshaft sensor | Signal |
| 3 | - | Not used | |
| 4 | output | VANOS control valve, advance | Signal |
| 5 | input | Intake camshaft sensor | Signal |
| 6 | input | Exhaust camshaft sensor | Signal |
| 7 | output | Mass air flow (MAF) sensor, throttle actuator (EDR) | Power supply |
| 8 | input | Crankshaft sensor | Signal |
| 9 | - | Not used | |
| 10 | input | Throttle potentiometer | Signal |
| 11 | - | Not used | |
| 12 | - | Not used | |
| 13 | input | Alternator, battery charge indicator light | Signal |
| 14 | output | Throttle potentiometer | Power supply |
| 15 | - | Not used | |
| 16 | - | Not used | |
| 17 | ground | Mass air flow (MAF) sensor | |
| 18 | - | Not used | |
| 19 | input | Tire pressure warning | |
| 20 | ground | Throttle actuator (EDR) | |
| 21 | ground | Crankshaft position sensor | |
| 22 | input | Intake air temperature (IAT) sensor | Signal |
| 23 | input | Throttle position sensor | Signal |
| 24 | input | Engine coolant temperature (ECT) sensor | Signal |
| 25 | ground | Engine sensors | |
| 26 | input | Engine oil pressure sensor | Signal |
| 27 | - | Not used | |
| 28 | - | Not used | |
| 29 | input | Knock sensor, cyl. 1 - 2 | |
| 30 | input | Knock sensor, cyl. 3 - 4 | |
| 31 | input | Knock sensor, cyl. 5 - 6 | |
| 32 | - | Not used | |
| 33 | input | Fuel injector control, cyl. 1 | Pulsed ground |
| 34 | input | Fuel injector control, cyl. 2 | Pulsed ground |
| 35 | input | Fuel injector control, cyl. 3 | Pulsed ground |
| 36 | input | Fuel injector control, cyl. 4 | Pulsed ground |
| 37 | input | Fuel injector control, cyl. 5 | Pulsed ground |

| Connector X60003 52-pin (engine signals) (continued) | | | |
|---|---|---|---|
| **Pin** | **Type** | **Component or function** | **Note** |
| 38 | input | Fuel injector control, cyl. 6 | Pulsed ground |
| 39 | input | Oil level sensor | Signal |
| 40 | - | Not used | |
| 41 | - | Not used | |
| 42 | output | Fuel tank venting valve | Signal |
| 43 | output | VANOS control valve, retard | Signal |
| 44 | output | VANOS control valve, advance | Signal |
| 45 | output | Tire pressure warning light | Signal |
| 46 | output | Idle speed control valve | Close signal |
| 47 | output | Idle speed control valve | Open signal |
| 48 | - | Not used | |
| 49 | - | Not used | |
| 50 | output | VANOS control valve, retard | Signal |
| 51 | output | Sucking jet pump | Signal |
| 52 | - | Not used | |

| Connector X60004 40-pin (vehicle signals) | | | |
|---|---|---|---|
| **Pin** | **Type** | **Component or function** | **Note** |
| 1 | output | Charge warning light | Instrument cluster |
| 2 | - | Not used | |
| 3 | output | Secondary air injection pump relay | Signal |
| 4 | input | Engine cooling fan | Signal |
| 5 | - | Not used | |
| 6 | input | Start signal | Instrument cluster |
| 7 | ground | Pedal position sensor | |
| 8 | input | Pedal position sensor | Signal |
| 9 | output | Pedal position sensor | Voltage supply |
| 10 | - | Not used | |
| 11 | input | Oil pressure warning | Instrument cluster |
| 12 | ground | Pedal position sensor | |
| 13 | input | Pedal position sensor 2 | Signal |
| 14 | output | Pedal position sensor | Voltage supply |
| 15 | - | Not used | |
| 16 | - | Not used | |
| 17 | output | Crankshaft speed signal | EWS module |
| 18 | output | Fuel tank leak detection module (DM-TL) | |
| 19 | input | Back-up lights | |

| Connector X60004 40-pin (vehicle signals) (continued) | | | |
|---|---|---|---|
| **Pin** | **Type** | **Component or function** | **Note** |
| 20 | output | Oil level sensor | |
| 21 | - | Not used | |
| 22 | input | Processed wheel speed, right rear | Signal from ABS / ASC or ABS / DSC module |
| 23 | - | Not used | |
| 24 | input | Brake light switch | Signal |
| 25 | - | Not used | |
| 26 | input | Terminal 15, fuse F26 | Switched power |
| 27 | output | Cruise control | |
| 28 | output | Fuel tank leak detection module (DM-TL) | |
| 29 | output | A/C compressor relay | Activation |
| 30 | output | Fuel tank leak detection module (DM-TL) | Activation |
| 31 | - | Not used | |
| 32 | input/output | Diagnosis link (TxD) | Signal to OBD II plug |
| 33 | input | Electronic immobilization (EWS) | Code signal |
| 34 | input | Brake light switch | Signal |
| 35 | ground | Cruise control | |
| 36 | input/output | CAN-bus high | Signal |
| 37 | input/output | CAN-bus low | Signal |
| 38 | - | Not used | |
| 39 | - | Not used | |
| 40 | - | Not used | |

| Connector X60001 9-pin (ignition signals) | | | |
|---|---|---|---|
| **Pin** | **Type** | **Component or function** | **Note** |
| 1 | output | Ignition coil control, cyl. 1 | Primary signal |
| 2 | output | Ignition coil control, cyl. 2 | Primary signal |
| 3 | output | Ignition coil control, cyl. 3 | Primary signal |
| 4 | - | Not used | |
| 5 | ground | ECM | |
| 6 | output | Ignition coil control, cyl. 5 | Primary signal |
| 7 | output | Ignition coil control, cyl. 6 | Primary signal |
| 8 | - | Not used | |
| 9 | output | Ignition coil control, cyl. 4 | Primary signal |

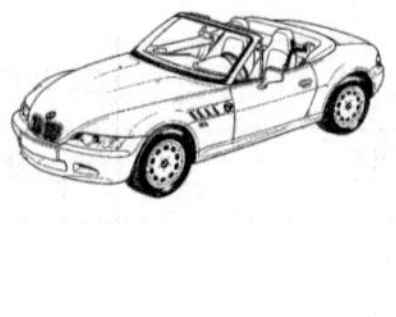

# 160 Fuel Tank and Fuel Pump





**TABLES**

## GENERAL

This repair group covers service information specifically for the fuel supply system. Information on the fuel injection system is covered in **130 Fuel Injection**.

In the descriptions and procedures given below, engine and engine management (DME) systems are referred to by manufacturer code. If necessary, see **100 Engine–General** for model year, engine code, and DME applications.

***NOTE—***
*Fuel filter replacement is covered in* **020 Maintenance**.

### Special tools

Some of the procedures in this group require the use of special tools.

◄ Automotive digital multimeter

> ***CAUTION—***
> *Only use a digital multimeter when testing automotive electrical components.*

1318

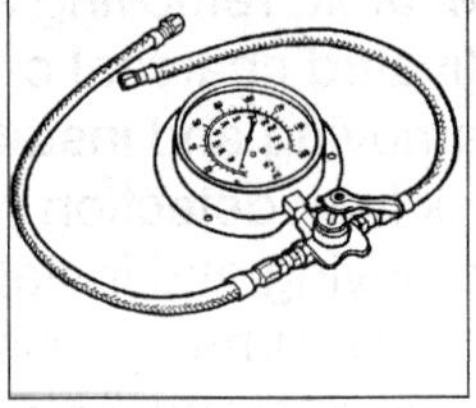

◄ Fuel pressure gauge
(Baum Tools Unlimited 1318)

13 3 010

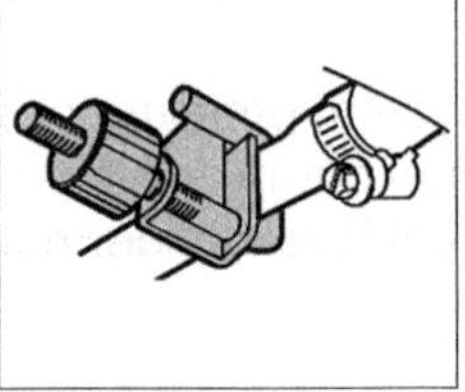

◄ Pinch clamp
(BMW special tool 13 3 010)

13 5 220

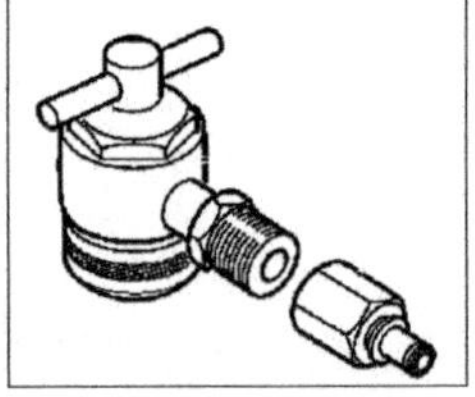

◄ Fuel pressure gauge adapter with shut-off valve
(BMW special tool 13 5 220)

13 5 281
13 5 282

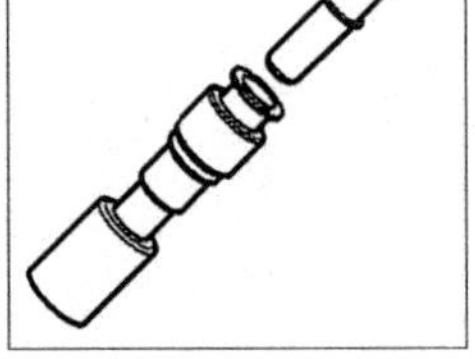

◄ Fuel line plugs
(BMW special tools 13 5 281, 13 5 282)

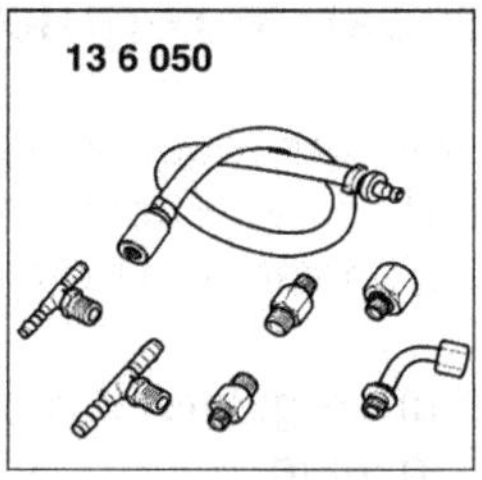


Fuel pressure gauge adapter fittings (BMW special tool 13 6 050)

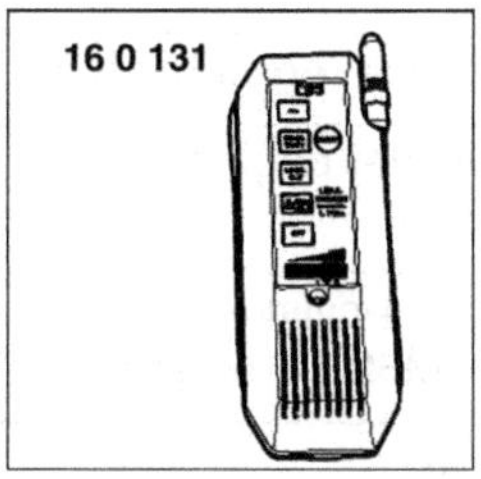


Fuel leak detector (BMW special tool 16 0 131)

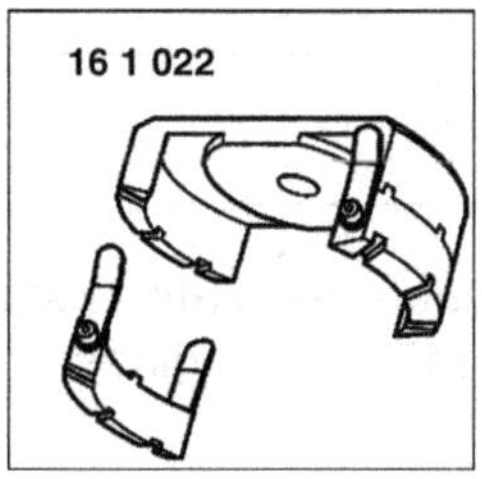


Fuel tank threaded collar spanner (BMW special tool 16 1 022)

Fuel tank cap pressure testing kit (BMW special tool kit 16 1 170)

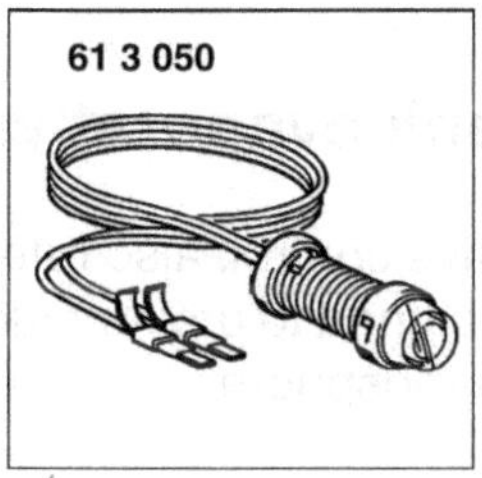


Fuel pump relay bypass switch (BMW special tool kit 61 3 050)

## Fuel pump

The electric fuel pump, mounted in the fuel tank in tandem with the fuel level sender, delivers pressurized fuel to the fuel injection system. A pressure regulator maintains system pressure in the fuel rail and at the fuel injectors. The quantity of fuel supplied exceeds demand, so excess fuel returns to the fuel tank via a return line. See **130 Fuel Injection** for more information.

***NOTE—***
- *The fuel rail in M54 and S54 engine consists of one fuel line, with no return line in the engine compartment.*
- *Fuel pump removal procedures are given under* **Fuel pump and fuel level sender, removing and installing.**

## Fuel pressure regulator

The fuel pressure regulator controls fuel pressure via a diaphragm and spring. Fuel pressure is modulated by engine load, based on vacuum from the intake manifold.

Fuel pressure regulator locations are summarized in the accompanying table.

| Fuel pressure regulator locations | |
|---|---|
| 1996 4-cylinder engine | In fuel injector rail, in engine compartment |
| 1997 - 1998 4-cylinder engine<br>M52, S52, M52 TU engine | In 3/2-way valve, below vehicle under driver seat |
| M54 engine | In fuel filter, below vehicle under driver seat |
| S54 engine | Ahead of fuel filter, below vehicle under driver seat |

Fuel pressure regulator

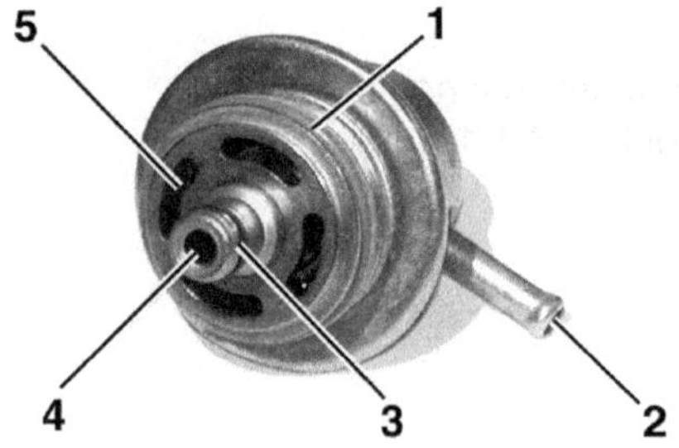


502160684

Fuel pressure regulator installed in some M44 and all M52, S52 and M52 TU engines is illustrated. Other models are similar.

1. Outer seal
2. Vacuum line connection
3. Inner seal
4. Fuel flow back to fuel tank
5. Fuel flow from fuel pump

## Fuel tank evaporative control system

Evaporative control, also referred to as running losses control, is designed to prevent fuel system fumes from venting into the atmosphere.

The components of this system allow control and monitoring of evaporative losses by the on-board diagnostic (OBD II) software incorporated into the engine control module (ECM).

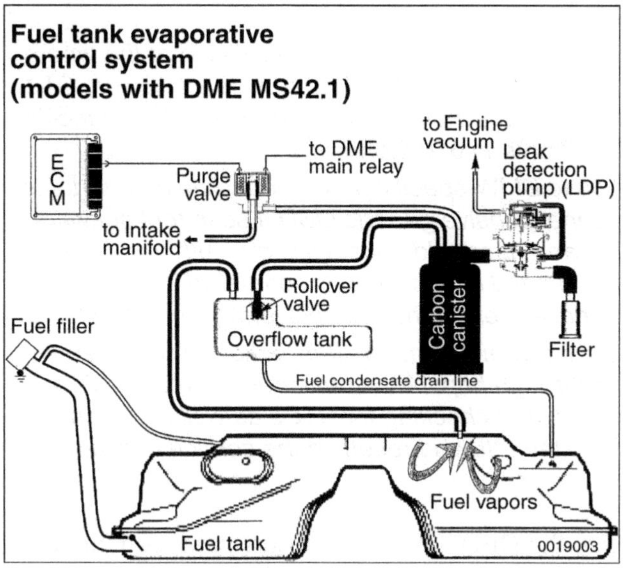


The main components of the evaporative control system and their functions:

- Fuel overflow tank acts as a liquid / vapor separator.
- Carbon canister stores evaporated fuel.
- Plumbing ducts vapors from fuel tank to canister and from canister to intake manifold.
- Carbon canister purge valve is controlled by engine control module (ECM).
- 3/2-way (or running losses) valve (not shown) shunts excess fuel volume directly back to fuel tank before it circulates through injector fuel rail (1997 - 1998 4-cylinder models, 6-cylinder models with Siemens DME MS 41.1, MS 42.1 engine management).
- Leak detection unit (LDP or DMTL) pressurizes fuel tank and evaporative system to monitor system leaks (all models except 1996 4-cylinder).

***NOTE—***
*3/2-way valve, running losses valve and fuel changeover valve are used interchangeably in BMW technical literature.*

## Evaporative system troubleshooting

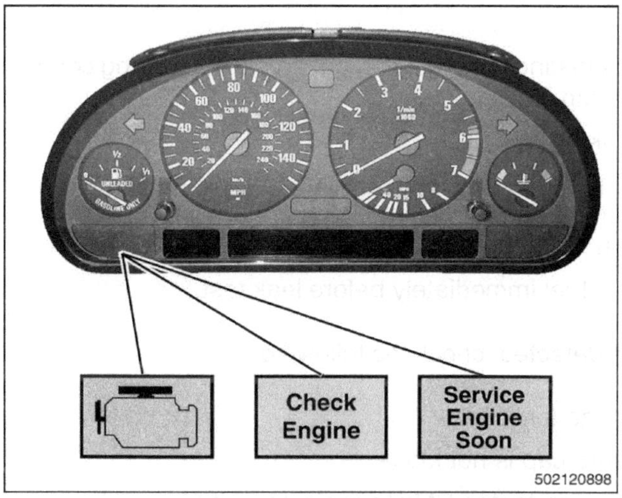


If vehicle emission levels exceed 1.5 times Federally mandated criteria, the OBD II system illuminates a malfunction indicator light (MIL) in the instrument cluster.

***NOTE—***
*BMW models covered in this manual are equipped with one of three versions of MIL (as shown).*

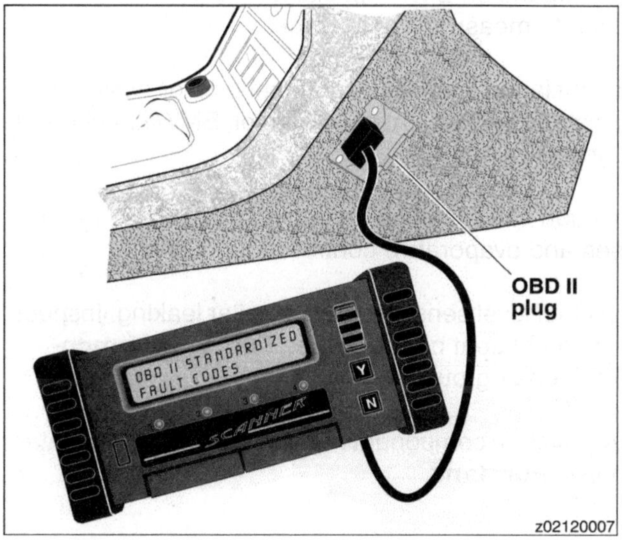


To begin fuel or evaporative system troubleshooting, connect a diagnostic scan tool (DISplus, GT1, MoDiC or other BMW-compatible tool) to OBD II plug on right side of center console.

Interrogate engine control module (ECM) for stored faults. The OBD II system detects ignition and fuel injection faults and stores a diagnostic trouble code (DTC) in the ECM along with other pertinent fault information.

Additional OBD II information, including a DTC lookup table, can be found in **OBD On-Board Diagnostics**.

***NOTE—***
- *Professional diagnostic scan tools available at the time of this printing include the BMW factory tools DISplus, GT1, MoDiC and a small number of aftermarket BMW-specific tools. See* **Tools** *in* **010 General**.
- *In addition to BMW-specific professional line of scan tools, there are inexpensive generic OBD II scan tool software programs and handheld units available. Although these tools have limited capabilities as compared to the dedicated tools, nevertheless they are powerful diagnostic tools.*
- *For the DIY owner, a simple aftermarket DTC reader is available through http://www.peakeresearch.com. This tool is capable of checking for DTCs as well as turning off the illuminated MIL and resetting the service indicator lights.*

In addition to the OBD II plug, BMW cars produced through 6 / 2000 are also equipped with a 20-pin data link connector (DLC) in the engine compartment. Engine management faults and other DTCs may be accessed and cleared and service lights reset via this connector as well as via the OBD II plug.

When leak testing the fuel tank, observe the following conditions to obtain plausible results:

- Fuel tank ¼ to ¾ full.
- Vehicle parked for at least 2 hours to allow fuel to reach room temperature. Ideal fuel temperature is 10°- 20°C (50° - 68°F).
- Do not refuel immediately before leak test.

If a leak is detected, check the following:

– Fuel filler cap leaking.
  - Make sure cap is not loose.
  - Use filler cap tester, BMW special tool kit 16 1 170 or equivalent, to measure cap leakage.

– Fuel tank ventilation lines leaking at fuel tank or activated carbon canister. Use fuel leakage detector, BMW special tool 16 0 131 or equivalent.

– Tank ventilation valve leaking (in engine compartment). Inspect hoses and evaporative control connections.

– Fuel pump / fuel level sensor threaded collar leaking. Inspect collar behind right seat backrest. See **Fuel tank components** in this repair group.

Evaporative system component replacement is covered later in this group in **Fuel tank**.

## Warnings and cautions

*WARNING —*

- *The fuel system is designed to retain pressure even when the ignition is off. When working with the fuel system, loosen fuel connections slowly to allow residual fuel pressure to dissipate. Avoid spraying fuel. Use shop rags to capture leaking fuel.*
- *Before beginning any work on the fuel system, place a fire extinguisher in the vicinity of the work area.*
- *Fuel is highly flammable. When working around fuel, do not disconnect any wires that could cause electrical sparks. Do not smoke or work near heaters or other fire hazards.*
- *Unscrew the fuel tank cap to release pressure in the tank before working on the tank or lines.*
- *When disconnecting a fuel hose, wrap shop rag around end of hose to prevent spray of fuel under pressure.*
- *Do not use a work light with an incandescent bulb near any fuel. Fuel may spray on the hot bulb causing a fire.*
- *Make sure the work area is properly ventilated.*

*CAUTION—*

- *Before making any electrical tests with the ignition turned on, disable the ignition system as described in* **120 Ignition System**. *Be sure the battery is disconnected when replacing components.*
- *To prevent damage to the ignition system or other DME components, including the engine control module (ECM), always connect and disconnect wires and test equipment with the ignition off.*
- *Prior to disconnecting the battery, read the battery disconnection cautions given in* **001 General Warnings and Caution**.
- *Cleanliness is essential when working with the fuel system. Thoroughly clean the fuel line unions before disconnecting any of the lines.*
- *Use only clean tools. Keep removed parts clean and sealed or covered with a clean, lint-free cloth, especially if completion of the repair is delayed.*
- *Do not move the car while the fuel system is open.*
- *Avoid using high pressure compressed air to blow out lines and components. High pressure can rupture internal seals and gaskets.*
- *Always replace seals, O-rings and hose clamps.*

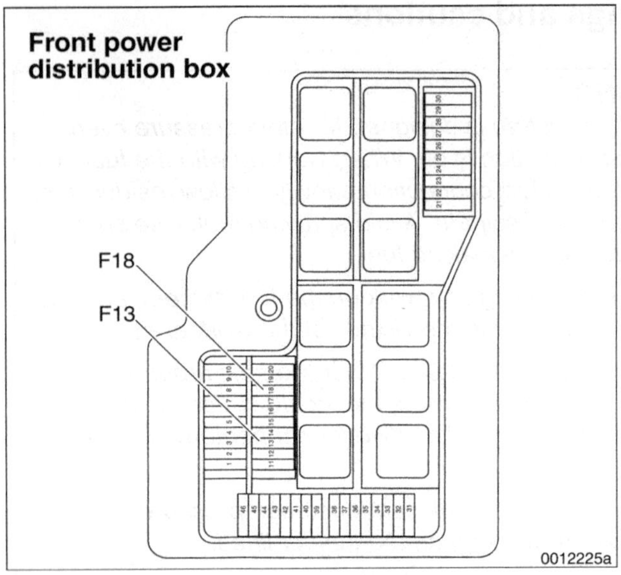


## FUEL SYSTEM ELECTRICAL TROUBLESHOOTING

### Fuel pump fuse and relay, locating

Begin troubleshooting fuel pump faults by locating the following:

Fuel pump fuses in front power distribution box, left rear of engine compartment.

- 1996 M44 engine: Fuse F18, 15A
- 1997 - 2002 M44, M52, S52, M2 TU, M54 engine:
  -Fuse F13, 5A / 7.5A
  -Fuse F18, 15A
- 2001 - 2002 S54 engine:
  -Fuse F13, 5A
  -Fuse F18, 20A

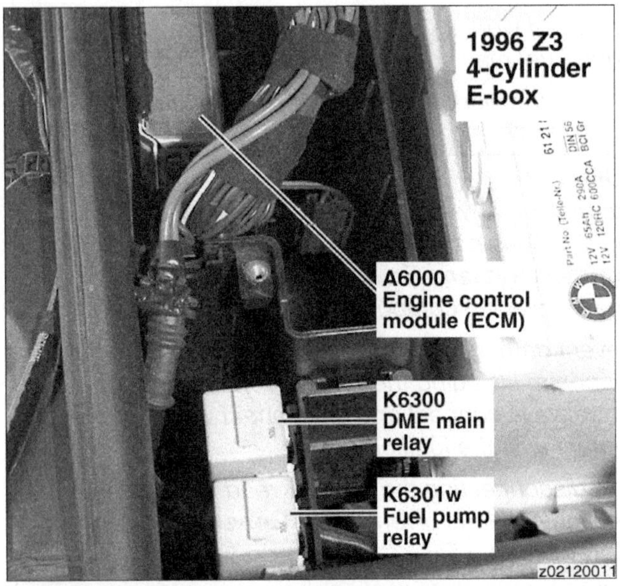


1996 4-cylinder: Fuel pump relay (K6301w) and DME main relay (K6000) in E-box, right rear of engine compartment.

> ***CAUTION—***
> *Relay positions can vary. Be sure to confirm relay position by identifying the wiring in the socket using the wiring diagrams in* **ELE Electrical Wiring Diagrams**.

1997 - 2002 Z3, 1998 - 2000 M version: Fuel pump relay (K6301w) in left kick panel.

2001 - 2002 M version: Fuel pump control module (K6301z) in center console, behind parking brake handle.

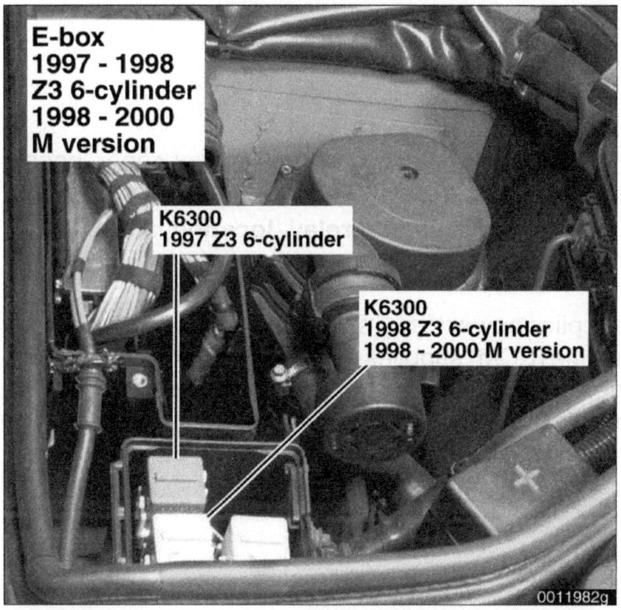


1997 - 1998 Z3, 1998 - 2000 M version: DME main relay (K6300) in E-box, right rear of engine compartment.

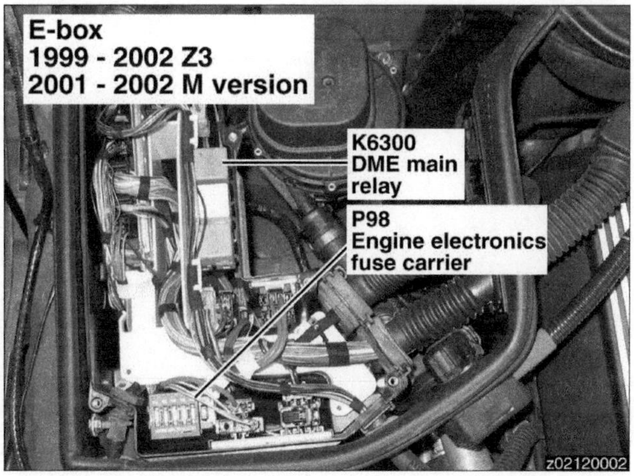


1999 - 2002 Z3, 2001 - 2002 M version: DME main relay (K6300) in E-box, right rear of engine compartment.

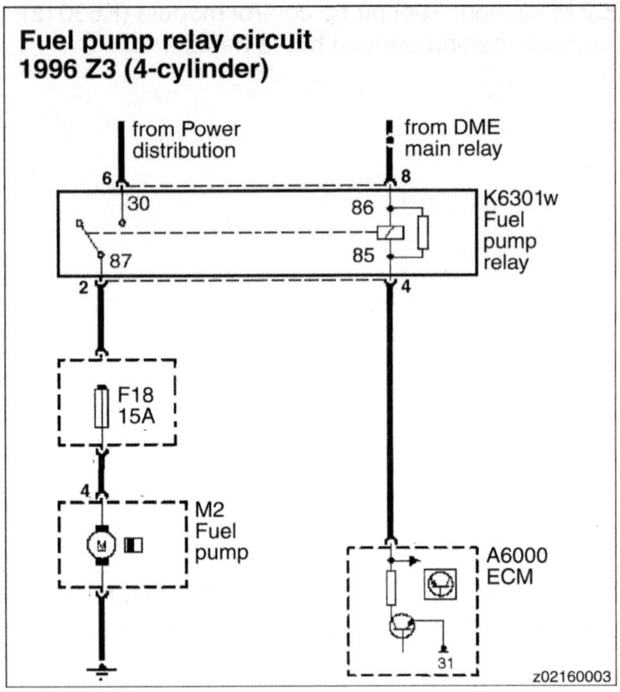


## Fuel pump relay circuit

1996 Z3 (4-cylinder) fuel pump wiring is unlike later models. Fuel pump fuse (F18) is between relay and pump.

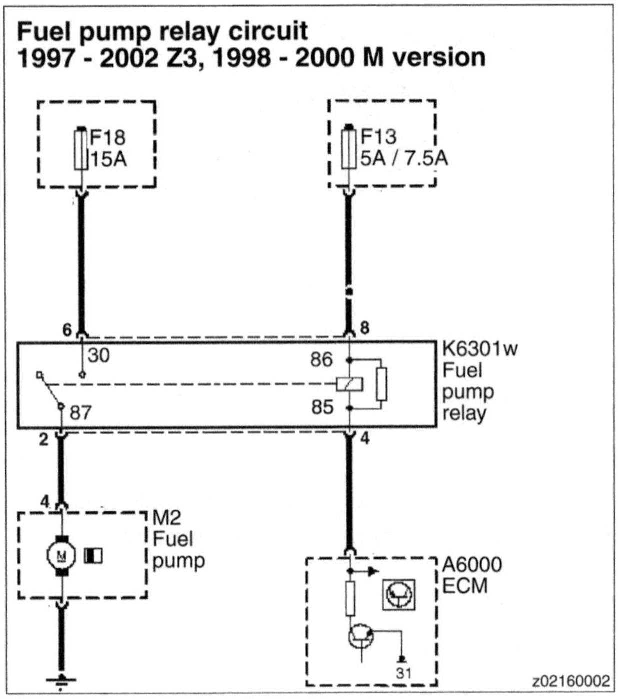


Fuel pump relay wiring for 1997 - 2002 Z3 and 1998 - 2000 M version. Fuses F13 and F18 are in front power distribution box. See **Fuel pump fuse and relay, locating** in this repair group.

Fuse 13 supplies power to coil side of fuel pump relay. Engine control module (ECM) supplies switched ground to relay. During starting, fuel pump runs as long as ignition switch is in START position and continues to run once engine starts. If electrical system fault interrupts power to fuel pump, engine does not run.

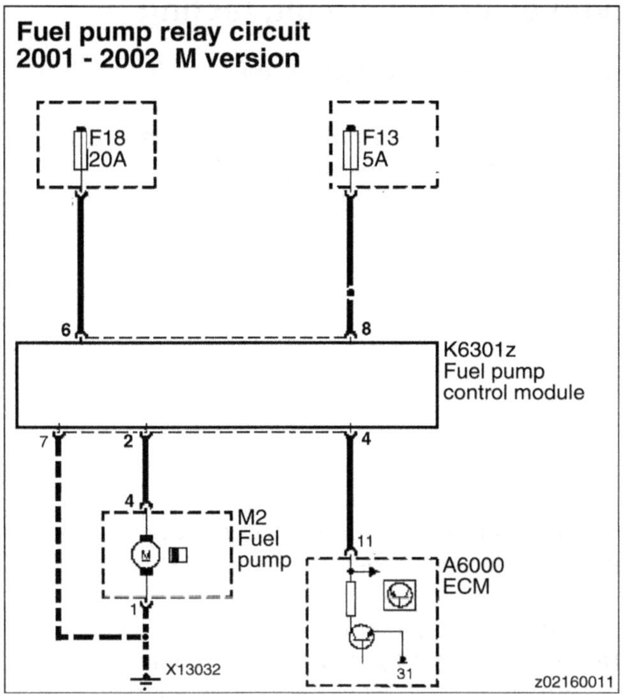


Fuel pump relay wiring for 2001 - 2002 M version. Fuses F13 and F18 are in front power distribution box. See **Fuel pump fuse and relay, locating** in this repair group.

***NOTE—***
*The fuel pump in the DME MS S54 system matches the fuel supply demands of the S54 engine. The ECM cycles a voltage signal to the fuel pump module, and the module cycles voltage to the fuel pump to control pump speed and pressure. The pump runs at low speed during idle and part load conditions, full speed during start-up and at full load.*

## Operating fuel pump for tests

To operate fuel pump for testing purposes without having to run engine, bypass fuel pump relay to power pump directly.

– To run fuel pump:
  - Make sure fuel pump fuses are intact.
  - Remove relay. See **Fuel pump fuse and relay, locating** in this repair group.
  - Connect socket terminal **6** (corresponds to relay terminal **30**) to socket terminal **2** (corresponds to relay terminal **87**) with BMW special tool 61 3 050 or a fused jumper wire. After completing tests, remove jumper harness.

***CAUTION—***
- *Relay locations vary. Use care when identifying relays and making electrical checks at the fuse / relay panel.*
- *Wire diameters and colors, where available from BMW, can be found in* ***ELE Electrical Wiring Diagrams****.*

***NOTE—***
- *The fuel pump runs as soon as the jumper wire is attached.*
- *To make a substitute for BMW special tool 61 3 050, make a jumper wire with a male spade connector at each end. The wire needs to be 1.5 mm$^2$ (14 ga.) and include an in-line fuse holder with a 15A fuse. To avoid relay connector damage from repeated connecting and disconnecting, also include a toggle switch in the jumper harness.*

– If pump does not run with jumper installed, fault could be in fuel pump, fuses F13 or F18, or wiring to pump. Check pump, fuses and pump wiring as described below.

## Fuel pump electrical circuit, testing

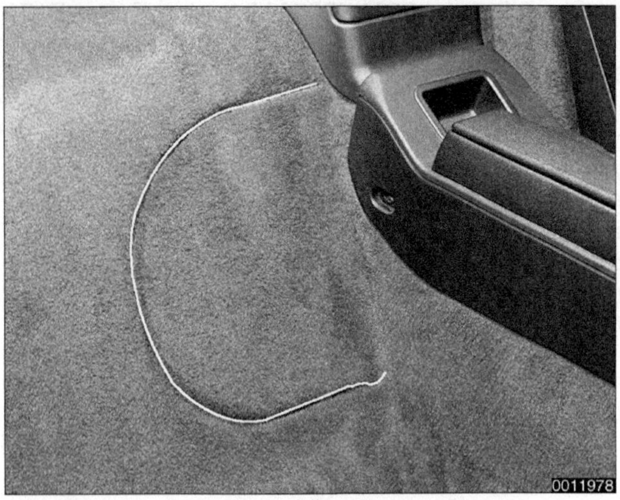


– Remove passenger seat. See **520 Seats**.

< Locate perforation mark on carpet behind front seat. Cut out carpet along perforation and fold back.

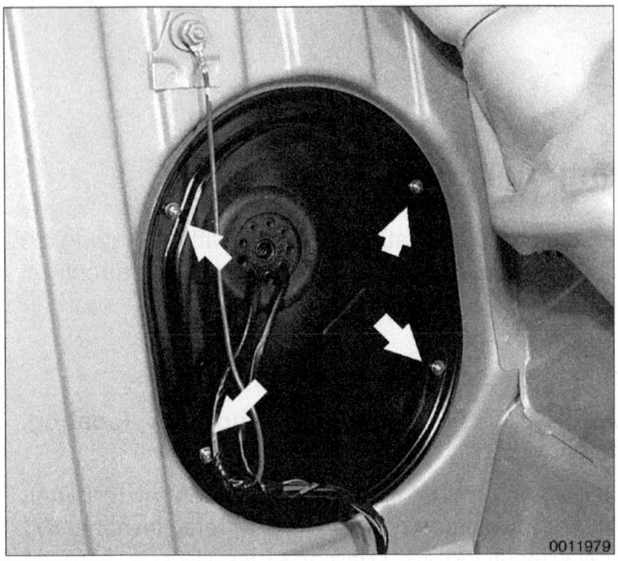


< Remove fuel pump / fuel level sender access cover mounting screws (**arrows**). Remove cover.

– Operate fuel pump using jumper at fuel pump relay socket. See **Operating fuel pump for tests** in this repair group. Listen for fuel pump operation. Remove jumper when finished.

– If pump does not run, disconnect harness connector from pump.

– Reconnect jumper at fuel pump relay socket.

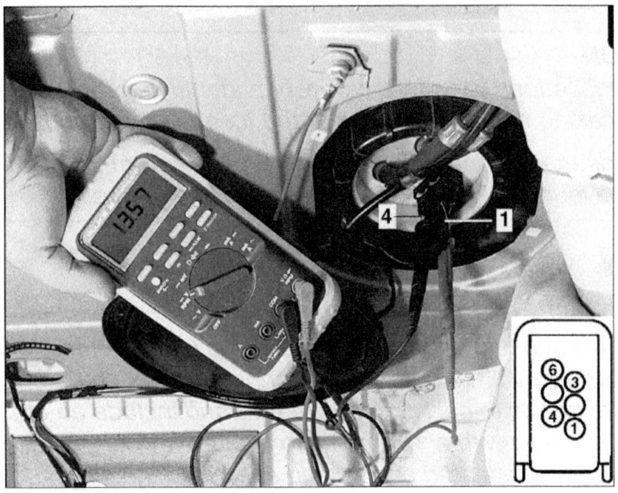


< Use digital multimeter to test for positive (+) battery voltage at green / violet wire at fuel pump connector **4**.

– Test for ground (-) at brown wire at fuel pump connector **1**.

– If voltage and ground are normal, the fuel pump is probably faulty.

– If there is no voltage, check wiring from fuel pump relay and make sure relay is functioning correctly.

## Fuel pump power consumption, testing

To achieve accurate test results, make sure battery voltage at fuel pump connector is at least 12.6 volts. Charge the battery as necessary.

– Remove passenger seat. See **520 Seats**.

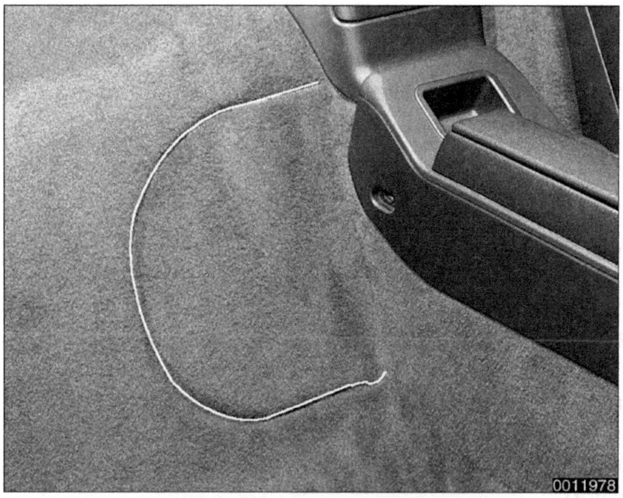

< Locate perforation mark on carpet behind front seat. Cut out carpet along perforation and fold back.

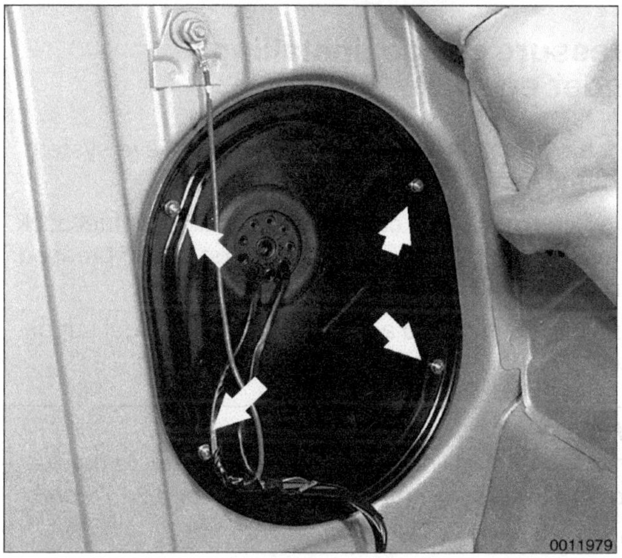

< Remove fuel pump / fuel level sender access cover mounting screws (**arrows**). Remove cover.

– Disconnect black 6-pin harness connector from fuel pump.

– Connect digital ammeter between green / violet wire (terminal 4) in connector and corresponding pump terminal.

– Use jumper to connect brown wire (terminal 1) in connector to terminal 1 on pump.

> ***CAUTION—***
> *Do not allow test leads to short to ground.*

– Run pump. See **Operating fuel pump for tests** in this repair group.

– Compare ammeter reading with specification listed below.

| Fuel pump current (typical) | |
|---|---|
| Current consumption (average) | 5.5 amps |

A higher than normal power consumption may indicate a worn fuel pump, causing intermittent fuel starvation due to pump overheating and seizure. The only remedy is pump replacement.

A lower than normal power consumption may indicate fuel blockage. Be sure to check that the return line and the pump pickup are not obstructed before replacing the pump.

## Fuel delivery troubleshooting

Checking fuel delivery is a fundamental part of troubleshooting and diagnosing the engine management system. Fuel pressure directly influences fuel delivery. Use an accurate fuel pressure gauge with a minimum range of 0 to 5 bar (0 to 75 psi) to make the tests.

There are three significant fuel delivery values to be measured:

- **System pressure** is created by the fuel pump and maintained by the fuel pressure regulator.
- **Fuel delivery volume** is created by the fuel pump and affected by restrictions, such as clogged fuel filter.
- **Residual pressure** is the pressure maintained in the closed system after the engine and fuel pump are shut off.

BMW specifies a special gauge and connecting adapters to measure fuel pressure. If the special tools or equivalent equipment is not available, a length of fuel line and a T-fitting can be installed to the fuel supply line and connected to a fuel gauge. On some vehicles, this is not possible without cutting into a fuel line.

### Fuel pressure gauge, installing (4-cylinder engine)

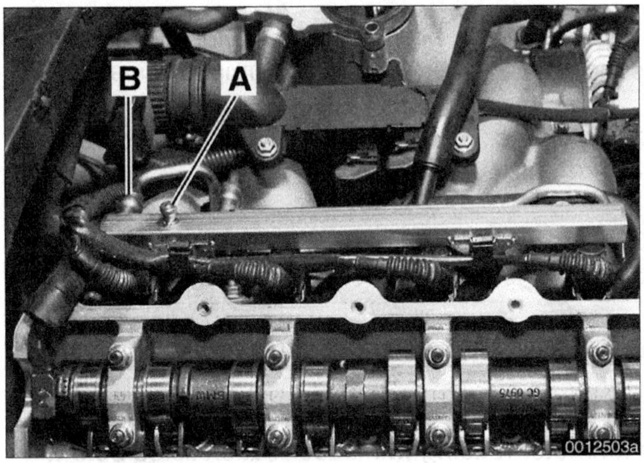


To prevent fuel from spraying on hot engine, relieve system fuel pressure before disconnecting fuel lines:

- Remove Schræder valve cap (**A**) at fuel rail. Blow fuel back to fuel tank through fitting using brief burst of compressed air (max. 3 bar / 43.5 psi).
- Alternatively, tightly wrap a shop towel around fuel supply line fitting (**B**) while loosening fitting.

***WARNING—***
- *When opening fuel lines, be prepared for fuel to be expelled under pressure (approx. 3 - 5 bar or 45 - 75 psi).*
- *Keep a fire extinguisher handy.*
- *Before disconnecting fuel hoses, wrap a cloth around fuel hoses to absorb any leaking fuel.*
- *Catch and dispose of escaped fuel.*
- *Plug all open fuel lines.*
- *Unscrew the fuel tank cap to release pressure in the tank before working on the tank or lines.*

***CAUTION—***
*Cleanliness is essential when working with fuel circuit components. Thoroughly clean the unions before disconnecting fuel lines.*

***NOTE—***
*Intake manifold top section and cylinder head cover are shown removed for purposes of illustration.*

– Attach fuel pressure gauge with minimum range of 0 to 5 bar (0 to 75 psi) to fuel rail:
  - Use standard tire valve tool to remove Schræder valve stem. Attach fuel line adapter, BMW special tool 13 5 220, to Schræder valve.
  - Alternatively, attach adapter kit and T-fitting, BMW special tool kit 13 6 050, between fuel supply line and rail.

***WARNING—***
*Make sure the fuel pressure gauge is securely connected to prevent it from coming loose under pressure.*

– Test system pressure and residual pressure. See **System pressure, testing** or **Residual pressure, testing** in this repair group.

– If disconnected, reattach fuel line using new sealing O-ring. Check for leaks with engine running.

– Check for fault codes and clear and reset ECM memory.

## Fuel pressure gauge, installing (M52, S52 engine)

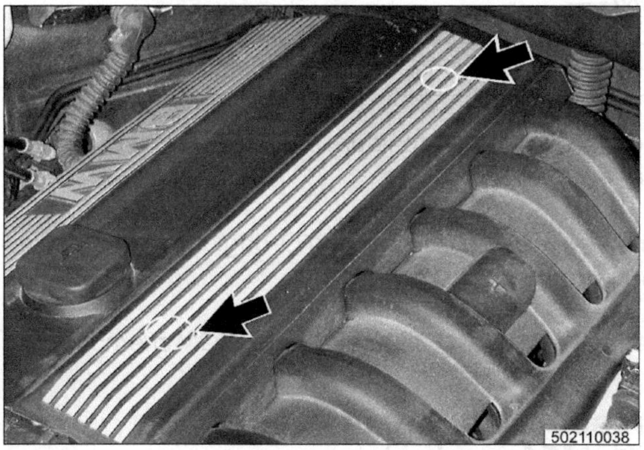


◄ To access fuel rail, remove plastic trim plugs (**arrows**), remove fasteners under them, and remove cover from above fuel rail and injectors.

Working at rear of intake manifold, use BMW special tool 16 1 050 to release locking clip at fuel supply line fitting. Disconnect line from fuel rail.

**WARNING —**
- *When opening fuel lines, be prepared for fuel to be expelled under pressure (approx. 3 - 5 bar or 45 - 75 psi).*
- *Do not smoke or work near heaters or other fire hazards. Keep a fire extinguisher handy.*
- *Before disconnecting fuel hoses, wrap a cloth around fuel hoses to absorb any leaking fuel.*
- *Catch and dispose of escaped fuel.*
- *Plug all open fuel lines.*
- *Unscrew the fuel tank cap to release pressure in the tank before working on the tank or lines.*

**CAUTION—**
*Cleanliness is essential when working with fuel circuit components. Thoroughly clean the unions before disconnecting fuel lines.*

– Attach adapter kit and T-fitting between fuel supply line and fuel rail.

– Attach fuel pressure gauge with minimum range of 0 to 5 bar (0 to 75 psi) to adapter:

**WARNING —**
*Make sure the fuel pressure gauge is securely connected to prevent it from coming loose under pressure.*

– Test system pressure and residual pressure as described in this repair group.

– Reattach fuel line using new sealing O-ring. Check for leaks with engine running.

– Check for fault codes and clear and reset ECM memory.

## Fuel pressure gauge, installing (M52 TU, M54 engine)

To access fuel rail, remove plastic trim plugs (**arrows**), remove fasteners under them, and remove cover from above fuel rail and injectors.

To prevent fuel from spraying on hot engine, relieve system pressure before disconnecting fuel lines:

- Remove Schræder valve cap (**arrow**) at fuel rail. Blow fuel back to fuel tank through fitting using brief burst of compressed air (max. 3 bar / 43.5 psi).
- Alternatively, tightly wrap a shop towel around fuel supply line fitting at rear of manifold while detaching fuel line.

***WARNING—***
- *When opening fuel lines, be prepared for fuel to be expelled under pressure (approx. 3 - 5 bar or 45 - 75 psi).*
- *Do not smoke or work near heaters or other fire hazards. Keep a fire extinguisher handy.*
- *Before disconnecting fuel hoses, wrap a cloth around fuel hoses to absorb any leaking fuel.*
- *Catch and dispose of escaped fuel.*
- *Plug all open fuel lines.*
- *Unscrew the fuel tank cap to release pressure in the tank before working on the tank or lines.*

***CAUTION—***
*Cleanliness is essential when working with fuel circuit components. Thoroughly clean the unions before disconnecting fuel lines.*

Attach fuel pressure gauge with minimum range of 0 to 5 bar (0 to 75 psi) to fuel rail:

- Use standard tire valve removal tool to remove Schræder valve stem. Attach fuel line adapter, BMW special tool 13 5 220 or equivalent, to Schræder valve.
- Alternatively, attach adapter kit and T-fitting, BMW special tool kit 13 6 050, between fuel supply line and rail.

***WARNING—***
*Make sure the fuel pressure gauge is securely connected to prevent it from coming loose under pressure.*

***CAUTION—***
*When using the BMW special tool, use care when turning the T-handle in to open the Schræder valve. Screw the handle in just until the valve opens. Turning the handle in too far will damage the valve.*

– Test system pressure and residual pressure. See **System pressure, testing** or **Residual pressure, testing** in this repair group.

– Reattach fuel line using new sealing O-ring. Check for leaks with engine running.

– Check for fault codes and clear and reset ECM memory.

## Fuel pressure gauge, installing (S54 engine)

< Working underneath driver seat, below car, remove protective cover from fuel filter and fuel pressure regulator assembly to access Schræder valve fuel pressure test port (**arrow**).

– Attach fuel pressure gauge with minimum range of 0 to 10 bar (0 to 150 psi):

- Use standard tire valve removal tool to remove Schræder valve stem and attach fuel gauge.
- Alternatively, attach fuel line adapter, BMW special tool 13 5 220 or equivalent, to Schræder valve. Attach fuel pressure gauge.

***WARNING—***
- *When opening fuel lines, be prepared for fuel to be expelled under pressure (approx. 5 bar or 75 psi).*
- *Keep a fire extinguisher handy.*
- *Before disconnecting fuel a fitting, wrap a cloth around the fitting to absorb any leaking fuel.*
- *Catch and dispose of escaped fuel.*
- *Plug all open fuel lines.*
- *Unscrew the fuel tank cap to release pressure in the tank before working on the tank or lines.*
- *Make sure the fuel pressure gauge is securely connected to prevent it from coming loose under pressure.*

***CAUTION—***
- *Cleanliness is essential when working with fuel circuit components. Thoroughly clean the unions before disconnecting fuel lines.*
- *When using the BMW special tool, use care when turning the T-handle in to open the Schræder valve. Screw the handle in just until the valve opens. Turning the handle in too far will damage the valve.*

– Test system pressure and residual pressure. See **System pressure, testing** in this repair group.

– Reattach fuel line using new sealing O-ring. Check for leaks with engine running.

– Check for fault codes and clear and reset ECM memory.

## System pressure, testing

System pressure is created by the fuel pump and maintained by the fuel pressure regulator. System pressure is not adjustable.

***NOTE—***
*A vacuum connection to the fuel pressure regulator is used to increase or decrease fuel pressure, depending on engine load. If you measure system pressure by running the engine instead of using the method used below, system pressure at idle (low load, high vacuum) will be 0.3 - 0.7 bar (4 - 10 psi) lower than the specifications given here. See* **Fuel pressure regulator** *in this repair group.*

– Remove fuel tank filler cap.

***NOTE—***
*Do not drive the car with the fuel filler cap removed. The OBD II system will illuminate the malfunction indicator light (MIL).*

0022146

< Attach fuel pressure gauge. See **Fuel pressure gauge, installing** in this repair group.

– Operate fuel pump. See **Operating fuel pump for tests** in this repair group. Compare system pressure to specifications in **Table a**.

| **Table a. System pressure specifications** | |
|---|---|
| 4-cylinder engine | 3.0 ± 0.2 bar (44 ± 3 psi) |
| M52, S52, M52 TU, M54 engine | 3.5 ± 0.2 bar (51 ± 3 psi) |
| S54 engine | 5.0 ± 0.2 bar (72.5 ± 3 psi) |

– If pressure is too low, repeat test while gradually pinching off return hose to fuel tank.

- If pressure does not rise rapidly fuel pump is most likely faulty.
- If pressure rises, fuel pressure regulator is most likely faulty.

– If pressure is too high, check return line from pressure regulator to tank. Check for kinks in hose. If no faults can be found, pressure regulator is most likely faulty.

***CAUTION—***
*The fuel pump is capable of developing a higher pressure than that regulated by the pressure regulator. In case the fuel pump check valve is faulty (stuck closed), make sure the system pressure does not rise above 6.0 bar (87 psi). Damage to the fuel lines or fuel system components could result.*

– Turn off fuel pump and reassemble engine, noting the following:
  - Remove fuel pressure gauge and fittings.
  - Replace Schræder valve stem, if removed.
  - Replace fuel line sealing O-rings and hose clamps, where applicable.
  - Reinstall fuel pump relay.
  - Replace fuel filler cap.

– Check for fault codes and clear and reset ECM memory.

## Fuel delivery volume, checking

– 4-cylinder, M52, S52 or M52 TU engine: Disconnect return line from fuel rail in engine compartment.

– M54 engine: Disconnect fuel supply line from fuel filter.

> ***WARNING—***
> - *When opening fuel lines, be prepared for fuel to be expelled under pressure (approx. 3 - 5 bar or 45 - 75 psi).*
> - *Do not smoke or work near heaters or other fire hazards. Keep a fire extinguisher handy.*
> - *Before disconnecting fuel hoses, wrap a cloth around fuel hoses to absorb any leaking fuel.*
> - *Catch and dispose of escaped fuel.*
> - *Plug all open fuel lines.*
> - *Unscrew the fuel tank cap to release pressure in the tank before working on the tank or lines.*

> ***CAUTION—***
> *Cleanliness is essential when working with fuel circuit components. Thoroughly clean the unions before disconnecting fuel lines.*

– Connect a length of hose to fuel supply line and place open end of hose in a suitable container for catching fuel (4 quart capacity).

– Run fuel pump for 1 minute and measure fuel collected. See **Operating fuel pump for tests** in this repair group. Compare results to specifications in **Table b**.

| **Table b. Fuel pump delivery volume (1 min. @12V tested with 3 bar counterpressure)** | |
|---|---|
| M44, M52, S52, M52 TU, M54 engine | > 2 liters (>2.2 US qt) |

***NOTE—***
*BMW does not provide fuel delivery volume information for the S54 engine.*

– When finished testing, reconnect fuel line. Tighten hose clamps where applicable.

– Check for fault codes and clear and reset ECM memory.

## Residual pressure, testing

For quick restarts and to avoid vapor lock when the engine is hot, the fuel injection system retains fuel pressure after the engine is shut off. This residual pressure is maintained by the fuel pressure regulator and a check valve at the fuel pump outlet.

0022146

< Attach fuel pressure gauge. See **Fuel pressure gauge, installing** in this repair group.

– Start engine and allow to run for approximately one minute. Note system pressure reading. Shut off engine.

– Observe fuel pressure reading after approximately 20 minutes. The pressure should not drop more than 0.5 bar (7 psi) from system pressure listed in accompanying table.

| **System pressure specifications** | |
|---|---|
| 4-cylinder engine | 3.0 ± 0.2 bar (44 ± 3 psi) |
| M52, S52, M52 TU, M54 engine | 3.5 ± 0.2 bar (51 ± 3 psi) |

***NOTE*** —
*BMW does not provide residual fuel pressure information for the S54 engine.*

– When finished, disconnect pressure gauge and fitting, and replace Schræder valve stem, if removed.

– If fuel system does not maintain pressure:
- Visually check for leaks in fuel lines and unions.
- Check for leaking injector(s).
- Check for faulty fuel pump check valve.

– To test fuel pump check valve:
- Repeat residual pressure test.
- Immediately after shutting off engine, pinch off fuel delivery line using a BMW special tool 13 3 010 or equivalent.
- If pressure is now maintained, fault is most likely at fuel pump check valve.
- If pressure again drops below specifications, fault is most likely in fuel pressure regulator.

– Check for fault codes and clear and reset ECM memory.

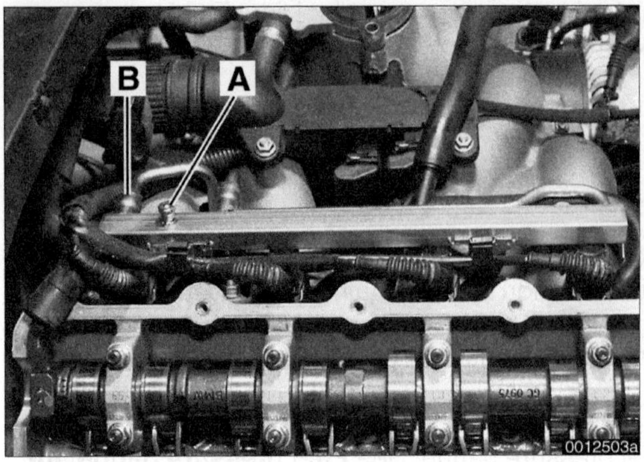


## FUEL PRESSURE REGULATOR

### Fuel pressure regulator, removing and installing (1996 4-cylinder engine)

On 1996 M44 engine, the fuel pressure regulator is mounted to the end of the fuel rail.

To prevent fuel from spraying on hot engine, relieve system pressure before disconnecting fuel lines:

- Remove Schræder valve cap (**A**) at fuel rail. Blow fuel back to fuel tank through fitting using brief burst of compressed air (max. 3 bar / 43.5 psi).
- Alternatively, tightly wrap a shop towel around fuel supply line fitting (**B**) while loosening fitting.

***WARNING—***
- *When opening fuel lines, be prepared for fuel to be expelled under pressure (approx. 3 - 5 bar or 45 - 75 psi).*
- *Do not smoke or work near heaters or other fire hazards. Keep a fire extinguisher handy.*
- *Catch and dispose of escaped fuel.*
- *Plug all open fuel lines.*
- *Unscrew the fuel tank cap to release pressure in the tank before working on the tank or lines.*

***CAUTION—***
*Cleanliness is essential when working with fuel circuit components. Thoroughly clean the unions before disconnecting fuel lines.*

***NOTE—***
*Intake manifold top section and cylinder head cover are shown removed for purposes of illustration.*

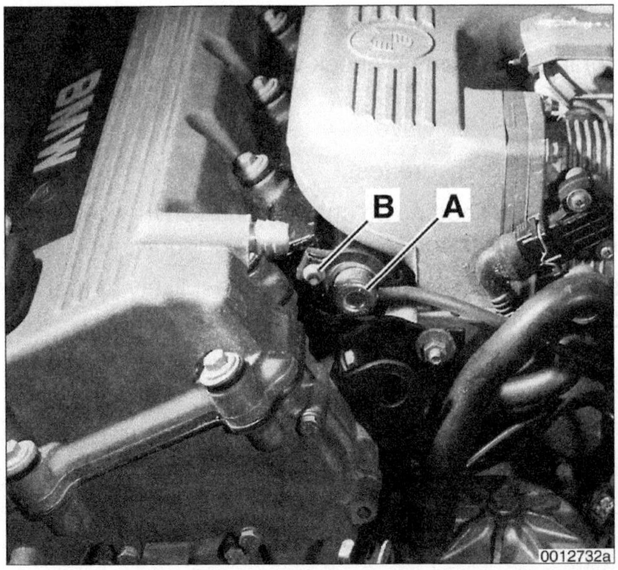


Working at front end of fuel rail:

- Disconnect vacuum hose (**A**) from fuel pressure regulator.
- Remove mounting bolt (**B**) from regulator clamp. Remove clamp. Mark orientation of vacuum hose nipple.

***NOTE—***
*Other versions of the fuel pressure regulator may be held in place by an internal circlip or retaining clip.*

– Wrap shop rag around regulator to catch dripping fuel, then remove regulator from fuel rail by pulling straight out.

***CAUTION—***
*Check nominal pressure rating stamped on fuel pressure regulator body. Replacement regulator must be rated with the same nominal pressure.*

– Installation is reverse of removal. Always replace sealing O-rings.

– Check for fault codes and clear and reset ECM memory.

## Fuel pressure regulator, removing and installing (1997 - 1998 4-cylinder engine, M52, S52 or M52 TU engine)

On 1997 - 1998 4-cylinder cars as well as cars with M52, S52 and M52 TU engine, the fuel pressure regulator is mounted underneath the car, below the driver's seat, in the 3/2-way valve assembly.

– Raise car and support in a safe manner.

> ***WARNING —***
> *Make sure the car is stable and well supported at all times. Use a professional automotive lift or jack stands designed for the purpose. A floor jack is not adequate support.*

– Working underneath driver seat, below car, remove protective cover from 3/2-way valve and fuel filter assembly.

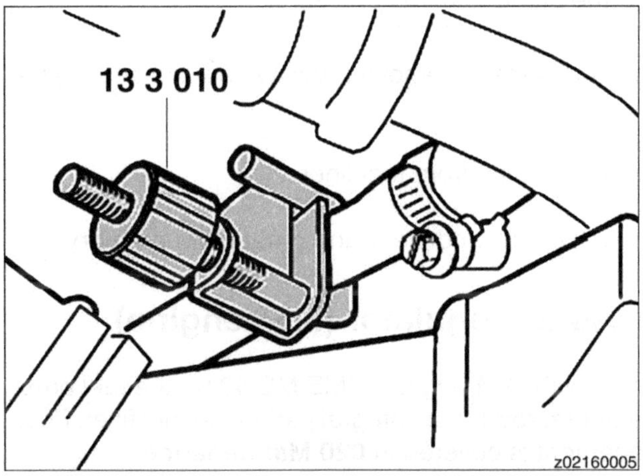


◄ Use BMW special tool 13 3 010 or equivalent to pinch off supply and return fuel lines. This will minimize fuel drips while pressure regulator is off.

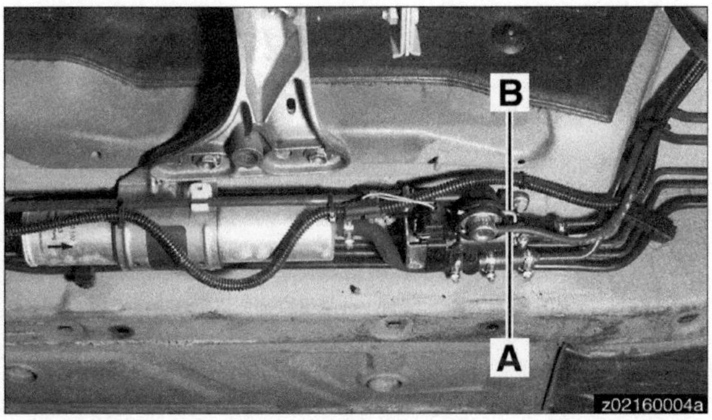


◄ Working at fuel pressure regulator:

- Pull vacuum hose (**A**) off regulator.
- Pry out regulator retaining clip (**B**).

– Wrap shop rag around regulator to catch spraying fuel, then remove regulator from 3/2 valve assembly by pulling straight down.

***WARNING—***
- *When opening fuel lines, be prepared for fuel to be expelled under pressure (approx. 3 - 5 bar or 45 - 75 psi).*
- *Do not smoke or work near heaters or other fire hazards. Keep a fire extinguisher handy.*
- *Before disconnecting fuel hoses, wrap a cloth around fuel hoses to absorb any leaking fuel.*
- *Catch and dispose of escaped fuel.*
- *Plug all open fuel lines.*
- *Unscrew the fuel tank cap to release pressure in the tank before working on the tank or lines.*

***CAUTION—***
- *Cleanliness is essential when working with fuel circuit components. Thoroughly clean the unions before disconnecting fuel lines.*
- *Check nominal pressure rating stamped on fuel pressure regulator body. Replacement regulator must be rated with the same nominal pressure.*

– Installation is reverse of removal. Always replace sealing O-rings.

– Remove pinch clamps from fuel lines.

– Check for fault codes and clear and reset ECM memory.

## Fuel pressure regulator (M54 engine)

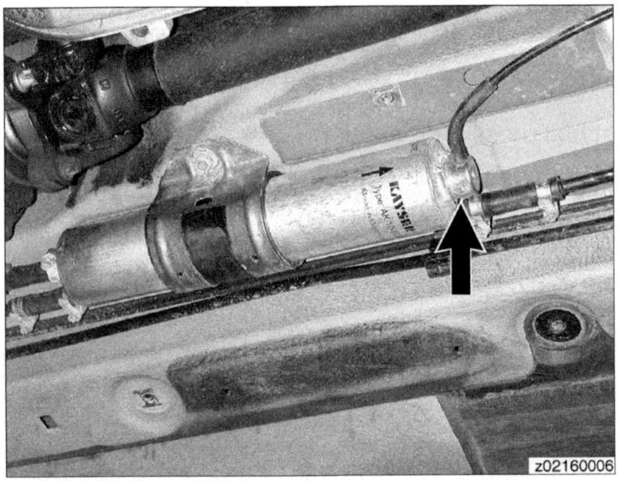


◄ In Z3 models with M54 engine (DME MS 43.0), the fuel pressure regulator (**arrow**) is an integral part of the fuel filter. Fuel filter replacement is covered in **020 Maintenance**.

## Fuel pressure regulator, removing and installing (S54 engine)

On cars with S54 engine, the fuel pressure regulator is mounted underneath the car, below the driver's seat, ahead of the fuel filter assembly.

– Raise car and support in a safe manner.

> ***WARNING —***
> *Make sure the car is stable and well supported at all times. Use a professional automotive lift or jack stands designed for the purpose. A floor jack is not adequate support.*

Working underneath driver seat, below car, remove protective cover to access fuel pressure regulator.

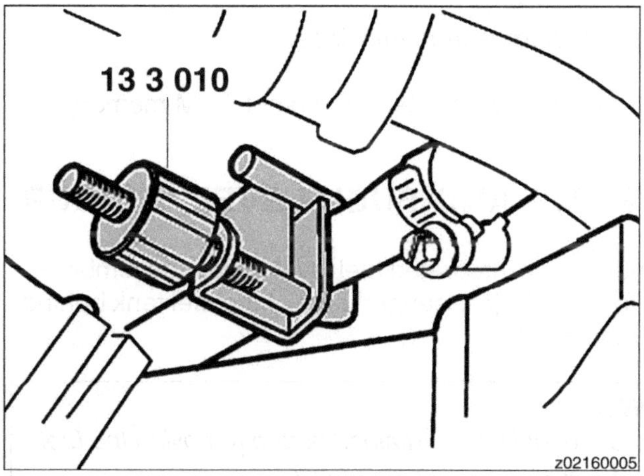


Use BMW special tool 13 3 010 or equivalent to pinch off supply and return fuel lines. This will minimize fuel drips while pressure regulator is off.

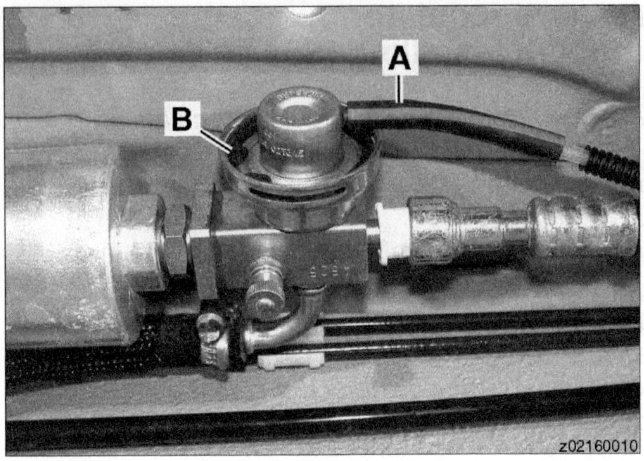


Working at fuel pressure regulator:
- Pull vacuum hose (**A**) off regulator.
- Pry out regulator retaining clip (**B**).

– Wrap shop rag around regulator to catch spraying fuel, then remove regulator from mounting.

> ***WARNING —***
> - *When opening fuel lines, be prepared for fuel to be expelled under pressure (approx. 5 bar or 75 psi).*
> - *Do not smoke or work near heaters or other fire hazards. Keep a fire extinguisher handy.*
> - *Before disconnecting fuel hoses, wrap a cloth around fuel hoses to absorb any leaking fuel.*
> - *Catch and dispose of escaped fuel.*
> - *Plug all open fuel lines.*
> - *Unscrew the fuel tank cap to release pressure in the tank before working on the tank or lines.*

> ***CAUTION—***
> *Cleanliness is essential when working with fuel circuit components. Thoroughly clean the unions before disconnecting fuel lines.*

– Installation is reverse of removal. Always replace sealing O-rings.

– Remove pinch clamps from fuel lines.

– Check for fault codes and clear and reset ECM memory.

## FUEL PUMP AND FUEL LEVEL SENDER

The fuel pump with integrated fuel level sender assembly is mounted in the right (passenger) side of the fuel tank behind the passenger seat.

> ***WARNING —***
> *Make sure the fuel tank is drained before removing the fuel pump and fuel level sender assembly.*

### Fuel pump and fuel level sender, removing and installing

– Drain fuel from tank. See **Fuel tank, draining** in this repair group.

– Remove passenger seat. See **520 Seats**.

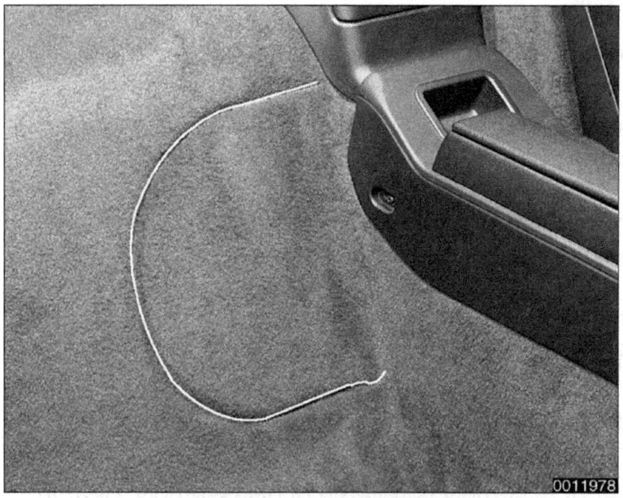


Locate perforation mark on carpet behind front seat. Cut out carpet along perforation and fold back.

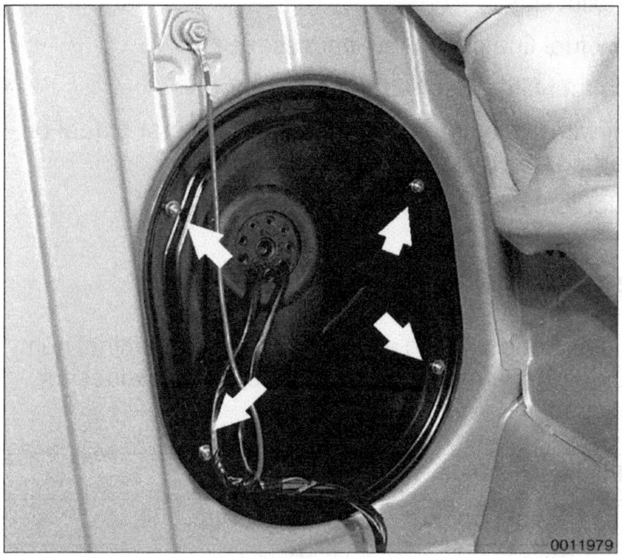


Remove fuel pump / fuel level sender access cover mounting screws (**arrows**). Remove cover.

– Label and remove hoses and harness connector from fuel pump / fuel level sender assembly.

> ***CAUTION—***
> - *Be careful to avoid dripping any fuel on interior carpeting or upholstery.*
> - *Cleanliness is essential when working with fuel circuit components. Thoroughly clean the unions before disconnecting fuel lines.*

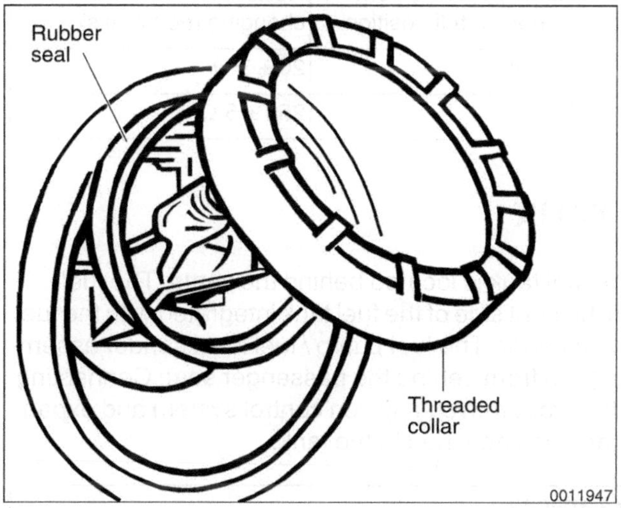


Use BMW special tool 16 1 020 to unscrew threaded collar from fuel tank. Slowly withdraw fuel pump / fuel level sender assembly, allowing fuel to drain off.

***NOTE—***
*If necessary, push level sender float arm towards fuel pump assembly to facilitate removal.*

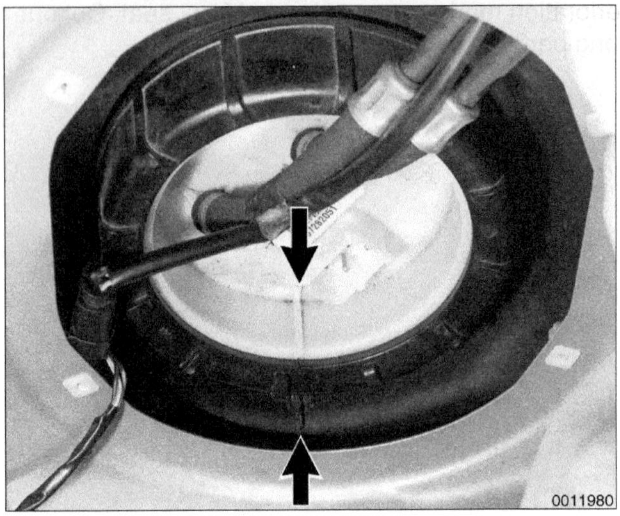

When installing:

- First install new rubber seal in fuel tank opening. Then install fuel pump / fuel level sender assembly.
- Align marks (**arrows**) on fuel pump / fuel level assembly with marks on fuel tank.

– Tighten new plastic threaded collar 2 turns by hand, then torque to specification using BMW special tool 16 1 020.

| Tightening torque | |
|---|---|
| Plastic threaded collar to fuel tank (new)<br>• Initial torque (by hand)<br>• Final torque (using tool 16 1 020) | <br>2 turns<br>55 Nm (41 ft-lb) |

– Remainder of assembly is reverse of disassembly. Bear in mind the following:

- Use new fuel line clamps as necessary.
- Replace fuel filter. See **020 Maintenance**.
- Fill tank with a minimum of 1.5 gallons (5 liters) of fuel before starting engine.
- Check for leaks by running engine.

## Fuel level sender, testing

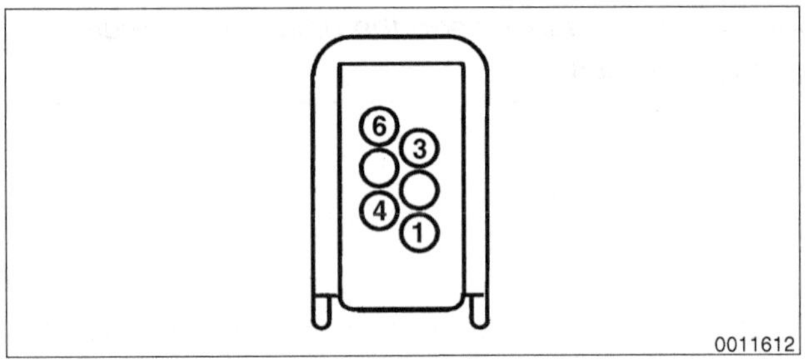


Test fuel level sender using an ohmmeter. Use terminal numbering guide illustrated. **Table c** lists resistance values for sender at specified float positions.

| **Table c. Fuel level sender resistances** | | |
|---|---|---|
| **Terminals** | **Float position** | **Test result** |
| **3** and **6** | Slowly moving from empty to full position | Resistance continuously changing (no breaks) |
| | Empty | 20 ± 2 Ω |
| | Full | 250 ± 5 Ω |

# FUEL TANK

The plastic fuel tank is located behind the seats. The fuel pump is in the right side of the fuel tank integrated with the fuel gauge level sender. The fuel pump / fuel level sender assembly is accessed from behind the passenger seat. Connecting lines for the evaporative emission control system and expansion tank are also attached to the tank.

| Fuel tank capacity | |
|---|---|
| Fuel tank | 51 liters (13.5 US gal) |
| Reserve | 8 liters (2.1 US gal) |

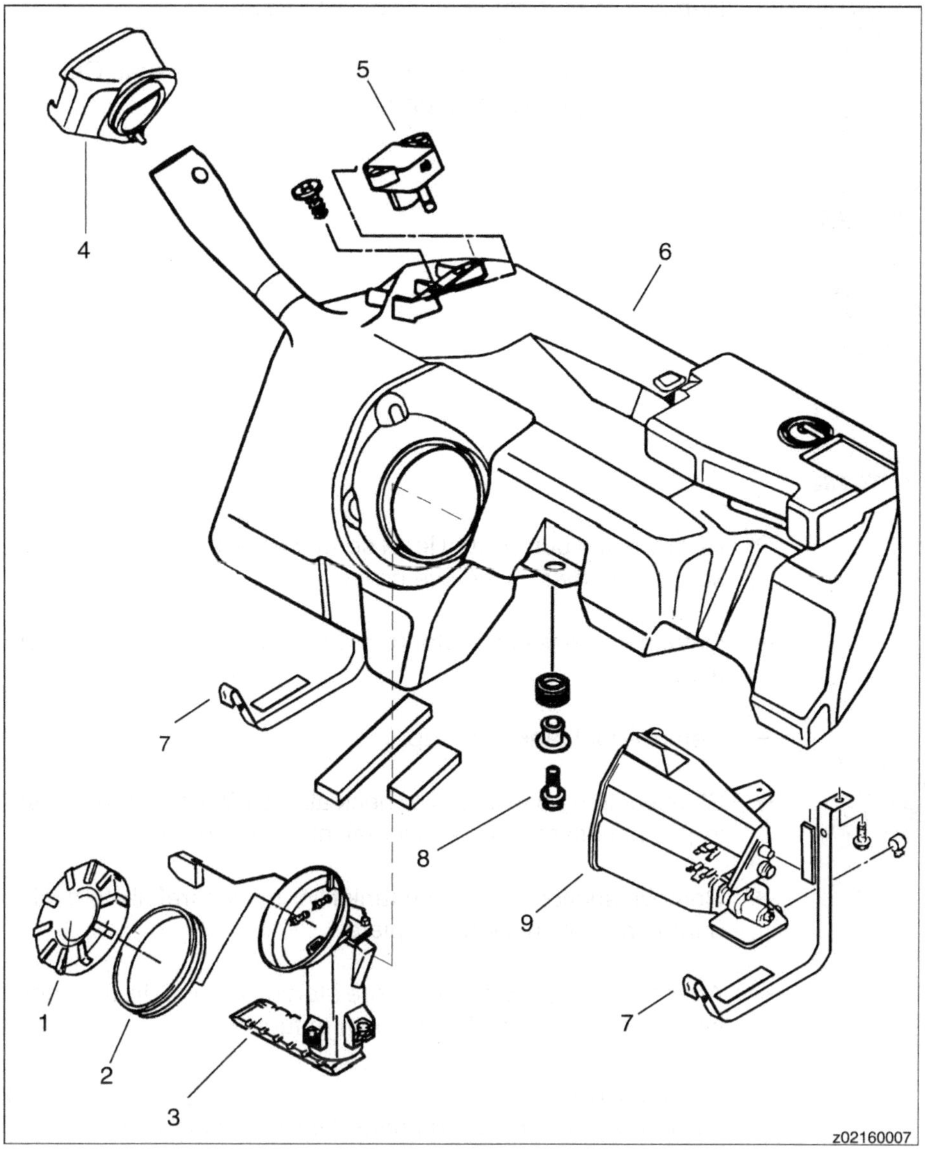


## Fuel tank components

1. **Fuel pump / fuel level sender plastic threaded collar**
2. **Rubber seal**
3. **Fuel pump / fuel level sender assembly**
4. **Fuel tank filler neck flange**
5. **Fuel tank pressure sensor**
6. **Plastic fuel tank**
7. **Mounting strap**
8. **Fuel tank center support bolt**
9. **Carbon canister**

## Fuel tank, draining

Drain fuel tank into a safe storage unit using an approved fuel pumping device.

> ***WARNING —***
> - *Fuel may be spilled. Do not smoke or work near heaters or other fire hazards.*
> - *Make sure work area is well-ventilated.*

– Disconnect negative cable from battery.

> ***CAUTION—***
> *Prior to disconnecting the battery, read the battery disconnection cautions given in* **001 General Warnings and Cautions**.

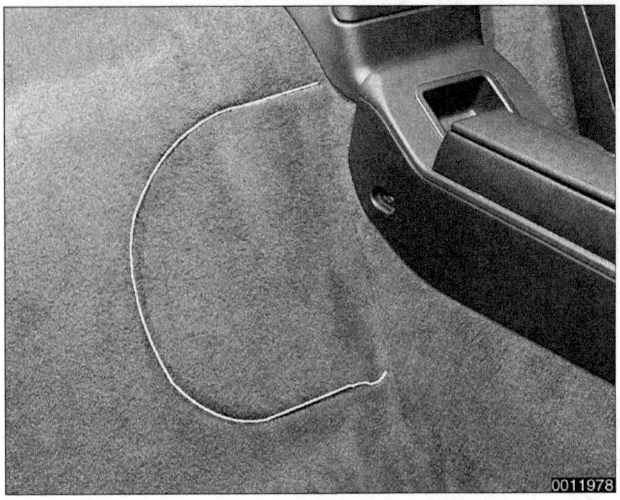


- Remove passenger seat. See **520 Seats**.

◄ Locate perforation mark on carpet behind front seat. Cut out carpet along perforation and fold back.

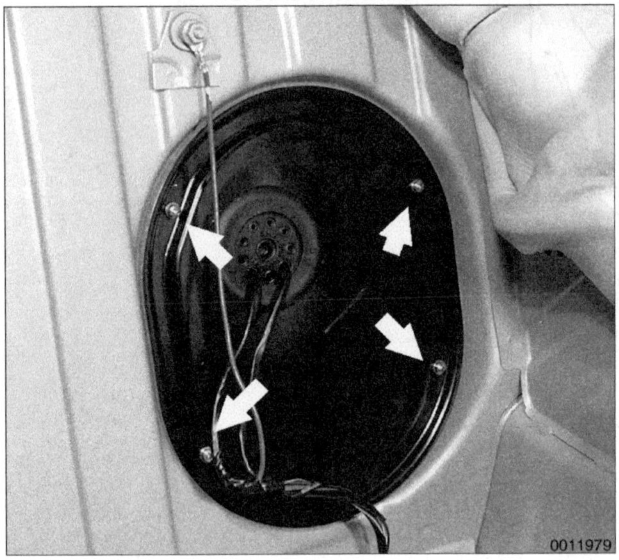


◄ Remove fuel pump / fuel level sender access cover mounting screws (**arrows**). Remove cover.

- Detach harness connector from fuel pump / fuel level sender assembly.

- Remove fuel tank filler cap.

- Slide suction hose into filler neck about 130 cm (51 in), twisting as necessary. Withdraw fuel into storage unit.

- Remove suction hose from tank filler neck carefully to avoid damaging filler neck baffle plate.

- After finishing repairs but before starting engine, fill fuel tank with at least 5 liters (1.5 gallons) of fuel.

> ***CAUTION—***
> *The fuel pump will be damaged if run without fuel.*

## Fuel tank, removing and installing

- Disconnect negative cable from battery.

> ***CAUTION—***
> *Prior to disconnecting the battery, read the battery disconnection cautions given in* **001 General Warnings and Cautions**.

- Drain fuel tank. See **Fuel tank, draining** in this repair group.

- Raise car and support in a safe manner.

> ***WARNING —***
> *Make sure the car is stable and well supported at all times. Use a professional automotive lift or jack stands designed for the purpose. A floor jack is not adequate support.*

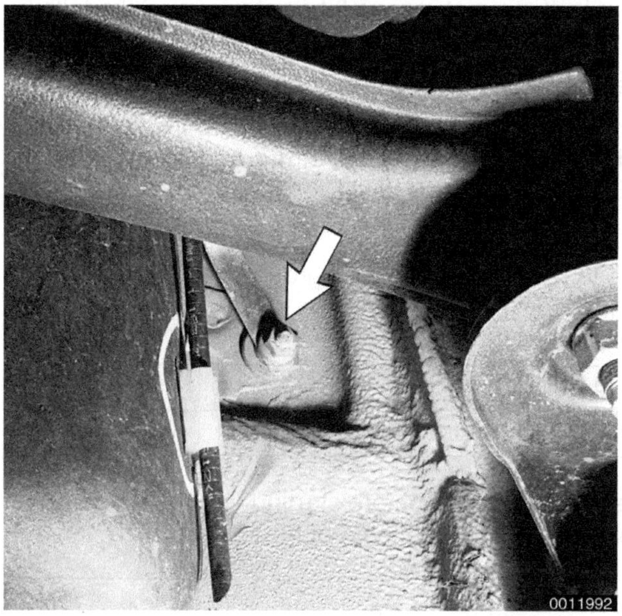


Working underneath car in right rear wheel well, remove fastener (**arrow**) and detach fuel tank filler ground wire from body.

– Unbolt and remove complete exhaust system. See **180 Exhaust System**. Remove exhaust system heat shield.

– Remove drive shaft. See **260 Driveshaft**.

– Remove complete rear axle. See **330 Rear Suspension**.

Working underneath left side of fuel tank at carbon canister:

- Remove left side fuel tank mounting strap.
- Disconnect vent lines (**A**).
- Detach harness connector (**B**).
- Loosen or cut evaporative line clamp (**C**). Disconnect hose from canister.

– Remove canister mounting screws and lower canister.

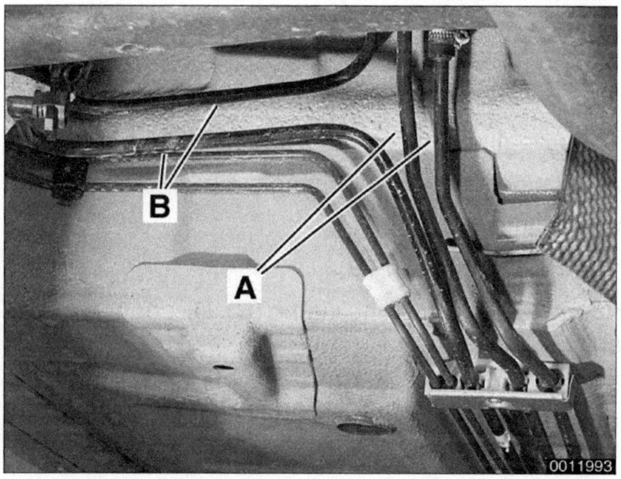


Working underneath fuel tank at fuel lines:

- Label, clamp off and disconnect fuel supply and return hoses (**A**) from lines going forward.
- Label, clamp off and disconnect fuel tank vent lines (**B**).

– Support tank from below and then remove bolt for right side tank mounting strap. Remove center support bolt from tank.

– Lower fuel tank approximately 2 - 4 inches. Disconnect harness connector from fuel tank pressure sensor at top and fuel pump / fuel level sender at front of tank.

– Carefully lower tank from car.

– Installation is reverse of removal, bearing in mind the following:
  - Use new hose clamps when reconnecting hoses.
  - Inspect rubber parts including hoses and buffers. Replace parts that are chafed, dried out or cracked.
  - Reconnect electrical connectors before raising tank into final position.
  - Be sure to reconnect filler neck ground strap.

| Tightening torques | |
|---|---|
| Fuel tank to body | 23 Nm (17 ft-lb) |
| Hose clamp (10 - 16 mm diameter) | 2 Nm (18 in-lb) |

– After finishing repairs but before starting engine, fill fuel tank with at least 5 liters (1.5 gallons) of fuel.

> ***CAUTION—***
> *The fuel pump will be damaged if run without fuel.*

– To make sure fuel tank is properly grounded, check electrical resistance between tank filler neck ground tab and wheel hub. The resistance should be no higher than 0.6Ω.

## Activated carbon canister, removing and installing

– Raise rear end of car and support safely.

> ***CAUTION—***
> *Make sure the car is stable and well supported at all times. Use a professional automotive lift or jack stands designed for the purpose. A floor jack is not adequate support.*

– Remove protective cover under left side of fuel tank, if applicable.

◄ Working underneath left side of fuel tank at carbon canister:
  - Remove left side fuel tank mounting strap.
  - Disconnect vent lines (**A**).
  - Detach harness connector (**B**).
  - Loosen or cut evaporative line clamp (**C**). Disconnect hose from canister.

– Remove canister mounting screws and lower canister.

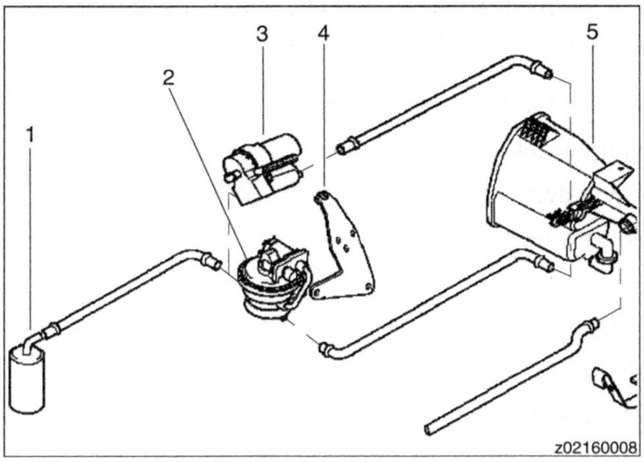


## Tank leak detection module, removing and installing

Evaporative loss control components at fuel tank:

1. Dust filter
2. Leak detection pump (LDP) (DME MS 42.1)
3. Detection module-tank leakage (DMTL) (DME MS 43.0)
4. Mounting bracket
5. Carbon canister

– Raise car and support safely.

> ***CAUTION—***
> *Make sure the car is stable and well supported at all times. Use a professional automotive lift or jack stands designed for the purpose. A floor jack is not adequate support.*

– Remove protective cover under left side of fuel tank, if applicable.

– Working underneath driver seat, below car, remove protective cover from 3/2-way valve and fuel filter assembly.

– Detach vent lines to and from leak detection module.

– Remove module and mounting bracket together. Separate on workbench.

– Installation is reverse of removal.

## 3/2–way (running losses) valve, removing and installing

The 3/2–way valve is located under the car underneath the driver seat, just ahead of the fuel filter. A protective cover shields fuel system components from road hazards and dirt.

The 3/2–way valve is controlled by the ECM. On engine start up, the valve solenoid is energized for 20 seconds in order to supply full fuel volume to the fuel rail. After 20 seconds, the solenoid is deactivated and sprung closed (fuel bypass to tank is opened). This reduces the amount of fuel circulating through the fuel rail and diverts the excess to return to the fuel tank through the fuel pressure regulator.

The fuel injectors are supplied with pressurized fuel, but excess fuel is diverted to the fuel tank without first passing through the fuel rail. This prevents the excess fuel from absorbing heat in the engine compartment, thus lowering the temperature and amount of vaporization in the fuel tank.

The 3/2–way valve is also activated briefly if an engine misfire is detected. This provides full fuel flow through the fuel rail to determine if the misfire was caused by a lean fuel condition. The valve is monitored by the ECM for faults.

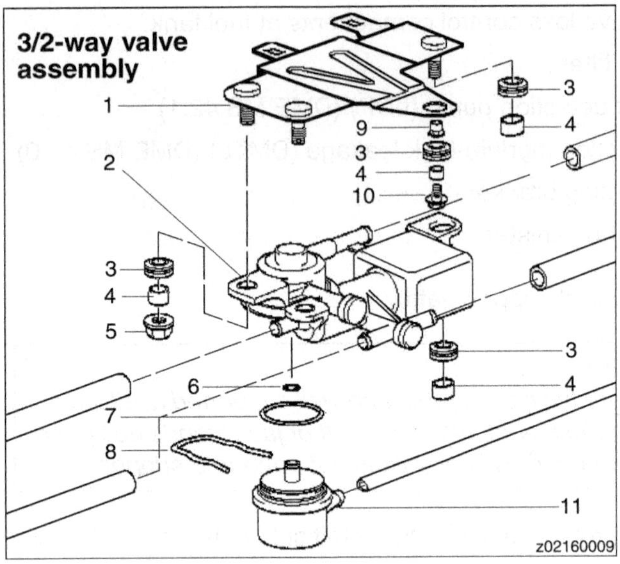


3/2-way valve assembly components:

1. Mounting bracket
2. 3/2-way valve
3. Rubber grommet
4. Spacer bushing
5. Nut (M6)
6. Sealing O-ring
7. Sealing O-ring
8. Fuel pressure regulator retaining clip
9. Blind rivet nut
10. Bolt (M6)
11. Fuel pressure regulator

– Raise car and support safely.

**CAUTION—**
*Make sure the car is stable and well supported at all times. Use a professional automotive lift or jack stands designed for the purpose. A floor jack is not adequate support.*

– Working underneath driver seat, remove protective panel from fuel filter and 3/2–way valve.

**WARNING—**
- *When opening fuel lines, be prepared for fuel to be expelled under pressure (approx. 3 - 5 bar or 45 - 75 psi).*
- *Do not smoke or work near heaters or other fire hazards. Keep a fire extinguisher handy.*
- *Before disconnecting fuel hoses, wrap a cloth around fuel hoses to absorb any leaking fuel.*
- *Catch and dispose of escaped fuel.*
- *Plug all open fuel lines.*
- *Unscrew the fuel tank cap to release pressure in the tank before working on the tank or lines.*

**CAUTION—**
*Cleanliness is essential when working with fuel circuit components. Thoroughly clean the unions before disconnecting fuel lines.*

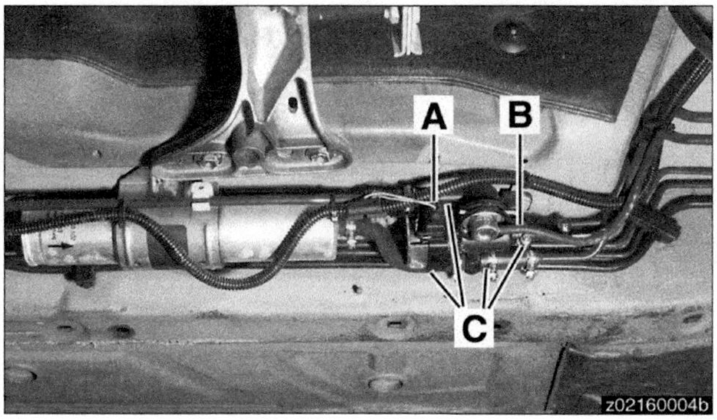


To remove 3/2-way valve:

- Disconnect electrical harness connector (**A**) from valve.
- Remove vacuum hose (**B**) from fuel pressure regulator.
- Loosen or remove hose clamps and disconnect fuel lines (**C**).
- Remove M6 mounting nuts and lower 3/2-way valve assembly.

> ***WARNING —***
> *Be prepared to catch dripping fuel in a pan.*

– Installation is reverse of removal. Use new hose clamps.

| Tightening torques | |
|---|---|
| Hose clamp 8 - 13 mm diameter | 2 Nm (18 in-lb) |
| Hose clamp 10 - 16 mm diameter | 2 Nm (18 in-lb) |

***NOTE —***
*Install protective cover and seals correctly to keep moisture and road dirt out of underbody fuel system components.*

– Check for fault codes and clear and reset ECM memory.

# 170 Radiator and Cooling System





**TABLES**

## GENERAL

This section covers troubleshooting and component repair information for the engine cooling system.

Procedures in this section are grouped by engine code. See **100 Engine–General** for application information.

Additional repair information for systems and items related to the cooling system may be found in the following repair groups:

- **020 Maintenance** for drive belt removal and installation
- **119 Radiator and Cooling System** for engine oil cooler replacement
- **130 Fuel Injection** for air filter housing and engine coolant temperature (ECT) sensor removal
- **510 Exterior Trim, Bumpers** for front bumper removal
- **640 Heating and Air Conditioning** for heater core replacement
- **OBD On-Board Diagnostics** for diagnostic trouble codes (DTCs)

### Special tools

Special tools are necessary for belt-driven fan clutch removal and also for pressure testing the cooling system.

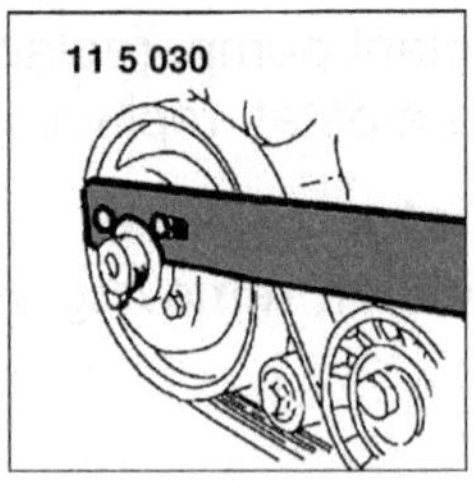


◄ Belt-driven fan counterhold wrench (BMW special tool 11 5 030)

◄ Belt-driven fan wrench (BMW special tool 11 5 040)

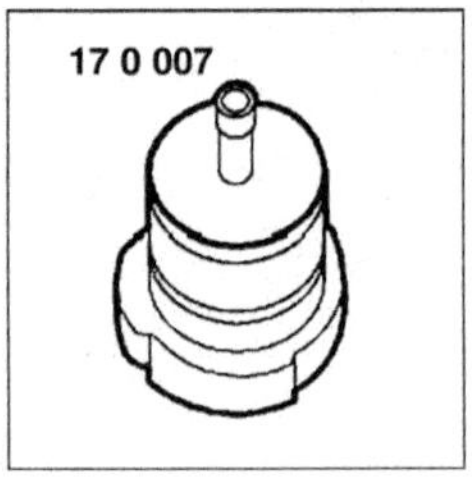


◄ Expansion tank cap test adaptor (BMW special tool 17 0 007)

◄ Cooling system pressure tester kit (BMW special tool 17 0 002 / 17 0 005)

## Coolant pump

A centrifugal-type coolant pump is mounted on the front of the cylinder block. The belt-driven pump circulates coolant through the system whenever the engine is running.

## Radiator and cooling fans

The radiator is a crossflow design. An expansion tank provides for coolant expansion at higher temperatures and easy monitoring of the coolant level.

The engine cooling fan system employs a combination of mechanical (belt-driven) and electrical fans, depending on engine installed.

**Mechanical (belt-driven) fan** is mounted to front of coolant pump through a fan clutch. A viscous fluid coupling controls speed of fan based on engine compartment temperature.

**Electric two-speed fan** is mounted behind the front grill on the *bumper side* of the radiator. This auxiliary fan is primarily used for the A/C system.

**Electric two-speed fan** is mounted on the *engine side* of the radiator of 4-cylinder model with manual transmission. The fan operates when the coolant temperature exceeds a predetermined level.

## Thermostat

The thermostat controls the flow of coolant into the radiator. When the engine is cold, the thermostat is closed so that coolant bypasses the radiator, recirculating from the engine directly back to the pump. When the engine reaches operating temperature, the thermostat opens and coolant circulates through the whole system, including the radiator.

Two types of thermostat are used in Z3 cars:

- **Mechanical thermostat** is used in 4-cylinder models, models with M52 engine built up to 9 / 1998 and models with S52 or S54 engine.
- **Electrically heated thermostat** is used in models with M52 TU or M54 engine from 9 / 1998.

### Mechanical thermostat

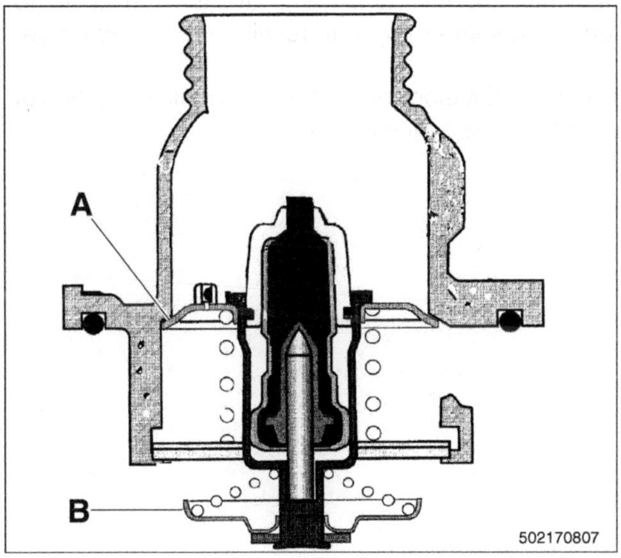


Mechanical thermostat relies on coolant temperature to operate. While coolant is cold, thermostat is closed. At higher coolant temperature, thermostat progressively opens to allow coolant flow to radiator, controlling engine temperature.

- Port **A** restricts coolant flow from radiator
- Port **B** restricts coolant flow to radiator

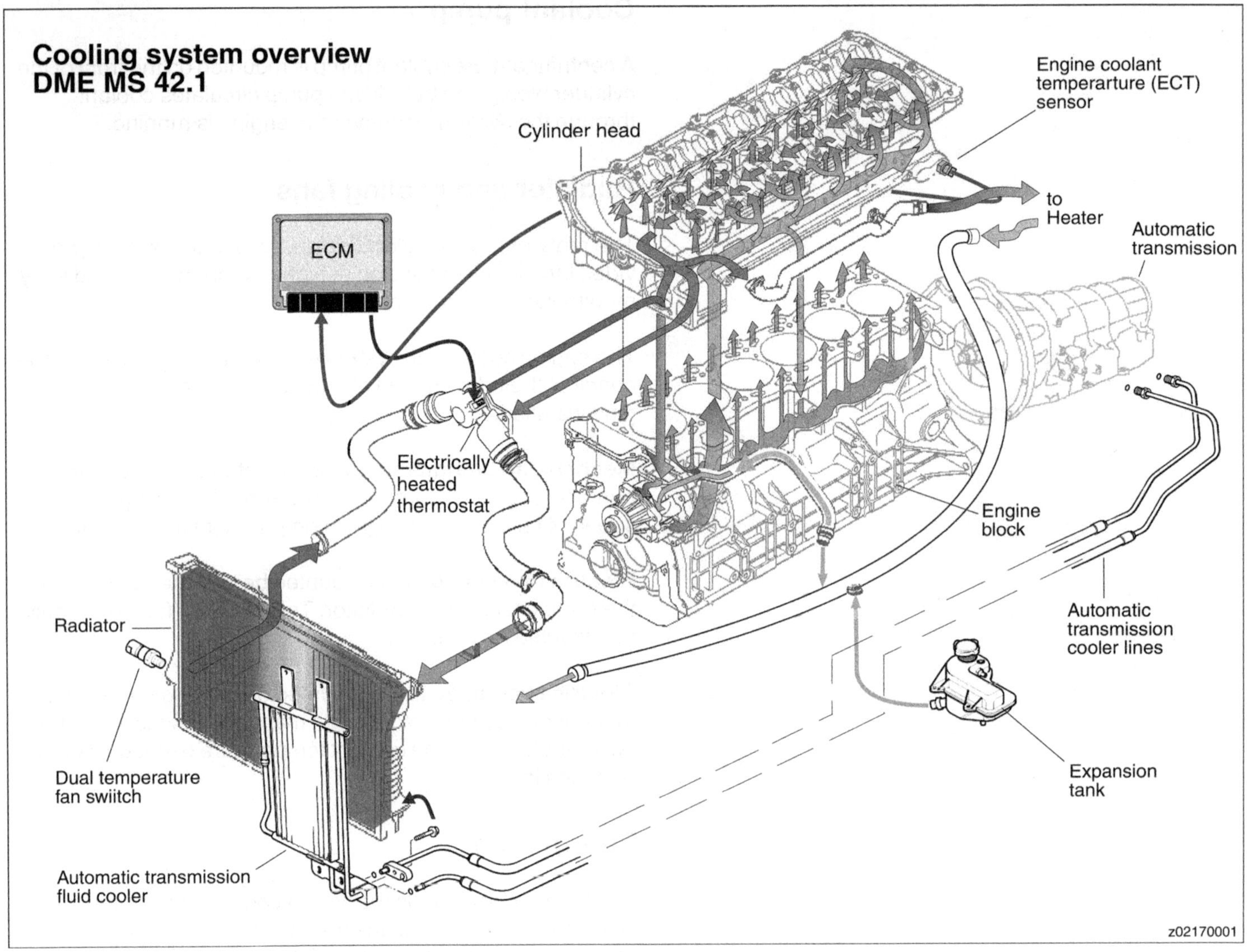


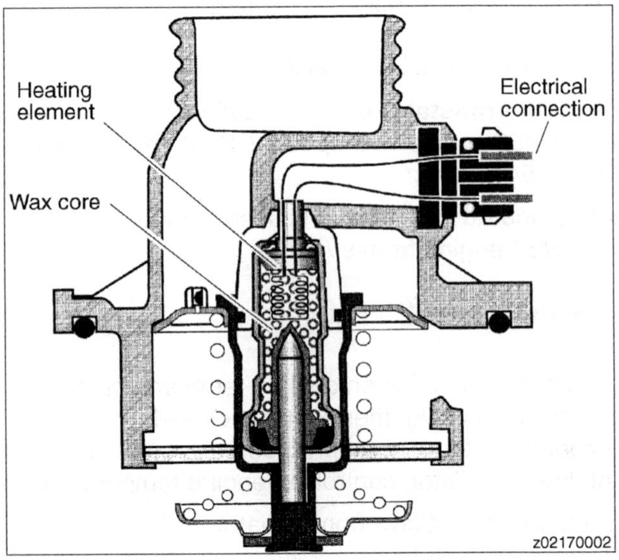


## Electrically heated thermostat

The electrically heated thermostat is DME map-controlled. The engine control module (ECM) activates the thermostat to maintain engine coolant temperature within a narrow range.

In case of failure of the electronics, the mechanical function of the thermostat acts as a fail-safe.

## Engine coolant temperature (ECT) sensor

The engine coolant temperature (ECT) sensor is a dual temperature sensor. One circuit provides coolant temperature information to the ECM while the other circuit provides coolant temperature information to the instrument cluster. See **Engine coolant temperature (ECT) sensor** entries in **130 Fuel Injection** for more information.

Engine coolant temperature (ECT) sensor (4-cylinder engine)

- 4-cylinder engine: The ECT sensor is located on the left front of the cylinder head above the oil filter.

M52, S52 engine: The ECT sensor is located below cylinder 1 intake runner.

Engine coolant temperature (ECT) sensor (M52 TU, M54 engine)

- M52 TU, M54 engine: The ECT sensor is located at the rear of the cylinder head, below cylinder 6 intake runner.

- S54 engine: The ECT sensor is located below cylinder 1 intake runner.

## Automatic transmission fluid cooler

Automatic transmission fluid lines circulate transmission fluid (ATF) to and from a heat exchanger in front of the radiator.

## Warnings and cautions

Observe the following warnings and cautions when working on the cooling system.

***WARNING—***

- *At normal operating temperature the cooling system is pressurized. Allow the system to cool thoroughly (a minimum of one hour), then release the cooling system pressure cap slowly to allow safe release of pressure.*
- *Releasing cooling system pressure lowers the coolant boiling point and the coolant may boil suddenly. Use heavy gloves and wear eye and face protection to guard against scalding.*
- *Use extreme care when draining and disposing of engine coolant. Coolant is poisonous and lethal to humans and pets. Pets are attracted to coolant because of its sweet smell and taste. Seek medical attention immediately if coolant is ingested.*

***CAUTION—***

- *Avoid adding cold water to the coolant while the engine is hot or overheated. If it is necessary to add coolant to a hot system, do so only with the engine running and coolant pump turning.*
- *To avoid excess silicate gel precipitation in the cooling system and loss of cooling capacity, use BMW coolant or equivalent low silicate antifreeze.*
- *If oil enters the cooling system, flush with cleaning agent. BMW recommends removal of the radiator and expansion tank to flush. See* **Cooling system, flushing** *in this repair group.*
- *When working on the cooling system, cover the alternator to protect it against coolant drips.*
- *Prior to disconnecting the battery, read the battery disconnection cautions given in* **001 General Warnings and Cautions**.

## TROUBLESHOOTING

Begin the diagnosis of cooling system problems with a thorough visual inspection. If no visual faults are found, use BMW scan tool DISplus, GT1, MoDiC or equivalent to check engine control module (ECM) fault memory for stored diagnostic trouble codes (DTCs).

Common cooling system faults can be grouped into one of 4 categories:

- Cooling system leaks
- Poor coolant circulation
- Radiator cooling fan faults
- Electrical or electronic faults

## Cooling system inspection

– Check that coolant pump drive belt tensioner is functioning properly and that belt tension is correct. Belt tensioner systems for different models are shown in **020 Maintenance**.

– Check condition of coolant pump drive belt.

– Check coolant hoses for cracks or softness. Check clamps for looseness. Check coolant level and check for evidence of coolant leaks from engine.

– Check that radiator fins are not blocked with dirt or debris. Clean radiator using low-pressure water or compressed air. Blow outward, from engine side out.

***NOTE—***
*Dead insects and other deposits may impede air flow through the radiator. The resulting increase in coolant temperature may lead to noticeably poorer engine performance.*

– To check coolant pump:

- Remove mechanical cooling fan. See **Mechanical (belt-driven) cooling fan, removing and installing**.
- Remove drive belt from coolant pump pulley. See **020 Maintenance**.
- Firmly grasp opposite sides of pulley and check for play in all directions.
- Spin pulley and check that shaft runs smoothly without play or wobble.

***NOTE—***
*Coolant provides lubrication for the pump shaft, so an occasional drop of coolant leaking from the pump is acceptable. If coolant drips steadily from pump vent hole, replace pump.*

– At normal engine operating temperature, the cooling system is pressurized. This raises the boiling point of coolant. Leaks may prevent the system from becoming pressurized. If visual evidence is inconclusive, pressure test the cooling system to help pinpoint hard-to-find leaks. See **Cooling system pressure test** in this repair group.

- If cooling system is full of coolant and holds pressure:
  - Use an appropriate scan tool to interrogate engine control module (ECM) for radiator fan or DME control circuit faults.
  - Check for loose or worn drive belt.
  - Test for failed thermostat or coolant pump impeller. Some pumps may be fitted with plastic impellers.
  - Check for clogged/plugged radiator or coolant passages.

### Thermostat test

If the engine overheats or runs too cool and no other cooling system tests indicate trouble, the thermostat may be faulty.

A thermostat that is stuck open causes the engine to warm up slowly and run below normal temperature at highway speed. A thermostat that is stuck closed restricts coolant flow to the radiator and causes overheating.

- To check if the thermostat is opening and coolant is circulating through the radiator, allow a cold engine to reach operating temperature (temperature gauge needle approximately centered).
  - Feel top of radiator and top radiator hose. If hot to the touch, coolant is probably circulating correctly.
  - If there are any cool areas in hose or radiator, coolant flow to radiator is probably restricted. Check for a faulty thermostat or a plugged radiator.

If the engine overheats and no other cooling system tests indicate trouble, the radiator may have some plugged passages that are restricting flow.

In engines equipped with the electrically heated thermostat, the fault may lie in the DME software or hardware, or it may lie in wiring to the thermostat.

### Cooling fan clutch, checking

An otherwise sound cooling system may still overheat, particularly with prolonged idling, due to a failure of the engine cooling fan viscous clutch coupling.

- Check fan clutch by spinning fan with engine off and cold. Make sure fan spins on clutch with some resistance.

- Check for signs of leaking fluid from clutch.

- If fan freewheels with no resistance, cannot be turned by hand, or there are signs of oil leakage, replace clutch.

## Electric cooling fan, checking

An auxiliary electric cooling fan provides additional radiator cooling when the air conditioning is on or if coolant temperature exceeds a predetermined level. The fan is powered by relays in the front power distribution box. Fan operation is controlled by the ECM.

If a faulty thermostat, trapped air or a restriction in the system is not allowing coolant to circulate through radiator, dual temperature switch does not close and cooling fan does not run.

– Make sure thermostat is operating correctly. See **Thermostat test** in this repair group.

– Check to make sure electric fan motor is not seized: Using wooden stick, reach into electric fan cowl and test freedom of movement of fan blades.

> ***WARNING —***
> - *When checking the electric cooling fan, keep hands and wires clear of the fan blades. The cooling fan can run at any time.*
> - *For greatest safety, check electric cooling fan when the engine is cold and the air conditioning OFF.*

< Dual temperature cooling fan switch (**arrow**) is mounted at right front of radiator. Check to make sure connector is on tightly.

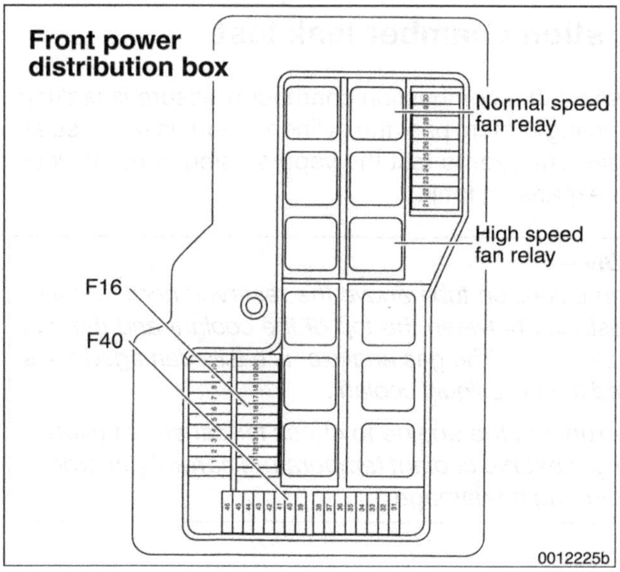


< Power supply for electric cooling fan:

- Fuses F16 (5A) and F40 (30A)
- Relays for normal speed and high speed fan operation

For additional electric cooling fan wiring information, see **ELE Electrical Wiring Diagrams**.

## Cooling system pressure test

Use cooling system pressure tester to look for coolant leaks, including internal ones. Common sources of internal coolant leaks are a faulty cylinder head gasket, a cracked cylinder head, or a cracked engine block.

> ***WARNING —***
> *At normal operating temperature the cooling system is pressurized. Allow the system to cool thoroughly (a minimum of one hour), then release the cooling system pressure cap slowly to allow safe release of pressure.*

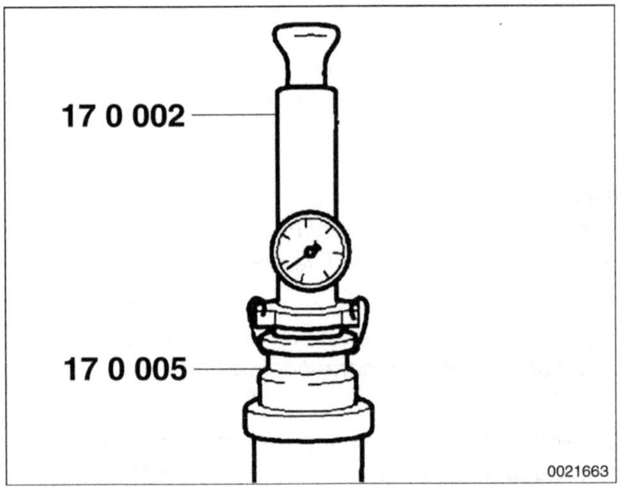


With engine cold, install pressure tester (BMW special tools 17 0 002 / 17 0 005 or equivalent) to expansion tank. Pressurize system to specification listed in **Table a**.

- Pressure should not drop more than 0.1 bar (1.45 psi) for at least two minutes.
- If pressure drops rapidly and there is no sign of external leakage, cylinder head gasket may be faulty. Perform compression and leak-down tests. See **113 Cylinder Head Removal and Installation**.
- Test expansion tank cap using pressure tester with correct adapter (BMW special tool 17 0 007 or equivalent). Replace faulty cap or cap gasket.

**Table a. Cooling system test pressures**

| Component | Test pressure |
|---|---|
| Cooling system pressure cap | 2 bar (29 psi) |
| Expansion tank or radiator | 1.5 bar (21.75 psi) |

> ***CAUTION—***
> *Exceeding the specified test pressure could damage the radiator or other system components.*

## Combustion chamber leak test

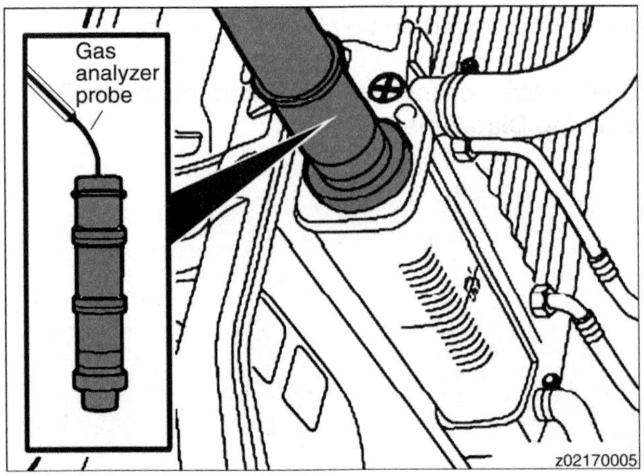


If you suspect that combustion chamber pressure is leaking into the cooling system past the cylinder head gasket, use an exhaust gas analyzer to test the vapors rising from the coolant at the expansion tank.

> ***CAUTION—***
> - *Use an extension tube above the reservoir neck to maintain distance between the top of the coolant and the gas analyzer probe. The gas analyzer is easily damaged if it is allowed to inhale liquid coolant.*
> - *While running the engine to check for causes of overheating, observe coolant temperature carefully in order to avoid engine damage.*

## ENGINE COOLANT SERVICE

Check antifreeze concentration in engine coolant as part of Inspections I and II. See **020 Maintenance**.

### Engine coolant

To maintain maximum coolant efficiency, fill cooling system with a 50 / 50 mixture of distilled water and ethylene glycol based antifreeze.

BMW requires the use of phosphate and nitrate free antifreeze. For this reason, use only BMW antifreeze or equivalent.

> ***CAUTION—***
> *BMW does not approve the use of coolant additives that provide additional protection against corrosion or are aimed at sealing hairline cracks. Due to the poorer transfer of heat from the cylinder head to the coolant, this can lead to heat accumulation and hot spots, which in turn can lead to cylinder head or cylinder head gasket failure.*

### Coolant recycling

Dispose of coolant properly by recycling. Follow instructions in the coolant manufacturer's safety data sheets and abide by local, state and country-specific regulations. Also see **Warnings and cautions** in this repair group.

### Coolant, draining and filling

Change coolant every 4 years (3 years for M version). Check antifreeze concentration in coolant each year before the cold season begins. Allow fresh coolant to circulate before checking coolant composition with a hydrometer.

– With engine cold, remove expansion tank cap.

> ***WARNING —***
> *At normal operating temperature the cooling system is pressurized. Allow the system to cool thoroughly (a minimum of one hour), then release the cooling system pressure cap slowly to allow safe release of pressure.*

> ***CAUTION—***
> *When working on the cooling system, cover the alternator to protect it against coolant drips.*

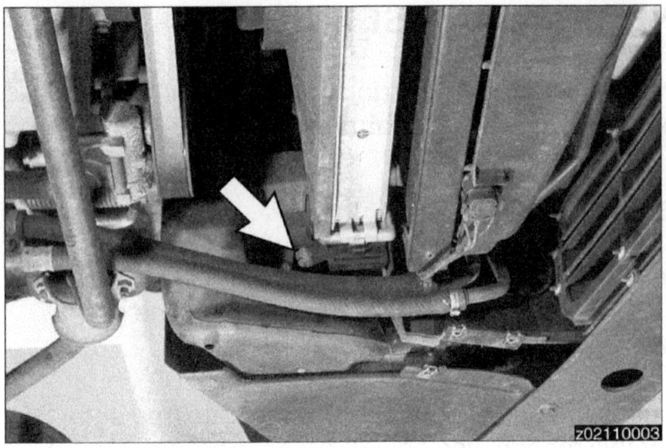


– Raise front of car and support safely.

> ***WARNING —***
> *Make sure the car is stable and well supported at all times. Use a professional automotive lift or jack stands designed for the purpose. A floor jack is not adequate support.*

– Remove splash shield from under engine.

< Place 3-gallon pail underneath radiator drain plug (**arrow**) and remove plug.

> ***WARNING —***
> *Coolant is poisonous and lethal, especially to pets. Clean up spills immediately and rinse the area with water.*

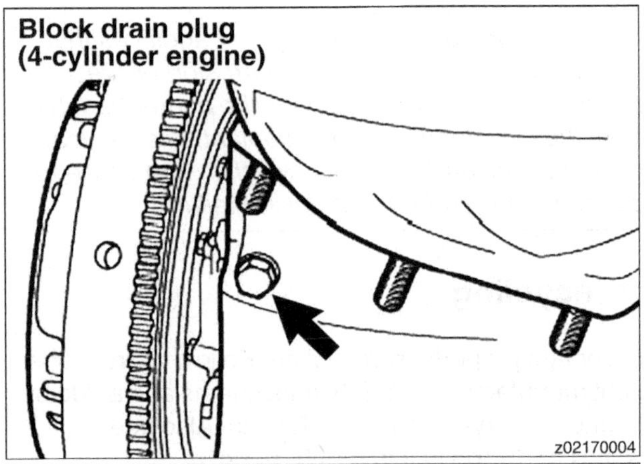


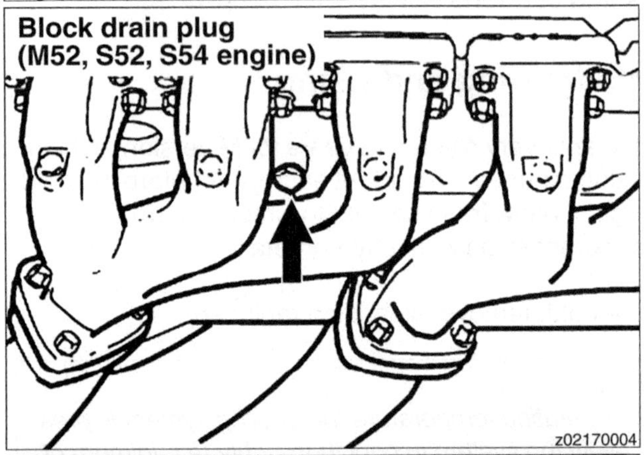

< Place 3-gallon pail underneath rear of engine block. Loosen and remove engine block coolant drain plug.

- 4-cylinder engine: Block drain plug is at rear of engine on exhaust side.
- M52, S52 or S54 engine: Block drain plug is on exhaust side of engine block between cylinders 3 and 4.
- M52 TU, M54 engine: Block drain plug is on exhaust side of engine block at cylinder 2.

– Reinstall radiator and engine block drain plugs using new sealing washers.

| Tightening torques | |
|---|---|
| Drain plug to engine block (M14) | 25 Nm (18 ft-lb) |
| Drain plug to radiator | 2 - 3 Nm (18 - 27 in-lb) |

– Before refilling radiator:

- Switch ignition to ON. (Do not start engine).
- Set temperature controls to full warm.
- Set blower control to low.

– Make sure cooling system bleed screws are open:

- Main bleed screw in all models is at top of coolant expansion tank.
- M52 engine: There is an additional bleed screw on coolant hose manifold next to oil filter housing.

– Using a coolant mixture of 50 / 50 BMW antifreeze or equivalent and water, fill expansion tank slowly. Continue until coolant emerges from bleed screw. Cooling system capacities are shown in **Table b**.

*NOTE—*

- *Be sure bleed screw(s) is loose when filling cooling system.*
- *Coolant can be reused provided it is clean and less than two years old. Do not reuse coolant after replacing damaged engine parts.*

**Table b. Cooling system capacity**

| Engine | Capacity |
|---|---|
| 4-cylinder | 6.9 liters (6.6 US qt) |
| 6-cylinder | 10.5 liters (11.1 US qt) |

– Bleed cooling system. See **Cooling system, bleeding** in this repair group.

## Cooling system, flushing

If oil or other impurities contaminate the coolant circuit, have the cooling system flushed at a BMW service center or other qualified repair shop equipped with coolant recycling and disposal equipment.

## Cooling system, bleeding

Air may become trapped in the cooling system during filling. Trapped air can prevent proper coolant circulation. Whenever the coolant is drained and filled, bleed the system of trapped air.

> ***WARNING—***
> *At normal operating temperature the cooling system is pressurized. Allow the system to cool thoroughly (a minimum of one hour), then release the cooling system pressure cap slowly to allow safe release of pressure.*

> ***CAUTION—***
> *When working on the cooling system, cover the alternator to protect it against coolant drips.*

With engine cold, add coolant mixture of 50 / 50 BMW antifreeze or equivalent and water to expansion tank until level reaches COLD mark on tank or on coolant level indicator.

*NOTE—*
*4-cylinder engine illustrated. Other models are similar.*

– Loosen bleed screw (**arrow**) on radiator expansion tank.

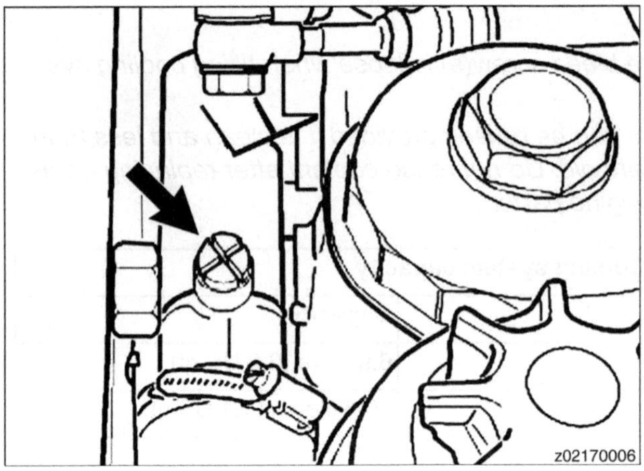


M52 engine: Loosen additional bleed screw (**arrow**) on coolant hose manifold next to oil filter housing.

– Add coolant until it spills from bleed screw(s). Tighten screw(s).

– Set temperature controls in passenger compartment to full warm and start engine.

– With engine idling at normal operating temperature and bleed screw(s) loose:
  - Slowly add coolant mixture to expansion tank until it spills from bleed screw(s).
  - When coolant spilling from bleed screw(s) is free of air bubbles, tighten screw(s) and replace cap on expansion tank.
  - Run engine until it reaches operating temperature.
  - Shut off engine and allow to cool. Recheck coolant level.
  - Top up so that coolant level indicator (if equipped) is at MAX.

– If trapped air cannot be bled manually, pressure-bleed system using cooling system pressure tester. See **Cooling system pressure test** in this repair group.

| **Tightening torque** | |
|---|---|
| Bleed screw to expansion tank | 2.5 Nm (22 in-lb) |

## Engine cooling fans

Z3 and M version cars are equipped with a belt-driven fan and 1 of 2 different versions of electric cooling fan.

### Mechanical (belt-driven) cooling fan, removing and installing

The belt-driven fan is attached to the coolant pump via a viscous fluid clutch.

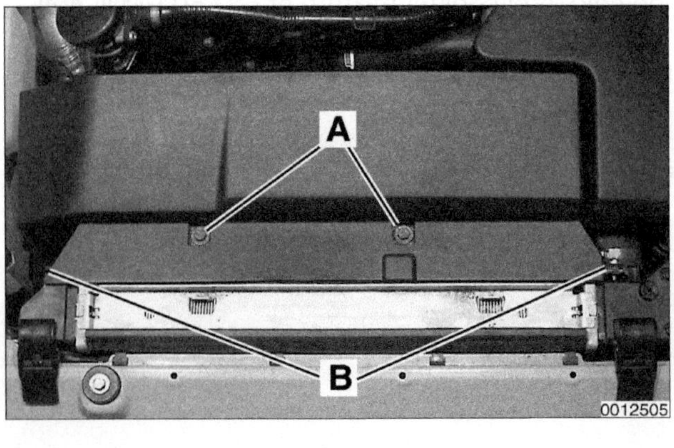


4-cylinder model:
- Remove engine intake air duct mounting screws (**A**) from top of radiator.
- Detach plastic rivets (**B**) at sides of radiator.
- Disconnect air duct and air filter housing top cover and remove from engine compartment.

Using a 32 mm wrench on cooling fan clutch nut (**arrow**), turn wrench quickly in a clockwise direction (working from front of car) to loosen. Spin fan off pump.

> ***CAUTION—***
> *The fan clutch nut has left-hand threads.*

***NOTE—***
*If necessary, use BMW special tools 11 5 030 and 11 5 040 to counterhold pulley and loosen nut.*

Working at top corners of radiator, pry out rivet pin (**arrow**) and remove rivet holding fan shroud to radiator. Remove fan and shroud together.

***NOTE—***
*Store the removed fan clutch assembly in an upright position to prevent loss of fan clutch fluid.*

– To replace fan clutch, remove fan mounting bolts and separate clutch from fan.

– Installation is reverse of removal.

| **Tightening torques** | |
|---|---|
| Clutch nut to coolant pump (left-hand thread)<br>• Without BMW special tool 11 5 040<br>• With BMW special tool 11 5 040 | <br>40 Nm (29 ft-lb)<br>30 Nm (22 ft-lb) |
| Fan to viscous clutch | 10 Nm (7 ft-lb) |

## Electric cooling fan, replacing (engine side fan)

The electric radiator cooling fan for 4-cylinder engine with manual transmission is mounted on the engine side of the radiator.

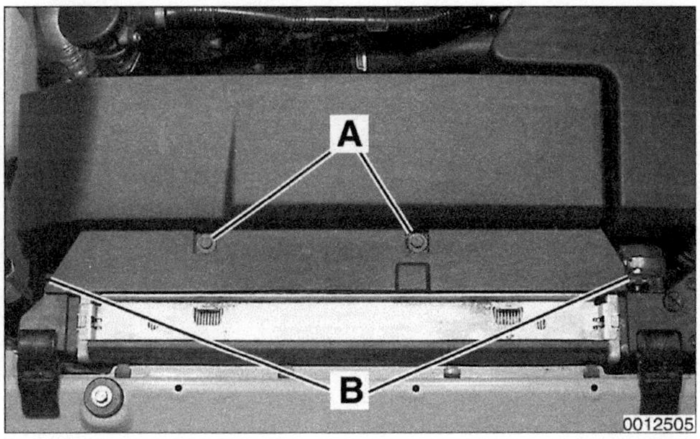


Remove engine intake air duct mounting screws (**A**) from top of radiator.

- Remove fan mounting screws (**B**) at sides of radiator.
- Disconnect air duct and air filter housing top cover and remove from engine compartment.

– Disconnect fan harness connector at bottom left of radiator.

– Lift out fan.

– Installation is reverse of removal. Be sure to align side tabs on fan housing with retaining tabs on radiator when reinstalling fan.

## Electric cooling fan, replacing (bumper side fan)

The electric radiator cooling fan for 6-cylinder engine is mounted behind the front bumper, on the front side of the A/C condenser. The fan is removed from below, necessitating removal of front bumper.

– Raise front of car and support safely.

> ***WARNING —***
> *Make sure the car is stable and well supported at all times. Use a professional automotive lift or jack stands designed for the purpose. A floor jack is not adequate support.*

– Remove splash shield from under engine.

– Remove front bumper. See **510 Exterior Trim, Bumpers**.

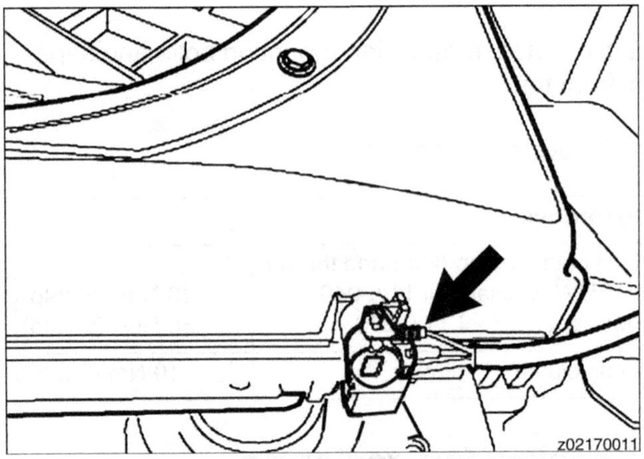


◄ Working underneath fan, detach harness connector (**arrow**).

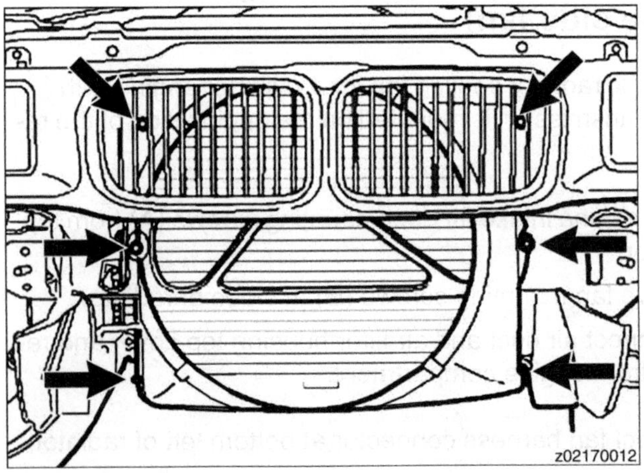


◄ Remove fan carrier mounting screws (**arrows**).

– Lower fan to remove.

– Installation is reverse of removal.

## Engine cooling fan switch, removing and installing

< dual temperature cooling fan switch (**arrow**) is mounted at right front of radiator.

- Detach electrical connector.
- Unscrew sensor from radiator.

*NOTE—*
- *Be prepared to catch dripping coolant.*
- *Remove old switch and install new switch quickly to avoid unnecessary coolant loss.*

- Install new switch, using new sealing O-ring.

| **Tightening torque** | |
|---|---|
| Dual temperature switch to radiator | 15 Nm (11 ft-lb) |

- Reattach electrical connector.
- Top up and bleed coolant. See **Cooling system, bleeding** in this repair group.
- Check for fault codes and clear and reset DME control module memory.

# COOLANT CIRCULATION COMPONENTS (4-CYLINDER ENGINE)

## Coolant pump, replacing (4-cylinder engine)

- With engine cold, drain cooling system. See **Coolant, draining and filling** in this repair group.

***WARNING—***
*At normal operating temperature the cooling system is pressurized. Allow the system to cool thoroughly (a minimum of one hour), then release the cooling system pressure cap slowly to allow safe release of pressure.*

***CAUTION—***
*When working on the cooling system, cover the alternator to protect it against coolant drips.*

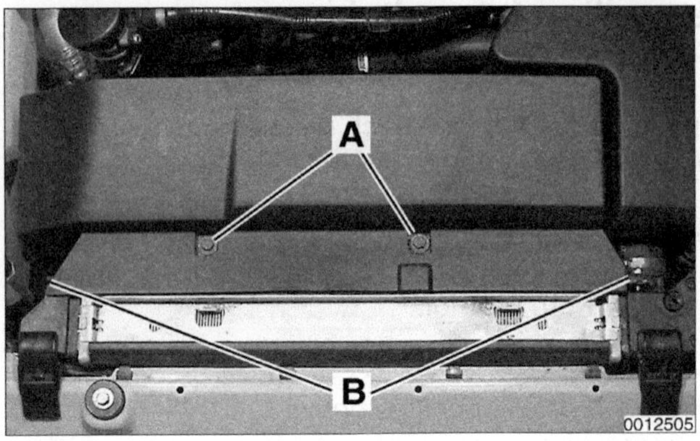


Working at top of radiator:

- Remove engine intake air duct mounting screws (**A**).
- Detach plastic rivets (**B**) at sides of radiator.
- Disconnect air duct and air filter housing top cover and remove from engine compartment.

– Remove belt-driven or engine side electric cooling fan. See **Engine cooling fans** in this repair group.

> ***CAUTION—***
> *The viscous fan clutch nut has left-hand threads.*

– Disconnect hoses from thermostat housing. Unbolt thermostat housing from front of cylinder head.

Working at front of engine:

- Pry off drive belt tensioner cover.
- Lever tensioner in counterclockwise direction (**arrow**) and slip belt off coolant pump pulley.
- Unbolt pulley from pump.

***NOTE—***
*Mark drive belt rotation direction if reusing belt.*

– Remove coolant pump mounting bolts from pump.

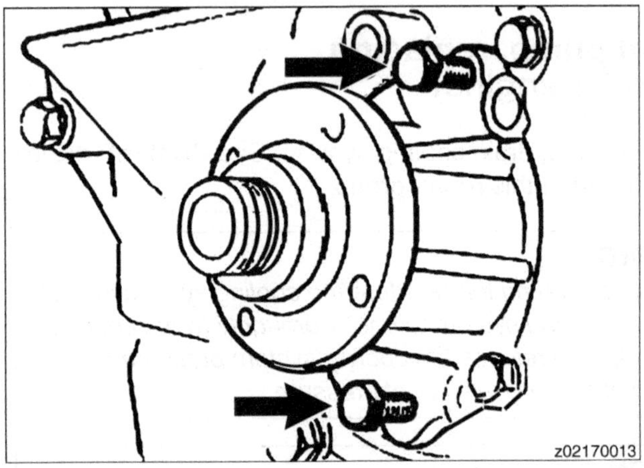


Insert two M6 bolts (**arrows**) in tapped bores and tighten uniformly until pump is freed from timing chain cover.

– Installation is the reverse of removal, noting the following

- Replace sealing O-ring and gaskets. Coat O-ring with water-based lubricant during installation.
- Fill and bleed cooling system. See **Cooling system, bleeding** in this repair group.

| **Tightening torques** | |
|---|---|
| Coolant pump pulley to coolant pump (M6) | 10 Nm (7 ft-lb) |
| Coolant pump to timing chain cover (M6) | 10 Nm (7 ft-lb) |

**Coolant thermostat housing (4-cylinder engine)**

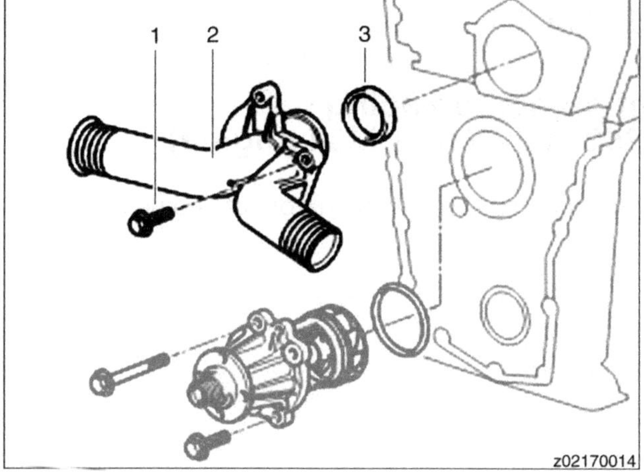


## Thermostat, replacing (4-cylinder engine)

On 4-cylinder engine, thermostat is integrated into thermostat housing. If thermostat requires replacement, replace complete thermostat housing.

1. M6 bolt
2. Thermostat housing with thermostat
3. Sealing O-ring

– With engine cold, partially drain cooling system. See **Coolant, draining and filling** in this repair group.

> ***WARNING —***
> *At normal operating temperature the cooling system is pressurized. Allow the system to cool thoroughly (a minimum of one hour), then release the cooling system pressure cap slowly to allow safe release of pressure.*

> ***CAUTION—***
> *When working on the cooling system, cover the alternator to protect it against coolant drips.*

– Remove belt-driven or engine side electric cooling fan. See **Engine cooling fans** in this repair group.

> ***CAUTION—***
> *The viscous fan clutch nut has left-hand threads.*

– Loosen hose clamps at thermostat housing and detach coolant hoses.

– Remove thermostat housing mounting bolts and remove housing with integral thermostat.

– Installation is the reverse of removal, noting the following:

- Replace sealing O-ring and hose clamps. Coat O-ring with water-based lubricant during installation.
- Fill and bleed cooling system. See **Cooling system, bleeding** in this repair group.

| **Tightening torques** | |
|---|---|
| Coolant hose clamp (32 - 48 mm) | 2.5 Nm (22 in-lb) |
| Thermostat housing to cylinder head | 10 Nm (7 ft-lb) |

# Coolant circulation components (M52, S52, M52TU or M54 engine)

## Coolant pump, replacing (M52, S52, M52 TU or M54 engine)

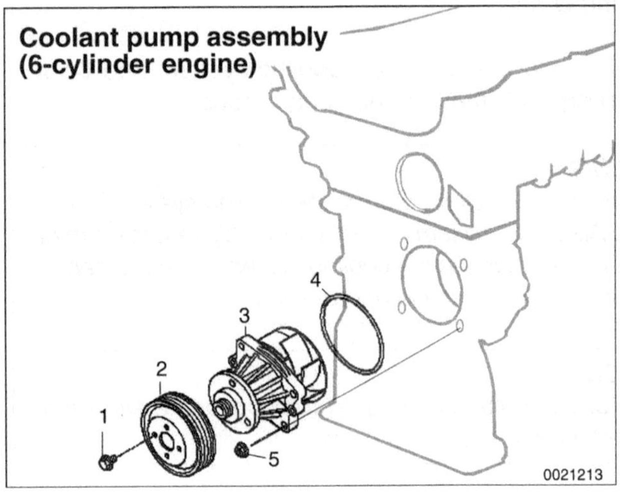


◄ Engine coolant pump is mounted in front of engine on timing chain cover.

1. M6 bolt
2. Pump pulley
3. Coolant pump
4. Sealing O-ring
5. M6 nut

– With engine cold, drain cooling system. See **Coolant, draining and filling** in this repair group.

> ***WARNING—***
> *At normal operating temperature the cooling system is pressurized. Allow the system to cool thoroughly (a minimum of one hour), then release the cooling system pressure cap slowly to allow safe release of pressure.*

> ***CAUTION—***
> *When working on the cooling system, cover the alternator to protect it against coolant drips.*

– Remove cooling fan and fan shroud. See **Mechanical (belt-driven) cooling fan, removing and installing** in this repair group.

> ***CAUTION—***
> *The viscous fan clutch nut has left-hand threads.*

– Remove coolant pump drive belt. See **020 Maintenance**.

***NOTE—***
*Mark drive belt rotation direction if reusing belt.*

◄ Remove coolant pump pulley bolts (**arrows**) and remove pulley from pump.

– Remove pump mounting nuts.

***NOTE—***
*The coolant pump is mounted on studs and retained by nuts.*

◄ Insert two M6 bolts (**arrows**) in tapped bores and tighten uniformly until pump is freed from timing chain cover (thermostat and hoses removed for clarity).

– Installation is reverse of removal, noting the following:

- Replace sealing O-ring and gaskets. Coat O-ring with water-based lubricant during installation.
- Fill and bleed cooling system. See **Cooling system, bleeding** in this repair group.

| **Tightening torques** | |
|---|---|
| Coolant pump pulley to coolant pump (M6) | 10 Nm (7 ft-lb) |
| Coolant pump to timing chain cover (M6) | 10 Nm (7 ft-lb) |

## Thermostat, replacing (M52, S52 engine)

The coolant thermostat is installed in a housing at the front of the engine.

– With engine cold, partially drain cooling system. See **Coolant, draining and filling** in this repair group.

> ***WARNING —***
> *At normal operating temperature the cooling system is pressurized. Allow the system to cool thoroughly (a minimum of one hour), then release the cooling system pressure cap slowly to allow safe release of pressure.*

> ***CAUTION—***
> *When working on the cooling system, cover the alternator to protect it against coolant drips.*

– Remove engine cooling fan and fan shroud. See **Mechanical (belt-driven) cooling fan, removing and installing** in this repair group.

> ***CAUTION—***
> *The viscous fan clutch nut has left-hand threads.*

– Unclip wiring duct retaining clips and move aside duct from above thermostat housing.

– Loosen hose clamps at thermostat housing and detach coolant hoses.

◄ Remove thermostat housing from front of engine. Remove thermostat. Note thermostat direction and orientation before removing.

– Installation is reverse of removal, noting the following:
  - Replace sealing O-ring and hose clamps. Coat O-ring with water-based lubricant during installation.
  - Fill and bleed cooling system. See **Cooling system, bleeding** in this repair group.

***NOTE —***
- *If the thermostat has a direction arrow or vent hole near the edge, install thermostat so that either the arrow or hole are at the top.*
- *Be sure to reconnect ground wire(s) at thermostat housing mounting bolt, where applicable.*

| **Tightening torques** | |
|---|---|
| Coolant hose clamp (32 - 48 mm) | 2.5 Nm (22 in-lb) |
| Engine lifting bracket to thermostat housing | 22 Nm (16 ft-lb) |
| Thermostat housing to cylinder head | 10 Nm (7 ft-lb) |

## Thermostat, replacing (M52 TU, M54 engine)

**Coolant thermostat housing (M52 TU, M54 engine)**

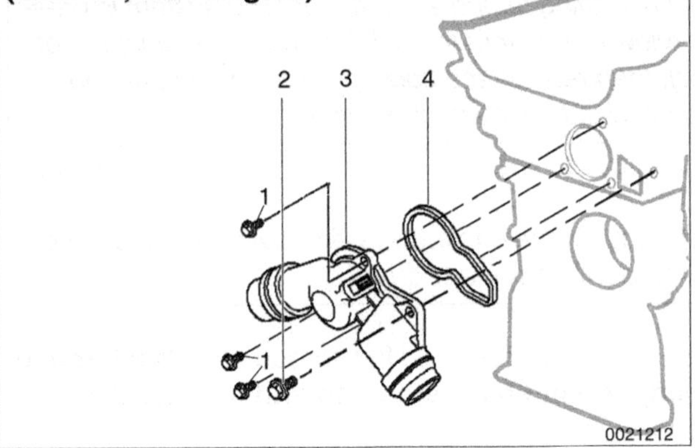


The electrically heated ECM-controlled ("characteristic map") thermostat is part of the thermostat housing at the front of the cylinder head. If the thermostat requires replacement, replace the complete thermostat housing.

1. M6 bolts
2. M8 bolt
3. Housing with integral thermostat
4. Seal

– If a faulty thermostat is suspected, use appropriate scan tool to interrogate ECM for stored fault codes.

– With engine cold, partially drain cooling system. See **Coolant, draining and filling** in this repair group.

> ***WARNING —***
> *At normal operating temperature the cooling system is pressurized. Allow the system to cool thoroughly (a minimum of one hour), then release the cooling system pressure cap slowly to allow safe release of pressure.*

> ***CAUTION—***
> *When working on the cooling system, cover the alternator to protect it against coolant drips.*

– Remove engine cooling fan and fan shroud. See **Mechanical (belt-driven) cooling fan, removing and installing** in this repair group.

> ***CAUTION—***
> *The viscous fan clutch nut has left-hand threads.*

– Disconnect electrical harness connector from thermostat housing.

0022355

< Lever out retaining clips (**arrows**) and pull hose fittings off thermostat housing.

– Unbolt and remove thermostat housing from front of engine. Loosen nut at top of engine lifting eye to facilitate removal.

– Installation is reverse of removal, noting the following:
  - Keep thermostat housing sealing faces free of oil.
  - Use new sealing gasket.
  - When reinstalling hose to thermostat housing, install clip on hose end fitting, then push hose onto housing until it clicks into place.
  - Fill and bleed cooling system. See **Cooling system, bleeding** in this repair group.

| **Tightening torques** | |
|---|---|
| Thermostat housing to cylinder head<br>• M6<br>• M8 | <br>10 Nm (7 ft-lb)<br>22 Nm (16 ft-lb) |

# Coolant circulation components (S54 engine)

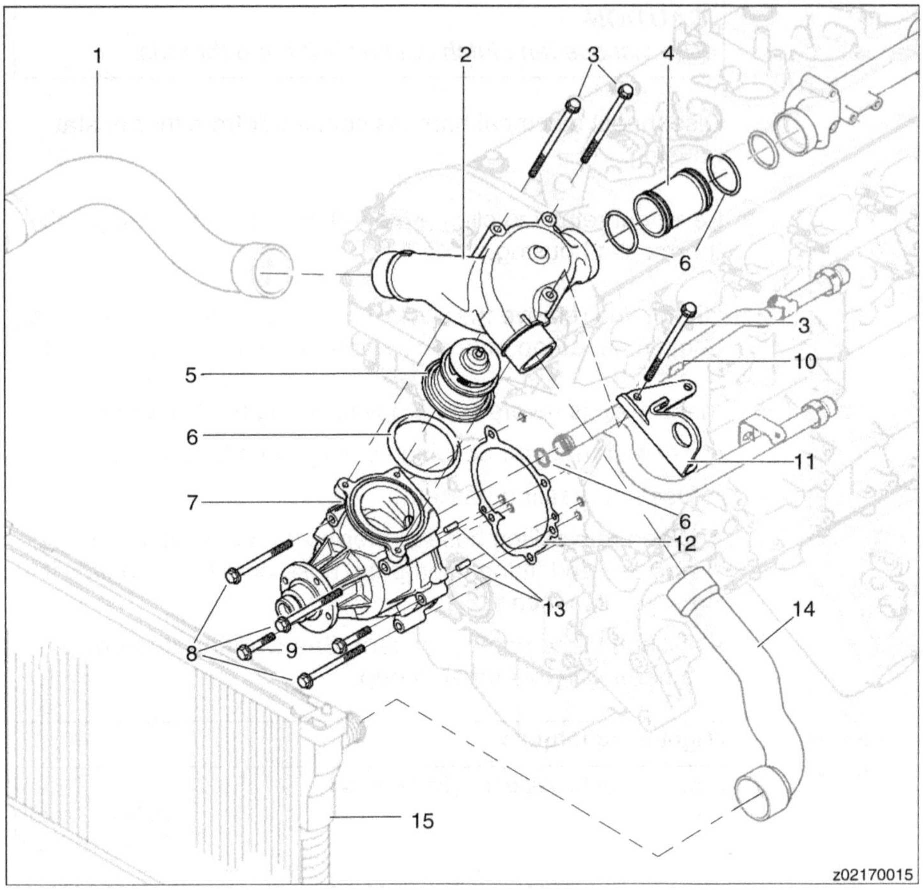


## Coolant pump components (S54 engine)

1. **Coolant hose from radiator**
2. **Thermostat housing cover**
3. **M6 x 85 bolt**
   - tighten to 10 Nm (7 ft-lb)
4. **Coolant return pipe**
5. **Thermostat**
6. **Sealing O-ring (replace)**
7. **Coolant pump housing**
8. **M6 x 70 bolt**
   - tighten to 10 Nm (7 ft-lb)
9. **M6 x 35 bolt**
   - tighten to 10 Nm (7 ft-lb)
10. **Coolant Y-pipe**
11. **Engine lifting bracket**
12. **Coolant pump gasket**
13. **Dowel pins**
14. **Coolant hose to radiator**
15. **Radiator**

## Coolant pump, replacing (S54 engine)

– With engine cold, drain cooling system. See **Coolant, draining and filling** in this repair group.

> ***WARNING —***
> *At normal operating temperature the cooling system is pressurized. Allow the system to cool thoroughly (a minimum of one hour), then release the cooling system pressure cap slowly to allow safe release of pressure.*

> ***CAUTION—***
> *When working on the cooling system, cover the alternator to protect it against coolant drips.*

– Remove cooling fan and fan shroud. See **Mechanical (belt-driven) cooling fan, removing and installing** in this repair group.

> ***CAUTION—***
> *The viscous fan clutch nut has left-hand threads.*

– Remove coolant pump drive belt. See **020 Maintenance**.

***NOTE—***
*If reusing belt, mark direction of rotation.*

– Remove coolant pump pulley bolts and remove pulley from pump.

– Remove thermostat. See **Thermostat, replacing (S54 engine)** in this repair group.

< Remove coolant pump mounting bolts (**arrows**).

– Remove pump from engine block. Pry gently to disengage from coolant Y-pipe.

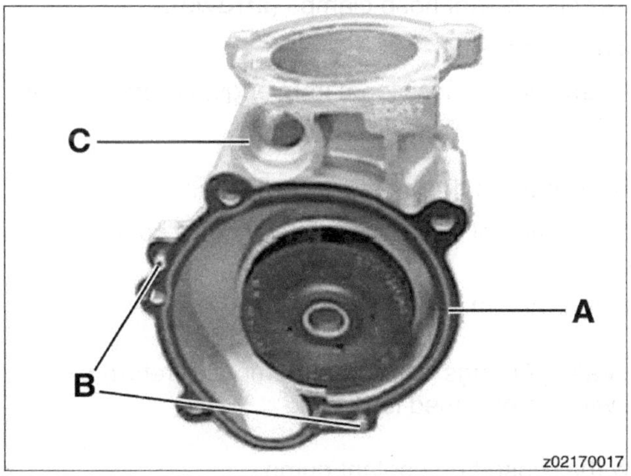


< When installing, make sure:

- New coolant pump gasket (**A**) is installed.
- Locating dowels (**B**) are undamaged and positioned correctly.
- Bore for Y-pipe connection (**C**) is clean. Install new sealing O-ring on end of Y-pipe and lubricate with water-based lubricant. Slide pump on pipe carefully.

> ***CAUTION—***
> *O-ring seal is easily damaged when installing pump.*

| **Tightening torque** | |
|---|---|
| Coolant pump to timing chain cover (M6) | 10 Nm (7 ft-lb) |

– Remainder of installation is reverse of removal. Fill and bleed cooling system. See **Cooling system, bleeding** in this repair group.

| **Tightening torques** | |
|---|---|
| Coolant pump pulley to coolant pump (M6) | 10 Nm (7 ft-lb) |
| Thermostat housing cover to coolant pump (M6) | 10 Nm (7 ft-lb) |

## Thermostat, replacing (S54 engine)

The coolant thermostat is installed in a housing at the top of the coolant pump, at the front of the engine.

– With engine cold, partially drain cooling system. See **Coolant, draining and filling** in this repair group.

> ***WARNING —***
> *At normal operating temperature the cooling system is pressurized. Allow the system to cool thoroughly (a minimum of one hour), then release the cooling system pressure cap slowly to allow safe release of pressure.*

> ***CAUTION—***
> *When working on the cooling system, cover the alternator to protect it against coolant drips.*

– Remove engine cooling fan and fan shroud. See **Mechanical (belt-driven) cooling fan, removing and installing** in this repair group.

> ***CAUTION—***
> *The viscous fan clutch nut has left-hand threads.*

– Remove air filter housing with mass air flow sensor. See **130 Fuel Injection**.

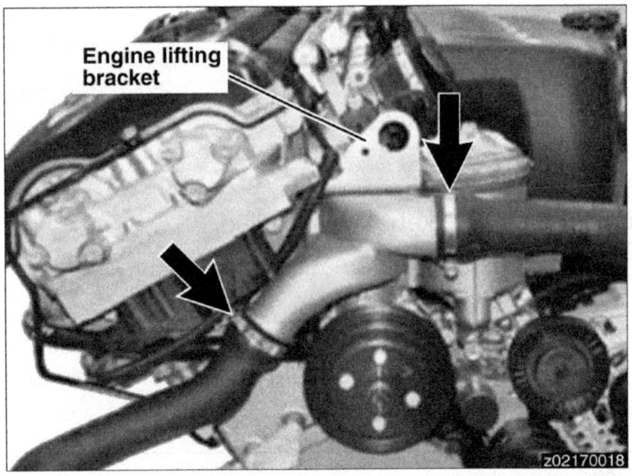


◄ Working at thermostat cover:

- Loosen or cut coolant hose clamps (**arrows**).
- Pull hoses off housing.
- Detach electrical harness loom from engine lifting bracket. Unbolt and remove bracket.

– Remove remainder of thermostat cover bolts.

- Remove thermostat cover, disengaging from coolant return pipe.
- Remove thermostat.

– Replace sealing O-rings on coolant pump and return pipe. Lubricate with water-based lubricant.

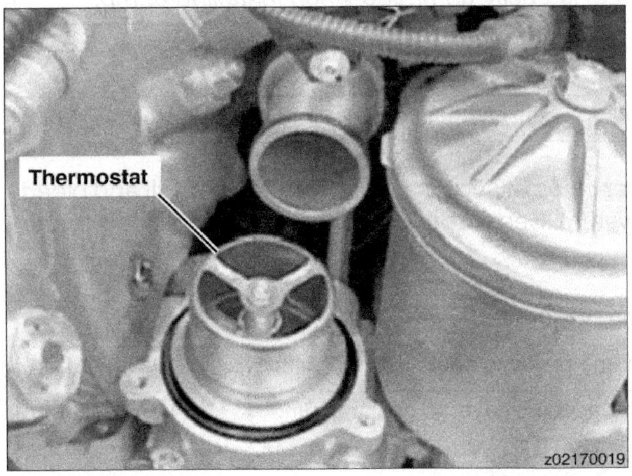


◄ Place new thermostat on coolant pump.

– Remainder of installation is reverse of removal. Fill and bleed cooling system. See **Cooling system, bleeding** in this repair group.

| **Tightening torques** | |
|---|---|
| Coolant hose clamp (32 - 48 mm) | 2.5 Nm (22 in-lb) |
| Thermostat housing to cylinder head | 10 Nm (7 ft-lb) |

## RADIATOR SERVICE

### Radiator, removing and installing

Radiator removal procedures for 4-cylinder and 6-cylinder models is similar.

– Raise front of car and support safely.

> ***WARNING —***
> *Make sure the car is stable and well supported at all times. Use a professional automotive lift or jack stands designed for the purpose. A floor jack is not adequate support.*

– Remove splash shield from under engine.

– With engine cold, drain cooling system. See **Coolant, draining and filling** in this repair group.

> ***WARNING —***
> *At normal operating temperature the cooling system is pressurized. Allow the system to cool thoroughly (a minimum of one hour), then release the cooling system pressure cap slowly to allow safe release of pressure.*

> ***CAUTION—***
> *When working on the cooling system, cover the alternator to protect it against coolant drips.*

– Remove air filter housing. See **130 Fuel Injection**.

– Remove belt-driven or engine side electric cooling fan. See **Engine cooling fans** in this repair group.

> ***CAUTION—***
> *The viscous fan clutch nut has left-hand threads.*

– Disconnect cooling fan harness connector (if applicable) from bottom of radiator.

0012506

◄ Detach harness connector from dual temperature cooling fan switch (**arrow**) at right front of radiator.

– Disconnect coolant hoses from radiator.

– 6-cylinder model with separate coolant expansion tank: Detach expansion tank hose from top left of radiator.

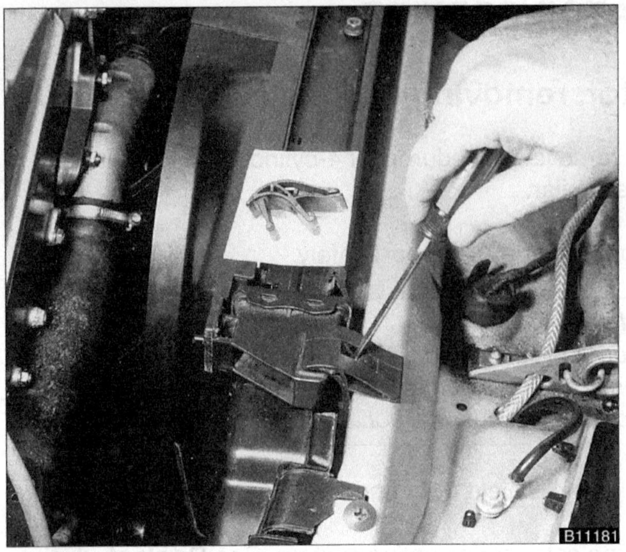


⮜ Use screwdriver to push down and pull forward to release radiator retaining clip.

– Slide top of radiator back toward engine and pull radiator up to remove.

***NOTE—***
*Radiator sits on two rubber mounts. Check to make sure the mounts do not stick to the bottom of the radiator.*

– Installation is reverse of removal, noting the following:
- Use new hose clamps as needed.
- Fill and bleed cooling system. See **Cooling system, bleeding** in this repair group.

# 180 Exhaust System

## General

This repair group covers exhaust system component replacement, exhaust manifold removal and installation and oxygen sensor replacement.

Procedures in this section are grouped by engine or engine management system code. See **100 Engine–General** for application information.

### Special tools

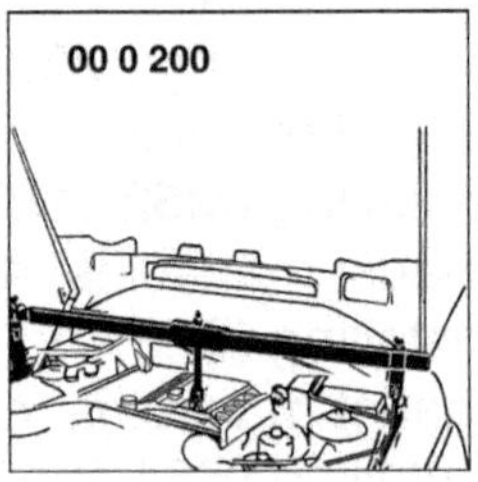


◀ Engine lifting tools
(BMW special tool kit 00 0 200)

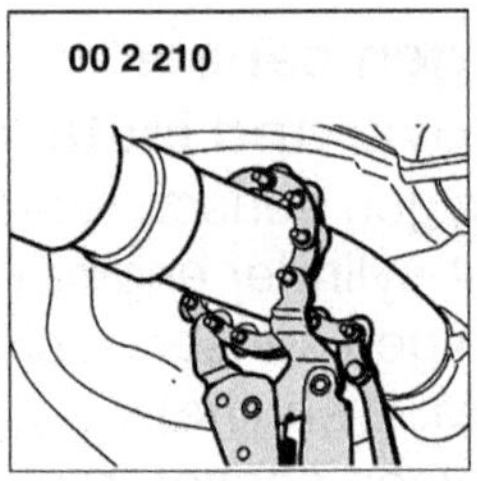


◀ Pipe cutter
(BMW special tool 00 2 210)

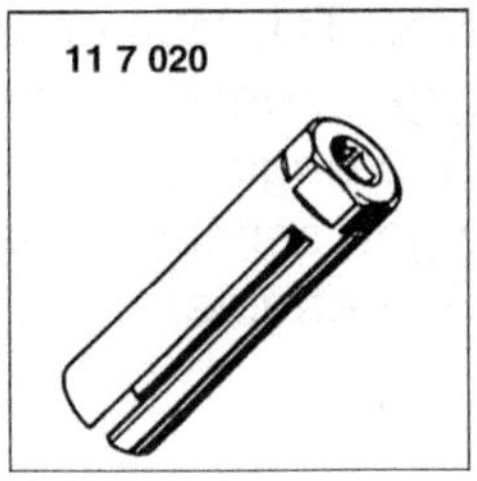


◀ Oxygen sensor socket
(BMW special tool 11 7 020)

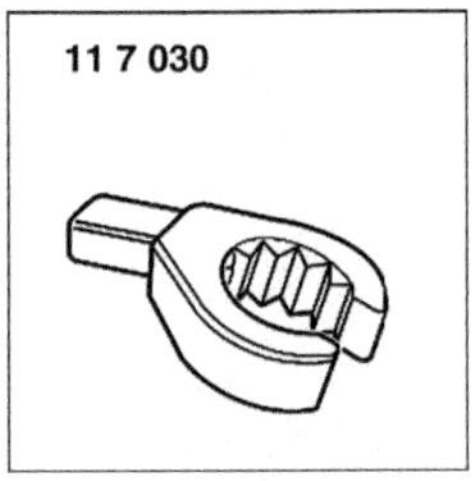


◀ Oxygen sensor crow-foot wrench
(BMW special tool 11 7 030)

◀ Universal extension (use with crow-foot wrench)
(BMW special tool 11 9 150)

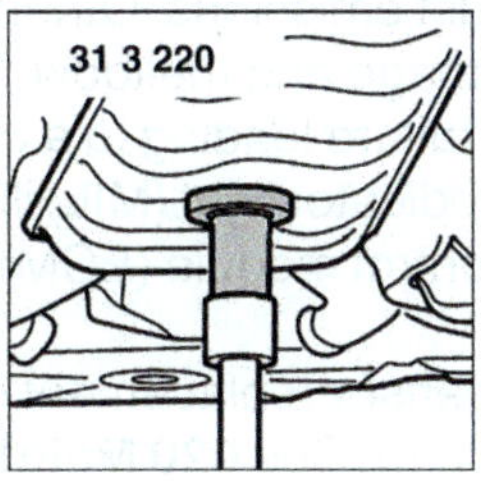


Exhaust system support (BMW special tool 31 3 220)

## Exhaust manifolds, catalytic converters and oxygen sensors

The exhaust system is designed to be maintenance free, although regular inspection is warranted due to the harsh operating conditions.

Exhaust manifolds and exhaust system components for all models are stainless steel.

- **1996 - 1998 4-cylinder Z3 (M44 engine)**: Catalytic converter is incorporated into front exhaust pipe, behind exhaust manifold. Oxygen sensors are threaded into front pipe fore and aft of catalytic converter.
- **1997 - 1998 6-cylinder Z3 (M52 engine)**: catalytic converter is incorporated into dual front exhaust pipe, behind exhaust manifolds. Oxygen sensors are threaded into front pipe fore and aft of catalytic converter.
- **1999 - 2002 Z3 (M52 TU** or **M54 engine)**: Exhaust manifolds incorporate catalytic converters. Two precatalyst and two post-catalyst oxygen sensors are mounted in exhaust manifold assembly.
- **1998 - 2000 M version (S52 engine)**: Exhaust manifolds identical to those in M52 models. However, two precatalyst oxygen sensors are screwed into tops of the exhaust manifolds. Dual front exhaust pipes are equipped with one catalytic converter in each pipe and post-catalyst oxygen sensor behind each catalyst.
- **2001 - 2002 M version (S54 engine)**: Exhaust manifolds incorporate catalytic converters. Two precatalyst and two post-catalyst oxygen sensors are mounted in exhaust manifold assembly.

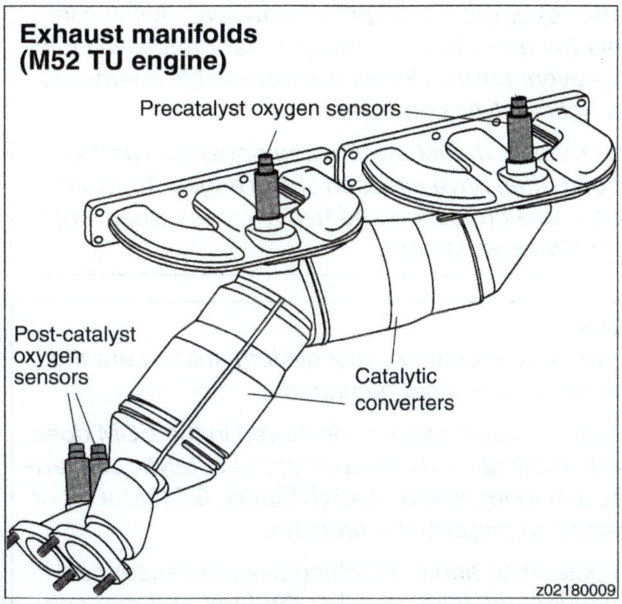


Mounting the catalytic converters close to the engine, as is done in the M52 TU, M54 or S54 engine, allows them to come up to operating temperature quickly.

See **Exhaust system diagrams** in this repair group for exhaust system components and details. See **100 Engine–General** and **130 Fuel Injection** for additional information on engine emissions, catalytic converters and oxygen sensors.

Under normal conditions, catalytic converters do not require replacement unless damaged. OBD II diagnostic software closely monitors fuel injection and ignition system to maintain precisely correct (calculated) fuel mixture. It is only under these precise conditions that the catalytic converter can control exhaust emissions. Otherwise the catalyst may sustain internal damage and meltdown. In case of hardware or software failure leading to a catalyst damaging fault, the malfunction indicator light (MIL) illuminates and a fault is set in the engine control module (ECM).

Oxygen sensor replacement is required every 100,000 miles (160,000 km). See **020 Maintenance**.

For more information on performance problems related to oxygen sensor failure, catalyst damaging faults, or catalyst failure, see **OBD On-Board Diagnostics**.

## Warnings and cautions

Observe the following warnings and cautions when servicing the exhaust system.

***WARNING—***

- *Exhaust gases are colorless, odorless, and very toxic. Run the engine only in a well-ventilated area. Immediately repair any leaks in the exhaust system. Also repair structural damage to the car body that might allow exhaust gases to enter the passenger compartment.*
- *The exhaust manifold(s), exhaust system and catalytic converter(s) operate at high temperatures. Allow components to cool before servicing. Wear protective clothing to prevent burns. Do not use flammable chemicals near a hot catalytic converter.*
- *Old, corroded exhaust system components crumble easily and often have exposed sharp edges. To avoid injury, wear eye protection and heavy gloves when working with old exhaust parts.*

***CAUTION—***

- *When working on the exhaust system, make sure the oxygen sensors are not damaged.*
- *A catalyst damaging fault code stored in the ECM does NOT automatically indicate a defective catalytic converter. Use a diagnostic scan tool (DISplus, GT1, MoDiC or equivalent) to pinpoint the problem.*
- *When detaching and reattaching oxygen sensor harness connectors, take care that the front and rear connectors are not mixed up.*

## Exhaust System Diagrams

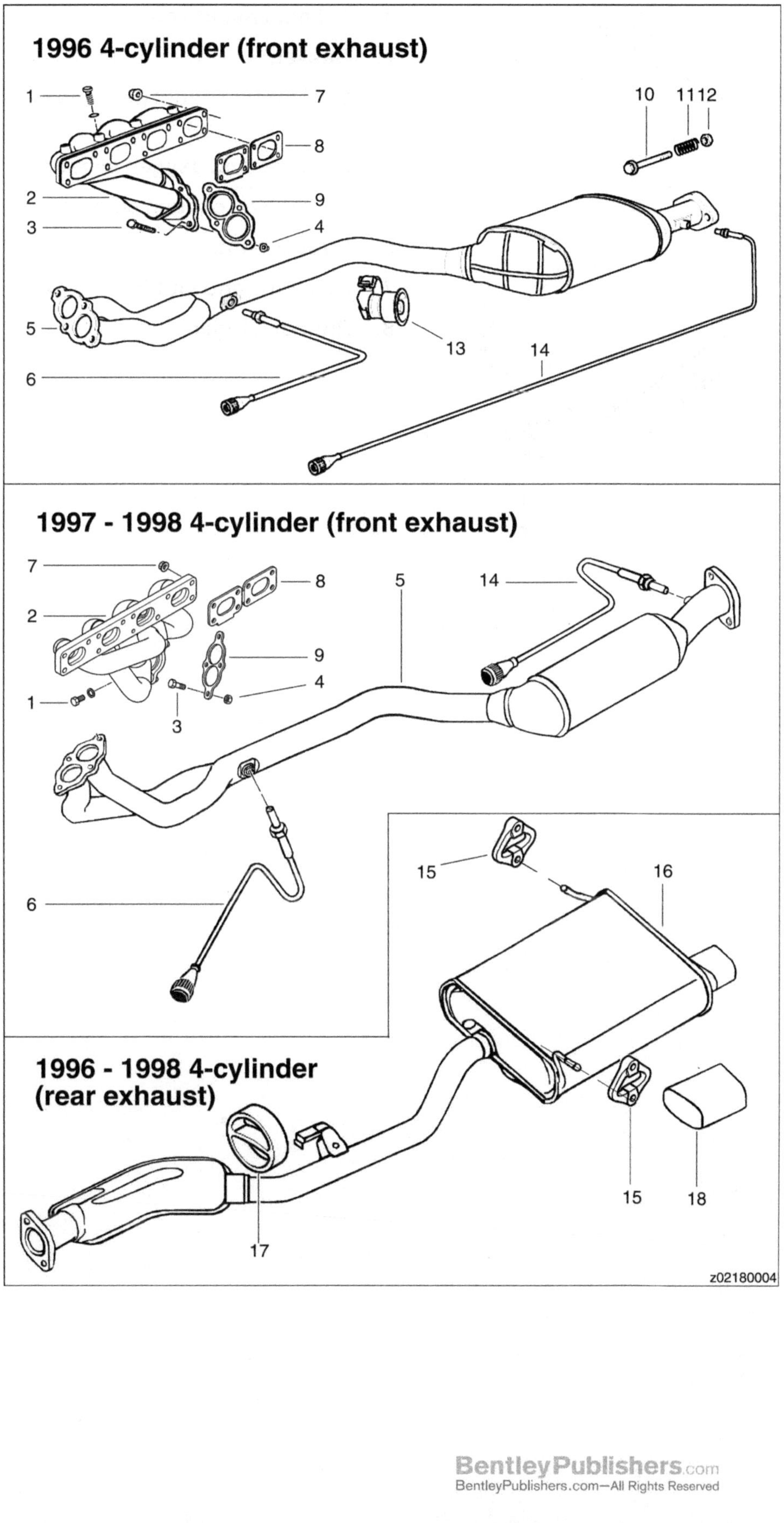


### 1996 - 1998 Z3 4-cylinder (M44 engine)

1. Plug (M8 bolt)
2. Exhaust manifold
3. M10 bolt
4. M10 nut
5. Front pipe and catalytic converter
6. Precatalyst oxygen sensor
7. M7 nut
8. Gasket
9. Gasket
10. M8 bolt
11. Spring
12. M8 nut
13. Vibration damper
14. Post-catalyst oxygen sensor
15. Rubber hanger
16. Rear pipe and muffler
17. Rubber hanger
18. Tailpipe extension

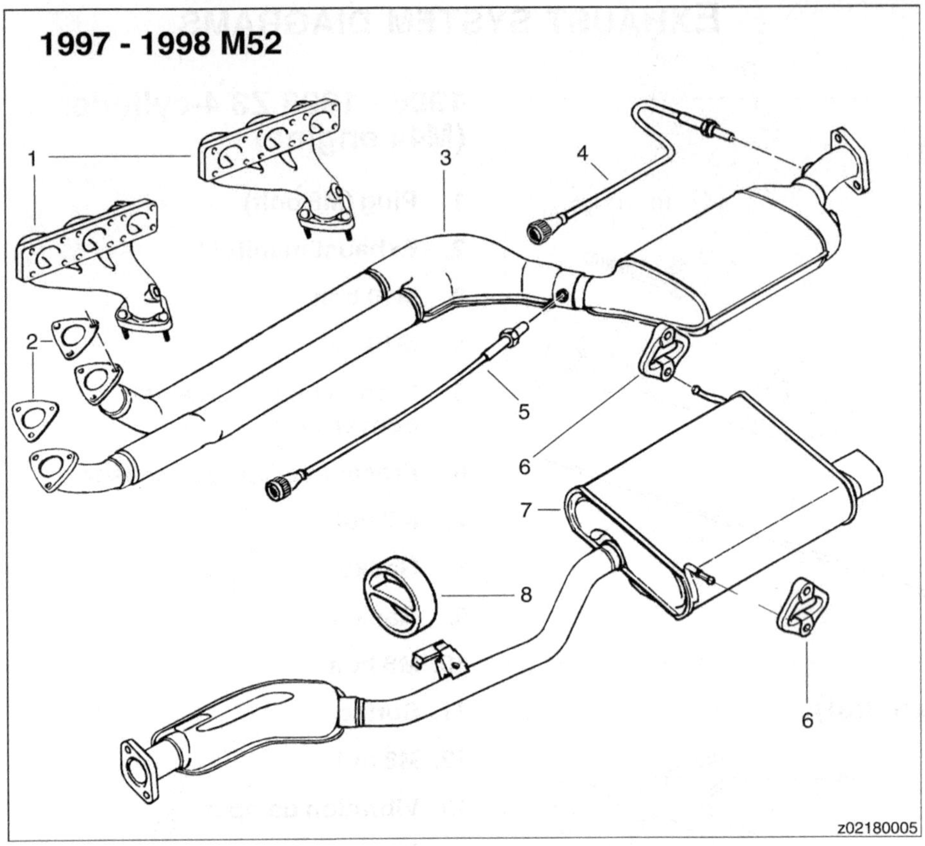


## 1997 - 1998 Z3 6-cylinder (M52 engine)

1. Exhaust manifolds
2. Gaskets
3. Dual front exhaust pipe and catalytic converter
4. Post-catalyst oxygen sensor
5. Precatalyst oxygen sensor
6. Rubber hanger
7. Rear pipe and muffler
8. Rubber hanger

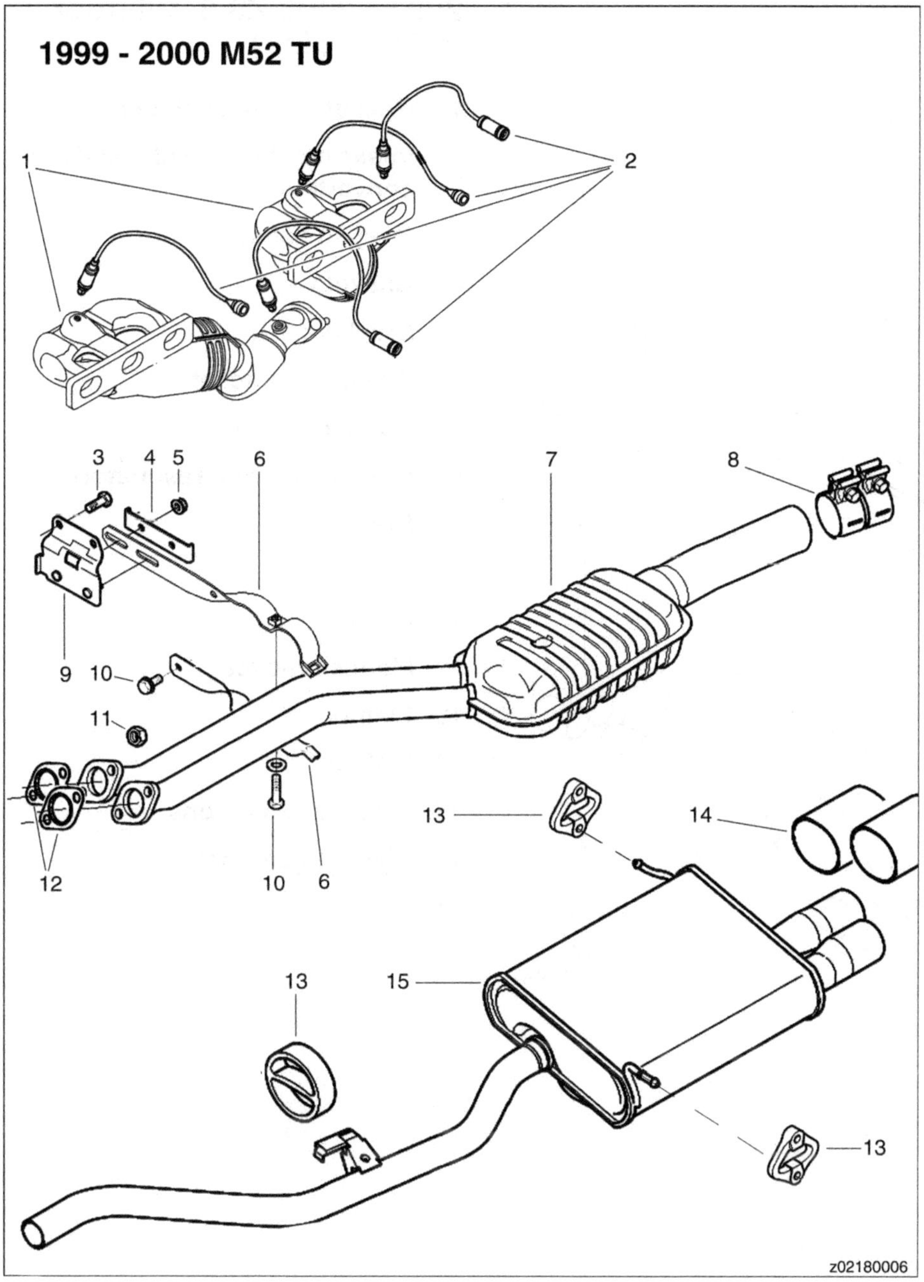


## 1999 - 2000 Z3 6-cylinder (M52 TU engine)

1. **Intake manifolds with catalytic converters**
2. **Oxygen sensors**
3. **M8 Torx bolt**
4. **Bracket**
5. **M8 nut**
6. **Clamping bracket**
7. **Front pipe and resonator**
8. **Clamp**
9. **Transmission bracket**
10. **M8 bolt**
11. **M10 nut**
12. **Gaskets**
13. **Rubber hanger**
14. **Tailpipe extension**
15. **Rear pipe and muffler**

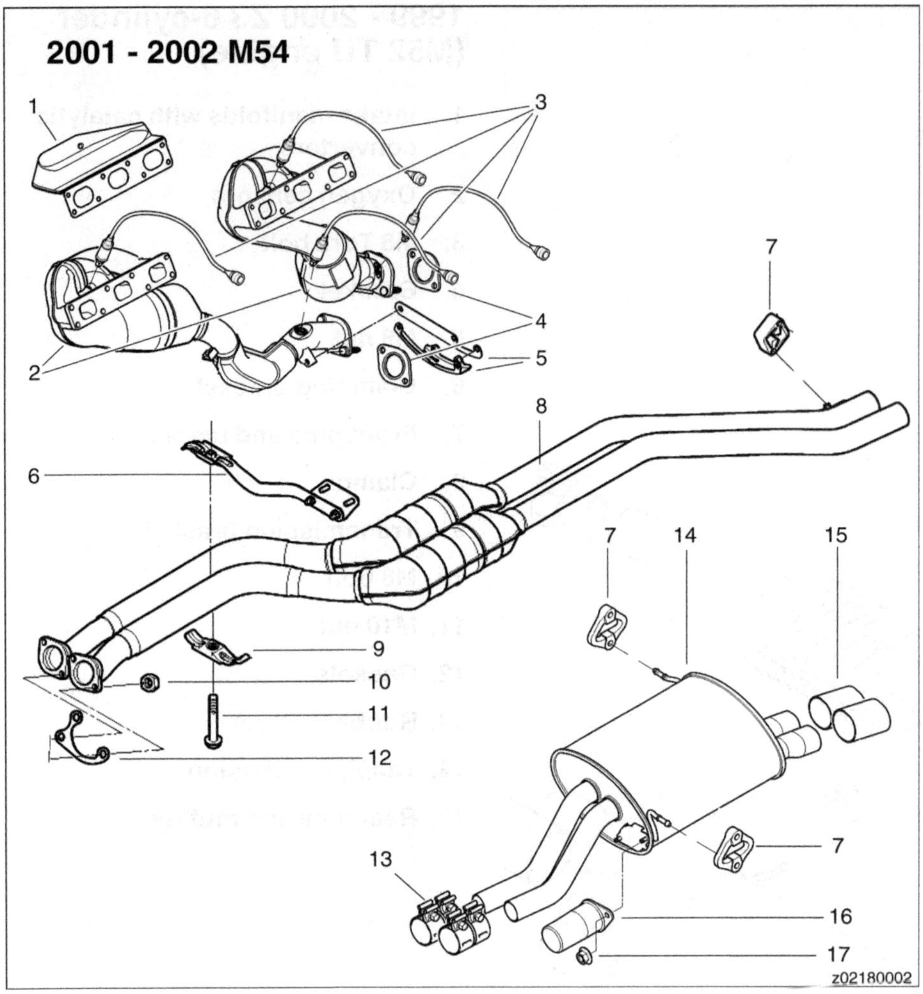


## 2001 - 2002 Z3 6-cylinder (M54 engine)

1. **Gasket with heat shield**
2. **Intake manifolds with catalytic converters**
3. **Oxygen sensors**
4. **Gaskets**
5. **Brackets**
6. **Clamping bracket**
7. **Rubber hanger**
8. **Front pipes with resonators**
9. **Bracket**
10. **M10 nut**
11. **M8 bolt**
12. **Flange bracket**
13. **Clamps**
14. **Muffler**
15. **Tailpipe extensions**
16. **Vibration damper**
17. **M8 nut**

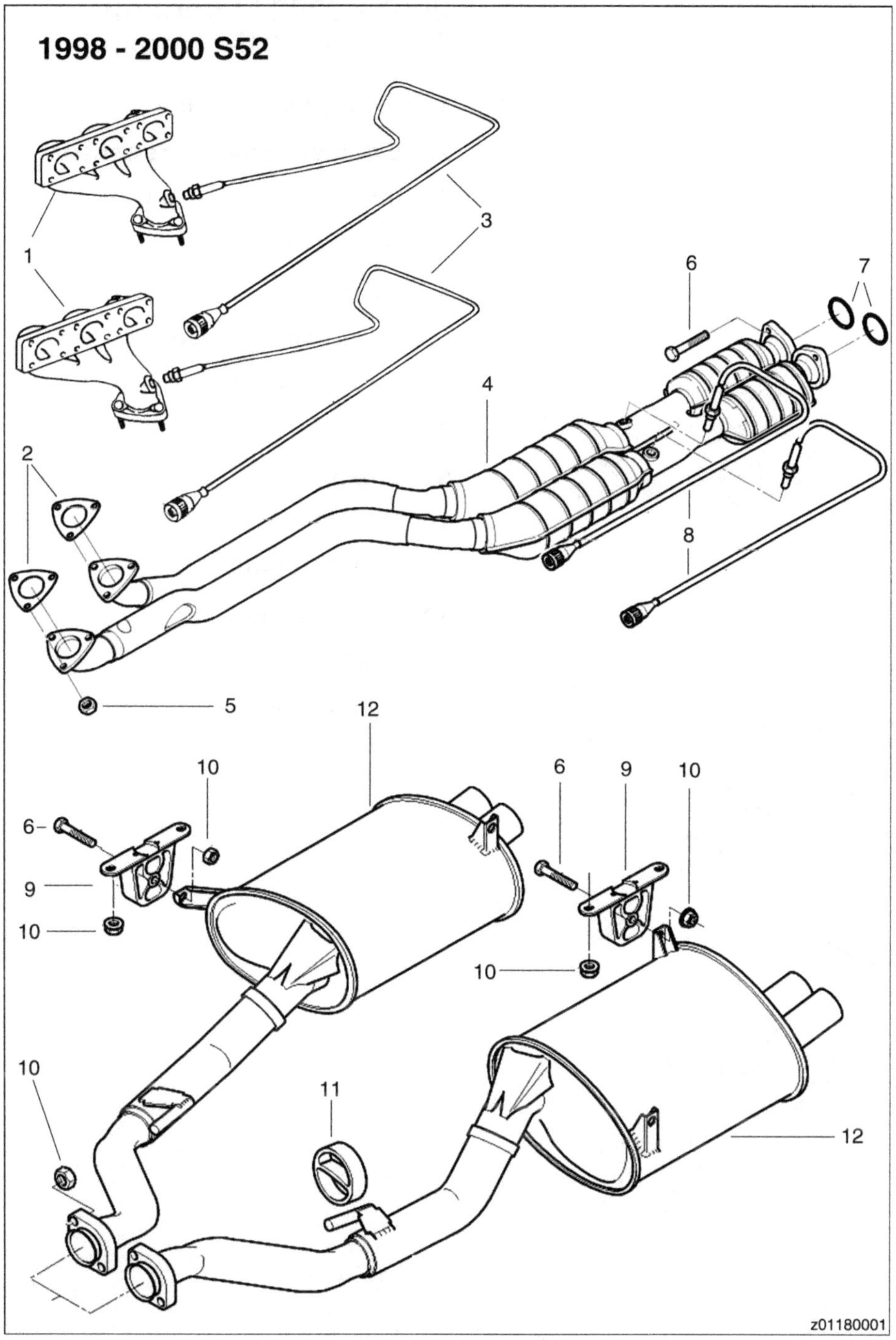


## 1998 - 2000 M version 6-cylinder (S52 engine)

1. Exhaust manifolds
2. Gaskets
3. Precatalyst oxygen sensors
4. Front pipes and catalytic converters
5. M10 nut
6. M8 bolt
7. O-ring seals
8. Post-catalyst oxygen sensors
9. Rubber hanger
10. M8 nut
11. Rubber hanger
12. Rear muffler

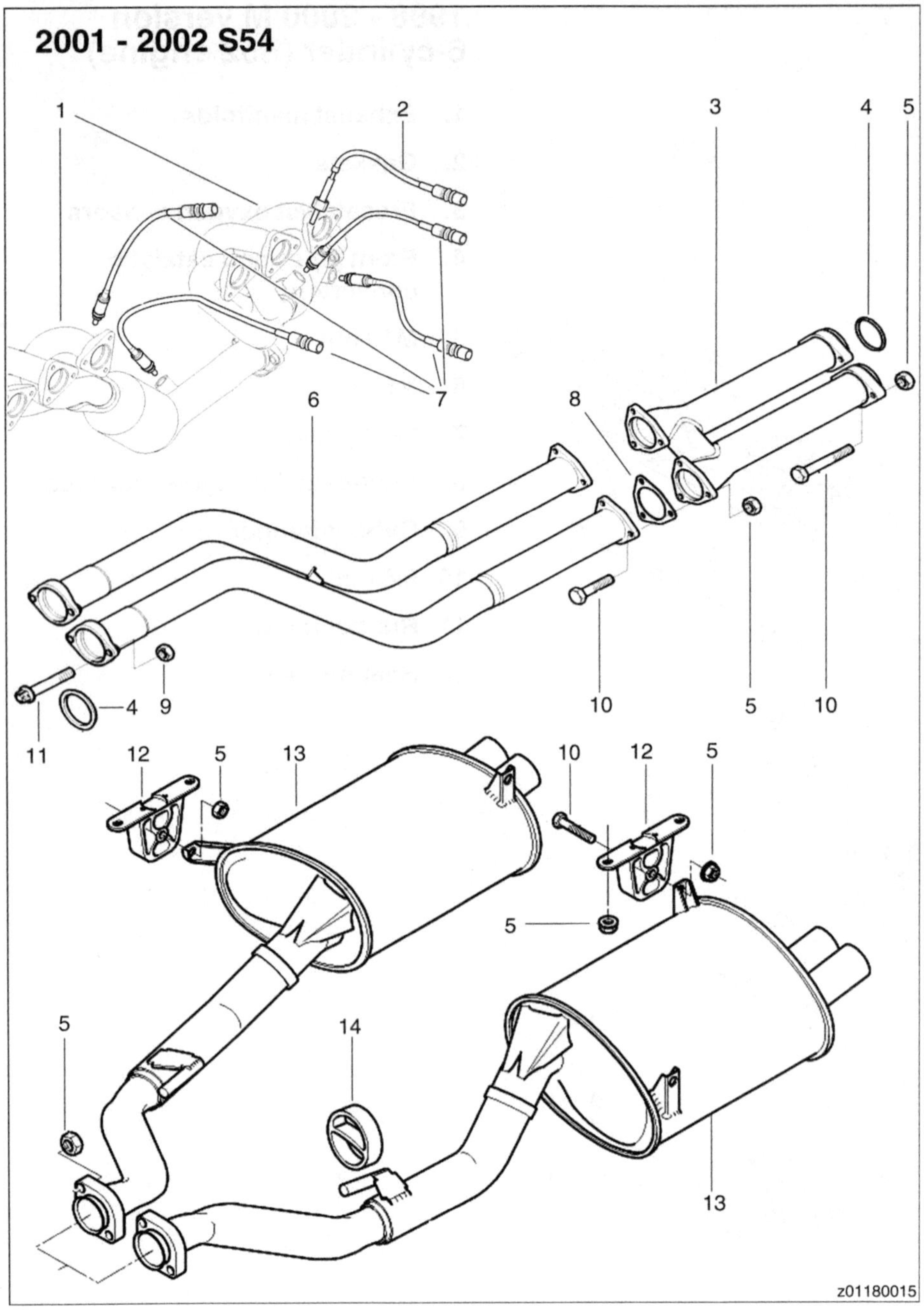


## 2001 - 2002 M version 6-cylinder (S54 engine)

1. **Exhaust manifolds**
2. **Exhaust temperature sensor**
3. **Intermediate pipe set**
4. **Gasket ring**
5. **M8 nut**
6. **Front exhaust pipe set**
7. **Oxygen sensors**
8. **Gasket**
9. **M10 nut**
10. **M8 bolt**
11. **M10 Torx bolt**
12. **Rubber mount**
13. **Muffler**
14. **Rubber hanger**

# EXHAUST SYSTEM REPLACEMENT

**Exhaust system removal:** Apply penetrating oil liberally to the exhaust system nuts and bolts in advance to make removal easier.

**Exhaust system installation:** Use new fasteners, clamps, rubber mounts and gaskets when replacing exhaust components.

Be sure to read **Warnings and cautions** in this repair group before starting work on the exhaust system.

## Exhaust system, removing and installing

The removal and installation procedures given here are general procedures for all models. Remove the exhaust system as a complete unit, then replace individual pipes and mufflers.

See **Exhaust system diagrams** in this repair group for components and layout for a specific model.

– With exhaust system fully cold, raise and support car for access to exhaust system.

> ***WARNING —***
> *Make sure the car is stable and well supported at all times. Use a professional automotive lift or jack stands designed for the purpose. A floor jack is not adequate support.*

– M44, M52, or S52 engine: Disconnect oxygen sensor connector(s).

– Remove body reinforcement braces from underneath exhaust system.

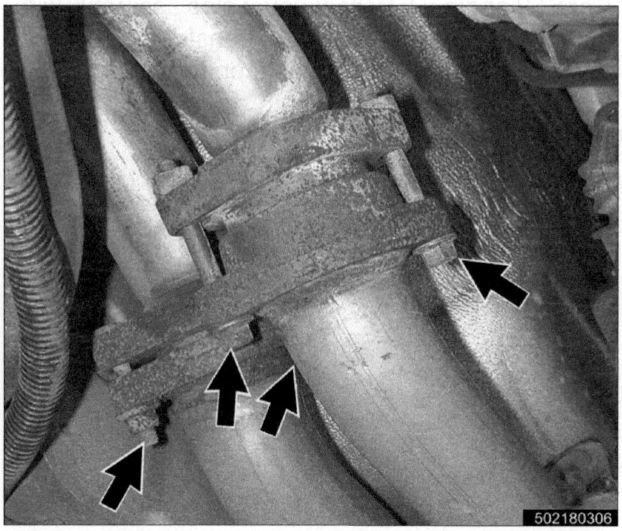
502180306

◄ Loosen and remove bolts (**arrows**) holding front exhaust pipe(s) to exhaust manifold(s).

– Support exhaust system.

> ***WARNING —***
> *The exhaust system is heavy. Do not loosen the exhaust system unless it is solidly supported.*

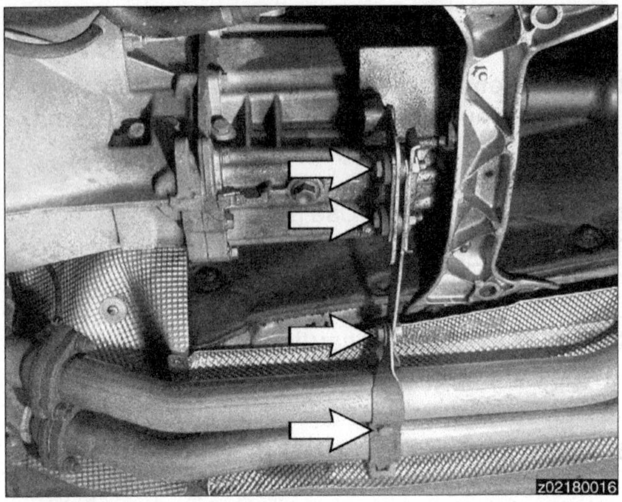

z02180016

< Remove exhaust support bracket mounting bolts (**arrows**) at transmission and set bracket aside.

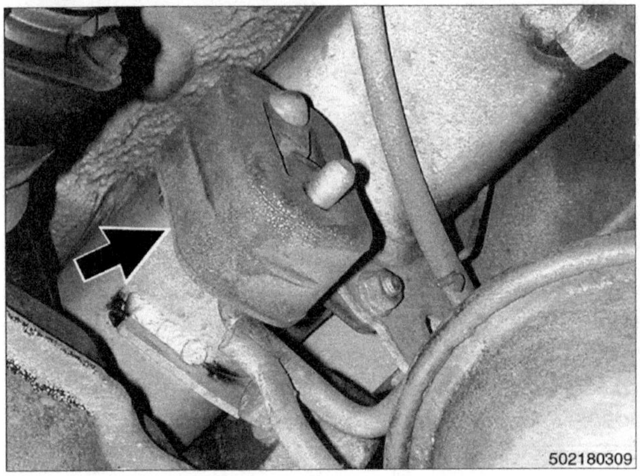

502180309

< Remove rubber hanger (**arrow**) by prying off bracket.

z02180017

< S52 or S54 engine: Detach rear muffler from underside of car by removing rubber hanger mounting nuts (**arrows**).

- Lower exhaust system as a single unit to ground.

- M44, M52, or S52 engine: If front exhaust pipe assembly with catalytic converter(s) is to be replaced, transfer oxygen sensor(s) to new pipe(s). Use copper paste on threads.

| Tightening torque | |
|---|---|
| Oxygen sensor to front pipe | 50 Nm (37 ft-lb) |

- Transfer muffler vibration damper to new muffler, if applicable.

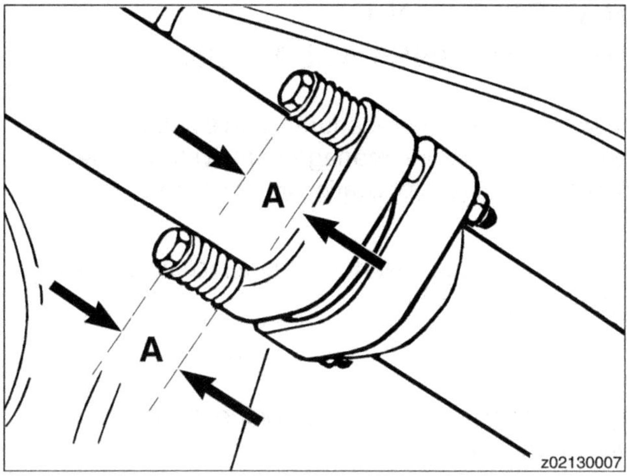


1996 4-cylinder model:

- Assemble front and rear pipes to each other using spring loaded fasteners.
- Tighten flange nuts evenly until correct compressed spring length is obtained.

| **Spring preload** | |
|---|---|
| Final spring length (**A**) | approx. 30 mm (1¼ in) |

– Install complete exhaust system to car, using new fasteners, gaskets and hangers. Bear in mind the following:

- Replace exhaust hanger bracket assembly components at transmission as necessary.
- Coat exhaust fasteners with copper paste before installing.
- Loosely install all exhaust fasteners and hangers before tightening fasteners to final torque.
- Make sure there is at least 20 mm (¾ in) clearance between exhaust system and car body at every point. Rubber hanger mounting points are adjustable for this purpose.

| **Tightening torque** | |
|---|---|
| Front exhaust pipe to exhaust manifold<br>• M10<br>• M10 Torx | <br>30 Nm (22 ft-lb)<br>24 Nm (18 ft-lb) |

## Muffler, replacing

BMW offers individual mufflers as replacement parts for the original one-piece assembly. Replacement of a rear muffler may be possible with the system installed on the car. Replacement of the front or middle muffler should only be done with the exhaust system removed from the car.

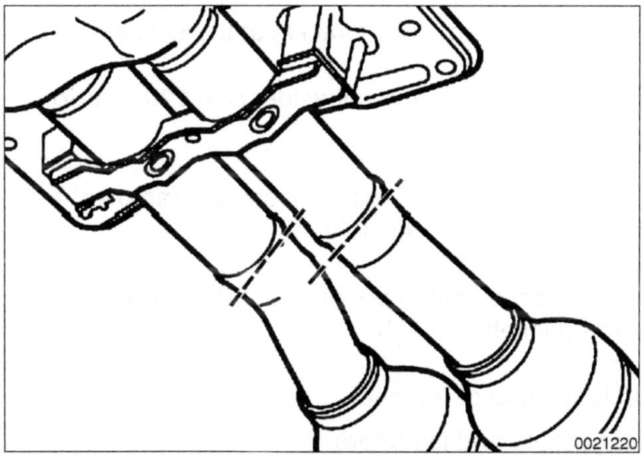


Mark length of new muffler pipe(s) on old system.

– Use exhaust pipe cutter to cut pipe(s) to proper length.

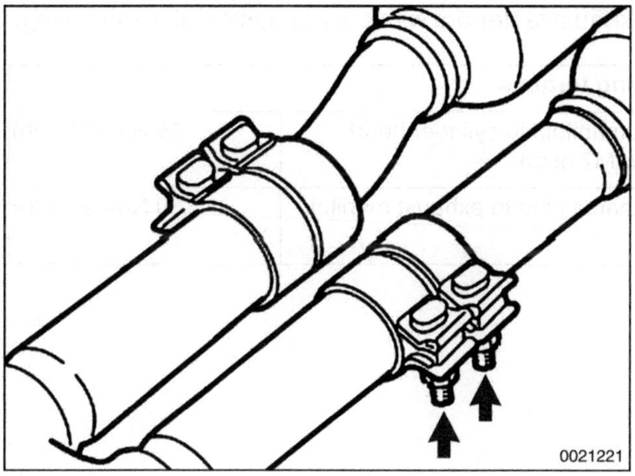


Use clamping sleeves to join new pipe(s) to old.

- Make sure pipe ends are centered inside sleeve.
- Point clamping sleeve bolt threads (**arrows**) down.

– Reinstall exhaust system, if removed.

## EXHAUST MANIFOLD REMOVAL AND INSTALLATION

Exhaust manifold removal procedures in this section are grouped by engine or engine management system code. See **100 Engine–General** for application information.

### Exhaust manifold, removing and installing (4-cylinder engine)

> ***WARNING —***
> *To avoid personal injury, be sure the engine is cold before starting the procedure.*

– Make sure ignition is OFF. Remove ignition key.

– Raise and support car for access to exhaust system.

> ***WARNING —***
> *Make sure the car is stable and well supported at all times. Use a professional automotive lift or jack stands designed for the purpose. A floor jack is not adequate support.*

– Unbolt front exhaust pipe from exhaust manifold. See **Exhaust system, removing and installing** in this repair group.

– Detach front exhaust pipe from bracket at transmission.

– Remove ignition coil pack and set aside. See **120 Ignition System**.

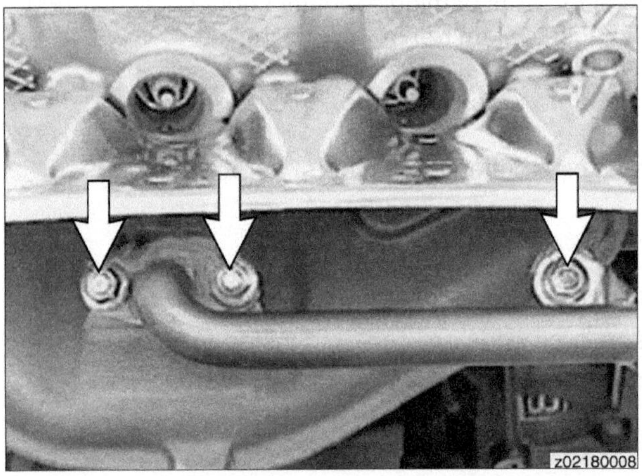


◄ Remove secondary air injection pipe fasteners (**arrows**) and detach pipe from exhaust manifold.

– Remove manifold mounting nuts and lift out manifold. Discard nuts and gaskets.

– Installation is reverse of removal.
- Use new gaskets and fasteners.
- Coat exhaust fasteners with copper paste before installing.

| Tightening torques | |
|---|---|
| Exhaust manifold to cylinder head (use new M7 nuts) | 15 Nm (11 ft-lb) |
| Front exhaust pipe to exhaust manifold (M10) | 30 Nm (22 ft-lb) |

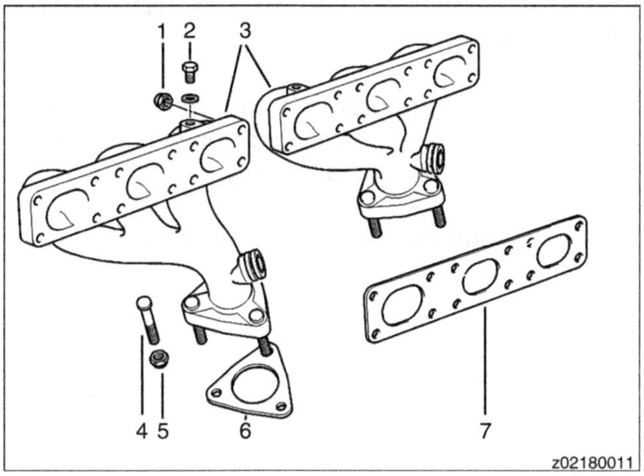


## Exhaust manifolds, removing and installing (M52 engine)

< Exhaust manifold components for M52 engine:

1. M7 nut
2. Plug
3. Exhaust manifolds
4. M10 stud
5. M10 nut
6. Gasket
7. Gasket

> ***WARNING —***
> *To avoid personal injury, be sure the engine is cold before starting the procedure.*

– Make sure ignition is OFF. Remove ignition key.

– Remove windshield washer fluid reservoir from right front of engine compartment. See **611 Wipers and Washers**.

– Raise and support car for access to exhaust system.

> ***WARNING —***
> *Make sure the car is stable and well supported at all times. Use a professional automotive lift or jack stands designed for the purpose. A floor jack is not adequate support.*

– Unbolt front exhaust pipe from exhaust manifolds. See **Exhaust system, removing and installing** in this repair group.

– Detach front exhaust pipe from bracket at transmission.

– Remove secondary air injection pipe and check valve fasteners and detach from exhaust manifold.

– Remove manifold mounting nuts and lift out manifolds. Discard nuts and gaskets.

– Installation is reverse of removal.

- Use new gaskets and fasteners.
- Coat exhaust fasteners with copper paste before installing.
- Install manifold gaskets with arched side facing manifolds.

| **Tightening torques** | |
|---|---|
| Exhaust manifold to cylinder head (use new M7 nuts) | 20 Nm (15 ft-lb) |
| Front exhaust pipe to exhaust manifold (M10) | 30 Nm (22 ft-lb) |

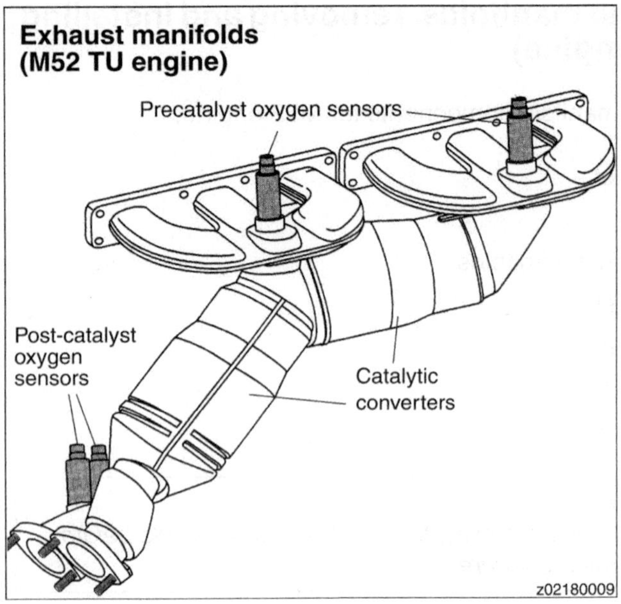


## Exhaust manifolds, removing and installing (M52 TU, M54 engine)

The exhaust manifolds incorporate the catalytic converters. Two precatalyst and two post-catalyst oxygen sensors are mounted in the exhaust manifold assembly.

Under normal conditions, the catalytic converters do not require replacement unless damaged.

> ***WARNING —***
> *To avoid personal injury, be sure the engine is cold before starting the procedure.*

- Make sure ignition is OFF. Remove ignition key.
- Remove windshield washer fluid reservoir from right front of engine compartment. See **611 Wipers and Washers**.
- Remove air filter housing and ducts. See **130 Fuel Injection**.

Remove fuel injector cover by removing plastic trim plugs (**arrows**) and removing fasteners under them. Lift off cover.

Disconnect post-catalyst oxygen sensor connectors (**arrows**) along left side of cylinder head cover.

***NOTE —***
*Mark connectors for later reassembly.*

Working at right side of engine above exhaust manifolds, remove precatalyst oxygen sensor harness connectors (**arrows**) from mounting brackets, then separate plugs.

– Remove oxygen sensors from intake manifolds and store safely. See **Oxygen sensors, replacing (M52 TU, M54 engine)** in this repair group.

– Remove secondary air system plumbing from right side of engine and from exhaust manifolds.

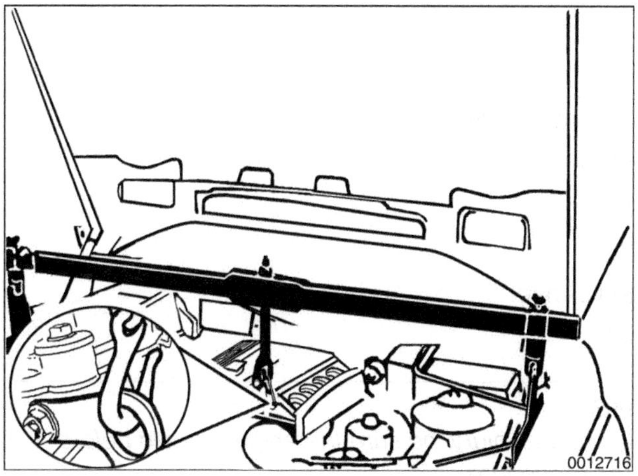


Connect engine lifting tool (BMW special tool set 00 0 200 or equivalent) to lifting eye on front of engine. Raise engine approx. 5 mm (¼ in).

– Raise and support car for access to exhaust system.

> ***WARNING —***
> *Make sure the car is stable and well supported at all times. Use a professional automotive lift or jack stands designed for the purpose. A floor jack is not adequate support.*

– Working underneath car, remove splash shield under engine compartment.

– Unbolt front exhaust pipe(s) from exhaust manifolds. See **Exhaust system, removing and installing** in this repair group.

– Detach front exhaust pipe(s) from bracket at transmission.

With engine supported from above, remove right engine mount and right engine mount bracket (**arrow**).

– Remove manifold mounting nuts and lift out manifolds. Discard nuts and gaskets.

***NOTE —***
*Remove front exhaust manifold first, before the rear manifold.*

- Installation is reverse of removal.
  - Coat oxygen sensor threads and exhaust fasteners with antiseize copper paste.
  - Use new gaskets and self-locking nuts.

| Tightening torques | |
|---|---|
| Exhaust manifold to cylinder head (use new M7 nuts) | 20 Nm (15 ft-lb) |
| Front exhaust pipe to exhaust manifold (M10) | 30 Nm (22 ft-lb) |
| Oxygen sensor to exhaust manifold | 50 Nm (37 ft-lb) |

## Exhaust manifolds, removing and installing (S52 engine)

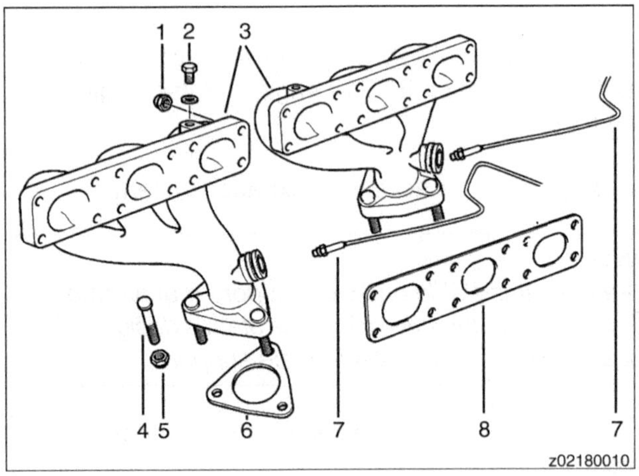


Exhaust manifold components for S52 engine:

1. M7 nut
2. Plug
3. Exhaust manifolds
4. M10 stud
5. M10 nut
6. Gasket
7. Oxygen sensor
8. Gasket

> ***WARNING —***
> *To avoid personal injury, be sure the engine is cold before starting the procedure.*

- Make sure ignition is OFF. Remove ignition key.
- Remove windshield washer fluid reservoir from right front of engine compartment. See **611 Wipers and Washers**.
- Remove secondary air injection pipe and check valve from exhaust manifold.
- Raise and support car for access to exhaust system.

> ***WARNING —***
> *Make sure the car is stable and well supported at all times. Use a professional automotive lift or jack stands designed for the purpose. A floor jack is not adequate support.*

- Unbolt front exhaust pipes from exhaust manifolds. See **Exhaust system, removing and installing** in this repair group.
- Detach front exhaust pipes from bracket at transmission.

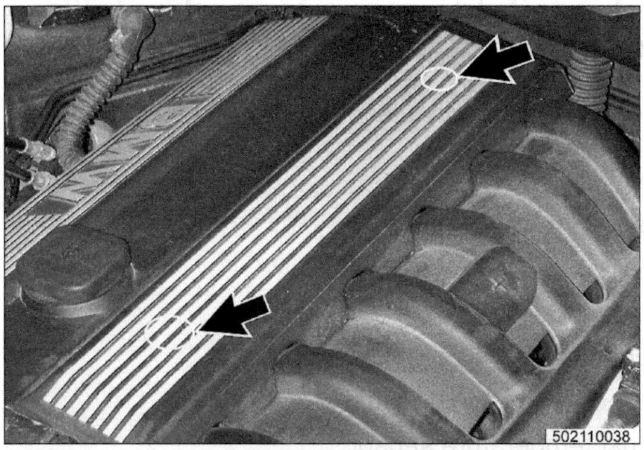


Remove fuel injector cover by removing plastic trim plugs (**arrows**) and removing fasteners under them. Lift off cover.

Disconnect oxygen sensor connectors (**arrows**) along left side of cylinder head cover.

***NOTE—***
*Mark connectors so that oxygen sensors and connectors can be reinstalled in their original positions.*

– Remove manifold mounting nuts and lift out manifolds. Discard nuts and gaskets.

***CAUTION—***
*To avoid damaging precatalyst oxygen sensors, use care when removing manifolds. To be safe, remove the sensors from the manifolds before removing manifolds. See* **Oxygen sensor, replacing (S52 engine)** *in this repair group.*

– Installation is reverse of removal.
- Use new gaskets and fasteners.
- Coat exhaust fasteners and oxygen sensors (if applicable) with copper paste before installing.
- Install manifold gaskets with arched side facing manifolds.

| **Tightening torques** | |
|---|---|
| Exhaust manifold to cylinder head (use new M7 nuts) | 20 Nm (15 ft-lb) |
| Front exhaust pipe to exhaust manifold (M10) | 30 Nm (22 ft-lb) |
| Oxygen sensor to exhaust manifold | 50 Nm (37 ft-lb) |

## Exhaust manifolds, removing and installing (S54 engine)

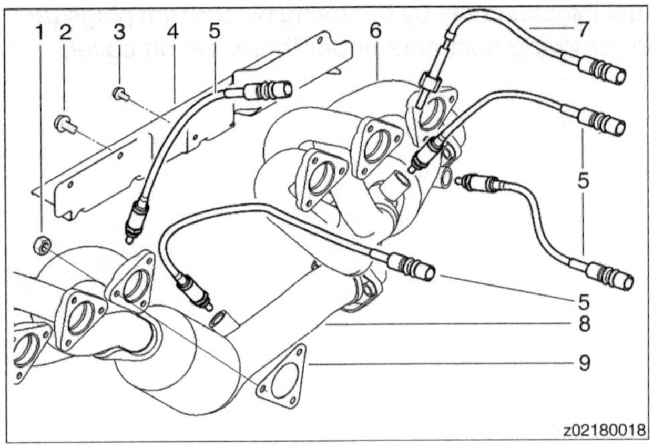


◀ Exhaust manifold components for S54 engine:

1. M7 nut
2. M8 bolt
3. M6 bolt
4. Heat shield
5. Oxygen sensor
6. Cylinder 4 - 6 exhaust manifold
7. Exhaust temperature sensor
8. Cylinder 1 - 3 exhaust manifold
9. Exhaust manifold gasket

> ***WARNING —***
> *To avoid personal injury, be sure the engine is cold before starting the procedure.*

– Make sure ignition is OFF. Remove ignition key.

– Working at rear of engine compartment, remove cable duct from bulkhead.

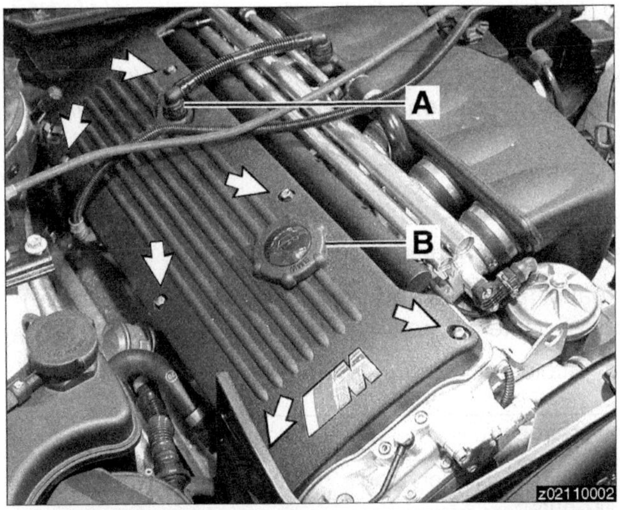


◀ Remove cylinder head top cover:

- Disconnect vent hose (**A**).
- Remove oil filler cover (**B**).
- Remove cover hold-down bolts (**arrows**) and lift off cover.

***NOTE —***
*Remove oil filler cap to allow removal of plastic cover. Be sure to replace it on the oil filler immediately to prevent debris from entering the engine.*

◀ Working at rear of cylinder head cover:

- Match mark oxygen sensor connectors (**arrows**) for correct reassembly.
- Separate connectors.

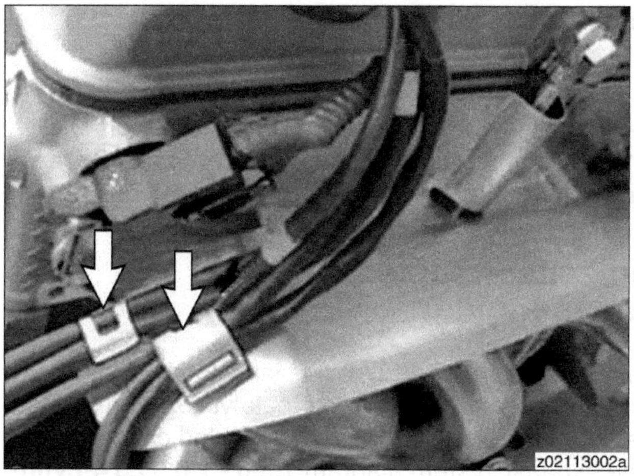


◀ Working at right rear of cylinder head, detach oxygen sensor wiring harnesses from exhaust heat shield brackets (**arrows**).

– Remove windshield washer fluid reservoir from right front of engine compartment. See **611 Wipers and Washers**.

– Remove exhaust manifold heat shield. Loosen A/C system refrigerant lines from mountings and move aside.

> ***CAUTION—***
> *Do not open refrigerant lines.*

– Raise and support car for access to exhaust system.

> ***WARNING —***
> *Make sure the car is stable and well supported at all times. Use a professional automotive lift or jack stands designed for the purpose. A floor jack is not adequate support.*

– Remove engine splash guard and body reinforcement cross-brace.

– Unbolt front exhaust pipes from exhaust manifolds. See **Exhaust system, removing and installing** in this repair group.

– Remove oxygen sensors and exhaust temperature sensor and set aside in a safe place. See **Oxygen sensor, replacing (S54 engine)** in this repair group.

– Detach front exhaust manifold (cylinder 1 - 3) from cylinder head and place on engine mount bracket.

– Detach rear exhaust manifold (cylinder 4 - 6) from cylinder head.

– Remove both exhaust manifolds together.

– Installation is reverse of removal.
  - Use new gaskets and fasteners.
  - Coat exhaust fasteners and oxygen sensors (if applicable) with copper paste before installing.

| **Tightening torques** | |
|---|---|
| Exhaust manifold to cylinder head (use new M7 nuts) | 20 Nm (15 ft-lb) |
| Front exhaust pipe to exhaust manifold (M10 Torx) | 24 Nm (18 ft-lb) |
| Oxygen sensor to exhaust manifold | 50 Nm (37 ft-lb) |

# Oxygen sensor removal and installation

Diagnose oxygen sensor faults using BMW scan tools DIS-plus, GT1, MoDiC or equivalent. Prior to oxygen sensor replacement, access and clear ECM fault memory.

## Oxygen sensor, replacing (4-cylinder engine)

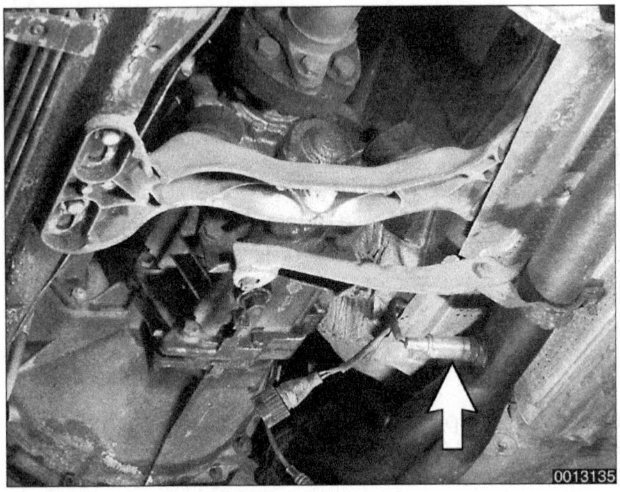


◄ Precatalyst oxygen sensor (**arrow**) is installed in front exhaust pipe next to transmission. Post-catalyst oxygen sensor is installed at rear of front exhaust pipe behind catalytic converter. See exhaust system diagram **1996 - 1998 Z3 4-cylinder (M44 engine)** in this repair group.

> ***WARNING —***
> *To avoid personal injury, be sure the engine is cold before starting the procedure.*

– Make sure ignition is OFF. Remove ignition key.

– Raise and support car for access to exhaust system.

> ***WARNING —***
> *Make sure the car is stable and well supported at all times. Use a professional automotive lift or jack stands designed for the purpose. A floor jack is not adequate support.*

– Disconnect oxygen sensor electrical connector. Unclip electrical harness from bracket.

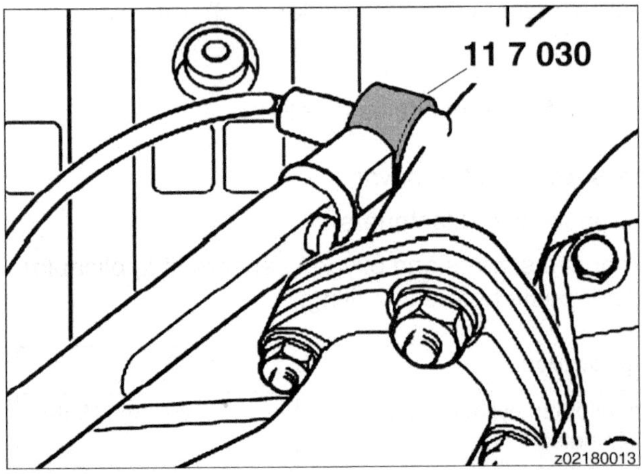


◄ Use BMW special tool 11 7 030 or equivalent to remove oxygen sensor from pipe.

– Installation is reverse of removal. Bear in mind the following:

- New oxygen sensor threads are coated at the factory with copper paste (antiseize compound).
- If an oxygen sensor is reused, apply a thin and even coat of copper paste to threads only.
- Reroute sensor electrical harness as before.

> ***CAUTION—***
> *Do not clean sensor tip which protrudes into the exhaust stream. Make sure tip does not contact any lubricant.*

| Tightening torque | |
|---|---|
| Oxygen sensor to exhaust pipe | 50 Nm (37 ft-lb) |

– Check for fault codes and clear and reset ECM memory.

## Oxygen sensor, replacing (M52 engine)

Precatalyst oxygen sensor is installed in front exhaust pipe ahead of catalytic converter. Post-catalyst oxygen sensor is installed in front exhaust pipe behind catalytic converter. See exhaust system diagram **1997 - 1998 Z3 6-cylinder (M52 engine)** in this repair group.

> ***WARNING —***
> *To avoid personal injury, be sure the engine is cold before starting the procedure.*

– Make sure ignition is OFF. Remove ignition key.

– Raise and support car for access to exhaust system.

> ***WARNING —***
> *Make sure the car is stable and well supported at all times. Use a professional automotive lift or jack stands designed for the purpose. A floor jack is not adequate support.*

– Disconnect oxygen sensor electrical connector. Unclip electrical harness from bracket.

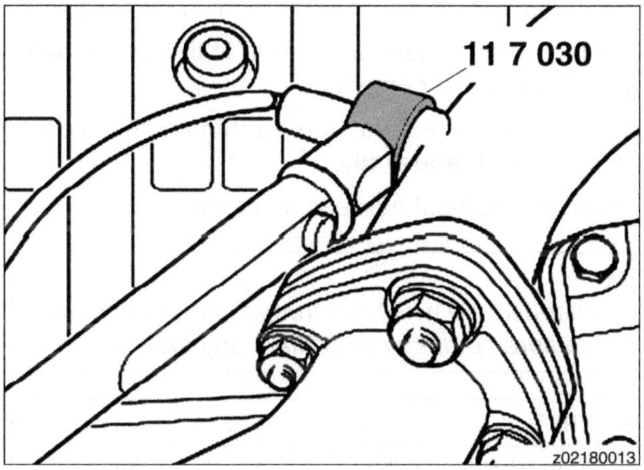


◄ Use BMW special tool 11 7 030 or equivalent to remove oxygen sensor from pipe.

– Installation is reverse of removal. Bear in mind the following:

- New oxygen sensor threads are coated at the factory with copper paste (antiseize compound).
- If an oxygen sensor is reused, apply a thin and even coat of copper paste to threads only.
- Reroute sensor electrical harness as before.

> ***CAUTION—***
> *Do not clean sensor tip which protrudes into the exhaust stream. Make sure tip does not contact any lubricant.*

| Tightening torque | |
|---|---|
| Oxygen sensor to exhaust pipe | 50 Nm (37 ft-lb) |

– Check for fault codes and clear and reset ECM memory.

## Oxygen sensor, replacing (M52 TU, M54 engine)

For oxygen sensor locations, see exhaust system diagrams **1999 - 2000 Z3 6-cylinder (M52 TU engine)** or **2001 - 2002 Z3 6-cylinder (M54 engine)** in this repair group.

### Precatalyst oxygen sensor, replacing

The precatalyst oxygen sensors are installed at the top of the exhaust manifolds.

> ***WARNING —***
> *To avoid personal injury, be sure the engine is cold before starting the procedure.*

– Make sure ignition is OFF. Remove ignition key.

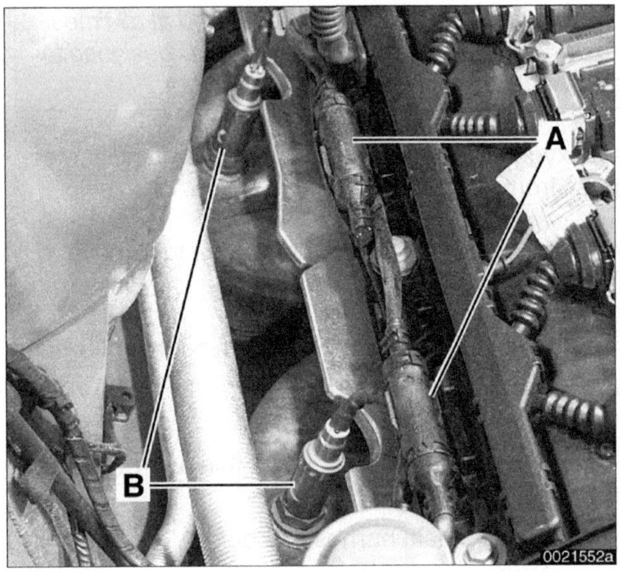


◀ Working at right side of engine above exhaust manifolds, remove front or rear oxygen sensor harness connector (**A**) from bracket, then separate plug.

– Remove front or rear precatalyst oxygen sensor (**B**) using BMW special tool 11 7 020 or equivalent.

– Installation is reverse of removal. Bear in mind the following:
  - New oxygen sensor threads are coated at the factory with copper paste (antiseize compound).
  - If an oxygen sensor is reused, apply a thin and even coat of copper paste to threads only.
  - Reroute sensor electrical harness as before.

> ***CAUTION—***
> *Do not clean sensor tip which protrudes into the exhaust stream. Make sure tip does not contact any lubricant.*

| Tightening torque | |
|---|---|
| Oxygen sensor to exhaust manifold | 50 Nm (37 ft-lb) |

– Check for fault codes and clear and reset ECM memory.

### Post-catalyst oxygen sensor, replacing

The post-catalyst oxygen sensors are installed at the rear of the catalytic converters near the bottom of the exhaust manifolds.

> ***WARNING —***
> *To avoid personal injury, be sure the engine is cold before starting the procedure.*

– Make sure ignition is OFF. Remove ignition key.

z02180012

◄ Remove fuel injector cover by removing plastic trim plugs (**arrows**) and removing fasteners under them. Lift off cover.

z02180014

◄ Disconnect post-catalyst oxygen sensor connectors (**arrows**) along left side of cylinder head cover.

***NOTE—***
*Mark connectors for later reassembly.*

– Raise and support car for access to exhaust system.

> ***WARNING—***
> *Make sure the car is stable and well supported at all times. Use a professional automotive lift or jack stands designed for the purpose. A floor jack is not adequate support.*

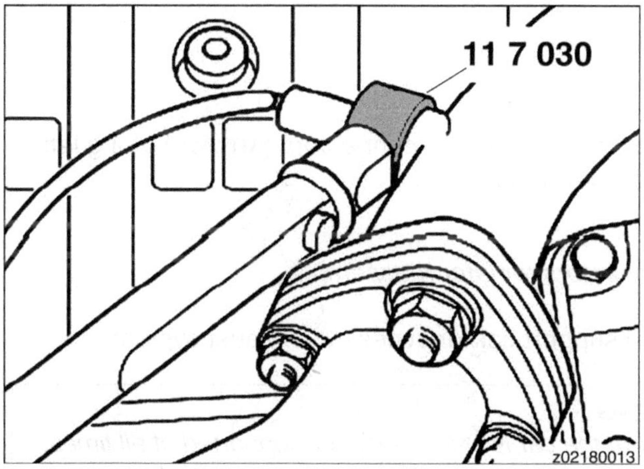


z02180013

– Working underneath car, remove splash shield under engine compartment.

◄ Working underneath rear of engine at exhaust manifolds, use BMW special too 11 7 030 or equivalent to remove oxygen sensor from manifold.

– Installation is reverse of removal. Bear in mind the following:
  - New oxygen sensor threads are coated at the factory with copper paste (antiseize compound).
  - If an oxygen sensor is reused, apply a thin and even coat of copper paste to threads only.
  - Reroute sensor electrical harness as before.

> ***CAUTION—***
> *Do not clean sensor tip which protrudes into the exhaust stream. Make sure tip does not contact any lubricant.*

| Tightening torque | |
|---|---|
| Oxygen sensor to exhaust manifold | 50 Nm (37 ft-lb) |

– Check for fault codes and clear and reset ECM memory.

## Oxygen sensor, replacing (S52 engine)

For oxygen sensor locations, see exhaust system diagram **1998 - 2000 M version 6-cylinder (S52 engine)** in this repair group.

### Precatalyst oxygen sensor, replacing

Precatalyst oxygen sensors are installed at base of exhaust manifolds.

> ***WARNING —***
> *To avoid personal injury, be sure the engine is cold before starting the procedure.*

– Make sure ignition is OFF. Remove ignition key.

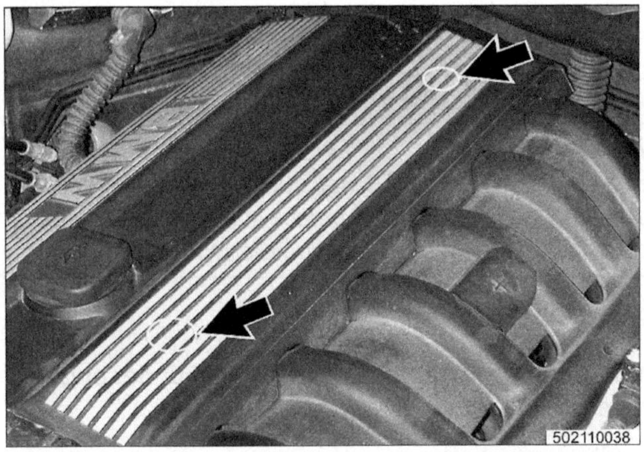
502110038

◀ Remove fuel injector cover by removing plastic trim plugs (**arrows**) and removing fasteners under them. Lift off cover.

502110207

◀ Disconnect oxygen sensor connectors (**arrows**) along left side of cylinder head cover.

> ***NOTE —***
> *Mark connectors for later reassembly.*

– Raise and support car for access to exhaust system.

> ***WARNING —***
> *Make sure the car is stable and well supported at all times. Use a professional automotive lift or jack stands designed for the purpose. A floor jack is not adequate support.*

– Working underneath car, remove splash shield under engine compartment.

– Use BMW special tool 11 7 020 or equivalent to remove oxygen sensor from intake manifold.

– Installation is reverse of removal. Bear in mind the following:
  - New oxygen sensor threads are coated at the factory with copper paste (antiseize compound).
  - If an oxygen sensor is reused, apply a thin and even coat of copper paste to threads only.
  - Reroute sensor electrical harness as before.

> ***CAUTION—***
> *Do not clean sensor tip which protrudes into the exhaust stream. Make sure tip does not contact any lubricant.*

| **Tightening torque** | |
|---|---|
| Oxygen sensor to exhaust manifold | 50 Nm (37 ft-lb) |

– Check for fault codes and clear and reset ECM memory.

## Post-catalyst oxygen sensor, replacing

Post-catalyst oxygen sensors are installed at rear of front exhaust pipes, behind catalytic converters and in front of resonators.

> ***WARNING—***
> *To avoid personal injury, be sure the engine is cold before starting the procedure.*

– Make sure ignition is OFF. Remove ignition key.

– Raise and support car for access to exhaust system.

> ***WARNING—***
> *Make sure the car is stable and well supported at all times. Use a professional automotive lift or jack stands designed for the purpose. A floor jack is not adequate support.*

– Disconnect oxygen sensor electrical connector. Unclip electrical harness from bracket.

– Use BMW special tool 11 7 030 or equivalent to remove oxygen sensor from pipe.

– Installation is reverse of removal. Bear in mind the following:
  - New oxygen sensor threads are coated at the factory with copper paste (antiseize compound).
  - If an oxygen sensor is reused, apply a thin and even coat of copper paste to threads only.
  - Reroute sensor electrical harness as before.

> ***CAUTION—***
> *Do not clean sensor tip which protrudes into the exhaust stream. Make sure tip does not contact any lubricant.*

| Tightening torque | |
|---|---|
| Oxygen sensor to exhaust pipe | 50 Nm (37 ft-lb) |

– Check for fault codes and clear and reset ECM memory.

## Oxygen sensor, replacing (S54 engine)

For oxygen sensor locations, see exhaust system diagram **2001 - 2002 M version 6-cylinder (S54 engine)** in this repair group.

### Precatalyst oxygen sensor, replacing

The precatalyst oxygen sensors are installed at the top of the exhaust manifolds.

> ***WARNING —***
> *To avoid personal injury, be sure the engine is cold before starting the procedure.*

– Make sure ignition is OFF. Remove ignition key.

– Working at rear of engine compartment, remove cable duct from bulkhead.

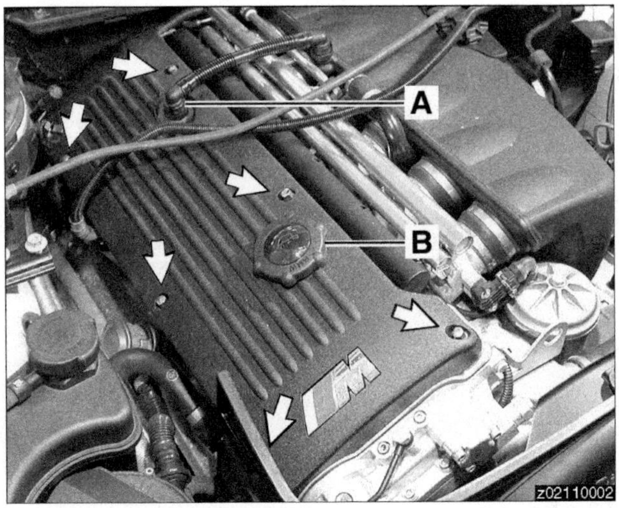


< Remove cylinder head top cover:
  - Disconnect vent hose (**A**).
  - Remove oil filler cover (**B**).
  - Remove cover hold-down bolts (**arrows**) and lift off cover.

***NOTE —***
*Remove oil filler cap to allow removal of plastic cover. Be sure to replace it on the oil filler immediately to prevent debris from entering the engine.*

z02110004

Working at rear of cylinder head cover:

- Match mark oxygen sensor connectors (**arrows**) for correct reassembly.
- Separate connectors.

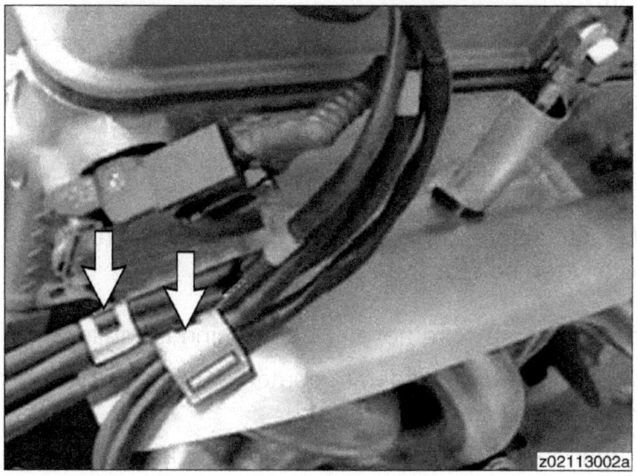
z02113002a

Working at right rear of cylinder head, detach oxygen sensor wiring harnesses from exhaust heat shield brackets (**arrows**).

– Raise and support car for access to exhaust manifolds from underneath.

> ***WARNING —***
> *Make sure the car is stable and well supported at all times. Use a professional automotive lift or jack stands designed for the purpose. A floor jack is not adequate support.*

– Use BMW special tools 11 7 030 and 11 9 150 or equivalent to remove oxygen sensor from pipe.

– Installation is reverse of removal. Bear in mind the following:

- New oxygen sensor threads are coated at the factory with copper paste (antiseize compound).
- If an oxygen sensor is reused, apply a thin and even coat of copper paste to threads only.
- Reroute sensor electrical harness as before.

> ***CAUTION—***
> *Do not clean sensor tip which protrudes into the exhaust stream. Make sure tip does not contact any lubricant.*

| **Tightening torque** | |
|---|---|
| Oxygen sensor to exhaust pipe | 50 Nm (37 ft-lb) |

– Check for fault codes and clear and reset ECM memory.

#### Post-catalyst oxygen sensor, replacing

Post-catalyst oxygen sensors are installed at base of exhaust manifolds.

> ***WARNING —***
> *To avoid personal injury, be sure the engine is cold before starting the procedure.*

– Make sure ignition is OFF. Remove ignition key.

– Separate oxygen sensor wiring harness.

– Raise and support car for access to exhaust manifolds from underneath.

> ***WARNING —***
> *Make sure the car is stable and well supported at all times. Use a professional automotive lift or jack stands designed for the purpose. A floor jack is not adequate support.*

– Use BMW special tools 11 7 030 and 11 9 150 or equivalent to remove oxygen sensor from pipe.

– Installation is reverse of removal. Bear in mind the following:

- New oxygen sensor threads are coated at the factory with copper paste (antiseize compound).
- If an oxygen sensor is reused, apply a thin and even coat of copper paste to threads only.
- Reroute sensor electrical harness as before.

> ***CAUTION—***
> *Do not clean sensor tip which protrudes into the exhaust stream. Make sure tip does not contact any lubricant.*

| Tightening torque | |
|---|---|
| Oxygen sensor to exhaust pipe | 50 Nm (37 ft-lb) |

– Check for fault codes and clear and reset ECM memory.

# 200 Transmission - General



**TABLES**

## GENERAL

This section of the manual covers repairs related to the clutch, manual and automatic transmissions, shift linkages and driveshaft. For repairs related to drive axles and the differential, see **331 Drive Axles, Differential**.

### Drivetrain

Z3 models are equipped with a longitudinal drivetrain. The transmission is bolted directly to the rear of the engine. A driveshaft connects the output shaft of the transmission to the differential. Individual drive axles with integrated constant velocity joints transfer rotational power from the differential to the rear wheels.

### Manual transmission

ZF manufactured manual transmission

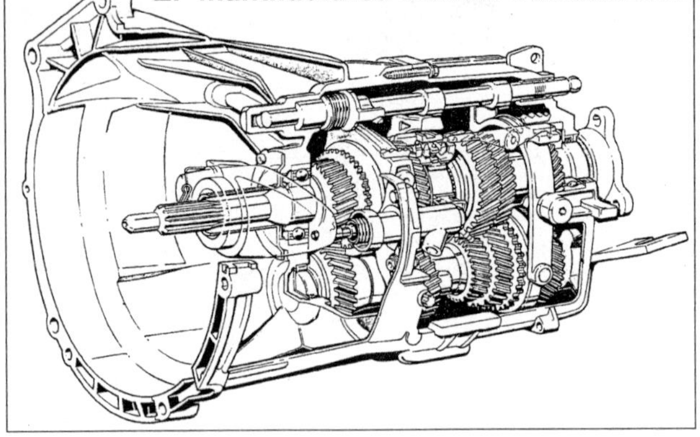

Due to different power characteristics and performance requirements, two different 5-speed manual transmissions are used in models covered by this manual. Manual transmission applications are given in **Table a**.

***NOTE —***
*For transmission gear ratios and repair information, see* **230 Manual Transmission**.

The manual transmissions use a single clutch disc with dual-mass flywheel. For further information, see **210 Clutch**.

Manual transmissions have a metal ID plate mounted on the side of the transmission. Do not rely on numbers cast on the transmission case for identification.

**Table a. Manual transmission applications**

| Model | Year | Transmission |
|---|---|---|
| Z3 1.9 | 1996 - 1998 | Getrag S5D 250 G |
| Z3 2.3 | 1999 - 2000 | Getrag S5D 250 G |
| Z3 2.5i | 2001 - 2002 | Getrag S5D 250 G |
| Z3 2.8 | 1997 - 2000 | ZF S5D 320 Z |
| Z3 3.0i | 2001 - 2002 | ZF S5D 320 Z |
| M 3.2 S52 | 1998 - 2000 | ZF S5D 320 Z |
| M 3.2 S54 | 2001 - 2002 | ZF S5D 320 Z |

## Automatic transmission

Two different automatic transmissions are used, depending on model and model year and/or production date. Both transmissions are electrohydraulically controlled by the transmission control module. Automatic transmission applications are given in **Table b**.

***NOTE—***
*M version cars are not equipped with automatic transmission.*

| Automatic transmission applications | | |
|---|---|---|
| **Model** | **Year** | **Transmission** |
| **Z3 1.9** | 1996 - 1998 | A4S 270 R / A4S 310 R (4-speed) |
| **Z3 2.3** | 1999 - 2000 | A4S 270 R / A4S 310 R (4-speed) |
| **Z3 2.5i** | 2001 - 2002 | A5S 360 R / A5S 390 R (5-speed Steptronic) |
| **Z3 2.8** | 1997 - 2000 | A4S 270 R / A4S 310 R (4-speed) |
| **Z3 3.0i** | 2001 - 2002 | A5S 360 R / A5S 390 R (5-speed Steptronic) |

Transmission Identification codes for automatic transmissions are located on metal ID tags mounted to the body of the transmission.

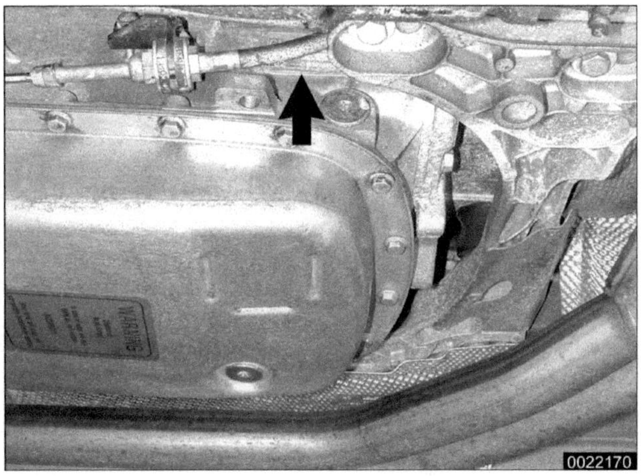
0022170

The transmissions, produced by GM, have an ID tag located on the left hand side of the transmission housing, behind the transmission selector cable, just above the transmission pan (**arrow**).

For automatic transmission repair information, see **240 Automatic Transmission**.

## Transmission lubricants

0022040

The lubricant used varies between transmissions and model years. Fluid application information is clearly marked on a label on the transmission.

On manual transmissions, either ATF (orange label reads ATF-Oil) or a special BMW lubricant (yellow label reads MTF-LT 1) is used. See **230 Manual Transmission**.

Automatic transmissions are filled with either Dexron III ATF (black transmission ID plate reads ATF-Oil), or a special life-time oil (green transmission ID plate reading Life-Time Oil). See **240 Automatic Transmission**.

***CAUTION—***
*If in doubt as to the type of fluid installed in a particular transmission, consult an authorized BMW dealer service department for the latest in operating fluids.*

# 210 Clutch

## GENERAL

This repair group covers replacement of the clutch mechanical and clutch hydraulic components. Special tools may be required for some of the service described in this repair group. Read the procedures through before beginning a job.

***NOTE—***
*See* **200 Transmission–General** *for transmission application information.*

The major components of the clutch system are shown in the **Clutch assembly and hydraulics** diagram in this section.

### Special tools

Some special tools are recommended for clutch repair. Be sure to read each procedure through before beginning a job.

Flywheel locking tool
(BMW special tool 11 2 170)

Pilot bearing removal tool
(BMW special tool 11 2 340)

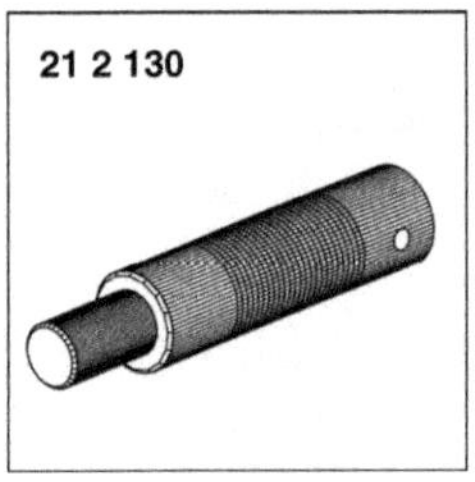


Clutch centering tool
(BMW special tool 21 2 130)

## CLUTCH HYDRAULICS

The clutch is hydraulically actuated by the master and slave cylinders. Clutch disc wear is automatically taken up by the pushrod travel of the slave cylinder, making periodic adjustments unnecessary.

A soft or spongy feel to the clutch pedal, long pedal free-play, or grinding noises from the gears while shifting can all indicate problems with the clutch hydraulics. In these circumstances it is best to start with a clutch fluid flush, followed, if necessary, by replacement of the hydraulic parts.

## Clutch hydraulic system, bleeding and flushing

If the clutch/brake fluid is murky or muddy, or has not been changed within the last 12 months, the system should be flushed. Flushing the old fluid from the clutch lines is done using a brake system pressure bleeder.

> ***WARNING—***
> - *Brake fluid is poisonous, highly corrosive and dangerous to the environment.*
> - *Wear safety glasses and rubber gloves when working with brake fluid.*
> - *Do not siphon brake fluid with your mouth.*
> - *Dispose of brake fluid properly.*

> ***CAUTION—***
> - *Immediately clean away any fluid spilled on painted surfaces and wash with water, as brake fluid dissolves paint.*
> - *Do not exceed 2 bars (29 psi) pressure at the fluid reservoir when bleeding or flushing the hydraulic system.*
> - *Always use new brake fluid from a fresh, unopened container. Brake fluid absorbs moisture from the air. This can lead to corrosion problems in the clutch and brake hydraulic systems, and also lowers the brake fluid boiling point.*

– Remove brake fluid reservoir cap. Using a clean syringe, remove brake fluid from reservoir. Refill reservoir with clean DOT 4 low viscosity brake fluid.

– Attach pressure brake bleeder to fluid reservoir.

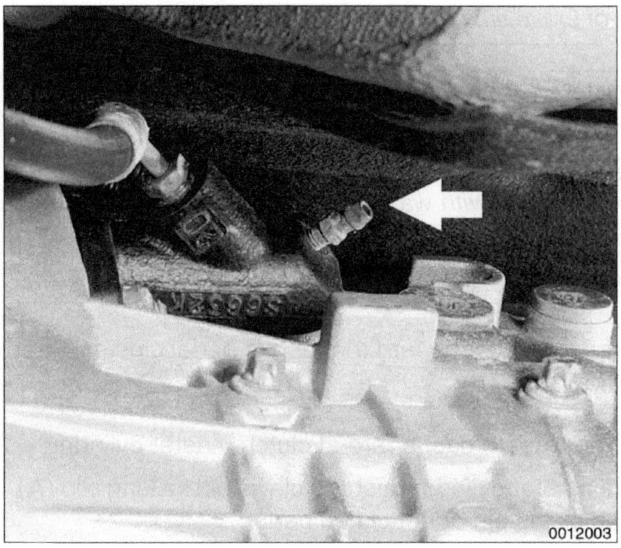
0012003

◄ Connect a length of hose from clutch slave cylinder bleeder valve on left side of transmission (**arrow**) to a container.

– Open bleeder valve and allow brake fluid to expel.

– Slowly operate clutch pedal several times until fluid runs out clear without bubbles.

– Close bleeder valve.

***NOTE—***
*If air is trapped in the lines, unbolt the slave cylinder from the transmission. Push the slave cylinder pushrod all the way in and slowly release it. This will force air in the system into the reservoir.*

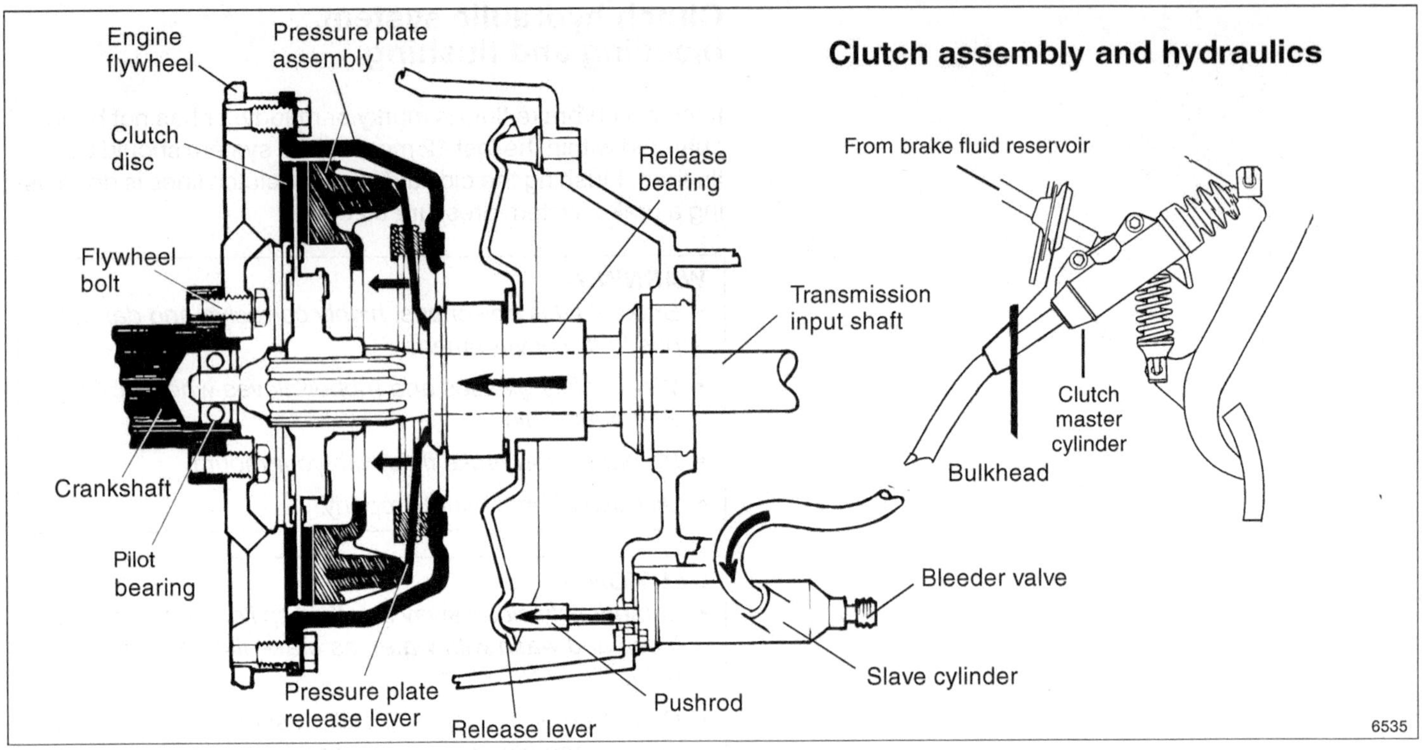


## Clutch master cylinder, replacing

The clutch master cylinder is mounted to the pedal assembly, directly above the clutch pedal.

– Working at left rear engine compartment, pinch off brake fluid supply hose from fluid reservoir to clutch master cylinder.

– Working in engine compartment, disconnect fluid line fitting from master cylinder.

> ***WARNING—***
> *Dispose of brake fluid properly.*

> ***CAUTION—***
> - *Be prepared to catch dripping fluid.*
> - *Immediately clean away any fluid spilled on painted surfaces and wash with water, as brake fluid dissolves paint.*
> - *Plug open brake fluid lines and ports.*

– Remove left lower dashboard trim above pedal cluster. See **513 Interior Trim**.

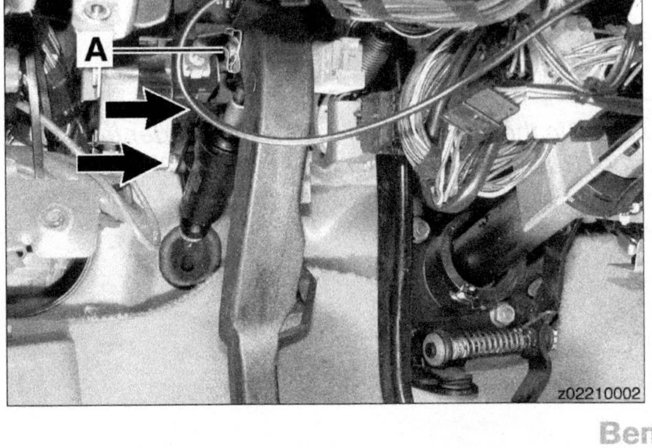


Working underneath dashboard at clutch master cylinder:

- Remove master cylinder pushrod clevis pin locking clip (**A**) and slide clevis pin out of clutch pedal.
- Remove master cylinder mounting bolts (**arrows**) and lift out master cylinder.

– Installation is reverse of removal. Remember to:
  - Fill fluid reservoir with clean fluid.
  - Bleed clutch hydraulics. See **Clutch hydraulic system, bleeding and flushing** in this repair group.

| Tightening torques | |
|---|---|
| Clutch master cylinder to body | 22 Nm (16 ft lb) |
| Fluid line to master cylinder | 17 Nm (12 ft lb) |

## Clutch slave cylinder, replacing

The clutch slave cylinder is mounted on the left side of the transmission.

– Remove brake fluid reservoir cap. Use a clean syringe to remove brake fluid from reservoir.

– Disconnect fluid hose from slave cylinder. Place a pan under hose to catch dripping fluid.

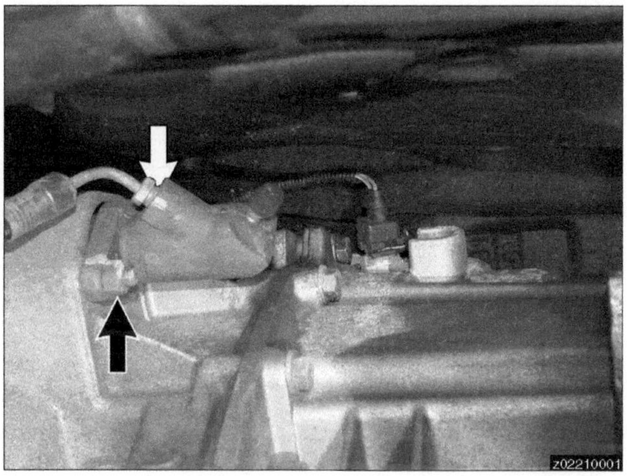


◄ Remove slave cylinder mounting fasteners (**arrows**) from transmission and lift off slave cylinder.

– Installation is reverse of removal, noting the following:
  - Check for wear on pushrod. Any wear except on tip is caused by misalignment.
  - Lightly coat pushrod tip with molybdenum disulfide grease (Molykote® Longterm 2 or equivalent).
  - During installation be sure pushrod tip engages recess in clutch release lever.
  - Fill fluid reservoir with clean fluid.
  - Bleed clutch hydraulics as described earlier.

| Tightening torques | |
|---|---|
| Clutch slave cylinder to transmission | 22 Nm (16 ft-lb) |
| Fluid line to slave cylinder | 17 Nm (12 ft-lb) |

# CLUTCH MECHANICAL

The transmission must be removed from the engine to access the clutch mechanical components. Replace the clutch disc, pressure plate and release bearing during a clutch overhaul.

Check the bottom of the bellhousing for oil. If oil is found, check for a faulty rear crankshaft oil seal. Crankshaft seal replacement is covered in **119 Lubrication System**.

## Clutch, removing

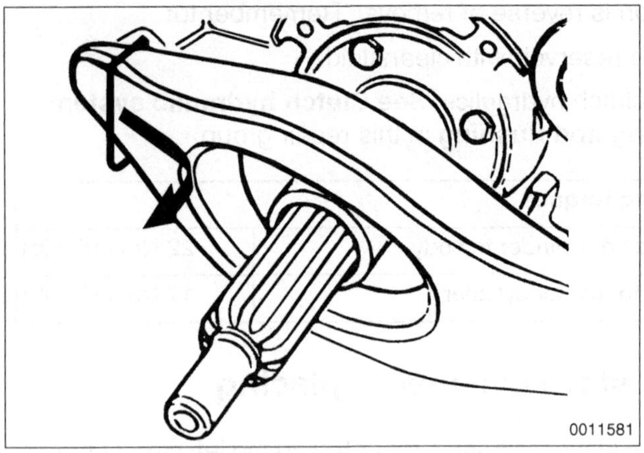

- Remove transmission from engine. See **230 Manual Transmission**.
- Remove release bearing from transmission input shaft.
- Remove clutch release lever by sliding it out from under spring clip (**arrow**).

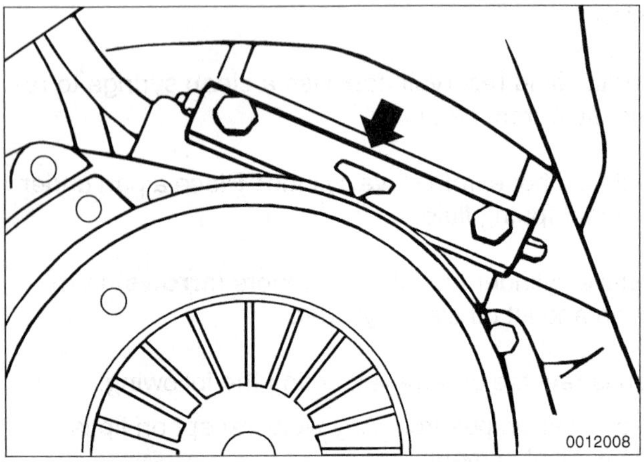

- Lock flywheel in position using BMW special tool 11 2 340 (**arrow**), or equivalent
- Install a clutch centering tool through clutch assembly to prevent clutch disc from falling out when removing pressure plate.

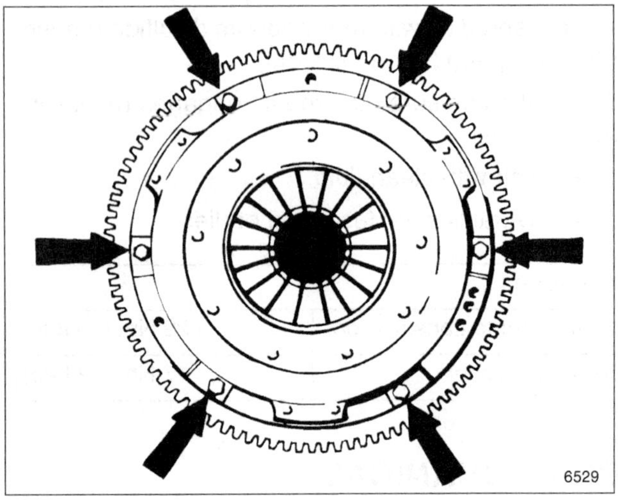

- Remove six bolts (**arrows**) from perimeter of pressure plate. Remove pressure plate and clutch disc.

## Clutch, inspecting and installing

- Inspect clutch disc for wear, cracks, loose rivets, contamination or excessive runout (warping). Replace if necessary.
- Inspect flywheel for scoring, hot spots, cracks or loose or worn guide pins. Replace flywheel if any faults are found.

***NOTE—***
*If flywheel is removed, install using new bolts.*

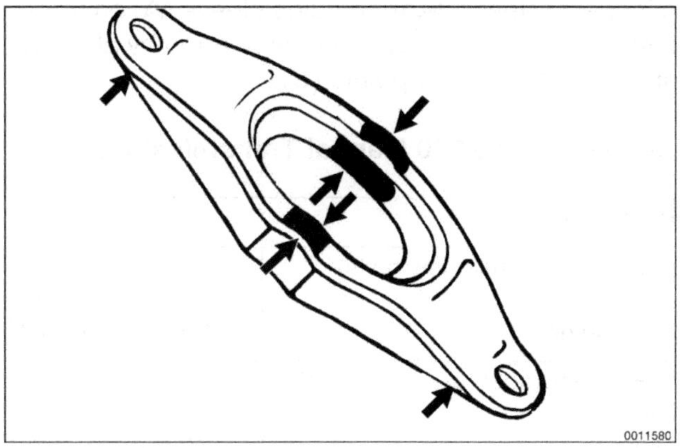


Inspect and clean release bearing lever. Apply a thin coat of grease to release bearing lever contact points (**arrows**). Also, lightly lubricate clutch disc splines and transmission input shaft splines.

| **Clutch and pressure plate wear limits** | |
|---|---|
| Minimum clutch disc lining thickness (at rivets) | 1.0 mm (0.04 in) |
| Maximum lateral runout (diaphragm springs to pressure plate) | 0.6 mm (0.02 in) |

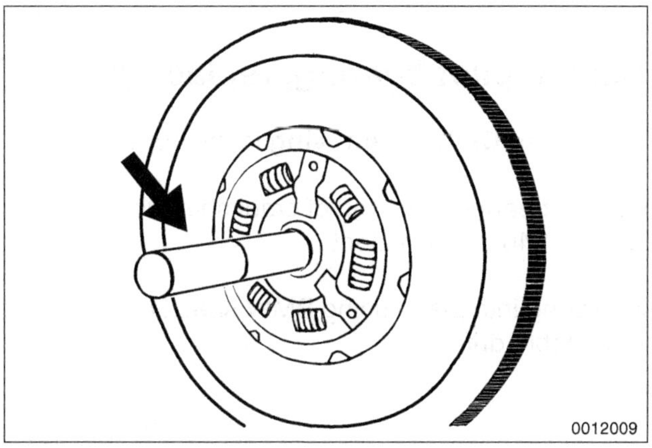


Center clutch disc on flywheel using clutch disc centering tool (**arrow**) (BMW special tool no. 21 2 130 or equivalent).

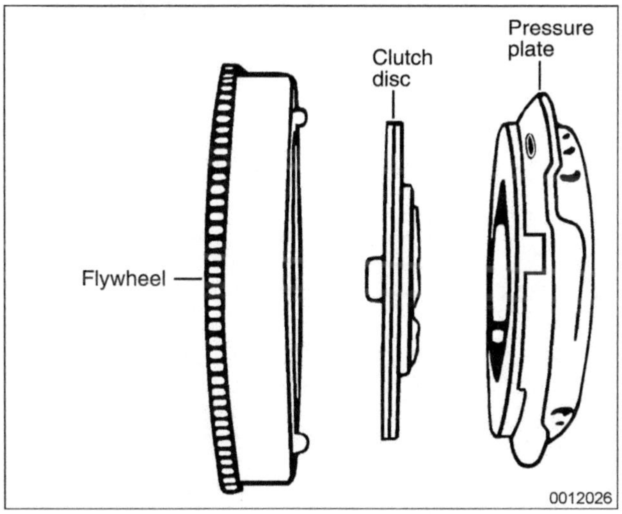


Be sure clutch disc is facing the correct way.

***NOTE—***
*The new disc should be marked* ***engine side*** *or* ***transmission side****.*

- Using new bolts, tighten them one turn at a time in a cross pattern until tight and then torque to specification.
- Remove clutch disc centering tool. Remove flywheel locking tool.

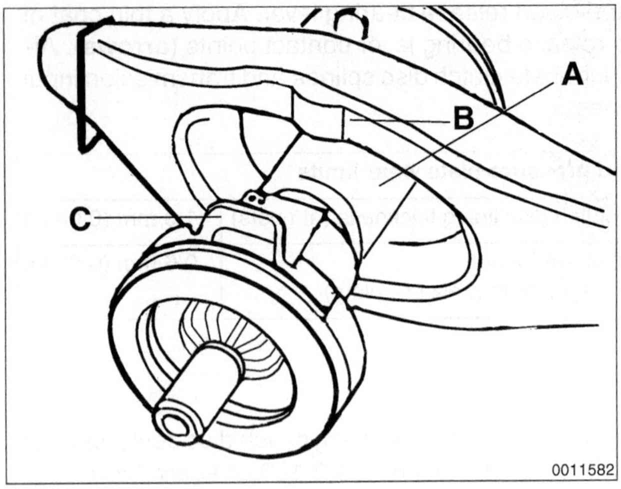


◄ Clean and inspect release bearing guide sleeve (**A**) on transmission. Install release lever and release bearing by aligning bearing tabs (**B**) with contact points (**C**).

– Install transmission. See **230 Manual Transmission**.

| **Tightening torques** | |
|---|---|
| Flywheel to crankshaft (use new bolts)<br>• 4 cylinder engine<br>• 6 cylinder engine | <br>120 Nm (89 ft-lb)<br>105 Nm (77 ft-lb) |
| Pressure plate to flywheel (use new bolts)<br>• 8.8 grade<br>• 10.9 grade | <br>24 Nm (18 ft-lb)<br>34 Nm (25 ft-lb) |

## Transmission pilot bearing, replacing

– Remove clutch. See **Clutch, removing** in this repair group.

– Remove release bearing from end of crankshaft using BMW special tool 11 2 340 or equivalent.

– Press new bearing into place using BMW special tool 11 2 350 or suitable drift.

# 230 Manual Transmission

## General

This repair group covers external transmission service, including removal and installation of the transmission unit. Internal transmission repair is not covered. Special press tools and procedures are required to disassemble and service the internal geartrain.

For crankshaft rear main seal replacement information, see **119 Lubrication System**.

### Special tools

BMW special tools are recommended for manual transmission service. If these tools are unavailable, equivalent tools may be substituted.

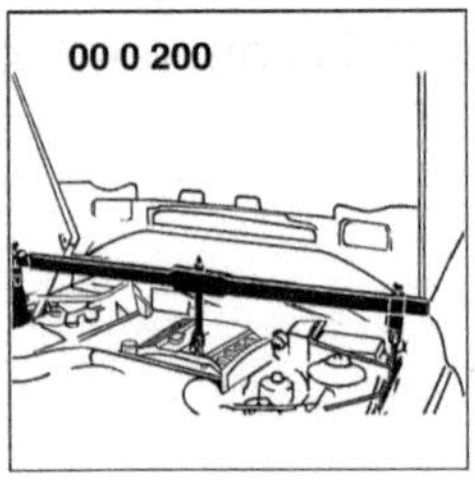


Engine support bracket
(BMW special tool set 00 0 200)

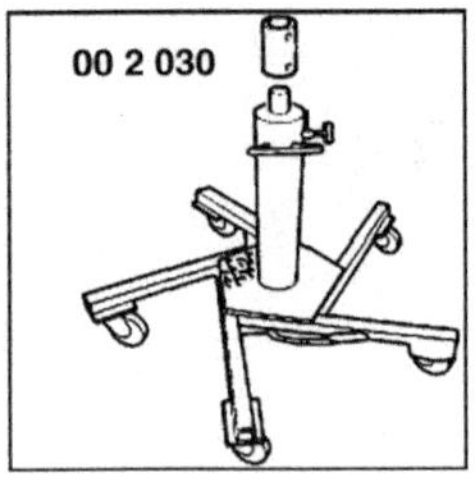


Transmission jack
(BMW special tool 00 2 030)

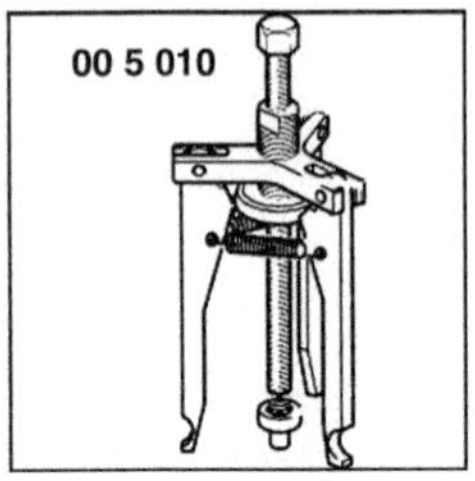


Seal puller
(BMW special tool 00 5 010)

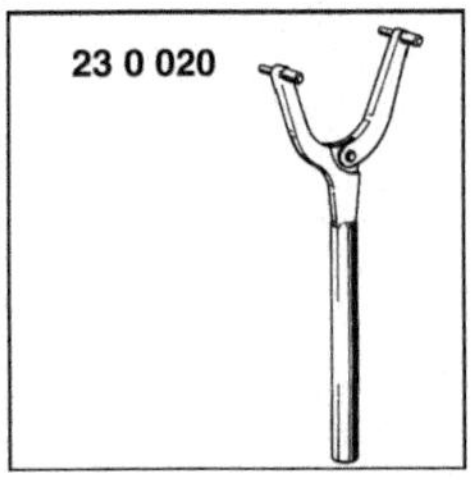


Transmission flange counterhold tool
(BMW special tool 23 0 020)

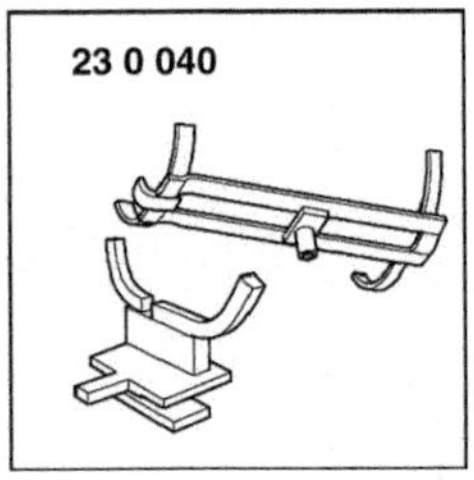


Transmission support (for use with transmission jack)
(BMW special tool 23 0 040)

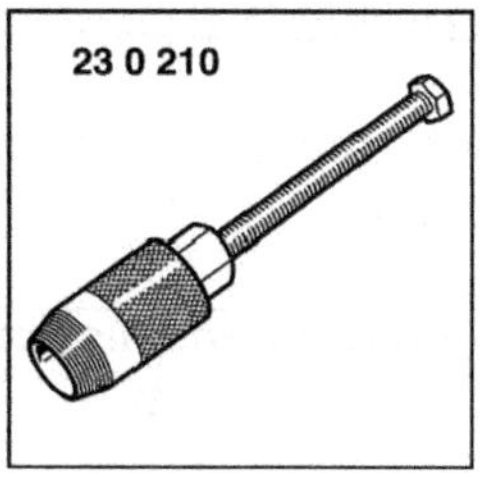


◀ Selector shaft seal extractor
(BMW special tool 23 0 210)

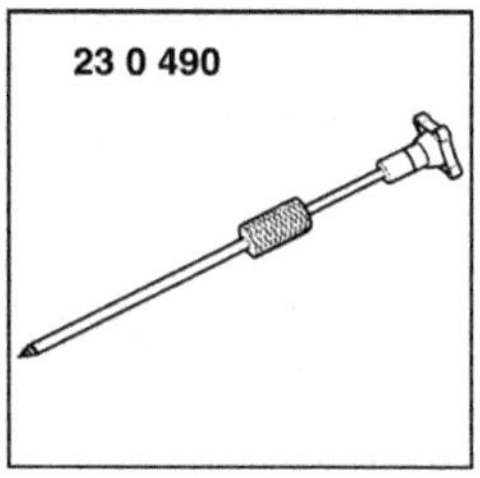


◀ Input shaft seal puller (slide hammer)
(BMW special tool 23 0 490)

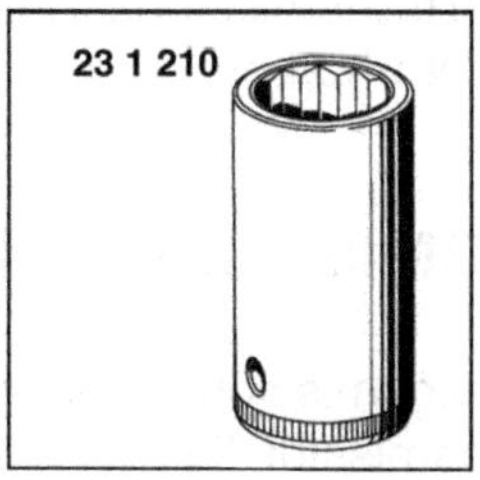


◀ Output shaft flange nut socket (27 mm)
(BMW special tool 23 1 210)

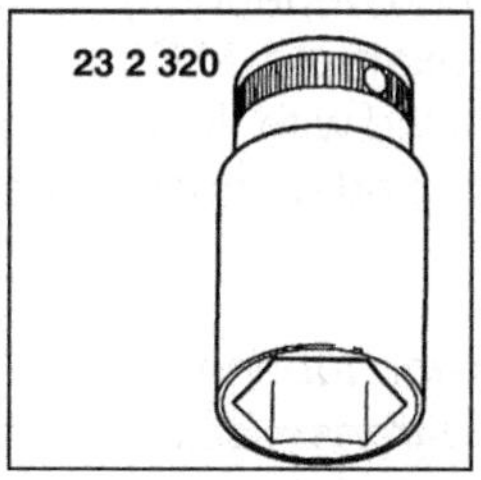


◀ Output shaft flange nut socket (36 mm)
(BMW special tool 23 2 320)

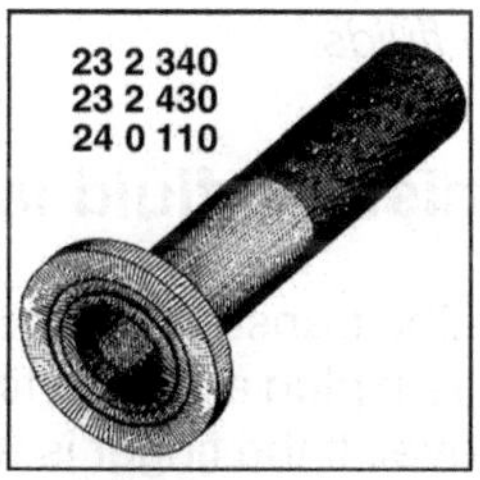


◀ Seal installation drift
(BMW special tool 023 2 340, 23 2 430, or 24 0 110)

◀ Input shaft seal installation sleeve
(BMW special tool 23 2 380 or 23 2 460)

### Transmission applications

Two 5-speed manual transmissions are used in Z3 / M Coupe and Roadsters.

**Table a. Manual transmission applications**

| Model | Year | Transmission |
|---|---|---|
| Z3 1.9 | 1996 - 1998 | Getrag S5D 250 G |
| Z3 2.3 | 1999 - 2000 | Getrag S5D 250 G |
| Z3 2.5i | 2001 - 2002 | Getrag S5D 250 G |
| Z3 2.8 | 1997 - 2000 | ZF S5D 320 Z |
| Z3 3.0i | 2001 - 2002 | ZF S5D 320 Z |
| M 3.2 S52 | 1998 - 2000 | ZF S5D 320 Z |
| M 3.2 S54 | 2001 - 2002 | ZF S5D 320 Z |

Manual transmissions have a metal ID plate mounted on the side of the transmissions. See **200 Transmission—General** for more information.

## TRANSMISSION FLUID SERVICE

Manual transmissions installed are normally filled with automatic transmission fluid (ATF), although alternative lubricants may be used depending on transmission type and model year. For example, M Coupe and Roadster transmissions are normally filled with "lifetime" manual transmission fluid (BMW MTF LT-1) and do not require periodic fluid change.

***NOTE—***
*Transmission lubricant type can be found on a label on the side of the transmission. Consult an authorized BMW dealer for the most up-to-date information regarding transmission operating fluids.*

### Transmission fluid level, checking

Checking the transmission oil level involves simply removing the side drain plug and inserting a finger into the hole to check the fluid level. If the finger is wetted by the transmission fluid, the level is correct.

### Transmission fluid, replacing

- Drive vehicle for a few miles to warm transmission.

- Raise and safely support vehicle.

***WARNING—***
*Make sure the car is stable and well supported at all times. Use a professional automotive lift or jack stands designed for the purpose. A floor jack is not adequate support.*

◀ Place a drain pan under transmission and remove drain plug (**2**) at bottom of transmission.

– Install and torque drain plug.

– Remove fill plug from side of transmission (**1**).

– Slowly fill with fluid until level is even with fill hole.

– Install and torque fluid fill plug.

| **Tightening torque** | |
|---|---|
| Transmission drain or fill plug to transmission | 50 Nm (37 ft-lb) |

| **Fluid capacity (use automatic transmission fluid)** | |
|---|---|
| S5D 250 G transmission | 1.0 liter (1.1 US qt) |
| S5D 320 Z transmission | 1.2 liter (1.3 US qt) |

## TRANSMISSION SERVICE

### Back-up light switch, replacing

– Raise and safely support vehicle.

> ***WARNING —***
> *Make sure the car is stable and well supported at all times. Use a professional automotive lift or jack stands designed for the purpose. A floor jack is not adequate support.*

◀ Disconnect harness connector (**arrow**) from back-up light switch.

***NOTE —***

- *Getrag transmission (S5D 250 G) from 2000 Z3 Roadster illustrated. The back-up light switch is located on the left of the transmission.*
- *The back-up light switch on ZF transmission (S5D 320 Z) is located on the right of the transmission.*

– Unscrew switch from transmission.

– Install new switch. Tighten switch to specification.

| **Tightening torque** | |
|---|---|
| Back-up light switch to transmission | 20 Nm (15 ft-lb) |

– Check transmission fluid level before lowering vehicle.

## Selector shaft seal, replacing

- Put transmission into second gear.
- Raise and safely support vehicle.

> ***WARNING—***
> *Make sure the car is stable and well supported at all times. Use a professional automotive lift or jack stands designed for the purpose. A floor jack is not adequate support.*

- Support transmission with transmission jack.
- Disconnect driveshaft at transmission output flange and hang driveshaft to side. See **260 Driveshaft**.
- Remove transmission crossmember and lower transmission at rear. Make sure cylinder head and manifold do not contact engine bulkhead or heater connections.

> ***CAUTION—***
> *Tilting the engine to lower the transmission can damage various components. Lower transmission slowly.*

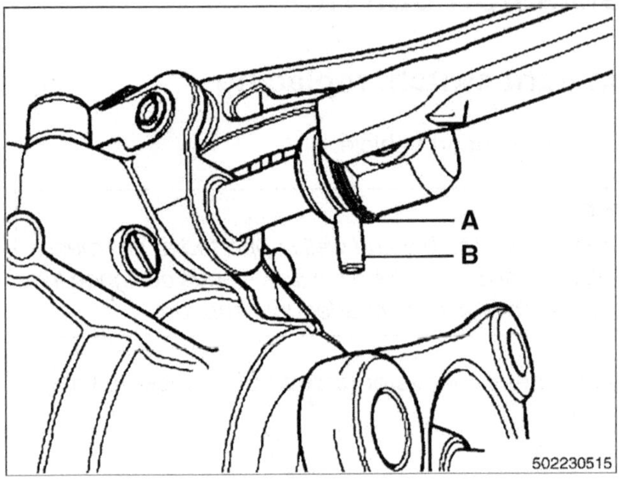


- Working at rear of selector shaft, disconnect shift linkage from selector shaft by prying back locking clip (**A**) and pushing locking pin (**B**) out.
- Detach shift rod from selector shaft and tie aside.

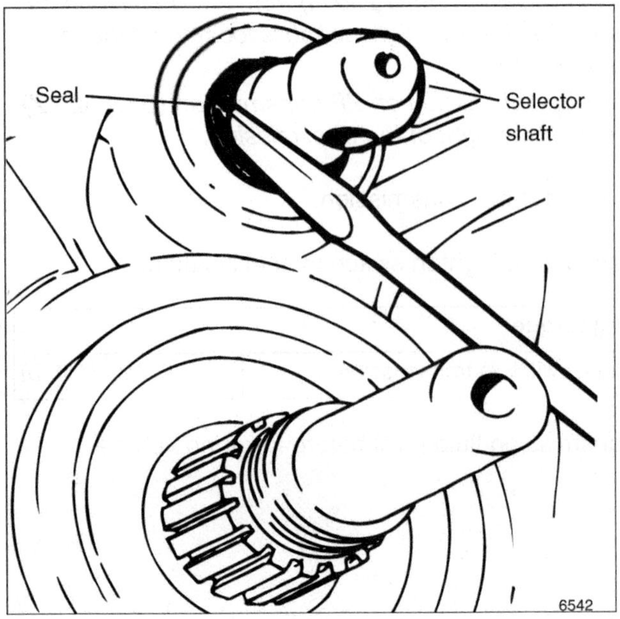


- Carefully pry seal out with BMW special tool 23 0 210, or a narrow seal remover or small screwdriver.

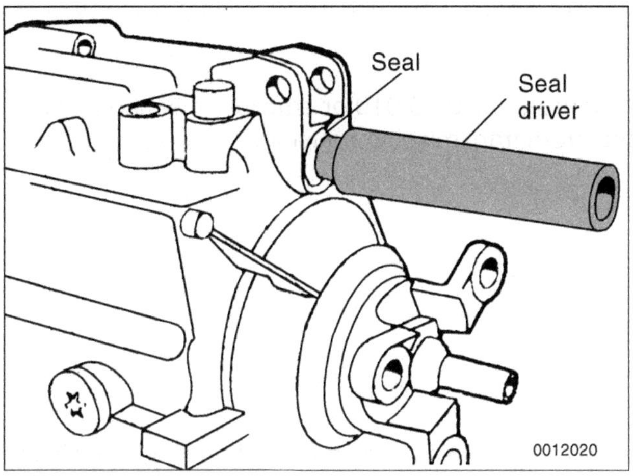


Coat new seal with transmission fluid. Drive new seal in flush with housing. Use BMW special tool 23 2 260 or equivalent and a soft-faced (plastic) hammer.

– Reinstall shift linkage to selector shaft.

– Raise transmission and reinstall crossmember.

– Install driveshaft using new nuts. See **260 Driveshaft**.

| Tightening torques | |
|---|---|
| Driveshaft to output flange | 48 Nm (35 ft-lb) |
| Transmission crossmember to chassis | 42 Nm (31 ft-lb) |

## Output shaft seal, replacing

– Raise and safely support vehicle.

> ***WARNING —***
> *Make sure the car is stable and well supported at all times. Use a professional automotive lift or jack stands designed for the purpose. A floor jack is not adequate support.*

– Support transmission with transmission jack.

– Disconnect driveshaft at transmission output flange and hang driveshaft to side. See **260 Driveshaft**.

– Remove transmission support bracket and lower transmission at the rear.

> ***CAUTION—***
> *When tilting engine and transmission assembly backward, the following components may be damaged through contact with the bulkhead or other components:*
> - *Ignition coil pack (4-cylinder engine). See* **120 Ignition System**.
> - *ASC throttle housing. See* **130 Fuel Injection**.
> - *Brake master cylinder. See* **340 Brakes**.

– Use screwdriver to pry out output flange collar nut retaining plate.

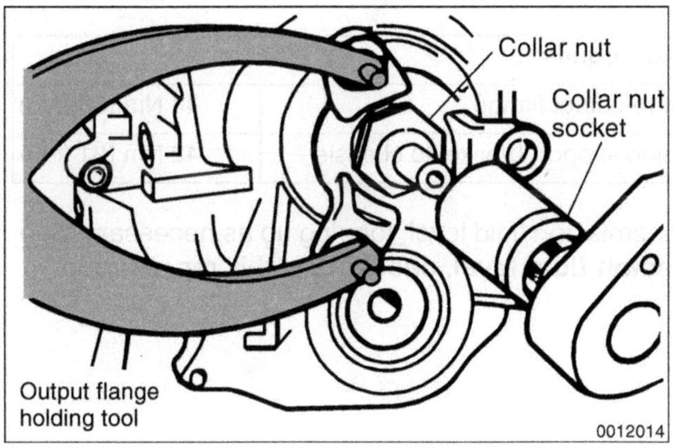


Use BMW special tool 23 0 020 or equivalent to counterhold output flange and remove collar nut with BMW special tool (see table below) or equivalent thin-walled deep socket.

| Collar nut sockets | |
|---|---|
| **Transmission** | **BMW special tool** |
| Getrag S5D 250G | 23 1 210 (27 mm) |
| ZF S5D 320Z | 23 2 320 (36 mm) |

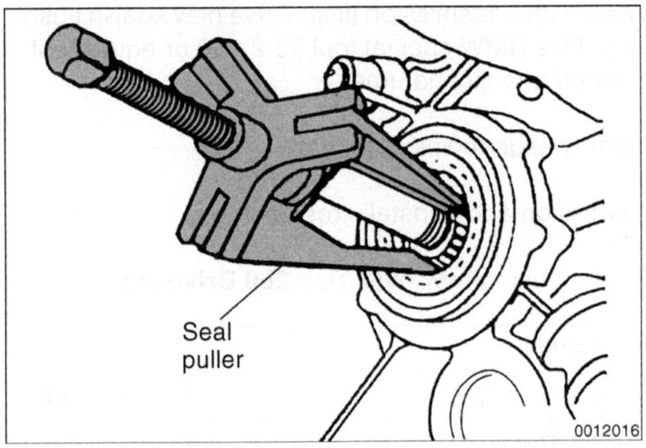


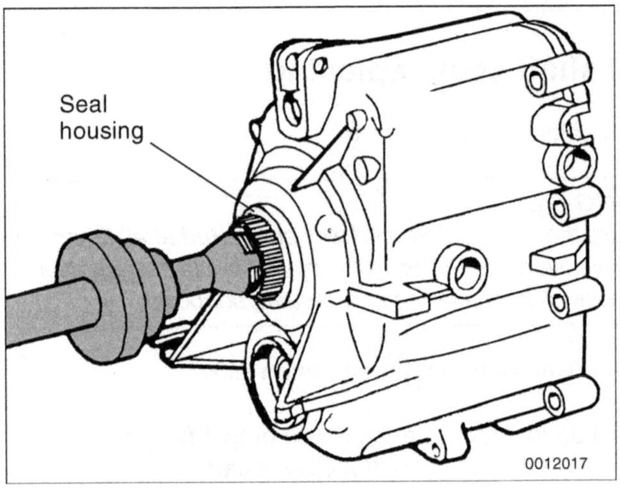


– Remove flange. If necessary, use a puller. Note placement of any shims.

◀ Use BMW special tool 00 5 010 or equivalent seal puller to remove seal from transmission housing.

◀ Coat new seal with transmission fluid and drive into position using BMW special tool 24 0 010 or suitable drift until seal is flush with housing.

– Reinstall flange to output shaft. Coat bearing surface of collar nut with sealer and install nut.

***NOTE—***
- *If any shims were removed, be sure to replace them in the correct position.*
- *BMW recommends the use of a sealer such as Loctite® 242 when installing the collar nut to prevent oil from leaking past the threaded shaft.*
- *On Getrag S5D 250 G transmission, BMW recommends heating the output flange to about 176°F (80°C) to aid in installation. This can be done by placing flange in hot water.*

– Tighten output flange collar nut then loosen and retighten to the correct torque specifications listed below. Install new lockplate. Bend tabs into flange grooves.

| **Tightening torque** | |
|---|---|
| Output flange to output shaft<br>• Stage I<br>• Stage II (after loosening) | <br>190 Nm (140 ft-lb)<br>120 Nm (89 ft-lb) |

– Raise transmission and reinstall crossmember.

– Install driveshaft using new nuts. See **260 Driveshaft.**

| **Tightening torques** | |
|---|---|
| Driveshaft to output flange | 48 Nm (35 ft-lb) |
| Transmission support bracket to chassis | 42 Nm (31 ft-lb) |

– Check transmission fluid level, topping up as necessary. See **Transmission fluid level, checking** in this repair group.

## Input shaft seal, replacing

Replacement of the input shaft seal requires the removal of the transmission from the vehicle as described in **Transmission Removal and Installation** in this repair group.

– With transmission on a workbench, remove input shaft seal cover from inside bellhousing.

– Remove clutch release bearing and release lever from inside bellhousing.

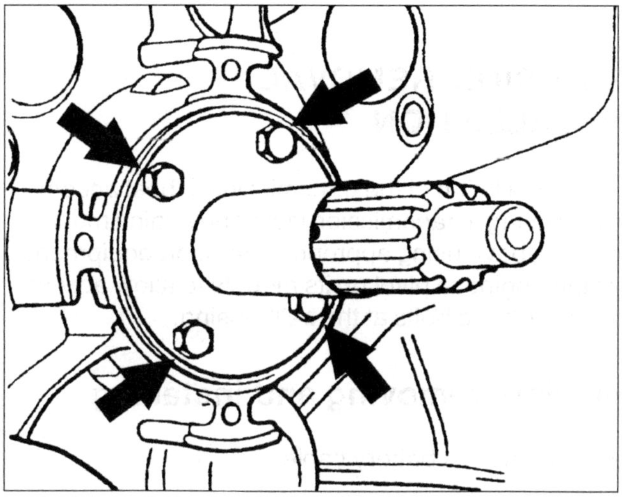

◀ Remove bolts for clutch release-bearing guide sleeve, noting bolt lengths. Remove sleeve and any shims. Note position of any shims.

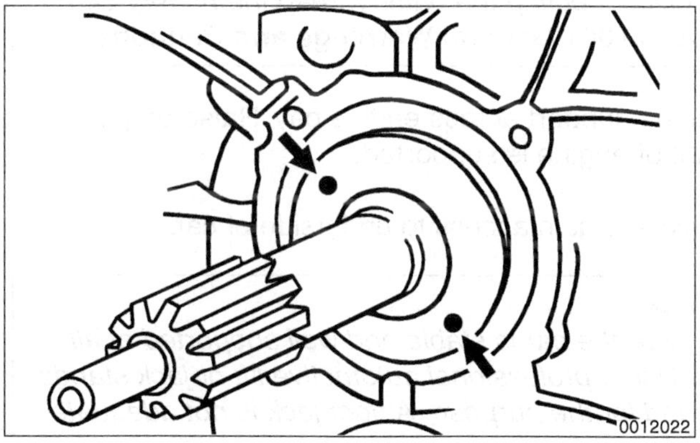


◀ Two covered removal holes (**arrows**) are present in seal. Thread BMW special tool 23 0 490 or equivalent slide hammer seal puller into one hole and remove seal.

– Getrag S5D 250G transmission:

- Place BMW special tool 23 2 460 or similar protective sleeve over input shaft.
- Coat sealing lip of new seal with transmission oil and push over protective sleeve up to transmission housing.
- Remove protective sleeve from input shaft.
- Use BMW special tool 23 2 430 or similar drift to drive seal until fully seated.

– ZF S5D 320Z transmission:

- Place BMW special tool 23 2 380 or similar protective sleeve over input shaft.
- Coat sealing lip of new seal with transmission oil and push over protective sleeve up to transmission housing.
- Remove protective sleeve from input shaft.
- Use BMW special tool 23 2 340 or similar drift to drive seal until fully seated.

- Thoroughly clean guide sleeve mounting bolts, sealing surfaces, and threads. Apply Loctite®243 or equivalent thread locking compound to guide sleeve bolts. Reinstall guide sleeve and spacer(s).

| Tightening torques | |
|---|---|
| Guide sleeve to transmission case | |
| • M6 | 10 Nm (7 ft-lb) |
| • M8 x 22 | 18 Nm (13 ft-lb) |
| • M8 x 30 | 25 Nm (18 ft-lb) |

- Reinstall transmission.

## TRANSMISSION REMOVAL AND INSTALLATION

Removal and installation of the transmission is best accomplished on a lift using a transmission jack. The engine must be supported from above using appropriate support equipment. This allows the engine to pivot on its mounts to allow access to the upper Torx-head bolts at the bellhousing.

### Transmission, removing and installing

- Disconnect negative (–) battery cable.

> ***CAUTION—***
> *Prior to disconnecting the battery, read the battery cautions given in* **001 General Warnings and Cautions**.

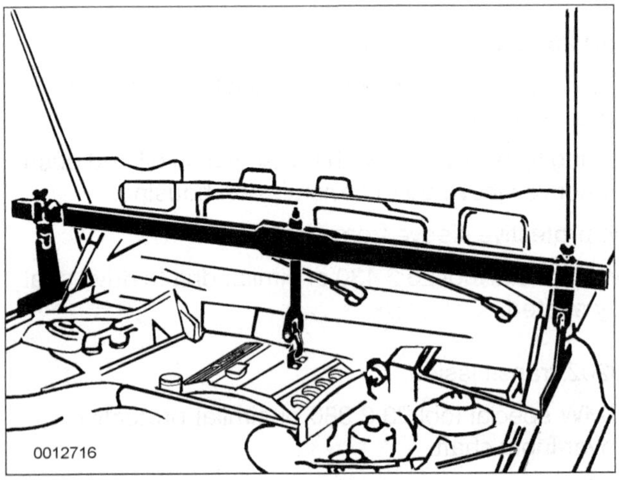


- Install engine support across engine bay. Raise engine so that weight of engine is supported.
- Raise vehicle to gain access to underside of car.

> ***WARNING—***
> - *Make sure the car is stable and well supported at all times. Use a professional automotive lift or jack stands designed for the purpose. A floor jack is not adequate support.*
> - *Removal of the transmission may upset the balance of the vehicle on a lift.*

- Disconnect harness connector from backup light switch on transmission.
- Remove complete exhaust system and heat shields. See **180 Exhaust System**.

***NOTE—***
*Disconnect oxygen sensor harness connector(s) before lowering exhaust system.*

- Remove driveshaft. See **260 Driveshaft**.
- Support transmission with transmission jack. If applicable, remove reinforcing brace from below engine/transmission.
- Remove rear transmission support bracket fasteners (**arrow**). Remove bracket and set aside.

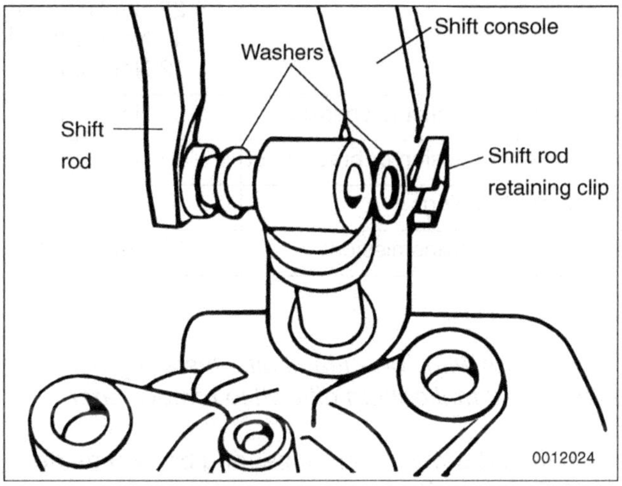


- Disconnect shift rod from transmission selector shaft. Disconnect shift console from top of transmission. See **250 Gearshift Linkage**.
- Unbolt clutch slave cylinder from side of transmission. Do not disconnect fluid hose. Suspend slave cylinder from body using stiff wire.

> ***CAUTION—***
> *Do not operate clutch with slave cylinder removed from transmission.*

- Lower transmission to access gearshift linkage.

> ***CAUTION—***
> *When tilting engine and transmission assembly backward, the following components may be damaged through contact with the bulkhead or other components:*
> - *Ignition coil pack (4-cylinder engine). See* **120 Ignition System**.
> - *ASC throttle housing. See* **130 Fuel Injection**.
> - *Brake master cylinder. See* **340 Brakes**.

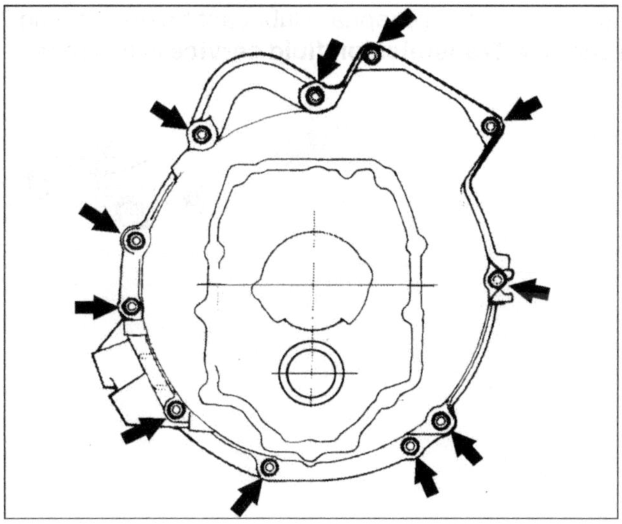

- Remove transmission mounting bolts (**arrows**) (Torx-head). Note length and location of bolts.

> ***CAUTION—***
> *Do not allow the transmission to be supported by the transmission input shaft.*

– Remove transmission by pulling backward until the transmission input shaft clears the clutch disc splines, then pull downwards. Lower jack and remove transmission.

> ***WARNING —***
> *Make sure car is stable and well supported at all times. Removal of transmission may upset balance of the vehicle.*

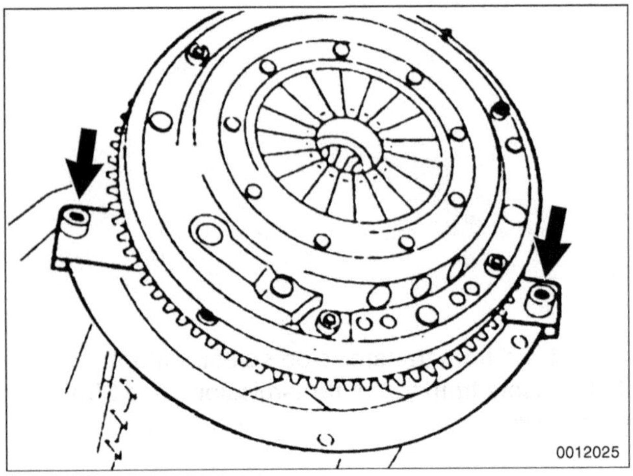

0012025

Installation is reverse of removal. Remember to:

- Thoroughly clean input shaft and clutch disc splines.
- Lightly lubricate transmission input shaft before installing.
- Make sure locating dowels (**arrows**) are undamaged.
- Center rear of transmission in driveshaft tunnel before tightening transmission support bracket.

| **Tightening torques** | |
|---|---|
| Transmission to engine (Torx-head)<br>• M8<br>• M10<br>• M12 | <br>22 Nm (16 ft-lb)<br>43 Nm (32 ft-lb)<br>72 Nm (53 ft-lb) |
| Transmission support bracket to chassis | 42 Nm (31 ft-lb) |
| Rubber mount to transmission or bracket nut | 42 Nm (31 ft-lb) |
| Transmission drain / fill plugs | 50 Nm (37 ft-lb) |
| Clutch slave cylinder to transmission | 22 Nm (16 ft-lb) |

***NOTE —***
*Always use washers with Torx-head bolts. Installing bolts without washers will make them difficult to remove later.*

– Install driveshaft and preload center bearing bracket as described in **260 Driveshaft**.

| **Tightening torque** | |
|---|---|
| Driveshaft to output flange | 48 Nm (35 ft-lb) |

– Refill transmission with appropriate lubricant before starting or towing car. See **Transmission fluid service** in this repair group.

# 240 Automatic Transmission

## General

This section covers maintenance and replacement of the automatic transmission for Z3 Coupe and Roadster models.

Automatic transmission internal repairs are not covered. Such repairs require special service equipment and knowledge. If it is determined that transmission internal service is required, consult an authorized BMW dealer about a factory reconditioned unit or a transmission rebuild.

> ***CAUTION—***
> *ATF does not circulate unless the engine is running. When towing an automatic transmission vehicle, use a flat-bed truck or raise the rear wheels off the ground.*

### Special tools

Some special tools are recommended for removal and installation of the automatic transmission. While these tools are not always required, they are especially helpful when dealing with the removal and replacement of the torque converter.

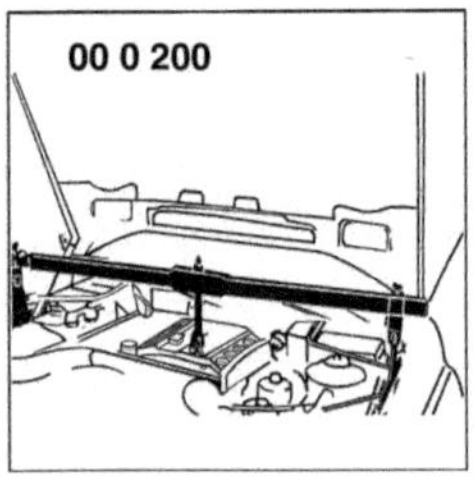


Engine support bracket
(BMW special tool set 00 0 200)

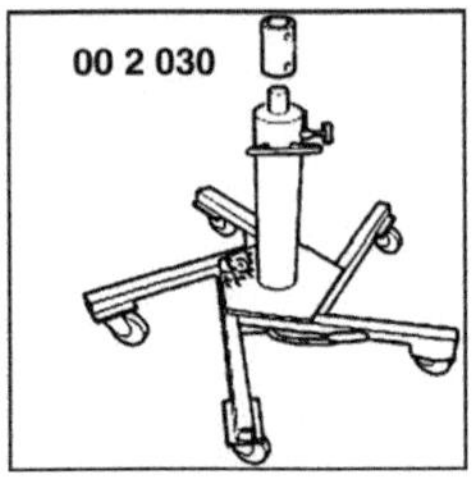


Transmission jack
(BMW special tool 00 2 030)

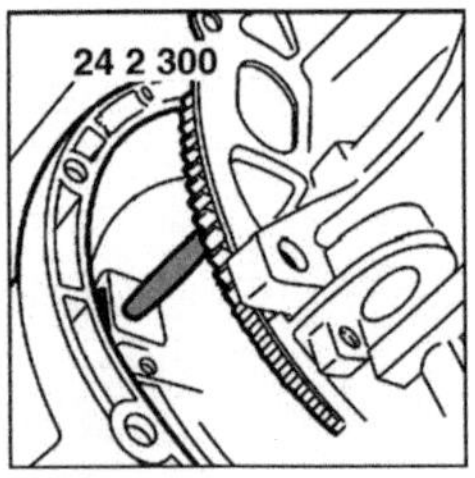


Spline bore alignment tool
(BMW special tool set 24 2 300)

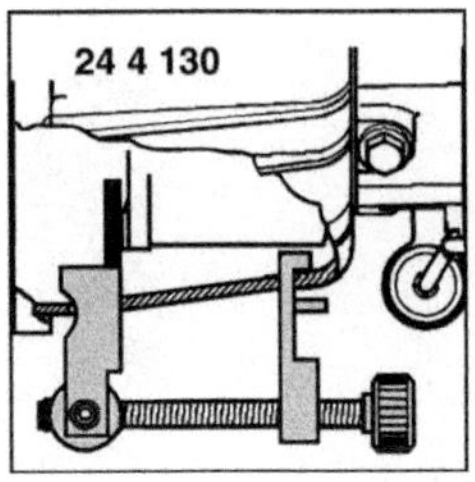


Torque converter clamp set
(BMW special tool set 24 4 130)

## Transmission applications

Depending on model year and engine size, Z3 cars are equipped with either a 4-speed automatic or a 5-speed Steptronic transmission.

M version cars are not equipped with automatic transmission.

| Automatic transmission applications | | |
|---|---|---|
| **Model** | **Year** | **Transmission** |
| **Z3 1.9** | 1996 - 1998 | A4S 270 R / A4S 310 R (4-speed) |
| **Z3 2.3** | 1999 - 2000 | A4S 270 R / A4S 310 R (4-speed) |
| **Z3 2.5i** | 2001 - 2002 | A5S 360 R / A5S 390 R (5-speed Steptronic) |
| **Z3 2.8** | 1997 - 2000 | A4S 270 R / A4S 310 R (4-speed) |
| **Z3 3.0i** | 2001 - 2002 | A5S 360 R / A5S 390 R (5-speed Steptronic) |

# TRANSMISSION SERVICE

## ATF level, checking

Automatic transmissions installed in Z3 models are normally filled with Dexron III automatic transmission fluid (ATF). Some transmissions are filled with special BMW "lifetime" ATF which does not require periodic fluid changes.

The transmission is not equipped with a dipstick. Checking the ATF level is a complicated procedure that requires measuring and maintaining a specified ATF temperature during the checking procedure.

***CAUTION—***
- *If the appropriate test equipment is not available, let an authorized BMW dealer check the ATF level. The checking procedure should be made on a lift with the engine running in neutral. In addition, the transmission fluid temperature must be maintained at a specified temperature at all times during the checking procedure.*
- *Be sure to use the correct type of ATF. If in doubt, consult an authorized BMW dealer for alternate fluid use and current information regarding transmission operating fluids.*

***NOTE—***
*ATF lubricant installed can be found on a type plate on the side of the transmission.*

– Place vehicle on a lift. Apply the parking brake.

***WARNING—***
*Make sure the car is stable and well supported at all times. Use a professional automotive lift or jack stands designed for the purpose. A floor jack is not adequate support.*

– Warm up drivetrain by running engine until the ATF reaches the specified temperature. Apply brake and shift through all gears several times. Turn engine off.

| ATF level checking | |
|---|---|
| Fluid temperature | 30 - 55°C (86 - 131°F) |

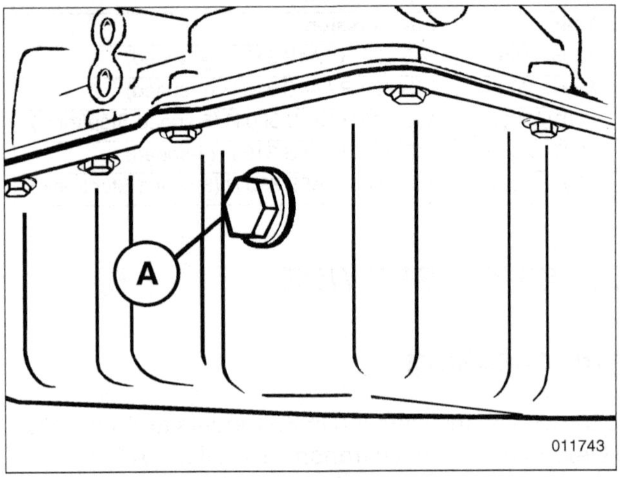


◄ Raise vehicle and remove ATF fill plug (**A**). Place transmission in PARK and restart engine.

***NOTE—***
*4-speed transmission shown. 5-speed is similar.*

– With ATF at the specified temperature, add fluid until it spills from filler hole. Reinstall fill plug.

> ***WARNING—***
> *Hot ATF can scald. Wear eye protection and protective clothing and gloves during the check. If the transmission was overfilled, hot ATF will spill from the filler hole when the fill plug is removed.*

| Tightening torque | |
|---|---|
| ATF fill plug to ATF sump<br>• A4S 270R/310R<br>• A5S 360R/390R | <br>33 Nm (24 ft-lb)<br>18 Nm (13 ft-lb) |

## ATF, draining and filling

The procedure given here includes removal and installation of the ATF fluid strainer (filter).

– Raise vehicle to gain access to transmission drain plug.

> ***WARNING—***
> *Make sure the car is stable and well supported at all times. Use a professional automotive lift or jack stands designed for the purpose. A floor jack is not adequate support.*

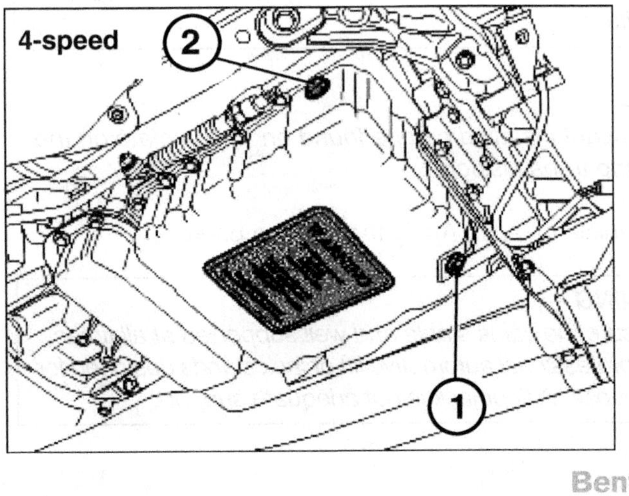


◄ Remove drain plug (**1**) and drain fluid into container.

> ***WARNING—***
> *The ATF should be warm when draining. Do not drain ATF if the engine and/or transmission is hot. Hot ATF can scald.*

– Remove mounting bolts from oil sump and remove sump and sump gasket.

***NOTE—***
*4-speed transmission has a front and rear sump. Draining procedure is similar for both.*

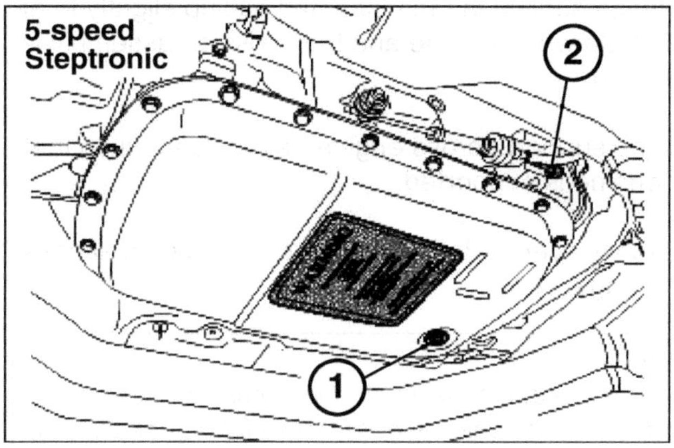


– A4S 270R/310R (4-speed): Remove ATF strainer mounting bolts and remove strainer together with gasket.

< A5S 360R/390R (5-speed): Pull ATF strainer and strainer sealing ring from transmission pump housing.

– Installation is the reverse of removal. Remember to:
  - Replace drain plug seal and all gasket(s).
  - Clean sump and sump magnet using a lint-free cloth.
  - Use new sump bolts and tighten bolts in-line, one after the other (i.e. do not tighten the bolts in a cross-wise pattern).
  - Fill transmission with the correct amount of fluid. See **ATF level, checking** in this repair group.

| **ATF capacity** | |
|---|---|
| A4S 270R/310R<br>• with torque converter<br>• without torque converter | <br>8.8 liters (9.3 US qt)<br>7.8 liters (8.2 US qt) |
| A5S 360R/390R<br>• with torque converter<br>• without torque converter | <br>9.0 liters (9.5 US qt)<br>8.0 liters (8.4 US qt) |

| **Tightening torques** | |
|---|---|
| ATF drain plug to ATF sump<br>• A4S 270R/310R<br>• A5S 360R/390R | <br>25 Nm (18 ft-lb)<br>18 Nm (13 ft-lb) |
| ATF strainer bolts to valve body | 5 Nm (3.7 ft-lb) |
| ATF sump to transmission | 12 Nm (9 ft-lb) |

## TRANSMISSION REMOVAL AND INSTALLATION

Removing and installing the transmission is best accomplished on a lift using a transmission jack. Use caution and safe workshop practices when working beneath car and lowering transmission.

Torx-head bolts are used to mount the transmission to the bellhousing. Be sure to have appropriate tools on hand before starting the job.

### Transmission, removing and installing

– Disconnect negative (–) battery cable.

> ***CAUTION—***
> *Prior to disconnecting the battery, read the battery cautions given in* **001 General Warnings and Cautions**.

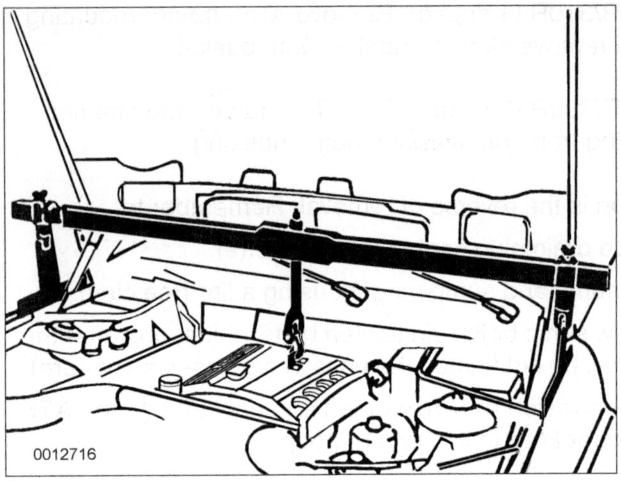


- Loosen engine cooling fan housing and pull up slightly to allow fan to tilt up when engine and transmission assembly is tilted backward.

< Install engine support across engine bay. Raise tool so that weight of engine is supported.

- Place gearshift lever in PARK. Raise vehicle to gain access to undercarriage.

> ***WARNING—***
> - *Make sure the car is stable and well supported at all times. Use a professional automotive lift or jack stands designed for the purpose. A floor jack is not adequate support.*
> - *Removal of the transmission may upset the balance of the vehicle on the lift.*

< Remove front crossbrace mounting bolts (**arrows**) and set crossbrace aside.

- Drain ATF from transmission. See **ATF, draining and filling** in this repair group.

- Remove exhaust system and heat shields. See **180 Exhaust System**.

***NOTE—***
*Disconnect oxygen sensor harness connector(s) before lowering exhaust system.*

- Disconnect driveshaft from transmission and tie aside. See **260 Driveshaft**.

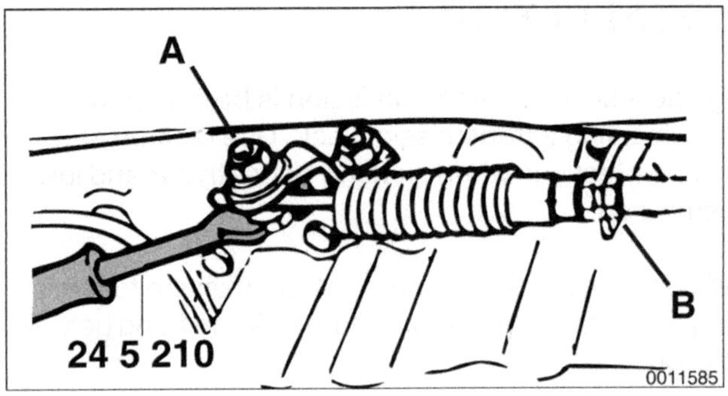


< With transmission in PARK, use BMW special tool 24 5 210 or equivalent thin wrench to counterhold selector shaft linkage clamping bolt at left side of transmission. This helps prevent twisting damage to cable end.

- Loosen clamping nut (**A**).
- Loosen cable clamping nut and remove cable from bracket (**B**).

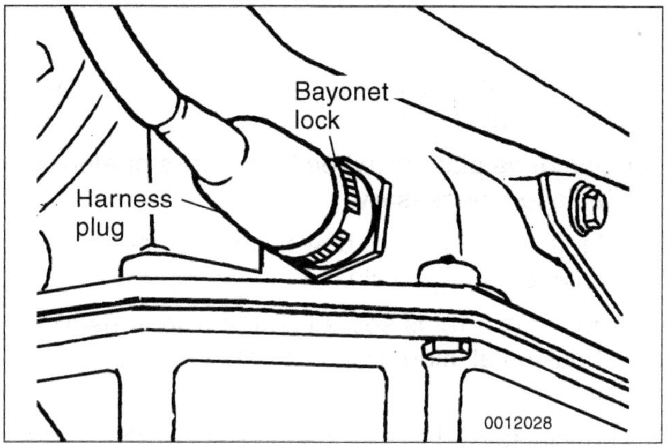


◄ Disconnect main harness connector from transmission by turning bayonet lock counterclockwise.

– Remove electrical connector from speed sensor on top of transmission.

– Disconnect transmission cooler line clamps from engine. Disconnect lines from transmission.

*NOTE —*
*Place pan under transmission lines to catch dripping fluid.*

– Support transmission with transmission jack. Remove transmission rear support bracket.

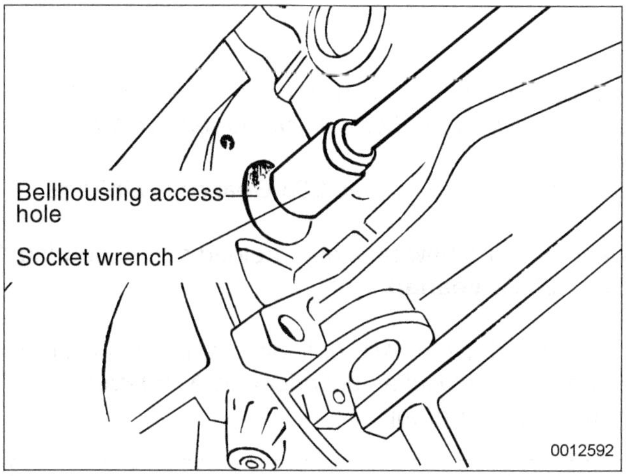


◄ Remove access plug in cover plate and remove torque converter bolts. Turn crankshaft to access converter bolts.

– Lower and tilt back engine and transmission assembly to access bellhousing-to-engine mounting bolts. Make sure cylinder head and manifold do not contact engine bulkhead or heater connections.

> ***CAUTION—***
> *When tilting engine and transmission assembly backward, the following components may be damaged through contact with the bulkhead or other components:*
> - *Ignition coil pack (4-cylinder engine). See* **120 Ignition System**.
> - *ASC throttle housing. See* **130 Fuel Injection**.
> - *Brake master cylinder. See* **340 Brakes**.

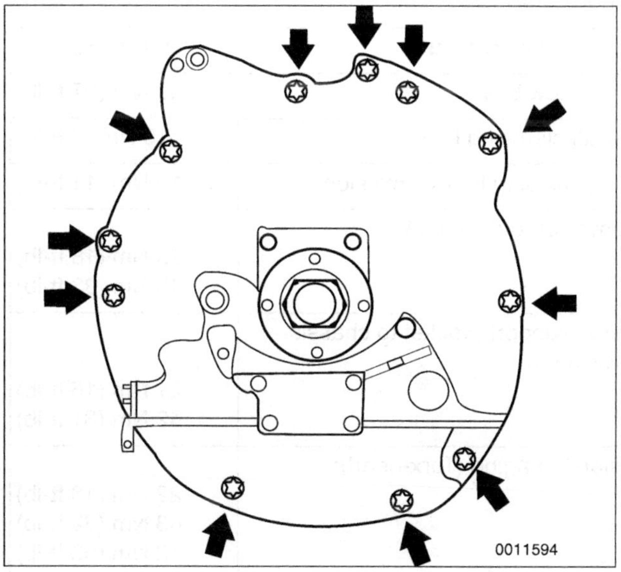


◄ Remove bellhousing-to-engine mounting bolts (**arrows**).

– Place a wedge of wood between engine oil pan and front subframe to prevent engine from dropping when transmission is removed.

– Remove transmission by pulling back and downwards. Lower jack to allow transmission to clear vehicle.

> ***CAUTION—***
> *Do not allow the torque converter to fall off the transmission input shaft. Use BMW special tool 24 4 130, or equivalent clamping tool, to hold toque converter.*

– If replacing faulty transmission, blow out oil cooler lines with low-pressure compressed air and flush cooler with clean ATF twice before reattaching lines to transmission.

> ***CAUTION—***
> - *Wear safety glasses when working with compressed air.*
> - *Do not reuse ATF used for flushing.*

– Automatic transmission installation is reverse of removal, noting the following:
  - Install new O-rings and sealing washers on transmission cooler lines and banjo bolts.
  - Inspect engine drive plate (flywheel) for cracks or elongated holes. Replace if necessary.
  - Be sure bellhousing alignment dowel sleeves are located correctly. Replace damaged sleeves.
  - Be sure torque converter is seated correctly in transmission during final installation. Use special torque converter mounting bolts.
  - Use BMW special tool 24 2 300 or equivalent to align three tabs on torque converter with indentations on engine drive plate.

***NOTE—***
*Always use washers with Torx-head bolts. Installing bolts without washers will make them difficult to remove later.*

– Adjust gearshift mechanism. See **250 Gearshift Linkage**.

– Install driveshaft using new nuts and preload center bearing bracket. See **260 Driveshaft**.

– Fill transmission with clean ATF until oil level is even with fill hole. Then check fluid level as described in **ATF level, checking** in this repair group.

| **Tightening torques** | |
|---|---|
| ATF drain plug to ATF sump<br>• A4S 270R/310R<br>• A5S 360R/390R | <br>25 Nm (18 ft-lb)<br>18 Nm (13 ft-lb) |
| ATF cooler line coupling nuts | 20 Nm (15 ft-lb) |
| ATF cooler hollow bolt | 37 Nm (27 ft-lb) |
| Selector shaft clamping bolt | 10 Nm (7 ft-lb) |
| Shifter cable housing to transmission | 15 Nm (11 ft-lb) |
| Torque converter to drive plate:<br>• M8<br>• M10 | <br>25 Nm (18 ft-lb)<br>45 Nm (33 ft-lb) |
| Transmission support bracket to chassis or transmission:<br>• M8<br>• M10 | <br><br>21 Nm (15 ft-lb)<br>42 Nm (31 ft-lb) |
| Transmission to engine (Torx-head):<br>• M8<br>• M10<br>• M12 | <br>22 Nm (16 ft-lb)<br>43 Nm (32 ft-lb)<br>72 Nm (53 ft-lb) |

# 250 Gearshift Linkage

## General

This repair group covers transmission gearshift and linkage service for both manual and automatic transmission equipped vehicles.

### Special tools

BMW recommends the use of one special tool for removal of the gearshift lever in manual transmission equipped vehicles.

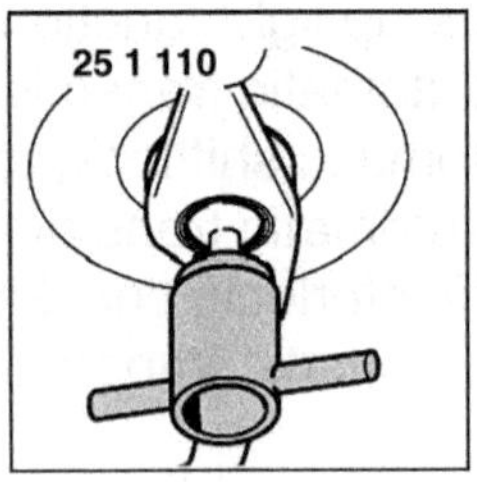


Shifter spanner (BMW special tool 25 1 110)

## Manual transmission gearshift

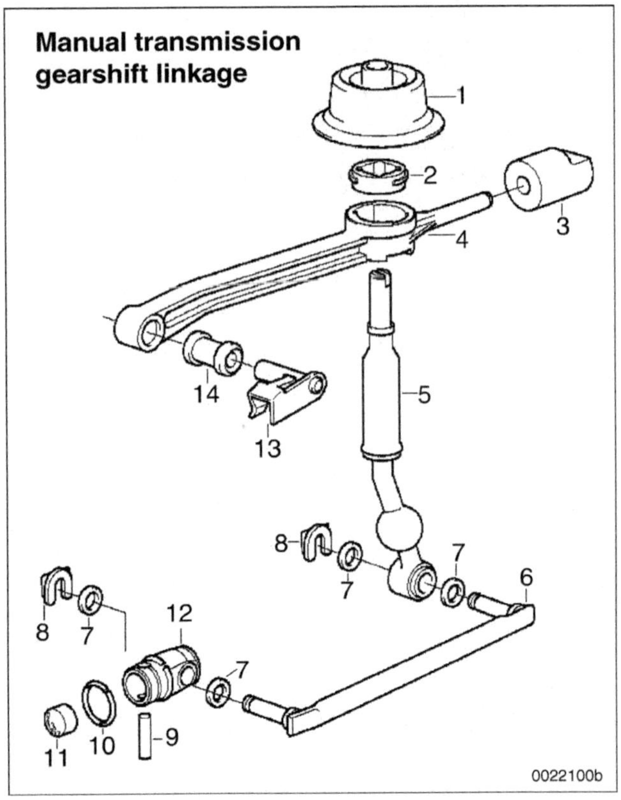


Gearshift linkage

1. Rubber boot
2. Shift lever bearing
3. Shift arm bearing
4. Shift arm
5. Shift lever
6. Selector link rod
7. Spacer
8. Circlip
9. Roll pin
10. Lock ring
11. Washer
12. Gear selector rod coupling
13. Locking pin
14. Bushing

### Gearshift linkage, removing (manual transmission)

To gain access to the complete gearshift mechanism it is necessary to remove the exhaust system and the driveshaft as described in **180 Exhaust System** and **260 Driveshaft**.

Use the exploded view illustration as a guide when removing and installing the manual gearshift.

– Pull straight up on shift knob to remove it from shift lever.

*NOTE—*
*Removal of the shift knob requires about 90 pounds of force. Do not twist knob or locating key will be damaged.*

– Pry up on rear of shift boot to unclip, then remove boot from front retainers. Pull boot up and off shift lever.

– Raise vehicle to gain access to undercarriage.

> ***WARNING—***
> *Make sure car is stable and well supported at all times. Use a professional automotive lift or jack stands designed for the purpose. A floor jack is not adequate support.*

– Disconnect oxygen sensor harness connectors.

– Remove complete exhaust system and heat shield. See **180 Exhaust System**.

– Remove driveshaft. See **260 Driveshaft**.

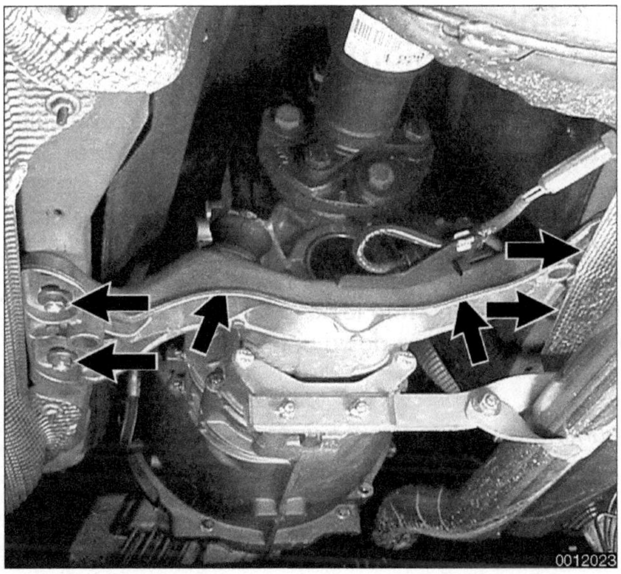


◄ Support transmission with transmission jack. Remove transmission rear support bracket fasteners (**arrows**). Set bracket aside.

– Lower transmission to access gearshift linkage.

> ***CAUTION—***
> *When tilting engine and transmission assembly backward, the following components may be damaged through contact with the bulkhead or other components:*
> - *Ignition coil pack (4-cylinder engine). See* **120 Ignition System**.
> - *ASC throttle housing. See* **130 Fuel Injection**.
> - *Brake master cylinder. See* **340 Brakes**.

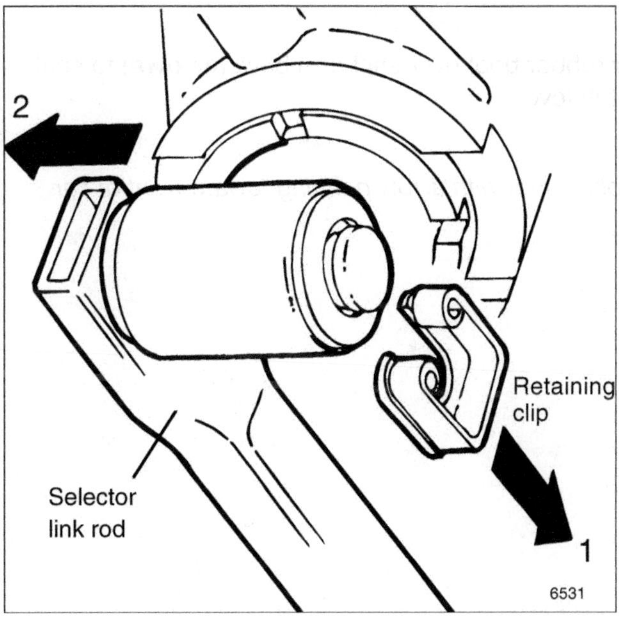


◄ Working above transmission, disconnect selector link rod from shift lever:

- Pull off shift rod retaining clip in direction of arrow (**1**).
- Disengage selector link rod from shift lever (**2**).
- Remove spacers on either side of shift lever end.

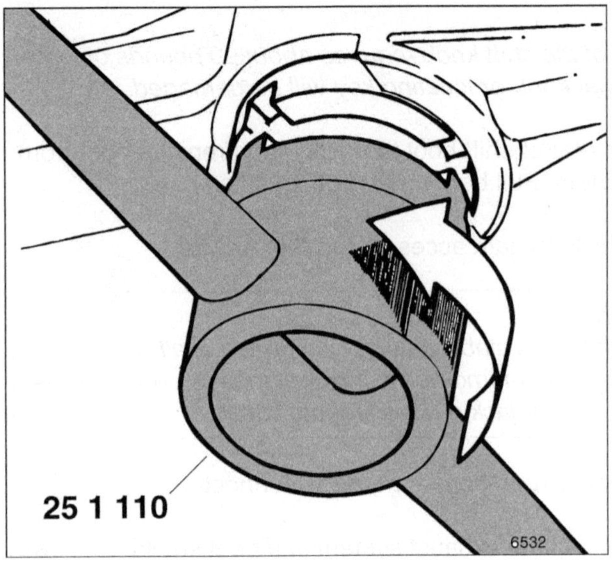


Release gearshift lever mounting ring using BMW special tool 25 1 110. Turn tool 90° (¼ turn) counterclockwise.

- Raise transmission and temporarily install transmission rear support bracket.
- Lower vehicle.
- Working from inside passenger compartment, pull up on gearshift lever to remove it together with inner rubber boot and retaining ring.

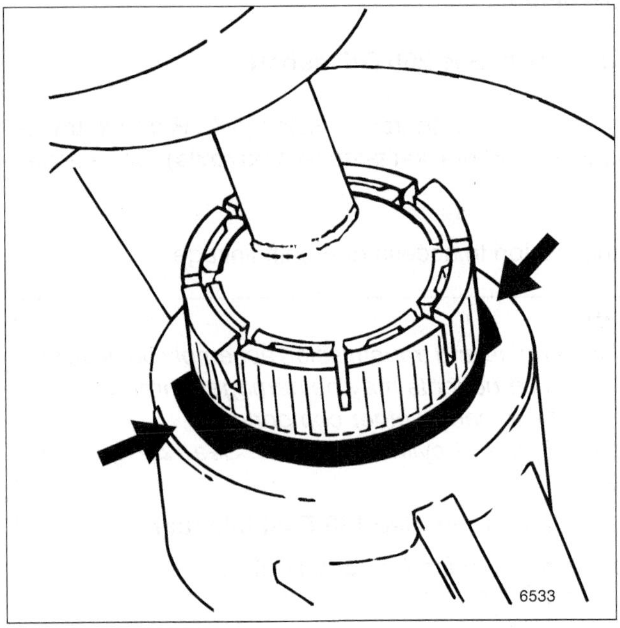


## Gearshift linkage, installing (manual transmission)

- Clean old grease from shift lever ball, mounting ring and shift arm bowl. Lubricate with molybdenum disulfide grease.
- Insert shift lever together with inner boot and mounting ring into shift arm bowl.

Align shift lever mounting ring locking tabs with slots (**arrows**) in shift arm. Press down until ring snaps into place.

***NOTE—***
*Install mounting ring with arrow pointing toward front of car.*

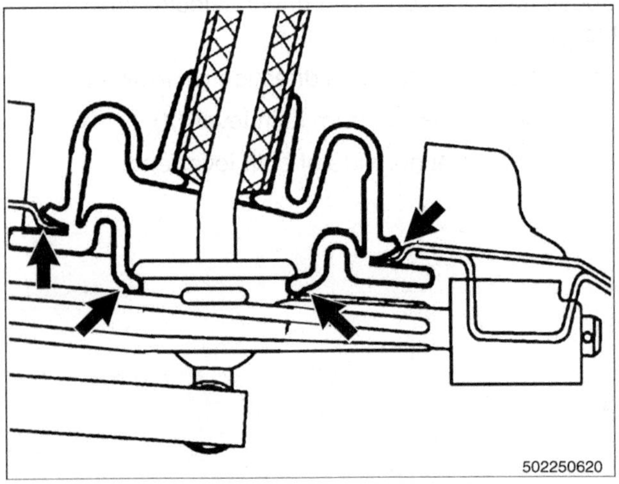


Pull inner rubber boot over shift arm bowl (**arrows**) to seal base of shift lever.

***NOTE—***
*Install rubber boot with arrow pointing toward front of car.*

– Raise car and support transmission. Remove transmission rear support bracket and lower transmission to access shift lever.

> ***CAUTION—***
> *When tilting engine and transmission assembly backward, the following components may be damaged through contact with the bulkhead or other components:*
> - *Ignition coil pack (4-cylinder engine). See* **120 Ignition System**.
> - *ASC throttle housing. See* **130 Fuel Injection**.
> - *Brake master cylinder. See* **340 Brakes**.

– Connect selector link rod to shift lever. Lift transmission, install rear support bracket and tighten mounting bolts.

- Install driveshaft and heat shield. See **260 Driveshaft**.
- Reinstall exhaust system. See **180 Exhaust System**.
- Lower vehicle. Install shift boot. Push shift knob on lever.

| **Tightening torques** | |
|---|---|
| Transmission rear support bracket<br>• To transmission<br>• To chassis | <br>21 Nm (15 ft-lb)<br>42 Nm (30 ft-lb) |

Automatic transmission gearshift (4-speed transmission)

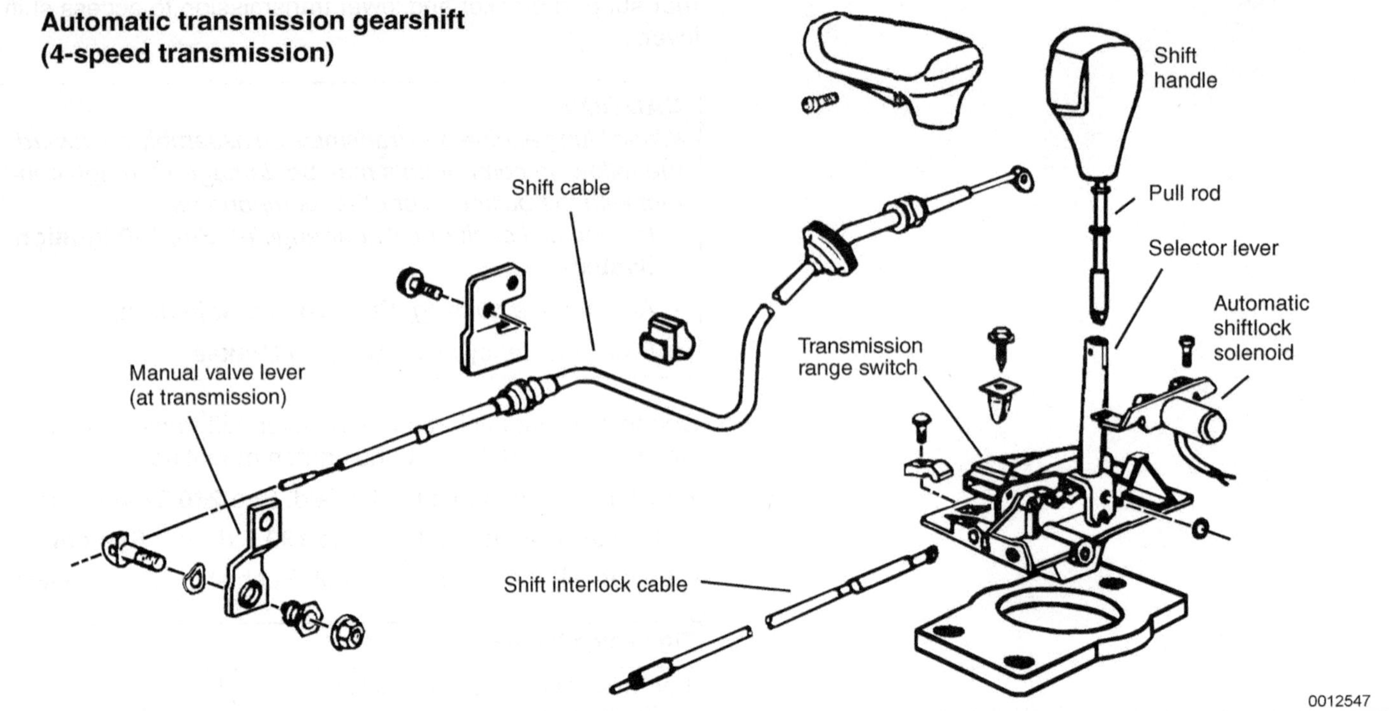


Steptronic transmission gearshift (5-speed)

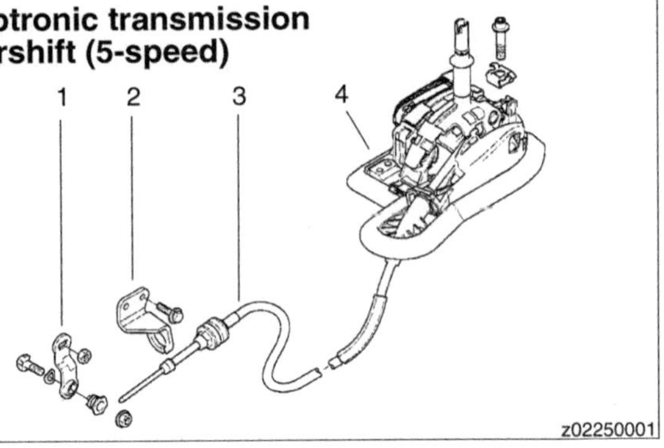


## Automatic Transmission Gearshift

Use the exploded-view illustration on this page as a guide when removing or installing the automatic transmission gearshift mechanism.

### Gearshift mechanism, adjusting (automatic transmission)

- Position selector lever in PARK.
- Raise vehicle to gain access to shift linkage.

> ***WARNING —***
> *Make sure car is stable and well supported at all times. Use a professional automotive lift or jack stands designed for the purpose. A floor jack is not adequate support.*

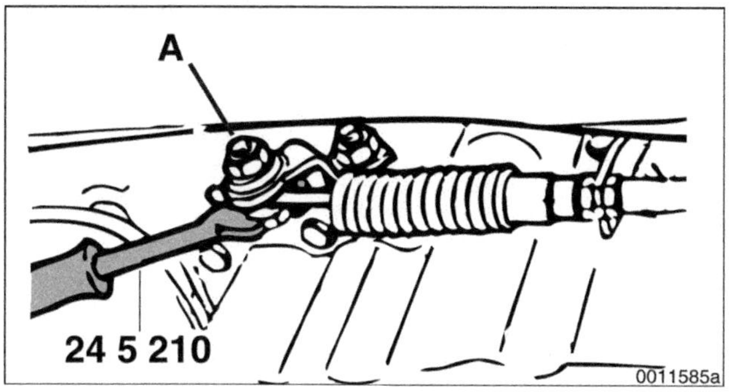


With transmission in PARK, use BMW special tool 24 5 210 or equivalent thin wrench to counterhold selector shaft linkage clamping bolt at left side of transmission. This helps prevent damaging cable end.

- Loosen clamping nut (**A**).
- Push shift lever forward toward engine (**PARK** position) while applying light rearward pressure on cable end. Tighten cable clamping nut.

***NOTE—***
*Do not overtighten nut so that it twists or distorts the cable.*

| Tightening torque | |
|---|---|
| Shift cable clamping nut | 10 Nm (7.5 ft-lb) |

## Transmission range switch, replacing (4-speed automatic transmission)

The transmission range switch functions as the neutral safety switch. Failure of the transmission range switch can cause a no-start condition. In addition, since it signals gearshift position to the automatic transmission control module, a faulty switch can cause erratic transmission operation.

– Remove shift knob:

- 1996 model: Remove hex-head screw (3 mm Allen wrench) from shift lever handle and remove handle. Carefully pry up cover from gearshift lever console.
- 1997 and later model: Pull up forcibly on shift knob to remove.

***NOTE—***
*Removal of the shift knob will require about 90 pounds of force. Do not twist knob or locating key will be damaged.*

– Pry up gently to remove gearshift boot and bezel.

- Use plastic prying tool or screwdriver with tip wrapped with tape.
- Disconnect window switches and any other electrical connectors on bezel.

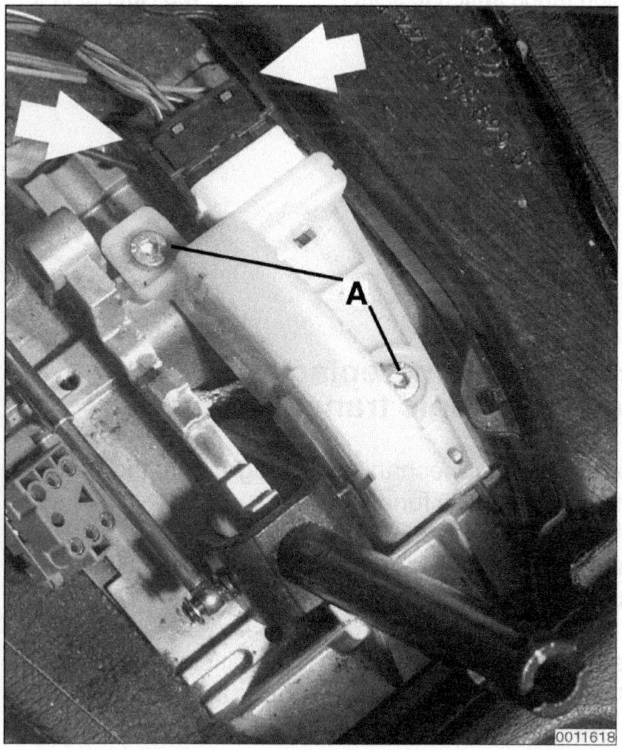


Disconnect range switch harness connector (**arrows**) and remove mounting screws (**A**).

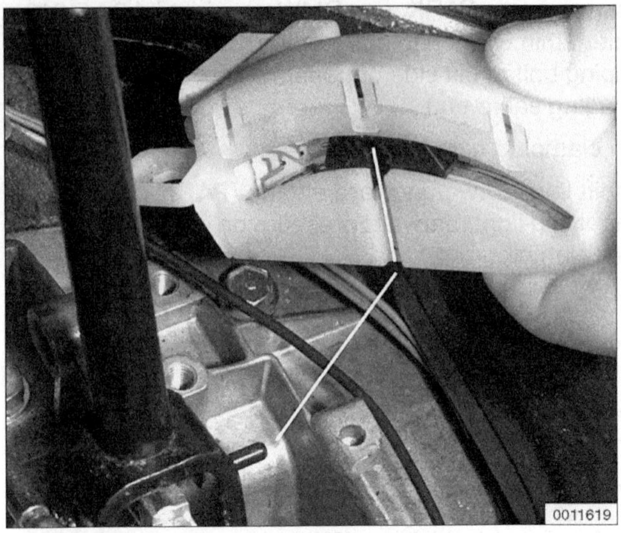


Remove range switch by aligning contact slide slot with drive pin on gearshift lever.

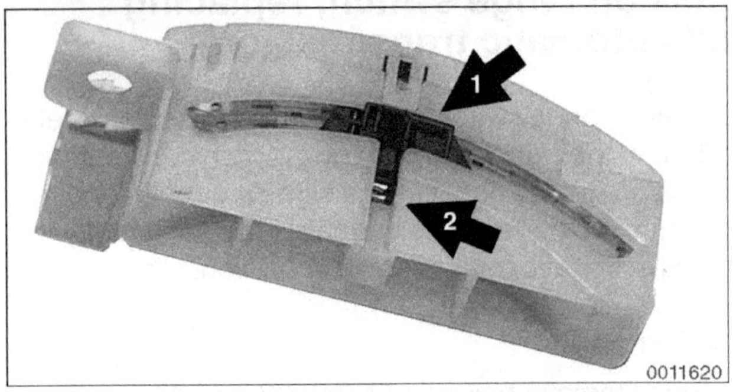


To reinstall switch, align plastic contact slide (**1**) with slot in switch (**2**).

- Install switch so that contact slide on switch aligns with drive pin on lever.

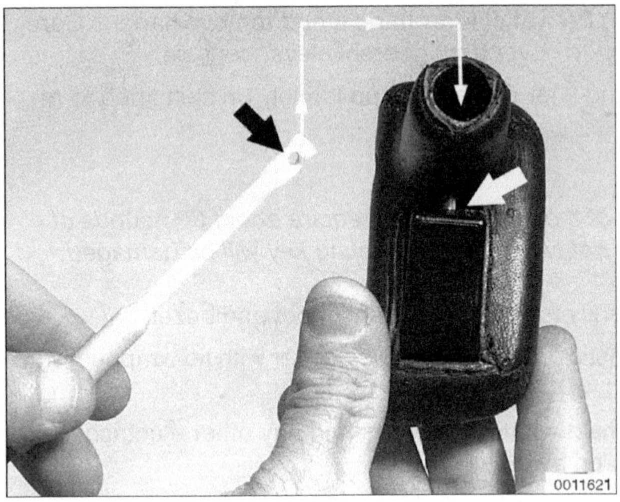


1996 model: When installing shift lever handle, engage pin on release button to hole in pull rod.

| **Tightening torques** | |
|---|---|
| Transmission range switch to shifter | 4.5 Nm (40 in-lb) |

## Shifter mechanism, replacing (5-speed Steptronic transmission)

The Steptronic shifter mechanism is integral with the transmission range switch. Its functions are:

**Neutral safety switch**. Transmission range switch failure can cause no-start condition.

**Gear position signal** to automatic transmission control module (EGS). Switch failure can cause erratic transmission operation.

Do not disassemble shifter mechanism. No repair parts are available separately.

– Place gearshift lever in PARK. Raise vehicle to gain access to undercarriage.

> ***WARNING —***
> *Make sure the car is stable and well supported at all times. Use a professional automotive lift or jack stands designed for the purpose. A floor jack is not adequate support.*

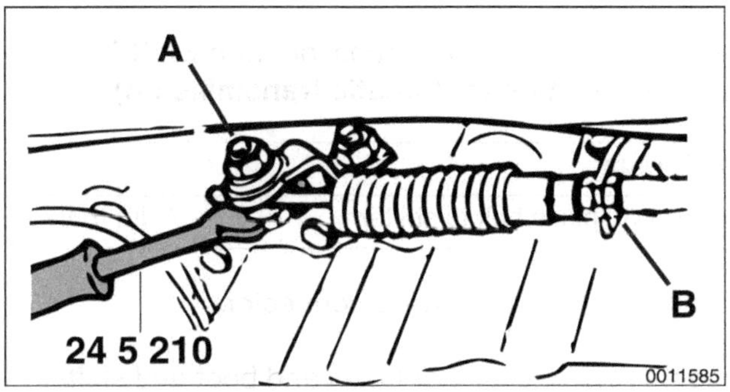


With transmission in PARK, use BMW special tool 24 5 210 or equivalent thin wrench to counterhold selector shaft linkage clamping bolt at left side of transmission. This helps prevent twisting damage to cable end.

- Loosen clamping nut (**A**).
- Loosen cable clamping nut and remove cable from bracket (**B**).

– Lower vehicle and remove shift knob. Pull up forcibly on shift knob to remove.

***NOTE —***
*Removal of the shift knob will require about 90 pounds of force. Do not twist knob or locating key will be damaged.*

– Pry up gently to remove gearshift boot and bezel.

- Use plastic prying tool or screwdriver with tip wrapped with tape.
- Disconnect window switches and any other electrical connectors on bezel.

– Remove center console. See **513 Interior Trim**.

– Disconnect shifter electrical connector.

– Remove interlock cable pinch bolt.

Remove shifter mounting bolts (**arrows**).

- Tilt shifter forward slightly to unhook interlock cable.
- Lift shifter out with shifter cable attached.

– When reinstalling, make sure shifter cable rubber seal (grommet) fits in center tunnel correctly.

| Tightening torque | |
|---|---|
| Shifter to center tunnel | 7 Nm (5 ft-lb) |

– Working underneath vehicle, attach shifter cable to transmission selector lever, then adjust. See **Gearshift mechanism, adjusting (automatic transmission)**.

> ***CAUTION—***
> *Wire tie shifter cable away from rotating drive shaft.*

| Tightening torques | |
|---|---|
| Selector shaft clamping bolt | 10 Nm (7 ft-lb) |
| Shifter cable housing to transmission | 15 Nm (11 ft-lb) |

– Adjust interlock cable for correct function. See **Shift interlock, checking function (automatic transmission)**.

| Tightening torque | |
|---|---|
| Interlock cable end pinch bolt to shifter | 7 Nm (5 ft-lb) |

– Reattach shifter electrical harness connector.

– Reinstall center console, shifter bezel and boot and shifter knob.

> ***CAUTION—***
> *Clip electrical harnesses to underside of shifter bezel to prevent scuffing damage from shifter.*

## Back-up light function (automatic transmission)

### 4-speed automatic transmission

Automatic transmission range switch triggers the back-up lights when placed in REVERSE.

### 5-speed Steptronic tranmission

Steptronic shifter sends REVERSE gear signal to automatic transmission control module (EGS). EGS triggers back-up light relay, K6325, in front power distribution box. See **610 Electrical Component Locations**.

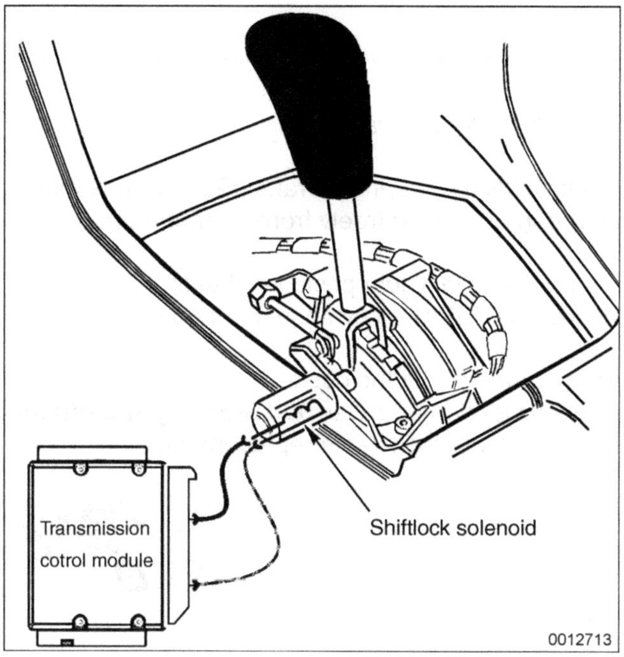


## Automatic shiftlock, checking function (automatic transmission)

◀ The automatic shiftlock uses an electric solenoid to lock the selector lever in PARK (P) or NEUTRAL (N). Depressing the foot brake with the ignition ON energizes the solenoid, allowing the lever to be moved into a drive gear. The solenoid is energized only when the engine speed is below 2,500 rpm and the vehicle speed is below 3 mph. The solenoid is mounted in the selector lever housing.

- With engine running and car stopped, place selector lever in P or N.
- Without depressing brake pedal, check that selector lever is locked in position (P or N).
- Depress brake pedal firmly. Solenoid should click audibly.
- Check that selector lever can now be moved out of P or N.

> ***WARNING —***
> *The next test should be performed in an open area with the parking brake on and with extreme caution.*

- With selector lever in P or N and brake pedal depressed, raise engine above 2,500 rpm. Check that selector lever cannot be moved out of P or N.
- If any faults are found check the electrical operation of shiftlock solenoid and check for wiring faults to or from transmission control module. See **ELE Electrical Wiring Diagrams**.

***NOTE —***
*The solenoid is controlled via the control module, using brake pedal position, engine speed and road speed as controlling inputs.*

## Shift interlock, checking function (automatic transmission)

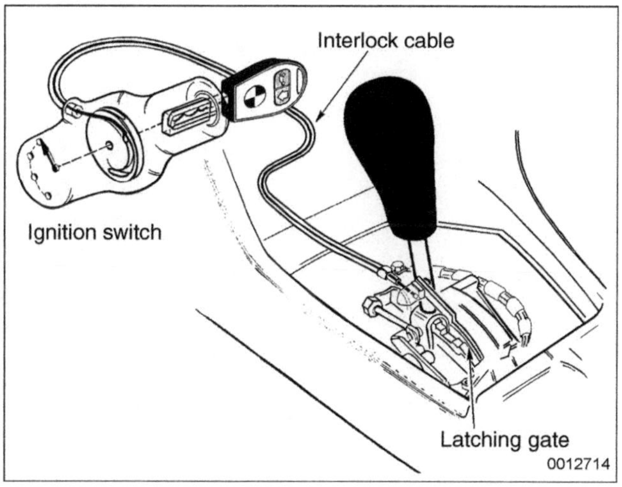


◀ The shift interlock uses a bowden cable between the ignition switch and the shift lever. The cable mechanically locks the shift lever in PARK (P) when the key is OFF or removed from the ignition lock. This feature also prevents the key from being removed from the ignition lock until selector lever is in P.

- Shift selector lever to P and turn ignition key OFF.
- Remove ignition key. Check that selector lever cannot be shifted out of P.

***NOTE—***
- *It must only be possible to remove ignition key with selector lever in P.*
- *Do not kink shift interlock cable.*

– Turn ignition key ON and depress brake pedal. Check that selector lever can be moved freely from gear to gear.

– With the selector lever in a gear position other than P, attempt to move the key to OFF. The key should not go to OFF.

– If any faults are found check the cable for kinks and the cable attachment points for damage. See **320 Steering and Wheel Alignment** for shift interlock cable replacement.

# 260 Driveshaft



**TABLE**



260

## GENERAL

This repair group covers repair and replacement of the driveshaft and driveshaft components.

Driveshaft repairs consist of replacement of the front flex-disc, center support assembly and centering guide. If the driveshaft universal joint is defective, replace that section of the driveshaft.

Drive axle repairs are covered in **331 Drive Axles, Differential**.

### Special tools

BMW recommends one special tool for servicing the driveshaft.

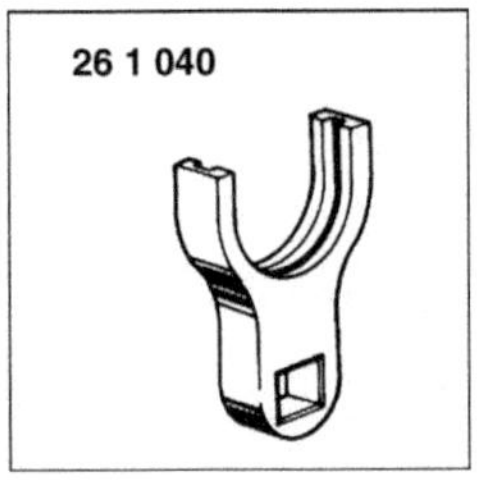


Driveshaft center slide nut spanner (BMW special tool 26 1 040)

### Driveshaft description

The two-piece driveshaft is joined in the center by a sliding splined coupling. This coupling compensates for slight fore and aft movement of the drive line. The driveshaft is connected to the transmission by a rubber flex-disc and to the differential by a universal joint. It is supported in the middle by a center support bearing. The bearing is mounted in rubber to isolate vibration.

### Troubleshooting

The source of driveline vibrations and noise can be difficult to pinpoint. Engine, transmission, rear axle, or wheel vibrations can be transmitted through the driveshaft to the car body. Noises from the rear of the car may be caused by differential problems, or by faulty wheel bearings, drive axles, or even worn or improperly inflated tires.

Driveshaft noise or vibration may be caused by worn or damaged components. In addition to inspecting for faulty driveshaft parts, the installed angles of the driveshaft should also be considered.

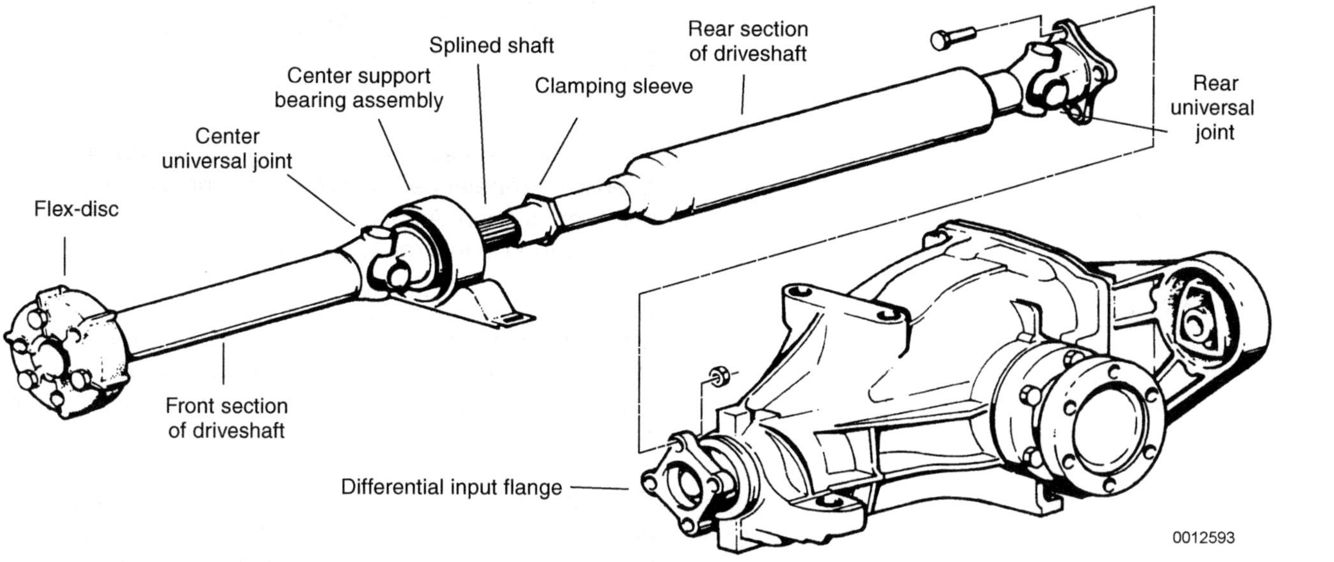


When troubleshooting driveshaft problems, begin with a visual inspection. Check the driveshaft for broken or missing balance weights. The weights are welded tabs on the driveshaft tubes. Check the torque of the fasteners at the flange connections, check the rubber of the flex-disc and center bearing for deterioration or tearing, and check for preload at the center bearing.

Check the universal joints for play. With the driveshaft installed, pull and twist the driveshaft while watching the joint. The BMW specification for play is very small, so almost any noticeable play could indicate a problem. If any universal joint is difficult to move or binds, replace the driveshaft section.

*With the driveshaft installed, the actual amount that the universal joints pivot is limited. For the most accurate test, check universal joints only in their normal range of movement.*

Further inspection requires removal of the driveshaft. Check the front centering guide on the transmission output flange for damage or misalignment. Also check runout at the transmission output flange and output shaft, and at the differential input flange. Check the bolt hole bores in the flange for wear and elongation.

| **Driveshaft components wear limits** | |
|---|---|
| Runout at transmission output flange (max. allowable):<br>• Axial play<br>• Radial play | <br><br>0.10 mm (0.004 in)<br>0.07 mm (0.003 in) |
| Differential input flange radial play (measured at driveshaft center lip) | 0.07 mm (0.003 in) |
| Universal joint play (max. allowable) | 0.15 mm (0.006 in) |

Spin the center bearing and check for smooth operation. Check that the splines of the sliding coupling move freely. Clean and lubricate the splines with molybdenum disulfide grease (Molykote® Longterm 2 or equivalent).

If inspection reveals nothing wrong with the driveshaft, it may need to be rebalanced. This can be done by a speciality driveshaft repair shop. Also, check the alignment of the driveshaft as described in this repair group.

***NOTE—***
*Minor driveshaft vibrations can often be corrected simply by disconnecting the driveshaft at the differential, and repositioning the driveshaft 90°, 180° or 270° in relation to the differential input flange.*

**Table a** lists symptoms of driveshaft problems and their probable causes. Most of the repair information is contained within this repair group. The bold type refers to repair information contained outside of this repair group.

## DRIVESHAFT SERVICE

Most components of the driveshaft are easily replaced once the driveshaft is removed from the car. Repair kits for the universal joints are not available, and it is not common practice to repair the universal joints on BMW driveshafts. Worn or damaged universal joints usually require replacement of that section of the driveshaft.

The driveshaft is balanced to close tolerances. Whenever it is to be removed or disassembled, the mounting flanges and driveshaft sections should be marked with paint before proceeding with work. This will ensure that the driveshaft can be reassembled or installed in exactly the same orientation.

**Table a. Driveshaft troubleshooting**

| Symptom | Probable cause | Corrective action |
|---|---|---|
| Vibration when starting off (forward or reverse) | Incorrect preload of center bearing | Check and adjust preload of center bearing. |
| | Center bearing rubber deteriorated | Inspect center bearing and rubber. Replace if necessary. |
| | Flex-disc damaged or worn | Inspect flex-disc. Replace if necessary. |
| | Engine or transmission mounts faulty | Inspect engine and transmission mounts. Align or replace, if necessary. |
| | Front centering guide worn, or driveshaft mounting flanges out of round | Check front centering guide and replace if necessary. Check runout of driveshaft flanges. |
| | Universal joint worn or seized | Check universal joint play and movement. Replace driveshaft section if necessary. |
| | Driveshaft misaligned | Check driveshaft alignment. |
| Noise during on/off throttle or when engaging clutch | Differential components worn or damaged (excessive pinion-to-ring-gear clearance) | Remove differential and repair. **331 Drive Axles, Differential**. |
| | Drive axle or CV joint faulty | Inspect drive axles and CV joints. Repair or replace as necessary. **331 Drive Axles, Differential**. |
| | Sliding coupling seized | Remove driveshaft and check movement of sliding coupling. Clean coupling splines and replace parts as necessary. |
| Vibration, audible rumble over 25 mph (40 kph) | Front centering guide worn, or driveshaft mounting flanges out of round or damaged | Check front centering guide and replace if necessary. Check runout of driveshaft mounting flanges. |
| | Universal joint worn or seized | Check universal joint play and movement. Replace driveshaft section if necessary. |
| | Flex-disc damaged or worn | Inspect flex-disc. Replace if necessary. |
| | Center bearing rubber deteriorated | Inspect center bearing rubber. Replace if necessary. |
| | Sliding coupling seized | Remove driveshaft and check movement of sliding coupling. Clean coupling splines and replace parts as necessary. |
| | Driveshaft misaligned | Check driveshaft alignment. |
| | Mounting flange bolts loose or holes worn | Remove driveshaft and check transmission output flange and differential input flange. Replace if necessary. |
| | Driveshaft unbalanced | Check driveshaft for loose or missing balance weights. Have driveshaft rebalanced or replace if necessary. |
| | Incorrect preload of center bearing | Check preload of center bearing. Readjust if necessary. |
| | Center bearing faulty | Replace center bearing. |
| | Differential rubber mount faulty | Inspect differential rubber bushing and replace if necessary. |

## Driveshaft, aligning

The alignment of the driveshaft does not normally need to be checked unless the engine/transmission or the differential have been removed and installed. If, however, all other parts of the driveshaft have been inspected and found to be okay, but there is still noise or vibration, driveshaft alignment should be checked.

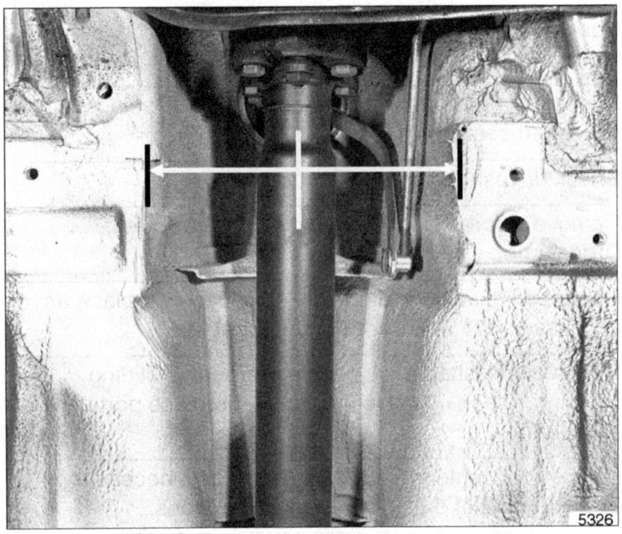
5326

There are two important driveshaft alignment checks. The first is to make sure that the driveshaft runs straight from the transmission to the differential, without any variation from side-to-side caused by misalignment of the engine/transmission in its mounts. Make a basic check by sighting along the driveshaft from back to front. Any misalignment should be apparent from the center bearing forward. To adjust the side-to-side alignment, loosen the transmission or engine mounts to reposition them, then retighten the mounts. The driveshaft should be exactly centered in the driveshaft tunnel.

The second important driveshaft alignment check is more complicated. It checks the amount the driveshaft is angled vertically at the joints. This angle is known as driveshaft deflection.

In general, there should be little deflection in the driveshaft between the engine, the center bearing, and the differential. Precise checks require the use of a large protractor or some other means of measuring the angle of the engine and the differential and comparing these angles to the angle of the driveshaft sections.

To change the deflection angle, shims can be placed between the center bearing and the body or between the transmission and its rear support. When using shims to change a deflection angle, keep in mind that the angle of adjacent joints will also change. Deflection angles should be as small as possible.

> ***CAUTION—***
> *Maximum allowable change in height of the center bearing or transmission support using shims is 3 mm (0.12 in.).*

## Driveshaft, removing

> ***WARNING —***
> - *Be sure the wheels are off the ground before removing the driveshaft. Set the parking brake before removing the driveshaft.*
> - *Once the driveshaft has been removed, the vehicle can roll regardless of whether the transmission is in gear or not.*
> - *The driveshaft is mounted to the transmission and differential with self-locking nuts. These nuts are designed to be used only once and should be replaced during reassembly.*

– Disconnect oxygen sensor electrical connectors.

– Remove complete exhaust system and heat shield (if necessary). See **180 Exhaust System**.

– Matchmark front and rear driveshaft connections at transmission, center bearing and differential.

5267

◄ Remove driveshaft mounting bolts. Counterhold with open end wrench (**arrow**). Discard old nuts.

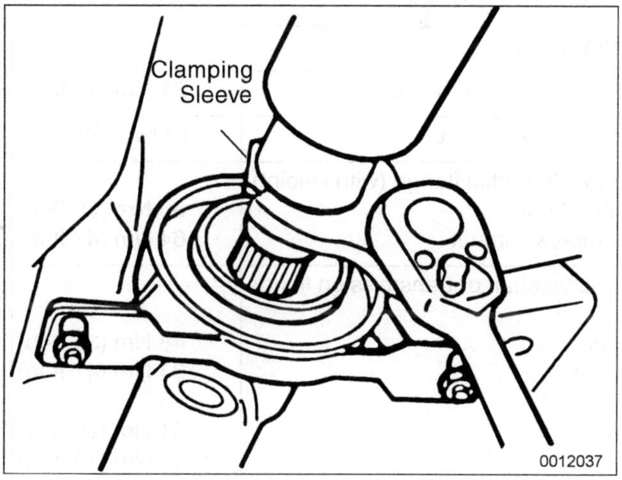

0012037

◄ Loosen threaded clamping sleeve on driveshaft a few turns using BMW special tool 26 1 040 or equivalent.

– Support driveshaft and remove center support bearing mounting bolts.

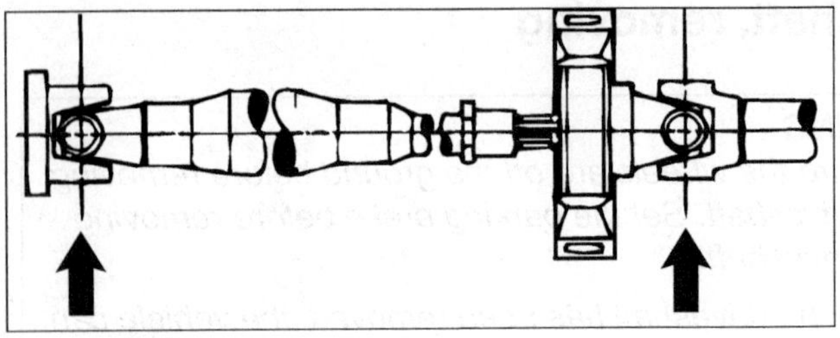

Remove driveshaft but do not separate two halves. If two halves were separate and not matchmarked, assemble driveshaft as shown in drawing. If a vibration occurs, disassemble driveshaft and rotate one section 180°. Centerlines of universal joints (**arrows**) must be in line with each other.

## Driveshaft, installing

- Align driveshaft matchmarks at differential and then at transmission flange. Use new self-locking nuts. Install nuts finger-tight.
- Position center support bearing. Use new self-locking nuts. Install nuts finger-tight.
- Tighten flange nuts while preventing bolts from twisting in coupling. Tighten differential flange first, then tighten coupling at transmission.

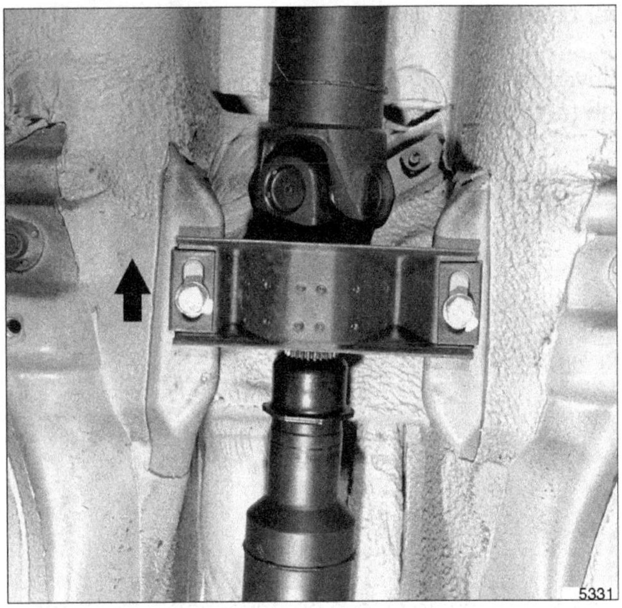


Preload center support bearing by pushing bearing forward 4 - 6 mm (0.16 - 0.24 in.) from center. Tighten attaching bolts.

- Tighten threaded sleeve on driveshaft to proper torque.
- Install heat shield.
- Install exhaust system. See **180 Exhaust System**. Connect wiring harness to oxygen sensors.
- Road test vehicle to check for noise or vibration.

> ***WARNING —***
> - *Do not reuse self-locking nuts. These nuts are designed to be used only once.*
> - *Avoid stressing the flex-disc when torquing the bolts. Do this by holding the bolts steady and turning the nuts on the flange side.*

| **Tightening torques** | |
|---|---|
| Driveshaft center mount to body | 21 Nm (16 ft-lb) |
| Driveshaft clamping sleeve | 10 Nm (89 in-lb) |
| Driveshaft to differential flange (with U-joint)<br>• M10 ribbed nut<br>• M10 compression nut | <br>80 Nm (59 ft-lb)<br>64 Nm (47 ft-lb) |
| Flex-disc to driveshaft or transmission flange<br>M10<br>• 8.8 grade<br>• 10.9 grade<br>M12<br>• 8.8 grade<br>• 10.9 grade | <br><br>48 Nm (35 ft-lb)<br>64 Nm (47 ft-lb)<br><br>81 Nm (60 ft-lb)<br>100 Nm (74 ft-lb) |

***NOTE—***
*Bolt grade is marked on the bolt head. When replacing bolts, only use bolts of the same strength and hardness as the originals installed.*

## Flex-disc, replacing

The flex-disc between the front section of the driveshaft and the output flange of the transmission should be checked for cracks, tears, missing pieces, or distortion. Check for worn bolt hole bores in the flange.

– Remove driveshaft. See **Driveshaft, removing** in this repair group.

***NOTE—***
*It is possible to only partially remove the driveshaft, leaving it connected to the differential. The driveshaft can be tilted down in the center and slid off the transmission flange once the clamping sleeve is loosened and the center bearing bracket is unbolted. Suspend the driveshaft using stiff wire in as close to the installed position as possible. If the driveshaft hangs unsupported, the rear universal joint may be damaged.*

– Unbolt flex-disc from driveshaft.

***NOTE—***
*Removal and installation of the bolts may be made easier by placing a large hose clamp around the flex-disc, and tightening the clamp slightly to compress the coupling.*

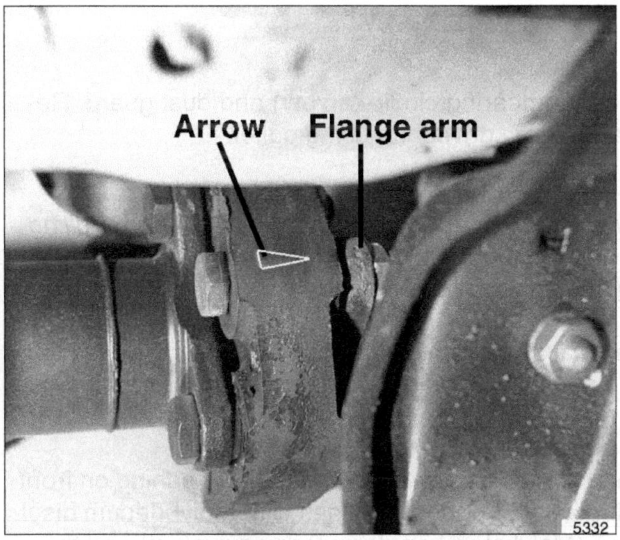


◄ Install new flex-disc using new self-locking nuts. Molded arrows on coupling should face flange arms.

– Reinstall driveshaft. See **Driveshaft, installing** in this repair group.

| Tightening torques | |
|---|---|
| Flex-disc to driveshaft or transmission flange | |
| M10 | |
| • 8.8 grade | 48 Nm (35 ft-lb) |
| • 10.9 grade | 64 Nm (47 ft-lb) |
| M12 | |
| • 8.8 grade | 81 Nm (60 ft-lb) |
| • 10.9 grade | 100 Nm (74 ft-lb) |

***NOTE—***
*Torque only the nuts while holding the bolt heads. This will prevent damaging or fatiguing the rubber.*

## Center bearing assembly, replacing

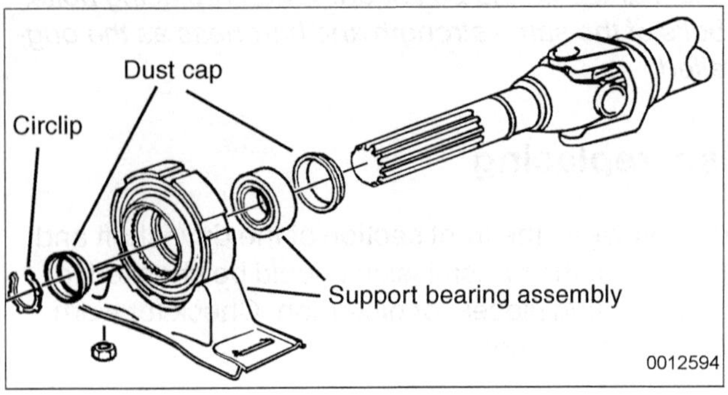


◀ To replace the center bearing assembly, the driveshaft must be removed from the car. The center bearing assembly consists of a grooved ball bearing in a rubber mount. The bearing assembly is pressed onto the front section of the driveshaft and secured by a circlip.

– Remove driveshaft. See **Driveshaft, removing** in this repair group.

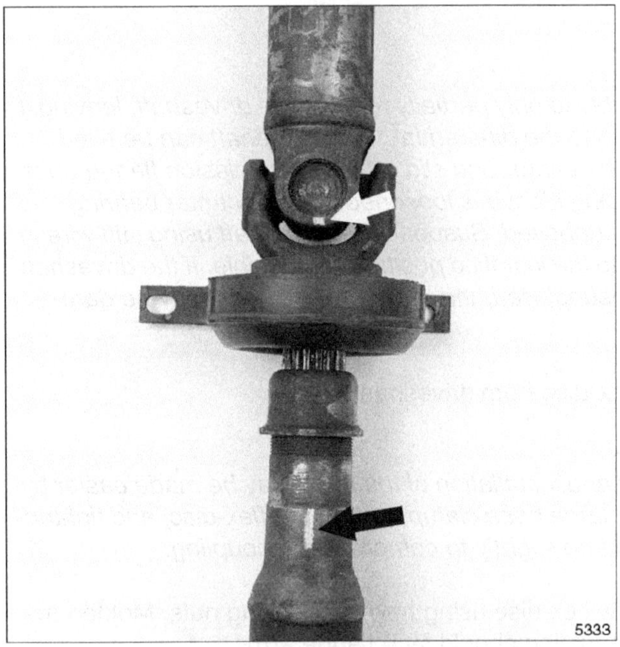


◀ Make matching marks (**arrows**) on front and rear driveshaft sections.

– Loosen clamping sleeve fully and pull driveshaft sections apart. Remove rubber bushing, washer, and clamping sleeve from front section.

– Inspect condition of rubber bushing for splined coupling. Replace worn or damaged parts.

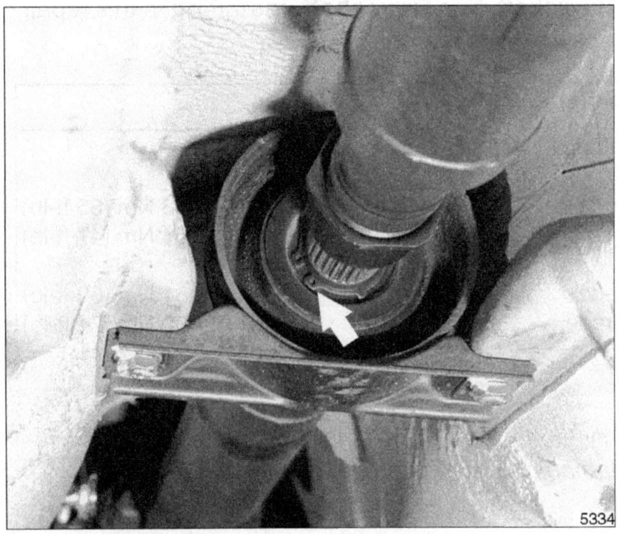


◀ Remove center bearing circlip (**arrow**) and dust guard. Remove bearing from driveshaft using puller.

***NOTE—***
*Install puller so that it pulls on inner hub of bearing. Pulling on outer ring of mount may tear rubber, and entire bearing assembly will need to be replaced.*

– Before reinstallation, make sure dust guard is on driveshaft, and then press center mount onto driveshaft flush with dust guard.

– Place clamping sleeve, washer, and rubber bushing on front driveshaft section. Lubricate splines with molybdenum disulfide grease (Molykote® Longterm 2 or equivalent) and then reassemble driveshaft.

***NOTE—***
*Do not retighten clamping sleeve until driveshaft is installed.*

– Reinstall driveshaft. See **Driveshaft, installing** in this repair group.

– Tighten clamping sleeve.

| **Tightening torque** | |
|---|---|
| Clamping sleeve to driveshaft | 10 Nm (7.5 ft-lb) |

## Front centering guide, replacing

The front centering guide centers the driveshaft in relation to the transmission. The guide is press-fit into a cavity in the front of the driveshaft and slides onto the transmission output shaft.

No specifications are given for wear of the guide, but generally the guide should fit snugly on the transmission output shaft.

***NOTE—***
*Some driveshafts have a dust cap installed on the end of the driveshaft, over the centering guide. The dust cap may become bent or distorted when the driveshaft is removed or installed. Damage to the dust cap should not affect the centering guide and should not be mistaken for guide wear.*

– Remove driveshaft. See **Driveshaft, removing** in this repair group.

– Pack cavity behind centering guide with heavy grease until grease is flush with bottom edge of guide.

– Insert 14 mm (approximately ½ in.) diameter mandrel or metal rod into guide. Strike guide with hammer to force centering guide out.

***NOTE—***
*The mandrel should fit snugly in the centering guide so that the grease cannot escape around the sides of the mandrel.*

– Remove old grease from driveshaft, lubricate new centering guide with molybdenum disulfide grease (Molykote® Longterm 2 or equivalent) and drive it into driveshaft.

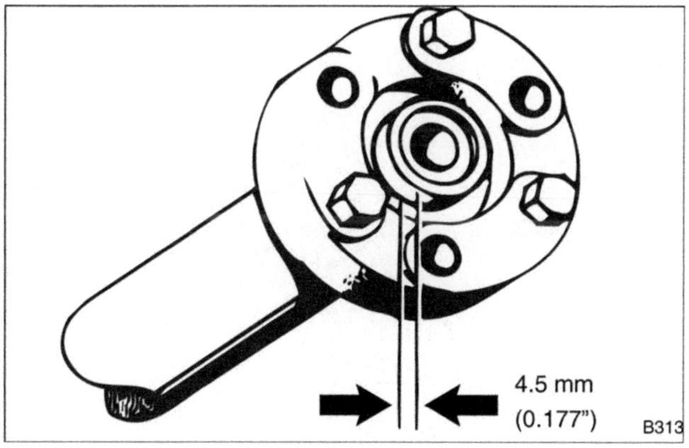


◄ The sealing lip of the guide should face outward and it should be driven into the driveshaft to a protrusion depth of 4.5 mm (0.177 in).

– Reinstall driveshaft. See **Driveshaft, installing** in this repair group.

# 300 Suspension, Steering and Brakes—General



**TABLES**

## General

This section covers general information for front and rear suspension, the steering system, and the braking and traction control systems.

The BMW Z3 suspension and steering systems are engineered to provide a compromise between taut, responsive handling and ride comfort. Although the front and rear suspension assemblies are independent subsystems, they work together to achieve BMW's overall combination of precise handling and comfort.

### Front suspension

The lower control arms connect the suspension struts to mounting points on the subframe crossmember and the body.

Each front strut assembly includes a shock absorber cartridge inside the tubular strut housing and a large coil spring. The upper strut mount bearing allows the strut to turn with the wheels. The lower end of the strut, including the stub axle for the front wheel, connects to a ball-joint on the control arm.

The front suspension is designed with minimum positive steering offset. This geometry contributes to stability when

**M version front suspension**

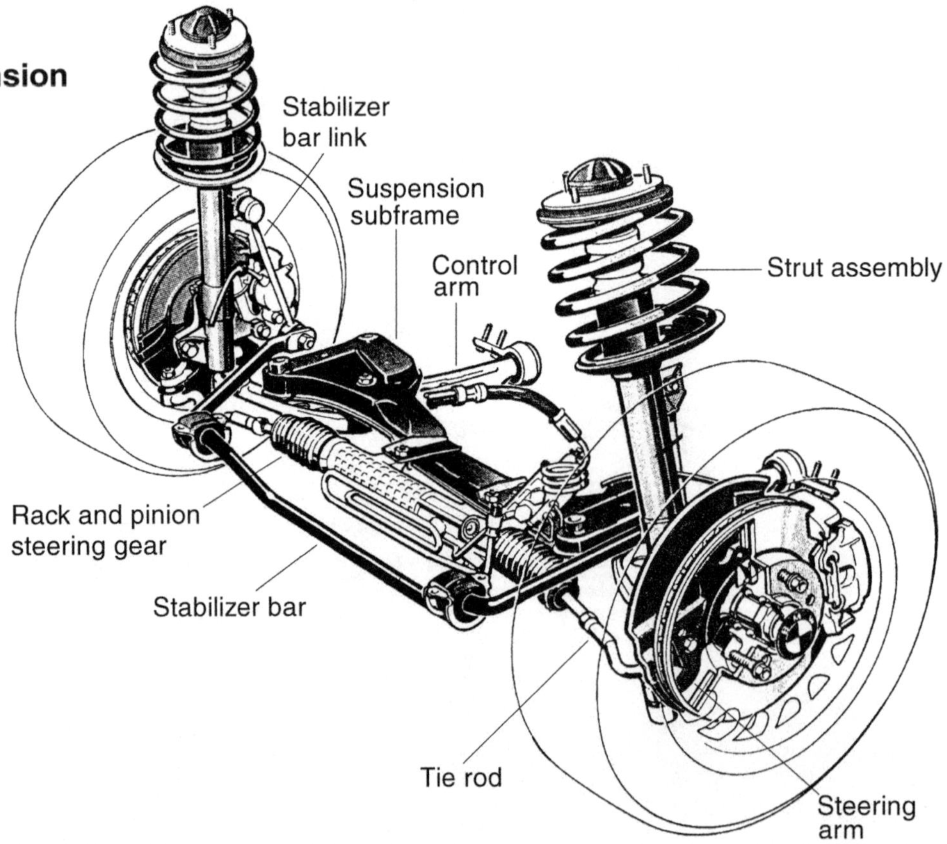

traction is unequal from side to side. Suspension travel is limited by rubber bump stops.

The three point mounting of each L-shaped control arm (**arrow**) precisely controls the front-to-rear and side-to-side position of the strut, while the flexibility of the joints and mounts also allows the movement necessary for suspension travel. The control arm mounting points are designed with anti-dive geometry. The suspension reduces the normal tendency for the front of the vehicle to dive under braking.

Control arm position is fixed, with no adjustment provisions on the control arms for altering front wheel alignment. A stabilizer bar mounted to the control arms or struts helps to reduce body roll when cornering.

## Steering

The variable-assist power steering system consists of an engine-driven hydraulic pump, a rack-and-pinion type steering gear, and connecting linkage to the road wheels. The Z3 utilizes an engine-speed dependent variable effort steering system. At low speeds, maximum power assist is provided to ease parking and city driving. At high speeds, assist is reduced to ensure stability.

The steering linkage connects the rack-and-pinion unit through tie rods to the wheels. The tie rod ends allow the wheels to pivot and react to suspension travel.

## Rear suspension

The rear axle carrier is the main mounting point for the final drive housing and the rear suspension components.

**Z3 rear suspension**

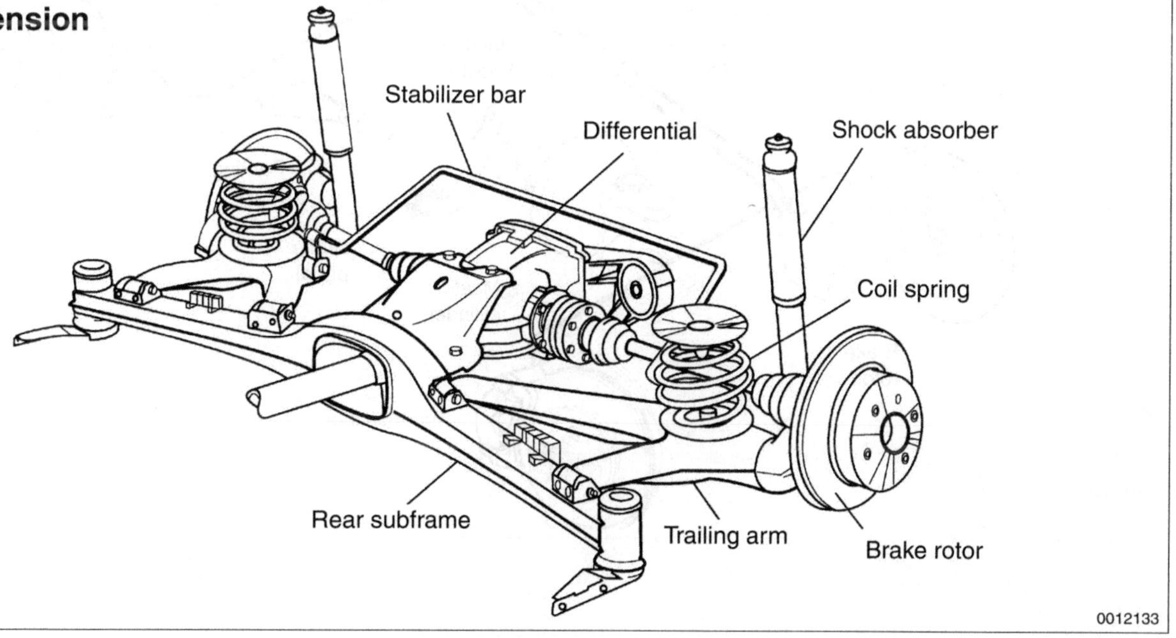

Trailing arms locate the rear wheels and anchor the springs, shocks and stabilizer bar. Drive axles with constant-velocity (CV) joints at both ends transfer power from the differential to the road wheels. The differential is mounted to the rear axle carrier through rubber mounts and bushings to help isolate drivetrain noise and vibration.

## Brakes

The Z3 is equipped with power disc brakes with an integral Antilock Brake System (ABS). The parking brake is a dual-drum system integrated with the rear brake rotors.

Power assist is provided by a vacuum booster when the engine is running. The brake pedal pushrod is connected directly to the master cylinder, so failure of the vacuum booster does not normally result in total brake failure.

Each disc brake uses a caliper with a single hydraulic cylinder. The dual-chamber master cylinder pressurizes two separate hydraulic circuits. If one circuit fails, the remaining circuit can still provide a portion of the braking force.

Brake pads in the left front and right rear contain wear sensors. When the pads need replacement, the sensors illuminate a light on the dashboard.

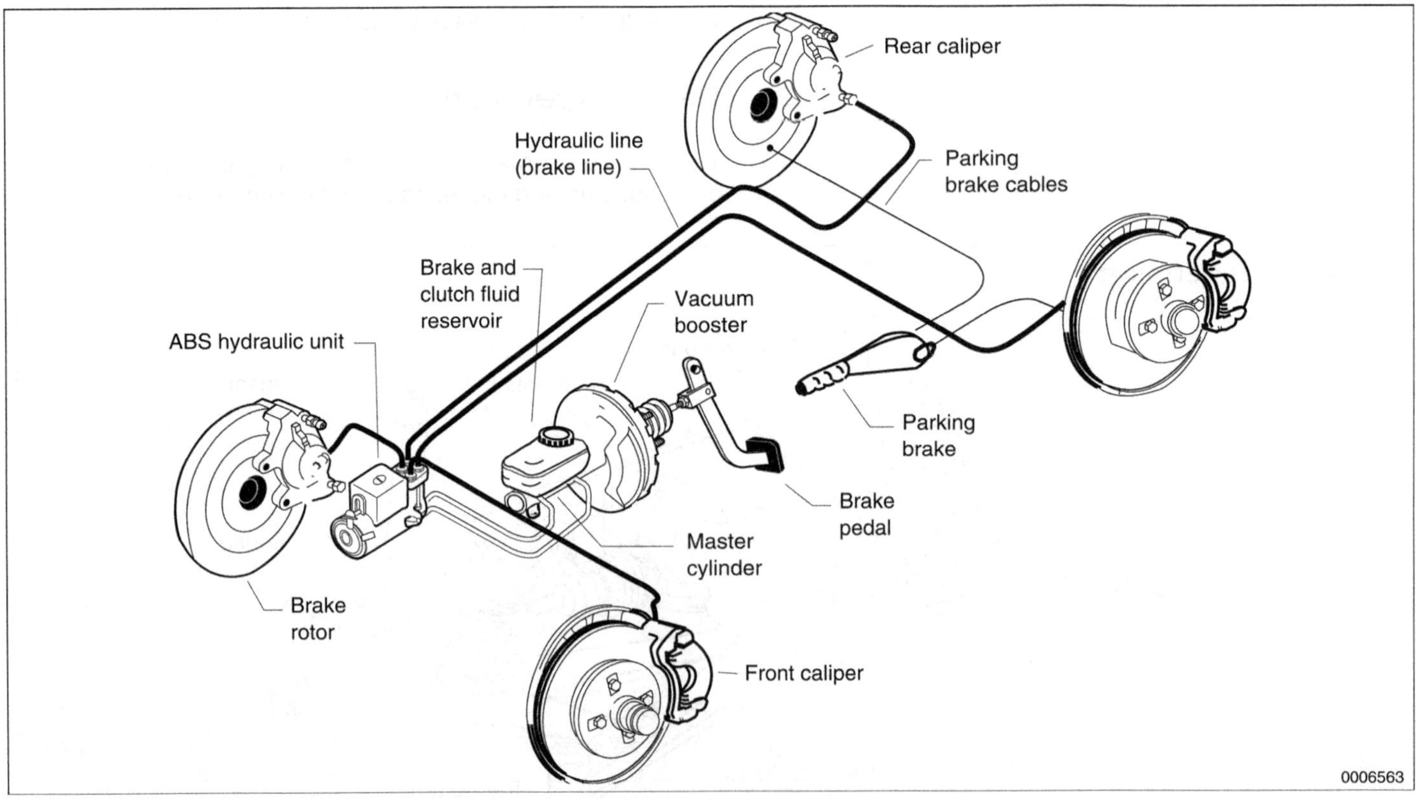

## Tires and wheels

Tire size is critical to proper ABS operation. Several different styles of wheels, in 15, 16 and 17 inch diameters, are available from an authorized BMW dealer.

***NOTE—***
*Select aftermarket wheels with care. Improperly fitted wheels can contact and damage suspension, brake or body components and may adversely affect vehicle stability.*

# ELECTRONIC BRAKE AND STABILITY CONTROL SYSTEMS

Z3 models are equipped with Antilock Braking System (ABS). Early models featured ABS with Automatic Stability Control (ABS / ASC). Later models came equipped with ABS and Dynamic Stability Control (ABS / DSC). DSC builds upon the existing ABS / ASC system to provide electronic control of drive and braking systems to insure vehicle stability.

This manual will refer to these systems as ABS. ASC or DSC will be specified when necessary.

## ABS system description

The electronically controlled ABS maintains vehicle stability and control during emergency braking by preventing wheel lock-up. ABS provides optimum deceleration and stability during adverse conditions. It automatically adjusts brake system hydraulic pressure at each wheel to prevent wheel lock-up.

The system's main components are:

- Wheel speed (pulse) sensors
- ABS / ASC or ABS / DSC electronic control module
- Hydraulic unit

Wheel speed sensors continuously send wheel speed signals to the electronic control module. The control module compares these signals to determine, in fractions of a second, whether any of the wheels are about to lock. If any wheel is nearing a lock-up condition, the module signals the hydraulic unit to maintain or reduce pressure at the appropriate wheel(s). Pressure is modulated by electrically-operated solenoid valves in the hydraulic unit.

## Automatic Stability Control (ASC)

***NOTE—***
*ASC may also be referred to as ASC+T (Automatic Stability Control+Traction).*

The Automatic Stability Control (ASC) system works in conjunction with ABS and the engine management system to enhance vehicle control. The main function of the ASC system is to maintain the rolling contact between the tires and the road surface under all driving conditions. This is achieved through exact application and management of braking and drivetrain forces.

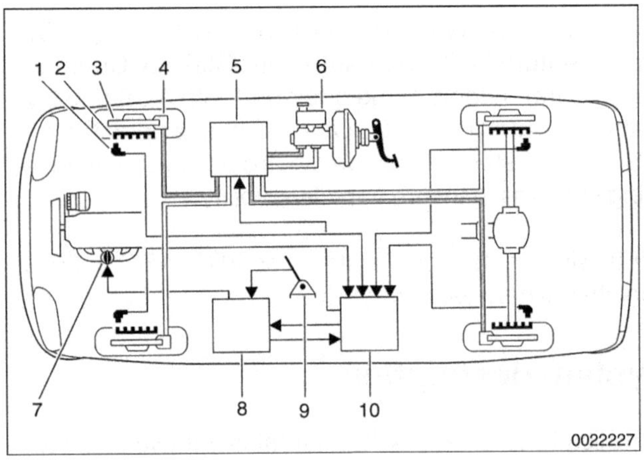


ABS / ASC traction control components

1. Wheel speed sensor
2. Wheel speed pulse wheel
3. Brake disk
4. Brake caliper
5. ABS / ASC hydraulic unit
6. Brake master cylinder
7. Throttle valve
8. Engine control module
9. Accelerator pedal
10. ABS / ASC control module

The ASC system improves traction by electronically applying the rear brakes when the rear drive wheels are spinning at a faster rate than the front wheels. The combined ABS / ASC control module, operating through the ABS hydraulic control unit, modulates braking force at the rear wheels.

In addition, ASC deactivates individual fuel injectors and controls throttle opening to reduce engine torque and maintain vehicle traction. The driver cannot increase the engine power output during ASC intervention regardless of how far the accelerator pedal is pushed.

The components that comprise the ASC system also function to replace the limited slip differential available in previous models. Even with the ASC system turned off, if the ASC control module senses a difference in wheel speed (one wheel spinning) the control module will apply modulated braking force to the slipping wheel until traction is regained, but will not override fuel injection function.

Traction control also comes into operation during deceleration. Decelerating on snowy or icy road surfaces can lead to rear wheel slip. If a rear wheel starts to drag or lock up, the ASC system can limit the problem by adjusting throttle, fuel injection and ignition timing.

A switch on the center console is used to toggle the ASC on or off.

ASC is maintenance free. There are no adjustments that can be made. Repair and troubleshooting of the ASC system requires special test equipment and knowledge and should be performed only by an authorized BMW dealer.

## Dynamic Stability Control (DSC)

Dynamic Stability Control (DSC) utilizes many principles and components of the ASC traction control system. DSC is active throughout the driving experience, unlike ASC which is only active during acceleration and braking. DSC helps stabilize the vehicle in cornering and avoidance maneuvers by adjusting engine controls such as throttle, ignition, fuel injection and the application of brake pressure to the wheels individually.

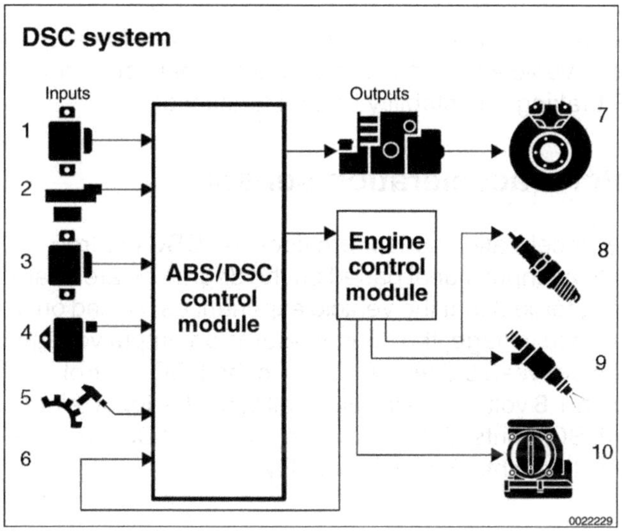


The DSC control module uses various inputs to determine vehicle instability during braking, cornering, or reduced traction situations. Based upon these inputs the ABS / DSC control module sends outputs to the engine control module and the ABS / DSC hydraulic unit to activate torque reduction protocols and braking intervention. DSC system inputs and outputs include:

1. Lateral acceleration sensor
2. Steering angle sensor
3. Rotational rate (yaw) sensor
4. Brake pressure sensor
5. ABS wheel speed sensor
6. Engine control module
7. ABS / DSC hydraulic module
8. Ignition (spark)
9. Fuel injection
10. Throttle valve

The DSC system can be toggled on and off by a switch mounted on the center console. Turning off the DSC system does not disable ABS or ASC functions.

## Control modules

Control module locations vary depending on year and system installed.

1996 - 1998 models with ASC:

- **ASC electronic control module** is located behind the glove compartment.
- **ABS / ASC hydraulic module** is located at the left side of the engine compartment.

1999 - 2000 models with ASC, all models with DSC:

- **ASC** or **DSC electronic control module** is integrated with **ABS / ASC** or **ABS / DSC hydraulic module** at left side of engine compartment.

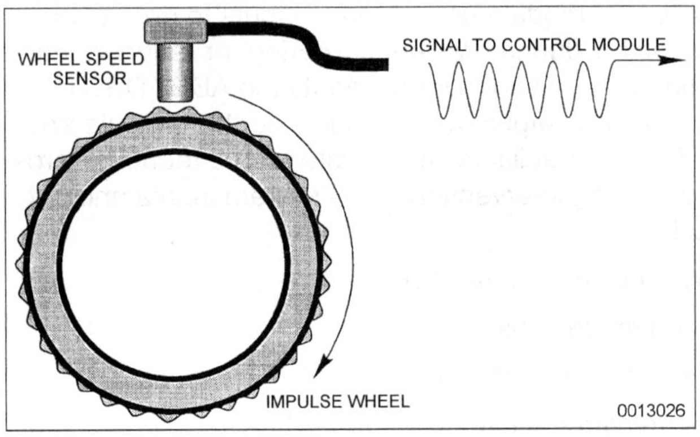


## Wheel speed sensors

Wheel speed sensors are a crucial component in every ABS system. Control modules use these sensor inputs to determine overall vehicle speed and individual wheel speed for both ABS braking and stability control functions.

## DSC lateral acceleration sensor

The lateral acceleration sensor provides the DSC control module with an input signal based on the degree of lateral acceleration (g forces) that the vehicle experiences. Based on a 5 volt reference voltage, the sensor returns an output voltage that ranges between 0.5 and 4.5 volts to the DSC control module, with 1.8 volts as a standing voltage. This input, along with other DSC inputs, determines the amount of DSC regulation needed to maintain vehicle stability.

## DSC rotational rate (yaw) sensor

The rotational rate sensor provides an analog voltage signal to the DSC control module to indicate the rotational speed (yaw) of the vehicle on its vertical axis. The control module supplies a 5 volt reference voltage to the sensor. The sensor returns a voltage between 0.25 and 4.65 volts based on the amount of yaw. If the vehicle's yaw exceeds preset parameters, the DSC control module activates a DSC regulation cycle to increase vehicle stability while cornering. In case of failure the sensor sends a constant voltage to the DSC control unit.

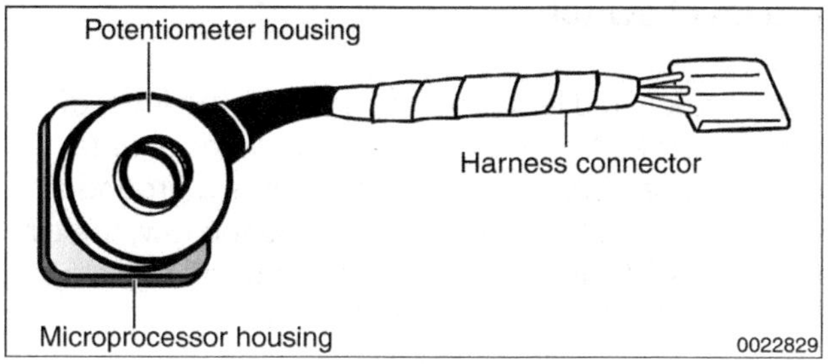


## DSC steering angle sensor

Using two potentiometers, the steering angle sensor determines the steering angle and the rate of steering change. The sensor processes the two potentiometer outputs and signals the DSC control unit. Sensor recalibration is required after steering angle sensor replacement or repairs to the steering column. Perform this recalibration using BMW scan tools DISplus, GT1 or MoDiC.

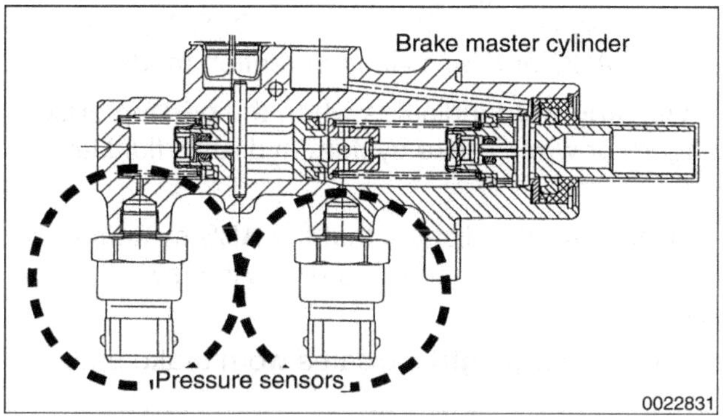


## DSC pressure sensor

The DSC pressure sensor(s) provides the ABS / DSC control module an analog voltage signal in proportion to brake fluid pressure in the master cylinder.

## DSC precharge pump

The DSC precharge pump provides the hydraulic unit with the necessary supply of hydraulic brake fluid. When the DSC system is activated, the precharge pump delivers brake fluid from the reservoir to the hydraulic unit at 10 bar (150 psi).

## Switches and indicators

### ASC or DSC control button

A control button on the dash is used to deactivate the ASC or DSC functions.

### Brake light switch

The brake light switch input signal is used by the control module to determine which stability control routine is necessary. The control module interrupts Automatic Stability Control (ASC) functions if the brake pedal is depressed during ASC intervention. On vehicles equipped with DSC, DSC operation is not cancelled during braking situations.

### Parking brake switch

The switch for the parking brake warning light is used to signal the control module if the parking brake is engaged. This signal is used in stability control system logic to cancel system functions.

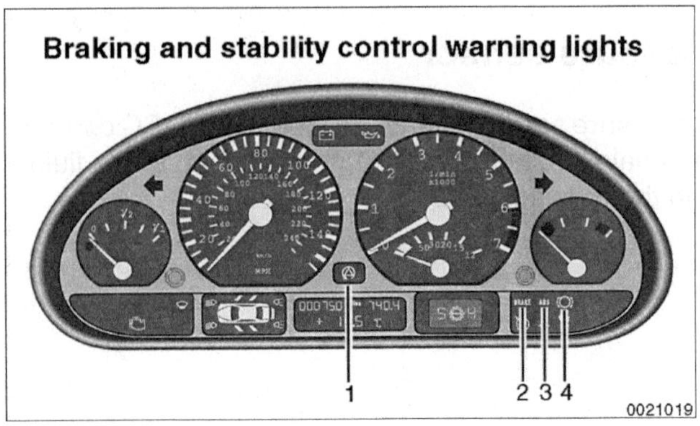


### Brake fluid level switch

The reed-type brake fluid level switch monitors the level of brake fluid in the brake fluid reservoir. When an adequate amount of fluid is present, the switch completes a ground circuit for the control module. When fluid level is too low, the circuit is broken and ASC or DSC functions are turned off. Normal braking and ABS remains unaffected.

### Warning lights

1. **ASC or DSC warning light** illuminates solidly when ASC or DSC system has been turned OFF. The light blinks when ASC or DSC is in intervention mode.
2. **Brake warning light** indicates when the parking brake is engaged, or when brake system hydraulic fluid level is low.
3. **ABS warning light** indicates that the ABS system has been deactivated, or when there is an ABS system malfunction.
4. **Brake pad warning light** indicates worn brake pads.

***CAUTION—***
*If the brake warning light, ABS warning light and ASC or DSC warning light are all illuminated at the same time, there is an ABS and stability control system failure. Do not drive vehicle without diagnosing the problem.*

## ABS troubleshooting

ABS is designed to be maintenance free. There are no adjustments that can be made to the system. Repair and troubleshooting of major ABS components requires special test equipment and knowledge and should be performed by an authorized BMW dealer.

ABS is self-tested by the ABS / ASC or ABS / DSC electronic control module each time the car is started. Once the test is complete, the ABS dashboard warning light turns off. If the light remains lit or comes on at any time during driving, a system fault has occurred and ABS is electronically disabled. The conventional braking system remains fully functioning.

When a system or component failure occurs in the electronic braking and stability control systems, either the brake warning light, ABS warning light, or the ASC or DSC warning light illuminates.

Use BMW scan tool DISplus, GT1 or MoDiC for:
- Diagnosis and troubleshooting
- Bleeding brakes
- Component coding and initialization

## ABS inspection

A visual inspection of the ABS system components may help to locate system faults. If no visual faults can be found and the ABS light remains on, have the system diagnosed by an authorized BMW dealer.

Carefully inspect the entire ABS wiring harness, particularly the wheel speed sensor harnesses and connectors near each wheel. Look for chafing or damage due to incorrectly routed wires.

Carefully remove the wheel speed sensors. Clean the sensor tips. Inspect toothed wheel on wheel hub. Check for missing, clogged or corroded teeth, or other damage that could alter the clearance between the sensor tip and toothed wheel.

# SUSPENSION AND STEERING TROUBLESHOOTING

Stable handling and ride comfort both depend on the integrity of the suspension and steering components. Any symptom of instability or imprecise road feel may be caused by worn or damaged suspension components.

When you are troubleshooting suspension and steering problems, in addition to inspecting for worn parts, consider also the condition of tires, wheels and their alignment. Tire wear and incorrect inflation pressures can dramatically affect handling. Subtle irregularities in wheel alignment angles also affect stability. Mixing different types or sizes of tires, particularly on the same axle, can affect alignment and may unbalance handling.

**Table a** lists the symptoms of common suspension and steering problems and their probable causes, and suggests corrective actions. **Bold type** indicates the **Repair Groups** where applicable test and repair procedures can be found.

**Table a. Suspension and steering troubleshooting**

| Symptom | Probable cause | Repairs (repair groups shown in bold) |
|---|---|---|
| Breaking away while braking | Worn struts or shock absorbers | Replace struts or shock absorbers. **310**, **330** |
| Car pulls to one side, wanders | Incorrect tire pressure | Check and correct tire pressures. **020** |
| | Incorrect wheel alignment | Check and adjust wheel alignment. **320** |
| | Faulty brakes (pulls only when braking) | Check for sticking / damaged front caliper. **340** |
| Front end or rear end vibration or shimmy | Worn struts or shock absorbers | Replace struts or shock absorbers. **310**, **330** |
| | Worn suspension bushings (control arm or trailing arm) | Replace worn bushings. **310**, **330** |
| | Worn front suspension ball joints (control arm, steering arm or steering tie-rod end) | Replace worn ball joints. **310** |
| | Unbalanced or bent wheels / tire | Balance tires. Check tires for uneven wear patterns. Check wheels for damage. |
| | Loose wheel lug bolts | Tighten lug bolts to proper torque. |

**Table a. Suspension and steering troubleshooting**

| Symptom | Probable cause | Repairs (repair groups shown in bold) |
|---|---|---|
| Poor handling, poor directional stability | Rear trailing arm or rear subframe bushings worn or damaged | Replace / replace rear suspension bushings. **330** |
| | Rear alignment incorrect | Check and adjust wheel alignment. **320** |
| Poor stability, repeated bouncing after bumps, suspension bottoms out easily | Worn struts or shock absorbers | Replace struts or shock absorbers. **310, 330** |
| Rear end hop with hard braking | Rear trailing arm bussing worn or damaged | Replace trailing arm bushing. **330** |
| Steering heavy, poor return-to-center | Worn upper strut mounts | Replace strut mounts. **310** |
| | Incorrect tire pressure | Check and correct tire pressures. **020** |
| | Power steering system faulty | Check power steering fluid level. **320** |
| Steering loose, imprecise | Incorrect tire pressure | Check and correct tire pressures. **020** |
| | Loose steering rack mounting bolt(s) | Inspect and tighten bolts. **320** |
| | Worn tie rod end(s) | Replace tie rod(s) and align wheels. **320** |
| | Faulty front wheel bearing | Replace wheel bearing. **310** |
| | Worn or damaged steering rack | Adjust or replace steering rack. **320** |
| | Worn tires | Replace tires. |
| Suspension noise, especially over bumps (drumming, rattling) | Worn front upper strut mounts | Replace upper strut mounts. **310** |
| | Worn suspension bushings (control arm or trailing arm) | Replace worn bushings. **310, 330** |
| | Worn stabilizer bar rubber mounts | Replace stabilizer bar rubber mounts. **310, 330** |
| | Loose suspension subframe | Check subframe for damage. Tighten mounting bolts. |
| Tail skid when braking | Rear trailing arm front bushing worn or damaged | Replace trailing arm bushing. **330** |
| Tire flat spots | Worn struts or shock absorbers | Replace struts or shock absorbers. **310, 330** |
| Uneven ride height | Incorrect coil springs | Measure ride height. **300** |
| | Bent or damaged suspension components | Inspect, repair / replace as necessary. **310, 330** |
| | Sagging coil springs | Replace springs as necessary. **310, 330** |
| Unsteady in curves, self-steering, poor rear end stability | Rear subframe bushings worn or damaged | Check / replace rear suspension bushings. |
| | Differential bushings worn or damaged | Check / replace rear differential bushings. **331** |
| | Rear shock absorbers worn | Replace rear shock absorbers. **330** |
| Wheel noise, continuous growling, may be more noticeable when turning | Worn wheel bearing | Replace wheel bearing. **310, 330** |
| Wheel-hop on normal road surface | Worn struts or shock absorbers | Replace struts or shock absorbers. **310, 330** |

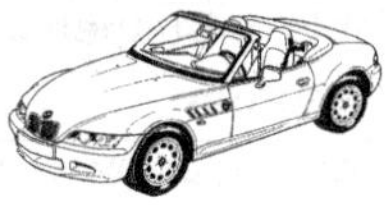

# 310 Front Suspension



**TABLE**

## General

This repair group covers front suspension repairs and front ride height. For a general description of the front suspension and components, see **300 Suspension, Steering and Brakes-General**.

### Special tools

Special service tools are required for most of the work described in this repair group. Read the procedures through before beginning any job.

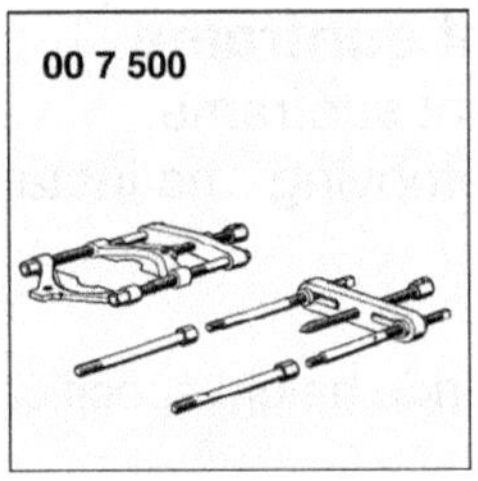


Puller set
(BMW special tool 00 7 500)

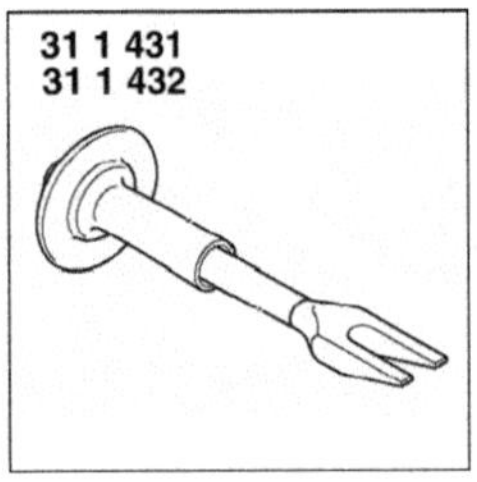


Ball joint fork
(BMW special tool 31 1 431 / 31 1 432)

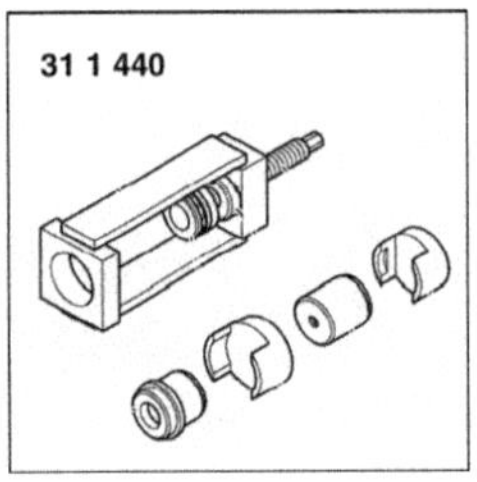


Ball joint press kit
(BMW special tool 31 1 440)

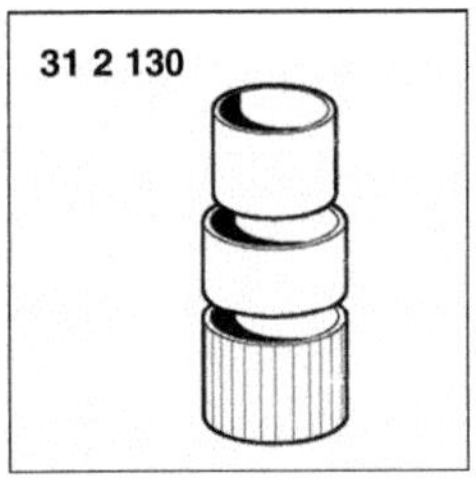


Bushing press tools
(BMW special tool 31 2 130)

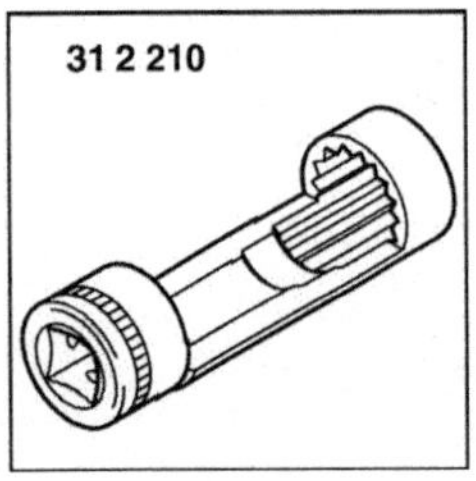


Strut nut socket
(BMW special tool 31 2 210)

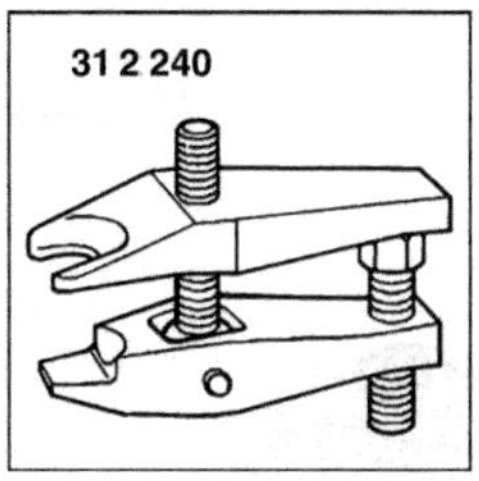


Ball joint press
(BMW special tool 31 2 240)

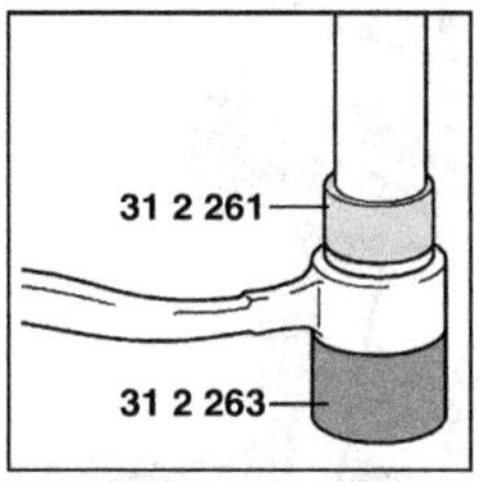


Bushing press tools
(BMW special tool 31 2 261 / 31 2 263)

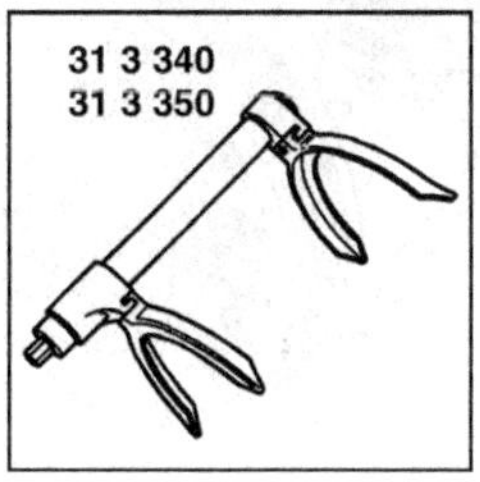


Spring compressor and spring holder arms
(BMW special tool 31 3 340 / 31 3 350)

***WARNING—***

- *Physical safety could be impaired if procedures described here are undertaken without the proper service tools and equipment. Be sure to have the right tools on hand before beginning any job.*
- *Do not reuse self-locking nuts or bolts. They are designed to be used only once and may fail if reused. Always replace self-locking fasteners any time they are loosened or removed.*
- *Do not install bolts and nuts coated with undercoating wax, as the correct tightening torque cannot be assured. Always clean the threads with solvent before installation, or install new parts.*
- *Do not attempt to weld or straighten any suspension components. Replace damaged parts.*

***CAUTION—***

- *Some models are equipped with aluminum suspension components. The following cautions apply:*
- *Due to the chemical and corrosion characteristics of aluminum, do not bring into contact with battery acid.*
- *Do not clean with wire brushes having brass or iron bristles. Only use brushes with stainless steel bristles.*
- *Do not expose to flying sparks from grinding / cutting operations.*
- *Do not subject to steel welding splashes.*
- *Do not expose to temperatures over 80°C (176°F), even for short periods. Temperatures in painting facilities are not a problem.*

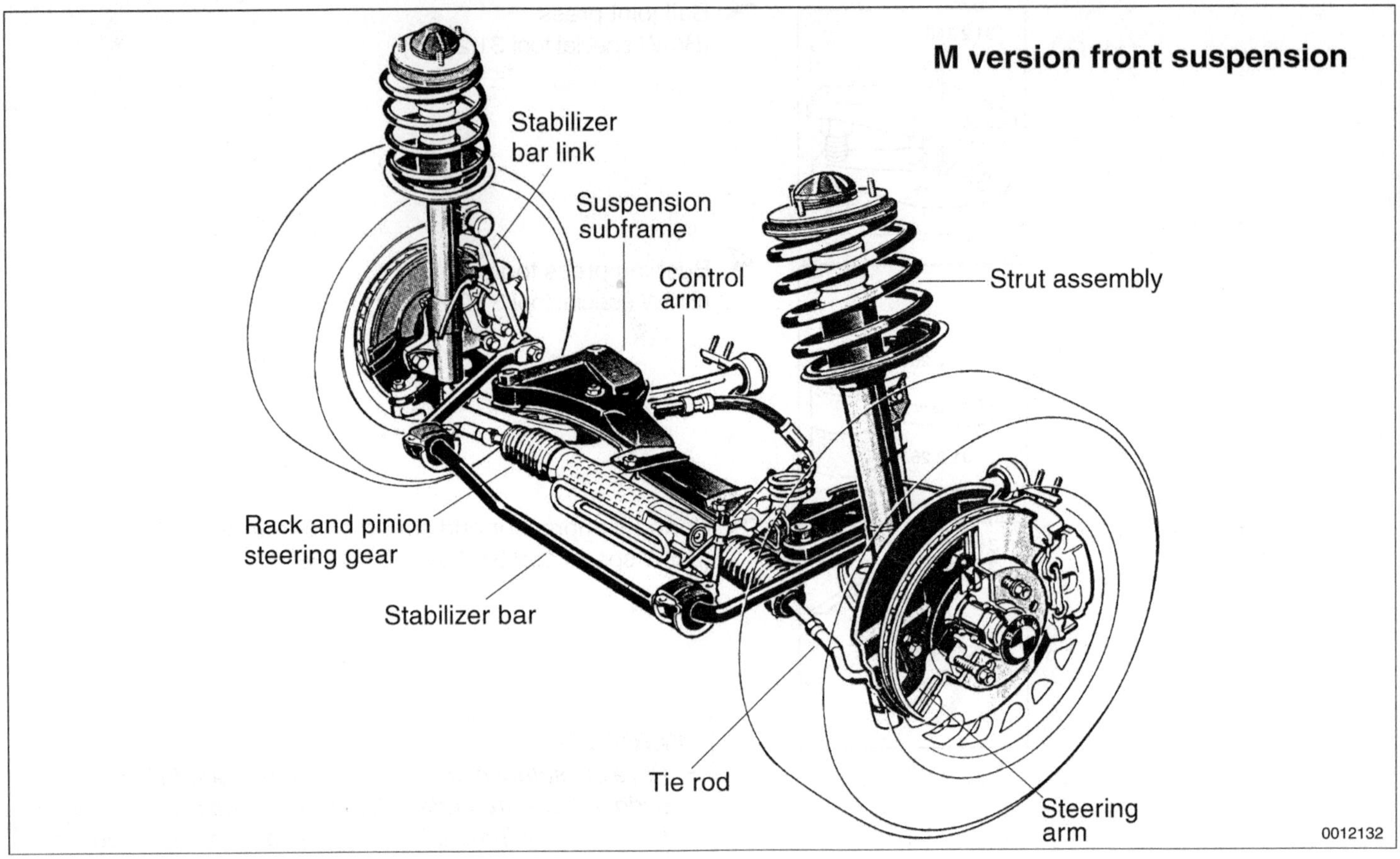


## Shock absorbers and springs

The front suspension shock absorbers are McPherson struts. The strut is a major component of the suspension and supports the spring. Most strut assembly components are available as replacement parts. Struts and springs should always be replaced in pairs.

Front strut, upper strut mount or spring replacement is a two step procedure:

- Removal of strut assembly from vehicle.
- Disassembly and replacement of components on work bench.

### Strut assembly, removing and installing

– Raise car and remove front wheel.

> ***WARNING —***
> *Make sure car is stable and well supported at all times. Use a professional automotive lift or jack stands designed for the purpose. A floor jack is not adequate support.*

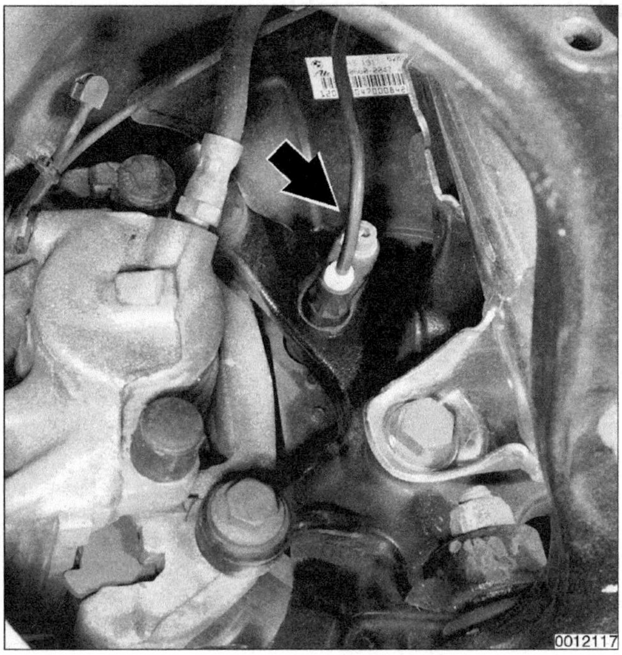


– Unbolt brake caliper assembly, keeping brake hose connected. Suspend brake caliper from chassis using stiff wire. Do not let it hang from brake hose. See **340 Brakes**.

< Remove ABS wheel speed sensor mounting bolt (**arrow**) and slide sensor out of steering arm.

– Unclip brake fluid line, ABS sensor wire harness, and brake pad wear sensor wire harness (left side only) from bracket on strut housing.

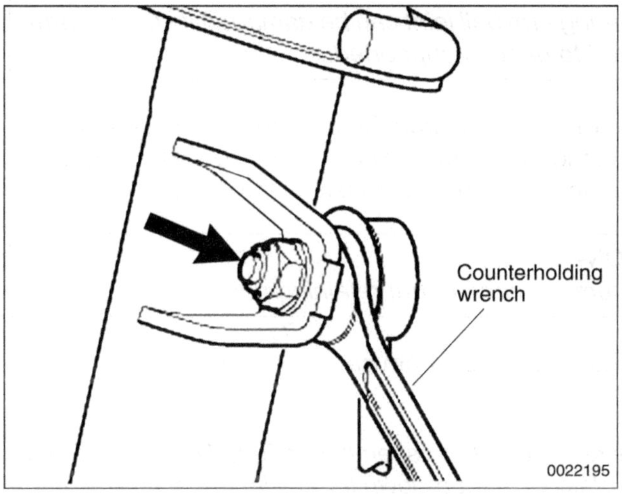


< M version: Loosen and remove stabilizer bar link mounting nut (**arrow**) from strut. Detach link from strut housing.

***NOTE—***
*Use a thin wrench to counterhold shaft of stabilizer link ball joint while removing mounting nut.*

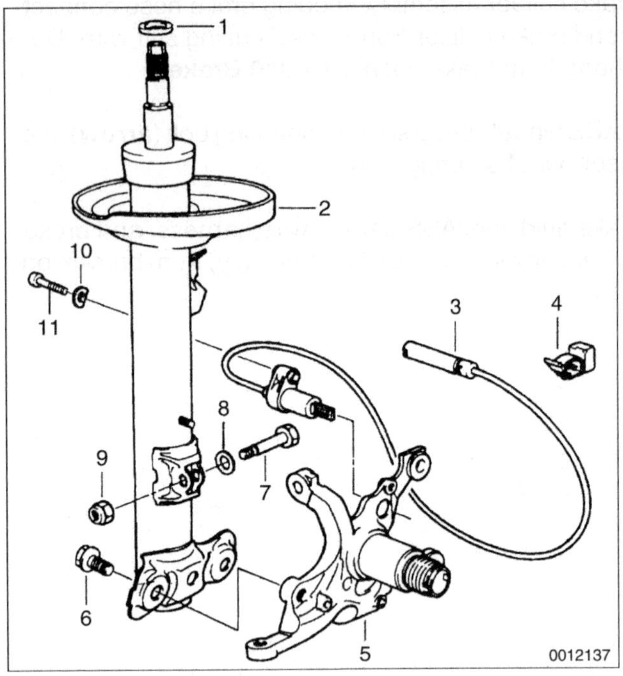


### Strut and steering arm assembly

1. Nut
2. Strut assembly
3. ABS wheel speed sensor
4. Cable holder
5. Steering arm
6. Bolt
7. Bolt
8. Flat washer
9. Self-locking nut
10. Washer
11. Bolt

– Remove bolts mounting strut to steering arm. Suspend steering arm assembly from chassis using stiff wire.

> ***CAUTION—***
> *The steering arm ball joint can be damaged if steering arm is allowed to hang unsupported.*

– Support strut assembly from below. From above, mark location of strut mounting studs in strut tower slots. Remove three strut mounting nuts on strut tower.

> ***CAUTION—***
> *Do not remove center strut retaining nut.*

– Remove strut assembly from car.

***NOTE—***
*Store the removed strut assembly upright. Do not allow it to lie horizontally or upside-down for any length of time. If incorrect storage has occurred, place the strut assembly upright with the rod fully extended for at least 24 hours.*

– Installation is reverse of removal. Remember to:
- Replace self-locking nuts and bolts.
- Align strut mounting studs to marks made previously on strut tower.
- Use new steering arm bolts, or clean bolts and use thread-locking compound.
- When attaching stabilizer link to strut, use a thin wrench to hold flats on link parallel to strut.
- Have car professionally aligned when job is complete.

| Tightening torques | |
|---|---|
| Brake caliper to strut housing | 110 Nm (81 ft-lb) |

| Tightening torques | |
|---|---|
| Road wheel to hub | 120±10 Nm (90 ±7 ft-lb) |
| Stabilizer bar link to strut | 59 Nm (44 ft-lb) |
| Steering arm to strut housing | 107 Nm (78 ft-lb) |
| Strut assembly to strut tower (self-locking nuts) | 22 Nm (16 ft-lb) |

## Strut assembly, disassembling and assembling

***NOTE—***
*Replace springs and struts in pairs.*

Replacing the upper strut mount, spring, and strut cartridge requires that the strut first be disassembled. This procedure requires a spring compressor specifically designed for the purpose.

> ***WARNING—***
> *Do not attempt to disassemble the struts without a spring compressor designed specifically for this job.*

- Remove strut assembly. See **Strut assembly, removing and installing** in this repair group.
- Using BMW special tool 31 3 340 / 31 3 350 or equivalent spring compressor, compress spring until spring force on upper mount is relieved.

**Z3 front strut assembly**

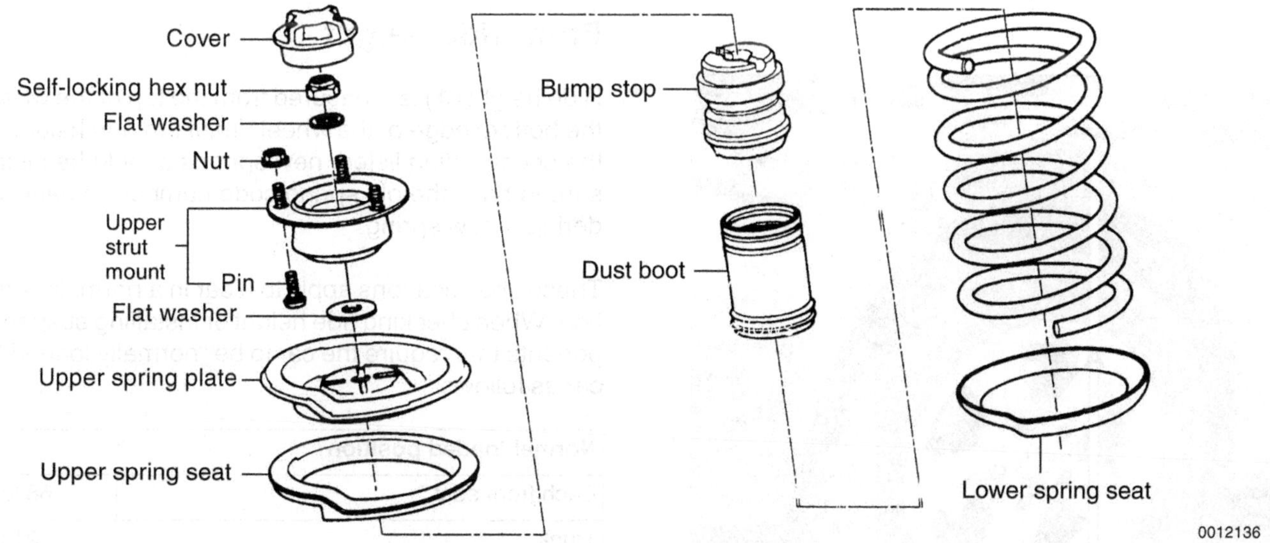

> ***WARNING —***
> *Make sure the spring compressor grabs the spring properly before compressing it.*

- Pry protective cover off top of strut assembly. Remove top (center) nut and washer while holding strut shaft stationary.
- Remove upper strut mount and related components.

***NOTE —***
*Store the removed strut assembly upright. Do not allow it to lie horizontally or upside-down for any length of time. If incorrect storage has occurred, place the strut assembly upright with the rod fully extended for at least 24 hours.*

- Assembly is reverse of disassembly noting the following:
  - Be sure upper spring plate is correctly installed to upper spring seat and spring ends are correctly located in the upper and lower spring seats.
  - Use a new self-locking top (center) nut. Tighten the nut fully before releasing spring compressor.
  - Release spring compressor carefully and evenly, allowing spring to expand slowly.
  - Have the car professionally aligned.

| **Tightening torques** | |
|---|---|
| Upper strut mount to strut cartridge shaft (self-locking nut)<br>• shaft with external hexagon<br>• shaft with internal hexagon | <br><br>65 Nm (48 ft-lb)<br>45 Nm (33 ft-lb) |
| Road wheel to hub | 120±10 Nm (90 ±7 ft-lb) |

## Front ride height

Ride height (**A**) is measured from the top of the wheel arch to the bottom edge of the wheel rim. If the ride height is outside the specification listed, new springs should be installed. Be sure to have the old spring code number on hand when ordering a new spring.

These specifications apply to a car in a normally loaded position. When checking ride height or installing suspension components that require the car to be "normally loaded," load the car as follows:

| **Normal loaded position** | |
|---|---|
| Each front seat | 68 kg (150 lbs) |
| Trunk | 21 kg (46 lbs) |
| Fuel tank | full |

| Front ride height specifications | |
|---|---|
| **Wheel size** | **Ride height** |
| **15 inch**<br>standard suspension<br>sport suspension<br>Coupe suspension | <br>647 mm (25.47 in)<br>632 mm (24.88 in)<br>642 mm (25.28 in) |
| **16 inch**<br>standard suspension<br>sport suspension<br>Coupe suspension | <br>660 mm (25.98 in)<br>645 mm (25.39 in)<br>655 mm (25.79 in) |
| **17 inch**<br>standard suspension<br>sport suspension<br>Z3 Coupe suspension<br>M Roadster / Coupe suspension | <br>632 mm (24.88 in)<br>660 mm (25.98 in)<br>670 mm (26.38 in)<br>664 mm (26.14 in) |
| **Maximum variation between sides: 10 mm (0.4 in)**<br>**Maximum deviation from specifications: 10 mm (0.4 in)** | |

## Control Arms

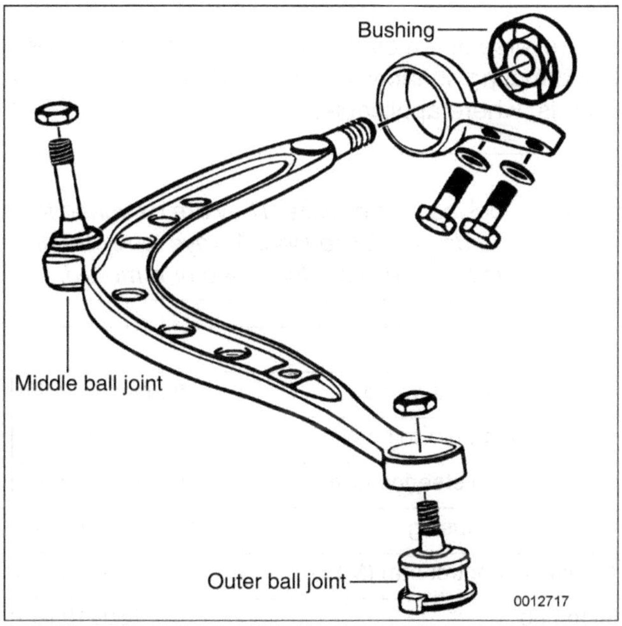


The outer mounting bushing and the outer ball joint on the control arm are available as replacement parts. If the control arm middle ball joint is worn, replace the complete control arm.

Special press tools are required to remove the control arm and to replace the control arm bushing. Read the procedure through before beginning the job.

### Control arm, removing and installing

– Raise car and remove wheel.

> ***WARNING —***
> *Make sure car is stable and well supported at all times. Use a professional automotive lift or jack stands designed for the purpose. A floor jack is not adequate support.*

Remove nut (**arrow**) attaching stabilizer bar link to control arm (non-M version).

– Suspend steering arm and brake assembly from strut spring using stiff wire. Be careful not to strain brake hose.

– Remove upper and two lower steering arm to strut mounting bolts. Separate steering arm from strut.

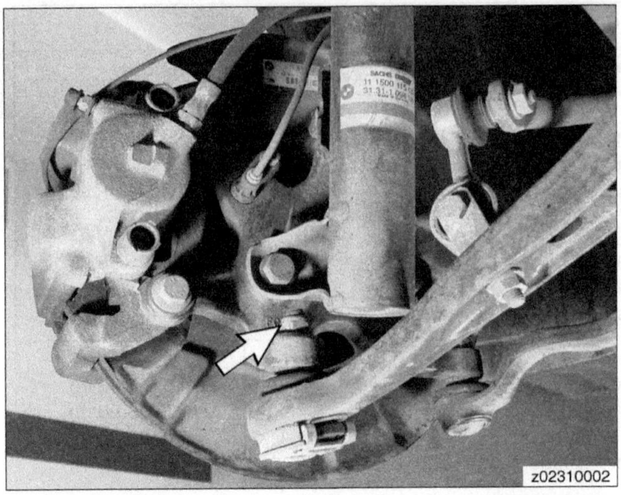


Remove control arm ball joint nut (**arrow**). Use BMW special tool 31 2 240 or equivalent to press ball joint out of steering arm.

– Working in middle of control arm, remove nut from middle ball joint. Knock ball joint loose with a plastic hammer.

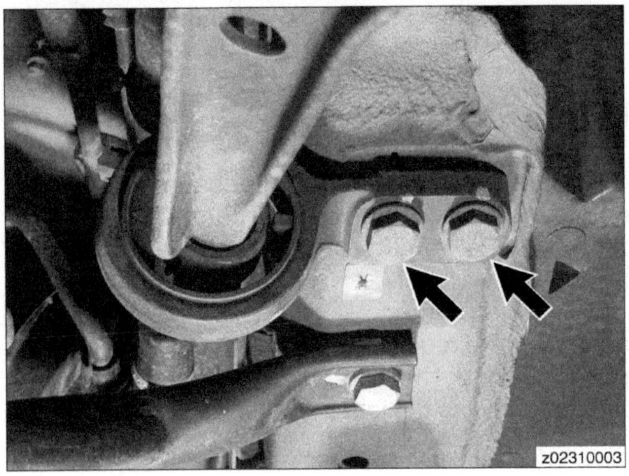


Working at other end of control arm, remove mounting bolts (**arrows**) attaching control arm bushing carrier to body and remove control arm.

– Installation is reverse of removal. Make sure all thread bores, bolts, nuts and mating surfaces are clean. Use new self-locking nuts or bolts, where applicable.

> ***WARNING —***
> *Do not reuse self-locking nuts or bolts. They are designed to be used only once and may fail if reused. Always replace self-locking fasteners any time they are loosened or removed.*

| **Tightening torques** | |
|---|---|
| Control arm ball joint to subframe | 85 Nm (62 ft-lb) |
| Control arm bushing carrier to body | 47 Nm (34 ft-lb) |
| Control arm ball joint to steering arm | 62 Nm (45 ft-lb) |
| Steering arm to strut housing | 107 Nm (78 ft-lb) |
| Stabilizer bar link to control arm (M10) | 59 Nm (44 ft-lb) |
| Road wheel to hub | 120±10 Nm (90±7 ft-lb) |

## Outer control arm ball joint, replacing

Only the outer ball joint is available as a replacement part. The ball joint can be replaced with the control arm installed.

– Raise car and remove wheel.

> ***WARNING —***
> *Make sure car is stable and well supported at all times. Use a professional automotive lift or jack stands designed for the purpose. A floor jack is not adequate support.*

– Loosen nut for ball joint until it contacts suspension strut.

– Use BMW special tool 31 1 430 or suitable ball joint fork to separate joint from control arm. Remove ball joint mounting nut.

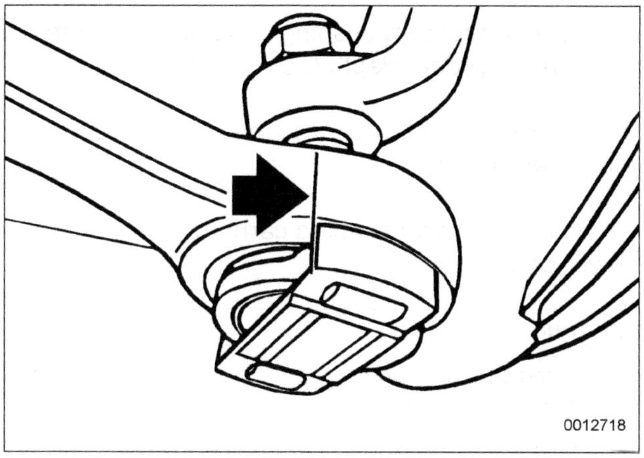


◀ Mark installed position (**arrow**) of ball joint guide to control arm.

– Press ball joint out of control arm using BMW special tool kit 31 1 440 or equivalent press.

– Make sure control arm ball joint taper bore is clean and free of any corrosion. Press new ball joint into position, making sure to align ball joint guide with mark made earlier. Verify that locking clip on new ball joint faces towards center of vehicle.

– Remainder of installation is reverse of removal.

| **Tightening torques** | |
|---|---|
| Control arm ball joint to steering arm | 62 Nm (45 ft-lb) |
| Road wheel to hub | 120±10 Nm (90±7 ft-lb) |

## Control arm bushing, replacing

Replace control arm bushings in pairs. Make sure the two bushings and bushing carriers have the same markings, indicating they come from the same manufacturer.

– Raise and safely support vehicle.

> ***WARNING —***
> *Make sure car is stable and well supported at all times. Use a professional automotive lift or jack stands designed for the purpose. A floor jack is not adequate support.*

– Unbolt control arm bushing carrier from body.

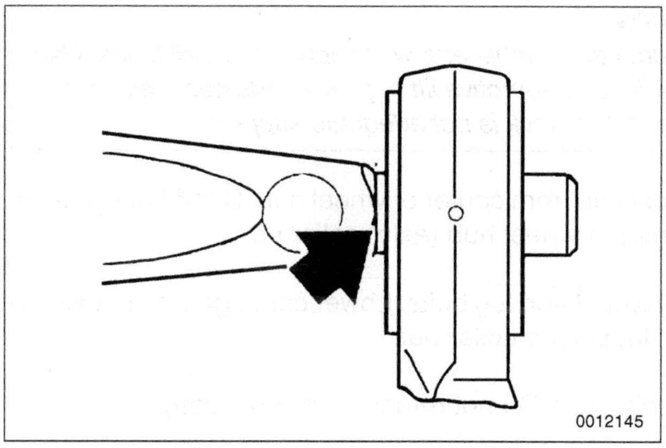


◀ Mark installed position (**arrow**) of bushing to control arm.

– Using appropriate press tools, pull rubber control arm bushing assembly off control arm.

– Coat end of control arm with BMW special lubricant (BMW part no. 81 22 9 407 284).

***NOTE —***
*Do not substitute a non-BMW approved lubricant. The kerosene-like lubricant is used only as a temporary aid in installing the bushing to the control arm stub. Work must be carried out in as short a time as possible. Within 30 minutes the control arm bushing will be fixed in position on the control arm.*

> ***CAUTION—***
> *BMW part numbers are provided for reference only and are subject to change.*

– Press new rubber control arm bushing assembly onto the end of control arm to proper depth (as marked).

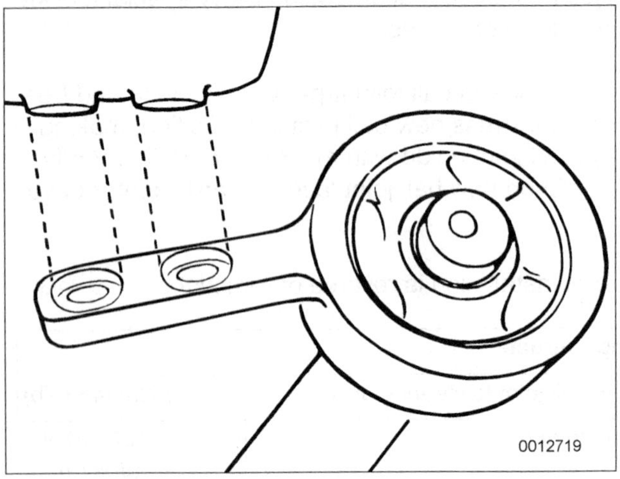

0012719

◄ Install bushing carrier to control arm so that carrier centering bores face up toward body.

– Immediately bolt control arm bushing carrier to body, torque bolts and lower car.

| **Tightening torque** | |
|---|---|
| Control arm bushing carrier to body | 47 Nm (34 ft-lb) |

– Lower the car to the ground and leave undisturbed in normal loaded position for a minimum of 30 minutes.

> ***CAUTION—***
> *Failure to follow the above procedure may impair the car's handling and stability.*

## FRONT WHEEL BEARINGS

The front wheel bearings are permanently sealed and require no maintenance. The bearing is integral with the wheel hub and pressed onto the steering arm stub axle.

Special press tools are required to replace the front wheel bearings. Read the procedure through before beginning the job.

### Front wheel bearing, replacing

– Raise car and remove wheel.

> ***WARNING—***
> *Make sure car is stable and well supported at all times. Use a professional automotive lift or jack stands designed for the purpose. A floor jack is not adequate support.*

0011199

◄ Pry off dust cap from center of wheel hub. Bend back staked part (**arrow**) of wheel hub (axle) collar nut.

– Remount wheel and lug bolts. Lower car to ground to gain leverage in loosening collar nut.

– Loosen collar nut. Do not remove nut completely.

*NOTE—*
*The wheel hub collar nut is tightened to a torque of 290 Nm (214 ft-lb). Make sure the car is firmly on the ground and use heavy-duty tools when loosening the nut.*

- Raise car and remove wheel.
- Remove ABS wheel speed sensor.
- Remove brake caliper assembly and rotor. See **340 Brakes**. Leave brake hose connected to caliper. Suspend caliper assembly from chassis using stiff wire.
- Remove wheel hub collar nut.

◀ Remove wheel hub with integral wheel bearing from steering arm using a puller.

| CAUTION— |
|---|
| *Never reuse bearing assembly after removing.* |

- If bearing inner race stays on steering arm, remove splash guard and use two-jaw puller to remove race.
- Install splash guard and new dust shield behind bearing. Press on new wheel hub/bearing assembly using appropriate press tools.

*NOTE—*
*When pressing the wheel hub/bearing on, apply force only to the bearing inner race.*

- Install new collar nut. Do not tighten nut to its final torque at this time.
- Install brake rotor and mount wheel and lug bolts. Lower car to ground to gain leverage.
- Tighten collar nut to final torque and stake nut to axle.
- Raise car and remove wheel.
- Install brake caliper as described in **340 Brakes**.
- Install a new grease cap, using Loctite® 638 sealant or equivalent. Install wheel and lower car.

| Tightening torques | |
|---|---|
| Brake rotor to wheel hub | 16 Nm (12 ft-lb) |
| Collar nut to stub axle | 290 Nm (214 ft-lb) |
| Brake caliper to strut housing | 110 Nm (81 ft-lb) |
| Road wheel to hub | 120 ± 10 Nm (90 ± 7 ft-lb) |

# FRONT STABILIZER BAR

## Front stabilizer bar, removing and installing

– Raise and safely support vehicle.

> ***WARNING—***
> *Make sure car is stable and well supported at all times. Use a professional automotive lift or jack stands designed for the purpose. A floor jack is not adequate support.*

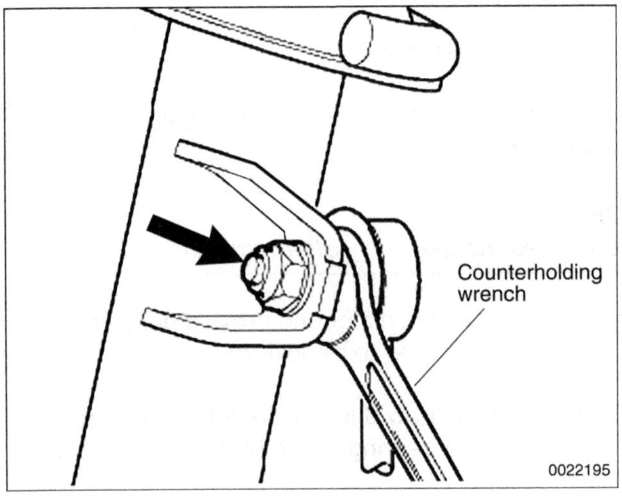


< M version: Loosen and remove stabilizer link mounting nut (**arrow**) from left and right struts.

– Other models: Loosen and remove stabilizer bar link from control arm.

***NOTE—***
*Use a thin wrench to counterhold shaft of stabilizer link ball joint while removing mounting nut.*

– Remove stabilizer bar connecting link from stabilizer bar on left and right sides.

– Remove two stabilizer bar bushing brackets on left and right sides. Remove bar.

– Installation is reverse of removal. Remember to:
- Clean grease from rubber mounts and stabilizer.
- Install open end of stabilizer bar bushings facing down.
- Use new self-locking nuts on connecting links.
- Perform final tightening of stabilizer bar with car level and in normal loaded position.

| Tightening torques | |
|---|---|
| Stabilizer bar bushing brackets to front subframe | 22 Nm (16 ft-lb) |
| Stabilizer bar link to control arm | 59 Nm (44 ft-lb) |
| Stabilizer bar link to stabilizer bar (M10) | 59 Nm (44 ft-lb) |
| Stabilizer bar link to strut housing | 59 Nm (44 ft-lb) |

## FRONT SUBFRAME

The front subframe provides rigid mounting points for the engine, suspension, and steering components. The subframe is not normally subject to wear and should only be replaced if structurally damaged.

Removing the subframe requires engine lifting equipment to support the weight of the engine so that the subframe can be removed from below.

> ***WARNING —***
> *Removal or replacement of the subframe may affect suspension and steering geometry, including front wheel alignment. Make appropriate matching marks during removal and have the front end aligned after installation.*

### Front subframe, removing and installing

0012716

- Using engine support equipment, raise engine until weight of engine is supported.

- Raise car and remove front wheels.

> ***WARNING —***
> *Make sure car is stable and well supported at all times. Use a professional automotive lift or jack stands designed for the purpose. A floor jack is not adequate support.*

- Working beneath car, remove nuts from engine mounts.

- Remove control arm bushing carrier from body.

- Remove control arm ball joint from subframe. Separate joint with a plastic hammer. Suspend control arm from chassis using stiff wire.

> ***WARNING —***
> *Do not allow the control arm to hang from outer ball joint. This can damage the ball joint.*

- Unbolt steering rack from subframe. See **320 Steering**. Suspend steering rack from chassis using stiff wire. Using a transmission jack or equivalent, support subframe from below.

- Remove reinforcing brace below oil pan.

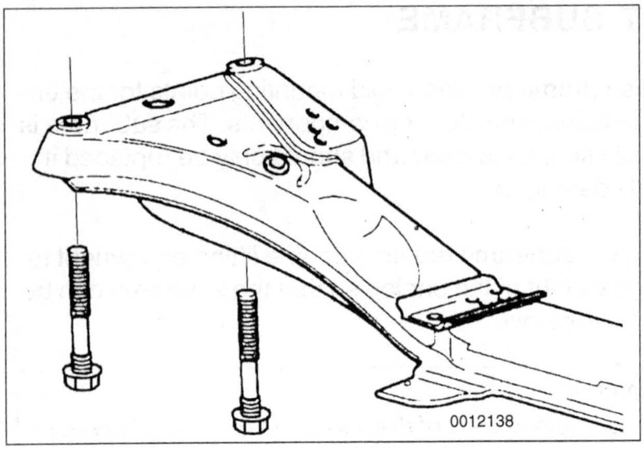


◄ Unbolt subframe from chassis on both sides.

– Slowly lower subframe, making sure all electrical leads and suspension components are clear during removal.

– Installation is reverse of removal, noting the following:
  - Make sure all bolts, bolt holes and mating surfaces are clean to ensure proper tightening and alignment.
  - Use new self-locking nuts or bolts, where applicable.
  - Lubricate subframe bolts with copper paste or equivalent antiseize compound before installing.
  - Lower engine onto engine mounts, allowing it to settle fully before tightening engine mount bolts.
  - When the job is completed have car professionally aligned.

| **Tightening torques** | |
|---|---|
| Front subframe to chassis (M12-12.9 grade bolt) | 105 Nm (77 ft-lb) |
| Steering gear to subframe (M10-10.9 grade self-locking bolt) | 42 Nm (31 ft-lb) |
| Control arm ball joint to front subframe | 85 Nm (62 ft-lb) |
| Control arm bushing to body | 47 Nm (34 ft-lb) |

# 320 Steering and Wheel Alignment



**TABLES**



320

## General

This repair group covers servicing of the mechanical and hydraulic components of the steering system, including wheel alignment information.

The variable-assist power steering system consists of an engine-driven hydraulic pump, a rack-and-pinion steering gear assembly with an integral hydraulic control valve, and connecting linkage to the road wheels.

At low speeds, maximum power assist is provided to ease parking and city driving. At high speeds, assist is reduced to ensure stability. The power steering system varies assist based on engine speed.

> ***WARNING —***
> - *Do not reuse self-locking nuts. They are designed to be used only once and may fail if reused. Always replace them with new locking nuts.*
> - *Do not install bolts and nuts coated with undercoating wax, as correct tightening torque cannot be assured. Always clean the threads with solvent before installation, or install new parts.*
> - *Do not attempt to weld or straighten any steering components. Always replace damaged parts.*

***NOTE —***
*For information on steering column-mounted switches and the ignition lock, see* ***612 Switches****.*

### Special tools

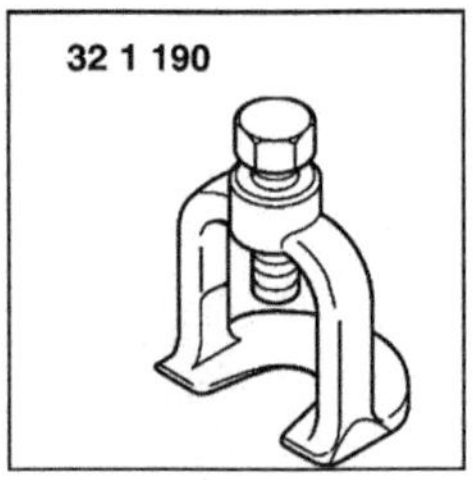


◄ Ball joint removal tool
(BMW special tool 32 1 190)

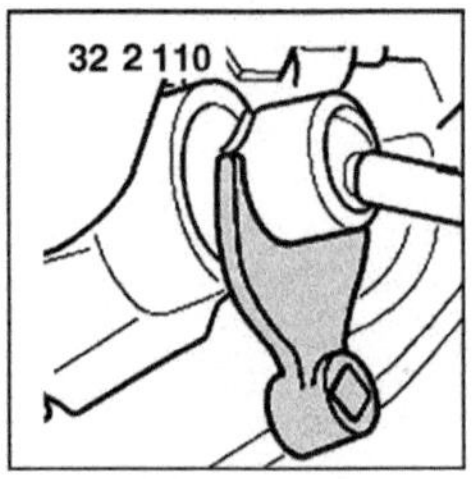


◄ Inner tie rod removal tool
(BMW special tool 32 2 210)

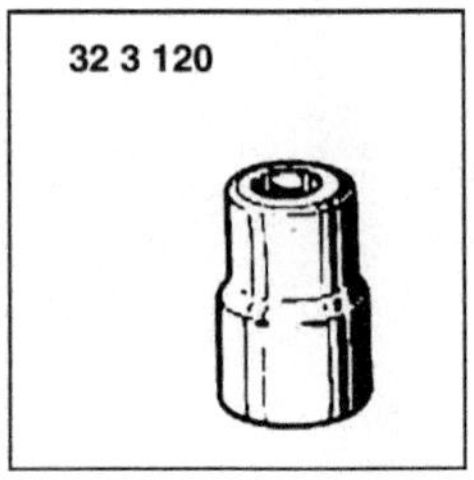


◄ Steering column shear bolt Torx socket (E8)
(BMW special tool 32 3 120)

## STEERING WHEEL

The BMW Z3 is equipped with an SRS airbag mounted in the steering wheel. Improper handling of the airbag could cause serious injury. The airbag is an explosive device and should be treated with extreme caution.

> ***WARNING —***
> - *The BMW Supplemental Restraint System (SRS) is complex and special precautions must be observed when servicing. Serious injury may result if system service is attempted by persons unfamiliar with the BMW SRS and its approved service procedures. BMW specifies that all inspection and service should be performed by an authorized BMW dealer.*
> - *BMW Supplemental Restraint System (SRS) is equipped with a back-up power supply inside the SRS control module. Wait at least 10 seconds after the battery cable has been disconnected before disturbing system. This will allow the reserve power supply to discharge.*

### Steering wheel, removing and installing

Use extreme caution when removing the airbag unit from the steering wheel. Always follow the airbag removal procedure as outlined in **721 Airbag System (SRS)**.

– Make sure front wheels are pointed straight ahead. Center steering wheel.

– Disconnect negative (–) battery cable and cover terminal with insulating material.

> ***CAUTION—***
> *Prior to disconnecting the battery, read the battery cautions in* **001 General Warnings and Cautions**.

***NOTE —***
*On 1996 cars, the battery is located in the right rear of the engine compartment. On later cars, the battery is located under the trunk floor mat.*

– Carefully remove airbag from front of steering wheel. See **721 Airbag System (SRS)**. Store airbag unit in a safe place.

> ***WARNING —***
> *Improper handling of the airbag could cause serious injury. Store the airbag with the horn pad facing up. If stored facing down, accidental deployment could propel it violently into the air, causing injury.*

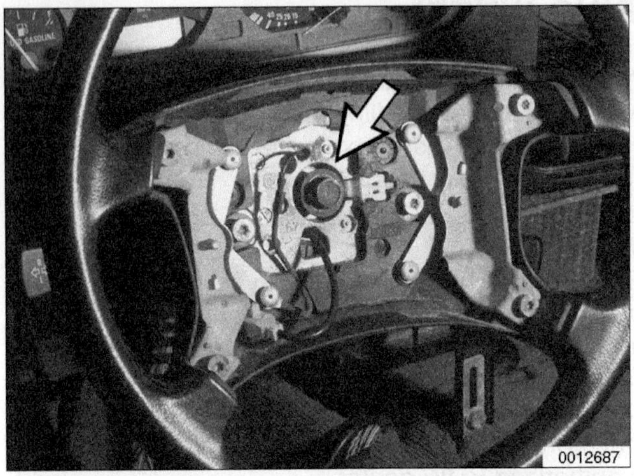


Remove steering wheel center bolt (**arrow**). Note position of steering wheel on steering column shaft. Lightly punch mark or scribe position, if necessary.

– Unlock steering wheel by turning ignition key ON. Remove steering wheel.

> ***CAUTION—***
> *The SRS contact reel is mounted to the rear of the steering wheel hub. The contact reel is a wound coil of wire that ensures continuous electrical contact for the air bag unit. Once the steering wheel mounting nut or bolt is removed the contact reel is locked in the center position and its position must not be altered.*

When installing steering wheel, align matching marks. Make sure airbag contact ring locking pin engages cut out in contact reel (**arrow**). Install steering column center nut. Do not over-torque.

***NOTE—***
*Depending on model, SRS contact reel may be different than that shown.*

– The remainder of installation is reverse of removal.

| **Tightening torque** | |
|---|---|
| Steering wheel to steering column spindle | 63 Nm (46 ft-lb) |

## STEERING COLUMN

### Steering lock, removing and installing

Steering lock and ignition switch components are shown in **Ignition switch components** diagram in **612 Switches**.

– Disconnect negative (–) battery cable and cover terminal with insulating material.

> ***CAUTION—***
> *Prior to disconnecting the battery, read the battery cautions in* **001 General Warnings and Cautions**.

– Carefully remove airbag from front of steering wheel. See **721 Airbag System (SRS)**. Store airbag unit in a safe place.

> ***WARNING—***
> *Improper handling of the airbag could cause serious injury. Store the airbag with the horn pad facing up. If stored facing down, accidental deployment could propel it violently into the air, causing injury.*

– Remove steering wheel. See **Steering wheel, removing and installing** in this repair group.

– Remove driver's side lower dashboard trim. See **513 Interior Trim**.

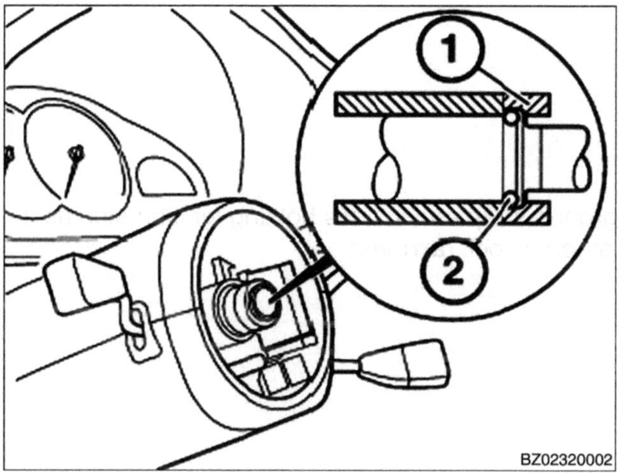

BZ02320002

◄ Remove collar (**1**) from steering shaft. Note position of snap ring (**2**) in collar recess.

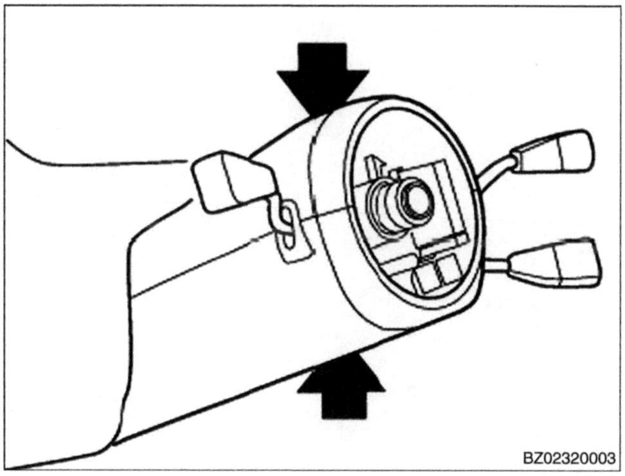
BZ02320003

◄ Remove upper and lower steering column trim screws (**arrows**).

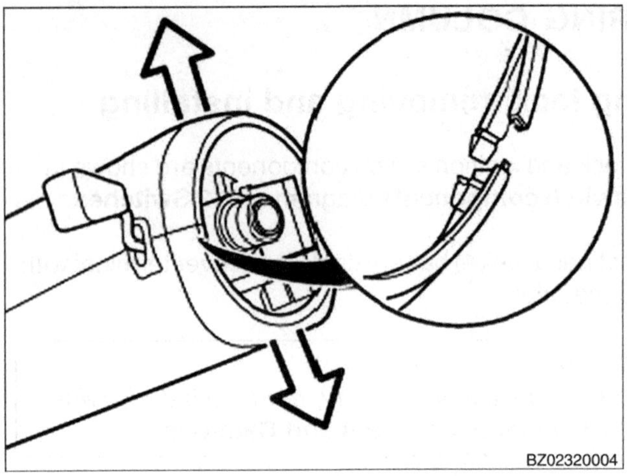
BZ02320004

Carefully separate upper and lower trim and remove.

- Remove ignition lock ring.

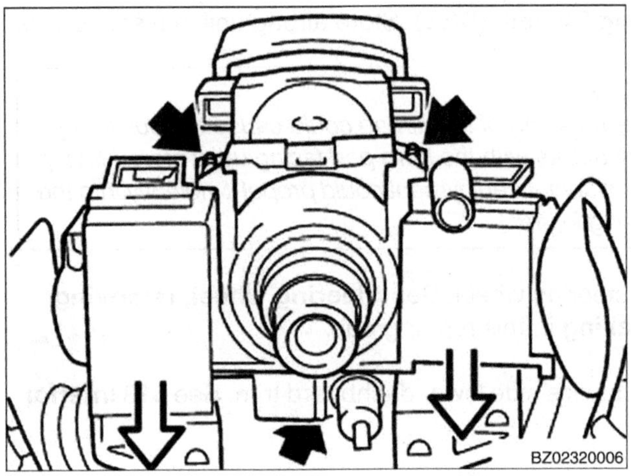
BZ02320006

Remove screws (**black arrows**) and remove clips (**white arrows**) from steering column combination switch. Remove combination switch.

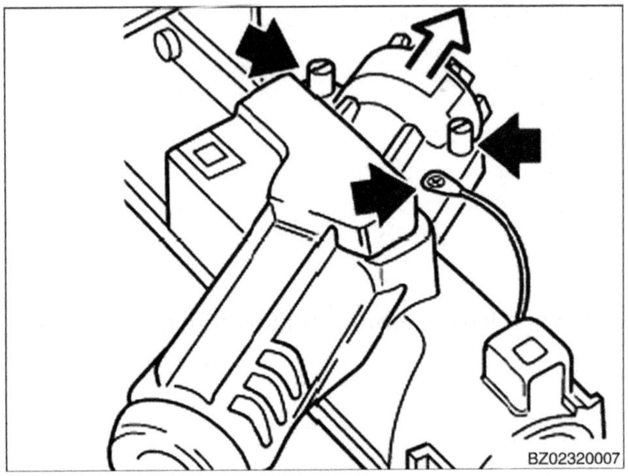
BZ02320007

Remove ground strap and screws holding ignition electrical switch to steering lock (**arrows**).

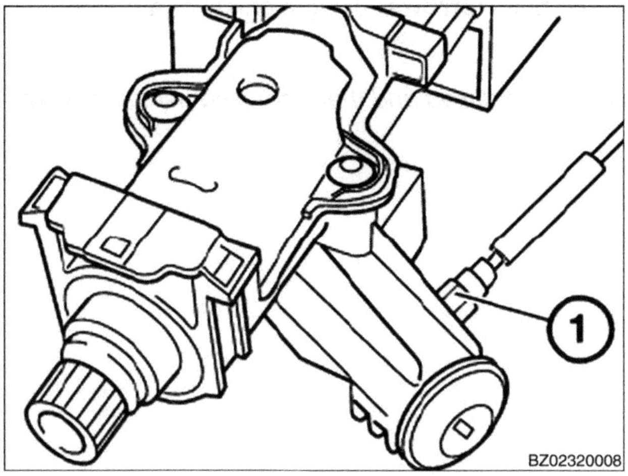


< Automatic transmission models: Loosen and disconnect interlock shift cable (**1**) from ignition lock assembly.

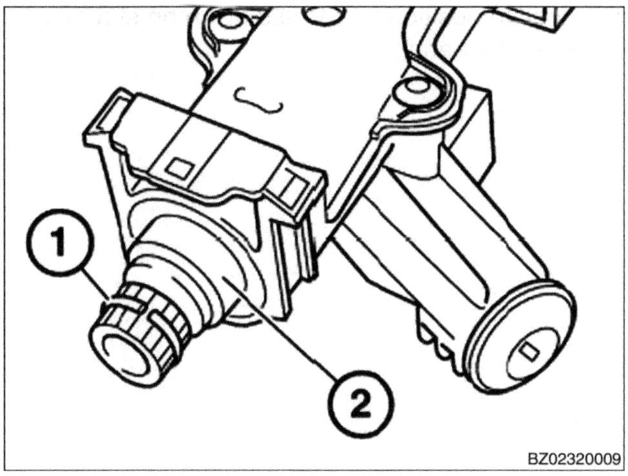


< Remove snap ring (**1**) and remove steering shaft sleeve (**2**).

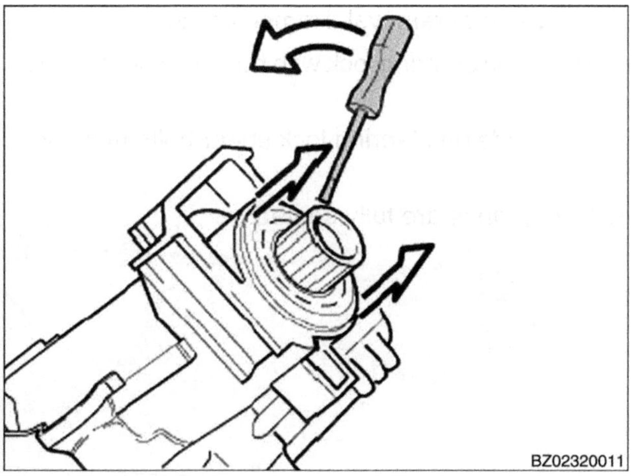


< Gently pry off steering shaft bearing.

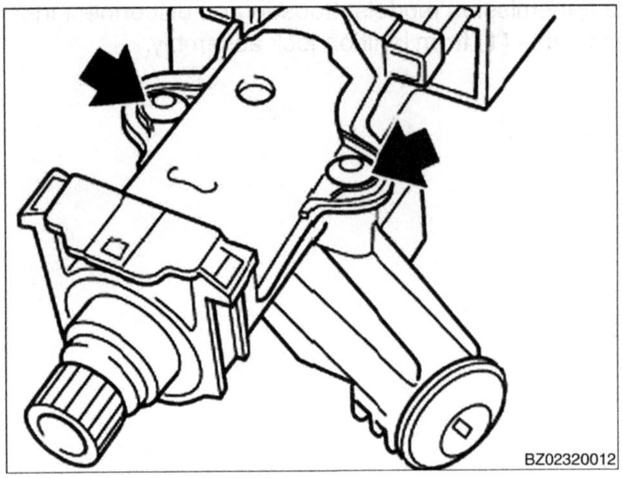


Remove shear bolts (**arrows**) using a chisel or a punch.

***NOTE —***
*Shear bolts can be drilled and then removed using an easy-out*

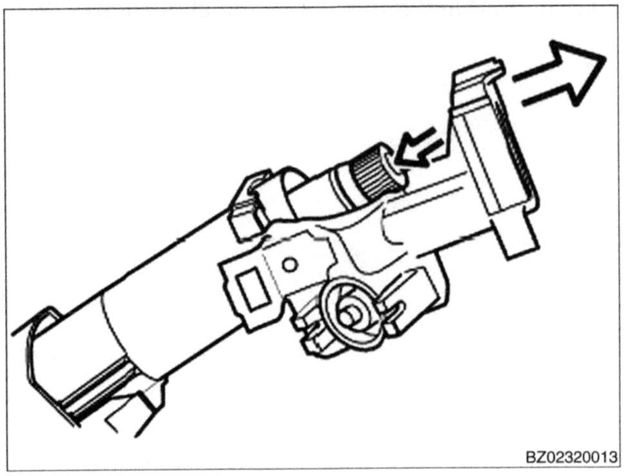


Turn steering lock to position "R". Press down on steering spindle and take off steering lock.

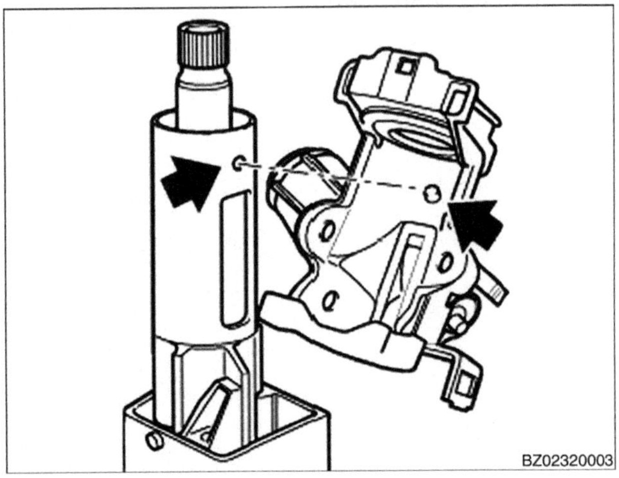


Installation is reverse of removal. Remember to:

- Match dowel pin on steering lock with hole on spindle cover (**arrow**).
- Tighten Torx heads on steering lock shear bolts until they shear.
- Make sure snap rings are fully seated.

## Steering column, removing and installing

– Disconnect negative (–) battery cable and cover terminal with insulating material.

> ***CAUTION—***
> *Prior to disconnecting the battery, read the battery cautions in* **001 General Warnings and Cautions**.

***NOTE—***
*On 1996 cars, the battery is located in the right rear of the engine compartment. On later cars, the battery is located under the trunk floor mat.*

– Carefully remove airbag from front of steering wheel. See **721 Airbag System (SRS)**. Store airbag unit in a safe place.

> ***WARNING—***
> *Improper handling of the airbag could cause serious injury. Store the airbag with the horn pad facing up. If stored facing down, accidental deployment could propel it violently into the air, causing injury.*

– Remove steering wheel. See **Steering wheel, removing and installing** in this repair group.

– Remove driver's side lower dashboard trim. See **513 Interior Trim**.

– Automatic transmission models: Loosen and disconnect interlock shift cable from ignition lock assembly.

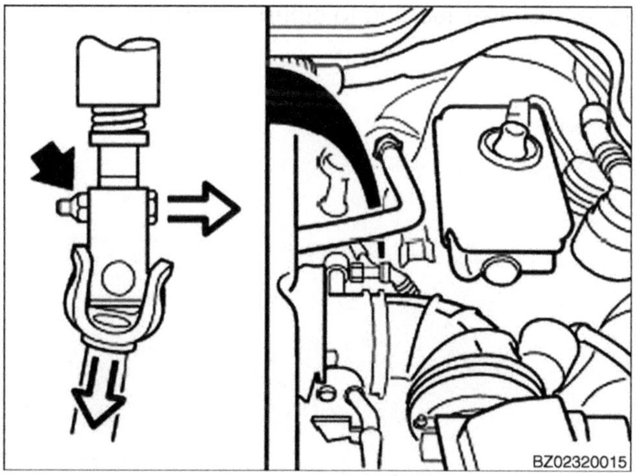


◄ Working in engine compartment, remove clamping bolt and pull lower steering spindle off splined end of upper steering spindle.

– When installing, replace self locking nut. Re-insert clamping bolt into slot in upper spindle.

| Tightening torque | |
|---|---|
| Spindle clamping bolt to upper spindle | 19 Nm (14 ft-lb) |

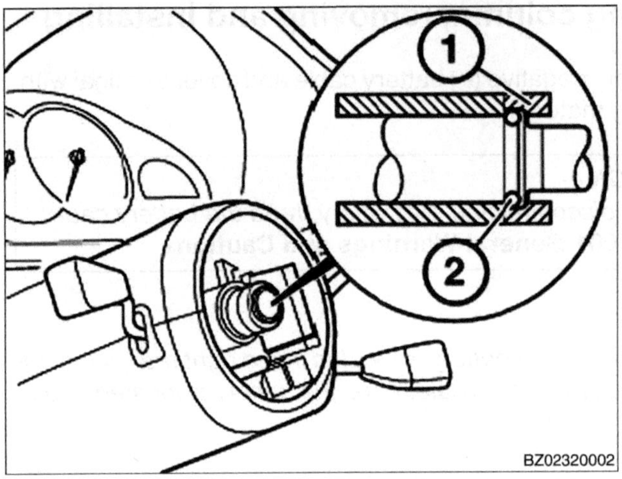


Remove collar (**1**) from steering shaft. Note position of snap ring (**2**) in collar recess.

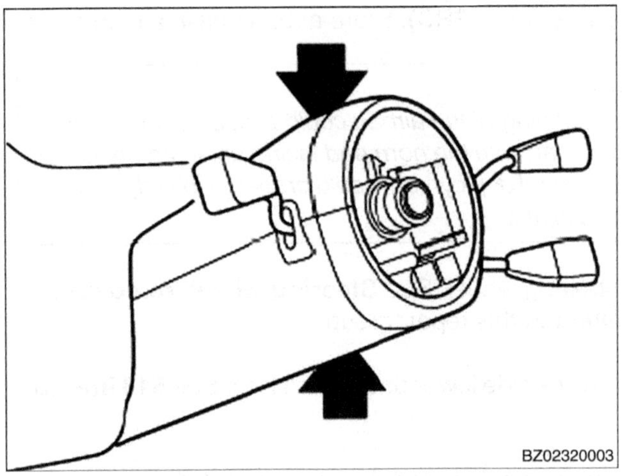


Remove upper and lower steering column trim screws (**arrows**).

Working under dash, turn steering spindle lock ring (**arrow**) at bulkhead counterclockwise to release. Release electrical connectors for all wires leading to steering column.

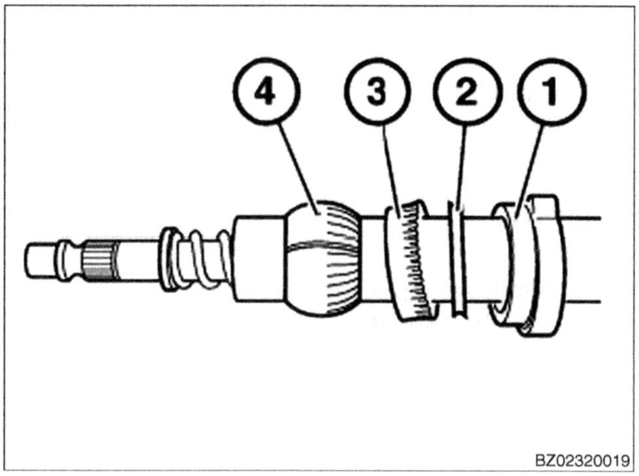

BZ02320019

To replace lock ring, note assembly sequence.

1. Lock ring
2. Rubber ring
3. Joint socket
4. Ball joint (with Teflon ring, if necessary)

***NOTE—***
*When installing, turn locking ring clockwise until it clicks into place.*

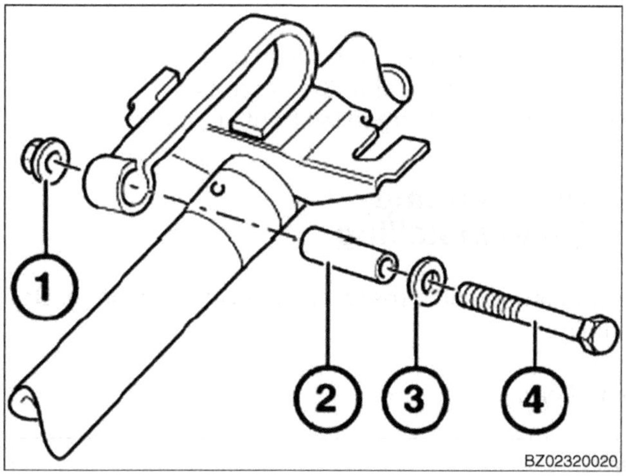

BZ02320020

Remove bolt holding steering column to dashboard reinforcement.

1. Nut
2. Spacer
3. Washer
4. Bolt (M8x80)

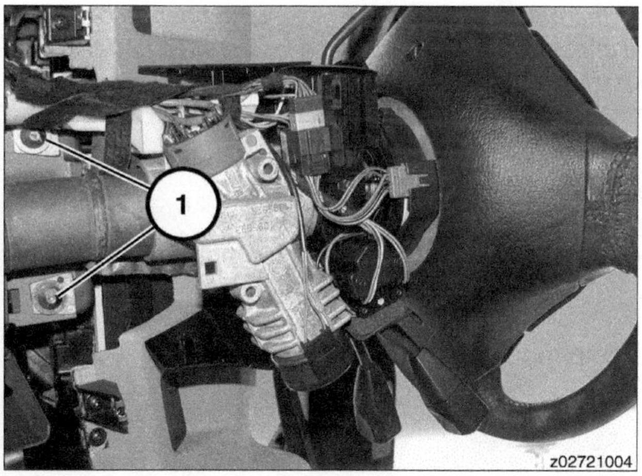

z02721004

Remove shear bolts (**1**) holding steering column to dash using a chisel or a punch and remove steering column.

***NOTE—***
*Shear bolts can be drilled and then removed using an easy-out.*

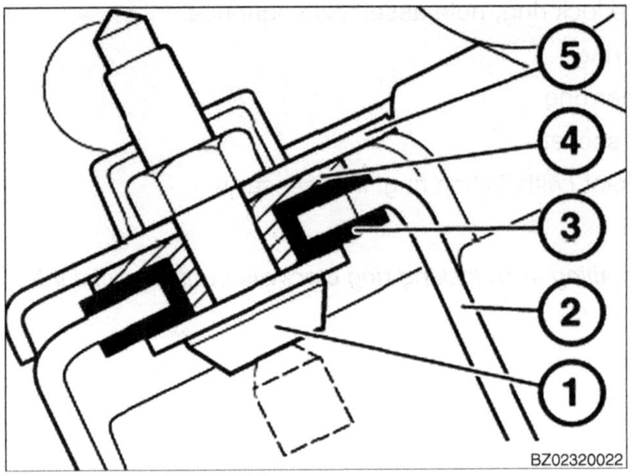


To replace column, note shear bolt installation sequence.

1. Shear bolt
2. Steering column
3. Spacer
4. Sleeve
5. Dashboard carrier

– Raise steering column and pull toward interior while tightening shear bolts until they shear.

– Remainder of installation is reverse of removal.

## POWER STEERING SYSTEM

Power assist is provided by a belt-driven pump on the lower left front of the engine, just below the alternator. The power steering fluid reservoir is located left front of engine.

### Power steering pump, removing and installing

– Empty power steering fluid reservoir using clean syringe. Do not reuse fluid.

– Raise front of car.

> ***WARNING —***
> *Make sure that the car is firmly supported on jack stands designed for the purpose. Place the jack stands beneath a structural chassis point. Do not place jack stands under suspension parts.*

– Remove fluid hoses from pump. Plug openings in pump and in hose ends.

– Remove drive belt from power steering pump. See **020 Maintenance**.

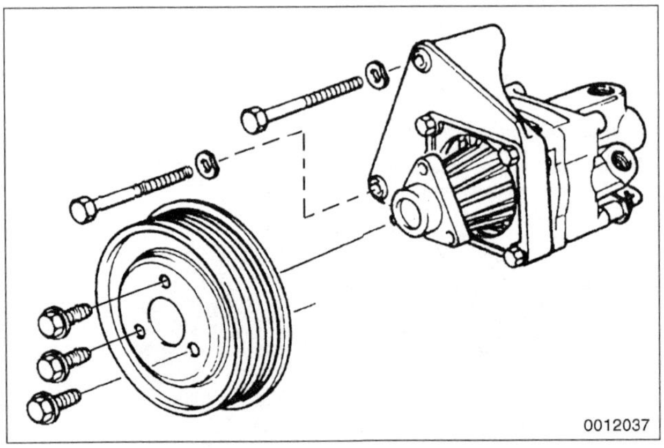


Remove pump mounting bolts and remove pump.

– Installation is reverse of removal, noting the following:
  - Make sure all thread bores, bolts, nuts, fluid couplings and mating surfaces are clean, and that drive belt is free of grease and dirt.
  - Use new sealing washers when reattaching power steering pressure lines. Do not over-torque banjo bolts.
  - Make sure hoses have adequate clearance from chassis.
  - Fill and bleed power steering system. See **Power steering system, bleeding and filling** in this repair group.

| Tightening torques | |
|---|---|
| Hydraulic hoses to power steering pump<br>• M10 banjo bolt<br>• M14 banjo bolt<br>• M16 banjo bolt | <br>12 Nm (9 ft-lb)<br>35 Nm (26 ft-lb)<br>40 Nm (30 ft-lb) |
| Power steering pump to bracket (self-locking nuts) | 22 Nm (16 ft-lb) |
| Power steering pump bracket to engine (self-locking nuts) | 22 Nm (16 ft-lb) |

## Power steering system, bleeding and filling

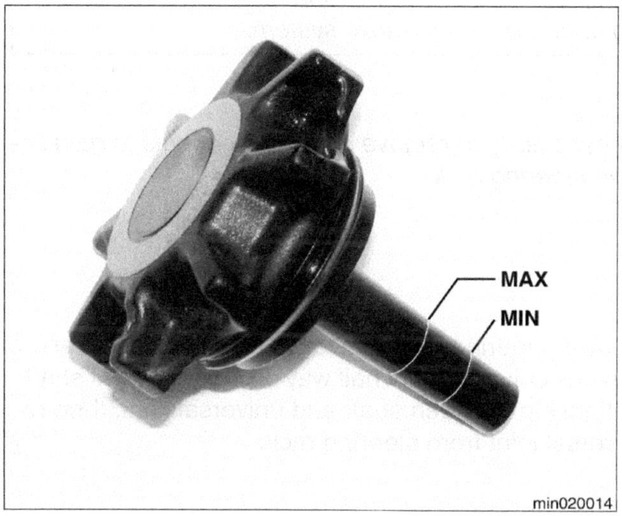

min020014

< With engine off, fill power steering fluid reservoir with clean fluid. Fill level to **MAX** mark on dipstick.

– Start engine. Slowly turn steering wheel from lock to lock a minimum of two times.

– Turn engine off and check fluid level, adding fluid if necessary.

| Power steering fluid | |
|---|---|
| Recommended fluid | Dexron III® ATF |

## STEERING RACK

The steering system is a power-assisted rack-and-pinion unit. The rack is maintenance free and non-adjustable.

The steering rack and linkage require no maintenance other than alignment and a periodic inspection for worn components.

## Steering rack, removing and installing

- Empty power steering fluid reservoir using clean syringe. Do not reuse fluid.

- Make sure front wheels are pointed straight ahead. Raise front of car.

> ***WARNING —***
> *Make sure the car is firmly supported on jack stands designed for the purpose. Place the jack stands beneath a structural chassis point. Do not place jack stands under suspension parts.*

- Remove front wheels.

- Remove outer tie rod ball joint nuts, then separate outer tie rod ends from steering arms using appropriate press tools.

- Disconnect power steering hose from fluid reservoir.

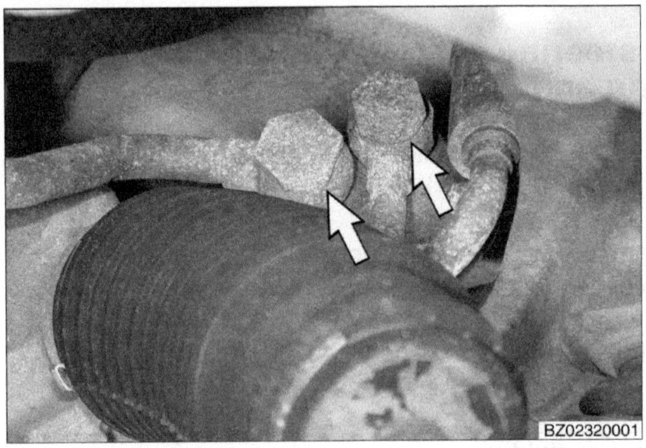

BZ02320001

< Remove fluid line banjo bolts (**arrows**) from steering rack. Be prepared to catch dripping fluid.

> ***CAUTION—***
> *Plug the openings in the pump and hose ends to prevent dirt from entering the hydraulic system.*

***NOTE —***
*It may be necessary to remove other components to gain access to the steering rack.*

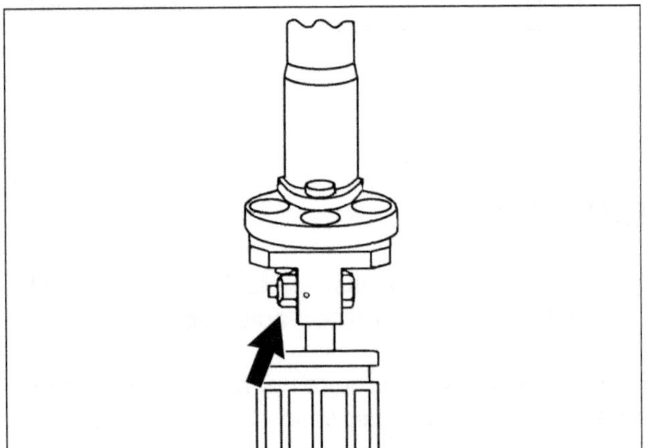

0012140

< Remove lower steering shaft universal joint pinch bolt (**arrow**), then pull universal joint half way off steering gear shaft. Mark relationship between shaft and universal joint, then remove universal joint from steering rack.

Remove steering rack mounting bolts (**arrows**), then remove steering rack from subframe crossmember by pulling it forward.

– Installation is reverse of removal, noting the following:
  - Make sure all thread bores, bolts, nuts, splines and mating surfaces are clean. Use new self-locking nuts wherever applicable.
  - Use marks made during removal to reassemble steering column universal joint to steering shaft.
  - Use a new sealing washer on banjo fitting. Do not over-torque banjo bolt.
  - Fill and bleed power steering system as described earlier.
  - Have car professionally aligned.

| Tightening torques | |
|---|---|
| Outer tie rod end to steering arm | 45 Nm (33 ft-lb) |
| Road wheel to hub | 120 ± 10 Nm (90 ± 7 ft-lb) |
| Steering column universal joint to steering shaft (self-locking nut) | 19 Nm (14 ft-lb) |
| Steering rack to subframe M10 bolt | 42 Nm (32 ft-lb) |
| Steering wheel to steering spindle | 63 Nm (46 ft-lb) |
| Hydraulic hoses to power steering pump<br>• M10 banjo bolt<br>• M14 banjo bolt<br>• M16 banjo bolt | <br>12 Nm (9 ft-lb)<br>35 Nm (26 ft-lb)<br>40 Nm (30 ft-lb) |

## Outer tie rod end, replacing

– Raise car and support safely.

> ***WARNING —***
> *Make sure the car is firmly supported on jack stands designed for the purpose. Place the jack stands beneath a structural chassis point. Do not place jack stands under suspension parts.*

– Remove wheel.

– Remove outer tie rod ball joint nut, and press out outer tie rod end ball joint using BMW special tool 32 1 190.

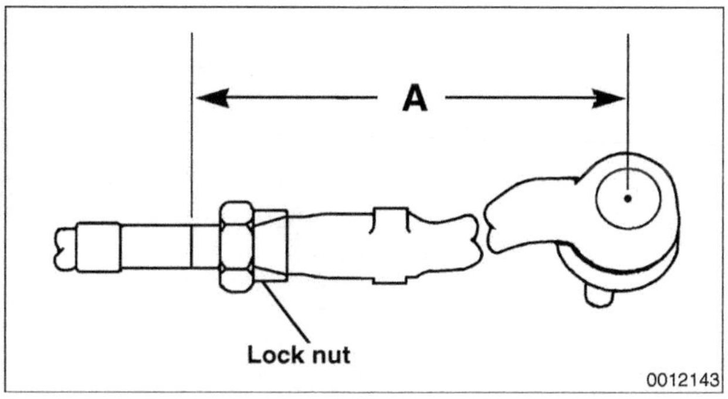


Make a reference measurement (**A**) of outer tie rod end to inner tie rod. Record measurement.

***NOTE —***
*Accurate measuring of the tie rod end in reference to the tie rod will help approximate correct wheel alignment when new parts are installed.*

– Loosen outer tie rod lock nut.

– Unscrew tie rod end from steering rack.

– Installation is reverse of removal. Remember to:
  - Make sure all threaded parts are clean.
  - Use new self-locking fasteners.
  - Have car professionally aligned.

| Tightening torques | |
|---|---|
| Outer tie rod end to steering arm | 45 Nm (33 ft-lb) |
| Outer tie rod end lock nut | 35 Nm (26 ft-lb) |

## Inner tie rod, replacing

– Raise car and support safely.

> ***WARNING —***
> *Make sure the car is firmly supported on jack stands designed for the purpose. Place the jack stands beneath a structural chassis point. Do not place jack stands under suspension parts.*

– Remove wheel.

– Remove outer tie rod end ball joint nut and press out tie rod end ball joint. See **Outer tie rod end, replacing** in this repair group.

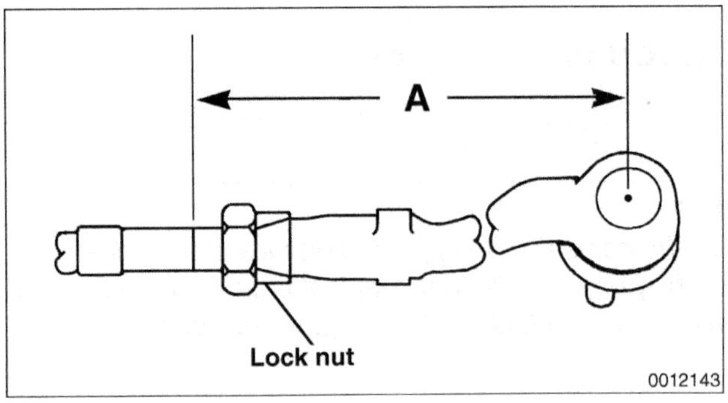


∢ Make a reference measurement (**A**) of outer tie rod end to inner tie rod. Record measurement.

***NOTE —***
*Accurate measuring of the tie rod end in reference to the tie rod will help approximate correct wheel alignment when new parts are installed.*

– Loosen outer tie rod lock nut.

– Unscrew outer tie rod end from tie rod.

– Loosen rack boot bellows clamp and slide bellows back. Inspect boot for any sign of damage. Replace if necessary.

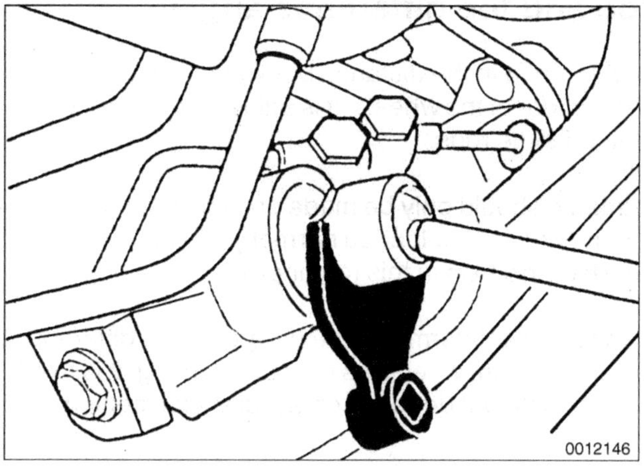


◄ Use BMW special tool 32 2 110 to unscrew tie rod.

– Installation is reverse of removal. Remember to:
  - Make sure all threaded parts are clean.
  - Use new self-locking fasteners.
  - Check outer tie rod end ball joints for wear.
  - Have car professionally aligned.

| Tightening torques | |
|---|---|
| Outer tie rod end to steering arm | 45 Nm (33 ft-lb) |
| Outer tie rod end lock nut | 35 Nm (26 ft-lb) |
| Inner tie rod to steering rack | 71 Nm (52 ft-lb) |

## WHEEL ALIGNMENT

Proper handling, stability, tire wear and driving ease depend upon the correct alignment of all four wheels. The BMW Z3 uses a sophisticated multi-link suspension at the front and rear of the car.

In what is known as a four wheel or thrust-axis alignment the front axle is aligned in relation to the rear axle, then the front wheels are aligned in relation to one another. Proper alignment requires computerized alignment equipment.

### Camber

Front and rear camber are fixed by the design of the car. No adjustments are possible. Any deviations are usually the result of worn or damaged suspension or body parts.

***NOTE—***
*Front upper strut mount replacement is covered in* **310 Front Suspension**.

### Caster

Front and rear caster is fixed by the design of the car and deviations are usually the result of worn or damaged suspension or body parts.

Check the upper strut mounts, control arms, lower ball joints and control arm bushing assemblies for wear, damage, and deformation if front caster problems are present. Check the trailing arms, rear suspension subframe and bushings if rear caster problems are present.

## Front toe and toe difference angle

Toe is the difference in the distance between the front of the wheels and the rear of the wheels. It is adjusted by altering the length of the outer tie rods.

Toe adjustments should only be made on alignment equipment and only with the car loaded normally. See the table **Normal loaded position** in this repair group.

Toe difference angle determines the progressively different paths taken by the front wheels as the car is turning. It is fixed by the steering rack, but depends on accurate toe setting.

## Rear toe

There is no provision for routine rear wheel toe adjustment. If any alignment angles deviate from specification, carefully inspect the rear trailing arms, rear suspension subframe and all associated bushings and flexible mounts for wear or damage. After replacing any parts that appear worn or damaged, recheck the rear toe measurements.

## Wheel alignment specifications

Alignment specifications for the Z3 are listed in **Table a** and **Table b**. The specifications only apply under the following conditions:

- Correct wheels and tires are installed, in good condition, and are at the correct inflation pressures.
- Steering and suspension parts and bushings are undamaged and show no signs of abnormal wear.
- Wheel bearings are in good condition.
- Ride height is in accordance with specifications. See **310 Front Suspension** and **330 Rear Suspension**.
- Car is in normal loaded position.

| **Normal loaded position** | |
|---|---|
| Each front seat | 68 kg (150 lb) |
| Trunk | 21 kg (46 lb) |
| Fuel tank | full |

| Table a. Front wheel alignment specifications | | |
|---|---|---|
| | **Base suspension** | **Sport suspension** |
| Toe angle (total)<br>• toe difference angle<br>• inside wheel @ 20° lock | <br>0° 18' ± 8'<br>– 1° 50'± 30' | <br>0° 18' ± 8'<br>– 2° 4' ± 30' |
| Camber | – 44' ± 30' | – 1° 4' ± 30' |
| Caster<br>• @ 10° wheel lock<br>• @ 20° wheel lock<br>• max deviation between sides | <br>3° 41' ± 30'<br>3° 48' ± 30'<br>30' | <br>3° 48' ± 30'<br>3° 56' ± 30'<br>30' |
| Front wheel displacement:<br>• Full lock<br>• inside wheel (approximate)<br>• outside wheel (approximate) | <br>0 ± 15'<br>41°<br>34° | <br>0 ± 15'<br>40°<br>33° |

| Table b. Rear wheel alignment specifications | | |
|---|---|---|
| | **Base suspension** | **Sport suspension** |
| **4-cylinder** | | |
| Toe angle (total) | 0° 34' ± 13' | 0° 44' ± 13' |
| Axis deviation (allowable range) | 0 ± 15' | 0 ± 15' |
| Camber | –2° 20' ± 30' | –2° 50' ± 30' |
| Max deviation between sides | 30' | 30' |
| **6-cylinder** | | |
| Toe angle (total) | 0° 20' ± 13' | 0° 30' ± 13' |
| Axis deviation (allowable range) | 0 ± 15' | 0 ± 15' |
| Camber | –2° 20' ± 30' | –2° 50' ± 30' |
| Max deviation between sides | 30' | 30' |

# 330 Rear Suspension



TABLE



330

## General

This repair group covers removal and replacement of rear suspension components.

For additional rear suspension information see the following:

- **300 Suspension, Steering and Brakes-General**
- **331 Drive Axles, Differential**

### Special tools

Special service tools are required for most of the work described in this repair group. Read the procedures through before beginning any job.

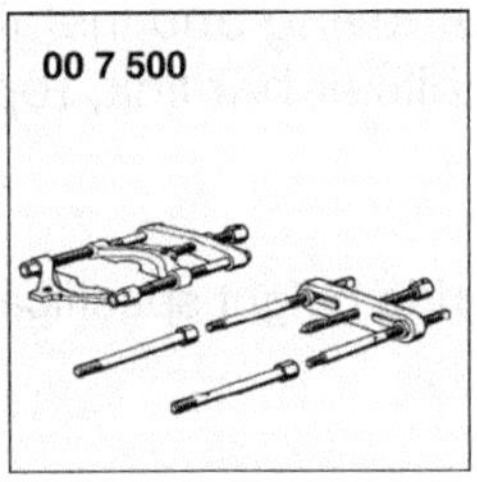


Puller
(BMW special tool 00 7 500)

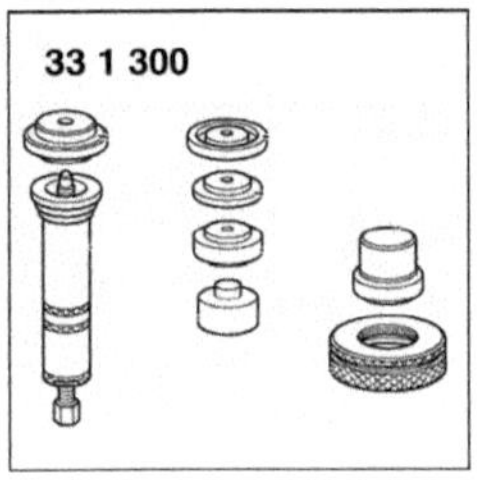


Tapered roller bearing extractor kit
(BMW special tool 33 1 300)

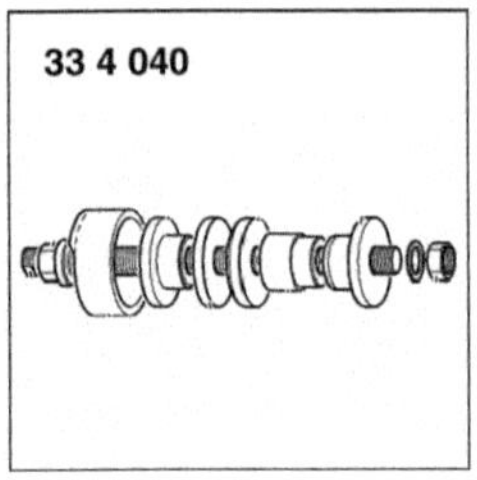


Rear wheel bearing removal and installation kit
(BMW special tool 33 4 040)

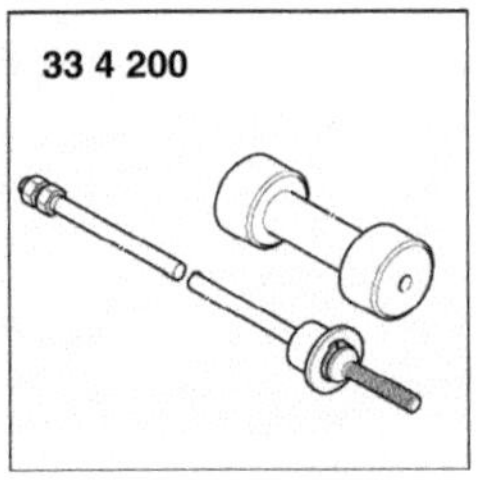


Slide hammer puller
(BMW special tool 33 4 200)

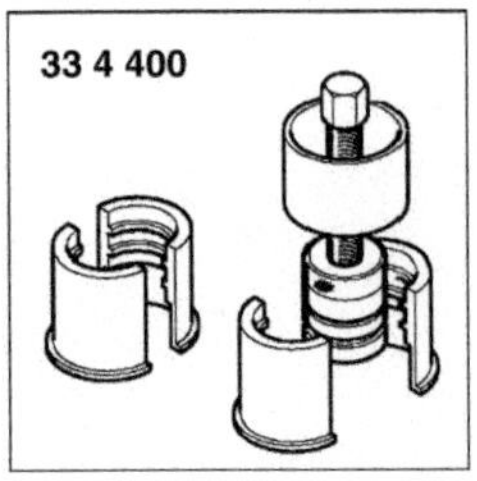


Inner race removal tool
(BMW special tool 33 4 400)

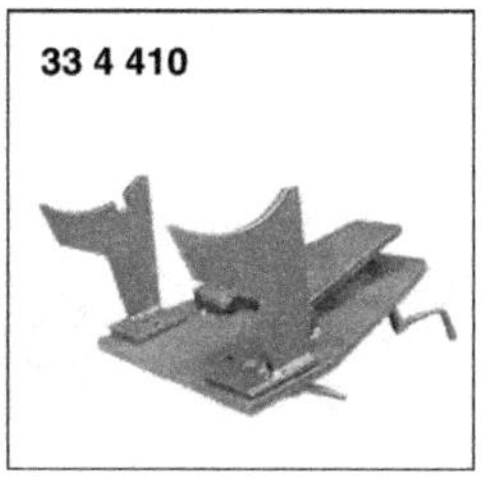


Differential support bracket (BMW special tool 33 4 410)

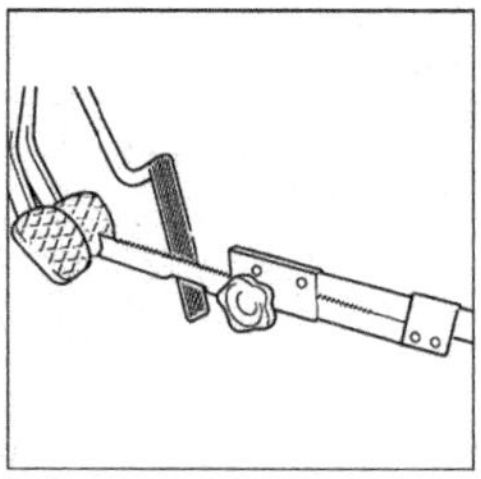

Pedal prop

> ***WARNING —***
> - *Physical safety could be impaired if procedures described here are undertaken without the proper service tools and equipment. Be sure to have the right tools on hand before beginning any job.*
> - *Do not reuse self-locking nuts or bolts. They are designed to be used only once and may fail if reused. Always replace self-locking fasteners any time they are loosened or removed.*
> - *Do not install bolts and nuts coated with undercoating wax, as the correct tightening torque cannot be assured. Always clean the threads with solvent before installation, or install new fasteners.*
> - *Do not attempt to weld or straighten any suspension components. Replace damaged parts.*

## SHOCK ABSORBERS AND SPRINGS

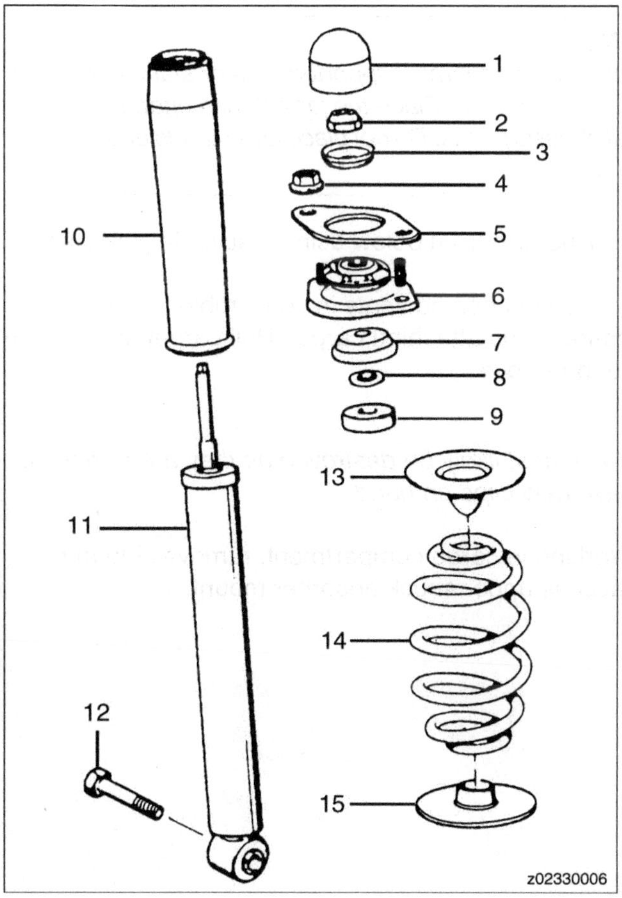


### Rear shock absorber and spring components

1. Dust cover
2. Self-locking hex nut (M10 x 1.0)
3. Plate
4. Nut (M8)
5. Gasket
6. Upper mount
7. Plate
8. Spacer ring
9. Bump stop
10. Dust cover
11. Shock absorber
12. Bolt (M12 x 1.5)
13. Spring seat, upper
14. Coil spring
15. Spring seat, lower

**Z3 rear suspension**

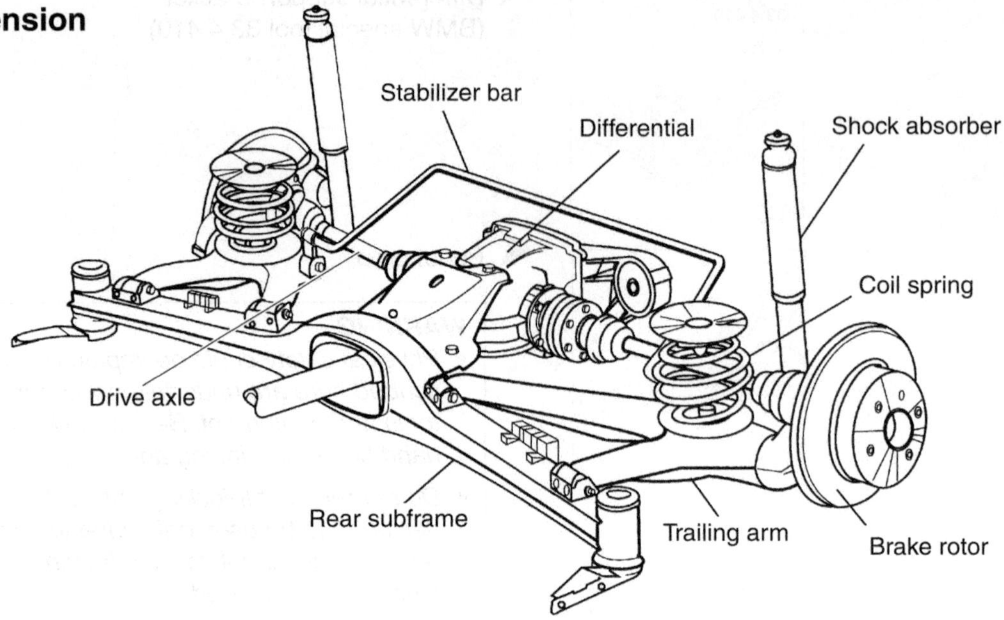


## Rear shock absorber, removing and installing

Replace shock absorbers and springs in pairs.

– Raise car and support safely.

> ***WARNING*** —
> *Make sure the car is firmly supported on jack stands designed for the purpose. Place the jack stands underneath structural chassis points. Do not place jack stands under suspension parts.*

– Support trailing arm from below using a suitable jack stand.

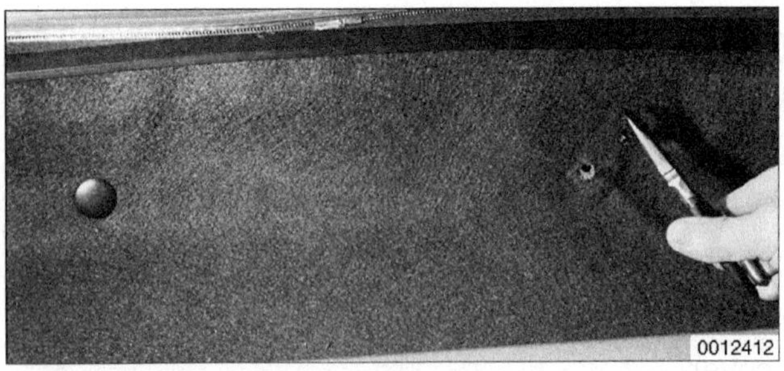


◄ Roadster: Working behind seats in convertible top compartment, unfasten clips attaching carpet. Pull carpet material downward to remove.

> ***NOTE*** —
> *The clips will most likely be destroyed as they are removed. Plan to have new clips on hand.*

– Coupe: Working in cargo compartment, remove side trim panel to access upper shock absorber mount.

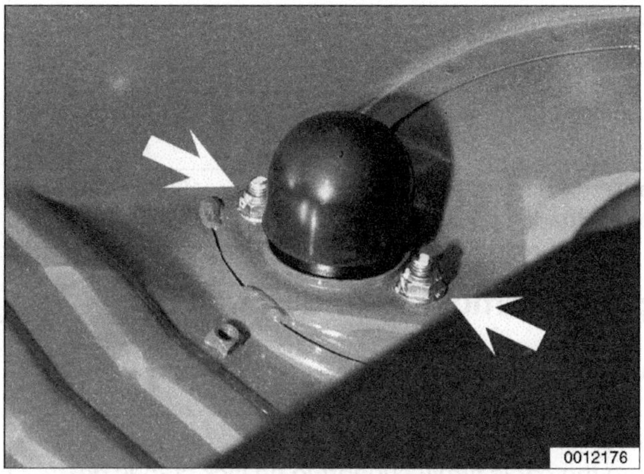


Remove shock absorber upper mount cover and mounting nuts (**arrows**).

> ***WARNING —***
> *The shock absorber prevents the drive axle and trailing arm from drooping too far. Never remove shock absorber without supporting trailing arm. The coil spring is under compression and is extremely dangerous.*

Working underneath car, remove shock absorber to trailing arm bolt (**arrow**).

– Carefully lower and remove shock absorber.

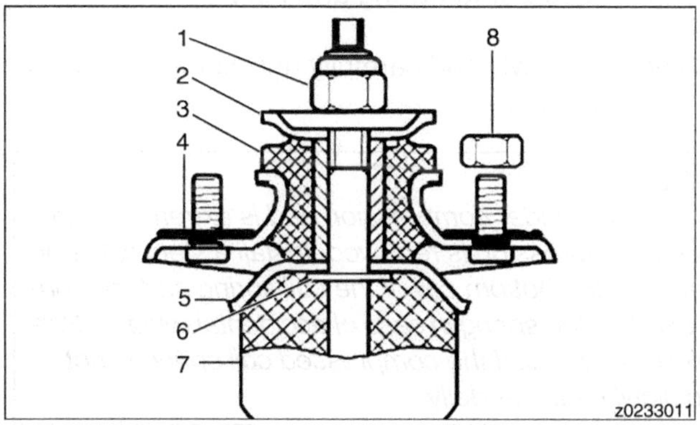


When installing, note order of upper shock absorber mount components:

1. Self-locking nut M10
   -tighten to 14 Nm (10 ft-lb)
2. Upper support plate
3. Top mount
4. Gasket
5. Lower support plate
6. Spacer ring
7. Rubber bump stop
8. Self-locking nut M8
   -tighten to 24 Nm (17 ft-lb)

– Installation is reverse of removal. Remember to:
  - Clean threaded bolts, nuts and mating surfaces.
  - Install shock absorber into shock tower using a new upper mounting gasket and new self-locking nuts.
  - Tighten lower bolt to its final torque once car is on ground.

| Tightening torques | |
|---|---|
| Shock absorber to upper mount | 14 Nm (10.3 ft-lb) |
| Shock absorber upper mount to body (self-locking nuts) | 24 Nm (18 ft-lb) |
| Shock absorber to trailing arm (car in normal loaded position) | 100 Nm (74 ft-lb) |

## Coil spring, removing and installing

– Raise car and remove rear wheel.

> ***WARNING —***
> *Make sure the car is firmly supported on jack stands designed for the purpose. Place the jack stands underneath structural chassis points. Do not place jack stands under suspension parts.*

– Support trailing arm from below using a suitable jack.

> ***WARNING —***
> *The shock absorber prevents the drive axle and trailing arm from drooping too far. Never disconnect shock absorber without supporting trailing arm.*

0012103

< Remove six bolts (**arrows**) holding drive axle to differential output flange. Suspend drive axle from chassis using stiff wire.

– Remove shock absorber to trailing arm bolt.

– Lower trailing arm slowly and carefully until compressed coil spring is fully unloaded.

> ***WARNING —***
> *The coil spring is under compression and is extremely dangerous until compression is removed. Install a steel cable or chain between the bottom coil of the coil spring and the control arm to retain the spring in case of accidental release. Personal injury can result if the compressed coil spring is not released slowly and carefully.*

– Remove retaining cable or chain and remove coil spring.

– If spring is to be reused, carefully inspect it for surface damage or corrosion. Inspect spring seats for signs of damage. Replace parts showing wear or damage.

– Fit coil spring to spring seats and install retaining cable or chain.

– Slowly lift trailing arm back into position, making sure coil spring is correctly seated in upper and lower coil spring seats.

– When trailing arm has been lifted sufficiently, attach shock absorber to trailing arm. Remove retaining cable or chain.

– Remainder of installation is reverse of removal.

| Tightening torques | |
|---|---|
| Drive axle to differential flange<br>• M8 x 40 bolt<br>• M10 x 20 Torx<br>• M10 x 46 black<br>• M10 x 46 silver (replace) | <br>64 Nm (47 ft-lb<br>83 Nm (61 ft-lb)<br>100 Nm (74 ft-lb)<br>80 Nm (59 ft-lb) |
| Road wheel to hub | 120 ± 10 Nm<br>(90 ± 7 ft-lb) |
| Shock absorber to trailing arm<br>(car in normal loaded position) | 100 Nm (74 ft-lb) |

## Rear ride height

Rear ride height (**A**) is measured from the lower edge of the wheel arch to the bottom edge of the wheel rim.

Rear ride height (**A**) is controlled by the rear springs. If rear ride height is not within the specifications listed, replace the rear springs.

Ride height specifications apply to a car in a normally loaded position, simulating a load of passengers, luggage and fuel. When checking ride height or installing suspension components, normal loaded position is as follows:

| Normal loaded position | |
|---|---|
| Each front seat | 68 kg (150 lb) |
| Trunk or cargo compartment | 21 kg (46 lb) |
| Fuel tank | full |

| Rear ride height specifications | |
|---|---|
| **Wheel installed** | **Ride height** |
| **15 inch**<br>• Standard suspension<br>• Sport suspension | <br>565 mm (22.24 in)<br>550 mm (21.65 in) |
| **16 inch**<br>• Standard suspension<br>• Sport suspension | <br>578 mm (22.76 in)<br>563 mm (22.17 in) |
| **17 inch**<br>• Standard suspension<br>• Sport suspension<br>• M Roadster suspension | <br>593 mm (23.35 in)<br>578 mm (22.76 in)<br>588 mm (23.15 in) |

## REAR SUBFRAME

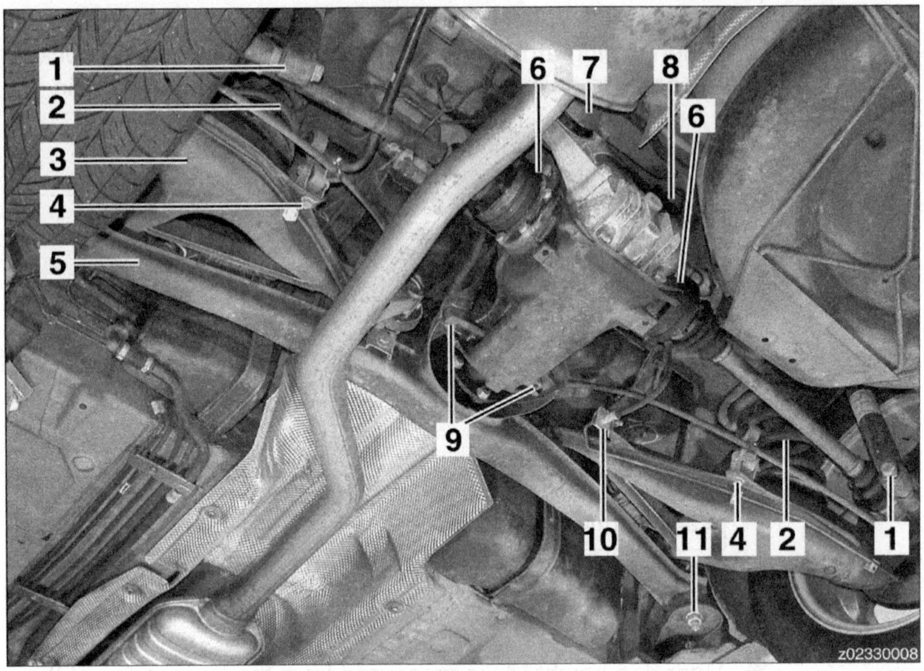


### Rear subframe components

1. **Shock absorber lower mounting bolt (M12)**
   - tighten with car in normal loaded position
   - tighten to 100 Nm (74 ft-lb)
2. **Coil spring**
3. **Trailing arm**
4. **Stabilizer bar link mounting nut (M8)**
   - tighten to 22 Nm (16 ft-lb)
5. **Rear subframe**
6. **Differential output flange**
7. **Differential rear mounting to body (M12)**
   - replace self-locking nut
   - tighten to 77 Nm (56 ft-lb)
8. **Vehicle speed sensor connector**
9. **Differential front mounting bolts to subframe (M12)**
   - tighten to 110 Nm (81 ft-lb)
10. **Brake line union**
    - tighten to 19 Nm (14 ft-lb)
11. **Rear subframe mounting fastener (M14)**
    - replace self-locking nut
    - tighten to 140 Nm (103 ft-lb)

### Rear subframe, removing and installing

The procedure given here describes removal of the complete rear suspension, including trailing arms with attached brake hardware, and differential with drive axles.

– Remove negative (-) battery cable from battery.

> *CAUTION—*
> *Prior to disconnecting the battery, read the battery cautions in* **001 General Warnings and Cautions**.

– Raise rear end of car and support it securely on jack stands. Remove rear wheels.

> ***WARNING—***
> *Make sure that the car is firmly supported on jack stands designed for the purpose. Place the jack stands beneath a structural chassis point. Do not place jack stands under suspension parts.*

– Remove complete exhaust system. See **180 Exhaust System**.

– Remove driveshaft. See **260 Driveshaft**.

– Remove parking brake cable from parking brake lever assembly. Working from below, remove cables from cable guides. See **340 Brakes**.

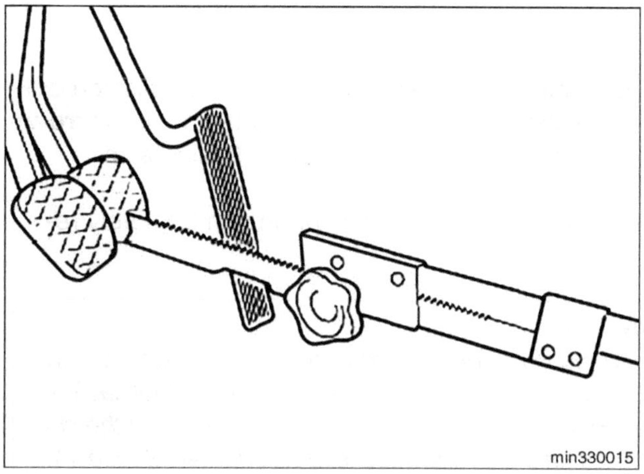
min330015

◄ Insert pedal prop and depress brake pedal slightly. This prevents brake fluid from escaping when brake lines are detached.

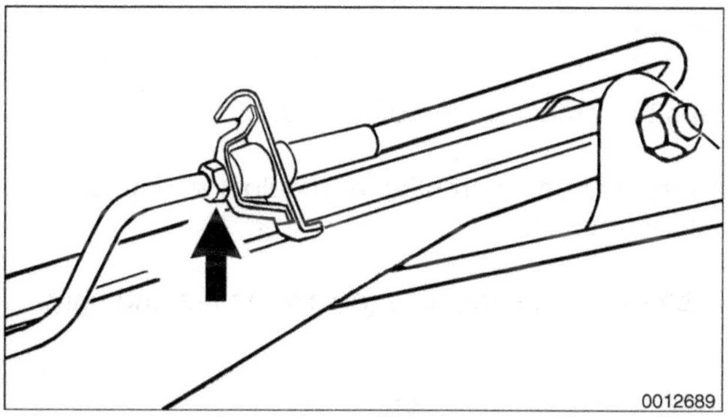
0012689

◄ Disconnect rear brake lines at left and right unions (**arrow**). Plug brake lines.

– Disconnect stabilizer bar at trailing arms.

– Disconnect harness connector at vehicle speed sensor on differential.

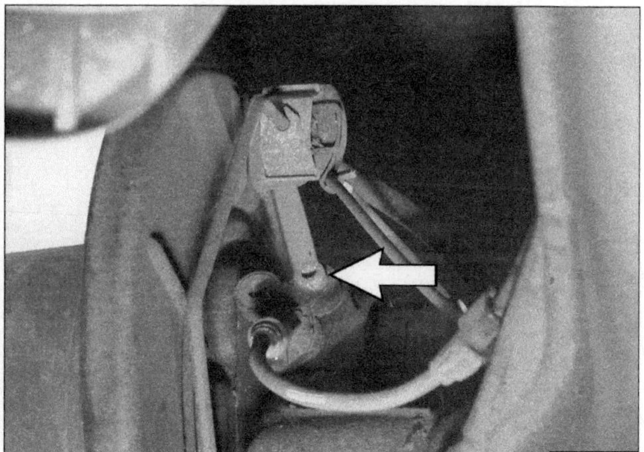
z02330007

◄ Remove ABS wheel speed sensor mounting bolts (**arrow**) at left and right trailing arms. Pull out sensors.

◀ Working above trailing arms, unclip wheel sensor connectors (**arrow**) from left and right trailing arms.

- Right side: Unclip brake pad wear sensor connector as well.
- Separate connectors.
- Set wheel speed sensors aside in a safe place.

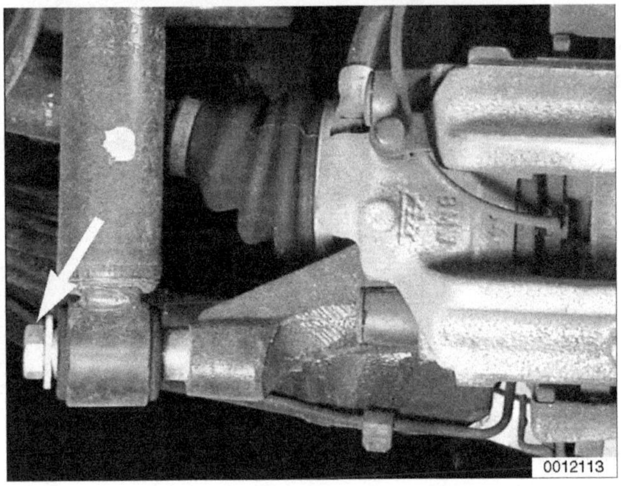


◀ Support right trailing arm from below with a transmission jack or equivalent. Remove shock absorber mounting bolt (**arrow**) from trailing arm. Pull shock absorber off trailing arm.

– Slowly lower trailing arm until coil spring can be safely removed.

> ***WARNING —***
> *The coil spring is under compression and is extremely dangerous until compression is removed. Install a steel cable or chain between the bottom coil of the coil spring and the control arm to retain the spring in case of accidental release. Personal injury can result if the compressed coil spring is not released slowly and carefully.*

– Repeat steps for removing left coil spring.

– Support subframe and differential from below with a transmission jack or equivalent.

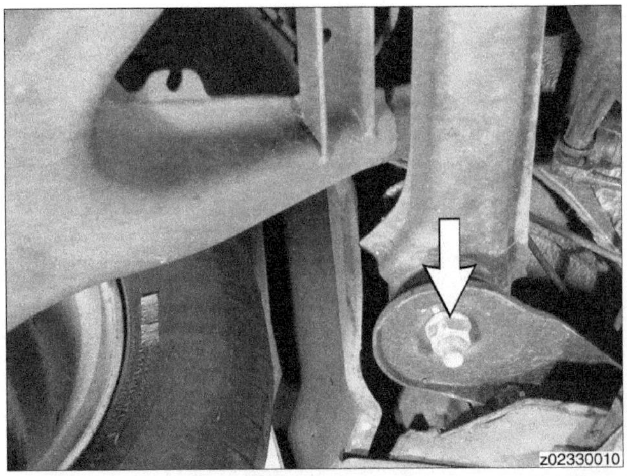


◀ Remove subframe mounting nuts (**arrow**) on left and right.

< Remove differential housing rear mounting fastener (**arrow**).

– Carefully lower complete subframe assembly.

– Installation is reverse of removal. Remember to:
  - Use new self-locking fasteners.
  - Check and adjust parking brake. See **340 Brakes**.
  - Bleed brake system. See **340 Brakes**.

| Tightening torques | |
|---|---|
| Brake line union | 19 Nm (14 ft-lb) |
| Differential rear mount to body (M12) (use new self-locking nut) | 77 Nm (56 ft-lb) |
| Rear subframe to outer rubber bushing (M14) (use new self-locking nut) | 140 Nm (103 ft-lb) |
| Road wheel to wheel hub | 120 ± 10 Nm (90 ± 7 ft-lb) |
| Shock absorber to trailing arm (car in normal loaded position) | 100 Nm (74 ft-lb) |
| Stabilizer bar link to trailing arm (M8) | 22 Nm (16 ft-lb |
| Subframe to differential (M12) | 110 Nm (81 ft-lb) |

## TRAILING ARMS

The trailing arms and their mounting bushings control the position of the rear wheels. A damaged trailing arm or worn bushings will change the rear wheel alignment and may adversely affect handling and stability.

> ***WARNING —***
> *Do not attempt to straighten a damaged trailing arm. Bending or heating may weaken the original part. If the trailing arm shows any signs of damage or excessive corrosion, replace it.*

### Trailing arm, removing and installing

– Raise rear end of car and support it securely on jack stands. Remove wheel.

> ***WARNING —***
> *Make sure that the car is firmly supported on jack stands designed for the purpose. Place the jack stands beneath a structural chassis point. Do not place jack stands under suspension parts.*

– Remove drive axle. See **311 Drive Axles, Differential**.

– Remove parking brake cable from parking brake lever assembly. See **340 Brakes**.

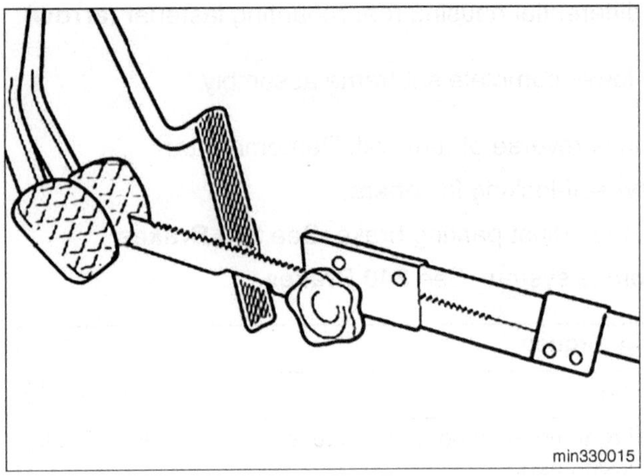
min330015

Insert pedal prop and depress brake pedal slightly. This prevents brake fluid from escaping when brake line is detached.

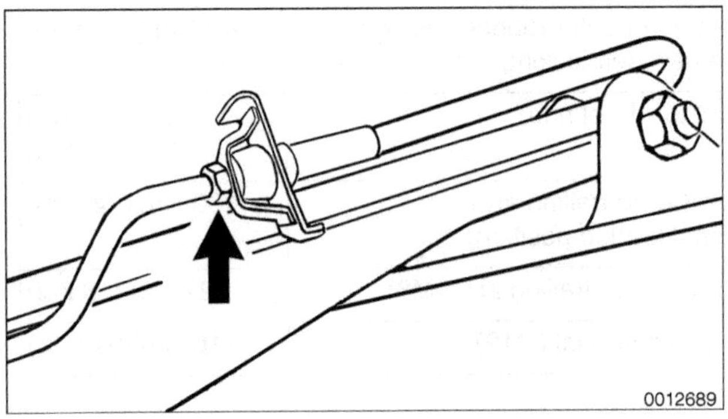
0012689

Disconnect rear brake line at union (**arrow**). Plug line openings.

- Support trailing arm from below with a transmission jack or equivalent.
- Remove shock absorber to trailing arm bolt. Slowly lower trailing arm until coil spring can be safely removed.

> ***WARNING —***
> *The coil spring is under compression and is extremely dangerous until compression is removed. Install a steel cable or chain between the bottom coil of the coil spring and the control arm to retain the spring in case of accidental release. Personal injury can result if the compressed coil spring is not released slowly and carefully.*

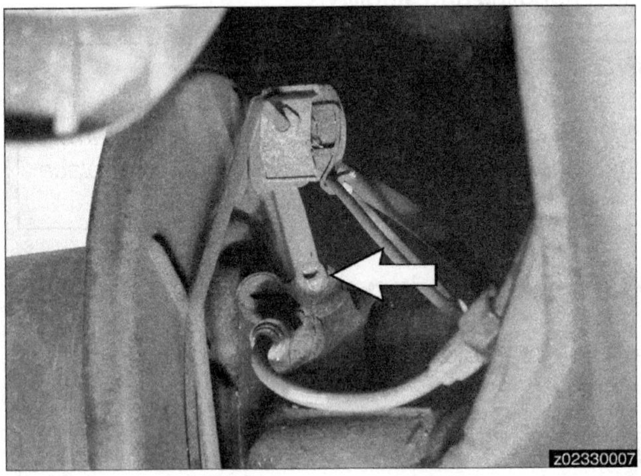
z02330007

Remove ABS wheel speed sensor mounting bolt (**arrow**) at trailing arm. Pull out sensor.

0012104

◄ Working above trailing arm, unclip wheel sensor wiring from trailing arm (**arrow**) and separate. Set sensor aside in a safe place.

– If removing right trailing arm, separate brake pad wear sensor connector as well.

– Detach stabilizer bar link from trailing arm.

– Remove two trailing arm through-bolts and nuts and remove trailing arm from car.

– If trailing arm is being replaced, transfer attached brake system components to new arm. See **340 Brakes**.

– Install new wheel bearing. See **Rear wheel bearing, replacing** in this repair group.

***NOTE—***
*BMW-supplied replacement trailing arms come with new rubber bushings installed. A new wheel bearing will have to be installed.*

– Installation is reverse of removal. Remember to:
  - Use new self-locking nuts.
  - Install inner trailing arm mounting bolts first.
  - Tighten trailing arm mounting bolts to final tightening torque with car on ground and loaded normally. See **Rear ride height** in this repair group.
  - Bleed brake system. See **340 Brakes.**

| **Tightening torques** | |
|---|---|
| Drive axle to differential flange<br>• M8 x 40 bolt<br>• M10 x 20 Torx<br>• M10 x 46 black<br>• M10 x 46 silver (replace) | <br>64 Nm (47 ft-lb<br>83 Nm (61 ft-lb)<br>100 Nm (74 ft-lb)<br>80 Nm (59 ft-lb) |
| Road wheel to wheel hub | 120 ± 10 Nm<br>(90 ± 7 ft-lb) |
| Shock absorber to trailing arm<br>(car in normal loaded position) | 100 Nm (74 ft-lb) |
| Stabilizer bar link to trailing arm (M8) | 22 Nm (16 ft-lb) |
| Trailing arm to rear subframe<br>(car in normal loaded position) | 67 Nm (49 ft-lb) |

**Trailing arm and rear subframe**

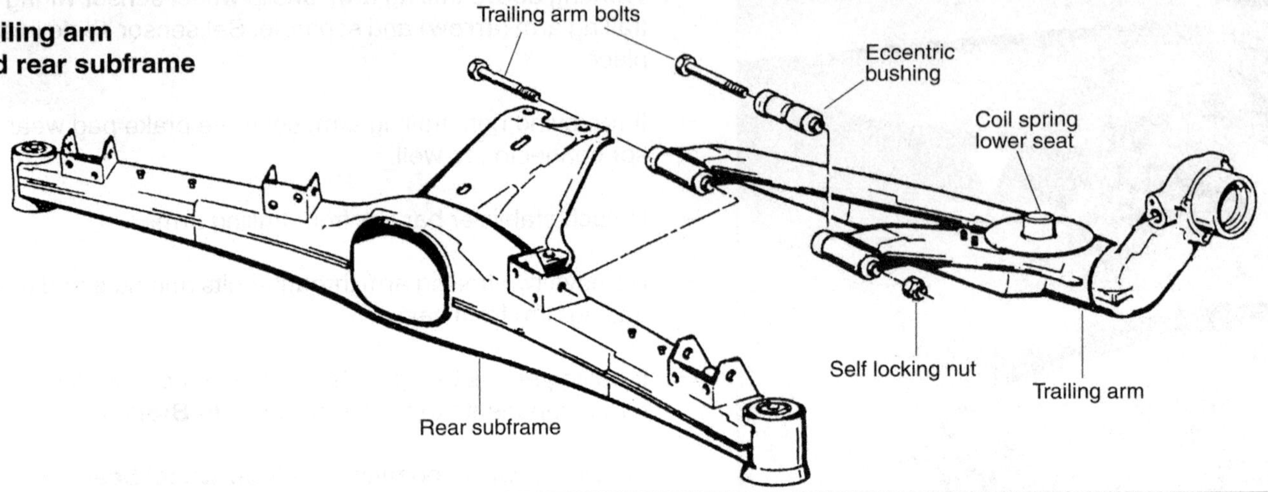


## Trailing arm bushings, replacing

***NOTE—***
*Special offset (eccentric) trailing arm bushings are available from BMW to correct for slight out-of-tolerance rear toe settings. See an authorized BMW dealer for additional parts information.*

– Remove trailing arm. See **Trailing arm, removing and installing** in this repair group.

– Mark installed position of eccentric bushing, then press out using appropriate press tools.

– Coat new bushing with Circolight® lubricant or equivalent.

– Draw new bushing into trailing arm using appropriate press tools.

– Install trailing arm. See **Trailing arm, removing and installing** in this repair group.

# REAR WHEEL BEARINGS

Special press tools are required to replace the wheel bearings. Read the procedure through before beginning the job.

## Rear wheel bearing, replacing

– Raise car and remove rear wheel.

> ***WARNING—***
> *Make sure the car is firmly supported on jack stands designed for the purpose. Place the jack stands underneath structural chassis points. Do not place jack stands under suspension parts.*

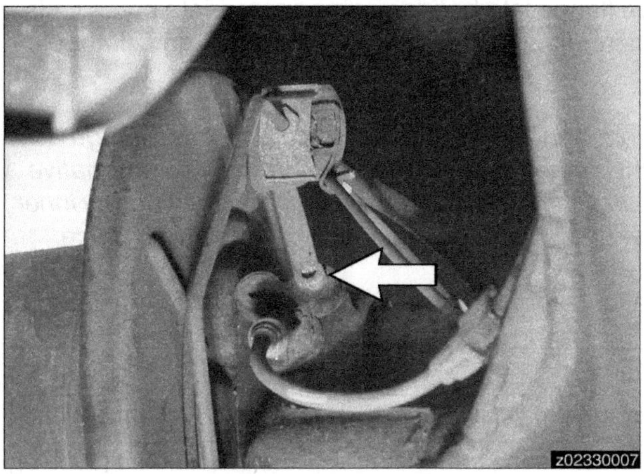


– Remove drive axle. See **311 Drive Axles, Differential**.

– Remove brake caliper assembly and rotor. See **340 Brakes**. Leave brake hose connected to caliper. Suspend caliper assembly from chassis using stiff wire.

◄ Remove ABS wheel speed sensor mounting bolt (**arrow**) at trailing arm. Pull out sensor and tie aside safely.

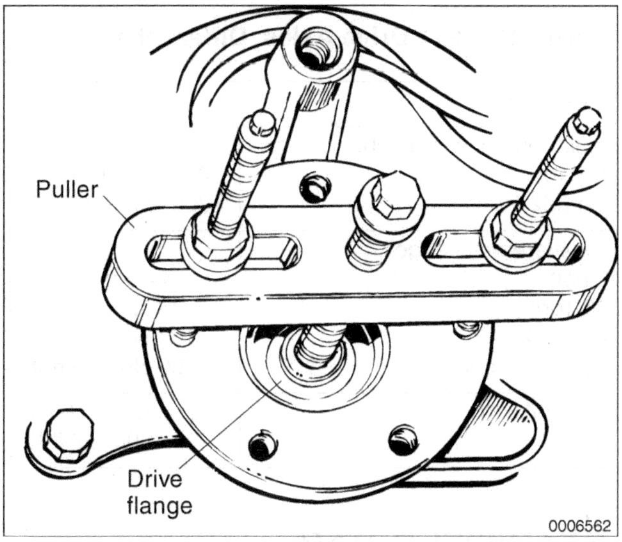


◄ Remove drive flange from bearing assembly. Use appropriate press tools or slide hammer to pull drive flange out of bearing assembly.

***NOTE —***
*The inner bearing race may remain attached to the drive flange. If this occurs, use BMW special tools 00 7 500 and 33 1 300 or equivalents to separate the inner race from the drive flange.*

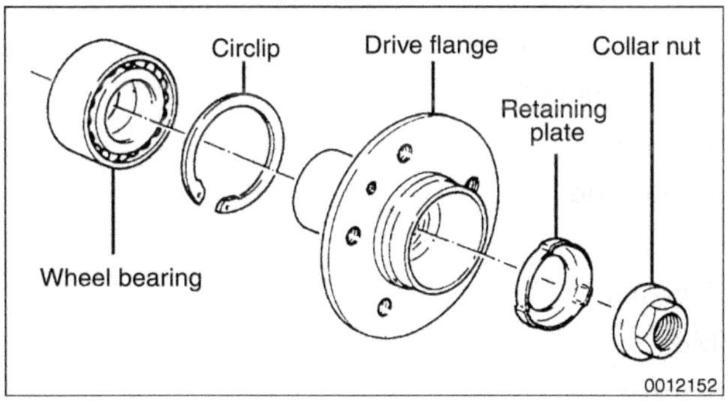


◄ Remove bearing retainer circlip from trailing arm bearing housing.

– Using appropriate press tools, pull bearing assembly out of trailing arm bearing housing.

– Inspect bearing housing for damage and contamination. Clean bearing housing bore thoroughly before installing new bearing. Make sure all thread bores, bolts, nuts and mating surfaces are clean.

– Install new bearing assembly using appropriate press tools. Apply force only to outer race when installing bearing.

– Install new circlip. Do not reuse old circlip.

***CAUTION—***
*Make sure the bearing is pressed in far enough to contact the shoulder at the back side of the housing and the circlip is fully seated in its groove.*

– Draw drive flange into bearing using appropriate press tools. Support bearing inner race when pressing in drive flange.

*NOTE—*
*BMW recommends special tools to pull the drive flange through the wheel bearing into position. If using alternative tools to install the flange, be sure to support the bearing inner race when pressing or pulling the drive flange into place.*

– Install brake caliper assembly and rotor. See **340 Brakes**.

| Tightening torques | |
|---|---|
| Brake rotor to drive flange | 16 Nm (12 ft-lb) |
| Brake caliper to trailing arm | 67 Nm (50 ft-lb) |

– Install drive axle. See **311 Drive Axles, Differential**.

| Tightening torques | |
|---|---|
| Drive axle collar nut to wheel hub:<br>• Z3 (M22 nut)<br>• M version (M27 nut) | <br>200 Nm (148 ft-lb)<br>300 Nm (221 ft-lb) |
| Drive axle to differential flange<br>• M8 x 40 bolt<br>• M10 x 20 Torx<br>• M10 x 46 black<br>• M10 x 46 silver (replace) | <br>64 Nm (47 ft-lb<br>83 Nm (61 ft-lb)<br>100 Nm (74 ft-lb)<br>80 Nm (59 ft-lb) |
| Road wheel to hub | 120 ± 10 Nm<br>(90 ± 7 ft-lb) |

## Rear stabilizer bar

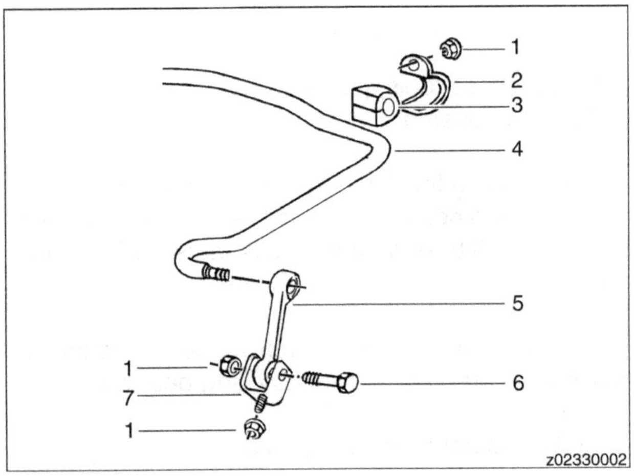


Rear stabilizer components:

1. Self-locking nut (M8)
2. Bushing bracket
3. Rubber bushing
4. Stabilizer bar
5. Stabilizer link
6. Bolt (M8)
7. Link bracket

### Rear stabilizer bar, removing and installing

– Raise and safely support vehicle.

*WARNING—*
*Make sure car is stable and well supported at all times. Use a professional automotive lift or jack stands designed for the purpose. A floor jack is not adequate support.*

– Loosen and remove stabilizer link mounting nuts at trailing arms.

– Remove two stabilizer bar bushing brackets on left and right sides. Remove bar.

– Installation is reverse of removal. Remember to:
  - Clean grease from rubber mounts and stabilizer.
  - Use new self-locking nuts.
  - Perform final tightening of stabilizer bar with car level and in normal loaded position.

| **Tightening torques** | |
|---|---|
| Stabilizer bar bushing bracket to body (M8) (replace self-locking nut) | 22 Nm (16 ft-lb) |
| Stabilizer bar link to trailing arm (M8) (replace self-locking nut) | 22 Nm (16 ft-lb) |

## Stabilizer link, replacing

– Remove stabilizer bar. See **Rear stabilizer bar, removing and installing** in this repair group.

– Clamp stabilizer link in a vice and pull bar off link with a twisting motion.

– Remove any trace of old bushing from stabilizer bar.

– Clamp new link in vice and coat inside of link bushing and end of stabilizer bar with Circolight® mounting agent.

– Using a twisting motion, press bar onto link bushing as far as possible.

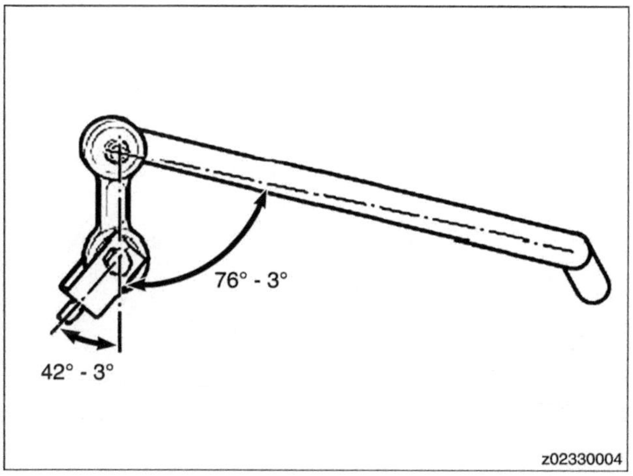


◄ Immediately rotate stabilizer link so that link to stabilizer bar orientation and link to link bracket orientation matches angles shown in illustration.

> ***CAUTION—***
> *Once mounting agent has evaporated, stabilizer link can no longer be moved.*

# 331 Drive Axles, Differential

## General

This repair group covers removal and repair information for the drive axle shafts, CV joints and CV joint boots. Also covered is differential (final drive) removal procedure and seal replacement information.

Internal repairs of the differential assembly are not covered in this manual.

### Special tools

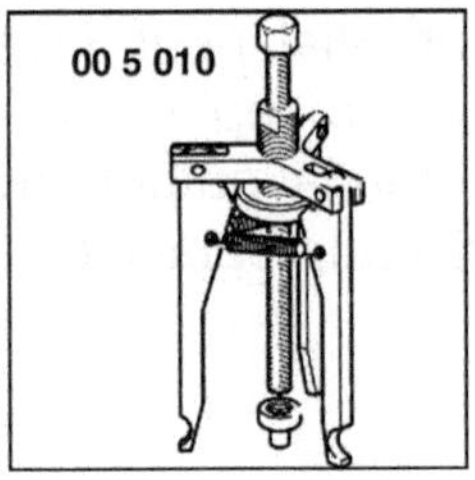


Radial seal puller (BMW special tool 00 5 010)

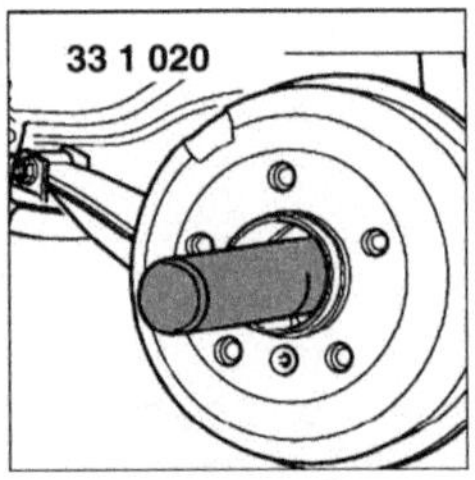


Wheel hub collar nut retaining plate installation tool (BMW special tool 33 1 020)

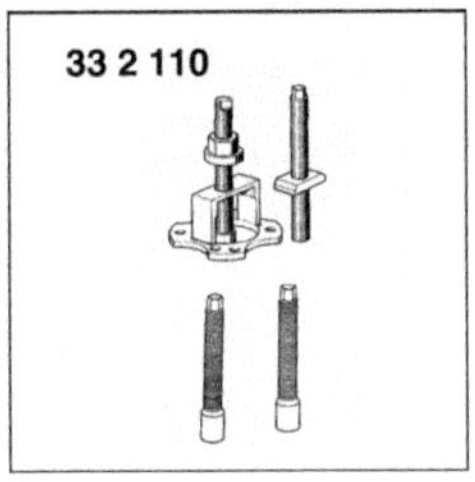


Drive axle removal and installation tool set (BMW special tool set 33 2 110)

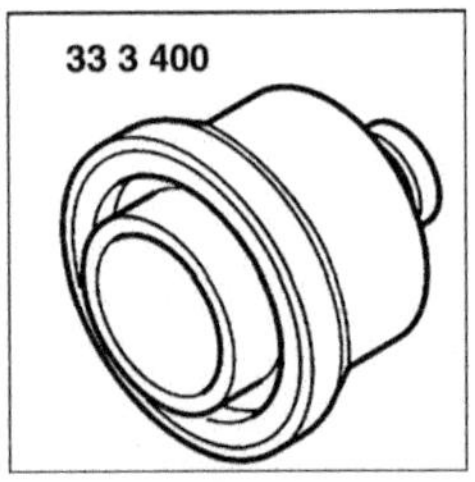


Output flange seal drift (BMW special tool 33 3 400)

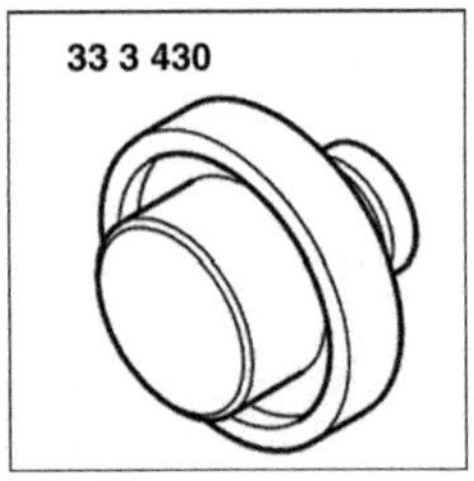


Input flange seal drift (BMW special tool 33 3 430)

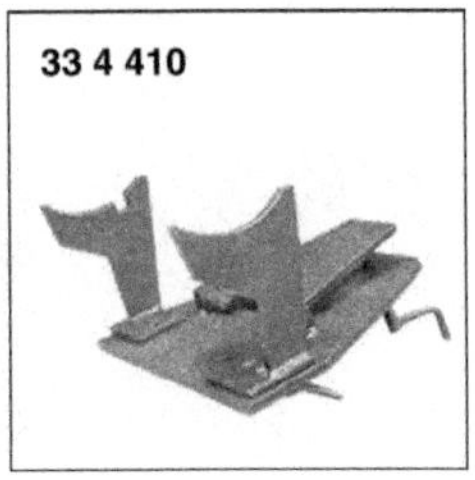


Differential support bracket (BMW special tool 33 4 410)

# Drive axles

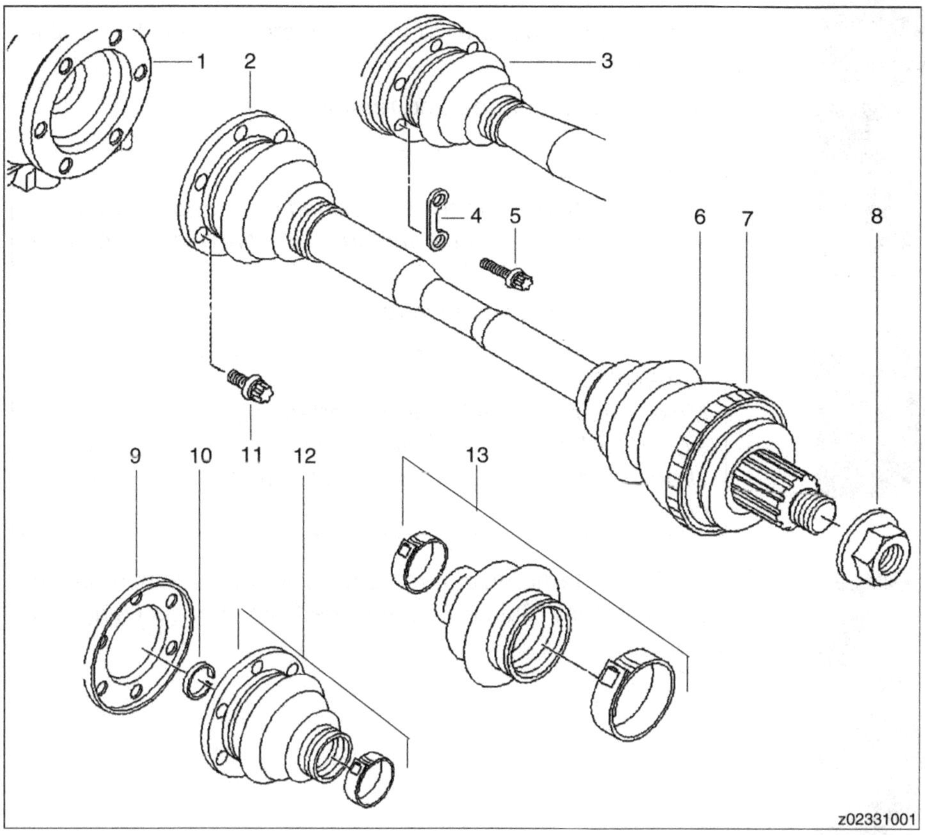


## Drive axle components

1. **Differential drive flange**
2. **Slip sleeve inner CV joint**
3. **Ball cage inner CV joint**
4. **Reinforcement**
5. **CV joint mounting bolt**
   - M8 x 40
     -tighten to 64 Nm (47 ft-lb)
   - M10 x 46 Torx black
     -tighten to 100 Nm (74 ft-lb)
   - M10 x 46 silver (replace)
     -tighten to 80 Nm (59 ft-lb)
6. **Outer CV joint**
7. **ABS wheel speed sensor pulse wheel**
8. **Wheel hub collar nut**
   - M22: tighten to 200 Nm (148 ft-lb)
   - M27: tighten to 300 Nm (221 ft-lb)
9. **Dust cover**
10. **Circlip**
11. **CV joint mounting bolt**
    - M10 x 20 Torx
      -tighten to 83 Nm (61 ft-lb)
12. **Inner CV joint boot kit**
13. **Outer CV joint boot kit**

## CV joint types

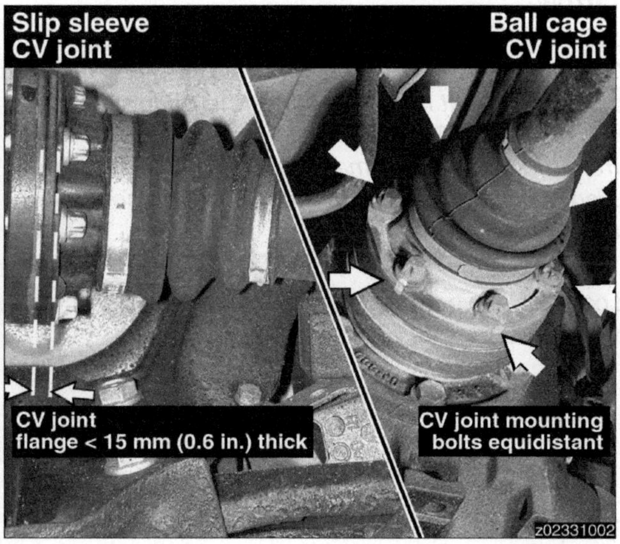


◄ The drive axles use constant-velocity (CV) joints on both ends. Two different styles of CV joints have been supplied:

- **Ball cage CV joint**. Traditional design CV joint with thick metal housing (approx. 40 mm / 1.6 in. thick). Balls slide in non-parallel grooves.
- **Slip sleeve CV joint**. Thin bolt-on flange (approx. 15 mm / 0.6 in. thick). CV joint has splined shaft that slides into end of drive axle.

For replacement parts, only CV joint boot kits or complete axles are offered by BMW. To replace a CV joint boot, remove the drive axle from the car.

## Drive axle, removing

◄ With an assistant applying brakes:

- If applicable, bend back locking plate at wheel hub collar nut (**arrow**).
- Break collar nut free and remove.

> ***WARNING—***
> *The wheel hub collar nut is tightened to a torque of over 200 Nm (148 ft-lb). Make sure the car is firmly on the ground.*

– Raise rear of car. Remove rear tire.

> ***WARNING—***
> *Make sure the car is firmly supported on jack stands designed for the purpose. Place the jack stands underneath structural chassis points. Do not place jack stands under suspension parts.*

◄ Working underneath car, remove six inner CV joint mounting bolts (**arrows**). Suspend drive axle from chassis using stiff wire.

– Support trailing arm from below, using a transmission jack or equivalent.

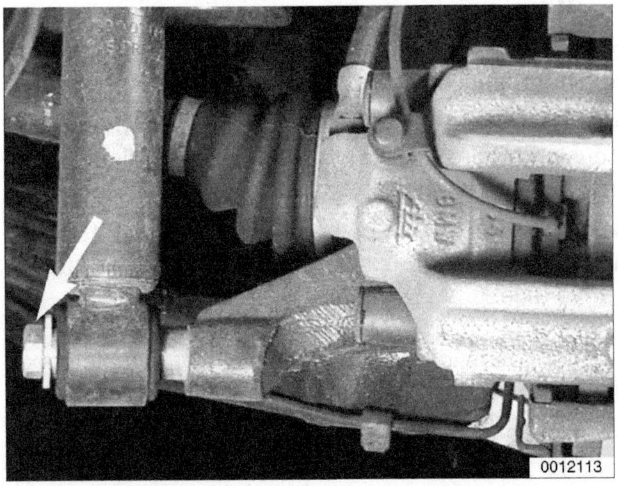


◄ Working underneath car, remove shock absorber to trailing arm bolt (**arrow**).

> ***CAUTION—***
> *The shock absorber at full extension is the bottom limit of the rear suspension. When the shock absorber is detached from the trailing arm, maintain compression on the suspension coil spring by maintaining support under the trailing arm.*

– Slow and carefully lower trailing arm sufficiently to gain clearance to remove drive axle.

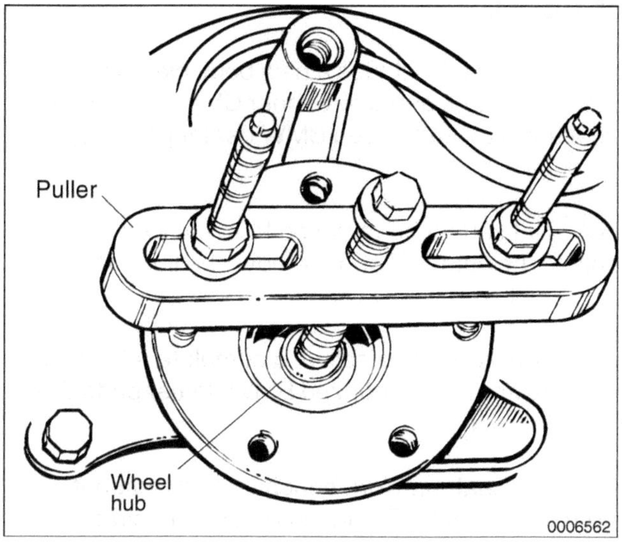


◄ If necessary, use BMW special tool 33 2 110 or equivalent puller to press drive axle out of wheel hub.

## Drive axle, installing

– Draw drive axle into wheel hub using BMW special tool set 33 2 110 or equivalent puller.

– Apply a light coating of oil to contact face of collar nut and install. Do not fully tighten yet.

– Raise trailing arm and install shock absorber to trailing arm.

| **Tightening torque** | |
|---|---|
| Shock absorber to trailing arm (car in normal loaded position) | 100 Nm (74 ft-lb) |

– Position drive axle on differential drive flange and install six CV joint bolts.

| **Tightening torques** | |
|---|---|
| Drive axle to differential flange<br>• M8 x 40 bolt<br>• M10 x 20 Torx<br>• M10 x 46 black<br>• M10 x 46 silver (replace) | <br>64 Nm (47 ft-lb<br>83 Nm (61 ft-lb)<br>100 Nm (74 ft-lb)<br>80 Nm (59 ft-lb) |

– Install road wheel and lower car to ground. Apply parking brake to prevent wheel from turning. Tighten wheel hub collar nut to its final torque.

> ***WARNING —***
> *The wheel hub collar nut is tightened to a torque of over 200 Nm (148 ft-lb). Make sure the car is firmly on the ground.*

| **Tightening torques** | |
|---|---|
| Drive axle collar nut to wheel hub:<br>• Z3 (M22 nut)<br>• M version (M27 nut) | <br>200 Nm (148 ft-lb)<br>300 Nm (221 ft-lb) |
| Road wheel to hub | 120 ± 10 Nm (90±7 ft-lb) |

***NOTE —***
- *Z3: M22 collar nut is locked in place using new retaining plate. Install using BMW special tool 33 1 020.*
- *M version: M27 collar nut is locked in place by staking lip on nut.*

## CV joint boots

A CV joint boot repair kit includes a new boot, clamping bands, special lubricant, and a new inner CV joint circlip. The kit is available from an authorized BMW dealer parts department.

The outer CV joint cannot be removed from the axle shaft. In order to replace the outer CV boot, remove the inner joint and boot first.

If the CV joints are worn or defective, a complete rebuilt axle shaft is available from an authorized BMW dealer parts department.

– Working with axle shaft on work bench, cut off old boot clamps and remove boots. Clean old grease off joints and shafts.

### Ball cage CV joint boot

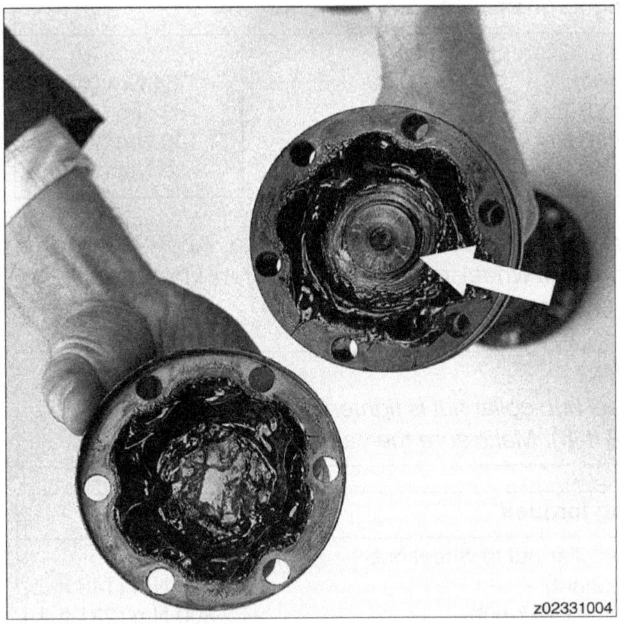


◄ Lift off dust cover from inner CV joint and remove circlip (**arrow**) retaining joint inner hub to axle shaft.

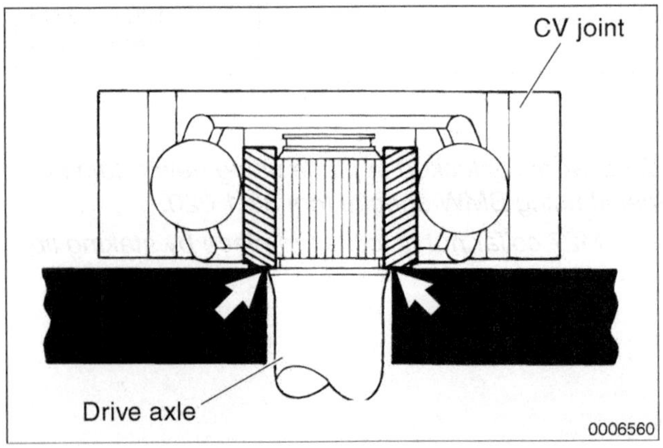


◄ Support inner hub at **arrows** when pressing axle shaft out of joint.

– Clean old lubricant off shaft splines and inner join splines.

> ***CAUTION—***
> *If the CV joint must be disassembled for cleaning and inspection, be sure to match mark inner and outer race and intermediate ball cage. This allows reassembly of parts in their original positions.*

– To inspect a CV joint, clean away the grease and look for galling, pitting and other signs of wear or physical damage.

- Polished surfaces or visible ball tracks alone are not necessarily cause for replacement.
- Discoloration due to overheating indicates lack of lubrication.

– Install new boots. Fill with requisite amount of grease.

| **CV joint lubricant capacity** | |
|---|---|
| Outer joint<br>• Z3<br>• M version | <br>80 grams (2.8 oz)<br>85 grams (3.0 oz) |
| Inner joint<br>• Z3<br>• M version | <br>85 grams (3.0 oz)<br>100 grams (3.5 oz) |

– Apply Loctite® 270 or equivalent thread-locking compound to drive axle splines. Position new CV joint on shaft so that raised or taller side of hub is facing shaft.

> ***CAUTION—***
> *Do not let the locking compound contact the balls in the joint. Apply only a thin coat to cover the splines.*

– While supporting axle shaft, press inner hub of CV joint onto shaft. Install new circlip.

– Clamp and secure boots to CV joints and axle shaft.

***NOTE—***
- *Do not let the inner CV joint ball hub pivot more than 20° in the outer ring of the joint. The balls may fall out if the hub is pivoted too far.*
- *Before installing each small boot clamp be sure to "burp" the boot by flexing the CV joint as far over as it will go. A small screwdriver inserted between the boot and the axle shaft will help the process.*

– Use sealing gel to seal dust cover to CV joint prior to reinstallation.

***NOTE—***
*BMW recommends Bostik®1513 or Epple®4851 adhesive, and Epple®39 or Curil®T sealer.*

### Slip sleeve CV joint boot

– Pull inner CV joint partially out of axle shaft. Match mark relative position of joint to shaft, then pull joint completely out of end of axle.

> ***CAUTION—***
> *If the joint and shaft are not reassembled in their original position the assembly may be out of balance and vibrate during operation.*

– Remove old boots and install new. Fill with grease.

***NOTE—***
*Before installing small boot clamp on outer CV joint, be sure to "burp" the boot by flexing the CV joint as far over as it will go. A small screwdriver inserted between the boot and the axle shaft will help the process.*

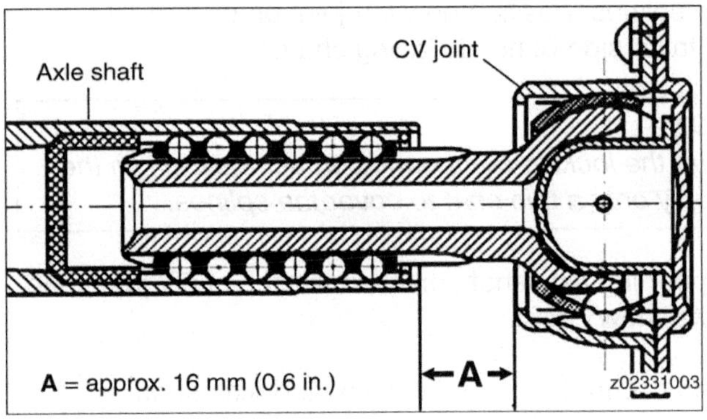


When reinstalling inner joint, push joint into shaft as far as possible, then pull it out to measurement **A** (16 mm / 0.6 in.).

***NOTE—***
*This step insures uniform lubrication of the CV joint.*

– Keep joint in position while installing inner (small) boot clamp.

– Make sure sealing lip of joint and boot are free of grease. The boot could otherwise slide off when boot clamp is tightened.

| Slip sleeve joint boot lengths | |
|---|---|
| Inner CV joint boot length | 65 mm (2.6 in) |
| Outer CV joint boot length | 55 mm (2.2 in) |

***NOTE—***
- *Position large boot clamp on the inner CV joint so that crimp in clamp faces rivet on CV joint flange.*
- *Install large clamps of inner and outer CV joints so that the crimps are offset by 180°.*

## DIFFERENTIAL

This section covers rear differential replacement and basic service. Internal differential repairs are not covered.

Low oil level caused by leaking oil seals may cause noisy differential operation. The output drive flange (side) and input drive flange (front) can be replaced without removing the differential.

*NOTE—*
*Do not mistake leaking CV joint grease for flange seal leaks. Degrease the differential to pinpoint the source of leaks prior to replacing seals.*

## Differential oil, checking

Z3 differentials for 1996 - 1997 and M version differentials for 1998 - 2002 contain Hypoid oil.

For Z3 differentials from 1998 on, BMW recommends using only specially formulated synthetic "lifetime" gear oils (SAF-XO), available through an authorized BMW dealer.

*NOTE—*
*For additional information on compatible lubricants, contact an authorized BMW dealer service department.*

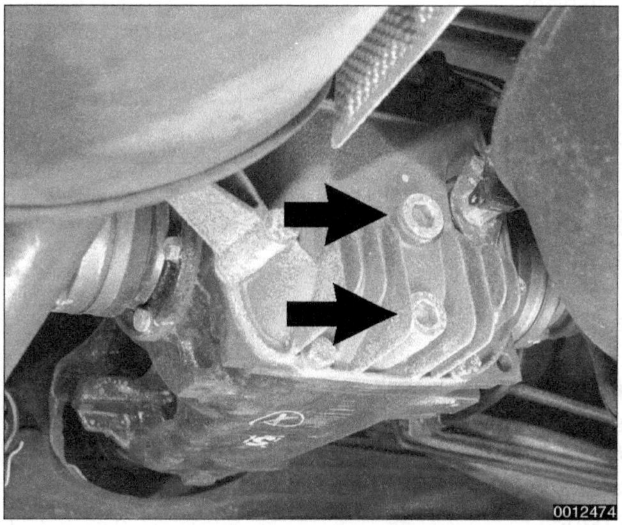
0012474

< Check differential oil level at filler plug (**upper arrow**).

– Check lubricant level with car on a level surface:
- Remove oil filler plug.
- Level is correct when fluid just reaches edge of filler hole.
- If necessary, top up fluid.
- Install and tighten oil filler plug when oil level is correct.

*NOTE—*
- *Use a 14 mm or 17 mm Allen socket to remove plugs.*
- *Make sure the car is level when checking fluid.*

| Tightening torque | |
|---|---|
| Drain and fill plugs to differential housing | 70 Nm (52 ft-lb) |

## Differential, removing and installing

– Raise rear end of car and support it safely on jack stands.

***WARNING—***
*Make sure the car is firmly supported on jack stands designed for the purpose. Place the jack stands underneath structural chassis points. Do not place jack stands under suspension parts.*

– Detach driveshaft from differential input flange. See **260 Driveshaft**.

0012103

< Remove inner CV joint mounting bolts (**arrows**) and detach both drive axles from differential. Suspend axles from chassis using stiff wire.

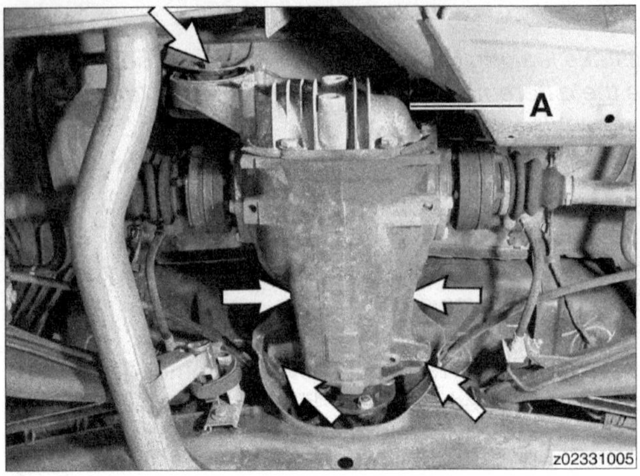


< Support differential with transmission jack. Working underneath differential:

- Disconnect vehicle speed sensor harness connector (**A**).
- Remove differential mounting bolts at front and rear (**arrows**).

– Slowly lower and tilt differential toward rear to remove.

– Installation is reverse of removal. Use new self-locking fasteners.

| Tightening torques | |
|---|---|
| Differential rear mounting to body (M12) | 77 Nm (57 ft-lb) |
| Differential to rear subframe (M12) | 110 Nm (81 ft-lb) |
| Drive axle to differential flange<br>• M8 x 40 bolt<br>• M10 x 20 Torx<br>• M10 x 46 black<br>• M10 x 46 silver (replace) | <br>64 Nm (47 ft-lb<br>83 Nm (61 ft-lb)<br>100 Nm (74 ft-lb)<br>80 Nm (59 ft-lb) |
| Driveshaft to differential (with U-joint)<br>• M10 ribbed nut<br>• M10 compression nut | <br>80 Nm (59 ft-lb)<br>64 Nm (47 ft-lb) |

## Differential input drive flange oil seal, replacing

– Raise rear of car and support safely.

> ***WARNING —***
> *Make sure the car is firmly supported on jack stands designed for the purpose. Place the jack stands underneath structural chassis points. Do not place jack stands under suspension parts.*

– Detach driveshaft from differential input shaft flange. See **260 Driveshaft**.

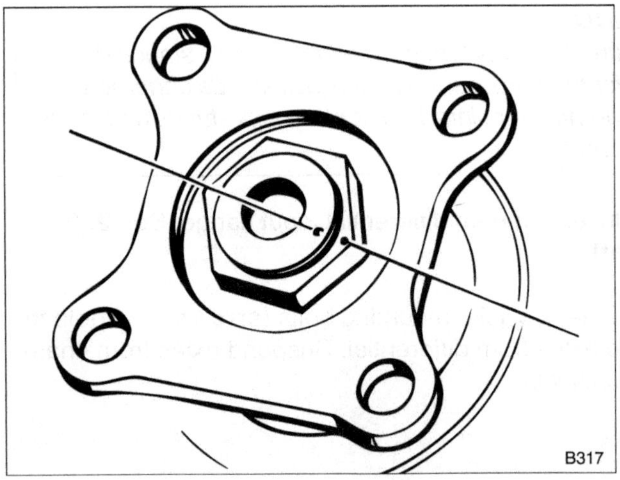


< Make matching marks on input shaft and input shaft collar nut.

– Pry lock plate from nut. Hold input flange using a suitable tool and remove collar nut. If necessary, use a puller to remove input flange.

***NOTE —***
*Be prepared to catch dripping fluid.*

– Inspect flange shaft sealing surface. Replace flange and shaft if shaft appears worn.

– Remove differential front dust cover.

– Pry seal from its recess using BMW special tool 00 5 010 or equivalent seal removal tool. Dip new seal in differential oil and drive it into position.

> ***CAUTION—***
> *Be careful not to mar the housing when removing seal.*

– Install new differential front dust cover.

– Lightly lubricate input shaft and press input flange back on. Install collar nut and slowly tighten until matching marks made previously line up. Come as close as possible to the specified torque.

| Tightening torques | |
|---|---|
| Input shaft flange to pinion (line up marks)<br>• 4-bolt side cover<br>• 6-bolt side covers | <br>175 Nm (129 ft-lb)<br>185 Nm (137 ft-lb) |

> ***CAUTION—***
> *If the input flange nut torque is exceeded, or the nut is tightened beyond the match marks, a crushable sleeve in the differential housing is damaged. Sleeve replacement requires differential disassembly.*

– Install a new lock plate and top off differential oil, if necessary. See **Differential oil, checking** in this repair group.

– Remainder of installation is reverse of removal.

| Tightening torques | |
|---|---|
| Drain and fill plugs to differential housing | 70 Nm (52 ft-lb) |
| Driveshaft to differential flange (with U-joint)<br>• M10 ribbed nut<br>• M10 compression nut | <br>80 Nm (59 ft-lb)<br>64 Nm (47 ft-lb) |

## Differential output drive flange oil seal, replacing

– Raise rear of car and support safely.

> ***WARNING—***
> *Make sure the car is firmly supported on jack stands designed for the purpose. Place the jack stands underneath structural chassis points. Do not place jack stands under suspension parts.*

0012103

◄ Working underneath car, remove six inner CV joint mounting bolts (**arrows**). Suspend drive axle from chassis using stiff wire.

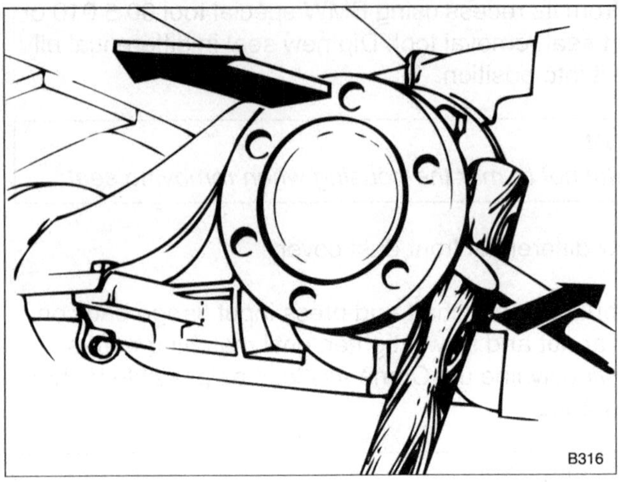


Pry output flange from differential. For leverage, use a wooden dowel as shown.

*NOTE—*
*Be prepared to catch dripping fluid.*

- Inspect flange shaft sealing surface. Replace flange and shaft if shaft appears worn.

- Remove snap ring from differential opening.

- Pry old seal from differential using BMW special tool 00 5 010 or equivalent seal removal tool.

***CAUTION—***
*Be careful not to mar the differential housing when removing the seal.*

- Coat new seal with differential lubricant and drive into place until fully seated.

- Install new snap ring in differential housing groove. Make sure both ends of ring are fully seated and that gap in ring is facing down.

- Install drive flange by pressing it in by hand until snap ring engages. Rotate drive flange slightly while pressing.

- Attach drive axle and tighten bolts.

| **Tightening torques** | |
|---|---|
| Drive axle to differential flange | |
| • M8 x 40 bolt | 64 Nm (47 ft-lb |
| • M10 x 20 Torx | 83 Nm (61 ft-lb) |
| • M10 x 46 black | 100 Nm (74 ft-lb) |
| • M10 x 46 silver (replace) | 80 Nm (59 ft-lb) |

- Top off differential with oil. See **Differential oil, checking** in this repair group.

# 340 Brakes



**TABLES**



340

## GENERAL

This repair group covers brake service:

- Brake pads, calipers, and disks
- Master cylinder, brake booster, and parking brake
- ABS/ASC and ABS/DSC component replacement

### Special tools

BMW requires the use of DISplus, GT1 or MoDiC scan tool to safely and completely bleed the braking and traction control systems. Read all procedures through before beginning a job.

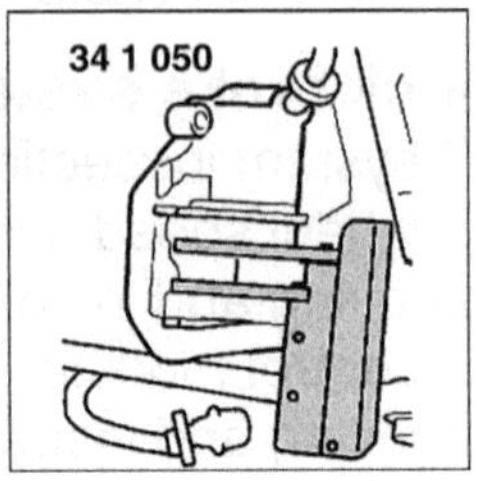


Brake caliper piston tool (BMW special tool 34 1 050)

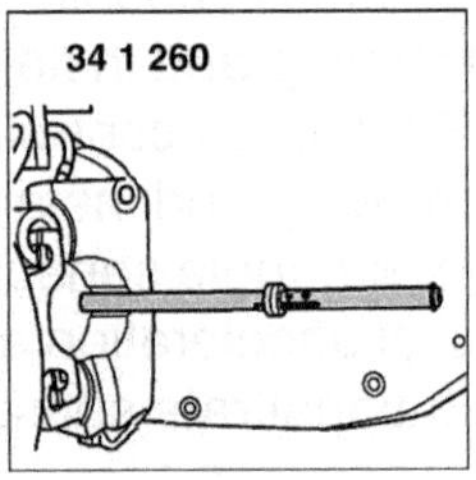


Brake pad lining gauge (BMW special tool 34 1 260)

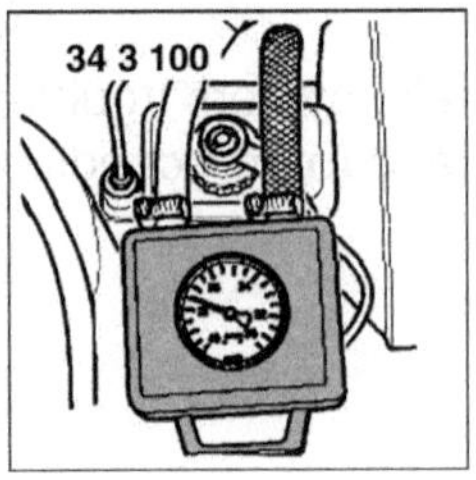


Brake booster vacuum tester (BMW special tool 34 3 100)

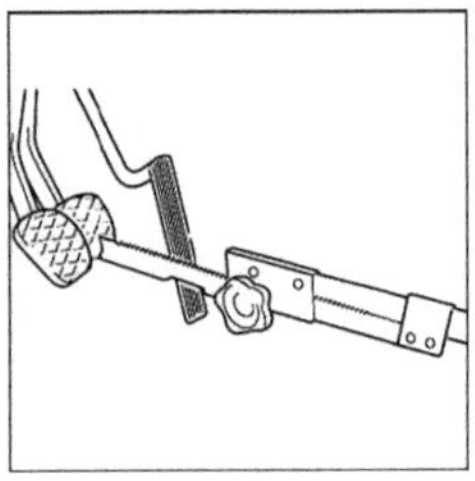

Pedal prop

### Brake system description

BMW Z3 and M version Roadster and Coupes are equipped with vacuum power-assisted four-wheel disc brakes with integral antilock brake system (ABS). Single-piston calipers act on solid or vented front rotors and solid or vented rear rotors. A brake pad wear sensor for each axle indicates when brake pads need replacement.

The dual drum-type parking brake system is integrated with the rear brake rotors.

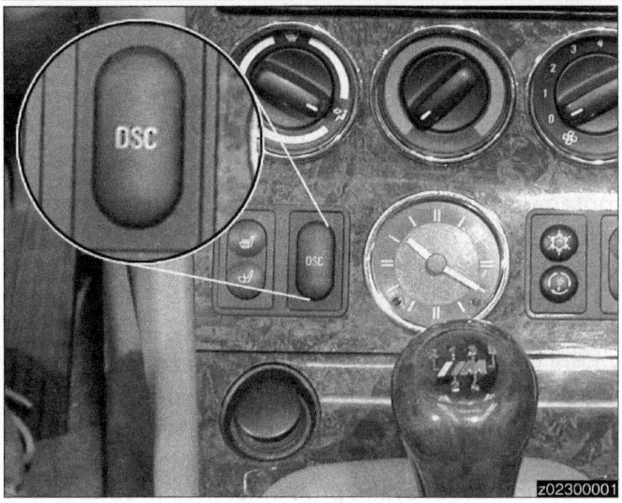


## Electronic braking and stability control

Z3 models were introduced with Automatic Stability Control (ASC). ASC is a computer controlled traction control system that uses the ABS system in conjunction with engine management system to control wheel spin and maintain vehicle stability while braking.

Dynamic Stability Control (DSC) was optionally available in Z3 models. This system uses ASC technology, but implements wheel speed modulation throughout all stages of driving. DSC is able to reduce understeer by applying differing amounts of braking force to each wheel, as well as overriding the engine management system during hard cornering.

For ABS / ASC and ABS / DSC system and component descriptions, see **300 Suspension, Steering and Brakes–General**.

> ***WARNING —***
> *A car with electronic stability control is still subject to normal physical laws. Avoid excessive speeds for the road conditions encountered.*

## Troubleshooting

Brake performance is mainly affected by three things:

- Brake fluid level and condition
- Brake system ability to create and maintain hydraulic pressure
- Condition of brake friction components

Air in brake fluid makes the brake pedal feel spongy during braking or increases the brake pedal force required to stop. Fluid contaminated by moisture or dirt corrodes system components. Inspect brake fluid inside the reservoir. If it is dirty or murky, or is over a year old, replace fluid.

Visually check hydraulic system starting at master cylinder. To check master cylinder function, hold brake pedal down hard with car stopped and engine running. The pedal should feel solid and stay solid. If pedal slowly falls to floor, either master cylinder is leaking internally, or fluid is escaping from other points in brake system. If no leaks can be found, replace master cylinder. Check all brake fluid lines and couplings for leaks, kinks, chafing and corrosion.

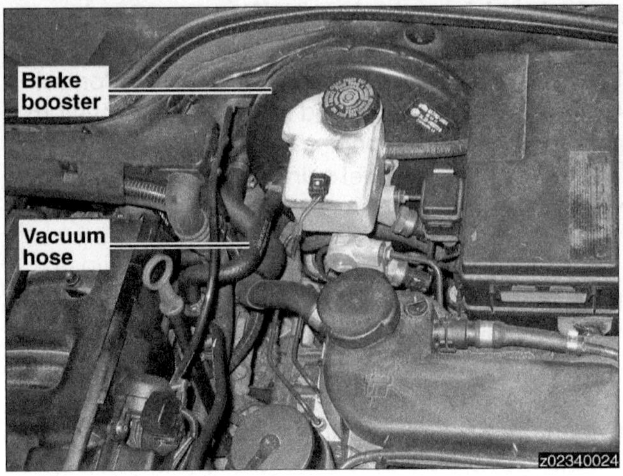


Check brake booster:

- Remove vacuum hose from brake booster and install BMW special tool 34 3 100 between hose and check valve at booster.
- Start engine and check build-up of partial vacuum. Switch engine off.
- Press brake pedal to set partial vacuum of no more than 0.8 bar (11.6 psi). Wait for vacuum value to stabilize.
- When pedal is released, partial vacuum should not drop by more then 0.06 bar (0.8 psi) over 1 hour.

– If vacuum values are not reached:

- Check vacuum hose junctions for tightness.
- Replace vacuum check valve.
- Check seal between brake booster and master cylinder.
- If values still cannot be reached, replace brake booster.

Worn or contaminated brake pads cause poor braking performance. Oil-contaminated or glazed pads cannot produce as much friction and stopping distances increase. Inspect brake rotors for glazing, discoloration and scoring.

Steering wheel vibration while braking at speed is often caused by warped rotors, but can also be caused by worn suspension components. See **310 Front Suspension**.

**Table a** lists symptoms of brake problems, their probable causes, and suggested corrective actions.

**Table a. Brake system troubleshooting**

| Symptom | Probable cause | Repairs |
|---|---|---|
| Brake squeal | Incorrectly installed brake pads or parking brake shoes | Check component installation. |
| | Brake pad carriers dirty or corroded | Remove brake pads and clean calipers. |
| | Brake pad anti-rattle springs faulty or missing | Install / replace anti-rattle springs. |
| | Brake pads heat-glazed or oil-soaked | Replace brake pads. Clean rotors. Replace leaking calipers as required. |
| | Wheel bearings worn (noise most pronounced when turning) | Replace worn bearings. See **310 Front Suspension** or **330 Rear Suspension**. |
| Pedal goes to floor when braking | Brake fluid level low due to system leaks | Check fluid level and inspect hydraulic system for signs of leakage. Fill and bleed system. |
| | Master cylinder faulty | Replace master cylinder. |
| Low pedal after system bleeding | Master cylinder faulty | Replace master cylinder. |

| Table a. Brake system troubleshooting | | |
|---|---|---|
| **Symptom** | **Probable cause** | **Repairs** |
| Pedal spongy or brakes work only when pedal is pumped | Air in brake fluid | Bleed system. |
| | Master cylinder faulty (internal return spring weak) | Replace master cylinder. |
| | Leaking line or hose unions | Repair or replace lines and hoses. Bleed system. |
| Excessive braking effort | Brake pads wet | Use light pedal pressure to dry pads while driving. |
| | Brake pads heat-glazed or oil-soaked | Replace brake pads. Clean rotors. Replace leaking calipers. |
| | Vacuum booster or vacuum hose connections to booster faulty | Inspect vacuum lines. Test vacuum booster and replace as required. Test vacuum check valve for one-way air flow. |
| Brakes pulsate, chatter or grab | Warped brake rotors | Resurface or replace rotors. |
| | Brake pads worn | Replace brake pads. |
| | Brake pads heat-glazed or oil-soaked | Replace brake pads. Clean rotors. Replace leaking calipers. |
| Uneven braking, car pulls to one side, rear brakes lock | Incorrect tire pressures or worn tires | Inspect tire condition. Check and correct tire pressures. |
| | Brake pads on one side of car heat-glazed or oil-soaked | Replace brake pads. Clean rotors. Replace leaking calipers. |
| | Caliper or brake pads binding | Clean and recondition brakes. |
| | Worn suspension components | Inspect for worn or damaged suspension components. See **310 Front Suspension** or **330 Rear Suspension**. |

## Warnings and cautions

***WARNING—***

- *Semi-metallic and metallic brake friction materials in brake pads or shoes produce dangerous dust. Avoid breathing in brake dust.*
- *Brake fluid is poisonous, corrosive and dangerous to the environment. Wear safety glasses and rubber gloves when working with brake fluid. Do not siphon brake fluid with your mouth. Dispose of brake fluid properly.*
- *Do not reuse self-locking nuts, bolts or fasteners. They are designed to be used only once and may fail if re-used. Replace with new self-locking fasteners.*

> ***CAUTION—***
> - *Models covered by this manual require special BMW service equipment to properly bleed the brake hydraulic system. See* **Bleeding brakes** *for more information.*
> - *Immediately clean brake fluid spilled on painted surfaces and wash with water, as brake fluid removes paint.*
> - *Use new brake fluid from an unopened container. Brake fluid absorbs moisture from the air. This can lead to corrosion problems in the braking system, and also lowers the fluid boiling point.*
> - *Plug open lines and brake fluid ports to prevent contamination.*
> - *Only tighten brake hoses on front wheels with wheels in straight ahead position.*
> - *If carrying out electric welding work, be sure to disconnect electrical harness connector from electronic control module.*
> - *Do not expose electronic control modules to high sustained heat, such as in a paint drying booth. Maximum heat exposure: 95°C (203°F) for short periods of time, 85°C (185°F) for long periods (approx. 2 hours).*

## BLEEDING BRAKES

Brake bleeding is usually done for one of two reasons; either to replace old brake fluid as part of routine maintenance or to expel trapped air in the system that resulted from opening the brake hydraulic system during repairs.

When adding or replacing brake fluid always add new brake fluid from an unopened container. It is important to bleed the entire system when any part of the hydraulic system has been opened.

If you are certain no air was introduced into the master cylinder or traction control hydraulic unit, brake bleeding can be done at the calipers using a pressure bleeder. See **Brake calipers, bleeding** in this repair group.

If air was introduced into the master cylinder or traction control hydraulic unit, or you are unsure, bleed brakes using a BMW service tester (DISplus / GT1 / MoDiC) or by an authorized BMW dealer. See **Brake system, bleeding** in this repair group.

When bleeding the brakes, start at the wheel farthest from the master cylinder and progress in the following order:

- right rear brake
- left rear brake
- right front brake
- left front brake

> ***WARNING —***
> - *The ASC and DSC systems use electronic controls and a sophisticated hydraulic unit. Once air enters the hydraulic unit, it is very difficult to remove using traditional methods. For this reason, pressure brake bleeding should be done in conjunction with the BMW service tester (DISplus/GT1/MoDiC). The tester electrically cycles the hydraulic unit solenoids while pressurized brake fluid is forced through the system.*
> - *When flushing brake fluid from the system, use extreme care to not let the brake fluid reservoir run dry. If air enters the hydraulic unit, the BMW service tester MUST be used to bleed the brake system before the vehicle is driven.*

## Brake calipers, bleeding

– Top off brake fluid in reservoir and connect pressure bleeder to reservoir. Connect bleeder hose and bottle to right rear caliper bleeder screw. Pressurize system to approximately 2 bar (29 psi).

> ***CAUTION—***
> *Do not exceed a pressure of 2 bar (29 psi) when pressure bleeding the brake system. Excessive pressure damages the brake fluid reservoir.*

– Have a helper hold brake pedal down.

< Open bleeder screw (**arrow**). Have helper slowly pump brakes about 10 times with bleeder screw open, holding pedal down on the last pump. When escaping fluid is free of air bubbles, close bleeder screw.

> ***CAUTION—***
> *Bleeder hose must always remain submersed in clean brake fluid whenever the bleeder valve is open.*

– Close bleeder screw and release brake pedal. Refill brake fluid reservoir and proceed to left rear wheel.

– Continue bleeding remaining wheels in the following order:
- left rear
- right front
- left front

| **Tightening torques** | |
|---|---|
| Bleeder to caliper:<br>• 7 mm screw<br>• 9 mm screw | <br>5 Nm (4 ft-lb)<br>6 Nm (5 ft-lb) |

## Brake system, bleeding

*NOTE—*
- *The procedure below requires a **BMW factory service tester**.*

– Determine type of traction control (ASC, DSC). Be sure to use correct specification brake fluid, according to application table below.

| Brake fluid applications | |
|---|---|
| ABS / ASC | DOT 4 brake fluid |
| ABS / DSC | DOT 4 low viscosity brake fluid |

– Top off brake fluid in reservoir and connect BMW service tester to Data Link Connector (DLC) under hood or 16 pin DLC / OBD II connector under dash.

– Call up service function Bleeding ABS / ASC or service function Bleeding ABS / DSC, depending on installed system.

– Fill pressure bleeding device with correct brake fluid and connect to brake fluid reservoir. Pressurize system.

> ***CAUTION—***
> *Do not exceed a pressure of 2 bar (29 psi) when pressure bleeding the brake system. Excessive pressure damages the brake fluid reservoir.*

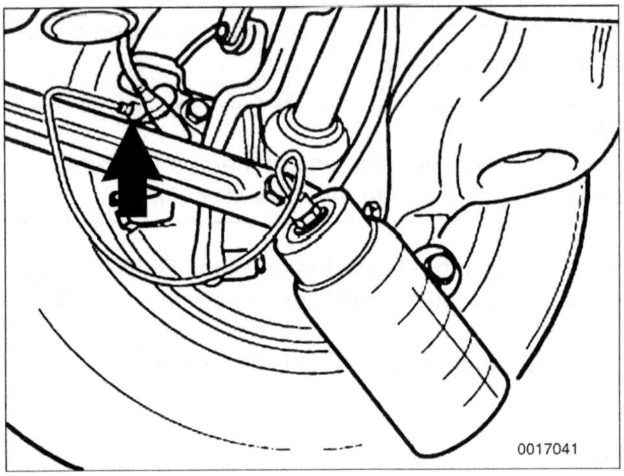


< Starting at **right rear** wheel, connect bleed hose and fluid receptacle to bleeder screw (**arrow**).

– Open bleeder screw. Carry out bleeding procedure with BMW service tester until clear, bubble-free fluid flows.
- Models with ASC: Close **right rear** bleeder.
- Models with DSC: After BMW service tester bleeding routine has finished, press brake pedal to floor 5 times until clear, bubble-free fluid flows. Close **right rear** bleeder.

> ***CAUTION—***
> *Bleeder hose must always remain submersed in clean brake fluid whenever the bleeder is open.*

- Repeat procedure at **left rear** brake.
- Repeat BMW service tester bleeding at **right front** brake.
  - Models with ASC: After BMW service tester bleeding routine has finished, press brake pedal to floor 12 times until clear, bubble-free fluid flows. Close **right front** bleeder.
  - Models with DSC: After BMW service tester bleeding routine has finished, press brake pedal to floor 5 times until clear, bubble-free fluid flows. Close **right front** bleeder.
- Repeat procedure at **left front** brake.
- Remove brake bleeding apparatus and disconnect BMW service tester. Top up brake fluid. Check to make sure brake fluid reservoir cap seal is intact.

| **Tightening torques** | |
|---|---|
| Bleeder to caliper:<br>• 7 mm screw<br>• 9 mm screw | <br>5 Nm (4 ft-lb)<br>6 Nm (5 ft-lb) |

## BRAKE PADS, CALIPERS AND ROTORS

Brake pads can be replaced without disconnecting the brake fluid hose from the caliper or having to bleed the brakes. The rotors can be replaced without disassembling the wheel hub and bearing. Always machine or replace rotors in pairs. Replace pads in sets.

***WARNING —***
- *Semi-metallic and metallic brake friction materials in brake pads or shoes produce dangerous dust.*
- *Brake fluid is poisonous, corrosive and dangerous to the environment. Wear safety glasses and rubber gloves when working with brake fluid. Do not siphon brake fluid by mouth. Dispose of brake fluid properly.*

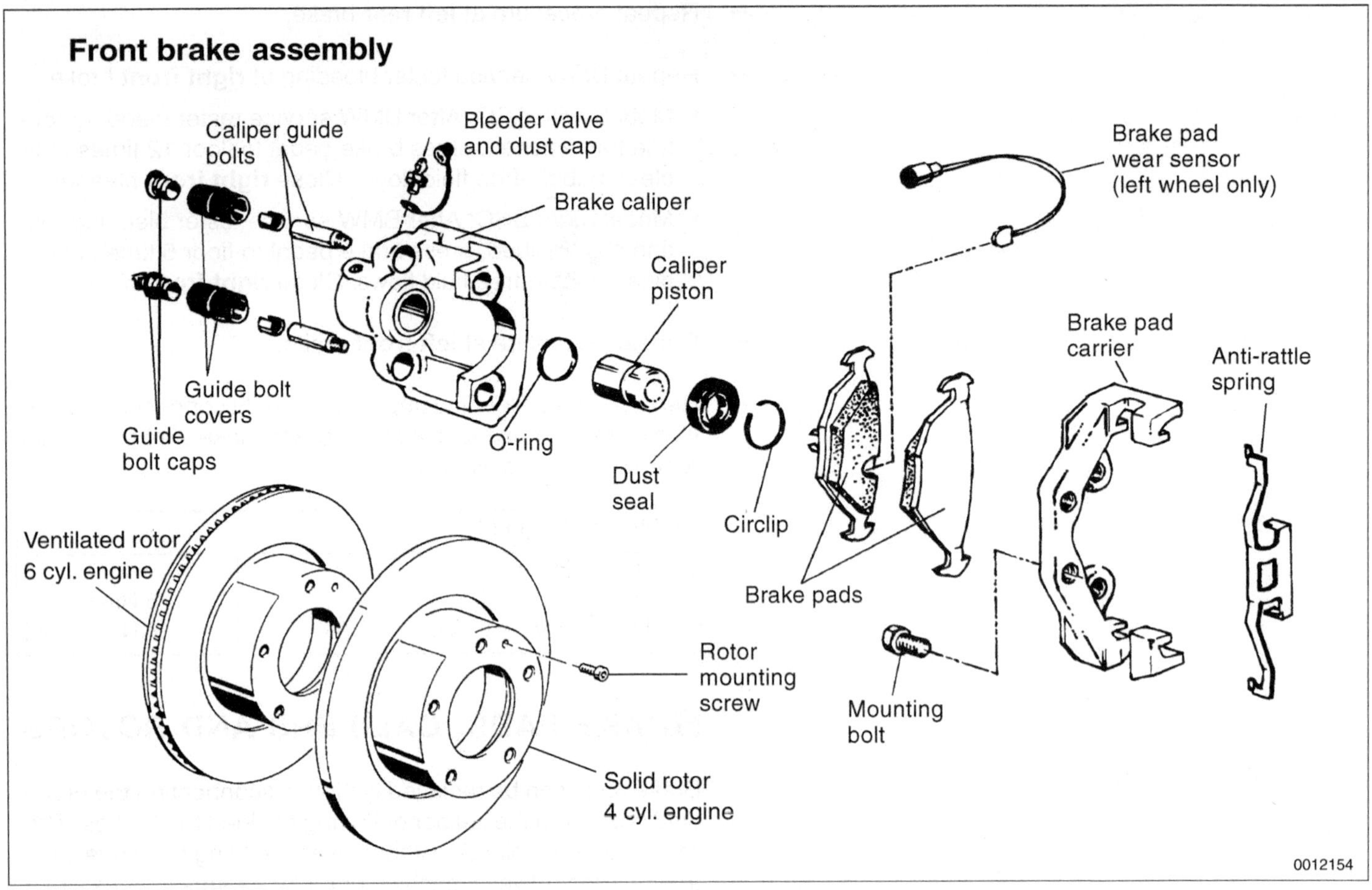


## Brake pads, replacing

This procedure is applicable to both front and rear brakes. Front and rear brake assemblies are similar, except that the rear brake rotors house the parking brake mechanism.

– Raise car and remove wheels.

> ***WARNING —***
> *Make sure the car is firmly supported on jack stands designed for the purpose. Place the jack stands underneath structural chassis points. Do not place jack stands under suspension parts.*

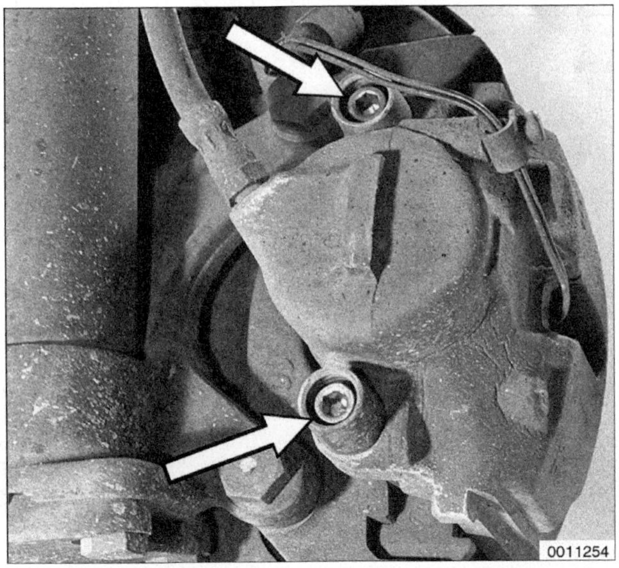


< Remove plastic caps from caliper guide bolts (**arrows**) and then remove guide bolts.

– If applicable, disconnect brake pad wear sensor connector and remove wiring from its holder.

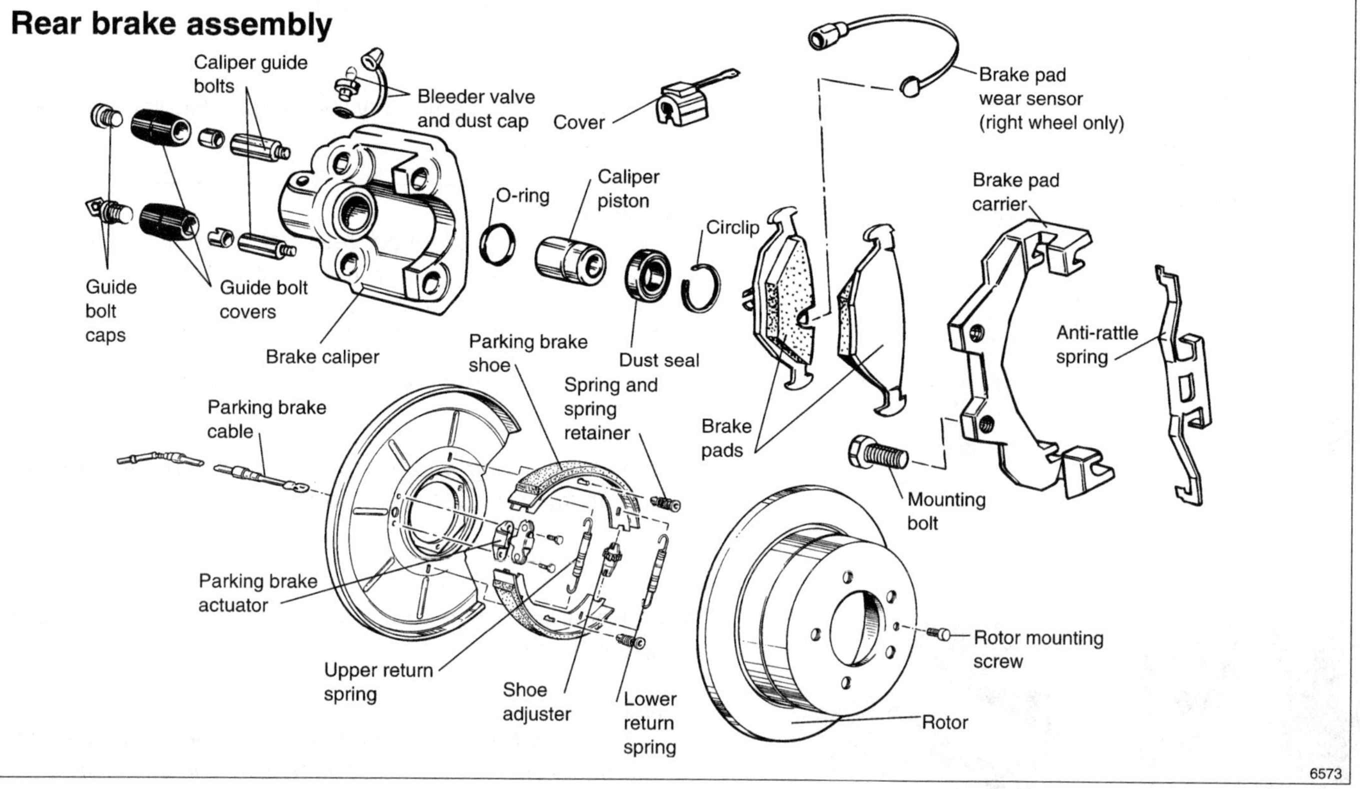


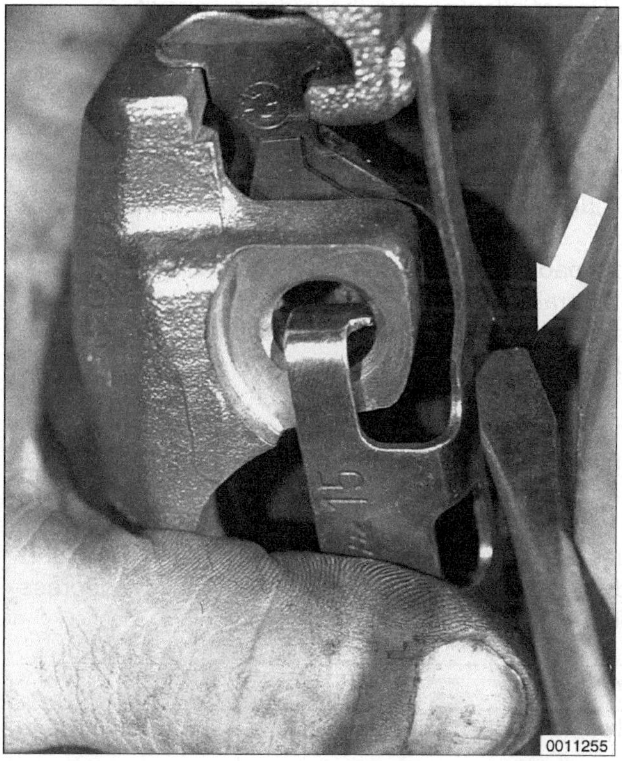


Remove anti-rattle clip from caliper by unhooking at top and bottom. Use screwdriver (**arrow**) as an aid.

– Pull caliper straight off rotor.

***CAUTION—***
*Do not let the brake caliper hang from the brake hose. Suspend it from the chassis using stiff wire.*

***NOTE—***
*If there is a ridge on the rotor edge, push the caliper piston back into the caliper before removing the caliper.*

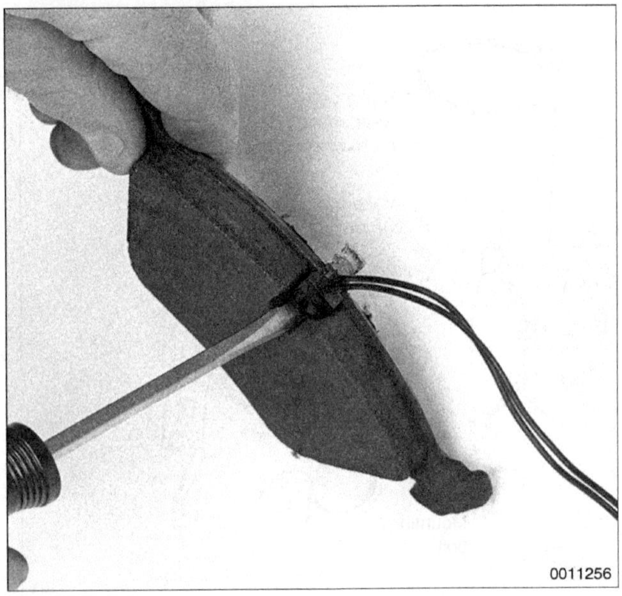
0011256

- Remove brake pads from caliper. Where applicable, carefully pry pad wear sensor from pad.

- Inspect brake caliper for signs of leakage. Check that the caliper piston slides smoothly into caliper. Replace caliper if any faults are found.

> ***CAUTION—***
> *With the pads removed, residual hydraulic pressure may cause the caliper piston to slide out. Use a wooden block to hold the piston in place.*

- If removed, replace brake pad wear sensor.

***NOTE—***
*If the brake lining wear warning light illuminated prior to brake pad, install new pad wear sensor.*

0011257

- Slowly press piston back into caliper. Use care not to damage piston dust seal.

***NOTE—***
- *To prevent old brake fluid from reentering the master cylinder, open the caliper bleeder screw and drain off some fluid when pressing the piston back into the caliper.*
- *Open the bleeder screw only when applying force to the piston. Do not allow air into the hydraulic circuit.*
- *Pressing the caliper piston in may cause the brake fluid reservoir to overflow. To prevent this, use a clean syringe to first remove some fluid from the reservoir.*

- Thoroughly clean all contact points on caliper and caliper carrier. Clean guide bolts and make sure bolts slide freely.

- Lightly coat pad contact points with brake anti-squeal paste. Place outer brake pad on caliper. Apply small amount of grease to top and bottom contact points where caliper contacts pad carrier.

- Place inner brake pad (with locating spring) onto brake caliper piston. Route pad wear sensor wiring through caliper opening and bleeder dust cap.

- Remainder of installation is reverse of removal. Do not lubricate guide bolts. Once brake assembly is installed, depress brake pedal several times before driving vehicle.

| **Tightening torques** | |
|---|---|
| Brake caliper to caliper carrier (guide bolt) | 32 Nm (23 ft-lb) |
| Road wheel to hub | 120 ± 10 Nm (90 ± 7 ft-lb) |

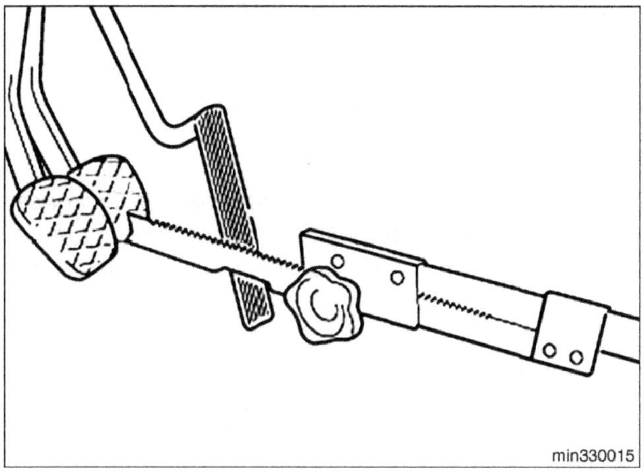


## Brake caliper and pad carrier, removing and installing

◀ Insert pedal prop and depress brake pedal slightly. This prevents brake fluid from escaping when brake lines are detached.

– Raise car and remove wheel.

> ***WARNING —***
> *Make sure the car is firmly supported on jack stands designed for the purpose. Place the jack stands underneath structural chassis points. Do not place jack stands under suspension parts.*

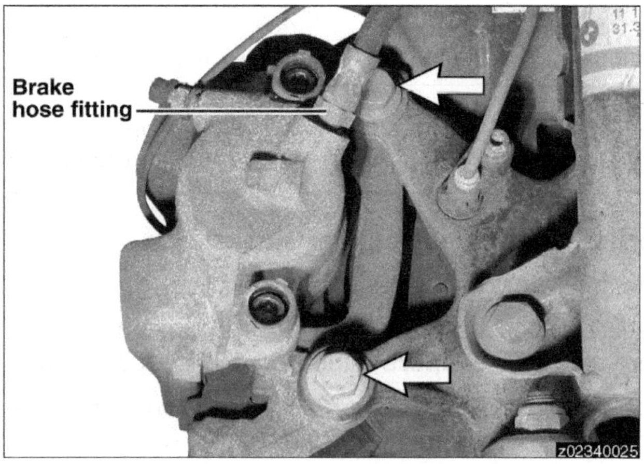


◀ Remove pad carrier and caliper mounting bolts (**arrows**) from steering arm (front caliper) or from trailing arm (rear caliper).

– Loosen brake hose fitting at caliper and spin caliper assembly off hose. Plug open fluid lines and ports.

– Inspect brake caliper for signs of leakage. Check that caliper piston slides smoothly into caliper. Replace caliper if any faults are found.

– Installation is reverse of removal. Bleed entire brake system before driving car. See **Brake calipers, bleeding** in this repair group.

| **Tightening torques** | |
|---|---|
| Brake caliper to pad carrier | 32 Nm (23 ft-lb) |
| Brake fluid hose to caliper<br>• M10 coupling<br>• M12 coupling | <br>14 Nm (10 ft-lb)<br>17 Nm (13 ft-lb) |
| Front pad carrier to strut assembly | 110 Nm (81 ft-lb) |
| Rear pad carrier to trailing arm | 67 Nm (50 ft-lb) |
| Road wheel to hub | 120 ± 10 Nm<br>(90 ± 7 ft-lb) |

## Brake rotor, removing and installing

Replace brake rotors in pairs.

– Without disconnecting brake fluid hose, remove brake pad carrier and caliper. See **Brake caliper and pad carrier, removing and installing** in this repair group.

> ***CAUTION—***
> *Do not let the brake caliper hang from the brake hose. Suspend it from the chassis using stiff wire.*

< Remove brake rotor mounting screw (**arrow**).

– Remove rotor and inspect for excessive grooving, cracks, and warping. Check rotor thickness. See **Table b**.

***NOTE —***
*On original equipment rotors, the minimum allowable thickness is stamped on the rotor hub. Measure the rotor braking surface with a micrometer at eight to ten different points and use the smallest measurement recorded.*

| Table b. Brake rotor reconditioning specifications | | |
|---|---|---|
| | **Front** | **Rear** |
| Vented rotor wear (min. thickness) | 20.4 mm (0.803 in) | 17.4 mm (0.685 in) |
| Solid rotor wear (min. thickness) | 10.4 mm (0.409 in) | 8.4 mm (0.331 in) |
| Axial runout (max.)<br>rotor removed<br>rotor installed | <br>0.05 mm (0.002 in)<br>0.20 mm (0.008 in) | <br>0.05 mm (0.002 in)<br>0.20 mm (0.008 in) |

– Installation is reverse of removal. Remember to:

- Clean rotor with brake cleaner before installing.
- If brake fluid hose was removed or loosened, bleed brake system. See **Brake calipers, bleeding** in this repair group.
- Once brake assembly is installed, depress brake pedal several times to adjust caliper and pads.

***NOTE —***
*When installing new rear brake rotors, adjust the parking brake. See* **Parking brake, adjusting** *in this repair group.*

| Tightening torques | |
|---|---|
| Brake rotor to wheel hub | 16 Nm (12 ft-lb) |
| Front pad carrier to strut assembly | 110 Nm (81 ft-lb) |
| Rear pad carrier to trailing arm | 67 Nm (50 ft-lb) |
| Road wheel to hub | 120 ± 10 Nm<br>(90 ± 7 ft-lb) |

## Brake Master Cylinder

The brake master cylinder is mounted to the front of the vacuum booster on the driver side engine compartment bulkhead.

### Brake master cylinder, removing and installing

– Using a clean syringe, empty brake fluid reservoir.

> ***WARNING—***
> *Brake fluid is poisonous, corrosive and dangerous to the environment. Dispose of it properly.*

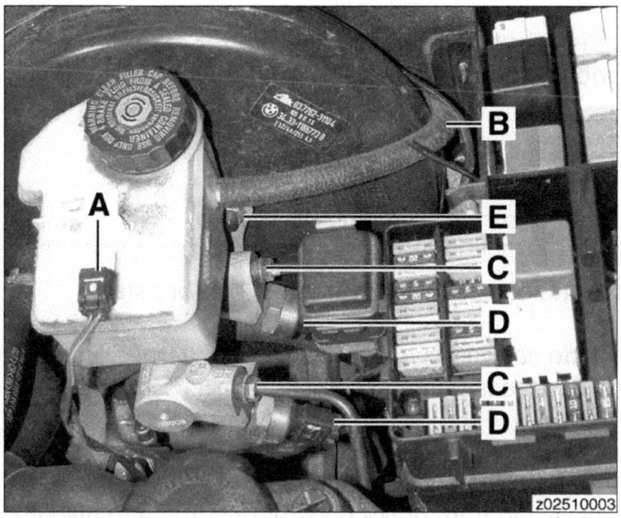


Working at brake master cylinder:

- Disconnect brake fluid level sender connector (**A**).
- Disconnect hydraulic clutch system supply line (**B**).
- Disconnect brake fluid lines (**C**).
- If applicable: Detach DSC pressure switch electrical connectors (**D**).
- Remove master cylinder mounting nuts (**E**) and remove master cylinder from brake booster.

> ***CAUTION—***
> - *Plug open brake lines.*
> - *Be prepared to catch dripping brake fluid.*

– When reinstalling, make sure all nuts, fluid couplings, thread bores, and mating surfaces are clean.

– If fluid reservoir was removed, install it carefully using new sealing grommets.

– Mount master cylinder to brake booster using a new O-ring and new self-locking nuts.

> ***CAUTION—***
> *Do not to over-torque the master cylinder mounting nuts. This could distort the brake booster and prevent proper vacuum build-up.*

– Connect all brake fluid lines. Connect hydraulic clutch hose.

– Reconnect fluid level sender and bleed entire brake system. See **Brake system, bleeding** in this repair group.

| **Tightening torques** | |
|---|---|
| Brake fluid line to master cylinder<br>• M10 coupling<br>• M12 coupling | <br>14 Nm (10 ft-lb<br>17 Nm (13 ft-lb) |
| Brake master cylinder to brake booster | 26 Nm (18 ft-lb) |

## Brake booster

The brake booster is mounted to the bulkhead on the driver's side of the engine compartment, directly behind the brake master cylinder. Intake manifold vacuum acts on a large diaphragm in the booster to reduce brake pedal effort.

### Brake booster, removing and installing

– Using a clean syringe, empty brake fluid reservoir.

> ***WARNING—***
> *Brake fluid is poisonous, corrosive and dangerous to the environment. Dispose of it properly.*

– Remove master cylinder. See **Brake master cylinder, removing and installing** in this repair group.

> ***CAUTION—***
> - *Plug open brake lines.*
> - *Be prepared to catch dripping brake fluid.*

– Disconnect engine vacuum hose from brake booster.

– Working inside car, remove panel(s) from underneath steering column to access pedal assembly. See **513 Interior Trim**.

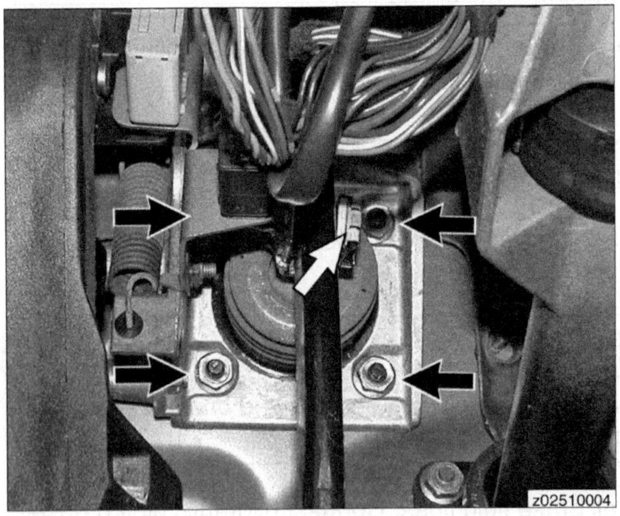

z02510004

◄ Working under left side of dashboard, remove clip (**white arrow**) and clevis pin from brake booster pushrod and disconnect it from brake pedal. Remove brake booster mounting nuts (**black arrows**).

– Working in engine compartment, remove brake booster.

– Make sure all nuts, fluid couplings, thread bores and mating surfaces are clean.

– Mount master cylinder to vacuum booster using a new O-ring and new self-locking nuts.

> ***CAUTION—***
> *Do not over-torque the master cylinder mounting bolts. This could distort the brake booster and prevent proper vacuum build-up.*

– Remainder of installation is reverse of removal. Remember to:
  - Reconnect all hydraulic lines.
  - Reconnect fluid level sender.
  - Bleed entire brake system. See **Brake system, bleeding** in this repair group.

| **Tightening torques** | |
|---|---|
| Brake booster to bulkhead | 22 Nm (16 ft-lb) |
| Brake fluid line to master cylinder<br>• M10 coupling<br>• M12 coupling | <br>14 Nm (10 ft-lb<br>17 Nm (13 ft-lb) |
| Brake master cylinder to brake booster | 26 Nm (18 ft-lb) |

***NOTE—***
*When replacing the brake booster check valve or the vacuum hose, install the valve so that the molded arrow on the valve is pointing towards the intake manifold. Use new hose clamps.*

## PARKING BRAKE

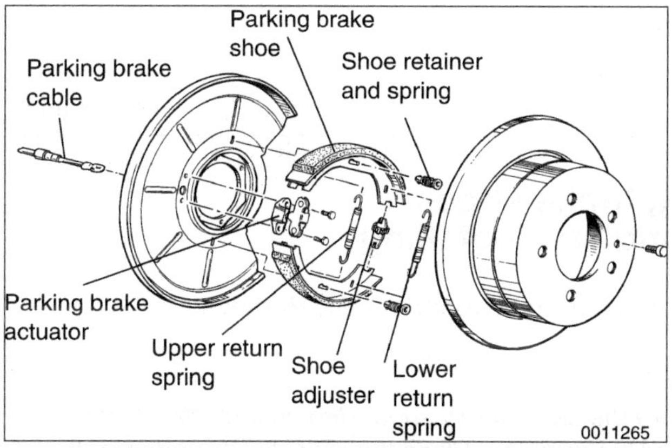


The parking brake is a drum system integrated into the rear wheel brake rotors. Adjustment of the shoes may be necessary from time to time to compensate for wear. The parking brake should also be adjusted any time cable, rear brake rotor, or parking brake shoes are replaced.

### Parking brake, adjusting

The parking brake can be adjusted with the wheels installed, but the rear wheels have to be raised off the ground.

Lift parking brake lever boot out of console. While holding cables stationary, loosen parking brake cable nuts (**arrows**) until they are completely slack.

– Raise rear of car and support safely.

***WARNING—***
*Make sure the car is firmly supported on jack stands designed for the purpose. Place the jack stands underneath structural chassis points. Do not place jack stands under suspension parts.*

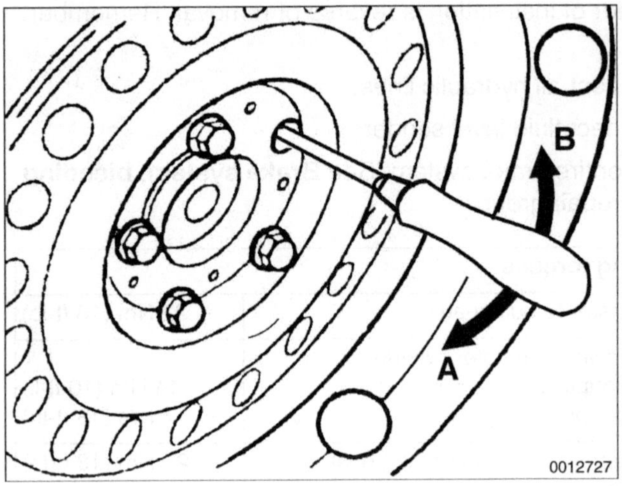


Remove one lug bolt from each rear wheel.

- Turn road wheel until lug bolt hole lines up with parking brake adjuster.
- Use flat bladed screwdriver to turn adjuster inside drum:
  -Left wheel: turn adjuster in direction **A** to expand shoes.
  -Right wheel: turn in direction **B** to expand shoes.
- Turn adjuster to expand brake shoes until road wheel can no longer turn, then back adjuster off about eighteen (18) notches.

– Working inside car, set parking brake several times to seat cable. Then pull parking brake lever up six notches. Tighten cable adjusting nuts until it is just possible to turn rear wheels with slight resistance.

– Release lever and make sure rear wheels turn freely.

– Turn on ignition. With parking brake lever released, make sure parking brake warning light in instrument panel is OFF. Check that light comes ON as soon as lever is pulled up.

– Install parking brake lever boot. Install road wheel lug bolts.

| Tightening torque | |
|---|---|
| Road wheel to hub | 120 ± 10 Nm<br>(90 ± 7 ft-lb) |

## Parking brake shoes, removing and installing

– Raise rear of car and support safely. Remove road wheels.

> ***WARNING —***
> *Make sure the car is firmly supported on jack stands designed for the purpose. Place the jack stands underneath structural chassis points. Do not place jack stands under suspension parts.*

– Without disconnecting brake fluid hose, remove rear brake calipers from trailing arms and support with stiff wire. Remove rear brake rotors. See **Brake rotor, removing and installing** in this repair group.

– Unhook upper return spring from brake shoes. Remove shoe retainers by pushing them in and rotating 90°. BMW special tool 34 4 000 can be used to remove retainers, if needed.

– Spread shoes apart at top and lift them out from below.

***NOTE —***
*When removing the brake shoes the shoe expander pivot pin may fall out.*

– Inspect shoe expander to make sure it functions properly. Apply a thin coat of grease to sliding parts and pins.

– Installation is reverse of removal. Be sure to adjust parking brake. See **Parking brake, adjusting** in this repair group.

## Parking brake cable, replacing

– Raise rear of car and support safely. Remove left or right road wheel.

> ***WARNING —***
> *Make sure the car is firmly supported on jack stands designed for the purpose. Place the jack stands underneath structural chassis points. Do not place jack stands under suspension parts.*

– Without disconnecting brake fluid hose, unbolt brake caliper and pad carrier from trailing arm and support with stiff wire. Remove brake rotor. See **Brake rotor, removing and installing** in this repair group.

– Remove parking brake shoes. See **Parking brake shoes, removing and installing** in this repair group.

◄ Release and remove parking brake cable from cable holder on trailing arm.

◄ Working inside car, lift parking brake lever boot out of console. Remove cable adjusting nut.

– Working underneath car, remove muffler and heat shield. See **180 Exhaust System**.

– Slide cable out of housing and install new cable.

– Remainder of installation is reverse of removal. Remember to adjust parking brake. See **Parking brake, adjusting** in this repair group.

# Antilock brake system (ABS)

Several versions of ABS / traction control are installed on cars covered by this manual.

1996 - 1998 models with ASC:

- **ASC electronic control module** is located behind the glove compartment.
- **ABS / ASC hydraulic module** is located at the left side of the engine compartment.

1999 - 2000 models with ASC, all models with DSC:

- **ASC** or **DSC electronic control module** is integrated with **ABS / ASC** or **ABS / DSC hydraulic module** at left side of engine compartment.

The ABS system is designed to be maintenance free. There are no adjustments that can be made to the system. Repair and troubleshooting of the major ABS system components requires special test equipment and knowledge and should be done by an authorized BMW dealer service department.

ABS is self-tested by the ABS control module each time the car is started. Once the test is complete, the ABS dashboard warning light turns OFF. If the light remains ON or comes ON at any time during driving, a system fault has occurred and the ABS system is electronically disabled. The conventional braking system remains fully functioning.

## ABS system inspection

A visual inspection of the ABS system may help to locate system faults. If no visual faults can be found and the ABS light remains ON, have the system diagnosed by an authorized BMW dealer.

Carefully inspect the entire ABS wiring harness, particularly the pulse sensor harnesses and connectors near each wheel. Look for chafing or damage due to incorrectly routed wires.

Carefully remove the wheel speed sensors. See **ABS wheel speed sensor, removing and installing** in this repair group. Clean the sensor tips. Inspect the toothed pulse wheel on the front wheel hub and the outer CV joint. Check for missing, clogged or corroded teeth, or other damage that could alter the clearance signal between the sensor tip and the pulse wheel.

***NOTE —***
- *If a front pulse wheel is damaged, replace the entire wheel hub. See* **310 Front Suspension**.
- *The rear pulse wheel is integral with the outer CV joint. See* **331 Differential, Drive Axles**.

## ABS wheel speed sensor, removing and installing

– Raise and safely support vehicle.

> ***WARNING —***
> *Make sure the car is firmly supported on jack stands designed for the purpose. Place the jack stands underneath structural chassis points. Do not place jack stands under suspension parts.*

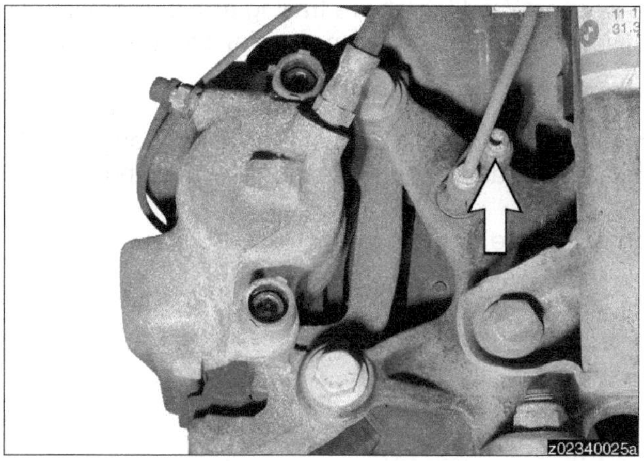


◀ Front ABS sensor: Unscrew wheel speed sensor mounting bolt (**arrow**). Remove ABS pulse sensor from its housing. Remove harness from retaining mounts.

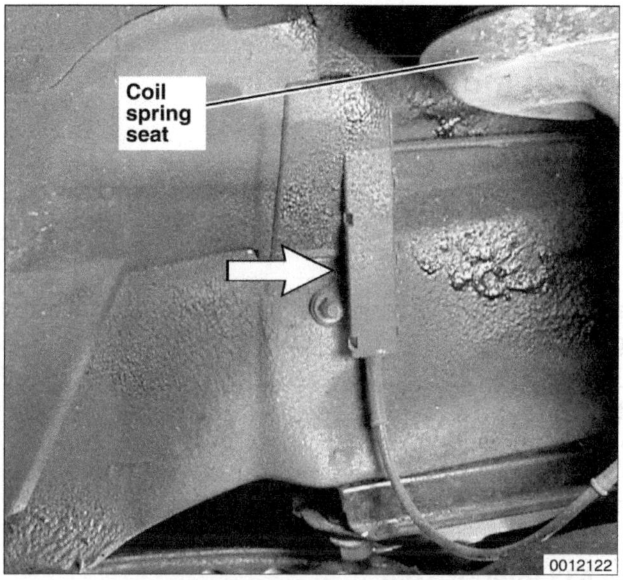


◀ Front ABS sensor: Working in front wheel housing, separate speed sensor connector (**arrow**) and remove sensor.

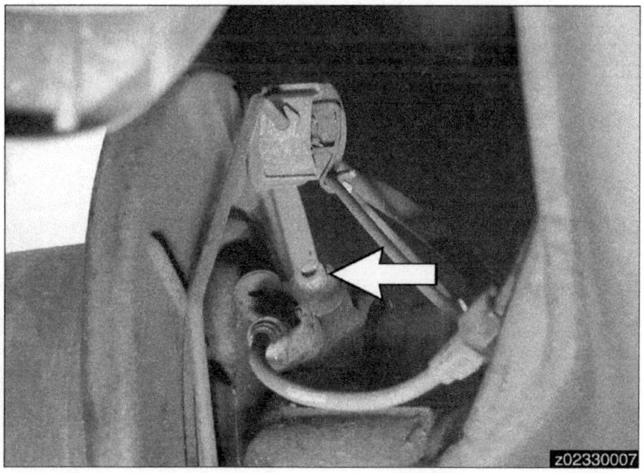


◀ Rear ABS sensor: Working above trailing arm, remove wheel speed sensor mounting bolt (**arrow**). Pull out sensor.

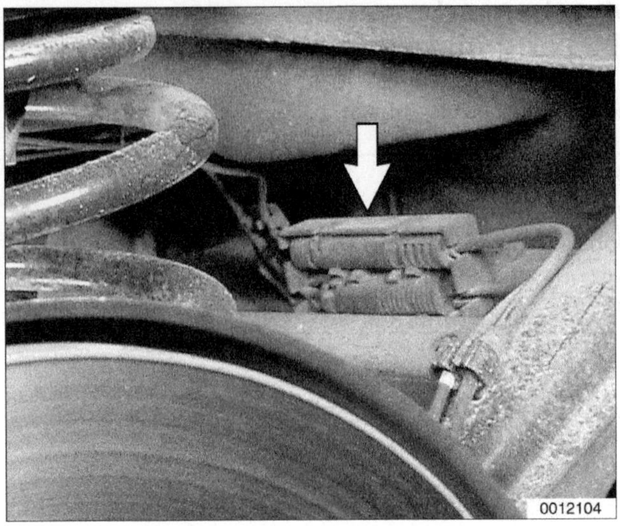
0012104

< Rear ABS sensor: Working above trailing arm in rear wheel housing, disconnect electrical connector (**arrow**) and remove sensor.

- During installation, apply thin coat of Molykote® Longterm 2 or an equivalent grease to pulse sensor and housing.
- Installation is reverse of removal.

## ABS control module, removing and installing (early version)

Early versions of the ABS / ASC control module are located under passenger side of dashboard to right of glove compartment.

- Disconnect negative (–) battery cable and cover terminal with insulating material.

> ***CAUTION—***
> *Prior to disconnecting the battery, read the battery cautions in* **001 General Warnings and Cautions**.

- Remove passenger side lower trim panel and glove compartment. See **513 Interior Trim**.

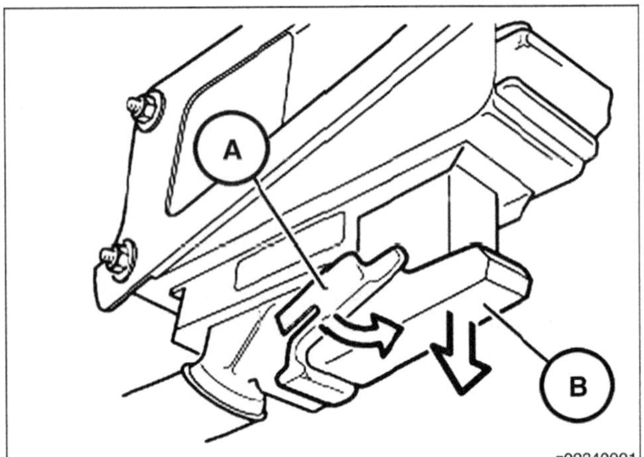

z02340001

< Release locking lever (**A**) on control module electrical connector. Pull connector (**B**) down to remove.

z02340002

< Remove control module mounting bolts (**arrows**).

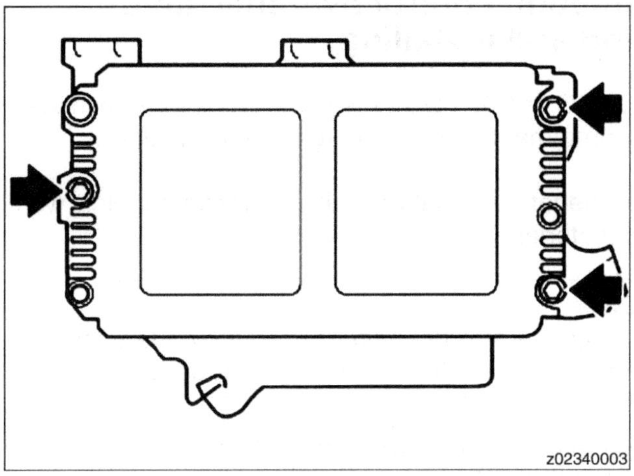
z02340003

If replacing control module, remove bolts (**arrows**) holding old module to bracket. Transfer bracket to new module.

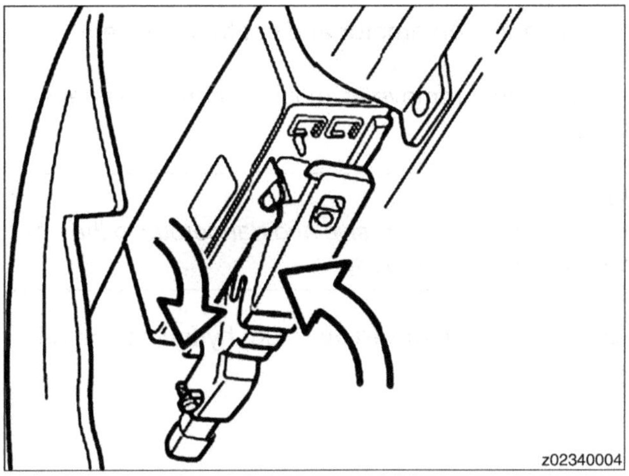
z02340004

To install electrical connector, attach left side first then press right side into place and fold down locking clamp.

- Use BMW service tester (DISplus / GT1 / MoDiC) to initialize control module.

- Remainder of installation is reverse of removal.

## ABS / traction control module, removing and installing (late version)

Later versions of the ABS control module are located in the engine compartment attached to the hydraulic unit.

- Disconnect negative (–) battery cable and cover terminal with insulating material.

> ***CAUTION—***
> *Prior to disconnecting the battery, read the battery cautions in* ***001 General Warnings and Cautions****.*

- Remove ABS hydraulic unit. See **ABS / traction control hydraulic unit, removing and installing** in this repair group.

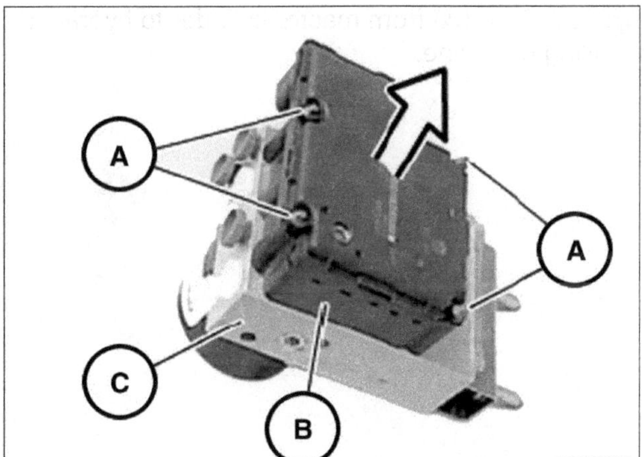

z02340010

Remove screws (**A**) and carefully pry control unit (**B**) away from hydraulic unit (**C**).

- Installation is reverse of removal. Remember to:
    - Replace self locking screws.
    - Bleed brakes. See **Brake system, bleeding** in this repair group.

## ABS / traction control hydraulic unit, removing and installing

The ABS / ASC or ABS / DSC hydraulic unit is located in the engine compartment near the brake master cylinder.

– Disconnect negative (–) battery cable and cover terminal with insulating material.

> ***CAUTION—***
> *Prior to disconnecting the battery, read the battery cautions in* **001 General Warnings and Cautions**.

– Remove air intake housing along with mass air flow sensor. See **130 Fuel Injection**.

– Unbolt cruise control servomotor and set to one side.

– Using a clean syringe, remove as much fluid from master cylinder as possible.

> ***WARNING —***
> *Brake fluid is poisonous, corrosive and dangerous to the environment. Dispose of it properly.*

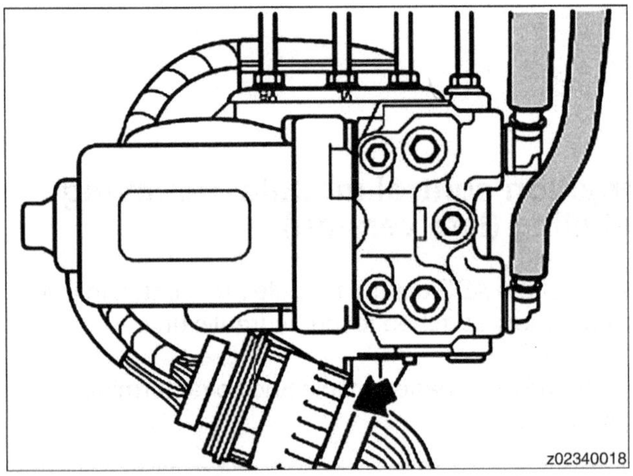
z02340018

Detach electrical connector (**arrow**) from hydraulic control unit harness.

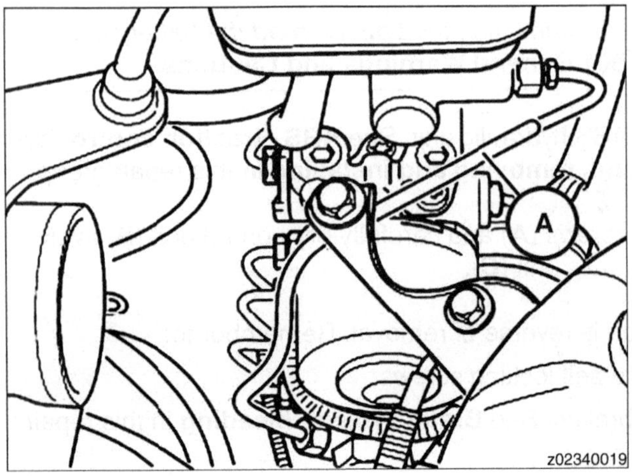

z02340019

Remove hydraulic line (**A**) from master cylinder to hydraulic control unit. Plug open line.

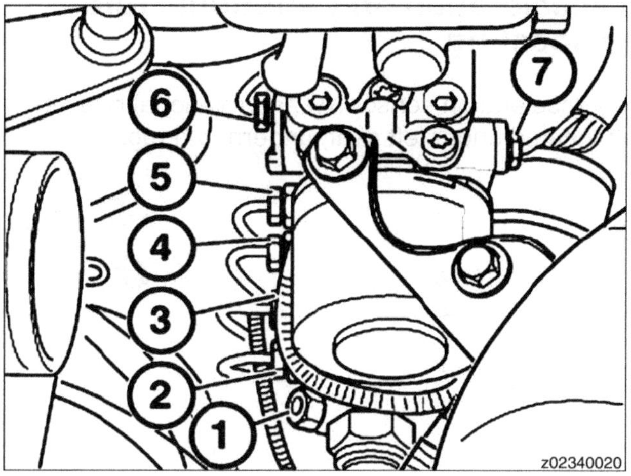

z02340020

Remove hydraulic lines (2 - 7) from hydraulic control unit. Mark location of hydraulic lines to aid assembly.

1. Brake master cylinder (rear axle)
2. Right front
3. Left front
4. Right rear
5. Left rear
6. Brake master cylinder (front axle)
7. Pressure accumulator

> ***WARNING—***
> *Do not mix hydraulic lines. Mark locations before removal.*

> ***CAUTION—***
> - *Plug open brake lines.*
> - *Be prepared to catch dripping brake fluid.*

| Tightening torque | |
|---|---|
| Hydraulic lines to hydraulic unit | 18 Nm (13 ft-lb) |

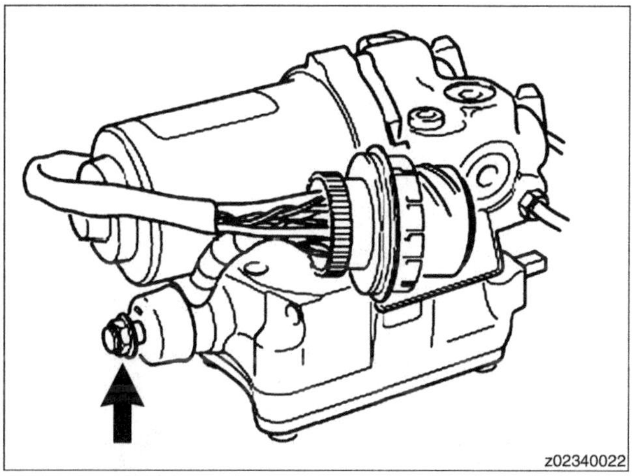
z02340022

Remove mounting bolt (**arrow**) and remove hydraulic unit.

***NOTE—***
*If necessary, remove carbon canister to aid removal of hydraulic unit.*

– Installation is reverse of removal. Remember to:

- Reconnect all hydraulic lines.
- Replace electrical connectors.
- Bleed entire brake system. See **Brake system, bleeding** in this repair group.

## Steering angle sensor, locating

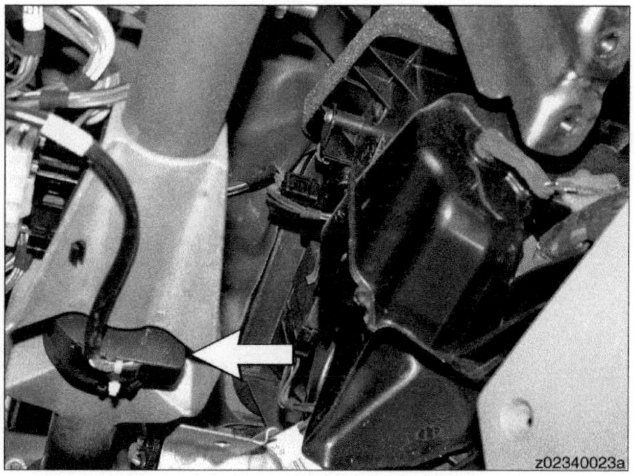
z02340023a

The steering angle sensor (**arrow**) is located on steering column under dashboard.

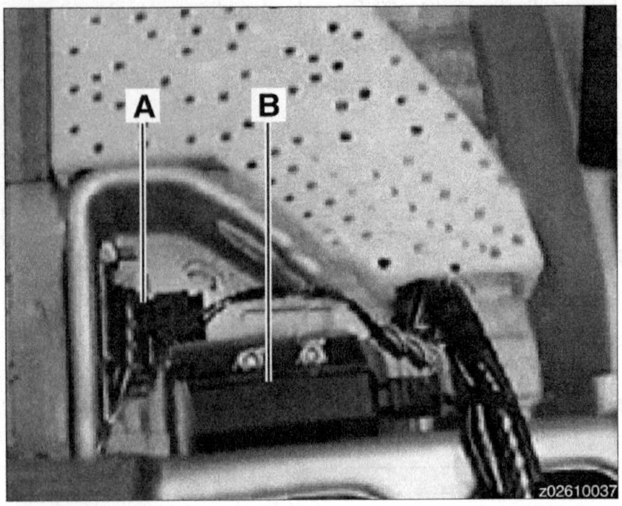


## Lateral acceleration and rotational rate sensors, locating

◄ DSC lateral acceleration sensor (**A**) and rotational rate sensor (**B**) are located under left seat, underneath rug.

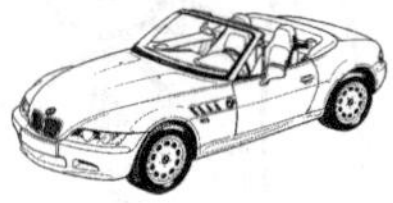

# 400 Body-General

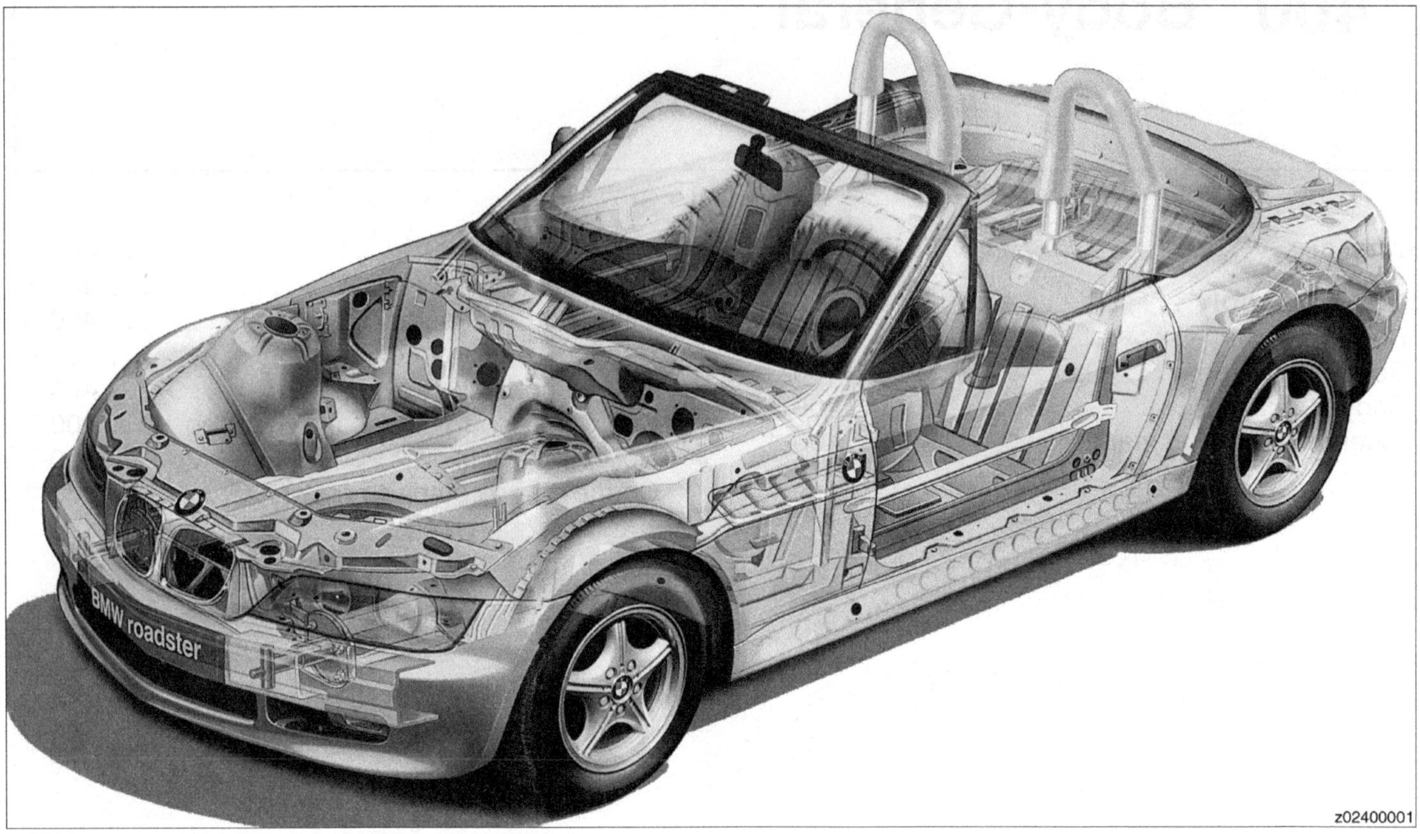


z02400001

## General

This section covers system descriptions and basic information for the repair groups found in **4 Body-General** and **5 Body Equipment**.

When working or servicing cars equipped with airbags, observe the following precautions to prevent personal injury.

> ***WARNING —***
> - *Airbags, pyrotechnic seat belt tensioners and battery safety terminal are deployed by explosive devices controlled by the MRS (SRS) control module. Handled improperly or without adequate safeguards, the MRS system can be very dangerous. Special precautions must be observed prior to any work at or near the steering wheel or steering column, dashboard, seats and seat belts or battery.*
> - *Always disconnect the battery and cover the negative (-) battery terminal with an insulator before starting diagnostic, troubleshooting or service work on the car and before doing any welding on the car.*

## Body assembly

### Body versions

The Z3 versions are:

- **Z3 Roadster**. 4-cylinder (1996 - 1998) and 6-cylinder (1997 - 2002)
- **M Roadster**. 6-cylinder (1998 - 2002)
- **Z3 Coupe**. 6-cylinder (1999 - 2002)
- **M Coupe**. 6-cylinder (1999 - 2002)

### Body dimensions

Body dimensions vary slightly between models and versions.

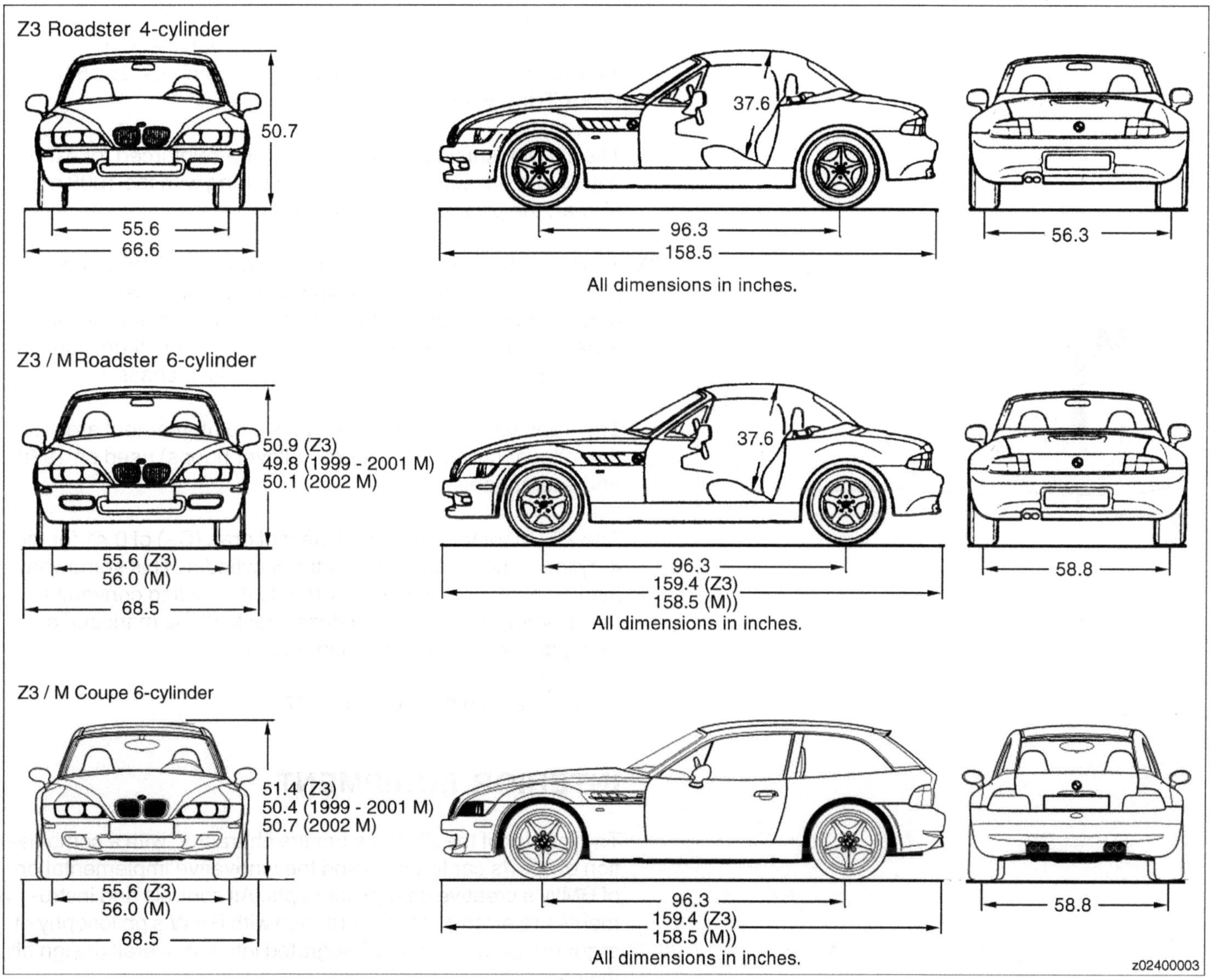

## Body structure

The BMW Z3 is a unibody design constructed of high strength steel. Attaching parts, such as the hood, front fenders, rear quarter panels, doors, trunk lid and hood are also constructed of steel.

## Exterior and aerodynamics

The Z3 exterior fender panels are corrosion-resistant zinc coated steel. Both the front and rear fenders (Roadster only) are bolted on and easily replaceable. See **410 Fenders, Engine Hood**.

The front and rear bumpers use an aluminum framework with overlaid high-impact, deformable plastic. The Roadster rear bumper mounting includes a pair of vibration dampers which utilize the mass of the bumper as an oscillating counterweight. This minimizes the feeling of body twist that is an inherent characteristic of an open-bodied vehicle. For information on removing and replacing bumpers, see **510 Exterior Trim, Bumpers**.

The Z3 windshield pillars are steel-tube-reinforced for extra safety and passenger protection. Starting with the 1998 model year, the pillar is finished in body color.

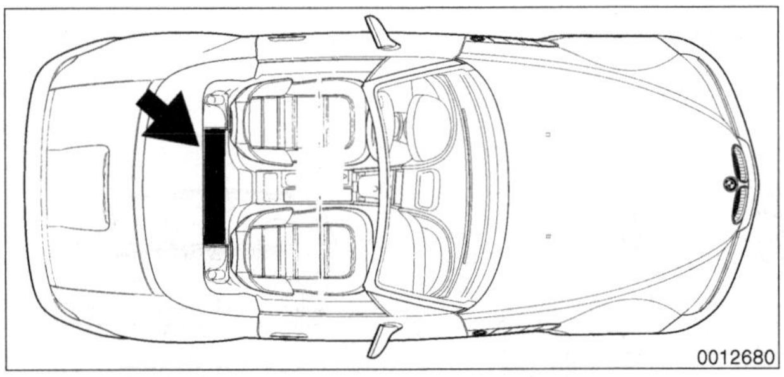


Starting with 1997 Roadster models, the horizontal surface of the body structure (**arrow**) between the upper seatbelt mounting posts was further reinforced. The reinforcement is a welded plate called a "transverse connector." It provides increased torsional rigidity for the convertible chassis.

The transverse connector provides a reinforcement area for the rollover protection system (rollover hoops) used on 1998 and later models.

The Z3 Roadster has a coefficient of drag ($C_D$) of 0.41 for the 4-cylinder model and 0.43 for the 6-cylinder. This somewhat inefficient number is primarily due to the folding convertible top. The vehicle is, nevertheless, quick off the mark and a superb performer in either configuration.

For Coupe models, the $C_D$ is 0.37.

## INTERIOR EQUIPMENT

The interior of the Z3 has a unique character, with a combination of sports car features and the innovative implementation of BMW's creative design concepts. All controls and instruments are arranged in accordance with BMW's philosophy of ergonomics and are fully integrated into the overall design of the car.

## Instruments and controls

BMW uses the term "cockpit" to describe its interior passenger compartment. Everything essential to the driver is logically grouped and easy to reach.

The padded dashboard houses the instrument cluster and the ventilation and heating system. The instrument cluster uses large easy-to-read analog instruments and is removable as a unit without removing the dashboard. See **620 Instruments**.

## Seats and seat belts

Anatomically-correct seats are constructed from polyurethane foam containing areas or "zones" of different firmness. The seat bases have steel springs for added support and strength.

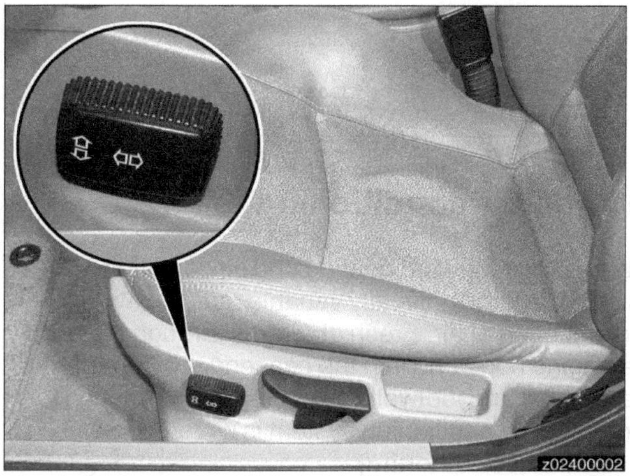


The 4-way power driver seat and 2-way or 4-way power passenger seat are provided with conveniently located control switches along the outside edge of the seat.

The Z3 is equipped with a dual airbag Supplemental Restraint System and automatic seat belt tensioners. The 1996 Roadster seatbelts used BMW's original-design spring tensioner. In 1997, the seat base design was modified to accommodate a pyrotechnic (explosive charge) seatbelt tensioner. Both types of tensioners are designed to automatically tension the belts by approximately 55 mm (about 2 inches) in the event of a collision.

***NOTE—***
*The seat base used on cars with pyrotechnic seatbelt tensioners is not interchangeable with the earlier seat base.*

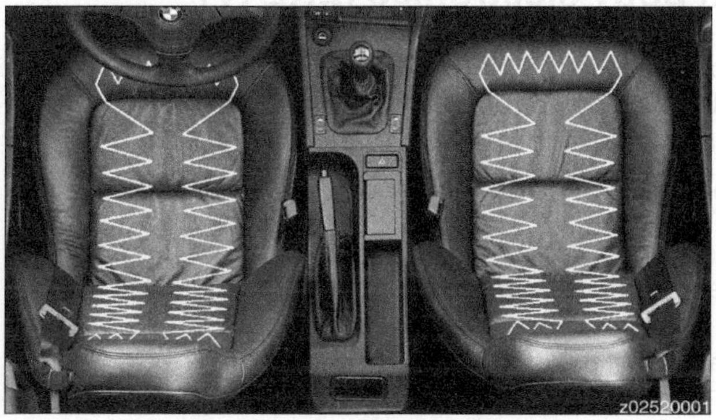


Seats heating is accomplished with a heating coil embedded in each seat cushion and backrest.

**Z3 heating and ventilation outlets**

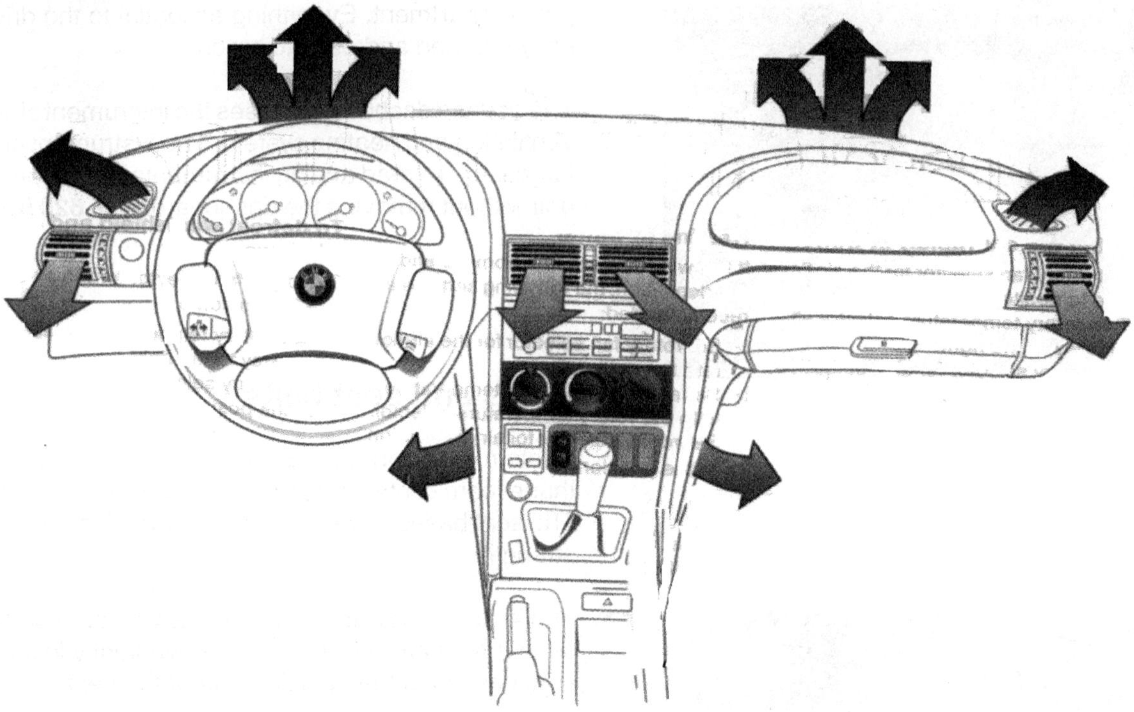

0012720

## Heating and air-conditioning

The electronically controlled heating and air conditioning system uses an extra large, infinitely variable radial blower motor for good distribution of air. Fresh air enters through the grille below the engine hood and into the passenger compartment via the dashboard and footwell vents. Repair information for the heating and air conditioning system is covered in **640 Heating and Air- conditioning**.

## Central body electronics (ZKE IV)

All Z3 models since the 1997 model year are equipped with ZKE IV. This is an electronic comfort and convenience package which includes the following features:

- One touch (up or down) power window operation
- Power window closure from driver or passenger door lock
- Vehicle locking or unlocking from the trunk lock

Other features of the system are discussed in **515 Central Locking and Anti-theft**. and **600 Electrical System–General**.

## Sunroof, convertible top and hardtop

Coupe models are equipped with a tilting sunroof. See **540 Sunroof** for details.

Opening and closing the convertible roof is particularly easy, involving just two rapid-action catches on the upper windshield frame. As of mid-1997, gas pressure springs were added to the convertible top frame, making it even easier to open and close the roof. And the top-of-the-line 6-cylinder model comes with electro-hydraulic operation of the roof as standard. The tonneau cover fits perfectly over the folded convertible top. With the 1998 model year, a body-colored hardtop was offered. Details about the convertible top and the hardtop can be found in **541 Convertible Top**.

## Rollover protection

z02541001

◄ Rollover protection bars are installed behind the seat headrests. These as well as the reinforced windshield pillars help provide protection in case of a rollover.

## Coupe cargo compartment

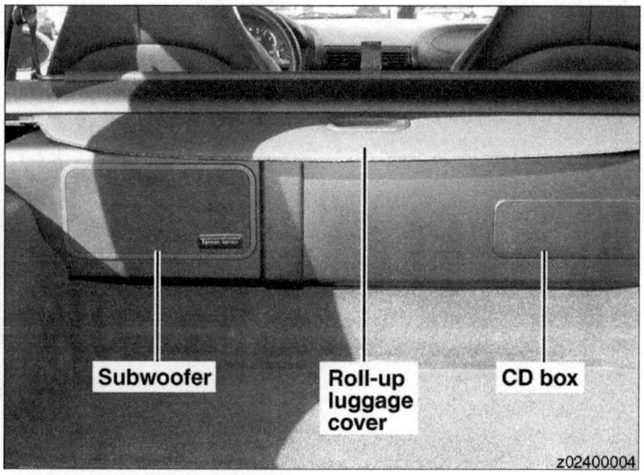


z02400004

◄ A CD player and subwoofer, as well as a forward-facing storage box, are located in the partition behind the Coupe seats.

The luggage cover rolls up, window-shade fashion, behind the seats.

The Coupe rear hatch window has an electric defogger and wiper and washer system.

# 410 Fenders, Engine Hood

## GENERAL

This repair group covers replacement of the front and rear fenders (side panels) and removal and installation of the engine hood.

The body is painted at the factory after assembly. Realignment of body panels may expose unpainted metal. To prevent corrosion, paint exposed metal.

## FENDERS

### Front fender, removing and installing

- Raise hood.

- Raise and safely support vehicle.

> ***WARNING —***
> *Make sure the car is stable and well supported at all times. Use a professional automotive lift or jack stands designed for the purpose. A floor jack is not adequate support.*

- Remove front wheel and remove inner liner from wheel well.

- Remove side turn signal light by pushing light housing backwards then carefully pulling light assembly out. Disconnect lamp socket by twisting counterclockwise.

- Working in wheel well, remove attaching bolt on inside front of fender.

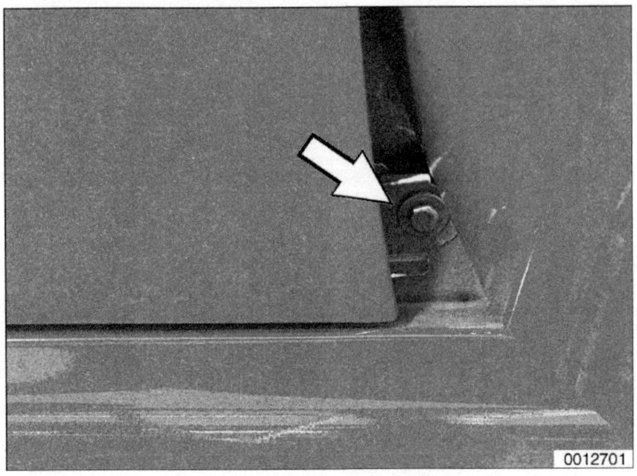


- Open door. Remove fender attaching screw (**arrow**) at bottom of fender. Open door all the way to gain access to screw.

◄ Open engine hood. Remove two attaching bolts (**arrows**) at top of fender.

– Remove fender.

– Installation is reverse of removal.

***NOTE —***
*Seam width between the fender and other body parts should be even at all points, approximately 5 mm (¼ in).*

## Rear fender (side panel), removing and installing

***NOTE —***
*This procedure applies to Roadster only. Coupe rear fenders are attached with electronic welds and are not normally replaceable. See your BMW dealer or a qualified body shop.*

– Remove trunk lid. See **412 Trunk Lid, Rear Hatch** for more information.

– Raise and safely support vehicle.

> ***WARNING —***
> *Make sure the car is stable and well supported at all times. Use a professional automotive lift or jack stands designed for the purpose. A floor jack is not adequate support.*

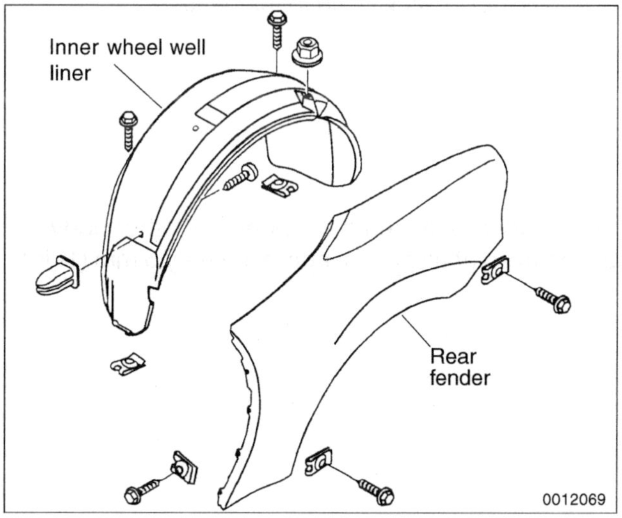


◄ Remove rear wheel and remove inner liner from wheel well.

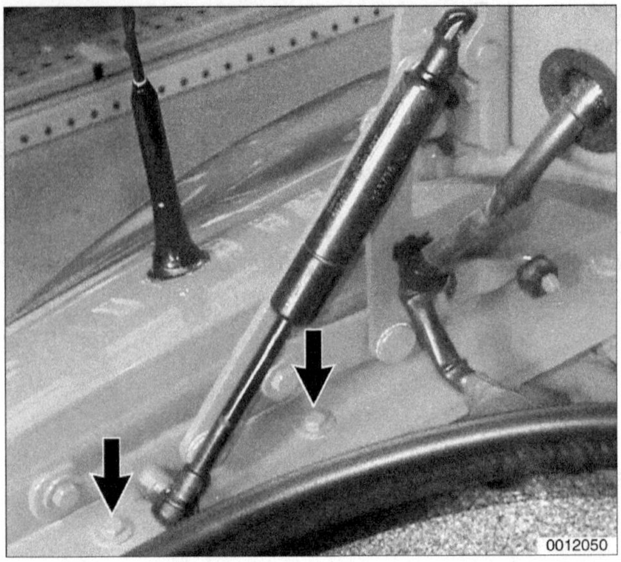
0012050

◀ Open trunk. Remove fender mounting bolts (**arrows**) from trunk sill.

– Remove retaining clip from bumper overlay.

0012706

◀ Open door. Carefully pry off finishing trim. Remove fender bolts from behind trim.

– Remove retainer screw on side panel of wheel arch.

– Carefully remove fender.

– Installation is reverse of removal. Remember to:
  - Align trunk lid. See **412 Trunk Lid, Rear Hatch**.
  - Replace damaged retaining clips.
  - Check door and trunk operation.

***NOTE—***
*Soak the trim clips in warm water for a short time before installing the trim panel. This will aid in installation.*

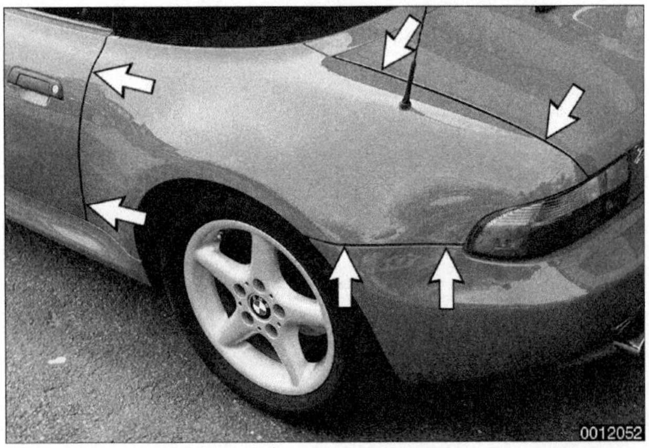
0012052

◀ Seam width (**arrows**) between the fender and other body parts should be even at all points, approximately 5 mm (¼ in).

# ENGINE HOOD

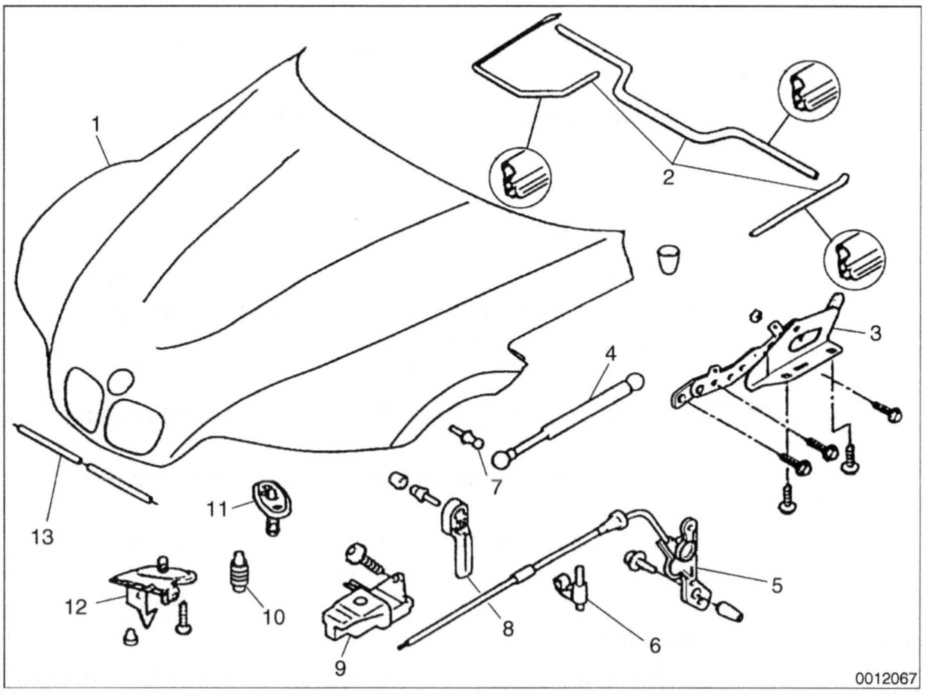


## Engine hood components

1. **Hood**
2. **Hood seal**
3. **Hinge assembly**
4. **Gas-charged piston with ball socket**
5. **Hood release and cable assembly**
6. **Hood release cable guide**
7. **Hood piston ball pivot**
8. **Hood release cable handle (inside car)**
9. **Hood lock catch**
10. **Buffer**
11. **Left hood latch**
12. **Right hood latch**
13. **Center hood release cable**

## Hood, raising to service position

When working in the engine compartment, it is often useful to raise the engine hood higher than it is normally allowed to lift. For this purpose, BMW has provided additional ball-socket attachments for the hood support pistons.

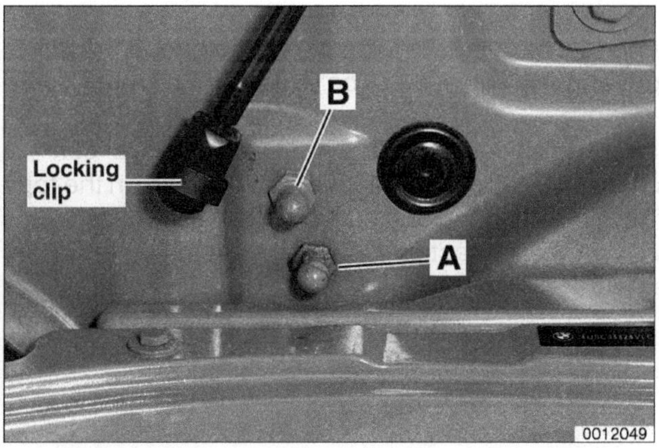


◄ Open hood. Remove locking clip from each support piston ball-socket.

– While an assistant supports hood, use a wooden stick to lever left and right hood support ball-sockets off pivot balls.

– Raise hood and press sockets on upper ball pivot.
  - Ball pivot **A** is normal position.
  - Ball pivot **B** is service position.

– When service work is complete, return supports to lower ball pivot.

## Hood, removing and installing

- Raise hood.
- Remove washer jet hoses from washer nozzles.
- Detach hood ground cable from hood.
- Detach gas-charged hood support pistons from hood.

> ***CAUTION—***
> *The hood is heavy. Before you remove the support pistons or retaining bolts at the hinges, be sure to have an assistant support the hood.*

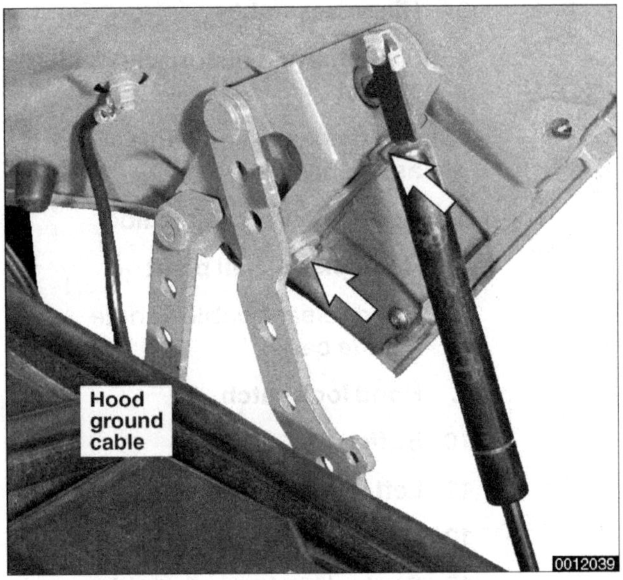


◄ Mark location of hinge mounting bolts (**arrows**) on hood.

- Remove hood hinge bolts and carefully remove hood.
- Installation is reverse of removal. Check hood alignment. See **Hood alignment** in this repair group.

> ***WARNING —***
> *Because of the hood's weight and size, do not rest it on top of the roof.*

## Hood alignment

When installing the hood, align the hinges as close to the original painted surface as possible. Movement of the hood on its attaching hardware may require touch up paint on exposed surfaces. Before shutting the hood, check the striker-to-lock alignment. Adjust the striker by loosening its attaching bolts and moving it, if necessary.

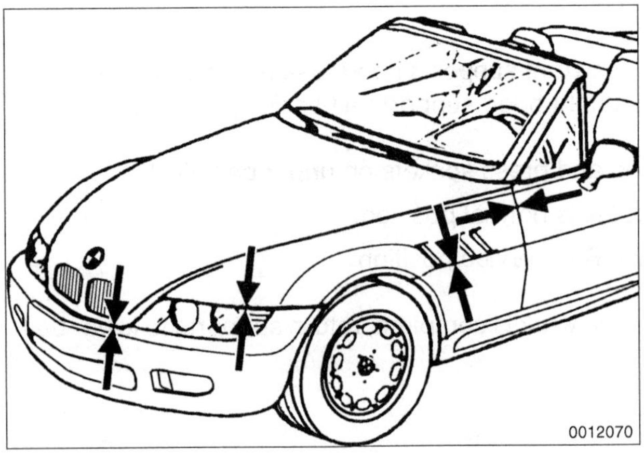


◄ Close the hood and check its alignment. The gap at the fenders, bumper and A-pillars should be even, about 5 mm (¼ in).

Check that the hood detent buffers are making proper contact. This is done by applying a light coat of grease on the top of the buffer. Shut the hood and then open it. The hood should have grease on the contact points. If not, turn the buffer in ¼-turn increments until contact is made.

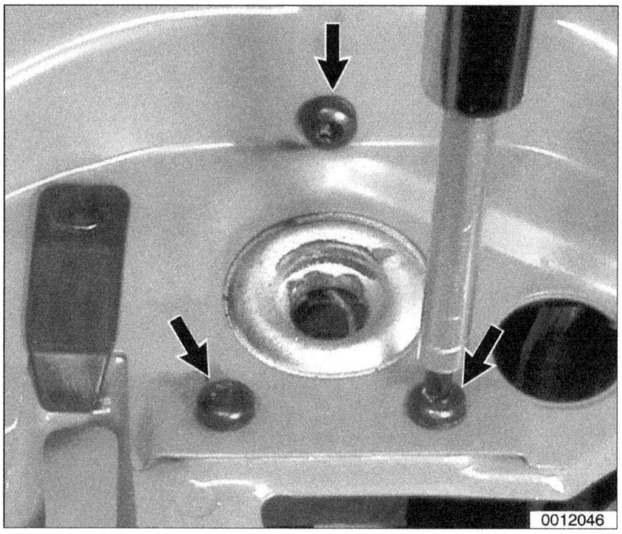


## Hood release cable, replacing

Use the engine hood assembly diagram earlier in this repair group as a guide when replacing the hood release cable.

– Raise hood.

< Remove three attaching screws (**arrows**) from right side hood catch. Remove catch from below.

– Remove attaching screws from left side hood catch. Remove catch from below.

– Disconnect center cable between two hood catches.

– Disconnect main cable from left side hood catch.

– From inside vehicle, remove hood release handle attaching screw. Remove left kick panel. Remove release cable clamping screw.

– Unplug rubber grommet where cable enters bulkhead.

– Disconnect cable from cable guides on left wheel well, and through inside of vehicle.

– Installation is reverse of removal.

– After assembly, check cable operation. Before shutting hood, be sure both lock catches function correctly.

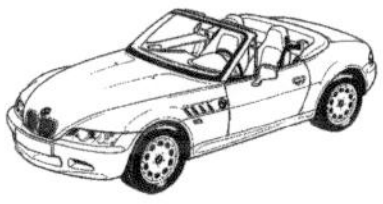

# 411 Doors

## GENERAL

This repair group covers door and interior door panel removal.

See also the following repair groups for related information:

- **512 Door Windows** for window repairs
- **515 Central Locking and Anti-theft** for door lock and handle repairs

> ***WARNING —***
> *Starting in 4 / 1998, Z3 models are equipped with side impact airbags installed in the doors. Read the warnings and cautions in* **721 Airbag System (SRS)** *prior to starting work on door.*

### Special tools

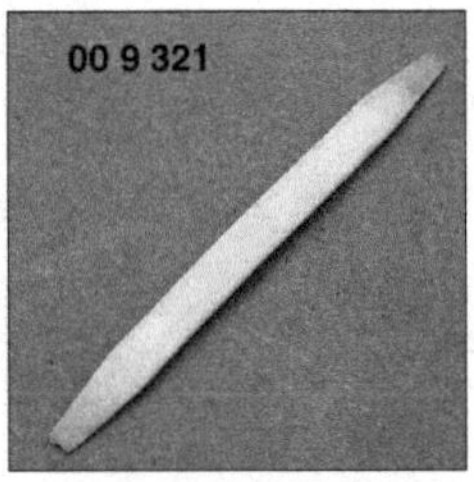


Plastic prying tool (BMW special tool 00 9 321)

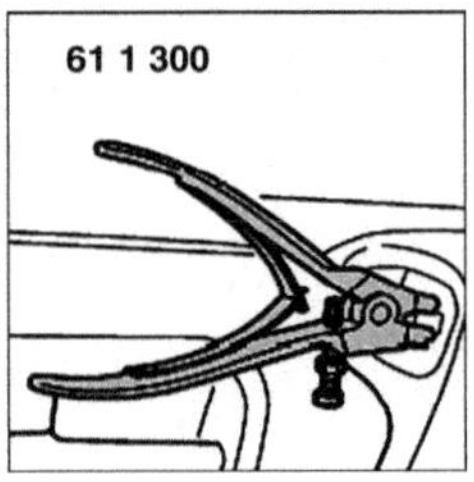


Mirror switch removal tool (BMW special tool 61 1 300)

## DOORS

### Door, removing and installing

- Make sure ignition is OFF. Remove key from ignition.
- Open door. Place floor jack under door for support. Raise jack just until door is supported. Use a rubber pad on jack to protect door finish.

> ***CAUTION—***
> *Do not raise the jack when the door is connected to its hinges. This could ruin the hinges and damage the unibody structure.*

Peel back rubber door check seal. Remove door check pin retaining clip (**A**). Tap pin out from below.

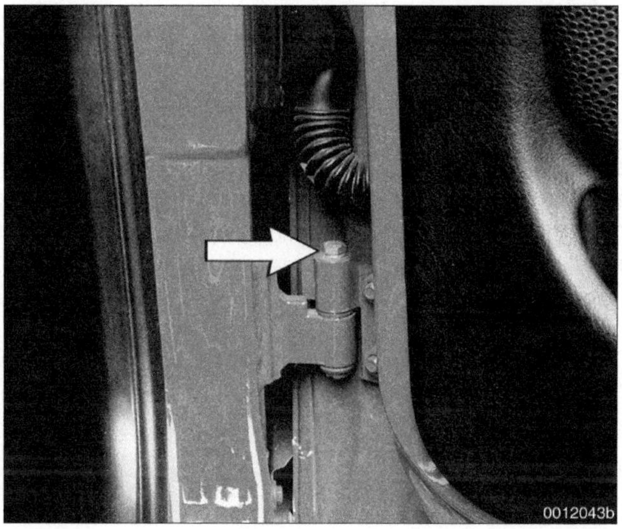


Loosen but do not remove threaded hinge pins (**arrow**) from upper and lower hinges.

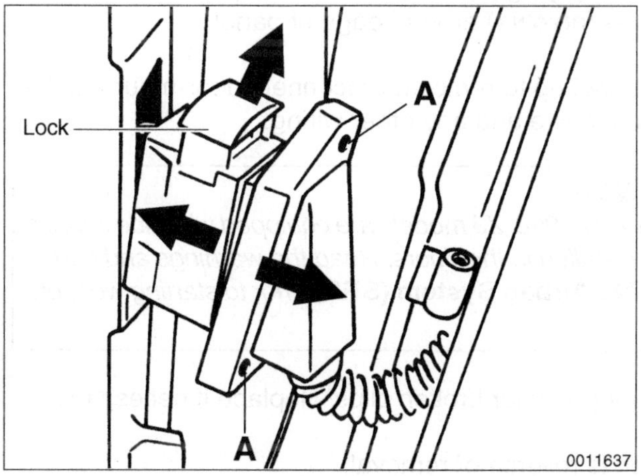


Disconnect wiring junction connector at door pillar by removing mounting screws (**A**). Separate connector by pulling up lock.

- Remove hinge pins while an assistant supports door. Remove door by lifting up off hinges.
- Installation is reverse of removal. No door alignment is required if hinge position was not disturbed.
- To replace door hinge:
  - Remove door.
  - Mark location of old hinge assembly on door and replace new one at the same spot.

***NOTE—***
*One half of the hinge is welded to the body. For replacement, consult a professional body shop.*

## Interior Door Panel

### Door panel, removing and installing

- Open door fully. Lower window.
- Make sure ignition is OFF. Remove key from ignition.

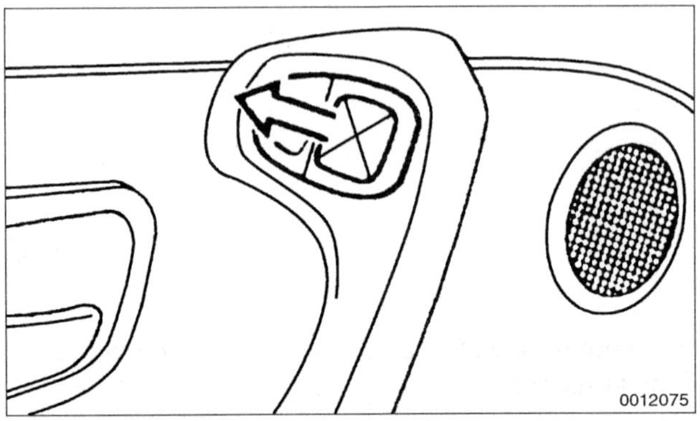

- Pry out power mirror switch or trim plug from top of door handle. If available, use BMW special tool 61 1 300.

- Remove panel mounting screw (**arrow**) from top of handle.
- Using a small flat-bladed screwdriver, carefully remove filler plate insert from inside door latch. Remove screw from behind filler plate. Remove door latch trim by sliding trim toward door handle.

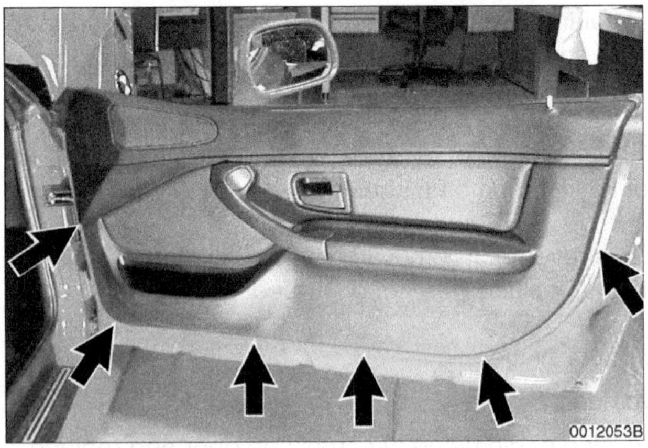

- Using plastic prying tool or door clip tool, pry door panel retaining clips (**arrows**) around edge of panel.
- Lift door panel up to remove. Disconnect door-mounted airbag, speaker wire and any other wiring.

> ***WARNING —***
> *Starting in 4 / 1998, Z3 models are equipped with side impact airbags installed in the doors. Read the warnings and cautions in* **721 Airbag System (SRS)** *prior to starting work on door.*

- Check door panel for broken clips. Replace if necessary.
- Installation is reverse of removal.

# 412 Trunk Lid, Rear Hatch

## GENERAL

This repair group covers removal and installation of trunk lid (Roadster) and rear hatch (Coupe).

> ***CAUTION—***
> - *In order to avoid damage to paintwork when installing a new trunk lid or rear hatch, make sure there is a sufficient gap between lid and body. One way to do this is to attach masking tape or fabric tape to points of contact on body.*
> - *The body is painted at the factory after assembly. Realignment of body panels may expose unpainted metal. To prevent corrosion, paint exposed metal.*

### Special tools

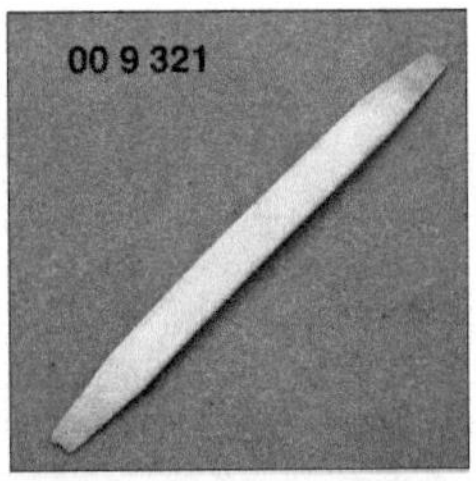


< Plastic prying tool (BMW special tool 00 9 321)

## TRUNK LID

### Trunk lid, removing and installing

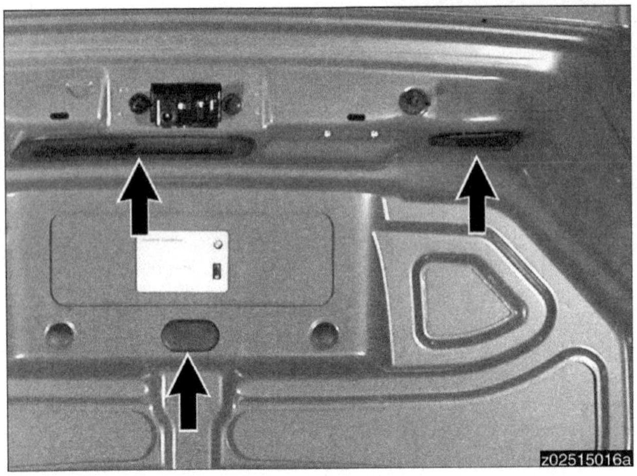


< Open trunk lid. Remove access covers (**arrows**) from inside lid.

– Disconnect wire leads from center brake light, trunk lock and trunk lock actuator. Remove wiring harness by snaking it through opening in upper left corner of trunk lid.

> ***NOTE—***
> *The trunk lid wiring harness does not have a single connector to disconnect. Each wire must be disconnected separately and then pulled out of the trunk lid opening.*

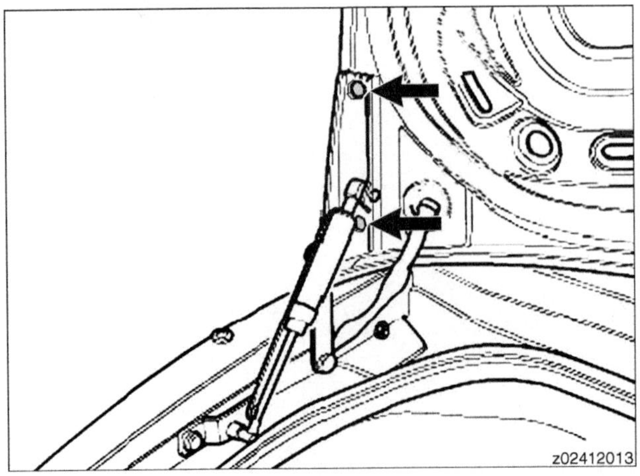


< With an assistant supporting trunk lid, remove attaching bolts (**A**) from each hinge.

– Remove trunk lid.

– Installation is reverse of removal. Make sure wiring loom does not chafe against trunk lid hinge.

– To align lid, see **Trunk lid, aligning** in this repair group.

## Trunk lid, aligning

When installing trunk lid, align hinges as close to original painted surface as possible.

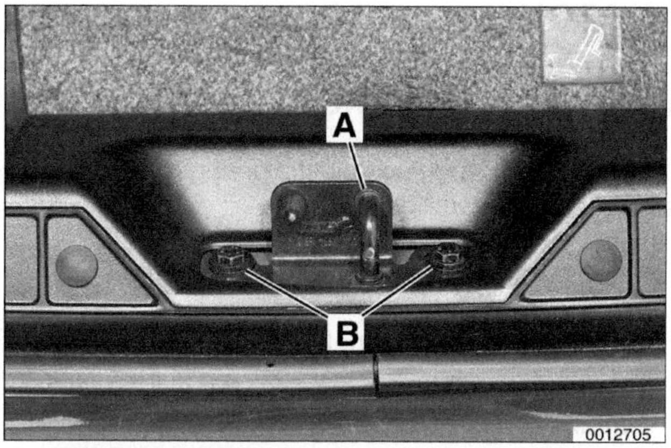


◄ Before shutting trunk lid, check striker-to-lock alignment. Adjust striker (**A**) by loosening attaching bolts (**B**) and moving striker.

◄ Close trunk lid. Check alignment and gap at front, sides and rear. The gap (**arrows**) should be even at all points, approximately 5 mm (¼ in).

◄ Check that detent buffers (**arrows**) make proper contact by applying a light coat of grease on top of buffer. Shut trunk and then open it. The panel should have grease on contact point. If adjustment is necessary, turn buffer in ¼ turn increments until contact is made.

## REAR HATCH

### Rear hatch, removing and installing

- Working at top of rear hatch, use plastic prying tool to gently pry center brake light assembly out.
- Detach center brake light electrical harness connector.
- Remove rear hatch inner trim panel:
  - Gently pry off plastic trim caps over screws behind latch mechanism and remove screws.
  - Unclip panel from hatch.
- Disconnect wire leads from rear wiper, hatch lock and hatch lock actuator. Remove wiring harness by snaking it through opening in upper left corner of hatch.

*The hatch lid wiring harness does not have a single connector to disconnect. Each wire must be disconnected separately and then pulled out of the lid opening.*

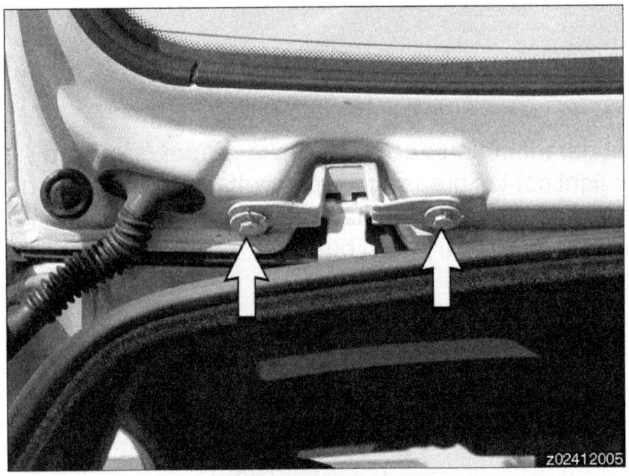


- Gently pry off plastic rear hatch hinge covers. Loosen hinge mounting bolts (**arrows**) on both sides.

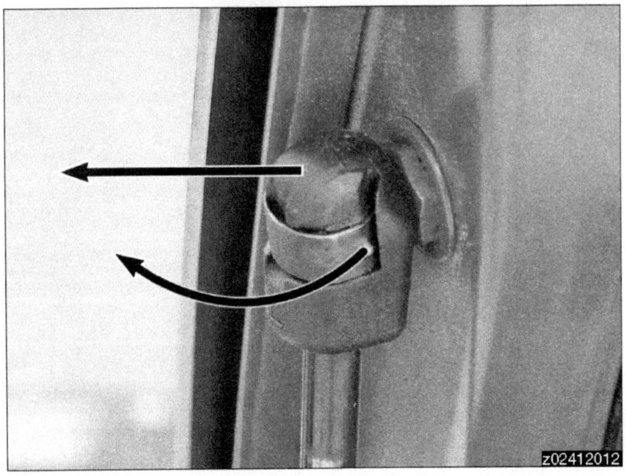


- With an assistant supporting hatch, unclip locking clip at upper end of hatch lid support strut and release strut from rear hatch.

> ***WARNING —***
> *Without hatch lid supports in place, the heavy lid can drop down and cause injury.*

- Remove hinge mounting bolts and lift off rear hatch.
- Installation is reverse of removal. Make sure wiring loom does not chafe.
- To align lid, see **Rear hatch alignment** in this repair group.

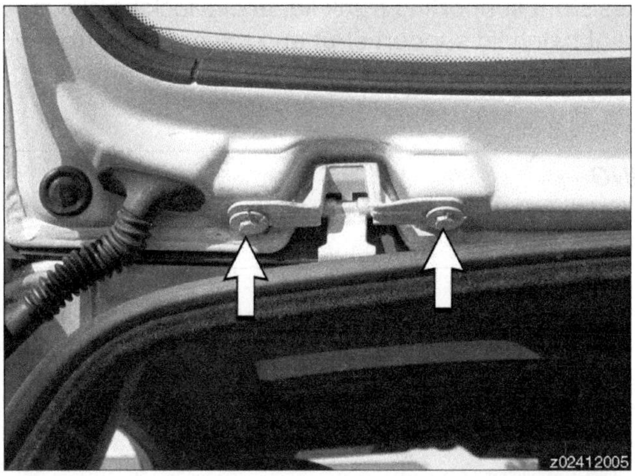


## Rear hatch alignment

⊲ Open rear hatch. Gently pry off plastic rear hatch hinge covers. Loosen hinge mounting bolts (**arrows**) on both sides.

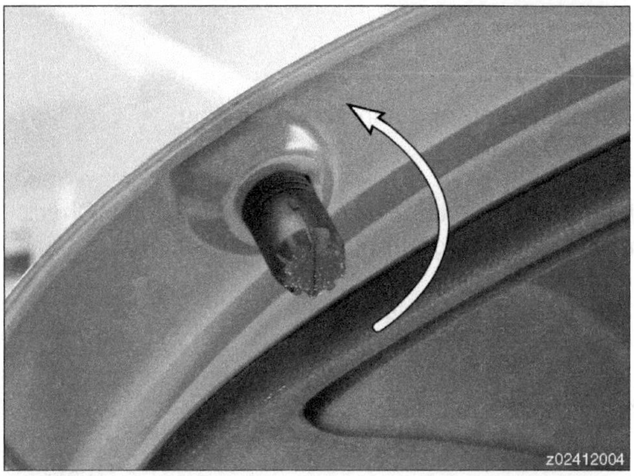


⊲ Unscrew and remove detent buffers.

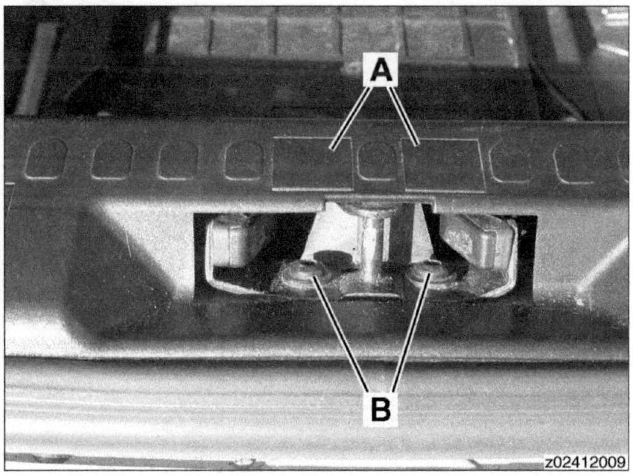


⊲ Working at rear hatch sill:

- Gently pry off plastic trim covers (**A**).
- Loosen rear hatch striker mounting bolts (**B**) so that striker can slide with resistance.

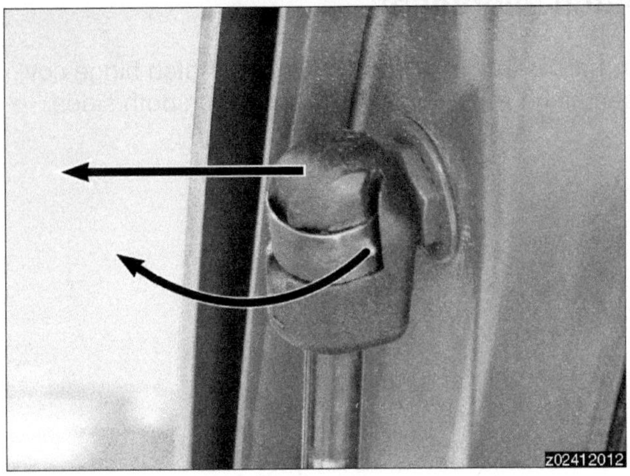


With an assistant supporting trunk lid, unclip locking clip at upper end of hatch lid support strut and release strut from rear hatch. Repeat step at lower end of strut and remove from car.

> ***WARNING —***
> *Without hatch lid supports in place, the heavy lid can drop down and cause injury.*

– Carefully lower hatch while holding lock release button in.

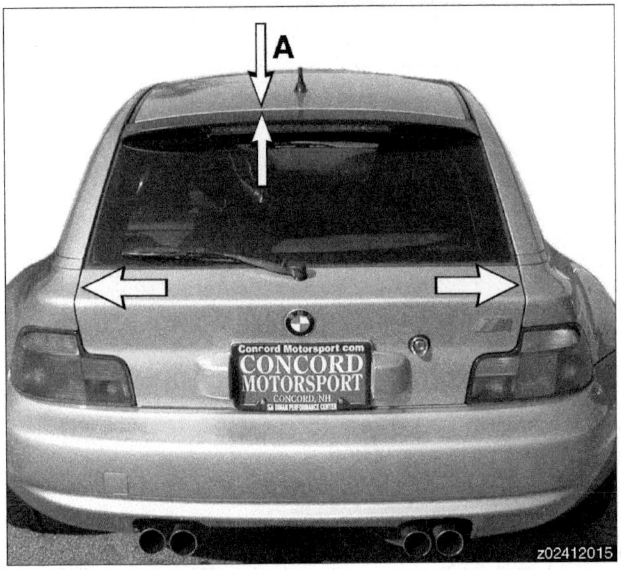


Correct hatch alignment:

- Even gaps (**arrows**) on sides.
- Maximum protrusion of hatch above roofline (**A**) = 1 mm (0.04 in).

– Open hatch and tighten latch striker and hinge bolts. Check alignment and repeat alignment steps if necessary.

– Reinstall hatch support struts.

– Reinstall detent buffers. Close hatch and make sure both sides of hatch rest on buffers.

# 510 Exterior Trim, Bumpers

## General

This repair group covers bumper removal and installation, as well as replacement of some exterior trim.

## Bumpers

The aluminum-framed front and rear bumpers are mounted on energy absorbing pistons and covered with a plastic cover. Removal of the cover requires removing the bumper from the vehicle. The removal and installation can be done with a T50 Torx socket and ordinary hand tools. If the bumpers are removed, follow the directions to adjust the bumpers to the proper position after installation.

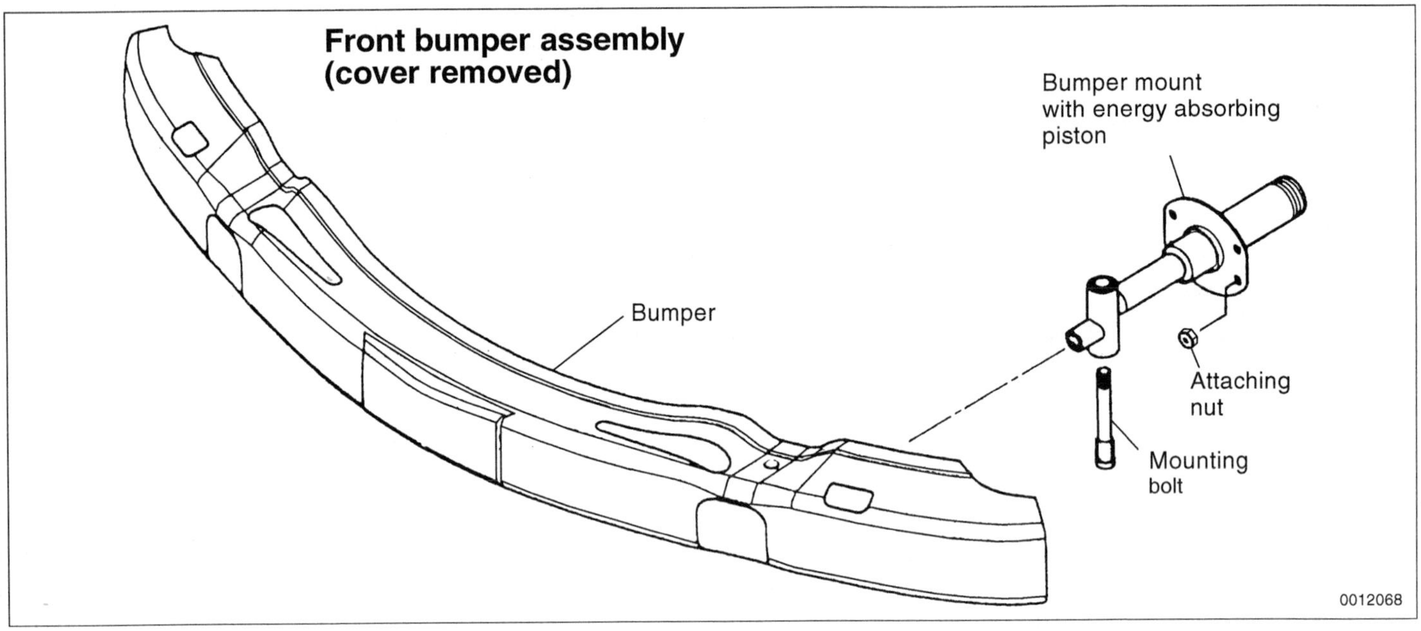


### Front bumper, removing and installing

The front bumper cover is held to the bumper by 16 expansions rivets. During removal of the cover, the rivets may become damaged. If removing the bumper cover, have new rivets on hand before beginning the job.

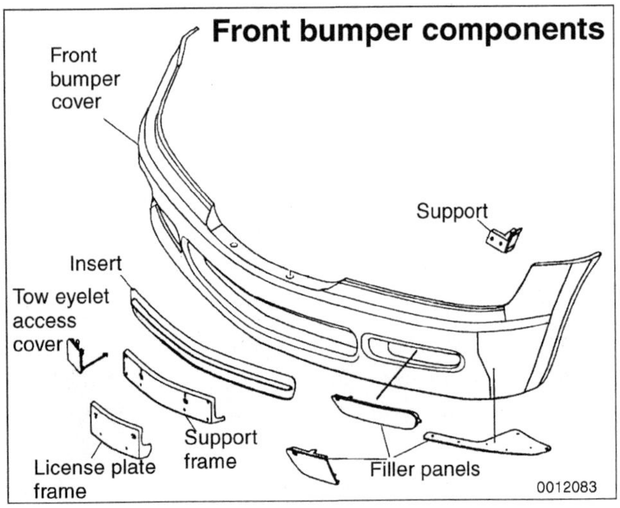


- ◄ Remove license plate frame.
- Raise and properly support vehicle.

◄ Remove engine splash shield mounting screws (**arrows**) under front bumpers and remove engine splash shield.

***NOTE—***
*4-cylinder splash shield shown. 6-cylinder models are similar.*

– Models with foglights: Disconnect foglight wiring connectors.

– Models with headlight washer: Disconnect headlight washer electrical connections and hoses.

– M version: Remove engine air intake hose from bumper.

– Using Torx 50 socket, remove bumper mounting bolts from left and right bumper mounts.

| **Tightening torque** | |
|---|---|
| Bumper to bumper mount | 42 Nm (31 ft-lb) |

– Pull bumper forward to remove from mounts.

◄ To remove bumper cover from bumper, remove expansion rivets (**arrows**) at top and bottom of bumper.

– Front bumper installation is reverse of removal. Check bumper for proper alignment at seams. There should be an even gap of approximately 5 mm (0.25 in).

***NOTE—***
*If the bumper seam gap is not within specification, the bumper can be adjusted by small amounts at the mounts, although a special tool is needed to make the adjustment. See* **Bumper height, adjusting**.

## Rear bumper, removing and installing

The rear bumper cover is held to the bumper by 16 expansions rivets. During removal of the cover, the rivets may become damaged. If removing the bumper cover, have new rivets on hand before beginning the job.

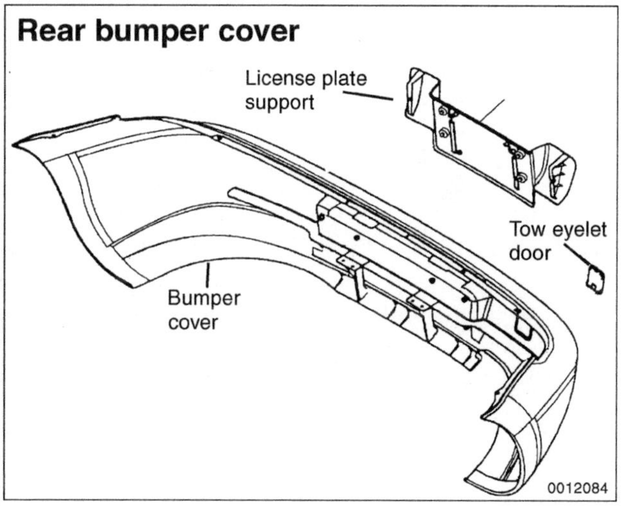


◄ Remove license plate frame screws. Disconnect license plate lamps from bumper.

– Raise and properly support rear of vehicle.

– Remove rear wheels.

– Remove spare wheel (if equipped).

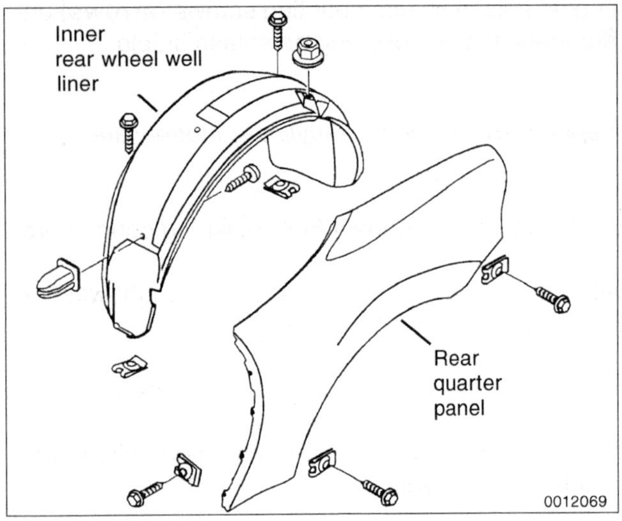


◄ Remove plastic liners from left and right wheel wells.

– 4-cylinder models: Disconnect bumper from rear wheel arch by removing expansion rivet.

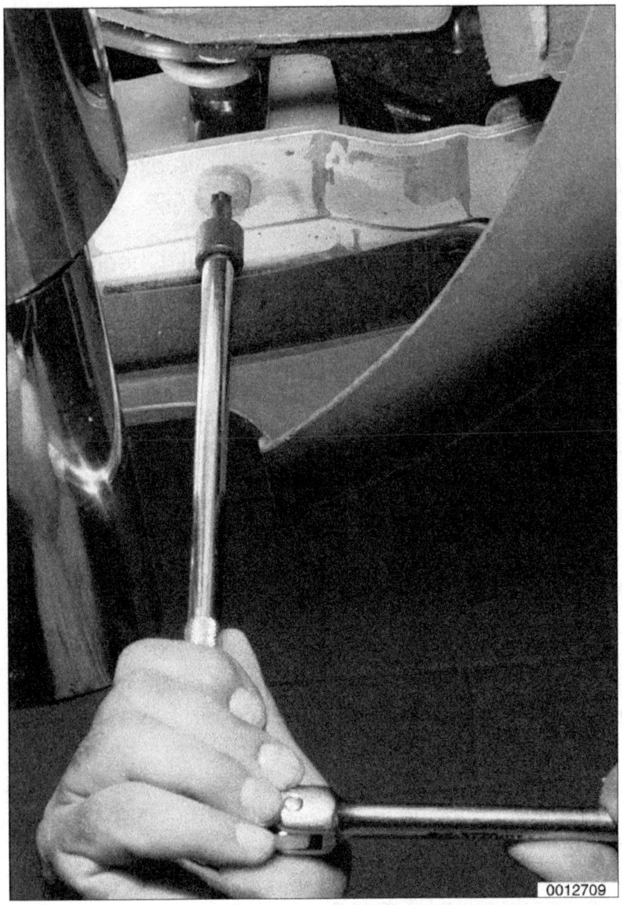


◄ Support bumper and remove two bumper bolts from underside of bumper using T50 socket.

– 6-cylinder models: Push bumper down and towards the rear to release bumper locating clips.

While an assistant pulls on bumper, press down on bumper cover and carefully slide bumper out from behind taillight assembly.

- To remove bumper cover from bumper, remove expansion rivets at top and bottom of bumper.

Rear bumper installation is reverse of removal. Check that corner of bumper cover is properly aligned with rear quarter panel. Check for proper gap at seams. There should be an even gap at all seams (**arrows**).

| Tightening torque | |
|---|---|
| Bumper to bumper mount | 42 Nm (31 ft-lb) |

***NOTE—***
*If unable to achieve an even seam gap, bumper can be adjusted by small amounts at the mounts, although a special tool is needed to make the adjustment. See* **Bumper height, adjusting**.

## Bumper height, adjusting

The adjustment procedure given below applies to both front and rear bumpers. A special tool is needed to adjust the bumper height, although bumper removal is not required to make the adjustment.

- Raise and properly support vehicle. Remove Torx 50 bumper mounting bolts.
- Attach BMW special tool no. 51 1 130 to a 12-inch extension.
- Feed socket up through bumper mount bore.

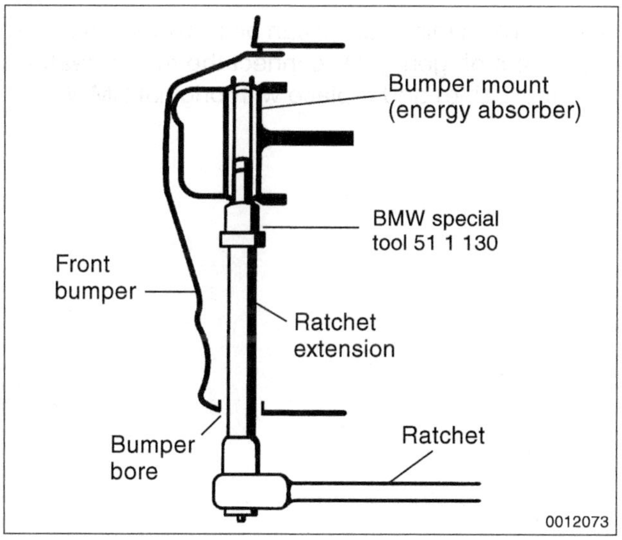


Turn tool clockwise or counterclockwise as needed to change height of adjusting collar.

- Reinstall mounting bolts and check bumper height.

| Tightening torque | |
|---|---|
| Bumper to bumper mount | 42 Nm (31 ft-lb) |

## Outside Rear View Mirror

The remote controlled rear view mirror assemblies can be replaced with simple hand tools. If just the mirror glass is broken, it can be replaced without removing the mirror assembly from the door.

### Outside mirror, removing and installing

- Make sure ignition switch is OFF. Remove key from ignition switch.
- Remove inside door panel. See **411 Doors**.

> ***WARNING —***
> *Starting in 4 / 1998, Z3 models are equipped with side impact airbags installed in the doors. Read the warnings and cautions in* **721 Airbag System (SRS)** *prior to starting work on door.*

- Disconnect cable or electrical connectors from mirror.

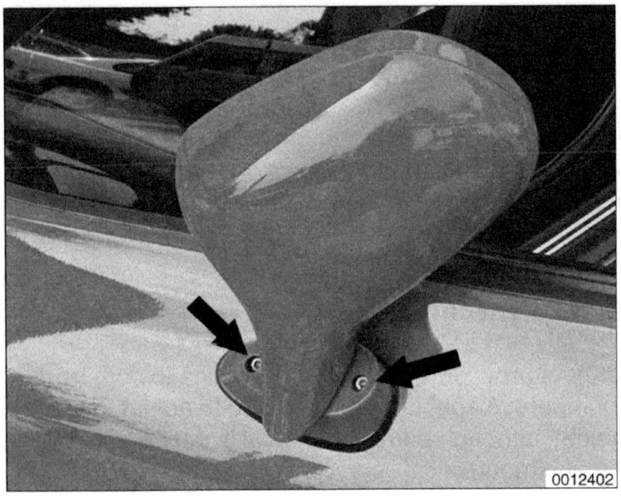
0012402

- ⮜ From outside vehicle, rotate mirror forward 90° to expose mounting screws (**arrows**). Remove screws.
- Carefully remove mirror assembly while feeding cable through opening in door.
- Installation is reverse of removal.

### Mirror glass, replacing

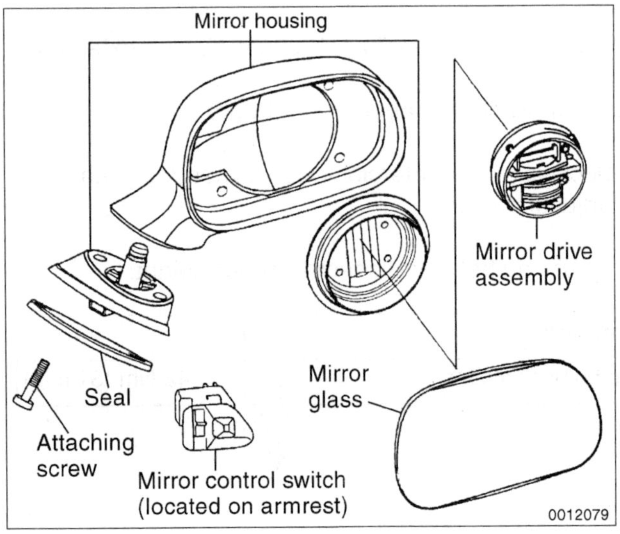

0012079

⮜ A discolored or broken mirror glass can be removed from the housing by prying it off gently. Disconnect the mirror heater wires (where applicable) and replace with original BMW parts.

# EXTERIOR TRIM

Exterior trim is fastened to the body with plastic clips and fasteners that may be damaged during removal. Emblems are glued to the body and cannot be reused once removed. Be sure to have necessary fasteners and emblems on hand for reinstallation.

## Emblems, removing and replacing

To aid removal, replace emblems at room temperature.

- Carefully work a nylon line coated with dish detergent behind emblem to remove.
- Remove excess adhesive using a rag dipped in benzene.

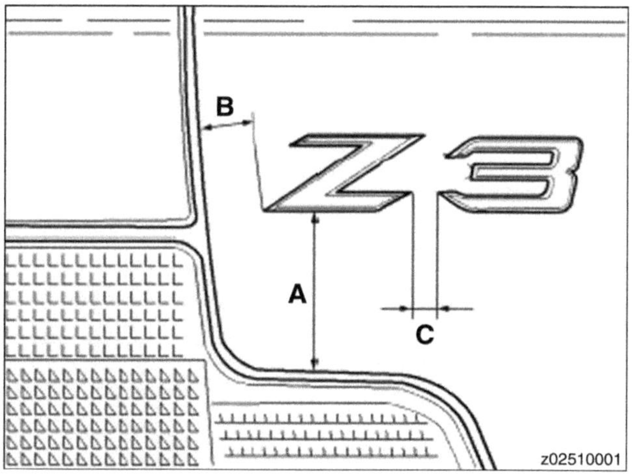


◀ Z3 emblem: apply masking tape to upper edge of dimension **A** and mark out remaining dimensions.

A = 65 mm (2.6 in)

B = 23 mm (0.9 in)

C = 10 mm (0.4 in)

- Peel back adhesive strip and firmly affix emblem.

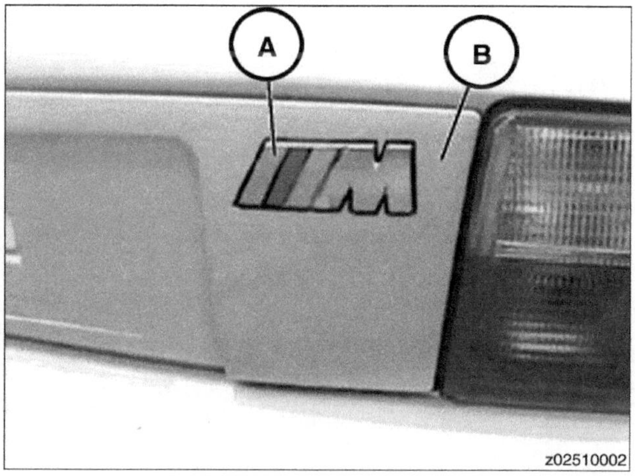


◀ M Roadster / Coupe emblem: Center emblem (**A**) in recess on rear panel (**B**).

- Peel back adhesive strip and firmly affix emblem.

0012047

## Grille inserts, removing and installing

< Raise hood. From inside hood depress locking tabs (arrows) of grille and push grille out from behind.

– Installation is reverse of removal.

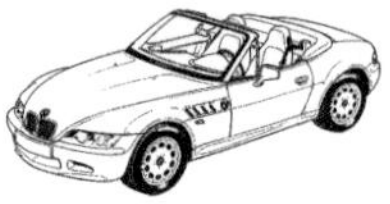

# 512 Door Windows

## General

This repair group covers replacement of door glass, window regulator and power window motor. For removal and installation of the doors and door interior panels, see **411 Doors**.

The stationary window glass mounted to the door and the door glass mounted to the window regulator are relatively easy to replace.

Replacing the windshield and fixed rear windows (coupe) requires special glue and tools. Replacement should be done by a professional glass installer.

> ***WARNING —***
> - *Always wear hand and eye protection when working with broken glass.*
> - *If the window is broken, vacuum all of the glass bits out of the door cavity. Use a blunt screwdriver to clean out any remaining glass pieces from the window guide rails.*

## Door windows

Coupe door windows are equipped with the BMW drop down feature:

- When the door is opened, the window is automatically lowered slightly to clear the door seal.
- When the door is closed, the window is raised fully up into the seal.

> ***CAUTION—***
> *The Coupe window shows approximately 15 mm (5/8 in) of glass when fully lowered. Do not attempt to adjust.*

### Door window, removing and installing

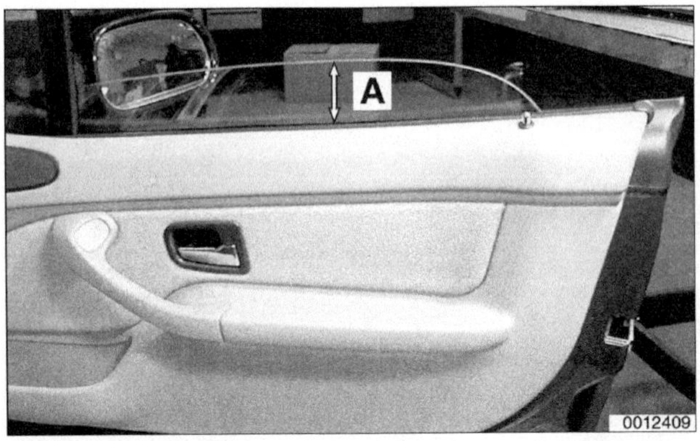


◄ Raise window (**A**) about 4 - 8 inches (100 - 200 mm) above fully lowered position.

– Turn ignition switch OFF and remove key from ignition switch.

– Remove inside door panel and interior vapor barrier. See **411 Doors**.

> ***WARNING —***
> *Starting in 4 / 1998, Z3 models are equipped with side impact airbags installed in the doors. Read the warnings and cautions in* **721 Airbag System (SRS)** *prior to starting work on door.*

– Remove fixed glass assembly from door. See **Fixed door glass, removing and installing** in this repair group.

– Working in large opening in door, slide window forward and out of horizontal window regulator channel at front.

– Pivot window up at front and out of door while at the same time sliding glass out of horizontal window regulator channel at rear. Remove glass from door cavity.

– Door window installation is the reverse of removal. To adjust window height, see **Door window, adjusting** in this repair group.

### Fixed door glass, removing and installing

– Remove inside door panel and interior vapor barrier. See **411 Doors**.

> ***WARNING —***
> *Starting in 4 / 1998, Z3 models are equipped with side impact airbags installed in the doors. Read the warnings and cautions in* **721 Airbag System (SRS)** *prior to starting work on door.*

– Working under front of door, remove bolt from front window channel.

– Remove two screws mounting bottom frame of fixed glass to door.

– Remove fixed window by lifting straight up.

– Installation is reverse of removal. Once installation is complete, check adjustment of window. See **Door window, adjusting** in this repair group.

## WINDOW REGULATOR SERVICE

### Window regulator, removing and installing

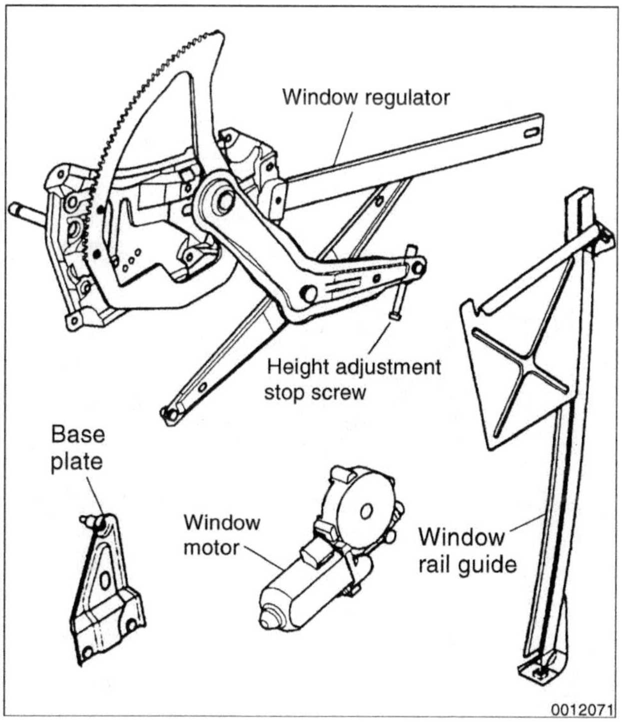


◄ Window regulator components.

– Remove inside door panel and interior vapor barrier. See **411 Doors**.

> ***WARNING —***
> *Starting in 4 / 1998, Z3 models are equipped with side impact airbags installed in the doors. Read the warnings and cautions in* **721 Airbag System (SRS)** *prior to starting work on door.*

– Remove door glass and fixed window glass. See **Door window, removing and installing** and **Fixed door glass, removing and installing** in this repair group.

- Disconnect wiring connector to power window motor.
- Remove fasteners mounting regulator to door.
- Remove two bolts mounting regulator to window track.
- Remove regulator assembly from door cavity. If necessary, operate motor to reposition regulator sector for additional clearance.
- Installation is reverse of removal.

> ***WARNING —***
> *Do not operate the window up or down with your hand anywhere in the door cavity.*

## Door window, adjusting

- Fully raise window. Close and latch convertible top. Close door.
- Remove inside door panel. See **411 Doors**.

> ***WARNING —***
> *Starting in 4 / 1998, Z3 models are equipped with side impact airbags installed in the doors. Read the warnings and cautions in* **721 Airbag System (SRS)** *prior to starting work on door.*

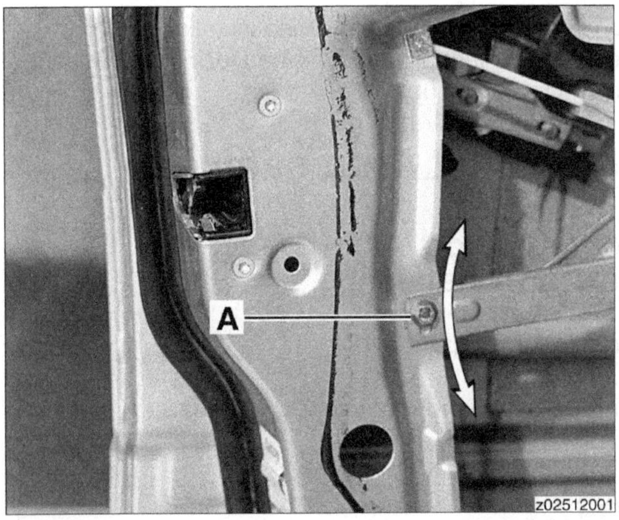


Check that window is parallel to its frame.

- If adjustment is necessary, loosen regulator rail mounting bolt (**A**) and reposition rail. Tighten bolt when adjustment is correct.

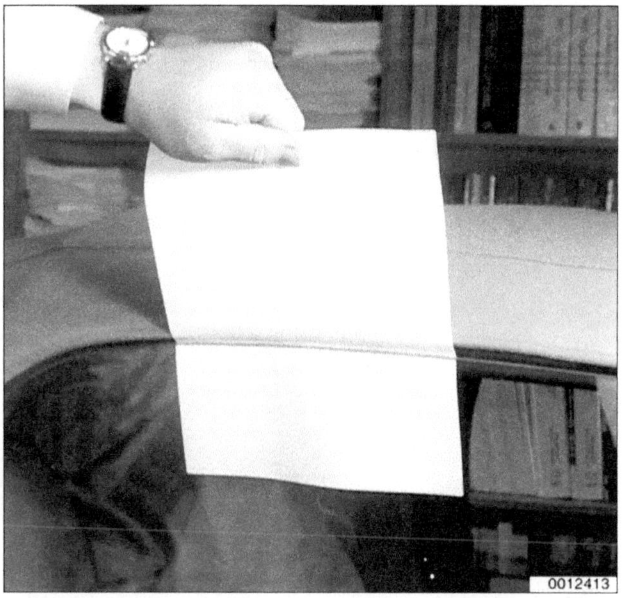


Check that window seats uniformly at all points on window seal/weatherstrip and that fixed glass is flush to A-pillar at windshield frame. Close window onto a piece of paper between window and weatherstrip. It should not be possible to pull paper out easily.

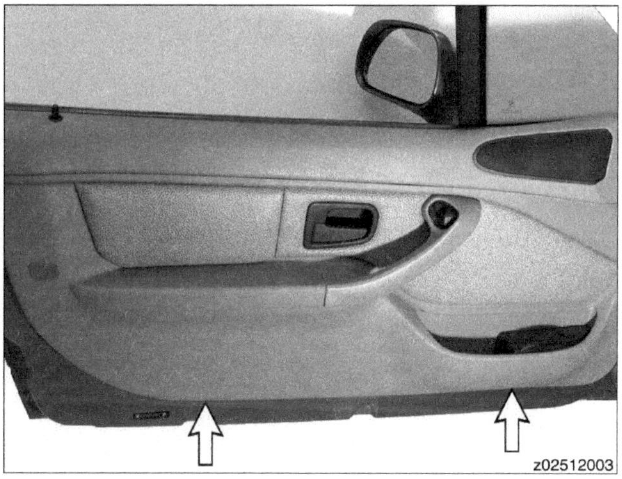


If adjustment is necessary, loosen bolts (**arrows**) at front and rear window channels and reposition channels as necessary. Tighten bolts when adjustment is correct.

***NOTE —***
*To check window-to-seal fit, spray window with a garden hose and inspect for leaks.*

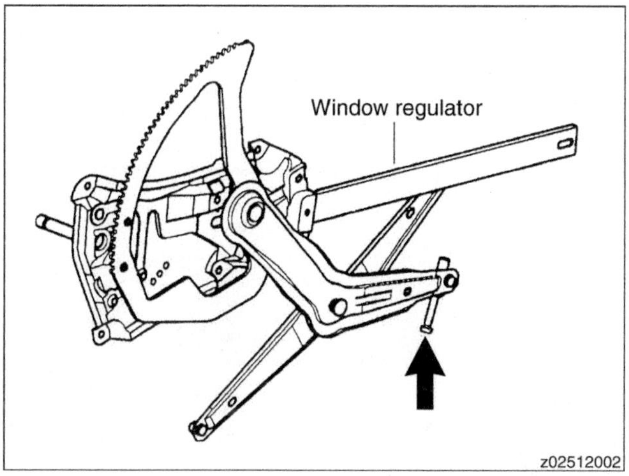


Check closed height of window to fixed glass. Window must transition smoothly to fixed glass.

- If height adjustment is necessary, turn height adjustment stop screw (**arrow**) on regulator.

# 513 Interior Trim



513

## GENERAL

This repair group covers interior trim removal and installation.

Additional information is in the following repair groups:

- **250 Gearshift Linkage** for shifter knob removal
- **411 Doors** for door trim panel removal
- **612 Switches**
- **620 Instruments** for instrument panel and Board Computer removal
- **640 Heating and Air-conditioning** for climate control panel removal
- **650 Radio**
- **721 Airbag System (SRS)**

### Special tool

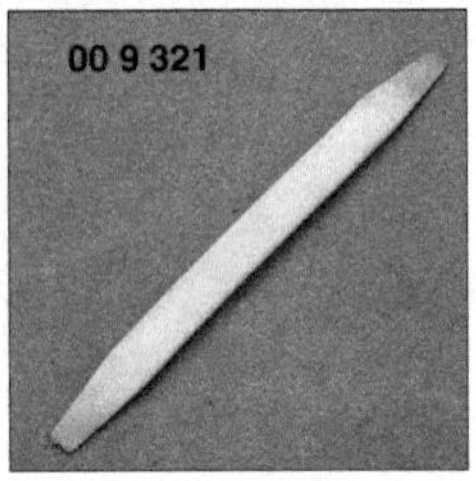


◄ Plastic prying tool
(BMW special tool 00 9 321)

### Interior trim repairs

Most of the interior trim and finish panels are clipped or screwed into place. Many of the trim clips are designed to be used only once. When removing trim that is held in place with clips, have spares on hand before beginning the job.

### Warnings and cautions

Z3 cars are equipped with airbags in the steering wheel and the right side dashboard. Be sure to read **Warnings and cautions** in this repair group and also in **721 Airbag System (SRS)**.

***WARNING—***
- *Serious injury may result if repairs are attempted by persons unfamiliar with the BMW SRS.*
- *BMW recommends that repairs involving SRS components be carried out by a qualified BMW technician.*
- *Disconnect the battery and cover the negative battery terminal with an insulator before starting service work on electrical components.*
- *After disconnecting the battery, wait 10 seconds before beginning work on pyrotechnical components.*

***CAUTION—***
- *Prior to disconnecting battery, read battery disconnection cautions given in* **001 General Warnings and Cautions**.
- *To prevent marring trim on interior components, work with a plastic prying tool or wrap a screwdriver tip with tape.*

# CENTER CONSOLE

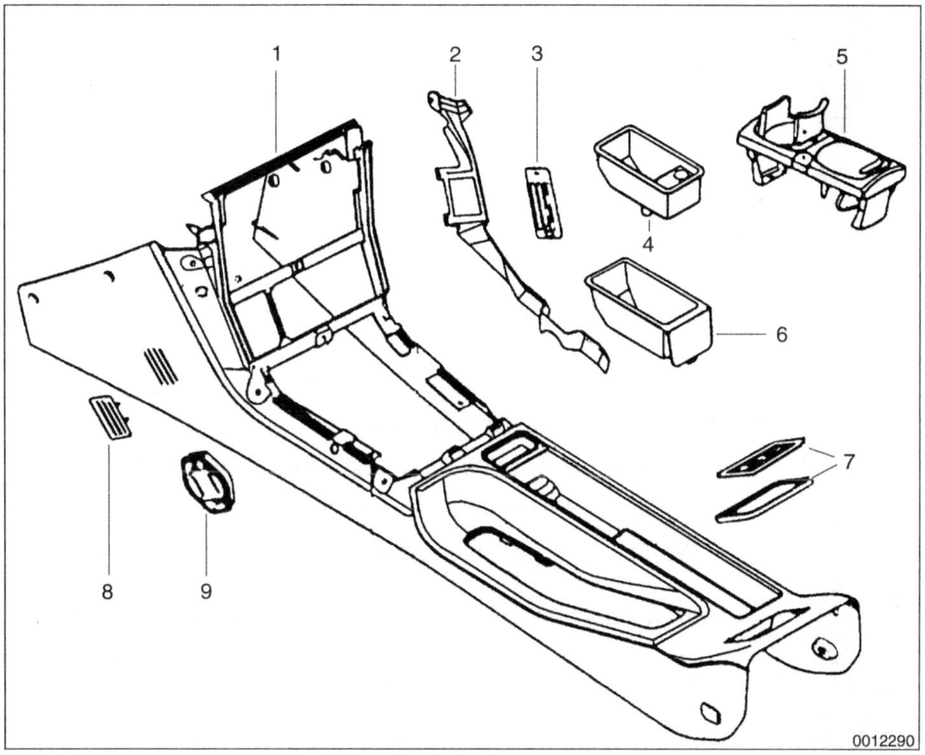


## Center console components

1. **Console**
2. **Reinforcement**
3. **Cover**
4. **Ashtray**
5. **Cup holder**
6. **Ashtray box**
7. **Convertible top switch and bezel**
8. **Cover**
9. **Frame**

## Center console, removing and installing

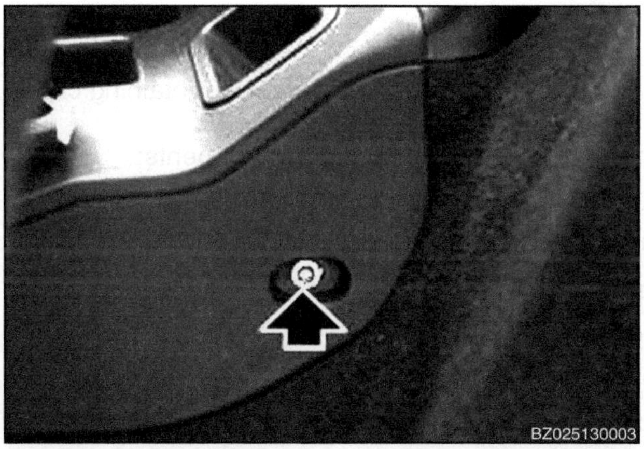


< Position seats as far forward as possible and fold backrests forward.

- Working behind seats, remove center console rear retaining screws (**arrow**) from left and right sides.
- Move seats back.

– Disconnect negative (–) battery cable and cover terminal with insulating material.

> ***CAUTION—***
> *Prior to disconnecting the battery, read the battery cautions in* **001 General Warnings and Cautions**.

– Remove radio. See **650 Radio**.

> ***CAUTION—***
> *Obtain anti-theft radio code before disconnecting radio.*

– Remove rubber boot around parking brake lever by prying it up and out. Pull small trim piece in front of and below parking brake lever backwards about 5 mm (¼ in), then lift to remove.

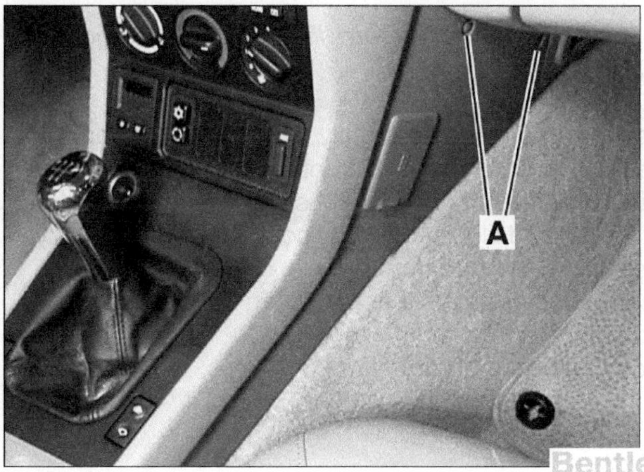


< Working in left and right footwell, lever off plastic screw covers (**A**) and remove center console front retaining screws.

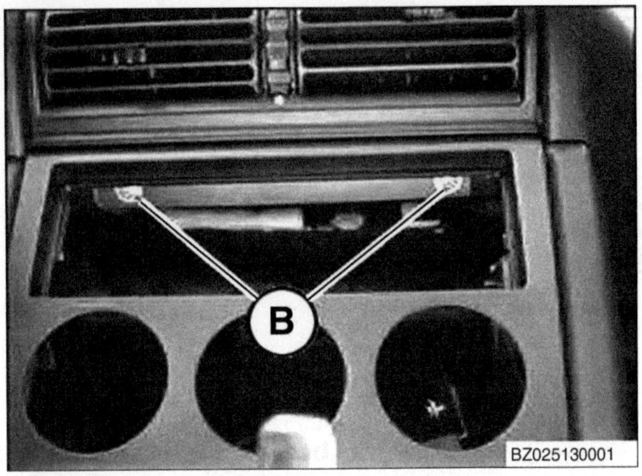


**<** Remove two screws (**B**) holding center console to upper dashboard.

**<** Working at center dashboard:

- Remove cover plate mounting screws (**outer arrows**).
- Unclip heater controls and slide inwards.
- Working behind cover plate, unplug switches and gauges.

– Remove shifter handle. See **250 Gearshift Linkage**.

– Using a plastic prying tool, remove shifter bezel and boot.

– Remove ash tray. Under ash tray, remove ash tray housing mounting screw.

– Remove storage box.

– Remove cup holder assembly by depressing retaining clips.

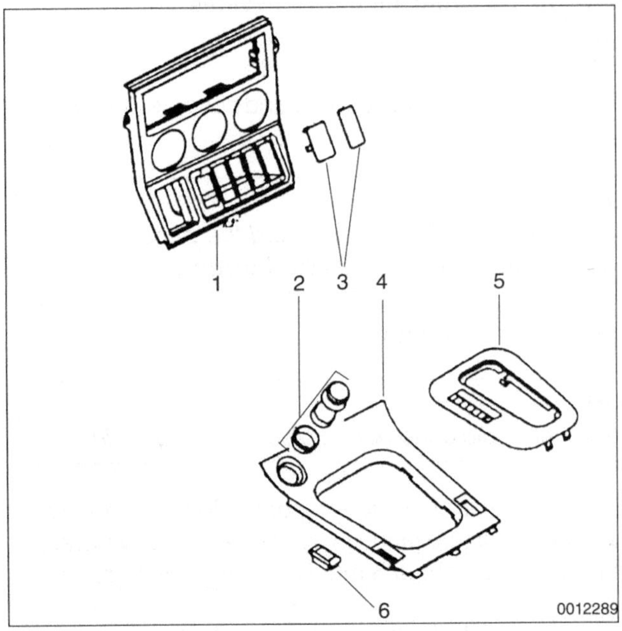


**<** Front console and center dashboard components:

1. Center dashboard cover plate
2. Lighter assembly
3. Cover bezels
4. Shifter panel
5. Automatic shifter bezel
6. Window switch

***NOTE—***
*1996 Roadster center dashboard illustrated. Other models are similar.*

– Raise rear of center console slightly, while pulling toward rear of vehicle. Pull console back until front locking clips disengage. While lifting console out, disconnect electrical connector for cigarette lighter and clock / onboard computer, if necessary. Carefully remove console from vehicle, checking for wiring, cables or mechanical parts that may snag during removal.

– Installation is reverse of removal.

## Dashboard

Dashboard removal and installation is a complicated job. The procedure involves the following:

- Center console removal. See **Center console, removing and installing** in this repair group.
- Left and right under-dash trim removal. See **Under-dash trim, left, removing and installing** and **Glove compartment, removing and installing** in this repair group.
- Driver and passenger airbag removal. See **721 Airbag System (SRS)**.

Read the complete procedures through before starting the job. Be sure to read **Warnings and cautions** in this repair group and also in **721 Airbag System (SRS)**.

### Under-dash trim, left, removing and installing

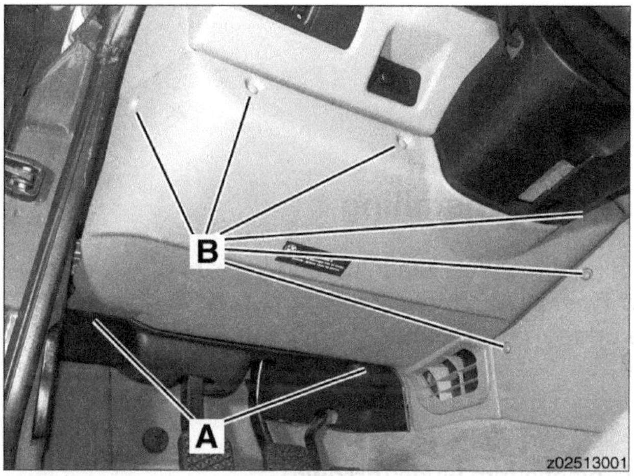

z02513001

◄ Working in left footwell:

- Twist plastic trim clips (**A**) ¼ turn each to disengage pedal cluster trim. Pull trim down and back to remove.
- Pry off plastic trim plugs (**B**) and remove trim mounting screws under plugs. Remove trim, disengaging any locking clips holding trim in place.

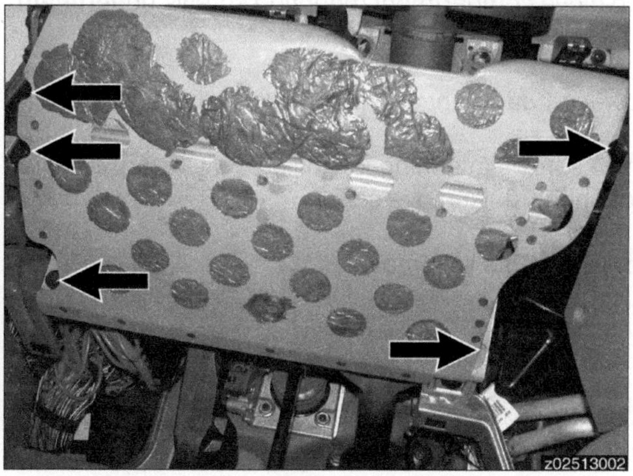
z02513002

◄ Remove knee bolster mounting bolts (**arrows**). Lift knee bolster out of vehicle.

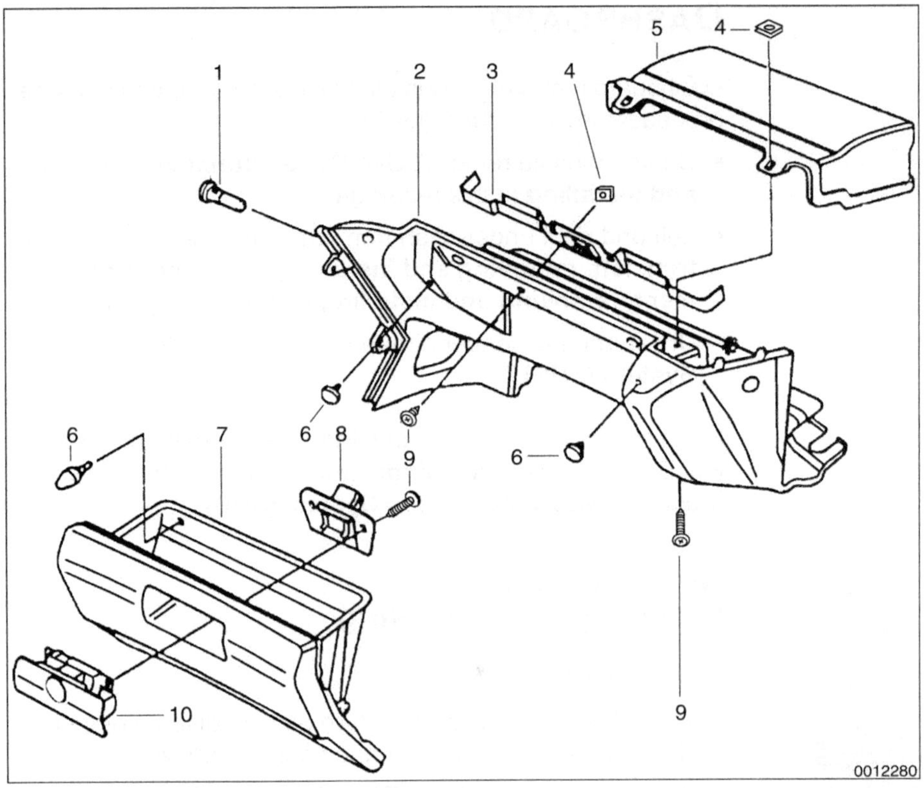


## Glove compartment components

1. **Pin**
2. **Frame**
3. **Lock catch**
4. **Nut**
5. **Top inside cover**
6. **Bump stop**
7. **Glove compartment**
8. **Lock brace**
9. **Screw**
10. **Lock**

## Glove compartment, removing and installing

– Working in right footwell, reach up and forward of glove compartment. Twist plastic trim clips ¼ turn each to disengage right side under-dash trim panel. Pull panel down and back to remove.

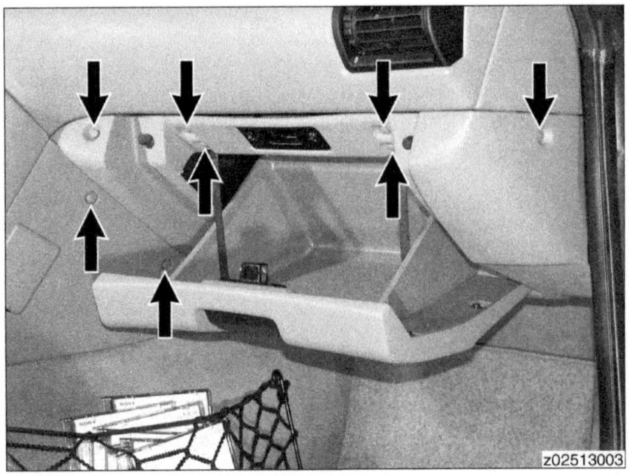


◄ Working at glove compartment:

- Remove plastic trim plugs (**arrows**).
- Remove glove compartment mounting screws underneath plugs.

– Close glove compartment and pull down and outward.

– Remove two screws attaching top inside glove compartment cover to dashboard. Remove cover.

– Installation is reverse of removal.

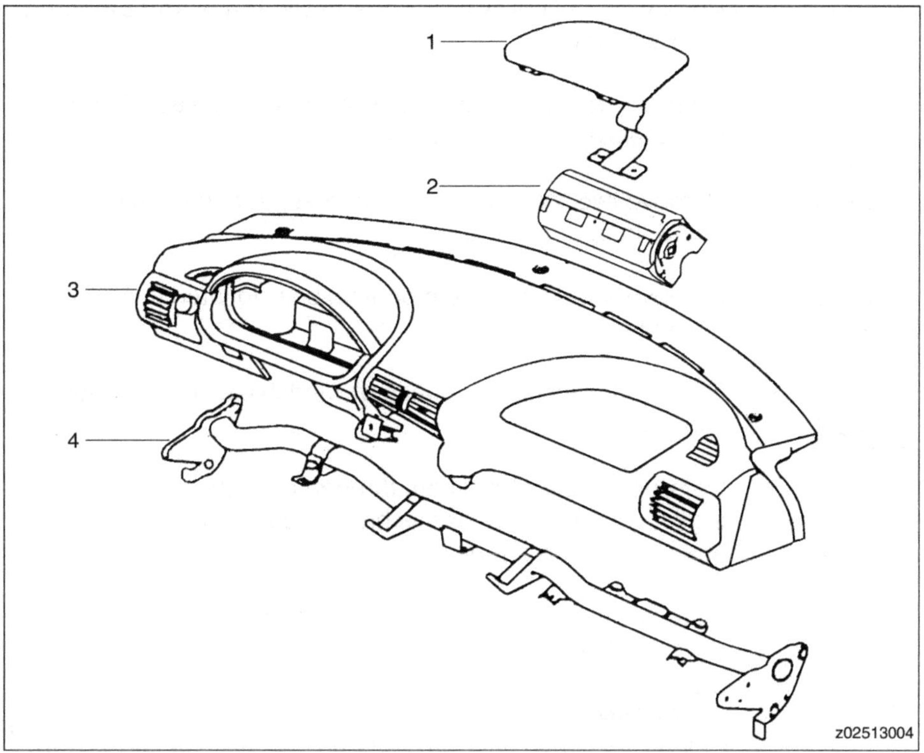


## Dashboard components

1. **Passenger airbag cover**
2. **Passenger airbag**
3. **Dashboard assembly**
4. **Dashboard reinforcement**

## Dashboard, removing and installing

– Disconnect negative (–) cable from battery and cover battery terminal with insulating material.

> ***CAUTION—***
> *Prior to disconnecting the battery, read the battery cautions in* **001 General Warnings and Cautions**.

– Remove trim and knee brace under left side of dashboard. See **Under-dash trim, left, removing and installing** in this repair group.

– Remove glove compartment and glove compartment inside cover. See **Glove compartment, removing and installing** in this repair group.

– Remove driver and passenger airbags. See **721 Airbag System (SRS)**.

> ***WARNING—***
> *Store the removed airbags in a safe place, facing up.*

– Remove steering wheel. See **320 Steering and Wheel Alignment**.

– Remove center console. See **Center console, removing and installing** in this repair group.

– Remove instrument cluster. See **620 Instrument Cluster**.

- Remove headlight switch. See **612 Switches and Electrical Accessories**.

- Remove steering column stalk switches. See **612 Switches and Electrical Accessories**.

z02513005

- Remove two mounting bolts (**arrows**) at each end of dashboard, in door jamb areas.

- Remove two dashboard mounting bolts located under center air vents.

- Remove 3 trim plugs at base of windshield. Then remove screws.

- Carefully lift dashboard out through passenger side, checking for wiring, cables or mechanical parts that might get snagged during removal.

- Installation is reverse of removal. Reconnect airbags before reconnecting battery.

> ***WARNING —***
> *Once the airbag units are installed and all other service procedures have been completed, start the engine and check that the airbag warning light goes out. If the warning light stays ON, the airbag system will not function as designed. Have the system diagnosed and repaired by an authorized BMW dealer.*

# 515 Central Locking and Anti-theft

## General

This section covers repair information for door, trunk or rear hatch locks and central locking. Also covered are the electronic immobilizer (EWS) and anti-theft alarm.

Z3 cars are equipped with sophisticated and self-diagnostic electrical systems. When experiencing malfunctions relating to central locking or anti-theft systems, use BMW service tester (DISplus, GT1, MoDiC) or equivalent for diagnosis. An advanced diagnostic scan tool can usually pinpoint electrical faults quickly and safely.

For additional electrical information, see the following repair groups:

- **600 Electrical System–General**
- **610 Electrical Component Locations**
- **ELE Electrical Wiring Diagrams.**

### Special tools

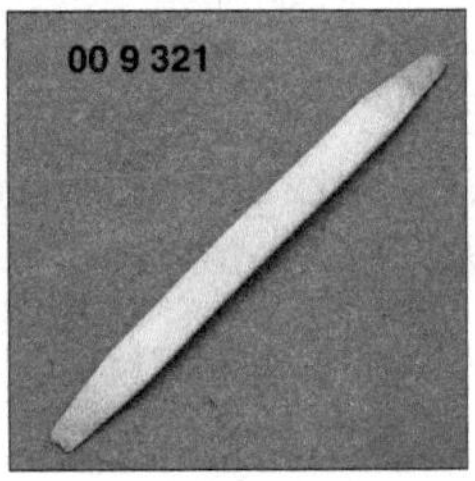


Plastic prying tool
(BMW special tool 00 9 321)

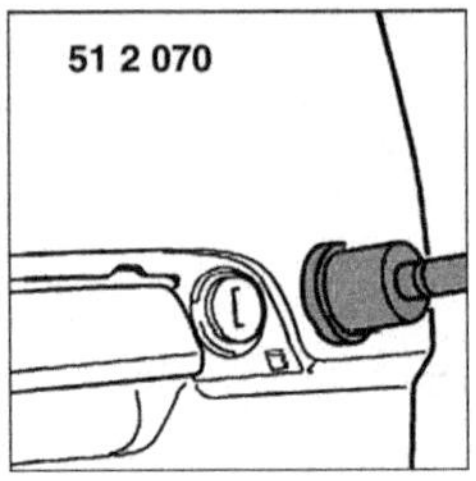


Door lock cylinder removal tool
(BMW special tool 51 2 070)

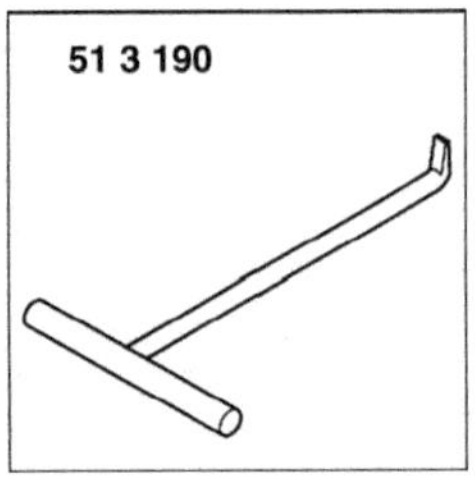


Door lock trim removal hook
(BMW special tool 51 3 190)

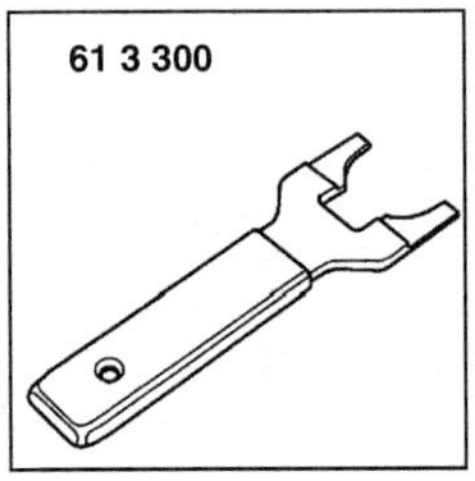


EWS ring antenna removal tool
(BMW special tool 61 3 300)

## Locking systems

The central locking system uses electrical lock actuators at each door and at the trunk lid or rear hatch. The locking system can be actuated by either front door lock or by the trunk lock / rear hatch. The remote control (keyless entry) key can also be used to activate the locks.

***NOTE—***
- *In Coupe models, only the UNLOCK option is available from the rear hatch lock.*
- *If the battery is discharged, the car can still be locked or unlocked using the key.*

When the key is turned to the lock position, a microswitch signals the locking control module to lock all doors, trunk or rear hatch, gas tank flap and, in Roadster models, the glove compartment lid. Where applicable, this also arms the anti-theft system.

In case of an accident, locking control responds by automatically unlocking the doors and turning the hazard warning lights and interior lights ON.

Two versions of central locking control were used in Z3 cars:
- 1996: ZVM
- 1997 - 2002: ZKE

## Central locking (ZVM)

The ZVM locking system utilizes a separate central locking module, located behind the glove compartment.

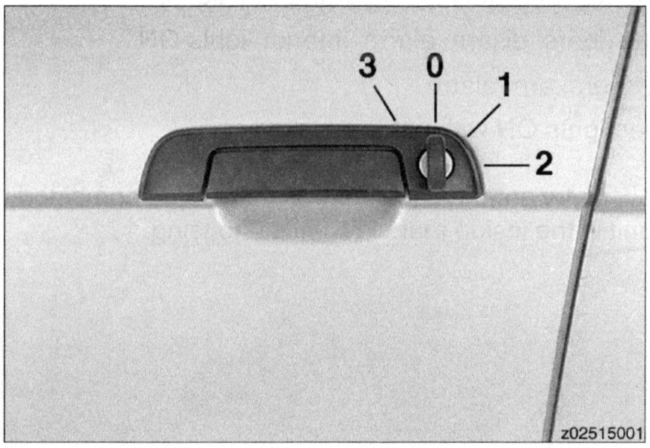


Three microswitches are operated via the door lock cylinder at each front door:

1. Position **1** actuates the door locking microswitch.
2. Position **2** actuates the double-lock microswitch.
3. Turning the key in the opposite direction (position **3)** actuates the unlocking microswitch and deactivates the alarm system.

ZVM responds to an accident via the inertia switch, mounted in the right footwell behind the speaker grille. When this function is activated, the doors are automatically unlocked and the hazard warning lights and interior lights are turned on.

## Central locking (ZKE)

In the ZKE system, central locking is integrated with a variety of comfort and safety features including window, sunroof or convertible top closure, interior lighting, wipers and washers and consumer cutoff function. See **600 Electrical System–General** for further details.

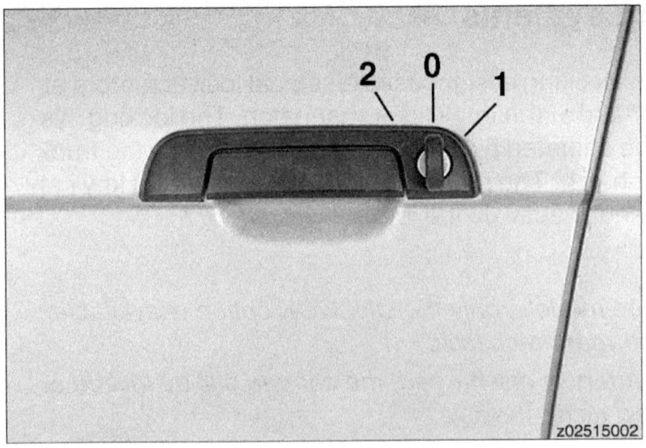

z02515002

The ZKE control module (referred to as the General Module or GM IV) is mounted behind the glove compartment.

In the ZKE locking system two microswitches are operated via the door lock cylinder at each front door:

1. Position **1** operates the door locking microswitch and activates the alarm system.
2. Turning the key approximately 45° in the opposite direction (position **2**) actuates the unlocking microswitch and deactivates the alarm system.

## Single lock, double lock

Door lock actuators incorporate two locking functions:

- **Single lock**. Vehicle is locked via the lock button.
- **Double lock** (also known as **central arrest**). Vehicle is locked from the outside using the driver's door key or a keyless entry system signal. This prevents the doors from being unlocked by any means except from an unlock request at the driver's door or via a keyless entry signal.

***CAUTION—***
*Do not engage double-lock with passengers in the car unless the master key is available. The door cannot be opened from outside or inside without the master key.*

## Keyless entry

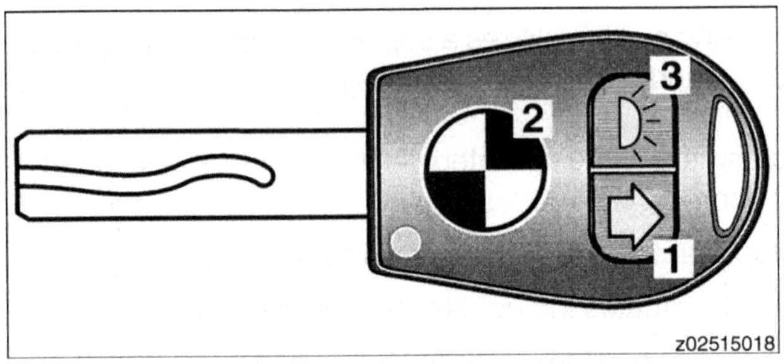

z02515018

The optional keyless entry system uses a tiny radio transmitter in the vehicle key to lock and unlock the doors and turn the interior lights on with the vehicle locked. Pressing the buttons on the key activates the following functions:

1. Unlock doors, disarm alarm, interior lights ON
2. Lock doors, arm alarm
3. Interior lights ON with vehicle locked

The keyless entry antenna and control module are combined into one unit in the inside rear view mirror housing.

## Warnings and cautions

Starting in 4 / 1988, Z3 cars are equipped with side impact door bags installed in the doors. Be sure to read the warnings and cautions that follow before starting work on the doors.

***WARNING—***
- *Serious injury may result if work on airbag components is attempted by persons unfamiliar with the BMW MRS and its approved service procedures. BMW advises that all such repairs be performed by an authorized BMW dealer.*
- *Airbags are deployed by explosive devices. Handled improperly or without adequate safeguards, these components are very dangerous.*
- *Always disconnect the battery and cover the negative battery terminal with an insulator before starting service work on electrical components.*
- *After disconnecting the battery, wait 10 seconds before beginning work on airbag components.*
- *Place an airbag unit that has been removed from the vehicle or taken out of its packaging with the padded side facing upward. Do not leave an airbag unit unattended.*

***CAUTION—***
- *Prior to disconnecting the battery, read the battery disconnection cautions given in* **001 General Warnings and Cautions**.
- *To prevent marring trim when working on interior components, work with a plastic prying tool or wrap a screwdriver tip with tape.*

# Door handles and locks

If just one of the central locking drives does not operate, the locking linkage could be frozen or stiff enough to prevent actuation of the drive. Remove the interior door panel and lubricate the locking linkage.

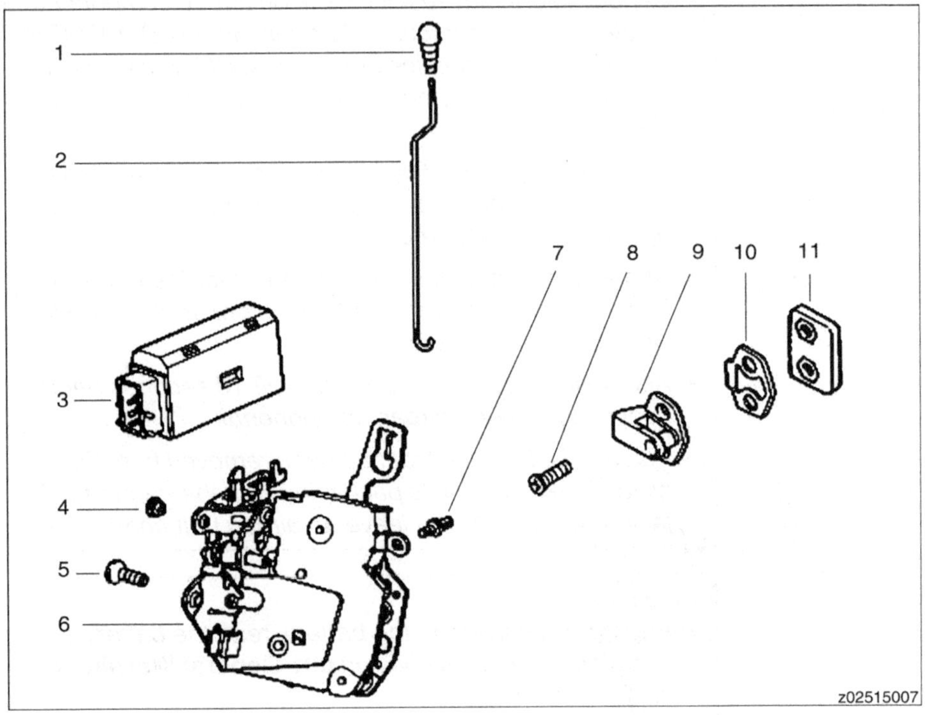


## Door lock components

1. **Lock button**
2. **Link rod**
3. **Central locking actuator**
4. **Grommet**
5. **Lock mounting bolt**
   - Torx M6
     tighten to 9 Nm (7 ft-lb)
6. **Door lock**
7. **Pin**
8. **Striker plate mounting bolt**
   - Torx M8
     tighten to 28 Nm (20 ft-lb)
9. **Door lock striker**
10. **Gasket**
11. **Striker back-up plate**

## Door lock striker, replacing

The door lock striker contains an electrical contact switch. If the switch is faulty, replace the striker.

◄ Remove striker mounting bolts (**arrows**).

***NOTE—***
*To prevent striker back-up plate from falling into B-pillar cavity, remove one bolt at a time.*

– Detach electrical connector to striker contact switch.

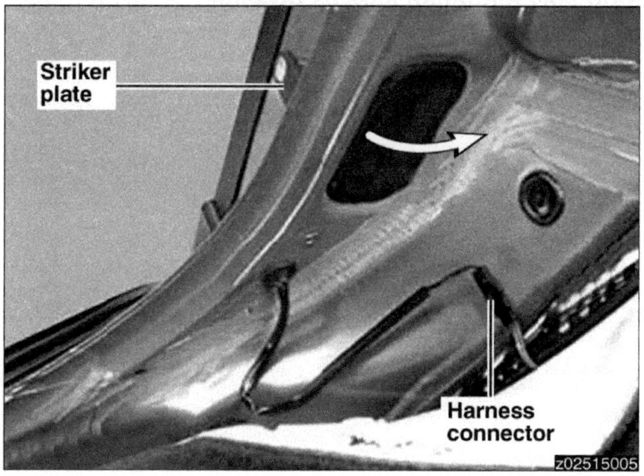


If striker backup plate falls into body cavity, or microswitch harness connector is not accessible from back of striker plate:

- Remove inside trim from the B-pillar (door-post pillar). Detach connector.
- Unclip plastic cover over striker back-up plate (**arrow**). Reach inside cavity to retrieve plate.

– Installation is reverse of removal.

| Tightening torque | |
|---|---|
| Striker plate to body pillar (Torx M8) | 28 Nm (20 ft-lb) |

## Door lock, removing and installing

> ***WARNING —***
> *Starting in 4 / 1998, Z3 models are equipped with side impact airbags installed in the doors. Read the* **Warnings and cautions** *in this repair group, as well the warnings and cautions in* **721 Airbag System (SRS)** *prior to starting work on the door.*

– Make sure ignition switch is OFF. Remove key from ignition switch.

– Remove door trim panel. See **411 Doors**.

– Model with door airbag:

- Unplug airbag carefully. Avoid damaging electrical connector.
- Remove airbag mounting screws.
- Remove airbag from door and store in safe place.

> ***WARNING —***
> *Place an airbag unit that has been removed from the vehicle with the padded side facing upward. Do not leave an airbag unit unattended.*

– Remove vapor barrier from door.

– Remove door glass. See **512 Windows**.

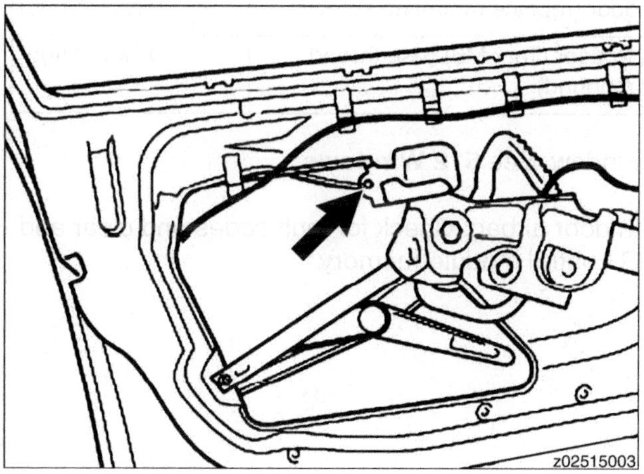


Remove inner door opener:

- Remove door opener mounting screw (**arrow**).
- Detach door opener from bowden cable and remove.

– Working inside door cavity, remove metal guard surrounding lock mechanism by removing mounting bolt.

◄ Remove door lock mounting screws (**arrows**), peeling back door seal as necessary.

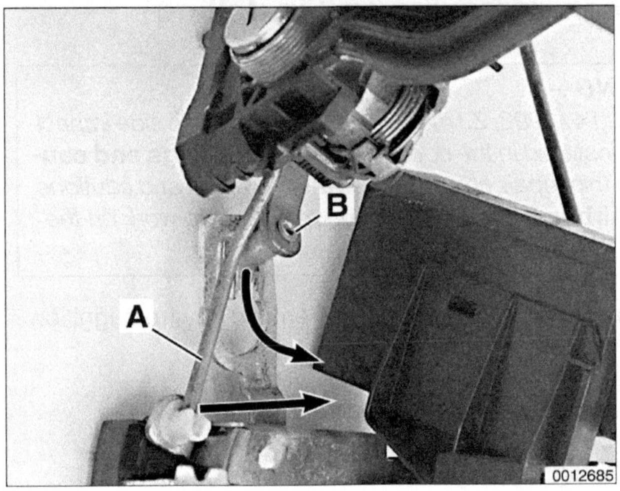


◄ Working inside door:

- Disconnect linkage and harness connector from exterior door handle, repositioning door lock as necessary.
- Pry out lock linkage (**A**) from plastic retainer.
- Lift door lock slightly to remove latching pin from lock gate (**B**).

***NOTE—***
*For clarity, door lock mechanism is shown removed from door.*

– Remove window guide rail mounting bolt at bottom outboard corner of door. Twist guide rail toward door hinge and remove from door cavity.

– Remove door lock from inside door cavity.

– Installation is reverse of removal. Remember to:

- If applicable, reinstall door airbag with new self-locking screws. Make sure airbag electrical connector snaps securely in place.
- Install lock mounting screws with Loctite®270 or equivalent thread locking compound.

| Tightening torques | |
|---|---|
| Airbag to door (replace mounting screws) | 9 Nm (7 ft-lb) |
| Door lock to door (Torx M6) (use thread locking compound) | 9 Nm (7 ft-lb) |

– Initialize window. See **512 Windows**.

– Model with door airbag: Check for fault codes and clear and reset MRS control module memory.

## Outside door handle with lock cylinder, removing and installing

> ***WARNING —***
> *Starting in 4/1998, Z3 models are equipped with side impact airbags installed in the doors. Read the* **Warnings and cautions** *in this repair group, as well the warnings and cautions in* **721 Airbag System (SRS)** *prior to starting work on the door.*

– Raise window to top.

– Make sure ignition switch is OFF. Remove key from ignition switch.

– Remove door trim panel. See **411 Doors**.

– Model with door airbag:
  - Unplug airbag carefully. Avoid damaging electrical connector.
  - Remove airbag mounting screws.
  - Remove airbag from door and store in safe place.

> ***WARNING —***
> *Place an airbag unit that has been removed from the vehicle with the padded side facing upward. Do not leave an airbag unit unattended.*

– Remove vapor barrier from door.

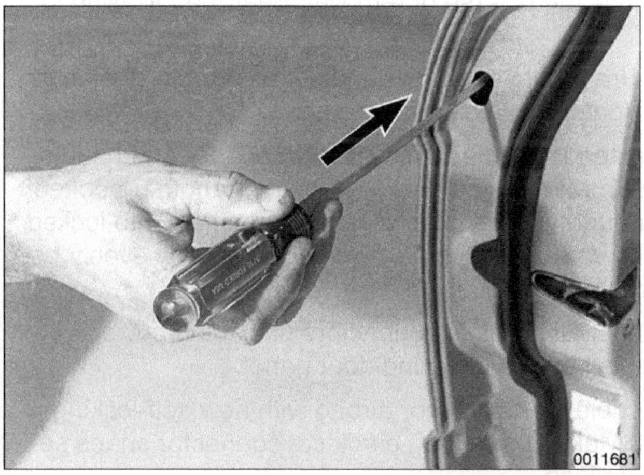


◄ Pry out access plug in side of door. Use BMW special tool 51 3 190 (or screwdriver) through access hole to release exterior trim lock plate. Carefully pry trim off door working from top.

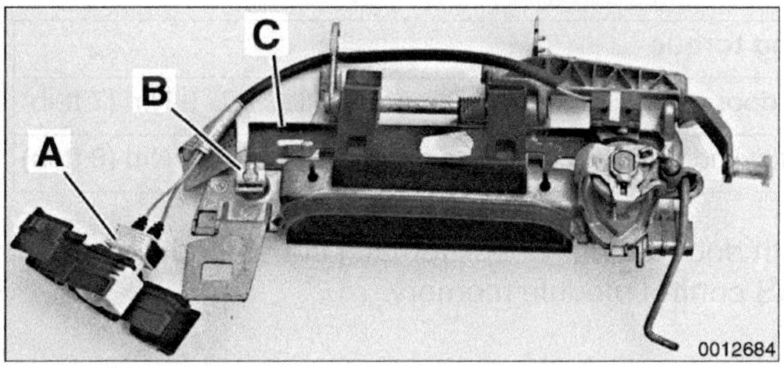


◄ Working inside door cavity:
- Twist main harness plastic connector (**A**) to remove from door handle.
- Remove retaining pin locking clip (**B**) and push pin out.

***NOTE —***
*Handle assembly shown removed from car to illustrate exterior trim locking plate (**C**).*

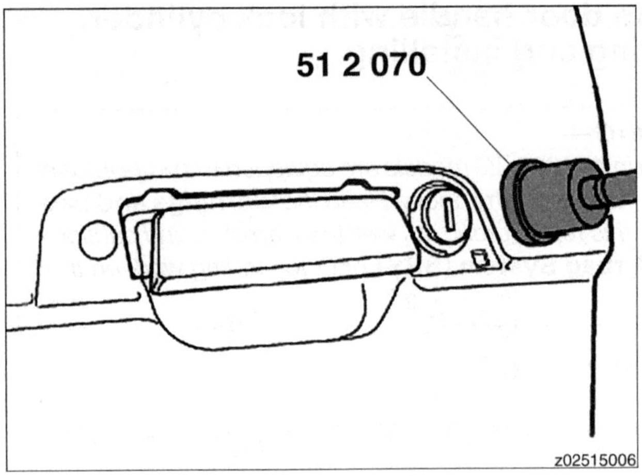


Working at door exterior, use special BMW spanner 51 2 070 or equivalent to remove lock cylinder collar nut while supporting handle.

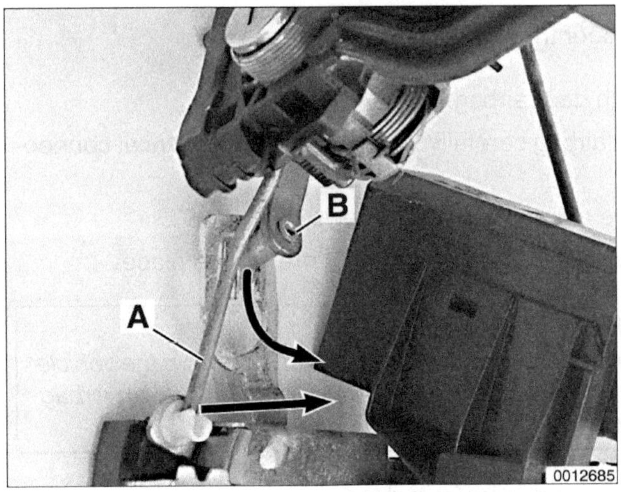


Working inside door:

- Pry out lock linkage (**A**) from plastic retainer at door lock.
- Lower door handle slightly to remove latching pin from lock gate (**B**).

***NOTE—***
*For clarity, door lock mechanism is shown removed from door.*

– Remove door handle from door by angling out of door cavity.

***NOTE—***
*Leave lock cylinder disassembly to a BMW technician or locksmith. Incorrect removal will allow the encoding plates to fall out of the lock cylinder.*

– Installation is the reverse of removal. Bear the following in mind:

- Make sure rubber seals for handle are correctly positioned before tightening into place.
- Place exterior handle trim over lock cylinder and handle, then use BMW special tool 51 3 190 (or another hooked tool) through access hole to pull locking plate into locked position. Make sure handle trim contacts body evenly before locking it in place.
- Reconnect all harness connectors (including power window motor) before installing door panel.
- If applicable, reinstall door airbag with new self-locking screws. Make sure airbag electrical connector snaps securely in place.

| Tightening torque | |
|---|---|
| Airbag to door (replace mounting screws) | 9 Nm (7 ft-lb) |
| Door lock cylinder collar nut to handle | 10 Nm (8 ft-lb) |

– Model with door airbag: Check for fault codes and clear and reset MRS control module memory.

# Trunk or Rear Hatch Lock

### Trunk lock (Roadster)

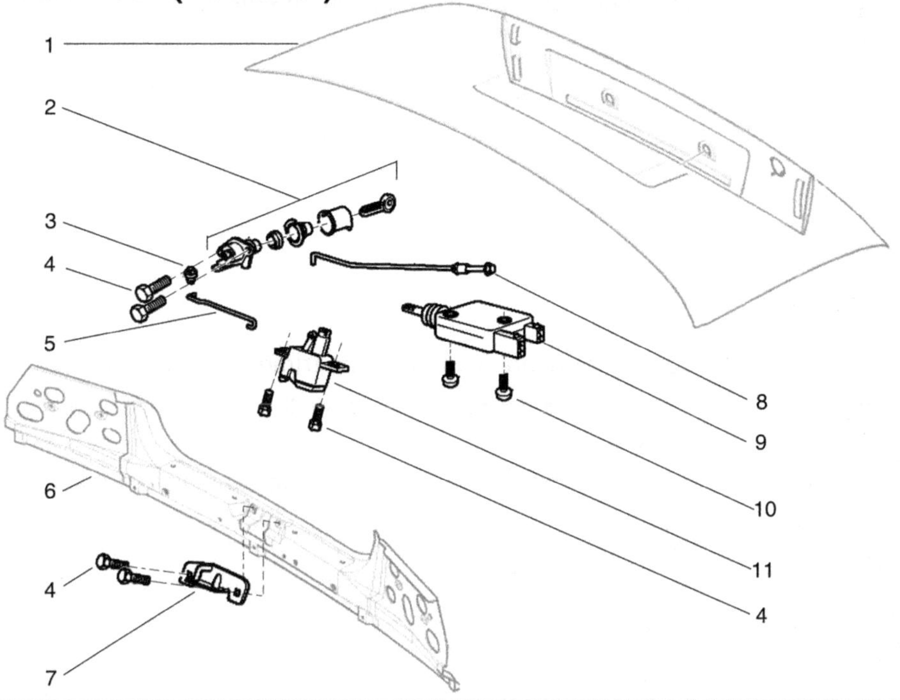


### Rear hatch lock (Coupe)

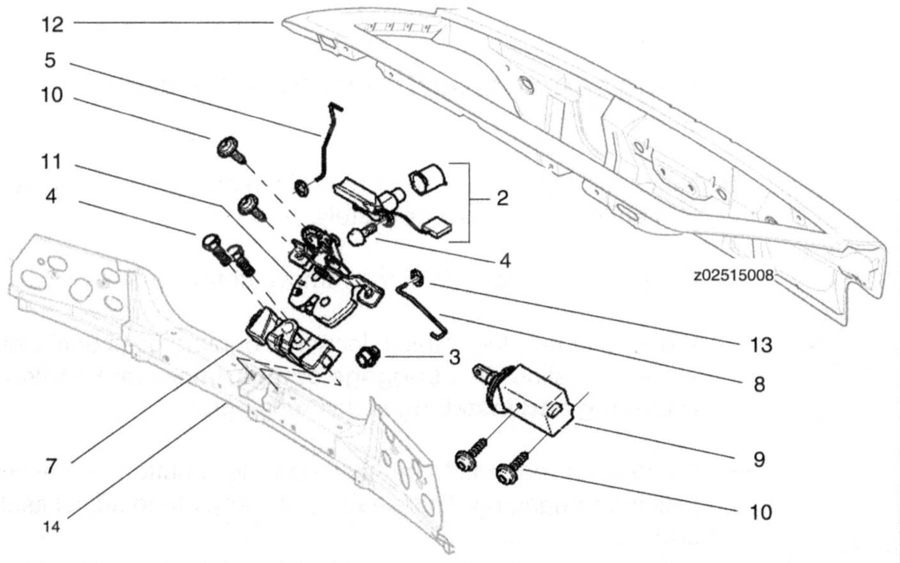


## Luggage compartment lock components

1. **Trunk lid**
2. **Lock cylinder assembly**
3. **Plastic clip**
4. **Mounting bolts (M6)**
5. **Actuating rod**
6. **Trunk rear panel**
7. **Lock striker**
8. **Link rod**
9. **Central locking actuator**
10. **Mounting screws (Torx)**
11. **Trunk lock**
12. **Rear hatch**
13. **Circlip**
14. **Cargo compartment rear panel**

## Luggage compartment lock striker, removing and installing

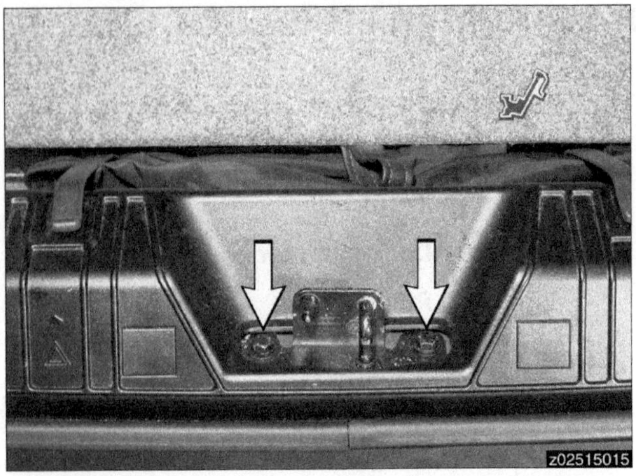


– Roadster: Working at trunk rear panel, remove lock striker mounting bolts (**arrows**) and lift off striker.

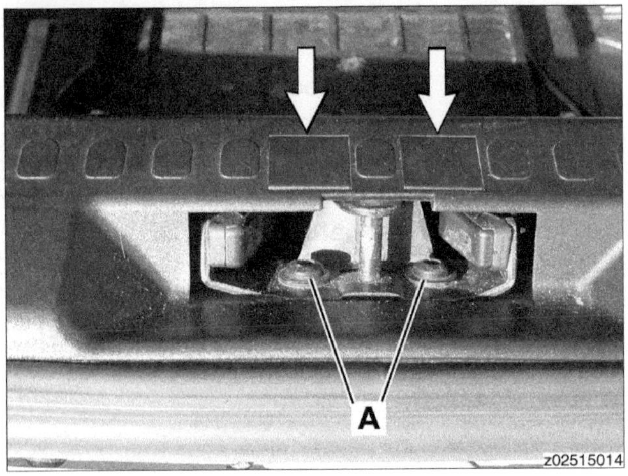


◄ Coupe: Working at cargo compartment rear panel:

- Use plastic prying tool to carefully pry out trim plugs (**arrow**).
- Remove lock striker mounting bolts and lift off striker.

## Luggage compartment lock striker, adjusting

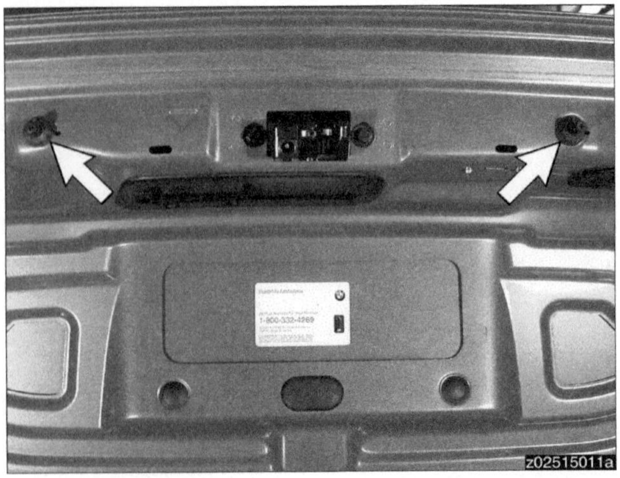


◄ Roadster: Open trunk lid or rear hatch and screw down detent buffers (**arrows**) completely.

– Coupe: Open rear hatch and remove detent buffers.

– Working at rear lock striker, loosen mounting bolts until striker can just slide. See **Luggage compartment lock striker, removing and installing** in this repair group.

– Close lid or rear hatch, but hold down lock button to prevent lock from engaging. This enables striker plate to adjust itself correctly.

– Open lid and tighten striker mounting bolts. Check adjustment by closing lid or hatch. Repeat adjustment if necessary.

– Unscrew or reinstall detent buffers and adjust so that closed lid or hatch is flush with sides of body.

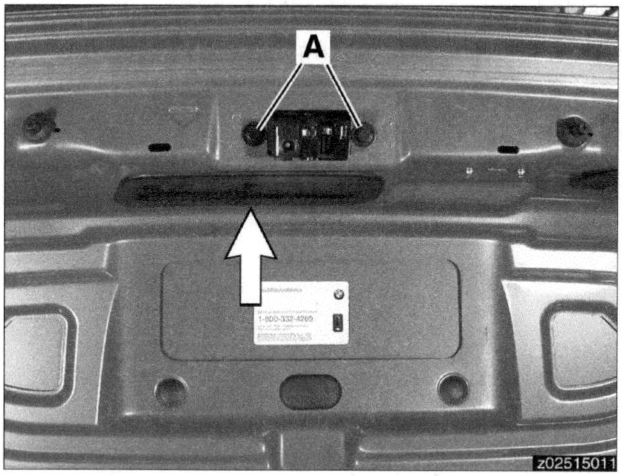


## Trunk lock, removing and installing (Roadster)

◀ Remove trunk lock plastic access cover (**arrow**) from inside trunk lid.

– Working inside trunk lid cavity, disengage lock actuating rod. See **Luggage compartment lock components** in this repair group.

– Remove lock mounting bolts (**A**).

– Working inside trunk lid cavity, disconnect electrical connector and remove lock.

– Installation is reverse of removal.
  - Check lock operation initially with trunk lid open.
  - Adjust striker location if necessary. See **Luggage compartment lock striker, adjusting** in this repair group.

## Rear hatch lock, removing and installing (Coupe)

– Remove rear hatch trim panel.

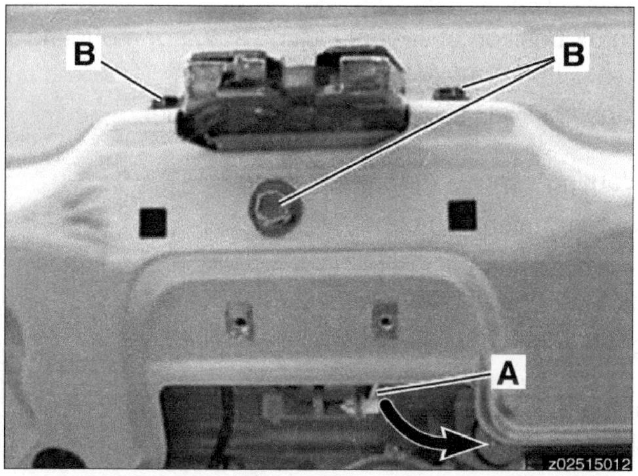


◀ Working at hatch lock:
  - Disengage lock actuating rod (**A**).
  - Remove lock mounting bolts (**B**).
  - Push lock into hatch cavity and to right.

– Working inside hatch cavity, disconnect electrical connector and remove lock.

– Installation is reverse of removal.
  - Check lock operation initially with hatch open.
  - Adjust striker location if necessary. See **Luggage compartment lock striker, adjusting** in this repair group.

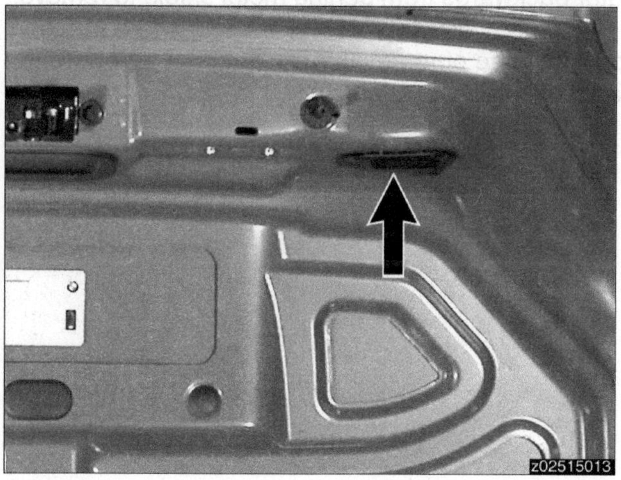


## Luggage compartment lock cylinder, removing and installing

◀ Roadster: Remove lock cylinder plastic access cover (**arrow**) from inside trunk lid.

– Coupe: Remove rear hatch trim panel.

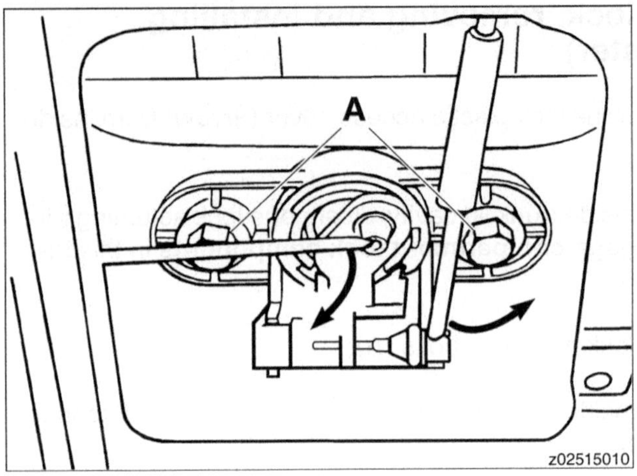


- Working inside luggage compartment lid cavity:
  - Disengage lock cylinder actuating rod and link rod (**arrows**).
  - Remove cylinder mounting bolts (**A**).
- Installation is reverse of removal.

## Luggage compartment locking actuator, removing and installing

> ***CAUTION—***
> *The actuator is adjustable. Match mark the position of screws before removal.*

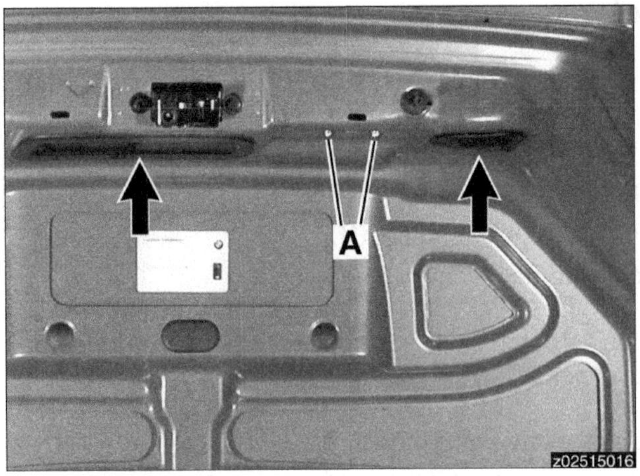


- Roadster: Working at open trunk lid:
  - Remove trunk lock and lock cylinder plastic access covers (**arrows**) from inside trunk lid.
  - Match mark position of central locking actuator mounting screws (**A**), then remove screws.
  - Disengage link rods from lock and lock cylinder.
- Coupe: Working at open rear hatch:
  - Remove rear hatch trim panel.
  - Match mark position of central locking actuator mounting screws (**A**), then remove screws.
  - Disengage link rods from lock and lock cylinder.

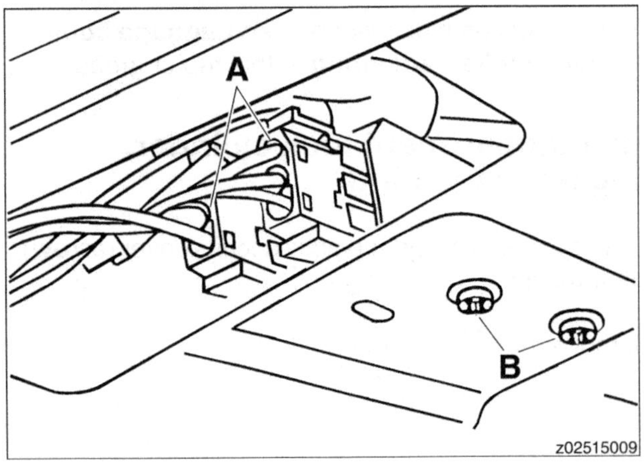


- Unplug actuator harness connectors (**A**). Actuator mounting screws shown at **B**.
- Pull actuator backwards from behind metal shield to remove.
- Installation is reverse of removal. Align actuator screws with match marks during installation.

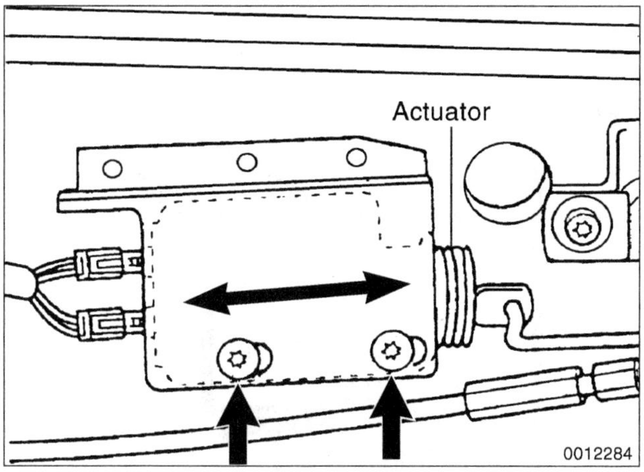


### Luggage compartment locking actuator, adjusting

- Be sure central locking system is in unlocked mode.
- Roadster: Remove access covers from inside trunk lid.
- Coupe: Remove rear hatch trim panel.

◀ Loosen rear central locking actuator mounting screws (**arrows**) just enough to allow actuator to slide.

- With luggage compartment lid open, operate driver's door lock with key, checking to make sure actuator plunger moves through its full range of motion.
- Set actuator position for maximum plunger motion. Tighten sliding screws and reinstall access covers or rear hatch trim.

## ELECTRONIC IMMOBILIZER (EWS)

Z3 vehicles are equipped with electronic immobilization (EWS) as a theft-prevention feature.

***NOTE —***
*At the time of introduction the system was referred to as Electronic Driveaway Protection which in German is* Electronische Wegfahrsperre *or EWS.*

The EWS ignition key is embedded with a computer chip and permanently encoded. A primary code is programmed into the key and into the vehicle itself. A secondary code is changed every time the vehicle is started. If the key code and EWS control module code do not match, the engine control module (ECM) and starter are disabled.

EWS ignition keys cannot be duplicated. The system is designed to have up to ten keys and only an authorized BMW dealer can provide replacement keys.

Two versions of EWS were used in Z3 cars:

- 1996 - 1998: EWS II
- 1999 - 2002 EWS III D

The primary difference between the two systems is that the function of the EWS transmitter-receiver in EWS II was merged into the EWS control module for the later system.

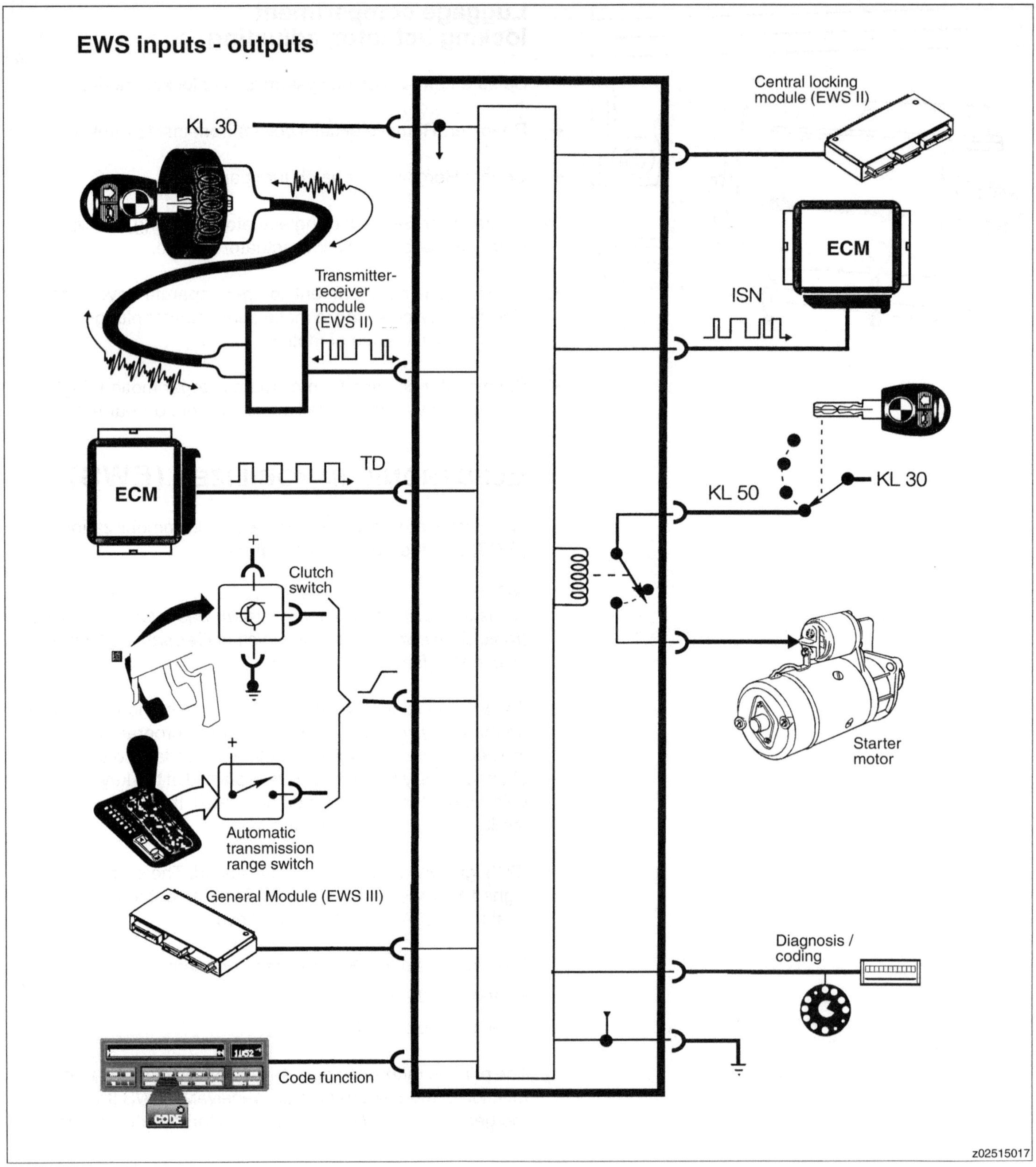
EWS inputs - outputs
KL 30
Transmitter-
receiver
module
(EWS II)
ECM
TD
Clutch
switch
Automatic
transmission
range switch
General Module (EWS III)
Code function
CODE
Central locking
module (EWS II)
ECM
ISN
KL 50
KL 30
Starter
motor
Diagnosis /
coding
z02515017

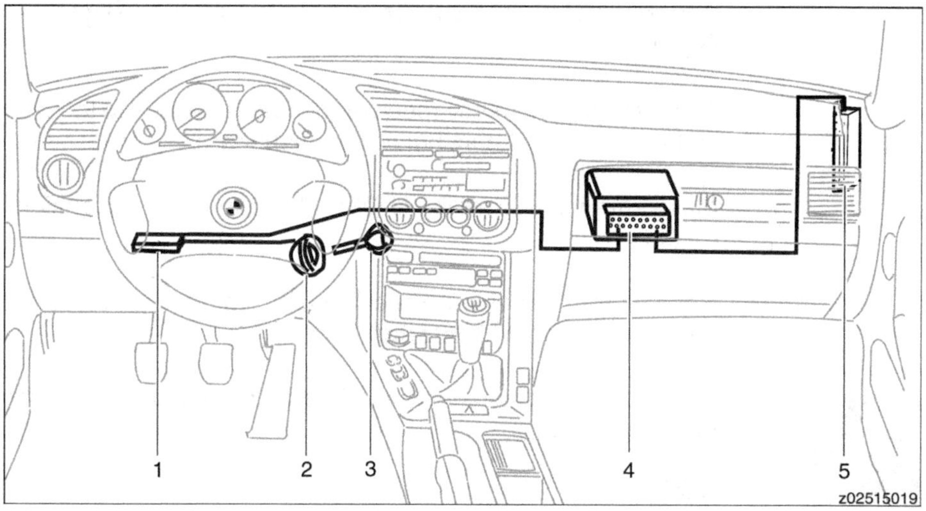


## EWS II components

1. **Transmitter-receiver module**
2. **Ring antenna**
3. **Ignition key**
4. **EWS II control module**
5. **ECM**

## EWS ring antenna, removing and installing

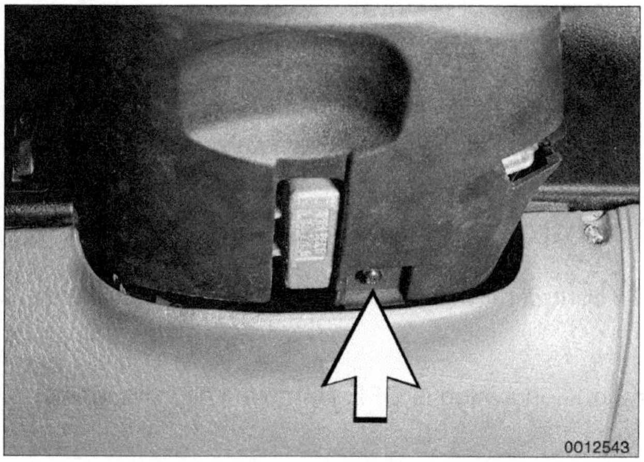


◄ Remove lower steering column retaining screw (**arrow**) and remove cover from steering column.

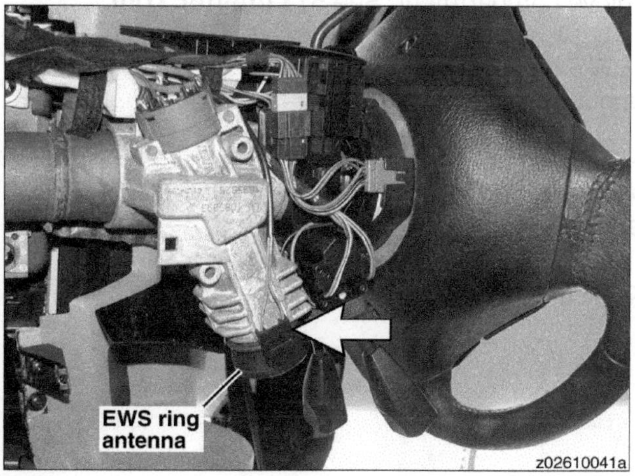


◄ Working at ignition switch:

- Detach ring antenna electrical connector (**arrow**).
- Use special tool 61 3 300 to pry ring antenna off ignition switch.

– Installation is reverse of removal.

## Transmitter-receiver module, replacing (EWS II)

The EWS II transmitter-receiver module is located to the left of the steering column, below the dashboard.

- Make sure ignition switch is OFF. Remove key from ignition switch.

- Remove trim panel and knee bolster from under left side of dashboard. See **513 Interior Trim**.

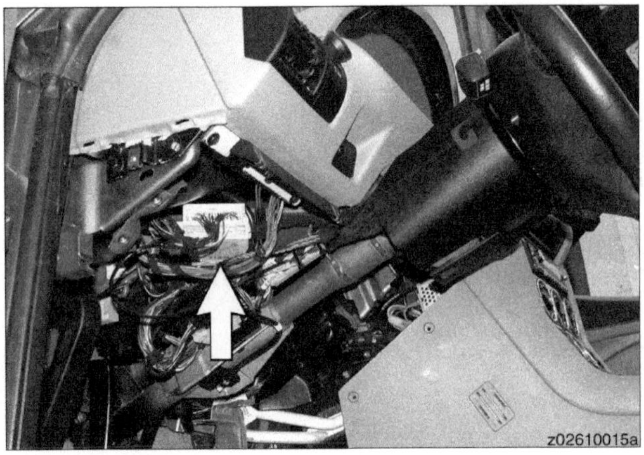
z02610015a

◄ Detach electrical connector and remove transmitter-receiver module (**arrow**).

*NOTE—*
*Check to make sure that the module is identified with EWS markings.*

- Installation is reverse of removal.

## EWS II control module, replacing

The EWS II control module is located behind the glove compartment.

- Make sure ignition switch is OFF. Remove key from ignition switch.

- Remove glove compartment. See **513 Interior Trim**.

z02610036a

◄ Remove EWS II control module harness connector (**arrow**) and pull module out of bracket.

*NOTE—*
*Check to make sure that the module is identified with EWS markings.*

- Installation is reverse of removal.

## EWS III control module, replacing

The EWS III control module is located to the left of the steering column, below the dashboard.

- Make sure ignition switch is OFF. Remove key from ignition switch.
- Remove trim panel and knee bolster from under left side of dashboard. See **513 Interior Trim**.

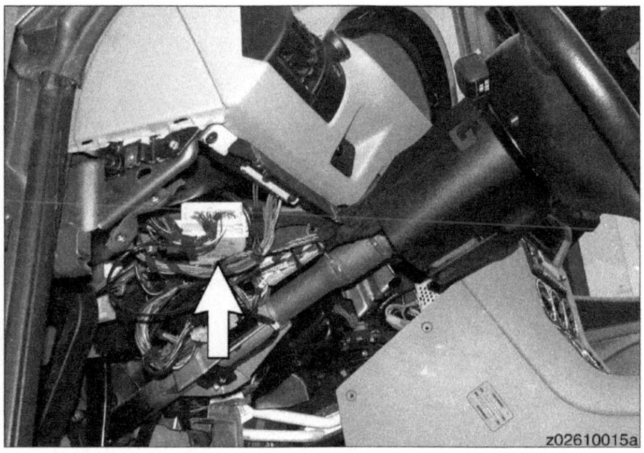


◄ Detach electrical connector and remove control module (**arrow**).

*NOTE—*
*Check to make sure that the module is identified with EWS markings.*

- Installation is reverse of removal.
- Use BMW service tester DISplus, GT1 or MoDiC to align new EWS module with ECM.

## ANTI-THEFT ALARM

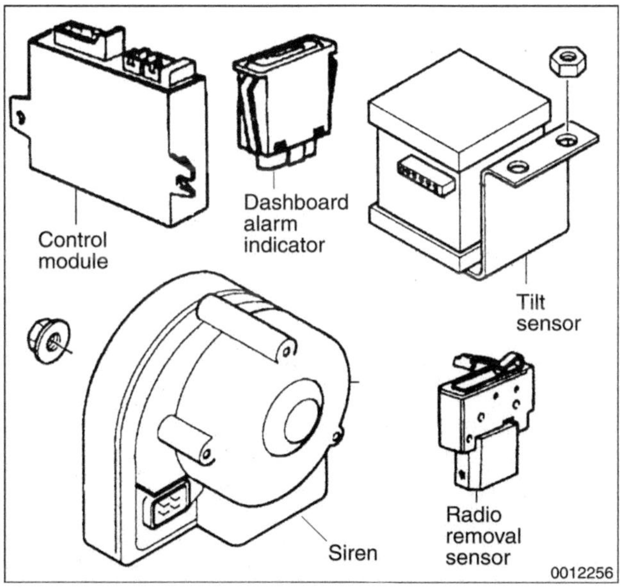


◄ The anti-theft alarm in Z3 cars is an optional dealer-installed system. Two versions of anti-theft alarm were used:

- Early productions cars: Alpine system
- Late production cars: DWA system

# 520 Seats

## GENERAL

This repair group covers seat removal and installation. Seat disassembly, upholstery repairs and diagnosis of electric seat problems is beyond the scope of this manual.

Additional information may be found in the following repair groups:

- **610 Electrical Component Locations** for fuse and relay positions
- **720 Seat Belts** for pyrotechnical seat belt tensioners
- **721 Airbag System (SRS)** for Multiple Restraint System (MRS)
- **ELE Electrical Wiring Diagrams**

### Special tool

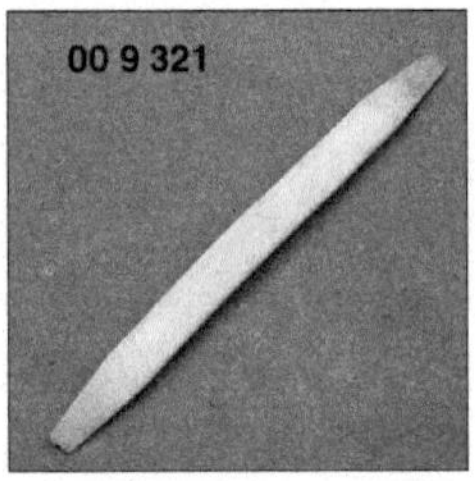


Plastic prying tool
(BMW special tool 00 9 321)

### Seats

The drivers seat is a four-way adjustable power seat. The passenger seat comes either as a two- or four-way power seat. Power seat adjustment is controlled by a switch mounted on the side of the seat. The seats are powered at all times. Both the driver and passenger seats are also equipped with a manual backrest adjustment controlled with a lever on the side of the seat.

Some Z3 models are equipped with heated seats, which can be individually adjusted to high or low temperature settings.

### Troubleshooting

> ***WARNING —***
> *Be sure to disable automatic seat belt tensioner before starting work on seat. See* **720 Seat Belts***.*

Troubleshooting the power seat can be done with the seat partially removed. Operation of power seat can be analyzed and tested without seat being connected to floor.

– If there is no movement or noise at seat, check:
  - Fuse 5 (30 A) in front power distribution box
  - Seat grounds. See **610 Electrical Component Locations**

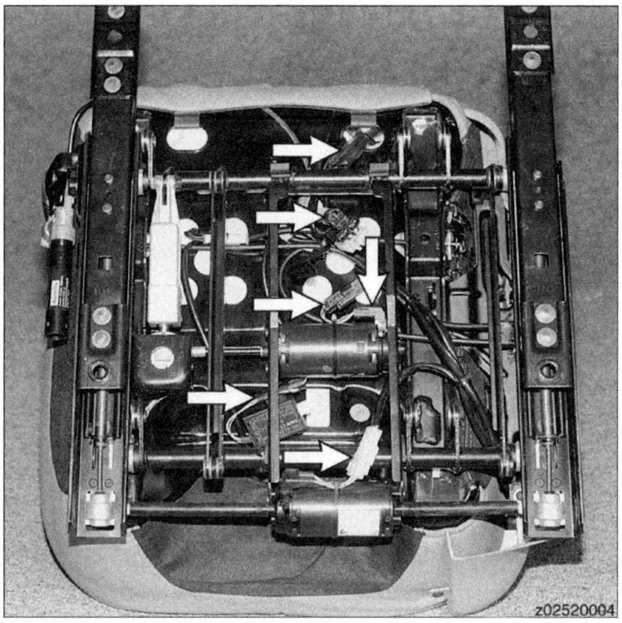
z02520004

Remove seat and place upside down on clean work surface.

- Use test light or a multimeter to check seat wiring and connectors (**arrows**). See **ELE Electrical Wiring Diagrams**.
- If checking seat heater, use multimeter to test heater element for continuity and heater timer operation.
- Check seat track movement at cable and power seat gear boxes located at top of each seat track.

## Warnings and cautions

Z3 cars are equipped with automatic seat belt tensioners. In 1997 and later models, the seat belt tensioners are pyrotechnical devices which deploy automatically via electrical signals from the MRS control module in case of an accident. The procedure for disabling the automatic tensioners is in **720 Seat Belts**.

Be sure to read the warnings and cautions in **721 Airbag System (SRS)** and also those which follow before starting work on the seats.

***WARNING —***
- *Automatic seatbelt tensioners are powerful devices. Handle with care.*
- *BMW recommends that repairs on pyrotechnical devices such seat belt tensioners be carried out by a qualified BMW technician.*
- *Disconnect the battery and cover the negative battery terminal with an insulator before starting service work on electrical components.*
- *After disconnecting the battery, wait 10 seconds before beginning work on pyrotechnical components.*

***CAUTION—***
- *Prior to disconnecting the battery, read the battery disconnection cautions given in* **001 General Warnings and Cautions**.
- *To prevent marring trim when working on interior components, work with a plastic prying tool or wrap a screwdriver tip with tape.*

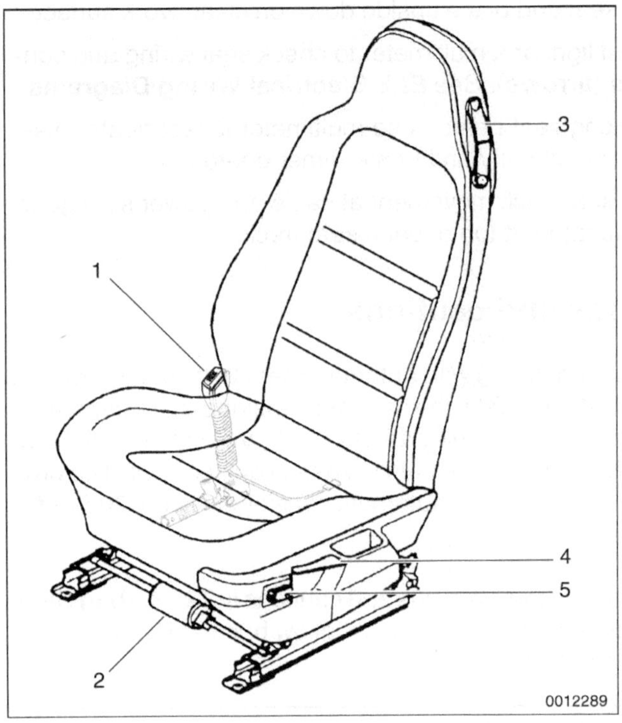

0012289

## Seat Removal and Installation

### Seat components

1. Automatic seat belt tensioner
2. Seat position control motor
3. Seat belt guide
4. Backrest release lever
5. Power seat adjustment control

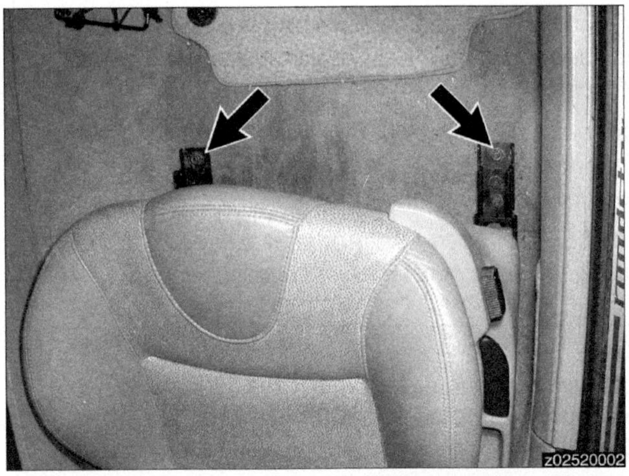
z02520002

### Seat, removing and installing

- Move seat all the way rearward. Remove front seat track mounting fasteners (**arrows**).

– Detach seat belt from guide at top of backrest.

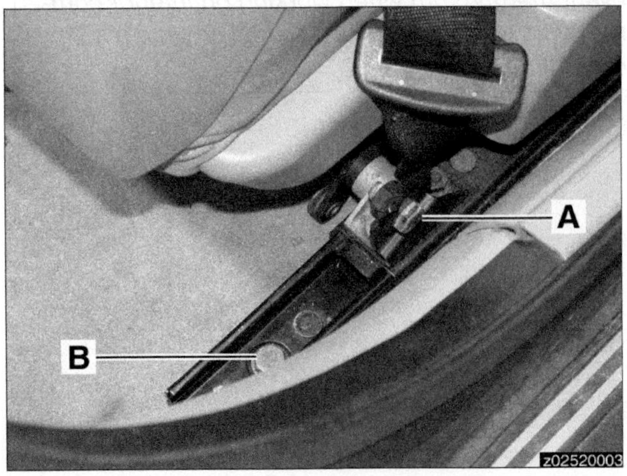

z02520003

- Move seat all the way forward.
  - Remove seat belt lower anchor bolt (**A**) below backrest. Note placement of washers and spacers.
  - Remove rear seat track mounting fasteners (**B**).

– Disconnect negative (–) battery cable and cover terminal with insulating material.

*NOTE—*
*On 1996 cars, the battery is located in the right rear of the engine compartment. On 1997 and later cars, the battery is located under the trunk floor mat.*

> ***CAUTION—***
> *Prior to disconnecting the battery, read the battery cautions in* **001 General Warnings and Cautions**.

– Tip seat back to access electrical connectors underneath. Separate connectors and remove front seat from interior of car.

> ***CAUTION—***
> *Cover door sill with blanket to prevent damage to trim.*

– Disarm automatic seat belt tensioner. See **720 Seat Belts**.

– Installation is reverse of removal.
  - Rearm seat belt tensioner before installing seat.
  - Avoid trapping floor carpet under seat tracks.
  - Note correct placement of washers and spacers on seat belt anchor bolt.
  - Tighten front track mounting fasteners first, then rear fasteners.

| **Tightening torques** | |
|---|---|
| Seat belt lower anchor to seat (use thread locking compound) | 48 Nm (35 ft-lb) |
| Seat belt tensioner to seat | 48 Nm (35 ft-lb) |
| Seat to floor<br>• Front (M8 nut)<br>• Rear (M10 bolt) | <br>22 Nm (16 ft-lb)<br>43 Nm (32 ft-lb) |

# 540 Sunroof

## GENERAL

This repair group covers the tilting sunroof assembly for Coupe models.

### Special tool

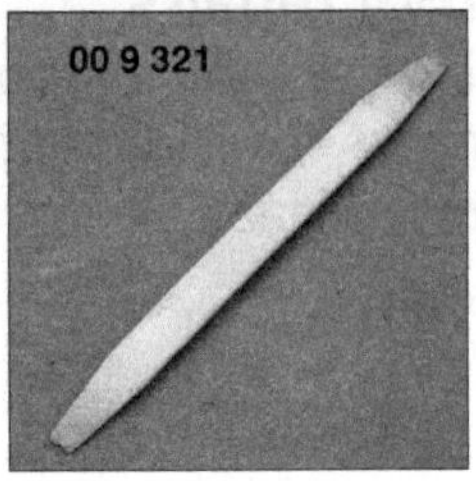


◄ Plastic prying tool (BMW special tool 00 9 321)

### Sunroof

The tilting sunroof is operated by an electric motor above the headliner.

In case of sluggish sunroof operation, lubricate guide channels with lithium based grease.

***NOTE—***
*Removal and repair of the sunroof frame, sunroof motor and other components inside the roof cavity is beyond the scope of this manual.*

## SUNROOF REPAIRS

### Tilting sunroof components

1. **Gasket**
2. **Sunroof frame**
3. **Glass**
4. **Sunroof motor**
5. **Circlip**
6. **Pin**
7. **Circlip**
8. **Torx bolt (M5)**
9. **Torx bolts (M5)**
10. **Sunroof tilting mechanism**
11. **Sunroof hinge mountings and fasteners**

## Sunroof panel, adjusting

Adjust the sunroof panel under the following circumstances:

- Top of closed sunroof misaligned with roof of car
- Closed sunroof not square
- Wind noises at speed (sunroof closed)
- Water leaks from headliner
- Sunroof removed and reinstalled

### Sunroof gap

z02540002

– Correct sunroof alignment with sunroof fully closed: Gap even all around edge of sunroof.

◀ Insert credit card to measure gap. Make sure card encounters equal resistance all around sunroof perimeter.

### Sunroof height

– Correct sunroof height:
  - Front: Flush to 1 mm (0.04 in.) below surface of roof.
  - Rear: Flush to 1 mm (0.04) above surface of roof.

***NOTE—***
*Upward adjustment of sunroof glass is possible at any time. Downward adjustment may be complicated due to the fact that the sunroof is glued in place.*

– To begin sunroof height adjustment, remove sunroof glass. See **Sunroof glass, removing and installing** in this repair group.

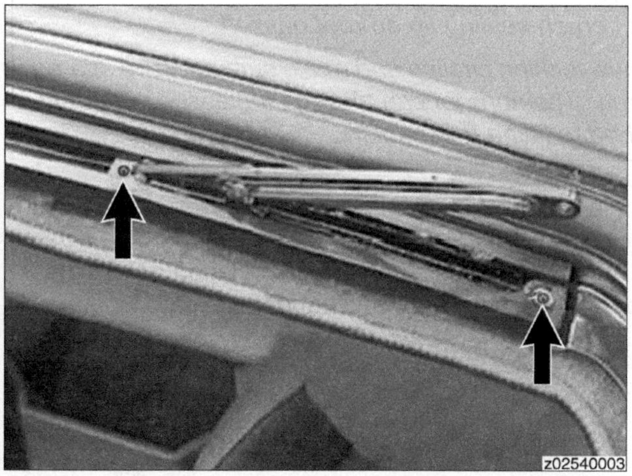
z02540003

◀ Rear height adjustment:
  - Press roof liner down a little and loosen tilting mechanism mounting screws (**arrows**). Do not remove.
  - At each screw position, insert U-shaped washer (1 mm thickness) between sunroof frame and tilting mechanism.
  - Tighten mounting screws.

| **Tightening torque** | |
|---|---|
| Tilting mechanism to sunroof frame | 3.5 ± 0.5 Nm<br>(31 ± 4 in-lb ) |

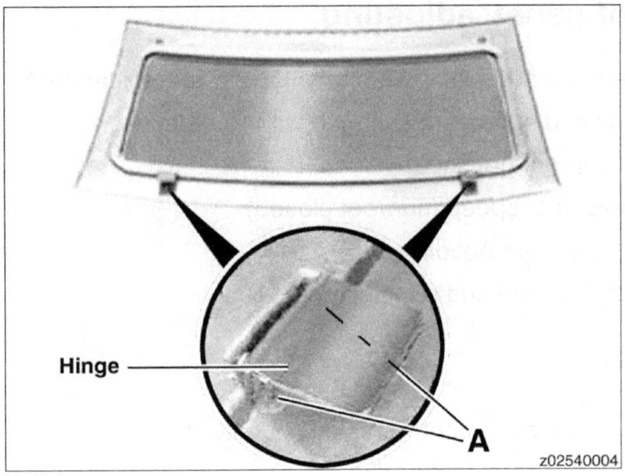


Front height adjustment:

- Loosen glass hinge mounting screws (**A**). Do not remove.
- Slide hinge up or down to adjust height.
- If correct adjustment cannot be attained, remove hinge and grind contact surface.

> ***CAUTION—***
> *Do not damage threaded bosses on hinge.*

– If ground down, prime hinge with paint to protect against rust. Reinstall hinge.

| Tightening torque | |
|---|---|
| Hinge to sunroof frame | 4 ± 0.5 Nm (35 ± 4 in-lb ) |

– Reinstall glass panel. Recheck alignment and height adjustments.

– When finished, test glass for secure installation: Press glass up from below with palms of hands and make sure it does not lift.

## Sunroof glass, removing and installing

– Open sunroof.

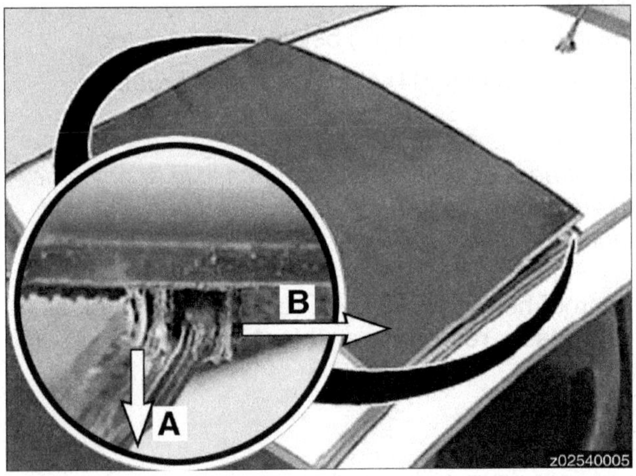


Working above roof at sunroof tilting mechanism:

- Release locking circlip (**A**).
- Slide out mounting pin (**B**).

– Tilt glass up and out of mounting. Store on a clean surface.

– When installing:

- Slide glass hinges into mountings first.
- Then attach glass to tilting mechanism.

> ***CAUTION—***
> - *Do not crush sealing lip on roof gasket.*
> - *Use new locking circlips and make sure they are seated securely. The sunroof can blow off if glass is not firmly attached to lifting mechanism.*

– Recheck glass alignment and height adjustments.

– When finished, test glass for secure installation: Press glass up from below with palms of hands and make sure it does not lift.

# 541 Convertible Top

## GENERAL

This repair group contains information on the manual and the electrohydraulic convertible tops for Roadster models.

Diagnosis and repair of the electrohydraulic top is beyond the scope of this manual.

### Special tool

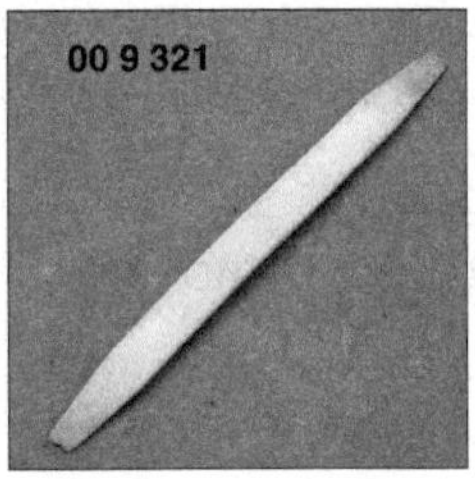


Plastic prying tool
(BMW special tool 00 9 321)

### Convertible top

The manual convertible top fitted to early Roadster models easily folds down with one operator.

An electrohydraulic power-operated version of the convertible top was fitted to later cars.

A removable hard top was available in 1998 as a dealer-installed option.

### Caution

**_CAUTION—_**
*To prevent marring trim when working on interior components, work with a plastic prying tool or wrap a screwdriver tip with tape.*

## Manual convertible top

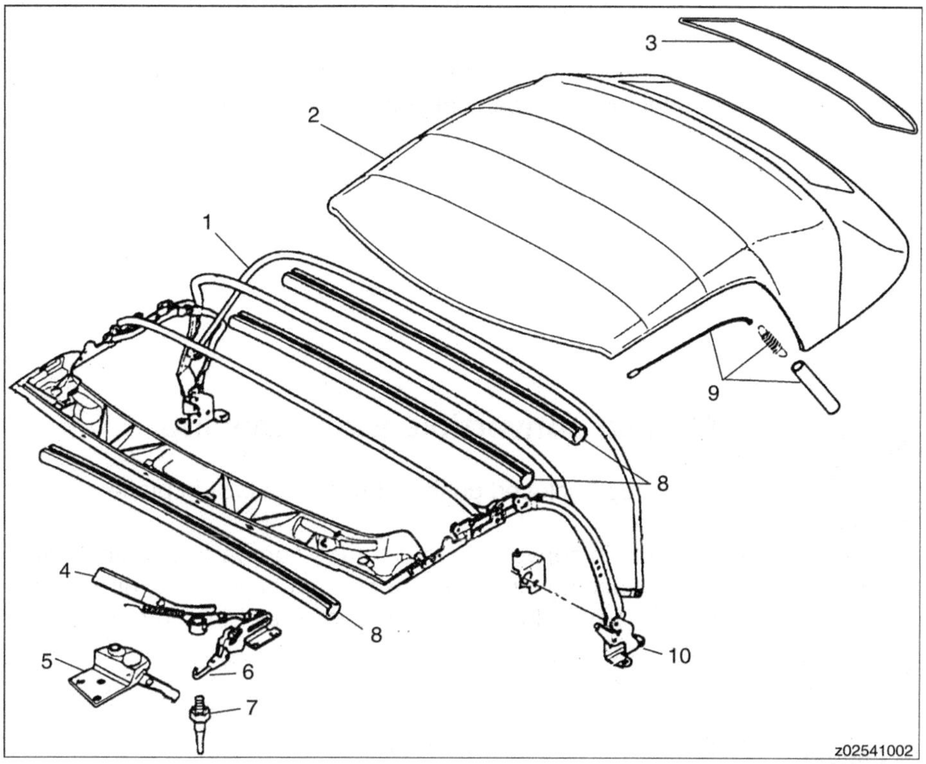


### Manual convertible top components

1. **Frame**
2. **Folding top**
3. **Rear window**
4. **Front latch**
5. **Visor lock plate**
6. **Visor hook**
7. **Centering pin**
8. **Padding**
9. **Cable assembly**
10. **Mounting plate**

### Manual convertible top, adjusting

- Lock and release convertible top at visor latch, checking for ease of operation.
- Lock top down and check that it is flush with front of windshield.

*NOTE—*
*Top must be no more than 2 mm (0.08 in.) lower than front of windshield.*

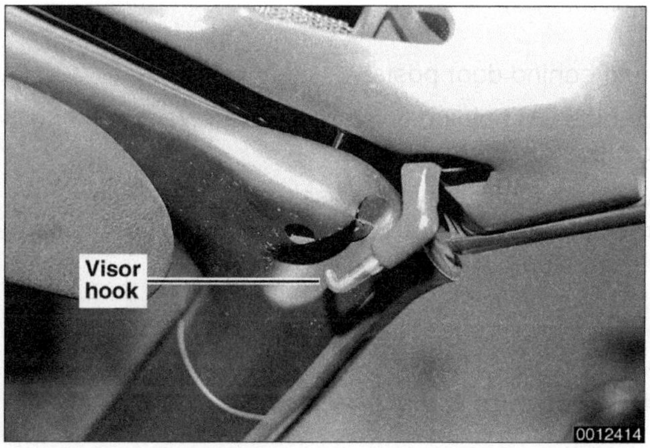


◄ If roof does not close tightly, is hard to release or alignment with windshield is incorrect:
- Loosen or tighten visor hooks as required using threaded section of rod.
- Lubricate visor latch if necessary.

- Check for front-to-rear fit by aligning centering pins at visor latch.

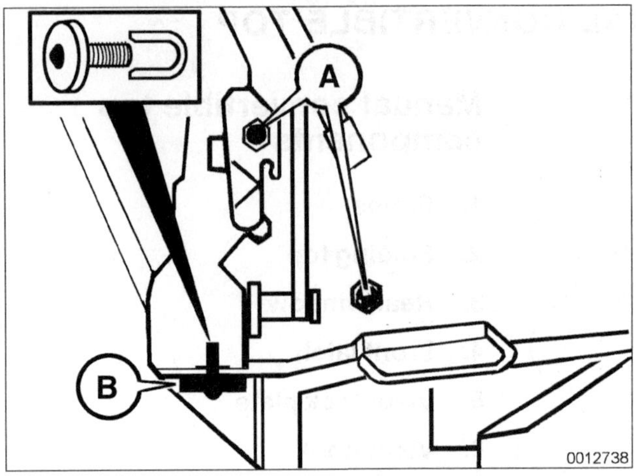


If alignment is necessary:

- Remove trim finisher pieces at tops of door posts and loosen nuts (**A**) on mounting plate studs at rear of door frame on each side.
- Remove or add U-shaped spacers at horizontal screws (**B**) until top is properly aligned. Tighten nuts and screws.

## Manual convertible top, removing

- Place top in closed position but do not lock in place.

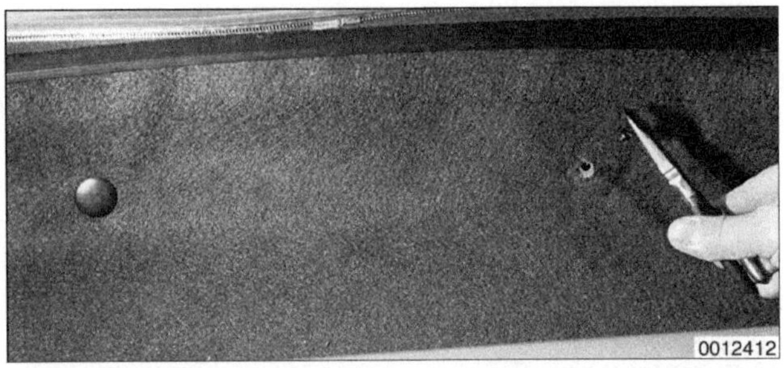


Working in convertible top storage compartment, unfasten clips attaching inner trim liner to rear bulkhead. Pull liner material (carpet) downward and peel away from bottom edge of convertible top.

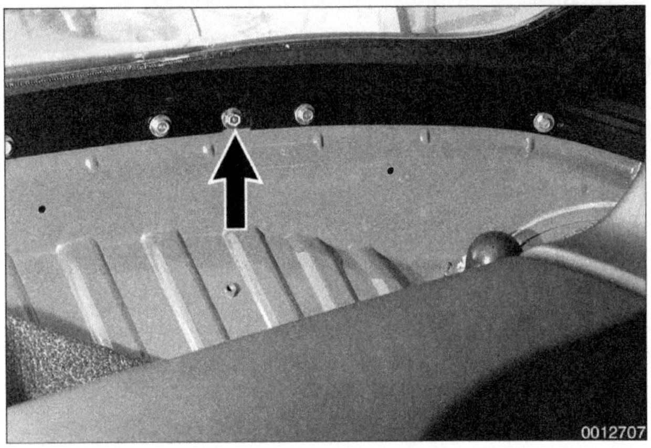


Peel back liner at rear of body.

- Remove all but two bolts from tack strip under liner.
- One bolt on each side (**arrow**) is recessed. Leave these bolts to hold tack strip to body.

***NOTE—***
*Note number of spacer discs on each bolt. Reinstall the same number of spacers at each fastener position of tack strip.*

- Fold top into storage compartment.
- Unclip and remove trim finisher piece from top of each door post.

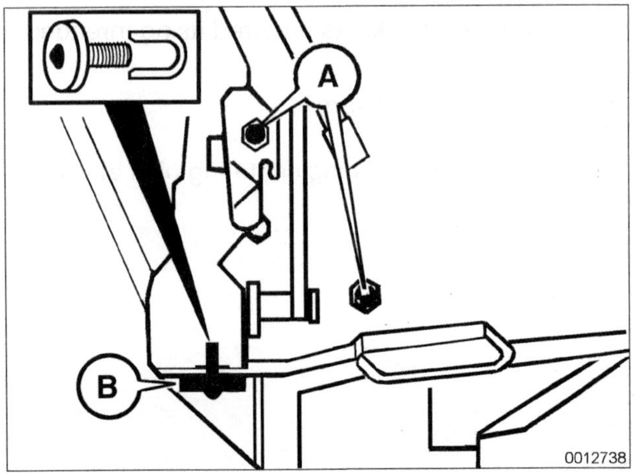


Working just behind door posts:

- Remove nuts (**A**) holding convertible top lower mounting plate to studs on car body.
- Remove horizontal screw with washers and U-shaped spacers (**B**) from each side of frame. Note number of spacers on each side.

- Press down on one corner of convertible top frame as assistant on other side of car raises other side. Lift top assembly out of car.

## Manual convertible top, installing

- With an assistant on opposite side of vehicle, lower convertible top onto one mounting plate and then on other while aligning mounting studs.

  ***NOTE —***
  *Be sure vertical studs for mounting plate each have a 1 mm spacer on them.*

- Install nuts on mounting plate studs. Tighten nuts and then loosen slightly.

- Insert horizontal screw with washer and 7 U-shaped spacers between lower mounting plate and body and tighten.

  ***NOTE —***
  *The 7 spacers are for the basic setting of the roof position. Note that some adjustment may be necessary at the spacers once the top is fully installed.*

- Retighten nuts on mounting plate studs on each side.

- Close convertible top but do not lock in place.

- Attach 2 mm thick butylene tape between tack strip and rear bulkhead.

- Fit back of convertible top into tack strip in rear bulkhead. Insert but do not tighten screw in center of tack strip.

- Insert screws in tack strip beside rear of door frame on each side of vehicle, but do not tighten.

- Install all remaining screws along tack strip.

- Tighten tack-strip center screw, then screws along door frames, and finally all other tack-strip screws.

- Check convertible top operation and alignment at latches. See **Manual convertible top, adjusting** in this repair group.

- Reinstall top storage compartment liner, using new plastic clips as necessary. Reinstall trim finisher pieces at tops of door posts.

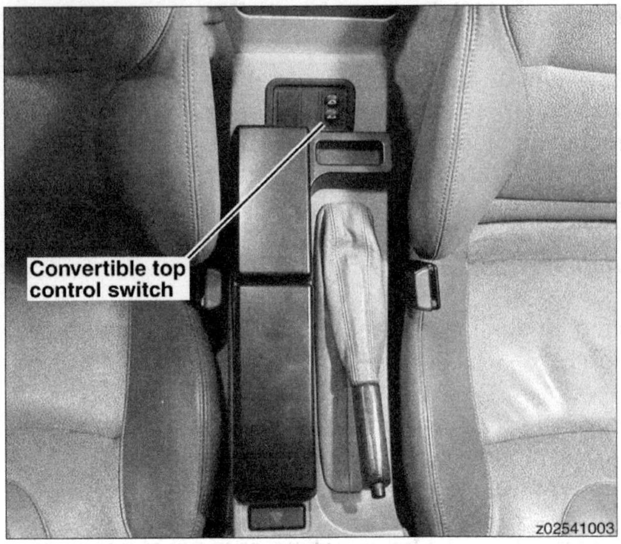


## Electrohydraulic Convertible Top

### Electrohydraulic top operation

To open or close convertible top:

- Stop vehicle. Lower windows slightly.
- Turn ignition to accessory position.
- Depress brake pedal.
- Manually unlock visor locks.
- Press control switch between seat backrests.

### Electrohydraulic top emergency closing

In case of electrical malfunction, follow this procedure to close the electrohydraulic convertible top manually.

– Open trunk and remove left side trim.

Open hydraulic pump emergency valve counterclockwise ½ turn. This depressurizes hydraulic top system and allows top to be closed manually.

### Electrohydraulic top fluid level

– Make sure top is closed.

– Open trunk and remove left side trim.

Use dimple (**arrows**) in side of fluid reservoir as level guide.

◀ If fluid needs topping up, open hydraulic pump emergency valve counterclockwise ½ turn.

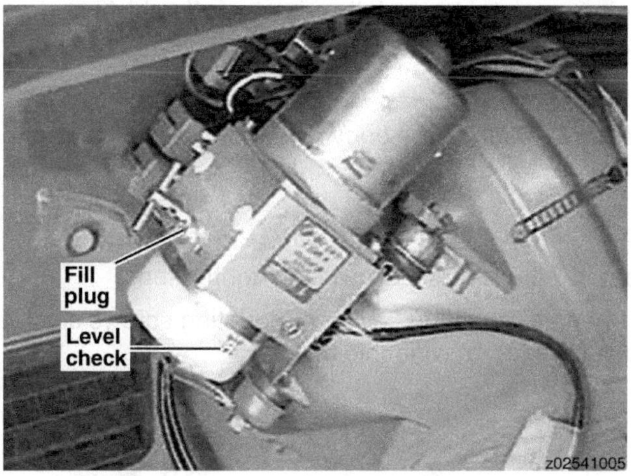


◀ Early version hydraulic pump.

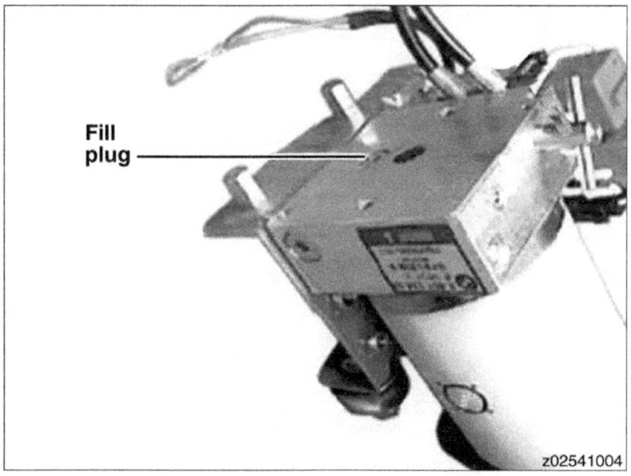


◀ Later version hydraulic pump, introduced early in 1999:

- Partially loosen pump on its mountings.
- Turn pump over to access fill plug on bottom.

– Replenish fluid as necessary through fill plug. Use new seal when replacing plug.

> ***CAUTION—***
> - *Cleanliness is essential when working with the hydraulic system. Thoroughly clean surface around fill plug before removing plug.*
> - *Spread shop rags below pump to avoid staining trunk trim.*

| **Fluid recommendation** | |
|---|---|
| Convertible top hydraulic fluid | BMW part no. 54 34 8 410 306 |

| **Tightening torque** | |
|---|---|
| Fill plug to hydraulic reservoir (use new seal) | 11 Nm (8 ft-lb) |

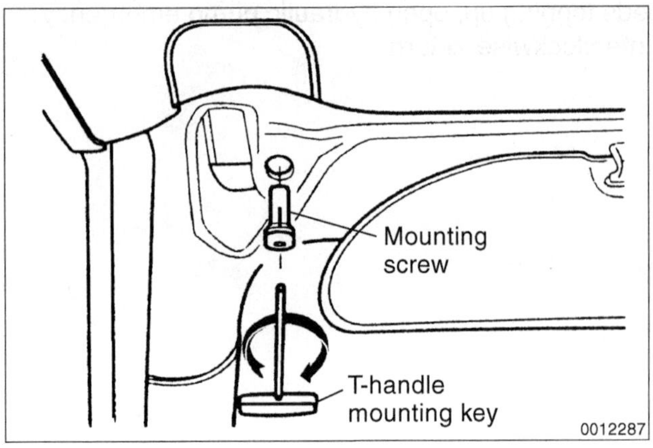


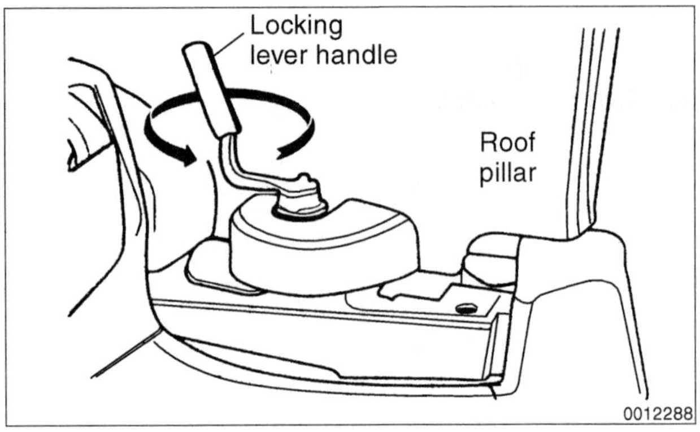


## Hardtop

### Hardtop, removing

- Use T-handle mounting key (from tool kit in trunk) to loosen two screws at front of hardtop, above windshield.

*NOTE —*
*The two screws remain in the windshield top frame.*

- Behind each seat at base of hardtop, fully loosen locking lever by turning counterclockwise.

– With aid of an assistant, carefully raise top off vehicle and store according to directions in owner's manual.

*NOTE —*
*Take care not to damage the body or paint work when removing the hardtop.*

– Installation is reverse of removal.
  - Turn rear locking levers clockwise to tighten top.
  - Using T-handle mounting key, turn mounting screws clockwise to secure top at windshield.

# 600 Electrical System–General



**TABLES**

## GENERAL

This repair group presents a brief description of the principal parts of the electrical system. Also covered here are basic electrical system troubleshooting tips.

For additional information on electrical systems and components, refer to the following repair groups:

- **121 Battery, Starter, Alternator**
- **400 Body–General**
- **512 Door Windows**
- **515 Central Locking and Anti-theft**
- **610 Electrical Component Locations**
- **ELE Electrical Wiring Diagrams**

### Electrical test equipment

For measuring voltage, current or resistance, use a digital multimeter. An analog meter (swing-needle) may draw enough current to damage sensitive electronic components.

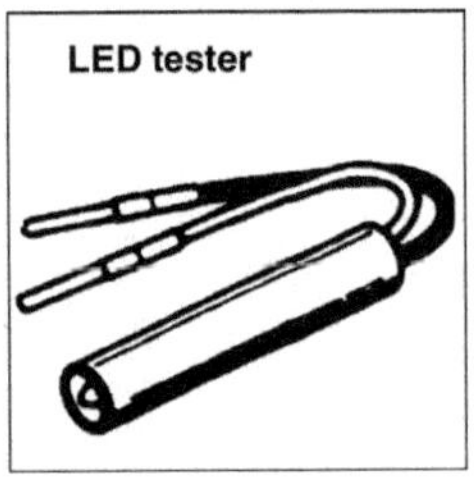


An LED test light is a safe, inexpensive tool that can be used to perform many simple electrical tests that would otherwise require a digital multimeter. The LED indicates when voltage is present between any two test-points in a circuit.

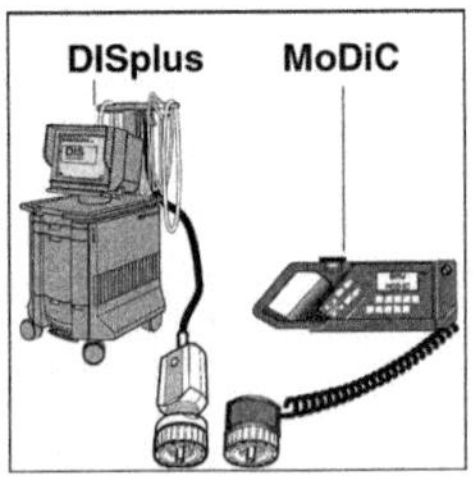


The quickest way to diagnose many electrical problems is to start out with a scan tool read out of diagnostic trouble codes (DTCs) using a scan tool such as BMW DISplus, GT1, MoDiC or equivalent. See **OBD On-Board Diagnostics** at rear of manual.

### Electrical wiring diagrams

Electrical schematics for selected models are in **ELE Electrical Wiring Diagrams** at the rear of this manual. Wiring schematics and electrical information are also available at the BMW subscription site:

**http://www.bmwtechinfo.com**

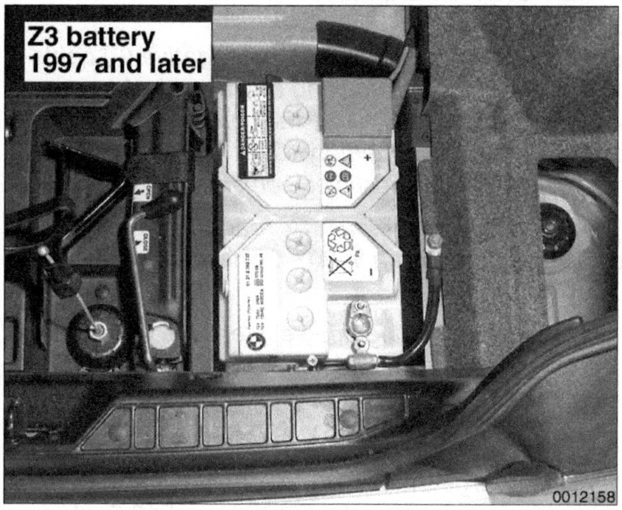


## Z3 electrical system

Z3 models have a 12-volt, direct current (dc), negative-ground electrical system. The voltage regulator maintains the voltage in the system at approximately the 12 vdc rating of the battery, and all circuits are grounded by direct or indirect connection to the negative (-) terminal of the battery.

In the 1996 model year, the Z3 battery was located in a small well in the right rear of the engine compartment.

In 1997, the battery was moved to the right side of the trunk, under the floor trim. It can be accessed by lifting up the carpeted panel that covers the tool kit.

Starting with models produced after 4 / 1999, the positive battery terminal was replaced by a pyrotechnic battery safety terminal (SBK). See **121 Battery, Starter, Alternator**.

All electrical circuits except those required for starting and operating the engine are protected by fuses. Fuses for individual components are primarily housed in the front power distribution box in the left rear engine compartment. For a full listing of fuses and fuse locations, see **610 Electrical Component Locations**.

## Electrical system safety precautions

***WARNING —***

- *Airbags, front seat belt tensioners and the battery safety terminal (SBK) are pyrotechnic (explosive charge) devices. Making repairs without the proper knowledge and special test equipment may cause serious personal injury. See* **721 Airbag System (SRS)**.
- *The ignition system of the car operates at lethal voltages. People with pacemakers or weak hearts should not expose themselves to the ignition system electric currents. Take extra precautions when working on the ignition system or when servicing the engine while it is running or the key is ON. See* **120 Ignition System** *for additional ignition system warnings and cautions.*
- *Keep hands, clothing and other objects clear of the electric radiator cooling fan when working on a warm engine. The fan may start at any time, even when the ignition is switched OFF.*

***CAUTION—***

- *Turn engine OFF and disconnect the negative (–) cable from the battery before removing any electrical components. Disconnecting the battery may erase fault code(s) stored in control module memory. Check for fault codes using special BMW diagnostic equipment.*
- *Prior to disconnecting the battery, read the battery disconnection cautions given in* **001 General Warnings and Cautions**.
- *Connect and disconnect ignition system wires, multiple connectors and ignition test equipment leads only while the ignition is OFF.*
- *Do not disconnect the battery with the engine running.*
- *Do not quick-charge the battery (for boost starting) for longer than one minute, and do not exceed 16.5 vdc at the battery with the boosting cables attached. Wait at least one minute before boosting the battery a second time.*
- *Many solid-state control modules are static sensitive. Static discharge will permanently damage them. Always handle the modules using proper static prevention equipment and techniques.*
- *Always switch a digital multimeter to the appropriate function and range before making test connections.*
- *To avoid damaging harness connectors or relay panel sockets, use jumper wires with flat-blade connectors that are the same size as the connector or relay terminals.*
- *Do not try to start the engine of a car which has been heated above 176°F (80°C) (for example, in a paint drying booth). Allow it to cool to normal temperature.*
- *Disconnect the battery before doing any electric welding on the car.*
- *Do not wash the engine while it is running, or any time the ignition is ON.*
- *Choose test equipment carefully. Use a digital multimeter with at least 10 MΩ input impedance, or an LED test light. An analog meter (swing-needle) or a test light with a normal incandescent bulb may draw enough current to damage sensitive electronic components.*
- *Do not use an ohmmeter to measure resistance on solid state components such as control modules or time delay relays.*
- *Always disconnect the battery before making resistance (ohm) measurements on a circuit.*

## BMW ACRONYMS

Acronyms used in wiring diagrams and in this manual are summarized in **Table a**.

**Table a. Common BMW acronyms**

| Acronym | Component or system |
|---|---|
| ABS | antilock brakes |
| A/C | air-conditioning |
| ADS | auxiliary throttle valve |
| ASC | traction control |
| DISA | dual resonance intake system |
| DLC | diagnostic link connector (20 pin) |
| DME | digital motor electronics |
| DMTL | diagnosis module—tank leakage |
| DSC | dynamic stability control |
| DTC | diagnostic trouble code |
| DWA | antitheft alarm |
| ECM | engine control module |
| ECT | engine coolant temperature (sensor) |
| EDK | electronic throttle (DME MS 43.0) |
| EDR | electronic throttle system (S54 engine) |
| EEPROM | flash programmable read-only memory |
| EGS | electronic transmission control |
| EPROM | erasable/programmable read-only memory |
| EWS | electronic immobilizer |
| GM | general module |
| IAT | intake air temperature (sensor) |
| IHKR | regulated heating and air conditioning system |
| KL 15 | battery positive, ignition switch on RUN, power |
| KL 30 | battery positive (B+), power |
| KL 31 | battery/chassis ground |
| KL 50 | ignition start position, power |
| LDP | fuel tank leak diagnosis pump |
| LED | light emitting diode |
| MAF | mass air flow sensor |
| MDK | motor driven throttle (DME MS 42.1) |
| MRS | multiple restraint system |
| MIL | malfunction indicator light |
| NTC | negative temperature coefficient resistor |
| OBC | on-board computer (board computer) |
| OBD II | second generation on-board diagnostics |
| PWG | accelerator pedal position sensor |
| RAM | random access memory |
| RXD | receive data line |
| SBK | battery safety terminal |
| SII | service interval indicator |
| SRS | supplemental restraint system |
| TCM | transmission control module |
| TDC | top dead center |
| TXD | transmit data line |
| ZKE | central body electronics |
| ZVM | central vehicle module |

## CENTRAL BODY ELECTRONICS (ZKE IV)

Central body electronics (ZKE IV) integrates and controls a large number of vehicle functions. The general module (GM IV) is the master controller for other ZKE system modules. It serves as the gateway for data communication with BMW diagnostic scan tool (DISplus, GT1, MoDiC). GM IV also communicates with the remote locking module for keyless entry operation.

The following functions are controlled directly by GM IV:

- Power windows. See **512 Door Windows**.
- Central locking, power trunk or rear hatch release, keyless entry, alarm system. See **515 Central Locking and Anti-theft**.
- Windshield and rear hatch wiper and washer system. See **611 Wipers and Washers**.
- Interior lighting. See **630 Lights**.
- Consumer sleep mode.

Other functions are not directly controlled by the GM IV but interconnected:

- Outside rear-view mirror control and heating. See **510 Exterior Trim, Bumpers**.
- Windshield washer jet heating. See **611 Wipers and Washers.**

## ELECTRICAL TROUBLESHOOTING

Four things are required for current to flow in any electrical circuit: a voltage source, wires or connections to transport the voltage, a load or device that uses the electricity, and a connection to ground.

Most problems can be found using a digital multimeter (volt/ohm/ammeter) to check the following:

- Voltage supply
- Breaks in the wiring (infinite resistance/no continuity)
- A path to ground that completes the circuit

Electric current is logical in its flow, always moving from the voltage source toward ground. Electrical faults can usually be located through a process of elimination. When troubleshooting a complex circuit, separate the circuit into smaller parts. General tests outlined below may be helpful in finding electrical problems. The information is most helpful when used with wiring diagrams.

Be sure to analyze the problem. Use wiring diagrams to determine the most likely cause. Get an understanding of how the circuit works by following the circuit from ground back to the power source.

When making test connections at connectors and components, use care to avoid spreading or damaging the connectors or terminals. Some tests may require jumper wires to bypass components or connections in the wiring harness. When connecting jumper wires, use blade connectors at the wire ends that match the size of the terminal being tested. The small internal contacts are easily spread apart, and this can cause intermittent or faulty connections that can lead to more problems.

## Voltage and voltage drops

Wires, connectors, and switches that carry current are designed with very low resistance so that current flows with a minimum loss of voltage. A voltage drop is caused by higher than normal resistance in a circuit. This additional resistance actually decreases or stops the flow of current. A voltage drop can be noticed by problems ranging from dim headlights to sluggish wipers. Some common sources of voltage drops are corroded or dirty switches, dirty or corroded connections or contacts, and loose or corroded ground wires and ground connections.

A voltage drop test is a good test to perform if current is flowing through the circuit but the circuit is not operating correctly. A voltage drop test will help pinpoint a corroded ground strap or a faulty switch. Normally, there should be less than 1 vdc drop across most wires or closed switches. A voltage drop across a connector or short cable should not exceed 0.5 vdc.

***NOTE—***
- *A voltage drop test is generally more accurate than a simple resistance check because the resistances involved are often too small to measure with most ohmmeters. For example, a resistance as small as 0.02 ohms would results in a 3 vdc drop in a typical 150 amp starter circuit. (150 amps x 0.02 Ω =3 vdc).*
- *Keep in mind that voltage with ignition key ON and voltage with engine running are not the same. With ignition ON and engine OFF (battery voltage), voltage should be approximately 12.6 vdc. With engine running (charging voltage), voltage should be approximately 14.0 vdc. Measure voltage at battery with ignition ON and then with engine running to get exact measurements.*

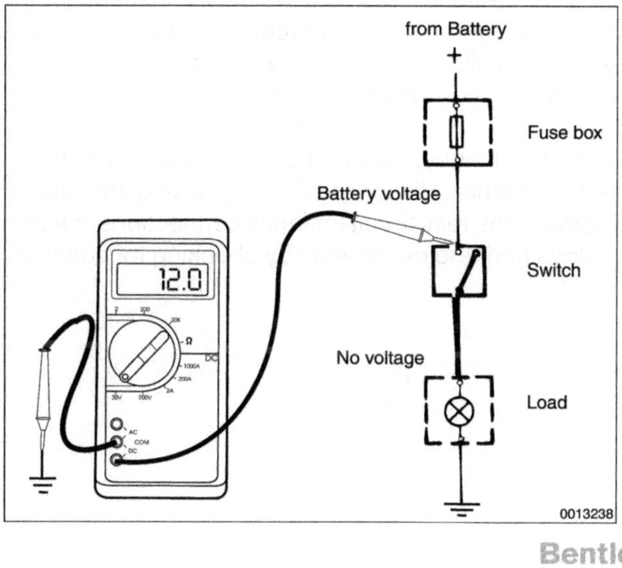


## Voltage, measuring

◄ Connect digital multimeter negative lead to a reliable ground point on car.

***NOTE—***
*The negative (-) battery terminal is always a good ground point.*

– Connect digital multimeter positive lead to point in circuit you wish to measure.

– Voltage reading should not deviate more than 1 vdc from voltage at battery. If voltage drop is more than this, check for a corroded connector or loose ground wire.

## Voltage drop, testing

Check voltage drop only when there is a load on the circuit, such as when operating the starter motor or turning on the headlights. Use a digital multimeter to ensure accurate readings.

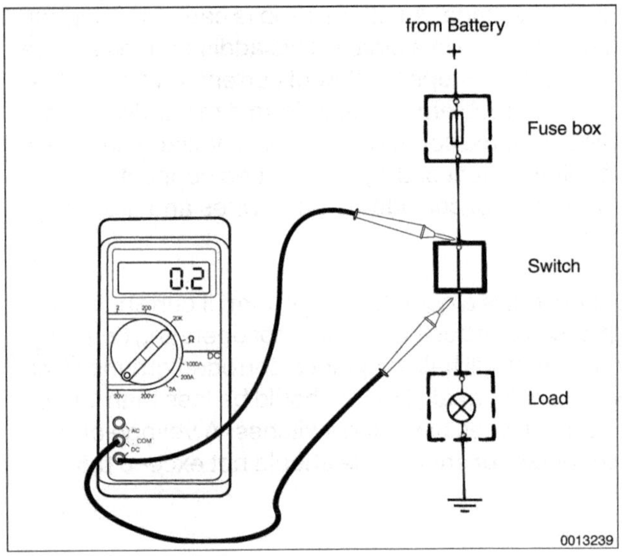


– Connect digital multimeter positive lead to positive (+) battery terminal or a positive power supply close to battery source.

– Connect digital multimeter negative lead to other end of cable or switch being tested.

– With power on and circuit working, meter shows voltage drop (difference between two points). This value should not exceed 1 vdc.

– Maximum voltage drop in an automotive circuit, as recommended by the Society of Automotive Engineers (SAE), is as follows:
  - 0 vdc for small wire connections
  - 0.1 vdc for high current connections
  - 0.2 vdc for high current cables
  - 0.3 vdc for switch or solenoid contacts

– On longer wires or cables, the drop may be slightly higher. In any case, a voltage drop of more than 1.0 vdc usually indicates a problem.

## Continuity, checking

Use continuity test to check a circuit or switch. Because most automotive circuits are designed to have little or no resistance, a circuit or part of a circuit can be easily checked for faults using an ohmmeter. An open circuit or a circuit with high resistance does not allow current to flow. A circuit with little or no resistance allows current to flow easily.

When checking continuity, turn ignition OFF. On circuits that are powered at all times, disconnect battery. Using the appropriate wiring diagram, test circuit for faulty connections, wires, switches, relays and engine sensors by checking for continuity.

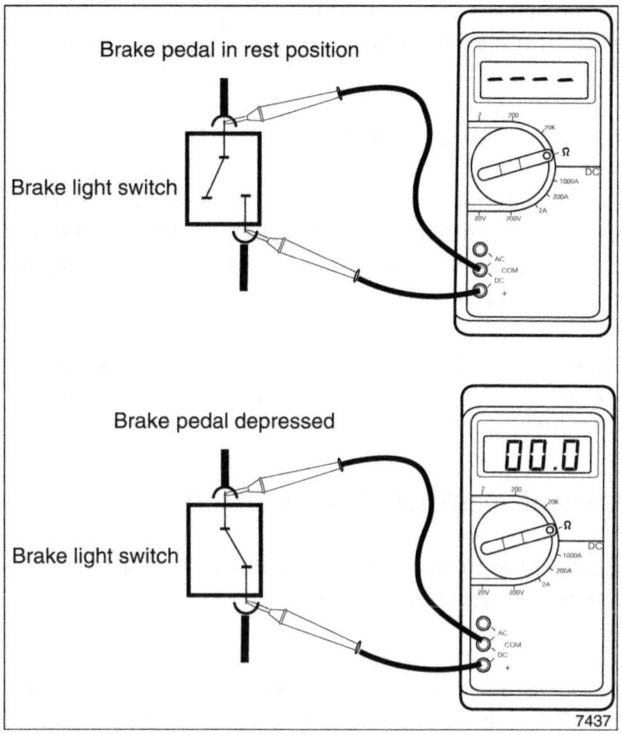


Example: Test brake light switch for continuity:

- With brake pedal in rest position (switch open) there is no continuity (infinite Ω).
- With pedal depressed (switch closed) there is continuity (0 Ω).

## Short circuits

Short circuits are exactly what the name implies. The circuit takes a shorter path than it was designed to take. The most common short that causes problems is a short to ground where the insulation on a positive (+) wire wears away and the metal wire is exposed. When the wire rubs against a metal part of the car or other ground source, the circuit is shorted to ground. If the exposed wire is live (positive battery voltage), a fuse blows and the circuit may be damaged.

Short circuits are often difficult to locate and may vary in nature. Short circuits can be found using a logical approach based on knowledge of the current path.

Use a digital multimeter to locate short circuits.

> ***CAUTION—***
> - *On circuits protected with high rating fuses (25 amp and greater), wires or circuit components may be damaged before the fuse blows. Check for wiring damage before replacing fuses of this rating.*
> - *When replacing blown fuses, confirm correct fuse rating by checking the fuse panel cover diagram.*

## Short circuit, testing with ohmmeter

– Remove blown fuse from circuit and disconnect cables from battery. Disconnect harness connector from circuit load or consumer.

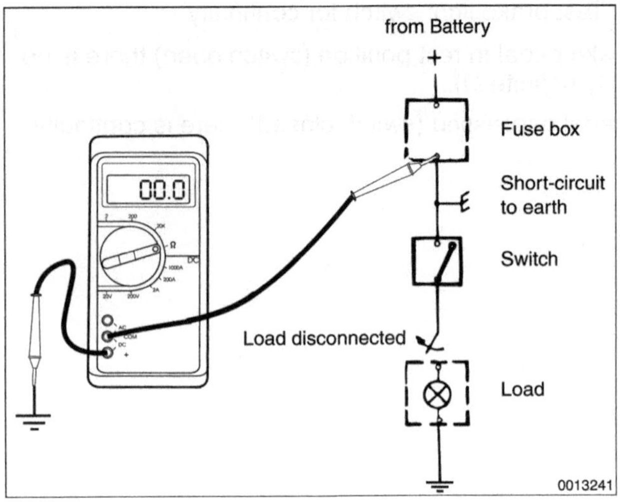


Using an ohmmeter, connect one test lead to load side of fuse terminal (terminal leading to circuit) and other test lead to ground.

– If there is continuity to ground, there is a short to ground.

– If there is no continuity, work from wire harness nearest to fuse/relay panel and move or wiggle wires while observing meter. Continue to move down harness until meter displays a reading. This is location of short to ground.

– Visually inspect wire harness at this point for any faults. If no faults are visible, carefully slice open harness cover or wire insulation for further inspection. Repair any faults found.

## Short circuit, testing with voltmeter

– Remove blown fuse from circuit. Disconnect harness connector from circuit load or consumer.

***NOTE —***
*Most fuses power more than one consumer. Be sure all consumers are disconnected when checking for a short circuit.*

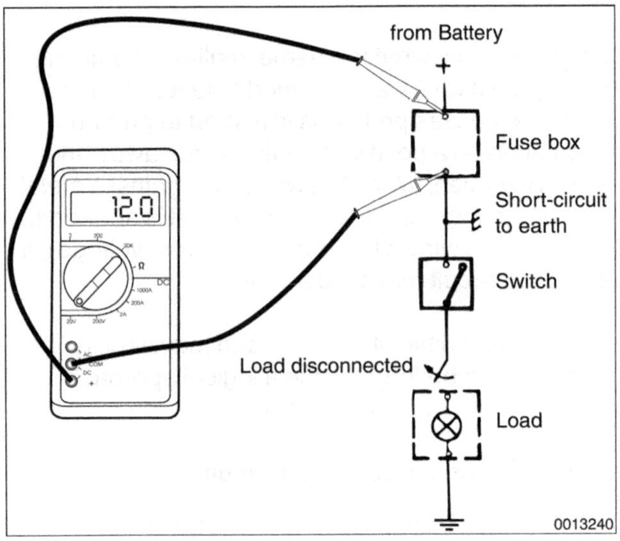


Using a digital multimeter, connect test leads across fuse terminals. Make sure power is present in circuit. If necessary turn key on.

– If voltage is present at voltmeter, there is a short to ground.

– If voltage is not present, work from wire harness nearest to fuse/relay panel and move or wiggle wires while observing meter. Continue to move down harness until meter displays a reading. This is location of short to ground.

– Visually inspect wire harness at this point for any faults. If no faults are visible, carefully slice open harness cover or wire insulation for further inspection. Repair any faults found.

# 610 Electrical Component Locations



**TABLES**

## GENERAL

This repair group covers fuse, relay, control module and ground location information, primarily via photos or illustrations.

The complex nature of Z3 electrical systems requires a large number of fuses and electrical components. Locating the correct component in this array of equipment is an important first step in any kind of electrical diagnosis. Electrical equipment and accessories installed vary depending on model and model year. Always confirm that the proper electrical component has been identified.

Investigating and correcting ground problems often clears mysterious and difficult to trace electrical problem. For ground information, see **Ground locations** in this repair group.

For additional electrical system information, see:

- **121 Battery, Starter, Alternator**
- **600 Electrical System–General**
- **ELE Electrical Wiring Diagrams**

***NOTE—***
*Acronyms commonly used in BMW technical literature are in* **600 Electrical System–General**.

### Special tools

Some special tools are needed to test and repair the electrical systems in the car.

Automotive digital multimeter

> ***CAUTION—***
> *Only use a digital multimeter when testing automotive electrical components.*

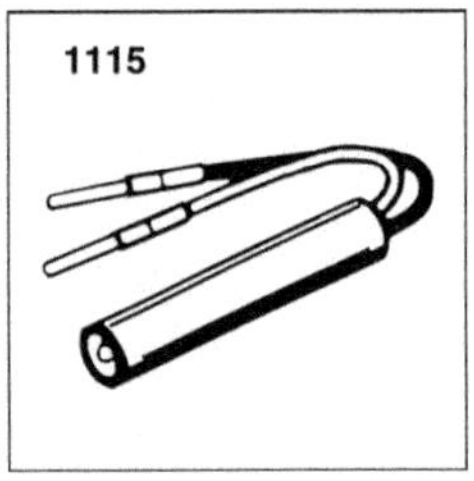


LED test light
(Baum tools 1115)

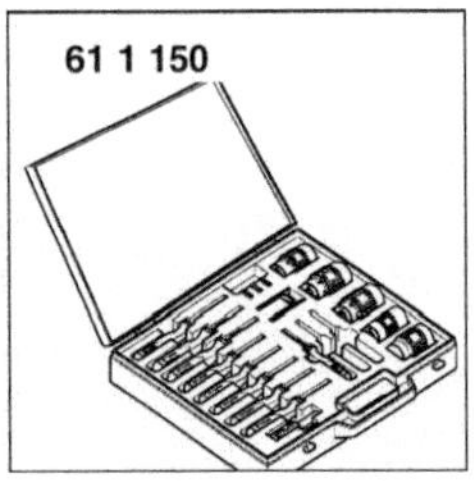


Wiring harness repair kit
(BMW special tool 61 1 150)

61 9 041

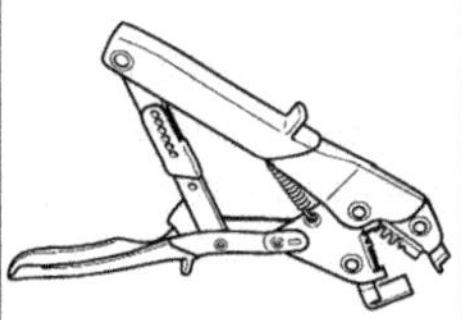

Wire end crimp tool
(BMW special tool 61 9 041)

## Electrical system safety precautions

Please read the following warnings and cautions before doing any work on your electrical system.

***WARNING—***
*The battery safety terminal, pyrotechnical seat belt tensioners, and airbags utilize explosive devices and must be handled with extreme care. Refer to the warnings and cautions in* **121 Battery, Alternator, Starter**; **720 Seat Belts**; *and* **721 Airbag System (SRS)**.

***CAUTION—***
- *Prior to disconnecting the battery, read the battery disconnection cautions in* **001 General Warnings and Cautions**.
- *Relay and fuse positions are subject to change and may vary from car to car. If questions arise, an authorized BMW dealer is the best source for the most accurate and up-to-date information.*
- *A good way to verify a relay position is to compare the wiring colors at the relay socket to the colors indicated on the wiring diagrams in* **ELE Electrical Wiring Diagrams**.
- *Always switch the ignition off and remove the negative (-) battery cable before removing any electrical components. Connect and disconnect ignition system wires, multiple connectors, and ignition test equipment leads only while the ignition is switched off.*
- *Only use a digital multimeter for electrical tests.*

## ELECTRICAL COMPONENTS TABLE

**Table a** is a cross-referenced listing of electrical components in Z3 and M version vehicles. Where available, photos of components in this repair group are referred to in the fourth column.

***NOTE —***
- *The* **Component identifier** *column in the table identifies the component by the alphanumeric code used in wiring diagrams.*
- *No Z3 vehicle is equipped with all components listed or illustrated.*

**Table a. Electrical component locations**

| Component | Component identifier | Location | Refer to |
|---|---|---|---|
| 3/2-way valve (running losses or fuel changeover valve) | Y6550 | Under car, left side, ahead of fuel filter, under protective cover | **Fig. 33** |
| A/C | | *see* Heating and A/C entries | |
| A/C compressor | Y2 | Right front lower engine | |
| A/C compressor relay | K19 | Front power distribution box, left rear engine compartment | **Fig. 13** |
| A/C control module | | *see* Heating and A/C control module | |
| A/C evaporator controller | S221 | Left side of center console, under dashboard, behind footwell duct | **Fig. 28** |
| A/C evaporator temperature sensor | B14 | Left side of center console, under dashboard, behind footwell duct | **Fig. 28** |
| A/C pressure switch | B8y, B9<br>B9508 | Right front of engine compartment at receiver / dryer | **Fig. 23** |
| A/C relay, DME, M version | K33a | Front power distribution box, left rear engine compartment | **Fig. 13** |
| ABS / ASC / DSC switch | S6, S6y | Center console below radio | |
| ABS / ASC / DSC warning lights | | In instrument cluster | |
| ABS / ASC diode | V7 | Front power distribution box, left rear engine compartment | |
| ABS / ASC electronic control module, 1996 - 1998 | A52 | Behind glove compartment | |
| ABS / ASC electronic control module, 1999 - 2002 | A52 | Left engine compartment, on ABS / ASC hydraulic unit | |
| ABS / ASC hydraulic unit | Y1w | Left engine compartment | |
| ABS / ASC pump relay | K75 | Front power distribution box, left rear engine compartment | **Fig. 13** |
| ABS / ASC relay, DME | K94 | Front power distribution box, left rear engine compartment | **Fig. 13** |
| ABS / ASC throttle valve position sensor | | Throttle position sensor | |
| ABS / ASC valve relay | K95 | Front power distribution box, left rear engine compartment | **Fig. 13** |
| ABS / DSC electronic control module | A65 | Left engine compartment on ABS / DSC hydraulic unit | **Fig. 11**<br>**Fig. 19** |
| ABS / DSC hydraulic unit | | Left engine compartment | |
| ABS / DSC precharge pump | M118 | Left rear engine compartment, under brake master cylinder | |
| ABS / DSC pressure sensor | B76 | Left rear engine compartment, on brake master cylinder | **Fig. 19** |
| ABS / DSC pressure transmitter | Y16z | Left rear engine compartment, on brake master cylinder | **Fig. 19** |
| ABS / DSC rotational (RPM) rate sensor | B75 | Under driver side carpet | **Fig. 31** |
| ABS / DSC steering angle sensor | R33 | Base of upper steering column | **Fig. 25**<br>**Fig. 26**<br>**Fig. 28** |

**Table a. Electrical component locations**

| Component | Component identifier | Location | Refer to |
|---|---|---|---|
| ABS / DSC transverse (lateral) acceleration sensor | B74 | Under driver side carpet | **Fig. 31** |
| ABS electronic control module, 1996 | A7 | Behind glove compartment | |
| ABS hydraulic unit | A202 | Left engine compartment | |
| ABS relay | K10 | Front power distribution box, left rear engine compartment | |
| ABS wheel speed sensor, front | B1, B2 | On steering arm | |
| ABS wheel speed sensor, rear | B3, B4 | On trailing arm | |
| Accelerator pedal sensor, DME MS 42.1 | | Integral part of throttle housing | |
| Accelerator pedal sensor, DME MS 43.0 | R10 | Above accelerator pedal | **Fig. 29** |
| Air flow sensor | | *see* Mass air flow sensor | |
| Air mass meter | | *see* Mass air flow sensor | |
| Airbag connector, passenger side | X782 | Right dashboard support, above glove compartment | **Fig. 30** |
| Airbag connector, steering wheel | X783 | Under steering wheel airbag | |
| Airbag contact spring | I01030 | Center of steering wheel under driver airbag, between steering wheel and column switches | |
| Airbag crash sensor, left | B78 | Under driver carpet | |
| Airbag crash sensor, right | B79 | Under passenger carpet | |
| Airbag module (SRS) | A12 | Below center console ahead of shifter | |
| Airbag warning light | | In instrument cluster | |
| Airbag, door | G14, G15 | In door trim panel | |
| Airbag, driver | G5 | Steering wheel center pad | |
| Airbag, passenger | G6 | Right dashboard | |
| Air-conditioning | | *see* A/C<br>*see also* Heating and A/C | |
| Alarm control module (Alpine) | N12w | Behind glove compartment | |
| Alarm indicator LED (DWA) | V5 | Center dash ahead of shifter | |
| Alarm siren, anti-theft (Alpine) | H1w | Left rear engine compartment | |
| Alarm siren, anti-theft (DWA) | H67 | Left rear engine compartment | |
| Alternator (generator) | G6509 | Left front of engine | |
| Amplifier, radio | | *see* Radio amplifier | |
| Antenna amplifier, Coupe | N8 | Above roof panel | |
| Antenna amplifier, Roadster | N8 | Left trunk, behind trim | **Fig. 37** |
| Anti-theft | | *see* Alarm | |
| ASC | | *see* ABS / ASC | |
| Automatic transmission control module (EGS) | A7000<br>A8500 | E-box, right rear engine compartment | |
| Automatic transmission position indicator light | E82 | Under shifter bezel | |
| Automatic transmission program switch | S8503 | Under shifter bezel | |
| Automatic transmission range switch | S3, S224z | Under shifter bezel | |
| Automatic transmission shiftlock | Y19 | Under shifter bezel | |
| Automatic transmission Steptronic switch | S224z | *see* Automatic transmission range switch | |
| Automatic transmission valve body | Y8505y<br>Y8521 | Inside transmission housing | |
| Automatic transmission valve body connector | X8505 | Right rear of transmission housing | **Fig. 34** |
| Auxiliary coolant pump | | *see* Heating and A/C auxiliary coolant pump | |

| Table a. Electrical component locations | | | |
|---|---|---|---|
| **Component** | **Component identifier** | **Location** | **Refer to** |
| Auxiliary fan | | *see* Engine cooling fan | |
| Auxiliary fuse box | | *see* **Fuse locations** in this repair group | |
| Auxiliary throttle control motor (ADS) | M93 | Left strut tower | **Fig. 18** |
| Auxiliary throttle position sensor, (ADS) | R31 | Left of intake manifold | **Fig. 11**<br>**Fig. 18** |
| B+ battery junction (high amperage fuse panel) | G7512 | Under left dashboard | **Fig. 2** |
| B+ battery junction (jumper point) | G6430 | Right rear engine compartment | **Fig. 12**<br>**Fig. 15**<br>**Fig. 16** |
| B+ jumper plug | B900 | Left footwell | |
| Back-up light relay, Steptronic transmission | K6325 | Front power distribution box, left rear engine compartment | **Fig. 13** |
| Back-up light switch, 4-speed automatic transmission | S3 | Automatic transmission range switch | |
| Back-up light switch, Getrag manual transmission | S8511 | Left of transmission housing | **Fig. 35** |
| Back-up light switch, Steptronic transmission | | Relay K6325 triggered by automatic transmission control module | |
| Back-up light switch, ZF manual transmission | S8511 | Right of transmission housing | |
| Battery safety terminal (SBK) | G19 | On positive battery terminal in trunk | **Fig. 40** |
| Battery, 1996 | G1 | Right rear engine compartment | **Fig. 14** |
| Battery, 1997 - 2002 | G1 | Right side trunk, behind trim panel | **Fig. 40** |
| Blower | | Heating and A/C blower motor | |
| Board computer | A35y | Center dashboard, below heating and A/C controls | |
| Body electronics control module | | General module | |
| Brake fluid level relay | K10485 | Right footwell behind trim | |
| Brake fluid level sensor | B18 | Top of brake fluid reservoir, left rear engine compartment | **Fig. 12**<br>**Fig. 18**<br>**Fig. 19** |
| Brake fluid pressure sensor | | *see* ABS / DSC pressure sensor | |
| Brake light switch | S30 | Above brake pedal | **Fig. 25** |
| Brake pad wear sensor connector, left front | X118 | In left front wheel housing | |
| Brake pad wear sensor connector, right rear | X713 | Ahead of right rear wheel housing | |
| Brake pad wear sensor, left front | B16 | Brake caliper, left front | |
| Brake pad wear sensor, right rear | B17 | Brake caliper, right rear | |
| Camshaft actuator (VANOS) | | *see* VANOS | |
| Camshaft sensor (cylinder identification sensor), 4-cylinder | B6209 | Upper timing chain cover, front of engine | |
| Camshaft sensor, exhaust, 6-cylinder | B6224 | Right front of cylinder head | **Fig. 12** |
| Camshaft sensor, intake, 6-cylinder | B6204 | Left front of cylinder head | |
| Carbon canister shut off valve | B6247 | Under left rear of car | |
| CD changer | N22 | Left trunk, behind trim | **Fig. 40** |
| Center brake light | H34 | | |
| Central body electronics (ZKE) control module | | *see* General module | |
| Central locking module (ZVM), 1996 | A30 | Behind glove compartment | |
| Charge pressure sensor | | see ASC / DSC pressure sensor | |
| Chime module | H10 | Dash panel above driver footwell | |
| Cigarette lighter | E28 | Center console near shifter | |

| Table a. Electrical component locations | | | |
|---|---|---|---|
| **Component** | **Component identifier** | **Location** | **Refer to** |
| Cigarette lighter light | E29 | Center console near shifter | |
| Climate control | | *see* A/C<br>*see also* Heating and A/C | |
| Clock, analog | N5 | Center dashboard | |
| Clock, digital | A110 | Center dashboard | |
| Clutch pedal position switches | S32, S235 | Above clutch pedal | **Fig. 25** |
| Combination (stalk) switch | S7 | On steering column | **Fig. 27** |
| Comfort relay | K78 | Left footwell, behind trim | |
| Compressor | | *see* A/C compressor | |
| Contact ring | | *see* Horn contact ring<br>*see also* Airbag contact spring | |
| Convertible top end-position microswitch | S164 | At convertible top linkage | |
| Convertible top motor | M101 | Left trunk, behind trim | **Fig. 37** |
| Convertible top relays | K18363<br>K18364 | Left trunk, behind trim | **Fig. 37** |
| Convertible top switch | S143 | Rear center console | |
| Coolant thermostat | | *see* Thermostat, characteristic map cooling | |
| Crankshaft sensor, 4-cylinder | B6203 | Left rear engine block under intake manifold | |
| Crankshaft sensor, 6–cylinder | B6203 | Left rear engine block below starter motor | **Fig. 21** |
| Crash control module,<br>1996 - 1999, late 2000 - 2002 | N17 | Driver footwell | **Fig. 24** |
| Crash control module, early 2000 Z3 | N17 | Behind glove compartment | |
| Crash sensor | | *see* Airbag crash sensor | |
| Cruise control actuator (Tempomat) | M25 | Left engine compartment | **Fig. 18** |
| Cruise control module (Tempomat), 1996 - 1998 | A8 | Behind glove compartment | |
| Cruise control module, 1999 - 2002 | A9008 | Under left dashboard, left of steering column | **Fig. 26** |
| Cruise control switch | S28 | Right of steering column | **Fig. 27** |
| Data link connector (DLC), 20–pin,<br>4-cylinder, M52, S52 engine | D100 | Right strut tower | **Fig. 11**<br>**Fig. 17** |
| Data link connector (DLC), 20–pin,<br>M52 TU, M54, S54 engine | D100 | Left strut tower | **Fig. 12**<br>**Fig. 19** |
| Data link connector (OBD II), 16–pin | D110 | Right side of center console | |
| DISA | | *see* Intake air resonance changeover valve (DISA) | |
| DME control module | | *see* Engine control module (ECM) | |
| DME main relay | K6300 | E-box, right rear engine compartment | **Fig. 14**<br>**Fig. 15**<br>**Fig. 16** |
| DMTL | | *see* Fuel tank leakage diagnosis module (LDP or DMTL) | |
| Door contact switch | S13, S14 | B pillar | |
| Door lock microswitch | S47, S49 | In door lock assembly | |
| Door lock motor | M12, M13 | In door | |
| Double temperature switch | | *see* Engine cooling fan switch | |
| DSC | | *see* ABS / DSC | |
| DWA | | *see* Alarm | |
| Dynamic stability control | | *see* ABS / DSC | |

**Table a. Electrical component locations**

| Component | Component identifier | Location | Refer to |
|---|---|---|---|
| E-box | | Right rear engine compartment, under plastic cover | **Fig. 12**<br>**Fig. 14**<br>**Fig. 15**<br>**Fig. 16** |
| E-box cooling fan | M6506 | Under E-box, right rear engine compartment | |
| E-box fuse holder | | *see* **Fuse locations** in this repair group | **Fig. 4**<br>**Fig. 16** |
| ECT sensor | | *see* Engine coolant temperature (ECT) sensor | |
| EGS | | *see* Automatic transmission control module (EGS) | |
| Electric coolant thermostat | | *see* Thermostat, characteristic map cooling | |
| Electrochromic rear view mirror | A22z | Top of windshield | |
| Electronic driveaway protection | | *see* EWS | |
| Electronic immobilizer | | *see* EWS | |
| Electronic throttle actuator (EDK), MS 43.0 | Y6390 | Throttle housing, left of intake manifold | |
| Electronics box | | *see* E-box | |
| Emergency flasher | | *see* Hazard flasher | |
| Engine compartment light switch | | *see* Engine hood contact switch | |
| Engine control module (ECM) | A6000 | E-box, right rear engine compartment | **Fig. 14**<br>**Fig. 15**<br>**Fig. 16** |
| Engine control module relay | | *see* DME main relay | |
| Engine coolant temperature (ECT) sensor, 4-cylinder engine | B6232 | left front of cylinder head | |
| Engine coolant temperature (ECT) sensor, M52 TU, M54 engine | B6236 | Under rear of intake manifold at rear of cylinder head | |
| Engine coolant temperature (ECT) sensor, M52, S52, S54 engine | B6256 | Below intake manifold runner 1 | |
| Engine cooling fan (electric) | M9 | Front of radiator | |
| Engine cooling fan high speed relay | K22 | Front power distribution box, left rear engine compartment | **Fig. 13** |
| Engine cooling fan normal speed relay | K21 | Front power distribution box, left rear engine compartment | **Fig. 13** |
| Engine cooling fan switch | S36y | Right side of radiator | **Fig. 22** |
| Engine electrical connectors | | Left rear engine compartment | **Fig. 11**<br>**Fig. 18** |
| Engine electronics fuse carrier | P98 | E-box, right rear engine compartment | **Fig. 4**<br>**Fig. 12**<br>**Fig. 16** |
| Engine hood contact switch | S21w | Near left headlight assembly | **Fig. 12** |
| Engine hood microswitch | S825 | Near left headlight assembly | |
| Evaporative emission control valve, 4-cylinder | Y6120 | Left strut tower | |
| Evaporative emission control valve, M52 TU, M54, S54 engine | Y6120 | Under intake manifold at throttle housing | **Fig. 20** |
| Evaporative emission control valve, M52, S52 engine, | Y6120 | Near mass air flow sensor | **Fig. 11** |
| Evaporator temperature sensor | | *see* A/C evaporator temperature sensor | |
| EWS II control module, 1997 - 1998 | A836 | Behind glove compartment | |
| EWS II transmitter module, 1997 - 1998 | A835 | Near steering column | |
| EWS III control module, 1999 - 2002 | A837 | Under left dashboard, left of steering column | **Fig. 26** |

| Table a. Electrical component locations | | | |
|---|---|---|---|
| **Component** | **Component identifier** | **Location** | **Refer to** |
| EWS ring antenna (toroidal ring coil) | L1 | Around ignition switch | **Fig. 27** |
| Exhaust camshaft sensor | | *see* Camshaft sensor | |
| Exhaust temperature sensor, S54 engine | B5351 | Exhaust pipe, right rear engine compartment | |
| Final stage, blower motor | | *see* Heating and A/C blower final stage | |
| Foglight relay, front | K47 | Front power distribution box, left rear engine compartment | **Fig. 13** |
| Foglight switch | S10 | Left dashboard | |
| Fresh air recirculation motor | | *see* Heating and A/C recirculation motor | |
| Front power distribution box | P90 | Left rear engine compartment | **Fig. 11**<br>**Fig. 12**<br>**Fig. 13** |
| Fuel changeover valve | | *see* 3/2–way valve | |
| Fuel injection valve (fuel injector) | Y6101 -<br>Y6106 | At intake manifold | |
| Fuel injector electrical harness | | Top of engine, under plastic cover | **Fig. 12**<br>**Fig. 18** |
| Fuel injector relay | K6327 | E-box, right rear engine compartment | **Fig. 16** |
| Fuel level sensor | B6y | Top of fuel tank, behind right seat, combined with fuel pump | |
| Fuel pump | M2 | Top of fuel tank, behind right seat, combined with fuel level sensor | |
| Fuel pump relay, 1996 | K6301w | E-box, right rear engine compartment | **Fig. 14** |
| Fuel pump relay,<br>1997 - 2002 Z3, 1998 - 2000 M version | K6301w | Underneath left dashboard | **Fig. 24** |
| Fuel pump relay, 2001 - 2002 M version | K6301z | Under rear of center console | |
| Fuel tank flap lock motor | M16 | Right rear wheel housing | |
| Fuel tank leakage diagnosis module (DMTL) | M119z | Under car, left side, ahead of fuel filter, under protective cover | |
| Fuel tank leakage diagnosis pump (LDP) | Y6139 | Under car, left side, ahead of fuel filter, under protective cover | **Fig. 33** |
| Fuel tank pressure sensor | B65 | Top of fuel tank | |
| Fuel tank vent valve | | *see* Evaporative emissions control valve | |
| Fuses and fuse panels | | *see* **Fuse locations** in this repair group | |
| Fuses, engine electronics | | *see* Engine electronics fuse pack | |
| Fusible link | P99 | Right trunk | **Fig. 1** |
| Gas generator, airbag | | *see* Airbag | |
| Gear position / neutral safety switch | | *see* Automatic transmission range switch | |
| Gear position indicator light | E96 | Under shifter bezel | |
| General module | A104 | Behind glove compartment | **Fig. 30** |
| Generator | | *see* Alternator | |
| Glove compartment light switch | S37 | Glove compartment top | |
| Grounds | | *see* **Ground locations** in this repair group. | |
| Hard top connector 1 | E99 | Near left rear wheel housing | |
| Hazard flasher | K16 | Front power distribution box, left rear engine compartment | **Fig. 13** |
| Hazard flasher switch | S18 | On center console behind shifter | |
| Headlight / foglight washer pump | M7 | On washer fluid reservoir, right front engine compartment | |
| Headlight dimmer / flasher | N3 | Combination (stalk) switch left of steering column | |
| Headlight relay, daylight (Canada) | K46z | Underneath left dashboard, behind kick panel | |
| Headlight relay, high beam | K46 | Underneath left dashboard, behind kick panel | **Fig. 24** |
| Headlight relay, low beam | K48 | Front power distribution box, left rear engine compartment | **Fig. 13** |

**Table a. Electrical component locations**

| Component | Component identifier | Location | Refer to |
|---|---|---|---|
| Heater | | *see also* A/C | |
| Heater blower | | *see* Heating and A/C blower motor | |
| Heater valve assembly, M version | Y4z | Left strut tower | |
| Heater valve assembly, Z3 | Y4z | Rear engine compartment on bulkhead | **Fig. 12** |
| Heating and A/C auxiliary coolant pump relay, M version | K104 | Front power distribution box, left rear engine compartment | **Fig. 13** |
| Heating and A/C auxiliary coolant pump, M version | M37a | Left strut tower | |
| Heating and A/C blower final stage (resistor pack), 1996 | R11 | Behind center of dashboard | |
| Heating and A/C blower final stage (resistor pack), 1997 - 2002 | R11 | Behind engine compartment bulkhead | |
| Heating and A/C blower motor | M30 | Behind engine compartment bulkhead | |
| Heating and A/C blower relay | K4 | Front power distribution box, left rear engine compartment | **Fig. 13** |
| Heating and A/C blower switch | S75 | Center dashboard | |
| Heating and A/C control panel illumination | E105, E106 | Behind panel in center dashboard | |
| Heating and A/C recirculation motor | M111 | Left side of center console, under dashboard, behind footwell duct | **Fig. 28** |
| Heating and A/C recirculation switch | S220 | Center dashboard | |
| Heating and A/C temperature wheel | S531 | Center dashboard | |
| High speed relay | | *see* Engine cooling fan high speed relay | |
| Hood switch | | *see* Engine hood switch | |
| Horn | | *see* Alarm siren | |
| Horn contact ring | C100, C101 | Top of steering column, under steering wheel | |
| Horn relay | K2 | Front power distribution box, left rear engine compartment | **Fig. 13** |
| Horn switch | S4 | Steering wheel, behind impact cushion | |
| Horn, left | H2 | Under left side front bumper | |
| Horn, right | H3 | Under right side front bumper | |
| Hot film mass air flow sensor | | *see* Mass air flow sensor | |
| Idle speed control valve, 4-cylinder engine | Y6130 | Left of intake manifold | **Fig. 18** |
| Idle speed control valve, M52, M52 TU, M54, S52 engine | Y6130 | Below center of intake manifold | |
| Idle speed control valve, S54 engine | Y6130 | Front of intake manifold | |
| Ignition coil circuit resistor | R8554 | Top of cylinder head | |
| Ignition coil pack, 4-cylinder engine | T6151 - T6154 | Right front engine compartment at strut tower | **Fig. 17** |
| Ignition coil, 6-cylinder engine | T6151 - T6156 | Top of cylinder head, above spark plugs, under cover | **Fig. 12** |
| Ignition switch | S2 | Right side steering column | **Fig. 27** |
| IHKR | | *see* A/C<br>*see also* Heating and A/C | |
| Impact sensor | | *see* Airbag crash sensor | |
| Individual control intake system valve | | *see* Intake manifold resonance valve | |
| Inertia switch | S27 | Right footwell behind speaker grill | |
| Instrument cluster module | A2w | Back of instrument cluster | |
| Intake air temperature (IAT) sensor, 4-cylinder | B6205 | Top of air filter housing | **Fig. 18** |

| Table a. Electrical component locations | | | |
|---|---|---|---|
| **Component** | **Component identifier** | **Location** | **Refer to** |
| Intake air temperature (IAT) sensor, M52 TU, M54 engine | B6205 | Top center of engine, between intake manifold and cylinder head | **Fig. 12** |
| Intake air temperature (IAT) sensor, M52, S52 engine | B6205 | Under intake manifold behind throttle housing | |
| Intake camshaft sensor | | *see* Camshaft sensor | |
| Intake manifold resonance changeover valve (DISA), 4-cylinder engine | Y6131 | Left of intake manifold | |
| Intake manifold resonance valve solenoid, DME MS 42.1, MS 43.0 | Y6167 | Left of intake manifold | **Fig. 12** |
| Interior light switch | S408 | Front of headliner at mirror | |
| Interior light, front | E34, E800 | Front of headliner at mirror | |
| Interior motion sensor and control module, Coupe | A121 | In headliner, center | |
| Interior motion sensor and control module, Roadster | A121 | Right center console | |
| Jet heater thermoswitch | | *see* Windshield washer jet heater thermoswitch | |
| Keyless entry antenna | | With keyless entry control module in interior rear view mirror | |
| Keyless entry control module | A194y | Interior rear view mirror | |
| Kickdown switch, automatic transmission | S8507 | Under accelerator pedal | |
| Knock sensor | B6240<br>B6241<br>B6241z<br>B6242<br>B6243<br>B62400 | Under intake manifold | |
| Lateral acceleration sensor | | *see* ABS / DSC transverse (lateral) acceleration sensor | |
| Leakage diagnosis pump (LDP) | | *see* Fuel tank leakage diagnosis module (LDP or DMTL) | |
| Light switch assembly | S8 | Left dashboard | |
| Load reduction relay | | *see* Unloader relay | |
| Luggage compartment light, Coupe | E32 | Rear headliner | |
| Mass air flow sensor | B6217<br>B6207 | Left engine compartment, in air intake duct | **Fig. 11**<br>**Fig. 12**<br>**Fig. 18** |
| Neutral safety switch | | *see* Automatic transmission range switch | |
| Normal speed relay | | *see* Engine cooling fan normal speed relay | |
| OBD II connector | | *see* Data link connector (OBD II), 16–pin | |
| Oil level sensor | B6254 | Bottom of oil pan | **Fig. 33** |
| Oil level sensor control module | A19007 | Behind glove compartment | **Fig. 30** |
| Oil pressure switch | B6231 | Left front of engine on oil filter housing | **Fig. 20** |
| Oil temperature / outside temperature gauge module | A3984 | Behind glove compartment | **Fig. 30** |
| Oil temperature gauge | B59 | Center dashboard | |
| Oil temperature sensor | B6238 | Left front of engine at oil filter housing | **Fig. 20** |
| On-board computer | | *see* Board computer | |
| Outside air temperature sensor, 4-cylinder engine | B21 | Left wheel housing | |
| Outside air temperature sensor, 6-cylinder engine | B21 | Under right corner of front bumper | |
| Outside mirror | Y5, Y6 | On door | |

**Table a. Electrical component locations**

| Component | Component identifier | Location | Refer to |
|---|---|---|---|
| Outside mirror control switch | S17 | Left door pull | |
| Oxygen sensor, 4-cylinder, M52, post-catalyst | B62002 | Pipe behind catalyst, underneath vehicle | |
| Oxygen sensor, 4-cylinder, M52, precatalyst | B62001 | Pipe in front of catalyst, underneath vehicle | |
| Oxygen sensors, M52 TU, M54, S54 precatalyst | B62101<br>B62201 | Top of exhaust manifolds, right engine compartment | |
| Oxygen sensors, M52 TU, M54, S54 post-catalyst | B62102<br>B62202 | Bottom of exhaust manifold, underneath vehicle | |
| Oxygen sensors, S52, post-catalyst | B62201<br>B62202 | Pipes behind catalysts, underneath vehicle | |
| Oxygen sensors, S52, precatalyst | B62101<br>B62102 | Bottom on exhaust manifolds, right engine compartment | |
| Park / neutral position switch | | *see* Automatic transmission range switch | |
| Parking brake switch | S31 | Under center console, rear of parking brake handle | |
| Parking light relay, left | K26 | Front power distribution box, left rear engine compartment | **Fig. 13** |
| Parking light relay, right, and license plate light relay | K25 | Front power distribution box, left rear engine compartment | **Fig. 13** |
| Pedal position sensor | | *see* Accelerator pedal sensor, DME MS 43.0 | |
| Power window | | *see* Window | |
| Precharge pump | | *see* ABS / DSC precharge pump | |
| Pressure sensor | | *see* ABS / DSC pressure sensor<br>*see also* A/C pressure switch | |
| Program switch | | *see* Automatic transmission program switch | |
| Pyrotechnical seat belt tensioner | | *see* Seat belt tensioner, pyrotechnical | |
| Radiator fan | | *see* Engine cooling fan (electric) | |
| Radio | N9 | Center dashboard | |
| Radio amplifier, 1996 | A18 | Left trunk behind trim | |
| Radio amplifier, 1997 - 2002 | A18 | Right trunk behind trim | **Fig. 40** |
| Radio amplifier, subwoofer | N44 | Left trunk behind trim | |
| Radio antenna | | *see* Antenna | |
| Radio anti-theft switch | S20w | Behind center dashboard | |
| Radio fuse holder | A605 | Behind center dashboard | |
| Rear window defogger switch, Coupe | S76 | Center dashboard | |
| Rear window defogger, Coupe | E9 | Rear window | |
| Rear wiper / washer | | *see* Washer / wiper motor, rear | |
| Recirculation switch | | *see* Heating and A/C recirculation switch | |
| Remote entry | | *see* Keyless entry | |
| Resonance valve solenoid, intake manifold | | *see* Intake manifold resonance valve solenoid | |
| Reversing light switch | | *see* Back-up light switch | |
| Ring coil (toroidal ring antenna) | | *see* EWS ring antenna | |
| Rotational (RPM) rate sensor | | *see* ABS / DSC rotational (RPM) rate sensor | |
| Running losses valve | | *see* 3/2 valve | |
| Seat adjustment motors | | Underside of seat | |
| Seat adjustment switch | S51, S56 | Seat base, outboard side | |
| Seat back heater | E56, E59 | In seat backrest | |
| Seat belt microswitch | S58, S59 | Seat belt buckle | |

**Table a. Electrical component locations**

| Component | Component identifier | Location | Refer to |
|---|---|---|---|
| Seat belt tensioner, pyrotechnical | G12, G23 | Under seat at inner seat rail | |
| Seat heater | E57, E58 | In front seat cushion | |
| Seat heater switch | S53, S54 | Center console, below IHKR control panel | |
| Seat limit switch | S910, S911 | In seat backrest | |
| Seat occupancy detector module | A113 | Front of passenger seat | |
| Seat occupancy sensor mat | I1008 | Passenger seat cushion | |
| Secondary air pump | Y6133 | E-box, right rear engine compartment | **Fig. 11**<br>**Fig. 12**<br>**Fig. 15**<br>**Fig. 16** |
| Secondary air pump relay, 4-cylinder engine | K6304 | Front power distribution box, left rear engine compartment | **Fig. 13** |
| Secondary air pump relay, 6-cylinder engine | K6304 | E-box, right rear engine compartment | **Fig. 15**<br>**Fig. 16** |
| Secondary air valve | Y6163 | Left rear engine, under intake manifold | |
| Shift interlock cable | | *see* Under shifter bezel | |
| Shiftlock solenoid | | *see* Automatic transmission shiftlock | |
| Siren | | *see* Alarm siren | |
| Slip control | | *see* ABS entries | |
| Slip ring, airbag contact | | *see* Airbag contact spring | |
| Sound system amplifier | | *see* Radio amplifier | |
| Speed sensor | | *see* ABS wheel speed sensor<br>*see also* Vehicle speed sensor | |
| Spiral spring | | *see* Airbag contact spring | |
| SRS control module | | *see* Airbag module (SRS) | |
| Starter | M6510 | Left rear of engine, under intake manifold | **Fig. 21** |
| Starter immobilization switch | | *see* Clutch pedal position switches | |
| Starting interlock relay, 1996 | K6324 | E-box, right rear engine compartment | **Fig. 14** |
| Steering angle sensor | | *see* ABS / DSC steering angle sensor | |
| Stop light switch | | *see* Brake light switch | |
| Sucking jet solenoid | Y6165 | Left rear engine compartment, right of brake fluid reservoir | |
| Sunroof control module, Coupe | A33 | In roof panel | |
| Sunroof motor, Coupe | M24y | In roof panel | |
| Sunroof switch, Coupe | S384 | In roof panel | |
| Supplemental restraint system (SRS) | | *see* Airbag | |
| Suppression capacitor, S54 | C6000 | Center cylinder head between ignition coils 2 and 3 | |
| Tailgate lock microswitch, Coupe | S117 | In tailgate | |
| Tailgate lock motor relay | K70 | Front power distribution box, left rear engine compartment | **Fig. 13** |
| Tailgate lock motor, Coupe | M17 | In tailgate | |
| Temperature switch | | *see* Engine cooling fan switch | |
| Thermostat, characteristic map cooling | Y6279 | Thermostat housing, front of engine | |
| Throttle position sensor, 4-cylinder, M52, S52 engine | R6252 | Left of engine at throttle housing | **Fig. 18** |
| Throttle position sensor, S54 engine | R6252 | Left of engine ahead of intake manifold | |
| Throttle valve actuator (EDK) (MS 43.0) | R6390 | Left of intake manifold | |
| Throttle valve actuator (MDK) (MS 42.1) | R6385 | Left of intake manifold | |

**Table a. Electrical component locations**

| Component | Component identifier | Location | Refer to |
|---|---|---|---|
| Tilt sensor | B28 | Left trunk behind trim | **Fig. 37** |
| Toroidal coil | | *see* EWS ring antenna (toroidal ring coil) | |
| Traction control | | *see* ABS entries | |
| Transmission connector | | *see* Automatic transmission valve body connector | |
| Transmission control module | | *see* Automatic transmission control module (EGS) | |
| Transmission range switch | | *see* Automatic transmission range switch | |
| Transverse acceleration sensor | | *see* ABS / DSC transverse (lateral) acceleration sensor | |
| Trunk lid lock microswitch | S117 | Underside of trunk lid, near license plate | |
| Trunk light switch | S826 | In trunk lid lock | |
| Trunk light, Roadster | E32 | In trim panel on trunk lid | |
| Trunk lock motor | M17 | In trunk lid | |
| Turn-signal / headlight dimmer switch | | Combination stalk switch on steering column | **Fig. 27** |
| Ultrasonic sensor (motion detector) | | *see* Interior motion sensor and control module | |
| Uncoupling diode, lights | V9, V10 | Front power distribution box, left rear engine compartment | |
| Unloader relay, terminal 15 | K6326 | E-box | **Fig. 14**<br>**Fig. 15** |
| Unloader relay, terminal 15 | K9 | Front power distribution box, left rear engine compartment | **Fig. 13** |
| Vanity mirror switch | S77, S78 | Sun visor | |
| VANOS control solenoid, exhaust camshaft | Y6276 | Right front of cylinder head | |
| VANOS control solenoid, intake camshaft | Y6275z | Left front of cylinder head | **Fig. 12** |
| VANOS valves, S54 engine | Y5255 | Front of cylinder head | |
| Vehicle speed sensor | B10 | On differential | **Fig. 36** |
| Voltmeter | B211 | Center dashboard | |
| Washer pump | M4 | Right front engine compartment on washer fluid reservoir | |
| Water valve | | *see* Heater valve assembly | |
| Wheel speed sensor | | *see* ABS wheel speed sensor | **Fig. 36** |
| Window motor | M21, M23 | Inside door | |
| Window switch | S126y<br>S127y | Center console, behind shifter | |
| Windshield washer jet heater | E51, E52 | Underside of engine hood at windshield washer nozzles | |
| Windshield washer jet heater thermoswitch, 1997 - 2002 | S198 | Right frame rail ahead of engine | |
| Windshield washer jet heater thermoswitch, 1999 | S198 | Left wheel housing | |
| Windshield wiper motor | M3 | Rear of engine compartment, behind bulkhead | |
| Wiper / washer control module | A51, A64 | Behind glove compartment | **Fig. 30** |
| Wiper / washer motor, rear, Coupe | A36 | In tailgate | |
| Wiper / washer switch (stalk) | S5 | Right side steering column | |
| Yaw sensor | | *see* ABS/DSC rotational (RPM) rate sensor | |
| ZKE control module | | *see* General module | |

## Fuse locations

Depending on year, model and equipment level, fuses are placed in a variety of locations throughout the vehicle. See **Table b**.

*NOTE—*
*In case of an isolated fuse the fuse rating is given in* **Table b**.

**Table b. Fuse panels in Z3 and M version cars**

| Fuse panel | Fuses | Rating in amps | Application | Location | Refer to |
|---|---|---|---|---|---|
| A605 | F953 | 2 | Heater control valve | Center of dashboard, behind radio | |
| G7512 | F101 - F108 | | High amperage fuses, 1999 - 2002 Z3, 2001 - 2002 M version | Under left side of dashboard | **Fig. 2** **Fig. 24** |
| P90 | F1 - F46 | | Front power distribution box 1996 - 2000 | Left rear engine compartment | **Fig. 3** |
| | F1 - F54 | | Front power distribution box 2001 - 2002 | Left rear engine compartment | |
| P95 | F49 | 50 | DME main relay, 1998 - 2000 all models | Under left side of dashboard | |
| P97 | F48 | 40 | IHKR (cooling fan relays) | Under left side of dashboard | |
| P98 | F201 - F205 | | Engine management fuse holder, 1999 - 2002 Z3, 2001 - 2002 M version | E-box, right rear engine compartment | **Fig. 4** **Fig. 11** **Fig. 12** **Fig. 16** |
| | F97 | 80 | Fusible link, 1997 - 1998 Z3 Fusible link, 1998 - 2000 M version | Near battery in trunk | |
| P99 | F100 | 200 | Fusible link, 1999 - 2002 Z3 Fusible link, 2001 - 2002 M version | Right front trunk, ahead of battery, under trim | **Fig. 1** |
| P900 | F51 - F53 | | DSC fuses, 1999 - 2000 Z3 | Left rear engine compartment, right of front power distribution box | **Fig. 3** |
| P901 | F201 | 15 | 2000 M version | Right side trunk | |

### High amperage fuses

**Fig. 1 P99 fusible link holder: F97 or F100**

In trunk near battery, under floor trim.

*NOTE—*
*2000 Z3 trunk illustrated. Other models are similar.*

**Table c. Fusible link holder P99 (in trunk)**

| Fuse | Rating in amps | Application |
|---|---|---|
| F97 | 80 | 1997 - 1998 Z3 1998 - 2000 M version |
| F100 | 200 | 1999 - 2002 Z3 2001 - 2002 M version |

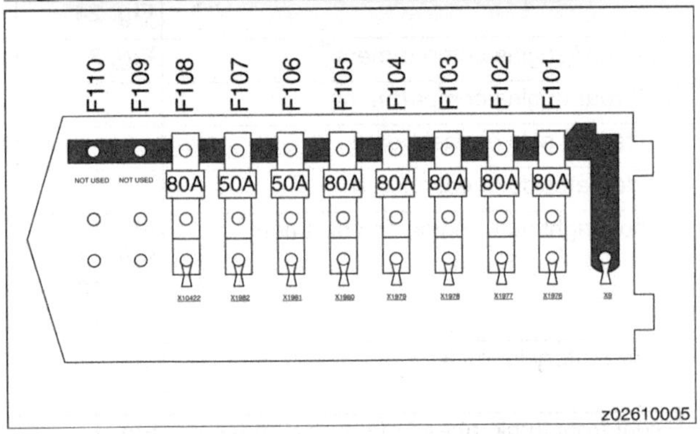


**Fig. 2** **G7512 fuse panel: F101 - F108**
**1999 - 2002 Z3, 2001 - 2002 M version**

Under left side of dash board:

- Remove trim panels under left side of dash. See **513 Interior Trim**.
- Remove left knee bolster.
- Remove power distribution fuse holder covers.

**Table d. Fuse panel G7512, fuses F101 - F108 (under left side of dash board)**

| Fuse | Rating in amps | Protected circuits |
|---|---|---|
| F101 | 80 | Fuses F31 - F36 |
| F102 | 50 | Fuses F14, F18, F20, F51 - F53 |
| F103 | 50 | Data link connector (DLC)<br>Fuses F1 - F2, F6, F19, F29 - F30 |
| F104 | 80 | Front fog light relay<br>Low beam light relay<br>Fuses F5, F7 - F12, F15 |
| F105 | 80 | Fuses F3, F37 - F40 |
| F106 | 50 | Ignition switch |
| F107 | | no information |
| F108 | 80 | B+ (battery jumper) at E-box<br>DME main relay<br>P98 engine electronics fuse panel in E-box<br>Engine control module (ECM)<br>Transmission control module (EGS) |

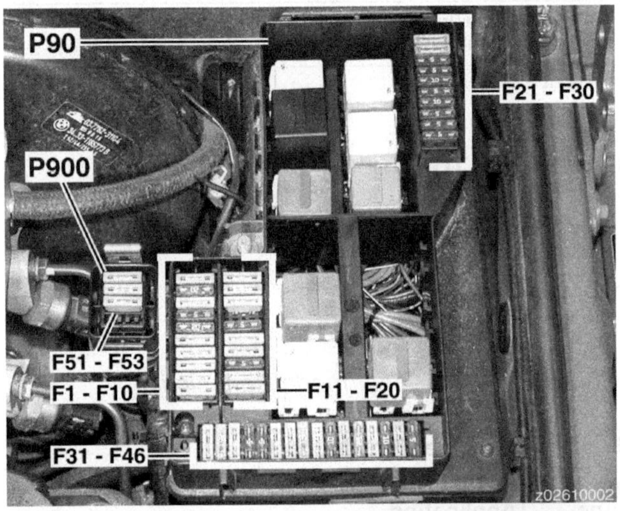


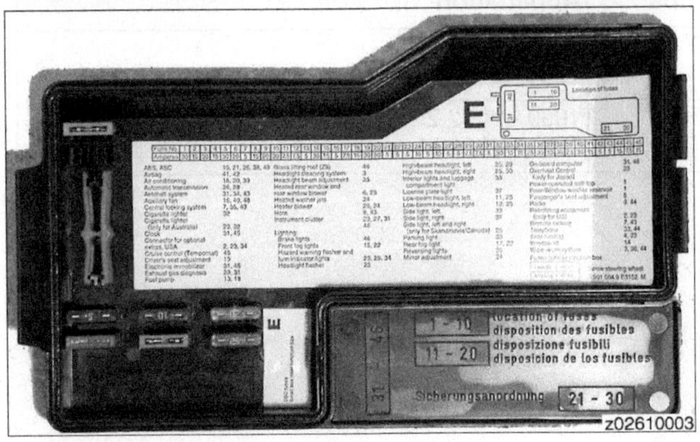


## Front power distribution box

**Fig. 3** **P90: fuses F1 - F46**
**P900: ABS / DSC fuses F51 - F53**

P90: Left rear engine compartment, under plastic cover
P900: Left rear engine compartment, to right of P90, under plastic cover

***NOTE—***

- *2000 Z3 front power distribution box illustrated. Earlier models are similar.*
- *In 2001 - 2002 models, fuses F51 - F54 are fitted in front power distribution box.*

P90: Fuse applications and capacities are listed on underside of front power distribution box cover.

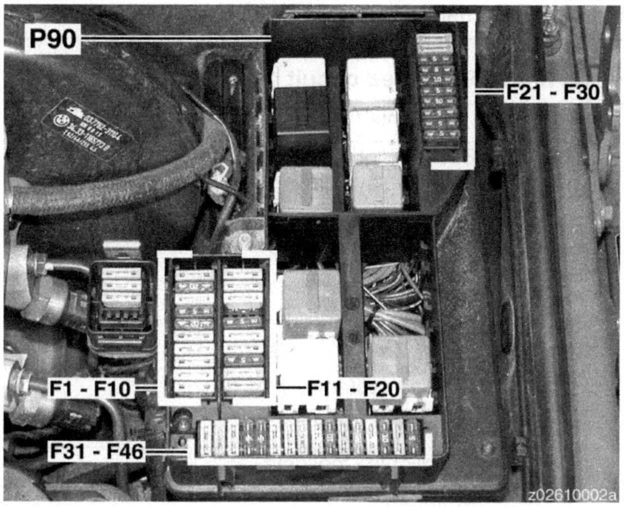


**Table e. 1996 fuse panel P90, fuses F1 - F46**

| Fuse | Rating in amps | Protected circuit |
|---|---|---|
| 1 - 3 | not used | |
| 4 | 15 | Seat heating |
| 5 | 30 | Seats |
| 6 | 20 | Rear window defogger |
| 7 | 5 | Central locking control |
| 8 | 15 | Horn |
| 9 | 20 | Radio / amplifier |
| 10 | 30 | ABS / ASC |
| 11 | 7.5 | Crash control module<br>Headlights / foglights |
| 12 | 7.5 | Crash control module<br>Headlights / foglights |
| 13 | 30 | Windows |
| 14 | not used | |
| 15 | 15 | Headlights / foglights |
| 16 | 5 | DME<br>IHKR |
| 17 | not used | |
| 18 | 15 | DME<br>E-box fan |
| 19 | 30 | Seats |
| 20 | 30 | Blower |
| 21 | 5 | ABS / ASC |
| 22 | 5 | Headlight switch |
| 23 | 5 | Headlight switch<br>Seat heating<br>Instrument cluster |
| 24 | 10 | Mirrors<br>Windshield washer jet heaters |
| 25 | 5 | Crash control module<br>Headlight switch |
| 26 | 10 | Back-up lights<br>EGS |
| 27 | 5 | ABS / ASC<br>Instrument cluster |
| 28 | 5 | Cruise control<br>EGS<br>EWS<br>Starting system |
| 29 | 7.5 | Headlights / foglights |
| 30 | 7.5 | Headlights / foglights |
| 31 | 5 | Board computer<br>Digital clock<br>EWS<br>Instrument cluster<br>Starting system |
| 32 | 30 | Interior accessories |
| 33 | 10 | Cellular telephone<br>Interior and exterior lights |
| 34 | 15 | Crash control module<br>Turn signals / hazard lights |
| 35 | 25 | Central locking module |
| 36 | 30 | Wipers / washers |
| 37 | 10 | Exterior lights<br>Headlight switch<br>Interior accessories |
| 38 | 30 | ABS / ASC |
| 39 | 7.5 | IHKR |
| 40 | 30 | IHKR |
| 41 | 15 | SRS |
| 42 | 5 | Instrument cluster<br>SRS |
| 43 | 5 | Central locking module, anti-theft |
| 44 | 15 | Cellular telephone<br>Radio / amplifier<br>Wipers / washers |
| 45 | 7.5 | Digital clock<br>EWS |
| 46 | 15 | ABS / ASC<br>Board computer<br>Brake lights<br>Cruise control<br>EGS<br>Instrument cluster |

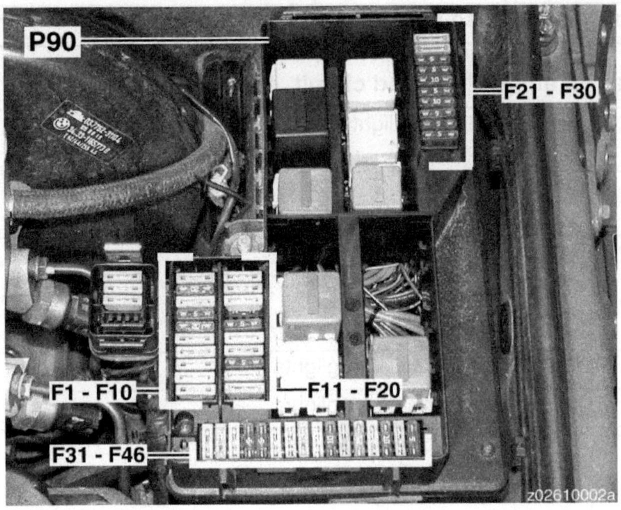


**Table f. 1997 - 1998 fuse panel P90, fuses F1 - F46**

| Fuse | Rating in amps | Protected circuit |
|---|---|---|
| 1 | 30 | Convertible top |
| 2 | not used | |
| 3 | 30 | Headlight washers<br>Wipers / washers |
| 4 | 15 | Seat heating |
| 5 | 30 | Seats |
| 6 | 20 | Rear window defogger |
| 7 | 5 | Central locking control |
| 8 | 15 | Horn |
| 9 | 20 | Radio / amplifier |
| 10 | 30 | ABS / ASC |
| 11 | 7.5 | Crash control module<br>Headlights / foglights |
| 12 | 7.5 | Crash control module<br>Headlights / foglights |
| 13 | 5 | DME |
| 14 | 30 | Windows |
| 15 | 15 | Headlights / foglights |
| 16 | 5 | DME<br>IHKR |
| 17 | not used | |
| 18 | 15 | DME |
| 19 | 30 | Seats |
| 20 | 30 | Blower |
| 21 | 5 | ABS / ASC |
| 22 | 5 | Headlights / foglights |
| 23 | 5 | Headlights / foglights<br>Turn signals / hazard lights |
| 24 | 10 | Mirrors<br>SRS<br>Windshield washer jet heaters |
| 25 | 5 | Headlights / foglights |
| 26 | 10 | Back-up lights<br>Charging system<br>EGS |
| 27 | 5 | Instrument cluster<br>Mirrors<br>SRS |
| 28 | 5 | EGS<br>EWS<br>Starting system |
| 29 | 7.5 | Headlights / foglights |
| 30 | 7.5 | Headlights / foglights |
| 31 | 5 | Anti-theft<br>Board computer<br>Digital clock<br>EWS<br>Instrument cluster<br>Starting system |
| 32 | 30 | Interior accessories |
| 33 | 10 | Cellular telephone<br>General module (ZKE IV)<br>Interior and exterior lights |
| 34 | 15 | Crash control module<br>Turn signals / hazard lights |
| 35 | 25 | Central locking module |
| 36 | 30 | Headlight washers<br>Wipers / washers |
| 37 | 10 | Interior and exterior lights<br>Interior accessories<br>Wipers / washers |
| 38 | 30 | ABS / ASC |
| 39 | 7.5 | DME<br>IHKR |
| 40 | 30 | IHKR |
| 41 | 15 | SRS |
| 42 | 5 | SRS |
| 43 | 5 | Central locking module, anti-theft<br>Interior lights |
| 44 | 15 | Cellular telephone<br>Headlight washers<br>Radio / amplifier<br>Wipers / washers |
| 45 | 7.5 | Digital clock<br>EWS |
| 46 | 15 | ABS / ASC<br>Board computer<br>Brake lights<br>Cruise control<br>EGS<br>Instrument cluster |

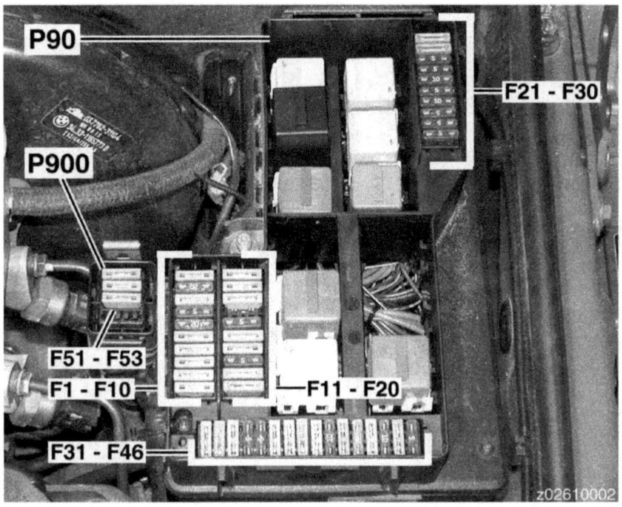


**Table g. 1999 - 2000 fuse panel P90, fuses F1 - F46**

| Fuse | Rating in amps | Protected circuit |
|---|---|---|
| 1 | 30 | Convertible top<br>Wipers / washers |
| 2 | 10 (Z3)<br>15 (M) | General module<br>Interior lights |
| 3 | 30 | Headlight washers<br>Wipers / washers |
| 4 | 15 | Seat heating<br>Instrument cluster |
| 5 | 30 | Seats |
| 6 | 20 | Rear window defogger |
| 7 | 5 | Central locking control<br>DSC III<br>Instrument cluster |
| 8 | 15 | Horn |
| 9 | 20 | Radio / amplifier |
| 10 | 30 | ABS / ASC |
| 11 | 7.5 | Crash control module<br>Headlights / foglights |
| 12 | 7.5 | Crash control module<br>Headlights / foglights |
| 13 | 7.5 (Z3)<br>5 (M) | ABS / ASC<br>Brake lights<br>Cruise control<br>DME |
| 14 | 30 | General module<br>Windows |
| 15 | 10 | Headlights / foglights |
| 16 | 5 | DME<br>IHKR |
| 17 | not used | |
| 18 | 15 | DME |
| 19 | 30 | Seats |
| 20 | 30 | Blower |
| 21 | 5 | ABS / ASC<br>DSC III |
| 22 | 5 | Headlights / foglights<br>Lighting switch |
| 23 | 5 | Headlights / foglights<br>Instrument cluster<br>Lighting switch<br>Rear window defogger<br>Turn signals / hazard lights |
| 24 | 10 | IHKR<br>Mirrors<br>Sunroof<br>SRS<br>Windshield washer jet heaters |
| 25 | 5 | Crash control module<br>Headlights / foglights<br>Lighting switch<br>2000: Parking lights<br>2000: Taillights<br>2000: Trunk illumination<br>2000: Under hood illumination |
| 26 | 10 | Back-up lights<br>Charging system<br>Cruise control<br>DME<br>E-box fan<br>EGS<br>Instrument cluster |
| 27 | 5 | ABS / ASC<br>Charging system<br>DME<br>Instrument cluster<br>SRS |
| 28 | 5 | EGS<br>EWS<br>Starting system |
| 29 | 7.5 | Headlights / foglights |
| 30 | 7.5 | Headlights / foglights |
| 31 | 5 | Anti-theft<br>Board computer<br>Clock<br>EWS<br>Instrument cluster<br>Starting system |
| 32 | 30 | Interior accessories |
| 33 | 10 | Cellular telephone<br>General module<br>Interior lights<br>License plate light<br>Lighting switch<br>Parking lights<br>Taillights<br>Trunk illumination<br>Under hood illumination |
| 34 | 15 | Crash control module<br>Turn signals / hazard lights |
| 35 | 25 | Central locking<br>General module |

| Table g. 1999 - 2000 fuse panel P90, fuses F1 - F46 | | |
|---|---|---|
| **Fuse** | **Rating in amps** | **Protected circuit** |
| 36 | 30 | Headlight washers<br>Wipers / washers |
| 37 | 10 | Interior accessories<br>License plate light<br>Lighting switch<br>Parking lights<br>Taillights<br>Trunk illumination<br>Under hood illumination<br>Wipers / washers |
| 38 | 30 | ABS / ASC |
| 39 | 7.5 | ECM<br>IHKR |
| 40 | 30 | IHKR |
| 41 | 15 | SRS |
| 42 | 5 | SRS |
| 43 | 5 | Anti-theft<br>Central locking<br>2000: Cruise control<br>2000: DME<br>General module<br>Horn<br>Interior lights<br>2000: Mirrors |
| 44 | 15 | Cellular telephone<br>Headlight washers<br>Radio / amplifier<br>Wipers / washers |
| 45 | 7.5 | Digital clock<br>EWS<br>Starting system |
| 46 | 15 | ABS / ASC<br>Brake lights<br>Convertible top<br>Cruise control<br>DME<br>EGS<br>Instrument cluster<br>2000: Board computer<br>2000: Sunroof |

| Table h. 1999 - 2000 fuse panel P900, fuses F51 - F53 | | |
|---|---|---|
| **Fuse** | **Rating in amps** | **Protected circuit** |
| 51 | 30 | DSC III |
| 52 | 30 | DSC III |
| 53 | 30 | DSC III |

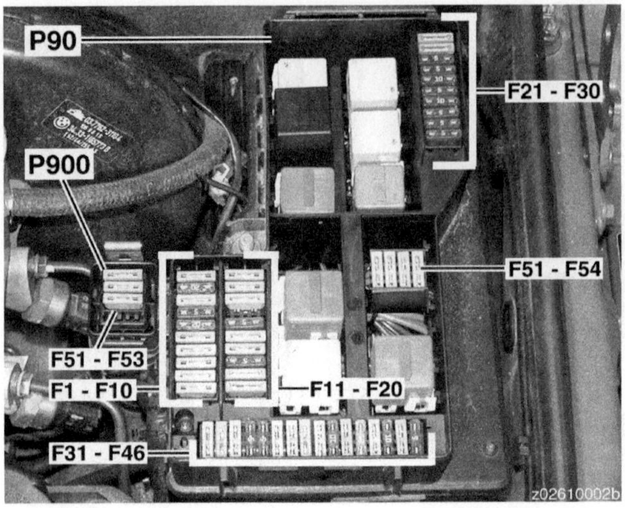


***NOTE —***
*The auxiliary fuse holder P900 was used through early production 2001 models. Locations for fuses F51 to F54 were added to the front power distribution box in late 2001 models.*

| Table i. 2001 - 2002 fuse panel P90, fuses F1 - F54 | | |
|---|---|---|
| **Fuse** | **Rating in amps** | **Protected circuit** |
| 1 | 30 | Convertible top<br>Wipers / washers |
| 2 | 15 | Anti-theft<br>General module<br>Interior lights |
| 3 | 30 | Headlight washers<br>Wipers / washers |
| 4 | 15 | Seat heating |
| 5 | 30 | Seats |
| 6 | 20 | Rear window defogger |
| 7 | 5 | Anti-theft<br>Central locking<br>DSC<br>2001: Instrument cluster<br>Keyless entry |
| 8 | 15 | Horn |
| 9 | 20 | Radio / amplifier |
| 10 | 30 | ABS / ASC |
| 11 | 7.5 | Crash control module<br>Headlights / foglights |
| 12 | 7.5 | Crash control module<br>Headlights / foglights |
| 13 | 5 | Brake lights<br>Cruise control<br>DME |
| 14 | 30 | General module<br>Windows |
| 15 | 15 (Z3)<br>5 (M) | DME<br>Headlights / foglights<br>Starting system |

| Table i. 2001 - 2002 fuse panel P90, fuses F1 - F54 | | |
|---|---|---|
| **Fuse** | **Rating in amps** | **Protected circuit** |
| 16 | 5 | Anti-theft<br>DME<br>IHKR |
| 17 | not used | |
| 18 | 15 (Z3)<br>20 (M) | DME |
| 19 | 30 | Seats |
| 20 | 30 | Blower |
| 21 | 5 | Traction control (ABS / ASC or DSC) |
| 22 | 5 | Headlights / foglights<br>Lighting switch |
| 23 | 5 | Headlights / foglights<br>Instrument cluster<br>Lighting switch<br>Rear window defogger<br>Turn signals / hazard lights |
| 24 | 10 | IHKR<br>Mirrors<br>Windshield washer jet heaters |
| 25 | 5 | Crash control module<br>Headlights / foglights<br>License plate light<br>Lighting switch<br>Parking lights<br>Taillights<br>Trunk illumination<br>Under hood illumination |
| 26 | 10 | Back-up lights<br>Charging system<br>Cruise control<br>DME<br>E-box fan<br>EGS<br>Instrument cluster |
| 27 | 5 | Charging system<br>DME<br>Instrument cluster<br>SRS<br>Tire pressure warning system<br>Traction control (ABS / ASC or DSC) |
| 28 | not used | |
| 29 | 7.5 | Headlights / foglights |
| 30 | 7.5 | Headlights / foglights |
| 31 | 5 | Anti-theft<br>Board computer<br>Clock<br>EWS<br>Instrument cluster<br>Starting system |
| 32 | 30 | Interior accessories |
| 33 | 10 | Cellular telephone<br>General module<br>Interior lights<br>Lighting switch<br>Parking lights / taillights / license plate light<br>Trunk illumination<br>Under hood illumination |

| Table i. 2001 - 2002 fuse panel P90, fuses F1 - F54 | | |
|---|---|---|
| **Fuse** | **Rating in amps** | **Protected circuit** |
| 34 | 15 | 2002: Anti-theft<br>Crash control module<br>Turn signals / hazard lights |
| 35 | 25 | Central locking<br>General module |
| 36 | 30 | Headlight washers<br>Wipers / washers |
| 37 | 10 | Interior accessories<br>License plate light<br>Lighting switch<br>Parking lights<br>Taillights<br>Trunk illumination<br>Under hood illumination<br>Wipers / washers |
| 38 | 30 | ABS / ASC |
| 39 | 7.5 | ECM<br>IHKR |
| 40 | 30 | IHKR |
| 41 | 7.5 | SRS |
| 42 | not used | |
| 43 | 5 | Anti-theft<br>Central locking<br>Cruise control<br>DME<br>General module<br>Horn<br>Interior lights<br>Keyless entry<br>Mirrors |
| 44 | 15 | Cellular telephone<br>Cruise control<br>DME<br>Headlight washers<br>Radio / amplifier<br>Wipers / washers |
| 45 | 7.5 | EWS<br>Starting system |
| 46 | 15 | Board computer<br>Brake lights<br>Convertible top<br>Cruise control<br>DME<br>EGS<br>Instrument cluster<br>Sunroof<br>Traction control (ABS / ASC or DSC) |
| 51 | 30 | DSC |
| 52 | 30 | DSC |
| 53 | 30 | DSC |
| 54 | 30 | IHKR |

## Engine electronics fuses

**Fig. 4 P98 engine electronics fuse carrier**

E-Box, right rear engine compartment

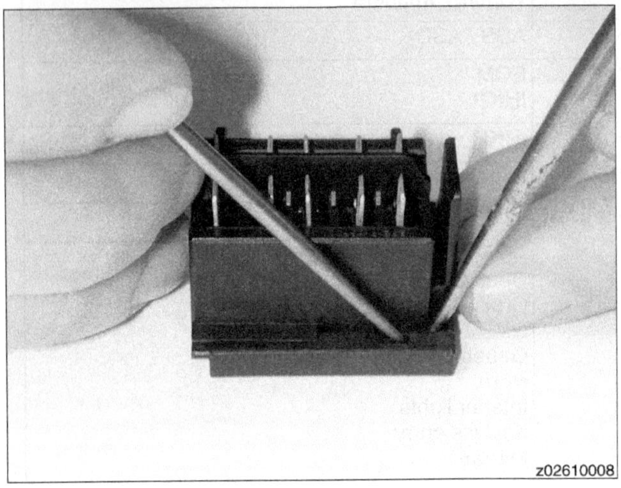


To remove an individual fuse:

- Lift out fuse carrier and disconnect harness.
- Slide cover off while prying gently with a pin tool to squeeze locking tabs.

> ***CAUTION—***
> *Once cover is off, fuses can fall out easily. Be sure to keep fuses in order while replacing defective fuse.*

**Table j. 1999 - 2000 fuse carrier P98, engine electronics**

| Fuse | Rating in amps | Protected circuits |
|---|---|---|
| F201 | 15 | Z3:<br>1999 - 2000 Canada model: daylight relay<br>1999 - 2000 US model: not used<br>M version: in fuse holder P901 in trunk |
| F202 | 30 | ECM<br>Evaporative emissions valve<br>Fuel injectors<br>Idle control valve<br>Intake resonance valve<br>Secondary air pump valve<br>VANOS solenoids |
| F203 | 20 | Back-up light relay<br>Camshaft sensors<br>Crankshaft sensor<br>Characteristic map thermostat<br>Mass air flow sensor<br>Transmission control (EGS)<br>Z3 (*via* F13):<br>Brake light switch<br>Fuel pump relay<br>M version (*via* F13):<br>ABS / ASC control module<br>ASC switch<br>Carbon canister filter shutoff valve |
| F204 | 30 | 3/2 (running losses) valve<br>Oxygen sensor heater<br>Secondary air pump relay<br>Tank leakage diagnosis pump (LDP)<br>Transmission control module (EGS) |
| F205 | 30 | Ignition coils *via* unloader relay |

| Table k. 2001 - 2002 Z3 fuse carrier P98, engine electronics | | |
|---|---|---|
| **Fuse** | **Rating in amps** | **Protected circuits** |
| F201 | 30 | Late production 2002: DME<br>Ignition coils |
| F202 | 30 | Early production: DME<br>Evaporative emissions valve<br>Idle control valve<br>Intake resonance valve<br>Secondary air pump valve<br>VANOS solenoids |
| F203 | 20 | Back-up light relay<br>Camshaft sensors<br>Crankshaft sensor<br>Characteristic map thermostat<br>E-box fan<br>Mass air flow sensor<br>*via* F13:<br>Brake light switch<br>Fuel pump relay<br>Secondary air pump relay<br>Tank leakage diagnosis pump (DMTL) |
| F204 | 30 | Oxygen sensor heater power<br>Secondary air pump relay<br>Transmission control module (EGS) |
| F205 | 30 | Fuel injectors relay |

| Table l. 2001 - 2002 M version fuse carrier P98, engine electronics | | |
|---|---|---|
| **Fuse** | **Rating in amps** | **Protected circuits** |
| F201 | 30 | Ignition coils<br>Suppression capacitor |
| F202 | 30 | DME<br>Evaporative emissions valve<br>Idle control valve<br>Sucking jet solenoid |
| F203 | 20 | Camshaft sensors<br>E-box fan<br>Mass air flow sensor<br>*via* F13:<br>Brake light switch<br>Fuel pump relay<br>Secondary air pump relay<br>Tank leakage diagnosis pump (DMTL) |
| F204 | 30 | Oxygen sensor heater |
| F205 | 30 | Fuel injectors relay |

## GROUND LOCATIONS

Grounds are distributed throughout the vehicle body. Many are found under the interior carpets or behind trim panels. Several components grounds are often ganged. Ground positions vary among models.

Lugs and connectors attached to ground are susceptible to damage and corrosion. Clean or renew as necessary.

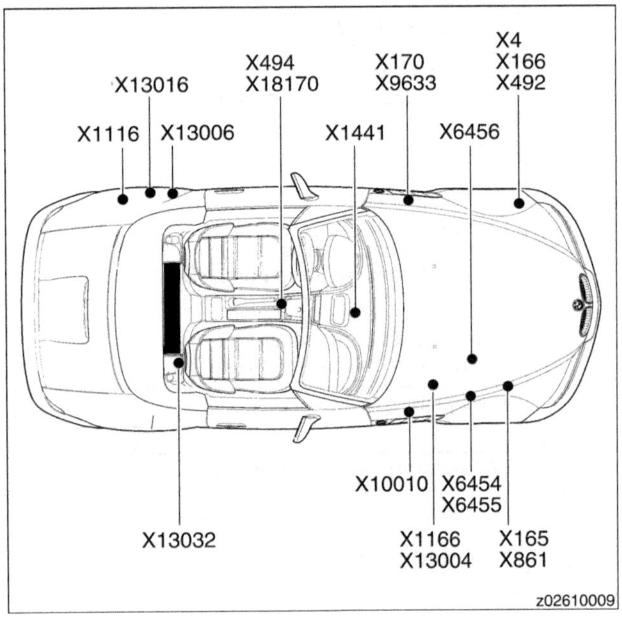


**Fig. 5 Grounds overview**

| Table m. Ground locations | | |
|---|---|---|
| **Designa tion** | **Location** | **Function / notes** |
| X4 | Left side engine compartment | ABS / ASC |
| X165 | Right side engine compartment, front of right strut tower | Engine cooling fan (M)<br>Exterior lights (right)<br>Headlight washers<br>Horn<br>Pressure switch<br>Radiator fan switch<br>Windshield washer jet heaters<br>Wiper / washers |
| X166 | Left side engine compartment | Aux. coolant pump (M)<br>Data link connector<br>Engine cooling fan (Z3)<br>Exterior lights (left)<br>Hood switch<br>IHKR<br>Instrument cluster |
| X170 | Behind left kick panel | ABS / ASC / DSC |
| X492 | Left side engine compartment | DSC |
| X494 | Below center console, behind shifter | Window switches |
| X861 | Right side engine compartment, front of right strut tower | Secondary air pump (M) |
| X1116 | Left side trunk, behind trim | Convertible top relay |
| X1166 | Behind glove compartment | ABS / ASC |
| X1441 | Below center console, ahead of shifter | Crash sensor<br>SRS module |
| X6454 | Right side engine compartment, rear of right strut tower | ECM<br>EGS<br>Fuel injectors<br>VANOS |
| X6455 | Right side engine compartment, rear of right strut tower | Secondary air pump (Z3) |
| X6456 | Top of cylinder head, under cover | Ignition coils |
| X9633 | Behind left footwell panel | Anti-theft and EWS<br>Blower relay<br>Brake fluid level sensor<br>Cruise control actuator<br>IHKR<br>Lights relay<br>Mirror (left)<br>Siren<br>Window motor (left) |

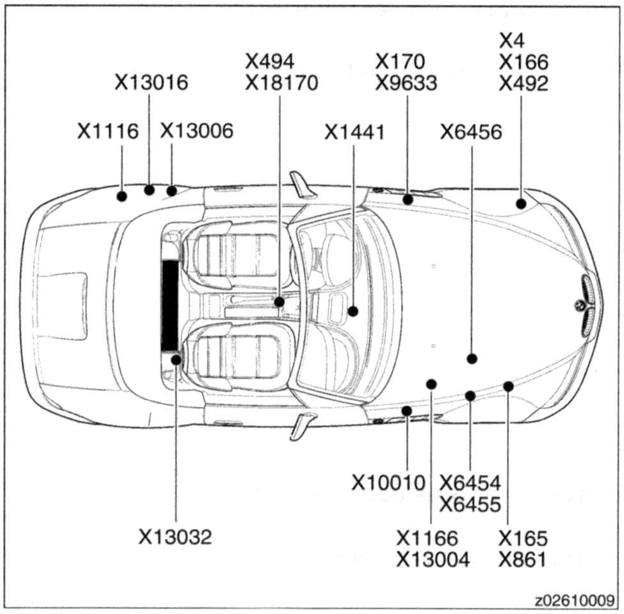


**Table m. Ground locations**

| Designation | Location | Function / notes |
|---|---|---|
| X10010 | Behind glove compartment | Anti-theft<br>Board computer<br>Clock<br>Clutch switch<br>Crash control module<br>Crash sensors<br>Cruise control<br>EWS transmitter<br>General module<br>Instrument cluster<br>Oil level sensor<br>Oil temperature gauge<br>Seat belt switch<br>Seat occupancy sensor<br>Wipers / washers |
| X13004 | Behind glove compartment | Instrument illumination<br>Interior lights<br>Mirror (right)<br>Seat (right)<br>Sunroof<br>Window motor (right) |
| X13006 | Left side trunk | Convertible top<br>Door jamb switch<br>Rear window defogger<br>Taillights and licence plate lights<br>Trunk light<br>Trunk lock<br>Wiper / washer, rear |
| X13016 | Left side trunk | Radio<br>Radio amplifier |
| X13032 | Behind right seat, behind trim | Fuel pump |
| X18170 | Below center console, behind shifter | ASC / DSC switch<br>Brake light switch<br>Chime module<br>Cigar lighter<br>Clock<br>Convertible top switch<br>EWS module<br>Gear position switch<br>General module<br>Hazard switch<br>Horn contact ring<br>Light switch<br>OBD II connector<br>Rear window defogger switch<br>Seat heaters<br>Secondary air pump relay<br>Steering column combination switch<br>Tire pressure control system |

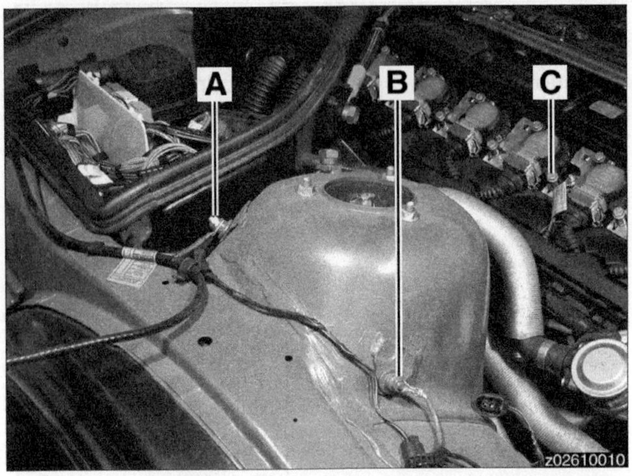


**Fig. 6 Right side engine compartment**

< Grounds at right strut tower; below engine top cover

- **A** X6454
  X6455
- **B** X165
  X861
- **C** X6456

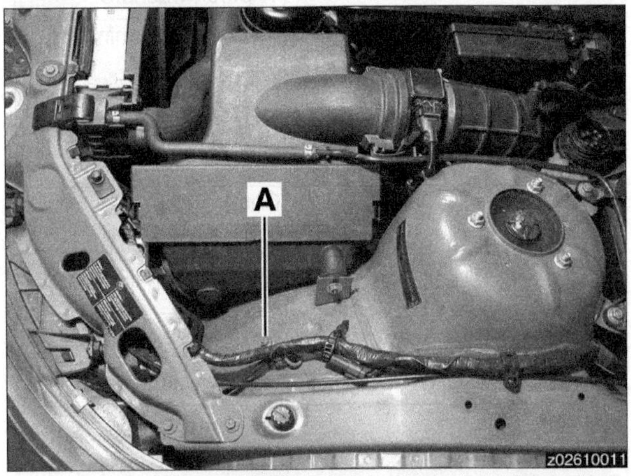


**Fig. 7 Left side engine compartment**

< Grounds next to air filter housing

- **A** X4
  X166
  X492

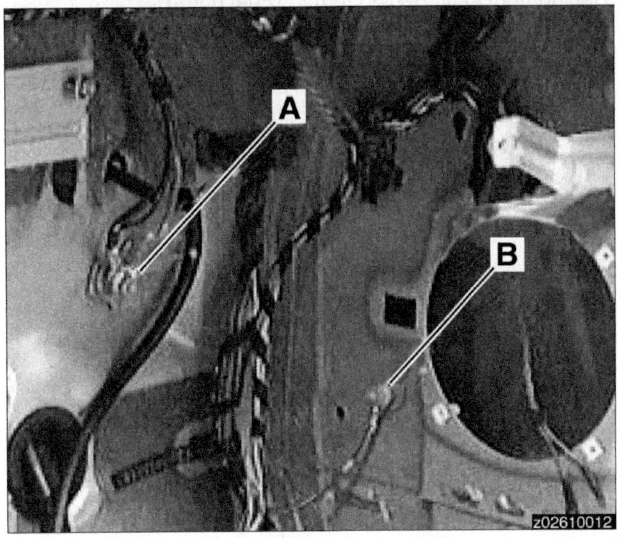


**Fig. 8 Behind glove compartment**

< Various body grounds

- **A** X1166
  X13004
- **B** X10010

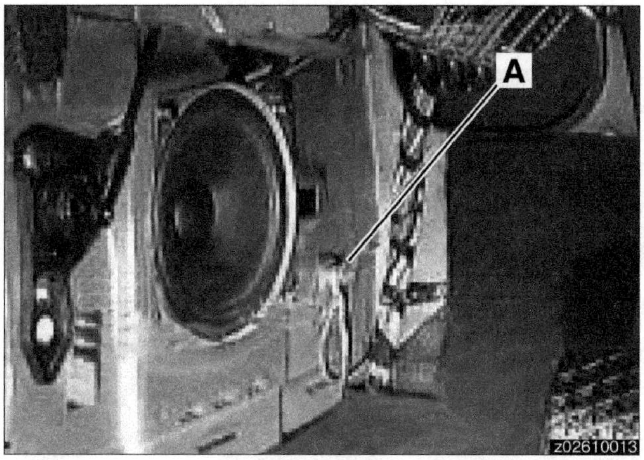


**Fig. 9 Behind left kick panel**

ABS / ASC / DSC and body grounds

**A** X170
X9633

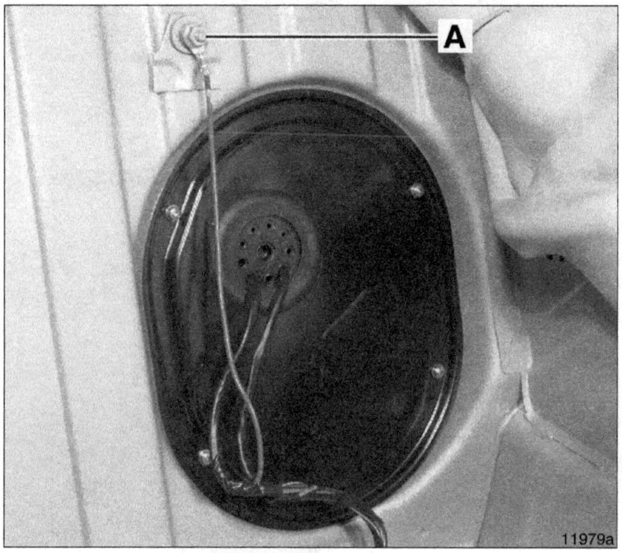


**Fig. 10 Behind right seat, behind trim**

Fuel pump ground

**A** X13032

## ELECTRICAL COMPONENT LOCATION PHOTOS

The following photos show the locations of major electrical components in Z3 vehicles, including M version cars.

***NOTE—***

- *Every component is not installed in every car.*
- *Due to changes in production, component locations may vary from what is illustrated. Consult your BMW dealer for the latest information.*
- *Confirm component identification by comparing wire colors to those shown in* **ELE Electrical Wiring Diagrams**.
- *The component location photos provided below are from a 2000 Z3 with standard transmission, unless otherwise noted.*
- *The automatic transmission range switch is also referred to as the neutral safety switch.*
- *The EWS system is also known as the drive-away protection system or the electronic immobilization system.*

## Engine compartment components

**Fig. 11 Engine compartment overview (1999 M Coupe)**

- **A** Grounds, X1166, X13004
- **B** Engine control module (ECM), A6000
- **C** Electronics box (E-box)
- **D** Ground jumper point
- **E** B+ battery jumper point, G6340
- **F** Secondary air pump, Y6133
- **G** Data link connector (20-pin DLC), D100
- **H** Brake fluid level sensor, B18
- **I** ABS / DSC electronic control module, A65
- **J** Engine harness connectors
- **K** Front power distribution box, P90
- **L** Auxiliary throttle control motor (ADS), M93
- **M** Grounds X165, X861
- **N** Throttle position sensor, R6252
- **O** Evaporative emission control valve, Y6120
- **P** Auxiliary throttle position sensor (ADS), R31
- **Q** Mass air flow sensor, B6207
- **R** Grounds X4, X166, X492
- **S** Washer fluid pump, M4
- **T** Intake camshaft VANOS solenoid, Y6275z
- **U** Alternator (generator), G6509
- **V** Engine hood contact switch, S21w

**Fig. 12 Engine compartment overview (2000 Z3)**

- **A** Grounds, X1166, X13004
- **B** Engine electronics fuse carrier, P98
- **C** Engine control module (ECM), A6000
- **D** Electronics box (E-box)
- **E** Secondary air pump, Y6133
- **F** B+ battery jumper point, G6340
- **G** Ignition coils, T5151 - T6156
- **H** Precatalyst oxygen sensor harness connectors
- **I** Post-catalyst oxygen sensor harness connectors
- **J** Ignition coils ground X6456
- **K** Fuel injector electrical harness
- **L** Intake air temperature (IAT) sensor, B6205
- **M** Heater valve, Y4z
- **N** Intake resonance control solenoid, Y6167
- **O** Brake fluid level sensor, B18
- **P** ABS / DSC pressure transmitter, Y162
- **Q** Data link connector (20-pin DLC), D100
- **R** Auxiliary fuse holder, P900
- **S** Front power distribution box, P90
- **T** Grounds X165, X861
- **U** Washer fluid pump, M4
- **V** Exhaust camshaft sensor, B6224
- **W** Characteristic map cooling thermostat
- **X** Intake camshaft VANOS solenoid, Y6275z
- **Y** Alternator (generator), G6509
- **Z** Mass air flow sensor, B6207
- **AA** Engine hood contact switch, S21w
- **AB** Grounds X4, X166, X492

**Fig. 13 Front power distribution box, P90 (2000 Z3)**

◄ Left rear engine compartment

- **A** Back-up lights relay, K6325 (Z3 with Steptronic transmission)
  Tailgate lock relay, K70 (M Coupe)
- **B** Left parking light relay, K26
- **C** Right parking light and license plate light relay, K25
  Secondary air pump relay, K6304 (4-cylinder)
- **D** Heating and A/C blower relay, K4
- **E** Hazard flasher, K16
- **F** Low beam light relay, K48
- **G** Front foglight relay, K47 (Z3)
  A/C relay, DME, K33a (M version)
- **H** Horn relay, K2
- **I** Engine cooling fan relay, normal speed, K21
- **J** A/C compressor relay, K19
- **K** Engine cooling fan relay, high speed, K22
- **L** ABS pump relay, K75 (1996 - 1998 Z3)
  ABS / ASC valve relay, K95 (1999 - 2000 M version)
- **M** ABS relay, K10 (1996 - 1998 Z3)
  ABS / ASC relay, DME, K94 (1998 - 2000 M version)
  Auxiliary coolant pump relay, K104 (2001 - 2002 M version)
- **N** Unloader relay terminal 15, K9

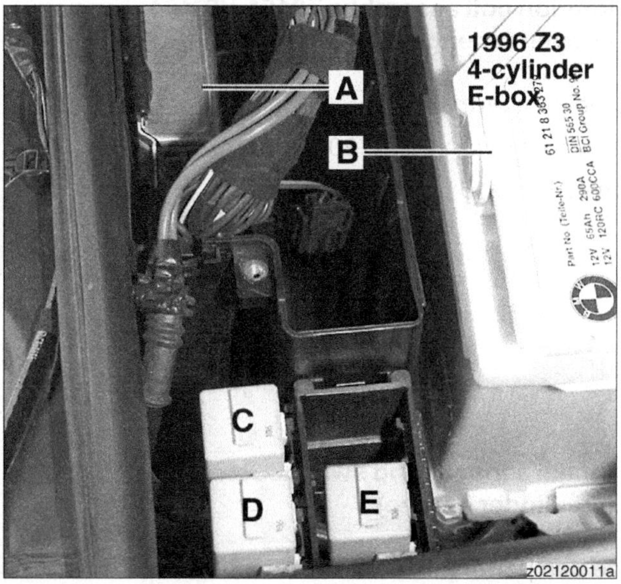


**Fig. 14 E-box (1996 4-cylinder engine)**

◄ Right rear engine compartment

- **A** Engine control module (ECM), A6000
- **B** Battery, G1
- **C** DME main relay, K6300
- **D** Fuel pump relay K6301w
- **E** Unloader relay, terminal 15, K6326
  Starter interlock relay, K6324

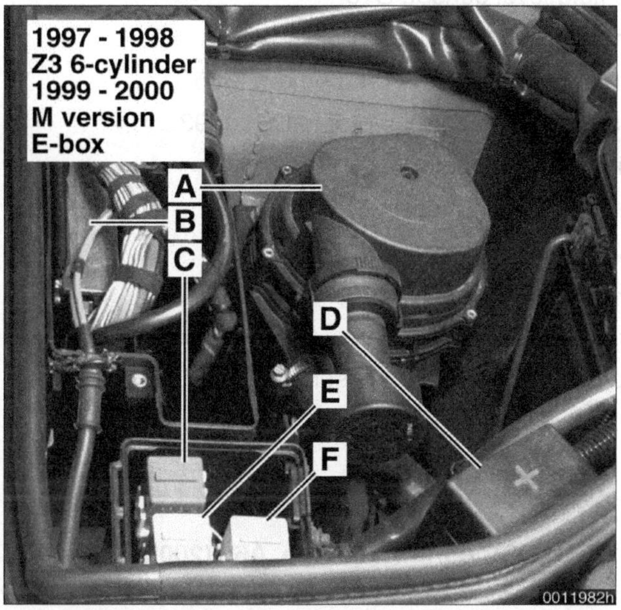


**Fig. 15 E-box (1997 - 1998 Z3 6-cylinder engine, 1998 - 2000 M version)**

◄ Right rear engine compartment

- **A** Secondary air pump, Y6133
- **B** Engine control module (ECM), A6000
- **C** DME main relay, K6300
- **D** B+ battery junction, G6340
- **E** Unloader relay terminal 15, K6325
- **F** Secondary air pump relay, K6304

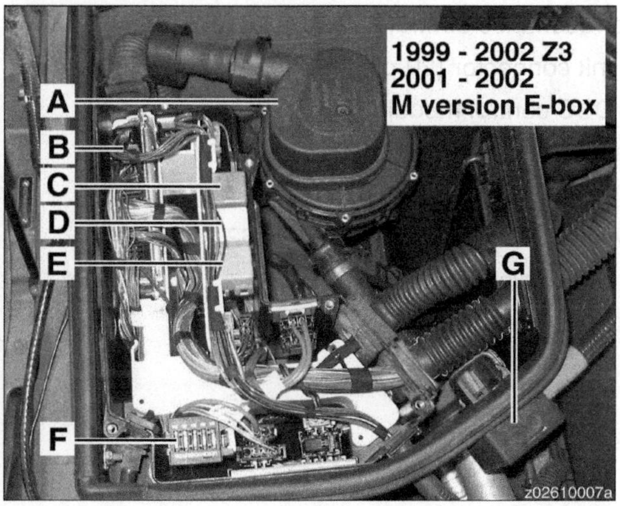


**Fig. 16 E-box (1999 - 2002 Z3, 2001 - 2002 M version)**

◄ Right rear engine compartment

- **A** Secondary air pump, Y6133
- **B** Engine control module (ECM), A6000
- **C** DME main relay, K6300
- **D** Fuel injector relay, K6327
- **E** Secondary air pump relay, K6304
- **F** Engine electronics fuse carrier, P98
- **G** B+ battery junction, G6340

**Fig. 17 Ignition coil assembly, T6151 - T6154 (4-cylinder engine)**

◄ Right engine compartment

- **A** Data link connector (DLC), 20-pin, D100
- **B** Ignition coils, T6151 - T6154

**Fig. 18 Intake sensors and switches (4-cylinder engine)**

◄ Left engine compartment

- **A** Brake fluid level sensor, B18
- **B** Engine electrical connectors
- **C** Cruise control actuator (Tempomat), M25
- **D** Auxiliary throttle control motor (ADS), M93
- **E** Fuel injector connector
- **F** Throttle position sensor, R6252
- **G** Idle speed control valve, Y6130
- **H** Auxiliary throttle position sensor (ADS), R31
- **I** Mass air flow sensor, B6207
- **J** Intake air temperature (IAT) sensor, B6205

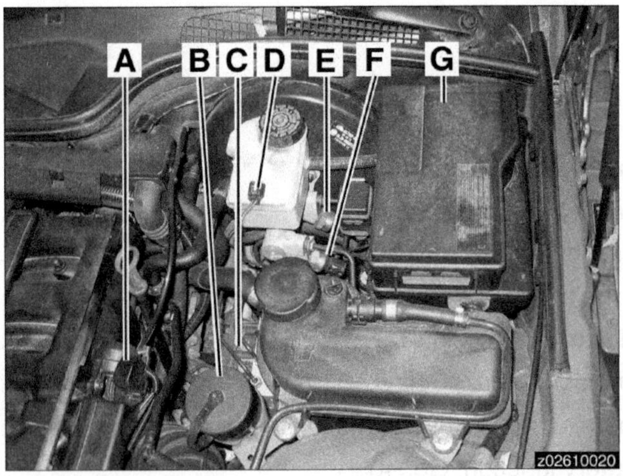


**Fig. 19 ABS / DSC control module, A65 (2000 Z3)**

◄ Left rear engine compartment

- **A** Intake resonance control solenoid,Y6167
- **B** Data link connector (20-pin DLC), D100
- **C** ABS/ DSC control module, A65
- **D** Brake fluid level sensor, B18
- **E** ABS/ DSC pressure sensor, B76
- **F** ABS / DSC pressure transmitter, Y162
- **G** Front power distribution box, P90

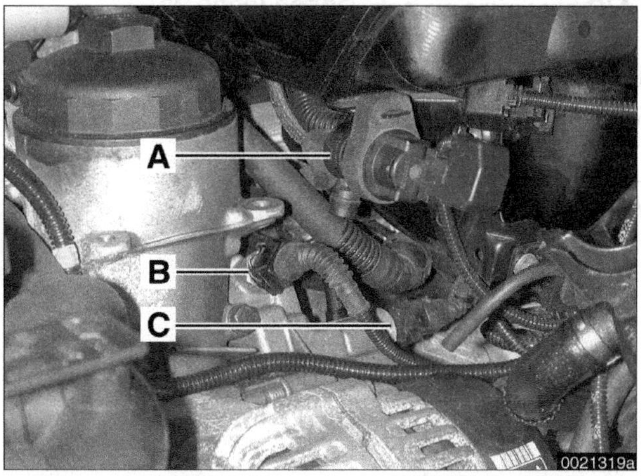


**Fig. 20 Evaporative emission control valve, Y6120 (M52 TU engine)**

< Left side of engine, underneath intake manifold

**A** Evaporative emission control valve, Y6120

**B** Oil temperature sender, B6238

**C** Oil pressure switch, B6231

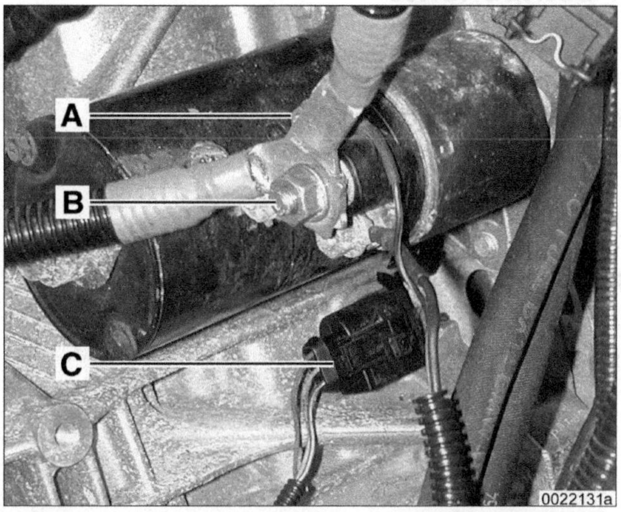


**Fig. 21 Starter motor, M6510 (M52 TU engine)**

< Left side of engine, underneath intake manifold

**A** Starter solenoid terminal 50

**B** Starter solenoid terminal 30

**C** Crankshaft sensor, B6203

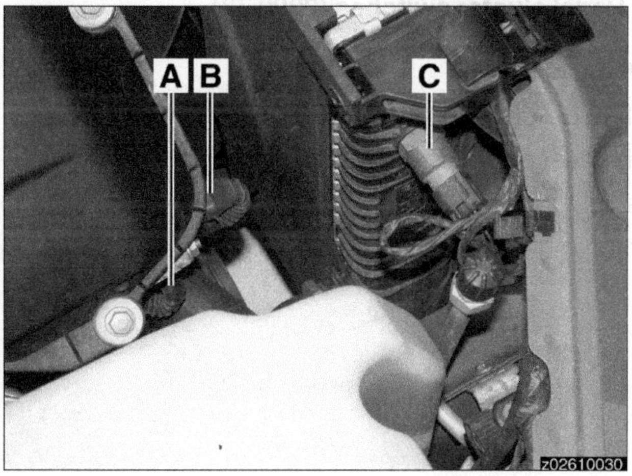


**Fig. 22 Engine cooling fan switch, S36 (2000 Z3)**

< Right front engine compartment

**A** Exhaust VANOS solenoid, Y6276

**B** Exhaust camshaft sensor, B6224

**C** Engine cooling fan switch, S36

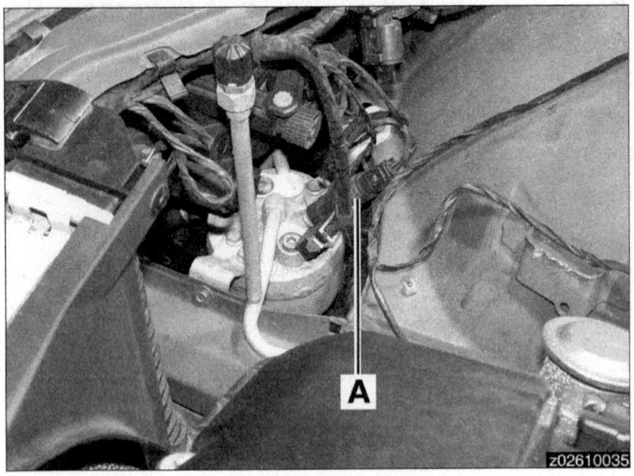


**Fig. 23 A/C pressure switch, B9508 (2000 Z3)**

Behind right headlight on A/C receiver/dryer

**A** A/C pressure switch, B9508

*NOTE —*
*Washer fluid reservoir removed for clarity.*

## Interior components

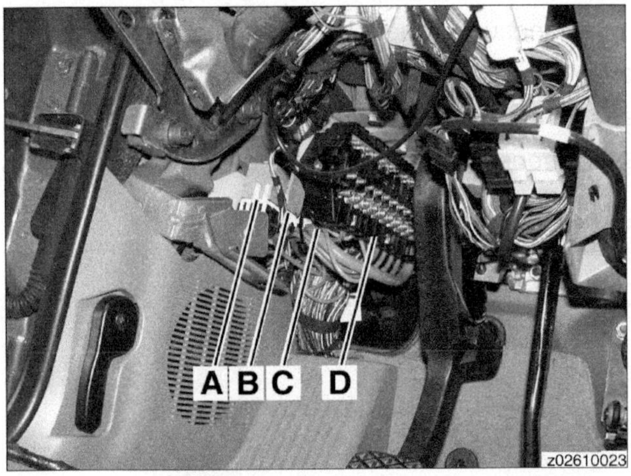


**Fig. 24 Fuel pump relay, K6301w (2000 Z3)**

Under left dashboard at kick panel

**A** Headlight relay, high beam, K46

**B** Crash control module, N17

**C** Comfort relay, K78 (1996)
Fuel pump relay, K6301w (1997 - 2002 Z3, 1998 - 2000 M version)

**D** High amperage fuse panel, G7512 (1999 - 2002 Z3, 2001 - 2002 M version)

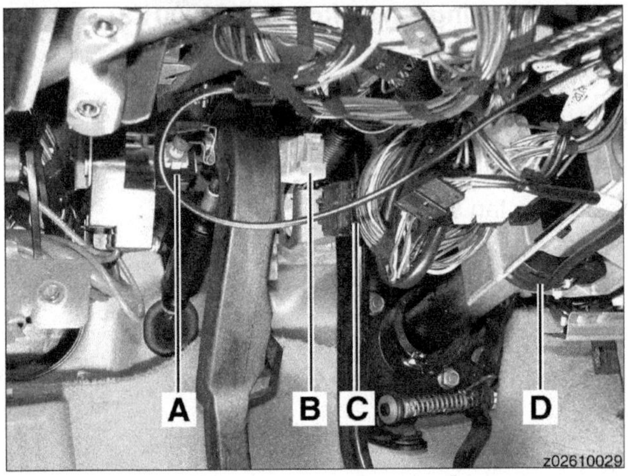


**Fig. 25 Pedal cluster switches (2000 Z3)**

Under left dashboard

**A** Clutch pedal starter immobilization switch, S235

**B** Clutch pedal position sensor, S32

**C** Brake light switch, S30

**D** Steering angle sensor, R33

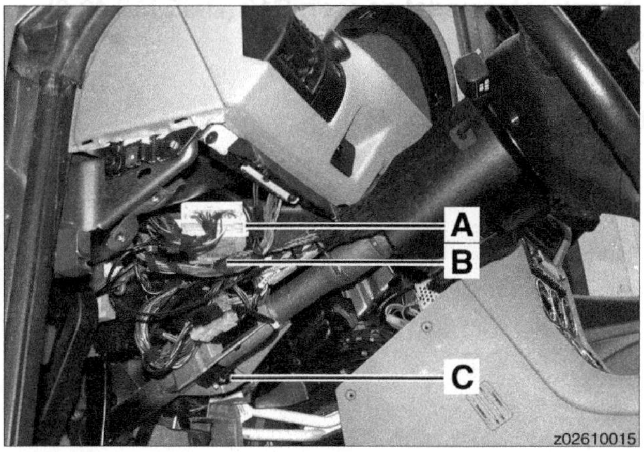


**Fig. 26 EWS control module, A837 (2000 Z3)**

Under left dashboard

- **A** Electronic immobilizer (EWS III) control module, A837 (1999 - 2002)
- **B** Cruise control module, A9008
- **C** Steering angle sensor, R33

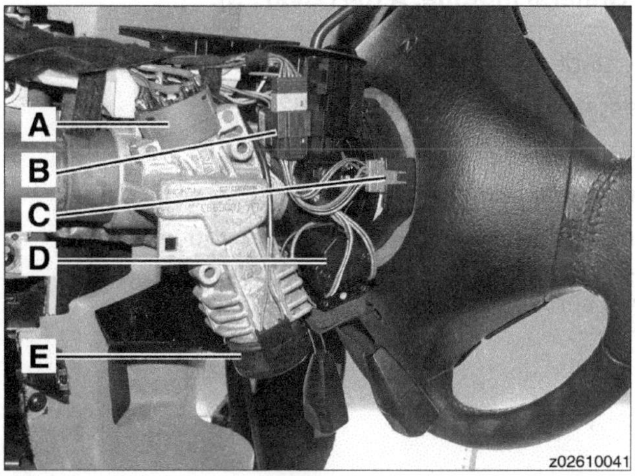


**Fig. 27 EWS ring antenna, L1 (2000 Z3)**

Under neath steering wheel

- **A** Ignition switch, S2
- **B** Combination switch connector
- **C** Horn and driver airbag connector
- **D** Cruise control switch, S28
- **E** EWS ring antenna, L1

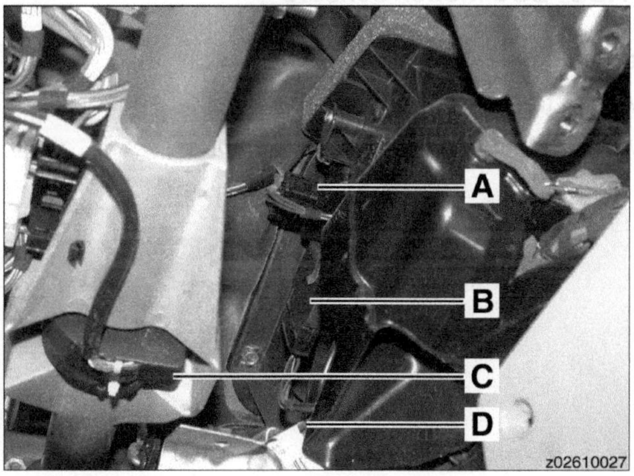


**Fig. 28 A/C and heating components (2000 Z3)**

Left side of center console, under dashboard

- **A** Fresh air / recirculation motor, M111
- **B** A/C evaporator controller, S221
- **C** Steering angle sensor, R33
- **D** A/C evaporator temperature sensor, B14

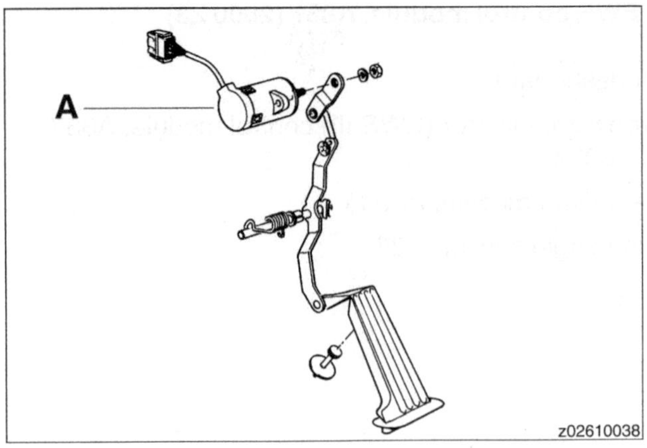


**Fig. 29 Accelerator pedal sensor, R10 (DME MS 43.0)**

Above accelerator pedal

- **A** Accelerator pedal sensor, R10

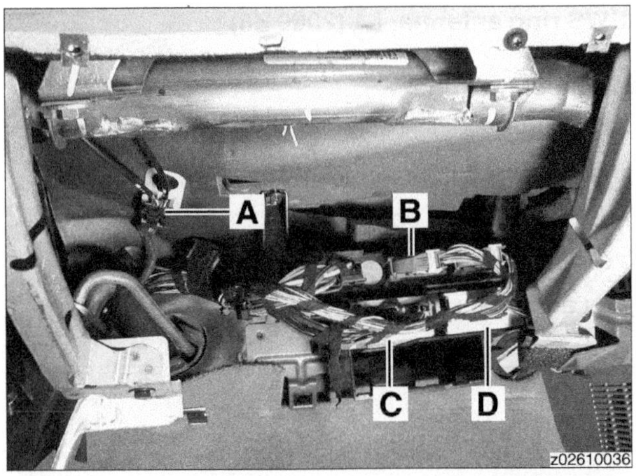


**Fig. 30 General module, A104 (2000 Z3)**

Right side under dashboard, behind glove compartment

- **A** Passenger airbag connector, X782
- **B** General module, A104
- **C** Wiper / washer control module, A51
- **D** Oil temperature / outside temperature gauge module, A3984
  Oil level sensor control module, A19007

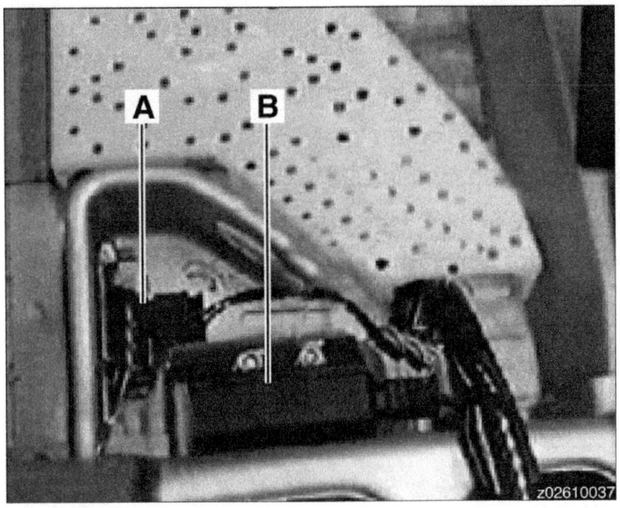


**Fig. 31 ABS / DSC motion sensors**

Under left seat, beneath rug

- **A** ABS / DSC transverse (lateral) acceleration sensor, B74
- **B** ABS / DSC rotational (RPM) rate sensor, B75

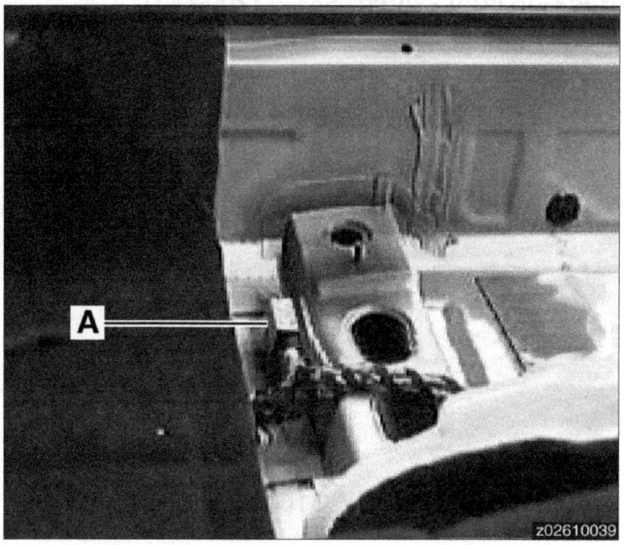


**Fig. 32 Airbag crash sensor, B78**

Under left seat, beneath rug

**A** Airbag crash sensor, left, B78

## Components underneath vehicle

**Fig. 33 Fuel system components (2000 Z3)**

Front left frame rail, under protective cover

**A** 3/2-way valve, Y6550

**B** Fuel tank leakage diagnosis pump (LDP), Y6139

**C** Engine oil level sensor, B6254

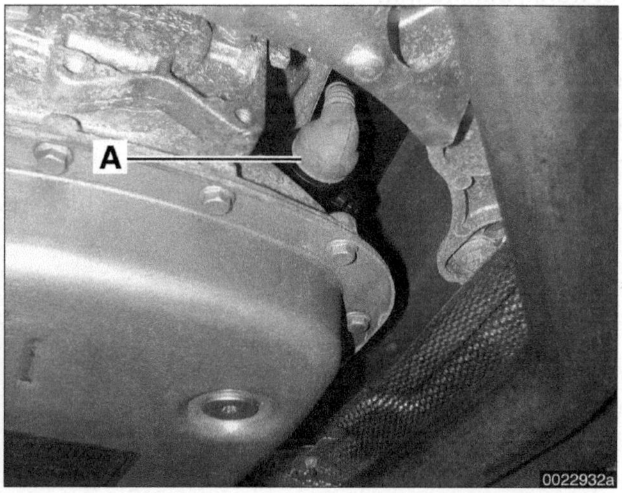


**Fig. 34 Automatic transmission connector, X8505 (GM transmission)**

Right rear of transmission housing

**A** Automatic transmission valve body connector, X8505

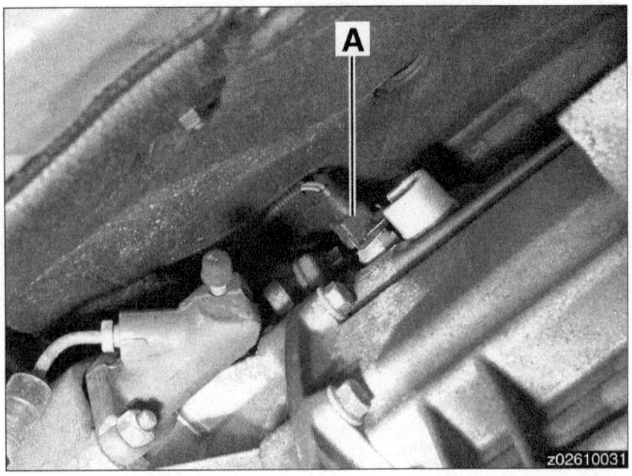


**Fig. 35 Back-up light switch, S8511 (2000 Z3)**

< Left side of transmission housing

**A** Back-up light switch, S8511 (Getrag manual transmission)

**Fig. 36 Vehicle speed sensor, B10 (2000 Z3)**

< Rear of differential housing

**A** Vehicle speed sensor, B10

## Components in trunk

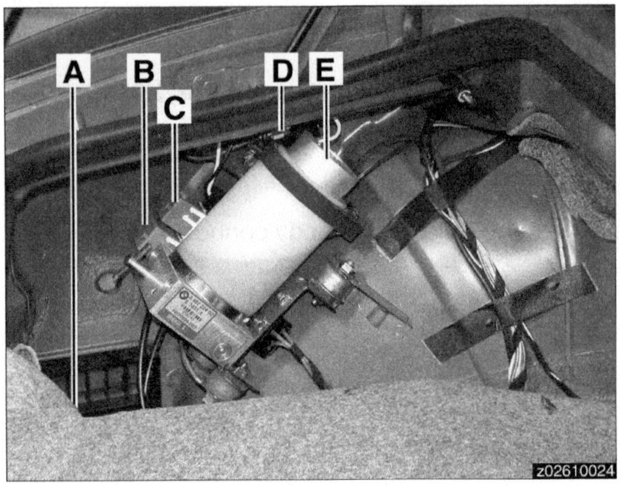


**Fig. 37 Convertible top components (2000 Z3)**

< Left side of trunk, behind trim

**A** Tilt sensor (anti-theft), B28

**B** Convertible top close relay, K18363

**C** Convertible top open relay, K18364

**D** Antenna amplifier connector, N8

**E** Convertible top hydraulic pump, M101

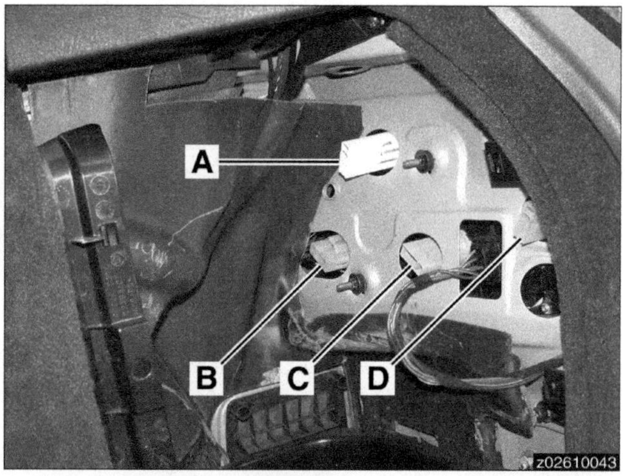


**Fig. 38 Right taillight, E17 (1999 M Coupe)**

Right rear cargo compartment, behind trim panel

- **A** Turn signal bulb
- **B** Brake light bulb
- **C** Taillight bulb
- **D** Back-up light bulb

***NOTE—***
*An additional bulb, the side marker, is behind rubberized trim.*

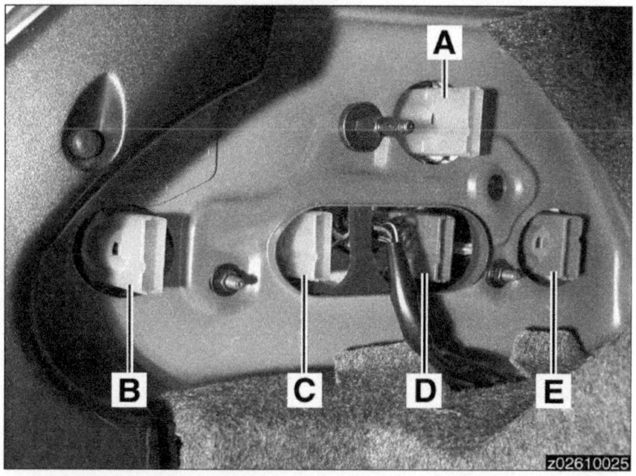


**Fig. 39 Left taillight, E16 (2000 Z3)**

Left rear trunk, behind plastic trim panel

- **A** Turn signal bulb
- **B** Taillight bulb
- **C** Back-up light bulb
- **D** Brake light bulb
- **E** Rear side marker bulb

**Fig. 40 Radio amplifier, A18 (2000 Z3)**

Right side trunk, behind trim

- **A** Radio amplifier, A18
- **B** Fusible link holder, P99
- **C** Battery safety terminal (SBK), G19
- **D** Battery, G1

# 611 Wipers and Washers

## General

This repair group covers front and rear wiper blade service. Also covered are repairs to front and rear wiper and washer assemblies.

For additional information see the following:

- **610 Electrical Component Locations**
- **612 Switches** for steering column stalk switch assembly replacement
- **ELE Electrical Wiring Diagrams**

### Special tools

Some BMW special tools are recommended for the procedures in this repair group.

Automotive digital multimeter

> ***CAUTION—***
> *Only use a digital multimeter when testing automotive electrical components.*

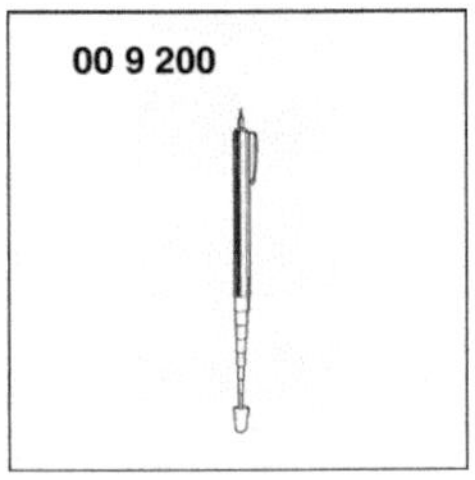


Washer nozzle adjusting tool (BMW special tool 00 9 200)

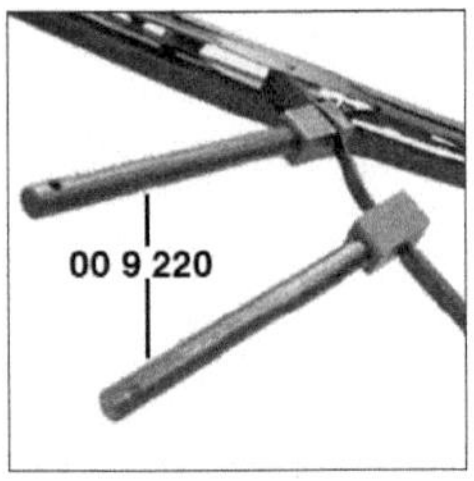


Wiper arm angle adjusting tool (BMW special tool 00 9 220)

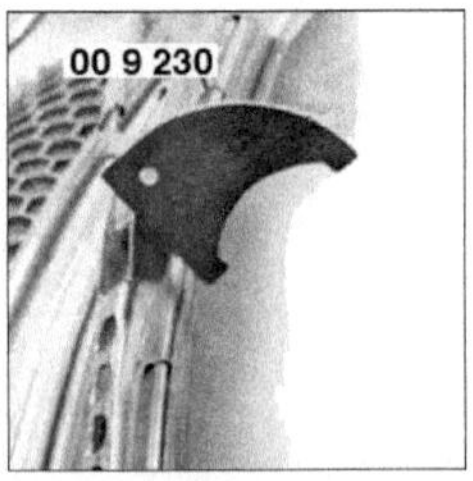


Wiper blade angle gauge (BMW special tool 00 9 230)

Wiper arm puller (BMW special tool 61 2 100)

## Cautions

*CAUTION—*
- *Switch the ignition off and remove the negative (-) battery cable before removing any electrical components.*
- *Prior to disconnecting the battery, read the battery disconnection cautions in* **001 General Warnings and Cautions**.
- *Prior to disconnecting the battery:*
  *-Obtain the radio operation code, if applicable.*
  *-Record radio presets.*
- *Only use a digital multimeter for electrical tests.*
- *Do not operate the windshield wipers with the engine hood open. The wiper arm may scrape the hood.*

## WIPER AND WASHER CONTROLS

Windshield wiping and washing functions are controlled by the wiper / washer control module (**arrow**) located behind the glove compartment.

### Wiper / washer switch

*NOTE—*
*Wiper / washer stalk switch replacement is covered in* **612 Switches**.

The wiper / washer switch communicates its setting to the wiper / washer module for the following:

**S (single)**. Momentarily holding the wiper switch down in the single position provides a ground signal to activate a single sweep at slow speed.

**I (slow)** and **II (fast)**. Stage I and stage II wiping speeds are road speed dependent. Stage I switches to intermittent when the vehicle is stopped and stage II switches to stage I when stopped.

**Windshield washing**. By pulling the stalk backward, a switched ground input to the wiper / washer module activates the windshield washer pump directly. The pump runs as long as the switch is held. Also, the wipers switch on and continue for 3 wiping cycles after the switch is released.

**Intermittent**. The interval time of the wipers is vehicle speed dependent. As vehicle speed increases, wiping interval times decrease. This timed interval is set in the control module and is not programmable. The programmed intermittent cycles intervals are shown in the table **Wiper interval program**.

| Wiper interval program | |
|---|---|
| **Vehicle speed** | **Programmed interval** |
| 0 - 10 mph | 14 sec |
| 10 - 25 mph | 8 sec |
| 25 - 50 mph | 7 sec |
| 50 - 69 mph | 6 sec |
| 69 - 93 mph | 5 sec |
| >93 mph | 4 sec |

## Wiper stall protection

The windshield wiper control module receives a feedback signal from the wiper motor when the wipers are switched on and operating. If the wipers stall for any reason (example: wiper blades frozen to the windshield, or excessive snow accumulation), the feedback signal is not generated.

If the signal is missing for several seconds after the wipers are switched on, the module switches the motor off. Be sure to clean off the windshield and free up the wipers.

If stall protection is activated several times, turn the ignition OFF and ON before the switching the wipers ON again.

## Rear window wiper / washer (Coupe)

The rear window of Coupe models has a wiper / washer. This system is activated through an additional set of contacts in the wiper / washer stalk switch. Operation is as follows:

**Intermittent wiping**. Push the wiper switch away from the steering wheel briefly to activate intermittent wipe. Pressing a second time switches the wiper OFF.

**Wash / wipe**. Push the lever away from the steering wheel and hold. This activates the washer pump and the rear wiper while the switch is held.

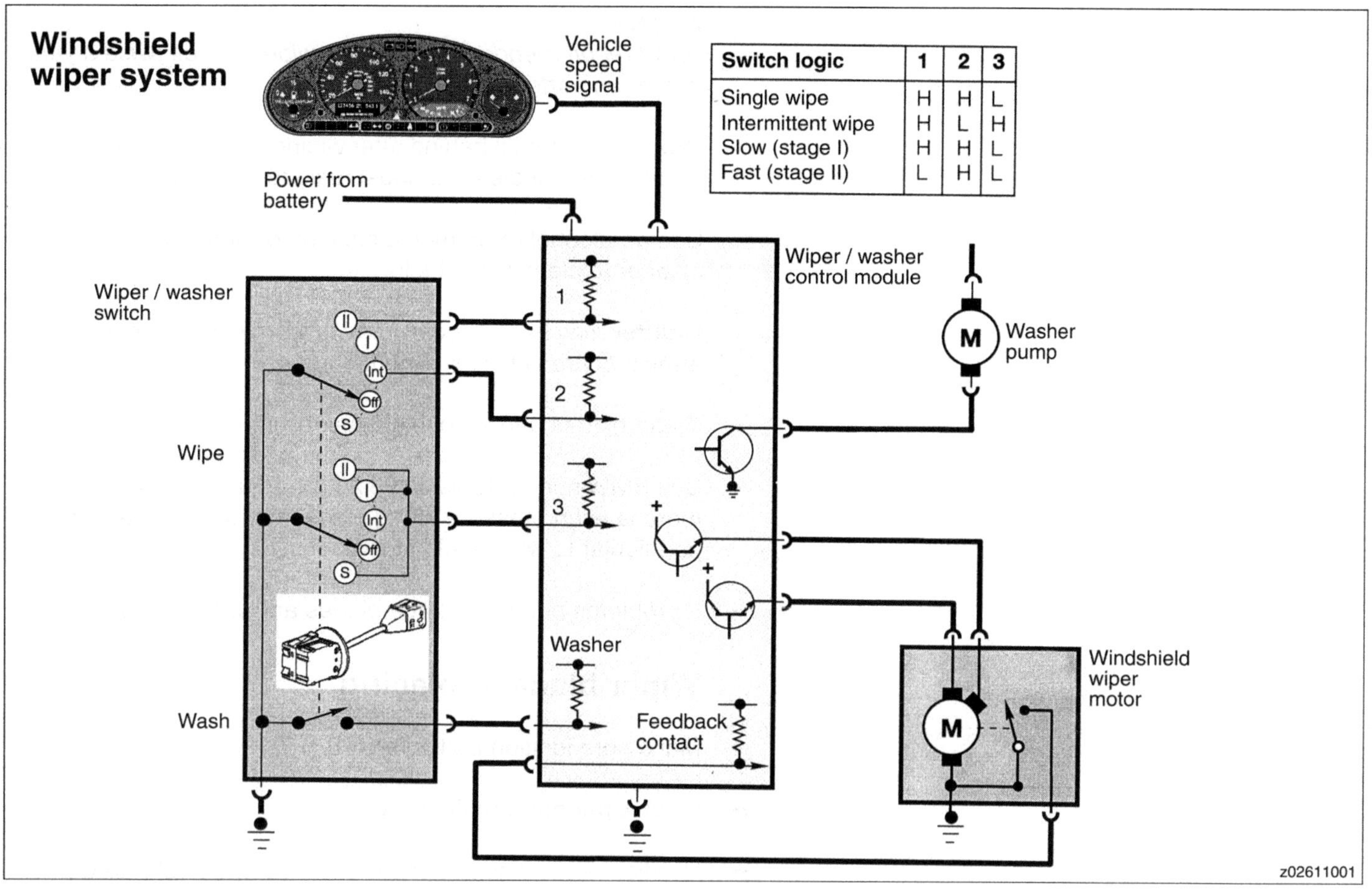


| Switch logic | 1 | 2 | 3 |
|---|---|---|---|
| Single wipe | H | H | L |
| Intermittent wipe | H | L | H |
| Slow (stage I) | H | H | L |
| Fast (stage II) | L | H | L |

## Troubleshooting

The wiper module is not connected to the diagnostic link for troubleshooting purposes. In case of problems, use the schematics in **ELE Electrical Wiring Diagrams** to diagnose the wiper / washer system.

> ***CAUTION—***
> *Only use a digital multimeter when testing automotive electrical components.*

# Wiper Arms and Blades

## Wiper blade cleaning problems

Common problems with the windshield wipers include streaking or sheeting, water drops after wiping, and blade chatter.

**Streaking** is usually caused when wiper blades are coated with road film or car wash wax.

– Clean blades using soapy water. If cleaning does not cure problem, replace blades.

***NOTE—***
*BMW recommends replacing the wiper blades twice a year, before and after the cold season.*

**Drops** that remain behind after wiping are usually caused by oil, road film, or diesel exhaust residue on the glass.

– Use an alcohol or ammonia solution or other non-abrasive cleaner to clean windshield.

**Chatter** may be caused by dirty or worn blades, or by wiper arms that are out of alignment.

– Clean blades and windshield as described above.

– Use BMW special tools 00 98 230 to adjust wiper arm so that there is even pressure along blade, and so that blade is perpendicular to windshield at rest.

– If problems persist, replace blades and wiper arms.

## Wiper blade, replacing

– Make sure ignition switch is OFF.

– Pivot wiper arm up off glass.

– Position wiper blade approximately perpendicular to wiper arm.

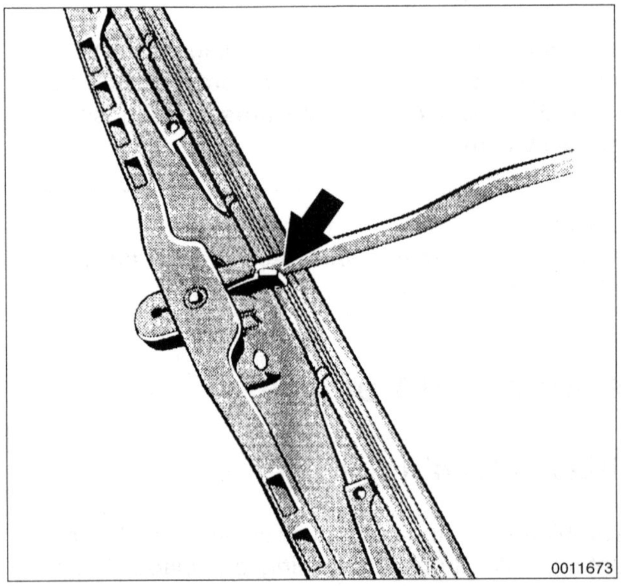


◄ Remove wiper blade from wiper arm by depressing retaining tab (**arrow**) and sliding blade out of arm.

***NOTE—***
*Some wiper blade versions may have two retaining tabs.*

– Installation is reverse of removal. Install wiper blade to wiper arm until it can be heard to click into position.

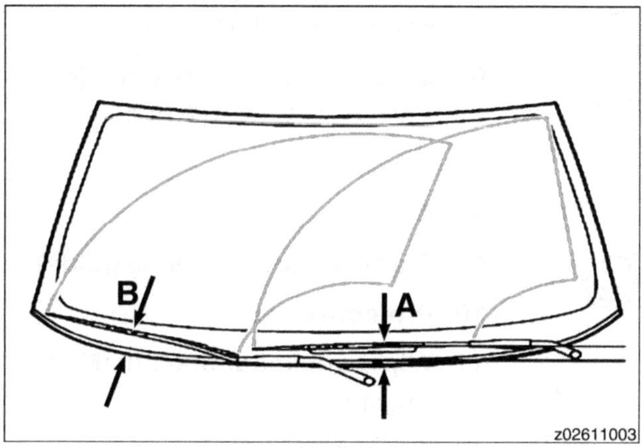


## Windshield wiper arm, removing and installing

- Make sure wipers are in PARK position and ignition switch is OFF. Open engine hood.
- Remove plastic trim plug from base of wiper arm.
- Remove wiper arm mounting nut (**arrow**).
- Mark position of wiper arm on shaft to aid installation.
- Remove wiper arm using BMW special tool 61 2 100 or equivalent puller.
- Installation is reverse of removal. Tighten mounting nut after adjusting position of arms.

| Wiper arm position | |
|---|---|
| Distance from wiper blade mounting to cowl frame:<br>• Driver wiper (**A**)<br>• Passenger wiper (**B**) | <br><br>25 mm (1 in)<br>45 mm (1.8 in) |

| Tightening torque | |
|---|---|
| Front wiper arm to shaft | 25 Nm (18 ft-lb) |

## Rear wiper arm, removing and installing (Coupe)

- Make sure wiper is in PARK position and ignition switch is OFF.
- Pivot plastic trim away from base of wiper arm.
- Pull rear window washer nozzle hose off wiper pivot.
- Remove wiper arm mounting nut.
- Mark position of wiper arm on shaft to aid installation.
- Remove wiper arm using BMW special tool 61 2 100 or equivalent puller.
- Installation is reverse of removal. Tighten mounting nut after adjusting position of arm. Be sure to reattach washer nozzle hose.

| Tightening torque | |
|---|---|
| Rear wiper arm to shaft | 12 Nm (9 ft-lb) |

## WIPER MOTOR AND LINKAGE

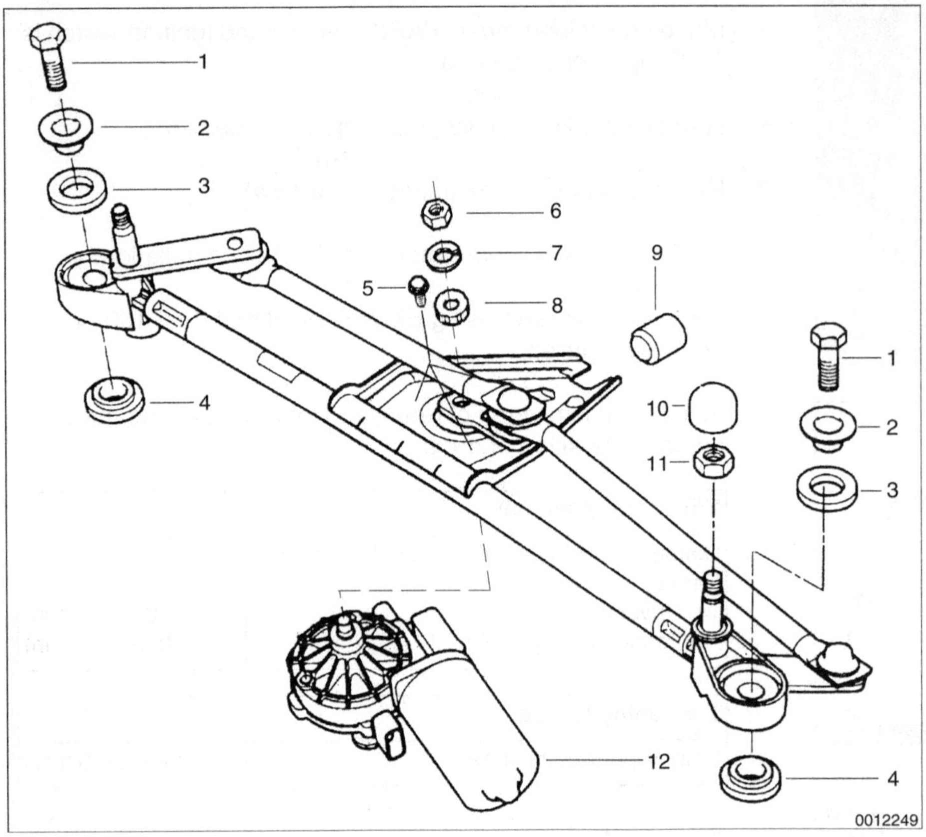


### Windshield wiper motor and linkage components

1. **Wiper assembly mounting bolt (M6)**
   - tighten to 8 Nm (6 ft-lb)
2. **Bushing**
3. **Upper isolation washer**
4. **Lower isolation washer**
5. **Motor mounting bolt (M6)**
   - tighten to 8 Nm (6 ft-lb)
6. **Motor crank mounting nut (M8)**
   - tighten to 27 Nm (20 ft-lb)
7. **Washer**
8. **Cover**
9. **Bushing (upper locking guide)**
10. **Protective cap**
11. **Wiper arm mounting nut (M10)**
    - tighten to 25 Nm (18 ft-lb)
12. **Wiper motor**

### Windshield wiper assembly, removing and installing

Remove the windshield wiper assembly (linkage and motor) as a single unit, then disassemble and repair as necessary.

– Make sure wipers are in PARK position and ignition switch is OFF.

> ***CAUTION—***
> *To avoid damaging the wiper arms and pivots, do not manually slide or force the wipers across the windshield.*

– Disconnect negative (–) battery cable and cover terminal with insulating material.

***NOTE—***
*On 1996 cars, the battery is located in the right rear of the engine compartment. On 1997 and later cars, the battery is located under the trunk floor mat.*

> ***CAUTION—***
> *Prior to disconnecting the battery, read the battery cautions in* **001 General Warnings and Cautions.**

– Remove windshield wiper arms. See **Windshield wiper arm, removing and installing** in this repair group.

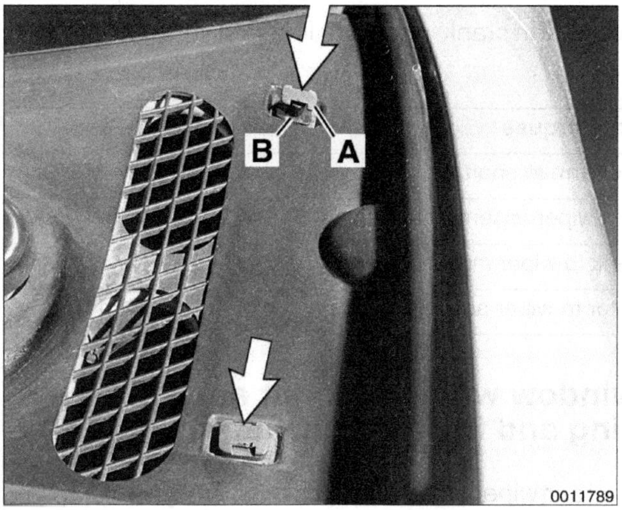


< Remove interior ventilation intake air grille at base of windshield.

- Peel up rubber seal from around protective grille.
- Remove expansion clips (**arrows**): First remove center piece (**A**), then slide out clip (**B**).
- Remove grille.

– Working below intake grille opening, remove left and right wiper assembly mounting bolts.

– Tilt assembly and pull toward right side of vehicle to disengage upper locking guide.

– Disconnect wiper motor harness connector.

– Pull assembly forward and up to remove from vehicle.

> ***CAUTION—***
> *When removing assembly, check that rubber isolation washers do not fall off.*

– Installation is reverse of removal. Be sure motor is in PARK position before installing assembly. If necessary, operate motor briefly so that motor stops in PARK position.

| Tightening torques | |
|---|---|
| Front wiper arm to shaft | 25 Nm (18 ft-lb) |
| Windshield wiper assembly to body (M6) | 10 Nm (7 ft-lb) |

## Windshield wiper motor, removing and installing

– Remove wiper assembly. See **Windshield wiper assembly, removing and installing** in this repair group.

– Detach wiper linkage rods from wiper motor crank.

– Mark position of crank to wiper motor shaft.

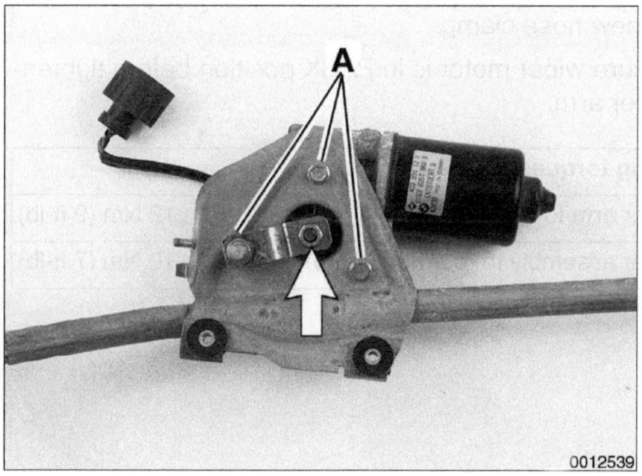


< Remove wiper crank mounting nut (**arrow**). Remove crank from motor shaft.

– Unscrew wiper motor mounting bolts (**A**) and remove motor from linkage assembly.

Installation is reverse of removal.

- If installing a new wiper motor, connect motor and operate wiper switch briefly so that motor stops in PARK position.
- Align wiper arm crank in straight line with linkage (**dashed line**).

| **Tightening torques** | |
|---|---|
| Front wiper arm to shaft | 25 Nm (18 ft-lb) |
| Windshield wiper assembly to body (M6) | 10 Nm (7 ft-lb) |
| Wiper crank to wiper motor shaft (M8) | 27 Nm (20 ft-lb) |
| Wiper motor to wiper assembly (M6) | 8 Nm (6 ft-lb) |

## Rear window wiper / washer assembly, removing and installing (Coupe)

– Make sure rear wiper is in PARK position and ignition switch is OFF.

> ***CAUTION—***
> *To avoid damaging the wiper arm and pivot, do not manually slide or force the wiper across the rear window.*

– Remove rear wiper arm. See **Rear wiper arm, removing and installing (Coupe)** in this repair group.

– Open rear hatch and remove hatch inner trim panel.

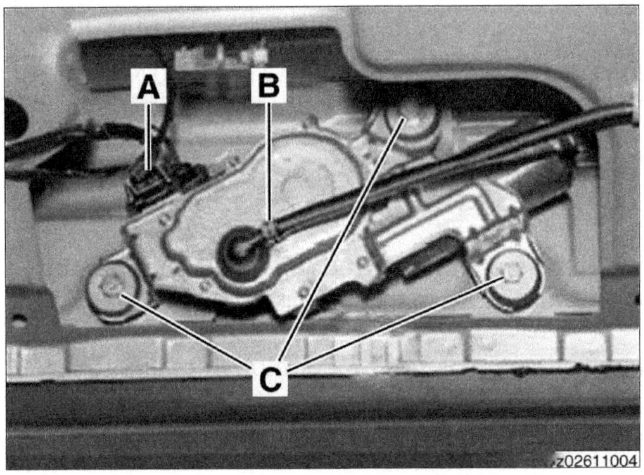


Working inside rear hatch cavity:

- Detach wiper module electrical connector (**A**).
- Loosen or cut hose clamp (**B**) and detach washer fluid hose.
- Remove wiper assembly mounting bolts (**C**).

– Remove wiper assembly.

– When installing:

- Replace porous or leaky rubber grommet as necessary.
- Install new hose clamp.
- Make sure wiper motor is in PARK position before tightening wiper arm.

| **Tightening torques** | |
|---|---|
| Rear wiper arm to shaft | 12 Nm (9 ft-lb) |
| Rear wiper assembly to rear hatch (M6) | 10 Nm (7 ft-lb) |

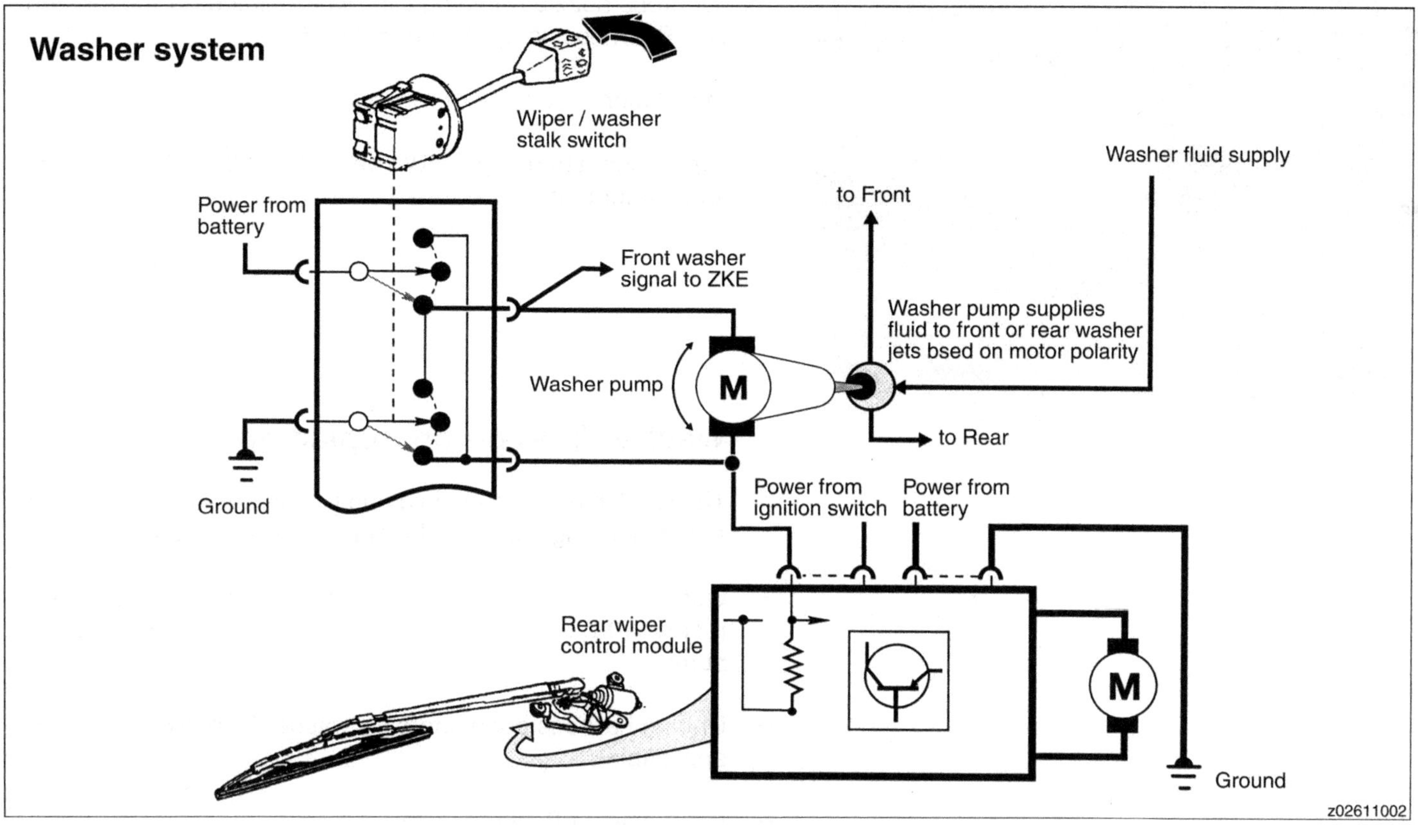


## WASHER SYSTEM

The washer system includes the windshield and rear window spray nozzles, the washer fluid pump, and the washer fluid reservoir.

The windshield spray nozzles, installed in the engine hood, are electrically heated. Heating is regulated by the outside air temperature sensor installed in the left wheel housing (4-cylinder models) or behind the right end of the front bumper (6-cylinder models).

The rear washer nozzle is part of the rear wiper arm.

A single washer pump, located in the washer fluid reservoir at right front of the engine compartment, is used to pump fluid to either the windshield or rear window nozzles.

### Windshield spray nozzle, removing and installing

- Open hood and remove hood insulating pad retaining clips until nozzle can be fully accessed.

- Carefully disconnect hose from nozzle. Detach nozzle electrical connector.

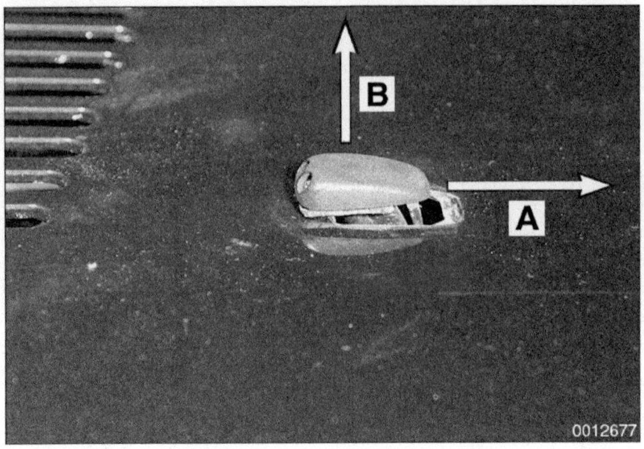


Depress nozzle retaining clips while sliding nozzle forward (**A**). Push nozzle up (**B**) out of hood.

– Installation is reverse of removal.

– Set nozzle aim using BMW special tool 00 9 200 or equivalent straight pin.

## Washer fluid pump, replacing

The windshield washer fluid pump is mounted in the fluid reservoir at the right front of the engine compartment.

– Open hood. Remove reservoir retaining screw(s). Loosen reservoir from mount and pull up slightly to access washer pump.

– Unplug harness connectors for washer fluid level switch (if applicable).

– Drain washer fluid reservoir.

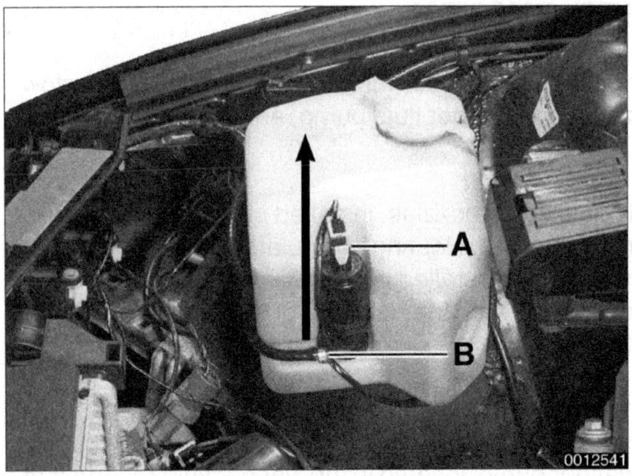


Working at reservoir:

- Pull pump out of reservoir (**arrow**).
- Detach pump electrical connector (**A**).
- Loosen or cut hose clamp (**B**) and disconnect fluid hose from pump.

– Installation is reverse of removal.

- Replace pump sealing grommet if damaged.
- Replace hose clamp.

# 612 Switches

# GENERAL

This section covers replacement information for interior switches and accessories.

For additional information, see:

- **119 Lubrication System** for oil pressure and oil level warning switch
- **170 Radiator and Cooling System** for engine cooling fan switch
- **230 Manual Transmission** for manual transmission back-up light switch
- **250 Gearshift Linkage** for automatic transmission back-up light switch function
- **610 Electrical Component Locations**
- **ELE Electrical Wiring Diagrams**

## Special tools

Some BMW special tools are recommended for the procedures in this repair group.

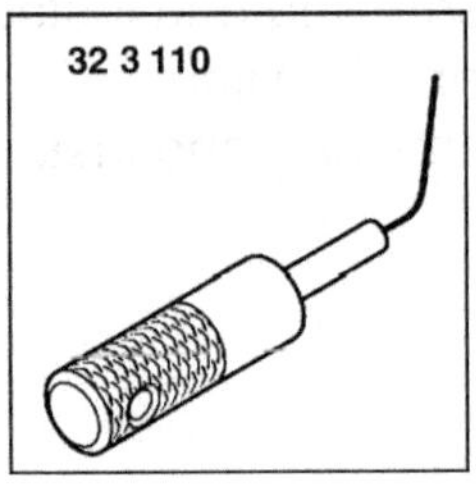


Ignition lock cylinder release tool (BMW special tool 32 3 110)

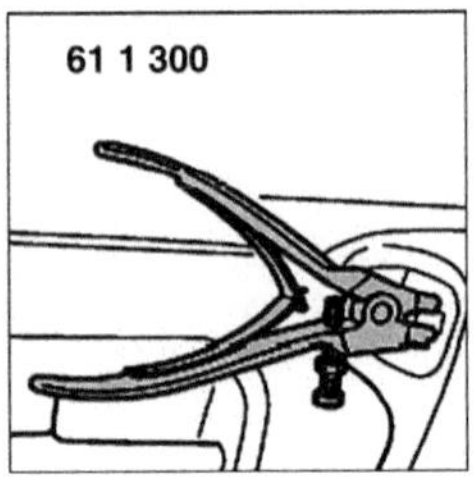


Mirror switch removal tool (BMW special tool 61 1 300)

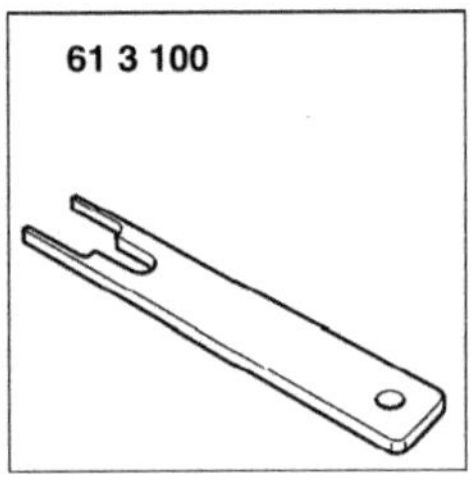


Headlight switch removal tool (BMW special tool 61 3 100)

## Cautions

> ***CAUTION—***
> - *Switch the ignition off and remove the negative (-) battery cable before removing any electrical components.*
> - *Prior to disconnecting the battery, read the battery cautions in* **001 General Warnings and Cautions**.
> - *To avoid marring the trim when working on interior components, use plastic prying tools or wrap the tips of screwdrivers and pliers with masking tape.*

## STEERING COLUMN SWITCHES

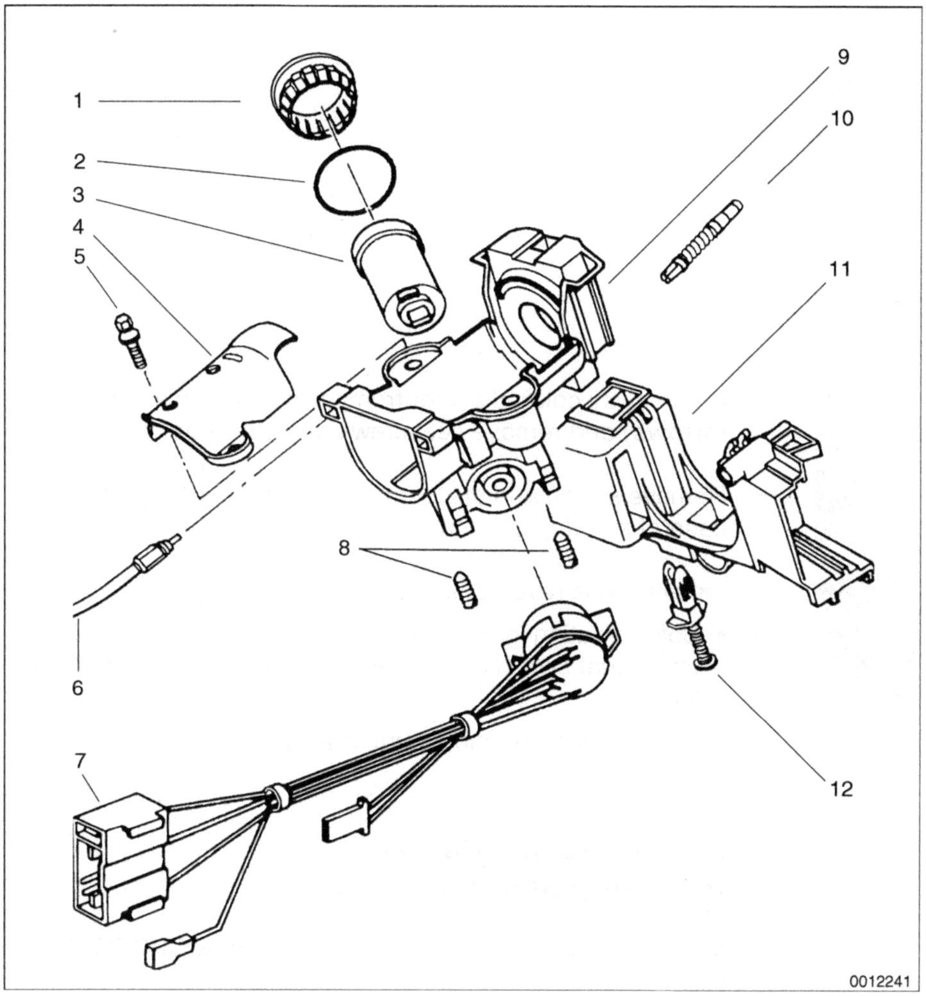


### Ignition switch components

1. **EWS ring antenna**
2. **O-ring**
3. **Ignition lock cylinder**
4. **cover**
5. **Shear bolt**
6. **Shift interlock cable**
7. **Ignition switch electrical connector**
8. **Set screws**
9. **Steering lock assembly**
10. **Airbag contact pin**
11. **Stalk switch bracket**
12. **Trim screw and plastic rivet**

The ignition electrical switch and the ignition lock cylinder can be removed with the ignition switch assembly attached to the steering column.

### Ignition electrical switch, removing and installing

– Disconnect negative (–) battery cable and cover battery terminal with insulating material.

> ***CAUTION—***
> *Prior to disconnecting the battery, read the battery cautions in* **001 General Warnings and Cautions**.

– Remove trim panel and knee bolster from lower left side of dashboard. See **513 Interior Trim**.

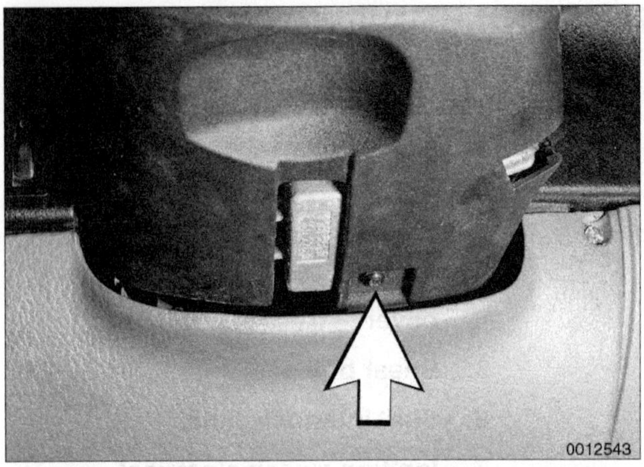
0012543

Remove lower steering column cover retaining screw (**arrow**) and remove cover from steering column.

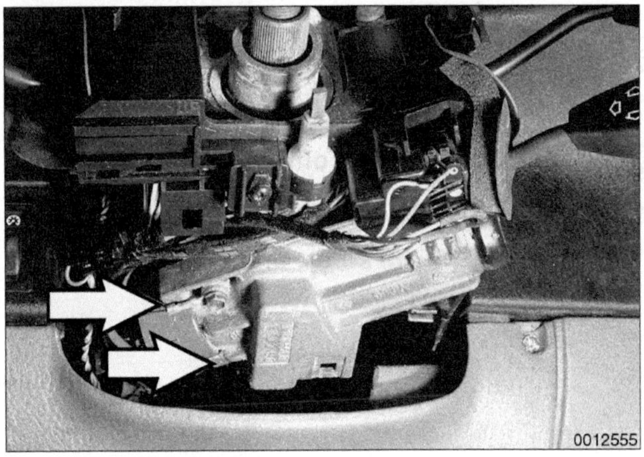
0012555

Clean away colored sealer from ignition switch set screws (**arrows**) and remove set screws on both sides of switch.

– Remove switch and disconnect electrical connector from base of steering column.

– Installation is reverse of removal.
  - Before mating ignition electrical switch to ignition lock, check that switch is in correct position relative to ignition lock.
  - Seal set screws using paint or lacquer.

## Ignition lock cylinder, removing and installing

– Insert key into lock cylinder and turn 60° to ACCESSORY position.

0011807

Insert BMW special tool 32 3 110 or equivalent stiff piece of wire into opening (**arrow**) in lock cylinder.
  - Pull out lock cylinder. If necessary rotate tool and key slightly from side to side.

***NOTE—***
*Use the valet key during this procedure for better access to the lock opening.*

– Installation is reverse of removal.

## Steering column stalk switches, removing and installing

The steering column stalk switches consist of the following:

- Left side: Headlight dimmer and flasher switch, combined with the turn signal switch
- Right side upper: Wiper / washer switch
- Right side lower: Cruise control switch

The stalk switches are all mounted to the stalk switch bracket. The following procedure is for removing the headlight dimmer / turn signal switch. Follow similar steps to remove the other stalk switches.

– Disconnect negative (–) battery cable and cover battery terminal with insulating material.

> ***CAUTION—***
> *Prior to disconnecting the battery, read the battery cautions in* **001 General Warnings and Cautions**.

– Remove airbag unit from steering wheel. See **721 Airbag System (SRS)**. Store airbag unit in a safe place.

> ***WARNING—***
> *The airbag is a vehicle safety system. Serious injury may result if system service is attempted by persons unfamiliar with the BMW SRS and its approved service procedures. BMW advises that all inspection and service be performed by an authorized BMW dealer.*

– With front wheels pointed straight ahead, remove steering wheel. See **320 Steering and Wheel Alignment**.

– Remove trim panel and knee bolster from lower left side of dashboard. See **513 Interior Trim**.

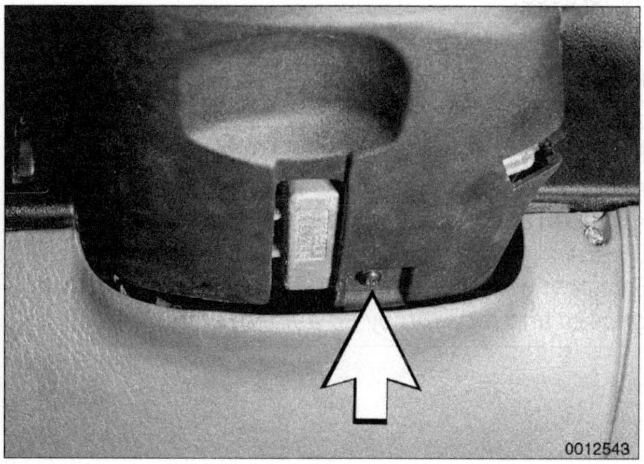


◀ Remove lower steering column cover retaining screw (**arrow**) and remove cover from steering column.

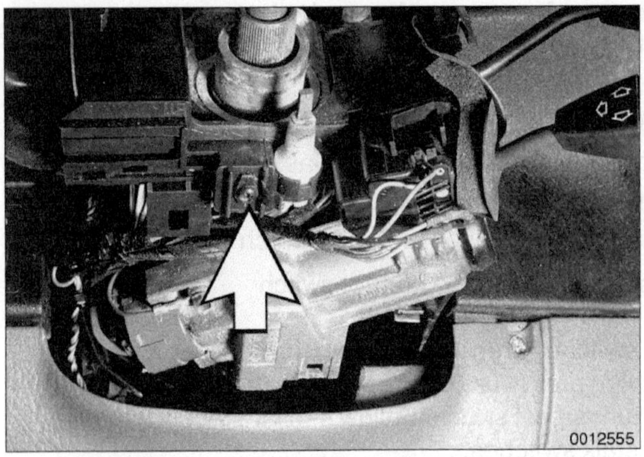


◄ Remove stalk switch bracket mounting screw (**arrow**).

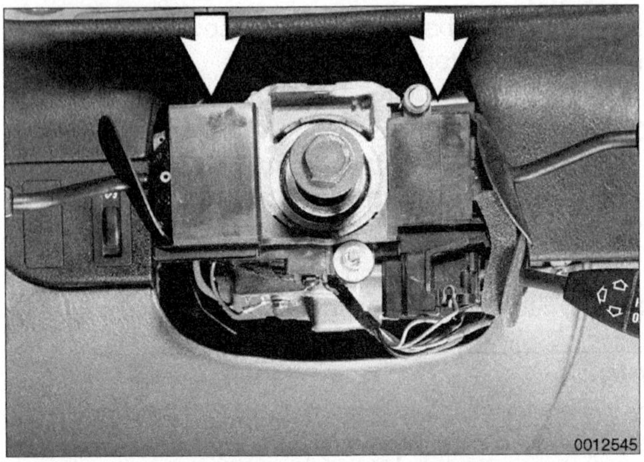


◄ Depress retaining tabs (**arrows**) to release stalk switch bracket. Push bracket downwards and off steering column.

◄ Simultaneously depress catches on headlight dimmer / turn signal switch and release from column.

– Cut wire ties retaining harness to steering column. Disconnect switch electrical connector located at bottom of steering shaft.

– Installation is reverse of removal.

| Tightening torques | |
|---|---|
| Steering wheel to steering column spindle (bolt) | 63 Nm (46 ft-lb) |

***WARNING —***
*Once the airbag unit is installed and all other service procedures have been completed, start the engine and check that the SRS warning light goes out. If the warning light stays on, the SRS system will not function as designed. Have the system diagnosed and repaired by an authorized BMW dealer.*

## PEDAL CLUSTER SWITCHES

### Brake light switch, removing and installing

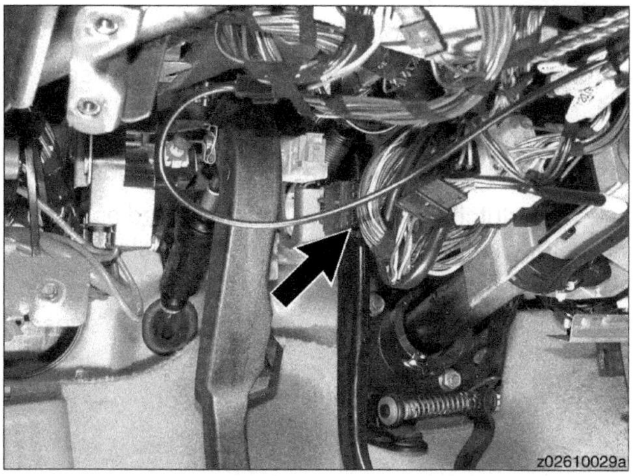


The brake light switch turns on the brake lights and is also used to deactivate cruise control when braking.

– Remove trim panel and knee bolster from lower left side of dashboard. See **513 Interior Trim**.

◄ Cut wire ties and detach brake light switch electrical connector (**arrow**).

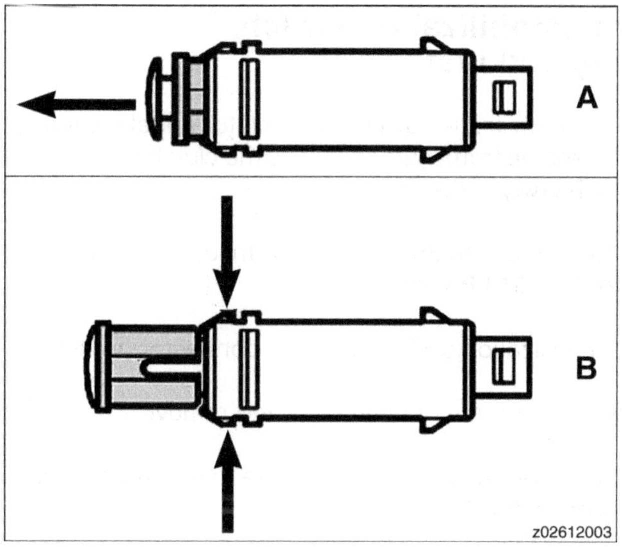


◄ Push brake pedal down as far as possible and lock in place using pedal stop. Working at brake light switch:

- **A** Pull switch tip forward completely (**arrow**).
- **B** Press in retaining clips (**arrows**) and pull out brake light switch backward.

– Installation is reverse of removal. Keep in mind the following:

- Make sure new brake light switch tip is as far forward as possible. Switch adjusts itself automatically as you slowly release brake pedal.
- Use new wire tie to reattach slack wiring harness to switch connector.
- Check function of brake light switch.
- Check cruise control cut-off function of switch.

### Clutch pedal position sensor, removing and installing

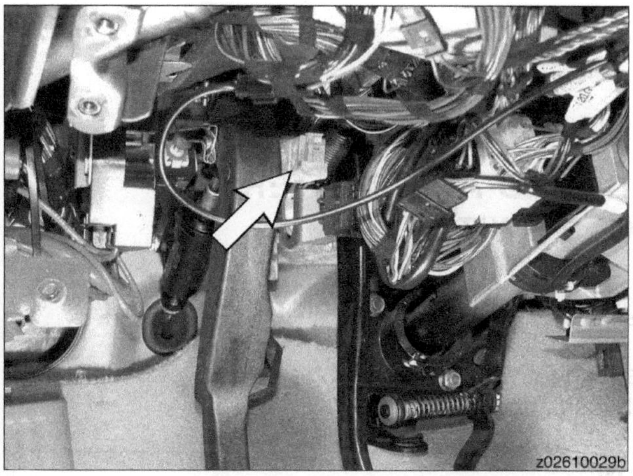


The clutch pedal position sensor is used to deactivate cruise control.

– Remove trim panel and knee bolster from lower left side of dashboard. See **513 Interior Trim**.

◄ Cut wire ties and detach clutch pedal position sensor electrical connector (**arrow**).

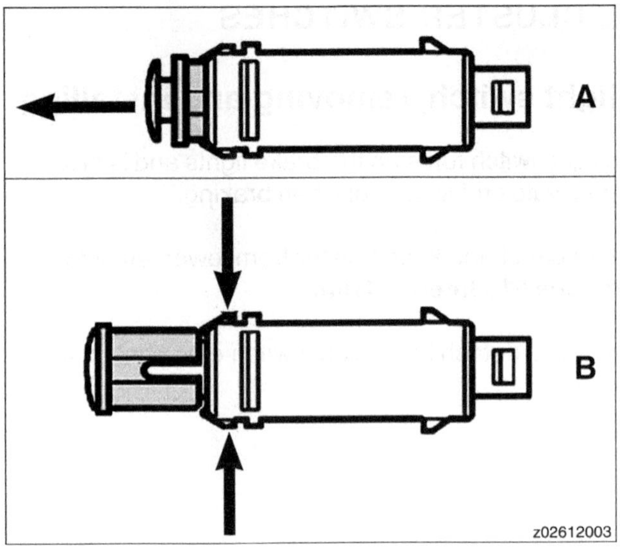

z02612003

Push clutch pedal down as far as possible and lock in place using pedal stop. Working at clutch pedal position sensor:

- **A** Pull switch tip forward completely (**arrow**).
- **B** Press in retaining clips (**arrows**) and pull out sensor backward.

– Installation is reverse of removal. Keep in mind the following:

- Make sure new clutch sensor tip is as far forward as possible. Sensor adjusts itself automatically as you slowly release clutch pedal.
- Use new wire tie to reattach slack wiring harness to sensor connector.
- Check cruise control cut-off function of sensor.

### Starter immobilization switch, removing and installing

The starter immobilization switch at the clutch pedal is used to prevent the engine from starting unless the clutch pedal is pressed all the way to the floor.

– Remove trim panel and knee bolster from lower left side of dashboard. See **513 Interior Trim**.

z02610029c

Detach immobilization switch electrical connector (**arrow**).

– Press together switch mounting clips and remove.

– Installation is reverse of removal. Be sure to test starting system after work is finished.

## DASHBOARD AND CENTER CONSOLE SWITCHES

### Headlight switch, removing and installing

– Disconnect negative (–) battery cable and cover battery terminal with insulating material.

> ***CAUTION—***
> *Prior to disconnecting the battery, read the battery cautions in* **001 General Warnings and Cautions**.

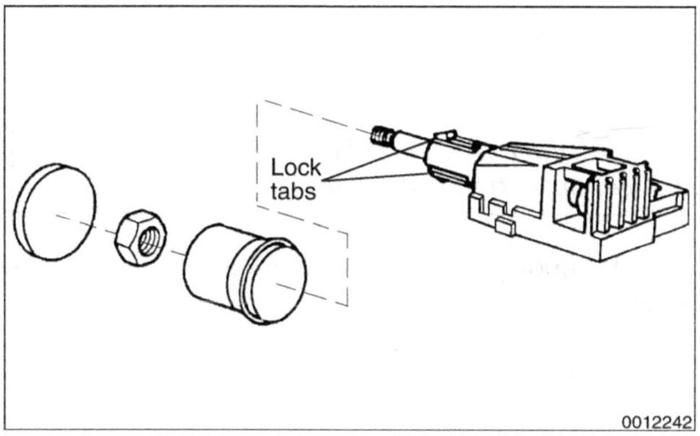


– Remove trim panel and knee bolster from lower left side of dashboard. See **513 Interior Trim**.

– Remove headlight switch knob center trim cap by carefully prying out with small flat bladed screwdriver.

> ***CAUTION—***
> *To avoid marring the trim, use a plastic prying tool or wrap the screwdriver tip with masking tape.*

– Remove lock nut inside headlight switch knob, then rotate knob counter-clockwise to remove.

< Using BMW special tool no. 61 3 100 or equivalent, press two headlight switch lock tabs inward and push switch forward into dashboard.

– Reach behind dashboard from below and pull switch down to access electrical connector. Detach connector.

– Installation is reverse of removal.

## Center dashboard switches, removing and installing

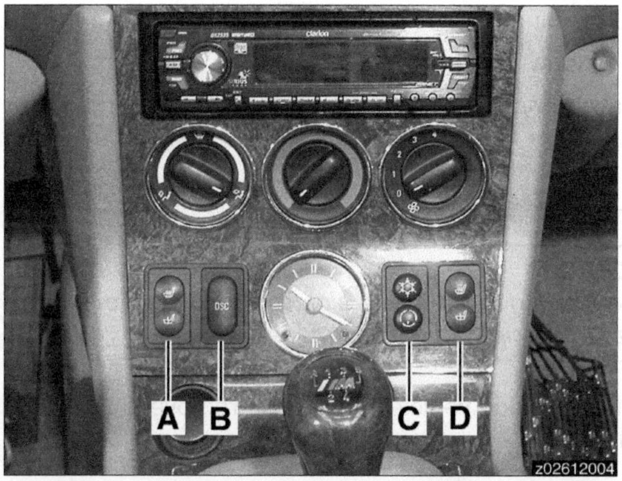


< Depending on year and model, the following switches may be installed in the center dashboard:

- Driver seat heater switch (**A**)
- ASC or DSC switch (**B**)
- A/C and recirculation switches (**C**)
- Passenger seat heater switch (**D**)

***NOTE—***
*2000 Z3 center dashboard illustrated.*

< Variant dashboard and center console:

- Driver seat heater switch (**A**)
- A/C and recirculation switches (**B**)
- Rear window defogger switch (**C**)
- ASC or DSC switch (**D**)
- Passenger seat heater switch (**E**)

***NOTE—***
*1999 M Coupe center dashboard and front console illustrated.*

These switch housings are externally identical, even though their internal construction and front logos are different. This means that the removal procedure for any of these switches is the same.

- Remove radio. See **650 Radio**.
- Reach through radio opening and push switch out.
- Alternatively, carefully pry switch out of console with small flat-bladed screwdriver.

> ***CAUTION—***
> *To avoid marring the trim, use a plastic prying tool or wrap the screwdriver tip with masking tape.*

- Pull switch out far enough to access electrical connector. Detach connector.
- Installation is reverse of removal.

## Shifter housing switches, removing and installing

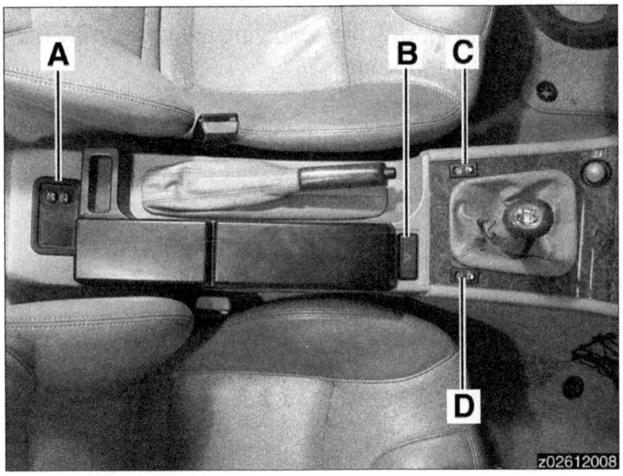


The following switches are installed in the shifter housing:

- Hazard warning switch (**B**)
- Power window switches (**C**, **D**)
- Automatic transmission program selector switch (not illustrated)

The removal procedure for these switches is similar.

> ***CAUTION—***
> *To avoid marring the trim, use a plastic prying tool or wrap the screwdriver tip with masking tape.*

- Manual transmission vehicle: Unclip and remove shift lever boot by lifting up.
- Automatic transmission vehicle: Remove shift lever bezel by carefully prying up on edge with a small flat screwdriver.
- If necessary, remove shifter knob. See **250 Gearshift Linkage**.
- Hazard warning switch: Push into shifter housing, then reach inside shift lever opening and pull switch out.
- Power window switch: Reach under switch and push up out of shifter housing.
- Automatic transmission program selector switch: Reach under switch and push up out of shifter housing.

- Detach switch electrical connector.
- Installation is reverse of removal.

### Convertible top switch, removing and installing

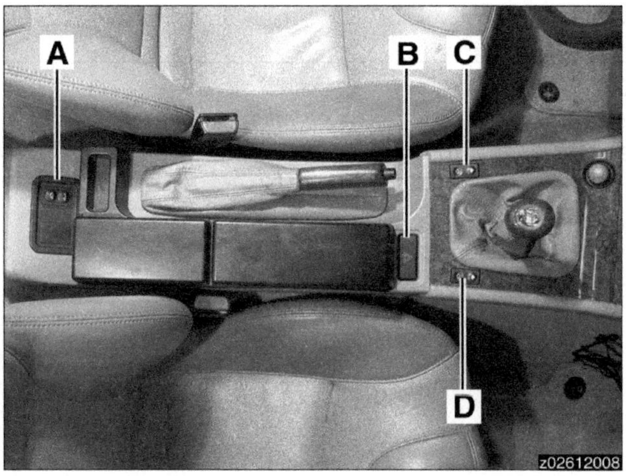


◄ The convertible top switch (**A**) is located in the rear of the center console, behind the parking brake handle.

- Working at switch panel edge, carefully pry convertible top switch out of console with small flat-bladed screwdriver.

> ***CAUTION—***
> *To avoid marring the trim, use a plastic prying tool or wrap the screwdriver tip with masking tape.*

- Detach electrical connector from back of switch.
- Installation is reverse of removal.

## OTHER INTERIOR SWITCHES

### Power mirror switch, removing and installing

◄ Working at left door pull, carefully pry off power mirror adjusting button (**arrow**).

> ***CAUTION—***
> *To avoid marring the trim, use a plastic prying tool or wrap the screwdriver tip with masking tape.*

- Using BMW special tool 61 1 300 or equivalent, reach in through mirror adjusting button opening and squeeze locking tabs to release switch.
- Pull out switch and detach electrical connector.
- Installation is reverse of removal.

## Sunroof switch, removing and installing

The sunroof operating switch (**arrow**) in Coupe models is located in the headliner toward the front.

– Carefully pry switch out of headliner, using a flat-bladed screwdriver.

> ***CAUTION—***
> *To avoid marring the trim, use a plastic prying tool or wrap the screwdriver tip with masking tape.*

– Pull out switch and detach electrical connector.

– Installation is reverse of removal.

# 620 Instruments



**TABLES**

## General

This repair group covers removal and installation of the instrument cluster. Instrument cluster self-test procedures are also covered.

### Special tools

Special BMW test harnesses and scan tool equipment are required to troubleshoot the cluster assembly. BMW dealer servicing is recommended.

Aftermarket warning light resetting tool (Peake Research)

Automotive digital multimeter

> ***CAUTION—***
> *Only use a digital multimeter when testing automotive electrical components.*

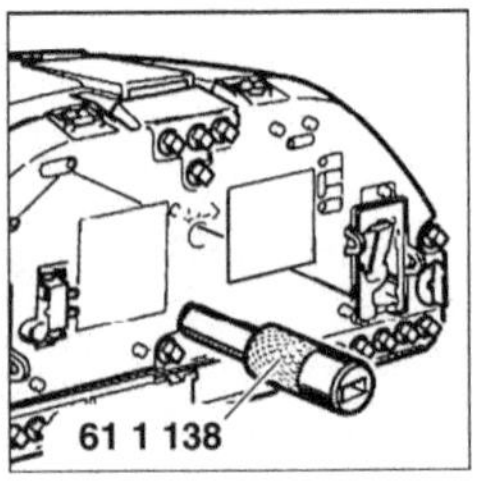


Instrument bulb removal tool (BMW special tool 61 1 138)

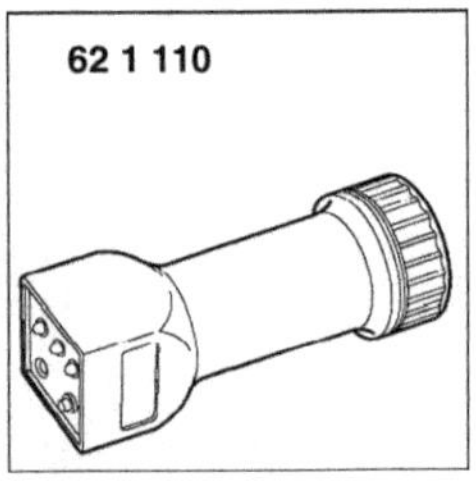


Service interval resetting tool (BMW special tool 62 1 110)

### Instrument cluster

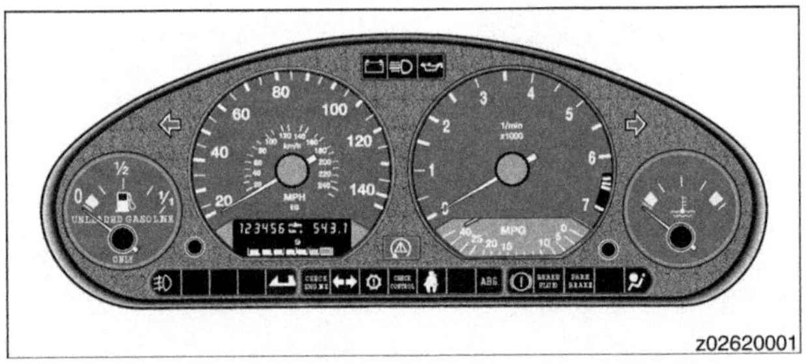


Analog gauges are used to display speed, engine revolutions per minute, engine coolant temperature and fuel tank level. A small LCD (liquid crystal display) readout inside the cluster assembly, at the base of the speedometer, displays the vehicle mileage and other information. Warning lights indicate when the engine oil pressure is too low, when the anti-lock brakes (ABS), the air bag system (SRS), or the engine control module (ECM) has detected a fault, or when the parking brake is engaged.

Although instrument clusters in Z3 and M version cars are similar, two different systems are used:

**1996 - 1998**. Cluster is self-contained. A coding plug is used for model-specific coding. Mileage and service information is kept in the coding plug.

**1999 - 2002**. In addition to the coding plug, the cluster is connected to the CAN-bus and the diagnostic link. Use BMW diagnostic scan tools (DISplus, GT1 or MoDiC) for diagnosis and coding functions.

## Board Computer (BC)

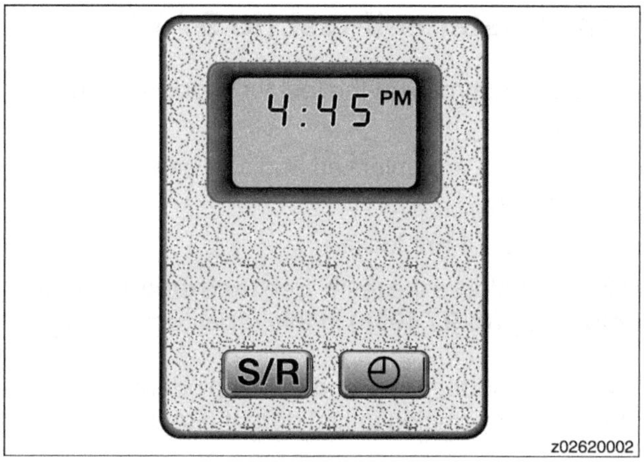

z02620002

A limited function Board Computer (BC) is offered as an option in Z3 models. It is located in the lower center dashboard, below the heating and A/C controls.

BC functions include:

- Digital clock
- Outside temperature
- Average fuel consumption
- Estimated range based on fuel left in tank
- Average road speed

BC functions are accessed via the turn signal stalk switch on the steering column.

## Additional instruments

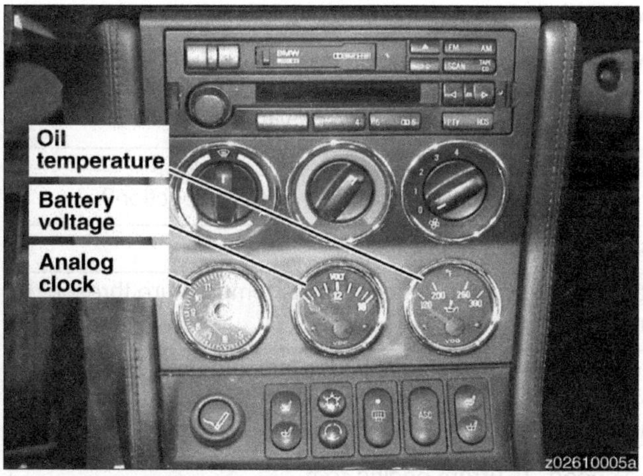

z02610005a

M version vehicles are equipped with additional gauges located in the lower center dash, below the radio and heating and A/C controls.

***NOTE—***
*1999 M Coupe lower dashboard is illustrated.*

## Cautions

***CAUTION—***
- *Switch the ignition off and remove the negative (-) battery cable before removing any electrical components.*
- *Prior to disconnecting the battery, read the battery disconnection cautions in* **001 General Warnings and Cautions**.
- *Prior to disconnecting the battery:*
  *-Obtain the radio operation code, if applicable.*
  *-Record radio presets.*
- *Only use a digital multimeter for electrical tests.*

## Instrument Cluster Self-Test

In early production Z3 cars, prior to 9 / 1998, the instrument cluster is not connected to the diagnostic link for troubleshooting. The cluster incorporates a sequence of 15 test functions which can be accessed through the LCD mileage display in the cluster.

Starting with 1999 models (production date 9 / 1998), the Z3 instrument cluster is capable of storing a fault memory, accessible *via* the diagnostic link. The cluster incorporates a sequence of 21 test functions which can be accessed through the LCD mileage display in the cluster.

### Instrument cluster self-test (to 9 / 1998)

**Table a** provides a summary of the test functions in early production Z3 models.

- Test functions 1 to 3 are unlocked and can be called up at any time for display.
- Test functions 4 to 14 are locked.
- Test function 15 unlocks the tests. See **Instrument cluster tests, unlocking (to 9 / 1998)** in this repair group.

To access self-test mode, press and hold trip odometer reset button while switching ignition to first detent (RADIO position).

**Table a. Instrument cluster test functions (to 9 / 1998)**

| Test | Function |
|---|---|
| 01: Self-test | Gauges and warning lights activated |
| 02: Engine data | Displays engine information from coding plug<br>Engine factor 2 = 4-cylinder<br>Engine factor 3 = 6-cylinder |
| 03: Elapsed distance | Interval since last service (in km)<br>Which service due at next inspection? |
| 04: Elapsed days | Not applicable in US |
| 05: Service interval evaluation | Engine speed, engine temperature threshold |
| 06: Fuel level and coolant temperature | Hexadecimal codes indicate gauge positions |
| 07: Engine speed | Current engine rpm |
| 08: Road speed | Current vehicle speed in kph |
| 09: Stored mileage | See **Instrument cluster mileage, resetting (to 9 / 1998)** |
| 10: Status bit | Digital input to cluster |
| 11: Status bit | Digital output from cluster |
| 12: Not used | |

| Table a. Instrument cluster test functions (to 9 / 1998) | |
|---|---|
| **Test** | **Function** |
| 13: Country code | Country version of cluster |
| 14: Software reset | Reset system, lock tests |
| 15: Unlock | See **Instrument cluster tests, unlocking (to 9 / 1998)** |

## Instrument cluster mileage, resetting (to 9 / 1998)

Vehicle mileage and service interval information are stored in an EEPROM on the instrument cluster main circuit board and also in the coding plug.

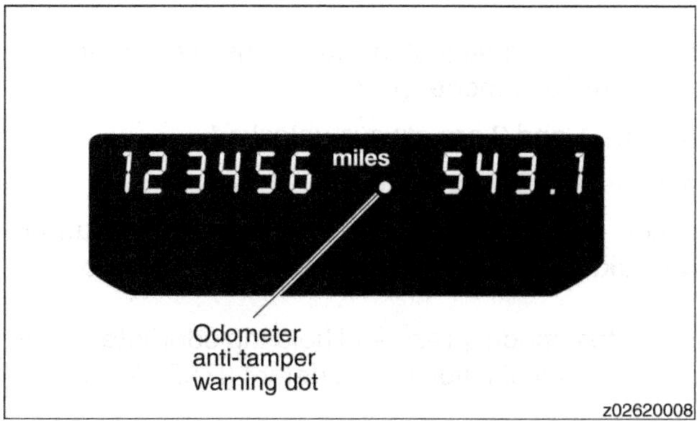


When the original coding plug is reinstalled in a new cluster, the mileage information stored in the new main circuit board differs from actual vehicle mileage stored in the coding plug. This causes an odometer anti-tamper warning dot to light up in the cluster.

Use test function 09, described below, to bring the two mileage figures into agreement and to extinguish the anti-tamper dot.

- Use test 15 to unlock test functions. See **Instrument cluster tests, unlocking (to 9 / 1998)** in this repair group.
- Use reset button to scroll to test 09. Display shows component with lower mileage:
  - Display **XXXXXX I** indicates that mileage in cluster circuit board EEPROM is lower than mileage stored in coding plug.
  - Display **XXXXXX E** indicates that mileage in coding plug is lower than mileage stored in cluster circuit board.
- Press and hold reset button for 4 seconds. This overwrites lower mileage with higher mileage and cancels anti-tamper warning.

***NOTE—***
*Service interval information is also transferred between storage sites by this procedure.*

z02620007

## Instrument cluster tests, unlocking (to 9 / 1998)

– Press and hold mileage reset button while switching ignition key to first detent (RADIO position). **tESt** appears in display.

< Press reset button to scroll through test functions until **tESt 15 L ON** is displayed.

– Press and hold reset button for 3 seconds until display changes to **tESt 15 L OFF**. Tests are now unlocked.

– Again press and hold reset button for 3 seconds. Display changes to **tESt 15 L ON**. Tests are now locked.

## Instrument cluster self-test (from 9 / 1998)

**Table b** provides a summary of test functions in the Z3 instrument cluster from 1999 model year.

- Test functions 1 and 2 are always unlocked.
- Test functions 3 to 21 are locked.
- Test function 19 unlock the tests. See **Instrument cluster tests, unlocking (from 9 / 1998)** in this repair group.

To access self-test mode, press and hold trip odometer reset button while switching ignition to first detent (RADIO position).

**Table b. Instrument cluster test functions (from 9 / 1998)**

| Menu | Submenu | Sample output | Meaning |
|---|---|---|---|
| Test 01: Vehicle specific data | 1.0 | 12345 | VIN |
| | 1.1 | 4663 | Body number |
| | 5_1.2 | 414061 | Cluster part number |
| | 1.3 | 029000 | Coding / diagnosis / bus index |
| | 1.4 | _2197 | Manufacturing data (calendar week / year) |
| | 1.5 | 01_700 | Hardware / software number of cluster |
| | 2_1.6 | FFFF_04 | Injection status, number of cylinders, engine factor |
| | 1.7 | _02_00 | CAN-bus index |
| Test 02: Cluster system test - Activates test for gauge drivers, indicators and LEDs to confirm function | | | |
| Test 03: Service interval data | 3.0 | 1500 | Consumption in liters since last reset |
| | 3.1 | | Elapsed days (not applicable in US) |
| Test 04: Fuel consumption data (current) | 4.0 | 0267 | 26.7 liters / 1000 km |
| | 4.1 | 0073 | 7.3 liters / hour |
| Test 05: Not used | | | |

**Table b. Instrument cluster test functions (from 9 / 1998)**

| Menu | Submenu | Sample output | Meaning |
|---|---|---|---|
| Test 06: Fuel level sensor input in liters | 6.0 | 0652 | 65.2 liters in fuel tank |
| | 6.1 | 0652 | 65.2 liters in fuel tank |
| | 1_6.2 | 0667 | Sensor diagnosis:<br>• 1_ = sensor ok<br>• 2_ = sensor faulty<br>• 3_ = implausible signal |
| Test 07: Temperature and speed | 7.0 | 032 | Coolant temperature 32°C (90°F) |
| | 7.1: Not used | | |
| | 7.2 | 5283 | Engine speed 5,283 rpm |
| | 7.3 | 058 | Vehicle speed 58 kph (36 mph) |
| Test 08: Input values in HEX code | 8.0 — 8.3 | XXX | Hex code, instrument cluster inputs |
| Test 09: Battery voltage | 9.0 | 125 | 12.5 volts |
| Test 10: Country coding | 10.0 | 02 | USA |
| Test 11: Cluster code | 11.0 | FFFF08 | Cluster code |
| Test 12: Not used | | | |
| Test 13: Gong test | 13.0 | Gong | Activate gong by pressing button (gong response is delayed). |
| Test 14: Fault memory (not for diagnosis) | | | |
| Test 15: Not used | | | |
| Test 16: Stored mileage | I_16.0 | 123459 | Internal EEPROM has lower stored mileage. |
| | E_16.0 | 000325 | External coding plug has lower stored mileage. |
| | 16.1 | COPY | Copy mileage data from higher to lower memory location. See **Instrument cluster mileage, resetting (from 9 / 1998)** in this repair group. |
| Test 17: Time | | | |
| Test 18: Not used | | | |
| Test 19: Unlock | 19.0 | L-ON L-OFF | Lock / unlock specific test functions. See **Instrument cluster tests, unlocking (from 9 / 1998)**. |
| Test 20: Average fuel consumption correction factor | 20.0 | | Press button to reset display to 1000. |
| | 20.1 | XXX9 | Press button when correct 1s position is reached. |
| | 20.2 | XX5X | Press button when correct 10s position is reached. |
| | 20.3 | 12XX | Press button when correct 100s position is reached. |
| Test 21: Software reset | 21.0 | reset | Reset software |

## Instrument cluster mileage, resetting (from 9 / 1998)

Vehicle mileage and service interval information are stored in an EEPROM on the instrument cluster main circuit board and also in the coding plug.

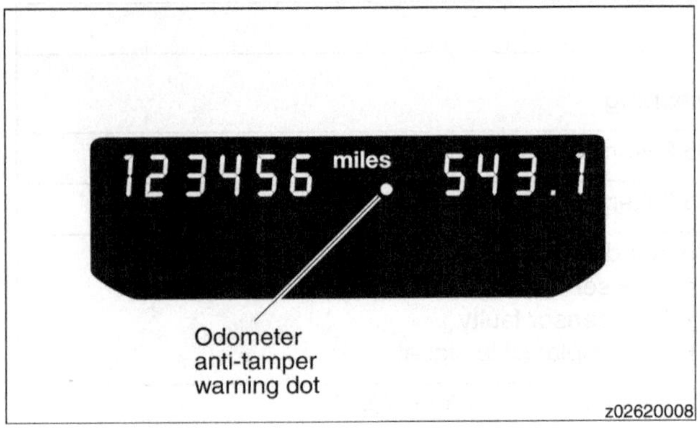


When the original coding plug is reinstalled in a new cluster, the mileage information stored in the new main circuit board differs from actual vehicle mileage stored in the coding plug. This causes an odometer anti-tamper warning dot to light up in the cluster.

Use test function 16, described below, to bring the two mileage figures into agreement and to extinguish the anti-tamper dot. Normal data exchange between two information storage sites is restored.

– Use test 19 to unlock test functions. See **Instrument cluster tests, unlocking (from 9 / 1998)** in this repair group.

– Use odometer reset button to scroll to test 16.
  - Display **XXXXXX I_16.0** indicates that mileage in cluster circuit board EEPROM is lower than mileage stored in coding plug.
  - Display **XXXXXX E_16.0** indicates that mileage in coding plug is lower than mileage stored in cluster circuit board.

– Press and hold odometer reset button for 4 seconds.
  - Display **COPY _16.1**overwrites lower mileage with higher mileage and extinguishes anti-manipulation warning dot.

***NOTE—***
- *Service interval information is also transferred by this procedure.*
- *If the battery is disconnected before 2 miles have accumulated on cluster, repeat test 16 to cancel anti-tamper warning.*

## Instrument cluster tests, unlocking (from 9 / 1998)

– Press and hold trip odometer reset button while switching ignition key to first detent (RADIO position). **tESt** appears in display.

Press reset button to scroll through test functions until **19 L ON** is displayed.

– Display switches from **L ON** (lock ON) to **L OFF** (lock OFF) every second.
  - Press reset button *immediately* when it changes to **L-OFF**.
  - Proceed to next test. (Example: test 16, odometer reset.)

***NOTE—***
*Cluster tests are automatically locked when exiting test functions.*

# INSTRUMENT COMPONENTS

## Instrument cluster, removing and installing

– Turn ignition switch ON, then center steering wheel. Turn ignition switch OFF.

– Disconnect negative (–) battery cable and cover terminal with insulating material.

***NOTE—***
*On 1996 cars, the battery is located in the right rear of the engine compartment. On 1997 and later cars, the battery is located under the trunk floor mat.*

> ***CAUTION—***
> *Prior to disconnecting the battery, read the battery cautions in* **001 General Warnings and Cautions**.

– Remove driver airbag. See **721 Airbag System (SRS)**.

> ***WARNING—***
> - *The BMW Supplemental Restraint System (SRS) is complex and special precautions must be observed when servicing. Serious injury may result if system service is attempted by persons unfamiliar with the BMW SRS and its approved service procedures. BMW specifies that all inspection and service should be performed by an authorized BMW dealer.*
> - *BMW Supplemental Restraint Systems (SRS) are equipped with a back-up power supply inside the SRS control module. Allow a 5 second discharge period after disconnecting the battery cable. This will allow the power supply to discharge, rendering the SRS system inoperative.*

– Remove steering wheel. See **320 Steering and Wheel Alignment**.

◄ Remove two trim screws (**arrows**) that mount instrument cluster assembly and bezel to dashboard.

– Pull cluster assembly and bezel straight out, then tilt cluster assembly down onto steering column.

> ***CAUTION—***
> *Cover upper steering column with shop rags or soft towel to avoid scratching the clear plastic cluster assembly cover.*

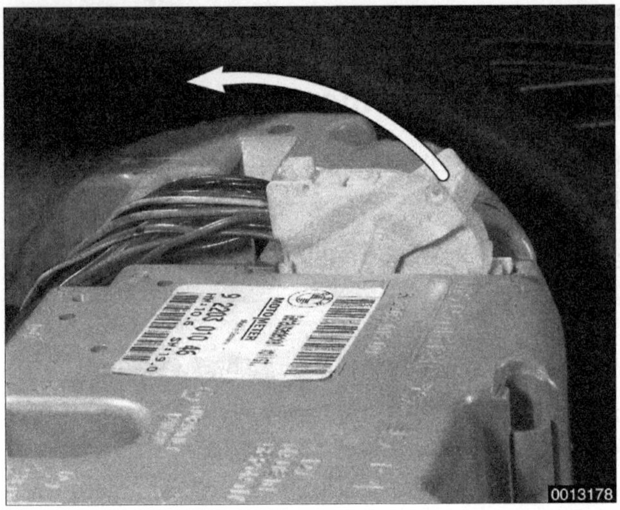
0013178

◄ Release instrument cluster assembly electrical connector locking levers (**arrow**).

> ***CAUTION—***
> *Attempting removal of the instrument cluster assembly electrical connectors without disengaging the locking levers will result in cluster and connector damage.*

– Disconnect cluster assembly electrical connectors, then remove cluster assembly.

– If replacing cluster, transfer coding plug. See **Coding plug, removing and installing** in this repair group.

– Replace cluster bulbs as necessary. See **Instrument cluster bulbs, removing and installing** in this repair group.

– When reinstalling, verify that electrical connector locking levers are in unlocked position. Attach and lock electrical connectors.

– Remainder of installation is reverse of disassembly.

***NOTE—***
*If installing new cluster, synchronize stored vehicle mileage and service information between coding plug and cluster. See* **Instrument cluster mileage, resetting (to 9 / 1998)** *or* **Instrument cluster mileage, resetting (from 9 / 1998)**.

## Coding plug, removing and installing

The coding plug stores one of two records of the vehicle mileage, as well as additional vehicle information.

– Remove instrument cluster. See **Instrument cluster, removing and installing** in this repair group.

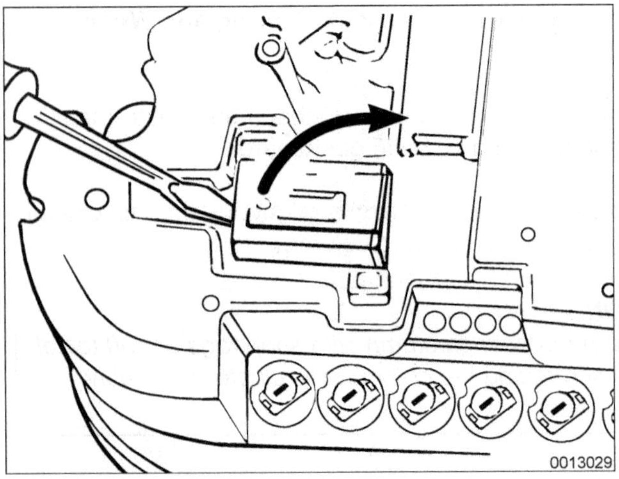
0013029

◄ Break open plastic retainer in back of instrument cluster and remove coding plug.

– Install coding plug and new retainer. Reinstall instrument cluster.

– Synchronize stored vehicle mileage and service information. See **Instrument cluster mileage, resetting (to 9 / 1998)** or **Instrument cluster mileage, resetting (from 9 / 1998)**.

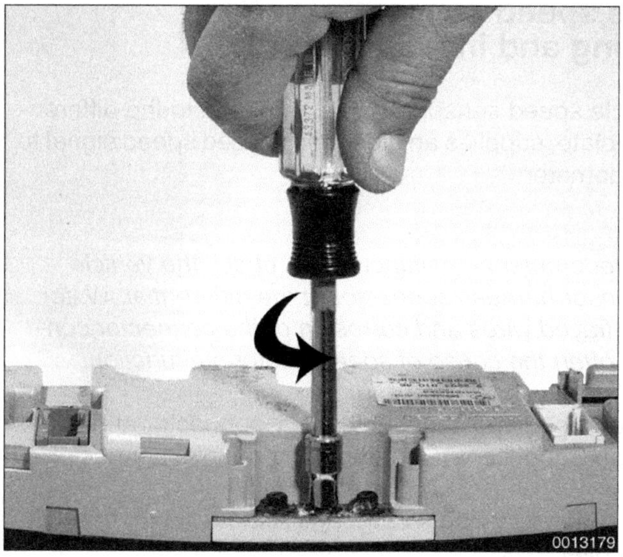


## Instrument cluster bulbs, removing and installing

– Remove instrument cluster. See **Instrument cluster, removing and installing** in this repair group.

< Turn faulty bulb 45° counterclockwise to remove. Use BMW special tool 61 1 138 or equivalent.

***CAUTION—***
*Bulbs with excessive wattage could melt cluster housing. Replace defective bulbs with bulbs of the same wattage.*

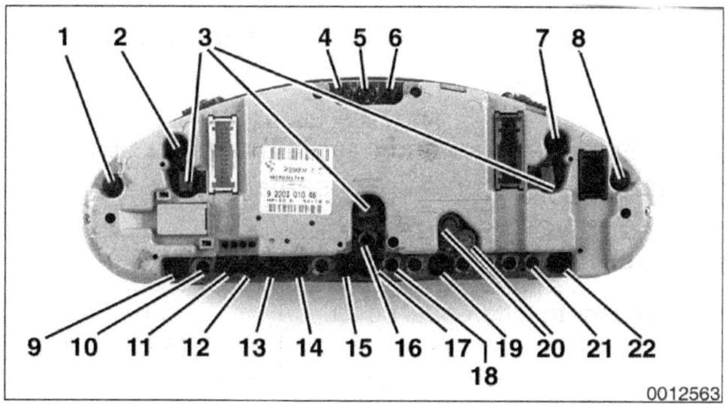


< Instrument cluster bulb assignment:

1. Temperature warning
2. Right turn signal
3. Instrument lighting
4. Oil pressure warning
5. High beam warning
6. Alternator warning
7. Left turn signal
8. Low fuel warning
9. Airbag warning
10. Catalytic converter overheating warning
11. Parking brake
12. Brake fluid level warning
13. Brake pad wear warning
14. ABS warning
15. Seat belt warning
16. ASC / DSC warning
17. Check control
18. Automatic transmission control (EGS) failure warning
19. Malfunction indicator light (MIL)
20. LCD illumination
21. Rear foglight (not used)
22. Front foglights

## Vehicle speed sensor, removing and installing

The vehicle speed sensor (**arrow**), mounted to the differential cover plate, supplies an electronic pulsed speed signal to the speedometer.

***NOTE—***
*If the speedometer is malfunctioning, check the vehicle speed sensor harness connector at the differential. Water damage, frayed wires and corrosion of the connector contacts are often the cause of speedometer malfunction.*

- Locate vehicle speed sensor electrical connector at back of differential. Squeeze retainer tabs and remove connector.
- Remove senor retaining screws and remove senor from differential cover.

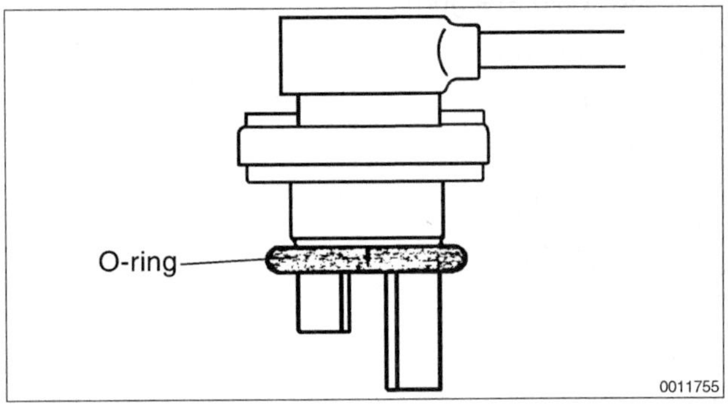


Prior to reinstallation, replace sensor O-ring and coat with oil (90W).

- Install sensor into differential and tighten mounting bolts evenly in stages until sensor is firmly seated.

| **Tightening torque** | |
|---|---|
| Vehicle speed sensor to differential cover | 6.6 - 8.4 Nm (53 - 74 in-lb) |

## Clock or Board Computer, removing and installing

The digital clock and the Board Computer (BC) are externally and visually similar. The same procedure is used to remove either component from the center lower dashboard.

- Remove radio from upper center dashboard. See **650 Radio**. Do not disconnect radio electrical harness.

Depress locking tabs (**arrows**) and remove clock or BC from console.

- Alternatively, remove center console. See **513 Interior Trim**. Push out BC or clock.
- Detach electrical connector.
- Installation is reverse of removal.

## Lower dashboard gauges, removing and installing (M version)

The gauges in the lower dashboard of M version vehicles are all the same size and installed the same way.

- Remove radio from upper center dashboard. See **650 Radio**. Do not disconnect radio electrical harness.

- Unfasten heater controls and press in.

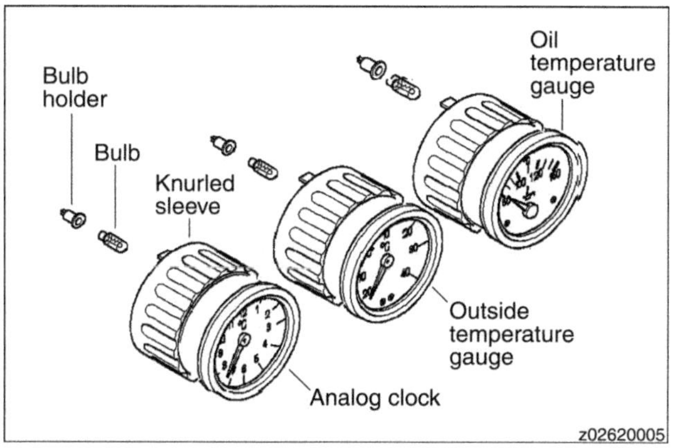

z02620005

< Reaching in through radio opening, loosen knurled sleeve in back of dash gauge and push gauge out.

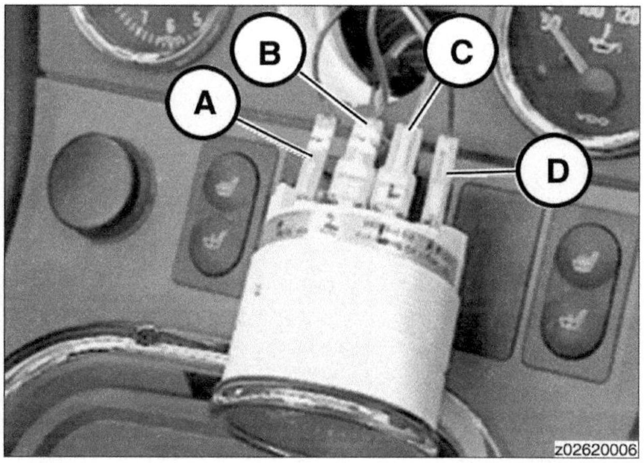

z02620006

< Mark electrical connectors at back of gauge (**A**, **B**, **C**, **D**) before disconnecting.

- Installation is reverse of removal.

# 630 Lights



TABLES

## General

This repair group covers exterior and interior lighting repair and bulb replacement for Z3 models.

For additional information, see:

- **230 Manual Transmission** for manual transmission back-up light switch
- **250 Gearshift Linkage** for automatic transmission back-up light switch function
- **610 Electrical Component Locations**
- **612 Switches**
- **ELE Electrical Wiring Diagrams**

### Special tools

< Automotive digital multimeter

> ***CAUTION—***
> *Only use a digital multimeter when testing automotive electrical components.*

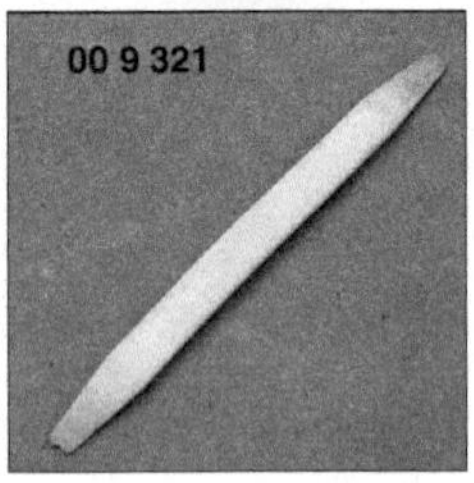


< Plastic prying tool
(BMW special tool 00 9 321)

### Lighting system

Headlight assembly

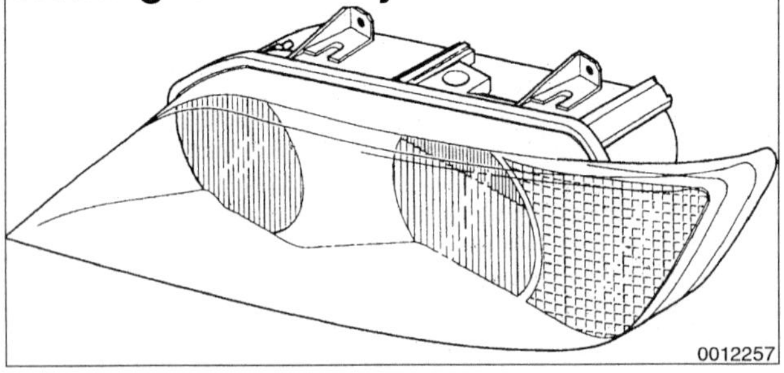


Z3 light bulbs consist of the following types:

- High and low beam headlights and foglights: High pressure halogen bulbs.
- Turn signals, side markers, taillights, parking lights, and license plate lights: Conventional tungsten-filament light bulbs. Turn signal bulbs are yellow.
- Interior and trunk lights: Conventional bulbs.
- Center brake light on Z3 Roadster from 4 / 1999: Integrated LED unit.

## Exterior bulb applications

For convenience, exterior bulb applications for Z3 cars are listed in **Table a**.

**Table a. Z3 exterior bulb applications**

| Location | Type & rating |
|---|---|
| Back-up light | 12v 21w |
| Brake light | 12v 21w |
| Center brake light<br>• Z3 Roadster 1996 - 4 / 1999<br>All M Roadster<br>• Z3 Roadster 4 / 1999 on<br>• Z3 and M Coupe | <br>12v 5w bulb without base<br><br>LED<br>Soldered bulb (no rating) |
| Foglight<br>• Front (halogen)<br>• Rear | <br>H7 12v 55w<br>12v 21w |
| Headlight (halogen)<br>• High beam<br>• Low beam | <br>HB3 12v 60w<br>HB4 12v 51w |
| Licence plate light | 12v 5w |
| Rear parking light | 12v 5w |
| Side marker light | 12v 5w |
| Turn signal<br>• Front / parking light<br>• Rear<br>• side | <br>12v 21/5w yellow<br>12v 21w yellow<br>12v 5w |

## Cautions

***CAUTION—***
- *Do not touch the glass of bulbs with bare skin. Dirt and skin oils may cause a bulb to fail prematurely. If necessary wipe bulb using a clean cloth dampened with rubbing alcohol.*
- *Use only original equipment replacement bulbs. Non-original bulbs may cause false failure readings on the Check Control display.*
- *To avoid marring car paint or trim, work with a plastic prying tool or wrap screwdriver tip with tape.*

# Headlights

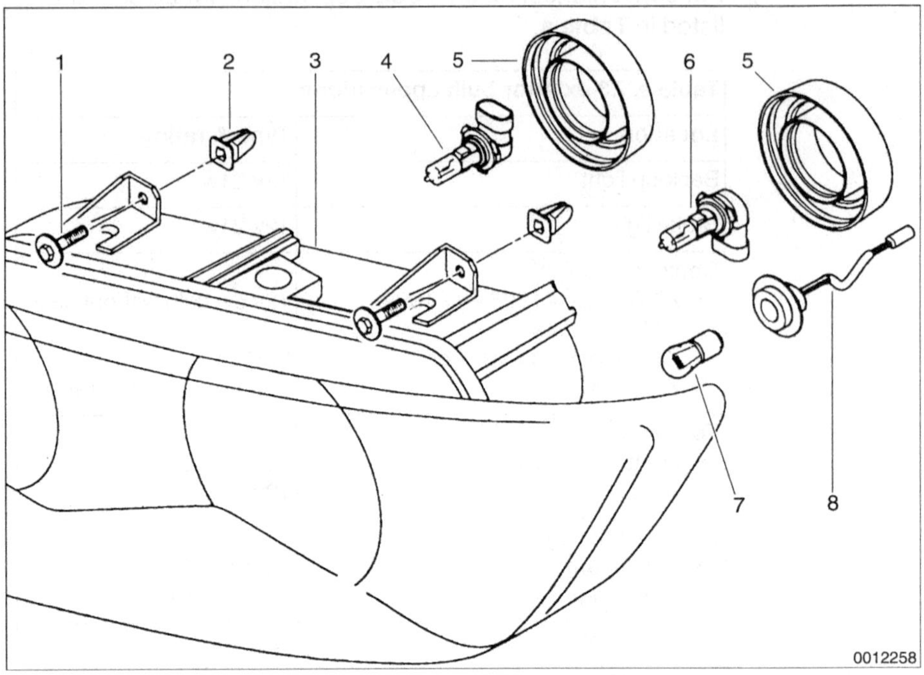


## Headlight components

1. **Mounting screw**
2. **Expanding rivet**
3. **Headlight housing**
4. **High beam bulb**
   - HB3 12v 60w
5. **Cap**
6. **Low beam bulb**
   - HB4 12v 51w
7. **Turn signal / parking light bulb**
   - 12v 21/5w yellow
8. **Turn signal / parking light bulb socket and harness**

## Headlight aim, adjusting

– Adjust headlight aim with tires at correct pressure, fuel tank full and weight of one person (approx. 75 Kg / 165 lb) in driver seat.

– Open engine hood.

– Left side: If necessary, remove air filter housing. See **130 Fuel Injection**.

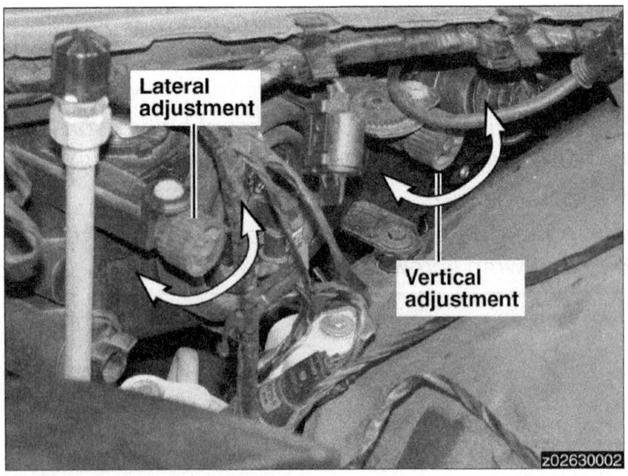


◄ Use knurled plastic screws in back of headlight to adjust aim.

***NOTE —***
*2000 Z3 Roadster right headlight illustrated. Left side and other models are similar.*

## Headlight bulb, replacing

– Verify headlight switch is OFF. Open engine hood.

> ***WARNING —***
> *If headlights have just been shut off, allow them sufficient time to cool. Halogen bulbs run hot and can cause burns.*

– Left side: If necessary, remove air filter housing. See **130 Fuel Injection**.

◄ Working behind headlight assembly, rotate headlight bulb to unlock it from its housing, then pull bulb straight out to remove.

– Detach electrical connector from headlight bulb, taking care not to damage connector locking clip.

– Insert new bulb into headlight housing assembly and rotate until it locks. Do not touch bulb with bare fingers. If necessary, clean bulb with soft, lint-free cloth and rubbing alcohol.

– Connect headlight electrical connector and check headlight operation.

## Headlight assembly, removing and installing

– Verify headlight switch is OFF. Open engine hood.

> ***WARNING —***
> *If headlights have just been shut off, allow them sufficient time to cool. Halogen bulbs run hot and can cause burns.*

– Left side: Remove air filter housing. See **130 Fuel Injection**.

– Remove bulbs and detach headlight electrical connectors. See **Headlight bulb, replacing** in this repair group. Store bulbs safely.

◄ Counterhold expanding rivets and remove headlight mounting screws (**arrows**). Tilt headlight forward to remove.

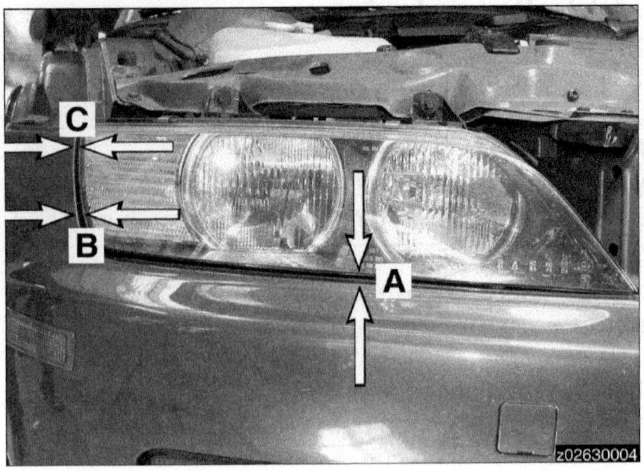


When installing:

- With headlight in place, screw in mounting screws as far as last thread. Counterhold expanding rivets.
- Rotate expanding rivets to adjust gaps **A**, **B** and **C**.

| Headlight assembly placement vs. bumper | |
|---|---|
| Lower gap (**A**) | approx. 5 mm (¼ in) |
| Side gap, lower (**B**) | approx. 5 mm (¼ in) |
| Side gap, upper (**C**) | approx. 2.5 mm (⅛ in) |

***NOTE—***
*If necessary, remove headlight to rotate expanding rivets.*

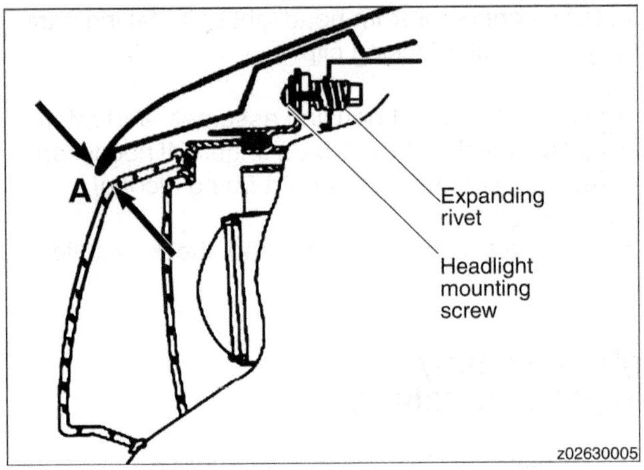


Close engine hood and check gap **A**.

| Headlight assembly placement vs. engine hood | |
|---|---|
| Headlight to hood gap (**A**) | approx. 0 mm |

- If necessary, open hood and rotate expanding rivets to adjust gap.
- Counterhold expanding nuts to tighten mounting screws.

– Reinstall bulbs and attach electrical connectors. Do not touch bulbs with bare fingers. If necessary, clean bulbs with soft, lint-free cloth and rubbing alcohol.

– Adjust headlight aim. See **Headlight aim, adjusting** in this repair group.

## EXTERIOR LIGHTS

### Front foglight bulb, replacing

– Verify headlight switch is OFF.

> ***CAUTION—***
> *If foglights have just been shut off, allow them sufficient time to cool. Halogen bulbs run hot and can cause burns.*

– Remove cover plate next to foglight by carefully prying on side closest to foglight lens. Use plastic prying tool or small flat-bladed screwdriver with tip wrapped with tape.

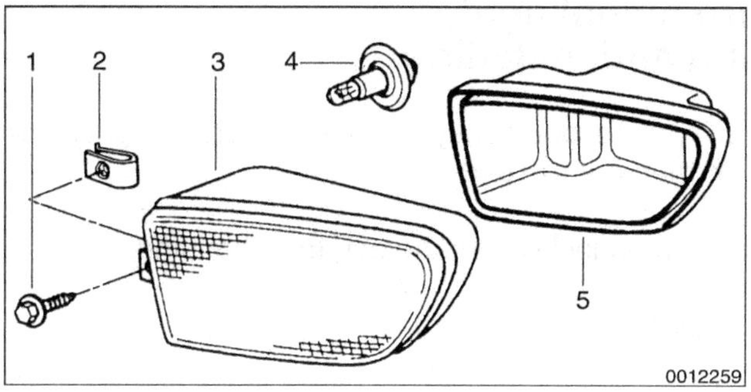


Remove foglight assembly mounting screw, pull out foglight socket and remove electrical connector.

1. Mounting screw
2. Body nut
3. Foglight lens
4. Foglight bulb
   H7 12v 55w
5. Foglight housing

- Disconnect wire spring clip that retains bulb and pull it out.
- Install new bulb and wire spring retaining clip. Do not touch bulb with bare fingers. If necessary, clean bulb with soft, lint-free cloth and rubbing alcohol.
- Remainder of installation is reverse of removal.

## Front turn signal / parking light bulb, replacing

- Verify headlight switch is OFF. Open engine hood.

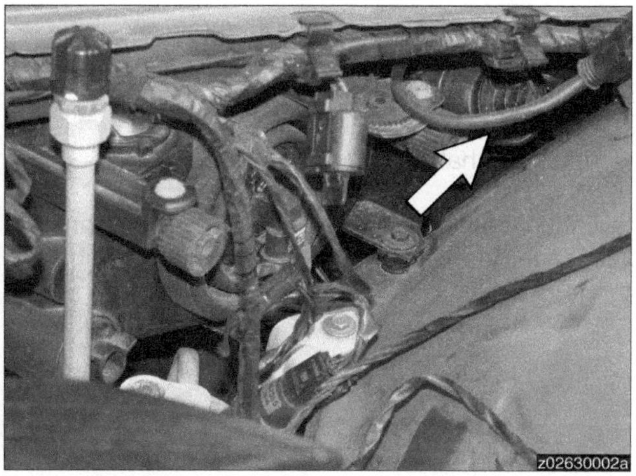


Working at corner of headlight assembly, rotate turn signal / parking light socket (**arrow**) counter-clockwise and pull straight back to remove.

***NOTE—***
- *2000 Z3 Roadster right headlight illustrated. Left side and other models are similar.*
- *If necessary for access, remove windshield washer fluid reservoir (right side) or air filter housing (left side).*

- Remove old bulb from socket and install new bulb. Do not touch bulb with bare fingers. If necessary, clean bulb with soft, lint-free cloth and rubbing alcohol.
- Reinstall bulb and socket in headlight assembly.

## Front side marker light, removing and installing

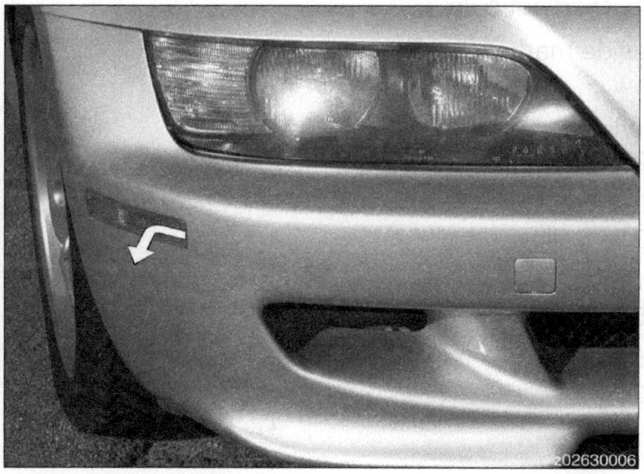


Using plastic prying tool, gently pry side marker light lens to side, then out (**arrow**).

- Twist bulb counterclockwise to detach from housing.
- Disconnect electrical connector.
- Installation is reverse of removal. Do not touch bulb with bare fingers. If necessary, clean bulb with soft, lint-free cloth and rubbing alcohol.

## Side turn signal light, removing and installing

< Using plastic prying tool, gently pry side turn signal light lens toward rear, then out (**arrow**).

– Twist bulb counterclockwise to detach from housing.

– Disconnect electrical connector.

– Installation is reverse of removal. Do not touch bulb with bare fingers. If necessary, clean bulb with soft, lint-free cloth and rubbing alcohol.

## Taillights, servicing

The Z3 rear light bulbs are mounted in a common taillight cluster assembly. Replacement of the parking, turn signal, brake and back-up light bulbs all follow the same procedure.

There are two versions of the taillight cluster.

### Version 1

Taillight bulbs are mounted to common cover.

– Verify headlight switch is OFF.

< Unscrew knurled nut that retains taillight assembly.

– Remove taillight assembly by carefully pulling it straight out. Taillight bulbs fit through stamped holes in rear body structure.

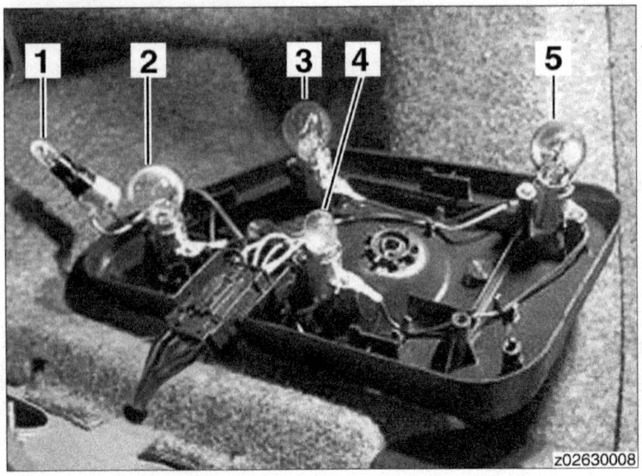


< Remove burned out bulb.

1. Rear side marker
   12v 5w
2. Brake light
   12v 21w
3. Turn signal
   12v 21w yellow
4. Taillight
   12v 5w
5. Back-up light
   12v 21w

– Install new bulb. Do not touch bulb with bare fingers. If necessary, clean bulb with soft, lint-free cloth and rubbing alcohol.

To replace socket assembly, disconnect electrical connector (**arrow**).

– Reinstall rear side marker light pig-tail into its socket. Verify side-marker pig-tail will not be damaged when knurled nut is tightened.

– Fully seat taillight assembly and tighten knurled nut.

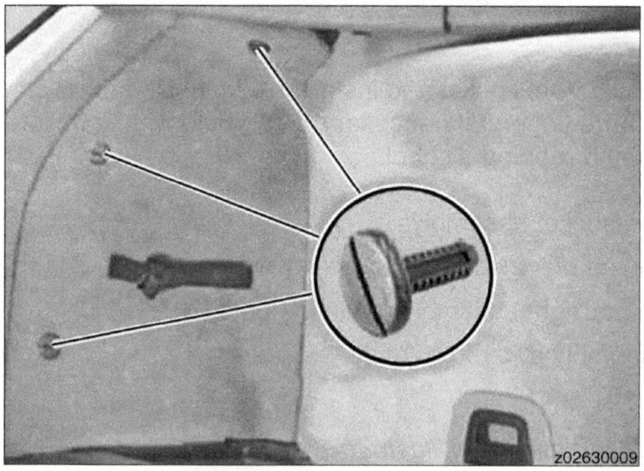


### Version 2

Remove taillight cover to access bulbs which are mounted to printed circuit board.

– Verify headlight switch is OFF.

– Unscrew knurled nut that retains taillight assembly cover.

Alternatively, release taillight assembly trim cover plastic retainers (**inset**).

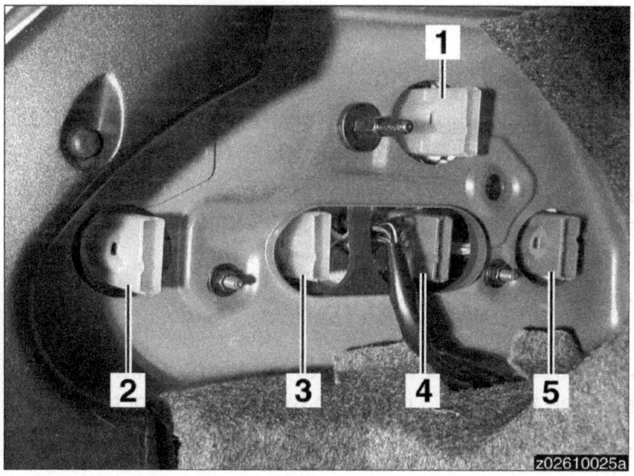


Z3 taillights: Twist bulb socket ¼ turn counterclockwise to remove burned out bulb with socket.

1. Turn signal
   12v 21w
2. Taillight
   12v 5w
3. Back-up light
   12v 21w
4. Brake light
   12v 21w
5. Rear side marker
   12v 5w

***NOTE—***
*2000 Z3 left taillight is illustrated.*

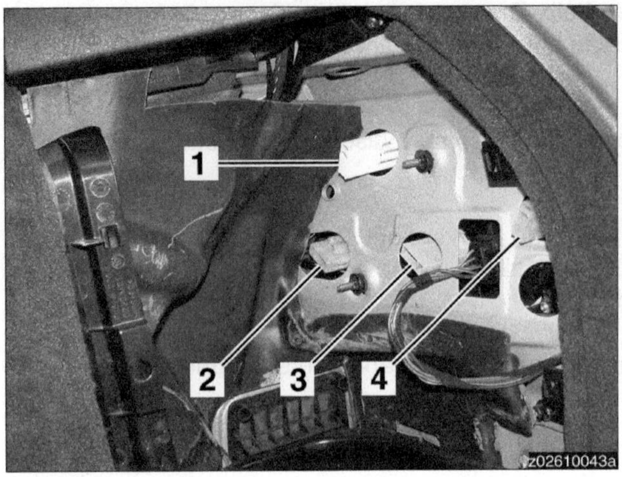


M Coupe taillights: Twist bulb socket ¼ turn counterclockwise to remove burned out bulb with socket.

1. Turn signal
   12v 21w yellow
2. Brake light
   12V 21W
3. Taillight
   12v 5w
4. Back-up light
   12v 21w

***NOTE—***
- *1999 M Coupe right taillight is illustrated.*
- *Rear side marker light is behind rubberized trim. Bulb is rated 12v 5w.*

– Remove bulb from socket and install new bulb. Do not touch bulb with bare fingers. If necessary, clean bulb with soft, lint-free cloth and rubbing alcohol.

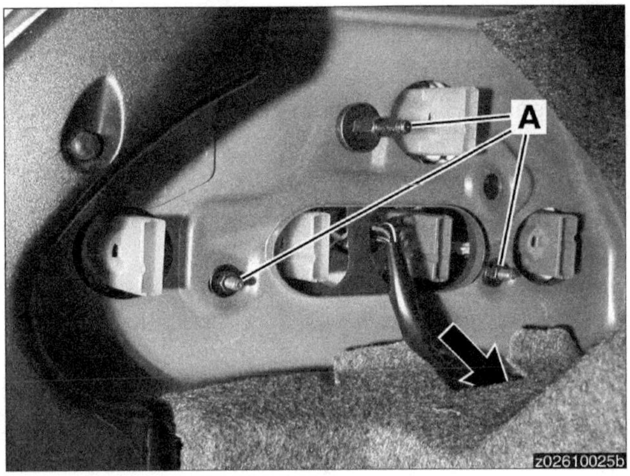


To replace socket assembly:
- Disconnect electrical connector (**arrow**).
- Remove taillight assembly mounting nuts (**A**).
- Pull assembly away from body.

***NOTE—***
- *2000 Z3 left taillight is illustrated.*
- *M Coupe electrical connector clips directly to taillight assembly.*

– Installation is reverse of removal. Keep in mind:
- When reinstalling assembly to body, make sure moisture seal is undamaged. Replace if necessary.
- Push electrical harness through body opening and route it as before.

## Center brake light, servicing

In Roadster models, the center brake light is mounted at the top center of the trunk lid. There are two versions of this light:

- Models to 4 / 1999: Light socket strip with multiple replaceable bulbs.
- Models from 4 / 1999: Light socket with integrated LEDs.

In Coupe models the center brake light is at the top center of the rear hatch. It consists of a strip with multiple replaceable bulbs soldered in place.

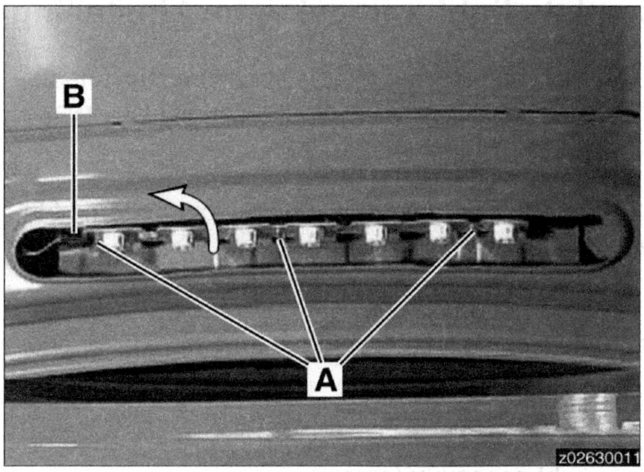


## Roadster version 1

– Open trunk lid. Carefully pry brake light plastic access cover off inside of trunk lid by lifting end with plastic prying tool or small, flat-bladed screwdriver.

◄ Rotate defective light bulb ¼ turn counterclockwise (**arrow**), then pull out.

– Pull old bulb out of socket, install new bulb and install light socket into light socket carrier. Do not touch bulb with bare fingers. If necessary, clean bulb with soft, lint-free cloth and rubbing alcohol.

– If center brake light assembly needs removal:
  - Remove mounting screws (**A**).
  - Detach electrical connector (**B**).
  - Unclip light assembly from trunk lid to remove.

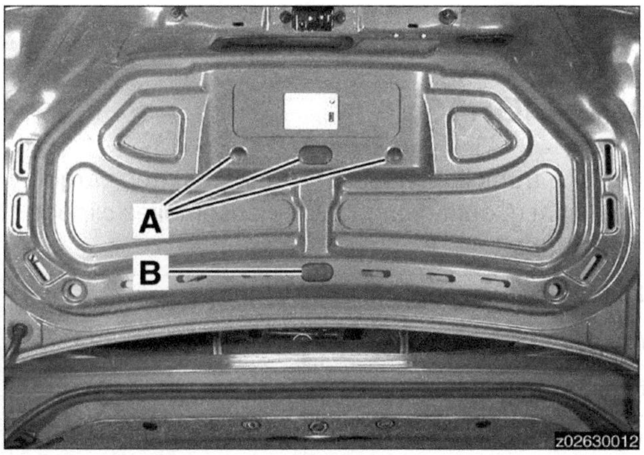


## Roadster version 2

◄ Open trunk lid. Working at center inside of trunk lid, pry off brake light assembly mount covers (**A**). Remove assembly mounting fasteners.

– Detach electrical connector behind cover **B** and remove integrated LED brake light assembly.

– When replacing, make sure moisture seals are undamaged. Replace seals if necessary.

## Coupe version

◄ Working at top of rear hatch, use plastic prying tool to gently pry center brake light assembly out.

– Detach electrical harness connector.

– Carefully disassemble unit. Pull printed circuit board bulb strip off housing, starting at one side.

– Use ohmmeter to diagnose defective bulb(s).

– Unsolder defective bulb(s) and replace. Do not touch bulbs with bare fingers. If necessary, clean bulbs with soft, lint-free cloth and rubbing alcohol.

## License plate light, removing and installing

License plate light sockets are either clipped into the sheet metal of the trunk lid or rear hatch or into a ridge in the rear bumper above the license plate.

z02630013

- **Version 1**. Use plastic prying tool to gently pry bottom of license plate light assembly (**arrow**).

– Replace bulb. Do not touch bulb with bare fingers. If necessary, clean bulb with soft, lint-free cloth and rubbing alcohol.

z02630014

- **Version 2**. Use plastic prying tool to gently pry left edge of license plate light assembly (**arrow**).

– Replace bulb. Do not touch bulb with bare fingers. If necessary, clean bulb with soft, lint-free cloth and rubbing alcohol.

# INTERIOR LIGHTS

The interior light, trunk light (Roadster versions) and cargo compartment light (Coupe versions) are assemblies clipped into the interior trim.

**Table b. Z3 interior bulb applications**

| Location | Type & rating |
|---|---|
| Interior or cargo compartment light | Festoon 12v 10w |
| Trunk light | Festoon 12v 10w |
| Glove compartment light | Festoon 12v 5w |
| Engine compartment light | Festoon 12v 10w |
| Vanity mirror light | Festoon 12v 5w |

## Interior light, removing and installing

Use this procedure to remove the interior light, cargo compartment light or trunk light.

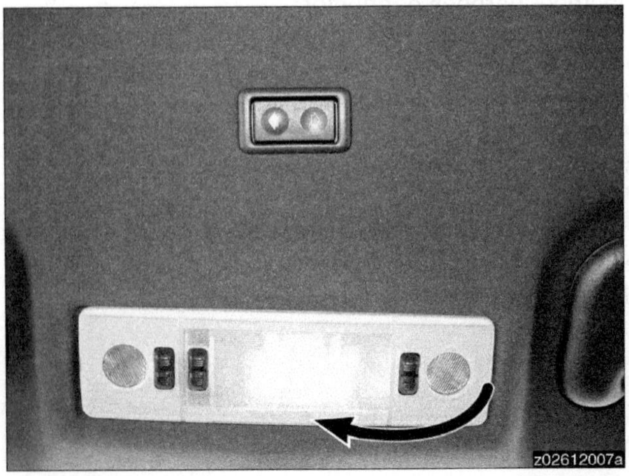

z02612007a

- Use plastic prying tool to gently pry edge of light assembly (**arrow**).

– Replace bulb. Do not touch bulb with bare fingers. If necessary, clean bulb with soft, lint-free cloth and rubbing alcohol.

# 640 Heating and Air-conditioning



**TABLES**

## GENERAL

This repair group covers the heating and air-conditioning (A/C) system.

The refrigerant used in Z3 vehicles is R134a. Many of the procedures require that the A/C refrigerant charge be evacuated. A/C system recharging procedures are beyond the scope of this manual. See **A/C system warnings and cautions** in this repair group.

For additional information, see:

- **020 Maintenance** for A/C drive belt
- **170 Radiator and Cooling System** for cooling system, radiator cooling fan and cooling fan switch

### Special tools

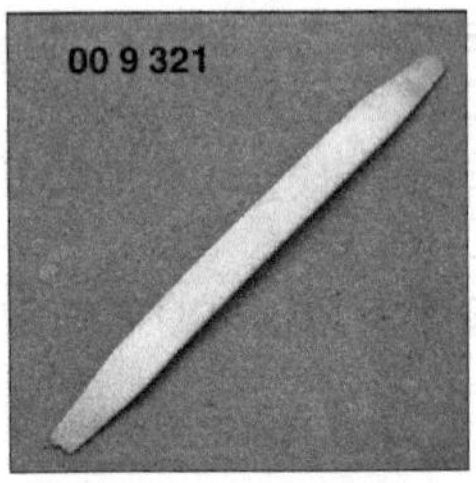


Plastic prying tool
(BMW special tool 00 9 321)

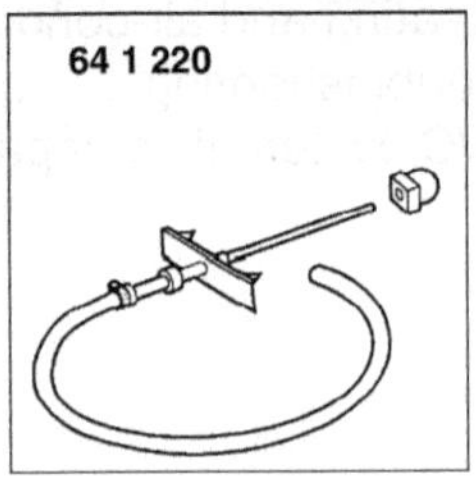


A/C evaporator spray-cleaning wand
(BMW special tool 64 1 220)

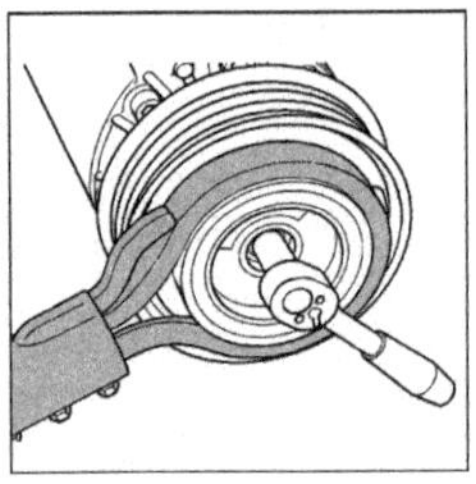

Compressor clutch counterholding strap

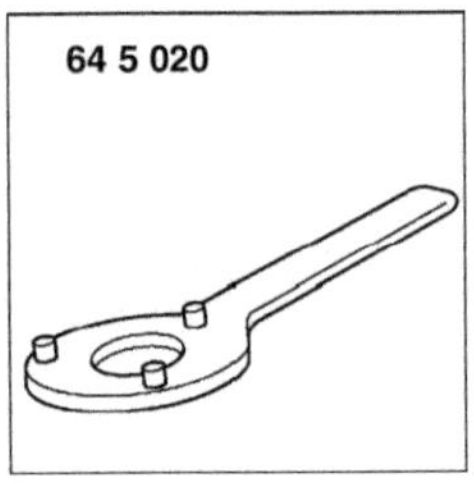


Seiko compressor clutch counterholding tool
(BMW special tool 64 5 020)

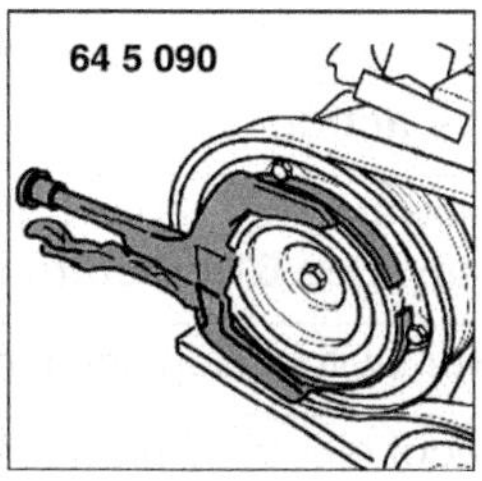


Nippondenso compressor clutch counterholding tool (BMW special tool 64 5 090)

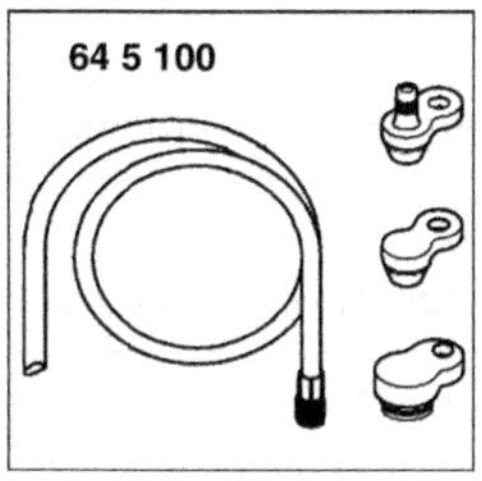


Refrigerant line plug kit (BMW special tool 64 5 100)

## Heating and air-conditioning system

The BMW Z3 uses a variant of BMW's climate control system with manual temperature and blower speed controls.

The manual controls are:

- **A** Air outlet selection knob
- **B** Air temperature control
- **C** Blower switch
- **D** Recirculation / compressor control switch

## Troubleshooting

Begin heating and air-conditioning system troubleshooting with an analysis of the problem. Begin by isolating the system involved, then attempt to track the problem to its logical conclusion, and repair the fault.

*Example:* If the heater is not supplying hot air, the problem could be a faulty engine coolant thermostat.

**Table a** provides common symptoms and possible corrective actions to aid in diagnosing simple faults.

**Table a. Heating and air-conditioning troubleshooting**

| Symptom | Probable cause | Corrective action |
|---|---|---|
| Blower does not work. | Blower fuse blown. | Inspect blower fuse (F20 in front power distribution box.) Replace if necessary. See **610 Electrical Component Locations**. |
| | Blower switch faulty. | Test blower switch. Replace if necessary. |
| | Blower motor wiring faulty. | Troubleshoot electrical harness. Repair as necessary. |
| Blower does not operate in lower three speeds (operates in highest position only). | Blower resistor pack faulty. | Test blower resistor pack. Replace if necessary. |
| | Blower switch faulty. | Test blower switch. Replace if necessary. |
| Air-conditioning system does not produce cold air. | A/C compressor not engaging. | Test A/C compressor and control system. Repair as necessary. |
| | Refrigerant charge too low or too high. | Test A/C system pressures. Repair as necessary. |
| | Heater temperature control system malfunction. | Verify that air temperature control knob is functioning correctly. (Test if hot to ambient air is available when knob is rotated from hot to cold position). Repair temperature control door if necessary. |
| Heater system does not produce hot air. | Engine coolant thermostat faulty. | Test engine coolant thermostat. Replace if necessary. See **170 Radiator and Cooling System**. |
| | Heater temperature control system malfunction. | Verify that air temperature control knob is functioning correctly. (Check if hot ambient air is available when knob is rotated from hot to cold position). Repair temperature control door if necessary. |
| | Heater control valve faulty. | Test heater control valve. Replace if necessary. |
| | Air recirculation flap malfunction. | Test air recirculation flap motor. Replace if necessary. |
| Air-conditioned air smells musty. | Mold growing in A/C evaporator vanes. | Apply disinfectant using evaporator cleaning wand. |

## HEATING AND A/C CONTROLS

### Climate controls, removing and installing

– Disconnect negative (–) battery cable and cover terminal with insulating material.

***NOTE—***
*On 1996 cars, the battery is located in the right rear of the engine compartment. On 1997 and later cars, the battery is located under the trunk floor mat.*

> ***CAUTION—***
> *Prior to disconnecting the battery, read the battery cautions in* **001 General Warnings and Cautions**.

– Remove center console. See **513 Interior Trim**.

– Remove three radio trim panel retaining screws and remove trim panel.

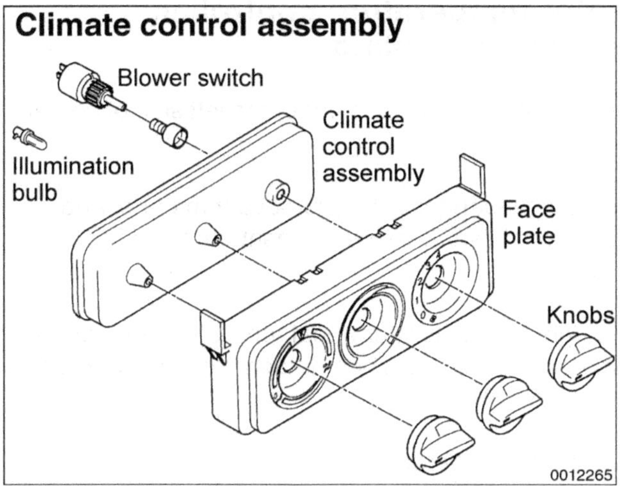


- Remove climate control assembly from center console.
- Remove clips retaining control cables to levers.
- Disconnect blower switch and heater valve control microswitch electrical connectors.
- Remove climate control assembly from vehicle.
- Installation is reverse of removal.

## Blower switch, removing and installing

- Remove radio. See **650 Radio**.
- Remove blower switch control knob.
- Remove blower switch retaining nut.
- Push blower switch into dashboard, then reach in through radio opening and pull switch through.
- Disconnect blower switch electrical connector. Installation is reverse of removal.

## Evaporator temperature sensor, removing and installing

The A/C evaporator temperature sensor is under the left side of center dashboard.

- Working inside passenger cabin, remove trim panel and knee bolster from lower left side of dashboard. See **513 Interior Trim**.

- Working to right of accelerator pedal, detach electrical harness (**arrow**) from evaporator sensor.
- Pull sensor out of heater housing.
- Installation is reverse of removal.

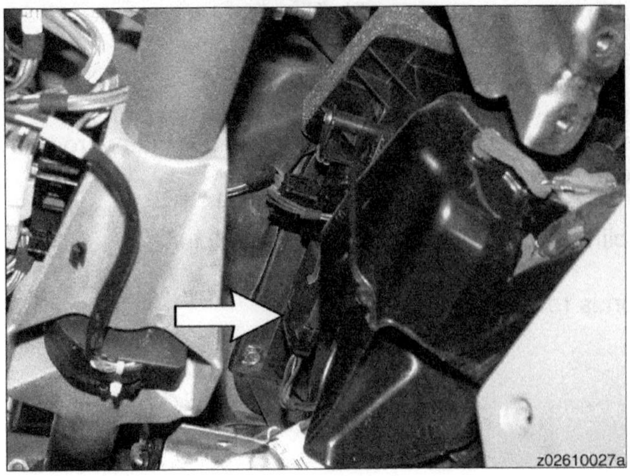


## Evaporator temperature controller, removing and installing

The A/C evaporator controller is under the left side of center dashboard.

– Working inside passenger cabin, remove trim panel and knee bolster from lower left side of dashboard. See **513 Interior Trim**.

< Working to right of accelerator pedal, unclip evaporator controller from heater housing. Then detach electrical harness (**arrow**) from controller.

– Installation is reverse of removal.

## Fresh air / recirculation flap motor, removing and installing

The fresh air / recirculation flap motor is under the left side of center dashboard, behind left footwell air duct.

– Working inside passenger cabin, remove trim panel and knee bolster from lower left side of dashboard. See **513 Interior Trim**.

< Working to right of accelerator pedal:

- Remove left footwell air duct mounting screw (**black arrow**).
- Swing air duct to left to remove (**white arrow**).

– Remove flap motor mounting screws. Remove electrical connector from flap motor.

– Pry flap motor slightly downward and unclip from actuating lever.

– When installing:

- Push actuating lever on shaft until it snaps into place.
- Mount flap motor with screws.

– Remainder of installation is reverse of removal.

## Blower components

The 4 speed blower is installed at the front of the heating and air-conditioning housing (heater box). It is accessible through the interior ventilation grille opening in the engine compartment, at the cowl.

The blower resistor is installed in the blower housing and must be removed when removing the blower.

The blower fuse (F20), rated 30A, is located in the front power distribution box. See **610 Electrical Component Locations**.

### Blower motor, removing and installing

– Disconnect negative (–) battery cable and cover terminal with insulating material.

*NOTE—*
*On 1996 cars, the battery is located in the right rear of the engine compartment. On 1997 and later cars, the battery is located under the trunk floor mat.*

> ***CAUTION—***
> *Prior to disconnecting the battery, read the battery cautions in* ***001 General Warnings and Cautions****.*

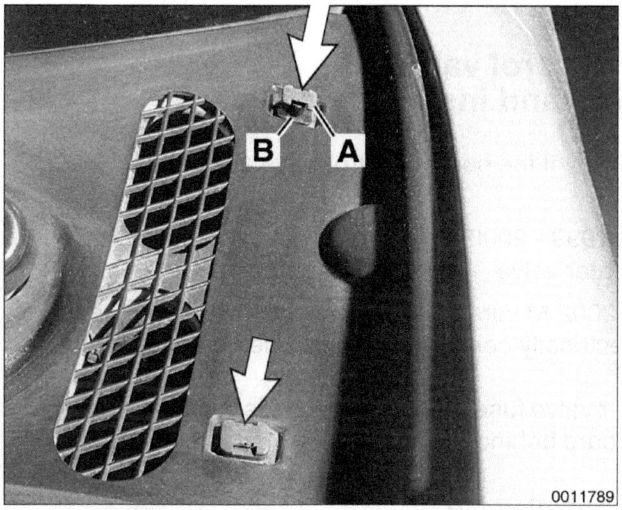


◀ Remove ventilation intake air grille at base of windshield.

- Remove windshield wiper arms.
- Peel up rubber seal from around protective grille.
- Remove expansion clips (**arrows**): First remove center piece (**A**), then slide out clip (**B**).
- Remove grille.

– Remove windshield wiper motor assembly. See **611 Wipers and Washers**.

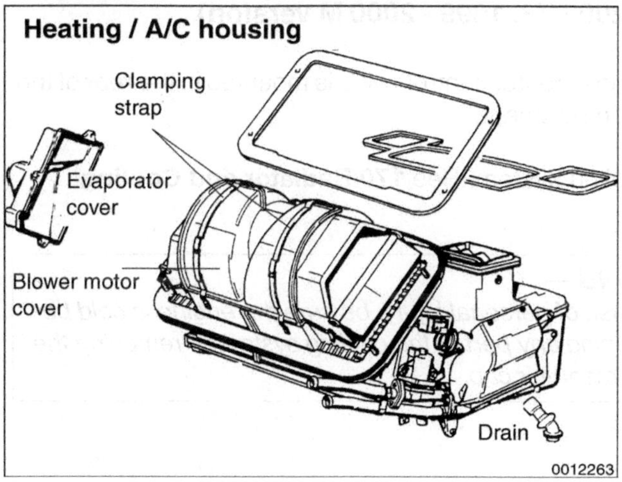


◀ Remove clamping straps from blower motor cover and remover cover.

– Remove fan cover, then disconnect blower motor resistor pack electrical connector.

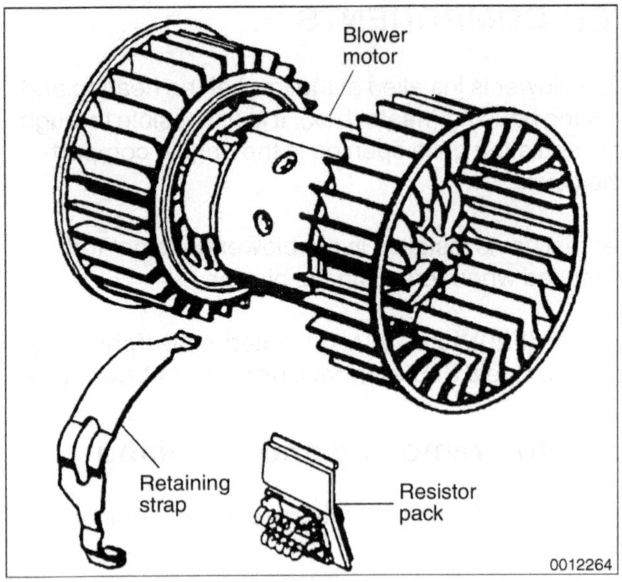


◄ Unlatch fan retaining strap and remove blower motor.

– Installation is reverse of removal.

## HEATER COMPONENTS

Engine coolant circulates through a heater core to heat the air passing into the interior of the vehicle. An electric heater control valve prevents coolant from passing through the heater core when the temperature control knob is set to COLD.

### Heater control valve, removing and installing

Two versions of the heater control valve are installed in Z3 models:

- Z3 and 1999 - 2000 M version: Electrically controlled solenoid heater valve
- 2001 - 2002 M version: Auxiliary coolant pump in tandem with electrically controlled solenoid heater valve

The heater valve fuse (F953), rated 2A, is installed in the center dashboard behind the radio.

#### Heater control valve (1996 - 2002 Z3, 1999 - 2000 M version)

The solenoid heater control valve is mounted in the rear of the engine compartment.

– Drain cooling system. See **170 Radiator and Cooling System**.

> ***WARNING —***
> *Due to risk of personal injury, be sure the engine is cold before opening any part of the cooling system or removing the coolant reservoir cap.*

– Detach wiring loom above rear of engine and stretch loom forward to expose heater valve.

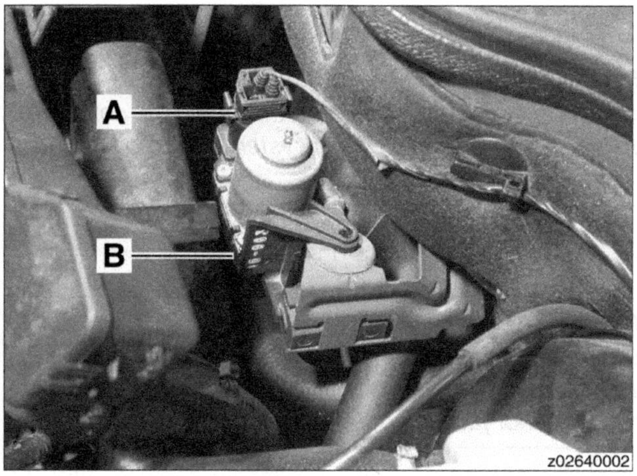


◄ Working at rear engine bulkhead:

- Remove heater control valve electrical connector (**A**).
- Loosen heater hose clamps (**B**) at valve. Label and then remove heater hoses from valve.

– Detach heater control valve from mounting brackets.

– Installation is reverse of removal. Bear in mind the following:

- Make sure heater valve rubber mounts are undamaged and seated correctly. Replace if damaged.
- Use new hose clamps.
- Fill and bleed cooling system. See **170 Radiator and Cooling System**.

### Heater control valve and auxiliary coolant pump (2001 - 2002 M version)

In M version models with S54 engine (model years 2001 - 2002), the auxiliary electric coolant pump in the heating circuit, integral with the heater control valve, is mounted at the left strut tower in the engine compartment.

– Drain cooling system. See **170 Radiator and Cooling System**.

> ***WARNING—***
> *Due to risk of personal injury, be sure the engine is cold before opening any part of the cooling system or removing the coolant reservoir cap.*

◄ Working at left strut tower:

- Remove auxiliary pump and heater control valve electrical connectors (**A**).
- Loosen heater hose clamps (**B**) at valve. Label and then remove heater hoses from valve.

– Detach pump and heater control valve from mounting brackets.

– Installation is reverse of removal. Bear in mind the following:

- Make sure coolant pump and heater valve rubber mounts are undamaged and seated correctly. Replace if damaged.
- Use new hose clamps.
- Fill and bleed cooling system. See **170 Radiator and Cooling System**.

## Heater core, removing and installing

- Drain cooling system. See **170 Radiator and Cooling System**.

> ***WARNING —***
> *Due to risk of personal injury, be sure the engine is cold before opening any part of the cooling system or removing the coolant reservoir cap.*

- Raise front of vehicle and support safely.

> ***WARNING —***
> *Make sure the car is stable and well supported at all times. Use a professional automotive lift or jack stands designed for the purpose. A floor jack is not adequate support.*

- Working underneath engine, loosen or remove heater hose clamps at engine bulkhead. Separate heater hoses from coolant pipes.

- Blow low pressure compressed air through heater core to remove any coolant still trapped.

> ***WARNING —***
> *Keep coolant out of eyes by using safety glasses.*

- Remove trim panel and knee bolster from under left side of dashboard. See **513 Interior Trim**.

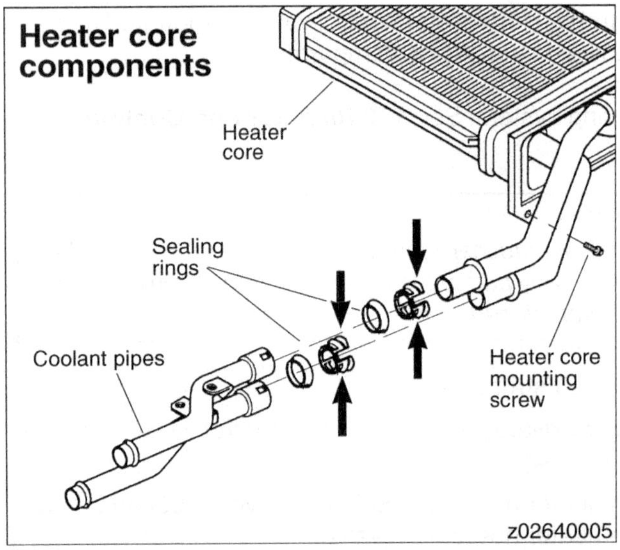


- Working to right of accelerator pedal under left dashboard, squeeze coolant pipe sealing ring clips together (**arrows**), then separate coolant pipes from heater core.

***NOTE —***
*Coolant pipes stay attached to bulkhead.*

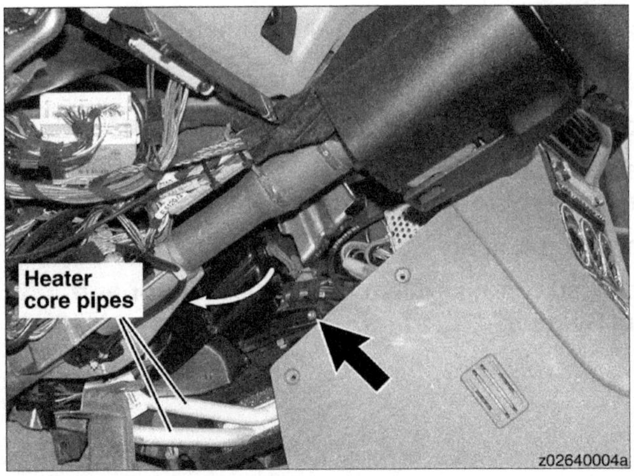


- Remove left footwell air duct mounting screw (**black arrow**).
  - Swing air duct to left to remove (**white arrow**).

- Remove heater core mounting screws, then pull heater core sideways to remove.

- Installation is reverse of removal. Bear in mind the following:
  - Use new hose clamps.
  - Fill and bleed cooling system. See **170 Radiator and Cooling System**.

## Heating and A/C housing, removing and installing

Heater housing removal and installation is a complicated job. The procedure involves the following:

- Draining of cooling system
- Discharging of air-conditioning refrigerant
- Removal of driver and passenger airbags
- Removal of steering wheel
- Removal of center console
- Removal of complete dashboard

Read through the complete procedure before starting the job. Be sure to read airbag warnings and cautions in **721 Airbag System (SRS)**.

– Disconnect negative (–) battery cable and cover terminal with insulating material.

***NOTE—***
*On 1996 cars, the battery is located in the right rear of the engine compartment. On 1997 and later cars, the battery is located under the trunk floor mat.*

***CAUTION—***
*Prior to disconnecting the battery, read the battery cautions in* **001 General Warnings and Cautions**.

– Drain cooling system. See **170 Radiator and Cooling System**.

***WARNING—***
*Due to risk of personal injury, be sure the engine is cold before opening any part of the cooling system or removing the coolant reservoir cap.*

– Raise front of vehicle and support safely.

***WARNING—***
*Make sure the car is stable and well supported at all times. Use a professional automotive lift or jack stands designed for the purpose. A floor jack is not adequate support.*

– Working underneath engine, loosen or remove heater hose clamps at engine bulkhead. Remove heater hoses from heater core fittings.

– Blow low pressure compressed air through heater core to remove any coolant still trapped in core.

***WARNING—***
*Keep coolant out of eyes by using safety glasses.*

– Following manufacturer's instructions, connect an approved refrigerant recovery / recycling / recharging unit to A/C system and discharge system.

> ***WARNING —***
> *Do not attempt to discharge or charge the A/C system without proper equipment and training. Damage to the vehicle and personal injury may result.*

⮜ Working at right rear of engine compartment at bulkhead, Remove bolts (**arrows**) mounting air-conditioning refrigerant lines to expansion valve double-line. Detach refrigerant lines.

> ***CAUTION—***
> *Plug open refrigerant lines and ports.*

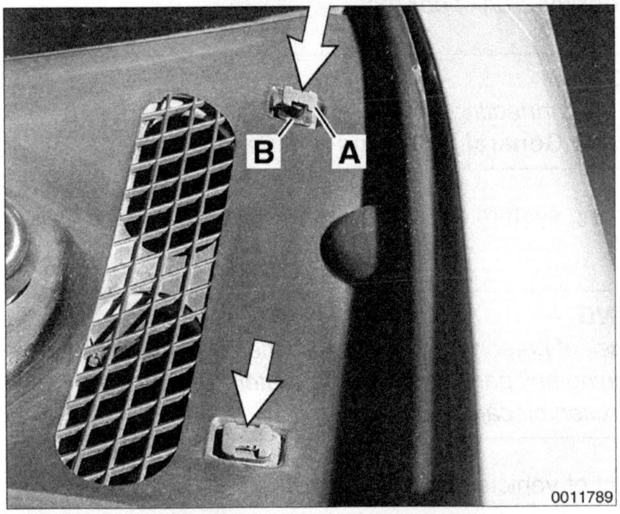


⮜ Remove ventilation intake air grille at base of windshield.

- Remove windshield wiper arms.
- Peel up rubber seal from around protective grille.
- Remove expansion clips (**arrows**): First remove center piece (**A**), then slide out clip (**B**).
- Remove grille.

– Remove windshield wiper motor assembly. See **611 Wipers and Washers**.

⮜ Working inside ventilation intake housing, remove heating and A/C housing mounting bolts (**arrows**).

– Working inside passenger cabin, remove center console. See **513 Interior Trim**.

– Remove glove compartment. See **513 Interior Trim**.

– Remove driver and passenger airbag. See **721 Airbag System (SRS)**.

– Remove steering wheel. See **320 Steering and Wheel Alignment**.

– Remove instrument cluster. See **620 Instrument Cluster**.

– Loosen steering column stalk switch. See **612 Switches**.

– Remove headlight switch. See **612 Switches**.

– Remove climate control panel from center dashboard. See **Climate controls, removing and installing** in this repair group.

– Remove complete dashboard and set aside. See **513 Interior Trim**.

◄ Remove dashboard reinforcement:

- Remove reinforcement center mounting bolts (**arrows**). Cut wire ties holding electrical harnesses to reinforcement.
- Remove reinforcement side bracket mounting bolts.
- Working at left and right kick panels, remove reinforcement side mounting bolts.
- Lift out reinforcement.

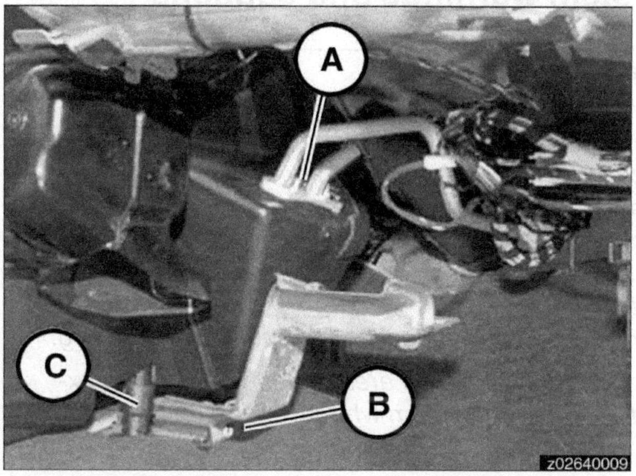


◄ Working at right side of heater housing:

- Remove refrigerant double-line mounting bolt (**A**). Detach double-line from expansion valve.
- Remove trim mounting bracket screws (**B**) and remove bracket.
- Detach heater housing drains (**C**) at right and left.

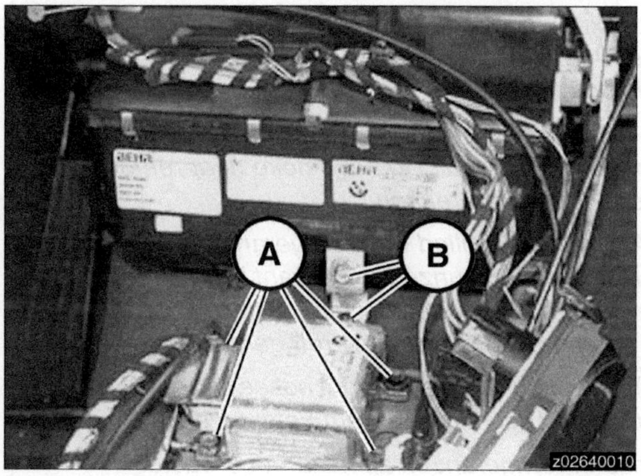


◄ Working at center tunnel:

- Remove SRS control module mounting bolts (**A**). Set module aside.
- Remove heater housing center mounting bracket bolts (**B**).

– Carefully lift out heating and A/C housing, checking for any wiring, hoses or mechanical parts that might get snagged.

– Installation is reverse of removal. Bear in mind the following:
  - Replace refrigerant line sealing O-rings.
  - Use new clamps at coolant hoses.
  - Reattach battery last, after airbags are installed.
  - Fill and bleed cooling system.
  - Recharge A/C system following equipment manufacturer's instructions. Be sure to use correct type and amount of refrigerant oil. See **Refrigerant and refrigerant oil** in this repair group.

| **Tightening torques** | |
|---|---|
| Heater housing to cowl | 3 Nm (26 in-lb) |
| Passenger airbag to dashboard reinforcement (M8) | 24 Nm (18 ft-lb) |
| Refrigerant double-line fasteners<br>• Nut<br>• Stud | <br>3.5 Nm (31 in-lb)<br>2 Nm (18 in-lb) |

## AIR-CONDITIONING SYSTEM

### A/C system warnings and cautions

***WARNING—***
- *Wear hand and eye protection (gloves and goggles) when working around the A/C system. If refrigerant oil comes in contact with your skin or eyes:*
  *-Remove contact lenses, if worn.*
  *-Do not rub skin or eyes.*
  *-Immediately flush skin or eyes with cool water for 15 minutes.*
  *-Rush to a doctor or hospital. Do not induce vomiting.*
- *Work in a well ventilated area. Switch on building exhaust / ventilation system when working on the A/C.*
- *If you inhale refrigerant oil, breathe fresh air immediately and then seek medical attention.*
- *Do not expose any component of the A/C system to high temperatures (above 80°C/176°F). Excessive heat could burst the system.*
- *Keep refrigerant away from open flames. Poisonous gas is produced if it burns. Do not smoke near refrigerant gases for the same reason.*
- *The A/C system is filled with refrigerant gas which is under pressure. Pressurized refrigerant in the presence of oxygen may form a combustible mixture. Do not introduce compressed air into any refrigerant container.*
- *Electric welding near refrigerant hoses causes R134a to decompose. Discharge system before welding.*
- *Dispose of refrigerant oil as hazardous waste.*

> ***CAUTION—***
> - *US law requires that any person who services a motor vehicle air-conditioner is properly trained and certified, and uses approved refrigerant recycling equipment. Technicians must complete an EPA-approved recycling course to be certified.*
> - *It is recommended that all A/C service be left to an authorized BMW dealer or other qualified A/C service facility.*
> - *State and local governments may have additional requirements regarding A/C servicing. Always comply with state and local laws.*
> - *Leak test and repair any A/C system which is known to discharge itself.*
> - *Do not top off a partially charged refrigerant system. Discharge system, evacuate and then recharge system.*
> - *Do not use R12 refrigerant, refrigerant oils or system components in R134a system. Component damage and system contamination results.*
> - *Refrigerant oil (PAG oil) is extremely hygroscopic (water-absorbent). Do not store in open container.*
> - *The mixture of refrigerant oil and refrigerant (R134a) attacks some metals and alloys (for example, copper) and breaks down certain hose materials. Use only hoses and pipes that are identified with a green mark (stripe) or the lettering "R134a".*
> - *Immediately plug open connections on A/C components and pipes to prevent dirt and moisture contamination.*
> - *Do not steam clean A/C condensers or evaporators. Use only cold water or compressed air.*
> - *Do not reuse refrigerant oil.*
> - *Do not store refrigerant oil in vicinity of flames, heat sources or strongly oxidizing materials.*
> - *In case of fire, use carbon dioxide ($CO_2$) extinguisher or sprayed water jet on A/C components.*
> - *After refilling or testing refrigerant lines, replace protective caps. They serve as additional seals.*

## A/C operation

Pressing the A/C switch signals the DME control module (ECM) and the evaporator control module that A/C operation is desired. The blower switch must be in one of the four operating speed positions. If evaporator temperature is within acceptable limits, the ECM supplies power to the A/C compressor clutch via the compressor control relay.

The ECM also turns on the radiator cooling fan at low speed whenever the air-conditioner is operating. If the second stage of the radiator temperature switch senses high engine coolant temperature, or if the refrigerant pressure switch, located in the high pressure-side refrigerant line, detects high refrigerant pressure, the radiator cooling fan runs at high speed.

The pressure switch also signals the evaporator control module if the refrigerant pressure is too high or too low for safe operation range. If this occurs, the ECM turns off the compressor.

Pressing the recirculation switch operates the fresh air recirculation motor. This moves the recirculation flap to the closed position and air intake from the outside is closed off.

## Refrigerant and refrigerant oil

The air-conditioning refrigerant in Z3 models is R134a. A synthetic oil, polyalkalene glycol (PAG), is used to lubricate the A/C compressor. This oil, being highly hygroscopic (water-absorbent) also functions as a system desiccant.

Refrigerant and refrigerant oil capacities, where available, are given in **Table b**.

A new A/C compressor is filled at the factory with 120 - 200 grams (4.2 - 7.2 oz) of refrigerant oil, depending on its manufacturer. If installing a new compressor, recharge with refrigerant, then follow instructions in **A/C compressor, initial break-in** in this repair group.

If the air-conditioning system is evacuated and recharged without a new compressor being installed, refrigerant oil is also extracted and collected in the oil separator of the recharging equipment. After evacuation, the refrigerant gasifies and only the refrigerant oil remains in the oil separator.

Measure and note down the quantity of refrigerant oil. When recharging pour the same amount of new oil plus an additional 10 grams (0.3 oz) into the compressor. Do not reuse the old oil.

| Tightening torque | |
|---|---|
| Compressor filler plug to compressor | 30 Nm (22 ft-lb) |

| Table b. A/C system fluid capacities | | |
|---|---|---|
| **Model** | **Refrigerant (R134a)** | **Refrigerant oil (PAG)** |
| Z3<br>• Nippondenso compressor<br>• Seiko compressor | <br>900 ± 25 grams (31.7 ± 0.9 oz)<br>900 ± 25 gram (31.7 ± 0.9 oz) | <br>120 ± 20 grams (4.1 ± 0.7 oz)<br>150 ± 10 grams (5.1 ± 0.3 oz) |
| M Coupe, M Roadster | 950 ± 25 grams (33.5 ± 0.9 oz) | not available |

## A/C compressor, initial break-in

When starting a new compressor for the first time, carry out the following break-in procedure:

- Switch on A/C system.
- Set all dashboard air vents to OPEN.
- Start engine and allow to idle.
- Set blower output to minimum 75% of maximum.
- Run A/C for at least 2 minutes at idle speed.

> ***CAUTION—***
> *The new compressor may be damaged if run at higher than idle speed at start-up.*

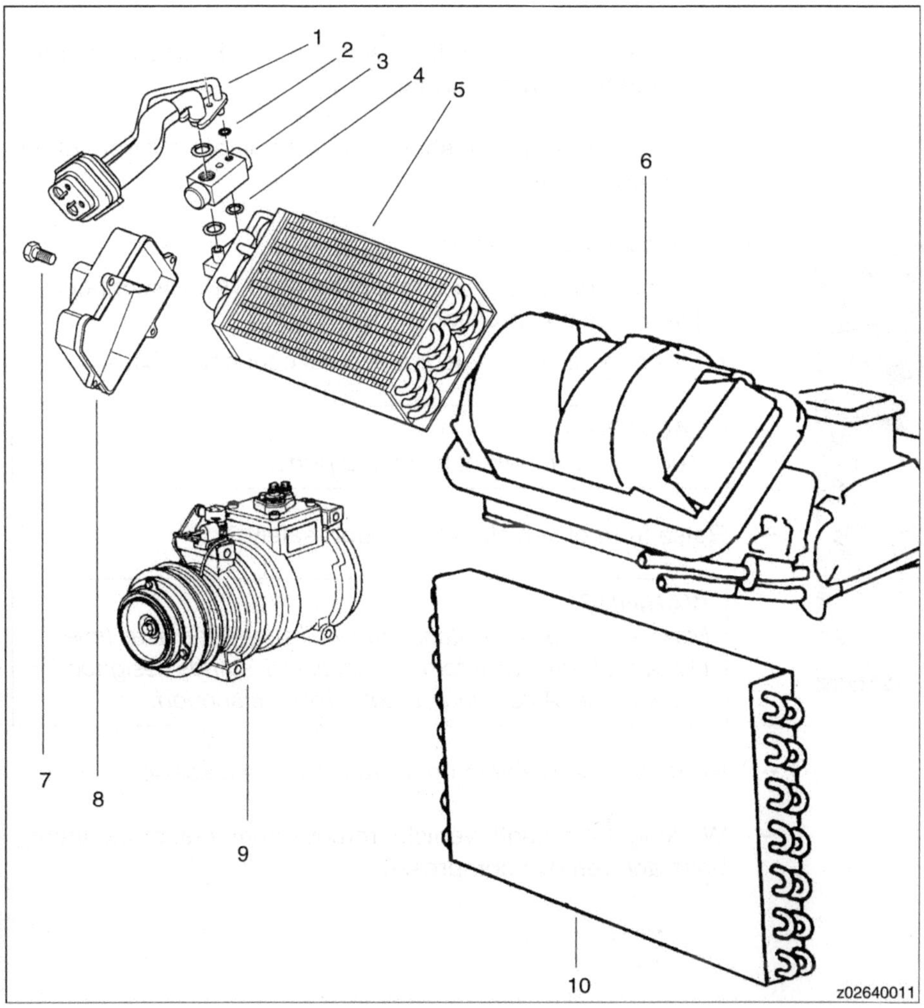


## A/C system components

1. **Refrigerant double-line**
2. **Sealing O-rings**
3. **Expansion valve**
4. **Sealing O-rings**
5. **A/C evaporator**
6. **Heating and A/C housing**
7. **Mounting bolt**
8. **Evaporator cover**
9. **A/C compressor**
10. **A/C condenser**

A/C compressor components

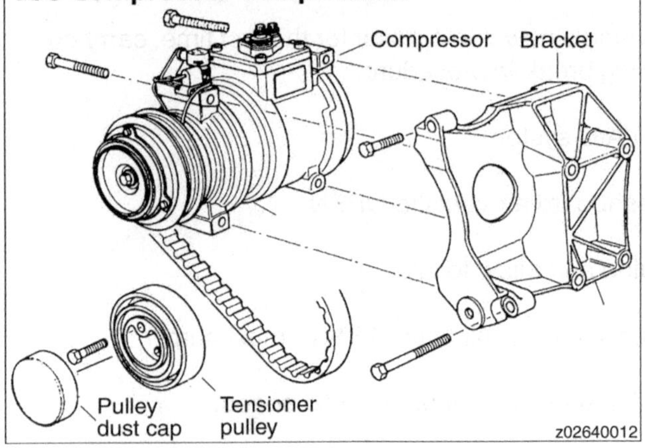


## A/C compressor, removing and installing

Z3 models may be equipped with A/C compressors manufactured by Seiko or Nippondenso. Look for the manufacturer identification on the compressor parts sticker.

Removal procedures for the two versions of the compressor are similar.

– Following manufacturer's instructions, connect an approved refrigerant recovery / recycling / recharging unit to A/C system and discharge system.

> ***WARNING —***
> *Do not attempt to discharge or charge the A/C system without proper equipment and training. Damage to the vehicle and personal injury may result.*

– Remove viscous clutch fan and shroud, if applicable. See **170 Radiator and Cooling System**.

– Remove A/C drive belt. If reusing belt, mark direction of rotation. See **020 Maintenance**.

– Remove windshield washer fluid reservoir at right front of engine compartment.

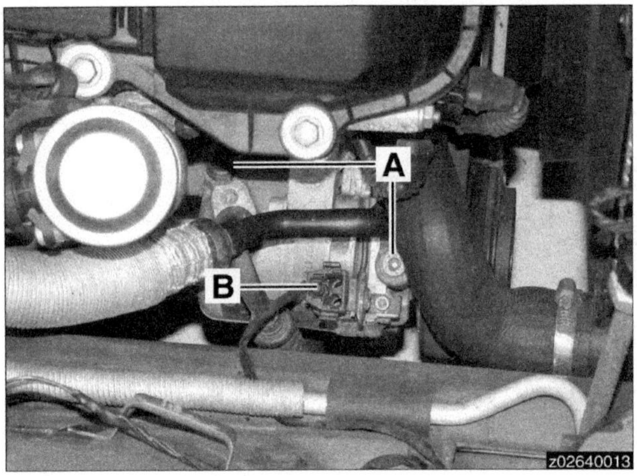


Working at compressor:

- Loosen refrigerant line mounting bolts (**A**). Disconnect lines from compressor.
- Disconnect compressor electrical connector (**B**).

> ***CAUTION—***
> *Plug open refrigerant lines and ports.*

– Raise front of vehicle and support safely.

> ***WARNING —***
> *Make sure the car is stable and well supported at all times. Use a professional automotive lift or jack stands designed for the purpose. A floor jack is not adequate support.*

– Remove splash shield under engine if necessary.

– Working underneath vehicle, remove compressor mounting bolts and remove compressor.

- Installation is reverse of removal.
  - Replace sealing O-rings when reconnecting refrigerant pipes.
  - If installing a new compressor, or if A/C system has been open for more than 24 hours, replace receiver / drier unit. See **A/C receiver / drier, removing and installing** in this repair group.
  - Install and tension A/C drive belt, noting previously made direction mark. See **020 Maintenance**.

| Tightening torques | |
|---|---|
| Compressor to mounting bracket (M8 bolt) | 22 Nm (18 ft-lb) |
| Refrigerant line to compressor (M8) | 20 Nm (15 ft-lb) |

- Recharge A/C system following equipment manufacturer's instructions. Be sure to use correct refrigerant oil. See **Refrigerant and refrigerant oil** in this repair group.

- Prior to starting new compressor, be sure to read and follow instructions in **A/C compressor, initial break-in** in this repair group.

## A/C compressor clutch, removing and installing

If the bearings in the compressor clutch become noisy the clutch will eventually seize. The compressor clutch may be available separately. Use diagrams and the procedure below when replacing compressor clutch.

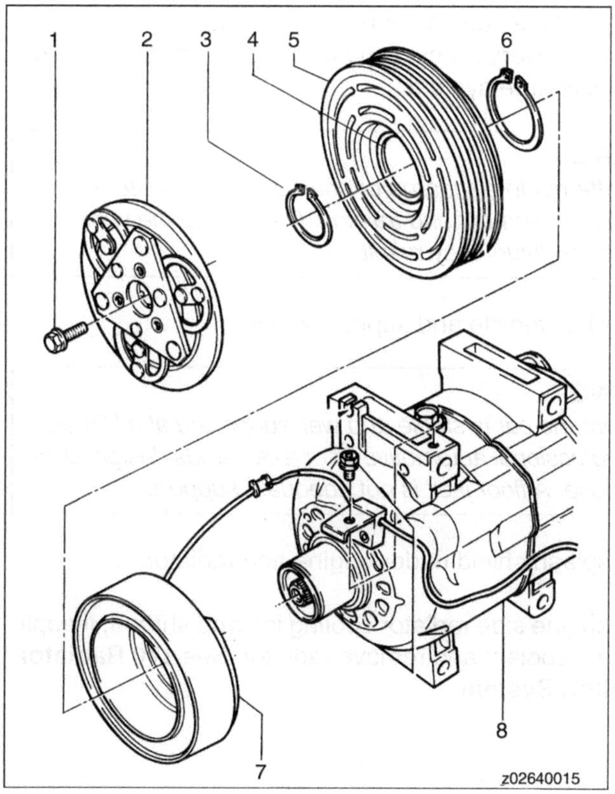


Compressor clutch components

1. Clutch plate retaining bolt
2. Clutch plate
3. Pulley retaining circlip
4. Clutch bearing
5. A/C compressor pulley
6. Clutch retaining circlip
7. Clutch (magnetic winding)
8. Compressor

The clutch can be replaced without disturbing the sealed refrigerant system, although the clearance in a 6-cylinder car is very tight. Detach the compressor from the compressor bracket to gain extra clearance.

- Remove viscous clutch fan and shroud, if applicable. See **170 Radiator and Cooling System**.

- Remove A/C drive belt. If reusing belt, mark direction of rotation. See **020 Maintenance**.

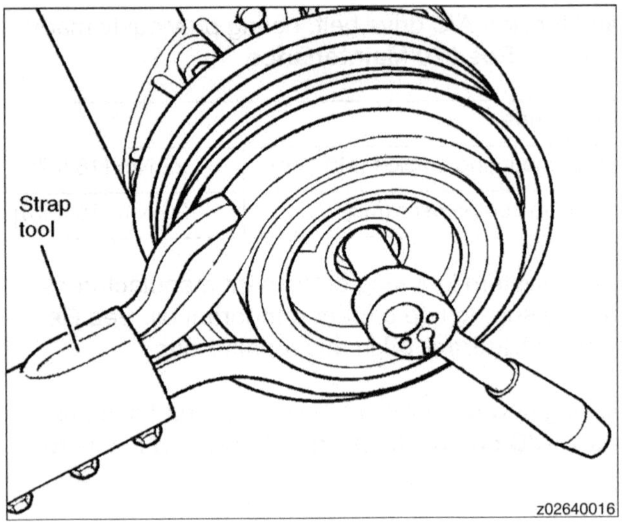


- Raise front of vehicle and support safely.

> ***WARNING—***
> *Make sure the car is stable and well supported at all times. Use a professional automotive lift or jack stands designed for the purpose. A floor jack is not adequate support.*

- Remove splash shield under engine if necessary.

- Working underneath vehicle, use compressor clutch counter-holding tool (BMW special tool 64 5 020, 64 5 090 or strap tool) to keep clutch steady while loosening mounting bolt.

- Remove clutch drive plate and pulley to access clutch magnetic winding. Use puller to remove clutch, if necessary.

- Installation is reverse of removal. Install and tension A/C drive belt, noting previously made direction mark. See **020 Maintenance**.

| Tightening torques | |
|---|---|
| Compressor clutch to compressor shaft (M8 bolt) | 20 Nm (15 ft-lb) |

## A/C condenser, removing and installing

The A/C condenser is mounted in front of the radiator and can be accessed once the radiator cooling fan and radiator are removed.

- Following manufacturer's instructions, connect an approved refrigerant recovery / recycling / recharging unit to A/C system and discharge system.

> ***WARNING—***
> *Do not attempt to discharge or charge the A/C system without proper equipment and training. Damage to the vehicle and personal injury may result.*

- Raise front of vehicle and support safely.

> ***WARNING—***
> *Make sure the car is stable and well supported at all times. Use a professional automotive lift or jack stands designed for the purpose. A floor jack is not adequate support.*

- Remove splash shield under engine and radiator.

- Remove engine side radiator cooling fan and shroud, if applicable. Drain coolant and remove radiator. See **170 Radiator and Cooling System**.

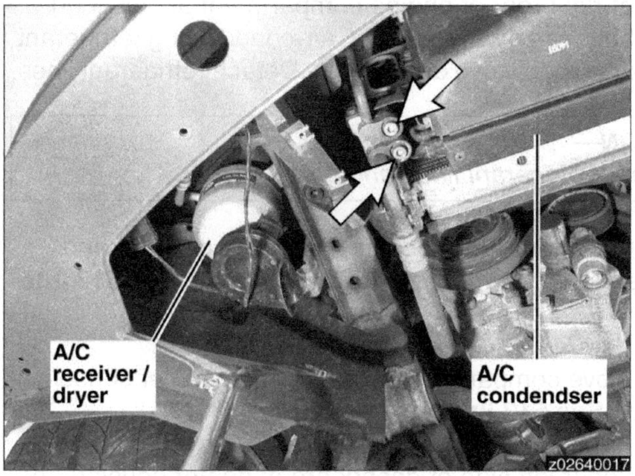


◄ Remove inlet and outlet refrigerant line mounting bolts (**arrows**) at A/C condenser. Detach lines.

> ***CAUTION—***
> *Plug open refrigerant lines and ports.*

– Remove condenser mounting bolts and remove condenser.

– Installation is reverse of removal.

- Replace sealing O-rings when reconnecting refrigerant lines.
- If A/C system has been open for more than 24 hours, replace receiver / drier unit. See **A/C receiver / drier, removing and installing** in this repair group.
- Fill and bleed cooling system. See **170 Radiator and Cooling System**.

| **Tightening torque** | |
|---|---|
| Refrigerant line to condenser (M8) | 20 Nm (15 ft-lb) |

– Recharge A/C system following equipment manufacturer's instructions. Be sure to use correct refrigerant oil. See **Refrigerant and refrigerant oil** in this repair group.

## A/C evaporator, removing and installing

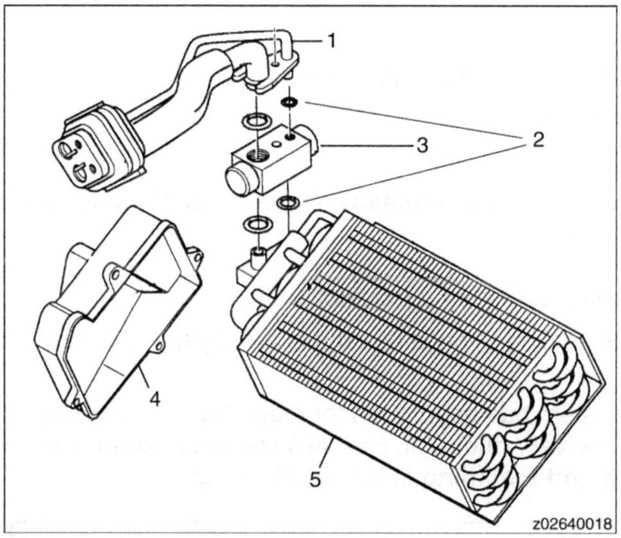


◄ The A/C evaporator and expansion valve are mounted in the right (passenger) side of the heating and A/C housing.

1. Refrigerant double-line
2. Sealing O-rings
3. Expansion valve
4. Evaporator cover
5. Evaporator

– Following manufacturer's instructions, connect an approved refrigerant recovery / recycling / recharging unit to A/C system and discharge system.

> ***WARNING —***
> *Do not attempt to discharge or charge the A/C system without proper equipment and training. Damage to the vehicle and personal injury may result.*

◄ Working at right rear of engine compartment at bulkhead, Remove bolts (**arrows**) mounting air-conditioning refrigerant lines to expansion valve double-line. Detach refrigerant lines.

> ***CAUTION—***
> *Plug open refrigerant lines and ports.*

– Working inside passenger cabin, remove trim panel and knee bolster from lower left side of dashboard. See **513 Interior Trim**.

– Remove glove compartment and right side center console trim panel. See **513 Interior Trim**.

– Working under left side of dashboard at heater housing, remove evaporator temperature sensor. See **Evaporator temperature sensor, removing and installing** in this repair group.

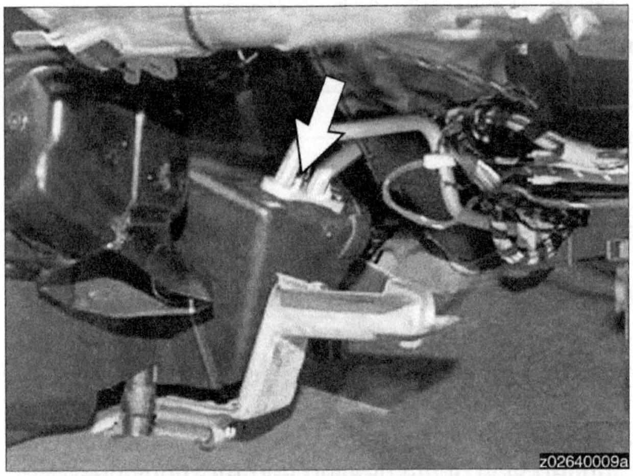


◄ Working under right side of dashboard, remove refrigerant double-line mounting bolt (**arrow**). Detach double-line from expansion valve.

– Remove A/C evaporator cover from heater housing.

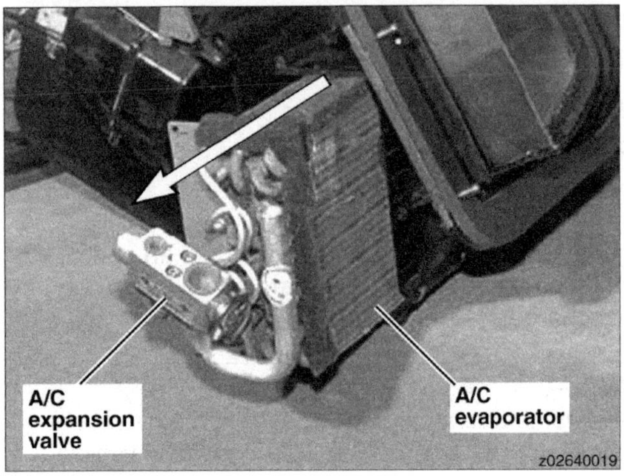


◄ Slide evaporator together with expansion valve out of heater housing (**arrow**).

***NOTE—***
*Heater housing is shown removed from the car for purpose of illustration.*

– Installation is reverse of removal.

- Replace sealing O-rings when reconnecting refrigerant lines.
- If A/C system has been open for more than 24 hours, replace receiver / drier unit. See **A/C receiver / drier, removing and installing** in this repair group.

| Tightening torque | |
|---|---|
| Refrigerant double-line fasteners<br>• Nut<br>• Stud | <br>3.5 Nm (31 in-lb)<br>2 Nm (18 in-lb) |

– Recharge A/C system following equipment manufacturer's instructions. Be sure to use correct refrigerant oil. See **Refrigerant and refrigerant oil** in this repair group.

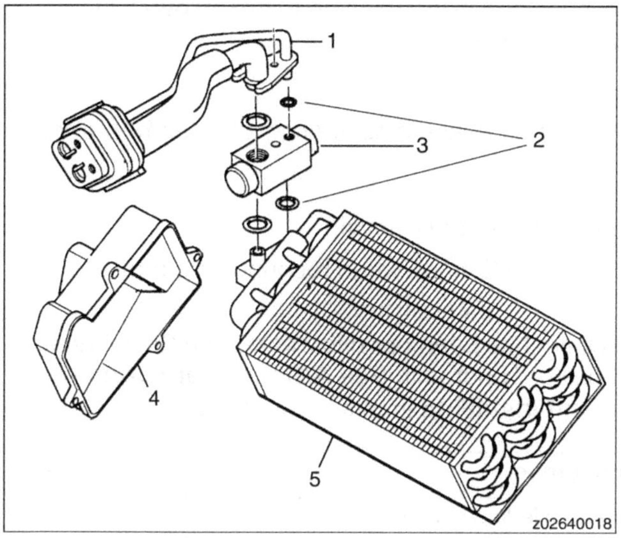


## A/C expansion valve, removing and installing

- The A/C evaporator and expansion valve are mounted in the right (passenger) side of the heating and A/C housing.

1. Refrigerant double-line
2. Sealing O-rings
3. Expansion valve
4. Evaporator cover
5. Evaporator

- Following manufacturer's instructions, connect an approved refrigerant recovery / recycling / recharging unit to A/C system and discharge system.

> ***WARNING —***
> *Do not attempt to discharge or charge the A/C system without proper equipment and training. Damage to the vehicle and personal injury may result.*

- Working at right rear of engine compartment at bulkhead, Remove bolts (**arrows**) mounting air-conditioning refrigerant lines to expansion valve double-line. Detach refrigerant lines.

> ***CAUTION—***
> *Plug open refrigerant lines and ports.*

- Working inside passenger cabin, remove glove compartment and right side center console trim panel. See **513 Interior Trim**.

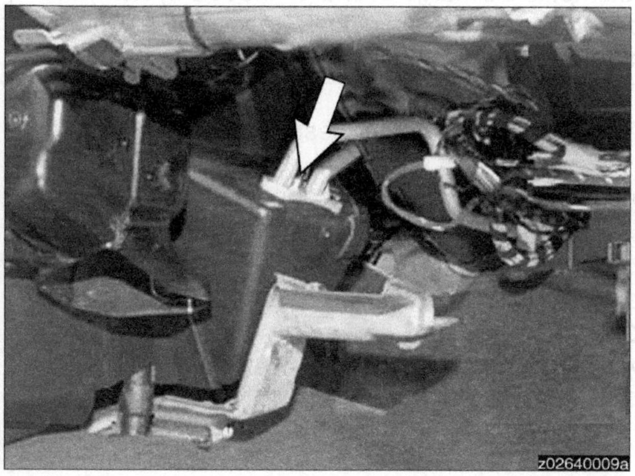


- Working under right side of dashboard, remove refrigerant double-line mounting bolt (**arrow**). Detach double-line from expansion valve.

- Remove A/C evaporator cover from heater housing.

z02640020

Remove expansion valve mounting bolts (**arrows**). Lift expansion valve off evaporator.

*NOTE—*
*Heater housing is shown removed from the car for purpose of illustration.*

– Installation is reverse of removal.
  - Replace sealing O-rings when reconnecting refrigerant lines.
  - If A/C system has been open for more than 24 hours, replace receiver / drier unit. See **A/C receiver / drier, removing and installing** in this repair group.

| Tightening torque | |
|---|---|
| Refrigerant double-line fasteners<br>• Nut<br>• Stud | <br>3.5 Nm (31 in-lb)<br>2 Nm (18 in-lb) |

– Recharge A/C system following equipment manufacturer's instructions. Be sure to use correct refrigerant oil. See **Refrigerant and refrigerant oil** in this repair group.

## A/C pressure switch, removing and installing

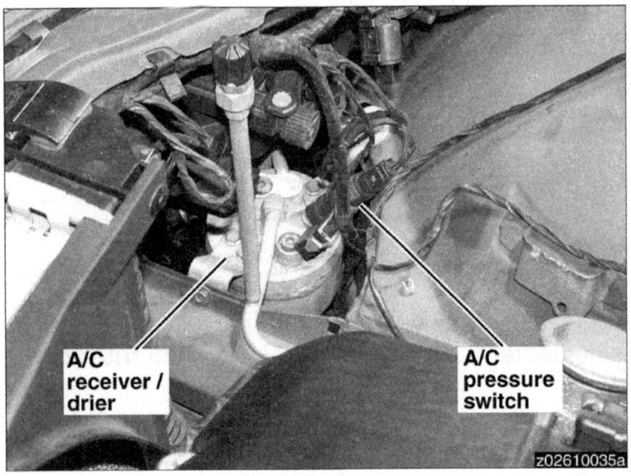

z02610035a

The A/C pressure switch is mounted to the receiver / drier in the right front (passenger side) corner of the engine compartment.

*NOTE—*
*2000 Z3 Roadster engine compartment is illustrated. Other models are similar.*

– Following manufacturer's instructions, connect an approved refrigerant recovery / recycling / recharging unit to A/C system and discharge system.

> ***WARNING—***
> *Do not attempt to discharge or charge the A/C system without proper equipment and training. Damage to the vehicle and personal injury may result.*

– Remove windshield washer fluid reservoir.

– Detach pressure switch electrical connector.

– Unscrew switch from receiver / drier.

> ***CAUTION—***
> *Plug open refrigerant ports.*

– Installation is reverse of removal.
  - Replace sealing O-rings as necessary.
  - If A/C system has been open for more than 24 hours, replace receiver / drier unit. See **A/C receiver / drier, removing and installing** in this repair group.

– Recharge A/C system following equipment manufacturer's instructions. Be sure to use correct refrigerant oil. See **Refrigerant and refrigerant oil** in this repair group.

## A/C receiver / drier, removing and installing

The A/C receiver / drier does not need to be replaced at regular service intervals in a functioning, leakproof A/C system. Replace the receiver / drier in the following situations:

- If the refrigerant is contaminated by compressor failure, resulting in metal shavings circulating in the system.
- If the A/C system is leaking.
- If refrigerant lines or ports have been open for more than 24 hours.

The A/C receiver / drier is mounted in the right front (passenger) corner of the engine compartment and is removed from below.

– Following manufacturer's instructions, connect an approved refrigerant recovery / recycling / recharging unit to A/C system and discharge system.

> ***WARNING —***
> *Do not attempt to discharge or charge the A/C system without proper equipment and training. Damage to the vehicle and personal injury may result.*

– Remove windshield washer fluid reservoir.

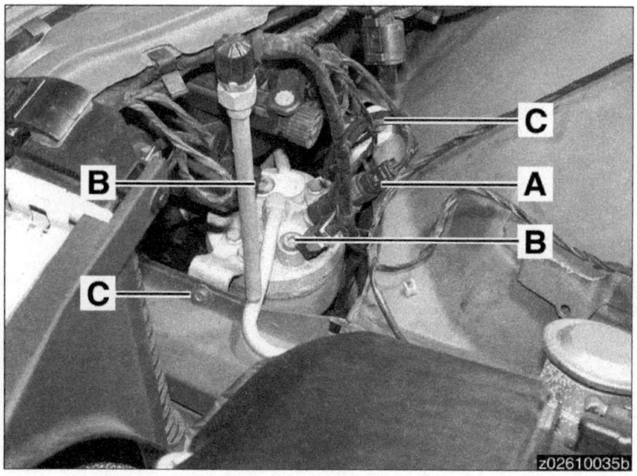


◀ Working at right front corner of engine compartment:
  - Detach pressure switch electrical connector (**A**).
  - Loosen and remove A/C refrigerant line mounting bolts (**B**).
  - Remove receiver / drier mounting bolts (**C**).

> ***CAUTION—***
> *Plug open refrigerant lines and ports.*

***NOTE —***
*2000 Z3 Roadster engine compartment is illustrated. Other models are similar.*

– Raise front of vehicle and support safely.

> ***WARNING —***
> *Make sure the car is stable and well supported at all times. Use a professional automotive lift or jack stands designed for the purpose. A floor jack is not adequate support.*

– Remove splash shield under engine and radiator.

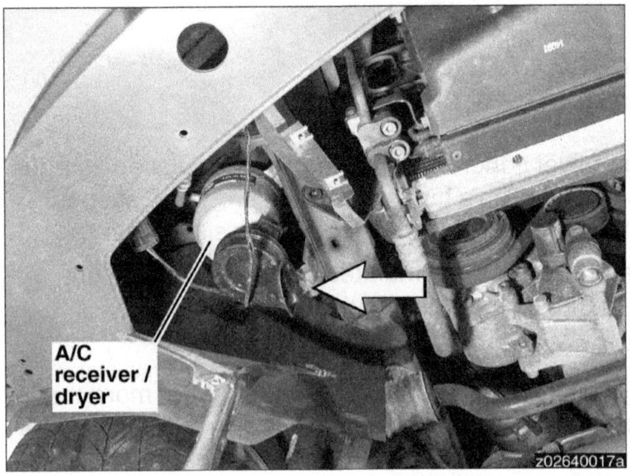


◀ Remove right side horn mounting bolt. Suspend horn out of the way.

– Carefully lower receiver / drier with pressure switch, checking for lines or electrical harnesses that night get snagged.

– Installation is reverse of removal. Replace sealing O-rings as necessary.

– Recharge A/C system following equipment manufacturer's instructions. Be sure to use correct refrigerant oil. See **Refrigerant and refrigerant oil** in this repair group.

# 650 Radio

## General

Basic BMW factory installed sound systems are covered in this repair group. A variety of radios, cassette players, cd players and amplifiers are used in Roadster and Coupe models. There are also many different speaker configurations.

After-market sound systems are beyond the scope of this manual.

For additional information, see:
- **610 Electrical Component Locations**
- **ELE Electrical Wiring Diagrams**

### Special tools

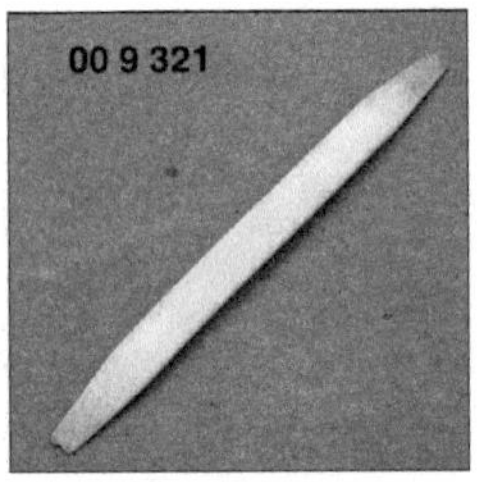


Plastic prying tool
(BMW special tool 00 9 321)

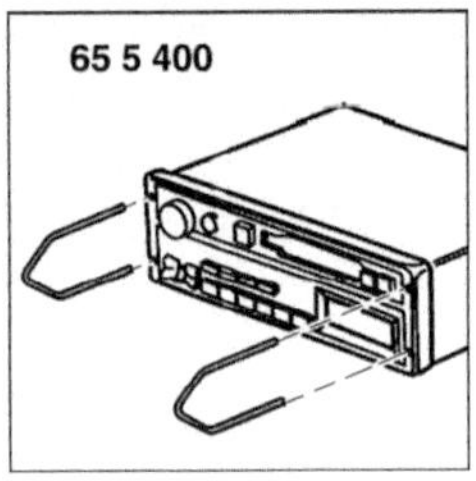


Radio removal tools (Bavaria radio only)
(BMW special tool set 65 5 400)

### Radio system

The Z3 radio, installed in the center of the lower dashboard, is equipped with an anti-theft system. If power is removed from the radio circuitry by detaching the radio electrical connectors, or by disconnecting the battery, the radio is rendered inoperable. The radio is not permanently damaged, and proper operation can be restored by entering the radio anti-theft code, or by bringing the vehicle and the radio to an authorized BMW dealer.

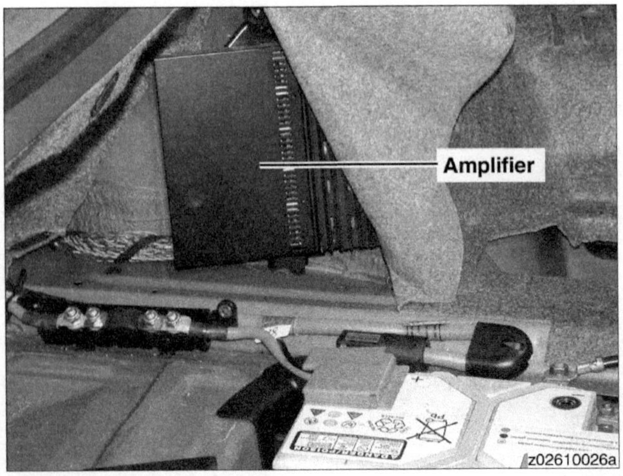


The sound system amplifier, if equipped, is in the trunk behind trim.

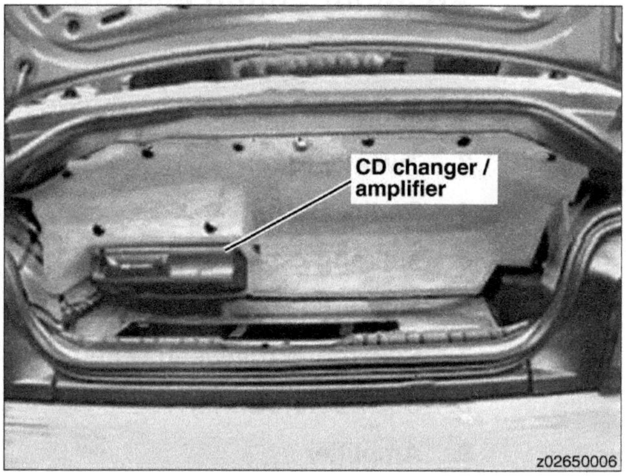

z02650006

An optional six-disc compact disc changer may be installed in the trunk. If installed, CD changer and amplifier are mounted to the same bracket in the trunk.

The vehicle is factory pre-wired for the CD changer.

## Speakers

Several speaker configurations are used in Z3 cars. The illustrations in **Roadster sound system** and **Coupe sound system** in this repair group depict sound system components and speakers for a variety of models. Bear in mind that not all components are found in every car.

In 1997 and later models the storage area behind the seats was eliminated to make room for a subwoofer and amplifier enclosure.

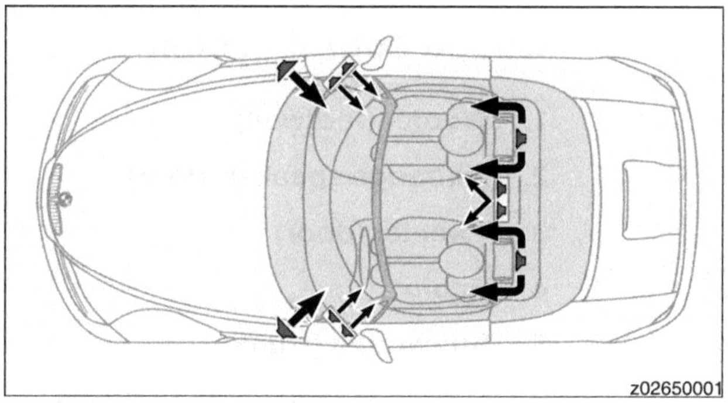
z02650001

On 1997 and later Roadster models, speakers are configured to surround the cockpit for optimum sound during top-down driving. Additionally, the bass and volume outputs automatically compensate for vehicle speed, making constant volume adjustments unnecessary.

## Antenna

In Roadster models, the antenna is mounted on the left rear fender. The antenna connections and antenna amplifier are in the left side of the trunk, behind trim.

In Coupe models there is a stub antenna screwed to the antenna base in the center rear of the roof. The antenna base and amplifier are above the headliner.

If equipped, the telephone utilizes the radio antenna.

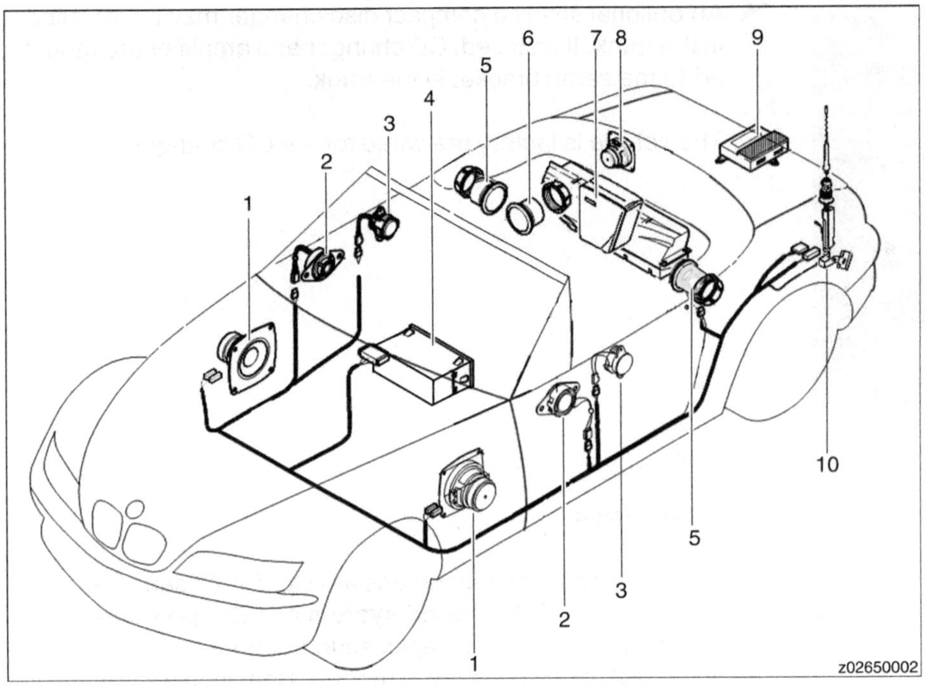


## Roadster sound system

1. **Woofer (footwell)**
2. **Midrange speaker (door)**
3. **Tweeter (door)**
4. **Radio**
5. **Midrange speaker (rear side)**
6. **Midrange speaker (behind seat)**
7. **Subwoofer (between seats)**
8. **Subwoofer (trunk)**
9. **Amplifier**
10. **Antenna**

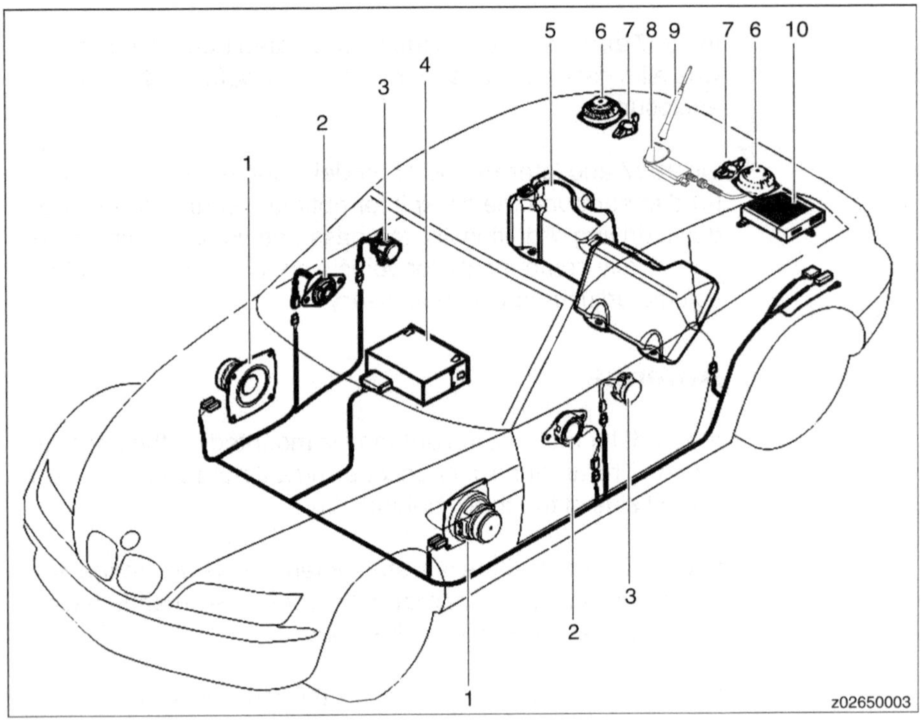


## Coupe sound system

1. **Woofer (footwell)**
2. **Midrange speaker (door)**
3. **Tweeter (door)**
4. **Radio**
5. **Subwoofer (trunk)**
6. **Midrange speaker (headliner)**
7. **Tweeter (headliner)**
8. **Antenna amplifier**
9. **Antenna mast**
10. **Amplifier**

***NOTE—***
*Not all components are found in every car.*

## Sound System Service

***CAUTION—***
- *Before beginning work on the radio or sound system verify that the radio is an original equipment BMW radio and that the wiring harness has not been modified.*
- *Aftermarket radios, amplifiers, speakers, etc. are beyond the scope of this manual. For information or service, refer to an authorized agent of the equipment in question.*
- *Do not remove the radio or disconnect the battery without obtaining the anti-theft code.*

### Radio, removing and installing

– Note and write down radio station presets.

– Obtain radio anti-theft activation code (if applicable) and write it down.

< Pry open two small access covers (**arrows**) on ends of radio faceplate.

– Loosen two lock tab retaining screws inside access pockets on sides of radio. If radio does not pull out easily, loosen screws further. Do not pull hard on radio chassis. The radio may be damaged.

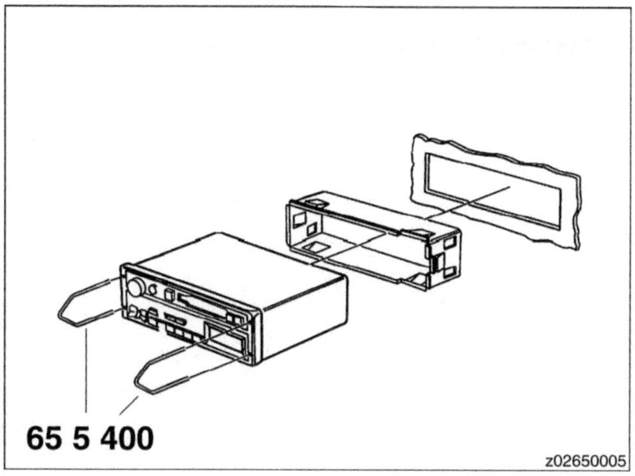


< On some models, BMW special tool set 65 5 400 is necessary to pull out radio.

– Pull radio far enough out to reach behind it and disconnect radio electrical and antenna connectors.

– When installing:
  - Attach radio electrical connectors first.
  - Push radio straight back into dash opening until it seats fully.
  - Tighten locking tab retaining screws until radio is secured firmly.
  - Snap shut access covers on sides of faceplate.
  - Enter anti-theft activation code.
  - Reprogram radio station pre-sets.

## Amplifier, removing and installing

*NOTE—*
*In models with cd changer, the amplifier and cd changer may be mounted on one bracket behind the driver seat.*

- Make sure radio and other sound system components are OFF.
- Roadster: Working in trunk, remove trim to access amplifier.
- Coupe: Working in cargo compartment, remove trim behind passenger seat to access amplifier.

< Disconnect amplifier electrical connectors (**arrows**).

- Remove amplifier bracket, then detach amplifier from bracket.
- Installation is reverse of removal.

## SPEAKERS

*CAUTION—*
*To avoid marring trim, work with a plastic prying tool or wrap screwdriver tip with tape.*

### Front speakers

The front speakers are a pair of woofers installed in the left and right footwells.

- Remove footwell trim.

< Remove speaker mounting screws (**arrows**).

- Pull out speaker and detach harness connector.

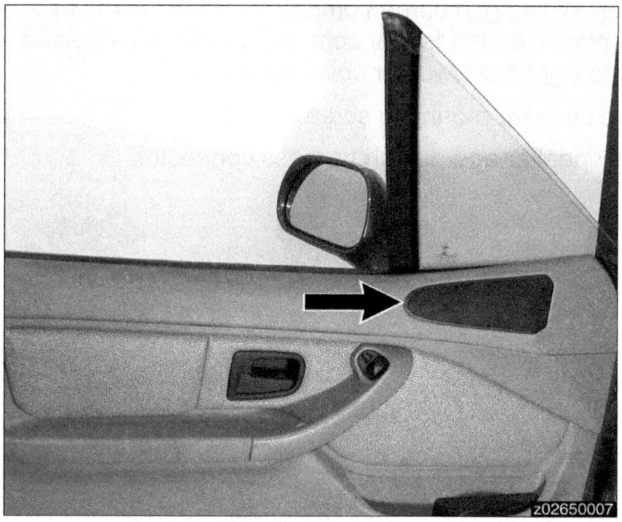


## Door speakers

< Use plastic prying tool or screwdriver with tip wrapped with tape to pry out door speaker trim cover (**arrow**).

– Remove speaker mounting screws.

– Pull out speaker and detach harness connector.

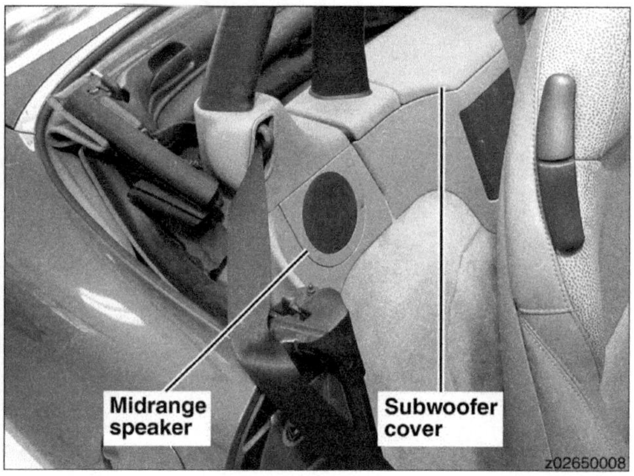


## Rear speakers, Roadster

< Midrange speaker: Use plastic prying tool or screwdriver with tip wrapped with tape to pry out speaker trim cover.

- Remove speaker mounting screws.
- Pull out speaker and detach harness connector.

– Subwoofer: Use plastic prying tool or screwdriver with tip wrapped with tape to pry up subwoofer cover.

- Remove speaker housing mounting screws.
- Pull housing up out of enclosure. Tilt housing forward to clear enclosure.
- Remove speaker from housing and detach harness connector.

> ***CAUTION—***
> *During installation, be sure that harness connector does not damage speaker diaphragm.*

## Rear speakers, Coupe

< Tweeter or midrange speaker: Working in cargo compartment at headliner, use plastic prying tool or screwdriver with tip wrapped with tape to pry off speaker cover.

- Remove speaker mounting screws.
- Pull out speaker and detach harness connector.

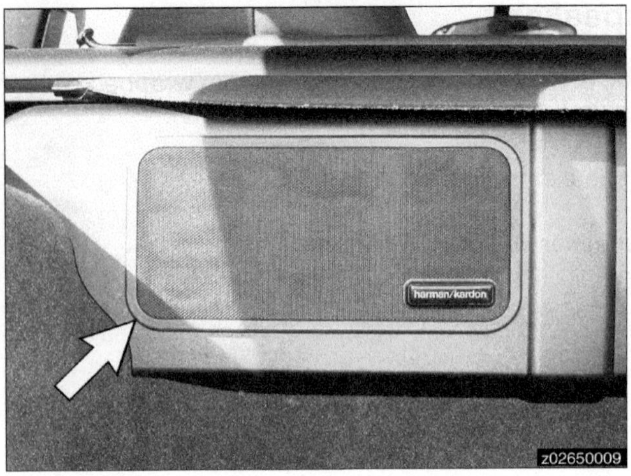


Subwoofer: Working in cargo compartment behind driver seat, use plastic prying tool or screwdriver with tip wrapped with tape to pry off subwoofer cover (**arrow**).

- Remove speaker mounting screws.
- Pull out speaker and detach harness connector.

# 720 Seat Belts



**TABLES**

## GENERAL

This repair group covers repairs to the seat belt system.

Operation of the seat belt system is linked to the airbag system. See **721 Airbag System (SRS)**.

### Special tools

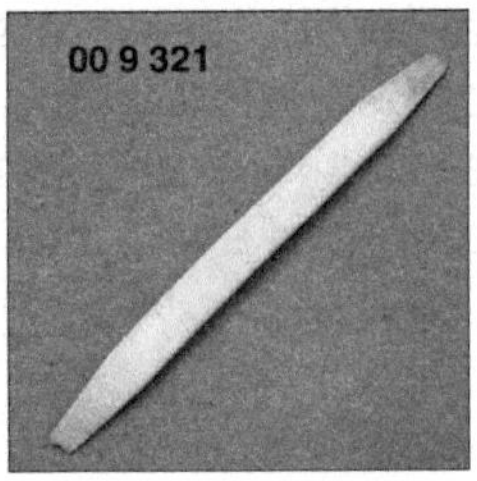


Plastic prying tool
(BMW special tool 00 9 321)

### Seat belt system

Z3 seats are equipped with 3-point shoulder belts.

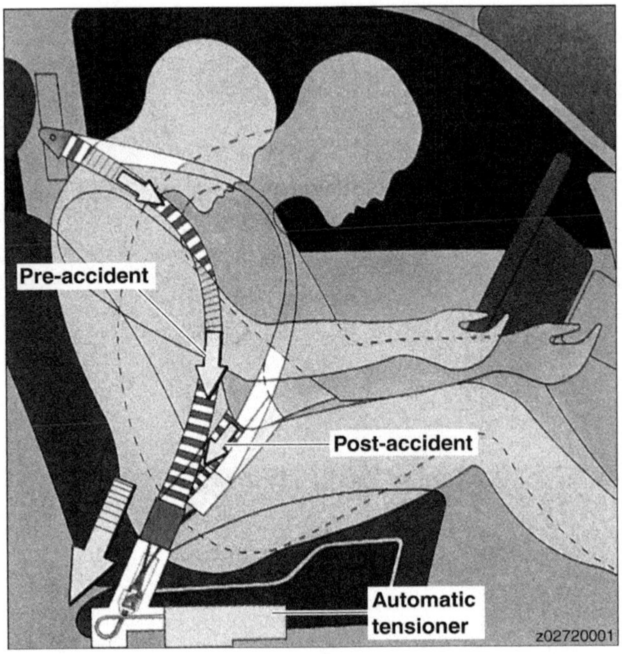


Automatic seat belt tensioners are designed to automatically retract and tension the seat belt by 55 mm (approximately 2 inches) in case of an accident.

Two types of automatic tensioner were used in Z3 models:

- 1996 models: Mechanical spring tensioner
- 1997 - 2002 models: Pyrotechnical (explosive charge) tensioner, electronically controlled by the Multiple Restraint System (MRS) control module. See **721 Airbag System (SRS)**.

### Troubleshooting

Make sure that the seat belt reel locks when pulled quickly and sharply. The automatic reel does not require any servicing and should never be opened.

**Table a** covers some common seat belt problems.

**Table a. Seat belt troubleshooting**

| Problem | Probable cause | Repairs |
|---|---|---|
| Damage to seat belt. | Accident in which bumpers were permanently deformed. | Replace complete seat belt. Also check and repair seat belt mounting on car body and seat belt mounting on seat runner. |
| Belt creased, unraveled, pinched, cut or melted. Belt buckle or plastic casing worn, damaged or missing. | Seat belts aged or worn. | Replace complete seat belt. |
| Seat belt buckle is not released when red button on seat belt lock is pressed. | Seat belt lock mechanism worn or damaged. | Replace seat belt lock and tensioner assembly. |
| Seat belt automatic reel does not lock when pulled out suddenly. | Automatic reel defective. | Replace reel assembly. |

| Table a. Seat belt troubleshooting | | |
|---|---|---|
| **Problem** | **Probable cause** | **Repairs** |
| Seat belt automatic reel jams when pulled out. | Automatic reel loose. | Tighten reel mounting bolt. |
| | Return spring broken inside reel. | Replace reel assembly. |
| Seat belt does not retract automatically. | Automatic reel loose. | Tighten reel mounting bolt. |
| | Return spring broken inside reel. | Replace reel assembly. |
| Seat belt squeaks when fastened or unfastened. | Excessive friction in belt guides. | Replace reel assembly. |
| | Automatic reel loose. | Tighten reel mounting bolt. |
| | Return spring broken inside reel. | Replace reel assembly. |
| Seat belt pyrotechnic tensioner triggered. | Accident triggered pyrotechnic deployment. | Replace complete seat belt. Also check belt tensioner bracket for damage. |

## Warnings and cautions

***WARNING —***

- *Mechanical and pyrotechnic seat belt tensioners are powerful devices and should be handled with extreme care. Incorrect handling can trigger the tensioner and cause injury.*
- *BMW recommends that all repair or replacement work on pyrotechnic devices must be carried out by a qualified BMW technician.*
- *Be sure to disconnect the battery and wait 5 seconds before attempting to work on pyrotechnic devices.*
- *Do not repair pyrotechnic devices. Always replace.*
- *Do not expose pyrotechnic components to temperatures above 75°C (167°F).*
- *Do not fire a pyrotechnic gas generator prior to disposal. It must be fired by a special disposal company or shipped back to the manufacturer in the packaging of the new component.*
- *During body straightening and welding with an electric arc welder, always disconnect the battery and the connection to the pyrotechnic gas generators.*
- *Replace seat belt if subject to occupant loading in a collision.*
- *Do not modify or repair seat belts. Do not modify seat belt mounting points.*
- *Do not bleach or dye seat belt webbing. Webbing that is severely faded or redyed does not meet strength requirements.*
- *Clean belts with a luke-warm soap solution only.*
- *Periodically inspect seat belts for webbing defects such as cuts or pulled threads.*
- *Immediately after replacing a damaged or worn seat belt, destroy old belt to prevent it from being reused.*

> ***CAUTION—***
> - *When working on electrical components, disconnect the negative (–) cable from the battery and insulate the cable end to prevent accidental reconnection.*
> - *Prior to disconnecting the battery, read the battery disconnection cautions given in* **001 General Warnings and Cautions**.
> - *To prevent marring trim when working on interior components, work with a plastic prying tool or wrap a screwdriver tip with tape.*

## SEAT BELTS

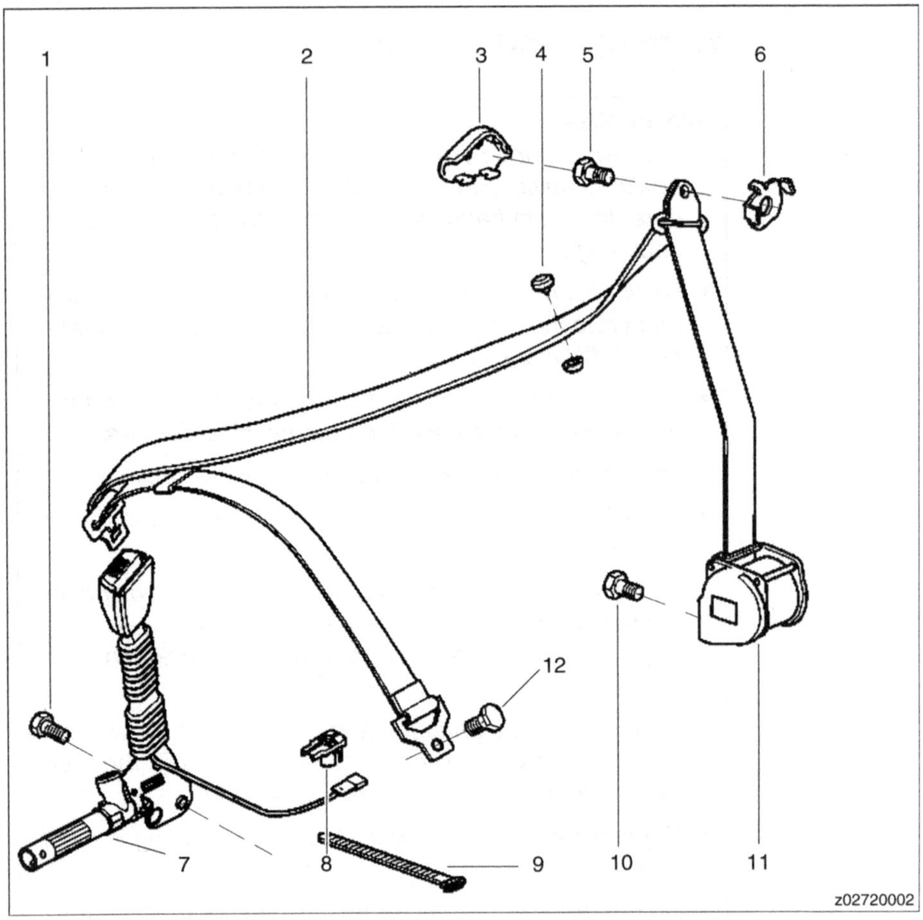


### Seat belt components, Roadster

1. **Tensioner bolt**
   - M10 Torx
     tighten to 48 Nm (35 ft-lb)
2. **Seat belt**
3. **Trim cover**
4. **Button**
5. **Seat belt upper anchor bolt**
   - 7/16" Torx
     tighten to 31 Nm (23 ft-lb)
6. **Support**
7. **Seat belt lock and pyrotechnic tensioner assembly**
8. **Electrical connector protector**
9. **Wire tie**
10. **Seat belt reel bolt**
    - 7/16" Torx
      tighten to 48 Nm (35 ft-lb)
11. **Seat belt reel**
12. **Seat belt lower anchor bolt**
    - 7/16" Torx
      tighten to 48 Nm (35 ft-lb)

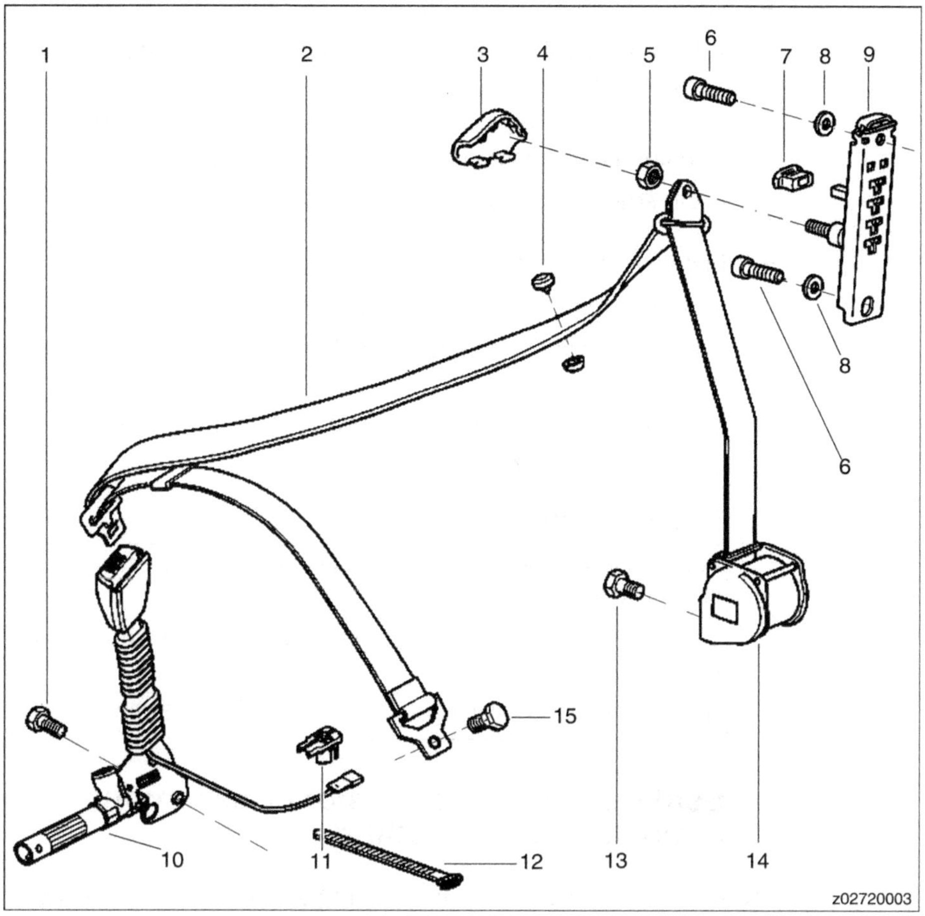


## Seat belt components, Coupe

1. **Tensioner bolt**
   - M10 Torx tighten to 48 Nm (35 ft-lb)
2. **Seat belt**
3. **Trim cover**
4. **Button**
5. **Seat belt upper anchor fastener**
   - 7/16" nut tighten to 31 Nm (23 ft-lb)
6. **Height adjuster bolt**
   - 7/16" or M8 Torx tighten to 24 Nm (18 ft-lb)
7. **Height adjusting lever**
8. **Washer**
9. **Height adjuster bracket**
10. **Seat belt lock and pyrotechnic tensioner assembly**
11. **Electrical connector protector**
12. **Wire tie**
13. **Seat belt reel bolt**
    - 7/16" Torx tighten to 48 Nm (35 ft-lb)
14. **Seat belt reel**
15. **Seat belt lower anchor bolt**
    - 7/16" Torx tighten to 48 Nm (35 ft-lb)

## Seat belt, removing and installing

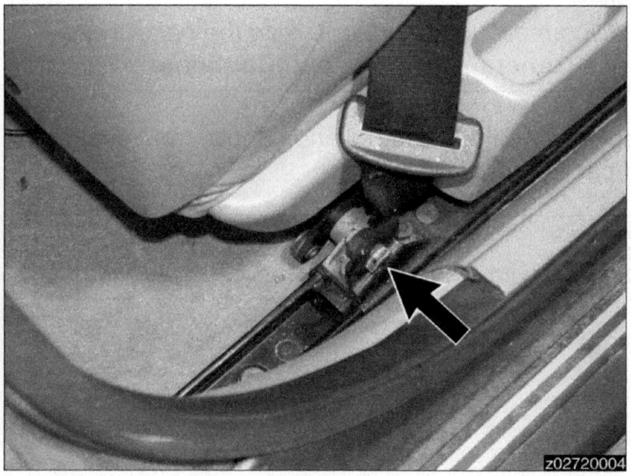


< Move seat forward. Remove seat belt lower anchor bolt (**arrow**) below backrest.

– Detach seat belt from guide at top of backrest.

– Roadster:
  - Remove side trim behind seat.
  - Remove seat belt upper anchor and seat belt reel mounting bolts, noting arrangement of washers and spacers.
  - Lift seat belt out of vehicle.

– Coupe: Remove B pillar trim.
  - Remove seat belt reel mounting bolt. Note position of locating lug.
  - Remove seat belt upper anchor fastener from height adjuster. Note position of washers and spacers.
  - Lift seat belt out of vehicle.

– Installation is reverse of removal. Bear in mind the following:
  - Use fasteners specifically designed for seat belt system.
  - Note correct placement of washers and spacers on fasteners.
  - Make sure locating lugs on seat belt anchors fit into recesses properly.

| Tightening torques | |
|---|---|
| Seat belt lower anchor to seat (use thread locking compound) | 48 Nm (35 ft-lb) |
| Seat belt upper anchor to height adjuster or to B-pillar | 31 Nm (23 ft-lb) |
| Seat belt reel to B-pillar | 48 Nm (35 ft-lb) |

## Seat belt lock and tensioner assembly, removing and installing

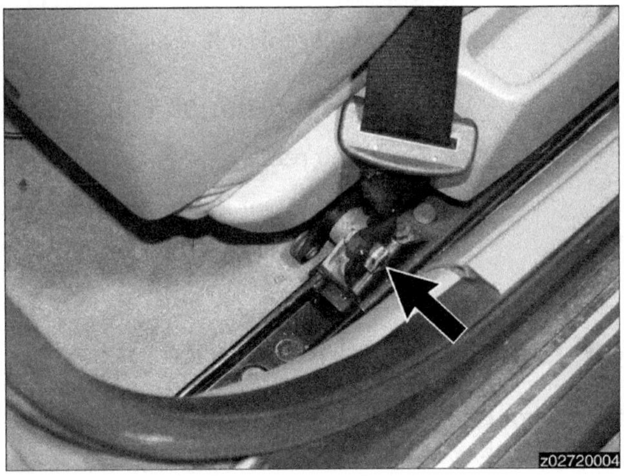


Move seat forward. Remove seat belt lower anchor bolt (**arrow**) below backrest. Note placement of washers and spacers.

– Disconnect negative (–) battery cable and cover terminal with insulating material.

***NOTE—***
*On 1996 cars, the battery is located in the right rear of the engine compartment. On 1997 and later cars, the battery is located under the trunk floor mat.*

> ***CAUTION—***
> *Prior to disconnecting the battery, read the battery cautions in* **001 General Warnings and Cautions**.

– Disconnect electrical connectors and remove front seat from interior of car. See **520 Seats**.

> ***CAUTION—***
> *Cover door sill with blanket to prevent damage to trim.*

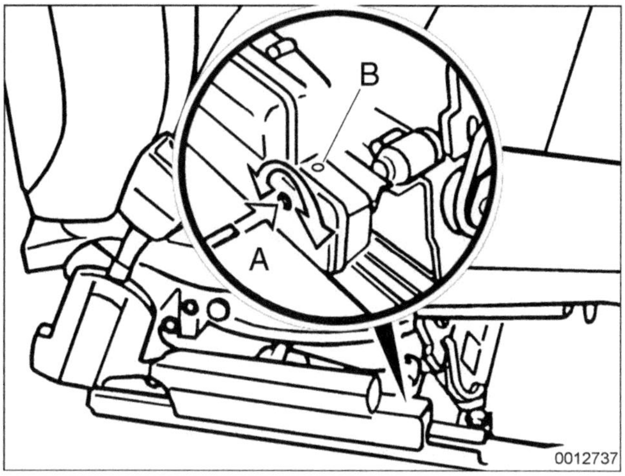


1996 model (mechanical tensioner): Working at side of seat, turn tensioner activating / deactivating screw (**A**) clockwise until display (**B**) changes from green to red.

***NOTE—***
- *Green is armed. Red is disarmed.*
- *The pyrotechnic tensioner in later models is disarmed when the electrical connector is detached.*

– Remove tensioner mounting bolt and detach seat belt lock and tensioner assembly from seat.

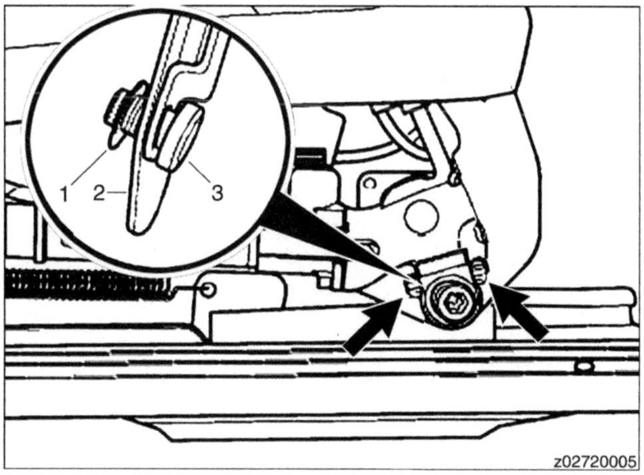


When installing assembly to seat:
- Note arrangement of fastener (**inset**):
  **1**. Washer
  **2**. Tensioner assembly
  **3**. Bolt
- Orient locating lugs (**arrows**) properly.

– Seat installation is reverse of removal.
- 1996 with mechanical tensioner: Reactivate tensioner before installing seat: Turn tensioner activating screw counterclockwise until green appears in window.
- Models with pyrotechnic tensioner: Make sure tensioner orange electrical connector is not under tension.
- Note correct placement of washers and spacers on seat belt lower anchor.

| **Tightening torques** | |
|---|---|
| Seat belt lower anchor to seat (use thread locking compound) | 48 Nm (35 ft-lb) |
| Seat belt tensioner to seat | 48 Nm (35 ft-lb) |

# 721 Airbag System (SRS)



721

## General

This repair group covers airbag components and replacement. Airbag system repair and fault diagnosis is not covered.

Individual airbag system components can only be tested electronically when installed in the car. Use BMW service testers DISplus, GT1, MoDiC or equivalent for diagnostic work.

> ***WARNING —***
> *If the airbag warning light is ON while driving, the airbags will not deploy in case of an accident. The warning light remains ON until problems are corrected and the fault memory cleared.*

For additional information, see the following repair groups:

- **121 Battery, Starter, Alternator** for battery safety terminal (SBK)
- **610 Electrical Component Locations**
- **720 Seat Belts** for pyrotechnic seat belt tensioners
- **ELE Electrical Wiring Diagrams**

### Special tool

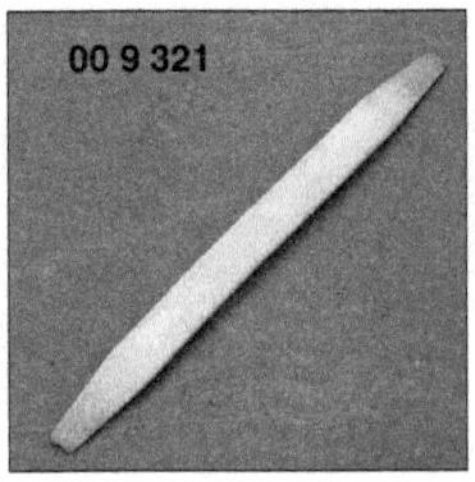


◄ Plastic prying tool
(BMW special tool 00 9 321)

### Z3 airbag systems

Airbags are also knows as the Supplemental Restraint System (SRS). In BMW vehicles the occupant safety system, including the airbags, is called the Multiple Restraint System (MRS).

Four versions of MRS were used in Z3 models.

| Multiple Restraint System versions | |
|---|---|
| **Date of manufacture** | **Restraint system** |
| 1996 to 8 / 1997 | ZAE |
| 9 / 1997 to 8 / 1998 | MRS |
| 9 / 1998 to 3 / 1999 | MRS II |
| 4 / 1999 to 2002 | MRS III |

z02721001

**ZAE** features the following:

- Centralized airbag control module with built-in capability for crash detection (electronic acceleration sensor)
- Driver airbag
- Passenger airbag
- Airbag warning light
- From 3 / 1997: Pyrotechnic seat belt tensioners

**MRS** and **MRS II** add the following features:

- Control module coded to vehicle using DISplus, GT1, MoDiC or equivalent
- Self-diagnostic capability
- Control module internal mechanical safety switch (must close in conjunction with acceleration sensor before airbag activation)
- Passenger seat occupancy sensor (SBE)
- Hall sensor seat belt microswitches

**MRS III** adds the following features:

- Side impact (door-mounted) airbags
- Side impact crash sensors
- Two-stage ("smart") airbags
- Battery safety terminal (SBK)
- Fuel pump safety cut off

## Airbag warning light

z02721003

The airbag system utilizes a self diagnostic control module to detect and store system faults. If a fault is detected, the airbag warning light in the instrument cluster comes on and stays on.

When the ignition key is turned on, the warning light illuminates for approximately 4 - 6 seconds. The illuminated light indicates the self-test of the electronic control/monitoring module, the ignitors and sensor circuits, and the system wiring. The airbag system should be inspected by an authorized BMW dealer if the warning light does not come on during self-test, if it does not go out in approximately 4 - 6 seconds after switching on the ignition, or if the light ever comes on during driving.

Special test equipment is required to retrieve SRS fault codes, diagnose system faults, and reset / turn off the SRS warning light. The SRS warning light remains ON until all problems have been corrected and the fault memory has been cleared.

Seat occupancy signal-converter module

z02712006

## Seat occupancy sensor (SBE) (MRS II, MRS III)

An occupancy sensor pad is incorporated in the passenger seat cushion. The pad is a foil contact sensor that is open when the seat is not occupied.

When the seat cushion is pressed by the weight of a passenger, the sensor circuit produces a resistance that varies with the weight and/or movement.

This variable resistance signal is converted by the seat occupancy module into a square wave modulated output signal that is sent to the MRS control module. The MRS control module uses this signal to determine passenger seat belt tensioner and airbag deployment thresholds.

## Side impact (door-mounted) airbags (MRS III)

z02721005

The door airbag, mounted on the door (**arrow**), is designed to protect the upper part of the occupant's body (chest or thorax) in case of a side impact. It may be referred to as a thorax airbag.

## Side impact crash sensors (MRS III)

Side impact sensors, mounted below the seats on the vehicle floor pan, detect the severity of side impacts. The pulse-modulated signal is used by the MRS III control module, along with its internal impact sensor signal, to determine level of deployment of passive safety devices (front airbags, side impact airbags, seat belt tensioners, battery safety terminal).

## Two-stage airbags (MRS III)

The two-stage airbag ensures that the force of airbag inflation is not greater than necessary to provide protection. The airbags are designed to provide "soft" deployment if the acceleration sensor detects a low-speed impact, and "hard" deployment in higher speed impacts.

## Warnings and cautions

Observe the following precautions to prevent personal injury.

***WARNING —***

- *If the airbag warning light is ON, the airbags will not deploy in case of an accident. Be sure to have the airbag system inspected and repaired immediately.*
- *Serious injury may result if MRS system service is attempted by persons unfamiliar with the BMW MRS and its approved service procedures. BMW advises that all inspection and service be performed by an authorized BMW dealer.*
- *Airbags, pyrotechnic seat belt tensioners and the battery safety terminal are deployed by explosive devices. Handled improperly or without adequate safeguards, these components are very dangerous.*
- *Always disconnect the battery and cover the negative (–) battery terminal with an insulator before starting diagnostic, troubleshooting or service work on electrical components, even if not associated with the airbags, and before doing any welding on the car.*
- *After disconnecting the battery, wait 10 seconds before beginning work on airbag components.*
- *If an airbag has been activated due to an accident, BMW specifies that airbag components be replaced. For more information on post-collision airbag service, see an authorized BMW dealer.*
- *Do not fire an airbag unit prior to disposal. It must be fired by a special disposal company or shipped back to BMW in the packaging of the new components.*
- *When removing a fired airbag unit, avoid contact with the skin; wear gloves. In case of skin contact, wash with water.*
- *Do not allow airbag system components to come in contact with cleaning solutions or grease. Never subject airbag components to temperatures above 167°F (75°C). When reconnecting the battery, no person should be inside the vehicle.*
- *Always place an airbag unit that has been removed from its packaging with the padded side facing upward. Do not leave an airbag unit unattended.*
- *If the airbag unit or airbag control module has been dropped from a height of ½ meter (1½ ft) or more, do not use the component.*

> ***CAUTION—***
> - *When working on electrical switches or lights, disconnect the negative (–) cable from the battery and insulate the cable end to prevent accidental reconnection.*
> - *Prior to disconnecting the battery, read the battery disconnection cautions given in* **001 General Warnings and Cautions**.
> - *To prevent marring trim when working on interior components, work with a plastic prying tool or wrap a screwdriver tip with tape.*

## AIRBAG ELECTRONICS

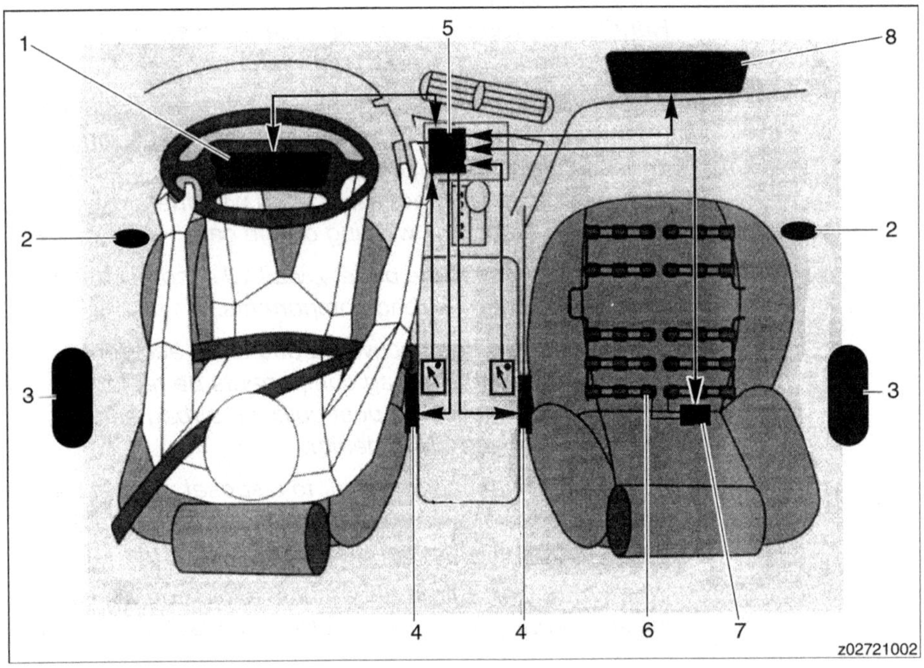


### MRS III schematic

1. **Driver airbag**
2. **Side impact sensor**
3. **Side impact (door-mounted) airbag**
4. **Pyrotechnic seat belt tensioner and seat belt microswitch (Hall sensor)**
5. **MRS control module**
6. **Seat occupancy sensor (SBE) pad**
7. **SBE module (analog- to-digital signal converter)**
8. **Passenger airbag**

### MRS control module, replacing

The MRS control module contains the electronic crash sensor for frontal impact. Sensors mounted on the driver and passenger floor are used for side impacts. The MRS control module is located under the center console ahead of the shifter.

– Disconnect negative (–) battery cable and cover terminal with insulating material.

***NOTE—***
*On 1996 cars, the battery is located in the right rear of the engine compartment. On 1997 and later cars, the battery is located under the trunk floor mat.*

> ***CAUTION—***
> *Prior to disconnecting the battery, read the battery cautions in* **001 General Warnings and Cautions**.

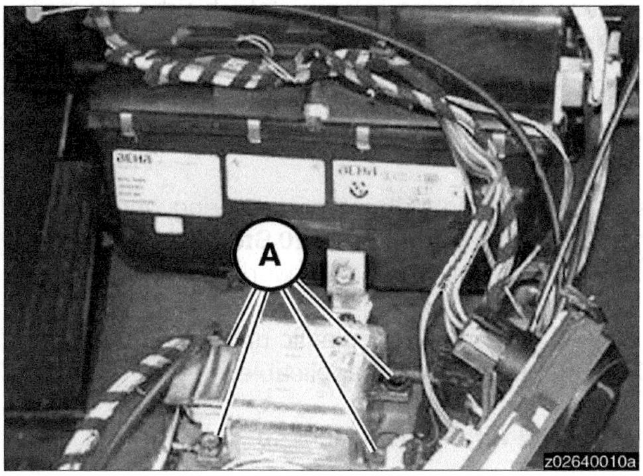


– Remove center console between front seats. See **513 Interior Trim**.

◀ Working in front of the shifter housing, remove module mounting fasteners (**A**) and lift up control module. Disconnect electrical harness connector.

– Installation is reverse of removal.

- Install module with arrow pointing to front of car.
- Be sure that module ground cable is firmly reattached.

| Tightening torque | |
|---|---|
| MRS control module to floor | 9 Nm (7 ft-lb) |

– Reconnect battery last.

– Check for fault codes and clear and reset MRS control module memory.

## Airbag contact ring, removing and installing

The airbag contact ring is mounted to the rear of the steering wheel and is made up of a wire ribbon that coils and uncoils as the steering wheel is turned. The contact ring assures continuous electrical contact to the driver airbag.

– Disconnect negative (–) battery cable and cover terminal with insulating material.

***NOTE—***
*On 1996 cars, the battery is located in the right rear of the engine compartment. On 1997 and later cars, the battery is located under the trunk floor mat.*

***CAUTION—***
*Prior to disconnecting the battery, read the battery cautions in* **001 General Warnings and Cautions**.

– Wait ten seconds for residual airbag power to discharge.

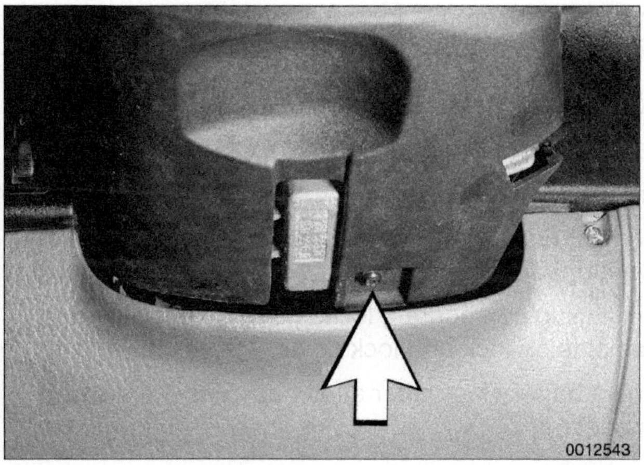


◀ Remove lower steering column trim mounting screw (**arrow**) and remove lower trim.

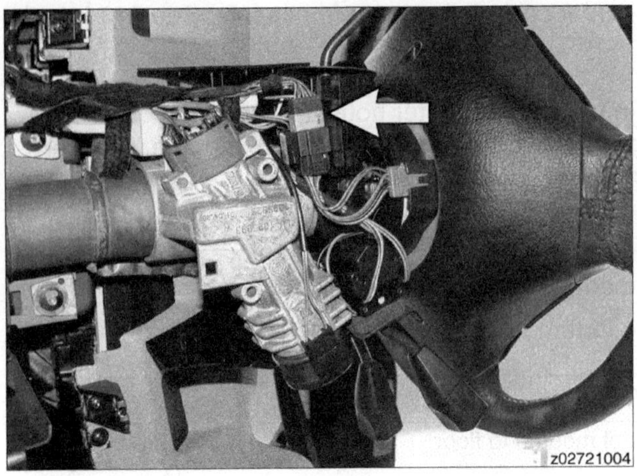


Working underneath steering column, detach airbag harness connector (**arrow**).

– Remove driver airbag from steering wheel. See **Driver airbag, removing and installing** in this repair group.

– With wheels facing straight ahead and steering wheel centered, remove steering wheel. See **320 Steering and Wheel Alignment**. Place steering wheel on work bench.

– Remove contact ring cover (flat plastic ring containing warning text) from steering wheel, if applicable.

– Disconnect ground wire, if applicable.

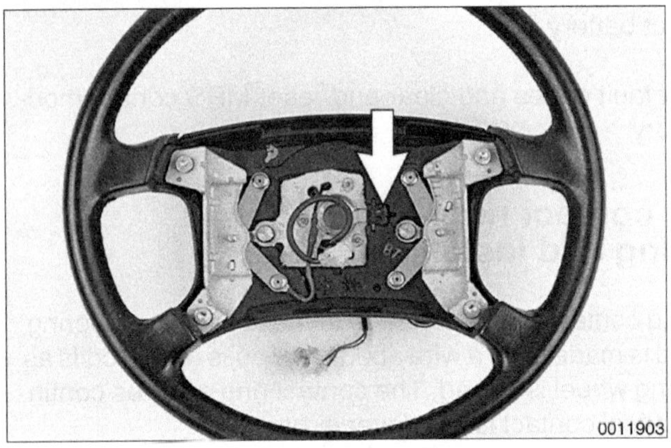


Pry out airbag contact ring lock (**arrow**) and remove spring from behind lock.

***NOTE —***
*Airbag steering wheel may not be equipped with contact ring lock.*

> ***CAUTION—***
> - *The contact ring lock holds the contact ring in the centered position. Once the lock is removed, the contact ring is free to rotate. Do not allow the contact ring to rotate once the lock is removed.*
> - *If the contact ring is accidentally rotated out of center, turn the ring either fully clockwise or counterclockwise, then turn in the opposite direction 3 complete revolutions until the alignment arrow can be seen through the lock opening.*

Remove contact ring mounting fasteners (**arrows**). Disconnect wire lead(s) and remove contact ring from rear of steering wheel.

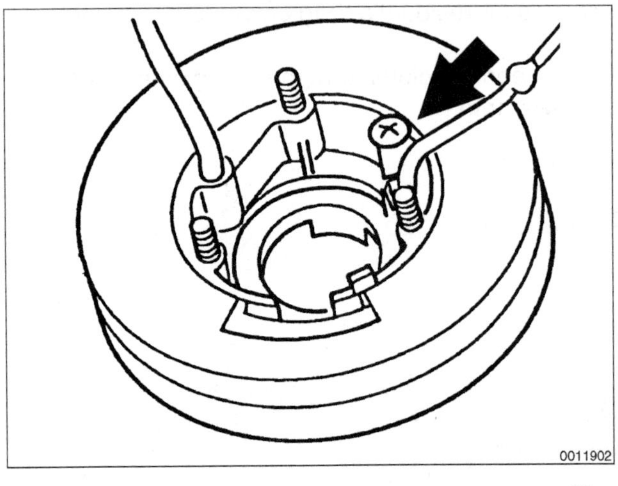


When installing a new contact ring from BMW, check for locking screw (**arrow**). Remove locking screw after installing contact ring to steering wheel, if applicable.

– Installation is reverse of removal, noting the following:
- Route wiring lead for horn through square opening in steering wheel.
- Lock contact ring mounting fasteners in place using lacquer paint or varnish.
- When installing contact ring lock, make sure lock spring engages pins on lock and lock cover.
- Reconnect battery last, after all airbag connections have been made.

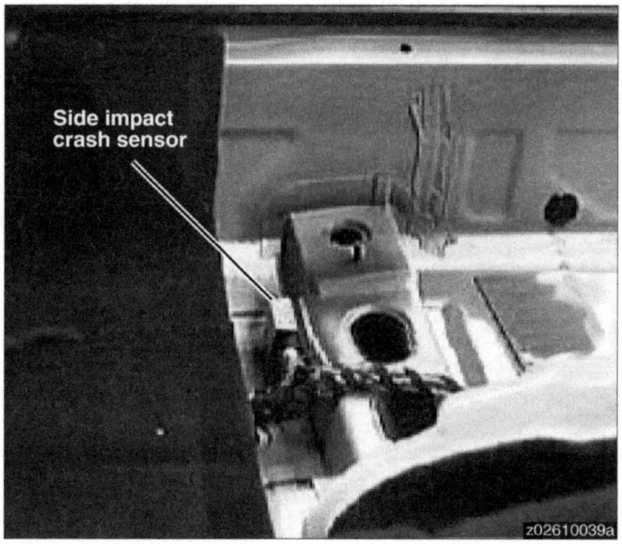


## Side impact crash sensor

The side impact crash sensor is mounted under the seat, one on each side of the vehicle near door sill. Remove seat and carpet to access.

| Tightening torque | |
|---|---|
| Side impact crash sensor to floor brace | 5.5 Nm (49 in-lb) |

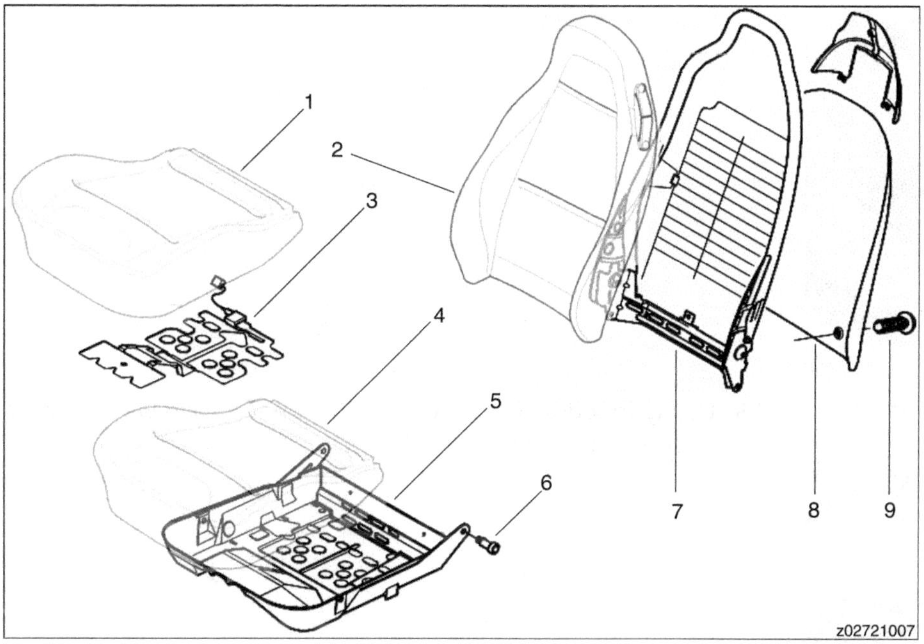


## Seat occupancy detection pad

1. **Seat cover**
2. **Backrest cover and cushion**
3. **Occupancy detection pad**
4. **Seat cushion**
5. **Seat bottom**
6. **Backrest mounting bolt**
7. **Backrest frame**
8. **Backrest shell**
9. **Backrest shell mounting rivet**

# AIRBAGS

## Driver airbag, removing and installing

– Disconnect negative (–) battery cable and cover terminal with insulating material.

***NOTE—***
*On 1996 cars, the battery is located in the right rear of the engine compartment. On 1997 and later cars, the battery is located under the trunk floor mat.*

> ***CAUTION—***
> *Prior to disconnecting the battery, read the battery cautions in* **001 General Warnings and Cautions**.

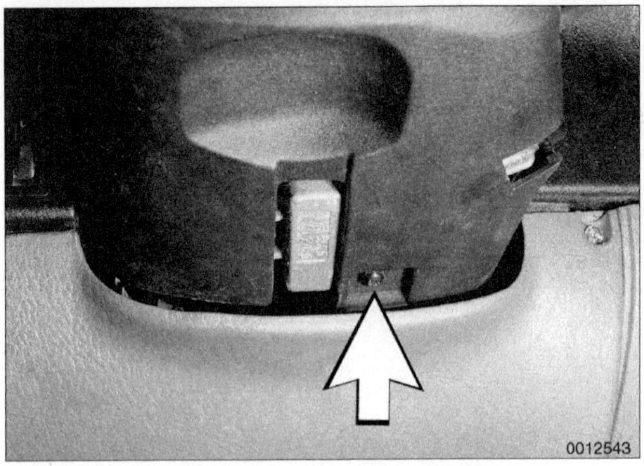
0012543

– Wait ten seconds for residual airbag power to discharge.

◄ Remove lower steering column trim mounting screw (**arrow**) and remove lower trim.

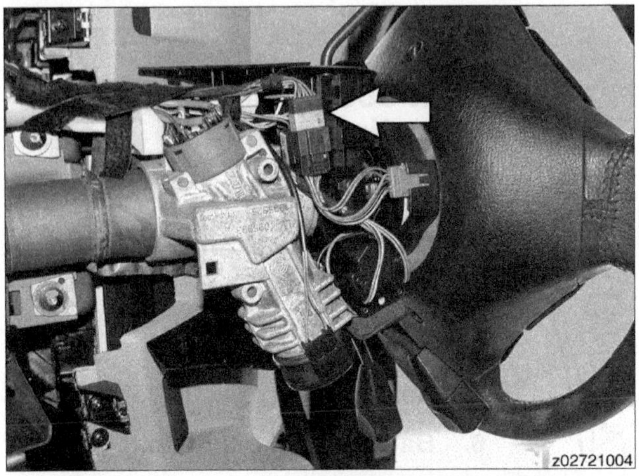
z02721004

◄ Working underneath steering column, detach airbag harness connector (**arrow**).

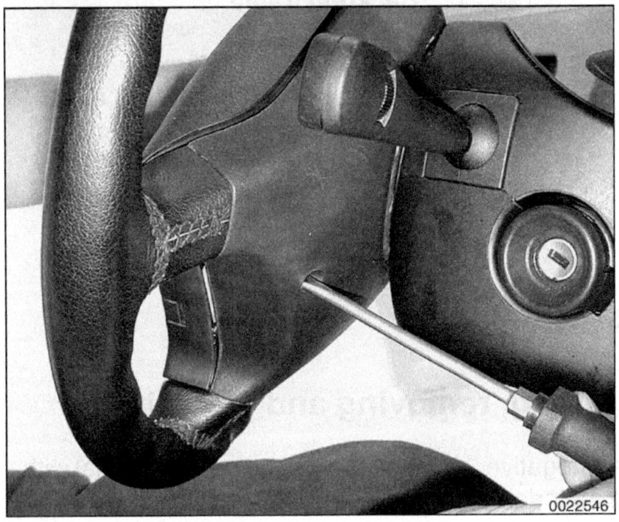
0022546

◄ Using Torx driver, remove airbag mounting screws from rear of steering wheel.

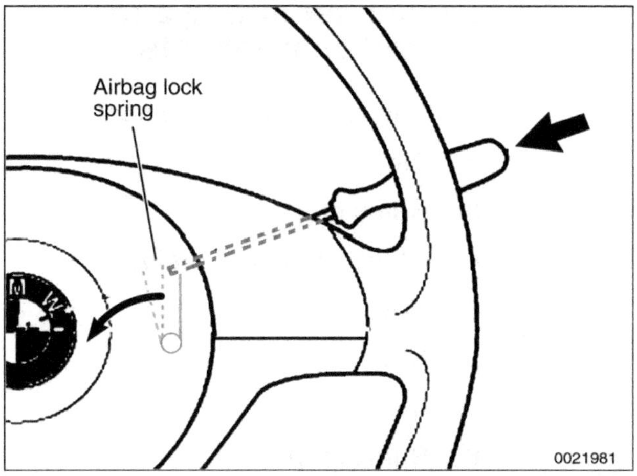


◀ Sport steering wheel:

- Insert screwdriver through opening in rear of steering wheel and push against spring tension to release airbag lock.
- Repeat on other side of steering wheel.

– Carefully lift airbag off front of steering wheel.

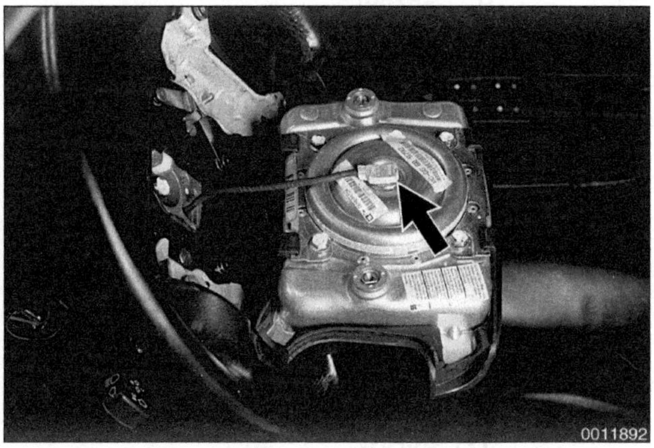


◀ Detach electrical harness connector (**arrow**).

***NOTE —***
*Early airbag connector is illustrated. "Smart"" airbag (two-stage) has two electrical connectors.*

> ***WARNING —***
> - *Store the removed airbag unit with the horn pad facing up. If stored facing down, accidental deployment could propel it violently into the air, causing injury.*
> - *Once the airbag is removed, do not drive the car.*
> - *Do not reconnect the battery with the airbag disconnected. A fault code will be stored, setting off the airbag warning light. Special tools are needed to reset the fault memory.*

– Installation is reverse of removal.

- Torque airbag using specification listed below.
- Sport steering wheel: Press airbag unit mounting pins into spring locks in steering wheel until they snap in firmly.

> ***CAUTION—***
> *Do not pinch airbag harness in center of steering wheel when installing airbag.*

| **Tightening torque** | |
|---|---|
| Airbag to steering wheel (M3) | 8 Nm (71 in-lb) |

– Reconnect airbag before reconnecting battery.

> ***WARNING —***
> *Once the airbag unit is installed and all other service procedures have been completed, start the engine and check that the airbag warning light goes out. If the warning light stays ON, the airbag system will not function as designed. Have the system diagnosed and repaired by an authorized BMW dealer.*

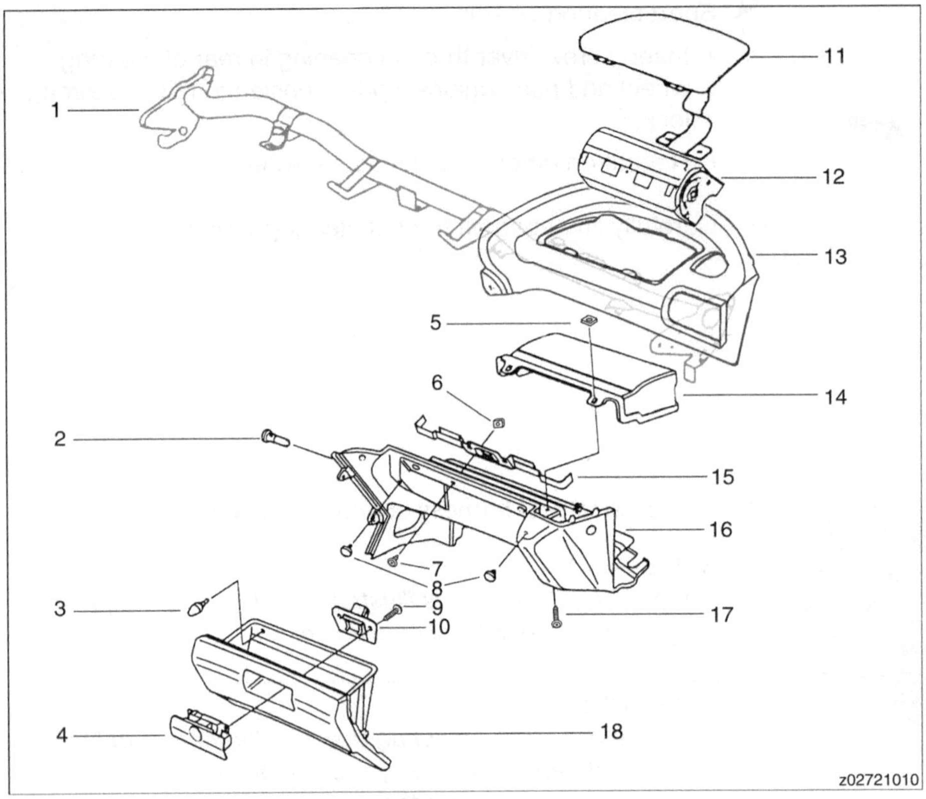


## Passenger airbag and related components

1. Dashboard reinforcement
2. Pin
3. Bump stop
4. Glove compartment lock
5. Nut
6. Nut
7. Screw
8. Bump stop
9. Screw
10. Glove compartment lock brace
11. Airbag cover with strap
12. Airbag
13. Right side dashboard
14. Top inside glove compartment cover
15. Glove compartment lock catch
16. Glove compartment frame
17. Screw
18. Glove compartment

## Passenger airbag, removing and installing

– Disconnect negative (–) battery cable and cover terminal with insulating material.

***NOTE—***
*On 1996 cars, the battery is located in the right rear of the engine compartment. On 1997 and later cars, the battery is located under the trunk floor mat.*

> ***CAUTION—***
> *Prior to disconnecting the battery, read the battery cautions in* **001 General Warnings and Cautions**.

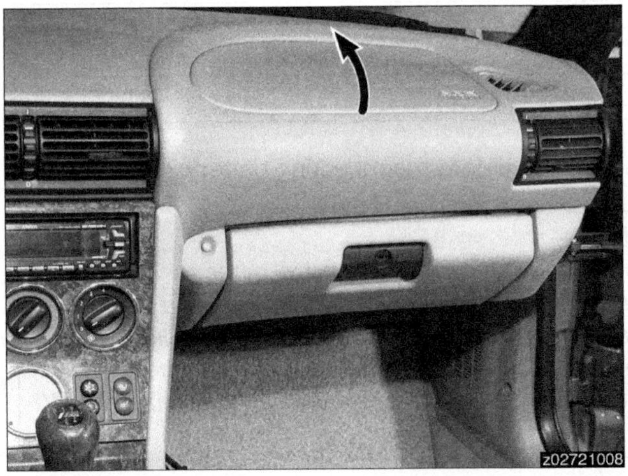


– Wait 10 seconds for residual airbag power to discharge.

◄ Using plastic prying tool or screwdriver with tip wrapped with tape, gently pry off passenger airbag cover (**arrow**).

– Remove trim under right side of dashboard. Remove glove compartment. See **513 Interior Trim**.

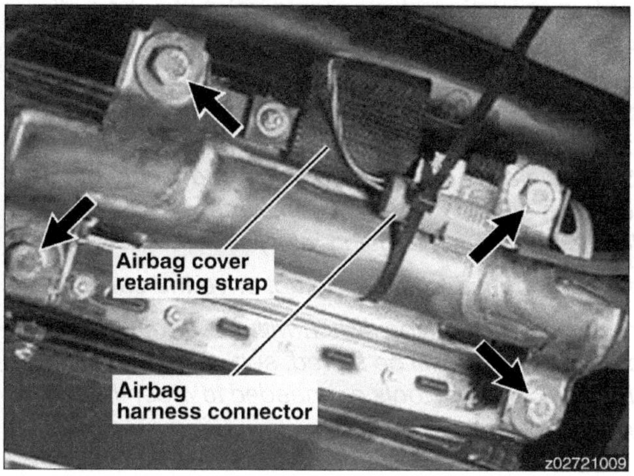


Working at airbag unit:

- Detach airbag harness connector.
- Remove airbag cover retaining strap mounting screws and detach cover.
- Remove airbag mounting bolts (**arrows**) at dashboard reinforcement and lift out airbag.

> ***WARNING —***
> - *Improper handling of the airbag could cause serious injury. Store the airbag with the outside pad facing up. If stored facing down, accidental deployment could propel it violently into the air, causing injury.*
> - *Once the airbag is removed, do not drive the car.*
> - *Do not reconnect the battery with the airbag disconnected. A fault code will be stored, setting off the airbag warning light. Special tools are needed to reset the fault memory.*

– Installation is reverse of removal.

> ***CAUTION—***
> - *If the airbag has been triggered, BMW recommends replacement of the airbag cover and strap.*
> - *Do not pinch the airbag harness when installing the unit in the dashboard.*

| Tightening torque | |
|---|---|
| Airbag to dashboard reinforcement (M8) | 24 Nm (18 ft-lb) |

– Reconnect airbag before reconnecting battery.

> ***WARNING —***
> *Once the airbag unit is installed and all other service procedures have been completed, start the engine and check that the airbag warning light goes out. If the warning light stays ON, the airbag system will not function as designed. Have the system diagnosed and repaired by an authorized BMW dealer.*

## Door airbag, removing and installing

– Make sure ignition switch is OFF. Remove key from ignition switch.

– Remove door trim panel. See **411 Doors**.

– Unplug airbag carefully. Avoid damaging electrical connector.

< Remove airbag mounting screws (**arrows**). Lift airbag off door and store in a safe manner.

> ***WARNING —***
> - *Improper handling of the airbag could cause serious injury. Store the airbag with the outside pad facing up. If stored facing down, accidental deployment could propel it violently into the air, causing injury.*
> - *Once the airbag is removed, do not drive the car.*
> - *Do not turn on the ignition switch with the airbag disconnected. A fault code will be stored, setting off the airbag warning light. Special tools are needed to reset the fault memory.*

– Installation is reverse of removal. Bear in mind:
- Reinstall airbag with new self-locking screws.
- Make sure airbag electrical connector snaps securely in place.

| Tightening torque | |
|---|---|
| Airbag to door (replace mounting screws) | 9 Nm (7 ft-lb) |

> ***WARNING —***
> *Once the airbag unit is installed and all other service procedures have been completed, start the engine and check that the airbag warning light goes out. If the warning light stays ON, the airbag system will not function as designed. Have the system diagnosed and repaired by an authorized BMW dealer.*

# ELE Electrical Wiring Diagrams



**Electrical Component Locations**

Relay and fuse positions, component and ground locations . . . . . . . . . . . . . . . . . . . . . . . . . see **Repair Group 610**

## GENERAL

The wiring schematics given in this section represent detailed electrical circuit information for BMW Z3 and M version cars for 1996, 1998, 2000 and 2002. In most cases 1997 cars are close to or identical to 1998 cars; similarly for 1999 / 2000 models and 2001 / 2002 models.

Each diagram shows the wiring, connectors, terminals, and electrical or electronic components of the circuit. It also identifies the wires by color and size.

Wiring diagrams are necessary to troubleshoot and repair electrical and electronic circuits. Remember that circuits often have more than one source of power and/or ground. In many cases there is a switched ground. Take time to study the schematics of the entire system to understand the circuit logic prior to circuit troubleshooting.

***NOTE—***
- *Wiring schematics and additional electrical information are also available at the BMW subscription site:* **http://www.bmwtechinfo.com.**
- *Common BMW acronyms are tabulated in* **600 Electrical System—General**.
- *Airbags systems, known by the acronym SRS (supplemental restraint system), are also referred to by the BMW acronym MRS (multiple restraint system).*
- *Every attempt has been made in this repair manual to standardize component names. However, in some cases, the same component may appear with different names. Regardless of the component name, the alphanumeric component designation used in the wiring diagrams is consistent throughout.*

## Special tools

Automotive digital multimeter

> ***CAUTION—***
> *Only use a digital multimeter when testing automotive electrical components.*

61 1 150

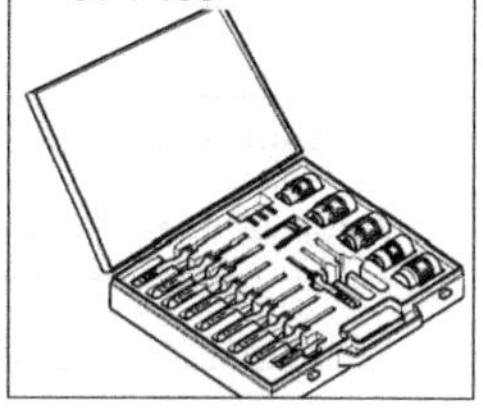

Wiring harness end repair tools
(BMW special tool BMW 61 1 150)

61 9 041

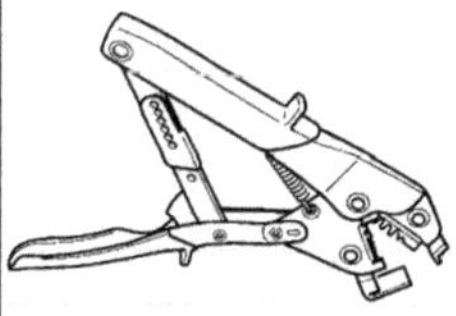

Wire end crimp tool
(BMW special tool BMW 61 9 041)

## Circuit and terminal descriptions

BMW identifies electrical circuits, including junctions and grounds, with unique designations which follow the German DIN standard. See **Table a**. For example, if a relay terminal is labeled '30', it tells you that positive (+) voltage is supplied to that terminal at all times directly from the battery.

Electrical components are identified in the schematics using a letter followed by a number. For example, A6000 is the DME control module. The letter A identifies the component as an electronic control module. **Table b** lists the letter code for common components.

| Table a. Electrical terminal designations (according to DIN 72 552 standard) | |
|---|---|
| **Number** | **Circuit description** |
| 1 | Low voltage switched coil terminal |
| 4 | High voltage center coil terminal |
| 15 | Originates at ignition switch. Supplies power when ignition switch is in RUN or START position. |
| 30 | Battery positive (+) voltage. Supplies power whenever battery is connected. (Not dependent on ignition switch position, unfused) |
| 31 | Ground, battery negative (-) terminal |
| 50 | Supplies power from battery to starter solenoid when ignition switch is in START position only. |
| +54 | Originates at ignition switch. Supplies power when ignition switch is in RUN position only. |
| 85 | Ground side (-) of relay coil |
| 86 | Power-in side (+) of relay coil |
| 87 | Power out when the relay coil is energized (usually terminal 87 connects to terminal 30 power when relay is energized) |
| B+ | Battery positive connection |
| D+ | Alternator warning light and field energizing circuit |

| Table b. BMW electrical component designation | | |
|---|---|---|
| **Letter** | **Function** | **Example** |
| A | Electronic control module | A6000 (DME engine control module) |
| B | Sensor, transducer | B6238 (oil temperature sensor) |
| E | Light, electric heater | E11 (left high beam) |
| F | Fuse | F3 (fuse 3 in fuse panel) |
| G | Power supply | G6509 (alternator), G1 (battery) |
| H | Warning light, signal indicator | H10 (chime module) |
| K | Relay | K6300 (DME relay) |
| M | Electric motor, actuator | M6510 (starter) |
| P | Fuse panel | P90 (front power distribution box) |
| S | Switch, coding plug | S235 (clutch pedal switch) |
| T | Ignition coil | T6151 (cylinder 1 ignition coil) |
| U | Radio / interference suppression | U4 (telephone) |
| X | Connector; splice connector; ground | X165 (right front engine compartment ground) |
| Y | Electromechanical component | Y6101 (cylinder 1 fuel injector) |
| Z | Interference suppression coil | Z400 (noise suppression coil) |

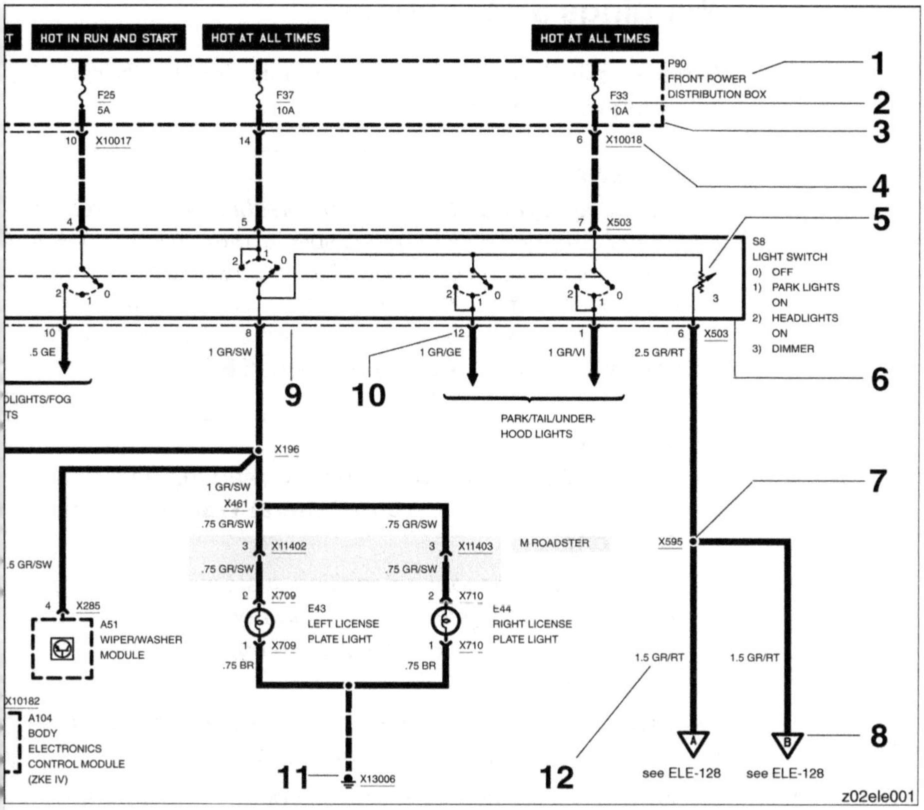


## Sample wiring schematic

1. **Alphanumeric designation and component name.**
2. **Fuse position and rating**
3. **Dashed border: Complete component is not in this diagram**
4. **Connector**
5. **Internal component**
6. **Solid border: Complete component in this diagram**
7. **Splice or connector**
8. **Circuit continues in another diagram**
9. **Dashed line: multipin connector**
10. **Connector pln number**
    - Example: Pin 12 in connector X503
11. **Ground**
12. **Wire gauge and color**
    - Example: 1.5 mm$^2$, grey-red

| Wiring color code | |
|---|---|
| BL | = blue |
| BR | = brown |
| GE | = yellow |
| GN | = green |
| GR | = grey |
| RT | = red |
| SW | = black |
| WS | = white |
| VI | = violet |

z02ele003

## Wire color and size

Wire insulation colors in this section are given with German color abbreviations.

Wire sizes follow the DIN (European) convention.

***NOTE*** —
*For example, an 0.5 wire is ½ mm$^2$ in cross-section area. This corresponds to approx. SAE 16 gauge wire.*

## Warnings and cautions

When working on electrical or electronic circuits, observe the following warnings and cautions.

> ***WARNING —***
> *On cars equipped with airbags, pyrotechnic seatbelt retractors and battery safety terminal, special precautions apply to any electrical testing or repair. These components are explosive devices and must be handled with extreme care. Before starting any work, refer to the warnings and cautions in* **720 Seat Belts** *and* **721 Airbag System (SRS)**.

> ***CAUTION—***
> - *Prior to disconnecting the battery, read the battery disconnection cautions in* **001 General Warnings and Cautions.**
> - *Connect and disconnect ignition system wires, multiple connectors and ignition test equipment leads only while ignition is switched off. Keep clothing, hands, and feet dry if possible.*
> - *Switch a test meter to the appropriate function and range* before *making test connections.*
> - *Switch the ignition OFF and disconnect the negative (–) battery cable* before *removing any electrical components.*

## Data link connections (DLC / OBD II), 1996

(1 of 5)

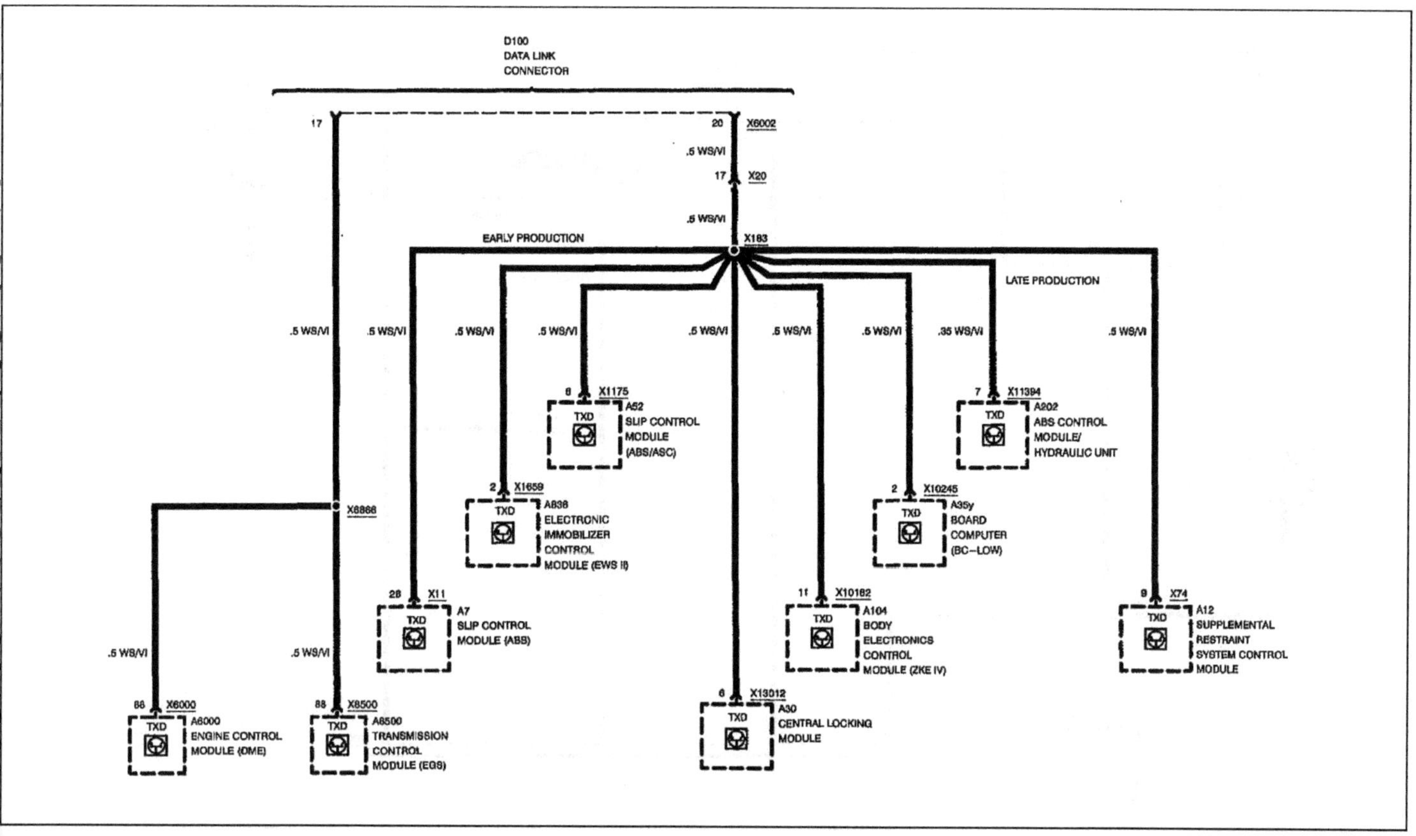


## Data link connections (DLC / OBD II), 1996

(2 of 5)

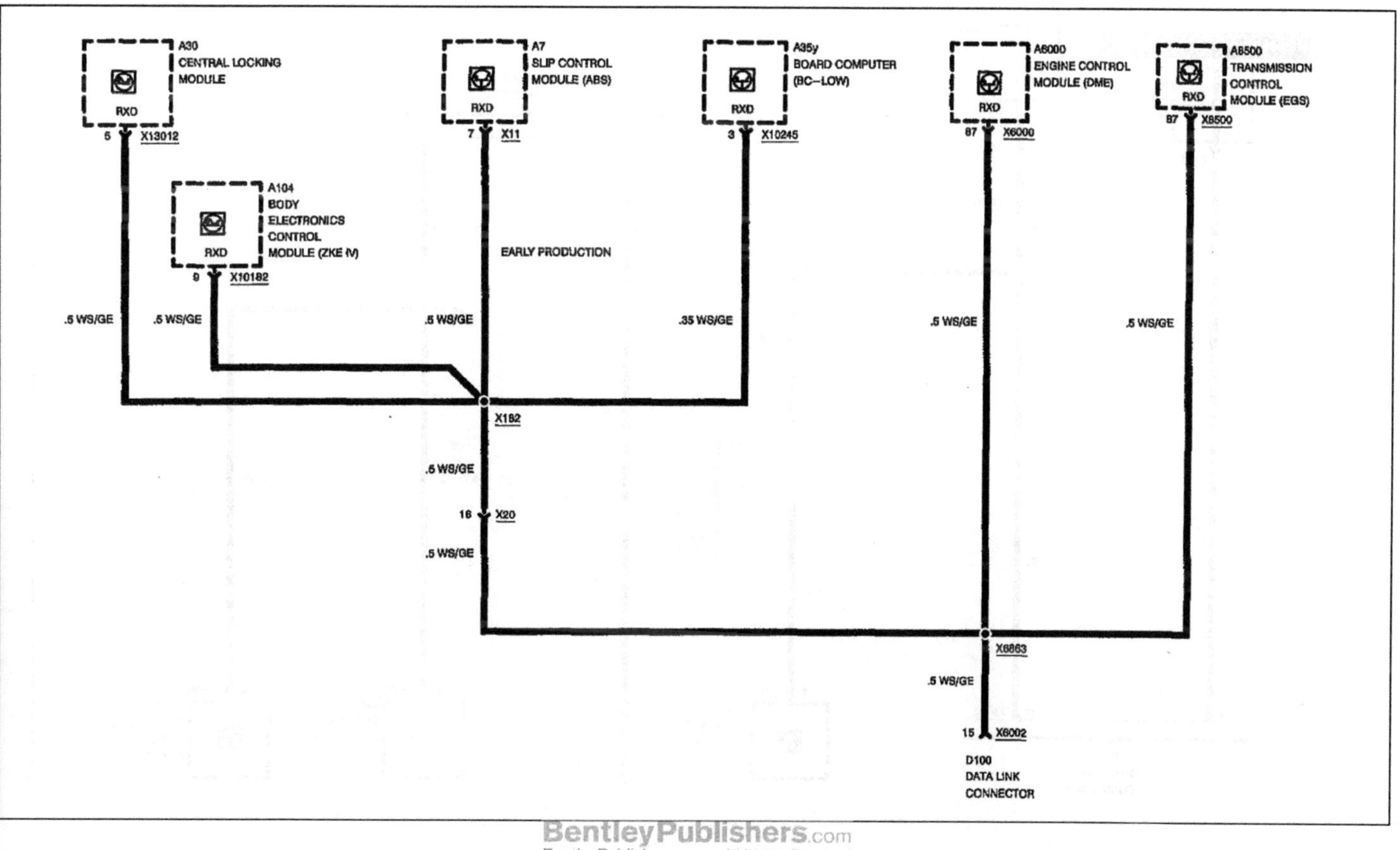

## Data link connections (DLC / OBD II), 1996
(3 of 5)

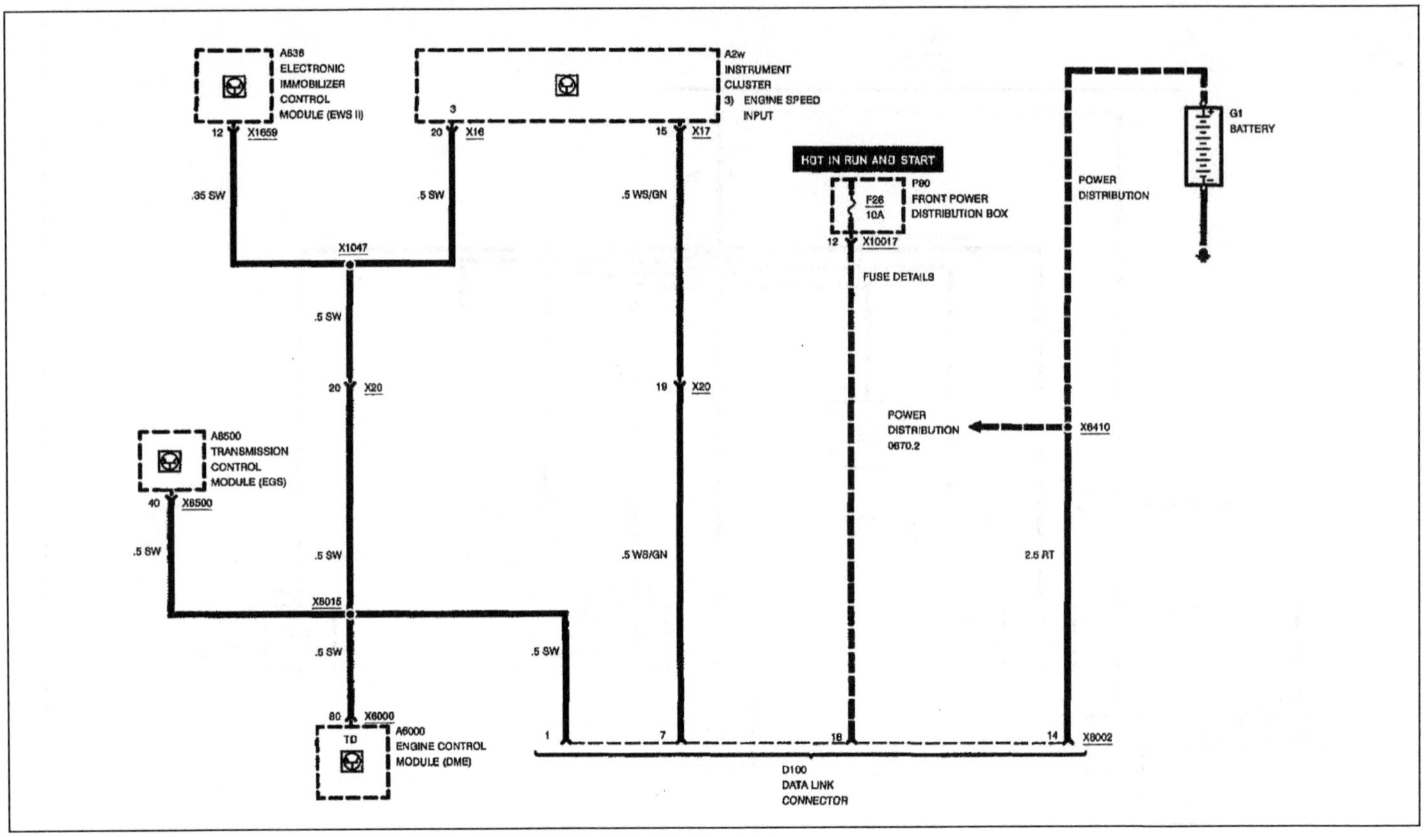


## Data link connections (DLC / OBD II), 1996
(4 of 5)

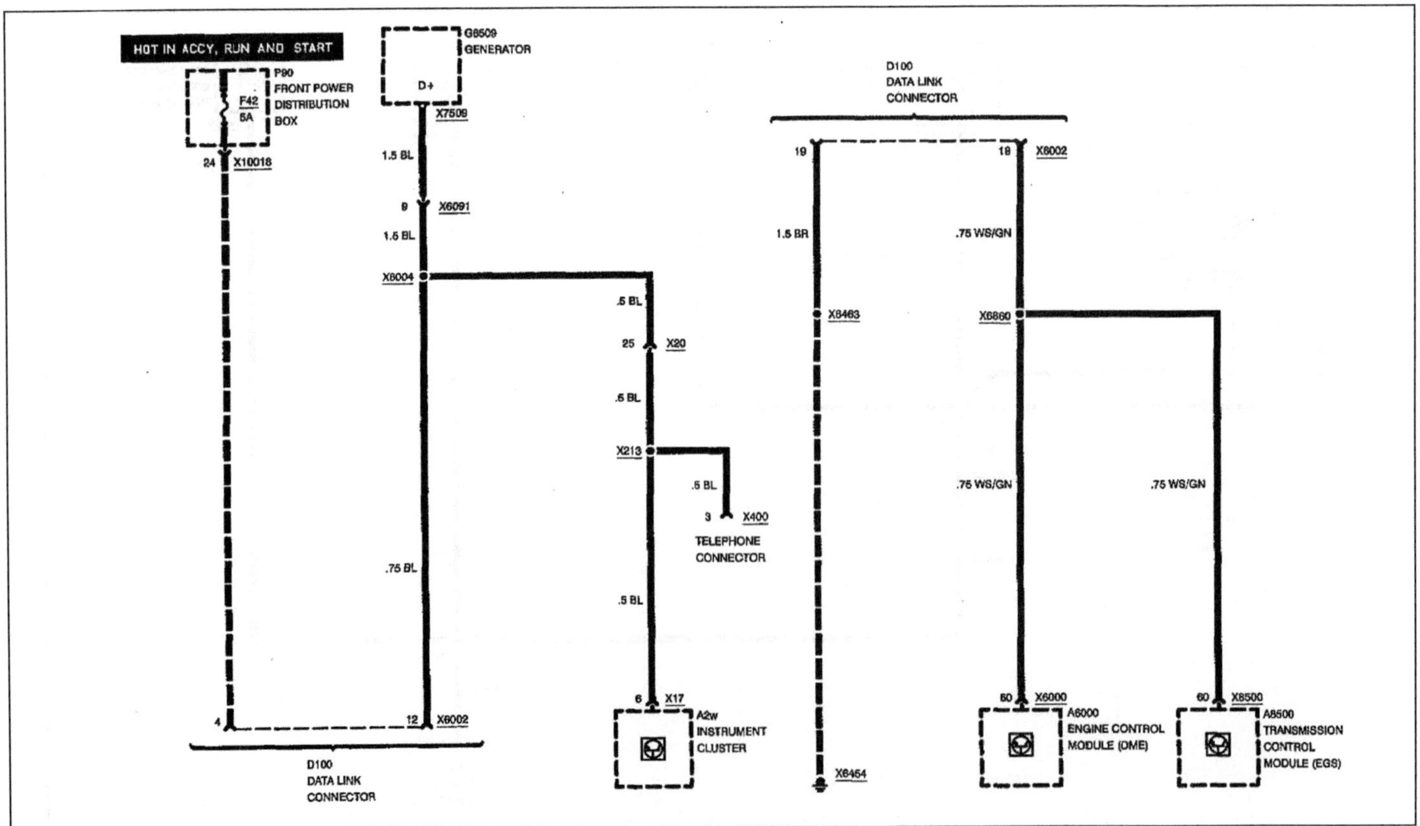

## Data link connections (DLC / OBD II), 1996

(5 of 5)

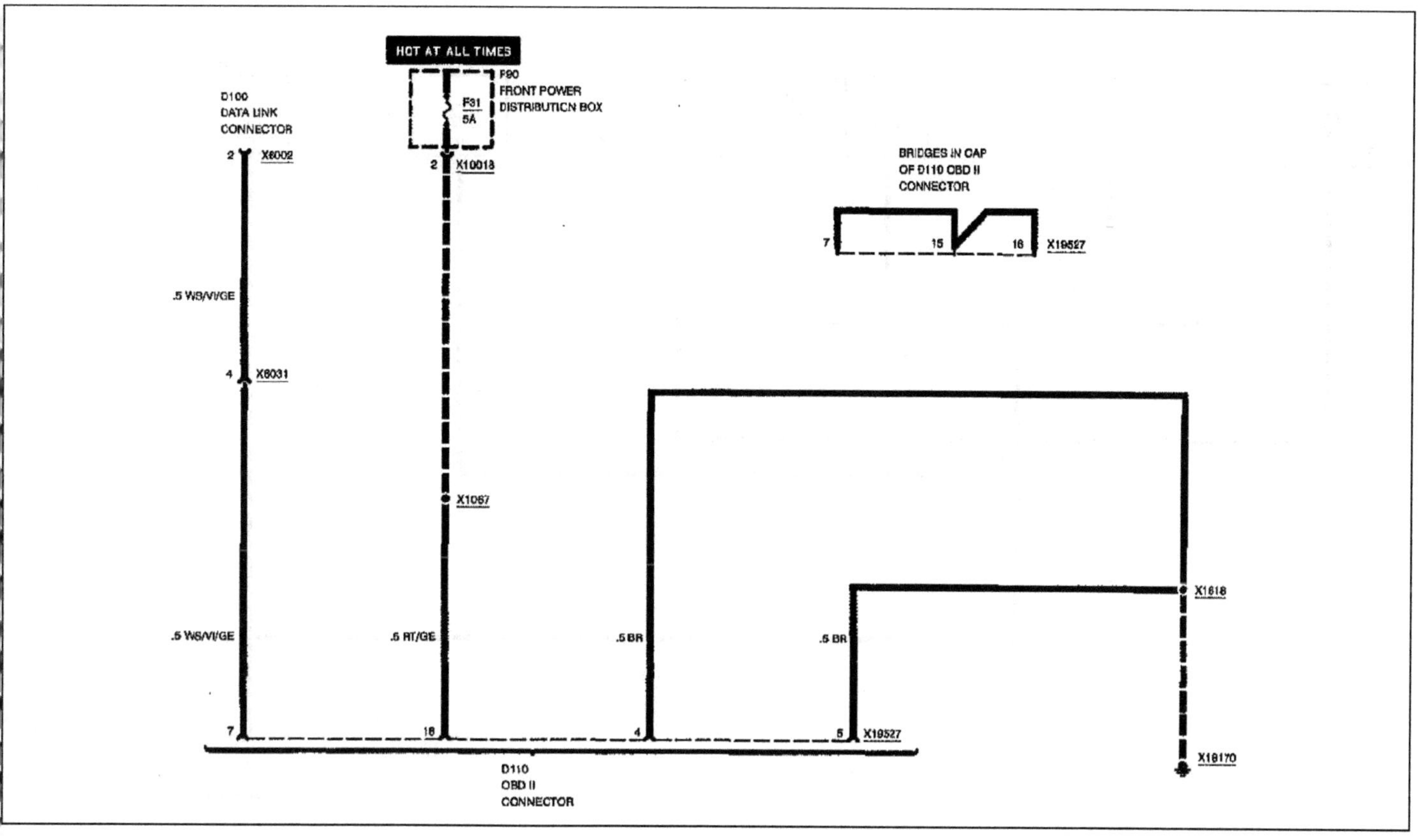


## Data link connections (DLC / OBD II), 1998

(1 of 5)

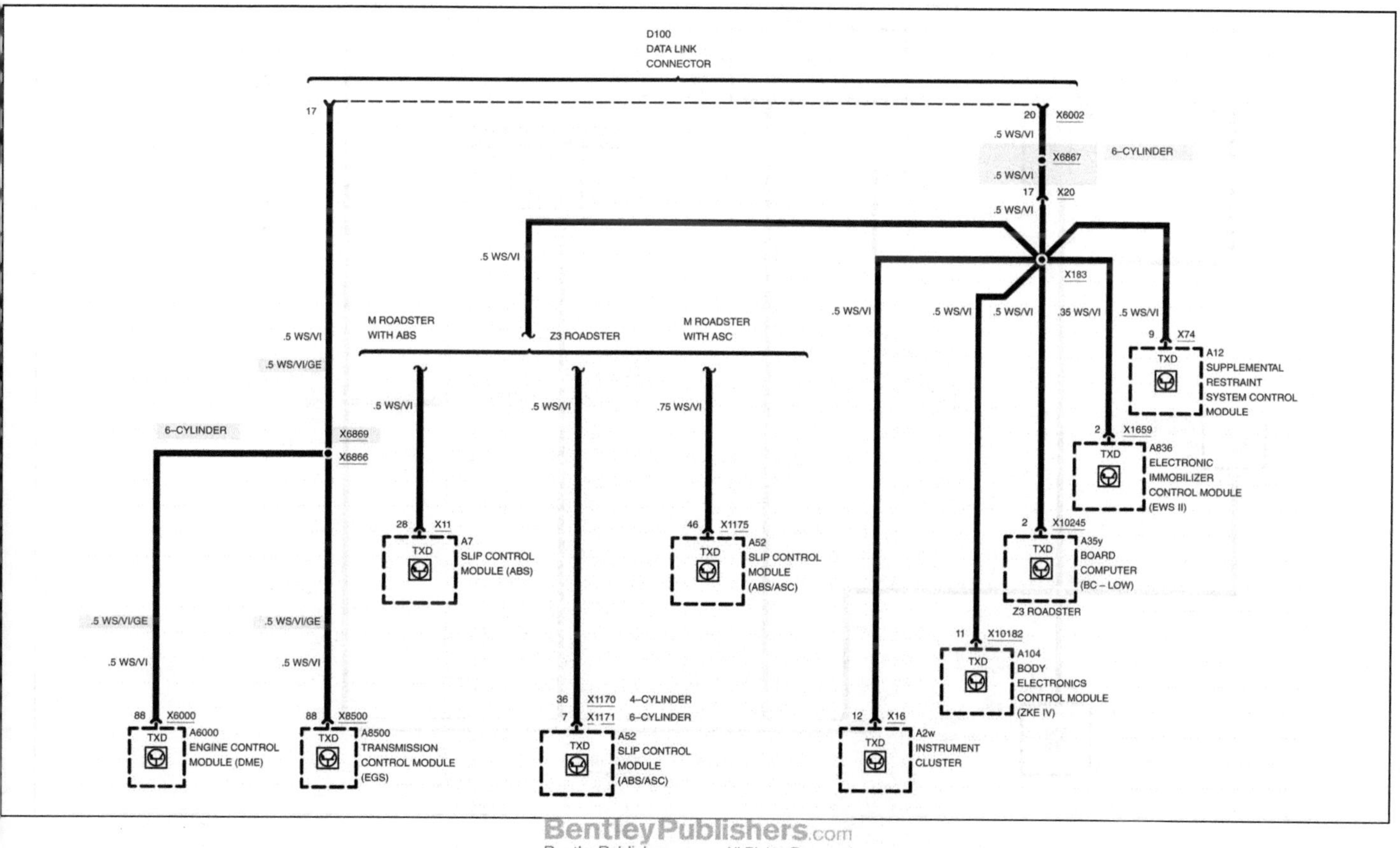

## Data link connections (DLC / OBD II), 1998
(2 of 5)

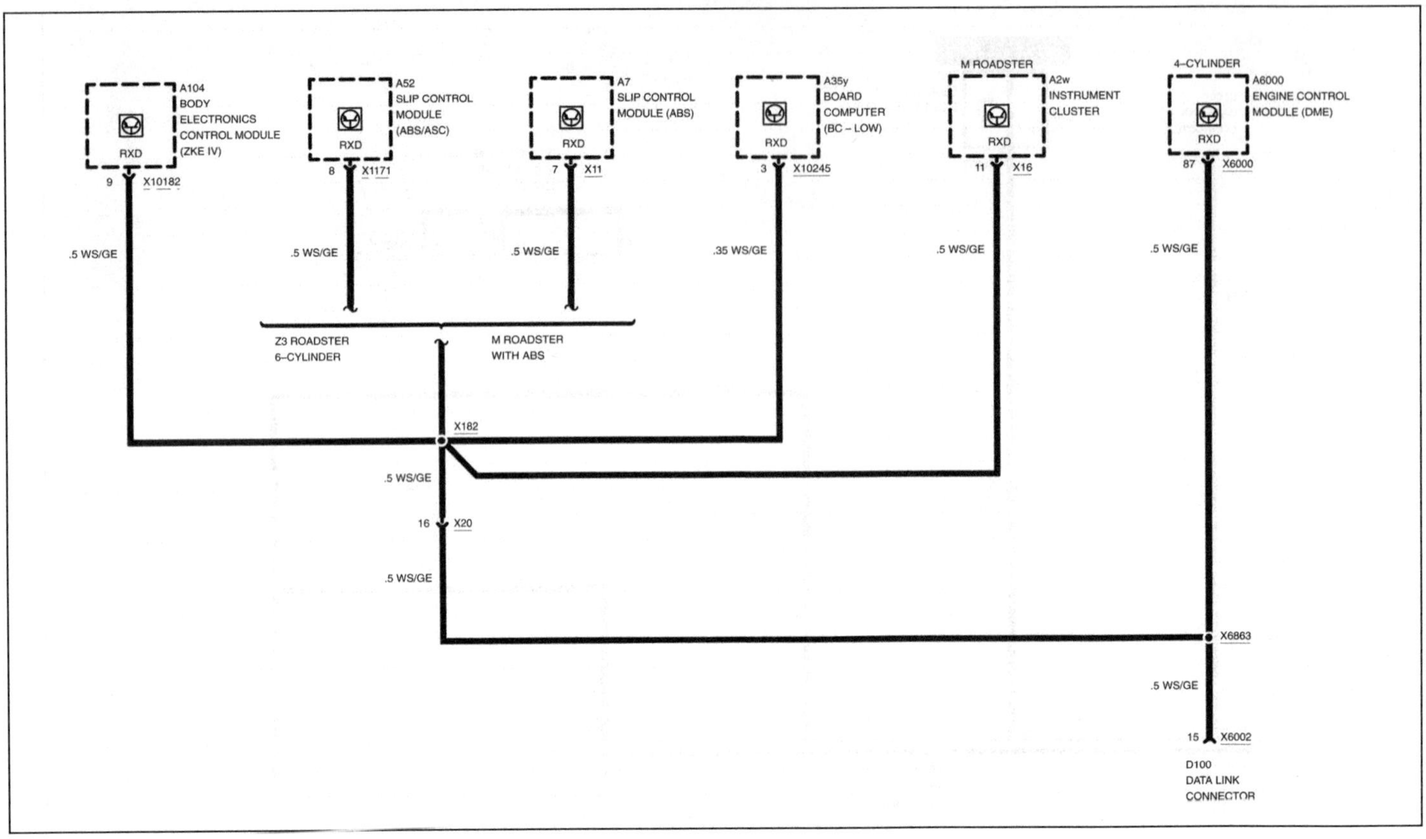


## Data link connections (DLC / OBD II), 1998
(3 of 5)

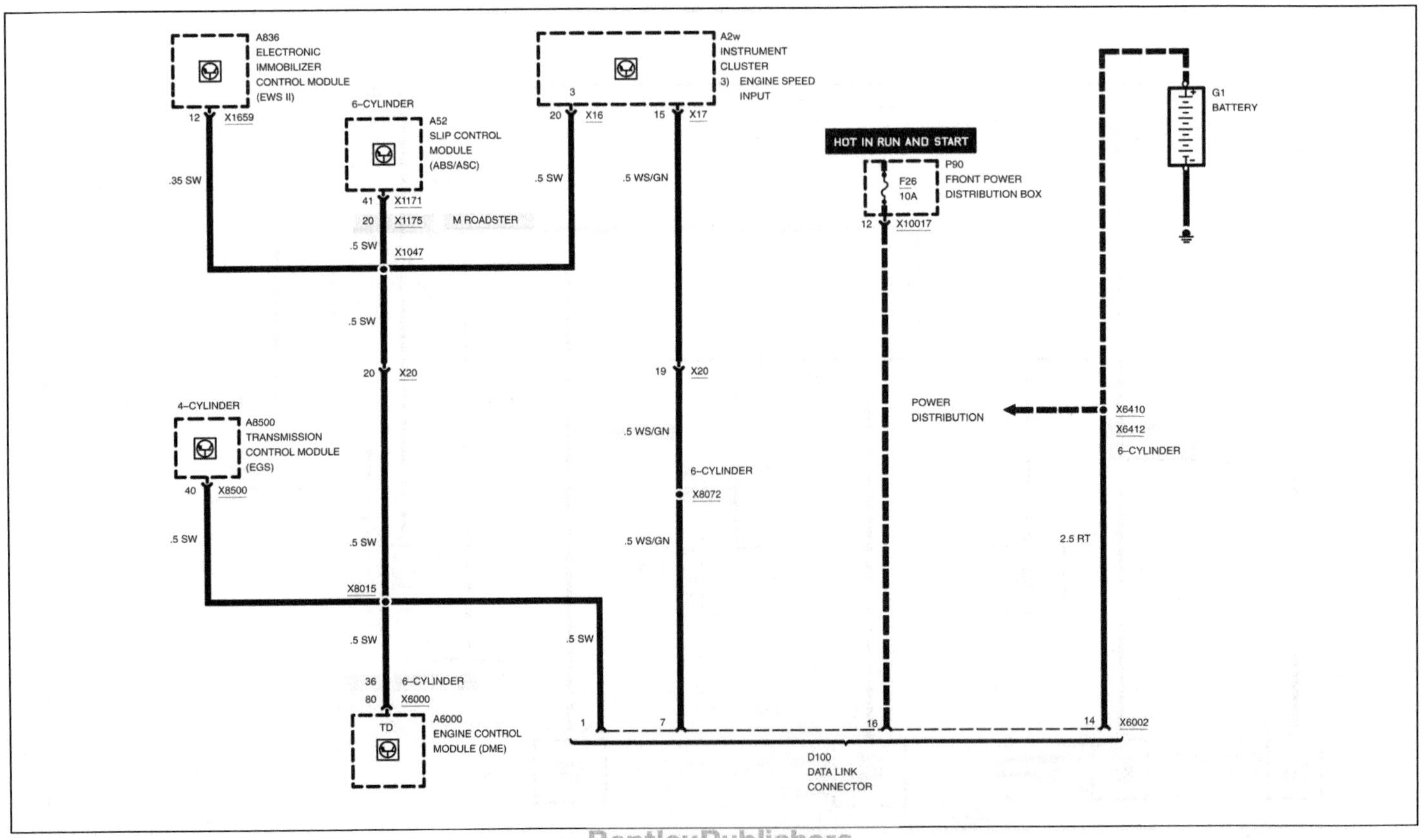

## Data link connections (DLC / OBD II), 1998
(4 of 5)

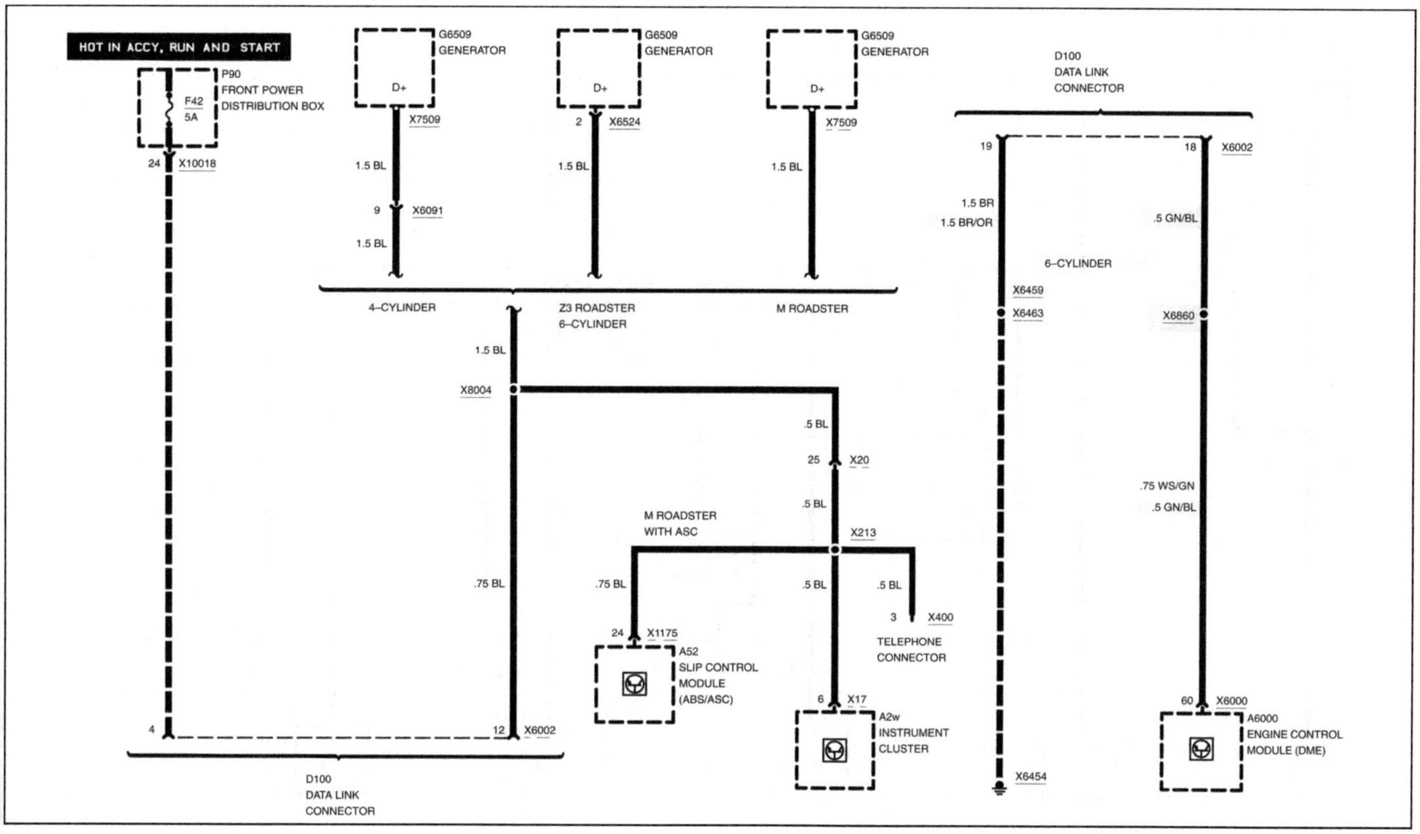


## Data link connections (DLC / OBD II), 1998
(5 of 5)

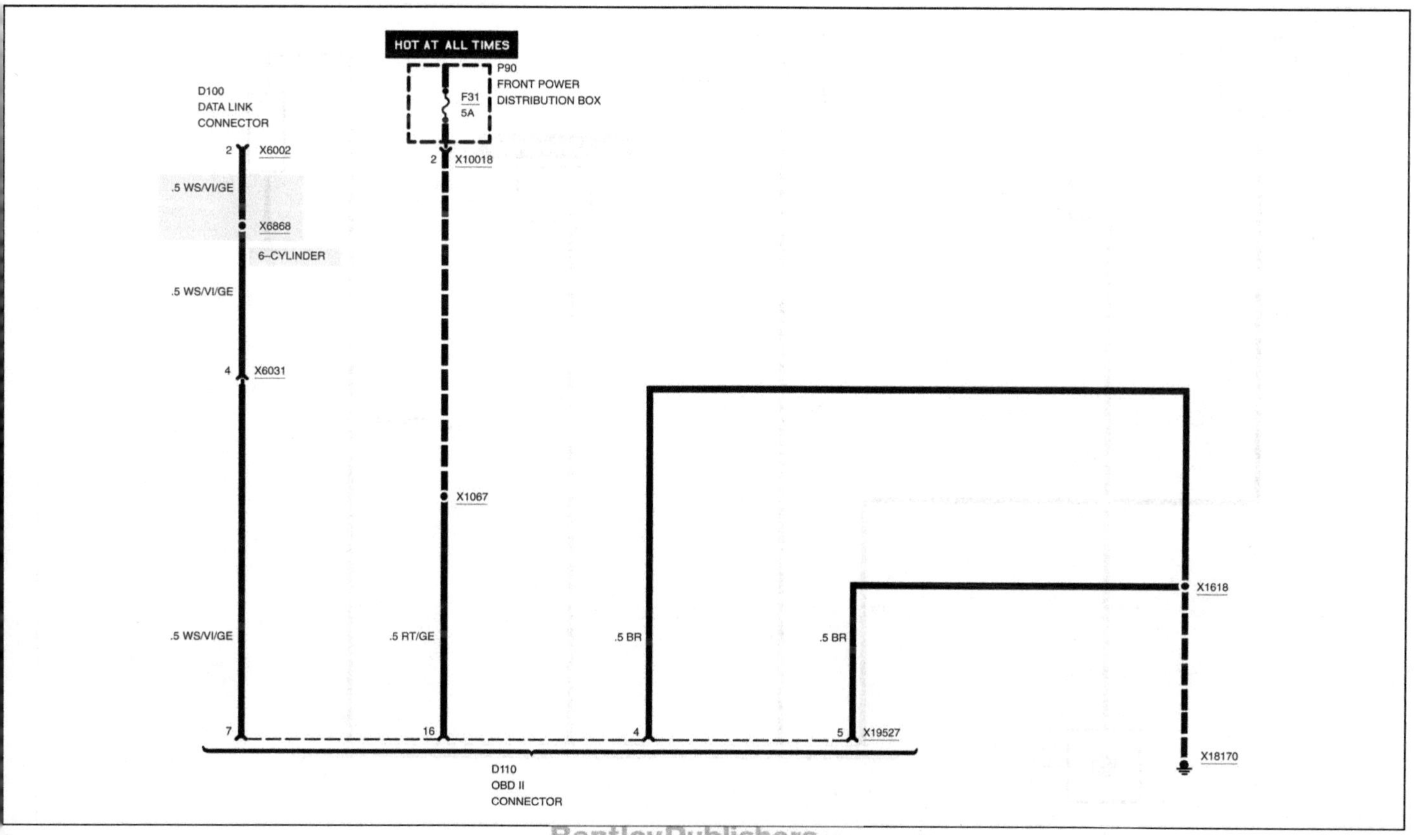

## Data link connections (DLC / OBD II), 2000
(1 of 6)

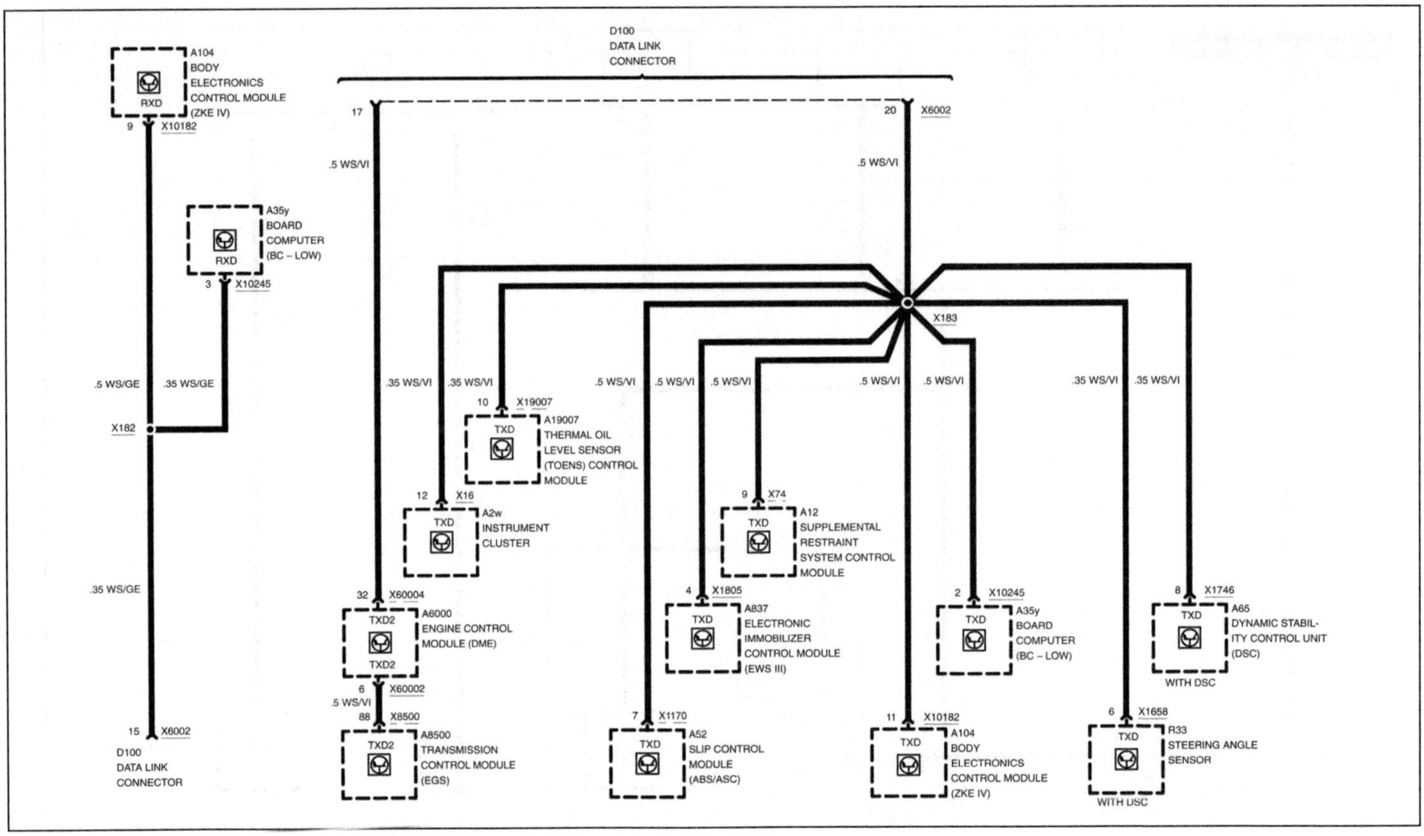


## Data link connections (DLC / OBD II), 2000
(2 of 6)

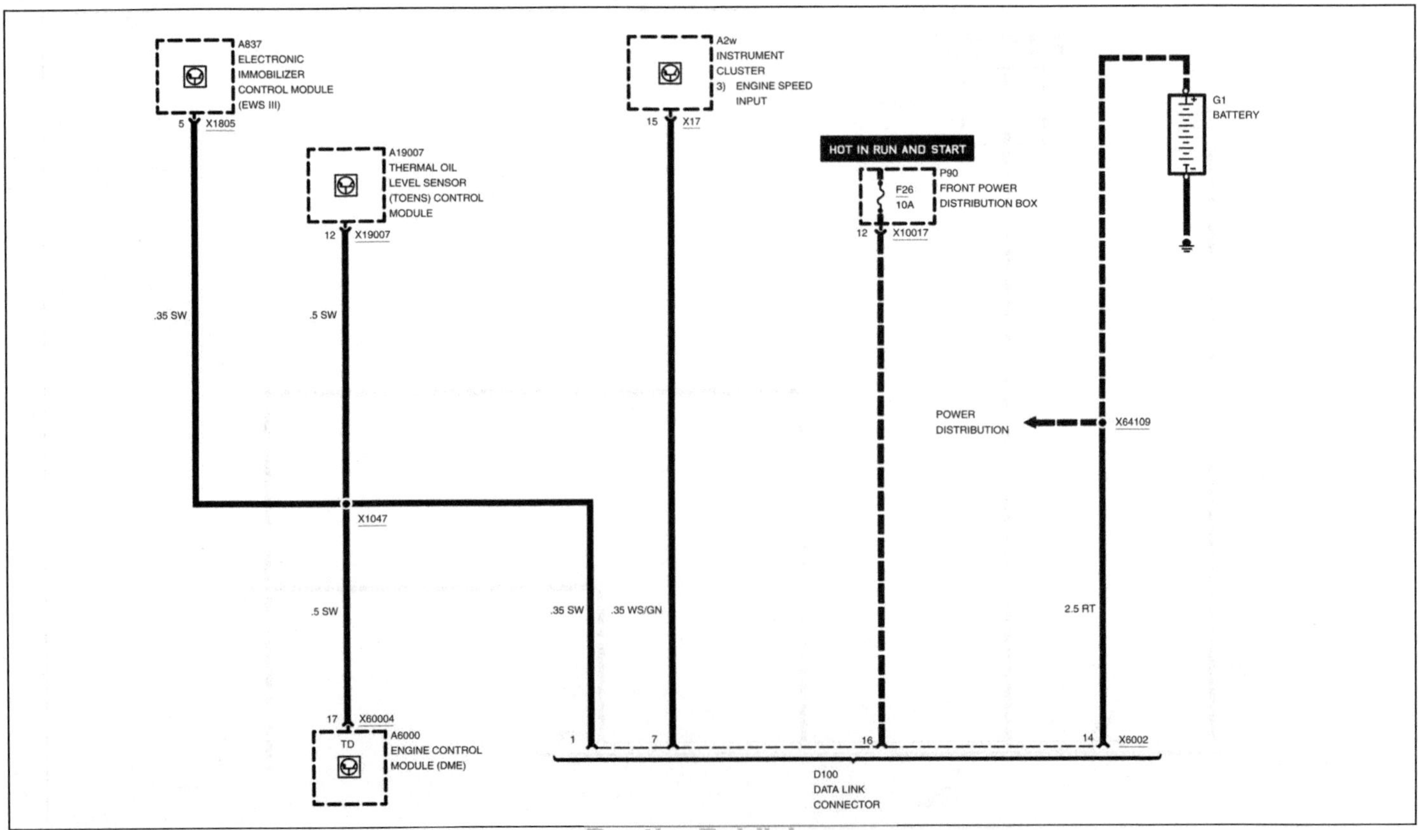

## Data link connections (DLC / OBD II), 2000
(3 of 6)

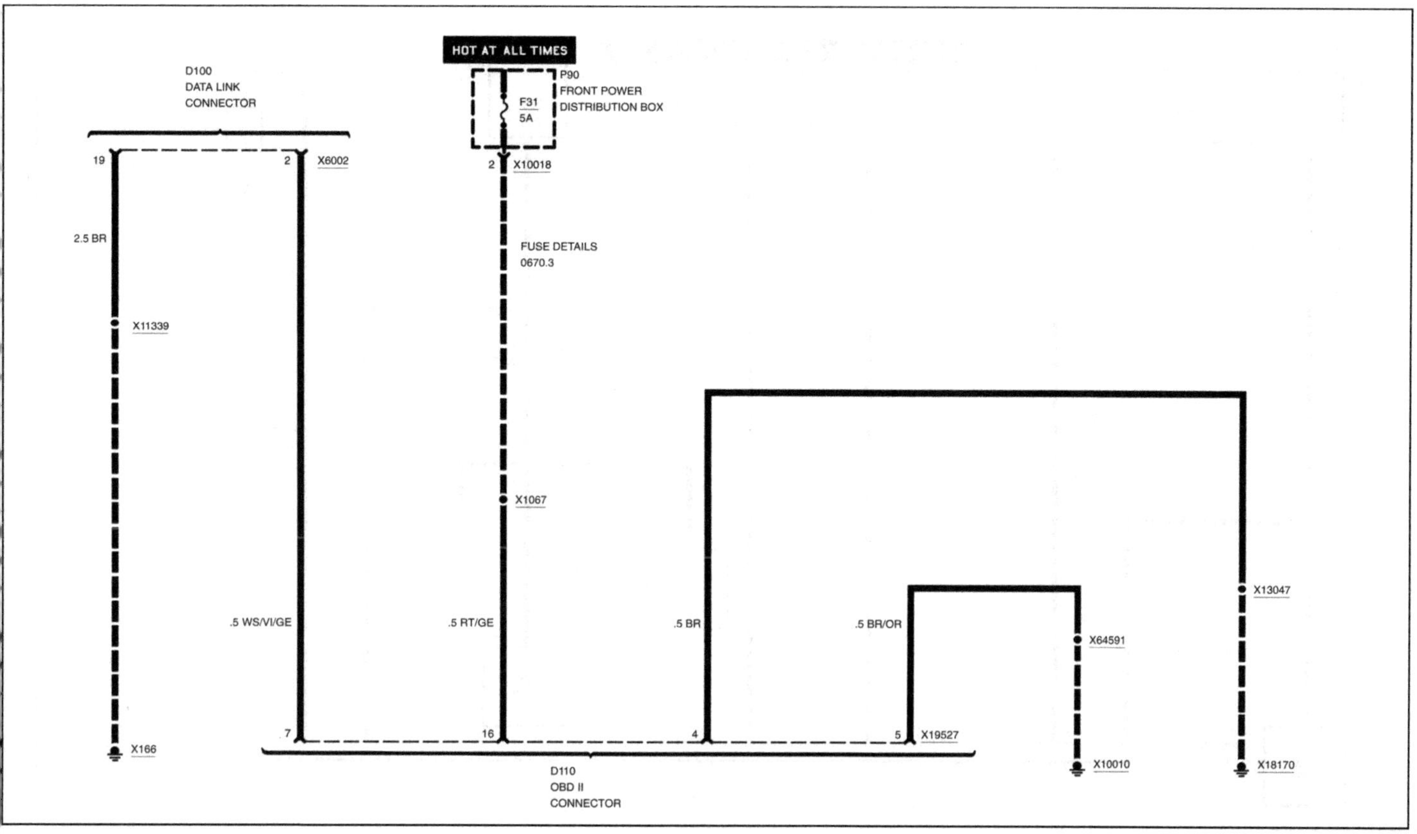


## Data link connections (DLC / OBD II), 2000
(4 of 6)

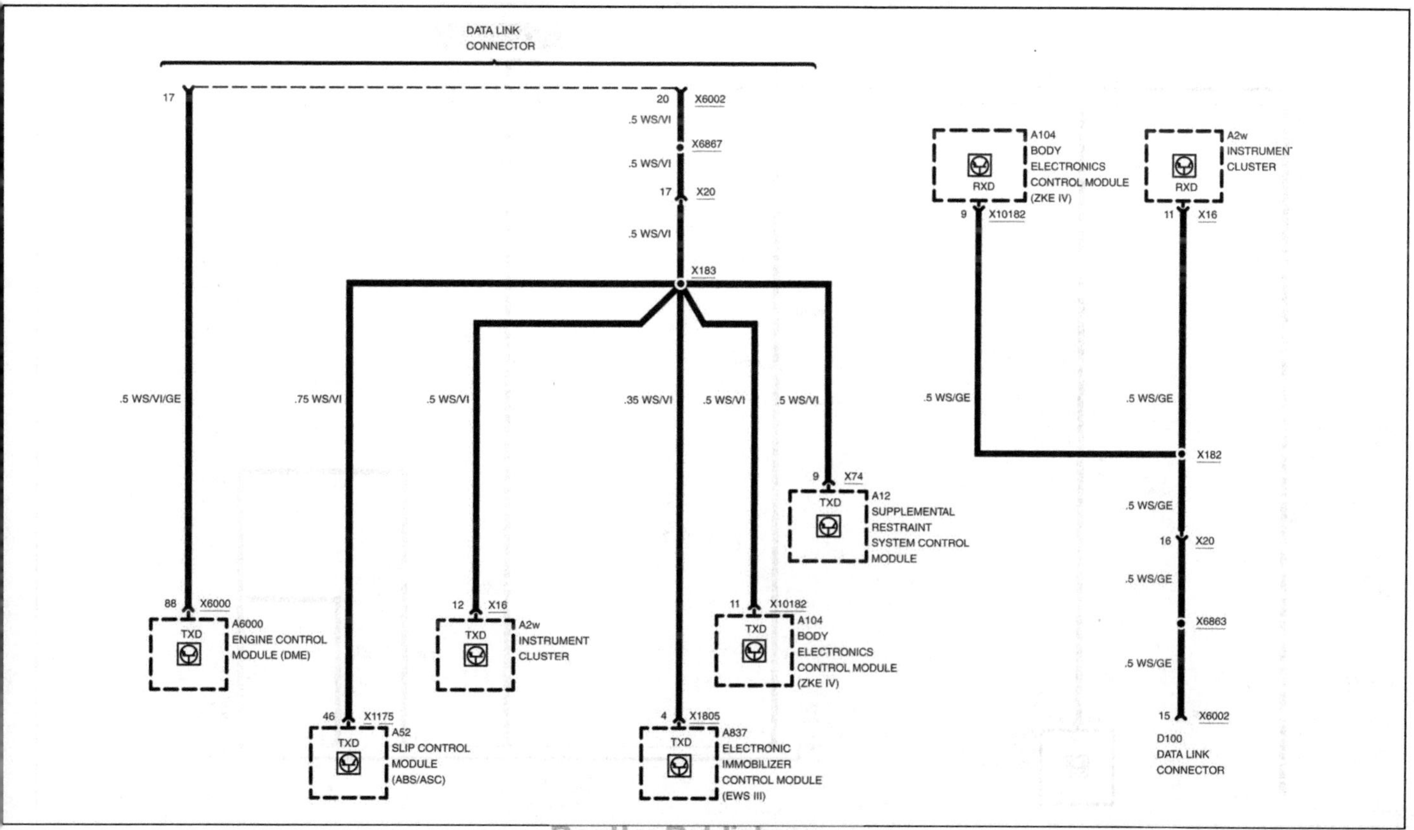

## Data link connections (DLC / OBD II), 2000
(5 of 6)

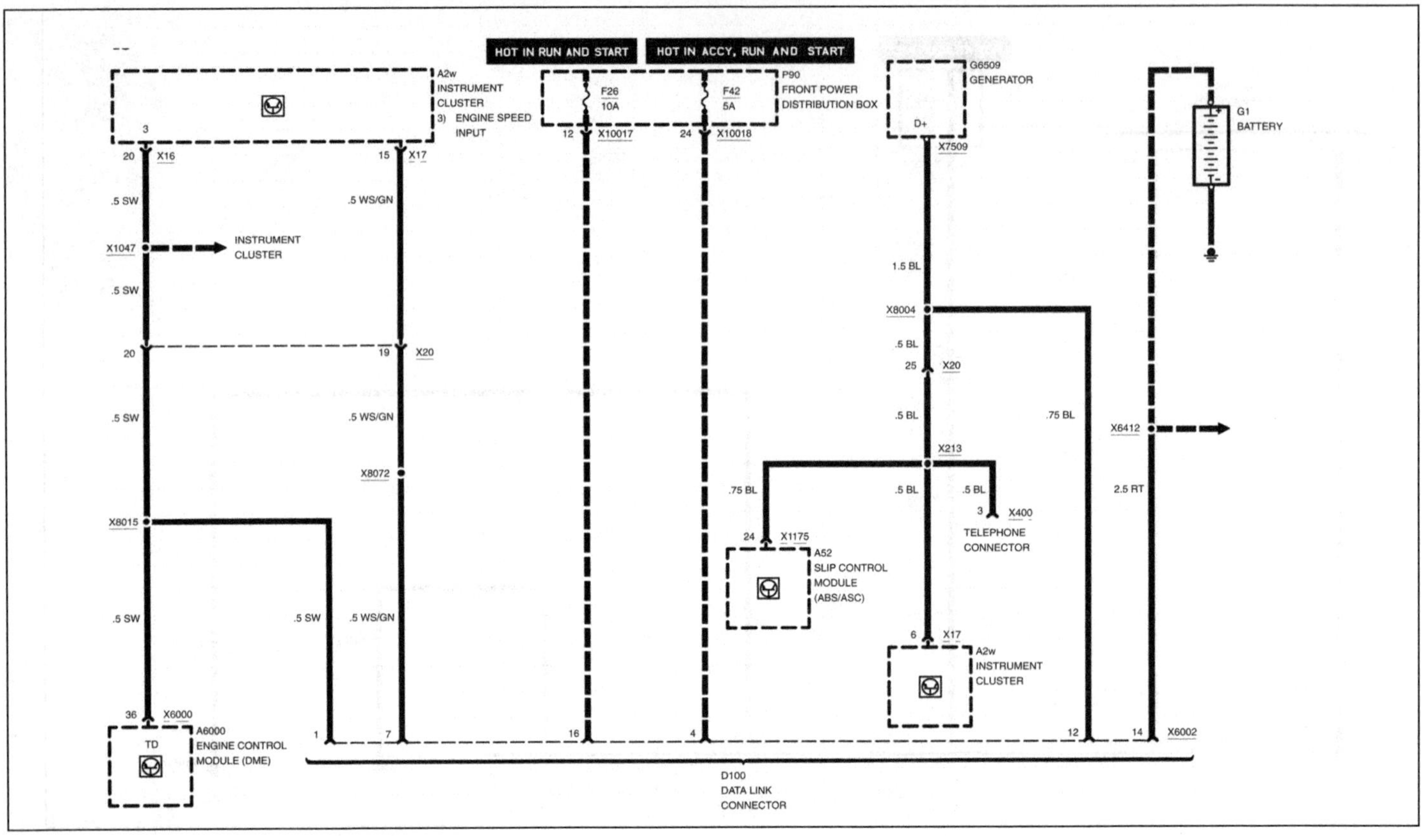


## Data link connections (DLC / OBD II), 2000
(6 of 6)

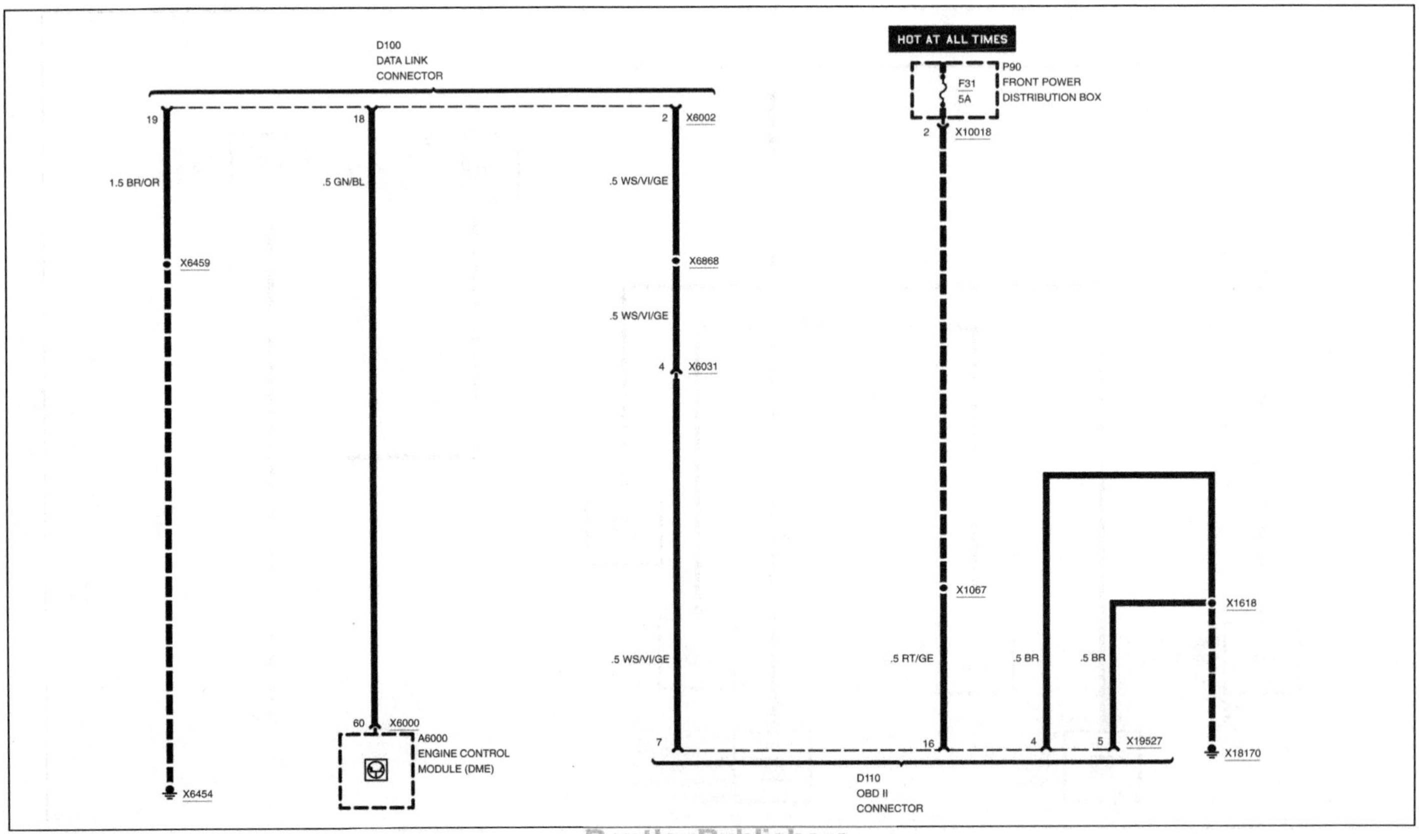

## Data link connections (DLC / OBD II), 2002
(1 of 3)

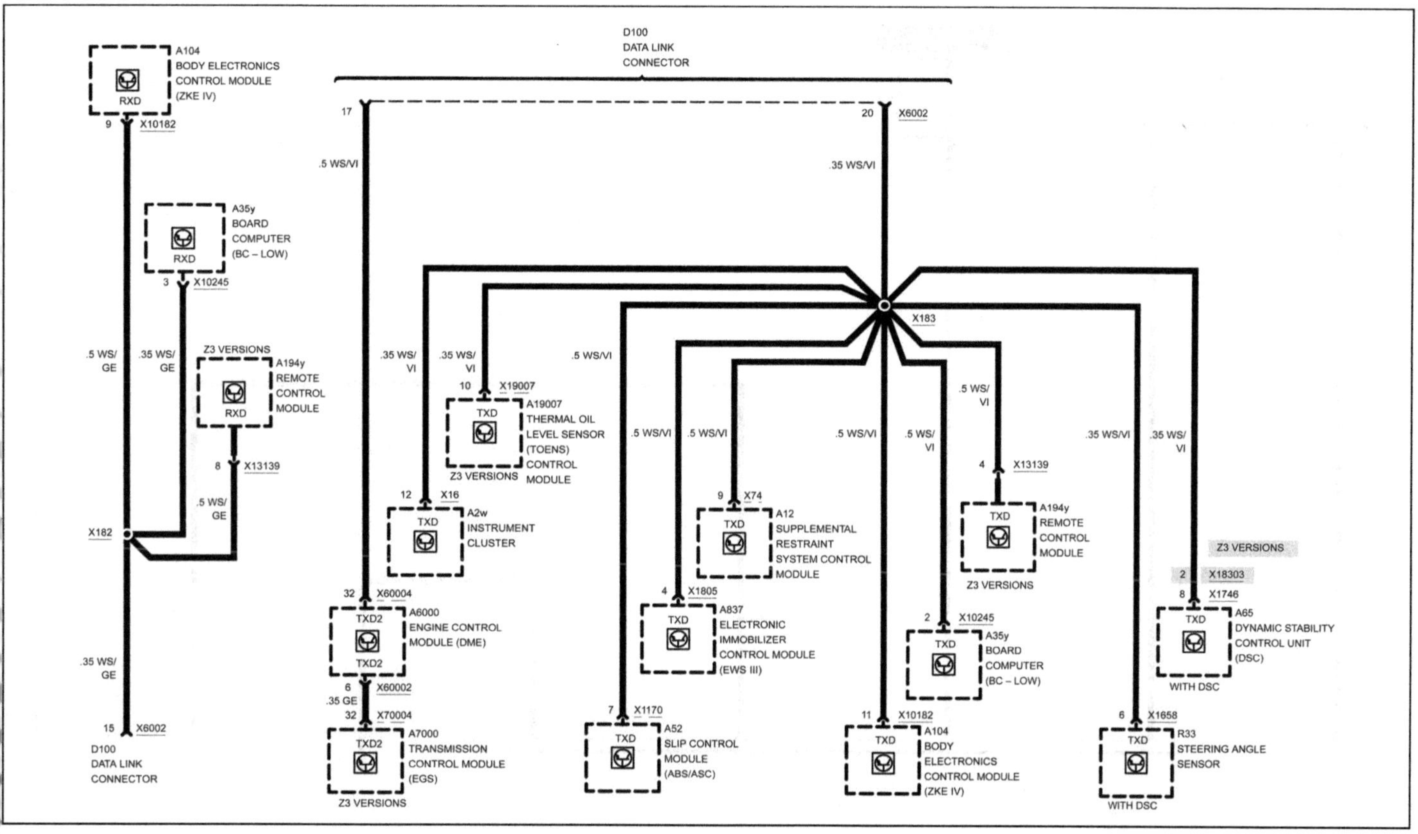


## Data link connections (DLC / OBD II), 2002
(2 of 3)

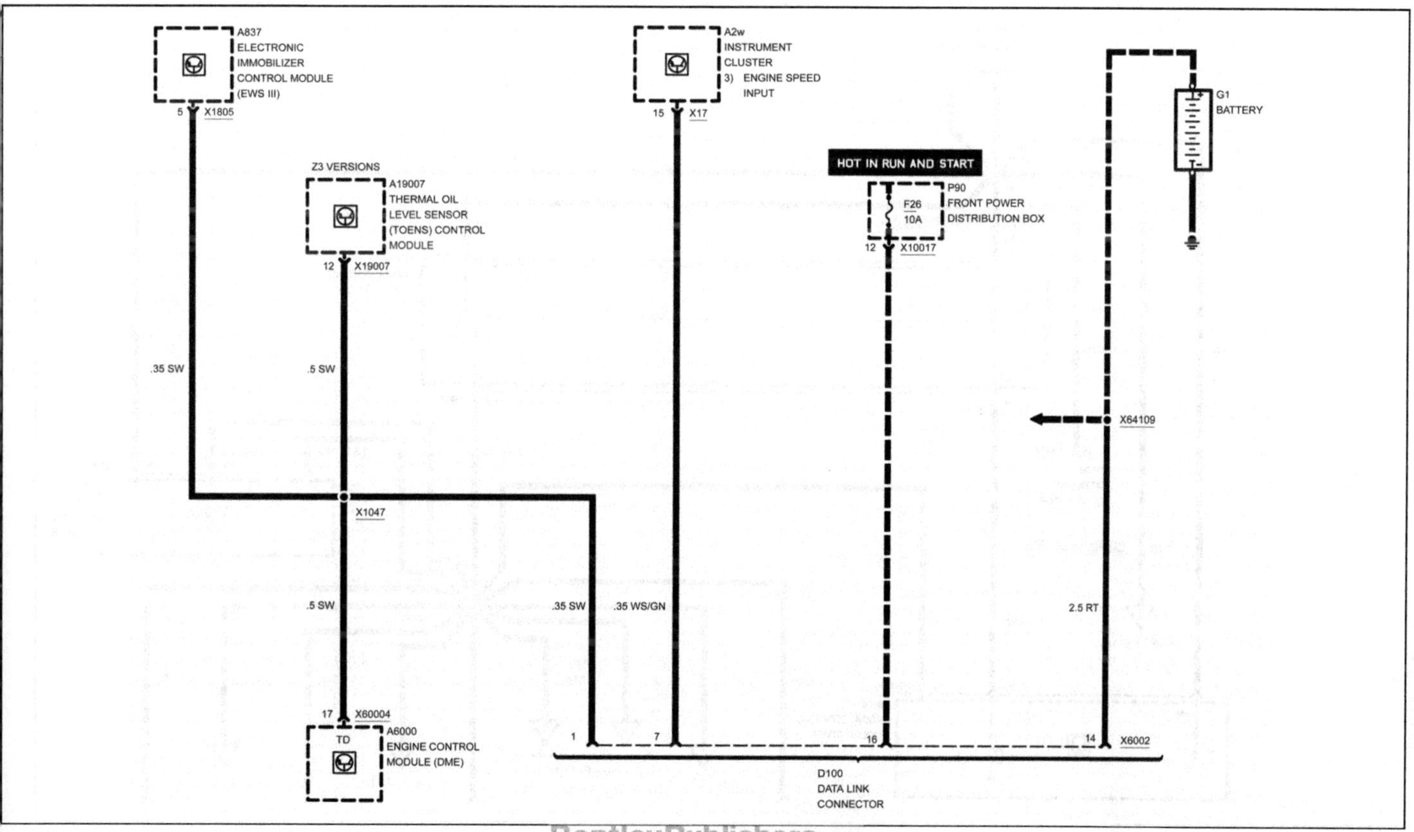

## Data link connections (DLC / OBD II), 2002
(3 of 3)

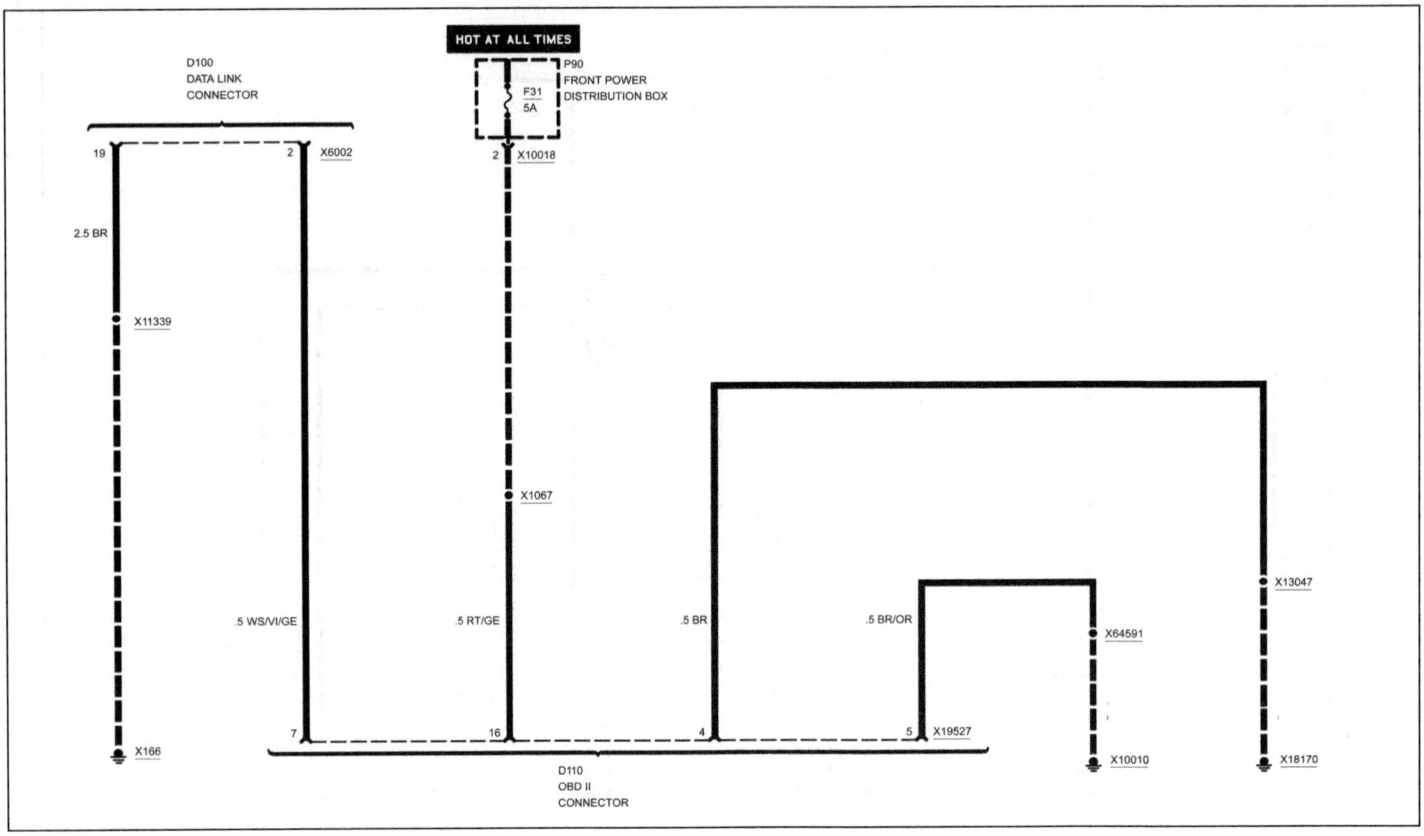


## Bosch DME M5.2 (M44 engine)
(1 of 8)

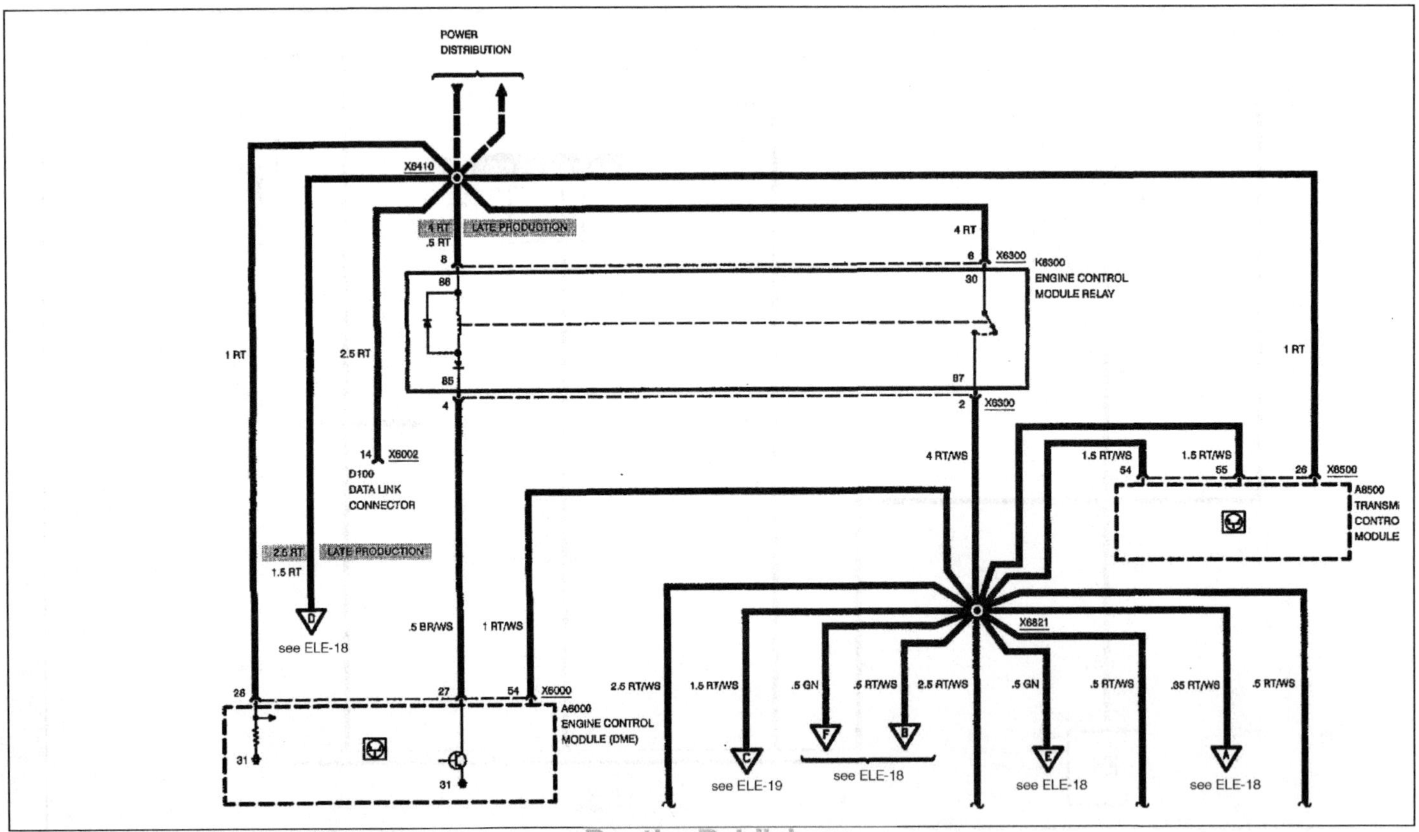

## Bosch DME M5.2 (M44 engine)

(2 of 8)

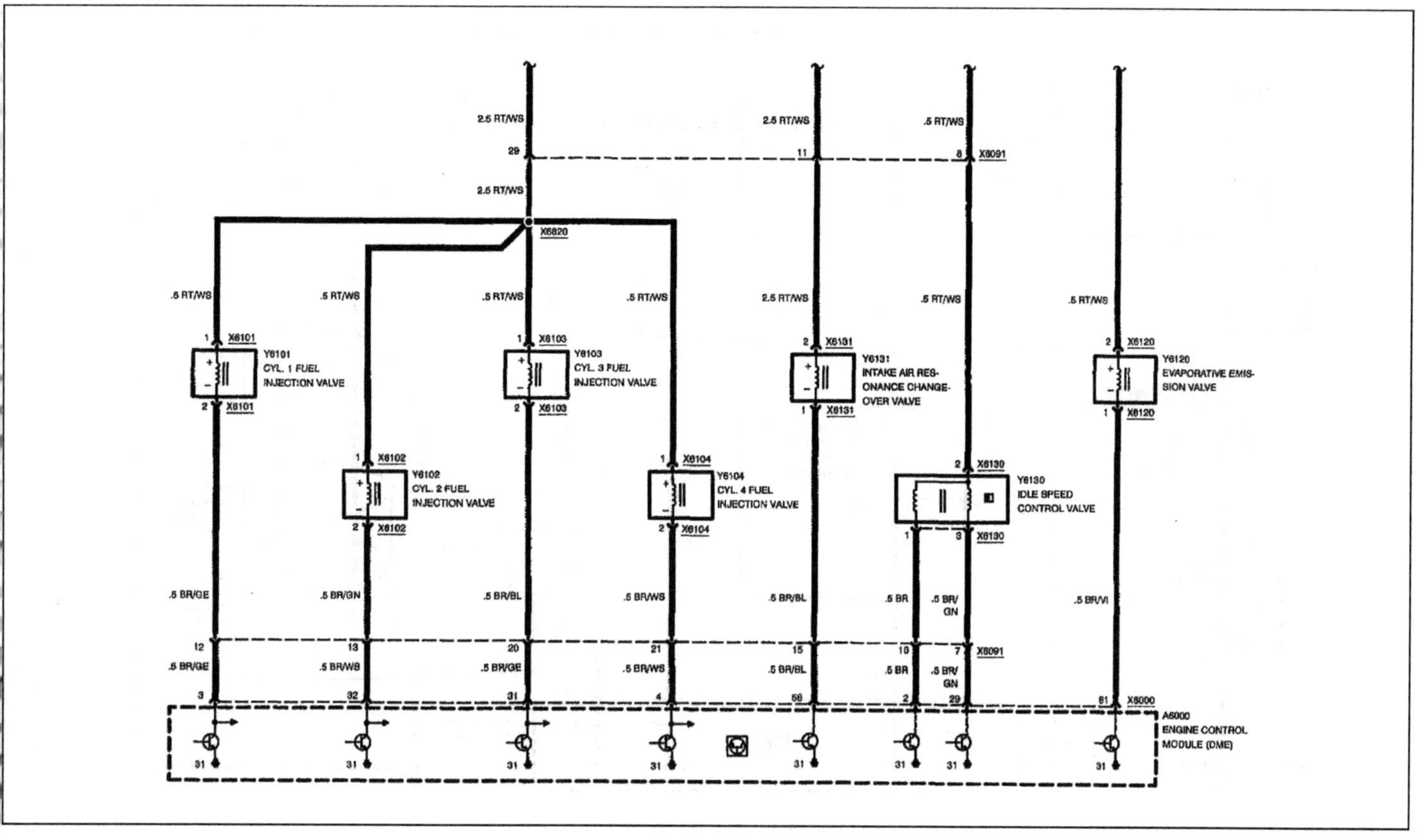


## Bosch DME M5.2 (M44 engine)

(3 of 8)

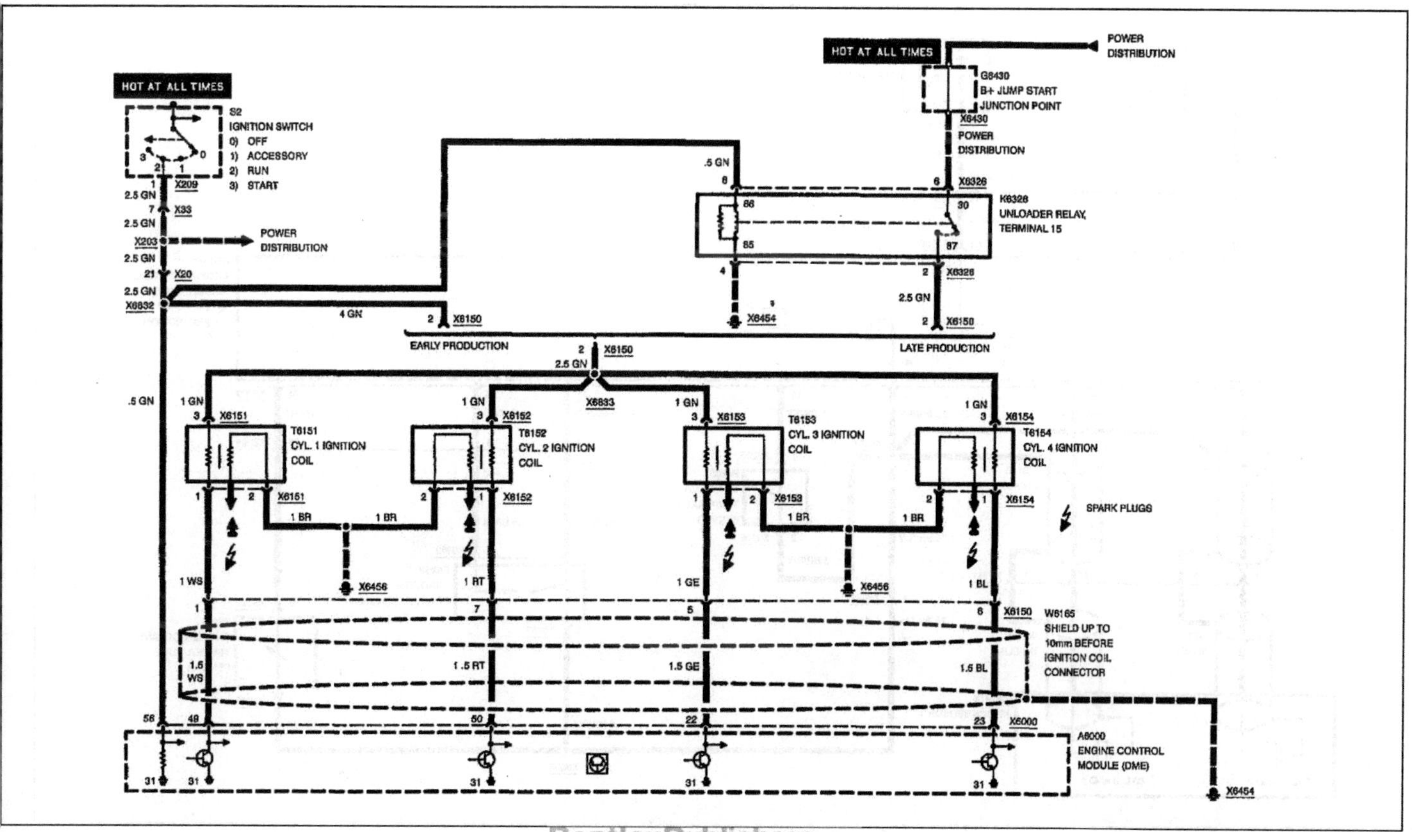

## Bosch DME M5.2 (M44 engine)

**(4 of 8)**

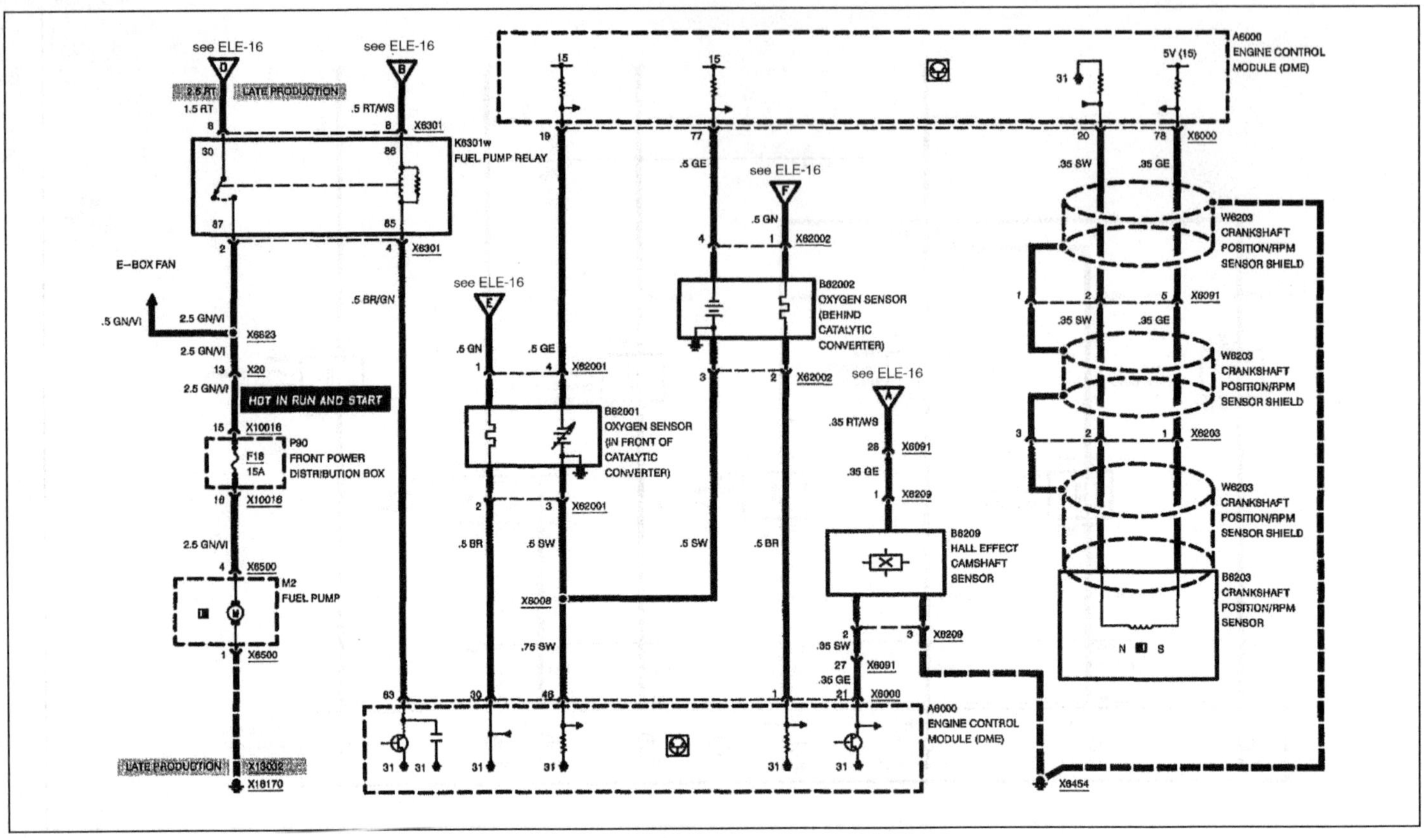


## Bosch DME M5.2 (M44 engine)

**(5 of 8)**

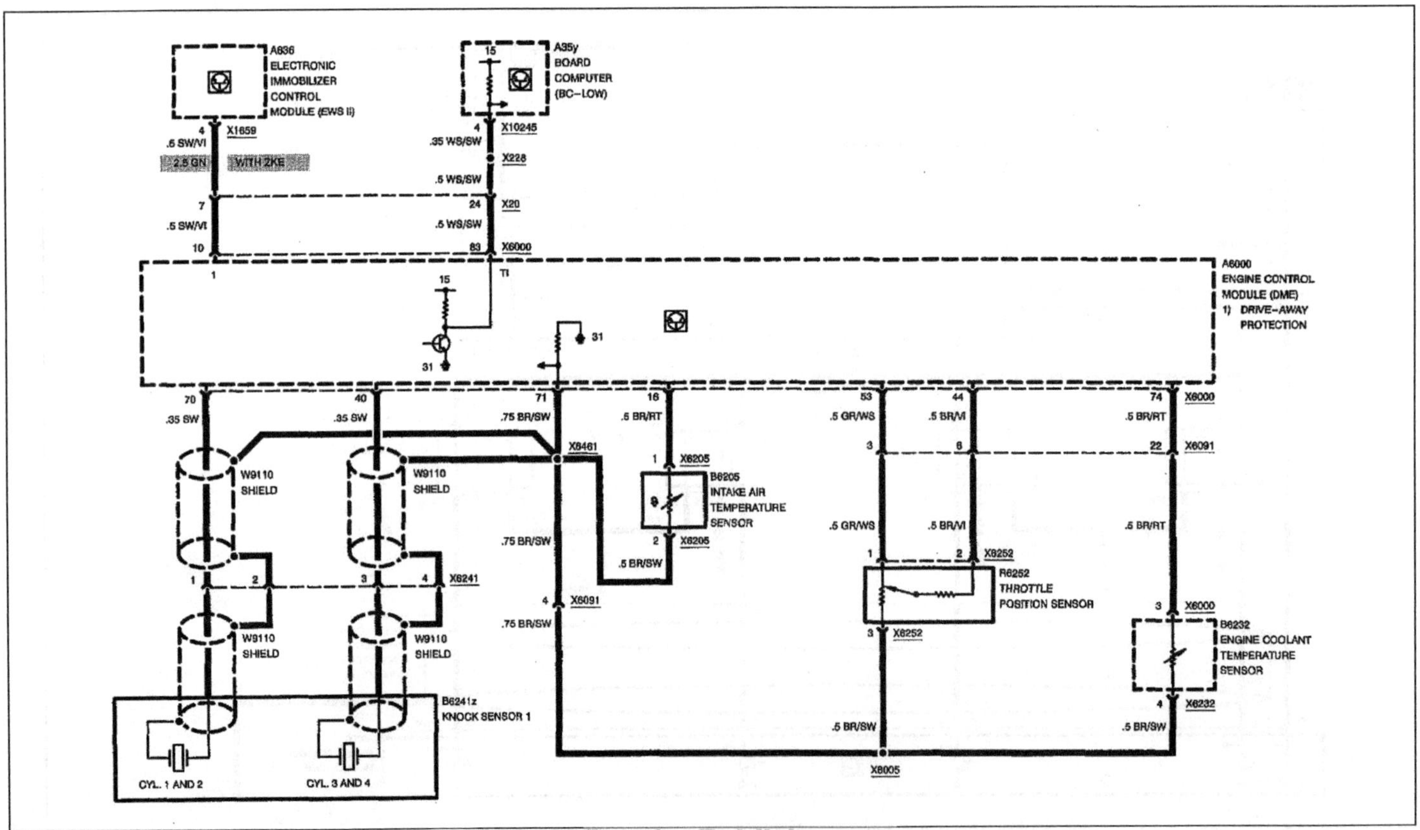

## Bosch DME M5.2 (M44 engine)

(6 of 8)

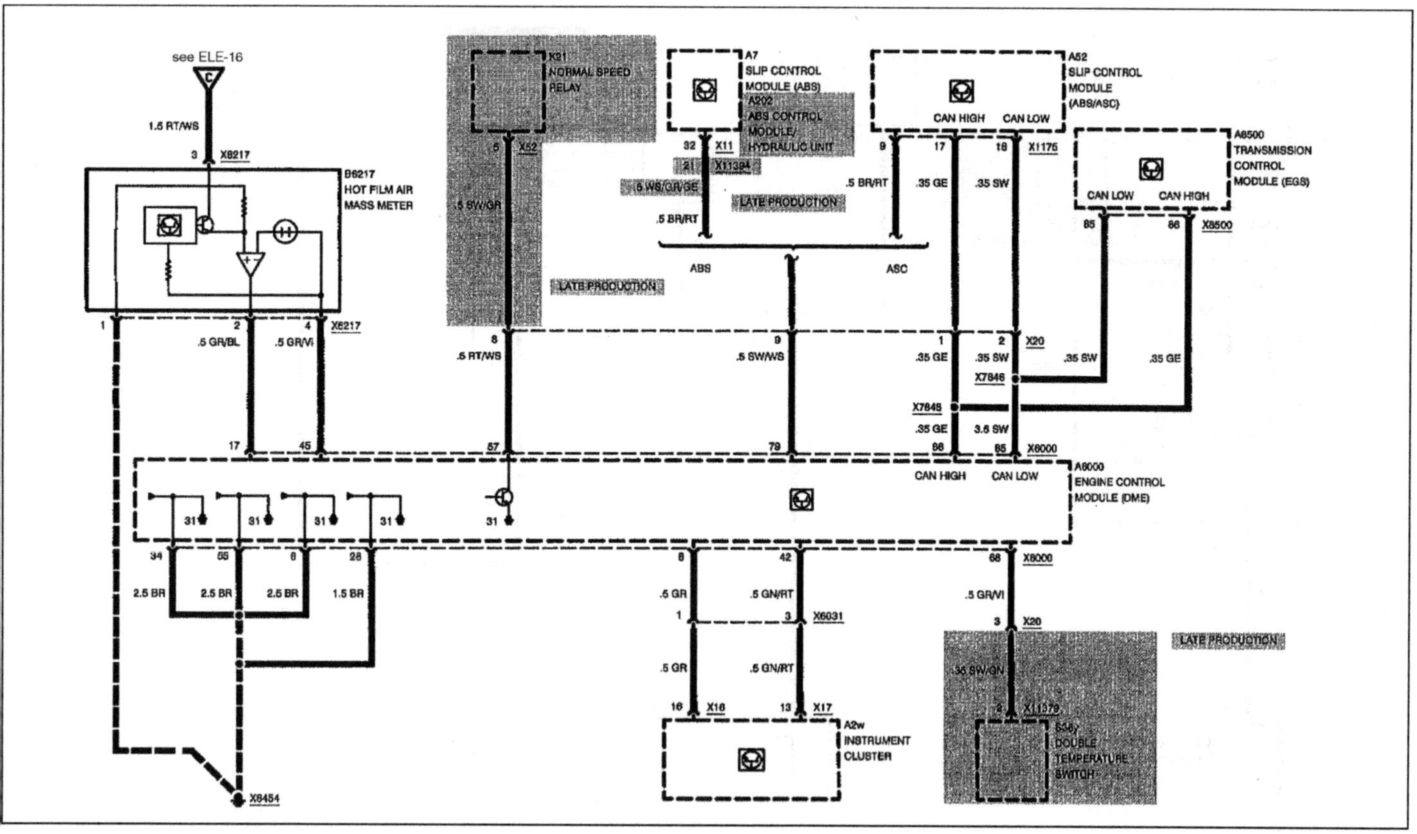


## Bosch DME M5.2 (M44 engine)

(7 of 8)

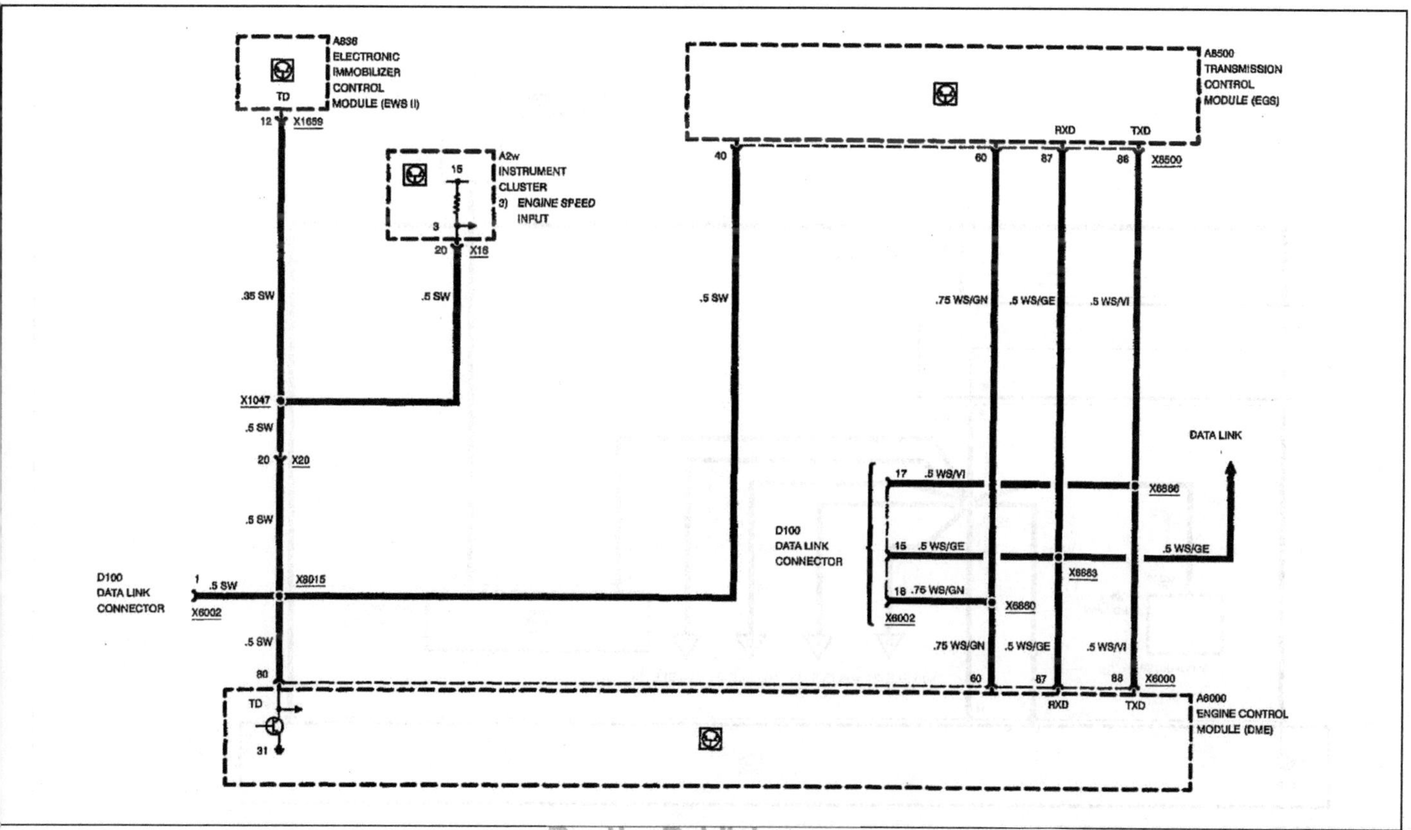

## Bosch DME M5.2 (M44 engine)

(8 of 8)

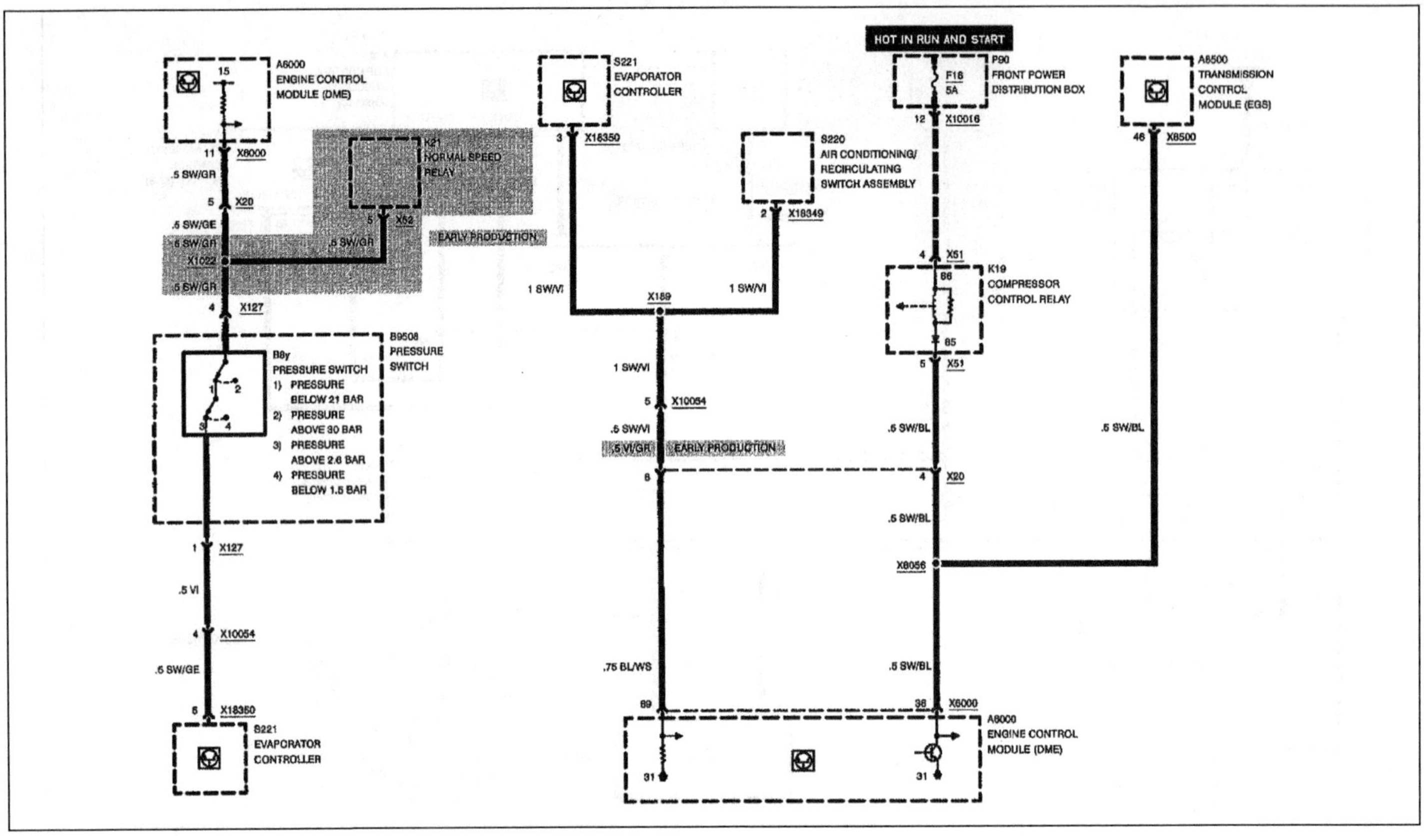


## Siemens DME MS 41.1 (M52, S52 engine)

(1 of 13)

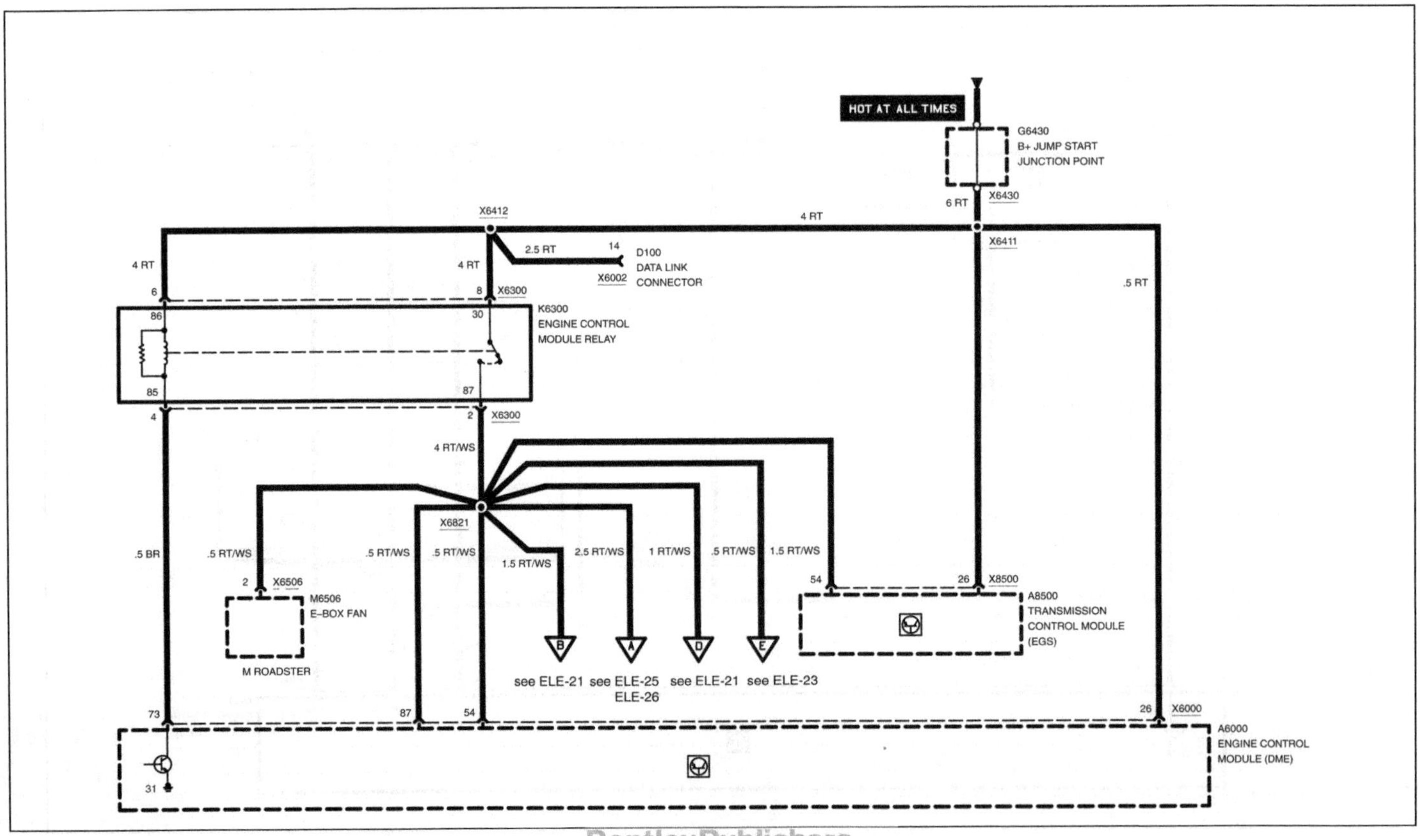

## Siemens DME MS 41.1 (M52, S52 engine)

**(2 of 13)**

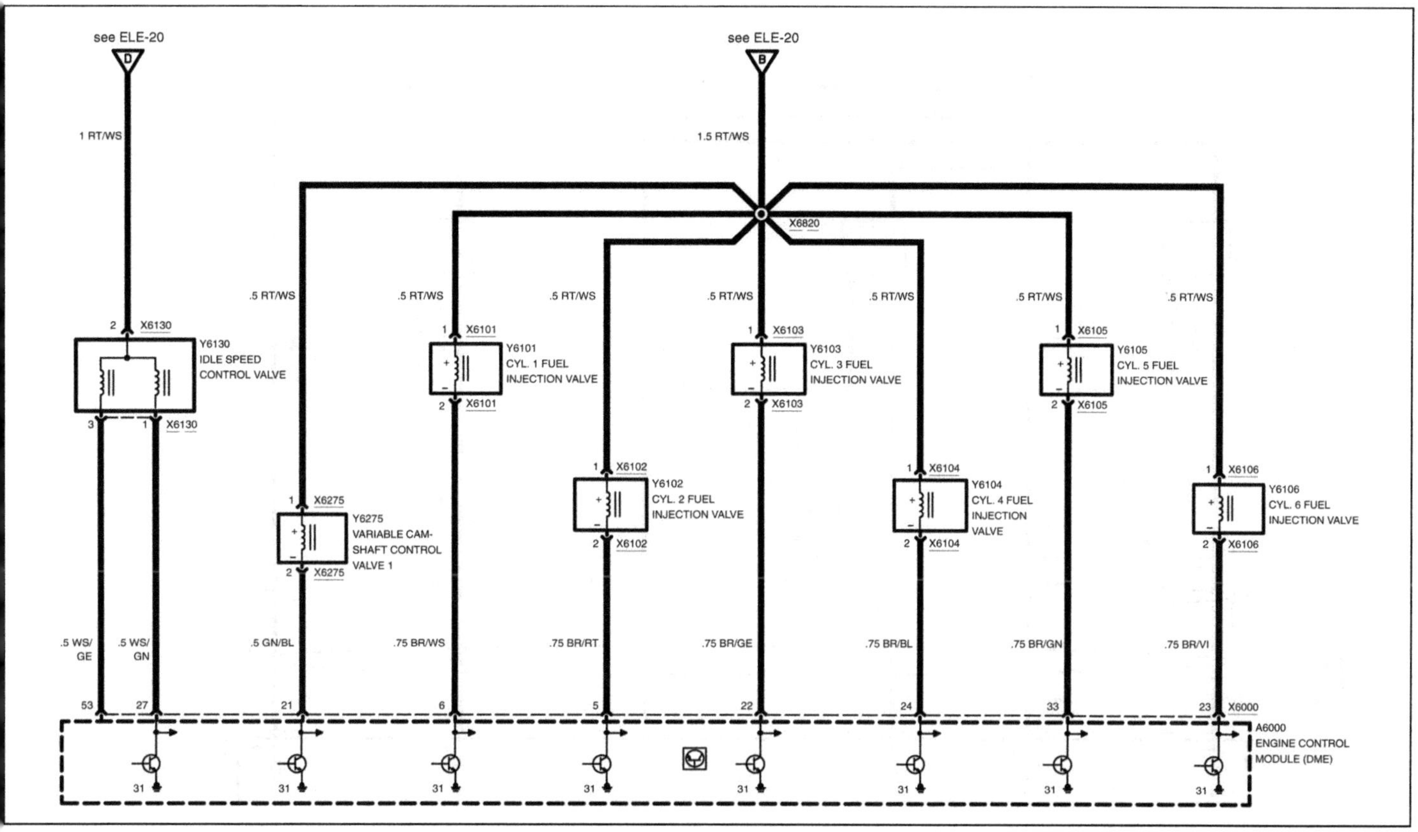


## Siemens DME MS 41.1 (M52, S52 engine)

**(3 of 13)**

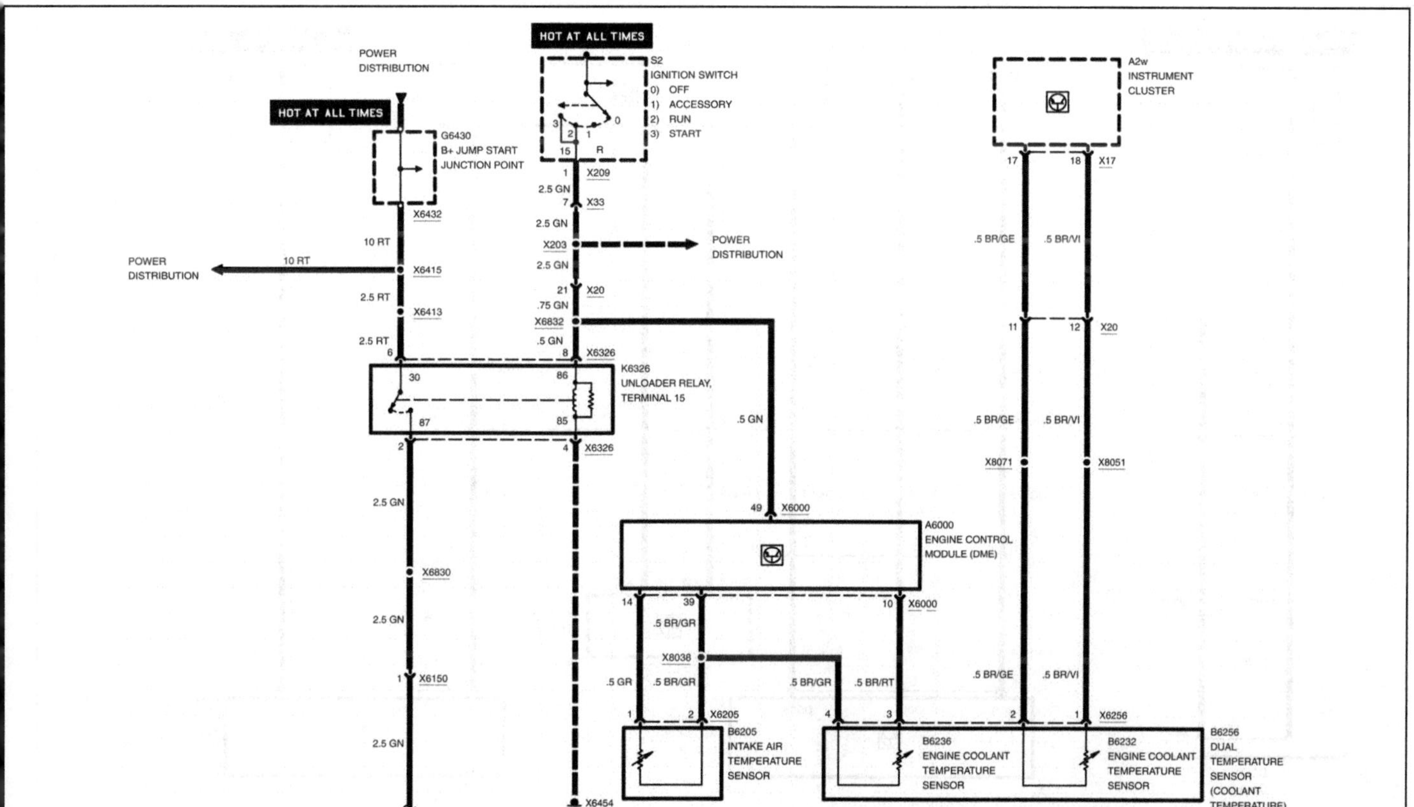

## Siemens DME MS 41.1 (M52, S52 engine)

(4 of 13)

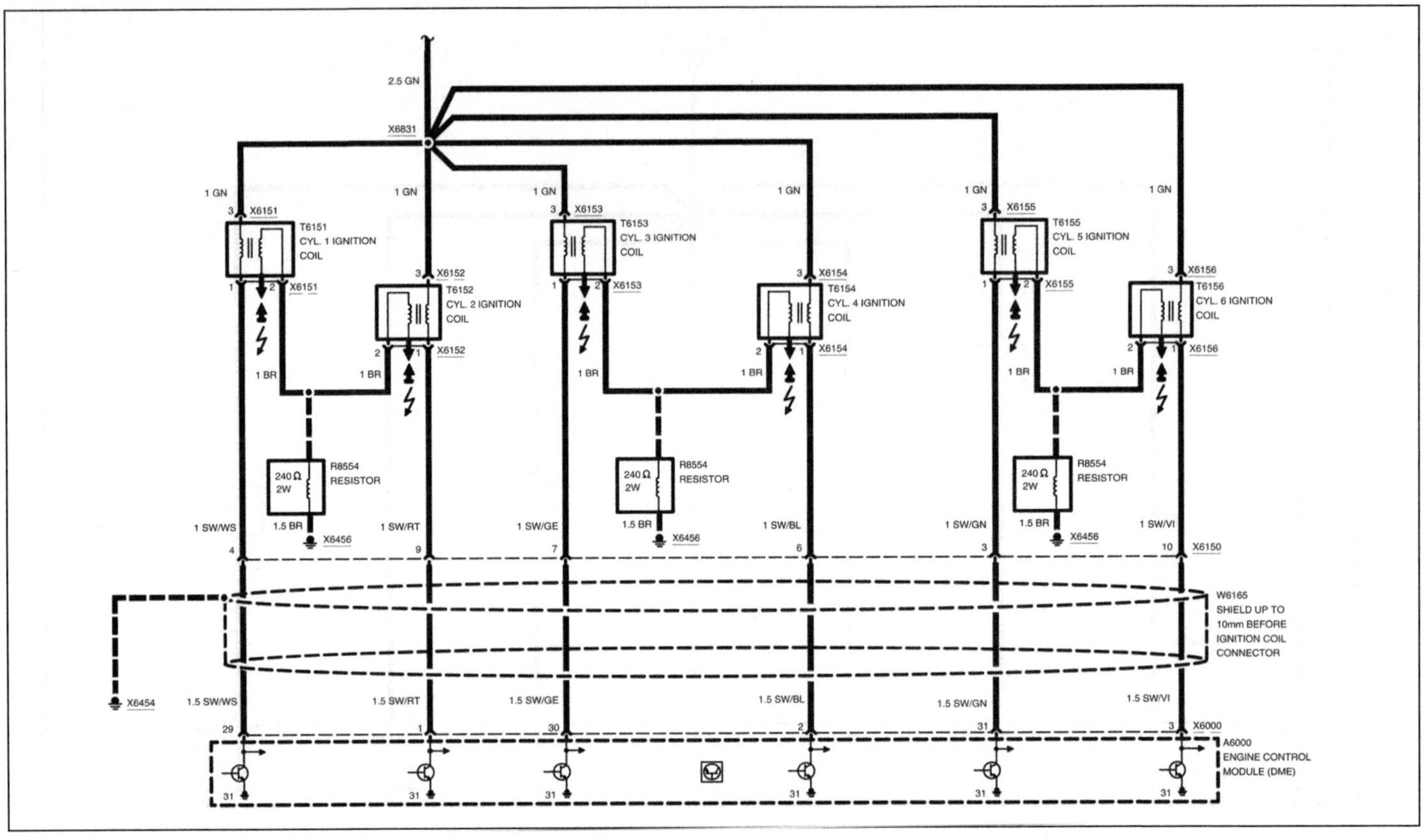


## Siemens DME MS 41.1 (M52, S52 engine)

(5 of 13)

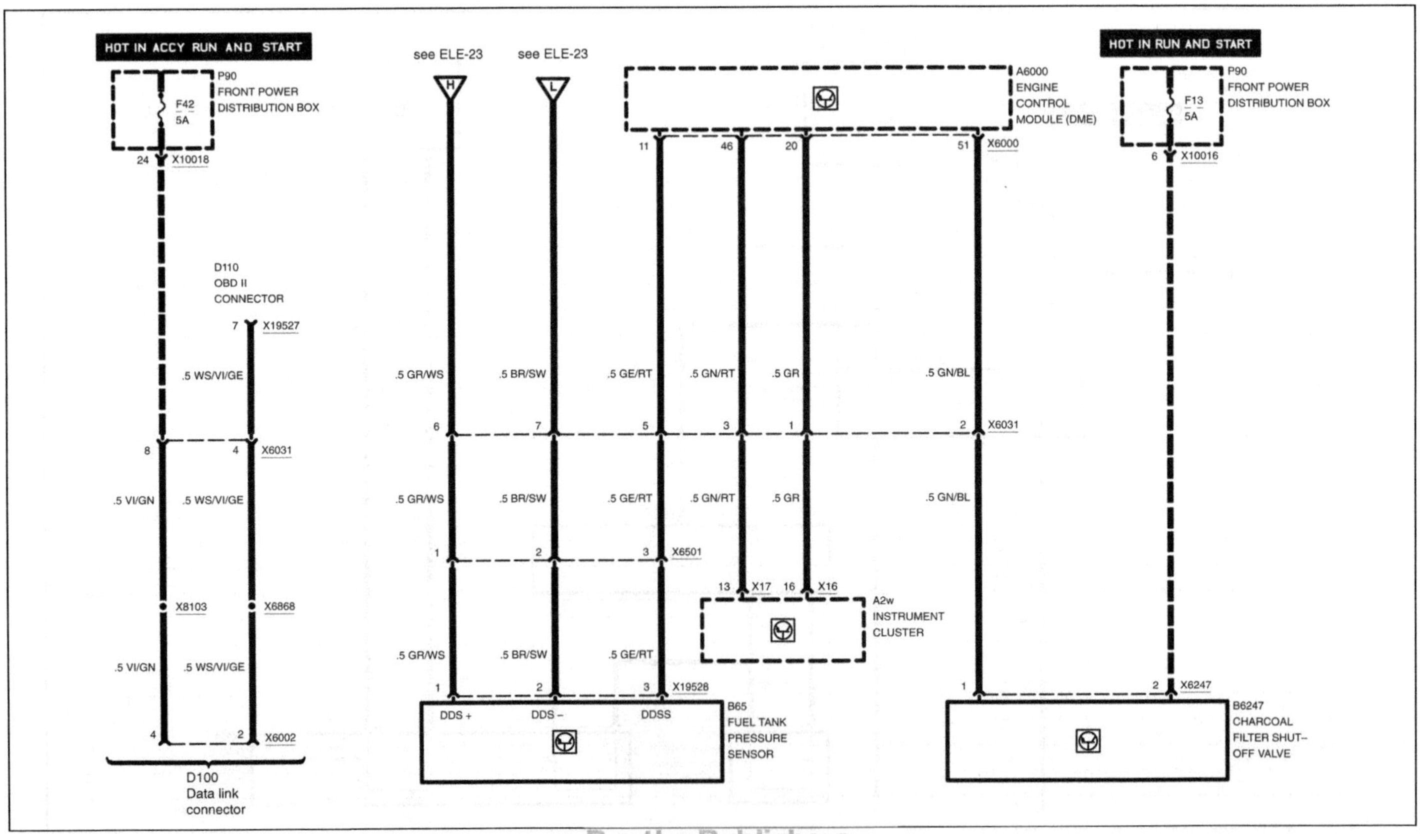

## Siemens DME MS 41.1 (M52, S52 engine)

**(6 of 13)**

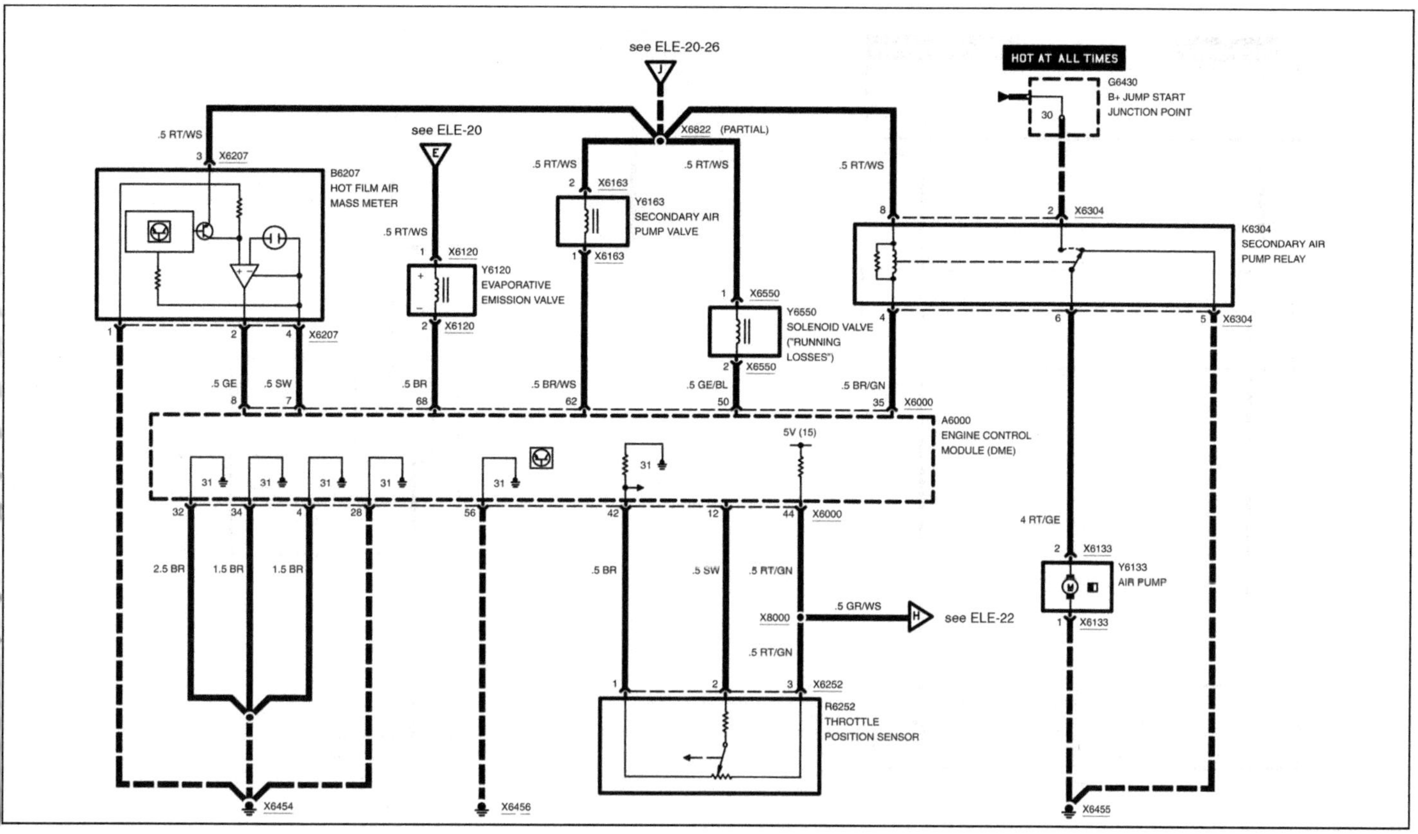


## Siemens DME MS 41.1 (M52, S52 engine)

**(7 of 13)**

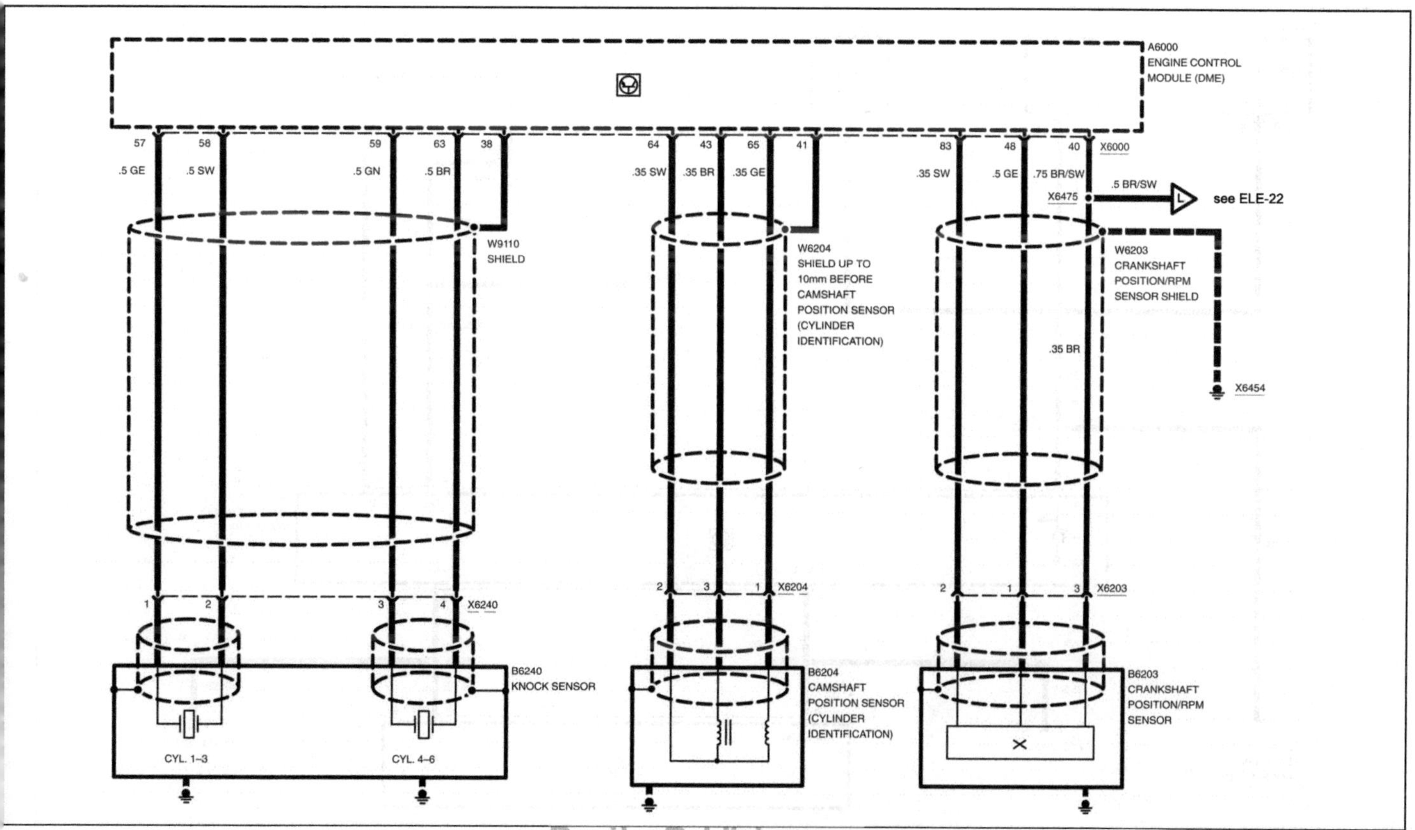

## Siemens DME MS 41.1 (M52, S52 engine)
(8 of 13)

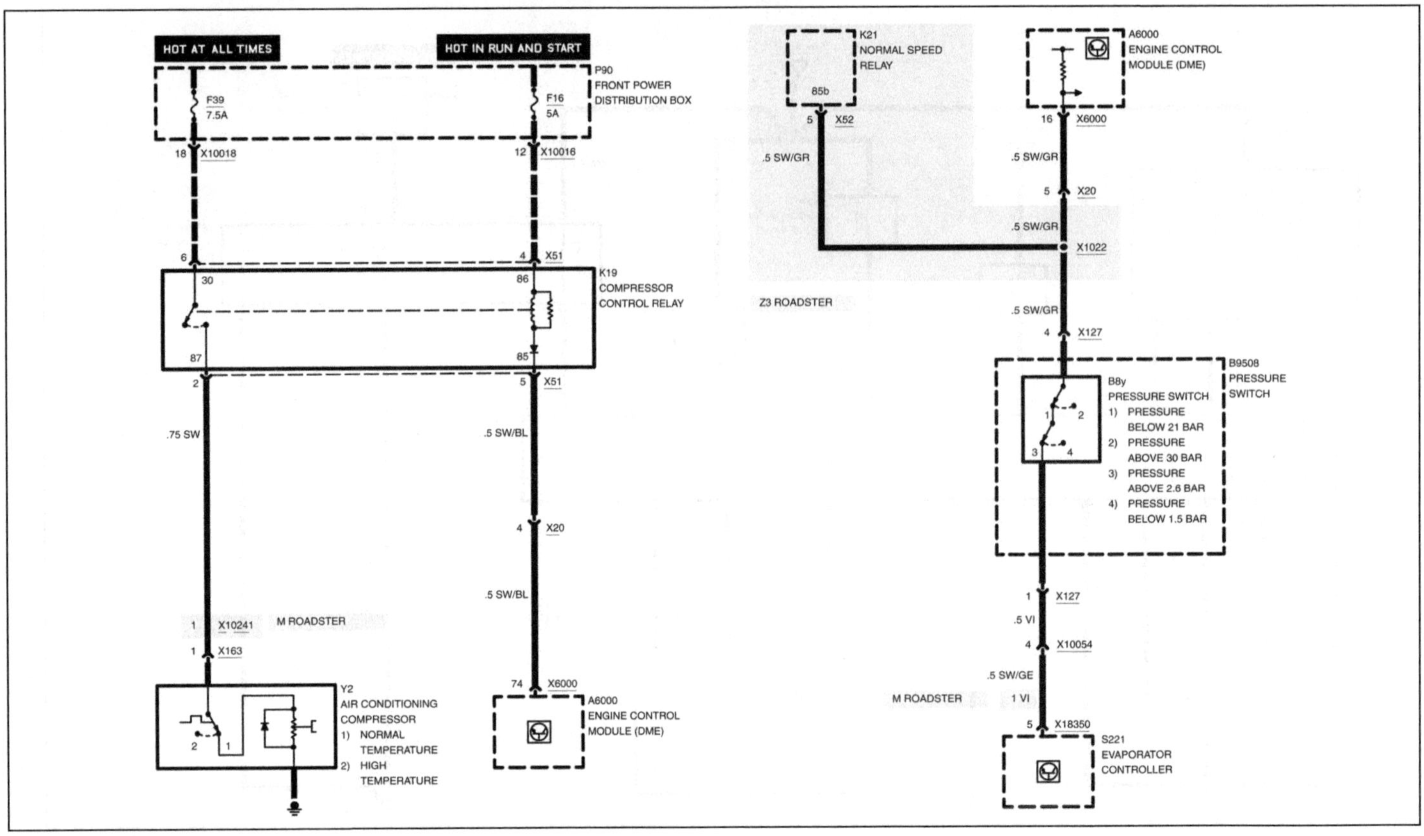


## Siemens DME MS 41.1 (M52, S52 engine)
(9 of 13)

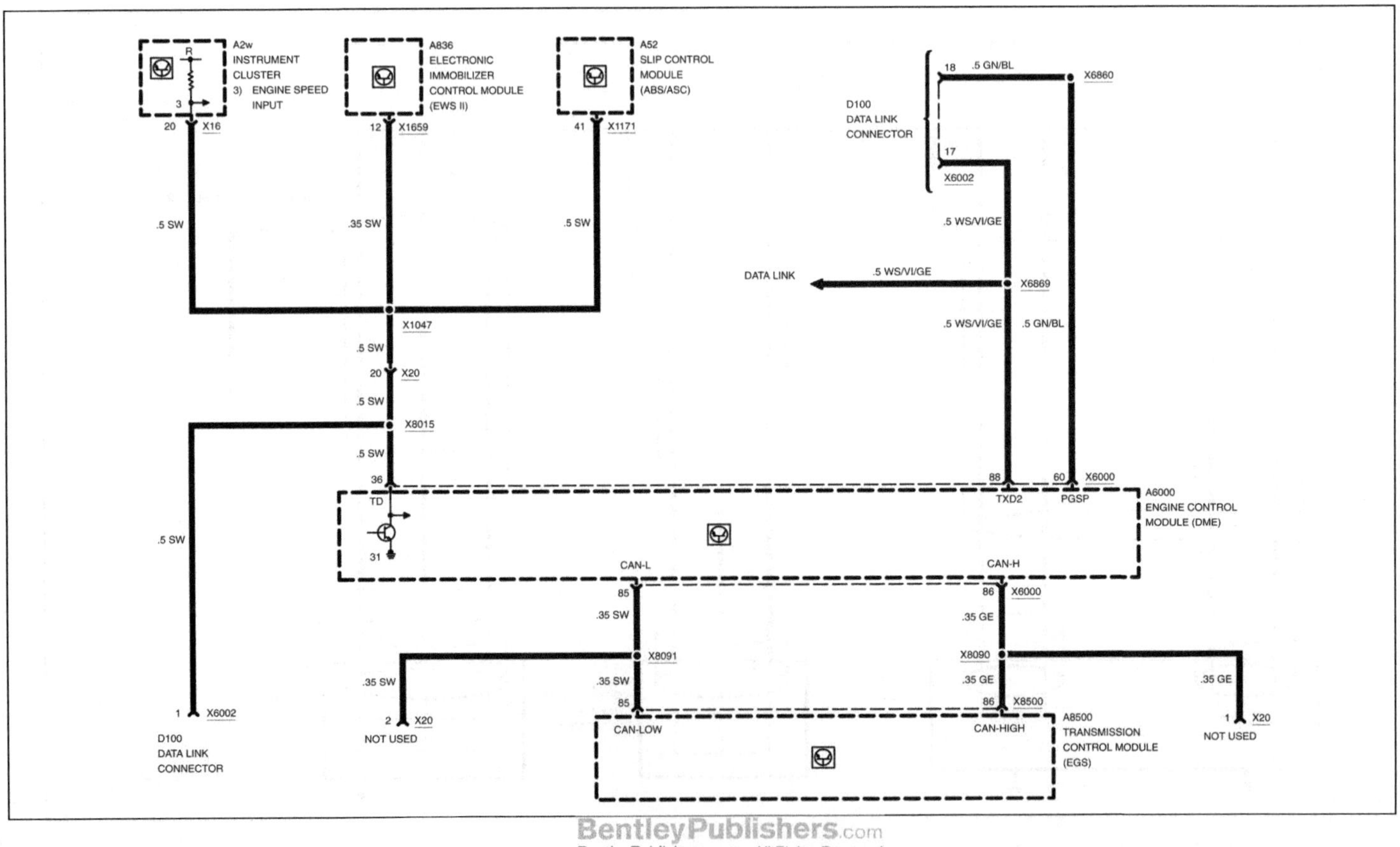

## Siemens DME MS 41.1 (M52, S52 engine)
(10 of 13)

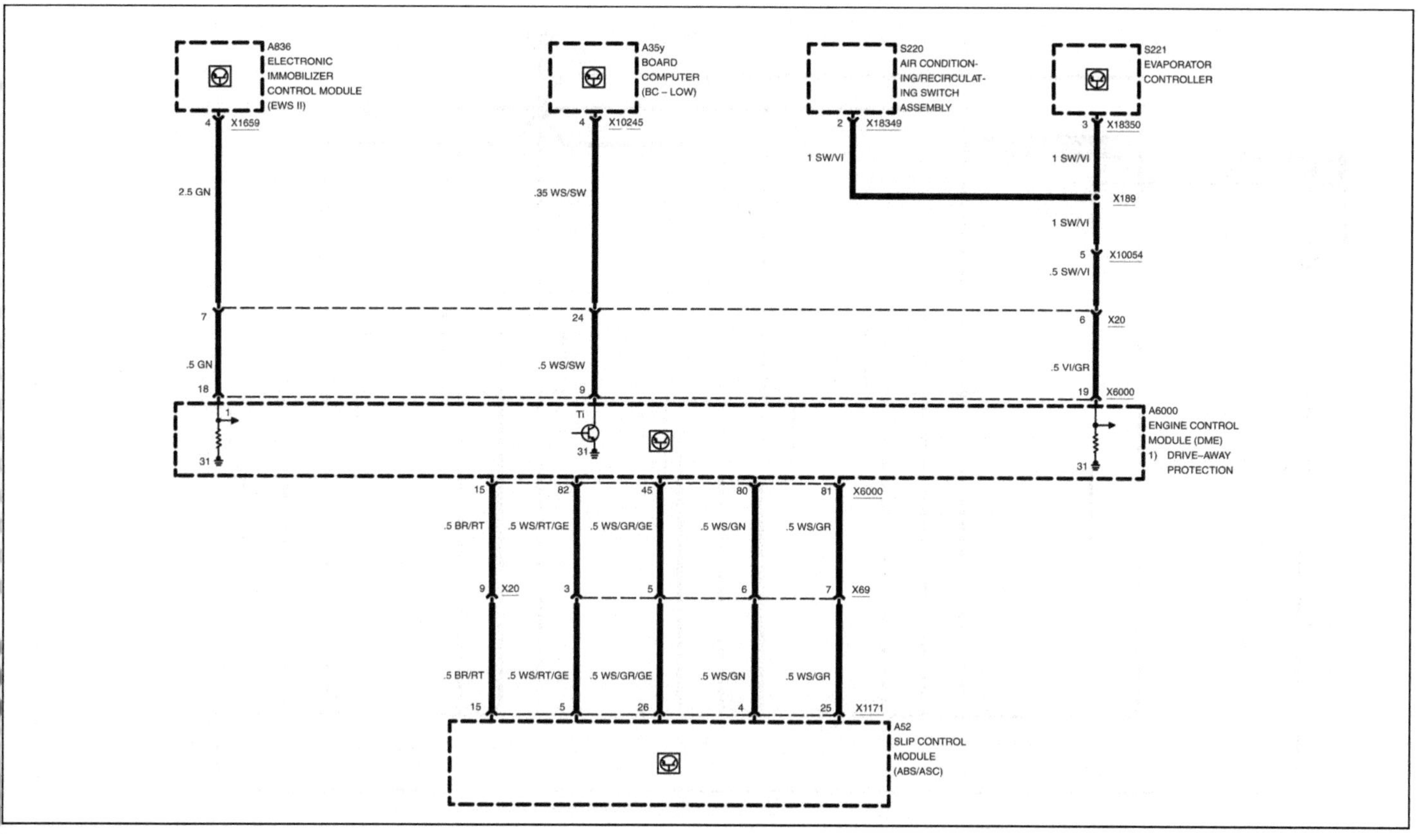


## Siemens DME MS 41.1 (M52, S52 engine)
(11 of 13)

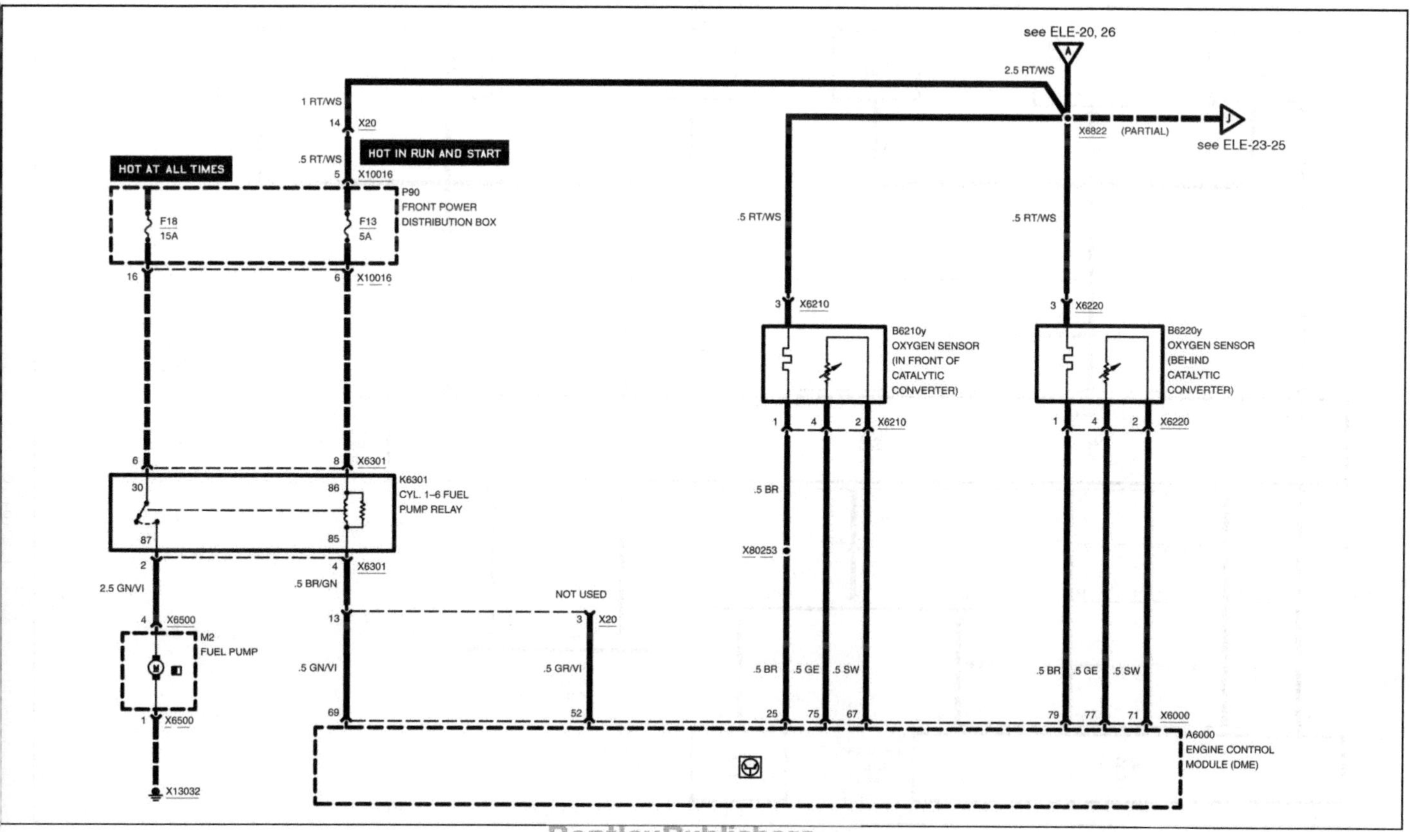

## Siemens DME MS 41.1 (M52, S52 engine)
### (12 of 13)

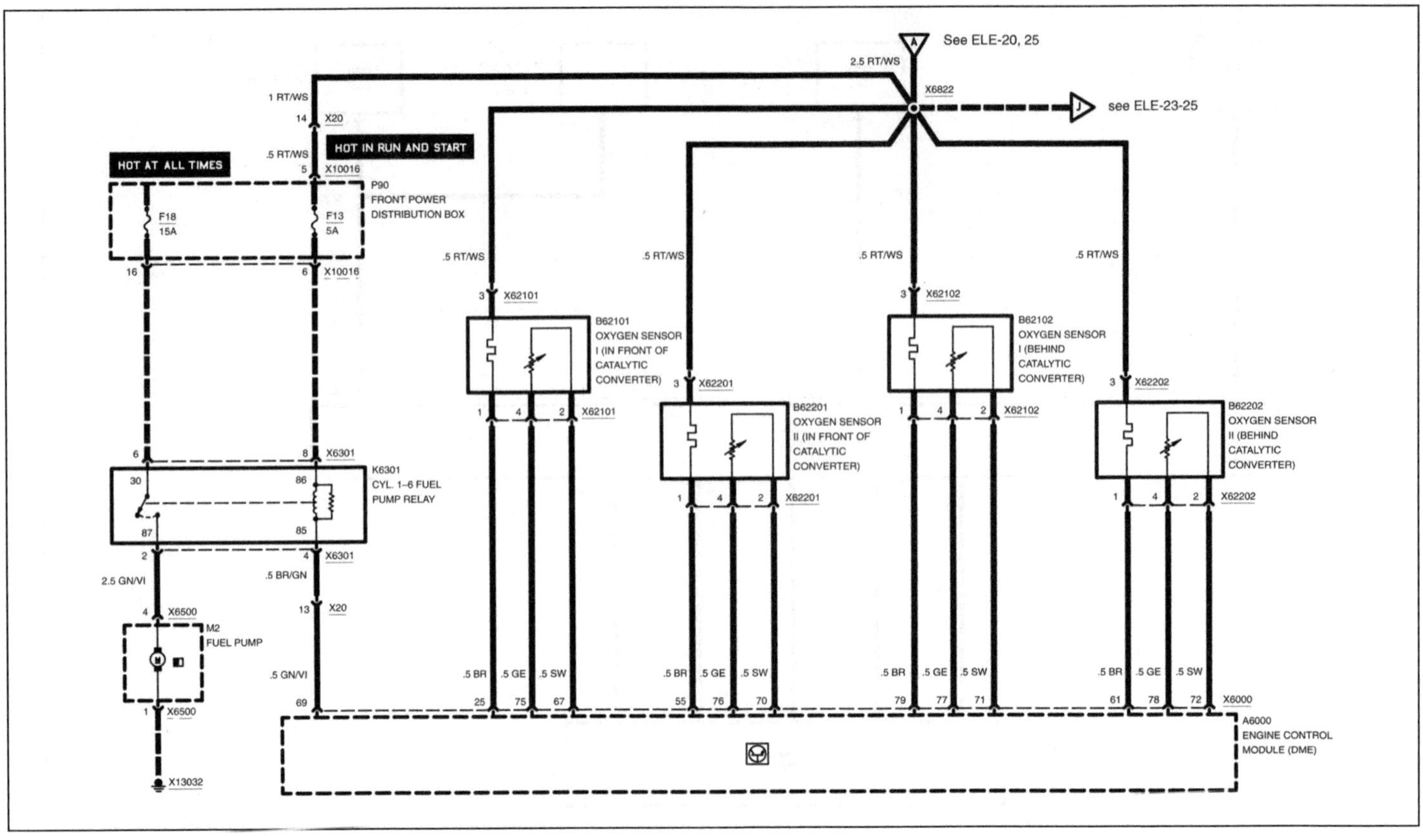


## Siemens DME MS 41.1 (M52, S52 engine)
### (13 of 13)

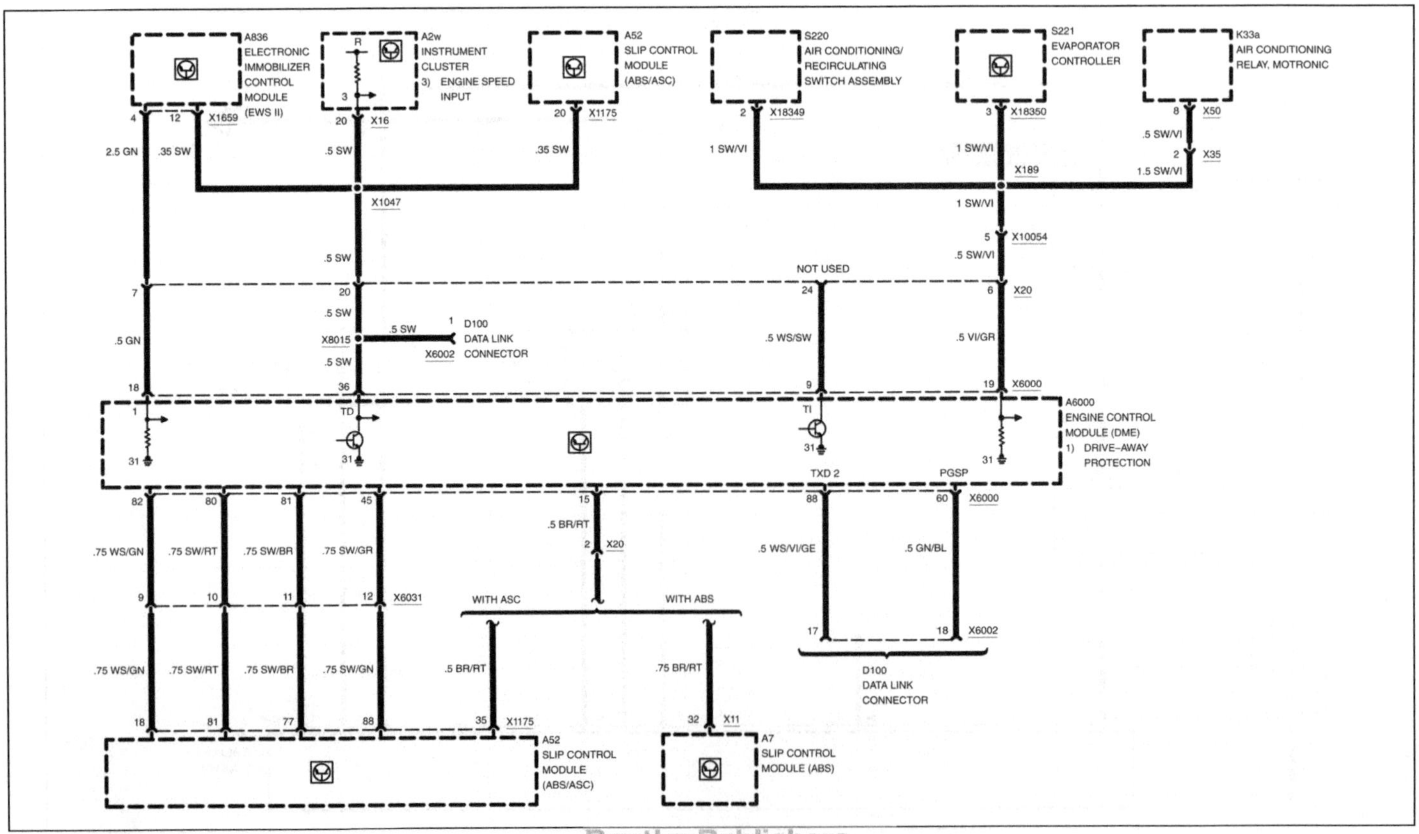

## Siemens DME MS 42.1 (M52 TU engine)
(1 of 11)

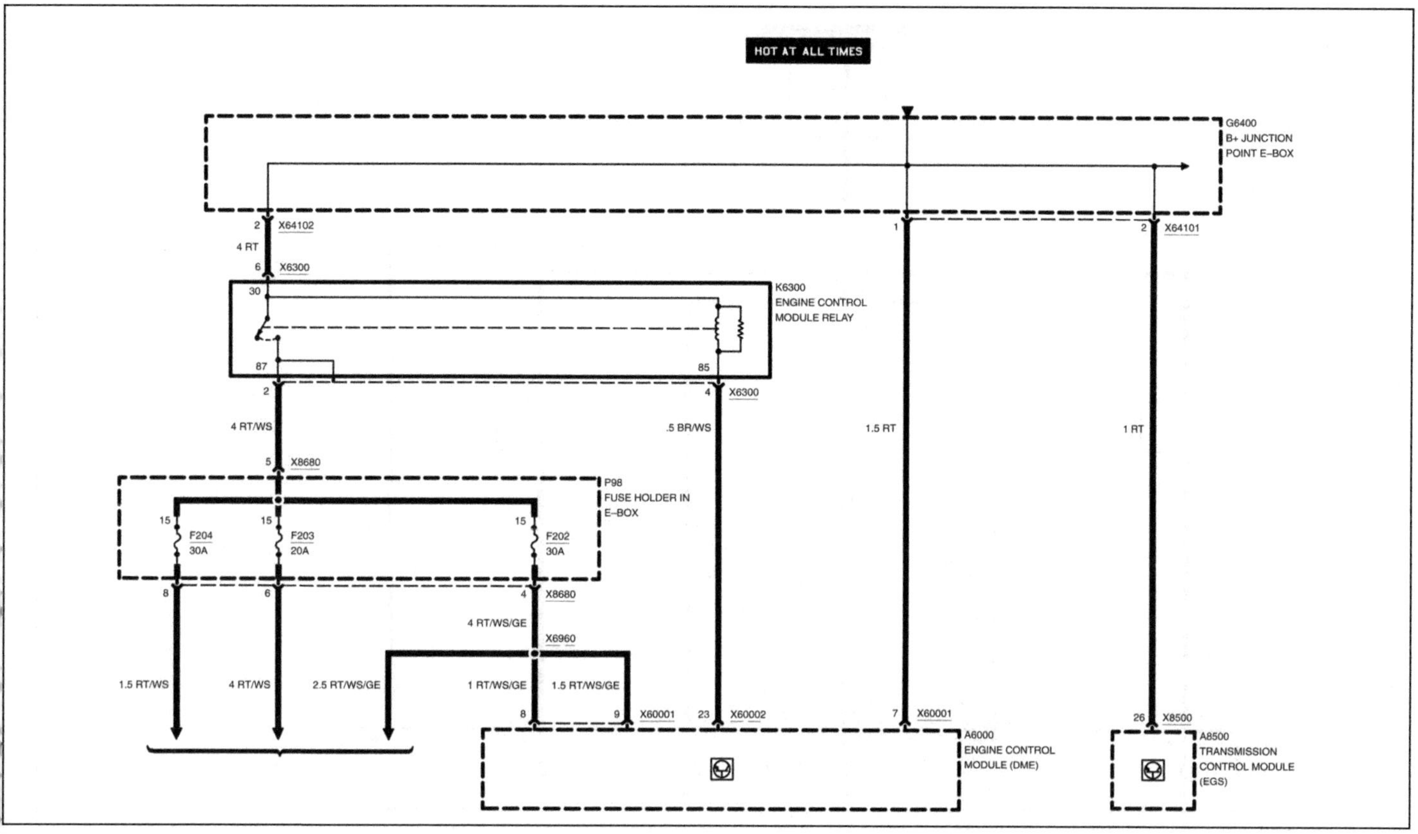


## Siemens DME MS 42.1 (M52 TU engine)
(2 of 11)

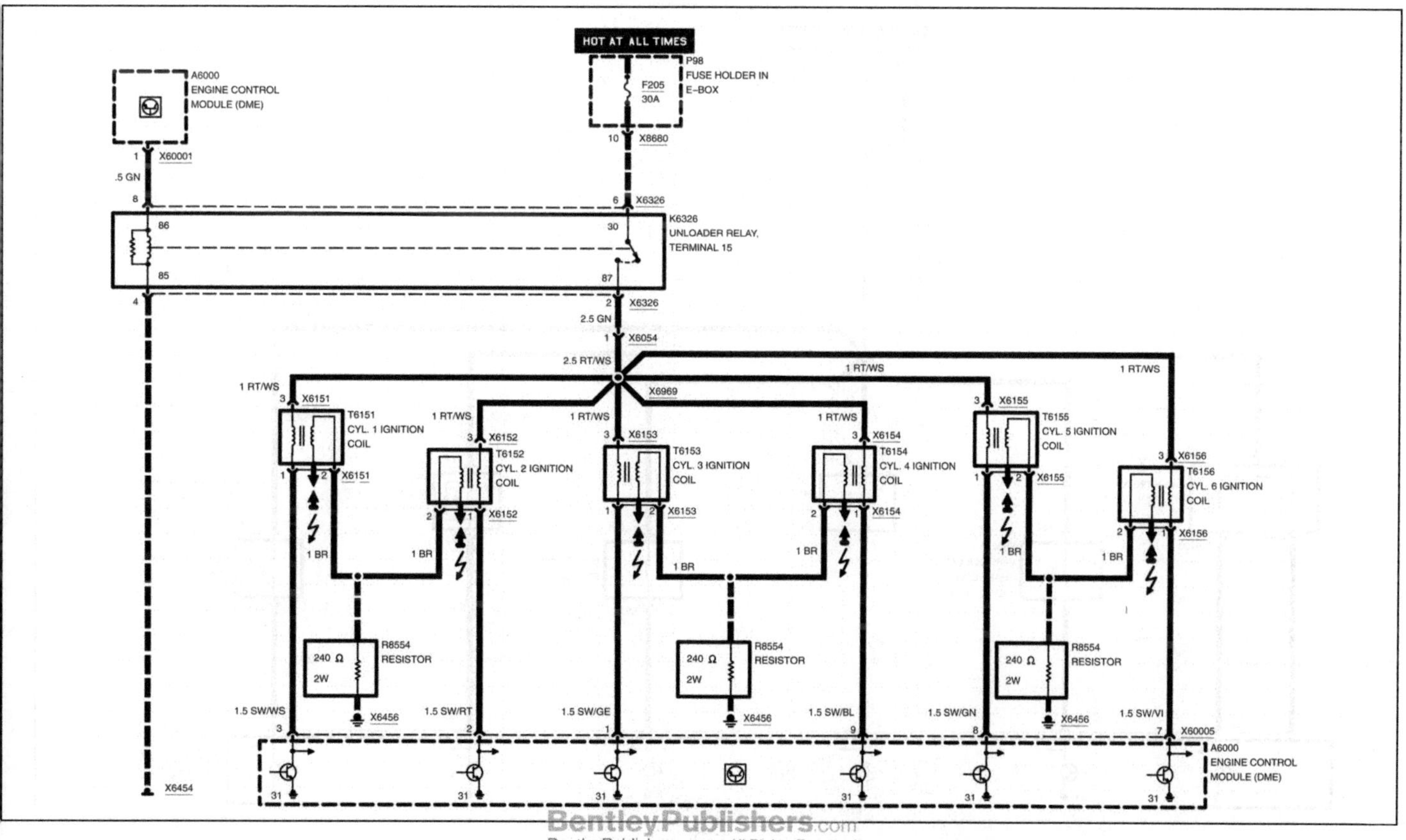

## Siemens DME MS 42.1 (M52 TU engine)
(3 of 11)

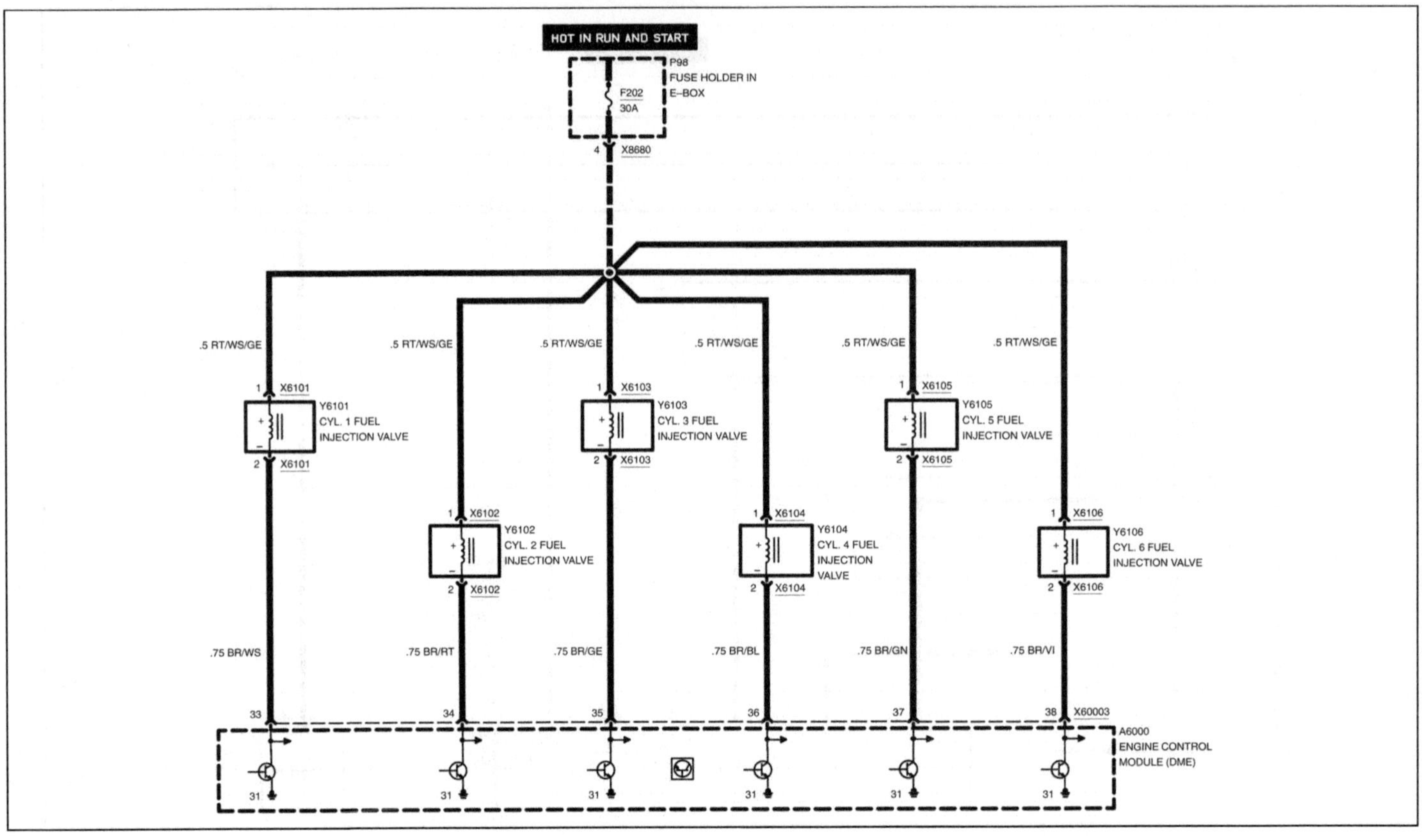


## Siemens DME MS 42.1 (M52 TU engine)
(4 of 11)

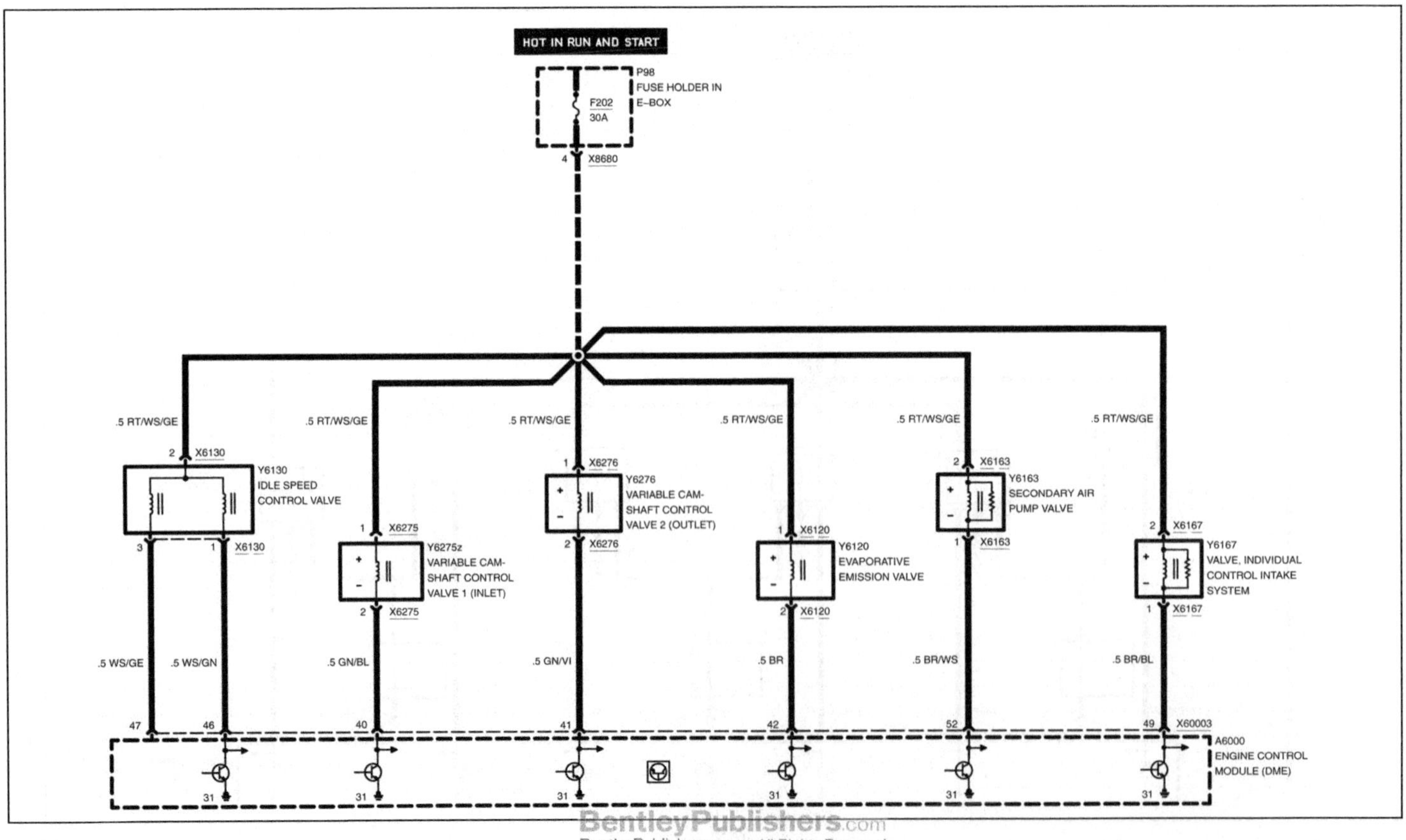

## Siemens DME MS 42.1 (M52 TU engine)

**(5 of 11)**

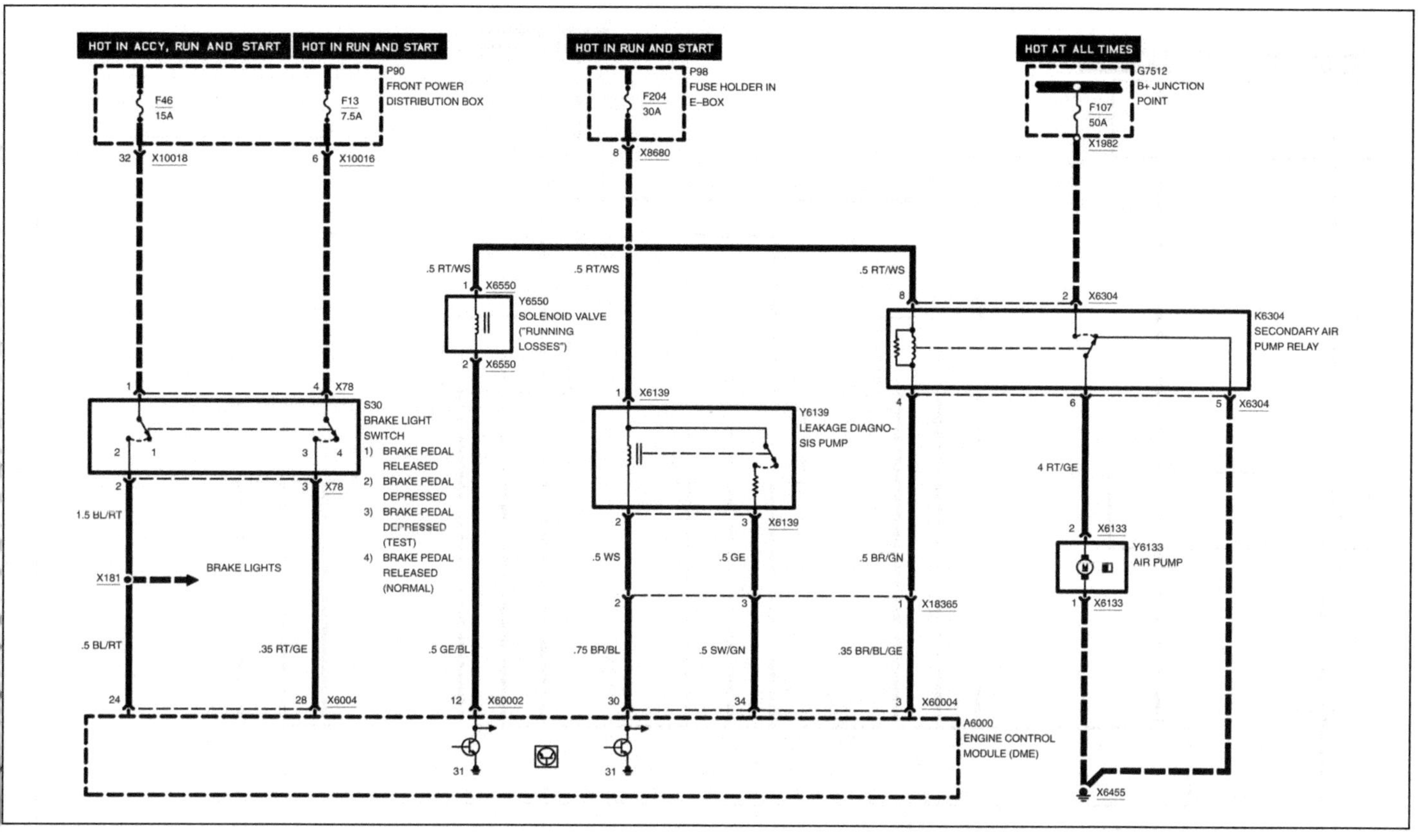


## Siemens DME MS 42.1 (M52 TU engine)

**(6 of 11)**

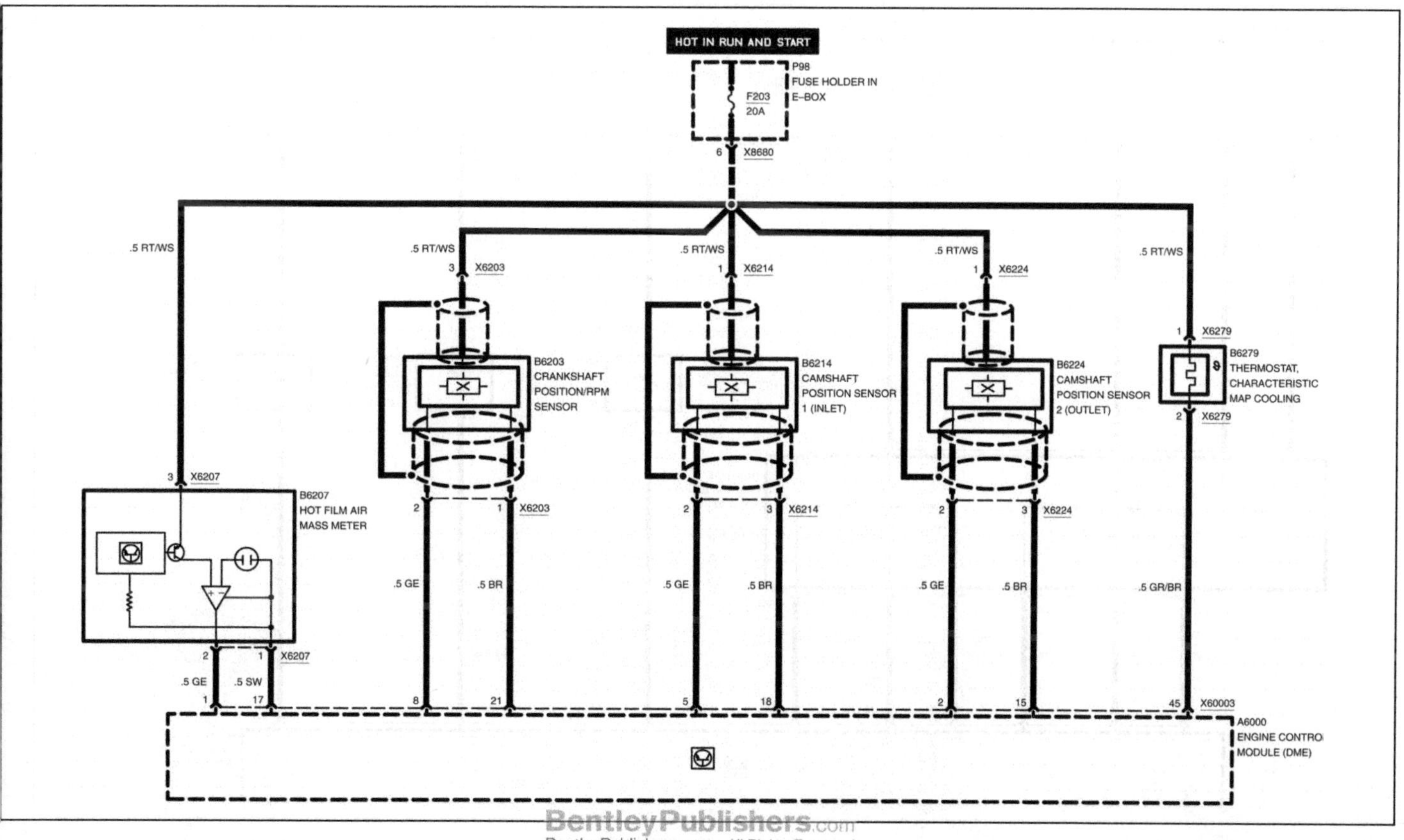

## Siemens DME MS 42.1 (M52 TU engine)

**(7 of 11)**

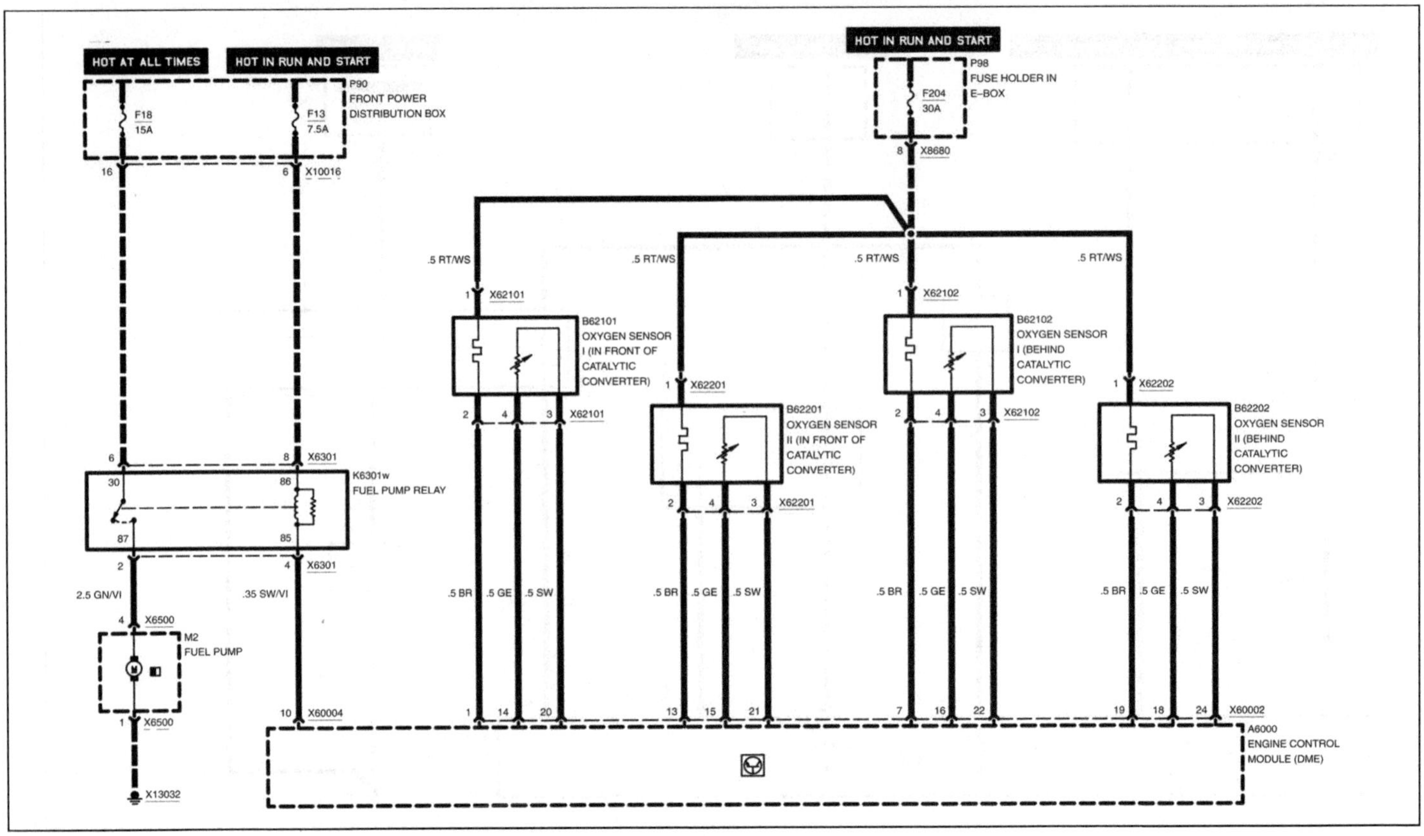


## Siemens DME MS 42.1 (M52 TU engine)

**(8 of 11)**

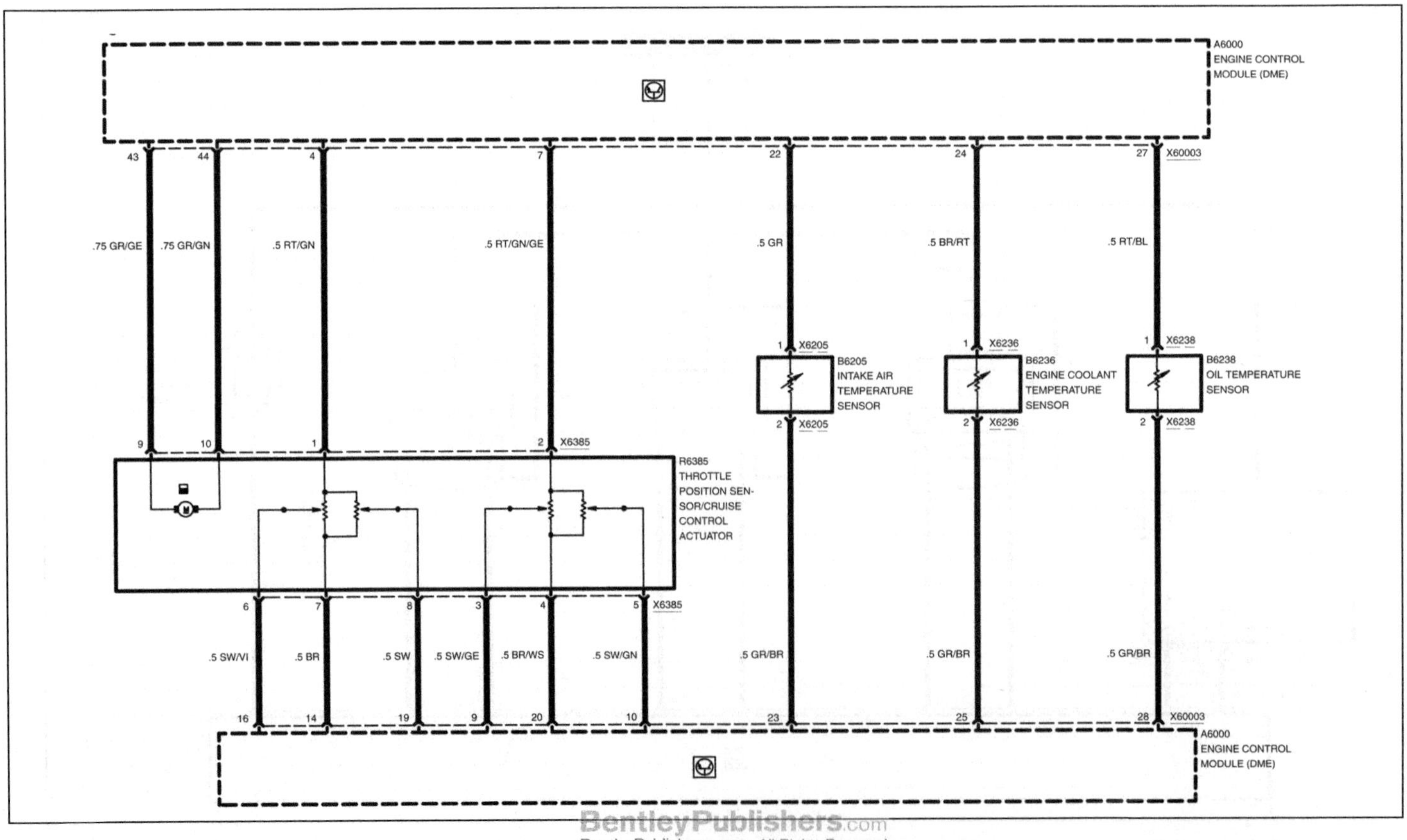

## Siemens DME MS 42.1 (M52 TU engine)
(9 of 11)

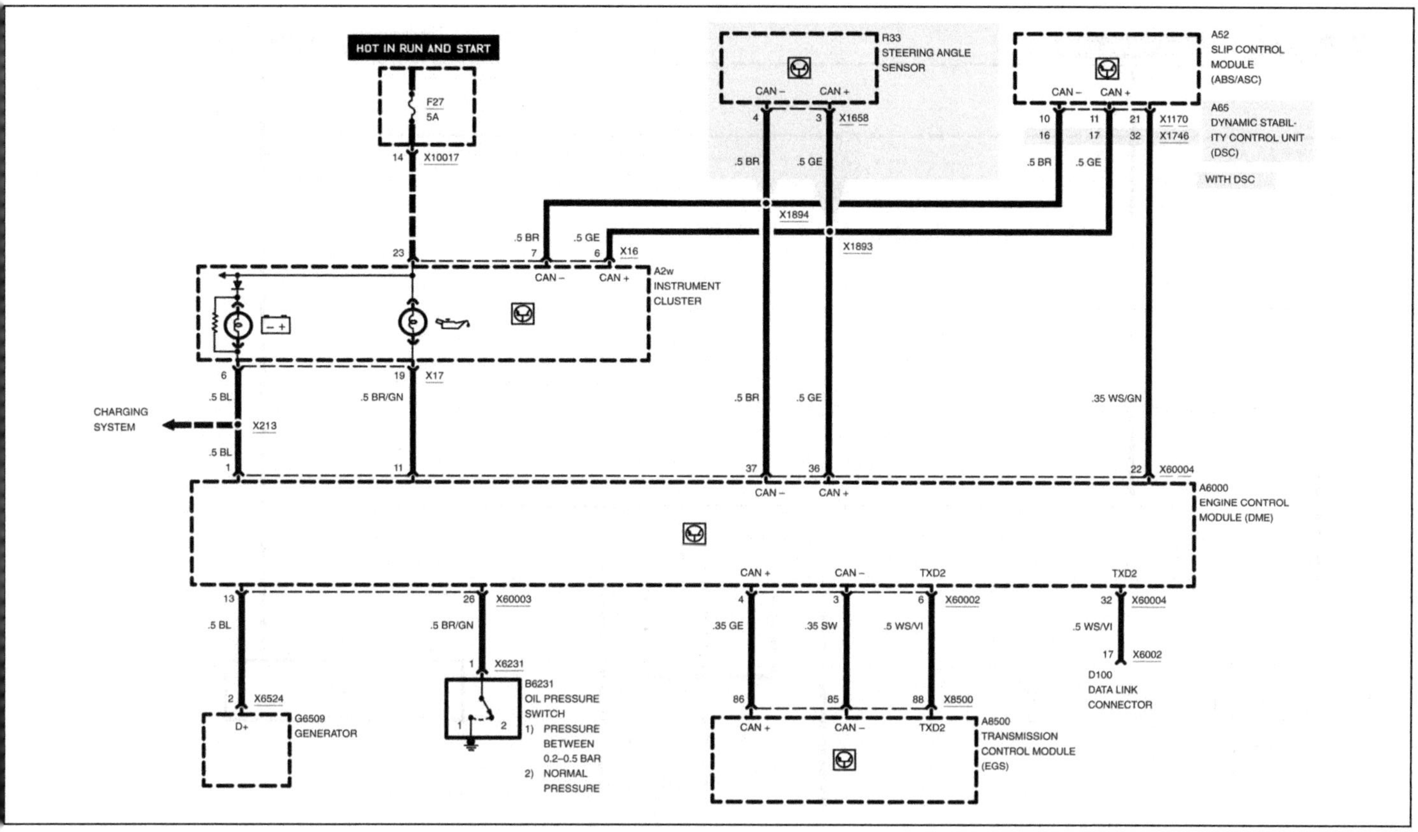


## Siemens DME MS 42.1 (M52 TU engine)
(10 of 11)

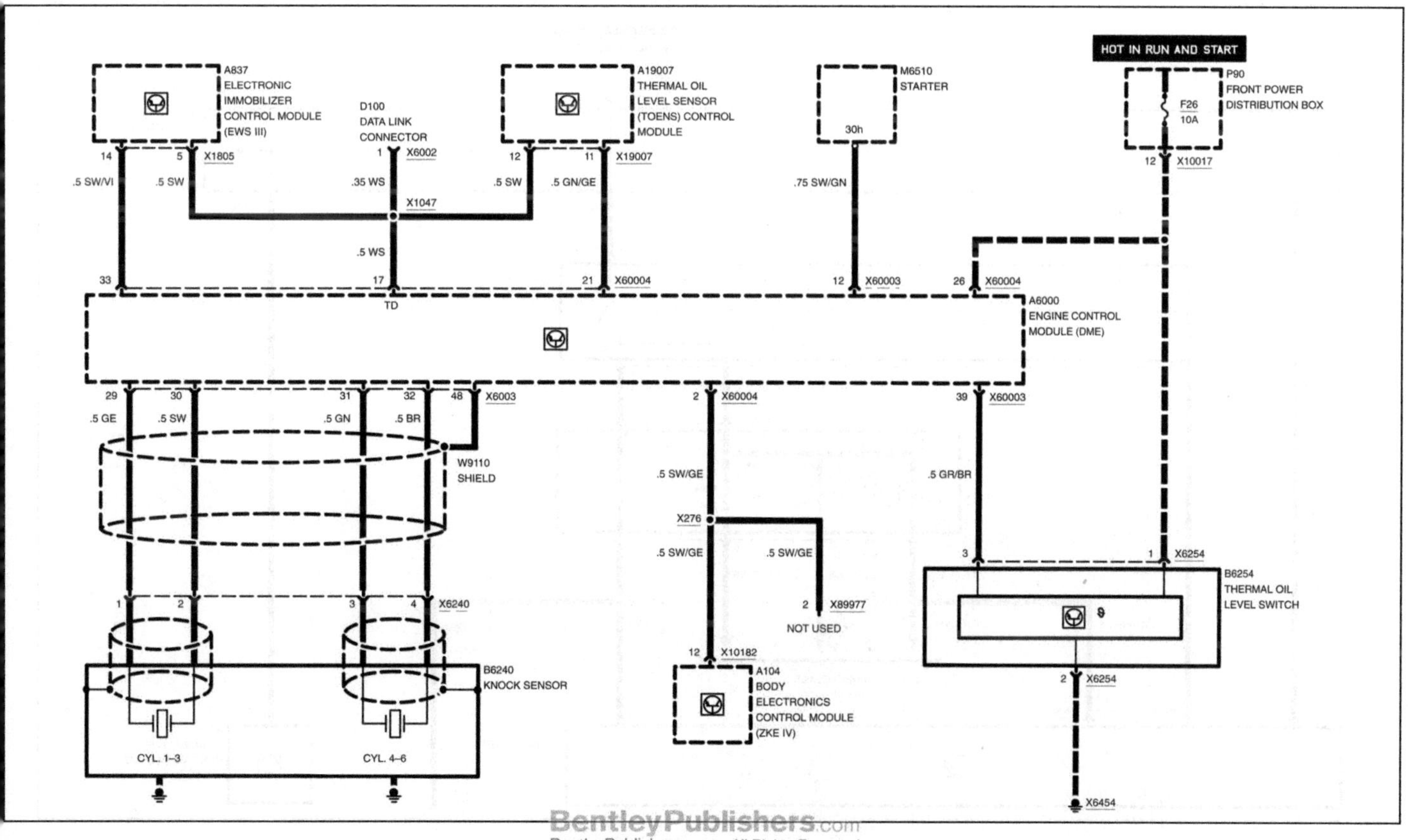

## Siemens DME MS 42.1 (M52 TU engine)
**(11 of 11)**

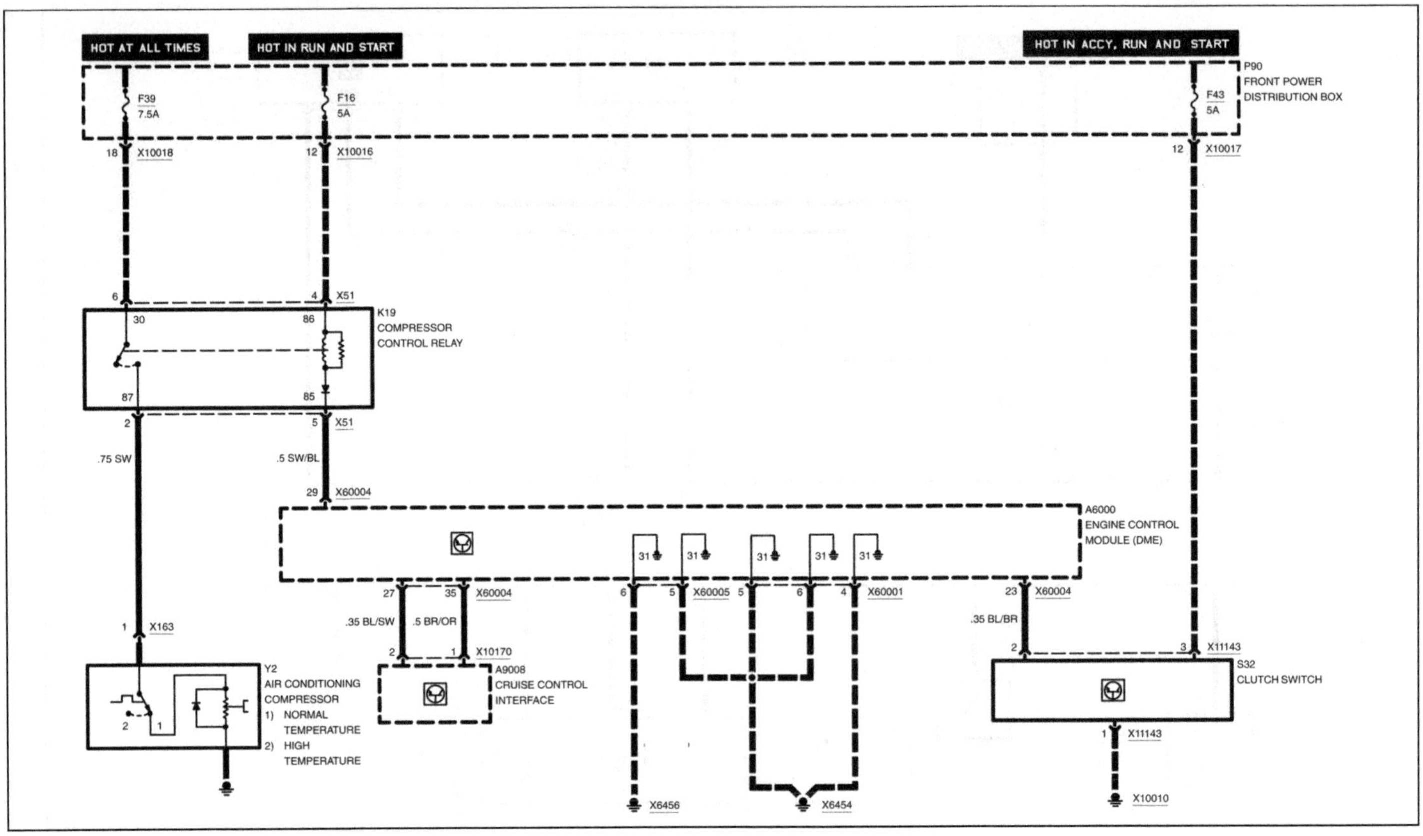


## Siemens DME MS 43.0 (M54 engine)
**(1 of 11)**

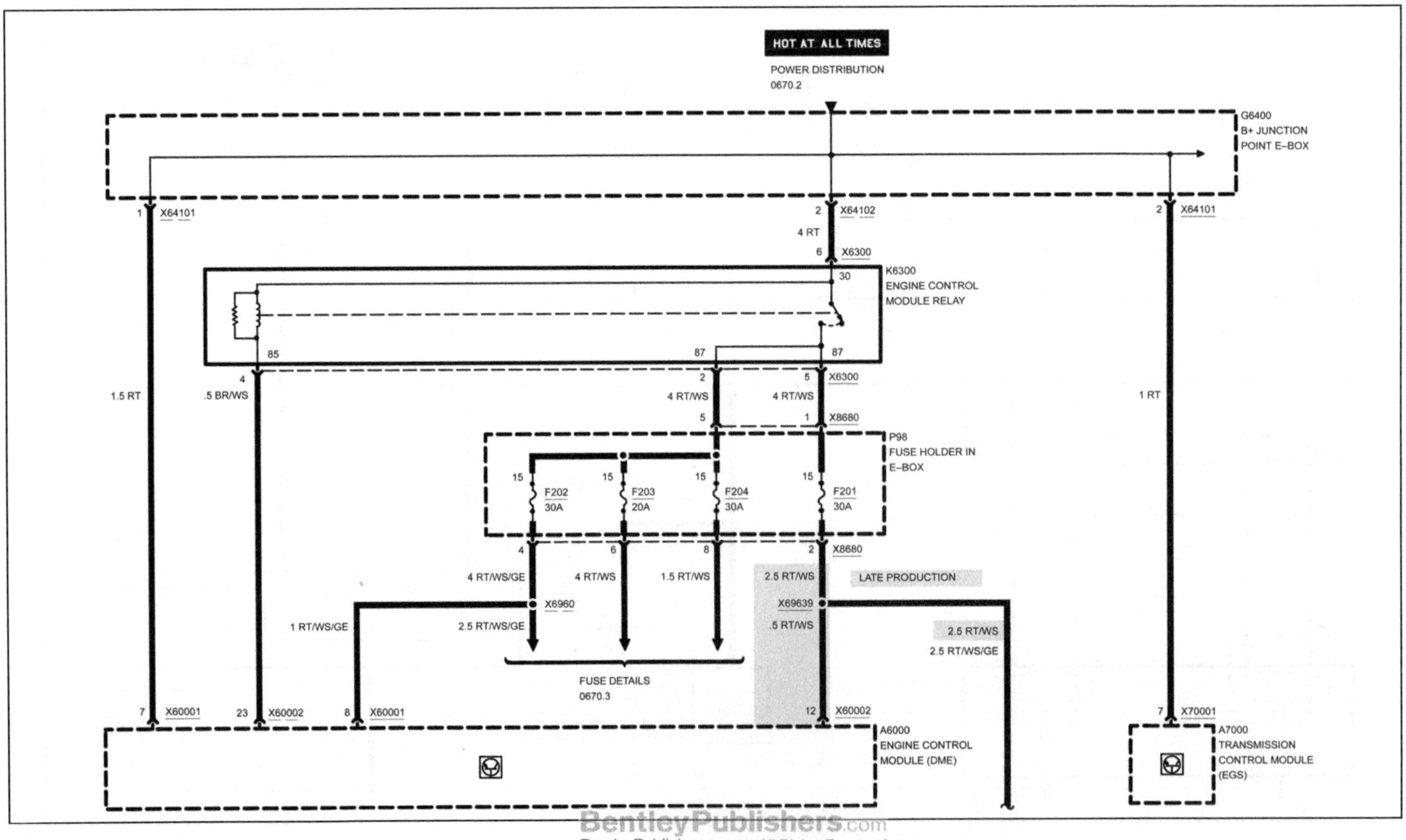

## Siemens DME MS 43.0 (M54 engine)
(2 of 11)

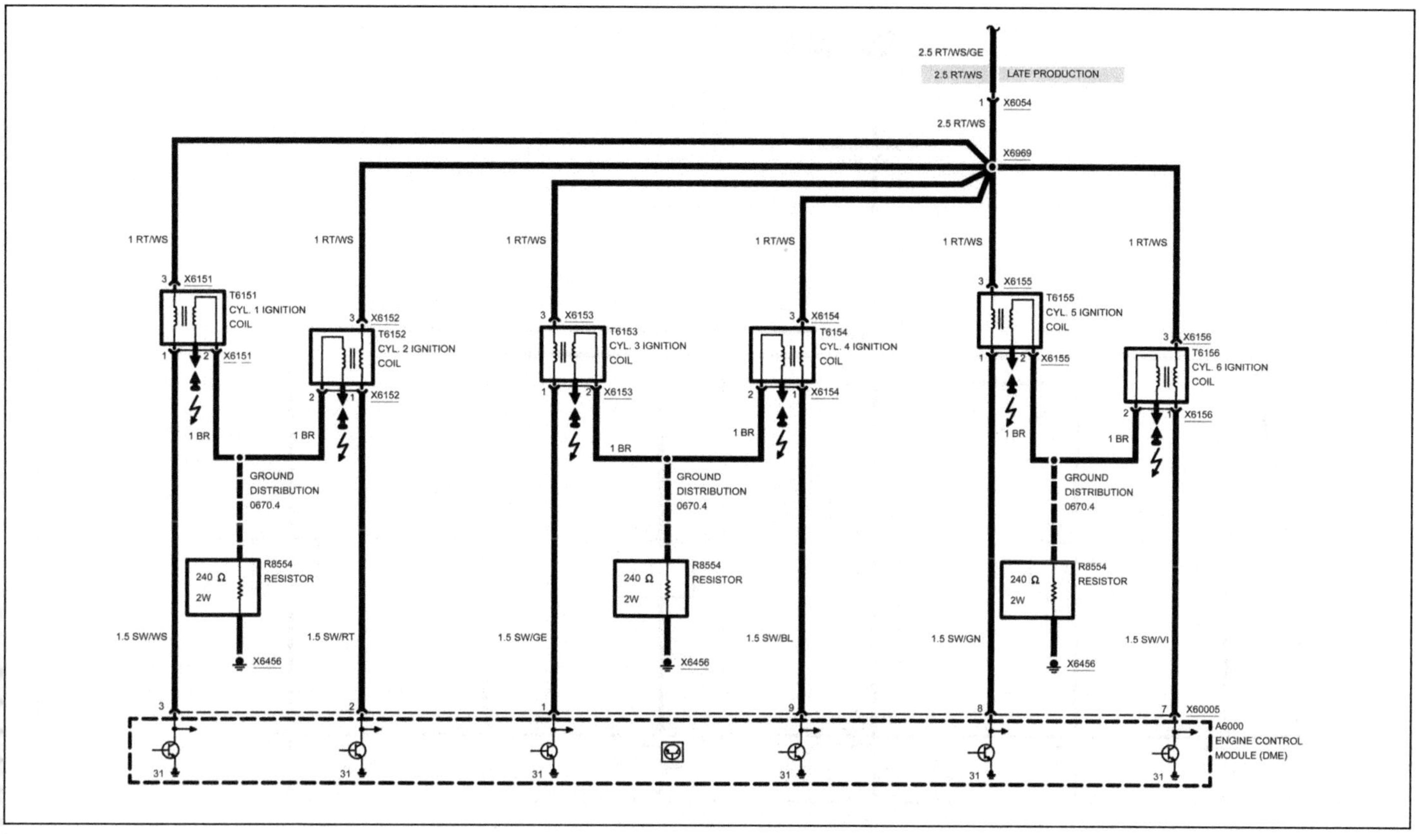


## Siemens DME MS 43.0 (M54 engine)
(3 of 11)

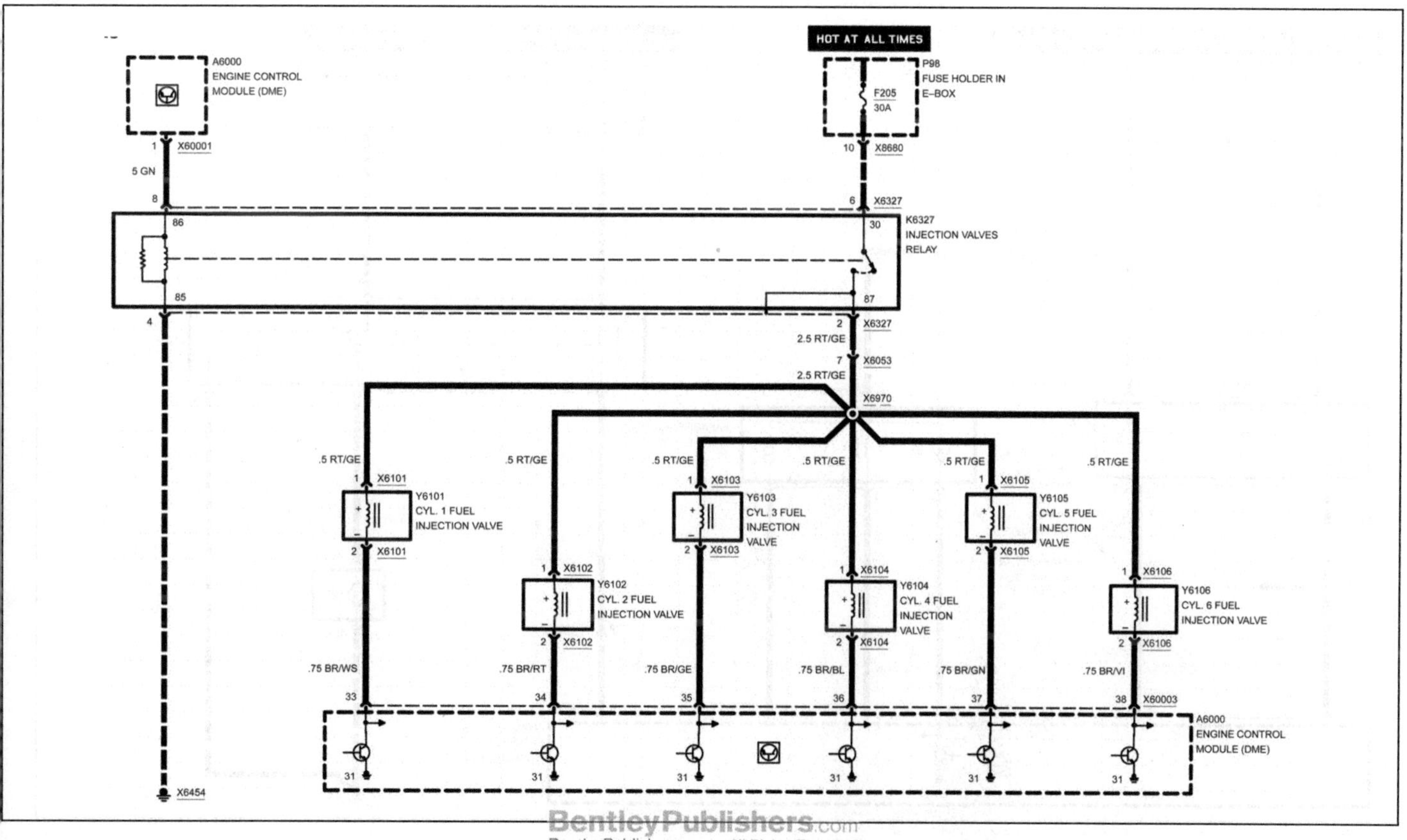

## Siemens DME MS 43.0 (M54 engine)
(4 of 11)

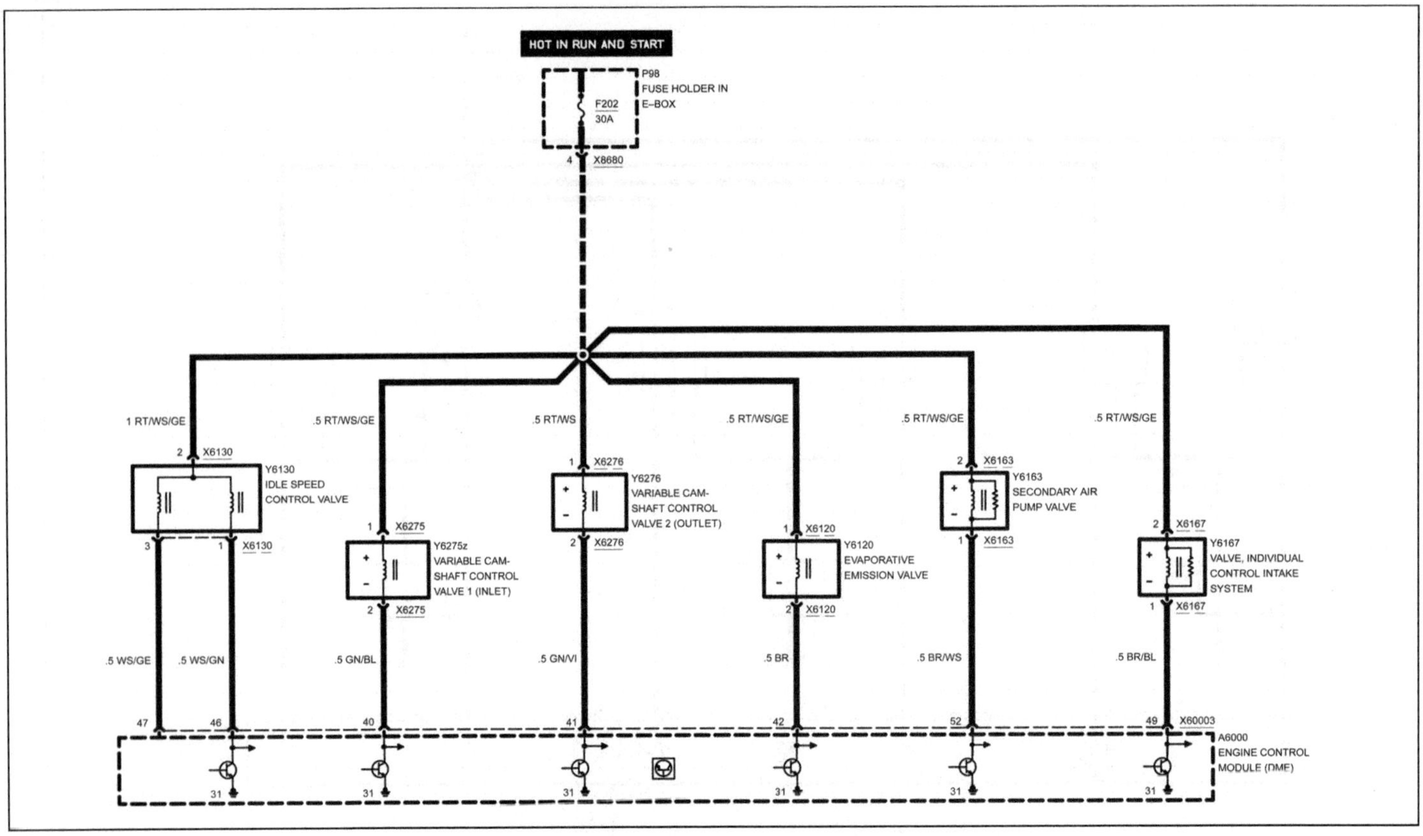


## Siemens DME MS 43.0 (M54 engine)
(5 of 11)

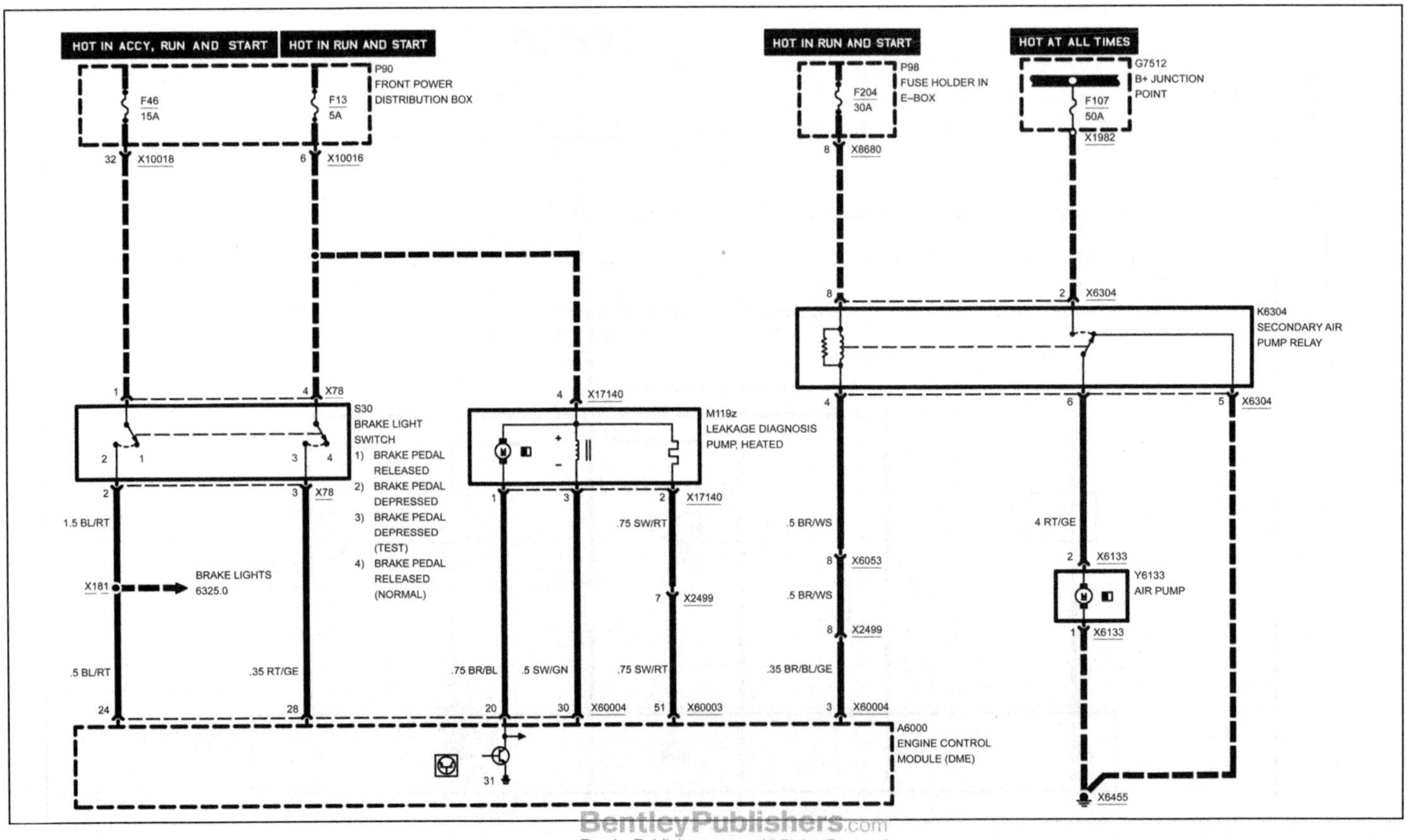

## Siemens DME MS 43.0 (M54 engine)
(6 of 11)

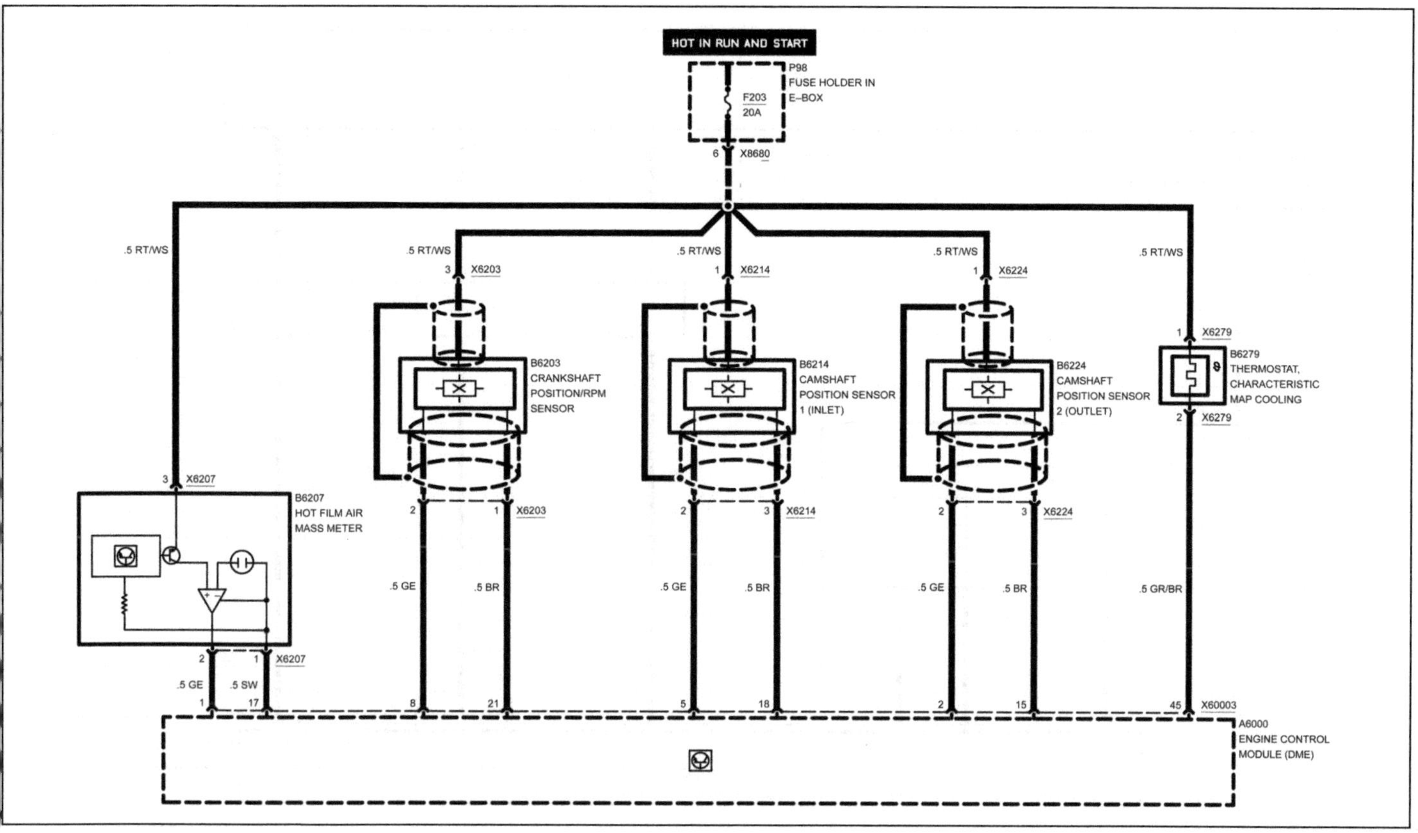


## Siemens DME MS 43.0 (M54 engine)
(7 of 11)

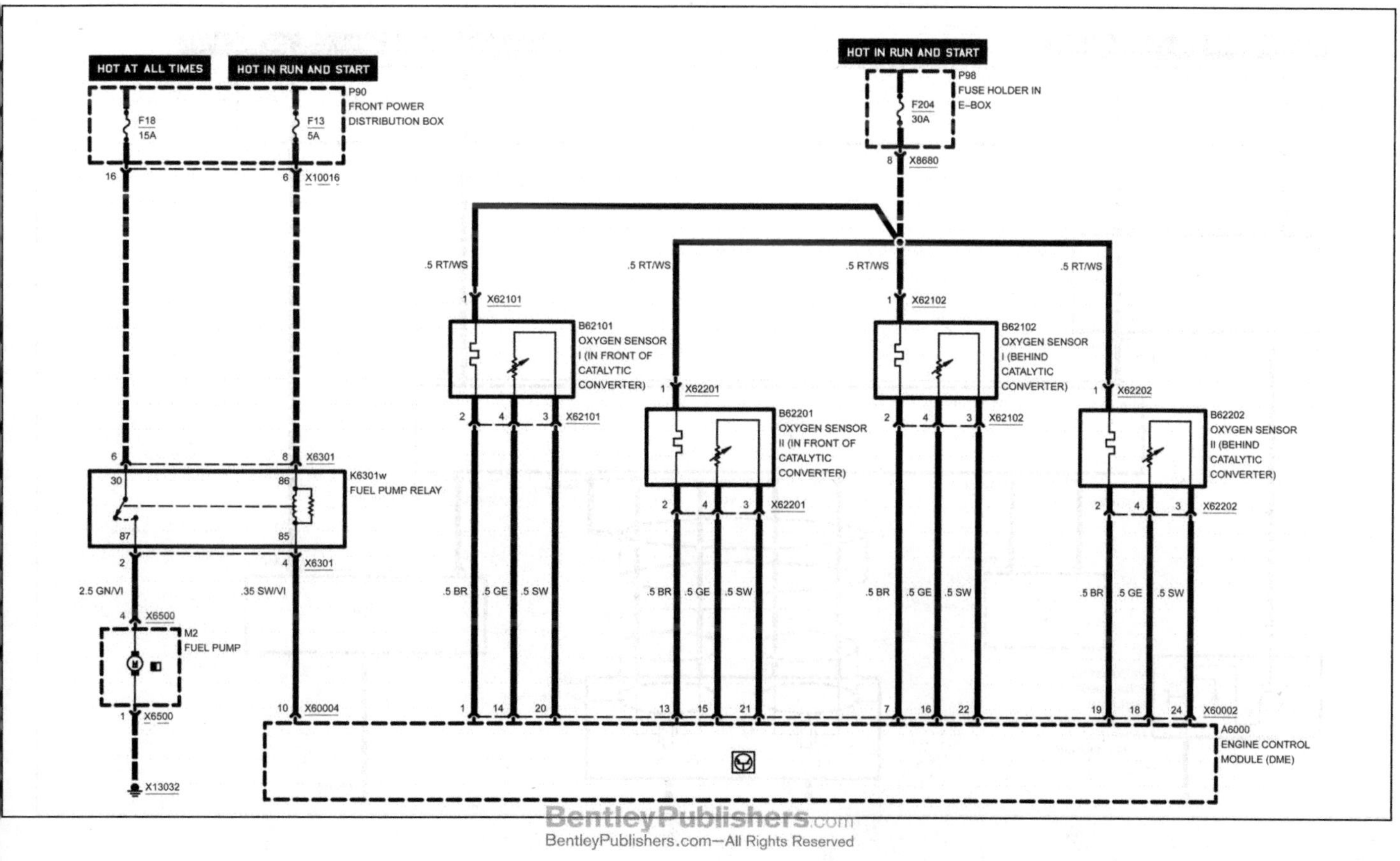

## Siemens DME MS 43.0 (M54 engine)
(8 of 11)

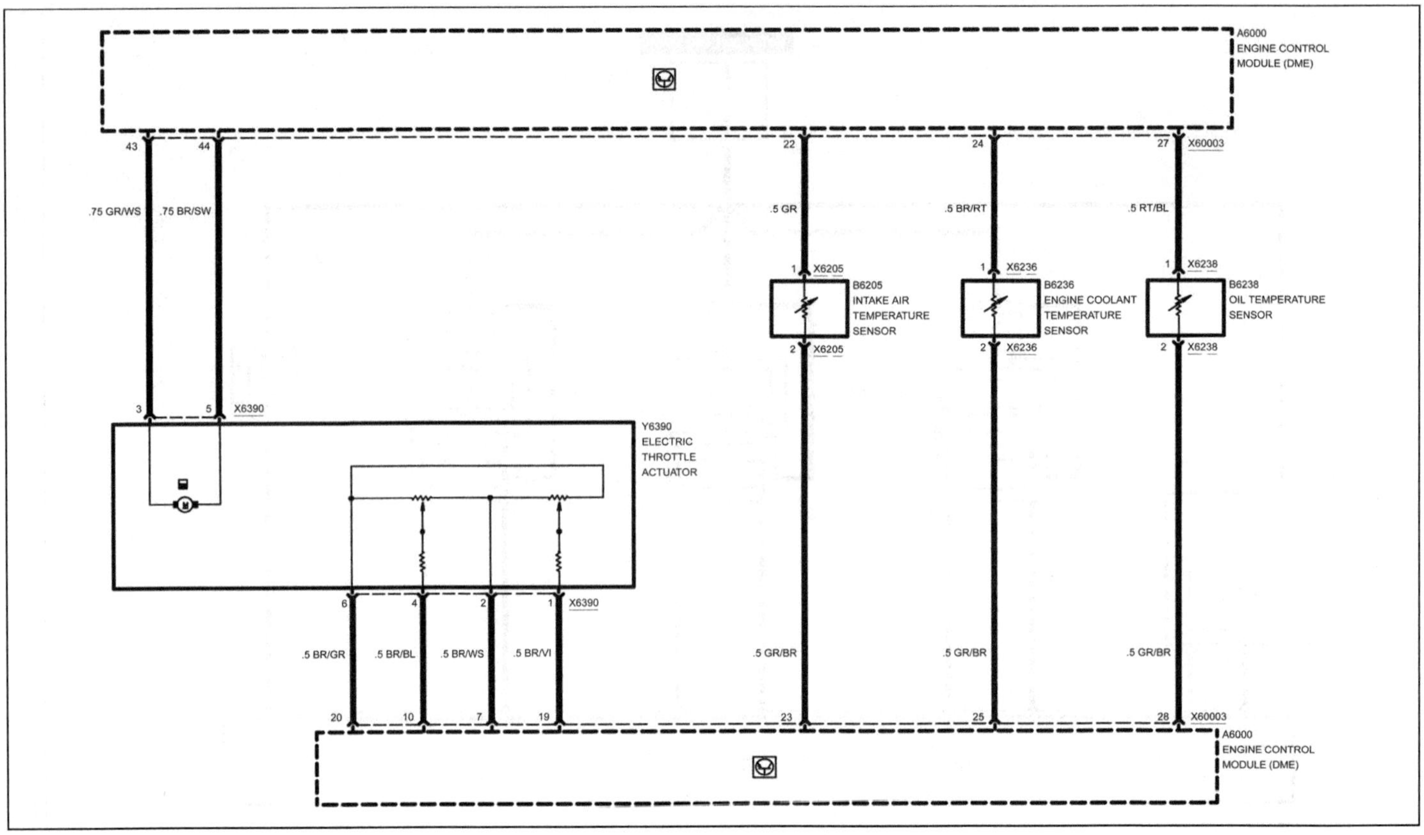


## Siemens DME MS 43.0 (M54 engine)
(9 of 11)

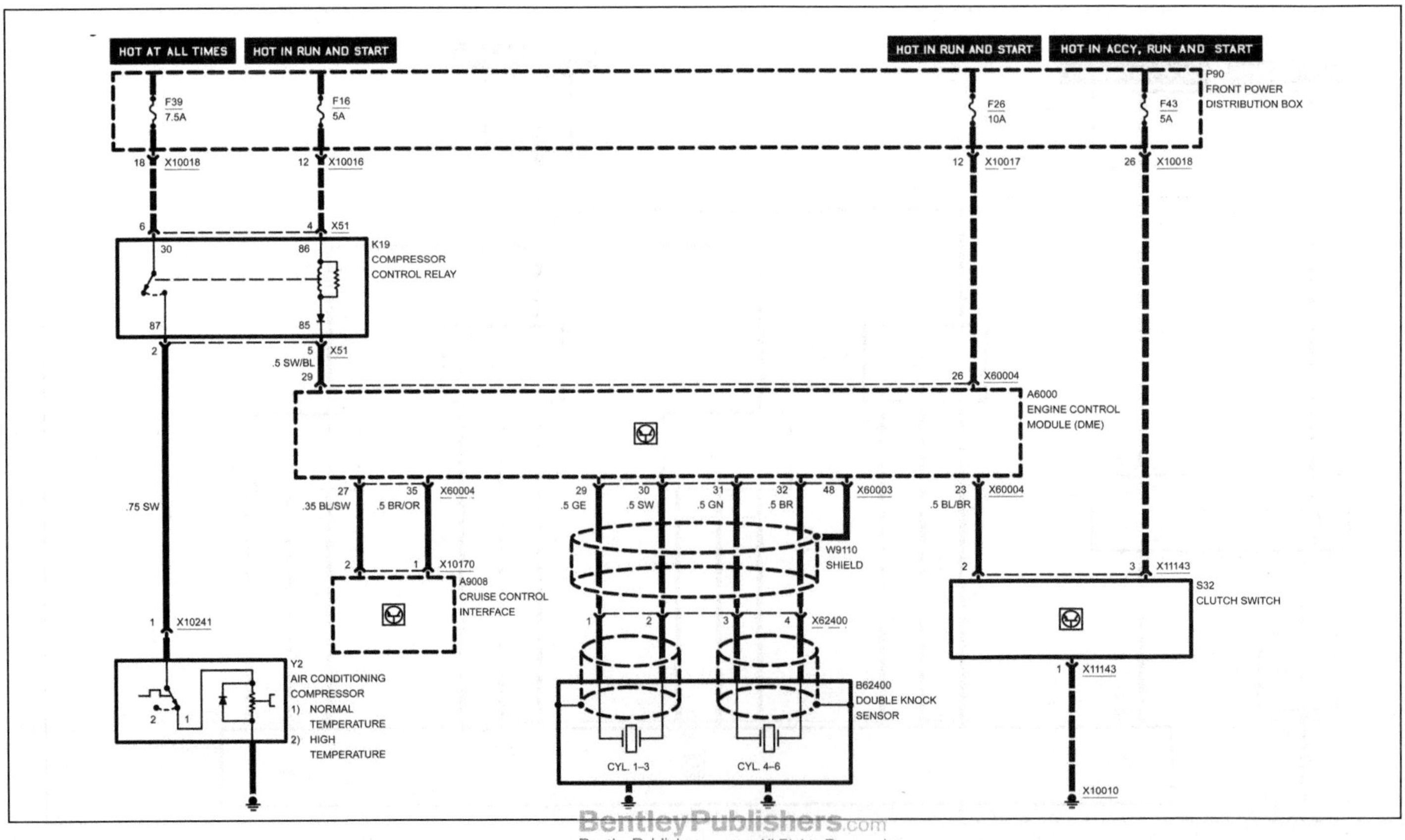

## wSiemens DME MS 43.0 (M54 engine)
(10 of 11)

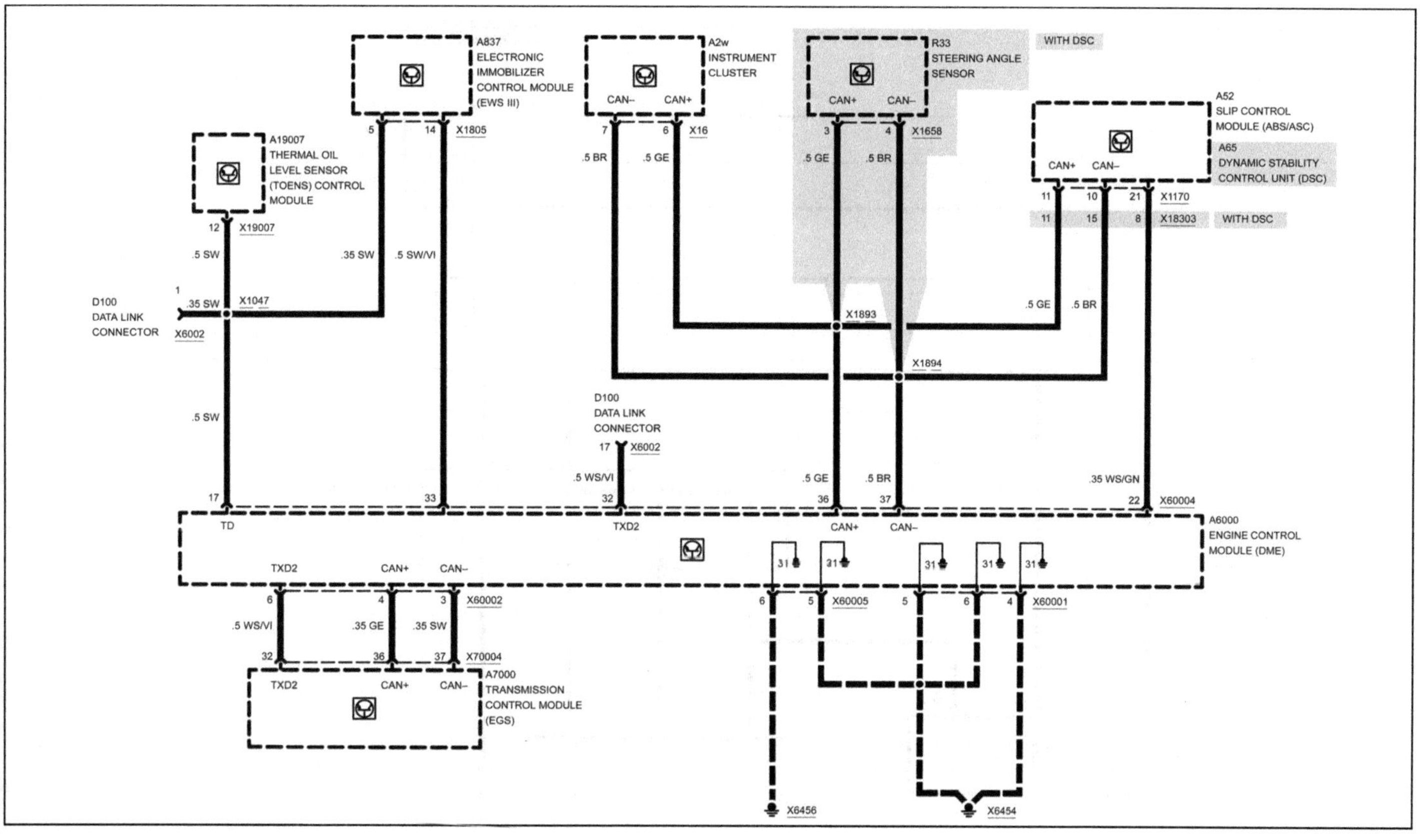


## Siemens DME MS 43.0 (M54 engine)
(11 of 11)

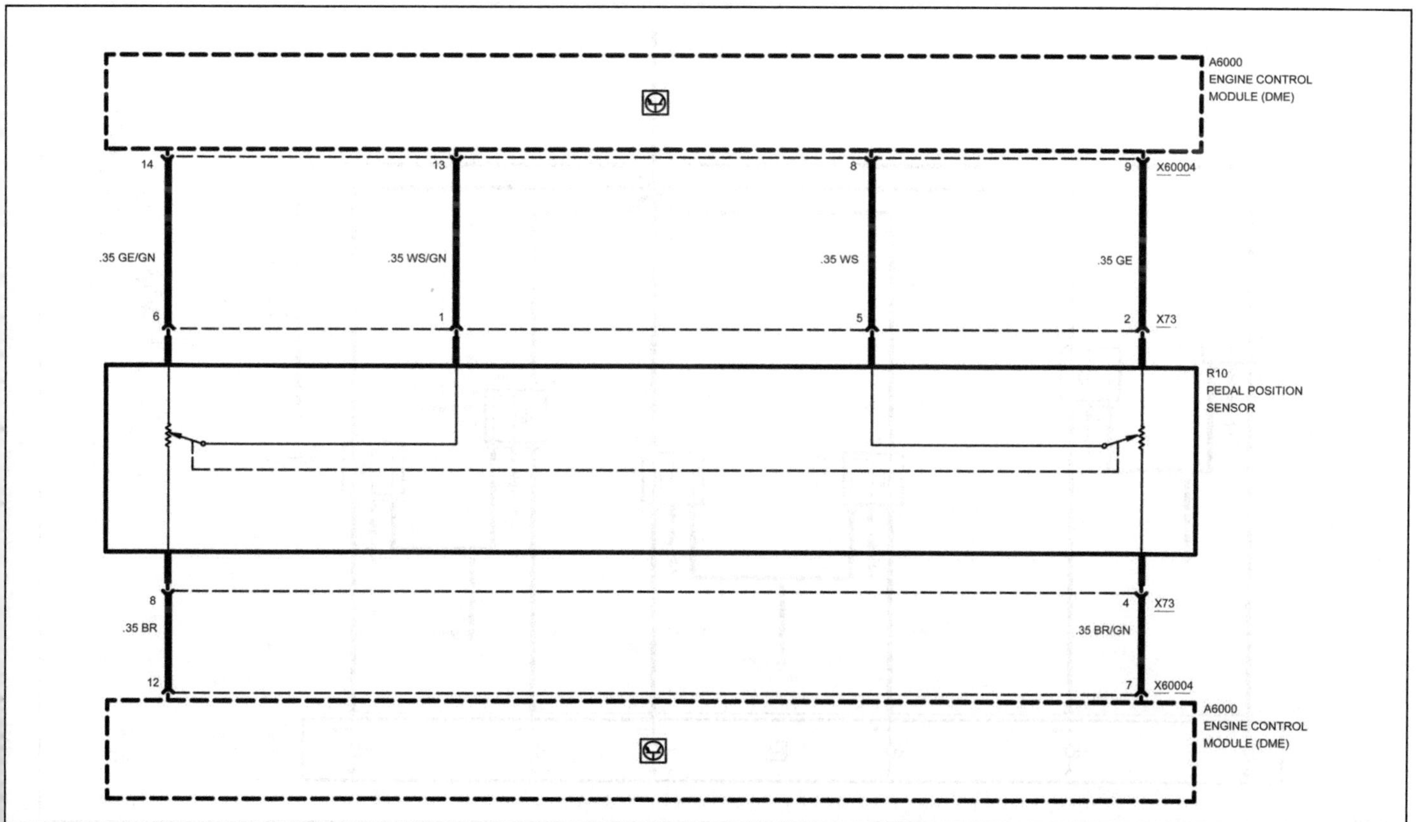

## Siemens DME MS S54 (S54 engine)
### (1 of 14)

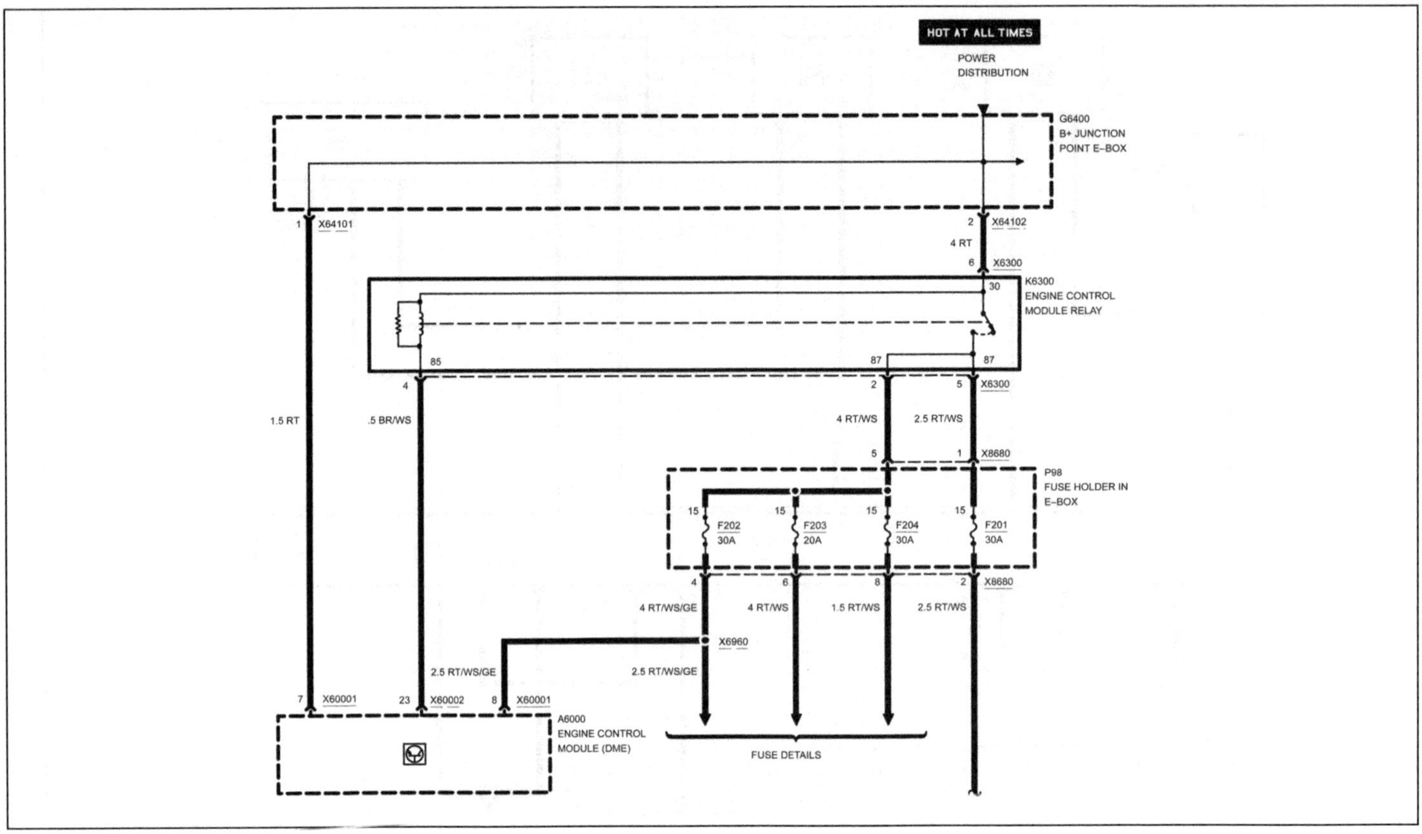


## Siemens DME MS S54 (S54 engine)
### (2 of 14)

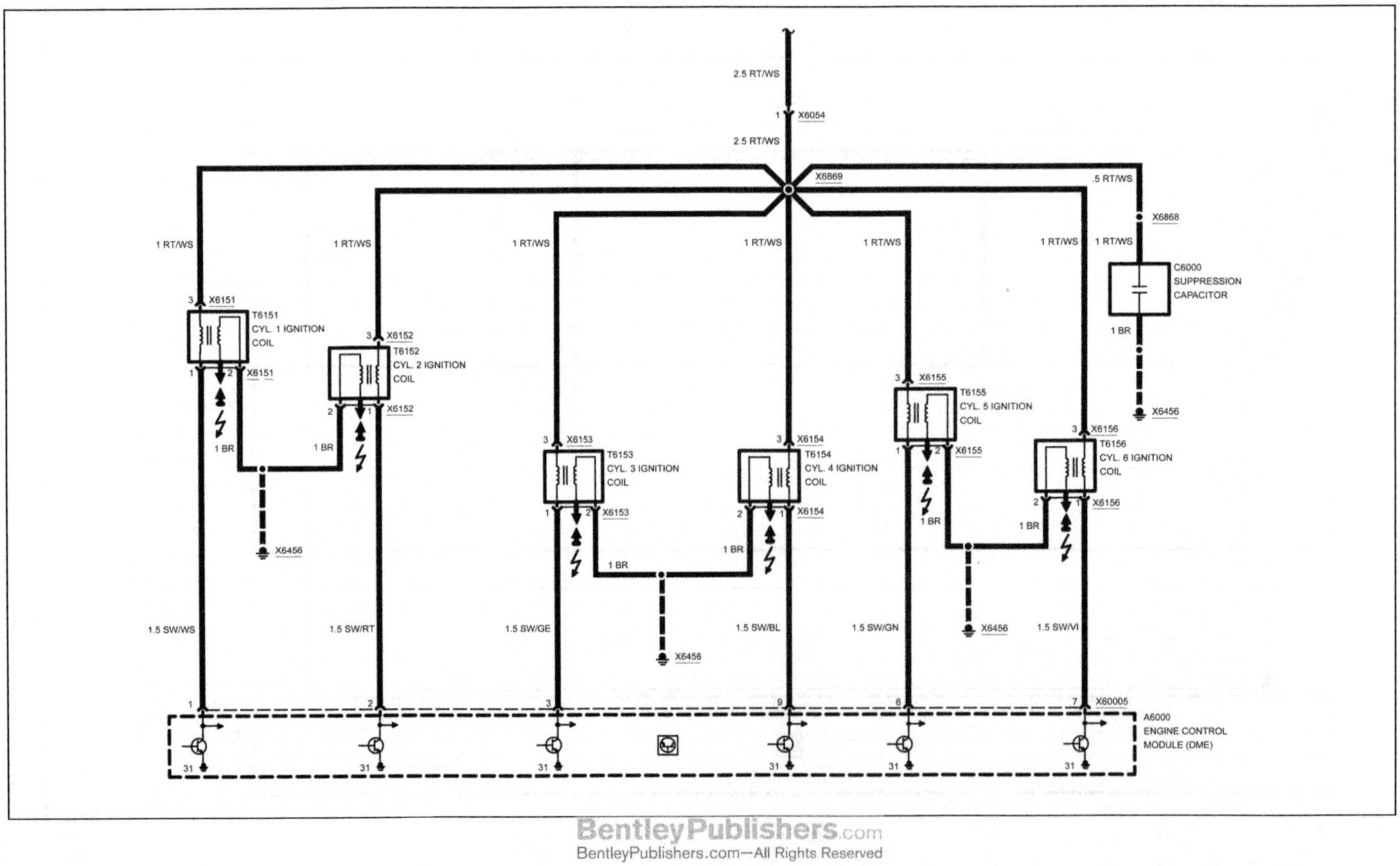

## Siemens DME MS S54 (S54 engine)

**(3 of 14)**

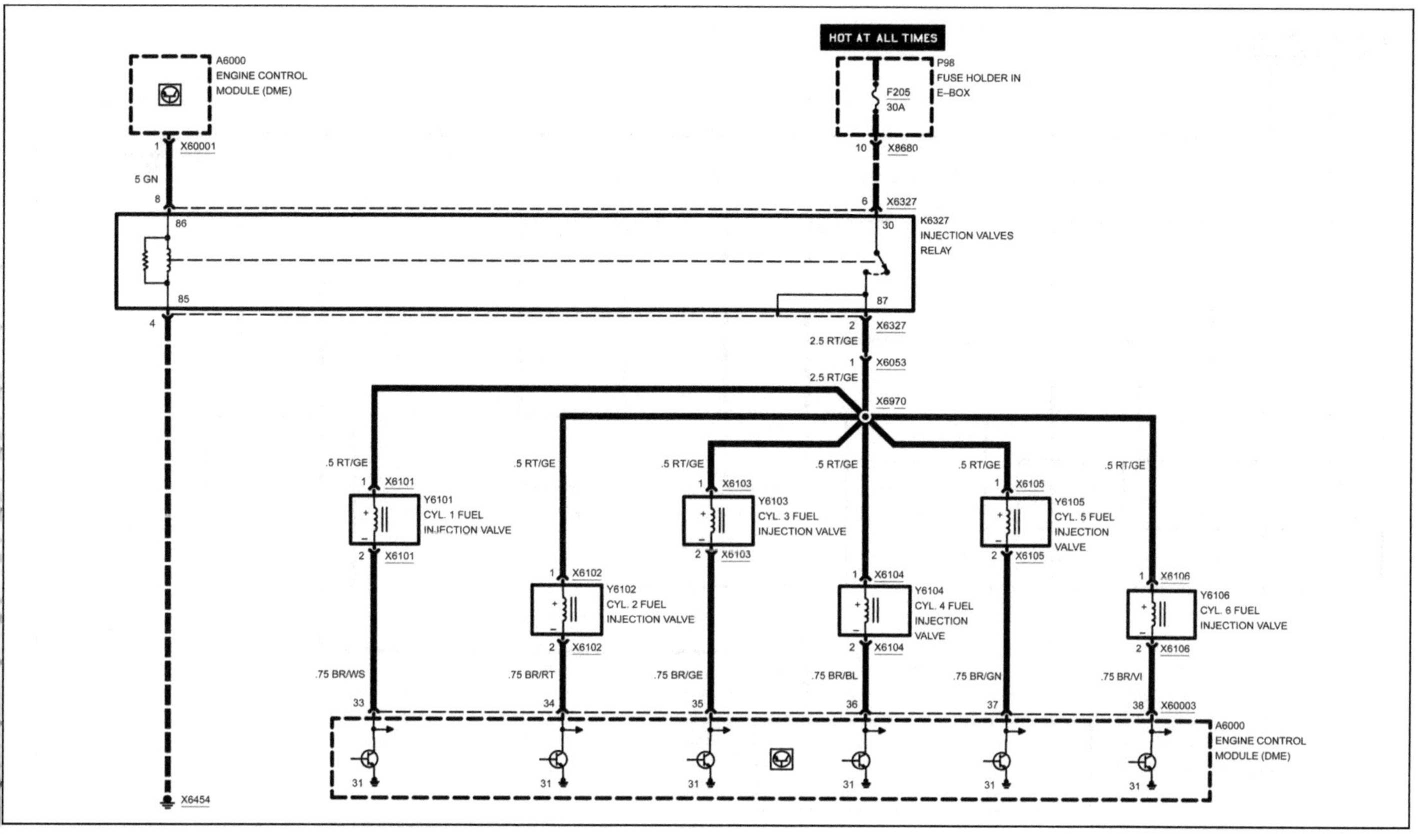


## Siemens DME MS S54 (S54 engine)

**(4 of 14)**

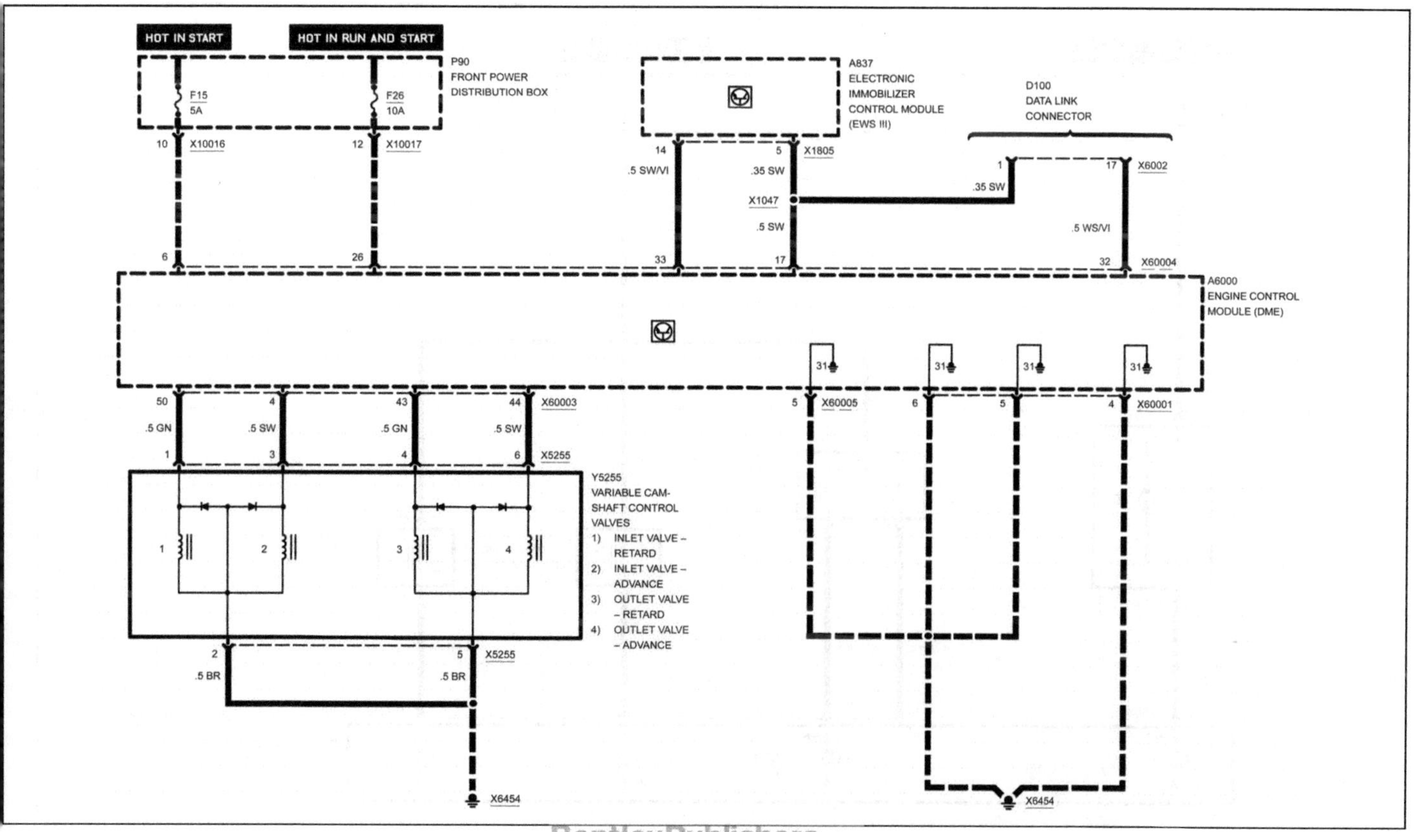

## Siemens DME MS S54 (S54 engine)
(5 of 14)

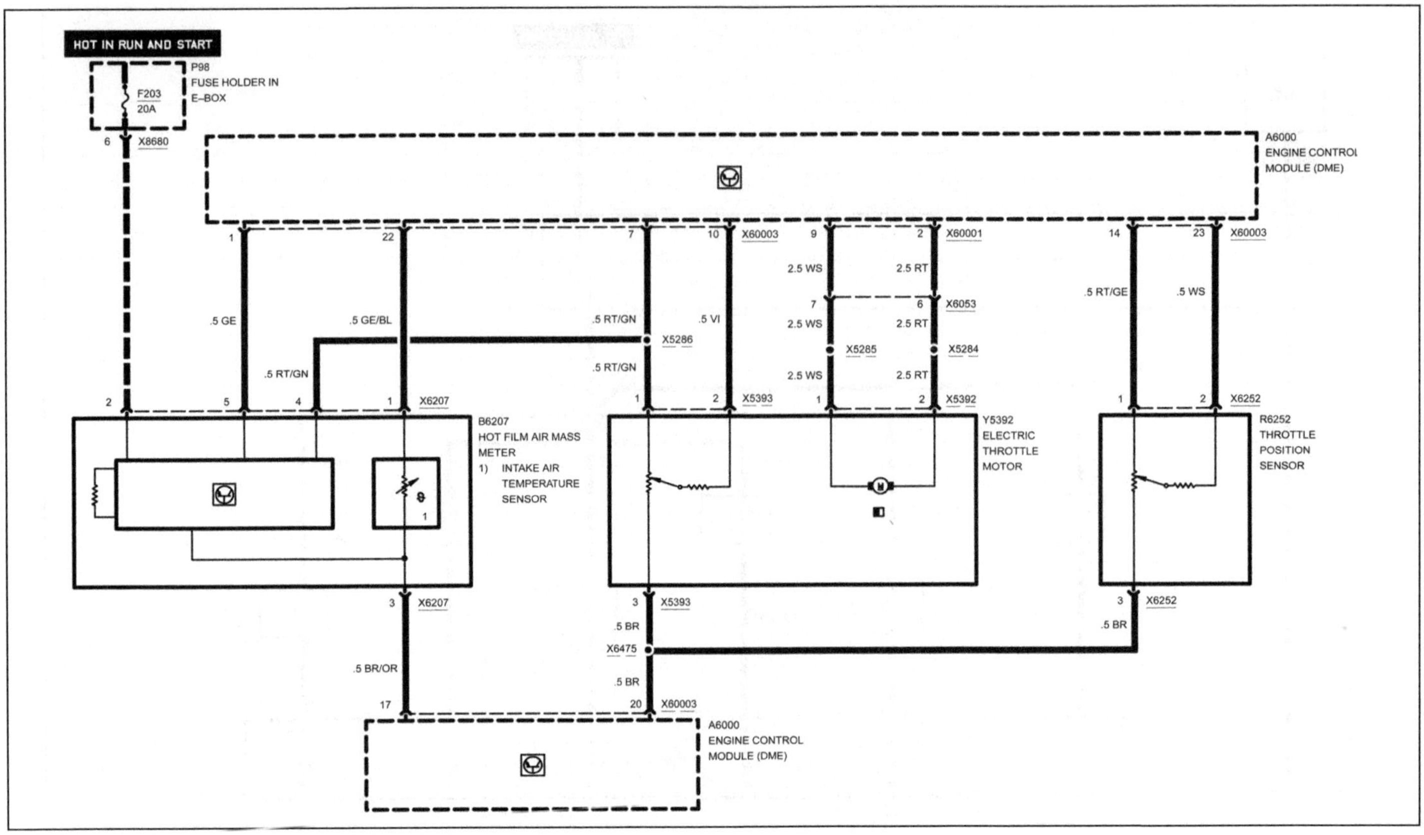


## Siemens DME MS S54 (S54 engine)
(6 of 14)

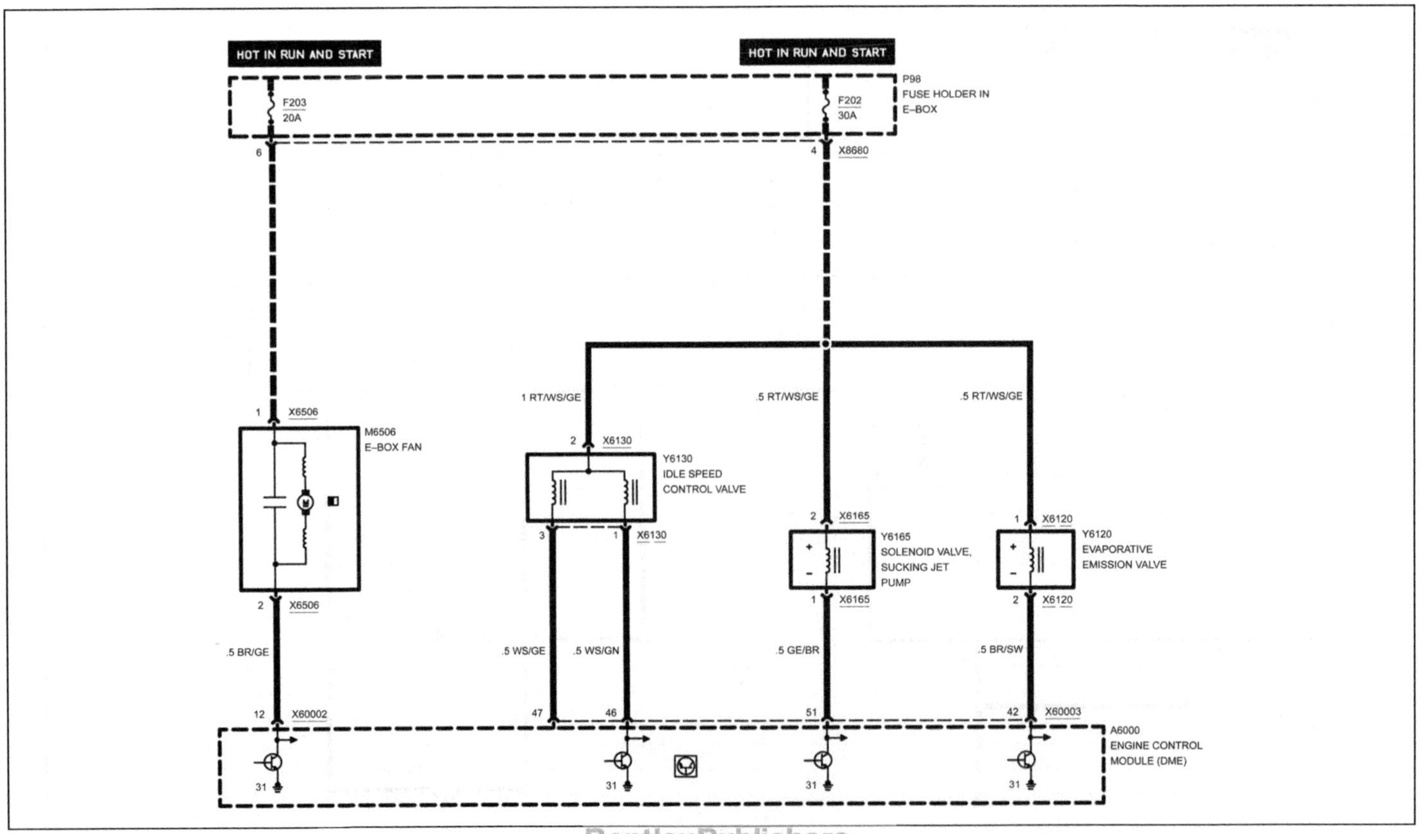

## Siemens DME MS S54 (S54 engine)

(7 of 14)

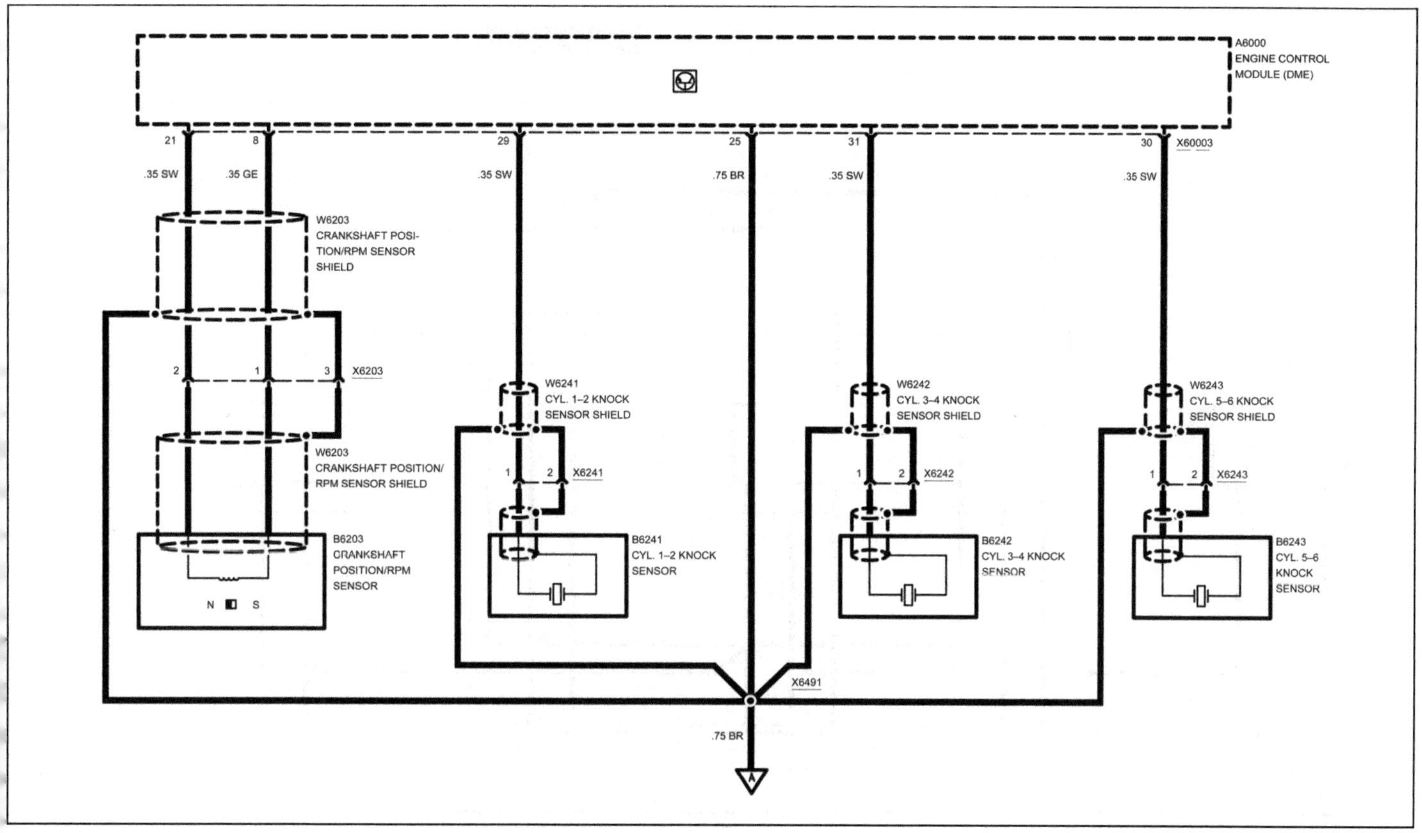


## Siemens DME MS S54 (S54 engine)

(8 of 14)

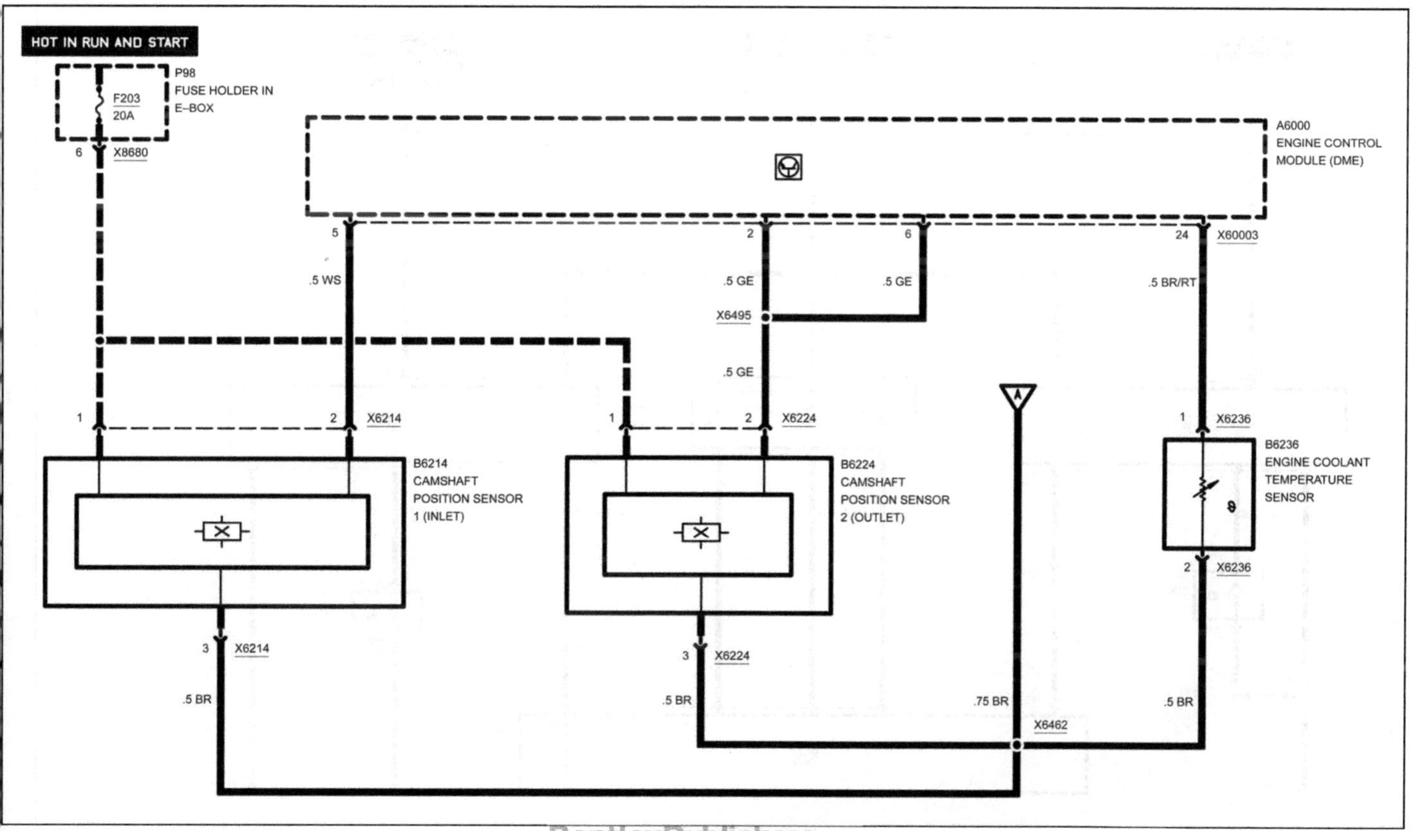

## Siemens DME MS S54 (S54 engine)
**(9 of 14)**

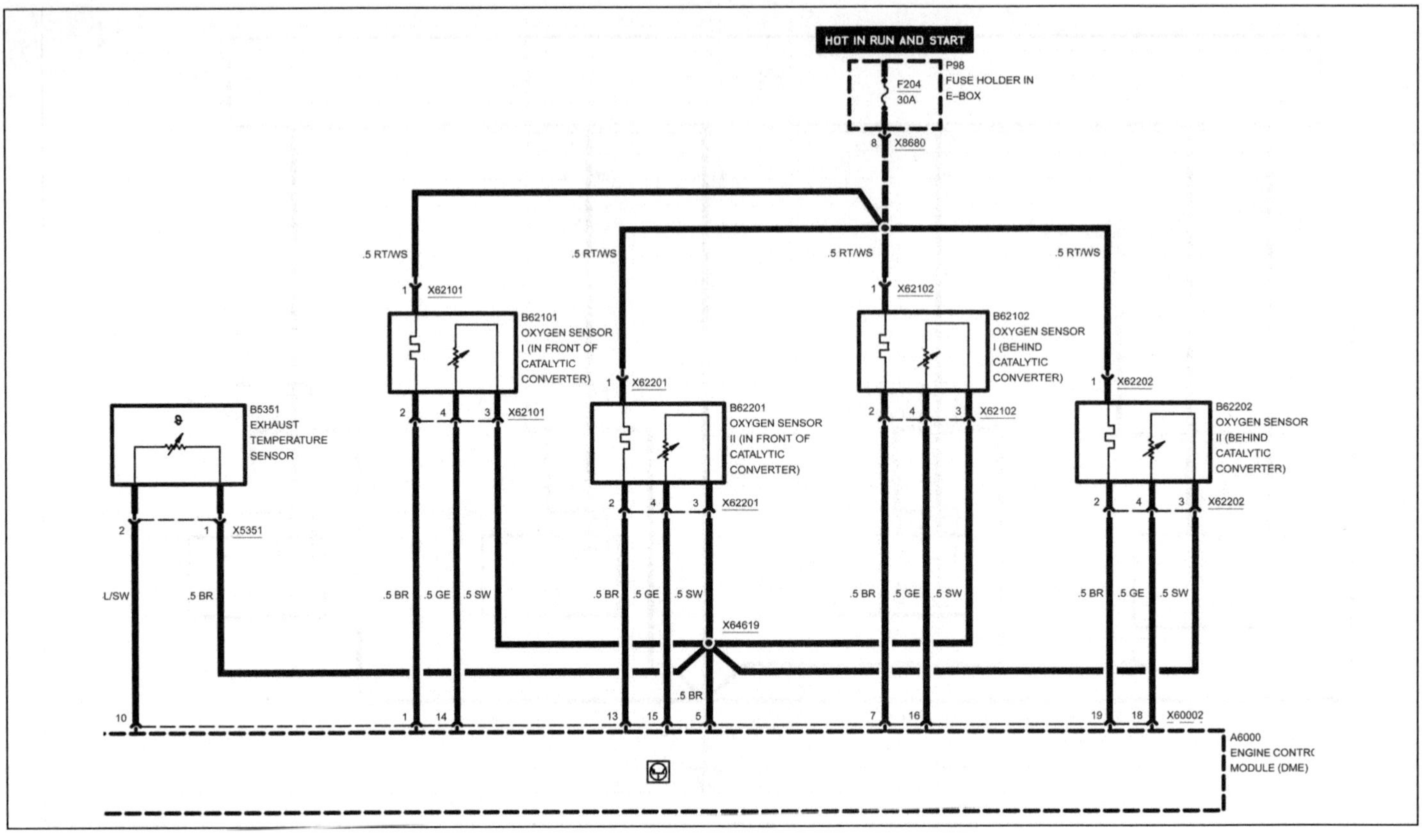


## Siemens DME MS S54 (S54 engine)
**(10 of 14)**

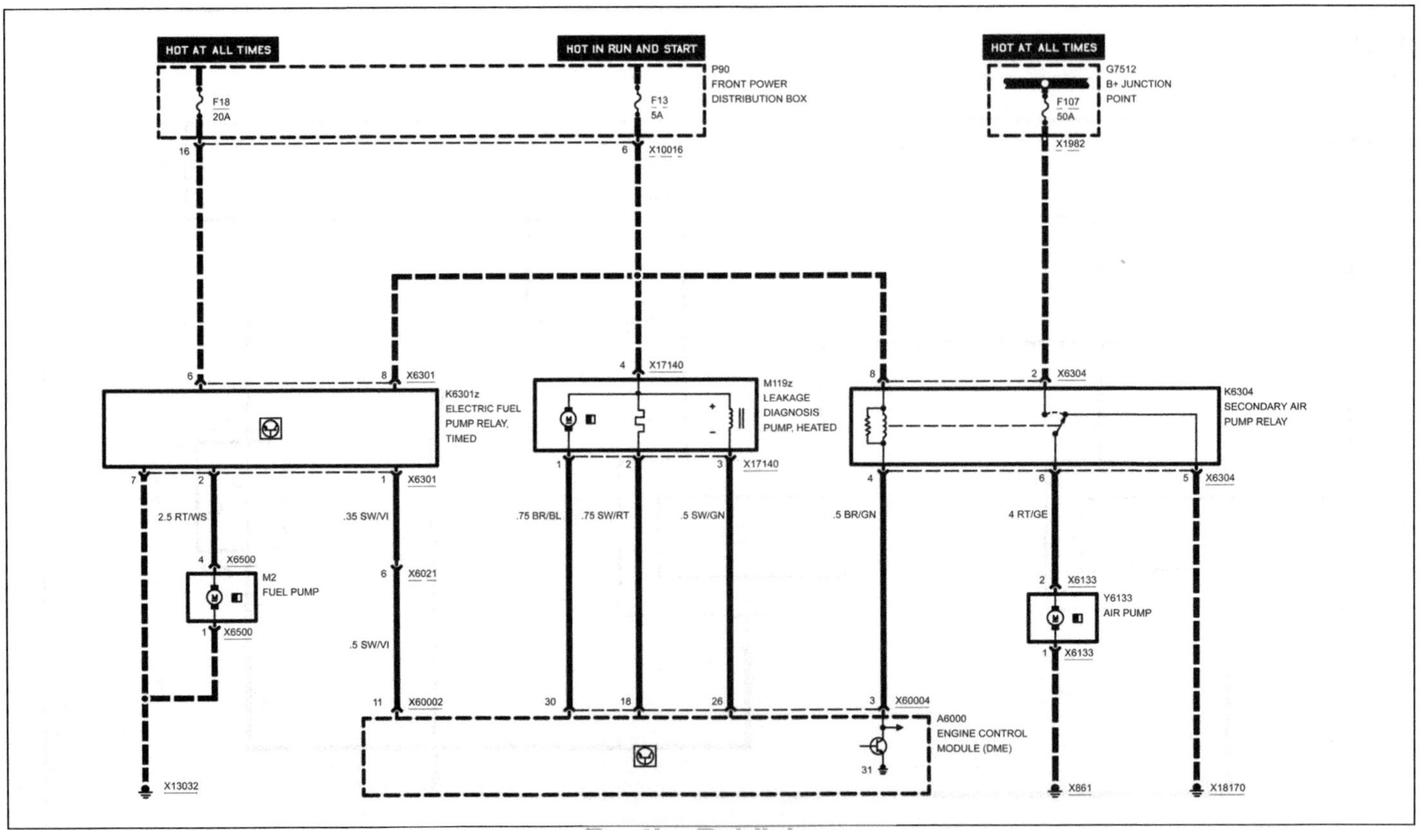

## Siemens DME MS S54 (S54 engine)

(11 of 14)

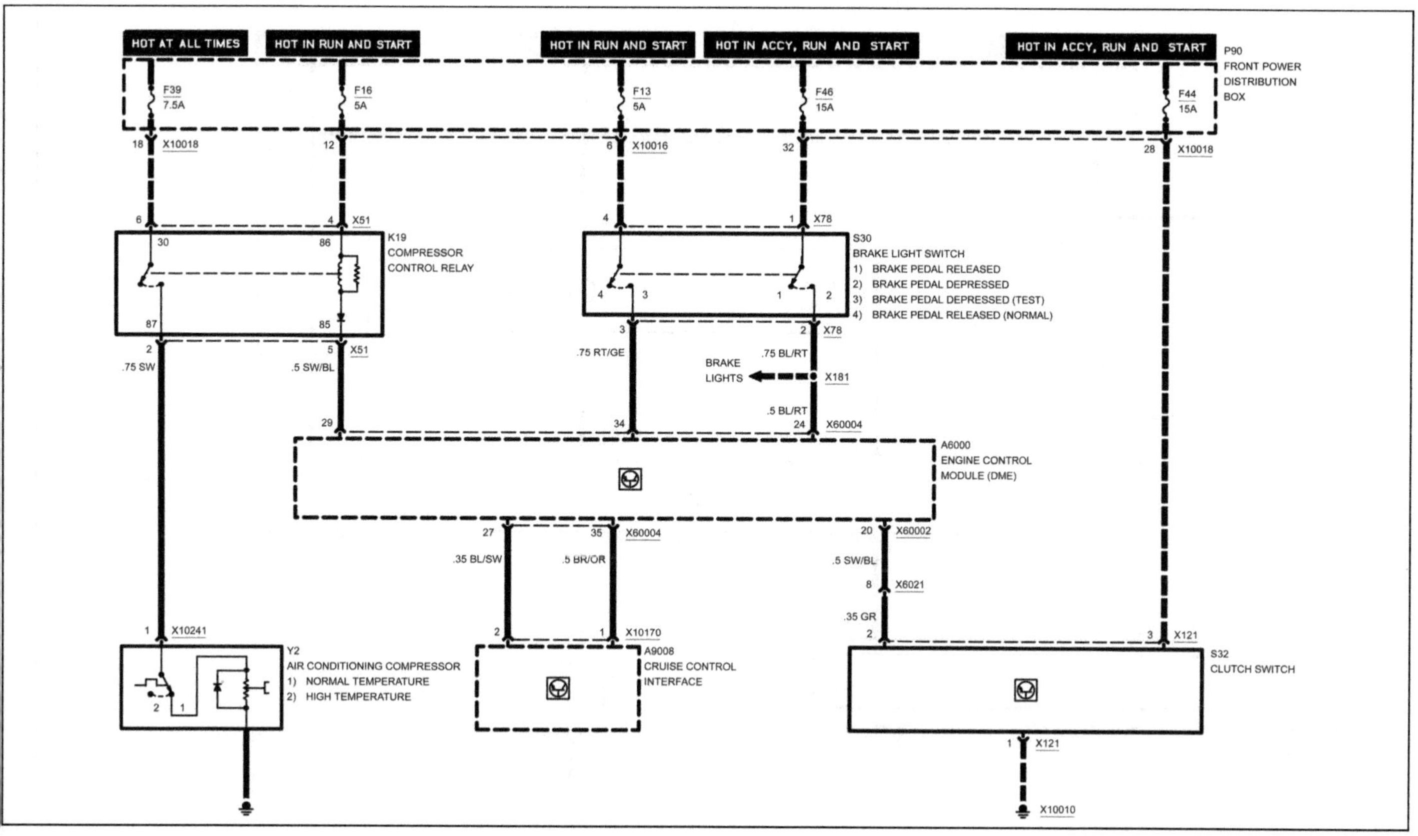


## Siemens DME MS S54 (S54 engine)

(12 of 14)

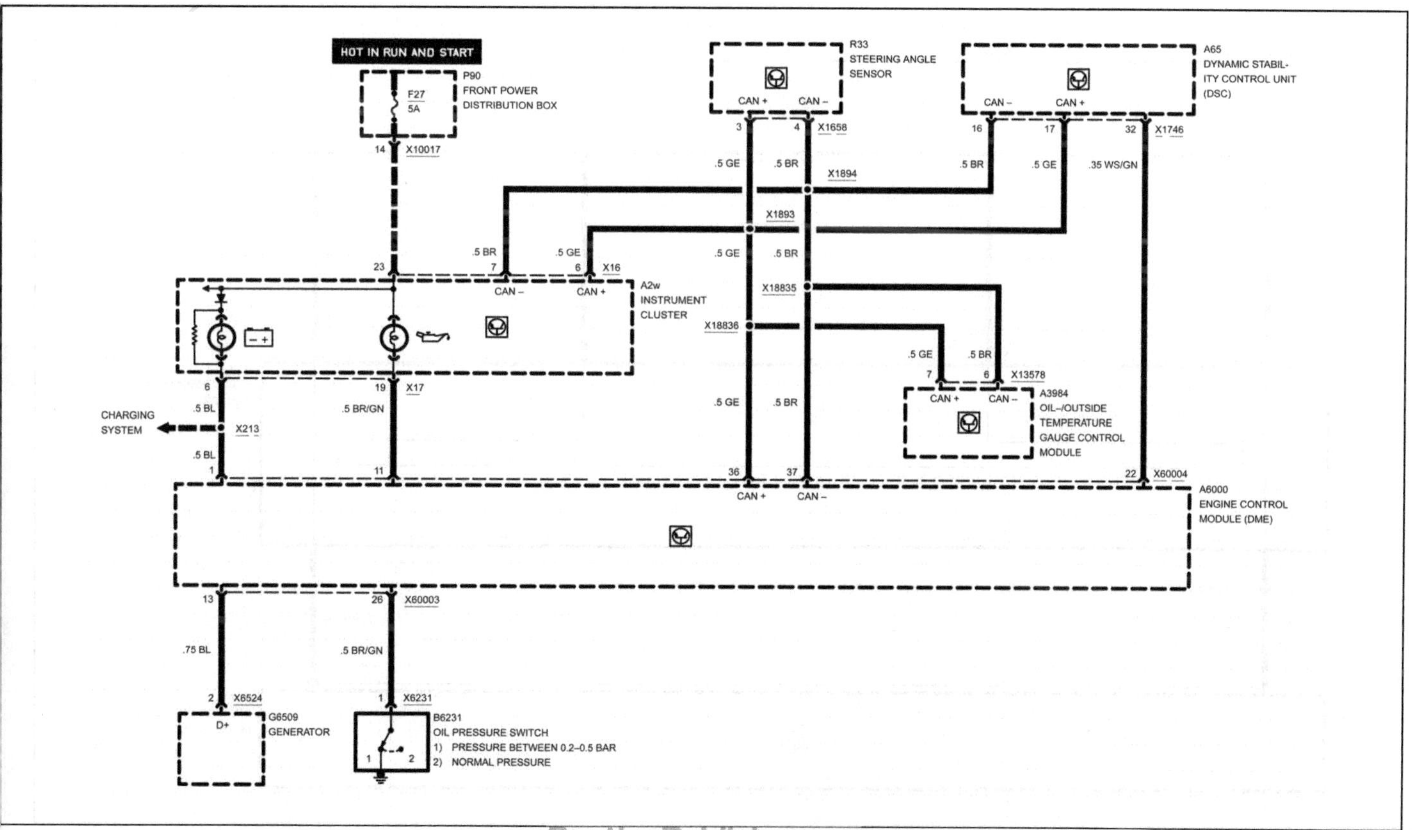

## Siemens DME MS S54 (S54 engine)
### (13 of 14)

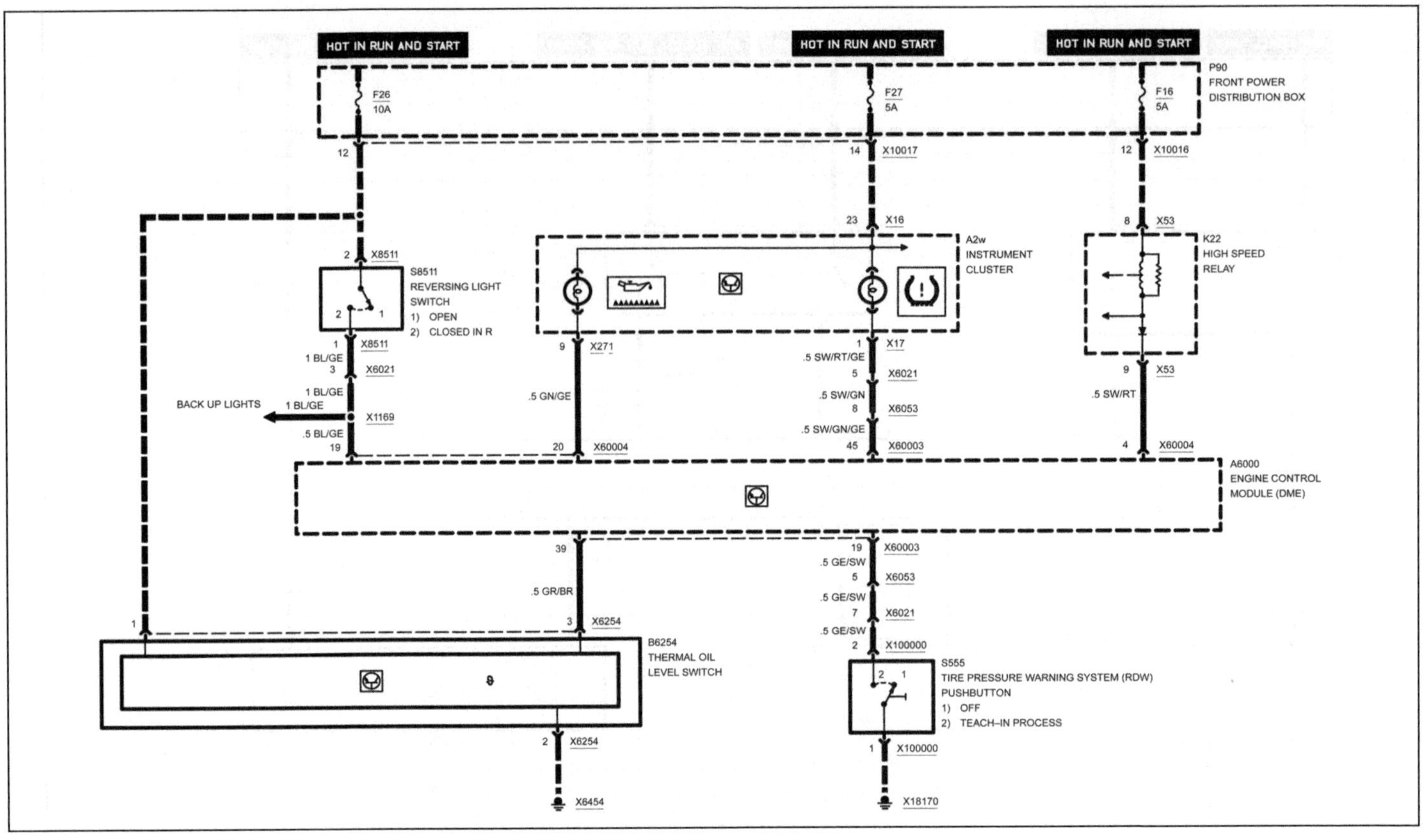


## Siemens DME MS S54 (S54 engine)
### (14 of 14)

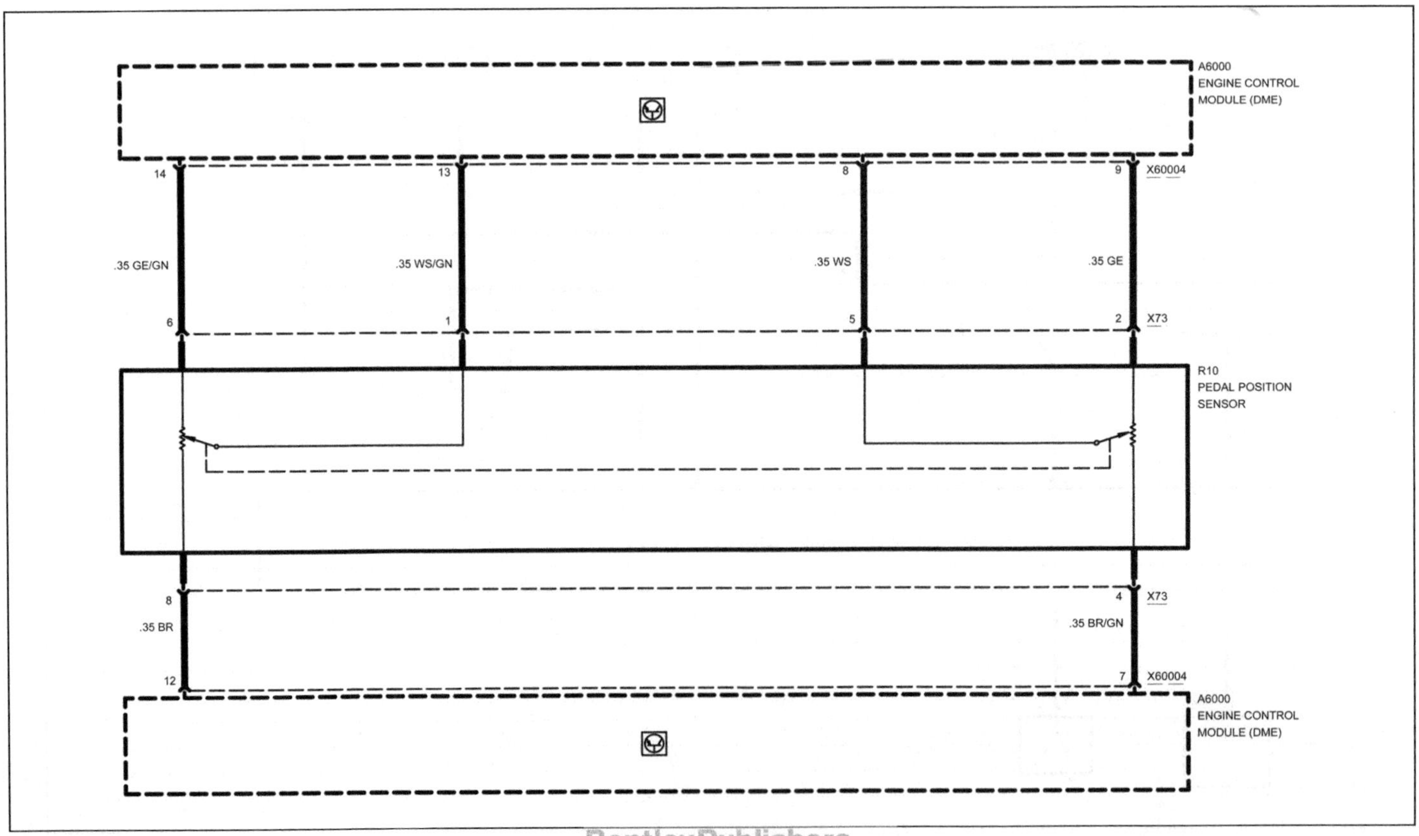

## Charging system, 1996 - 1998, 2000 M
**(1 of 2)**

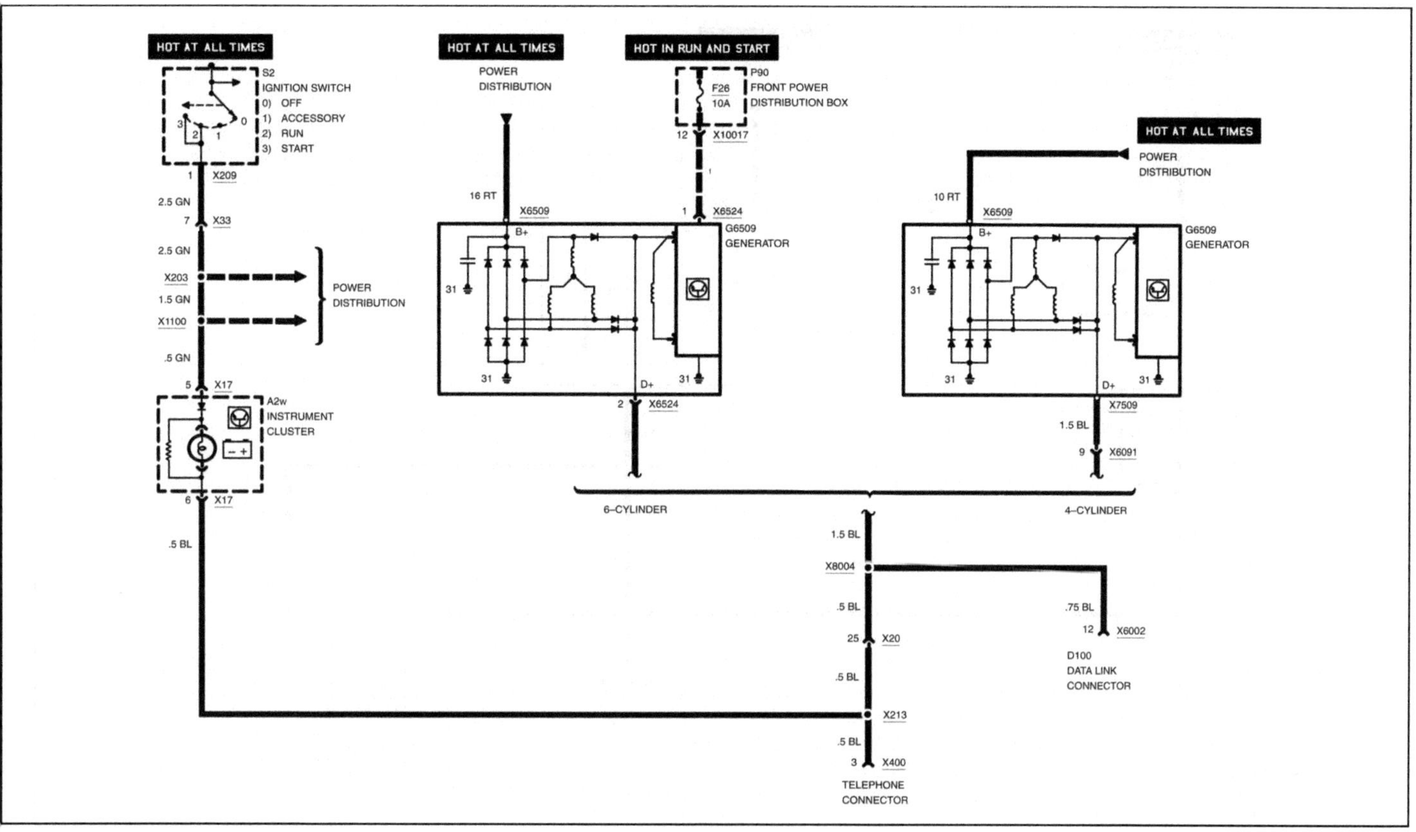


## Charging system, 1996 - 1998, 2000 M
**(2 of 2)**

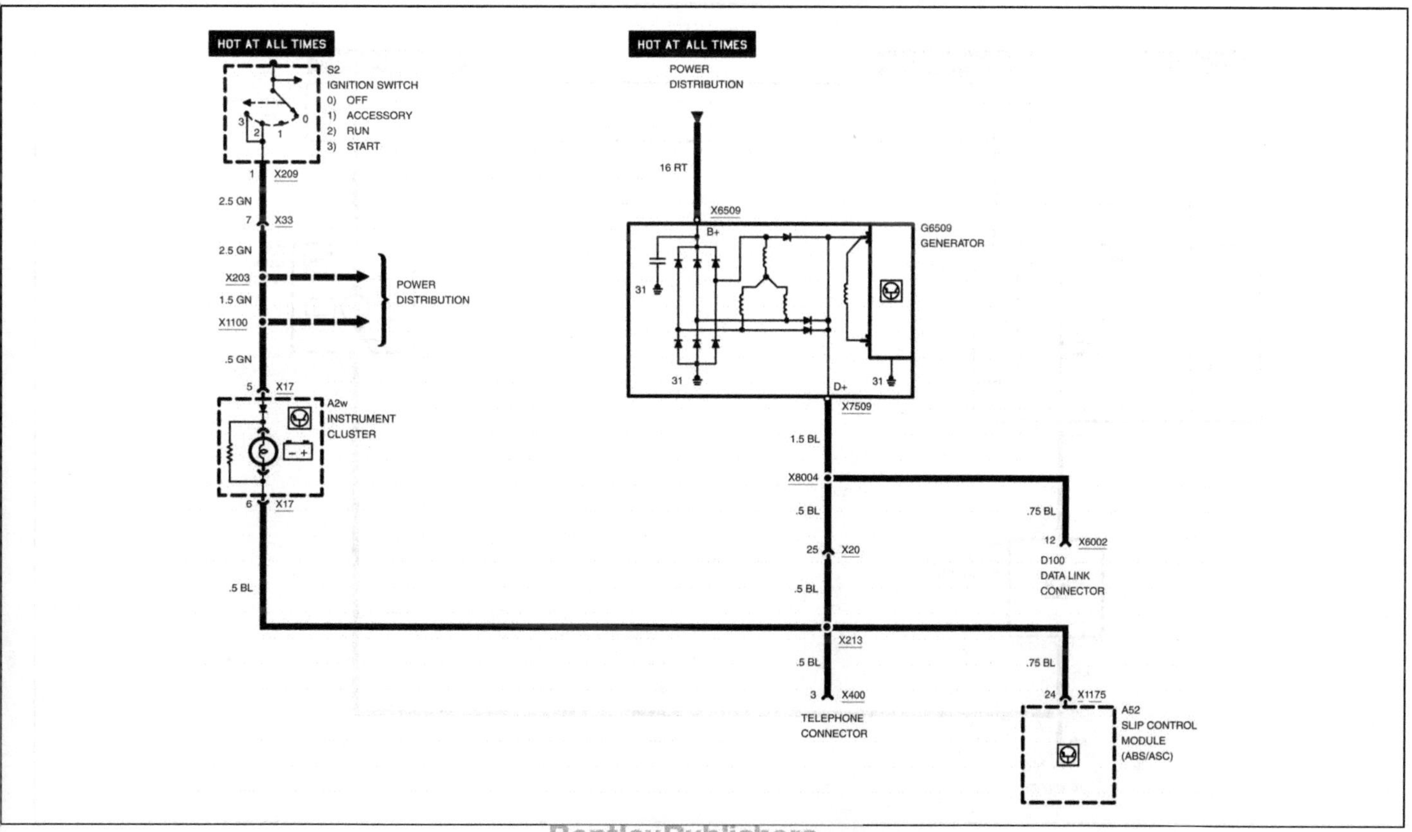

## Charging system, 2000 - 2002
(1 of 2)

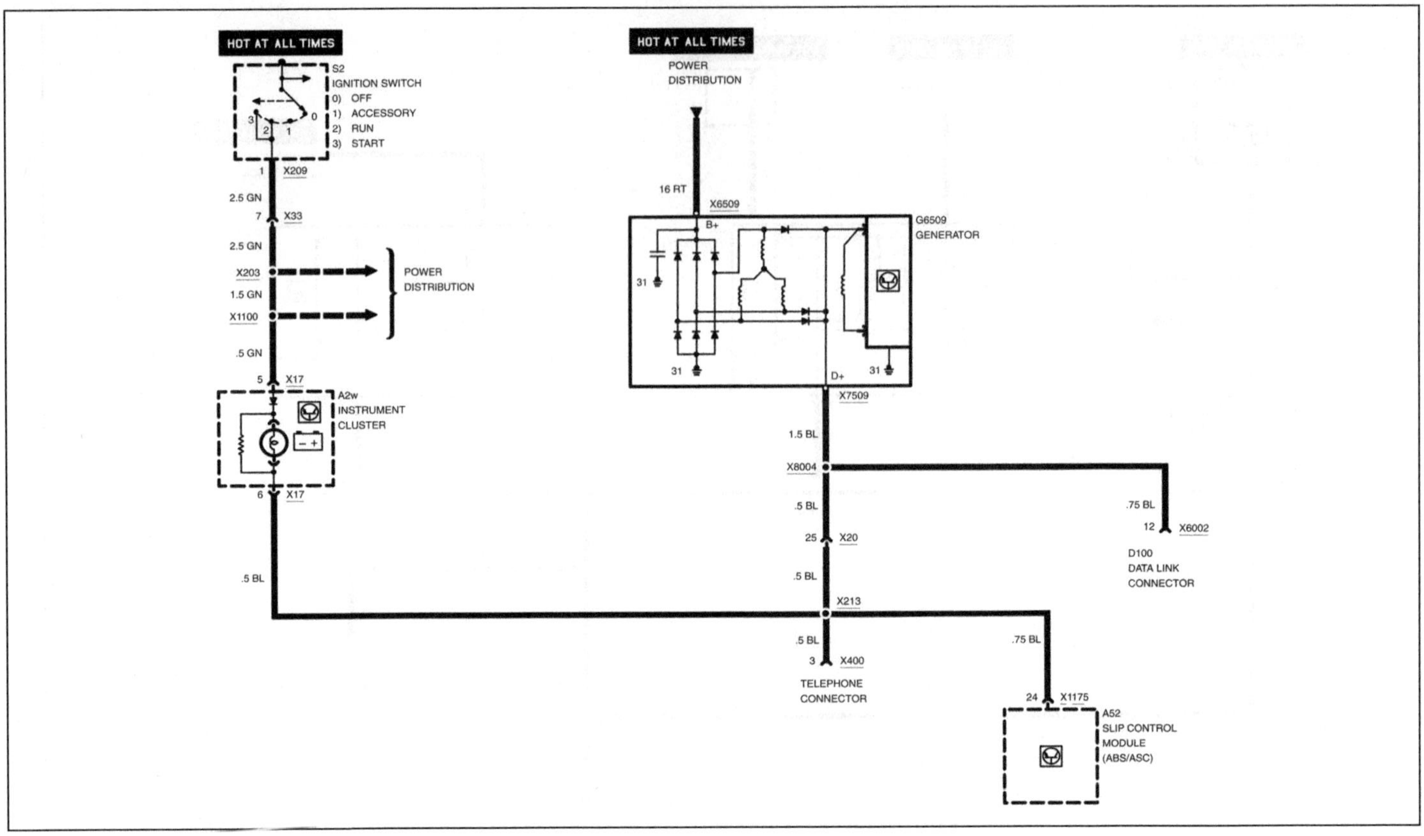


## Charging system, 2000 - 2002
(2 of 2)

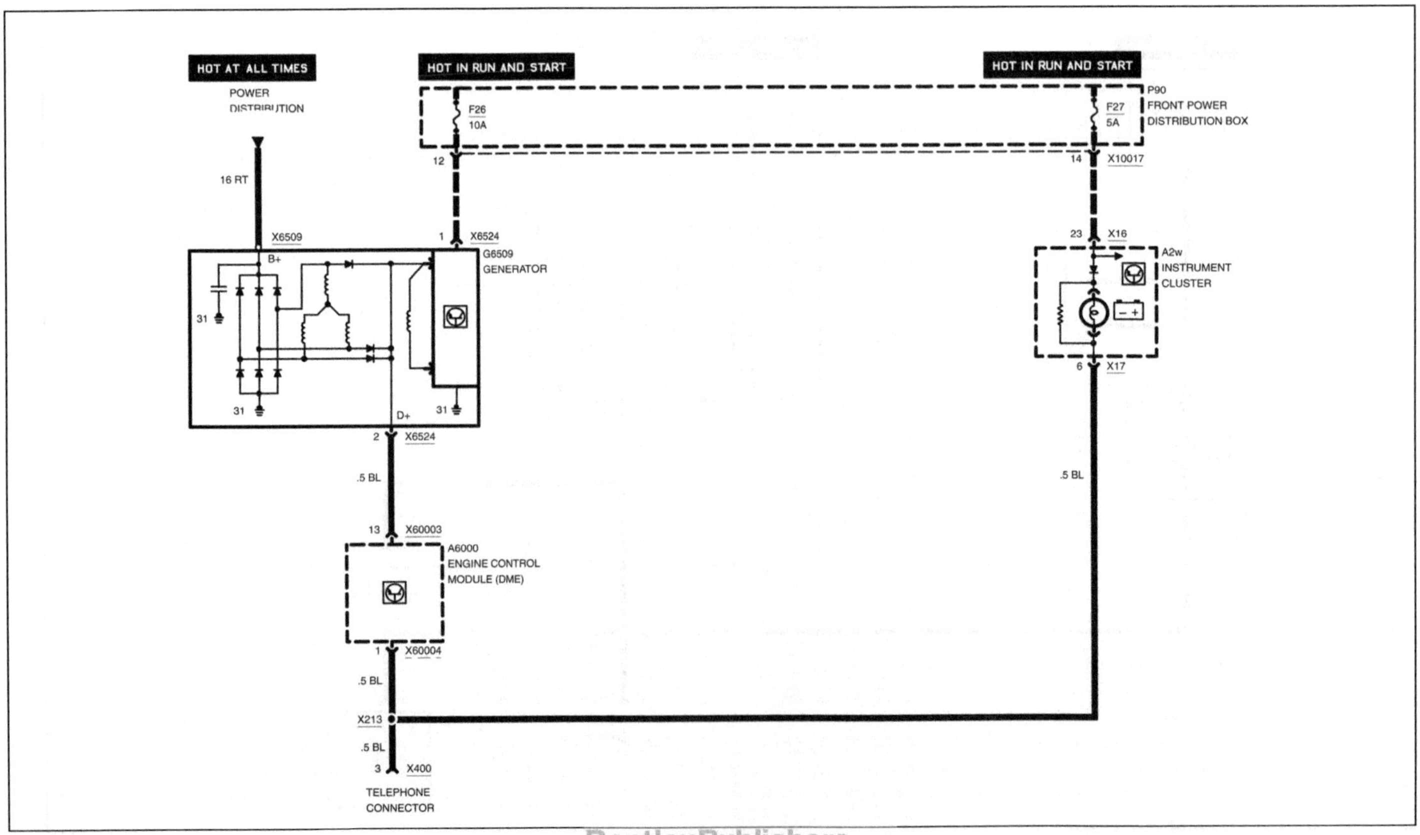

## Starting system, 1996 - 1998, 2000 M
(1 of 2)

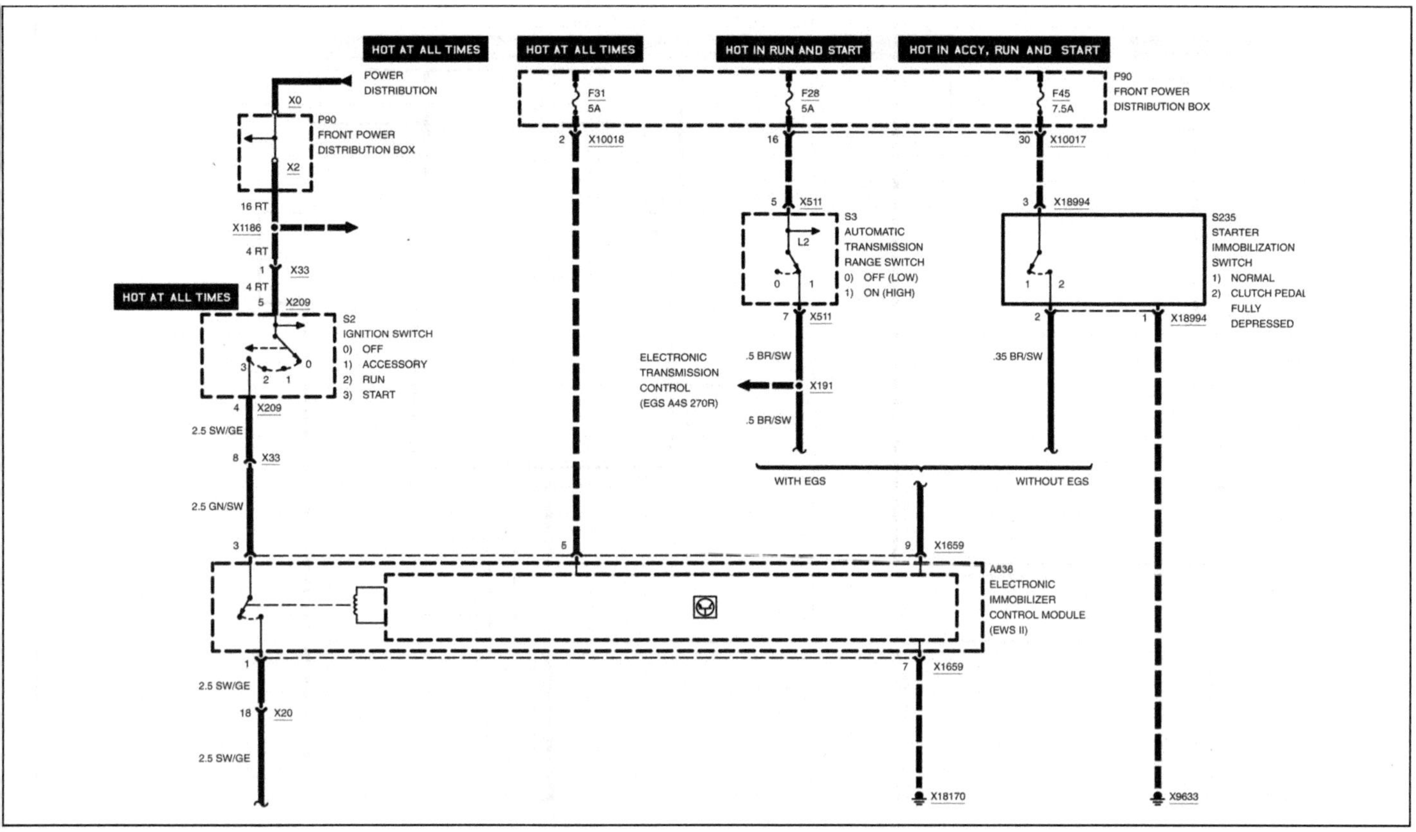


## Starting system, 1996 - 1998, 2000 M
(2 of 2)

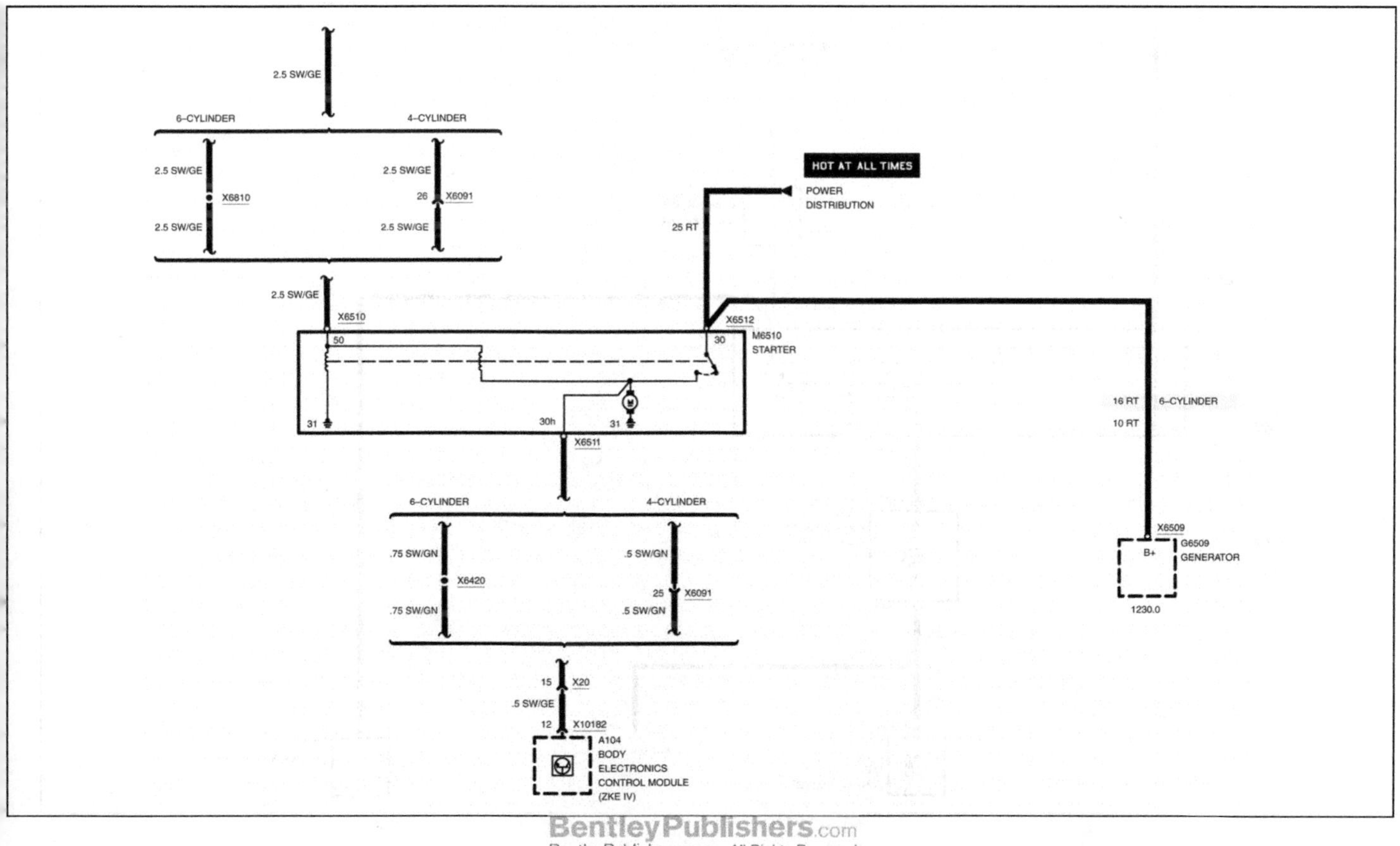

## Starting system, 2000 - 2002 Z
(1 of 2)

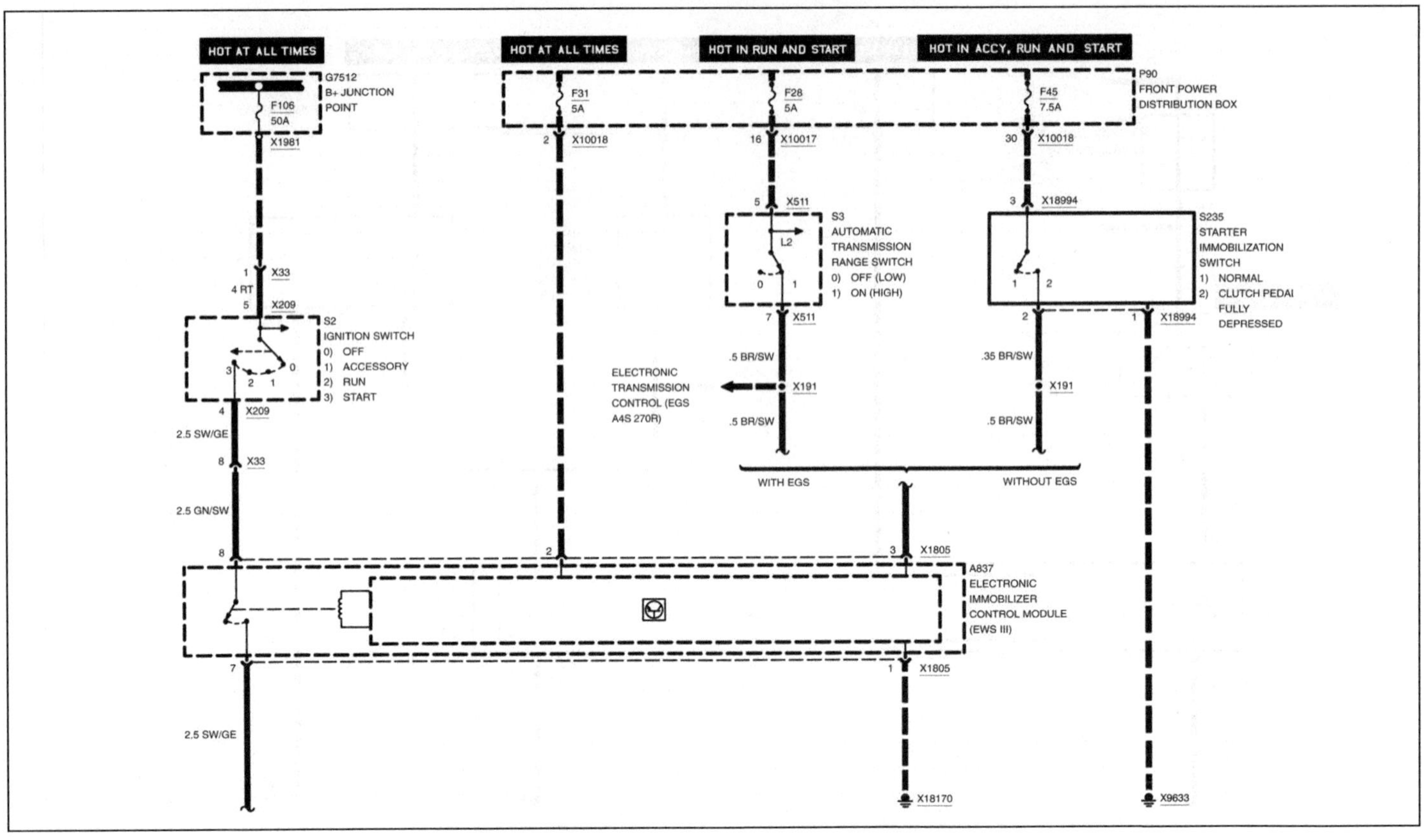


## Starting system, 2000 - 2002 Z
(2 of 2)

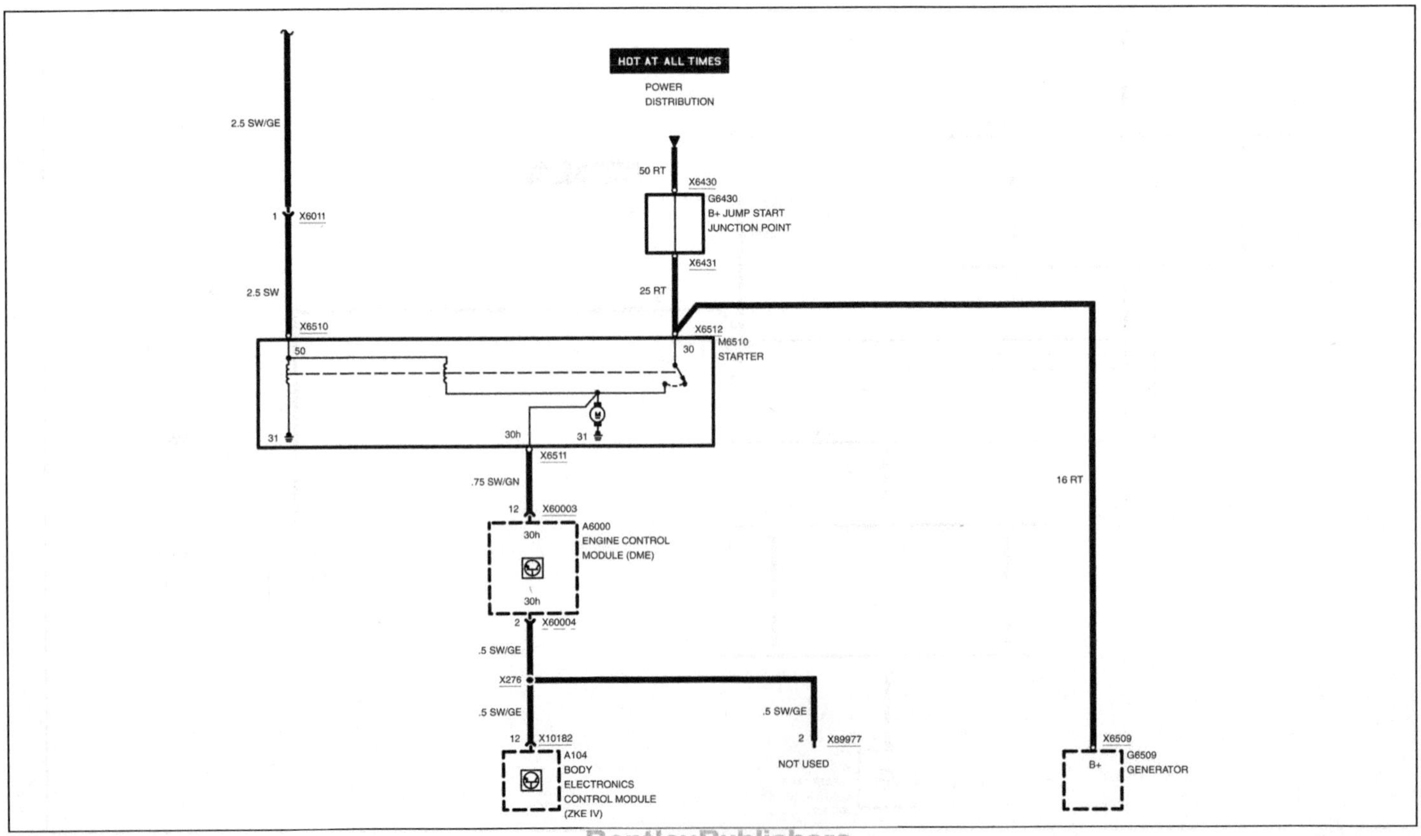

## Starting system, 2000 - 2002 M

**(1 of 2)**

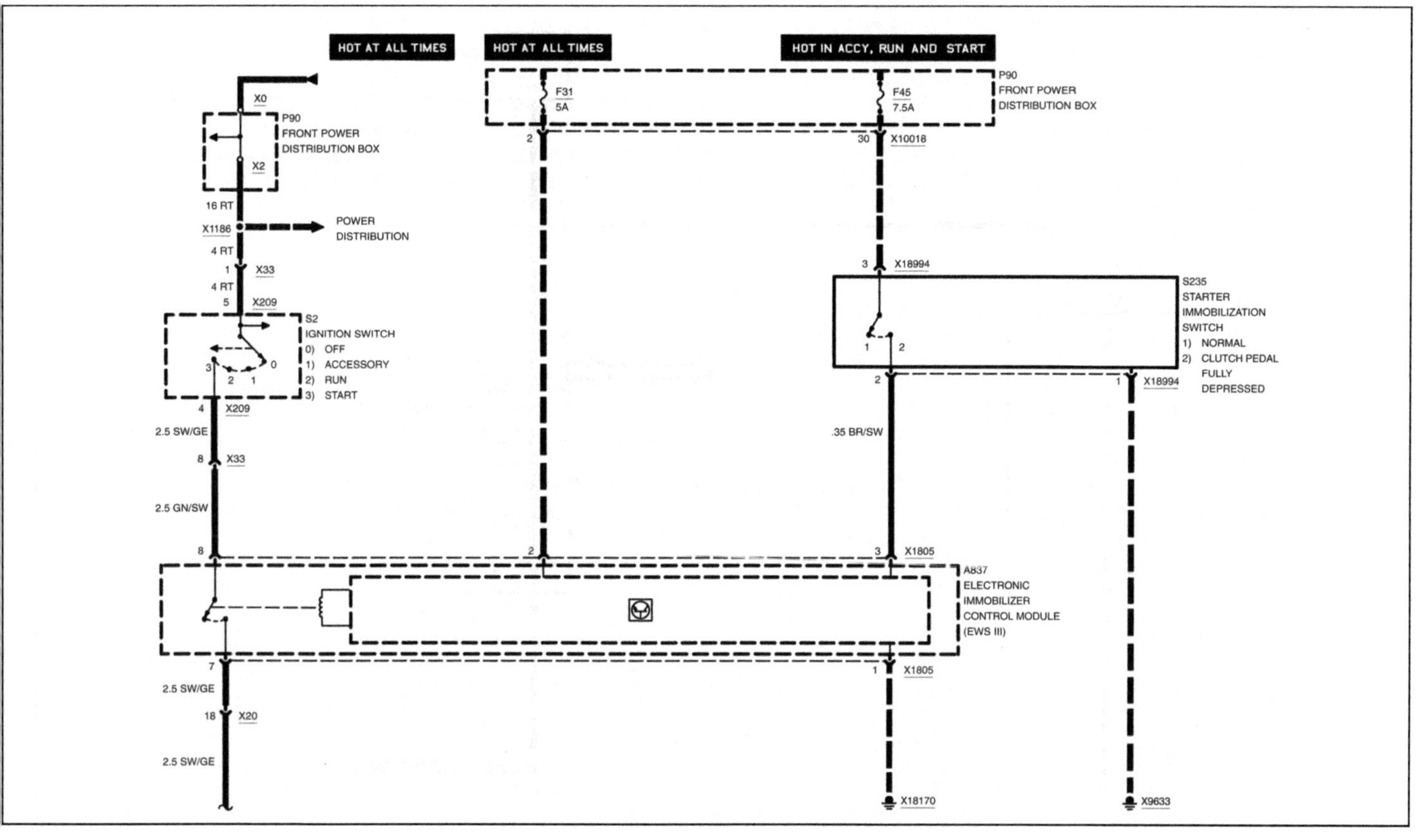


## Starting system, 2000 - 2002 M

**(2 of 2)**

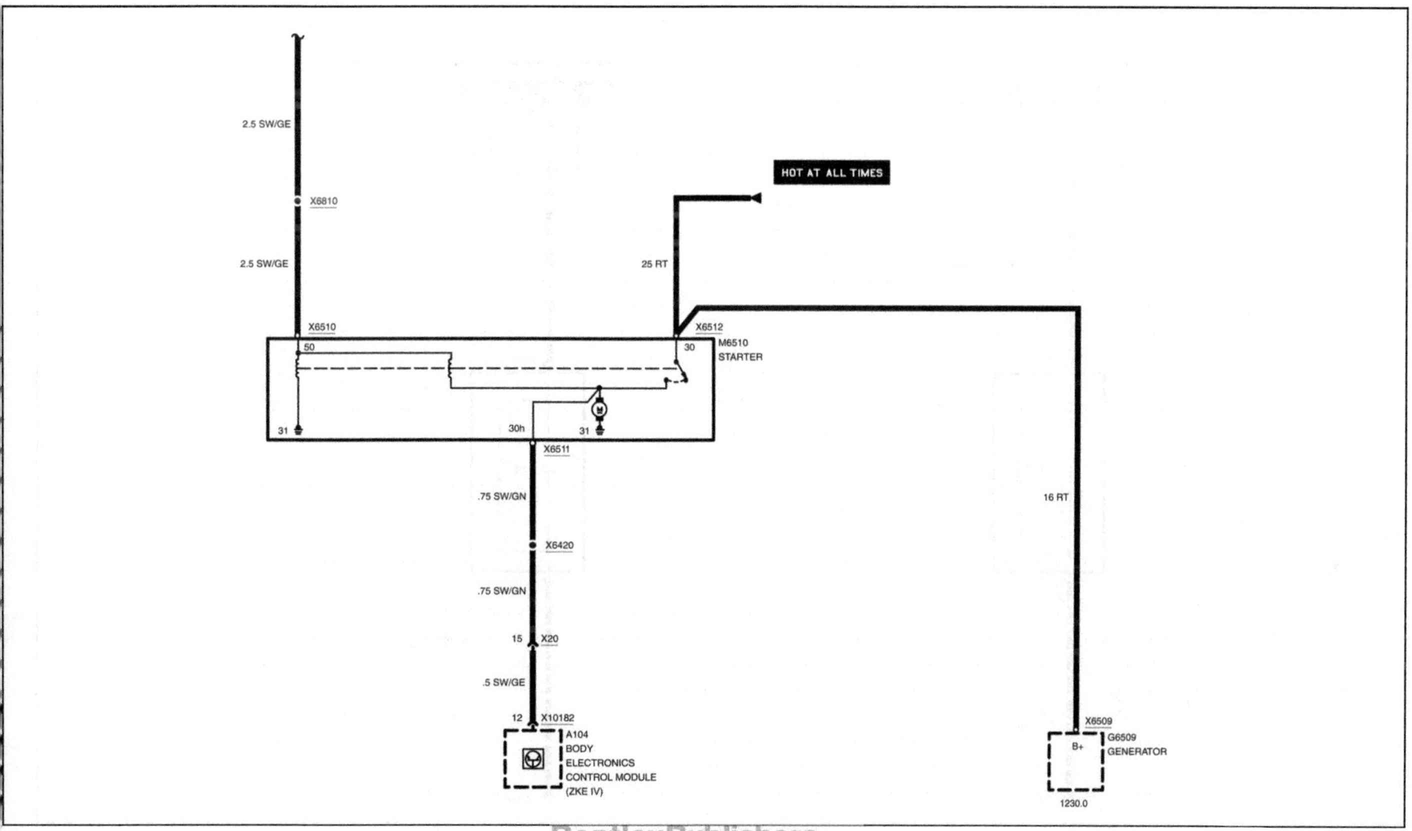

## E-box fan, 1996

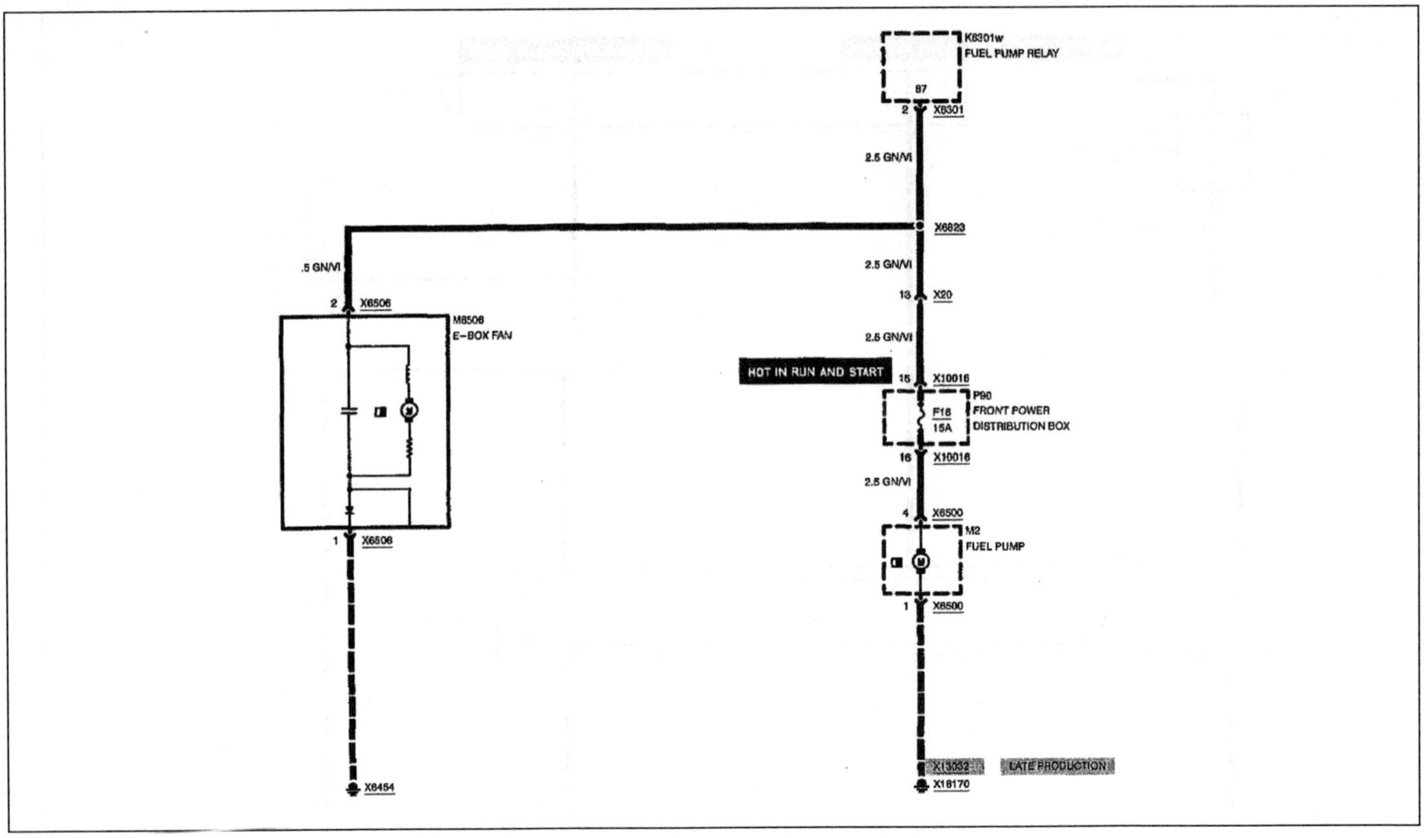


## E-box fan, 1998 - 2000

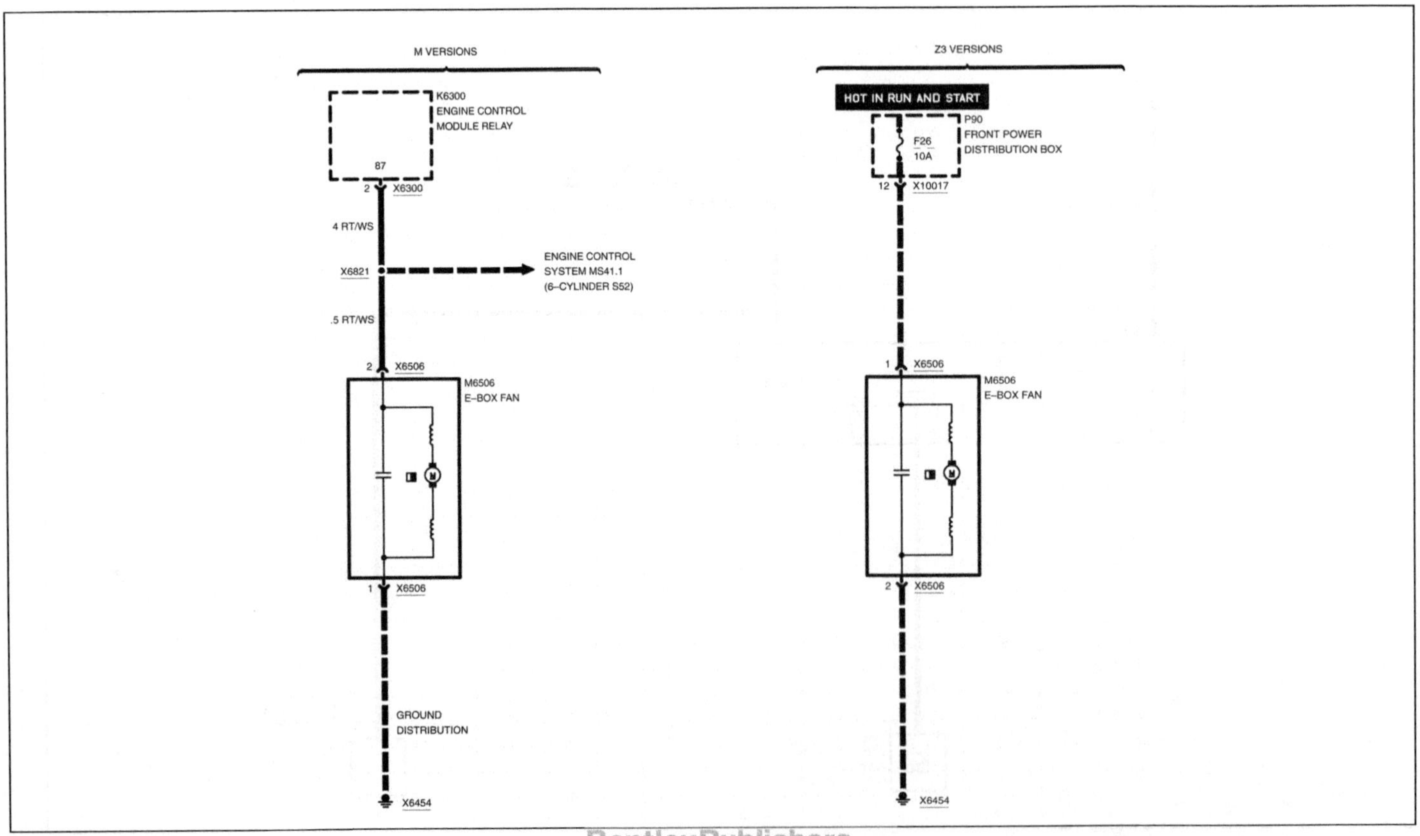

## E-box fan, 2002

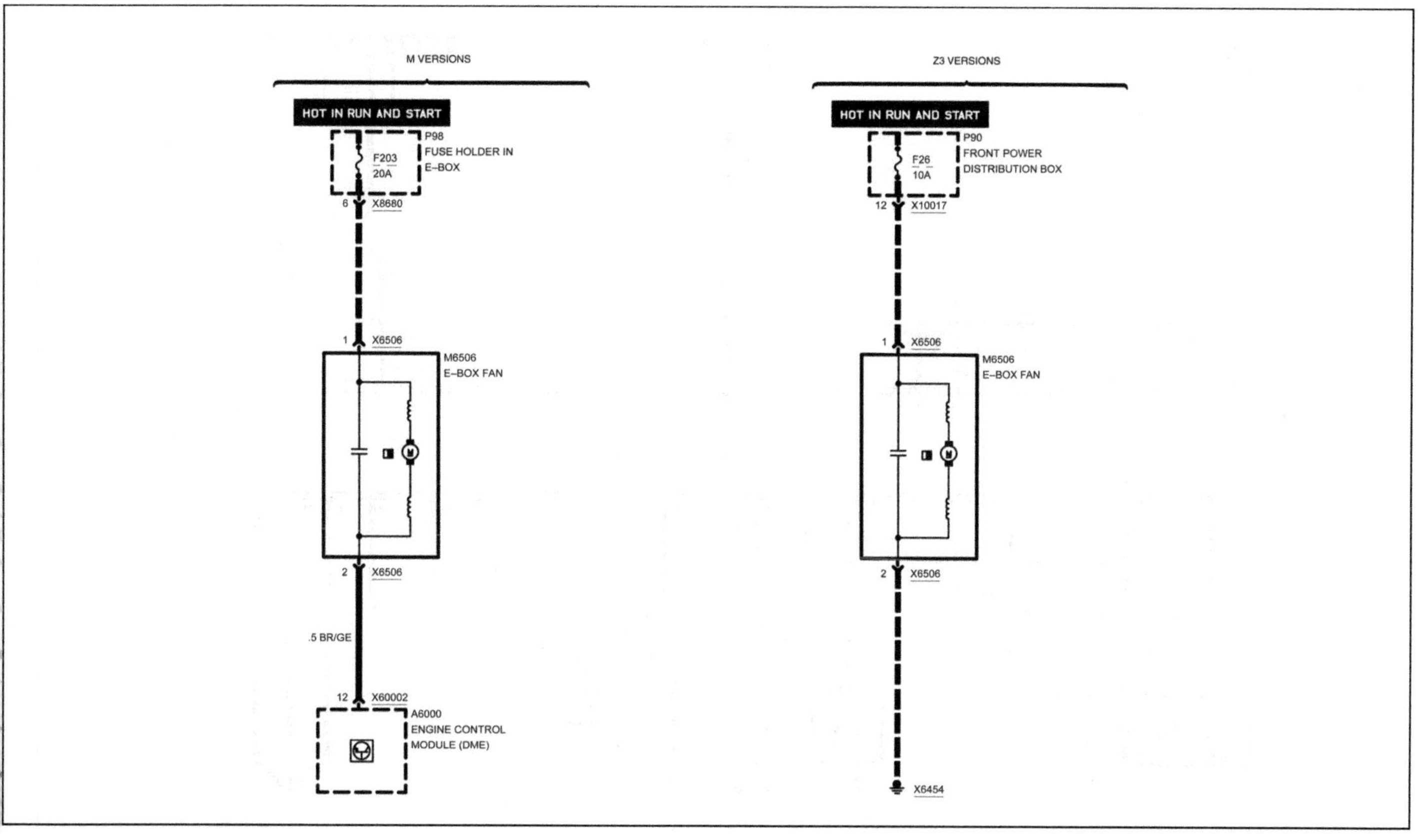


## Automatic transmission control (EGS) A4S 310R (4-cylinder)

**(1 of 5)**

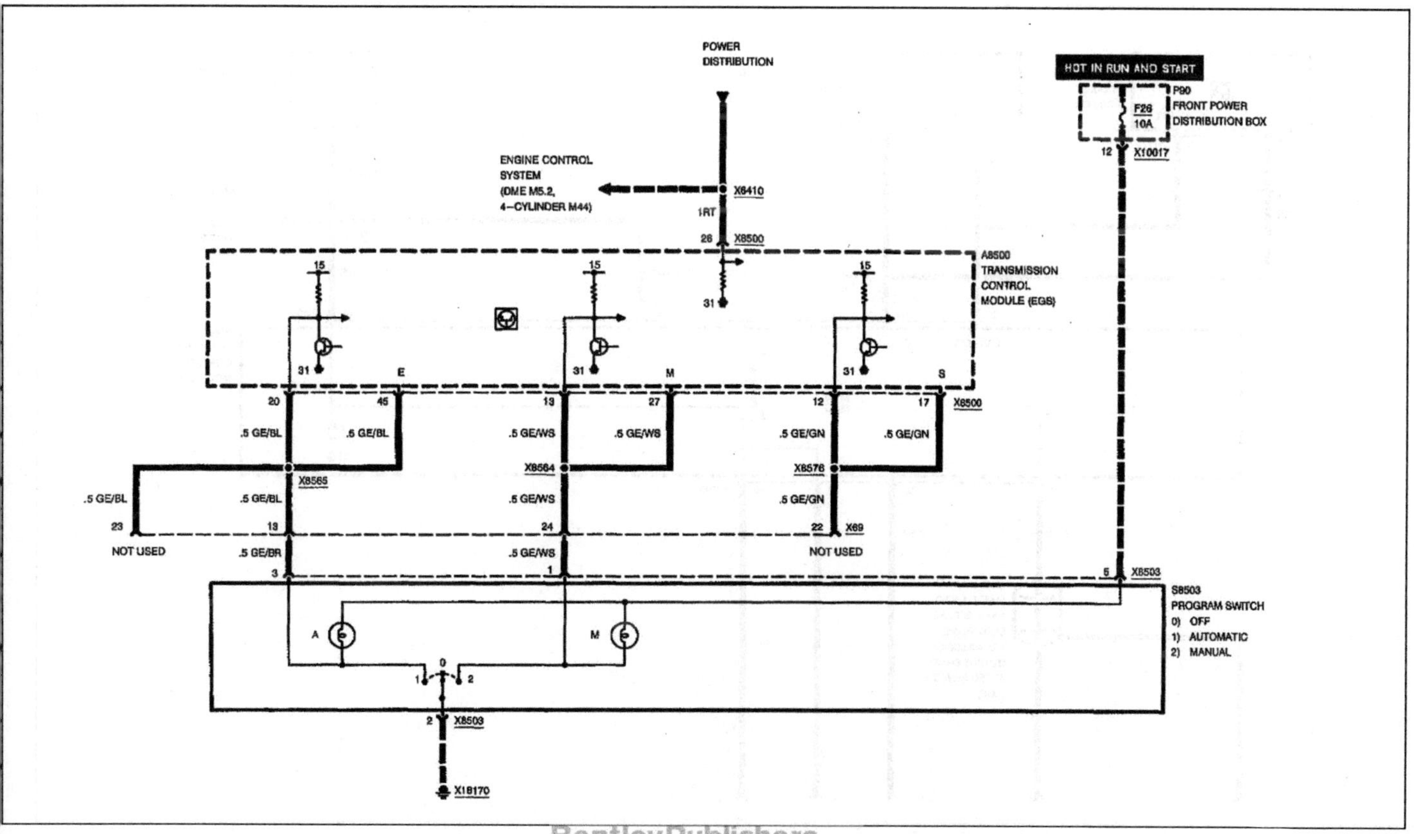

## Automatic transmission control (EGS) A4S 310R (4-cylinder)

(2 of 5)

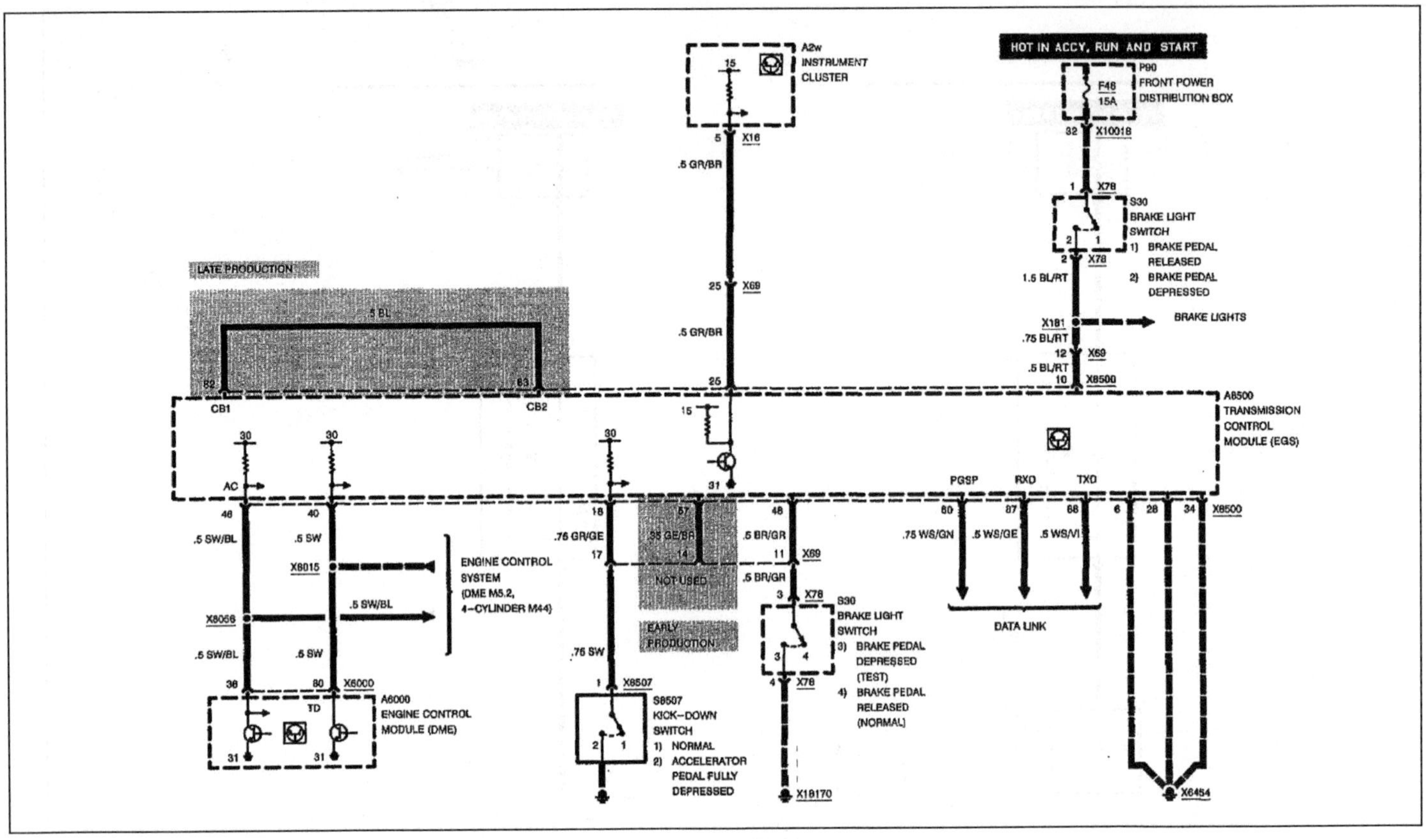


## Automatic transmission control (EGS) A4S 310R (4-cylinder)

(3 of 5)

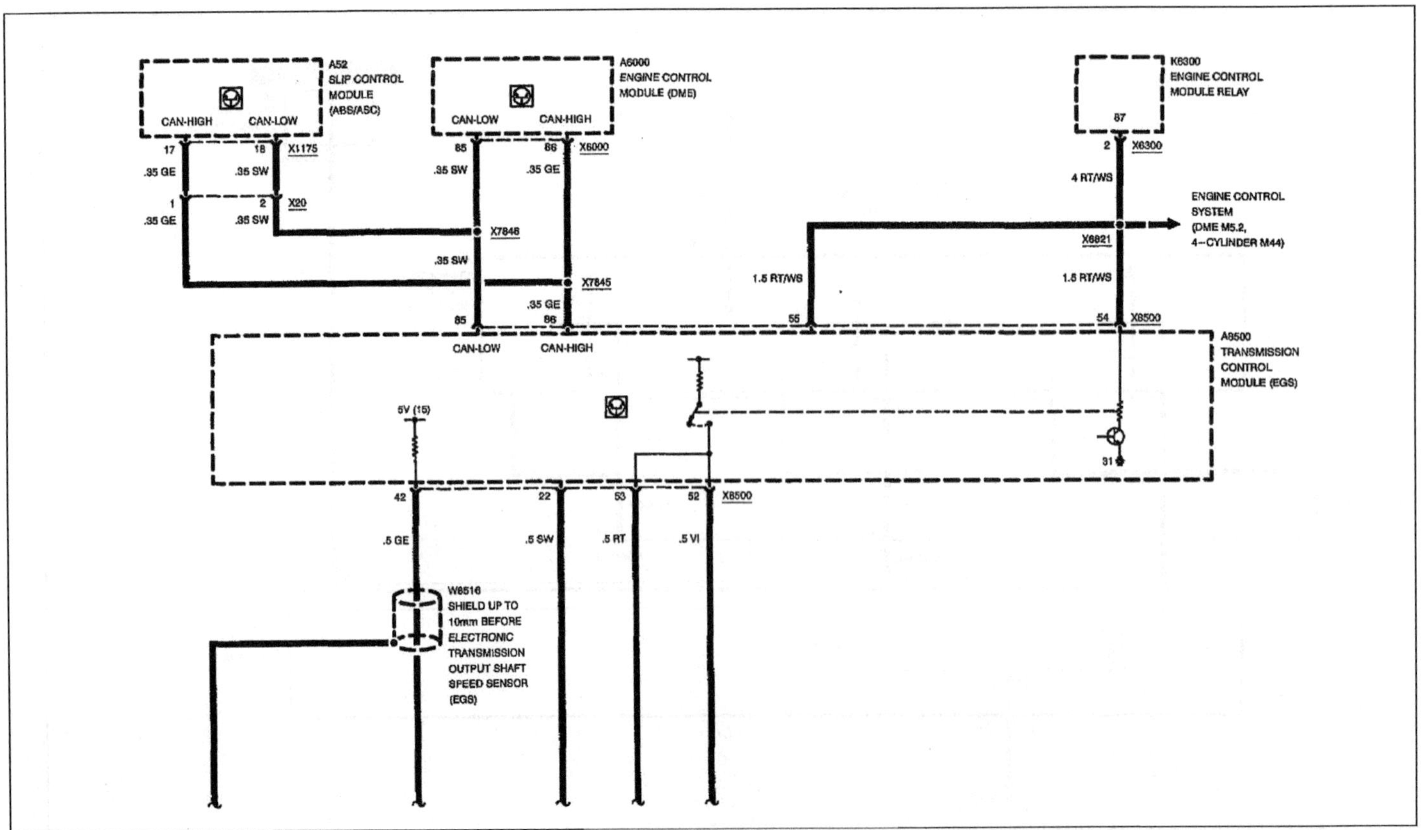

## Automatic transmission control (EGS) A4S 310R (4-cylinder)

(4 of 5)

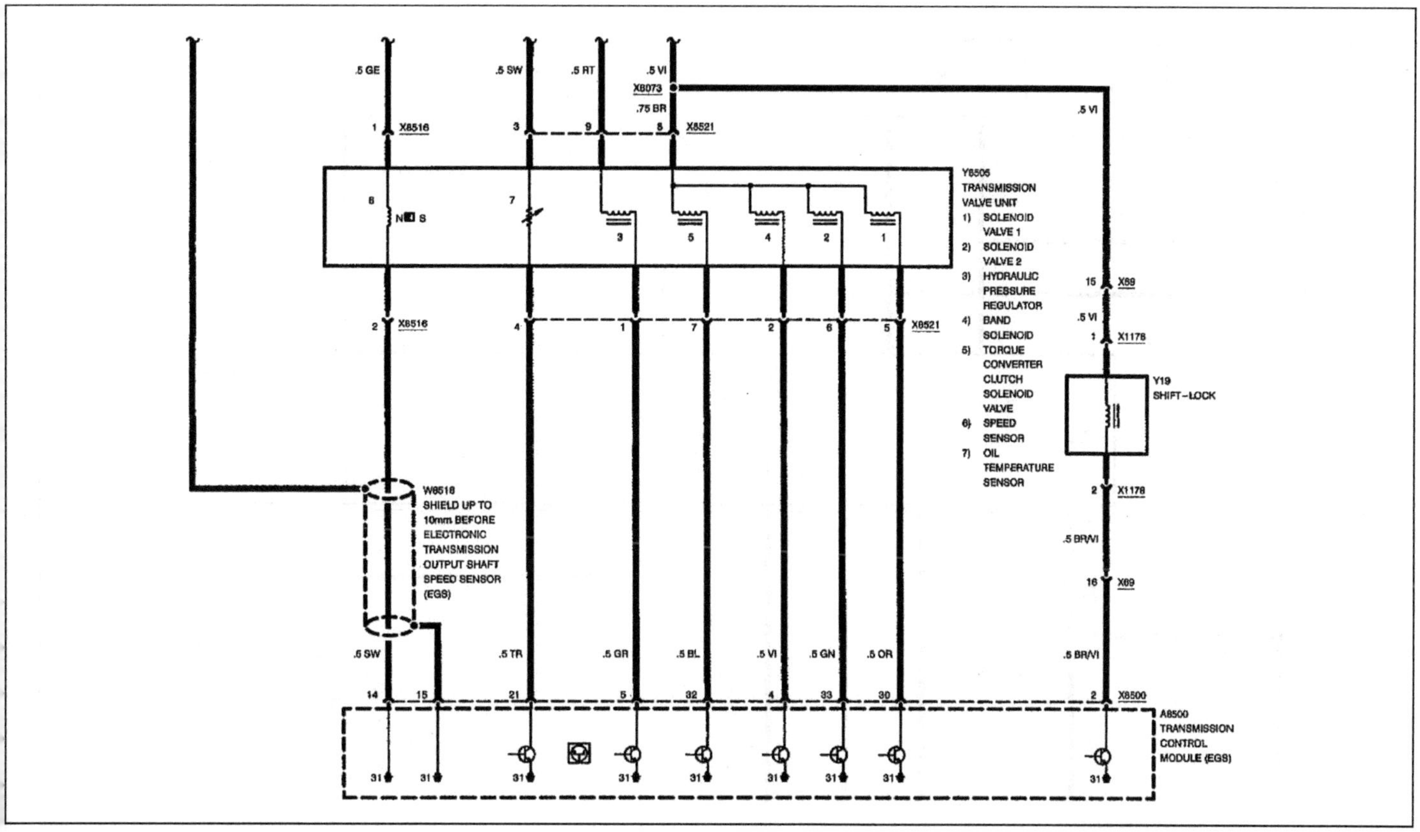


## Automatic transmission control (EGS) A4S 310R (4-cylinder)

(5 of 5)

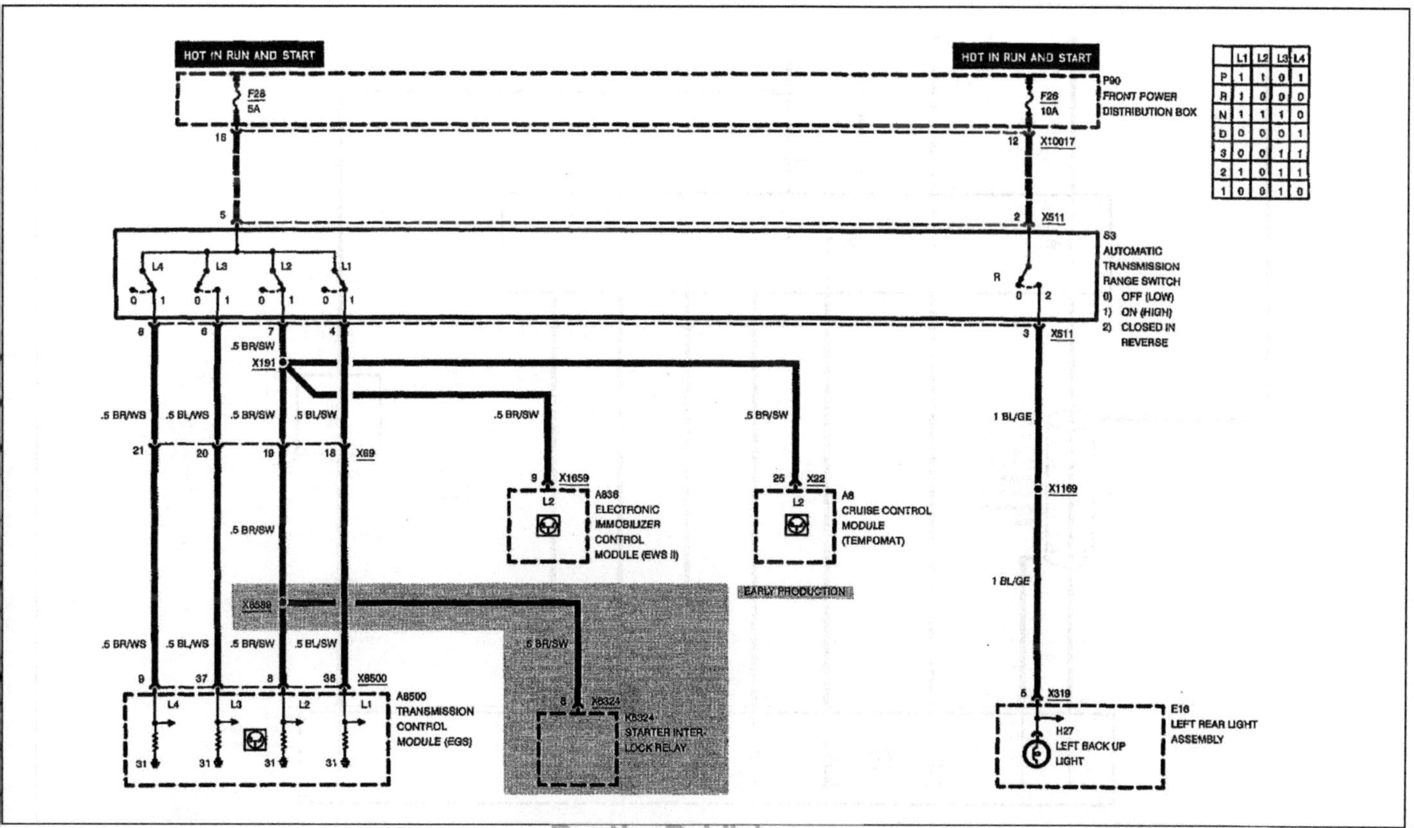


| | L1 | L2 | L3 | L4 |
|---|---|---|---|---|
| P | 1 | 1 | 0 | 1 |
| R | 1 | 0 | 0 | 0 |
| N | 1 | 1 | 1 | 0 |
| D | 0 | 0 | 0 | 1 |
| 3 | 0 | 0 | 1 | 1 |
| 2 | 1 | 0 | 1 | 1 |
| 1 | 0 | 0 | 1 | 0 |

## Automatic transmission control (EGS) A4S 270R (4-cylinder)
### (1 of 5)

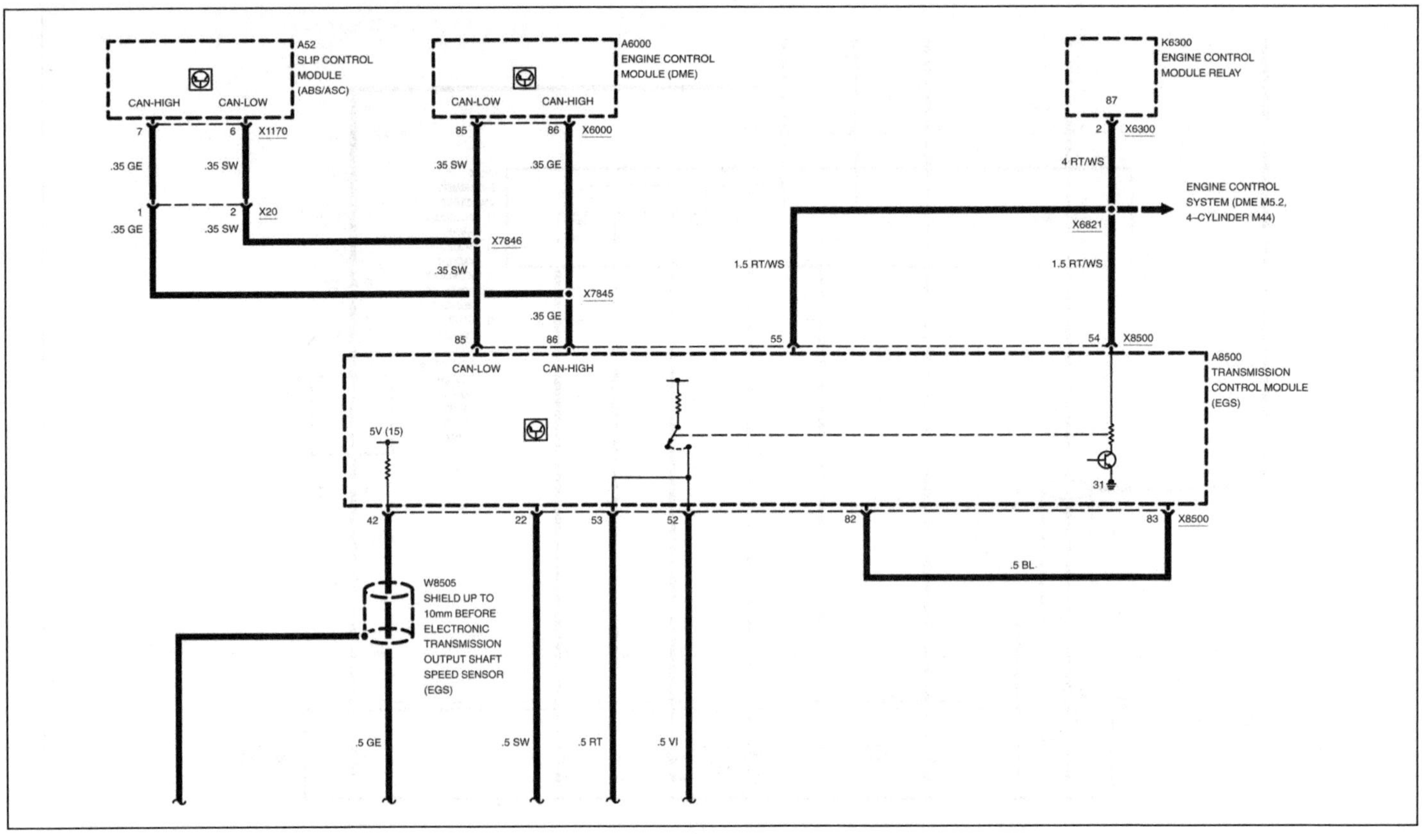


## Automatic transmission control (EGS) A4S 270R (4-cylinder)
### (2 of 5)

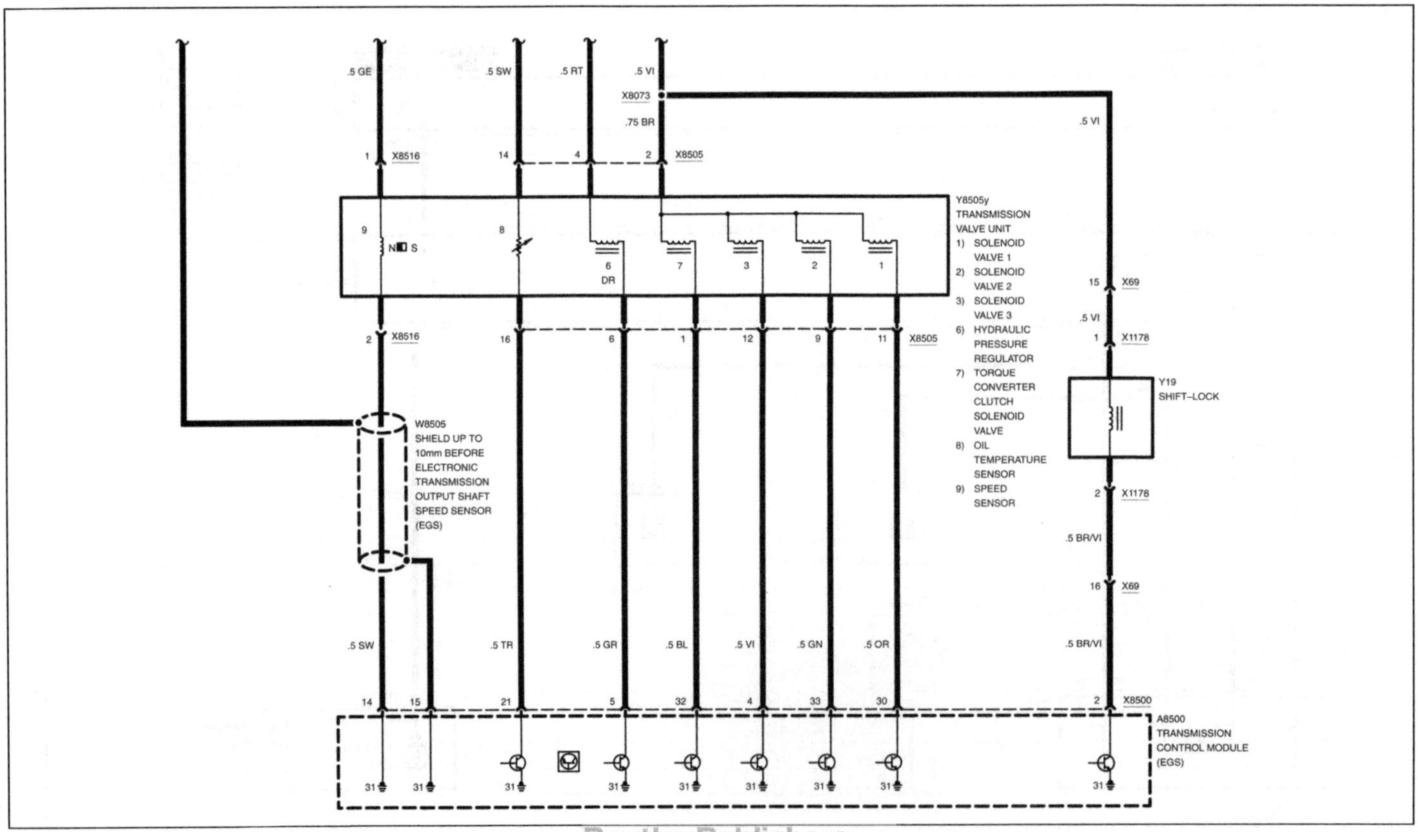

## Automatic transmission control (EGS) A4S 270R (4-cylinder)

(3 of 5)

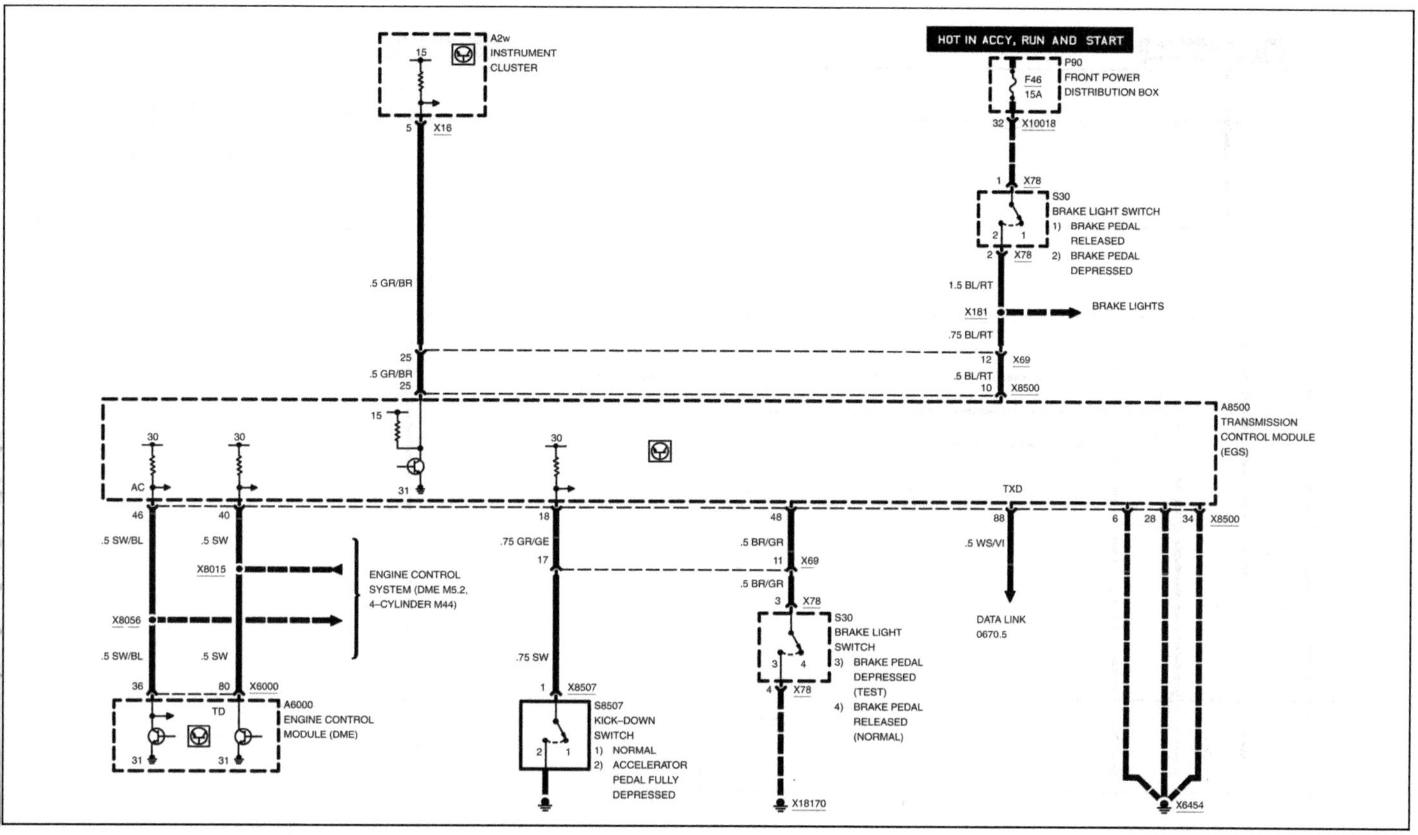


## Automatic transmission control (EGS) A4S 270R (4-cylinder)

(4 of 5)

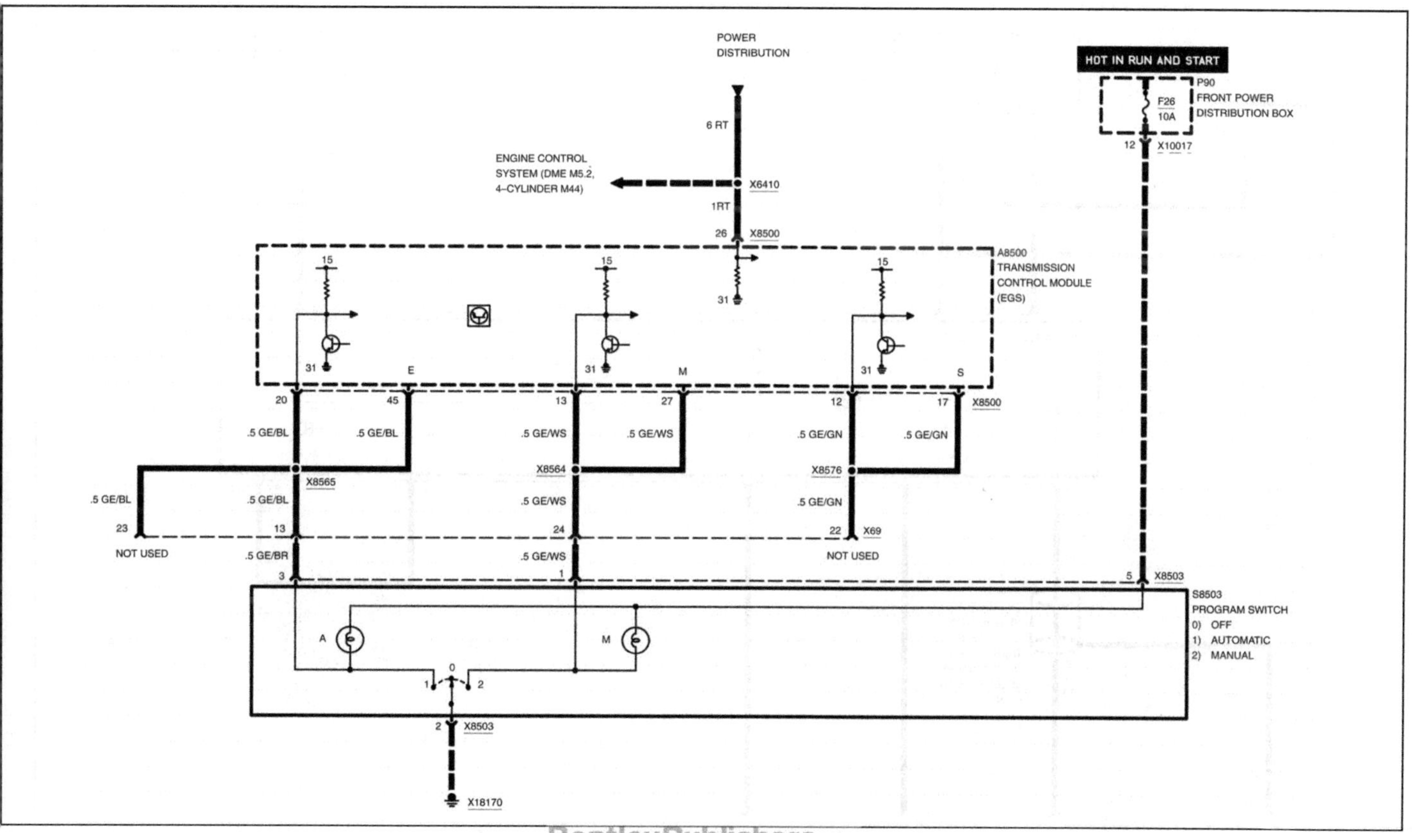

## Automatic transmission control (EGS) A4S 270R (4-cylinder)
### (5 of 5)

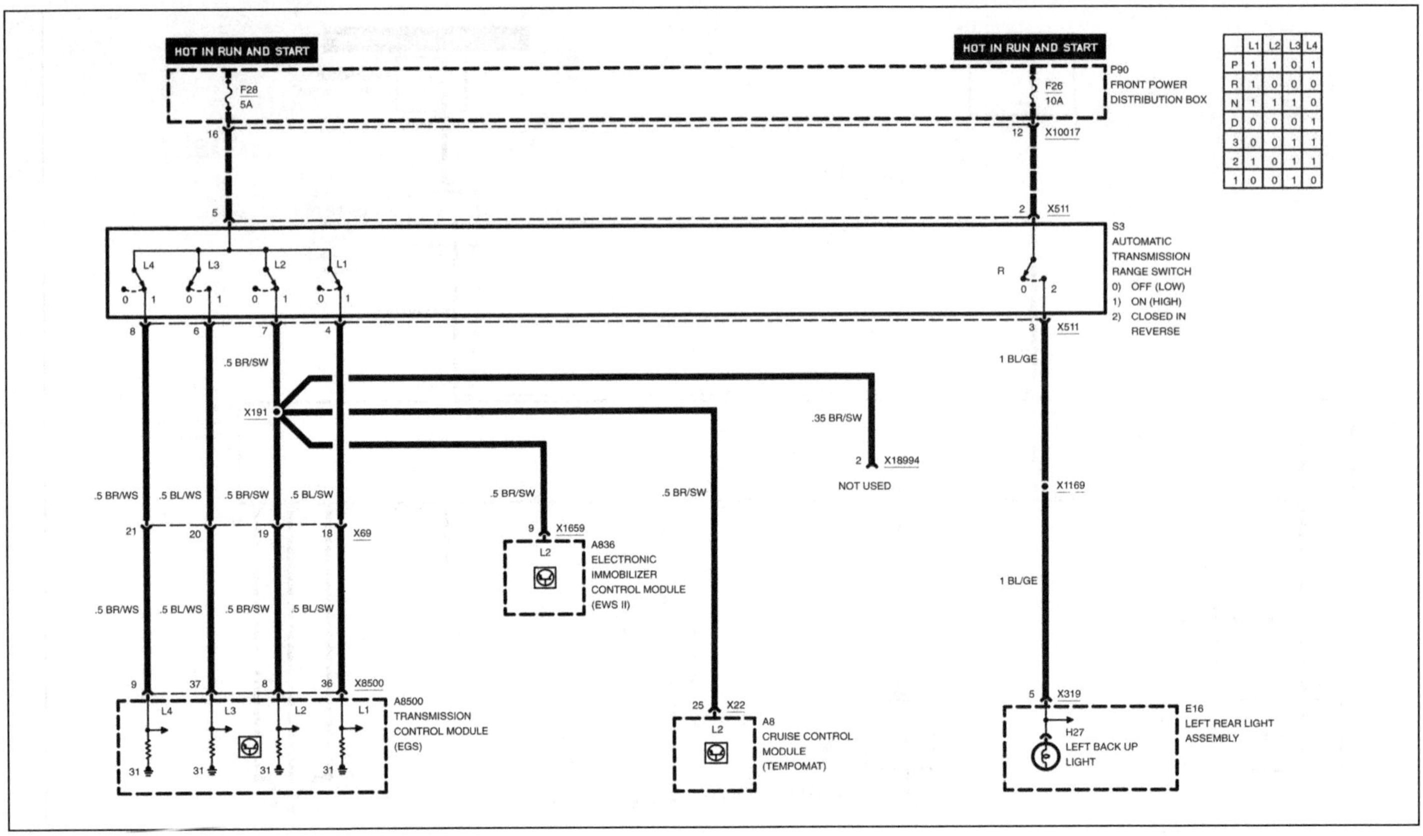


| | L1 | L2 | L3 | L4 |
|---|---|---|---|---|
| P | 1 | 1 | 0 | 1 |
| R | 1 | 0 | 0 | 0 |
| N | 1 | 1 | 1 | 0 |
| D | 0 | 0 | 0 | 1 |
| 3 | 0 | 0 | 1 | 1 |
| 2 | 1 | 0 | 1 | 1 |
| 1 | 0 | 0 | 1 | 0 |

## Automatic transmission control (EGS) A4S 270R (6-cylinder)
### (1 of 4)

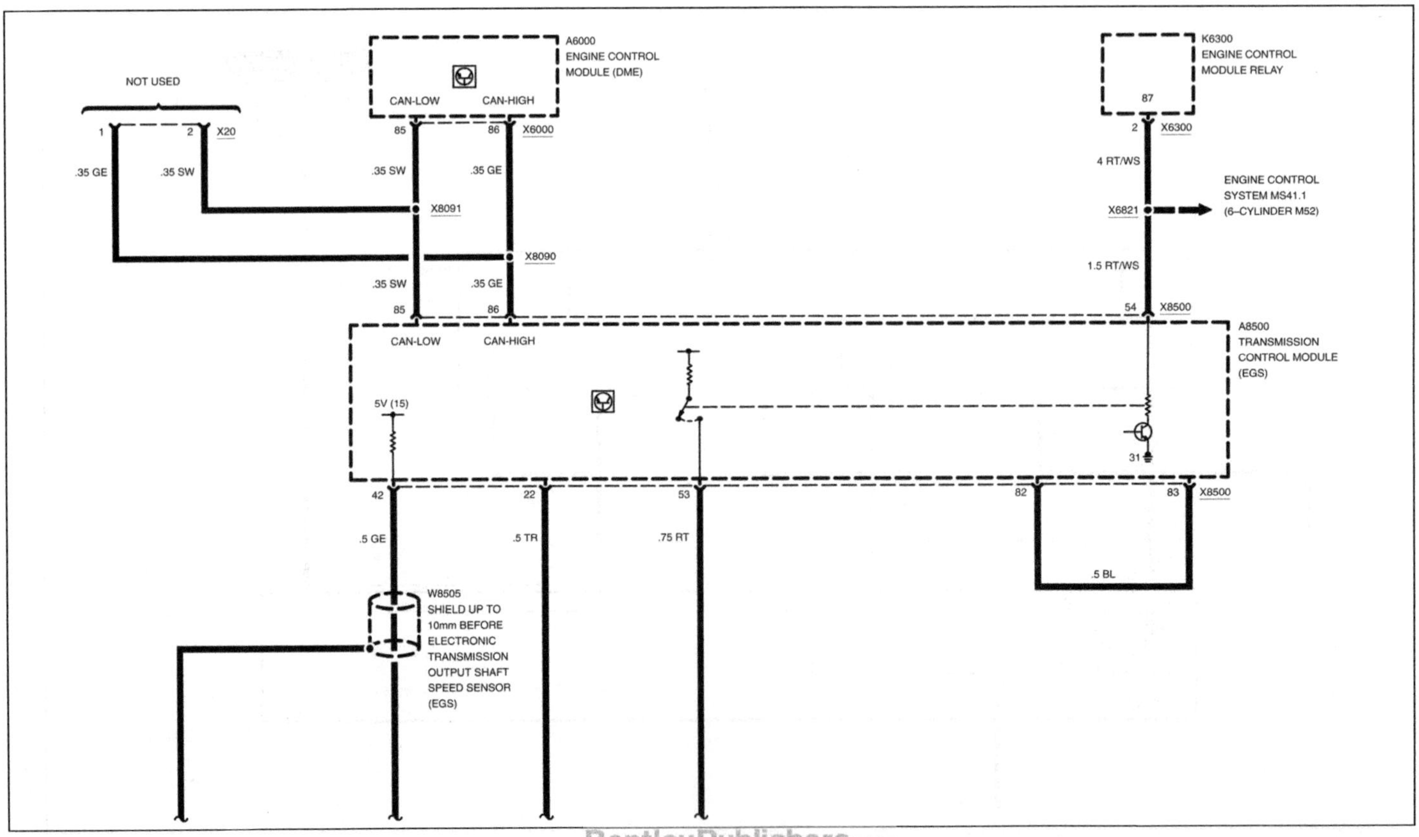

## Automatic transmission control (EGS) A4S 270R (6-cylinder)

**(2 of 4)**

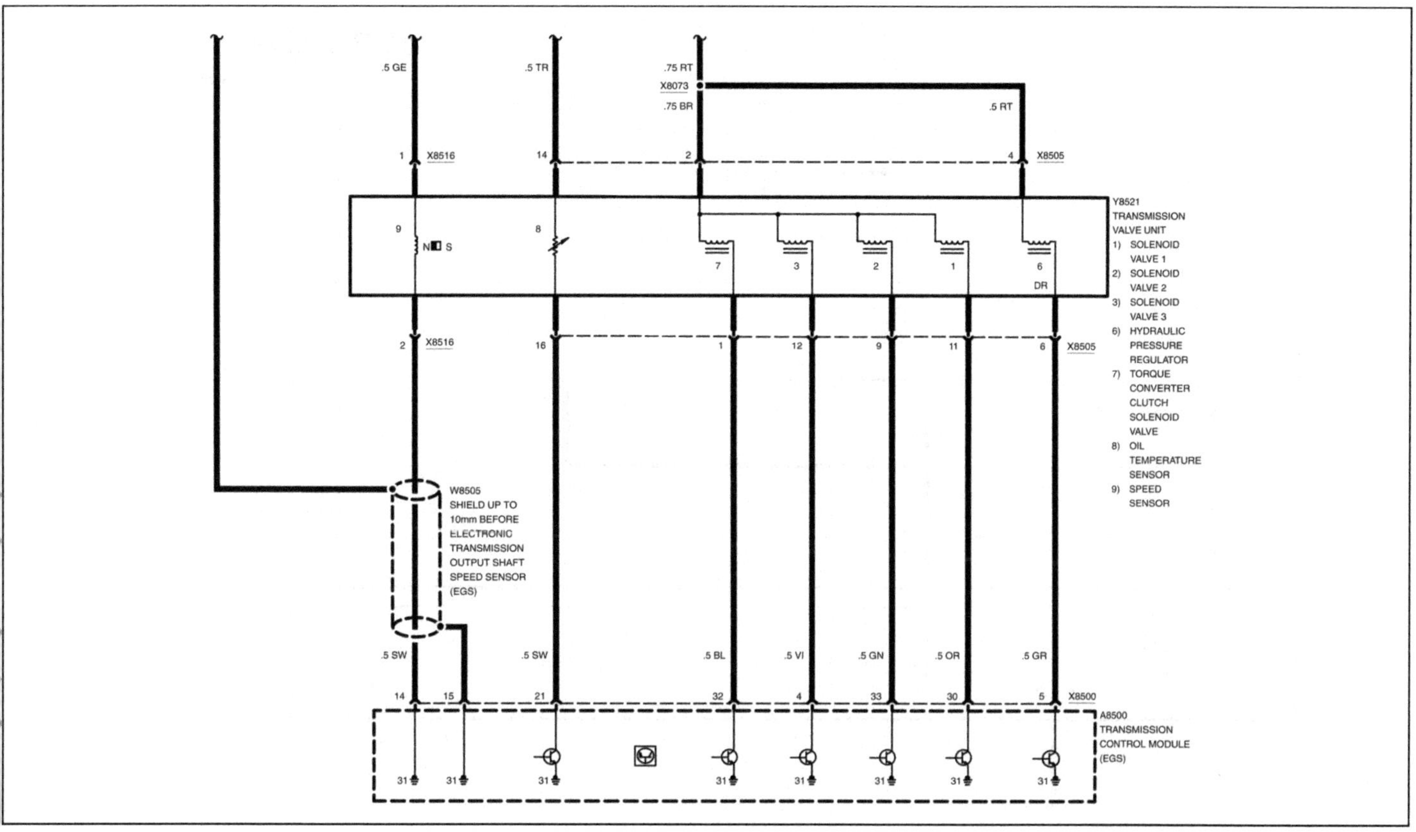


## Automatic transmission control (EGS) A4S 270R (6-cylinder)

**(3 of 4)**

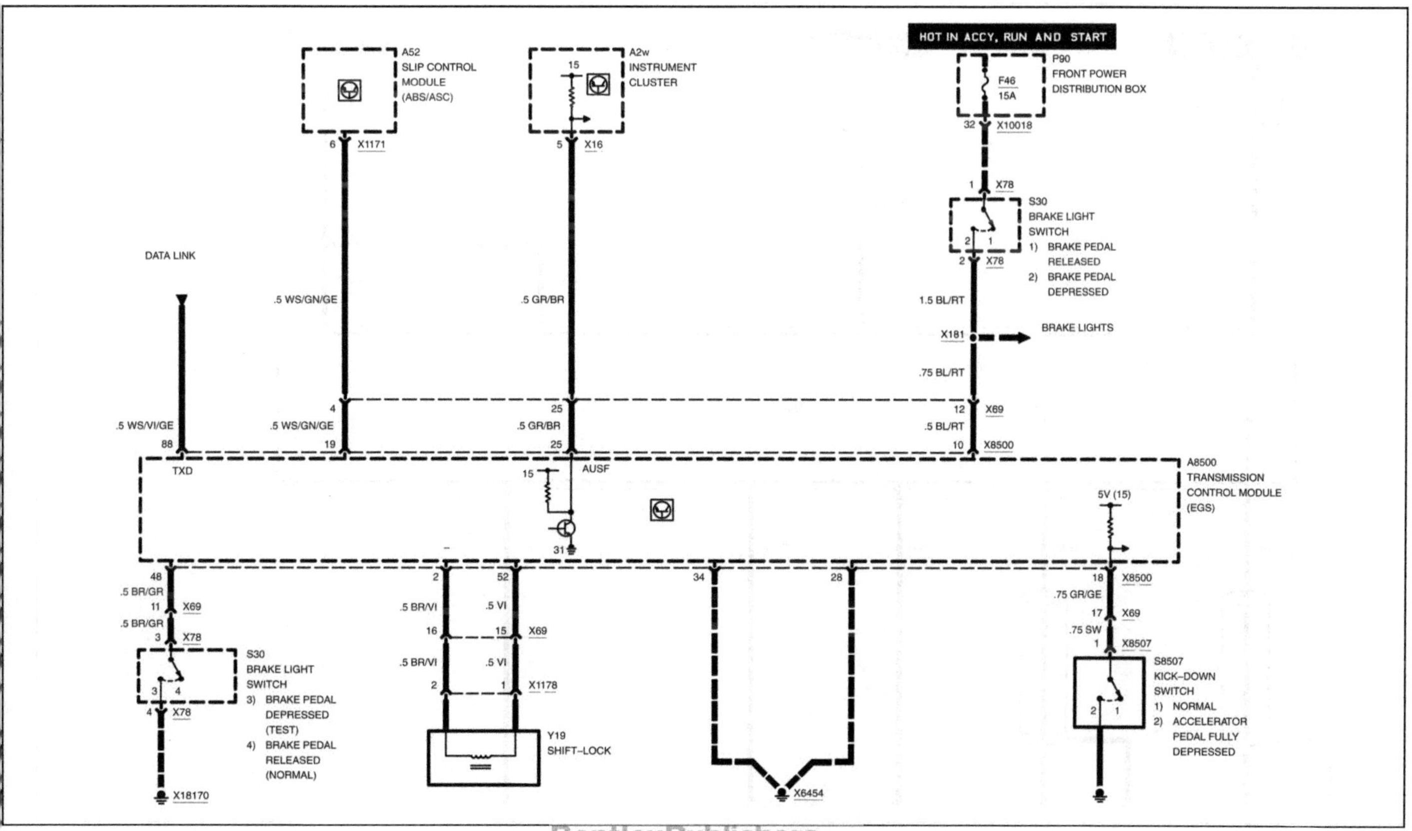

## Automatic transmission control (EGS) A4S 270R (6-cylinder)

(4 of 4)

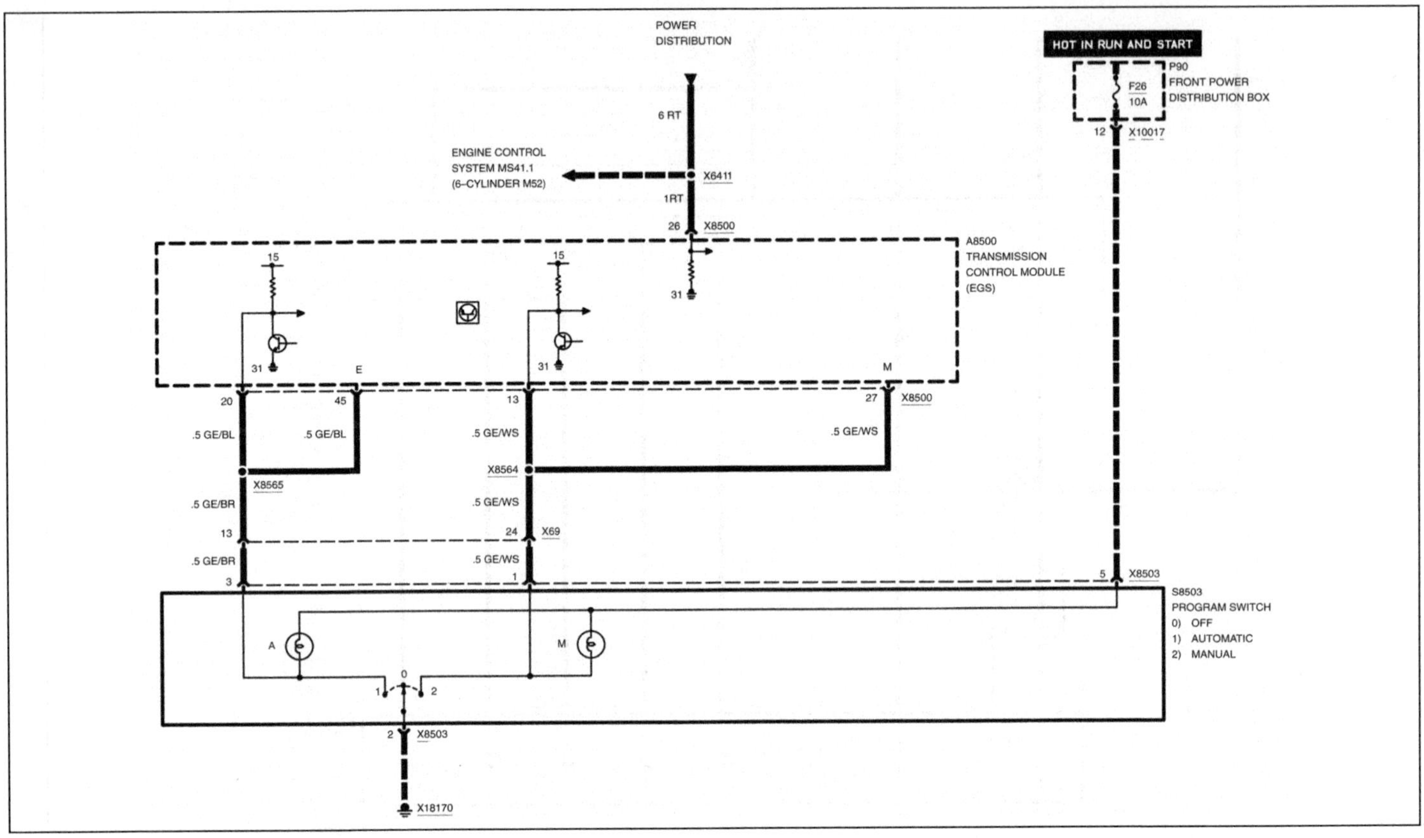


## Automatic transmission control (EGS) A5S 360R

(1 of 5)

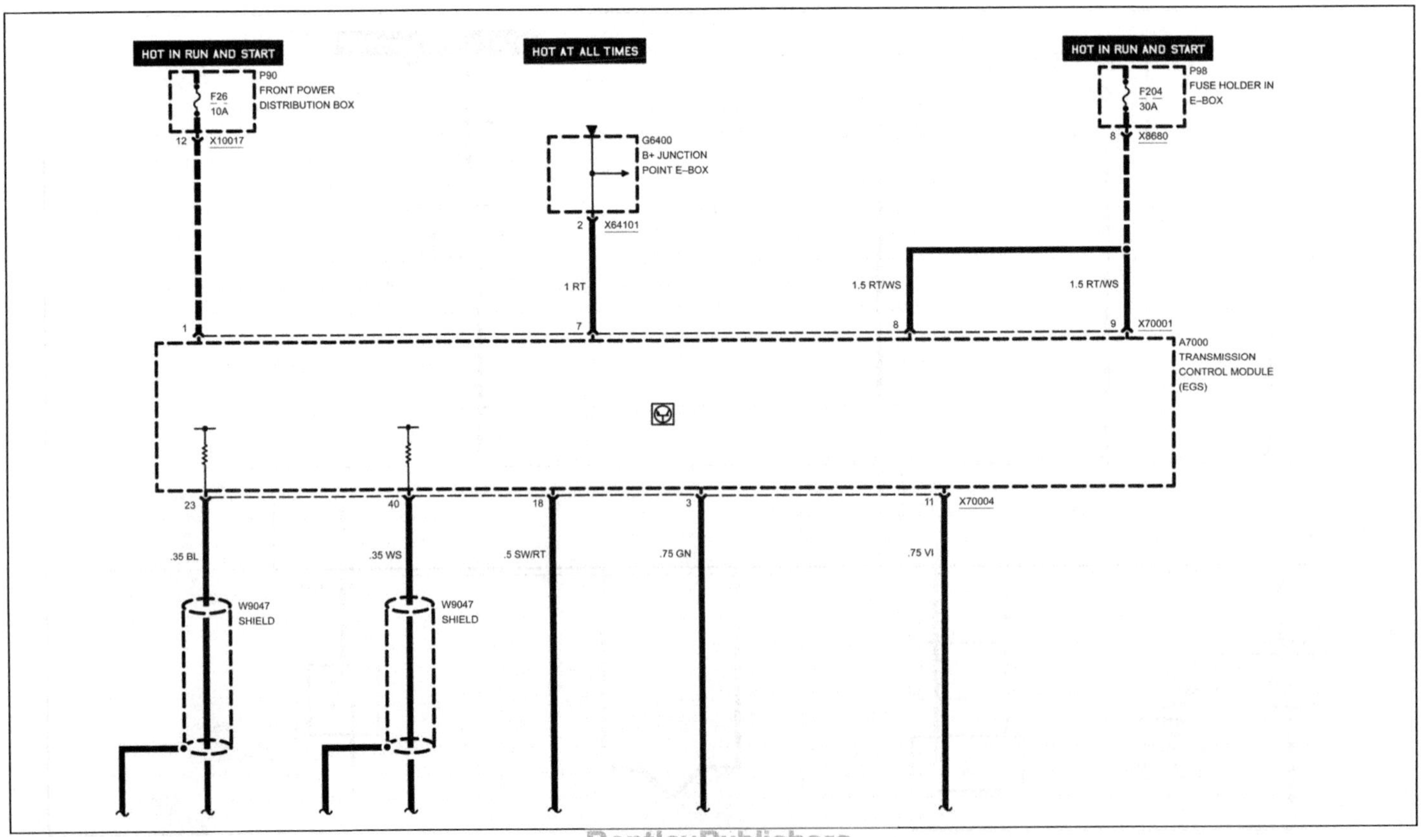

## Automatic transmission control (EGS) A5S 360R

(2 of 5)

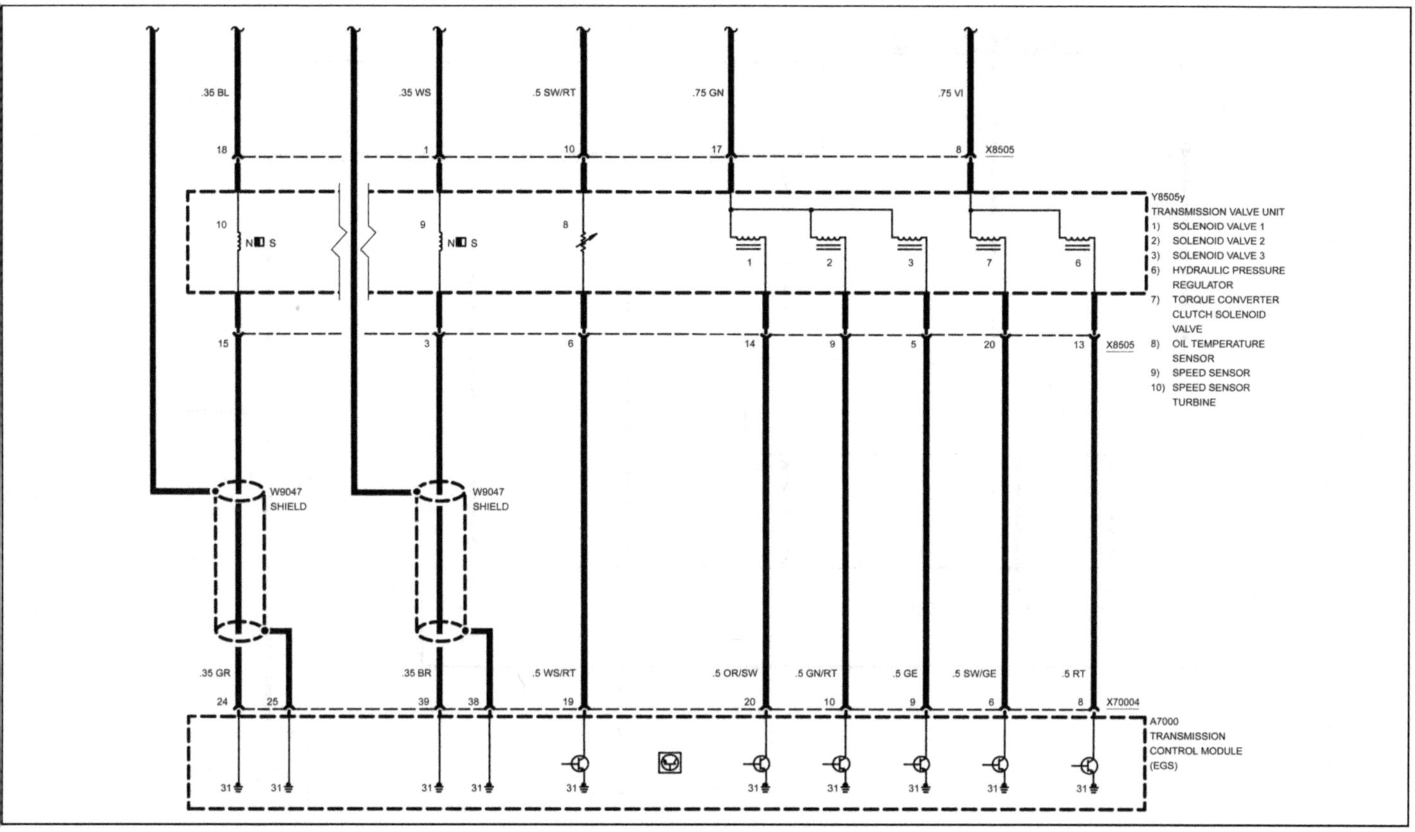


## Automatic transmission control (EGS) A5S 360R

(3 of 5)

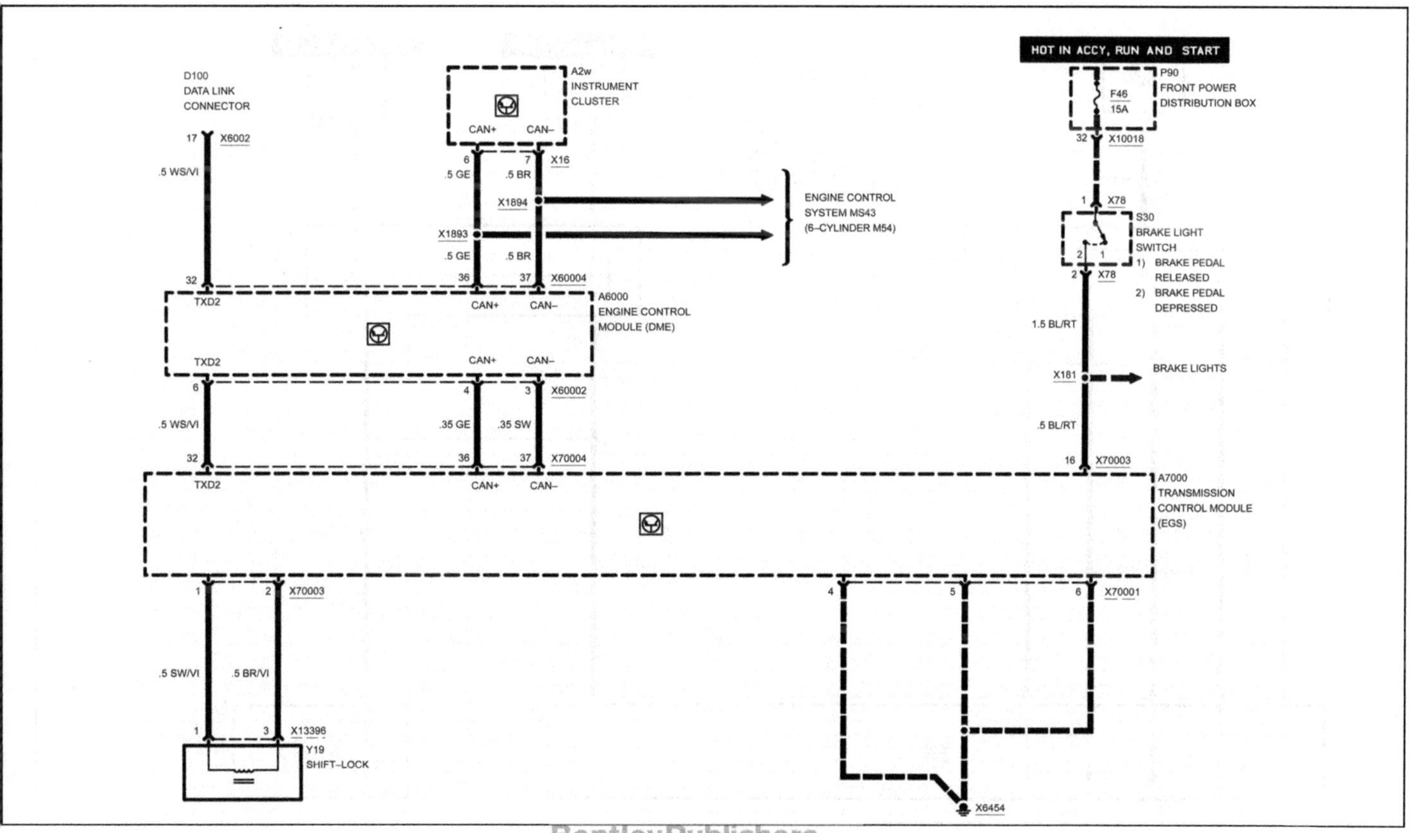

## Automatic transmission control (EGS) A5S 360R
(4 of 5)

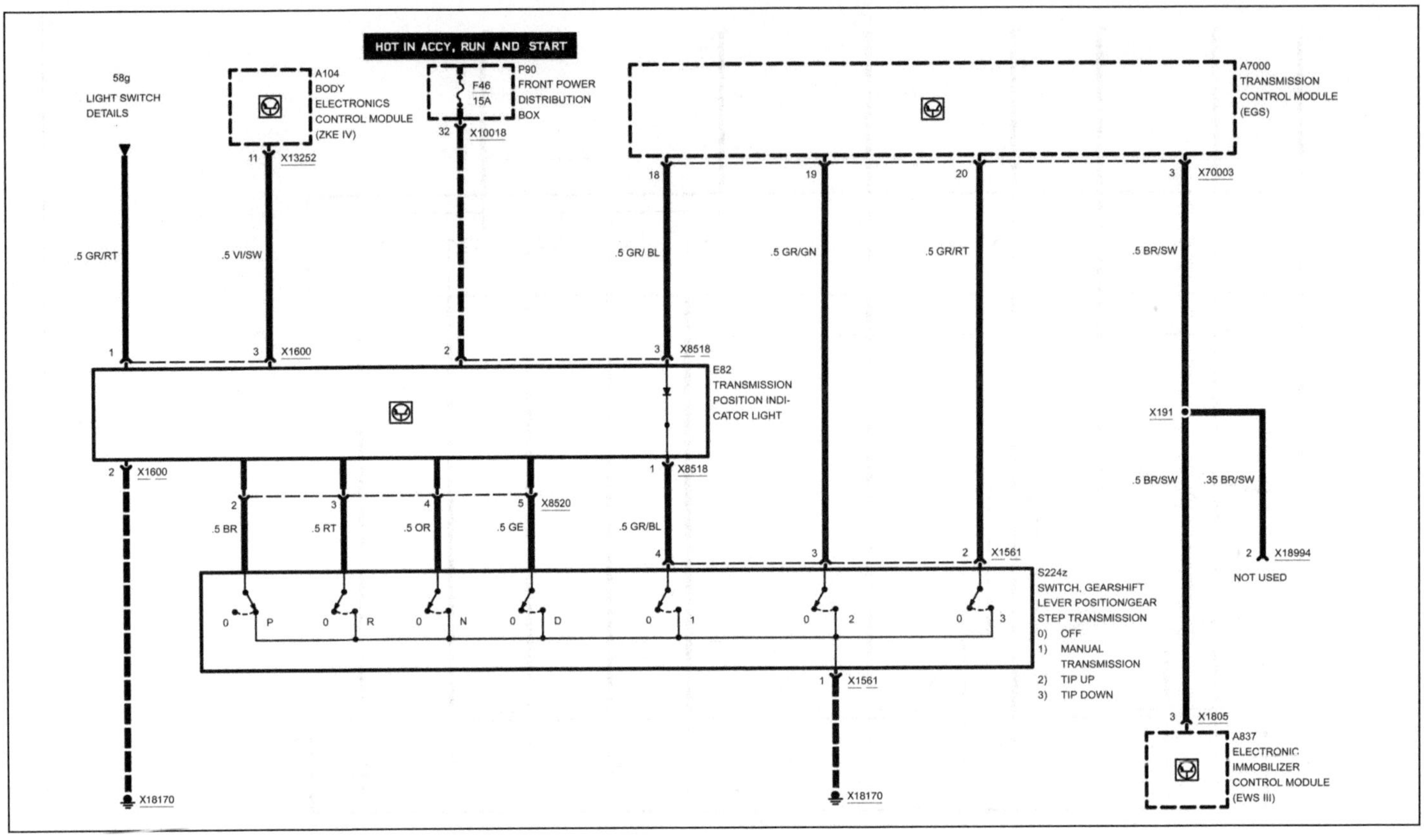


## Automatic transmission control (EGS) A5S 360R
(5 of 5)

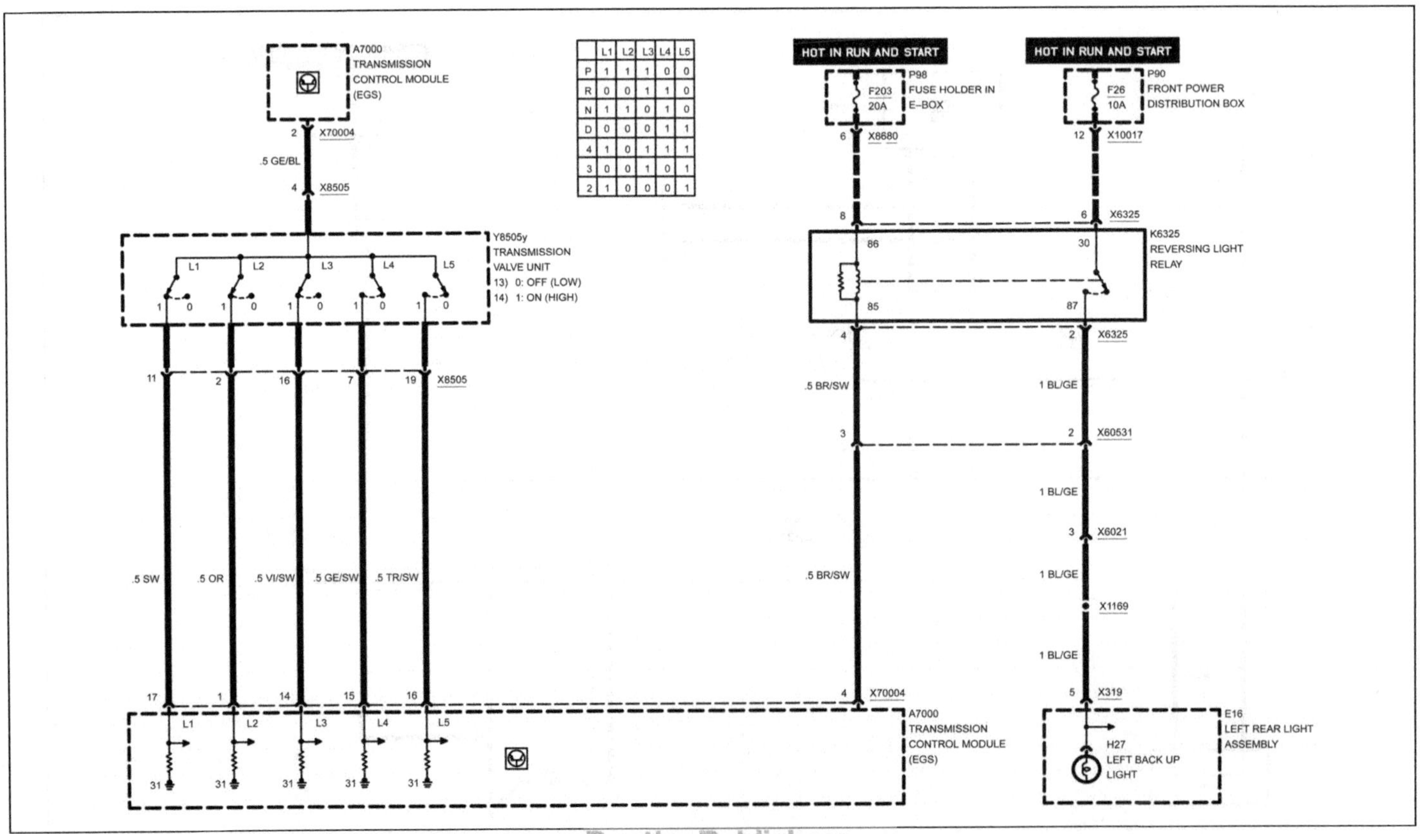


| | L1 | L2 | L3 | L4 | L5 |
|---|---|---|---|---|---|
| P | 1 | 1 | 1 | 0 | 0 |
| R | 0 | 0 | 1 | 1 | 0 |
| N | 1 | 1 | 0 | 1 | 0 |
| D | 0 | 0 | 0 | 1 | 1 |
| 4 | 1 | 0 | 1 | 1 | 1 |
| 3 | 0 | 0 | 1 | 0 | 1 |
| 2 | 1 | 0 | 0 | 0 | 1 |

## Airbag system (SRS), 1996

(1 of 2)

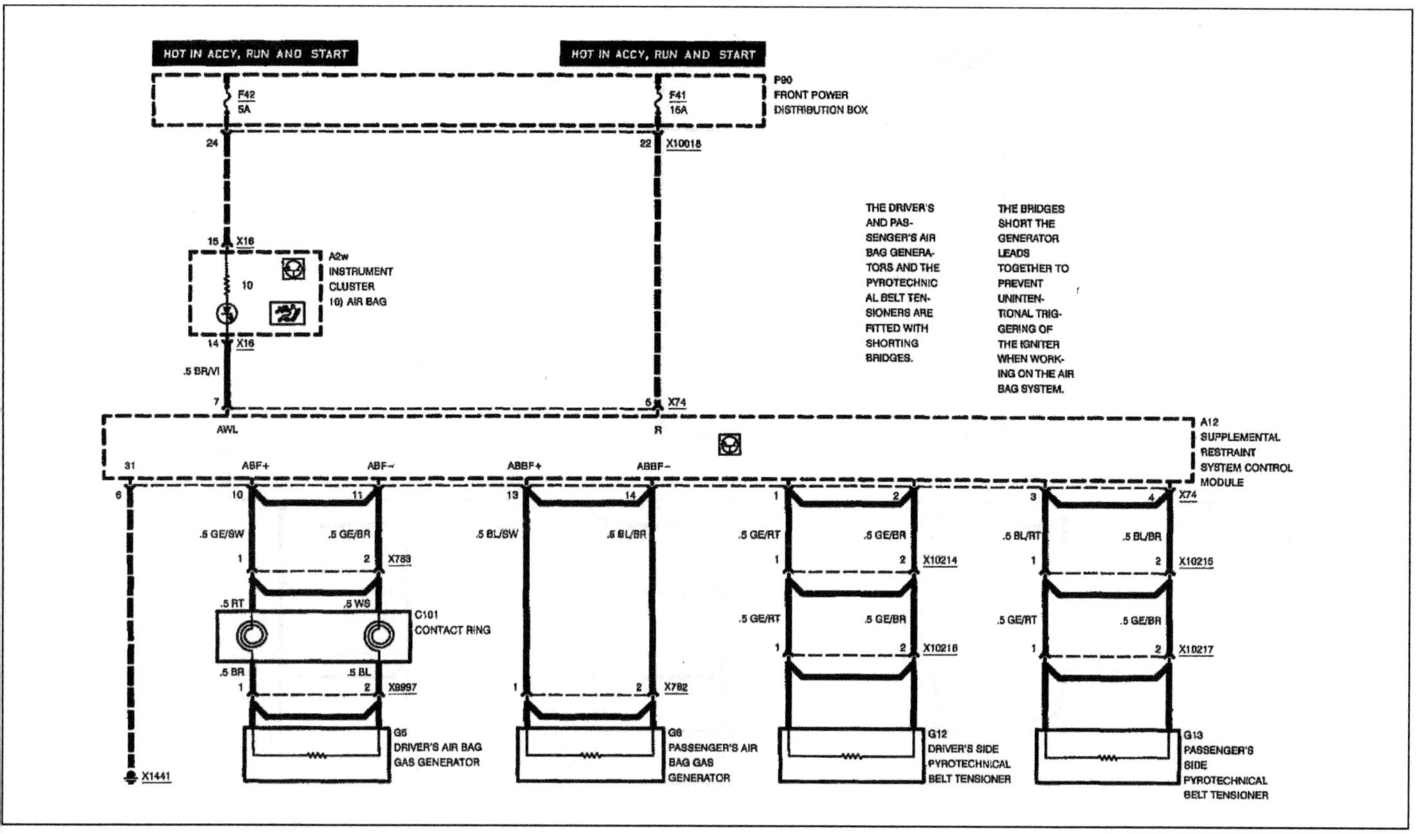


## Airbag system (SRS), 1996

(2 of 2)

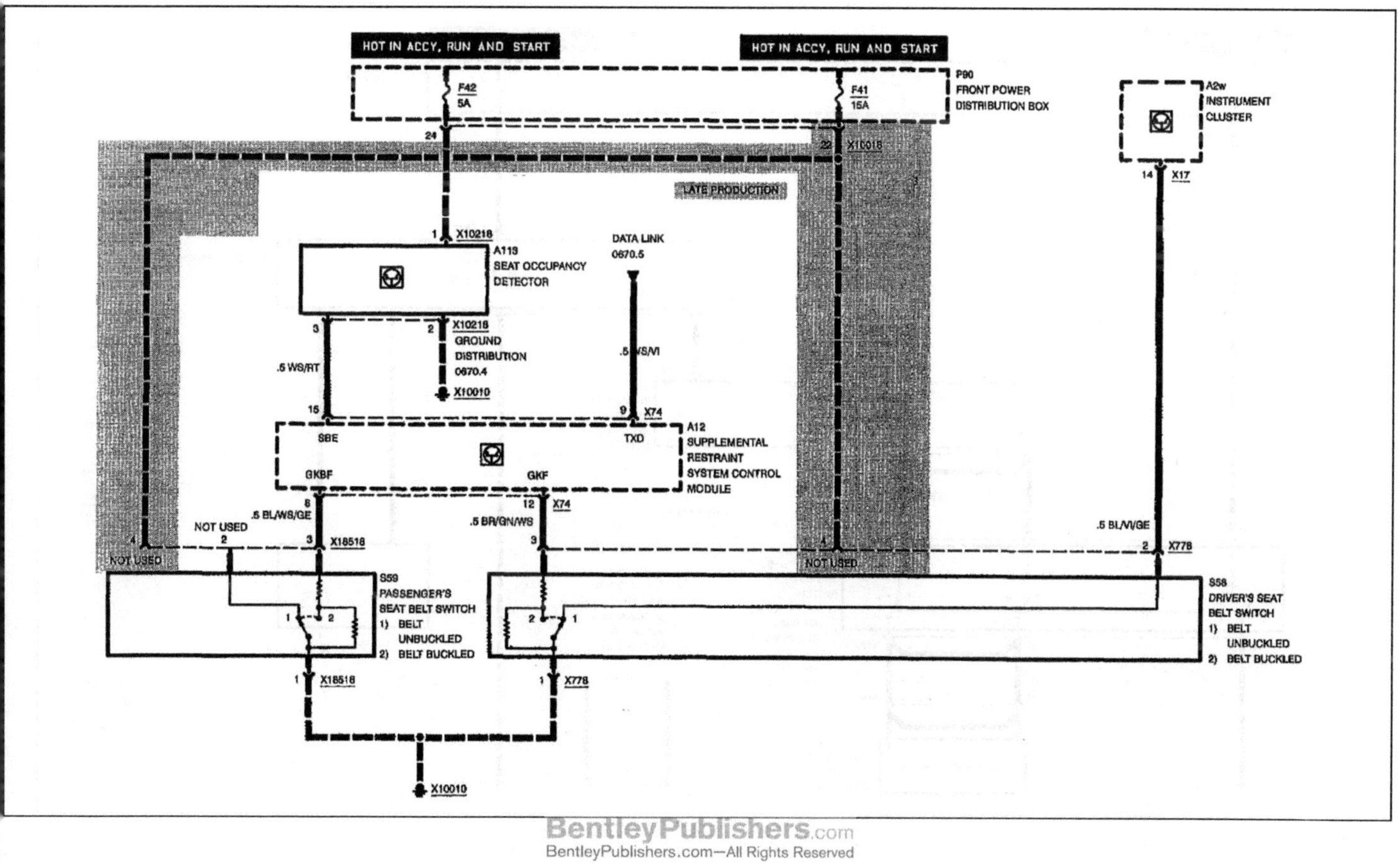

## Airbag system (SRS), 1998

(1 of 4)

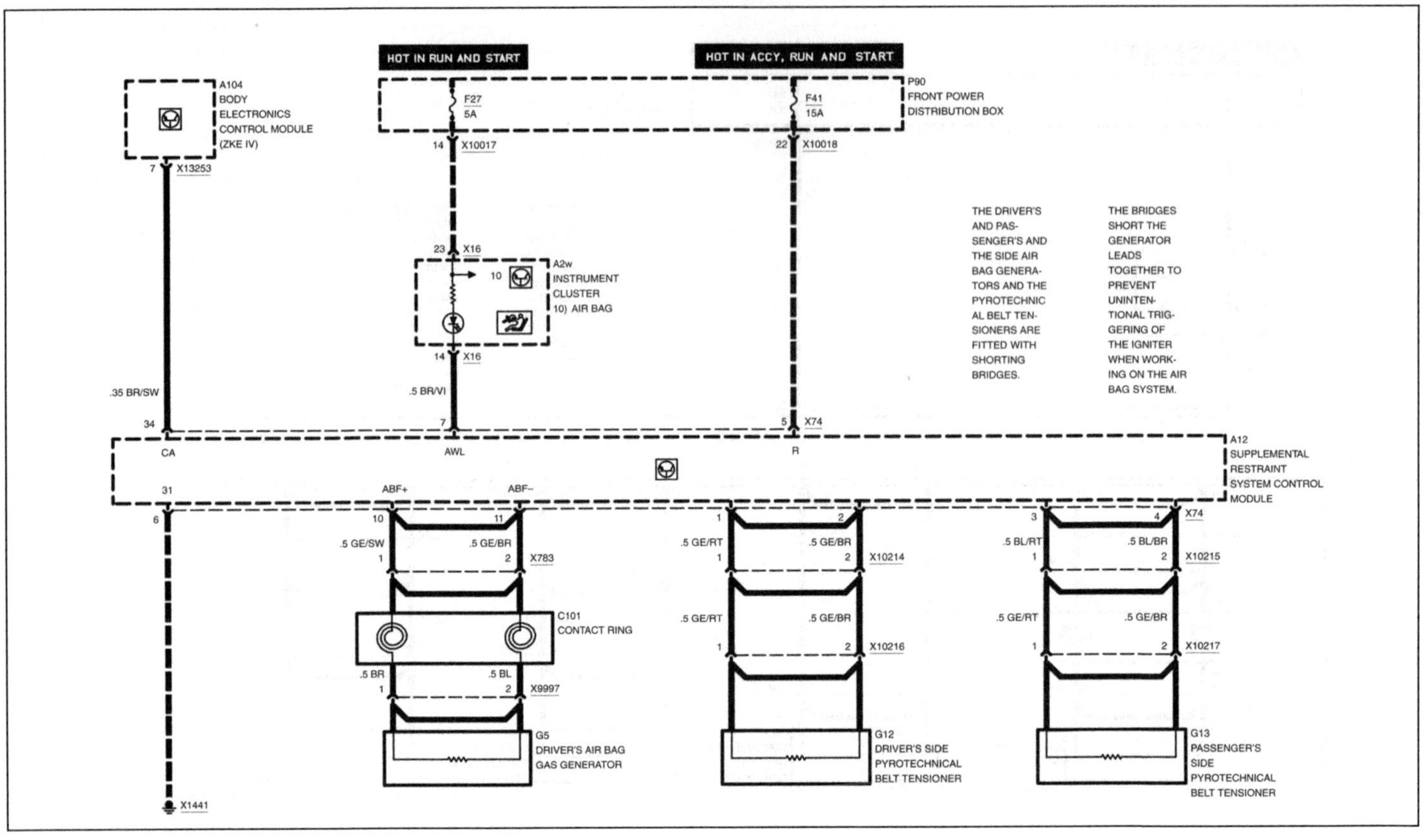


## Airbag system (SRS), 1998

(2 of 4)

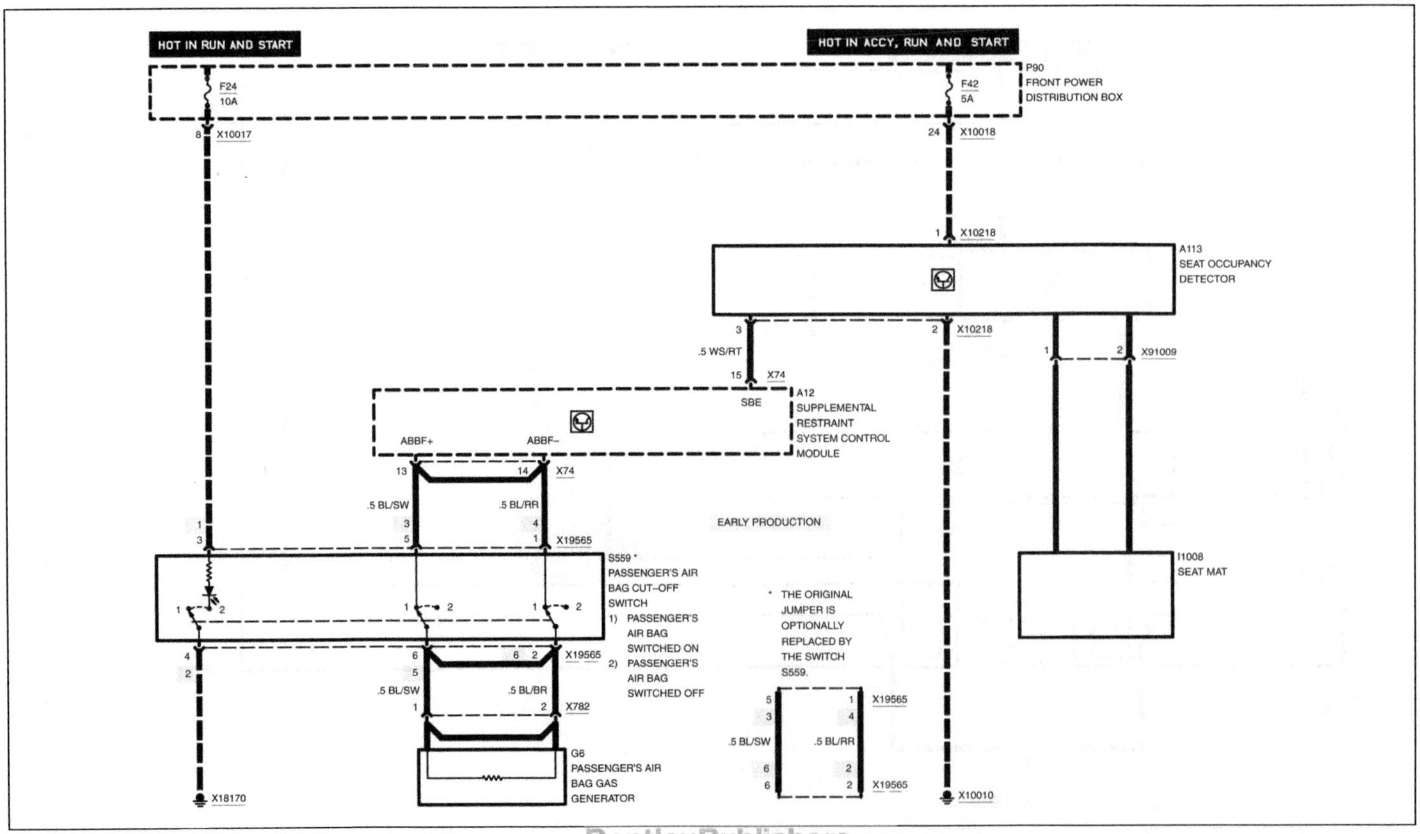

## Airbag system (SRS), 1998
### (3 of 4)

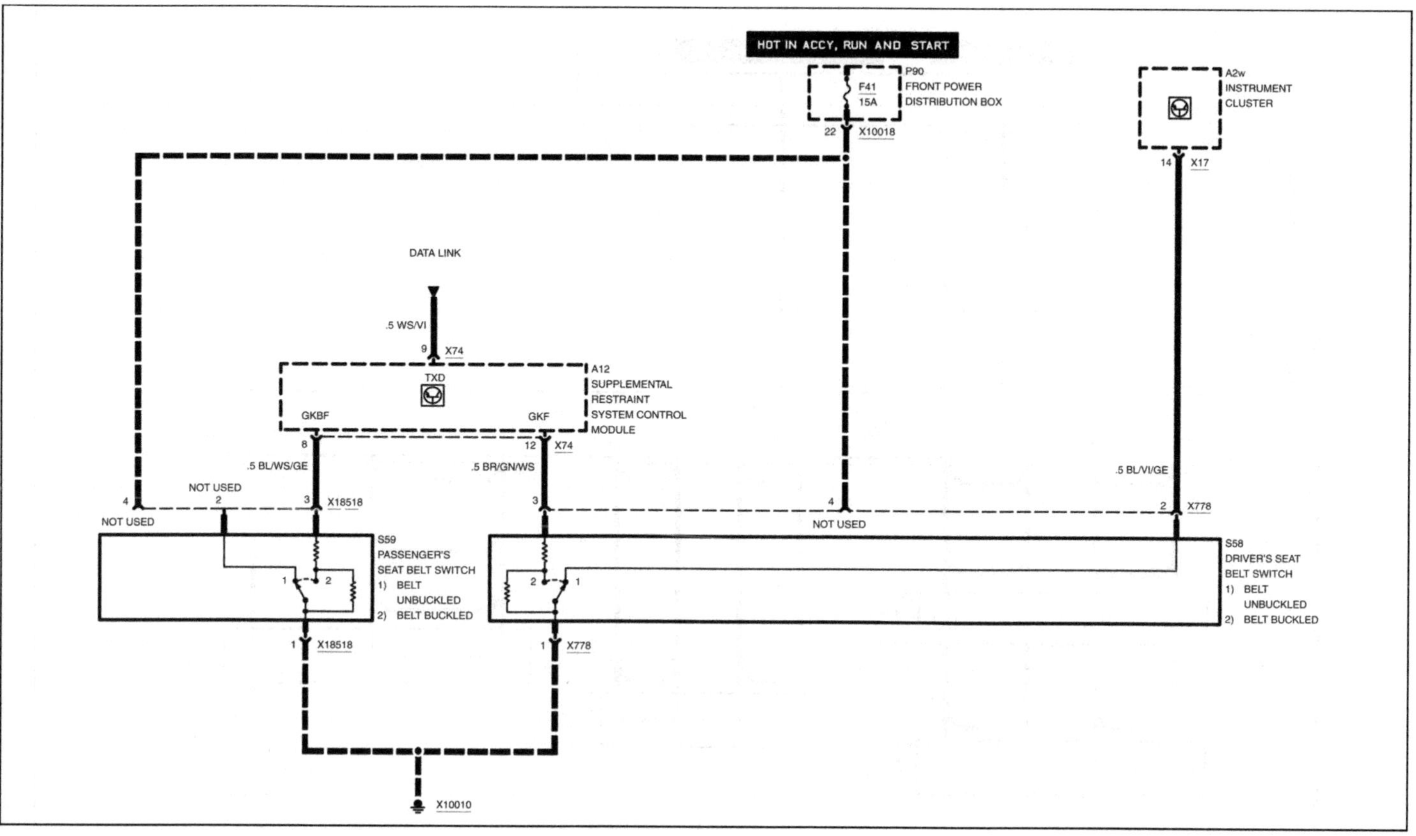


## Airbag system (SRS), 1998
### (4 of 4)

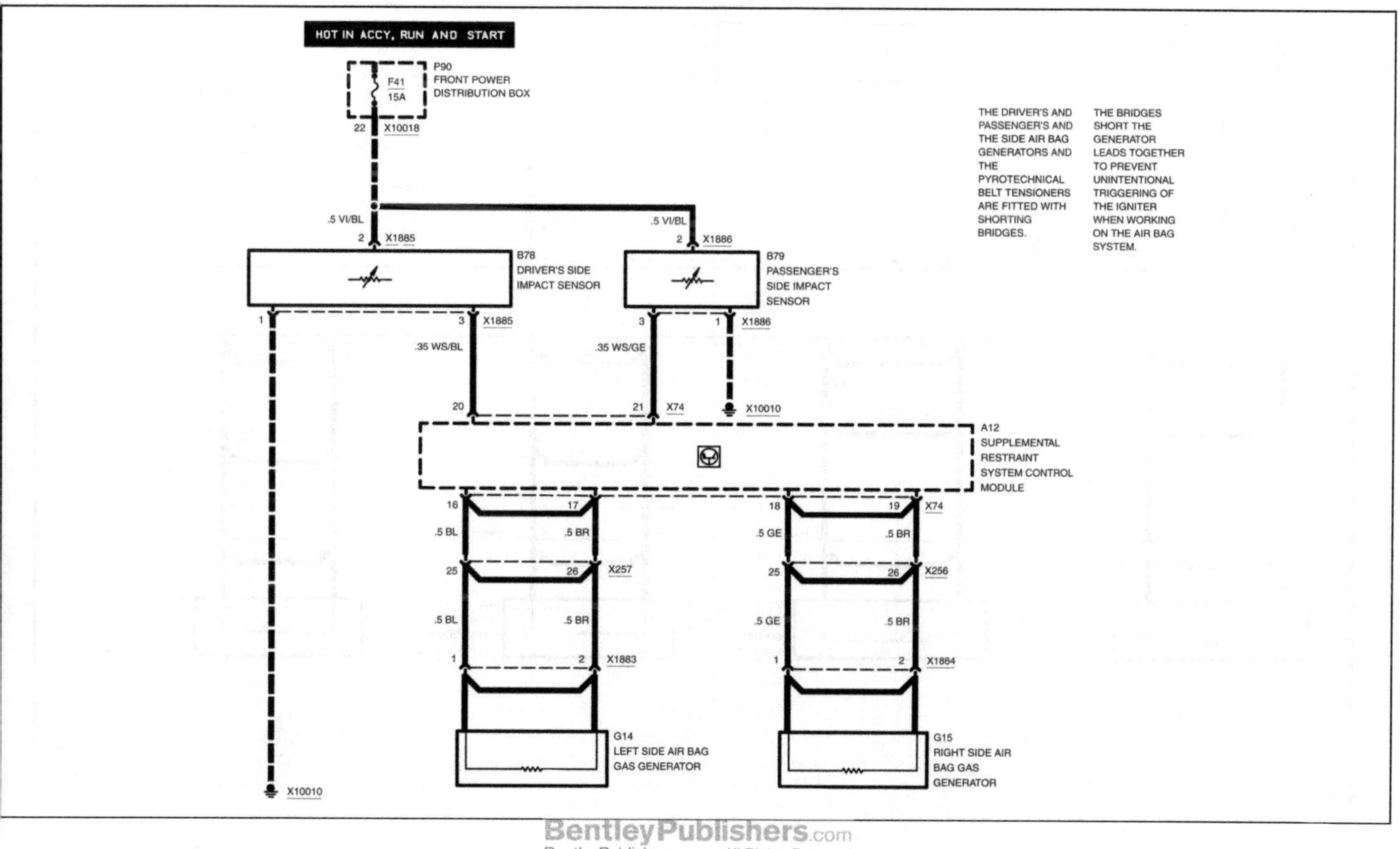

## Airbag system (SRS), 2000
(1 of 4)

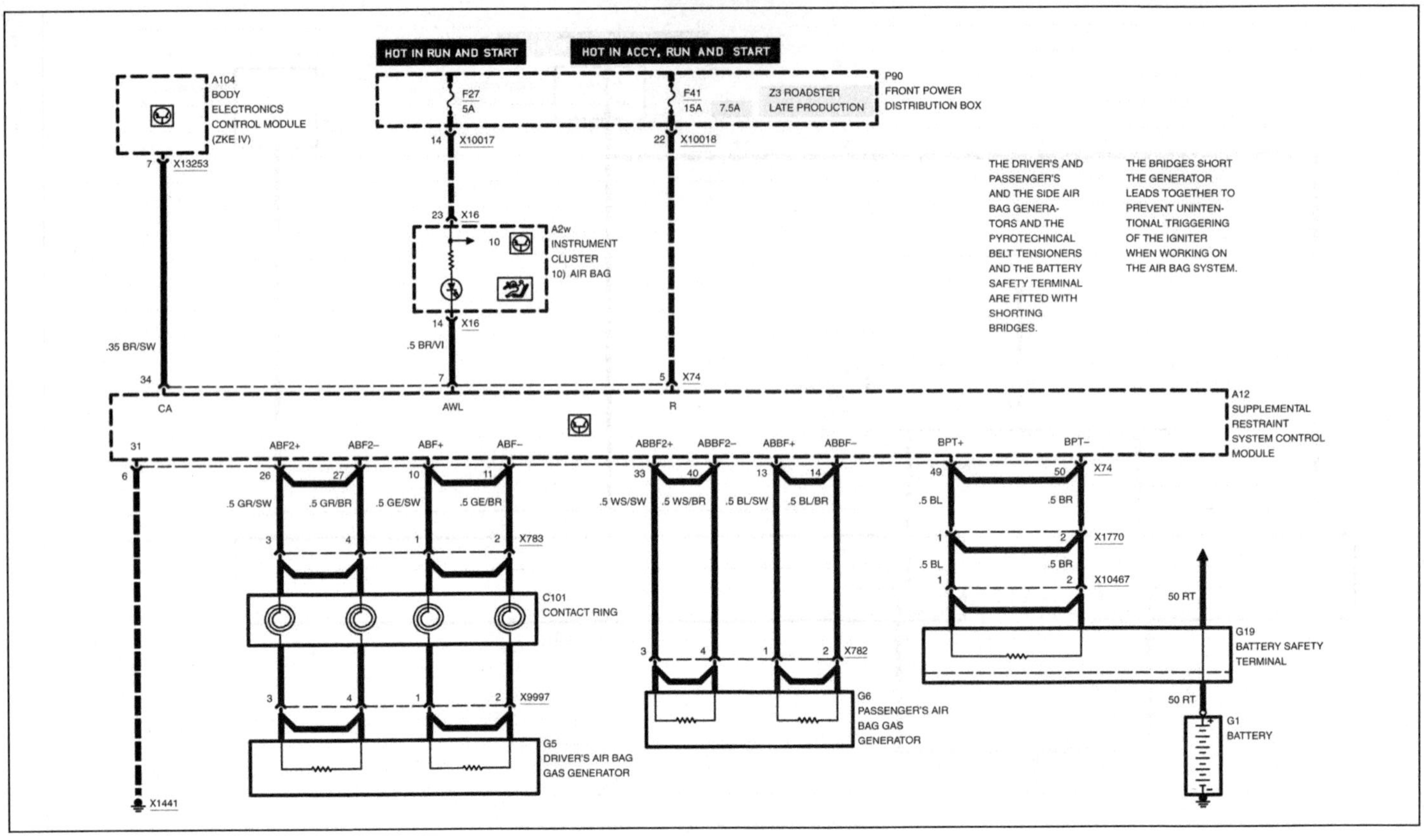


## Airbag system (SRS), 2000
(2 of 4)

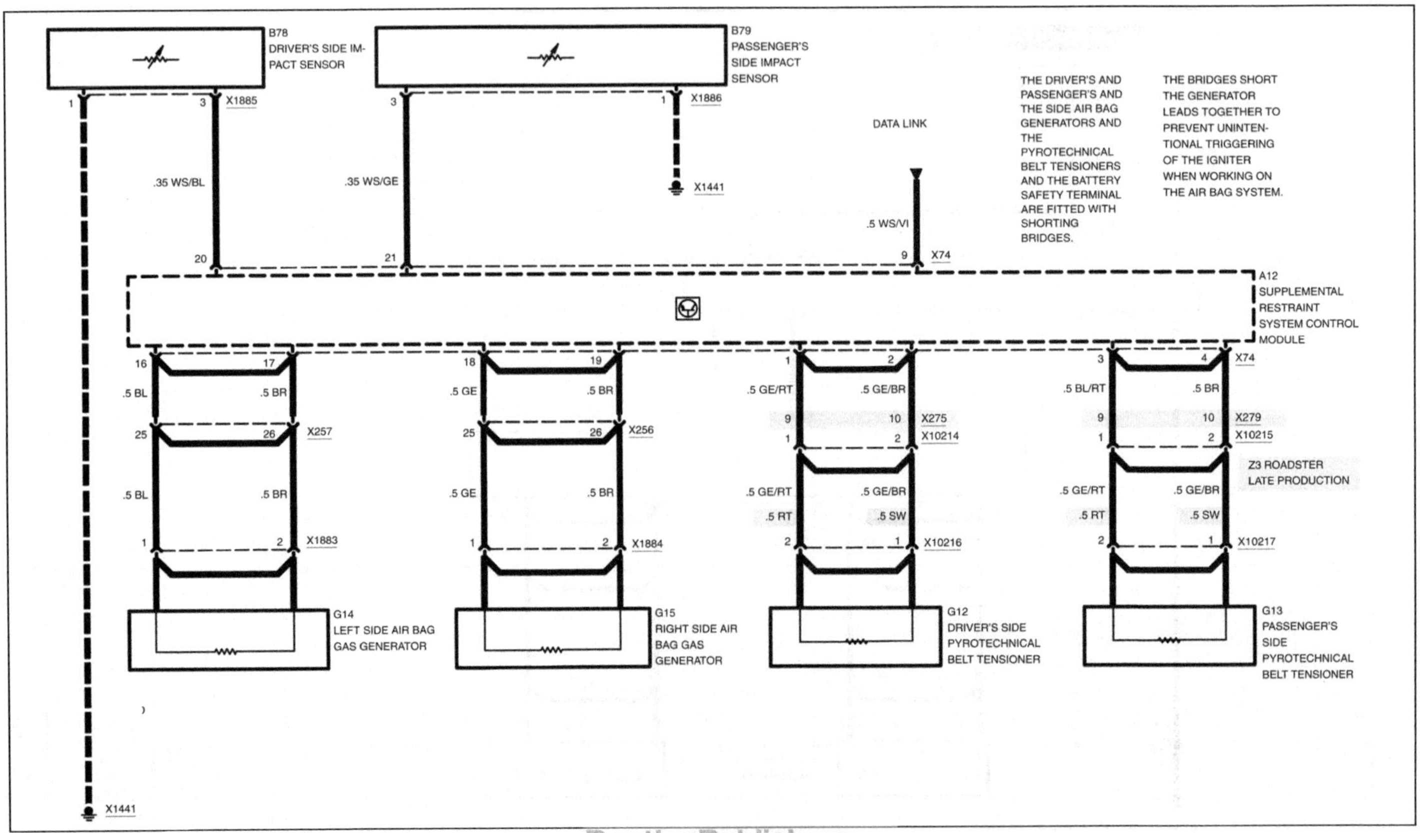

## Airbag system (SRS), 2000
(3 of 4)

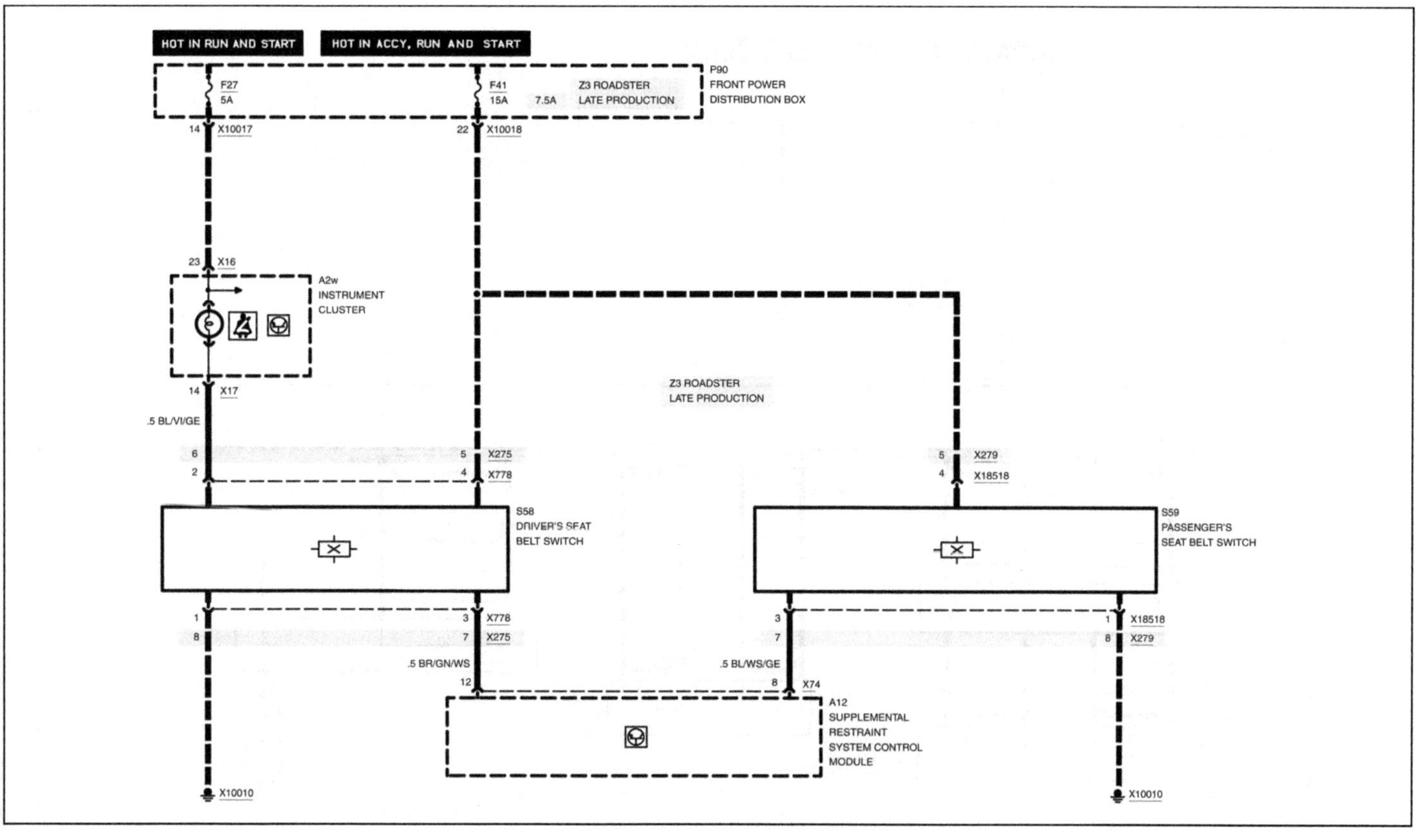


## Airbag system (SRS), 2000
(4 of 4)

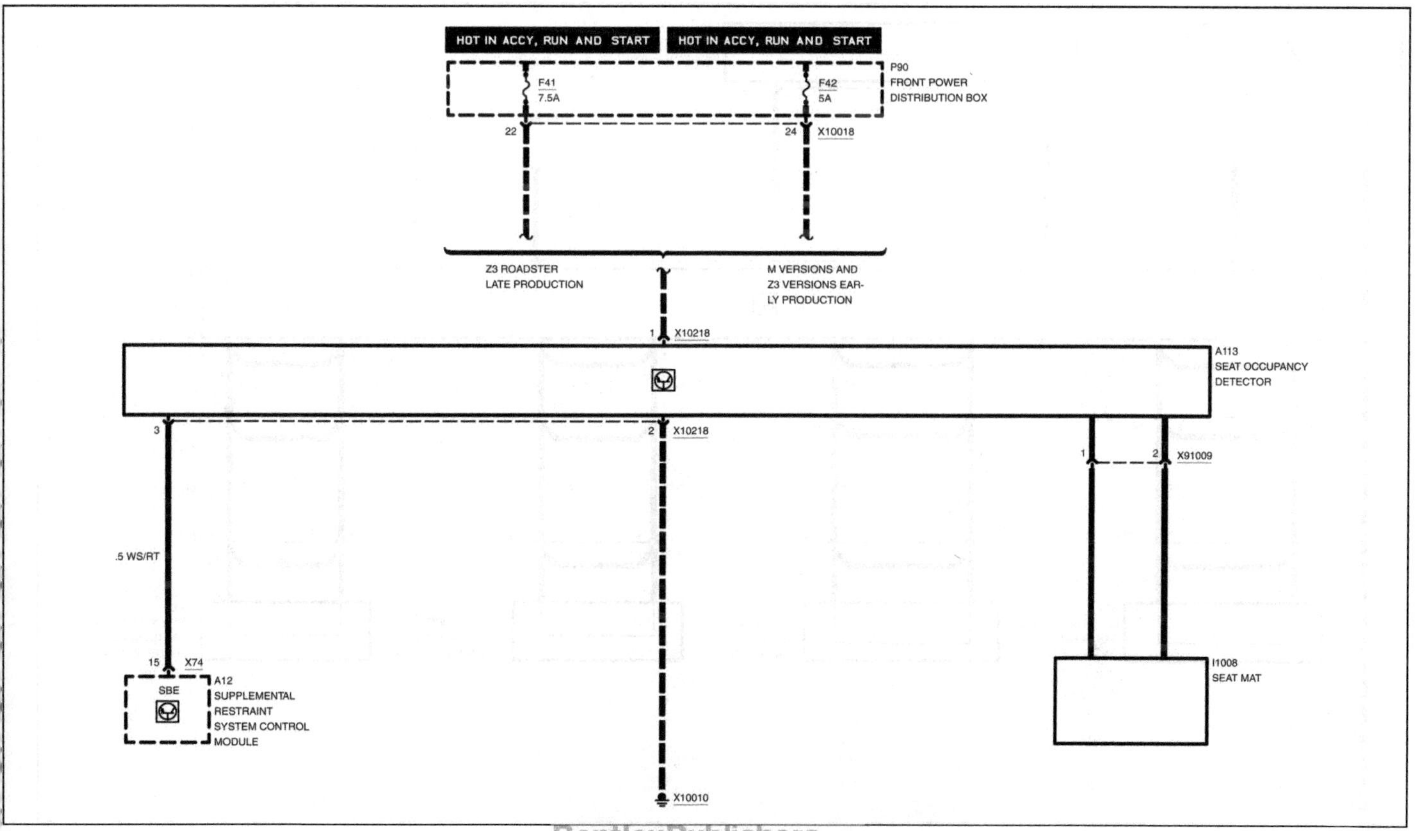

## Airbag system (SRS), 2002
(1 of 4)

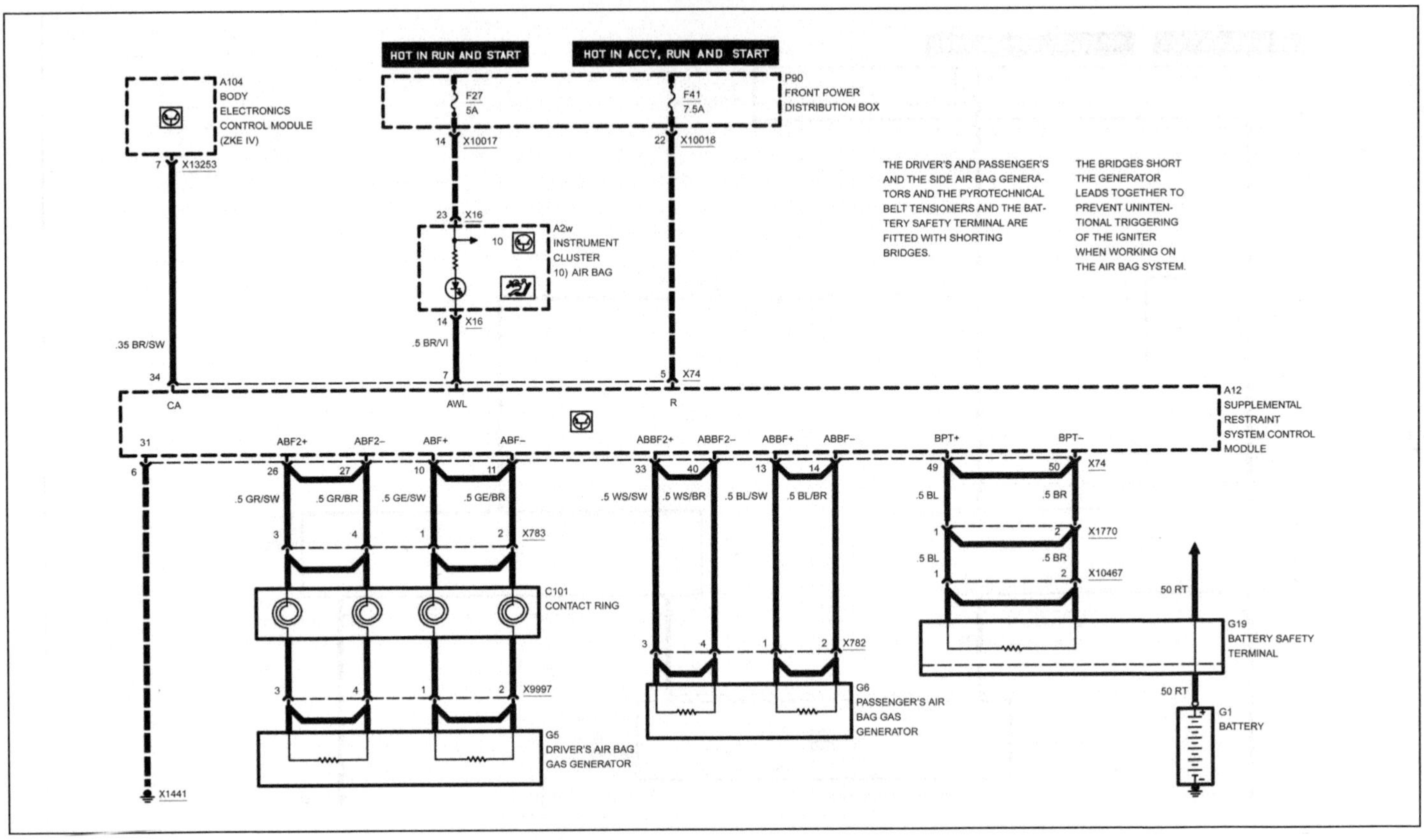


## Airbag system (SRS), 2002
(2 of 4)

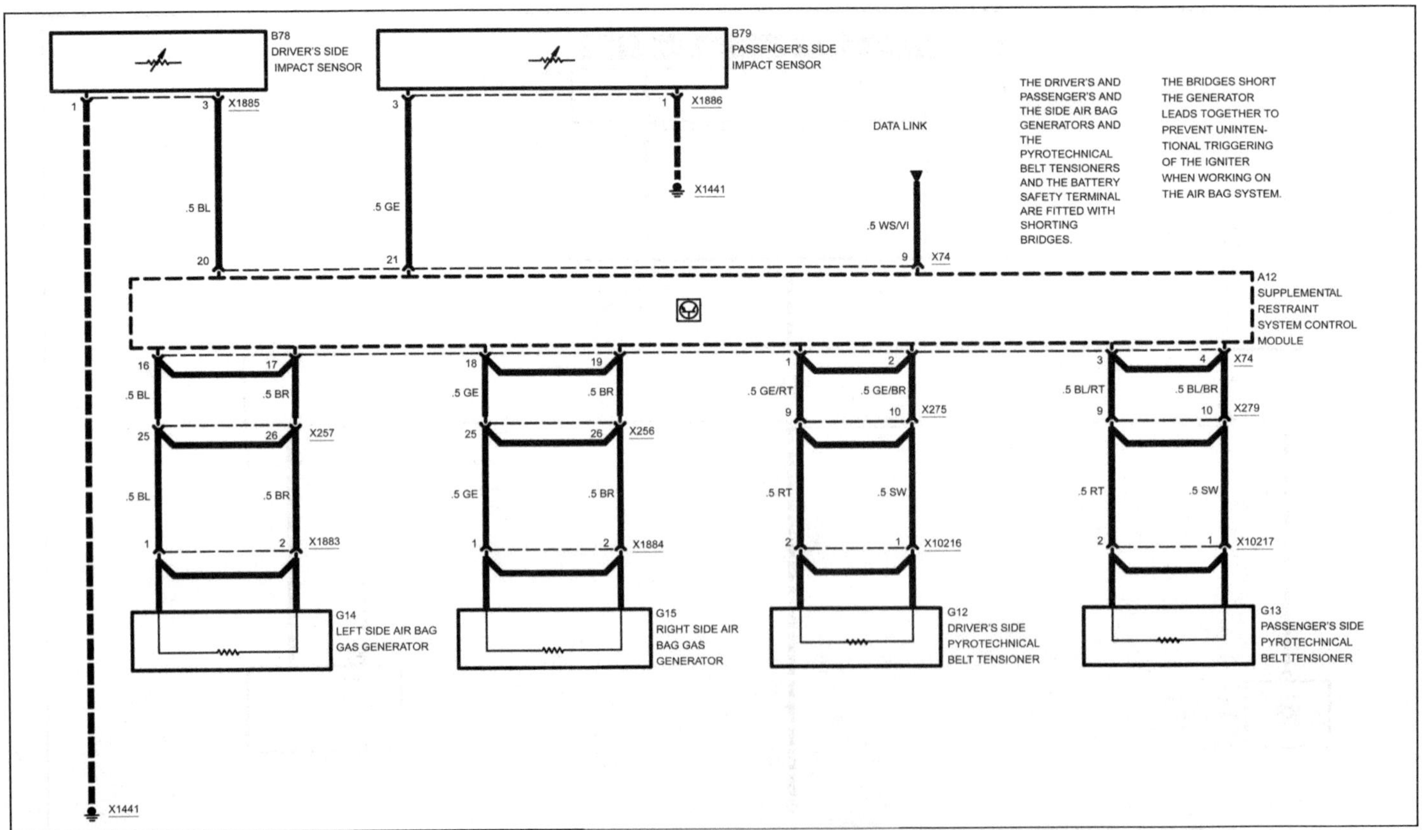

## Airbag system (SRS), 2002

(3 of 4)

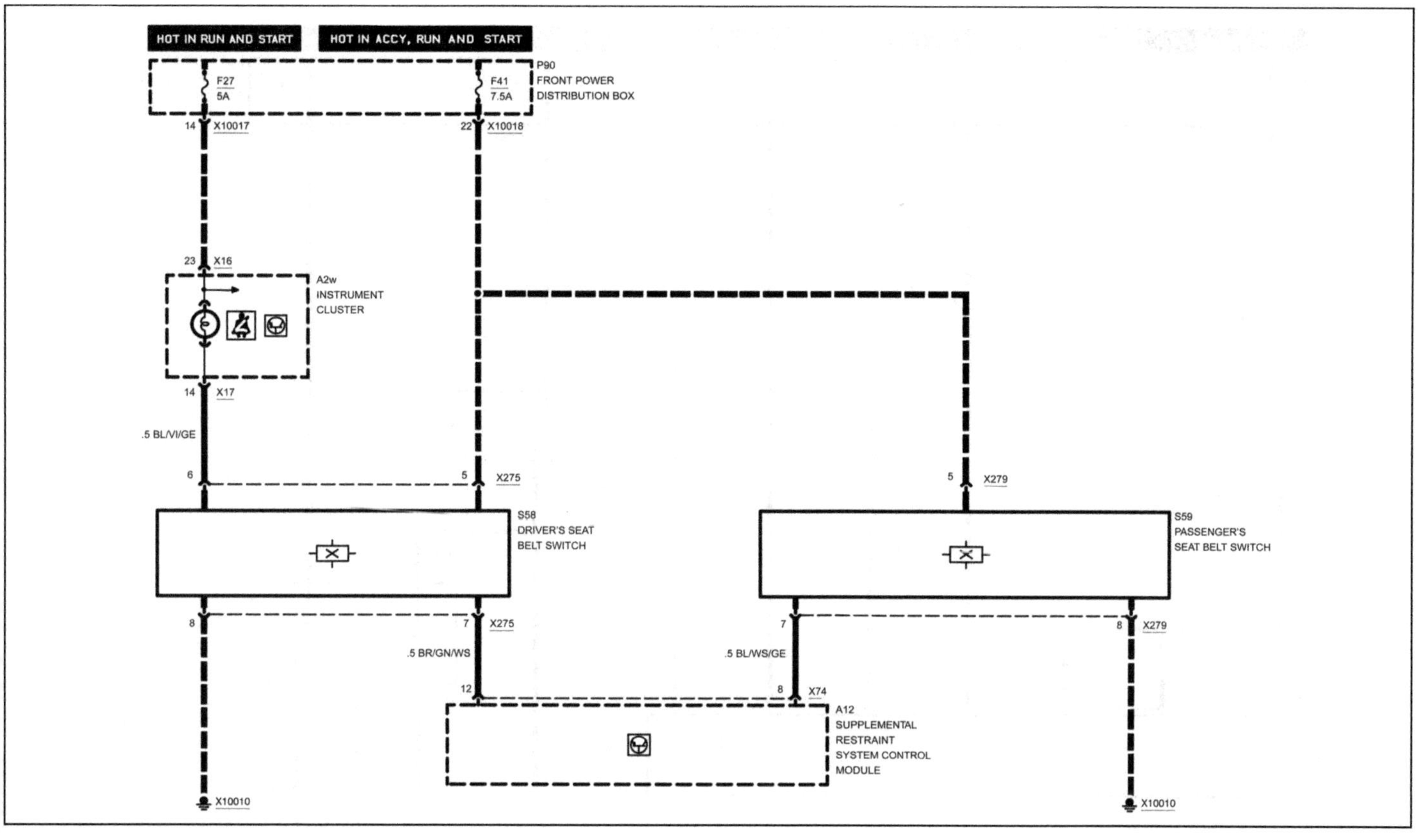


## Airbag system (SRS), 2002

(4 of 4)

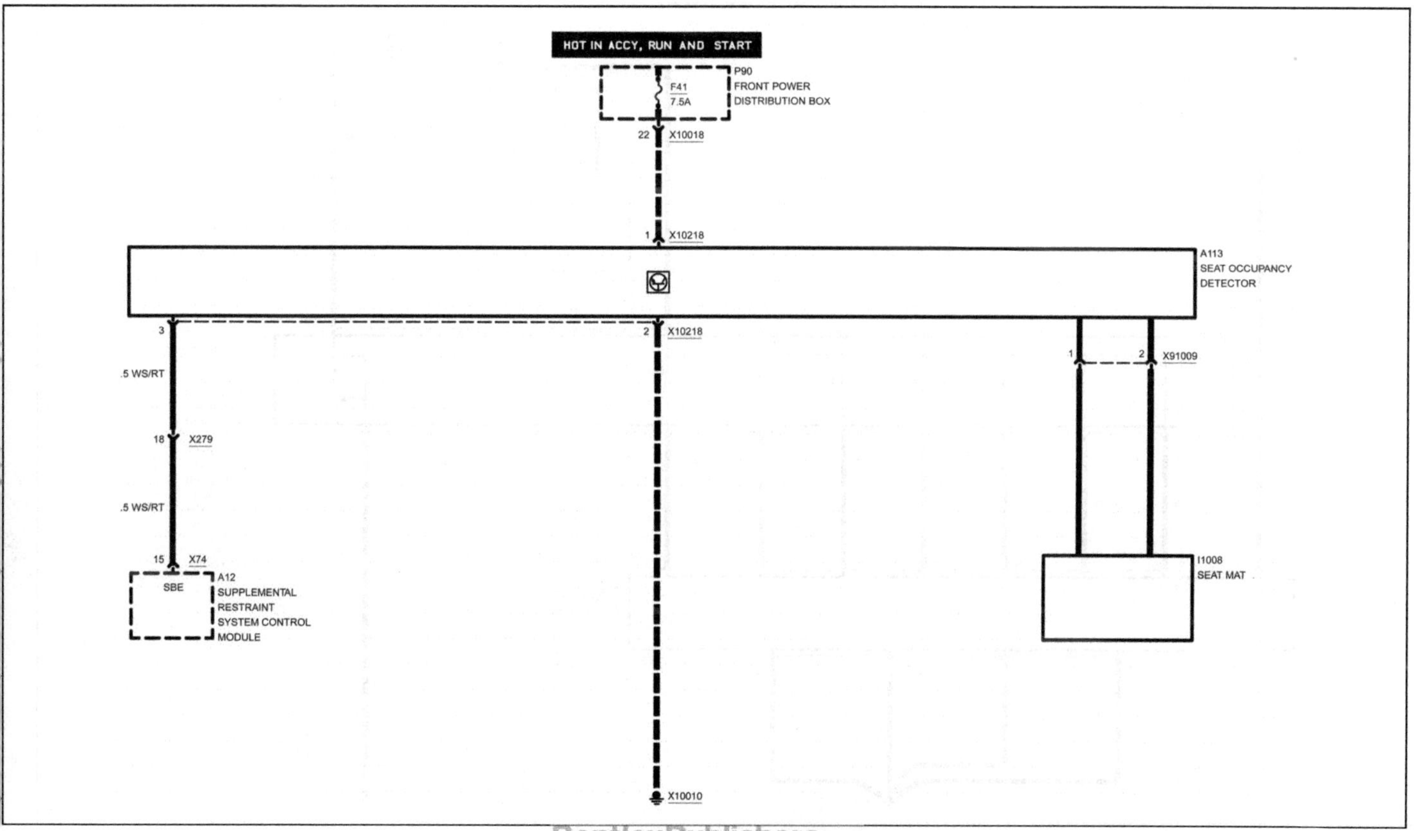

## ABS / ASC, 1996, 1998 M
(1 of 3)

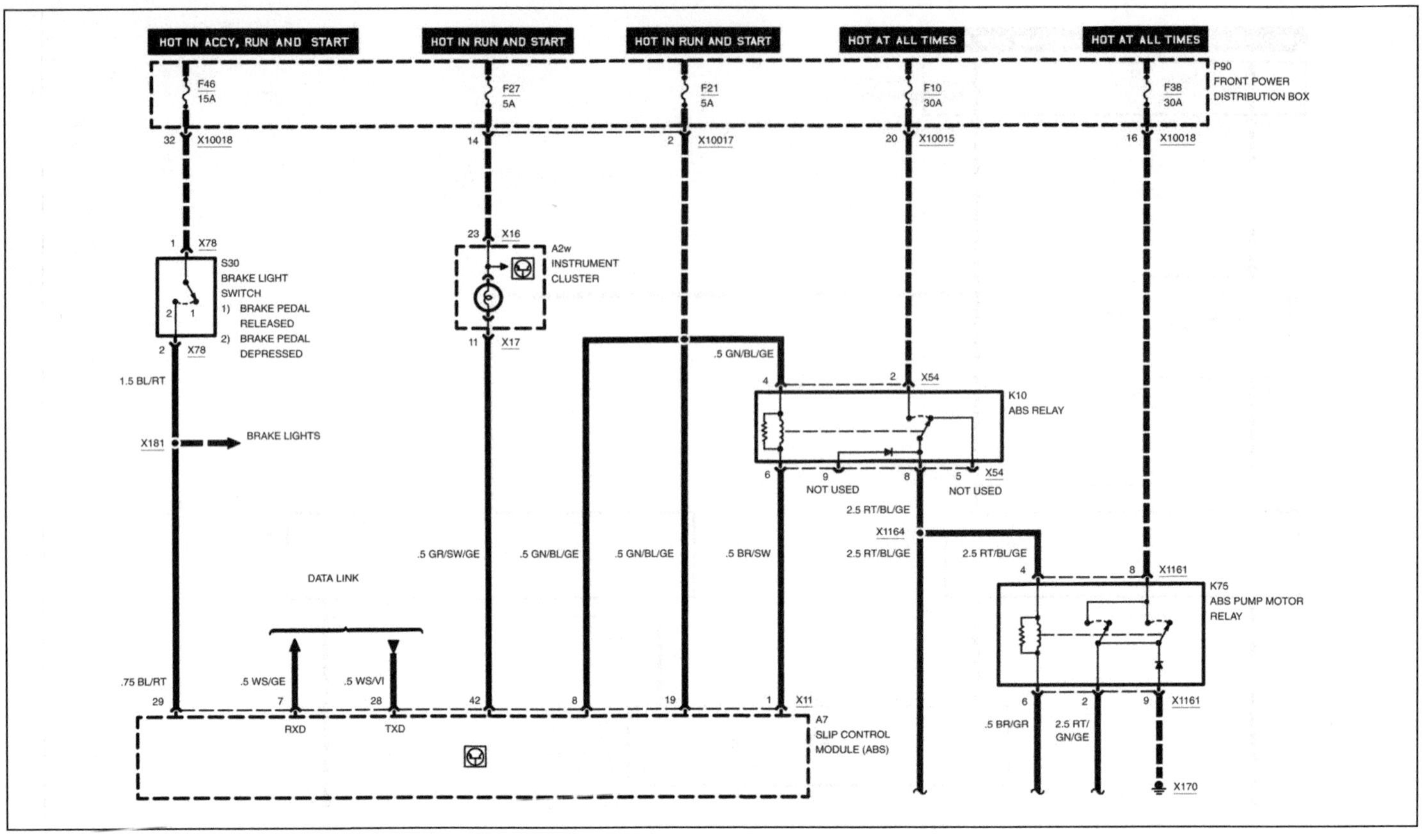


## ABS / ASC, 1996, 1998 M
(2 of 3)

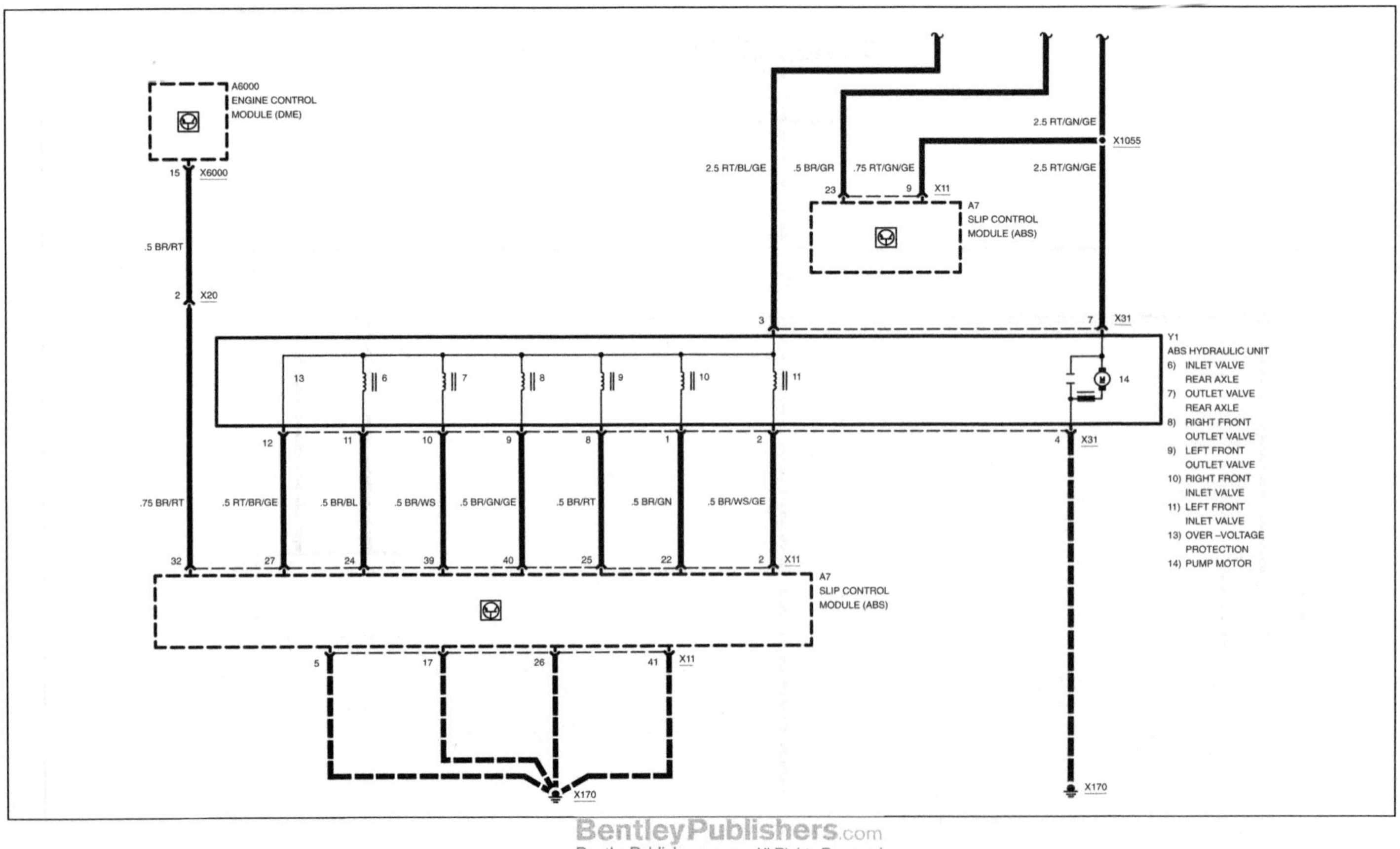

## ABS / ASC, 1996, 1998 M
(3 of 3)

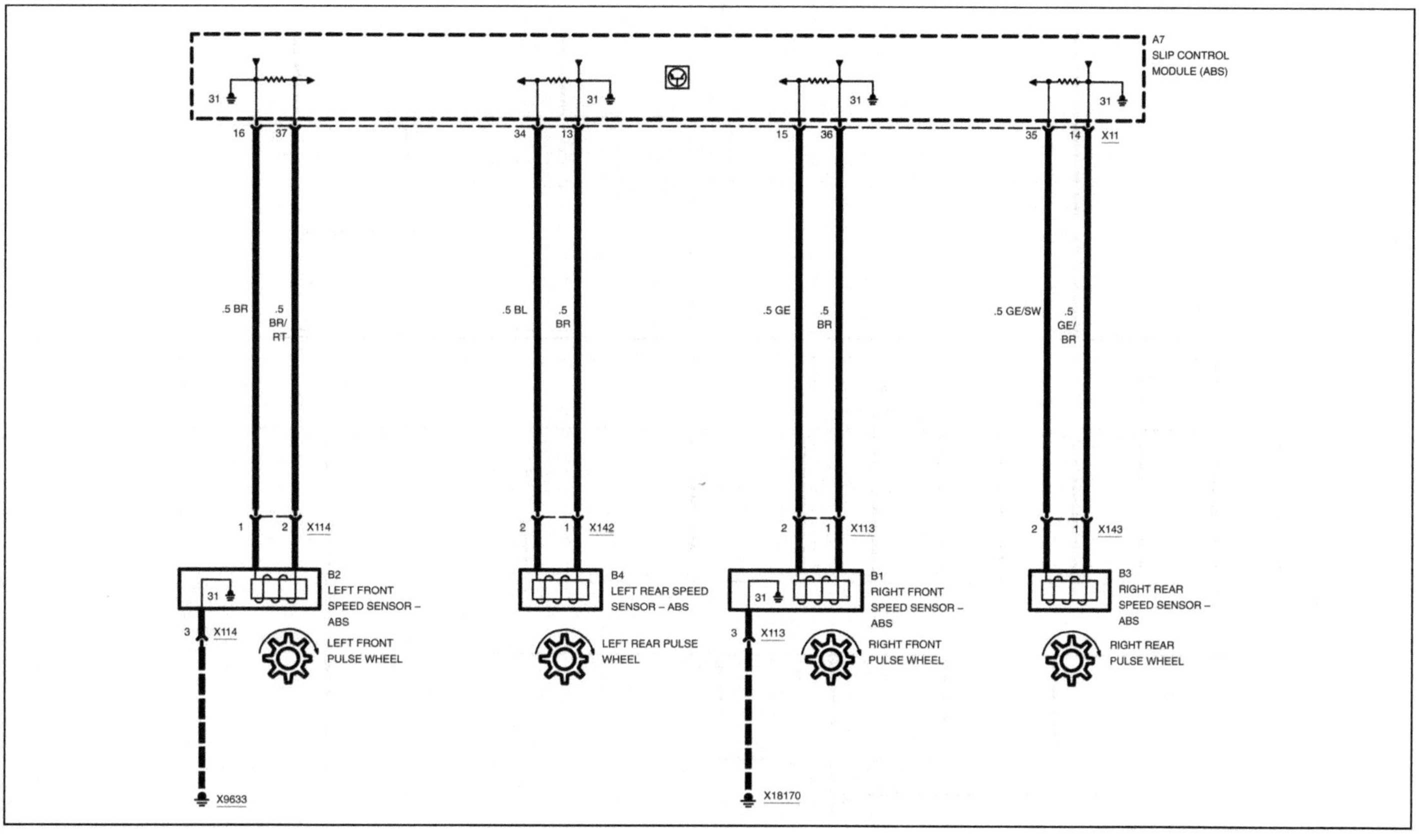


## ABS / ASC, 1998 Z3 (4-cylinder)
(1 of 4)

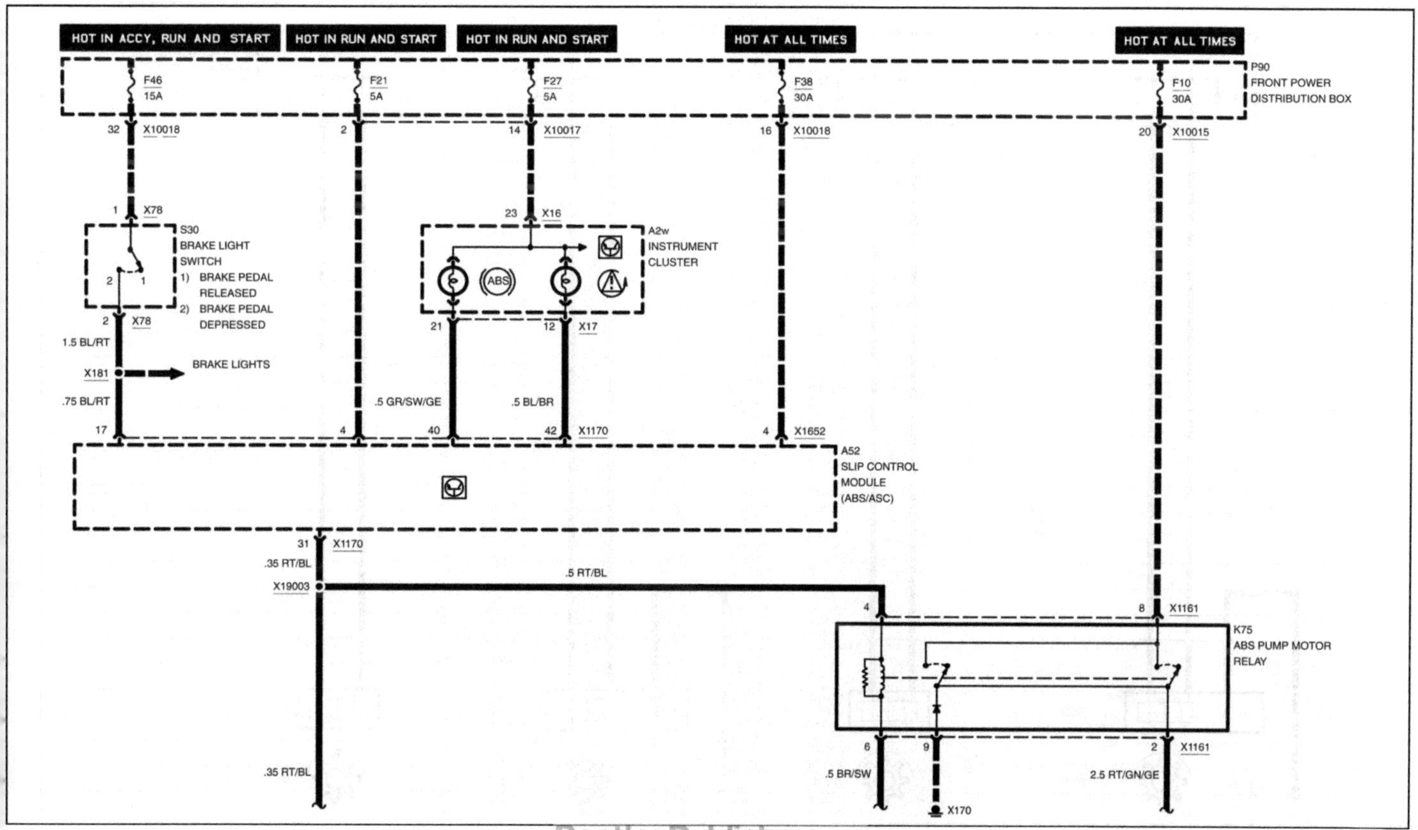

## ABS / ASC, 1998 Z3 (4-cylinder)
### (2 of 4)

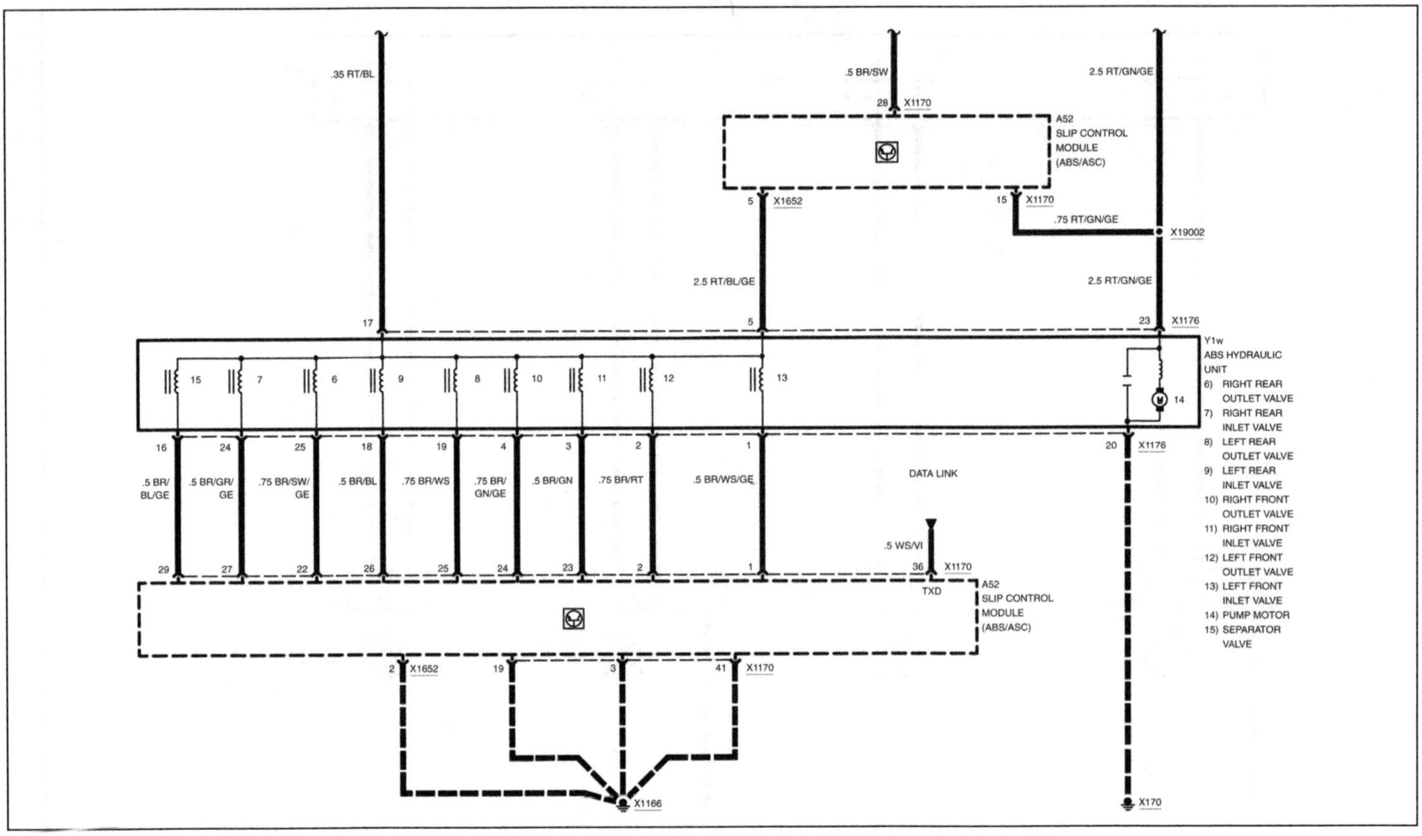


## ABS / ASC, 1998 Z3 (4-cylinder)
### (3 of 4)

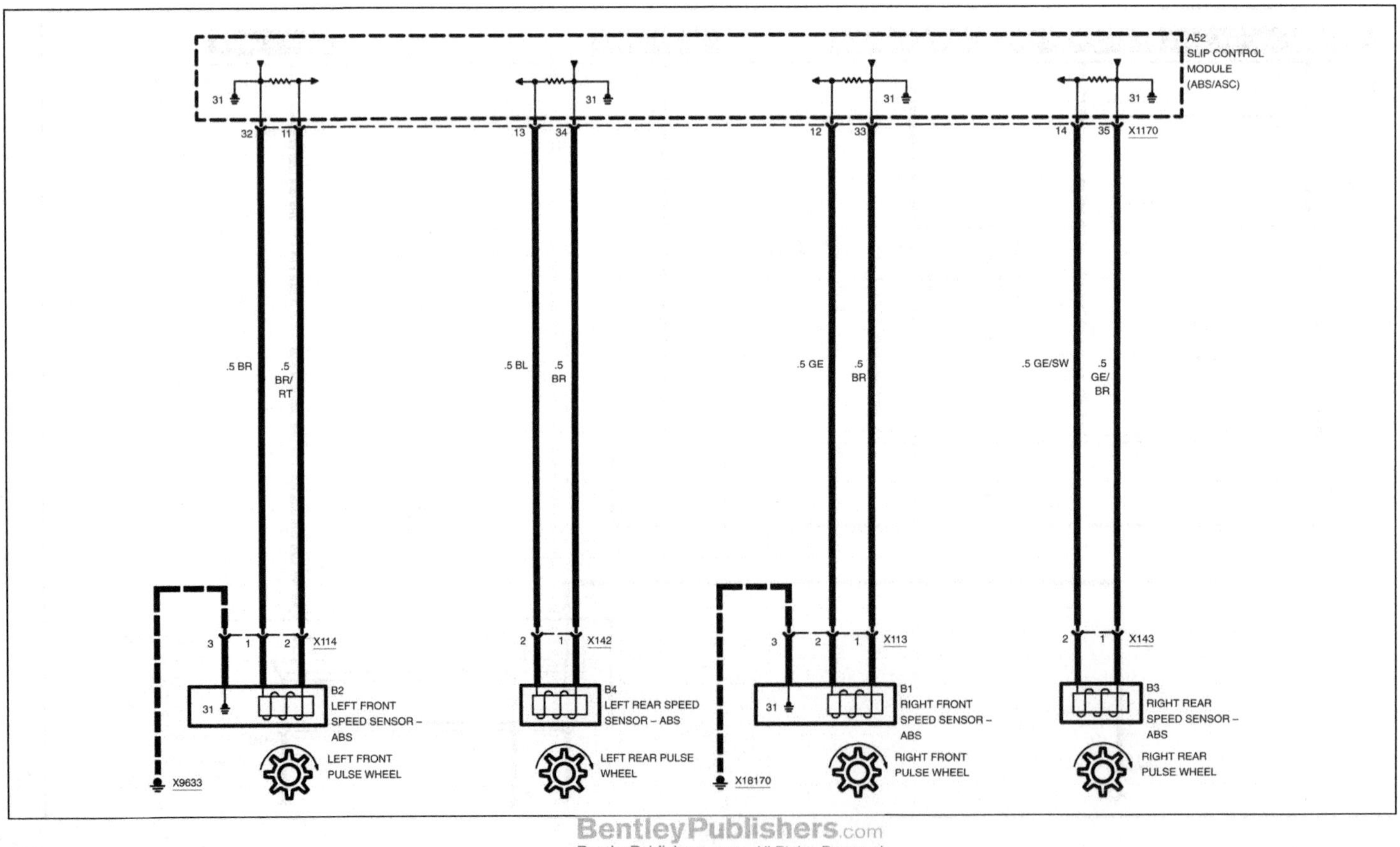

## ABS / ASC, 1998 Z3 (4-cylinder)

(4 of 4)

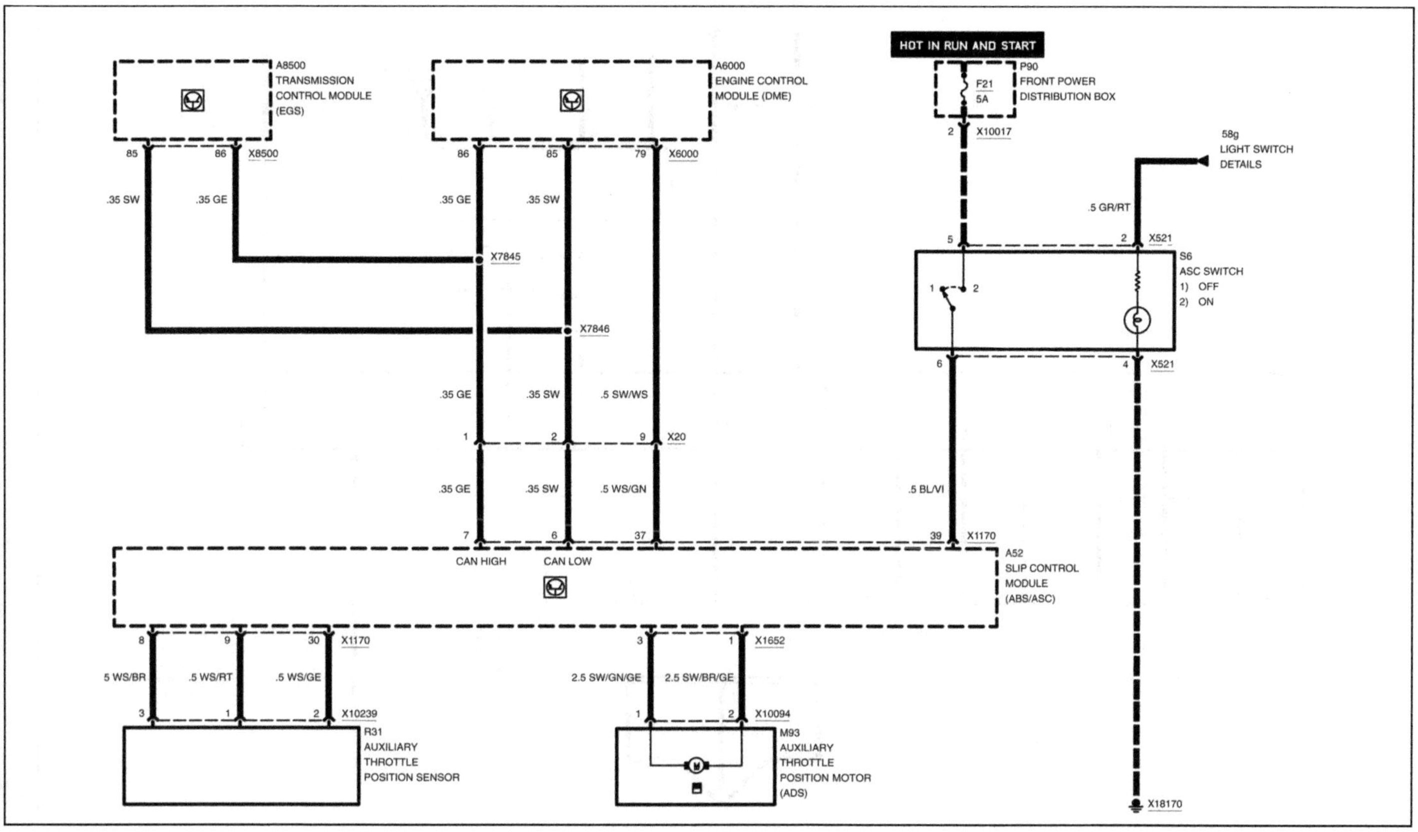


## ABS / ASC, 1998 Z3 (6-cylinder)

(1 of 5)

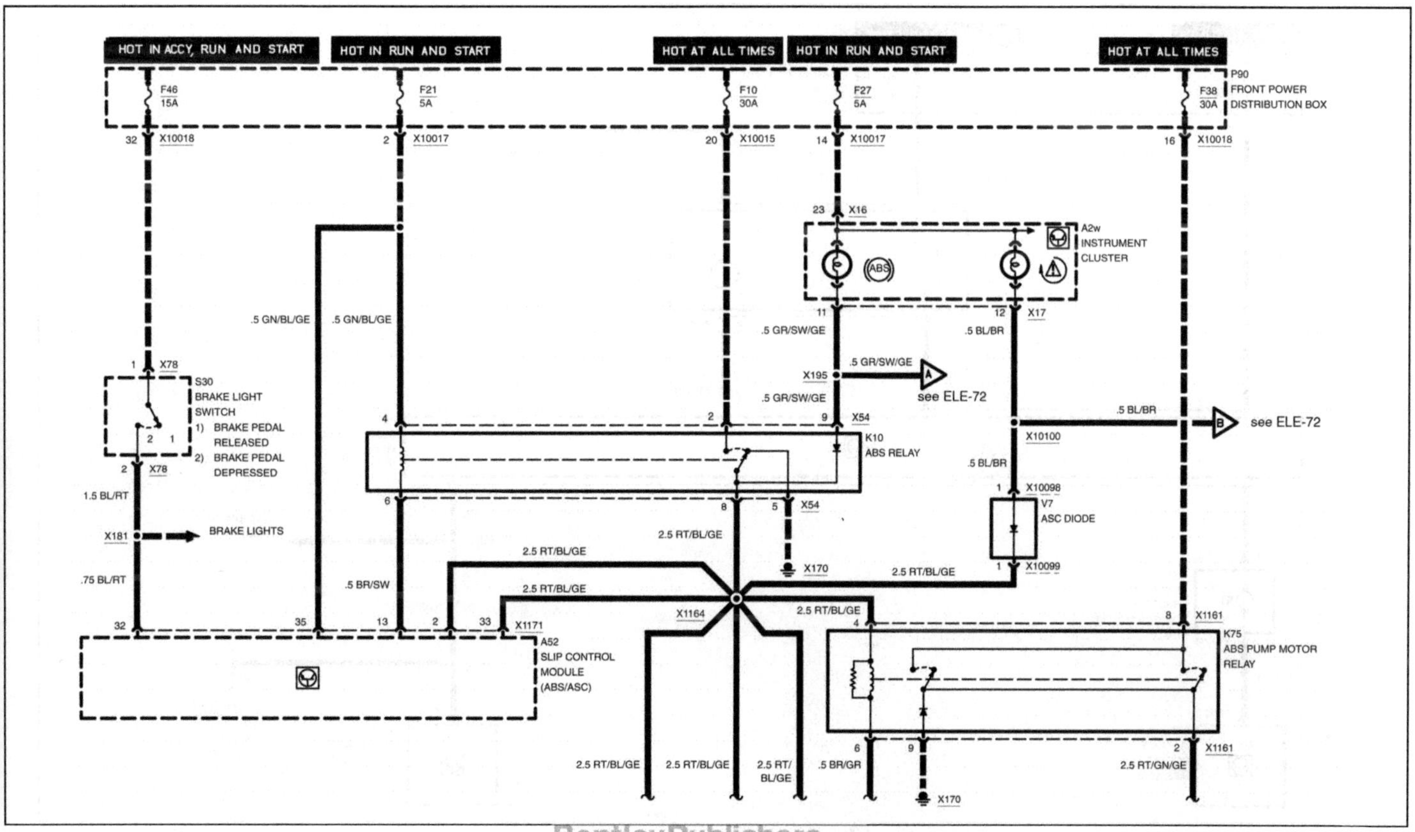

## ABS / ASC, 1998 Z3 (6-cylinder)

**(2 of 5)**

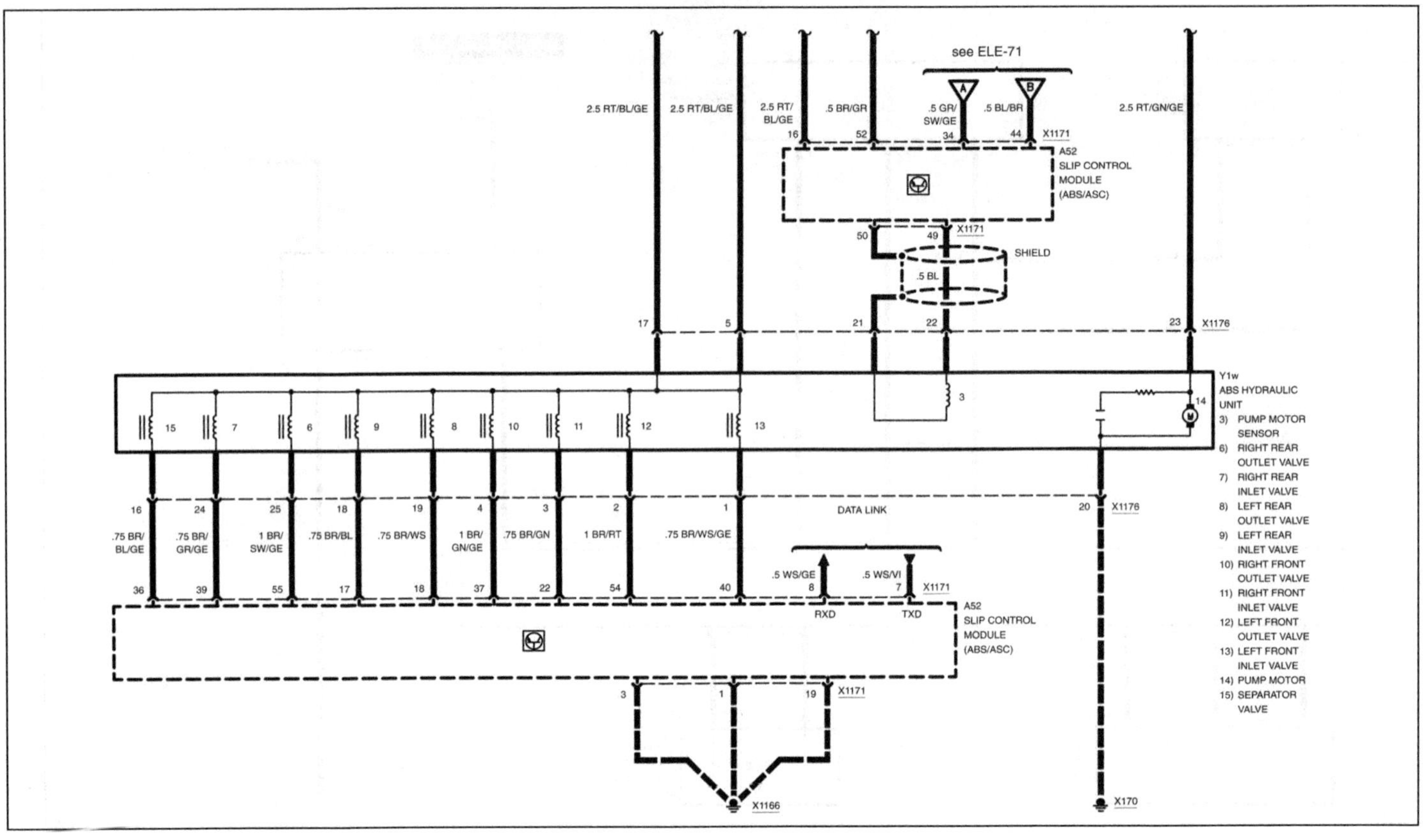


## ABS / ASC, 1998 Z3 (6-cylinder)

**(3 of 5)**

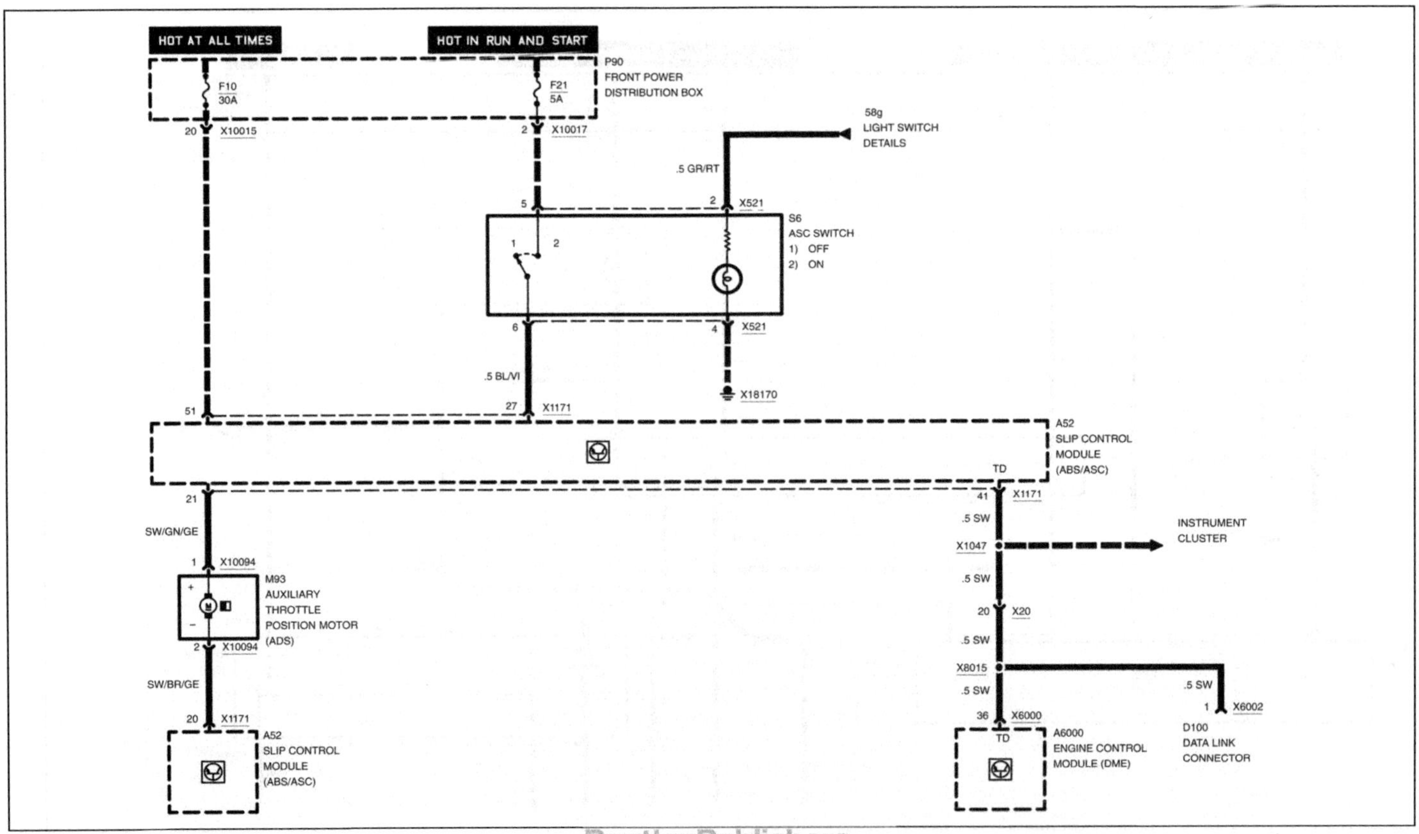

## ABS / ASC, 1998 Z3 (6-cylinder)

(4 of 5)

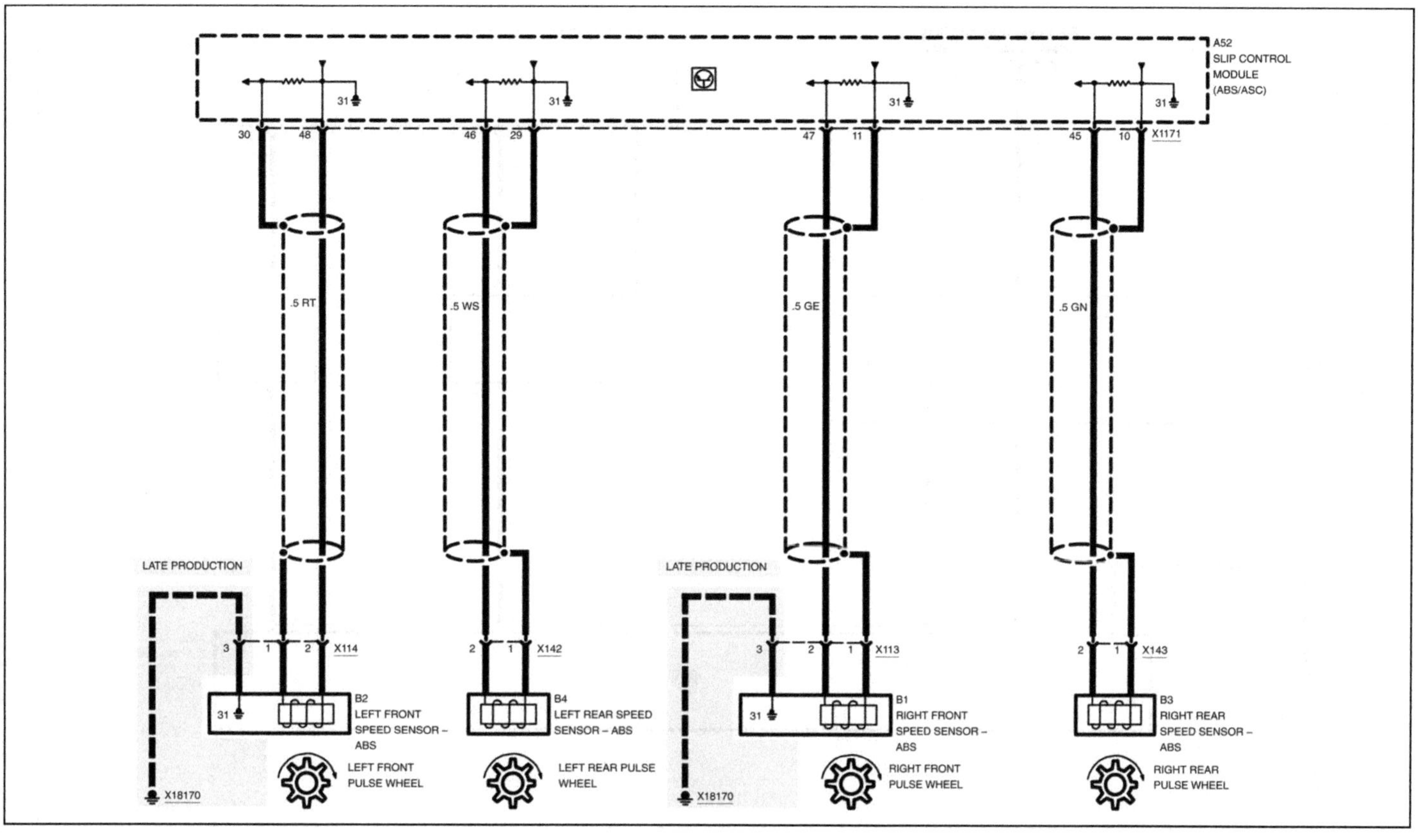


## ABS / ASC, 1998 Z3 (6-cylinder)

(5 of 5)

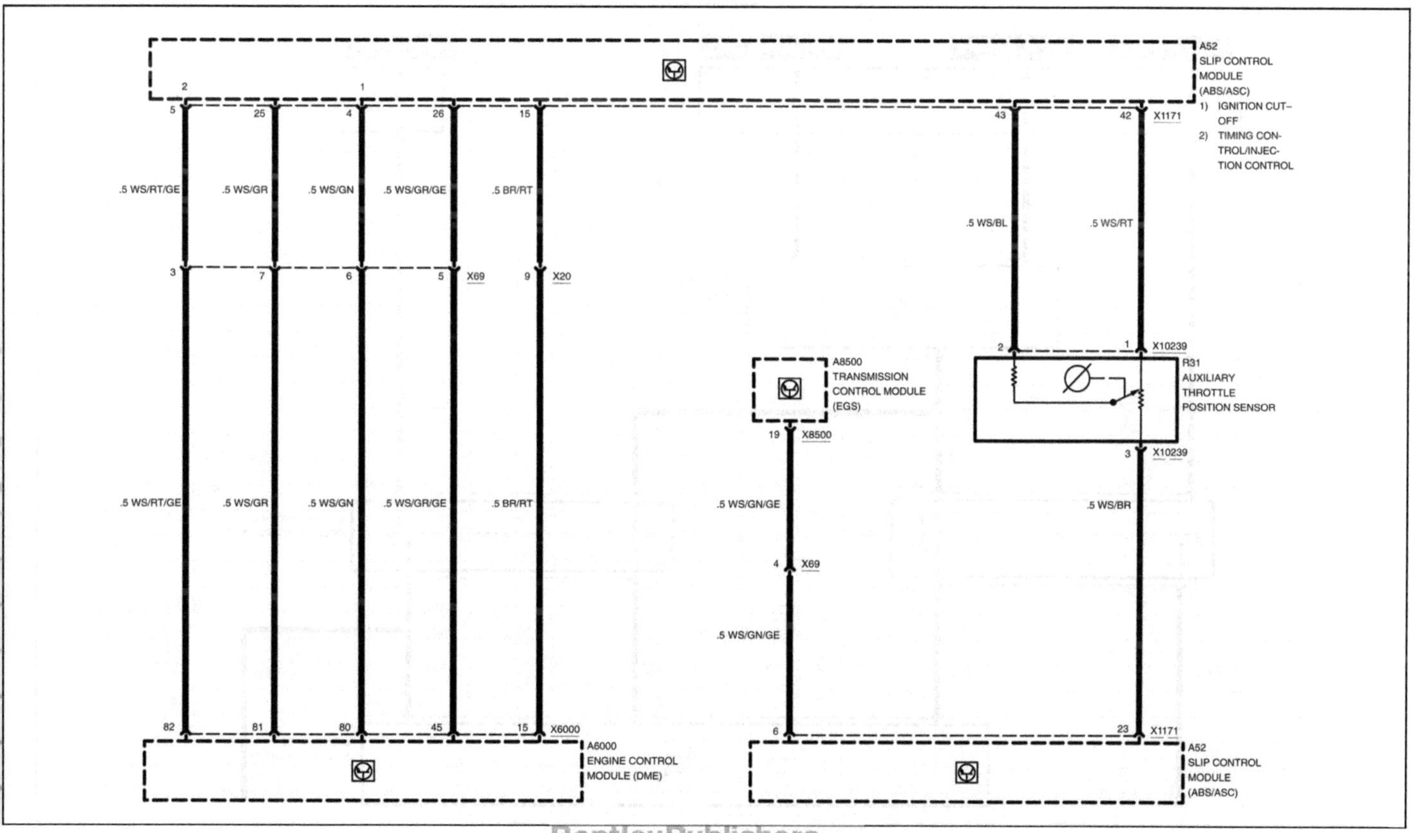

## ABS / ASC, 1998 M roadster, 2000 M

(1 of 5)

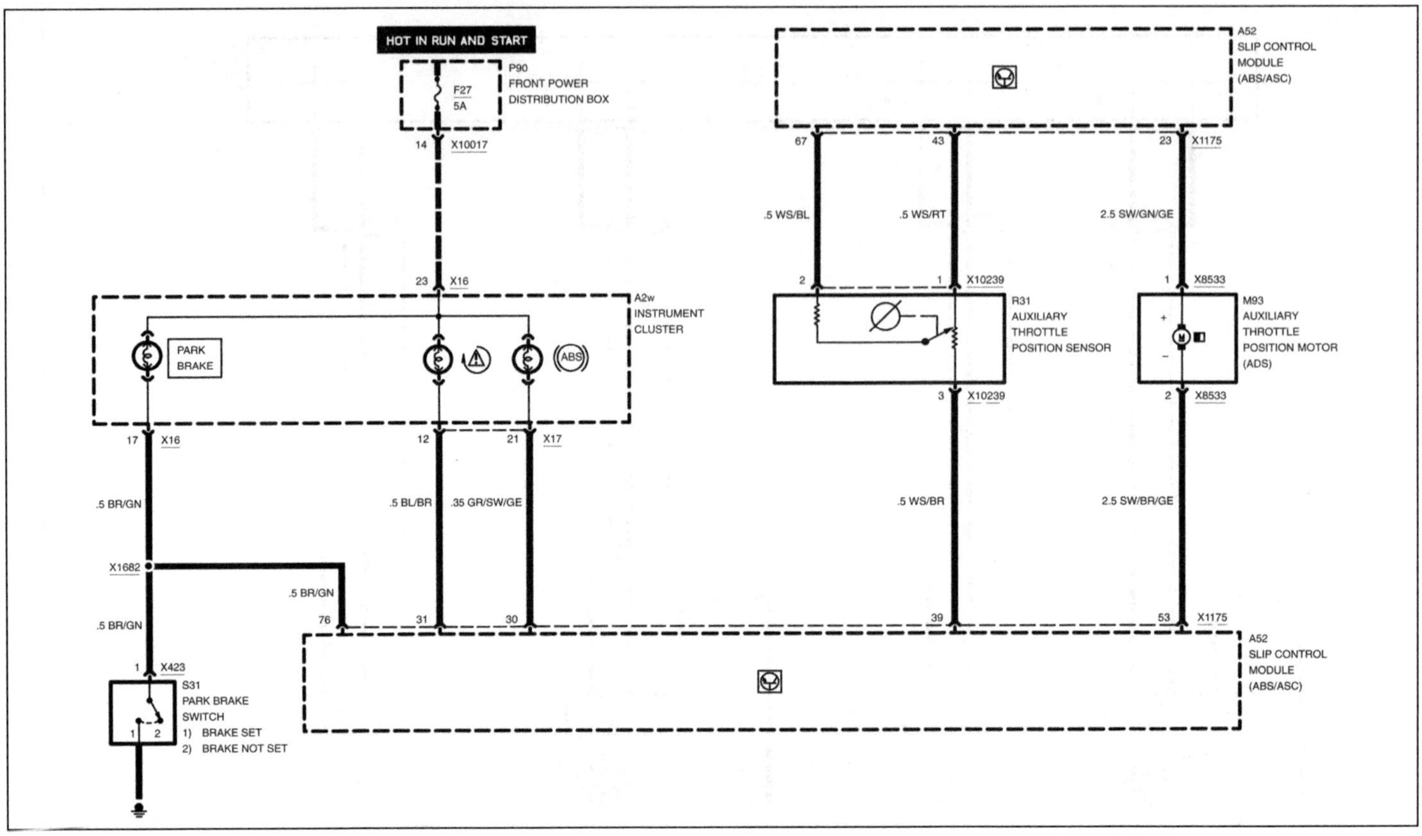


## ABS / ASC, 1998 M roadster, 2000 M

(2 of 5)

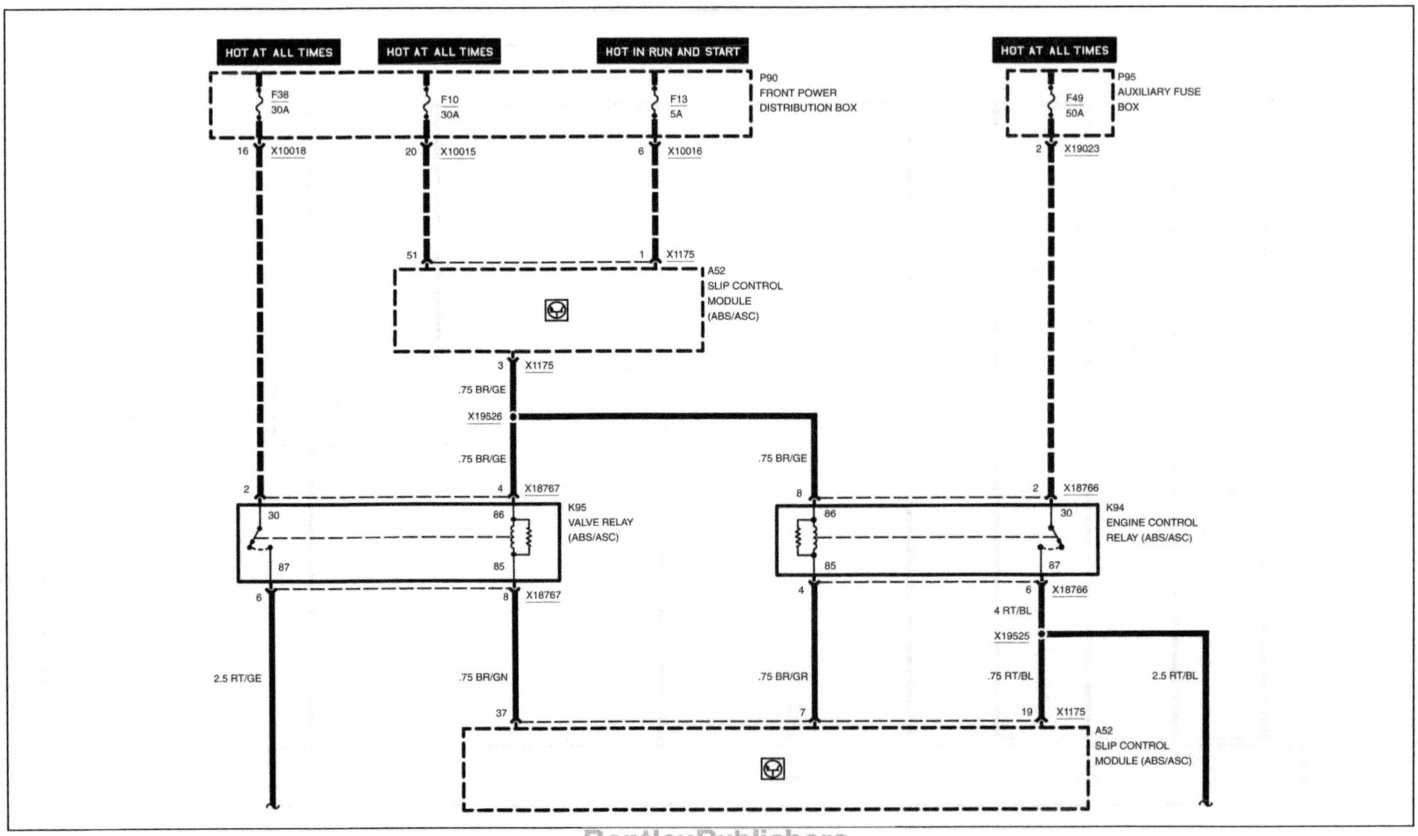

## ABS / ASC, 1998 M roadster, 2000 M

(3 of 4)

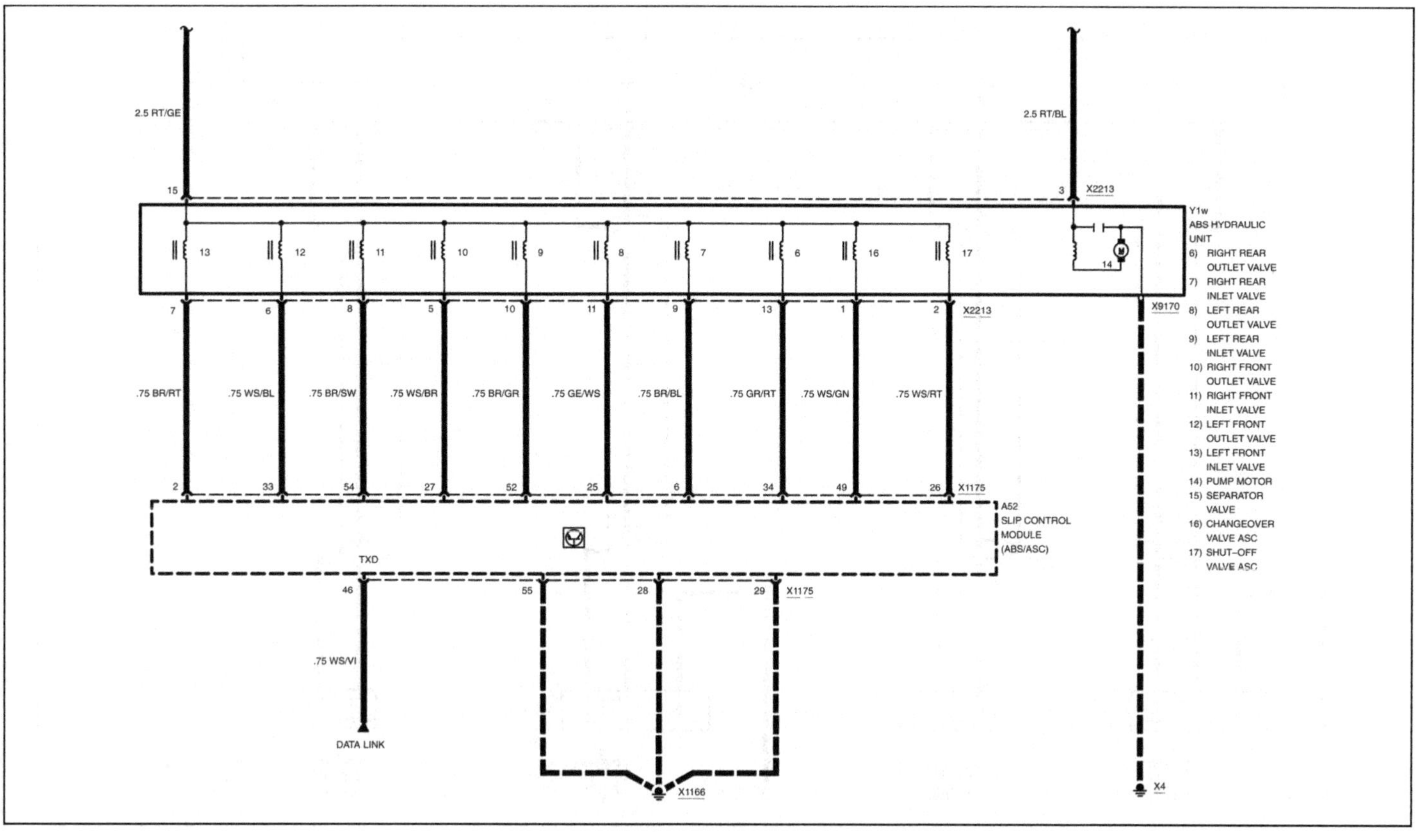


## ABS / ASC, 1998 M roadster, 2000 M

(4 of 5)

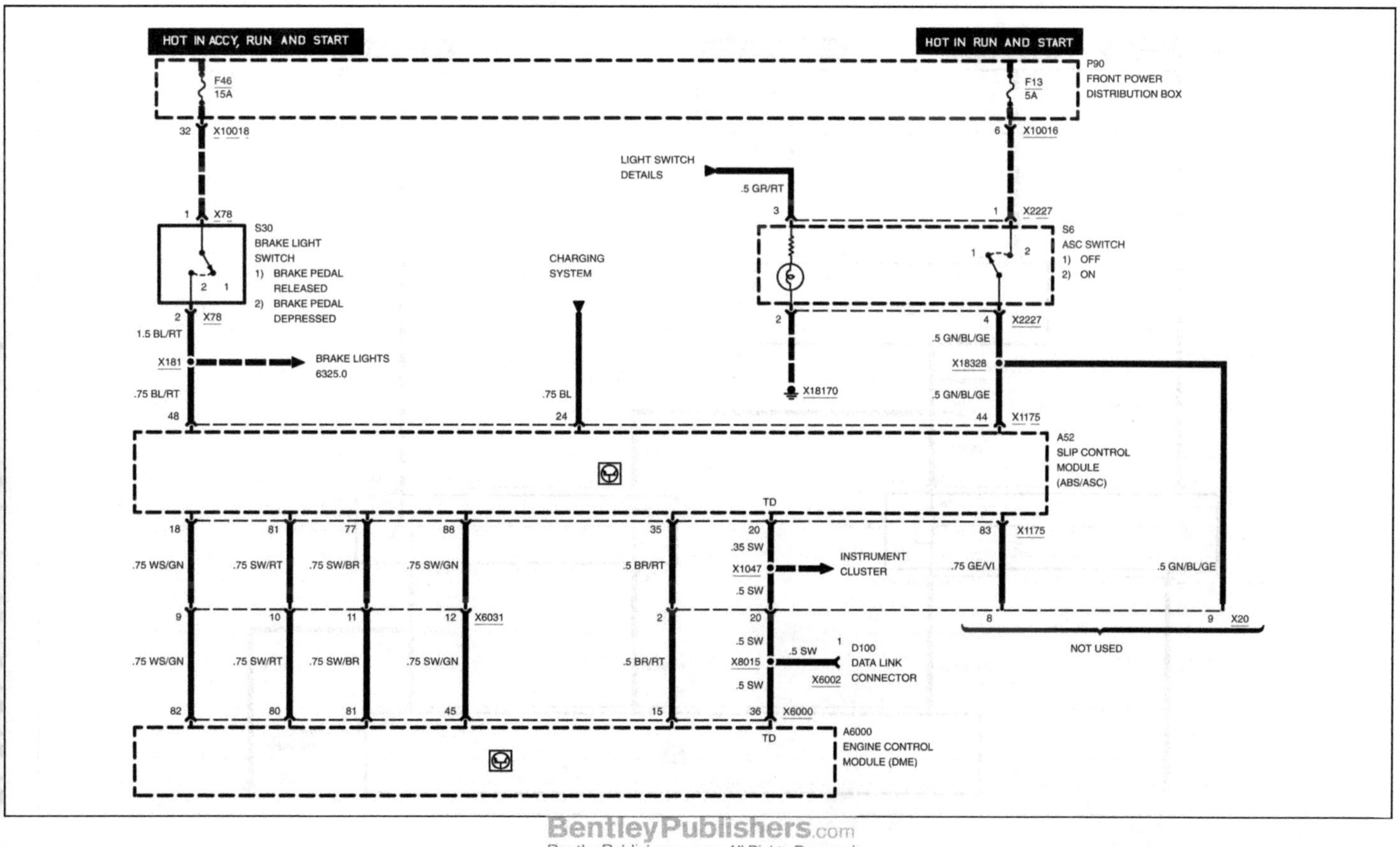

## ABS / ASC, 1998 M roadster, 2000 M
(5 of 5)

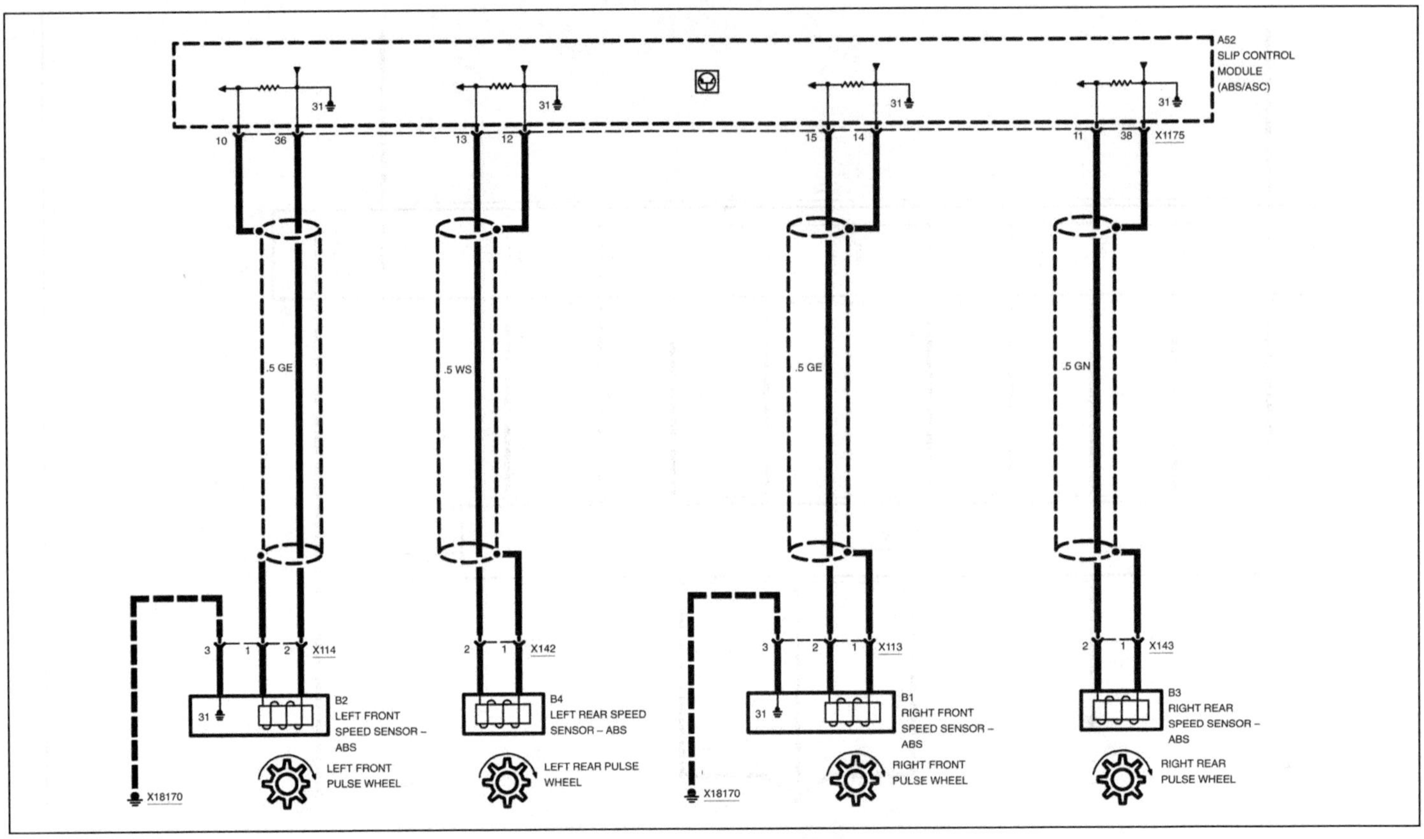


## ABS / ASC, 2000 - 2002 Z3
(1 of 8)

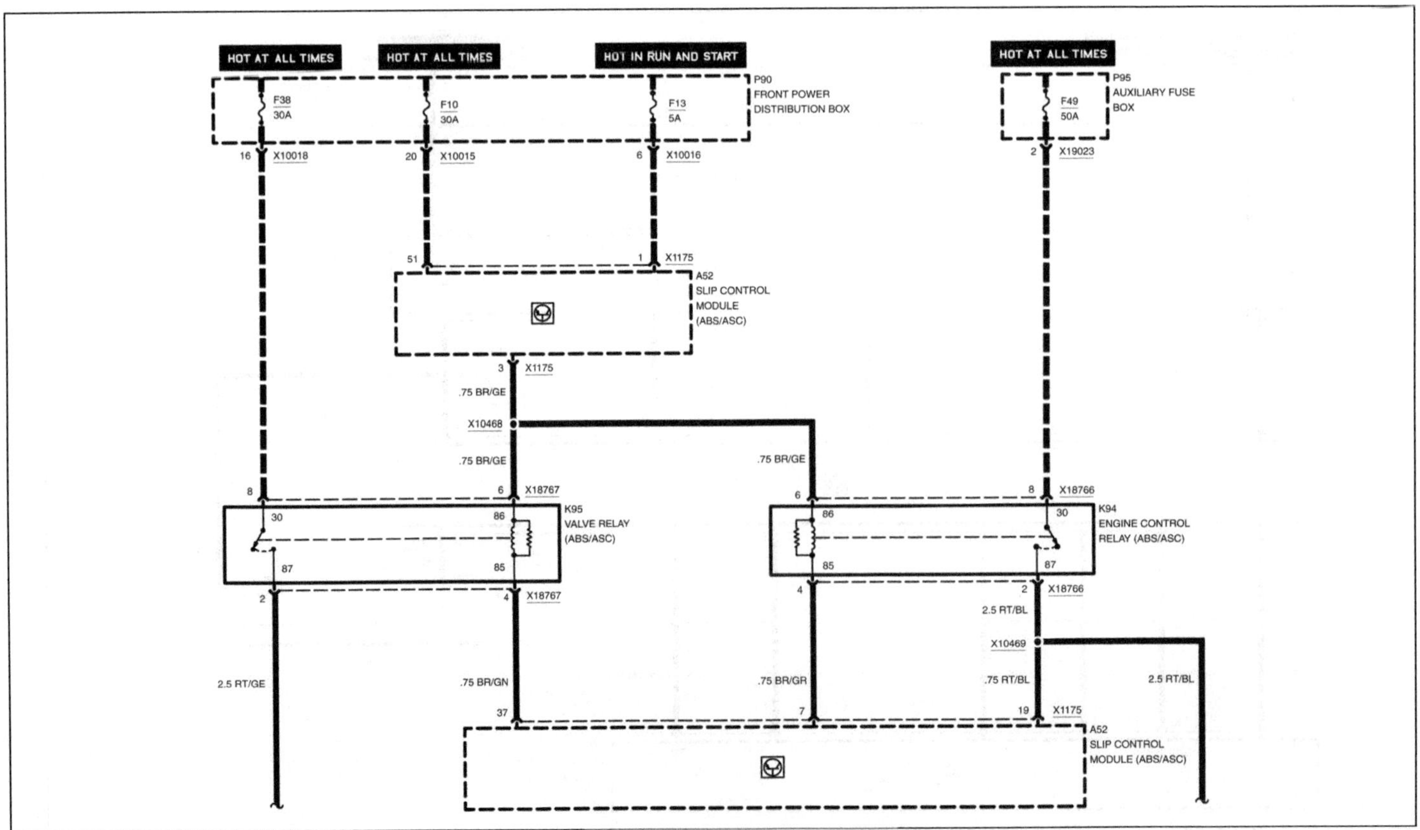

## ABS / ASC, 2000 - 2002 Z3
(2 of 8)

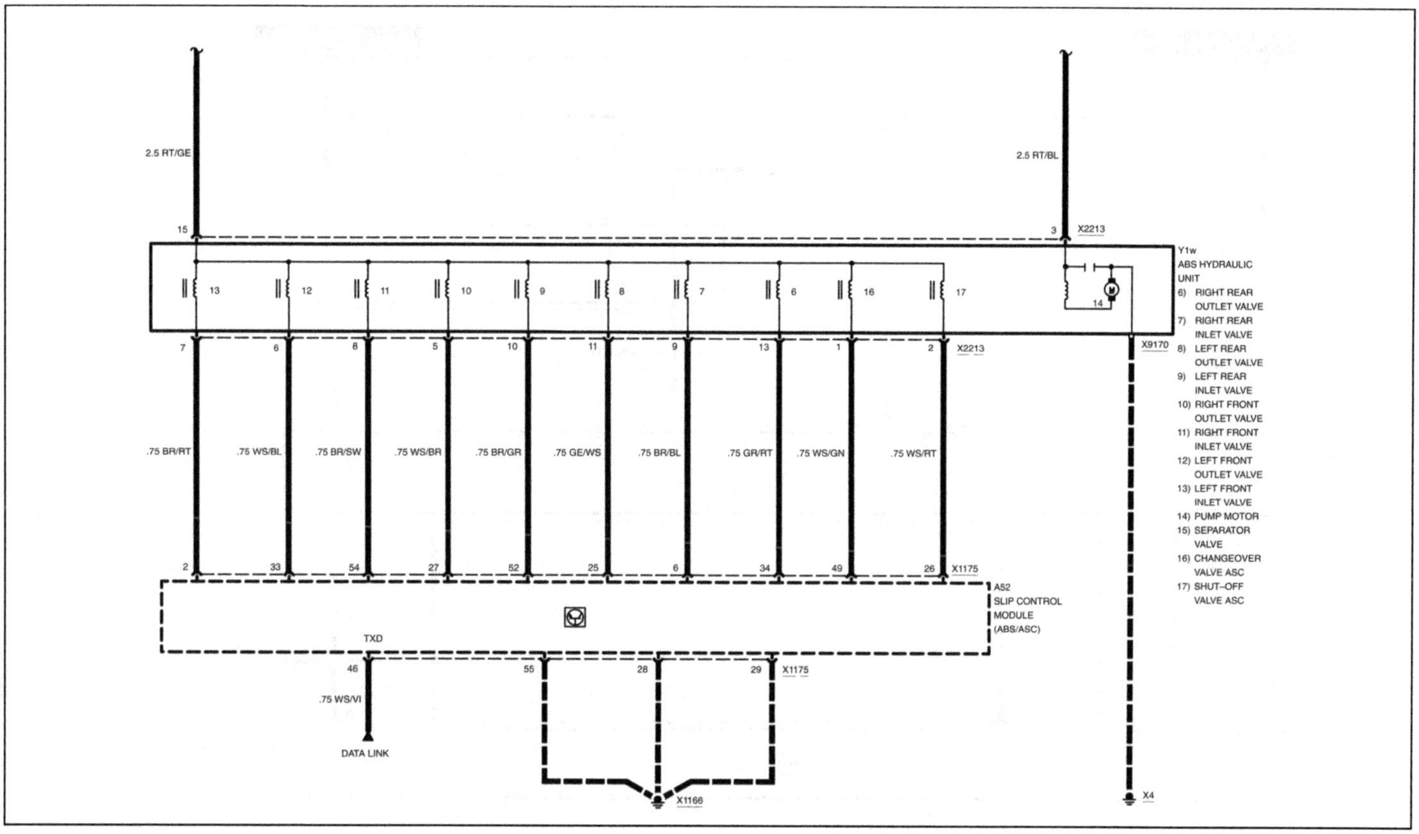


## ABS / ASC, 2000 - 2002 Z3
(3 of 8)

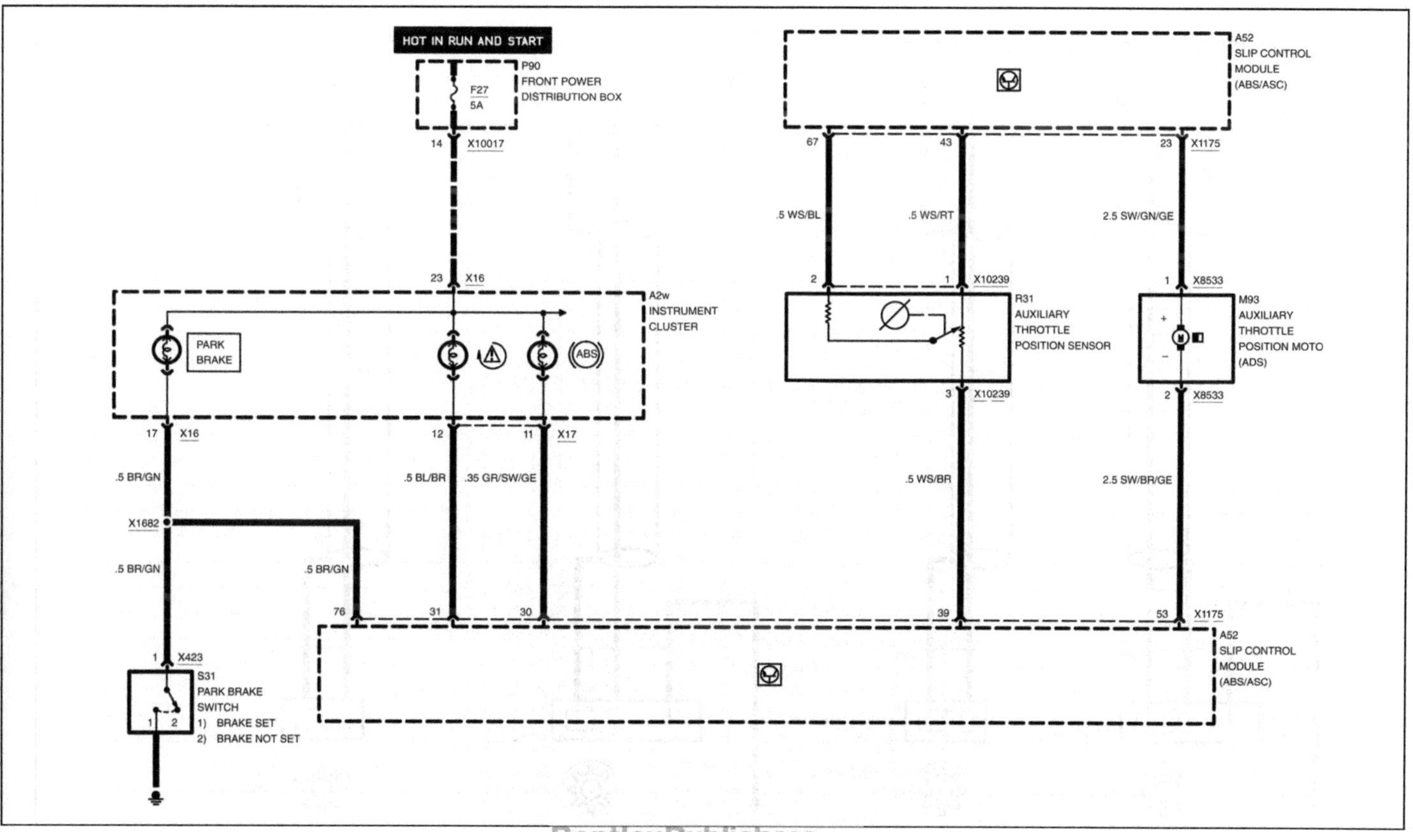

## ABS / ASC, 2000 - 2002 Z3
(4 of 8)

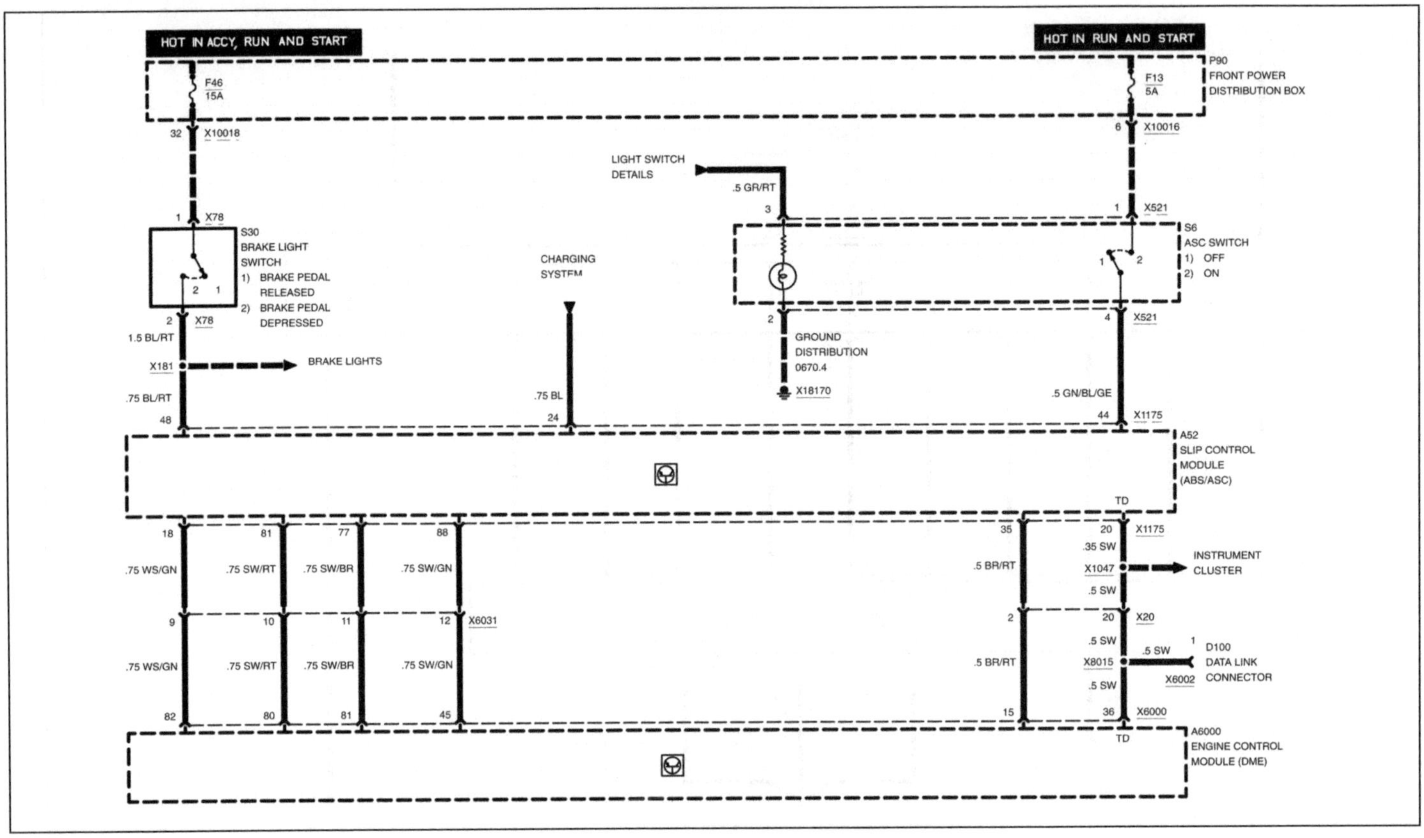


## ABS / ASC, 2000 - 2002 Z3
(5 of 8)

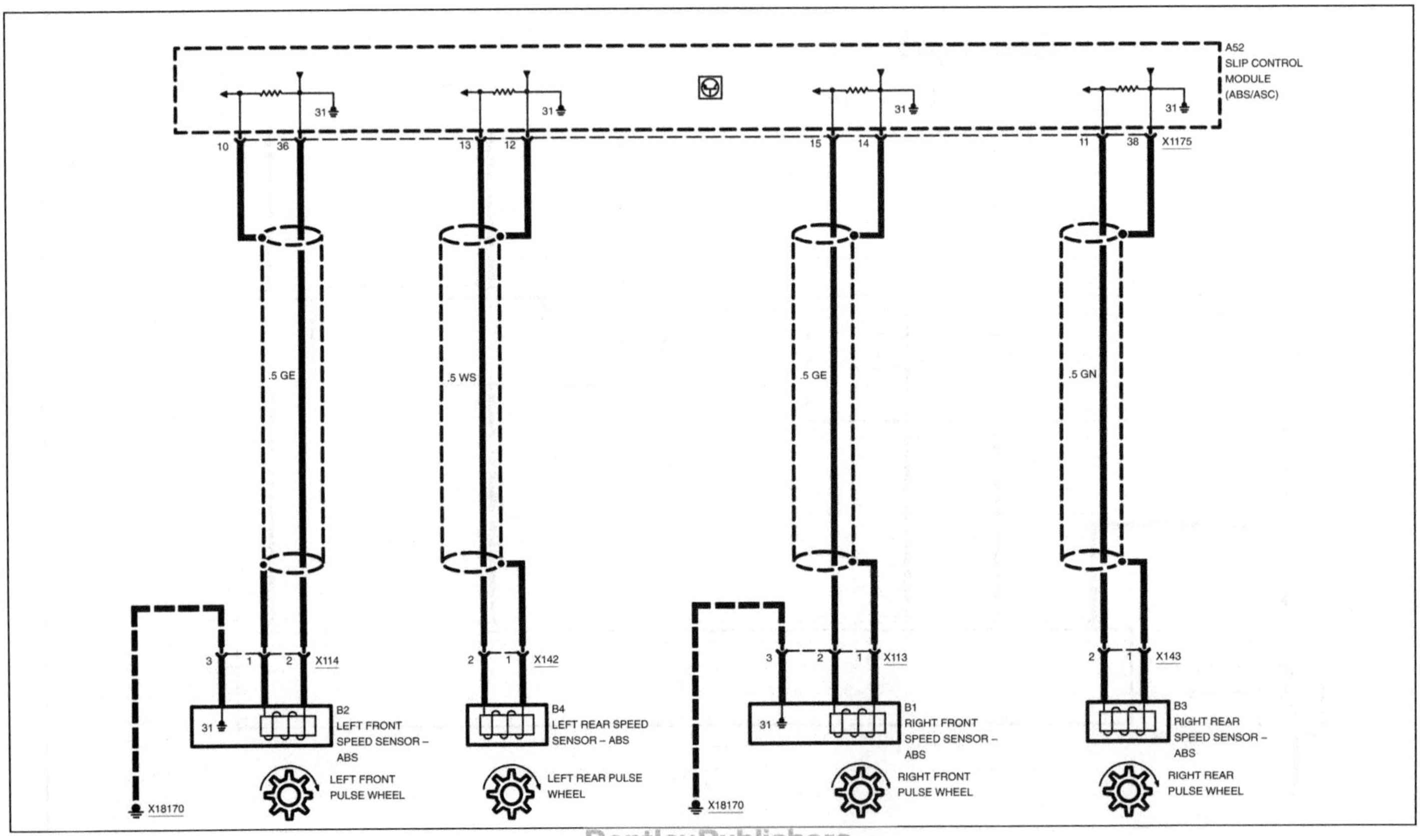

## ABS / ASC, 2000 - 2002 Z3
(6 of 8)

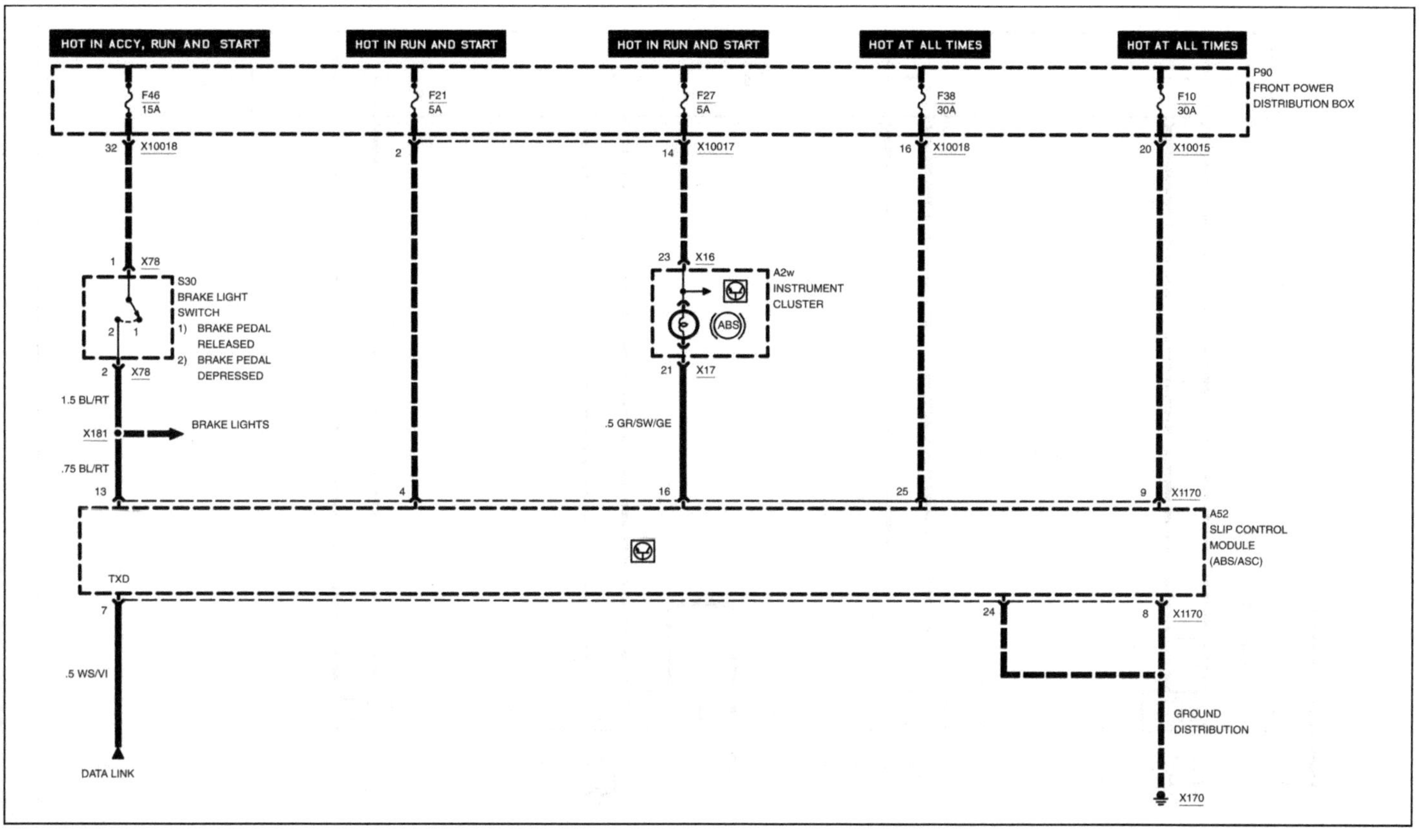


## ABS / ASC, 2000 - 2002 Z3
(7 of 8)

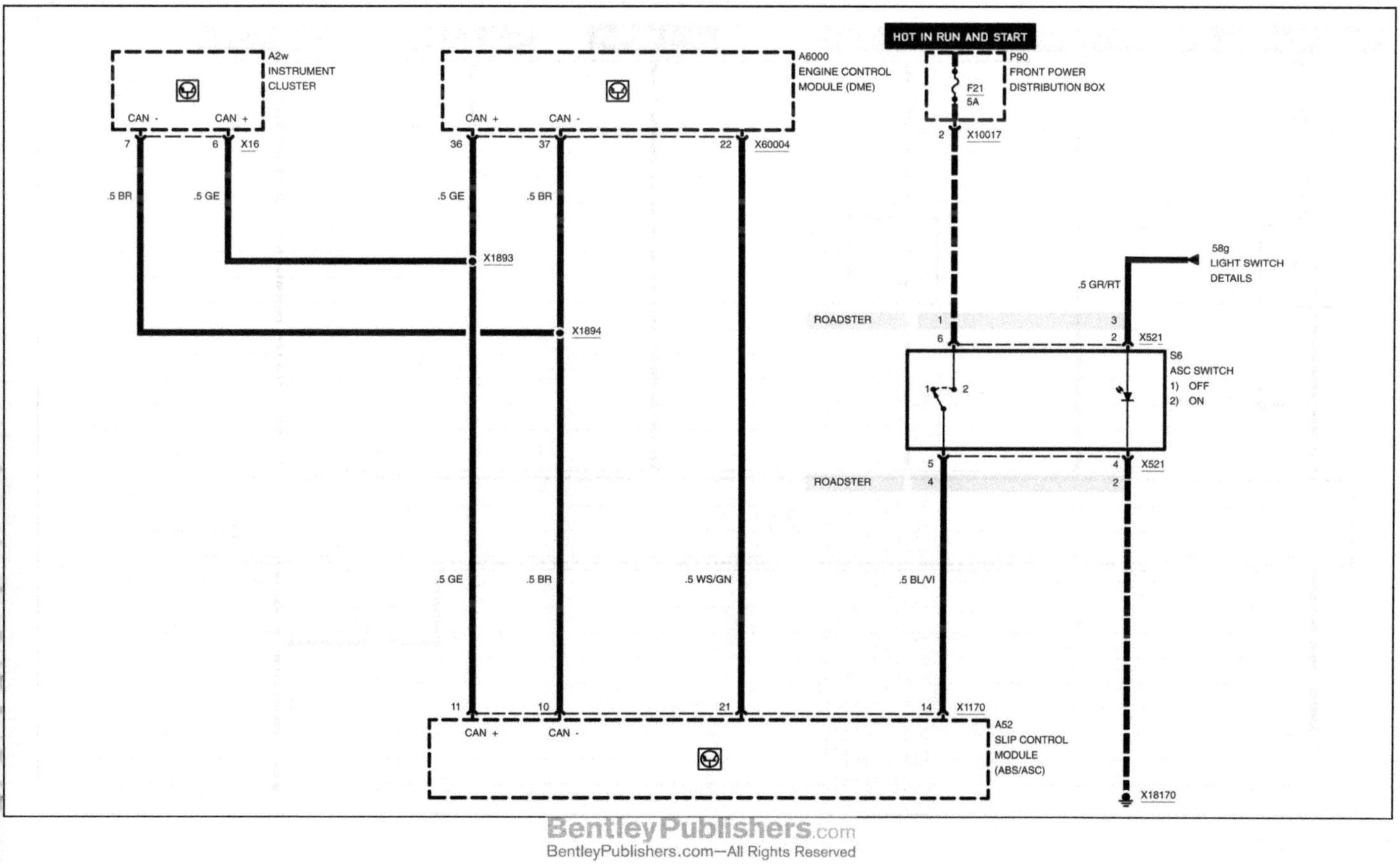

## ABS / ASC, 2000 - 2002 Z3

(8 of 8)

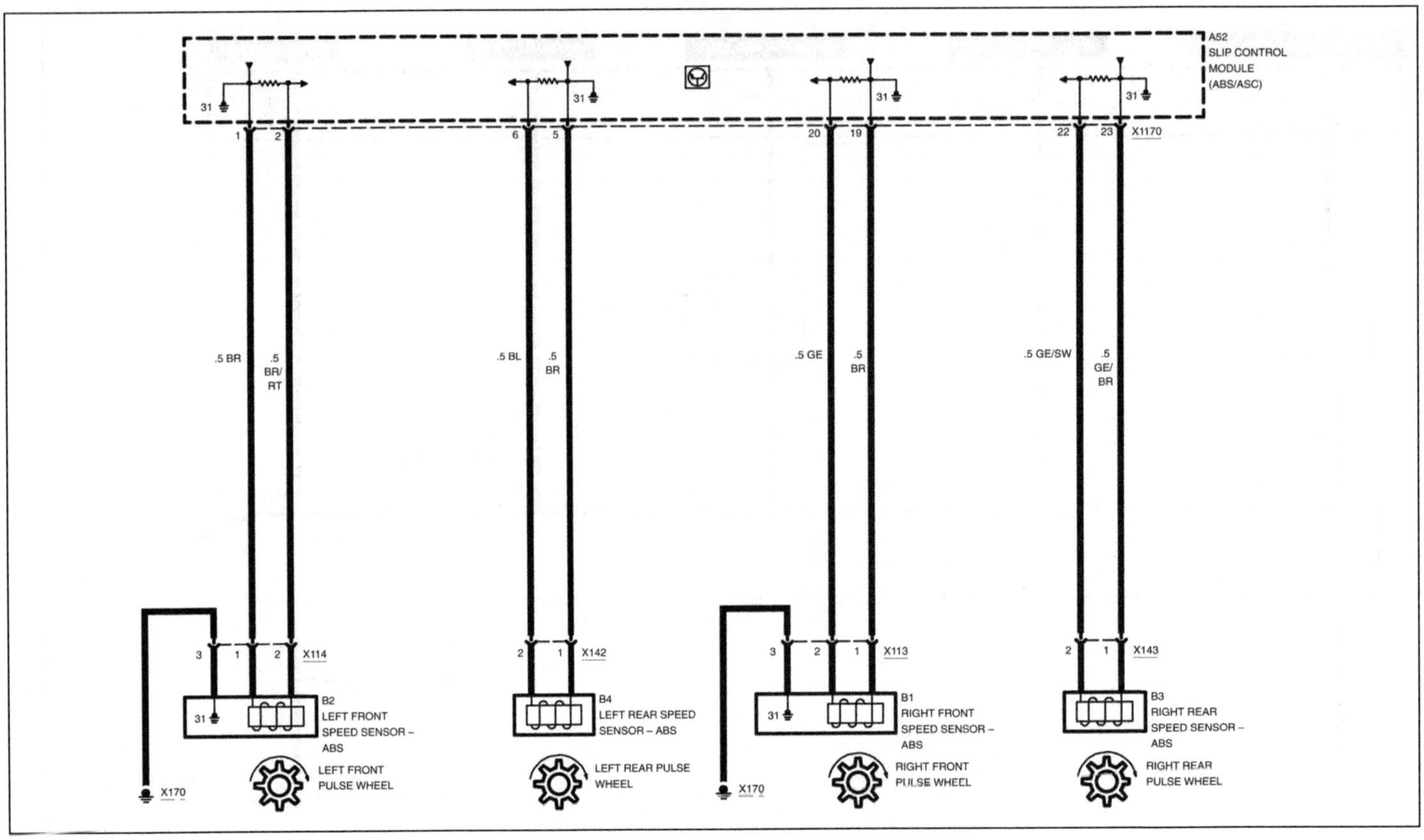


## ABS / DSC III, 2000, 2002 M

(1 of 6)

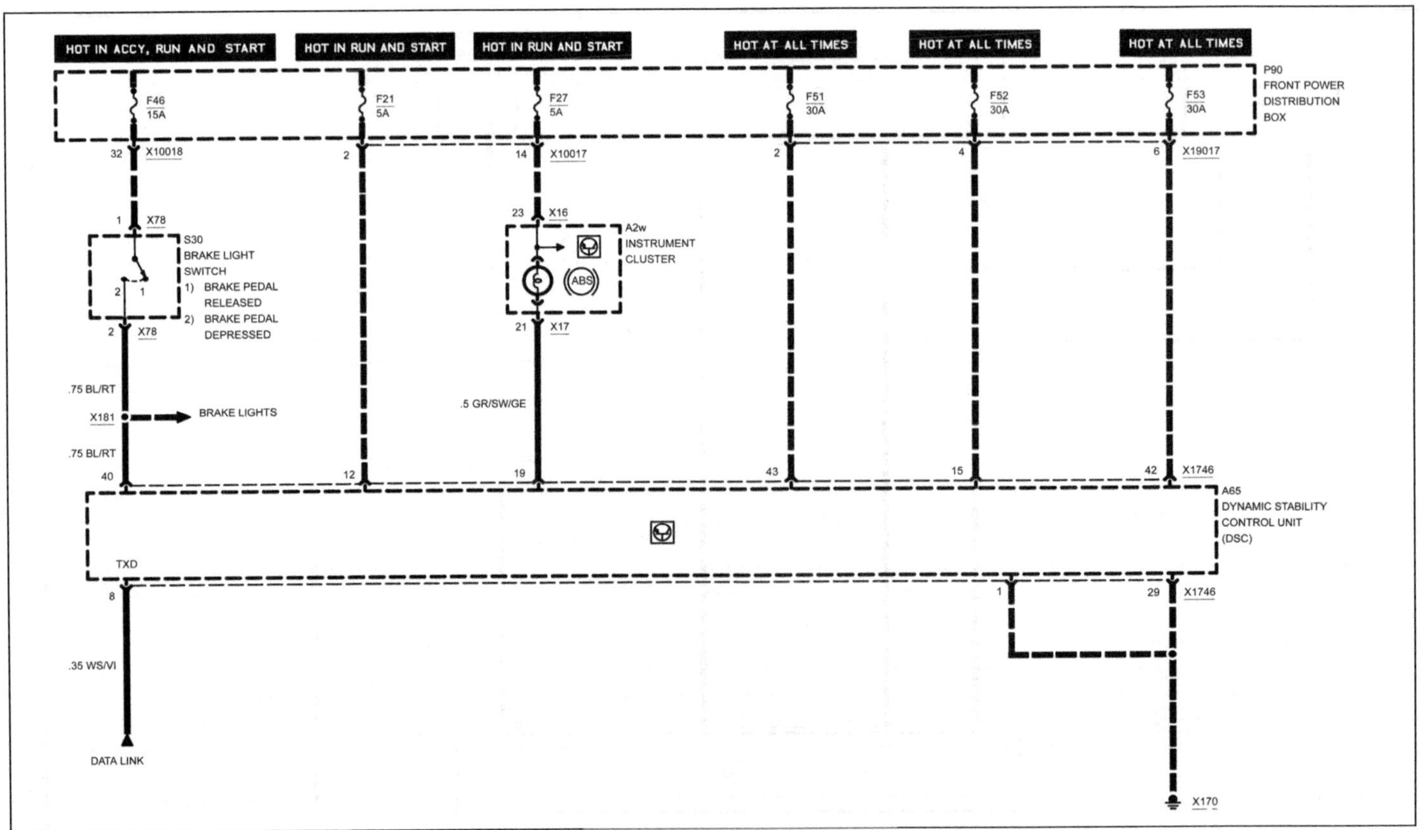

## ABS / DSC III, 2000, 2002 M
(2 of 6)

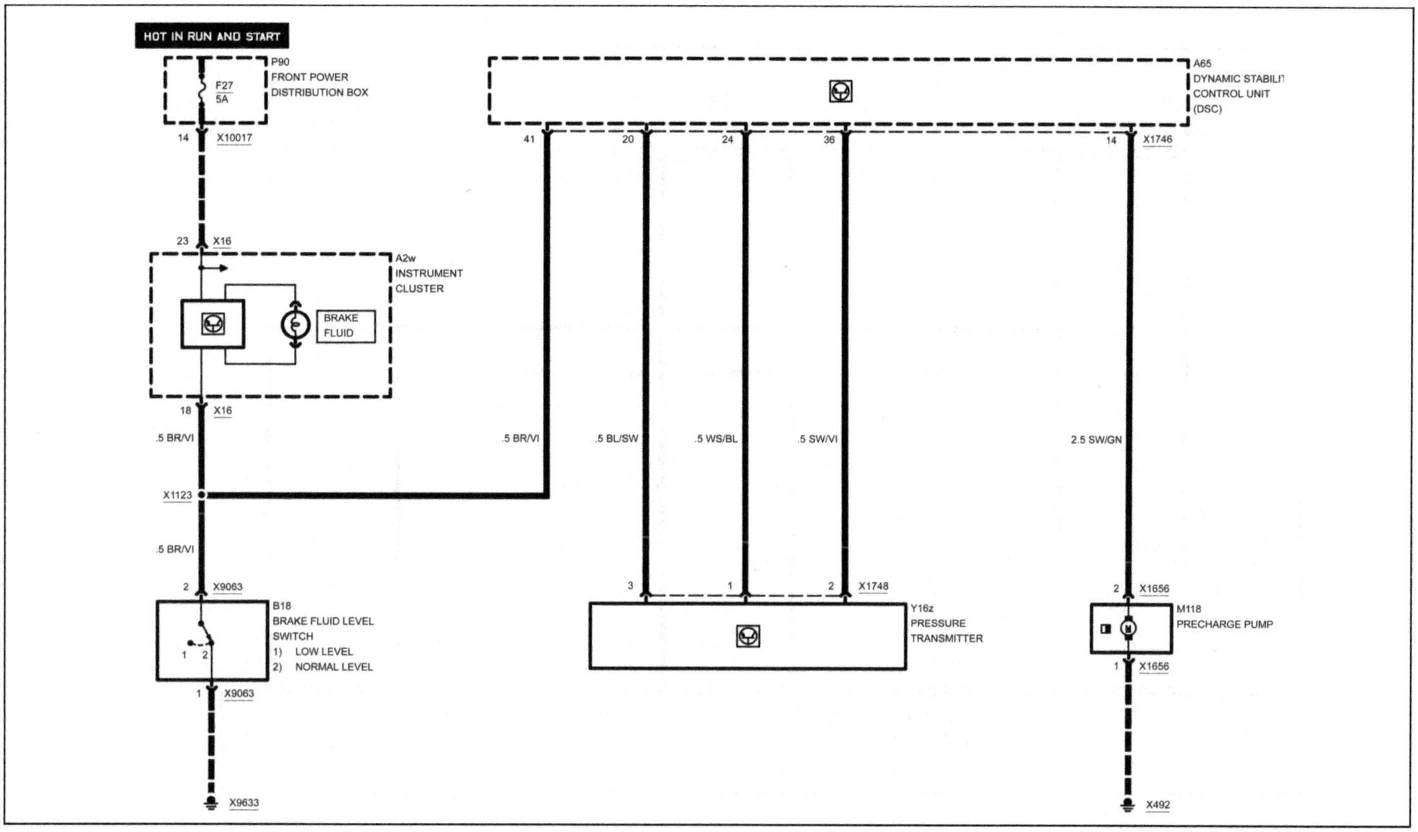


## ABS / DSC III, 2000, 2002 M
(3 of 6)

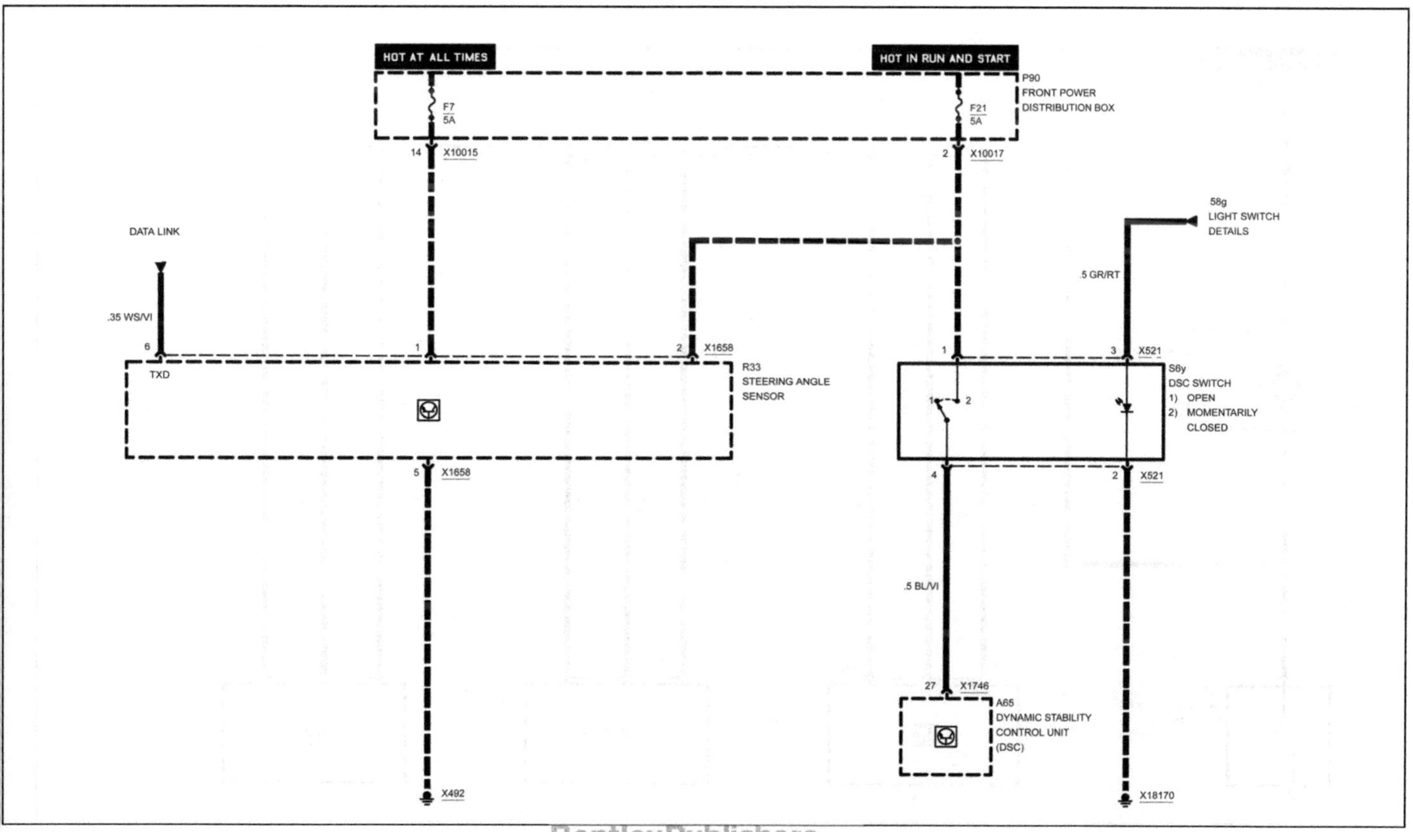

## ABS / DSC III, 2000, 2002 M
(4 of 6)

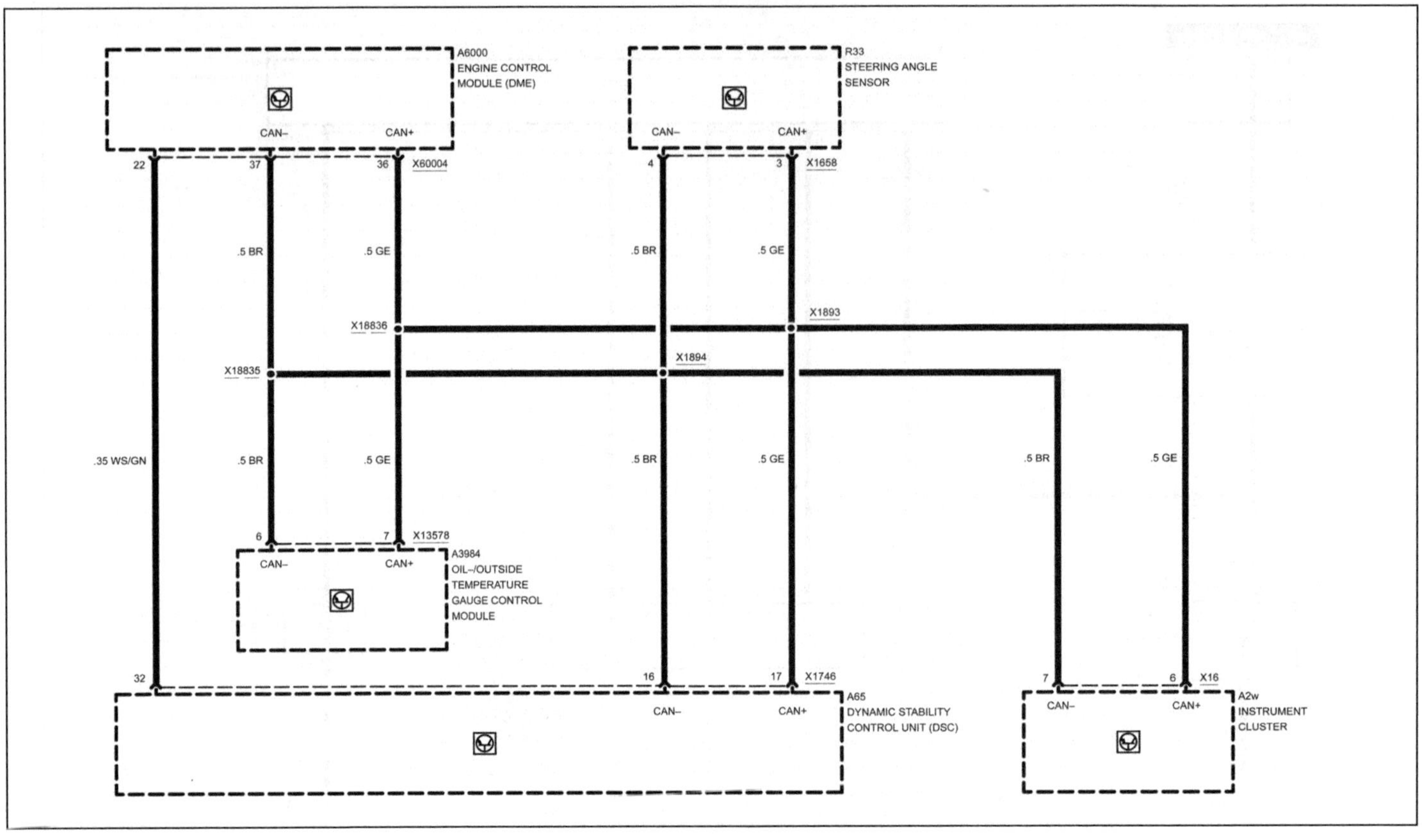


## ABS / DSC III, 2000, 2002 M
(5 of 6)

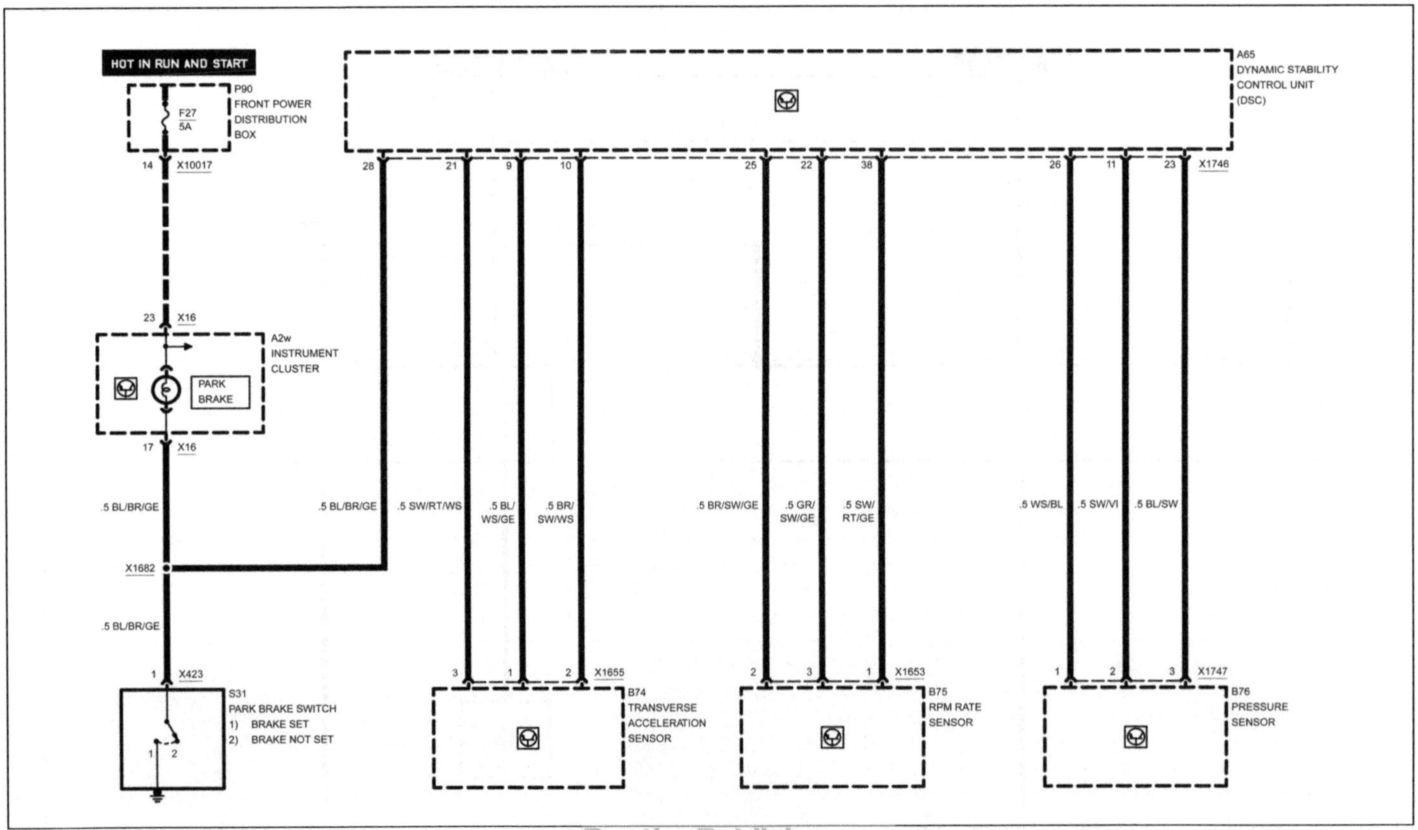

## ABS / DSC III, 2000, 2002 M

(6 of 6)

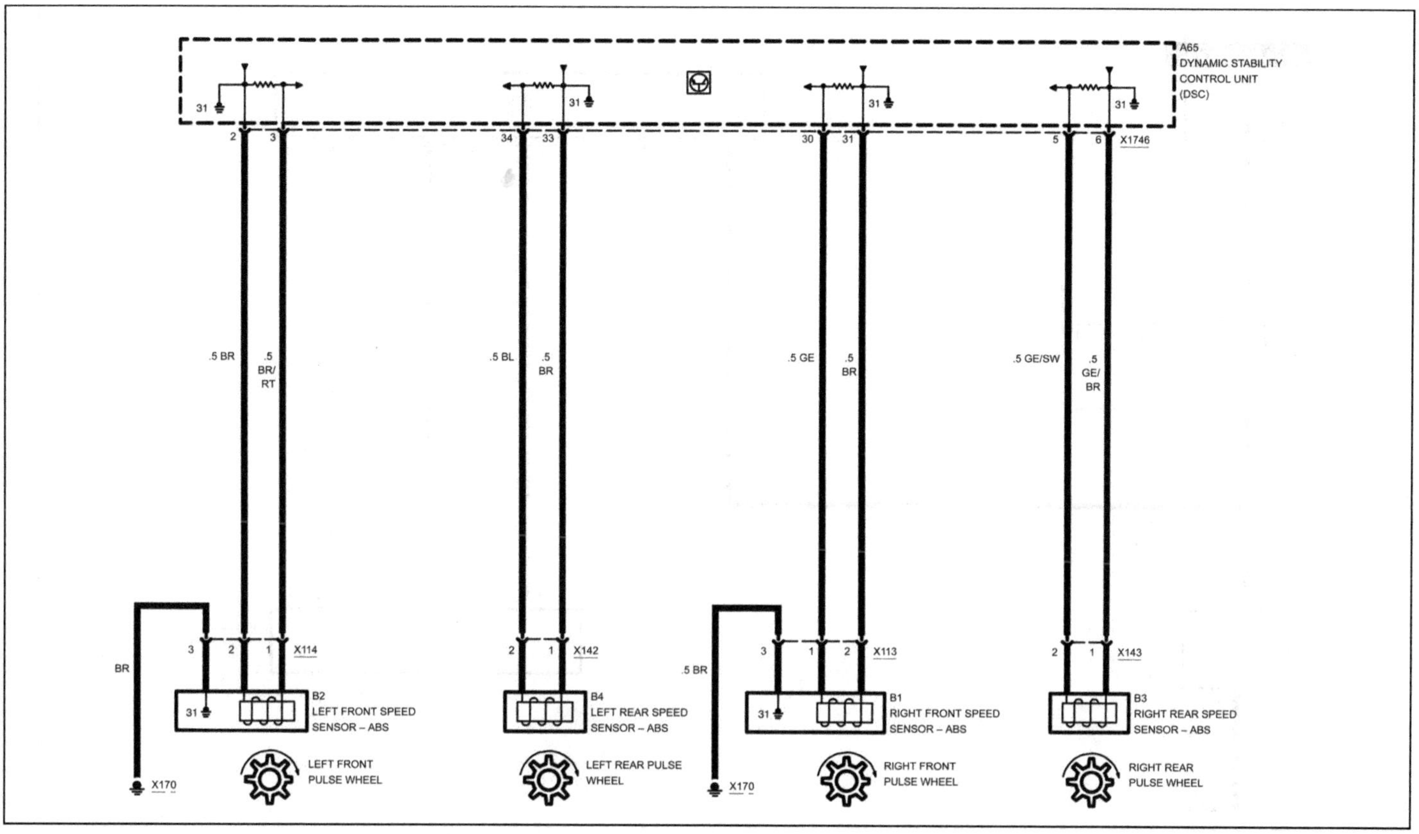


## ABS / DSC MK 60, 2002 Z3

(1 of 6)

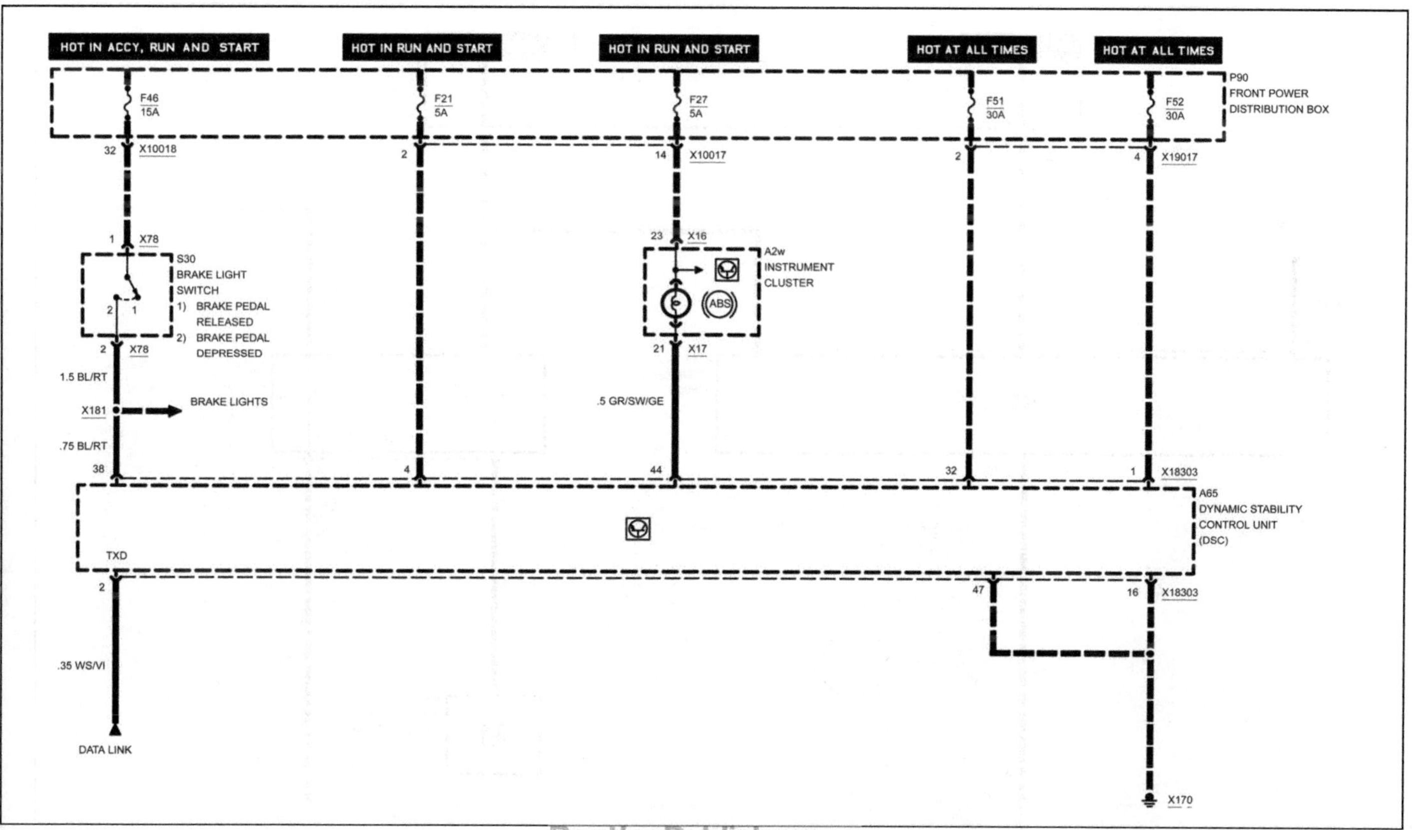

## ABS / DSC MK 60, 2002 Z3

(2 of 6)

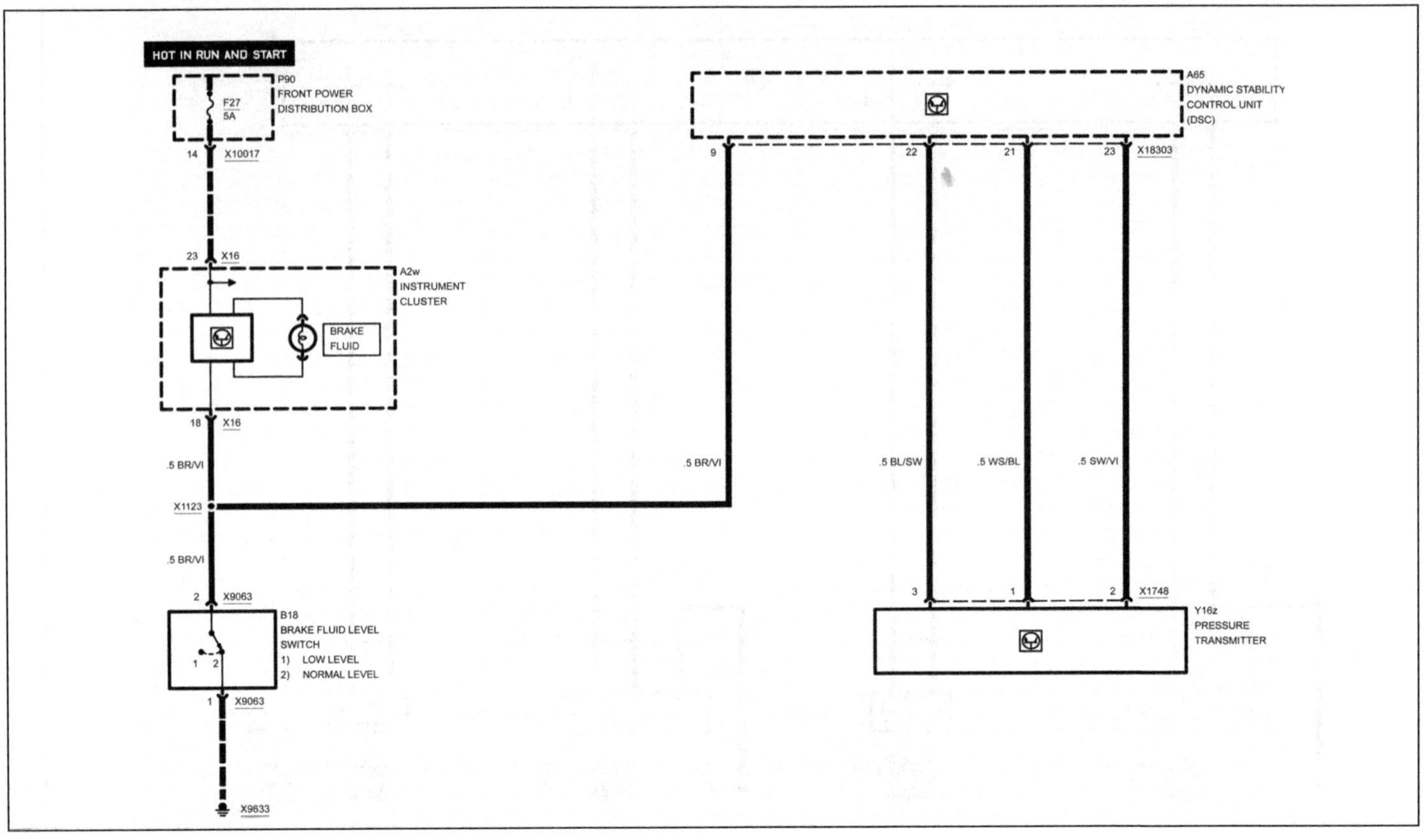


## ABS / DSC MK 60, 2002 Z3

(3 of 6)

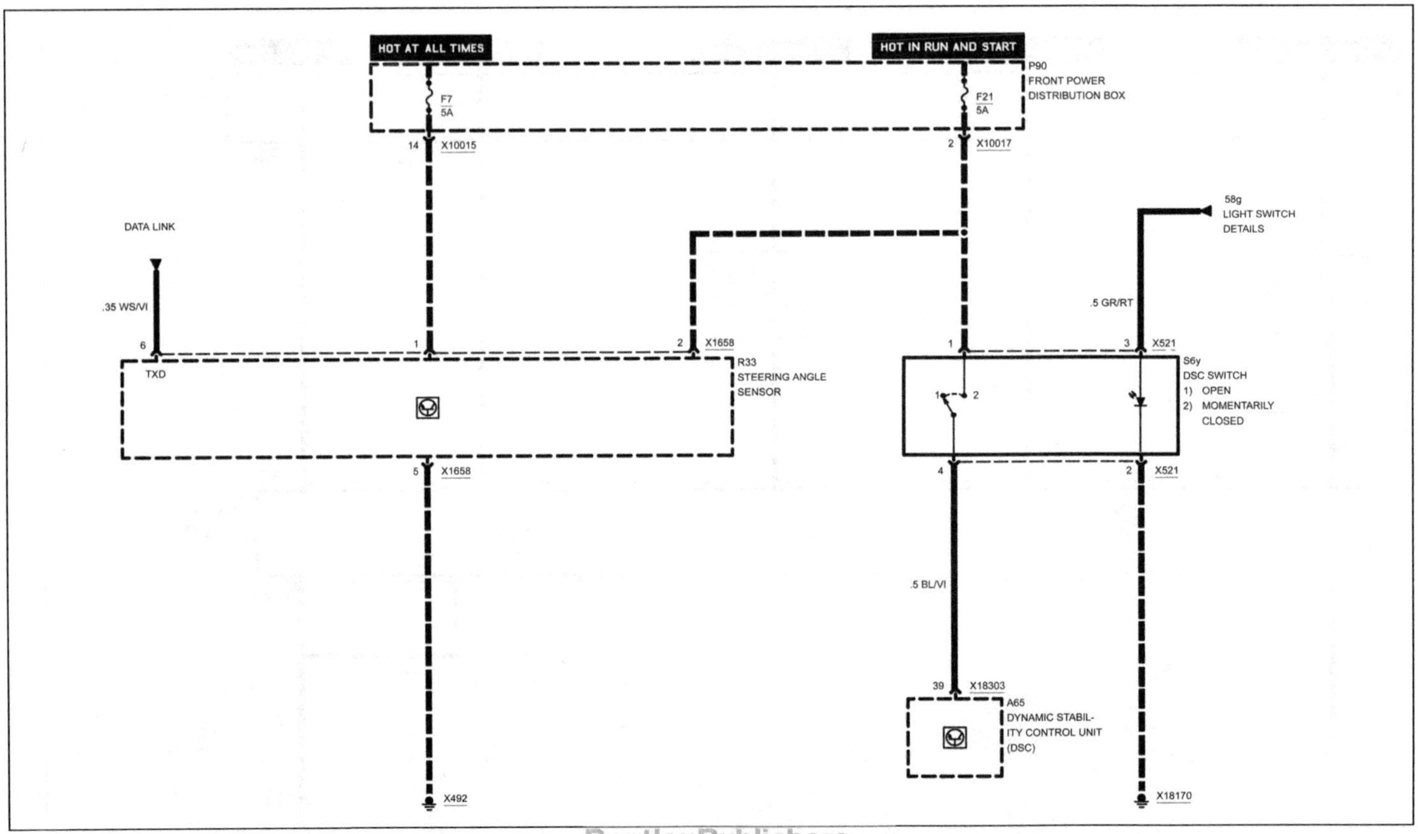

## ABS / DSC MK 60, 2002 Z3
(4 of 6)

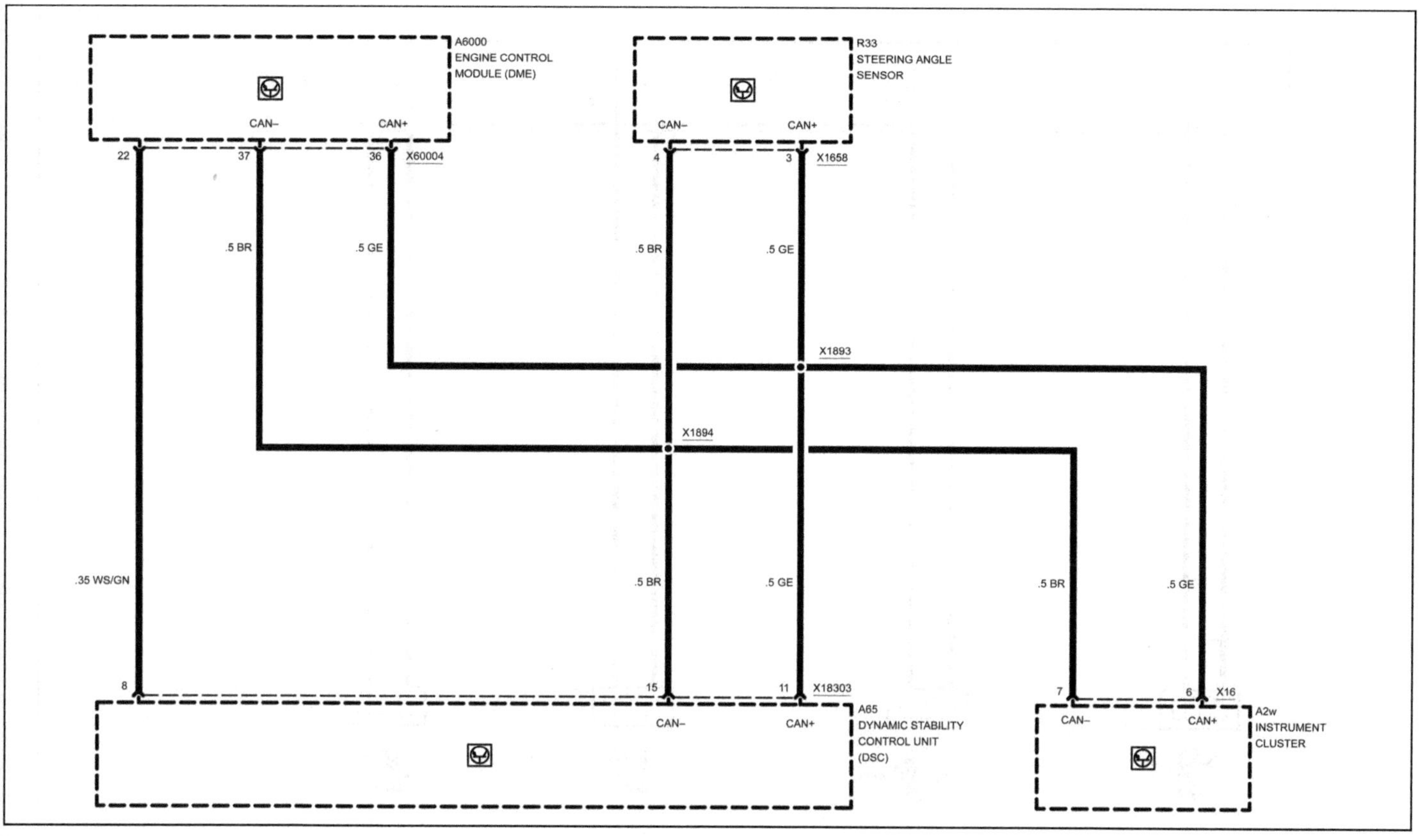


## ABS / DSC MK 60, 2002 Z3
(5 of 6)

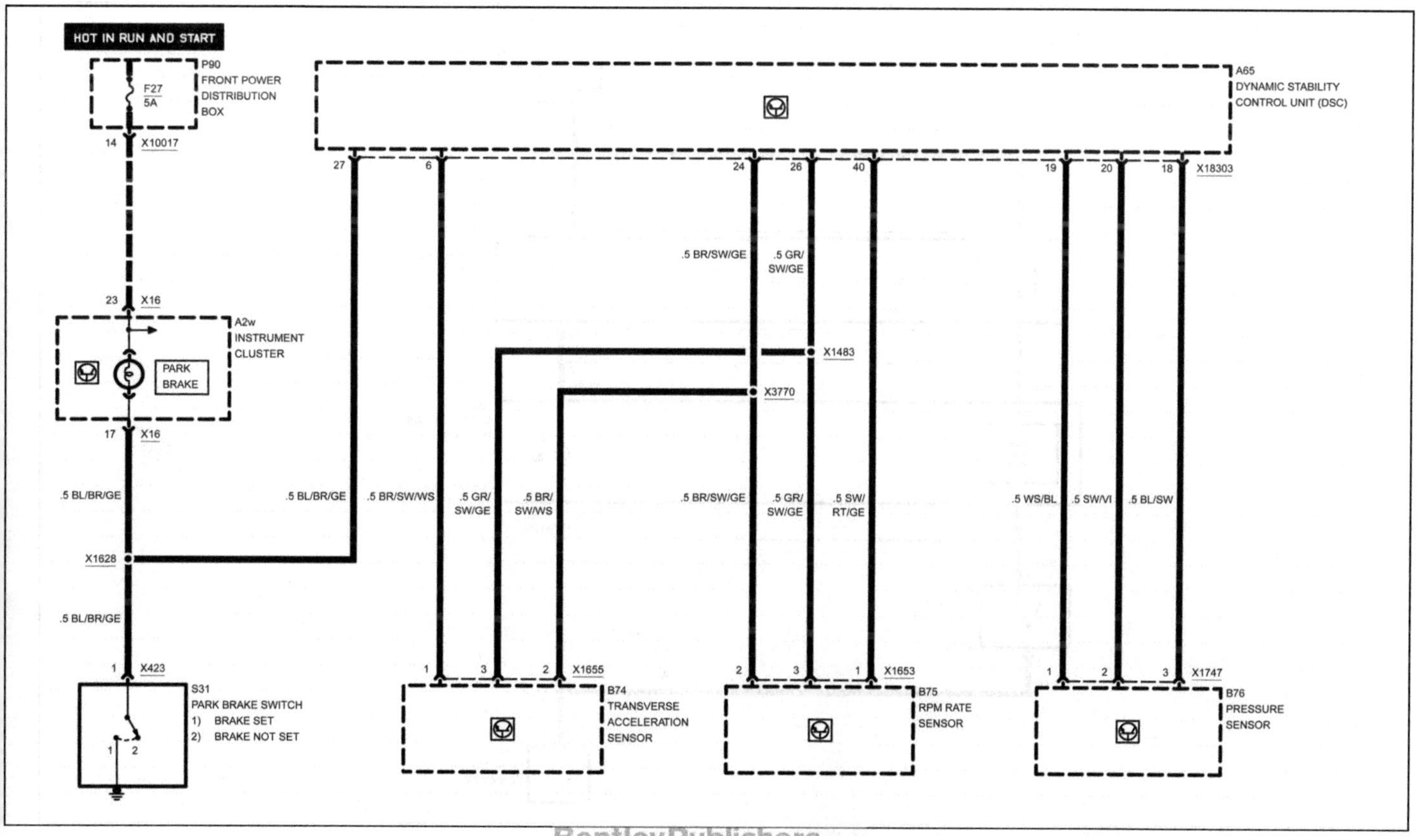

## ABS / DSC MK 60, 2002 Z3

(6 of 6)

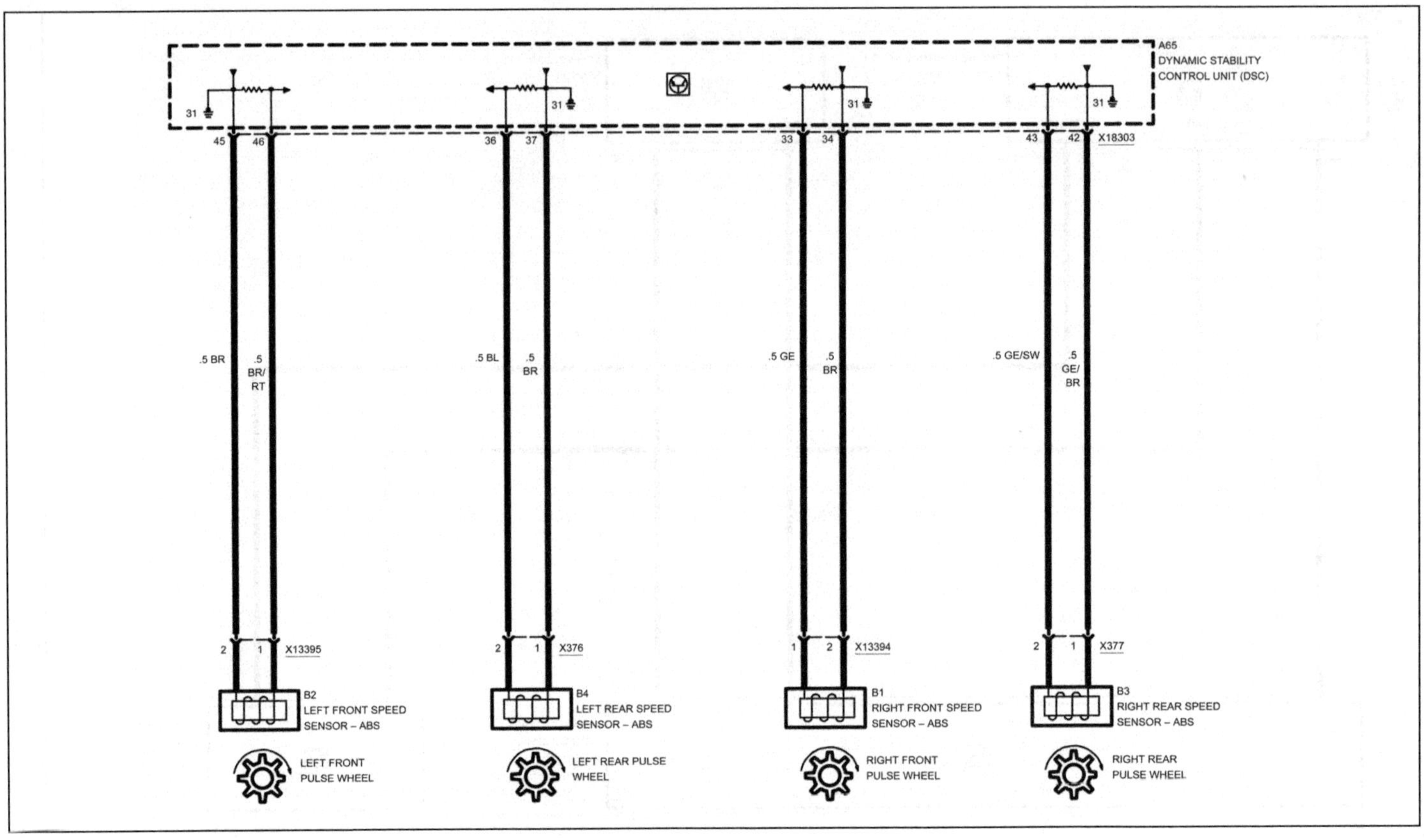


## Tire inflation, 1998 M Roadster, 2000 - 2002

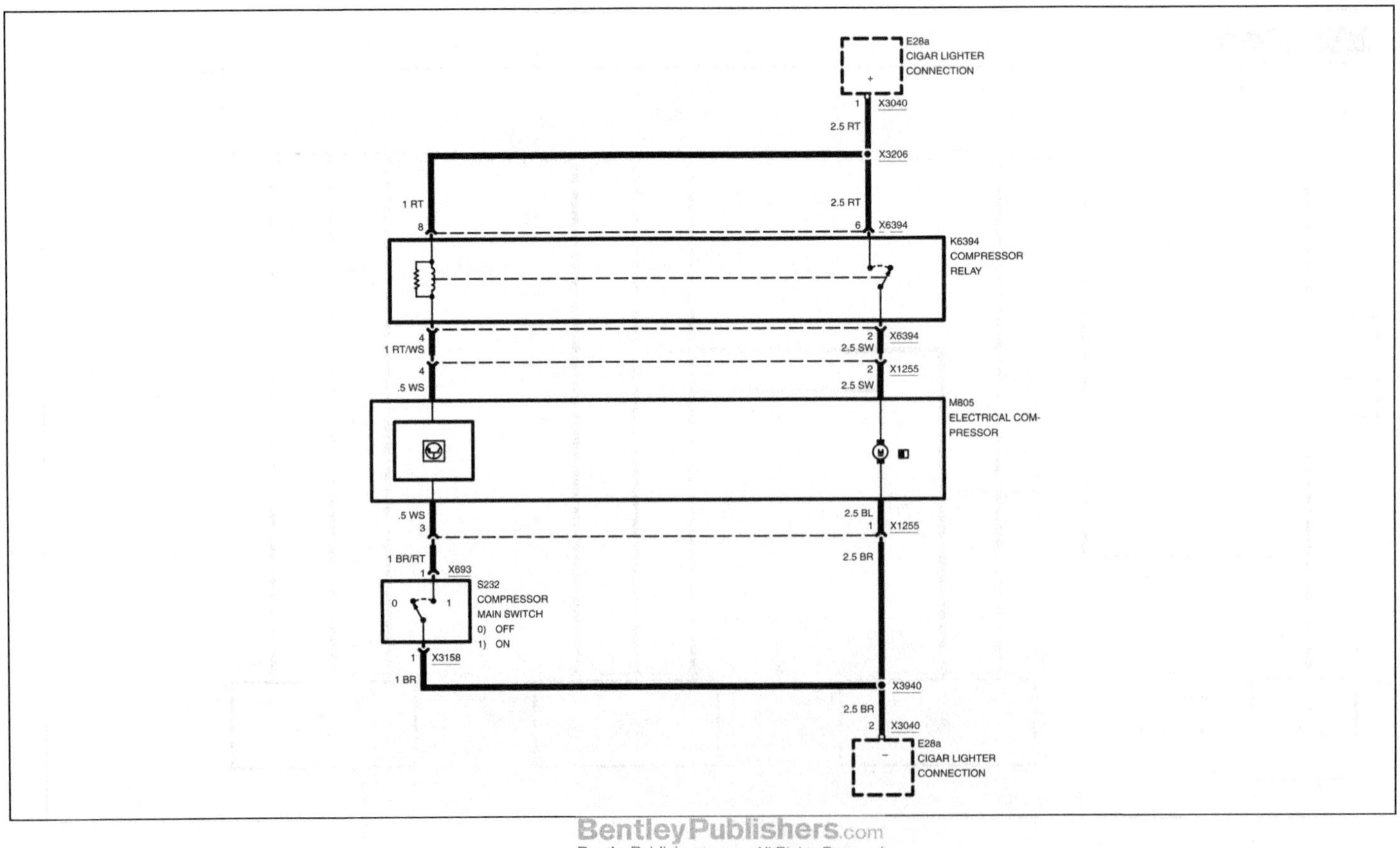

## Tire pressure warning system (RDW)

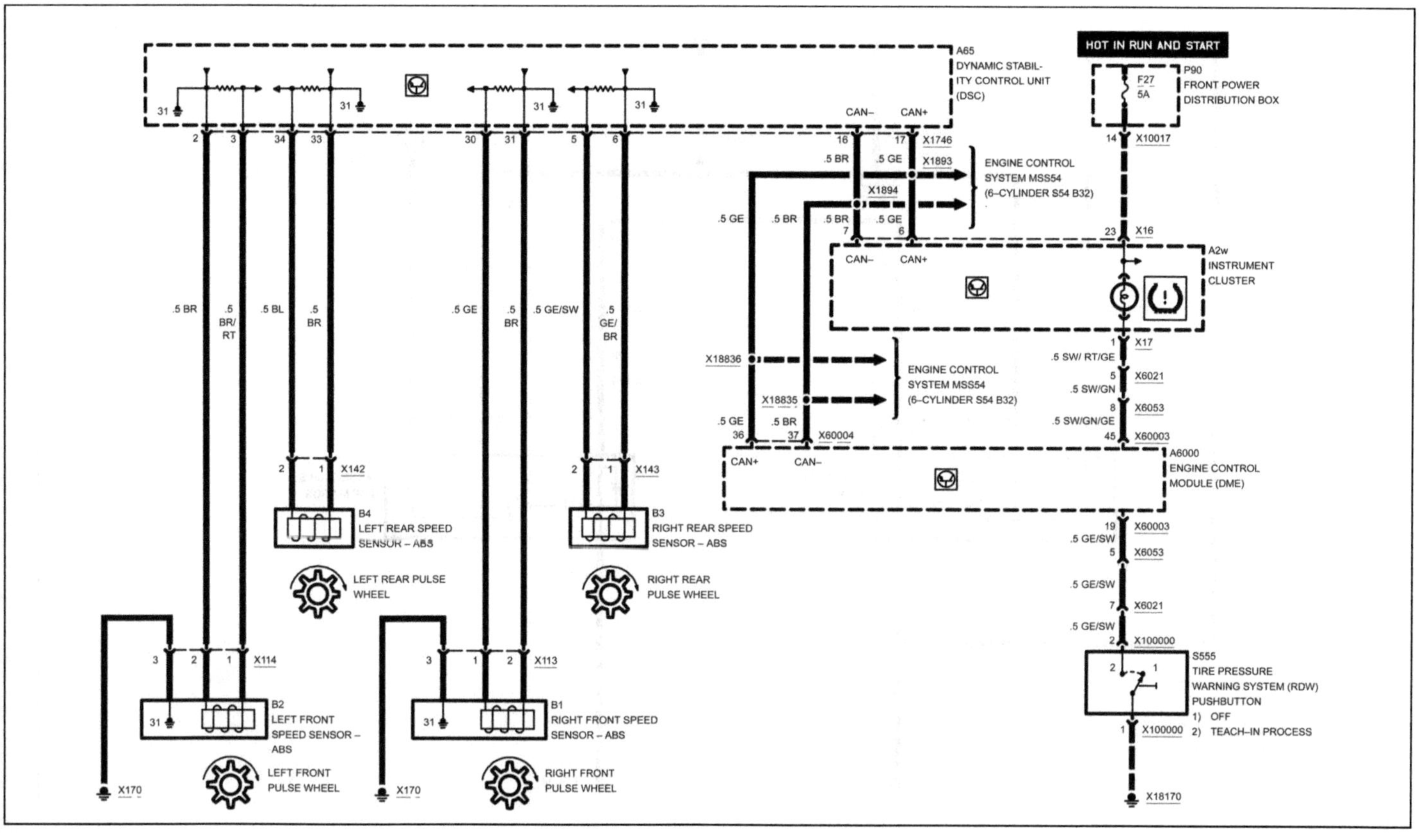


## Mirrors
(1 of 2)

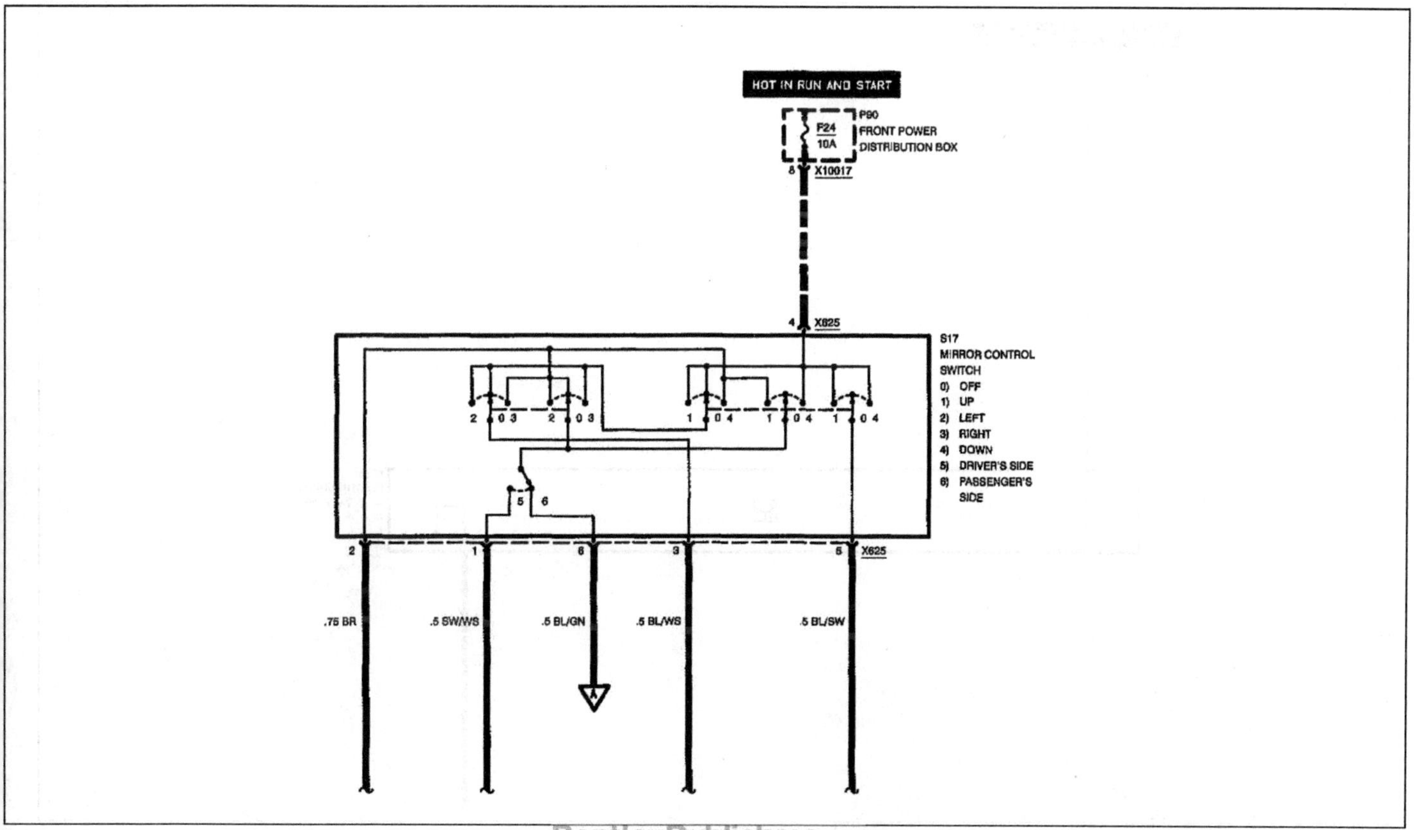

## Mirrors
(2 of 2)

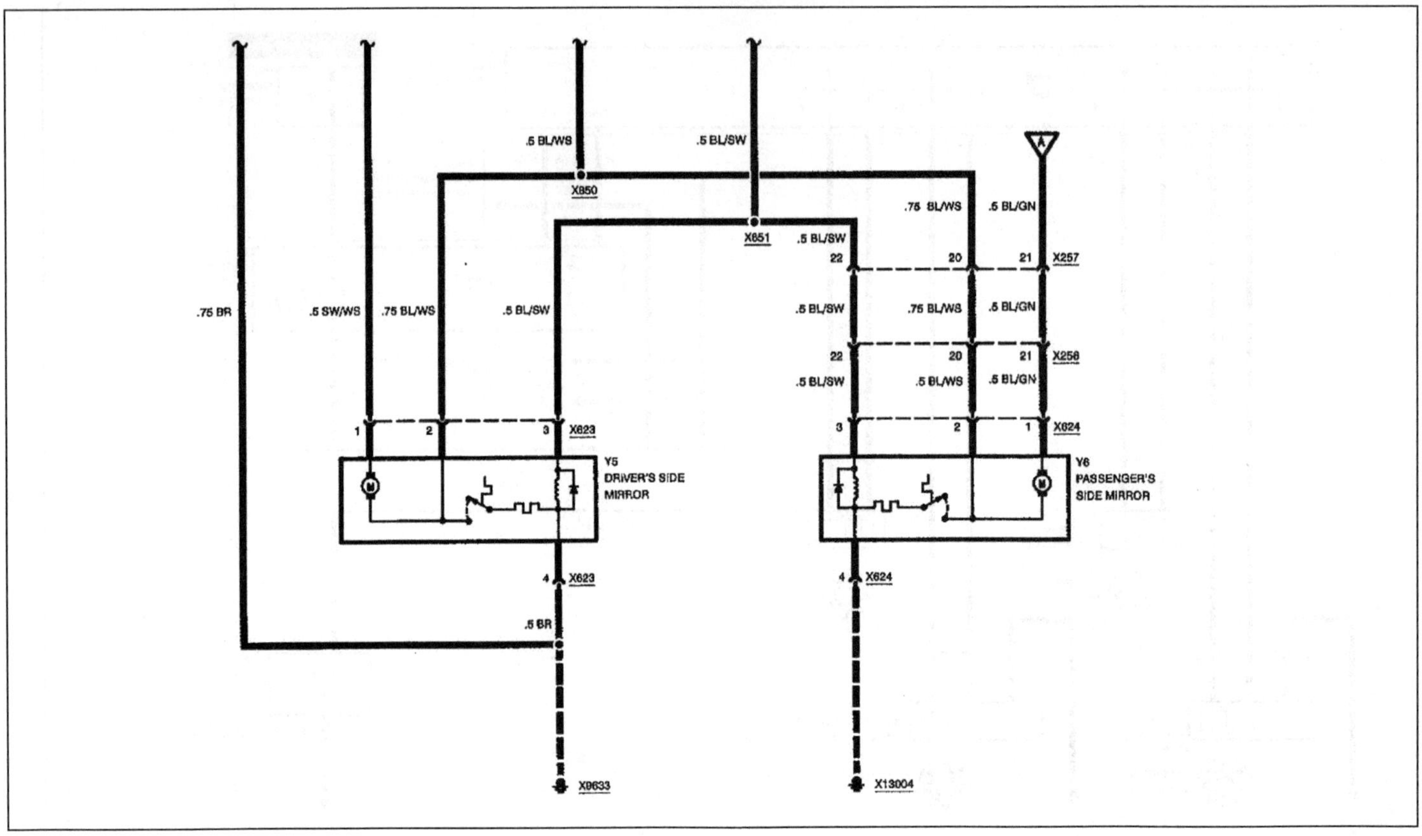


## Mirror, electrochromic rear view

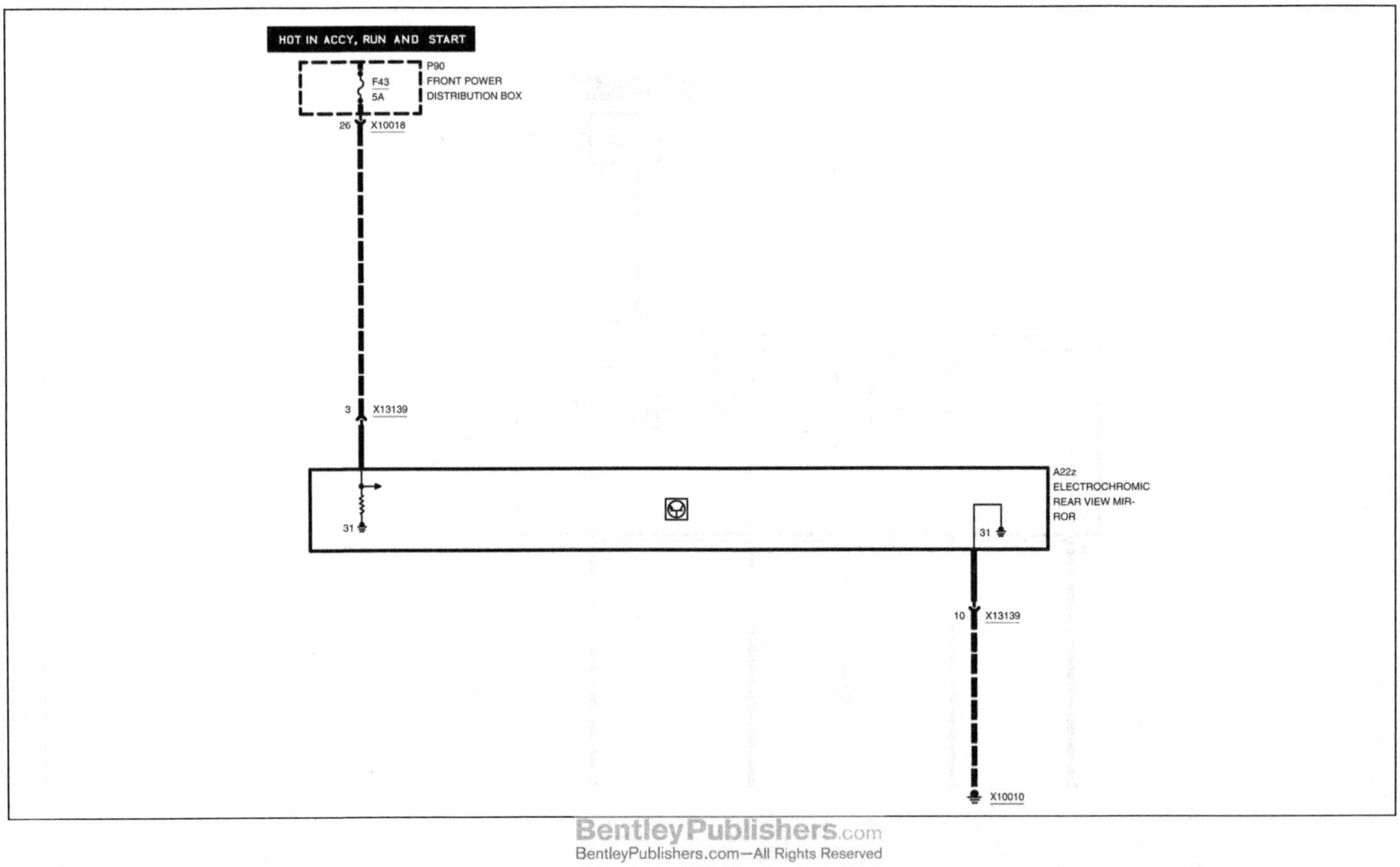

## Central locking (ZKE)
(1 of 6)

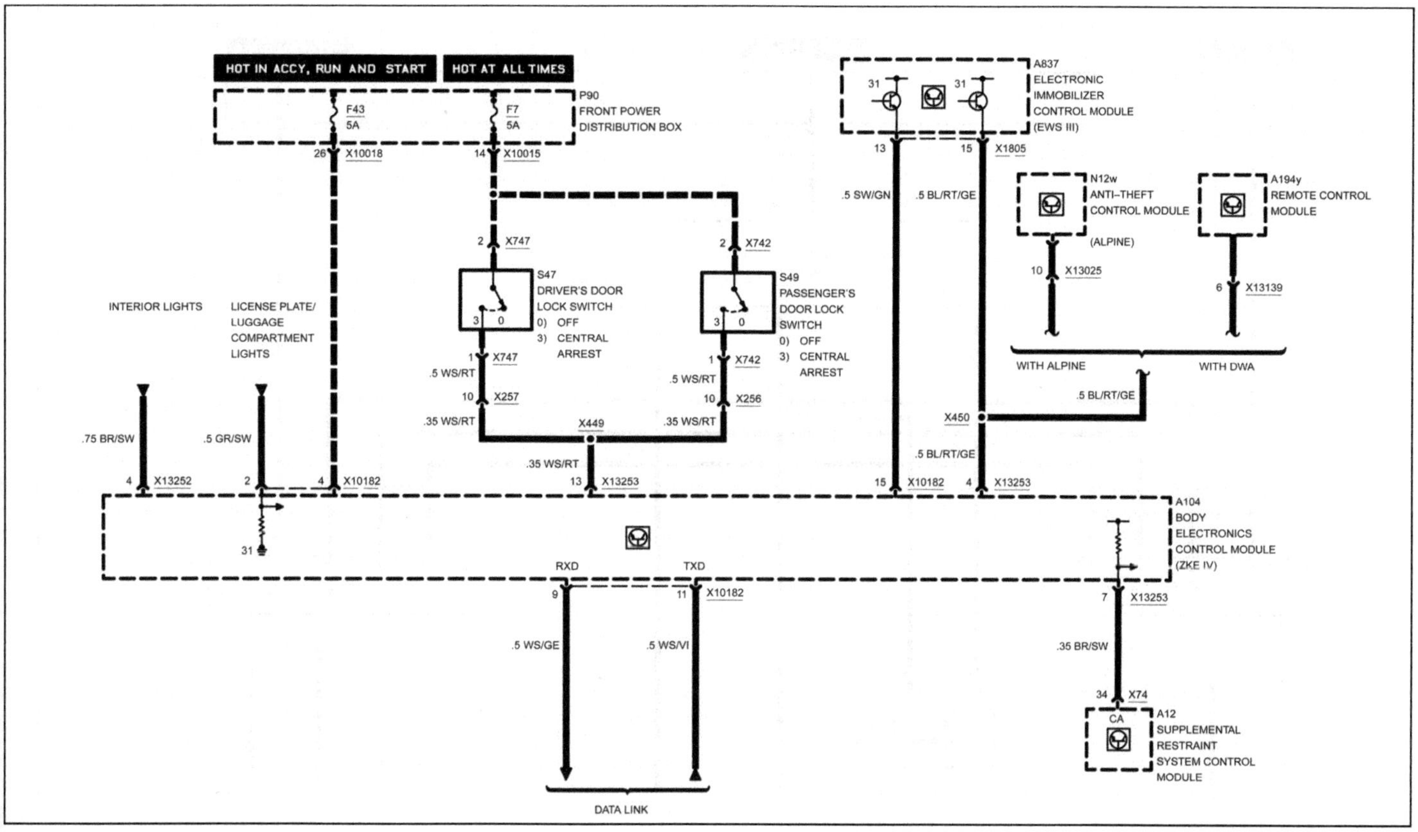


## Central locking (ZKE)
(2 of 6)

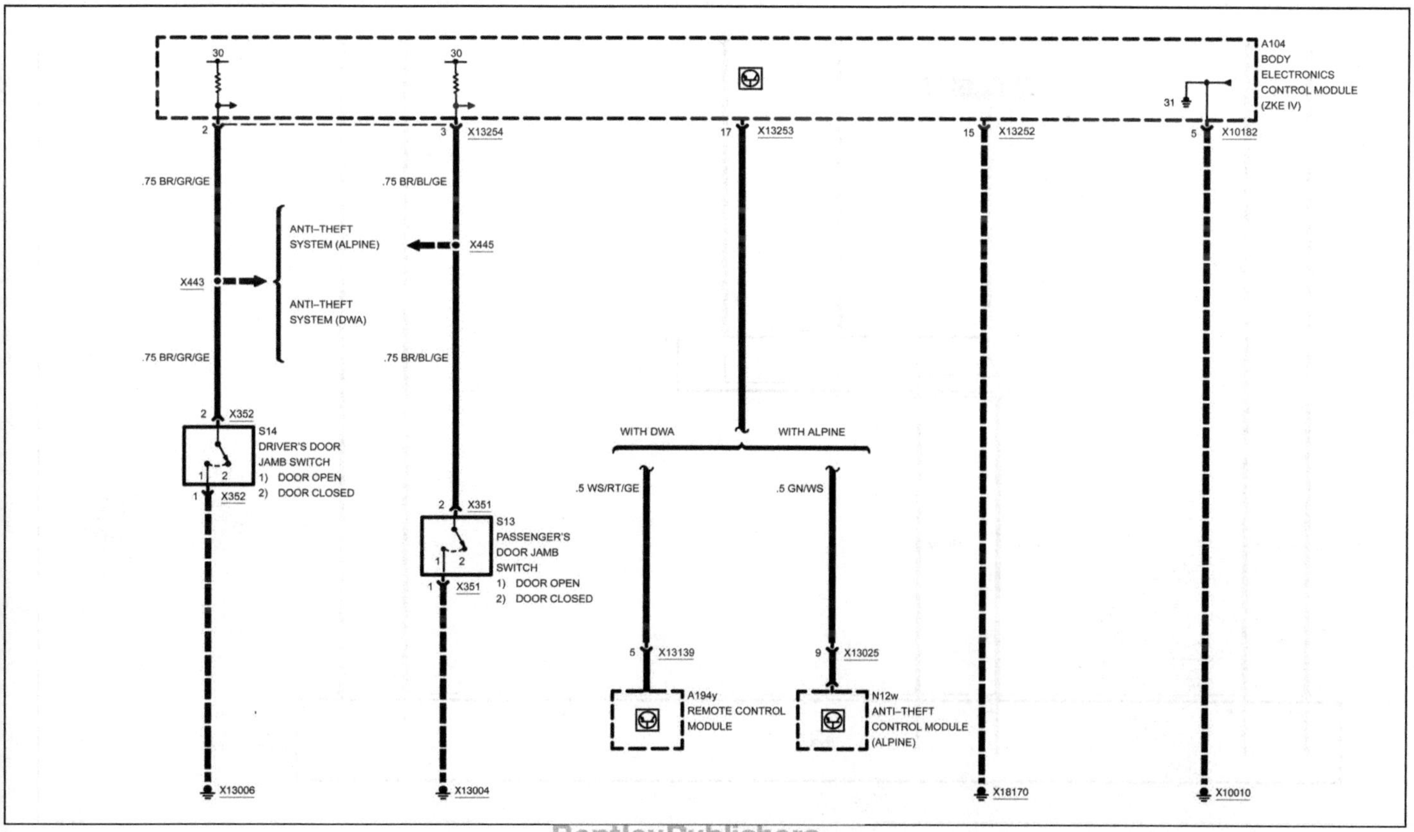

## Central locking (ZKE)
### (3 of 6)

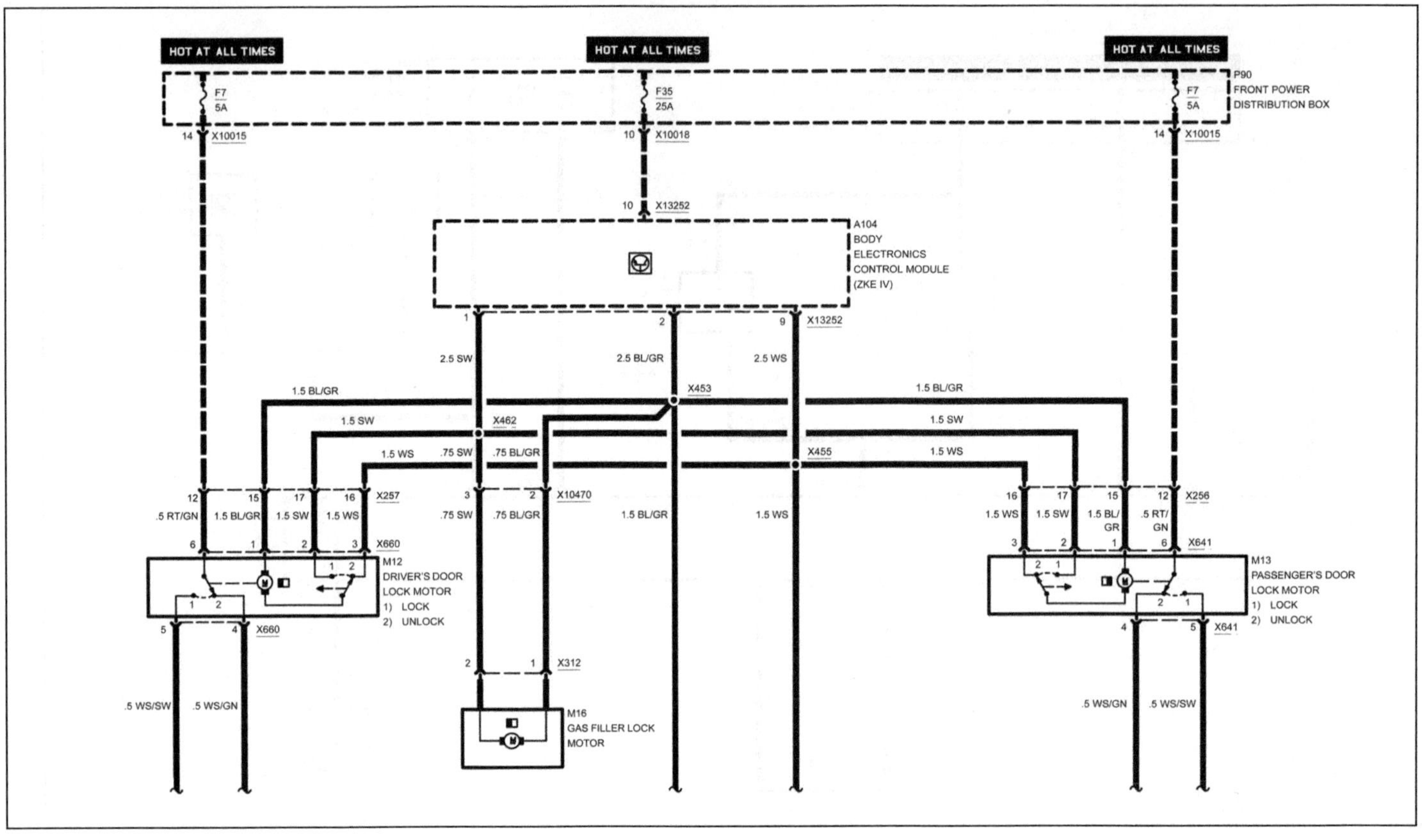


## Central locking (ZKE)
### (4 of 6)

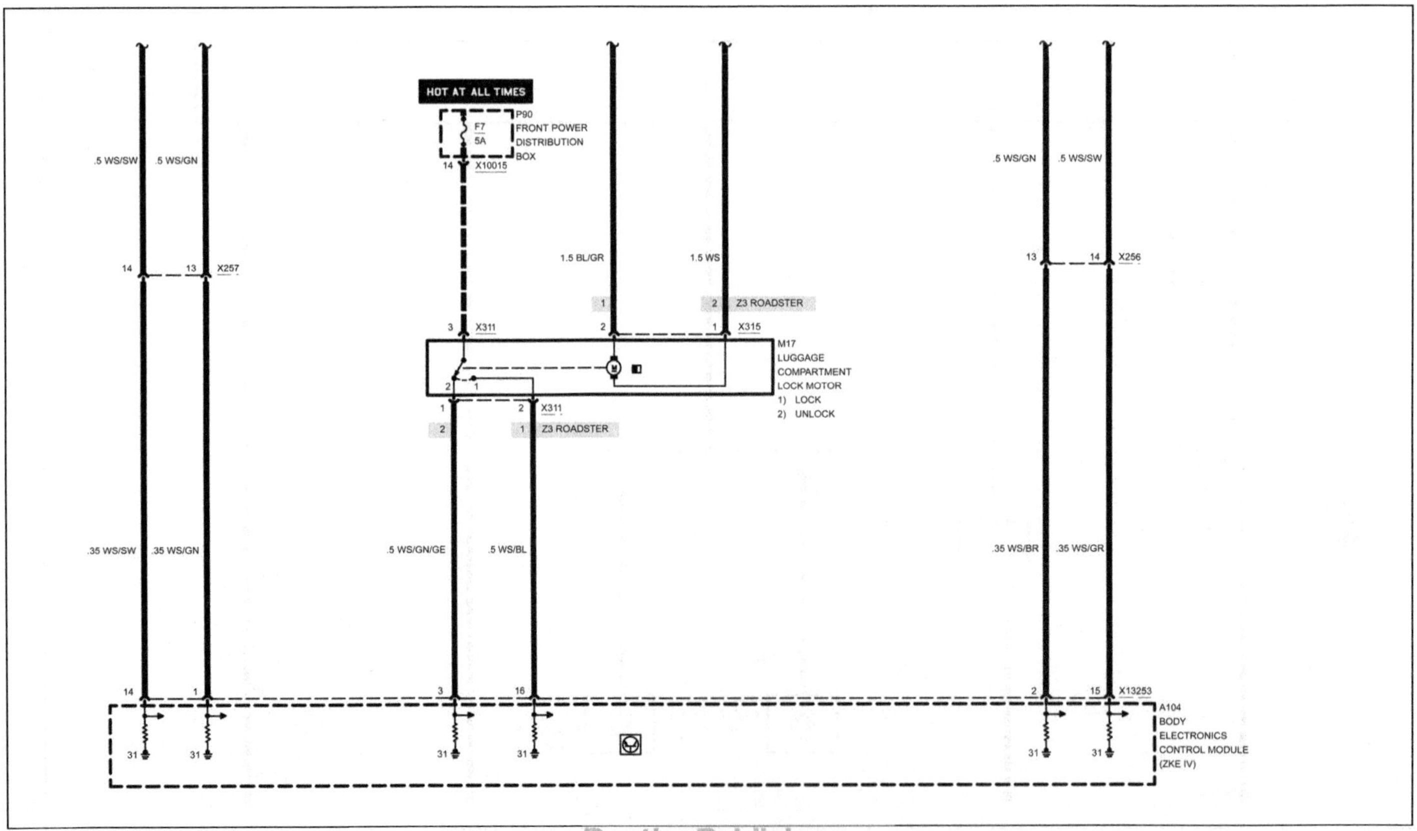

# Central locking (ZKE)

(5 of 6)

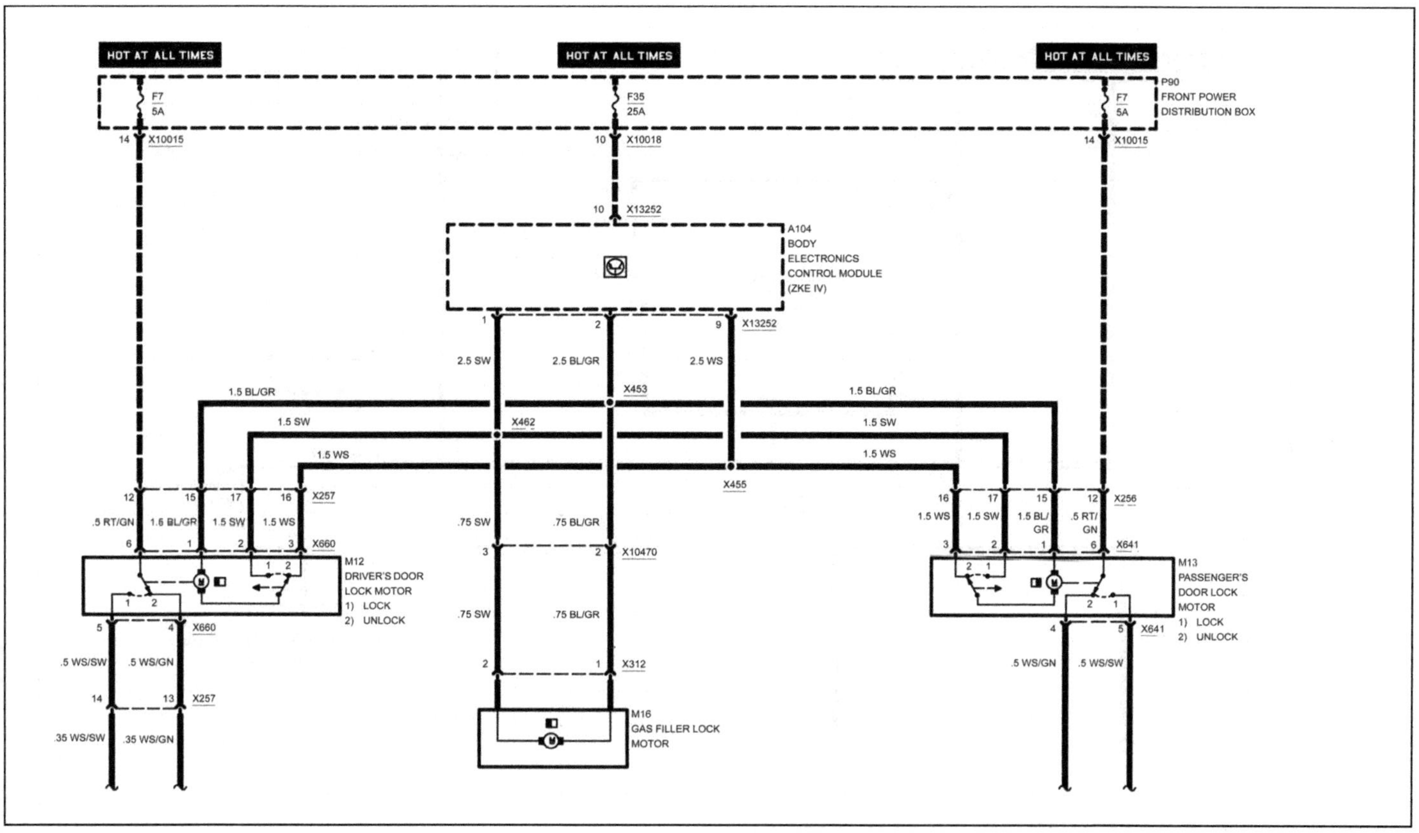


# Central locking (ZKE)

(6 of 6)

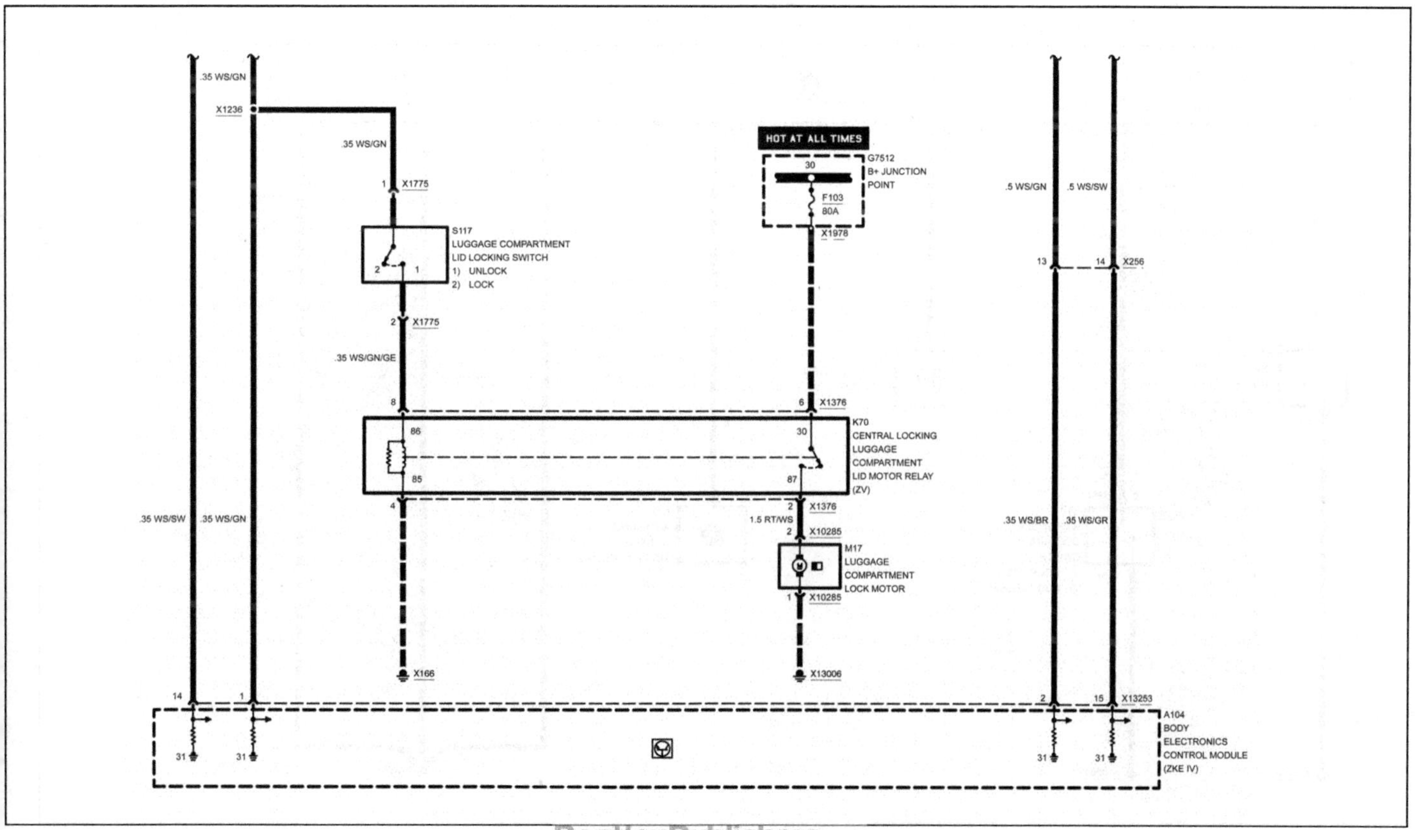

## Central locking (ZVM), 1996
(1 of 4)

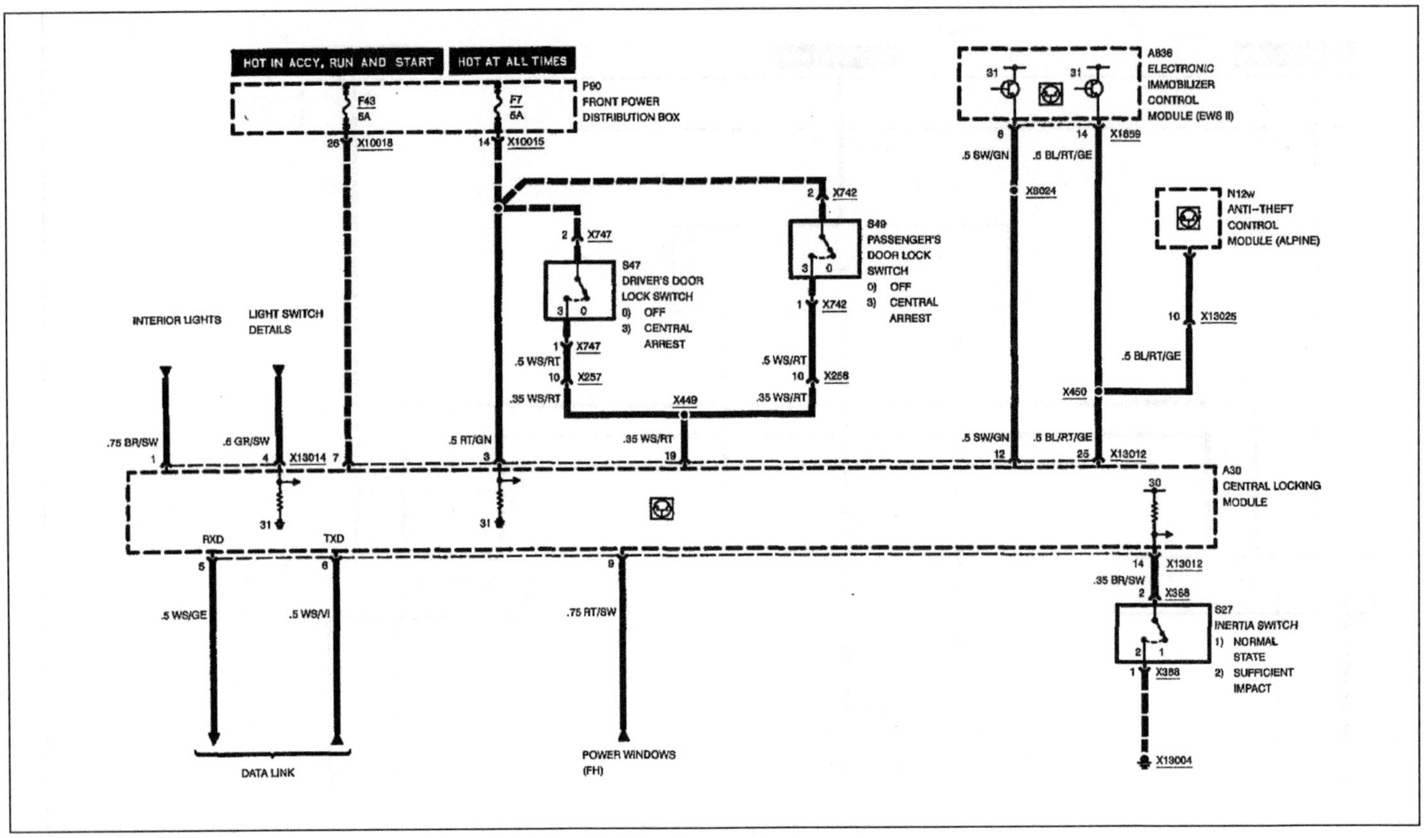


## Central locking (ZVM), 1996
(2 of 4)

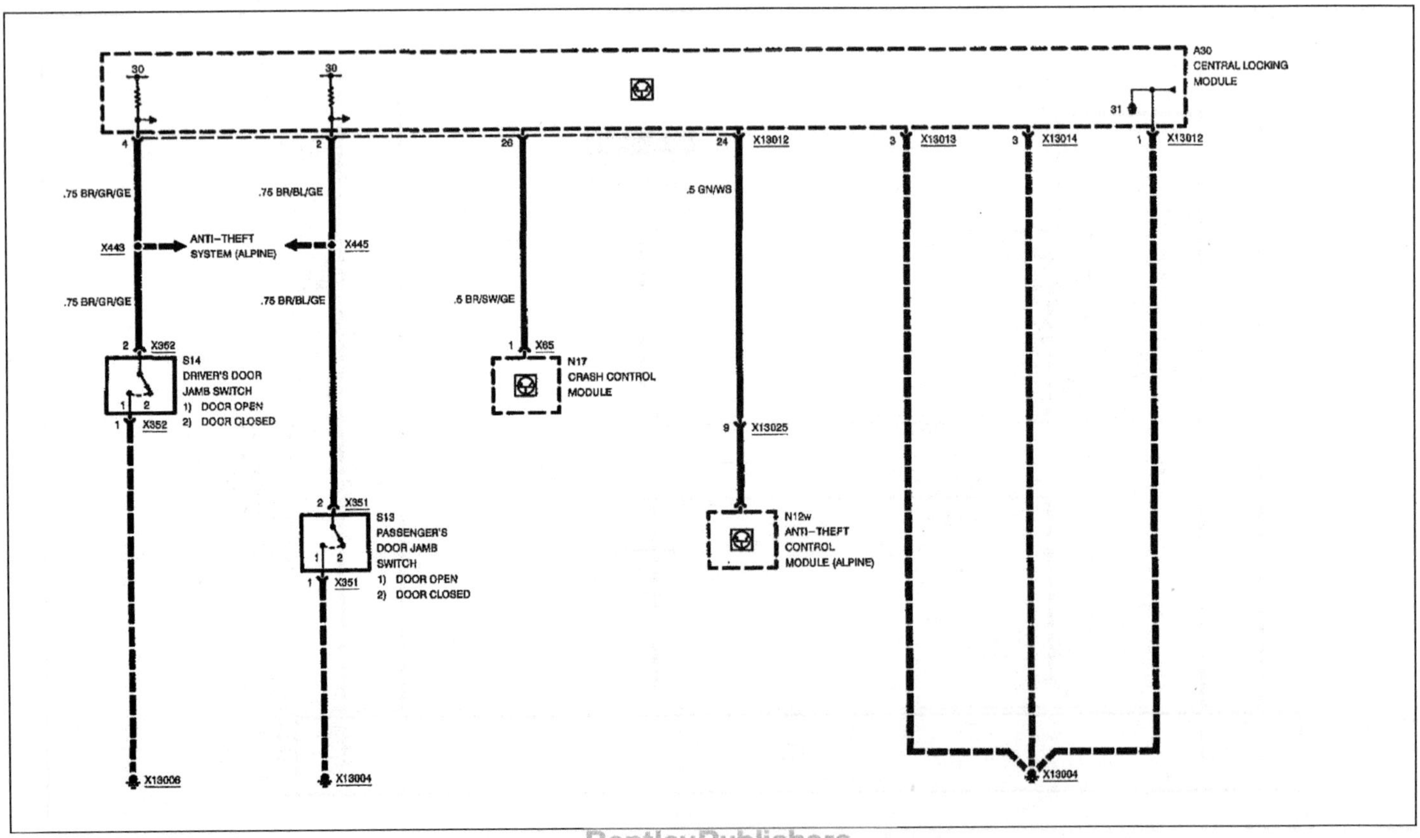

# Central locking (ZVM), 1996

(3 of 4)

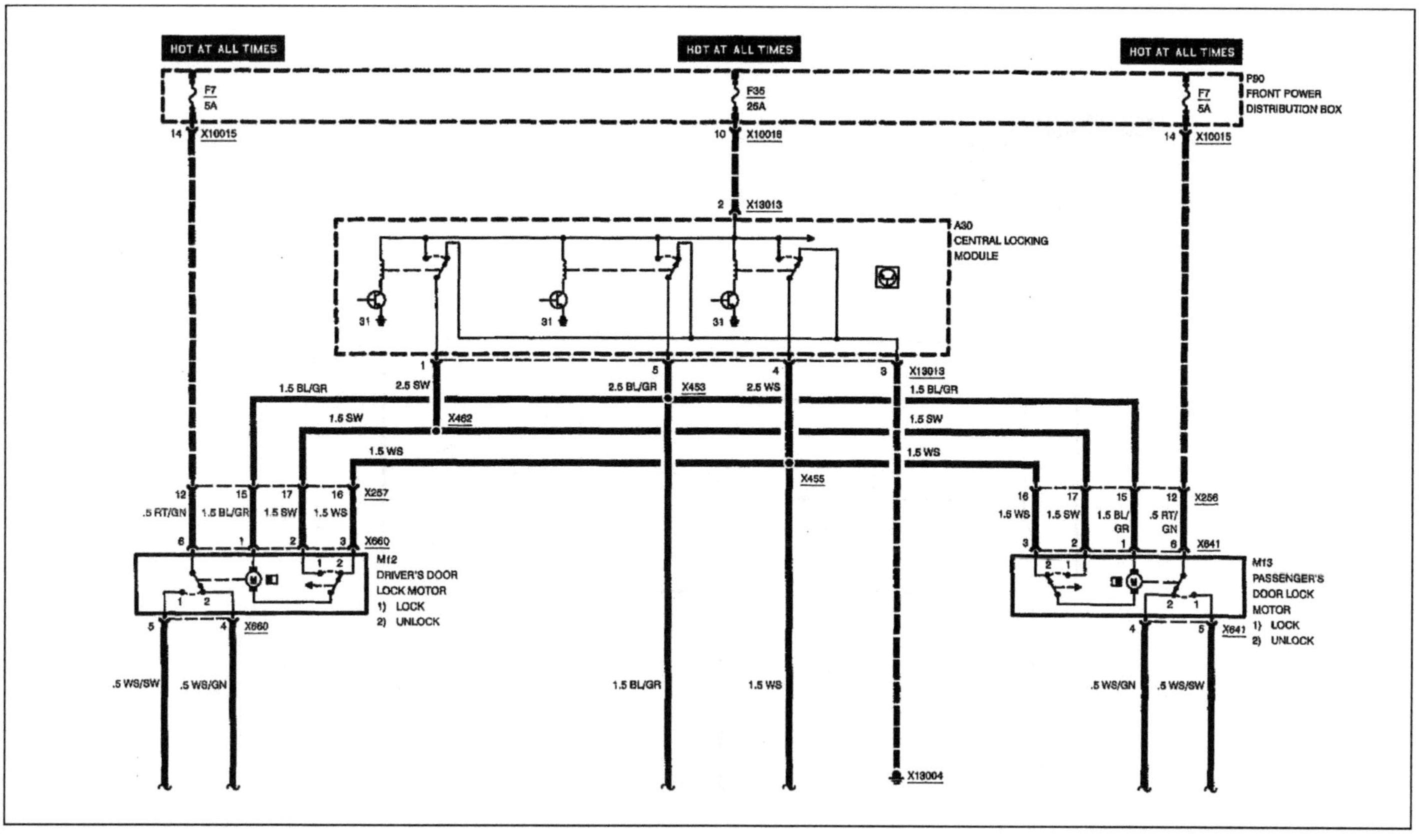


# Central locking (ZVM), 1996

(4 of 4)

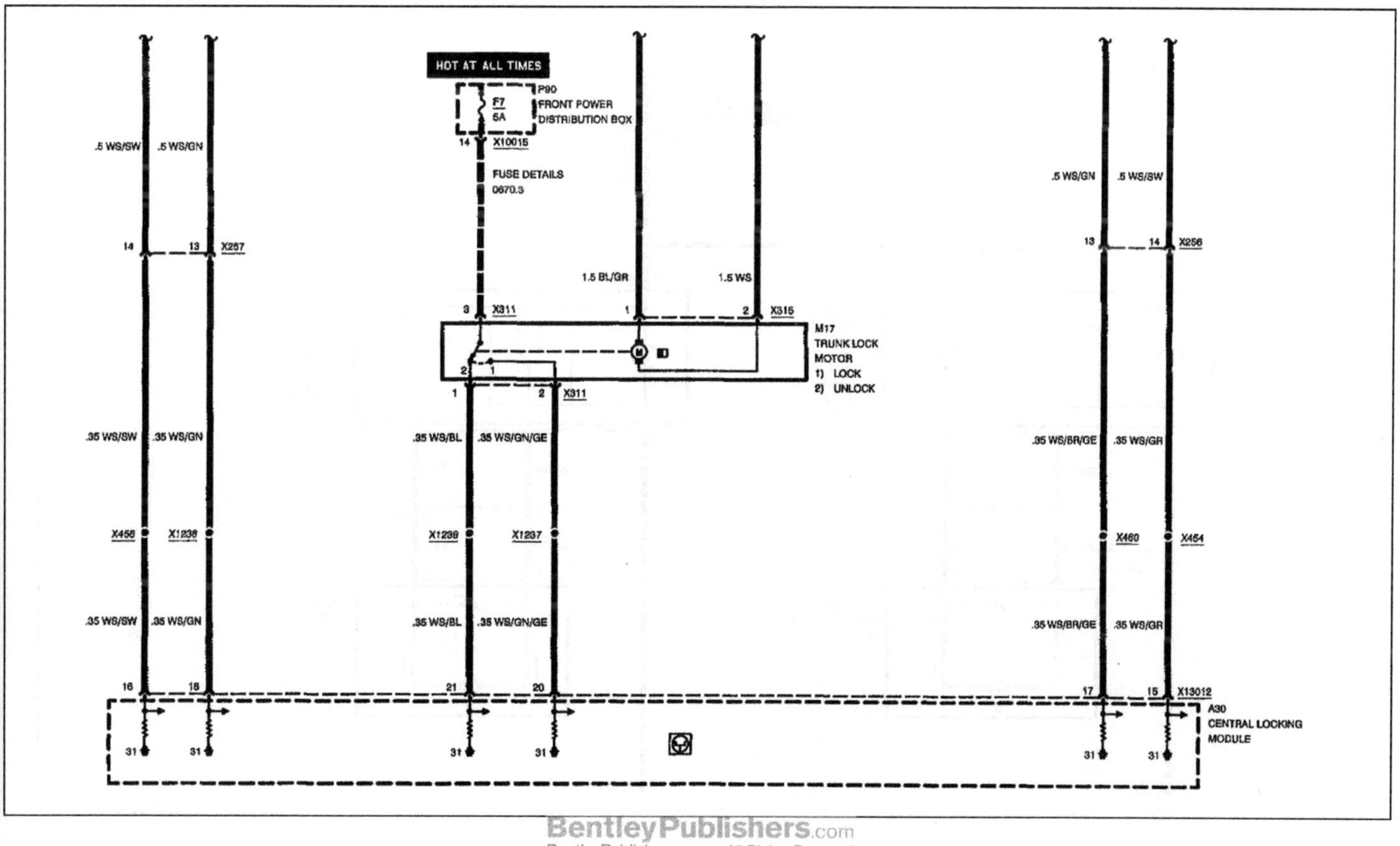

## Windows (ZVM)

(1 of 2)

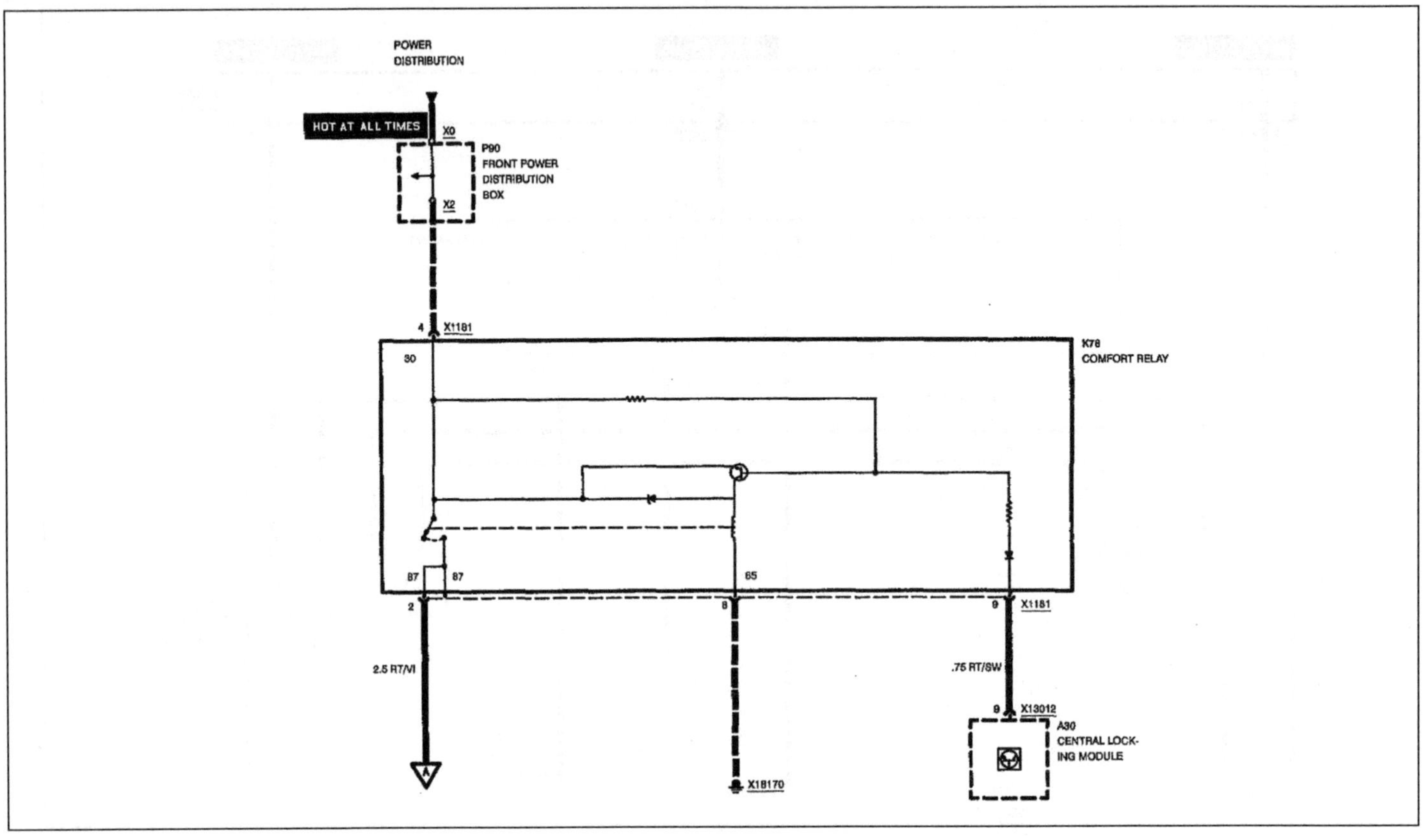


## Windows (ZVM)

(2 of 2)

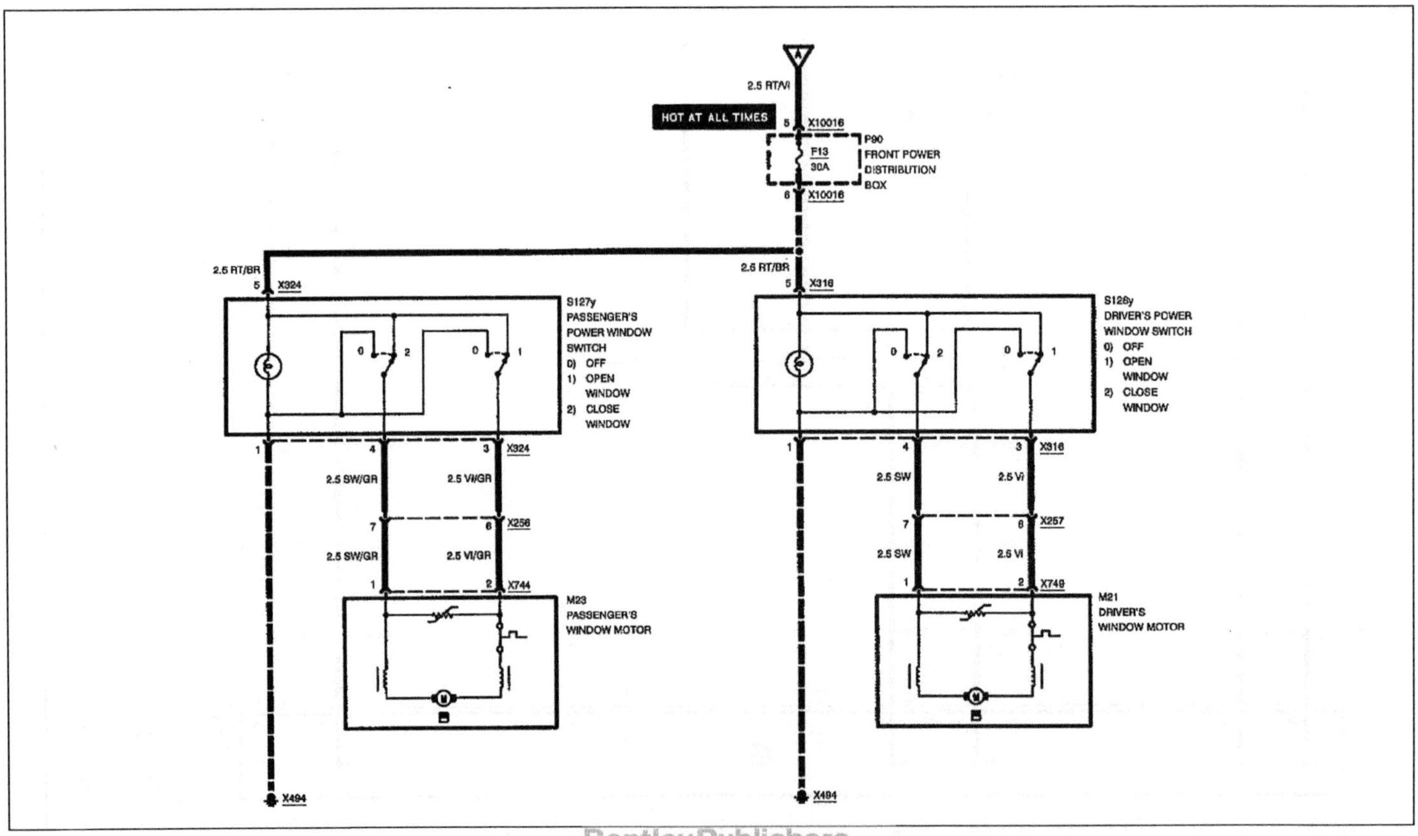

## Windows (ZKE)

(1 of 3)

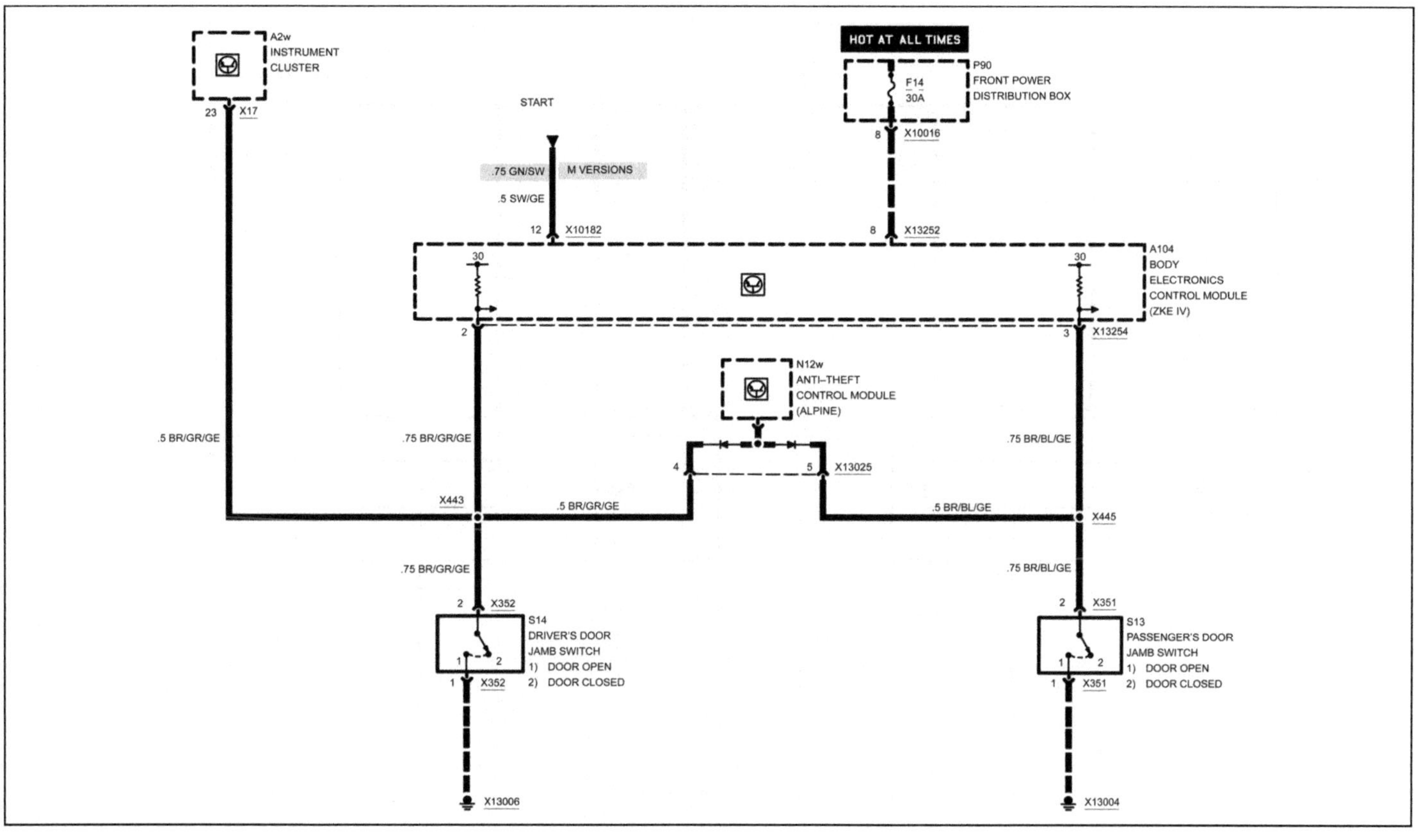


## Windows (ZKE)

(2 of 3)

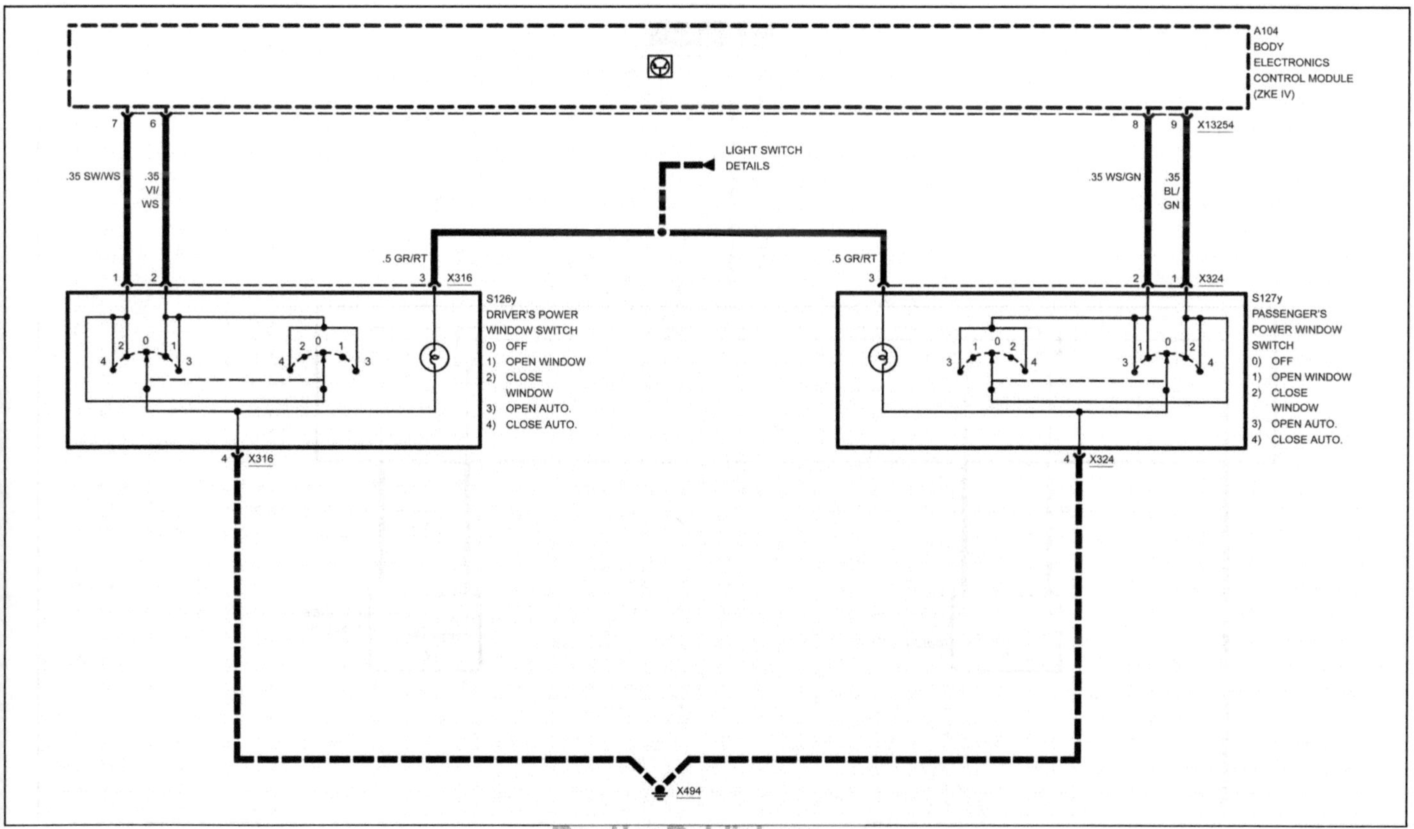

## Windows (ZKE)
(3 of 3)

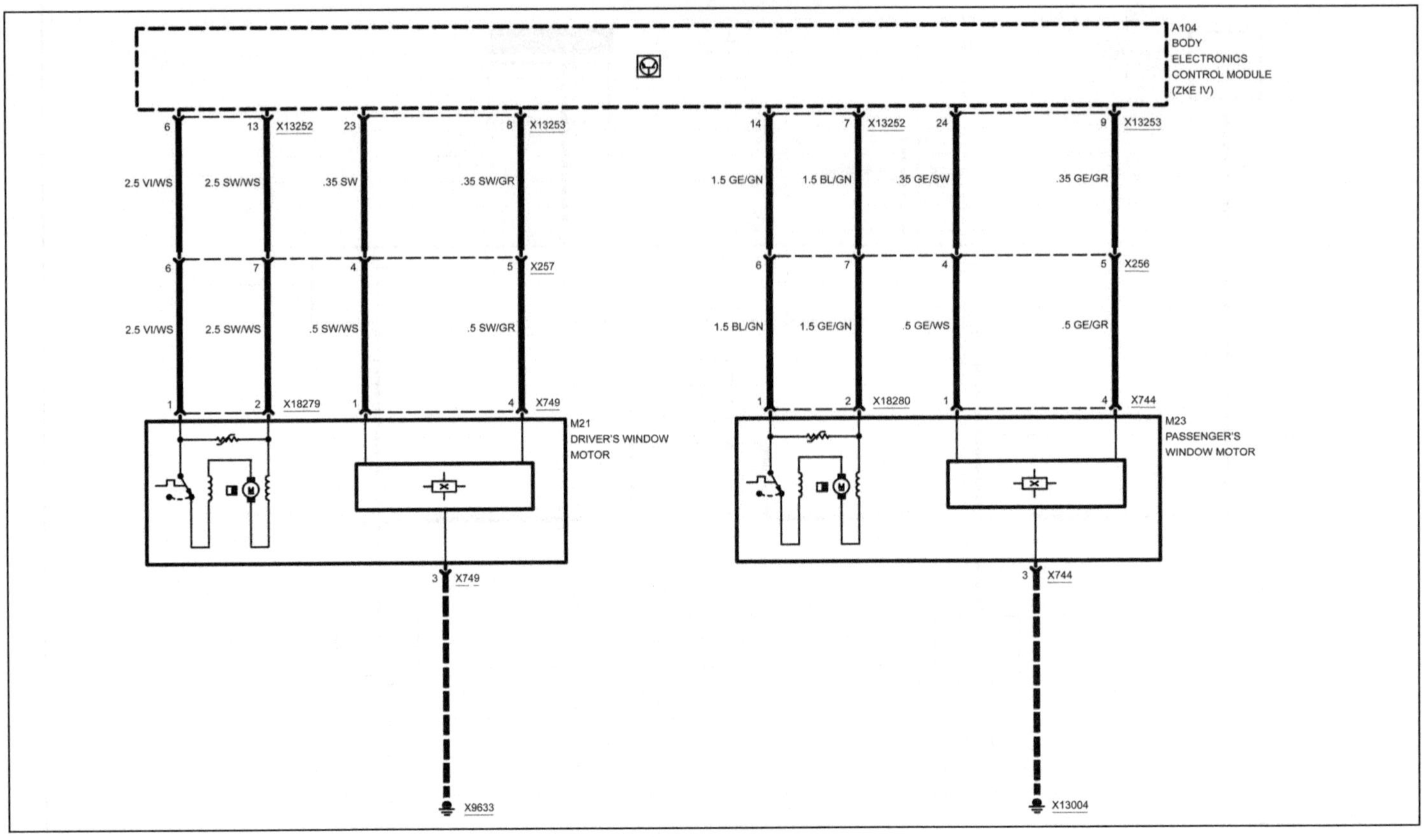


## Seats
(1 of 4)

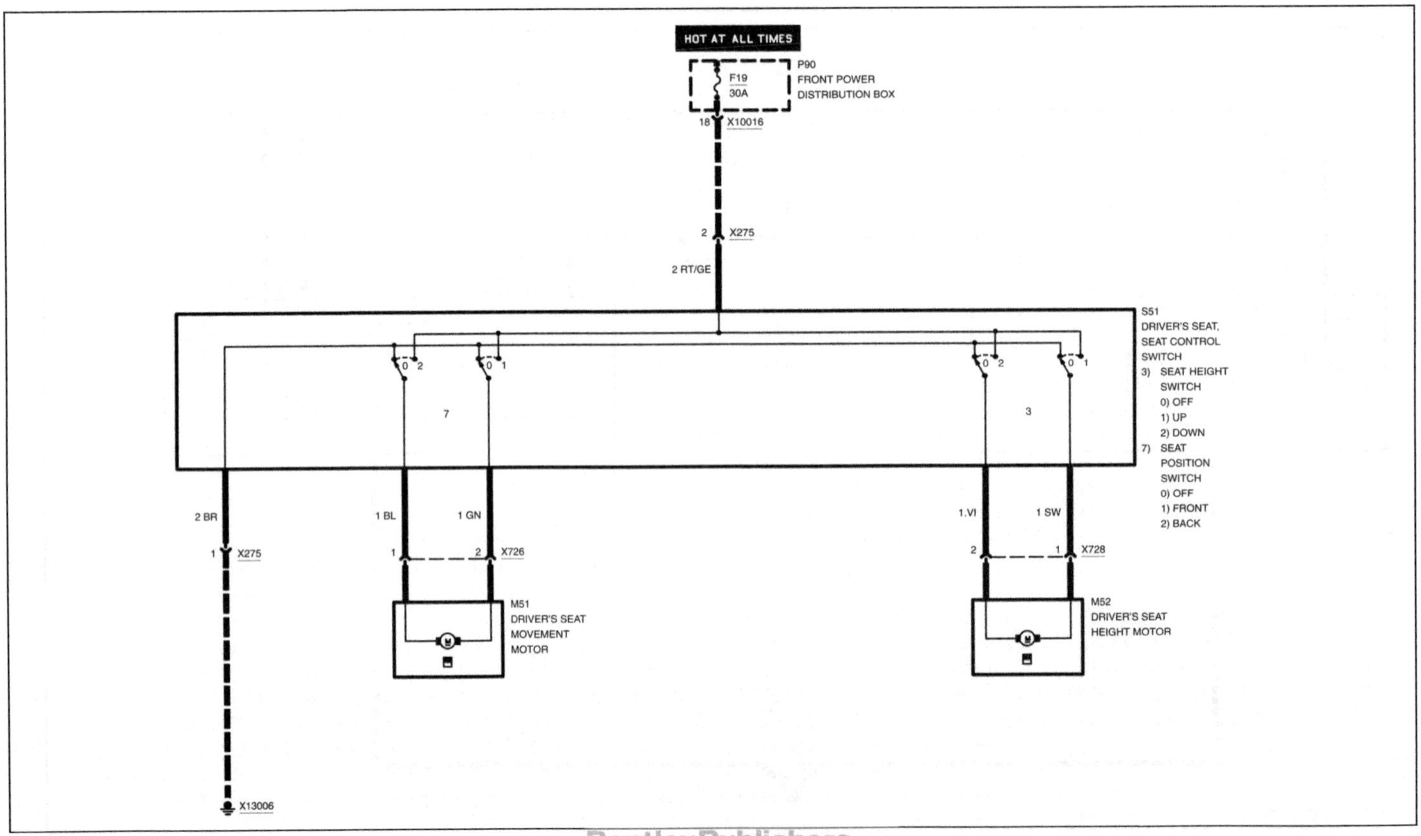

# Seats
## (2 of 4)

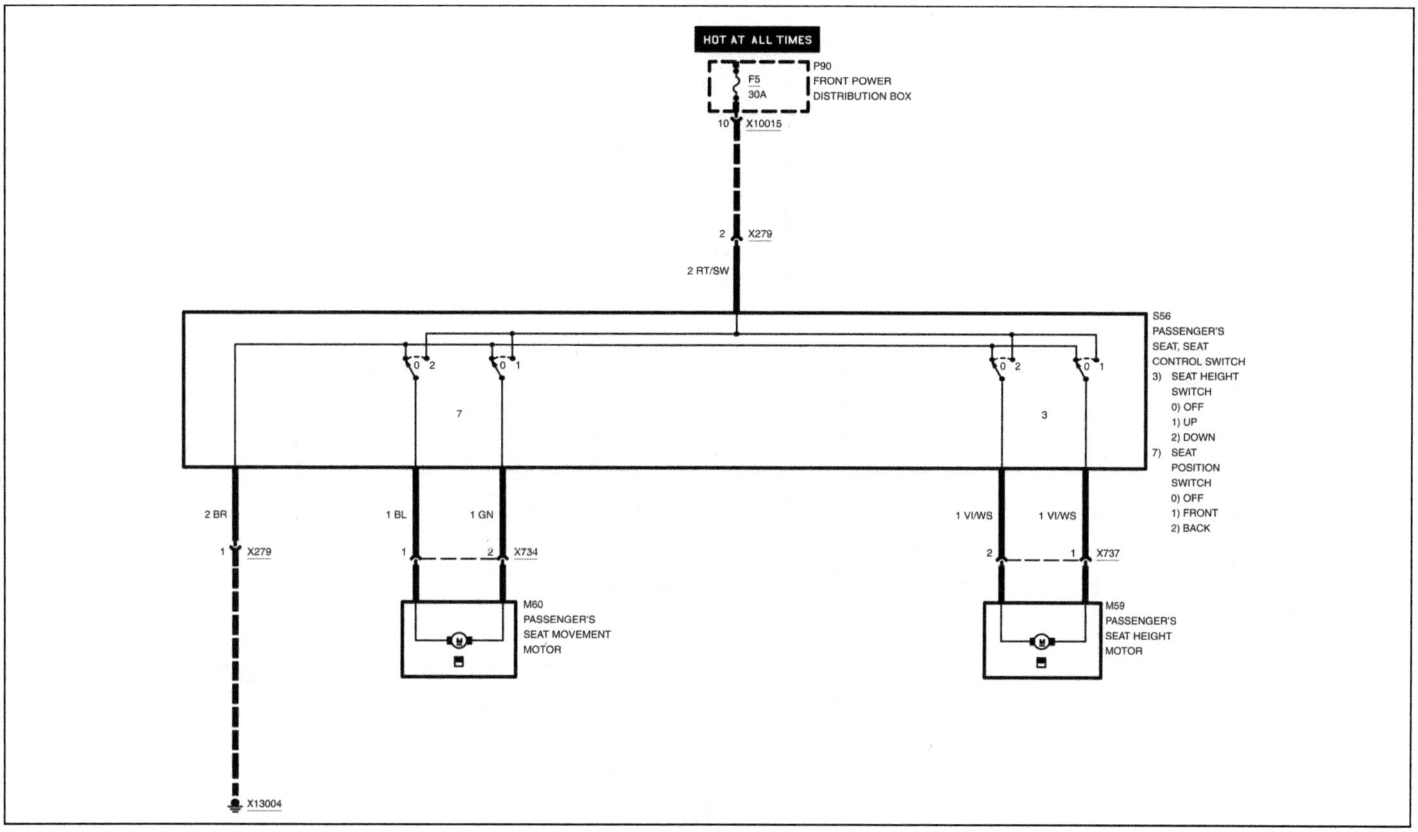


# Seats
## (3 of 4)

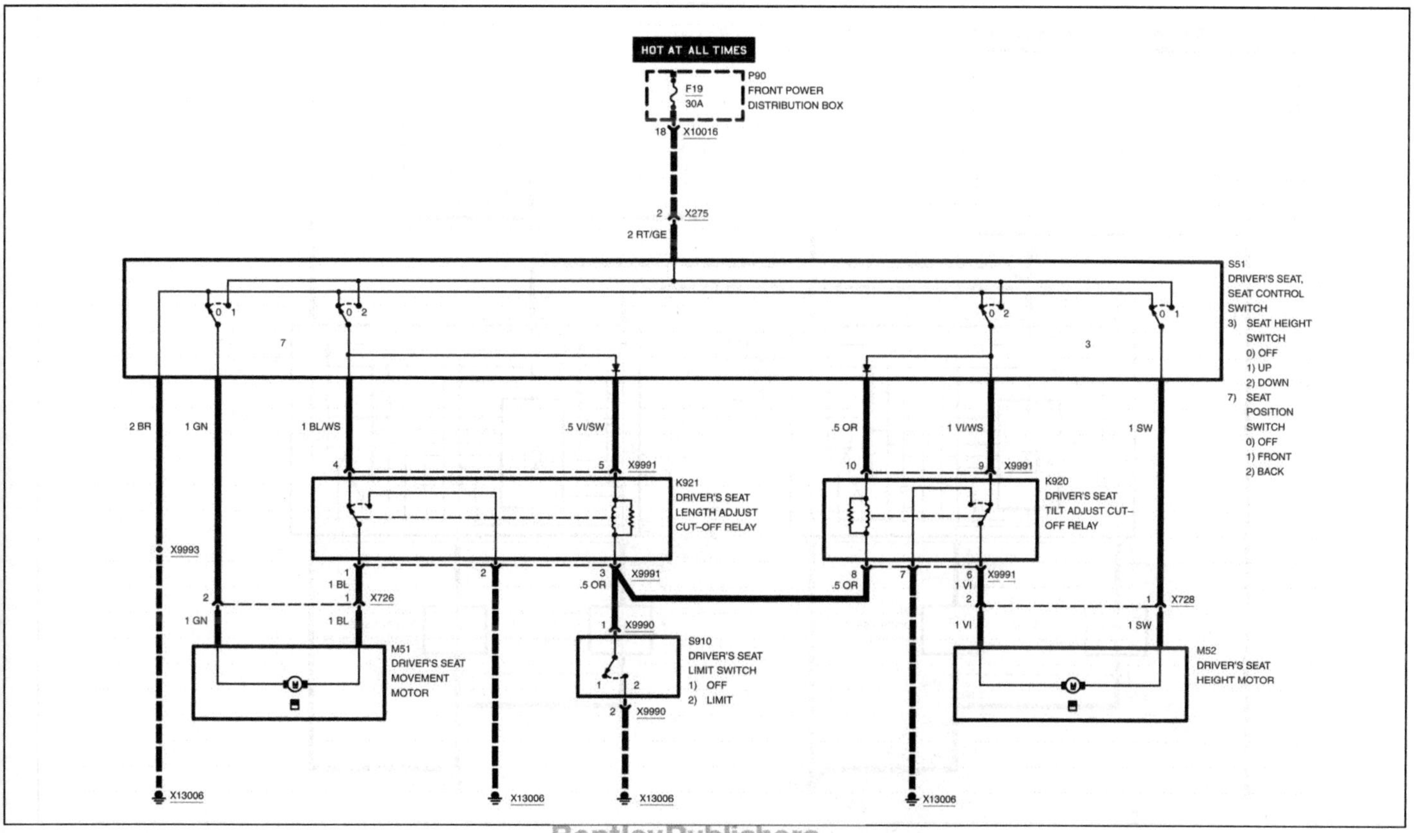

## Seats
(4 of 4)

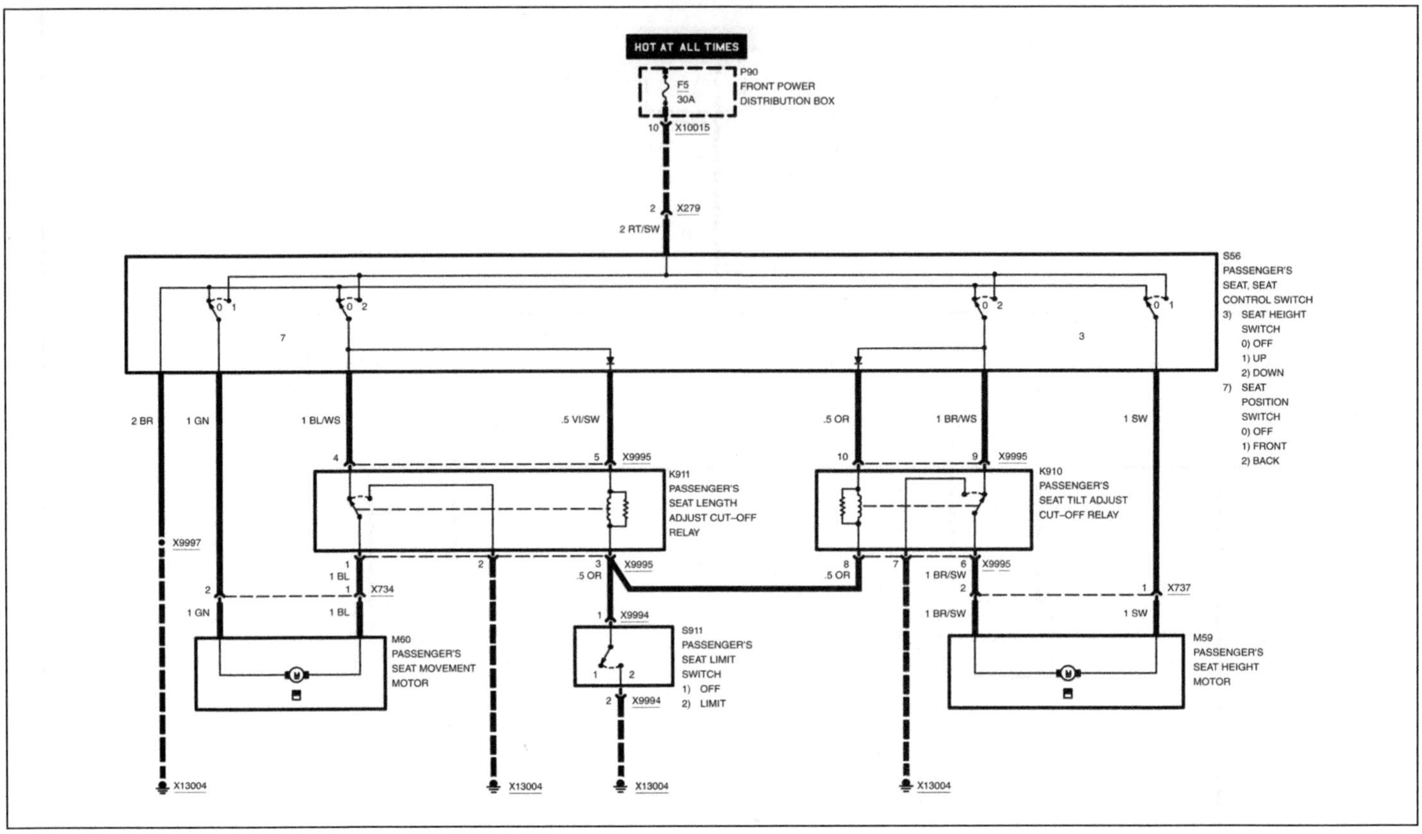


## Seat heating, 1996 - early 1998

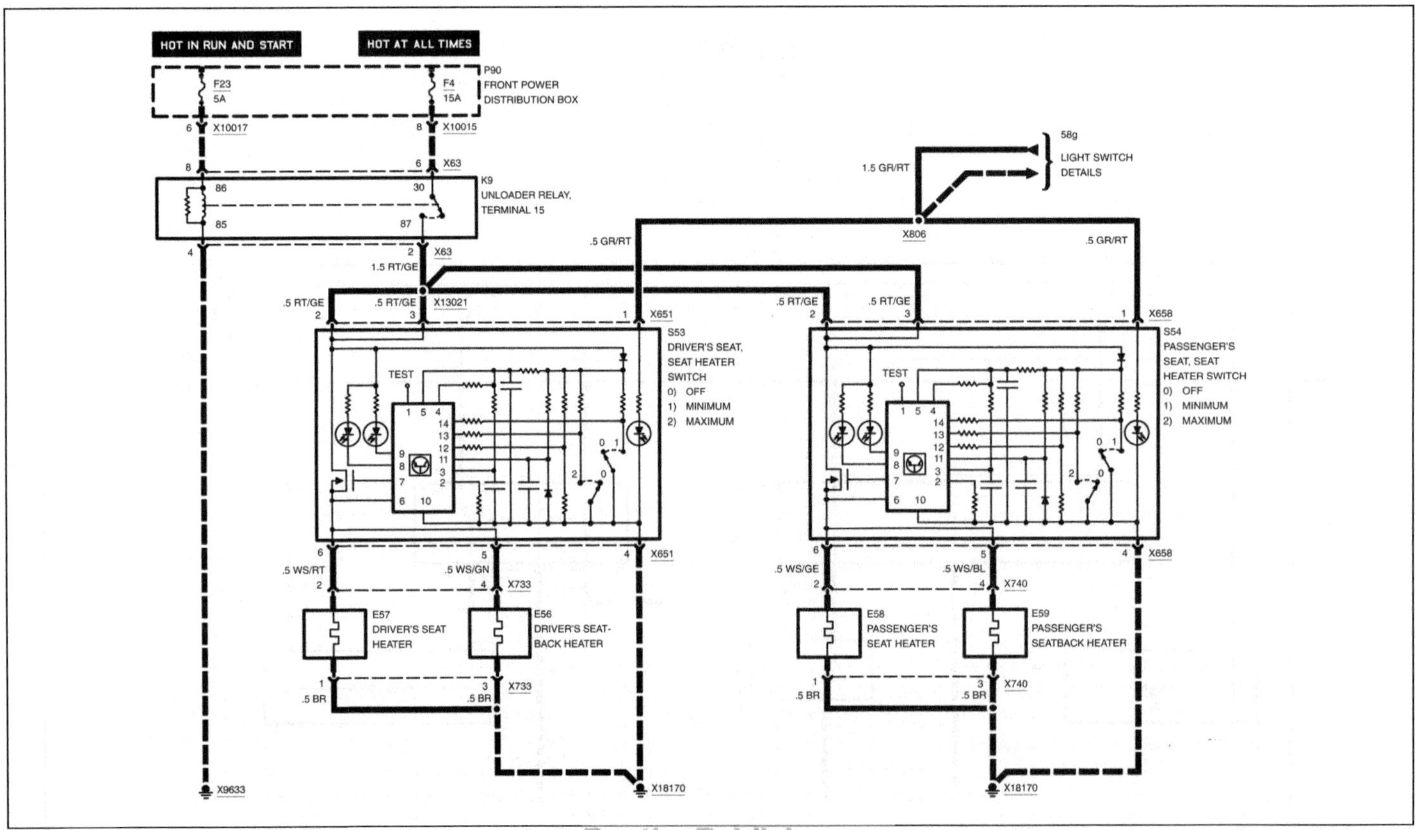

## Seat heating, late 1998 - 2002

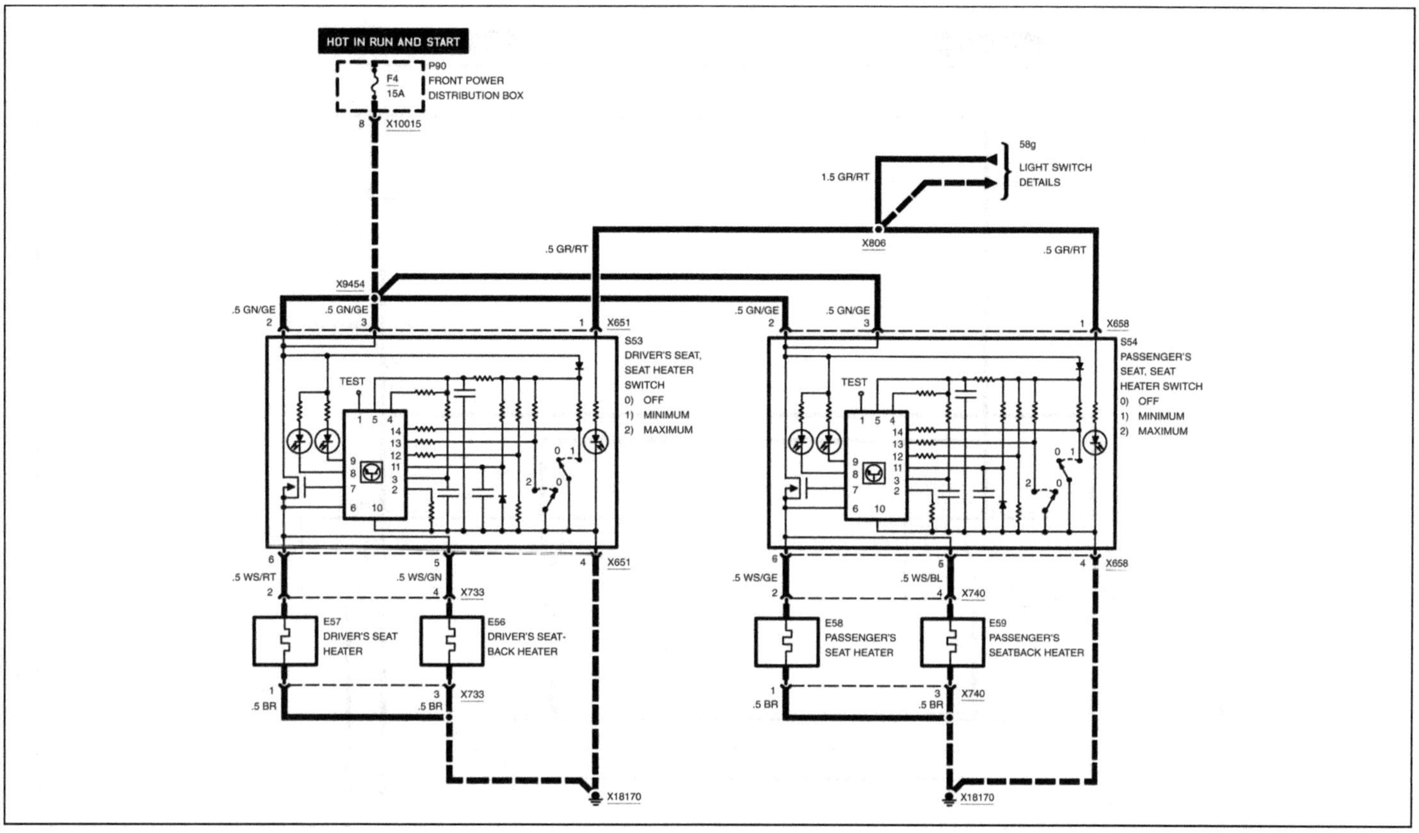


## Sunroof

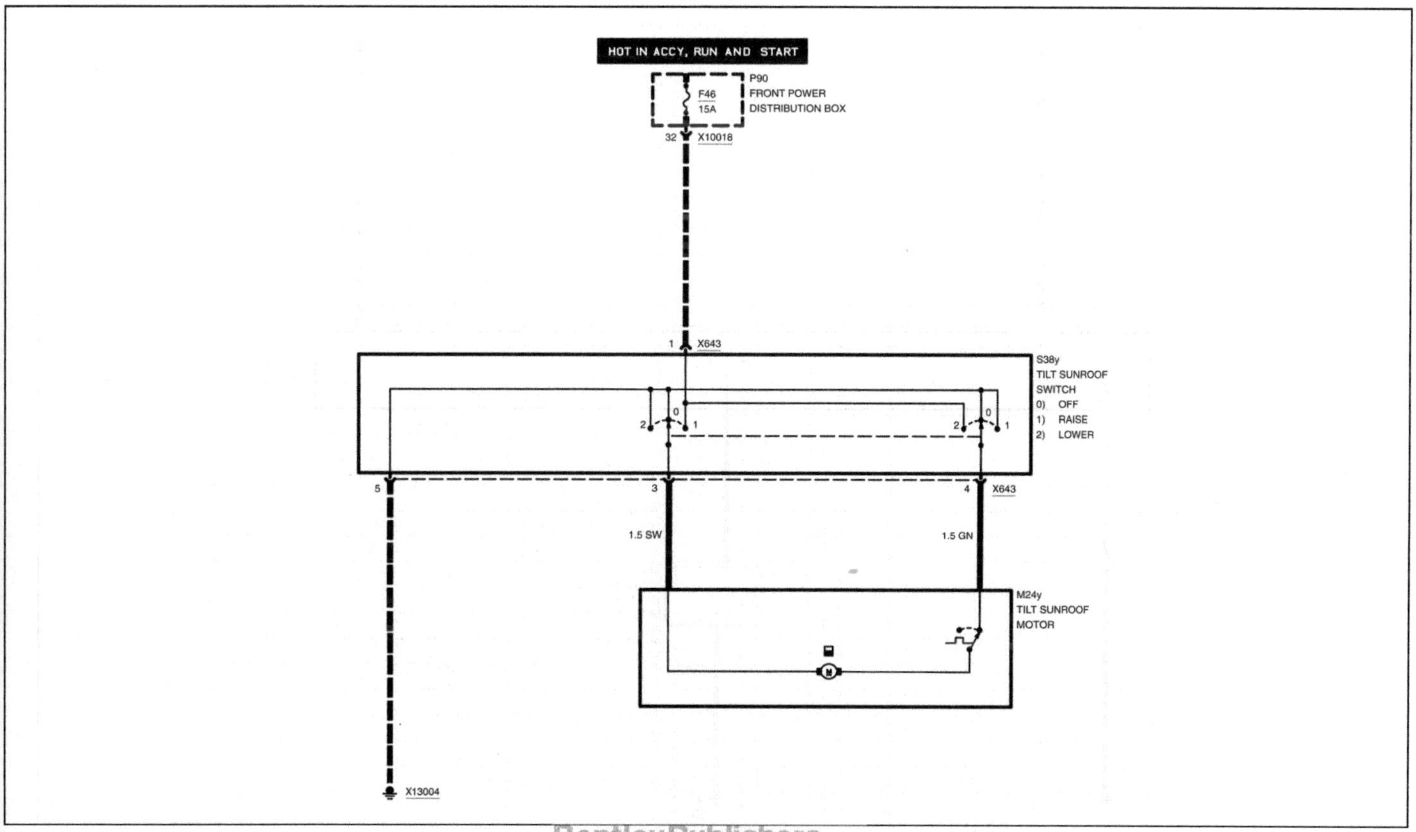

## Convertible top
(1 of 2)

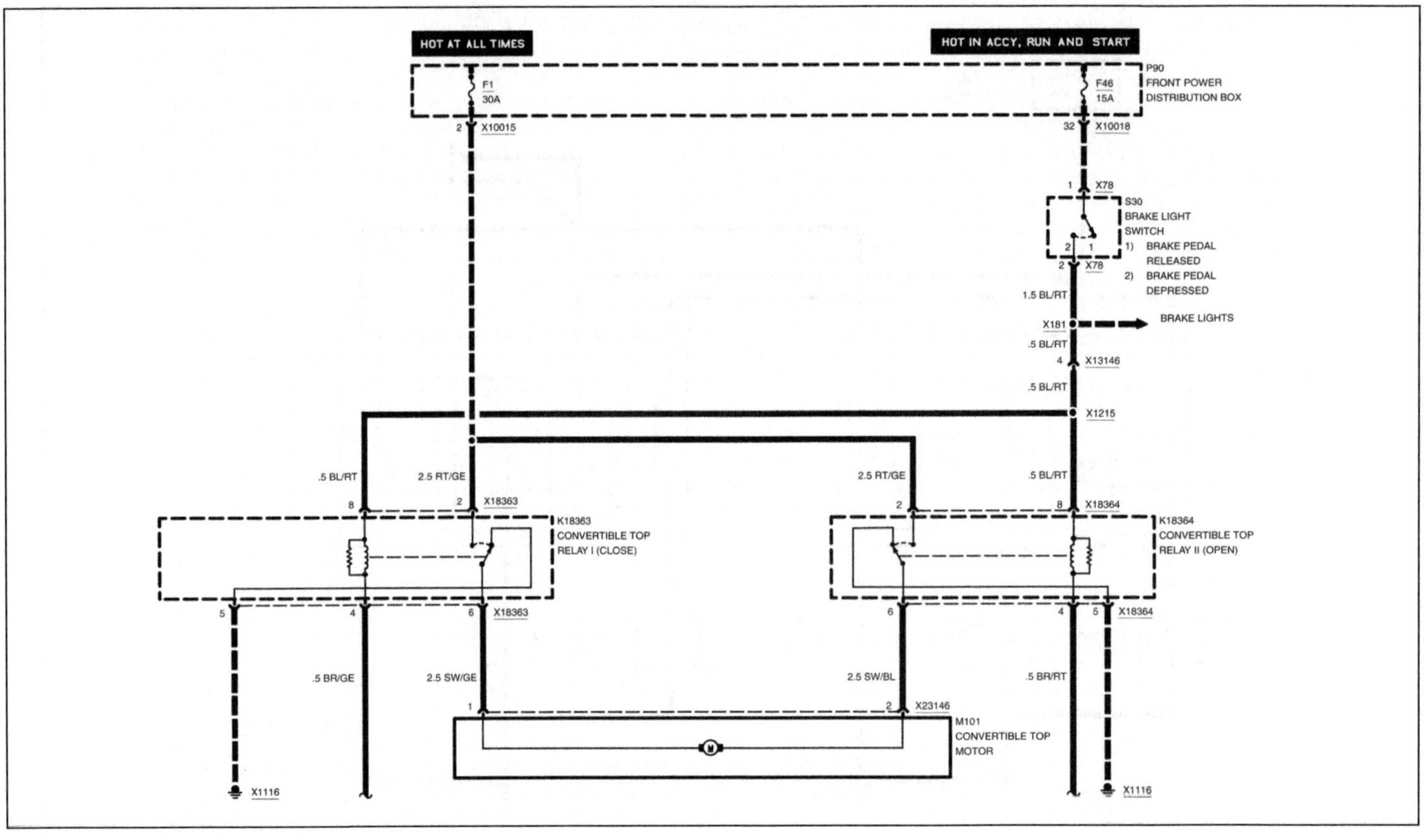


## Convertible top
(2 of 2)

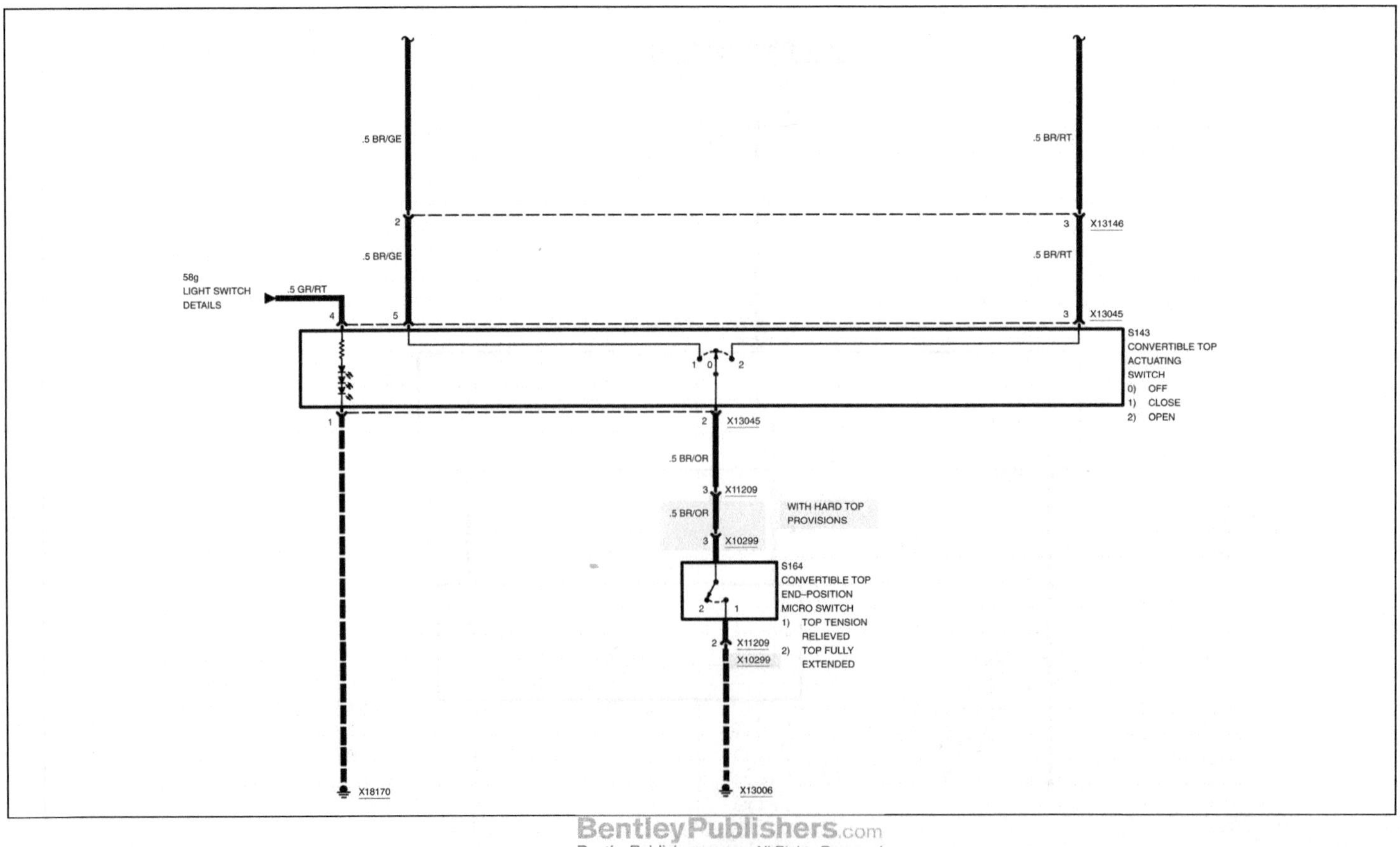

## ZKE IV, 1996 - 1998

(1 of 2)

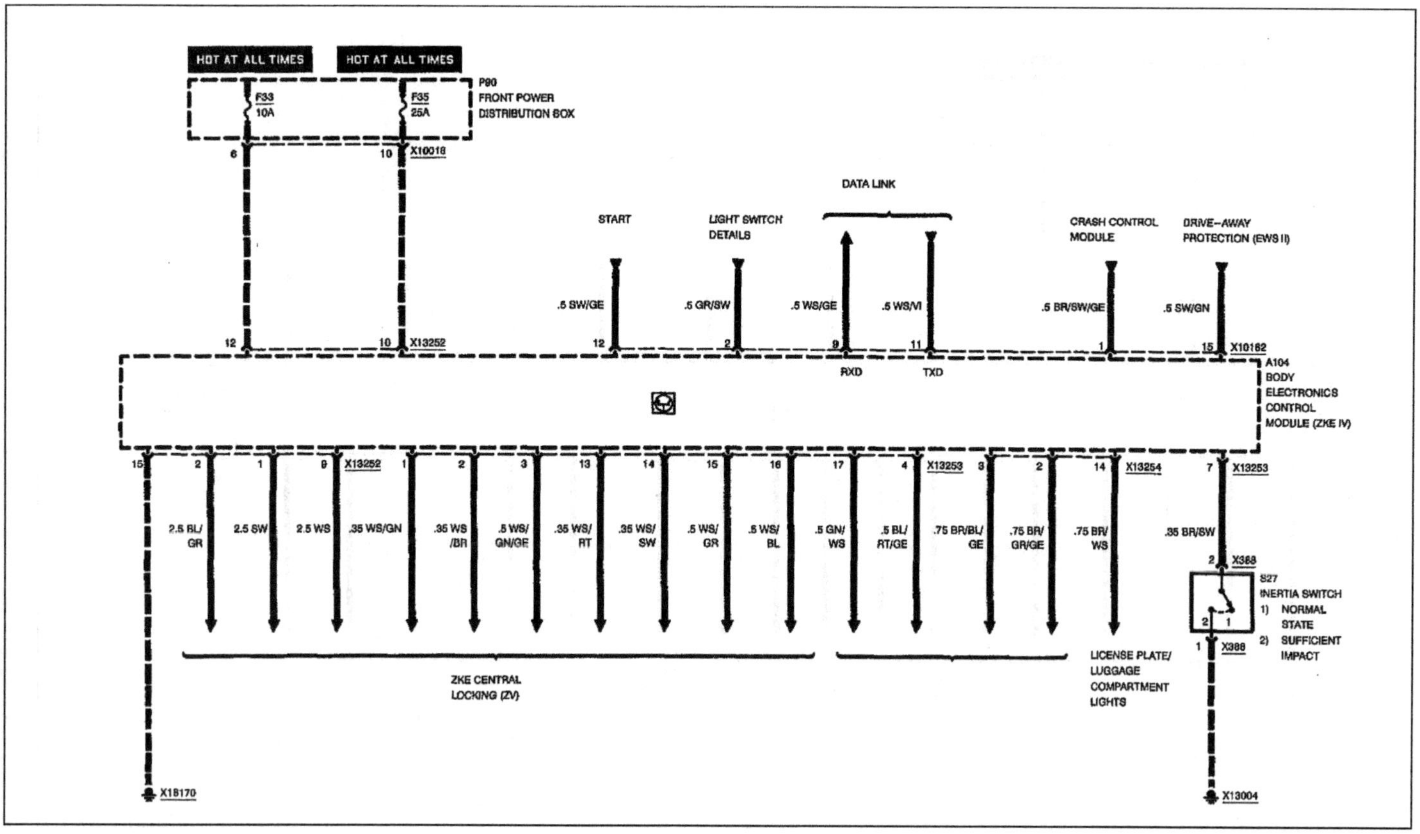


## ZKE IV, 1996 - 1998

(2 of 2)

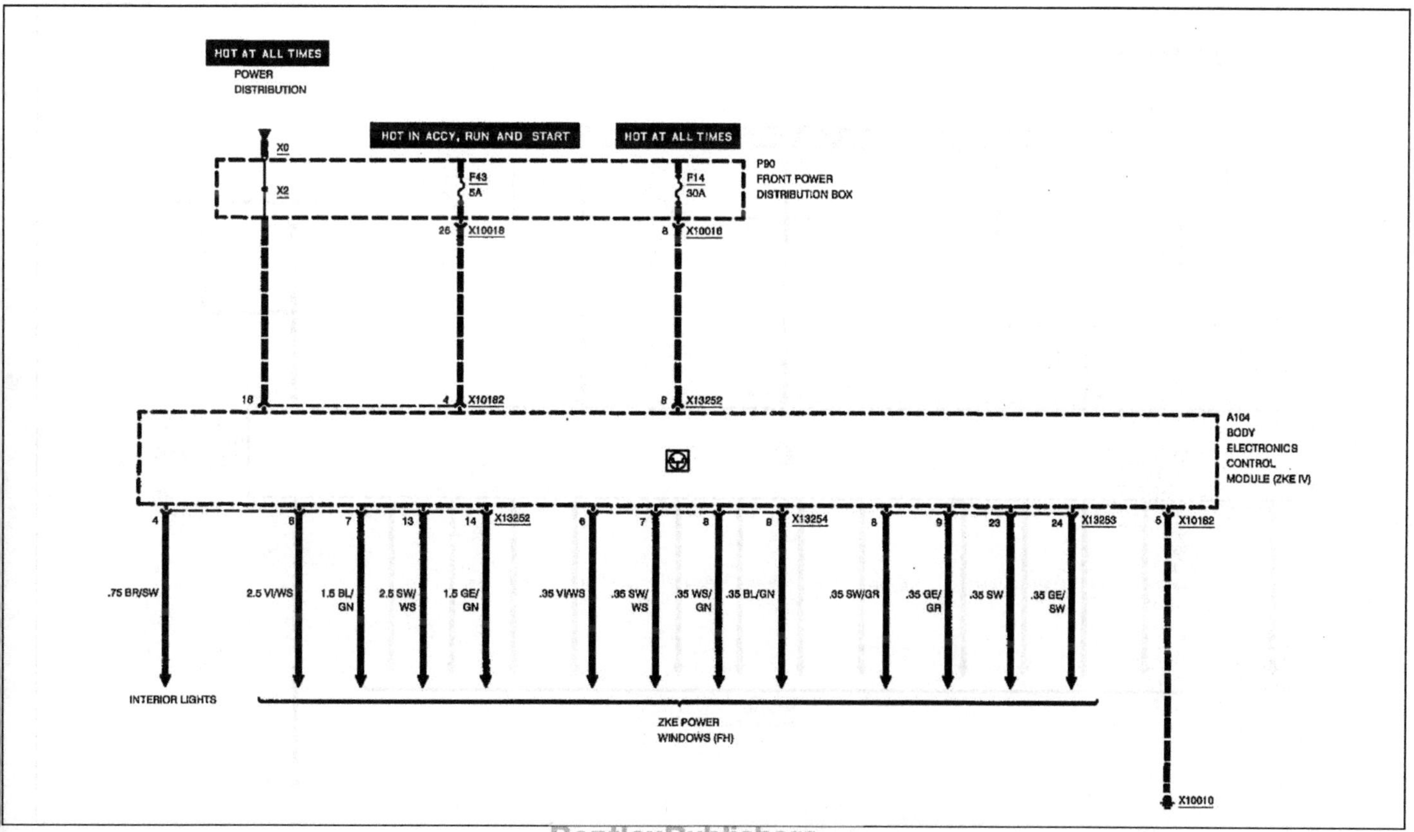

## ZKE IV, 2000
(1 of 2)

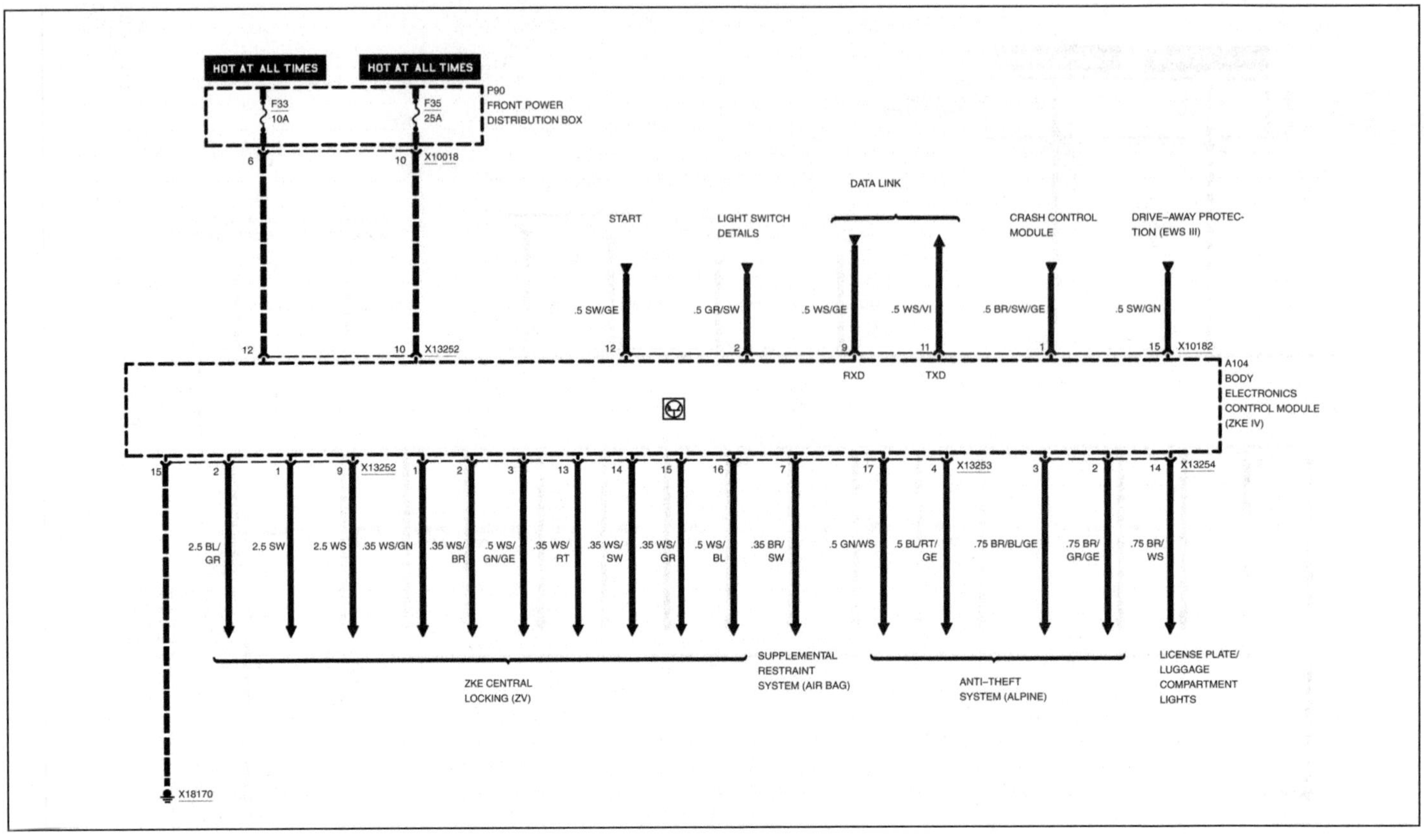


## ZKE IV, 2000
(2 of 2)

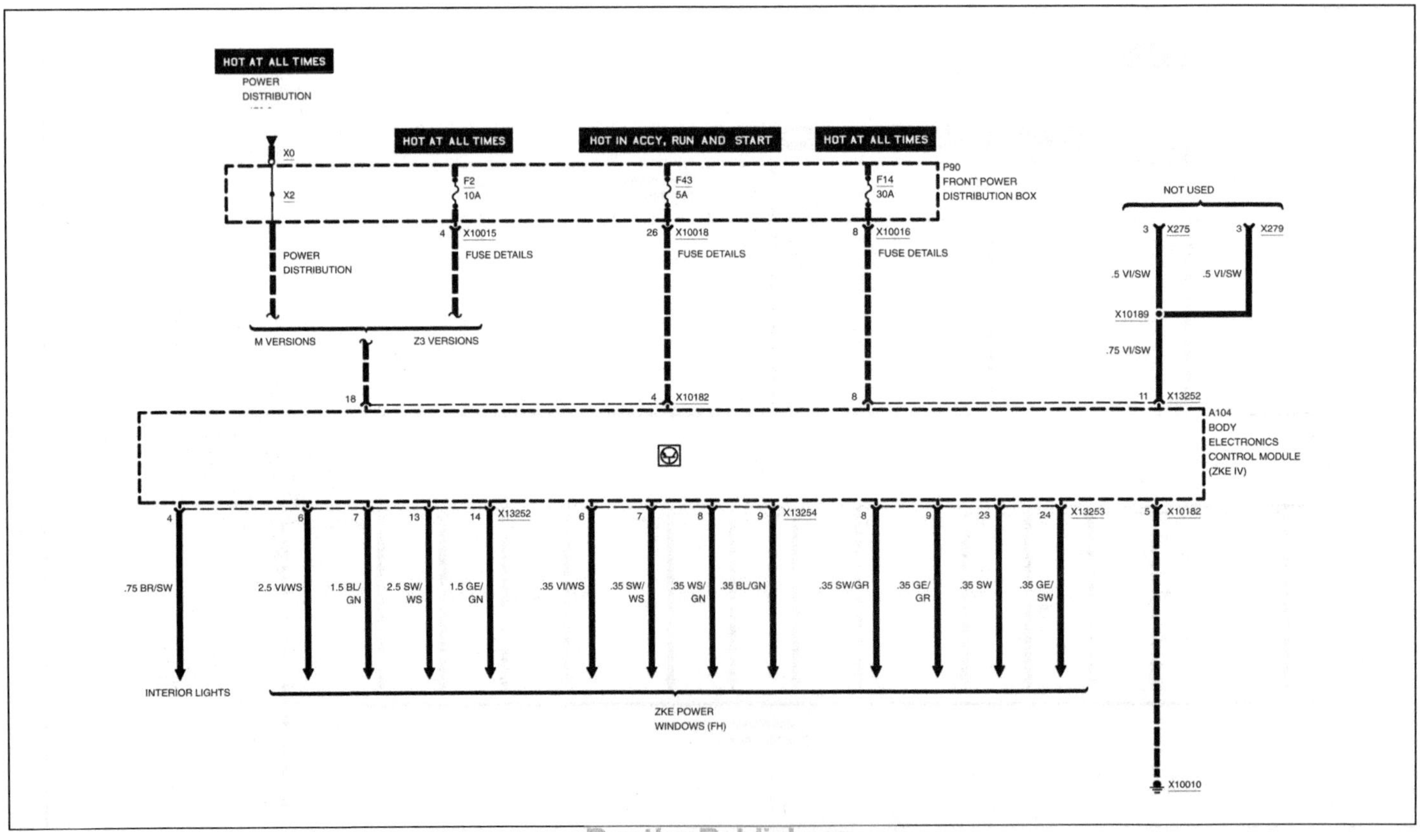

## ZKE IV, 2002
(1 of 2)

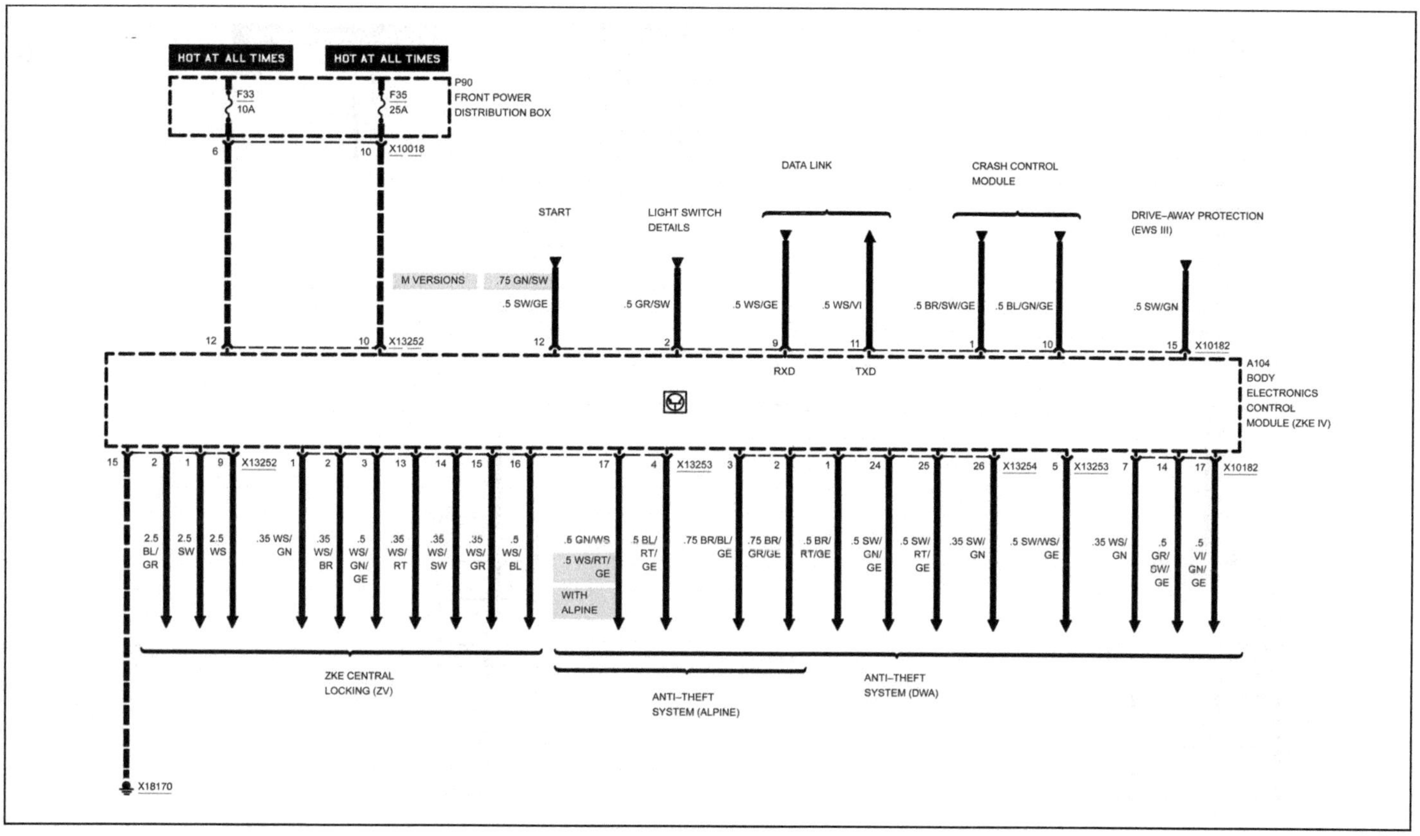


## ZKE IV, 2002
(2 of 2)

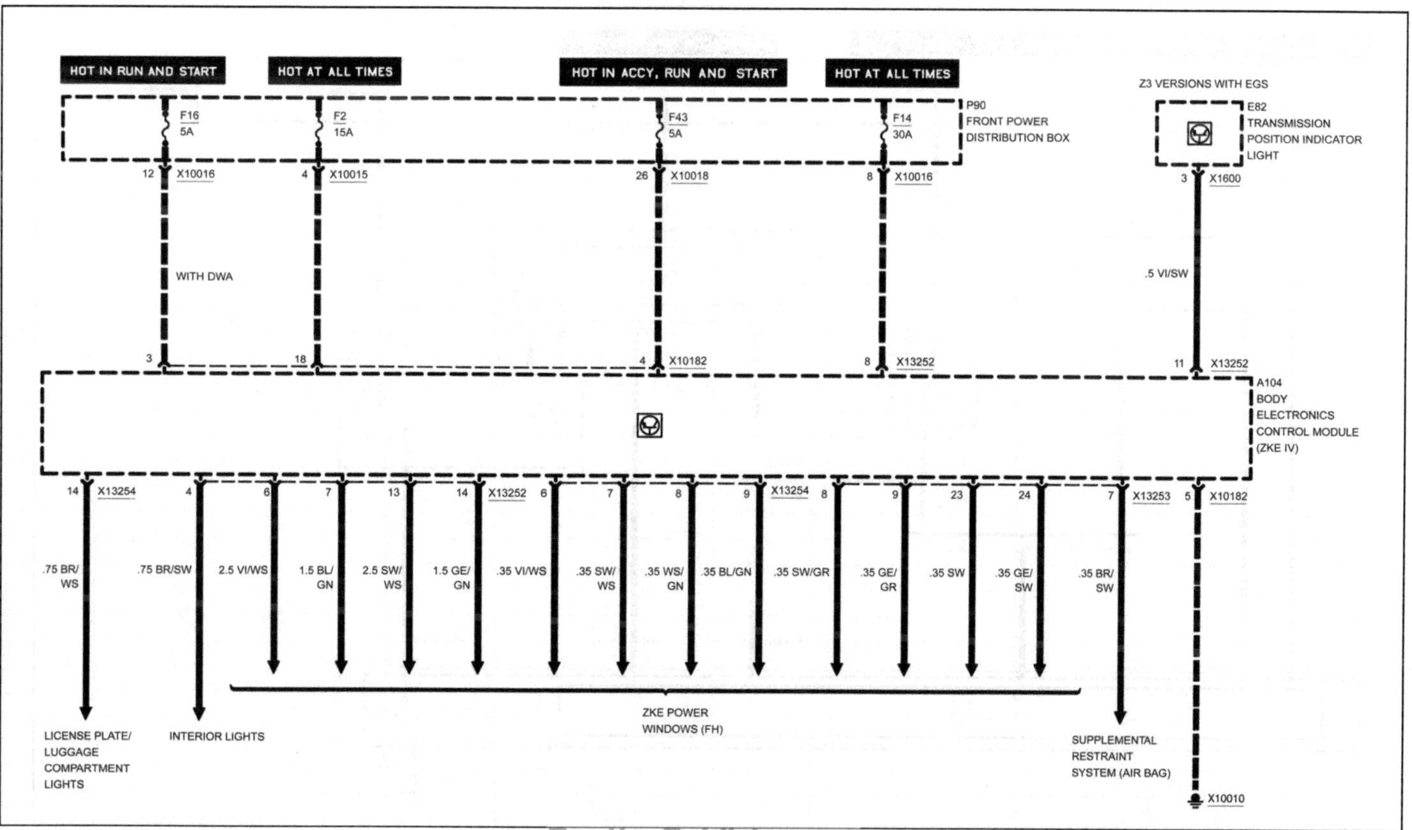

## Horn

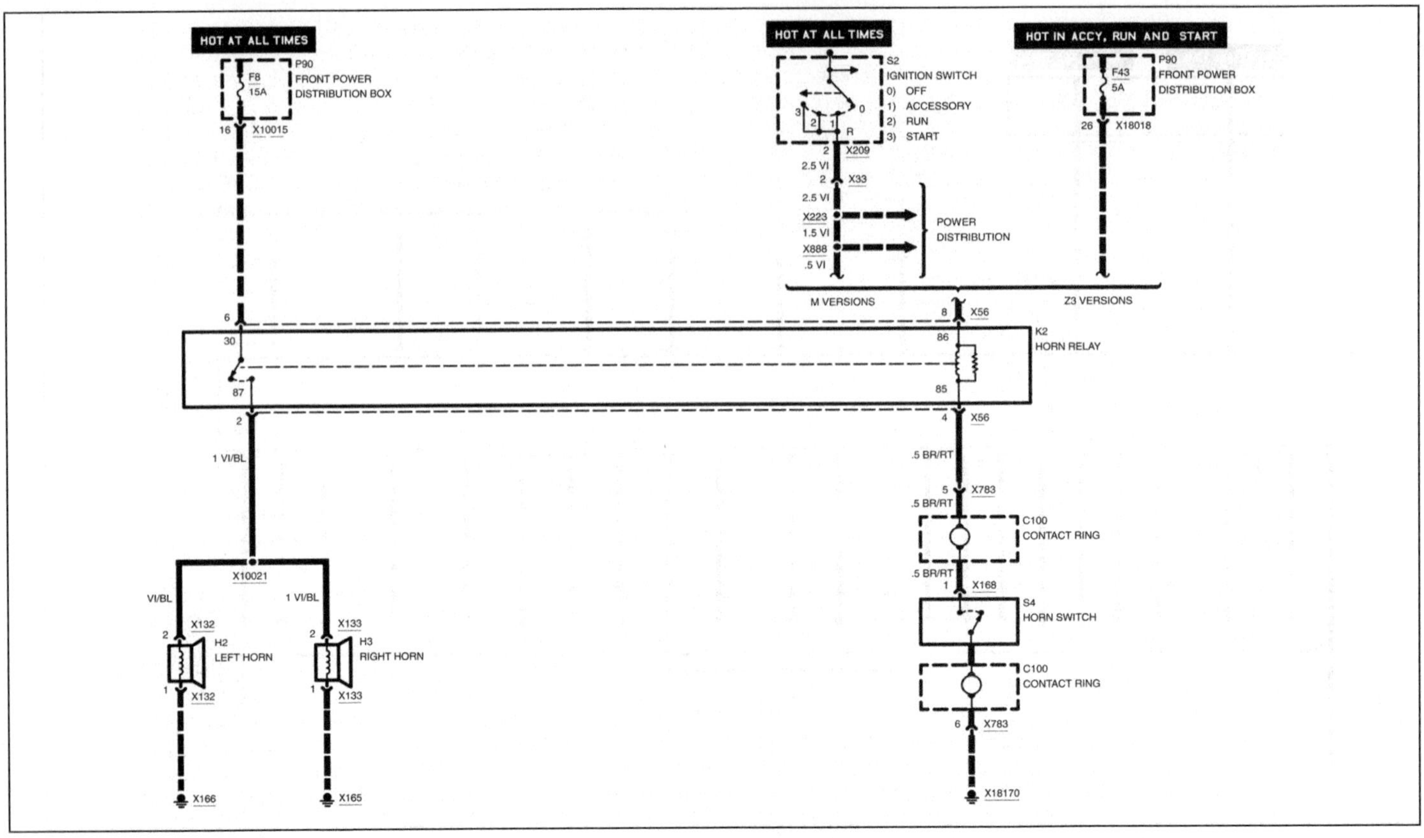


## Electronic immobilization (EWS II) 1996 - 1998

(1 of 2)

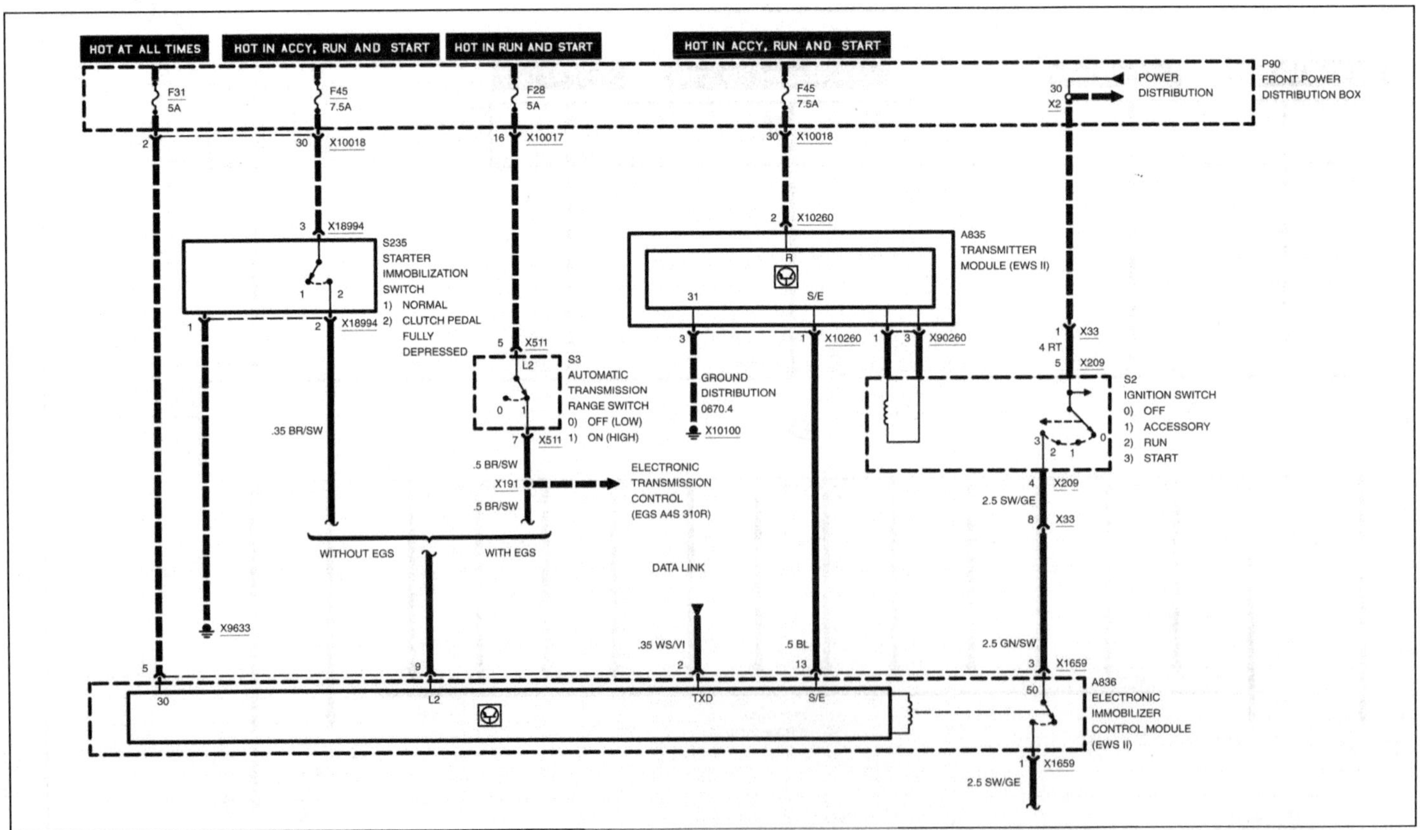

## Electronic immobilization (EWS II) 1996 - 1998

(2 of 2)

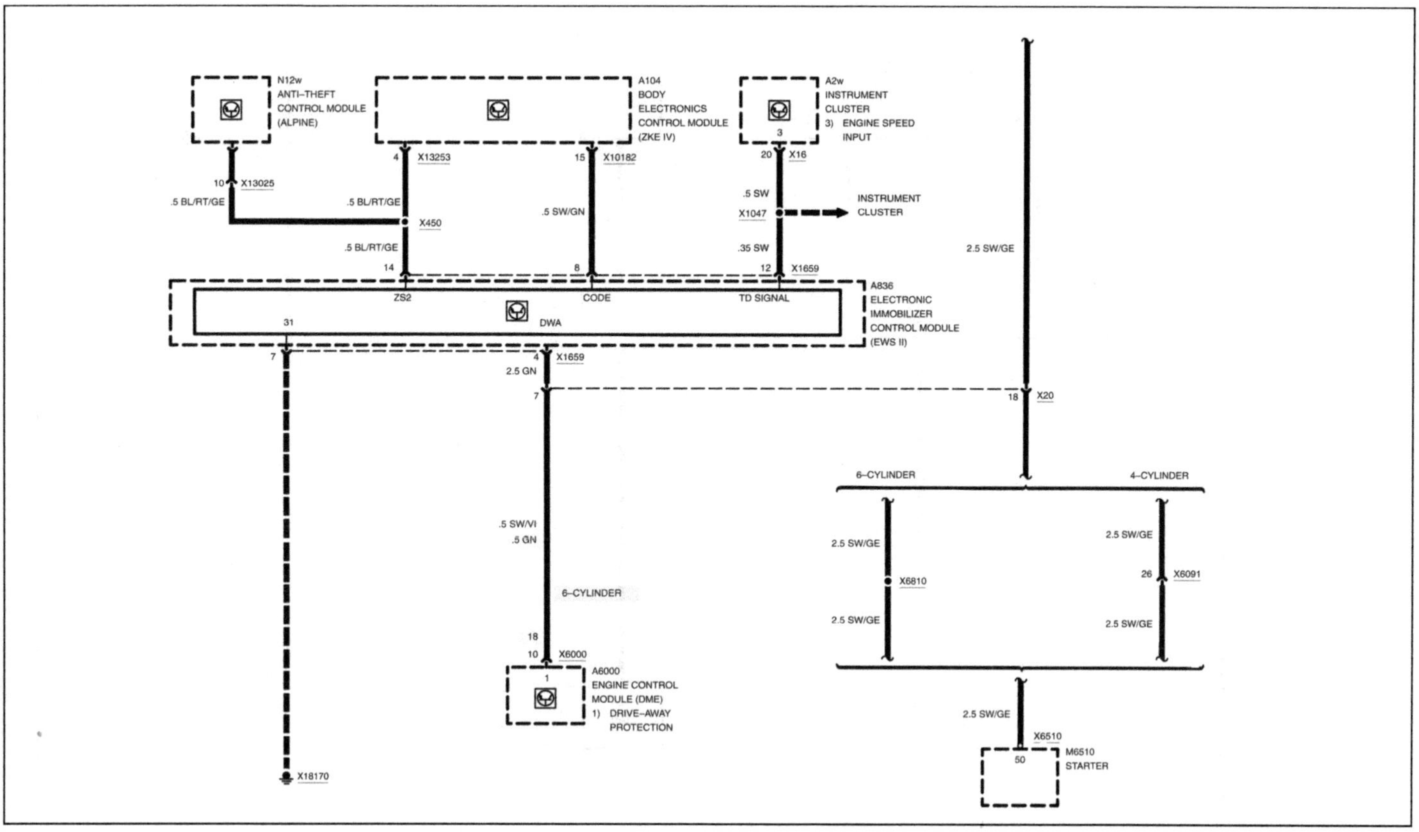


## Electronic immobilization (EWS III) 2000 Z3

(1 of 2

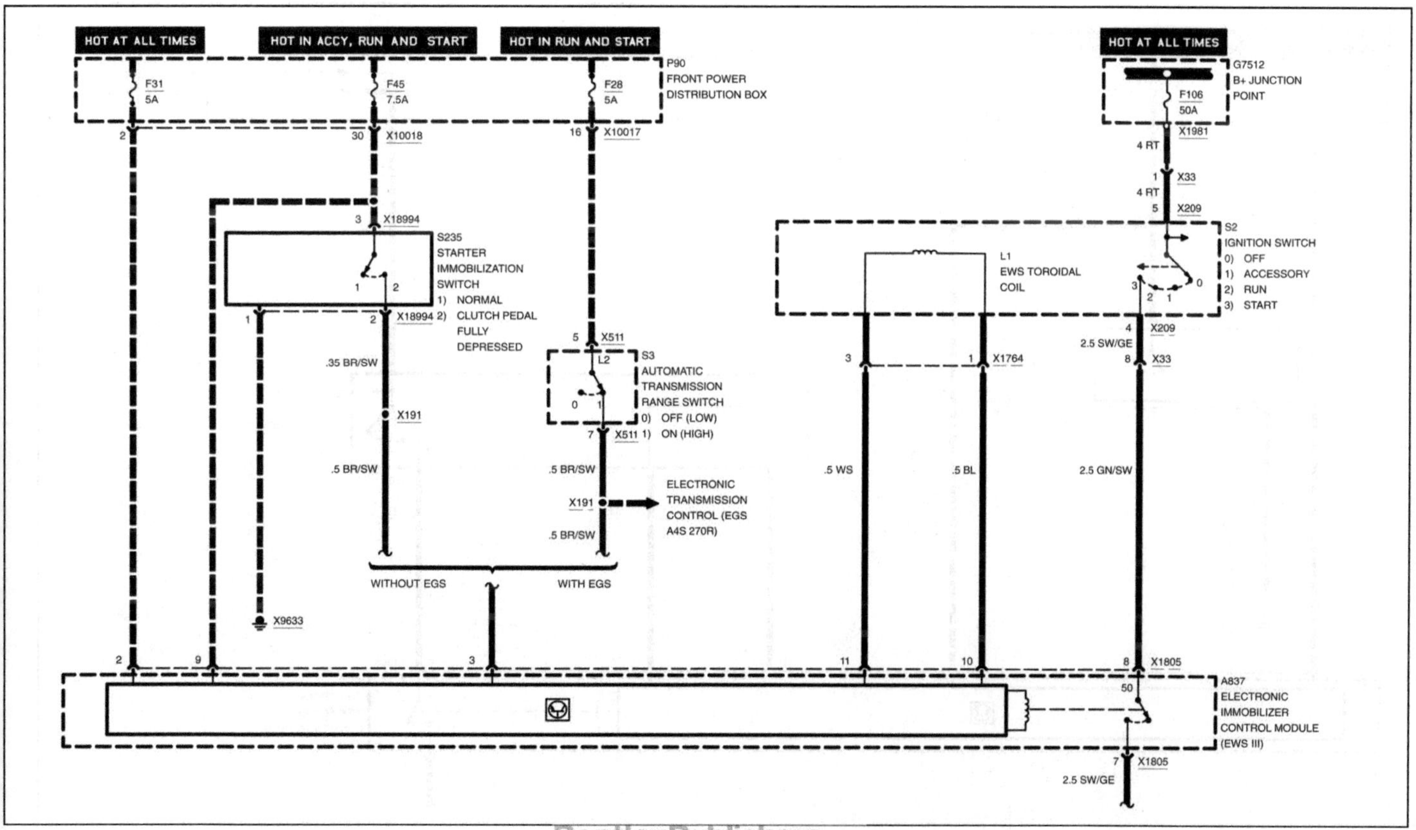

## Electronic immobilization (EWS III) 2000 Z3

(2 of 2)

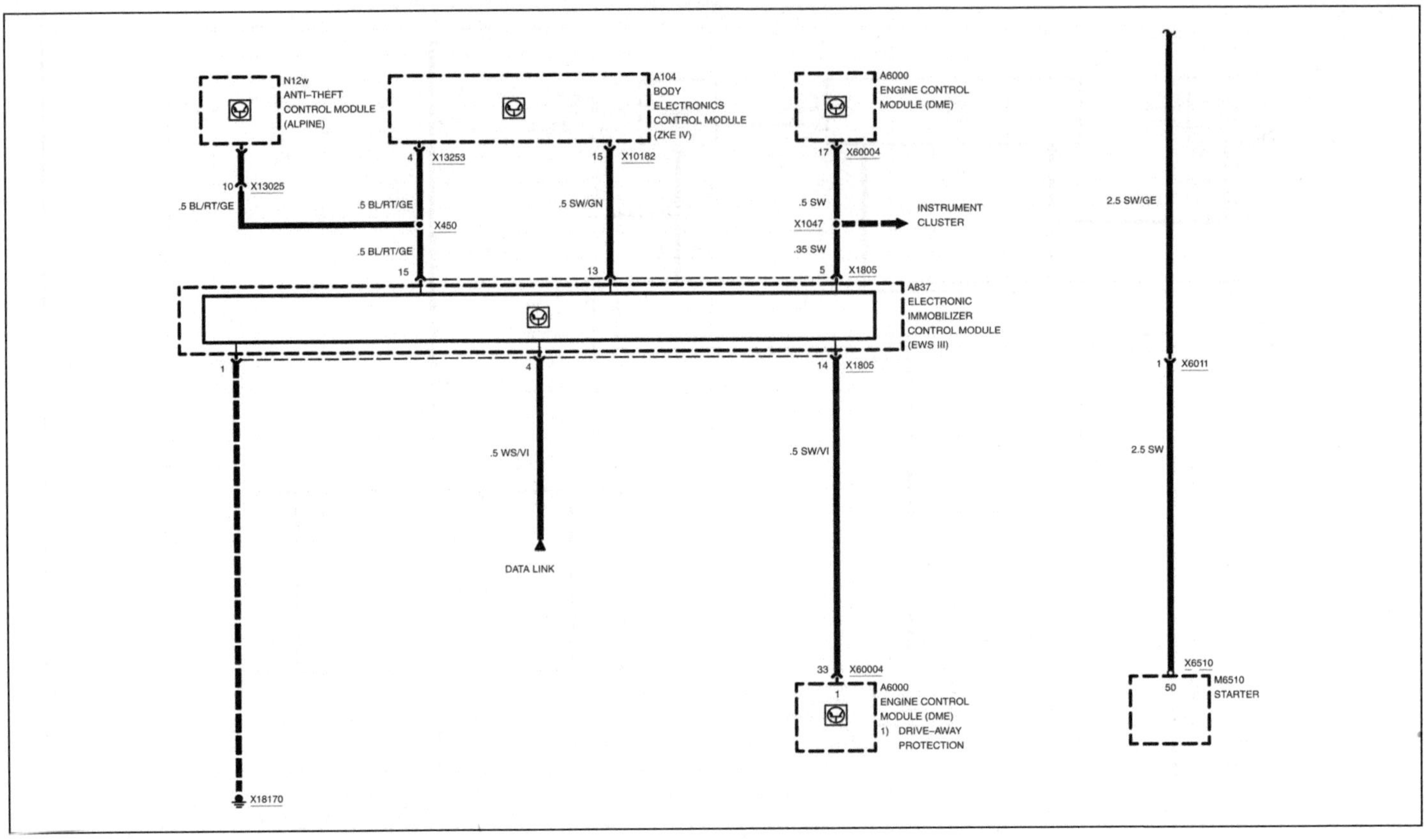


## Electronic immobilization (EWS III) 2000 M

(1 of 2)

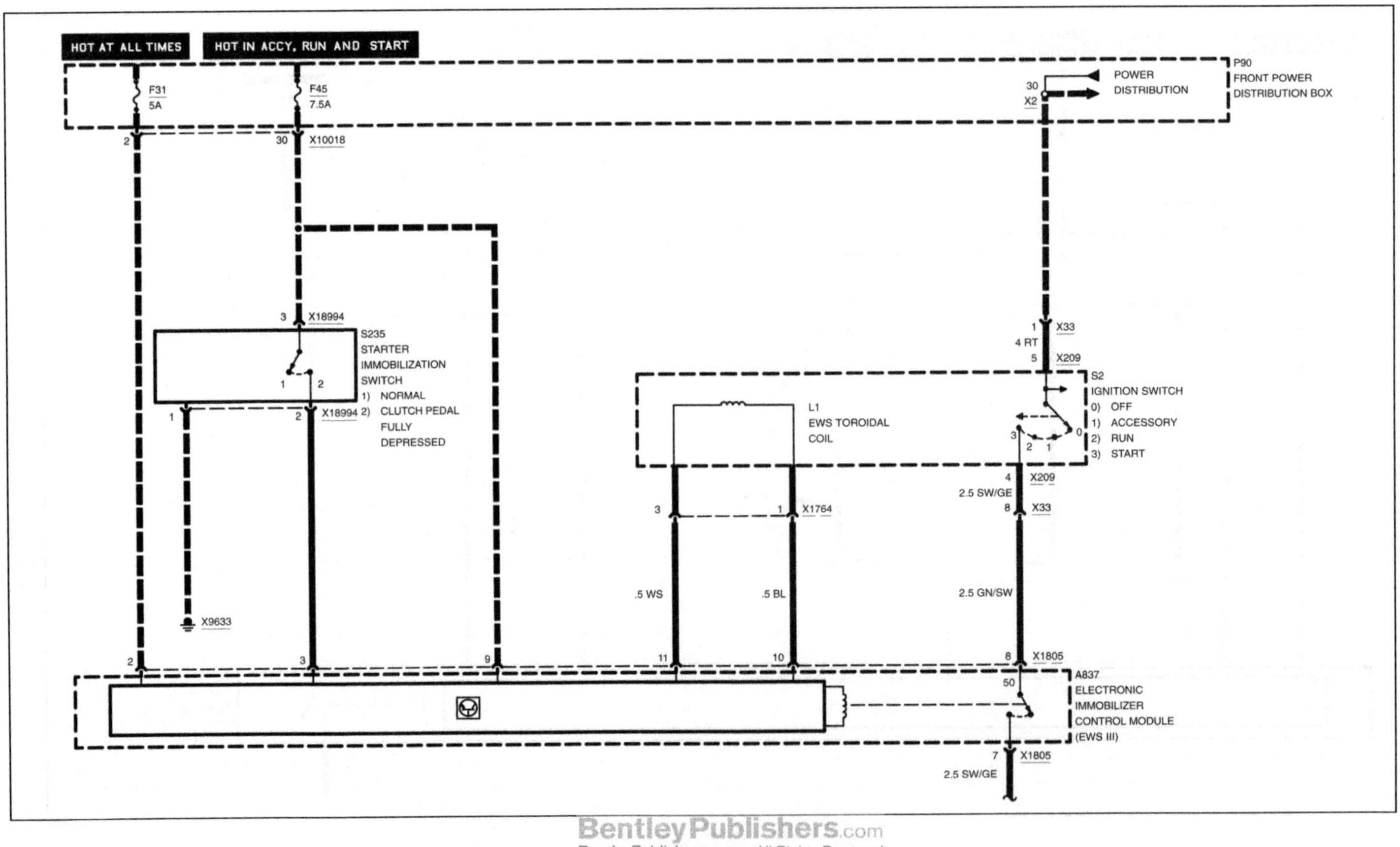

## Electronic immobilization (EWS III) 2000 M

(2 of 2)

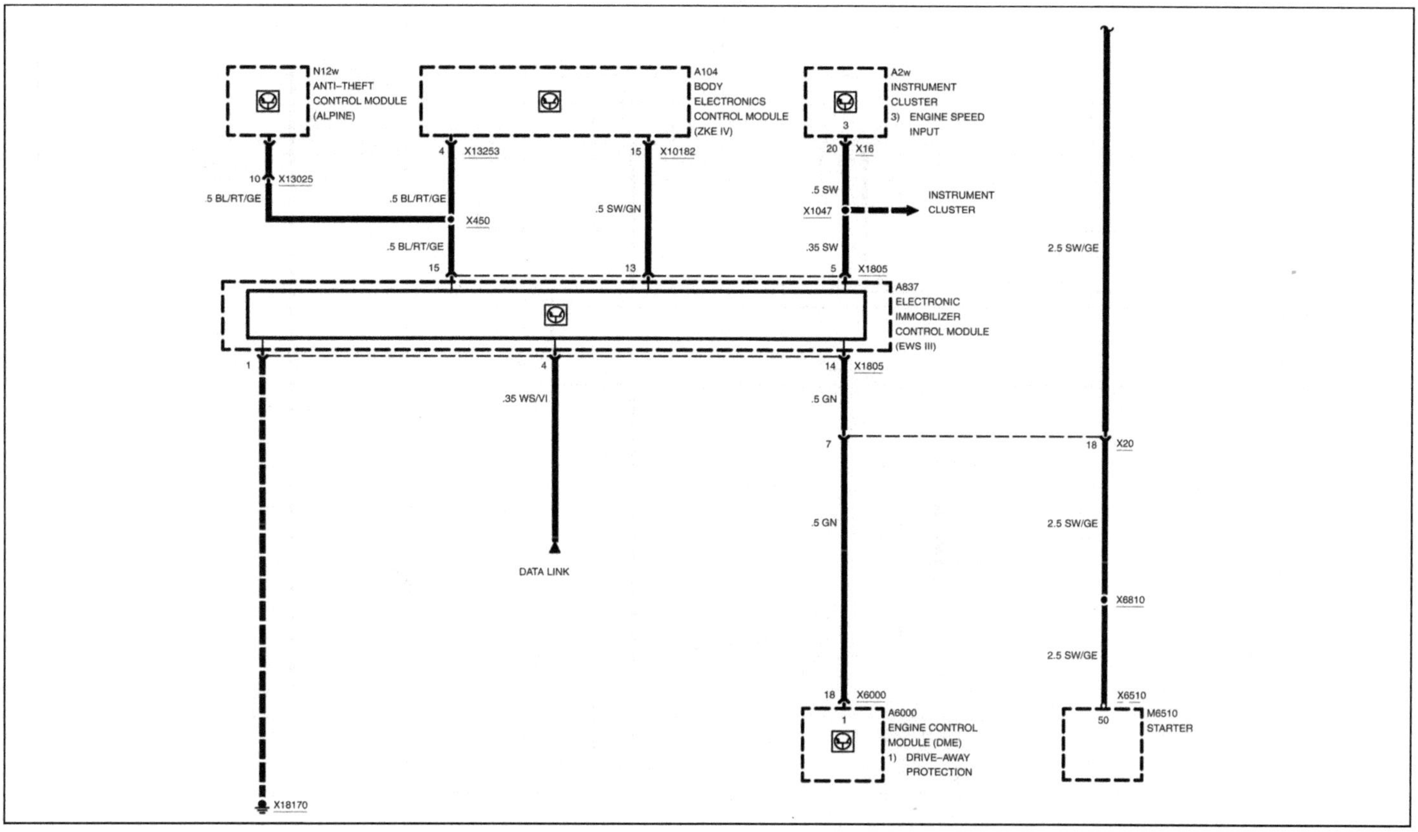


## Electronic immobilization (EWS III) 2002 M

(1 of 2)

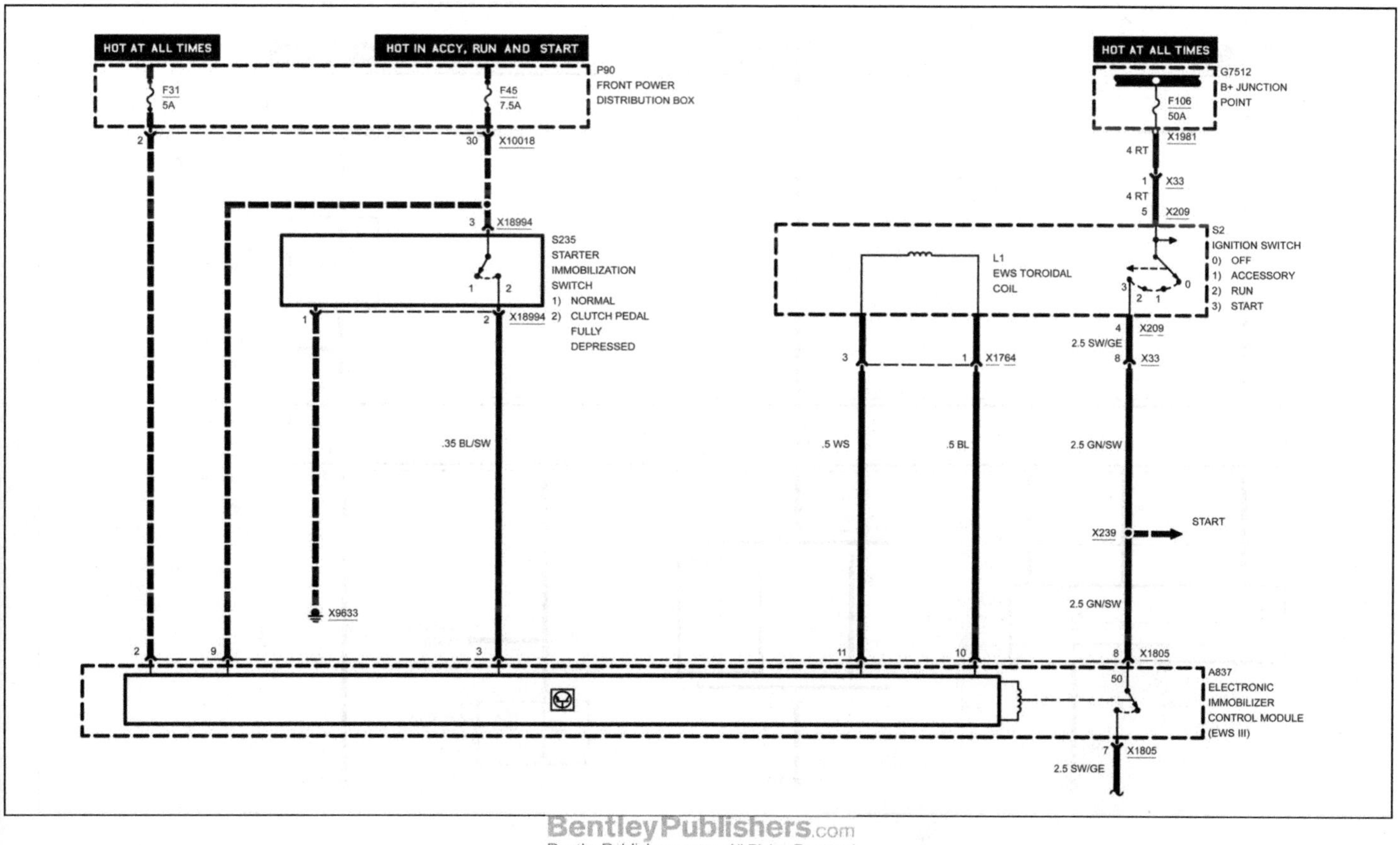

## Electronic immobilization (EWS III) 2002 M

(2 of 2)

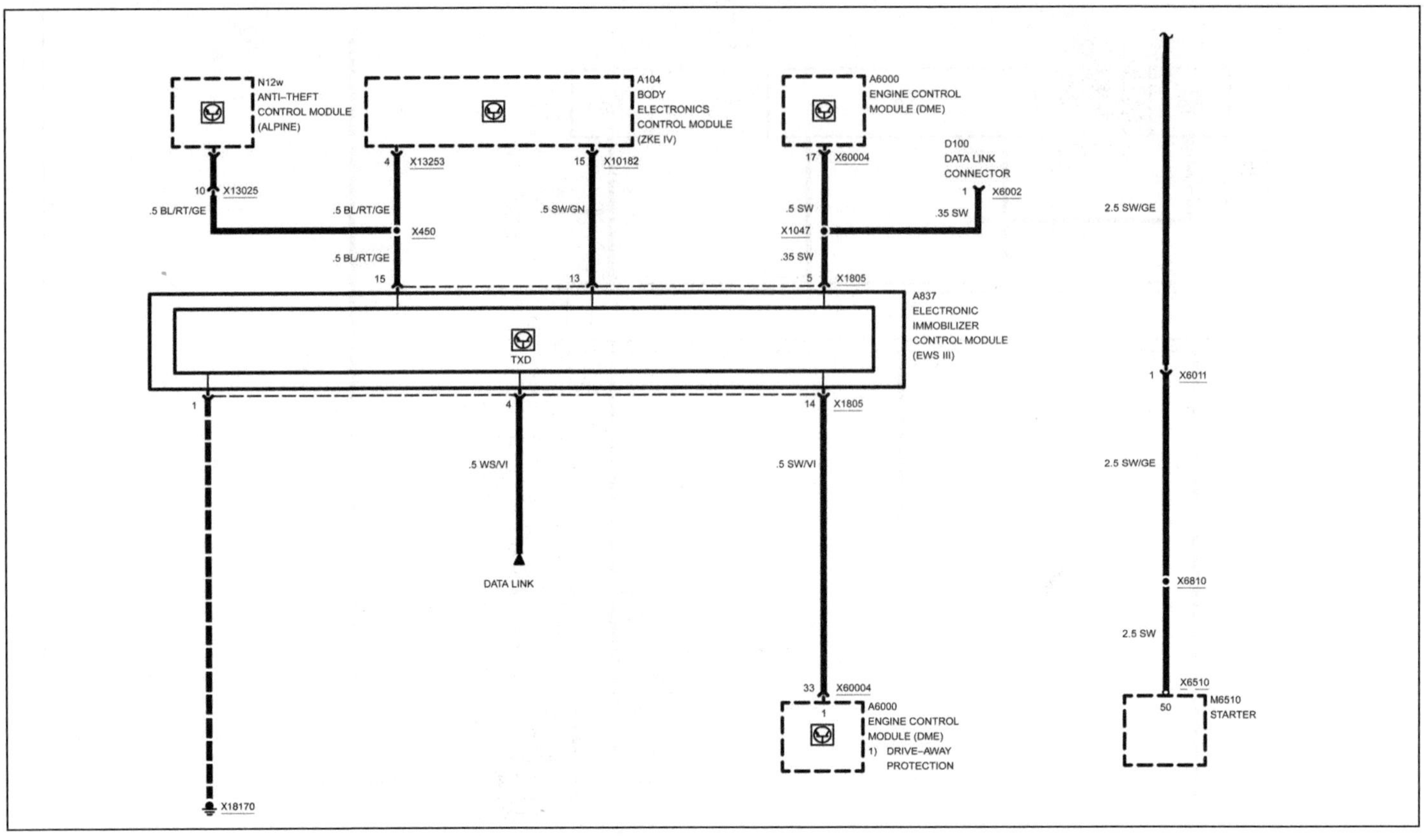


## Wiper / washer, 1996, 1998, 2000 - 2002 Roadster

(1 of 2)

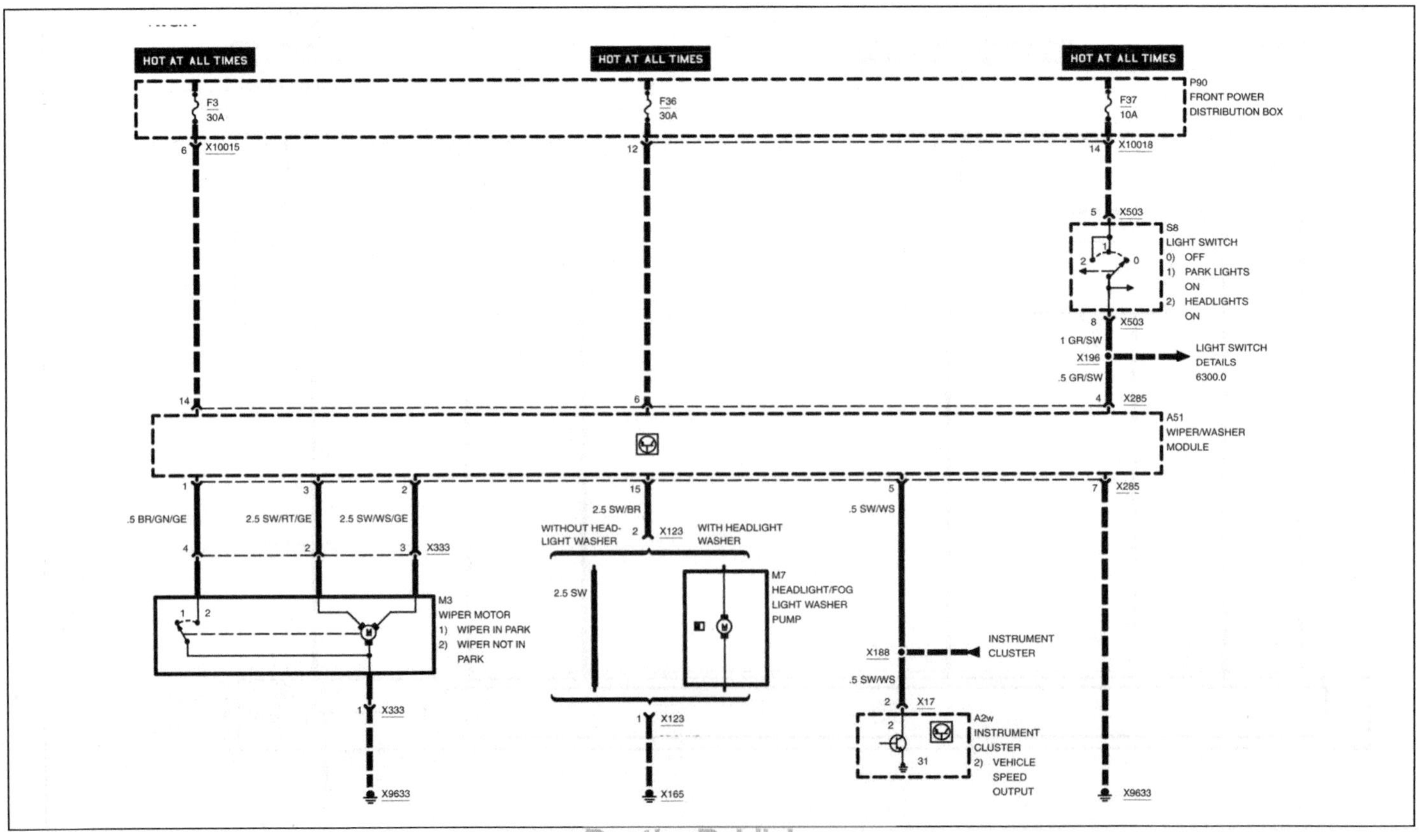

## Wiper / washer, 1996, 1998, 2000 - 2002 Roadster
(2 of 2)

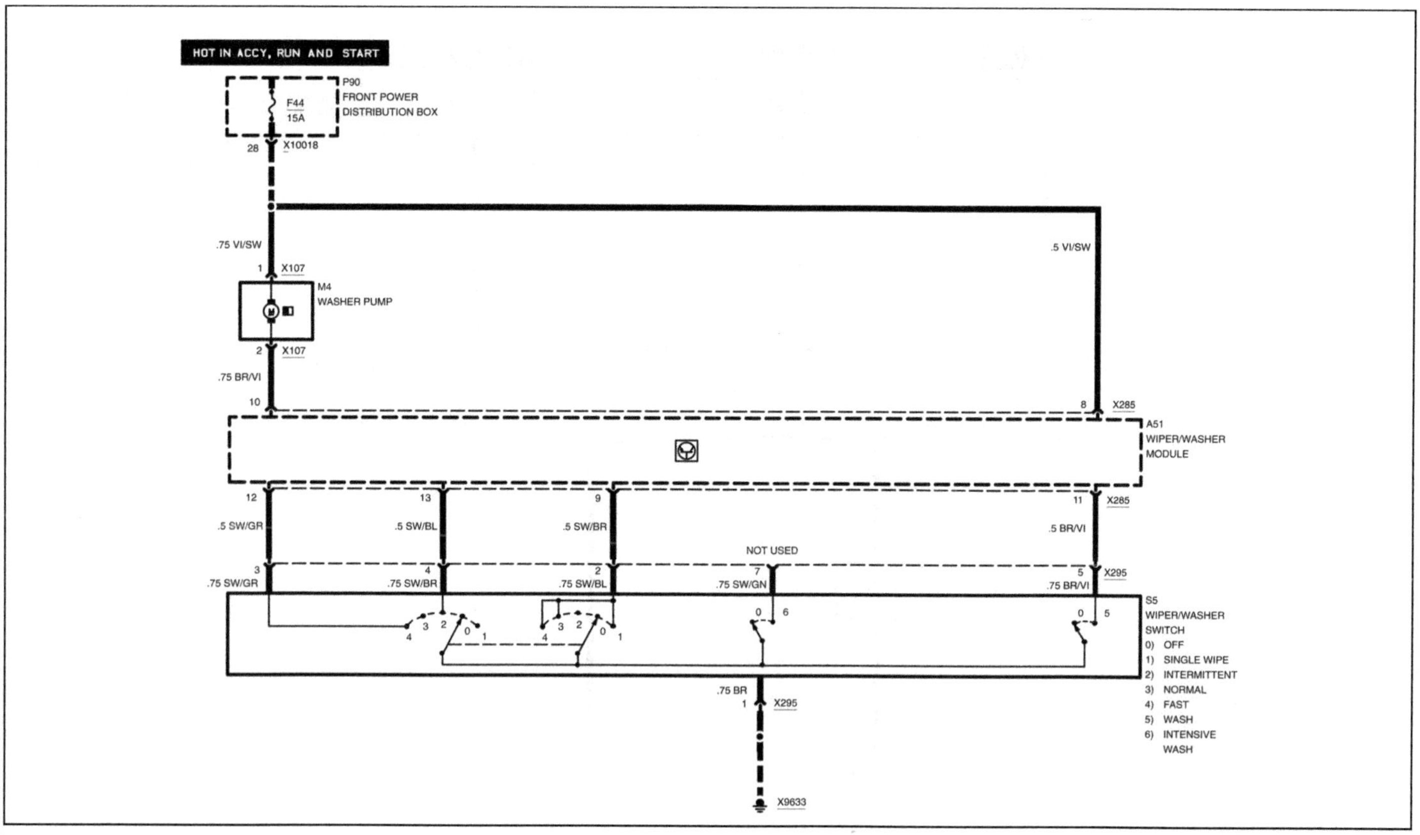


## Wiper / washer, high, 2000 - 2002 Coupe

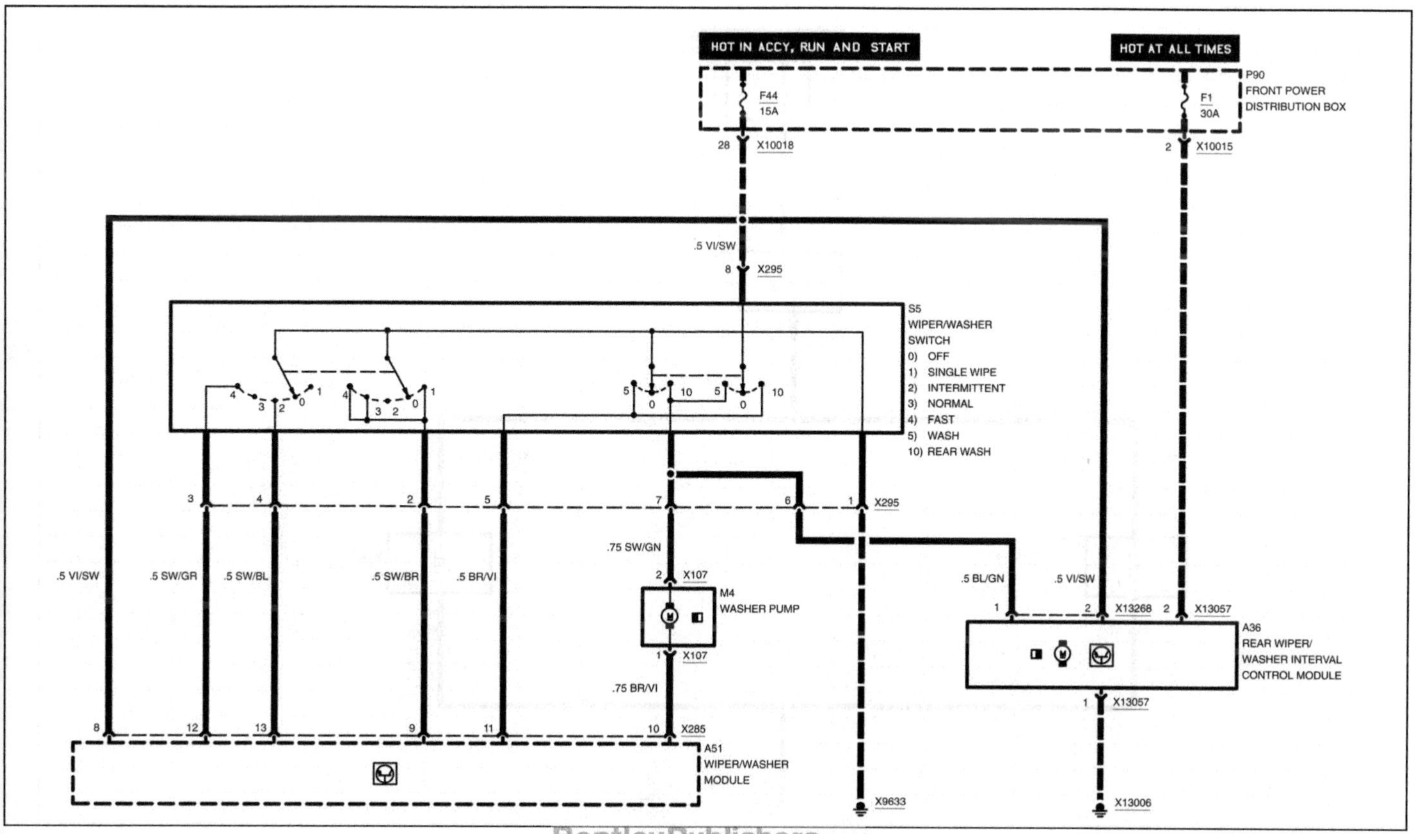

## Headlight washer

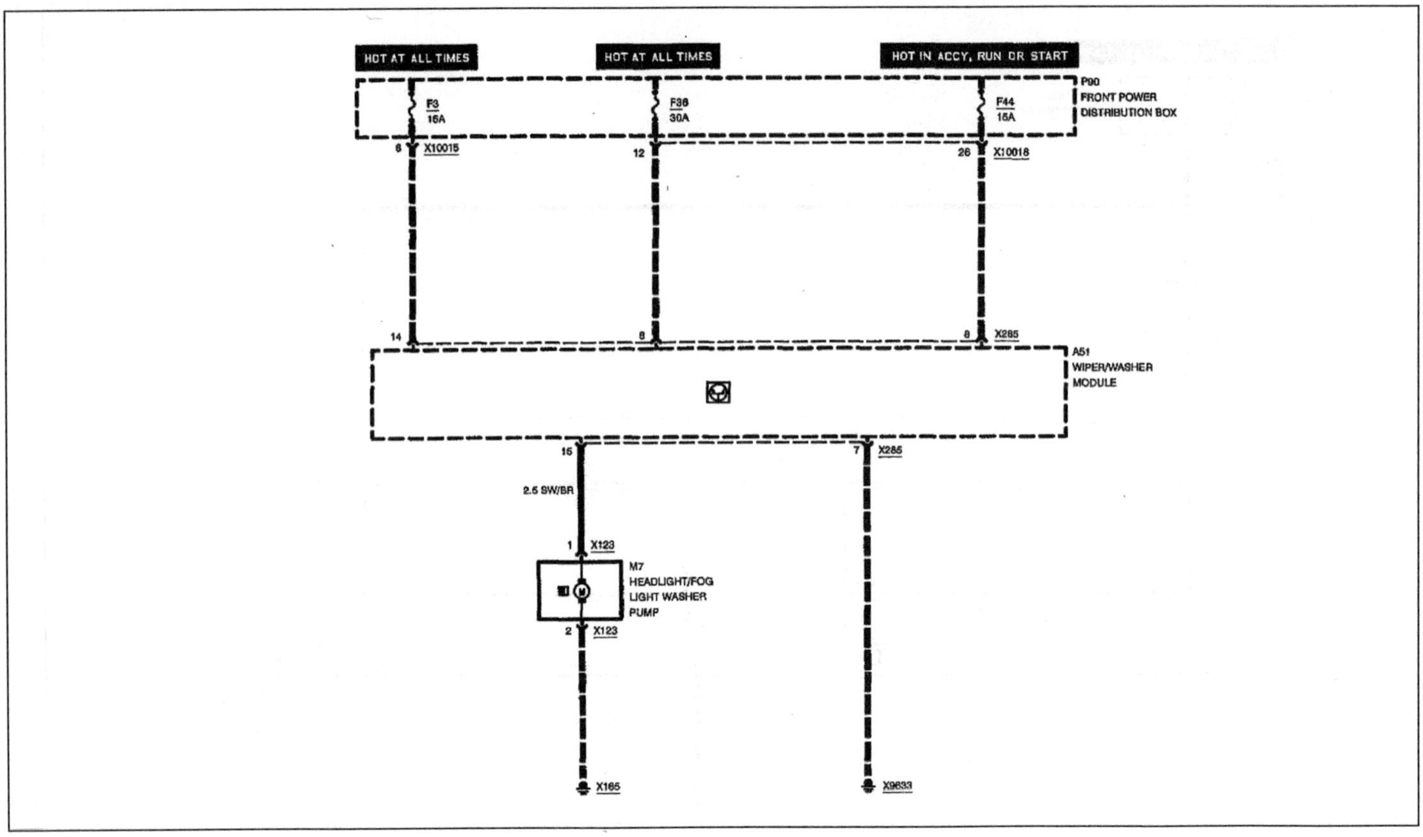


## Windshield washer jet heaters

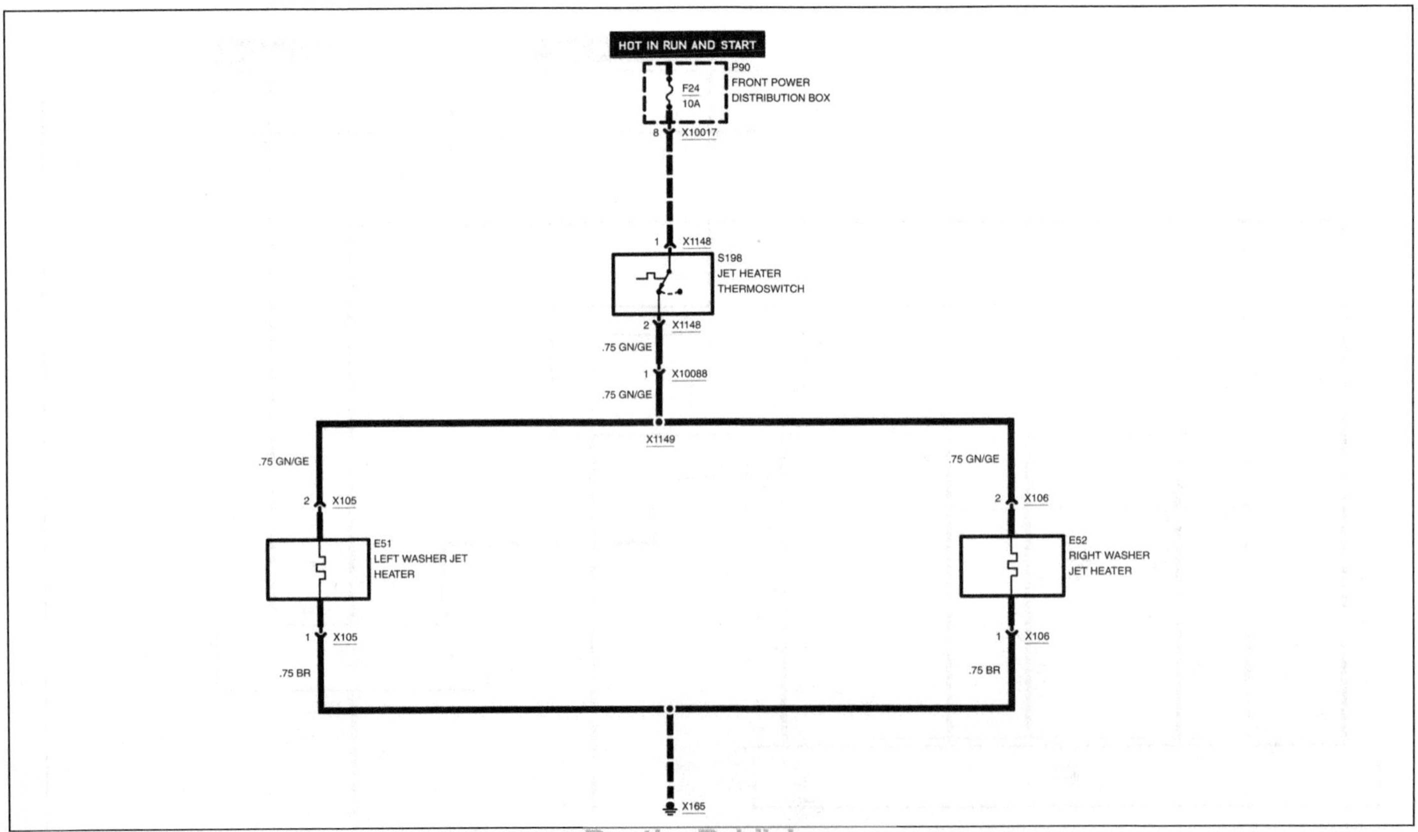

## Instrument cluster, 1996
(1 of 6)

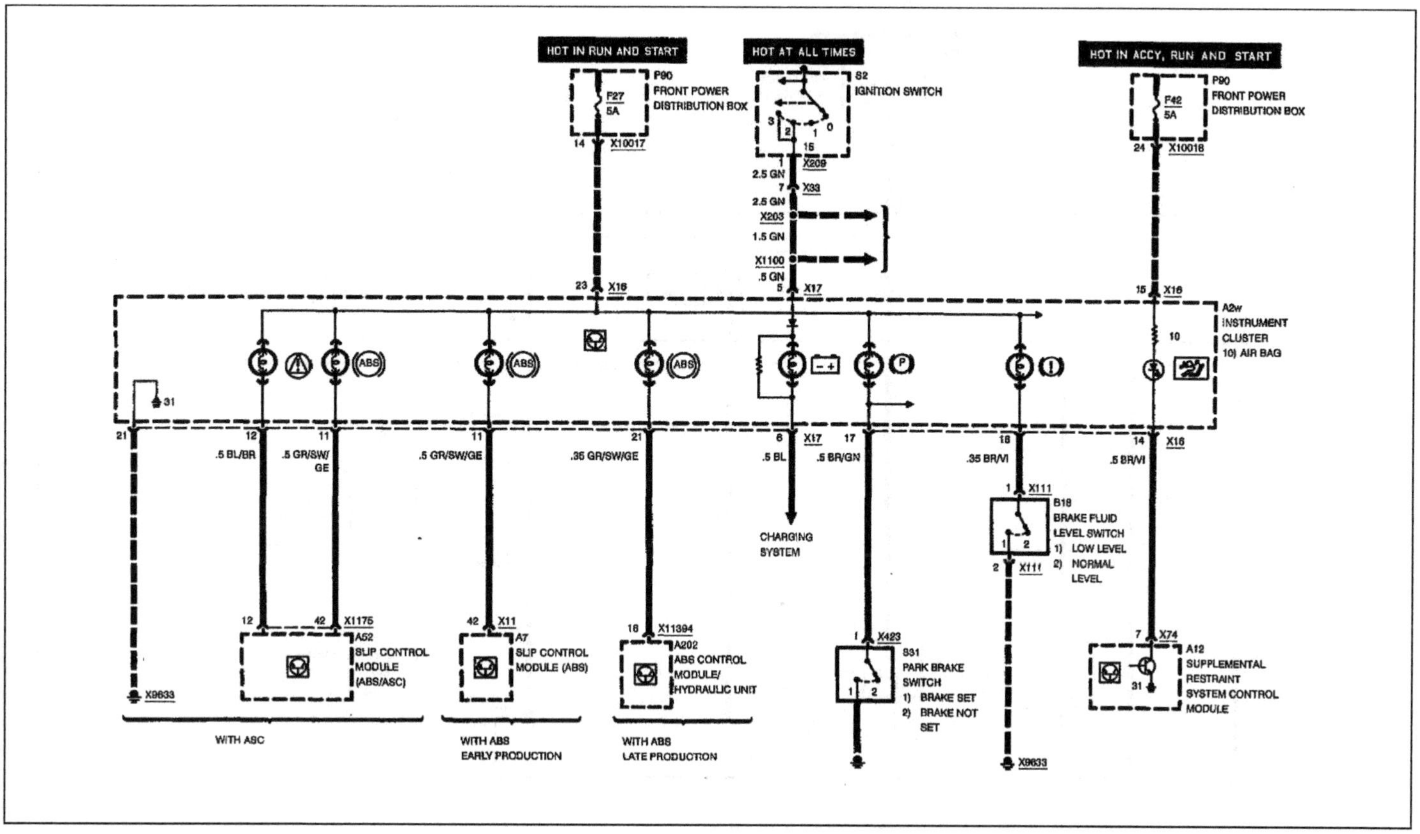


## Instrument cluster, 1996
(2 of 6)

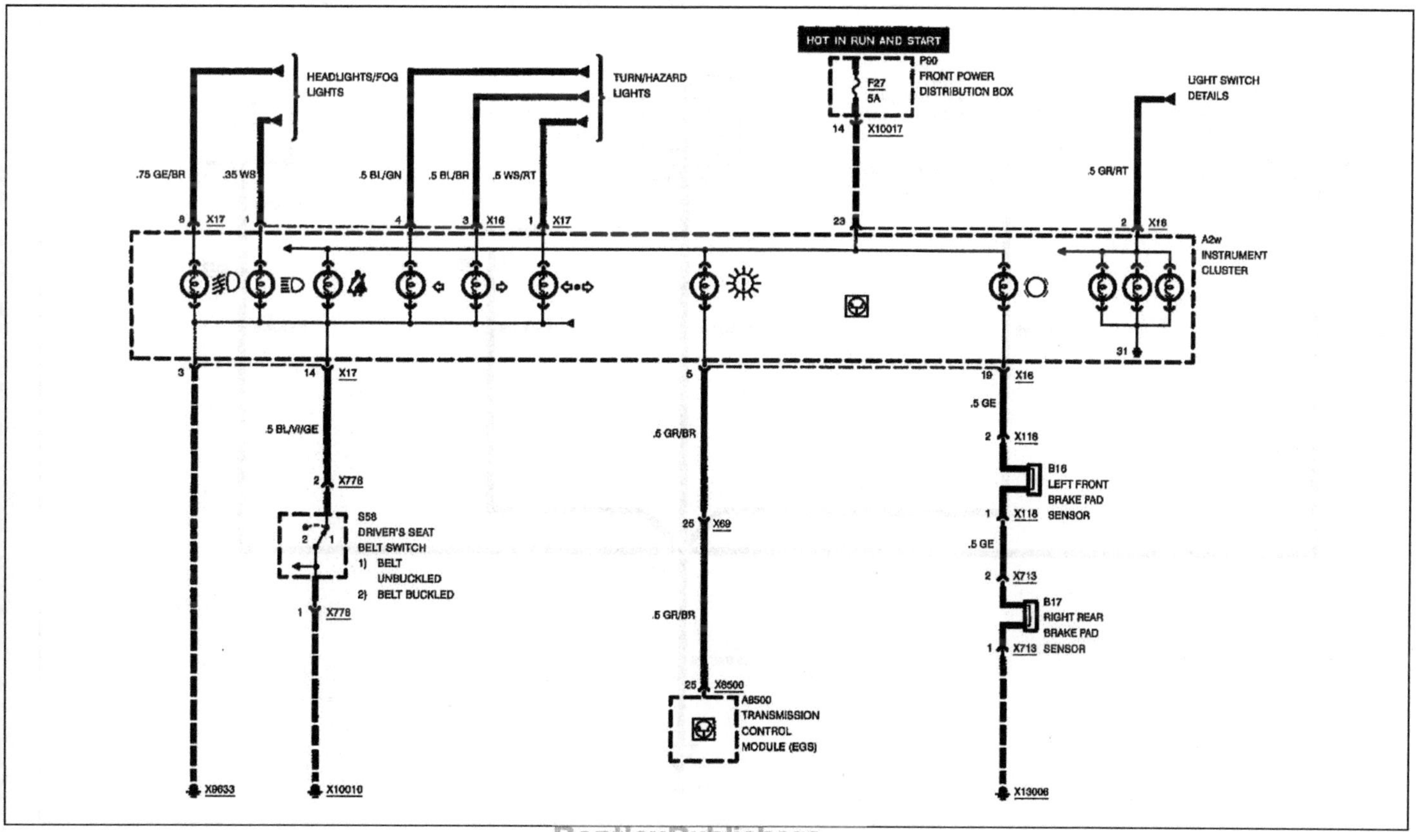

## Instrument cluster, 1996
(3 of 6)

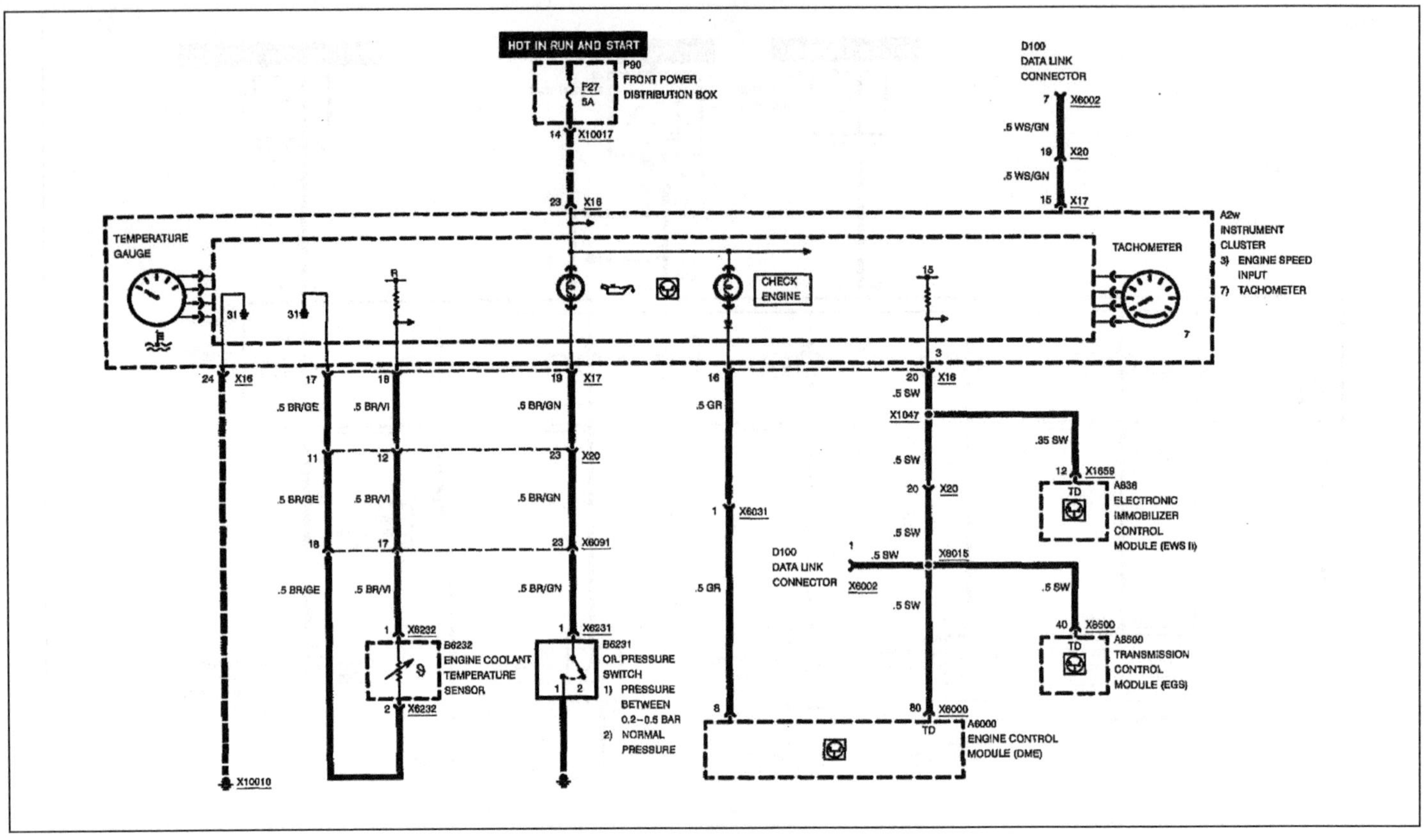


## Instrument cluster, 1996
(4 of 6)

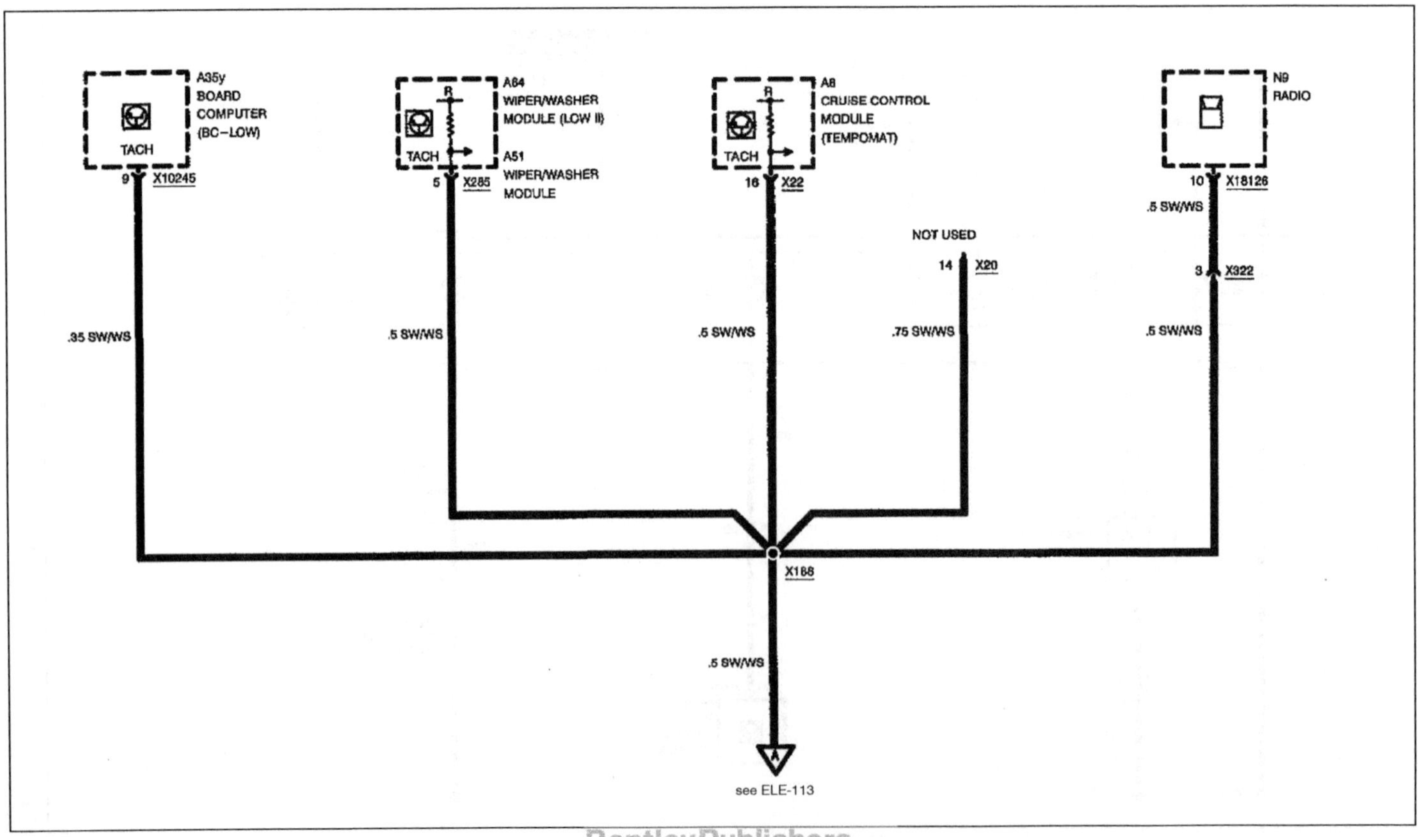

see ELE-113

## Instrument cluster, 1996
(5 of 6)

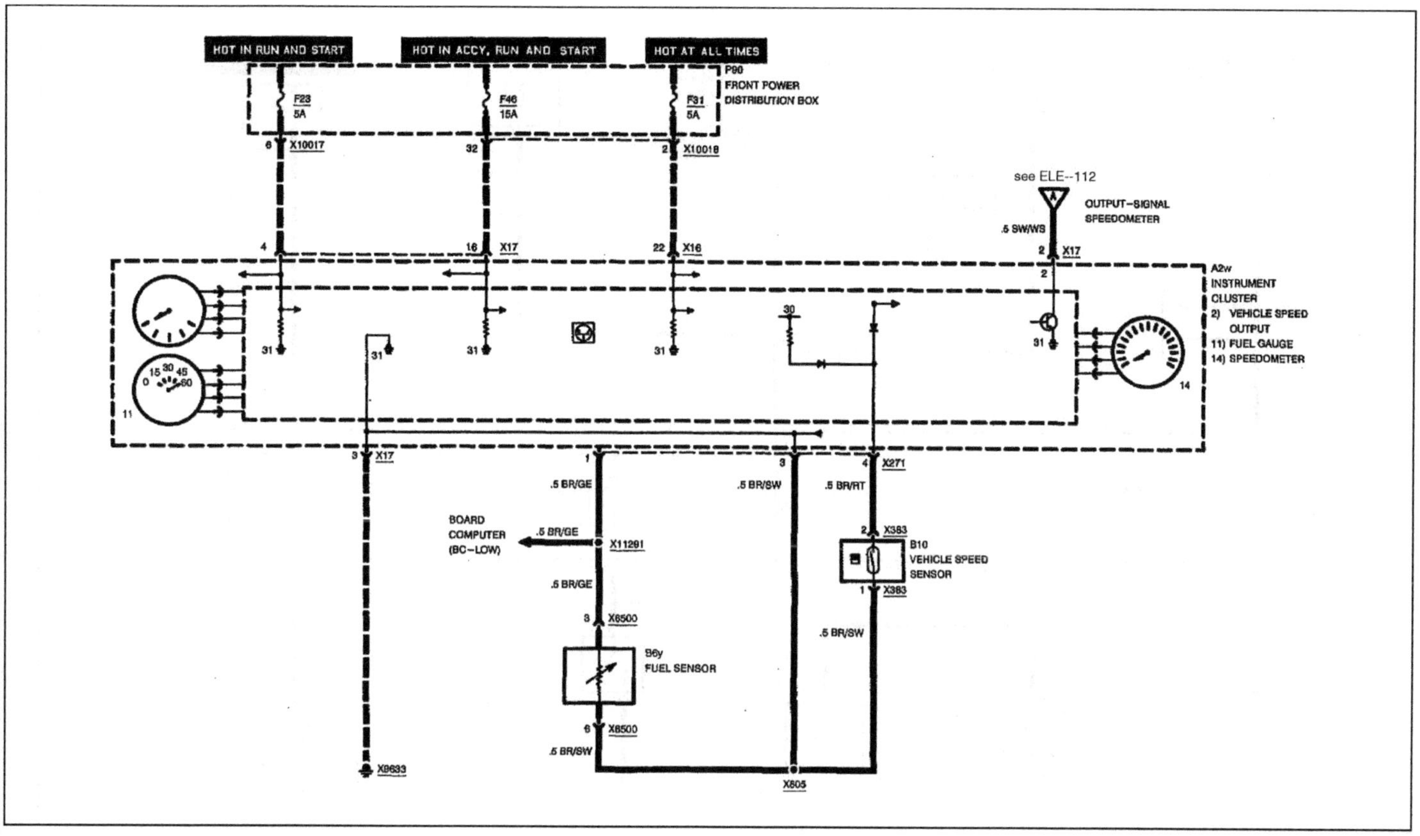


## Instrument cluster, 1996
(6 of 6)

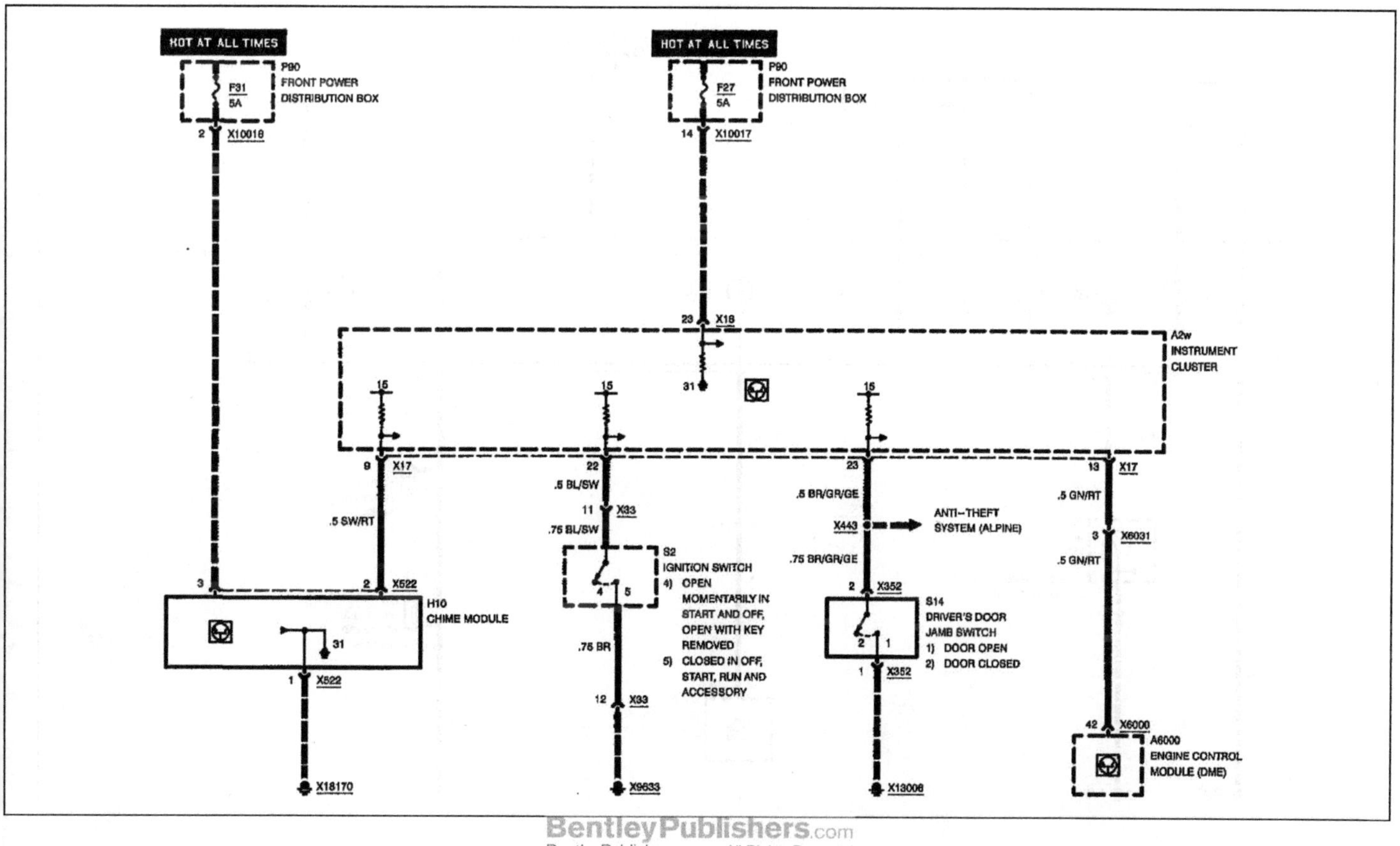

## Instrument cluster, 1998
(1 of 8)

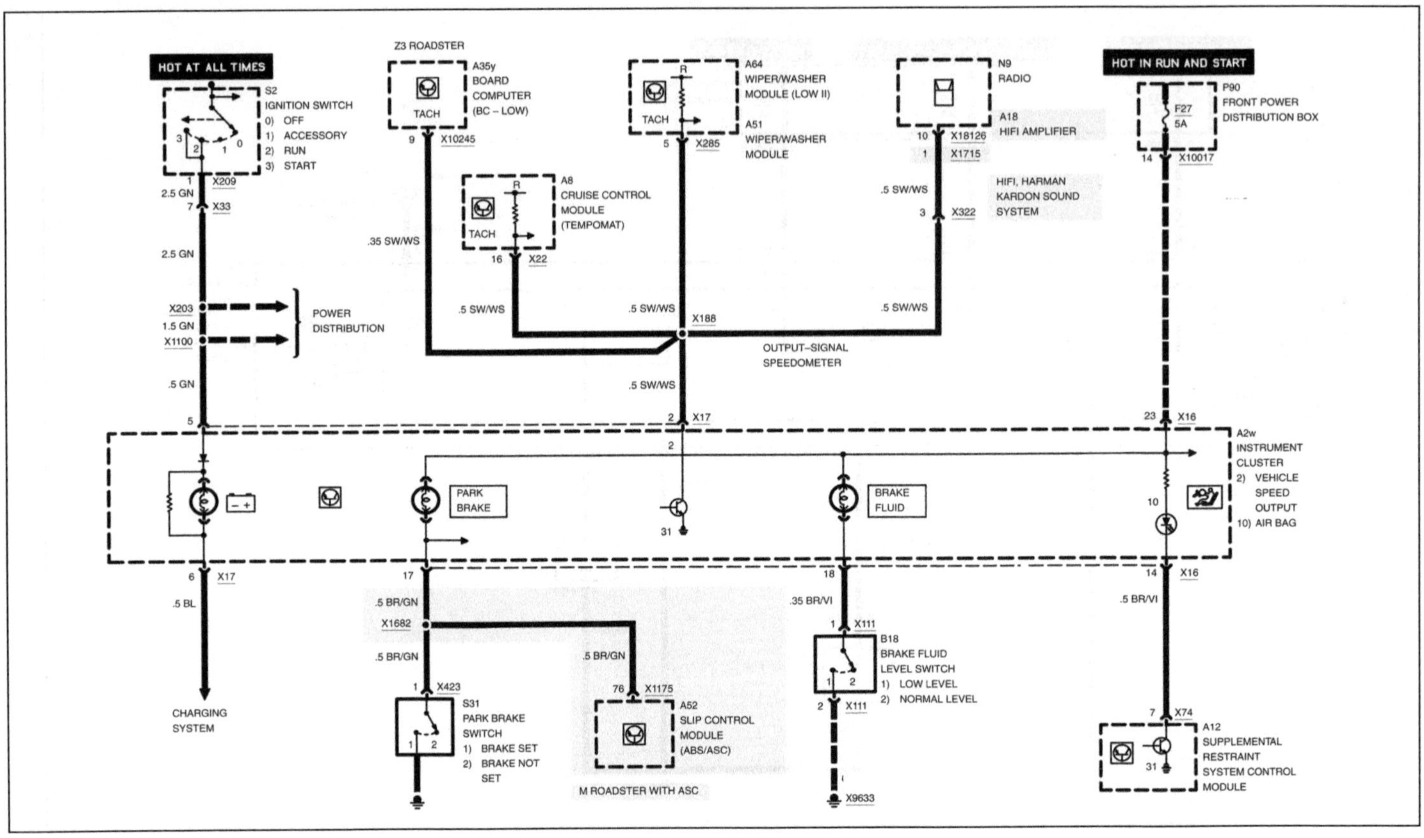


## Instrument cluster, 1998
(2 of 8)

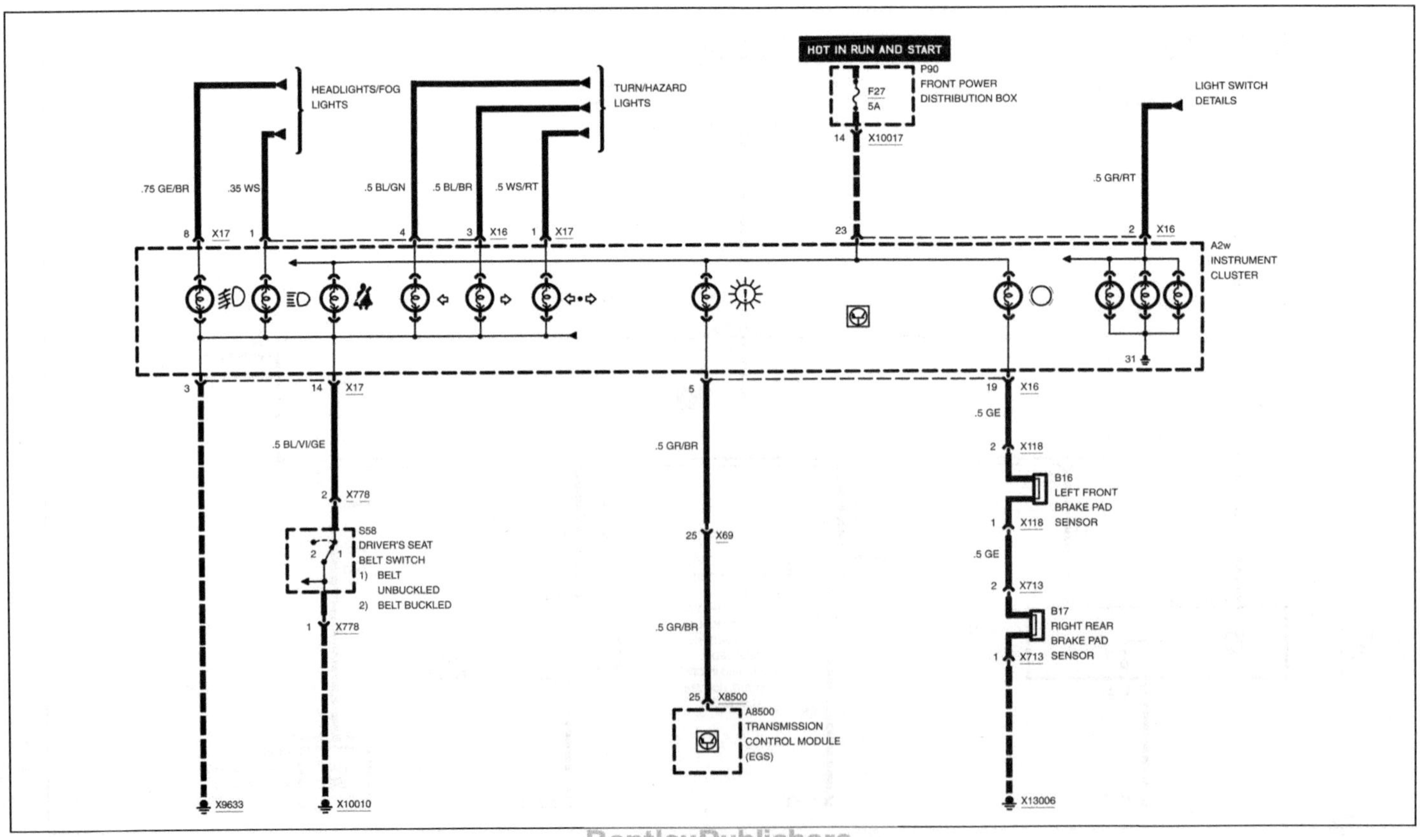

## Instrument cluster, 1998

**(3 of 8)**

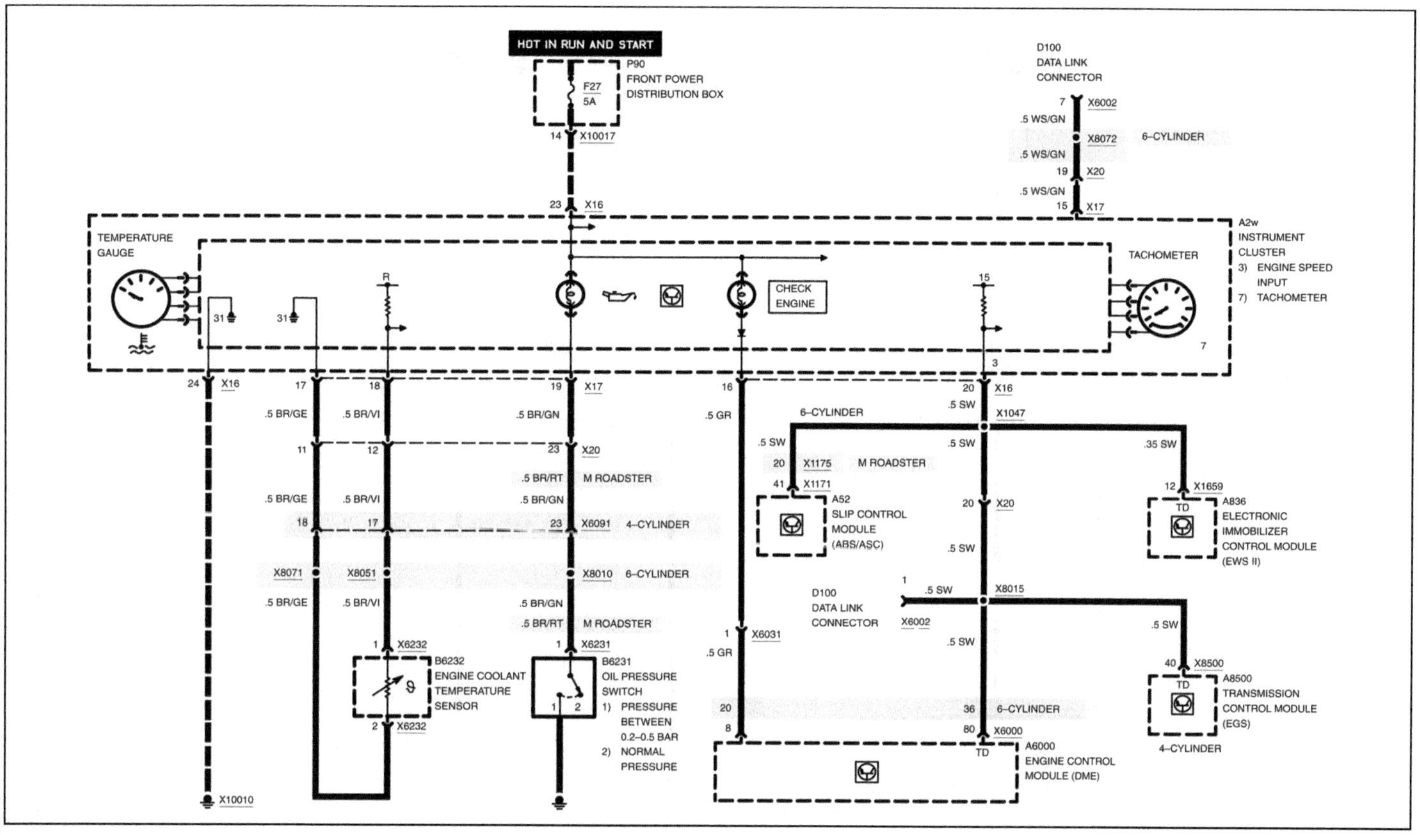


## Instrument cluster, 1998

**(4 of 8)**

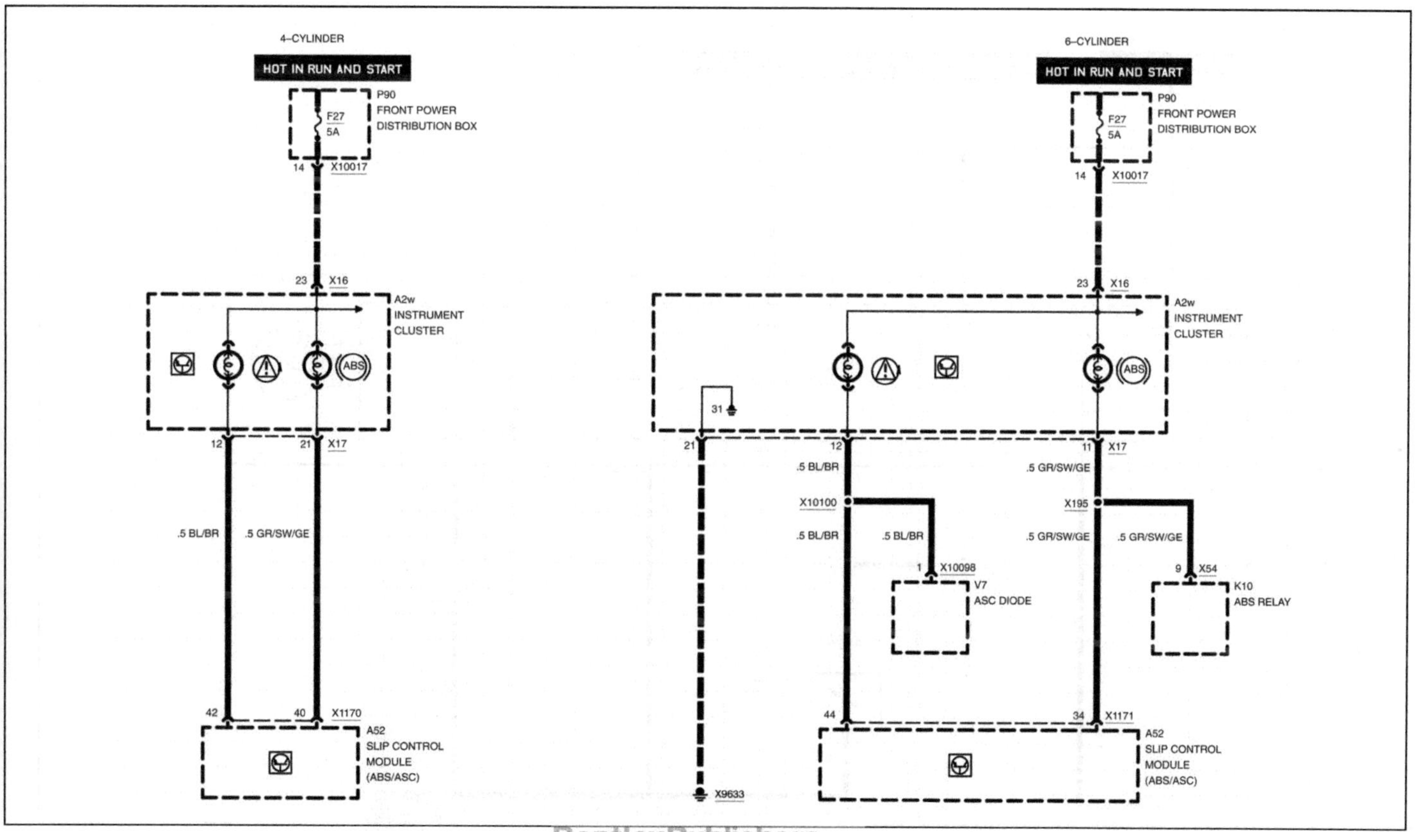

## Instrument cluster, 1998
(5 of 8)

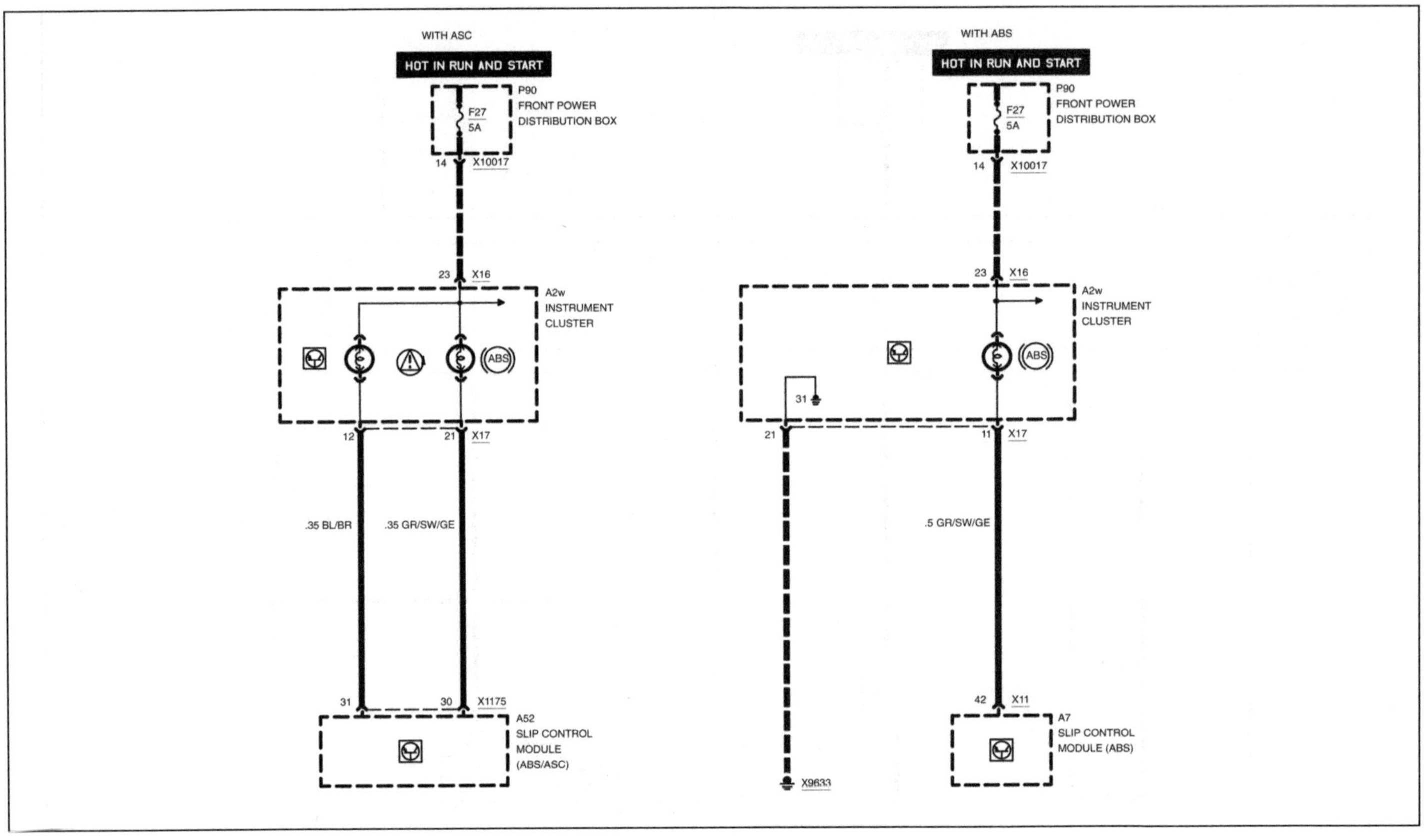


## Instrument cluster, 1998
(6 of 8)

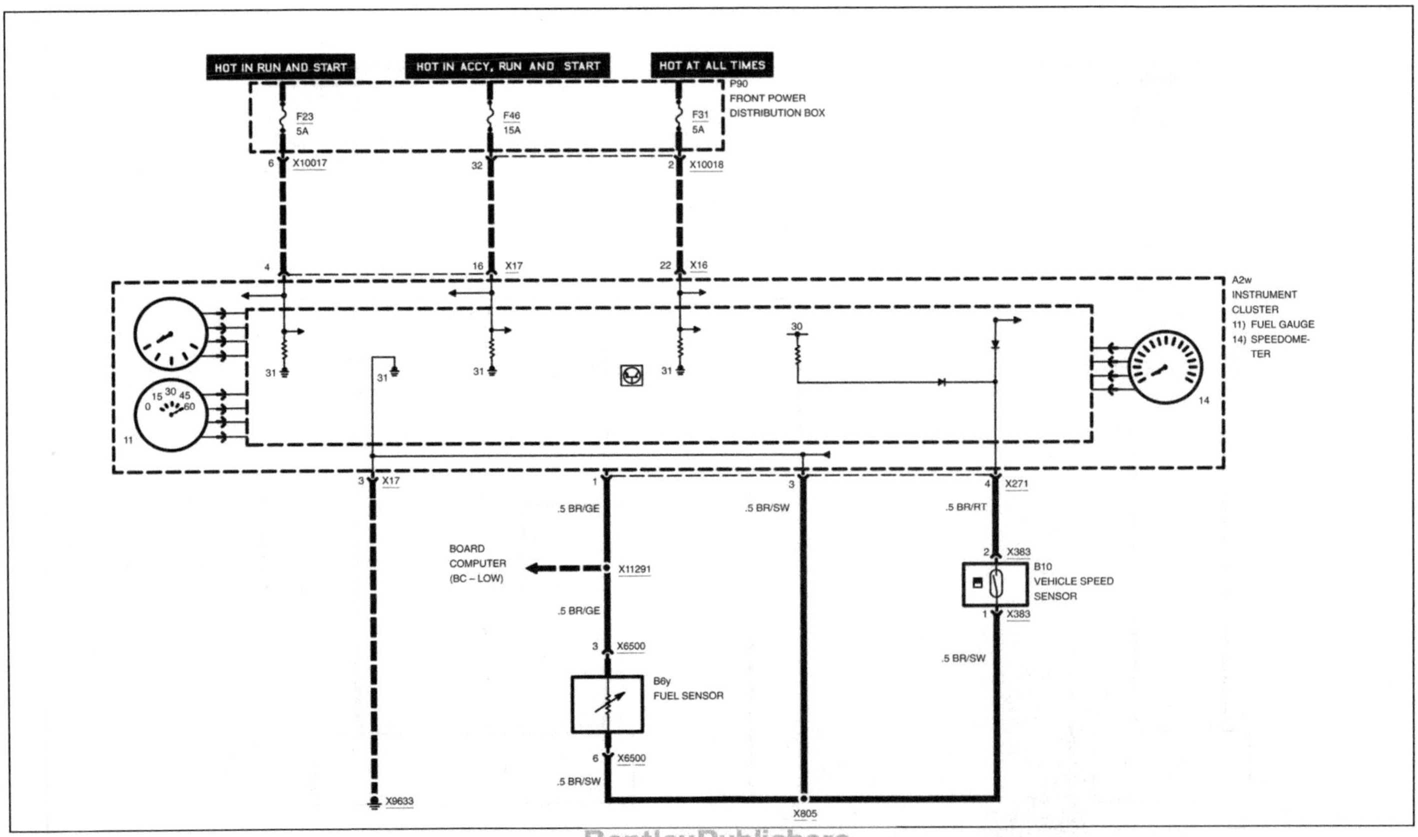

## Instrument cluster, 1998
(7 of 8)

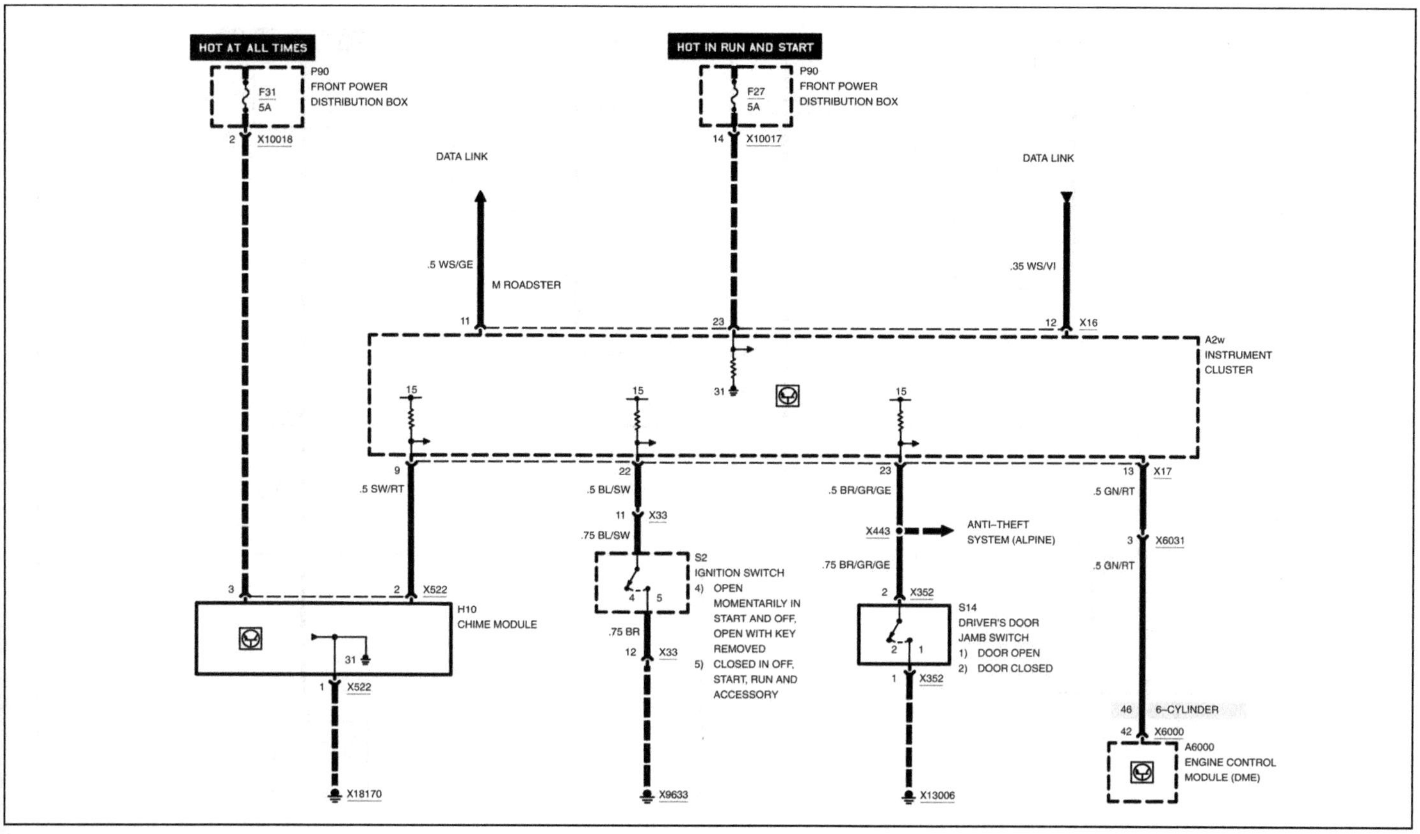


## Instrument cluster, 1998
(8 of 8)

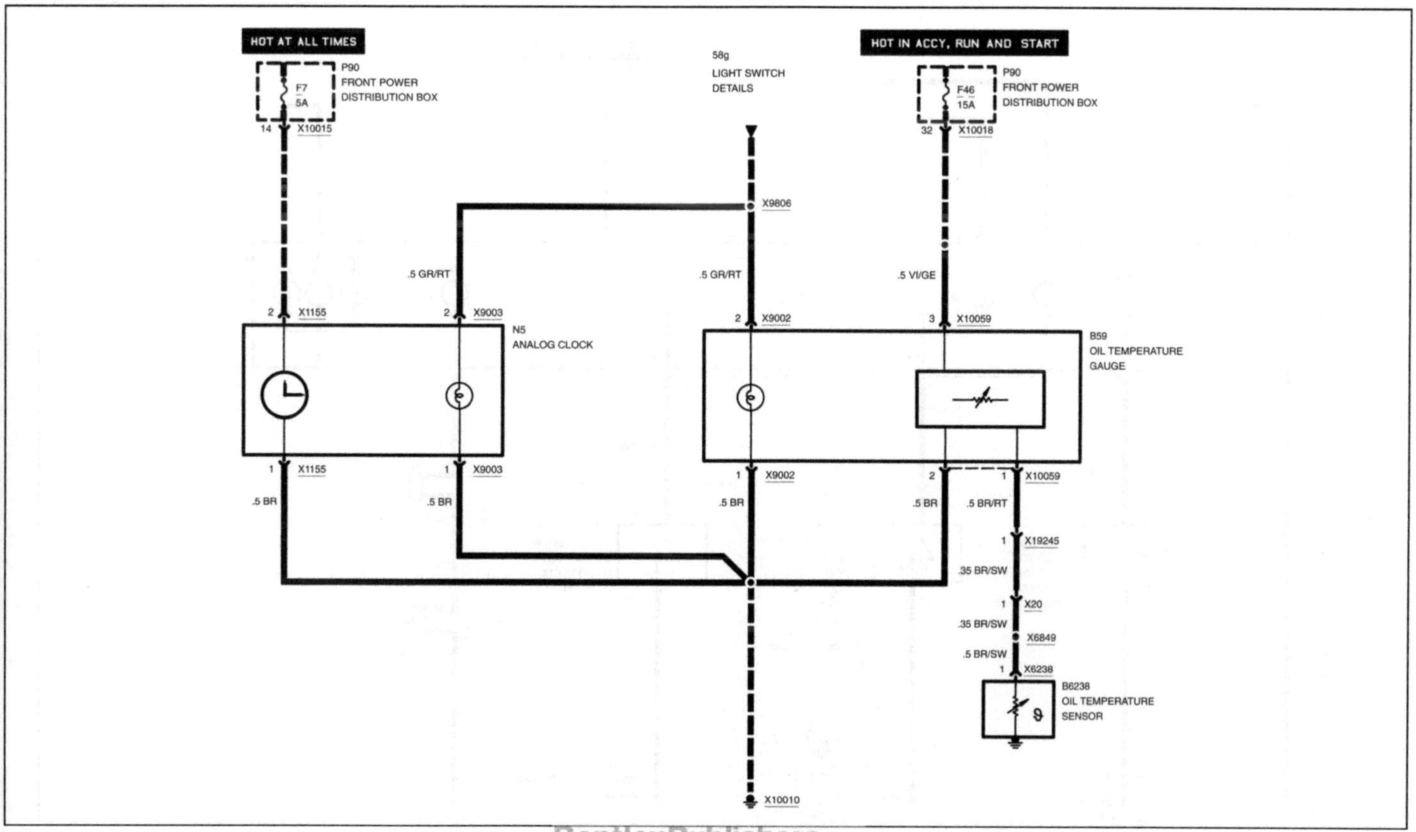

## Instrument cluster, 2000

(1 of 10)

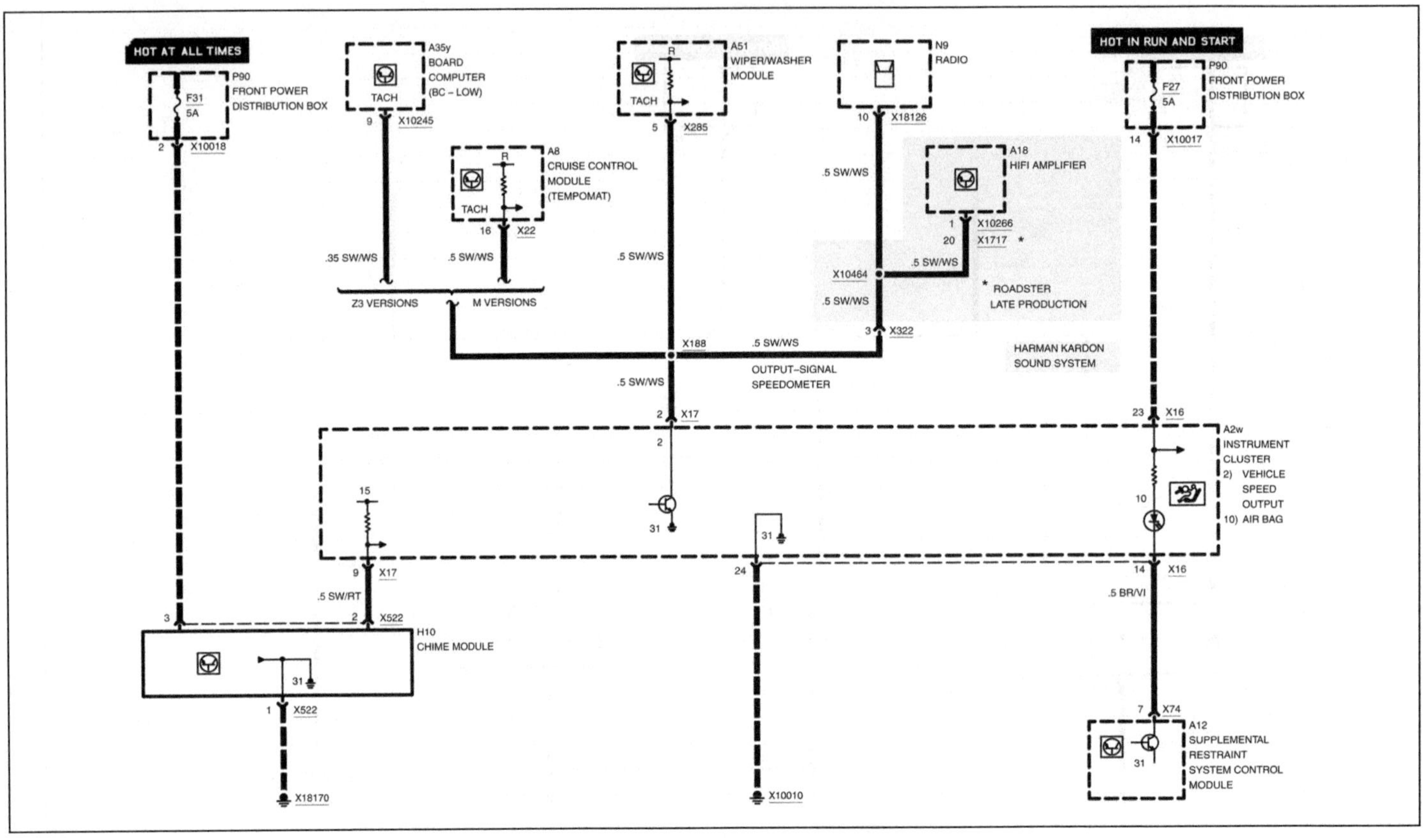


## Instrument cluster, 2000

(2 of 10)

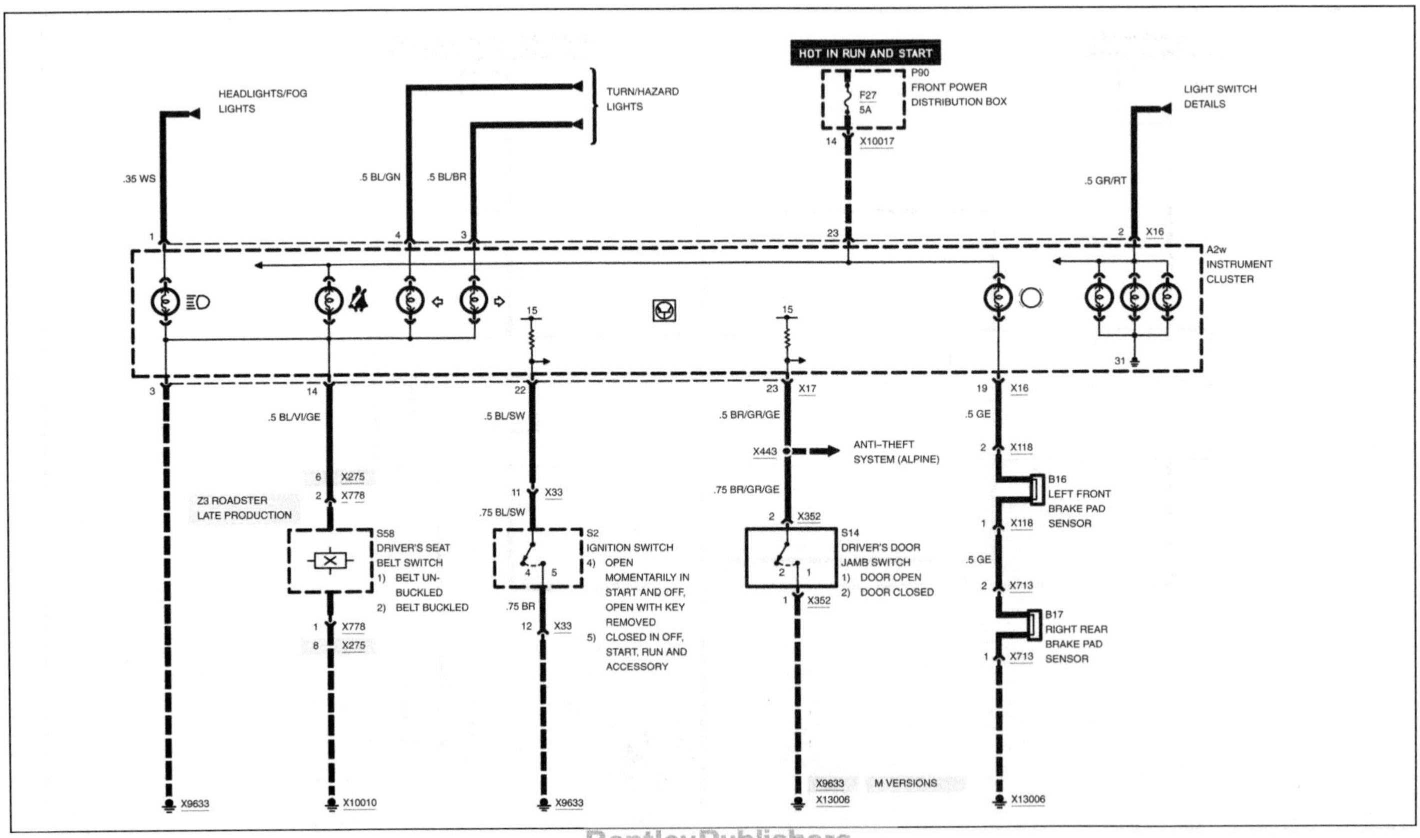

## Instrument cluster, 2000

### (3 of 10)

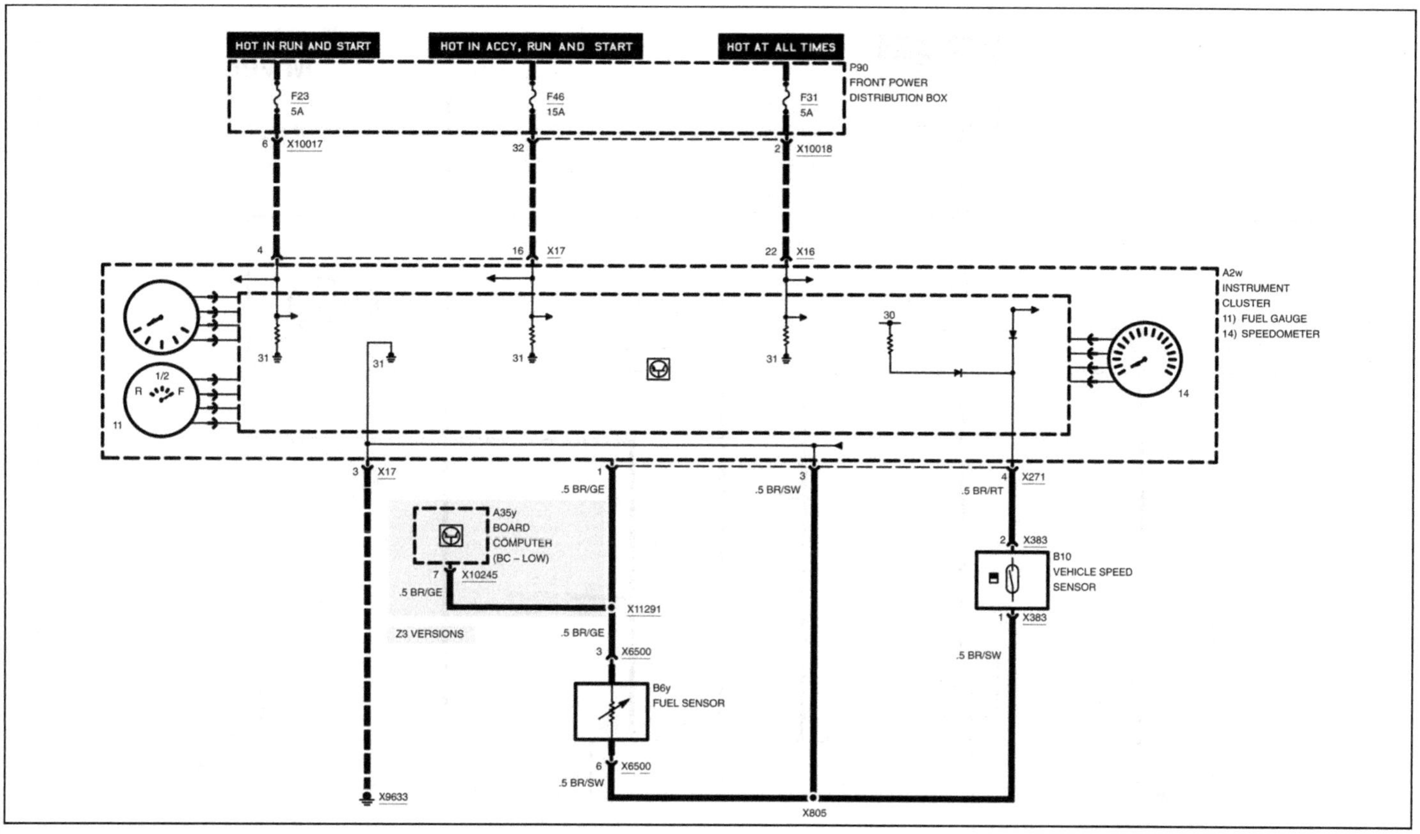


## Instrument cluster, 2000

### (4 of 10)

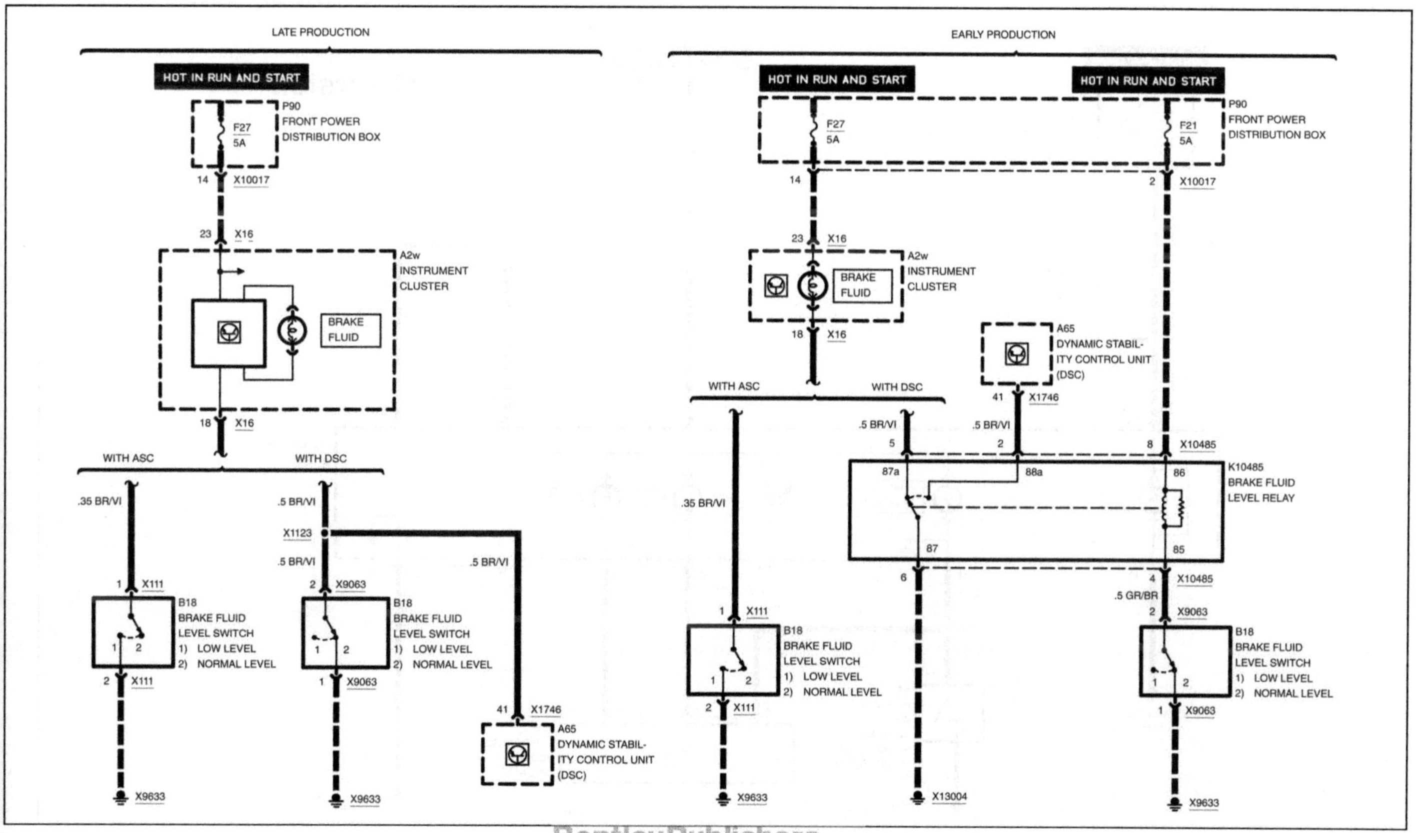

## Instrument cluster, 2000
(5 of 10)

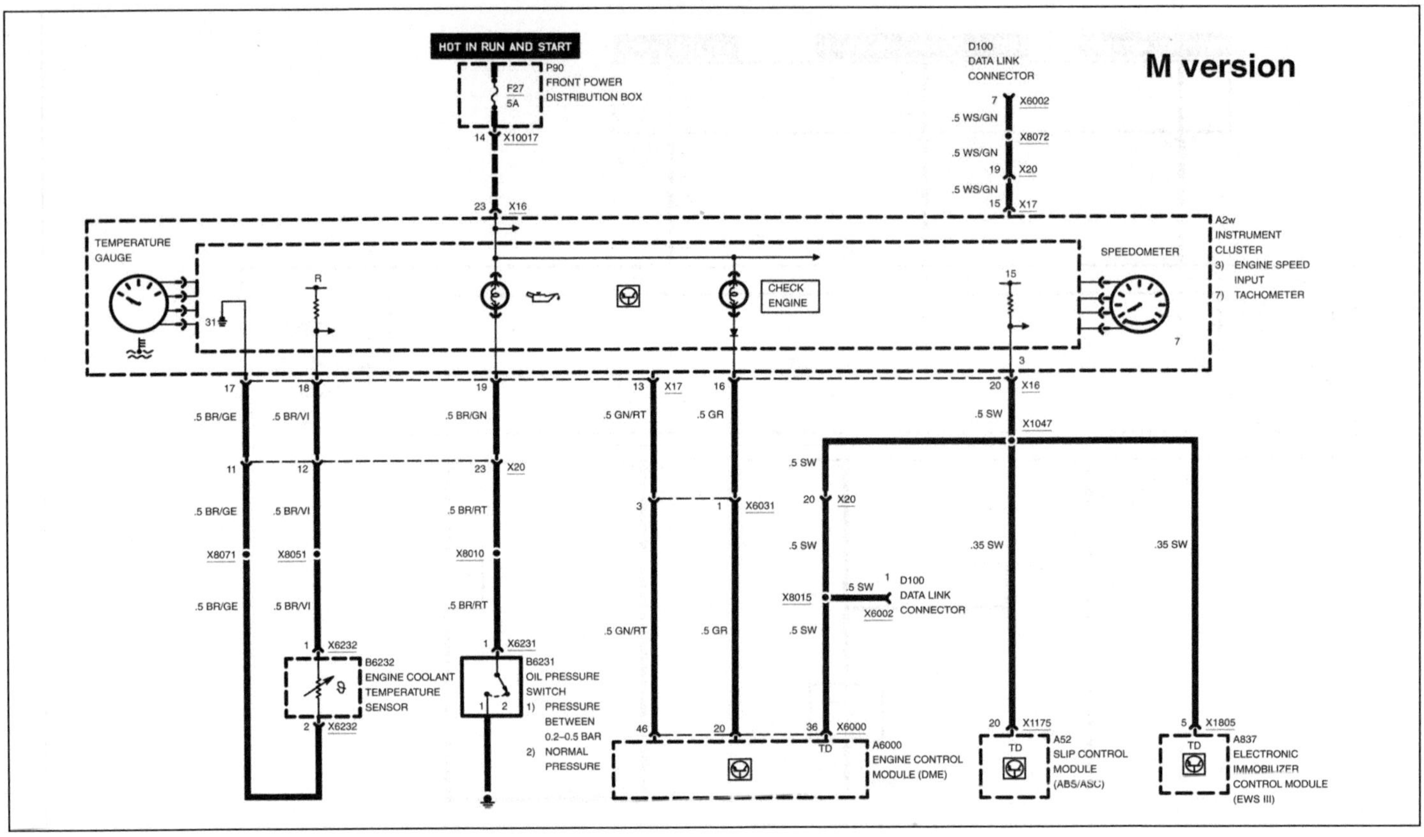


## Instrument cluster, 2000
(6 of 10)

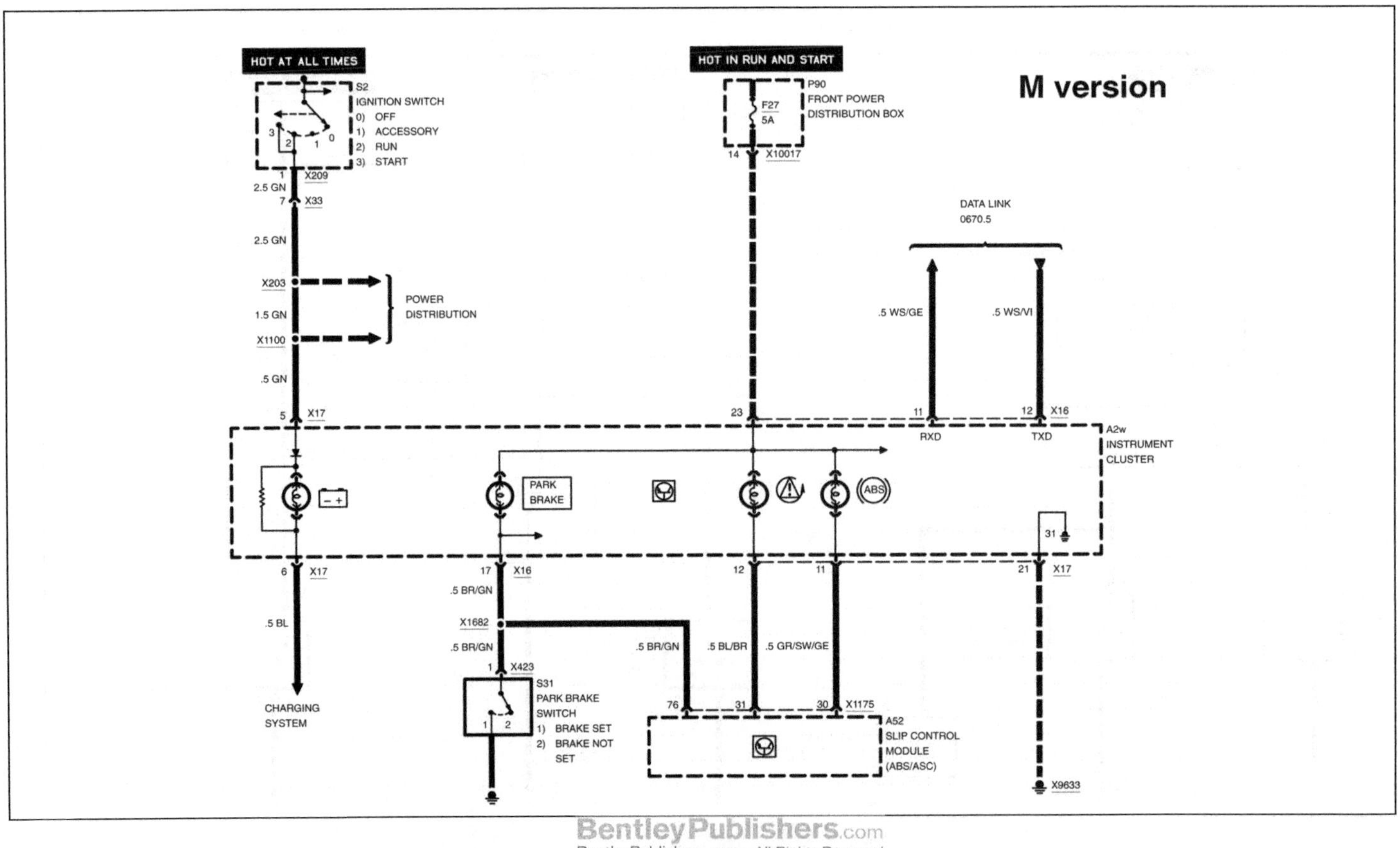

## Instrument cluster, 2000

(7 of 10)

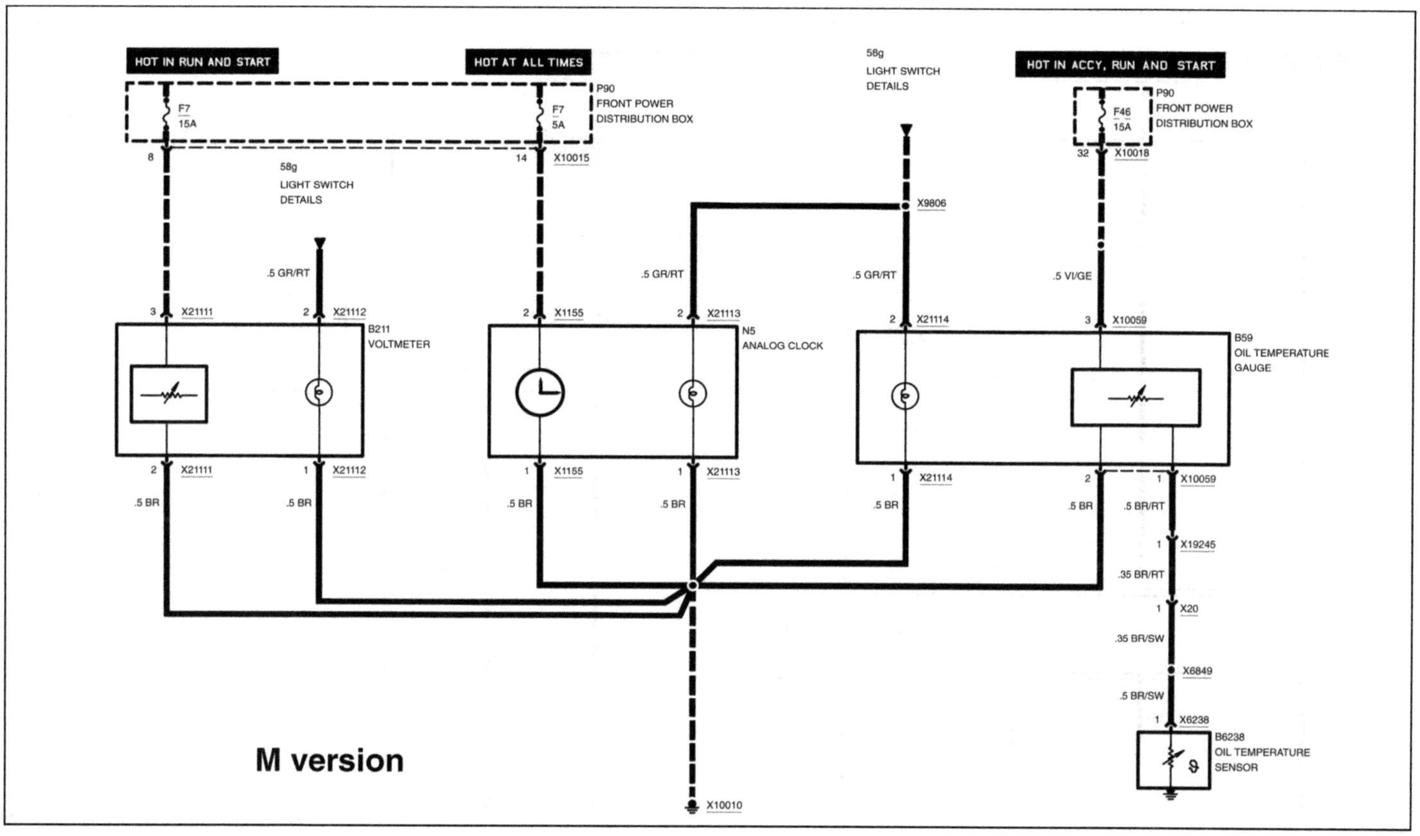


## Instrument cluster, 2000

(8 of 10)

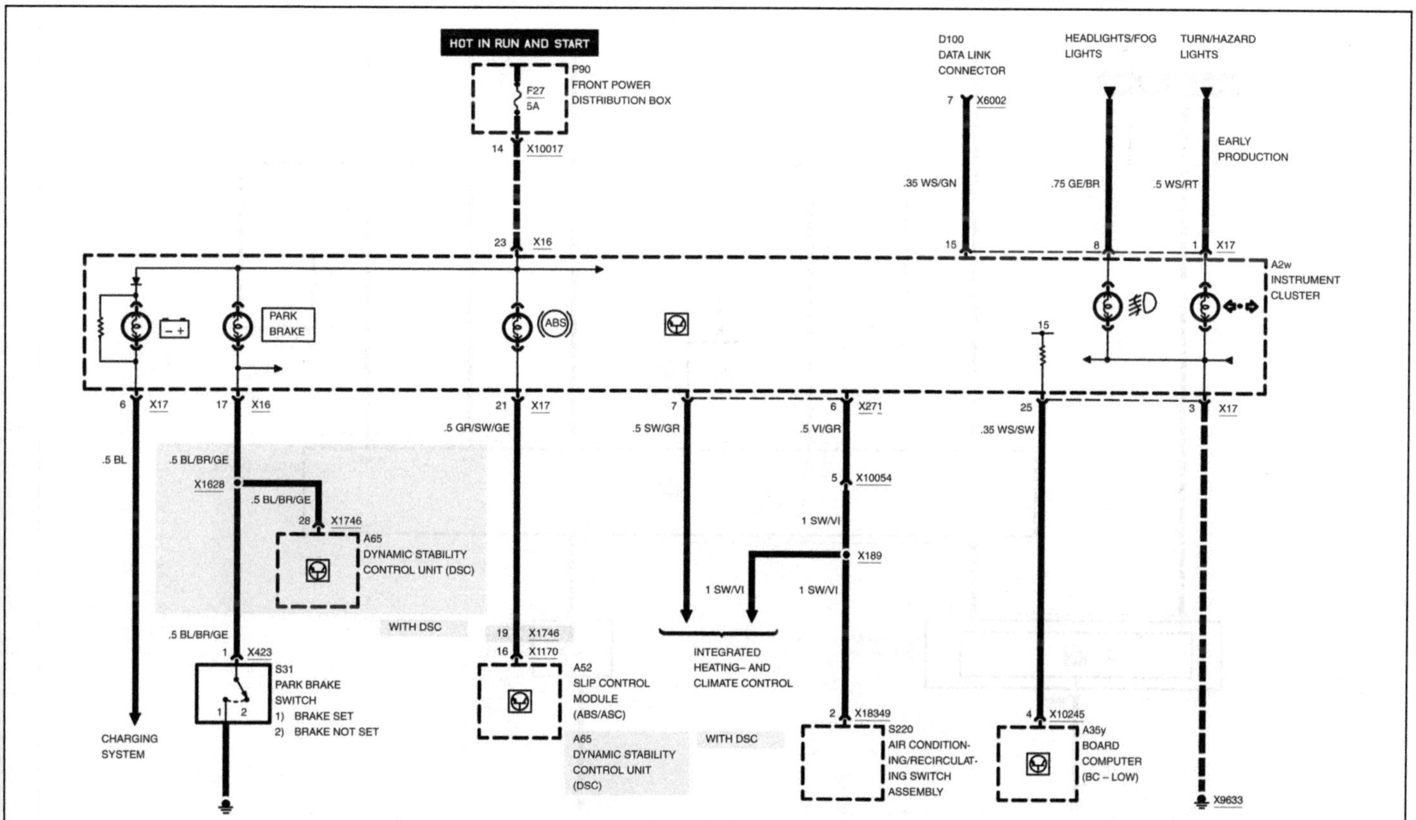

## Instrument cluster, 2000
(9 of 10)

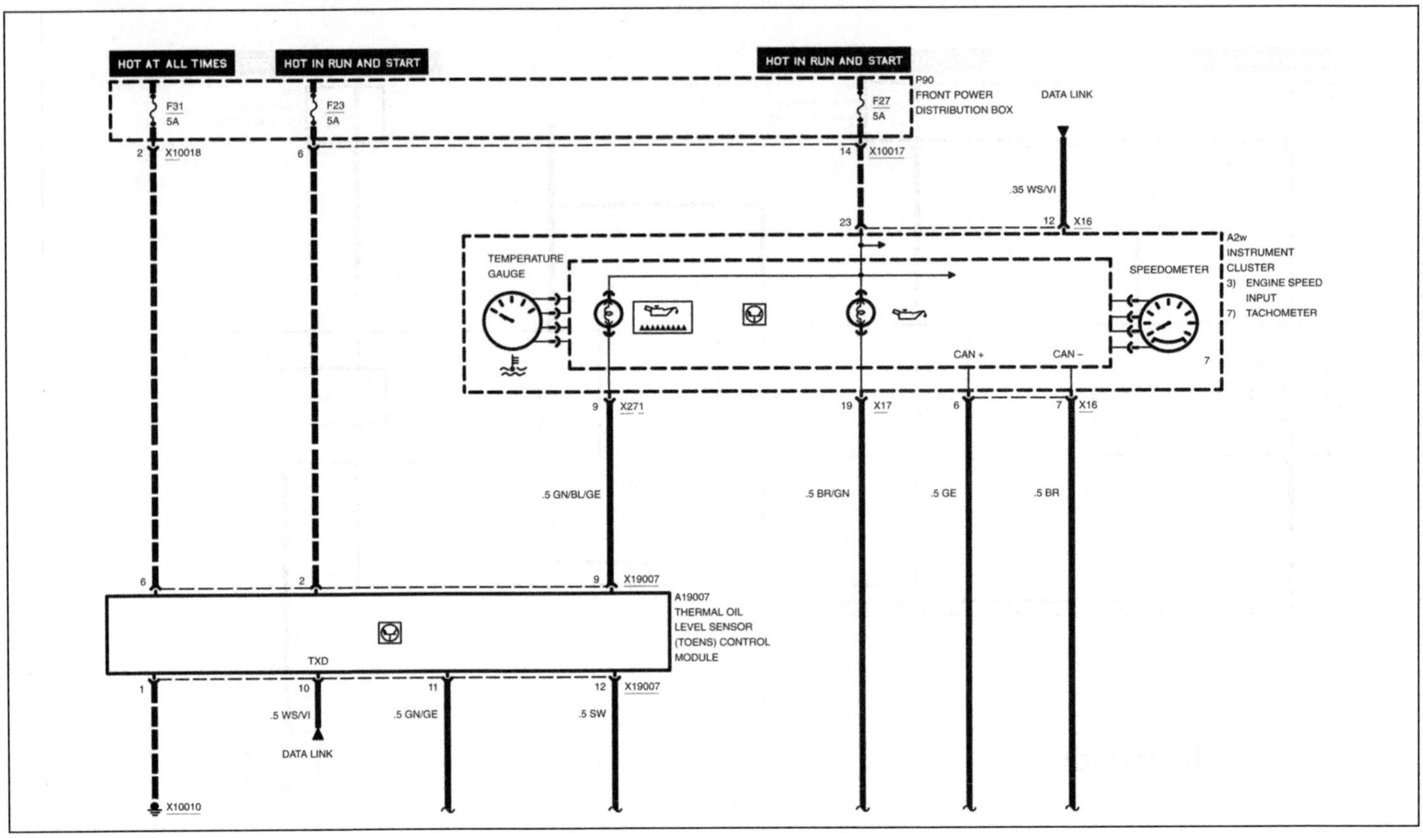


## Instrument cluster, 2000
(10 of 10)

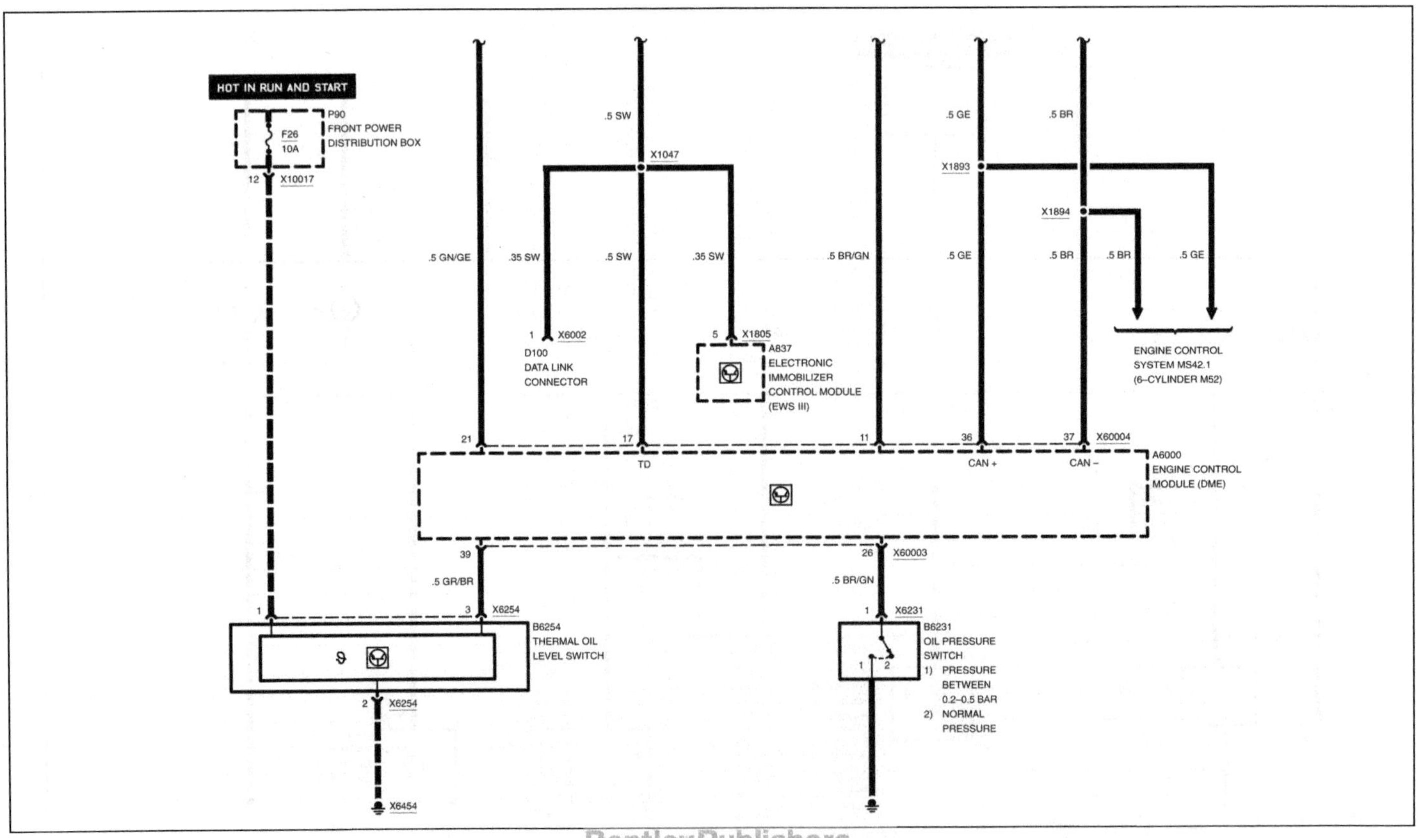

## Instrument cluster, 2002
(1 of 7)

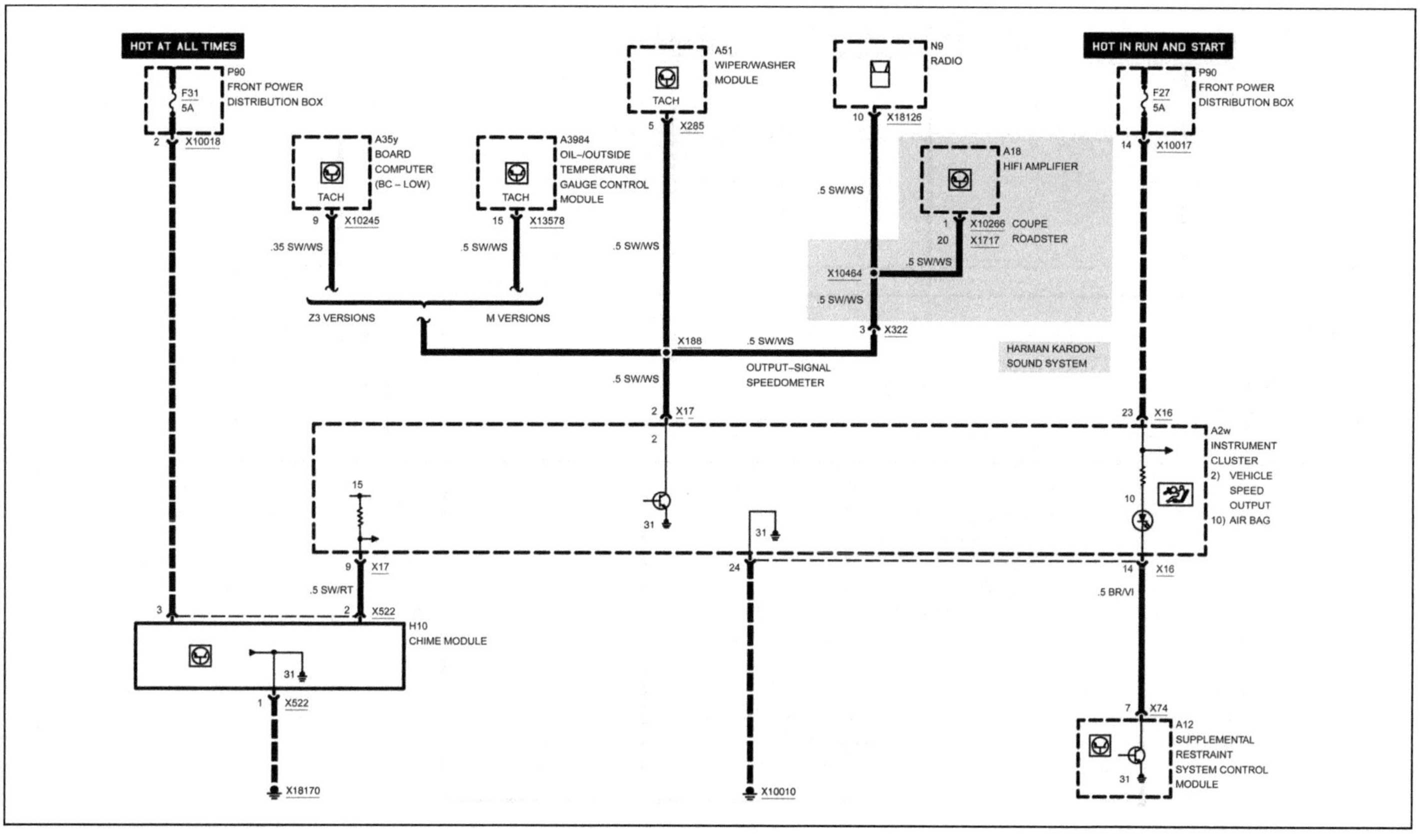


## Instrument cluster, 2002
(2 of 7)

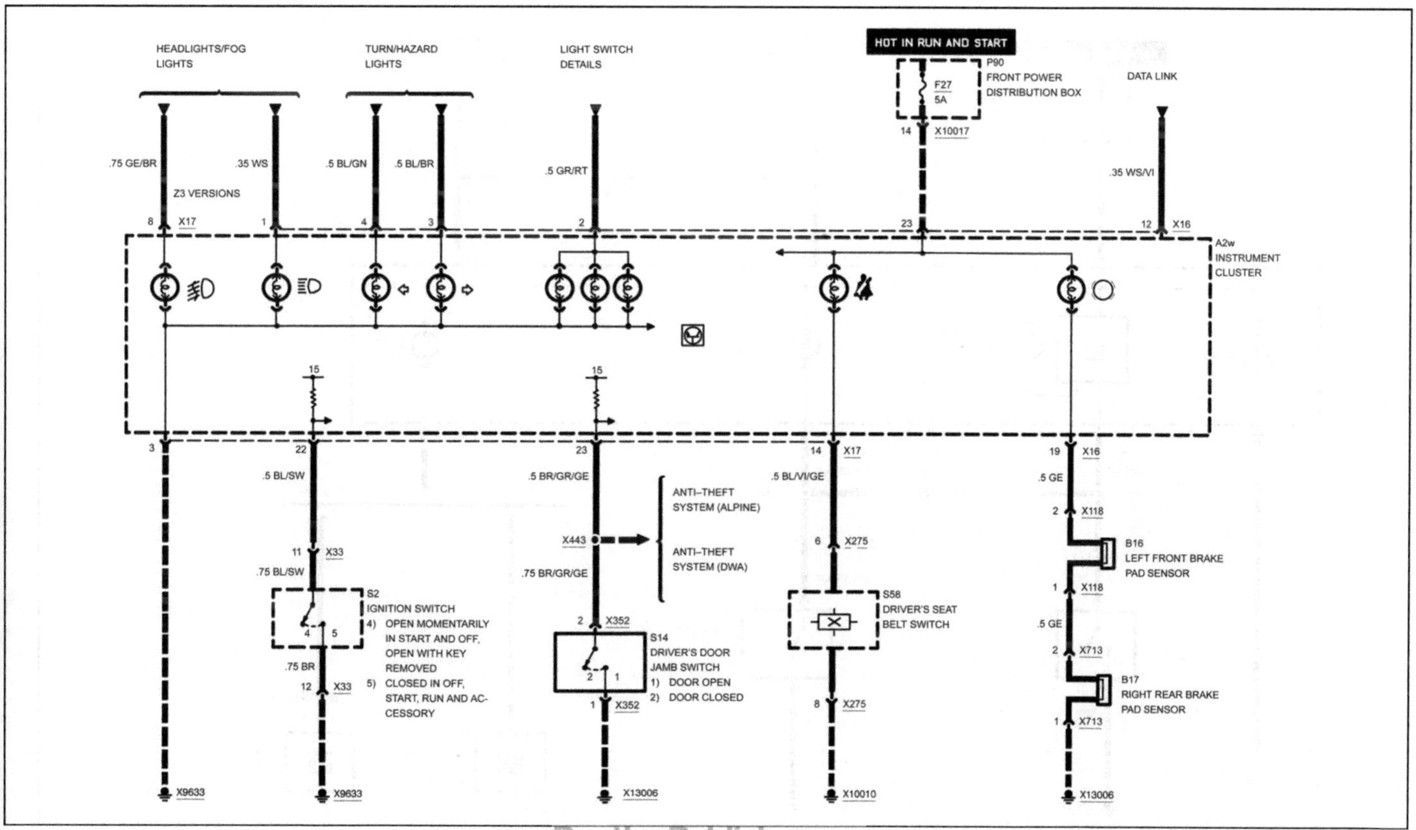

## Instrument cluster, 2002
### (3 of 7)

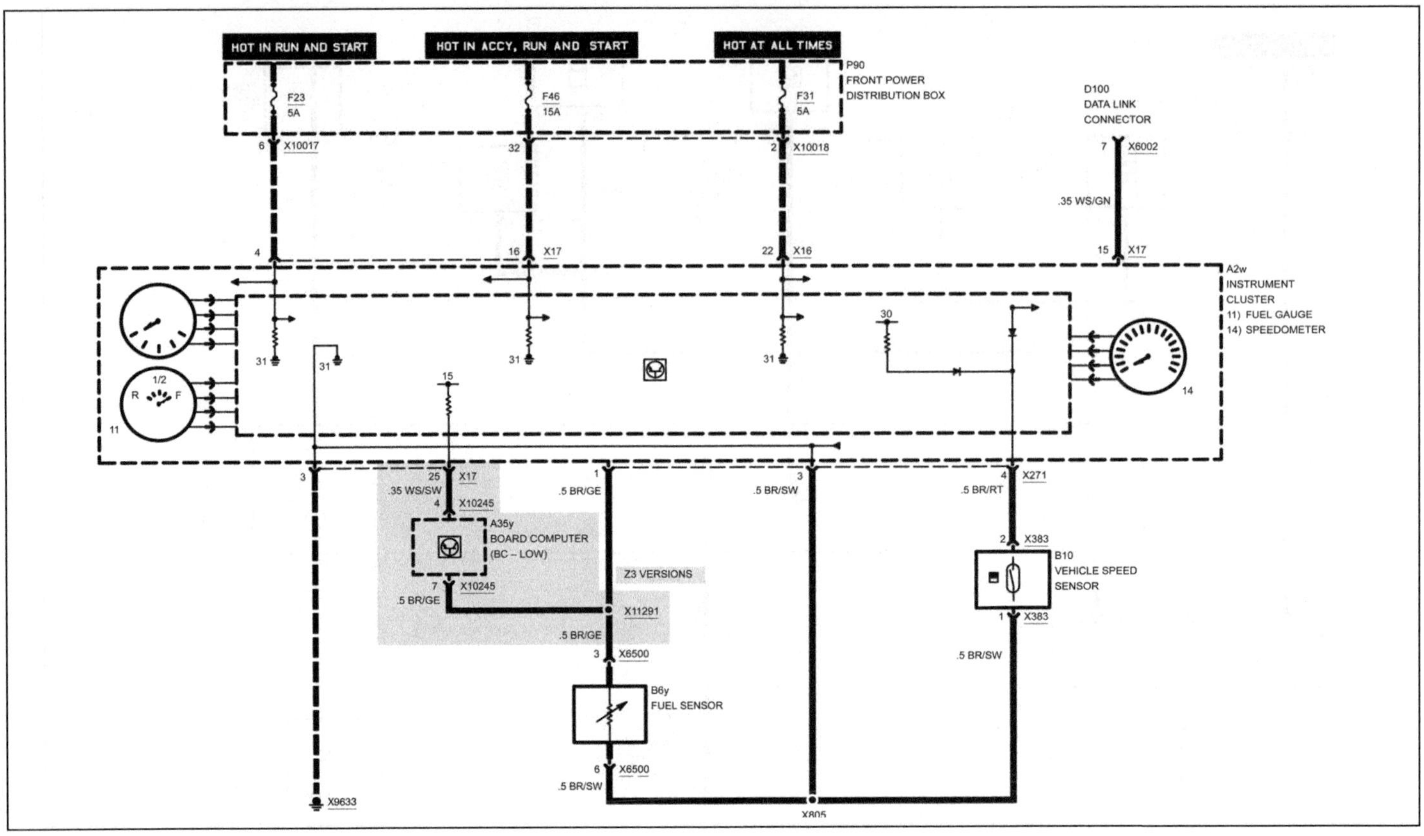


## Instrument cluster, 2002
### (4 of 7)

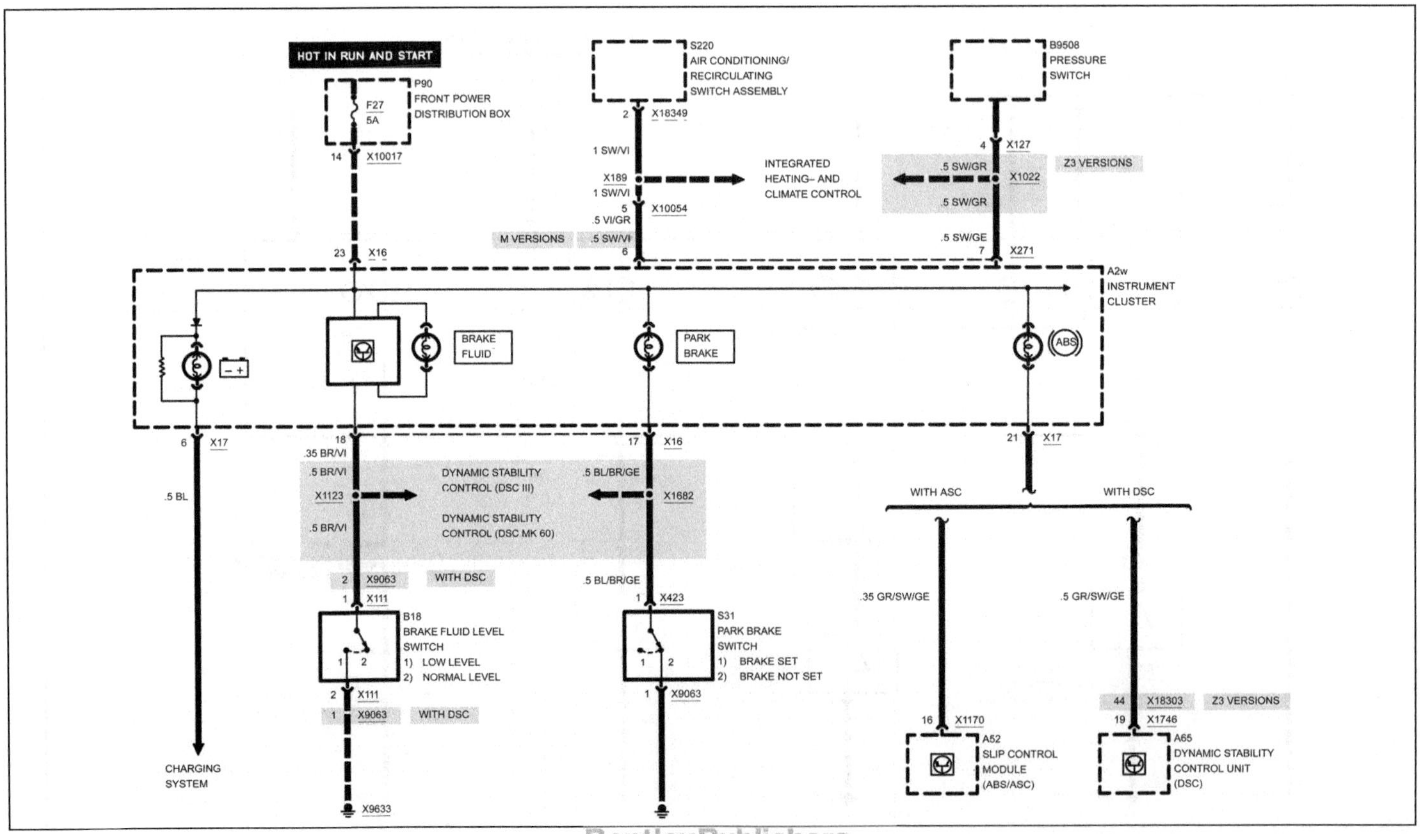

## Instrument cluster, 2002
(5 of 7)

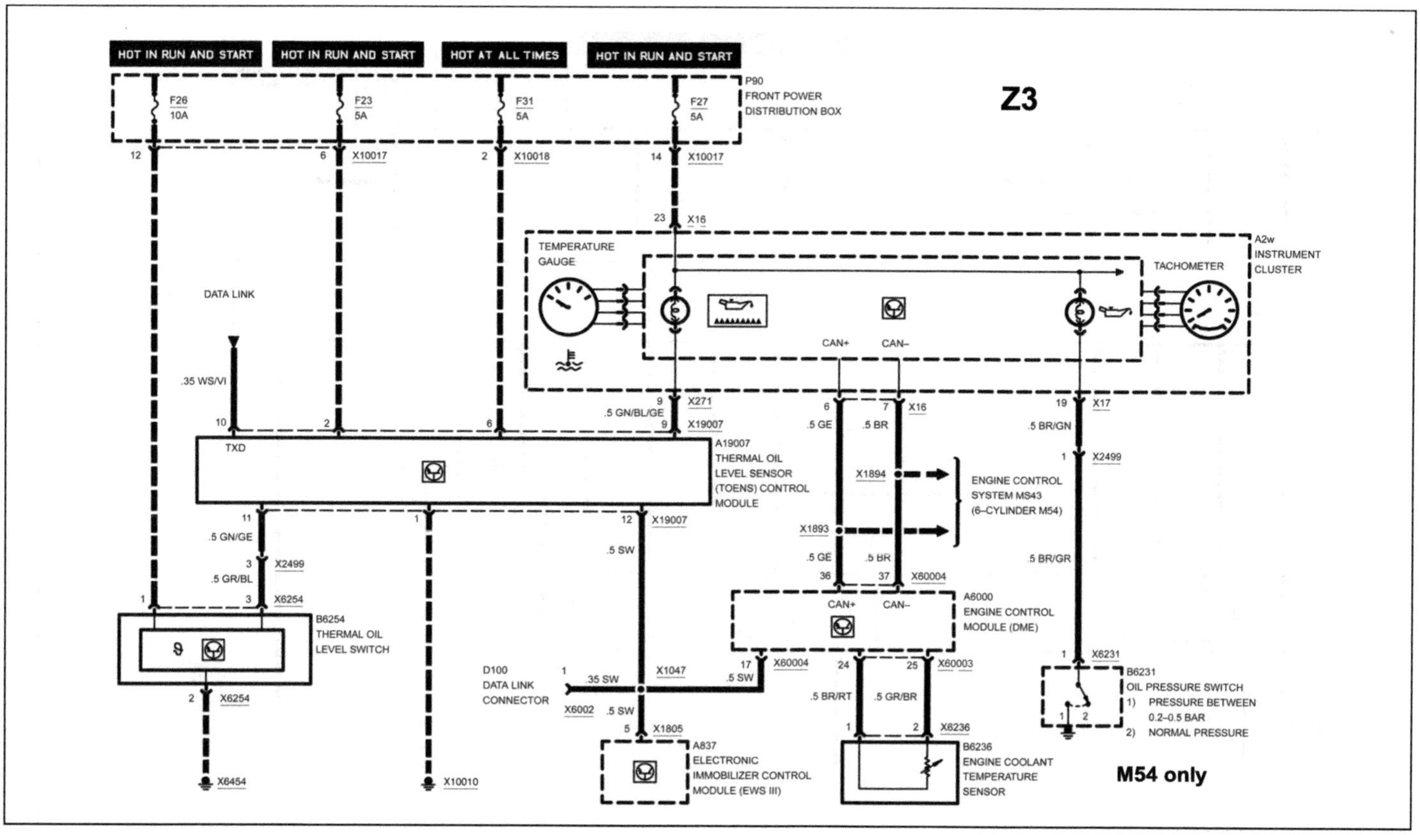


## Instrument cluster, 2002
(6 of 7)

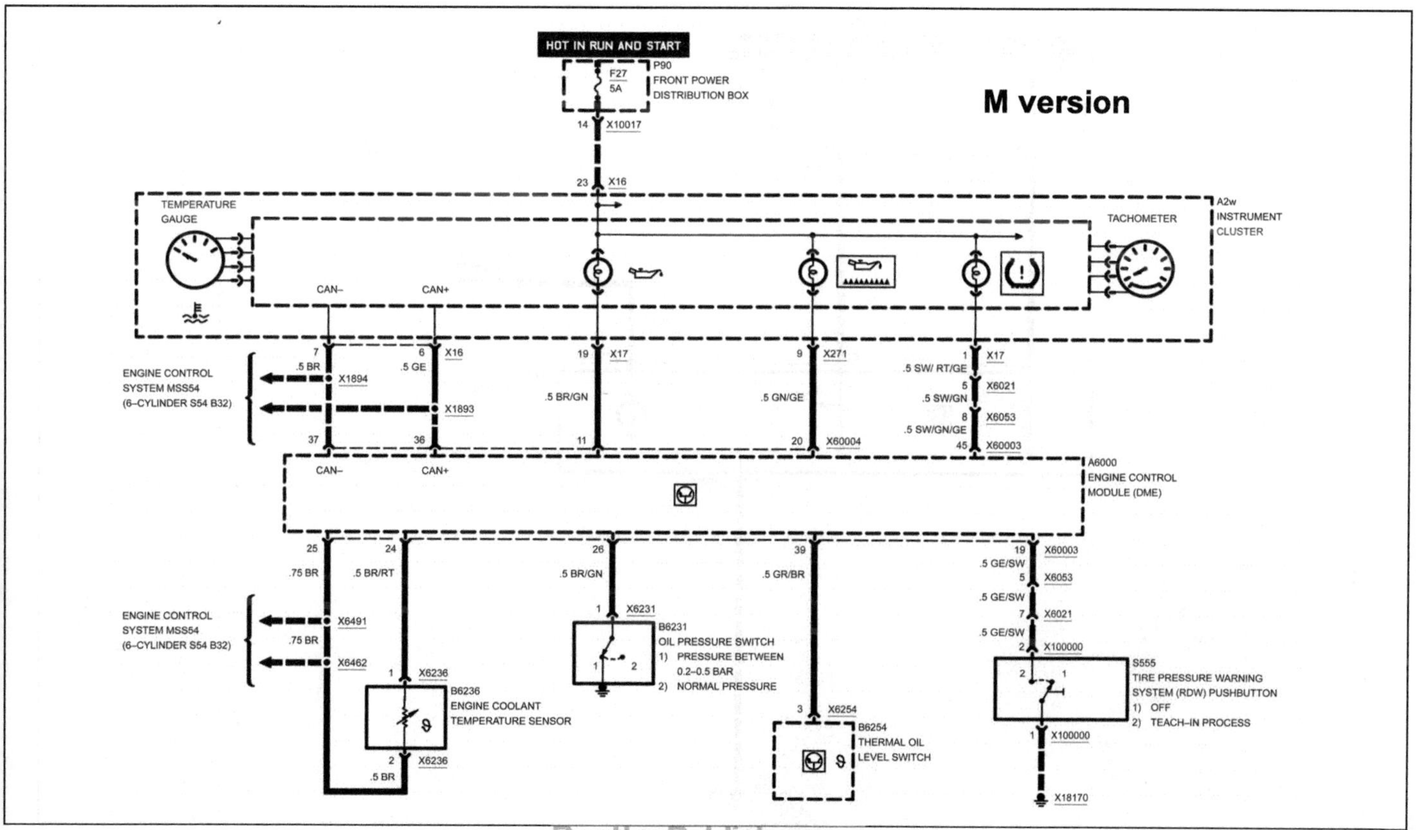

## Instrument cluster, 2002
(7 of 7)

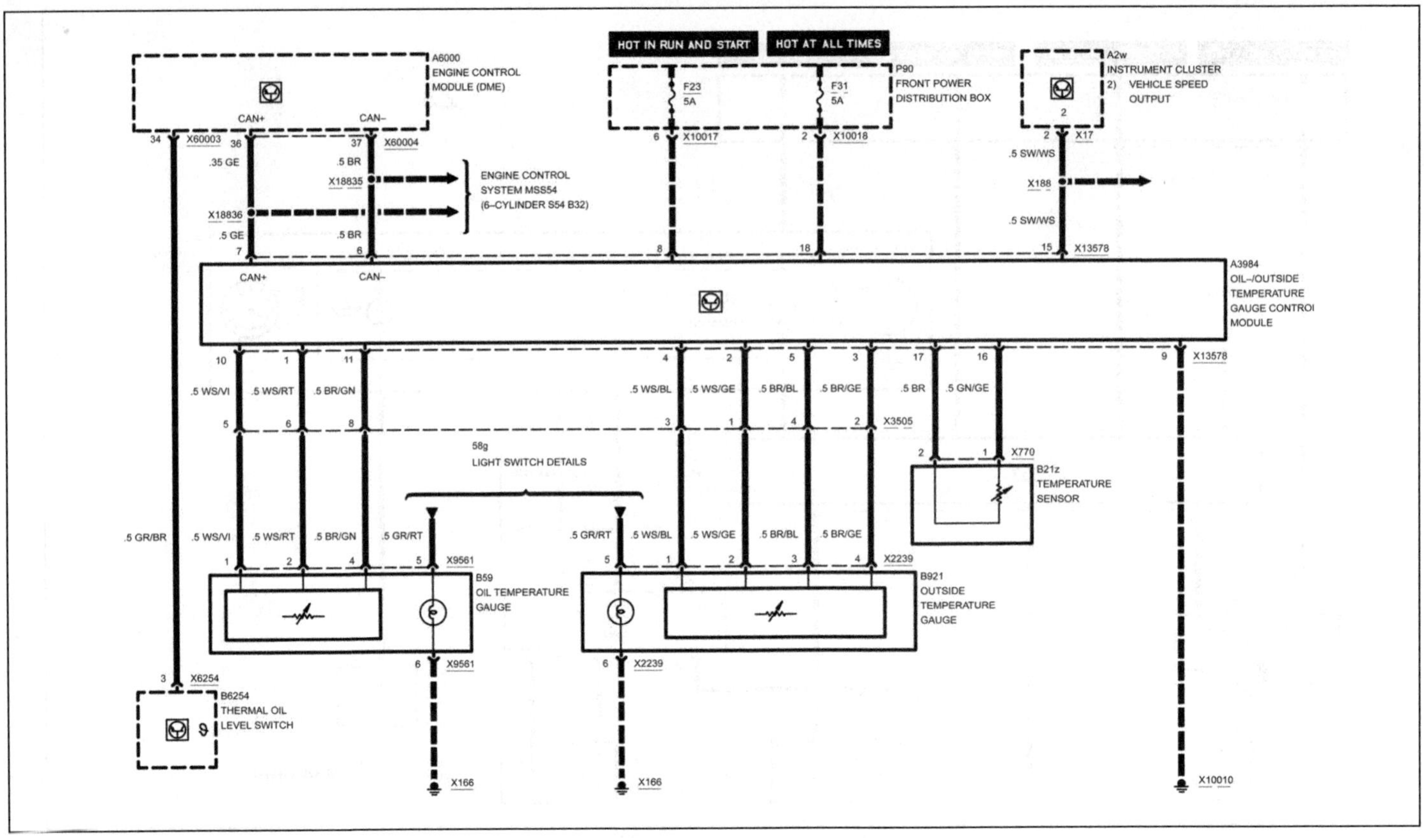


## Clock, digital

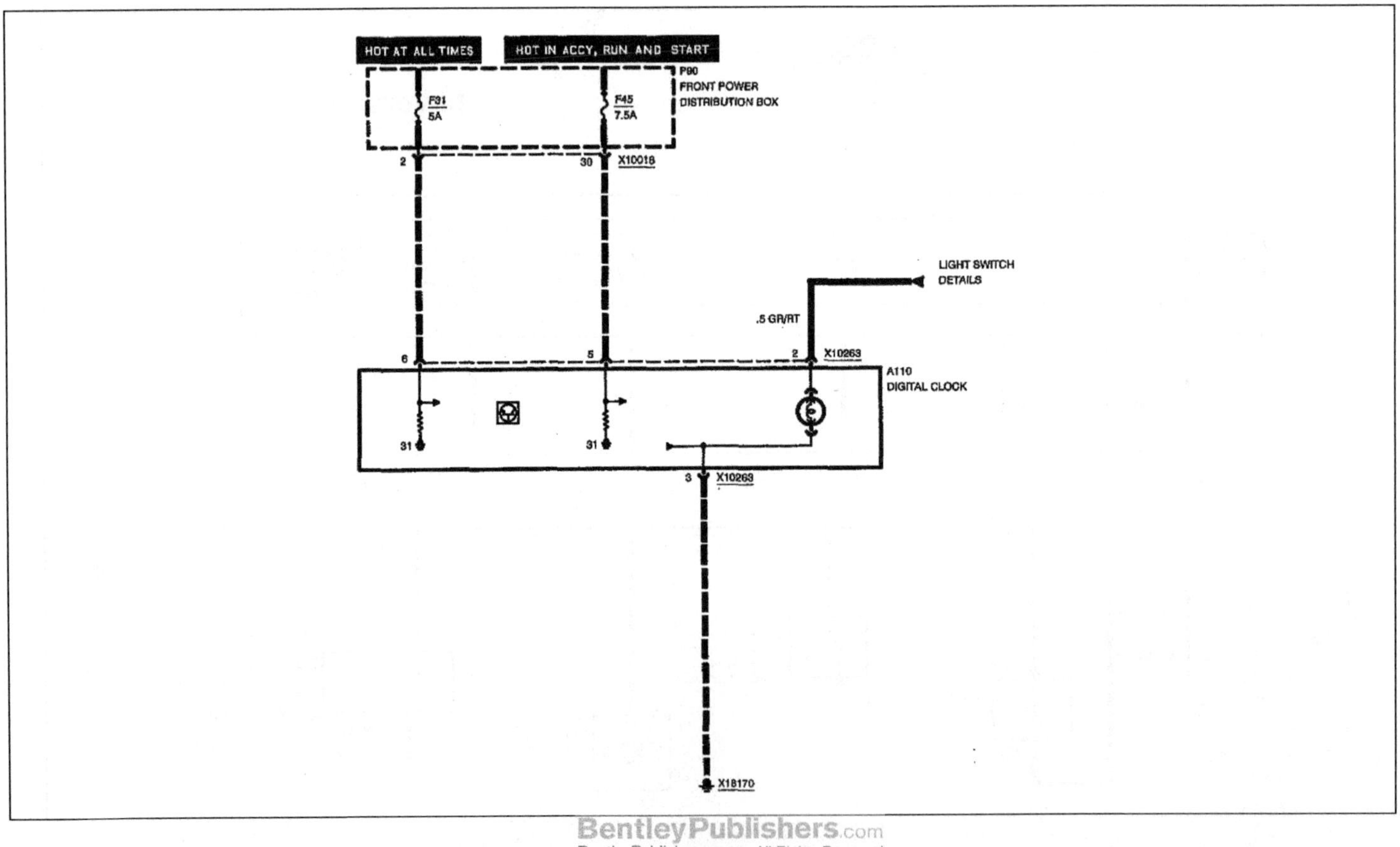

# Clock, analog

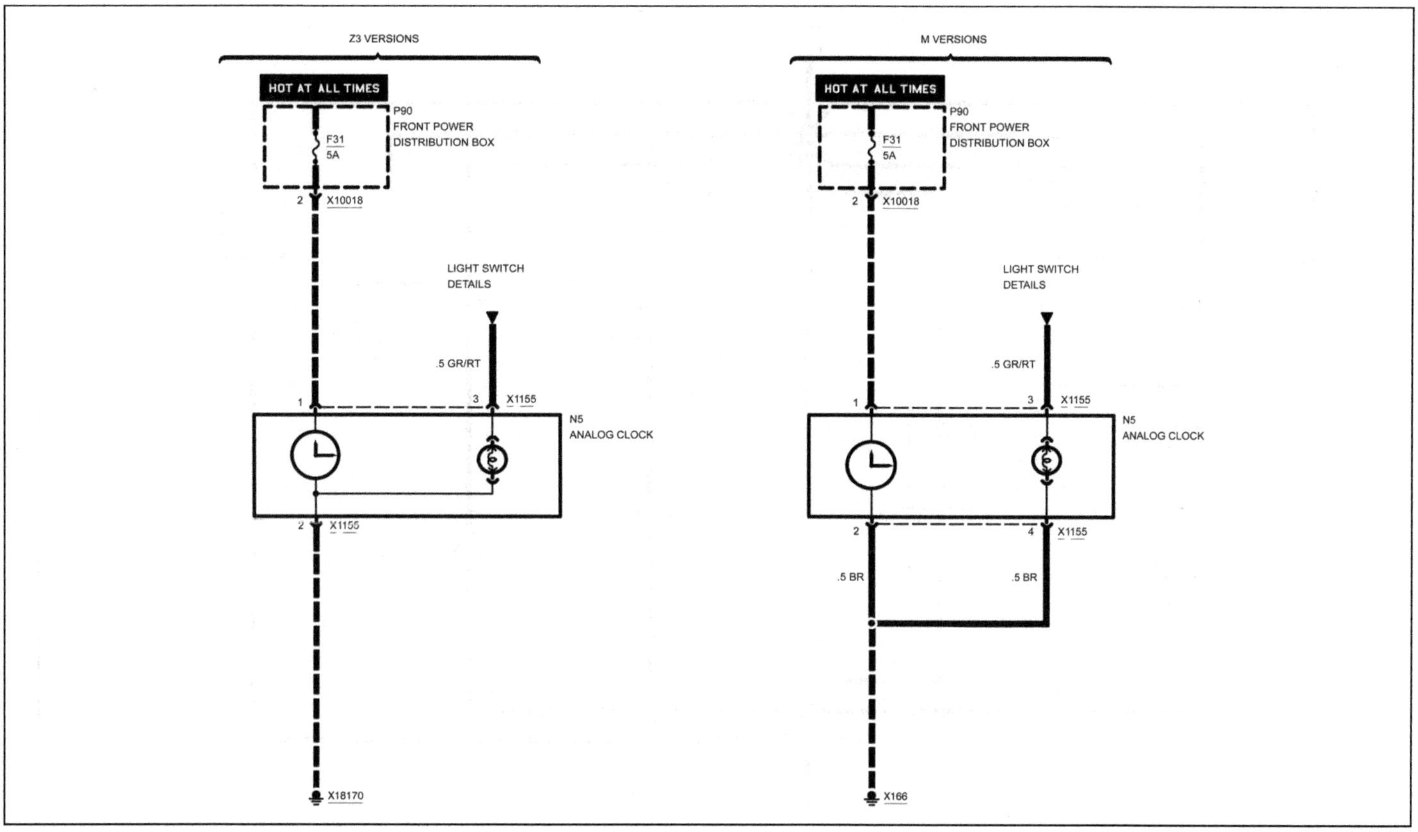


# Light switch details, 1996 - 1998

**(1 of 4)**

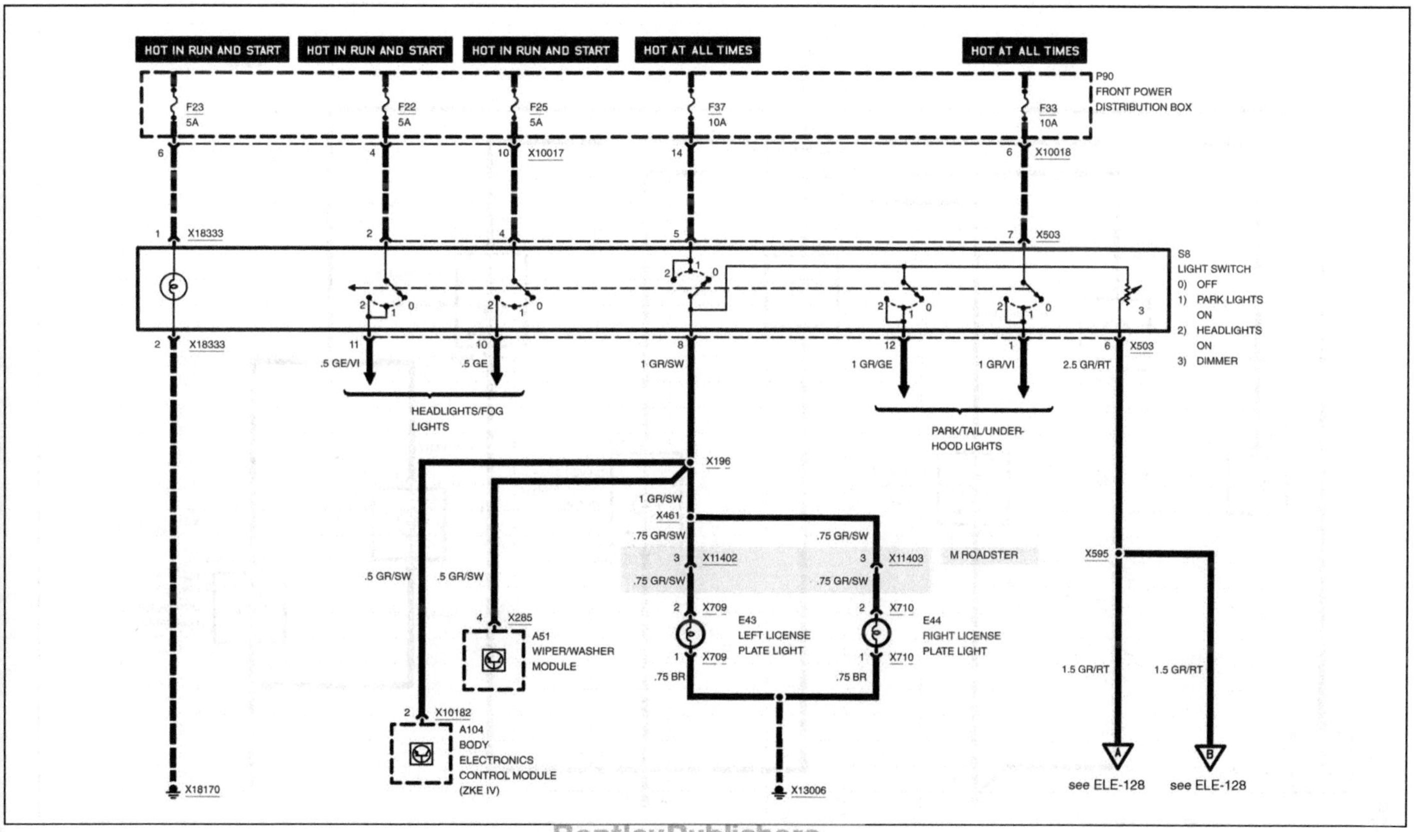

## Light switch details, 1996 - 1998
(2 of 4)

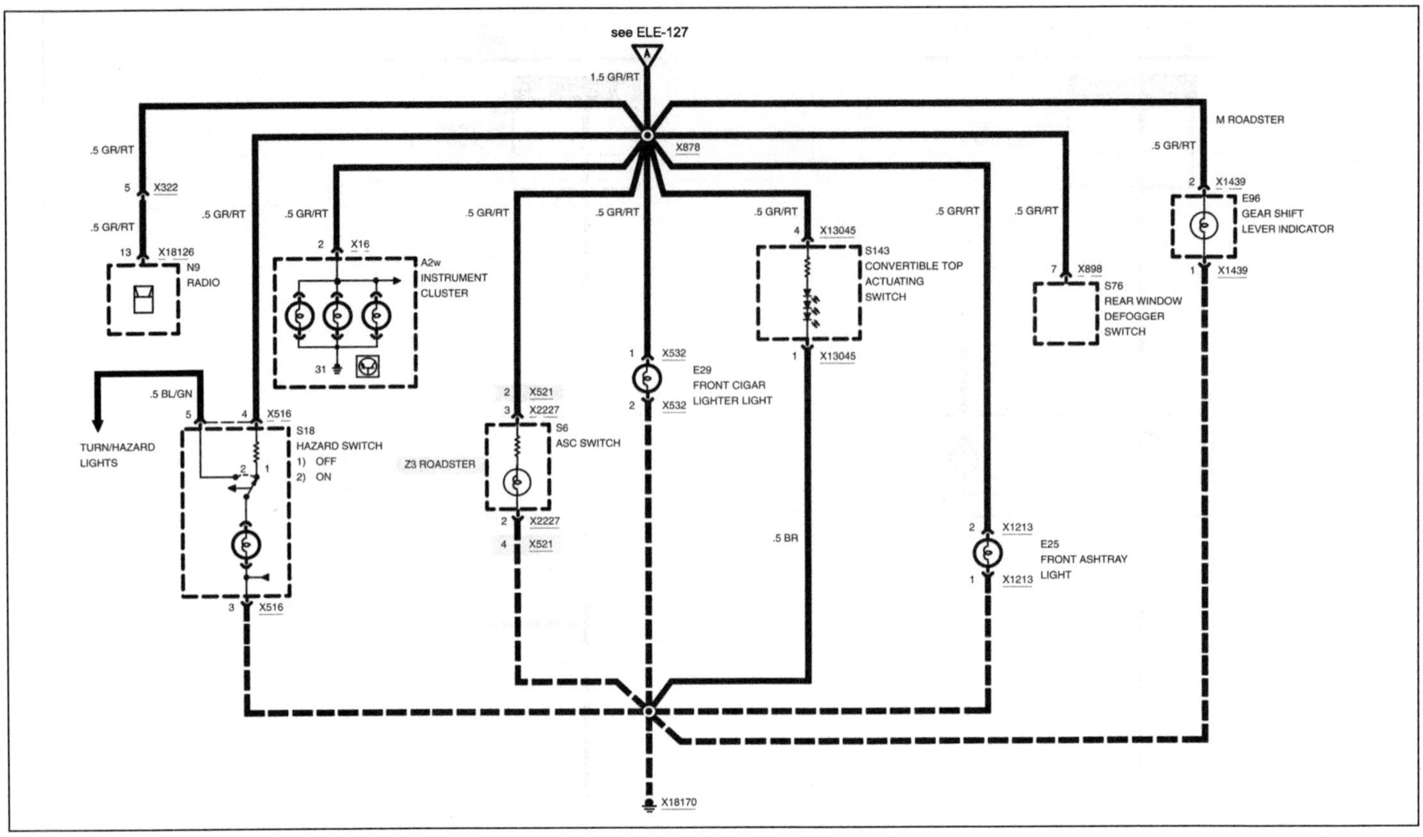


## Light switch details, 1996 - 1998
(3 of 4)

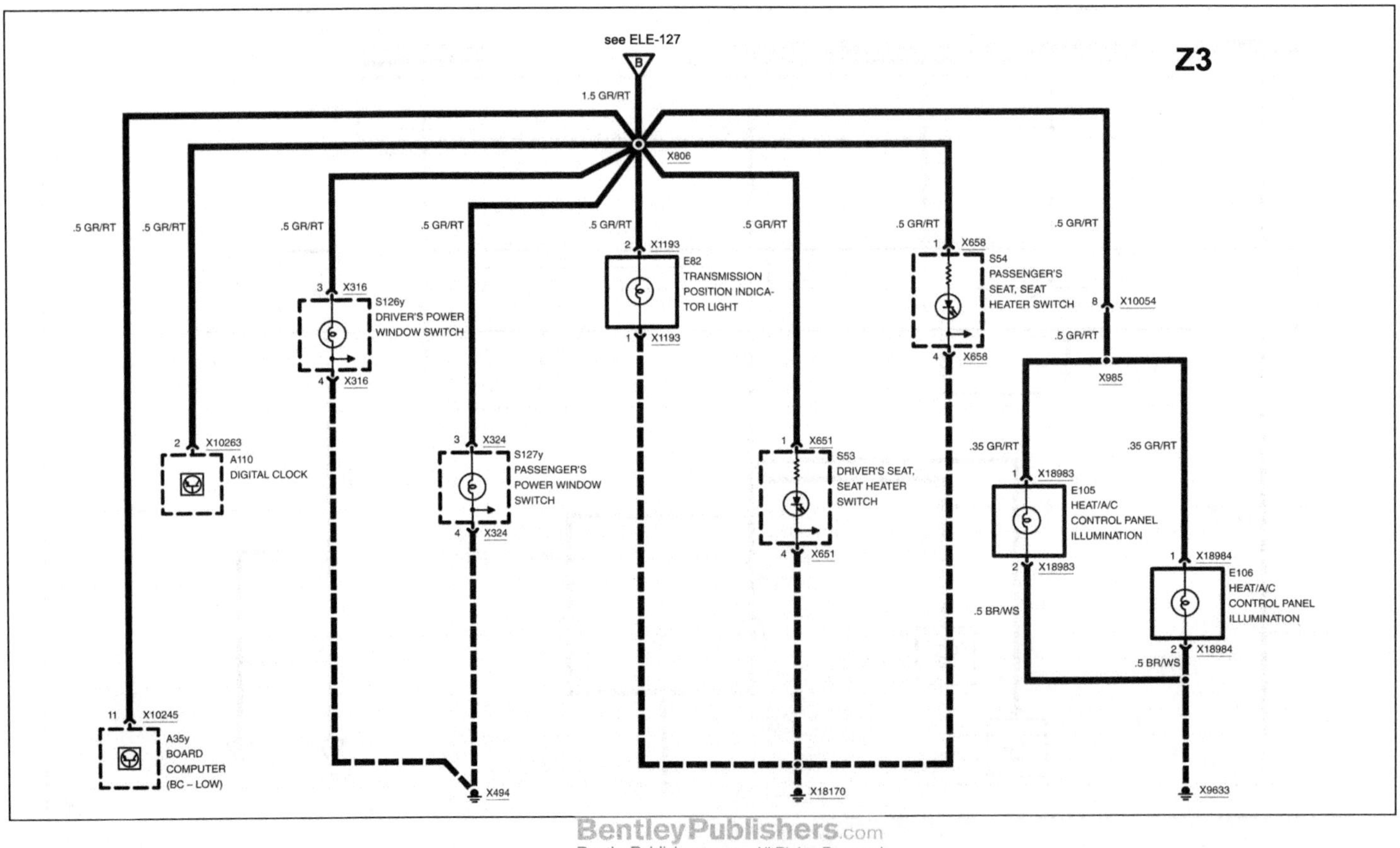

## Light switch details, 1996 - 1998
**(4 of 4)**

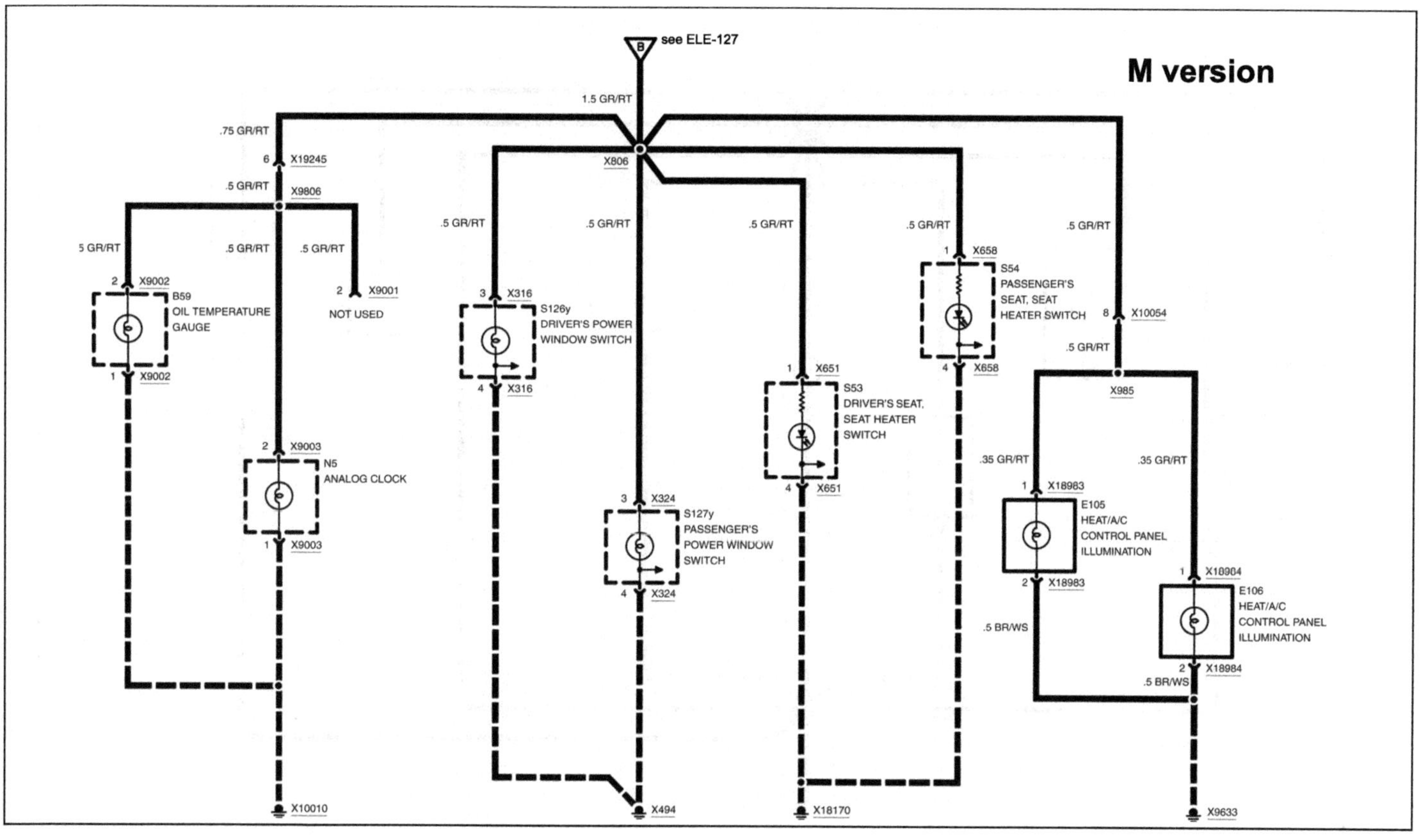


## Light switch details, 2000 - 2002
**(1 of 4)**

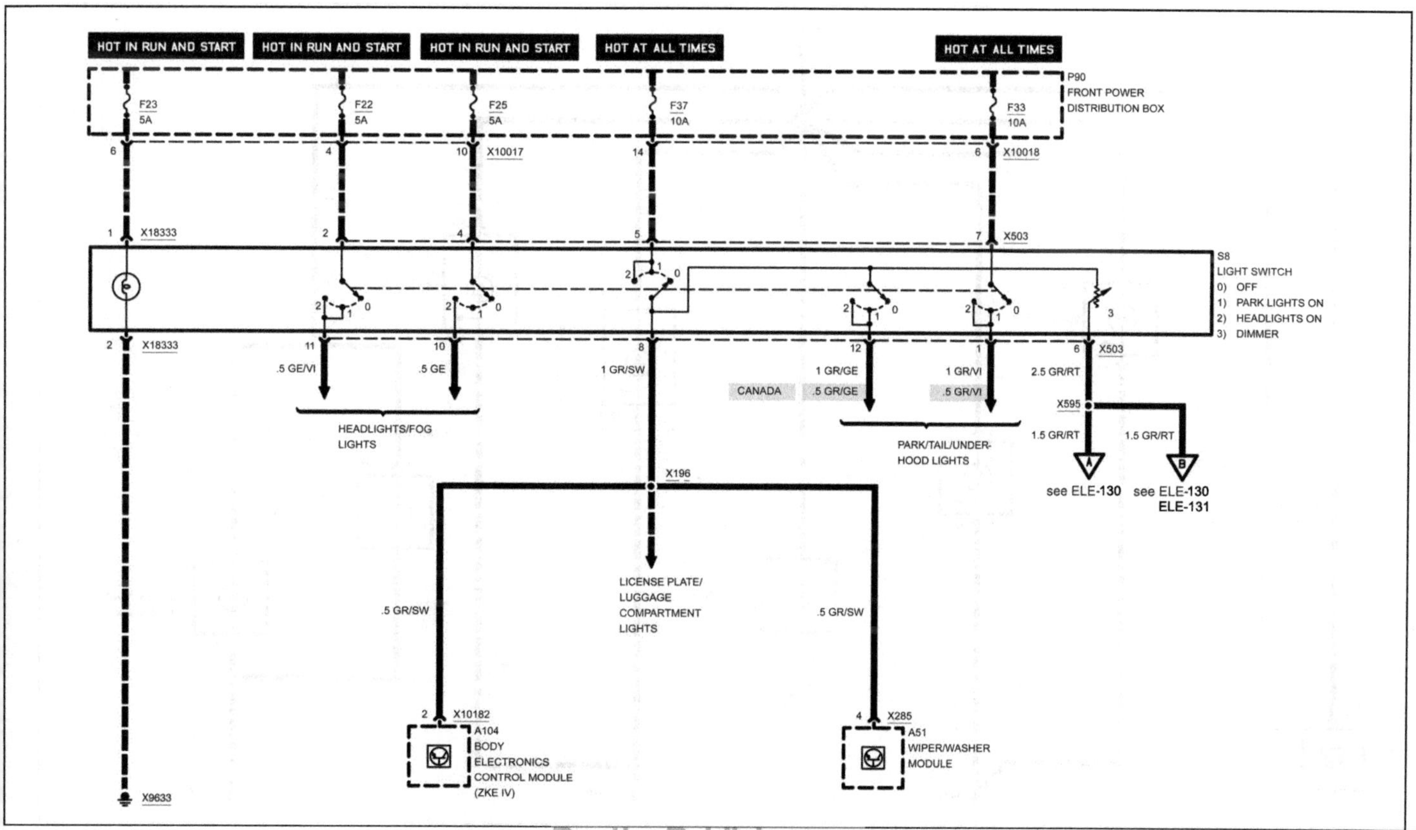

## Light switch details, 2000 - 2002
(2 of 4)

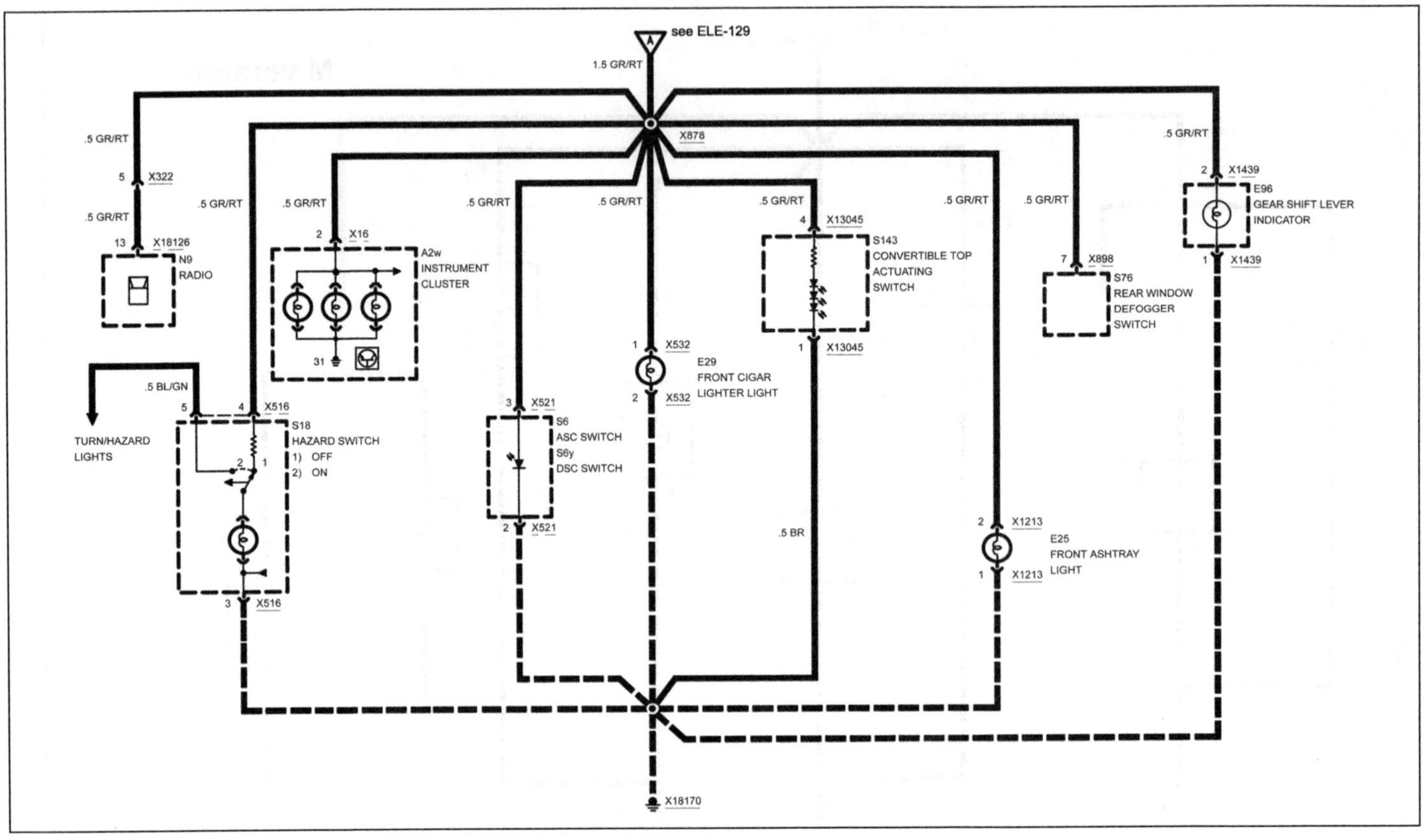


## Light switch details, 2000 - 2002
(3 of 4)

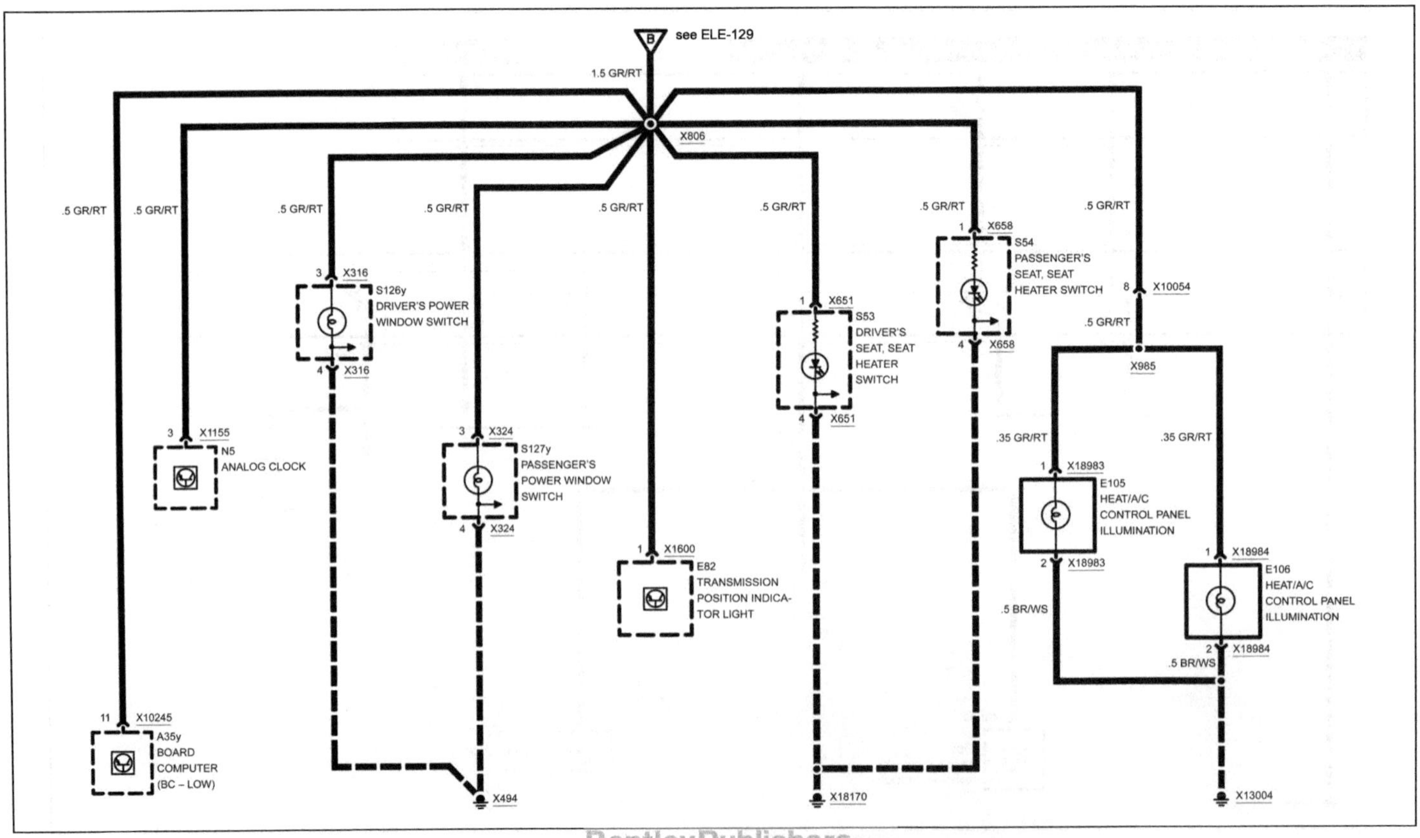

## Light switch details, 2000 - 2002
(4 of 4)

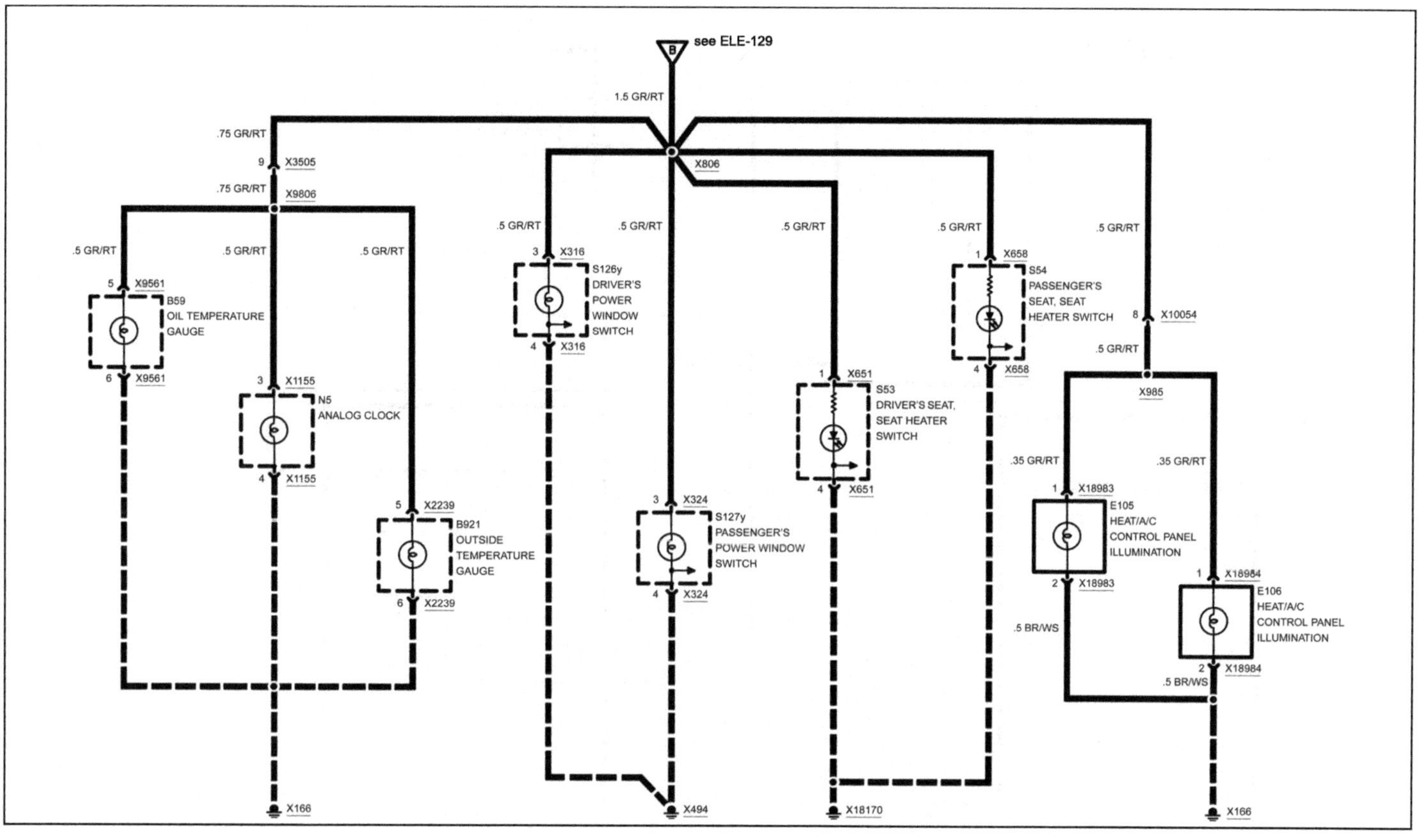


## Headlights / foglights, 1996 - 1998
(1 of 2)

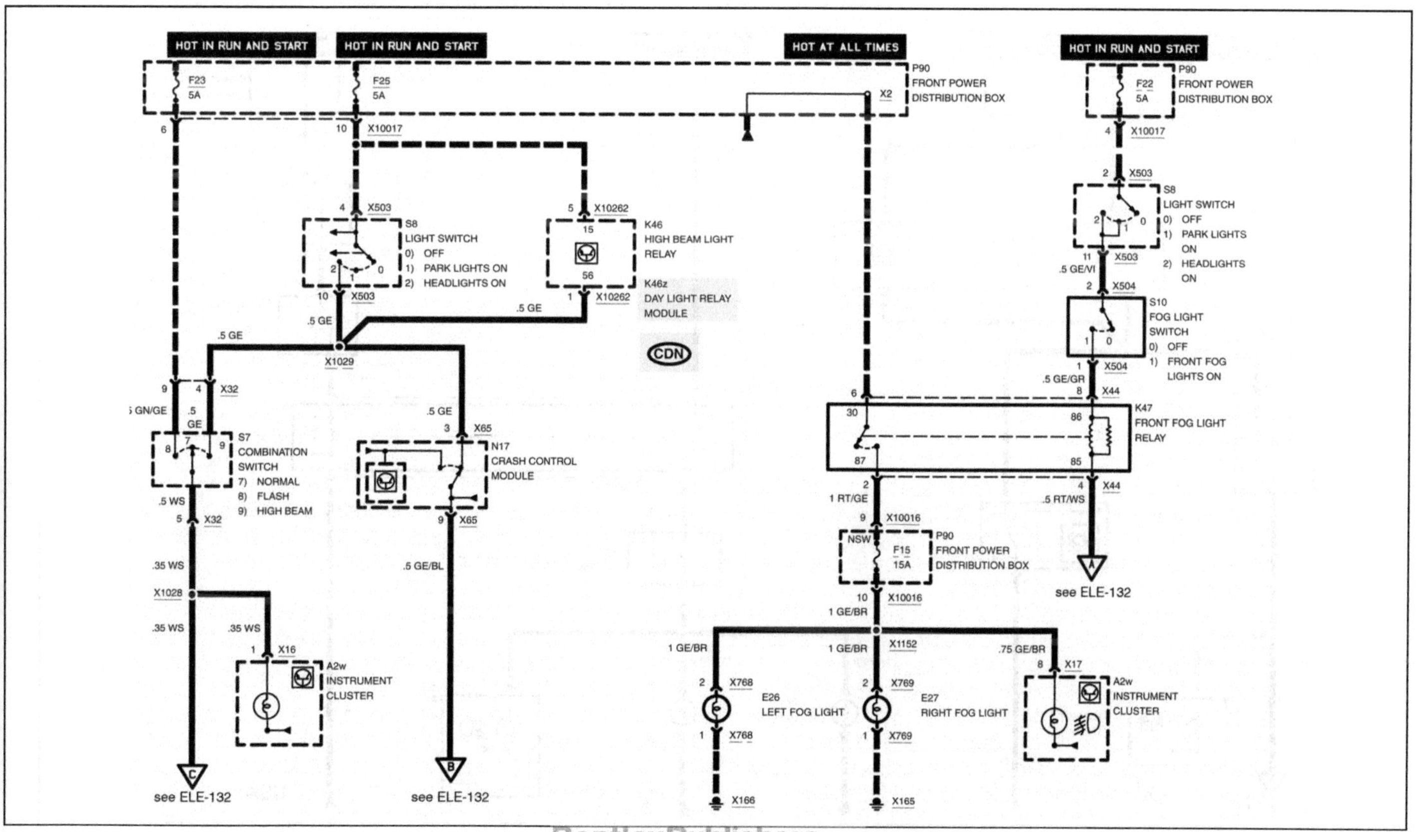

## Headlights / foglights, 1996 - 1998
(2 of 2)

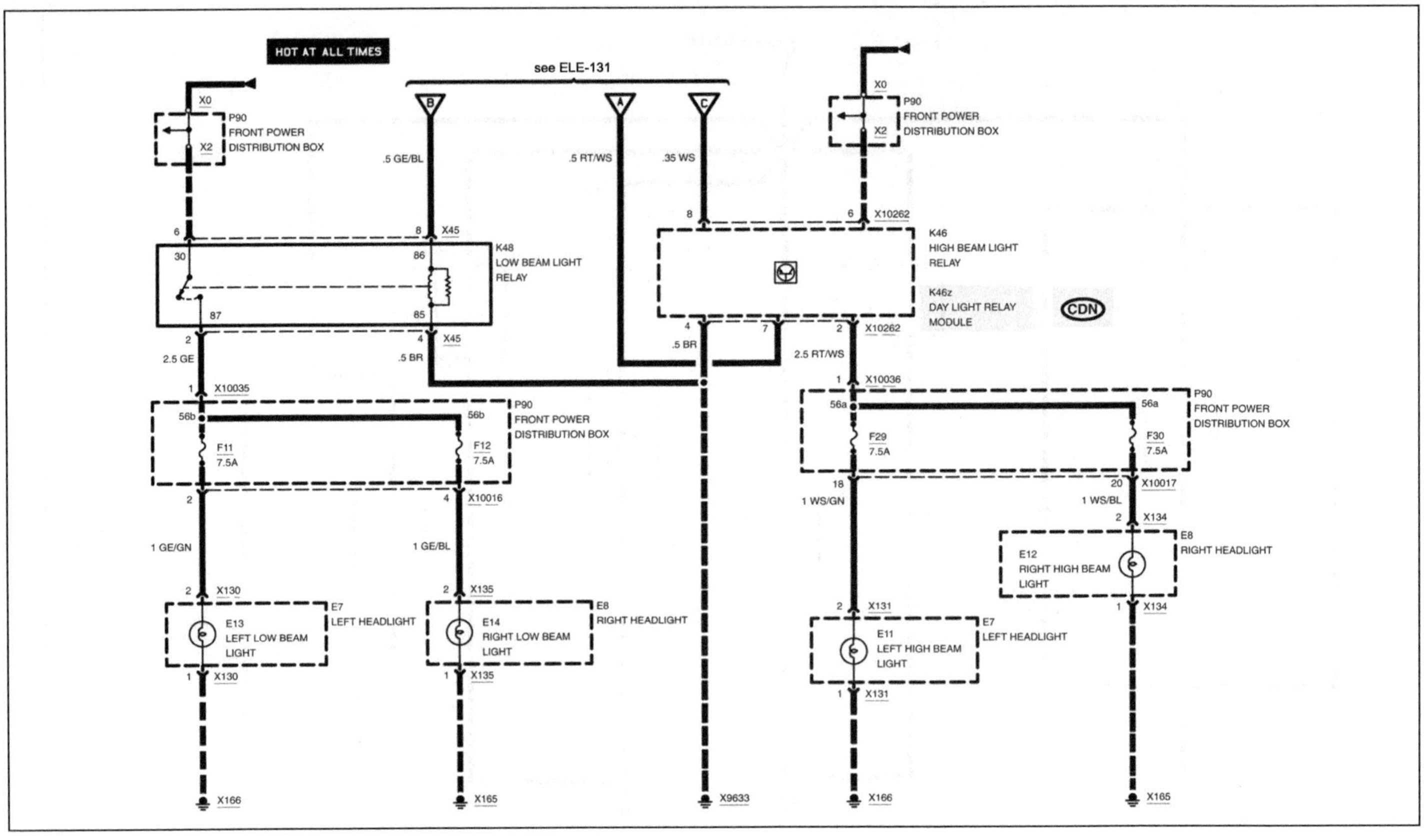


## Headlights / foglights, 2000 - 2002
(1 of 4)

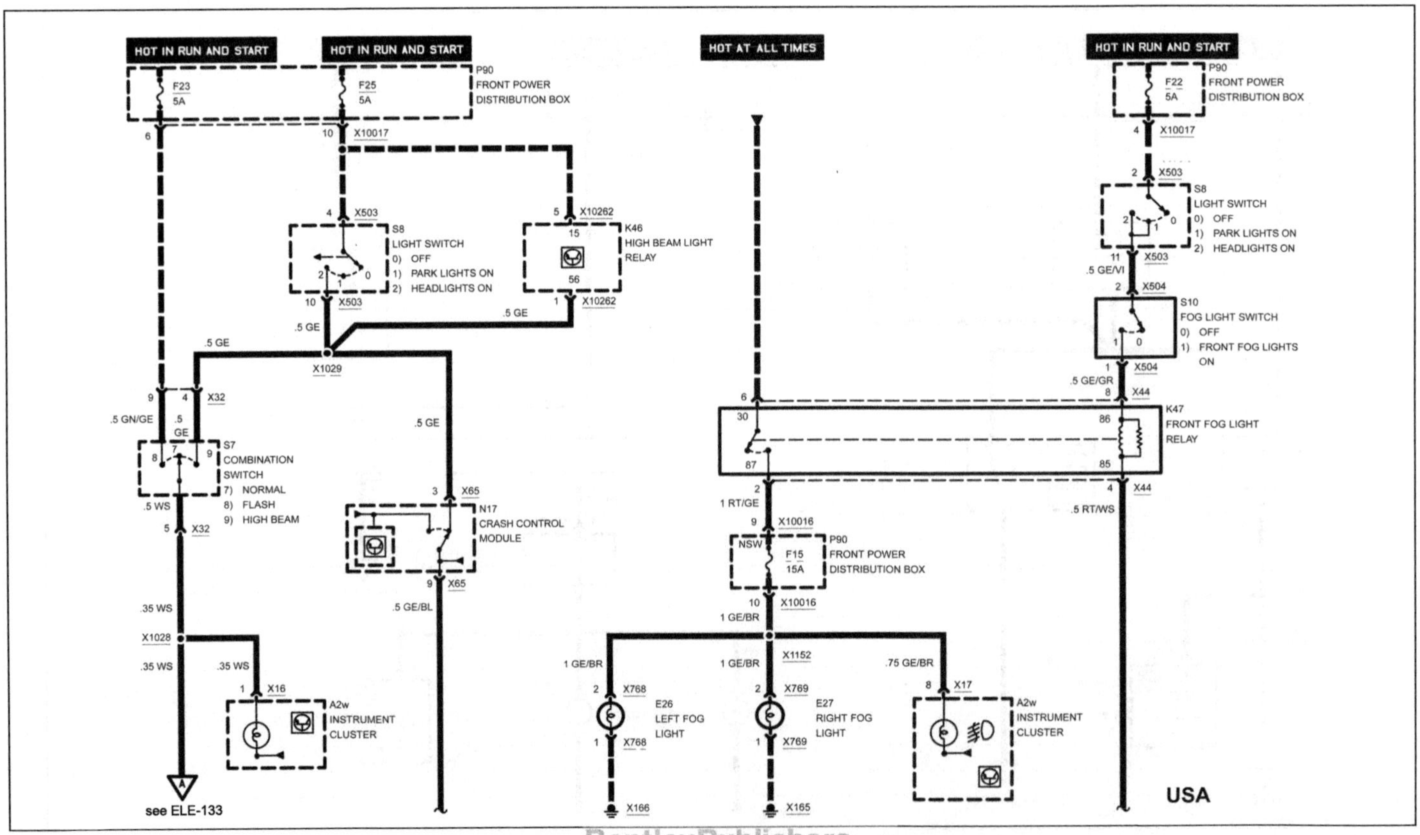

## Headlights / foglights, 2000 - 2002
(2 of 4)

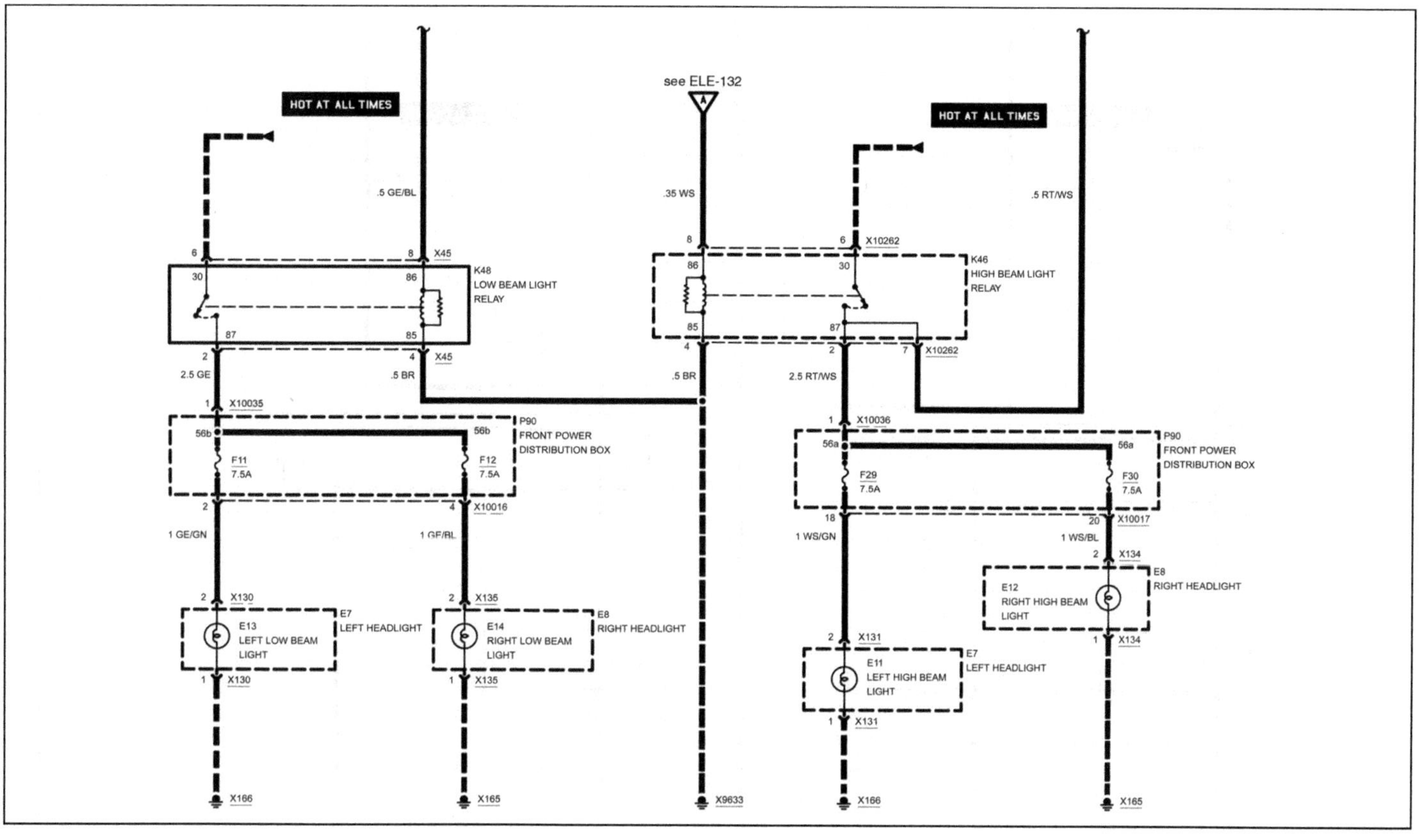


## Headlights / foglights, 2000 - 2002
(3 of 4)

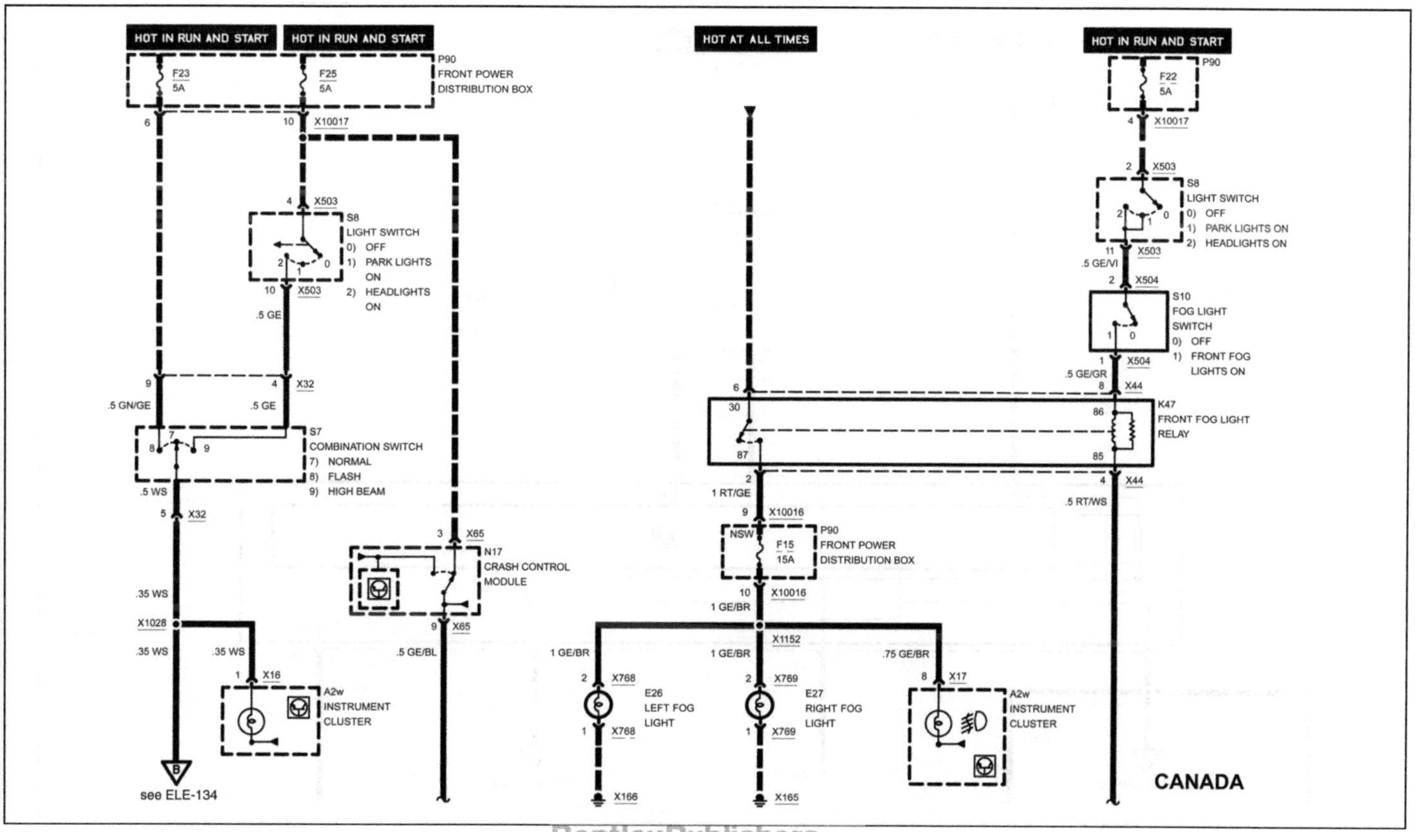

## Headlights / foglights, 2000 - 2002
(4 of 4)

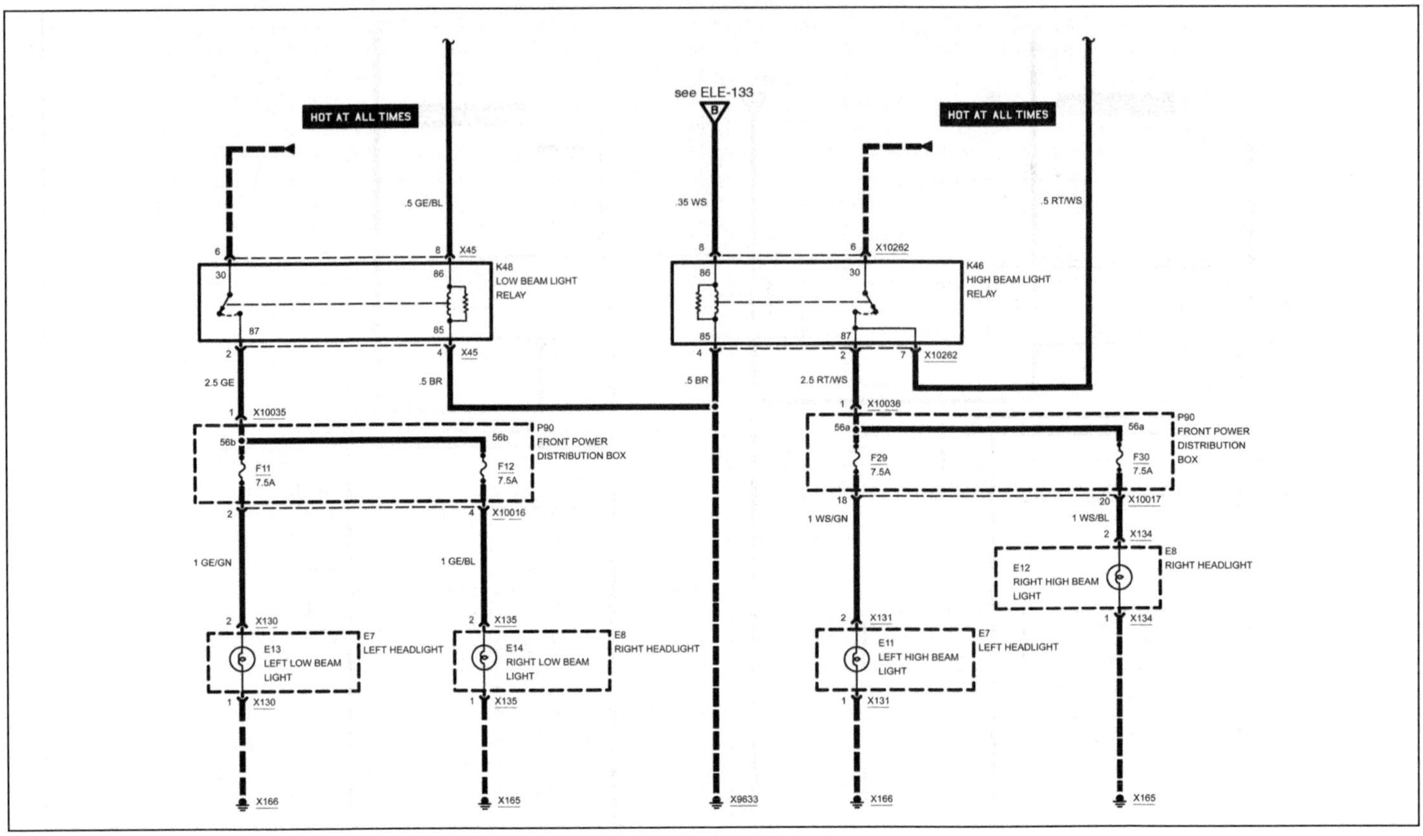


## Turn signals / hazard lights
(1 of 4)

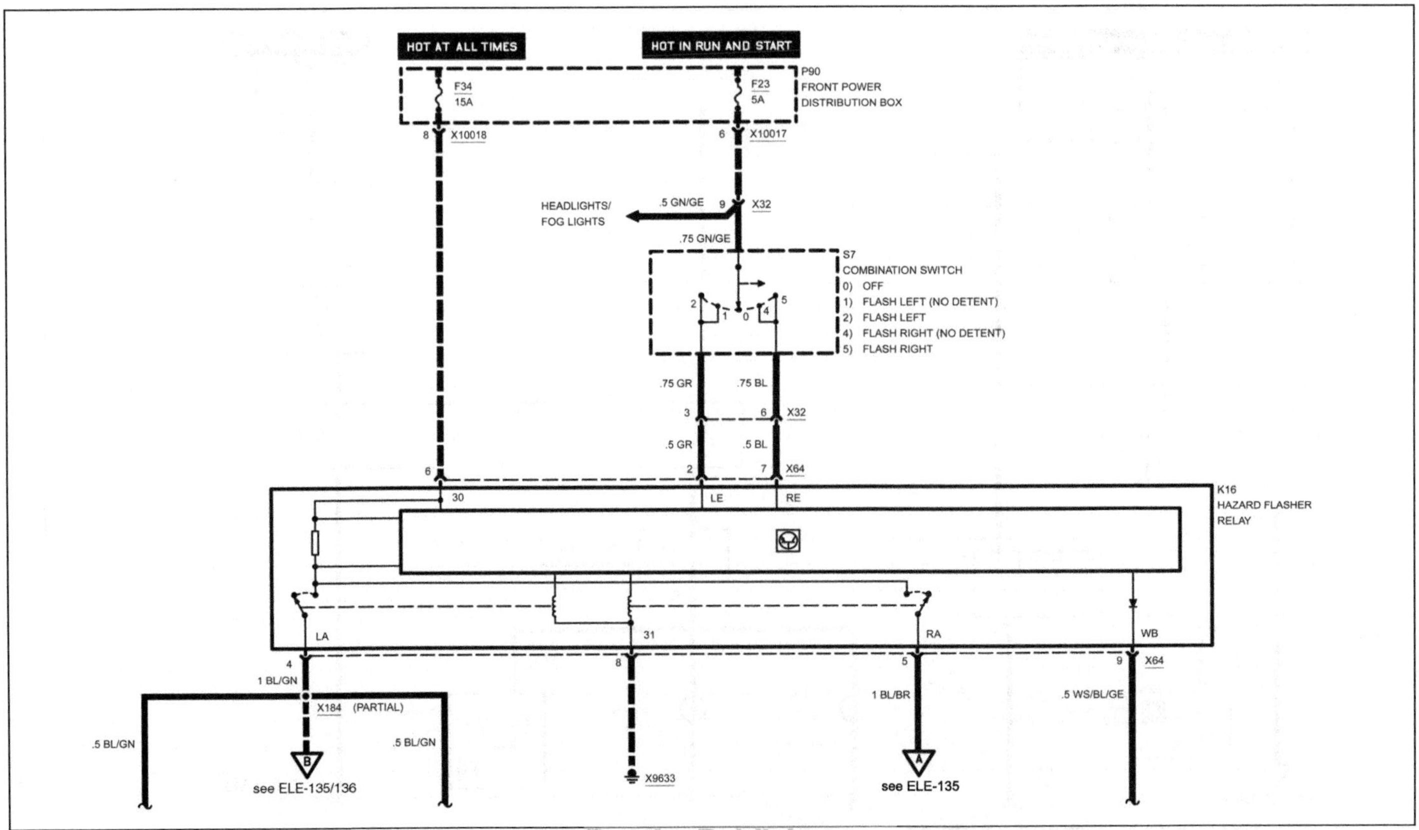

## Turn signals / hazard lights
(2 of 4)

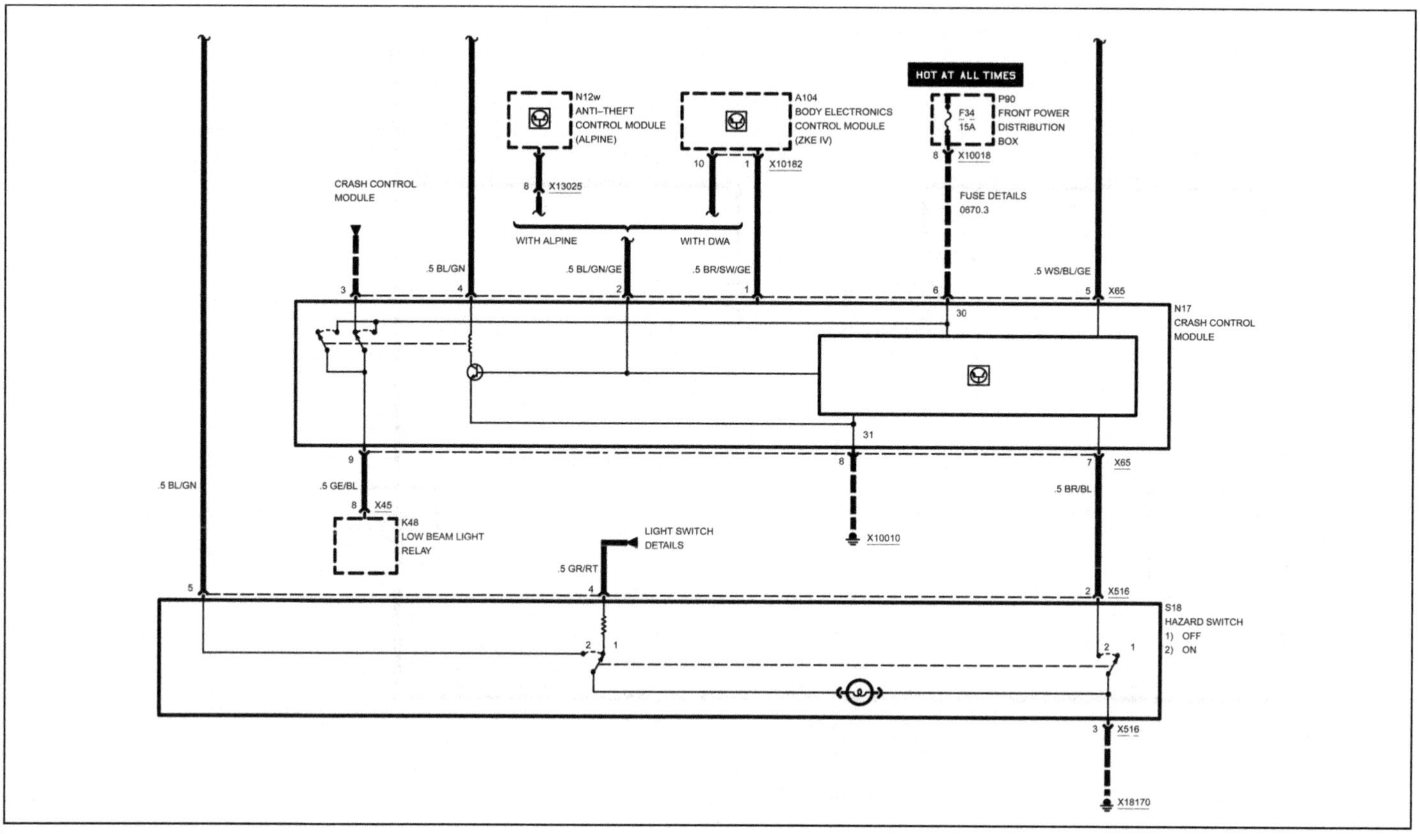


## Turn signals / hazard lights
(3 of 4)

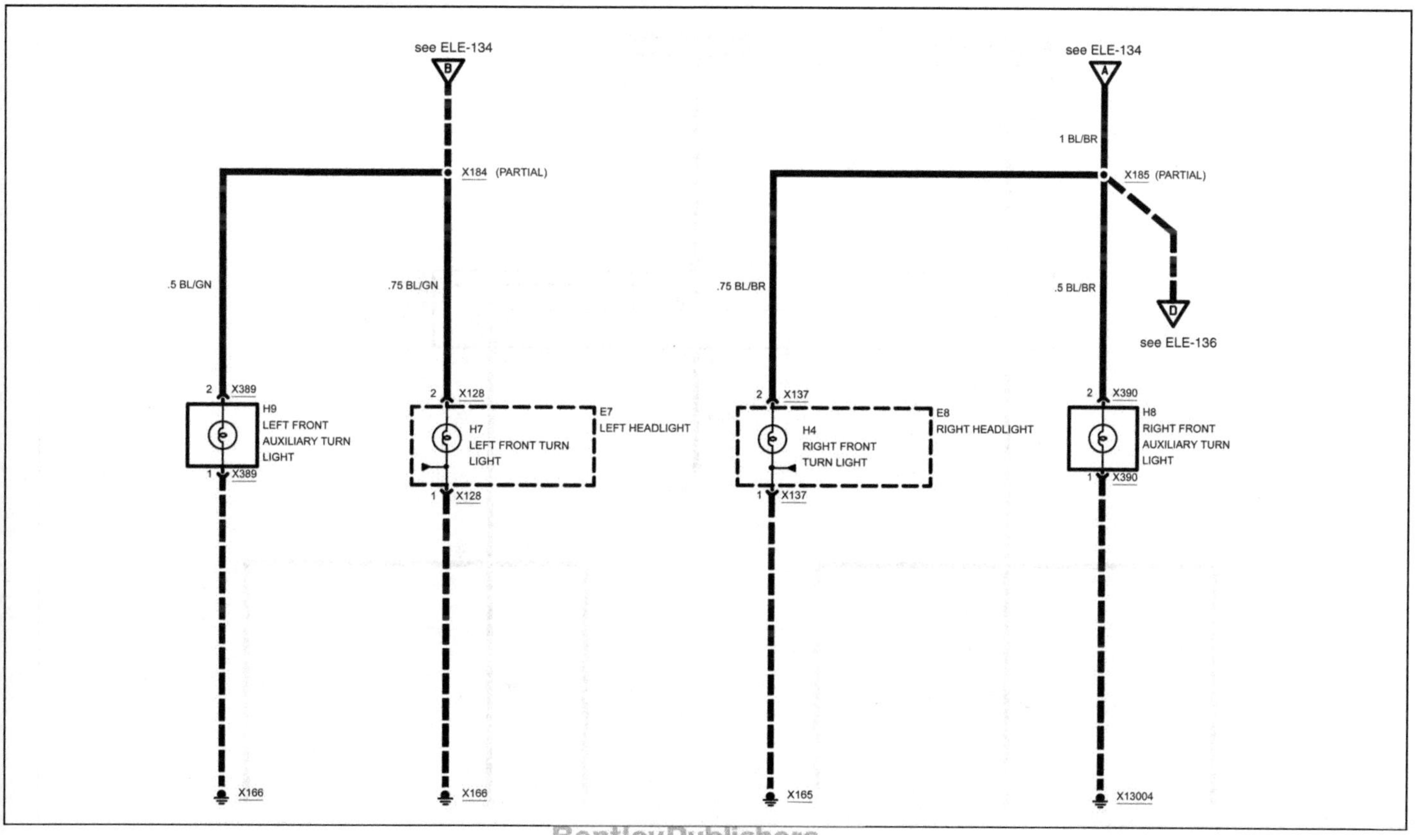

## Turn signals / hazard lights

**(4 of 4)**

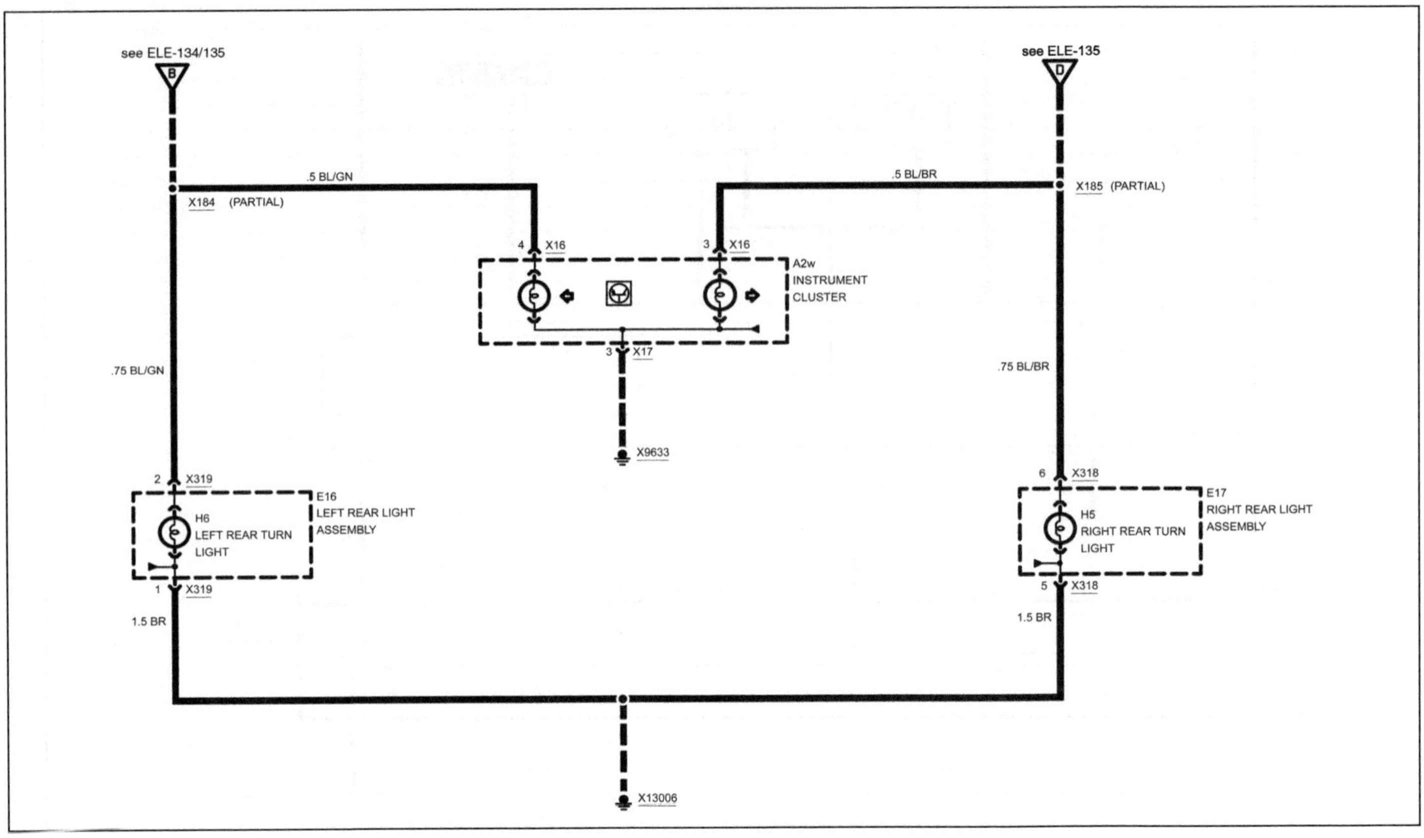


## Park / tail / underhood lights, 1996 - 1998

**(1 of 2)**

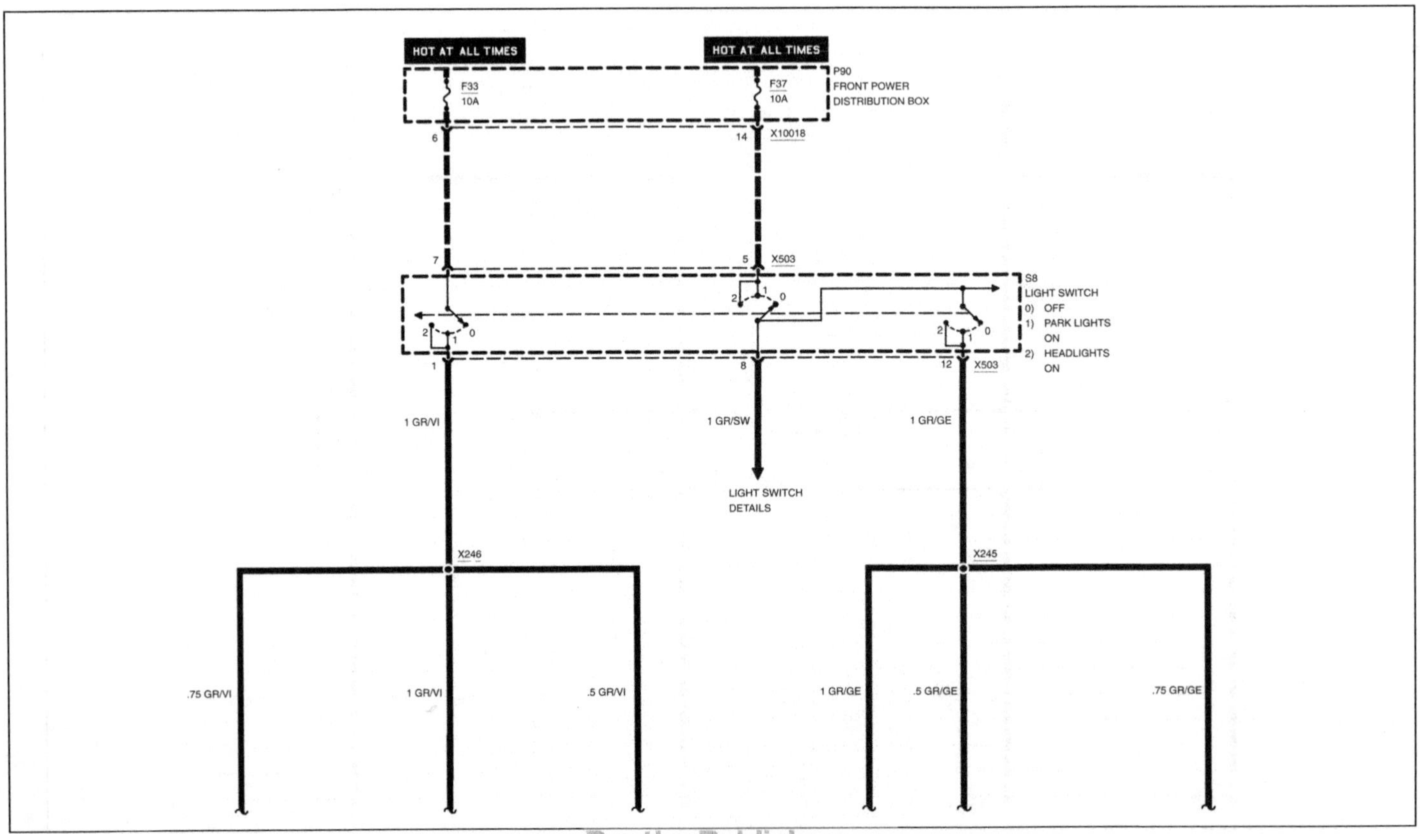

## Park / tail / underhood lights, 1996 - 1998

(2 of 2)

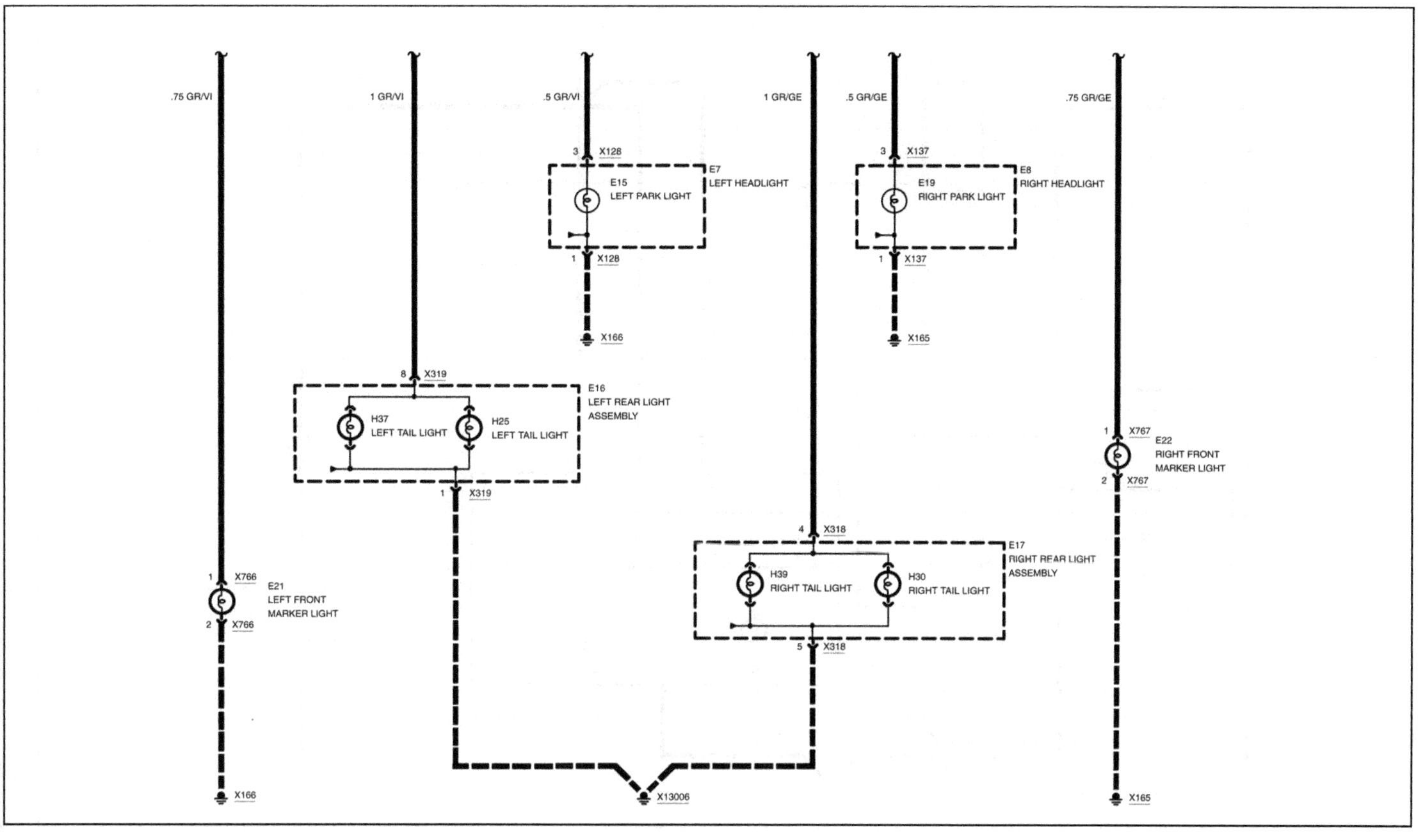


## Park / tail / underhood lights, 2000 - 2002

(1 of 4)

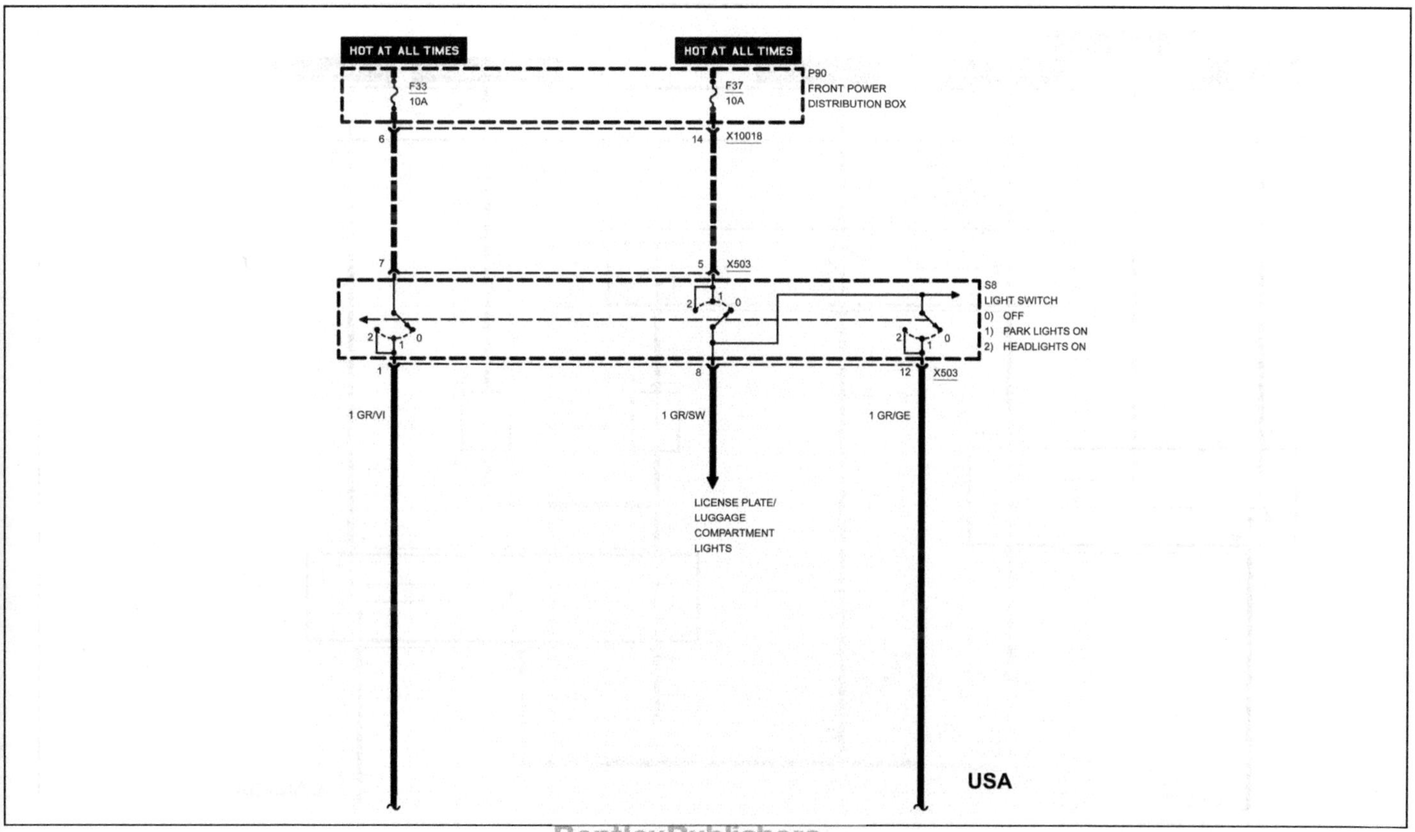

## Park / tail / underhood lights, 2000 - 2002
(2 of 4)

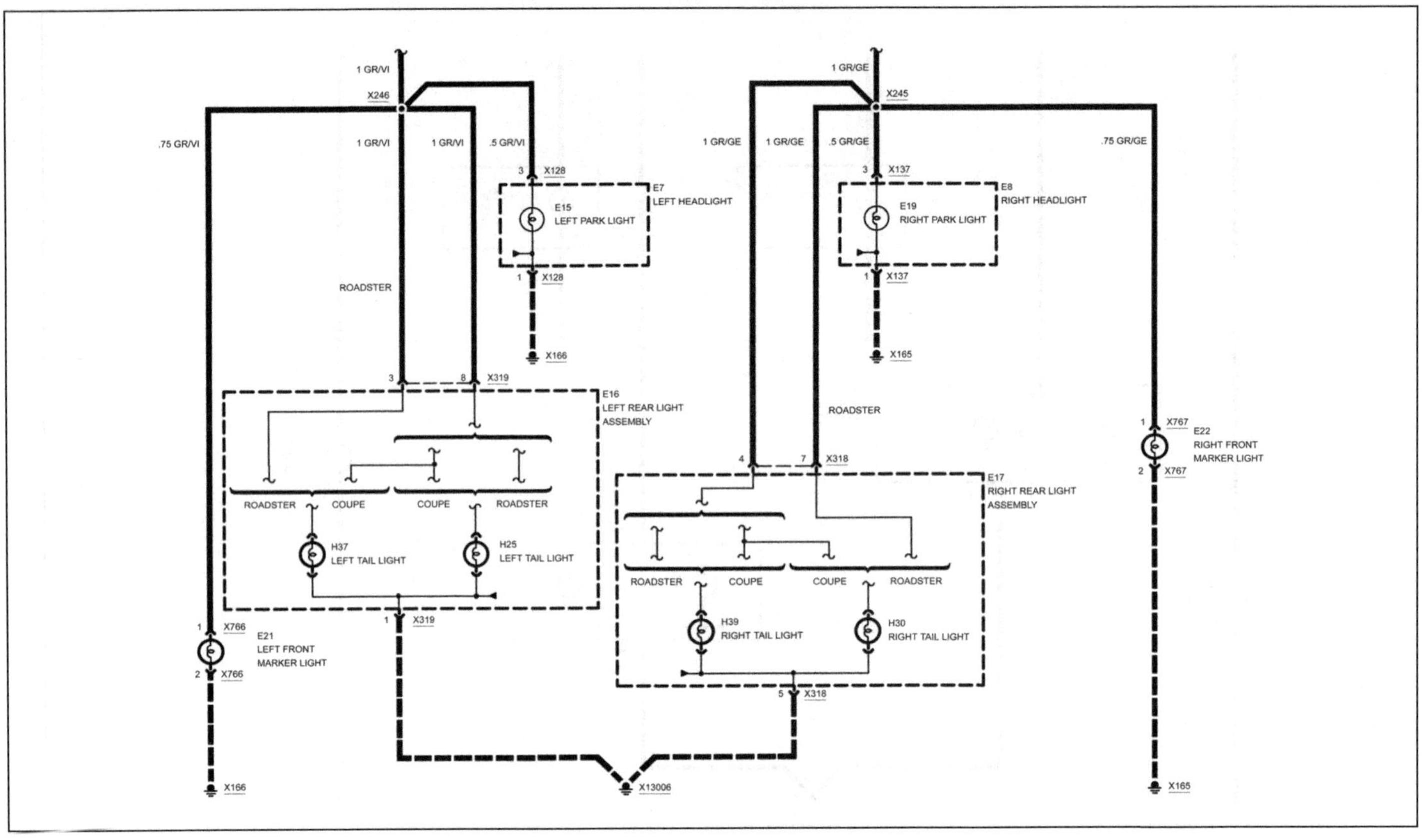


## Park / tail / underhood lights, 2000 - 2002
(3 of 4)

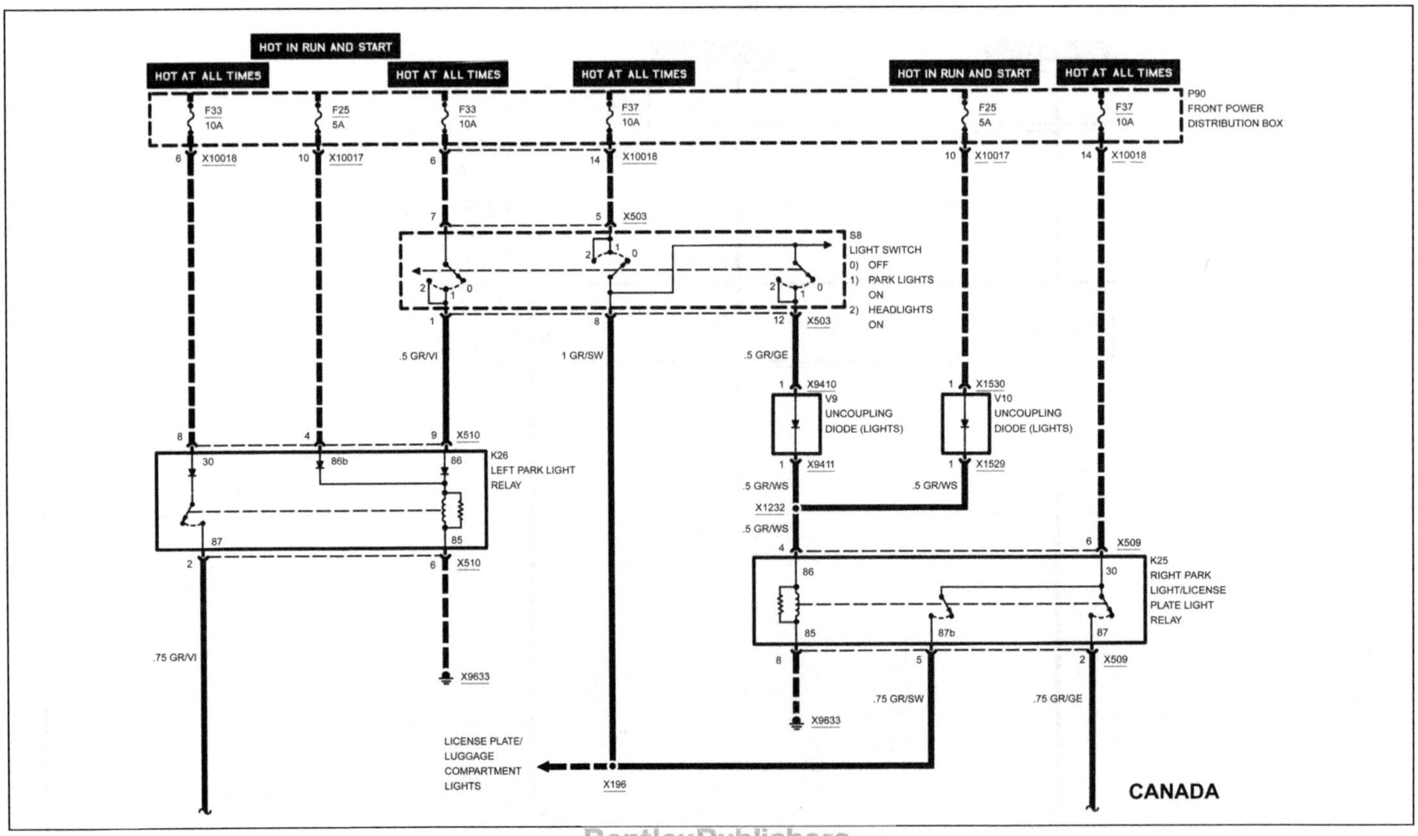

## Park / tail / underhood lights, 2000 - 2002

(4 of 4)

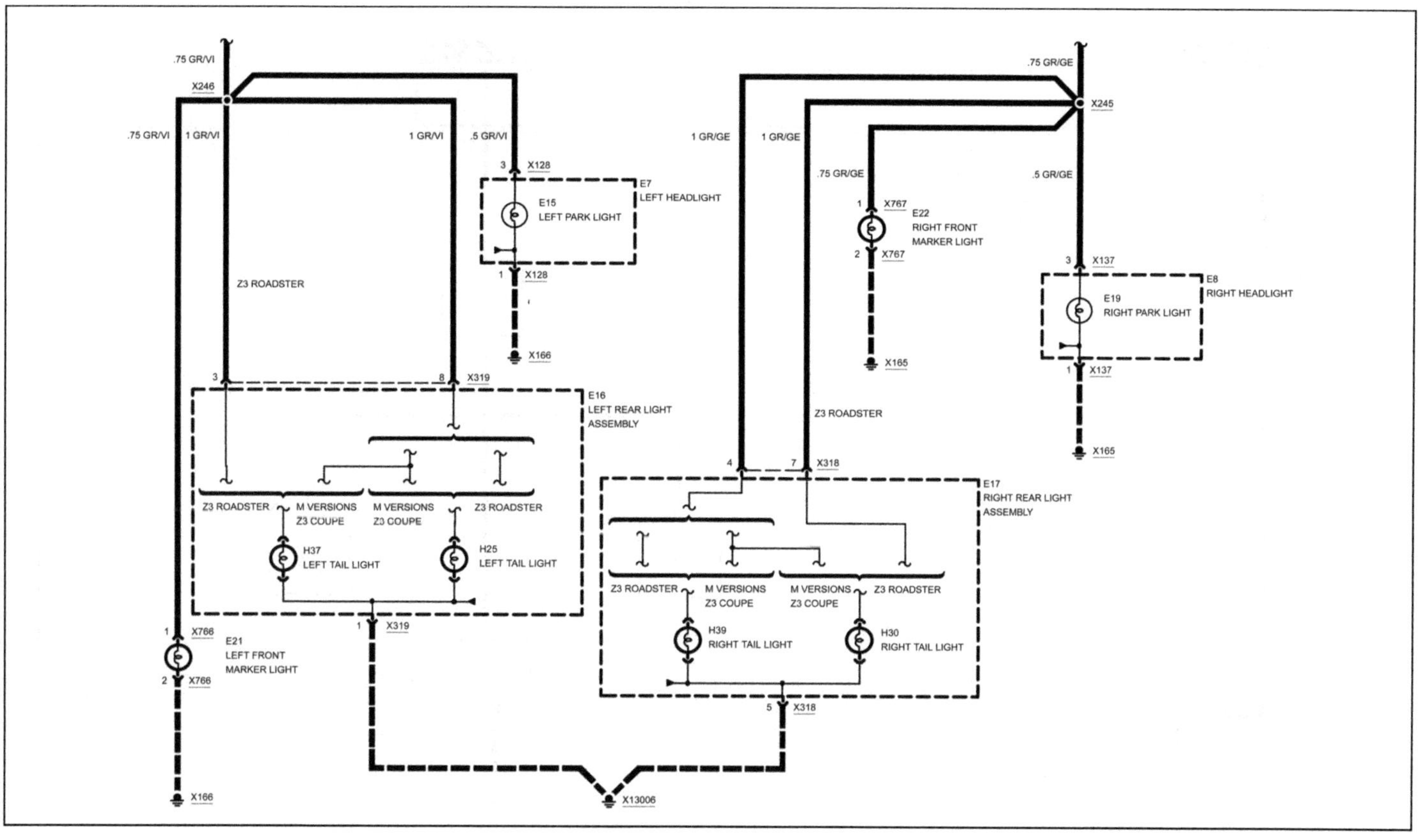


## License plate / trunk lights, 1996

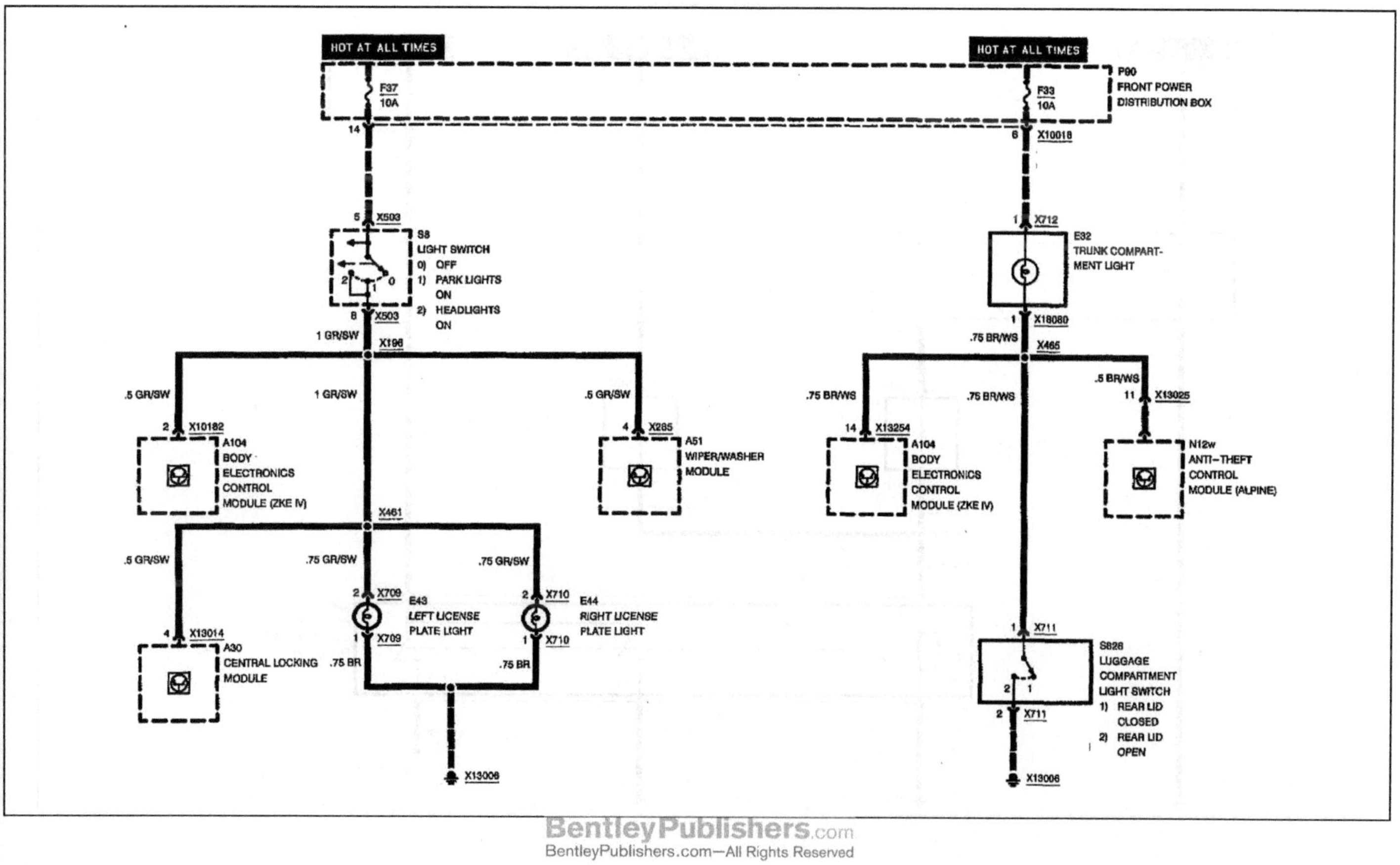

## License plate / trunk lights, 1998

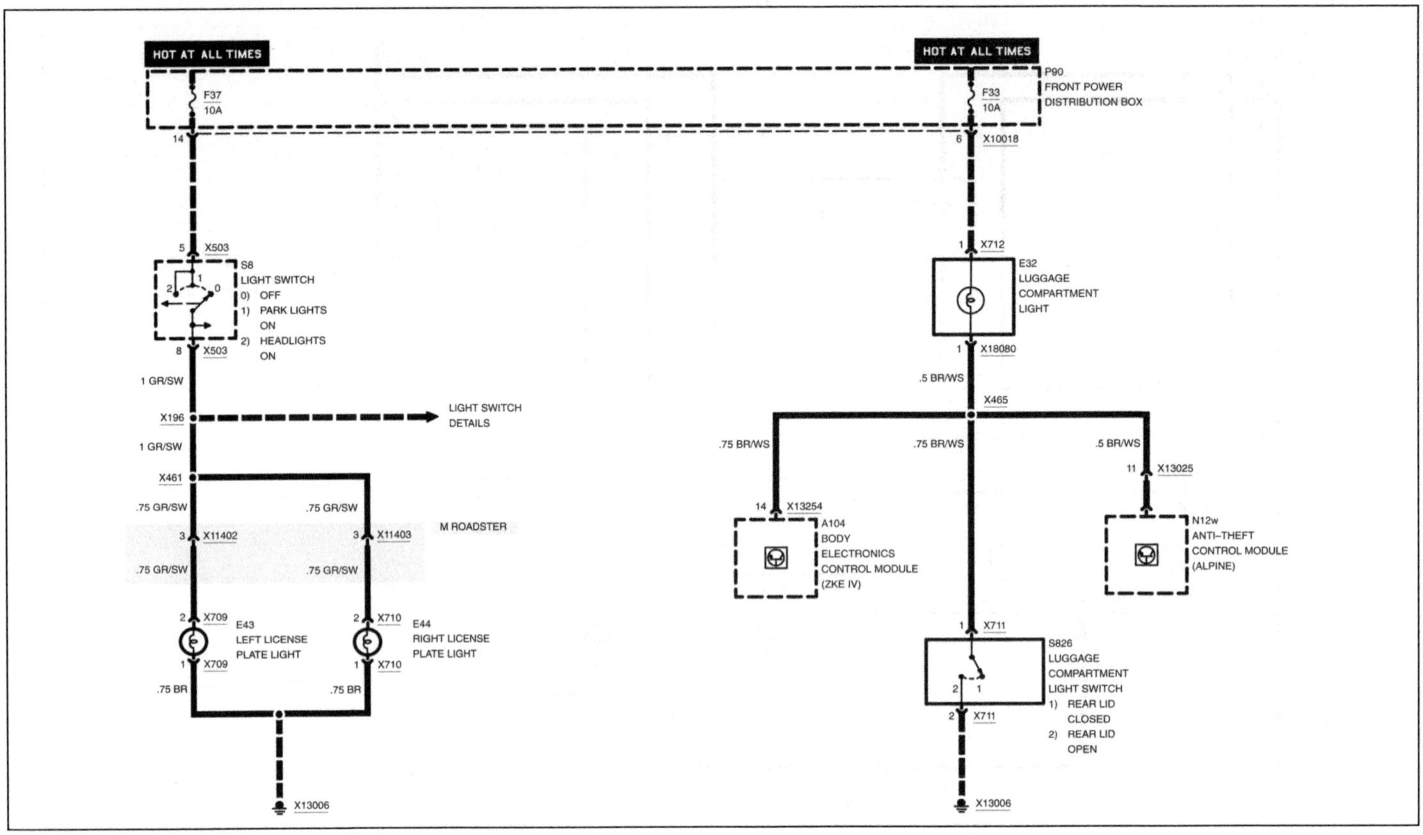


## License plate / trunk lights, 2000 - 2002
(1 of 4)

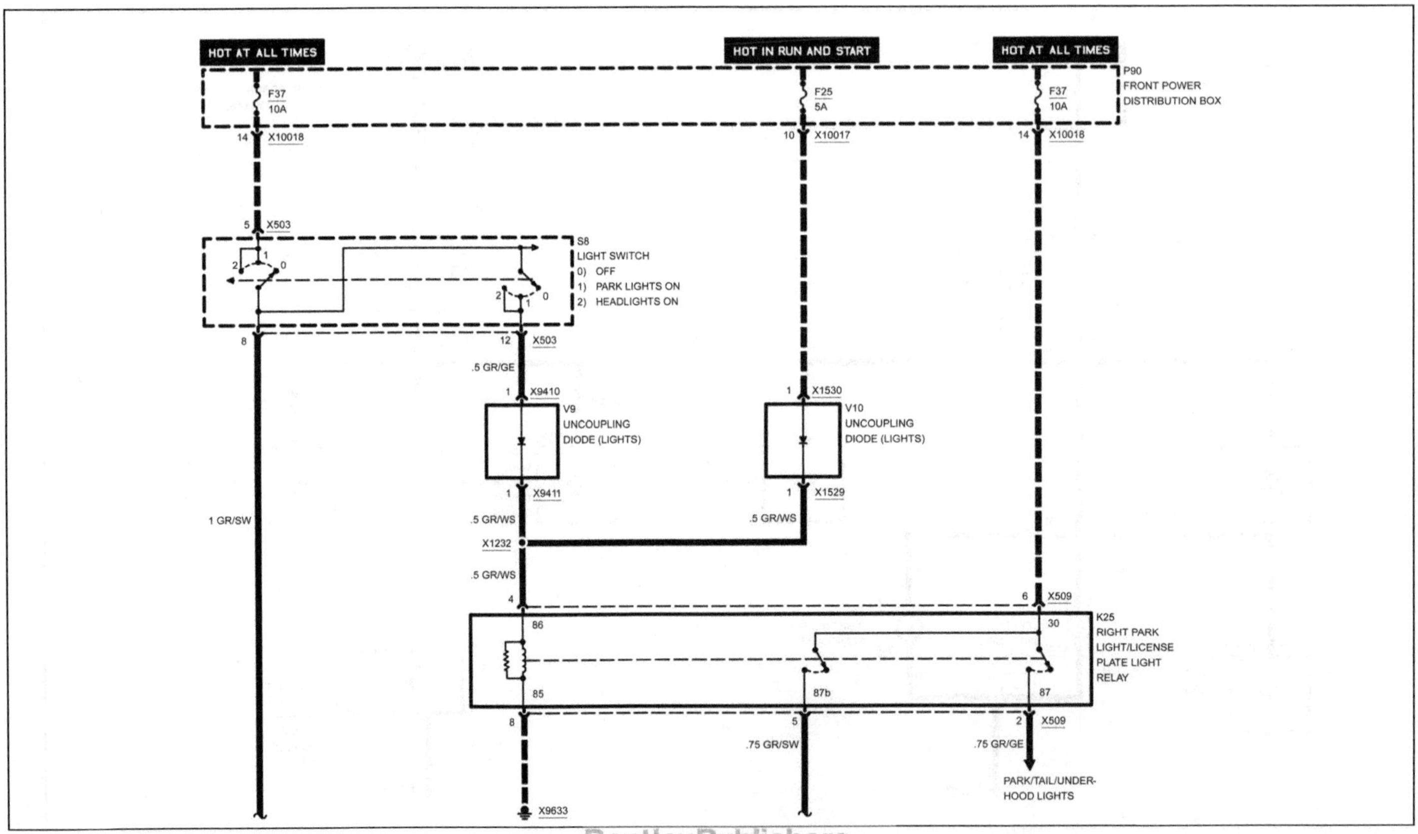

## License plate / trunk lights, 2000 - 2002
(2 of 4)

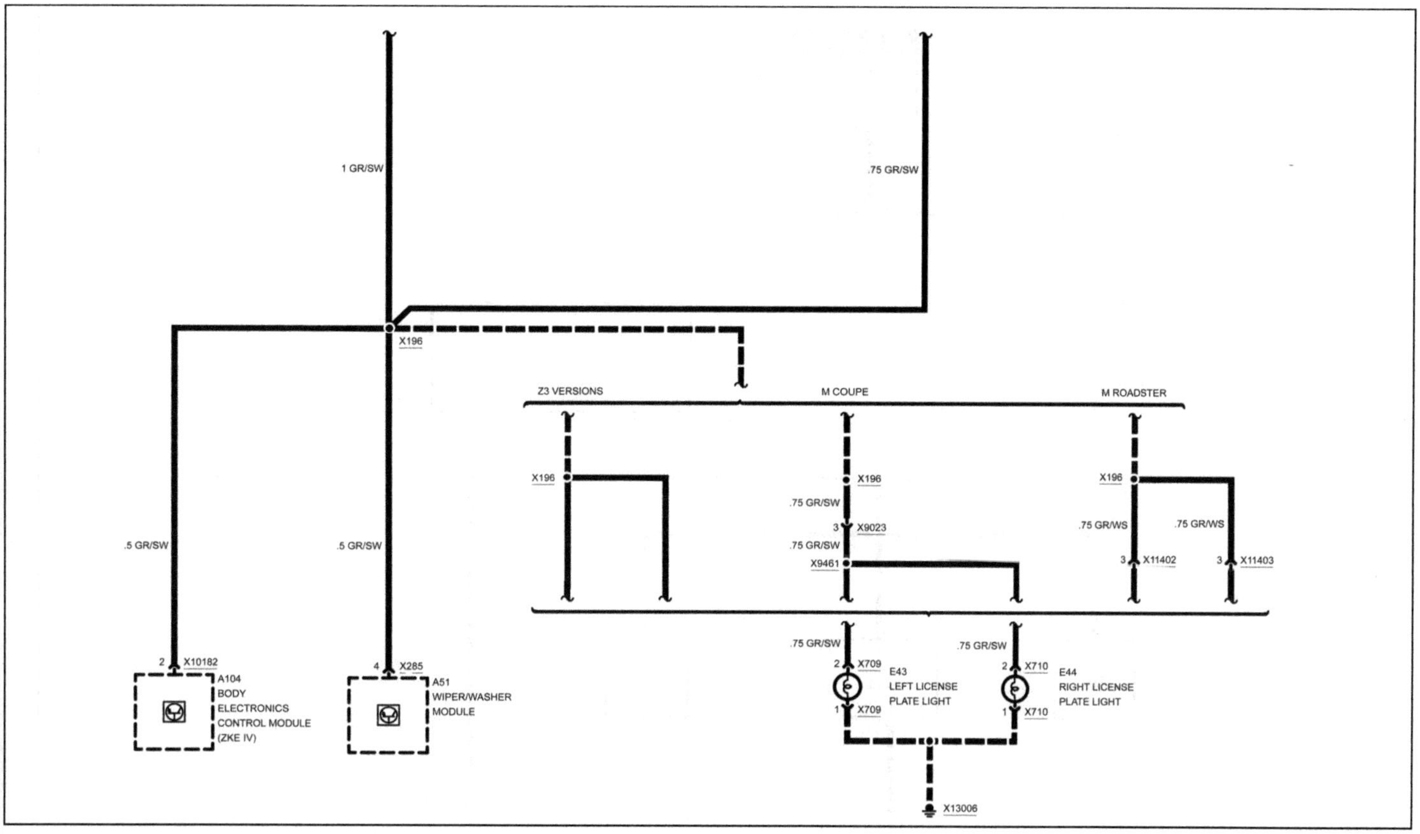


## License plate / trunk lights, 2000 - 2002
(3 of 4)

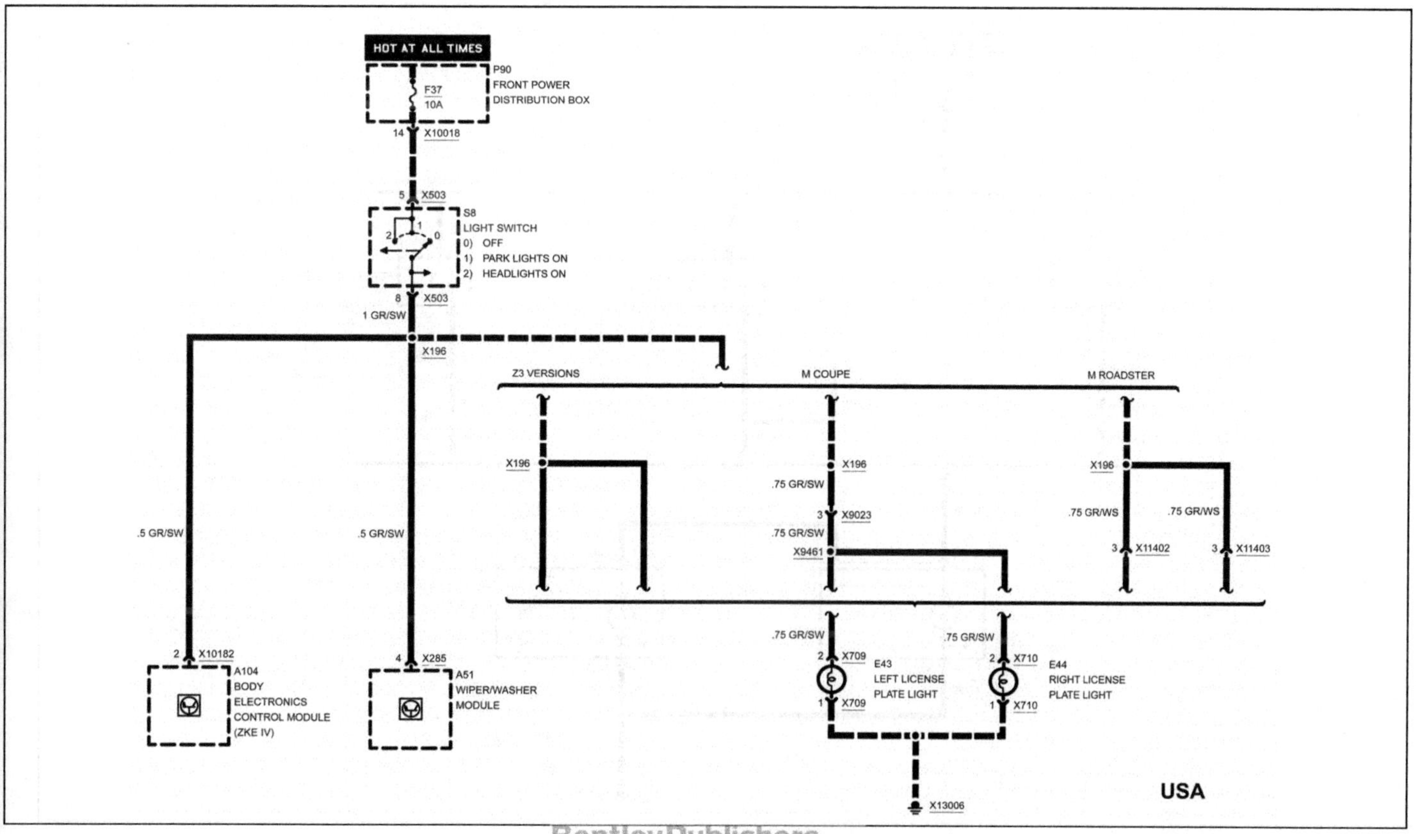

## License plate / trunk lights, 2000 - 2002
(4 of 4)

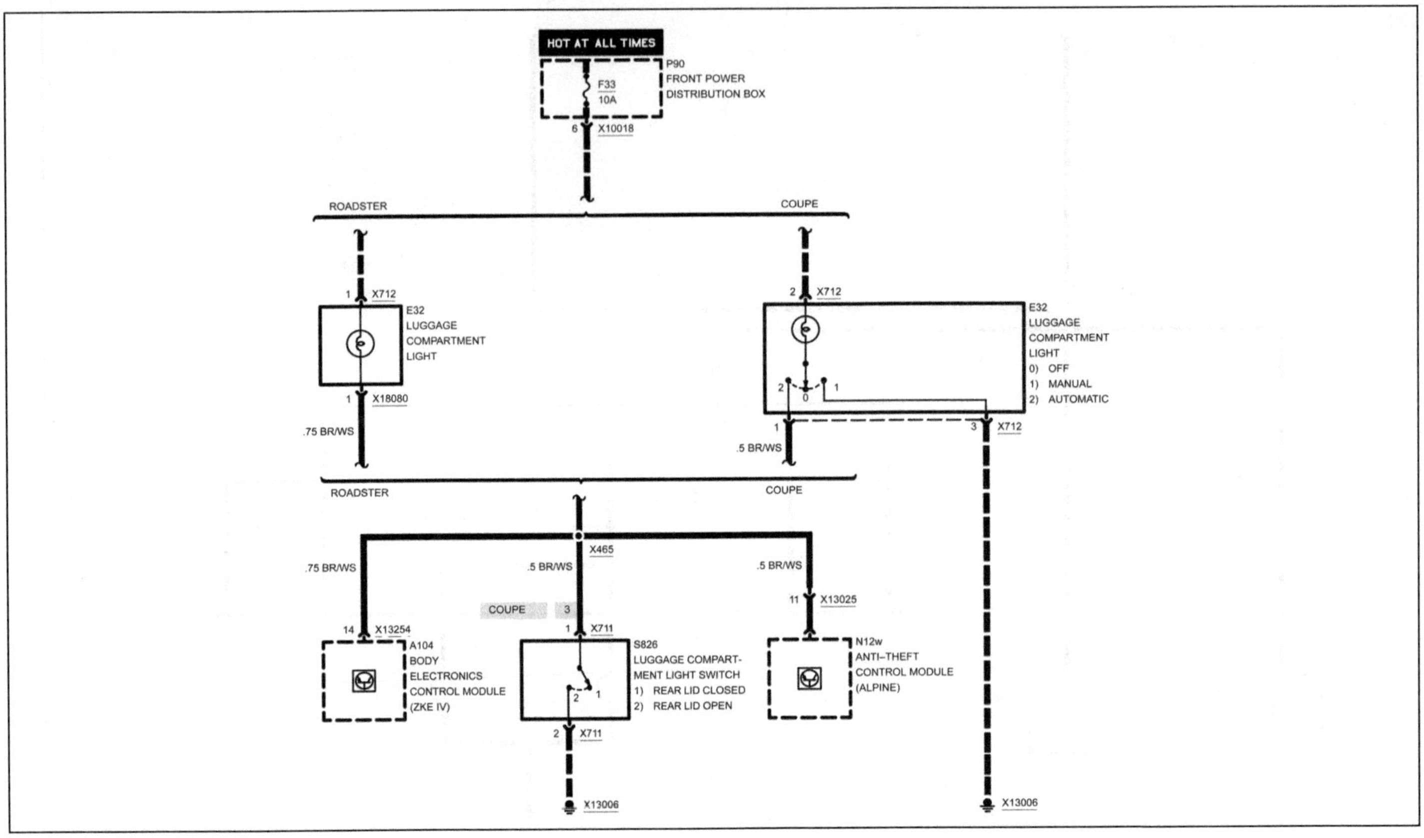


## Back-up lights, 1996 - 2000

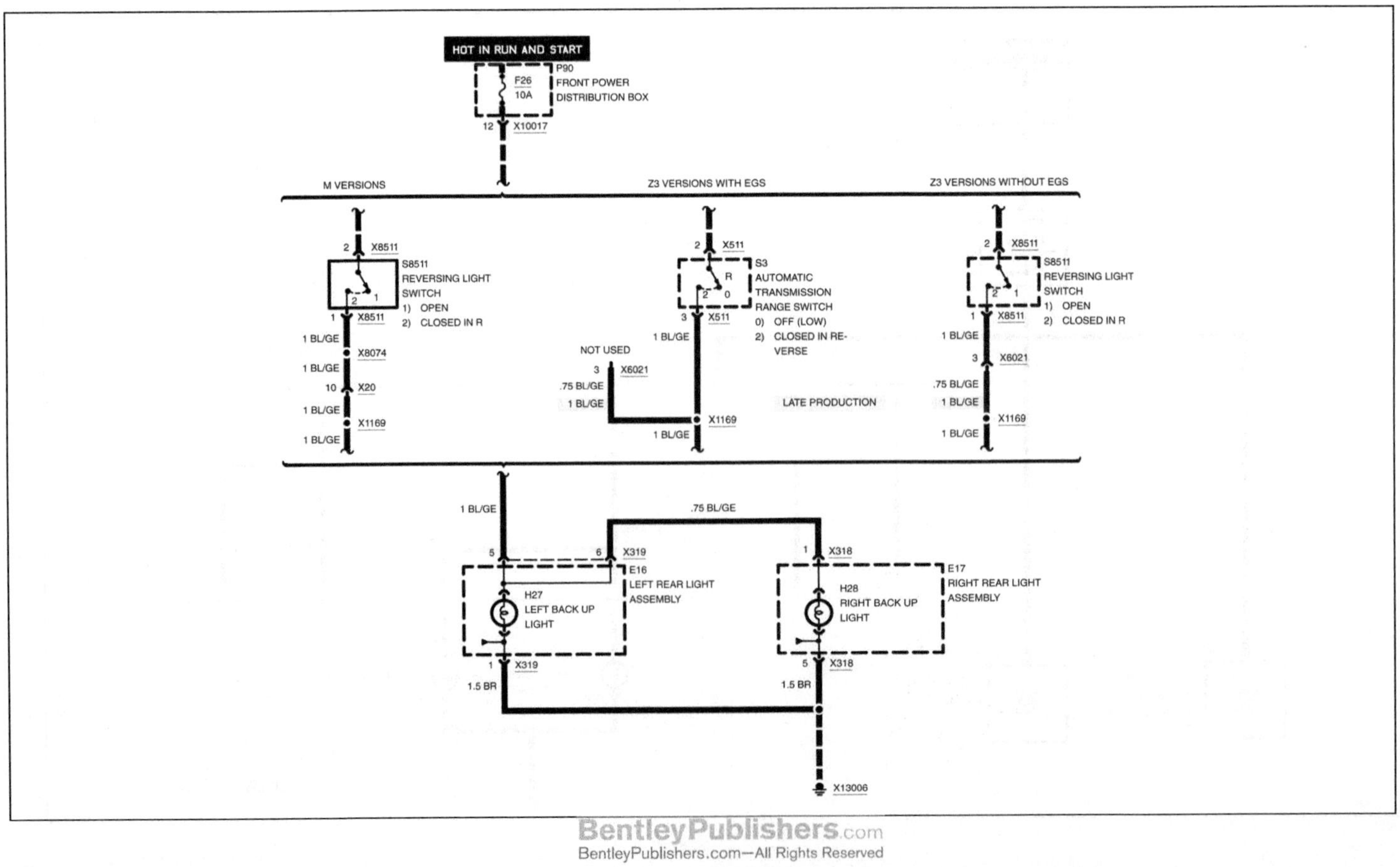

## Back-up lights, 2002

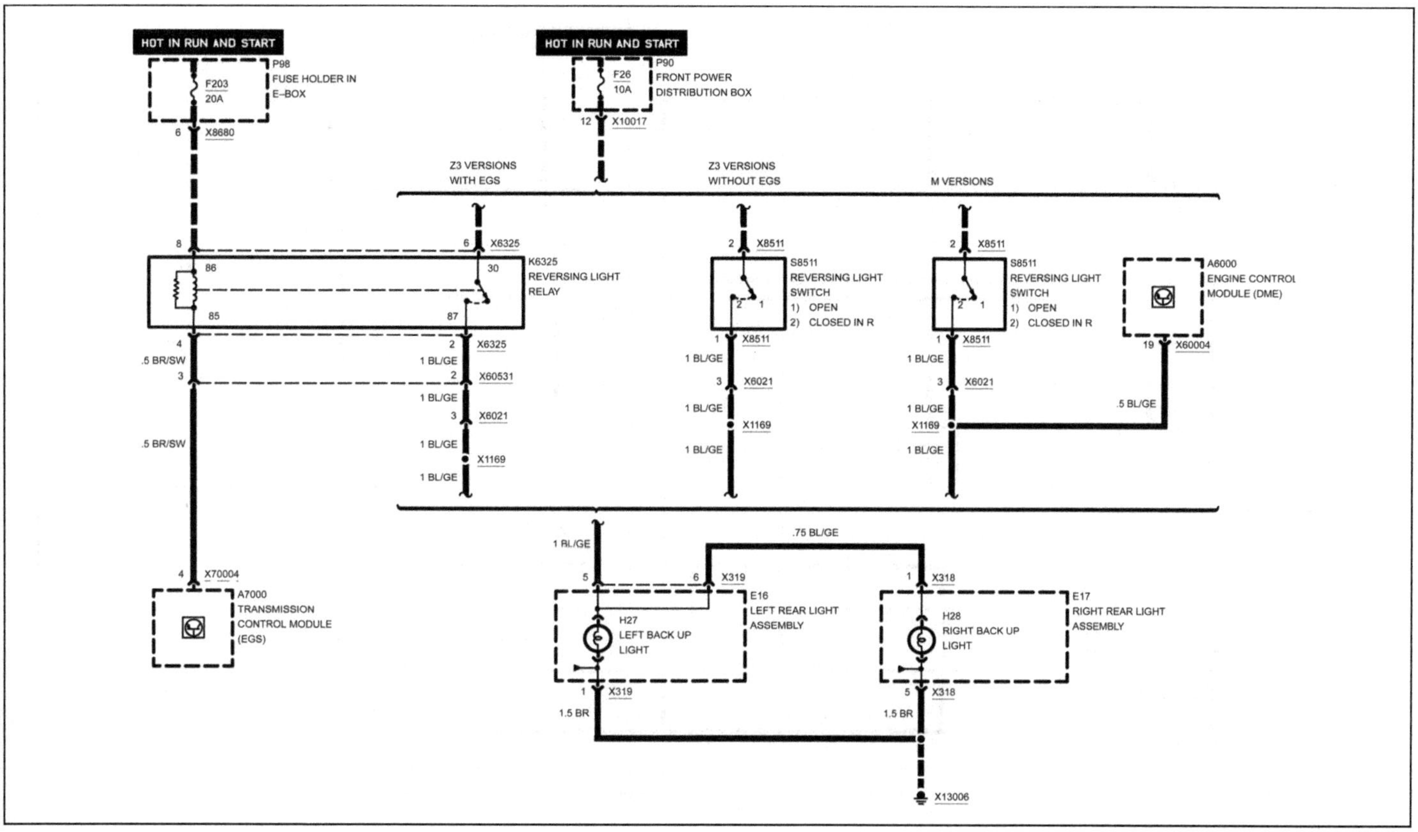


## Brake lights, 1996

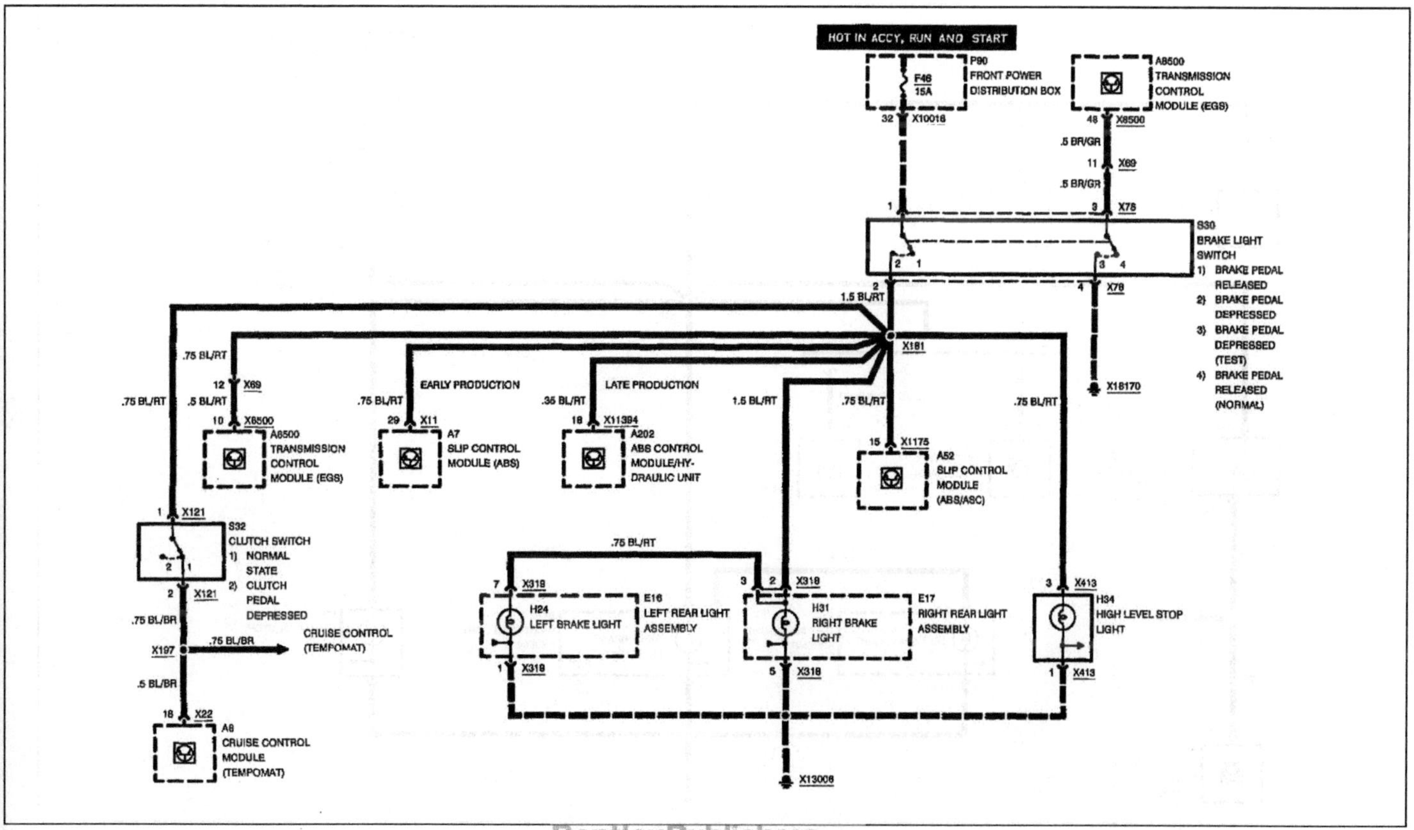

## Brake lights, 1998 Z3

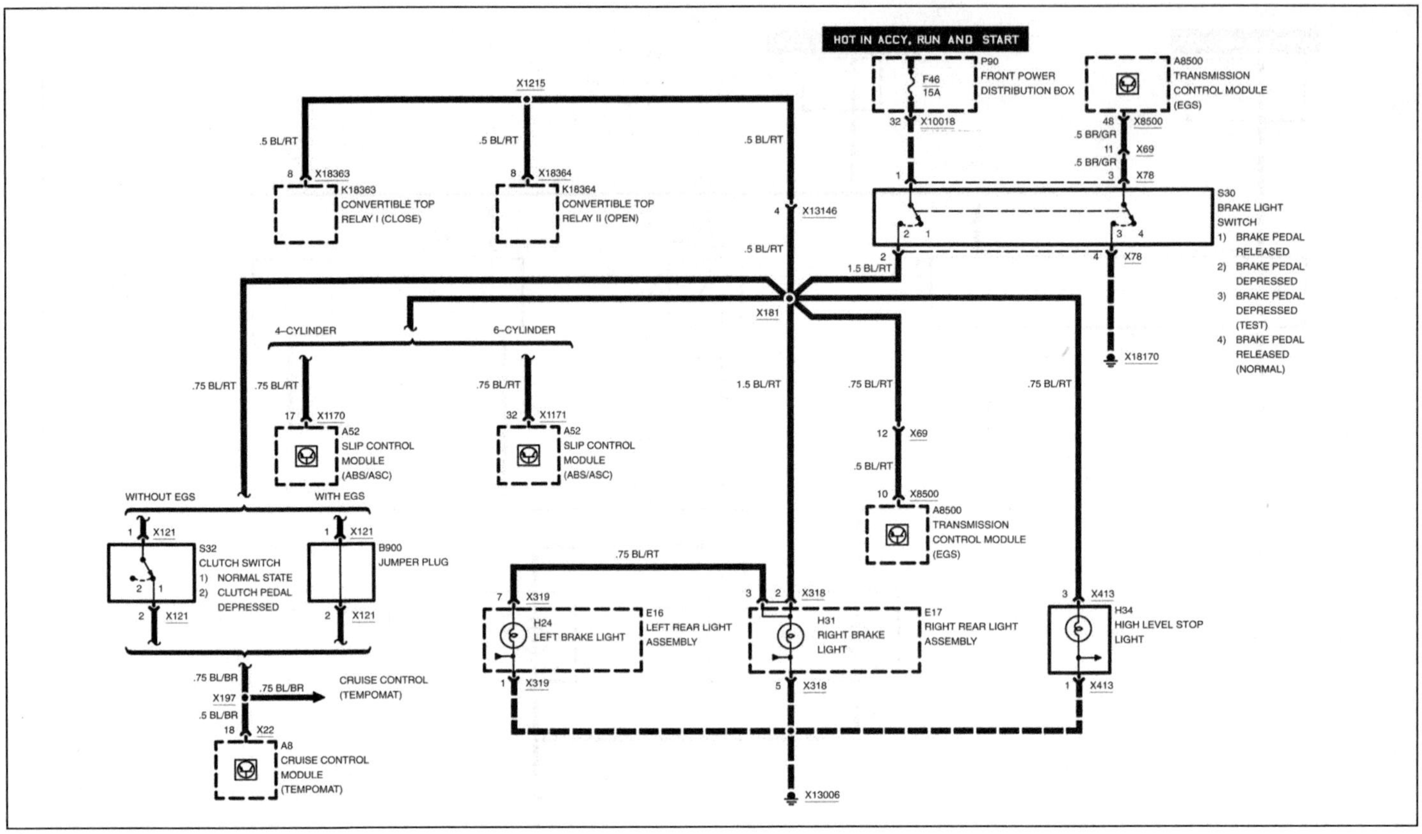


## Brake lights, 1998 M

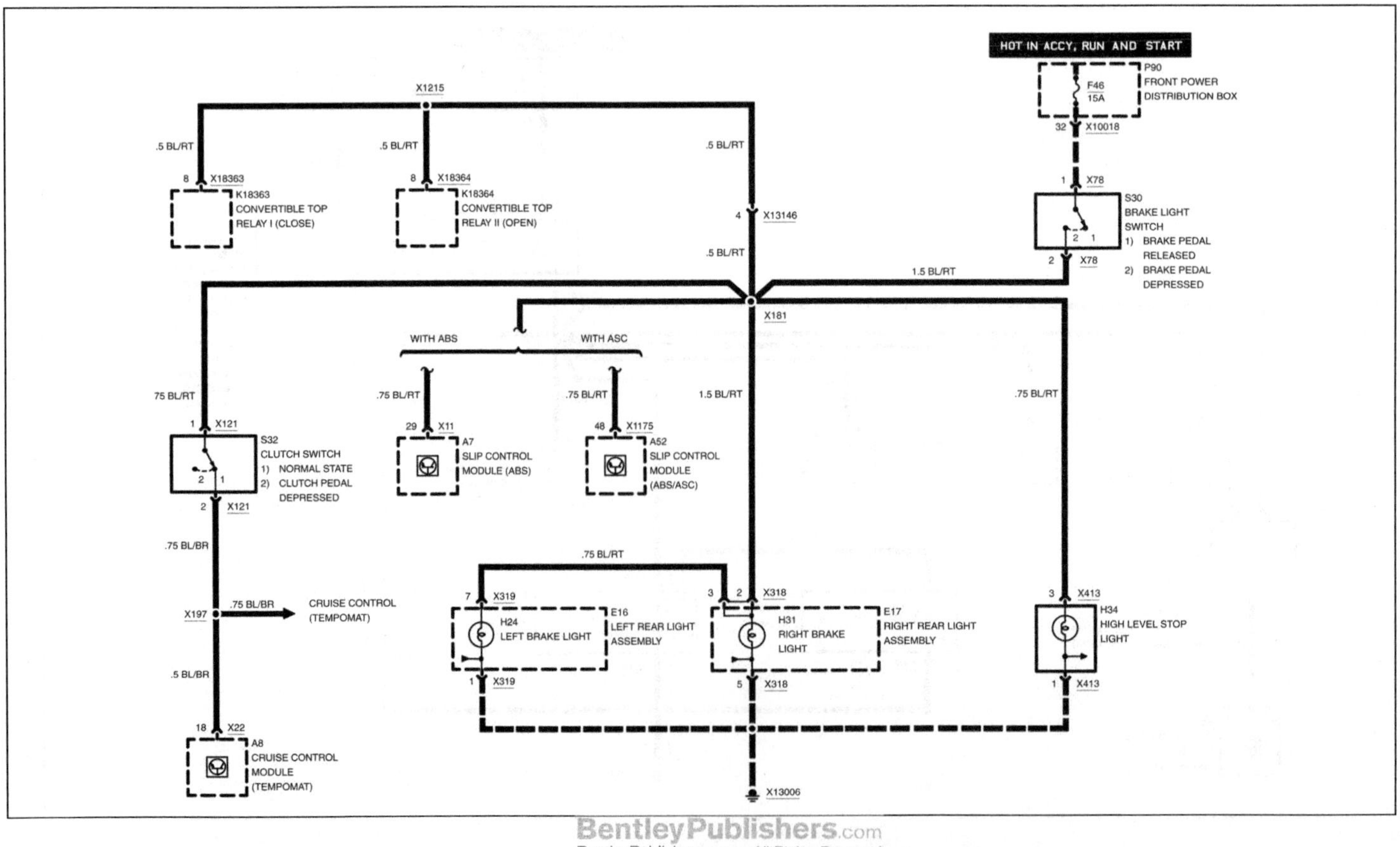

## Brake lights, 2000 Z3

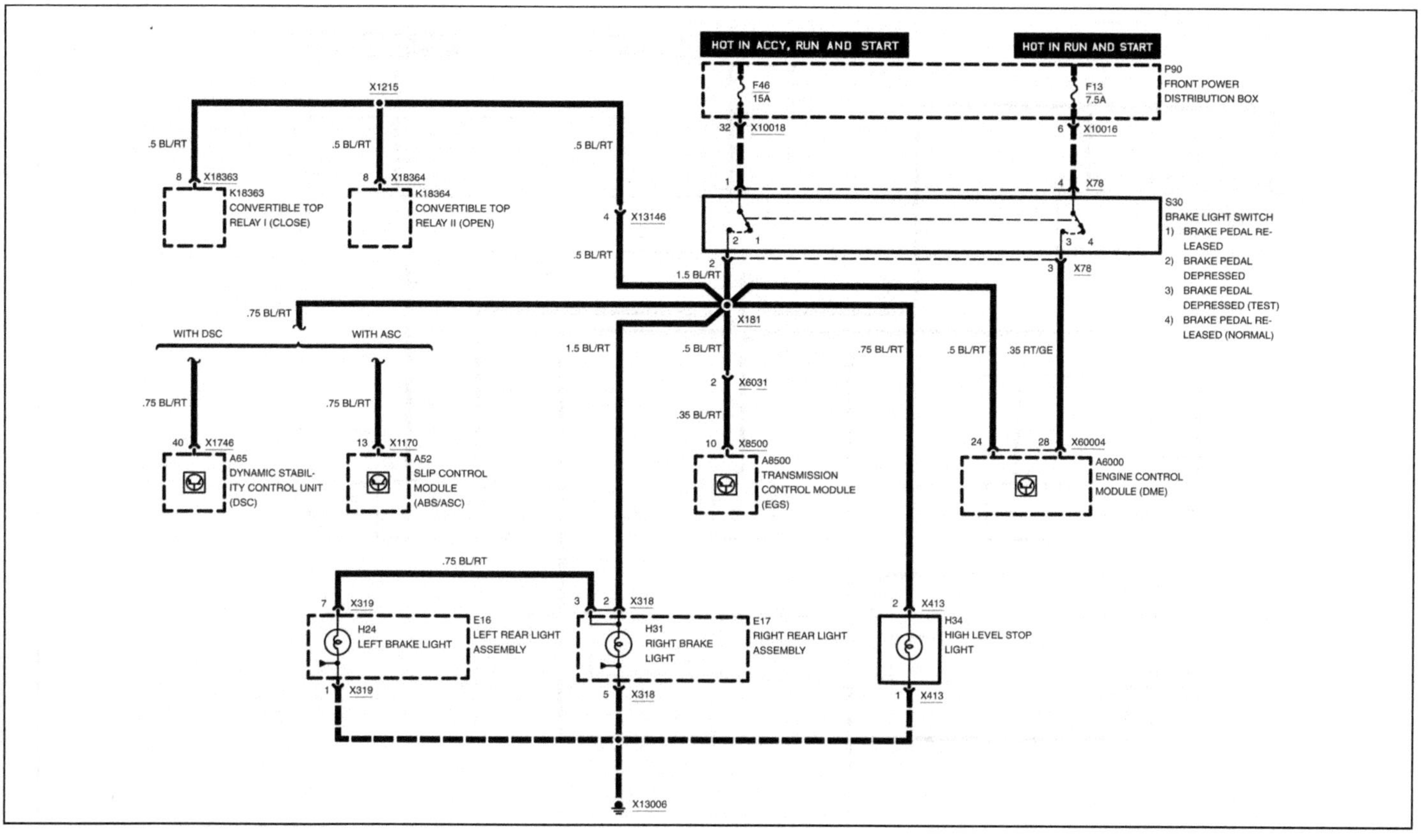


## Brake lights, 2000 M

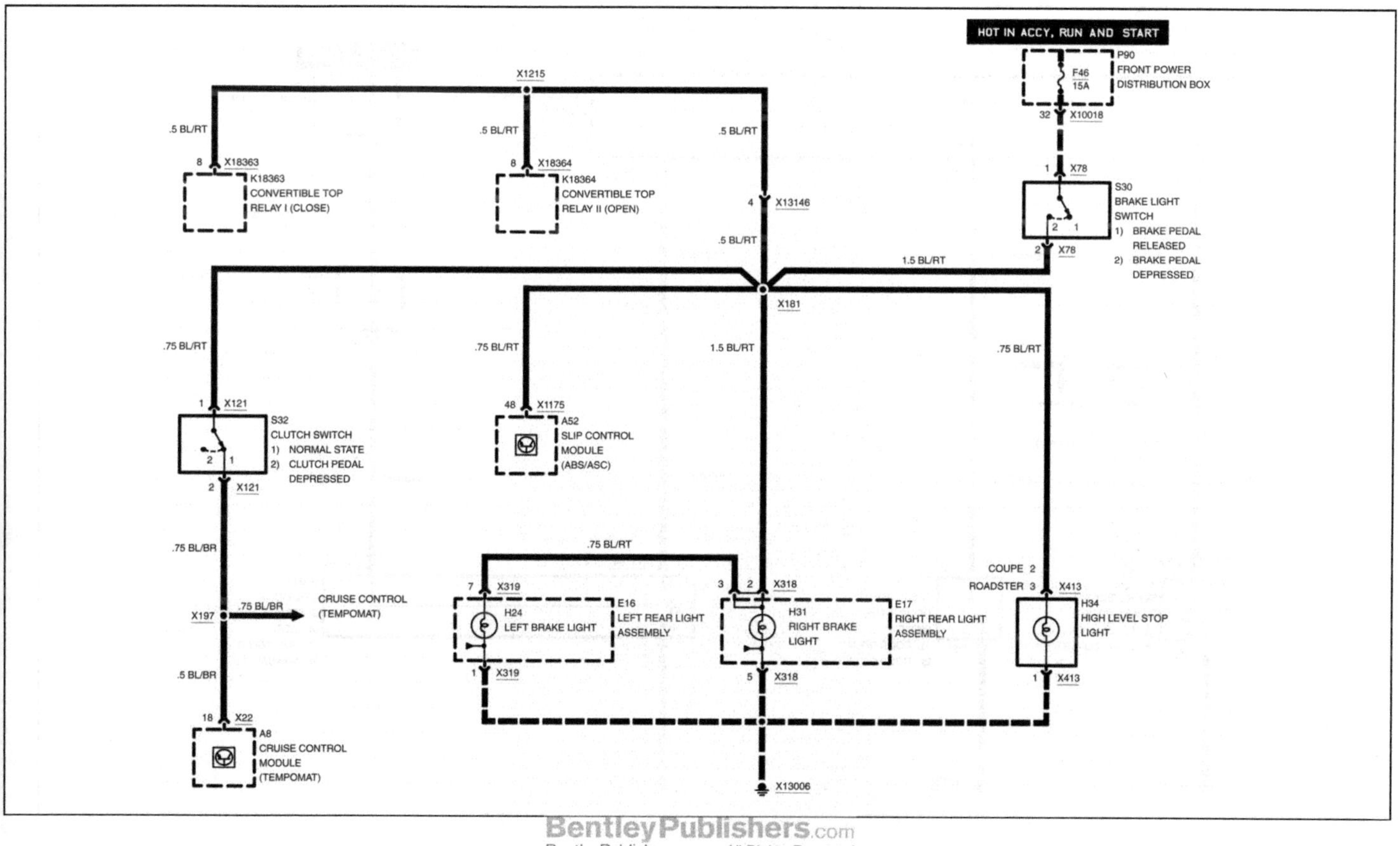

## Brake lights, 2002

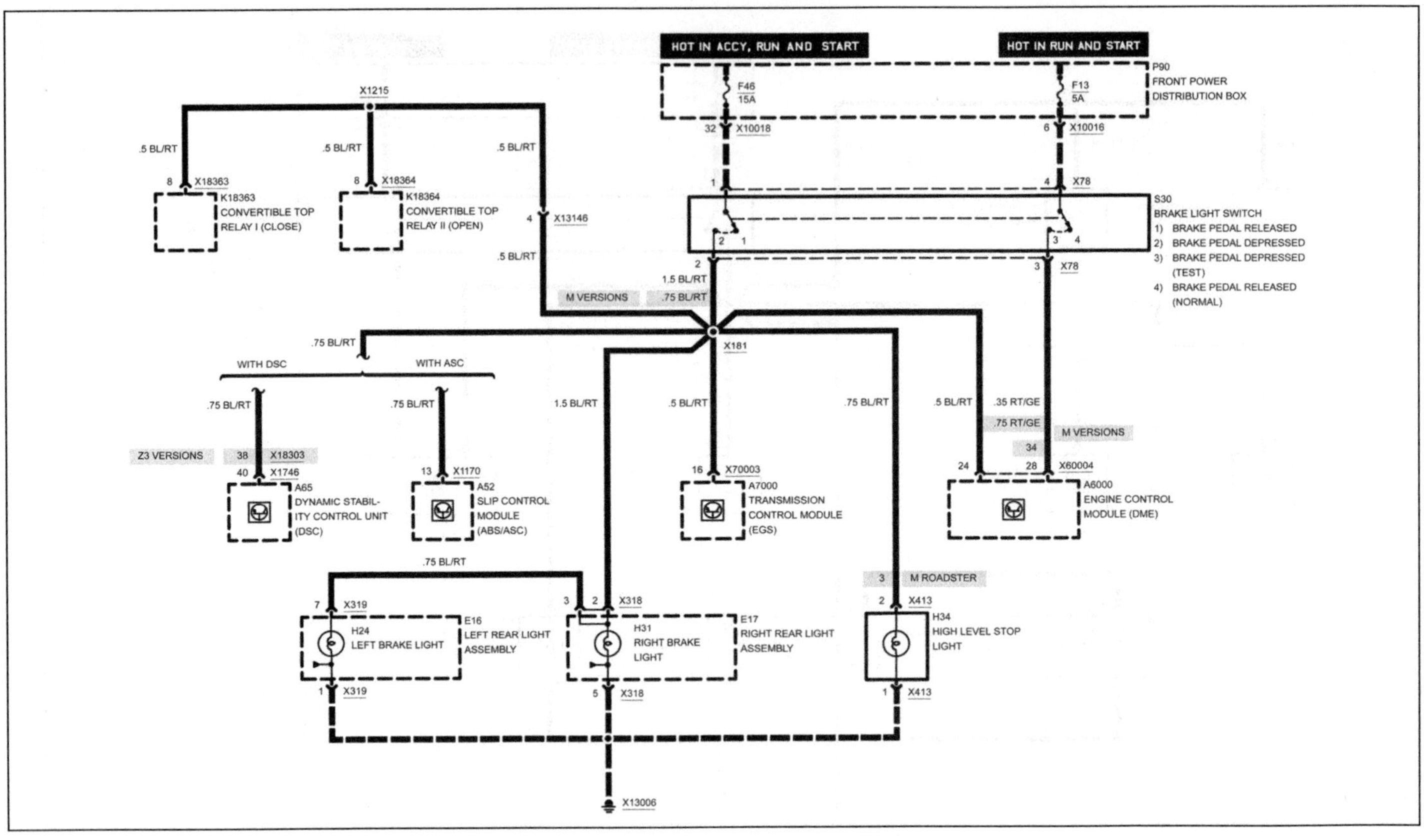


## Interior lights (ZVM), 1996

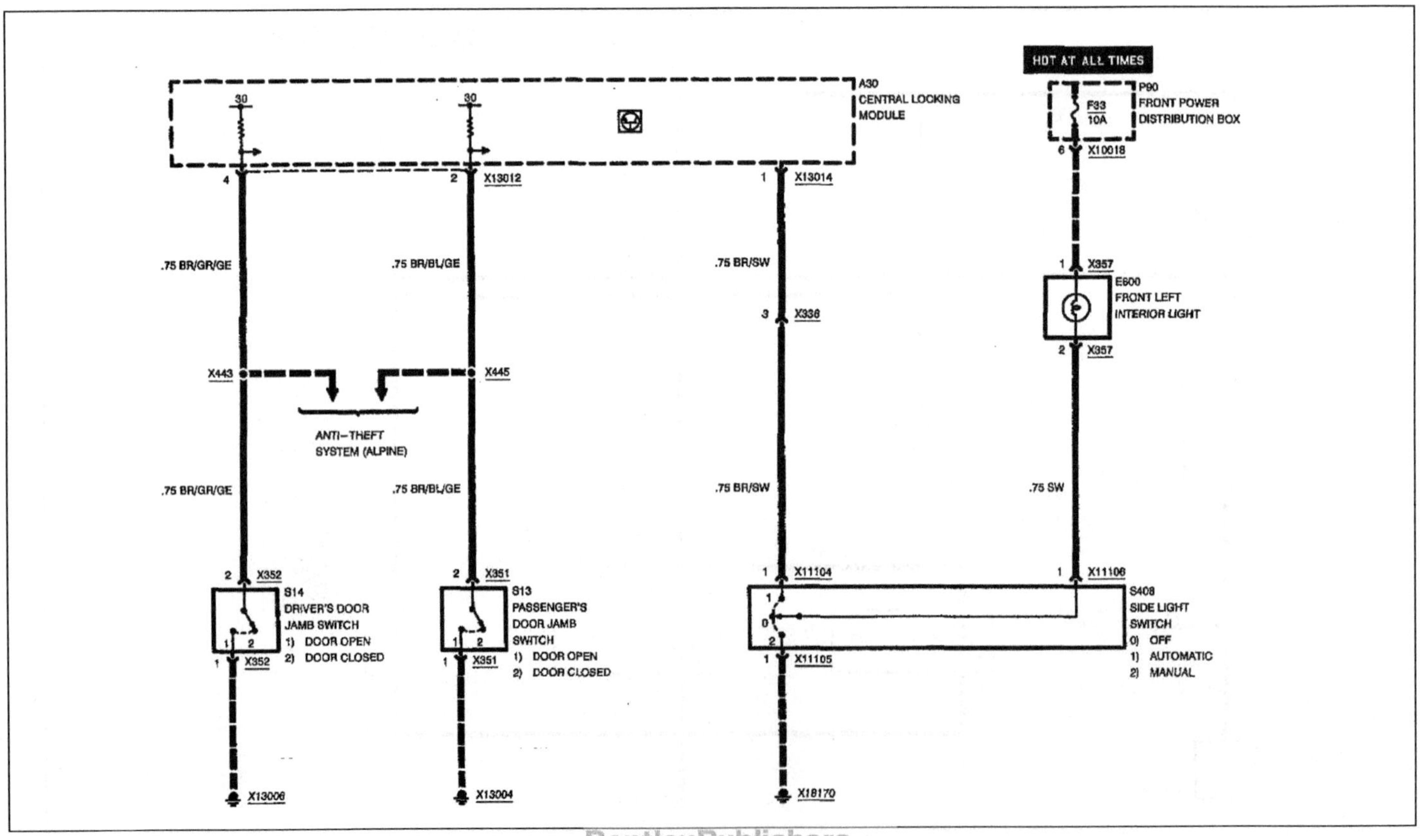

## Interior lights, 1996 with ZKE, 1998

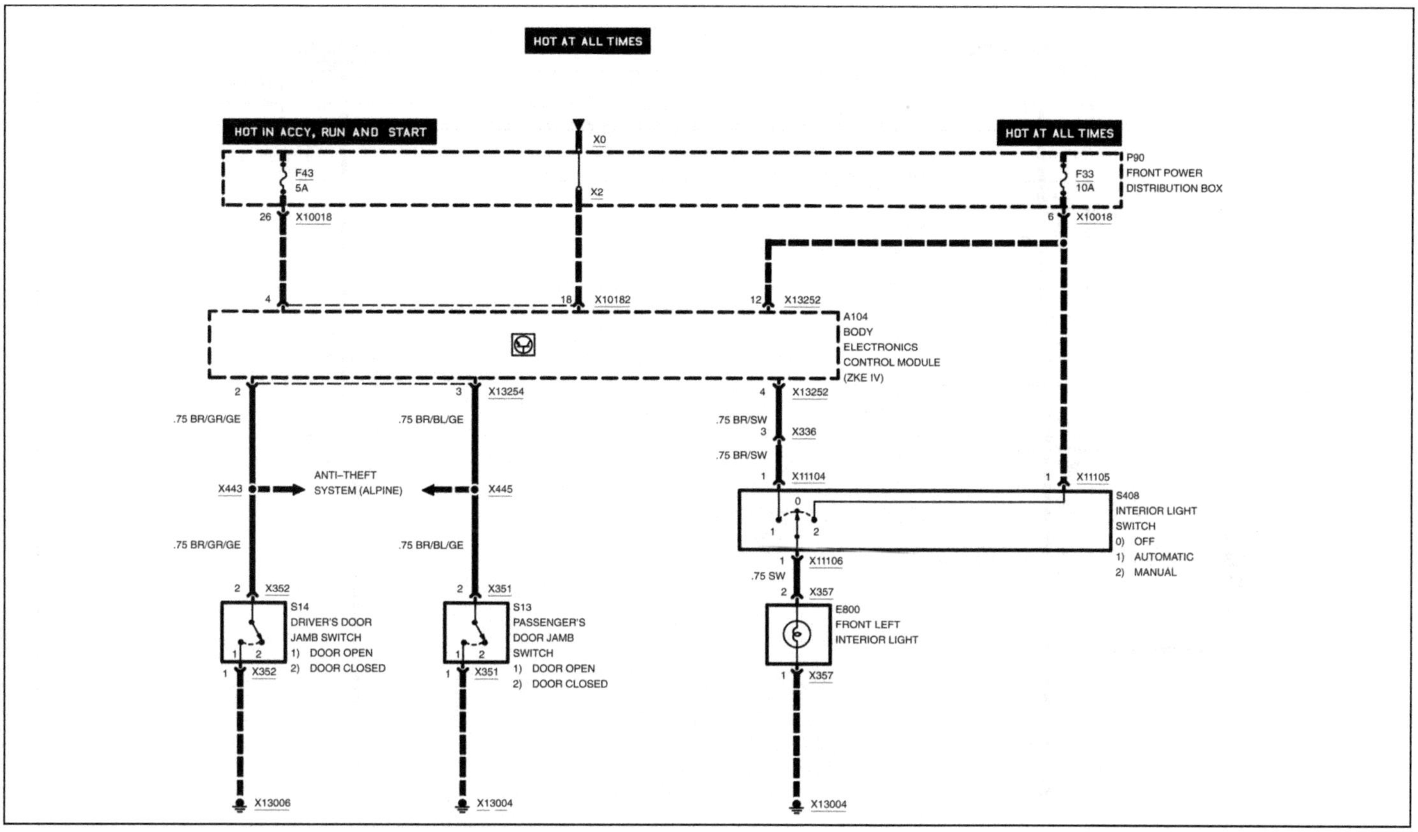


## Interior lights, 2000 - 2002

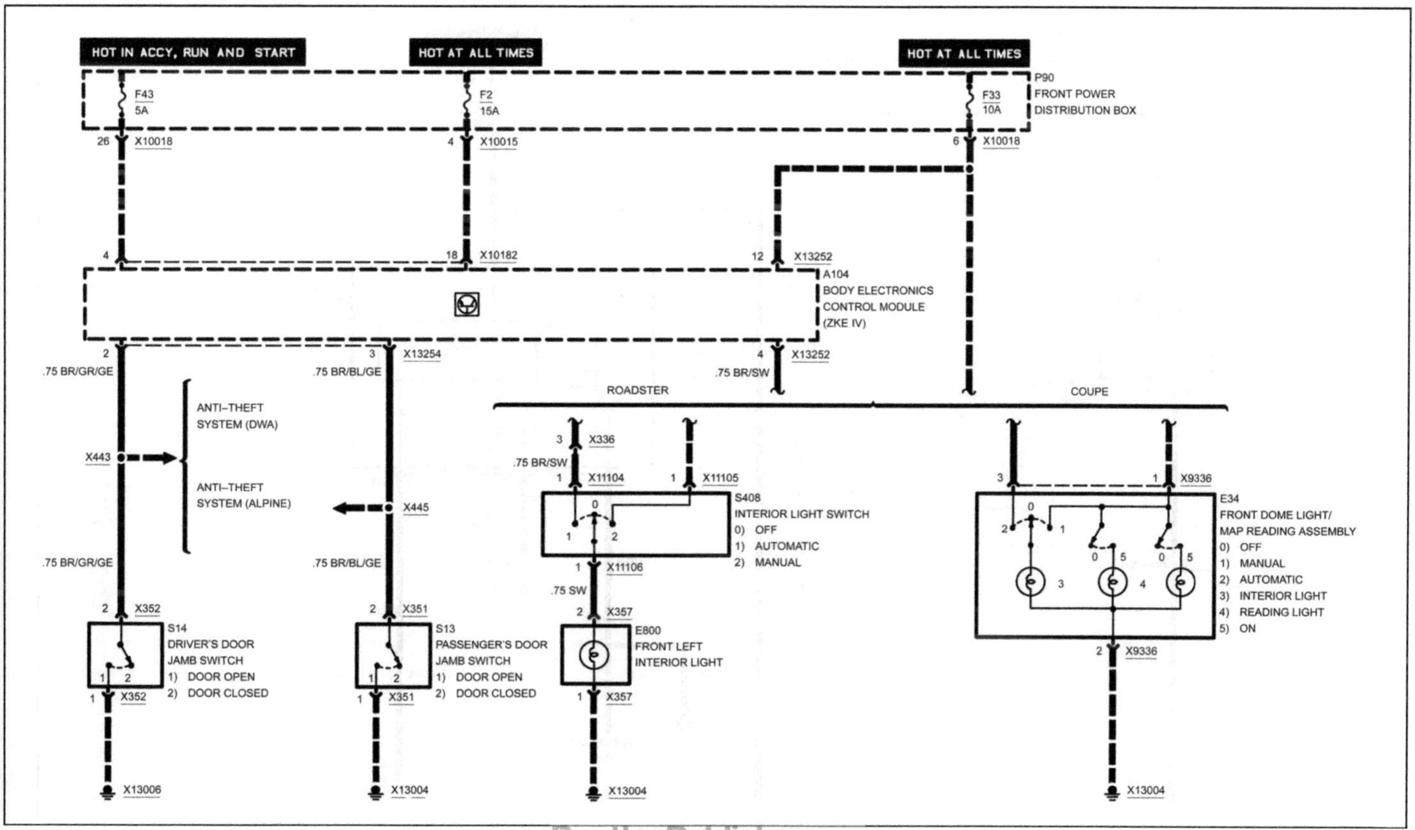

## Interior accessories

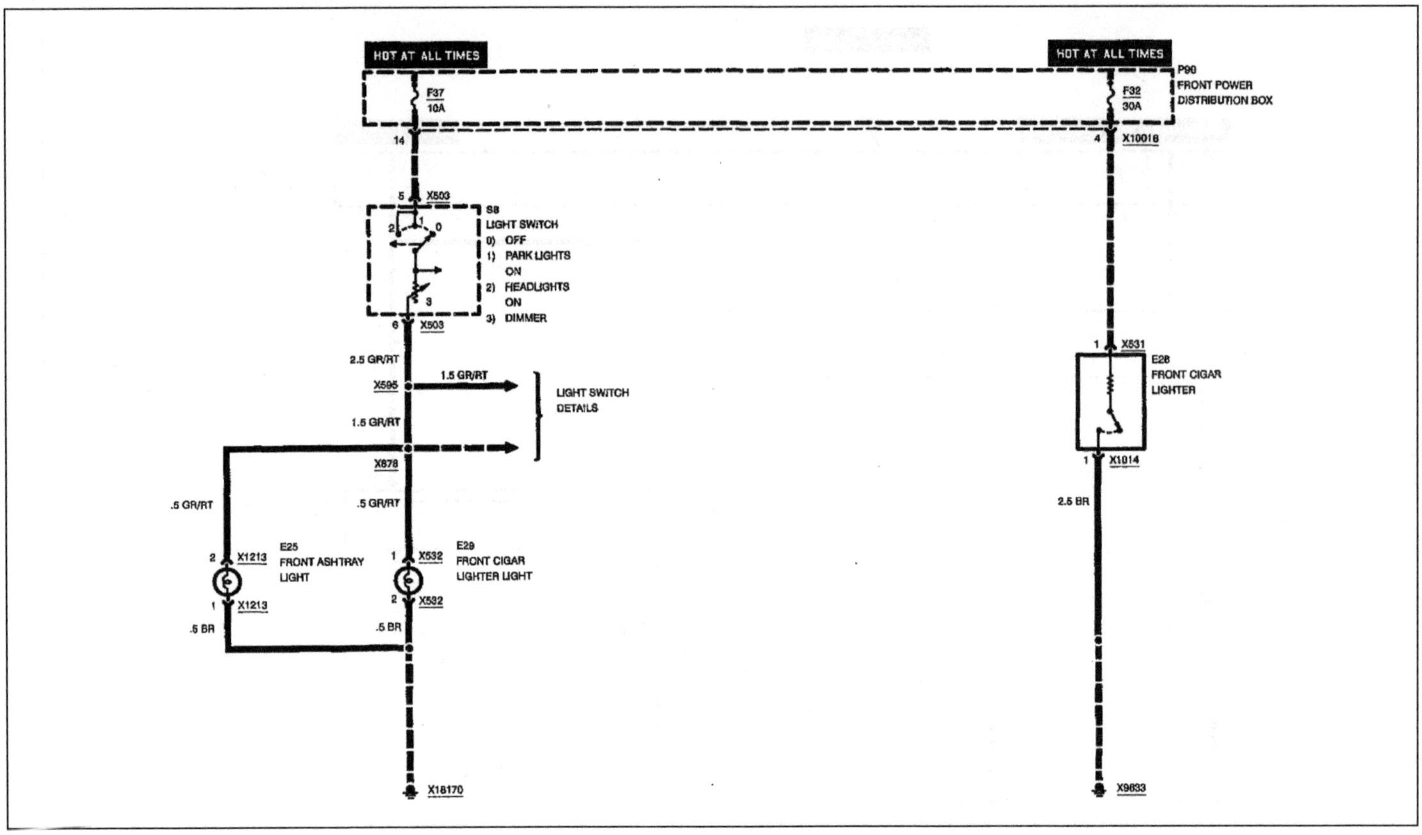


## Crash control module, 1996 - 1998

(1 of 3)

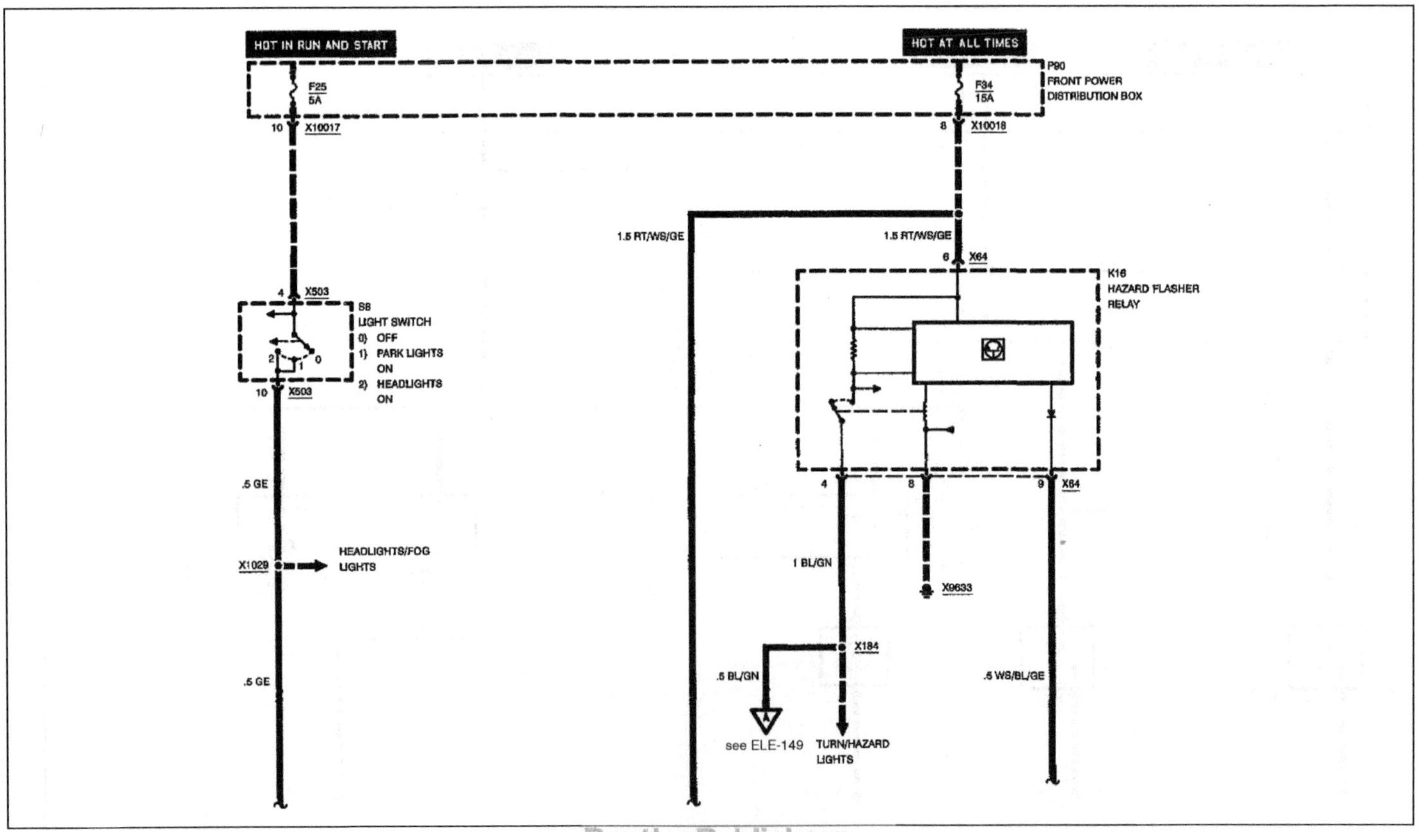

## Crash control module, 1996 - 1998

**(2 of 3)**

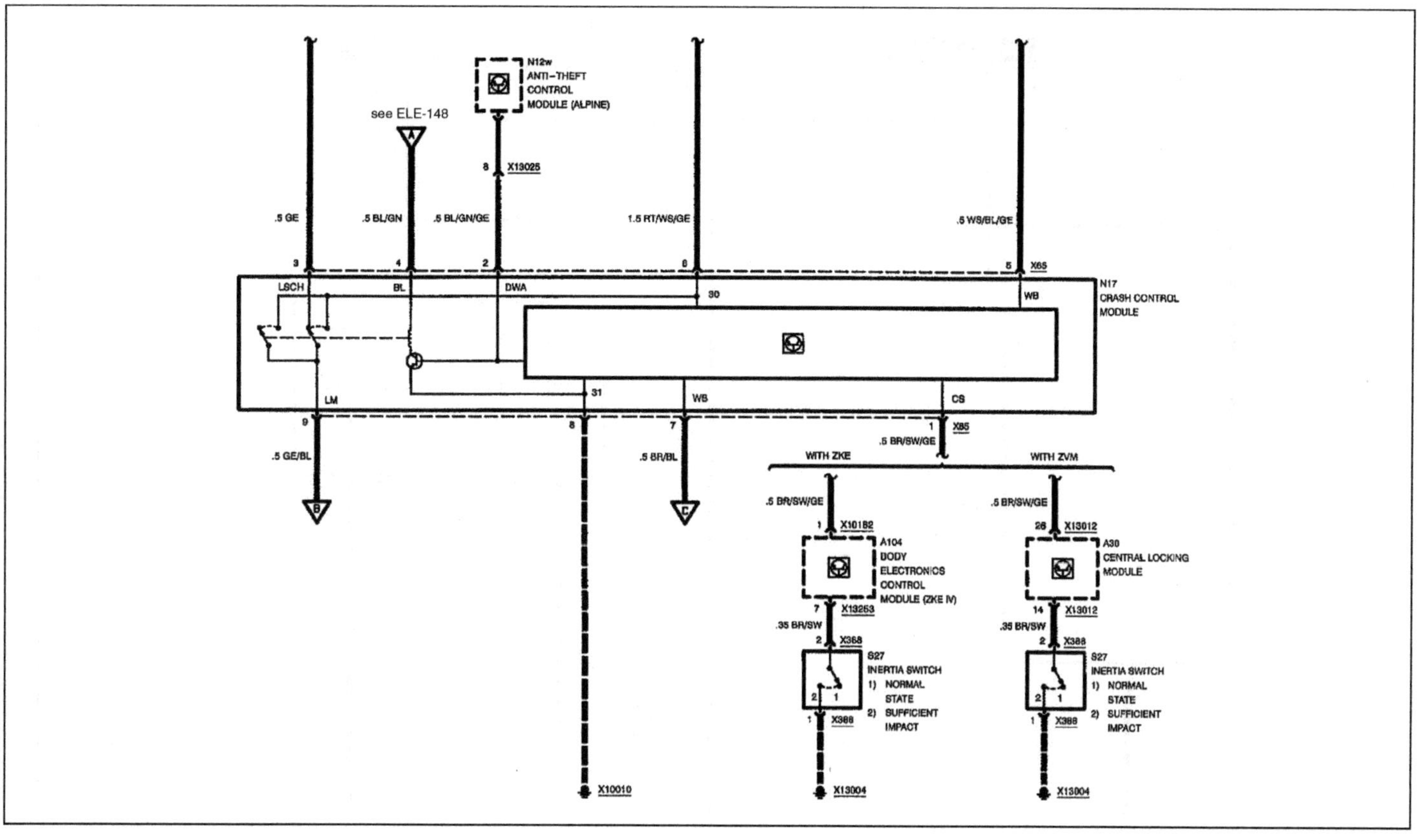


## Crash control module, 1996 - 1998

**(3 of 3)**

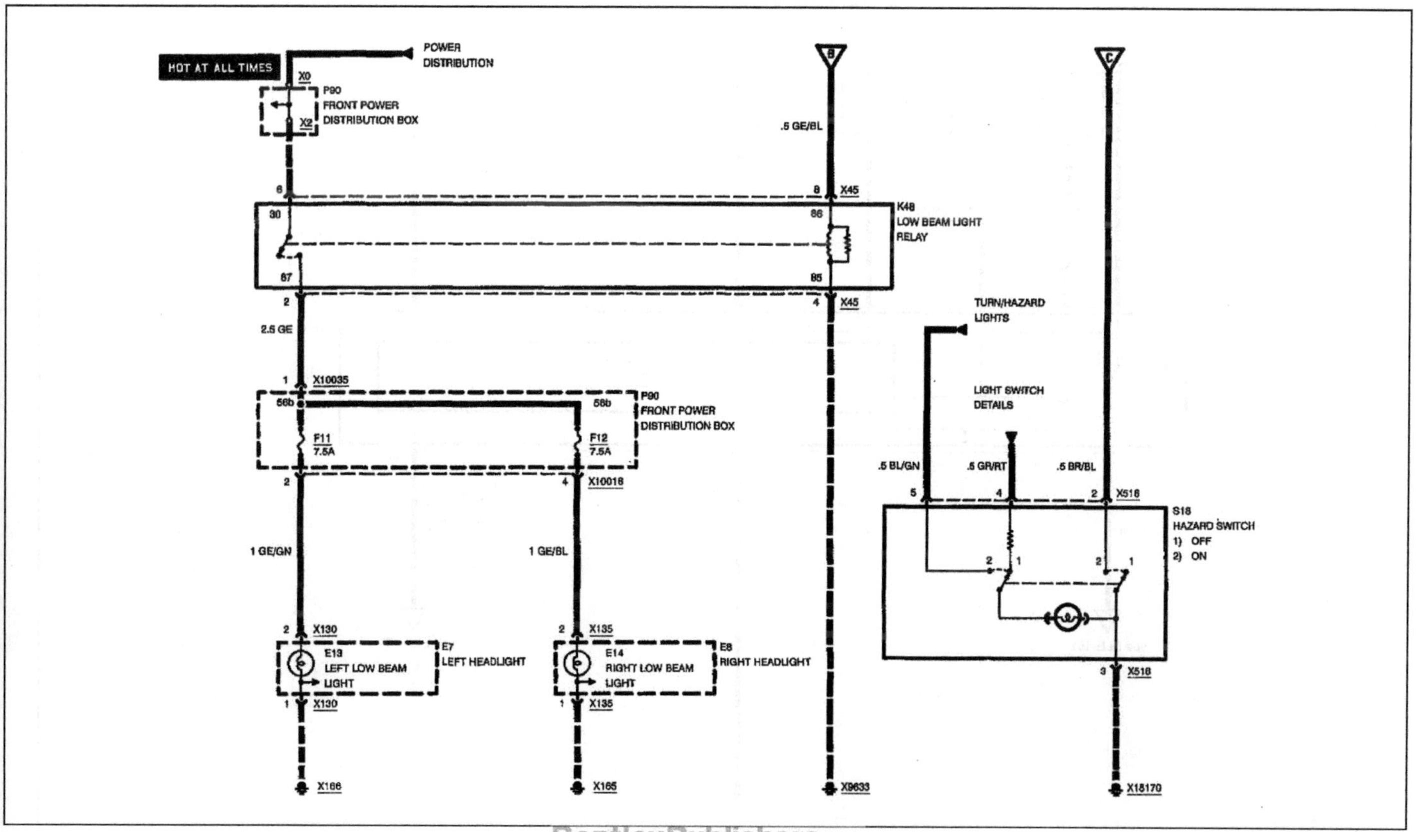

## Crash control module, 2000 - 2002
(1 of 3)

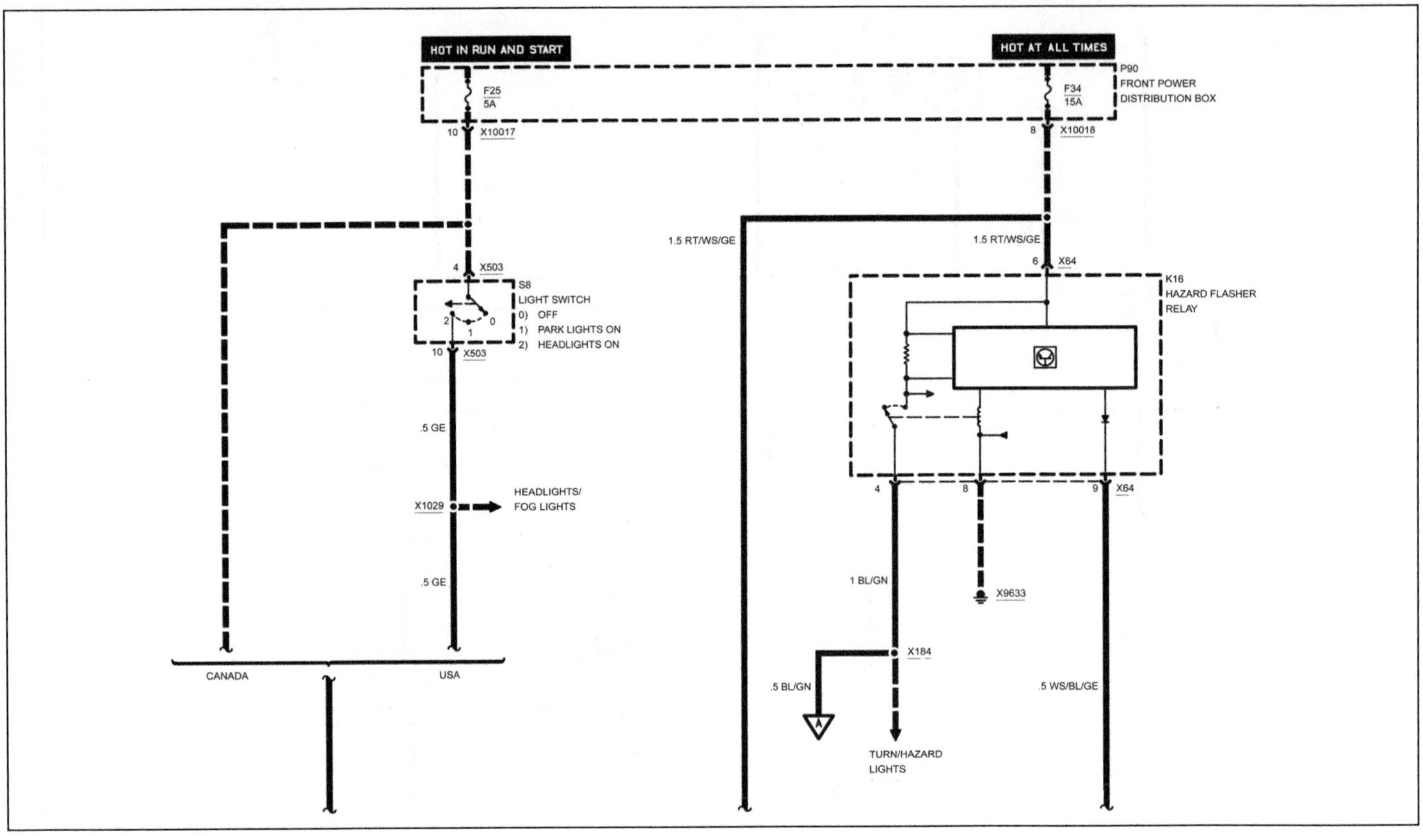


## Crash control module, 2000 - 2002
(2 of 3)

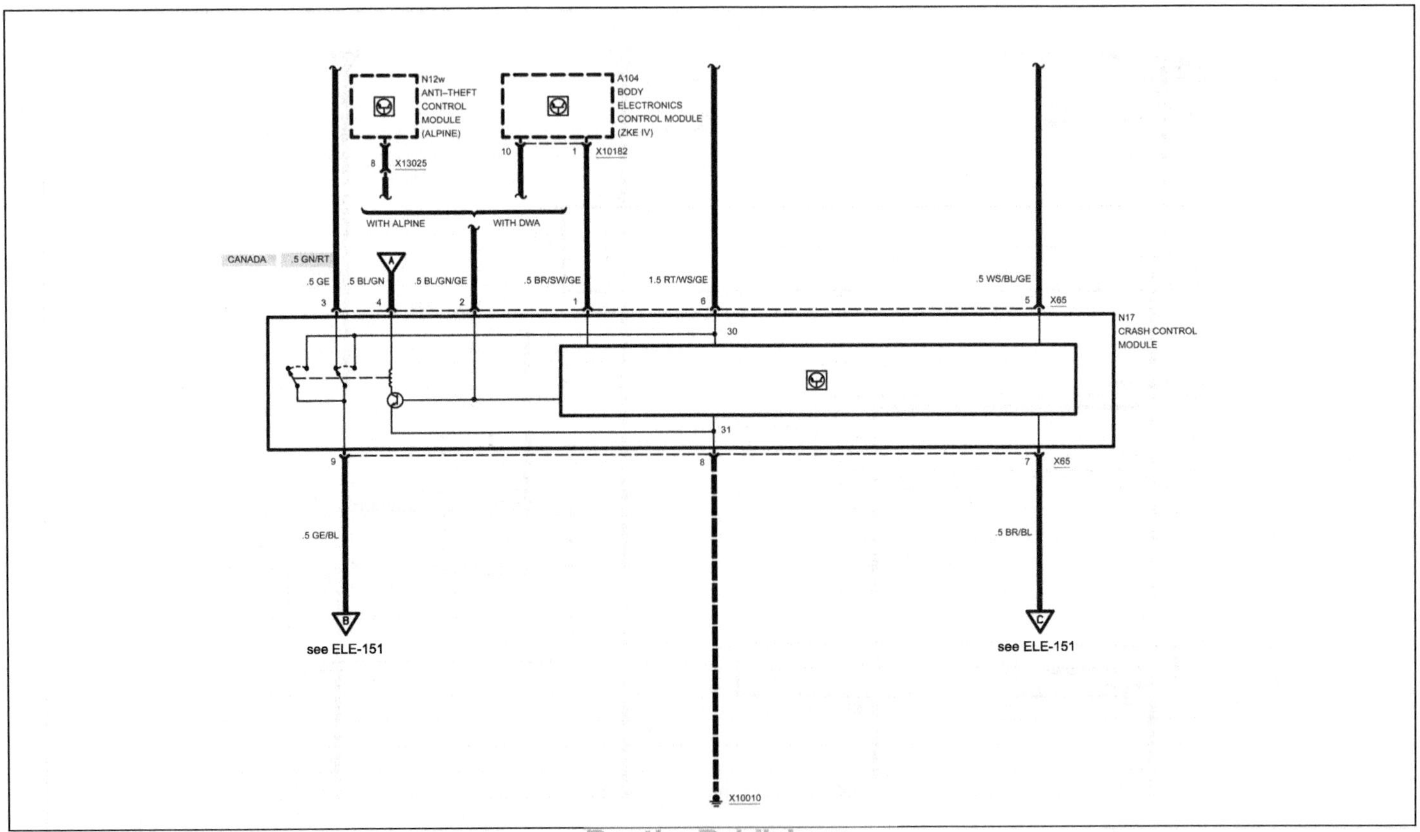

## Crash control module, 2000 - 2002
**(3 of 3)**

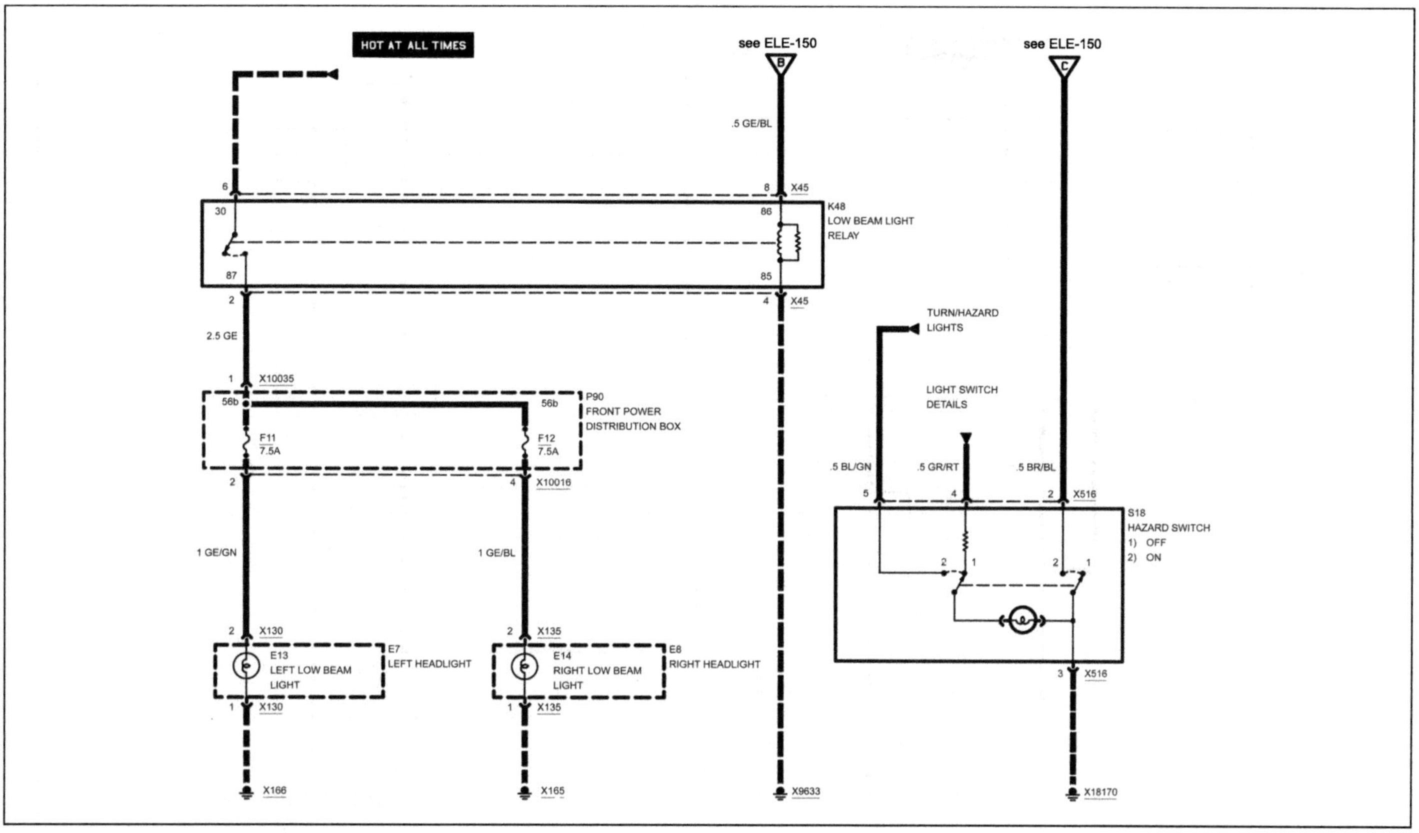


## Rear window defogger, 1998
**(1 of 2)**

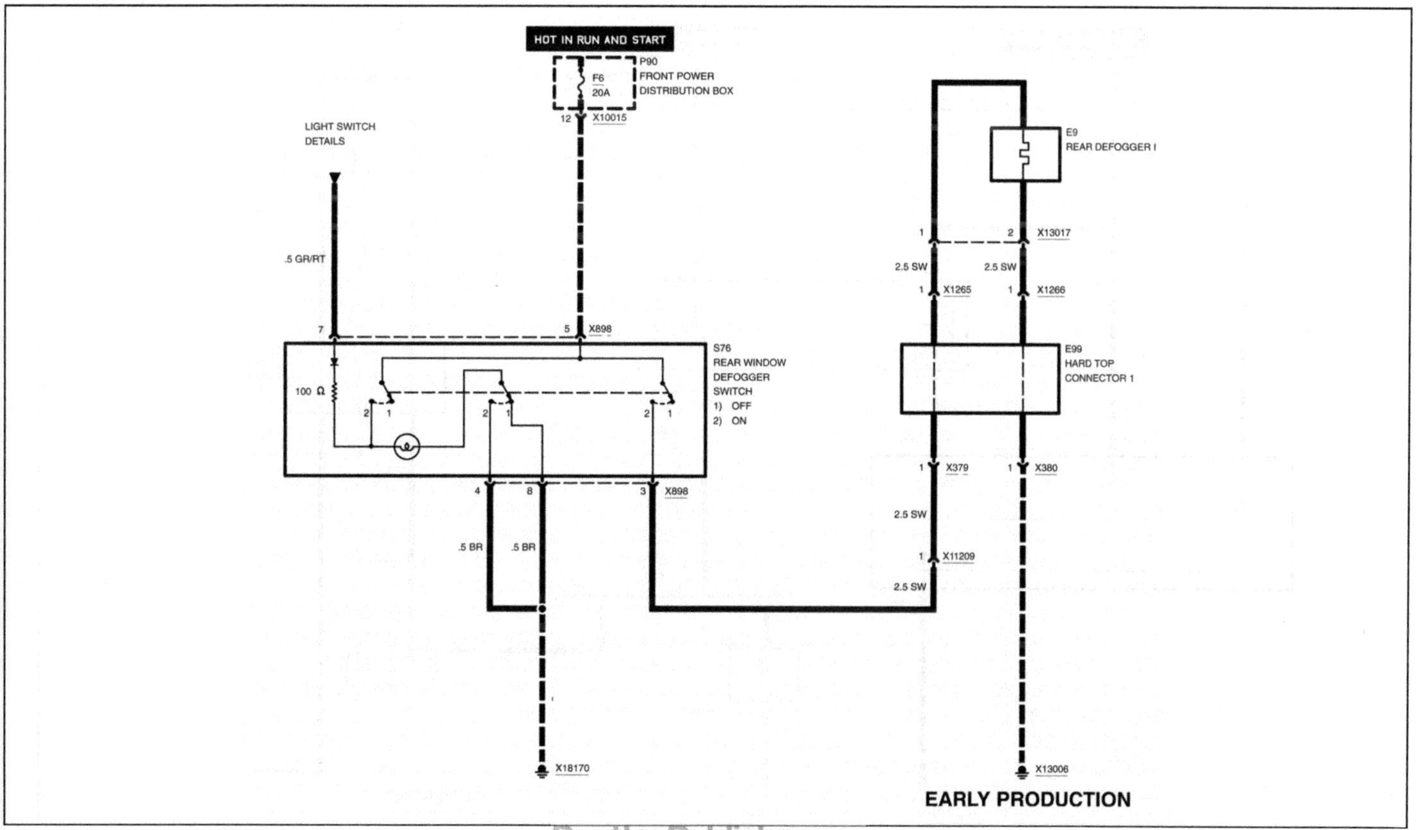


**EARLY PRODUCTION**

## Rear window defogger, 1998
(2 of 2)

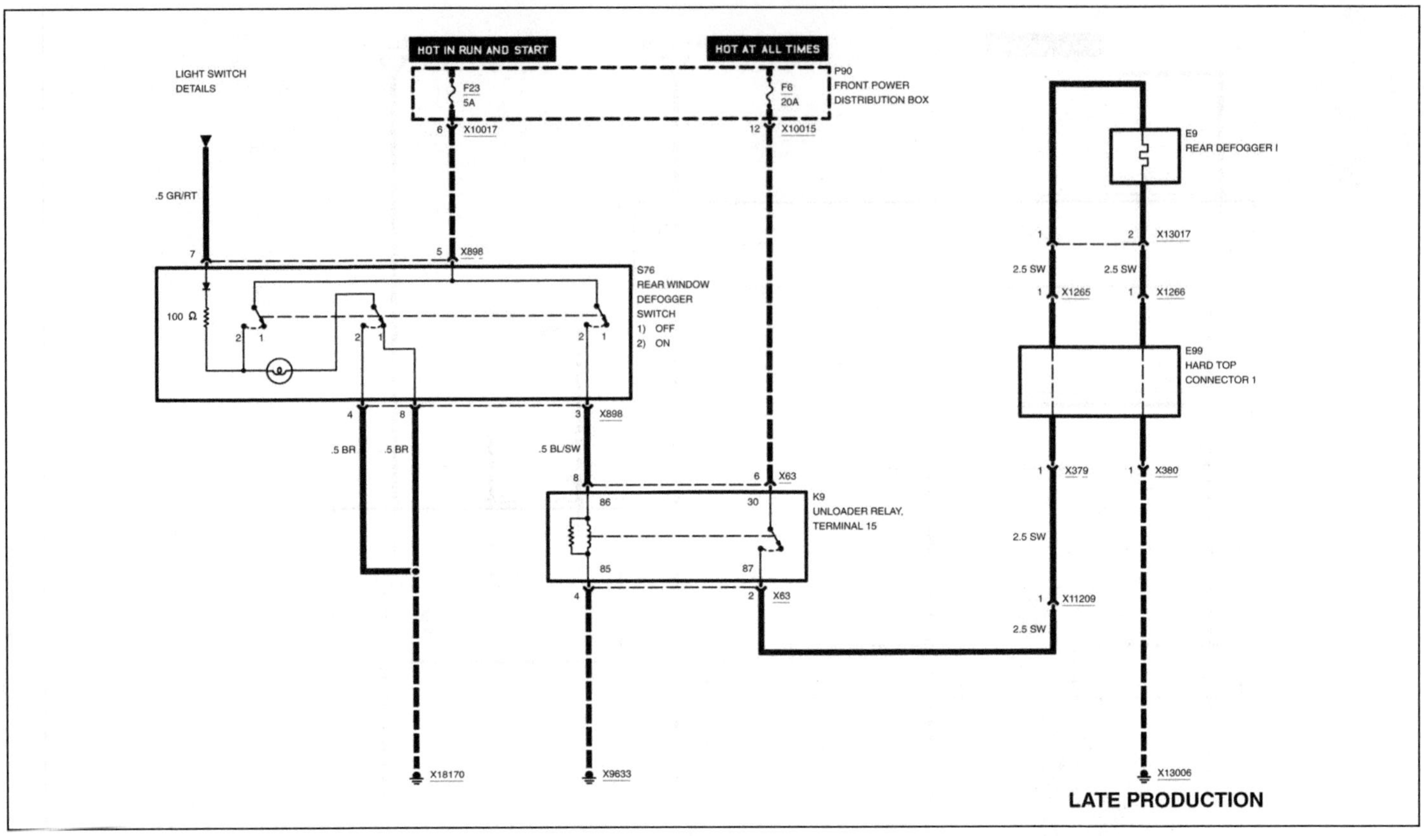


## Rear window defogger, 2000
(1 of 2)

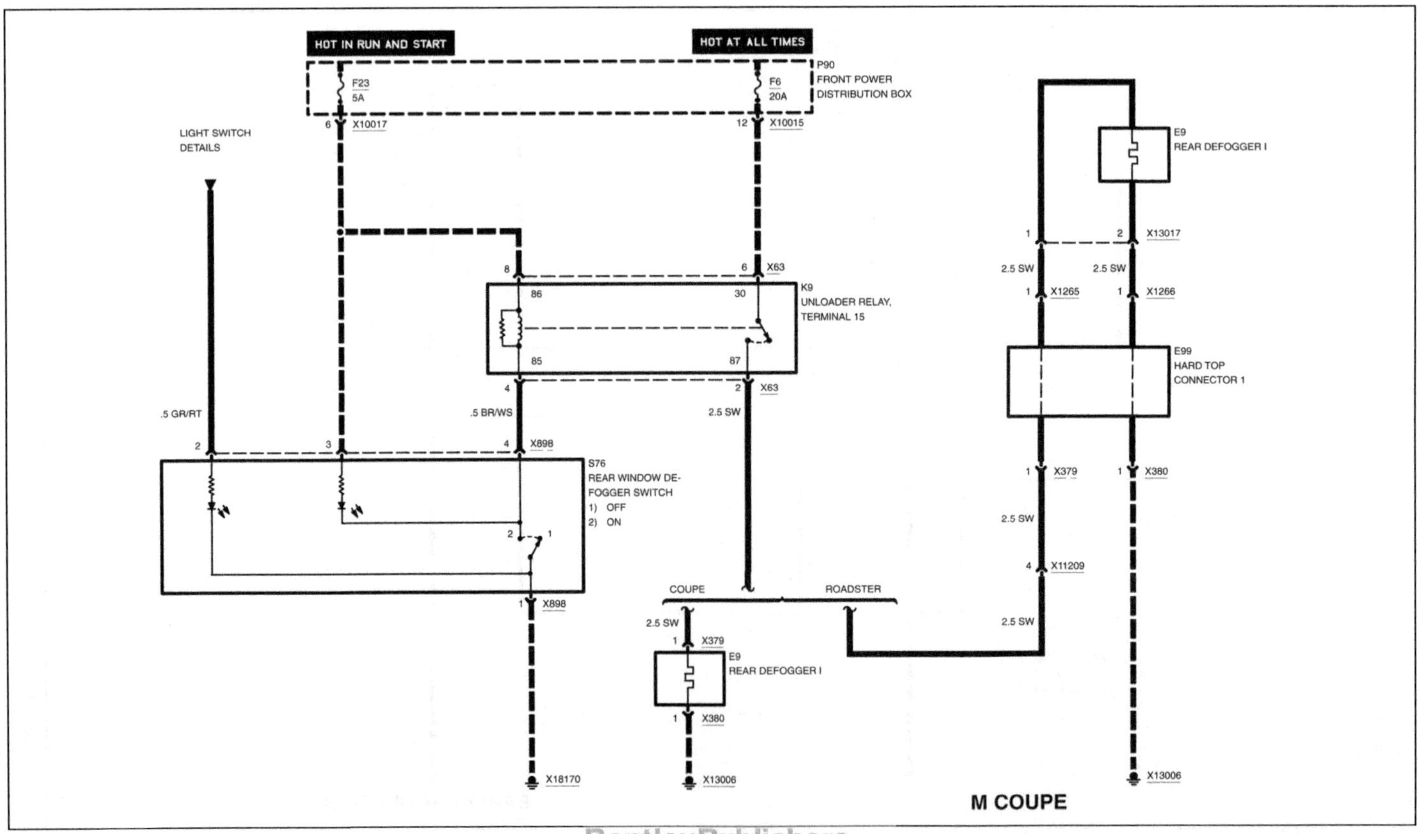

## Rear window defogger, 2000

(2 of 2)

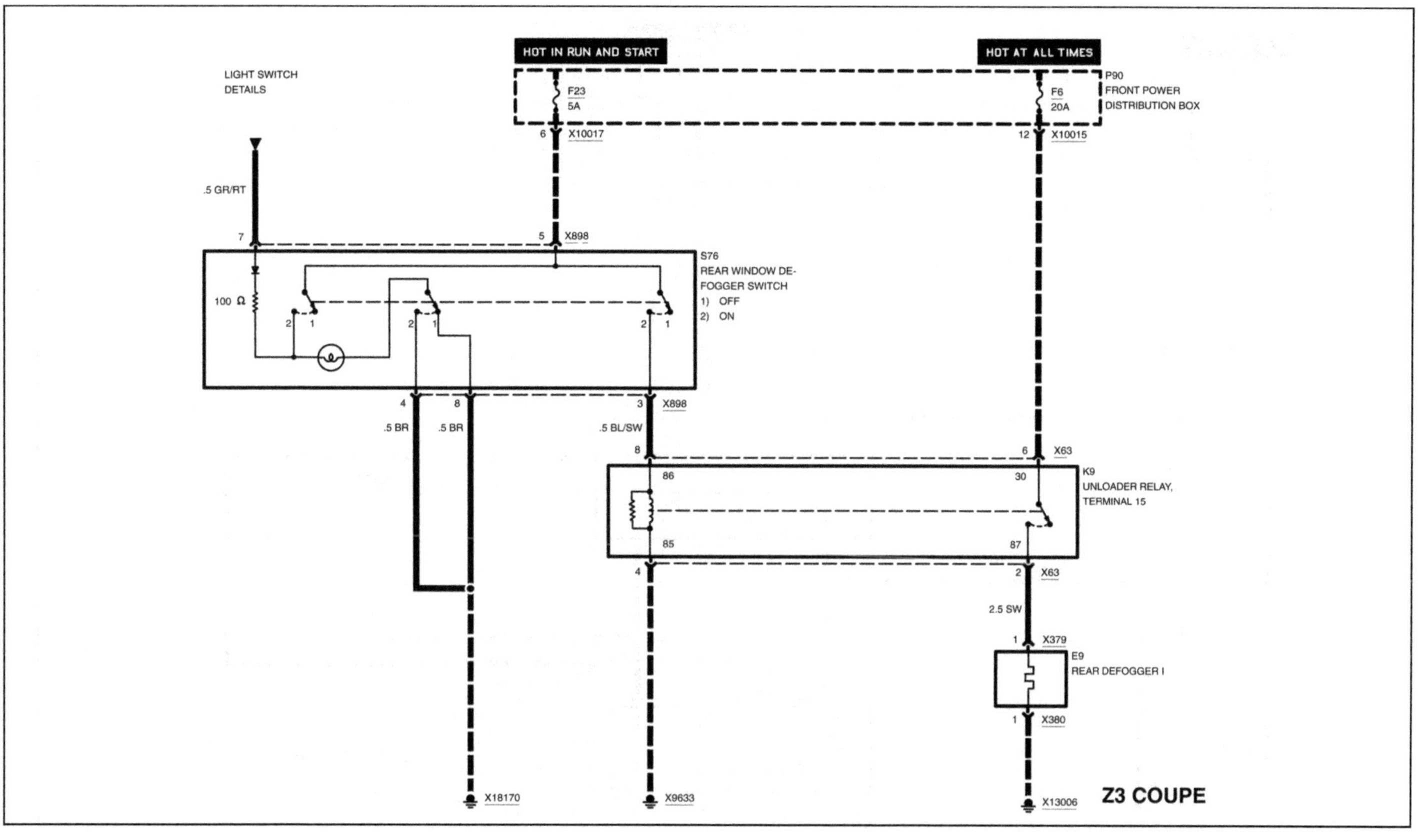


## Rear window defogger, 2002

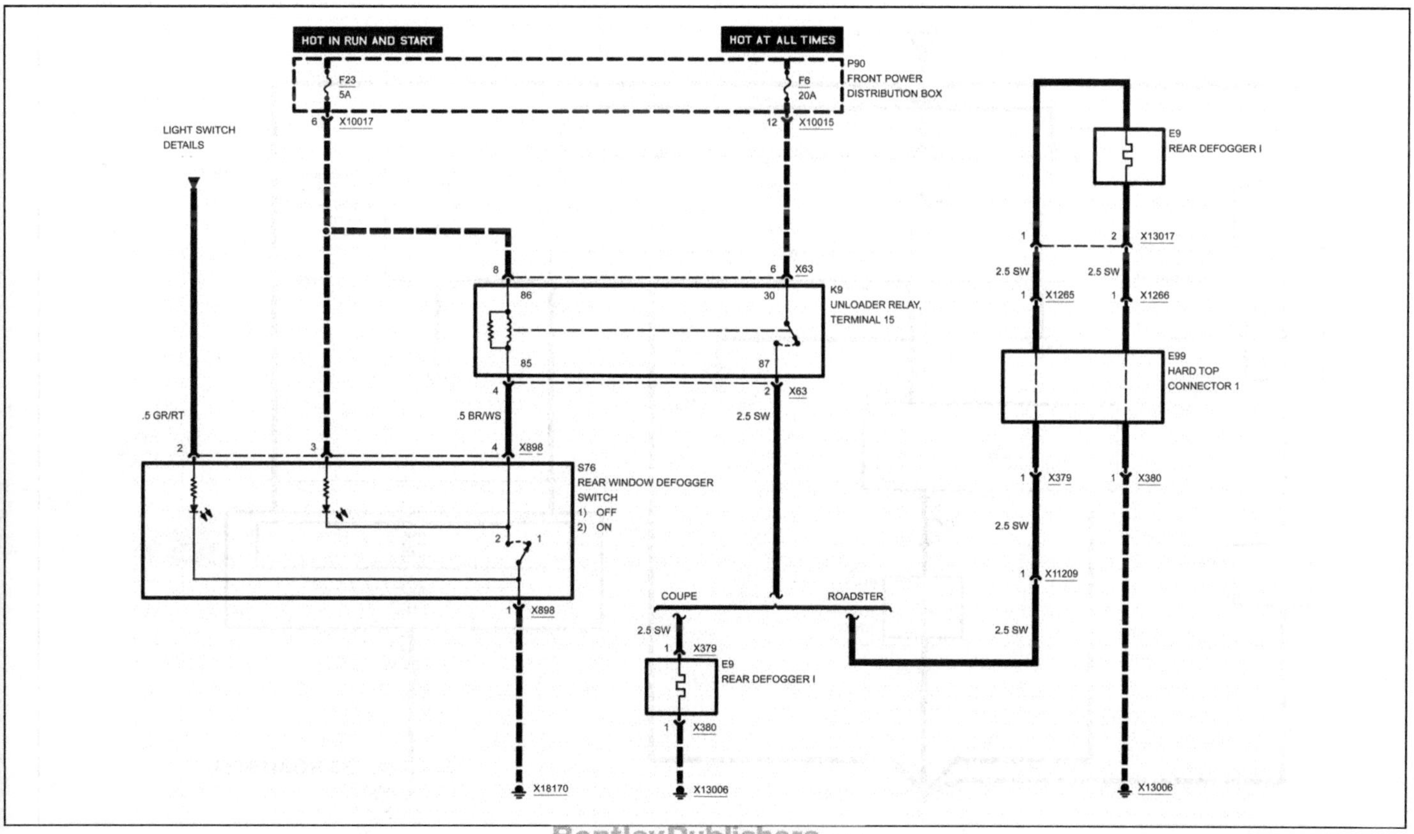

## Heating and air-conditioning
(1 of 9)

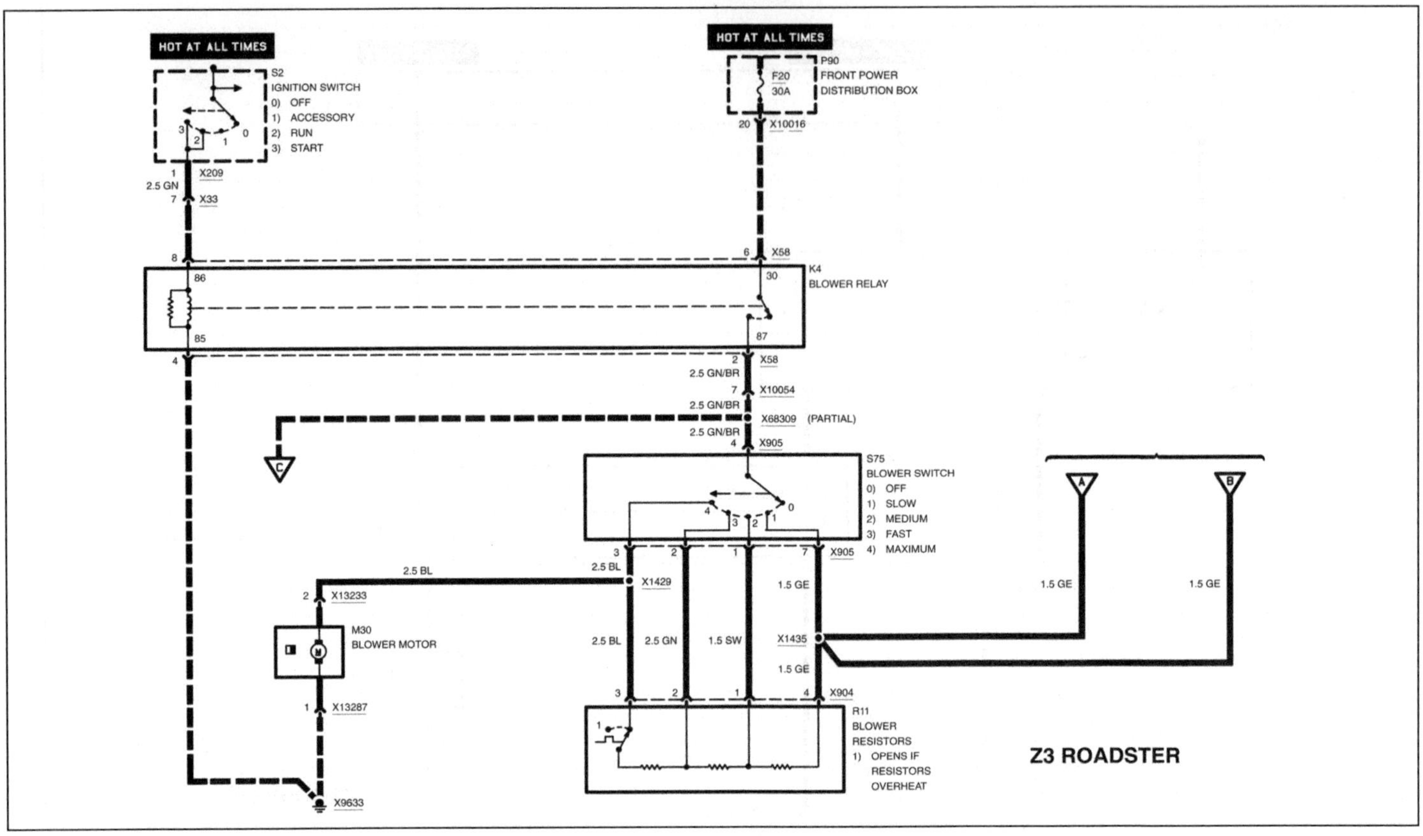


## Heating and air-conditioning
(2 of 9)

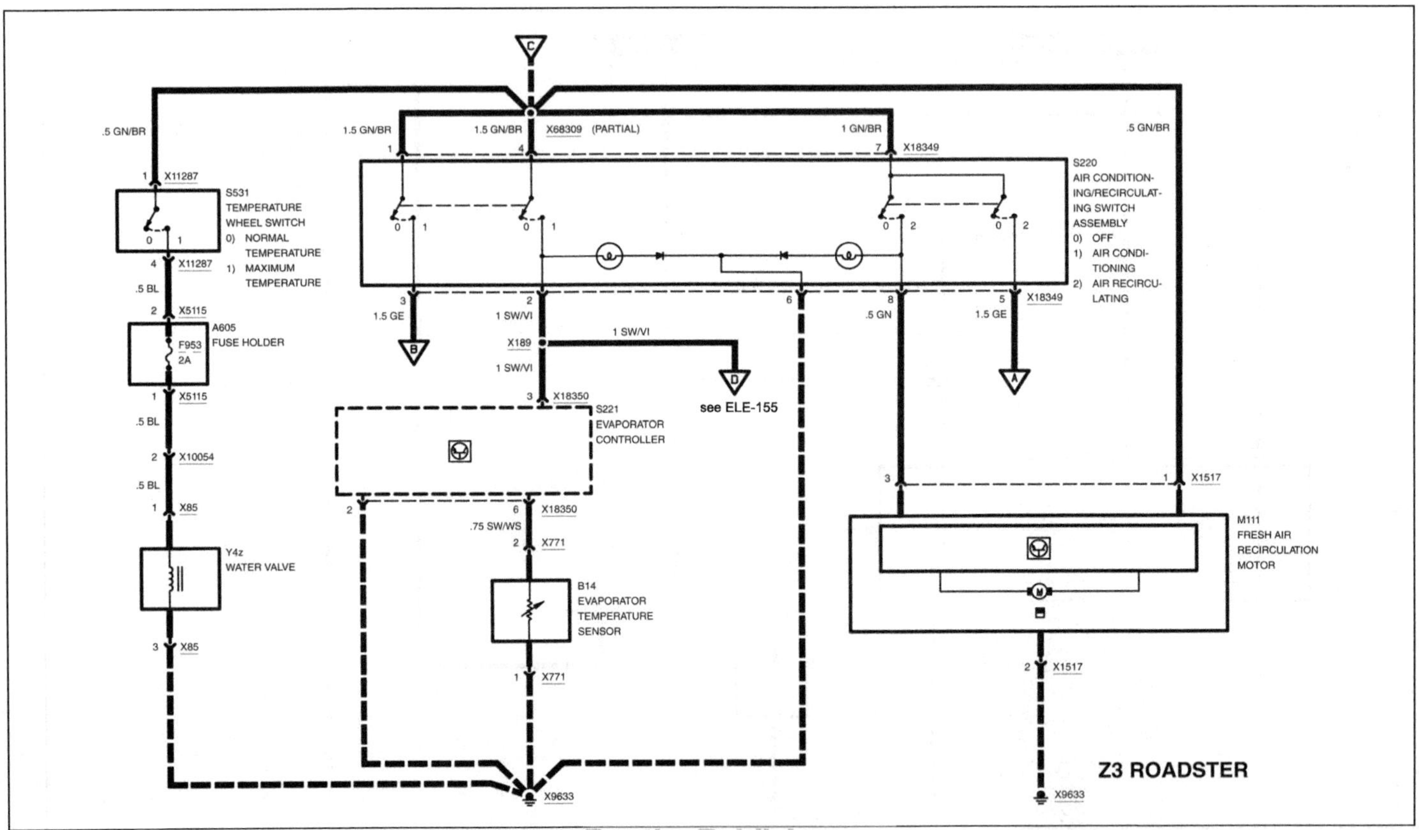

## Heating and air-conditioning
(3 of 9)

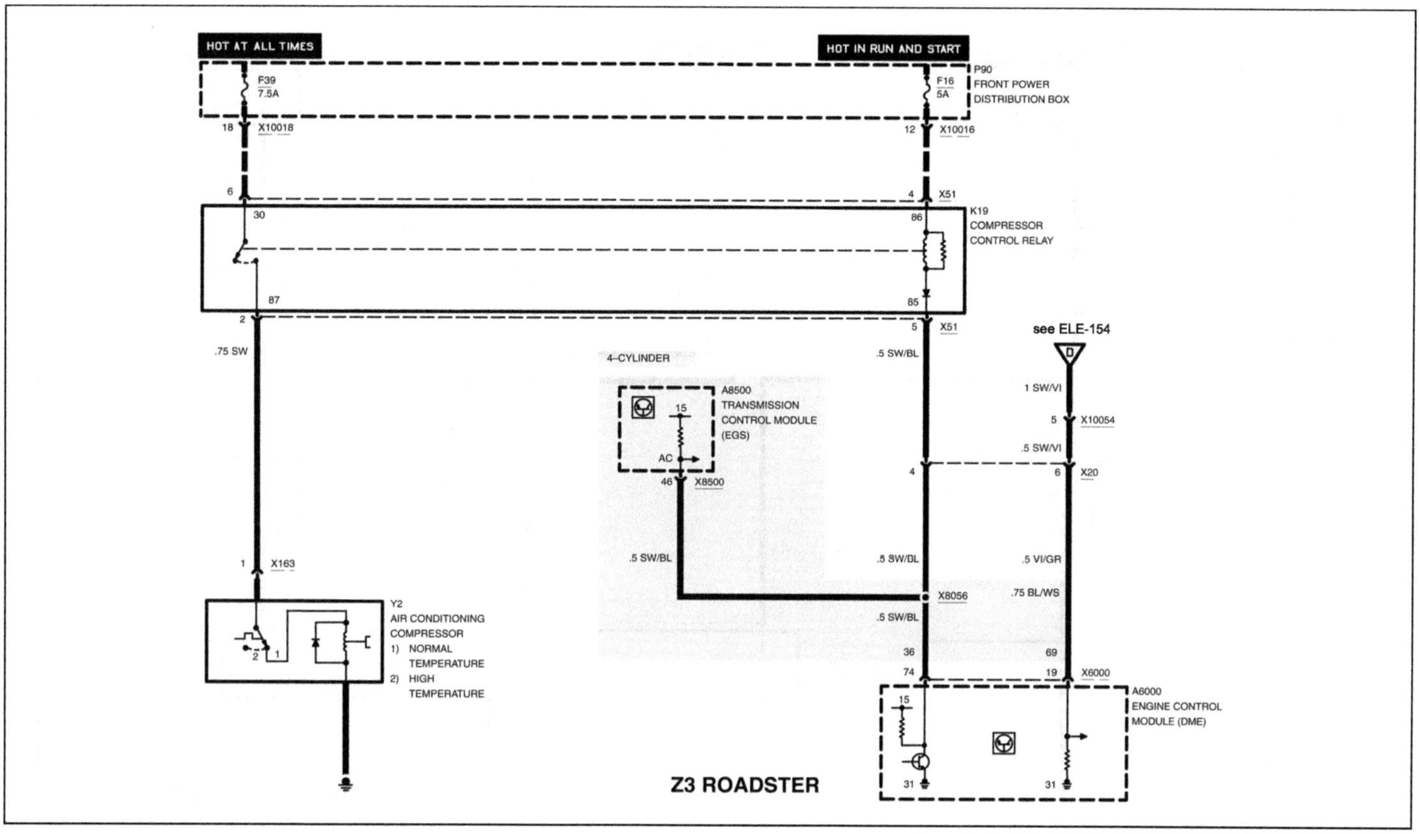


## Heating and air-conditioning
(4 of 9)

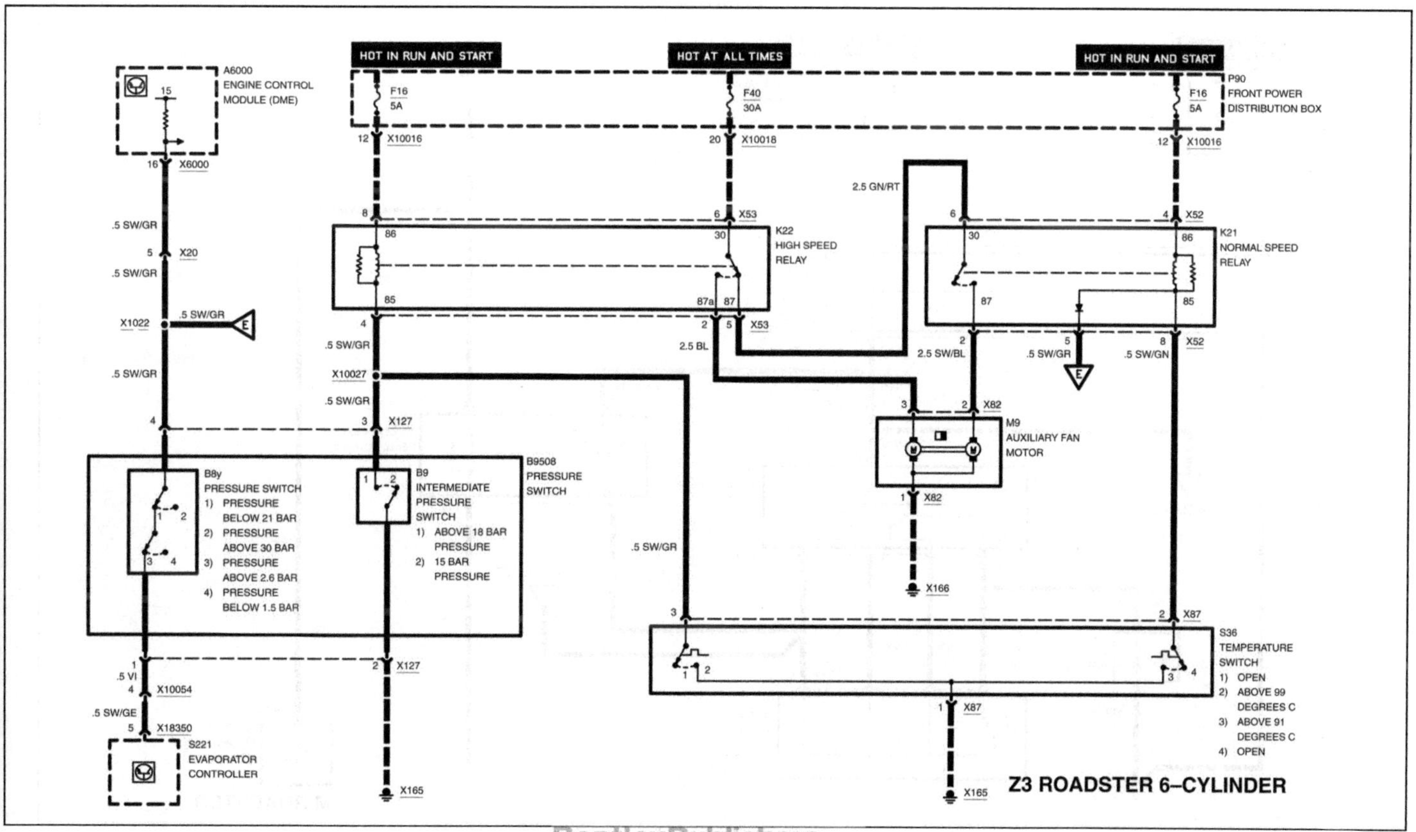

## Heating and air-conditioning
(5 of 9)

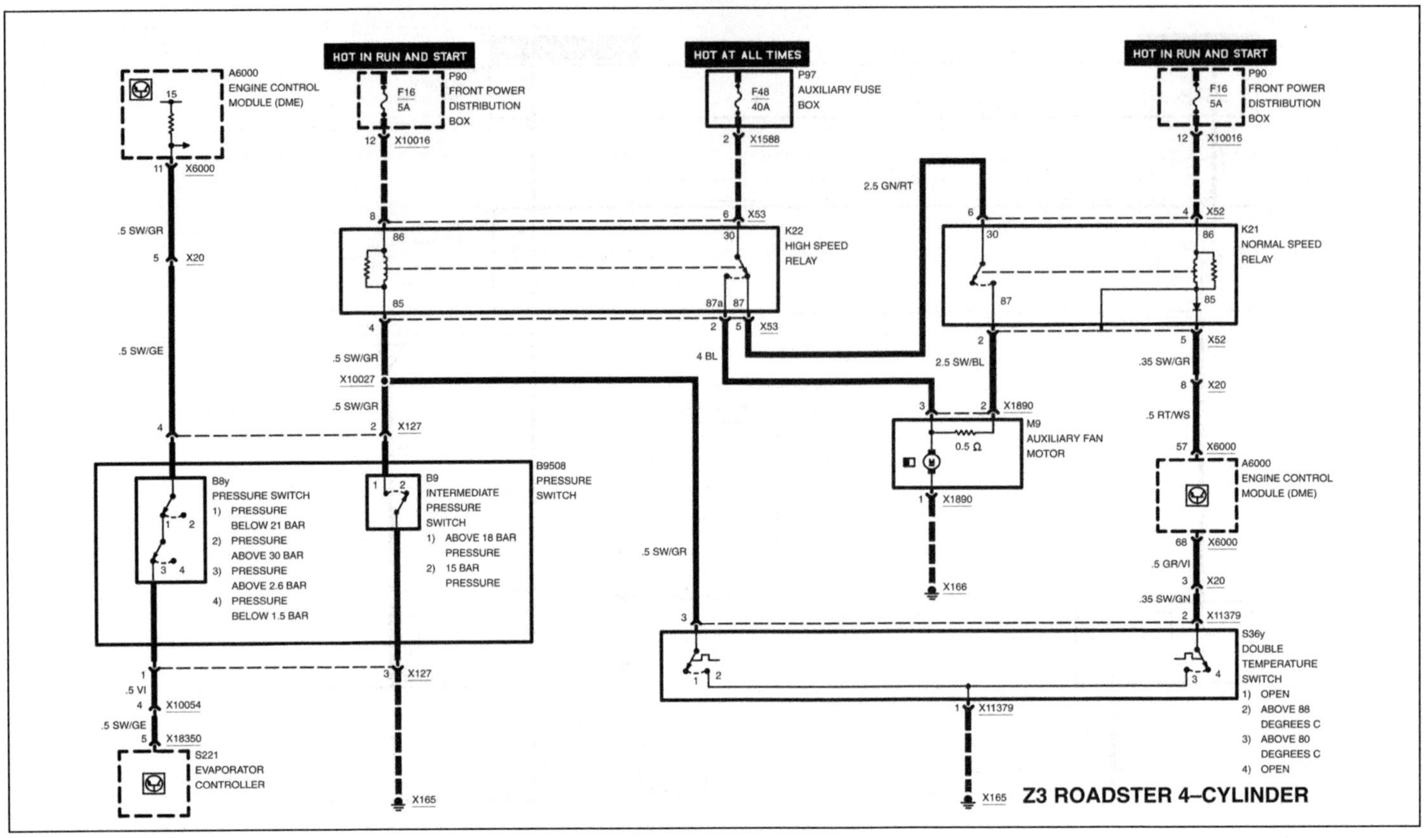


## Heating and air-conditioning
(6 of 9)

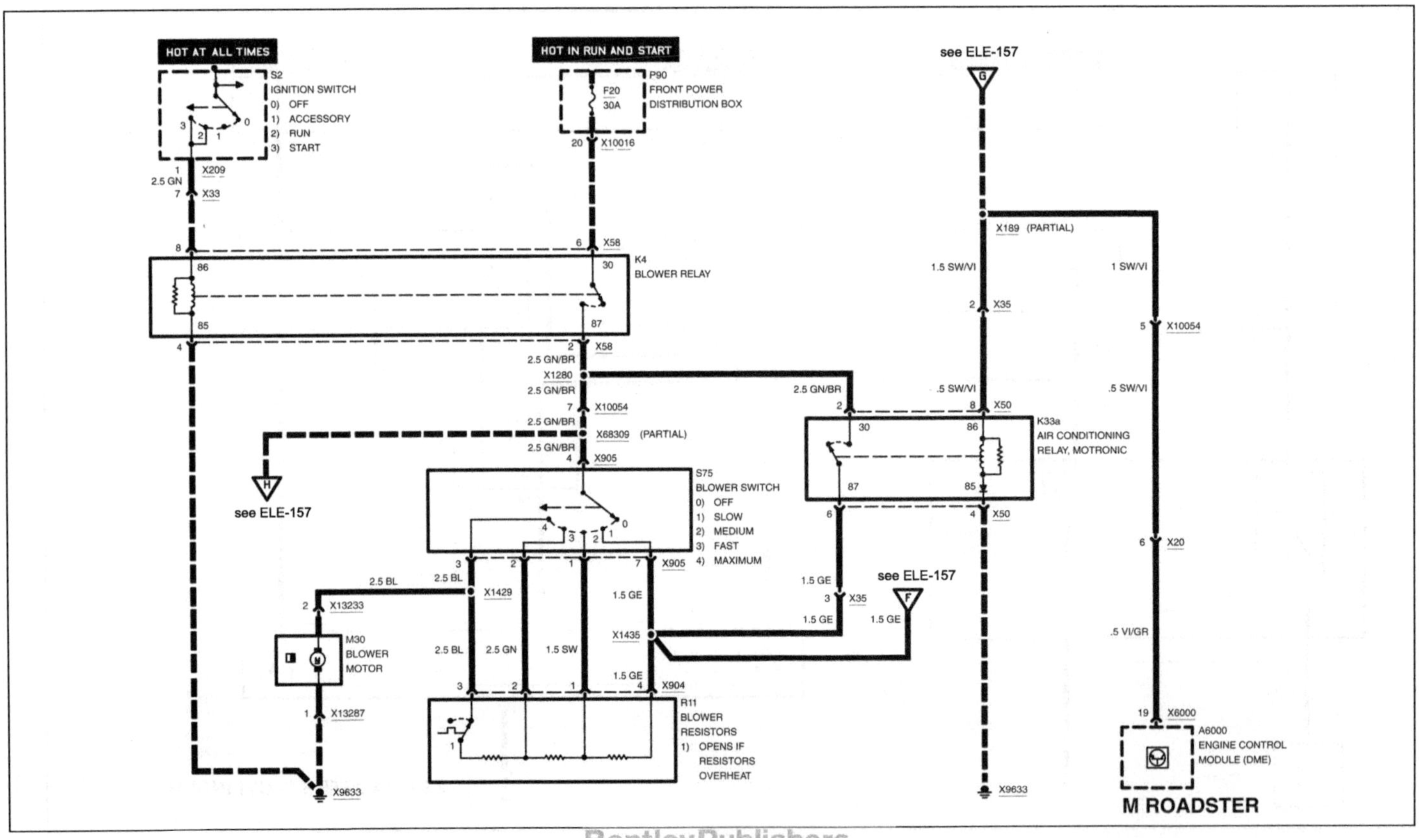

## Heating and air-conditioning
(7 of 9)

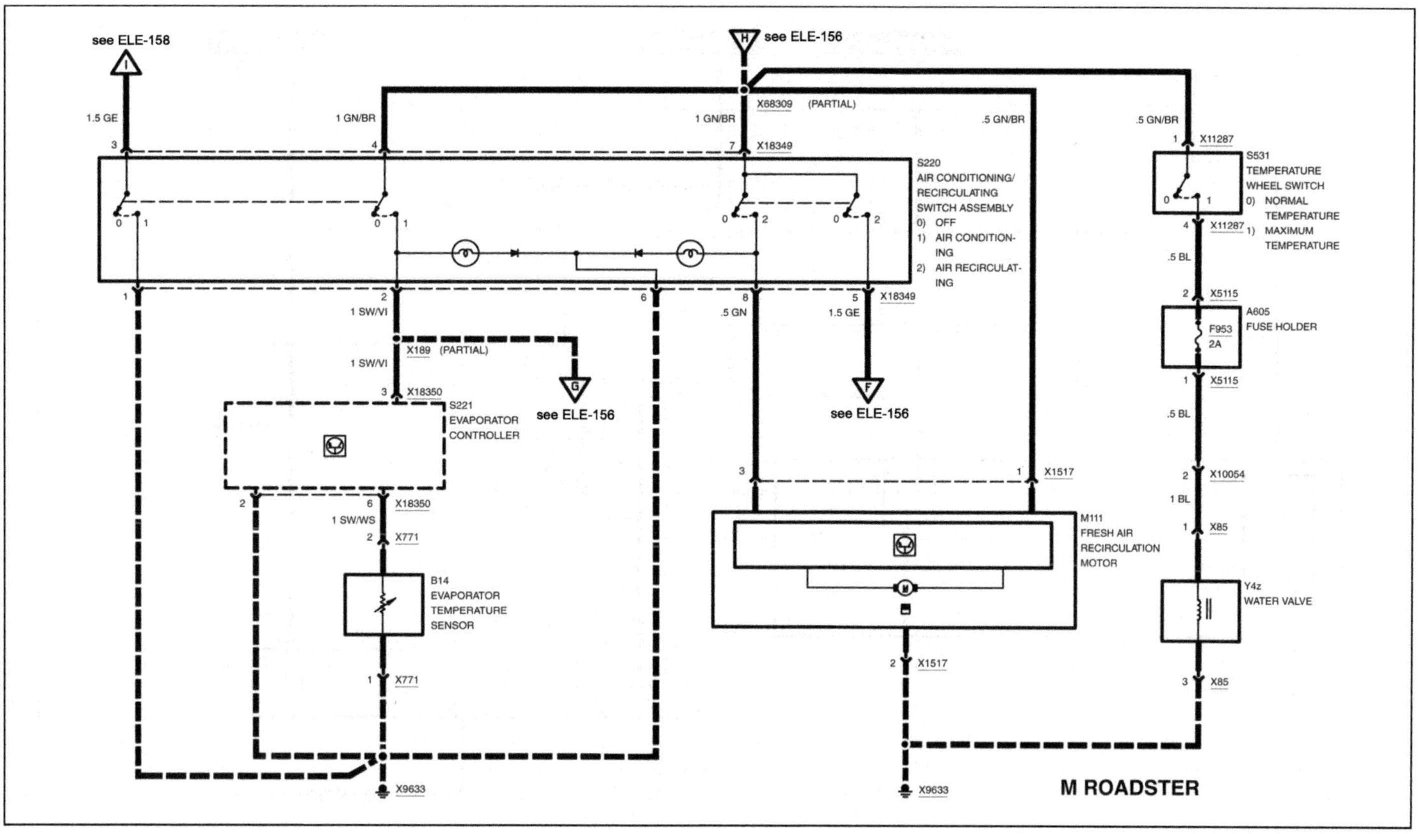


## Heating and air-conditioning
(8 of 9)

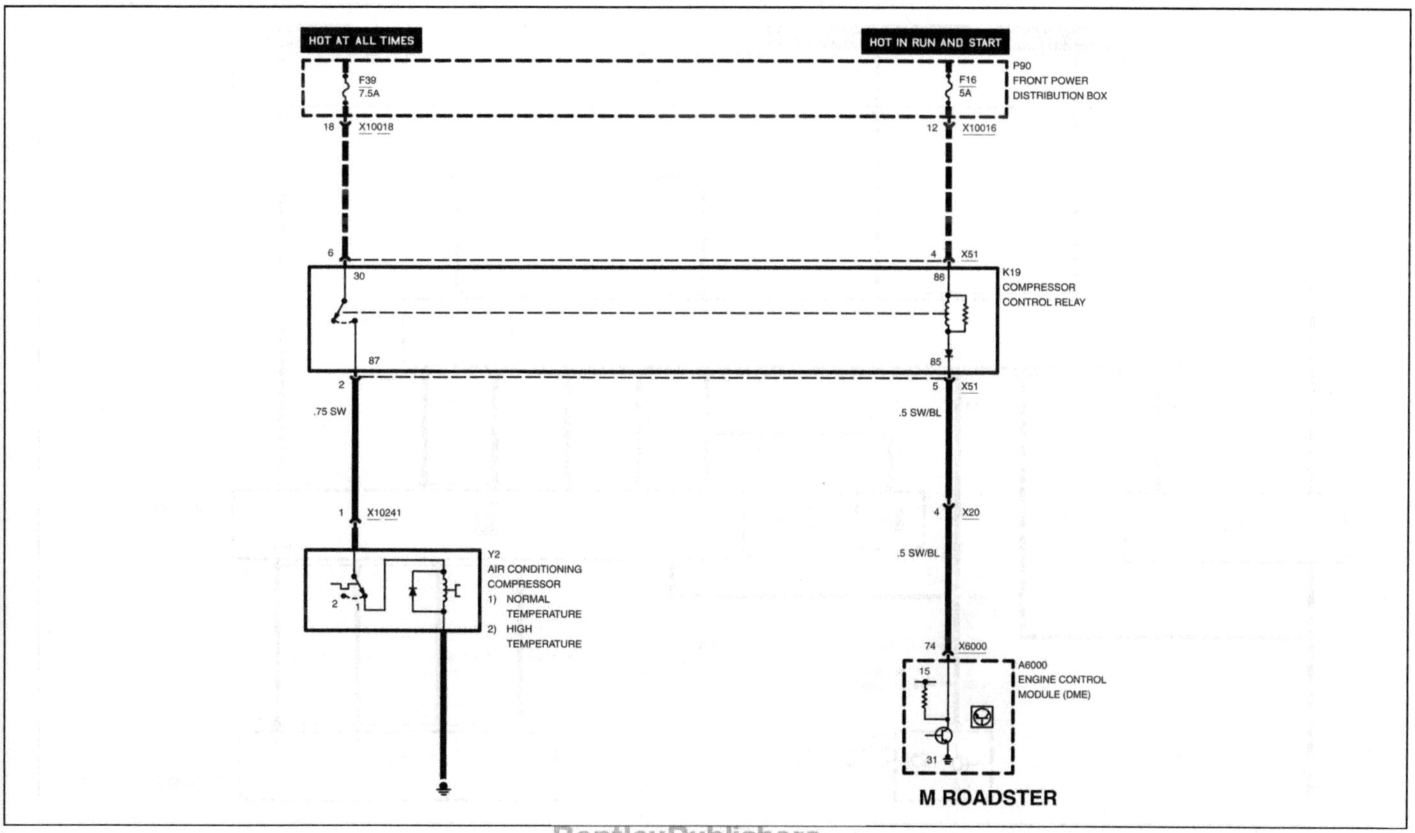

## Heating and air-conditioning
(9 of 9)

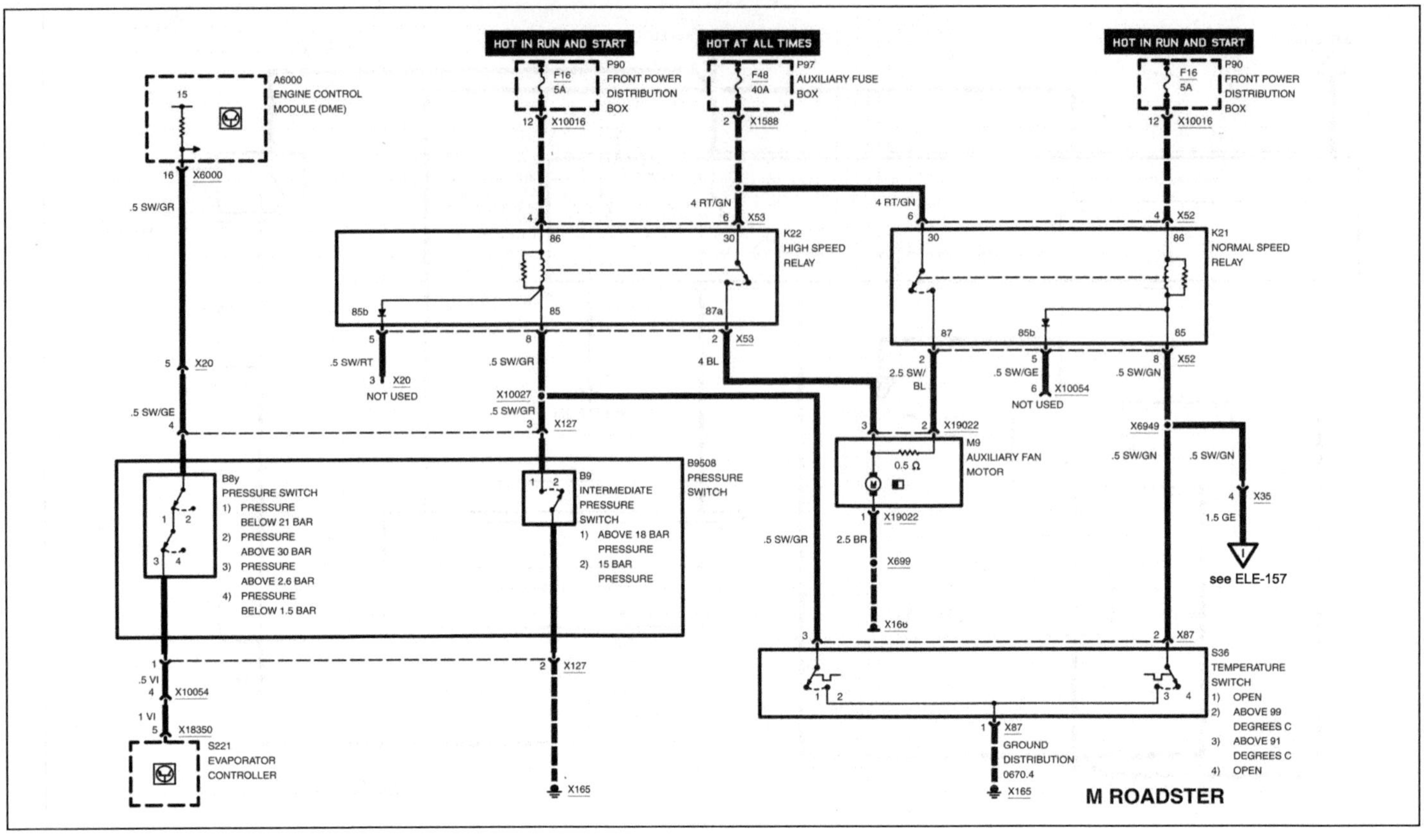


## Radio
(1 of 11)

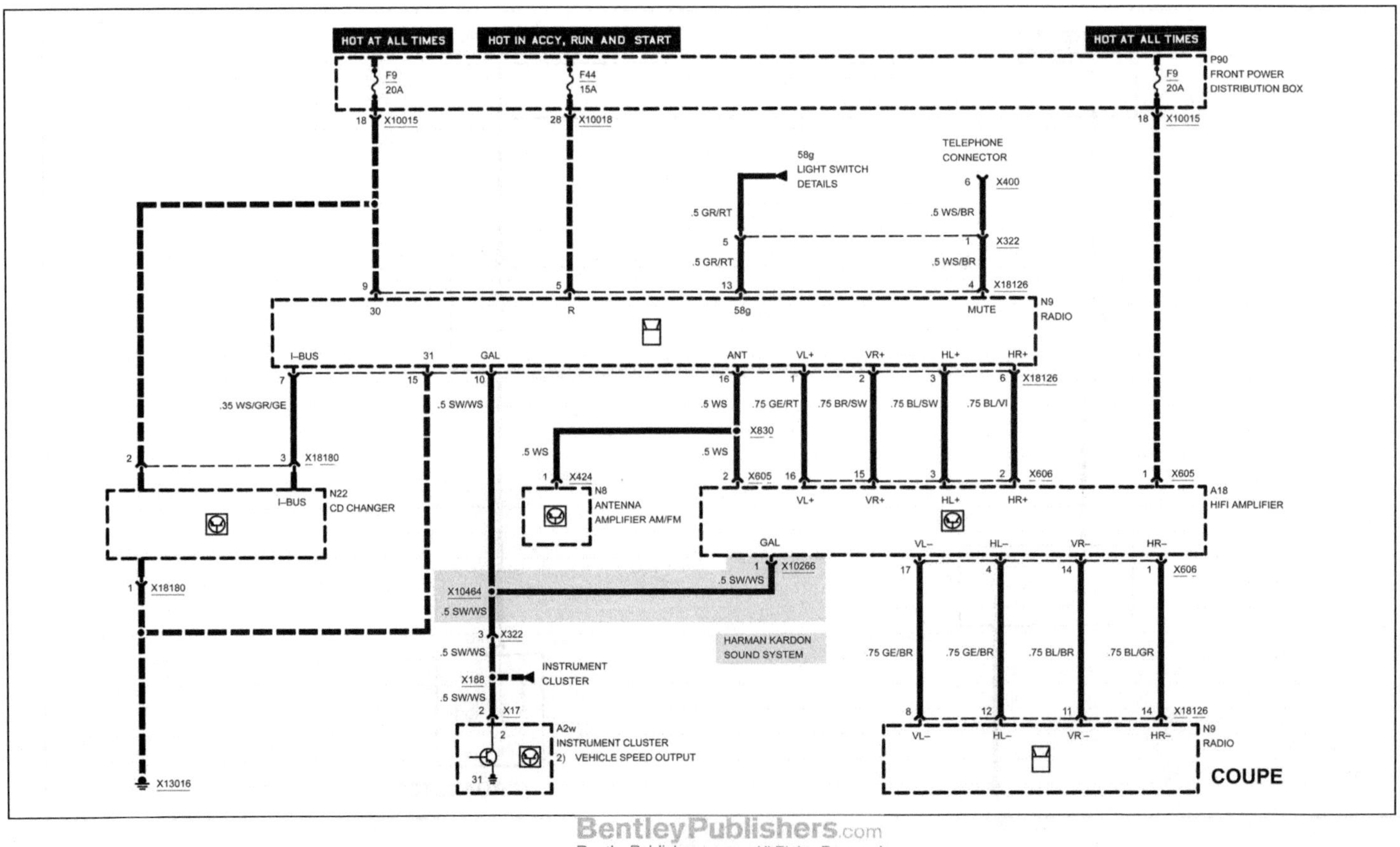

# Radio
(2 of 11)

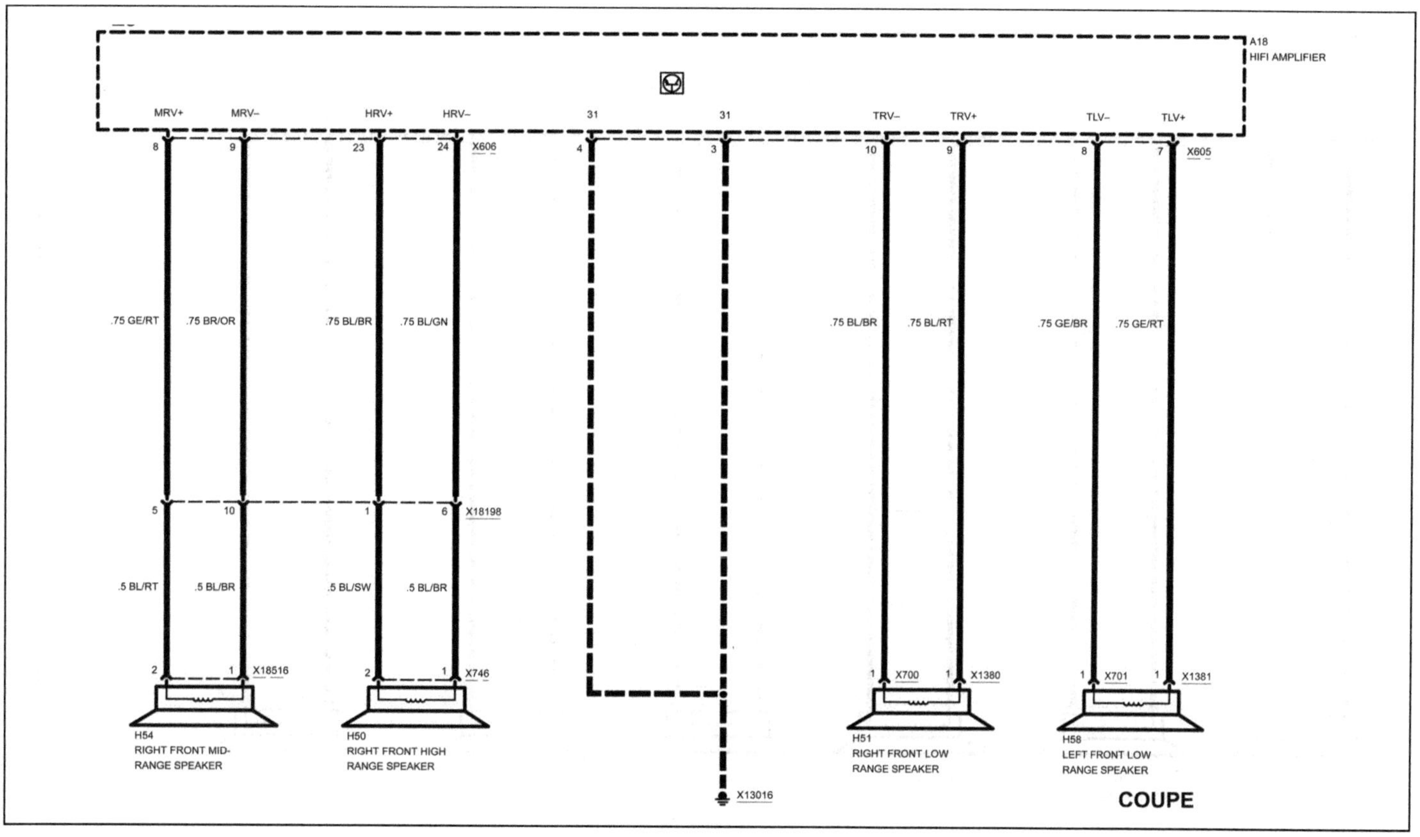


# Radio
(3 of 11)

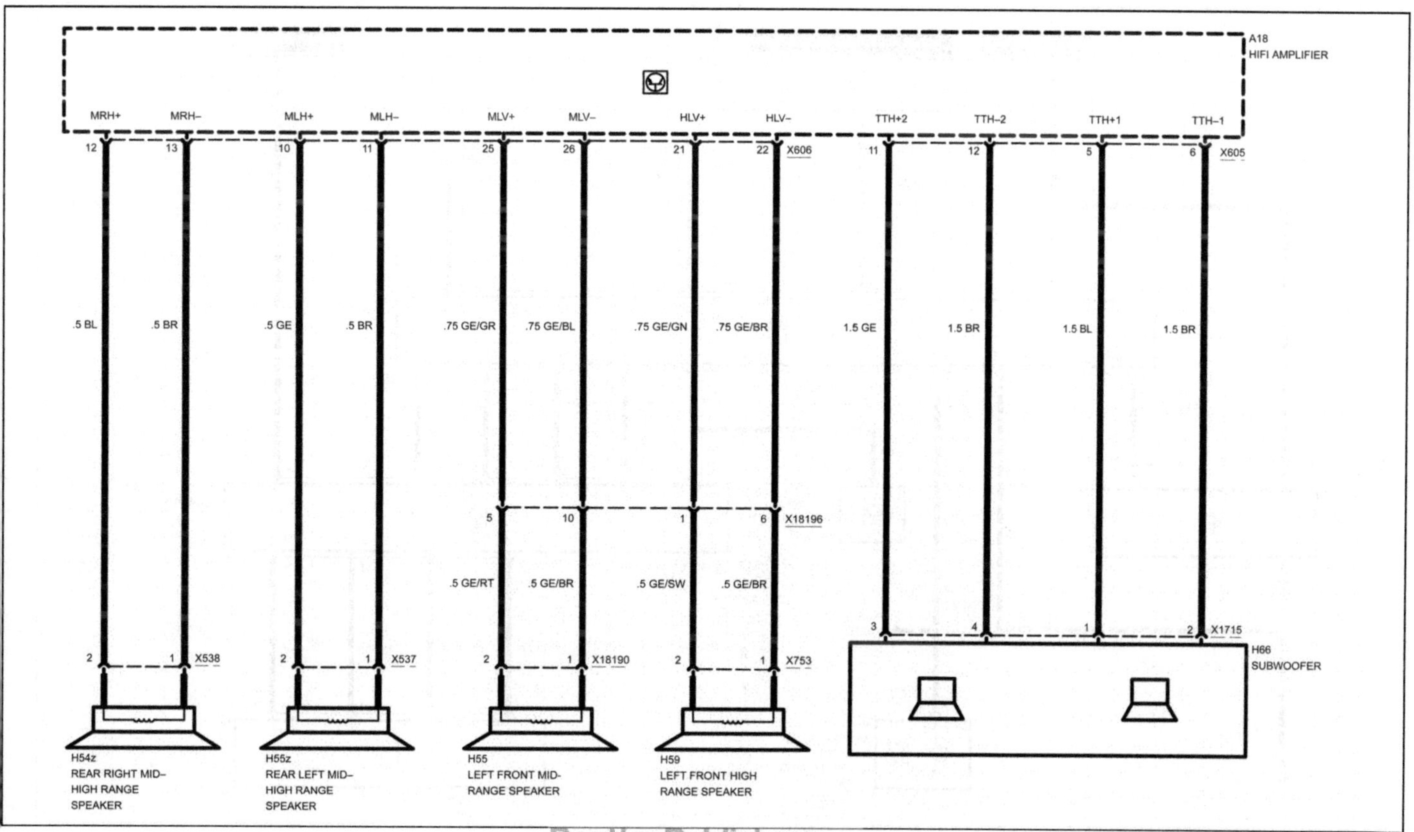

## Radio
(4 of 11)

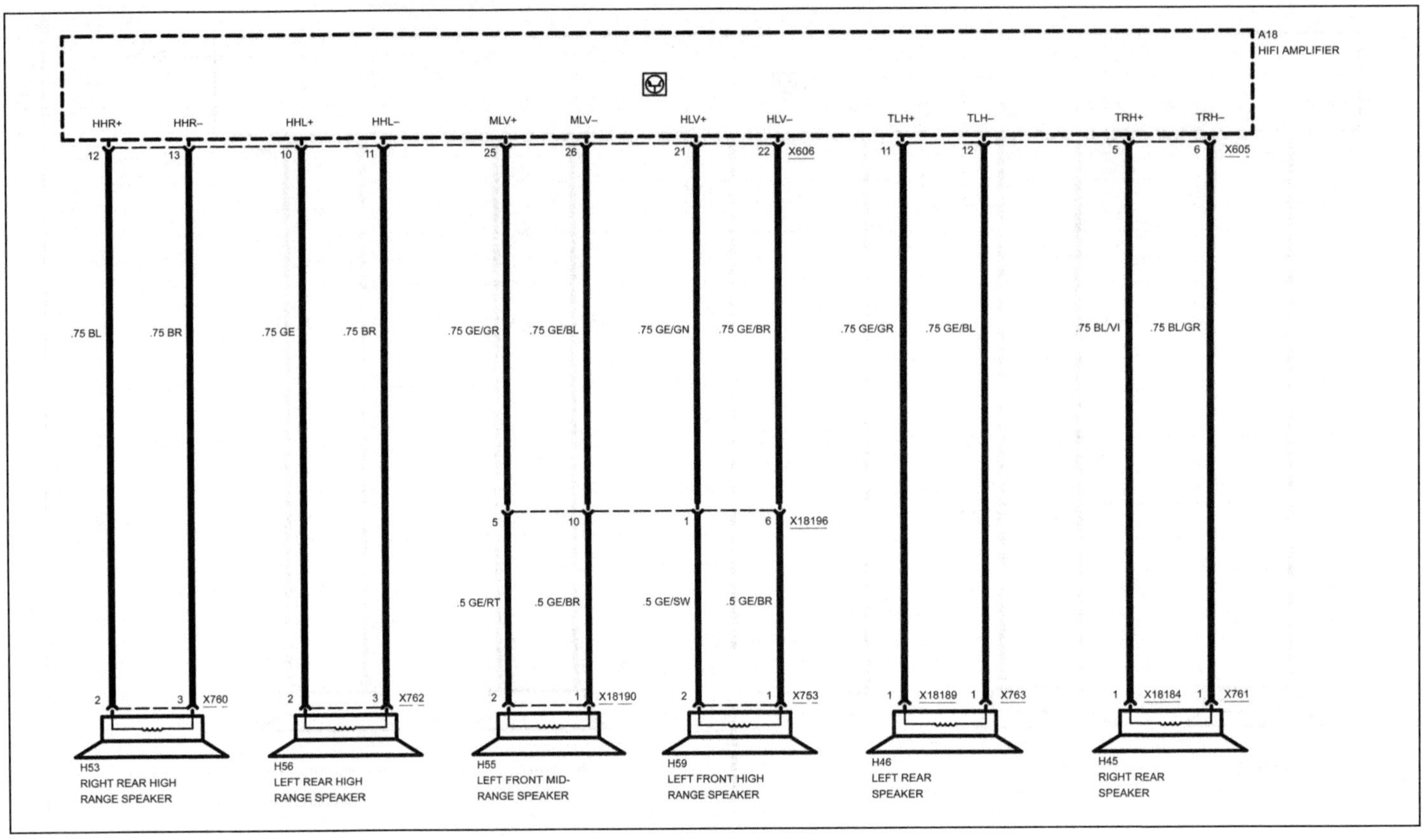


## Radio
(5 of 11)

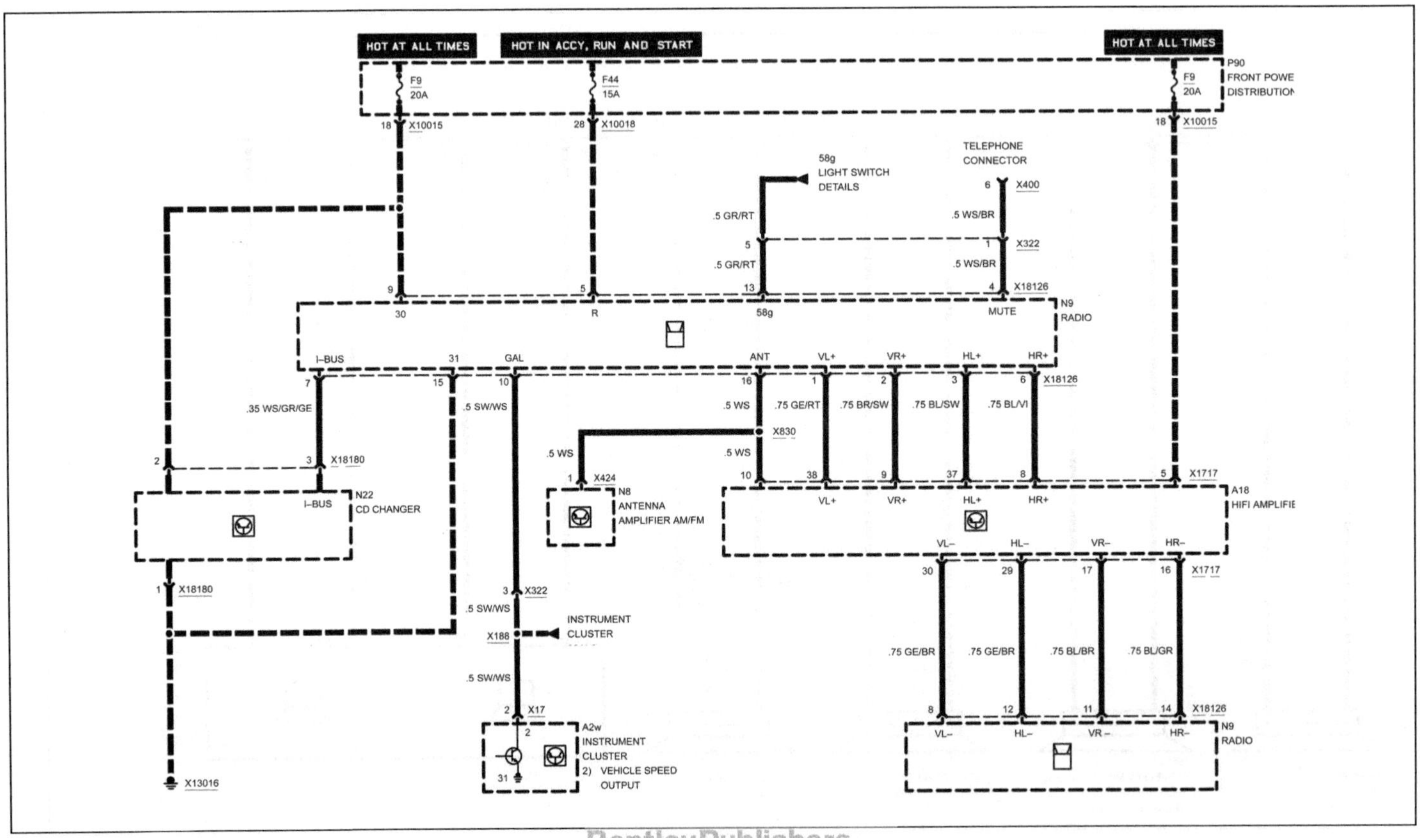

# Radio
(6 of 11)

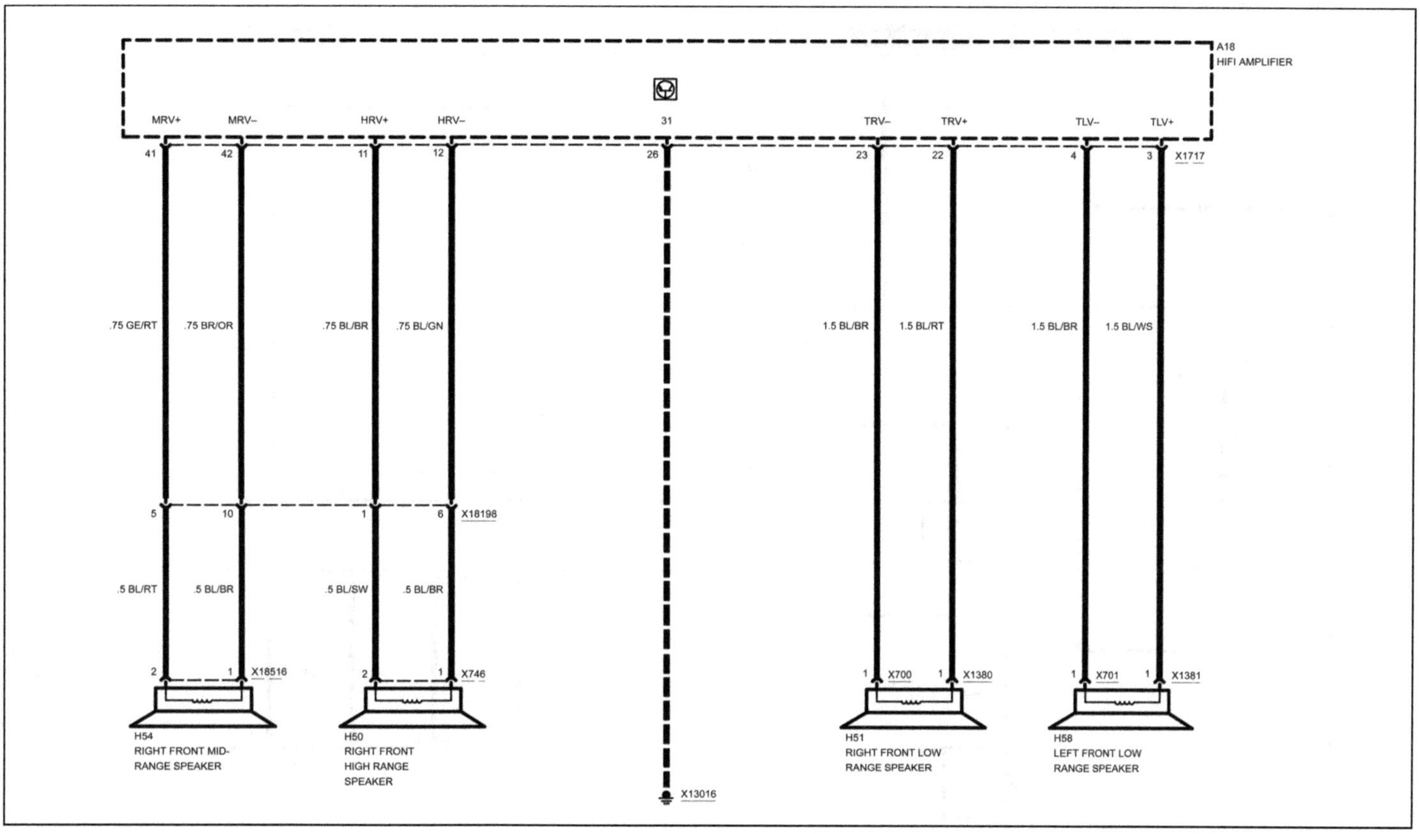


# Radio
(7 of 11)

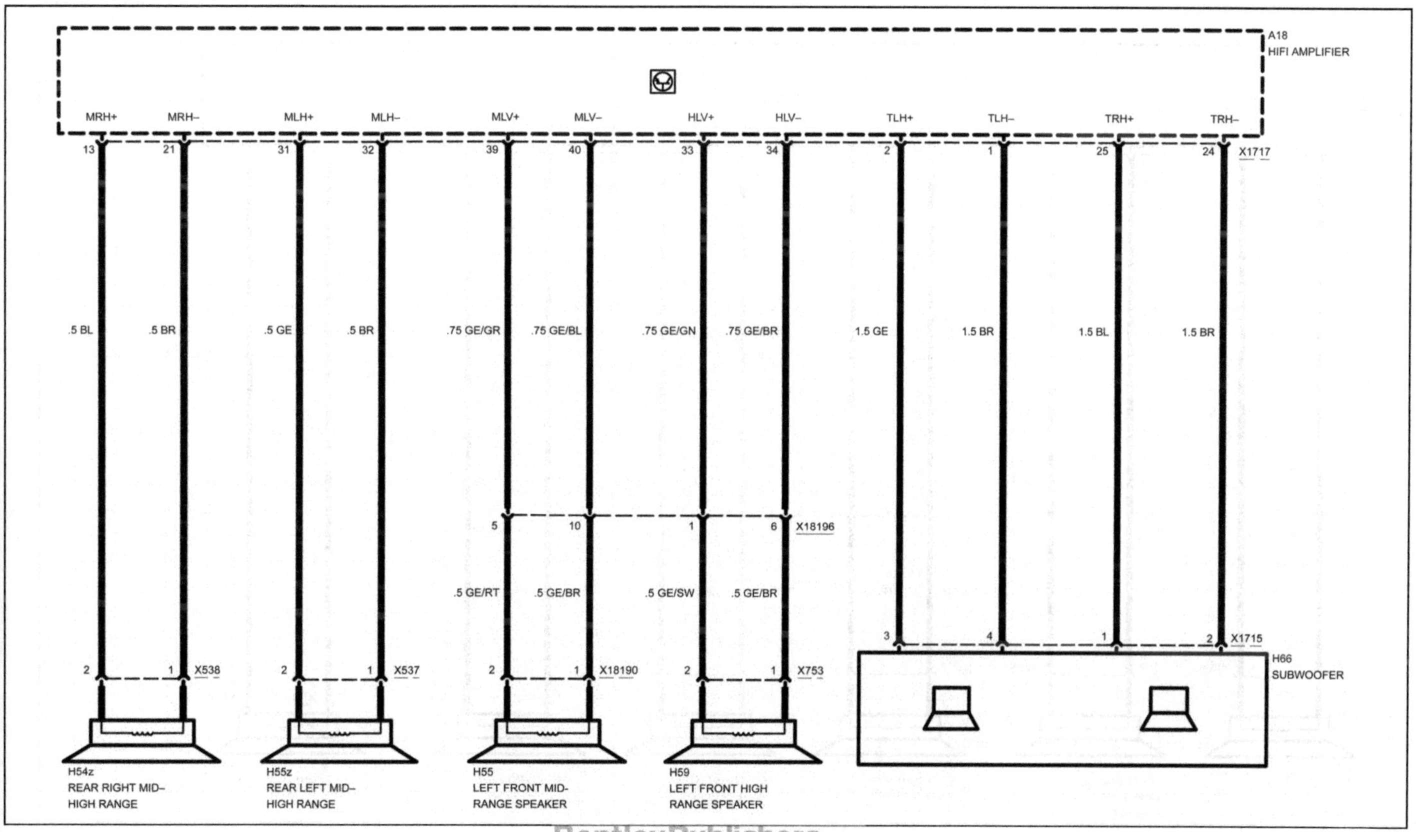

## Radio
(8 of 11)

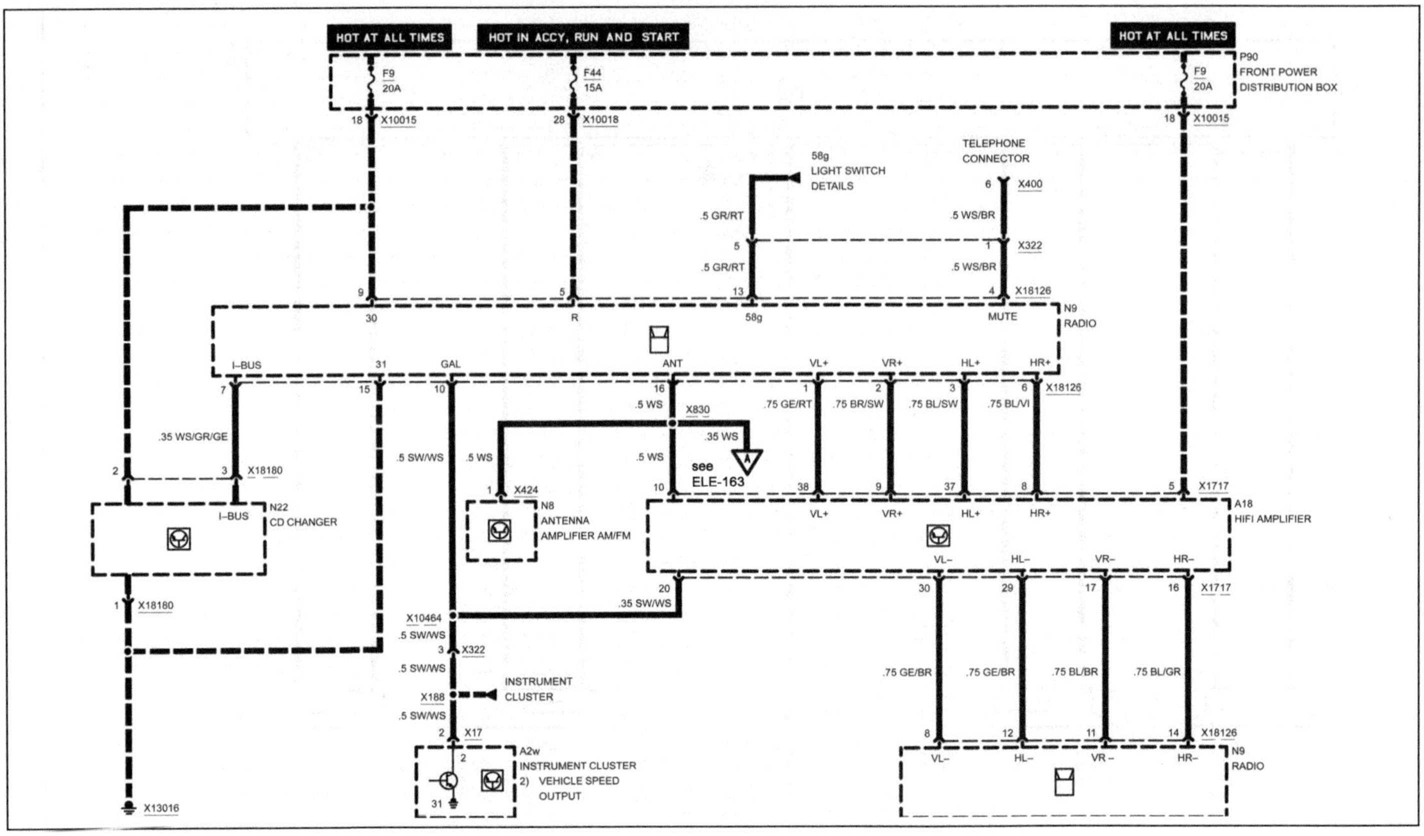


## Radio
(9 of 11)

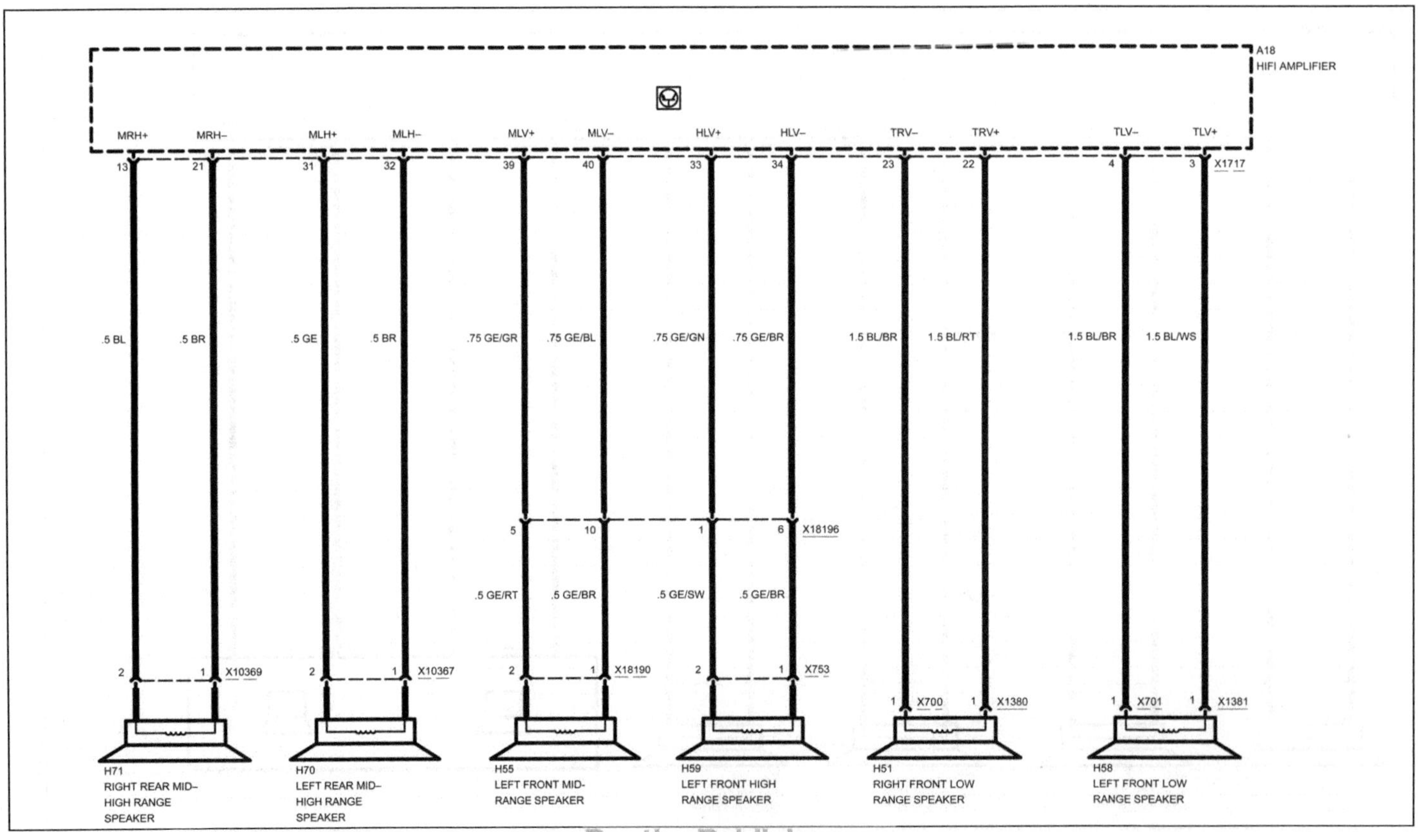

# Radio
**(10 of 11)**

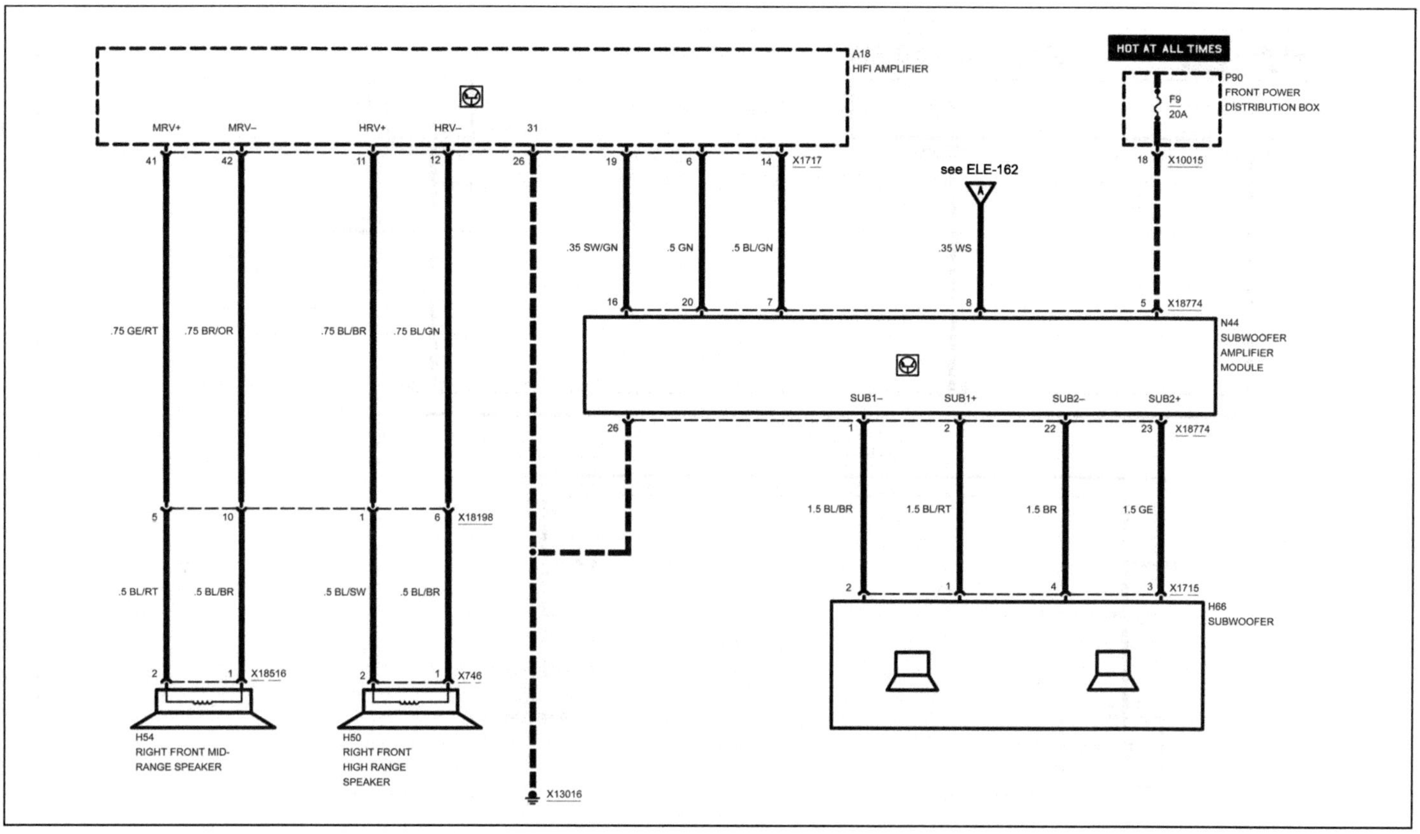


# Radio
**(11 of 11)**

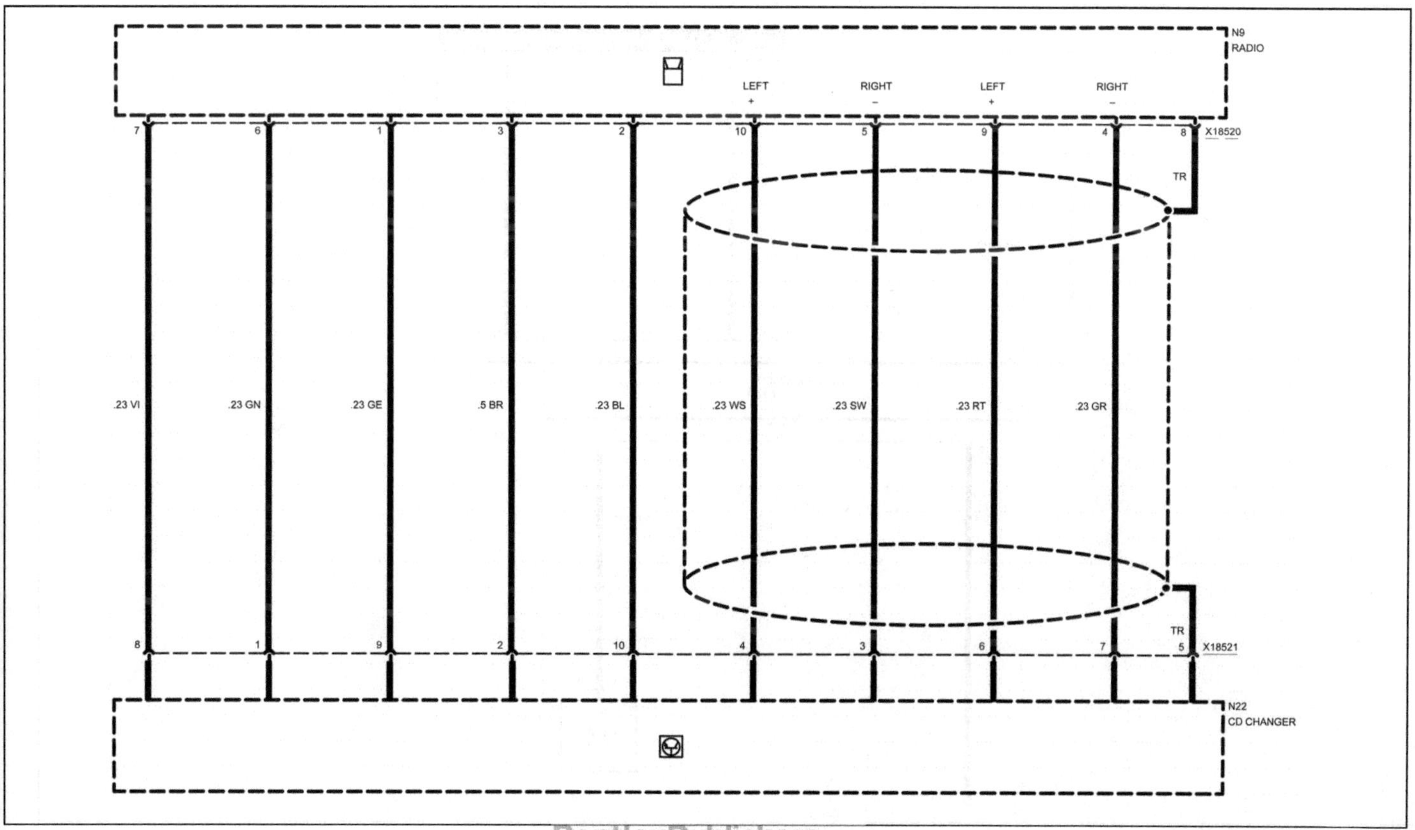

## Radio connections, 1996

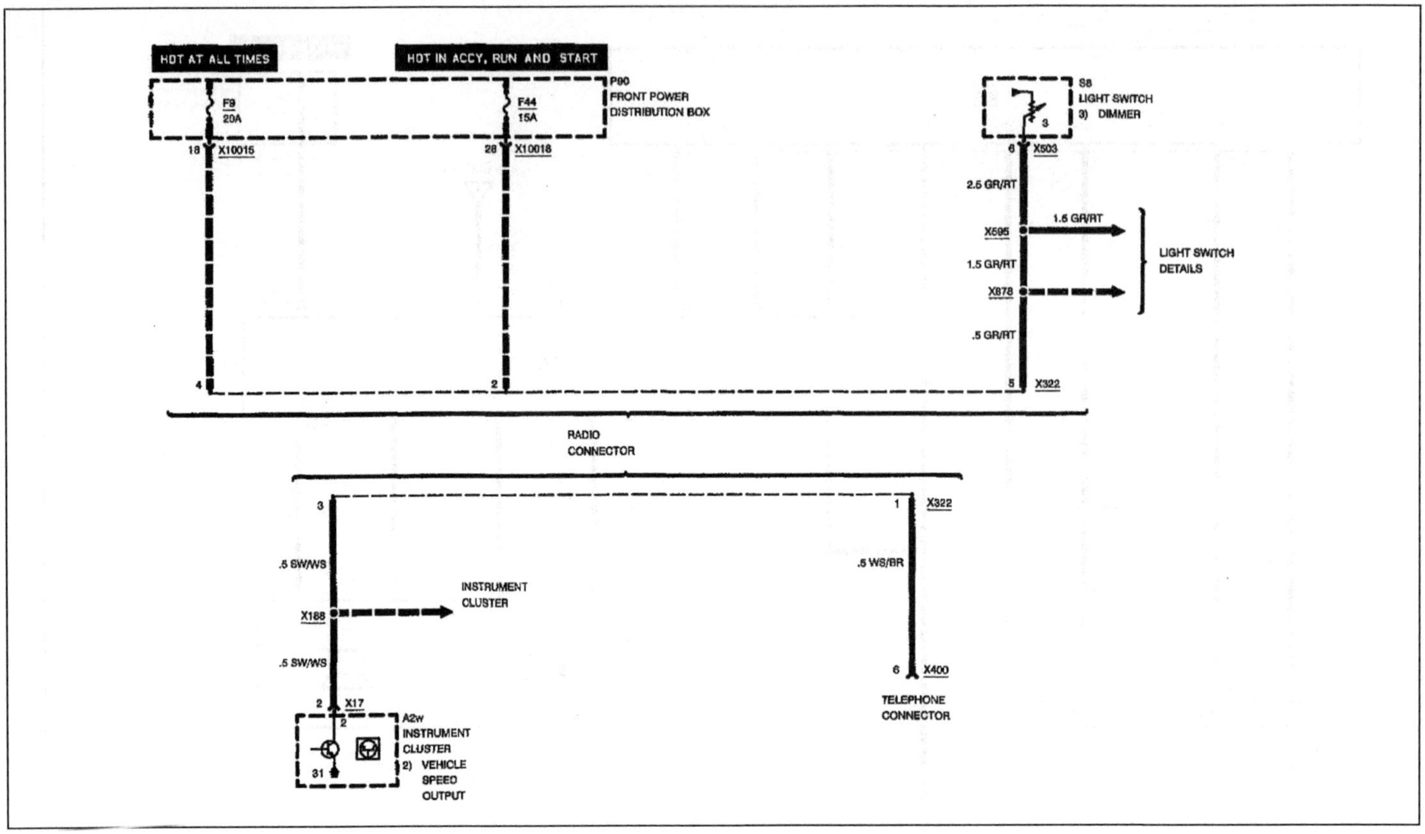


## Telephone connections

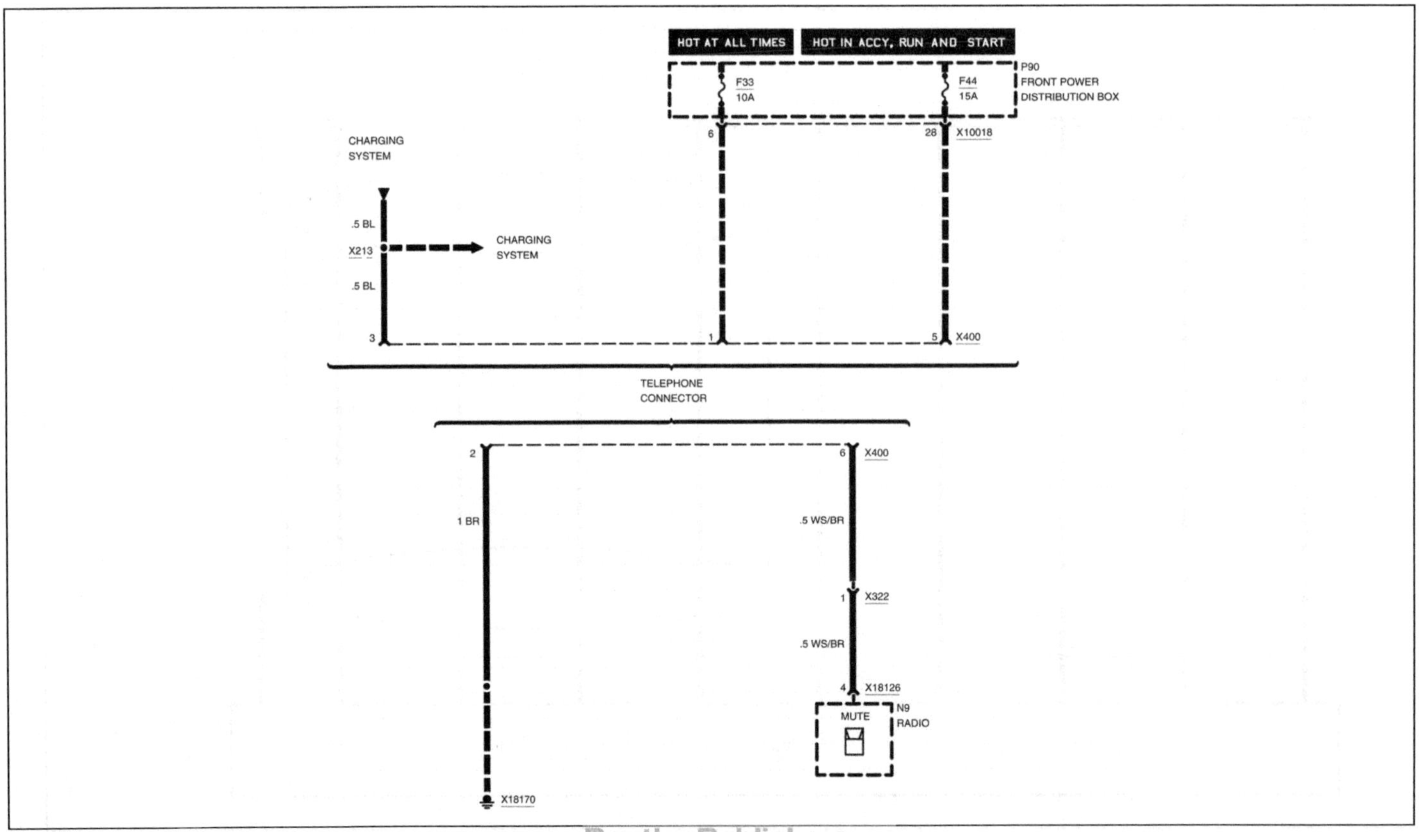

# Cruise control, 1996 - 1998
(1 of 2)

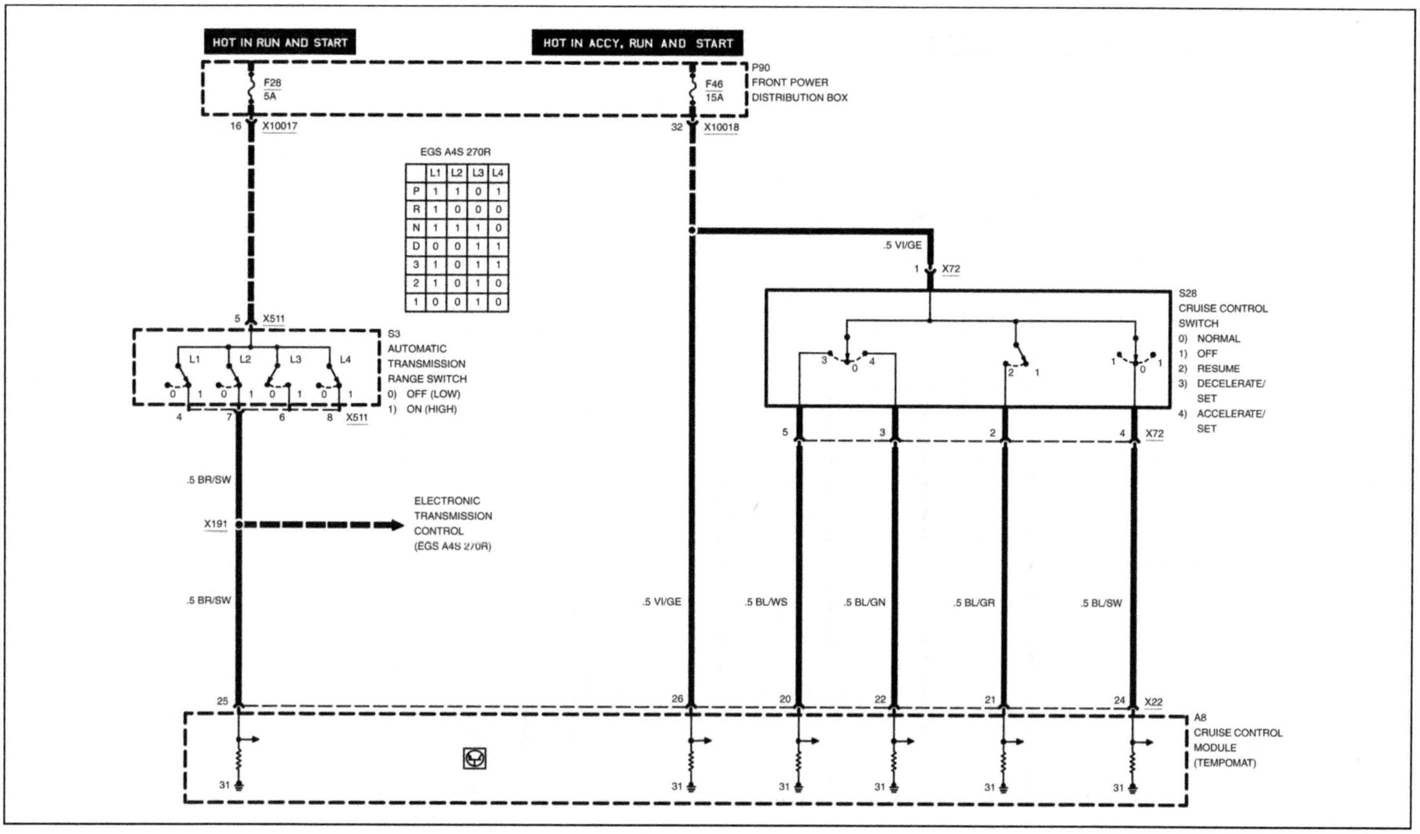


| | L1 | L2 | L3 | L4 |
|---|---|---|---|---|
| P | 1 | 1 | 0 | 1 |
| R | 1 | 0 | 0 | 0 |
| N | 1 | 1 | 1 | 0 |
| D | 0 | 0 | 1 | 1 |
| 3 | 1 | 0 | 1 | 1 |
| 2 | 1 | 0 | 1 | 0 |
| 1 | 0 | 0 | 1 | 0 |

# Cruise control, 1996 - 1998
(2 of 2)

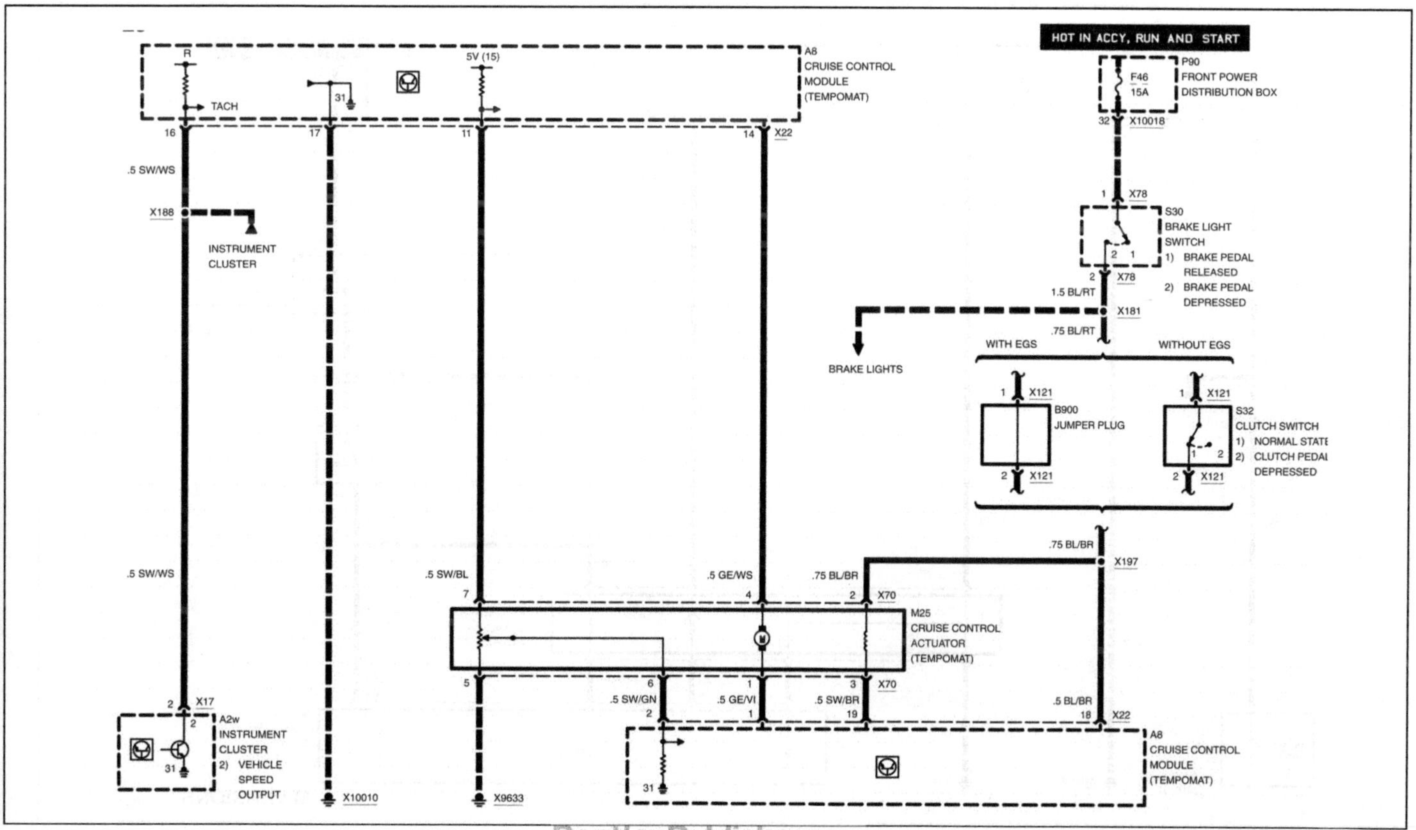

## Cruise control, 2000
(1 of 4)

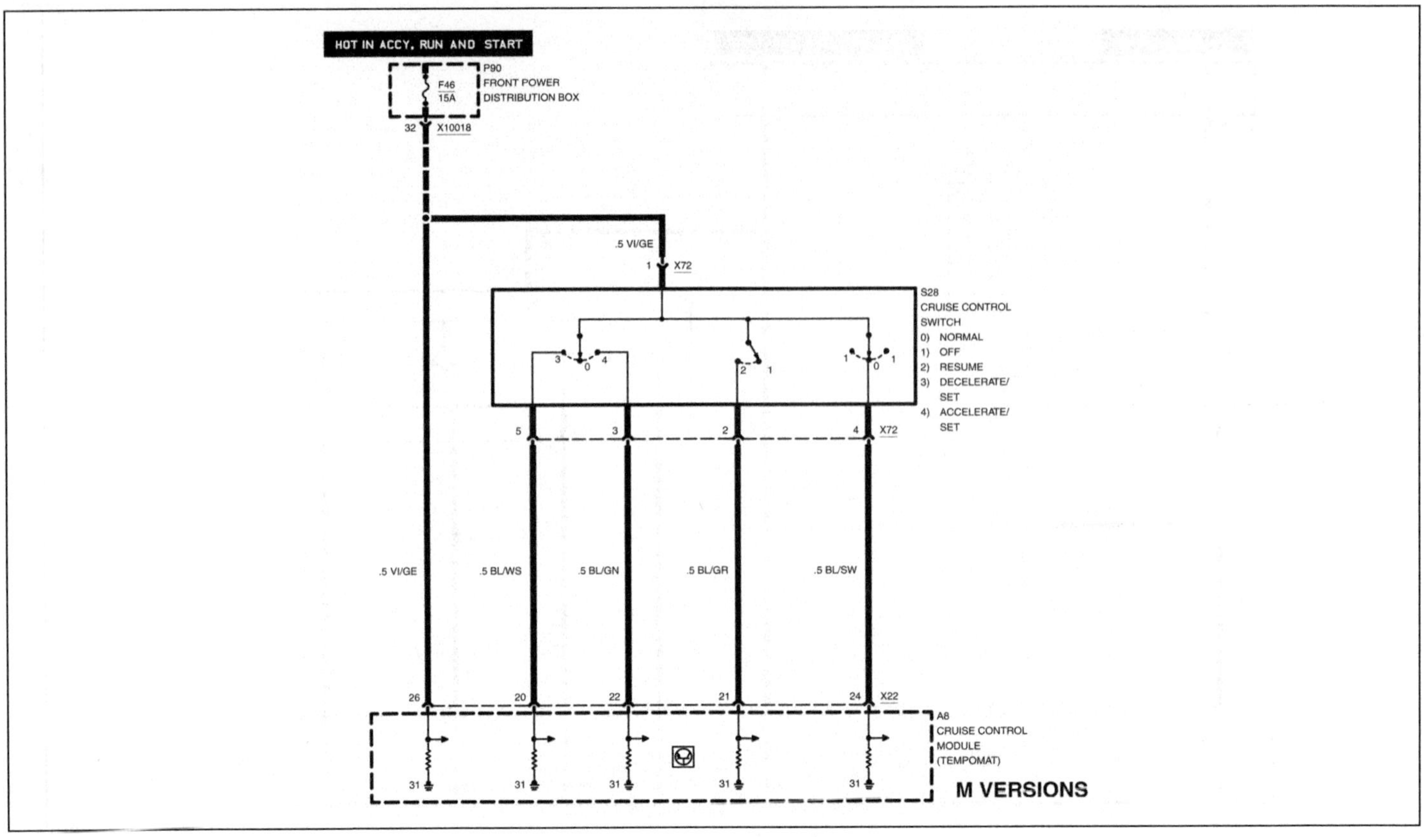


## Cruise control, 2000
(2 of 4)

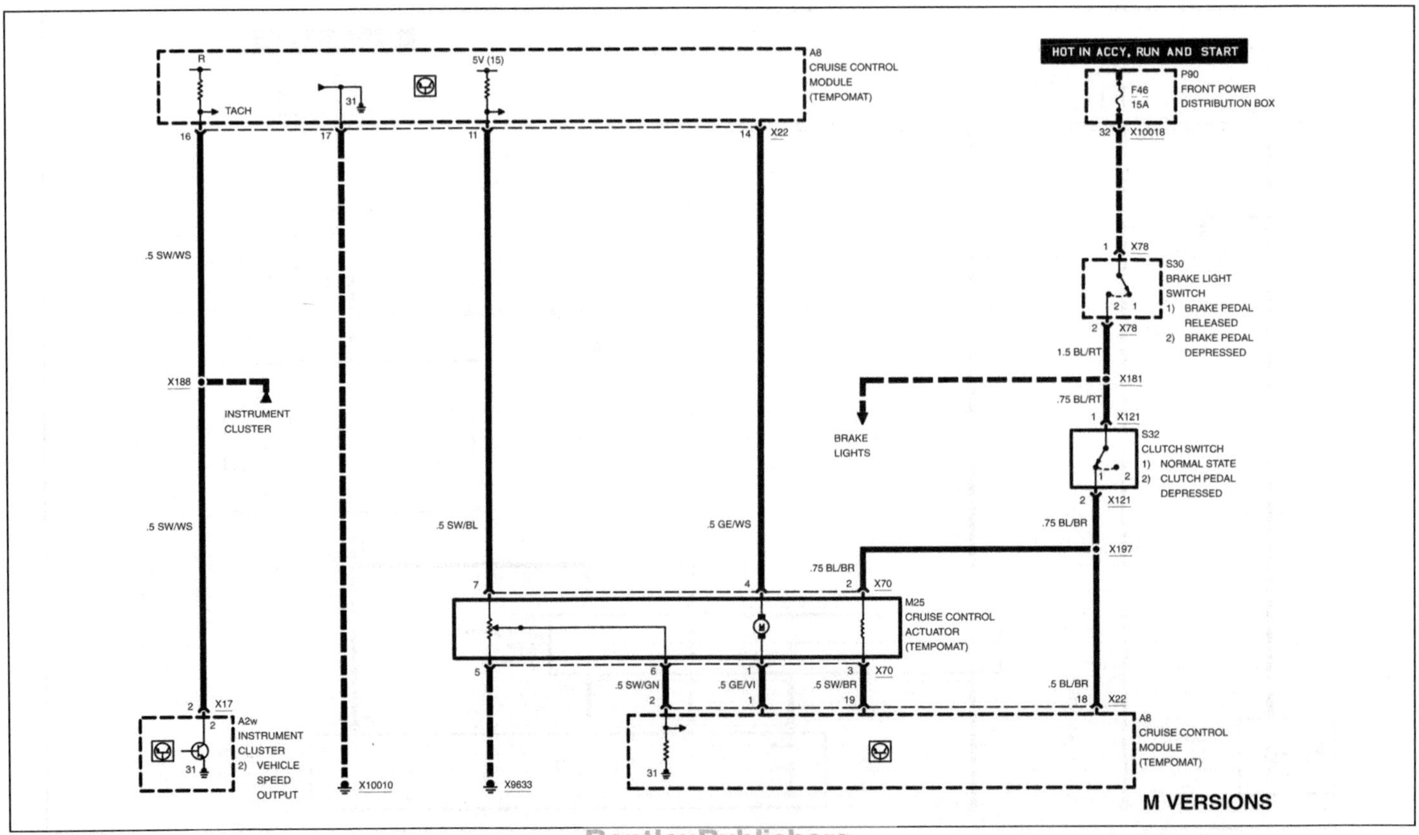

## Cruise control, 2000
(3 of 4)

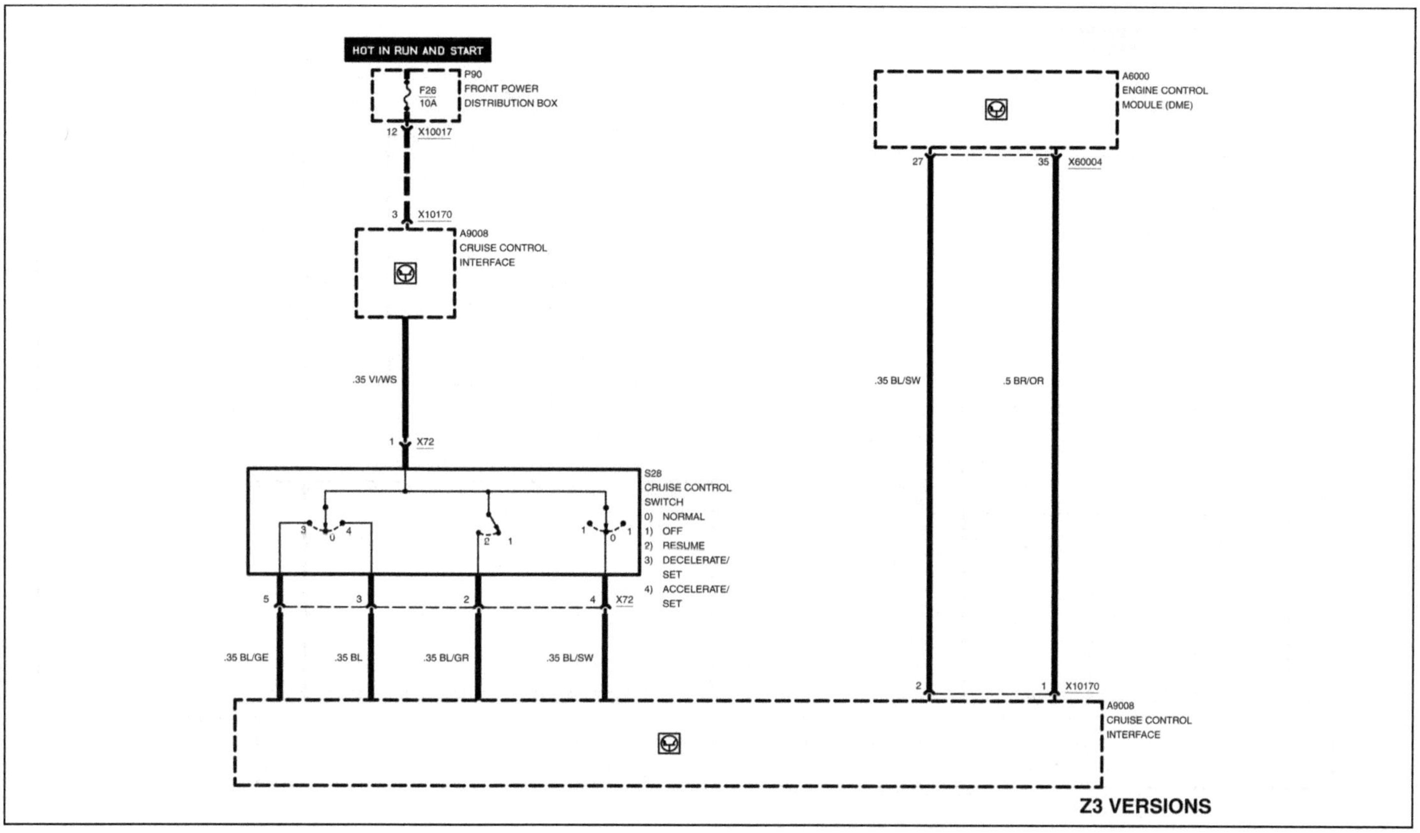


## Cruise control, 2000
(4 of 4)

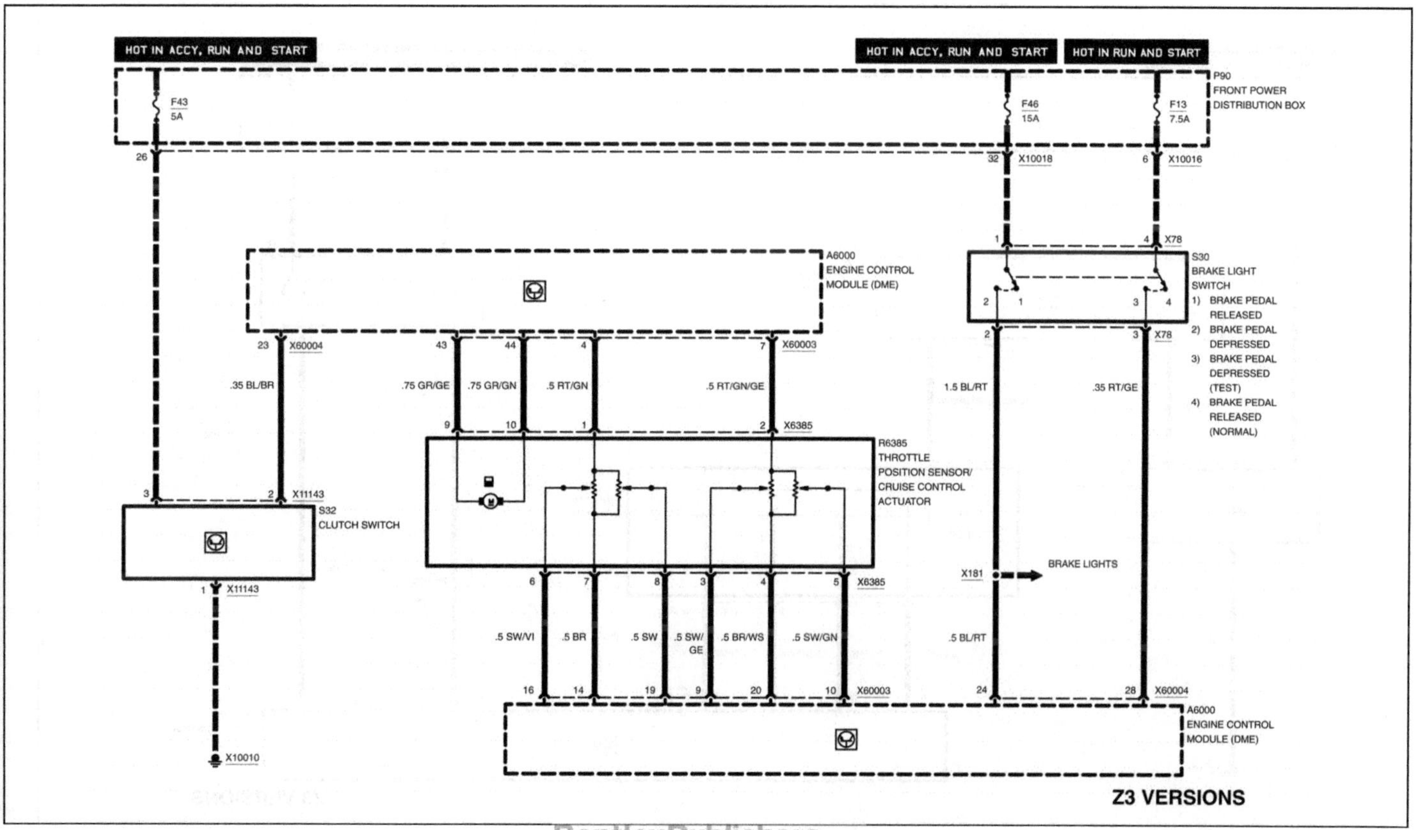

## Cruise control, 2002

(1 of 3)

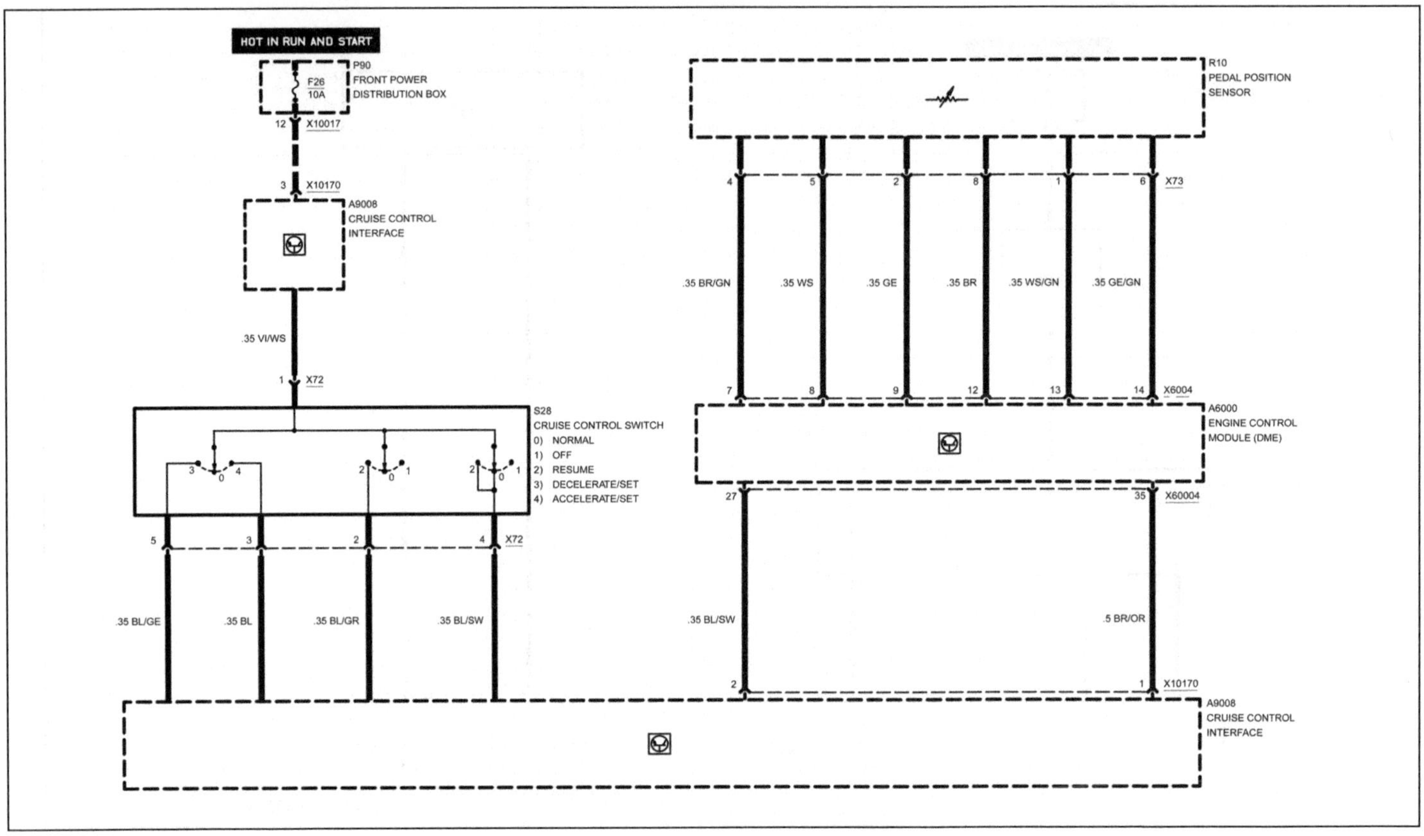


## Cruise control, 2002

(2 of 3)

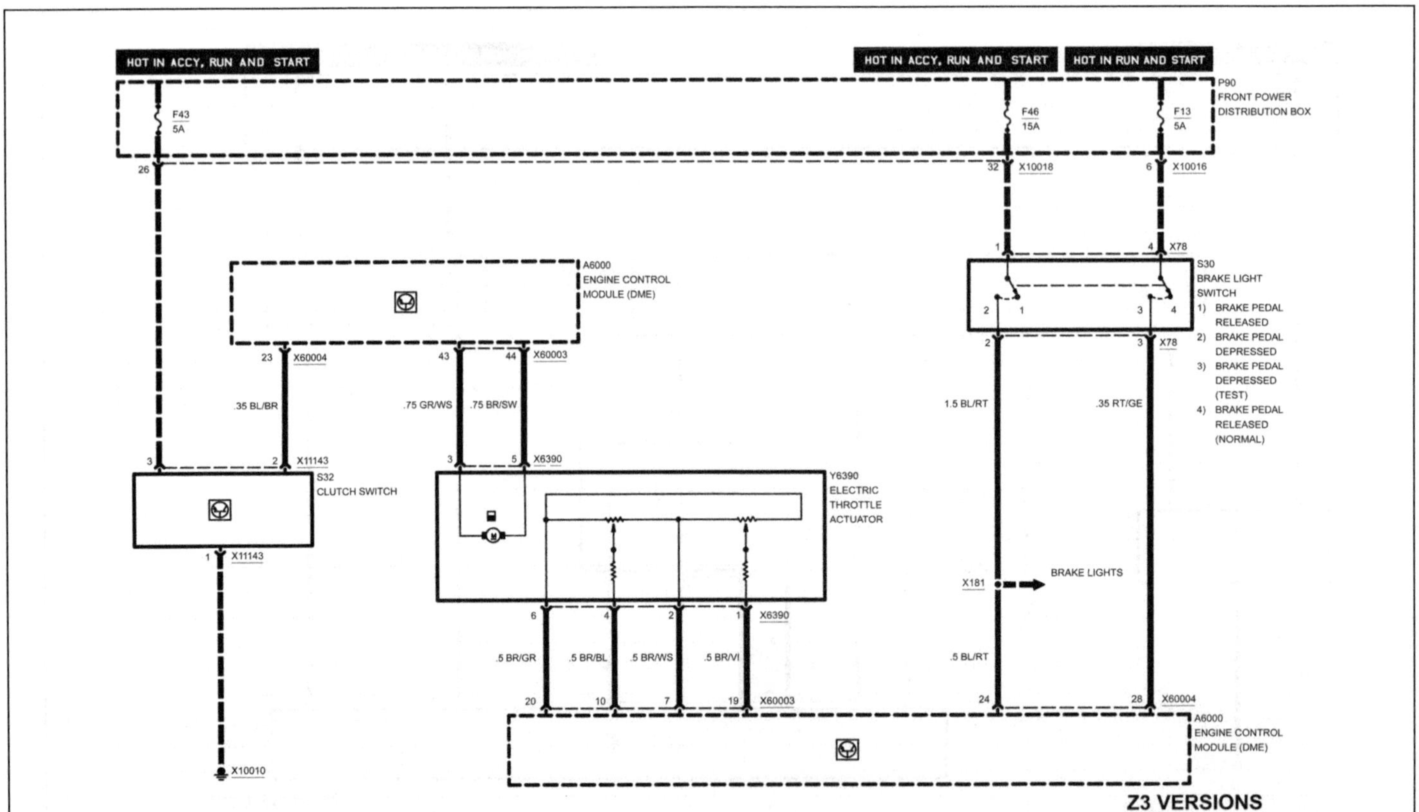


**Z3 VERSIONS**

## Cruise control, 2002

(3 of 3)

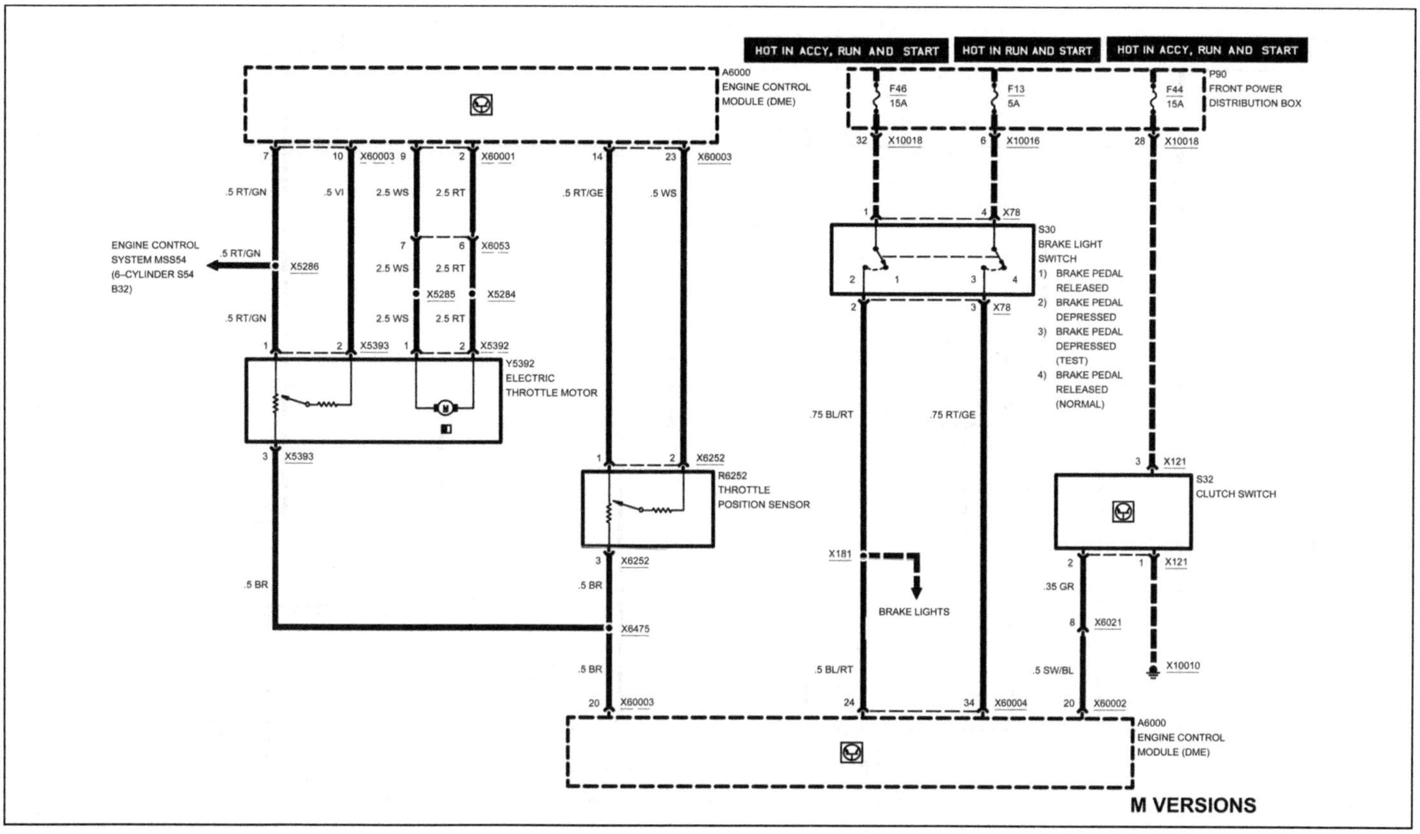


## Anti-theft system (Alpine), 1996 - 2002

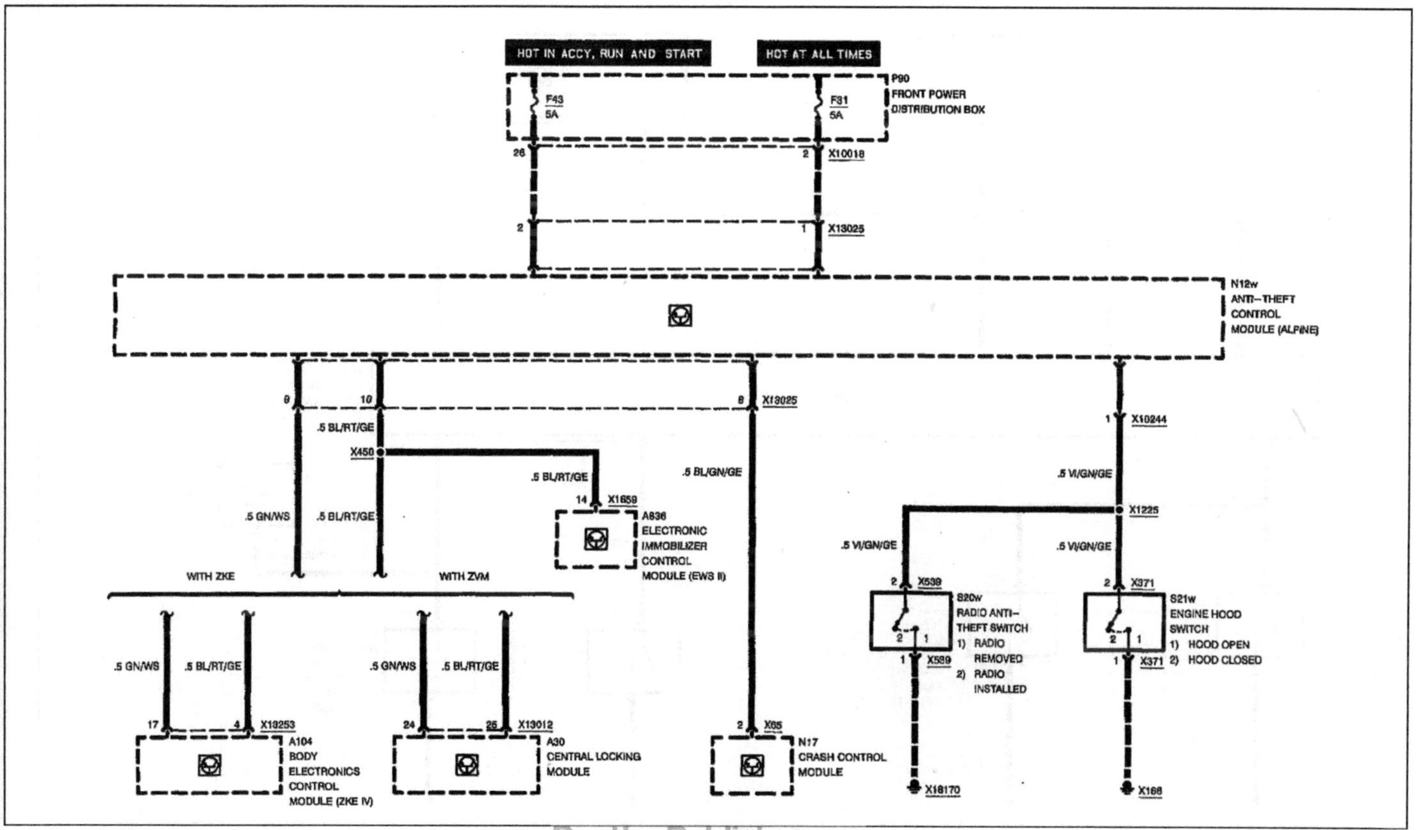

## Anti-theft system (DWA), 2002

(1 of 4)

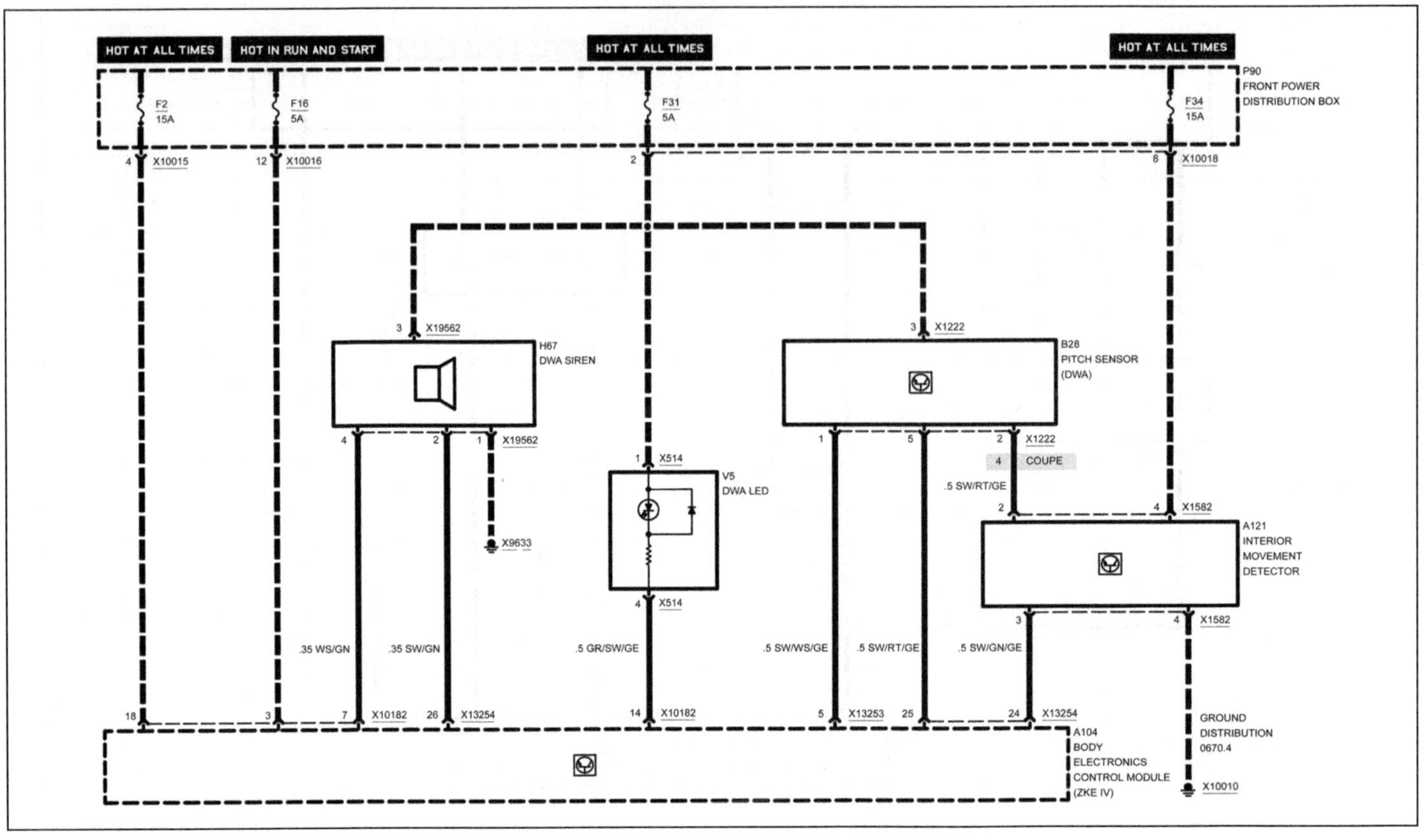


## Anti-theft system (DWA), 2002

(2 of 4)

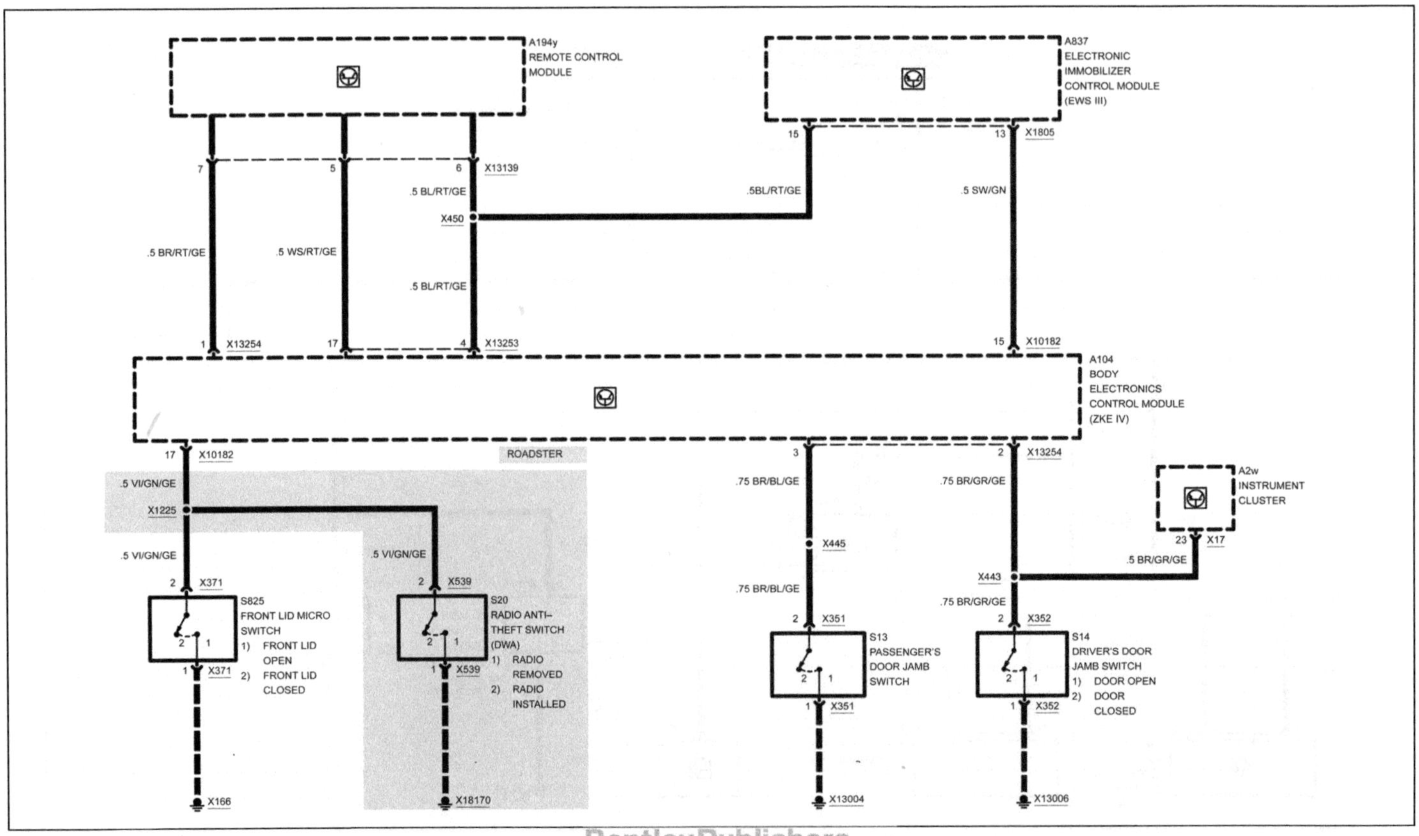

## Anti-theft system (DWA), 2002

(3 of 4)

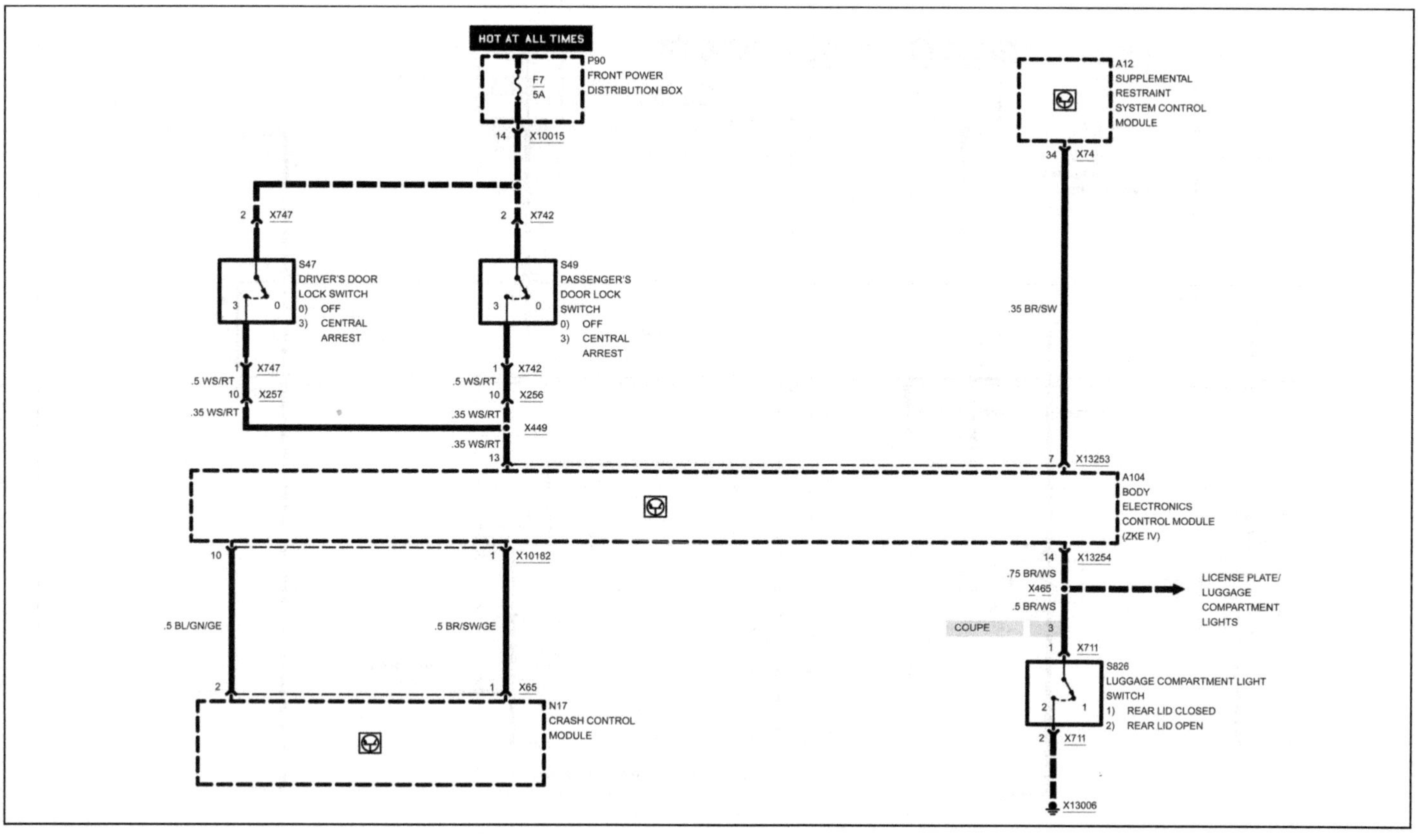


## Anti-theft system (DWA), 2002

(4 of 4)

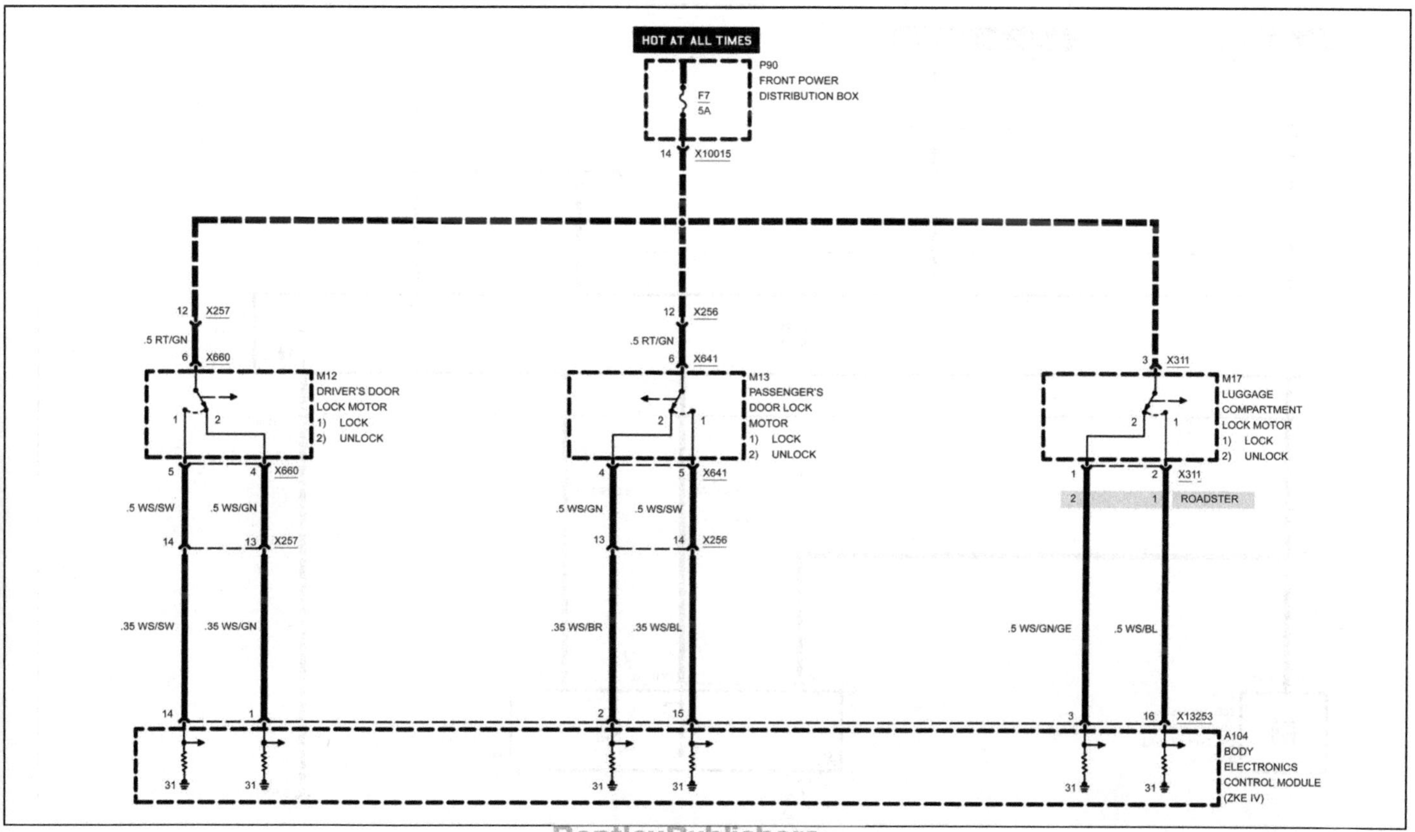

## Board computer

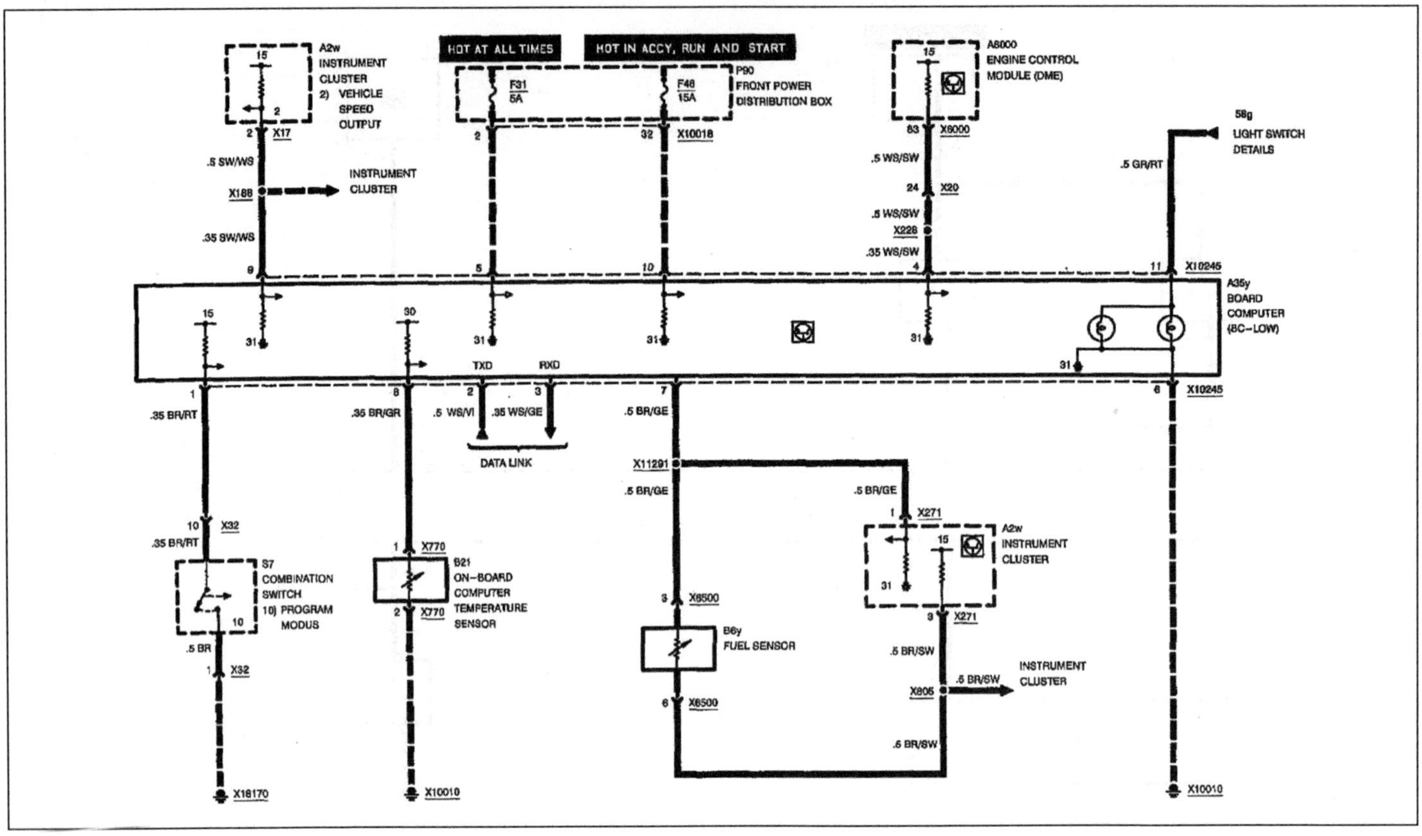


## Keyless entry

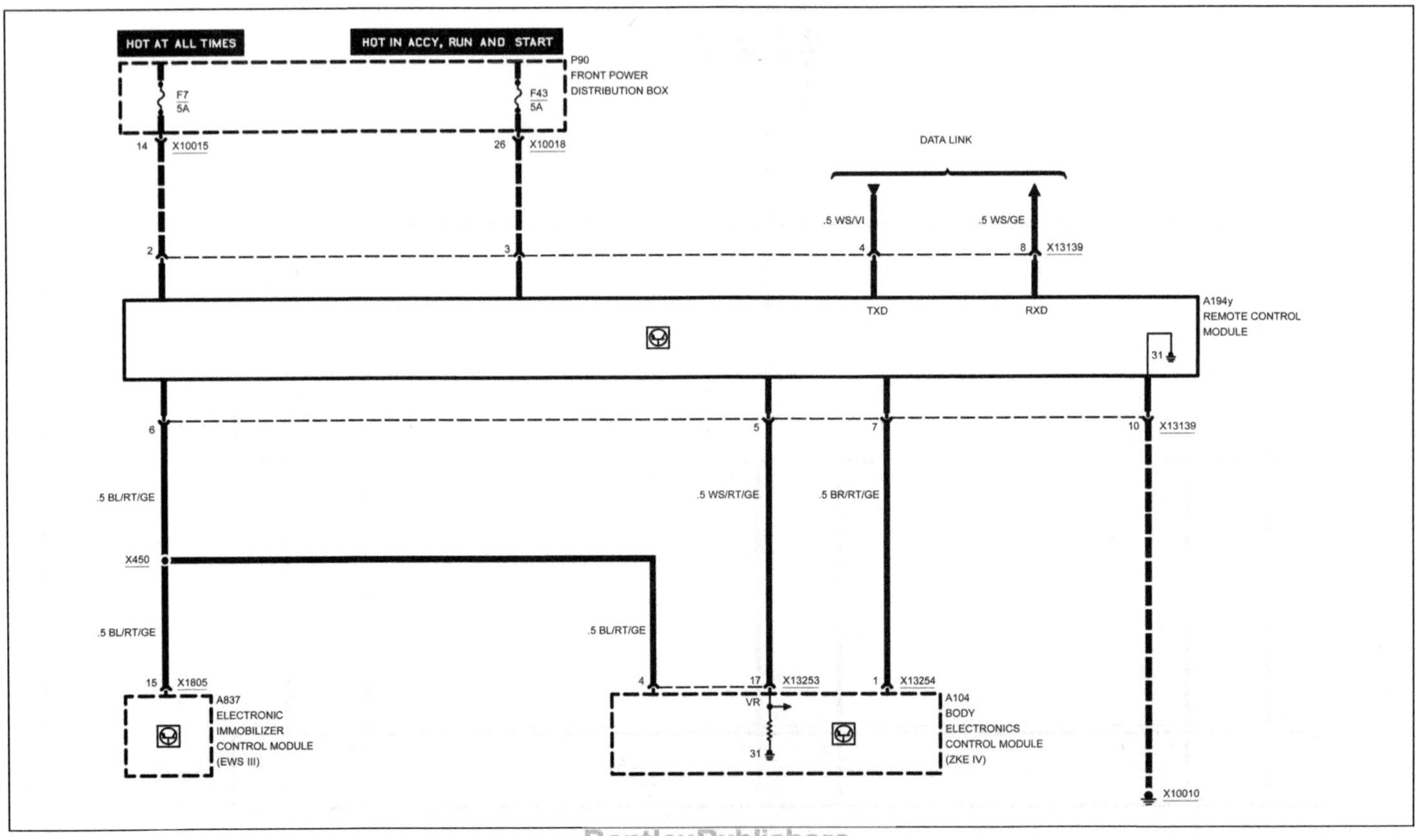

# OBD On-Board Diagnostics



**DTC TABLES**

## General

This chapter outlines the fundamentals and equipment requirements of the SAE (Society of Automotive Engineers) and CARB (California Air Resource Board) On-Board Diagnostics II (OBD II) standard as it applies to BMW vehicles. Also covered here is a comprehensive listing of Z3 and M version OBD II diagnostic trouble codes (DTCs).

## On-Board Diagnostics II (OBD II)

OBD II standards have been applied to all passenger vehicles sold in the United States from model year 1996.

OBD capabilities are incorporated into engine software to monitor virtually every component that can affect vehicle emissions. Each emission-influencing component is checked by a diagnostic routine to verify that it is functioning properly. If a problem or malfunction is detected, the OBD II system illuminates the malfunction indicator light (MIL) on the instrument panel.

The OBD II system also stores important information about the detected malfunction in the engine control module (ECM) so that a repair technician can find the problem and repair it accurately.

***NOTE—***
- *Specialized OBD II scan tool equipment is needed to access the fault memory and OBD II data.*
- *The OBD II fault memory (including the MIL) can only be reset using a special scan tool. Removing the connector from the engine control module (ECM) or disconnecting the battery will not erase the fault memory.*

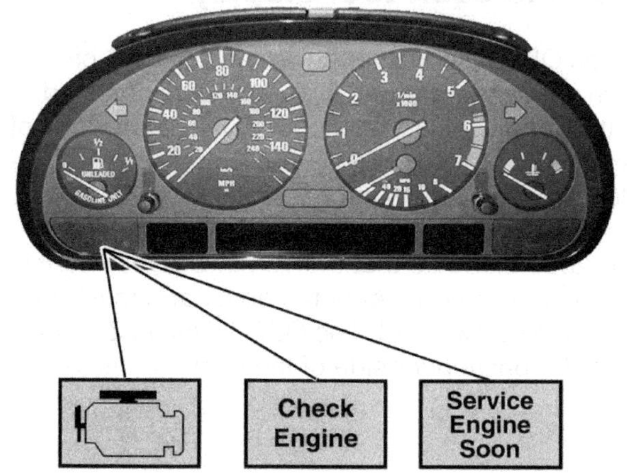


502120898

## Malfunction indicator light (MIL)

BMW models covered in this manual are equipped with one of three versions of MIL (as shown).

A bulb check function illuminates the malfunction indicator light (MIL) when the ignition key is in the ON position before cranking the engine. The OBD II system illuminates the MIL when emission levels exceed 1.5 times the Federal standards or when an emission-related system malfunction occurs.

Conditions that illuminate the MIL:

- Engine management system fault detected for **two** consecutive OBD II drive cycles. (See **Drive cycle** in this repair group.)
- Catalyst damaging fault.
- Malfunction in component(s) monitoring emissions.
- Manufacturer-defined specifications exceeded.
- Implausible sensor input signal.
- Misfire.
- Leak detected in fuel tank evaporative system.
- No purge flow detected from purge valve / evaporative system.
- Failure of DME to enter closed-loop operation within specified time.
- Engine management system enters "limp home" mode.

Additional information on MIL:

- A fault code is stored within the engine control module (ECM) upon the first occurrence of a fault being identified.
- Two **complete** consecutive drive cycles with the fault present illuminate the MIL. The exception to the two-fault requirement is a catalyst-damaging fault which turns the light on immediately.
- If the second drive cycle is not completed and the fault not checked, the engine control module (ECM) counts the third drive cycle as the next consecutive drive cycle. The MIL is illuminated if the fault is still present.
- Once the MIL is illuminated, it remains illuminated unless the system is tested and found to be without faults through three complete consecutive drive cycles.
- In all cases except catalyst-damaging fault(s), the fault code is cleared from memory automatically if the system is checked through **40** consecutive drive cycles without the fault being detected.
- In case of a catalyst-damaging fault, the fault code is cleared from memory automatically if the system is checked through **80** consecutive drive cycles without the fault being detected.

## Scan tool and scan tool display

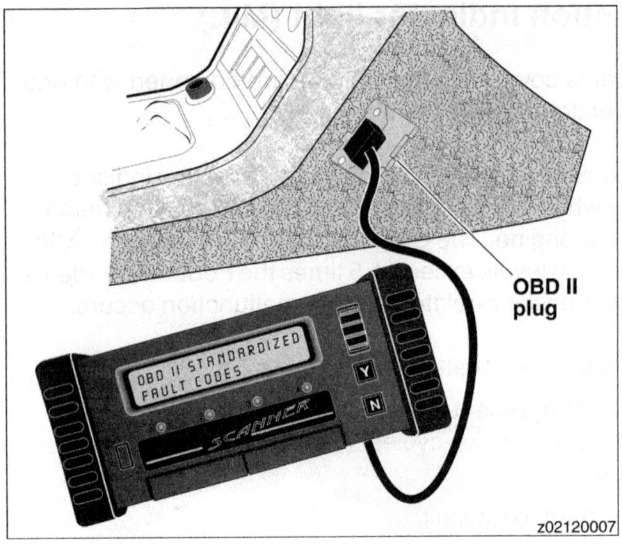


With a generic OBD II scan tool connected to the 16-pin OBD II plug, diagnostic trouble codes (DTCs) can be obtained, along with the conditions associated with the illumination of the MIL. A more advanced or BMW-dedicated scan tool accesses additional proprietary information.

◀ OBD II standards mandate that the 16-pin OBD II plug must be located within three (3) feet of the driver and must not require any tools to be exposed. The OBD II plug in Z3 models is located on the passenger side of the center console.

◀ The OBD II plug is mandated by federal law from 1996 on and is found in all Z3 and M version cars. However, BMW cars produced through 6 / 2000 are also equipped with a 20 pin data link connector (DLC) in the engine compartment. Engine management faults and other DTCs may be accessed and cleared, and service lights reset via this connector as well as via the OBD II plug.

***NOTE***—
- *In addition to the BMW factory scan tools, Baum Tools, Snap-on, and Assenmacher Tool are specialized tool companies that may offer BMW-specific scan tools. Check with the tool manufacturer for availability.*
- *Generic OBD II scan tool software programs and handheld units are readily available. Though limited in capabilities compared to the factory tools, they are nonetheless powerful diagnostic devices which read live data streams, freeze-frame data and other valuable diagnostic information.*
- *A simple aftermarket DTC reader is available from Peake Research. This tool is capable of checking for DTCs as well as turning off the illuminated MIL.*

## Diagnostic monitors

Diagnostic monitors are software routines which run tests and checks on specific systems, components, or functions.

A complete drive cycle (see **Drive cycle** in this repair group) is required for the tests to be valid. The diagnostic monitor signals the engine control module (ECM) of the loss or impairment of the signal or component and determines if a signal or sensor is faulty based on 3 conditions:

- Signal or component shorted to ground
- Signal or component shorted to B+
- Signal or component missing (open circuit)

The OBD II system monitors all emission control systems in the vehicle. Not all vehicles have a full complement of emission control systems. For example, a vehicle may not be equipped with secondary air injection, so naturally no secondary air readiness code would be present.

The OBD II protocol requires monitoring of the following:

- Oxygen sensor(s)
- Catalytic converter(s)
- Engine misfire
- Fuel tank evaporative control system (EVAP)
- Secondary air
- Fuel system

**Oxygen sensor monitoring**: When drive conditions allow, response rate and switching time of each oxygen sensor is monitored. The oxygen sensor heater function is also monitored. The OBD II system differentiates between precatalyst and post-catalyst oxygen sensors. In order for the oxygen sensors to be effectively monitored, the system must be in closed loop operation.

**Catalyst monitoring**: Comparing precatalyst and post-catalyst oxygen sensor output yields information about oxygen storage in the catalyst. This, in turn, indicates catalyst efficiency.

**Misfire detection**: This strategy monitors crankshaft speed fluctuations to determine if an engine misfire occurs. The system identifies a misfire, as well as other pertinent misfire information such as:

- Specific cylinder(s) misfiring
- Severity of misfire event
- Relevance to vehicle emissions
- Potential of damage to catalyst

Misfire detection is an on-going process only disabled under certain limited conditions.

**Fuel system monitoring**: This monitor checks fuel delivery for proper engine operation based on programmed data (long and short term fuel trim). If too much or not enough fuel is delivered over a predetermined time, a DTC is set and the MIL is turned on.

***NOTE—***
*Fuel trim refers to adjustments to programmed fuel delivery. Long-term fuel trim refers to gradual adjustments to fuel delivery calibration. Long term fuel trim adjustments compensate for gradual changes that occur over time.*

Fuel system monitoring monitors the calculated injection time (ti) in relation to engine speed, load, and precatalyst oxygen sensor(s) signals.

**Evaporative system (EVAP) monitoring**: This monitor checks the fuel storage system and related fuel lines for leaks. It can detect very small leaks anywhere in the system. A pressure test is performed on the EVAP system on a continuous basis if the drive cycle allows.

## Drive cycle

The OBD II drive cycle is a specific driving routine used by the EPA to test vehicle emissions. The purpose of the drive cycle is to run all of the emission-related on-board diagnostics over a broad range of driving conditions.

A drive cycle is considered complete when all of the diagnostic monitors have run their tests without interruption. For a drive cycle to be initiated, the vehicle must be started cold and brought up to an oil temperature of 160°F, at least 40°F above its original starting temperature.

Once a drive cycle is completed, the system status or inspection/maintenance (I/M) readiness codes are set to "Yes." When all codes are set to "Yes" the system is described as having established "readiness".

System status codes will be set to "No" in the following cases:

- Battery or ECM is disconnected.
- DTCs have been erased after completion of repairs but drive cycle has not be completed.

A scan tool can be used to determine I/M readiness.

## Readiness codes

The inspection/maintenance (I/M) readiness codes are mandated as part of OBD II. The readiness code is stored after diagnostic monitoring of all installed components and systems has been carried out. The readiness code function was put into place to prevent manipulation of I/M emission test procedures by clearing faults codes or disconnecting the ECM or battery. Once set, the readiness code remains set until the battery is disconnected or DTCs are erased.

**I/M readiness codes**

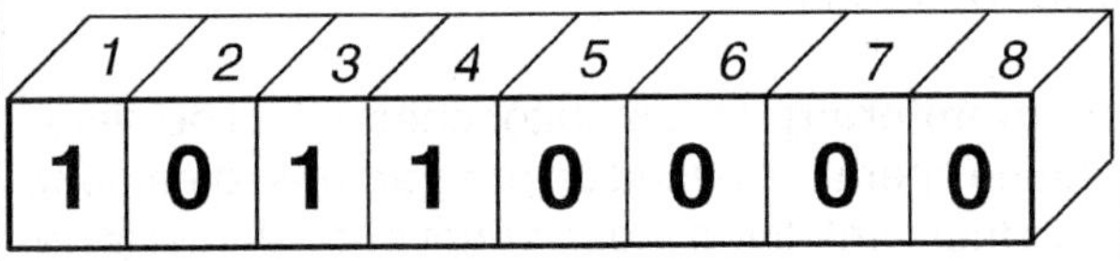


1. EGR
2. Oxygen sensor heater
3. Oxygen sensor
4. A/C refrigerant (R12 only)
5. Secondary air injectionr
6. EVAP monitoring
7. Catalyst heating
8. Catalyst efficiency

The readiness code can be displayed using an aftermarket scan tool. The code is binary:

- 0 means "yes", the function is ready.
- 1 means "no", the function is not ready.

When all zeros are displayed, the system has established its readiness.

## Diagnostic trouble codes (DTCs)

The SAE mandates a 5-digit diagnostic trouble code (DTC) standard. Emission related DTCs start with the letter "P" for power train and are commonly referred to as "P-codes". When the malfunction indicator light (MIL) is illuminated it indicates that a DTC has been stored:

- DTCs are stored as soon as they occur, whether or not the MIL illuminates.
- DTCs store and display a time stamp.
- DTCs record the current fault status.

| DTC digit interpretation | |
|---|---|
| 1st digit | |
| P | powertrain |
| B | body |
| C | chassis |
| 2nd digit | |
| 0 | SAE |
| 1 | BMW |
| 3rd digit (in P-codes) | |
| 0 | total system |
| 1 | air/fuel induction |
| 2 | fuel injection |
| 3 | ignition system or missfire |
| 4 | auxiliary emission control |
| 5 | vehicle speed & idle control |
| 6 | ECM inputs/outputs |
| 7 | transmission |
| 4th - 5th digits | individual circuits or components |

### DTC example: P 0 3 0 6

- **P:** A powertrain problem
- **0**: SAE sanctioned or generic
- **3:** Related to an ignition system/misfire
- **06** Misfire has been detected at cylinder #6

**Freeze frame data:** DTCs provide a "freeze frame" or snapshot of a vehicle performance or emissions fault at the moment that the fault first occurs. This information is accessible through generic scan tools.

Freeze frame data contains, but is not limited to, the following information:

- Engine load (calculated)
- Engine RPM
- Short and long-term fuel trim
- Vehicle speed
- Coolant temperature
- Intake manifold pressure
- Open/closed loop operation
- Fuel pressure (if available)
- DTC

## P-CODES (DTCs) AND BMW FAULT CODES

BMW fault codes (FC) expand on the SAE sanctioned DTCs and are accessible primarily through BMW (or BMW specific) diagnostic scan tools. Below is a comprehensive listing of SAE P-codes, the corresponding BMW FCs and an explanation of their meanings.

**Table a. M44 engine: Bosch DME M5.2 from 01/01/1996 to 08/31/1996**

| PCode | BMW-FC | PCode text |
|---|---|---|
| P0100 | 115 | Mass or Volume Air Flow Circuit |
| P0111 | 124 | Intake Air Temperature Sensor 1 Circuit Range/Performance |
| P0116 | 123 | Engine Coolant Temperature Circuit Range/Performance |
| P0120 | 117 | Throttle/Pedal Position Sensor/Switch 'A' Circuit |
| P0130 | 10 | O2 Sensor Circuit (Bank 1 Sensor 1) |
| P0133 | 15 | O2 Sensor Circuit Slow Response (Bank 1 Sensor 1) |
| P0135 | 13 | O2 Sensor Heater Circuit (Bank 1 Sensor 1) |
| P0136 | 12 | O2 Sensor Circuit (Bank 1 Sensor 2) |
| P0139 | 17 | O2 Sensor Circuit Slow Response (Bank 1 Sensor 2) |
| P0170 | 26 | Fuel Trim (Bank 1) |
| P0201 | 150 | Injector Circuit/Open - Cylinder 1 |
| P0202 | 151 | Injector Circuit/Open - Cylinder 2 |
| P0203 | 152 | Injector Circuit/Open - Cylinder 3 |
| P0204 | 153 | Injector Circuit/Open - Cylinder 4 |
| P0300 | 62 | Random/Multiple Cylinder Misfire Detected |
| P0300 | 75 | Random/Multiple Cylinder Misfire Detected |
| P0301 | 50 | Cylinder 1 Misfire Detected |
| P0301 | 63 | Cylinder 1 Misfire Detected |
| P0302 | 51 | Cylinder 2 Misfire Detected |

| Table a. M44 engine: Bosch DME M5.2 from 01/01/1996 to 08/31/1996 | | |
|---|---|---|
| **PCode** | **BMW-FC** | **PCode text** |
| P0100 | 115 | Mass or Volume Air Flow Circuit |
| P0302 | 64 | Cylinder 2 Misfire Detected |
| P0303 | 52 | Cylinder 3 Misfire Detected |
| P0303 | 65 | Cylinder 3 Misfire Detected |
| P0304 | 53 | Cylinder 4 Misfire Detected |
| P0304 | 66 | Cylinder 4 Misfire Detected |
| P0325 | 210 | Knock Sensor 1 Circuit (Bank 1 or Single Sensor) |
| P0330 | 211 | Knock Sensor 2 Circuit (Bank 2) |
| P0335 | 111 | Crankshaft Position Sensor 'A' Circuit |
| P0340 | 112 | Camshaft Position Sensor 'A' Circuit (Bank 1 or Single Sensor) |
| P0420 | 40 | Catalyst System Efficiency Below Threshold (Bank 1) |
| P0440 | 93 | Evaporative Emission System |
| P0443 | 98 | Evaporative Emission System Purge Control Valve Circuit |
| P0500 | 120 | Vehicle Speed Sensor 'A' |
| P0505 | 32 | Idle Air Control System |
| P0600 | 236 | Serial Communication Link |
| P0601 | 103 | Internal Control Module Memory Check Sum Error |
| P0603 | 102 | Internal Control Module Keep Alive Memory (KAM) Error |
| P0604 | 101 | Internal Control Module Random Access Memory (RAM) Error |
| P0604 | 104 | Internal Control Module Random Access Memory (RAM) Error |
| P1140 | 121 | Mass or Volume Air Flow Circuit Range/Performance Problem |
| P1174 | 27 | Fuel Trim Adaptation Malfunction (Bank 1) |
| P1176 | 16 | O2 Sensor Slow Response (Bank 1) |
| P1186 | 14 | O2 Sensor Heater Control Circuit (Bank 1 Sensor 2) |
| P1386 | 220 | Control Module Self-Test, Knock Control Circuit Baseline Test (Bank 1) |
| P1509 | 168 | Idle-Speed Control Valve Opening Solenoid Control Circuit Signal Electrical |
| P1511 | 175 | DISA (Differentiated Intake Manifold) Control Circuit Electrical |
| P1550 | 169 | Idle-Speed Control Valve Closing Coil Electrical |
| P1585 | 8 | Misfire Detected with Low Fuel |
| P1589 | 222 | Control Module Self-Test, Knock Control Test Pulse (Bank 1) |
| P1690 | 165 | Malfunction Indicator Lamp (MIL) Control Circuit Electrical |

| Table b. M44 engine: Bosch DME M5.2 from 09/01/1996 to 12/31/1996 | | |
|---|---|---|
| **PCode** | **BMW-FC** | **PCode text** |
| P0100 | 115 | Mass or Volume Air Flow Circuit |
| P0111 | 124 | Intake Air Temperature Sensor 1 Circuit Range/Performance |

| Table b. M44 engine: Bosch DME M5.2 from 09/01/1996 to 12/31/1996 | | |
|---|---|---|
| **PCode** | **BMW-FC** | **PCode text** |
| P0116 | 123 | Engine Coolant Temperature Circuit Range/Performance |
| P0120 | 117 | Throttle/Pedal Position Sensor/Switch 'A' Circuit |
| P0130 | 10 | O2 Sensor Circuit (Bank 1 Sensor 1) |
| P0133 | 15 | O2 Sensor Circuit Slow Response (Bank 1 Sensor 1) |
| P0135 | 13 | O2 Sensor Heater Circuit (Bank 1 Sensor 1) |
| P0136 | 12 | O2 Sensor Circuit (Bank 1 Sensor 2) |
| P0139 | 17 | O2 Sensor Circuit Slow Response (Bank 1 Sensor 2) |
| P0170 | 26 | Fuel Trim (Bank 1) |
| P0201 | 150 | Injector Circuit/Open - Cylinder 1 |
| P0202 | 151 | Injector Circuit/Open - Cylinder 2 |
| P0203 | 152 | Injector Circuit/Open - Cylinder 3 |
| P0204 | 153 | Injector Circuit/Open - Cylinder 4 |
| P0300 | 62 | Random/Multiple Cylinder Misfire Detected |
| P0300 | 75 | Random/Multiple Cylinder Misfire Detected |
| P0301 | 50 | Cylinder 1 Misfire Detected |
| P0301 | 63 | Cylinder 1 Misfire Detected |
| P0302 | 51 | Cylinder 2 Misfire Detected |
| P0302 | 64 | Cylinder 2 Misfire Detected |
| P0303 | 52 | Cylinder 3 Misfire Detected |
| P0303 | 65 | Cylinder 3 Misfire Detected |
| P0304 | 53 | Cylinder 4 Misfire Detected |
| P0304 | 66 | Cylinder 4 Misfire Detected |
| P0325 | 210 | Knock Sensor 1 Circuit (Bank 1 or Single Sensor) |
| P0330 | 211 | Knock Sensor 2 Circuit (Bank 2) |
| P0335 | 111 | Crankshaft Position Sensor 'A' Circuit |
| P0340 | 112 | Camshaft Position Sensor 'A' Circuit (Bank 1 or Single Sensor) |
| P0420 | 40 | Catalyst System Efficiency Below Threshold (Bank 1) |
| P0440 | 93 | Evaporative Emission System |
| P0443 | 98 | Evaporative Emission System Purge Control Valve Circuit |
| P0500 | 120 | Vehicle Speed Sensor 'A' |
| P0505 | 32 | Idle Air Control System |
| P0600 | 236 | Serial Communication Link |
| P0601 | 103 | Internal Control Module Memory Check Sum Error |
| P0603 | 102 | Internal Control Module Keep Alive Memory (KAM) Error |
| P0604 | 101 | Internal Control Module Random Access Memory (RAM) Error |
| P0604 | 104 | Internal Control Module Random Access Memory (RAM) Error |
| P1140 | 121 | Mass or Volume Air Flow Circuit Range/Performance Problem |

| Table b. M44 engine: Bosch DME M5.2 from 09/01/1996 to 12/31/1996 | | |
|---|---|---|
| **PCode** | **BMW-FC** | **PCode text** |
| P1174 | 27 | Fuel Trim Adaptation Malfunction (Bank 1) |
| P1176 | 16 | O2 Sensor Slow Response (Bank 1) |
| P1186 | 14 | O2 Sensor Heater Control Circuit (Bank 1 Sensor 2) |
| P1386 | 220 | Control Module Self-Test, Knock Control Circuit Baseline Test (Bank 1) |
| P1509 | 168 | Idle-Speed Control Valve Opening Solenoid Control Circuit Signal Electrical |
| P1511 | 175 | DISA (Differentiated Intake Manifold) Control Circuit Electrical |
| P1550 | 169 | Idle-Speed Control Valve Closing Coil Electrical |
| P1585 | 8 | Misfire Detected with Low Fuel |
| P1589 | 222 | Control Module Self-Test, Knock Control Test Pulse (Bank 1) |
| P1690 | 165 | Malfunction Indicator Lamp (MIL) Control Circuit Electrical |

| Table c. M44 engine: Bosch DME M5.2 from 01/01/1997 to 08/31/1997 | | |
|---|---|---|
| **PCode** | **BMW-FC** | **PCode text** |
| P0100 | 115 | Mass or Volume Air Flow Circuit |
| P0111 | 124 | Intake Air Temperature Sensor 1 Circuit Range/Performance |
| P0116 | 123 | Engine Coolant Temperature Circuit Range/Performance |
| P0120 | 117 | Throttle/Pedal Position Sensor/Switch 'A' Circuit |
| P0130 | 10 | O2 Sensor Circuit (Bank 1 Sensor 1) |
| P0133 | 15 | O2 Sensor Circuit Slow Response (Bank 1 Sensor 1) |
| P0135 | 13 | O2 Sensor Heater Circuit (Bank 1 Sensor 1) |
| P0136 | 12 | O2 Sensor Circuit (Bank 1 Sensor 2) |
| P0170 | 26 | Fuel Trim (Bank 1) |
| P0201 | 150 | Injector Circuit/Open - Cylinder 1 |
| P0202 | 151 | Injector Circuit/Open - Cylinder 2 |
| P0203 | 152 | Injector Circuit/Open - Cylinder 3 |
| P0204 | 153 | Injector Circuit/Open - Cylinder 4 |
| P0300 | 62 | Random/Multiple Cylinder Misfire Detected |
| P0300 | 75 | Random/Multiple Cylinder Misfire Detected |
| P0301 | 50 | Cylinder 1 Misfire Detected |
| P0301 | 63 | Cylinder 1 Misfire Detected |
| P0302 | 51 | Cylinder 2 Misfire Detected |
| P0302 | 64 | Cylinder 2 Misfire Detected |
| P0303 | 52 | Cylinder 3 Misfire Detected |
| P0303 | 65 | Cylinder 1 Misfire Detected |
| P0304 | 53 | Cylinder 4 Misfire Detected |
| P0304 | 66 | Cylinder 4 Misfire Detected |

| Table c. M44 engine: Bosch DME M5.2 from 01/01/1997 to 08/31/1997 | | |
|---|---|---|
| **PCode** | **BMW-FC** | **PCode text** |
| P0325 | 210 | Knock Sensor 1 Circuit (Bank 1 or Single Sensor) |
| P0330 | 211 | Knock Sensor 2 Circuit (Bank 2) |
| P0335 | 111 | Crankshaft Position Sensor 'A' Circuit |
| P0340 | 112 | Camshaft Position Sensor 'A' Circuit (Bank 1 or Single Sensor) |
| P0420 | 40 | Catalyst System Efficiency Below Threshold (Bank 1) |
| P0440 | 93 | Evaporative Emission System |
| P0442 | 97 | Evaporative Emission System Leak Detected (small leak) |
| P0443 | 98 | Evaporative Emission System Purge Control Valve Circuit |
| P0451 | 99 | Evaporative Emission System Pressure Sensor/Switch Range/Performance |
| P0455 | 94 | Evaporative Emission System Leak Detected (large leak) |
| P0500 | 120 | Vehicle Speed Sensor 'A' |
| P0505 | 32 | Idle Air Control System |
| P0600 | 236 | Serial Communication Link |
| P0601 | 103 | Internal Control Module Memory Check Sum Error |
| P0603 | 102 | Internal Control Module Keep Alive Memory (KAM) Error |
| P0604 | 101 | Internal Control Module Random Access Memory (RAM) Error |
| P0604 | 104 | Internal Control Module Random Access Memory (RAM) Error |
| P1140 | 121 | Mass or Volume Air Flow Circuit Range/Performance Problem |
| P1145 | 2 | Solenoid Valve Running Losses Control Circuit Electrical |
| P1174 | 27 | Fuel Trim Adaptation Malfunction (Bank 1) |
| P1186 | 14 | O2 Sensor Heater Control Circuit (Bank 1 Sensor 2) |
| P1386 | 220 | Control Module Self-Test, Knock Control Circuit Baseline Test (Bank 1) |
| P1423 | 80 | Secondary Air System (Bank 1) |
| P1453 | 84 | Secondary Air Injection Pump Relay Control Circuit Electrical |
| P1509 | 168 | Idle-Speed Control Valve Opening Solenoid Control Circuit Signal Electrical |
| P1511 | 175 | DISA (Differentiated Intake Manifold) Control Circuit Electrical |
| P1550 | 169 | Idle-Speed Control Valve Closing Coil Electrical |
| P1585 | 8 | Misfire Detected with Low Fuel |
| P1589 | 222 | Control Module Self-Test, Knock Control Test Pulse (Bank 1) |
| P1690 | 165 | Malfunction Indicator Lamp (MIL) Control Circuit Electrical |

| Table d. M44 engine: Bosch DME M5.2 from 09/01/1997 to 08/31/1998 | | |
|---|---|---|
| **PCode** | **BMW-FC** | **PCode text** |
| P0100 | 115 | Mass or Volume Air Flow Circuit |
| P0111 | 124 | Intake Air Temperature Sensor 1 Circuit Range/Performance |
| P0116 | 123 | Engine Coolant Temperature Circuit Range/Performance |

| Table d. M44 engine: Bosch DME M5.2 from 09/01/1997 to 08/31/1998 | | |
|---|---|---|
| **PCode** | **BMW-FC** | **PCode text** |
| P0120 | 117 | Throttle/Pedal Position Sensor/Switch 'A' Circuit |
| P0130 | 10 | O2 Sensor Circuit (Bank 1 Sensor 1) |
| P0133 | 15 | O2 Sensor Circuit Slow Response (Bank 1 Sensor 1) |
| P0135 | 13 | O2 Sensor Heater Circuit (Bank 1 Sensor 1) |
| P0136 | 12 | O2 Sensor Circuit (Bank 1 Sensor 2) |
| P0170 | 26 | Fuel Trim (Bank 1) |
| P0201 | 150 | Injector Circuit/Open - Cylinder 1 |
| P0202 | 151 | Injector Circuit/Open - Cylinder 2 |
| P0203 | 152 | Injector Circuit/Open - Cylinder 3 |
| P0204 | 153 | Injector Circuit/Open - Cylinder 4 |
| P0300 | 62 | Random/Multiple Cylinder Misfire Detected |
| P0300 | 75 | Random/Multiple Cylinder Misfire Detected |
| P0301 | 50 | Cylinder 1 Misfire Detected |
| P0301 | 63 | Cylinder 1 Misfire Detected |
| P0302 | 51 | Cylinder 2 Misfire Detected |
| P0302 | 64 | Cylinder 2 Misfire Detected |
| P0303 | 52 | Cylinder 3 Misfire Detected |
| P0303 | 65 | Cylinder 3 Misfire Detected |
| P0304 | 53 | Cylinder 4 Misfire Detected |
| P0304 | 66 | Cylinder 4 Misfire Detected |
| P0325 | 210 | Knock Sensor 1 Circuit (Bank 1 or Single Sensor) |
| P0330 | 211 | Knock Sensor 2 Circuit (Bank 2) |
| P0335 | 111 | Crankshaft Position Sensor 'A' Circuit |
| P0340 | 112 | Camshaft Position Sensor 'A' Circuit (Bank 1 or Single Sensor) |
| P0420 | 40 | Catalyst System Efficiency Below Threshold (Bank 1) |
| P0440 | 93 | Evaporative Emission System |
| P0442 | 97 | Evaporative Emission System Leak Detected (small leak) |
| P0443 | 98 | Evaporative Emission System Purge Control Valve Circuit |
| P0451 | 99 | Evaporative Emission System Pressure Sensor/Switch Range/Performance |
| P0455 | 94 | Evaporative Emission System Leak Detected (large leak) |
| P0500 | 120 | Vehicle Speed Sensor 'A' |
| P0505 | 32 | Idle Air Control System |
| P0600 | 236 | Serial Communication Link |
| P0601 | 103 | Internal Control Module Memory Check Sum Error |
| P0603 | 102 | Internal Control Module Keep Alive Memory (KAM) Error |
| P0604 | 101 | Internal Control Module Random Access Memory (RAM) Error |
| P0604 | 104 | Internal Control Module Random Access Memory (RAM) Error |

| Table d. M44 engine: Bosch DME M5.2 from 09/01/1997 to 08/31/1998 | | |
|---|---|---|
| **PCode** | **BMW-FC** | **PCode text** |
| P1140 | 121 | Mass or Volume Air Flow Circuit Range/Performance Problem |
| P1145 | 2 | Solenoid Valve Running Losses Control Circuit Electrical |
| P1174 | 27 | Fuel Trim Adaptation Malfunction (Bank 1) |
| P1186 | 14 | O2 Sensor Heater Control Circuit (Bank 1 Sensor 2) |
| P1386 | 220 | Control Module Self-Test, Knock Control Circuit Baseline Test (Bank 1) |
| P1423 | 80 | Secondary Air System (Bank 1) |
| P1453 | 84 | Secondary Air Injection Pump Relay Control Circuit Electrical |
| P1509 | 168 | Idle-Speed Control Valve Opening Solenoid Control Circuit Signal Electrical |
| P1511 | 175 | DISA (Differentiated Intake Manifold) Control Circuit Electrical |
| P1550 | 169 | Idle-Speed Control Valve Closing Coil Electrical |
| P1585 | 8 | Misfire Detected with Low Fuel |
| P1589 | 222 | Control Module Self-Test, Knock Control Test Pulse (Bank 1) |
| P1690 | 165 | Malfunction Indicator Lamp (MIL) Control Circuit Electrical |

| Table e. M52 engine: Siemens DME MS41.1 from 09./01/1996 to 08/31/1997 | | |
|---|---|---|
| **PCode** | **BMW-FC** | **PCode text** |
| P0101 | 8 | Mass or Volume Air Flow Circuit Range/Performance |
| P0111 | 14 | Intake Air Temperature Sensor 1 Circuit Range/Performance |
| P0116 | 10 | Engine Coolant Temperature Circuit Range/Performance |
| P0120 | 112 | Throttle/Pedal Position Sensor/Switch 'A' Circuit |
| P0120 | 113 | Throttle/Pedal Position Sensor/Switch 'A' Circuit |
| P0120 | 118 | Throttle/Pedal Position Sensor/Switch 'A' Circuit |
| P0120 | 173 | Throttle/Pedal Position Sensor/Switch 'A' Circuit |
| P0120 | 174 | Throttle/Pedal Position Sensor/Switch 'A' Circuit |
| P0125 | 222 | Insufficient Coolant Temperature for Closed Loop Fuel Control |
| P0130 | 150 | O2 Sensor Circuit (Bank 1 Sensor 1) |
| P0130 | 151 | O2 Sensor Circuit (Bank 1 Sensor 1) |
| P0130 | 152 | O2 Sensor Circuit (Bank 1 Sensor 1) |
| P0133 | 229 | O2 Sensor Circuit Slow Response (Bank 1 Sensor 1) |
| P0134 | 186 | O2 Sensor Circuit No Activity Detected (Bank 1 Sensor 1) |
| P0135 | 25 | O2 Sensor Heater Circuit (Bank 1 Sensor 1) |
| P0136 | 156 | O2 Sensor Circuit (Bank 1 Sensor 2) |
| P0136 | 157 | O2 Sensor Circuit (Bank 1 Sensor 2) |
| P0136 | 215 | O2 Sensor Circuit (Bank 1 Sensor 2) |
| P0141 | 61 | O2 Sensor Heater Circuit (Bank 1 Sensor 2) |
| P0150 | 153 | O2 Sensor Circuit (Bank 2 Sensor 1) |

| Table e. M52 engine: Siemens DME MS41.1 from 09./01/1996 to 08/31/1997 | | |
|---|---|---|
| **PCode** | **BMW-FC** | **PCode text** |
| P0150 | 154 | O2 Sensor Circuit (Bank 2 Sensor 1) |
| P0150 | 155 | O2 Sensor Circuit (Bank 2 Sensor 1) |
| P0153 | 230 | O2 Sensor Circuit Slow Response (Bank 2 Sensor 1) |
| P0154 | 187 | O2 Sensor Circuit No Activity Detected (Bank 2 Sensor 1) |
| P0155 | 55 | O2 Sensor Heater Circuit (Bank 2 Sensor 1) |
| P0156 | 159 | O2 Sensor Circuit (Bank 2 Sensor 2) |
| P0156 | 160 | O2 Sensor Circuit (Bank 2 Sensor 2) |
| P0156 | 216 | O2 Sensor Circuit (Bank 2 Sensor 2) |
| P0161 | 79 | O2 Sensor Heater Circuit (Bank 2 Sensor 2) |
| P0170 | 202 | Fuel Trim (Bank 1) |
| P0173 | 203 | Fuel Trim (Bank 2) |
| P0201 | 6 | Injector Circuit/Open - Cylinder 1 |
| P0202 | 5 | Injector Circuit/Open - Cylinder 2 |
| P0203 | 22 | Injector Circuit/Open - Cylinder 3 |
| P0204 | 24 | Injector Circuit/Open - Cylinder 4 |
| P0205 | 33 | Injector Circuit/Open - Cylinder 5 |
| P0206 | 23 | Injector Circuit/Open - Cylinder 6 |
| P0301 | 238 | Cylinder 1 Misfire Detected |
| P0302 | 239 | Cylinder 2 Misfire Detected |
| P0303 | 240 | Cylinder 3 Misfire Detected |
| P0304 | 241 | Cylinder 4 Misfire Detected |
| P0305 | 242 | Cylinder 5 Misfire Detected |
| P0306 | 243 | Cylinder 6 Misfire Detected |
| P0325 | 57 | Knock Sensor 1 Circuit (Bank 1 or Single Sensor) |
| P0330 | 59 | Knock Sensor 2 Circuit (Bank 2) |
| P0335 | 83 | Crankshaft Position Sensor 'A' Circuit |
| P0340 | 65 | Camshaft Position Sensor 'A' Circuit (Bank 1 or Single Sensor) |
| P0412 | 62 | Secondary Air Injection System Switching Valve A Circuit |
| P0420 | 233 | Catalyst System Efficiency Below Threshold (Bank 1) |
| P0430 | 234 | Catalyst System Efficiency Below Threshold (Bank 2) |
| P0440 | 250 | Evaporative Emission System |
| P0441 | 144 | Evaporative Emission System Incorrect Purge Flow |
| P0442 | 145 | Evaporative Emission System Leak Detected (small leak) |
| P0443 | 68 | Evaporative Emission System Purge Control Valve Circuit |
| P0500 | 214 | Vehicle Speed Sensor 'A' |
| P0505 | 204 | Idle Air Control System |
| P0600 | 217 | Serial Communication Link |

| Table e. M52 engine: Siemens DME MS41.1 from 09./01/1996 to 08/31/1997 | | |
|---|---|---|
| **PCode** | **BMW-FC** | **PCode text** |
| P0601 | 100 | Internal Control Module Memory Check Sum Error |
| P0601 | 170 | Internal Control Module Memory Check Sum Error |
| P0601 | 171 | Internal Control Module Memory Check Sum Error |
| P1140 | 149 | Mass or Volume Air Flow Circuit Range/Performance Problem |
| P1145 | 50 | Solenoid Valve Running Losses Control Circuit Electrical |
| P1161 | 122 | Fuel Trim Adaptation Additive High (Bank 2) (M52: Engine Oil Temperature Sensor Circuit) |
| P1178 | 231 | O2 Sensor Signal Circuit Slow Switching from Rich to Lean (Bank 1 Sensor 1) |
| P1179 | 232 | O2 Sensor Signal Circuit Slow Switching from Rich to Lean (Bank 2 Sensor 1) |
| P1186 | 190 | O2 Sensor Heater Control Circuit (Bank 1 Sensor 2) |
| P1187 | 191 | O2 Sensor Heater Control Circuit (Bank 2 Sensor 2) |
| P1188 | 227 | Fuel Control (Bank 1 Sensor 1) |
| P1189 | 228 | Fuel Control (Bank 2 Sensor 1) |
| P1190 | 235 | Pre Catalyst Fuel Trim System (Bank 1) |
| P1191 | 236 | Pre Catalyst Fuel Trim System (Bank 2) |
| P1192 | 225 | Post Catalyst Fuel Trim System (Bank 1) |
| P1193 | 226 | Post Catalyst Fuel Trim System (Bank 2) |
| P1397 | 18 | Camshaft Position Sensor 'B' Circuit (Bank 1) |
| P1421 | 246 | Secondary Air System (Bank 2) |
| P1423 | 245 | Secondary Air System (Bank 1) |
| P1432 | 247 | Secondary Air Injection System Incorrect Flow Detected |
| P1453 | 35 | Secondary Air Injection Pump Relay Control Circuit Electrical |
| P1509 | 53 | Idle-Speed Control Valve Opening Solenoid Control Circuit Signal Electrical |
| P1519 | 103 | Oil-Condition Sensor Temperature Measurement (M62/M52/S52: 'A' Camshaft Position Actuator Bank 1) |
| P1525 | 21 | 'A' Camshaft Position Actuator Control Open Circuit (Bank 1) |
| P1550 | 27 | Idle-Speed Control Valve Closing Coil Electrical |
| P1593 | 124 | DISA (Differentiated Intake Manifold) Control Circuit Electrical |
| P1622 | 123 | Map Cooling Thermostat Control Circuit Electrical |
| P1624 | 168 | Pedal Position Sensor Potentiometer Supply Channel 1 Electrical (M52: Coolant Thermostat (Coolant Temperature Below Thermostat Regulating Temperature)) |

| Table f. M52 engine: Siemens DME MS41.1 from 09/01/1997 to 08/31/1998 | | |
|---|---|---|
| **PCode** | **BMW-FC** | **PCode text** |
| P0101 | 8 | Mass or Volume Air Flow Circuit Range/Performance |
| P0111 | 14 | Intake Air Temperature Sensor 1 Circuit Range/Performance |
| P0116 | 10 | Engine Coolant Temperature Circuit Range/Performance |
| P0120 | 112 | Throttle/Pedal Position Sensor/Switch 'A' Circuit |
| P0120 | 113 | Throttle/Pedal Position Sensor/Switch 'A' Circuit |

| Table f. M52 engine: Siemens DME MS41.1 from 09/01/1997 to 08/31/1998 | | |
|---|---|---|
| **PCode** | **BMW-FC** | **PCode text** |
| P0120 | 118 | Throttle/Pedal Position Sensor/Switch 'A' Circuit |
| P0120 | 173 | Throttle/Pedal Position Sensor/Switch 'A' Circuit |
| P0120 | 174 | Throttle/Pedal Position Sensor/Switch 'A' Circuit |
| P0125 | 222 | Insufficient Coolant Temperature for Closed Loop Fuel Control |
| P0130 | 150 | O2 Sensor Circuit (Bank 1 Sensor 1) |
| P0130 | 151 | O2 Sensor Circuit (Bank 1 Sensor 1) |
| P0130 | 152 | O2 Sensor Circuit (Bank 1 Sensor 1) |
| P0133 | 229 | O2 Sensor Circuit Slow Response (Bank 1 Sensor 1) |
| P0134 | 186 | O2 Sensor Circuit No Activity Detected (Bank 1 Sensor 1) |
| P0135 | 25 | O2 Sensor Heater Circuit (Bank 1 Sensor 1) |
| P0136 | 156 | O2 Sensor Circuit (Bank 1 Sensor 2) |
| P0136 | 157 | O2 Sensor Circuit (Bank 1 Sensor 2) |
| P0136 | 215 | O2 Sensor Circuit (Bank 1 Sensor 2) |
| P0141 | 61 | O2 Sensor Heater Circuit (Bank 1 Sensor 2) |
| P0150 | 153 | O2 Sensor Circuit (Bank 2 Sensor 1) |
| P0150 | 154 | O2 Sensor Circuit (Bank 2 Sensor 1) |
| P0150 | 155 | O2 Sensor Circuit (Bank 2 Sensor 1) |
| P0153 | 230 | O2 Sensor Circuit Slow Response (Bank 2 Sensor 1) |
| P0154 | 187 | O2 Sensor Circuit No Activity Detected (Bank 2 Sensor 1) |
| P0155 | 55 | O2 Sensor Heater Circuit (Bank 2 Sensor 1) |
| P0156 | 159 | O2 Sensor Circuit (Bank 2 Sensor 2) |
| P0156 | 160 | O2 Sensor Circuit (Bank 2 Sensor 2) |
| P0156 | 216 | O2 Sensor Circuit (Bank 2 Sensor 2) |
| P0161 | 79 | O2 Sensor Heater Circuit (Bank 2 Sensor 2) |
| P0170 | 202 | Fuel Trim (Bank 1) |
| P0173 | 203 | Fuel Trim (Bank 2) |
| P0201 | 6 | Injector Circuit/Open - Cylinder 1 |
| P0202 | 5 | Injector Circuit/Open - Cylinder 2 |
| P0203 | 22 | Injector Circuit/Open - Cylinder 3 |
| P0204 | 24 | Injector Circuit/Open - Cylinder 4 |
| P0205 | 33 | Injector Circuit/Open - Cylinder 5 |
| P0206 | 23 | Injector Circuit/Open - Cylinder 6 |
| P0301 | 238 | Cylinder 1 Misfire Detected |
| P0302 | 239 | Cylinder 2 Misfire Detected |
| P0303 | 240 | Cylinder 3 Misfire Detected |
| P0304 | 241 | Cylinder 4 Misfire Detected |
| P0305 | 242 | Cylinder 5 Misfire Detected |

| Table f. M52 engine: Siemens DME MS41.1 from 09/01/1997 to 08/31/1998 | | |
|---|---|---|
| **PCode** | **BMW-FC** | **PCode text** |
| P0306 | 243 | Cylinder 6 Misfire Detected |
| P0325 | 57 | Knock Sensor 1 Circuit (Bank 1 or Single Sensor) |
| P0330 | 59 | Knock Sensor 2 Circuit (Bank 2) |
| P0335 | 83 | Crankshaft Position Sensor 'A' Circuit |
| P0340 | 65 | Camshaft Position Sensor 'A' Circuit (Bank 1 or Single Sensor) |
| P0412 | 62 | Secondary Air Injection System Switching Valve A Circuit |
| P0420 | 233 | Catalyst System Efficiency Below Threshold (Bank 1) |
| P0430 | 234 | Catalyst System Efficiency Below Threshold (Bank 2) |
| P0440 | 250 | Evaporative Emission System |
| P0441 | 144 | Evaporative Emission System Incorrect Purge Flow |
| P0442 | 145 | Evaporative Emission System Leak Detected (small leak) |
| P0443 | 68 | Evaporative Emission System Purge Control Valve Circuit |
| P0500 | 214 | Vehicle Speed Sensor 'A' |
| P0505 | 204 | Idle Air Control System |
| P0600 | 217 | Serial Communication Link |
| P0601 | 100 | Internal Control Module Memory Check Sum Error |
| P0601 | 170 | Internal Control Module Memory Check Sum Error |
| P0601 | 171 | Internal Control Module Memory Check Sum Error |
| P1140 | 149 | Mass or Volume Air Flow Circuit Range/Performance Problem |
| P1145 | 50 | Solenoid Valve Running Losses Control Circuit Electrical |
| P1161 | 122 | Fuel Trim Adaptation Additive High (Bank 2) (M52: Engine Oil Temperature Sensor Circuit) |
| P1178 | 231 | O2 Sensor Signal Circuit Slow Switching from Rich to Lean (Bank 1 Sensor 1) |
| P1179 | 232 | O2 Sensor Signal Circuit Slow Switching from Rich to Lean (Bank 2 Sensor 1) |
| P1186 | 190 | O2 Sensor Heater Control Circuit (Bank 1 Sensor 2) |
| P1187 | 191 | O2 Sensor Heater Control Circuit (Bank 2 Sensor 2) |
| P1188 | 227 | Fuel Control (Bank 1 Sensor 1) |
| P1189 | 228 | Fuel Control (Bank 2 Sensor 1) |
| P1190 | 235 | Pre Catalyst Fuel Trim System (Bank 1) |
| P1191 | 236 | Pre Catalyst Fuel Trim System (Bank 2) |
| P1192 | 225 | Post Catalyst Fuel Trim System (Bank 1) |
| P1193 | 226 | Post Catalyst Fuel Trim System (Bank 2) |
| P1421 | 246 | Secondary Air System (Bank 2) |
| P1423 | 245 | Secondary Air System (Bank 1) |
| P1432 | 247 | Secondary Air Injection System Incorrect Flow Detected |
| P1453 | 35 | Secondary Air Injection Pump Relay Control Circuit Electrical |
| P1470 | 126 | Leakage Diagnostic Pump Control Circuit Electrical |
| P1475 | 140 | Leakage Diagnostic Pump Reed Switch Did not Close |

| Table f. M52 engine: Siemens DME MS41.1 from 09/01/1997 to 08/31/1998 | | |
|---|---|---|
| **PCode** | **BMW-FC** | **PCode text** |
| P1476 | 143 | Leakage Diagnostic Pump Clamped Tube (M52 MY99/00: Leakage Diagnostic Pump Reed Switch Circuit) |
| P1509 | 53 | Idle-Speed Control Valve Opening Solenoid Control Circuit Signal Electrical |
| P1519 | 103 | Oil-Condition Sensor Temperature Measurement (M62/M52/S52: 'A' Camshaft Position Actuator Bank 1) |
| P1525 | 21 | A' Camshaft Position Actuator Control Open Circuit (Bank 1) |
| P1550 | 27 | Idle-Speed Control Valve Closing Coil Electrical |
| P1593 | 124 | DISA (Differentiated Intake Manifold) Control Circuit Electrical |
| P1622 | 123 | Map Cooling Thermostat Control Circuit Electrical |
| P1624 | 168 | Pedal Position Sensor Potentiometer Supply Channel 1 Electrical (M52: Coolant Thermostat (Coolant Temperature Below Thermostat Regulating Temperature)) |

| Table g. S52 engine: Siemens DME MS41.1 from 09/01/1998 to 04/30/1999 | | |
|---|---|---|
| **PCode** | **BMW-FC** | **PCode text** |
| P0101 | 8 | Mass or Volume Air Flow Circuit Range/Performance |
| P0111 | 14 | Intake Air Temperature Sensor 1 Circuit Range/Performance |
| P0116 | 10 | Engine Coolant Temperature Circuit Range/Performance |
| P0121 | 12 | Throttle/Pedal Position Sensor/Switch 'A' Circuit Range/Performance |
| P0125 | 222 | Insufficient Coolant Temperature for Closed Loop Fuel Control |
| P0130 | 75 | O2 Sensor Circuit (Bank 1 Sensor 1) |
| P0133 | 229 | O2 Sensor Circuit Slow Response (Bank 1 Sensor 1) |
| P0134 | 200 | O2 Sensor Circuit No Activity Detected (Bank 1 Sensor 1) |
| P0135 | 25 | O2 Sensor Heater Circuit (Bank 1 Sensor 1) |
| P0136 | 77 | O2 Sensor Circuit (Bank 1 Sensor 2) |
| P0141 | 79 | O2 Sensor Heater Circuit (Bank 1 Sensor 2) |
| P0150 | 76 | O2 Sensor Circuit (Bank 2 Sensor 1) |
| P0153 | 230 | O2 Sensor Circuit Slow Response (Bank 2 Sensor 1) |
| P0154 | 201 | O2 Sensor Circuit No Activity Detected (Bank 2 Sensor 1) |
| P0155 | 55 | O2 Sensor Heater Circuit (Bank 2 Sensor 1) |
| P0156 | 78 | O2 Sensor Circuit (Bank 2 Sensor 2) |
| P0161 | 61 | O2 Sensor Heater Circuit (Bank 2 Sensor 2) |
| P0170 | 202 | Fuel Trim (Bank 1) |
| P0173 | 203 | Fuel Trim (Bank 2) |
| P0201 | 6 | Injector Circuit/Open - Cylinder 1 |
| P0202 | 5 | Injector Circuit/Open - Cylinder 2 |
| P0203 | 22 | Injector Circuit/Open - Cylinder 3 |
| P0204 | 24 | Injector Circuit/Open - Cylinder 4 |
| P0205 | 33 | Injector Circuit/Open - Cylinder 5 |
| P0206 | 23 | Injector Circuit/Open - Cylinder 6 |

| Table g. S52 engine: Siemens DME MS41.1 from 09/01/1998 to 04/30/1999 | | |
|---|---|---|
| **PCode** | **BMW-FC** | **PCode text** |
| P0300 | 0 | Random/Multiple Cylinder Misfire Detected |
| P0301 | 238 | Cylinder 1 Misfire Detected |
| P0302 | 239 | Cylinder 2 Misfire Detected |
| P0303 | 240 | Cylinder 3 Misfire Detected |
| P0304 | 241 | Cylinder 4 Misfire Detected |
| P0305 | 242 | Cylinder 5 Misfire Detected |
| P0306 | 243 | Cylinder 6 Misfire Detected |
| P0325 | 57 | Knock Sensor 1 Circuit (Bank 1 or Single Sensor) |
| P0330 | 59 | Knock Sensor 2 Circuit (Bank 2) |
| P0335 | 83 | Crankshaft Position Sensor 'A' Circuit |
| P0340 | 65 | Camshaft Position Sensor 'A' Circuit (Bank 1 or Single Sensor) |
| P0412 | 62 | Secondary Air Injection System Switching Valve A Circuit |
| P0420 | 233 | Catalyst System Efficiency Below Threshold (Bank 1) |
| P0430 | 234 | Catalyst System Efficiency Below Threshold (Bank 2) |
| P0440 | 250 | Evaporative Emission System |
| P0440 | 255 | Evaporative Emission System |
| P0441 | 194 | Evaporative Emission System Incorrect Purge Flow |
| P0441 | 252 | Evaporative Emission System Incorrect Purge Flow |
| P0442 | 195 | Evaporative Emission System Leak Detected (small leak) |
| P0442 | 251 | Evaporative Emission System Leak Detected (small leak) |
| P0443 | 68 | Evaporative Emission System Purge Control Valve Circuit |
| P0446 | 253 | Evaporative Emission System Vent Control Circuit |
| P0455 | 254 | Evaporative Emission System Leak Detected (large leak) |
| P0500 | 214 | Vehicle Speed Sensor 'A' |
| P0505 | 204 | Idle Air Control System |
| P0600 | 217 | Serial Communication Link |
| P0601 | 100 | Internal Control Module Memory Check Sum Error |
| P1145 | 50 | Solenoid Valve Running Losses Control Circuit Electrical |
| P1178 | 231 | O2 Sensor Signal Circuit Slow Switching from Rich to Lean (Bank 1 Sensor 1) |
| P1179 | 232 | O2 Sensor Signal Circuit Slow Switching from Rich to Lean (Bank 2 Sensor 1) |
| P1186 | 235 | O2 Sensor Heater Control Circuit (Bank 1 Sensor 2) |
| P1187 | 236 | O2 Sensor Heater Control Circuit (Bank 2 Sensor 2) |
| P1188 | 227 | Fuel Control (Bank 1 Sensor 1) |
| P1189 | 228 | Fuel Control (Bank 2 Sensor 1) |
| P1396 | 244 | Crankshaft Position Sensor Segment Timing Plausibility |
| P1403 | 51 | Carbon Canister Shutoff Valve Control Circuit Electrical (M73: Heated Catalyst Battery Voltage or Current Too Low during Heating (Bank 2)) |

| Table g. S52 engine: Siemens DME MS41.1 from 09/01/1998 to 04/30/1999 | | |
|---|---|---|
| **PCode** | **BMW-FC** | **PCode text** |
| P1421 | 246 | Secondary Air System (Bank 2) |
| P1423 | 245 | Secondary Air System (Bank 1) |
| P1432 | 247 | Secondary Air Injection System Incorrect Flow Detected |
| P1453 | 35 | Secondary Air Injection Pump Relay Control Circuit Electrical |
| P1470 | 196 | Leakage Diagnostic Pump Control Circuit Electrical |
| P1475 | 191 | Leakage Diagnostic Pump Reed Switch Did not Close |
| P1476 | 193 | Leakage Diagnostic Pump Clamped Tube (M52 MY99/00: Leakage Diagnostic Pump Reed Switch Circuit) |
| P1477 | 190 | Leakage Diagnostic Pump Reed Switch Did Not Open |
| P1509 | 53 | Idle-Speed Control Valve Opening Solenoid Control Circuit Signal Electrical |
| P1519 | 212 | Oil-Condition Sensor Temperature Measurement (M62/M52/S52: 'A' Camshaft Position Actuator Bank 1) |
| P1525 | 21 | 'A' Camshaft Position Actuator Control Open Circuit (Bank 1) |
| P1550 | 27 | Idle-Speed Control Valve Closing Coil Electrical |

| Table h. S52 engine: Siemens DME MS41.1 from 05/01/1999 to 08/31/2000 | | |
|---|---|---|
| **PCode** | **BMW-FC** | **PCode text** |
| P0101 | 8 | Mass or Volume Air Flow Circuit Range/Performance |
| P0111 | 14 | Intake Air Temperature Sensor 1 Circuit Range/Performance |
| P0116 | 10 | Engine Coolant Temperature Circuit Range/Performance |
| P0121 | 12 | Throttle/Pedal Position Sensor/Switch 'A' Circuit Range/Performance |
| P0125 | 222 | Insufficient Coolant Temperature for Closed Loop Fuel Control |
| P0130 | 75 | O2 Sensor Circuit (Bank 1 Sensor 1) |
| P0133 | 229 | O2 Sensor Circuit Slow Response (Bank 1 Sensor 1) |
| P0134 | 200 | O2 Sensor Circuit No Activity Detected (Bank 1 Sensor 1) |
| P0135 | 25 | O2 Sensor Heater Circuit (Bank 1 Sensor 1) |
| P0136 | 77 | O2 Sensor Circuit (Bank 1 Sensor 2) |
| P0141 | 79 | O2 Sensor Heater Circuit (Bank 1 Sensor 2) |
| P0150 | 76 | O2 Sensor Circuit (Bank 2 Sensor 1) |
| P0153 | 230 | O2 Sensor Circuit Slow Response (Bank 2 Sensor 1) |
| P0154 | 201 | O2 Sensor Circuit No Activity Detected (Bank 2 Sensor 1) |
| P0155 | 55 | O2 Sensor Heater Circuit (Bank 2 Sensor 1) |
| P0156 | 78 | O2 Sensor Circuit (Bank 2 Sensor 2) |
| P0161 | 61 | O2 Sensor Heater Circuit (Bank 2 Sensor 2) |
| P0170 | 202 | Fuel Trim (Bank 1) |
| P0173 | 203 | Fuel Trim (Bank 2) |
| P0201 | 6 | Injector Circuit/Open - Cylinder 1 |
| P0202 | 5 | Injector Circuit/Open - Cylinder 2 |

| Table h. S52 engine: Siemens DME MS41.1 from 05/01/1999 to 08/31/2000 | | |
|---|---|---|
| PCode | BMW-FC | PCode text |
| P0203 | 22 | Injector Circuit/Open - Cylinder 3 |
| P0204 | 24 | Injector Circuit/Open - Cylinder 4 |
| P0205 | 33 | Injector Circuit/Open - Cylinder 5 |
| P0206 | 23 | Injector Circuit/Open - Cylinder 6 |
| P0300 | 0 | Random/Multiple Cylinder Misfire Detected |
| P0301 | 238 | Cylinder 1 Misfire Detected |
| P0302 | 239 | Cylinder 2 Misfire Detected |
| P0303 | 240 | Cylinder 3 Misfire Detected |
| P0304 | 241 | Cylinder 4 Misfire Detected |
| P0305 | 242 | Cylinder 5 Misfire Detected |
| P0306 | 243 | Cylinder 6 Misfire Detected |
| P0325 | 57 | Knock Sensor 1 Circuit (Bank 1 or Single Sensor) |
| P0330 | 59 | Knock Sensor 2 Circuit (Bank 2) |
| P0335 | 83 | Crankshaft Position Sensor 'A' Circuit |
| P0340 | 65 | Camshaft Position Sensor 'A' Circuit (Bank 1 or Single Sensor) |
| P0412 | 62 | Secondary Air Injection System Switching Valve A Circuit |
| P0420 | 233 | Catalyst System Efficiency Below Threshold (Bank 1) |
| P0430 | 234 | Catalyst System Efficiency Below Threshold (Bank 2) |
| P0440 | 250 | Evaporative Emission System |
| P0440 | 255 | Evaporative Emission System |
| P0441 | 194 | Evaporative Emission System Incorrect Purge Flow |
| P0441 | 252 | Evaporative Emission System Incorrect Purge Flow |
| P0442 | 195 | Evaporative Emission System Leak Detected (small leak) |
| P0442 | 251 | Evaporative Emission System Leak Detected (small leak) |
| P0443 | 68 | Evaporative Emission System Purge Control Valve Circuit |
| P0446 | 253 | Evaporative Emission System Vent Control Circuit |
| P0455 | 254 | Evaporative Emission System Leak Detected (large leak) |
| P0500 | 214 | Vehicle Speed Sensor 'A' |
| P0505 | 204 | Idle Air Control System |
| P0600 | 217 | Serial Communication Link |
| P0601 | 100 | Internal Control Module Memory Check Sum Error |
| P1145 | 50 | Solenoid Valve Running Losses Control Circuit Electrical |
| P1178 | 231 | O2 Sensor Signal Circuit Slow Switching from Rich to Lean (Bank 1 Sensor 1) |
| P1179 | 232 | O2 Sensor Signal Circuit Slow Switching from Rich to Lean (Bank 2 Sensor 1) |
| P1186 | 235 | O2 Sensor Heater Control Circuit (Bank 1 Sensor 2) |
| P1187 | 236 | O2 Sensor Heater Control Circuit (Bank 2 Sensor 2) |
| P1188 | 227 | Fuel Control (Bank 1 Sensor 1) |

| Table h. S52 engine: Siemens DME MS41.1 from 05/01/1999 to 08/31/2000 | | |
|---|---|---|
| PCode | BMW-FC | PCode text |
| P1189 | 228 | Fuel Control (Bank 2 Sensor 1) |
| P1396 | 244 | Crankshaft Position Sensor Segment Timing Plausibility |
| P1403 | 51 | Carbon Canister Shutoff Valve Control Circuit Electrical (M73: Heated Catalyst Battery Voltage or Current Too Low during Heating (Bank 2)) |
| P1421 | 246 | Secondary Air System (Bank 2) |
| P1423 | 245 | Secondary Air System (Bank 1) |
| P1432 | 247 | Secondary Air Injection System Incorrect Flow Detected |
| P1453 | 35 | Secondary Air Injection Pump Relay Control Circuit Electrical |
| P1470 | 196 | Leakage Diagnostic Pump Control Circuit Electrical |
| P1475 | 191 | Leakage Diagnostic Pump Reed Switch Did not Close |
| P1476 | 193 | Leakage Diagnostic Pump Clamped Tube (M52 MY99/00: Leakage Diagnostic Pump Reed Switch Circuit) |
| P1477 | 190 | Leakage Diagnostic Pump Reed Switch Did Not Open |
| P1509 | 53 | Idle-Speed Control Valve Opening Solenoid Control Circuit Signal Electrical |
| P1519 | 212 | Oil-Condition Sensor Temperature Measurement (M62/M52/S52: 'A' Camshaft Position Actuator Bank 1) |
| P1525 | 21 | 'A' Camshaft Position Actuator Control Open Circuit (Bank 1) |
| P1550 | 27 | Idle-Speed Control Valve Closing Coil Electrical |

| Table i. M52 TU engine: Siemens DME MS42.1 from 06/01/1998 to 05/31/1999 | | |
|---|---|---|
| PCode | BMW-FC | PCode text |
| P0101 | 8 | Mass or Volume Air Flow Circuit Range/Performance |
| P0111 | 14 | Intake Air Temperature Sensor 1 Circuit Range/Performance |
| P0116 | 10 | Engine Coolant Temperature Circuit Range/Performance |
| P0120 | 112 | Throttle/Pedal Position Sensor/Switch 'A' Circuit |
| P0120 | 113 | Throttle/Pedal Position Sensor/Switch 'A' Circuit |
| P0120 | 118 | Throttle/Pedal Position Sensor/Switch 'A' Circuit |
| P0120 | 173 | Throttle/Pedal Position Sensor/Switch 'A' Circuit |
| P0120 | 174 | Throttle/Pedal Position Sensor/Switch 'A' Circuit |
| P0125 | 222 | Insufficient Coolant Temperature for Closed Loop Fuel Control |
| P0128 | 168 | Coolant Thermostat (Coolant Temperature Below Thermostat Regulating Temperature) |
| P0130 | 150 | O2 Sensor Circuit (Bank 1 Sensor 1) |
| P0130 | 151 | O2 Sensor Circuit (Bank 1 Sensor 1) |
| P0130 | 152 | O2 Sensor Circuit (Bank 1 Sensor 1) |
| P0133 | 229 | O2 Sensor Circuit Slow Response (Bank 1 Sensor 1) |
| P0134 | 186 | O2 Sensor Circuit No Activity Detected (Bank 1 Sensor 1) |
| P0135 | 25 | O2 Sensor Heater Circuit (Bank 1 Sensor 1) |
| P0136 | 156 | O2 Sensor Circuit (Bank 1 Sensor 2) |
| P0136 | 157 | O2 Sensor Circuit (Bank 1 Sensor 2) |

| Table i. M52 TU engine: Siemens DME MS42.1 from 06/01/1998 to 05/31/1999 | | |
|---|---|---|
| **PCode** | **BMW-FC** | **PCode text** |
| P0136 | 215 | O2 Sensor Circuit (Bank 1 Sensor 2) |
| P0139 | 220 | O2 Sensor Circuit Slow Response (Bank 1 Sensor 2) |
| P0141 | 61 | O2 Sensor Heater Circuit (Bank 1 Sensor 2) |
| P0150 | 153 | O2 Sensor Circuit (Bank 2 Sensor 1) |
| P0150 | 154 | O2 Sensor Circuit (Bank 2 Sensor 1) |
| P0150 | 155 | O2 Sensor Circuit (Bank 2 Sensor 1) |
| P0153 | 230 | O2 Sensor Circuit Slow Response (Bank 2 Sensor 1) |
| P0154 | 187 | O2 Sensor Circuit No Activity Detected (Bank 2 Sensor 1) |
| P0155 | 55 | O2 Sensor Heater Circuit (Bank 2 Sensor 1) |
| P0156 | 159 | O2 Sensor Circuit (Bank 2 Sensor 2) |
| P0156 | 160 | O2 Sensor Circuit (Bank 2 Sensor 2) |
| P0156 | 216 | O2 Sensor Circuit (Bank 2 Sensor 2) |
| P0159 | 221 | O2 Sensor Circuit Slow Response (Bank 2 Sensor 2) |
| P0161 | 79 | O2 Sensor Heater Circuit (Bank 2 Sensor 2) |
| P0170 | 202 | Fuel Trim (Bank 1) |
| P0173 | 203 | Fuel Trim (Bank 2) |
| P0201 | 6 | Injector Circuit/Open - Cylinder 1 |
| P0202 | 5 | Injector Circuit/Open - Cylinder 2 |
| P0203 | 22 | Injector Circuit/Open - Cylinder 3 |
| P0204 | 24 | Injector Circuit/Open - Cylinder 4 |
| P0205 | 33 | Injector Circuit/Open - Cylinder 5 |
| P0206 | 23 | Injector Circuit/Open - Cylinder 6 |
| P0301 | 238 | Cylinder 1 Misfire Detected |
| P0302 | 239 | Cylinder 2 Misfire Detected |
| P0303 | 240 | Cylinder 3 Misfire Detected |
| P0304 | 241 | Cylinder 4 Misfire Detected |
| P0305 | 242 | Cylinder 5 Misfire Detected |
| P0306 | 243 | Cylinder 6 Misfire Detected |
| P0325 | 57 | Knock Sensor 1 Circuit (Bank 1 or Single Sensor) |
| P0330 | 59 | Knock Sensor 2 Circuit (Bank 2) |
| P0335 | 83 | Crankshaft Position Sensor 'A' Circuit |
| P0340 | 65 | Camshaft Position Sensor 'A' Circuit (Bank 1 or Single Sensor) |
| P0412 | 62 | Secondary Air Injection System Switching Valve A Circuit |
| P0420 | 233 | Catalyst System Efficiency Below Threshold (Bank 1) |
| P0430 | 234 | Catalyst System Efficiency Below Threshold (Bank 2) |
| P0440 | 250 | Evaporative Emission System |
| P0442 | 145 | Evaporative Emission System Leak Detected (small leak) |

| Table i. M52 TU engine: Siemens DME MS42.1 from 06/01/1998 to 05/31/1999 | | |
|---|---|---|
| PCode | BMW-FC | PCode text |
| P0443 | 68 | Evaporative Emission System Purge Control Valve Circuit |
| P0500 | 214 | Vehicle Speed Sensor 'A' |
| P0505 | 204 | Idle Air Control System |
| P0600 | 217 | Serial Communication Link |
| P0601 | 100 | Internal Control Module Memory Check Sum Error |
| P0601 | 170 | Internal Control Module Memory Check Sum Error |
| P0601 | 171 | Internal Control Module Memory Check Sum Error |
| P1132 | 188 | O2 Sensor Heater Control Circuit (Bank 1 Sensor 1) |
| P1133 | 189 | O2 Sensor Heater Control Circuit (Bank 2 Sensor 1) |
| P1140 | 149 | Mass or Volume Air Flow Circuit Range/Performance Problem |
| P1145 | 50 | Solenoid Valve Running Losses Control Circuit Electrical |
| P1161 | 122 | Fuel Trim Adaptation Additive High (Bank 2) (M52: Engine Oil Temperature Sensor Circuit) |
| P1180 | 223 | O2 Sensor Signal Circuit Slow Switching from Rich to Lean (Bank 1 Sensor 2) |
| P1181 | 224 | O2 Sensor Signal Circuit Slow Switching from Rich to Lean (Bank 2 Sensor 2) |
| P1186 | 190 | O2 Sensor Heater Control Circuit (Bank 1 Sensor 2) |
| P1187 | 191 | O2 Sensor Heater Control Circuit (Bank 2 Sensor 2) |
| P1188 | 227 | Fuel Control (Bank 1 Sensor 1) |
| P1189 | 228 | Fuel Control (Bank 2 Sensor 1) |
| P1190 | 235 | Pre Catalyst Fuel Trim System (Bank 1) |
| P1191 | 236 | Pre Catalyst Fuel Trim System (Bank 2) |
| P1192 | 225 | Post Catalyst Fuel Trim System (Bank 1) |
| P1193 | 226 | Post Catalyst Fuel Trim System (Bank 2) |
| P1397 | 18 | Camshaft Position Sensor 'B' Circuit (Bank 1) |
| P1421 | 246 | Secondary Air System (Bank 2) |
| P1423 | 245 | Secondary Air System (Bank 1) |
| P1432 | 247 | Secondary Air Injection System Incorrect Flow Detected |
| P1453 | 35 | Secondary Air Injection Pump Relay Control Circuit Electrical |
| P1470 | 126 | Leakage Diagnostic Pump Control Circuit Electrical |
| P1475 | 140 | Leakage Diagnostic Pump Reed Switch Did not Close |
| P1476 | 143 | Leakage Diagnostic Pump Clamped Tube (M52 MY99/00: Leakage Diagnostic Pump Reed Switch Circuit) |
| P1478 | 146 | Evaporative Emission System Leak Detected (very small leak) |
| P1509 | 53 | Idle-Speed Control Valve Opening Solenoid Control Circuit Signal Electrical |
| P1519 | 103 | Oil-Condition Sensor Temperature Measurement (M62/M52/S52: 'A' Camshaft Position Actuator Bank 1) |
| P1520 | 104 | Oil-Condition Sensor Level Measurement (M52: 'B' Camshaft Position Actuator Bank 1) |
| P1522 | 105 | Oil-Condition Sensor Permeability Measurement (M62: 'A' Camshaft Position Actuator Bank 2; M52: 'A' Camshaft Position Actuator Tight or Jammed) |
| P1523 | 106 | 'A' Camshaft Position Actuator Signal Low (Bank 1) (M52: 'B' Camshaft Position Actuator Tight or Jammed) |

| Table i. M52 TU engine: Siemens DME MS42.1 from 06/01/1998 to 05/31/1999 | | |
|---|---|---|
| **PCode** | **BMW-FC** | **PCode text** |
| P1525 | 21 | 'A' Camshaft Position Actuator Control Open Circuit (Bank 1) |
| P1529 | 19 | 'B' Camshaft Position Actuator Control Circuit Signal Low (Bank 1) |
| P1542 | 110 | Pedal Position Sensor Electrical |
| P1542 | 111 | Pedal Position Sensor Electrical |
| P1542 | 117 | Pedal Position Sensor Electrical |
| P1542 | 120 | Pedal Position Sensor Electrical |
| P1542 | 172 | Pedal Position Sensor Electrical |
| P1542 | 175 | Pedal Position Sensor Electrical |
| P1542 | 176 | Pedal Position Sensor Electrical |
| P1550 | 27 | Idle-Speed Control Valve Closing Coil Electrical |
| P1580 | 109 | Throttle Valve Mechanically Stuck (M73: Throttle Valve 1 Spring Test) |
| P1580 | 114 | Throttle Valve Mechanically Stuck (M73: Throttle Valve 1 Spring Test) |
| P1580 | 119 | Throttle Valve Mechanically Stuck (M73: Throttle Valve 1 Spring Test) |
| P1593 | 124 | DISA (Differentiated Intake Manifold) Control Circuit Electrical |
| P1622 | 123 | Map Cooling Thermostat Control Circuit Electrical |
| P1623 | 115 | Pedal Position Sensor Potentiometer Supply |
| P1623 | 116 | Pedal Position Sensor Potentiometer Supply |

| Table j. M52 TU engine: Siemens DME MS42.1 from 06/01/1999 to 10/31/2000 | | |
|---|---|---|
| **PCode** | **BMW-FC** | **PCode text** |
| P0101 | 8 | Mass or Volume Air Flow Circuit Range/Performance |
| P0111 | 14 | Intake Air Temperature Sensor 1 Circuit Range/Performance |
| P0116 | 10 | Engine Coolant Temperature Circuit Range/Performance |
| P0120 | 112 | Throttle/Pedal Position Sensor/Switch 'A' Circuit |
| P0120 | 113 | Throttle/Pedal Position Sensor/Switch 'A' Circuit |
| P0120 | 118 | Throttle/Pedal Position Sensor/Switch 'A' Circuit |
| P0120 | 173 | Throttle/Pedal Position Sensor/Switch 'A' Circuit |
| P0120 | 174 | Throttle/Pedal Position Sensor/Switch 'A' Circuit |
| P0125 | 222 | Insufficient Coolant Temperature for Closed Loop Fuel Control |
| P0128 | 168 | Coolant Thermostat (Coolant Temperature Below Thermostat Regulating Temperature) |
| P0130 | 150 | O2 Sensor Circuit (Bank 1 Sensor 1) |
| P0130 | 151 | O2 Sensor Circuit (Bank 1 Sensor 1) |
| P0130 | 152 | O2 Sensor Circuit (Bank 1 Sensor 1) |
| P0133 | 229 | O2 Sensor Circuit Slow Response (Bank 1 Sensor 1) |
| P0134 | 186 | O2 Sensor Circuit No Activity Detected (Bank 1 Sensor 1) |
| P0135 | 25 | O2 Sensor Heater Circuit (Bank 1 Sensor 1) |

| Table j. M52 TU engine: Siemens DME MS42.1 from 06/01/1999 to 10/31/2000 | | |
|---|---|---|
| PCode | BMW-FC | PCode text |
| P0136 | 156 | O2 Sensor Circuit (Bank 1 Sensor 2) |
| P0136 | 157 | O2 Sensor Circuit (Bank 1 Sensor 2) |
| P0136 | 215 | O2 Sensor Circuit (Bank 1 Sensor 2) |
| P0141 | 61 | O2 Sensor Heater Circuit (Bank 1 Sensor 2) |
| P0150 | 153 | O2 Sensor Circuit (Bank 2 Sensor 1) |
| P0150 | 154 | O2 Sensor Circuit (Bank 2 Sensor 1) |
| P0150 | 155 | O2 Sensor Circuit (Bank 2 Sensor 1) |
| P0153 | 230 | O2 Sensor Circuit Slow Response (Bank 2 Sensor 1) |
| P0154 | 187 | O2 Sensor Circuit No Activity Detected (Bank 2 Sensor 1) |
| P0155 | 55 | O2 Sensor Heater Circuit (Bank 2 Sensor 1) |
| P0156 | 159 | O2 Sensor Circuit (Bank 2 Sensor 2) |
| P0156 | 160 | O2 Sensor Circuit (Bank 2 Sensor 2) |
| P0156 | 216 | O2 Sensor Circuit (Bank 2 Sensor 2) |
| P0161 | 79 | O2 Sensor Heater Circuit (Bank 2 Sensor 2) |
| P0170 | 202 | Fuel Trim (Bank 1) |
| P0173 | 203 | Fuel Trim (Bank 2) |
| P0201 | 6 | Injector Circuit/Open - Cylinder 1 |
| P0202 | 5 | Injector Circuit/Open - Cylinder 2 |
| P0203 | 22 | Injector Circuit/Open - Cylinder 3 |
| P0204 | 24 | Injector Circuit/Open - Cylinder 4 |
| P0205 | 33 | Injector Circuit/Open - Cylinder 5 |
| P0206 | 23 | Injector Circuit/Open - Cylinder 6 |
| P0301 | 238 | Cylinder 1 Misfire Detected |
| P0302 | 239 | Cylinder 2 Misfire Detected |
| P0303 | 240 | Cylinder 3 Misfire Detected |
| P0304 | 241 | Cylinder 4 Misfire Detected |
| P0305 | 242 | Cylinder 5 Misfire Detected |
| P0306 | 243 | Cylinder 6 Misfire Detected |
| P0325 | 57 | Knock Sensor 1 Circuit (Bank 1 or Single Sensor) |
| P0330 | 59 | Knock Sensor 2 Circuit (Bank 2) |
| P0335 | 83 | Crankshaft Position Sensor 'A' Circuit |
| P0340 | 65 | Camshaft Position Sensor 'A' Circuit (Bank 1 or Single Sensor) |
| P0412 | 62 | Secondary Air Injection System Switching Valve A Circuit |
| P0420 | 233 | Catalyst System Efficiency Below Threshold (Bank 1) |
| P0430 | 234 | Catalyst System Efficiency Below Threshold (Bank 2) |
| P0440 | 250 | Evaporative Emission System |
| P0442 | 145 | Evaporative Emission System Leak Detected (small leak) |

| Table j. M52 TU engine: Siemens DME MS42.1 from 06/01/1999 to 10/31/2000 | | |
|---|---|---|
| **PCode** | **BMW-FC** | **PCode text** |
| P0443 | 68 | Evaporative Emission System Purge Control Valve Circuit |
| P0500 | 214 | Vehicle Speed Sensor 'A' |
| P0505 | 204 | Idle Air Control System |
| P0600 | 217 | Serial Communication Link |
| P0601 | 100 | Internal Control Module Memory Check Sum Error |
| P0601 | 170 I | Internal Control Module Memory Check Sum Error |
| P0601 | 171 | Internal Control Module Memory Check Sum Error |
| P1132 | 188 | O2 Sensor Heater Control Circuit (Bank 1 Sensor 1) |
| P1133 | 189 | O2 Sensor Heater Control Circuit (Bank 2 Sensor 1) |
| P1140 | 149 | Mass or Volume Air Flow Circuit Range/Performance Problem |
| P1145 | 50 | Solenoid Valve Running Losses Control Circuit Electrical |
| P1161 | 122 | Fuel Trim Adaptation Additive High (Bank 2) (M52: Engine Oil Temperature Sensor Circuit) |
| P1180 | 223 | O2 Sensor Signal Circuit Slow Switching from Rich to Lean (Bank 1 Sensor 2) |
| P1181 | 224 | O2 Sensor Signal Circuit Slow Switching from Rich to Lean (Bank 2 Sensor 2) |
| P1184 | 220 | HO2S Sensor Voltage Excursion Electrical (Bank 1 Sensor 1) |
| P1185 | 221 | HO2S Sensor Voltage Excursion Electrical (Bank 2 Sensor 1) |
| P1186 | 190 | O2 Sensor Heater Control Circuit (Bank 1 Sensor 2) |
| P1187 | 191 | O2 Sensor Heater Control Circuit (Bank 2 Sensor 2) |
| P1188 | 227 | Fuel Control (Bank 1 Sensor 1) |
| P1189 | 228 | Fuel Control (Bank 2 Sensor 1) |
| P1190 | 235 | Pre Catalyst Fuel Trim System (Bank 1) |
| P1191 | 236 | Pre Catalyst Fuel Trim System (Bank 2) |
| P1192 | 225 | Post Catalyst Fuel Trim System (Bank 1) |
| P1193 | 226 | Post Catalyst Fuel Trim System (Bank 2) |
| P1397 | 18 | Camshaft Position Sensor 'B' Circuit (Bank 1) |
| P1421 | 246 | Secondary Air System (Bank 2) |
| P1423 | 245 | Secondary Air System (Bank 1) |
| P1432 | 247 | Secondary Air Injection System Incorrect Flow Detected |
| P1453 | 35 | Secondary Air Injection Pump Relay Control Circuit Electrical |
| P1470 | 126 | Leakage Diagnostic Pump Control Circuit Electrical |
| P1475 | 140 | Leakage Diagnostic Pump Reed Switch Did not Close |
| P1476 | 143 | Leakage Diagnostic Pump Clamped Tube (M52 MY99/00: Leakage Diagnostic Pump Reed Switch Circuit) |
| P1478 | 146 | Evaporative Emission System Leak Detected (very small leak) |
| P1509 | 53 | Idle-Speed Control Valve Opening Solenoid Control Circuit Signal Electrical |
| P1519 | 103 | Oil-Condition Sensor Temperature Measurement (M62/M52/S52: 'A' Camshaft Position Actuator Bank 1) |
| P1520 | 104 | Oil-Condition Sensor Level Measurement (M52: 'B' Camshaft Position Actuator Bank 1) |

| Table j. M52 TU engine: Siemens DME MS42.1 from 06/01/1999 to 10/31/2000 | | |
|---|---|---|
| **PCode** | **BMW-FC** | **PCode text** |
| P1522 | 105 | Oil-Condition Sensor Permeability Measurement (M62: 'A' Camshaft Position Actuator Bank 2; M52: 'A' Camshaft Position Actuator Tight or Jammed) |
| P1523 | 106 | 'A' Camshaft Position Actuator Signal Low (Bank 1) (M52: 'B' Camshaft Position Actuator Tight or Jammed) |
| P1525 | 21 | 'A' Camshaft Position Actuator Control Open Circuit (Bank 1) |
| P1529 | 19 | 'B' Camshaft Position Actuator Control Circuit Signal Low (Bank 1) |
| P1542 | 110 | Pedal Position Sensor Electrical |
| P1542 | 111 | Pedal Position Sensor Electrical |
| P1542 | 117 | Pedal Position Sensor Electrical |
| P1542 | 120 | Pedal Position Sensor Electrical |
| P1542 | 172 | Pedal Position Sensor Electrical |
| P1542 | 175 | Pedal Position Sensor Electrical |
| P1542 | 176 | Pedal Position Sensor Electrical |
| P1550 | 27 | Idle-Speed Control Valve Closing Coil Electrical |
| P1580 | 109 | Throttle Valve Mechanically Stuck (M73: Throttle Valve 1 Spring Test) |
| P1580 | 114 | Throttle Valve Mechanically Stuck (M73: Throttle Valve 1 Spring Test) |
| P1580 | 119 | Throttle Valve Mechanically Stuck (M73: Throttle Valve 1 Spring Test) |
| P1593 | 124 | DISA (Differentiated Intake Manifold) Control Circuit Electrical |
| P1622 | 123 | Map Cooling Thermostat Control Circuit Electrical |
| P1623 | 115 | Pedal Position Sensor Potentiometer Supply |
| P1623 | 116 | Pedal Position Sensor Potentiometer Supply |

| Table k. M54 engine: Siemens DME MS43.0 from 04/01/2000 to 08/31/2001 | | |
|---|---|---|
| **PCode** | **BMW-FC** | **PCode text** |
| P0011 | 103 | 'A' Camshaft Position Timing Over-Advanced or System Performance (Bank 1) |
| P0012 | 105 | 'A' Camshaft Position Timing Over-Retarded (Bank 1) |
| P0014 | 104 | 'B' Camshaft Position Timing Over-Advanced or System Performance (Bank 1) |
| P0015 | 106 | 'B' Camshaft Position Timing Over-Retarded (Bank 1) |
| P0101 | 149 | Mass or Volume Air Flow Circuit Range/Performance |
| P0102 | 8 | Mass or Volume Air Flow Circuit Low Input |
| P0103 | 8 | Mass or Volume Air Flow Circuit High Input |
| P0107 | 197 | Manifold Absolute Pressure/Barometric Pressure Circuit Low Input |
| P0108 | 197 | Manifold Absolute Pressure/Barometric Pressure Circuit High Input |
| P0112 | 14 | Intake Air Temperature Sensor 1 Circuit Low |
| P0113 | 14 | Intake Air Temperature Sensor 1 Circuit High |
| P0117 | 10 | Engine Coolant Temperature Circuit Low |
| P0118 | 10 | Engine Coolant Temperature Circuit High |
| P0121 | 118 | Throttle/Pedal Position Sensor/Switch 'A' Circuit Range/Performance |

| Table k. M54 engine: Siemens DME MS43.0 from 04/01/2000 to 08/31/2001 | | |
|---|---|---|
| **PCode** | **BMW-FC** | **PCode text** |
| P0122 | 112 | Throttle/Pedal Position Sensor/Switch 'A' Circuit Low |
| P0123 | 112 | Throttle/Pedal Position Sensor/Switch 'A' Circuit High |
| P0125 | 222 | Insufficient Coolant Temperature for Closed Loop Fuel Control |
| P0128 | 168 | Coolant Thermostat (Coolant Temperature Below Thermostat Regulating Temperature) |
| P0131 | 150 | O2 Sensor Circuit Low Voltage (Bank 1 Sensor 1) |
| P0132 | 150 | O2 Sensor Circuit High Voltage (Bank 1 Sensor 1) |
| P0133 | 229 | O2 Sensor Circuit Slow Response (Bank 1 Sensor 1) |
| P0133 | 231 | O2 Sensor Circuit Slow Response (Bank 1 Sensor 1) |
| P0134 | 150 | O2 Sensor Circuit No Activity Detected (Bank 1 Sensor 1) |
| P0135 | 188 | O2 Sensor Heater Circuit (Bank 1 Sensor 1) |
| P0137 | 152 | O2 Sensor Circuit Low Voltage (Bank 1 Sensor 2) |
| P0138 | 152 | O2 Sensor Circuit High Voltage (Bank 1 Sensor 2) |
| P0139 | 215 | O2 Sensor Circuit Slow Response (Bank 1 Sensor 2) |
| P0139 | 248 | O2 Sensor Circuit Slow Response (Bank 1 Sensor 2) |
| P0140 | 152 | O2 Sensor Circuit No Activity Detected (Bank 1 Sensor 2) |
| P0141 | 190 | O2 Sensor Heater Circuit (Bank 1 Sensor 2) |
| P0151 | 151 | O2 Sensor Circuit Low Voltage (Bank 2 Sensor 1) |
| P0152 | 151 | O2 Sensor Circuit High Voltage (Bank 2 Sensor 1) |
| P0153 | 230 | O2 Sensor Circuit Slow Response (Bank 2 Sensor 1) |
| P0153 | 232 | O2 Sensor Circuit Slow Response (Bank 2 Sensor 1) |
| P0154 | 151 | O2 Sensor Circuit No Activity Detected (Bank 2 Sensor 1) |
| P0155 | 189 | O2 Sensor Heater Circuit (Bank 2 Sensor 1) |
| P0157 | 153 | O2 Sensor Circuit Low Voltage (Bank 2 Sensor 2) |
| P0158 | 153 | O2 Sensor Circuit High Voltage (Bank 2 Sensor 2) |
| P0159 | 216 | O2 Sensor Circuit Slow Response (Bank 2 Sensor 2) |
| P0159 | 249 | O2 Sensor Circuit Slow Response (Bank 2 Sensor 2) |
| P0160 | 153 | O2 Sensor Circuit No Activity Detected (Bank 2 Sensor 2) |
| P0161 | 191 | O2 Sensor Heater Circuit (Bank 2 Sensor 2) |
| P0171 | 202 | System Too Lean (Bank 1) |
| P0171 | 227 | System Too Lean (Bank 1) |
| P0172 | 202 | System Too Rich (Bank 1) |
| P0172 | 227 | System Too Rich (Bank 1) |
| P0174 | 203 | System Too Lean (Bank 2) |
| P0174 | 228 | System Too Lean (Bank 2) |
| P0175 | 203 | System Too Rich (Bank 2) |
| P0175 | 228 | System Too Rich (Bank 2) |
| P0197 | 122 | Engine Oil Temperature Sensor Low |

| Table k. M54 engine: Siemens DME MS43.0 from 04/01/2000 to 08/31/2001 | | |
|---|---|---|
| **PCode** | **BMW-FC** | **PCode text** |
| P0198 | 122 | Engine Oil Temperature Sensor High |
| P0201 | 6 | Injector Circuit/Open - Cylinder 1 |
| P0202 | 5 I | Injector Circuit/Open - Cylinder 2 |
| P0203 | 22 | Injector Circuit/Open - Cylinder 3 |
| P0204 | 24 | Injector Circuit/Open - Cylinder 4 |
| P0205 | 33 | Injector Circuit/Open - Cylinder 5 |
| P0206 | 23 | Injector Circuit/Open - Cylinder 6 |
| P0221 | 119 | Throttle/Pedal Position Sensor/Switch 'B' Circuit Range/Performance |
| P0222 | 113 | Throttle/Pedal Position Sensor/Switch 'B' Circuit Low |
| P0223 | 113 | Throttle/Pedal Position Sensor/Switch 'B' Circuit High |
| P0261 | 6 | Cylinder 1 Injector Circuit Low |
| P0262 | 6 | Cylinder 1 Injector Circuit High |
| P0264 | 5 | Cylinder 2 Injector Circuit Low |
| P0265 | 5 | Cylinder 2 Injector Circuit High |
| P0267 | 22 | Cylinder 3 Injector Circuit Low |
| P0268 | 22 | Cylinder 3 Injector Circuit High |
| P0270 | 24 | Cylinder 4 Injector Circuit Low |
| P0271 | 24 | Cylinder 4 Injector Circuit High |
| P0273 | 33 | Cylinder 5 Injector Circuit Low |
| P0274 | 33 | Cylinder 5 Injector Circuit High |
| P0276 | 23 | Cylinder 6 Injector Circuit Low |
| P0277 | 23 | Cylinder 6 Injector Circuit High |
| P0301 | 238 | Cylinder 1 Misfire Detected |
| P0302 | 239 | Cylinder 2 Misfire Detected |
| P0303 | 240 | Cylinder 3 Misfire Detected |
| P0304 | 241 | Cylinder 4 Misfire Detected |
| P0305 | 242 | Cylinder 5 Misfire Detected |
| P0306 | 243 | Cylinder 6 Misfire Detected |
| P0313 | 238 | Misfire Detected with Low Fuel |
| P0313 | 239 | Misfire Detected with Low Fuel |
| P0313 | 240 | Misfire Detected with Low Fuel |
| P0313 | 241 | Misfire Detected with Low Fuel |
| P0313 | 242 | Misfire Detected with Low Fuel |
| P0313 | 243 | Misfire Detected with Low Fuel |
| P0327 | 57 | Knock Sensor 1 Circuit Low (Bank 1 or Single Sensor) |
| P0332 | 59 | Knock Sensor 2 Circuit Low (Bank 2) |
| P0335 | 83 | Crankshaft Position Sensor 'A' Circuit |

| Table k. M54 engine: Siemens DME MS43.0 from 04/01/2000 to 08/31/2001 | | |
|---|---|---|
| **PCode** | **BMW-FC** | **PCode text** |
| P0339 | 83 | Crankshaft Position Sensor 'A' Circuit Intermittent |
| P0340 | 65 | Camshaft Position Sensor 'A' Circuit (Bank 1 or Single Sensor) |
| P0344 | 65 | Camshaft Position Sensor 'A' Circuit Intermittent (Bank 1 or Single Sensor) |
| P0365 | 18 | Camshaft Position Sensor 'B' Circuit (Bank 1) |
| P0369 | 18 | Camshaft Position Sensor 'B' Circuit Intermittent (Bank 1) |
| P0413 | 62 | Secondary Air Injection System Switching Valve A Circuit Open |
| P0414 | 62 | Secondary Air Injection System Switching Valve A Circuit Shorted |
| P0420 | 233 | Catalyst System Efficiency Below Threshold (Bank 1) |
| P0430 | 234 | Catalyst System Efficiency Below Threshold (Bank 2) |
| P0441 | 250 | Evaporative Emission System Incorrect Purge Flow |
| P0443 | 68 | Evaporative Emission System Purge Control Valve Circuit |
| P0444 | 68 | Evaporative Emission System Purge Control Valve Circuit Open |
| P0445 | 68 | Evaporative Emission System Purge Control Valve Circuit Shorted |
| P0455 | 143 | Evaporative Emission System Leak Detected (large leak) |
| P0456 | 143 | Evaporative Emission System Leak Detected (very small leak) |
| P0491 | 245 | Secondary Air Injection System Insufficient Flow (Bank 1) |
| P0500 | 214 | Vehicle Speed Sensor 'A' |
| P0505 | 204 | Idle Air Control System |
| P0600 | 217 | Serial Communication Link |
| P0604 | 100 | Internal Control Module Random Access Memory (RAM) Error |
| P1111 | 11 | Engine Coolant Temperature Sensor Radiator Outlet Low Input |
| P1112 | 11 | Engine Coolant Temperature Sensor Radiator Outlet High Input |
| P1120 | 114 | Pedal Position Sensor Circuit |
| P1121 | 117 | Pedal Position Sensor 1 Range/Performance Problem |
| P1122 | 110 | Pedal Position Sensor 1 Low Input |
| P1123 | 110 | Pedal Position Sensor 1 High Input |
| P1134 | 25 | O2 Sensor Heater Circuit Signal Intermittent (Bank 1 Sensor 1) |
| P1135 | 25 | O2 Sensor Heater Circuit Low Voltage (Bank 1 Sensor 1) |
| P1136 | 25 | O2 Sensor Heater Circuit High Voltage (Bank 1 Sensor 1) |
| P1137 | 79 | O2 Sensor Heater Circuit Signal Intermittent (Bank 1 Sensor 2) |
| P1138 | 79 | O2 Sensor Heater Circuit Low Voltage (Bank 1 Sensor 2) |
| P1139 | 79 | O2 Sensor Heater Circuit High Voltage (Bank 1 Sensor 2) |
| P1151 | 55 | O2 Sensor Heater Circuit Signal Intermittent (Bank 2 Sensor 1) |
| P1152 | 55 | O2 Sensor Heater Circuit Low Voltage (Bank 2 Sensor 1) |
| P1153 | 55 | O2 Sensor Heater Circuit High Voltage (Bank 2 Sensor 1) |
| P1155 | 61 | O2 Sensor Heater Circuit Signal Intermittent (Bank 2 Sensor 2) |
| P1156 | 61 | O2 Sensor Heater Circuit Low Voltage (Bank 2 Sensor 2) |

| Table k. M54 engine: Siemens DME MS43.0 from 04/01/2000 to 08/31/2001 | | |
|---|---|---|
| **PCode** | **BMW-FC** | **PCode text** |
| P1157 | 61 | O2 Sensor Heater Circuit High Voltage (Bank 2 Sensor 2) |
| P1190 | 235 | Pre Catalyst Fuel Trim System (Bank 1) |
| P1191 | 236 | Pre Catalyst Fuel Trim System (Bank 2) |
| P1192 | 225 | Post Catalyst Fuel Trim System (Bank 1) |
| P1193 | 226 | Post Catalyst Fuel Trim System (Bank 2) |
| P1222 | 111 | Pedal Position Sensor 2 Low Input |
| P1223 | 111 | Pedal Position Sensor 2 High Input |
| P1342 | 238 | Misfire During Start Cylinder 1 |
| P1343 | 238 | Misfire Cylinder 1 with Fuel Cut-Off |
| P1344 | 239 | Misfire During Start Cylinder 2 |
| P1345 | 239 | Misfire Cylinder 2 with Fuel Cut-Off |
| P1346 | 240 | Misfire During Start Cylinder 3 |
| P1347 | 240 | Misfire Cylinder 3 with Fuel Cut-Off |
| P1348 | 241 | Misfire During Start Cylinder 4 |
| P1349 | 241 | Misfire Cylinder 4 with Fuel Cut-Off |
| P1350 | 242 | Misfire during Start Cylinder 5 |
| P1351 | 242 | Misfire Cylinder 5 with Fuel Cut-Off |
| P1352 | 243 | Misfire during Start Cylinder 6 |
| P1500 | 211 | Idle-Speed Control Valve Stuck Open |
| P1501 | 211 | Idle-Speed Control Valve Stuck Closed |
| P1502 | 27 | Idle-Speed Control Valve Closing Solenoid Control Circuit Signal High |
| P1503 | 27 | Idle-Speed Control Valve Closing Solenoid Control Circuit Signal Low |
| P1504 | 27 | Idle-Speed Control Valve Closing Solenoid Control Open Circuit |
| P1506 | 53 | Idle-Speed Control Valve Opening Solenoid Control Circuit Signal High |
| P1507 | 53 | Idle-Speed Control Valve Opening Solenoid Control Circuit Signal Low |
| P1508 | 53 | Idle-Speed Control Valve Opening Solenoid Control Open Circuit |
| P1353 | 243 | Misfire Cylinder 6 with Fuel Cut-Off |
| P1413 | 35 | Secondary Air Injection Pump Relay Control Circuit Signal Low |
| P1414 | 35 | Secondary Air Injection Pump Relay Control Circuit Signal High |
| P1444 | 142 | Diagnostic Module Tank Leakage (DM-TL) Pump Control Open Circuit |
| P1445 | 140 | Diagnostic Module Tank Leakage (DM-TL) Pump Control Circuit Signal Low |
| P1446 | 140 | Diagnostic Module Tank Leakage (DM-TL) Pump Control Circuit Signal High |
| P1447 | 142 | Diagnostic Module Tank Leakage (DM-TL) Pump Current Too High during Switching Solenoid Test |
| P1448 | 142 | Diagnostic Module Tank Leakage (DM-TL) Pump Current Too Low |
| P1449 | 142 | Diagnostic Module Tank Leakage (DM-TL) Pump Current Too High |
| P1451 | 126 | Diagnostic Module Tank Leakage (DM-TL) Switching Solenoid Control Circuit Signal Low |
| P1452 | 126 | Diagnostic Module Tank Leakage (DM-TL) Switching Solenoid Control Circuit Signal High |

| Table k. M54 engine: Siemens DME MS43.0 from 04/01/2000 to 08/31/2001 | | |
|---|---|---|
| **PCode** | **BMW-FC** | **PCode text** |
| P1512 | 124 | DISA (Differentiated Intake Manifold) Control Circuit Signal Low |
| P1513 | 124 | DISA (Differentiated Intake Manifold) Control Circuit Signal High |
| P1523 | 21 | 'A' Camshaft Position Actuator Signal Low (Bank 1) (M52: 'B' Camshaft Position Actuator Tight or Jammed) |
| P1524 | 21 | 'A' Camshaft Position Actuator Control Circuit Signal High (Bank 1) |
| P1525 | 21 | 'A' Camshaft Position Actuator Control Open Circuit (Bank 1) |
| P1529 | 19 | 'B' Camshaft Position Actuator Control Circuit Signal Low (Bank 1) |
| P1530 | 19 | 'B' Camshaft Position Actuator Control Circuit Signal High (Bank 1) (S54 to 09/00: Throttle Valve Position Control, Control Deviation) |
| P1531 | 19 | 'B' Camshaft Position Actuator Control Open Circuit (Bank 1) |
| P1602 | 48 | Control Module Self-Test, Control Module Defective |
| P1602 | 58 | Control Module Self-Test, Control Module Defective |
| P1602 | 63 | Control Module Self-Test, Control Module Defective |
| P1602 | 66 | Control Module Self-Test, Control Module Defective |
| P1602 | 67 | Control Module Self-Test, Control Module Defective |
| P1602 | 70 | Control Module Self-Test, Control Module Defective |
| P1602 | 71 | Control Module Self-Test, Control Module Defective |
| P1602 | 72 | Control Module Self-Test, Control Module Defective |
| P1603 | 49 | Control Module Self-Test, Torque Monitoring |
| P1604 | 50 | Control Module Self-Test, Speed Monitoring |
| P1604 | 51 | Control Module Self-Test, Speed Monitoring |
| P1619 | 123 | Map Cooling Thermostat Control Circuit Signal Low |
| P1620 | 123 | Map Cooling Thermostat Control Circuit Signal High |
| P1624 | 146 | Pedal Position Sensor Potentiometer Supply Channel 1 Electrical (M52: Coolant Thermostat (Coolant Temperature Below Thermostat Regulating Temperature)) |
| P1625 | 147 | Pedal Position Sensor Potentiometer Supply Channel 2 Electrical |
| P1632 | 115 | Throttle Valve Adaptation Conditions Not Met |
| P1633 | 115 | Throttle Valve Adaptation Limp-Home Position Unknown |
| P1634 | 115 | Throttle Valve Adaptation Spring Test Failed |
| P1634 | 135 | Throttle Valve Adaptation Spring Test Failed |
| P1634 | 136 | Throttle Valve Adaptation Spring Test Failed |
| P1635 | 115 | Throttle Valve Adaptation Lower Mechanical Stop not Adapted |
| P1636 | 109 | Throttle Valve Control Circuit |
| P1637 | 162 | Throttle Valve Position Control, Control Deviation |
| P1638 | 160 | Throttle Valve Position Control Throttle Stuck Temporarily |
| P1639 | 161 | Throttle Valve Position Control Throttle Stuck Permanently |

| Table I. M54 engine: Siemens DME MS43.0 from 09/01/2001 to 08/31/2002 | | |
|---|---|---|
| **PCode** | **BMW-FC** | **PCode text** |
| P0011 | 103 | 'A' Camshaft Position Timing Over-Advanced or System Performance (Bank 1) |
| P0012 | 105 | 'A' Camshaft Position Timing Over-Retarded (Bank 1) |
| P0014 | 104 | 'B' Camshaft Position Timing Over-Advanced or System Performance (Bank 1) |
| P0015 | 106 | 'B' Camshaft Position Timing Over-Retarded (Bank 1) |
| P0030 | 25 | HO2S Heater Control Circuit (Bank 1 Sensor 1) |
| P0031 | 25 | HO2S Heater Control Circuit Low (Bank 1 Sensor 1) |
| P0032 | 25 | HO2S Heater Control Circuit High (Bank 1 Sensor 1) |
| P0036 | 79 | HO2S Heater Control Circuit (Bank 1 Sensor 2) |
| P0037 | 79 | HO2S Heater Control Circuit Low (Bank 1 Sensor 2) |
| P0038 | 79 | HO2S Heater Control Circuit High (Bank 1 Sensor 2) |
| P0050 | 55 | HO2S Heater Control Circuit (Bank 2 Sensor 1) |
| P0051 | 55 | HO2S Heater Control Circuit Low (Bank 2 Sensor 1) |
| P0052 | 55 | HO2S Heater Control Circuit High (Bank 2 Sensor 1) |
| P0056 | 61 | HO2S Heater Control Circuit (Bank 2 Sensor 2) |
| P0057 | 61 | HO2S Heater Control Circuit Low (Bank 2 Sensor 2) |
| P0058 | 61 | HO2S Heater Control Circuit High (Bank 2 Sensor 2) |
| P0101 | 149 | Mass or Volume Air Flow Circuit Range/Performance |
| P0102 | 8 | Mass or Volume Air Flow Circuit Low Input |
| P0103 | 8 | Mass or Volume Air Flow Circuit High Input |
| P0107 | 197 | Manifold Absolute Pressure/Barometric Pressure Circuit Low Input |
| P0108 | 197 | Manifold Absolute Pressure/Barometric Pressure Circuit High Input |
| P0112 | 14 | Intake Air Temperature Sensor 1 Circuit Low |
| P0113 | 14 | Intake Air Temperature Sensor 1 Circuit High |
| P0117 | 10 | Engine Coolant Temperature Circuit Low |
| P0118 | 10 | Engine Coolant Temperature Circuit High |
| P0121 | 118 | Throttle/Pedal Position Sensor/Switch 'A' Circuit Range/Performance |
| P0122 | 112 | Throttle/Pedal Position Sensor/Switch 'A' Circuit Low |
| P0123 | 112 | Throttle/Pedal Position Sensor/Switch 'A' Circuit High |
| P0125 | 222 | Insufficient Coolant Temperature for Closed Loop Fuel Control |
| P0128 | 168 | Coolant Thermostat (Coolant Temperature Below Thermostat Regulating Temperature) |
| P0131 | 150 | O2 Sensor Circuit Low Voltage (Bank 1 Sensor 1) |
| P0132 | 150 | O2 Sensor Circuit High Voltage (Bank 1 Sensor 1) |
| P0133 | 229 | O2 Sensor Circuit Slow Response (Bank 1 Sensor 1) |
| P0134 | 150 | O2 Sensor Circuit No Activity Detected (Bank 1 Sensor 1) |
| P0135 | 188 | O2 Sensor Heater Circuit (Bank 1 Sensor 1) |
| P0137 | 152 | O2 Sensor Circuit Low Voltage (Bank 1 Sensor 2) |
| P0138 | 152 | O2 Sensor Circuit High Voltage (Bank 1 Sensor 2) |

| Table I. M54 engine: Siemens DME MS43.0 from 09/01/2001 to 08/31/2002 | | |
|---|---|---|
| **PCode** | **BMW-FC** | **PCode text** |
| P0140 | 152 | O2 Sensor Circuit No Activity Detected (Bank 1 Sensor 2) |
| P0141 | 190 | O2 Sensor Heater Circuit (Bank 1 Sensor 2) |
| P0151 | 151 | O2 Sensor Circuit Low Voltage (Bank 2 Sensor 1) |
| P0152 | 151 | O2 Sensor Circuit High Voltage (Bank 2 Sensor 1) |
| P0153 | 230 | O2 Sensor Circuit Slow Response (Bank 2 Sensor 1) |
| P0154 | 151 | O2 Sensor Circuit No Activity Detected (Bank 2 Sensor 1) |
| P0155 | 189 | O2 Sensor Heater Circuit (Bank 2 Sensor 1) |
| P0157 | 153 | O2 Sensor Circuit Low Voltage (Bank 2 Sensor 2) |
| P0158 | 153 | O2 Sensor Circuit High Voltage (Bank 2 Sensor 2) |
| P0160 | 153 | O2 Sensor Circuit No Activity Detected (Bank 2 Sensor 2) |
| P0161 | 191 | O2 Sensor Heater Circuit (Bank 2 Sensor 2) |
| P0171 | 227 | System Too Lean (Bank 1) |
| P0172 | 227 | System Too Rich (Bank 1) |
| P0174 | 228 | System Too Lean (Bank 2) |
| P0175 | 228 | System Too Rich (Bank 2) |
| P0197 | 122 | Engine Oil Temperature Sensor Low |
| P0198 | 122 | Engine Oil Temperature Sensor High |
| P0201 | 6 | Injector Circuit/Open - Cylinder 1 |
| P0202 | 5 | Injector Circuit/Open - Cylinder 2 |
| P0203 | 22 | Injector Circuit/Open - Cylinder 3 |
| P0204 | 24 | Injector Circuit/Open - Cylinder 4 |
| P0205 | 33 | Injector Circuit/Open - Cylinder 5 |
| P0206 | 23 | Injector Circuit/Open - Cylinder 6 |
| P0221 | 119 | Throttle/Pedal Position Sensor/Switch 'B' Circuit Range/Performance |
| P0222 | 113 | Throttle/Pedal Position Sensor/Switch 'B' Circuit Low |
| P0223 | 113 | Throttle/Pedal Position Sensor/Switch 'B' Circuit High |
| P0261 | 6 | Cylinder 1 Injector Circuit Low |
| P0262 | 6 | Cylinder 1 Injector Circuit High |
| P0264 | 5 | Cylinder 2 Injector Circuit Low |
| P0265 | 5 | Cylinder 2 Injector Circuit High |
| P0267 | 22 | Cylinder 3 Injector Circuit Low |
| P0268 | 22 | Cylinder 3 Injector Circuit High |
| P0270 | 24 | Cylinder 4 Injector Circuit Low |
| P0271 | 24 | Cylinder 4 Injector Circuit High |
| P0273 | 33 | Cylinder 5 Injector Circuit Low |
| P0274 | 33 | Cylinder 5 Injector Circuit High |
| P0276 | 23 | Cylinder 6 Injector Circuit Low |

| Table I. M54 engine: Siemens DME MS43.0 from 09/01/2001 to 08/31/2002 | | |
|---|---|---|
| **PCode** | **BMW-FC** | **PCode text** |
| P0277 | 23 | Cylinder 6 Injector Circuit High |
| P0301 | 238 | Cylinder 1 Misfire Detected |
| P0302 | 239 | Cylinder 2 Misfire Detected |
| P0303 | 240 | Cylinder 3 Misfire Detected |
| P0304 | 241 | Cylinder 4 Misfire Detected |
| P0305 | 242 | Cylinder 5 Misfire Detected |
| P0306 | 243 | Cylinder 6 Misfire Detected |
| P0313 | 238 | Misfire Detected with Low Fuel |
| P0313 | 239 | Misfire Detected with Low Fuel |
| P0313 | 240 | Misfire Detected with Low Fuel |
| P0313 | 241 | Misfire Detected with Low Fuel |
| P0313 | 242 | Misfire Detected with Low Fuel |
| P0313 | 243 | Misfire Detected with Low Fuel |
| P0327 | 57 | Knock Sensor 1 Circuit Low (Bank 1 or Single Sensor) |
| P0332 | 59 | Knock Sensor 2 Circuit Low (Bank 2) |
| P0335 | 83 | Crankshaft Position Sensor 'A' Circuit |
| P0339 | 83 | Crankshaft Position Sensor 'A' Circuit Intermittent |
| P0340 | 65 | Camshaft Position Sensor 'A' Circuit (Bank 1 or Single Sensor) |
| P0344 | 65 | Camshaft Position Sensor 'A' Circuit Intermittent (Bank 1 or Single Sensor) |
| P0365 | 18 | Camshaft Position Sensor 'B' Circuit (Bank 1) |
| P0369 | 18 | Camshaft Position Sensor 'B' Circuit Intermittent (Bank 1) |
| P0413 | 62 | Secondary Air Injection System Switching Valve A Circuit Open |
| P0414 | 62 | Secondary Air Injection System Switching Valve A Circuit Shorted |
| P0420 | 233 | Catalyst System Efficiency Below Threshold (Bank 1) |
| P0430 | 234 | Catalyst System Efficiency Below Threshold (Bank 2) |
| P0441 | 250 | Evaporative Emission System Incorrect Purge Flow |
| P0443 | 68 | Evaporative Emission System Purge Control Valve Circuit |
| P0444 | 68 | Evaporative Emission System Purge Control Valve Circuit Open |
| P0445 | 68 | Evaporative Emission System Purge Control Valve Circuit Shorted |
| P0455 | 143 | Evaporative Emission System Leak Detected (large leak) |
| P0456 | 143 | Evaporative Emission System Leak Detected (very small leak) |
| P0491 | 245 | Secondary Air Injection System Insufficient Flow (Bank 1) |
| P0492 | 246 | Secondary Air Injection System Insufficient Flow (Bank 2) |
| P0500 | 214 | Vehicle Speed Sensor 'A' |
| P0505 | 204 | Idle Air Control System |
| P0600 | 217 | Serial Communication Link |
| P0604 | 100 | Internal Control Module Random Access Memory (RAM) Error |

| Table I. M54 engine: Siemens DME MS43.0 from 09/01/2001 to 08/31/2002 | | |
|---|---|---|
| **PCode** | **BMW-FC** | **PCode text** |
| P1083 | 202 | Fuel Control Limit Mixture Too Lean (Bank 1 Sensor 1) |
| P1084 | 202 | Fuel Control Limit Mixture Too Rich (Bank 1 Sensor 1) |
| P1085 | 203 | Fuel Control Limit Mixture Too Lean (Bank 2 Sensor 1) |
| P1086 | 203 | Fuel Control Limit Mixture Too Rich (Bank 2 Sensor 1) |
| P1087 | 229 | O2 Sensor Circuit Slow Response in Lean Control Range (Bank 1 Sensor 1) |
| P1088 | 229 | O2 Sensor Circuit Slow Response in Rich Control Range (Bank 1 Sensor 1) |
| P1089 | 230 | O2 Sensor Circuit Slow Response in Lean Control Range (Bank 1 Sensor 2) |
| P1090 | 235 | Pre Catalyst Fuel Trim System Too Lean (Bank 1) |
| P1091 | 236 | Pre Catalyst Fuel Trim System Too Lean (Bank 2) |
| P1092 | 235 | Pre Catalyst Fuel Trim System Too Rich (Bank 1) |
| P1093 | 236 | Pre Catalyst Fuel Trim System Too Rich (Bank 2) |
| P1094 | 230 | O2 Sensor Circuit Slow Response in Rich Control Range (Bank 2 Sensor 1) |
| P1111 | 11 | Engine Coolant Temperature Sensor Radiator Outlet Low Input |
| P1112 | 11 | Engine Coolant Temperature Sensor Radiator Outlet High Input |
| P1120 | 114 | Pedal Position Sensor Circuit |
| P1121 | 117 | Pedal Position Sensor 1 Range/Performance Problem |
| P1122 | 110 | Pedal Position Sensor 1 Low Input |
| P1123 | 110 | Pedal Position Sensor 1 High Input |
| P1143 | 215 | O2 Sensor Activity Check Signal Too High (Bank 1 Sensor 2) |
| P1144 | 215 | O2 Sensor Activity Check Signal Too Low (Bank 1 Sensor 2) |
| P1149 | 216 | O2 Sensor Activity Check Signal Too High (Bank 2 Sensor 2) |
| P1150 | 216 | O2 Sensor Activity Check Signal Too Low (Bank 2 Sensor 2) |
| P1171 | 196 | Ambient Pressure Sensor Variant Recognition Value in Boot Range Implausible |
| P1172 | 196 | Ambient Pressure Sensor Variant Recognition Error Value Stored in Boot Range |
| P1173 | 196 | Ambient Pressure Sensor Variant Recognition Learning Failed |
| P1222 | 111 | Pedal Position Sensor 2 Low Input |
| P1223 | 111 | Pedal Position Sensor 2 High Input |
| P1314 | 202 | Fuel Mixture Deviation Detected with Low Fuel |
| P1314 | 203 | Fuel Mixture Deviation Detected with Low Fuel |
| P1314 | 227 | Fuel Mixture Deviation Detected with Low Fuel |
| P1314 | 228 | Fuel Mixture Deviation Detected with Low Fuel |
| P1342 | 238 | Misfire During Start Cylinder 1 |
| P1343 | 238 | Misfire Cylinder 1 with Fuel Cut-Off |
| P1344 | 239 | Misfire During Start Cylinder 2 |
| P1345 | 239 | Misfire Cylinder 2 with Fuel Cut-Off |
| P1346 | 240 | Misfire During Start Cylinder 3 |
| P1347 | 240 | Misfire Cylinder 3 with Fuel Cut-Off |

| Table I. M54 engine: Siemens DME MS43.0 from 09/01/2001 to 08/31/2002 | | |
|---|---|---|
| **PCode** | **BMW-FC** | **PCode text** |
| P1348 | 241 | Misfire During Start Cylinder 4 |
| P1349 | 241 | Misfire Cylinder 4 with Fuel Cut-Off |
| P1350 | 242 | Misfire during Start Cylinder 5 |
| P1351 | 242 | Misfire Cylinder 5 with Fuel Cut-Off |
| P1352 | 243 | Misfire during Start Cylinder 6 |
| P1353 | 243 | Misfire Cylinder 6 with Fuel Cut-Off |
| P1413 | 35 | Secondary Air Injection Pump Relay Control Circuit Signal Low |
| P1414 | 35 | Secondary Air Injection Pump Relay Control Circuit Signal High |
| P1445 | 140 | Diagnostic Module Tank Leakage (DM-TL) Pump Control Circuit Signal Low |
| P1446 | 140 | Diagnostic Module Tank Leakage (DM-TL) Pump Control Circuit Signal High |
| P1447 | 142 | Diagnostic Module Tank Leakage (DM-TL) Pump Current Too High during Switching Solenoid Test |
| P1448 | 142 | Diagnostic Module Tank Leakage (DM-TL) Pump Current Too Low |
| P1449 | 142 | Diagnostic Module Tank Leakage (DM-TL) Pump Current Too High |
| P1451 | 126 | Diagnostic Module Tank Leakage (DM-TL) Switching Solenoid Control Circuit Signal Low |
| P1452 | 126 | Diagnostic Module Tank Leakage (DM-TL) Switching Solenoid Control Circuit Signal High |
| P1500 | 211 | Idle-Speed Control Valve Stuck Open |
| P1501 | 211 | Idle-Speed Control Valve Stuck Closed |
| P1502 | 27 | Idle-Speed Control Valve Closing Solenoid Control Circuit Signal High |
| P1503 | 27 | Idle-Speed Control Valve Closing Solenoid Control Circuit Signal Low |
| P1504 | 27 | Idle-Speed Control Valve Closing Solenoid Control Open Circuit |
| P1506 | 53 | Idle-Speed Control Valve Opening Solenoid Control Circuit Signal High |
| P1507 | 53 | Idle-Speed Control Valve Opening Solenoid Control Circuit Signal Low |
| P1508 | 53 | Idle-Speed Control Valve Opening Solenoid Control Open Circuit |
| P1512 | 124 | DISA (Differentiated Intake Manifold) Control Circuit Signal Low |
| P1513 | 124 | DISA (Differentiated Intake Manifold) Control Circuit Signal High |
| P1523 | 21 | 'A' Camshaft Position Actuator Signal Low (Bank 1) (M52: 'B' Camshaft Position Actuator Tight or Jammed) |
| P1524 | 21 | 'A' Camshaft Position Actuator Control Circuit Signal High (Bank 1) |
| P1525 | 21 | 'A' Camshaft Position Actuator Control Open Circuit (Bank 1) |
| P1529 | 19 | 'B' Camshaft Position Actuator Control Circuit Signal Low (Bank 1) |
| P1530 | 19 | 'B' Camshaft Position Actuator Control Circuit Signal High (Bank 1) (S54 to 09/00: Throttle Valve Position Control, Control Deviation) |
| P1531 | 19 | 'B' Camshaft Position Actuator Control Open Circuit (Bank 1) |
| P1602 | 48 | Control Module Self-Test, Control Module Defective |
| P1602 | 58 | Control Module Self-Test, Control Module Defective |
| P1602 | 63 | Control Module Self-Test, Control Module Defective |
| P1602 | 66 | Control Module Self-Test, Control Module Defective |
| P1602 | 67 | Control Module Self-Test, Control Module Defective |

| Table l. M54 engine: Siemens DME MS43.0 from 09/01/2001 to 08/31/2002 | | |
|---|---|---|
| **PCode** | **BMW-FC** | **PCode text** |
| P1602 | 70 | Control Module Self-Test, Control Module Defective |
| P1602 | 71 | Control Module Self-Test, Control Module Defective |
| P1602 | 72 | Control Module Self-Test, Control Module Defective |
| P1603 | 49 | Control Module Self-Test, Torque Monitoring |
| P1604 | 50 | Control Module Self-Test, Speed Monitoring |
| P1619 | 123 | Map Cooling Thermostat Control Circuit Signal Low |
| P1620 | 123 | Map Cooling Thermostat Control Circuit Signal High |
| P1624 | 146 | Pedal Position Sensor Potentiometer Supply Channel 1 Electrical (M52: Coolant Thermostat (Coolant Temperature Below Thermostat Regulating Temperature)) |
| P1625 | 147 | Pedal Position Sensor Potentiometer Supply Channel 2 Electrical |
| P1632 | 115 | Throttle Valve Adaptation Conditions Not Met |
| P1633 | 115 | Throttle Valve Adaptation Limp-Home Position Unknown |
| P1634 | 115 | Throttle Valve Adaptation Spring Test Failed |
| P1635 | 115 | Throttle Valve Adaptation Lower Mechanical Stop not Adapted |
| P1636 | 109 | Throttle Valve Control Circuit |
| P1637 | 162 | Throttle Valve Position Control, Control Deviation |
| P1638 | 160 | Throttle Valve Position Control Throttle Stuck Temporarily |
| P1639 | 161 | Throttle Valve Position Control Throttle Stuck Permanently |
| P1675 | 135 | Throttle Valve Actuator Start Test Re-Adaptation Required |
| P1694 | 136 | Throttle Valve Actuator Start Test Spring Test and Limp-Home Position Failed |
| P2096 | 225 | Post Catalyst Fuel Trim System Too Lean (Bank 1) |
| P2098 | 226 | Post Catalyst Fuel Trim System Too Lean (Bank 2) |

| Table m. S54 engine: Siemens DME MS S54 from 03/01/2001 to 08/31/2001 | | |
|---|---|---|
| **PCode** | **BMW-FC** | **PCode text** |
| P0011 | 184 | 'A' Camshaft Position Timing Over-Advanced or System Performance (Bank 1) |
| P0012 | 184 | 'A' Camshaft Position Timing Over-Retarded (Bank 1) |
| P0014 | 185 | 'B' Camshaft Position Timing Over-Advanced or System Performance (Bank 1) |
| P0015 | 185 | 'B' Camshaft Position Timing Over-Retarded (Bank 1) |
| P0101 | 73 | Mass or Volume Air Flow Circuit Range/Performance |
| P0102 | 41 | Mass or Volume Air Flow Circuit Low Input |
| P0103 | 41 | Mass or Volume Air Flow Circuit High Input |
| P0111 | 77 | Intake Air Temperature Sensor 1 Circuit Range/Performance |
| P0112 | 77 | Intake Air Temperature Sensor 1 Circuit Low |
| P0113 | 77 | Intake Air Temperature Sensor 1 Circuit High |
| P0116 | 105 | Engine Coolant Temperature Circuit Range/Performance |
| P0117 | 78 | Engine Coolant Temperature Circuit Low |
| P0118 | 78 | Engine Coolant Temperature Circuit High |

| Table m. S54 engine: Siemens DME MS S54 from 03/01/2001 to 08/31/2001 | | |
|---|---|---|
| **PCode** | **BMW-FC** | **PCode text** |
| P0121 | 118 | Throttle/Pedal Position Sensor/Switch 'A' Circuit Range/Performance |
| P0121 | 120 | Throttle/Pedal Position Sensor/Switch 'A' Circuit Range/Performance |
| P0121 | 121 | Throttle/Pedal Position Sensor/Switch 'A' Circuit Range/Performance |
| P0122 | 118 | Throttle/Pedal Position Sensor/Switch 'A' Circuit Low |
| P0123 | 118 | Throttle/Pedal Position Sensor/Switch 'A' Circuit High |
| P0128 | 105 | Coolant Thermostat (Coolant Temperature Below Thermostat Regulating Temperature) |
| P0131 | 13 | O2 Sensor Circuit Low Voltage (Bank 1 Sensor 1) |
| P0132 | 13 | O2 Sensor Circuit High Voltage (Bank 1 Sensor 1) |
| P0133 | 214 | O2 Sensor Circuit Slow Response (Bank 1 Sensor 1) |
| P0134 | 13 | O2 Sensor Circuit No Activity Detected (Bank 1 Sensor 1) |
| P0134 | 218 | O2 Sensor Circuit No Activity Detected (Bank 1 Sensor 1) |
| P0135 | 37 | O2 Sensor Heater Circuit (Bank 1 Sensor 1) |
| P0137 | 87 | O2 Sensor Circuit Low Voltage (Bank 1 Sensor 2) |
| P0138 | 87 | O2 Sensor Circuit High Voltage (Bank 1 Sensor 2) |
| P0139 | 92 | O2 Sensor Circuit Slow Response (Bank 1 Sensor 2) |
| P0140 | 87 | O2 Sensor Circuit No Activity Detected (Bank 1 Sensor 2) |
| P0141 | 39 | O2 Sensor Heater Circuit (Bank 1 Sensor 2) |
| P0151 | 12 | O2 Sensor Circuit Low Voltage (Bank 2 Sensor 1) |
| P0152 | 12 | O2 Sensor Circuit High Voltage (Bank 2 Sensor 1) |
| P0153 | 215 | O2 Sensor Circuit Slow Response (Bank 2 Sensor 1) |
| P0154 | 12 | O2 Sensor Circuit No Activity Detected (Bank 2 Sensor 1) |
| P0154 | 219 | O2 Sensor Circuit No Activity Detected (Bank 2 Sensor 1) |
| P0155 | 38 | O2 Sensor Heater Circuit (Bank 2 Sensor 1) |
| P0157 | 88 | O2 Sensor Circuit Low Voltage (Bank 2 Sensor 2) |
| P0158 | 88 | O2 Sensor Circuit High Voltage (Bank 2 Sensor 2) |
| P0159 | 93 | O2 Sensor Circuit Slow Response (Bank 2 Sensor 2) |
| P0160 | 88 | O2 Sensor Circuit No Activity Detected (Bank 2 Sensor 2) |
| P0161 | 40 | O2 Sensor Heater Circuit (Bank 2 Sensor 2) |
| P0171 | 144 | System Too Lean (Bank 1) |
| P0171 | 174 | System Too Lean (Bank 1) |
| P0172 | 144 | System Too Rich (Bank 1) |
| P0172 | 174 | System Too Rich (Bank 1) |
| P0174 | 145 | System Too Lean (Bank 2) |
| P0174 | 175 | System Too Lean (Bank 2) |
| P0175 | 145 | System Too Rich (Bank 2) |
| P0175 | 175 | System Too Rich (Bank 2) |
| P0201 | 3 | Injector Circuit/Open - Cylinder 1 |

| Table m. S54 engine: Siemens DME MS S54 from 03/01/2001 to 08/31/2001 | | |
|---|---|---|
| **PCode** | **BMW-FC** | **PCode text** |
| P0202 | 5 | Injector Circuit/Open - Cylinder 2 |
| P0203 | 4 | Injector Circuit/Open - Cylinder 3 |
| P0204 | 33 | Injector Circuit/Open - Cylinder 4 |
| P0205 | 31 | Injector Circuit/Open - Cylinder 5 |
| P0206 | 32 | Injector Circuit/Open - Cylinder 6 |
| P0221 | 119 | Throttle/Pedal Position Sensor/Switch 'B' Circuit Range/Performance |
| P0222 | 119 | Throttle/Pedal Position Sensor/Switch 'B' Circuit Low |
| P0223 | 119 | Throttle/Pedal Position Sensor/Switch 'B' Circuit High |
| P0261 | 3 | Cylinder 1 Injector Circuit Low |
| P0262 | 3 | Cylinder 1 Injector Circuit High |
| P0264 | 5 | Cylinder 2 Injector Circuit Low |
| P0265 | 5 | Cylinder 2 Injector Circuit High |
| P0267 | 4 | Cylinder 3 Injector Circuit Low |
| P0268 | 4 | Cylinder 3 Injector Circuit High |
| P0270 | 33 | Cylinder 4 Injector Circuit Low |
| P0271 | 33 | Cylinder 4 Injector Circuit High |
| P0273 | 31 | Cylinder 5 Injector Circuit Low |
| P0274 | 31 | Cylinder 5 Injector Circuit High |
| P0276 | 32 | Cylinder 6 Injector Circuit Low |
| P0277 | 32 | Cylinder 6 Injector Circuit High |
| P0300 | 213 | Random/Multiple Cylinder Misfire Detected |
| P0301 | 205 | Cylinder 1 Misfire Detected |
| P0302 | 206 | Cylinder 2 Misfire Detected |
| P0303 | 207 | Cylinder 3 Misfire Detected |
| P0304 | 208 | Cylinder 4 Misfire Detected |
| P0305 | 209 | Cylinder 5 Misfire Detected |
| P0306 | 210 | Cylinder 6 Misfire Detected |
| P0335 | 16 | Crankshaft Position Sensor 'A' Circuit |
| P0340 | 7 | Camshaft Position Sensor 'A' Circuit (Bank 1 or Single Sensor) |
| P0365 | 10 | Camshaft Position Sensor 'B' Circuit (Bank 1) |
| P0369 | 162 | Camshaft Position Sensor 'B' Circuit Intermittent (Bank 1) |
| P0394 | 15 | Camshaft Position Sensor 'B' Circuit Intermittent (Bank 2) |
| P0411 | 170 | Secondary Air Injection System Incorrect Flow Detected |
| P0413 | 63 | Secondary Air Injection System Switching Valve A Circuit Open |
| P0414 | 63 | Secondary Air Injection System Switching Valve A Circuit Shorted |
| P0418 | 19 | Secondary Air Injection System Control 'A' Circuit |
| P0420 | 178 | Catalyst System Efficiency Below Threshold (Bank 1) |

| Table m. S54 engine: Siemens DME MS S54 from 03/01/2001 to 08/31/2001 | | |
|---|---|---|
| **PCode** | **BMW-FC** | **PCode text** |
| P0430 | 179 | Catalyst System Efficiency Below Threshold (Bank 2) |
| P0440 | 232 | Evaporative Emission System |
| P0442 | 180 | Evaporative Emission System Leak Detected (small leak) |
| P0443 | 36 | Evaporative Emission System Purge Control Valve Circuit |
| P0444 | 36 | Evaporative Emission System Purge Control Valve Circuit Open |
| P0445 | 36 | Evaporative Emission System Purge Control Valve Circuit Shorted |
| P0491 | 170 | Secondary Air Injection System Insufficient Flow (Bank 1) |
| P0492 | 170 | Secondary Air Injection System Insufficient Flow (Bank 2) |
| P0500 | 135 | Vehicle Speed Sensor 'A' |
| P0506 | 128 | Idle Air Control System RPM Lower Than Expected |
| P0507 | 128 | Idle Air Control System RPM Higher Than Expected |
| P0601 | 150 | Internal Control Module Memory Check Sum Error |
| P0601 | 157 | Internal Control Module Memory Check Sum Error |
| P0604 | 157 | Internal Control Module Random Access Memory (RAM) Error |
| P0605 | 150 I | Internal Control Module Read Only Memory (ROM) Error |
| P0605 | 157 | Internal Control Module Read Only Memory (ROM) Error |
| P0606 | 30 | ECM/PCM Processor |
| P0606 | 107 | ECM/PCM Processor |
| P0606 | 158 | ECM/PCM Processor |
| P0606 | 163 | ECM/PCM Processor |
| P1121 | 111 | Pedal Position Sensor 1 Range/Performance Problem |
| P1122 | 60 | Pedal Position Sensor 1 Low Input |
| P1123 | 60 | Pedal Position Sensor 1 High Input |
| P1134 | 37 | O2 Sensor Heater Circuit Signal Intermittent (Bank 1 Sensor 1) |
| P1135 | 37 | O2 Sensor Heater Circuit Low Voltage (Bank 1 Sensor 1) |
| P1136 | 37 | O2 Sensor Heater Circuit High Voltage (Bank 1 Sensor 1) |
| P1137 | 39 | O2 Sensor Heater Circuit Signal Intermittent (Bank 1 Sensor 2) |
| P1138 | 39 | O2 Sensor Heater Circuit Low Voltage (Bank 1 Sensor 2) |
| P1139 | 39 | O2 Sensor Heater Circuit High Voltage (Bank 1 Sensor 2) |
| P1151 | 38 | O2 Sensor Heater Circuit Signal Intermittent (Bank 2 Sensor 1) |
| P1152 | 38 | O2 Sensor Heater Circuit Low Voltage (Bank 2 Sensor 1) |
| P1153 | 38 | O2 Sensor Heater Circuit High Voltage (Bank 2 Sensor 1) |
| P1155 | 40 | O2 Sensor Heater Circuit Signal Intermittent (Bank 2 Sensor 2) |
| P1156 | 40 | O2 Sensor Heater Circuit Low Voltage (Bank 2 Sensor 2) |
| P1157 | 40 | O2 Sensor Heater Circuit High Voltage (Bank 2 Sensor 2) |
| P1178 | 216 | O2 Sensor Signal Circuit Slow Switching from Rich to Lean (Bank 1 Sensor 1) |
| P1179 | 217 | O2 Sensor Signal Circuit Slow Switching from Rich to Lean (Bank 2 Sensor 1) |

| Table m. S54 engine: Siemens DME MS S54 from 03/01/2001 to 08/31/2001 | | |
|---|---|---|
| **PCode** | **BMW-FC** | **PCode text** |
| P1221 | 112 | Pedal Position Sensor 2 Range/Performance Problem |
| P1222 | 61 | Pedal Position Sensor 2 Low Input |
| P1223 | 61 | Pedal Position Sensor 2 High Input |
| P1340 | 213 | Multiple Cylinder Misfire During Start |
| P1341 | 204 | Multiple Cylinder Misfire with Fuel Cut-Off |
| P1342 | 205 | Misfire During Start Cylinder 1 |
| P1503 | 2 I | Idle-Speed Control Valve Closing Solenoid Control Circuit Signal Low |
| P1343 | 196 | Misfire Cylinder 1 with Fuel Cut-Off |
| P1344 | 206 | Misfire During Start Cylinder 2 |
| P1345 | 197 | Misfire Cylinder 2 with Fuel Cut-Off |
| P1346 | 207 | Misfire During Start Cylinder 3 |
| P1347 | 198 | Misfire Cylinder 3 with Fuel Cut-Off |
| P1348 | 208 | Misfire During Start Cylinder 4 |
| P1349 | 199 | Misfire Cylinder 4 with Fuel Cut-Off |
| P1350 | 209 | Misfire during Start Cylinder 5 |
| P1351 | 200 | Misfire Cylinder 5 with Fuel Cut-Off |
| P1352 | 210 | Misfire during Start Cylinder 6 |
| P1353 | 201 | Misfire Cylinder 6 with Fuel Cut-Off |
| P1447 | 127 | Diagnostic Module Tank Leakage (DM-TL) Pump Current Too High during Switching Solenoid Test |
| P1448 | 127 | Diagnostic Module Tank Leakage (DM-TL) Pump Current Too Low |
| P1449 | 127 | Diagnostic Module Tank Leakage (DM-TL) Pump Current Too High |
| P1450 | 27 | Diagnostic Module Tank Leakage (DM-TL) Switching Solenoid Control Open Circuit |
| P1451 | 27 | Diagnostic Module Tank Leakage (DM-TL) Switching Solenoid Control Circuit Signal Low |
| P1452 | 27 | Diagnostic Module Tank Leakage (DM-TL) Switching Solenoid Control Circuit Signal High |
| P1504 | 2 | Idle-Speed Control Valve Closing Solenoid Control Open Circuit |
| P1507 | 29 | Idle-Speed Control Valve Opening Solenoid Control Circuit Signal Low |
| P1508 | 29 | Idle-Speed Control Valve Opening Solenoid Control Open Circuit |
| P1509 | 29 | Idle-Speed Control Valve Opening Solenoid Control Circuit Signal Electrical |
| P1525 | 67 | 'A' Camshaft Position Actuator Control Open Circuit (Bank 1) |
| 1525 | 72 | 'A' Camshaft Position Actuator Control Open Circuit (Bank 1) |
| P1531 | 22 | 'B' Camshaft Position Actuator Control Open Circuit (Bank 1) |
| P1550 | 2 | Idle-Speed Control Valve Closing Coil Electrical |
| P1585 | 149 | Misfire Detected with Low Fuel |
| P1602 | 122 | Control Module Self-Test, Control Module Defective |
| P1602 | 231 | Control Module Self-Test, Control Module Defective |
| P1634 | 107 | Throttle Valve Adaptation Spring Test Failed |
| P1636 | 45 | Throttle Valve Control Circuit |

| Table m. S54 engine: Siemens DME MS S54 from 03/01/2001 to 08/31/2001 | | |
|---|---|---|
| PCode | BMW-FC | PCode text |
| P1637 | 107 | Throttle Valve Position Control, Control Deviation |
| P1637 | 230 | Throttle Valve Position Control, Control Deviation |
| P1639 | 107 | Throttle Valve Position Control Throttle Stuck Permanently |
| P1640 | 155 | Internal Control Module RAM/ROM Error |
| P1640 | 156 | Internal Control Module RAM/ROM Error |
| P1663 | 156 | EWS (Electronic Immobilizer) Rolling Code Faulty Storage in EEPROM |

| Table n. S54 engine: Siemens DME MS S54 from 09/01/2001 to 08/31/2002 | | |
|---|---|---|
| PCode | BMW-FC | PCode text |
| P0011 | 184 | 'A' Camshaft Position Timing Over-Advanced or System Performance (Bank 1) |
| P0012 | 184 | 'A' Camshaft Position Timing Over-Retarded (Bank 1) |
| P0014 | 185 | 'B' Camshaft Position Timing Over-Advanced or System Performance (Bank 1) |
| P0015 | 185 | 'B' Camshaft Position Timing Over-Retarded (Bank 1) |
| P0030 | 37 | HO2S Heater Control Circuit (Bank 1 Sensor 1) |
| P0031 | 37 | HO2S Heater Control Circuit Low (Bank 1 Sensor 1) |
| P0032 | 37 | HO2S Heater Control Circuit High (Bank 1 Sensor 1) |
| P0036 | 39 | HO2S Heater Control Circuit (Bank 1 Sensor 2) |
| P0037 | 39 | HO2S Heater Control Circuit Low (Bank 1 Sensor 2) |
| P0038 | 39 | HO2S Heater Control Circuit High (Bank 1 Sensor 2) |
| P0050 | 38 | HO2S Heater Control Circuit (Bank 2 Sensor 1) |
| P0051 | 38 | HO2S Heater Control Circuit Low (Bank 2 Sensor 1) |
| P0052 | 38 | HO2S Heater Control Circuit High (Bank 2 Sensor 1) |
| P0056 | 40 | HO2S Heater Control Circuit (Bank 2 Sensor 2) |
| P0057 | 40 | HO2S Heater Control Circuit Low (Bank 2 Sensor 2) |
| P0058 | 40 | HO2S Heater Control Circuit High (Bank 2 Sensor 2) |
| P0101 | 73 | Mass or Volume Air Flow Circuit Range/Performance |
| P0102 | 41 | Mass or Volume Air Flow Circuit Low Input |
| P0103 | 41 | Mass or Volume Air Flow Circuit High Input |
| P0111 | 77 | Intake Air Temperature Sensor 1 Circuit Range/Performance |
| P0112 | 77 | Intake Air Temperature Sensor 1 Circuit Low |
| P0113 | 77 | Intake Air Temperature Sensor 1 Circuit High |
| P0116 | 105 | Engine Coolant Temperature Circuit Range/Performance |
| P0117 | 78 | Engine Coolant Temperature Circuit Low |
| P0118 | 78 | Engine Coolant Temperature Circuit High |
| P0121 | 118 | Throttle/Pedal Position Sensor/Switch 'A' Circuit Range/Performance |
| P0121 | 120 | Throttle/Pedal Position Sensor/Switch 'A' Circuit Range/Performance |
| P0121 | 121 | Throttle/Pedal Position Sensor/Switch 'A' Circuit Range/Performance |

| Table n. S54 engine: Siemens DME MS S54 from 09/01/2001 to 08/31/2002 | | |
|---|---|---|
| PCode | BMW-FC | PCode text |
| P0122 | 118 | Throttle/Pedal Position Sensor/Switch 'A' Circuit Low |
| P0123 | 118 | Throttle/Pedal Position Sensor/Switch 'A' Circuit High |
| P0128 | 105 | Coolant Thermostat (Coolant Temperature Below Thermostat Regulating Temperature) |
| P0131 | 13 | O2 Sensor Circuit Low Voltage (Bank 1 Sensor 1) |
| P0132 | 13 | O2 Sensor Circuit High Voltage (Bank 1 Sensor 1) |
| P0133 | 214 | O2 Sensor Circuit Slow Response (Bank 1 Sensor 1) |
| P0134 | 13 | O2 Sensor Circuit No Activity Detected (Bank 1 Sensor 1) |
| P0134 | 218 | O2 Sensor Circuit No Activity Detected (Bank 1 Sensor 1) |
| P0135 | 37 | O2 Sensor Heater Circuit (Bank 1 Sensor 1) |
| P0137 | 87 | O2 Sensor Circuit Low Voltage (Bank 1 Sensor 2) |
| P0138 | 87 | O2 Sensor Circuit High Voltage (Bank 1 Sensor 2) |
| P0139 | 92 | O2 Sensor Circuit Slow Response (Bank 1 Sensor 2) |
| P0140 | 87 | O2 Sensor Circuit No Activity Detected (Bank 1 Sensor 2) |
| P0141 | 39 | O2 Sensor Heater Circuit (Bank 1 Sensor 2) |
| P0151 | 12 | O2 Sensor Circuit Low Voltage (Bank 2 Sensor 1) |
| P0152 | 12 | O2 Sensor Circuit High Voltage (Bank 2 Sensor 1) |
| P0153 | 215 | O2 Sensor Circuit Slow Response (Bank 2 Sensor 1) |
| P0154 | 12 | O2 Sensor Circuit No Activity Detected (Bank 2 Sensor 1) |
| P0154 | 219 | O2 Sensor Circuit No Activity Detected (Bank 2 Sensor 1) |
| P0155 | 38 | O2 Sensor Heater Circuit (Bank 2 Sensor 1) |
| P0157 | 88 | O2 Sensor Circuit Low Voltage (Bank 2 Sensor 2) |
| P0158 | 88 | O2 Sensor Circuit High Voltage (Bank 2 Sensor 2) |
| P0159 | 93 | O2 Sensor Circuit Slow Response (Bank 2 Sensor 2) |
| P0160 | 88 | O2 Sensor Circuit No Activity Detected (Bank 2 Sensor 2) |
| P0161 | 40 | O2 Sensor Heater Circuit (Bank 2 Sensor 2) |
| P0171 | 144 | System Too Lean (Bank 1) |
| P0171 | 174 | System Too Lean (Bank 1) |
| P0172 | 144 | System Too Rich (Bank 1) |
| P0172 | 174 | System Too Rich (Bank 1) |
| P0174 | 145 | System Too Lean (Bank 2) |
| P0174 | 175 | System Too Lean (Bank 2) |
| P0175 | 145 | System Too Rich (Bank 2) |
| P0175 | 175 | System Too Rich (Bank 2) |
| P0201 | 3 | Injector Circuit/Open - Cylinder 1 |
| P0202 | 5 | Injector Circuit/Open - Cylinder 2 |
| P0203 | 4 | Injector Circuit/Open - Cylinder 3 |
| P0204 | 33 | Injector Circuit/Open - Cylinder 4 |

| Table n. S54 engine: Siemens DME MS S54 from 09/01/2001 to 08/31/2002 | | |
|---|---|---|
| **PCode** | **BMW-FC** | **PCode text** |
| P0205 | 31 | Injector Circuit/Open - Cylinder 5 |
| P0206 | 32 | Injector Circuit/Open - Cylinder 6 |
| P0221 | 119 | Throttle/Pedal Position Sensor/Switch 'B' Circuit Range/Performance |
| P0222 | 119 | Throttle/Pedal Position Sensor/Switch 'B' Circuit Low |
| P0223 | 119 | Throttle/Pedal Position Sensor/Switch 'B' Circuit High |
| P0261 | 3 | Cylinder 1 Injector Circuit Low |
| P0262 | 3 | Cylinder 1 Injector Circuit High |
| P0264 | 5 | Cylinder 2 Injector Circuit Low |
| P0265 | 5 | Cylinder 2 Injector Circuit High |
| P0267 | 4 | Cylinder 3 Injector Circuit Low |
| P0268 | 4 | Cylinder 3 Injector Circuit High |
| P0270 | 33 | Cylinder 4 Injector Circuit Low |
| P0271 | 33 | Cylinder 4 Injector Circuit High |
| P0273 | 31 | Cylinder 5 Injector Circuit Low |
| P0274 | 31 | Cylinder 5 Injector Circuit High |
| P0276 | 32 | Cylinder 6 Injector Circuit Low |
| P0277 | 32 | Cylinder 6 Injector Circuit High |
| P0300 | 213 | Random/Multiple Cylinder Misfire Detected |
| P0301 | 205 | Cylinder 1 Misfire Detected |
| P0302 | 206 | Cylinder 2 Misfire Detected |
| P0303 | 207 | Cylinder 3 Misfire Detected |
| P0304 | 208 | Cylinder 4 Misfire Detected |
| P0305 | 209 | Cylinder 5 Misfire Detected |
| P0306 | 210 | Cylinder 6 Misfire Detected |
| P0335 | 16 | Crankshaft Position Sensor 'A' Circuit |
| P0340 | 7 | Camshaft Position Sensor 'A' Circuit (Bank 1 or Single Sensor) |
| P0365 | 10 | Camshaft Position Sensor 'B' Circuit (Bank 1) |
| P0369 | 162 | Camshaft Position Sensor 'B' Circuit Intermittent (Bank 1) |
| P0394 | 15 | Camshaft Position Sensor 'B' Circuit Intermittent (Bank 2) |
| P0411 | 170 | Secondary Air Injection System Incorrect Flow Detected |
| P0413 | 63 | Secondary Air Injection System Switching Valve A Circuit Open |
| P0414 | 63 | Secondary Air Injection System Switching Valve A Circuit Shorted |
| P0418 | 19 | Secondary Air Injection System Control 'A' Circuit |
| P0420 | 178 | Catalyst System Efficiency Below Threshold (Bank 1) |
| P0430 | 179 | Catalyst System Efficiency Below Threshold (Bank 2) |
| P0441 | 232 | Evaporative Emission System Incorrect Purge Flow |
| P0442 | 180 | Evaporative Emission System Leak Detected (small leak) |

| Table n. S54 engine: Siemens DME MS S54 from 09/01/2001 to 08/31/2002 | | |
|---|---|---|
| **PCode** | **BMW-FC** | **PCode text** |
| P0443 | 36 | Evaporative Emission System Purge Control Valve Circuit |
| P0444 | 36 | Evaporative Emission System Purge Control Valve Circuit Open |
| P0445 | 36 | Evaporative Emission System Purge Control Valve Circuit Shorted |
| P0491 | 170 | Secondary Air Injection System Insufficient Flow (Bank 1) |
| P0492 | 170 | Secondary Air Injection System Insufficient Flow (Bank 2) |
| P0500 | 135 | Vehicle Speed Sensor 'A' |
| P0506 | 128 I | Idle Air Control System RPM Lower Than Expected |
| P0507 | 128 | Idle Air Control System RPM Higher Than Expected |
| P0601 | 150 | Internal Control Module Memory Check Sum Error |
| P0601 | 157 | Internal Control Module Memory Check Sum Error |
| P0604 | 150 | Internal Control Module Random Access Memory (RAM) Error |
| P0604 | 157 | Internal Control Module Random Access Memory (RAM) Error |
| P0605 | 150 | Internal Control Module Read Only Memory (ROM) Error |
| P0605 | 157 | Internal Control Module Read Only Memory (ROM) Error |
| P1502 | 2 | Idle-Speed Control Valve Closing Solenoid Control Circuit Signal High |
| P1503 | 2 | Idle-Speed Control Valve Closing Solenoid Control Circuit Signal Low |
| P0606 | 30 | ECM/PCM Processor |
| P0606 | 107 | ECM/PCM Processor |
| P0606 | 158 | ECM/PCM Processor |
| P0606 | 163 | ECM/PCM Processor |
| P1121 | 111 | Pedal Position Sensor 1 Range/Performance Problem |
| P1122 | 60 | Pedal Position Sensor 1 Low Input |
| P1123 | 60 | Pedal Position Sensor 1 High Input |
| P1178 | 216 | O2 Sensor Signal Circuit Slow Switching from Rich to Lean (Bank 1 Sensor 1) |
| P1179 | 217 | O2 Sensor Signal Circuit Slow Switching from Rich to Lean (Bank 2 Sensor 1) |
| P1221 | 112 | Pedal Position Sensor 2 Range/Performance Problem |
| P1222 | 61 | Pedal Position Sensor 2 Low Input |
| P1223 | 61 | Pedal Position Sensor 2 High Input |
| P1340 | 213 | Multiple Cylinder Misfire During Start |
| P1341 | 204 | Multiple Cylinder Misfire with Fuel Cut-Off |
| P1342 | 205 | Misfire During Start Cylinder 1 |
| P1343 | 196 | Misfire Cylinder 1 with Fuel Cut-Off |
| P1344 | 206 | Misfire During Start Cylinder 2 |
| P1345 | 197 | Misfire Cylinder 2 with Fuel Cut-Off |
| P1346 | 207 | Misfire During Start Cylinder 3 |
| P1347 | 198 | Misfire Cylinder 3 with Fuel Cut-Off |
| P1348 | 208 | Misfire During Start Cylinder 4 |

| Table n. S54 engine: Siemens DME MS S54 from 09/01/2001 to 08/31/2002 | | |
|---|---|---|
| **PCode** | **BMW-FC** | **PCode text** |
| P1349 | 199 | Misfire Cylinder 4 with Fuel Cut-Off |
| P1350 | 209 | Misfire during Start Cylinder 5 |
| P1351 | 200 | Misfire Cylinder 5 with Fuel Cut-Off |
| P1352 | 210 | Misfire during Start Cylinder 6 |
| P1353 | 201 | Misfire Cylinder 6 with Fuel Cut-Off |
| P1434 | 176 | Diagnostic Module Tank Leakage (DM-TL) |
| P1447 | 127 | Diagnostic Module Tank Leakage (DM-TL) Pump Current Too High during Switching Solenoid Test |
| P1448 | 127 | Diagnostic Module Tank Leakage (DM-TL) Pump Current Too Low |
| P1449 | 127 | Diagnostic Module Tank Leakage (DM-TL) Pump Current Too High |
| P1450 | 27 | Diagnostic Module Tank Leakage (DM-TL) Switching Solenoid Control Open Circuit |
| P1451 | 27 | Diagnostic Module Tank Leakage (DM-TL) Switching Solenoid Control Circuit Signal Low |
| P1452 | 27 | Diagnostic Module Tank Leakage (DM-TL) Switching Solenoid Control Circuit Signal High |
| P1504 | 2 | Idle-Speed Control Valve Closing Solenoid Control Open Circuit |
| P1506 | 29 | Idle-Speed Control Valve Opening Solenoid Control Circuit Signal High |
| P1507 | 29 | Idle-Speed Control Valve Opening Solenoid Control Circuit Signal Low |
| P1508 | 29 | Idle-Speed Control Valve Opening Solenoid Control Open Circuit |
| P1525 | 67 | 'A' Camshaft Position Actuator Control Open Circuit (Bank 1) |
| P1525 | 72 | 'A' Camshaft Position Actuator Control Open Circuit (Bank 1) |
| P1531 | 21 | 'B' Camshaft Position Actuator Control Open Circuit (Bank 1) |
| P1531 | 22 | 'B' Camshaft Position Actuator Control Open Circuit (Bank 1) |
| P1585 | 149 | Misfire Detected with Low Fuel |
| P1602 | 122 | Control Module Self-Test, Control Module Defective |
| P1602 | 231 | Control Module Self-Test, Control Module Defective |
| P1634 | 107 | Throttle Valve Adaptation Spring Test Failed |
| P1636 | 45 | Throttle Valve Control Circuit |
| P1637 | 107 | Throttle Valve Position Control, Control Deviation |
| P1637 | 230 | Throttle Valve Position Control, Control Deviation |
| P1639 | 107 | Throttle Valve Position Control Throttle Stuck Permanently |
| P1640 | 155 | Internal Control Module RAM/ROM Error |
| P1640 | 156 | Internal Control Module RAM/ROM Error |
| P1663 | 155 | EWS (Electronic Immobilizer) Rolling Code Faulty Storage in EEPROM |
| P1663 | 156 | EWS (Electronic Immobilizer) Rolling Code Faulty Storage in EEPROM |

**WARNING**

*Your common sense, good judgement and general alertness are crucial to safe and successful service work. Before attempting any work on your BMW, be sure to read* **001 General Warnings and Cautions** *and the copyright page at the front of the manual. Review these warnings and cautions each time you prepare to work on your car. Please also read any warnings and cautions that accompany the procedures in the manual.*

> ***WARNING***
>
> *Your common sense, good judgement and general alertness are crucial to safe and successful service work. Before attempting any work on your BMW, be sure to read* **001 General Warnings and Cautions** *and the copyright page at the front of the manual. Review these warnings and cautions each time you prepare to work on your car. Please also read any warnings and cautions that accompany the procedures in the manual.*

> ***WARNING***
>
> *Your common sense, good judgement and general alertness are crucial to safe and successful service work. Before attempting any work on your BMW, be sure to read* **001 General Warnings and Cautions** *and the copyright page at the front of the manual. Review these warnings and cautions each time you prepare to work on your car. Please also read any warnings and cautions that accompany the procedures in the manual.*

> ***WARNING***
>
> *Your common sense, good judgement and general alertness are crucial to safe and successful service work. Before attempting any work on your BMW, be sure to read* **001 General Warnings and Cautions** *and the copyright page at the front of the manual. Review these warnings and cautions each time you prepare to work on your car. Please also read any warnings and cautions that accompany the procedures in the manual.*

**WARNING**

*Your common sense, good judgement and general alertness are crucial to safe and successful service work. Before attempting any work on your BMW, be sure to read* **001 General Warnings and Cautions** *and the copyright page at the front of the manual. Review these warnings and cautions each time you prepare to work on your car. Please also read any warnings and cautions that accompany the procedures in the manual.*

**Notes**

## Notes

## Notes

## Notes

## Notes

## Notes

## Notes